10 0567988 8
the biology of
ANCER

D0303303

the biology of CANCER

Robert A. Weinberg

Vice President: Denise Schanck
Editorial Assistants: Sigrid Masson and Alan Grose
Production Editor, Proofreader, and Layout: Emma Jeffcock
Text Editor: Elizabeth Zayatz
Copy Editor: Richard K. Mickey
Illustrator: Nigel Orme
Designer: Matthew McClements, Blink Studio Ltd.
Indexer: Merrall-Ross International Ltd.
Media Production and Design: Michael Morales

About the Author

Dr. Robert A.Weinberg is a founding member of the Whitehead Institute for Biomedical Research. He is also the Daniel K. Ludwig Professor for Cancer Research at the Massachusetts Institute of Technology (MIT) and American Cancer Society Research Professor.

This book contains information obtained from authentic and highly regarded sources. Reprinted material is quoted with permission, and sources are indicated. Every attempt has been made to source the figures accurately. Reasonable efforts have been made to publish reliable data and information, but the author and the publisher cannot assume responsibility for the validity of all materials or for the consequences of their use.

Library of Congress Cataloging-in-Publication Data

Weinberg, Robert A. (Robert Allan)
The biology of cancer / by Robert A. Weinberg.
p. ; cm.
Includes bibliographical references and index.
ISBN 0-8153-4078-8 (hardcover) -- ISBN 0-8153-4076-1 (softcover)
1. Cancer--Molecular aspects. 2. Cancer--Genetic aspects.
3. Cancer cells. I. Title.
[DNLM: 1. Neoplasms--genetics. 2. Cell Transformation, Neoplastic.
3. Genes, Neoplasm. QZ 202 W423b 2007]
RC268.4.W45 2007
616.99'4071--dc22

2006001825

Published by Garland Science, Taylor & Francis Group, LLC, an informa business
270 Madison Avenue, New York, NY 10016, USA, and
2 Park Square, Milton Park, Abingdon, OX14 4RN, UK

Printed in the United States of America

15 14 13 12 11 10 9 8 7 6 5 4 3 2 1

Front Cover

Carcinoma cells at the outer edge of an island of these tumor cells often undergo an epithelial-mesenchymal transition, which enables them to become motile and to invade the adjacent stroma. Seen here are experimentally transformed human mammary epithelial cells, which express cytokeratins *(red),* which are typical of epithelial cells. However, human carcinoma cells that are in contact with the surrounding mouse stroma (revealed by *blue* nuclei) have shed their cytokeratin expression and express instead vimentin *(green),* a marker of mesenchymal cells. These cells have also changed shape and some have already invaded deep into the surrounding stroma. (Courtesy of Kimberly A. Hartwell and Tan A. Ince.)

Dedication

This book is dedicated to my wife, Amy Shulman Weinberg, who lived through the long hours when my attentions were focused on nothing else but this book, and who patiently listened to my claims that it would be finished in short order, knowing full well that I had no idea of how long this project would really take. Her constant support and love made the writing of this possible.

Preface

This textbook describes the development of a relatively new area of biology. The beginnings of this field of biomedical research can be traced, with some precision, to the discoveries in 1975 concerning the proto-oncogene, made in the laboratory of H.E. Varmus and J. M. Bishop in San Francisco, California. Research conducted since that time has yielded a rich harvest of information about the molecules and genes responsible for cancer in human beings.

Prior to this pioneering research, we knew almost nothing about the molecular and cellular mechanisms that create cancer. However, some intriguing clues lay before the community of cancer researchers: We knew that carcinogenic agents often, but not always, operate as mutagens; this suggested that mutant genes are involved in some fashion in programming the abnormal proliferation of cancer cells. We knew that the development of cancer is often a long, protracted process. And we knew that individual cancer cells extracted from tumors behave very differently than their counterparts in normal tissues.

Now, more than three decades later, we have an abundance of information about the root causes of cancer—almost too much. The genes and proteins that have been implicated in the causation of human cancer can be numbered in the hundreds. And cancer itself has been found to be not one disease but many, with the above-noted agents being involved in creating the approximately 110 distinct types of human cancers cataloged so far. The techniques of genetic analysis, which were quite primitive at the beginning of this period, have now advanced to the stage that we can sequence entire mammalian genomes and catalog hundreds of mutant alleles present in the genomes of human cancer cells. In the first years of the new millennium, we are deluged with a vast amount of genetic, biochemical, and cell biological information about cancer development, far more almost than any human mind can assimilate and comprehend.

This textbook has been written to lay out the underlying laws and principles that are emerging from the vast collection of facts that grow in number with each passing year. Such laws and principles explain many of the complex behaviors of human tumors. By providing context and perspective, they can be used to help us understand most and possibly all types of human cancer. This field of biological research is still in active ferment, and new, fascinating discoveries are being reported almost every month. And so, many of the principles of cancer development are also still being uncovered. Nevertheless, we can already perceive the outlines of some lasting truths, and to the extent possible, these are laid out in this book.

In part, this book has been written as a recruiting pamphlet, as new generations of researchers are needed to move cancer research forward. The campaign to conquer cancer has only begun. The principles that have been elucidated over these past three decades, and which are the subject of this book, have still not been successfully applied to make major inroads into the prevention, diagnosis, and cure of human cancers.

And yes, there are still major questions that remain murky and poorly resolved. We still do not understand how cancer cells learn to invade and to create the metastases that are responsible for 90% of cancer mortality. We have only a very imperfect understanding of the role of the immune system in preventing cancer development. And while we know much about the individual signaling molecules operating inside human cells, we lack a clear understanding of how the complex signaling circuitry that they form is able to operate to make the life-and-death decisions that determine the fate of individual cells within our body. Those decisions ultimately determine whether or not one of our cells will begin a journey down the long road leading to cancerous proliferation and, finally, to a life-threatening tumor.

Contemporary cancer research has enriched numerous other areas of modern biomedical research. Consequently, much of what you will learn from this book will be useful in understanding many aspects of immunology, neurobiology, developmental biology, and a dozen other biomedical research fields. Enjoy the ride!

Robert A. Weinberg
Cambridge, Massachusetts
May 2006

A Note to the Reader

This book is organized into 16 chapters of quite different lengths. The chapters are meant to be read in the order that they appear, in that each builds on the ideas that have been presented in the chapters before it. The first chapter is a condensed refresher course for undergraduate biology majors and graduate students; it lays out the background concepts that are assumed in the subsequent chapters.

The main purpose is to illustrate the concepts that together constitute our current thinking about how cancer arises and how this disease should be treated in the future. Some experiments are described in detail to indicate the logic supporting many of these concepts. You will find numerous schematic drawings, often coupled with micrographs, that will help you to appreciate how experimental results have been assembled, piece-by-piece, generating the syntheses that underlie modern cancer research.

Scattered about the text are numerous "Sidebars," which consist of commentary that detours slightly from the main thrust of the discussion. Often these Sidebars contain anecdotes or elaborate on ideas presented in the main text. Read them if you are interested, or skip over them if you find them too distracting. They are presented to provide additional interest—a bit of extra seasoning in the rich stew of ideas that constitutes contemporary research in this area.

There are extensive cross-references whenever topics under discussion have been introduced or described in other chapters. Many of these have been inserted in the event that you read the chapters in an order different from their presentation here. These cross-references should not induce you to continually leaf through other chapters in order to track down cited sections or figures. If you feel that you will benefit from earlier introductions to a topic, use these cross-references; otherwise, ignore them.

Each chapter ends with a forward-looking summary entitled "Synopsis and Prospects." This section synthesizes the main concepts of the chapter and addresses ideas that remain matters of contention. It also considers where research might go in the future. This overview is extended by a list of key concepts and a set of questions. Some of the questions are deliberately challenging and we hope they will provoke you to think more deeply about many of the issues and concepts developed. Finally, most chapters have an extensive list of articles from research journals. These will be useful if you wish to explore a particular topic in detail. Most of the references are review articles, and many contain detailed discussions of various subfields of research as well as recent findings. In addition, there are occasional references to older publications that will clarify how certain lines of research developed.

Perhaps the most important goal is to enable you to move beyond the textbook level and jump directly into the primary research literature. Therefore, some of the text is directed toward teaching the elaborate, specialized vocabulary of the cancer research literature, and many of its terms are defined in the glossary. Boldface type has been used throughout to highlight key terms that you should understand. Cancer research, like most areas of contemporary biomedical research, is plagued by numerous abbreviations and acronyms that pepper the text of many published reports. The book provides a key to deciphering this alphabet soup by defining these acronyms. You will find a list of such abbreviations at the back.

The CD-ROM packaged contains a variety of media for students and instructors. All the figures, tables, and micrographs are available on the CD and have been pre-loaded into PowerPoint® presentations, one for each chapter. A separate folder contains individual versions of each figure, table, and micrograph in JPEG format. Additional "Sidebars" are available on the CD-ROM to complement material presented in the text. These are cross-referenced throughout and each is highlighted with a blue icon. There are also several movies that will aid in understanding some of the processes described. The author has recorded mini-lectures on the following topics for students and instructors: Mutations and the Origin of Human Cancer, Metastasis, Growth Factors, Cancer Therapies, Immunology and Cancer, and p53 and Apoptosis. These are available on the CD in MP3 format and can be transferred to an iPod™ or other MP3 player, as well as listened to on your computer. Many of these media items, together with future media updates, will also be available to students and instructors at: http://www.garlandscience.com.

All of the teaching supplements on the CD-ROM are available to qualified instructors online at the Garland Science Classwire™ Web site. Garland Science Classwire™ offers access to instructional resources from all of the Garland Science textbooks and provides free online course management tools. For additional information, please visit http://www.classwire.com/garlandscience or e-mail science@garland.com.

The poster entitled "The Pathways of Human Cancer" summarizes many of the signaling pathways implicated in tumor development. This poster has been produced by Cell Signaling Technology.

Because this book describes an area of research in which new and exciting findings are being announced all the time, some of the details and interpretations presented here may become outdated (or, equally likely, proven to be wrong) once this book is in print. Still, the primary concepts presented here will remain, as they rest on solid foundations of experimental results.

The author and the publisher would greatly appreciate your feedback. Every effort has been made to minimize errors. Nonetheless, you may find them here and there, and it would be of great benefit if you took the trouble to communicate them. Even more importantly, much of the science described herein will require reinterpretation in coming years as new discoveries are made. Please email us at tboc@taylorandfrancis.com with your suggestions, which will be considered for incorporation into future editions.

Acknowledgments

The science described in this book is the opus of a large, highly interactive research community stretching across the globe. Its members have moved forward our understanding of cancer immeasurably over the past generation. The colleagues listed below have helped the author in countless ways, large and small, by providing sound advice, referring me to critical scientific literature, and analyzing complex and occasionally contentious scientific issues. They are representatives of this community, whose members are, virtually without exception, ready and pleased to provide a helping hand to those who request it. I am most grateful to them. Not listed below are the many colleagues who generously provided high quality versions of their published images; they are acknowledged through the literature citations in the figure legends.

Joan Abbott; Eike-Gert Achilles; Jerry Adams; Kari Alitalo; James Allison; David Alpers; Fred Alt; Carl Anderson; Andrew Aprikyan; Jon Aster; Laura Attardi; Frank Austen; Joseph Avruch; Sunil Badve; William Baird; Frances Balkwill; Allan Balmain; Alan Barge; J. Carl Barrett; David Bartel; Renato Baserga; Richard Bates; Philip Beachy; Camille Bedrosian; Anna Belkina; Robert Benezra; Thomas Benjamin; Yinon Ben-Neriah; Ittai Ben-Porath; Bradford Berk; René Bernards; Anton Berns; Kenneth Berns; Monica Bessler; Neil Bhowmick; Marianne Bienz; Line Bjørge; Harald von Boehmer; Gareth Bond; Thierry Boon; Dorin-Bogdan Borza; Chris Boshoff; Noël Bouck; Thomas Brabletz; Douglas Brash; Cathrin Brisken; Garrett Brodeur; Patrick Brown; Richard Bucala; Patricia Buffler; Tony Burgess; Suzanne Bursaux; Randall Burt; Stephen Bustin; Janet Butel; Lisa Butterfield; Blake Cady; John Cairns; Judith Campisi; Harvey Cantor; Robert Cardiff; Peter Carroll; Arlindo Castelanho; Bruce Chabner; Ann Chambers; Howard Chang; Andrew Chess; Ann Cheung; Lynda Chin; Francis Chisari; Yunje Cho; Margaret Chou; Karen Cichowski; Michael Clarke; Hans Clevers; Brent Cochran; Robert Coffey; John Coffin; Samuel Cohen; Graham Colditz; Kathleen Collins; Dave Comb; John Condeelis; Suzanne Cory; Christopher Counter; Sara Courtneidge; Sandra Cowan-Jacob; John Crispino; John Crissman; Carlo Croce; Tim Crook; Christopher Crum; Marcia Cruz-Correa; Gerald Cunha; George Daley; Riccardo Dalla-Favera; Alan D'Andrea; Chi Dang; Douglas Daniels; James Darnell, Jr.; Robert Darnell; Galina Deichman; Titia de Lange; Hugues de Thé; Chuxia Deng; Edward Dennis; Lucas Dennis; Ronald DePinho; Theodora Devereaux; Tom DiCesare; Jules Dienstag; John DiGiovanni; Peter Dirks; Ethan Dmitrovsky; Daniel Donoghue; John Doorbar; G. Paolo Dotto; William Dove; Julian Downward; Glenn Dranoff; Thaddeus Dryja; Raymond DuBois; Nick Duesbery; Michel DuPage; Harold Dvorak; Nicholas Dyson; Michael Eck; Walter Eckhart; Argiris Efstratiadis; Robert Eisenman; Klaus Elenius; Steven Elledge; Elissa Epel; John Eppig; Raymond Erikson; James Eshleman; John Essigmann; Gerard Evan; Mark Ewen; Guowei Fang; Juli Feigon; Andrew Feinberg; Stephan Feller; Bruce Fenton; Stephen Fesik; Isaiah Fidler; Gerald Fink; Alain Fischer; Zvi Fishelson; David Fisher; Richard Fisher; Richard Flavell; Riccardo Fodde; M. Judah Folkman; David Foster; Uta Francke; Emil Frei; Errol Friedberg; Peter Friedl; Stephen Friend; Jonas Frisen; Elaine Fuchs; Margaret Fuller; Yuen Kai (Teddy) Fung; Kyle Furge; Amar Gajjar; Joseph Gall; Donald Ganem; Judy Garber; Frank Gertler; Charlene Gilbert; Richard Gilbertson; Robert Gillies; Doron Ginsberg; Edward Giovannucci; Inna Gitelman; Steve Goff; Lois Gold; Alfred Goldberg; Mitchell Goldfarb; Richard Goldsby; Joseph Goldstein; Susanne Gollin; Mehra Golshan; Todd Golub; Jeffrey Gordon; Michael Gordon; Siamon Gordon; Martin Gorovsky; Arko Gorter; Joe Gray; Douglas Green; Yoram Groner; John Groopman; Steven Grossman; Wei Gu; David Guertin; Piyush Gupta; Barry Gusterson; Daniel Haber; James Haber; William Hahn; Kevin Haigis; Senitiroh Hakomori; Alan Hall; Dina Gould Halme; Douglas Hanahan; Philip Hanawalt; Adrian Harris; Curtis Harris; Lyndsay Harris; Stephen Harrison; Kimberly Hartwell; Leland Hartwell; Harald zur Hausen; Carol Heckman; Ruth Heimann; Samuel Hellman; Brian Hemmings; Lothar Hennighausen; Meenhard Herlyn; Glenn Herrick; Avram Hershko; Douglas Heuman; Richard Hodes; Jan Hoeijmakers; Robert Hoffman; Robert Hoover; David Hopwood; Gabriel Hortobagyi; H. Robert Horvitz; Marshall Horwitz; Alan Houghton; Peter Howley; Robert Huber; Tim Hunt; Tony Hunter; Stephen Hursting; Nancy Hynes; Richard Hynes; Antonio Iavarone; J. Dirk Iglehart; Tan Ince; Max Ingman; Mark Israel; Kurt Isselbacher; Tyler Jacks; Rudolf Jaenisch; Rakesh Jain; Bruce Johnson; David Jones; Richard Jones; William Kaelin, Jr.; Raghu Kalluri; Alexander Kamb; Barton Kamen; Manolis Kamvysselis; Yibin Kang; Philip Kantoff; Paul Kantrowitz; Jan Karlsreder; Michael Kastan; Michael Kauffman; William Kaufmann; Robert Kerbel; Scott Kern; Khandan Keyomarsi; Marc Kirschner; Christoph Klein; George Klein; Yoel Kloog; Alfred Knudson; Frederick Koerner; Anthony Komaroff; Kenneth Korach; Alan Korman; Eva Kramarova; Jackie Kraveka; Wilhelm Krek; Charlotte Kuperwasser; James Kyranos; Carole LaBonne; Peter Laird; Sergio Lamprecht; Eric Lander; Laura Landweber; Lewis Lanier; Andrew Lassar; Robert Latek; Lester Lau; Derek Le Roith; Chung Lee; Keng Boon Lee; Richard Lee; Jacqueline Lees; Rudolf Leibel; Mark Lemmon; Christoph Lengauer;

Jack Lenz; Gabriel Leung; Arnold Levine; Beth Levine; Jay Levy; Ronald Levy; Fran Lewitter; Frederick Li; Siming Li; Frank Lieberman; Elaine Lin; Joachim Lingner; Martin Lipkin; Joe Lipsick; David Livingston; Harvey Lodish; Lawrence Loeb; Edward Loechler; Michael Lotze; Lawrence Lum; Vicky Lundblad; David MacPherson; Sendurai Mani; Alberto Mantovani; Sandy Markowitz; Larry Marnett; G. Steven Martin; Seamus Martin; Joan Massagué; Patrice Mathevet; Paul Matsudaira; Andrea McClatchey; Frank McCormick; Patricia McManus; Mark McMenamin; U. Thomas Meier; Matthew Meyerson; George Miller; Nathan Miselis; Randall Moon; David Morgan; Rebecca Morris; Simon Conway Morris; Robert Moschel; Bernard Moss; Paul Mueller; Anja Mueller-Homey; William A. Muller; Gregory Mundy; Karl Münger; Lance Munn; Ruth Muschel; Lee Nadler; David G. Nathan; Jeremy Nathans; Sergei Nedospasov; Benjamin Neel; David Neuhaus; Donald Newmeyer; Leonard Norkin; Lloyd Old; Kenneth Olive; Tamer Onder; Moshe Oren; Terry Orr-Weaver; Barbara Osborne; Michele Pagano; David Page; Asit Parikh; Chris Parker; William Paul; Amanda Paulovich; Tony Pawson; Mark Peifer; David Pellman; David Phillips; Jacqueline Pierce; Malcolm Pike; John Pintar; Maricarmen Planas-Silva; Roland Pochet; Daniel Podolsky; Beatriz Pogo; Roberto Polakiewicz; Jeffrey Pollard; Nicolae Popescu; Christoph Poremba; Richmond Prehn; Carol Prives; Vito Quaranta; Peter Rabinovitch; Al Rabson; Priyamvada Rai; Klaus Rajewsky; Sridhar Ramaswamy; Anapoorni Rangarajan; Jeffrey Ravetch; Ilaria Rebay; John Reed; Steven Reed; Alan Rein; Ee Chee Ren; Elizabeth Repasky; Jeremy Rich; Andrea Richardson; Dave Richardson; Darrell Rigel; James Roberts; Diane Rodi; Clifford Rosen; Jeffrey Rosen; Neal Rosen; Naomi Rosenberg; Michael Rosenblatt; Theodora Ross; Martine Roussel; Steve Rozen; Jeffrey Ruben; José Russo; David Sabatini; Julien Sage; Ronit Sarid; Edward Sausville; Charles Sawyers; David Scadden; David Schatz; Christina Scheel; Joseph Schlessinger; Anja Schmidt; Stuart Schnitt; Robert Schoen; Robert Schreiber; Edward Scolnick; Ralph Scully; Harold Seifried; William Sessa; Jeffrey Settleman; Fergus Shanahan; Jerry Shay; James Sherley; Charles Sherr; Ethan Shevach; Chiaho Shih; Frank Sicheri; Peter Sicinski; Sandy Simon; Dinah Singer; Arthur Skarin; Jonathan Skipper; Judy Small; Gilbert Smith; Lauren Sompayrac; Holger Sondermann; Gail Sonenshein; Deborah Spector; Michael Sporn; Eric Stanbridge; E. Richard Stanley; Louis Staudt; Philipp Steiner; Ralph Steinman; Gunther Stent; Sheila Stewart; Charles Stiles; Jonathan Stoye; Michael Stratton; Bill Sugden; Takashi Sugimura; John Sullivan; Nevin Summers; Calum Sutherland; Clifford Tabin; John Tainer; Jussi Taipale; Shinichiro Takahashi; Martin Tallman; Steven Tannenbaum; Susan Taylor; Margaret Tempero; Masaaki Terada; Satvir Tevethia; Jean Paul Thiery; William Thilly; David Thorley-Lawson; Jay Tischfield; Robertus Tollenaar; Stephen Tomlinson; Dimitrios Trichopoulos; Elaine Trujillo; James Umen; Alex van der Eb; Wim van Egmond; Diana van Heemst; Laura van't Veer; Harold Varmus; Alexander Varshavsky; Anna Velcich; Ashok Venkitaraman; Björn Vennström; Inder Verma; Shelia Violette; Bert Vogelstein; Peter Vogt; Olga Volpert; Evan Vosburgh; Geoffrey Wahl; Graham Walker; Gernot Walter; Jack Wands; Elizabeth Ward; Jonathan Warner; Randolph Watnick; I. Bernard Weinstein; Robin Weiss; Irving Weissman; Danny Welch; H. Gilbert Welch; Zena Werb; Forest White; Michael White; Raymond White; Max Wicha; Walter Willet; Owen Witte; Richard Wood; Andrew Wyllie; John Wysolmerski; Michael Yaffe; Yukiko Yamashita; George Yancopoulos; Jing Yang; Moshe Yaniv; Chun-Nan Yeh; Richard Youle; Richard Young; Stuart Yuspa; Claudio Zanon; David Zaridze; Patrick Zarrinkar; Bruce Zetter; Drazen Zimonjic; Leonard Zon; Weiping Zou

Reviewers: The following provided great help by reviewing individual chapters and providing much-appreciated critiques. Their scientific expertise and their insights into pedagogical clarity have proven to be invaluable.

Barbara Danowski (Union College); Bob Duronio (University of North Carolina, Chapel Hill); George Edick (Rensselaer Polytechnic Institute); Hung Y. Fan (University of California, Irvine); Paul G. Greenwood (Colby College); Yoji Ikawa (RIKEN, Japan); Helen James (University of East Anglia, UK); John Haanen (Netherlands Cancer Institute, The Netherlands); Momna Hejmadi (University of Bath, UK); Richard Karp (University of Cincinnati); Zhi-Chun Lai (Pennsylvania State University); Richard McIntosh (University of Colorado, Boulder); Satya Narayan (University of Florida); Peter J. Peters (Netherlands Cancer Institute, The Netherlands); Stephanie Richards (Bates College); Jeffrey E. Segall (Albert Einstein College of Medicine); A.D. Sharrocks (University of Manchester, UK); Robert Sikes (University of Delaware); Joanne Willey (Hofstra University); Jolene Windle (Virginia Commonwealth University).

Class Testers: The ultimate test of a textbook is its performance in the classroom, and the following instructors and their students were gracious and generous, using chapters of this book in their classes on cancer biology. The feedback that they provided was extremely useful in improving the clarity of the exposition of many parts of this book.

Iswar Hariharan (University of California, Berkeley); Kunxin Luo (University of California, Berkeley); G. Steven Martin (University of California, Berkeley); Charles Rogler (Albert Einstein School of Medicine, Yeshiva University); Jeffrey E. Segall (Albert Einstein School of Medicine, Yeshiva University); Paula M. Vertino (Emory University School of Medicine); Jean Wang (University of California, San Diego).

Readers: Through their careful reading of the text, these two graduate students provided extraordinarily useful feedback in improving many sections of this book and in clarifying sections that were, in their original versions, poorly written and confusing.

David Kashatus (University of North Carolina Medical School); Mark Schramp (University of California, Berkeley).

Whitehead Institute/MIT: Christine Hickey was responsible over several years' time in organizing the enormous collection of visual materials that were acquired and provided help that can only be considered far beyond the call of duty and truly extraordinary.

José Torradas and Ji Zhang, two MIT undergraduates, provided great help in the cataloging and processing of many of the images.

Dave Richardson of the Whitehead Institute library helped on countless occasions to retrieve papers from obscure corners of the vast scientific literature.

Garland: While this book has a single recognized author, it really is the work of many hands. The prose was edited by Elizabeth Zayatz and Richard K. Mickey, two editors who are nothing less than superb. To the extent that this book is clear and readable, much of this is a reflection of their dedication to clarity, precision of language, graceful syntax, and the use of images that truly serve to enlighten rather than confound. I have been most fortunate to have two such extraordinary people looking over my shoulder at every step of the writing process. And, to be sure, I have learned much from them.

Many of the figures are the work of Nigel Orme, an illustrator of great talent, whose sense of design and dedication to precision and detail are, once again, nothing less than extraordinary.

Garland Publishing determined the structure and design and provided unfaltering support and encouragement through every step of the three year-long process that was required to bring this project to fruition. Denise Schanck gave guidance and cheered me on every step of the way. Unfailingly gracious, she is, in every sense, a superb publisher, whose instincts for design and standards of quality publishing are a model. All textbook authors should be as fortunate as I have been to have someone of her qualities at the helm!

The editorial and logistical support required to organize and assemble a book of this complexity was provided first by Alan Grose and then by Sigrid Masson, both of whom are multitalented and exemplars of ever-cheerful competence, thoroughness, and helpfulness. The truly Herculean task of procuring permissions for publication of the myriad figures fell on the shoulders of Gina Pujols and Sigrid Masson. This remains a daunting task, even in this age of internet and email. Without their help, it would have been impossible to share with the reader many of the images that have created the field of modern cancer research.

The layout is a tribute to the talents of Emma Jeffcock, once again an exemplar of competence, who has an unerring instinct for how to make images and the pages that hold them accessible and welcoming to the reader; she also provided much-valued editorial help that resulted in many improvements of the prose.

The electronic media associated with this book are the work of Michael Morales, whose ability to organize clear and effective visual presentations are indicated by the electronic files that are carried in the accompanying CD. He is recognized and thanked for his dedication to detail, thoroughness, and his great talents in providing accessible images that inform the reader and complement the written text.

Additional, highly valuable input into the organization and design were provided by Adam Sendroff. Savita Poornam also helped in developing ideas for supplementary materials. Their help is gratefully acknowledged.

Contents

List of Headings

Chapter 8

pRb and Control of the Cell Cycle Clock 255

Chapter 9

p53 and Apoptosis: Master Guardian and Executioner 307

Chapter 10

Eternal Life: Cell Immortalization and Tumorigenesis 357

Chapter 11

Multi-Step Tumorigenesis 399

Chapter 12

Maintenance of Genomic Integrity and the Development of Cancer 463

Chapter 13

Dialogue Replaces Monologue: Heterotypic Interactions and the Biology of Angiogenesis 527

Chapter 14

Chapter 15

Chapter 16

The Rational Treatment of Cancer 725

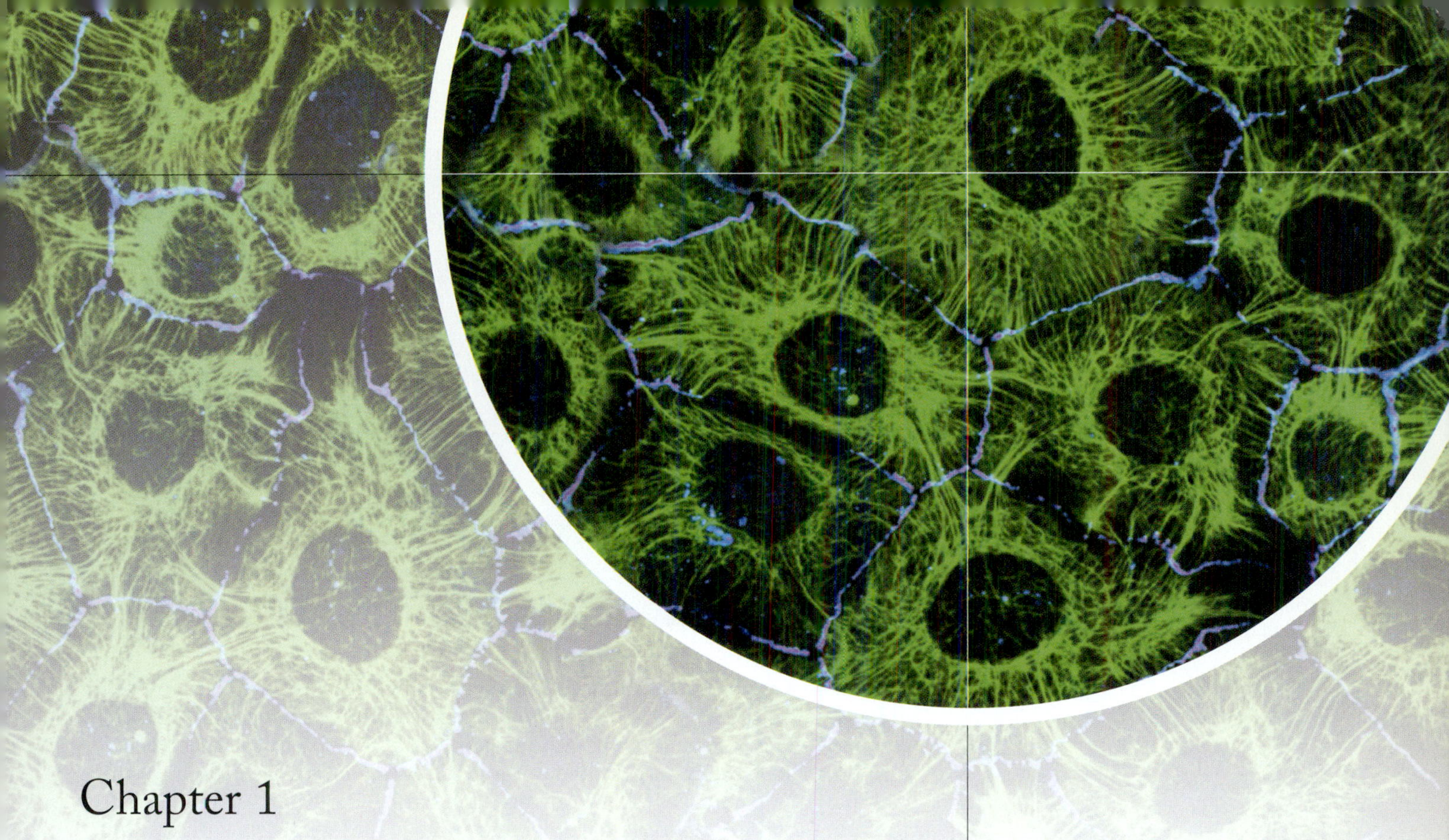

Chapter 1

The Biology and Genetics of Cells and Organisms

> Protoplasm, simple or nucleated, is the formal basis of all life... Thus it becomes clear that all living powers are cognate, and that all living forms are fundamentally of one character. The researches of the chemist have revealed a no less striking uniformity of material composition in living matter.
>
> Thomas Henry Huxley, evolutionary biologist, 1868

> Anything found to be true of *E. coli* must also be true of elephants.
>
> Jacques Monod, pioneer molecular biologist, 1954

The biological revolution of the twentieth century totally reshaped all fields of biomedical study, cancer research being only one of them. The fruits of this revolution were revelations of both the outlines and the intimate details of genetics and heredity, of how cells grow and divide, how they assemble to form tissues, and how the tissues develop under the control of specific genes. Everything that follows in this text draws directly or indirectly on this new knowledge.

This revolution, which began in mid-century and was triggered by Watson and Crick's discovery of the DNA double helix, continues to this day. Indeed, we are still too close to this breakthrough to properly understand its true importance and its long-term ramifications. The discipline of molecular biology, which grew up following this discovery, delivered solutions to the most profound problem of twentieth-century biology—how does the genetic constitution of a cell and organism determine its appearance and function?

(A)

(B)

Figure 1.1 Darwin and Mendel (A) Charles Darwin's 1859 publication of *On the Origin of Species by Means of Natural Selection* exerted a profound effect on thinking about the origin of life, the evolution of organismic complexity, and the relatedness of species. (B) Darwin's theory of evolution lacked a genetic rationale until the work of Gregor Mendel. The synthesis of Darwinian evolution and Mendelian genetics is the foundation for much of modern biological thinking. (A, from the Grace K. Babson Collection of the Works of Sir Isaac Newton on permanent deposit at the Dibner Institute and Burndy Library, Cambridge, MA; B, courtesy of the Mendelianum Museum Moraviae, Brno, Czech Republic.)

Without this molecular foundation, modern cancer research, like many other biological disciplines, would have remained a descriptive science that cataloged diverse biological phenomena without being able to explain the mechanics of how they occur. Today, our understanding of how cancers arise is being continually enriched by discoveries in diverse fields of biological research, most of which draw on the sciences of molecular biology and genetics. Perhaps unexpectedly, many of our insights into the origins of malignant disease are not coming from the laboratory benches of cancer researchers. Instead, the study of diverse organisms, ranging from yeast to worms to flies, provides us with much of the intellectual capital that fuels the forward thrust of the rapidly moving field of cancer research.

Those who fired up this biological revolution stood on the shoulders of nineteenth-century giants, specifically, Darwin and Mendel (Figure 1.1). Without the concepts established by these two, which influence all aspects of modern biological thinking, molecular biology and contemporary cancer research would be inconceivable. So, throughout this chapter, we frequently make reference to evolutionary processes as proposed by Charles Darwin and genetic systems as conceived by Gregor Mendel.

1.1 Mendel establishes the basic rules of genetics

Many of the basic rules of genetics that govern how genes are passed from one complex organism to the next were discovered in the nineteenth century and have come to us basically unchanged. Working in the 1860s, Gregor Mendel laid out these rules and described how they affect the appearance and behavior of individual organisms. Mendel's work focused largely on the genetics of pea plants. His results and conclusions were soon forgotten, only to be rediscovered independently by three researchers in 1900. During the decade that followed, it became clear that these rules—we now call them Mendelian genetics—apply to virtually all sexual organisms, including **metazoa** (multicellular animals), as well as **metaphyta** (multicellular plants).

Mendel's most fundamental insight came from his realization that genetic information is passed in particulate form from an organism to its offspring. This implied that the entire repertoire of genetic information carried by an organism, now called its "genome," is organized as a collection of discrete, separable information packets, which later came to be called *genes*. Only in recent years have we begun to know with any precision how many distinct genes are present in the genomes of mammals; many current analyses of the human genome—the best

	Seed shape	Seed color	Flower color	Flower position	Pod shape	Pod color	Plant height
One form of trait (dominant)	round	yellow	violet-red	axial	inflated	green	tall
A second form of trait (recessive)	wrinkled	green	white	terminal	pinched	yellow	short

studied of these—place the number in the range of 22,000, barely more than the 19,000 genes identified in the genome of the fruit fly, *Drosophila melanogaster*.

Mendel's work also implied that the constitution of an organism, including its physical and chemical makeup, could be divided into a series of discrete, separable entities. (Of course, this separability is implicit in the way we use language to describe anatomical forms and functions—distinct words to describe the nose and the arm and the stomach.) Mendel went further by showing that distinct anatomical parts are controlled by distinct genes. He found that the heritable material controlling the smoothness of peas behaved independently of the material governing plant height or flower color. In effect, each observable trait of an individual might be traceable to a separate gene that served as its blueprint. Thus, Mendel's research implied that the genetic constitution of an organism (its **genotype**) could be divided into hundreds, perhaps thousands of discrete information packets; in parallel, its observable, outward appearance (its **phenotype**) could be subdivided into a large number of discrete physical or chemical traits (Figure 1.2).

Mendel's thinking launched a century-long research project among geneticists, who applied his principles to studying thousands of traits in a variety of experimental animals, including flies (*D. melanogaster*), worms (*Caenorhabditis elegans*), and mice (*Mus musculus*). In the mid-twentieth century, geneticists also began to apply Mendelian principles to study the genetic behavior of single-celled organisms, such as the bacterium *Escherichia coli* and baker's yeast, *Saccharomyces cerevisiae*. The principle of genotype governing phenotype was directly transferable to these simpler organisms and their genetic systems.

While Mendelian genetics represents the foundation of contemporary genetics, it has been adapted and extended in myriad ways since its embodiments of 1865 and 1900. For example, the fact that single-celled organisms often reproduce asexually, that is, without mating, created the need for adaptations of Mendel's original rules. Moreover, the notion that each attribute of an organism could be traced to instructions carried in a single gene was realized to be simplistic. The

Figure 1.2 Genetic information is organized in discrete parcels One of Gregor Mendel's fundamental insights was that the genetic content of an organism consists of discrete parcels of information, each responsible for a distinct observable trait of the organism. Shown here are seven of the traits of the pea plant whose behavior Mendel studied through breeding experiments. In each case, a given trait could have two alternative phenotypic manifestations, which we now know to be specified by the alternative versions of genes that we call alleles. When the two alternative alleles coexisted within a single plant, one of these traits (which we call "dominant") was always observed while the other (the "recessive" trait) was never observed. (Courtesy of J. Postlethwait.)

great majority of observable traits of an organism are traceable to the cooperative interactions of a number of genes. Conversely, almost all the genes carried in the genome of a complex organism play roles in the development and maintenance of multiple organs, tissues, and physiologic processes.

Mendelian genetics revealed for the first time that genetic information is carried redundantly in the genomes of complex plants and animals. Mendel deduced that there were two copies of a gene for flower color and two for pea shape. Today we know that this twofold redundancy applies to the entire genome with the exception of the genes carried in the sex chromosomes. Hence, the genomes of higher organisms are termed **diploid**.

Mendel's observations also indicated that the two copies of a gene could convey different, possibly conflicting information. Thus, one gene copy might specify rough-surfaced and the other smooth-surfaced peas. In the twentieth century, these different versions of a gene came to be called **alleles**. An organism may carry two identical alleles of a gene, in which case, with respect to this gene, it is said to be **homozygous**. Conversely, the presence of two different alleles of a gene in an organism's genome renders this organism **heterozygous** with respect to this gene.

Because the two alleles of a gene may carry conflicting instructions, our views of how genotype determines phenotype become more complicated. Mendel found that in many instances, the voice of one allele may dominate over that of the other in deciding the ultimate appearance of a trait. For example, a pea genome may be heterozygous for the gene that determines the shape of peas, carrying one smooth and one rough allele. However, the pea plant carrying this pair of alleles will invariably produce smooth peas. This indicates that the smooth allele is **dominant**, and that it will invariably overrule its **recessive** counterpart allele (rough) in determining phenotype (see Figure 1.2). (Strictly speaking, using proper genetic parlance, we would say that one phenotype encoded by a gene is dominant with respect to another, the latter being recessive.)

Twentieth-century genetic research revealed an additional subtlety in the ways that the alleles of a gene interact with one another. Often the "voices" of two conflicting alleles blend in determining outcome—they seek a compromise halfway between their disparate messages. For example, a coexisting pair of red and white alleles in the gene determining flower color in snapdragons will yield a pink phenotype (Figure 1.3). This represents an example of **incomplete dominance**. In other situations, the voices of both alleles can be clearly heard—the phenomenon of **co-dominance**. Thus, individuals who are heterozygous at the genetic locus determining blood type (e.g., A and B alleles) will have red cells that express both the A and the B antigens.

1.2 Mendelian genetics helps to explain Darwinian evolution

In the early twentieth century, it was not apparent how the distinct allelic versions of a gene arise. At first, this variability in information content seemed to have been present in the collective gene pool of a species from its earliest evolutionary beginnings. This perception changed only later, beginning in the 1920s and 1930s, when it became apparent that genetic information is corruptible; the information content in genetic texts, like that in all texts, can be altered. **Mutations** were found to be responsible for changing the information content of a gene, thereby converting one allele into another or creating a new allele from one previously widespread within a species. An allele that is present in the great majority of individuals within a species is usually termed **wild type**, the term implying that that such an allele, being naturally present in large numbers of apparently healthy organisms, is compatible with normal structure and function.

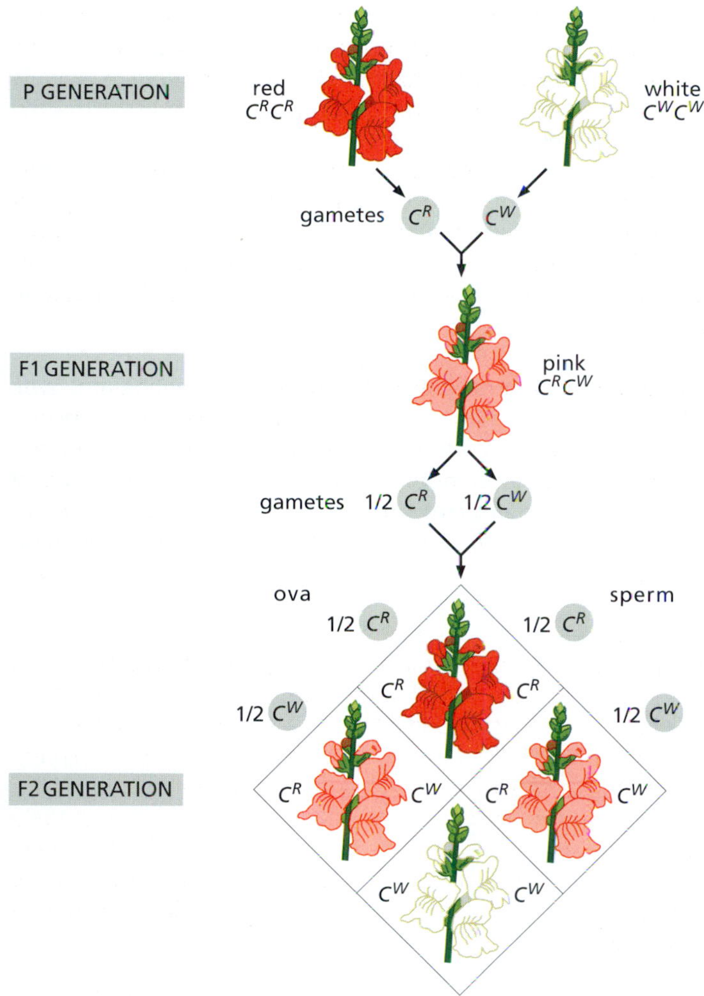

Figure 1.3 Incompletely dominant alleles While Mendel studied a series of genes that carry only dominant and recessive alleles, many genes function in a more complex fashion, so that both alleles contribute to phenotype—the phenomenon of incomplete dominance. As indicated here, when snapdragons having red and white flowers are bred with one another, the initial offspring in the F1 generation develop pink flowers. These pink snapdragons are genetically heterozygous, and when, as indicated here, they are bred with one another, homozygotes will emerge in the second generation that, once again, bear either red or white flowers. (Adapted from N. Campbell, Biology, 4th ed. San Francisco, CA: Benjamin Cummings, 1996.)

Mutations alter genomes continually throughout the evolutionary life span of a species, which usually extends over millions of years. They strike the genome and its constituent genes randomly. Mutations provide a species with a method for continually tinkering with its genome, for trying out new versions of genes that offer the prospect of novel, possibly improved phenotypes. The result of the continuing rain of mutations on the genome is a progressive increase during the evolutionary history of a species in the genetic diversity of its members. Thus, the collection of alleles present in the genomes of all members of a species—the **gene pool** of this species—becomes progressively more heterogeneous as the species grows older.

This means that older species carry more distinct alleles in their genomes than younger ones. Humans, belonging to a relatively young species (<150,000 years old), have one-half as many alleles and genetic diversity as chimpanzees, which therefore have been around as a species twice as long as we have.

The continuing diversification of alleles in a species' genome, occurring over millions of years, is countered to some extent by the forces of natural selection that Charles Darwin first described. Some alleles of a gene may confer more advantageous phenotypes than others, and so individual organisms carrying these advantageous alleles have a greater probability of leaving numerous descendants than do other members of the same species that happen not to carry these alleles. Consequently, natural selection results in a continual discarding of many of the alleles that have been generated by random mutations. In the long run, all things being equal, disadvantageous alleles are lost from the pool of alleles carried by the members of a species, advantageous alleles increase in number, and the overall fitness of the species improves incrementally.

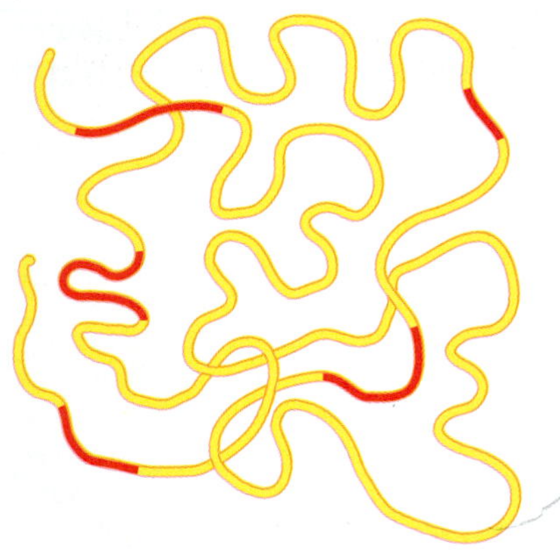

Figure 1.4 Biologically important sequences in the human genome The human genome can be characterized as a collection of relatively small islands of biologically important sequences (~3.5% of the total genome; *red*) floating amid a sea of "junk" DNA (*yellow*). The proportion of sequences carrying biological information has been greatly exaggerated for the sake of illustration.

Now, a century after Mendel was rediscovered and Mendelian genetics revived, we have come to realize that the great bulk of the genetic information in our own genome—indeed, in the genomes of all mammals—does not seem to specify phenotype and is often not associated with specific genes. Reflecting the discovery in 1944 that genetic information is encoded in DNA molecules, these "non-coding" stretches in the genome are often called **junk DNA** (Figure 1.4). Only about 1.5% of a mammal's genomic DNA carries sequence information that encodes the structures of proteins. Recent sequence comparisons of human, mouse, and dog genomes suggest that another ~2% encodes important information regulating gene expression and mediating other, still-poorly understood functions.

Because mutations act randomly on a genome, altering true genes and junk DNA indiscriminately, the great majority of mutations alter genetic information—DNA sequences—that have no effect on cellular or organismic phenotype. These mutations remain silent phenotypically and are said, from the point of view of natural selection, to be **neutral mutations**, being neither advantageous nor disadvantageous (Figure 1.5). Since the alleles created by these muta-

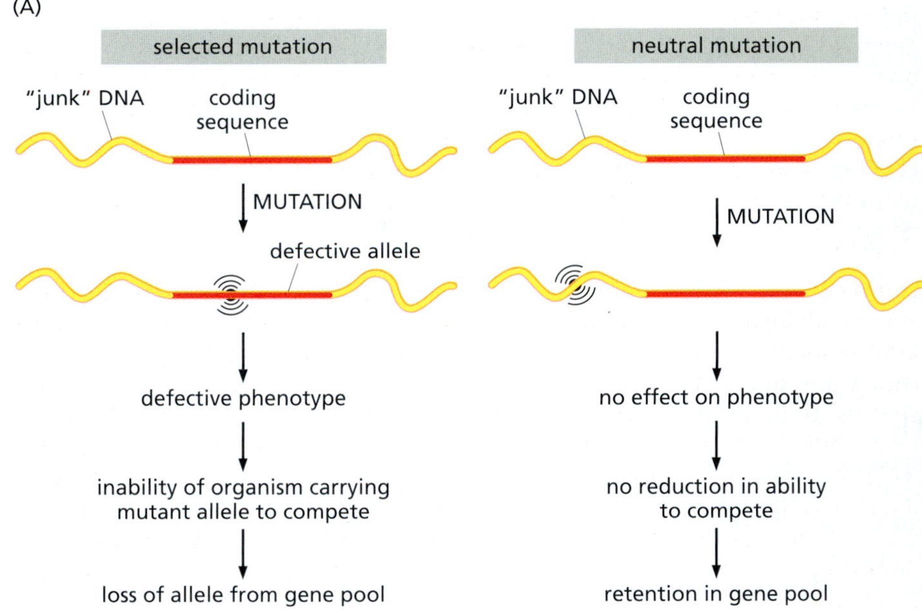

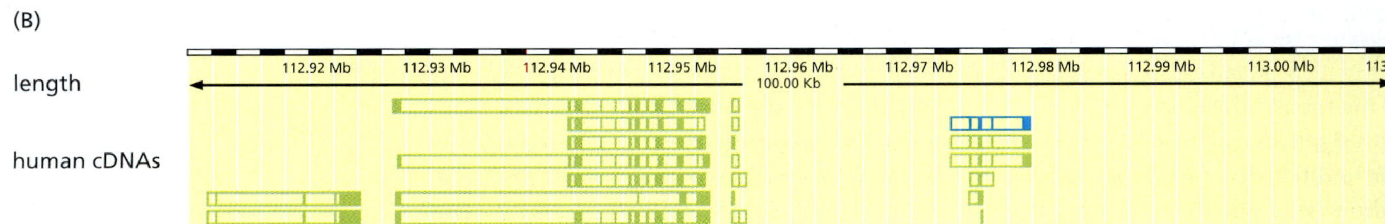

Figure 1.5 Neutral mutations and evolution (A) The coding sequences (*red*) of most genes have already been optimized over extended evolutionary time periods. Hence, many mutations affecting amino acid sequence and thus protein structure (*left*) create alleles that compromise the organism's ability to survive. For this reason, these mutant alleles are likely to be eliminated from the species' gene pool. In contrast, mutations striking "junk" DNA (*yellow*) will have no effect on phenotype and are therefore often preserved in the species' gene pool. This explains why, over extended periods of evolutionary time, coding DNA sequences change slowly, while noncoding DNA sequences change far more rapidly. (B) Depicted is a physical map of a randomly chosen 0.1-megabase segment of human Chromosome 1 from base pair 112,912,286 to base pair 113,012,285. Four distinct genes have been found here. Each is split up into a small number of islands (*small solid rectangles*) that are known or likely to specify segments of mRNA molecules (i.e., exons) and large stretches of intervening sequences (i.e., introns) that do not appear to specify biological information (see Figure 1.17). As is apparent, large stretches of DNA sequence between genes have not been associated with any particular biological function. (B, courtesy of The Wellcome Trust Sanger Institute.)

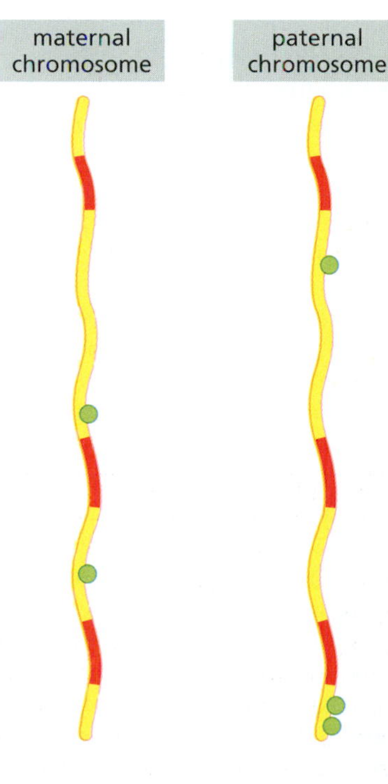

Figure 1.6 Polymorphic diversity in the human gene pool Because the great majority of human genomic DNA does not encode biologically important information (*yellow*), it has evolved relatively rapidly and has accumulated many subtle differences in sequences—polymorphisms—that are phenotypically silent (see Figure 1.5). Such polymorphisms are transmitted like Mendelian alleles, but their presence in a genome can be ascertained only by molecular techniques such as DNA sequencing. The dots (*green*) indicate where the sequence on this chromosome differs from the sequence that is most common in the human gene pool. For example, the prevalent sequence in one stretch of chromosomal DNA may be TAACTGG, while the variant sequence TTACTGG may be carried by a minority of humans and constitute a polymorphism. The presence of a polymorphism in one chromosome but not the other represents a region of heterozygosity, even though a nearby gene (*red*) may be present in identical allelic version on both chromosomes and therefore be in a homozygous configuration.

tions are silent, their existence could not be discerned by early geneticists whose work depended on gauging phenotypes. However, with the advent of DNA sequencing techniques, it became apparent that hundreds of thousands, even a million functionally silent mutations can be found scattered throughout the genomes of organisms such as humans. The genome of each human carries its own unique array of these functionally silent genetic alterations. The term *polymorphism* was originally used to describe variations in shape and form that distinguish normal individuals within a species from each other. These days, geneticists use the term **genetic polymorphisms** to describe the inter-individual, functionally silent differences in DNA sequence that make each human genome unique (Figure 1.6).

During the course of evolution, the approximately 3.5% of the genome that does encode biological function behaves much differently from the junk DNA. Junk DNA sequences suffer mutations that have no effect on the viability of an organism. Consequently, countless mutations in the noncoding sequences of a species' genome survive in its gene pool and accumulate progressively during its evolutionary history. In contrast, mutations affecting the coding sequences usually lead to loss of function and, as a consequence, loss of organismic viability; hence, these mutations are weeded out of the gene pool by the hand of natural selection, explaining why genetic sequences that do specify biological phenotypes generally change very slowly over long evolutionary time periods (see Sidebar 1.1).

1.3 Mendelian genetics governs how both genes and chromosomes behave

In the first decade of the twentieth century, Mendel's rules of genetics were found to have a striking parallel in the behavior of the chromosomes that were then being visualized under the light microscope. Both Mendel's genes and the chromosomes were found to be present in pairs. Soon it became clear that an identical set of chromosomes is present in almost all the cells of a complex organism. This chromosomal array, often termed the **karyotype**, was found to be duplicated each time a cell went through a cycle of growth and division.

The parallels between the behaviors of genes and chromosomes led to the speculation, soon validated in hundreds of different ways, that the mysterious information packets called genes were carried by the chromosomes. Each chromosome was realized to carry its own unique set of genes in a linear array. Today, we know that as many as several thousand genes may be arrayed along a mammalian chromosome. (Human Chromosome 1—the largest of the set—has been

Sidebar 1.1 Evolutionary forces dictate that certain genes are highly conserved Many genes encode cellular traits that are essential for the continued viability of the cell. These genes, like all others in the genome, are susceptible to the ever-tinkering hand of mutation, which is continually creating new gene sequences by mutating existing ones. Natural selection tests these novel sequences and determines whether they specify phenotypes that are more advantageous than the preexisting ones. Almost invariably, the sequences in genes required for cell and therefore organismic viability were already optimized hundreds of millions of years ago. Consequently, almost all subsequently occurring changes in the sequence information of these genes would have been deleterious and would have compromised the viability of the cell and, in turn, the organism. These mutant alleles were soon lost, because the mutant organisms carrying them failed to leave descendants. This dynamic explains why the sequences of many genes have been highly conserved over vast evolutionary time periods. Stated more accurately, the structures of their encoded proteins have been highly conserved.

In fact, the great majority of the proteins that are present in our own cells and are required for cell viability were first developed during the evolution of single-cell **eukaryotes** (nucleated cells). This is indicated by numerous observations showing that many of our proteins have clearly recognizable counterparts in single-cell eukaryotes, such as baker's yeast. Another large repertoire of highly conserved genes and proteins is traceable to the appearance of the first multicellular animals (metazoa); these genes enabled the development of distinct organs and of organismic physiology. Hence, another large group of our own genes and proteins is present in counterpart form in worms and flies (Figure 1.7).

By the time the ancestor of all mammals first appeared more than 150 million years ago, virtually all the biochemical and molecular features present in contemporary mammals had already developed. The fact that they have changed little in the intervening time points to their optimization long before the appearance of the various mammalian orders. This explains why the embryogenesis, physiology, and biochemistry of all mammals is very similar, indeed, so similar that lessons learned through the study of laboratory mice are almost always transferable to an understanding of human biology.

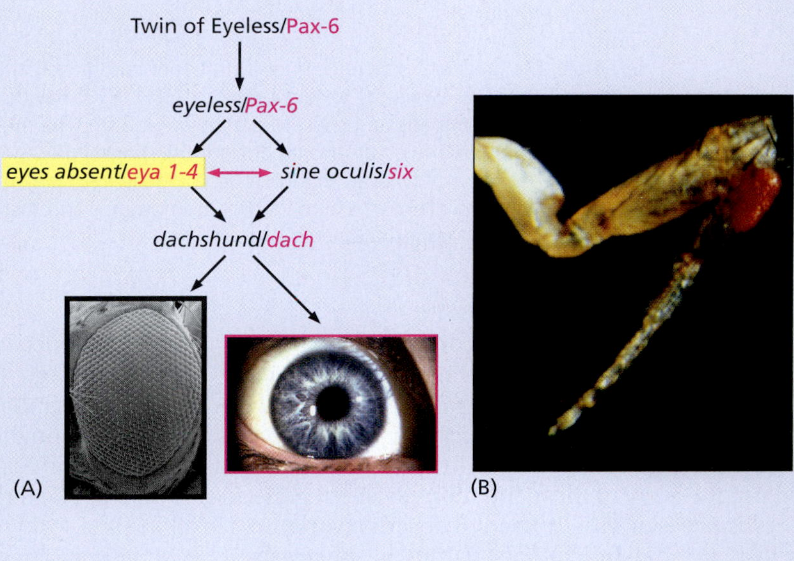


```
human          FGLARAFGIPIRVYTHEVVTLWYRSPEVLLGS
S. pombe       FGLARSFGVPLRNYTHEIVTLWYRAPEVLLGS
S. cerevisiae  FGLARAFGVPLRAYTHEIVTLWYRAPEVLLGG

human          ARYSTPVDIWSIGTIFAELATKLPLFHGDSEI
S. pombe       RHYSTGVDIWSVGCIFAENIRRSPLFPGDSEI
S. cerevisiae  KQYSTGVDTWSIGCIFAEHCNRLPIFSGDSEI

human          DQLFRIPRALGTPNNEVWPEVESLQDYKNTFP
S. pombe       DEIFKIPQVLGTPNEEVWPGVTLLQDYKSTFP
S. cerevisiae  DQIFKIPRVLGTPNEAIWPDIVYLPDFKPSFP
```
(C)

Figure 1.7 Extraordinary conservation of gene function The last common ancestor of flies and mammals lived more than 600 million years ago. Moreover, fly (i.e., arthropod) eyes and mammalian eyes show totally different architectures. Nevertheless, the genes that orchestrate eye development (*eyeless* in the fly, *Pax-6/small eye* in the mouse) are interchangeable. (A) Thus, the components of the signal transduction cascades that operate downstream of these master regulators to trigger eye development (*black* for flies, *pink* for mice) are also highly conserved and interchangeable. (B) The expression of the mouse *Pax-6/small eye* gene, like the *Drosophila eyeless* gene, in an inappropriate (*ectopic*) location in a fly embryo results in the fly developing an eye on its leg, demonstrating the interchangeability of the two genes. (C) The conservation of genetic function over vast evolutionary distances is often manifested in the amino acid sequences of homologous proteins. Here, the amino acid sequence of a human protein is given together with the sequences of the corresponding proteins from two yeast species, *S. pombe* and *S. cerevisiae*. (A, courtesy of I. Rebay; B, courtesy of Walter Gehring; C, from B. Alberts et al., Essential Cell Biology, 2nd ed. New York: Garland Science, 2004.)

estimated at one stage of human genome sequencing to hold 3041 distinct genes.) Indeed, the length of a chromosome, as viewed under the microscope, is roughly proportional to the number of genes that it carries.

Each gene was found to be localized to a specific site along the length of a specific chromosome. This site is often termed a genetic **locus**. Much effort was expended by geneticists throughout the twentieth century to map the sites of genes—genetic loci—along the chromosomes of a species (Figure 1.8).

The diploid genetic state that reigns in most cells throughout the body was found to be violated in one important class of cells—the *germ cells*, sperm and egg. These cells carry only a single copy of each chromosome and gene and thus are said to be **haploid**. During the formation of germ cells in the testes and ovaries, each pair of chromosomes is separated and one of the two chromosomes (and thus associated genes) is chosen at random for incorporation into the sperm or egg. When sperm and egg combine subsequently during the process of fertilization, the two haploid genomes fuse to yield the new diploid genome of the fertilized egg. All cells in the organism descend directly from this diploid cell and, if all goes well, inherit precise replicas of its diploid genome. In a large multicellular organism like the human, this means that a complete copy of the genome is present in almost all of the approximately 3×10^{13} cells throughout the body!

With the realization that genes reside in chromosomes, and that a complete set of chromosomes is present in almost all cell types in the body, came yet another conclusion that was rarely noted: genes act to create the phenotypes of an organism through their ability to act locally by influencing the behavior of its individual cells. The alternative—that a single set of genes residing at some unique anatomical site in the organism controls the entire organism's development and physiology—was now discredited.

The rule of paired, similarly appearing chromosomes was found to be violated by some of the sex chromosomes. In the cells of female placental mammals (in which the embryo is sustained through much of its development by the placenta), there are two similarly appearing X chromosomes, and these behave like the other chromosomes in the female cells—the nonsex chromosomes, often called **autosomes**. But in males, an X chromosome is paired with a Y chromosome, which is smaller and carries a much smaller repertoire of genes. In humans, the X chromosome is thought to carry about 900 genes, compared with the 78 distinct genes on the Y chromosome, which, because of redundancy, specify only 27 distinct proteins (Figure 1.9).

This asymmetry in the configuration of the sex chromosomes puts males at a biological disadvantage. Many of the 900 or so genes on the X chromosome are vital to normal organismic development and function. The twofold redundancy created by the paired X chromosomes guarantees more robust biology. If a gene

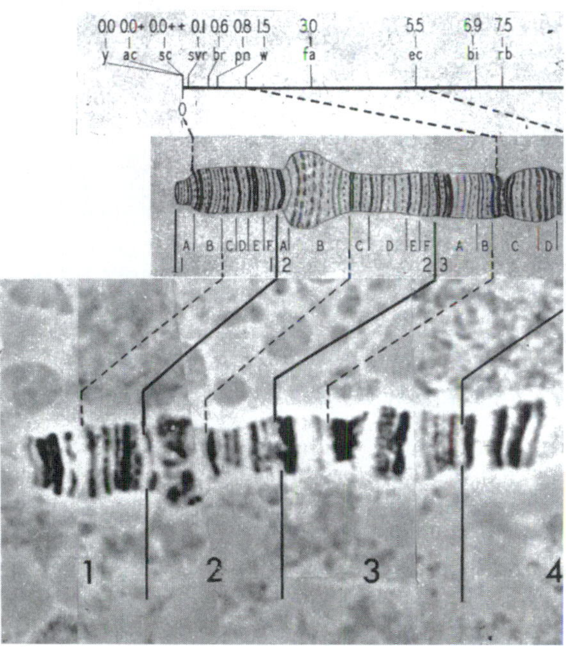

(A)

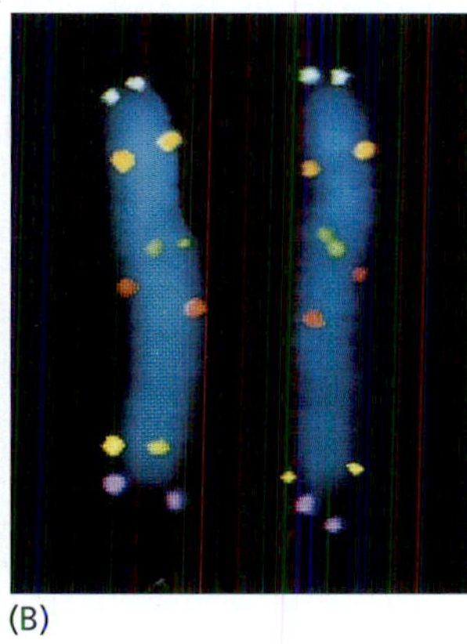
(B)

Figure 1.8 Localization of genes along chromosomes (A) The physical structure of the *Drosophila* chromosomes could be mapped by taking advantage of the fly's salivary gland chromosomes, which exhibit banding patterns resulting from alternating light (sparse) and dark (condensed) chromosomal regions (*bottom*). Independently, genetic crosses yielded the relative chromosomal locations of hundreds of distinct genetic loci along linear genetic maps (*top*). The locations of these loci were then aligned with physical banding maps, like the one shown here for the beginning of the left arm of *Drosophila* Chromosome 1. (B) The availability of DNA probes that hybridize specifically to various genes now makes it possible to localize genes along a chromosome by tagging each probe with a specific fluorescent dye or combination of dyes. Here, using this procedure of fluorescence *in situ* hybridization (FISH) on metaphase chromosomes, six distinct genes have been localized to various sites along human Chromosome 5. (There are two dots for each gene because chromosomes are present in duplicate form during metaphase of mitosis.) (A, courtesy of M. Singer, and P. Berg, Genes and Genomes, Mill Valley, CA: University Science Books, 1991, as taken from C.B. Bridges, *J. Heredity* 26:60, 1935; B, courtesy of David C. Ward.)

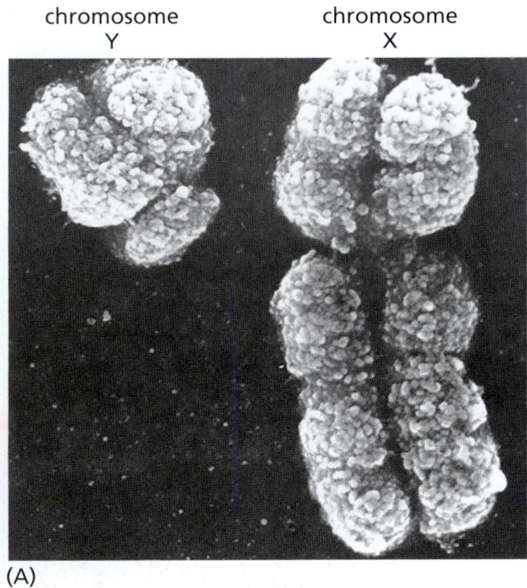

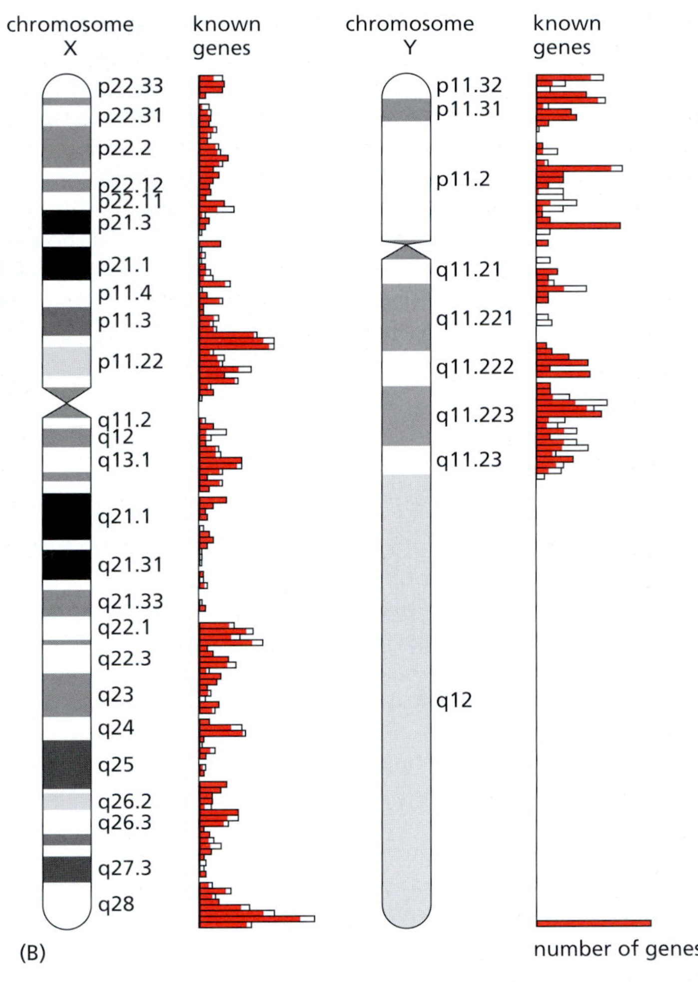

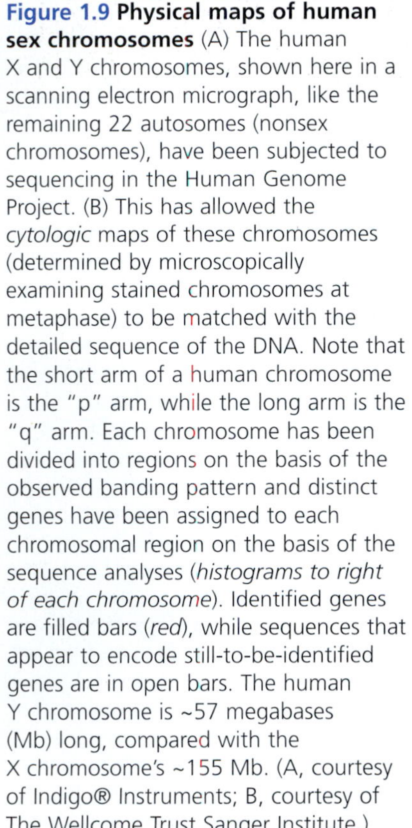
Figure 1.9 Physical maps of human sex chromosomes (A) The human X and Y chromosomes, shown here in a scanning electron micrograph, like the remaining 22 autosomes (nonsex chromosomes), have been subjected to sequencing in the Human Genome Project. (B) This has allowed the *cytologic* maps of these chromosomes (determined by microscopically examining stained chromosomes at metaphase) to be matched with the detailed sequence of the DNA. Note that the short arm of a human chromosome is the "p" arm, while the long arm is the "q" arm. Each chromosome has been divided into regions on the basis of the observed banding pattern and distinct genes have been assigned to each chromosomal region on the basis of the sequence analyses (*histograms to right of each chromosome*). Identified genes are filled bars (*red*), while sequences that appear to encode still-to-be-identified genes are in open bars. The human Y chromosome is ~57 megabases (Mb) long, compared with the X chromosome's ~155 Mb. (A, courtesy of Indigo® Instruments; B, courtesy of The Wellcome Trust Sanger Institute.)

copy on one of the X chromosomes is defective (i.e., a nonfunctional mutant allele), chances are that the second copy of the gene on the other X chromosome can continue to carry out the task of the gene, ensuring normal biological function. Males lack this genetic fail-safe system in their sex chromosomes. One of the more benign consequences of this is color blindness, which strikes males frequently and females infrequently, due to the localization of the genes encoding the color-sensing proteins of the retina on the X chromosome. However, this disparity between the genders is mitigated somewhat by nature (Sidebar 1.2).

Color blindness reveals the virtues of having two redundant gene copies around to ensure that biological function is maintained. If one copy is lost through mutational inactivation, the surviving gene copy is often capable of specifying a wild-type phenotype. Such functional redundancy operates for the great majority of genes carried by the autosomes. As we will see later, this dynamic plays an important role in cancer development, since virtually all of the genes that operate to prevent runaway proliferation of cells are present in two redundant copies in our cells, both of which must be lost before their growth-suppressing functions are lost and malignant cell proliferation occurs.

1.4 Chromosomes are altered in most types of cancer cells

Individual genes are far too small to be seen with a light microscope, and subtle mutations within a gene are smaller still. Consequently, the great majority of the mutations that play a part in cancer cannot be visualized through microscopy. However, the examination of chromosomes through the light microscope can

Sidebar 1.2 Female cells can access information only on single X chromosomes In the simplest depiction of sex chromosome behavior, each male cell relies on the information carried by its single X chromosome, while each female cell is able to consult both of its X chromosomes. Given the large number of genes present on the X chromosome, this disparity would create substantial physiologic differences between male and female cells, since female cells would, in general, express twice as much of each of the products specified by the genes on their X chromosomes. This problem is solved by the mechanism of X-inactivation. Early in embryogenesis, one of the two X chromosomes is randomly inactivated in each of the cells of a female embryo. This inactivation causes the functional silencing of almost all of the genes on this chromosome and the physical shrinkage of this chromosome into a small particle termed the **Barr body** (Figure 1.10). Subsequently, all descendants of a cell that chose to inactivate a particular X chromosome will inherit this pattern of chromosomal inactivation and will therefore continue to carry the same inactivated X chromosome. Accordingly, the female advantage of carrying redundant copies of X chromosome–associated genes is only a partial one.

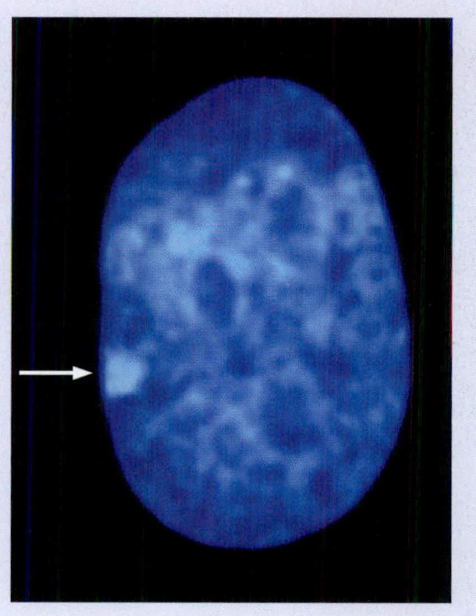

Figure 1.10 X chromosomes as Barr bodies One of the two X chromosomes in each cell is randomly inactivated in the cells of early female embryo and remains inactivated in all linear descendants. It is visible in interphase nuclei, where it remains condensed and associated with the nuclear membrane (*white arrow*). (From B.P. Chadwick et al., *Semin. Cell Dev. Biol.* 14:359–367, 2003.)

give evidence of large-scale alterations of the cell genome. Indeed, such alterations were noted as early as 1914, specifically in cancer cells.

We now know that cancer cells often exhibit aberrantly structured chromosomes of various sorts, the loss of entire chromosomes, the presence of extra copies of others, and the fusion of the arm of one chromosome with part of another. These changes in overall chromosomal configuration expand our conception of how mutations can affect the genome: since alterations of overall chromosomal structure and number also constitute types of genetic change, these changes must be considered to be the consequences of mutations (see Sidebar 1.3). And importantly, the aberrant chromosomes seen initially in cancer cells provided the first clue that these cells might be genetically aberrant, that is, that they were mutants.

The normal configuration of chromosomes is often termed the **euploid** karyotypic state. Euploidy implies that each of the autosomes is present in normally structured pairs and that the X and Y chromosomes are present in the numbers appropriate for the sex of the individual carrying them. Deviation from the euploid karyotype—the state termed **aneuploidy**—is seen, as mentioned above, in many cancer cells. Often this aneuploidy is merely a consequence of the general chaos that reigns within a cancer cell. However, this connection between aneuploidy and malignant cell proliferation also hints at a theme that we will return to repeatedly in this book: the acquisition of extra copies of one chromosome or the loss of another can create a genetic configuration that somehow benefits the cancer cell and its agenda of runaway proliferation.

1.5 Mutations causing cancer occur in both the germ line and the soma

Because mutations alter the information content of genes, mutant alleles of a gene can be passed from parent to offspring. This transmission from one generation to the next, made possible by the germ cells (sperm and egg), is said to occur via the **germ line**. Importantly, the germ-line transmission of a recently

Sidebar 1.3 Cancer cells are often aneuploid The presence of abnormally structured chromosomes and changes in chromosome number provided the first clue, early in the twentieth century, that changes in cell genotype often accompany and perhaps cause the uncontrolled proliferation of malignant cells. These deviations from the normal euploid karyotype can be placed into a number of categories. Chromosomes that seem to be structurally normal may accumulate in extra copies, leading to three, four, or even more copies of these chromosomes per cancer cell nucleus (Figure 1.11); such deviations from normal chromosome number are manifestations of aneuploidy.

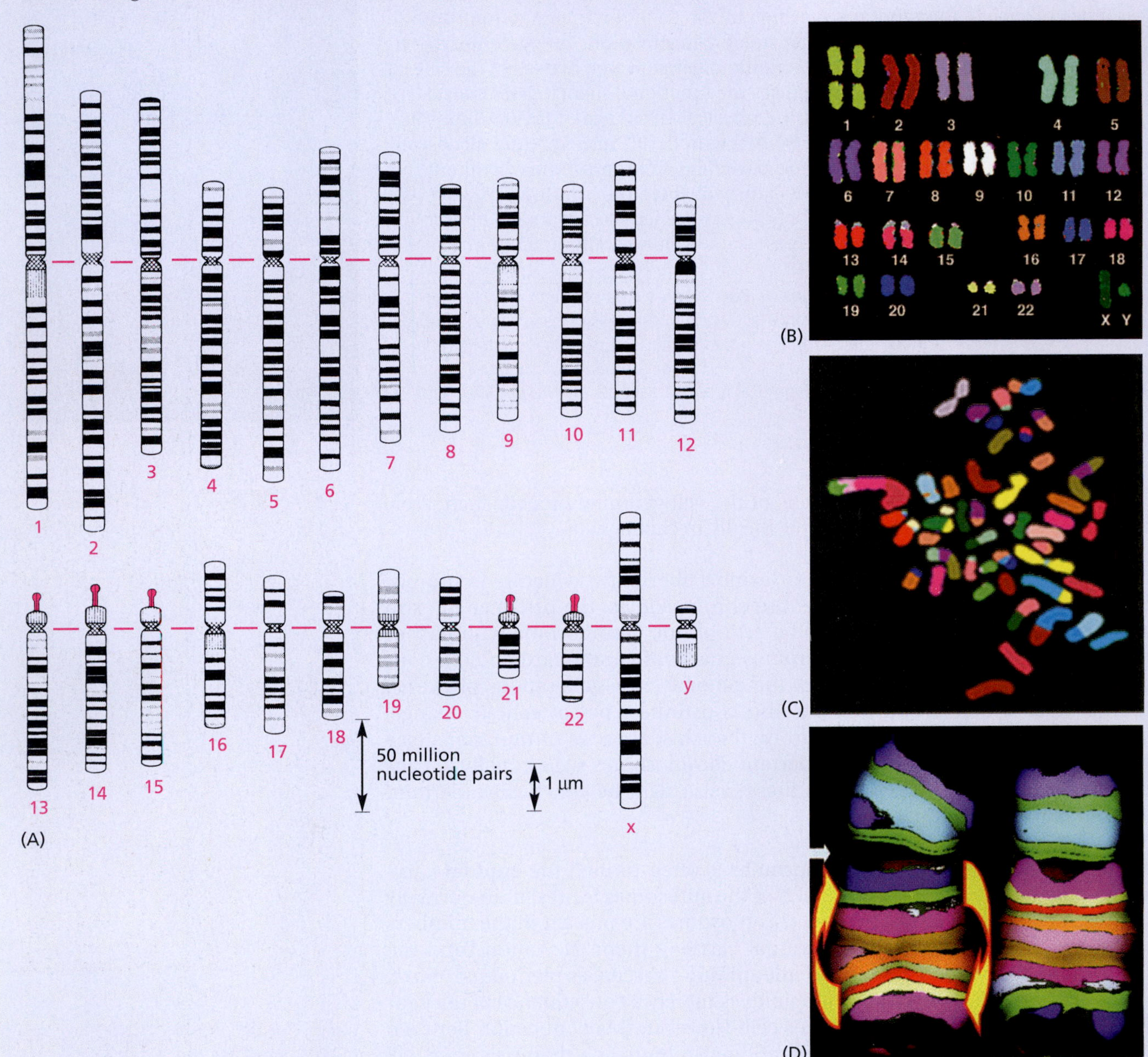

Figure 1.11 Normal and abnormal chromosomal complements (A) Staining of metaphase chromosomes reveals a characteristic light and dark banding pattern for each chromosome. The full array of human chromosomes is indicated here schematically. (B) The techniques of spectral karyotype (SKY) analysis and M-band fluorescence *in situ* hybridization (mFISH) allow an experimenter to "paint" each individual metaphase human chromosome with a distinct color (by hybridizing chromosome-specific DNA probes labeled with various fluorescing dyes to the chromosomes). The actual colors in images such as these are generated by computer. The diploid karyotype of a normal human cell is presented. (C) The aneuploid karyotype of a human breast cancer cell, in which some chromosomes are present in inappropriate numbers and in which translocations (exchanges of segments between chromosomes) are apparent. (D) Here, mFISH was used to label intrachromosomal subregions with specific fluorescent dyes, revealing that a large portion of an arm of normal human Chromosome 5 (*right*) has been inverted (*left*); these were in a cell of a worker who had been exposed to plutonium in the nuclear weapons industry of the former Soviet Union. (A, adapted from U. Francke, *Cytogenet. Cell Genet.* 31:24–32, 1981; B, from E. Schröck et al., *Science* 273:494–497, 1996. © AAAS; C, courtesy of J.M. Davidson and P.A.W. Edwards; D, from M.P. Hande et al, *Am. J. Hum. Genet.* 72:1162–1170, 2003.)

Alternatively, chromosomes may undergo changes in their structure. A segment may be broken off one chromosomal arm and become fused to the arm of another chromosome, resulting in a chromosomal **translocation** (Figure 1.11C). Alternatively, chromosomal segments may be exchanged between chromosomes from different chromosome pairs, resulting in **reciprocal translocations**. A chromosomal segment may also become inverted, which may affect the regulation of genes that are located near the breakage-and-fusion points (Figure 1.11D).

A segment of a chromosome may be copied many times over, and the resulting extra copies may be fused head-to-tail in long arrays within a chromosomal segment that is termed an HSR (**homogeneously staining region**; Figure 1.12A). A segment may also be cleaved out of a chromosome, replicate as an autonomous, extrachromosomal entity, and increase to many copies per nucleus, resulting in the appearance of subchromosomal fragments termed DMs (**double minutes**; Figure 1.12B). These latter two changes cause increases in the copy number of genes carried in such segments, resulting in **gene amplification**. Sometimes, both types of amplification coexist in the same cell (Figure 1.12C). Gene amplification can favor the growth of cancer cells by increasing the copy number of growth-promoting genes.

On some occasions, certain growth-inhibiting genes may be discarded by cancer cells during their development. For example, when a segment in the middle of a chromosomal arm is discarded and the flanking chromosomal regions are joined, this results in an **interstitial deletion** (Figure 1.12D).

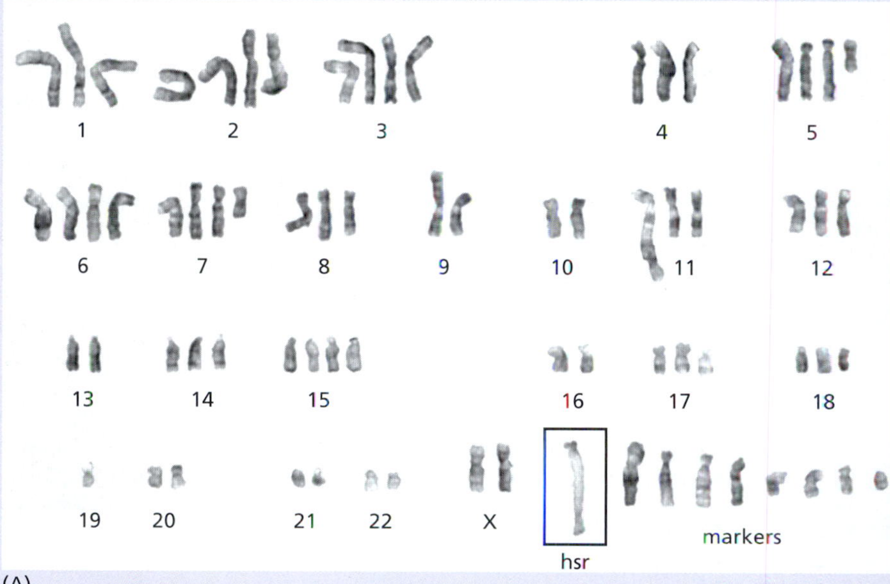

(A)

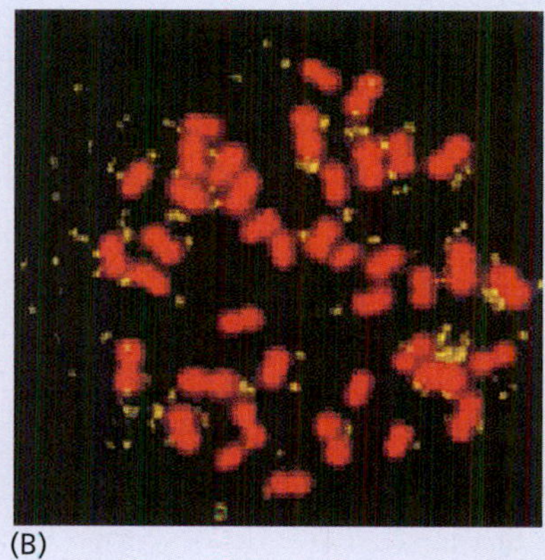
(B)

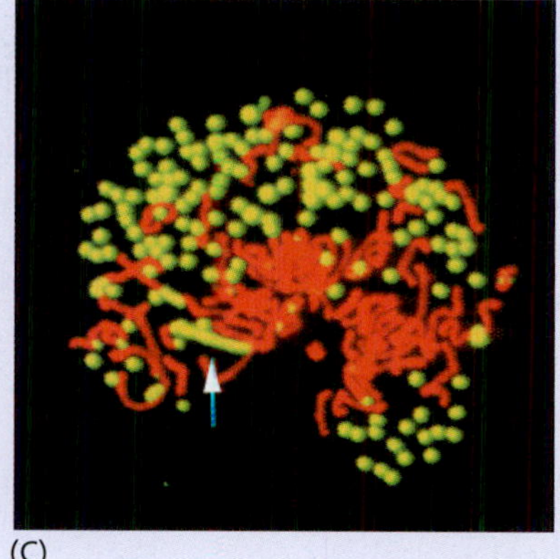
(C)

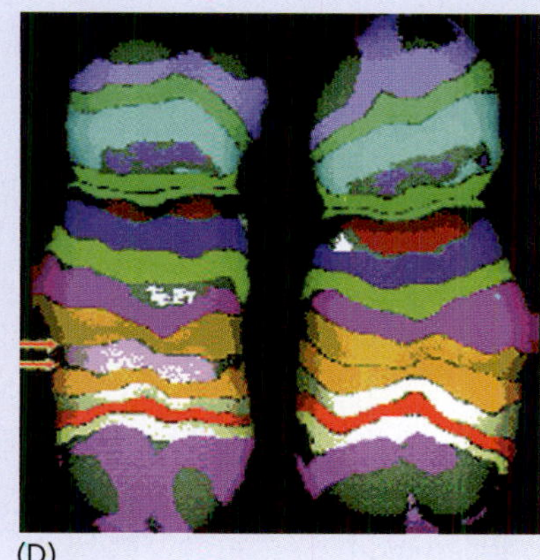
(D)

Figure 1.12 Increases and decreases in copy number of chromosomal segments (A) Giemsa stain may reveal changes in the normal banding patterns of chromosomal regions. A homogeneously staining region (HSR) occurs because of repeated rounds of reduplication of a chromosomal segment, yielding many tandem copies of this segment and a concomitant amplification (increase) in the copy number of its genes. Seen here is an HSR observed in a human hepatocellular carcinoma cell line. (B) Double minute chromosomes (DMs) derive from chromosomal segments that have broken loose from their original sites and have been replicated repeatedly as extrachromosomal elements; like the chromatids of normal chromosomes, these structures are doubled during the metaphase of mitosis. Shown are mouse breast cancer cells that exhibit amplified copies of the *HER2/Neu* oncogene borne on DMs (*yellow spots*) that are not associated with the chromosomes (*red*) in these cells. The DMs were detected using FISH with a probe reactive against this oncogene. (C) Occasionally, an amplified gene may be found both in a homogeneously staining region (within a chromosome) and in double minutes. Here, analysis of COLO320 cells, a line of human neuroendocrinal tumors cells, reveals multiple copies of the *myc* oncogene (*yellow*), which we will study in Chapter 8, amid the chromosomes (*red*). One HSR is indicated by the arrow, while many dozens of DMs are apparent. (D) The use of mFISH revealed that a segment within normal human Chromosome 5 (*right*) has been deleted (an interstitial deletion) following extensive exposure to radiation from plutonium (paired arrows, left). (A, from J-M. Wen et al., *Cancer Genet. Cytogenet.* 135:91–95, 2002; B, from C. Montagna et al., *Oncogene* 21:890–898, 2002; C, from N. Shimizu et al., *J. Cell Biol.* 140:1307–1320, 1998; D, from M.P. Hande et al., *Am. J. Hum. Genet.* 72:1162–1170, 2003.)

created mutant allele from one organism to its offspring can occur only if a precondition has been met: the responsible mutation must strike a gene carried in the genome of sperm or egg or in the genome of one of the cell types that are immediate precursors of the sperm or egg within the gonads. Mutations affecting the genomes of cells everywhere else in the body—which constitute the **soma**—may well affect the particular cells in which such mutations strike but will have no prospect of being transmitted to the offspring of an organism. Such **somatic mutations** cannot become incorporated into the vehicles of generation-to-generation genetic transmission—the chromosomes of sperm or eggs.

Somatic mutations are of central importance to the process of cancer formation. As described repeatedly throughout this book, a somatic mutation can affect the behavior of the cell in which it occurs and, through repeated rounds of cell growth and division, can be passed on to all descendant cells within the tissue. These direct descendants of a single progenitor cell, which may ultimately number in the millions or even billions, are said to constitute a cell **clone**, in that all members of this group of cells trace their ancestry directly back to the single cell in which the mutation originally occurred.

An elaborate cellular repair apparatus within each cell continuously monitors the cell's genome and, with great efficiency, removes and erases mutant sequences, replacing them with appropriate wild-type sequences (see Sidebar 1.4; in Chapter 12, we will examine this repair apparatus in depth). This apparatus

Sidebar 1.4 Reproductive cloning demonstrates the extraordinary efficiency of the repair apparatus The efficiency of the complex apparatus that repairs DNA (see Chapter 12), the resulting suppression of somatic mutations, and the consequent integrity of the genomes of somatic cells have all been illustrated dramatically by the successes of animal cloning in recent years. In the much-celebrated case of the sheep Dolly, cells were taken from the breast tissue of her "mother," the nuclei of these cells were removed, and these nuclei were then implanted in egg cells (oocytes) whose own nuclei had previously been eliminated. The resulting cells, containing donor nuclei and recipient egg cytoplasm, were then induced to proliferate and to form embryos (Figure 1.13). The fact that an essentially normal sheep (Dolly) was born many months later, having developed from one of these embryos, demonstrated that the genome in one of her mother's mammary gland cells carried no mutations that compromised normal embryologic development.

The implied intactness of the genome in the breast cells of Dolly's mother focuses attention on the mother's life, specifically, how she developed from an embryo. During this development, the genome of the fertilized egg that was destined to develop into Dolly's mother was copied and recopied dozens of times during the cycles of cell growth and division that intervened between the initial formation of the fertilized egg and the formation, many months later, of the mother's breast tissue. This recopying process, involving DNA replication and a repair apparatus that operated to weed out mutant DNA sequences, was highly effective in preserving an intact somatic cell genome that was capable of programming normal organismic development.

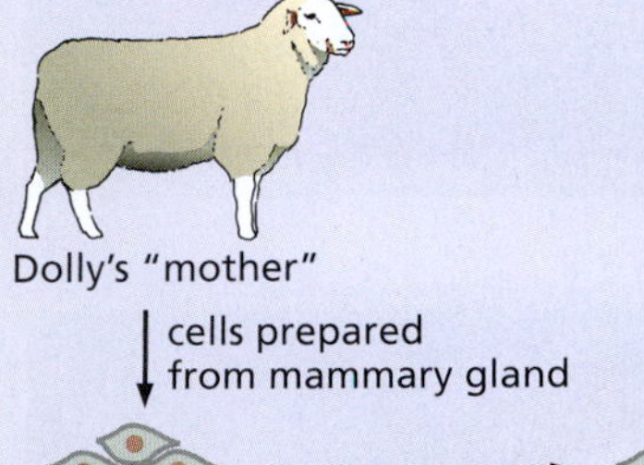

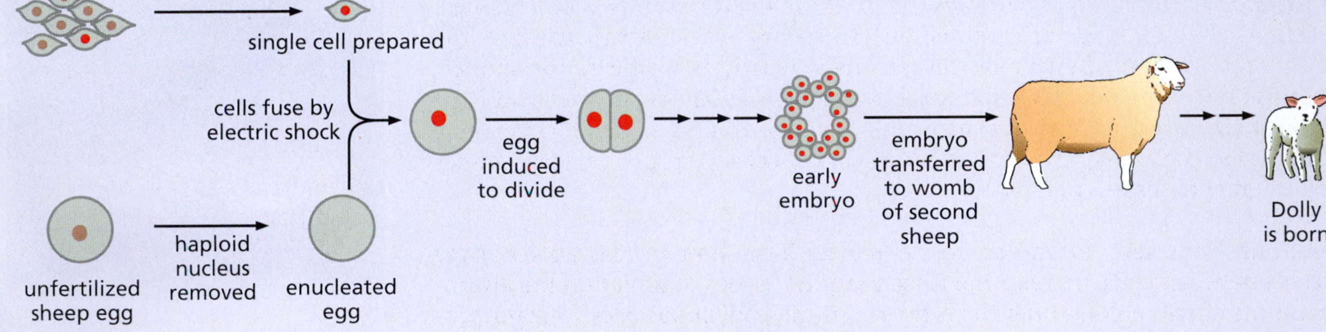

Figure 1.13 Cloning and genomic integrity In the reproductive cloning procedure, the nucleus of an unfertilized egg is removed and the nucleus from a somatic cell is introduced in its stead. The resulting diploid cell, formally equivalent to a fertilized egg, is then induced to divide, generating an embryo that can then be implanted in the womb of a foster mother and may develop into a newborn. Successful generation of a healthy newborn reveals that the donor somatic cell carried the genome of the species in an essentially intact, functional form.

maintains genomic integrity by minimizing the number of mutations that strike the genome and are then perpetuated by transmission to descendant cells. However, no system of damage detection and repair is infallible. Some mistakes in genetic sequence survive its scrutiny, become fixed in the cell genome, are copied into new DNA molecules, and are then passed on as mutations to progeny cells. In this sense, many of the mutations that accumulate in the genome represent the consequences of occasional oversights made by the repair apparatus. Yet others are the results of catastrophic damage to the genome that exceeds the capacities of the repair apparatus.

1.6 Genotype embodied in DNA sequences creates phenotype through proteins

The genes studied in Mendelian genetics are essentially mathematical abstractions. Mendelian genetics explains their transmission, but it sheds no light on how genes create cellular and organismic phenotypes. Phenotypic attributes can range from complex, genetically templated behavioral traits to the *morphology* (shape, form) of cells and subcellular organelles to the biochemistry of cell metabolism. This mystery of how genotype creates phenotype represented the major problem of twentieth-century biology. Indeed, attempts at forging a connection between these two became the obsession of many molecular biologists during the second half of the twentieth century and continue as such into the twenty-first, if only because we still possess an incomplete understanding of how genotype influences phenotype.

Molecular biology has provided the basic conceptual scaffold for understanding this connection. In 1944, DNA was proven to be the chemical entity in which the genetic information of cells is carried. Nine years later, Watson and Crick elucidated the double-helical structure of DNA. A dozen years after that, in 1965, it became clear that the sequences in the bases of the DNA double helix determine precisely the sequence of amino acids in proteins. The unique structure and function of each type of protein in the cell is determined by its sequence of amino acids. Therefore, the specification of amino acid sequence, which is accomplished by base sequences in the DNA, provides almost all the information that is required to construct a protein.

Once synthesized within cells, proteins proceed to create phenotype, doing so in a variety of ways. Proteins can assemble within the cell to create complex structures, which are often termed components of the cellular **cytoarchitecture**, or more specifically, the **cytoskeleton** (Figure 1.14A,B). When secreted into the space between cells, the complex structure is often termed the **extracellular matrix** (ECM); it ties cells together, enabling them to form complex tissues (Figure 1.14C,D). As we will see later, the structure of the ECM is often disturbed by malignant cancer cells, enabling them to migrate to sites within a tissue and organism that are normally forbidden to them.

Many proteins function as enzymes that catalyze the thousands of biochemical reactions that together are termed **intermediary metabolism**; without the active intervention of enzymes, few of these reactions would occur spontaneously. Proteins can also contract and create cellular movement (**motility**; Figure 1.15) as well as muscle contraction. Cellular motility also appears to play a role in cancer development by allowing cancer cells to spread through tissues and migrate to distant organs.

And most important, for the process of cancer formation, proteins can convey signals between cells, thereby enabling complex tissues to maintain the appropriate numbers of constituent cell types. Within individual cells, certain proteins receive signals from one source, process these signals, and pass them on to

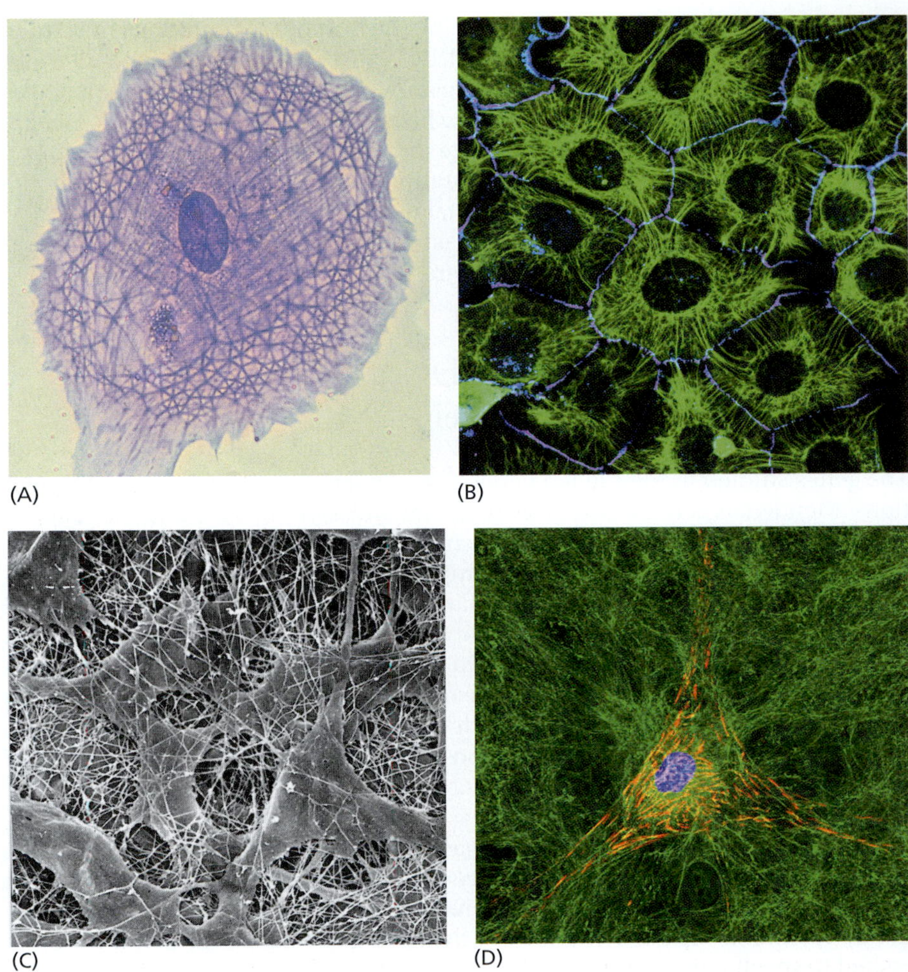

Figure 1.14 Intracellular and extracellular scaffolding (A) The cytoskeleton is assembled from complex networks of intermediate filaments, actin, and microtubules. Together, they generate the shape of a cell and enable its motion. (B) Here, an important intermediate filament of epithelial cells—keratin—is detected using an anti-keratin-specific antibody (*green*). The boundaries of cells are labeled with a second antibody that reacts with a plasma membrane protein (*blue*). (C) Cells secrete a diverse array of proteins, which are assembled into the extracellular matrix (ECM). A scanning electron micrograph reveals the complex meshwork of collagen fibers, glycoproteins, hyaluronan, and proteoglycans, in which fibroblasts (connective tissue cells) are embedded. (D) A cell of the NIH 3T3 cell line, which is used extensively in cancer cell biology, is shown here amid an ECM network of fibronectin fibers (*green*). The points of cellular attachment to the fibronectin are mediated by integrin receptors on the cell surface *(orange, yellow)*. (A, courtesy of Colin Smith; B, courtesy of Kathleen Green and Evangeline Amargo; C, from T. Nishida et al., *Invest. Ophthalmol. Vis. Sci.* 29:1887–1890, 1988. © Association for Research in Vision and Ophthalmology; D, from E. Cukierman et al., *Curr. Opin. Cell Biol.* 14:633–639, 2002.)

another protein within the cell; such signal-processing functions, often termed intracellular signal **transduction**, are also central to the creation of cancers, since many of the abnormal-growth phenotypes of cancer cells are the result of aberrantly functioning intracellular signal-transducing molecules.

The functional versatility of proteins makes it apparent that almost all aspects of cell and organismic phenotype can be created by the actions of proteins. Once we realize this, we can depict genotype and phenotype in the simplest of molecular terms: genotype resides in the sequences of bases in DNA, while phenotype derives from the actions of proteins. (In fact, this depiction is simplistic, because it ignores the important role of RNA molecules as intermediaries between DNA sequences and protein structure and the recently discovered abilities of certain RNA molecules to function as enzymes and as regulators of the expression of certain genes.)

In the complex eukaryotic cells of animals, as in the simpler **prokaryotic** cells of bacteria, DNA sequences are copied into RNA molecules in the process termed **transcription**; a gene that is being transcribed is said to be actively **expressed**, while a gene that is not being transcribed is often considered to be **repressed**. In the simplest version of transcription, the transcription of a gene yields an RNA molecule of length comparable to the gene itself. Once synthesized, the base sequences in the RNA molecule are **translated** by the protein-synthesizing factories in the cell, its **ribosomes**, into a sequence of amino acids. The resulting macromolecule, which may be hundreds, even thousands of amino acids long, folds up into a unique three-dimensional configuration and becomes a functional protein (Figure 1.16).

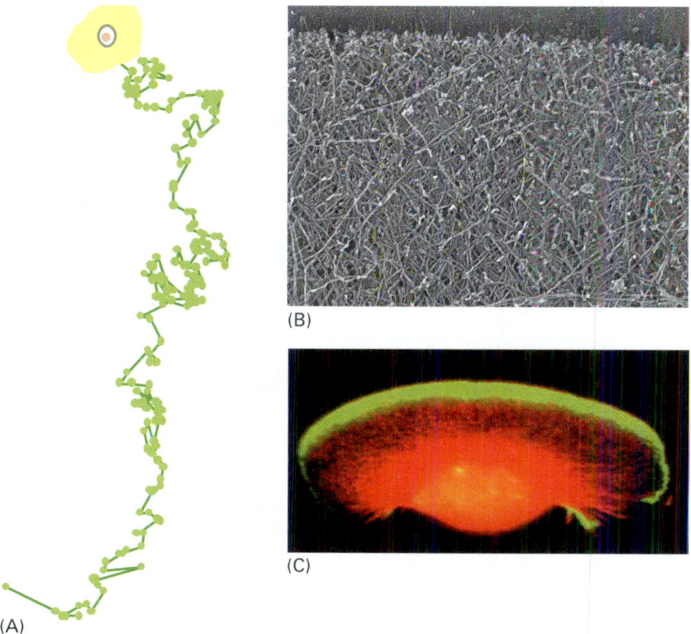

Figure 1.15 Cell motility (A) The movement of individual cells can be observed in a culture dish, in which their locations are plotted at intervals and scored electronically. This image traces the movement of a human vascular endothelial cell (the cell type that forms the lining of blood vessels) toward two attractants located at the bottom—vascular endothelial growth factor (VEGF) and sphingosine-1-phosphate (S1P). Such locomotion is presumed to be critical to the formation of new blood vessels within a tumor. Each point represents a position plotted at 10-minute intervals. This motility is made possible by complex networks of proteins that form the cells' cytoskeletons. (B) Seen here is the network of actin filaments that is assembled at the leading edge of a motile cell. (C) The advancing cell is a fish keratocyte; its leading edge (*green*) is pushed forward by the actin filament network. (A, courtesy of C. Furman and F. Gertler; B and C, from T. Svitkina and G. Borisy, *J. Cell Biol.* 145:1009–1026, 1999. © The Rockefeller University Press.)

Post-translational modification of the initially synthesized protein may result in the covalent attachment of certain chemical groups to specific amino acid residues in the protein chain; included among these modifications are, notably, phosphates, complex sugar side chains, and acetate groups (see Sidebar 1.5). An equally important post-translational modification involves the cleavage of one protein by a second protein termed a **protease**, which has the ability to cut amino acid chains at certain sites. Accordingly, the final, matured form of a protein chain may include far fewer amino acid residues than were present in the initially synthesized protein. Following their synthesis, many proteins are dispatched to specific sites within the cell or are exported from the cell through the process of secretion.

In eukaryotic cells—the main subject of this book—the synthesis of RNA is itself a complex process. An RNA molecule transcribed from its parent gene

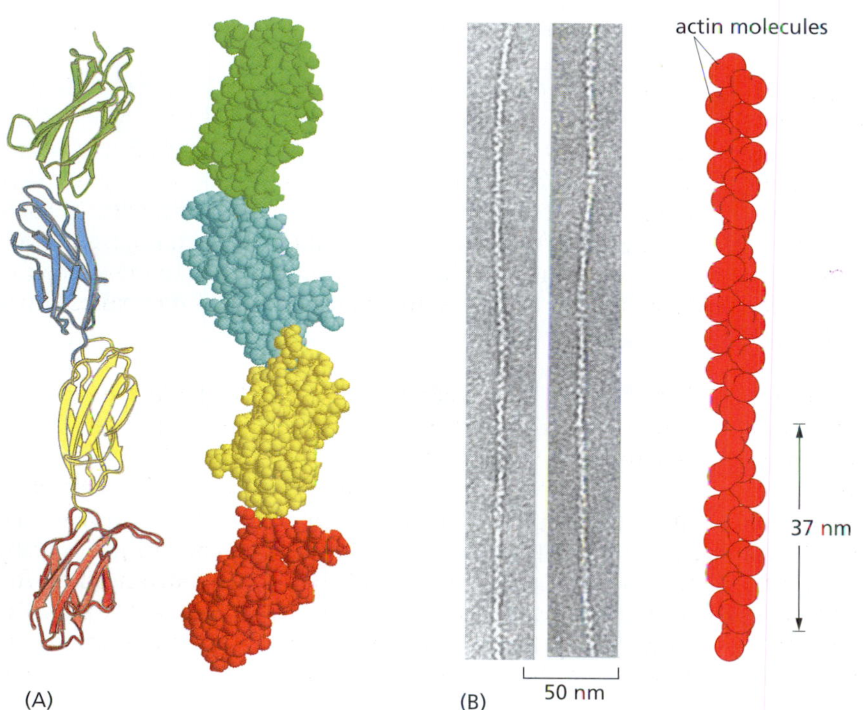

Figure 1.16 Structures of proteins and multiprotein assemblies (A) The three-dimensional structure of part of fibronectin, an important extracellular matrix protein (see Figure 1.14), can be depicted as a ribbon diagram (*left*), which illustrates the path taken by the amino acid chain comprising this protein; alternatively, in a space-filling model (*right*), the positions of the individual atoms are indicated. One domain of fibronectin is composed of four distinct, similarly structured domains, which are shown here with different colors. (B) The actin fibers (*left*), which constitute an important component of the cytoskeleton (see Figures 1.14 and 1.15), are composed of assemblies of individual protein molecules, each of which is illustrated here as a distinct two-lobed body (*right*). (A, adapted from D.J. Leahy, I. Aukhil, and H.P. Erickson, *Cell* 84:155–164, 1996; B, (*left*) courtesy of Roger Craig, (*right*) from B. Alberts et al., Molecular Biology of the Cell, 4th ed. New York: Garland Science, 2002.)

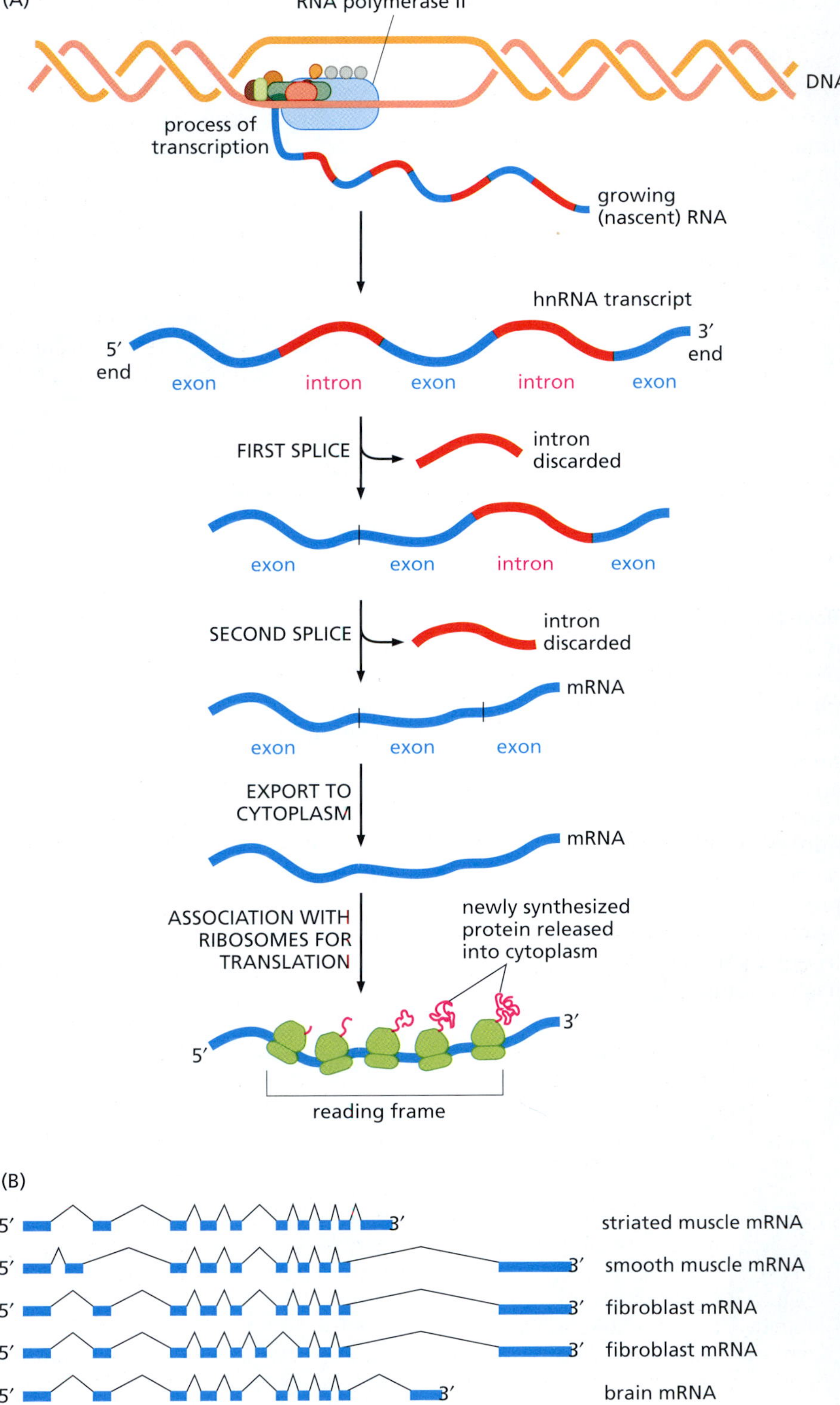

Figure 1.17 Processing of pre-mRNA (A) By synthesizing a complementary RNA copy of one of the two DNA strands of a gene, RNA polymerase II creates a molecule of heterogeneous nuclear RNA (hnRNA) (*red and blue*). Those hnRNA molecules that are processed into mRNAs are termed pre-mRNA. The progressive removal of the introns (*red*) leads to a processed mRNA containing only exons (*blue*). (B) A given pre-mRNA molecule may be spliced in a number of alternative ways, yielding distinct mRNAs that may encode distinct protein molecules. Illustrated here are the tissue-specific alternative splicing patterns of the α-tropomyosin pre-mRNA molecule, whose mRNA products specify important components of cell (and thus muscle) contractility. In this case, the introns are indicated as *black carets* while the exons are indicated as *blue rectangles*. (B, adapted from B. Alberts et al., Molecular Biology of the Cell, 4th ed. New York: Garland Science, 2002.)

may initially be almost as long as that gene. However, soon after its synthesis, segments of the RNA molecule, some very small and others enormous, will be cleaved out of the initially synthesized molecule. These segments, termed **introns**, are soon discarded and consequently have no impact on the subsequent coding ability of the RNA molecule (Figure 1.17). (In fact, this editing of the sequences of an RNA molecule may occur while it is still being synthesized on its DNA template.)

Flanking each intron are two retained sequences, the **exons**, which are fused together during this process of **splicing**. The initially synthesized RNA molecule and its derivatives found at various stages of splicing, together with nuclear RNA transcripts being processed from other genes, collectively constitute the **hnRNA** (heterogeneous nuclear RNA). The end product of these post-transcriptional modifications may be an RNA molecule that is only a small fraction of the length of its initially synthesized, hnRNA precursor. This final matured RNA molecule is likely to be exported into the cytoplasm, where, as an **mRNA** (messenger RNA) molecule, it serves as the template on which ribosomes assemble the amino acids that form the proteins. (The term **pre-mRNAs** is often used to designate those hnRNAs that are known precursors of cytoplasmic mRNAs.) Some mature mRNAs may be less than 1% of the length of their pre-mRNA precursor. The complexity of post-transcriptional modification of RNA and post-translational modification of proteins yields an enormous array of distinct protein species within the cell (Sidebar 1.5).

1.7 Gene expression patterns also control phenotype

The 22,000 or so genes in the mammalian genome, acting combinatorially within individual cells, are able to create the extraordinarily complex organismic phenotypes of the mammalian body. A central goal of twenty-first-century biology is to relate the functioning of this large repertoire of genes to organismic physiology, developmental biology, and disease development. The complexity of this problem is illustrated by the fact that there are at least several hundred distinct cell types within the mammalian body, each with its own behavior, its own distinct metabolism, and its own physiology.

This complexity is acquired during the process of organismic development, and its study is the purview of developmental biologists. They wrestle with a problem

Sidebar 1.5 How many distinct proteins can be found in the human body? While some have ventured to provide estimates of the total number of human genes (a bit more than 22,000), it is difficult to extrapolate from this number to the total number of distinct proteins encoded in the human genome. The simplest estimate comes from the assumption that each gene encodes the structure of a single protein. But this assumption is naive, because it ignores the fact that the pre-mRNA transcript deriving from a single gene may be subjected to several **alternative splicing** patterns, yielding multiple, distinctly structured mRNAs, many of which may in turn encode distinct proteins (see Figure 1.17). Thus, in some cells, splicing may include certain exons in the final mRNA molecule made from a gene, while in other cells, these exons may be absent. Such alternative splicing patterns can generate mRNAs having greatly differing structures and protein-encoding sequences. In one, admittedly extreme case, a single *Drosophila* gene has been found to be capable of generating 38,016 distinct mRNAs and thus proteins through various alternative splices of its pre-mRNA; genes having similarly complex alternative splicing patterns are likely to reside in our own genome.

An additional dimension of complexity derives from the post-translational modifications of proteins. The proteins that are exported to the cell surface or released in soluble form into the extracellular space are usually modified by the attachment of complex trees of sugar molecules during the process of **glycosylation**. Intracellular proteins often undergo other types of chemical modifications. Proteins involved in transducing the signals that govern cell proliferation often undergo **phosphorylation** through the covalent attachment of phosphate groups to serine, threonine, or tyrosine amino acid residues. Many of these phosphorylations affect some aspect of the functioning of these proteins. Similarly, the histone proteins that wrap around DNA and control its access by the RNA polymerases that synthesize hnRNA are subject to **acetylation**, as well as more complex post-translational modifications.

The polypeptide chains that form proteins may also undergo cleavage at specific sites following their initial assembly, often yielding small proteins showing functions that were not apparent in the uncleaved precursor proteins. Later, we will describe how certain signals may be transmitted through the cell via a cascade of the protein-cleaving enzymes termed proteases. In these cases, protein A may cleave protein B, activating its previously latent protease activity; thus activated, protein B may cleave protein C, and so forth. Taken together, alternative splicing and post-translational modifications of proteins generate vastly more distinct protein molecules than are apparent from counting the number of genes in the human genome.

that is inherent in the organization of all multicellular organisms. All of the cells in the body of an animal are the lineal descendants of a fertilized egg. Moreover, almost all of these cells carry genomes that are reasonably accurate copies of the genome that was initially present in this fertilized egg (see Sidebar 1.4). The fact that cells throughout the body are phenotypically quite distinct from one another (e.g., a skin cell versus a brain cell) while being genetically identical creates this central problem of developmental biology: how do these various cell types acquire distinct phenotypes if they all carry identical genetic templates? The answer, documented in thousands of ways over the past three decades, lies in the selective reading of the genome by different cell types (Figure 1.18).

As cells in the early embryo pass through repeated cycles of growth and division, the cells located in different parts of the embryo begin to assume distinct phenotypes, this being the process of **differentiation**. Differentiating cells become committed to form one type of tissue rather than another, for example, gut as opposed to nervous system. All the while, they retain the same set of genes. This discrepancy leads to a simplifying conclusion: sooner or later, differentiation must be understood in terms of the sets of genes that are expressed (i.e., transcribed) in some cells but not in others.

By being expressed in a particular cell type, a suite of genes dictates the synthesis of a cohort of proteins that collaborate to create a specific cell phenotype.

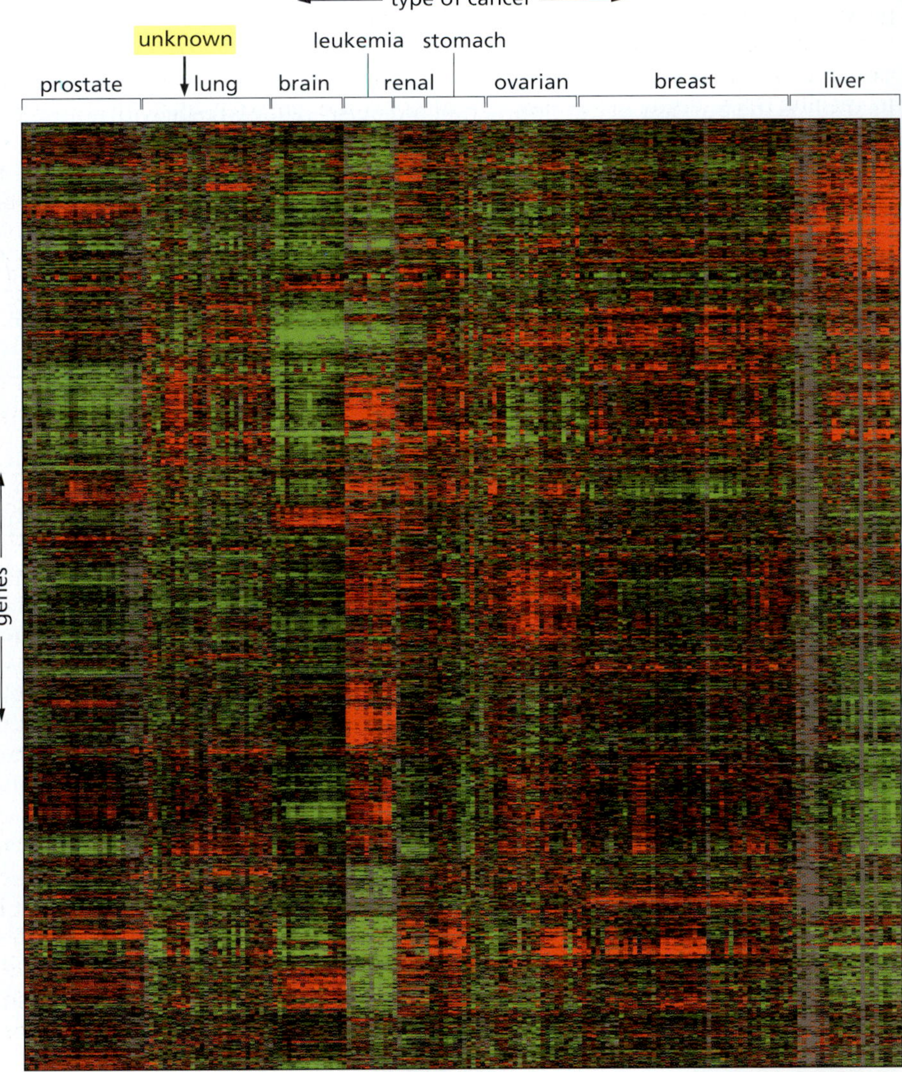

Figure 1.18 Global surveys of gene expression arrays The development of gene expression microarrays has made it possible to survey the expression levels of hundreds, even thousands of genes within a given type of cell. In this image, higher-than-average levels of expression are indicated as small *red* bars, while lower-than-average levels are indicated by small *green* bars. The mRNAs from 142 different human tumors (*arrayed left to right*) were analyzed. In each case, the expression levels of 1800 selected human genes were measured (*top to bottom*). Each class of tumors has its characteristic spectrum of expressed genes. In this case, a tumor of unknown type (*yellow label*) was judged to be a lung cancer because its pattern of gene expression was similar to those of a series of identified lung cancers. (Courtesy of P.O. Brown, D. Botstein and The Stanford Expression Collaboration.)

Accordingly, the phenotype of each kind of differentiated cell in the body should, in principle, be understandable in terms of the specific subset of genes that is expressed in that cell type.

The genes within mammalian cells can be grouped into two broad classes—the **housekeeping** and the **tissue-specific** genes. Many genes encode proteins that are required universally to maintain viability of all cell types throughout the body or to carry out certain biological functions common to all cell types. These commonly expressed genes are classified as housekeeping genes. Within a given differentiated cell type, housekeeping genes represent the great majority of expressed genes.

A minority of genes within a differentiated cell—the tissue-specific genes—are dedicated to the production of proteins and thus phenotypes that are associated specifically with this differentiated cell. It may be, for example, that within a given differentiated cell type, 10,000–15,000 housekeeping genes are expressed while far fewer than 1000 tissue-specific genes are responsible for the distinguishing, differentiated characteristics of the cell. By implication, in each type of differentiated cell, a significant proportion of the 20,000 or so genes in the genome are unexpressed, since they are not required either for the cell's specific differentiation program or for general housekeeping purposes.

1.8 Transcription factors control gene expression

The foregoing description of differentiation makes it clear that large groups of genes must be coordinately expressed while other genes must be repressed in order for cells to display complex, tissue-specific phenotypes. Such coordination of expression is the job of **transcription factors** (TFs). These proteins bind to specific DNA sequences present in the control region of each gene and determine whether or not the gene will be transcribed. The specific stretch of nucleotide sequence to which the TFs bind, often called a **sequence motif**, is usually quite short, being typically 5–10 nucleotides long. In ways that are still incompletely understood at the molecular level, some TFs provide the RNA polymerase enzyme, which is responsible for carrying out transcription, with access to a gene. Yet other TFs may block such access and thereby ensure that a gene is transcriptionally repressed.

The control region within a gene contains a series of short DNA nucleotide sequences that are recognized by some specialized transcription factors that proceed to bind to them and then to exercise control over the transcription process (Figure 1.19). The presence or absence of these short DNA sequences (often termed **enhancers**) therefore determines whether or not a transcription factor can bind to the control region of a gene and impose control. Often the control sequences lie immediately upstream of the site within the gene where transcription begins. In addition, a defined region around the start site of transcription, termed the **promoter**, is bound by RNA polymerase II and an assembly of general transcription factors. Hence, a gene can be separated into two functionally significant regions—the nontranscribed control sequences, and the transcribed sequences that are represented in pre-mRNA and mRNA molecules. In many genes, however, enhancer sequences may also be scattered within the transcribed region of the gene, frequently in introns.

Transcription factors can exercise great power, since a single type of TF can simultaneously affect the expression of a large cohort of downstream responder genes, each of which carries the recognition sequences (an enhancer) for this factor in its promoter. This ability of a single TF (or a single gene that specifies this TF) to elicit multiple changes within a cell or organism is often termed **pleiotropy**. In the case of cancer cells, a malfunctioning, pleiotropically acting TF may simultaneously orchestrate the expression of a large cohort of responder genes that together proceed to create the cancer cell phenotype.

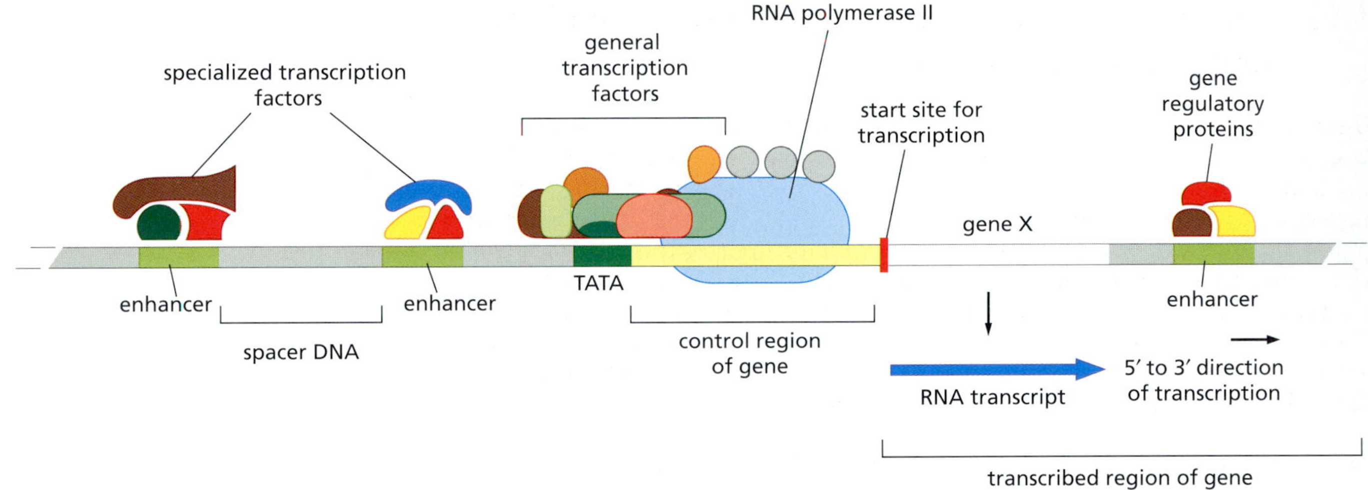

Figure 1.19 Regulation of gene expression The control region of a gene includes specific segments of DNA (*light green*) to which gene regulatory proteins known as transcription factors (TFs) bind, often as multiprotein complexes. In addition, the *promoter* of the gene (*darker green, yellow*) contains sequences to which RNA polymerase II (pol II) can bind together with associated general transcription factors. The bound transcription factors can influence the structure of chromatin (notably the histone proteins that package chromosomal DNA), creating a localized environment that enables pol II to initiate and elongate an RNA transcript (*blue arrow*). (The general transcription factors are involved in initiating the transcription of many genes throughout the genome, while the specialized transcription factors are involved in regulating the expression of small subsets of genes.) Some of the regulatory sequences (sometimes termed *enhancers*) may be located within the transcribed region of a gene. (From B. Alberts et al., Molecular Biology of the Cell, 4th ed. New York: Garland Science, 2002.)

The transcription of most genes is dependent upon the actions of several distinct TFs that must sit down together, each at its appropriate sequence site (i.e., enhancer) in or near the gene promoter, and collaborate to activate gene expression. These interactions inject another element into our thinking—that the expression of a gene is most often the result of the combinatorial actions of several TFs. Therefore, the coordinated expression of multiple genes within a cell, often called its gene **expression program**, is dependent on the actions of multiple TFs acting in combination on large numbers of gene promoters.

1.9 Metazoa are formed from components conserved over vast evolutionary time periods

These descriptions of cell biology, genetics, and evolution are informed in part by our knowledge of the history of life on Earth. Metazoa probably arose only once during the evolution of life on this planet, perhaps 700 million years ago. Once the principal mechanisms governing their genetics, biochemistry, and embryonic development were developed, these mechanisms remained largely unchanged in the descendant organisms up to the present. This sharing of conserved traits among various animal phyla has profound consequences for cancer research, since many lessons learned from the study of more primitive but genetically malleable organisms, such as flies and worms, have proven to be directly transferable to our understanding of how mammalian tissues, including those of humans, develop and function.

Upon surveying the diverse organisms grouped within the mammalian class, one finds that the differences in biochemistry and cell biology are minimal. For this reason, throughout this book we will move effortlessly back and forth between mouse biology and human biology, treating them as if they are essentially identical. On occasion, where species-specific differences are important, these will be pointed out.

The complex signaling circuits operating within cells seem to be organized in virtually identical fashion in the cells of all types of mammals. Even more stunning is the interchangeability of the component parts. It is rare that a human protein cannot function in place of its counterpart *orthologous* protein (see Sidebar 1.6) in mouse cells. In the case of many types of proteins, this conservation of both function and structure is so profound that proteins can be swapped between organisms that are separated by far greater evolutionary distances. A striking example of this, noted earlier (see Figure 1.7), is provided by the gene and thus protein that specifies eye formation in mammals and in flies.

1.10 Gene cloning techniques revolutionized the study of normal and malignant cells

Until the mid-1970s, the molecular analysis of mammalian genes was confined largely to the genomes of DNA tumor viruses, indeed the viruses described later in Chapter 3. These viruses have relatively simple genomes that accumulate to a high copy number (i.e., number of molecules) per cell. This made it possible for biologists to readily purify and study the detailed structure and functioning of viral genes that operate much like the genes of the host cells in which these viruses multiplied. In contrast, molecular analysis of cellular genes was essentially impossible, since there are so many of them (tens of thousands per haploid genome) and they are embedded in a genome of daunting complexity (~3.2 billion base pairs of DNA per haploid cellular genome).

All this changed with the advent of gene cloning. Thereafter, cellular genomes could be fragmented and used to create the collections of DNA fragments known as genomic **libraries**. Various DNA hybridization techniques could then be used to identify the genomic fragments within these libraries that were of

Sidebar 1.6 Orthologs and homologs All higher vertebrates (birds and mammals) seem to have comparable numbers of genes—in the range of 25,000. Moreover, almost every gene present in the bird genome seems to have a closely related counterpart in the human genome. The correspondence between mouse and human genes is even stronger, given the closer evolutionary relatedness of these two mammalian species.

Within the genome of any single species, there are genes that are clearly related to one another in their information content and in the related structures of the proteins they specify. Such genes form a **gene family**. For example, the group of genes in the human genome encoding globins constitutes such a group. It is clear that these related genes arose at some point in the evolutionary past through repeated cycles of the process in which an existing gene is duplicated followed by the divergence of the two duplicated nucleotide sequences from one another (Figure 1.20).

Genes that are related to one another within a single species' genome or genes that are related to one another in the genomes of two distinct species are said to be **homologous** to one another. Often the precise counterpart of a gene in a human can be found in the genome of another species. These two closely related genes are said to be **orthologs** of one another. Thus, the precise counterpart—the ortholog—of the c-*myc* gene in humans is the c-*myc* gene in chickens. To the extent that there are other *myc*-like genes harbored by the human genome (i.e., N-*myc* and L-*myc*), the latter are members of the same gene family as c-*myc* but are not orthologs of one another or of the c-*myc* gene in chickens.

Throughout this book we will often refer to genes without making reference to the species from which they were isolated. This is done consciously, since in the great majority of cases, the functioning of a mouse gene (and encoded protein) is indistinguishable from that of its human or chicken ortholog.

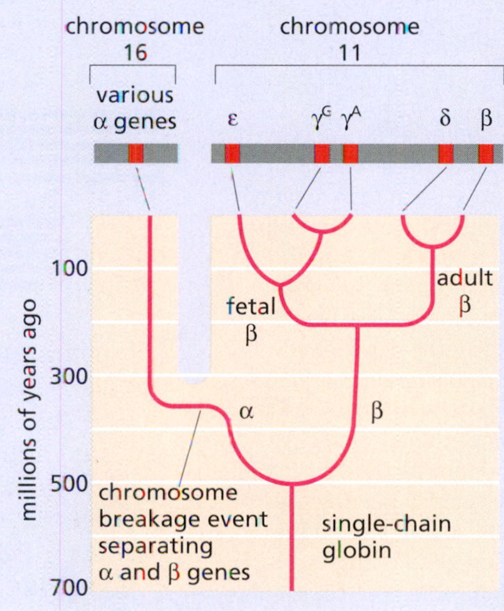

Figure 1.20 Evolutionary development of gene families The evolution of organismic complexity has been enabled, in part, by the development of increasingly specialized proteins. New proteins are "invented" largely through a process of gene duplication, followed by the separate, diverging evolutions of the two resulting genes. Repeated cycles of such gene duplications followed by divergence have led to the development of large numbers of multi-gene families. During vertebrate evolution, an ancestral globin gene, shown here, which encoded the protein component of hemoglobin, was duplicated repeatedly, leading to the large number of distinct globin genes in the modern mammalian genome that are present on two human chromosomes. Because these globins have distinct amino acid sequences, each can subserve a specific physiologic function. (From B. Alberts et al., Essential Cell Biology, 2nd ed. New York: Garland Science, 2004.)

special interest to the experimenter, in particular the DNA fragment that carried part or all of a gene under study. The retrieval of such a fragment from the library and the amplification of this retrieved fragment into millions of identical copies yielded a purified, **cloned** fragment of DNA and thus a cloned gene (Figure 1.21).

Yet other techniques were used to generate DNA copies of the mRNAs that are synthesized in the nucleus and exported to the cytoplasm, where they serve as the templates for protein synthesis. Discovery of the enzyme **reverse transcriptase** (RT; see Sidebar 3.1) was of central importance here. Use of this enzyme made it possible to synthesize ***in vitro*** (i.e., in the test tube) complementary DNA copies of mRNA molecules. These DNA molecules, termed **cDNAs**, carry the sequence information that is present in an mRNA molecule after the process of splicing has removed all introns. While we will refer frequently throughout this book to DNA clones of the genomic (i.e., chromosomal) versions of genes and to cDNAs generated from the mRNA transcripts of such genes, space limitations preclude any detailed descriptions of the cloning procedures per se.

For cancer researchers, gene cloning arrived just at the right time. As we will see in the next chapters, research in the 1970s diminished the candidacy of tumor viruses as the cause of most human cancers. As these viruses moved off center stage, cellular genes took their place as the most important agents responsible for the formation of human tumors. Study of these genes would have been impossible without the newly developed gene cloning technology, which became widely available in the late 1970s, just when it was needed by the community of scientists intent on finding the root causes of cancer.

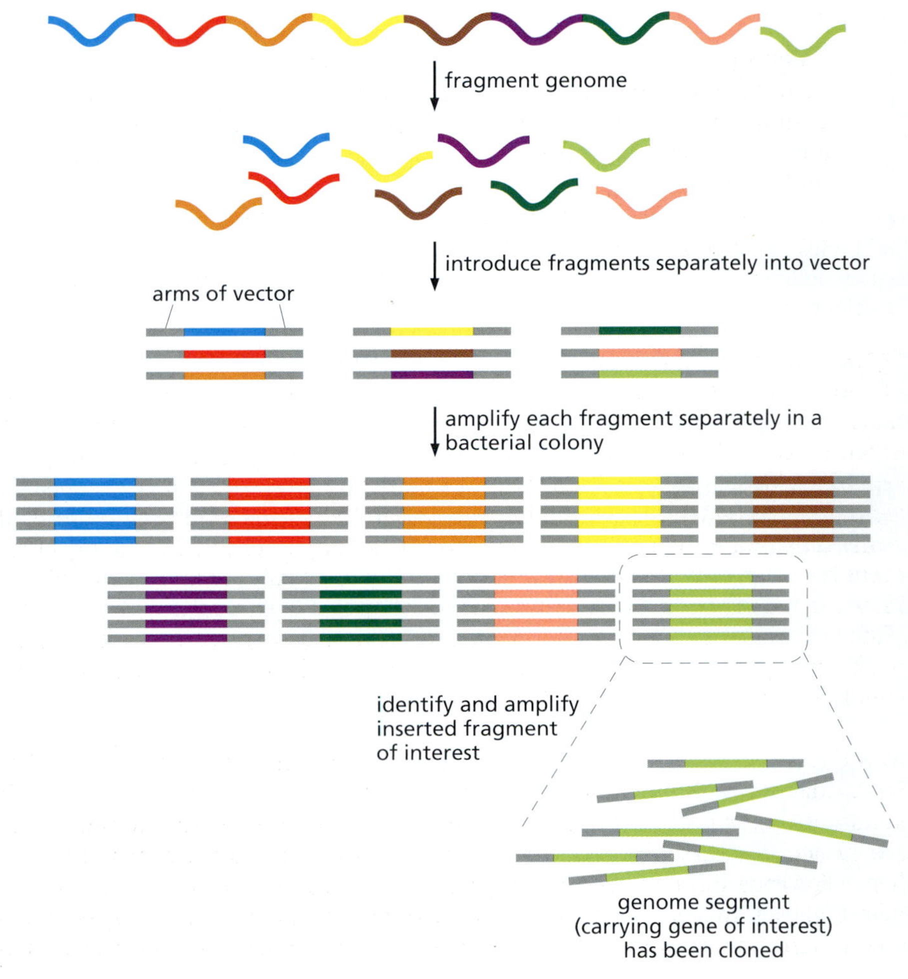

Figure 1.21 Molecular cloning of genes Many versions of the gene cloning procedure have been developed since this technology was first invented in the early and mid-1970s. Pictured here is an outline of one version of this process, in which a specific DNA segment (e.g., containing a gene of interest) is isolated from a complex genome like that of humans. The genome is fragmented into relatively small segments, often several tens of kilobases long (*top*); each DNA segment is inserted individually into a vector (*gray arms*) by linking it to the DNA of the vector, yielding a recombinant vector; each of the resulting vector + insert DNAs is then inserted independently into a bacterium. Each bacterium carrying a vector + inserted gene segment is then expanded into a bacterial colony containing many thousands of bacteria, and the bacterial colony bearing the gene of interest is identified and the vector bearing the gene of interest is retrieved, yielding millions of copies of this genomic fragment (*bottom right*).

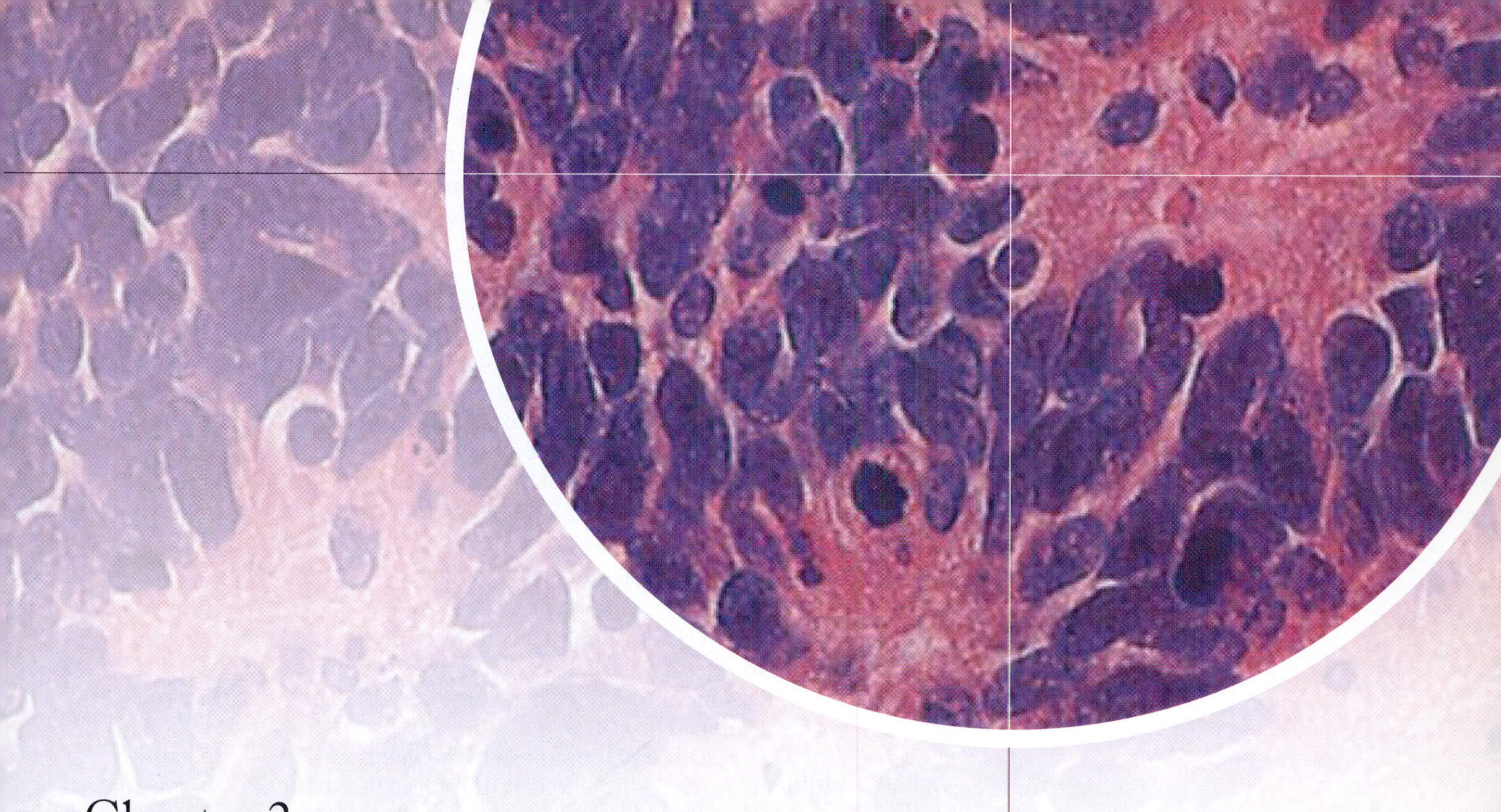

Chapter 2

The Nature of Cancer

> When I published the results of my experiments on the development of double-fertilized sea-urchin eggs in 1902, I added the suggestion that malignant tumors might be the result of a certain abnormal condition of the chromosomes, which may arise from multipolar mitosis. … So I have carried on for a long time the kind of experiments I suggested, which are so far without success, but my conviction remains unshaken .
>
> Theodor Boveri, pathologist, 1914

The cellular organization of metazoan tissues has made possible the evolution of an extraordinary diversity of anatomical designs. Much of this plasticity in design can be traced to the fact that the building blocks of tissue and organ construction—individual cells—are endowed with great autonomy and versatility. Most types of cells in the metazoan body carry a complete organismic genome—far more information than any one of these cells will ever require. And many cells retain the ability to grow and divide long after organismic development has been completed. This retained ability to proliferate and to participate in tissue **morphogenesis** (the creation of shape) makes possible the maintenance of adult tissues throughout the life span of an organism. Such maintenance may involve the repair of wounds and the replacement of cells that have suffered attrition after extended periods of service.

At the same time, this versatility and autonomy poses a grave danger, in that individual cells within the organism may gain access to information in their genomes that is normally denied to them. Moreover, their genomic sequences are subject to corruption by various mechanisms that alter the structure and hence information content of the genome. The resulting mutated genes may divert cells into acquiring novel, often highly abnormal phenotypes. Such changes may be incompatible with the normally assigned roles of these cells in organismic structure and physiology. Among these inappropriate changes may be alterations in cellular growth programs, and these in turn can lead to the appearance of large populations of cells that no longer obey the rules governing normal tissue construction and maintenance.

When portrayed in this way, the renegade cells that form a tumor are the result of normal development gone awry. In spite of extraordinary safeguards taken by the organism to prevent their appearance, cancer cells somehow learn to thrive. Normal cells are carefully programmed to participate in constructing the diverse tissues that make possible organismic survival. Cancer cells have a quite different and more focused agenda. They appear to be motivated by only one consideration: making more copies of themselves.

2.1 Tumors arise from normal tissues

A confluence of discoveries in the mid- and late nineteenth century led to our current understanding of how tissues and complex organisms arise from fertilized eggs. The most fundamental of these was the discovery that all tissues are composed of cells and cell products, and that all cells arise through the division of pre-existing cells. Taken together, these two revelations led to the deduction, so obvious to us now, that all the cells in the body of a complex organism are members of cell lineages that can be traced back to the fertilized egg. Conversely, the fertilized egg is able to spawn all the cells in the body, doing so through repeated cycles of cell growth and division.

These realizations had a profound impact on how tumors were perceived. Previously, many had portrayed tumors as foreign bodies that had somehow taken root in the body of an afflicted patient. Now, tumors, like normal tissues, could be examined under the microscope by researchers in the then-new science of **histology** (or **histopathology**). These examinations, first of normal tissue **sections** (thin slices) and later of sections made from tumor masses, revealed that tumors, like normal tissues, were composed of masses of cells (Figure 2.1).

In addition, evidence accumulated that tumors of various types, rather than invading the body from the outside world, often derive directly from the normal tissues in which they are first discovered. However, tumors did seem to be capable of moving within the confines of the human body: In many patients, multiple tumors were discovered at anatomical sites quite distant from where their

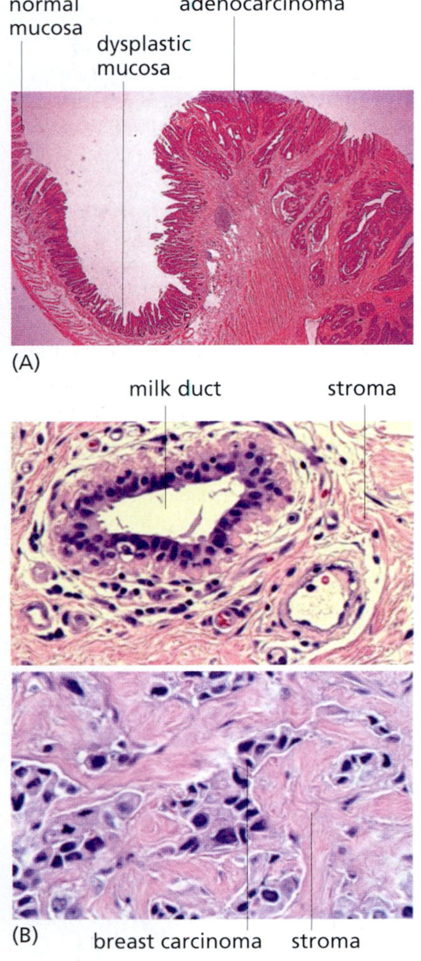

Figure 2.1 Normal versus neoplastic tissue (A) This histological section of the lining of the ileum in the small intestine, viewed at low magnification, reveals the continuity between normal and cancerous tissue. To the left is the normal epithelial lining, termed the mucosa. In the middle is mucosal tissue that has become highly abnormal, being termed "dysplastic." To the right is a frank tumor—an adenocarcinoma—which has begun to invade underlying tissues. (B) This pair of sections of tissue from the human breast, viewed at high magnification, shows how normal tissue architecture becomes deranged in tumors. In the normal mammary gland *(upper panel),* a milk duct is lined by epithelial cells *(dark purple nuclei)*. These ducts are surrounded by mesenchymal tissue termed "stroma," which consists of connective tissue cells, such as fibroblasts and adipocytes, and collagen matrix *(pink)*. In an invasive ductal breast carcinoma *(lower panel),* the cancer cells, which arise from the epithelial cells lining the normal ducts, have abnormally large nuclei *(purple)*, no longer form well-structured ducts, and have invaded the stroma *(pink)*. (A, from A.T. Skarin, Atlas of Diagnostic Oncology, 3rd ed. Philadelphia: Elsevier Science Ltd., 2003; B, courtesy of A. Orimo.)

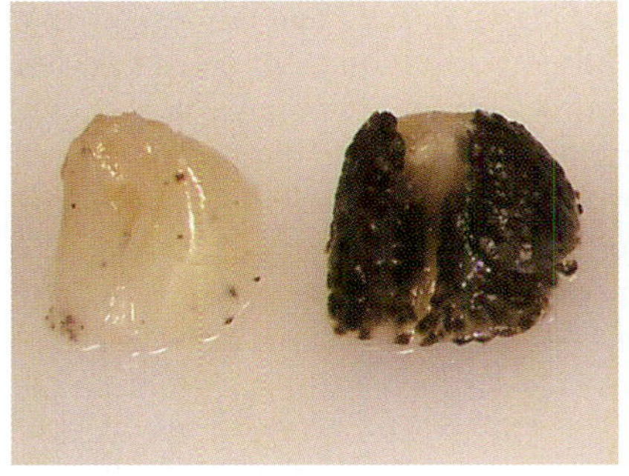

(A)

(B)

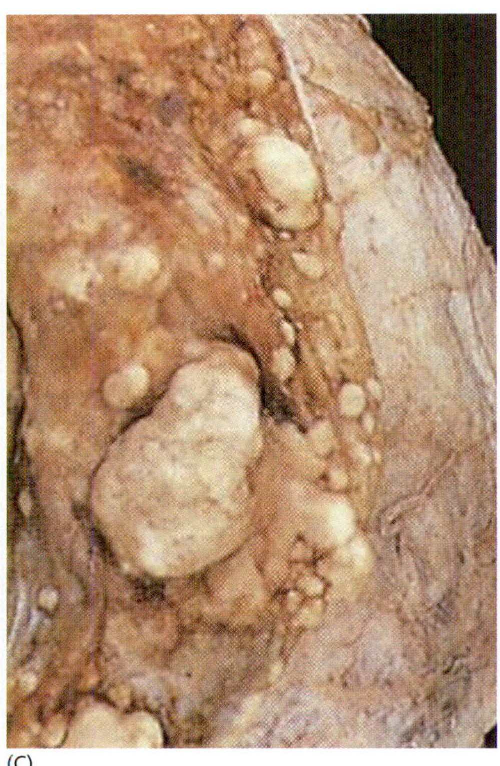

(C)

Figure 2.2 Metastasis of cancer cells to distant sites Many types of tumors eventually release cancer cells that migrate to distant sites in the body, where they form the secondary tumors known as metastases. (A) Metastasis can be studied with facility in the mouse, in which the location of melanoma cells can be pinpointed because of their distinctive dark pigment. Seen here are the lungs of a mouse in which the formation of metastases has been almost entirely blocked *(left)* and one in which hundreds of metastases *(black spots)* were allowed to form, as observed two weeks after B16 mouse melanoma cells were introduced, via injection, into the tail vein of a mouse *(right)*. (B) Metastases *(white)* in the liver often arise in patients with advanced colon carcinomas. The portal vein, which drains blood from the colon into the liver, provides a route for metastasizing colon cancer cells to migrate directly into the liver. (C) Breast cancer often metastasizes to the brain. Here, large metastases are revealed post mortem in the right side of a brain where the dura (membrane covering) of the brain *(right)* has been removed. (A, from F. Nimmerjahn et al. *Immunity* 23:41–51,2005; B, courtesy of Peter Isaacson; C, from A.T. Skarin, Atlas of Diagnostic Oncology, 3rd ed. Philadelphia: Elsevier Science Ltd., 2003.)

disease first began, a consequence of the tendency of cancers to spread throughout the body and to found new colonies of cancer cells (Figure 2.2). These new settlements, termed **metastases**, were often traceable directly back to the site where the disease of cancer had begun—the founding or **primary tumor**.

Invariably, detailed examination of the organization of cells within tumor masses gave evidence of a tissue architecture that was less organized and structured than the architecture of nearby normal tissues (see Figure 2.1). These histopathological comparisons provided the first seeds of an idea that would take the greater part of the twentieth century to prove: tumors are created by cells that have lost the ability to assemble and create tissues of normal form and function. Stated more simply, cancer came to be viewed as a disease of malfunctioning cells.

It followed logically that all tumors should, in principle, be traceable back to the specific tissue or organ site in which they first arose, often using the histopathological analyses of tumor sections to provide critical clues. This simple idea led for the first time to a new way of classifying these growths, which depended on their presumed tissues of origin. The resulting classifications often united under one roof cancers that arise in tissues and organs that have radically different functions in the body but share common types of tissue organization.

The science of histopathology also made it possible to understand the relationship between the clinical behavior of a tumor (i.e., the effects that the tumor had on the patient) and its microscopic features. Most important here were the criteria that segregated tumors into two broad categories depending on their degree of aggressive growth. Those that grew locally without invading adjacent tissues were classified as **benign**. Others that invaded nearby tissues and spawned metastases were termed **malignant**.

In fact, the great majority of primary tumors arising in humans are benign and are harmless to their hosts, except in the rare cases where the expansion of these localized masses causes them to press on vital organs or tissues. Some benign tumors, however, may cause clinical problems because they release dangerously high levels of hormones that create physiologic imbalances in the body. For example, thyroid **adenomas** (premalignant epithelial growths) may cause excessive release of thyroid hormone into the circulation, leading to hyperthyroidism; pituitary adenomas may release growth hormone into the circulation, causing excessive growth of certain tissues—a condition known as **acromegaly**. Nonetheless, deaths caused by benign tumors are relatively uncommon. The vast majority of cancer-related mortality derives from malignant tumors. More specifically, it is the metastases spawned by these tumors that are responsible for some 90% of deaths from cancer.

2.2 Tumors arise from many specialized cell types throughout the body

The majority of human tumors arise from epithelial tissues. **Epithelia** are sheets of cells that line the walls of cavities and channels or, in the case of skin, serve as the outside covering of the body. By the first decades of the twentieth century, detailed histological analyses had revealed that normal tissues containing epithelia are all structured similarly. Thus, beneath the epithelial cell layers in each of these tissues lies a **basement membrane** (sometimes called a **basal lamina**); it separates the epithelial cells from the underlying layer of supporting connective tissue cells, termed the **stroma** (Figure 2.3).

The basement membrane is a specialized type of extracellular matrix (ECM) and is assembled from proteins secreted largely by the epithelial cells. Yet other types of basement membrane are present in other kinds of tissue types. For example,

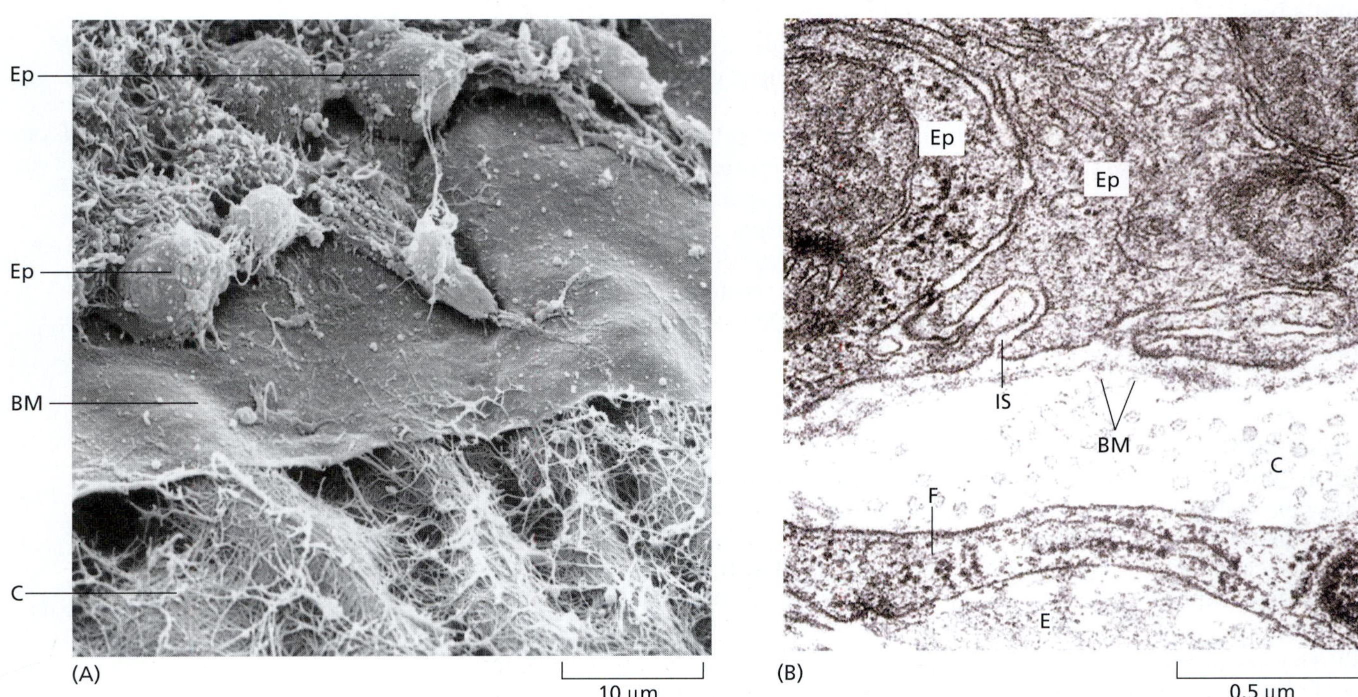

Figure 2.3 Basement membranes (A) This scanning electron micrograph of a chick corneal epithelium illustrates the basic plan of epithelial tissues, in which epithelial cells (Ep) are tethered to one side of the basement membrane. The basement membrane (BM), sometimes termed "basal lamina", seen here as a continuous sheet, is formed as meshwork of extracellular matrix proteins. Below this are seen stromal cells and a network of collagen fibers (C) that anchors the underside of the basement membrane to the ECM of the stroma. (B) When the epithelium of the mouse trachea is viewed in section at far higher magnification through a transmission electron microscope, the basement membrane (BM) can be visualized. Several epithelial cells (Ep) are seen above the BM, while below are collagen fibrils (C), a fibroblast (F), and elastin fibers (E). Note that the basement membrane is not interrupted at the intercellular space (IS). (A, courtesy of Robert Trelstad; B, from B. Young et al., Wheater's Functional Histology, 4th ed. Edinburgh: Churchill Livingstone, 2003.)

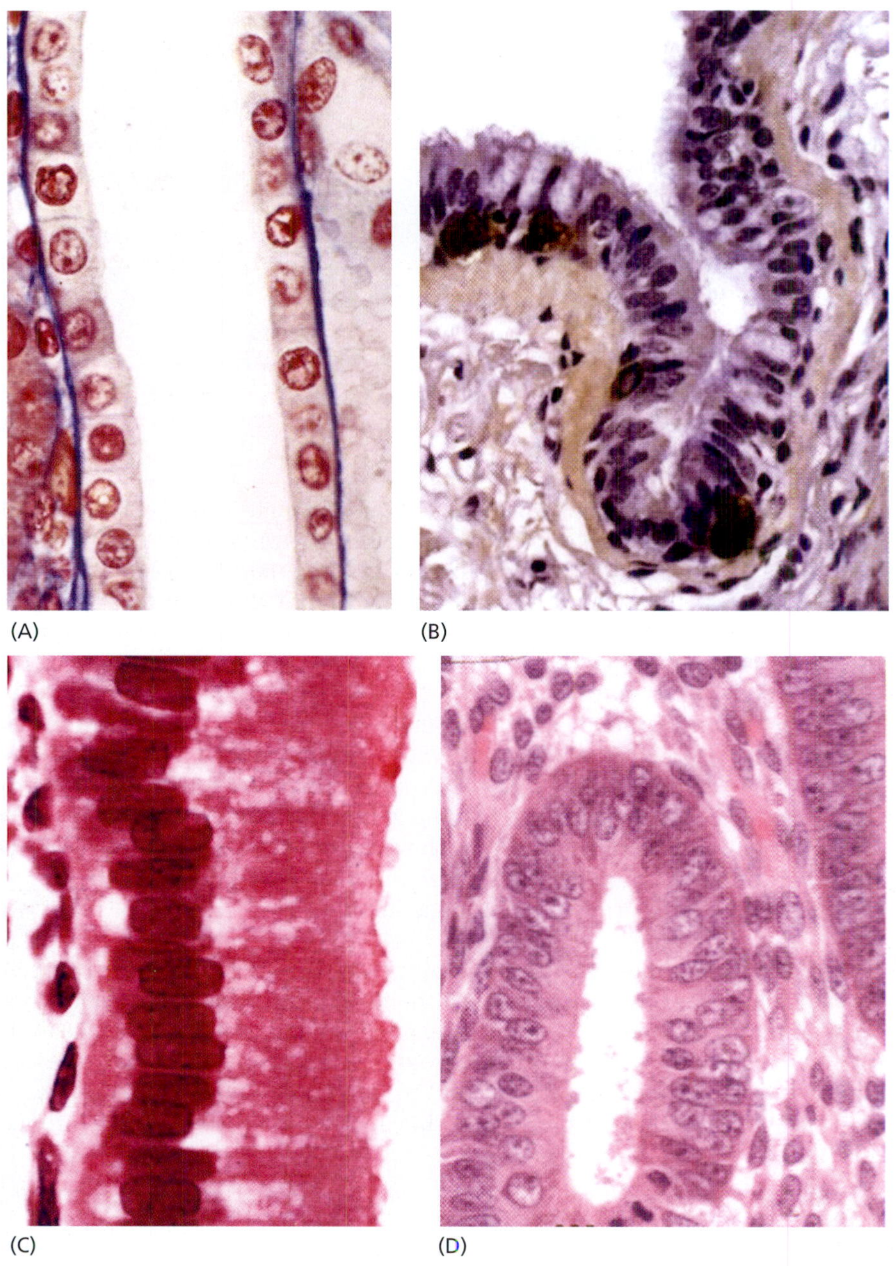

Figure 2.4 Architecture of epithelial tissues A common organizational plan describes most of the epithelial tissues in the body: The mature, differentiated epithelial cells are at the exposed surface of an epithelium. In many tissues, underlying these epithelia are less differentiated epithelial cells, not seen in this figure. Beneath the epithelial cell layer lies a basement membrane (see Figure 2.3), which is usually difficult to visualize in the light microscope. Shown here are epithelia of (A) a collecting tubule of the kidney, (B) the bronchiole of the lung, (C) the columnar epithelium of the gallbladder, and (D) the endometrium of the uterus. In each case, the epithelial cells protect the underlying tissue from the contents of the cavity that they are lining. (From B. Young et al., Wheater's Functional Histology, 4th ed. Edinburgh: Churchill Livingstone, 2003.)

the **endothelial** cells, which form the inner linings of capillaries and larger vessels, rest on a specialized basement membrane that separates them from an outer layer of specialized smooth muscle cells. In all cases, these basement membranes serve as a structural scaffolding of the tissue. In addition, as we will learn later, cells attach a variety of biologically active signaling molecules to basement membranes.

Epithelia are of special interest here, because they spawn the most common human cancers—the **carcinomas**. These tumors are responsible for more than 80% of the cancer-related deaths in the Western world. Included among the carcinomas are tumors arising from the epithelial cell layers of the gastrointestinal tract—which includes mouth, esophagus, stomach, and small and large intestines—as well as the skin, mammary gland, pancreas, lung, liver, ovary, gallbladder, and urinary bladder. Examples of normal epithelial tissues are presented in Figure 2.4.

This group of tissues embraces cell types that arise from all three of the primitive cell layers in the early vertebrate embryo. Thus, the epithelia of the lungs,

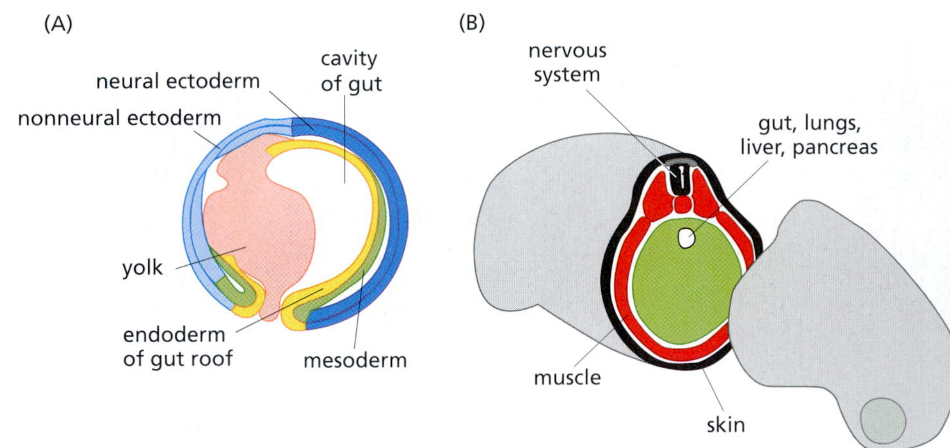

Figure 2.5 Embryonic cell layers (A) The tissues of more complex metazoa develop from three embryonic cell compartments—ectoderm *(blue)*, mesoderm *(green)*, and endoderm *(yellow)*. Each of the three embryonic cell layers is precursor to distinct types of differentiated cells. (B) In an early-stage tadpole, the skin and nervous system develop from the ectoderm *(gray, black)*, while the connective tissue, including bone, muscle, and blood-forming cells, develops from the mesoderm *(red)*. The gut and derived outpouchings, including lung, pancreas, and liver, develop from the endoderm *(white)*. The development of all vertebrates follows this plan. (Adapted from T. Mohun et al., *Cell* 22:9–15, 1980.)

liver, gallbladder, pancreas, esophagus, stomach, and intestines all derive from the inner cell layer, the **endoderm**. Skin arises from the outer embryonic cell layer, termed the **ectoderm**, while the ovaries originate embryologically from the middle layer, the **mesoderm** (Figure 2.5). Therefore, in the case of carcinomas, histopathological classification is not informed by the developmental history of the tissue of origin.

The epithelial and stromal cells of these various tissues collaborate in forming and maintaining the epithelial sheets. When viewed from the perspective of evolution, it now seems that the embryologic mechanisms for organizing and structuring epithelial tissues were invented early in metazoan evolution, likely more than 600 million years ago, and that these mechanistic principles have been exploited time and again during metazoan evolution to construct tissues and organs having a wide array of physiologic functions.

Most of the carcinomas fall into two major categories that reflect the two major biological functions associated with epithelia (Table 2.1). Some epithelial sheets serve largely to seal the cavity or channel that they line and to protect the underlying cell populations (Figure 2.6). Tumors that arise from epithelial cells forming these protective cell layers are said to be **squamous cell carcinomas**. For example, the epithelial cells lining the skin and the esophagus spawn tumors of this type.

Many epithelia also contain specialized cells that secrete substances into the ducts or cavities that they line. This class of epithelial cells generates **adenocarcinomas**. Often these secreted products are used to protect the epithelial cell layers from the contents of the cavities that they surround (see Figure 2.6). Thus, some epithelial cells lining the lung and stomach secrete mucus layers that protect them, respectively, from the air (and airborne particles) and from the corrosive effects of high concentrations of acid. The epithelia in some organs such as the lung, uterus, and cervix have the capacity to give rise to pure adenocarcino-

Table 2.1 Carcinomas

(A) Tissue sites of more common types of adenocarcinoma	(B) Tissue sites of more common types of squamous cell carcinoma	(C) Other types of carcinoma
lung	skin	small-cell lung carcinoma
colon	nasal cavity	large-cell lung carcinoma
breast	oropharynx	hepatocellular carcinoma
pancreas	larynx	renal cell carcinoma
stomach	lung	transitional-cell carcinoma (of urinary bladder)
esophagus	esophagus	
prostate	cervix	
endometrium		
ovary		

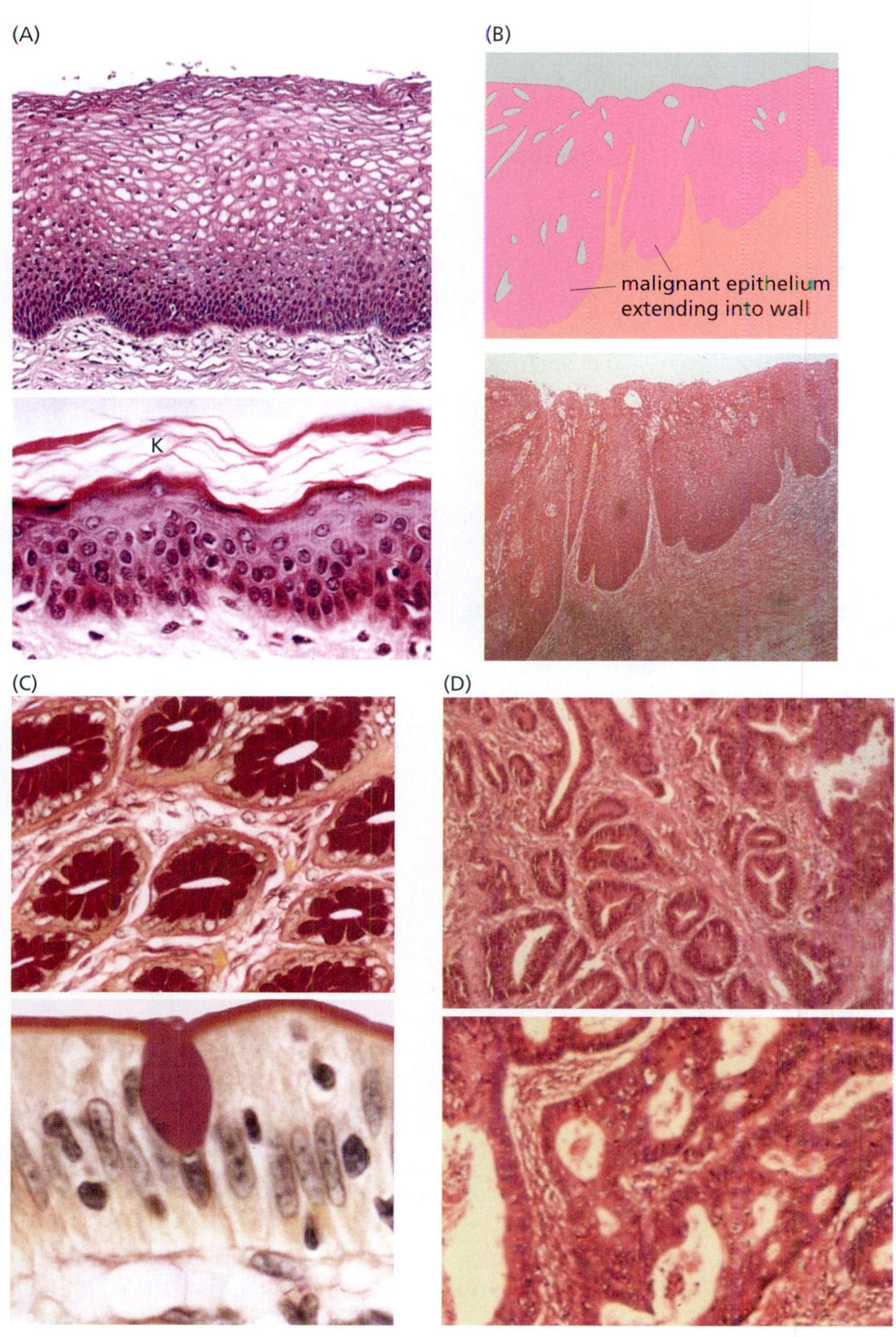

Figure 2.6 Epithelia and derived carcinomas Epithelia can be classified into subtypes depending on the shape and function of the normal epithelial cells and the carcinomas arising from them. The origins of squamous cell carcinomas and adenocarcinomas are seen here. (A) Normal squamous cells are often flattened and function to protect the epithelium and underlying tissue from the contents of the lumen or, in the case of skin, from the outside world. The squamous epithelia of the cervix of the uterus *(above)* and the skin *(below)* are organized quite similarly, with mature flattened cells at the surface being continually shed [e.g., the dead keratinocytes (K) of the skin] and replaced by less mature cells produced below. (B) In this carcinoma of the esophagus, large tongues of malignant squamous epithelial cells are invading the underlying stromal/mesenchymal tissue. (C) In some tissues, the glandular cells within epithelia secrete mucopolysaccharides to protect the epithelium; in other tissues, they secrete proteins that function within the *lumina* (cavities) of ducts or are distributed to distant sites in the body. Pits in the stomach wall are lined by mucus-secreting cells *(dark red, upper panel)*. In the epithelium of the small intestine *(lower panel)* a single mucus-secreting goblet cell *(purple)* is surrounded by epithelial cells of a third type—columnar cells, which are involved in the absorption of water. (D) These adenocarcinomas of the stomach *(upper panel)* and colon *(lower panel)* show multiple ductal elements, which are clear indications of their derivation from secretory epithelia such as those in panel C. (A and C, from B. Young et al., Wheater's Functional Histology, 4th ed. Edinburgh: Churchill Livingstone, 2003; B and D, from A.T. Skarin, Atlas of Diagnostic Oncology, 3rd ed. Philadelphia: Elsevier Science Ltd., 2003.)

mas or pure squamous cell carcinomas; quite frequently, however, tumors in these organs are found in which both types of carcinoma cell coexist.

The remainder of malignant tumors arise from nonepithelial tissues throughout the body. The first major class of nonepithelial cancers derive from the various connective tissues throughout the body, which do share a common origin in the mesoderm of the embryo (Table 2.2). These tumors, the **sarcomas**, constitute only about 1% of the tumors encountered in the oncology clinic. Sarcomas derive from a variety of **mesenchymal** cell types. Included among these are **fibroblasts** and related connective tissue cell types that secrete collagen, the major structural component of the extracellular matrix of tendons and skin; **adipocytes**, which store fat in their cytoplasm; **osteoblasts**, which assemble calcium phosphate crystals within matrices of collagen to form bone; and **myocytes**, which assemble to form muscle (Figure 2.7). A relatively unusual tumor, an **angiosarcoma**,

Table 2.2 Various types of more common sarcomas

osteosarcoma
liposarcoma
leiomyosarcoma
rhabdomyosarcoma
malignant fibrous histiocytoma
fibrosarcoma
synovial sarcoma
angiosarcoma
chondrosarcoma

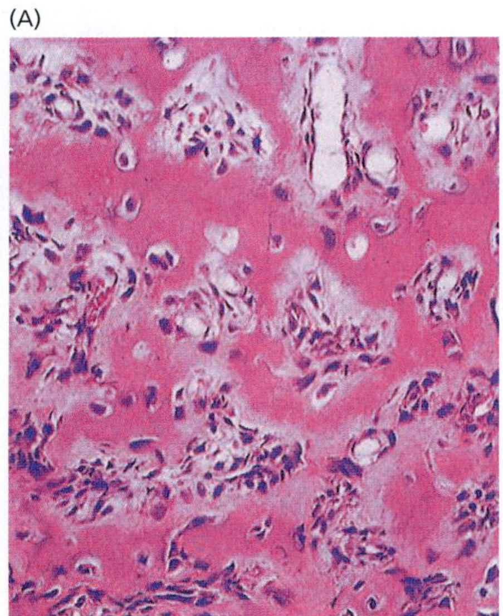

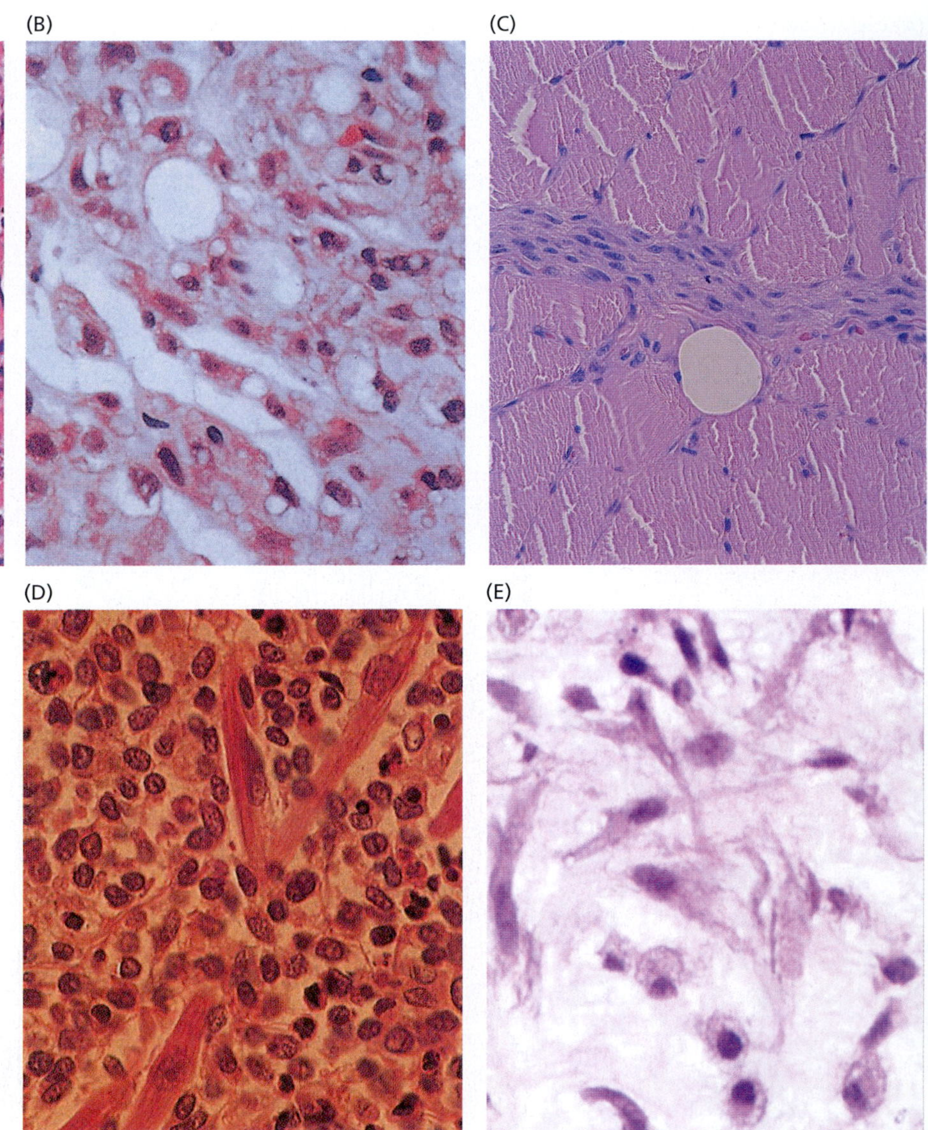

Figure 2.7 Mesenchymal tumors
(A) The osteosarcoma seen here has malignant bone-forming cells—osteoblasts *(dark purple nuclei)*—growing amid mineralized bone *(pink)* that has been constructed by these osteoblasts in the surrounding extracellular matrix. (B) A liposarcoma arises from cells closely related to adipocytes, which are responsible for storing lipid globules in various tissues. The presence of these globules throughout this tumor gives it a foamy appearance. (C) This leiomyosarcoma *(center, dark purple nuclei)*, which arises in the cells that form smooth muscle *(light purple)*, is dispatching individual tumor cells to grow among the muscle fibers *(light purple)*. (D) Rhabdomyosarcomas arise from the cells forming striated skeletal muscles; the cancer cells *(dark red nuclei)* are seen here amid several normal muscle cells *(red)*. (E) This particular sarcoma arose in an unusual anatomic location—the meninges, which form the protective covering of the brain. (A to D, from A.T. Skarin, Atlas of Diagnostic Oncology, 3rd ed. Philadelphia: Elsevier Science Ltd., 2003; E, from B. Young et al., Wheater's Functional Histology, 4th ed. Edinburgh: Churchill Livingstone, 2003.)

arises from precursors of the endothelial cells. The stromal layers of epithelial tissues include many of these mesenchymal cell types.

The second group of nonepithelial cancers arise from the various cell types that constitute the blood-forming (**hematopoietic**) tissues, including the cells of the immune system (Table 2.3 and Figure 2.8). Among these are cells destined to form **erythrocytes** (red blood cells), antibody-secreting (**plasma**) cells, as well as T and B **lymphocytes**. The term **leukemia** (literally "white blood") refers to malignant derivatives of several of these hematopoietic cell lineages that move freely through the circulation and, unlike the red blood cells, are nonpigmented. **Lymphomas** include tumors of the **lymphoid** lineages (B and T lymphocytes) that aggregate to form solid tumor masses, most frequently found in lymph nodes, rather than the dispersed, single-cell populations of tumor cells seen in leukemias.

The third and last major grouping of nonepithelial tumors arises from cells that form various components of the central and peripheral nervous system (Table 2.4). These are often termed **neuroectodermal** tumors to reflect their origins in the outer cell layer of the early embryo. Included here are **gliomas**, **glioblastomas**, **neuroblastomas**, **schwannomas**, and **medulloblastomas** (Figure 2.9). While comprising only 1.3% of all diagnosed cancers, these are responsible for about 2.5% of cancer-related deaths.

Table 2.3 Various types of more common hematopoietic malignancies

acute lymphocytic leukemia
acute myelogenous leukemia
chronic myelogenous leukemia
chronic lymphocytic leukemia
multiple myeloma
non-Hodgkin's lymphoma[a]
Hodgkin's disease

[a]The non-Hodgkin's lymphoma types, also known as lymphocytic lymphomas, can be placed in as many as 15–20 distinct subcategories, depending upon classification system.

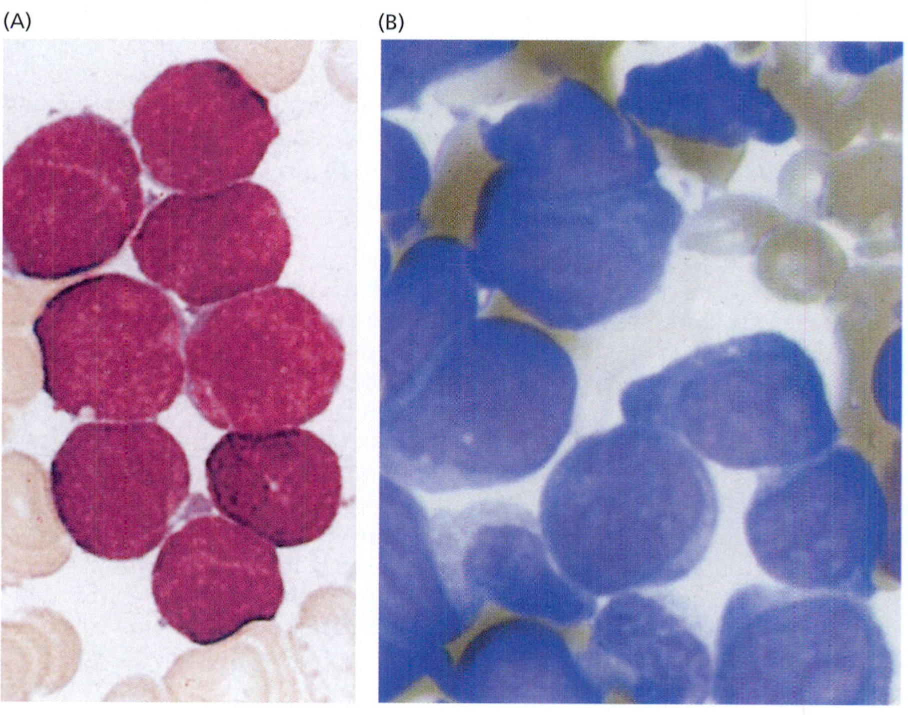

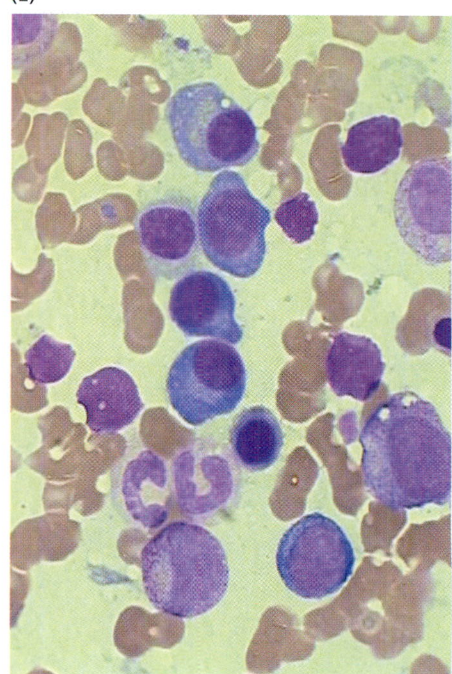

Figure 2.8 Hematopoietic malignancies Hematopoietic malignancies take a variety of forms. (A) Acute lymphocytic leukemias (ALLs) arise from both the B-cell (80%) and T-cell (20%) lineages of lymphocytes (see Section 15.1). The cells forming this particular tumor exhibited the antigenic markers indicating origin from pre-B cells. (B) As in many hematopoietic malignancies, these acute myelogenous leukemia (AML) cells have only a small rim of cytoplasm around their large nuclei. They derive from precursor cells of the lineage that forms various types of granulocytes as well as monocytes, the latter developing into macrophages. (C) The erythroblasts in this erythroleukemia closely resemble the precursors of differentiated red blood cells—erythrocytes. (D) In chronic myelogenous leukemia (CML), a variety of ostensibly differentiated leukemic cells of the myeloid (marrow) lineage of cells are apparent, suggesting the differentiation of myeloid stem cells into several distinct cell types. (E) Multiple myeloma (MM) is a malignancy of the plasma cells of the B-cell lineage, which are responsible for producing and secreting antibody molecules, hence, their relatively large cytoplasms. Seen here are plasma cells of MM at various stages of differentiation *(purple nuclei)*. In all of these micrographs, numerous lightly staining erythrocytes (red blood cells) are seen in the background. (From A.T. Skarin, Atlas of Diagnostic Oncology, 3rd ed. Philadelphia: Elsevier Science Ltd., 2003.)

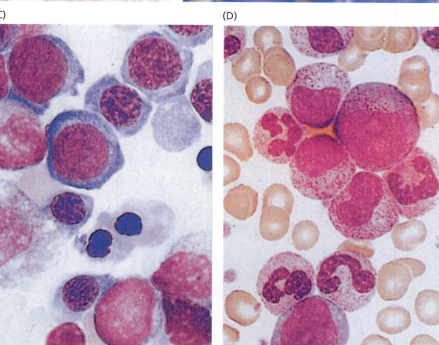

Table 2.4 Various types of neuroectodermal malignancies

glioblastoma multiforme
astrocytoma
meningioma
neurinoma
retinoblastoma
neuroblastoma
ependymoma
oligodendroglioma
medulloblastoma

2.3 Some types of tumors do not fit into the major classifications

Not all tumors fall neatly into one of these four large groups. For example, **melanomas** derive from melanocytes, the pigmented cells of the skin and the retina. The melanocytes, in turn, arise from a primitive embryonic structure termed the **neural crest**. While having an embryonic origin close to that of the neuroectodermal cells, the melanocytes end up during development as wanderers that settle in the skin and the eye, provide pigment to these tissues, and acquire no direct connections with the nervous system (Figure 2.10).

Small-cell lung carcinomas (SCLCs) contain cells having many attributes of **neurosecretory** cells, such as those of neural crest origin in the **adrenal** glands that sit above the kidneys. Such cells, often in response to neuronal signaling, secrete biologically active peptides. It remains unclear whether the SCLCs, frequently seen in tobacco users, arise from neuroectodermal cells that have insinuated themselves during normal development into the developing lung. According to a more likely alternative, these tumors originate in endodermal cell populations of the lung that have shed some of their epithelial characteristics and taken on those of a neuroectodermal lineage.

This switching of tissue lineage and resulting acquisition of an entirely new set of differentiated characteristics is often termed **transdifferentiation**. The term implies that the commitments cells have made during embryogenesis to enter into one or another tissue and cell lineage are not irreversible, and that under certain conditions, cells can move from one differentiation lineage to another. Such a change in phenotype may affect both normal and cancer cells. For example, at the borders of many carcinomas, epithelial cancer cells often change shape and gene expression programs and take on attributes of the nearby stromal cells of mesenchymal origin. This dramatic shift in cell phenotype, termed the **epithelial-mesenchymal transition** or simply EMT, implies great plasticity on the part of cells that normally seem to be fully committed to behaving like epithelial cells. As described later (Chapters 13 and 14), this transition may often accompany and enable the invasion by carcinoma cells into adjacent normal tissues.

Still, even with these occasional rule-breaking exceptions in mind, one major biological principle seems to govern the vast majority of cancers. While cancer cells deviate substantially in behavior from their normal cellular precursors, they almost always retain some of the distinctive attributes of the normal cell types from which they have arisen. These attributes provide critical clues about the origins of most tumors; they enable pathologists to examine tumor biopsies under the microscope and assign a tissue of origin and tumor classification, even without prior knowledge of the anatomical site from which these biopsies were prepared.

In a small minority of cases, (1–2%), the tumors given to pathologists for analysis have shed virtually all of the tissue-specific, differentiated traits of their normal precursor tissues. The cells in such tumors are said to have **dedifferentiated**, and the tumors as a whole are **anaplastic**, in that it is no longer possible to use histopathological criteria to identify the tissues from which they have arisen (Figure 2.11).

2.4 Cancers seem to develop progressively

Between the two extremes of fully normal and highly malignant tissue architectures lies a broad spectrum of tissues of intermediate appearance. The different gradations of abnormality may well reflect cell populations that are evolving

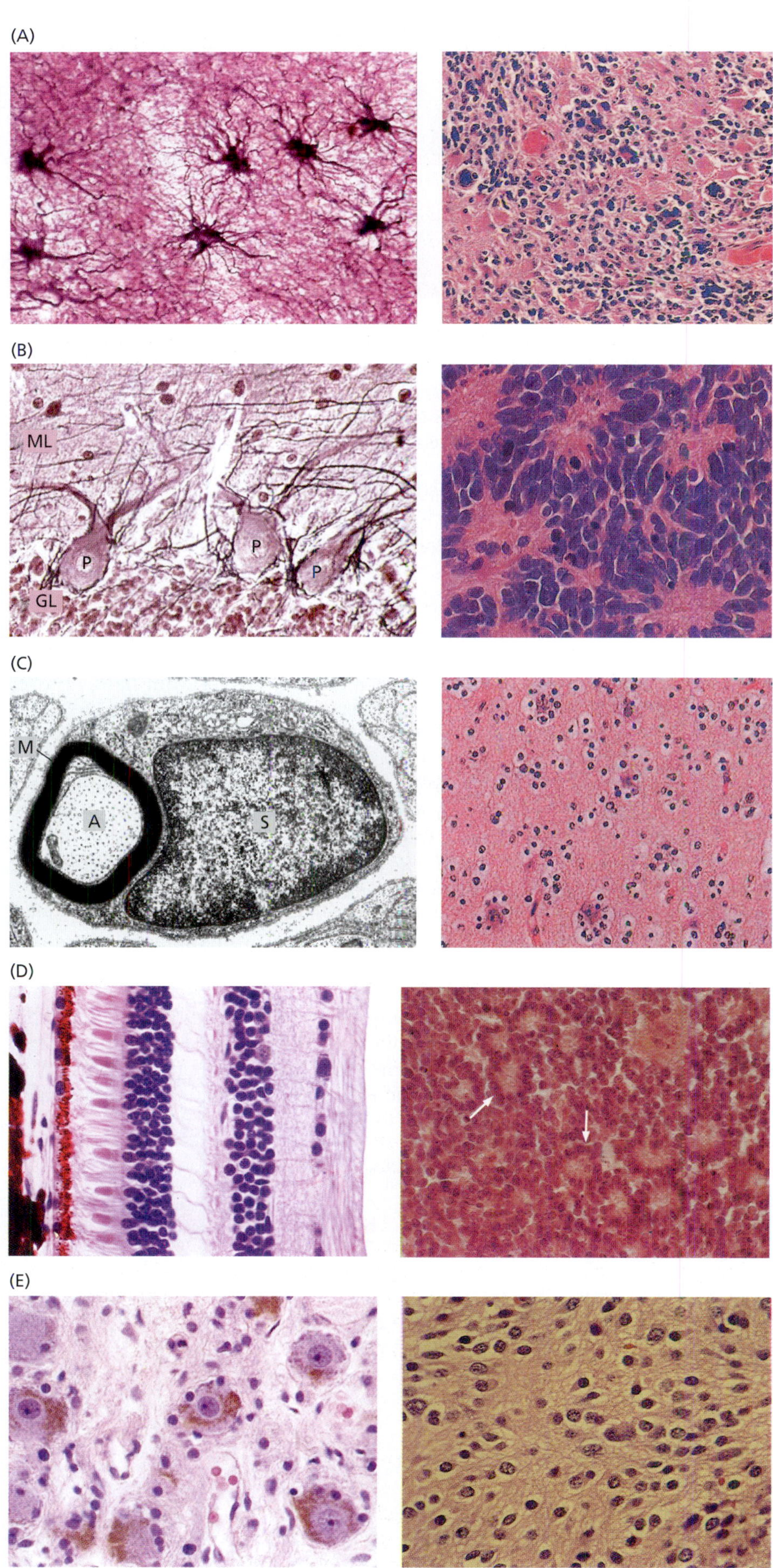

Figure 2.9 Neuroectodermal tumors Various cellular components of the central and peripheral nervous systems can become malignant. (A) Astrocytes—nonneuronal, supporting cells of the brain *(dark purple, left panel)*—are the presumed precursors of astrocytomas and glioblastomas *(right panel)*. Glioblastoma multiforme takes its name from the multiple distinct neuroectodermal cell types that constitute the tumor. The tumor cells are seen to have nuclei of various sizes *(purple)*. (B) Cells of the granular layer (GL) of the cerebellum *(left panel)* reside below Purkinje cells (P) and cells of the molecular layer (ML) in the cortex of the cerebellum. The precursors of granular cells yield medulloblastomas *(right panel)*, the cells of which are notable for their ability to differentiate into neurons, glial cells, and pigmented neuroepithelial cells *(purple nuclei, pink cytoplasms)*. About one-third of these tumors show the rosettes of cells seen here. (C) The Schwann cells (S, *left panel)*, one of which is seen in this electron micrograph, normally wrap multiple membrane layers (M) around axons (A) to provide the latter with a sheathing that aids in conductance. They are precursors of the oligodendrogliomas *(right panel)*. Each of the neoplastic cell nuclei here has a halo around it, which is characteristic of this tumor. (D) Rods, cones and other neuronal cell types *(left panel)* constitute important components of the normal retina. Retinoblastomas *(right panel)* arise from the cells that are the common precursors of the cells forming the normal retina. Retinoblastomas often show the characteristic rosettes, indicated here with *arrows*. (E) Cells of the sympathetic ganglia of the peripheral nervous system *(larger cells, left panel)* give rise to neuroblastomas *(right panel)*, which are usually seen in children. The individual tumor cells here are surrounded by dense fibrillary webs, which are derived from neurites—cytoplasmic processes used by neurons to communicate with one another. (A to E *left panels*, from B. Young et al., Wheater's Functional Histology, 4th ed. Edinburgh: Churchill Livingstone, 2003; A to E *right panels*, from A.T. Skarin, Atlas of Diagnostic Oncology, 3rd ed. Philadelphia: Elsevier Science Ltd., 2003.)

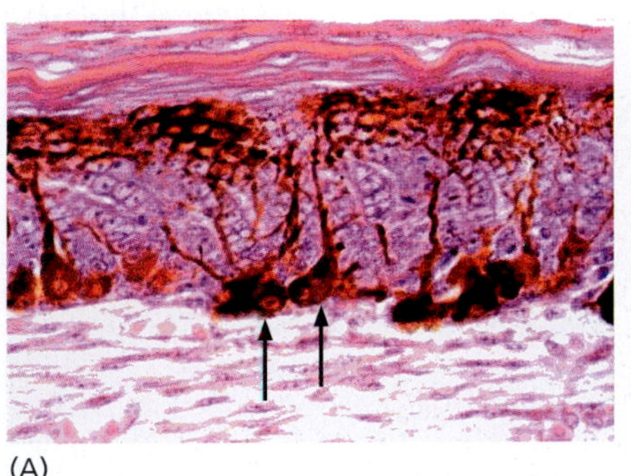
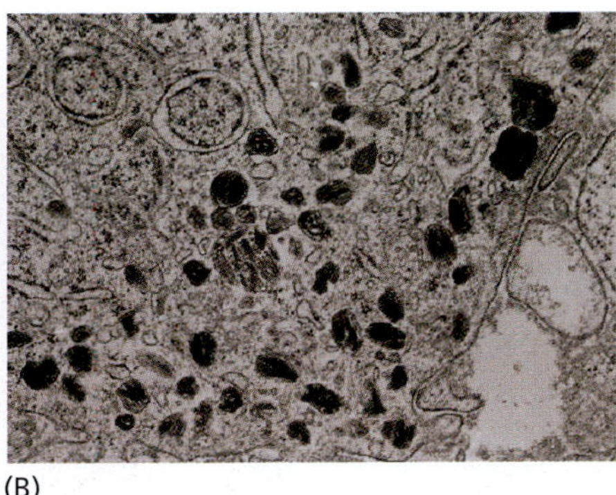
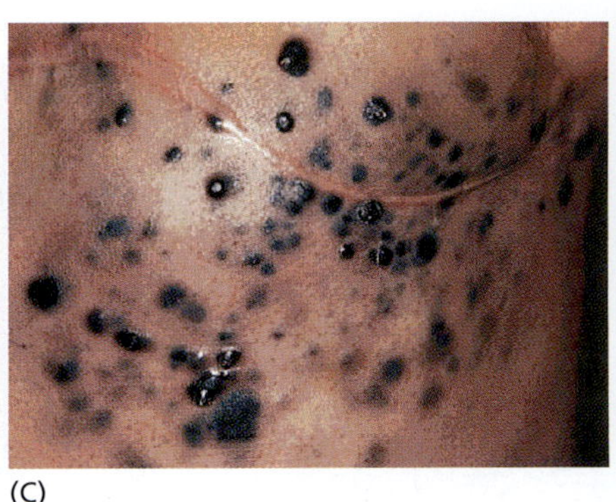

(A) (B) (C)

Figure 2.10 Melanocytes and melanomas (A) Melanocytes *(arrows)*, which form pigment granules, are normally scattered among the basal keratinocytes of the skin (see also Figure 2.6A). They extend long, thin cytoplasmic processes through which they deposit these granules in the cytoplasms of the keratinocytes, which form the bulk of the epithelium. Layers of dead keratinocytes at the surface of the skin *(above)* and stroma cells *(below)* are also apparent. (B) The pigment granules, visualized here by electron microscopy, have made melanomas favored objects of research because of the readily detected metastases that they often form (e.g., see Figure 2.2A). Once melanomas have begun to invade vertically from the superficial layers of the skin into the underlying stroma, they have a high tendency to metastasize to distant tissues sites. (C) This case of cutaneous melanoma dramatizes the metastatic nature of the disease and the readily observed, pigmented metastases. (A, from W.J. Bacha Jr. et al., Color Atlas of Veterinary Histology, 2nd ed. Philadelphia: Lippincott, Williams and Wilkins, 2000; B and C, from A.T. Skarin, Atlas of Diagnostic Oncology, 3rd ed. Philadelphia: Elsevier Science Ltd., 2003.)

progressively away from normal and toward greater degrees of aggressive and invasive behavior. Thus, each type of growth may represent a distinct step along this evolutionary pathway. If so, these architectures suggest, but hardly prove, that the development of tumors is a complex, multi-step process, a subject that is discussed in great detail in Chapter 11.

Some growths contain cells that deviate only minimally from those of normal tissues but may nevertheless be abnormal in that they contain excessive *numbers* of cells. Such growths are termed **hyperplastic** (Figure 2.12). In spite of their apparently deregulated proliferation, the cells forming hyperplastic growths have retained the ability to assemble into tissues that appear reasonably normal.

An equally minimal deviation from normal is seen in **metaplasia**, where one type of normal cell layer is displaced by cells of another type that are not normally encountered in this site within a tissue. These invaders, although present in the wrong location, often appear completely normal under the microscope. Metaplasia is most frequent in epithelial transition zones where one type of epithelium meets another. Transition zones like these are found at the junction of the cervix with the uterus and the junction of the esophagus and the stomach.

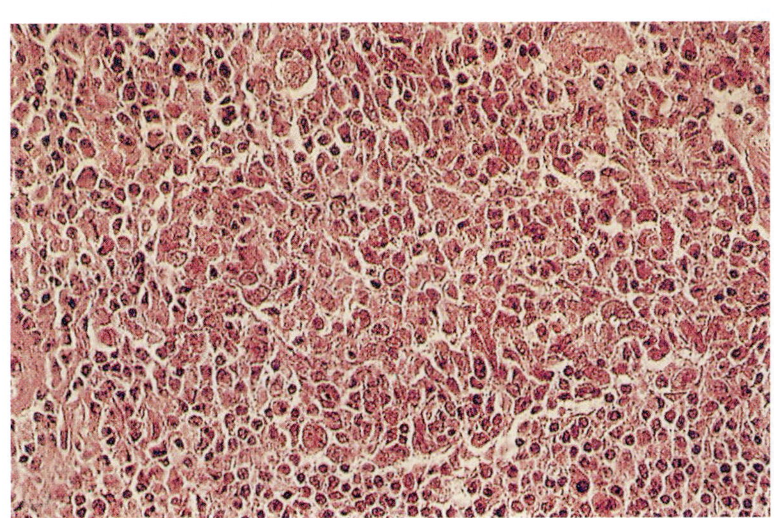

Figure 2.11 Anaplastic tumors of obscure origin The histological appearance of an anaplastic tumor, such as that shown here, gives little indication of its tissue of origin. Attempts to determine the origin of these cells with an antibody stain that specifically recognizes one or another tissue-specific protein marker may also prove uninformative. (From A.T. Skarin, Atlas of Diagnostic Oncology, 3rd ed. Philadelphia: Elsevier Science Ltd., 2003.)

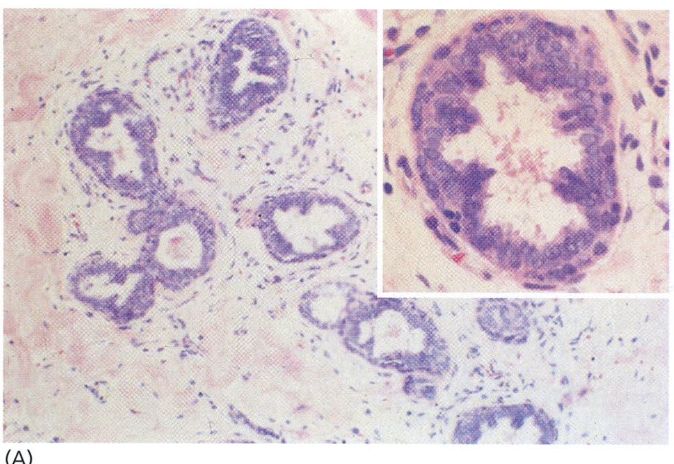

(A)

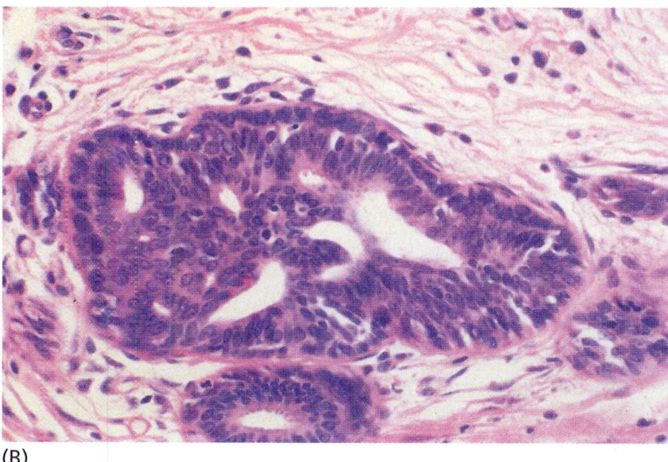

(B)

Figure 2.12 Normal versus hyperplastic epithelium The morphology of the normal ductal epithelium of the mammary gland (see Figure 2.1B) can be compared with different degrees of hyperplasia. (A) In these mildly hyperplastic milk ducts, shown at low magnification and high magnification *(inset)*, mammary epithelial cells have begun to pile up and to protrude into the lumina. (B) A more advanced hyperplastic mammary duct shows epithelial cells that are crowded together and almost completely fill the lumen. However, they have not penetrated the basement membrane *(not visible)* and invaded the surrounding stroma. (From A.T. Skarin, Atlas of Diagnostic Oncology, 3rd ed. Philadelphia: Elsevier Science Ltd., 2003.)

In both locations, a squamous epithelium normally undergoes an abrupt transition into a mucus-secreting epithelium. For example, an early indication of premalignant change in the esophagus is a metaplastic condition termed **Barrett's esophagus**, in which the normally present squamous epithelium is replaced by secretory epithelial cells of a type usually found within the stomach (Figure 2.13). Even though these gastric cells have a quite normal appearance, this metaplasia is considered an early step in the development of esophageal carcinomas. Indeed, patients suffering from Barrett's esophagus have a thirty-fold increased risk of developing these highly malignant tumors.

A slightly more abnormal tissue is said to be **dysplastic**. Cells within a dysplasia are usually abnormal **cytologically**; that is, the appearance of individual cells is no longer normal. The cytological changes include variability in nuclear size and shape, increased nuclear staining by dyes, increased ratio of nuclear versus cytoplasmic size, increased mitotic activity, and lack of the cytoplasmic features associated with the normal differentiated cells of the tissue (Figure 2.14). In dysplastic growths, the relative numbers of the various cell types seen in the normal tissue are no longer observed. Together, these changes in individual cells and in cell numbers have major effects on the overall tissue architecture. Dysplasia is considered to be a transitional state between completely benign growths and those that are premalignant.

Even more abnormal are the growths that are seen in epithelial tissues and termed variously adenomas, **polyps**, adenomatous polyps, **papillomas**, and, in skin, warts (Figure 2.15). These are often large growths that can be readily detected with the naked eye. They contain all the cell types found in the normal

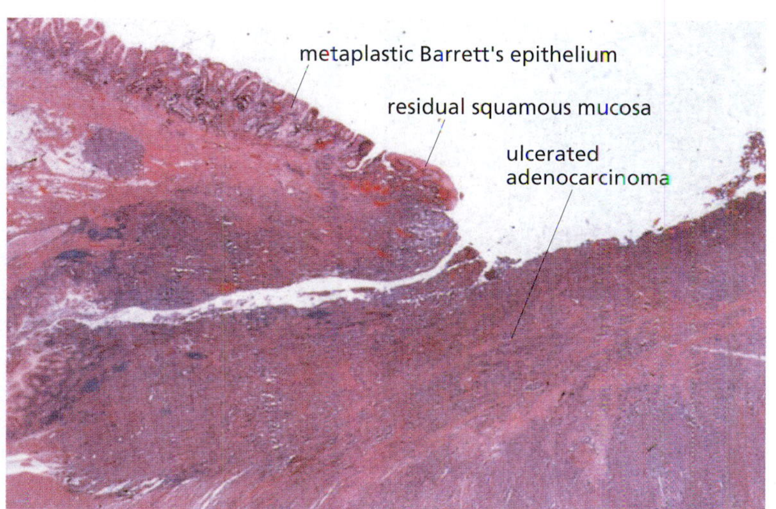

Figure 2.13 Metaplastic conversion of epithelia In certain precancerous conditions, the normally present epithelium is replaced by an epithelium from a nearby tissue—the process of metaplasia. For example, in Barrett's esophagus (sometimes termed Barrett's esophagitis), the squamous cells that normally line the wall of the esophagus *(residual squamous mucosa)* are replaced by secretory cells that migrate from the lining of the stomach *(metaplastic Barrett's epithelium)*. This particular metaplasia, which is provoked by chronic acid reflux from the stomach, can become a precursor lesion to an esophageal carcinoma, which has developed here from cells of gastric origin (*ulcerated adenocarcinoma*). (Adapted from A.T. Skarin, Atlas of Diagnostic Oncology, 3rd ed. Philadelphia: Elsevier Science Ltd., 2003.)

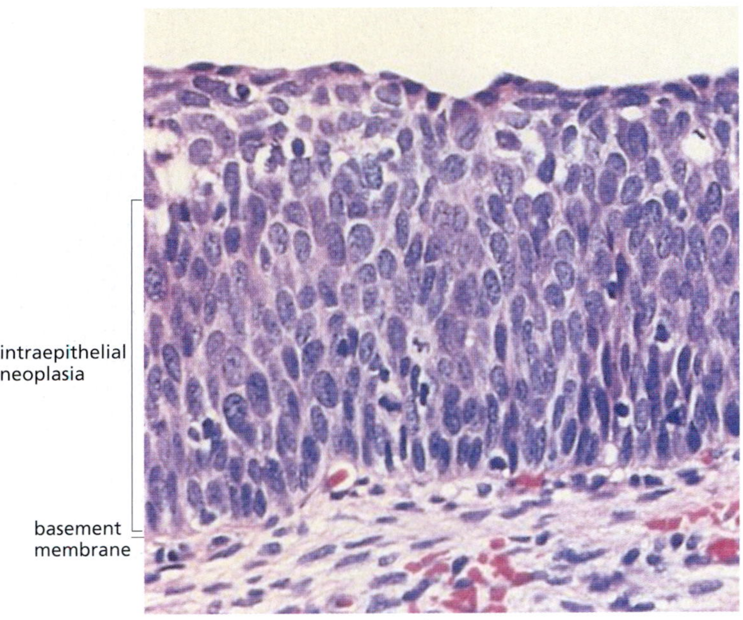

Figure 2.14 Formation of dysplastic epithelium In this high-grade intraepithelial squamous neoplasia of the cervix, the epithelial cells have not broken through the basement membrane *(not visible)* and invaded the underlying stroma. However, only the top layer of cells retains the flattened appearance of squamous cells, while the cell layers below *(purple)* have lost the differentiated, flattened appearance (see Figure 2.6A), have slightly enlarged nuclei, and have accumulated in extra cell layers. (From A.T. Skarin, Atlas of Diagnostic Oncology, 3rd ed. Philadelphia: Elsevier Science Ltd., 2003.)

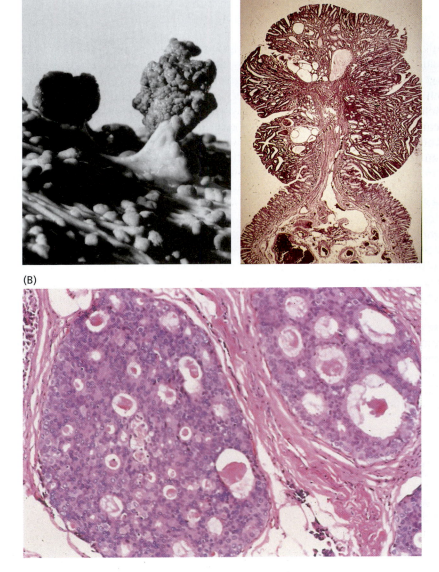

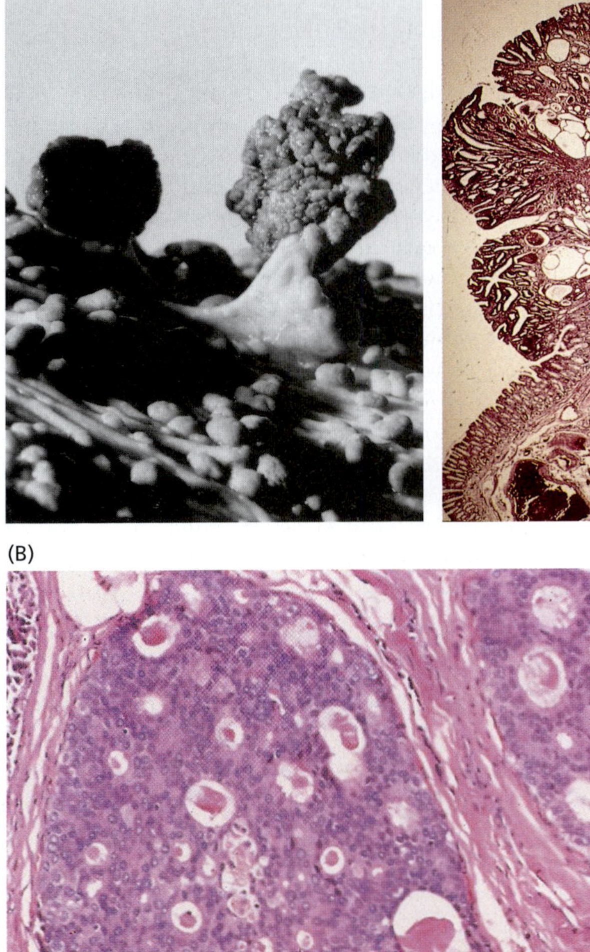

Figure 2.15 Pre-invasive adenomas and carcinomas Adenomatous growths, termed polyps in certain organs, have a morphology that sets them clearly apart from normal and dysplastic epithelium. (A) In the colon, pre-invasive growths appear as either flat thickenings of the colonic wall (sessile polyps, *not shown*) or as the stalk-like growths (pedunculated polyps) shown here. These growths, also termed "adenomas," have not penetrated the basement membrane and invaded the underlying stroma. Polyps are seen here in a photograph *(left)* and a micrograph section *(right)*. (B) In this intraductal carcinoma of the breast, the epithelial cancer cells have almost completely filled two ducts *(left, right)* and have expanded them to great size but have not yet broken through the surrounding basement membrane and invaded the stroma. (A, *left*, courtesy of John Northover and Cancer Research, UK; *right*, courtesy of Anne Campbell; B, from A.T. Skarin, Atlas of Diagnostic Oncology, 3rd ed. Philadelphia: Elsevier Science Ltd., 2003.)

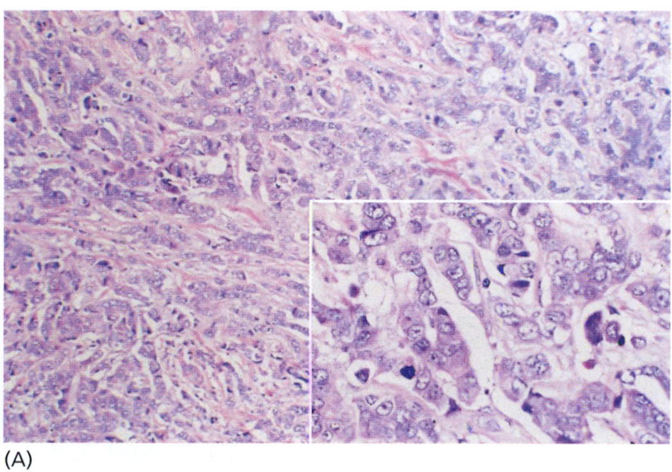
(A)

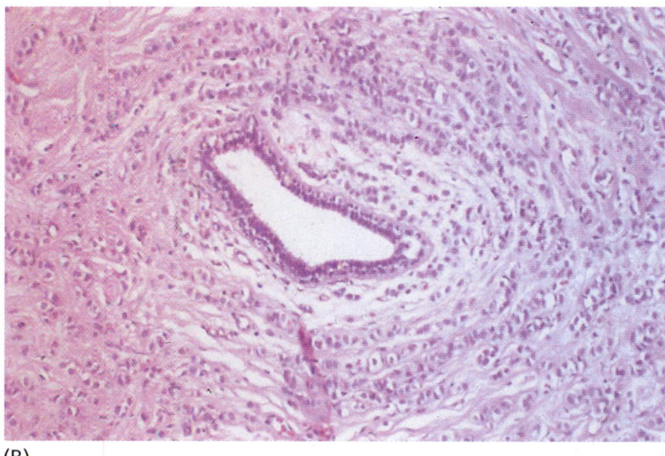
(B)

Figure 2.16 Invasive carcinomas Tumors are considered malignant only after they have breached the basement membrane and invaded the nearby surrounding stroma. (A) In this invasive ductal carcinoma of the breast, islands of epithelial cancer cells are intermingled with stromal cells. The ductal nature of this carcinoma is revealed at higher magnification *(inset)*, where the carcinoma cells are seen to form rudimentary ducts *(e.g., middle of inset)* with lumina. (B) In this invasive lobular carcinoma of the breast, individual carcinoma cells have ventured out into the stroma, often doing so in single-file formation *(numerous small purple nuclei)*. They surround a single duct—a carcinoma *in situ*—in which the epithelial cancer cells are growing facing the lumen of the duct and have not penetrated the basement membrane and invaded the surrounding stroma. (From A.T. Skarin, Atlas of Diagnostic Oncology, 3rd ed. Philadelphia: Elsevier Science Ltd., 2003.)

epithelial tissue, but this mass of cells has launched a program of substantial expansion, creating a macroscopic mass. Under the microscope, the tissue within these adenomatous growths is seen to be dysplastic. They usually grow to a certain size and then stop growing, and they respect the boundary created by the basement membrane, which continues to separate them from underlying epithelium. Since adenomatous growths do not penetrate the basement membrane and invade underlying tissues, they are considered to be benign.

A further degree of abnormality is represented by growths that do invade underlying tissues (Figure 2.16). Here, for the first time, we encounter malignant tumors that have a substantial potential of threatening the life of the individual who carries them. Clinical oncologists and surgeons often reserve the word "cancer" for these and even more abnormal growths. However, in this book, as in much of contemporary cancer research, the word cancer is used more loosely to include all types of abnormal growths. (In the case of epithelial tissues, the term "carcinoma" is usually applied to growths that have acquired this degree of invasiveness.) This disparate collection of growths—both benign and malignant—are called collectively **neoplasias**, i.e., new types of tissue. (Some reserve the term "neoplasia" for malignant tumors.)

As mentioned above, cells in an initially formed primary tumor may seed new tumor colonies at distant sites in the body through the process of metastasis. This process is itself extraordinarily complex, and it depends upon the ability of cancer cells to invade adjacent tissues, to enter into blood and lymph vessels, to migrate through these vessels to distant anatomical sites, to leave the vessels and invade underlying tissue, and to found a new tumor cell colony at the distant site. These steps are the subject of detailed discussion in Chapter 14.

Because the various growths cataloged here represent increasing degrees of tissue abnormality, it would seem likely that they are distinct stopping points along the road of **tumor progression**, in which a normal tissue evolves progressively into one that is highly malignant. However, the precursor–product relationships of these various growths (i.e., normal → hyperplastic → dysplastic → neoplastic → metastatic) is only suggested by the above descriptions but by no means proven.

2.5 Tumors are monoclonal growths

Even if we accept the notion that tumors arise through the progressive alteration of normal cells, another question remains unanswered: how many normal cells are the ancestors of those that congregate to form a tumor (Figure 2.17)? Do the tumor cells descend from a single ancestral cell that crossed over the boundary

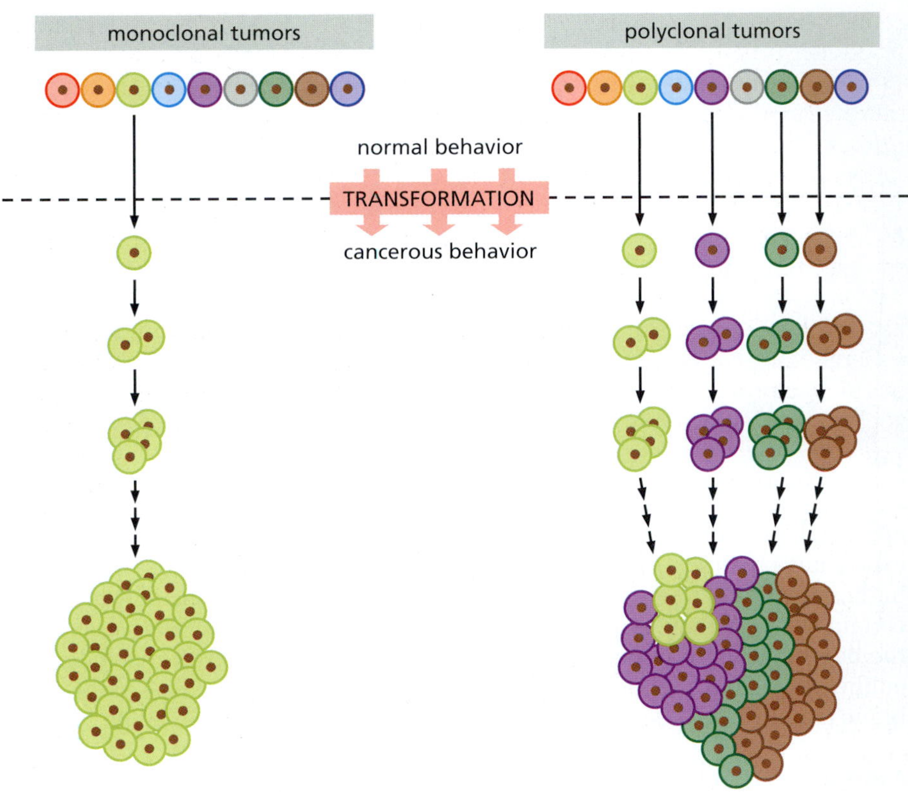

Figure 2.17 Monoclonality versus polyclonality of tumors In theory, tumors may be polyclonal or monoclonal in origin. In a polyclonal tumor *(right),* multiple cells cross over the border from normalcy to malignancy to become the ancestors of several, genetically distinct subpopulations of cells within a tumor mass. In a monoclonal tumor *(left),* only a single cell is transformed from normal to cancerous behavior to become the ancestor of the cells in a tumor mass.

from normal to abnormal growth? Or did a large cohort of normal cells undergo this change, each becoming the ancestor of a distinct subpopulation of cells within a tumor mass?

The most effective way of addressing this issue is to determine whether all the cells in a tumor mass share a common, highly unique genetic or biochemical marker. For example, a randomly occurring somatic mutation might mark a cell in a very unusual way. If this particular genetic marker were present in all cells within a tumor, this would suggest that they all descend from the initially mutated cell and that they have all inherited the marker from this common progenitor. Such a population of cells, all of which derive from a common ancestral cell, is said to be **monoclonal**. Alternatively, if the tumor mass is composed of a series of genetically distinct subpopulations of cells that give no indication of a common origin, it can considered to be **polyclonal**.

The first experiments designed to measure the clonality of tumor cell populations actually relied on a naturally occurring, nongenetic (**epigenetic**) marking event. As described earlier (Sidebar 1.2; Figure 1.10), in the somatic cells of early embryos of female placental mammals, one of the two X chromosomes in each cell is selected randomly for silencing. This silencing causes almost all genes on one X chromosome in a cell to be repressed transcriptionally and is manifested karyotypically through the condensation of the silenced X chromosome into a small particle termed the Barr body. Once the decision to inactivate an X chromosome (of maternal or paternal origin) has occurred in an embryonic cell, all descendant cells in adult tissues appear to respect the decision made by their ancestor in the embryo and thus continue to inactivate the same X chromosome.

The gene for glucose-6-phosphate dehydrogenase (G6PD) is located on the X chromosome, and more than 30% of African American women are heterozygous at this locus. Thus, they carry two alleles specifying forms of this enzyme that can be distinguished either by starch gel electrophoresis or by the fact that one form is more resistant to heat inactivation than the other. Because of X-chromosome silencing, each of the cells in these heterozygous women will express only one or the other allele of the *G6PD* gene, which is manifested in turn in the variant of the G6PD protein that these cells synthesize (Figure 2.18). In most of their tissues, half of the cells make one variant enzyme, while the other half make the other variant. In 1965, observations were reported on a number of **leiomyomas** (benign tumors of the uterine wall) in African American heterozygotes. Each leiomyoma invariably expressed either one or the other variant form of the G6PD enzyme. This meant that, with great likelihood, its component cells all descended from a single common progenitor that expressed only that particular allele.

This initial demonstration of the monoclonality of human tumors was followed by many other confirmations of this concept. One proof came from observations of **myelomas**, which derive from the B-cell precursors of antibody-producing plasma cells. Normally, the pool of these B-cell precursors consists of hundreds of thousands, likely millions of distinct subpopulations, each expressing its own specific antibody molecules as a consequence of a particular **immunoglobulin** gene rearrangement. In contrast, all the myeloma cells in a patient produce the identical antibody molecule, indicating their descent from a single, common ancestor that was present years earlier in this complex, heterogeneous cell population (Figure 2.19A).

Perhaps the most vivid demonstrations of tumor monoclonality have come from observations of cancer cells that have a variety of chromosomal aberrations that can be visualized microscopically when chromosomes are condensed during the metaphase of mitosis. Often, a very peculiar chromosomal abnormality—the clear result of a rare genetic accident—is seen in all the cancer cells within a tumor mass (Figure 2.19B). This observation makes it obvious that all

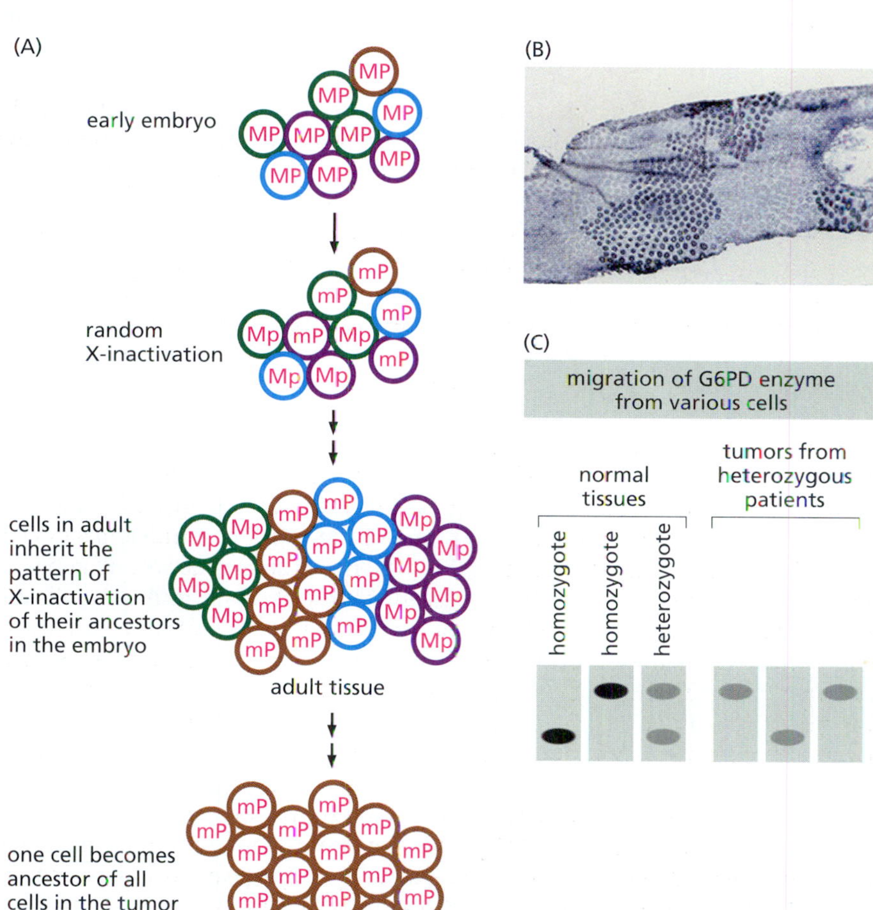

Figure 2.18 X-chromosome inactivation patterns and the monoclonality of tumors (A) While the female embryo begins with both X chromosomes in an equally active state, either the the X chromosome inherited from the mother (M) or the X chromosome inherited from the father (P) soon undergoes inactivation at random. Such inactivation silences expression of almost all genes residing on that chromosome. In the adult, all of the lineal descendants of a particular embryonic cell continue to inactivate the same X chromosome. Hence, the adult female body is made of patches (clones) of cells of the type Mp and patches of the type mP, where the *lowercase letter* denotes an inactivated state. (B) The two allelic forms of glucose-6-phosphate dehydrogenase (G6PD), which is encoded by a gene on the X chromosome, have differing sensitivities to heat inactivation. Hence, gentle heating of tissue from a heterozygote—in this case a section of intestine—reveals patches of cells that carry the heat-resistant, still-active enzyme variant *(dark blue spots)* among patches that do not. The cells in each patch are the descendants of an embryonic cell that decided early in embryogenesis to inactivate either its maternal or paternal X chromosome. (C) Use of starch gel electrophoresis to resolve the two forms of G6PD showed that all of the cancer cells in a tumor arising in a *G6PD* heterozygous patient express either one or the other form of this enzyme. This indicated that these cells all descended from a common, founding ancestor that already exhibited this particular pattern of X-inactivation. This finding suggested that the cancer cells within a tumor mass constitute a monoclonal growth. (B, from M. Novelli et al., *Proc. Natl. Acad. Sci. USA* 100:3311–3314, 2003; C, adapted from P.J. Fialkow, *N. Engl. J. Med.* 291:26–35, 1974.)

the malignant cells within this tumor descend from the single ancestral cell in which this chromosomal restructuring originally occurred.

While such observations seem to provide compelling proof that tumor populations are monoclonal, tumorigenesis may actually be more complex. Let us imagine, as a counterexample, that ten normal cells in a tissue simultaneously crossed over the border from being normal to being malignant (or at least premalignant) and that each of these cells, and its descendants in turn, proliferated uncontrollably (see Figure 2.17). Each of these founding cells would spawn a large monoclonal population, and the tumor mass, as a whole, consisting of a mixture of these ten cell populations, would be polyclonal.

It is highly likely that each of these ten clonal populations varies subtly from the other nine in a number of characteristics, among them the time required for cells in this population to double. Simple mathematics indicates that a cell population that exhibits a slightly shorter doubling time will, sooner or later, outgrow all the others, and that the descendants of these cells will dominate in the tumor mass, creating what will appear to be a monoclonal tumor. In fact, many tumors seem to require decades to develop, which is plenty of time for one clonal subpopulation to dominate in the overall tumor cell population. Hence, the monoclonality of the cells in a large tumor mass hardly proves that this tumor was strictly monoclonal during its early stages of development.

A second confounding factor derives from the genotypic and phenotypic instability of tumor cell populations. As we will discuss in great detail in Chapter 11, the population of cells within a tumor may begin as a relatively homogeneous collection of cells (thus constituting a monoclonal growth), but soon this population may become quite heterogeneous because of the continual acquisition of new mutant alleles by different cells in this population. The resulting genetic

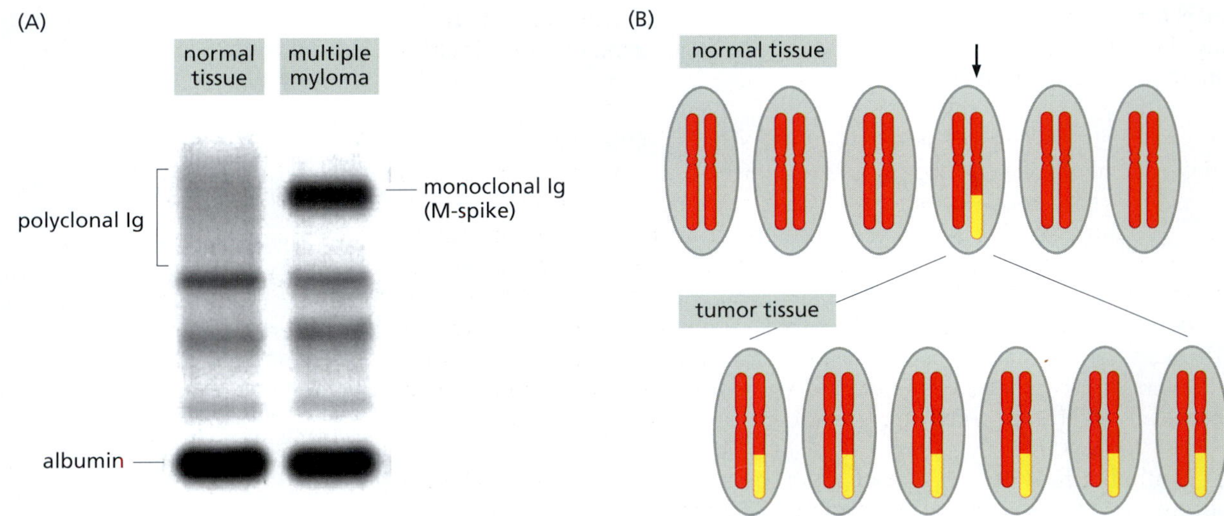

Figure 2.19 Additional proofs of tumor monoclonality (A) In normal plasma, the immunoglobulin (Ig) molecules (i.e., antibodies) migrate as a heterogeneous collection of molecules upon gel electrophoresis (polyclonal Ig, *top of left channel*); this heterogeneity is indicative of the participation of a diverse spectrum (a polyclonal population) of plasma cells in antibody production. However, in the disease of multiple myeloma, this heterogeneous population of Ig molecules is replaced by a single antibody species (termed an M-spike) that is produced by a single clonal population of antibody-secreting tumor cells. (B) Illustrated here is an unusual translocation *(arrow)* that involves exchange of segments between two separate (nonhomologous) chromosomes—a *red* and a *yellow* chromosome; the translocation affects only one of the paired homologous chromosomes. This translocation event, occurring among a population of karyotypically normal cells *(top row)*, creates a characteristic "signature" of the particular tumor being studied. (Only one of the two chromosomal products of the translocation is shown here.) Since all of the cancer cells within a given tumor carry the identical, rare translocation *(bottom row)*, this indicates their descent from a common progenitor in which this translocation initially occurred. (A, courtesy of S. Chen-Kiang and S. Ely.)

heterogeneity may mask the true monoclonal origin of this cell population, since many of the genetic markers in these descendant cells will be present only in specific subpopulations of cells within the tumor mass.

These caveats complicate our assessment of the monoclonal origins of tumors. Nonetheless, it is a widespread consensus that the vast majority of human tumors are monoclonal growths descended from single progenitor cells that took the first small steps to becoming cancerous.

2.6 Cancers occur with vastly different frequencies in different human populations

The nature of cancer suggests that it is a disease of chaos, a breakdown of existing biological order within the body. More specifically, the disorder seen in cancer appears to derive directly from malfunctioning of the controls that are normally responsible for determining when and where cells throughout the body will multiply. In fact, there is ample opportunity for the disorder of cancer to strike a human body. Most of the more than 10^{13} cells in the body continue to carry the genetic information that previously allowed them to come into existence and might, in the future, allow them to multiply once again. This explains why the risk of uncontrolled cell proliferation in countless sites throughout the body is substantial throughout the lives of mammals like ourselves.

To be more accurate, the risk of cancer is far greater than the $>10^{13}$ population size would suggest, since this number represents the average, steady-state population of cells in the body at any point in time during adulthood. The aggregate number of cells that are formed during an average human lifetime is about 10^{16}, a number that testifies to the enormous amount of cell turnover—involving cell death and replacement (almost 10^7 events per second)—that occurs continuously in many tissues in the body. As discussed in Chapters 9 and 12, each time a new cell is formed by the complex process of cell growth and division, there are many ways for things to go awry. Hence, the chance for disaster to strike, including the inadvertent formation of cancer cells, is great.

Since a normal biological process (incessant cell division) is likely to create a substantial risk of cancer, it would seem logical that human populations throughout the world would experience similar frequencies of cancer. However, when cancer **incidence** rates (that is, the rates with which the disease is diagnosed) are examined in various countries, we learn that the risks of many types of cancer vary dramatically (Table 2.5), while other cancers (not indicated in Table 2.5) do indeed show comparable incidence rates across the globe. So, our speculation that all cancers should strike different human populations at comparable rates is simply wrong. Some do and some don't. This realization forces us to reconsider our thinking about how cancers are formed.

Some of the more than 100 types of human cancers do seem to have a high proportion of tumors that are caused by random, unavoidable accidents of nature and thus occur with comparable frequencies in various human populations. This seems to be true for certain pediatric tumors. In addition to this relatively constant "background rate" of some specific cancers, yet other factors appear to intervene in certain populations to increase dramatically the total number of cancer cases. The two obvious contributory factors here are heredity and environment. Different human populations may carry cancer-susceptibility alleles at greatly different frequencies. Alternatively, the environments in which people live may contribute enormously to disease incidence rates. The environment, in the broadest sense, includes both the air and the water that enter into their bodies, as well as aspects of lifestyle, such as dietary choices, reproductive habits, and tobacco usage.

Table 2.5 Geographic variation in cancer incidence and death rates

Countries showing highest and lowest incidence of specific types of cancer[a]			
Cancer site	Country of highest risk	Country of lowest risk	Relative risk H/L[b]
Skin (melanoma)	Australia (Queensland)	Japan	155
Lip	Canada (Newfoundland)	Japan	151
Nasopharynx	Hong Kong	United Kingdom	100
Prostate	U.S. (African American)	China	70
Liver	China (Shanghai)	Canada (Nova Scotia)	49
Penis	Brazil	Israel (Ashkenazic)	42
Cervix (uterus)	Brazil	Israel (non-Jews)	28
Stomach	Japan	Kuwait	22
Lung	U.S. (Louisiana, African American)	India (Madras)	19
Pancreas	U.S. (Los Angeles, Korean American)	India	11
Ovary	New Zealand (Polynesian)	Kuwait	8

Geographic areas showing highest and lowest death rates from specific types of cancer[c]			
Cancer site	Area of highest risk	Area of lowest risk	Relative risk H/L[b]
Lung, male	Eastern Europe	West Africa	33
Esophagus	Southern Africa	West Africa	16
Colon, male	Australia, New Zealand	Middle Africa	15
Breast, female	Northern Europe	China	6

[a]See C. Muir, J. Waterhouse, T. Mack et al., eds., Cancer Incidence in Five Continents, vol. 5, Lyon: International Agency for Research on Cancer, 1987; excerpted by V.T. DeVita, S. Hellman, and S.A. Rosenberg, Cancer: Principles and Practice of Oncology, Philadelphia: Lippincott, 1993.
[b]Relative risk: age-adjusted incidence or death rate in highest country or area (H) divided by age-adjusted incidence or death rate in lowest country or area (L). These numbers refer to age-adjusted rates, e.g., the relative risk of a 60-year-old dying from a specific type of tumor in one country compared with a 60-year-old in another country.
[c]See P. Pisani, D.M. Parkin, F. Bray and J. Ferlay, *Int. J. Cancer* 83:18–29, 1999. This survey divided the human population into 23 geographic areas and surveyed the relative mortality rates of various cancer types in each area.

Which of these two alternatives—heredity or environment—is the dominant determinant of the country-to-country variability of cancer incidence? While many types of disease-causing alleles are distributed unequally in the gene pools of different human populations, these alleles do not seem to explain the dramatically different incidence rates of various cancers throughout the world. This point is demonstrated most dramatically by measuring cancer rates in migrant populations. For example, Japanese experience rates of stomach cancer that are 6 to 8 times higher than those of Americans (Figure 2.20). However, when Japanese settle in the United States, within a generation their offspring exhibit a stomach cancer rate that is comparable to that of the surrounding population. For the great majority of cancers, disease risk therefore seems to be "environmental," where this term is understood to include both physical environment and lifestyle.

As indicated in Table 2.5, the incidence of some types of cancer may vary enormously from one population to the next. Thus, breast cancer in China is about one-sixth as common as in the United States or Northern Europe. Having excluded genetic contributions to this difference, we might then conclude that as much as 85% of the breast cancers in the United States might in theory be avoidable, if only American women were to experience an environment and lifestyle comparable to those of their Chinese counterparts. Even within the American population, there are vast differences in cancer mortality: the

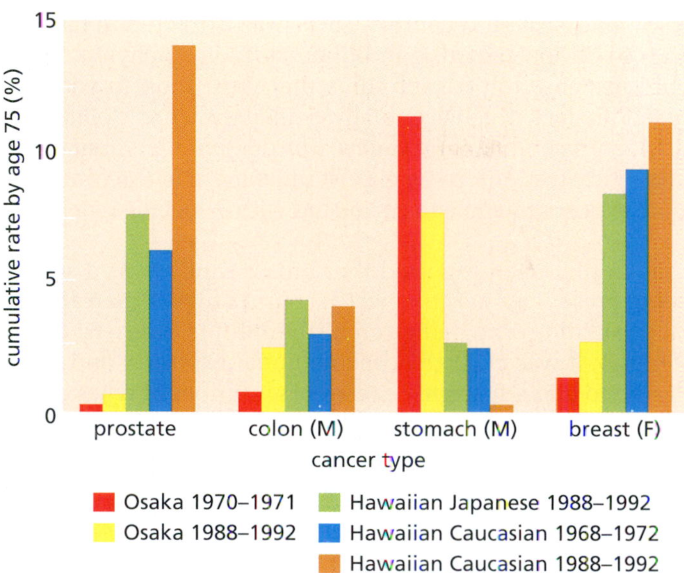

Figure 2.20 Country-to-country comparisons of cancer incidence Public health records reveal dramatic differences in the incidence of certain cancers in different countries. Here, the relative incidence of a group of cancers in Japan and in the American island of Hawaii are presented. Invariably, after Japanese have immigrated to Hawaii, within a generation their cancer rates approach those of the population that settled there before them. This indicates that the differing cancer rates are not due to genetic differences between the Japanese and the American populations. (From J. Peto, *Nature* 411:390–395, 2001.)

Seventh-Day Adventists, whose religion discourages smoking, heavy drinking, and the consumption of meat, die from cancer at a rate that is only about half that of the general population.

For those who wish to understand the **etiologic** (causative) mechanisms of cancer, these findings lead to an inescapable conclusion: the great majority of the commonly occurring cancers are caused by factors or agents that are external to the body, enter into the body, and somehow attack and corrupt its tissues. In a minority of cancers, substantial variations in cancer risk may be attributable to differences in reproductive behavior and the resulting dramatic effects on the hormonal environment within the human female body.

Let us imagine, for the sake of argument, that avoidance of certain obvious cancer-causing factors in diet and lifestyle resulted in a 50% reduction in the risk of dying from cancer in the West, leaving the disease of cancer as the cause of about 10% of overall mortality in this population. Under these conditions, given the approximately 10^{16} mitoses occurring in each human body during a normal life span, we calculate that only 1 in 10^{17} cell divisions would lead directly or indirectly to a clinically detectable cancer. Now, we become persuaded that in spite of the enormous intrinsic risk of developing cancer, the body must be able to mount highly effective defenses that usually succeed in holding off the disease for the 70 or 80 years that most of us spend on this planet. These defenses are the subject of many discussions throughout this book.

2.7 The risks of cancers often seem to be increased by assignable influences including lifestyle

Evidence that certain kinds of cancers are associated with specific exposures or lifestyles is actually quite old, predating modern epidemiology by more than a century. The first known report comes from the observations of the English physician John Hill, who in 1761 noted the connection between the development of nasal cancer and the excessive use of tobacco snuff. Fourteen years later, Percivall Pott, a surgeon in London, reported that he had encountered a substantial number of skin cancers of the scrotum in adolescent men who, in their youth, had worked as chimney sweeps. Within three years, the Danish sweepers guild urged its members to take daily baths to remove the apparently cancer-causing material from their skin. This practice was likely to be the cause of the markedly lower rate of scrotal cancer in continental Europe when compared with Britain even a century later.

Beginning in the mid-sixteenth century, silver was extracted in large quantities from the mines in St. Joachimsthal in Bohemia, today Jáchymov in the Czech Republic. (The silver Joachimsthaler coins that were soon in wide circulation came to be called "thaler," which eventually yielded the word "dollar"!) By the first half of the nineteenth century, lung cancer was documented at high rates in the miners, a disease that was otherwise almost unheard of at the time. Once again, an occupational exposure had been correlated with a specific type of cancer.

In 1839, an Italian physician reported that breast cancer was a scourge in the nunneries, being present at rates that were six times higher than among women in the general population who had given birth multiple times. By the end of the nineteenth century, it was clear that occupational exposure and lifestyle were closely connected to and apparently causes of a number of types of cancer.

The range of agents that might trigger cancer was expanded with the discovery in the first decade of the twentieth century that physicians and others who experimented with the then-recently invented X-ray tubes experienced increased rates of cancer, often developing tumors that arose at the site of irradiation. These observations led, many years later, to an understanding of the lung cancer in the St. Joachimsthaler miners: their greatly increased lung cancer incidence could be attributed to the high levels of radioactivity in the ores coming from these mines.

Perhaps the most compelling association between environmental exposure and cancer incidence was forged in 1949 and 1950 when two groups of epidemiologists reported that individuals who were heavy cigarette smokers ran a lifetime risk of lung cancer that was more than twentyfold higher than that of nonsmokers. The initial results of one of these landmark studies are given in Table 2.6. These various epidemiologic correlations proved to be critical for subsequent cancer research, since they suggested that cancers often had specific, assignable causes, and that a chain of causality might one day be traced between these ultimate causes and the cancerous changes observed in certain human tissues. Indeed, in the half century that followed the 1949–1950 reports, epidemiologists identified a variety of environmental and lifestyle factors that were strongly correlated with the incidence of certain cancers (Table 2.7); in some of these cases, researchers have been able to discover the specific biological mechanisms through which these factors act to cause increased incidence of some of these cancers.

2.8 Specific chemical agents can induce cancer

Coal tar condensates, much like those implicated in cancer causation by Percivall Pott's work, were used in Japan at the beginning of the twentieth century to induce skin cancers in rabbits. Repeated painting of localized areas of the skin of their ears resulted, after many months, in the outgrowth of carcinomas.

Table 2.6 Relative risk of lung cancer as a function of the number of cigarettes smoked per day[a]

	Lifelong nonsmoker	Smokers			
Most recent number of cigarettes smoked (by subjects) per day before onset of disease	—	≥1, <5	≥5, <15	≥15, <25	≥25
Relative risk	1	8	12	14	27

[a]The relative risk indicates the risk of contracting lung cancer compared with that of a nonsmoker, which is set at 1.

From R. Doll and A.B. Hill, *BMJ* 2:739–748, 1950.

Table 2.7 Known or suspected causes of human cancers

Environmental and lifestyle factors known or suspected to be etiologic for human cancers in the United States[a]	
Type	**% of total cases[b]**
Cancers due to occupational exposures	1–2
Lifestyle cancers	
Tobacco-related (sites: e.g., lung, bladder, kidney)	34
Diet (low in vegetables, high in nitrates, salt) (sites: e.g., stomach, esophagus)	5
Diet (high fat, lower fiber, broiled/fried foods) (sites: e.g., bowel, pancreas, prostate, breast)	37
Tobacco and alcohol (sites: mouth, throat)	2

Specific carcinogenic agents implicated in the causation of certain cancers[c]	
Cancer	**Exposure**
Scrotal carcinomas	chimney smoke condensates
Liver angiosarcoma	vinyl chloride
Acute leukemias	benzene
Nasal adenocarcinoma	hardwood dust
Osteosarcoma	radium
Skin carcinoma	arsenic
Mesothelioma	asbestos
Vaginal carcinoma	diethylstilbestrol
Oral carcinoma	snuff

[a]Adapted from Cancer Facts and Figures, American Cancer Society, 1990.
[b]A large number of cancers are thought to be provoked by a diet high in calories acting in combination with many of these lifestyle factors.
[c]Adapted from S. Wilson, L. Jones, C. Coussens and K. Hanna, eds., Cancer and the Environment: Gene-Environment Interaction, Washington, DC: National Academy Press, 2002.

This work, first reported by Katsusaburo Yamagiwa in 1915, was little noticed in the international scientific community of the time (Figure 2.21). In retrospect, it represented a stunning advance, because it directly implicated chemicals (those in coal tar) in cancer causation. Equally important, Yamagiwa's work, together with that of Peyton Rous (to be described in Chapter 3), demonstrated that cancer could be induced at will in laboratory animals. Before these breakthroughs, researchers had been forced to wait for tumors to

(A)

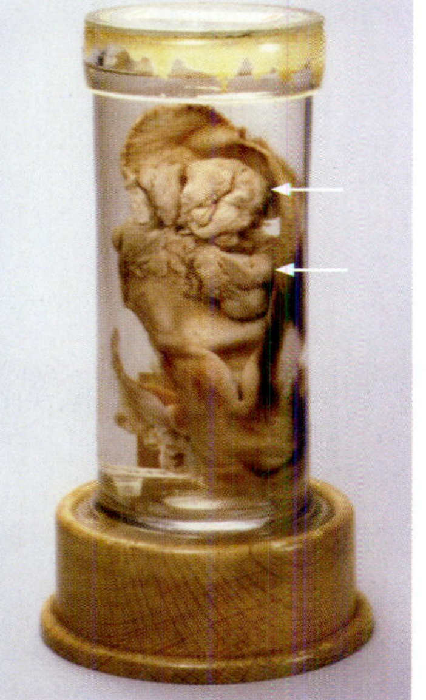
(B)

Figure 2.21 The first induction of tumors by chemical carcinogens (A) In 1915, Katsusaburo Yamagiwa reported the first successful induction of cancer by repeated treatment of rabbit ears with a chemical carcinogen, in this case coal tars. (B) The skin carcinomas *(arrows)* that he induced on the ears of these rabbits are preserved to this day in the medical museum of the University of Tokyo. This particular carcinoma was harvested and fixed following 660 days of painting with coal tar. (Courtesy of T. Taniguchi.)

appear spontaneously in wild or domesticated animals. Now, cancers could be produced according to a predictable schedule, often involving many months of experimental treatment of animals.

By 1940, British chemists had purified several of the components of coal tar that were particularly **carcinogenic** (i.e., cancer-causing), as demonstrated by the ability of these compounds to induce cancers on the skin of laboratory mice. Compounds such as 3-methylcholanthrene, benzo[*a*]pyrene, and 1,2,4,5-dibenz[*a,h*]anthracene were common products of combustion, and some of these hydrocarbons, notably benzo[*a*]pyrene, were subsequently found in the condensates of cigarette smoke as well (Figure 2.22). These findings suggested that certain chemical species that entered into the human body could perturb tissues and cells and ultimately provoke the emergence of a tumor. The same could be said of X-rays, which were also able to produce cancers, ostensibly through a quite different mechanism of action.

While these discoveries were being reported, an independent line of research developed that portrayed cancer as an infectious disease. As described in detail in Chapter 3, researchers in the first decade of the twentieth century found that viruses could cause leukemias and sarcomas in infected chickens. By mid-century, a wide variety of viruses had been found able to induce cancer in rabbits, chickens, mice, and rats. As a consequence, those intent on uncovering the origins of human cancer were pulled in three different directions, since the evidence of cancer causation by chemical, viral, and physical (i.e., radiation) agents had become compelling.

2.9 Both physical and chemical carcinogens act as mutagens

The confusion caused by the three competing theories of carcinogenesis was reduced significantly by discoveries made in the field of fruit fly genetics. In 1927, Hermann Muller discovered that he could induce mutations in the genome of *Drosophila melanogaster* by exposing these flies to X-rays. Most important, this discovery revealed that the genome of an animal was mutable, that is, that its information content could be changed through specific treatments, notably irradiation. At the same time, it suggested at least one mechanism by which X-rays could induce cancer: perhaps radiation was able to mutate the genes of normal cells, thereby creating mutant cells that grew in a malignant fashion.

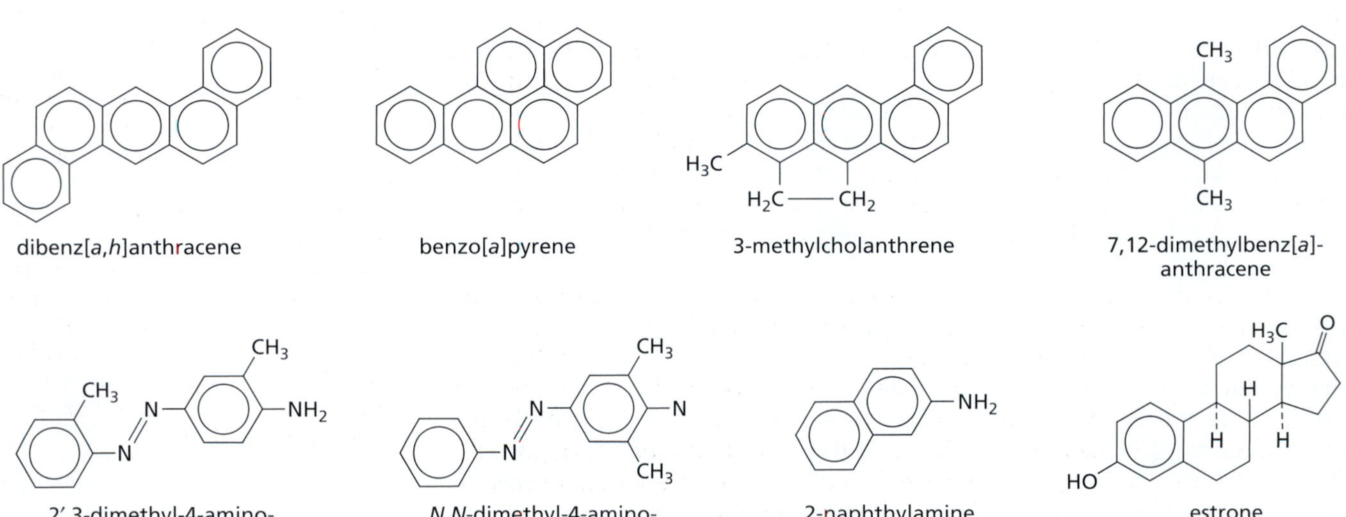

Figure 2.22 Structures of carcinogenic hydrocarbons These chemical species arise from the incomplete combustion of organic (i.e., carbon-containing) compounds. Each of the chemical structures shown here, which were already determined before 1940, represents a chemical species that was found, following purification, to be potently carcinogenic. The four compounds shown in the top row are all polycyclic aromatic hydrocarbons (PAHs). (From E.C. Miller, *Cancer Res.* 38:1479–1496, 1978.)

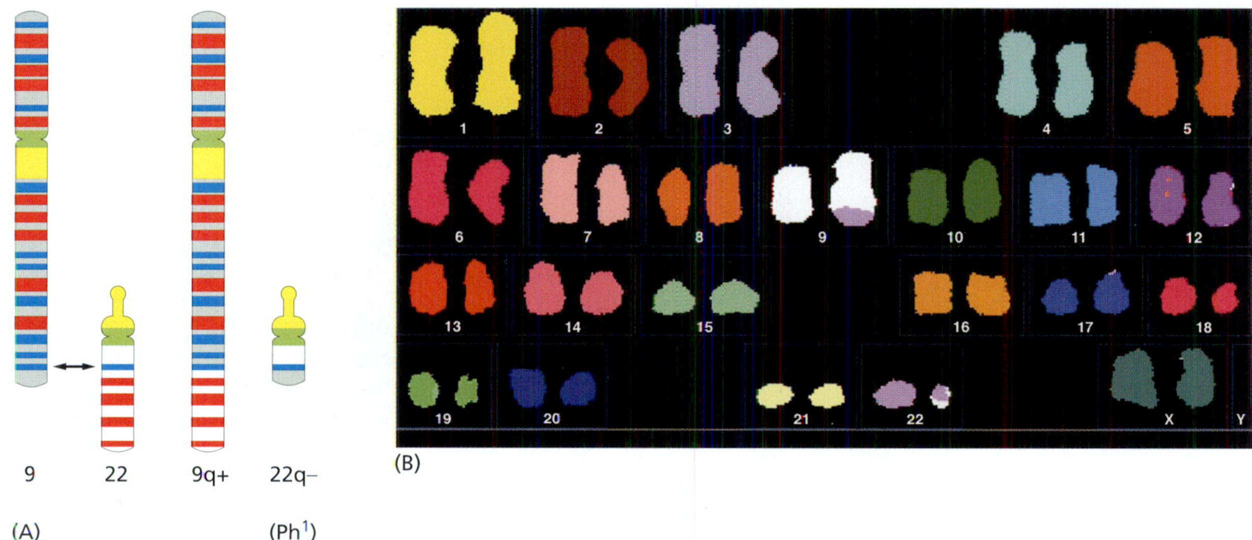

Figure 2.23 Structure of the Philadelphia chromosome Analyses of the banding patterns of metaphase chromosomes of chronic myelogenous leukemia (CML) cells first revealed the characteristic tiny chromosome (called the "Philadelphia chromosome" or Ph^1) that is present in the great majority of CML cells. (A) This banding pattern, determined through light-microscopic surveys of stained metaphase chromosomes, is illustrated here schematically. While the chromosomal translocation generating the two altered chromosomes (9q+,22q–) is *reciprocal* (i.e., involving a loss and a gain by each of the two chromosomes), the sizes of the exchanged chromosomal arms are unequal, leading to the greatly truncated Chromosome 22 (i.e., 22q–). The arrow indicates the point of crossing over, known as the translocation *breakpoint*. (B) The relatively minor change to the tumor cell karyotype that is created by the CML translocation is apparent in this SKY analysis, in which chromosome-specific probes are used, together with fluorescent dyes and computer-generated coloring, to visualize the entire chromosomal complement of CML cells. As is apparent, one of the two Chromosomes 9 has acquired a *light purple* segment (a color assigned to Chromosome 22) at the end of its long arm. Reciprocally, one of the two Chromosomes 22 has acquired a white region (characteristic of Chromosome 9) at the end of its long arm. (A, from B. Alberts et al., Molecular Biology of the Cell, 4th ed. New York: Garland Science, 2002; B, courtesy of Thomas Ried and Nicole McNeil.)

By the late 1940s, a series of chemicals, many of them alkylating agents of the type that had been used in World War I mustard gas warfare, were also found to be **mutagenic** for fruit flies. Soon thereafter, some of these same compounds were shown to be carcinogenic in laboratory animals. These findings caused several geneticists to speculate that cancer was a disease of mutant genes, and that carcinogenic agents, such as X-rays and certain chemicals, succeeded in inducing cancer through their ability to mutate genes.

These speculations were hardly the first ones of this sort. As early as 1914, the German biologist Theodor Boveri, drawing on yet older observations of others, suggested that chromosomes, which by then had been implicated as carriers of genetic information, were aberrant within cancer cells, and that cancer cells might therefore be mutants. Boveri's notion, along with many other speculations on the origin of cancer, gained few adherents, however, until the discovery in 1960 of an abnormally configured chromosome in a large proportion of cases of chronic myelogenous leukemia (CML). This chromosome, soon called the Philadelphia chromosome after the place of its discovery, was clearly a distinctive characteristic of this type of cancer (Figure 2.23). Its reproducible association with this class of tumor cells suggested, but hardly proved, that it played a causal role in tumorigenesis.

In 1975 Bruce Ames, a bacterial geneticist working at the University of California in Berkeley, reported experimental results that lent great weight to the theory that carcinogens can function as mutagens. Decades of experiments with laboratory mice and rats had demonstrated that chemical carcinogens acted with vastly different potencies, differing by as much as 1 million-fold in their ability to induce cancers. Such experiments showed, for example, that one microgram of aflatoxin, a compound produced by molds growing on peanuts and wheat, was as potently carcinogenic as was a 10,000 times greater weight of the synthetic compound benzidine. Ames posed the question whether these various compounds were also mutagenic, more specifically, whether compounds that were potent carcinogens also happened to be potent mutagens.

The difficulty that Ames faced in his initial attempts to address this question was a simple one: there were no good ways of measuring the relative mutagenic potencies of various chemical species. So Ames set out to devise his own method for quantifying mutagenic potency. He developed an experimental protocol that consisted of applying various carcinogenic chemicals to a population of *Salmonella* bacteria growing in Petri dishes and then scoring for the abilities of these carcinogens to mutate the bacteria. The readout here was the number of colonies of *Salmonella* that grew out following exposure to one or another chemical.

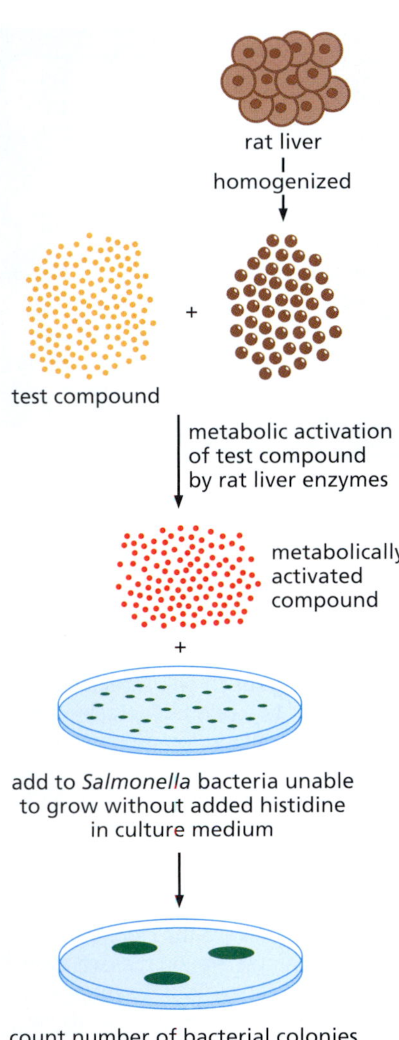
rat liver
homogenized
test compound
+
metabolic activation of test compound by rat liver enzymes
metabolically activated compound
+
add to *Salmonella* bacteria unable to grow without added histidine in culture medium
count number of bacterial colonies that have undergone mutation enabling them to grow without added histidine

Figure 2.24 The Ames test for gauging mutagenicity The Ames test makes it possible to quantitatively assess the mutagenic potency of a test compound. To begin, the liver of a rat (or other species) is homogenized, releasing the enzymes that can metabolically activate a chemical to its mutagenic form. The rat liver homogenate *(brown dots)* is then mixed with the test compound *(orange)*, which often results in the conversion of the test compound to a chemically activated state *(red)*. This mixture (still containing the liver homogenate, *not shown*) is applied to a dish of mutant *Salmonella* bacteria *(small green dots)* that require the amino acid histidine in their culture medium in order to grow. Since histidine is left out of the medium, only those bacteria that are mutated to a histidine-independent genotype (and phenotype) will be able to grow, and each of these will yield a large colony *(green)* that can be counted with the naked eye, indicating how many mutant bacteria (and thus mutant alleles) were generated by the brief exposure to the activated compound.

In detail, Ames used a strain of *Salmonella* that was already mutant and therefore unable to grow in medium lacking the amino acid histidine. The mutant allele that caused this phenotype was susceptible to back-mutation to a wild-type allele. Once the wild-type allele was formed in response to exposure to a mutagen, a bacterium carrying this allele became capable of growing in Ames's selective medium, multiplying until it formed a colony that could be scored by eye (Figure 2.24).

In principle, Ames needed only to introduce a test compound into a Petri dish containing his special *Salmonella* strain. By counting the number of bacterial colonies that appeared later on, he could gauge the mutagenic potency of this compound. But there remained one substantial obstacle to the success of this mutagenesis assay. Detailed studies of a number of chemical carcinogens had shown that after carcinogenic molecules entered into the tissues of laboratory animals, they were metabolized into yet other chemical species. In many cases, the resulting products of metabolism, rather than the initially introduced chemicals, seemed to be the agents that were directly responsible for the observed cancer induction. These metabolized compounds were found to be highly reactive chemically and able to form covalent bonds with the various macromolecules known to be present in cells—DNA, RNA, and protein.

The original, unmodified compounds that were introduced into laboratory animals came to be called **procarcinogens** to indicate their ability to become converted into actively carcinogenic compounds, which were labeled **ultimate carcinogens**. This chemical conversion complicated the design of Ames's mutagenesis assay. If many compounds required metabolic activation before their carcinogenicity was apparent, it seemed plausible that their mutagenic powers would also be evident only after such metabolic conversion. Given the radically different metabolisms of bacteria and mammalian cells, it was highly unlikely that Ames's *Salmonella* bacteria would be able to accomplish the metabolic activation of procarcinogens that occurred in the tissues of laboratory animals.

This information forced Ames to add an extra step to his mutagenesis assay, a step suggested by earlier work of others. It was known that a great many chemicals introduced into the body undergo metabolic conversion in the liver. Many of these conversions could be achieved in the test tube simply by mixing such chemicals with homogenized liver. So Ames mixed rat liver homogenates with his test compounds and then introduced this mixture into the Petri dishes carrying *Salmonella*. (We now know that the metabolic activation of procarcinogens in the liver is often mediated by enzymes that are normally involved, paradoxically, in the **detoxification** of compounds introduced into the body; see Section 12.6.)

When Ames added liver homogenate to his *Salmonella* cultures, his assay of mutagenic potency worked beautifully. It revealed that a number of known carcinogens were also actively mutagenic. Even more important were the correlations that Ames found. Chemicals that were potently mutagenic were also powerful carcinogens. Those that were weakly mutagenic induced cancer poorly. These correlations, as plotted by others, extended over five orders of magnitude of potency (Figure 2.25).

As we have read, the notion that carcinogens are mutagens predated Ames's work by a quarter of a century. Nonetheless, the results of his analyses galvanized researchers interested in the origins of cancer, since they addressed the carcinogen–mutagen relationship so directly. Their reasoning went like this: Ames had demonstrated the mutagenic powers of certain chemical compounds in bacteria. Since the genomes of bacterial and animal cells are both made of the same chemical substance—double-stranded DNA—it was likely that the compounds that induced mutations in the *Salmonella* genome were similarly capable of inducing mutations in the genomes of animal cells. Hence, the "Ames test," as it came to be known, should be able to predict the mutagenicity of these compounds after they had been introduced into the body of a mammal. And in light of the correlation between mutagenic and carcinogenic potency, the Ames test could be employed to screen various substances for their carcinogenic powers, and thus for their threat to human health. By 1976, Ames and his group reported on the mutagenic potencies of 300 distinct organic compounds. Yet other tests for mutagenic potency were developed in the years that followed (Sidebar 2.1).

Ames's results led to the next deduction, really more of a speculation: if, as Ames argued, carcinogens are mutagens, then it followed that the carcinogenic powers of various agents derived directly from their ability to induce mutations in the cells of target tissues. As a further deduction, it seemed inescapable that the cancer cells created by chemical carcinogens carry mutated genes. These mutated genes, whatever their identity, must in some way be responsible for the aberrant growth phenotypes of such cancer cells.

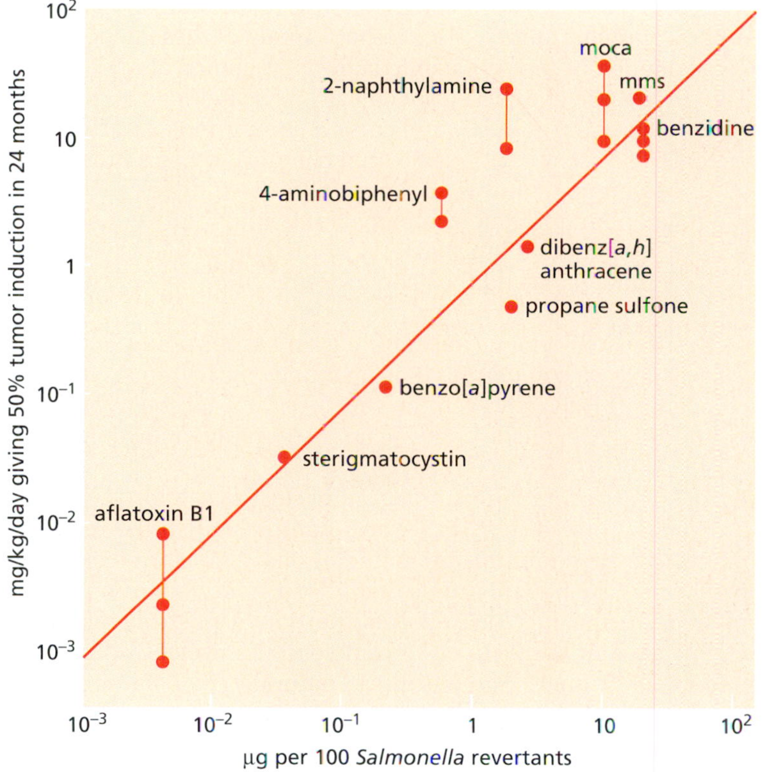

Figure 2.25 Mutagenic versus carcinogenic potency On this log–log plot, the relative carcinogenic potencies of a group of chemicals (*ordinate*) that have been used to treat laboratory animals (rats and mice) are plotted as a function of their mutagenic potencies (*abscissa*) as gauged by the Ames test (see Figure 2.24). Since both the ordinate and abscissa are plotted as the amount of compound required to elicit an observable effect (yielding tumors in 50% of treated animals or 100 colonies of mutant *Salmonella* bacteria, termed here "revertants"), the compounds that are the most potent mutagens and most potent carcinogens appear in the lower left of this graph. Note that both parameters vary by five orders of magnitude. (Adapted from M. Meselson et al., in H.H. Hiatt et al., eds., Origins of Human Cancer, Book C: Human Risk Assessment. Cold Spring Harbor, NY: Cold Spring Harbor Laboratory Press, 1977.)

Sidebar 2.1 Other tests for mutagenicity help assess possible carcinogenicity The Ames test is only one of a number of biological assay systems that can be used to assess the mutagenic potency of suspected carcinogenic chemicals. Many of these other assays depend upon exposing mammalian cells directly to the chemical compounds being tested and the subsequent use of a diverse array of biological readouts. For example, a test for sister chromatid exchange (SCE) depends on the ability to measure crossingover between the two paired chromatids that together constitute a chromosome in the late (i.e., G_2) phase of a cell's growth-and-division cycle. Many mutagenic agents have been shown to provoke this SCE. Mutagenic agents may also register as being capable of inducing the formation of fragmented cell nuclei, that is, **micronuclei**. Use of genetics has made it possible to test mammalian cells for their mutation-induced loss of thymidine kinase enzyme or the HGPRT enzyme (hypoxanthine guanine phosphoribosyl transferase). The ability to examine under the light microscope the chromosomal array (i.e., the karyotype) of cells in the metaphase of mitosis makes it possible to screen for chromosomal aberrations inflicted by test compounds. Yet another assay gauges the degree of DNA labeling in those cells that are in the G_1 or G_2 phase of the cell cycle (described in Chapter 8); such non-S-phase labeling, sometimes referred to as "unscheduled DNA synthesis," has also been shown to be a good indicator of the genomic damage that has been inflicted on a cultured cell, since this type of DNA synthesis represents one key step in the process used by cells to repair damaged DNA.

None of these tests has proven to be ideal as a predictor of the carcinogenicity of a test substance. The Ames test, as an example, has been found by some to have a sensitivity (% of established carcinogens identified as mutagens) of about 54% and a specificity (% of noncarcinogens identified as nonmutagens) of 70%.

This logic was transferable to X-ray carcinogenesis as well. Since X-rays were mutagens and carcinogens, it followed that they also induced cancer through their ability to mutate genes. This convergence of cancer research with genetics had a profound effect on researchers intent on puzzling out the origins of cancer. Though still unproven, it appeared likely that the disease of cancer could be understood in terms of the mutant genes carried by cancer cells.

2.10 Mutagens may be responsible for some human cancers

The connection between carcinogenesis and mutagenesis seemed to shed light on how human tumors arise. Perhaps many of these neoplasms were the direct consequence of the mutagenic actions of chemical and physical carcinogens. The mutagenic chemicals, specifically, procarcinogens, need not derive exclusively from the combustion of carbon compounds and the resulting formation of coal tars. It seemed plausible that chemical species present naturally in foodstuffs or generated during cooking could also induce cancer. Even if many foods did not contain ultimate carcinogens, chemical conversions carried out by liver cells or by the abundant bacteria in the colon might well succeed in creating actively mutagenic and thus carcinogenic chemical species.

As this research on the causes of human cancer proceeded, it became apparent that virtually all compounds that are mutagenic in human cells are likely to be carcinogenic as well. However, the converse does not seem to hold: chemical compounds that are carcinogenic are not necessarily mutagenic (see Sidebar 2.2).

Sidebar 2.2 Not all carcinogens are mutagenic By the early 1990s, it became apparent that the carcinogen–mutagen equivalence no longer held. By then, a more extensive use of the Ames test showed that as many as 40% of the compounds that were known to be carcinogenic in rodents showed no obvious mutagenicity in the *Salmonella* mutation assay. So the conclusions drawn earlier from the initial applications of Ames's test required major revision: some carcinogens act through their ability to mutate DNA, while others promote the appearance of tumors through nongenetic mechanisms. We will encounter these nonmutagenic carcinogens, often called tumor **promoters**, again in Chapter 11.

By 1991, Ames and others had used his test to catalog the mutagenic powers of a diverse group of chemicals and natural foodstuffs, including many of the plants that are common and abundant in the Western diet. As Ames argued, the presence of such compounds in foodstuffs derived from plants was hardly surprising, since plants have evolved thousands, possibly millions of distinct toxic chemical compounds in order to defend themselves from predation by insects and larger animals. Some of these naturally toxic compounds, initially developed as anti-predator defenses, might also, as an unintended side-effect, be mutagenic (Table 2.8).

A diverse set of discoveries led to the model, which remains unproven in many of its aspects to this day, that a significant proportion of human cancer is attributable directly to the consumption of foodstuffs that are mutagenic and hence carcinogenic. Included among these foodstuffs is, for example, red meat, which upon cooking at high temperatures generates compounds such as heterocyclic amines, which are potently mutagenic (see Section 12.6).

The difficulties in proving this model derive from several sources. Each of the plant and animal foodstuffs in our diet is composed of thousands of diverse chemical species present in vastly differing concentrations. Almost all of these compounds undergo metabolic conversions once they are inside our bodies, first in the gastrointestinal tract and often thereafter in the liver. Accordingly, the number of distinct chemical species that are introduced into our bodies is incalculable. Each of these introduced compounds, once it is present in the body, may then be concentrated in some cells or quickly metabolized and excreted. This creates a further dimension of complexity.

Moreover, the actual mutagenicity of various compounds in different cell types may vary enormously because of metabolic differences in these cells. For example, some cells, such as **hepatocytes** in the liver, express high levels of biochemical species designed to scavenge and inactivate mutagenic compounds, while others, such as fibroblasts, express far lower levels. In sum, the ability to relate the mutagenicity of foodstuffs to actual rates of mutagenesis and carcinogenesis in the human body is far beyond our reach at present—a problem of intractable complexity (see also Sidebar 2.3).

Table 2.8 A sampling of Bruce Ames's roster of carcinogens identified in the normal diet[a]

Foodstuff	Compound	Concentration in foodstuff
Black pepper	piperine	100 mg/g
Common mushroom	agaritine	3 mg/g
Celery[b]	furocoumarins, psoralens	1 μg/g, 0.8 μg/g
Rhubarb	anthraquinones	varies
Cocoa powder	theobromine	20 mg/g
Mustard, horseradish	allyl isothiocyanate	varies
Alfalfa sprouts	canavanine[c]	15 mg/g
Burnt materials[d]	large number	varies
Coffee	caffeic acid	11.6 mg/g

[a]Ames has cited 37 naturally occurring compounds that have registered as carcinogens in laboratory animals; one or more have been found in each of the following foodstuffs: absinthe, allspice, anise, apple, apricot, banana, basil, beet, broccoli, Brussels sprouts, cabbage, cantaloupe, caraway, cardamom, carrot, cauliflower, celery, cherries, chili pepper, chocolate, cinnamon, cloves, coffee, collard greens, comfrey herb tea, coriander, corn, currants, dill, eggplant, endive, fennel, garlic, grapefruit, grapes, guava, honey, honeydew melon, horseradish, kale, lemon, lentils, lettuce, licorice, lime, mace, mango, marjoram, mint, mushrooms, mustard, nutmeg, onion, orange, paprika, parsley, parsnip, peach, pear, peas, black pepper, pineapple, plum, potato, radish, raspberries, rhubarb, rosemary, rutabaga, sage, savory, sesame seeds, soybean, star anise, tarragon, tea, thyme, tomato, turmeric, and turnip
[b]The levels of these can increase 100-fold in diseased plants.
[c]Canavanine is indirectly genotoxic because of oxygen radicals that are released, perhaps during the inflammatory reactions associated with elimination of canavanine-containing proteins.
[d]On average, several grams of burnt material are consumed daily in the form of bread crusts, burnt toast, and burnt surfaces of meats cooked at high temperature.

Adapted from B.N. Ames, *Science* 231:1256–1264, 1983; B.N. Ames and L.S. Gold, *Proc. Natl. Acad. Sci. USA* 87:7777–7781, 1990; and L.S. Gold, B.N. Ames and T.H. Slone, Misconceptions about the causes of cancer, in D. Paustenbach, ed., Human and Environmental Risk Assessment: Theory and Practice, New York: John Wiley & Sons, 2002, pp. 1415–1460.

Sidebar 2.3 The search for elusive human carcinogens Ideally, the identification of important human carcinogens should have been aided by the use of *in vitro* assays, such as the Ames test (Section 2.9), and *in vivo* tests—exposure of laboratory animals to agents suspected of causing cancer (Section 2.8). In truth, however, these various types of laboratory tests have failed to register important human carcinogens. Instead, we have learned about their carcinogenicity because of various epidemiologic studies. For example, the most important known human carcinogen—tobacco smoke—would likely have escaped detection because it is a relatively weak carcinogen in laboratory rodents; and another known human carcinogen—asbestos—would have eluded detection by both *in vitro* and *in vivo* laboratory tests. Conversely, some frequently used drugs, such as phenobarbital and isoniazid, register positively in the Ames test, and saccharin registers as a carcinogen in male laboratory rats (see Sidebar 11.16), but epidemiologic evidence indicates conclusively that none of these is actually associated with increased cancer risk in humans who have been exposed to these compounds over long periods of time. Hence, the development of truly useful, predictive tests of human carcinogens still lies in the future.

2.11 Synopsis and prospects

The descriptions of cancer and cancer cells developed during the second half of the nineteenth century and the first half of the twentieth indicated that tumors were nothing more than normal cell populations that had run amok. Moreover, many tumors seemed to be composed largely of the descendants of a single cell that had crossed over the border from normalcy to malignancy and proceeded to spawn the billions of descendant cells constituting these neoplastic masses. This scenario drew attention to the nature of the cells that founded tumors and to the mechanisms that led to their transformation into cancer cells. It seemed that each tumor mass was composed largely of replicas of a founding, transformed cell. If one could understand why this cell multiplied uncontrollably, somehow other pieces of the cancer puzzle were likely to fall into place.

Still, existing observations and experimental techniques offered little prospect of revealing precisely why a cell altered its behavior, transforming itself from a normal into a malignant cell. The carcinogen = mutagen theory seemed to offer some clarification, since it implicated mutant cellular genes as the agents responsible for disease development and, therefore, for the aberrant behavior of cancer cells. Perhaps there were mutant genes operating inside cancer cells that programmed the runaway proliferation of these cells, but the prospects for discovering such genes and understanding their actions seemed remote. No one knew how many genes were present in the human genome and how to analyze them. If mutant genes really did play a major part in cancer causation, they were likely to be small in number and dwarfed by the apparently vast number of genes present in the genome as a whole. They seemed to be the proverbial needles in the haystack, in this case a vast haystack of unknown size.

This theorizing about cancer's origins was further complicated by two other important considerations. First, many apparent carcinogens failed the Ames test, providing strong suggestion that they were nonmutagenic. Second, certain viral infections seemed to be closely connected to the incidence of a small but significant subset of human cancer types. Somehow, their carcinogenic powers had to be reconciled with the actions of mutagenic carcinogens and mutant cellular genes.

By the mid-1970s, recombinant DNA technology, including gene cloning, began to influence a wide variety of biomedical research areas. While appreciating the powers of this new technology to isolate and characterize genes, cancer researchers remained frustrated as to how they should proceed to exploit it to track down the elusive mutant genes that were responsible for cancer. One thing was clear, however. Sooner or later, the process of cancer **pathogenesis** (disease development) needed to be explained and understood in molecular terms. Somehow, the paradigm of DNA, RNA, and proteins, so powerful in elucidating a vast range of biological processes, would need to be brought to bear on the cancer problem.

In the end, the breakthrough that opened up this logjam came from study of the tumor viruses, which by most accounts were minor players in human cancer development. Tumor viruses were genetically simple, and yet they possessed potent carcinogenic powers. To understand these viruses and their import, we need to move back, once again, to the beginning of the twentieth century and confront another of the ancient roots of modern cancer research. This we do in Chapter 3.

Key concepts

- The nineteenth-century discovery that all cells of an organism descend from the fertilized egg led to the realization that tumors are not foreign bodies but growths derived from normal tissues. The comparatively disorganized tissue architecture of tumors pointed toward cancer being a disease of malfunctioning cells.
- Tumors can be either benign (localized, noninvasive) or malignant (invasive, metastatic). The metastases spawned by malignant tumors are responsible for almost all deaths from cancer.
- Tumors are classified into four major groups according to their origin (epithelial, mesenchymal, hematopoietic, and neuroectodermal).
- Virtually all cell types in the body can give rise to cancer, but the most common human cancers are of epithelial origin—the carcinomas. Most carcinomas fall into two categories: squamous cell carcinomas arise from epithelia that form protective cell layers, while adenocarcinomas arise from secretory epithelia.
- Nonepithelial malignant tumors include (1) sarcomas, which originate from mesenchymal cells; (2) hematopoietic cancers, which arise from cells of the circulatory and the immune systems; and (3) neuroectodermal tumors, which originate from components of the nervous system.
- Some tumors do not fit this classification scheme. Moreover, occasionally the origin of a tumor cannot be identified because its cells have dedifferentiated (shed all tissue-specific traits); such tumors are said to be anaplastic.
- Cancers seem to develop progressively, with tumors demonstrating different gradations of abnormality along the way from benign to metastatic.
- Benign tumors may be hyperplastic or metaplastic. Hyperplastic tissues appear normal except for an excessive number of cells, whereas metaplastic tissues show displacement of normal cells by normal cell types not usually encountered at that site. Metaplasia is most frequent in epithelial transition zones.
- Dysplastic tumors contain cells that are cytologically abnormal. Dysplasia is a transitional state between completely benign and premalignant. Adenomatous growths (adenomas, polyps, papillomas, and warts) are dysplastic epithelial tumors that are considered to be benign because they respect the boundary created by the basement membrane.
- Tumors that breach the basement membrane and invade underlying tissue are malignant. An even further degree of abnormality is metastasis, the seeding of tumor colonies to different sites in the body. Metastasis requires not only invasiveness but also such newly acquired traits as motility and adaptation to foreign environments.
- Biochemical and genetic markers were used to determine that human tumors are monoclonal (descended from one ancestral cell) rather than polyclonal (descended from different subpopulations of cells), although confounding factors may mask the true nature of a tumor's origins.
- Although the incidence of some cancers (e.g., some pediatric ones) is comparable worldwide, many vary dramatically by country and therefore cannot be

due simply to a normal biologic process gone awry by chance. Differences in heredity or environment might well explain these differences; in fact, epidemiologic studies have shown that environment is the dominant determinant of the country-by-country variations in cancer incidence.

- Laboratory research supported the epidemiologic studies by directly implicating chemical and physical agents (tobacco, coal dust, X-rays) as causes of cancers. However, the possibility of cancer as an infectious disease arose when viruses were found to cause leukemias and sarcomas in chickens.
- A possible mechanism that supported carcinogenesis by physical and chemical agents surfaced in 1927 when mutations were induced in fruit flies by exposing them to X-rays. By 1950, a series of chemicals also were found to be mutagenic for fruit flies and carcinogenic in lab animals. This led to the speculation that cancer was a disease of mutant genes and that carcinogenic agents induced cancer through their ability to mutate genes.
- In 1975 the Ames test provided support for this idea by showing that many carcinogens can act as mutagens. However, additional research showed that almost all compounds that are mutagenic are likely to be carcinogens, but the converse does not hold true. So, some carcinogens act through their ability to mutate DNA, while others promote tumorigenesis through nongenetic mechanisms. Such nonmutagenic carcinogens are called tumor promoters.
- The Ames test combined with other discoveries led to the model, still unproven, that a significant portion of human cancers is attributable directly to the consumption of foodstuffs that are mutagenic and hence carcinogenic.

Thought questions

1. What types of observation allow a trained pathologist to identify the tissue of origin of a tumor? And why are certain tumors (5–10%) extremely difficult to assign to a specific tissue of origin?
2. Under certain circumstances, all tumors of a class can be traced to a specific embryonic cell layer, while in other classes of tumors, no such association can be made. What tumors would fit into each of these two groupings?
3. What evidence persuades us that a cancer arises from the native tissues of an individual rather than invading the body from outside and thus being of foreign origin?
4. How compelling are the arguments for the monoclonality of tumor cell populations and what logic and observations undermine the conclusion of monoclonality?
5. How can we estimate what percentage of cancers in a population that are avoidable (through virtuous life styles) and what percentage occur because of an unavoidable background incidence that strikes a population independent of the specifics of its life style?
6. What limitations does the Ames test have in predicting the carcinogenicity of various agents?
7. In the absence of being able to directly detect mutant genes within cancer cells, what types of observation allow one to infer that cancer is a disease of mutant cells?

Additional reading

Ames BN (1983) Dietary carcinogens and anticarcinogens. *Science* 231, 1256–1264.

Bickers DR & Lowy DR (1989) Carcinogenesis: a fifty-year historical perspective *J. Invest. Dermatol.* 92, 121S–131S.

Greenlee RT, Murray T, Bolden S & Wingo PA (2000) Cancer statistics 2000. *CA Cancer J. Clin.* 50, 7–33.

Peto J (2001) Cancer epidemiology in the last century and the next decade. *Nature* 411, 390–395.

Pisani P, Parkin DM, Bray F & Ferlay J (1999) Estimates of worldwide mortality from 25 cancers in 1990. *Int. J. Cancer* 83, 18–29.

Preston-Martin S, Pike MC, Ross RK et al. (1990) Increased cell division as a cause of human cancer. *Cancer Res.* 50, 7415–7421.

Wilson S, Jones L, Coussens C & Hanna K (eds.) (2002) Cancer and the Environment: Gene-Environment Interaction. Washington, DC: National Academy Press.

Chapter 3

Tumor Viruses

> A tumor of the chicken ... has been propagated in this laboratory since October, 1909. The behavior of this new growth has been throughout that of a true neoplasm, for which reason the fact of its transmission by means of a cell-free filtrate assumes exceptional importance.
>
> Peyton Rous, cancer biologist, 1911

Viruses are capable of causing a wide variety of human diseases, ranging from rabies to smallpox to the common cold. The great majority of these infectious agents do harm through their ability to multiply inside infected host cells, to kill these cells, and to release progeny virus particles that proceed to infect other hosts nearby. The **cytopathic** (cell-killing) effects of viruses, together with their ability to spread rapidly throughout a tissue, enable these agents to leave a wide swath of destruction in their wake.

But the peculiarities of certain viral replication cycles may on occasion yield quite another outcome. Rather than killing infected cells, some viruses may, quite paradoxically, force their hosts to thrive, indeed, to proliferate uncontrollably. In so doing, such viruses—often called tumor viruses—can create cancer.

At one time, in the 1970s, tumor viruses were studied intensively because they were suspected to be the cause of many common human cancers. This notion was not borne out by the evidence subsequently gathered during that decade, which indicated that virus-induced cancers represent only a minority of the cancer types afflicting humans. Nonetheless, this line of research proved to be invaluable for cancer biologists: study of various tumor viruses provided the key for opening many of the long-hidden secrets of human cancers, including the great majority of cancers that have no connection with tumor virus infections.

As we will see, tumor virus research had a highly variable history over the course of the last century. These infectious agents were discovered in the first decade of the twentieth century and then retreated from the center stage of science. Half a century later, interest in these agents revived, culminating in the frenetic pace of tumor virus research during the 1970s.

The cancer-causing powers of tumor viruses drove many researchers to ask precisely how they succeed in creating disease. Most of these viruses possess relatively simple genomes containing small numbers of viral genes, yet some were found able to overwhelm an infected cell and its vastly more complex genome and to redirect cell growth in new directions. Such behavior indicated that tumor viruses have developed extremely potent genes to perturb the complex regulatory circuitry of the host cells that they infect.

By studying tumor viruses and their mechanisms of action, researchers changed the entire mindset of cancer research. Cancer became a disease of genes and thus a condition that was susceptible to analysis by the tools of molecular biology and genetics. When this story began, no one anticipated how obscure tumor viruses would one day revolutionize the study of human cancer pathogenesis.

3.1 Peyton Rous discovers a chicken sarcoma virus

In the last two decades of the nineteenth century, the research of Louis Pasteur and Robert Koch uncovered the infectious agents that were responsible for dysentery, cholera, rabies, and a number of other diseases. By the end of the century, these agents had been placed into two distinct categories, depending on their behavior upon filtration. Solutions of infectious agents that were trapped in the pores of filters were considered to contain bacteria. The other agents, which were small enough to pass through the filters, were classified as viruses. On the basis of this criterion, the agents for rabies, foot-and-mouth disease, and smallpox were categorized as viruses.

Cancer, too, was considered a candidate infectious disease. As early as 1876, a researcher in Russia reported the transmission of a tumor from one dog to another: chunks of tumor tissue from the first dog were implanted into the second, whereupon a tumor appeared several weeks later. This success was followed by many others using rat and mouse tumors.

The significance of these early experiments remained controversial. Some researchers interpreted these outcomes as proof that cancer was a transmissible disease. Yet others dismissed these transplantation experiments, since in their eyes, such work showed only that tumors, like normal tissues, could be excised from one animal and forced to grow as a graft in the body of a second animal.

In 1908, two researchers in Copenhagen reported their success in extracting a filterable agent from chicken leukemia cells and transmitting this agent to other birds, which then contracted the disease. The two Danes did not follow up on their initial discovery, and it remained for Peyton Rous, working at the Rockefeller Institute in New York, to found the discipline of tumor virology (Figure 3.1).

In 1909, Rous began his study of a sarcoma that had appeared in the breast muscle of a hen. In initial experiments, Rous succeeded in transmitting the tumor by preparing small fragments of tumor and implanting these into other birds of the same breed. Later, as a variation of this experimental plan, he ground up a fragment of a sarcoma in sand and filtered the resulting homogenate (Figure 3.2). When he injected the resulting filtrate into young birds, they too developed tumors, sometimes within several weeks. He subsequently found that these

Figure 3.1 The young and the old Peyton Rous Peyton Rous began his work in 1910 that led to the discovery of Rous sarcoma virus (RSV). More than 50 years later (1966), he received the Nobel Prize in Medicine and Physiology for this seminal work—a tribute to his persistence and longevity. (Courtesy of the Rockefeller University Archives.)

induced tumors could also be homogenized to yield, once again, an infectious agent that could be transmitted to yet other birds, which also developed sarcomas at the sites of injection.

These serial passages of the sarcoma-inducing agent from one animal to another yielded a number of conclusions that are obvious to us now but at the time were nothing less than revolutionary. The carcinogenic agent, whatever its nature, was clearly very small, since it could pass through a filter. Hence, it was a virus (see Sidebar 3.1). This virus could cause the appearance of a sarcoma in an injected chicken, doing so on a predictable timetable. Such an infectious agent offered researchers the unique opportunity to induce cancers at will rather than relying on the spontaneous and unpredictable appearance of tumors in animals or humans. In addition to its ability to induce cancer, this agent, which came to be called Rous sarcoma virus (RSV), was capable of multiplying within the tissues of the chicken; far more virus could be recovered from an infected tumor tissue than was originally injected.

In 1911, when Rous finally published his work, yet another report appeared on a transmissible virus of rabbit tumors, called myxomas. Soon thereafter, Rous

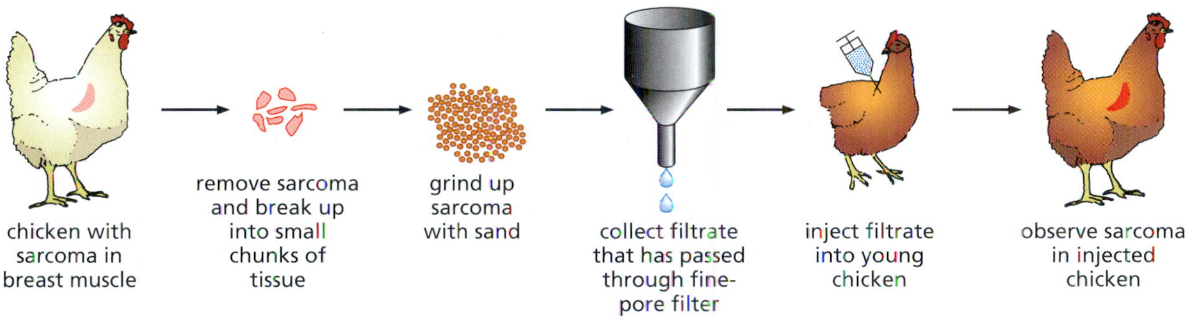
chicken with sarcoma in breast muscle
remove sarcoma and break up into small chunks of tissue
grind up sarcoma with sand
collect filtrate that has passed through fine-pore filter
inject filtrate into young chicken
observe sarcoma in injected chicken

Figure 3.2 Rous's protocol for inducing sarcomas in chickens Rous removed a sarcoma from the breast muscle of a chicken, ground it with sand, and passed the resulting homogenate through a fine-pore filter. He then injected the filtrate (the liquid that passed through the filter) into the wing web of a young chicken and observed the development of a sarcoma many weeks later. He then ground up this new sarcoma and repeated the cycle of homogenization, filtration, and injection, once again observing a tumor in another young chicken. These cycles could be repeated indefinitely; after repeated serial passaging, the virus was able to produce sarcomas far more rapidly than the original viral isolate.

Sidebar 3.1 Viruses have simple life cycles The term "virus" refers to a diverse array of infectious particles that infect and multiply within a wide variety of cells, ranging from bacteria to the cells of plants and metazoa. Relative to the cells that they infect, individual virus particles, often termed **virions**, are tiny. Virions are generally simple in structure, with a nucleic acid (DNA or RNA) genome wrapped in a protein coat (a **capsid**) and, in some cases, a lipid membrane surrounding the capsid. In isolation, viruses are metabolically inert. They can multiply only by infecting and parasitizing a suitable host cell. The viral genome, once introduced into the cell, provides instructions for the synthesis of progeny virus particles. The host cell, for its part, provides the low–molecular-weight precursors needed for the synthesis of viral proteins and nucleic acids, the protein-synthetic machinery, and, in many cases, the polymerases required for replicating and transcribing the viral genome.

The endpoint of the resulting infectious cycle is the production of hundreds, even thousands of progeny virus particles that can then leave the infected cell and proceed to infect other susceptible cells. The interaction of the virus with the host cell can either be a **virulent** one, in which the host cell is destroyed during the infectious cycle, or a **temperate** one, in which the host cell survives for extended periods, all the while harboring the viral genome and releasing progeny virus particles.

Many viruses carrying double-stranded DNA (dsDNA) replicate in a fashion that closely parallels the macromolecular metabolism of the host cell (Figure 3.3). This allows them to use host-cell DNA polymerases to replicate their DNA, host-cell RNA polymerases to transcribe the viral mRNAs from double-stranded viral DNA templates, and host ribosomes to translate the viral mRNAs. Once synthesized, viral proteins are used to coat (**encapsidate**) the newly synthesized viral genomes, resulting in the assembly of complete progeny virions, which then are released from the infected cell.

Since cells do not express enzymes that can replicate RNA molecules, many RNA-containing virus particles encode their own RNA-dependent RNA polymerases to replicate their genomes. Poliovirus, as an example, makes such an enzyme, as does rabies virus. RNA tumor viruses like Rous sarcoma virus, as we will learn later in this chapter, follow a much more circuitous route for replicating their viral RNA.

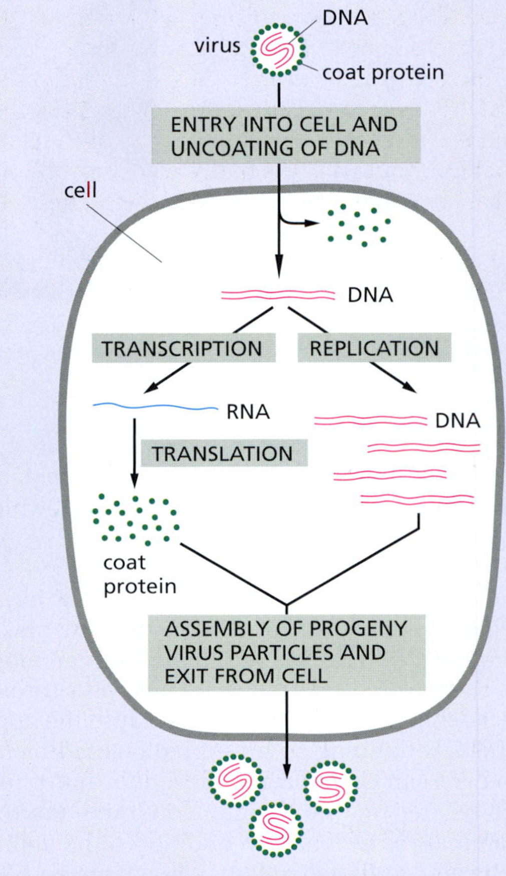

Figure 3.3 Life cycle of viruses with dsDNA genomes The life cycle of viruses with double-stranded DNA (dsDNA) genomes closely parallels that of the host cell. Almost all of the synthetic steps leading to the synthesis of viral DNA, RNA, and proteins can be achieved by using the synthetic machinery provided by the infected host cell. (From B. Alberts et al., Molecular Biology of the Cell, 4th ed. New York: Garland Science, 2002.)

and his collaborators found two other chicken viruses, and yet another chicken sarcoma virus was reported by others in Japan. Then, there was only silence for another two decades until other novel tumor viruses were discovered. The molecular nature of viruses and the means by which they multiplied would remain mysteries for more than half a century after Rous's initial discovery.

Still, his finding of a sarcoma virus reinforced the convictions of those who believed that virtually all human diseases were provoked by infectious agents. In their eyes, cancer could be added to the lengthening list of diseases, such as cholera, tuberculosis, rabies, and sepsis, whose causes could be associated with a specific microbial agent. By 1913, the Dane Johannes Grib Fibiger reported that stomach cancers in rats could be traced to spirochete worms that they harbored. His work, for which he received the 1926 Nobel Prize, represented direct and strong validation of the idea, first indicated by Rous's work, that cancer was yet another example of an infectious disease.

Within months of Fibiger's 1926 Nobel award, he passed away and his scientific opus began to disintegrate. The stomach tumors that he had described were not tumors at all. Instead, they were found to be metaplastic stomach epithelia. As it turned out, these lesions were present in the rats' stomachs because of the profound vitamin deficiencies that these animals suffered; they lived in sugar refineries and ate sugar cane almost exclusively. Fibiger's Nobel Prize became an embarrassment to the still-small community of cancer researchers. They threw the proverbial baby out with the bathwater, discrediting both his work and the notion that cancer could ever be caused by infectious agents.

Interest in the origins of cancer shifted almost totally to chemically induced cancers. Chemicals had been discovered in the early twentieth century that were clearly carcinogenic (see Section 2.8). Study of Rous sarcoma virus and the other tumor viruses languished and entered into a deep sleep for several decades.

3.2 Rous sarcoma virus is discovered to transform infected cells in culture

The rebirth of Rous sarcoma virus research began largely at the California Institute of Technology in Pasadena, in the laboratory of Renato Dulbecco. Dulbecco's postdoctoral fellow Harry Rubin found that when stocks of RSV were introduced into Petri dishes carrying cultures of chicken embryo fibroblasts (CEFs), the RSV-infected cells survived, apparently indefinitely. It seemed that RSV parasitized these cells, forcing them to produce a steady stream of progeny virus particles for many days, weeks, even months (Figure 3.4). Most other viruses, in contrast, were known to enter into host cells, multiply, and quickly kill their hosts; the multitude of progeny virus particles released from dying cells could then proceed to infect yet other susceptible cells in the vicinity, repeating the cycle of infection, multiplication, and cell destruction.

Most important, the RSV-infected cells in these cultures displayed many of the traits that had been previously associated with cancer cells. Thus, after RSV particles were applied to chicken fibroblasts in a culture dish, **foci** (clusters) of infected cells appeared. Under the microscope, these cells strongly resembled the cells isolated from chicken sarcomas (Figure 3.5). Thus, such cells showed a rounded **morphology** (shape, form) and had a metabolism reminiscent of that seen in cells isolated from tumors. This resemblance led Rubin, Howard Temin (a student in Dulbecco's laboratory; Figure 3.6), and others to conclude that the process of cell **transformation**—conversion of a normal cell into a tumor cell—could be accomplished within the confines of a Petri dish, not just in the complex and difficult-to-study environment of a living tissue.

These simple observations radically changed the course of twentieth-century cancer research, because they clearly demonstrated that cancer formation could be studied at the level of individual cells whose behavior could be tracked closely under the microscope. This insight suggested the further possibility that the entire complex biology of tumors could one day be understood by studying the transformed cells forming tumor masses. So, an increasing number of biologists began to view cancer as a disease of malfunctioning cells rather than abnormally developing tissues.

Temin and Rubin, soon followed by many others, used this experimental model to learn some basic principles about cell transformation. They were interested in the fate of a cell that was initially infected by RSV. How did such a cell proliferate

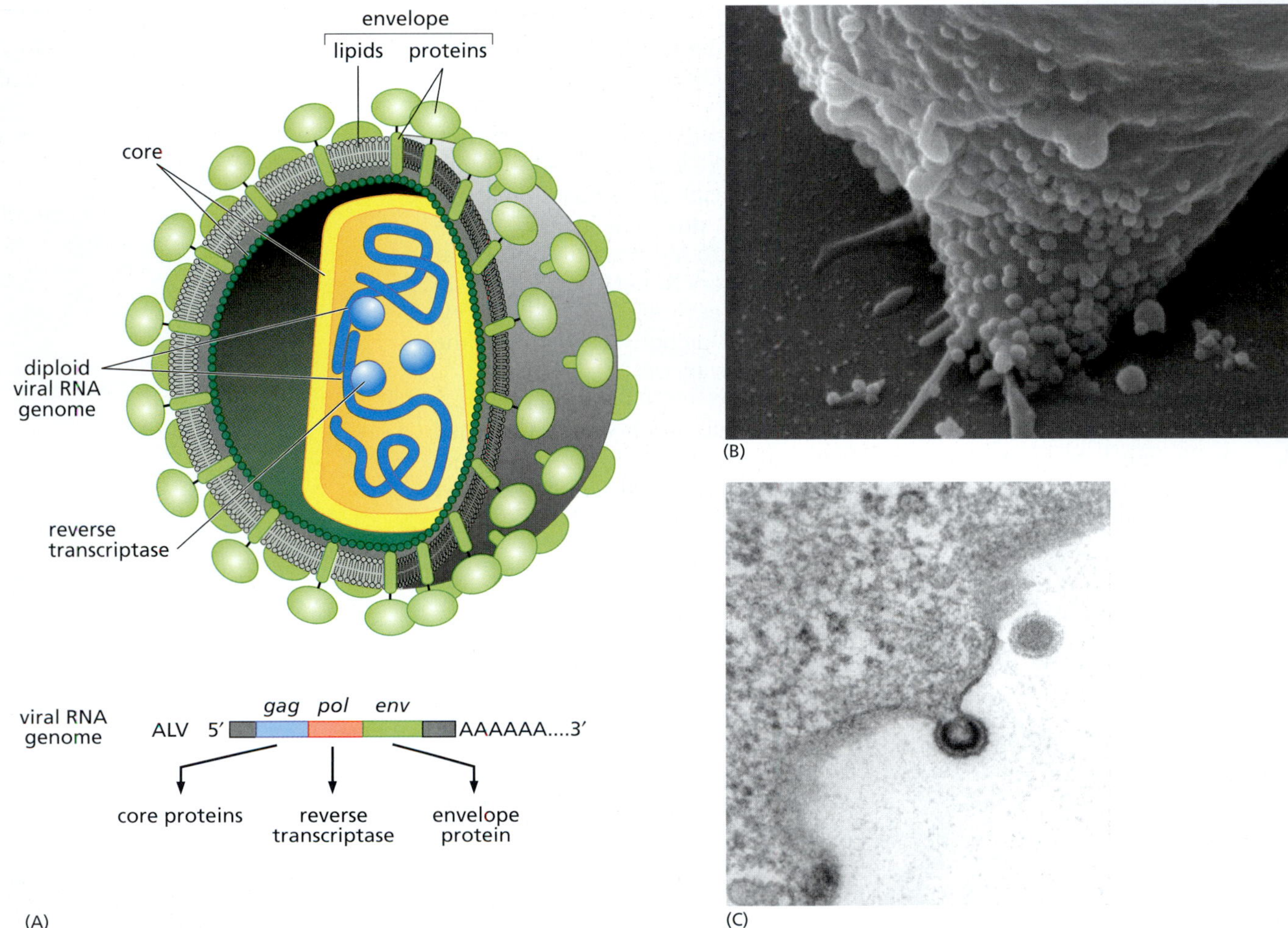

Figure 3.4 The virion of RSV and related viruses (A) This schematic drawing of the structure of a *retrovirus* virion, such as that of Rous sarcoma virus, indicates three major types of viral proteins. The glycoprotein spikes (encoded by the viral *env* gene) protrude from the lipid bilayer that surrounds the virion; these spikes enable the virion to *adsorb* (attach) to the surface of a cell and to introduce the internal contents of the virion into its cytoplasm. These include a complex protein coat formed by the several core proteins encoded by the viral *gag* gene. Within this protein shell are found two identical copies of the viral genomic RNA and a number of reverse transcriptase molecules specified by the viral *pol* gene. (B) Scanning electron micrograph and (C) transmission electron micrograph showing murine leukemia virus (MLV) particles budding from the surface of an infected cell. As the nucleocapsids (containing the gag proteins, the virion RNA, and the reverse transcriptase) leave the cell, they wrap themselves with a patch of lipid bilayer taken from the plasma membrane of the infected cell. (A, adapted from H. Fan et al., The Biology of AIDS. Boston, MA: Jones and Bartlett Publishers, 1989; B, courtesy of Albert Einstein College of Medicine; C, courtesy of Laboratoire de Biologie Moleculaire.)

when compared with uninfected neighboring cells? After exposure of cells to a solution of virus particles (often called a **virus stock**), the two researchers would place a layer of agar above the cell layer growing at the bottom of the Petri dish, thereby preventing virus particles from spreading from initially infected cells to uninfected cells in other parts of the dish. Hence, any changes in cell behavior were the direct result of the initial infection by virus particles of chicken embryo fibroblasts.

The foci that Temin and Rubin studied revealed the dramatic differences in the behavior of normal versus transformed cells. When first introduced into a Petri dish, normal cells formed islands scattered across the bottom of the dish. They then proliferated and eventually filled up all the space in the bottom of the dish, thereby creating **confluent** cultures. Once they reached confluence, however, these normal cells stopped proliferating, resulting in a one-cell-thick (or slightly thicker) layer of cells, often called a cell **monolayer** (Figure 3.7).

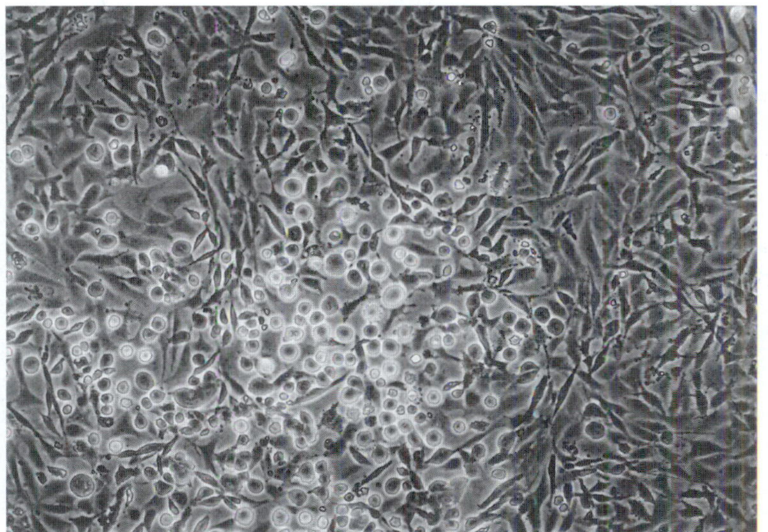

Figure 3.5 An RSV-induced focus This phase-contrast microscope reveals a focus of Rous sarcoma virus–transformed chicken embryo fibroblasts surrounded by a monolayer of uninfected cells. The focus stands out because it is multiple cell layers thick and because of the rounded morphology of the transformed cells and their refractility, which contrasts with the flattened morphology of the normal cells. (Courtesy of P.K. Vogt.)

The cessation of growth of these normal cells after forming confluent monolayers was the result of a process that came to be called **contact inhibition, density inhibition**, or **topoinhibition**. Somehow, high cell density or contact with neighbors caused these cells to stop dividing. This behavior of normal cells contrasted starkly with that of the **transformants** within the RSV-induced foci. The latter clearly had lost contact (density) inhibition and consequently continued to proliferate, piling up on top of one another and creating multilayered clumps of cells so thick that they could often be seen with the naked eye.

Under certain experimental conditions, it could be shown that all the cells within a given focus were the descendants of a single progenitor cell that had been infected and presumably transformed by an RSV particle. Today, we would term such a flock of descendant cells a cell clone and the focus as a whole, a clonal outgrowth, implying in both cases the descent of these cells from a common progenitor.

The behavior of these foci gave support to one speculation about the possible similarities between the cell transformation triggered by RSV in the Petri dish and the processes that led to the appearance of tumors in living animals, including humans: maybe all the cells within a spontaneously arising human tumor mass also constitute a clonal outgrowth and therefore are the descendants of a single common progenitor cell that somehow underwent transformation and then launched a program of replication that led eventually to the millions, even billions, of descendant cells that together formed the mass. As discussed earlier (see Section 2.5), detailed genetic analyses of human tumor cells were required, in the end, to test this notion in a truly definitive way.

Figure 3.6 Howard Temin Howard Temin, pictured here in 1964, began his work by demonstrating the ability of Rous sarcoma virus to transform cells *in vitro*, and showed the persistence of the virus in the infected, transformed cells. He postulated the existence of a DNA provirus and subsequently shared the 1975 Nobel Prize in Medicine and Physiology with David Baltimore for their simultaneous discoveries of the enzyme reverse transcriptase. (Courtesy of the University of Wisconsin Archives.)

3.3 The continued presence of RSV is needed to maintain transformation

The behavior of the cells within an RSV-induced focus indicated that the transformation phenotype was transmitted from an initially infected, transformed chicken cell to its direct descendants. This transmission provoked another set of questions: Did an RSV particle infect and transform the progenitor cell of the focus and, later on, continue to influence the behavior of all of its direct descendants, ensuring that they also remained transformed? Or, as an alternative, did RSV act in a "hit-and-run" fashion by striking the initially infected progenitor cell,

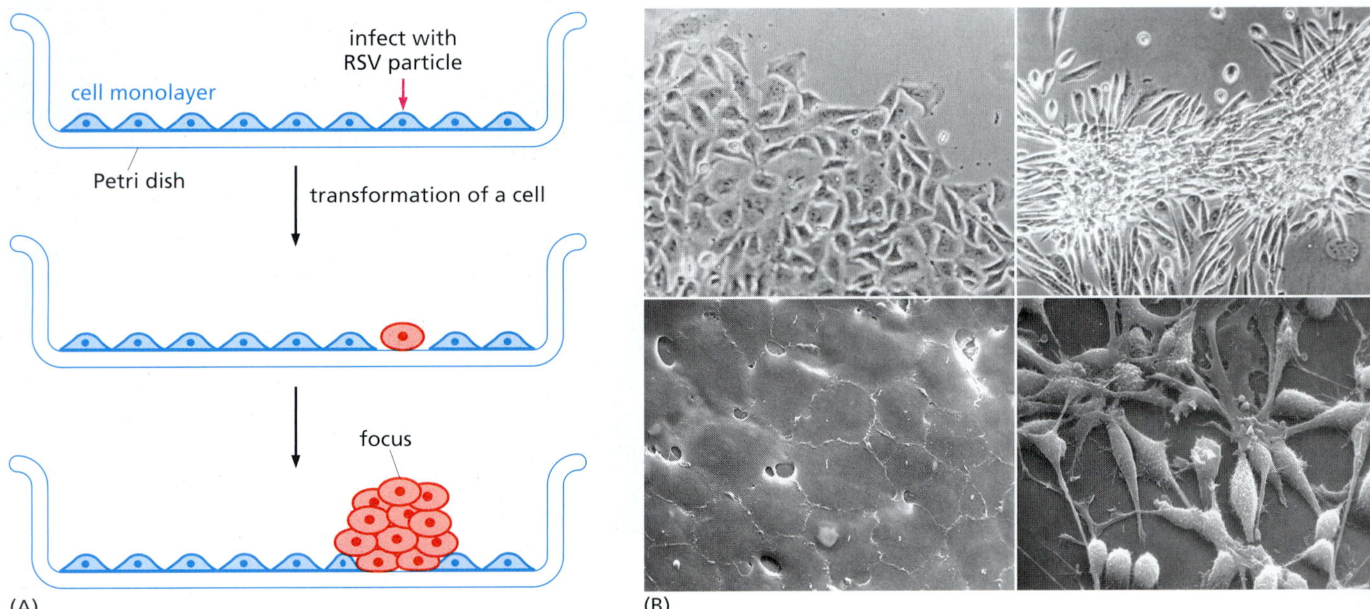

Figure 3.7 Transformed cells forming foci (A) Normal chicken embryo fibroblasts (*blue*) growing on the bottom of a Petri dish form a layer one cell thick—a monolayer. This is because when these cells touch one another, they cease proliferating—the behavior termed "contact inhibition." However, when one of these cells is infected by RSV, this cell (*red*) and its descendants acquire a rounded morphology and lose contact inhibition. As a consequence, they continue to proliferate in spite of touching one another, and eventually accumulate in a thick clump of cells (a focus) many cells thick, which can be seen with the naked eye. (B) The effects of RSV infection and transformation can be observed through both phase-contrast microscopy (*upper panels*) and scanning electron microscopy (*lower panels*). Normal chicken embryo fibroblasts are spread out and form a continuous monolayer (*left panels*). However, upon transformation by RSV, they round up, become refractile (*white halos, upper right panel*), and pile up upon one another (*right panels*). (B, courtesy of L.B. Chen.)

altering its behavior, and then leaving the scene of the crime? According to this second scenario, the progenitor cell could somehow transmit the phenotype of cancerous growth to its descendants without the continued presence of RSV.

Temin and Rubin's work made it clear that the descendants of an RSV-infected cell continued to harbor copies of the RSV genome, but that evidence, on its own, settled little. The real question was, Did the transformed state of the descendant cells actually *depend* on some continuing influence exerted by the RSV genomes that they carried?

An experiment performed in 1970 at the University of California, Berkeley, settled this issue unambiguously. A mutant of RSV was developed that was capable of transforming chicken cells when these cells were cultured at 37°C but not at 41°C (the latter being the normal temperature at which chicken cells grow). **Temperature-sensitive** (*ts*) mutants like this one were known to encode partially defective proteins, which retain their normal structure and function at one temperature and lose their function at another temperature, presumably through thermal **denaturation** of the structure of the mutant protein.

After the chicken embryo fibroblasts were infected with the *ts* mutant of RSV, these cells became transformed if they were subsequently cultured at the lower (**permissive**) temperature of 37°C, as anticipated. Indeed, these cells could be propagated for many cell generations at this lower temperature and continued to grow and divide just like cancer cells, showing their characteristic transformed morphology (see Figure 3.7B). But weeks later, if the temperature of these infected cultures was raised to 41°C (the *nonpermissive* temperature), these cells lost their transformed shape and quickly reverted to the shape and growth pattern of cells that had never experienced an RSV infection (Figure 3.8).

The Berkeley experiments led to simple and yet profoundly important conclusions. Since the cells that descended from a *ts* RSV–infected cell continued to show the temperature-sensitive growth trait, it was obvious that copies of the genome of the infecting virus persisted in these cells for weeks after the initial infection. These copies of the RSV genome in the descendant cells continued to make some temperature-sensitive protein (whose precise identity was not known). Most important, the *continuing actions* of this protein were required in order to maintain the transformed growth phenotype of the RSV-infected cells.

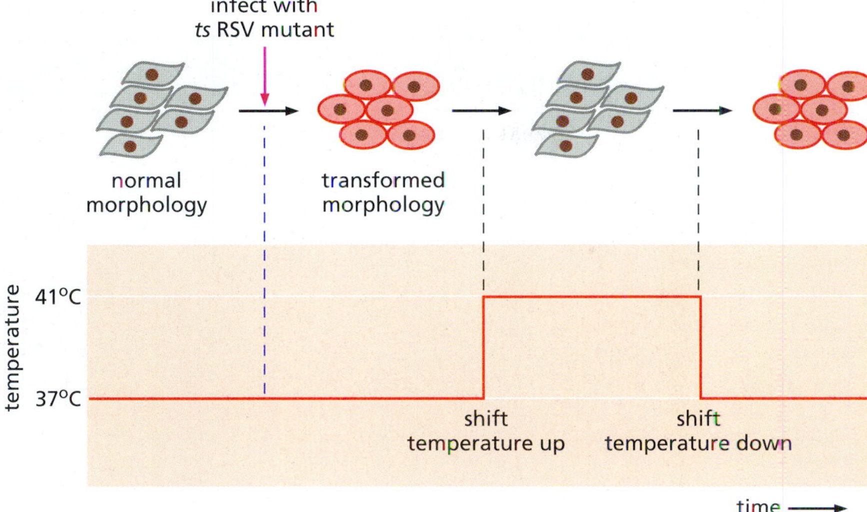

Figure 3.8 Temperature-sensitive mutant and the maintenance of transformation by RSV When chicken embryo fibroblasts were infected at the *permissive* temperature (37°C), at which the viral transforming function can be expressed, the cells became transformed. When the cultures containing these infected cells were shifted to the *nonpermissive* temperature (41°C), the viral transforming function was inactivated and the cells reverted to a normal, nontransformed morphology. Later, when the temperature of the culture was shifted back to the permissive temperature, the viral transforming function was regained and the cells once again exhibited a transformed morphology. The loss of the transformed phenotype upon temperature shift-up demonstrated that the continuous action of some temperature-sensitive viral protein was required in order to maintain this phenotype. The reacquisition of the transformed phenotype after temperature shift-down indicated that the viral genome continued to be present in these cells at the high temperature in spite of their normal appearance.

This work showed that cell transformation, at least that induced by RSV, was not a hit-and-run affair. In the language of the tumor virologists, the viral transforming gene was required to both *initiate* and *maintain* the transformed phenotype of virus-infected cells.

3.4 Viruses containing DNA molecules are also able to induce cancer

RSV was only one of a disparate group of viruses that were found able to induce tumors in infected animals. By 1960, four other classes of tumor viruses had become equally attractive agents for study by cancer biologists. A new type of tumor virus was uncovered almost a quarter century after Rous's pioneering work. This virus, discovered by Richard Shope in rabbits, caused **papillomas** (warts) on the skin. These were really benign lesions, which on rare occasions progressed to true tumors—squamous cell carcinomas of the skin.

By the late 1950s, it became clear that Shope's virus was constructed very differently from RSV. The papillomavirus particles carried DNA genomes, whereas RSV particles were known to carry RNA molecules. Also, the Shope virus particles were sheathed in a protein coat, whereas RSV clearly had, in addition, a lipid membrane coating on the outside. In the decades that followed, more than 100 distinct human papillomavirus (HPV) types, all related to the Shope virus, would be discovered (Figure 3.9).

Research begun in the 1930s in Bar Harbor, Maine, yielded, three decades later, the agent (mouse mammary tumor virus, MMTV) that was responsible for the mother-to-offspring, milk-borne transmission of breast cancer susceptibility in certain strains of mice. The MMTV genome, like that of RSV, was found to be composed of RNA molecules and was therefore quite different from the papillomavirus genome.

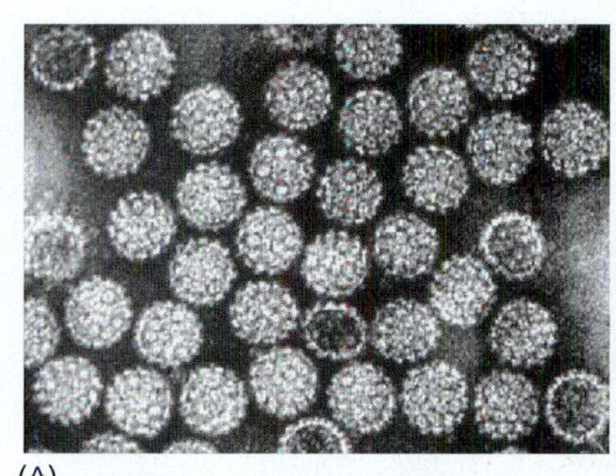
(A)

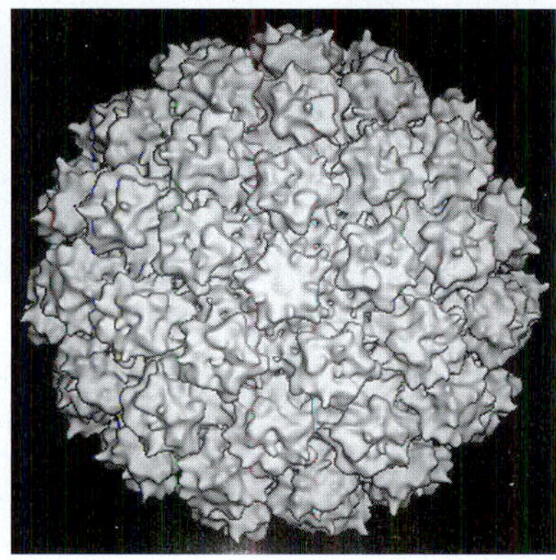
(B)

Figure 3.9 Shope papillomavirus (A) Electronic micrograph of Shope papillomavirus. (B) This cryo-electron micrograph, together with image enhancement, reveals the structure of a human papillomavirus particle. (A, courtesy of D. DiMaio; B, from B.L. Trus et al., *Nat. Struct. Biol.* 4:413–420, 1997.)

Table 3.1 Tumor virus genomes

	Virus family	Approximate size of genome (kb)
DNA viruses		
Hepatitis B virus (HBV)	hepadna	3
SV40/polyoma	papova	5
Human papilloma 16 (HPV)	papova	8
Human adenovirus 5	adenovirus	35
Human herpesvirus 8 (HSV-8; KSHV)	herpesviruses	165
Shope fibroma virus	poxviruses	160
RNA viruses		
Rous sarcoma virus (RSV)	retrovirus	9
Human T-cell leukemia virus (HTLV-I)	retrovirus	9

Adapted in part from G.M. Cooper, Oncogenes, 2nd ed. Boston: Jones and Bartlett Publishers, 1995.

In the 1950s and 1960s, various other DNA tumor viruses were discovered (Table 3.1). Polyomavirus, named for its ability to induce a variety of distinct tumor types in mice, was discovered in 1953. Closely related to polyomavirus in its size and chemical makeup was SV40 virus (the 40th *s*imian *v*irus in a series of isolates). This monkey virus had originally been discovered as a contaminant of the poliovirus vaccine stocks prepared in the mid- and late 1950s (Figure 3.10A). Clever virological sleuthing revealed that SV40 particles often hid out in cultures of the rhesus and cynomolgus monkey kidney cells used to propagate poliovirus during the preparation of vaccine. In fact, the presence of SV40 was not initially apparent in these cell cultures. However, when poliovirus stocks that had been propagated in these monkey cells were later used to infect African green monkey kidney (AGMK) cells, SV40 revealed itself by inducing a very distinctive cytopathic effect—numerous large **vacuoles** (fluid-filled bubble-like structures) in the cytoplasm of infected cells (Figure 3.10B). Within a day after the vacuoles formed in an SV40-infected cell, this cell would lyse, releasing tens of thousands

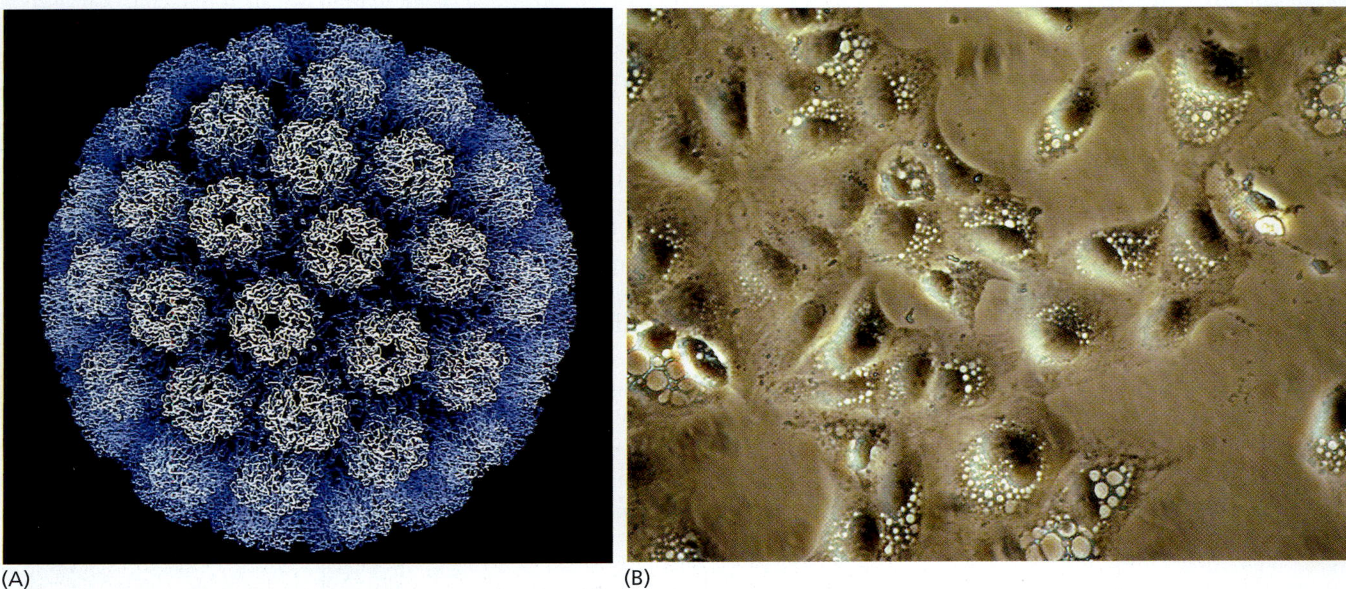

Figure 3.10 SV40 virus (A) As determined by X-ray diffraction, the protein capsid of the SV40 DNA tumor virus consists of three virus-encoded proteins that are assembled into pentamers and hexamers with icosahedral symmetry. The dsDNA genome of SV40 is carried within this capsid. (B) SV40 launches a lytic cycle in permissive host cells, such as the kidney cells of several monkey species. The resulting cytopathic effect seen here, involves the formation of large cytoplasmic vacuoles prior to the death of the cell and the release of tens of thousands of progeny virus particles. (A, courtesy of Robert Grant, Stephen Crainic, and James M. Hogle; B, from A. Gordon-Shaag et al. *J. Virol.* 77:4273–4282, 2003.)

of progeny virus particles. (Because of SV40 contamination, during the course of poliovirus vaccine production, some poliovirus-infected cell cultures yielded far more SV40 virus particles than poliovirus particles!)

This **lytic cycle** of SV40 contrasted starkly with its behavior in cells prepared from mouse, rat, or hamster embryos. SV40 was unable to replicate in these cells, which were therefore considered to be *nonpermissive* hosts. But on occasion, in one cell out of thousands in an infected population, a transformant grew out that shared many characteristics with RSV-transformed cells, i.e., a cell that had undergone changes in morphology and loss of contact inhibition, and had acquired the ability to seed tumors *in vivo*. On the basis of this, SV40 was classified as a tumor virus.

By some estimates, between one-third and two-thirds of the polio vaccines—the oral, live Sabin vaccine and the inactivated, injected Salk vaccine—administered between 1955 and 1963 contained SV40 virus as a contaminant, and between 10 and 30 million people were exposed to this virus through vaccination. In 1960, the fear was first voiced that the SV40 contaminant might trigger cancer in many of those who were vaccinated; reassuringly, epidemiologic analyses conducted over the succeeding four decades indicated little, if any, increased risk of cancer among those exposed to these two vaccines (Sidebar 3.2).

Shope's papillomavirus, the mouse polyomavirus, and SV40 were grouped together as the **papovavirus** class of DNA tumor viruses, the term signifying *pa*pilloma, *po*lyoma, and the *va*cuoles induced by SV40 during its lytic infection. By the mid-1960s, it was apparent that the genomes of the papovaviruses were all formed from circular double-stranded DNA molecules (Figure 3.11). This represented a great convenience for experimenters, since there were several techniques in use at the time that made it possible to separate these relatively

Sidebar 3.2 Is SV40 responsible for the mesothelioma plague? The mesothelium is the membranous outer lining of many internal organs and derives directly from the embryonic mesoderm. Mesotheliomas—tumors of various mesothelial surfaces—were virtually unknown in the United States in the first half of the twentieth century. Beginning in about 1960, however, their incidence began to climb steeply. By the end of the century, the annual incidence had approached 2500 in the United States. Much of this increased incidence has been attributed to exposure to asbestos, specifically a subtype termed crocidolite. Thus, the disease is seen frequently among those who have worked with asbestos, which was used as a heat-resisting insulating material until its use was finally banned in the last two decades of the century. However, 20% of mesothelioma patients have no documented exposure to asbestos, a fact that has provoked a search for other etiologic (causative) agents.

SV40 is a plausible etiologic agent of mesothelioma. By 2003, forty-one laboratories across the world had reported SV40 DNA, RNA, or protein in mesothelioma cells. Traces of the virus are otherwise rarely found in human tumors, with the exception of certain types of brain tumors. Cultured human mesothelioma cells are readily infected by SV40, and this infection leads rapidly to their **immortalization**, that is, to the ability of these cells, which usually have limited proliferative potential in culture, to multiply indefinitely.

The presence of SV40 contamination of poliovirus vaccine stocks has raised concerns that mesothelioma may have been induced in many individuals as an unintended side effect of poliovirus vaccination. In fact, there are many instances of mesotheliomas in individuals who are highly unlikely to have been exposed to these vaccines. Moreover, the viral T-antigen protein, whose presence is invariably observed in SV40-transformed cells (see Section 3.6), is rarely detected in mesothelioma tumors. And attempts at demonstrating viral DNA have usually yielded DNA segments that are indicative of laboratory artifacts, for example, contamination of mesothelioma tumor samples by laboratory stocks of SV40 or by recombinant DNA plasmids that had previously been engineered to include segments of the viral genome. The discoveries of such contaminations provide increasing ammunition for those who are skeptical of SV40's role in mesothelioma pathogenesis.

As many as 85% of humans are known to be infected with two viruses that are closely related to SV40—JC and BK—and antisera that recognize the capsids of these viruses cross-react with the capsid of SV40, explaining many of the claims that SV40 is often present in human tissues. In immunosuppressed individuals, notably AIDS patients, JC virus can cause a fatal brain degeneration—progressive multifocal leukoencephalopathy (PML)—but there is no evidence that it is also carcinogenic.

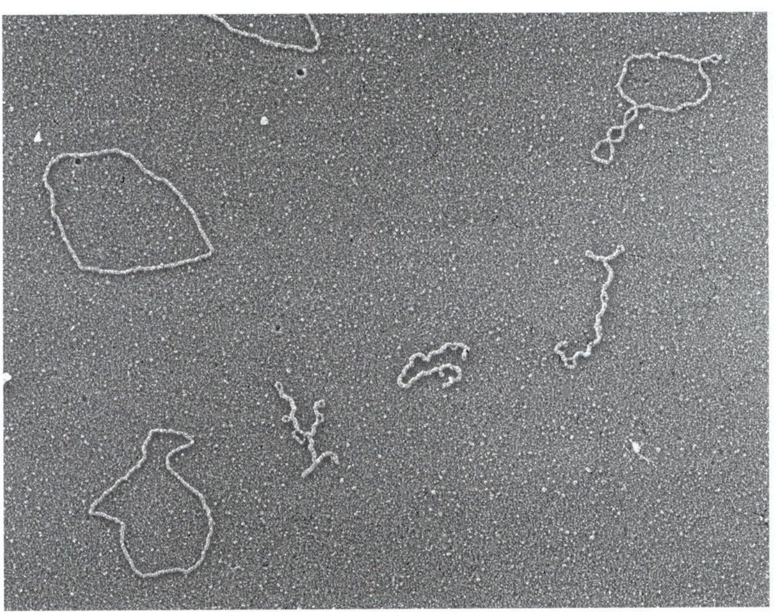

Figure 3.11 Papovavirus genomes Electron-microscopic analyses revealed the structure of DNA genomes of papovaviruses, including that of the SV40 genome shown here, to be double-stranded and circular. Shown here are the closed-circular, supercoiled form of SV40 DNA (form I; kinked DNAs) and the relaxed, nicked circular form (form II; open, spread circles). Form III, which is a linear form of the viral DNA resulting from a double-strand DNA break, is not shown shown. (Courtesy of J.D. Griffith.)

small, circular DNA molecules [about 5–8 kilobases (kb) in length] from the far larger, linear DNA molecules present in the chromosomes of infected host cells.

The group of DNA tumor viruses grew further with the discovery that human adenovirus, known to be responsible for upper respiratory infections in humans, was able to induce tumors in infected hamsters. Here was a striking parallel with the behavior of SV40. The two viruses could multiply freely in their natural host cells, which were therefore considered to be permissive. During the resulting lytic cycles of the virus, permissive host cells were rapidly killed in concert with the release of progeny virus particles. But when introduced into nonpermissive cells, both adenovirus and SV40 failed to replicate and instead left behind, albeit at very low frequency, clones of transformants.

Other entrants into the class of DNA tumor viruses were members of the herpesvirus group. While human herpesvirus types 1 and 2 were apparently not **tumorigenic** (capable of inducing tumors), a distantly related herpesvirus of *Saimiri* monkeys provoked rapid and fatal lymphomas when injected into monkeys from several other species. Another distantly related member of the herpesvirus family—Epstein–Barr virus (EBV)—was discovered to play a causal role in provoking Burkitt's lymphomas in young children in Equatorial Africa and New Guinea as well as nasopharyngeal carcinomas in Southeast Asia. Finally, at least two members of the poxvirus class, which includes smallpox virus, were found to be tumorigenic: Shope fibroma virus and Yaba monkey virus cause benign skin lesions in rabbits and rhesus monkeys, respectively. The tumorigenic powers of these viruses, which have very large genomes (135–160 kb), remain poorly understood to this day.

Researchers found that adenovirus and herpesvirus particles contain long, linear, double-stranded DNA (dsDNA) molecules, which, like the genomes of RSV, carry the information required for both viral replication and virus-induced cell transformation. Compared with the papovaviruses, the herpesviruses had genomes of enormous size (see Table 3.1), suggesting that they carried a proportionately larger number of genes. In the end, it was the relatively small sizes of papovavirus genomes that made them attractive objects of study by those interested in the molecular origins of cancer.

Most of the small group of genes in a papovavirus genome were apparently required to program viral replication; included among these were several genes specifying the proteins that form the capsid coat of the virus particle.

This dictated that papovaviruses could devote only a small number of their genes to the process of cell transformation.

This realization offered the prospect of greatly simplifying the cancer problem by reducing the array of responsible genes and causal mechanisms down to a very small number. Without such simplification, cancer biologists were forced to study the cancer-causing genes that were thought to be present in the genomes of cells transformed by nonviral mechanisms. Cellular genomes clearly harbored large arrays of genes, possibly more than a hundred thousand. At the time, the ability to analyze genomes of such vast complexity and to isolate individual genes from these genomes was a distant prospect.

3.5 Tumor viruses induce multiple changes in cell phenotype including acquisition of tumorigenicity

Like RSV-transformed cells, the cells transformed by SV40 showed profoundly altered shape and piled up on one another. This loss of contact inhibition was only one of a number of changes exhibited by virus-transformed cells (Table 3.2). As discussed later in this book, normal cells in culture will not proliferate unless they are provided with serum and serum-associated growth-stimulating factors in their culture medium. Cells transformed by a variety of tumor viruses were often found to have substantially reduced requirements for these factors in their culture medium.

Yet another hallmark of the transformed state is an ability to proliferate in culture for longer time periods than normal cells. Researchers discovered that normal cells have a limited proliferative potential in culture and ultimately stop multiplying after a certain, apparently predetermined number of cell divisions. Cancer cells seemed to be able to proliferate indefinitely in culture, and hence were described as being *immortalized* (as discussed in Chapter 10).

When transformed cells were suspended in an agar gel, these cells were able to proliferate into spherical colonies containing dozens, even hundreds of cells (Figure 3.12). This ability to multiply without attachment to the solid substrate provided by the bottom of the Petri dish was termed the trait of **anchorage independence**. Normal cells, in contrast, demonstrated an absolute requirement for tethering to a solid substrate before they would grow and were therefore considered to be **anchorage-dependent**. This ability of cells to grow in an anchorage-independent fashion *in vitro* usually served as a good predictor of their ability to form tumors *in vivo* following injection into appropriate host animals.

This tumor-forming ability—the phenotype of tumorigenicity—represented the acid test of whether cells were fully transformed, that is, had acquired the full

Table 3.2 Properties of transformed cells

Altered morphology (rounded shape, refractile in phase-contrast microscope)
Loss of contact inhibition (ability to grow over one another)
Ability to grow without attachment to solid substrate (anchorage independence)
Ability to proliferate indefinitely (immortalization)
Reduced requirement for mitogenic growth factors
High saturation density (ability to accumulate large numbers of cells in culture dish)
Inability to halt proliferation in response to deprivation of growth factors
Increased transport of glucose
Tumorigenicity

Adapted in part from S.J. Flint, L.W. Enquist, R.M. Krug et al., Principles of Virology. Washington, DC: ASM Press, 2000.

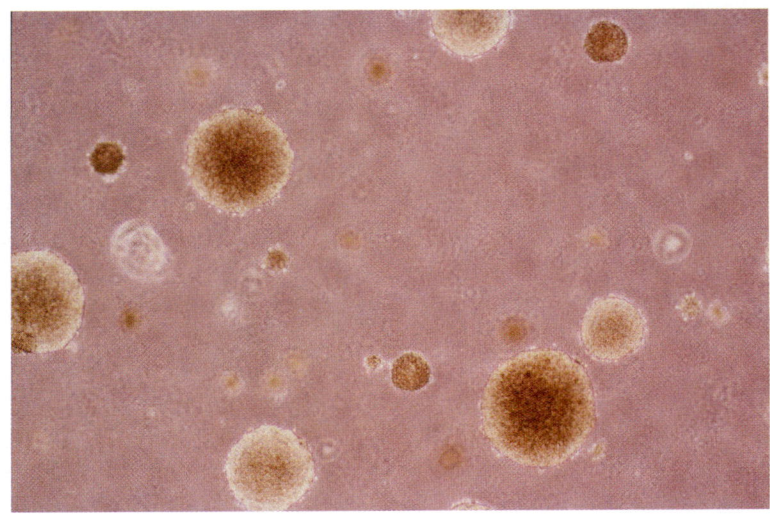

Figure 3.12 Anchorage-independent growth A photomicrograph of colonies of cells growing in an anchorage-independent fashion. (Each of the larger colonies seen here may contain several hundred cells.) This test is usually performed by suspending cells in a semi-solid medium, such as agarose or methylcellulose. By holding cells in suspension, these media ensure that cells are prevented from attaching to a solid substrate, specifically, the bottom of the Petri dish. The ability of cells to proliferate while held in suspension—the phenotype of anchorage independence—is usually a good (but hardly infallible) predictor of the ability of such cells to form tumors *in vivo*. (Courtesy of A. Orimo.)

repertoire of neoplastic traits. Tests for tumorigenicity could be performed when the cells used in an *in vitro* transformation experiment were prepared from a strain of mice and later injected into host mice of the same strain (Sidebar 3.3). Since the host and injected cells came from the same genetic strain, the immune systems of such **syngeneic** host mice would not recognize the transformed cells as being foreign bodies and therefore would not attempt to eliminate them—the process of **tumor rejection** (a process to which we will return in Chapter 15). This allowed injected cells to survive in their animal hosts, enabling them to multiply into large tumors if, indeed, they had acquired the tumorigenic phenotype.

Often, it was impossible to test the tumorigenicity of tumor virus-infected cells in a syngeneic host animal, simply because the cells being studied came from a species in which inbred syngeneic hosts were not available. This forced the use of **immunocompromised** hosts whose immune systems were tolerant of a wide variety of foreign cell types, including those from other species (see Sidebar 3.3). Mice of the Nude strain soon became the most commonly used hosts to test the tumorigenicity of a wide variety of cells, including those of human origin. Quite

Sidebar 3.3 The powers of the immune system dictate whether transformed cells can form tumors A number of distinct strains of mice and rats have been developed through repeated cycles of inbreeding over the past century. Because of this inbreeding, an individual mouse from an inbred strain is genetically identical to all other members of the strain (aside from the gender-specific differences created by the Y chromosome). Included among these mouse strains are, for example, the BALB/c, C3H, and C57/Bl6 strains. The genetic identity of all individuals within a strain means that tissue fragments (including tumors) from one animal of a strain can be transplanted to other animals of that strain and become established in the recipient animals.

This ability to transplant tissues from one individual to another does not exist in outbred populations, such as humans and natural populations of mice. Except for the rare instances of identical twins, tissues that are introduced from one animal (or human) to another are rapidly recognized as being foreign by the immune system of the recipient. The immune system then undertakes to eliminate the foreign tissue and usually succeeds in doing so rapidly. This means that true tests of the tumorigenicity of *in vitro* transformed mouse cells can be undertaken only when such cells derive from the same strain of mice as the hosts into which these cells are introduced. The lack of inbred strains of chickens severely limited the type of cancer research that could be undertaken with these birds.

This barrier to transplantation can be circumvented in several ways. Often the immune system of a very young animal is more tolerant of genetically foreign tissue than the more robust, well-developed immune system of an adult. This allowed Rous to carry out some of his experiments. More common and practical, however, is the use of **immunocompromised** strains of mice, which lack fully functional immune systems and therefore tolerate engrafted tissues of foreign genetic origin, including cells and tissues originating in other species (termed **xenografts**). Frequently used strains of immunocompromised mice include the Nude, RAG, and NOD/SCID mice, each of which has a defect in one or more key components of its immune system.

Figure 3.13 Nude mice Mice of the Nude strain have two advantages in tests of tumorigenicity. Because they lack a thymus, they are highly immunocompromised, and therefore are relatively receptive to engrafted cells from genetically unrelated sources, including cells from a foreign species. In addition, because these mice are hairless, it is easy to monitor closely the progress of tumor formation after transformed cells have been injected under the skin. (Courtesy of Harlan Sprague Dawley.)

frequently, candidate tumor cells are injected **subcutaneously**, i.e., directly under the skin of these animals. Since they also lack the ability to grow hair, nude mice provide the additional advantage of allowing the experimenter to closely monitor the progress of implanted tumor cells (Figure 3.13).

As mentioned, the small size of the genomes of RNA tumor viruses (e.g., RSV) and papovaviruses dictated that each of these use only a small number of genes (perhaps as few as one) to elicit multiple changes in the cells that they infected and transformed. Recall that the ability of a gene to concomitantly induce a number of distinct alterations in a cell is termed pleiotropy. Accordingly, though little direct evidence about gene number was yet in hand, it seemed highly likely that the genes used by tumor viruses to induce cell transformation were acting pleiotropically on a variety of molecular targets within cells.

3.6 Tumor virus genomes persist in virus-transformed cells by becoming part of host cell DNA

The Berkeley experiments (see Section 3.3) provided strong evidence that the continued actions of the RSV genome were required to maintain the transformed state of cells, including those that were many cell generations removed from an initially infected progenitor cell. This meant that some or all of the viral genetic information needed to be perpetuated in some form, being passed from a transformed mother cell to its two daughters and, further on, to descendant cells through many cycles of cell growth and division. Conversely, a failure to transmit these viral genes to descendant cells would result in their reversion to cells showing normal growth behavior.

Paralleling the behavior of RSV, cell transformation achieved by two intensively studied DNA tumor viruses—SV40 and polyomavirus—also seemed to depend on the continued presence of viral genomes in the descendants of an initially transformed cell. The evidence proving this came in a roundabout way, largely from the discovery of tumor-associated proteins (T antigens) that were found in cancers induced by these two viruses. For example, sera prepared from mice carrying an SV40-induced tumor showed strong reactivity with a nuclear protein that was present characteristically in tumors triggered by SV40 and absent in tumors induced by polyomavirus or by other carcinogenic agents. The implication was that the viral genome residing in tumor cells encoded a protein (e.g., the SV40 T antigen) that induced a strong immunological response in the tumor-bearing mouse or rat host (Figure 3.14).

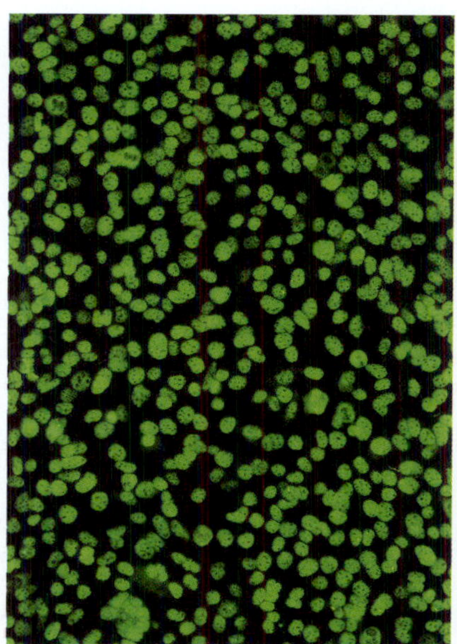

Figure 3.14 Expression of the SV40 T antigen When SV40-transformed cells are introduced into a syngeneic host, these cells provoke a strong immune response against the SV40-encoded T antigen (tumor antigen), which is usually termed "large T" (LT). Anti-LT serum, together with appropriate immunofluorescent tags, can be used to stain virus-transformed cells, revealing the nuclear location of large T. Shown here is a version of this in which an anti-LT monoclonal antibody (rather than anti-tumor serum) has been used to stain mouse cells that have been transformed by SV40. (Courtesy of S.S. Tevethia.)

The display of the virus-induced T antigen correlated directly with the transformed state of these cells. Therefore, cells that lost the T antigen would also lose the transformation phenotype induced by the virus. This correlation suggested, but hardly proved, that the viral sequences responsible for transformation were associated with or closely linked to viral sequences encoding the T antigen.

The cell-to-cell transmission of viral genomes over many cell generations represented a major conceptual problem. Cellular genes were clearly transmitted with almost total fidelity from mother cells to daughter cells through the carefully programmed processes of chromosomal DNA replication and mitosis that occur during each cellular growth-and-division cycle. How could viral genomes succeed in being replicated and transmitted efficiently through an unlimited number of cell generations? This was especially puzzling, since viral genomes seemed to lack the genetic elements that were thought to be required for proper allocation of chromosomes to daughter cells during mitosis.

Adding to this problem was the fact that the DNA metabolism of papovaviruses, such as SV40 and polyomavirus, was very different from that of the host cells that they preyed upon. When SV40 and polyomavirus infected permissive host cells, the viral DNAs were replicated as autonomous, extrachromosomal molecules. Both viruses could form many tens of thousands of circular, double-stranded DNA genomes of about 5 kb in size from a single viral DNA genome initially introduced by infection (see Figure 3.11). While the viral DNA replication exploited a number of host-cell DNA replication enzymes, it proceeded independently of the infected cells' chromosomal DNA replication. This nonchromosomal replication occurring during the lytic cycles of SV40 and polyomavirus shed no light on how these viral genomes were perpetuated in populations of virus-transformed cells. The latter were, after all, nonpermissive and therefore prevented these viruses from replicating their DNA.

A solution to this puzzle came in 1968, when it was discovered that the viral DNA in SV40-transformed cells was tightly associated with their chromosomal DNA. Using centrifugation techniques to gauge the molecular weights of DNA molecules, it became clear that the SV40 DNA in these cells no longer sedimented like a small (~5 kb) viral DNA genome. Instead, the SV40 DNA sequences in virus-transformed cells co-sedimented with the high molecular weight (>50 kb) chromosomal DNA of the host cells (Figure 3.15). In fact, the viral DNA in these

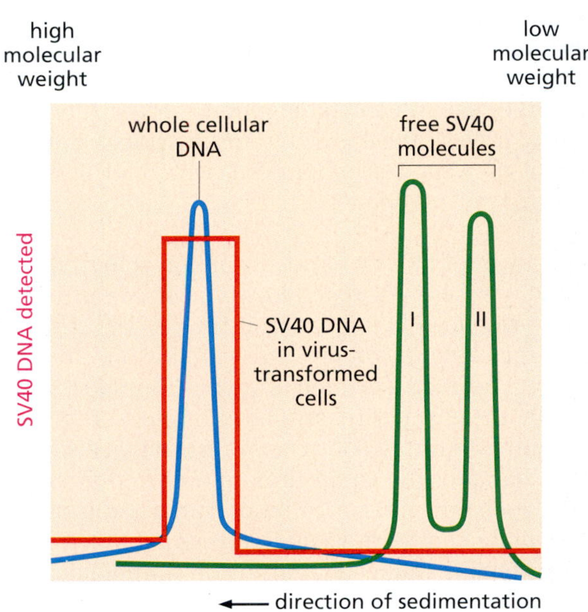

Figure 3.15 Integration of SV40 DNA DNA molecules from SV40-transformed cells were isolated and sedimented by centrifugation through an alkaline solution stabilized by a sucrose gradient (used to prevent mixing of different fluid strata within the centrifuge tube). Under these conditions, the high–molecular-weight cellular DNA (*blue*) sedimented a substantial distance down the sucrose gradient (*left side of graph*). In contrast, the SV40 DNA isolated from virus particles (*green*) sedimented more slowly, indicative of its lower molecular weight. Forms I and II viral DNA refer to the closed-circular and nicked-circular DNAs of SV40, respectively (Figure 3.11). Use of nucleic acid hybridization revealed that the SV40 DNA sequences in SV40 virus–transformed cells co-sedimented with the high–molecular-weight chromosomal DNA of the virus-transformed cells (*red*). (Adapted from J. Sambrook et al., *Proc. Natl. Acad. Sci. USA* 60:1288–1295, 1968.)

transformed cells could not be separated from the cells' chromosomal DNA by even the most stringent methods of dissociation, including the harsh treatment of exposure to alkaline pH.

These results indicated that SV40 DNA in virus-transformed cells had become covalently linked to the chromosomal DNA. Such **integration** of viral genome solved an important problem in viral transformation: transmission of viral DNA sequences from a mother cell to its offspring could be guaranteed, since the viral DNA would be co-replicated with the cell's chromosomal DNA during the S (DNA synthesis) phase of each cell cycle. In effect, by integrating into the chromosome, the viral DNA sequences became as much a part of a cell's genome as the cell's own native genes.

Some years later, this ability of papovavirus genomes to integrate into host-cell genomes became highly relevant to the pathogenesis of one common form of human cancer—cervical carcinoma. Almost all (>99.7%) of these tumors were found to carry fragments of human papillomavirus (HPV) genomes integrated into their chromosomal DNA. Provocatively, intact viral genomes were rarely discovered to be present in integrated form in cancer cell genomes. Instead, only the portion of the viral genome that contains **oncogenic** (cancer-causing) information was found in the chromosomal DNA of these cancer cells, while the portion that enables these viruses to replicate and construct progeny virus particles was almost always absent or present in only fragmentary form.

3.7 Retroviral genomes become integrated into the chromosomes of infected cells

The ability of SV40 and polyomavirus to integrate copies of their genomes into host-cell chromosomal DNA solved one problem but created another that seemed much less solvable: how did RSV succeed in transmitting its genome through many generations within a cell lineage? The genome of RSV is made of single-stranded RNA (Figure 3.16), which clearly could not be integrated directly into the chromosomal DNA of an infected cell. Still, RSV succeeded in transmitting its genetic information through many successive cycles of cell growth and division (Sidebar 3.4).

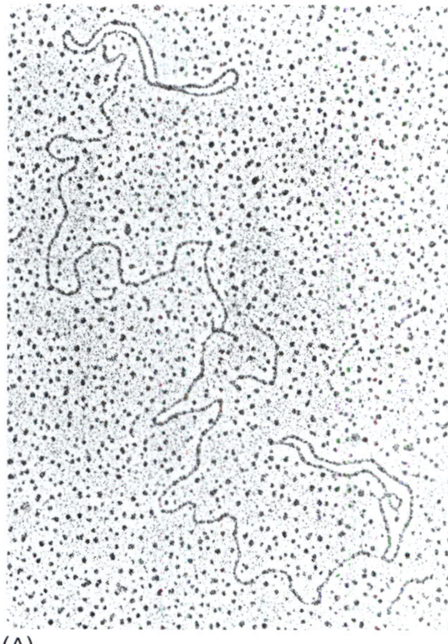

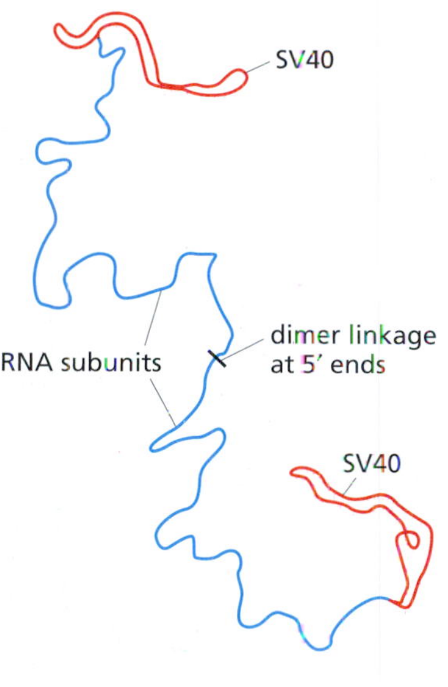

Figure 3.16 Genome structure of RNA tumor viruses Viruses like RSV carry single-stranded RNA (ssRNA) genomes. Uniquely, the genomes of this class of viruses were discovered to be diploid, i.e., to carry two identical copies of the viral genetic sequence. The organization of such genomes was revealed by, among other techniques, electron microscopy. (A) This electron micrograph shows the genome of a distant relative of RSV termed "baboon endogenous virus." (B) As is indicated in the schematization, circular SV40 dsDNA molecules (*red, top, bottom*), to which oligo dT (oligodeoxythymidine) tails were attached enzymatically, have been used to visually label the 3′ termini of the two single-stranded viral RNA molecules (*blue*). The viral RNA molecules (like cellular mRNA molecules) contains polyadenylate at their 3′ termini, and therefore anneals to the oligo dT tails attached to the circular SV40 molecules. As is also indicated, the two ssRNAs are associated at their 5′ ends. (Adapted from W. Bender and N. Davidson, *Cell* 7:595–607, 1976.)

Sidebar 3.4 Re-infection could not explain the stable transmission of RSV genomes The transmission of RSV genomes to the descendants of initially infected cells could be explained, in principle, by a mechanism in which the virus-infected cells continually release virus particles that subsequently infect daughter cells. Such repeated cycles of re-infection can ensure that an RSV genome can be perpetuated indefinitely in a population of cells without the need for chromosomal integration of viral genetic information, as is the case with SV40 virus (Section 3.6). In fact, stable transformation of cell populations by RSV was also observed using strains of RSV that could infect and transform a cell but were unable to replicate in that cell. In addition, certain cell populations were found that could be transformed by an initial RSV infection but resisted subsequent cycles of re-infection. Together, such observations effectively ruled out re-infection as the means by which RSV perpetuated itself in lineages of cells over multiple cell generations.

This puzzle consumed Temin in the mid- and late-1960s and caused him to propose a solution so unorthodox that it was ridiculed by many, landing him in the scientific wilderness. Temin argued that after RSV particles (and those of related viruses) infected a cell, they made double-stranded DNA (dsDNA) copies of their RNA genomes. It was these dsDNA versions of the viral genome, he said, that became established in the chromosomal DNA of the host cell. Once established in this way, the DNA version of the viral genome—which he called a **provirus**—then assumed the molecular configuration of a cellular gene and would be replicated each time the cell replicated its chromosomal DNA. In addition, the proviral DNA, once established in the genome, could serve as a template for transcription by cellular RNA polymerase, thereby yielding RNA molecules that could be incorporated into progeny virus particles or, alternatively, could function as messenger RNA (mRNA) that was used for the synthesis of viral proteins (Figure 3.17).

The process of **reverse transcription** that Temin proposed—making DNA copies of RNA—was without precedent in the molecular biology of the time, which recognized information flow only in a single direction, specifically, DNA→RNA→proteins. But the idea prevailed, receiving strong support from Temin's and David Baltimore's simultaneous discoveries in 1970 that RSV and related virus particles carry the enzyme reverse transcriptase. As both research groups discovered, this enzyme has the capacity to execute the key copying step that Temin had predicted—the step required in order for RSV to transmit its genome through many cycles of cell growth and division.

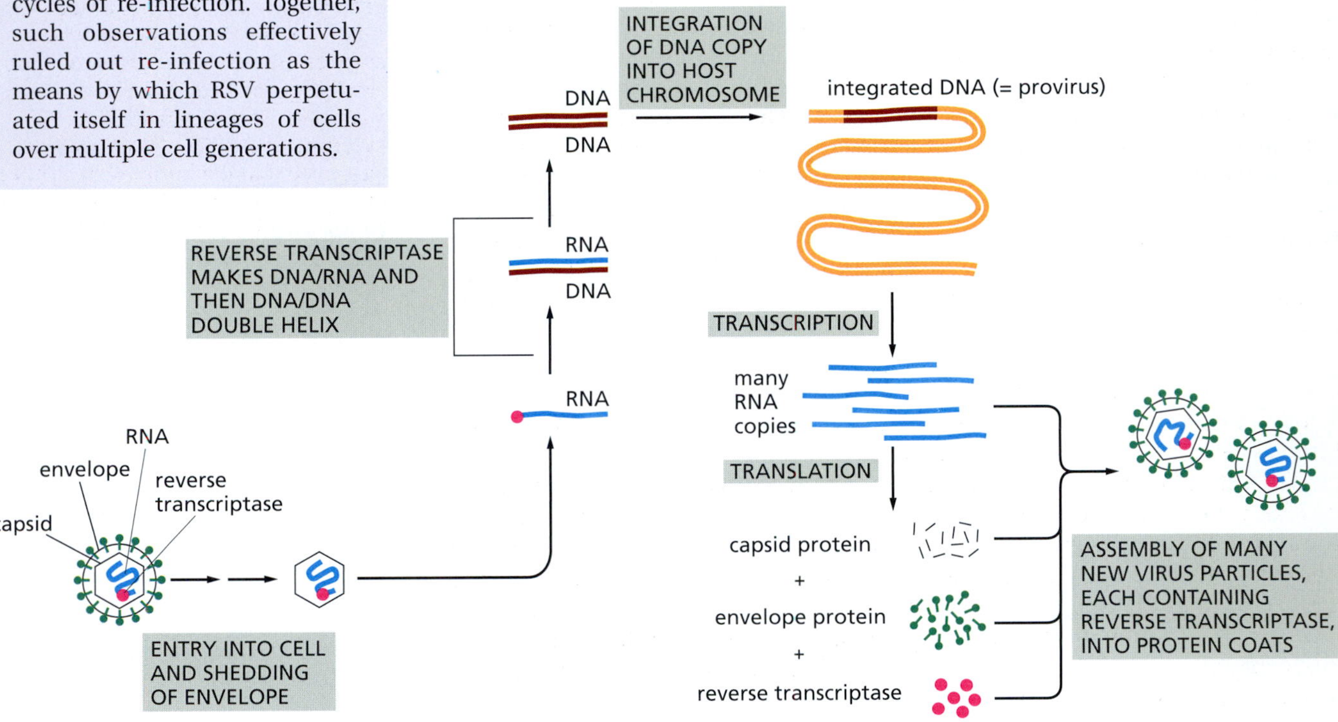

Figure 3.17 The life cycle of an RNA tumor virus like RSV The retrovirus replication cycle was proposed in outline by Temin; later, the molecular details were uncovered, resulting in the scheme depicted here. An infecting virion introduces its single-stranded (ss) RNA genome (*blue*) into the cytoplasm of a cell together with the reverse transcriptase enzyme (RT, *red*). The RT makes a double-stranded (ds) DNA copy (*brown*) of the viral RNA using the viral RNA as template. The reverse-transcribed DNA then moves into the nucleus, where it becomes integrated into the cellular chromosomal DNA (*orange*). The resulting integrated provirus is transcribed into ssRNA molecules (*blue*) by the RNA polymerase II (pol II) of the host cell. The progeny RNA molecules are exported to the cytoplasm, where they either serve as mRNAs to make viral proteins or are packaged into progeny virus particles that leave the cell and initiate a new round of infection. (From B. Alberts et al., Essential Cell Biology, 2nd ed. New York: Garland Science, 2004.)

It soon became apparent that RSV was only one of a large group of similarly constructed viruses, which together came to be called **retroviruses** to reflect the fact that their cycle of replication depends on information flowing "backward" from RNA to DNA. Within a year of the discovery of reverse transcriptase, the presence of proviral DNA was detected in the chromosomal DNA of RSV-infected cells. Hence, like SV40 and polyomavirus, retroviruses rely on integration of their genomes into the chromosome to ensure the stable retention and transmission of their genomes.

There is, however, an important distinction between the integration mechanisms used by retroviruses like RSV and the DNA tumor viruses such as SV40 and polyomavirus. Integration is a normal, essential part of the replication cycle of retroviruses. But in the case of DNA tumor viruses, chromosomal integration of their genomes is a very rare accident (<<1 per 1000 infections) that enables the perpetuation of viral genomes in the descendants of an initially infected cell; the rare SV40 genomes that do succeed in becoming established in chromosomal DNA are found integrated in a haphazard fashion that often includes only fragments of the wild-type genome (Figure 3.18).

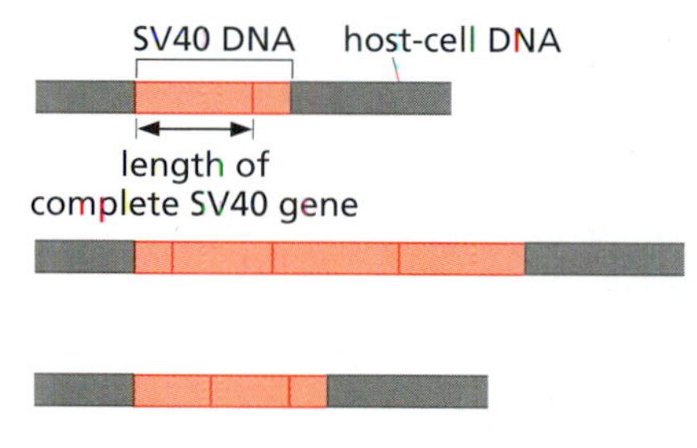

Figure 3.18 Configuration of integrated SV40 genomes Use of restriction enzyme cleavage-site mapping revealed the configurations of SV40 genomes integrated into the chromosomal DNA of virus-transformed cells. In some cells, only a portion of the viral genome was present; in yet others, there were multiple, head-to-tail tandem arrays present. These integration events involved a linearization of the circular SV40 DNA and some type of random, nonhomologous ("illegitimate") recombination between viral DNA and chromosomal DNA, with no specific sites in either DNA being involved in the recombination events. The only feature shared in common by the integrated genomes was the presence of the "early region" of the SV40 genome, which encodes the viral transforming proteins, specifically, the large T and small T antigens. Shown here are four examples of how SV40 DNA becomes integrated into the chromosomal DNA of cells. (Adapted from M. Botchan, W. Topp, and J. Sambrook, *Cell* 9:269–287, 1976.)

3.8 A version of the *src* gene carried by RSV is also present in uninfected cells

Because the genomes of retroviruses, like those of papovaviruses, were quite small (<10 kb), it seemed likely that the coding capacity of the retroviral genomes was limited to a small number of genes, probably far fewer than ten. Using this small repertoire of genes, retroviruses nevertheless succeeded in specifying some viral proteins needed for viral genome replication, others required for the construction and assembly of progeny virus particles, and yet other proteins used to transform infected cells.

In the case of RSV, the use of mutant viruses revealed that the functions of viral replication (including reverse transcription and the construction of progeny virions) required one set of genes, while the function of viral transformation required another. Thus, some mutant versions of RSV could replicate perfectly well in infected cells, producing large numbers of progeny virus particles, yet such mutants lacked transforming function. Conversely, other mutant derivatives of RSV could transform cells but had lost the ability to replicate and make progeny virions in these transformed cells.

At least three retroviral genes were implicated in viral replication. Two of these encode structural proteins that are required for assembly of virus particles; a third specifies the reverse transcriptase (RT) enzyme, which copies viral RNA into DNA shortly after retrovirus particles enter into host cells. A comparison of the RNA genome of RSV with the genomes of related retroviruses lacking transforming ability suggested that there was rather little information in the RSV genome devoted to encoding the remaining known viral function—transformation. Consequently, geneticists working with RSV speculated that all the viral transforming functions resided in a single gene, which they termed *src* (pronounced "sark"), to indicate its role in triggering the formation of sarcomas in infected chickens (Figure 3.19).

In 1974, the laboratory run jointly by J. Michael Bishop and Harold Varmus at the University of California, San Francisco, undertook to make a DNA **probe** that specifically recognized the transformation-associated (i.e., *src*) sequences of the RSV genome in order to understand its origins and functions (see Sidebar 3.5). This *src*-specific probe was then used to follow the fate of the *src* gene after cells were infected with RSV. The notion here was that uninfected chicken cells would carry no *src*-related DNA sequences in their genomes. However, following RSV

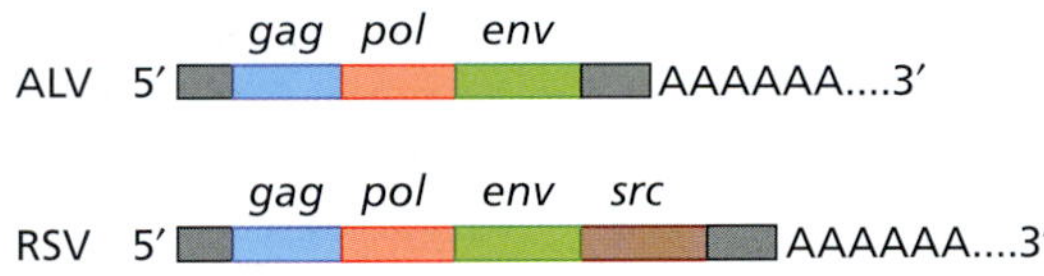

Figure 3.19 Structure of the Rous sarcoma virus genome The RNA genome of RSV is closely related to that of a relatively common infectious agent of chickens, avian leukosis virus (ALV). Note that the ALV genome encodes the *gag* proteins that, together with the viral RNA, form the nucleoprotein core of these particles (see Figure 3.4); the *pol* gene, which encodes the reverse transcriptase; and the *env* gene, which specifies the glycoprotein spikes of the virion. The RSV genome carries, in addition, a gene (*src*, *brown*) that specifies the Src protein, which causes cell transformation. Not indicated here is the specialized cap structure at the 5′ end of these RNA molecules.

infection, *src* sequences would become readily detectable in cells, having been introduced by the infecting viral genome.

The actual outcome of this experiment was, however, totally different from expectation. In 1975, this research group, using their *src*-specific probe, found that *src* sequences were clearly present among the DNA sequences of uninfected chicken cells. These *src* sequences were present as single-copy cellular genes; that is, two copies of the *src*-related DNA sequences were present per diploid chicken cell genome—precisely the representation of the great majority of genes in the cellular genome.

The presence of *src* sequences in the chicken cell genome could not be dismissed as some artifact of the hybridization procedure used to detect them. Moreover, careful characterization of these *src* sequences made it unlikely that they had been inserted into the chicken genome by some retrovirus. For example, *src*-related DNA sequences were readily detectable in the genomes of several related bird species, and, more distantly on the evolutionary tree, in the

Sidebar 3.5 The making of a *src*-specific DNA probe In order to make a *src*-specific DNA probe, a researcher working in Bishop and Varmus's laboratory exploited two types of RSV virus strains: a wild-type RSV genome that carried all of the sequences needed for viral replication and transformation, and a mutant RSV genome that was able to replicate but had lost, because of a major deletion of genetic sequences, the *src* sequences required for transformation (Figure 3.20). Using reverse transcriptase, he made a DNA copy of the wild-type sequences, yielding single-stranded DNA molecules complementary to the viral RNA. He then fragmented this DNA, and hybridized it to the RNA genome of the RSV deletion mutant that lacked *src* sequences, creating DNA–RNA hybrid molecules. He then retrieved the ssDNA molecules that failed to form DNA–RNA hybrids (discarding those that did form the hybrids). The result was ssDNA fragments that specifically recognized sequences contained within the deleted portion of the mutant RSV genome, that is, those lying within the *src* gene.

Because the initial reverse transcription of the wild-type RSV RNA was carried out in the presence of radiolabeled deoxyribonucleoside triphosphates, the *src*-specific DNA fragments (which constituted the *src* "probe") were also labeled with radioisotope. This made it possible to discover whether DNAs of interest (such as the DNAs prepared from virus-infected or uninfected cells) also carried *src* sequences by determining whether the *src* probe (with its associated radioactivity) was able to hybridize to these cellular DNAs.

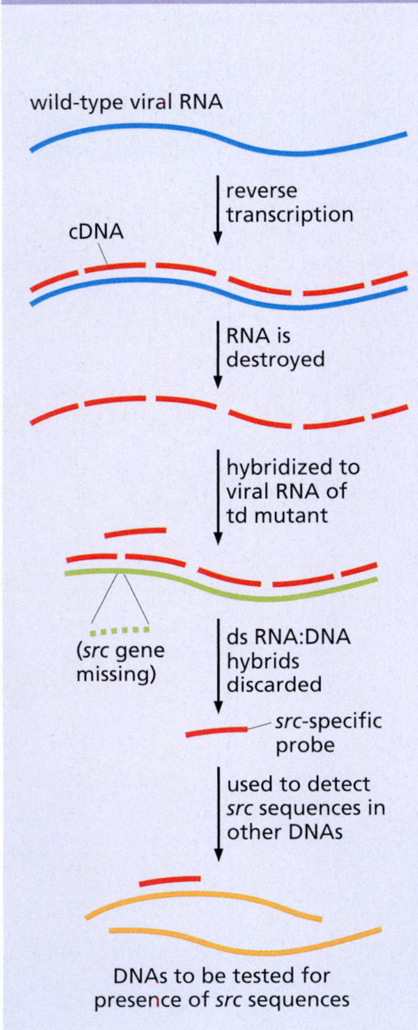

Figure 3.20 The construction of a *src*-specific DNA probe Wild-type (wt) RSV RNA (*blue*) was reverse-transcribed under conditions where only a single-stranded (ss) complementary DNA molecule (a cDNA; *red*) was synthesized. The wt single-stranded viral DNA was then annealed (hybridized) to viral RNA (*green*) of the transformation-defective (td) mutant of RSV, which had lost its transforming function and apparently deleted its *src* gene. The resulting ds RNA:DNA hybrids were discarded, leaving behind the ssDNA fragment of the wtDNA that failed to hybridize to the RNA of the td mutant. This surviving DNA fragment, if radiolabeled, could then be used as a *src*-specific probe in order to detect *src*-related sequences in various cellular DNAs (*orange*).

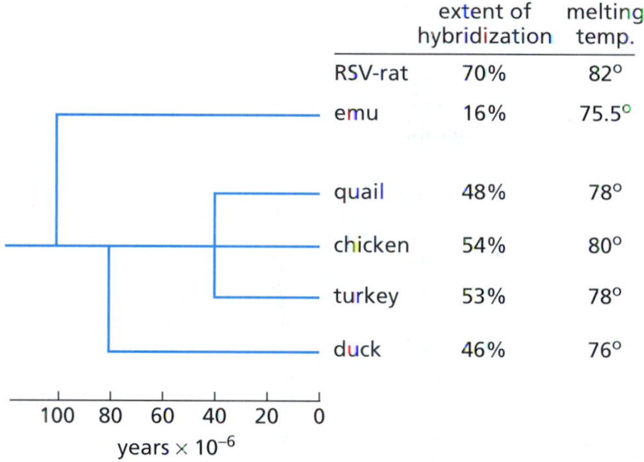

extent of hybridization
melting temp.
RSV-rat 70% 82°
emu 16% 75.5°
quail 48% 78°
chicken 54% 80°
turkey 53% 78°
duck 46% 76°
100 80 60 40 20 0
years × 10^{-6}

Figure 3.21 Evolutionary tree of the *src* gene The *src*-specific cDNA probe (see Figure 3.20) was annealed with the DNAs of a variety of avian species—whose relatedness to the chicken is indicated on this evolutionary tree—as well as to a rat cell that had recently been infected and transformed by Rous sarcoma virus (RSV). The time in the distant evolutionary past when the last common ancestor shared by various pairs of species lived is indicated by the abscissa below. The best hybridization between the RSV *src* probe was with the DNA of an RSV-transformed rat cell carrying an RSV provirus (used as a positive control), as indicated by the percentage of the probe radioactivity that annealed to this cellular DNA and the melting temperature (T_m) of the resulting hybrid. (High melting temperature indicates a high degree of sequence complementarity between two strands of a hybrid; low melting temperature indicates numerous sequence mismatches between the two annealing DNA strands.) Chicken DNA annealed to the *src* probe to a lesser extent and the resulting hybrids were denatured at a slightly lower temperature, indicating that the sequence of the *src* gene of RSV deviates slightly from the related gene in the chicken genome. The *src* proto-oncogenes of species that were further removed evolutionarily from chickens showed lesser extents of hybridization, and the hybrids that were formed melted at lower temperatures. Subsequent work, not shown here, revealed the presence of *src* sequences in the DNA of organisms belonging to other metazoan phyla, such as arthropods (e.g., *Drosophila melanogaster*) and even a sponge. (Courtesy of H.E. Varmus.)

DNAs of several mammals (Figure 3.21). The more distant the evolutionary relatedness of a species was to chickens, the weaker was the reactivity of the *src* probe with its DNA. This was precisely the behavior expected of a cellular gene that had been present in the genome of a common ancestral species and had acquired increasingly divergent DNA sequences as descendant species evolved progressively away from one another over the course of millions of years.

The evidence converged on the idea that the *src* sequences present in the genome of an uninfected chicken cell possessed all the properties of a normal cellular gene, being present in a single copy per haploid genome, evolving slowly over tens of millions of years, and being present in vertebrate species that were ancestral to both birds and mammals. This realization created a revolution in thinking about the origins of cancer.

3.9 RSV exploits a kidnapped cellular gene to transform cells

The presence of a highly conserved *src* gene in the genome of a normal organism implied that this cellular version of *src*, sometimes termed c-*src* (i.e., cell *src*), played some role in the life of this organism (the chicken) and its cells. How could this role be reconciled with the presence of a transforming *src* gene carried in the genome of RSV? This viral transforming gene (v-*src*) was closely related to the c-*src* gene of the chicken, yet the two genes had drastically different effects and apparent functions. When ensconced in the cellular genome, the actions of c-*src* were apparently compatible with normal cellular behavior and normal organismic development. In contrast, the very similar v-*src* gene borne by the RSV genome acted as a potent **oncogene**—a gene capable of transforming a normal chicken cell into a tumor cell.

One solution to this puzzle came from considering the possibility that perhaps the *src* gene of RSV was not naturally present in the genome of the retrovirus ancestral to RSV. This hypothetical viral ancestor, while lacking *src* sequences, was perfectly capable of replicating in chicken cells. In fact, such a *src*-negative retrovirus—avian leukosis virus (ALV)—was common in chickens and was capable of infectious spread from one chicken to another (see Sidebar 3.6). This suggested that during the course of infecting a chicken cell, an ancestral virus, similar to this common chicken virus, had acquired sequences from the host-cell genome (Figure 3.22), doing so through some genetic trick. The acquired cellular sequences (e.g., the *src* sequences) were then incorporated into the viral genome, thereby adding a fourth gene to the existing three genes that this retrovirus (like similar retroviruses) used for its replication in infected cells (see Figure 3.19).

Sidebar 3.6 Viruses like RSV have very short lives Circumstantial evidence indicates that infection of a chicken cell by a virus like avian leukosis virus (ALV) resulted in introduction of a copy of the chicken c-*src* gene into the ALV genome, creating the hybrid virus that we term RSV. This recombination event seems to have occurred only once in the twentieth century; that is, all *src*-containing chicken retroviruses appear to be the descendants of the RSV that arose on one occasion in a chicken coop in about 1909. The discovery of RSV depended on a fortuitous circumstance: Peyton Rous was able to study the virus only because a Long Island, New York, chicken farmer brought him a prized, tumor-bearing hen with the hope that Rous would be able to cure the bird of its cancer.

The presence of the acquired *src* gene in the viral genome confers no obvious growth advantage on this virus, and RSV is not naturally spread from animal to animal. Moreover, similarly configured retroviruses described later, which also carry both viral and cellular sequences in their genomes, spread poorly from one host animal to another, often because they have deleted essential viral replication genes from their genomes in order to accommodate acquired oncogenes of cellular origin. For these reasons, it seems clear that hybrid retroviruses such as RSV arise through rare genetic recombination events within infected animals, create tumors in these animals, and disappear when these animal die, unless an alert chicken farmer or slaughterer happens to discover the tumors and pass them on to an interested virologist. The latter may or may not succeed in isolating a transforming retrovirus from the tumor.

Once present in the genome of RSV, the kidnapped *src* gene could then be altered and exploited by this virus to transform subsequently infected cells.

This scheme attributed great cleverness to retroviruses by implying that these viruses had the ability to pick up and exploit preexisting cellular genes for their own purposes. Such behavior is most unusual for a virus, since virtually all other types of viruses carry genes that have little if any relatedness to DNA sequences native to the cells that they infect (see Sidebar 3.7).

But there was an even more important lesson to be learned here, this one concerning the c-*src* gene. This cellular gene, one among tens of thousands in the chicken cell genome, could be converted into a potent viral oncogene following some slight remodeling by a retrovirus such as RSV. Because it was a precursor to an active oncogene, c-*src* was called a **proto-oncogene**. The very concept of a proto-oncogene was revolutionary: it implied that the genomes of normal vertebrate cells carry a gene that has the potential, under certain circumstances, to induce cell transformation and thus cancer.

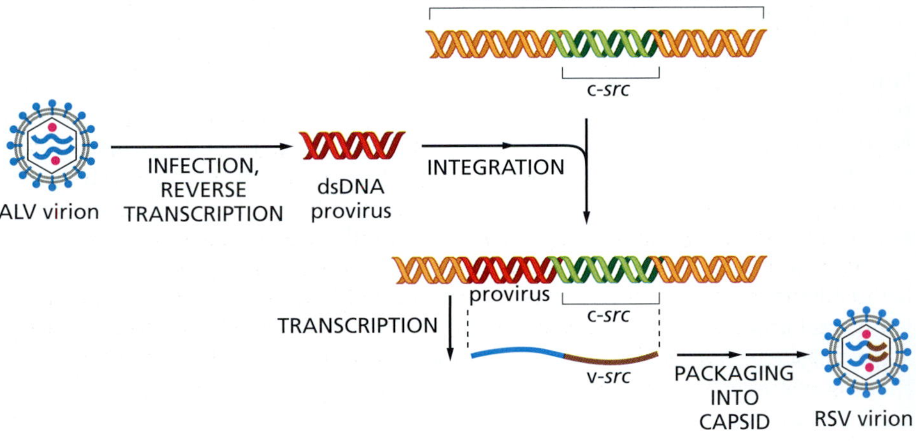

Figure 3.22 Capture of *src* by avian leukosis virus The precise mechanism by which an avian leukosis virus (ALV) captured a cellular *src* gene (c-*src*) is not known. A plausible scenario is indicated here. Thus, ALV proviral DNA (*red*) became integrated (by chance) next to a c-*src* proto-oncogene (*green*) in an infected chicken cell. The ALV provirus and adjacent c-*src* gene were co-transcribed into a single hybrid RNA transcript (*blue and brown*). After splicing out c-*src* introns (not shown), this hybrid viral RNA was packaged into a virus particle that became the ancestor of Rous sarcoma virus (RSV). (Not shown is the attachment of ALV sequences at the 3′ side of v-*src*, with the result that this viral oncogene became flanked on both sides by ALV sequences.)

The structures of the c-*src* proto-oncogene and the v-*src* oncogene were worked out rapidly in the years that followed these discoveries in 1975–1976. Just as the viral geneticists had speculated, all of the viral transforming sequences resided in a single viral oncogene. Within the RSV RNA genome, the v-*src* gene was found at the 3′ end of the genome, added to the three preexisting retroviral genes that were involved in viral replication (see Figure 3.19).

This scenario of acquisition and activation of c-*src* by a retrovirus led to three further ideas. First, if retroviruses could activate this proto-oncogene into a potent oncogene, perhaps other types of mutational mechanisms might operate to yield a similar outcome through their ability to reshape the normal c-*src* gene. Maybe these other mutational mechanisms caused the activation of a cellular proto-oncogene into an oncogene without removing the normal gene from its normal roosting site on the cellular chromosome. Maybe the information for inducing cancer was already present in the normal cell genome, waiting to be unmasked.

Second, it became clear that all of the transforming powers of RSV derived from the presence of a single gene—v-*src*—in its genome. This was of great importance, because it implied that a single oncogene could, as long suspected, elicit a large number of changes in the shape, metabolism, and growth behavior of a cell. More generally, this suggested that other cancer-causing genes could also act pleiotropically. Accordingly, if a transformed, tumorigenic cell differed from a normal cell in 20 or 30 distinct traits, perhaps these multiple changes were not dependent on the alteration of 20 or 30 different genes; instead, maybe a small number of genes would suffice to transform a normal into a tumorigenic cell.

Third, RSV and its v-*src* oncogene might represent a model for the behavior of other types of retroviruses that were similarly capable of transforming infected

Sidebar 3.7 Where do viruses and their genes originate? The discovery by Varmus and Bishop's laboratory of the cellular origins of the RSV *src* gene provoked the question of where most viral genes come from and how viruses originate. Viral genome replication—in both DNA and RNA virus replication cycles—is executed with far less fidelity than is the replication of cellular DNA. The consequence is that viral genome sequences evolve far more rapidly than the genomic sequences of metazoan cells. Hence, any traces of precursor gene sequences that may have been present in other viruses or in cells were erased by this rapid evolution hundreds of millions of years ago.

We will probably never know where and how most viral genomes originated. Presumably, viruses have been around since the cells that they parasitize first appeared. Some types of viruses may have begun as renegade cellular genes that broke away from cellular genomes and struck out on their own. RNA viruses may bear vestiges of early cellular life forms that used RNA molecules as genomes and, in the case of retroviruses, of a later stage of cellular evolution when a reverse transcriptase-like cellular enzyme enabled the transition from RNA to DNA genomes.

Whatever their origins, the great majority of viral genes have no obvious relatedness with the genes of their hosts. This highlights the uniqueness of the Varmus and Bishop finding that retroviruses can apparently acquire and **transduce** (carry in their genomes) cellular genes, seemingly promiscuously. This flexibility of their genomes suggested a novel application of retroviruses, which began to flourish in the mid-1980s: a variety of interesting genes were introduced into retroviral genomes using recombinant DNA techniques. The resulting viral genomes were then used as **vectors** to transduce these genes into cultured cells *in vitro* and into living tissues *in vivo* (the latter application often being referred to as "gene therapy").

cells *in vitro* (in the culture dish) and inducing tumors *in vivo* (in living tissue). Perhaps these other transforming retroviruses had acquired other cellular genes unrelated to *src*. While c-*src* was certainly the first cellular proto-oncogene to be discovered, maybe other cellular proto-oncogenes were hiding in the vertebrate cellular genome, waiting to be picked up and activated by some passing retrovirus.

3.10 The vertebrate genome carries a large group of proto-oncogenes

An accident of history—an encounter in 1909 between a Long Island chicken farmer and Peyton Rous—made RSV the first tumorigenic retrovirus to be isolated and characterized in detail. Consequently, RSV was favored initially with the most detailed molecular and genetic analysis. However, in the 1950s and 1960s, a group of other chicken and rodent tumor viruses were found that were subsequently realized to be members, like RSV, of the retrovirus class.

The diversity of these other transforming retroviruses and the diseases that they caused suggested that they, like RSV, might be carrying kidnapped cellular proto-oncogenes and might be using these acquired genes to transform infected cells. So, within a year of the discovery of v-*src* and c-*src*, the race began to find yet other viruses that had traveled down a similar genetic path and picked up other, potentially interesting proto-oncogenes.

Another chicken retrovirus—the MC29 myelocytomatosis virus—which was known to be capable of inducing a bone marrow malignancy in chickens, was one of this class. MC29 was also found to carry an acquired cellular gene in its genome, termed the v-*myc* oncogene, which this virus exploited to induce rapidly growing tumors in infected chickens. As was the case with v-*src*, the origin of the v-*myc* gene could be traced to a corresponding proto-oncogene residing in the normal chicken genome. Like *src*, *myc* underwent some remodeling after being incorporated into the retroviral genome. This remodeling imparted potent oncogenic powers to a gene that previously had played a benign and possibly essential role in the life of normal chicken cells.

Of additional interest was the discovery that MC29, like RSV, descended from avian leukosis virus (ALV). This reinforced the notion that retroviruses like ALV were adept at acquiring random pieces of a cell's genome. The biological powers of the resulting hybrid viruses would presumably depend on which particular cellular genes had been picked up. In the case of the large majority of acquired cellular genes, hybrid viruses carrying such genes would show no obvious phenotypes such as tumor-inducing potential. Only when a growth-promoting cellular gene, that is, a proto-oncogene, was acquired might the hybrid virus exhibit a cancer-inducing phenotype that could lead to its discovery and eventual isolation by a virologist.

Mammals were also found to harbor retroviruses that are distantly related to ALV and, like ALV, are capable of acquiring cellular proto-oncogenes and converting them into potent oncogenes. Among these is the feline leukemia virus, which acquired the *fes* oncogene in its genome, yielding feline sarcoma virus, and a hybrid rat–mouse leukemia virus, which on separate occasions acquired two distinct proto-oncogenes: the resulting transforming retroviruses, Harvey and Kirsten sarcoma viruses, carry the H-*ras* and K-*ras* oncogenes, respectively, in their genomes. Within a decade, the repertoire of retrovirus-associated oncogenes had increased to more than two dozen, many named after the viruses in which they were originally discovered (Table 3.3). By now, more than thirty distinct vertebrate proto-oncogenes have been discovered through this route.

Table 3.3 Acutely transforming retroviruses and the oncogenes that they have acquired[a]

Name of virus	Viral oncogene	Species	Major disease	Nature of oncoprotein
Rous sarcoma	*src*	chicken	sarcoma	non-receptor TK
Y73/Esh sarcoma	*yes*	chicken	sarcoma	non-receptor TK
Fujinami sarcoma	*fps*[b]	chicken	sarcoma	non-receptor TK
UR2	*ros*	chicken	sarcoma	RTK; unknown ligand
Myelocytomatosis 29	*myc*	chicken	myeloid leukemia[c]	transcription factor
Mill Hill virus 2	*mil*[d]	chicken	myeloid leukemia	ser/thr kinase
Avian myeloblastosis E26	*myb*	chicken	myeloid leukemia	transcription factor
Avian myeloblastosis E26	*ets*	chicken	myeloid leukemia	transcription factor
Avian erythroblastosis ES4	*erbA*	chicken	erythroleukemia	thyroid hormone receptor
Avian erythroblastosis ES4	*erbB*	chicken	erythroleukemia	EGF RTK
3611 murine sarcoma	*raf*[e]	mouse	sarcoma	ser/thr kinase
SKV770	*ski*	chicken	endothelioma (?)	transcription factor
Reticuloendotheliosis	*rel*	turkey	immature B-cell lymphoma	transcription factor
Abelson murine leukemia	*abl*	mouse	pre-B-cell lymphoma	non-receptor TK
Moloney murine sarcoma	*mos*	mouse	sarcoma, erythroleukemia	ser/thr kinase
Harvey murine sarcoma	H-*ras*	rat, mouse	sarcoma	small G protein
Kirsten murine sarcoma	K-*ras*	mouse	sarcoma	small G protein
FBJ murine sarcoma	*fos*	mouse	osteosarcoma	transcription factor
Snyder–Theilen feline sarcoma	*fes*[f]	cat	sarcoma	non-receptor TK
McDonough feline sarcoma	*fms*	cat	sarcoma	CSF-1 RTK
Gardner–Rasheed feline sarcoma	*fgr*	cat	sarcoma	non-receptor TK
Hardy–Zuckerman feline sarcoma	*kit*	cat	sarcoma	steel factor RTK
Simian sarcoma	*sis*	woolly monkey	sarcoma	PDGF
AKT8	*akt*	mouse	lymphoma	ser/thr kinase
Avian virus S13	*sea*	chicken	erythroblastic leukemia[g]	RTK; unknown ligand
Myeloproliferative leukemia	*mpl*	mouse	myeloproliferation	TPO receptor
Regional Poultry Lab v. 30	*eyk*	chicken	sarcoma	RTK; unknown ligand
Avian sarcoma virus CT10	*crk*	chicken	sarcoma	SH2/SH3 adaptor
Avian sarcoma virus 17	*jun*	chicken	sarcoma	transcription factor
Avian sarcoma virus 31	*qin*	chicken	sarcoma	transcription factor[h]
AS42 sarcoma virus	*maf*	chicken	sarcoma	transcription factor
Cas NS-1 virus	*cbl*	mouse	lymphoma	SH2-dependent ubiquitylation factor

Abbreviations: CSF, colony-stimulating factor; EGF, epidermal growth factor; G, GTP-binding; PDGF, platelet-derived growth factor; RTK, receptor tyrosine kinase; ser/thr, serine/threonine; SH, src-homology segment; TK, tyrosine kinase; TPO, thrombopoietin.

[a]Not all viruses that have yielded these oncogenes are indicated here.
[b]Ortholog of the mammalian *fes* oncogene.
[c]Also causes carcinomas and endotheliomas.
[d]Ortholog of the mammalian *raf* oncogene.
[e]Ortholog of the avian *mil* oncogene.
[f]Ortholog of the avian *fps* oncogene.
[g]Also causes granulocytic leukemias and sarcomas.
[h]Functions as a transcriptional repressor.

Adapted in part from S.J. Flint, L.W. Enquist, R.M. Krug et al., Principles of Virology. Washington, DC: ASM Press, 2000. Also in part from G.M. Cooper, Oncogenes. Boston: Jones and Bartlett Publishers, 1995.

In each case, a proto-oncogene found in the DNA of a mammalian or avian species was readily detectable in the genomes of all other vertebrates. There were, for example, chicken, mouse, and human versions of c-*myc*, and these genes seemed to function identically in their respective hosts. The same could be said of all the other proto-oncogenes that were uncovered. Soon it became clear that this large repertoire of proto-oncogenes must have been present in the genome of the vertebrate that was the common ancestor of all mammals and birds, and that this group of genes, like most others in the vertebrate genome, was inherited by all of the modern descendant species.

Now, a quarter of a century later, we realize that these transforming retroviruses provided cancer researchers with a convenient window through which to view the cellular genome and its cohort of proto-oncogenes. Without these retroviruses, the discovery of proto-oncogenes would have been exceedingly difficult. By fishing these genes out of the cellular genome and revealing their latent powers, these viruses catapulted cancer research forward by decades.

3.11 Slowly transforming retroviruses activate proto-oncogenes by inserting their genomes adjacent to these cellular genes

As described above, each of the various tumorigenic retroviruses arose when a nontransforming retrovirus, such as avian leukosis virus (ALV) or murine leukemia virus (MLV), acquired a proto-oncogene from the genome of an infected host cell. In fact, the "nontransforming" precursor viruses could also induce cancers, but they were able to do so only on a much more extended timetable; often months passed before these viruses succeeded in producing cancers. The oncogene-bearing retroviruses, in contrast, often induced tumors within days or weeks after they were injected into host animals.

When the rapidly transforming retroviruses were used to infect cells in culture, the cells usually responded by undergoing the changes in morphology and growth behavior that typify the behavior of cancer cells (see Table 3.2). In contrast, when the slowly tumorigenic viruses, such as ALV and MLV, infected cells, these cells released progeny virus particles but did not show any apparent changes in shape or growth behavior. This lack of change in cell phenotypes was consistent with the fact that these viruses lacked oncogenes in their genomes.

These facts, when taken together, represented a major puzzle. How could viruses like MLV or ALV induce a malignancy if they carried no oncogenes? In 1981 this puzzle was solved.

The solution came from study of the leukemias that ALV induced in chickens, more specifically, from detailed analysis of the genomic DNAs of the leukemic cells. These cells invariably carried copies of the ALV provirus integrated into their genomes. Some researchers undertook to map the sites in the chromosomal DNA where the ALV proviruses had integrated. By the time these experiments began, a decade after Temin and Baltimore's discovery, it had become clear that the integration of proviruses occurs at random sites throughout the chromosomal DNA of infected host cells. Given the size of the chicken genome, there might be many millions of distinct chromosomal sites used by ALV to integrate its provirus.

But molecular analysis of a series of ALV-induced leukemias, each arising independently in a separate chicken, turned up a major surprise. In a substantial portion (>80%) of these leukemia cell genomes, the ALV provirus was found to be integrated into the chromosomal DNA immediately adjacent to the c-*myc* proto-oncogene (Figure 3.23)! This observation, on its own, was difficult to reconcile with the notion that provirus integration occurs randomly at millions of sites throughout the genomes of infected cells.

It soon became clear that the close physical association of the integrated viral genomes and the cellular *myc* gene led to a functional link between these two genetic elements. The viral transcriptional promoter, nested within the ALV provirus, disrupted the control mechanisms that normally govern expression of the c-*myc* gene (see Figure 3.23). Now, instead of being regulated by its own native gene promoter, the cellular *myc* gene was placed directly under viral transcriptional control. As a consequence, rather than being regulated up and down

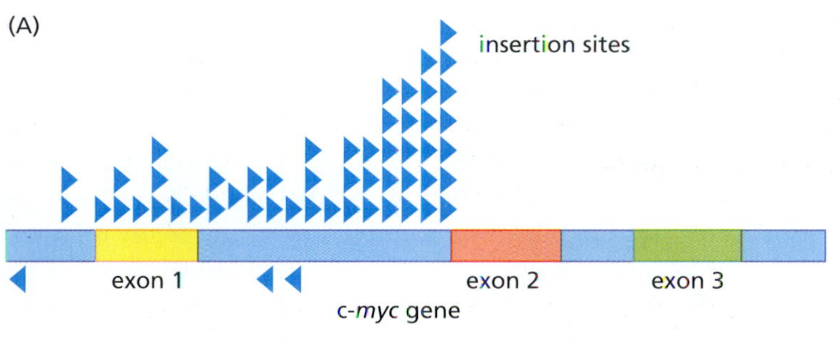

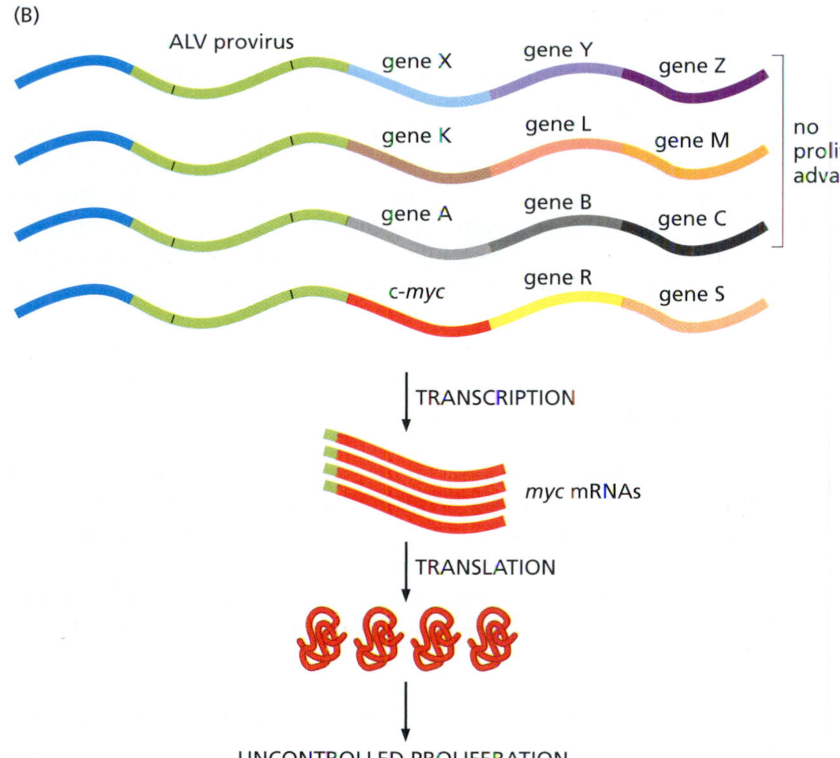

Figure 3.23 Insertional mutagenesis The oncogenic actions of viruses, such as avian leukosis virus (ALV), that lack acquired oncogenes could be explained by the integration of their proviral DNA adjacent to a cellular proto-oncogene. (A) Analysis of numerous B-cell lymphomas that were induced by ALV revealed that a large proportion of the ALV proviruses were integrated into the chromosomal DNA segment carrying the c-*myc* proto-oncogene; the majority were integrated between the first noncoding exon of c-*myc* and the second exon, in which the *myc* reading frame begins. The integration sites are shown here by the *filled triangles*. As indicated, most but not all of the proviruses were integrated in the same transcriptional orientation as that of the c-*myc* gene. (B) This behavior could be rationalized as follows. In the course of ALV infection of chicken lymphocytes, ALV proviruses (*green*) become integrated randomly at millions of different sites in the chromosomal DNA of the lymphocytes. On rare occasions, an ALV provirus becomes integrated (by chance) within the c-*myc* proto-oncogene (*red*). This may then cause transcription of the c-*myc* gene to be driven by the strong, constitutively acting ALV promoter. Because high levels of the Myc protein are potent in driving cell proliferation, the cell carrying this particular integrated provirus and activated c-*myc* gene will now multiply uncontrollably, eventually spawning a large host of descendants that will constitute a lymphoma. (Adapted from S.J. Flint, L.W. Enquist, R.M. Krug et al., Principles of Virology. Washington, DC: ASM Press, 2000.)

by the finely tuned control circuitry of the cell, c-*myc* expression was taken over by a foreign usurper that drove its expression unceasingly and at a high rate. In essence, this hybrid viral–cellular gene arising in the chromosomes of leukemic cells now functioned much like the v-*myc* oncogene carried by avian myelocytomatosis virus.

Suddenly, all the clues needed to solve the puzzle of **leukemogenesis** (leukemia formation) by ALV fell into place. The solution went like this. During the course of infecting a chicken, ALV spread to thousands, then millions of cells in the hematopoietic system of this bird. Soon, the infection was so successful that the bird would become **viremic**, that is, its bloodstream carried high concentrations of virus particles.

Each of these tens of millions of infections resulted in the insertion of an ALV provirus at some random location in the genome of an infected cell. Almost always, this provirus integration had no effect on the infected host cell, aside from forcing the host to produce large numbers of progeny virus particles. But on rare occasions, perhaps in 1 out of 10 million infections, a provirus became integrated by chance next to the c-*myc* gene. This jackpot event led to an explosive outcome—conversion of the c-*myc* gene into a potent oncogene. The cell carrying this deregulated *myc* gene then began uncontrolled proliferation, and within weeks, some of the progeny cells evolved further into more aggressive cancer cells that constituted a leukemia.

This scenario explains the slow kinetics with which these leukemias arise after initial viral infection of a bird. Since activation of the c-*myc* gene through provirus integration is a low-probability event, many weeks and many millions of infectious events are required before these malignancies are triggered. This particular mechanism of proto-oncogene activation came to be called **insertional mutagenesis**; it explains, as well, the leukemogenic powers of other slowly acting retroviruses, such as MLV. By now, study of avian and murine retrovirus-induced infections has demonstrated integration events next to more than 25 distinct cellular proto-oncogenes. Indeed, insertional mutagenesis can be used as a powerful strategy to find new proto-oncogenes (Sidebar 3.8).

Unvoiced by those who uncovered insertional mutagenesis was another provocative idea: Maybe it was possible that nonviral carcinogens could achieve the same end result as ALV did. Perhaps these other carcinogens, including X-rays and mutagenic chemicals, could alter cellular proto-oncogenes while these genes resided in their normal sites in cellular chromosomes. The result might be a disruption of cellular growth control that was just as destabilizing as the events that led to ALV-induced leukemias.

3.12 Some retroviruses naturally carry oncogenes

The descriptions of retroviruses provided in this chapter indicate that there are essentially two classes of these viruses. Some, such as ALV and MLV, carry no oncogenes but can induce tumors that erupt only after a long latent period (i.e., many weeks) following the initial infection of a host animal. Other viruses, such as RSV, can induce cancer rapidly (i.e., in days or several weeks), having acquired an oncogene from a cellular proto-oncogene precursor.

In reality, there is a third class of retroviruses that conforms to neither of these patterns. Human T-cell leukemia virus (HTLV-I) infects about 1% of the inhabitants of Kyushu, the south island of Japan. An endemic infection is also present, albeit at a lower rate, in some islands of the Caribbean. Lifelong HTLV-I infection carries a 3–4% risk of developing adult T-cell leukemia, and the virus seems to be maintained in the population via milk-borne, mother-to-infant transmission.

There are no indications, in spite of extensive molecular surveys, that HTLV-I provirus integration sites are clustered in certain chromosomal regions. Accordingly, it appears highly unlikely that HTLV-I uses insertional mutagenesis to incite leukemias. Instead, its leukemogenic powers seem to be traceable to one or more viral proteins that are naturally encoded by the viral genome. The best understood of these is the viral *tax* gene, whose product is responsible for activating transcription of proviral DNA sequences, thereby enabling production of progeny RNA genomes. At the same time, the *tax* gene product appears to activate transcription of two cellular genes that specify important growth-stimulating proteins—IL-2 (interleukin-2) and GM-CSF (granulocyte macrophage colony-stimulating factor). These "growth factors," to which we will return in Chapter 5, are released by virus-infected cells and proceed to stimulate the proliferation of several types of hematopoietic cells. While such induced proliferation, on its own, does not directly create a leukemia, it seems that populations of these HTLV-I–stimulated cells may progress at a low but predictable frequency to spawn variants that are indeed neoplastic. In this instance, the expression of certain viral oncogenes, notably *tax*, appears to be an intrinsic and essential component of a retroviral replication cycle within host animals, rather than the consequence of rare genetic accidents that yield unusual hybrid genomes, such as the genome of RSV.

Sidebar 3.8 Insertional mutagenesis uncovers novel proto-oncogenes As described earlier (see Section 3.10), the analysis of the genomes of rapidly transforming retroviruses enabled investigators to identify a large cohort of proto-oncogenes. The phenomenon of insertional mutagenesis, first discovered through the insertion of an ALV genome adjacent to the c-*myc* proto-oncogene, offered an alternative strategy for discovering these cellular genes. Thus, a researcher could study a series of independently arising tumors, all of which had been induced by a retrovirus, such as ALV or MLV, that was known to lack its own oncogene. More specifically, this researcher could analyze the locations of the host-cell sequences that lay immediately adjacent to the integrated proviruses in the chromosomal DNA of tumor cells. The hope was that the proviruses might be found to be integrated repeatedly next to a (possibly still-unknown) cellular gene whose activation was triggered by the transcriptional promoter of the provirus. The adjacent gene could be readily cloned, since it was effectively tagged by the closely linked proviral DNA.

The initial fruits of this strategy came from studying the breast cancers induced by mouse mammary tumor virus (MMTV), another retrovirus. Researchers mapped the integration sites of MMTV proviruses in the genomes of mouse breast cancers that had been induced by this virus. Most of the proviruses were found to be integrated in one of three alternative genomic locations, clustering next to cellular genes that were then called *int-1*, *int-2*, and *int-3* (Table 3.4). Each of these genes was later discovered to encode a protein involved in stimulating cell proliferation in one way or another. The deregulated expression of each of these genes, resulting from nearby MMTV provirus integration, seemed to be responsible for triggering the cell proliferation that led to the appearance of mammary tumors.

The *int-1* gene, which was found to be homologous to the *wingless* gene of *Drosophila*, was renamed *Wnt-1*, and was the forerunner of a whole series of *Wnt* genes that have proven to be important vertebrate mitogens and **morphogens**, i.e., factors important for controlling morphogenesis. More recently, this search strategy has been used to uncover a large group of other cellular genes, each of which, when activated by insertional mutagenesis mediated by MLV, triggers leukemias in mice (Table 3.4).

Table 3.4 Examples of cellular genes found to be activated by insertional mutagenesis

Gene	Insertional mutagen	Tumor type	Species	Type of oncoprotein
myc	ALV	B-cell lymphoma	chicken	transcription factor
myc	ALV, FeLV	T-cell lymphoma	chicken, cat	transcription factor
nov	ALV	nephroblastoma	chicken	growth factor
erbB	ALV	erythroblastosis	chicken	receptor TK
mos	IAP	plasmacytoma	mouse	ser/thr kinase
int-1[a]	MMTV	mammary carcinoma	mouse	growth factor
int-2[b]	MMTV	mammary carcinoma	mouse	growth factor
int-3	MMTV	mammary carcinoma	mouse	receptor[c]
int-H/int-5	MMTV	mammary carcinoma	mouse	enzyme[d]
pim-1	Mo-MLV	T-cell lymphoma	mouse	ser/thr kinase
pim-2	Mo-MLV	B-cell lymphoma	mouse	ser/thr kinase
bmi-1	Mo-MLV	T-cell lymphoma	mouse	transcription repressor
tpl-2	Mo-MLV	T-cell lymphoma	mouse	non-receptor TK
lck	Mo-MLV	T-cell lymphoma	mouse	non-receptor TK
p53	Mo-MLV	T-cell lymphoma	mouse	transcription factor
GM-CSF	IAP	myelomonocytic leukemia	mouse	growth factor
IL2	GaLV	T-cell lymphoma	gibbon ape	cytokine[e]
IL3	IAP	T-cell lymphoma	mouse	cytokine
K-*ras*	F-MLV	T-cell lymphoma	mouse	small G protein
CycD1	F-MLV	T-cell lymphoma	mouse	G1 cyclin
CycD2	Mo-MLV	T-cell lymphoma	mouse	G1 cyclin

[a]Subsequently renamed *Wnt-1*.
[b]Subsequently identified as a gene encoding a fibroblast growth factor (FGF).
[c]Related to notch receptors.
[d]Enzyme that converts androgens to estrogens.
[e]Cytokines are GFs that largely regulate various types of hematopoietic cells.

Abbreviations: ALV, avian leukosis virus; FeLV, feline leukemia virus; F-MLV, Friend murine leukemia virus; GaLV, gibbon ape leukemia virus; GF, growth factor; IAP, intracisternal A particle (a retrovirus-like genome that is endogenous to cells); Mo-MLV, Moloney murine leukemia virus; MMTV, mouse mammary tumor virus; ser/thr, serine/threonine; TK, tyrosine kinase.

Adapted in part from J. Butel, *Carcinogenesis* 21:405–426, 2000; and from N. Rosenberg and P. Jolicoeur, in J.M. Coffin, S.H. Hughes and H.E. Varmus (eds.), Retroviruses. Cold Spring Harbor, NY: Cold Spring Harbor Laboratory Press, 1997. Also in part from G.M. Cooper, Oncogenes, 2nd ed. Boston: Jones and Bartlett Publishers, 1995.

3.13 Synopsis and prospects

By studying tumors in laboratory and domesticated animals, cancer biologists discovered a wide array of cancer-causing viruses during the twentieth century. Many of these viruses, having either DNA or RNA genomes, were found able to infect cultured cells and transform them into tumorigenic cells. These transforming powers pointed to the presence of powerful oncogenes in the genomes of the viruses, indeed, oncogenes that were potent enough to induce many of the phenotypes associated with cancer cells (see Table 3.2). Moreover, the ability of these viruses to create transformed cells in the culture dish shed light on the mechanisms by which these viruses could induce cancers in the tissues of infected host animals.

A major conceptual revolution came from the detailed study of RNA tumor viruses, specifically, Rous sarcoma virus (RSV). Its oncogene, termed v-*src*, was found to have originated in a normal cellular gene, c-*src*. This discovery revealed the ability of nontransforming, slowly tumorigenic retroviruses, such as ALV (avian leukosis virus), to acquire normal cellular genes and convert these captured genes into potently transforming oncogenes. The hybrid viruses that arose following these genetic acquisitions were now able to rapidly induce tumors in infected hosts.

Even more important were the implications of finding the c-*src* gene. Its presence in a normal cellular genome demonstrated that the cellular genome carries a proto-oncogene that can be converted into an oncogene following alterations to the sequences of the normal gene. (The details of these alterations will be described in the next several chapters.) Soon a number of retroviruses of both avian and mammalian origin were discovered to carry other oncogenes that had been acquired in similar fashion from the genomes of infected cells. While each of these proto-oncogenes was found initially in the genome of one or another vertebrate species, we now know that all of these genes are represented in the genomes of all vertebrates. Consequently, the generic vertebrate genome carries dozens of such normal genes, each of which has the potential to become converted into an active oncogene.

Yet other proto-oncogenes were discovered by studying the integration sites of proviruses in the genomes of tumors that had been induced by nontransforming retroviruses, such as murine leukemia virus (MLV) and ALV. The random integration of these proviruses into chromosomal DNA occasionally yielded, through the process of insertional mutagenesis, the conversion of a proto-oncogene into an activated oncogene that could readily be isolated because of its close linkage to the provirus. On many occasions, insertional mutagenesis led to rediscovery of a proto-oncogene that was already known because of its presence in an acutely transforming retrovirus; *myc* and avian myelocytomatosis virus (AMV) exemplify this situation. On other occasions, truly novel proto-oncogenes were discovered through study of provirus integration sites; the *int-1* gene activated by MMTV provides a striking example of this route of discovery. In fact, the process of insertional mutagenesis remained little more than a laboratory curiosity, of interest to only a small cadre of cancer biologists, until it was reported to lead to tumors in several patients being treated by gene therapy (Sidebar 3.9).

These discoveries of proto-oncogenes and oncogenes, as profound as they were, provoked as many questions as they answered. It remained unclear how the retrovirus-encoded oncogene proteins (called **oncoproteins**) differed functionally from the proteins encoded by corresponding proto-oncogenes. The biochemical mechanisms used by these oncoproteins to transform cells were also obscure.

The molecular mechanisms used by DNA tumor viruses to transform infected cells were even more elusive, since these viruses seemed to specify oncoproteins that were very different from the proteins made by their host cells. Such

Sidebar 3.9 Gene therapy can occasionally have tragic consequences Gene therapy has been found to be most applicable to diseases of the hematopoietic system. Thus, children who are born with a severe immunodeficiency due to a germ-line–specified defect in one or another critical component of the immune system can, in principle, be cured if the missing gene is transduced into their bone marrow stem cells using retrovirus vectors (see Sidebar 3.7). After infection *in vitro* by a gene-transducing retroviral vector, stem cells are introduced into the afflicted children, in whose bone marrow these cells become stably engrafted. The differentiated progeny of these engrafted, genetically altered stem cells are then able to supply missing immune functions, thereby reversing the congenital immunodeficiency.

Just such a therapeutic approach was launched in France, in which bone marrow stem cells from ten children suffering from a congenital immunodeficiency—X-linked severe combined immunodeficiency—were infected with a Moloney murine leukemia virus (MLV)–derived retroviral vector that transduced a gene specifying the gene product that they lacked—the γc protein. Nine of these children responded by showing a dramatic reconstitution of their immunological function. However, as reported in 2003, the two youngest children in the trial developed T-cell leukemia 30 and 34 months after the initiation of the gene therapy trial. A third case was reported later.

In these cases, analyses of the DNA of leukemic cells revealed proviruses derived from the viral vector that had become integrated within several kilobase pairs of the first exon of the *LMO2* gene, a proto-oncogene that was previously known to be activated in human T-cell leukemias. Given the known role of the *LMO2* oncogene in leukemogenesis, these inserted proviruses were almost certainly responsible for triggering the three leukemias. Hence, insertional mutagenesis leading to oncogenesis, which had long been feared as a possible but remote risk incurred by such a gene therapy strategy, became a grim reality and may ultimately limit the options available to those interested in correcting inborn defects through gene therapy.

differences suggested that these viral oncoproteins could not insinuate themselves into the cellular growth-regulating machinery in any easy, obvious way. Only in the mid-1980s, ten years after this research began, did their transforming mechanisms become apparent, as we will see in Chapters 8 and 9.

For many cancer researchers, and for the public that supported this research, there was a single overriding issue that had motivated much of this work in the first place: did any of these viruses and the proto-oncogenes that they activated play key roles in causing human cancers? As we will learn, about one-fifth of the human cancer burden worldwide is associated with infectious agents. Hepatitis B and C viruses (HBV, HCV), as well as human papillomaviruses (HPVs), play key roles in triggering commonly occurring cancers. Indeed, even infrequently occurring human tumors that seem to be familial have been traced in recent years to viral infections (Sidebar 3.10). So the recognized role of viruses in cancer pathogenesis is substantial and growing.

Still, even if RNA and DNA tumor viruses were not responsible for inciting a single case of human cancer, the research into their transforming mechanisms would have been justified. This research opened the curtain on the genes in our genome that play central roles in all types of human cancer. It accelerated by decades our understanding of cancer pathogenesis at the level of genes and molecules. It catapulted cancer research from a descriptive science into one where complex phenomena could finally be understood and explained in precise, mechanistic terms.

Sidebar 3.10 Classic Kaposi's sarcoma appears to be a familial disease Some virus-induced malignancies are limited largely to small subpopulations and thus resemble familial cancer syndromes. We will encounter many of these syndromes in Chapters 7, 8, 9, and 12. There, we will learn that inheritance of mutant alleles of tumor suppressor genes or DNA repair genes can create a strong, inborn cancer predisposition.

Prior to the onset of the AIDS epidemic, the disease of Kaposi's sarcoma (KS)— apparently a malignancy of cells related to those forming the endothelial lining of lymph ducts—was confined largely to small subpopulations, notably men of Mediterranean and Jewish descent. This resembled a familial cancer, in which cancer-predisposing alleles were present only in the gene pools of certain ethnic subpopulations. After the onset of the AIDS epidemic, however, KS became 1000-fold more common, and at least this form of KS could be associated with an infectious agent—human herpesvirus-8 (HHV-8), also known as KSHV (KS herpesvirus). This virus, along with a number of other infectious agents, is an opportunistic pathogen that thrives in the bodies of those lacking a functional immune system. Because of the AIDS epidemic in Africa, KS has now become the fourth most common infection-induced cancer worldwide.

The virology of HHV-8 failed to explain how the "classic," pre-AIDS KS is transmitted in immunocompetent populations. Indeed, these tumors could also be associated with HHV-8 infections. Examination of the various substrains of HHV-8, as defined by sequence polymorphisms in viral DNAs, has recently revealed that different substrains of the virus are present in different subpopulations of Jews; one substrain predominates among Ashkenazic Jews (of recent European descent), while a second is common among Sephardic Jews (of North African and Middle Eastern descent). Both populations have infection rates that are as much as 10 to 20 times higher than in non-AIDS Western populations. Provocatively, within these subgroups, the transmission of specific HHV-8 substrains is correlated with inheritance of certain mitochondrial DNA polymorphisms far more strongly than with inheritance of Y-chromosome polymorphic markers. Mitochondrial DNA is transmitted maternally, indicating that maternal transmission of virus (occurring possibly via saliva) has played a major role in creating pockets of disease in family lineages that are likely to extend back to founder populations that existed more than 2000 years ago. [Another maternal transmission route may explain the high incidence of adult T-cell leukemia (caused by the HTLV-I retrovirus) in southern Japan (see Section 3.12).]

Hence, certain geographically and ethnically localized malignancies, such as classic KS and adult T-cell leukemia, are actually due to viruses that spread poorly "horizontally" (i.e., from one adult to another) but can be transmitted "vertically" (between parent and offspring) through long-term, intimate contact. This echoes the behavior of certain strains of mice that have high rates of breast cancer. As first shown in Bar Harbor, Maine, in 1933, transmission of disease from parent to offspring occurred when females of high-incidence strains were mated to males of low-incidence strains, but not following the reverse matings. Also, when female pups of high-incidence strains were transferred to low-incidence foster nursing mothers within 24 hours of birth, only 8% eventually developed breast cancer, compared with a 92% incidence exhibited by mice that had been nursed by mothers from a high-incidence strain. This led to the conclusion that this breast cancer susceptibility was transmitted from one generation to the next by a milk-borne infectious agent, which was later identified as mouse mammary tumor virus (MMTV).

Key concepts

- The notion that cancer might be an infectious disease gained favor with Peyton Rous's 1910 discovery that a virus—the Rous sarcoma virus (RSV)—could induce tumors in chickens, but the idea lost credibility in 1926 when the stomach lesions of Fibiger's rats were found to result from vitamin deficiency and not spirochete infection.
- Decades later, Howard Temin and Harry Rubin's discovery that cultured cells infected with RSV were transformed into tumor cells resurrected tumor virus research and led to the realization that cancer could be studied at the level of the cell.
- Transformed cells in culture show several characteristics: (1) unlike normal cells, transformed cells lack contact inhibition and consequently manifest as a multilayered clump known as a focus; and (2) the focus is a clonal outgrowth, with all its cells descended from a single, common progenitor. (3) The cells in a focus can often grow in an anchorage-independent fashion.
- Tumorigenicity in a host animal is the acid test for full cellular transformation.
- The transformation phenotype induced by RSV infection was found to be transmitted to progeny cells, and experiments using a temperature-sensitive *(ts)* mutant of RSV showed that retention of the transformed state depends on the continued activity of an RSV gene product.

- In addition to RNA viruses like RSV, several classes of DNA viruses—including papovavirus, human adenovirus, herpesvirus, and poxvirus—were found to induce cancers.
- While the genomes of RNA viruses consist of single-stranded RNA, the genomes of DNA viruses consist of double-stranded DNA (dsDNA). The papovaviruses—which include Shope papillomavirus, mouse polyomavirus, and SV40—have circular dsDNA genomes, while adenoviruses and herpesviruses have long, linear dsDNA genomes. Like SV40, human adenovirus induces cell lysis after replicating in its natural, permissive host, but cannot replicate in nonpermissive host cells and instead may transform these cells.
- Since replication of viral DNA genomes occurs independently of the host cells' DNA and since viral genomes lack the elements to properly segregate during mitosis, the transmission of DNA tumor virus genomes from one cell generation to the next posed a conceptual problem, until it was discovered that DNA tumor virus genomes integrate into host-cell chromosomal DNA.
- Since the genomes of RNA viruses consist of single-stranded RNA that cannot be incorporated into host DNA and re-infection does not explain the persistence of the transformed state in descendant cells, Temin postulated that RNA viruses make double-stranded DNA copies of their genomes—the heretofore unheard-of process of reverse transcription—and that these DNA copies are integrated into the host's chromosomal DNA as a part of a normal viral replication cycle. This is a major distinction from DNA tumor viruses, for which integration is a very rare and haphazard event that is not an integral part of viral replication.
- Because their replication cycle depends on information flowing backward (i.e., from RNA to DNA), RNA viruses came to be called retroviruses and the DNA version of their viral genomes was called a provirus.
- Working with RSV, researchers found that viral replication and cell transformation were specified by separate genes, with the transforming function residing in a single gene called *src*.
- Using a DNA probe that specifically recognized the transformation-associated (i.e., *src*) sequences of the RSV genome, researchers made the unexpected discovery that *src*-related sequences were present in the DNA of chicken cells not infected by RSV. Further research indicated that the *src* gene was a normal, highly conserved gene of all vertebrate species (as later proved true of many other such genes).
- The difference between the actions of the cellular version of *src* (c-*src*), which supports normal cell function, and the viral version (v-*src*), which acts as an oncogene, can be explained if v-*src* were altered after it was plucked from a cellular genome by an ancestor of RSV.
- Because it can serve as a precursor to an oncogene, c-*src* was called a proto-oncogene, a term that carried the implication that normal vertebrate cells contain genes that have the intrinsic potential to induce cancer.
- The acquisition and activation of *src* by a retrovirus indicated that a single oncogene could act pleiotropically to evoke a multiplicity of changes in cellular traits, as well as the possibility that other mutation mechanisms might activate proto-oncogenes that continued to reside in their normal sites in cellular chromosomes.
- Some retroviruses can induce cancer although they do not carry oncogenes in their genomes; these viruses work much more slowly to induce cancer than those bearing oncogenes. Such nontransforming, slowly tumorigenic retroviruses activate proto-oncogenes by inserting their genomes adjacent to these cellular genes in cellular chromosomes, a process called insertional mutagenesis. This chance occurrence places the proto-oncogene under the control of the viral transcriptional promoter, which deregulates the gene's expression

and leads to uncontrolled cell proliferation. Insertional mutagenesis can be exploited to find new proto-oncogenes.

- In addition to the nontransforming retroviruses (which work via insertional mutagenesis) and the acutely transforming ones (which work via acquired oncogenes), retroviruses exist whose carcinogenic powers are traceable to their own normal gene products. A case in point is human T-cell leukemia virus (HTLV-I), whose *tax* gene encodes a protein that activates transcription of proviral DNA and, as a side effect, also stimulates expression of cellular growth factors that induce cell proliferation.

Thought questions

1. What observations favor or argue against the notion that cancer is an infectious disease?
2. How can one prove that tumor virus genomes must be present in order to maintain the transformed state of a virus-induced tumor? What genetic mechanisms, do you imagine, might enable this process to become "hit-and-run," in which the continued presence of a tumor virus is not required to maintain the tumorigenic phenotype after a certain time?
3. Why are oncogene-bearing viruses like Rous sarcoma virus so rarely encountered in wild populations of chickens?
4. What evidence suggests that the phenotypes of cells transformed by tumor viruses *in vitro* reflect comparable phenotypes of tumor cells *in vivo*?
5. What logic suggests that the chromosomal integration of tumor virus genomes is an intrinsic, obligatory part of the replication cycle of RNA tumors viruses but an inadvertent side-product of DNA tumor virus replication?
6. What evidence suggests that a proto-oncogene like *src* is actually a normal cellular gene rather than a gene that has been inserted into the germ line by an infecting retrovirus?
7. How do you imagine that DNA tumor viruses and retroviruses like avian leukosis virus arose in the distant evolutionary past?
8. Why do retroviruses like avian leukosis virus take so long to induce cancer?

Additional reading

Bishop JM (1985) Viral oncoproteins. *Cell* 42, 23–38.

Boshoff C (2004) Kaposi virus scores cancer coup. *Nat. Med.* 9, 261–262.

Bruix J et al., (2004) Focus on hepatocellular carcinoma. *Cancer Cell* 5, 215–219.

Bute J (2000) Viral carcinogenesis: revelation of molecular mechanisms and etiology of human disease. *Carcinogenesis* 21, 405–426.

Carbone M, Pass HI, Miele L and Bocchetta M (2003) New developments about the association of SV40 with human mesothelioma. *Oncogene* 22, 5173–5180.

Coffin JM, Hughes SH and Varmus HE (eds.) (1997) Retroviruses. Cold Spring Harbor, NY: Cold Spring Harbor Laboratory Press.

Evans AS and Mueller NE (1990) Viruses and cancer: causal associations. *Ann. Epidemiol.* 1, 71–92.

Fields BN, Knipe DM, Howley PM et al. (eds.) (1996) Field's Virology, 3rd ed. Philadelphia: Lippincott–Raven.

Flint SJ, Enquist LW, Krug RM et al. (2000) Principles of Virology. Washington, DC: ASM Press.

Kung HJ, Boerkoel C and Cater TH (1991) Retroviral mutagenesis of cellular oncogenes: a review with insights into the mechanisms of insertional activation. *Curr. Top. Microbiol. Immunol.* 171, 1–15.

López-Rios F, Illei PB, Rusch V and Ladanyi M (2004) Evidence against a role for SV40 infection in human mesotheliomas and high-risk, false-positive PCR results owing to presence of SV40 sequences in common laboratory plasmids. *Lancet* 364, 1157–1166.

Newton R, Beral V and Weiss R (1999) Human immunodeficiency virus infection and cancer. *Cancer Surv.* 33, 237–262.

Parsonnet J (ed.) (1999) Microbes and Malignancy: Infection as a Cause of Human Cancers. London, UK: Oxford University Press.

Phillips AC and Vousden KH (1999) Human papillomavirus and cancer: the viral transforming genes. *Cancer Surv.* 33, 55–74.

Stehelin D, Varmus HE, Bishop JM and Vogt PK (1976) DNA related to the transforming gene(s) of avian sarcoma viruses is present in normal avian DNA. *Nature* 260,170–173.

Weiss R, Teich N, Varmus H and Coffin J (eds.) (1985) Molecular biology of tumor viruses: RNA tumor viruses, 2nd ed. Cold Spring Harbor, NY: Cold Spring Harbor Laboratory Press.

zur Hausen H (1991) Viruses in human cancers. *Science* 254, 1167–1173.

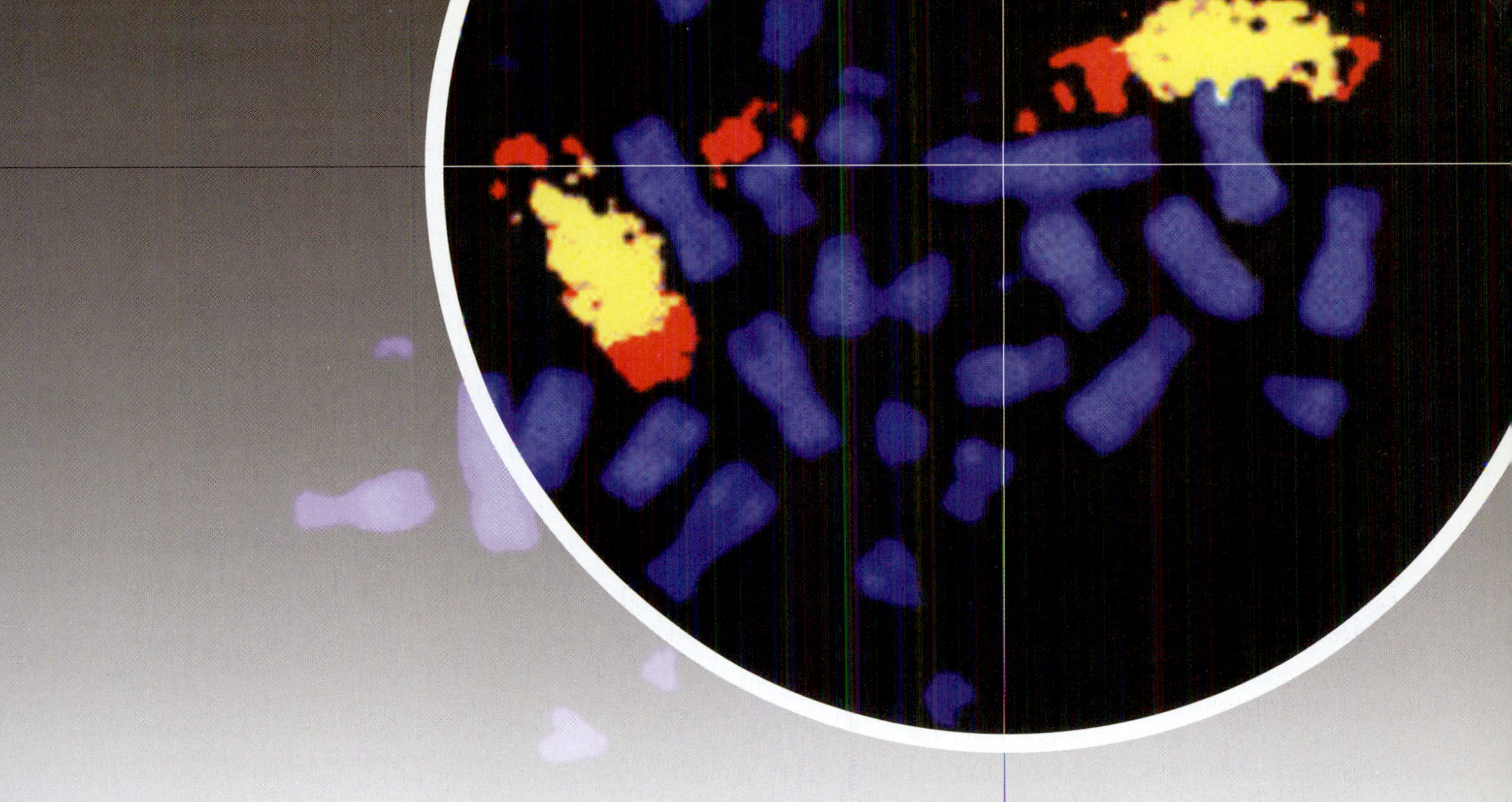

Chapter 4

Cellular Oncogenes

> The viral origin of the majority of all malignant tumors ... has now been documented beyond any reasonable doubt. It ... would be rather difficult to assume a fundamentally different etiology for human tumors.
>
> Ludwik Gross, tumor virologist, 1970

The DNA and RNA tumor viruses characterized in the 1970s provided cancer biologists with a simple and powerful theory of how human tumors could arise. Viruses that occurred commonly in the human population might, with some frequency, infect susceptible tissues and cause the transformation of infected cells. These cells, in turn, would begin to multiply and, sooner or later, form the large cell masses that were encountered frequently in the oncology clinic. Since tumor viruses succeeded in transforming normal rodent and chicken cells into tumor cells with only a small number of introduced genes, these viruses might have similar powers in transforming human cells as well.

With the passage of time, this scenario, attractive as it was, became increasingly difficult to reconcile with the biology and epidemiology of human cancer. Most types of human cancer clearly did not spread from one individual to another as an infectious disease. Significant clusters of cancer cases—mini-epidemics of disease—were hard to find. Even more important, attempts undertaken during the 1970s to isolate viruses from most types of human tumors were unsuccessful. Of the hundred and more tumor types encountered in the oncology clinic, only two commonly occurring tumor types in the Western world—cervical carcinomas and **hepatomas** (liver carcinomas)—could clearly be tied to specific viral causative agents.

These realizations evoked two responses. Those who hung tenaciously to tumor viruses as causative agents of all human cancers argued that chemical and physical carcinogens interacted with viruses that normally hid within the body's cells, activating their latent cancer-causing powers. Other researchers responded by jettisoning viruses entirely and began looking at another potential source of the genes responsible for human cancers—the cellular genome with its tens of thousands of genes. This second tack eventually triumphed, and by the late 1980s, the cell genome was recognized to be a rich source of the genes that drive human cancer cell proliferation.

So, tumor viruses, once viewed as the key agents triggering all human cancers, failed to live up to these high expectations. Ironically, however, tumor virus research proved to be critical in uncovering the cellular genes that are indeed responsible for the neoplastic cell phenotype. The large catalog of cellular cancer-causing genes assembled over the ensuing decades—oncogenes and tumor suppressor genes—derives directly from these early efforts to find infectious cancer-causing agents in human populations.

4.1 Can cancers be triggered by the activation of endogenous retroviruses?

Research begun in Japan by Katsusaburo Yamagiwa in the first decade of the twentieth century revealed that chemical agents could induce cancers in laboratory animals (see Section 2.8). As mentioned earlier, his work showed that repeated painting of coal tars on the ears of rabbits yielded skin carcinomas after several months' time. By the middle of the following decade, a Ph.D. thesis in Paris documented more than a hundred cases of human cancer, largely of the skin, in individuals who had worked with X-ray tubes. In both cases, it was clear that the agents that directly provoked the tumors were nonbiological, being either organic chemicals or radiation (see Sections 2.8 and 2.9).

These discoveries were well known to all cancer researchers by the mid-twentieth century and were hard to reconcile with the theory that all cancers are triggered in one way or another by the actions of infectious agents, that is, tumor viruses. Responding to this, some adherents of the virus theory of cancer, especially those working with retroviruses, proposed a new mechanism in the early 1970s. Their model explained how tumor viruses could participate in the formation of the many cancers that had no outward signs of viral infection.

This new scheme derived from the peculiar biology of retroviruses. On occasion, retrovirus genomes become integrated into the germ-line chromosomes of various vertebrate species, and the resulting proviruses are then transmitted like Mendelian alleles from one generation to the next (Sidebar 4.1). More often than not, these **endogenous proviruses** are transcriptionally silent, and their presence in all of the cells of an organism is not apparent. On rare occasions, however, it is possible to awaken the expression of such latent endogenous proviruses, which often retain the ability to encode infectious retrovirus particles.

Activation of an endogenous retroviral (ERV) genome in fibroblasts prepared from certain strains of mice can be accomplished by culturing these cells in the presence of the thymidine analog bromodeoxyuridine (BrdU). In response, these connective tissue cells, which were ostensibly free of retroviral infection, suddenly begin to release retrovirus particles, due to the transcriptional derepression of their normally silent, endogenous proviruses. Similarly, latent endogenous proviruses may be activated spontaneously *in vivo* in a small number of cells in a mouse. Once infectious virus particles are released from these few cells, they can multiply by cell-to-cell infection, spread rapidly throughout the body, and induce leukemias in these animals.

Knowing this behavior of endogenous retroviruses, some suspected that human cancers might arise in a similar fashion. For example, mutagenic carcinogens, such as those present in tobacco tar, might provoke the activation of previously latent endogenous retroviruses. The resulting virus particles would then begin multiplication by spreading throughout the body of an individual and, like the endogenous retroviruses in some mouse strains, cause cancers to form in one or another susceptible tissue. At the same time, while capable of spreading throughout a person's tissues, such endogenous viruses might be unable to spread horizontally to another individual, explaining the repeated observations that cancer does not behave like a communicable disease. Another related scheme postulated that retroviruses had inserted viral oncogenes into the germ lines of various species, and these latent viral oncogenes became activated by various types of carcinogens.

While attractive in concept, these models of human cancer causation soon collapsed because supportive evidence was not forthcoming. Reports of infectious retroviral particles in human tumors could not be verified. Even reverse transcriptase–containing virus particles were difficult to find in human tumor samples.

It became clear that most endogenous retroviral genomes present in the human genome are relics of ancient germ-line infections that occurred 5 million years ago and earlier in ancestral primates. Since that time, these proviruses mutated progressively into sequences that were no longer capable of specifying infectious retrovirus particles and in this way joined the ranks of the junk DNA sequences that form the great bulk of our genome. Even though as much as 8% of the human genome derives from endogenous retroviral genomes, only several of the approximately 40,000 retrovirus-derived segments have ever been shown to be genetically intact and capable, in principle, of specifying infectious retrovirus particles. One subfamily of these viruses, termed HERV-K, has entered into the human germ line relatively recently, and several of its proviruses are seemingly intact, but to date, even these have not been found to produce infectious viruses or to become mobilized in cancer cells. (It remains unclear why our germ line has not continued to acquire new, functional endogenous proviruses during recent evolutionary times, while the germ lines of other mammalian species, such as that of the mouse, harbor recently acquired endogenous viruses that remain genetically intact and hence retain biological function.) So cancer researchers began to look elsewhere for the genetic elements that might be triggering human cancer formation.

4.2 Transfection of DNA provides a strategy for detecting nonviral oncogenes

For those researchers intent on understanding nonviral carcinogenesis, the demise of the endogenous retrovirus theory left one viable theory on the table. According to this theory, carcinogens function as mutagens (Section 2.9). Whether physical (e.g., X-rays) or chemical (e.g., tobacco tars), these agents induce cancer through their ability to mutate critical growth-controlling genes in the genomes of susceptible cells. Such growth-controlling genes might be, for example, normal cellular genes, such as the proto-oncogenes discovered by the retrovirologists. Once these genes were mutated, the resulting mutant alleles might function as active oncogenes, driving the cancerous growth of the cells that carried them.

Stated differently, this model—really a speculation—predicted that chemically transformed cells carried mutated genes and that these genes were responsible for programming the aberrant growth of these cells. It was impossible to predict the number of such mutated genes present in the genomes of these cells. More

Sidebar 4.1 Endogenous retroviruses can explain tumor development in the absence of infectious viral spread The exposure of a mouse or chicken to a retrovirus often results in the infection of a wide variety of cell types in the body including, on occasion, the cells in the gonads—ovaries or testes. Infections of cells in these organs can result in the integration of a retroviral provirus (see Section 3.7) into the chromosomes of cells that serve as precursors to either sperm or egg. Such proviruses become established in a genetic configuration that is equivalent to that of the cellular genes carried by the sperm or egg. Therefore, when these gametes participate in fertilization, the provirus can be transmitted to a fertilized egg and thus to all of the cells of a resulting embryo and adult (Figure 4.1). This provirus might now become ensconced in the genome of all animals that descend from the initially infected animal. Such a provirus would be termed an *endogenous virus* to distinguish it from viruses that are transmissible from one individual to the next via infection.

Endogenous proviruses that are readily transcribed in an animal's tissues are likely to create a viremia and may induce cancer in that animal early in its life. Such disease-inducing endogenous proviruses are therefore disadvantageous and will be rapidly eliminated from a species' gene pool. This explains why the endogenous proviruses found in the germ lines of most species are, with rare exception, transcriptionally silent.

Careful examinations of the genomes of a variety of mammalian and avian species demonstrate the presence of numerous endogenous retroviral genomes, most of which are clearly relics of germ-line infections that occurred in the distant evolutionary past. Having resided for millions of years in a species' germ line, most have suffered so many mutations that they are no longer able to specify infectious virus particles. However, a small subset of endogenous viral genomes, notably those that have been recently inserted into a species' germ line, remain genetically intact. Given the proper stimulus, these previously latent proviruses may suddenly be transcribed in one or another cell, spread from this cell throughout the body, and eventually launch some type of malignancy, usually of hematopoietic cells (see Section 3.11). For example, the high rate of leukemia in the AKR mouse strain is attributable to the frequent spontaneous activation of an endogenous murine leukemia provirus in an AKR mouse, the subsequent infectious spread of virus throughout the mouse, viremia, and, finally, via insertional mutagenesis, the activation of a proto-oncogene and the eruption of a leukemia.

Figure 4.1 Origin of endogenous retroviruses (A) These viral genomes arise when retroviruses *(red dots)* succeed in establishing a systemic infection in an organism (e.g., a mouse, *top)* and infect, among other cells, a precursor cell of gametes—sperm or eggs. Once a resulting provirus *(green rectangle)* becomes integrated in the genome of a gamete (in this case sperm), it can be introduced into the genome of a fertilized egg and thereafter be distributed to all cells of the organism arising from this zygote *(green dots)*. This organism can in turn transmit the provirus to its descendants via the normal route of sexual reproduction. Activation of expression of the endogenous provirus in an animal *(below)* can lead to infectious spread throughout the body, viremia, and eventually leukemia. (B) The presence of endogenous retrovirus (ERV) genomes can be detected by probing the genomic DNAs of an organism with the DNA of an infectious retrovirus. Shown here are the ERV genomes present in the DNAs of a variety of mouse strains and subspecies as visualized by the Southern blotting procedure (see Figure 4.4). In this case, only the subclass of ERV genomes related to *xenotropic* murine retroviruses is being probed. Each band in a gel channel represents a restriction fragment in a cell genome that carries part or all of an ERV genome. The variability of ERV integration sites from one lab strain to another indicates that numerous ERVs have been integrated into the mouse germ line since the speciation of *Mus musculus,* the species from which all these strains derive. (C) In contrast to the mouse ERV genomes, those detected [using as a probe a fragment of the clone of a human ERV *(above)*] in a collection of human DNAs show rather similar integration sites across the species, indicating their germ-line integration long before the emergence of the human species; the polymorphic differences *(black arrows)* are largely the results of recombination between the terminal LTR sequences at the ends of individual ERV proviruses and resulting deletion of the intervening stretches of proviral DNA. (B, from K. Tomonaga and J.M. Coffin, *Virol.* 73:4327–4340, 1999; C, from J.F. Hughes and J.M. Coffin, *Proc. Natl. Acad. Sci. USA* 101:1668–1672, 2004.)

important, experimental proofs of the existence of these cancer-causing genes represented a daunting challenge. If they were really present in the genomes of chemically transformed cells, including perhaps human tumor cells, how could they possibly be found? If these genes were mutant versions of normal cellular genes, then they were embedded in cancer cell genomes together with tens of thousands, perhaps even a hundred thousand other genes, each present in at least one copy per haploid genome. These cancer genes, if they existed, were clearly tiny needles buried in very large haystacks.

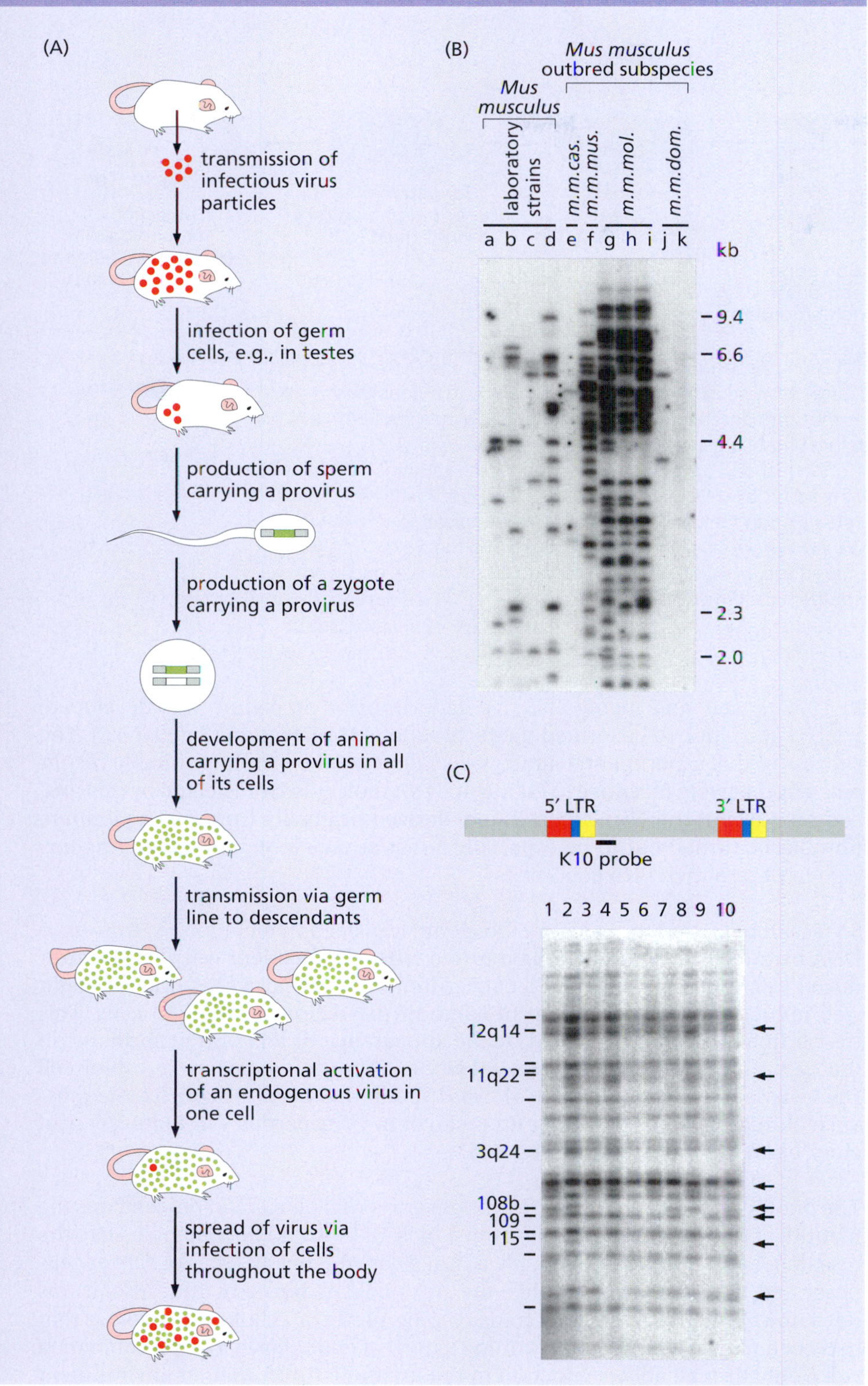

This difficulty caused some to craft a novel experimental strategy to search for oncogenes in the genomes of various types of chemically transformed cells. In outline, this strategy involved introducing DNA (and thus the genes) of cancer cells into normal recipient cells, and then determining whether the recipient cells became transformed in response to the introduced tumor cell DNA. This strategy depended on several experimental advances, including (1) the development of an effective gene transfer procedure, (2) the finding of appropriate cancer cells from which to extract DNA, and (3) the choice of appropriate recipient cells into which this DNA could be introduced (Figure 4.2).

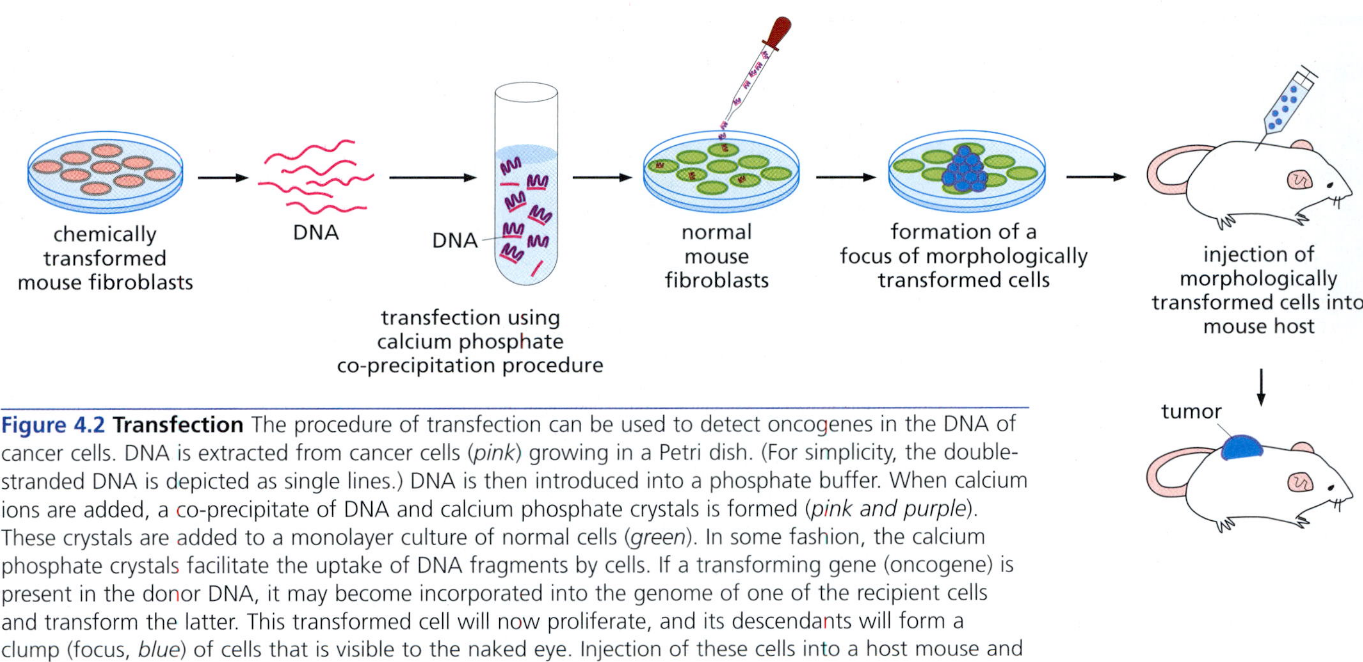

Figure 4.2 Transfection The procedure of transfection can be used to detect oncogenes in the DNA of cancer cells. DNA is extracted from cancer cells (*pink*) growing in a Petri dish. (For simplicity, the double-stranded DNA is depicted as single lines.) DNA is then introduced into a phosphate buffer. When calcium ions are added, a co-precipitate of DNA and calcium phosphate crystals is formed (*pink and purple*). These crystals are added to a monolayer culture of normal cells (*green*). In some fashion, the calcium phosphate crystals facilitate the uptake of DNA fragments by cells. If a transforming gene (oncogene) is present in the donor DNA, it may become incorporated into the genome of one of the recipient cells and transform the latter. This transformed cell will now proliferate, and its descendants will form a clump (focus, *blue*) of cells that is visible to the naked eye. Injection of these cells into a host mouse and resulting tumor formation can be used to confirm the transformed state of these cells.

In 1972, a new and highly effective gene transfer procedure was developed, which soon came to be termed the technique of **transfection** (Sidebar 4.2). The success of this experimental strategy also depended on finding suitable recipient cells that were receptive to taking up DNA molecules transfected by this procedure. Cells of the NIH 3T3 cell line, derived originally from mouse embryo fibroblasts, turned out to be especially adept at taking up and integrating foreign DNA into their own genomes.

So researchers used the calcium phosphate transfection technique to introduce DNA extracted from tumor cells into the NIH 3T3 recipient cells. If the introduced tumor cell DNA carried a cancer-inducing gene or genes, then it might well induce the transformation of some of the recipient NIH 3T3 cells. This transformation could be scored by the appearance of foci of transformants in the cultures of NIH 3T3 cells several weeks after their exposure to tumor cell DNA—essentially the assay that Howard Temin had used to score for the presence of infectious transforming Rous sarcoma virus particles in monolayers of chick embryo fibroblasts (Section 3.2).

The final issue to be settled before this experimental plan could proceed was the identity of the donor cancer cells from which DNA would be prepared. Here the researchers were working blind. It was not clear whether all types of cancer cells possessed transforming genes like the *src* oncogene borne by RSV. Also, it was not known whether a cellular transforming gene—a cellular oncogene—that had been responsible for transforming a normal epithelial cell into a carcinoma cell would also be able to function in the unfamiliar intracellular environment of connective tissue (fibroblastic) cells, like that of the NIH 3T3 cells. There were yet other possible problems. For example, an oncogene that was responsible for transforming normal human cells into cancer cells might fail to transform normal mouse cells because of some interspecies incompatibilities.

With these concerns in mind, researchers chose donor tumor cells derived from mouse fibroblasts. These particular cancer cells originated with mouse fibroblasts of the C3H10T1/2 mouse cell line that had been treated repeatedly with the potent carcinogen and mutagen 3-methylcholanthrene (3-MC), a known component of coal tars. Importantly, these cells bore no traces of either tumor virus

Sidebar 4.2 Transfection represents a highly useful gene transfer technique The development in 1972 of a transfection procedure was originally motivated by the need to introduce naked viral RNA and DNA molecules directly into mammalian cells. By doing so, experimenters could circumvent the usual route through which viral genomes enter cells—being carried in by virions. Transfection of adenovirus, SV40, and even retrovirus DNAs into appropriate recipient cells was found to result in viral replication cycles that were indistinguishable from those initiated by infectious virus particles. These successes demonstrated the ability of the transfection procedure to introduce relatively long DNA molecules, often larger than 20 kilobases, into recipient cells. Later, these procedures were adapted to transfer cellular genes from one cell to another.

In order to carry out a transfection (see Figure 4.2), a purified DNA of interest is suspended in a phosphate buffer. The addition of calcium to this solution causes the precipitation of calcium phosphate crystals and the co-precipitation of DNA that may be present in the solution. When these calcium phosphate crystals are placed on monolayers of recipient cells grown in culture, they facilitate the introduction of the DNA molecules into these cells, doing so through a mechanism that remains unclear. Once inside recipient cells, a small portion of the transfected DNA enters into the nucleus and somehow becomes integrated into the chromosomal DNA of these cells, thereby gaining the ability to be transmitted to the progeny of these cells together with all of their native genes. Other gene transfer techniques have been developed since this technique was invented, but the calcium phosphate procedure is still widely used.

infection or activated endogenous retroviral genomes. Hence, any transforming oncogenes detected in the genome of these cells would, with great likelihood, be of cellular origin, that is, mutant versions of normal cellular genes.

In 1978–1979, DNAs extracted from several such 3-MC-transformed mouse cell lines were transfected into cultures of NIH 3T3 recipient cells, yielding large numbers of foci after several weeks. The cells plucked from the resulting foci were later found to be both anchorage-independent and tumorigenic. This simple experiment proved that the donor tumor DNA carried one or several genetic elements that were able to convert a non-tumorigenic NIH 3T3 recipient cell into a cell that was strongly tumorigenic.

DNA extracted from normal, untransformed C3H10T1/2 cells was unable to induce foci in the NIH 3T3 cell monolayers. This difference made it highly likely that previous exposure of normal C3H10T1/2 cells to the 3-MC carcinogen had altered the genomes of these cells in some way, resulting in the creation of novel genetic sequences that possessed transforming powers. In other words, it seemed likely that the 3-MC carcinogen had converted a previously normal C3H10T1/2 gene (or genes) into a mutant allele that now could function as a transforming oncogene when introduced into NIH 3T3 cells.

At first, it seemed to be quite difficult to determine whether the donor tumor cells carried a single oncogene in their genomes or several distinct transforming oncogenes that acted in concert to transform the recipient cells. Careful analysis of the transfection procedure soon resolved this issue. Researchers discovered that when cellular DNA was applied to a recipient cell, only about 0.1% of a cell genome's worth of donor DNA became established in the genome of each transfected recipient cell. The probability of two independent, genetically unlinked donor genes both being introduced into a single recipient cell was therefore $10^{-3} \times 10^{-3} = 10^{-6}$, that is, a highly unlikely event. From this calculation they could infer that only a single gene was responsible for the transformation of NIH 3T3 cells following transfection of donor tumor cell DNA. This led, in

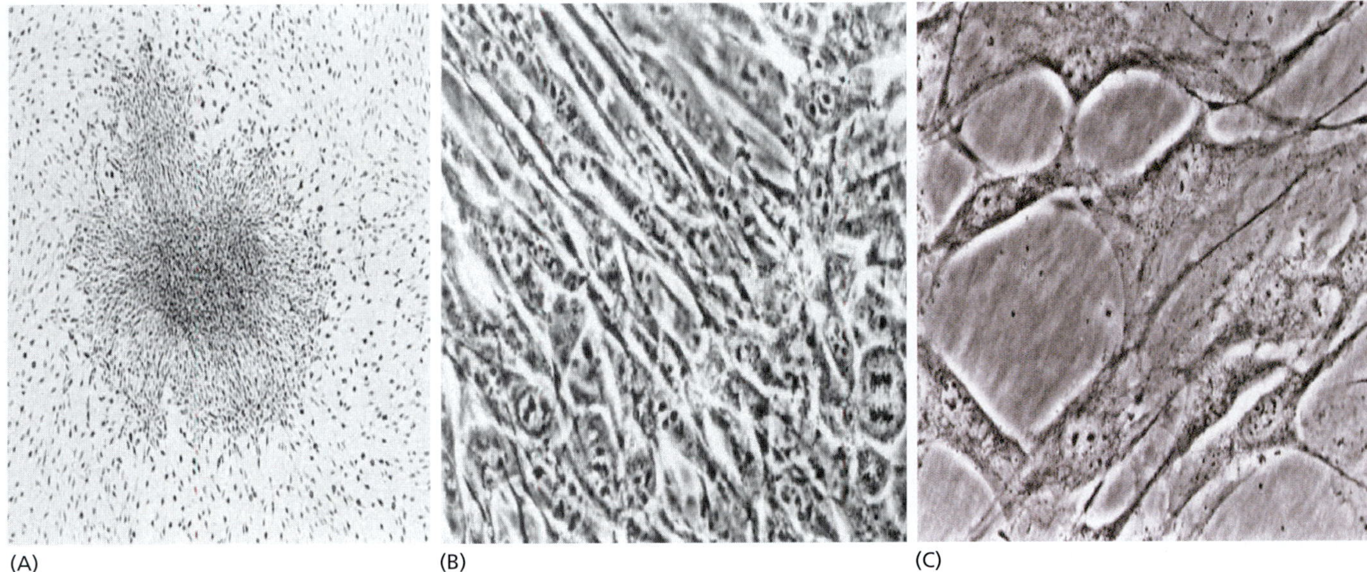

Figure 4.3 Transformation of mouse cells by human tumor DNA The introduction via transfection of various human tumor DNAs into NIH 3T3 cells yielded foci of transformants. (A) A focus generated by transfection of DNA from the T24 human bladder carcinoma cell line. (B) High-magnification image of the transformed cells within this focus. Like many types of transformed fibroblasts, these are spindle-shaped, refractile, and piled up densely on one another. (C) High-magnification image of the NIH 3T3 cells in the surrounding untransformed cell monolayer. Like normal fibroblasts, these have wide, extended cytoplasms and are not piled on one another. (From M. Perucho et al., *Cell* 27:467–476, 1981.)

turn, to the conclusion that years earlier exposure of normal C3H10T1/2 mouse cells to the 3-MC carcinogen had caused the formation of a single mutant oncogenic allele; this allele was able, on its own, to transform both the C3H10T1/2 cells and, later on, the recipient NIH 3T3 cells into which this allele was introduced by gene transfer.

These transfection experiments were highly important, in that they provided strong indication that oncogenes can arise in the genomes of cells through mechanisms that have no apparent connection with viral infection. Perhaps human tumor cells, which also appeared to arise via nonviral mechanisms, also carried transfectable oncogenes. Would human oncogenes, if present in the genomes of these cells, also be able to alter the behavior of mouse cells?

Both of these questions were soon answered in the affirmative. DNAs extracted from cell lines derived from human bladder, lung, and colon carcinomas, as well as DNA from a human promyelocytic leukemia, were all found capable of transforming recipient NIH 3T3 cells (Figure 4.3). This meant that the oncogenes in these cell lines, whatever their nature, were capable of acting across species and tissue boundaries to induce cell transformation.

4.3 Oncogenes discovered in human tumor cell lines are related to those carried by transforming retroviruses

The oncogenes detected by transfection in the genomes of various human tumor cells were ostensibly derived from preexisting normal cellular genes that lacked oncogenic function. This seemed to parallel the process that led to the appearance of transforming retroviruses (Section 3.9). Recall that during the formation of these viruses, preexisting normal cellular genes—proto-oncogenes—became activated into potent oncogenes, albeit through an entirely different genetic mechanism.

These apparent parallels led to an obvious question: could the same group of cellular proto-oncogenes become activated into oncogenes by marauding retroviruses in one context and by nonviral mutagens in another? Or did the retrovirus-associated oncogenes and those activated by nonviral mechanisms arise from two very distinct groups of cellular proto-oncogenes?

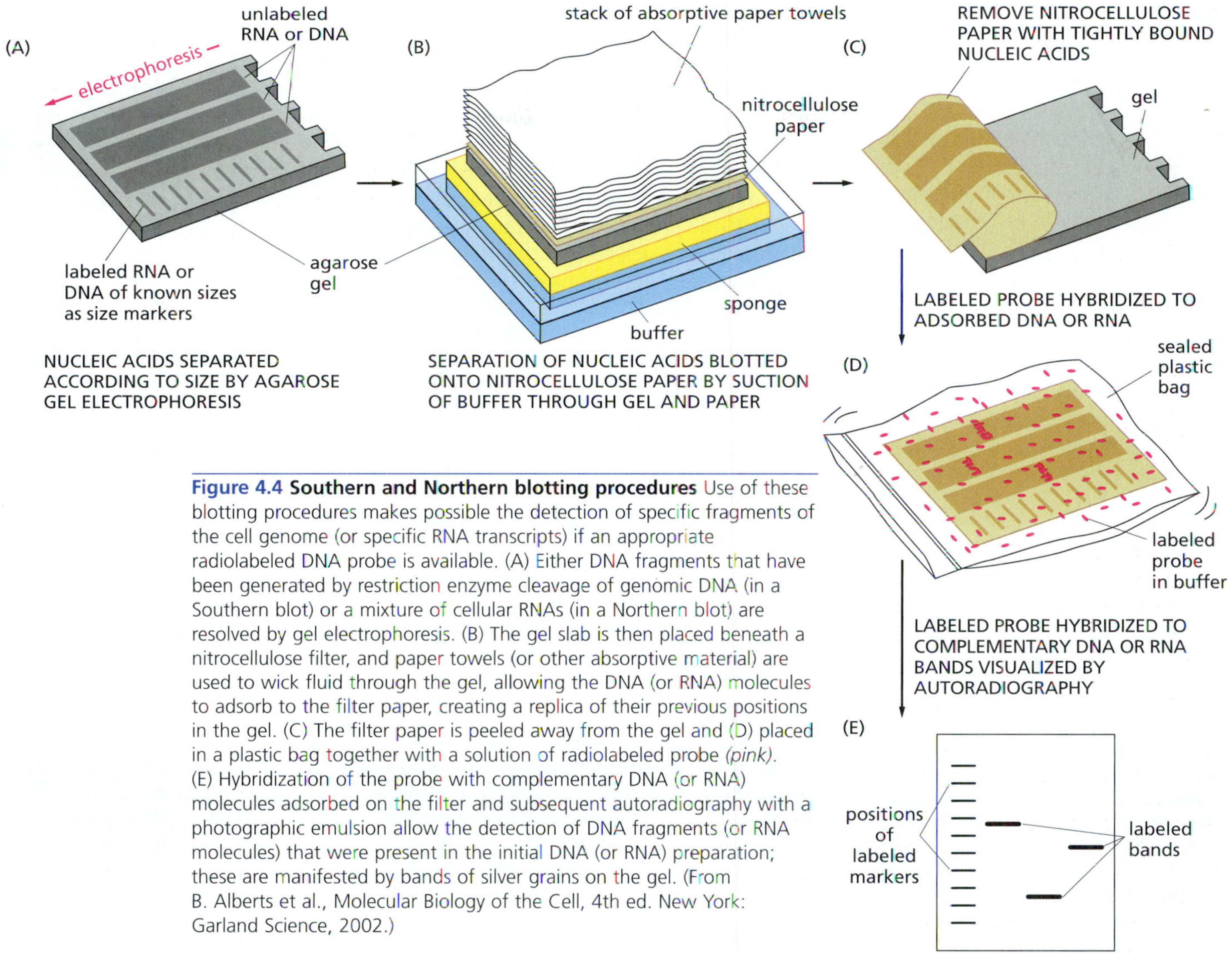

(A)
electrophoresis
unlabeled RNA or DNA
labeled RNA or DNA of known sizes as size markers
agarose gel
NUCLEIC ACIDS SEPARATED ACCORDING TO SIZE BY AGAROSE GEL ELECTROPHORESIS

(B)
stack of absorptive paper towels
nitrocellulose paper
sponge
buffer
SEPARATION OF NUCLEIC ACIDS BLOTTED ONTO NITROCELLULOSE PAPER BY SUCTION OF BUFFER THROUGH GEL AND PAPER

(C)
REMOVE NITROCELLULOSE PAPER WITH TIGHTLY BOUND NUCLEIC ACIDS
gel
LABELED PROBE HYBRIDIZED TO ADSORBED DNA OR RNA

(D)
sealed plastic bag
labeled probe in buffer
LABELED PROBE HYBRIDIZED TO COMPLEMENTARY DNA OR RNA BANDS VISUALIZED BY AUTORADIOGRAPHY

(E)
positions of labeled markers
labeled bands

Figure 4.4 Southern and Northern blotting procedures Use of these blotting procedures makes possible the detection of specific fragments of the cell genome (or specific RNA transcripts) if an appropriate radiolabeled DNA probe is available. (A) Either DNA fragments that have been generated by restriction enzyme cleavage of genomic DNA (in a Southern blot) or a mixture of cellular RNAs (in a Northern blot) are resolved by gel electrophoresis. (B) The gel slab is then placed beneath a nitrocellulose filter, and paper towels (or other absorptive material) are used to wick fluid through the gel, allowing the DNA (or RNA) molecules to adsorb to the filter paper, creating a replica of their previous positions in the gel. (C) The filter paper is peeled away from the gel and (D) placed in a plastic bag together with a solution of radiolabeled probe *(pink)*. (E) Hybridization of the probe with complementary DNA (or RNA) molecules adsorbed on the filter and subsequent autoradiography with a photographic emulsion allow the detection of DNA fragments (or RNA molecules) that were present in the initial DNA (or RNA) preparation; these are manifested by bands of silver grains on the gel. (From B. Alberts et al., Molecular Biology of the Cell, 4th ed. New York: Garland Science, 2002.)

Use of DNA probes specific for the retrovirus-associated oncogenes provided the answers in short order. Using the Southern blot procedure (Figure 4.4), a DNA probe derived from the H-*ras* oncogene present in Harvey rat sarcoma virus was able to recognize and form hybrids with the oncogene detected by transfection in the DNA of a human bladder carcinoma cell (Figure 4.5). A related oncogene, termed K-*ras* from its presence in the genome of Kirsten sarcoma virus, was able to anneal with the oncogene detected by transfection of DNA from a human colon carcinoma cell line.

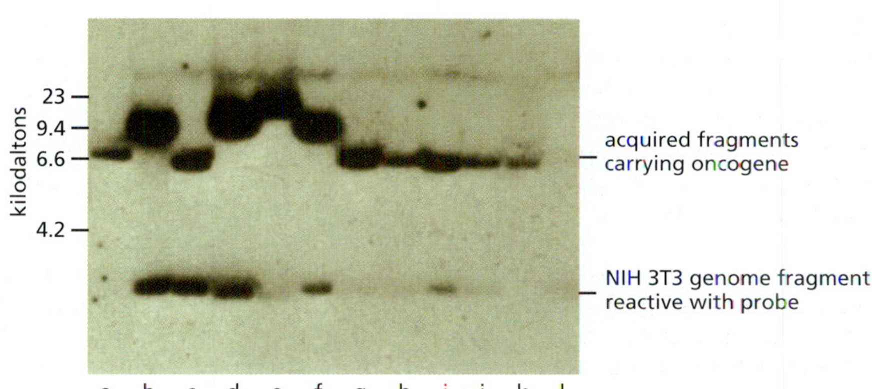

Figure 4.5 Homology between transfected oncogenes and retroviral oncogenes The Southern blot procedure (see Figure 4.4) was used to determine whether there was any relatedness between retrovirus-associated oncogenes and those discovered by transfection of tumor cell DNA. Cloned retroviral oncogene DNAs were used to make radiolabeled probes, while the restriction enzyme–cleaved genomic DNAs from transfected cells were analyzed by the Southern blot procedure. Shown here is the annealing between a radiolabeled H-*ras* oncogene probe (cloned from the genome of Harvey murine sarcoma virus) and the genomic DNAs from a series of 11 lines of NIH 3T3 cells *(channels a through k)* that had been transformed by transfection of DNA extracted from a human bladder carcinoma cell line; the DNA of untransfected NIH 3T3 cells was analyzed in channel l. (From L.F. Parada et al., *Nature* 297:474–478, 1982.)

The list of connections between the retrovirus-associated oncogenes and oncogenes present in non-virally induced human tumors soon grew by leaps and bounds (Table 4.1). In these cases, the connections were often forged following the discovery that the retrovirus-associated oncogenes were present in increased copy number in human tumor cell genomes. The *myc* oncogene, originally known from its presence in avian myelocytomatosis virus (AMV; see Section 3.10) was found to be present in multiple copies in the DNA of the HL-60 human promyelocytic leukemia cell line. These extra copies of the *myc* gene (about 10–20 per diploid genome) were the result of the process of **gene amplification** and suggested that the multiple copies of this gene caused proportionately increased levels of its protein product; this somehow favored the proliferation of the cancer cells. The *erbB* gene, first discovered through its presence in the genome of avian erythroblastosis virus (AEV; refer to Table 3.3), was discovered to be present in increased copy number in the DNAs of human stomach, breast, and brain tumor cells. (**Erythroblastosis** is a malignancy of red blood cell precursors.) Elevated expression of the homolog of the *erbB* gene is now thought to be present in the majority of human carcinomas.

In 1987, amplification of the *erbB*-related gene known variously as *erbB2, neu*, or *HER2*, was reported in many breast cancers (Figure 4.6A). Increases in gene copy number of more than five copies per cancer cell were found to correlate with a decreased survival of patients bearing these tumors (Figure 4.6B). (This figure shows a **Kaplan–Meier plot**, in which the percentage of patients surviving is plotted on the ordinate as a function of the time after initial diagnosis or treatment, which is plotted on the abscissa. We will use this graphing convention

Table 4.1 Examples of retrovirus-associated oncogenes that have been discovered in altered form in human cancers

Name of virus	Species	Oncogene	Type of oncoprotein	Homologous oncogene found in human tumors
Rous sarcoma	chicken	*src*	receptor TK	colon carcinoma[a]
Abelson leukemia	mouse	*abl*	nonreceptor TK	CML
Avian erythroblastosis	mouse	*erbB*	receptor TK	gastric, lung, breast[b]
McDonough feline sarcoma	cat	*fms*	receptor TK	AML[c]
H-Z feline	cat	*kit*	receptor TK[d]	gastrointestinal stromal
Murine sarcoma 3611	mouse	*raf*	Ser/Thr kinase[e]	bladder carcinoma
Simian sarcoma	monkey	*sis*	growth factor (PDGF)	many types[f]
Harvey sarcoma	mouse/rat	H-*ras*[g]	small G protein	bladder carcinoma
Kirsten sarcoma	mouse/rat	K-*ras*[g]	small G protein	many types
Avian erythroblastosis	chicken	*erbA*	nuclear receptor[h]	liver, kidney, pituitary
Avian myeloblastosis E26	chicken	*ets*	transcription factor	leukemia[i]
Avian myelocytoma	chicken	*myc*[j]	transcription factor	many types
Reticuloendotheliosis	turkey	*rel*[k]	transcription factor	lymphoma

[a]Mutant forms found in a small number of these tumors.
[b]Receptor for EGF; the related erbB2/HER2/Neu protein is overexpressed in 30% of breast cancers.
[c]Fms, the receptor for colony-stimulating factor (CSF-1), is found in mutant form in a small number of AMLs; the related Flt3 (Fms-like tyrosine kinase-3) protein is frequently found in mutant form in these leukemias.
[d]Receptor for stem cell factor.
[e]The closely related B-Raf protein is mutant in the majority of melanomas.
[f]Protein is overexpressed in many types of tumors.
[g]The related N-*ras* gene is found in mutant form in a variety of human tumors.
[h]Receptor for thyroid hormone.
[i]27 distinct members of the Ets family of transcription factors are encoded in the human genome. Ets-1 is overexpressed in many types of tumors; others are involved in chromosomal translocations in AML and in Ewing sarcomas.
[j]The related N-*myc* gene is overexpressed in pediatric neuroblastomas and small-cell lung carcinomas.
[k]Rel is a member of a family of proteins that constitute the NF-κB transcription factor, which is constitutively activated in a wide range of human tumors.

Abbreviations: AML, acute myelogenous leukemia; CML, chronic myelogenous leukemia.

Adapted in part from J. Butel, *Carcinogenesis* 21:405–426, 2000; and G.M. Cooper, Oncogenes, 2nd ed. Boston and London: Jones and Bartlett, 1995.

repeatedly throughout this book.) Significantly, the observed amplification of the *erbB2/HER2* gene was correlated with an increased expression of its encoded protein (Figure 4.6C). Among a large group of breast cancer patients, those whose tumors expressed normal levels of this protein showed a median survival of six to seven years after diagnosis, while those patients whose tumors expressed elevated levels had a median survival of only three years. This inverse correlation between *erbB2/neu* expression levels and long-term patient survival provided a strong indication that this gene, in amplified form, was causally involved in driving the malignant growth of the breast cancer cells (but see Sidebar 4.3).

Ironically, mutant alleles of the *src* oncogene, the first cellular oncogene to be discovered, proved to be elusive in human tumor cell genomes. Finally, in 1999—almost a quarter of a century after the *src* gene was first cloned—mutant forms of the *src* gene were found in the genomes of human tumor cells, specifically, in the genomes of 12% of advanced human colon carcinomas.

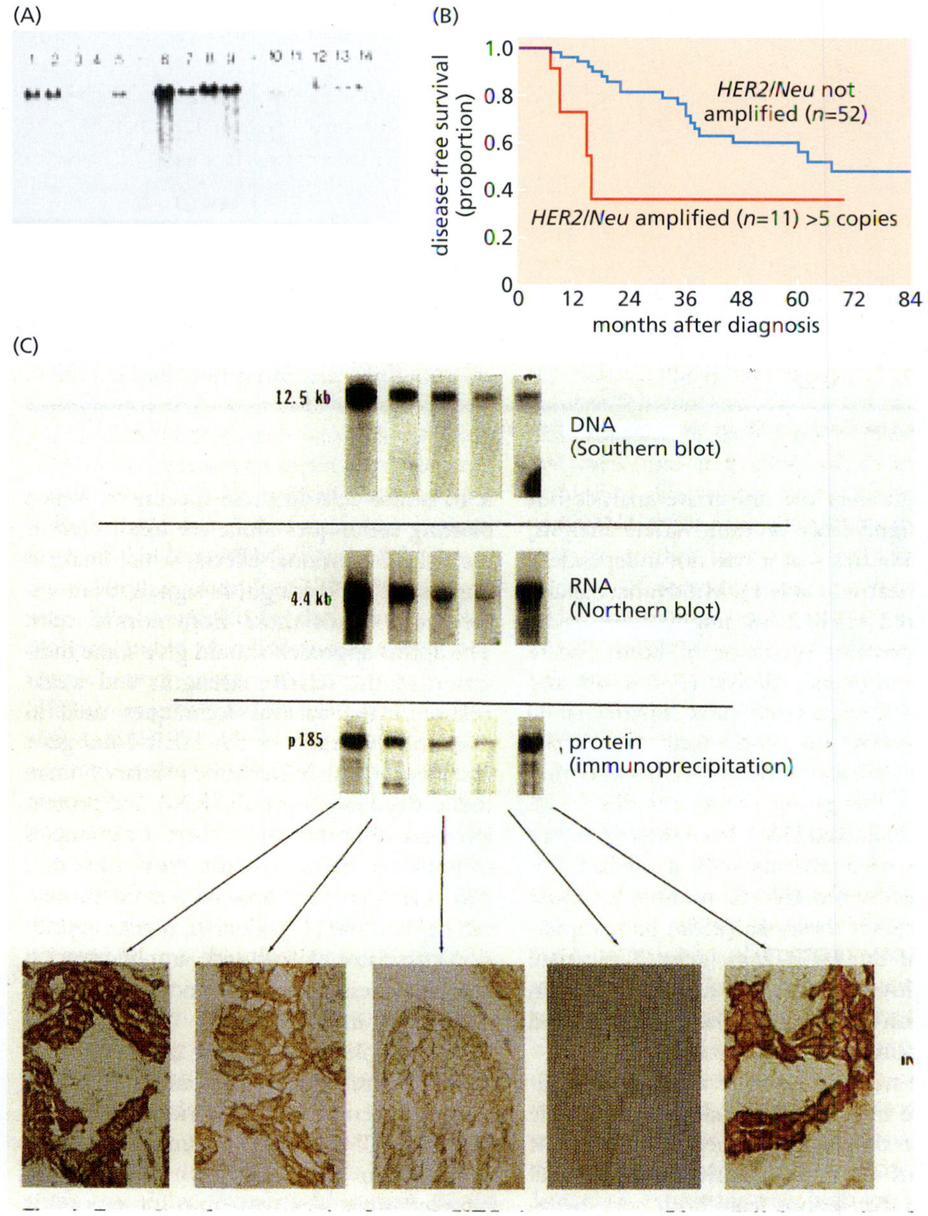

Figure 4.6 Amplification of the *erbB2/neu* oncogene in breast cancers (A) The Southern blotting procedure was used to determine whether the DNA of human breast carcinomas carried extra (i.e., amplified) copies of the *erbB2/neu* oncogene (also termed *HER2*), a close relative of the *erbB* oncogene. As indicated by the dark bands representing restriction enzyme fragments, some human breast carcinomas carried extra copies of this gene. (B) This gene amplification is correlated with a poor prognosis for the breast cancer patient, as indicated by this Kaplan–Meier plot. Those patients whose tumors carried more than five copies of the *erbB2/neu* gene were far more prone to experience a relapse in the first 18 months after diagnosis and treatment than were those patients whose tumors lacked this amplification. (All patients included in this study had breast cancer cells in the lymph nodes draining the involved breast.) (C) Subsequent work indicated that while the *erbB2/neu* oncogene was amplified in some tumors ("DNA"), others overexpressed the mRNA without gene amplification ("RNA"), and yet others expressed increased levels of the protein without indications of gene amplification or elevated transcription ("protein"). Increased ErbB2/Neu protein could also be demonstrated by staining tissue sections with an antibody that reacted with the protein and produced dense staining *(brown)* in some tumors but not in others ("immunohistochemistry"). (A and B, from D.J. Slamon et al., *Science* 235:177–182, 1987; C, courtesy of D.J. Slamon.)

Sidebar 4.3 Gene amplifications may be difficult to interpret The discovery that the *erbB2/neu/HER2* gene is amplified in about 30% of human breast cancers, and that this amplification is correlated with poor prognosis (see Figure 4.6), would seem to explain how many highly malignant breast cancers acquire their aggressive phenotypes. It is known that elevated signaling by this protein drives cells into endless rounds of growth and division and also protects them from programmed cell death—apoptosis. However, analyses of gene expression patterns (Figure 4.7) yield more complex interpretations. In the expression array analysis shown here, the expression levels of a cohort of 160 genes that flank this gene (labeled here ERBB2) on both sides along human chromosome 17q, together with the expression of this gene itself, were monitored in a series of 360 human breast cancers. Elevated expression is indicated in red while normal expression is indicated in green. As is apparent, in about one-fourth of these breast cancers *(right quarter of array)*, expression of *erbB2/neu/HER2* RNA was elevated, as might be expected from the amplification that this gene had undergone in many of these tumors. At the same time, in many of these tumors, expression of closely linked genes mapping to both sides of this gene was also elevated. This reflects the fact that the unit of DNA amplification—the **amplicon**—almost always included a stretch of chromosomal DNA that was far longer than the *erbB2/neu/HER2* gene itself, leading to co-amplification of these neighboring genes. Among these genes are several that may also positively influence cell proliferation and survival, including *GRB7* and *PPARB*, whose protein products interact with ErbB2 (see Chapter 5) and with the apoptosis circuitry (see Chapter 9), respectively. Hence, in such cases, a number of co-amplified genes may be collaborating to orchestrate the malignant phenotype of human breast cancer cells, and it becomes difficult to ascribe specific cancer cell phenotypes to the elevated expression of only a single gene, such as the *erbB2/neu/HER2* discussed here.

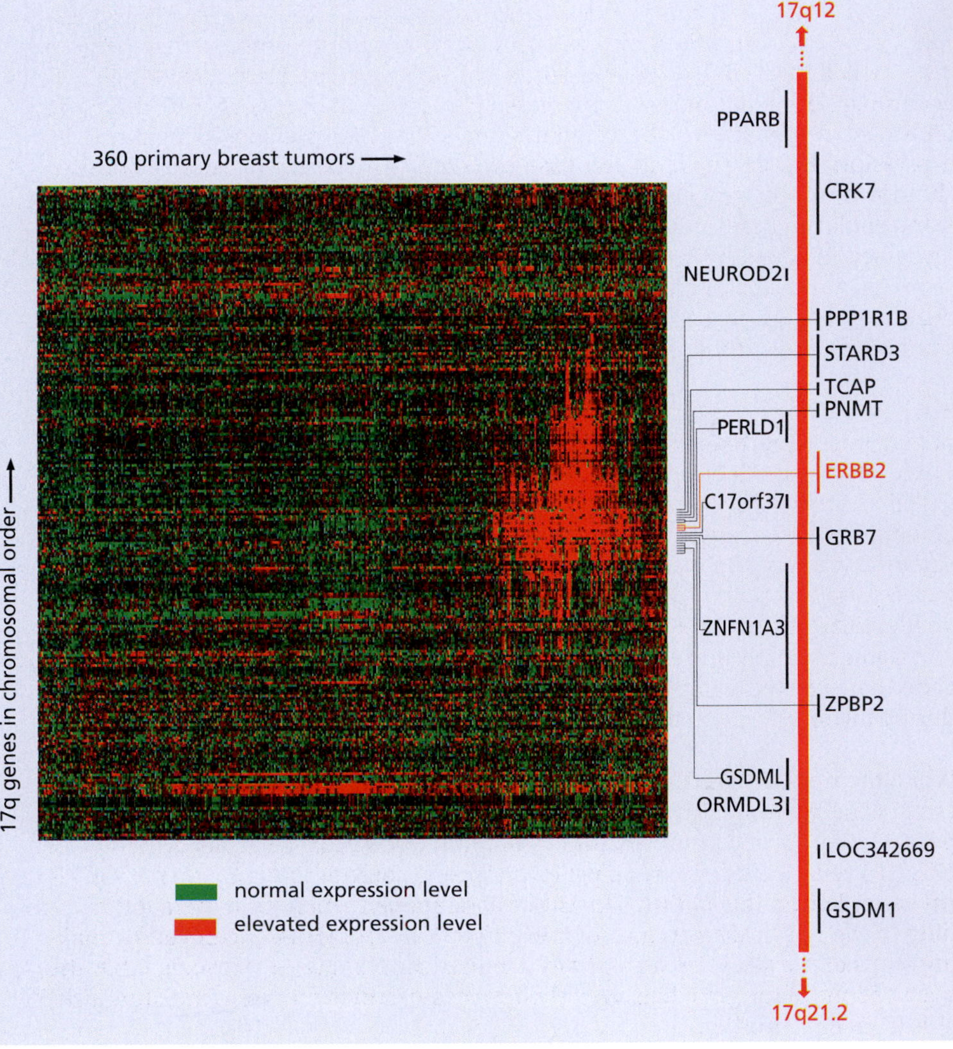

Figure 4.7 Elevation of expression of 17q genes together with overexpression of *HER2/Neu/erbB2* The amplification of a gene, such as *HER2/Neu/erbB2* (i.e., *HER2*), occurs as a consequence of the amplification of an entire chromosomal segment—an amplicon—that usually extends beyond this gene on both sides for several megabases. Because the amplicon encompasses additional genes, these other genes will also be amplified and may affect tumor cell phenotype (in this case that of breast cancer cells). The map of some of the genes identified that flank *HER2/Neu/erbB2* on both sides is provided *(red vertical bar, right)*. In this case, RNA samples from 360 primary breast tumors were analyzed *(columns, left to right)*, while probes for 160 distinct genes in this chromosomal region were arrayed in the order of their location along human Chromosome 17q (i.e., the long arm of Chromosome 17) *(rows, top to bottom)*. Those tumors with similar patterns of gene expression, including elevated *HER2* expression, were clustered together by a computer and are grouped on the *right*. As is apparent, genes flanking *HER2* were also overexpressed in a number of these tumors. (Courtesy of L.D. Miller.)

The lesson taught by these numerous cross connections was simple and clear: many of the oncogenes originally discovered through their association with avian and mammalian retroviruses could be found in a mutated, activated state in human tumor cell genomes. This meant that a common set of cellular proto-oncogenes might be activated either by retroviruses (in animals) or, alternatively, by nonviral mutational mechanisms operating during the formation of human cancers.

4.4 Proto-oncogenes can be activated by genetic changes affecting either protein expression or structure

While a number of proto-oncogenes were found in activated, oncogenic form in human tumor genomes, the precise genetic alterations that led to many of these activations remained unclear. In the case of retrovirus-associated oncogenes, one mechanism became obvious once the organization of the transforming retrovirus genomes was known. In the normal cell, the expression of each proto-oncogene seemed to be regulated by its own transcriptional promoter—the DNA sequence that controls the level of its transcription. The promoter of each proto-oncogene enabled the gene to respond to a variety of physiologic signals. Often the needs of the cell, communicated through these signals, caused a proto-oncogene to be expressed at very low levels. On other occasions, when required by the cell, expression of the gene might be strongly induced.

A quite different situation pertained after a proto-oncogene was acquired by a retrovirus. After insertion into the retrovirus genome, expression of this captured gene was controlled by a retroviral transcriptional promoter (see Figure 3.19), which invariably drove the gene's expression unceasingly and at high levels. Transcription of this virus-associated gene, now an oncogene, was therefore no longer responsive to the cellular signals that had previously regulated its expression. For example, in the case of c-*myc,* expression or **repression** (i.e., shutdown) of this gene is normally tightly controlled by the changing levels of extracellular signals, such as those conveyed by mitogenic growth factors (to be discussed in Chapter 5). Once present in the genome of avian myelocytomatosis virus (AMV), expression of this gene (now called v-*myc*) is found to be at far higher levels than are seen normally in cells, and this expression occurs at a constant (sometimes termed **constitutive**) level.

But how did a normal human H-*ras* proto-oncogene become converted into the potent oncogene that was detected by transfection of human bladder carcinoma DNA (see Section 4.2)? Gene amplification could not be invoked to explain its activation, since this oncogene seemed to be present in bladder carcinoma DNA as a single-copy gene. The puzzle grew when this H-*ras* bladder carcinoma oncogene was isolated by molecular cloning (Sidebar 4.4). It was localized to a genomic DNA fragment of 6.6 kilobases in length. Provocatively, an identically sized DNA fragment was found in normal human DNAs. The latter fragment clearly represented the human H-*ras* proto-oncogene—the normal gene that suffered some type of mutation that converted it into an oncogene during the formation of the bladder carcinoma.

While their overall DNA structures were very similar, these two versions of the H-*ras* gene performed in dramatically different ways. The oncogene that had been cloned from human bladder carcinoma cells caused transformation of NIH 3T3 cells, while its normal proto-oncogene counterpart (i.e., the normal H-*ras* gene) lacked this ability. The mystery deepened when more detailed mapping of the physical structures of these two DNA segments—achieved by making maps of the cleavage sites of various restriction enzymes—revealed that the two versions of the gene had overall physical structures that were indistinguishable from one another.

Sidebar 4.4 Cloning of transfected oncogenes The oncogene of the T24/EJ human bladder carcinoma cell line was cloned by two research groups before its relationship with the H-*ras* oncogene of Harvey sarcoma virus was known. These groups faced the challenge of isolating a gene without knowing anything about its sequence or structure. One group of researchers transfected DNA of the human bladder carcinoma cells into NIH 3T3 mouse cells (Figure 4.8). They used Southern blotting (see Figure 4.4) to detect the donor DNA by exploiting probes that were specific for the ***Alu* repeat** sequences, which are scattered randomly about the human genome in almost 1 million locations but are absent from the mouse genome. (More precisely, distantly related mouse repeat sequences are not recognized by DNA probes that are specific for the human *Alu* repeats.) Thus, *Alu* sequences are present in the human genome, on average, about every 5 kb or so. Accordingly, it was likely that the human bladder carcinoma oncogene carried with it some linked human *Alu* sequences into recipient mouse cells during the transfection procedure.

Indeed, the researchers found a relatively small number of human *Alu* sequences (about 0.1% of a total human genome's worth of *Alu* sequences; see Section 4.2) in the genomes of transfected, transformed mouse cells. They then prepared whole genomic DNA from these transformed NIH 3T3 cells and transfected it once again into fresh NIH 3T3 cells and isolated transformed cells that arose following this second cycle of transfection. Once again, only about 0.1% of the donor DNA was transferred from donor to recipient (see Figure 4.8). In these secondarily transfected cells, the only human *Alu* sequences that had survived the two cycles of transfection were those that were closely linked to the oncogene responsible for the observed transformation (i.e., they were "carried along for the ride" together with the human oncogene whose phenotype was being selected). These researchers then used a human *Alu*-specific sequence probe to identify the presence of an *Alu*-containing DNA fragment in a collection (a genomic library) of DNA fragments prepared from the genomic DNA of the secondarily transfected cells. They then retrieved this fragment using standard gene cloning procedures. The cloned *Alu*-containing DNA segment was found to carry, in addition, the long-sought bladder carcinoma oncogene.

The other research group used an elegant procedure that caused the bladder carcinoma gene to become closely linked to a bacterial gene during the initial transfection event. They then followed the fate of this bacterial segment through another cycle of transfection and, by using a probe specific for it, were able to isolate both this bacterial segment and the linked bladder carcinoma cell from transfected cells by molecular cloning.

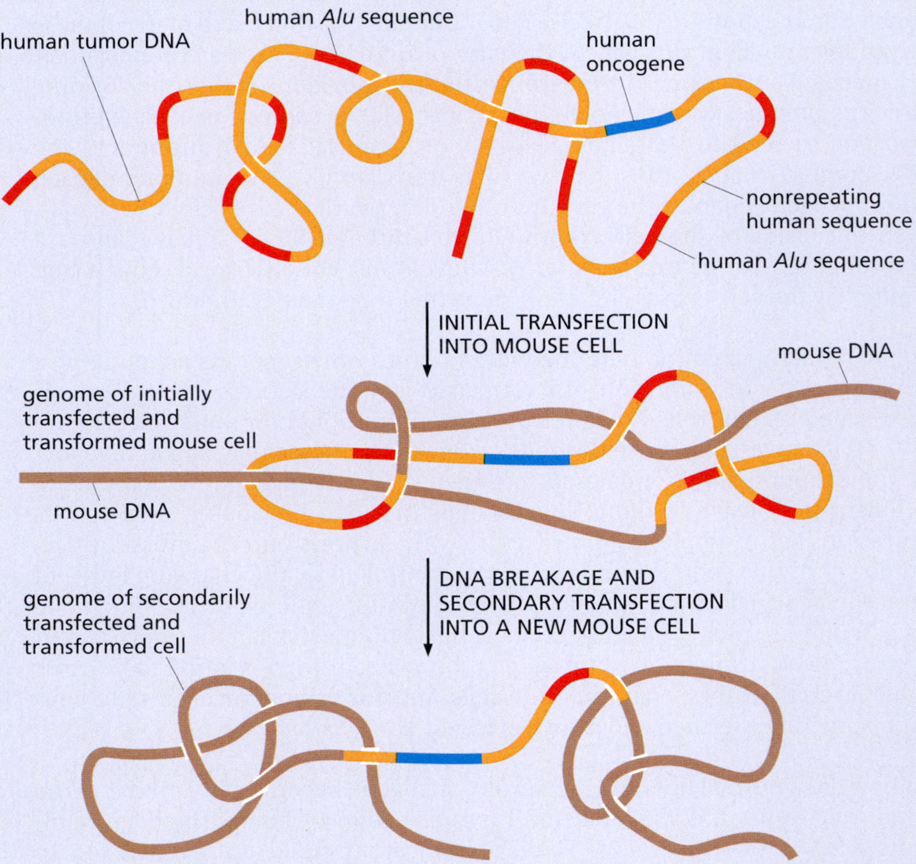

Figure 4.8 Cloning of transfected human oncogenes This strategy for cloning the oncogene present in a human bladder carcinoma depended on the presence of *Alu* sequences *(red segments),* which are present in almost a million copies scattered throughout the human genome *(orange segments)*. As a consequence, virtually all human genes are closely linked to one or more of the *Alu* sequences. If the genomic DNA of a human tumor cell bearing an oncogene (*blue segment*) is transfected into a mouse cell [whose DNA (*light brown line*) lacks sequences closely related to human sequences], the introduced human DNA can be detected by use of an *Alu*-specific probe in the Southern blotting procedure (see Figure 4.4). Because so many human *Alu* sequences were co-introduced together with the human oncogene into an initially transformed mouse cell, the DNA of this transformant was extracted, fragmented, and used in a second cycle of transfection, once again into mouse cells. The only human DNA and associated *Alu* sequences that were present in the secondarily transformed cells were those that were closely linked to the human oncogene (whose presence was selected because of the transformed phenotype that it caused in these cells). A genomic library could then be made from the DNA of these secondary transformants, and the DNA clone containing the oncogene could be identified (using an *Alu*-specific probe) and retrieved, resulting in the cloning of the bladder carcinoma oncogene.

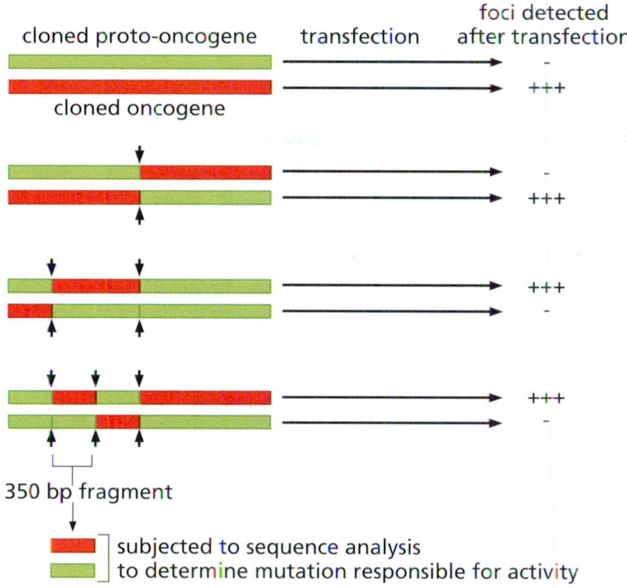

Figure 4.9 Localization of an oncogene-activating mutation The cloned DNAs of a human bladder carcinoma oncogene *(red segments)* and the closely related human H-*ras* proto-oncogene *(green segments)* were cleaved by restriction enzymes at the sites indicated *(vertical arrows)* and recombinant genes were made by ligating (linking) the resulting DNA fragments from the two sources and testing the hybrid DNA molecules for their transforming activity using the transfection-focus assay (see Figure 4.2). This made it possible to progressively localize the mutation responsible for oncogene activation to a small 350-base-pair segment, which could then be subjected to sequence analysis in order to determine the precise sequence change that distinguished the two allelic versions of this gene. (From C.J. Tabin et al., *Nature* 300:143–149, 1982.)

Yet clearly, the two versions of the H-*ras* gene had some significant difference in their sequences, because they functioned so differently. The critical sequence difference was initially localized by recombining segments of the cloned proto-oncogene with other segments deriving from the oncogene (Figure 4.9). This made it possible to narrow down the critical difference to a segment only 350 base pairs long.

The puzzle was finally solved when the corresponding 350-bp segments from the proto-oncogene and oncogene were subjected to DNA sequence analysis. The critical difference was extraordinarily subtle—a single base substitution in which a G (guanosine) residue in the proto-oncogene was replaced by a T (thymidine) in the oncogene. This single base-pair replacement—a **point mutation**—appeared to be all that was required to convert the normal gene into a potent oncogene (Figure 4.10)! This important discovery was made simultaneously in three laboratories, eliminating all doubt about its correctness.

The discovery of this point mutation represented a significant milestone in cancer research. It was the first time that a mutation was discovered in a gene that contributed causally to the neoplastic growth of a human cancer. Equally important, it seemed that this genetic change arose as a somatic mutation.

With this information in hand, researchers could devise a likely explanation for the origin of the bladder carcinoma and, by extension, other similar tumors. The particular bladder carcinoma from which the H-*ras* oncogene had been cloned was said to have arisen in a middle-aged man who had been smoking for four

```
                                            CCCGGG CCGCAGGCCC TTGAGGAGCG
                                       ↓
                                      gly ⁄ proto-oncogene
met thr glu tyr lys leu val val val gly ala GGC gly val gly lys ser ala leu thr
ATG ACG GAA TAT AAG CTG GTG GTG GTG GGC GCC GTC GGT GTG GGC AAG AGT GCG CTG ACC
                                      val ⁀ oncogene
                                                             splice
                                                               ↓
ile gln leu ile gln asn his phe val asp glu tyr asp pro thr ile glu
ATC CAG CTG ATC CAG AAC CAT TTT GTG GAC GAA TAC GAC CCC ACT ATA GAG GTGAGCCTGC

GCCGCCGTCC AGGTGCCAGC AGCTGCTGCG GGCGAGCCCA GGACACAGCC AGGATAGGGC TGGCTGCAGC

CCCTGGTCCC CTGCATGGTG CTGTGGCCCT GTCTCCTGCT TCCTCTAGAG GAGGGGAGTC CCTCGTCTCA

GCACCCCAGG AGAGGAGGGG GCATGAGGGG CATGAGAGGT ACC
```

Figure 4.10 Mutation responsible for H-*ras* oncogene activation As indicated in Figure 4.9, the critical difference between the human bladder carcinoma oncogene and its proto-oncogene could be localized to a subgenic fragment of 350 base pairs. The sequences of the two 350-nucleotide-long DNA fragments from the oncogene and proto-oncogene were then determined. The two differed at a single nucleotide, which affected the 12th codon of the H-*ras* reading frame (*arrow*), converting the normally present glycine-encoding codon to one specifying valine. (From C.J. Tabin et al., *Nature* 300:143–149, 1982.)

decades. During this time, carcinogens present in cigarette smoke were introduced in large amounts into his lungs, and passed from there through the bloodstream to his kidneys, which excreted these chemical species with the urine. While in the bladder, some of the carcinogen molecules present in the urine had entered cells lining the bladder and attacked their DNA. On one occasion, a mutagenic carcinogen introduced a point mutation in the H-*ras* proto-oncogene of an epithelial cell. Thereafter, this mutant cell and its descendants proliferated uncontrollably, being driven by the potent transforming action of the H-*ras* oncogene that they carried. The result, years later, was the large tumor mass that was eventually diagnosed in this patient.

Importantly, this base-pair substitution occurred in the **reading frame** of the H-*ras* gene—the portion of the gene dedicated to encoding amino acid sequence (see Figure 4.10). In particular, this point mutation caused the substitution of a glycine residue present in the normal H-*ras*–encoded protein by a valine residue. The effects of this amino acid substitution on the function of the H-*ras* oncoprotein will be discussed later in Chapters 5 and 6.

The discovery of this point mutation established a mechanism for oncogene activation that was quite different from that responsible for the creation of *myc* oncogenes. In the case of H-*ras,* a change in the structure of the encoded protein appeared to be critical. In the contrasting case of *myc,* deregulation of its expression seemed to be important for imparting oncogenic powers to this gene.

Within a decade, a large number of human tumors were found that carried point mutations in one of the three *ras* genes present in the mammalian genome: H-*ras,* K-*ras,* and N-*ras.* Significantly, in each of these tumors, the point mutation that was uncovered was present in one of three specific codons in the reading frame of a *ras* gene. Consequently, all Ras oncoproteins (whether made by the H-, K-, or N-*ras* gene) were found to carry amino acid substitutions in residues 12, 61, or (less frequently) 13. Taken together, more than 20% of human tumors arising in a variety of tissues carry such point-mutated *ras* genes (Table 4.2).

Both activation mechanisms—regulatory and structural—might collaborate to create an active oncogene. In the case of the *myc* oncogene carried by avian myelocytomatosis virus, for example, expression of this gene was found to be strongly deregulated by the viral transcription promoter. At the same time, some subtle alterations in the reading frame of the *myc* oncogene (and thus changes in the structure of its encoded oncoprotein, Myc) further enhanced the already-

Table 4.2 A list of point-mutated *ras* oncogenes carried by a variety of human tumor cells

Tumor type	Proportion (%) of tumors carrying a point-mutated *ras* gene[a]
Pancreas	90 K
Thyroid (papillary)	60 (H, K, N)
Thyroid (follicular)	55 (H, K, N)
Colorectal	45 (K)
Seminoma	45 (K, N)
Myelodysplasia	40 (N, K)
Lung (non-small-cell)	35 (K)
Acute myelogenous leukemia	30 (N)
Liver	30 (N)
Melanoma	15 (K)
Bladder	10 (K)
Kidney	10 H

[a]H, K, and N refer to the human *H-RAS, K-RAS,* and *N-RAS* genes, respectively.

Adapted from J. Downward, *Nat. Rev. Cancer* 3:11–22, 2003.

potent transforming powers of this oncogene. Similarly, the H-*ras* oncogene carried by Harvey sarcoma virus was found to carry a point mutation in its reading frame (like that discovered in the bladder carcinoma oncogene); at the same time, this gene was greatly overexpressed, being driven by the retroviral transcriptional promoter.

4.5 Variations on a theme: the *myc* oncogene can arise via at least three additional distinct mechanisms

The observation that the v-*myc* oncogene of avian myelocytomatosis virus (AMV) arose largely through deregulation of its expression only hints at the diverse mechanisms that are capable of creating this oncogene. As cited in Section 4.3, in some human tumors, expression of the *myc* gene continues to be driven by its own natural transcriptional promoter, but the copy number of this gene is found to be elevated to levels many times higher than the two copies present in the normal human genome. In 30% of childhood neuroblastomas, a close relative of c-*myc*, termed N-*myc*, has also been found to be amplified, specifically in the more aggressive tumors of this type (Sidebar 4.5). In both instances, the increased gene copy numbers result in corresponding increases in the level of gene products—the Myc and N-Myc proteins. As we will discuss later in Chapter 8, proteins of the Myc family possess potent growth-promoting powers. Consequently, when present at excessive levels, these proteins seem to drive uncontrolled cell proliferation.

Sidebar 4.5 N-*myc* amplification and childhood neuroblastomas Amplification of the N-*myc* gene occurs in about 40% of advanced pediatric neuroblastomas, which are tumors of the peripheral nervous system. This amplification, which is associated with the formation of either double minutes (DMs) or homogeneously staining regions (HSRs), represents a bad prognosis for the patient (Figure 4.11). The HSRs, which contain multiple copies of the genomic region encompassing the N-*myc* gene, are often found to have broken away from the normal chromosomal mapping site of N-*myc* and, in one study, to have become associated with at least 18 other different chromosomal regions. For unknown reasons, amplification of the N-*myc* gene leads to a bimodal distribution of gene copies, with some tumors having 10 to 30 gene copies while others carry 100 to 150 copies of this gene. While N-*myc* amplification was originally thought to be a peculiarity of neuroblastomas (and thus a specific diagnostic marker for this particular disease), it has now been found in a variety of neuroectodermal tumors, including astrocytomas and retinoblastomas; in addition, small-cell lung carcinomas, which have neuroendocrinal traits, also often exhibit amplified N-*myc* genes.

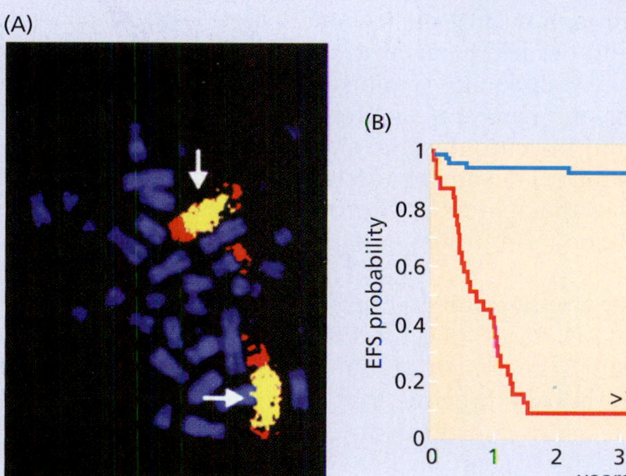

Figure 4.11 N-*myc* amplification and neuroblastoma prognosis (A) The N-*myc* gene is often amplified in human childhood neuroblastomas. Multiple copies of this gene have been detected *(yellow)* through the technique of fluorescence *in situ* hybridization (FISH). The fact that these N-*myc* gene copies are present as tandem arrays within chromosomes *(yellow)* means that they constitute homogeneous staining regions (HSRs) rather than extrachromosomal particles—double minutes (DMs)—which are also frequently seen in these tumors. (B) This Kaplan–Meier plot illustrates the event-free survival (EFS) of children suffering from neuroblastoma, i.e., no clinically significant cancer-related observations or occurrences in the indicated years following initial diagnosis and treatment. Those who have minimal or no N-*myc* amplification have a very good prognosis and minimal clinical events, while those who have extensive N-*myc* amplification have a dramatically poorer prognosis and therefore short survival times after diagnosis. (A, from C. Lengauer et al., *Nature* 396:643–649, 1998; B, from M.L. Schmidt et al., *J. Clin. Oncol.* 18:1260–1268, 2000.)

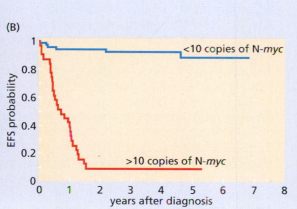

Table 4.3 Some frequently amplified chromosomal regions and the genes they are known to carry

Name of oncogene[a]	Human chromosomal location	Human cancers	Nature of protein
erbB1	7q12–13	glioblastomas (50%); squamous cell carcinomas (10–20%)	RTK
cab1–erbB2–grb7	17q12	gastric, ovarian, breast carcinomas (10–25%)	RTK, adaptor protein
k-sam	7q26	gastric, breast carcinomas (10–20%)	RTK
FGF-R1	8p12	breast carcinomas (10%)	RTK
met	7q31	gastric carcinomas (20%)	RTK
K-*ras*	6p12	lung, ovarian, bladder carcinomas (5–10%)	small G protein
N-*ras*	1p13	head and neck cancers (30%)	TF
c-*myc*	8q24	various leukemias, carcinomas (10–50%)	TF
L-*myc*	1p32	lung carcinomas (10%)	TF
N-*myc–DDX1*	2p24–25	neuroblastomas, lung carcinomas (30%)	TF
akt-1	14q32–33	gastric cancers (20%)	ser/thr kinase
cyclin D1–exp1–hst1–ems1	(11q13)	breast and squamous cell carcinomas (40–50%)	G1 cyclin
cdk4–mdm2–sas–gli	12q13	sarcomas (40%)	CDK, p53 antagonist
cyclin E	19q12	gastric cancers (15%)	cyclin
akt2	(19q13)	pancreatic, ovarian cancers (30%)	ser/thr kinase
AIB1, BTAK	(20q12–13)	breast cancers (15%)	receptor co-activator
cdk6	(19q21–22)	gliomas (5%)	CDK
myb	6q23–24	colon carcinoma, leukemias	TF
ets-1	11q23	lymphoma	TF
gli	12q13	glioblastomas	TF
FGFR2	10q26	breast carcinomas	RTK

[a]The listing of several genes indicates the frequent co-amplification of a number of closely linked genes; only the products of the most frequently amplified genes are described in the right column.
Courtesy of M. Terada, Tokyo, and adapted from G.M. Cooper, Oncogenes, 2nd ed. Boston and London: Jones and Bartlett, 1995.

Note, by the way, the notations that are used here and throughout this text. Nonhuman oncogenes are usually written as uncapitalized three-letter words in italics (e.g., *myc*) while their protein products are written in roman font with an initial capital (e.g., Myc). The *myc* proto-oncogene itself is often termed c-*myc* to distinguish it from its two cousin genes, N-*myc* and L-*myc*. To make matters more confusing, human genes follow a different nomenclature, so that the human *myc* gene is denoted as *MYC* and its protein product is written as MYC. We will generally use the nonhuman acronym conventions in this book.

The gene amplification process, which is responsible for increases in *myc* copy number, occurs through the preferential replication of a limited region of chromosomal DNA, leaving the more distantly located chromosomal regions unaffected (see Figure 4.7). Since the region of chromosomal DNA that undergoes amplification—the amplicon—usually includes a stretch of DNA far longer than the c-*myc* or N-*myc* gene (e.g., typically including 0.5 to 10 megabases of DNA), the amplified chromosomal regions are often large enough to be observed at the metaphase of mitosis through a light microscope. Often gene amplification yields large, repeating end-to-end linear arrays of the chromosomal region, which appear as homogeneously staining regions (HSRs) in the microscope (see Figure 1.12). Alternatively, the chromosomal region carrying a *myc* or N-*myc* gene may break away from the chromosome and can be seen as small, independently replicating, extrachromosomal particles (double minutes; see Figure 1.12). Indeed, we now know that a number of proto-oncogenes can be found in amplified gene copy number in various types of human tumors (Table 4.3).

An even more unusual way of deregulating *myc* expression levels has already been cited (see Section 3.11). Recall that the insertional mutagenesis mechanism causes the expression of the c-*myc* proto-oncogene to be placed under the transcriptional control of an ALV provirus that has integrated nearby in the chromosomal DNA. The resulting constitutive overexpression of c-*myc* RNA and thus Myc protein results, once again, in flooding of the cell with excessive growth-promoting signals.

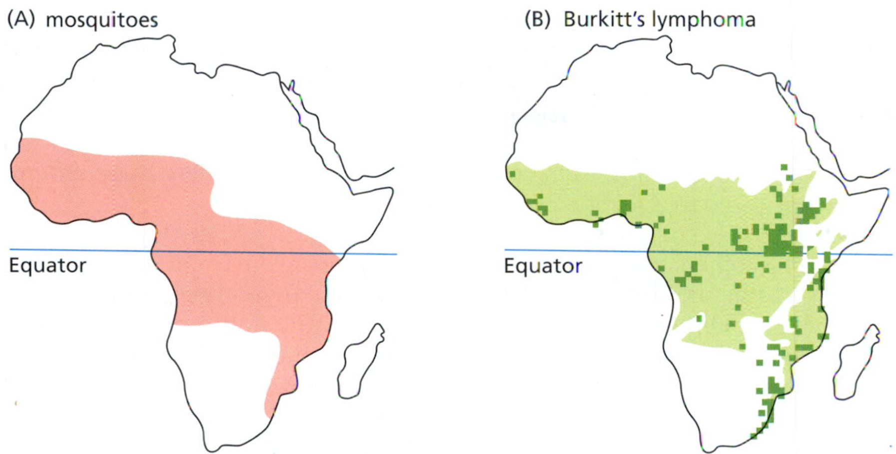

Figure 4.12 Burkitt's lymphoma incidence in Africa (A) The geographic distribution of the *Aedes simpsoni* mosquito throughout Africa, which was known to be a vector in the transmission of malaria. (B) The geographic distribution of childhood Burkitt's lymphoma (BL), as originally documented by Dennis Burkitt. Because these two distribution maps are roughly congruent, this suggested that malarial infection was likely to be an etiologic factor in this disease. In addition, the invariable presence of the Epstein–Barr virus (EBV) genome in the BL tumor cells indicates a second etiologic factor. As the tumor cells evolve, a chromosomal translocation involving the c-*myc* gene develops, which contributes to the neoplastic proliferation of the EBV-infected B lymphocytes. (A and B, from A.J. Haddow, in D.P. Burkitt and D.H. Wright (eds.), Burkitt's Lymphoma. Edinburgh and London: E. & S. Livingstone Co. and Baltimore: Williams and Wilkins.)

Such activation by provirus integration hinted at a more general mode of activation of the c-*myc* proto-oncogene: even while continuing to reside at its normal chromosomal site, c-*myc* can become involved in cancer development if it happens to come under the control of foreign transcriptional promoters. In the disease of Burkitt's lymphoma (BL), this principle was validated in a most dramatic way. This tumor occurs with some frequency among young children in East and Central Africa (Figure 4.12). The etiologic agents of this disease include chronic infections both by Epstein–Barr virus (EBV, a distant relative of human herpesviruses; Section 3.4) and by malarial parasites.

Neither of these etiologic factors shed light on the nature of a critical genetic change inside Burkitt's lymphoma cells that is responsible for the runaway proliferation of these cells. Careful examination of metaphase chromosome spreads of these tumor cells did, however, uncover a striking clue: the tumor cells almost invariably carried chromosomal **translocations** (see also Figure 2.23). Such alterations fuse a region from one chromosome with another region from a second, unrelated chromosome (Figure 4.13). Translocations such as these are

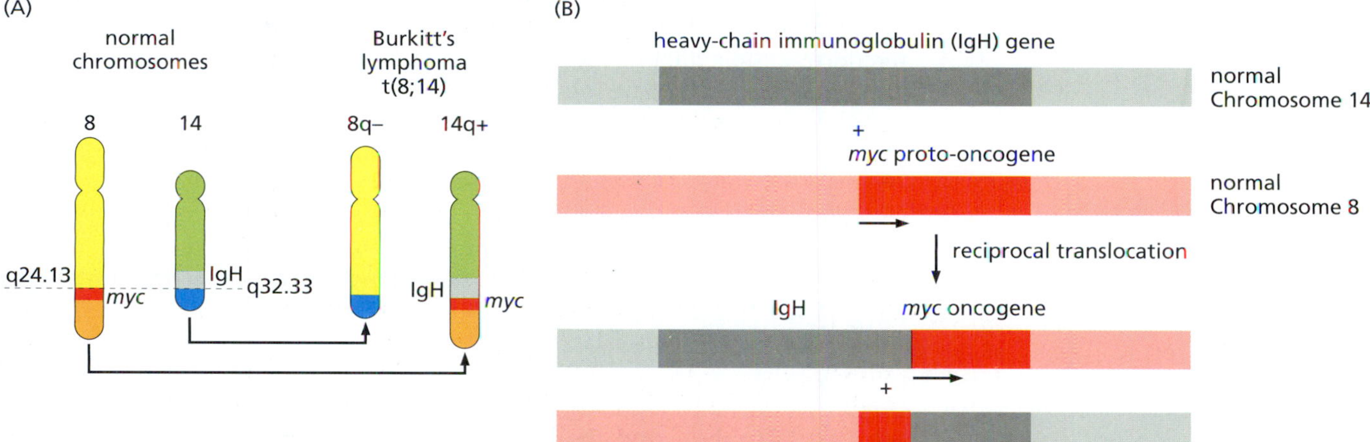

Figure 4.13 Chromosomal translocations in Burkitt's lymphomas (A) In the genomes of Burkitt's lymphoma (BL) cells, the expression of the c-*myc* gene is placed under control of the transcription-controlling enhancer sequences of an immunoglobulin gene as a direct consequence of reciprocal chromosomal translocations. These translocations juxtapose immunoglobulin genes on Chromosomes 2, 14, or 22 with the *myc* gene on Chromosome 8. [These occur in 9% (κ chain), 16% (λ chain), and 75% (heavy chain) of BLs, respectively.] The most common translocation, t (8; 14), is shown here. (B) Depicted is a genetic map of the translocation event that places the c-*myc* gene *(red rectangle)* on Chromosome 8 under the control of the immunoglobulin heavy-chain sequences (IgH; *gray rectangle*) present on human Chromosome 14. Because the immunoglobulin enhancer sequences direct high, constitutive expression, the normal modulation of *myc* expression in response to physiologic signals is abrogated. The resulting *myc* oncogene initially makes structurally normal Myc protein but in abnormally high amounts. (Subsequently occurring point mutations in the *myc* reading frame may further potentiate the function of the Myc oncoprotein.) (From P. Leder et al., *Science* 222:765–771, 1983.)

often found to be *reciprocal*, in the sense that a chromosomal region from chromosome A lands on chromosome B, while the displaced segment of chromosome B ends up being linked to chromosome A. In the case of Burkitt's lymphomas, three distinct, alternative chromosomal translocations were found, involving human Chromosomes 2, 14, or 22. The three translocations were united by the fact that in each case, a region from one of these three chromosomes was fused to a section of Chromosome 8.

In 1983 researchers realized that the *myc* proto-oncogene could be found in the region on human Chromosome 8 that is involved in these three distinct types of translocation. On the other side of the fusion site (often termed the chromosomal **breakpoint**) were found the transcription-promoting sequences from any one of three distinct **immunoglobulin** (antibody) genes. Thus, the immunoglobulin heavy-chain gene cluster is found on Chromosome 14, the κ antibody light-chain gene is located on Chromosome 2, and the λ antibody light-chain gene is found on Chromosome 22. (There is clear evidence that the enzymes responsible for rearranging the sequences of antibody genes during the development of the immune system occasionally lose specificity and, instead of creating a rearranged antibody gene, inadvertently fuse part of an antibody gene with the *myc* proto-oncogene.) Parenthetically, none of this explains the role of EBV in the pathogenesis of Burkitt's lymphoma (Sidebar 4.6).

Suddenly, the grand design underlying these complex chromosomal changes became clear, and it was a simple one: these translocations separate the *myc* gene from its normal transcriptional promoter and place it, instead, under the control of one of three highly active transcriptional regulators, each from an immunoglobulin gene (see Figure 4.13). Once its expression is subjugated by the antibody gene promoters, *myc* becomes a potent oncogene and drives the relentless proliferation of lymphoid cells in which these transcriptional promoters are highly active. Hence, the proliferation of the rare cell that happens to acquire such a deregulated *myc* gene will be strongly favored.

Since these discoveries, almost a dozen distinct chromosomal translocations have been found to cause deregulated expression of known proto-oncogenes; most of these genes remain poorly characterized (Table 4.4). Altogether, more than 300 distinct recurring translocations (i.e., those that have been encountered in multiple, independently arising human tumors) have been cataloged, and more than 100 of the novel hybrid genes created by these translocations have been isolated by molecular cloning.

Sidebar 4.6 How does Epstein–Barr virus (EBV) cause cancer? The 1982 discovery of the Burkitt's lymphoma-associated chromosomal translocations shed little light on the precise mechanism by which the etiologic factors (chronic EBV and malarial parasite infections) favor the formation of these tumors. The contributions of each of these agents to lymphoma pathogenesis are still not totally clear. It appears that chronic malarial infection can compromise the immune defenses of children and thereby render them susceptible to runaway EBV infections. This can lead, in turn, to the accumulation of large pools of EBV-immortalized B lymphocytes that are driven to proliferate continuously by the virus. In some of these cells, the enzymatic machinery dedicated to organizing normal immunoglobulin gene rearrangement misfires and, in so doing, occasionally creates inappropriate juxtapositions of segments of antibody genes with the c-*myc* proto-oncogene. The resulting *myc* oncogenes, working together with several genes (possible oncogenes) expressed by EBV, then drive the cell proliferation that leads eventually to the eruption of lymphomas. An additional puzzle comes from the involvement of EBV in nasopharyngeal carcinomas (NPCs) in Southeast Asia, where this viral infection, together with certain lifestyle factors (possibly the consumption of Chinese-style salted fish early in life), has been implicated as an etiologic agent. Even more puzzling is the fact that in Western populations, EBV infection is common and causes mononucleosis in **immunocompetent** individuals (i.e., those with a fully functional immune system), triggering malignancies only in rare instances.

Table 4.4 Translocations in human tumors that deregulate proto-oncogene expression and thereby create oncogenes

Oncogene	Neoplasm
myc	Burkitt's lymphoma; other B- and T-cell malignancies
bcl-2	follicular B-cell lymphomas
bcl-3	chronic B-cell lymphomas
bcl-6	diffuse B-cell lymphomas
hox1	acute T-cell leukemia
lyl	acute T-cell leukemia
rhom-1	acute T-cell leukemia
rhom-2	acute T-cell leukemia
tal-1	acute T-cell leukemia
tal-2	acute T-cell leukemia
tan-1	acute T-cell leukemia

Adapted from G.M. Cooper, Oncogenes, 2nd ed. Boston and London: Jones and Bartlett, 1995.

In summary, the three alternative ways of activating the c-*myc* proto-oncogene—through provirus integration, gene amplification, or chromosomal translocation—all converge on a common mechanistic theme. Invariably, the gene is deprived of its normal physiologic regulation and is forced instead to be expressed at high, constitutive levels.

In general, the mechanisms leading to the overexpression of genes in cancer cells remain poorly understood. Some overexpression, as indicated here, is achieved through gene amplification and chromosomal translocation. But even more frequently, genes that are present in normal configuration and at normal copy number are transcribed at excessively high levels in cancer cells through the actions of deregulated transcription factors; the latter are largely uncharacterized. To complicate matters, other genes that are found to be amplified in cancer cell genomes are not always overexpressed (Sidebar 4.7); this observation indicates that gene amplification per se does not constitute evidence proving that a gene plays a key role in driving cancer cell proliferation.

Sidebar 4.7 Gene amplification does not always lead to overexpression Recent systematic analyses of chromosomal regions in tumor cell genomes and the expression levels of the genes within these regions indicate that the copy number of a gene does not always predict its level of expression. Instead, these analyses demonstrate that even though a set of genes may undergo measurable increases in copy number—gene amplification—only 40 to 60% of these genes will show corresponding increases in their RNA transcripts (and thus proteins). Such observations indicate that the expression levels of many genes are regulated by complex negative-feedback mechanisms; these function to ensure physiologically appropriate levels of expression, even in the presence of excessive copies of these genes. Such feedback mechanisms, which often operate by controlling the activities of various transcription factors, dictate that the overexpression of certain genes can only be achieved through deregulation of the transcription factors or alterations of the gene promoters to which these factors bind.

4.6 A diverse array of structural changes in proteins can also lead to oncogene activation

The point mutation discovered in *ras* genes was the first of many mutations that were found to affect the structures of proto-oncogene–encoded proteins, converting them into active oncoproteins. As an important example, the formation of certain human tumors, such as gastric and mammary carcinomas and glioblastoma brain tumors (also called gliomas), involves the protein that serves as the cell surface **receptor** for epidermal growth factor (EGF). As we will discuss in detail in the next chapter, this receptor protein extends from the extracellular space through the plasma membrane of cells into their cytoplasm. Normally, the EGF receptor, like almost 60 similarly structured receptors, recognizes the presence of its cognate **ligand** (i.e., EGF) in the extracellular space and, in response, informs the cell interior of this encounter. In about one-third of glioblastomas examined, however, the EGF receptor has been found to be decapitated, lacking most of its extracellular domain (Figure 4.14). We now know that such truncated receptors send growth-stimulatory signals into cells, even in the absence of any EGF. In so doing, they act as oncoproteins to drive cell proliferation. In Chapter 5, we will describe more precisely how structural alterations of growth factor receptors convert them into potent oncoproteins.

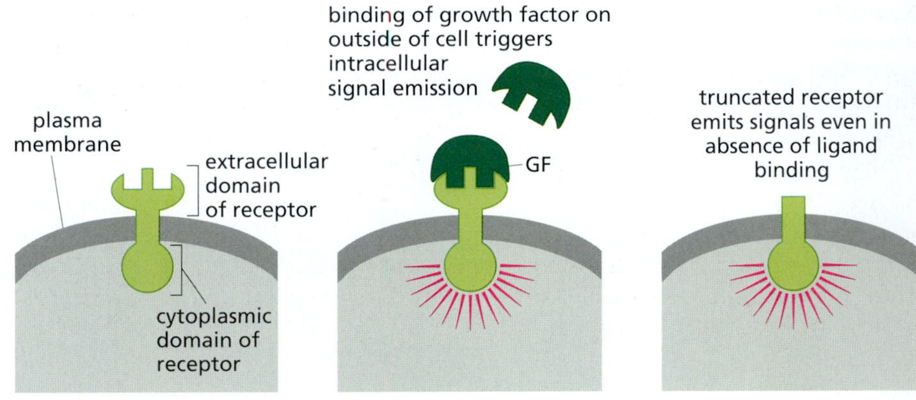

Figure 4.14 Deregulated firing of growth factor receptors Normally, growth factor receptors displayed on the plasma membrane of a cell release signals into the cell interior only when the extracellular domain of the receptor has bound the appropriate growth factor (GF; i.e., the "ligand" of the receptor). However, if the extracellular domain of certain receptors is deleted because of a mutation in the receptor-encoding gene or alternative splicing of the receptor pre-mRNA, the resulting truncated receptor protein then emits signals into the cell without binding its growth factor ligand.

A type of chromosomal translocation quite different from that found in Burkitt's lymphoma cells is seen in the great majority (>95%) of cases of chronic **myelogenous** leukemia (CML), a bone marrow–derived malignancy. Unlike the case of the *myc* gene in Burkitt's lymphomas, these translocations cause the fusion of two distinct reading frames—a discovery made in the same year (1982) that the Burkitt's lymphoma genes were first characterized at the molecular level. The consequence of this translocation is a now-enlarged reading frame that encodes the structure of a hybrid protein (see Figure 2.23).

On one side of the translocation breakpoint are found the sequences that encode the protein made by the *abl* proto-oncogene; the *abl* gene was originally discovered by virtue of its presence as an acquired oncogene in **Ab**elson murine **l**eukemia virus, a rapidly tumorigenic retrovirus (refer to Table 3.3). The *abl* gene, which maps to Chromosome 9q34 (i.e., the fourth band of the third region of the long arm of human Chromosome 9), was found, upon examination of various tumors, to be fused with sequences that are clustered in a narrow region in 22q11 (Figure 4.15). [The standard notation used here is t(9; 22)(q34; q11), where t signifies a translocation, q the long arm, and p the short arm of a chromosome.] This area on Chromosome 22 was called the **b**reakpoint **c**luster **r**egion; hence, *bcr*. Subsequently, all of these breakpoints were found to lie within the *bcr* gene. The resulting fusion of Abl with Bcr amino acid sequences deregulates the normally well-controlled Abl protein, causing it to emit growth-promoting signals in a strong, deregulated fashion.

Since the discovery of the *bcr–abl* translocation, dozens of other, quite distinct translocations have been documented that also result in the formation of hybrid proteins (e.g., Table 4.5). Almost all of these have been found in hematopoietic malignancies, notably leukemias and lymphomas. Because each of these types of translocation is found only rarely, the function of many of the resulting fused proteins remains obscure.

4.7 Synopsis and prospects

By the late 1970s, disparate lines of evidence concerning cancer genes coalesced into a relatively simple idea. The genomes of mammals and birds contain a cohort of proto-oncogenes, which function to regulate normal cell proliferation and differentiation. Alterations of these genes that affect either the control of their expression or the structure of their encoded proteins can lead to overly active growth-promoting genes, which appear in cancer cells as activated oncogenes. Once formed, such oncogenes proceed to drive cell multiplication and, in so doing, play a central role in the pathogenesis of cancer.

Many of these cellular genes were originally identified because of their presence in the genomes of rapidly transforming retroviruses, such as Rous sarcoma virus, avian erythroblastosis virus, and Harvey sarcoma virus. Subsequently, transfection experiments revealed the presence of potent transforming genes in the genomes of cells that had been transformed by exposure to chemical carcinogens and cells derived from spontaneously arising human tumors. These tumor cells had no associations with retrovirus infections. Nonetheless, the oncogenes that they carried were found to be related to those carried by transforming retroviruses. This meant that a common repertoire of proto-oncogenes could be activated by two alternative routes: retrovirus acquisition or somatic mutation.

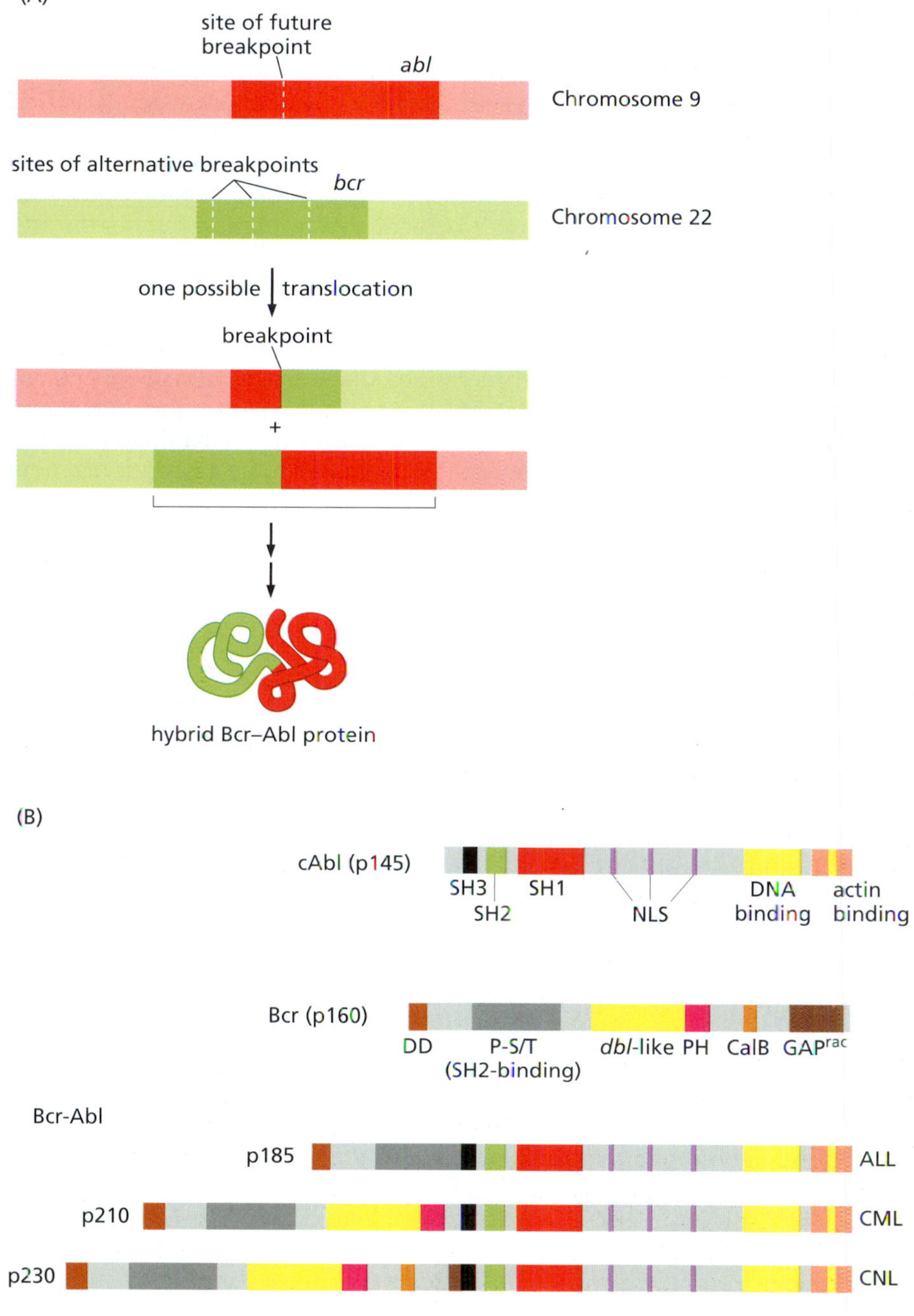

Figure 4.15 Formation of the *bcr-abl* oncogene (A) Reciprocal chromosomal translocations between human Chromosomes 9 and 22, which carry the *abl* and *bcr* genes, respectively, result in the formation of fused, hybrid genes that encode hybrid Bcr–Abl proteins. (B) As is indicated here, different breakpoints in *bcr* are observed in the chromosomal translocations encountered in different types of human leukemia—ALL, acute lymphocytic leukemia; CML, chronic myelogenous leukemia; CNL, chronic neutrophilic leukemia. (The normal structures that encode these proteins are shown at top, for comparison.) The cAbl and the Bcr protein are each multidomain, multifunction proteins, as indicated by the labels attached to the colored areas; and the fusion proteins therefore carry a large number of distinct signaling activities. (B, adapted from A.M. Pendergast, in A.M. Carella, Chronic Myeloid Leukaemia: Biology and Treatment. London: Martin Dunitz, 2001.)

Table 4.5 Translocations in human tumors that cause the formation of oncogenic fusion proteins of novel structure and function

Oncogene	Neoplasm
bcr/abl	chronic myelogenous leukemia; acute lymphocytic leukemia
dek/can	acute myeloid leukemia
E2A/pbx1	acute pre-B-cell leukemia
PML/RAR	acute promyelocytic leukemia
?/erg	myeloid leukemia
irel/urg	B-cell lymphoma
CBFβ/MYH11	acute myeloid leukemia
aml1/mtg8	acute myeloid leukemia
ews/fli	Ewing sarcoma
lyt-10/Cα1	B-cell lymphoma
hrx/enl	acute leukemias
hrx/af4	acute leukemias
NPM/ALK	large-cell lymphomas

Adapted from G.M. Cooper, Oncogenes, 2nd ed. Boston and London: Jones and Bartlett, 1995.

The somatic mutations that caused proto-oncogene activation could be divided into two categories—those that caused changes in the structure of encoded proteins and those that led to elevated, deregulated expression of these proteins. Mutations affecting structure included the point mutations affecting *ras* proto-oncogenes and the chromosomal translocations that yielded hybrid genes such as *bcr–abl.* Elevated expression could be achieved in human tumors through gene amplification or chromosomal translocations, such as those that place the *myc* gene under the control of immunoglobulin enhancer sequences.

These revelations about the role of mutant cellular genes in cancer pathogenesis eclipsed for some years the observations that certain human cancers are associated with and likely caused by infectious agents, notably viruses and bacteria. With the passage of time, however, the importance of infections in human cancer pathogenesis became clear. As is indicated by Table 4.6, viral infections are now known to be involved in causing a variety of human cancers, some of them quite common. About one-fifth of deaths from cancer worldwide are associated in one way or another with infectious agents. Thus, the 9% of worldwide

Table 4.6 Viruses implicated in human cancer causation

Virus[a]	Virus family	Cells infected	Human malignancy	Transmission route
EBV	Herpesviridae	B cells	Burkitt's lymphoma	saliva
		oropharyngeal epithelial cells	nasopharyngeal carcinoma	saliva
		lymphoid	lymphoma[b]	Hodgkin's disease
HTLV-I	Retroviridae	T cells	non-Hodgkin's lymphoma	parenteral, venereal[c]
HHV-8[d]	Herpesviridae	endothelial cells	Kaposi's sarcoma, body cavity lymphoma	venereal
HBV	Hepadnaviridae	hepatocytes	hepatocellular carcinoma	parenteral, venereal
HCV	Flaviviridae	hepatocytes	hepatocellular carcinoma	parenteral
HPV	Papillomaviridae	cervical epithelial	cervical carcinoma	venereal
JCV[e]	Polyomaviridae	central nervous system	astrocytoma, glioblastoma	?

[a]Most of the viruses carry one or more potent growth-promoting genes/oncogenes in their genomes. However, such genes have not been identified in the genomes of HBV and HCV.
[b]These tumors, which bear copies of EBV genomes, appear in immunosuppressed patients.
[c]Parenteral, blood-borne; venereal, via sexual intercourse.
[d]Also known as KSHV, Kaposi's sarcoma herpesvirus.
[e]JCV (JC virus, a close relative of SV40) infects more than 75% of the human population by age 15, but the listed virus-containing tumors are not common. Much correlative evidence supports the role of JCV in the transformation of human central nervous system cells but evidence of a causal role in tumor formation is lacking.

Adapted in part from J. Butel, *Carcinogenesis* 21:405–426, 2000.

cancer mortality caused by stomach cancer is associated with long-term infections by *Helicobacter pylori.* Six percent of cancer mortality is caused by carcinomas of the liver (hepatomas), almost all of which are associated with chronic hepatitis B and C virus infections. And the 5% of cancer mortality from cervical carcinomas is largely attributable to human papillomavirus (HPV) infections.

Somehow, we must integrate the carcinogenic mechanisms activated by these various infectious agents into the larger scheme of how human cancers arise—the scheme that rests largely on the discoveries, described in this chapter, of cellular oncogenes. In some instances, the infecting viruses introduce viral oncogenes into cells that contribute to the transformation phenotype. Viruses like Epstein–Barr virus (EBV) and human herpesvirus-8 (HHV-8) come to mind here. In other cases, viruses like hepatitis B and hepatitis C (HBV, HCV) contribute in a less direct way to tumorigenesis. Because the latter two viruses lack oncogenes and do not participate in insertional mutagenesis, a third mechanism must be invoked: as we will learn in Chapter 11, these viruses act through their abilities to induce chronic tissue damage and associated inflammation.

We focus in the next two chapters on the cellular genes that trigger cancer as a consequence of somatic mutations. At first, the discoveries of cellular oncogenes seemed, on their own, to provide the definitive answers about the origins and growth of human tumors. But soon it became clear that cellular oncogenes and their actions could explain only a part of human cancer development. Yet other types of genetic elements were clearly involved in programming the runaway proliferation of cancer cells. Moreover, a definitive understanding of cancer development depended on understanding the complex machinery inside cells that enables them to respond to the growth-promoting signals released by oncoproteins. So, the discovery of cellular oncogenes was a beginning, a very good one, but still only a beginning. Two decades of further work were required before a more complete understanding of cancer pathogenesis could be assembled.

Key concepts

- The inability to find tumor viruses in the majority of human cancers in the mid-1970s left researchers with one main theory of how most human cancers arise: that carcinogens act as mutagens and function by mutating normal growth-controlling genes into oncogenes.
- To verify this model's prediction that transformed cells carry mutated genes functioning as oncogenes, a novel experimental strategy was devised: DNA from chemically transformed cells was introduced into normal cells—the procedure of transfection—and the recipient cells were then monitored to determine whether they too had become transformed.
- Cultures of NIH 3T3 cells that had been transfected with DNA from chemically transformed mouse cells yielded numerous transformants, which proved to be both anchorage-independent and tumorigenic; this indicated that the chemically transformed cells carried genes that could function as oncogenes and that oncogenes could arise in the genomes of cells independently of viral infections.
- Further experiments using human tumor donor cells transfected into murine cells showed that the oncogenes could act across species and tissue boundaries to induce cell transformation.
- The oncogenes detected by transfection in human tumor cells and the oncogenes of transforming retroviruses were both found to derive from preexisting normal cellular genes. DNA hybridization experiments proved that the same group of normal proto-oncogenes served as precursors of these two types of oncogenes. This meant that many of the oncogenes originally discovered in retroviruses could be found in a mutated state in human tumor cell genomes.

- Retrovirus-associated oncogenes were often found to be present in increased copy number in human tumor cell genomes, which suggested that gene amplification resulted in increased levels of protein products that favored the proliferation of cancer cells. An example is the amplification of the *erbB2/neu/HER2* gene and increased expression of its encoded protein. This particular amplification, when detected in breast cancers, was correlated with decreased patient survival, suggesting that it played a causal role in driving the growth of these tumors.
- While one important activation mechanism for retrovirus-associated genes was apparent (deregulation by placing their expression under the control of viral transcriptional promoters), the mechanism(s) by which normal human proto-oncogenes became converted into oncogenes in the absence of viruses was not evident.
- Sequencing of a transfected H-*ras* oncogene revealed that it differed from its corresponding proto-oncogene in a single base that was located in the gene's reading frame. This resulted in an amino acid substitution in the product, yielding a protein with aberrant behavior.
- The discovery of the point mutation in H-*ras* established a new mechanism for oncogene activation based on a change in the structure of an oncogene-encoded protein.
- Both activation mechanisms—regulatory and structural—might collaborate to create an active oncogene.
- The *myc* oncogene was found to arise through several different mechanisms: provirus integration, gene amplification, and chromosomal translocation.
- Gene amplification occurs through preferential replication of a segment (the amplicon) of chromosomal DNA. The result may be repeating end-to-end linear arrays of the segment, which appear as homogeneously staining regions (HSRs) of a chromosome when viewed under the light microscope. Alternatively, the region carrying the amplified segment may break away from the chromosome and can be seen as small, independently replicating, extrachromosomal particles (double minutes). Gene amplification does not always result in overexpression of the gene.
- Translocation involves the fusion of a region from one chromosome to a nonhomologous chromosome. Translocation can place a gene under the control of a foreign transcriptional promoter and lead to its overexpression, as is the case with the *myc* oncogenes in Burkitt's lymphomas. In these tumors, the transcription-promoting sequences from any one of three different immunoglobulin genes are relocated next to the section of Chromosome 8 that carries the *myc* proto-oncogene. By separating the *myc* gene from its normal transcriptional promoter and placing it under the control of a highly active transcriptional regulator, this translocation results in the overexpression of *myc*.
- A variety of structural changes in proteins can also lead to oncogene activation. Examples include alterations in the structure of growth factor receptors (such as the EGF receptor) and translocations that fuse two distinct reading frames to yield a hybrid protein (such as Bcr-Abl). Both types of alterations deregulate the proteins, causing them to emit growth-promoting signals in a strong, unremitting fashion.

Thought questions

1. What evidence do we have that suggests that endogenous retrovirus genomes are not responsible for a portion of human cancers?
2. Why might an assay like the transfection-focus assay fail to detect certain types of human tumor-associated oncogenes?
3. What mechanisms might cause a certain region of chromosomal DNA to accidentally undergo amplification?
4. How many distinct molecular mechanisms might be responsible for converting a proto-oncogene into a potent oncogene?
5. How many distinct molecular mechanisms might allow chromosomal translocations to produce activate proto-oncogenes into oncogenes?
6. What experimental search strategies would you propose if you wished to launch a systematic screening of a vertebrate genome in order to enumerate all of the proto-oncogenes that it harbors?
7. Since proto-oncogenes represent distinct liabilities for an organism, in that they can incite cancer, why have these genes not been eliminated from the genomes of chordates?

Additional reading

Alitalo K and Schwab M (1986) Oncogene amplification in tumor cells. *Adv. Cancer Res.* 47, 235–281.

Barbacid M (1987) *ras* genes. *Annu. Rev. Biochem.* 56, 779–827.

Bishop JM (1991) Molecular themes in oncogenesis. *Cell* 64, 235–248.

Bos JL (1989) *Ras* oncogenes in human cancer: a review. *Cancer Res.* 49, 4682–4689.

Brodeur G (2003) Neuroblastoma: biological insights into a clinical enigma. *Nat. Rev. Cancer* 3, 203–216.

Butel J. (2000) Viral carcinogenesis: revelation of molecular mechanisms and etiology of human disease. *Carcinogenesis* 21, 405–426.

Cleary ML (1991) Oncogenic conversion of transcription factors by chromosomal translocations. *Cell* 66, 619–622.

Cooper GM (1995) Oncogenes, 2nd ed. Boston and London: Jones and Bartlett.

Lowy DR & Willumsen BM (1993) Function and regulation of ras. *Annu. Rev. Biochem.* 62, 851–891.

Martin GS (2001) The hunting of the Src. *Nat. Rev. Mol. Cell Biol.* 2, 467–475.

Mitelman F, Johansson B & Mertens F (eds) (2004) Mitelman Database of Chromosome Aberrations in Cancer. http://cgap.nci.nih.gov/Chromosomes/Mitelman

Rabbitts TH (1991) Translocations, master genes, and differences between the origins of acute and chronic leukemias. *Cell* 67, 641–644.

Reddy EP, Skalka AM & Curran T (eds) (1988) The Oncogene Handbook. Amsterdam: Elsevier.

Rosenberg N & Jolicoeur P (1997) Retroviral pathogenesis. In Retroviruses (JM Coffin, SH Hughes, HE Varmus eds), pp 475–585. Cold Spring Harbor, NY: Cold Spring Harbor Laboratory Press.

Rowley JD (2001) Chromosome translocations: dangerous liaisons revisited. *Nat. Rev. Cancer* 1, 245–250.

Solomon E, Borrow J & Goddard AD (1991) Chromosome aberrations and cancer. *Science* 254, 1153–1160.

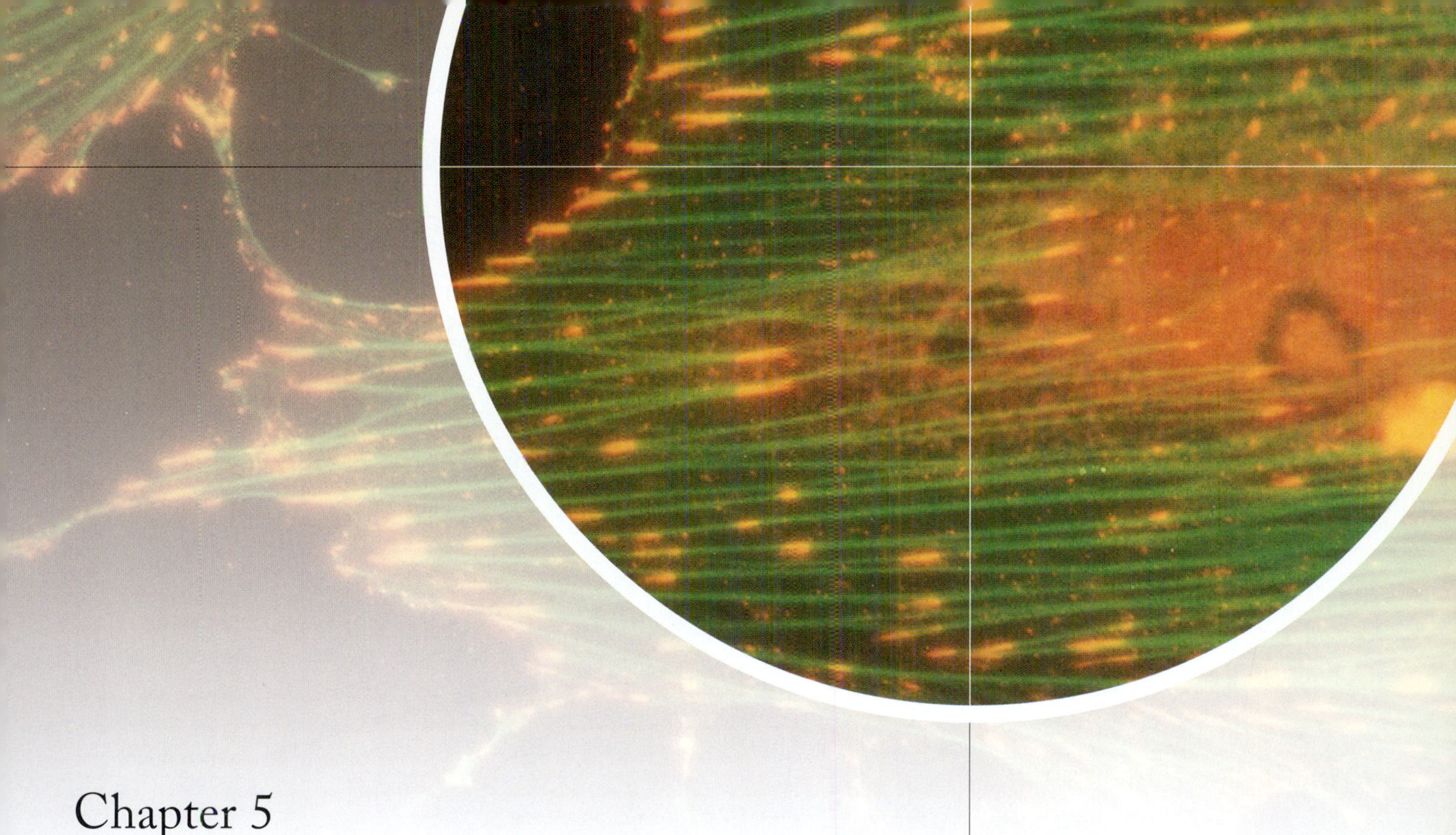

Chapter 5

Growth Factors, Receptors, and Cancer

> The ability of certain fetal serums to stimulate cell growth and the decreased requirement for such factors by transformed cells may be due to the fact that these serum factors are the same or similar to the transforming factors synthesized by some embryonic or neoplastic cells.
>
> David E. Comings, geneticist, 1973

The discovery of oncogenes and their precursors, the proto-oncogenes, stimulated a variety of questions. The most pointed of these centered on the issue of how the oncogenes, acting through their encoded protein products, succeed in perturbing cell behavior so profoundly. A variety of cell phenotypes were concomitantly altered by the actions of oncoproteins such as Src and Ras, the products of the *src* and *ras* oncogenes. How could a single protein species succeed in changing so many different cellular regulatory pathways at the same time?

The vital clues about oncoprotein functioning came from detailed studies of how normal cells regulate their growth and division. Normal cells receive growth-stimulatory signals from their surroundings. These signals are processed and integrated by complex circuits within the cell, which decide whether cell growth and division is appropriate or not (Figure 5.1).

This need to receive extracellular signals at the cell surface and to transfer them into the cytoplasm creates a challenging biochemical problem. The extra- and intracellular spaces are separated by the lipid bilayer that is the plasma membrane. This membrane is a barrier that effectively blocks the

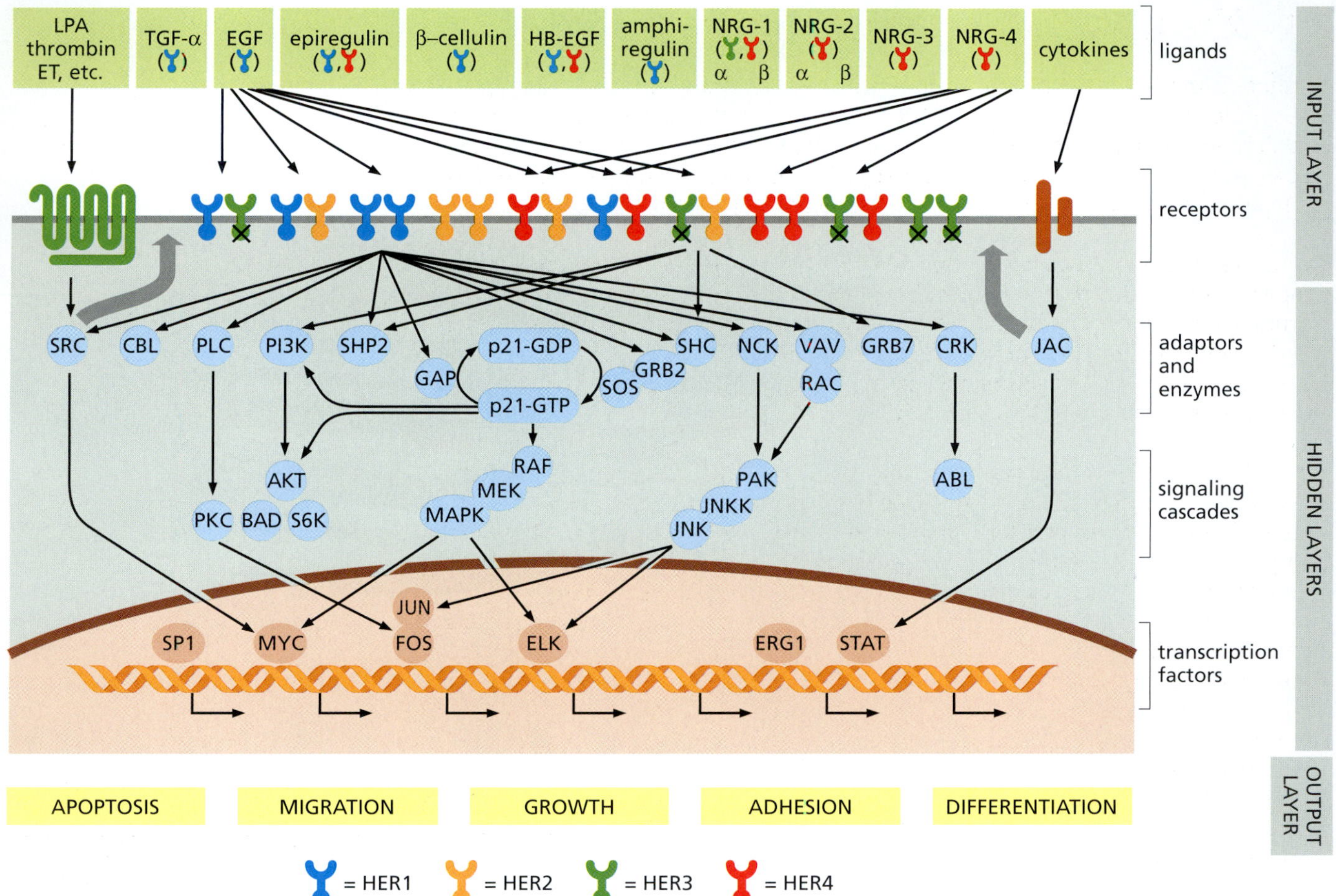

Figure 5.1 The ErbB signaling network: how cells communicate with their surroundings This cartoon is one representation of how cells communicate with their surroundings—the topic of much of this chapter. As is indicated here, a variety of protein messengers (ligands, *green squares, top)* interact with a complex array of cell surface receptors, which transduce signals across the plasma membrane *(gray)* into the cytoplasm, where a complex network of signal-transducing proteins processes these signals, funnels signals into the nucleus *(bottom)*, and ultimately evokes a variety of biological responses ("output layer," *yellow rectangles, bottom)*. Many of the components of this circuitry, both at the cell surface and in the cell interior, are involved in cancer pathogenesis. This cartoon focuses on a small subset of the receptors that are displayed on the surfaces of mammalian cells—the main topic of this chapter. The adaptors and signaling cascades will be covered in the next chapter. (From Y. Yarden and M.X. Sliwkowski, *Nat. Rev. Mol. Cell Biol.* 2:127–137, 2001.)

movement of virtually all but the smallest molecules through it, resulting in dramatically differing concentrations of many types of molecules (including ions) on its two sides. How have cells managed to solve the problem of passing (**transducing**) signals through a membrane that is almost impervious? And given this barrier, how can the inside of the cell possibly know what is going on in the surrounding extracellular space?

These signaling processes are part of the larger problem of cell-to-cell communication. This problem needed to be addressed and solved at the time when the first multicellular animals (metazoa) arose 600 to 700 million years ago. Without an effective means of intercellular communication, the behavior of individual cells could not be coordinated, and the formation of architecturally complex tissues and organisms was inconceivable. Obviously, such communication depended on the ability of some cells to emit signals and of others to receive them and respond in specific ways.

In very large part, the signals passed between cells are carried by proteins. Hence, signal emission requires an ability by some cells to release proteins into the extracellular space. Such release—the process of protein secretion—is also complicated by the imperviousness of the plasma membrane. After these signaling proteins are released into the extracellular space, the designated recipient cells must be able to sense the presence of these proteins in their surroundings. Much of this chapter is focused on this second problem—how normal cells receive signals from the environment that surrounds them. As we will see, the deregulation of this signaling is central to the formation of cancer cells.

5.1 Normal metazoan cells control each other's lives

As already implied (Section 4.6), the normal versions of oncogene-encoded proteins often serve as components of the machinery that enables cells to receive and process biochemical signals regulating cell proliferation. Therefore, to truly appreciate the complexities of oncogene and oncoprotein function, we need to understand the details of how normal cell proliferation is governed.

Our entrée into this discussion comes from a basic and far-reaching principle. Proper tissue architecture depends absolutely on maintaining appropriate proportions of different constituent cell types within a tissue, on the replacement of missing cells, and on discarding extra, unneeded cells (Figure 5.2). Wounds must be repaired, and attacks by foreign infectious agents must be warded off through the concerted actions of many cells within the tissue.

All of these functions depend upon cooperation among large groups of cells. This explains why cells in a living tissue are constantly talking to one another. Much of this incessant chatter is conveyed by **growth factors (GFs)**. These are relatively small proteins that are released by some cells, make their way through intercellular space, and eventually impinge on yet other cells, carrying with them specific biological messages. Growth factors convey many of the signals that tie the cells within a tissue together into a single community, all members of which are in continuous communication with their neighbors.

Decisions about growth versus no-growth must be made for the welfare of the entire tissue and whole organism, not for the benefit of its individual component cells. For this reason, no single cell within the condominium of cells that is a living tissue can be granted the autonomy to decide on its own whether it should proliferate or remain in a nongrowing, quiescent state. This weighty decision can be undertaken only after consultation with other cells within the tissue. These neighbors may provide a particular cell with growth factors that stimulate its proliferation or release growth-inhibitory factors that discourage it.

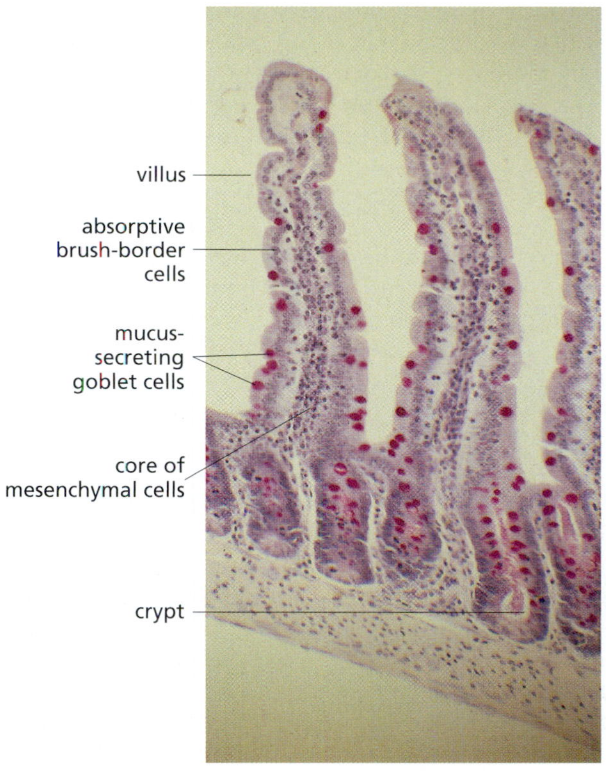

Figure 5.2 Maintenance of tissue architecture The epithelial lining of the small intestine shown here illustrates the fact that a number of distinct cell types coexist within a tissue in order to create normal tissue structure and function. The relative numbers and positions of each cell type must be tightly controlled. This control is largely achieved via the exchange of signals between neighboring cells within the tissue. In this particular epithelium, cell-to-cell signaling also ensures that new epithelial cells—in the intestine termed *enterocytes*—are continually being generated in the crypts (at the bases of fingerlike villi) in order to replace others that have migrated up the sides of the villi to the tips, where they are sloughed off. Included among the epithelial cells are mucus-secreting goblet cells *(dark red)* and absorptive cells, as well as enteroendocrine cells and Paneth cells *(not shown)*. In addition, in the middle of each villus is a core of mesenchymal cell types, which together constitute the stroma and include fibroblasts, endothelial cells, pericytes, and macrophages. The proper proportions of these various distinct cell types must be maintained in order to ensure proper tissue structure and function. (From B. Alberts et al., Molecular Biology of the Cell, 4th ed. New York: Garland Science, 2002.)

In the end, all the decisions made by an individual cell about its proliferation must, by necessity, represent a consensus decision shared with the cells that reside in its neighborhood.

The dependence of individual cells on their surroundings is illustrated nicely by the behavior of normal cells when they are removed from living tissue and propagated in a Petri dish. Even though the liquid medium placed above the cells contains all the nutrients required to sustain their growth and division, including amino acids, vitamins, glucose, and salts, such medium, on its own, does not suffice to induce these cells to proliferate. Instead, this decision depends upon the addition to the medium of serum, usually prepared from the blood of calves or fetal calves. Serum contains the growth factors that persuade cells to multiply.

Serum is produced when blood is allowed to clot. The blood platelets adhere to one another and form a matrix that gradually contracts and traps most of the cellular components of the blood, including both the white and red cells. In the context of a wounded tissue, this clot formation is designed to stanch further bleeding. The clear fluid that remains after clot formation and retraction constitutes the serum.

While the platelets in a wound site are in the process of aggregating as part of clot formation, they also initiate the wound-healing process, doing so through the release of growth factors, notably platelet-derived growth factor (PDGF), into the medium around them (Figure 5.3). PDGF is a potent stimulator of fibroblasts, which form much of the connective tissue including the cell layers beneath epithelia (refer to Figure 2.3). Growth-stimulating factors such as PDGF are often termed **mitogens**, to indicate their ability to induce cells to proliferate. More specifically, PDGF attracts fibroblasts into the wound site and then stimulates their proliferation (e.g., see Figure 5.4A). [Other growth factors, such as epidermal growth factor (EGF), can cause dramatic changes in cell shape as well (Figure 5.4B).] Without stimulation by serum-derived PDGF, cultured fibroblasts will remain viable and metabolically active for weeks in a Petri dish, but they will not grow and divide.

This dependence of fibroblasts on growth-stimulating signals released by a second cell type, in this case blood platelets, mirrors hundreds of similar cell-to-cell communication routes that operate within living tissues to encourage or discourage cell proliferation. Moreover, as discussed in greater detail below, PDGF and EGF are only two of a large and disparate group of growth factors that help to convey important growth-controlling messages from one cell to another.

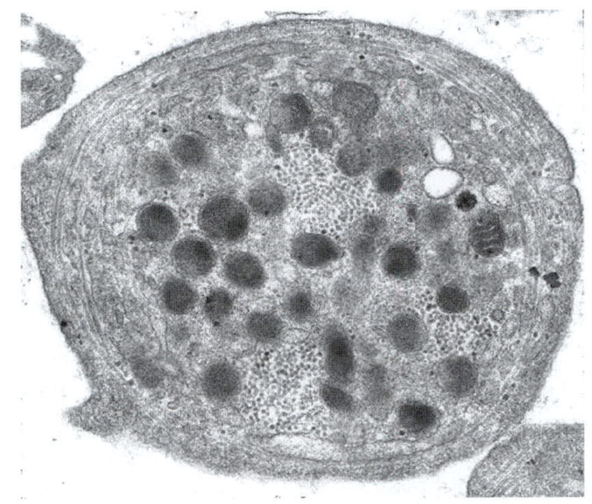

Figure 5.3 Storage of PDGF by platelets Platelets carry large numbers of secretory (*exocytotic*) vesicles termed α-granules *(large dark spots)*, which contain a number of important signaling molecules, prominent among them being several types of platelet-derived growth factor (PDGF). When platelets become activated during clot formation, these vesicles fuse with the plasma membranes, releasing previously stored mitogens and survival factors into the extracellular space. (Courtesy of S. Israels.)

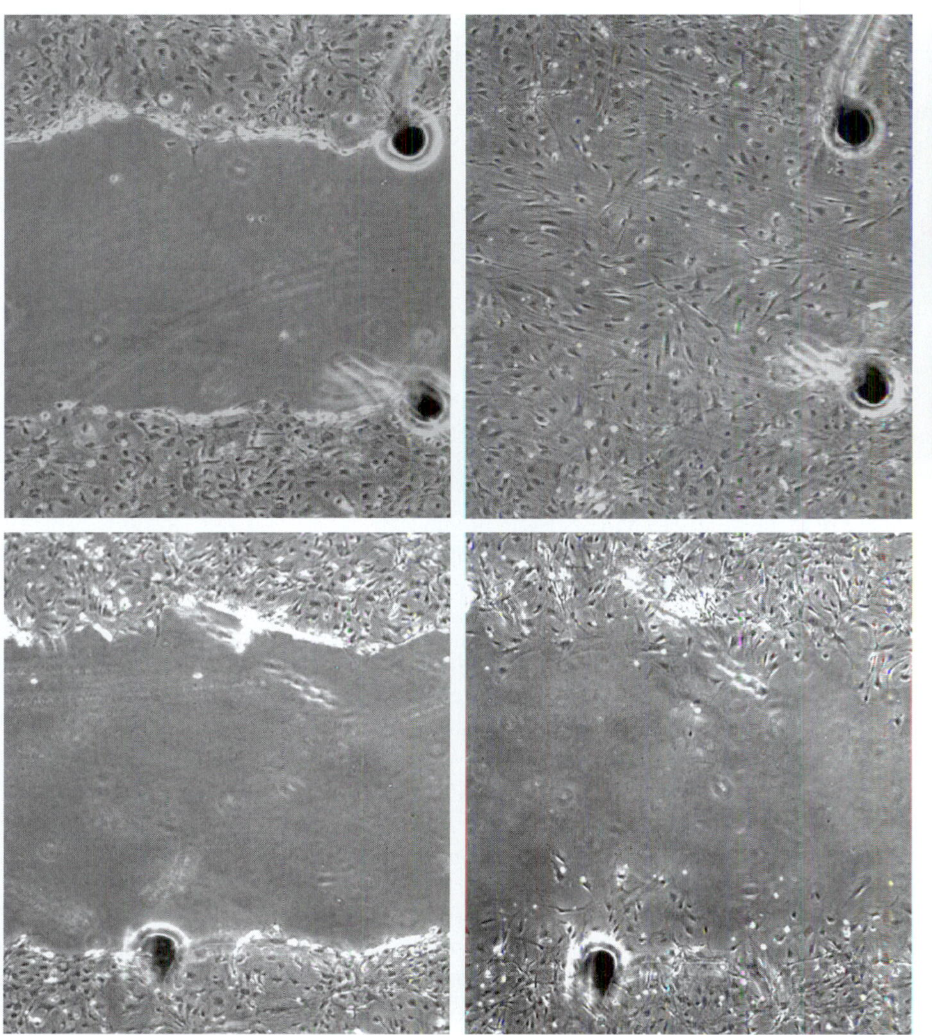

(B)

Figure 5.4 Effects of growth factors on cells (A) The PDGF released by platelets upon clotting is a potent attractant and mitogen for fibroblasts, which swarm into a wound site in order to reconstruct the tissue. In this *in vitro* model of wound healing, two monolayers of fibroblasts in Petri dishes have been wounded by the tip of a pipette (*left panels*). In the presence of one form of PDGF (PDGF-BB) added to their growth medium, large numbers of wild-type fibroblasts *(top left)* migrate into the wound site within 48 hours (*top right*). In contrast, mutant fibroblasts that have been rendered unresponsive to PDGF (through the loss of a PDGF cell surface receptor; *lower left)* have failed to migrate into the wound site after the same interval *(lower right)*. (B) Five minutes after epidermal growth factor (EGF) that has been immobilized on a bead *(green dot)* is applied to a mouse mammary carcinoma cell, the cell has reorganized its actin cytoskeleton *(red)* and extended an arm of cytoplasm toward the growth factor *(above,* top view; *below,* side view*)*. (A, from Z. Gao et al., *J. Biol. Chem.* 280:9375–9389, 2005; B, from S.J. Kempiak et al., *J. Cell Biol.* 162:781–787, 2003.)

As developed in the rest of this chapter, the signaling molecules that enable cells to sense the presence of growth factors in their surroundings, to convey this information into the cell interior, and to process this information, are usurped by oncogene-encoded proteins. By taking charge of the natural growth-stimulating machinery of the cell, oncoproteins are able to delude a cell into believing that it has encountered growth factors in its surroundings. Once this deception has taken place, the cell responds slavishly by beginning to proliferate, just as it would have done if abundant growth factors were indeed present in the medium around it.

5.2 The Src protein functions as a tyrosine kinase

The first clues to how cell-to-cell signaling via growth factors operates came from biochemical analysis of the v-*src* oncogene and the protein that it specifies. The trail of clues led, step-by-step, from this protein to the receptors used by cells to detect growth factors and in turn to the intracellular signaling pathways that control normal and malignant cell proliferation.

The initial characterization of the v-*src*–encoded oncoprotein attracted enormous interest. This was the first cellular oncoprotein to be studied and therefore had the prospect of giving cancer researchers their first view of the biochemical mechanisms of cell transformation. After the molecular cloning of the *src* oncogene, the amino acid sequence of its protein product was deduced directly from

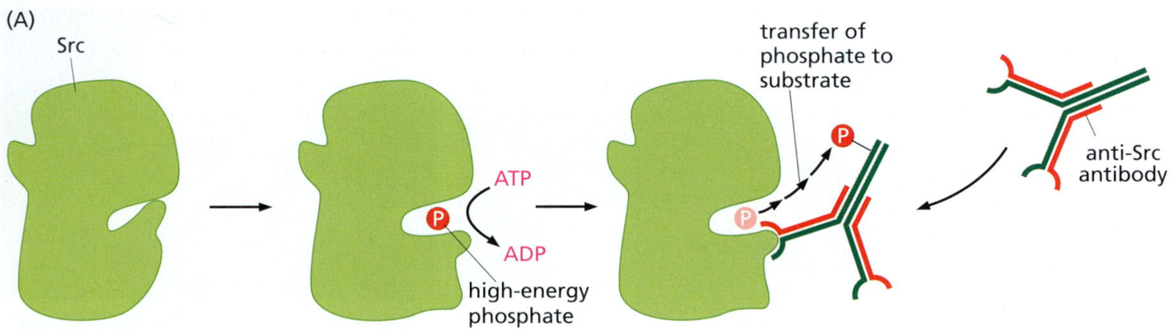

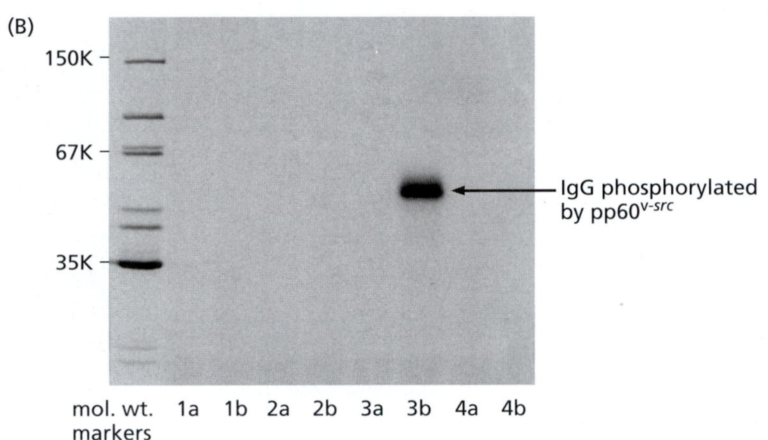

Figure 5.5 Phosphorylation of a precipitating antibody molecule by Src Protein kinases operate by removing the high-energy γ phosphate from ATP and attaching it to the hydroxyl groups in the side chains of serine, threonine, or tyrosine residues of substrate proteins. (A) The antibody molecule that was used to immunoprecipitate Src molecules (see Sidebar 5.1) also happened to serve as a substrate for phosphorylation by this kinase. Its phosphorylation is indicated by the presence of an attached phosphate group *(red)*. (B) This experiment revealed, for the first time, the biochemical activity associated with an oncoprotein. Normal rabbit serum *(tracks 1a, 2a, 3a, 4a)* and serum from a rabbit bearing an RSV-induced tumor *(tracks 1b, 2b, 3b, 4b)* were used to immunoprecipitate cell lysates that were incubated with ^{32}P-radiolabeled ATP. Lysates were prepared from uninfected chicken embryo fibroblasts (CEFs, *tracks 1a, b*), avian leukosis virus (ALV)–infected CEFs *(tracks 2a, b)*, wild-type (wt) RSV-infected CEFs *(tracks 3a, b)*, and CEFs infected by a transformation-defective mutant of RSV *(tracks 4a, b)*. Only the combination of tumor-bearing rabbit serum and wt RSV infection yielded a strongly ^{32}P-labeled protein that co-migrated with the precipitating antibody, indicating that the antibody molecule had become phosphorylated by the Src oncoprotein (whose molecule weight is not apparent in this analysis). (B, from M.S. Collett and R.L. Erikson, *Proc. Natl. Acad. Sci. USA* 75:2021–2024, 1978.)

the nucleotide sequence of the cloned gene. The encoded protein was made as a polypeptide chain of 533 amino acid residues and had a mass of almost precisely 60 kilodaltons (kD).

The amino acid sequence of Src provided few clues as to how this oncoprotein functions to promote cell proliferation and transformation. Avian and mammalian cells transformed by the v-*src* oncogene were known to exhibit a radically altered shape, to pump in glucose from the surrounding medium more rapidly than normal cells, to grow in an anchorage-independent fashion, to lose contact inhibition, and to form tumors. Whatever the precise mechanisms of action, it was obvious that Src affected, directly or indirectly, a wide variety of cellular targets.

Clever biochemical sleuthing in 1977–1978 solved the puzzle of how Src operates. Antibodies were produced that specifically recognized and bound Src (Sidebar 5.1). The antibody molecules were found to become phosphorylated when they were incubated in solution with both Src and adenosine triphosphate (ATP), the universal phosphate donor of the cell. Phosphorylation was known to involve the covalent attachment of phosphate groups to the side chains of specific amino acid residues. Hence, it was clear that Src operates as a protein **kinase**—an enzyme that removes a high-energy phosphate group from ATP and transfers it to a suitable protein substrate, in this instance, an antibody molecule (Figure 5.5). While Src does not normally phosphorylate antibody molecules, its ability to do so in these early experiments suggested that its usual mode of action also involved the phosphorylation of certain target proteins within cells.

Independent of this kinase activity, Src was itself a **phosphoprotein**, that is, it carried phosphate groups attached covalently to one or more of its amino acid side chains. This indicated that Src served as a substrate for phosphorylation by a protein kinase—either phosphorylating itself (**autophosphorylation**) or serving as the substrate of yet another kinase. Its molecular weight and phosphoprotein status caused it initially to be called pp60^{src}, although hereafter we will refer to it simply as Src.

Sidebar 5.1 Making anti-Src antibodies presented a major challenge Thorough biochemical analyses of thousands of cellular proteins have depended upon the ability to specifically detect and analyze the protein of interest (e.g., Src) amid the complex soup of proteins present in cells. This detection frequently relies on the use of antibodies that specifically recognize and bind a protein such as Src while ignoring all other cellular proteins around it. The production of such antibodies involves the injection of an intact protein (e.g., the entire Src protein) or a chemically synthesized **oligopeptide** (e.g., a short protein whose amino acid sequence reflects the sequence of a small segment of the Src protein) into an animal, usually a mouse, rabbit, or goat. The hope is that the injected material will be **immunogenic** (will elicit an immune response) in the animal, causing it to form antibodies that specifically bind and precipitate this protein from a cell **lysate** (a preparation of disrupted cells).

The immune systems of mammals often react vigorously to a foreign **antigen**, i.e., a protein carrying peptide sequences that are unfamiliar and novel for the immune system. Conversely, the immune system often mounts no immune response at all (and thus fails to make antibody) against an injected protein that closely resembles one of the proteins normally present in the animal. This lack of immunoreactivity against familiar proteins is termed immunological **tolerance**, a phenomenon that we will return to in Chapter 15.

Those who first produced anti-Src antibodies used a strain of RSV that can infect and transform mammalian cells and can therefore induce tumors in mammals. They anticipated that the resulting tumors would express the chicken Src protein, and that dying tumor cells would expose this protein to the immune systems of the tumor-bearing mammalian hosts and thus provoke an immune response. In fact, the Src protein proved to be poorly immunogenic in most animals. The reason for this was discovered much later: the structure of the Src protein is highly conserved evolutionarily. Accordingly, the chicken Src protein is almost identical to the corresponding protein of most mammalian species and is therefore not readily recognized as a foreign protein by these species' immune systems. In the end, these investigators succeeded (for unknown reasons) in producing anti-Src antiserum only from rabbits bearing RSV-induced tumors. The factors that determine the immunogenicity of a protein such as Src in one or another species are complex and often obscure. Once in hand, the anti-Src antiserum could be used to precipitate (i.e., **immunoprecipitate**) Src protein from lysates of a wide variety of RSV-transformed cells and from normal, untransformed cells (Figure 5.6). The long-sought Src protein was detected as a polypeptide migrating as a 60-kD protein in channels 8 and 9 of Figure 5.6.

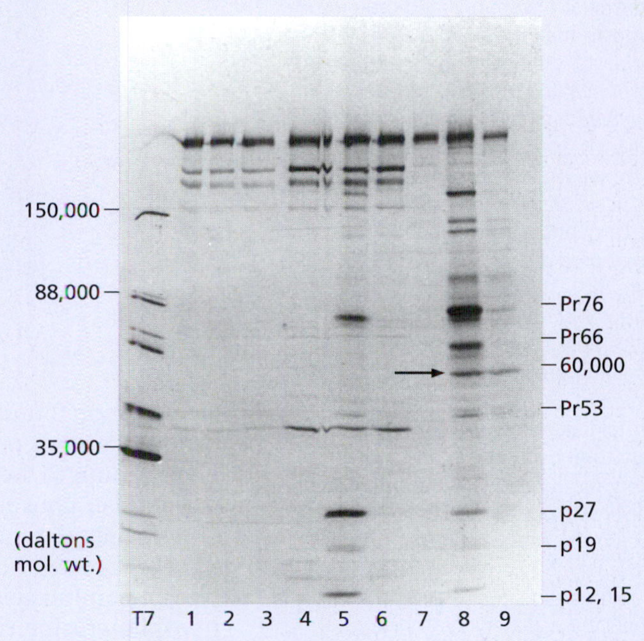

Figure 5.6 Immunoprecipitate of Src protein When the Src protein was first analyzed, the only means of studying it depended upon *immunoprecipitation* of this protein from the mixture of tens of thousands of other cellular proteins. In this case, many attempts at making an anti-Src serum finally resulted in an antibody preparation that could be used to immunoprecipitate this protein from cells. Molecular weight markers (in daltons) are in leftmost channel T7; channels 1–3: lysate of normal, uninfected cells; channels 4–6: lysate of cells infected with RSV mutant with deleted *src* gene; channels 7–9: lysate of cells infected with wild-type RSV. Channel 7: normal rabbit serum; channel 8: tumor-bearing rabbit serum; channel 9: same as channel 8 except preincubated with lysate of RSV particles. The band at 60 kD in channels 8 and 9 (*arrow*) represented the first detection of the RSV oncoprotein. (p12, p15, p19, p27, Pr53, Pr66, and Pr76 are RSV virion proteins.) (From A.F. Purchio et al., *Proc. Natl. Acad. Sci. USA* 75:1567–1571, 1978.)

The fact that Src functions as a kinase was a major revelation. In principle, a protein kinase can phosphorylate multiple, distinct substrate proteins within a cell. (In the case of Src, more than 50 distinct substrates have been enumerated.) Once phosphorylated, each of these substrate proteins may be functionally altered and proceed, in turn, to alter the functions of its own set of downstream targets. This mode of action (Figure 5.7) seemed to explain how a protein like Src could act pleiotropically to perturb multiple cell phenotypes. At the time of this discovery, other protein kinases had already been found able to regulate complex circuits, notably those involved in carbohydrate metabolism.

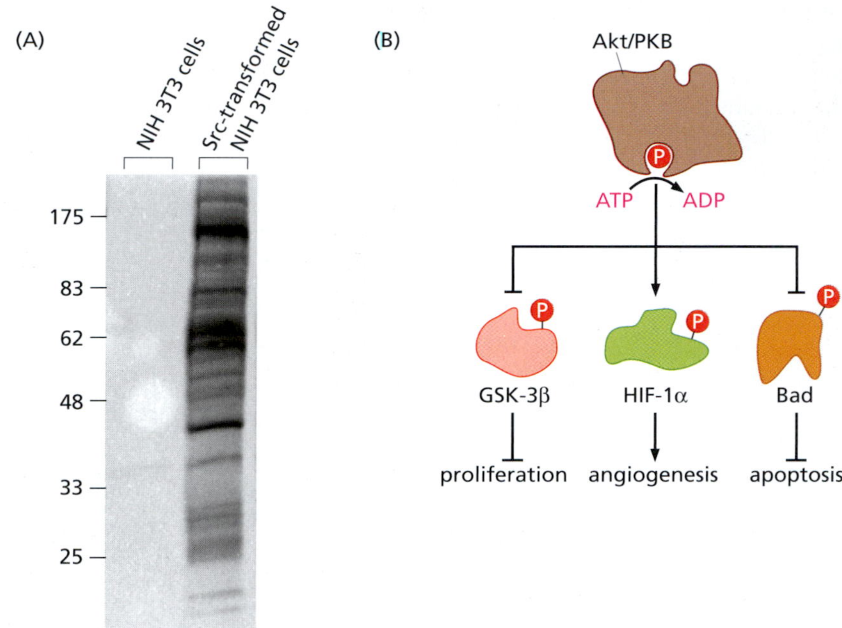

Figure 5.7 Actions of protein kinases (A) This autoradiograph of proteins separated by gel electrophoresis illustrates the fact that the phosphorylation state of a number of distinct protein species within a cell can be altered by the actions of the Src kinase expressed in cells by Rous sarcoma virus. The left channel shows the array of phosphorylated proteins in untransformed mouse NIH 3T3 fibroblasts; the right channel shows the phosphorylated proteins in cells expressing an active Src kinase protein. Molecular weight markers in kilodaltons are shown on left. In fact, this analysis specifically detects the protein species that carry phosphotyrosine residues, the known products of Src kinase action (see Figure 5.8). The left channel in this figure is essentially blank, indicating the relative absence of phosphotyrosine-containing proteins in cells that lack Src expression. (B) The pleiotropic actions of a protein kinase usually derive from its ability to phosphorylate and thereby modify the functional state of a number of distinct substrate proteins. Illustrated here are the actions of the Akt/PKB kinase, which happens to be a serine/threonine kinase and can influence a wide variety of biological processes through phosphorylation of the indicated major control proteins. Thus, Akt/PKB can inactivate the antiproliferative actions of GSK-3β and the pro-apoptotic powers of Bad, and can activate the angiogenic (blood vessel–inducing) powers of HIF-1α. (A, from S.M. Ulrich et al., *Biochemistry* 42:7915–7921, 2003.)

Src was soon found to be quite different from all other protein kinases that had been uncovered previously. These other kinases were known to attach phosphate groups to the side chains of serine and threonine amino acid residues. Src, in contrast, phosphorylated certain tyrosine residues of its protein substrates (Figure 5.8). Careful quantification of the phosphorylated amino acid residues in cells revealed how unusual this enzymatic activity is. More than 99% of the phosphoamino acids in normal cells are phosphothreonine or phosphoserine; phosphotyrosine constitutes as little as 0.05 to 0.1% of these cells' total phosphoamino acids.

After transformation of cells by the v-*src* oncogene, the level of phosphotyrosine was found to rise dramatically, becoming as much as 1% of the total phosphoamino acids in these cells. (This helps to explain the dramatic differences between the two channels of Figure 5.7A). When the same cells were transformed by other oncogenes, such as H-*ras*, their phosphoamino acid content hardly changed at all. Therefore, the creation of phosphotyrosine residues was a specific attribute of Src and was not associated with all mechanisms of cell transformation. Subsequent studies yielded a second, related conclusion: signaling through tyrosine phosphorylation is a device that is used largely by mitogenic signaling pathways in mammalian cells, whereas the kinases involved in thousands of other signaling processes rely almost exclusively on serine and threonine phosphorylation to convey their messages.

Mutant forms of Src that had lost the ability to phosphorylate substrate proteins also lost their transforming powers. So the accumulating evidence converged on the idea that Src succeeds in transforming cells through its ability to act as a tyrosine kinase (TK) and phosphorylate a still-uncharacterized group of substrate proteins within these cells. While an important initial clue, these advances only allowed a reformulation of the major question surrounding this research area. Now the problem could be restated as follows: How did the phosphorylation of tyrosine residues result in cell transformation?

5.3 The EGF receptor functions as a tyrosine kinase

After the cloning and sequencing of the v-*src* oncogene, the oncogenes of a number of other acutely transforming retroviruses were isolated by molecular cloning and subjected to DNA sequencing. Unfortunately, in the great majority

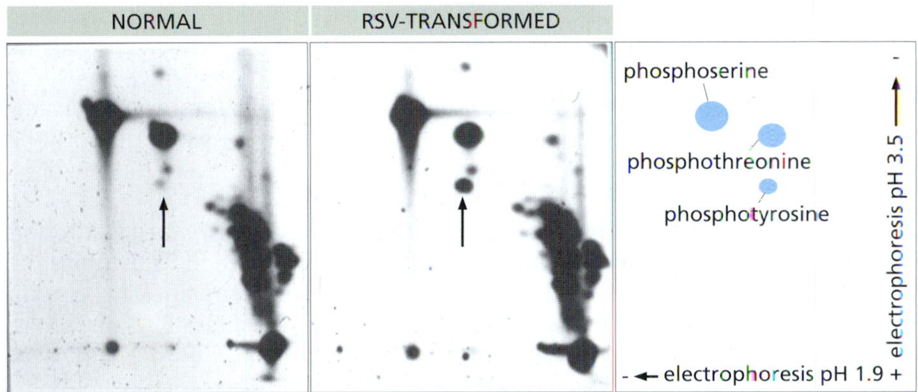

Figure 5.8 Substrate specificity of the Src kinase Electrophoresis can be used to resolve the three types of phosphoamino acids—phosphotyrosine, phosphothreonine, and phosphoserine. In the images shown here, electrophoresis was performed toward the cathode at pH 1.9 *(right to left)* and toward the cathode at pH 3.5 *(bottom to top)*. The right panel shows the expected locations of these three phosphoamino acids following electrophoresis. In normal chicken embryo fibroblasts (CEFs; *left panel)*, phosphoserine and phosphothreonine are present in abundance, while almost no phosphotyrosine is evident *(arrow)*. However, in *src*-transformed CEFs (center panel), the level of phosphotyrosine rises dramatically. Unrelated phosphorylated compounds and inorganic phosphate are seen at the lower right of each electophoresis. (Courtesy of T. Hunter.)

of cases, the amino acid sequences of their protein products provided few clues about biochemical function. The next major insights came instead from another area of research. Cell biologists interested in how cells exchange signals with one another isolated a variety of proteins involved in cell-to-cell signaling and determined the amino acid sequences of these proteins. Quite unexpectedly, close connections were uncovered between these signaling proteins and the protein products of certain oncogenes.

This line of research began with epidermal growth factor (EGF), the first of the growth factors to be discovered. EGF was initially characterized because of its ability to provoke premature eye opening in newborn mice. Soon after, EGF was found to have mitogenic effects when applied to a variety of epithelial cell types.

EGF was able to bind to the surfaces of the cells whose growth it stimulated; other cells to which EGF was unable to bind were unresponsive to its mitogenic effects. Taken together, such observations suggested the involvement of a cell surface protein, an EGF receptor (EGF-R), which was able to specifically recognize EGF in the extracellular space, bind to it, and inform the cell interior that an encounter with EGF had occurred. In the language of biochemistry, EGF served as the ligand for its cognate receptor—the still-hypothetical EGF receptor (EGF-R).

Isolation of the EGF-R protein proved challenging, because this receptor, like many others, is usually expressed at very low levels in cells. Researchers circumvented this problem by taking advantage of a human tumor cell line (an epidermoid carcinoma of the uterus) that expresses the EGF-R at elevated levels—as much as 100-fold higher than normal. These cells yielded an abundance of the receptor protein. It could then be purified biochemically and subjected to amino acid sequence analysis.

The sequence of the EGF-R protein provided important insights into its overall structural features and how this structure enables it to function. Its large N-terminal domain of 621 amino acid residues, which protrudes into the extracellular space (and is therefore termed its **ectodomain**), was clearly involved in recognizing and binding the EGF ligand. The EGF receptor possessed a second distinctive domain that is common to many cell surface glycoproteins; this **transmembrane** domain of 23 amino acid residues threads its way from outside the cell through the lipid bilayer of the plasma membrane into the cytoplasm. The existence of this transmembrane domain could be deduced from the presence of a continuous stretch of **hydrophobic** amino acid residues located in the middle of the protein's sequence; this domain of EGF-R protein resides comfortably in the highly hydrophobic environment of the lipid bilayer that forms the plasma membrane. Finally, a third domain of 542 residues at the C-terminus (i.e., the end of the chain synthesized last) of the EGF receptor was found to extend into the cytoplasm (Figure 5.9).

The overall structure of the EGF receptor suggested in outline how it functions. After its ectodomain binds EGF, a signal is transmitted through the plasma membrane to activate the cytoplasmic domain of the receptor; once activated, the latter emits signals that induce a cell to grow and divide. Importantly, examination of this cytoplasmic domain revealed a clear sequence similarity with the already-known sequence of the Src protein (Figure 5.9A).

Suddenly, it became clear how the EGF receptor emits signals in the cell interior: once its ectodomain binds EGF, somehow the Src-like kinase in its cytoplasmic domain becomes activated, proceeds to phosphorylate tyrosines on certain cytoplasmic proteins, and thereby causes a cell to proliferate. Subsequent sequencing efforts revealed overall sequence similarities among a variety of tyrosine kinases, many of which can function as oncoproteins (Figure 5.9B).

Since these initial analyses of the EGF receptor structure, a large number of other, similarly structured receptors have been characterized (Figure 5.10). As discussed later, each of these receptors has its own growth factor ligand or set of ligands (Table 5.1). Depending on the particular growth factor–receptor pair, the binding of a ligand to its receptor can trigger multiple biological responses in the cell in addition to stimulating growth and division. Important among these are changes in cell shape, cell survival, and cell **motility** (movement). Interestingly, these receptors and their ligands represent relatively recent evolutionary inventions (Sidebar 5.2).

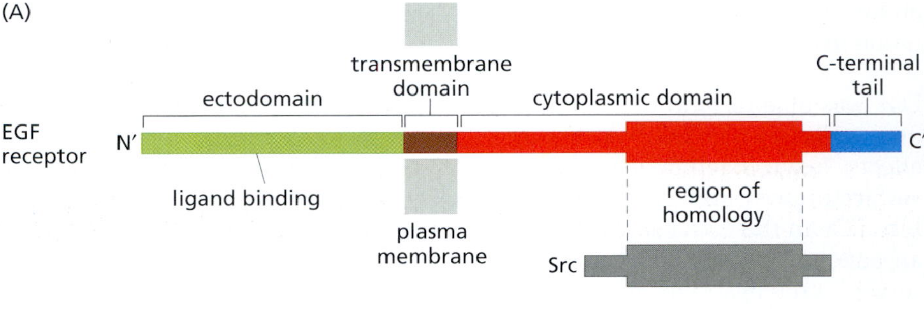

(B)

```
Src   RESLRLEAKLGGGCFGEVWMGTWNDTTR-----VAIKTLKPGTMSPE--AFLQEAQVMKK
Abl   RTDITMKHKLGGGQYGEVYEGVWKKYSLT----VAVKTLKEDTMEVE--EFLKEAAVMKE
Fes   HEDLVLGEQIGRGNFGEVFSGRLRADNTL----VAVKSCRETLPPDIKAKFLQEAKILKQ
ErbB  ETEFKKVKVLGSGAFGTIYKGLWIPEGEKVKIPVAIKELREATSPKANKEILDEAYVMAS
Raf   ASEVMLSTRIGSGSFGIVYKGKWHGD-------VAVKILKVVDPTPEQLQAFRNEVAVLR
Mos   WEQVCLMHRLGSGGFGSVYKATYHGVP------VAIKQVNKCTEDLRASQRSFWAELNIA

Src   LR-HEKLVQLYAVVSEEP-IYIVIEYMSK-----GSLLDFLKGEMGKY--------LRLP
Abl   IK-HPNLVQLLGVCTREPPFYIITEFMTY-----GNLLDYLRECNRQE--------VSAV
Fes   YS-HPNIVRLIGVCTQKQPIYIVMELVQG-----GDFLTFLRTEGAR---------LRMK
ErbB  VD-NPHVCRLLGICLTST-VQLITQLMPY-----GCLLDYIREHKDN---------IGSQ
Raf   KTRHVNI--LLFMGYMTKDNLAIVTQWCEGSSLYKHLHVQETK-------------FQMF
Mos   GLRHDNIVRVVAASTRTPEDSNSLGTIIMEFGGNVTLHQVIYDATRSPEPLSCRKQLSLG

Src   QLVDMAAQIASGMAYVERMNYVHRDLRAANILVGENLVCKVADFGLARLIEDNEYTARQG
Abl   VLLYMATQISSAMEYLEKKNFIHRDLAARNCLVGENHLVKVADFGLSRLMTGDTYTAHAG
Fes   TLLQMVGDAAAGMEYLESKCCIHRDLAARNCLVTEKNVLKISDFGMSREEADGVYAASGG
ErbB  YLLNWCVQIAKGMNYLEERRLVHRDLAARNVLVKTPQHVKITDFGLAKLLGADEKEYHAE
Raf   QLIDIARQTAQGMDYLHAKNIIHRDMKSNNLFLHEGLTVKIGDFGLATVKSRWSGSQQVE
Mos   KCLKYSLDVVNGLLFLHSQSILHLDLKPANILISEQDVCKISDFGCSQKLQDLRGRQASP

Src   AKF--PIKWTAPEAALYGR---FTIKSDVWSFGILLTELTTKGRVPYPGMVNREVLDQVE
Abl   AKF--PIKWTAPESLAYNK---FSIKSDVWAFGVLLWEIATYGMSPYPGIDLSQVYELLE
Fes   LRLV-PVKWTAPEALNYGR---YSSESDVWSFGILLWETFSLGASPYPNLSNQQTREFVE
ErbB  GGKV-PIKWMALESILHRI---YTHQSDVWSYGVTVWELMTFGSKPYDGIPASEISSVLE
Raf   QPTG-SVLWMAPEVIRMQDDNPFSFQSDVYSYGIVLYELMA-GELPYAHINNRDQIIFMV
Mos   PHIGGTYTHQAPEILKGEI---ATPKADIYSFGITLWQMTT-REVPYSGEPQYVQYAVVA

Src   R--GYRMPCPPECPESLHD----LMCQCWRKDPEERPTFKYLGAQLLPA
Abl   K--DYRMERPEGCPEKVYE----LMRACWQWNPSDRPSFAEIHQAFETM
Fes   K--GGRLPCPELCPDAVFR----LMEQCWAYEPGQRPSFSAFYQELQSI
ErbB  K--GERLPQPPICTIDVYM----IMVKCWMIDADSRPKFRELIAEFSKM
Raf   GR-GYASPDLSRLYKNCPKAIKRLVADCVKKVKEERPLFPQILSSIELL
Mos   YNLRPSLAGAVFTASLTGKALQNIIQSCWEARGLQRPSAELLGRDLKAF
```

Figure 5.9 Structure of the EGF receptor (A) The receptor for epidermal growth factor (EGF-R) is a complex protein with an extracellular domain (ectodomain, *green*), a transmembrane domain that threads its way through the plasma membrane *(brown)*, and a cytoplasmic domain *(red, blue)*. (B) Comparison of the amino acid sequences (using the single-letter amino acid code) of the cytoplasmic domain of the EGF-R with those of Src revealed areas of sequence identity *(green)*, suggesting that the EGF-R (labeled here as "ErbB"), like Src, emits signals by functioning as a tyrosine kinase. Yet other viral oncoproteins, such as Abl and Fes, were found to share sequence similarity with these two. While the sequence identities seem to be quite scattered (e.g., between Src and ErbB), this relatively small number of shared residues nonetheless indicates clear evolutionary relatedness (homology) between these proteins. Shown as well are the sequences of two other viral oncoproteins—those of the *raf* and *mos* oncogenes—which are more distantly related (being serine/threonine kinases) and share even fewer sequences with Src and the EGF-R. (Adapted from G.M. Cooper, Oncogenes, 2nd ed. Boston and London: Jones and Bartlett, 1995.)

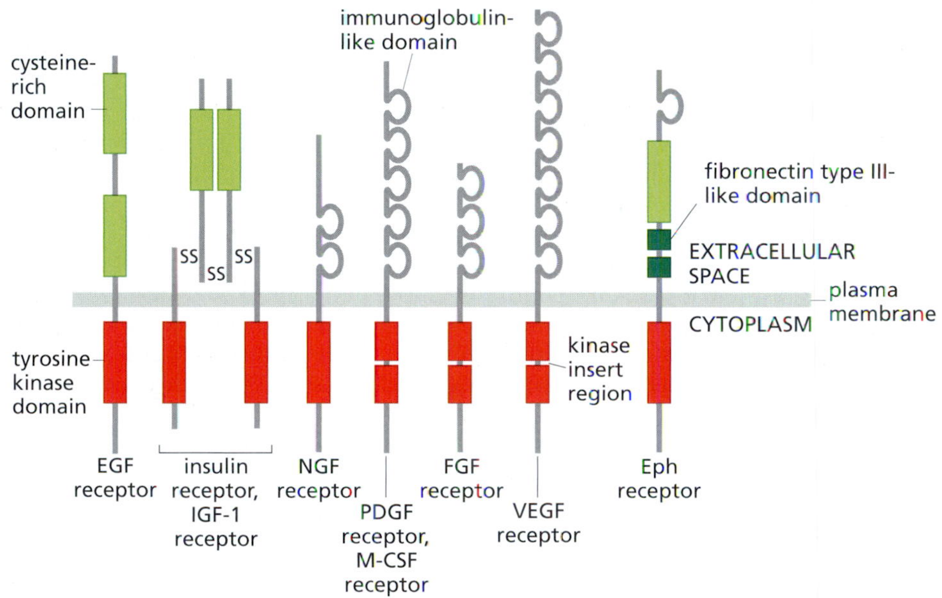

Figure 5.10 Structure of tyrosine kinase receptors The EGF receptor (Figure 5.9) is only one of a large number of similarly structured receptors that are encoded by the human genome. These tyrosine kinase receptors (RTKs) can be placed into distinct families, depending on the details of their structure. Representatives of most of these families are shown here. All of these share in common quite similar cytoplasmic tyrosine kinase domains *(red)*, although in some cases (e.g., the PDGF receptor), the tyrosine kinase domains are interrupted in the middle by small "insert" regions. The ectodomains of these receptors (which protrude into the extracellular space, *green, gray*) have highly variable structures, reflecting the fact that they recognize and bind a wide variety of extracellular ligands. (From B. Alberts et al., Molecular Biology of the Cell, 4th ed. New York: Garland Science, 2002.)

5.4 An altered growth factor receptor can function as an oncoprotein

A real bombshell fell in 1984 when the sequence of the EGF receptor was recognized to be closely related to the sequence of a known oncoprotein specified by the *erbB* oncogene. This oncogene had been discovered originally in the genome of avian erythroblastosis virus (AEV), a transforming retrovirus that rapidly induces a leukemia of the red blood cell precursors (an **erythroleukemia**). In

Table 5.1 Growth factors (GFs) and tyrosine kinase receptors that are often involved in tumor pathogenesis

Name of GF	Name of receptor	Cells responding to GF
PDGF[a]	PDGF-R	endothelial, VSMCs, fibroblasts, other mesenchymal cells, glial cells
EGF[b]	EGF-R[c]	many types of epithelial cells, some mesenchymal cells
NGF	Trk	neurons
FGF[d]	FGF-R[e]	endothelial, fibroblasts, other mesenchymal cells, VSMCs, neuroectodermal cells
HGF/SF	Met	various epithelial cells
VEGF[f]	VEGF-R[g]	endothelial cells in capillaries, lymph ducts
IGF[h]	IGF-R1	wide variety of cell types
GDNF	Ret	neuroectodermal cells
SCF	Kit	hematopoietic, mesenchymal cells

[a]PDGF is represented by four distinct polypeptides, PDGF-A, -B, -C, and -D. The PDGF-Rs consist of at least two distinct species, α and β, that can homodimerize or heterodimerize and associate with these ligands in different ways.
[b]The EGF family of ligands, all of which bind to the EGF-R (ErbB1) and/or heterodimers of erbB1 and one of its related receptors (footnote c), includes—in addition to EGF—TGF-α, HB-EGF, amphiregulin, betacellulin, and epiregulin.
[c]The EGF-R family of receptors consists of four distinct proteins, ErbB1 (EGF-R), ErbB2 (HER2, Neu), ErbB3 (HER3), and ErbB4 (HER4). They often bind ligands as heterodimeric receptors, for example, ErbB1 + ErbB3, ErbB1 + ErbB2, or ErbB2 + ErbB4; ErbB3 is devoid of kinase activity and is phosphorylated by ErbB2 when the two form heterodimers. ErbB3 and ErbB4 bind neuregulins, a family of more than 15 ligands that are generated by alternative splicing. Because ErbB3 has no intrinsic kinase activity, it becomes phosphorylated in heterodimeric complexes by ErbB2, which has no ligand of its own but does have strong tyrosine kinase activity.
[d]FGFs constitute a large family of GFs. The prototypes are acidic FGF (aFGF) and basic FGF (bFGF); in addition there are other known members of this family.
[e]There are four well-characterized FGF-Rs.
[f]There are four known VEGFs. VEGF-A and -B are involved in angiogenesis, while VEGF-C and -D are involved predominantly in lymphangiogenesis.
[g]There are three known VEGF-Rs: VEGF-R1 (also known as Flt-1) and VEGF-R2 (also known as Flk-1/KDR), involved in angiogenesis; and VEGF-R3, involved in lymphangiogenesis.
[h]The two known IGFs, IGF-1 and IGF-2, both related in structure to insulin, stimulate cell growth (i.e., increase in size) and survival; they also appear to be mitogenic.

Abbreviation: VSMC, vascular smooth muscle cell.

Adapted in part from B. Alberts et al., Molecular Biology of the Cell, 4th ed. New York: Garland Science, 2002.

Sidebar 5.2 Old and new cellular hardware The mechanisms of cell-to-cell communication are a relatively recent invention in the history of life on Earth. The earliest living cells seem to have arisen about 3.5 billion years ago. For the next 2.8 billion years or so, life on Earth was represented exclusively by single-cell organisms, first in the form of prokaryotes and then eukaryotes. Only in the last 600 to 700 million years have cells learned to talk to one another in ways that allow them to form complex tissues and create metazoan body plans.

This relatively recent acquisition of intercellular communication skills has implications for the proteins inside our cells and those on their surfaces. The great majority of intracellular proteins found in metazoa are of very ancient lineage. Many of the activities that these proteins carry out, such as energy metabolism, macromolecular synthesis, and cell growth and division, were already central to the lives of unicellular eukaryotes 1.5 billion years ago. For this reason, the functions of many of our intracellular proteins can be studied profitably by examining the behavior of their homologs in the cells of baker's yeast or fission yeast.

The cellular proteins on the surfaces of cells, as well as the proteins exchanged between cells, are relative newcomers. In the great majority of cases, their antecedents cannot be found among the proteins made by single-cell eukaryotes. Many of these newer proteins were invented when metazoan life first arose and have been retained in clearly recognizable form in most and possibly all modern metazoa. This explains why our understanding of growth factors and how they signal, as laid out in this chapter, has been greatly enriched by the powerful genetic techniques that allow researchers to dissect the cell-to-cell communication systems of simpler metazoan organisms, such as worms and flies.

one stroke, two areas of cell biology were united. A protein used by the cell to sense the presence of a growth factor in its surroundings had been appropriated (in its avian form) and converted into a potent retrovirus-encoded oncoprotein.

Once examined in detail, the oncoprotein specified by the *erbB* oncogene of avian erythroblastosis virus was found to lack sequences present in the N-terminal ectodomain of the EGF receptor (Figure 5.11). Without these N-terminal sequences, the ErbB oncoprotein clearly cannot recognize and bind EGF, and yet it functions as a potent stimulator of cell proliferation. This realization led to an interesting speculation that was soon validated: somehow, deletion of the ectodomain enables the resulting truncated EGF receptor protein to send growth-stimulating signals into cells in a constitutive fashion, fully independent of EGF. As indicated in Figure 5.12A, mutations in the genes encoding growth factor receptors, including those specifying truncated receptors, may trigger ligand-independent firing by these receptors. Indeed, a variety of mutations, including those creating amino acid substitutions in the transmembrane domain (shown), or in the ectodomain and cytoplasmic domains (not shown), of some receptors can provoke ligand-independent firing. (In Chapter 16, we will see how such structurally altered receptors influence the responsiveness of human tumors to anti-cancer therapeutic drugs.)

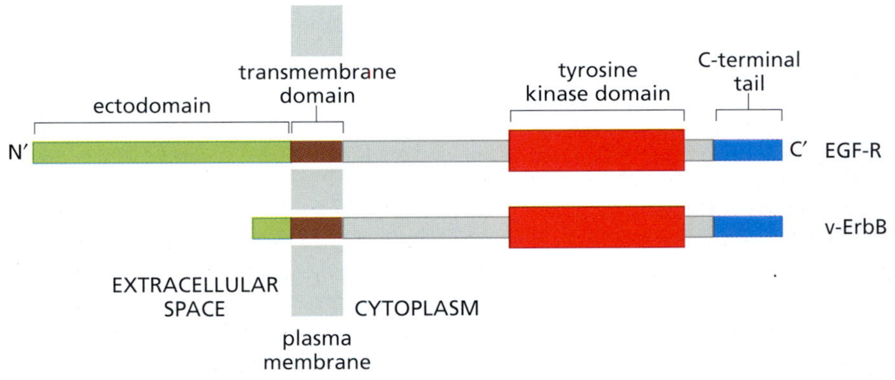

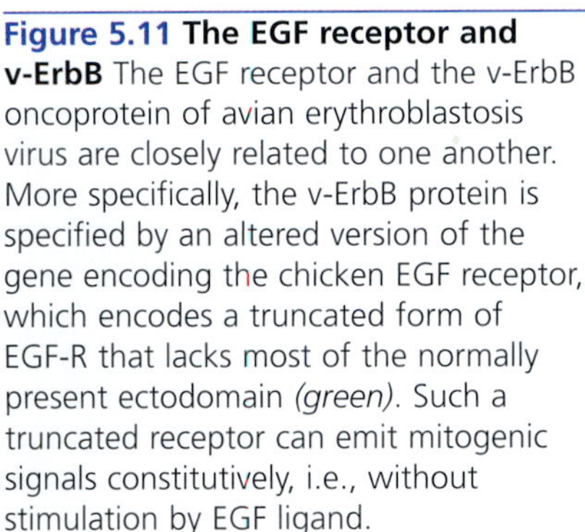
Figure 5.11 The EGF receptor and v-ErbB The EGF receptor and the v-ErbB oncoprotein of avian erythroblastosis virus are closely related to one another. More specifically, the v-ErbB protein is specified by an altered version of the gene encoding the chicken EGF receptor, which encodes a truncated form of EGF-R that lacks most of the normally present ectodomain *(green)*. Such a truncated receptor can emit mitogenic signals constitutively, i.e., without stimulation by EGF ligand.

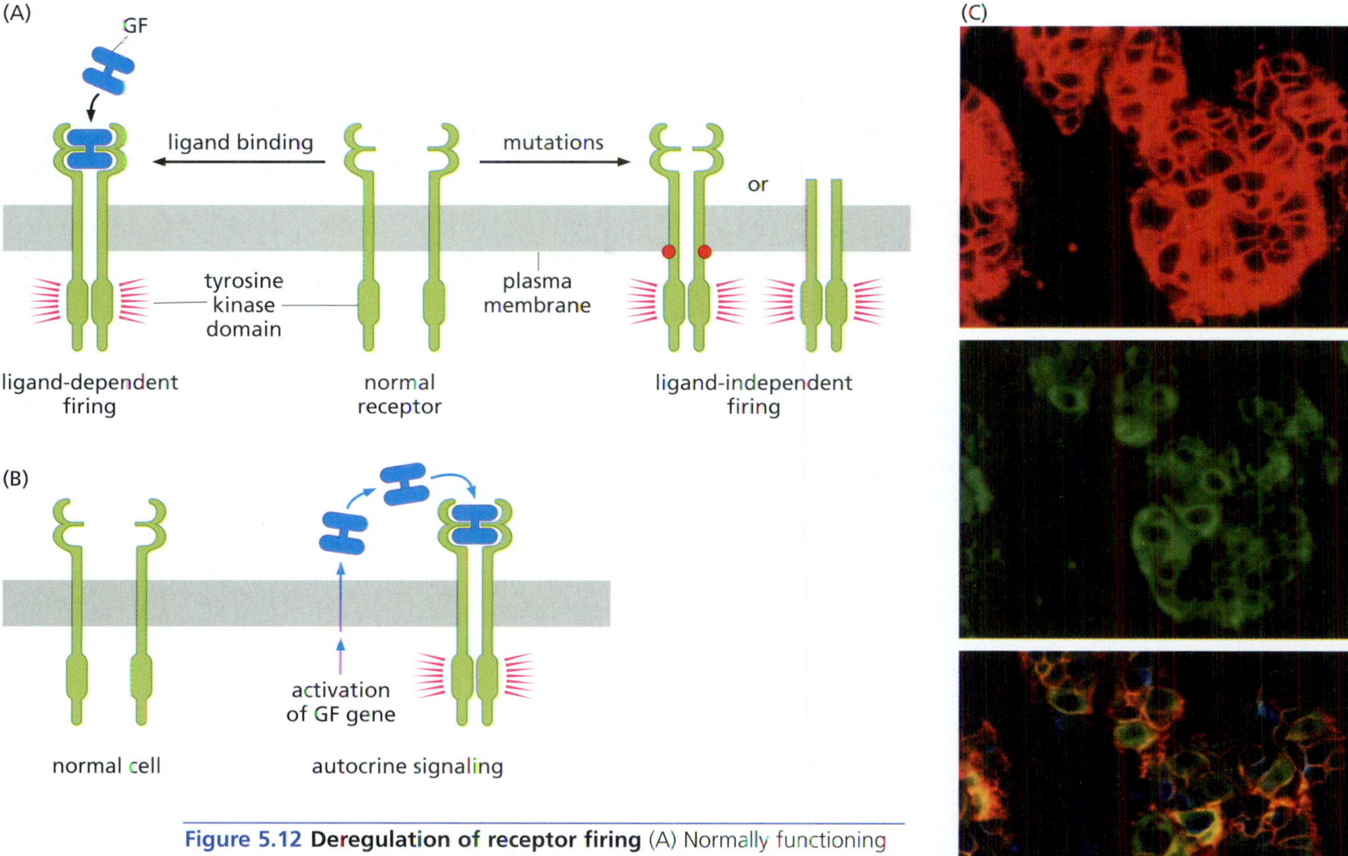

Figure 5.12 Deregulation of receptor firing (A) Normally functioning growth factor receptors emit cytoplasmic signals *(red)* in response to binding ligand *(blue, left)*. However, mutations in the genes encoding the receptor molecules *(right)* can cause subtle alterations in protein structure, such as amino acid substitutions *(red dots)*, that cause ligand-independent firing. More drastic alterations in receptor structure, including truncation of the ectodomain (e.g., Figure 5.11), may also yield such deregulated signaling. (B) In general, cells do not synthesize and release a growth factor ligand whose cognate receptor they also display. However, in many types of cancer, tumor cells acquire the ability to express a growth factor *(blue)* that is not normally expressed by cells of their lineage, whose cognate receptor may already be expressed by these cells. This creates an auto-stimulatory or autocrine signaling loop. (C) An example of this is seen in these successive sections of an invasive human breast carcinoma, in which islands of cancer cells are surrounded by non-staining stroma *(black)*. The top section has been immunostained for expression of the EGF receptor *(red)*, while the middle section has been stained for expression of TGF-α, a ligand of the EGF receptor *(green)*. When these two images are superimposed *(bottom panel)*, cells that express both receptor and ligand are stained *yellow*, while nuclei are stained *blue*. (C, from J.S. de Jong et al., *J. Pathol.* 184:44-52, 1998.)

These insights into receptor function provided one solution to a long-standing problem in cancer cell biology. As mentioned earlier, normal cells had long been known to require growth factors in their culture medium in order to grow. Cancer cells, in contrast, were known to have a greatly reduced dependence on growth factors for their growth and survival. The discovery of the ErbB–EGF-R connection yielded a simple, neat explanation of this particular trait of cancer cells: The ErbB oncoprotein releases signals very similar to those emitted by a ligand-activated EGF receptor. However, unlike the EGF receptor, the ErbB oncoprotein can send a constant, unrelenting stream of growth-stimulating signals into the cell, thereby persuading the cell that substantial amounts of EGF are present in its surroundings when none might be there at all.

Sidebar 5.3 Receptor overexpression results from a variety of molecular mechanisms The overexpression of growth factor receptors at the cell surface is achieved through a number of distinct molecular mechanisms. In some cancer cells, receptors are overexpressed because the encoding gene is transcribed at an elevated rate, resulting in correspondingly increased levels of mRNA and protein. Alternatively, in a number of types of human cancers, these receptor genes are often amplified, leading once again to proportionately increased levels of mRNA transcript and receptor protein.

Yet other, more subtle mechanisms may also be responsible for increased numbers of receptor molecules at the cell surface. One of these derives from the mechanisms that govern the *lifetime* of receptor molecules at the cell surface. Many receptors are displayed at the surface for only a limited amount of time before they are internalized via the mechanism known as **endocytosis**, in which a patch of plasma membrane together with associated proteins is pulled into the cytoplasm, where it forms a vesicle. Thereafter, the contents of this vesicle may be dispatched to **lysosomes**, in which they are degraded, or they may be recycled back to the cell surface.

One protein, termed huntingtin-interacting protein-1 (HIP1), is part of a complex of proteins that participate in and facilitate endocytosis. HIP1 has been found to be overexpressed in a variety of human carcinomas. For reasons that are poorly understood, overexpression of HIP1 prevents the normal endocytosis of a number of cell surface proteins, including the EGF receptor (EGF-R). As a consequence, cells overexpressing HIP1 accumulate excessive levels of EGF-R at the cell surface and, in the case of NIH 3T3 mouse fibroblasts, become hyper-responsive to EGF. This explains why such cells can grow vigorously in the presence of very low levels of serum (0.1%) that cause normal NIH 3T3 cells to retreat into a nongrowing, quiescent state. Moreover, these cells become transformed in the presence of 10% serum in the tissue culture medium—the serum concentration that is routinely used to propagate untransformed NIH 3T3 cells.

Another protein, called cyclin G-associated kinase (GAK), is a strong promoter of EGF-R endocytosis. When its expression is suppressed, levels of cell surface EGF-R increase by as much as fiftyfold. Its expression levels in human tumors have not yet been measured. A third protein, called c-Cbl, is responsible for tagging ligand-activated EGF-R, doing so via a covalent modification termed monoubiquitylation (to be described in Sidebar 7.8). This tagging causes the endocytosis of the EGF-R and its subsequent degradation in lysosomes. The v-Cbl viral oncoprotein, as well as several cellular proteins (Sts-1, Sts-2), can block this endocytosis, resulting, once again, in accumulation of elevated levels of EGF-R at the cell surface. In a more general sense, the hyper-responsiveness to growth factors resulting from receptor overexpression is likely to drive the proliferation of a variety of human cancer cell types *in vivo*.

As mentioned earlier in Section 4.6, we now realize that truncated versions of the EGF receptor are found in a number of human tumor cell types. In many lung cancers, for example, the EGF-R mRNA lacks the coding sequences carried by exons 2 through 7. This deletion, which removes sequences specifying much of the ectodomain of the receptor, often results from alternative splicing of the precursor of this mRNA. More generally, a variety of growth factor receptors that are configured much like the EGF receptor have been found in human tumors to be overexpressed (Sidebar 5.3) or synthesized in a structurally altered form (Table 5.2).

5.5 A growth factor gene can become an oncogene: the case of *sis*

The notion that oncoproteins can activate mitogenic signaling pathways received another big boost when the platelet-derived growth factor (PDGF) protein was isolated and its amino acid sequence determined. In 1983, the B chain of PDGF was found to be closely related in sequence to the oncoprotein encoded by the v-*sis* oncogene of simian sarcoma virus. Once again, study of the oncogenes carried by rapidly transforming retroviruses paid off handsomely.

The PDGF protein was discovered to be unrelated in structure to EGF and to stimulate proliferation of a different set of cells. PDGF stimulates largely mesenchymal cells, such as fibroblasts, adipocytes, smooth muscle cells, and endothelial cells, while the mitogenic activities of EGF are focused largely (but not entirely) on epithelial cells. This specificity of action could be understood

Table 5.2 Tyrosine kinase GF receptors altered in human tumors[a]

Name of receptor	Main ligand	Type of alteration	Types of tumor
EGF-R/ErbB1	EGF, TGF-α	overexpression	non-small cell lung cancer; breast, head and neck, stomach, colorectal, esophageal, prostate, bladder, renal, pancreatic, and ovarian carcinomas; glioblastoma
EGF-R/ErbB1		truncation of ectodomain	glioblastoma, lung and breast carcinomas
ErbB2/HER2/Neu	NRG, EGF	overexpression	30% of breast adenocarcinomas
ErbB3, 4	various	overexpression	oral squamous cell carcinoma
Flt-3	FL	tandem duplication	acute myelogenous leukemia
Kit	SCF	amino acid substitutions	gastrointestinal stromal tumor
Ret		fusion with other proteins, point mutations	papillary thyroid carcinomas, multiple endocrine neoplasias 2A and 2B
FGF-R3	FGF	overexpression; amino acid substitutions	multiple myeloma, bladder and cervical carcinomas

[a]See also Figure 5.17.

once the PDGF receptor was isolated: the PDGF-R was found to be expressed on the surfaces of mesenchymal cells and is not displayed by epithelial cells, while the EGF-R largely shows the opposite pattern of expression. (Like the EGF receptor, the PDGF-R uses a tyrosine kinase in its cytoplasmic domain to broadcast signals into the cell; see Figure 5.10.)

The connection between PDGF and the *sis*-encoded oncoprotein suggested another important mechanism by which oncoproteins could transform cells: Simian sarcoma virus infects a cell. Its *sis* oncogene then causes the infected cell to release copious amounts of a PDGF-like Sis protein into the surrounding medium. Once present in the extracellular space, the PDGF-like molecules attach to the PDGF-R displayed by the same cell that has just synthesized and released them. The result is strong activation of this cell's PDGF receptors and, in turn, a flooding of the cell with a constant flux of the growth-stimulating signals released by the ligand-activated PDGF-R.

These discoveries also resolved a long-standing puzzle. Most types of acutely transforming retroviruses are able to transform a variety of infected cell types. Simian sarcoma virus, however, was known to be able to transform fibroblasts, but it failed to transform epithelial cells. The cell type–specific display of the PDGF-R explained the differing susceptibilities to transformation by simian sarcoma virus.

Once again, a close connection had been forged between a protein involved in mitogenic signaling and a viral oncoprotein. In this instance, a virus-infected cell was forced to make and release a growth factor to which it could also respond. Rather than a growth factor signal sent from one type of cell to another cell type located nearby (often termed **paracrine** signaling), or sent via the circulation from cells in one tissue in the body to others in a distant tissue (**endocrine** signaling), this represented an auto-stimulatory, or **autocrine**, signaling loop in which a cell manufactured its own mitogens (Figure 5.12B, C). (An autocrine signaling mechanism also explains how Friend leukemia virus, another retrovirus of mice, provokes disease; see Sidebar 5.4.)

In fact, a variety of human tumor cells are known to produce and release substantial amounts of growth factors to which these cells can also respond (Table 5.3). Some human cancers, such as certain lung cancers, produce as many as three distinct growth factors (tumor growth factor-α, or TGF-α; stem cell factor, or SCF; and insulin-like growth factor, or IGF) and at the same time express the receptors for these three ligands, thereby establishing three autocrine signaling loops simultaneously. These signaling loops seem to be functionally important for the growth of tumors. For example, in one study of small-cell lung cancer (SCLC) patients, those whose tumors expressed Kit, the receptor for stem cell factor

Sidebar 5.4 Autocrine signaling solves the mystery of Friend leukemia virus leukemogenesis At one time, the biological powers of the Friend murine leukemia virus (FLV) represented a major mystery. This virus was found to induce erythroleukemias rapidly in mice, yet careful examination of the viral genome failed to reveal any apparent oncogene that had been acquired from the cellular genome, as was the case with all other rapidly oncogenic retroviruses. Thus, FLV seemed to contain only the *gag*, *pol*, and *env* genes that are required for replication by all retroviruses (see Figure 3.19).

A clue came from a smaller retroviral genome that coproliferated together with FLV. This smaller genome, which apparently resulted from deletion of most of the genes of a wild-type retrovirus genome, carried only the gene for a viral glycoprotein, that is, a gene similar to and derived from the retroviral *env* gene. (Usually the *env*-encoded glycoproteins are used by retrovirus virions to attach to target cells and to fuse the virion membrane with the plasma membrane of these cells.) Detailed characterization of this glycoprotein, termed gp55, solved the puzzle of FLV-induced erythroleukemias. gp55 was found to act as a mimic of the growth factor **erythropoietin** (EPO). Normally, when oxygen tension in the blood is less than normal, EPO is released from the kidneys and binds to the EPO receptors displayed by cells in the bone marrow that are the immediate precursors of erythroblasts (precursors of the mature red blood cells). This activation of the EPO-R causes these cells to increase in number and stimulates them to differentiate into **erythrocytes** (red blood cells). In mice infected by FLV, the gp55 released from an infected erythroblast acts in an autocrine manner on the infected cell by binding to and stimulating its EPO receptors, thereby driving its proliferation. (The resulting erythrocyte precursor cells then accumulate in large numbers and suffer additional mutations that cause them to become neoplastic.) It is still unclear how Friend leukemia virus has succeeded in remodeling a retroviral glycoprotein into an effective mimic of EPO.

(SCF), survived for an average of only 71 days after diagnosis of their disease, while those whose tumors lacked Kit expression survived an average of 288 days.

Possibly the champion autocrine tumor is Kaposi's sarcoma, a tumor of cells closely related to the endothelial cells that form lymph ducts (see Sidebar 3.10). To date, Kaposi's tumors have been documented to produce PDGF, TGF-β, IGF-1, Ang2, CCl8/14, CXCL11, and endothelin—all ligands of cellular origin—as well as the receptors for these ligands. At the same time, the causal agent of this disease, the human herpesvirus-8 (HHV-8) genome that is present in Kaposi's tumor cells, produces two additional ligands—vIL6 and vMIP—whose cognate receptors are also expressed by the endothelial cell precursors that generate these tumors.

In most of these cases, expression of the cellular genes encoding various mitogenic growth factors has somehow become deregulated, resulting in the production of mitogens in cells that do not normally express significant levels of these cellular proteins. This pattern of behavior is compounded by HHV-8, which forces infected cells to synthesize two novel growth factors that are not encoded in the cellular genome.

Table 5.3 Examples of human tumors making autocrine growth factors

Ligand	Receptor	Tumor type(s)
HGF	Met	miscellaneous endocrinal tumors, invasive breast and lung cancers, osteosarcoma
IGF-2	IGF-1R	colorectal
IL-6	IL-6R	myeloma, HNSCC
IL-8	IL-8R A	bladder cancer
NRG	ErbB2[a]/ErbB3	ovarian carcinoma
PDGF-BB	PDGF-Rα/β	osteosarcoma, glioma
PDGF-C	PDGF-α/β	Ewing's sarcoma
PRL	PRL-R	breast carcinoma
SCF	Kit	Ewing's sarcoma, SCLC
VEGF-A	VEGF-R (Flt-1)	neuroblastoma, prostate cancer, Kaposi's sarcoma
TGF-α	EGF-R	squamous cell lung, breast and prostate adenocarcinoma, pancreatic, mesothelioma
GRP	GRP-R	small-cell lung cancer

[a]Also known as HER2 or Neu receptor.

Sidebar 5.5 Autocrine signaling influences the development of anti-cancer therapies Testing of the efficacy of a novel anti-cancer drug usually begins with analyses of the ability of the drug to affect the growth and/or survival of cancer cells grown in culture (*in vitro*). These tests are then followed up by the seeding of such cancer cells in host mice, where they are allowed to form tumors (*in vivo*). The responsiveness of these tumor grafts to drug treatment of the tumor-bearing mice is often used to predict whether the drug under study will have clinical efficacy, that is, an ability to stop further growth of a tumor or to cause its shrinkage in patients.

On some occasions, the *in vitro* drug responsiveness of a tumor cell population may predict its behavior *in vivo*; in other cases, it will not. The experience of some drug developers indicates that these contrasting behaviors correlate with the production of autocrine growth factors by tumor cell lines. Those cancer cells that produce abundant autocrine growth factors usually respond to drug treatment similarly *in vitro* and *in vivo*, while those that do not secrete growth factors behave very differently under these two conditions. It seems that the tumor cells that secrete autocrine factors create their own growth factor environment both *in vitro* and *in vivo*. Conversely, those that do not secrete these factors depend on serum-associated mitogens *in vitro* and on the growth factors supplied by nearby host cells *in vivo*; the latter may differ dramatically from the spectrum of mitogens present in serum.

In a more general sense, autocrine signaling loops seem to represent potential perils for tissues and organisms. In normal tissues, the proliferation of individual cells almost always depends on signals received from other cells; such interdependence ensures the stability of cell populations and the constancy of tissue architecture. A cell that has gained the ability to control its own proliferation (by making its own mitogens) therefore creates an imminent danger, since self-reinforcing, positive feedback loops often lead to gross physiologic imbalances.

The activation of these autocrine signaling loops yields an outcome very similar to that occurring when a structurally altered receptor protein such as ErbB/EGF-R is expressed by a cell. In both cases, the cell generates its own mitogenic signals and its dependence upon exogenous mitogens is greatly reduced. Interestingly, these dynamics also have important implications for the development of anti-cancer drugs (Sidebar 5.5).

5.6 Transphosphorylation underlies the operations of receptor tyrosine kinases

The actions of oncogenes such as *sis* and *erbB* provide a satisfying biological explanation of how a cell can become transformed. By supplying cells with a continuous flux of growth-stimulatory signals, oncoproteins are able to drive the repeated rounds of cell growth and division that are needed in order for large populations of cancer cells to accumulate and for tumors to form. Still, this biological explanation sidesteps an important biochemical question lying at the heart of mitogenic signaling: How do growth factor receptors containing tyrosine kinases (often called receptor tyrosine kinases or RTKs) succeed in transducing signals from the extracellular space into the cytoplasm of cells?

Knowing the presence of the tyrosine kinase enzymatic activity borne by the cytoplasmic domains of these proteins allows us to rephrase this question: How do growth factor receptors use their tyrosine kinase domains to emit signals in response to ligand binding?

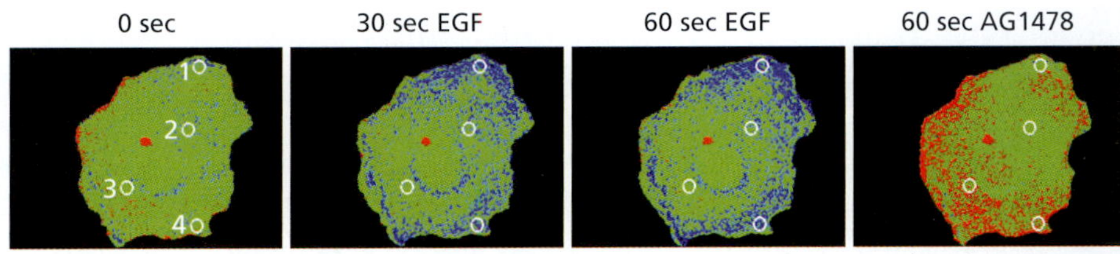

Figure 5.13 Formation of phosphotyrosine on the EGF-R following ligand addition The use of a fluorescent reagent that binds specifically to a phosphotyrosine residue on the EGF-R enables the visualization of receptor activation following ligand binding. Here, receptor activation is measured on a monkey kidney cell at a *basal* level (0 second), as well as 30 and 60 seconds after EGF addition. In addition, following a 2-minute stimulation by EGF the effects of a 60-second treatment by a chemical inhibitor of the EGF-R kinase (AG1478) are shown *(right)*. Receptor activity above the basal level is indicated in *blue*, while activity below the basal level is indicated in *red*. The response to AG1478 treatment indicates that a significant basal level of EGF-R activity was present (at 0 second) even before EGF addition. (From M. Offterdinger et al., *J. Biol. Chem.* 279:36972–36981, 2004.)

The solution to this problem came from a detailed examination of the proteins that become phosphorylated in cells within seconds after a growth factor such as EGF is applied to cells expressing its cognate receptor, in this case the EGF-R (Figure 5.13). It would seem reasonable that a variety of cytoplasmic proteins become phosphorylated on their tyrosine residues following ligand binding to a growth factor receptor. Indeed, there are a number of such proteins. But the most prominent among these phosphotyrosine-bearing proteins is often the receptor molecule itself (Figure 5.14)! Hence, these receptors seem to be capable of autophosphorylation.

Another clue came from the structure of many growth factor ligands; they were often found to be **dimeric**, being composed either of two identical protein subunits **(homodimers)** or of very similar but nonidentical subunits (**heterodimers**). A third clue derived from observations that many transmembrane proteins constructed like the EGF and PDGF receptors have lateral mobility in the plane of the plasma membrane. That is, as long as the transmembrane domains of these receptor proteins remain embedded within the lipid bilayer of this membrane, they are relatively free to wander back and forth across the surface of the cell.

Taken together, these facts led to a simple model. In the absence of ligand, a growth factor receptor exists in a **monomeric** (single-subunit) form, embedded as always in the plasma membrane. When presented with its growth factor ligand, which is a homodimer, a receptor molecule will bind to one of the two subunits of this ligand. Thereafter, the ligand–receptor complex will wander around the plasma membrane until it encounters another receptor molecule, on which occasion the second, still-unengaged subunit of the ligand will bind this second receptor molecule. The result is a cross-linking of the two receptor molecules achieved by the dimeric ligand that forms a bridge between them (Figure 5.15). X-ray crystallographic studies of growth factors bound to the ectodomains of their receptors have subsequently revealed other mechanisms by which these ligands are able to induce receptor dimerization (Sidebar 5.6).

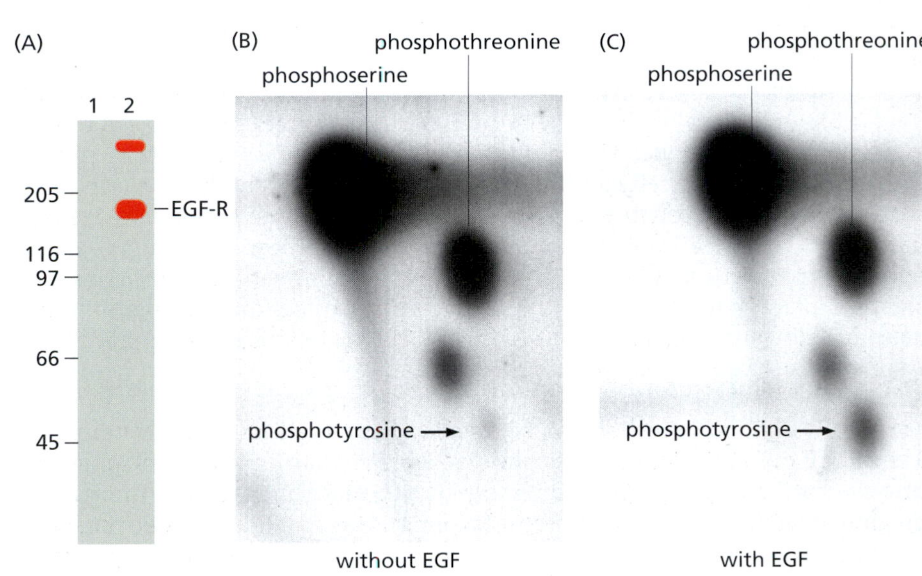

Figure 5.14 Apparent autophosphorylation of the EGF receptor (A) When human A431 epidermoid carcinoma cells, which greatly overexpress the EGF-R, are incubated in $^{32}PO_4$-containing medium and then exposed to EGF, a radiolabeled protein can be immunoprecipitated with an anti-phosphotyrosine antiserum *(lane 2)*; this protein co-migrates upon gel electrophoresis with the EGF-R and is not detectable in the absence of prior EGF treatment *(lane 1)*. (B) The $^{32}PO_4$-labeled phosphoamino acids borne by the proteins in such cells in the absence of EGF (as resolved by electrophoresis; see Figure 5.8) are seen here. (C) Following the addition of EGF to A431 cells, a spot in the lower right becomes darker; internal markers indicate that this is phosphotyrosine rather than phosphoserine or phosphothreonine. (A, from A.B. Sorokin et al., *FEBS Lett.* 268:121–124, 1990; B and C, courtesy of A.R. Frackelton Jr.)

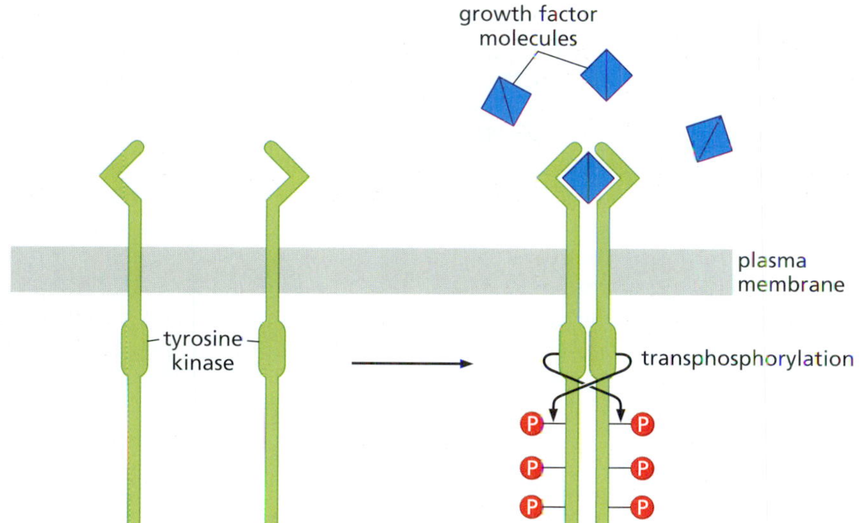

Figure 5.15 Receptor dimerization following ligand binding This scheme indicates how many growth factor receptors are thought to become activated to emit signals. In the absence of ligand, receptor molecules *(green)* are free to move laterally in the plane of the plasma membrane. In the presence of growth factor ligand, two receptor molecules are brought together to form a dimer. Once the receptor is dimerized, the tyrosine kinase domain of each receptor subunit is able to phosphorylate the C′-terminal cytoplasmic tail of the other subunit—the process of transphosphorylation *(arrows)*.

Once dimerization of the ectodomains of two receptor molecules has been achieved by ligand binding, the cytoplasmic portions are also pulled together. Now each kinase domain phosphorylates tyrosine residues present in the cytoplasmic domain of the other receptor. So the term "autophosphorylation" (see Figure 5.14) is actually a misnomer. This bidirectional, reciprocal phosphorylation is best described as a process of **transphosphorylation** (see Figure 5.15). (Indeed, it is unclear whether a single isolated kinase molecule is ever able to phosphorylate itself.)

The phosphorylation of these tyrosine residues can have at least two consequences. The catalytic cleft of a kinase—the region of the protein where its enzymatic function occurs—may normally be partially obstructed by a loop of the protein, preventing the kinase from interacting effectively with its substrates. Transphosphorylation of a critical tyrosine in the obstructing "activation loop" causes it to swing out of the way, thereby providing the catalytic cleft with direct access to substrate molecules. In addition, transphosphorylation results in the phosphorylation of an array of tyrosine residues present in cytoplasmic portions of the growth factor receptor outside the kinase domain, as indicated in Figure 5.15. As we will learn later, these phosphorylation events enable the resulting phosphorylated receptor molecule to activate a diverse array of downstream signaling pathways.

The receptor dimerization model explains how growth factor receptors can participate in the formation of cancers when the receptor molecules are overexpressed, that is, displayed on the cell surface at levels that greatly exceed those seen in normal cells (see Sidebar 5.3). Since these receptors are free to move laterally in the plane of the plasma membrane, their high numbers cause them to collide frequently, and these encounters, like the dimerization events triggered by ligand binding, can result in transphosphorylation, receptor activation, and signal emission. For instance, as Table 5.2 indicates, the EGF receptor is overexpressed in a wide variety of human tumors, mostly carcinomas. In such tumors, this overexpression may result in ligand-independent receptor dimerization and firing. Alternatively, excessive EGF receptor expression may make some cancer cells hyper-responsive to the low levels of EGF and TGF-α ligands that may be present in their surroundings.

It is usually difficult to distinguish between these two alternative mechanisms. One set of experiments, however, provides unambiguous indication that receptor overexpression can act via the first of these mechanisms—ligand-independent dimerization and firing. In these experiments, the human Met receptor (whose ligand is HGF/SF) was overexpressed in liver cells of a genetically altered

Sidebar 5.6 Growth factors induce receptor dimerization in various ways In reality, the simple model of receptor dimerization depicted in Figure 5.15 represents the behavior of only a subset of growth factor receptors. This model applies, for example, to nerve growth factor (NGF) and its Trk receptor, and to vascular endothelial growth factor (VEGF) and its Flt-1 receptor (Figure 5.16A). However, in the case of fibroblast growth factor (FGF), the growth factor molecules are monomers that do not contact each other directly. Instead, a pair of FGF molecules are linked together by binding to a common heparin molecule; each of the FGF monomers then attracts an FGF-R molecule, causing its dimerization (Figure 5.16B).

Other growth factors function as single polypeptide chains. Thus, a single molecule of erythropoietin (EPO) is able to simultaneously contact two receptor molecules, thereby causing dimerization *(not shown)*. Yet other ligands may function as heterodimers. In the case of the EGF receptor, two ligand molecules (such as EGF or a similar EGF-R ligand termed TGF-α) bind as monomers to monomeric receptors; this binding causes two such ligand-activated receptors to expose dimerization domains on the faces of the ectodomains opposite to the ligand-binding sites and associate with one another without any connection whatsoever between the two ligand monomers (Figure 5.16C).

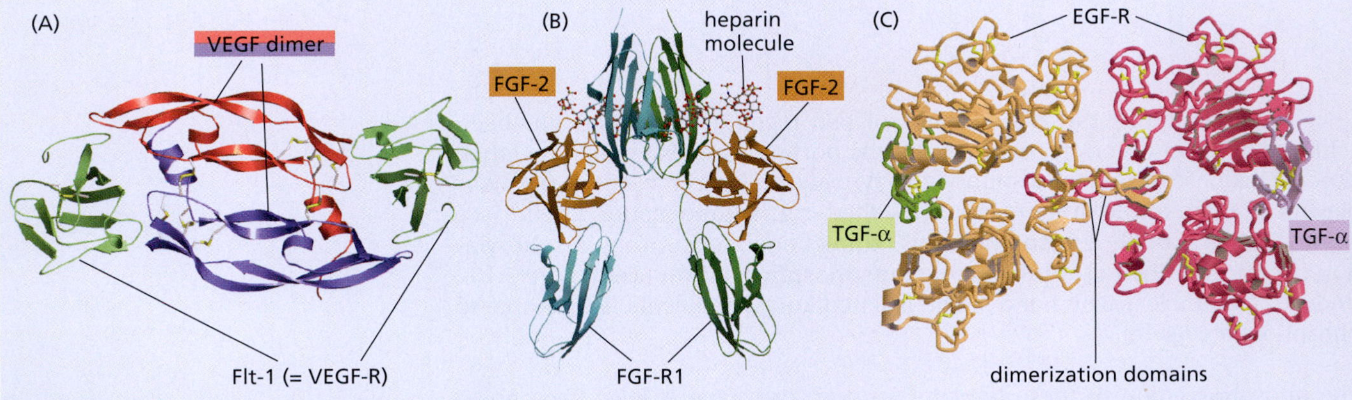

Figure 5.16 Alternative mechanisms of growth factor–induced receptor dimerization X-ray crystallography has revealed how a number of growth factor ligands cause receptor dimerization. (A) Vascular endothelial growth factor (VEGF) binds as a dimer *(red, blue)* to two monomers of its receptor, called Flt-1 *(green)*, thereby bringing them together. This complex is seen as a bottom view, i.e., from the plasma membrane looking outward. (B) In the case of the fibroblast growth factor receptor-1 (FGF-R1), each of the FGF-2 ligands *(dark orange)* binds to the ligand-binding domain of a receptor monomer *(green, aquamarine)*. Each ligand-binding domain of the receptor subunits, seen here in side view, is composed of two subdomains. (The two *green domains* are lying atop one another in this view.) Receptor dimerization occurs only if the two FGF ligand molecules are bound, in turn, to a common heparin molecule *(stick figure)*, the latter being a glycosaminoglycan component of the extracellular matrix. (C) In the case of the EGF-R, molecules of tumor growth factor-α (TGF-α), one of the alternative ligands of this receptor, bind individually *(green, left; purple, right)* to the "front sides" of two receptor subunits *(dark yellow, dark pink)*. As seen in a *top view* facing the plasma membrane, receptor monomers respond to ligand binding by exposing domains that mediate their "back-to-back" dimerization. Hence, each growth factor molecule is actually far removed from the region of the receptor monomer that is directly involved in dimerization. (A, from C. Wiesmann et al., *Cell* 91:695–704, 1997; B, from A. Plotnikov et al., *Cell* 98:641–650, 1999; C, from T.P.J. Garrett et al., *Cell* 110:763–773, 2002.)

mouse strain; this led, in turn, to the development of hepatocellular carcinomas (HCCs). Since the human Met receptor cannot bind the mouse HGF/SF ligand, this dictated that the observed receptor activation and cell transformation could only be ascribed to a ligand-independent process, specifically spontaneous dimerization of overexpressed Met receptor molecules.

A diverse array of structural alterations can also cause ligand-independent receptor dimerization, doing so in the absence of receptor overexpression. This lesson is driven home most dramatically by a mutant version of the HER2/ErbB2/Neu receptor, a cousin of the EGF receptor that is found in certain rat tumors. Exposure of pregnant rats to the mutagen ethylnitrosourea (ENU) results in the birth of pups that often succumb to neuroectodermal tumors 3 to 6 months later. The tumors almost invariably carry a point mutation in the sequence specifying the transmembrane domain of this receptor (see Figure 5.12A). The resulting substitution of a valine by a glutamic acid residue favors

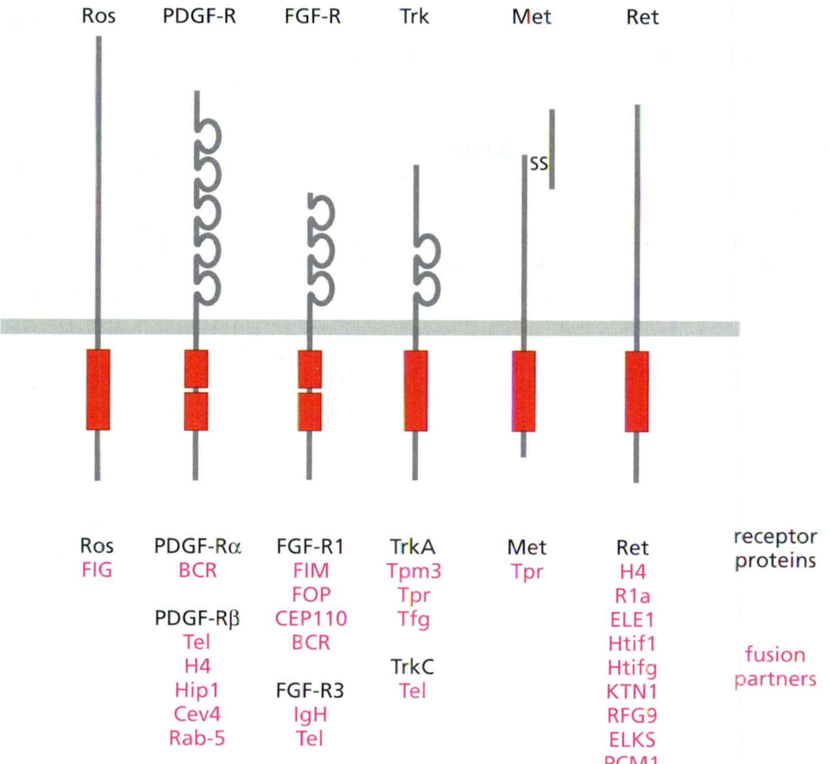

Figure 5.17 Gene fusion causing constitutively dimerized receptors In a number of malignant tumors, receptor dimerization occurs when the genes encoding growth factor receptors *(black type)* become fused to unrelated genes that happen to specify proteins that normally dimerize or form higher-order oligomers *(red type)*. As a consequence, the receptor portions of these hybrid fusion proteins (usually their ectodomains) are dragged together by the oligomerizing proteins to which they have become joined. Listed here are a number of such mutant, fused receptors found in human tumors; in certain cases, a given receptor may be found fused to multiple alternative oligomerizing proteins. (Courtesy of A. Charest.)

the constitutive dimerization of the two receptors, even in the absence of ligand binding, thereby creating a potent oncoprotein.

Like HER2/ErbB2/Neu, the EGF and FGF receptors may be affected by point mutations or small deletions that generate constitutively active receptors; the responsible somatic mutations alter the cytoplasmic domains of these receptors, perturbing the domains that regulate the receptor-associated tyrosine kinases. In a series of non-small-cell lung carcinomas, less than 10% were found to have mutant EGF receptors, and these turned out to be particularly responsive to therapeutic drugs directed against the receptor molecule, as we will learn in Section 16.12. Structurally altered receptors have been found to be far more common in bladder carcinomas: at least 60% of these tumors have been found to harbor subtly altered versions of the FGF receptor-3. The precise mechanism by which truncated, mutant versions of the EGF receptor lacking ectodomains become constitutively activated (see Figure 5.11) is less well understood.

Other types of genetic alterations of tyrosine kinase receptors may also give rise to oncoproteins. For example, Met, which functions as the receptor for hepatocyte growth factor (HGF), and TrkA, which serves as the receptor for nerve growth factor (NGF), have both been found to have become converted into oncoproteins in some tumors as a consequence of gene fusion events. These fusions have resulted in truncation of the sequences encoding the ectodomains of these receptors and a fusion of the remaining portions of these receptors with other proteins that are normally prone to dimerize or oligomerize (Figure 5.17). This results in a constitutive, ligand-independent dimerization of these receptors, explaining the oncogenic powers of these fusion proteins. Yet other, quite subtle changes in receptor structure can also encourage ligand-independent firing in diverse cell types (Sidebar 5.7).

Although we have placed great emphasis here on receptor dimerization, it represents only one of the molecular changes that tyrosine kinase receptor molecules can undergo after they bind their ligands. There is increasing evidence that, in addition to this dimerization, the ligand-binding ectodomain undergoes some type of rotation or stereochemical shift, and that this shift is transmitted

Sidebar 5.7 Mutant forms of a single tyrosine kinase receptor may play a causal role in very different types of cancer The Kit protein was originally discovered as an oncoprotein encoded by a feline sarcoma retrovirus. The *kit* proto-oncogene encodes a tyrosine kinase receptor whose ligand is stem cell factor (SCF), a growth factor that is important in stimulating the formation of various types of cells in the blood (the process of **hematopoiesis**), as well as the development of a variety of nonhematopoietic cell types, including melanocytes and the cells mediating gut motility. Mutant forms of the Kit receptor, which fire constitutively in a ligand-independent fashion, are found in a diverse array of malignancies. In many leukemias of dogs, mutant Kit receptors carrying amino acid substitutions in the cytoplasmic **juxtamembrane** residues—those close to the transmembrane domain—are common.

The pacemaker cells of the intestine, which control the peristaltic contractions of the smooth muscles of this organ, also rely on the Kit receptor for their proper development, and indeed, these cells are absent if the *kit* gene is inactivated in the germ line (and thus in all cells) of a mouse. In gastrointestinal stromal tumors (GISTs) of humans, which seem to arise from these same cells, juxtamembrane amino acid substitutions similar to those seen in dog leukemias are commonly observed. Analyses of the structure of the Kit receptor and related tyrosine kinase receptors indicate that the juxtamembrane domain of this receptor plays an *inhibitory* role in regulating the receptor, and that its ability to suppress receptor firing is compromised by the amino acid substitutions and deletions in this domain that are commonly seen in GISTs (Figure 5.18). In one group of GIST patients, those whose tumors carried subtle (missense) mutations affecting exon 11 of the *Kit* gene were all alive four years after initial diagnosis and treatment, while only half of those whose GISTs carried deletions of this entire exon were still alive at this time.

Figure 5.18 Multiple structural alterations affect Kit receptor firing (A) As seen here, the Kit receptor may be found in a constitutively active, mutant form in a number of different types of human tumors *(red letters).* (B) The structure of the cytoplasmic domain of the Flt-3 receptor (determined by X-ray crystallography), which is very closely related to that of Kit, is shown here as a ribbon structure. Like almost all protein kinases, this one is composed of an N-terminal lobe *(red)* and a C-terminal lobe *(blue),* with a catalytic cleft in between. The catalytic cleft is normally obstructed by both an "activation loop" *(green)* and a juxtamembrane domain (*yellow*). Note the tyrosine residue in the activation loop, whose phosphorylation causes this loop to swing out of the way, allowing the catalytic cleft access to substrates. (C) The Kit receptor, which is found in mutant form in almost all gastrointestinal stromal tumors (GISTs), often carries alterations affecting the juxtamembrane (JM) domain, depicted here as a yellow cylinder that sits above the N-terminal lobe of the two-lobed Kit tyrosine kinase *[side view, plasma membrane (not shown) above].* Ligand binding to the Kit receptor results initially in transphosphorylation of tyrosine residues in the JM domain, which causes it to dissociate from the side of the N-terminal lobe; the resulting phosphorylated structure *(yellow zigzag)* is depicted here as sitting above the kinase domain, out of the way. After this has occurred, transphosphorylation of a key tyrosine residue in the activation loop *(green),* which normally obstructs the catalytic cleft (see panel B), also contributes to activation of kinase function. Mutant Kit receptors with defective JM domains are therefore partially activated even in the absence of ligand binding. (A, courtesy of D. Emerson; B, from J. Griffith, *J. Mol. Cell* 13:169–178, 2004; C, from P.M. Chan et al., *Mol. Cell. Biol.* 23:3067–3078, 2003.)

through the plasma membrane to the cytoplasmic kinase domains of these receptors, which themselves are moved into new configurations that favor transphosphorylation. In fact, in some tyrosine kinase receptors, constitutive dimerization may exist even without ligand binding, and the ligand works only to effect a concurrent rotation of both the ectodomains and cytoplasmic domains of the receptor molecules that facilitates signaling by their kinase domains.

The sequencing of the human genome has revealed the full repertoire of tyrosine kinase receptors that are likely to be expressed in human tissues. Of the approximately 22,000 genes in our genome, at least 58 encode proteins having the general structures of the EGF and PDGF receptors, that is, a ligand-binding ectodomain, a transmembrane domain, and a tyrosine kinase cytoplasmic domain. These receptors are further grouped into subclasses according to their structural similarities (see Figure 5.10). To date, only a minority of these receptors have been implicated as agents contributing to human cancer formation. It is likely that the involvement of other receptors of this class in human cancer will be demonstrated in the future. Significantly, mutant alleles of certain tyrosine kinase receptor genes may also be transmitted in the human germ line. This explains the origins of a number of familial cancer syndromes, in which affected family members show greatly increased risks of contracting certain types of cancer (Sidebar 5.8).

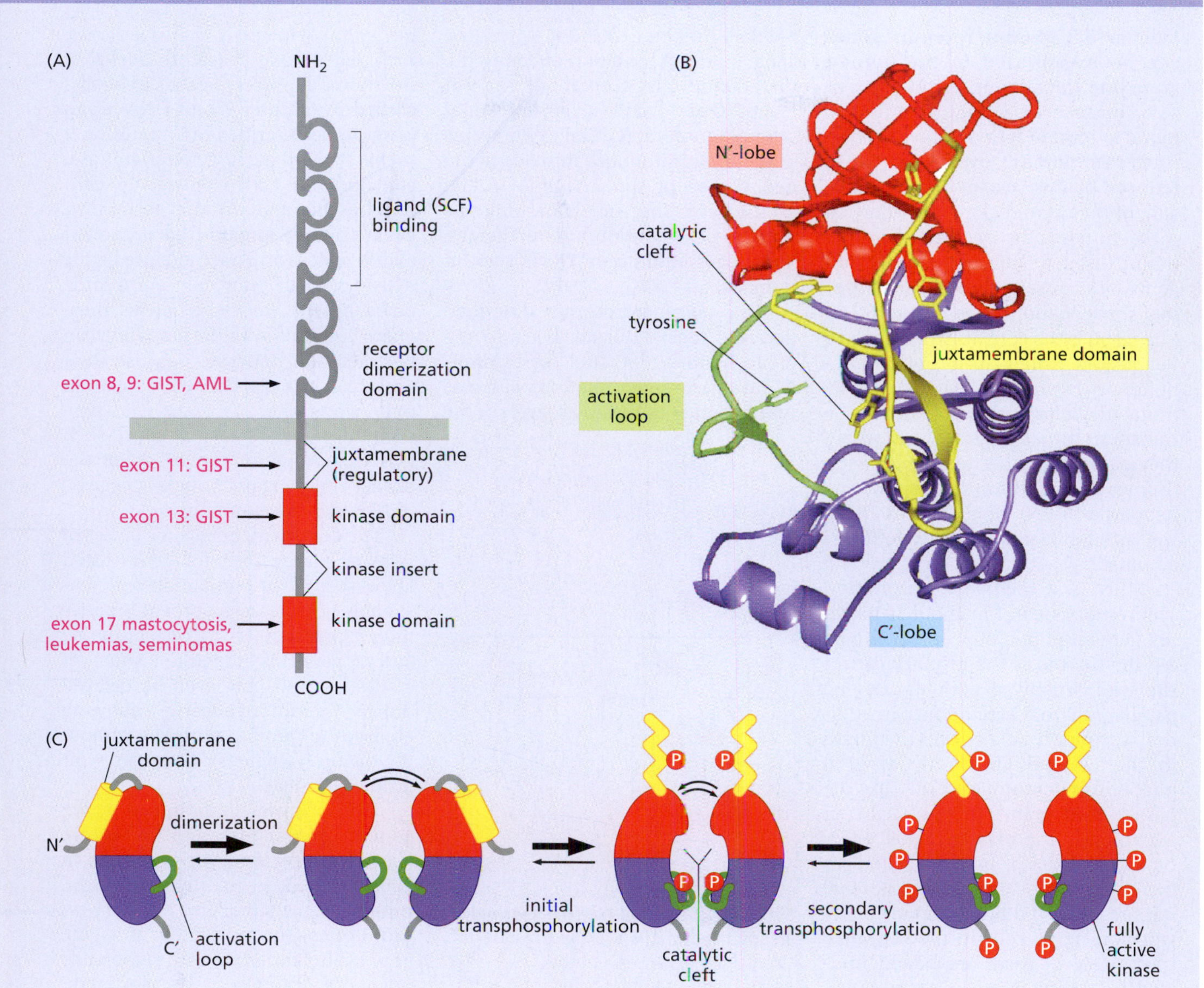

5.7 Yet other types of receptors enable mammalian cells to communicate with their environment

By the 1980s, a number of other distinct types of metazoan receptors had been discovered. By then, it had become clear that the tyrosine kinase receptors (RTKs) described above represent only one class of receptors among a far larger group of structurally diverse receptors, all designed to enable cells to sense the biochemical state of their surroundings. As is the case with tyrosine kinase receptors, each of these other receptor types is specialized to detect the presence of specific extracellular ligands or groups of related ligands and to respond to this encounter by transducing signals into the cytoplasm.

One class of receptors, which is especially important in receiving signals controlling the development of various hematopoietic cell types, also uses tyrosine kinases. In this case, however, the responsible enzymes, termed Jaks (Janus kinases), are separate polypeptides that associate with the cytoplasmic domains of these receptors through noncovalent links. Included here is the receptor for erythropoietin (EPO), which, as mentioned above, regulates the development of erythrocytes (red blood cells), and the **thrombopoietin** (TPO) receptor, which controls the development of the precursors of blood platelets,

Sidebar 5.8 Mutant receptor genes can be transmitted in the human germ line The *ret* gene encodes a tyrosine kinase receptor that is frequently found in mutant form in papillary thyroid carcinomas, especially those induced by inadvertent clinical exposure of the thyroid gland to X-rays. In these tumors, the *ret* gene is often found fused to one of several other genes as a consequence of chromosomal translocations (see Figure 5.17). The result is the constitutive firing of a Ret fusion protein in a ligand-independent manner. More subtly mutated alleles of the *ret* gene can be transmitted in the germ line, where they cause the inherited cancer syndromes known as multiple endocrine neoplasia (MEN) types 2A and 2B, as well as familial medullary thyroid carcinoma.

MEN is a relatively uncommon syndrome, seen in 1 in about 30,000 in the general population. In MEN type 2A, the thyroid is the primary cancer site, with virtually all patients carrying papillary thyroid carcinomas. In addition, secondary cancer sites are found in the adrenal glands (leading to pheochromocytomas) and in the parathyroid gland (hyperplasias or adenomas). The closely related MEN type 2B disease is manifested as increased risk of thyroid and adrenal tumors and, in addition, tumors of the ganglion nerve cells in the intestinal tract. MEN-2A cases are usually initiated by an inherited point mutation that causes the substitution of a cysteine residue in the extracellular domain of the Ret tyrosine kinase receptor. MEN-2B cases are caused by a substitution in the tyrosine kinase domain of this receptor.

Hereditary papillary renal cancer is due to inherited mutant alleles in the *met* gene, which encodes the receptor for hepatocyte growth factor (HGF). These mutant germ-line alleles of *met* usually carry point mutations that cause amino acid substitutions in the tyrosine kinase domain of Met that result in constitutive, ligand-independent firing by the receptor.

Members of a Japanese family (Figure 5.19) have been reported to transmit a mutant allele of the Kit receptor in their germ line. Like the somatically mutated forms of *kit* (see Sidebar 5.7), the mutant germ-line allele carries a mutation affecting the juxtamembrane domain of the Kit receptor and yields gastrointestinal stromal tumors (GISTs). When this mutation is introduced into one of the two copies of the *kit* gene in the mouse germ line, mice inheriting this mutant allele develop tumors that are indistinguishable from GISTs seen in humans.

The cancer syndromes described here are most unusual, because the great majority of such syndromes involve germ-line mutations of genes that result, at the cellular level, in the formation of inactive, recessive alleles. Included among these are alleles of the tumor suppressor genes to be discussed in Chapter 7 and DNA repair genes to be described in Chapter 12. It seems likely that, in general, mutant, constitutively active oncogenes cannot be tolerated in the germ line, because they function as dominant alleles in cells and are therefore highly disruptive of normal embryonic development. Recessive alleles present in single copy in the germ line can be tolerated, however, because their presence in cells becomes apparent only when the surviving wild-type dominant allele is lost. Because such loss is infrequent, expression of the cancer-inducing phenotype is delayed until late in development.

The dominant, oncogenic alleles of tyrosine kinase receptor genes listed above seem to be compatible with reasonably normal development because their expression (in the case of the *ret* and *kit* genes) is limited to a small set of tissues, and may even be delayed during embryogenesis, allowing embryos to develop normally without the disruptive effects of these growth-promoting alleles.

This logic was undermined by the genetic characterization of a rare familial cancer syndrome reported in 2005: Contrary to preconception, mutant, activated alleles of the H-*ras* proto-oncogene present in a zygote are not incompatible with reasonably normal human development. Individuals inheriting these alleles are afflicted with Costello syndrome, which is associated with mental retardation, high birth weight, **cardiomyopathy** (pathological defects in the heart muscle), and a tendency to develop tumors; together these pathologies prevent these individuals from surviving to adulthood. (This explains why the 12 cases that have been described all represented *de novo* mutations, that is, new mutations that arose during the formation of a sperm or egg rather than being inherited from a similarly afflicted parent.) Strikingly, most organ systems in these individuals develop relatively normally in spite of the presence of a highly penetrant allele of an activated H-*ras* oncogene (see Section 4.4) in all of their cells.

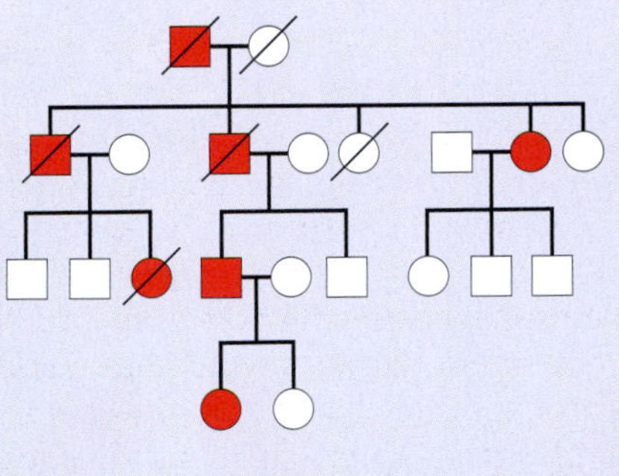

Figure 5.19 Familial gastrointestinal stromal tumors (GISTs) Gastrointestinal stromal tumors arise from the mesenchymal pacemaker cells of the lower gastrointestinal tract and contrast with the far more frequent colon carcinomas, which derive from the epithelial cell layer lining the lumen of the gut. In the case of the kindred depicted here, an allele encoding a mutant, constitutively active Kit tyrosine kinase receptor has been carried by at least four generations, afflicting them with GISTs and/or intestinal obstructions. The transmission of this mutant allele, which carries a deletion of a single amino acid residue in the juxtamembrane domain of the receptor protein (see Figure 5.18), through so many generations indicates that phenotypic expression of this mutant germ-line allele is often delayed until relatively late in life after childbearing years. Affected individuals *(red);* males *(squares);* females *(circles);* deceased *(strikethroughs)*. (From T. Nishida et al., *Nat. Genet.* 19:323–324, 1998.)

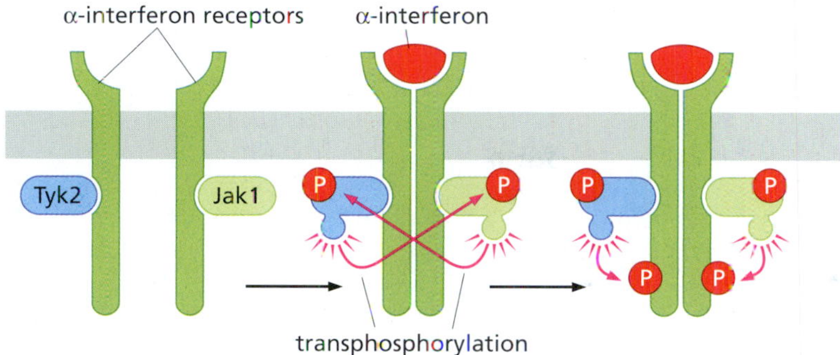

Figure 5.20 Structure of cytokine receptors The interferon receptor (IFN-R, *dark green*), like the receptors for erythropoietin and thrombopoietin, carries noncovalently attached tyrosine kinases (TKs) of the Jak family (in this case, Tyk2 and Jak1). Upon ligand binding (in this case α-interferon), the two TKs transphosphorylate and thereby activate one another. Subsequently, they phosphorylate the C-terminal tails of the receptor subunits, placing the receptor in an actively signaling configuration, much like the receptor tyrosine kinases described in Section 5.6.

called **megakaryocytes**. The receptors for the important immune regulator interferon are also members of this class. When receptor molecules of this class dimerize in response to ligand binding, the associated Jaks phosphorylate and activate each other (Figure 5.20); the activated Jaks then proceed to phosphorylate the C-terminal tails of the receptor molecules, thereby creating receptors that are activated to emit signals, much like the tyrosine kinase receptors described in the previous section. We will return to the details of Jak signaling in the next chapter.

Transforming growth factor-β (TGF-β) and related ligands have receptors that are superficially similar to the tyrosine kinase receptors, in that they have an extracellular ligand-binding domain, a transmembrane domain, and a cytoplasmic kinase domain. However, these receptors invariably function as heterodimers, such as the complex of the type I and type II TGF-β receptors (Figure 5.21). Importantly, their kinase domains phosphorylate serine and threonine rather than tyrosine residues. TGF-βs play centrally important roles in cancer pathogenesis, since they suppress the proliferation of normal epithelial cells while promoting the acquisition of invasive properties by already-transformed cells. Upon ligand binding, the type II TGF-β receptor subunit, which has a constitutively active serine/threonine kinase, is brought into close proximity with the type I TGF-β receptor subunit, which it then phosphorylates. The kinase belonging to the type I TGF-β receptor subunit becomes activated as a consequence, and proceeds to phosphorylate cytosolic proteins that migrate to the nucleus, where they trigger expression of certain target genes (see Figure 5.21). (A series of TGF-β–related factors, including activin and bone morphogenetic proteins, or BMPs, use similar receptors for signaling; they are not discussed further here, as their role in cancer development has not been documented.) We postpone more detailed discussion of the TGF-β signaling system until Chapter 8.

A far more primitive form of a transmembrane signaling system is embodied in the Notch receptor (termed simply Notch) and its multiple alternative ligands (NotchL, Delta, Jagged; Figure 5.22). After binding ligand, Notch is cleaved successively by two proteases. One of the resulting proteolytic Notch fragments, deriving from its cytoplasmic domain, then migrates to the nucleus, where it

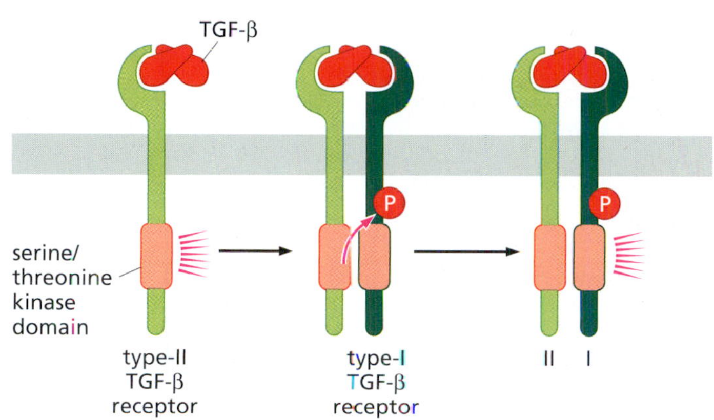

Figure 5.21 Structure of the TGF-β receptor The structure of the TGF-β receptor is superficially similar to that of tyrosine kinase receptors (RTKs), in that both types of receptors signal through cytoplasmic kinase domains. However, the kinase domains of TGF-β receptors specifically phosphorylate serine and threonine, rather than tyrosine residues. After the type II receptor (TGF-βRII) is brought into contact with the type I receptor (TGF-βRI) through the binding of TGF-β ligand, it phosphorylates and activates the kinase carried by the type I receptor. The activated type I receptor kinase then emits signals by phosphorylating certain cytoplasmic substrates.

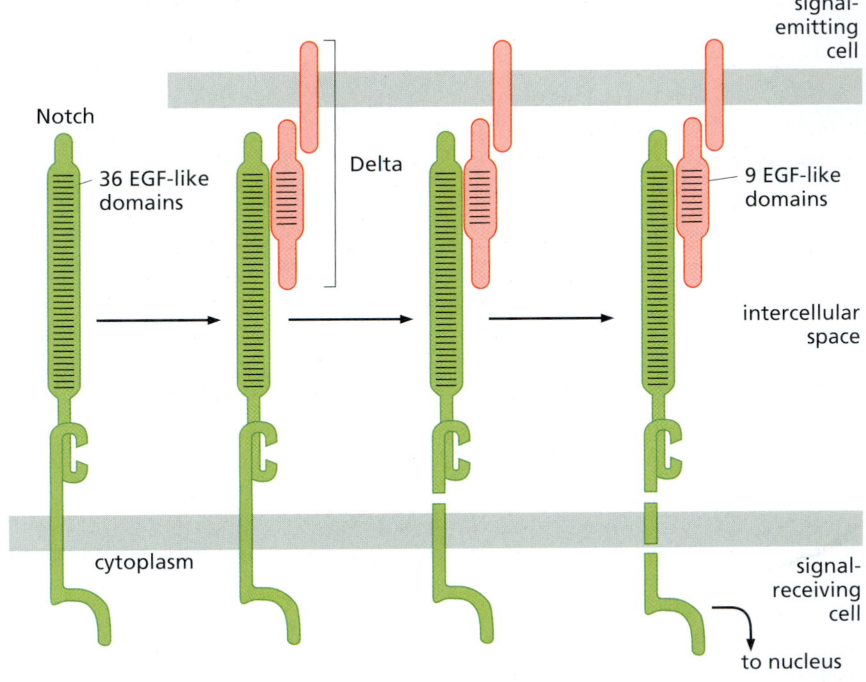

Figure 5.22 Structure of the Notch receptor The Notch receptor *(green)* appears to embody a very primitive type of signaling. After it has bound a ligand (e.g., NotchL or, as shown here, Delta, *pink*), which is being displayed on the surface of an adjacent cell, Notch undergoes two successive proteolytic cleavage events, and the resulting C-terminal cytoplasmic fragment is thereby freed, allowing it to migrate to the cell nucleus, where it alters the expression of certain genes.

functions as part of a complex of transcription factors that activate expression of a cohort of responder genes. The Notch signaling pathway appears to contribute to Ras-mediated cell transformation and is involved in a number of morphogenetic processes. Mutant, constitutively active forms of Notch, which fire in a ligand-independent fashion, have been found in half of adult T-cell leukemias.

Yet another receptor, termed Patched, is constructed from multiple transmembrane domains that weave their way back and forth through the plasma membrane (Figure 5.23). When the ligands of Patched—proteins of the Hedgehog (Hh) class—bind to Patched (Ptc), the latter moves away from a second membrane-

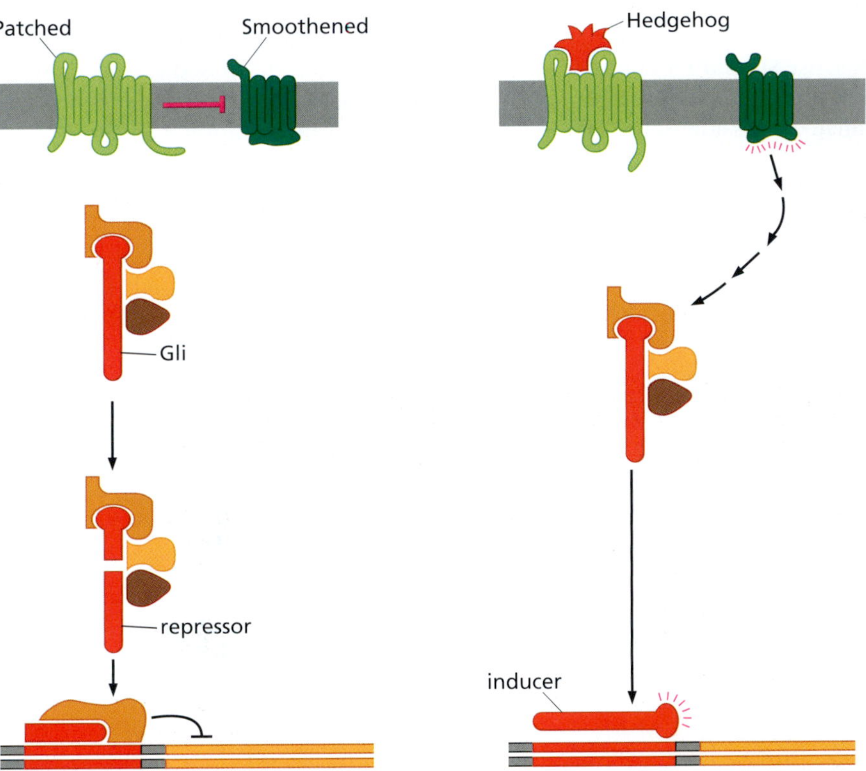

Figure 5.23 The Patched–Smoothened signaling system The Hedgehog pathway utilizes its own, unique signaling system. Normally, Smoothened, a seven-membrane-spanning surface receptor *(dark green)*, is functionally inert, being inhibited by Patched *(light green)*, which contains 12 membrane-spanning domains *(top left)*. Under these conditions, Gli *(red)* is cleaved into a protein that moves into the nucleus, where it operates as a repressor of transcription *(bottom left)*. However, when the Hedgehog ligand *(red, top right)* binds to Patched, the latter releases its inhibitory grip on Smoothened. Smoothened then proceeds, in some unknown fashion, to prevent cleavage of Gli, enabling Gli to move into the nucleus, where it can act as an inducer of transcription *(bottom right)*.

Sidebar 5.9 How do growth factors travel through intercellular space? The discussions of growth factors in this chapter imply that these proteins, once secreted by cells, diffuse freely through the intercellular fluid until they reach their intended targets, that is, other cells displaying cognate cell-surface receptors. In truth, many growth factor molecules, perhaps the great majority of them, become attached to components of the extracellular matrix (ECM) soon after their release from the cells that have synthesized them. The ECM is a complex meshwork of secreted glycoproteins and **proteoglycans** that fills most of the intercellular space and provides much of the structural rigidity to tissues; we will read about the ECM in greater detail shortly (Section 5.8).

Many types of secreted growth factors seem to decorate the ECM much as ornaments are used to decorate a Christmas tree. Such growth factors may remain tethered in inert form to the ECM for extended periods of time, and may then be released from the ECM and converted into soluble, activated factors. The cells that release these tethered growth factors from the ECM behave much like animals that browse on shrubs in search of food. These browsing cells usually utilize secreted proteases to mobilize the tethered growth factors. These cells may also display the receptors that enable them to respond to the growth factors that they have mobilized.

spanning protein called Smoothened (Smo). Smoothened then signals to a cytoplasmic complex that releases a transcription factor, and the latter translocates to the nucleus, where it activates the target genes of Hedgehog. Mutant alleles of both *Ptc* and *Smo* have been found in the common basal cell carcinomas of the skin, and mutant alleles of *Ptc* have been found in many medulloblastomas; we return to this class of tumors in Chapter 16.

The Wnt factors represent yet another, independent signaling system. As described earlier, their discovery is traced back to a *Drosophila* mutant termed ***Wingless*** and a related gene termed ***int-1*** in mice; the latter was discovered because of insertional mutagenesis by mouse mammary tumor virus (see Sidebar 3.8). These Wnt proteins are tethered tightly to the extracellular matrix (ECM) and, through a lipid tail, to cell membranes, and thus do not appear to be freely diffusible like several of the growth factors that we discussed earlier (see Sidebar 5.9). Growth factors of the Wnt class activate receptors of the Frizzled (Frz) family, which, like Patched, are complex receptors that weave back and forth through the plasma membrane multiple times (Figure 5.24). Without Wnt signaling, the cytoplasmic enzyme glycogen synthase kinase-3β (GSK-3β) phosphorylates several key proliferation-promoting proteins, including β-catenin,

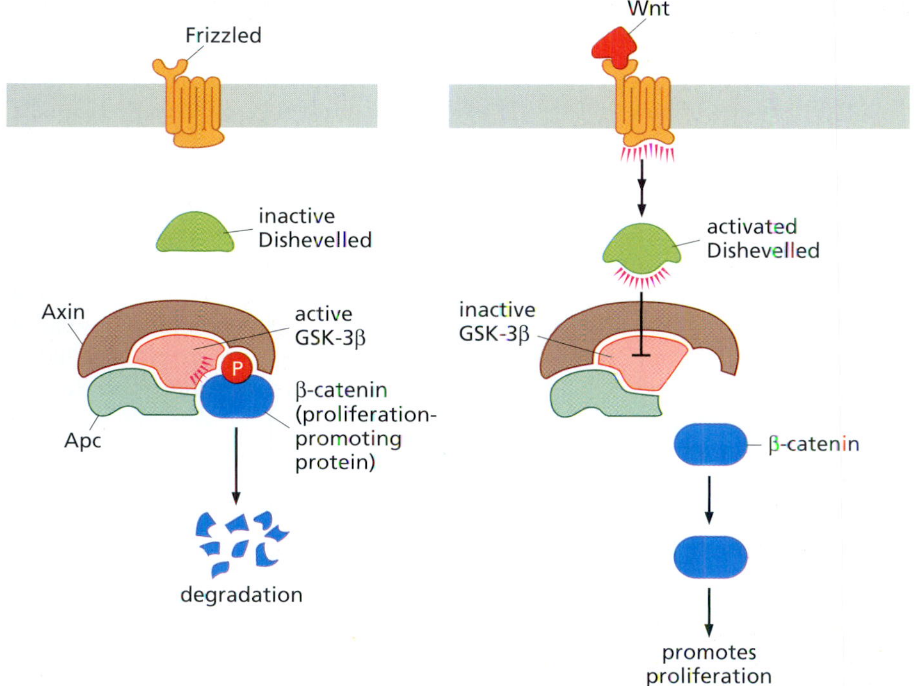

Figure 5.24 Signaling by the Frizzled receptors Receptors of the Wnt proteins are members of the Frizzled (Frz) family of transmembrane proteins. In the absence of ligand binding *(left)*, a complex of Axin *(brown)* and Apc *(light green)* allows glycogen synthase kinase-3β (GSK-3β, *pink*) to phosphorylate β-catenin *(blue)*. This marks β-catenin for rapid destruction by *proteolysis*. However, when a Wnt ligand binds to a Frizzled receptor *(right)*, the resulting activated Frizzled receptor, acting via the Dishevelled protein *(light green)*, causes inhibition of GSK-3β. This spares β-catenin, which may accumulate and promote cell proliferation.

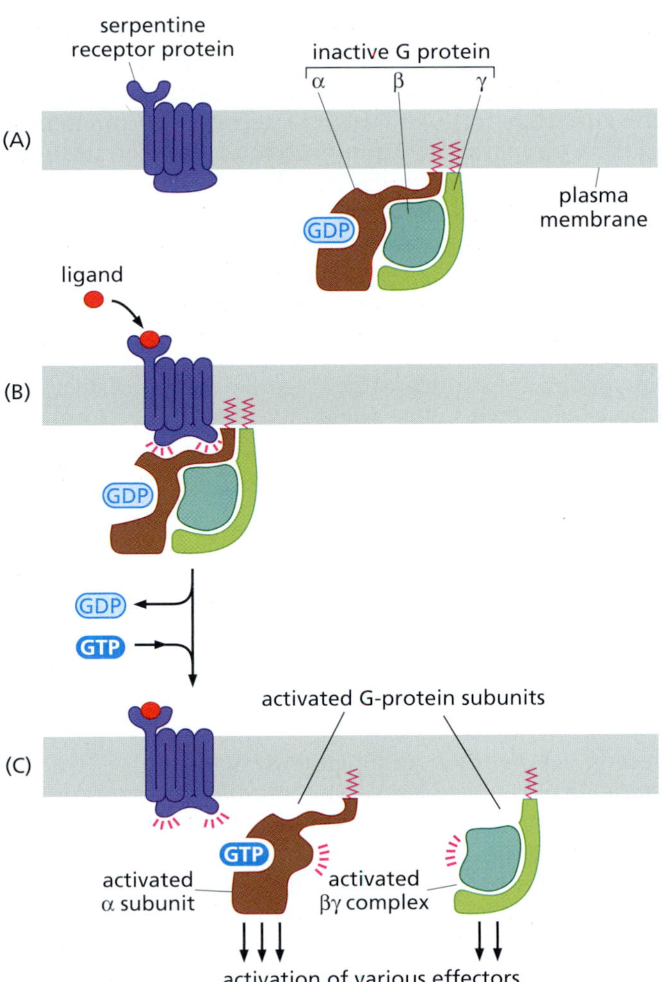

Figure 5.25 Signaling by serpentine receptors (A) The seven-membrane-spanning (serpentine) receptors *(left)* are associated with heterotrimeric G proteins, which are composed of α, β, and γ subunits *(right)*. (B) Once a receptor has bound its cognate ligand *(red)*, it stimulates the α subunit of the G protein to release its GDP (guanosine diphosphate) and bind GTP (guanosine triphosphate) instead *(light blue)*. (C) As part of this response, the α subunit dissociates from the β + γ pair. Now, the α subunit *(brown)* and, independently, the β + γ complex *(blue, light green)*, can each proceed to regulate enzymes that evoke a variety of downstream responses. This signaling is halted when the α subunit hydrolyzes its bound GTP, causing it to reassociate with the β + γ complex *(not shown)*.

tagging them for destruction. However, the binding of Wnts to Frizzled receptors triggers a cascade of steps that shut down GSK-3β firing, allowing these proteins to escape degradation and promote cell proliferation. This pathway also plays a critical role in cancer pathogenesis, and we will discuss it in great detail in the several chapters that follow.

A recent survey of the human genome sequence indicates that it encodes, in addition to the tyrosine kinase receptors, eleven TGF-β-like receptors, four Notch receptors, one or two Patched receptors, and ten Frizzled receptors. Far more numerous than these are a class of receptors that, like the Patched and Frizzled proteins, weave back and forth through the plasma membrane. Like the Frizzled proteins, this class of receptors, sometimes called "**serpentine**" receptors because of their imagined snakelike configuration, traverse the plasma membrane precisely seven times (Figure 5.25). Following ligand binding, they signal through their ability to activate G proteins (shorthand for guanine nucleotide-binding proteins) in the cytoplasm, and are therefore often called G-protein–coupled receptors (GPCRs). There are more than a thousand distinct types of GPCRs encoded in the mammalian genome. These versatile receptors are involved in functions as diverse as olfaction in the nose, taste in the mouth, detection of photons in the retina, and detection of neurotransmitters in the brain. Together, the genes encoding these receptors represent almost 5% of the total roster of genes present in the human genome.

The G proteins stimulated by GPCRs act as binary switches, flipping back and forth between GDP- and GTP-bound states. These G proteins are structurally complex; they are heterotrimers composed of α, β, and γ subunits. As indicated in Figure 5.25, once a GPCR binds its ligand, a poorly understood shift in the configuration of the receptor allows it to trigger the release of GDP by the α subunit of the G protein and the binding of GTP instead. The α subunit then dissociates from its β + γ partners and proceeds to interact with other **effector** proteins that transmit signals further into the cell interior; the β + γ complex may also interact with its own downstream effectors. Eventually, the α subunit turns itself off by hydrolyzing its bound GTP (converting it to GDP) and reassociates with its β + γ partners. These heterotrimeric complexes then retreat into their inactive, nonsignaling state, awaiting signals once again from a ligand-activated GPCR. While vast in number, these receptors have to date been found to contribute to the pathogenesis of only a relatively small number of human cancers (to be described in Chapter 6).

5.8 Integrin receptors sense association between the cell and the extracellular matrix

The biology of normal and transformed cells provides hints of yet another type of cell surface receptor, one dedicated to sensing a quite different class of molecules in the extracellular space. Recall that an important attribute of transformed cells is their ability to grow in an anchorage-independent fashion, i.e., to proliferate without attachment to a solid substrate such as the bottom of a Petri dish (see Section 3.5). This behavior contrasts with that of normal cells, which require attachment in order to proliferate. Indeed, in the absence of such attachment, many types of normal cells will activate a version of their death program (**apoptosis**) that is often termed **anoikis**. These cell death processes will be explored in detail in Chapter 9.

Biochemical analyses of the solid substrate to which cells adhere at the bottom of Petri dishes have revealed that, in very large part, cells are not anchored directly to the glass or plastic surface of these dishes. Instead, they attach to a complex network of molecules that closely resembles the extracellular matrix (ECM) usually found in the spaces between cells within most tissues. The ECM is composed of a series of glycoproteins, including collagens, laminins, proteoglycans, and fibronectin (Figure 5.26). After cells are introduced into a Petri dish, they secrete ECM components that adhere to the glass or plastic; once an ECM is constructed in this way, the cells attach themselves to this matrix. Consequently, anchorage dependence really reflects the need of normal cells to be tethered to ECM components in order to survive and proliferate.

Figure 5.26 Extracellular matrix This scanning electron micrograph illustrates fibroblasts embedded in the specialized extracellular matrix (ECM) present in the cornea (of a rat). Many of the components of this ECM, including glycoproteins, hyaluronan, and proteoglycans, have been removed in order to highlight the major component of this ECM—collagen fibers. (From T. Nishida et al., *Invest. Ophthalmol. Vis. Sci.* 29:1887–1890, 1988.)

The trait of anchorage dependence make it obvious that cells are able to sense whether or not they have successfully attached to the ECM. As was learned in the mid-1980s, such sensing depends on specialized receptors that inform cells about the extent of tethering to the ECM and about the identities of specific molecular components of the ECM (e.g., collagens, laminins, fibronectin) to which tethering has occurred. Much, if not all, of this sensing is accomplished by a specialized class of cell surface receptors termed **integrins**. In effect, the molecular components of the ECM serve as the ligands of the integrin receptors. At the same time, the integrins create mechanical stability in tissues by tethering cells to the scaffolding formed by the ECM.

Integrins constitute a large family of heterodimeric transmembrane cell surface receptors composed of α and β subunits. At least eighteen α and eight β subunits have been enumerated; together, 24 distinct heterodimers are known. The ectodomains of these receptors bind specific ECM components (Figure 5.27). Table 5.4 indicates that each integrin heterodimer shows specificity for binding a specific ECM molecule or a small subset of ECM components. The much-studied α5/β1 integrin, for example, is the main receptor for fibronectin, an important glycoprotein component of the ECM found in vertebrate tissues. Laminins,

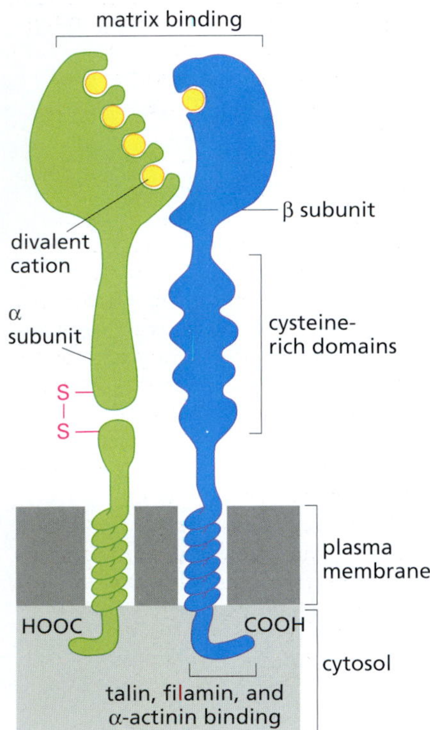

Figure 5.27 Structure of integrins Integrin molecules function as heterodimeric cell surface receptors, each composed of an α plus a β subunit *(green, blue)*. The ectodomains of these receptors bind specific components of the extracellular matrix (ECM). At the same time, the cytoplasmic domain (largely that of the β subunit) is linked, via intermediary proteins, to the cytoskeleton (largely that constructed by actin fibers); in addition, the cytoplasmic domains may attract a variety of signal-transducing proteins that become activated when the ectodomain binds an ECM ligand.

which are large, multidomain ECM molecules, have been reported to be bound by as many as 12 distinct integrin heterodimers.

Having bound their ECM ligands through their ectodomains, integrins cluster to form *focal adhesions* (Figure 5.28A). This clustering affects the organization of the cytoskeleton underlying the plasma membrane, since some integrins are linked directly or indirectly via their cytoplasmic domains to important components of the cytoskeleton, such as actin, vinculin, talin, and paxillin (Figure 5.28B). The formation of focal adhesions may also cause the cytoplasmic domains of integrins to activate signaling pathways that evoke a variety of cellular responses, including cell migration, proliferation, and survival. For example, by triggering the release of anti-apoptotic signals, integrins reduce the likelihood of anoikis. These integrin functions continue to be critical during the development of tumors (Sidebar 5.10).

Integrins are most unusual in one other respect. Normally we think of receptors as passing information from outside the cell into the cytoplasm. Integrins surely do this. But in addition, it is clear that signals originating in the cytoplasm are used to control the binding affinities of integrins for their ECM ligands. Such "inside-out" signaling enables cells to modulate their associations with various types of ECM or with various points of contact with an ECM, breaking existing contacts and forging new ones in their place. Rapid modulation of extracellular contacts enables cells to free themselves from one microenvironment within a tissue and move into another, and to traverse a sheet of ECM *in vitro*. Cultured fibroblasts lacking focal adhesion kinase (FAK), one of the signaling molecules that associate with the cytoplasmic domains of integrins, are unable to remodel their focal contacts and lack motility, indicating that the signals transduced throughout the cytoplasm by FAK are important for reconfiguring the cytoskeleton—the structure that enables cells to change shape and to move. Later, in Chapter 14, we will see how cell motility is critical to the ability of cancer cells to invade and metastasize.

These various descriptions of tyrosine kinase receptors and integrins reveal that mammalian cells use specialized cell surface receptors to sense two very distinct

Table 5.4 Examples of integrins and their extracellular matrix ligands

Integrin	ECM ligand
α1β1	collagens, laminin
α1β1	vitronectin, fibronectin
αvβ3	vitronectin, fibrinogen, thrombospondin
α5β1	fibronectin
α6β1	laminin
α7β1	laminin
α2β3	fibrinogen
α6β4	laminin–epithelial hemidesmosomes

Adapted in part from B. Alberts et al., Molecular Biology of the Cell, 4th ed. New York: Garland Science, 2002; and from H. Lodish et al., Molecular Cell Biology. New York: W.H. Freeman, 1995.

(A)

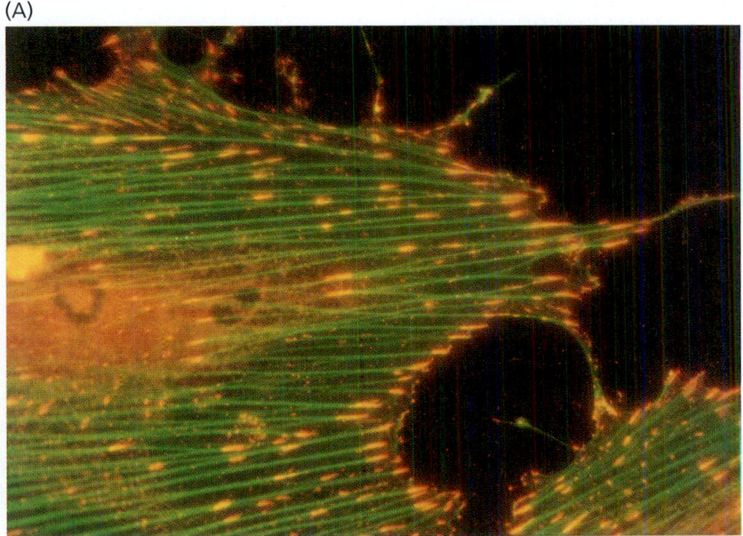

(B)

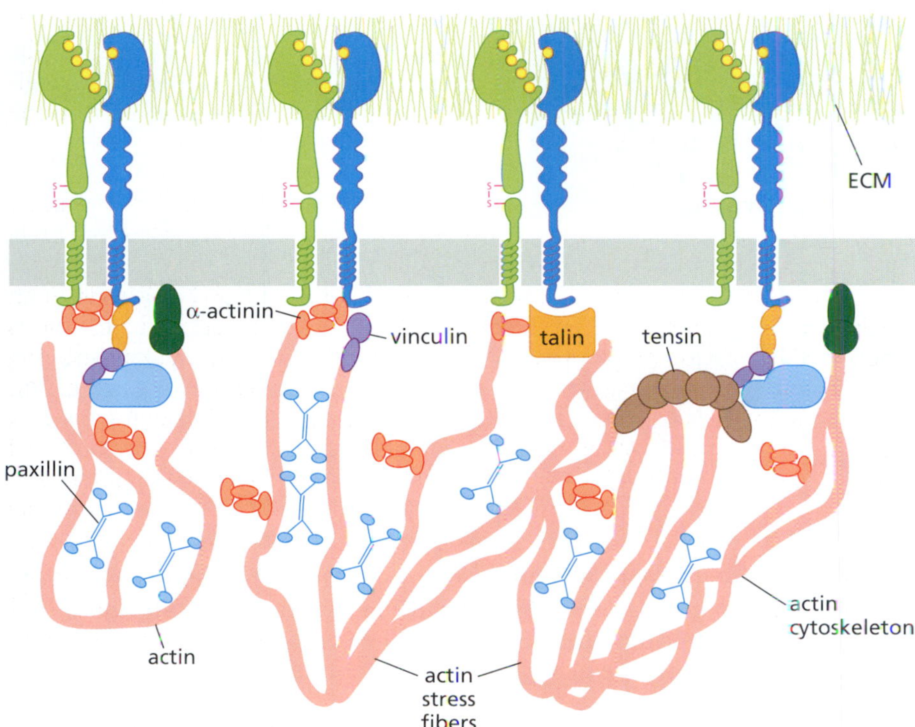

Figure 5.28 Integrin tethering to the ECM and cytoskeleton (A) This fluorescence micrograph illustrates the discrete foci *(yellow/orange)* on the cell surface, termed focal adhesions, at which cells use integrins to tether themselves to the extracellular matrix (ECM). This clustering also affects the organization of the actin fibers *(green)* forming the cytoskeleton. (B) This schematic figure of the organization of integrins *(green, blue)* indicates their association with the ECM *(green fibers, above)* through their ectodomains and their association with the actin *cytoskeleton (pink chains)* through their cytoplasmic domains via the β subunit of each heterodimer. A series of intermediary proteins, such as actinin, vinculin, and talin, allows these linkages to be formed. (A, courtesy of Keith Burridge; B, adapted from C. Miranti and J. Brugge, *Nat. Cell Biol.* 4:E83–E90, 2002.)

types of extracellular molecules. Some receptors, such as the EGF and PDGF receptors, sense soluble (or solubilized) growth factors, while others, notably the integrins, sense attachment to the essentially insoluble scaffolding of the ECM. Together, these receptors enable a normal cell to determine whether two preconditions have been satisfied before this cell undertakes growth and division: the cell must sense the presence of adequate levels of mitogenic growth factors in its surroundings and the existence of adequate anchoring to the ECM.

Both of these requirements for cell proliferation are known to be abrogated if a cell carries an activated *ras* oncogene. Thus, *ras*-transformed cells can grow in the presence of relatively low concentrations of serum and serum-associated mitogenic growth factors; in addition, many types of *ras*-transformed cells can proliferate in an anchorage-independent fashion. These behaviors suggest that, in some way, the Ras oncoprotein is able to mimic the signals introduced into the cell by ligand-activated growth factor receptors and by integrins that have

Sidebar 5.10 Some integrins are essential for tumorigenesis The ability to selectively inactivate ("knock out") genes within targeted tissues has made it possible to evaluate the contributions of various signaling proteins to tumorigenesis. For example, inactivation of the gene encoding the β1 integrin in the mouse germ line leads to embryonic lethality; however, introduction of a "conditional" allele into the mouse germ line has enabled the inactivation of this gene specifically in the epithelial cells of the mouse mammary gland (Figure 5.29A, B). Such inactivation of both copies of the β1 integrin gene has minimal effects on mammary gland development. But if a transgene specifying the middle T oncogene of polyomavirus is expressed in the breast tissue of these mice, the normally observed premalignant hyperplastic nodules are reduced by more than 75% and the β1 integrin–negative cells are fully unable to develop further to form mammary carcinomas (Figure 5.29C). Detailed analyses indicate that this loss of β1 integrin expression permits the oncogene-expressing mammary epithelial cells to survive but precludes their active proliferation. This suggests, in a speculative vein, that interfering with β1 integrin expression or function may be useful in blocking the development of certain types of human tumors.

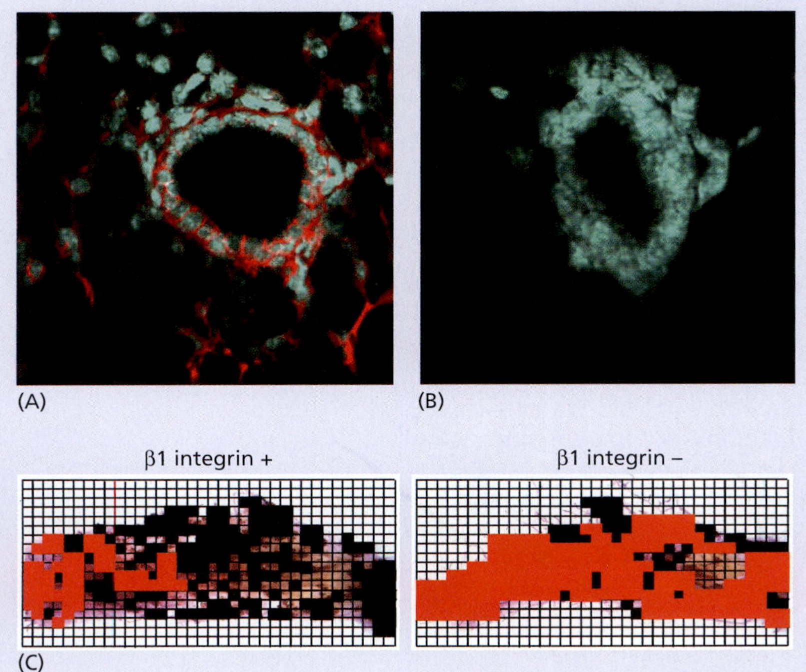

Figure 5.29 β1 integrin and mammary tumorigenesis The selective inactivation of the gene encoding β1 integrin in the mouse mammary gland allows a test of its importance in tumorigenesis. (A) The immunostaining of β1 integrin *(red)* amid the cells forming a mouse mammary duct *(blue)*. (B) In the absence of β1 integrin, ductal morphogenesis still proceeds normally. Moreover, the overall development of the mammary gland is normal *(not shown)*. (C) In a transgenic mouse prone to mammary tumorigenesis *(left)*, premalignant hyperplastic nodules, scored by microscopic examination of whole mammary glands, are present in abundance *(black squares)* amid the normal epithelium of these glands *(red squares)*. However, the great majority of these nodules fail to form in mammary glands in which most epithelial cells have been deprived of β1 integrin expression *(right)*. (From D.E. White et al., *Cancer Cell* 6:159–170, 2004.)

engaged ECM components. An understanding of the biochemical basis of these various signals demands some insight into the structure and function of the normal and oncogenic Ras proteins. So, we will move back in history to the early 1980s to see how the puzzle of Ras function was solved.

5.9 The Ras protein, an apparent component of the downstream signaling cascade, functions as a G protein

The discoveries that two oncogenes—*erbB* and *sis*—encoded components of the mitogenic growth factor signaling machinery (Sections 5.4 and 5.5) sparked an intensive effort to relate other known oncoproteins, notably the Ras oncoprotein, to this signaling machinery. The *ras* oncogene clearly triggered many of the same changes in cells that were seen when cells were transformed by either *src*, *erbB*, or *sis*. Was there some type of signaling cascade—a molecular bucket brigade—operating in the cell, in which protein A transferred a signal to protein B, which in turn signaled protein C? And if such a cascade did exist, could the Ras oncoprotein be found somewhere downstream of *erbB* and *sis*? Did the signals emitted by these various proteins all converge on some common target at the bottom of this hypothesized signaling cascade?

At the biochemical level, it was clear that growth factor ligands activated tyrosine kinase receptors, and that these receptors responded by activating their cytoplasmic tyrosine kinase domains. But which proteins were affected thereafter by the resulting receptor phosphorylation events? And how did this phosphorylation lead further to a mitogenic response by the cell—its entrance into a phase of active growth and division? In the early 1980s, progress on these key issues ground to a halt because biochemical experiments offered no obvious way to move beyond the tyrosine kinase receptors to the hypothesized downstream signaling cascades.

All the while, substantial progress was being made in understanding the biochemistry of the Ras protein. In fact, the three distinct *ras* genes in mammalian cells (Table 4.2) encode four distinct Ras proteins (since K-*ras* specifies a second protein via alternative splicing of an mRNA precursor). Because all four Ras proteins have almost identical structures and function similarly, we refer to them simply as "Ras" in the discussions that follow. At their C-termini, they all carry covalently attached lipid tails, composed of farnesyl, palmitoyl, or geranylgeranyl groups (or combinations of several of these groups). These lipid moieties enable the Ras proteins, all of about 21 kD mass, to become anchored to cytoplasmic membranes, largely to the cytoplasmic face of the plasma membrane.

(For many years, it appeared that this membrane anchoring is achieved through the insertion of the various lipid tails directly into the hydrophobic environments within various lipid bilayer membranes. However, more recent studies indicate greater complexity. In the case of H-Ras, for example, its palmitate moiety is indeed inserted directly into cytoplasmic membranes, while its farnesyl group is inserted into a hydrophobic pocket of a specialized farnesyl-binding protein; the latter, in turn, facilitates the interactions of H-Ras with other partner proteins and also strengthens the binding of H-Ras to various membranes.)

Like the heterotrimeric G proteins (Section 5.7), Ras was found to bind and hydrolyze (i.e., cleave) guanosine nucleotides. This action as a GTPase indicated a mode of action quite different from that of the Src and erbB tyrosine kinases, yet intriguingly, all three oncoproteins had very similar effects on cell behavior.

In further analogy to the G proteins, Ras was found (1) to bind a GDP molecule when in its quiescent, inactive state; (2) to jettison its bound GDP after receiving some stimulatory signal from upstream in a signaling cascade; (3) to acquire a GTP molecule in place of the recently evicted GDP; (4) to shift into an activated, signal-emitting configuration while binding this GTP; and (5) to cleave this GTP after a short period by using its own intrinsic GTPase function, thereby placing itself, once again, in its non-signal-emitting configuration (Figure 5.30). In

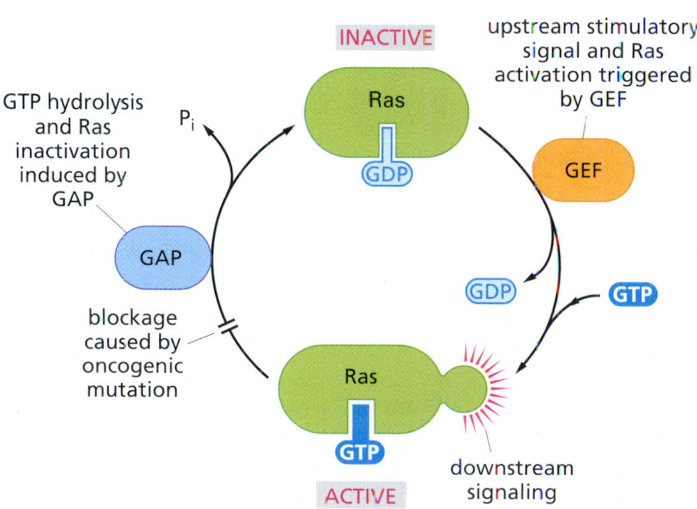

Figure 5.30 The Ras signaling cycle Detailed study of the biochemistry of Ras proteins revealed that they, like the heterotrimeric G proteins (see Figure 5.25), operate as binary switches, binding GDP in their inactive state *(top)* and GTP in their active, signal-emitting state *(bottom)*. Thus, an inactive, GDP-binding Ras protein is stimulated by a GEF (guanine nucleotide exchange factor, *orange*) to release its GDP and acquire a GTP in its stead, placing Ras in its active, signaling configuration. This period of signaling is halted after a short period by the actions of a GTPase activity intrinsic to Ras, which hydrolyzes GTP to GDP *(left)*. This GTPase activity is strongly stimulated by GAPs (GTPase-activating proteins) that Ras might encounter while in its activated state. Amino acid substitutions caused by oncogenic point mutations block this cycle by inactivating the intrinsic GTPase activity of Ras; this traps Ras in its activated, signal-emitting state.

effect, the Ras molecule seemed to behave like a light switch that automatically turns itself off after a certain predetermined time.

In more detail, the hypothesized chain of events went like this. Mitogenic signals, perhaps transduced in some way by tyrosine kinase receptors, activated a guanine nucleotide exchange factor (GEF) for Ras. This GEF induced an inactive Ras protein to shed its GDP and bind GTP instead. The resulting activated Ras would emit signals to a (still-unknown) downstream target or group of targets. This period of signaling would be terminated, sooner or later, when Ras decided to hydrolyze its bound GTP. In doing so, Ras would revert to its nonsignaling state. As was learned later, a group of other proteins known as GAPs (GTPase-activating proteins) could actively intervene and encourage Ras to undertake this hydrolysis (Sidebar 5.11). Implied in this model (see Figure 5.30) is the notion that Ras acts as a signal-transducing protein, in that it receives signals from upstream in a signaling cascade and subsequently passes these signals on to a downstream target.

Sidebar 5.11 GTPase-activating proteins and cancer When purified Ras proteins are examined in isolation, they show a very low level of GTPase activity. However, within living cells, it is clear that the GTP-bound form of a Ras can interact with a second class of proteins termed GAPs (GTPase-activating proteins). This interaction causes an increase of as much as 1000-fold in the GTPase activity of the Ras protein and thus results in the rapid conversion of active, GTP-bound Ras into its inactive, GDP-bound form. (The Ras-GAP protein, one of two known GAPs that act on Ras, has been found to insert an arginine "finger" into the GTPase cleft of Ras; in doing so, it actively participates in the catalysis that converts GTP to GDP.)

The lifetime of the activated, GTP-bound form of Ras may therefore be governed by the time required for its encounter with a GAP molecule. Since the GAPs themselves may be under the control of yet other signals, this means that the GTPase activity of the Ras proteins is modulated indirectly by other signaling pathways. Moreover, as described later (Section 7.10), the loss from a cell of the other Ras GAP, termed NF1, leaves the GTPase activity of Ras at an abnormally low level in this cell, resulting in the accumulation of GTP-loaded Ras, the hyperactivity of Ras signaling, and ultimately the formation of a neuroectodermal tumor.

A striking discovery was made in the course of detailed biochemical analysis of a Ras oncoprotein made by a point-mutated *ras* oncogene, specifically, the Ras oncoprotein encoded by Harvey sarcoma virus. Like the normal Ras protein, the Ras oncoprotein could bind GTP. However, the oncoprotein was found to have lost virtually all GTPase activity (see Figure 5.30). In such condition, it could be pushed into its active, signal-emitting configuration by some upstream stimulatory signal and guanine nucleotide exchange factors (GEFs). But once in this activated state, the Ras oncoprotein was unable to turn itself off! The negative-feedback loop that is so essential to the operations of the normal Ras protein had been sabotaged by this point mutation and resulting amino acid substitution (Figure 5.31).

This provided a clear explanation of how Ras can operate as an oncoprotein: rather than sending out short, carefully rationed pulses of growth-stimulating signals, the oncoprotein emits these signals for a long, possibly indefinite period of time, thereby flooding the cell with these signals, whatever their nature. Subsequent analysis of Ras protein biochemistry has revealed additional levels of control that may go awry in cancer cells (see Sidebar 5.11).

These findings provided a solution to a puzzle created by the sequencing of the mutant *ras* oncogenes present in many human tumors. As we learned (Section 4.4), invariably the point mutations creating these oncoproteins are **missense** mutations (which cause an amino acid substitution) rather than **nonsense** mutations (which cause premature termination of the growing protein chain). And invariably these mutations strike either the 12th or the 61st codon of the reading frame of the *ras* genes (or, on rare occasion, the 13th codon).

We might imagine that these particular nucleotides represent sites in the genomic DNA that are particularly susceptible to attack and alteration by mutagenic carcinogens. However, detailed study of the structure of the Ras proteins, as generated by X-ray crystallography (see Figure 5.31), has led to a far more compelling model. Examination of the 12th, 13th, and 61st amino acid residues of these proteins reveals that these residues are located around the cavity in the Ras protein where the GTPase catalytic activity operates. Consequently, almost all substitutions of these three amino acids, such as the glycine-to-valine substitution that we encountered earlier, compromise GTPase function. [To be more precise, they compromise the ability of GTPase-activating proteins (GAPs; see Figure 5.30) to trigger hydrolysis of the GTP bound by Ras.]

Knowing this, we can immediately understand why point mutations affecting only a limited number of amino acid residues are found in the *ras* oncogenes carried by human tumor cell genomes. Large-scale alterations of the *ras* proto-oncogenes, such as deletions, are clearly not productive for cancer, since they

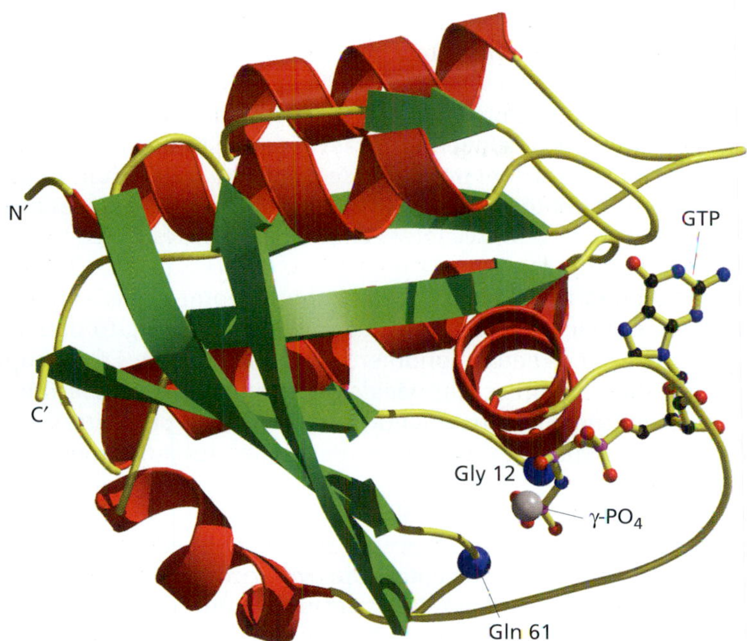

Figure 5.31 The structure of the Ras protein This diagram of the structure of a Ras protein, as determined by X-ray crystallography, depicts the arrangement of the polypeptide backbone of Ras and its α-helical *(red)* and β-pleated sheet *(green)* domains. GTP is indicated as a stick figure, and the two most frequently altered amino acid residues found in human tumor oncoproteins—glycine 12 and glutamine 61—are shown as *blue balls*. As is apparent, both of these residues are closely associated with the γ-phosphate of GTP *(gray ball)*, helping to explain why substitutions of these residues affect the GTPase activity of Ras, and therefore why the codons specifying these residues are preferentially mutated in human tumor cell genomes. (Courtesy of A. Wittinghofer.)

would result in the elimination of Ras protein function rather than an enhancement of it. Similarly, the vast majority of point mutations striking *ras* proto-oncogenes will yield mutant Ras proteins that have lost rather than gained the ability to emit growth-stimulatory signals. Only when the signal-emitting powers of Ras are left intact and its GTPase negative-feedback mechanism is inactivated selectively (by amino acid substitutions at one of these three sites in the protein) does the Ras protein gain *increased* power to drive cell proliferation and transform the cell.

Only those rare cells that happen to have acquired mutations affecting one of the three amino acid residues of Ras (residues 12, 61, or, rarely, 13) gain proliferative advantage over their wild-type counterparts and thereby stand the chance of becoming the ancestors of the cells in a tumor mass. So, even though point mutations affecting all the amino acid residues of a Ras protein are likely to occur with comparable frequencies, only those few conferring substantial growth advantage will actually be found in tumor cells. Other cells that happen to acquire point mutations affecting other residues in the Ras protein will retain a normal phenotype or may even lose proliferative ability.

5.10 Synopsis and prospects

In the early 1980s, the discovery of oncogenes such as *erbB* and *sis* revealed the intimate connections between growth factor signaling and the mechanisms of cell transformation. More specifically, these connections suggested that cell transformation results from hyperactivation of the mitogenic signaling pathways. This notion galvanized researchers interested in the biochemical mechanisms of carcinogenesis. Many began to study the growth factor receptors and their biochemical mechanisms of action.

A major problem surrounding growth factor receptors was the mechanism used by them to transfer signals from outside a cell into its interior. A simple model was soon developed to explain how this occurs: The growth factor ligands of receptors bind to their ectodomains. By doing so, they encourage receptor molecules to come together to form dimers. The dimerization of the receptor ectodomains that is initially achieved by the growth factors then encourages dimerization of the cytoplasmic domains of these receptors. Somehow, this enables the cytoplasmic domains to emit signals.

This model shed no light on how, at the biochemical level, receptors are able to emit signals. Here, research on the biochemistry of the Src protein proved to be telling. It was found to be a tyrosine kinase (TK), an enzyme that transfers phosphate groups to the side chains of tyrosine residues of various substrate proteins. Once the amino acid sequences of several growth factor receptors were analyzed, the cytoplasmic domains of these receptors were found to be structurally (and thus functionally) related to Src. Hence, tyrosine kinases create the means by which growth factor receptors emit biochemical signals.

This information was then integrated into the receptor dimerization model: when growth factor receptors are induced by ligand binding to dimerize, the tyrosine kinase of each receptor monomer transphosphorylates the cytoplasmic domain of the other monomer. This yields phosphorylated cytoplasmic tails of receptor molecules that, in some unknown fashion, allow signaling to proceed. Much of the next chapter is focused on these downstream signaling mechanisms.

Hyperactive signaling by these receptors is encountered in many types of human cancer cells. Often, the receptors are overexpressed, resulting in ligand-independent firing. Even more potent ligand-independent firing is achieved by various types of structural alterations of receptors. Most of these occur as the consequences of somatic mutations, but a number of mutant alleles of receptor genes have been found in the human gene pool; they are associated with a variety of inborn cancer susceptibility syndromes.

These tyrosine kinase growth factor receptors (RTKs) provide only one of many ways by which cells sense their surroundings. The TGF-β receptors, for example, are superficially similar to the RTKs, in that they have a ligand-binding ectodomain and a signal-emitting kinase domain in their cytoplasmic portion. However, the kinases of the TGF-β receptors are serine/threonine kinases and, as we will learn later, signal through much different mechanisms.

A disparate array of other signal-transducing receptors have been uncovered, including those of the Notch, Patch–Smoothened, and Frizzled classes. These receptors use a diversity of signal-transducing mechanisms to release signals into the cytoplasm. As we will see throughout this book, aberrant signaling by these receptors plays a key role in the pathogenesis of many types of human cancers.

Cells must also sense their contacts with the extracellular matrix (ECM)—the cage of secreted glycoproteins and proteoglycans that surrounds each cell within a tissue. Here, yet another class of receptors, the integrins, play a central role. Having bound components of the extracellular matrix, the heterodimeric integrin receptors transduce signals into cells that stimulate proliferation and suppress cells' apoptotic suicide program.

A distinct line of research, pursued in parallel, elucidated the biochemistry of the Ras protein, another key player in cancer pathogenesis. Ras operates like a binary switch, continually flipping between active, signal-emitting and quiescent states. The mutant alleles found in cancer cells cause amino acid substitutions in the GTPase pockets of Ras proteins, thereby disabling the mechanism that these proteins use to shut themselves off. This traps the Ras proteins in their active, signal-emitting configuration for extended periods of time, causing cells to be flooded with an unrelenting stream of mitogenic signals.

While these investigations revealed how isolated components of the cellular signaling machinery (e.g., RTKs, Ras) function, they provided no insights into how these proteins communicate with one another. That is, the *organization* of the signaling circuitry remained a mystery. Cell physiology hinted strongly that ErbB (i.e., the EGF receptor) and other tyrosine kinase receptors operated in a common signaling pathway with the Ras oncoprotein. Src fit in somewhere as well. These various oncoproteins, while having diverse biochemical functions, were found to exert very similar effects on cells. They all caused the rounding up of cells in monolayer culture, the loss of contact inhibition, the acquisition of anchorage independence, and reduction of a cell's requirement for mitogenic growth factors in its culture medium. This commonality of function suggested participation in a common signaling cascade.

One possible connection between Ras and the tyrosine kinase receptors came from the discovery that an activated *ras* oncogene causes many types of cells to produce and release growth factors. Prominent among these was transforming growth factor-α (TGF-α), an EGF-like growth factor that was found to be released by a variety of oncogene-transformed cells. Like EGF, TGF-α binds to and activates the EGF receptor (see Figure 5.16C). This suggested the following scenario: Once released by a *ras*-transformed cell, TGF-α might function in an autocrine manner (see Figure 5.12B) to activate EGF receptors displayed on the surface of that cell. This, in turn, would evoke a series of responses quite similar to those created by a mutant, constitutively activated EGF receptor. When viewed from this perspective, Ras appeared to operate *upstream* of a growth factor receptor rather than being an important component of the signaling cascade lying *downstream* of the receptor (Figure 5.32A).

In the end, this autocrine scheme was able to explain only a small part of *ras* function, since *ras* was found to be able to transform cells that lacked receptors for the growth factors that it induced. In addition, ample biochemical evidence accumulated that ligand binding of growth factor receptors led rapidly to acti-

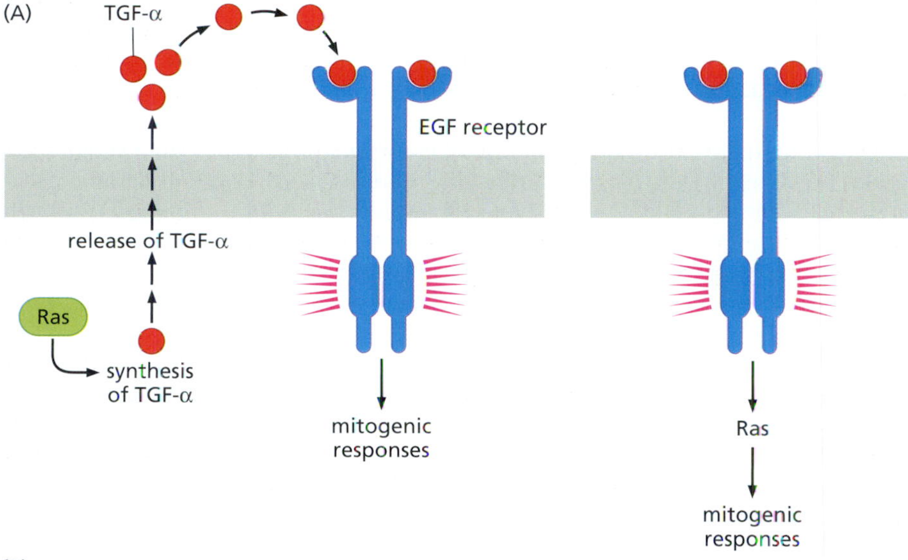

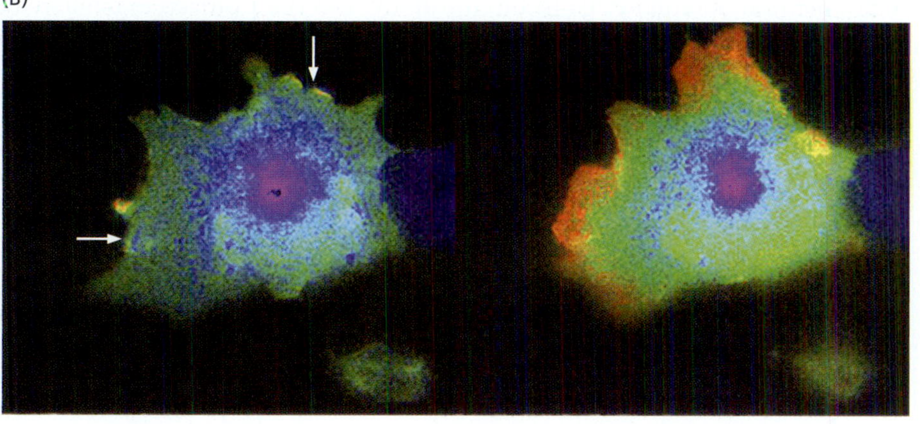

Figure 5.32 Alternative mechanisms of transformation by Ras (A) One depiction *(left)* of how Ras *(green)* normally operates to promote cell proliferation was inspired by the observation that *ras*-transformed cells release a number of growth factors into their surroundings, such as the TGF-α shown here *(red circles)*. Once secreted, growth factors like TGF-α might act in an autocrine fashion to activate a receptor, such as the EGF receptor shown here, thereby promoting cell proliferation. In this depiction, Ras operates "upstream" of the EGF receptor. An alternative scheme *(right)* placed Ras in the midst of a signaling cascade that operates "downstream" of growth factor receptors such as the EGF receptor. (B) Extensive evidence favoring the second scheme accumulated. This graphic example, produced years later, shows that 5 minutes after EGF is added to cells, previously inactive Ras protein *(blue shading)* becomes activated near the cell surface *(red shading)*. (This response is so rapid that it indicates close communication between the EGF receptor and Ras.) Interestingly, this activation occurred only on the two sides of a cell *(white arrows)* that were not in contact with nearby cells. (B, from N. Mochizuki et al., *Nature* 411:1065–1068, 2001.)

vation of Ras (Figure 5.32B). This left open only one alternative: direct intracellular signaling between ligand-activated growth factor receptors and Ras proteins. So the search was on to find the elusive biochemical links between tyrosine kinase receptors and the proteins like Ras that appeared to form components of the hypothesized downstream signaling cascade.

The use of biochemistry to discover these connections had serious limitations. As this chapter has made clear, biochemistry provided researchers with powerful tools to elucidate the functioning of individual, isolated cellular proteins. However, it had limited utility in revealing how different proteins might talk to one another. If there really were signaling cascades in the cell of the form A → B → C, biochemistry might well reveal how A or C functioned in isolation (e.g., as a kinase or GTPase) but would shed little light on the organization of the signaling pathway that lay between them.

Detailed biological characterization of human tumor cells overexpressing certain growth factor receptors also suggested another dimension of complexity. In the case of human breast cancers, overexpression of the HER2/Neu receptor (see Section 4.3) was found to be correlated with a bewildering array of phenotypes displayed by the associated cancer cells. Those cells expressing elevated levels of this protein showed increased rates of DNA synthesis, better anchorage-independent growth, greater efficiency in forming tumors when implanted into host mice (i.e., tumorigenicity), greater tendency to metastasize, and less dependence on estrogen for their growth. Hence, these receptor proteins were acting pleiotropically to confer a number of distinct changes on cancer cells. Such action seemed to be incompatible with a simple linear signaling cascade operating downstream of activated receptors. Instead, it appeared more likely that a number of distinct downstream signaling pathways were radiating out from these receptors, each involved in evoking a distinct cancer cell phenotype. But once again, there seemed to be little prospect of mapping these pathways and their signaling components.

This stalemate could be broken only by exploiting new types of experimental tools and a new way of thinking about the organization of complex signaling circuits. The science of genetics provided the new tool kit and mindset. However, genetics could be practiced with facility only on relatively simple experimental organisms for which powerful genetics techniques had been developed, specifically, those associated with the study of flies, worms, and yeast.

As it turned out, the successful elucidation of these signaling cascades also depended on one more factor. Much had already been learned from genetically dissecting the simpler metazoans and single-cell eukaryotes. Were lessons learned about their signaling systems transferable to the signaling cascades operating in mammalian cells? Here, fortune favored the cancer biologists. Because these intracellular signaling cascades were of very ancient lineage and had changed little over many hundreds of millions of years, discoveries about their organization and design in the worm, fly, and yeast proved to be directly applicable to mammalian cells. The consequence was the enrichment of cancer cell biology by research fields that had appeared, at least superficially, to be far removed from and irrelevant to an understanding of the problem of human cancer. As was often the case in cancer research, the important advances came from unexpected quarters, from the benches of researchers who never thought of themselves as foot soldiers in the war against cancer.

Key concepts

- Because a multicellular organism can exist only if its individual cells work in a coordinated fashion, the problem of cell-to-cell communication must have been solved by the time the first metazoa arose. Deregulation of such cell signaling is central to the formation of cancer.

- Src provided the first insight into how oncoproteins function when it was found to be a protein kinase—an enzyme that transfers phosphates from ATP to other proteins in the cell. Since multiple distinct protein substrates could be phosphorylated, and since each of these proteins could affect its own set of downstream targets, this explained how Src could act pleiotropically to yield the numerous phenotypic changes observed in RSV-transformed cells.
- Src attaches phosphates to tyrosine residues of proteins instead of the threonine or serine residues typical of most protein kinases. Subsequent research showed that tyrosine phosphorylation is primarily used by mitogenic signaling pathways.
- Epidermal growth factor (EGF) was found to have mitogenic effects on a variety of epithelial cell types, but only if it was able to bind to their surfaces. This suggested the involvement of a cell surface receptor, later isolated from a human tumor cell line.
- Once the N-terminal ectodomain of the EGF receptor binds EGF, a signal is transmitted across the plasma membrane that activates a Src-like kinase present in the C-terminal cytoplasmic domain. The kinase then phosphorylates tyrosines on the EGF receptor itself, which emit signals that induce cell proliferation.
- The EGF receptor and the ErbB oncoprotein are closely related. The ErbB oncoprotein strongly stimulates cell proliferation by emitting growth-stimulatory signals in a constitutive fashion. Hyperactive signaling of growth factor receptors (generally due to overexpression or structural alterations) is encountered in many types of human cancer cells.
- The close relationship between the B chain of PDGF and the Sis oncoprotein of simian sarcoma virus suggested an additional mechanism for cell transformation. The *sis* oncogene of the virus causes a virus-infected cell to release copious amounts of the PDGF-like oncoproteins, which attach to the PDGF receptors of the same cell. This creates an autocrine signaling loop in which a cell manufactures its own mitogen.
- The binding of ligand to a tyrosine kinase receptor induces its dimerization. This results in transphosphorylation of tyrosine residues located in the cytoplasmic domain of a receptor outside the immediate kinase domain. This phosphorylation causes activation of downstream signaling pathways.
- The dimerization model explains how overexpression of growth factor receptors favors cancer formation: because of their high numbers, the receptor molecules collide frequently as they move around the plasma membrane, resulting in dimerization, transphosphorylation, receptor activation, and emission of mitogenic signals.
- A variety of mutations affecting any of the three RTK domains may create ligand-independent firing by receptors. Gene fusion events that truncate the ectodomains while fusing the remaining portions of these receptors with proteins that are prone to dimerize or oligomerize yield potent oncoproteins. Amino acid substitutions or deletions in the cytoplasmic domain are also found in some cancers.
- There are other classes of receptors besides RTKs that are important in cancer pathogenesis. The receptors involved in controlling the development of hematopoietic cells rely on separate tyrosine kinases called Jaks that are associated noncovalently with their cytoplasmic domains. Receptors for TGF-β are cytoplasmic kinase domains that phosphorylate serine and threonine rather than tyrosine. When Notch binds one of its ligands, this receptor is cleaved twice, liberating a cytoplasmic domain fragment that can then join a nuclear complex that activates gene expression. The ligand-bound Patched receptor moves away from a second membrane-spanning protein called Smoothened; the latter then induces a cytoplasmic complex to release a transcription factor that activates target genes. Binding of Wnt factors to a

Frizzled receptor triggers a cascade of steps that prevents a cytoplasmic kinase from tagging several growth-promoting proteins for destruction. G-protein–coupled receptors (GPCRs) can activate G proteins that flip from an inactive GDP-bound state to a signaling GTP-bound state when activated by GPCRs and back again when they hydrolyze the GTP.

- Integrins constitute a large family of heterodimeric transmembrane cell surface receptors that have components of the ECM as their ligands. Integrins pass information both into and out of the cell. Upon ligand binding, integrins form focal adhesions, which link some integrins via their cytoplasmic domains to components of the cytoskeleton.
- Like G proteins, Ras binds GDP when quiescent, trades the GDP for GTP upon receiving a signal to become activated, emits a signal, and with the help of GTPase-activating proteins (GAPs) functions as a GTPase to hydrolyze the GTP and inactivate itself.
- The difference between a normal Ras protein and its oncogenic counterpart is generally a missense point mutation that obliterates GTPase activity, leading to a protein that, once activated, cannot turn itself off.

Thought questions

1. Why is autocrine signaling an intrinsically destabilizing force for a normal tissue?
2. Each growth factor elicits its own, quite characteristic set of biological responses in cells. How might you alter a cell so that its biological responses to one growth factor (e.g., EGF) are characteristic instead of the responses that it usually makes after being exposed to another growth factor (e.g., PDGF)?
3. The responsiveness of a cell to exposure to a growth factor are usually attenuated after a period of time (e.g., half an hour), after which time it loses this responsiveness. Given what you have already learned about growth factor receptors, what mechanisms might be employed by a cell to reduce its responsiveness to a growth factor?
4. What lines of evidence can you cite to support the notion that growth factor receptor firing following ligand binding is often dependent on the dimerization of a receptor (rather than some other molecular change in the receptor)?
5. In what ways are the heterodimeric G protein similar to the low molecular weight G proteins (e.g., Ras) and in what way do they differ fundamentally?
6. Why has the study of lower organisms, notably yeast, flies and worms, proven to be highly revealing in characterizing the intracellular signaling cascades and less useful in studying the receptors at the cell surface?
7. Integrins exhibit a novel type of control, termed "inside-out signaling", in which intracellular signals dictate the affinity of these receptors for their various extracellular ligands (i.e., components of the extracellular matrix). Why is such signaling necessitated by the phenomenon of cell motility?

Additional reading

Cross M & Dexter TM (1991) Growth factors in development, transformation and tumorigenesis. *Cell* 64, 271–280.

Giancotti FG & Ruoslahti E (1999) Integrin signaling. *Science* 285, 1028–1032.

Goldfarb M (1996) Functions of fibroblast growth factors in vertebrate development. *Cytokine Growth Factor Rev.* 7, 311–325.

Massagué J (2000) How cells read TGF-β signals. *Nat. Rev. Mol. Cell Biol.* 1, 169–178.

Ostman A & Heldin CH (2001) Involvement of platelet-derived growth factor in disease: development of specific antagonists. *Adv. Cancer Res.* 80, 1–38.

Schlapefer DD & Mitra SJ (2004) Multiple contacts link FAK to cell motility and invasion. *Curr. Opin. Genet. Dev.* 14, 92–101.

Schlessinger J (2002) Ligand-induced, receptor-mediated dimerization and activation of EGF receptor. *Cell* 110, 669–672.

Schwartz MA & Ginsberg MH (2002) Networks and crosstalk: integrin signalling spreads. *Nat. Cell Biol.* 4, E65–E68.

Ullrich A & Schlessinger J (1990) Signal transduction by receptors with tyrosine kinase activity. *Cell* 61, 203–212.

Yarden Y & Sliwkowski MX (2001) Untangling the signalling network. *Nat. Rev. Mol. Cell Biol.* 2, 127–137.

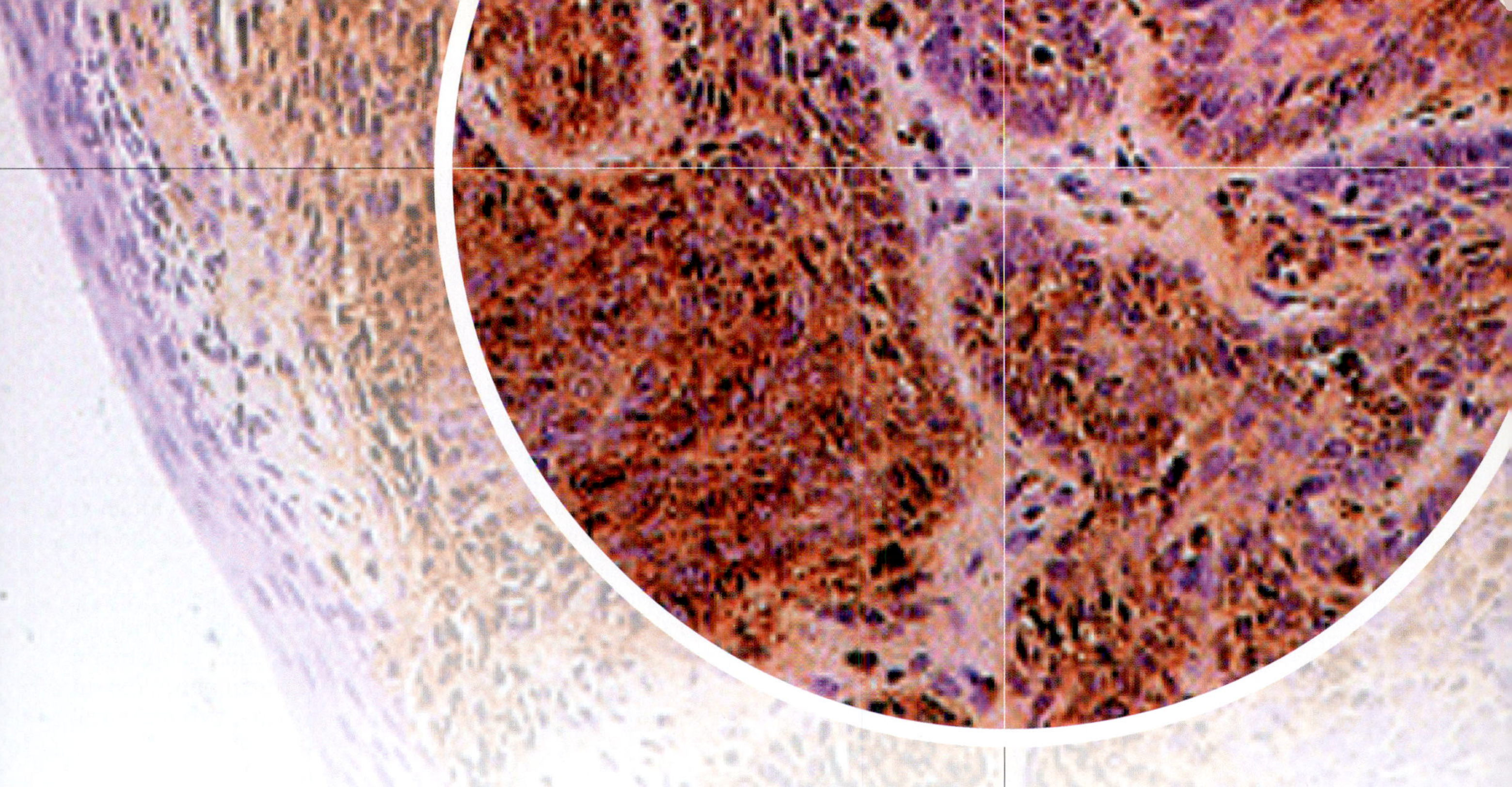

Chapter 6

Cytoplasmic Signaling Circuitry Programs Many of the Traits of Cancer

> Any living cell carries with it the experiences of a billion years of experimentation by its ancestors. You can expect to explain so wise and old bird in a few simple words.
>
> Max Delbrück, geneticist, 1966

Cancer is a disease of uncontrolled cell proliferation. Because the proliferative behavior of cancer cells is so aberrant, we might imagine that cancer cells invent entirely new ways of programming their growth and division—that the control circuitry within cancer cells is organized quite differently from that of normal, healthy cells. Such thinking greatly exaggerates the actual differences between normal and neoplastic cells. In truth, the two types of cells utilize control circuitry that is almost identical. Cancer cells discover ways of making relatively minor modifications of the control machinery operating inside cells. They tweak existing controls rather than demolishing the entire machinery and assembling a new version from the remnants of the original.

The present chapter is about this control machinery—really about the signal-processing circuitry that operates within the cell cytoplasm and governs cell proliferation. An electronic circuit board is assembled from complex arrays of hard-wired components that function as resistors, capacitors, diodes, and transistors. The cell also uses circuits assembled from arrays of intercommunicating components, but these are, almost without exception, proteins. While individual proteins and their functions are relatively simple, the operating systems that can be assembled from these components are often extraordinarily complex.

In Chapter 5, we read about the receptors displayed at the cell surface that gather a wide variety of signals and funnel them into the cytoplasm. Here, we study how these signals, largely those emitted by growth factor receptors, are processed and integrated in the cytoplasm. Many of the outputs of this signal-processing circuitry are then transmitted to the nucleus, where they provide critical inputs to the central machinery that governs cell proliferation. Our discussion of this nuclear governor—the *cell-cycle clock*—will follow in Chapter 8.

A single cell may express 20,000 or more distinct proteins, many of which are actively involved in the cytoplasmic regulatory circuits that are described here. These regulatory proteins are found in various concentrations and in different locations throughout the cytoplasm. Rather than floating around in dilute concentrations in the intracellular water, they form a thick soup. Indeed, as much as 30% of the volume of the cell is taken up by proteins rather than aqueous solvent.

These proteins must be able to talk to one another and to do so with great specificity and precision. Consequently, a signaling protein operating in a linear signaling cascade must recognize only those signals that come from its upstream partner protein(s) and pass them on to its intended downstream partner(s). In so doing, it will largely ignore the thousands of other proteins within the cell (Figure 6.1).

This means that each protein component of a signaling circuit must actually solve two problems. The first is one of specificity: how can it exchange signals only with the small subset of cellular proteins that are its intended signaling partners in the circuit? Second, how can this protein acquire rapid, almost instantaneous access to these signaling partners, doing so while operating in the viscous soup that is present in the cytoplasm and nucleus?

We study this circuitry because its design and operations provide key insights into how cancer cells arise. Thus, many of the oncoproteins described in previous chapters create cancer through their ability to generate signaling imbalances in this normally well-regulated system. At one level, cancer is surely a disease of inappropriate cell proliferation, but as we bore deep into the cancer cell, we will come to understand cancer at a very different level: cancer is really a disease of aberrant signal processing.

The present chapter will perhaps be the most challenging of all chapters in this book. The difficulty comes from the sheer complexity of signal transduction

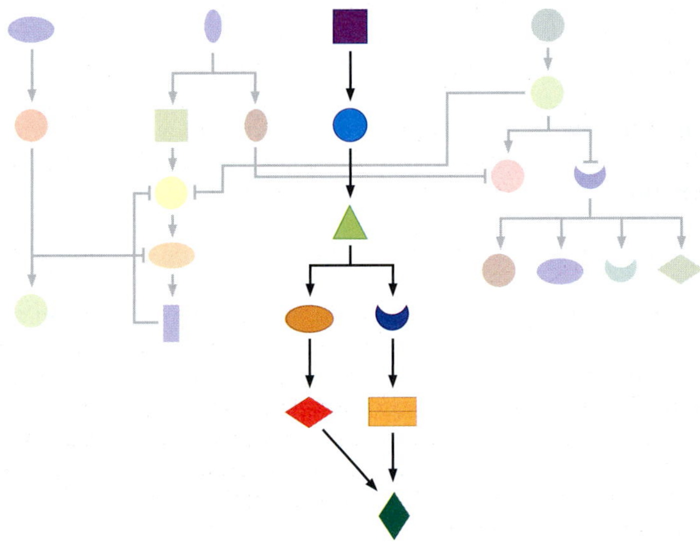

Figure 6.1 Signaling cascades This diagram of imaginary signaling circuitry illustrates how a signal transduction cascade—a series of signaling proteins that operates much like a molecular bucket brigade—passes signals from an upstream source *(purple square)* to its intended downstream target *(dark green diamond)* and, at the same time, avoids inadvertent activation of dozens of other signaling proteins in the cell *(faintly drawn symbols)*.

biochemistry, a field that is afflicted with many facts and blessed with only a small number of unifying principles. So, absorb this material in pieces; the whole is far too much for one reading.

6.1 A signaling pathway reaches from the cell surface into the nucleus

The growth and division that cells undertake following exposure to mitogens clearly represent complex regulatory programs involving the coordinated actions of hundreds, even thousands of distinct cellular proteins. How can growth factors succeed in evoking all of these changes simply by binding and activating their receptors? And how can the deregulation of these proliferative programs lead to cell transformation? And where does Ras fit into this circuitry, if indeed it does?

A major insight into these issues came from examination conducted in 1981 of the behavior of cultured normal cells, specifically fibroblasts, that had been deprived of serum-associated growth factors for several days. During this period of "serum starvation," the cells within a culture were known to enter reversibly into the nongrowing, quiescent state termed G_0 ("G zero"). After remaining in this G_0 state for several days, serum-starved cells were then exposed to fresh serum and thus to abundant amounts of mitogenic growth factors. The intent of this experiment was to induce all the cells in such cultures to enter into a state of active growth and division **synchronously** (coordinately). Indeed, within a period of 9 to 12 hours, the great majority of these previously quiescent cells would begin to replicate their DNA and, a number of hours later, initiate cell division.

These readily observed changes in cell behavior provided no hint of the large number of less obvious but nonetheless important molecular changes that occurred in these cells, often within minutes of exposure to fresh growth factors. For example, within less than an hour, transcription of an ensemble of more than 100 cellular genes was induced (Figure 6.2). Expression of these genes, which we now call **immediate early genes** (IEGs), increased rapidly in the half hour after growth factor stimulation. As is suggested by the contents of Table 6.1, the products of some of these genes help cells in various ways to emerge from the quiescent G_0 state into their active growth-and-division cycle. In fact, experiments involving several of these genes showed that blocking their expression prevents cells' emergence from the G_0 state.

A variation of this 1981 experiment was undertaken in which fresh serum was added together with **cycloheximide**, a drug that shuts down all cellular protein synthesis. In spite of the inhibition of protein synthesis, induction of the immediate early genes proceeded quite normally. This indicated that all of the proteins required to activate transcription of these genes were already in place at the moment when the serum was added to the cell. Stated differently, the induction of these genes did not require ***de novo*** (new) protein synthesis.

These results also demonstrated that, in addition to the growth factor receptors at the cell surface, an array of other proteins was present within the cell that could rapidly convey mitogenic signals from the receptors located at the cell surface to transcription factors (TFs) operating in the nucleus. Clearly, the functional activation of such cytoplasmic signal-transducing proteins did not depend upon increases in their concentration (since cycloheximide had no apparent effect on this signaling). Instead, changes in the proteins' structure, configuration, and intracellular localization appeared to play a dominant role in their functional activation soon after the stimulation of tyrosine kinase receptors by their growth factor ligands.

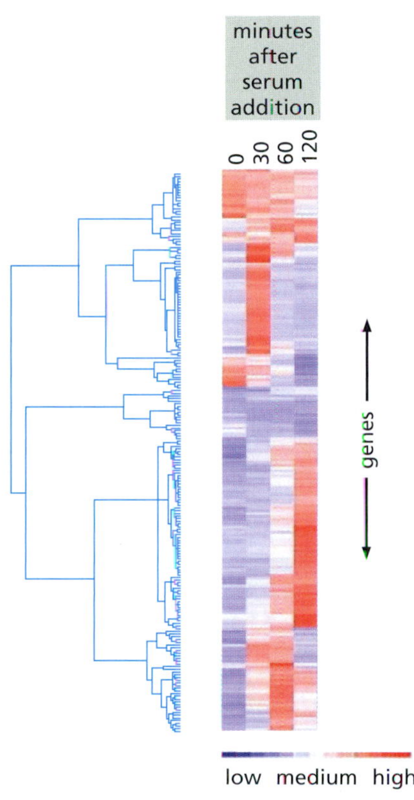

Figure 6.2 Expression of immediate early genes Expression of a large array of "immediate early genes" (IEGs) is induced within the first two hours after serum-starved cells are exposed to fresh serum. This expression occurs even when protein synthesis is blocked during this period, indicating that the transcription factors required for this expression preexist in the serum-starved cells. The gene expression array used here surveyed the expression of 14,824 genes at 30, 60, and 120 minutes following serum addition. Analyses of resulting expression patterns led to the identification of 229 genes *(unlabeled, arrayed top to bottom)* that behave like IEGs. Based on the various schedules of gene expression, these could be grouped into 10 subclasses (hierarchical clustering diagram, *left*) each of which showed its own characteristic schedule of induction and, in some cases, subsequent repression. *Red* indicates high level of RNA expression, while *blue* indicates low expression. Note that some IEGs are expressed for only a brief period of time before being repressed even though the serum and associated mitogens continue to be present in the culture medium. (From A. Selvaraj and R. Prywes, *BioMed Central Mol. Biol.* 5:13, 2004.)

Table 6.1 A sampling of immediate early genes[a]

Name of gene	Location of gene product	Function of gene product
fos[b]	nucleus	component of AP-1 TF
junB	nucleus	component of AP-1 TF
egr-1	nucleus	zinc finger TF
nur77	nucleus	related to steroid receptors
Srf-1[c]	nucleus	TF
myc	nucleus	bHLH TF
β-actin	cytoplasm	cytoskeleton
γ-actin	cytoplasm	cytoskeleton
tropomyosin	cytoplasm	cytoskeleton
fibronectin	extracellular	extracellular matrix
glucose transporter	plasma membrane	glucose import
JE	extracellular	cytokine
KC	extracellular	cytokine

[a]The total number of distinct immediate early genes is variously estimated to be between 50 and 100.
[b]Expression of a group of fos-related genes is also induced as IEGs. These include *fosB, fra-1,* and *fra-2.*
[c]Srf is a TF that binds to the promoters of other immediate early genes such as *fos, fosB, junB, egr-1* and *egr-2, nur77,* and cytoskeletal genes such as actins and myosins.

Adapted in part from H.R. Herschman, *Annu. Rev. Biochem.* 60:381–319, 1991; and from B.H. Cochran, in R. Grzanna and R. Brown (eds.), Activation of Immediate Early Genes by Drugs of Abuse. Rockville, MD: National Institutes of Health, 1993, pp. 3–24.

The immediate early genes encode a number of interesting proteins (see Table 6.1). Some of them specify transcription factors which, once synthesized, help to induce a second wave of gene expression. Included here are the *myc, fos,* and *jun* genes, which were originally identified through their association with transforming retroviruses. Yet other immediate early genes encode proteins that are secreted growth factors ("cytokines") or help to construct the cellular cytoskeleton.

The levels of *myc* mRNA were found to rise steeply following mitogen addition and to collapse quickly following mitogen removal. Moreover, the Myc protein itself turns over rapidly, having a **half-life** ($T_{1/2}$) of only 25 minutes. Such observations indicated that the levels of the Myc protein serve as an intracellular marker or indicator of the amount of mitogens present in the nearby extracellular space.

As an aside, we might note that the activity of Myc operating as a signaling protein derives in large part from major changes in its concentration within the cell nucleus. This contrasts starkly with the behavior of cytoplasmic signal-transducing proteins such as Ras and Src. These cytoplasmic proteins respond to mitogenic signals by undergoing noncovalent and covalent alterations in their structure rather than exhibiting significant increases in concentration. This difference is mirrored by the changes occurring when the normal *myc, ras,* and *src* genes are converted into oncogenes: the *levels* of Myc protein become deregulated (being expressed constitutively rather than being modulated in response to physiologic signals), whereas the *structures* of the Ras and Src proteins undergo alteration but the amounts of these proteins do not increase.

Within an hour after the immediate early mRNAs appear, a second wave of gene induction was found to occur. Significantly, the induction of these **delayed early genes** was largely blocked by the presence of cycloheximide, indicating that their expression did indeed depend on *de novo* protein synthesis (Figure 6.3). (In fact, expression of these delayed early genes seems to depend on transcription factors that are synthesized in the initial wave of immediate early gene expression.)

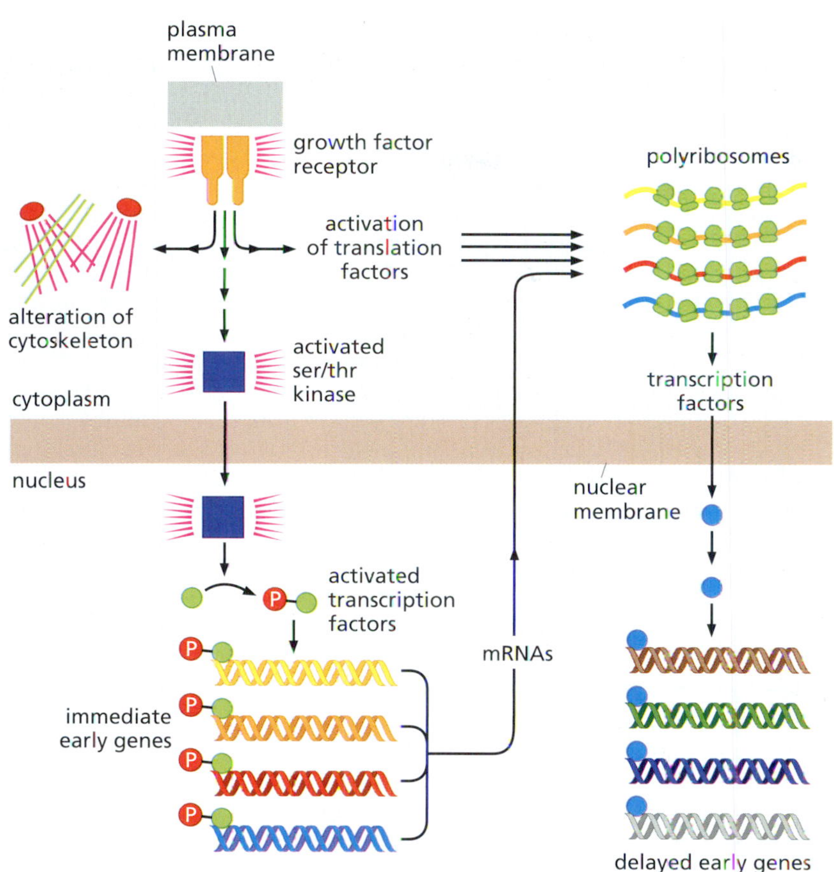

Figure 6.3 Immediate vs. delayed early genes The profound changes in cell physiology following the application of serum-associated mitogenic growth factors to a cell indicate that growth factor receptors can release a diverse array of biochemical signals once they are activated by ligand binding. The transcription of some genes (immediate early genes) is observed within minutes of growth factor stimulation *(lower left)*, while the transcription of others (delayed early genes) occurs with a lag, because it requires the synthesis of transcription factors *(lower right)*.

Addition of growth factors to quiescent cells was found to provoke yet other changes in cell physiology besides the rapid induction of nuclear genes. Following exposure to serum, the rate of cellular protein synthesis increases significantly, this being achieved through functional activation of the proteins that enable ribosomes to initiate translation of cellular mRNAs. Some growth factors were discovered to induce motility in cells, as indicated by the movement across the bottom surfaces of Petri dishes. Yet others can provoke a reorganization of actin fibers that help to construct the cell's cytoskeleton—the scaffolding that defines cell shape (Figure 6.4). Subsequently, many growth factors were discovered to provide survival signals to cells, thereby protecting them from inadvertent activation of the cell suicide program known as apoptosis.

These diverse responses indicated that a variety of distinct biochemical signals radiate from ligand-activated growth factor receptors, and that these signals

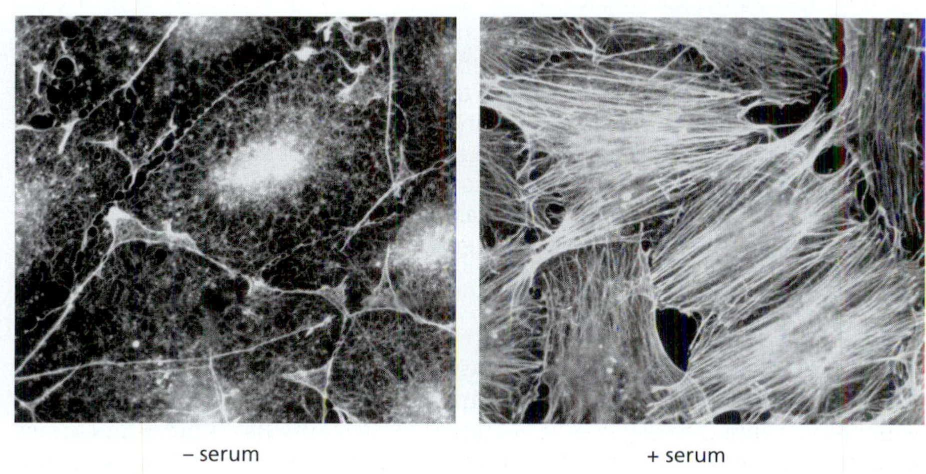

Figure 6.4 Serum-induced alterations of cell shape Like many other cell types, serum-starved mouse Swiss 3T3 cells undergo profound changes in their actin cytoskeleton following serum stimulation. In addition, the serum stimulation, much of it mediated by platelet-derived growth factor (PDGF), also induces greatly increased cell motility *(not shown)*. (Courtesy of A. Hall.)

Sidebar 6.1 Transcriptional responses to mitogens provide no indication of the gene expression in continuously growing cells Simple logic would dictate that the transcriptional responses observed shortly after growth factor stimulation of a serum-starved cell can be used to predict the transcriptional state of cells in which mitogenic signaling occurs continuously because of ongoing exposure to growth factors. Such signaling also seems to operate in cancer cells, in which certain oncoproteins release mitogenic signals continuously. Accordingly, the spectrum of immediate and delayed early genes induced within the first hour of growth factor stimulation should also be expressed continuously and at a high rate in transformed cells.

In fact, some genes that are members of the immediate and delayed early gene expression groups are found to be expressed at high levels in cancer cells, while others are not. This discordance derives from the operations of negative-feedback controls that force the shutdown of some genes shortly after they have been expressed. For example, expression of the *fos* gene, itself a proto-oncogene, increases rapidly in response to serum stimulation, peaks, and then declines dramatically within an hour or so after initial serum stimulation (e.g., see Figure 6.2). This shutdown is due to the ability of the Fos protein, once synthesized, to act as a repressor of the synthesis of more *fos* mRNA. Numerous other negative-feedback mechanisms operate to ensure that many of the initial biochemical responses evoked by growth factors operate only transiently following initial exposure of cells to these growth-stimulating factors.

impinge on a diverse array of cellular targets. Some of these signals seemed to be channeled directly into the nucleus, where they altered gene expression programs,while others were clearly directed toward cytoplasmic targets, including the protein-synthesizing machinery and the proteins that organize the structure of the cytoskeleton.

An understanding of how these signaling cascades operate clearly was very relevant to the cancer problem: if it were true that some oncoproteins flood cells with a continuous stream of mitogenic signals (Chapter 5), then the transformed state of cancer cells might well represent an exaggerated version of the responses that normal cells exhibit following exposure to growth factors. In fact, many of the traits of cancer cells can indeed be traced to responses evoked by growth factors, while others cannot be predicted by the initial responses of cells to these factors (Sidebar 6.1).

6.2 The Ras protein stands in the middle of a complex signaling cascade

The disparate responses of cells to growth factors represented a challenge to those interested in intracellular signal transduction, since almost nothing was known about the organization and function of the communication channels operating within the cell. Over a period of a decade (the 1980s), these circuits were slowly pieced together, much like a jigsaw puzzle. The clues came from many sources. The story started with Ras and then moved up and down the signaling cascades until the links in the signaling chains were finally connected to one another.

The advances in Ras biochemistry that we discussed in the last chapter (Section 5.9) seemed to explain much about how it operates as a binary switch, but provided no insight into its context—how this protein is connected with the overall signaling circuitry. Indeed, it remained possible that the Ras protein functioned in a signaling pathway that was independent of the one controlled by growth factors and their receptors.

In the end, the solution to the Ras problem came from a totally unexpected quarter. The genetics of eye development in the fruit fly *Drosophila melanogaster* revealed a series of genes whose products were essential for the normal development of the ommatidia, the light-sensing units forming the compound eye. One important gene here came to be called *sevenless*; in its absence, the seventh cell in each ommatidium failed to form (Figure 6.5). Provocatively, after cloning and sequencing, the *sevenless* gene was found to encode a protein with the structural features of the EGF receptor.

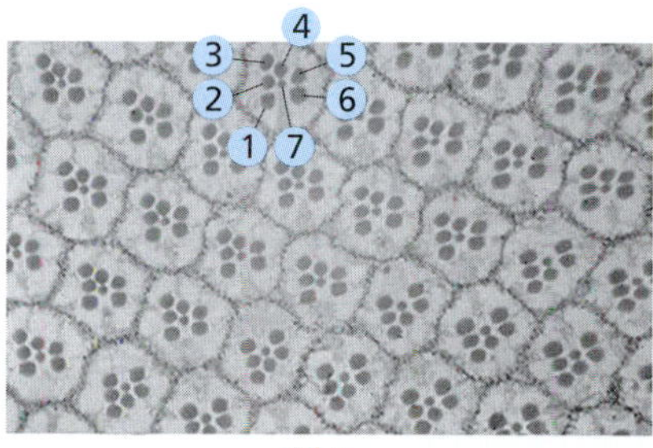

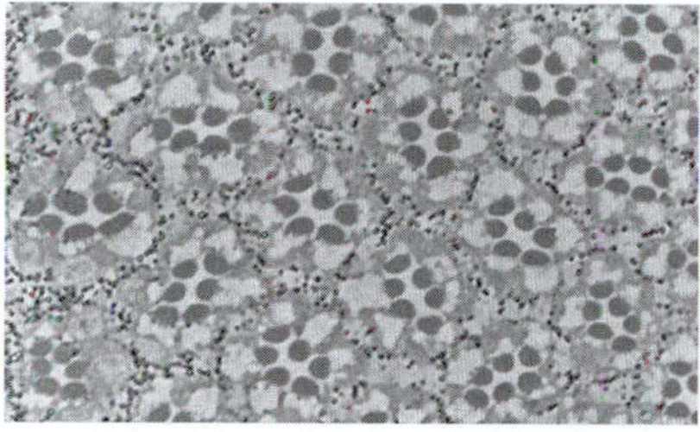

Figure 6.5 Structure of the fly eye Study of the light-sensing units in the *Drosophila melanogaster* eye—the ommatidia (*upper panel*)—revealed that each ommatidium is formed from a series of seven cells, with six outer ones surrounding a seventh in the center, as seen in this electron micrograph section (*numbering in top middle*). A fly carrying the mutation in the gene that came to be called *sevenless* produced ommatidia lacking the seventh, central cell (*lower panel*). Later, *sevenless* was found to encode a homolog of the epidermal growth factor (EGF) receptor. (Courtesy of E. Hafen.)

Yet other mutations mimicked the *sevenless* mutation. Genetic complementation tests revealed that these mutations affected genes whose products operate downstream of *sevenless*, ostensibly in a linear signal-transduction cascade. One of these downstream proteins was encoded by a gene that was named *son of sevenless* or simply *sos*. Close examination of the Sos protein showed it to be related to proteins known from yeast biochemistry to be involved in provoking nucleotide exchange by G (guanine nucleotide–binding) proteins such as Ras. The yeast proteins, often called guanine nucleotide exchange factors (GEFs), were known to induce G proteins to release their bound GDPs, thereby making room for GTPs to jump aboard (see Figure 5.30). The consequence was an activation of these G proteins from their inactive to their active, signal-emitting configuration. This was precisely the effect that Sos exerted on Ras proteins (Figure 6.6). Therefore, Sos was the long-sought upstream stimulator that kicked Ras into its active configuration, and a new, critical component could be placed in the scheme of Ras regulation.

Soon other intermediates in the signaling cascade were uncovered. Two of these, Shc (pronounced "shick") and Grb2 (pronounced "grab two") were discovered through genetic and biochemical screens for proteins that interacted with phosphorylated receptors or derived peptides. It became apparent that these proteins function as adaptors, forming physical bridges between growth factor receptors and Sos; these bridging proteins will be discussed in greater detail later. A third such adaptor protein, called Crk (pronounced "crack"), was identified as the oncoprotein encoded by the CT10 avian sarcoma virus.

These discoveries demonstrated the very ancient origins of this signaling pathway, which was already well developed in the common ancestor of all contemporary metazoan phyla that lived more than 600 million years ago. Once in place, the essential components of this cascade remained relatively unchanged in the cells of descendant organisms. Indeed, the individual protein components of these cascades have been so conserved that in many cases the protein components from the cells of one phylum (e.g., chordates) can be exchanged with those of another (e.g., arthropods) to reconstitute a functional signaling pathway.

Together, the genetic and biochemical data from these distantly related animal phyla could be combined in a scheme that suggested a linear signaling cascade organized like this: tyrosine kinase receptor → Shc → Grb → Sos →Ras. While this cascade provided the outline of a signaling channel, it gave little insight into the biochemical interactions that enabled these various proteins to exchange signals with one another.

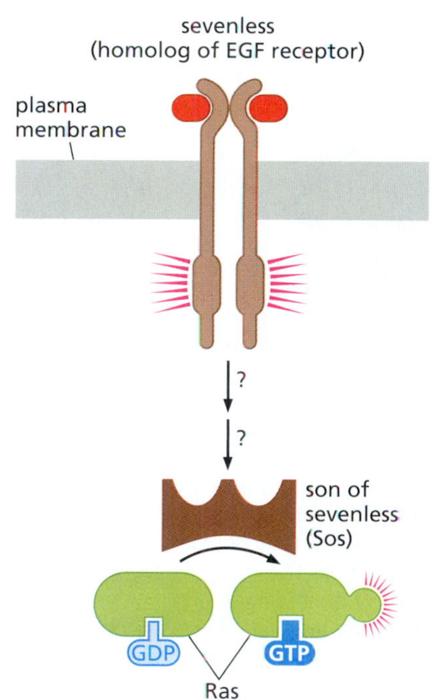

Figure 6.6 Son of sevenless Genetic studies of *Drosophila* eye development revealed another gene whose product appeared to lie downstream of Sevenless (Figure 6.5) in a signaling cascade and specified a guanine nucleotide exchange factor (GEF) (*brown*), capable of activating the Ras protein. This gene was named *son of sevenless* or simply *sos*.

6.3 Tyrosine phosphorylation controls the location and thereby the actions of many cytoplasmic signaling proteins

Among the major biochemical issues left unanswered, perhaps the most critical were the still-mysterious actions of the kinases carried by many growth factor receptors. Was the phosphorylation of these receptors on tyrosine residues critical to their ability to signal, or was it a distraction? And if this phosphorylation was important, how could it activate the complex signaling circuitry that lay downstream?

Two alternative mechanisms seemed plausible. The first predicted that a receptor-associated tyrosine kinase would phosphorylate a series of target proteins in the cytoplasm. Such covalent modification would alter the three-dimensional *conformation* of the proteins (i.e., their **stereochemistry**), thereby placing each of them in an actively signaling state that allowed them to transfer signals to a partner one step further down in a signaling cascade. This model implied, among other things, that the phosphorylation of the receptor molecule itself was of secondary importance to signaling.

According to the alternative model, the phosphorylation of the cytoplasmic tail of a receptor following growth factor binding affected the *physical location* of its downstream signaling partners without necessarily changing their intrinsic activity. Once relocalized to new sites within the cytoplasm, these downstream partners could then proceed with their task of emitting signals to yet other targets in the cell.

As it turned out, the second model, involving protein relocalization, proved to be more important. The salient insights came from analyzing the detailed structure of the Src protein. Researchers noted three distinct amino acid sequence domains, each of which appeared both in Src and in a number of other, otherwise unrelated proteins. These sequences, called Src homology domains 1, 2, and 3 (SH1, SH2, and SH3), provided the key to the puzzle of receptor signaling (Figure 6.7).

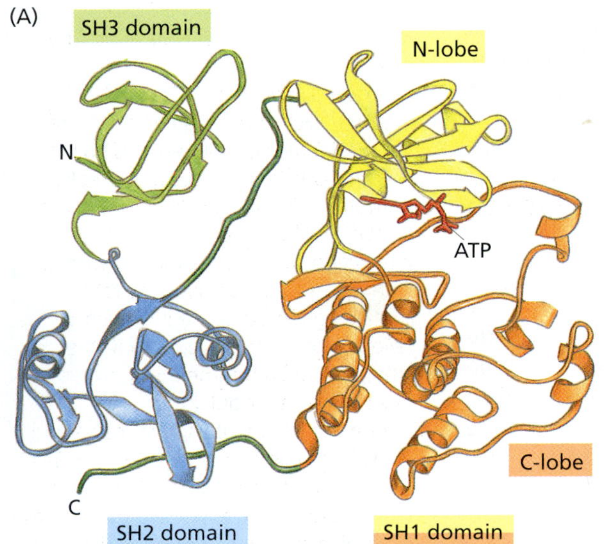

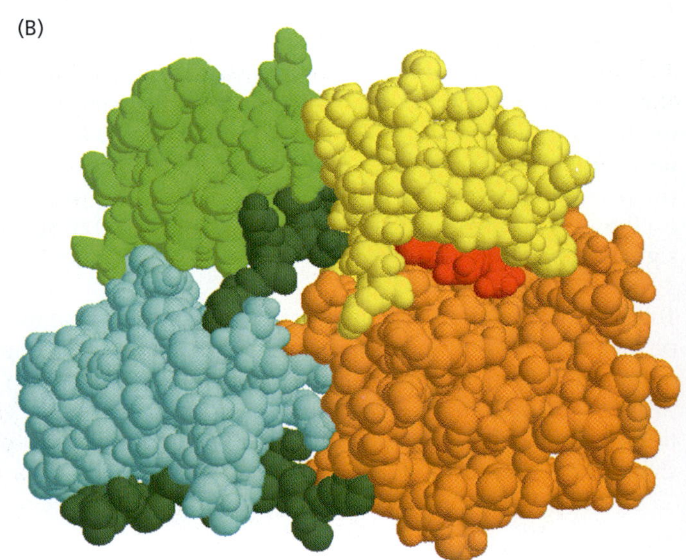

Figure 6.7 Domain structure of the Src protein X-ray crystallographic analyses of the Src protein have revealed three distinct structural domains that are present as well in a large number of other proteins. (A) The detailed structure of Src is shown in this ribbon diagram, in which the helical domains indicate α helices, while the flattened ribbons indicate β pleated sheets. Like many other subsequently analyzed protein kinases, the catalytic SH1 domain is seen to be composed of two domains *(right),* labeled N- and C-lobe kinase domains (*yellow, orange*). Between these lobes is the ATP-binding site where catalysis takes place *(red stick figure)*. The SH2 and SH3 domains (*blue, light green*) are to the left and are involved in substrate recognition and regulation of catalytic activity. Connecting sequences are shown in *dark green*. (B) Space-filling model of Src. (A, from B. Alberts et al., Molecular Biology of the Cell, 4th ed. New York: Garland Science, 2002; B, courtesy of M.J. Eck, see W. Xu et al., *Mol. Cell* 3:629–638, 1999.)

The SH1 domain of Src represents its catalytic domain, and indeed similar sequence domains are present in all receptor tyrosine kinases, as well as in the other, nonreceptor tyrosine kinases that are configured like Src. This shared sequence relatedness testifies to the common evolutionary origin of all these kinases.

Clever biochemical sleuthing conducted in the late 1980s uncovered the role of the SH2 domain present in Src and other signaling proteins. This relatively small structural domain of about 100 amino acid residues (Figure 6.8) acts as an intracellular "receptor." The "ligand" for this SH2 receptor is a short oligopeptide sequence that contains both a phosphorylated tyrosine and a specific oligopeptide sequence 3 to 6 residues long that flanks the phosphotyrosine on its C-terminal side.

Soon, it became apparent that there are dozens of distinct SH2 domains, each carried by a different protein and each having an affinity for a specific phosphotyrosine-containing oligopeptide sequence that functions as its ligand. By now, the sequence specificities of a substantial number of SH2 groups have been cataloged. The human genome is estimated to encode at least 117 distinct SH2 groups, each constituting a domain of a larger protein and each apparently having an affinity for binding a particular phosphotyrosine together with a flanking oligopeptide sequence.

Most commonly, an SH2 domain enables the protein carrying it to associate with a partner protein that is displaying a specific phosphotyrosine plus flanking amino acid sequence, thereby forming a physical complex between these two proteins. Some SH2-bearing proteins carry no enzymatic activity at all. Other such proteins, in addition to their SH2 domains, carry catalytic sites that are quite different from the tyrosine kinase (TK) activity present in Src itself. For example, one form of phospholipase C carries an SH2 domain, as does the p85 subunit of the enzyme phosphatidylinositol 3-kinase (PI3K). The SHP1 protein, to be discussed shortly, carries an SH2 domain linked to a **phosphatase** catalytic domain, that is, an enzymatic activity that removes phosphate groups, thereby reversing the actions of kinases.

These findings indicated that the catalytic domains of the various proteins and their SH2 groups function as independent structural modules. These modules have been pasted together in various combinations by evolution. The SH2 group

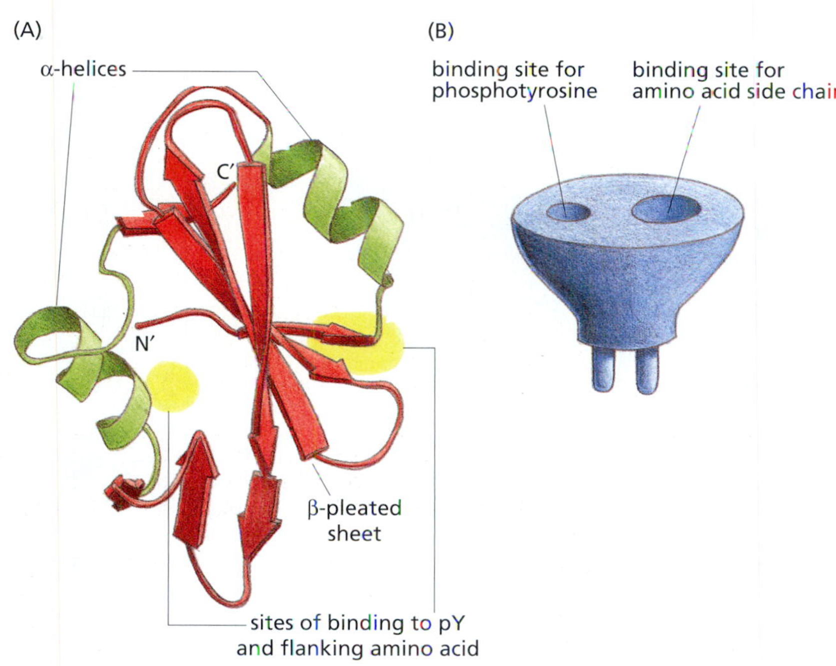

Figure 6.8 Structure and function of SH2 groups (A) This ribbon diagram demonstrates the three-dimensional structure of a typical SH2 domain as revealed by X-ray crystallography. The SH2 domain is seen in a different orientation from that in Figure 6.7. A typical SH2 domain is composed of about 100 amino acid residues and is assembled from a pair of anti-parallel β pleated sheets *(red)* surrounded by a pair of α helices *(green)*. The sites of interaction by the SH2 domain with its "ligand"—a phosphotyrosine (pY)-containing peptide—are indicated by the *yellow* spots. (B) As indicated in the schematic drawing, the SH2 domain functions as a modular plug: it recognizes both a phosphotyrosine and the side-chains of amino acids that flank this phosphotyrosine on its C-terminal side. These flanking amino acid residues determine the specificity of binding, i.e., the identities of the particular phosphotyrosinated oligopeptides that are recognized and bound by a particular SH2 domain. (A, based on data from G. Waksman et al., *Cell* 72:779–790, 1993; B, from B. Alberts et al., Molecular Biology of the Cell, 4th ed. New York: Garland Science, 2002.)

borne by each of these proteins allows it to become localized to certain sites within a cell, specifically to those sites that contain particular phosphotyrosine-bearing proteins to which the SH2-containing protein can become tethered.

The discoveries about SH2 groups finally solved the puzzle of how tyrosine kinase receptors are able to emit signals. The story is illustrated schematically in Figure 6.9. As a consequence of ligand-induced transphosphorylation, a receptor molecule, such as the EGF receptor, will acquire and display a characteristic array of phosphotyrosine residues on its cytoplasmic tail; likewise, the PDGF receptor will display its own spectrum of phosphotyrosines on its cytoplasmic tail. The unique identity of each phosphotyrosine residue borne by these tails is dictated by the sequence of amino acid residues flanking the phosphotyrosine on its C-terminal side. These phosphotyrosines become attractive homing sites for various SH2-containing cytoplasmic proteins, specifically, proteins that are normally found in the soluble portion of the cytoplasm (the **cytosol**) and are therefore free to move from one location to another in the cytoplasm. Consequently, shortly after becoming activated by ligand binding, a growth factor receptor becomes decorated with a specific set of these SH2-containing partner proteins that are attracted to its various phosphotyrosines.

For example, one ligand-activated form of the PDGF receptor attracts Src, PI3K, Ras-GAP (which stimulates Ras GTPase activity; refer to Sidebar 5.11), SHP2 (SH2-containing tyrosine phosphatase 2), and PLC-γ (phospholipase C-γ). Each of these proteins carries at least one SH2 domain, and all of them flock to the PDGF-R after its cytoplasmic tail and a short segment in the middle of its kinase domain (called a "kinase insert") become tyrosine-phosphorylated. Once tethered to the PDGF-R, some of these SH2-containing proteins then become substrates for tyrosine phosphorylation by the PDGF-R kinase. (It remains unclear whether all of these SH2-containing proteins and yet others can bind simultaneously to a single ligand-activated PDGF receptor molecule.)

The biochemical diversity of these associated proteins provides us with at least one solution to the question of how a growth factor is able to elicit a diversity of

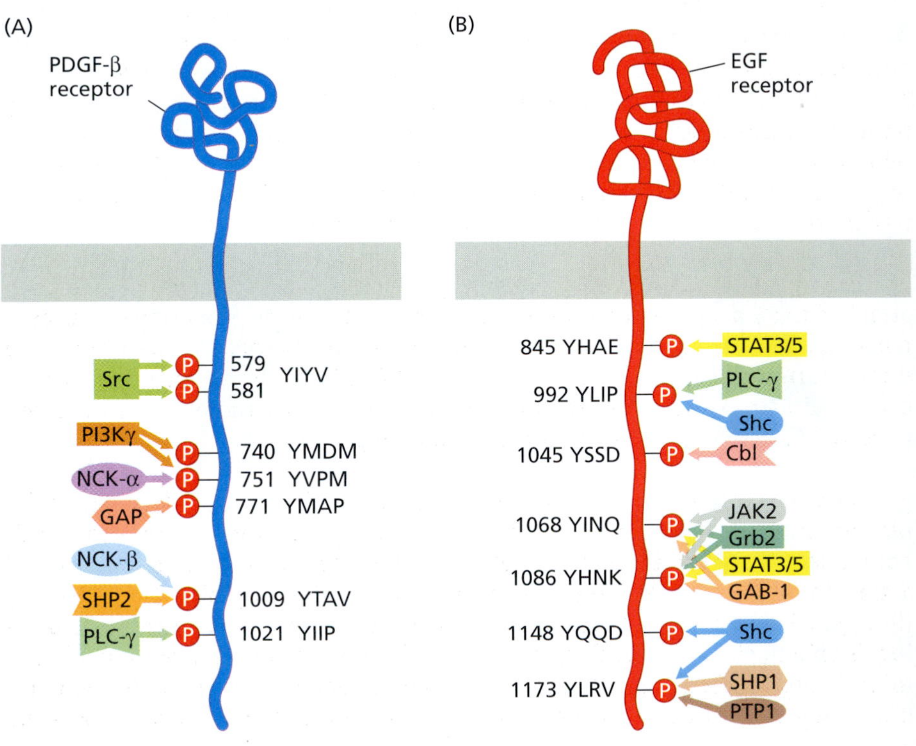

Figure 6.9 Attraction of signal-transducing proteins by phosphorylated receptors (A) This schematic diagram of a molecule of the platelet-derived growth factor-β (PDGF-β) receptor, which omits its tyrosine kinase domain, reveals a large number of tyrosine residues in the C-terminal portion of the receptor that undergo phosphorylation following ligand binding and receptor activation. (The positions of these tyrosine residues in the receptor polypeptide chain are indicated by the numbers.) Listed to the left are seven distinct cytoplasmic proteins, each of which can bind to one or more of the phosphotyrosines of the PDGF-β receptor via its SH2 domain(s). The amino acid sequence environment of each of these phosphotyrosines is indicated in the single-letter amino acid code. As is implied, the three amino acid residues that flank each tyrosine (Y) residue on its C-terminal side define the unique binding site recognized by the various SH2 domains carried by these seven associated proteins. (B) Similarly, a constellation of phosphotyrosines can be formed after the EGF receptor binds its ligand, often forming heterodimers with the HER2 receptor protein. However, the spectrum of SH2-containing proteins that associates with the EGF-R is quite different, allowing this receptor to activate a different set of downstream signaling pathways. (A, adapted from T. Pawson, *Adv. Cancer Res.* 64:87–110, 1994; B, adapted from R. Sordella et al., *Science* 305:1163–1167, 2004.)

Sidebar 6.2 An SH2-containing molecule determines outcomes of Olympic competitions One of the SH2-containing partners of tyrosine kinase receptors is a phosphotyrosine phosphatase (PTP), an enzyme that removes the phosphate group from phosphotyrosine. Once tethered via its SH2 group to a ligand-activated, tyrosine-phosphorylated receptor, this particular phosphatase, termed SHP1, begins to chew away the phosphate groups on tyrosine residues that have enabled the receptor to attract other downstream signaling partners. In this fashion, an enzyme such as SHP1 creates a negative-feedback loop to shut down further receptor firing. In fact, the SHP1 phosphatase is responsible for shutting down signaling by a variety of receptors, among them the receptor for the erythropoietin (EPO) growth factor, which normally responds to EPO by triggering **erythropoiesis**—red blood cell formation (to be described in Figure 6.22).

Some individuals inherit a mutant receptor gene encoding an EPO receptor that lacks the usual docking site for the SHP1 phosphatase. In individuals whose cells express this mutant receptor, the resulting defective negative-feedback control leads to a hyperactive EPO receptor, and, in turn, to higher-than-normal levels of red cells and thus oxygen-carrying capacity in their blood. In the case of several members of a Finnish family with this inherited genetic defect, this condition seems to have played a key role in their winning Olympic medals in cross-country skiing! (Provocatively, expression of SHP1 is shut down in many leukemias and non-Hodgkin's lymphomas; this permits multiple mitogen receptor molecules to fire excessively in the tumor cells.)

biochemical and biological responses from a cell: each of these proteins controls its own downstream signaling cascade, and each of these cascades, in turn, influences a different aspect of cellular behavior. By becoming tethered to a receptor molecule, these SH2-containing proteins are poised to activate their respective cascades. Some of these SH2-bearing proteins function to activate negative-feedback loops that ultimately shut down receptor signaling (Sidebar 6.2).

Because the cytoplasmic domains of tyrosine kinase receptors are found near the inner surface of the plasma membrane, the SH2-containing partner proteins attracted to the receptor are brought into close proximity to the plasma membrane and thus close to other molecules that are, for their own reasons, already tethered to this membrane. This close juxtaposition enables various SH2-containing proteins to interact directly with the membrane-associated proteins and phospholipids, thereby generating a variety of biochemical signals that can be transmitted to various downstream signal-transducing cascades.

For example, the Ras GTPase-activating protein (Ras-GAP), having become attached to a phosphorylated receptor, is tethered in close proximity to Ras proteins, many of which are permanently anchored to the plasma membrane via their C-terminal lipid tails. This allows Ras-GAP to interact with nearby Ras molecules, stimulating them to hydrolyze their bound GTPs, thereby causing them to convert from an active to an inactive signaling state.

Similarly, when phosphatidylinositol 3-kinase (PI3K) becomes tethered near the plasma membrane, it is able to reach over and phosphorylate inositol lipids embedded in the membrane; we return to this enzyme and its mechanism of action in Section 6.6. Yet another example is provided by phospholipase C-γ (PLC-γ), which, when juxtaposed to the plasma membrane via receptor tethering, is able to cleave a membrane-associated phospholipid [phosphatidylinositol (4,5)-diphosphate, or PIP_2] into two products, each of which, on its own, has potent signaling powers. The point here is that many reactions become possible (or can proceed at much higher rates) when enzymes and their substrates are brought into close proximity through tethering of these enzymes to phosphorylated receptors.

The SH3 domain, the third of the sequence motifs present in Src (see Figure 6.7), binds specifically to certain proline-rich sequence domains in partner proteins; these proline-rich sequences thus serve as the ligands of the SH3 domains. (A small subset of SH3 groups seem to recognize ligands of a different structure.) These SH3 domains are a relatively ancient invention (apparently originating more than 1.5 billion years ago), since there are at least 28 of them present in various proteins of the yeast *S. cerevisiae*. The SH2 domain, in contrast, seems to be of more recent vintage, having been invented possibly at the same time as tyrosine phosphorylation came into widespread use—perhaps 600 million years ago when metazoa arose (Figure 6.10A).

In the specific case of the SH3 domain of Src, there is evidence that this domain is used to recognize and bind to certain substrates that carry proline-rich domains and, once bound to Src, can then be phosphorylated by Src's kinase

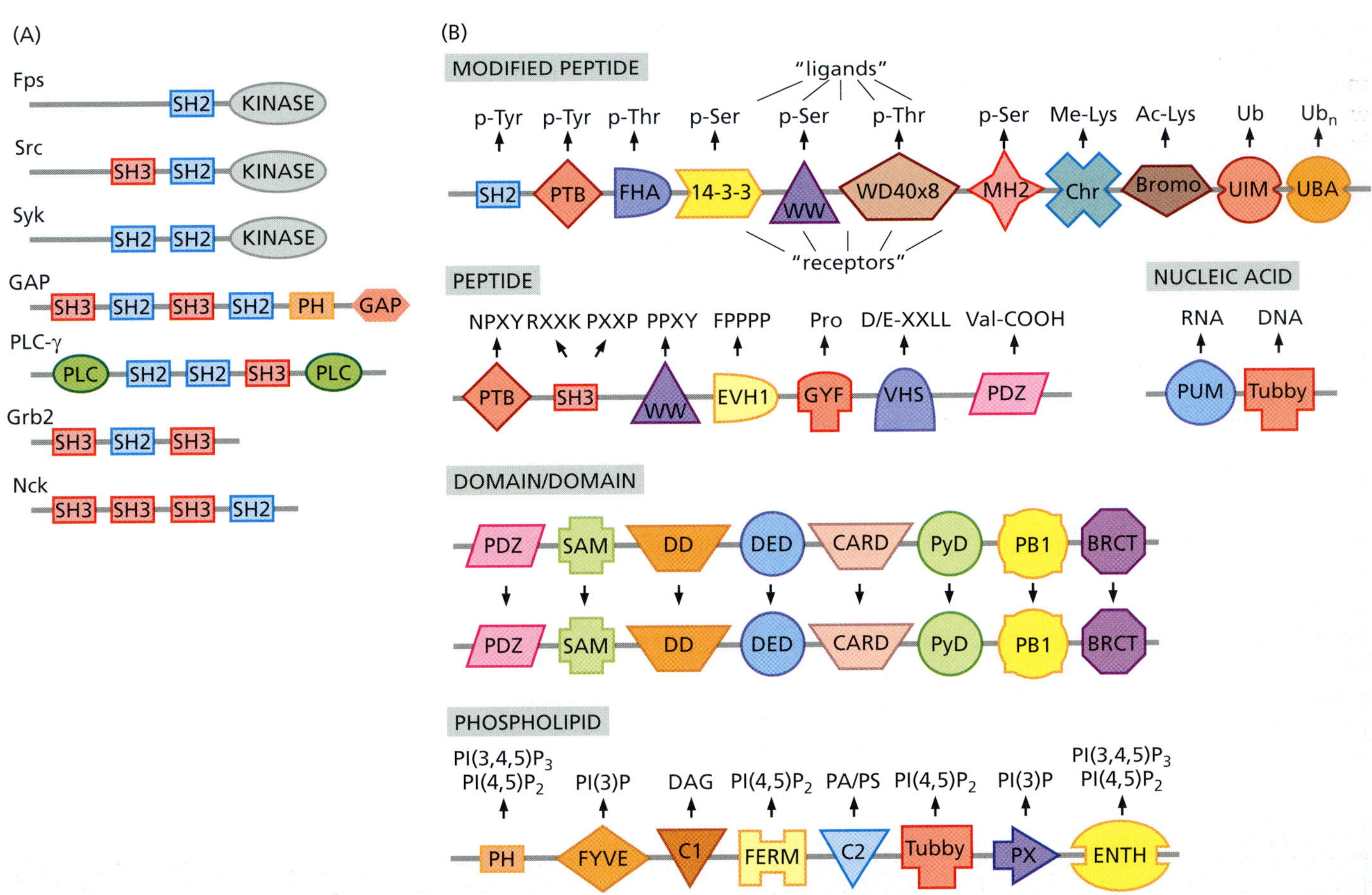

Figure 6.10 Protein interaction domains as modular units of protein structure Specific multiprotein complexes are formed using domains such as the SH2, SH3, and PH domains, which bind phosphotyrosines, proline-rich sequences, and PIP_3, respectively. As is indicated here, evolution has manipulated these domains in a modular fashion, attaching them to tyrosine kinase catalytic domains in the case of the first three tyrosine kinases (Fps, Src, and Syk), to a GAP (GTPase-activating protein) domain, or to a PLC (phospholipase C) domain. At the bottom are shown two adaptor or "bridging" proteins that contain only SH2 and SH3 domains and lack catalytic domains; each of these adaptors serves as a linker between pairs of other proteins, one of which carries a certain phosphotyrosine (to which an SH2 domain can bind) and the other of which carries a proline-rich domain (to which an SH3 domain can bind). (B) The spectrum of molecular "ligands" that can be bound by complementary "receptor" domains in order to facilitate intermolecular interactions is far larger than indicated in panel (A). As shown, a variety of modified amino acid residues, peptide sequences, nucleic acids, and phospholipids can serve as ligands of the indicated domains. In addition, certain domains ("domain-domain") favor the homodimerization of two protein molecules. (A, adapted from T. Pawson, *Adv. Cancer Res.* 64:87–110, 1994; B, from T. Pawson and P. Nash, *Science* 300:445-452, 2003.)

activity. Yet other evidence points to a proline-rich linker domain within Src itself, to which Src's SH3 domain associates, creating intramolecular binding. Through analyses of the human genome sequence, 253 distinct SH3 domains, each part of a larger protein, have been uncovered.

The SH2 and SH3 domains were the first of a series of specialized protein domains to be recognized and cataloged (Figure 6.10B). Each of these specialized domains is able to recognize and bind a specific sequence or structure present in a partner molecule. Some of these specialized domains recognize and bind phosphoserine or phosphothreonine on partner proteins, while others bind to phosphorylated forms of certain membrane lipids (Table 6.2). Another specialized domain, termed PTB, is able, like SH2 groups, to recognize and bind phosphotyrosine residues; in this instance, however, the flanking amino acid residues on the N-terminal side of the phosphotyrosine define the unique identity of the phosphotyrosine ligand.

Taken together, these various domains illustrate how the lines of communication within a cell are restricted, since the ability of signaling molecules to physically associate in highly specific ways with target molecules ensures that signals are passed only to these intended targets and not to other proteins within the cell. In addition, examination of an array of signaling proteins has revealed that these various domains have been used as modules, being assembled in different ways through evolution to ensure the specificity of a variety of intermolecular interactions (see Figure 6.10B).

The discovery of SH2 groups also explained how Src, the original oncoprotein, operates within the normal cell, and how Rous sarcoma virus has reconfigured the structure of Src, making it into an oncoprotein (Sidebar 6.3).

6.4 SH2 groups explain how growth factor receptors activate Ras and acquire signaling specificity

The discovery of SH2 groups and their mechanisms of action made it possible to solve several important problems. The biochemical clues were now in place to explain precisely how the receptor → Sos → Ras cascade operates.

The key was provided by the structure of the Grb2 protein. It contains two SH3 groups and one SH2 group. Its SH3 domains have an affinity for two distinct proline-rich sequences present in Sos, while its SH2 sequence associates with a phosphotyrosine present on the C-terminus of many ligand-activated receptors. Consequently, Sos, which seems usually to float freely in the cytoplasm, now becomes tethered via the Grb2 linker to the receptor (Figure 6.12).

Table 6.2 Binding domains that are carried by various proteins[a]

Name of domain	Ligand	Example of proteins carrying this domain
SH2	phosphotyrosine	Src (tyrosine kinase), Grb2 (adaptor protein), Shc (scaffolding protein), SHP2 (phosphatase), Cbl (ubiquitylation)
PTB	phosphotyrosine	Shc (adaptor protein), IRS-1 (adaptor for insulin RTK signaling), X11 (neuronal protein)
SH3	proline-rich	Src (tyrosine kinase), Crk (adaptor protein), Grb2 (adaptor protein)
14-3-3	phosphoserine	Cdc25 (CDK phosphatase), Bad (apoptosis regulator), Raf (ser/thr kinase), PKC (protein kinase C ser/thr kinase)
Bromo	acetylated lysine	P/CAF (transcription co-factor), chromatin proteins
PH[b]	phosphorylated inositides	PLC-δ (phospholipase C), Akt/PKB (ser/thr kinase), BTK

[a]At least 32 distinct types of binding domains have been identified. This table presents six of these that are often associated with transduction of mitogenic signals.
[b]The phosphoinositide-binding groups include, in addition to the PH domain, the Fab1, YOTB, Vac1, EEA1 (FYVE), PX, ENTH, and FERM domains.

Sidebar 6.3 The SH2 domain of Src has two alternative functions We have portrayed SH2 as a protein domain that enables its carrier to recognize and bind to other phosphotyrosine-containing partner proteins, thereby forming bimolecular complexes. In tyrosine kinases such as Src, however, the SH2 group can also play a quite different role. Like most signaling molecules, the Src kinase is usually held in a functionally silent configuration. It is kept in this inactive state by a negative regulatory domain of Src—a stretch of polypeptide that lies draped across the catalytic site of Src and blocks access to the substrates of this kinase. This obstructing domain (sometimes called an "activation loop"; see Figure 5.18) is held in place, in part, through the actions of the Src SH2 group, which recognizes and binds a phosphotyrosine in position 527 of Src itself. Hence, this SH2 group is forming an *intra*molecular rather than an *inter*molecular bridge (Figure 6.11A).

Some receptors, such as the PDGF receptor, may become activated by ligand binding and display phosphotyrosines in their cytoplasmic domains that are also attractive binding sites for the SH2 group of Src. As a consequence, the fickle Src SH2 group breaks its intramolecular bond and re-forms a new intermolecular bond with the PDGF receptor; this results in the tethering of Src to the PDGF receptor (Figure 6.11B). At the same time, the catalytic cleft of the Src protein has been freed from the obstructive effects of this negative regulatory domain, and therefore Src can begin to phosphorylate its own clientele of substrates, thereby aiding the broadcasting of signals by the PDGF receptor (Figure 6.11C).

The Src oncoprotein specified by Rous sarcoma virus is a mutant form of the normal c-Src protein, in which the C-terminal oligopeptide domain containing the tyrosine 527 residue has been replaced by another, unrelated amino acid sequence. As a consequence, a phosphotyrosine can never be formed at this site and the intramolecular inhibitory loop cannot form and obstruct the Src catalytic site. The viral oncoprotein (v-Src) is therefore given a free hand to fire constitutively. A similar mutant Src, in which the C-terminal six amino acids are deleted (including tyrosine 527), has been reported in liver metastases of human colon carcinomas. While the Src kinase is hyperactive in many colon carcinomas, this is one of the very rare examples of a mutant, structurally altered Src in human cancers.

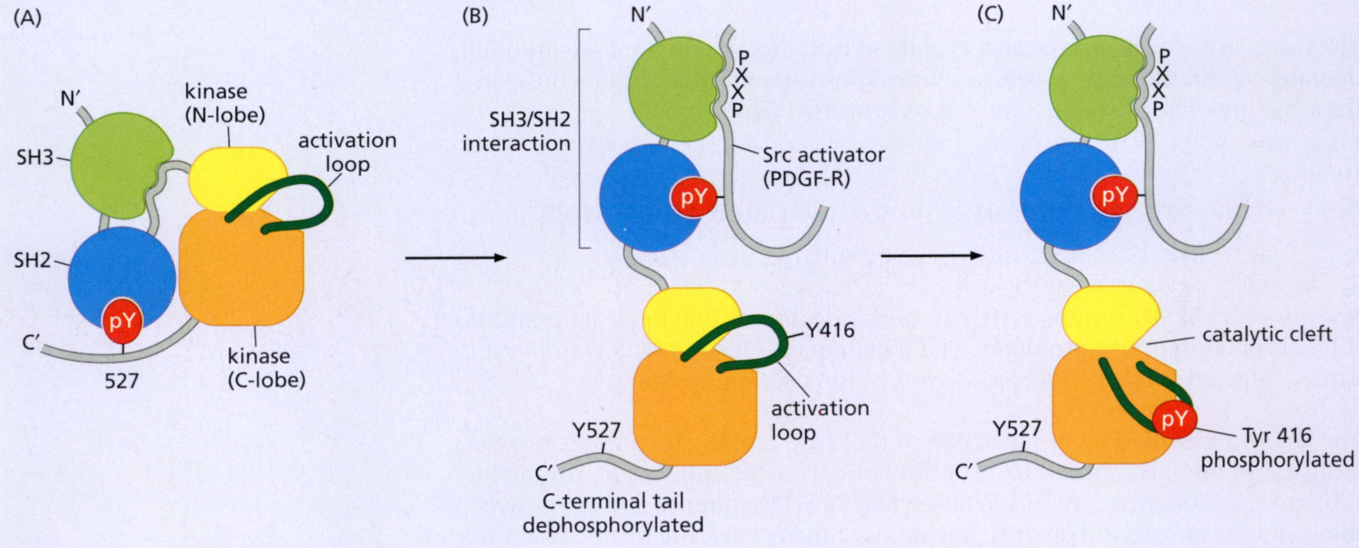

Figure 6.11 Structure and function of Src (A) The SH2 group of Src *(blue)* normally binds in an intramolecular fashion to a phosphotyrosine residue (pY) at position 527 near the C-terminus of Src *(red)*. This binding causes the catalytic cleft of Src, which is located between the N-lobe and the C-lobe of the kinase domain, to be obstructed. At the same time, the SH3 *(light green)* domain binds a proline-rich segment in the N-lobe of the kinase domain. (B) However, when the PDGF receptor (indicated here as a "Src activator") becomes phosphorylated, one of its phosphotyrosines *(red)* serves as a ligand for this SH2 group of Src, causing the SH2 to switch from its intramolecular binding to an intermolecular binding. Concomitantly, the SH3 group of Src detaches from its intramolecular binding to bind a proline-rich domain (PXXP) on the receptor. These changes place the catalytic cleft of Src in a configuration that enables it to fire. (C) A final phosphorylation of tyrosine 416 moves an obstructing oligopeptide activation loop *(dark green)* out of the way of the catalytic cleft, yielding the full tyrosine kinase activity of Src. (Adapted from W. Xu et al., *Mol. Cell* 3:629–638, 1999.)

Detailed analysis of the structures of Grb2 and similarly configured proteins (e.g., Crk) indicate that they carry no other functional domains beyond their SH2 and SH3 domains. Apparently, these adaptor proteins are nothing more than bridge builders designed specifically to link other proteins to one another. (An alternative bridging, also shown in Figure 6.12, can also be achieved when a growth factor receptor associates with Sos via Grb2 and Shc.)

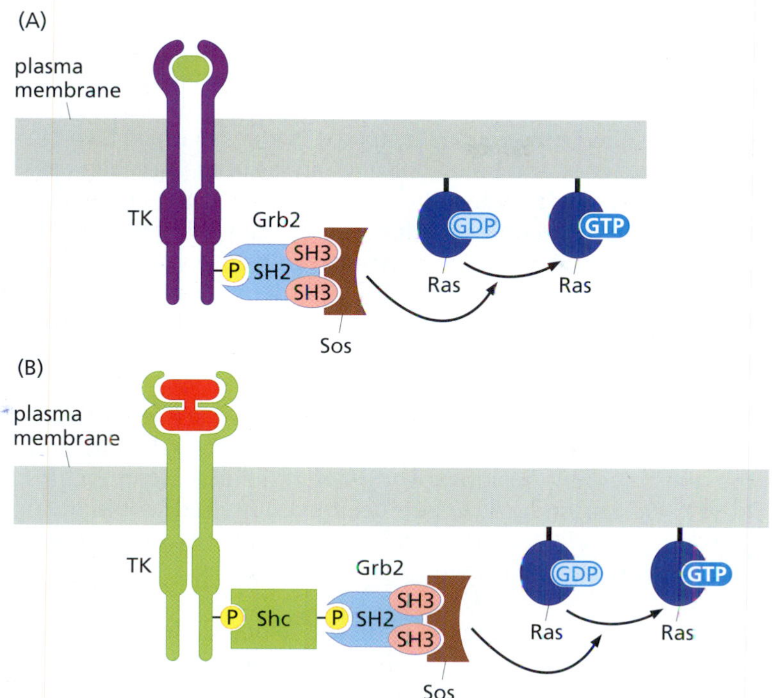

Figure 6.12 Intermolecular links forged by the Grb2 and Shc bridging proteins (A) The association of Sos, the Ras guanine nucleotide exchange factor (GEF), with ligand-activated growth factor receptors is achieved by bridging proteins, such as Grb2 shown here. The two SH3 domains of Grb2 bind to proline-rich domains of Sos, while its SH2 domain binds to a phosphotyrosine located on the C-terminal tail of a receptor. (B) An alternative mode of association can be achieved through a collaboration between two bridging proteins, Grb2 and Shc. Grb2 uses its SH3 domains to bind Sos and its SH2 domain to bind a phosphotyrosine on Shc; the latter uses its own SH2 domain to bind a phosphotyrosine on a ligand-activated receptor. (Adapted from G.M. Cooper, Oncogenes, 2nd ed. Boston and London: Jones and Bartlett, 1995.)

Once Sos becomes anchored via Grb2 (or via Grb2 + Shc) to a receptor, it is brought into close proximity with Ras proteins, most of which seem to be permanently tethered to the inner face of the plasma membrane. Sos is then physically well positioned to interact directly with these Ras molecules, inducing them to release GDPs and bind instead GTPs. This guanine nucleotide exchange causes the activation of Ras protein signaling. Hence, the biochemically defined pathway could now be drawn like this:

Receptor → Grb2 → Sos → Ras

or

Receptor → Shc → Grb2 → Sos →Ras

Because a diverse group of signaling proteins becomes attracted to ligand-activated receptors (see, for example, Figure 6.9), the Sos–Ras pathway represents only one of a number of signaling cascades radiating from growth factor receptors.

6.5 A cascade of kinases forms one of three important signaling pathways downstream of Ras

The multiple cellular responses that a growth factor elicits could now be explained by the fact that its *cognate* receptor (i.e., the receptor that specifically binds it) is able to activate a specific combination of downstream signaling pathways. Each of these pathways might be responsible for inducing, in turn, one or another of the biological changes occurring after cells are stimulated by this particular growth factor. Moreover, exaggerated forms of this signaling might well be operating in cancer cells that experience continuous growth factor stimulation because of an autocrine signaling loop or a mutant, constitutively activated receptor.

This branching of downstream pathways did not, however, explain how the mutant Ras oncoprotein (which represents only one of the multiple pathways downstream of growth factor receptors) is able to evoke a number of distinct changes in cells. Here, once again, signal transduction biochemistry provided critical insights. We now know that at least three major downstream signaling cascades radiate from the Ras protein. Their diverse actions help to explain how the Ras oncoprotein causes so many changes in cell behavior.

The layout of these downstream signaling pathways was mapped through a combination of yeast genetics and biochemical analyses of mammalian cells. These approaches made it clear that when Ras binds GTP, a domain of the Ras protein, termed its **effector loop**, is able to interact physically with several alternative downstream signaling partners that are known collectively as Ras *effectors*, i.e., the proteins that carry out the actual work of Ras (Figure 6.13). Each of these effectors binds quite tightly to the effector loop of the GTP-bound form of Ras protein while having little affinity for the loop presented by its GDP-bound form.

The first of these Ras effectors to be discovered was the Raf kinase. Like the great majority of protein kinases in the cell, Raf phosphorylates substrate proteins on their serine and threonine residues. Long before its association with Raf was uncovered, the Raf kinase had already been encountered as the oncoprotein specified by both a rapidly transforming murine retrovirus and, in homologous form, a chicken retrovirus (see Table 3.3).

The activation of Raf by Ras depends upon the relocalization of Raf within the cytoplasm. Recall that Ras proteins are always anchored to a membrane, usually the inner surface of the plasma membrane, through their C-terminal hydrophobic tails. Once it has bound GTP, activated Ras attracts and binds Raf via its effector loop. Before this association, Raf is found in the cytosol; thereafter, Raf becomes tethered via Ras to the plasma membrane (or, in certain cases, to another cytoplasmic membrane).

The association of Raf with Ras seems to shift the three-dimensional configuration of the Raf kinase molecule. In addition, during the time it is anchored to Ras, Raf may become phosphorylated by still-poorly understood kinases, possibly including Src. Raf now acquires active signaling powers and proceeds to phosphorylate and thereby activate a second kinase known as MEK (MAPKK). MEK is actually a "dual-specificity kinase," that is, one that can phosphorylate

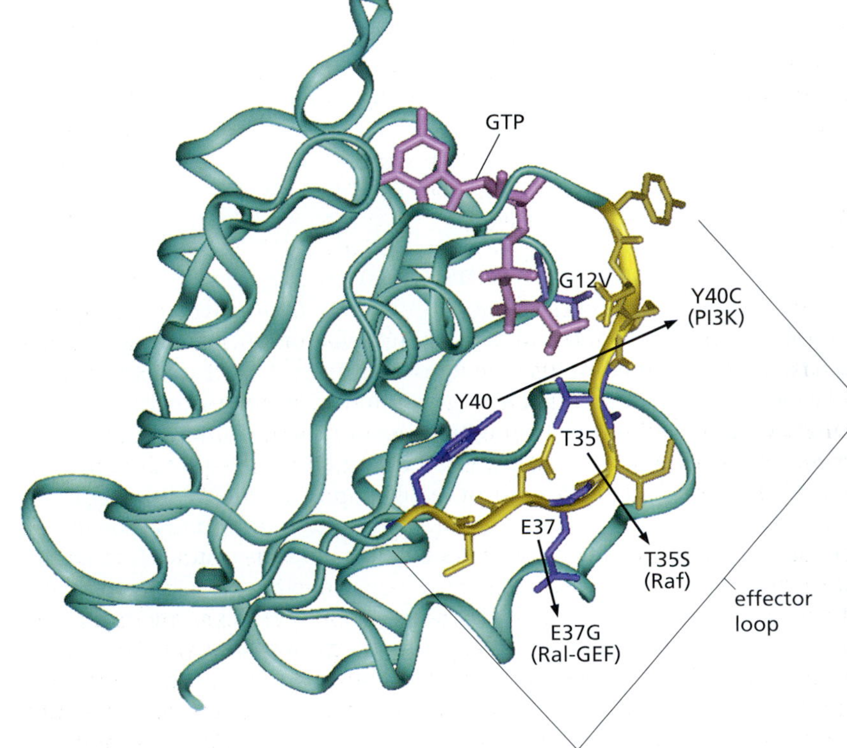

Figure 6.13 The Ras effector loop The crystallographic structure of the Ras protein, viewed from a different angle than in Figure 5.31, is depicted here in a ribbon diagram *(largely aquamarine)*. The guanine nucleotide *(purple)* is shown together with the nearby glycine residue, which, upon replacement by a valine, causes inactivation of the GTPase activity (G12V). The effector loop *(yellow)* interacts with at least three important downstream effectors of Ras. Amino acid substitutions *(arrows)* affecting three residues in the effector loop create three alternative forms of the Ras protein that interact preferentially with either PI3 kinase, Raf, or Ral-GEF. (Courtesy of R. Latek and A. Rangarajan, adapted from data in U. Krengel et al., *Cell* 62:539–548, 1990.)

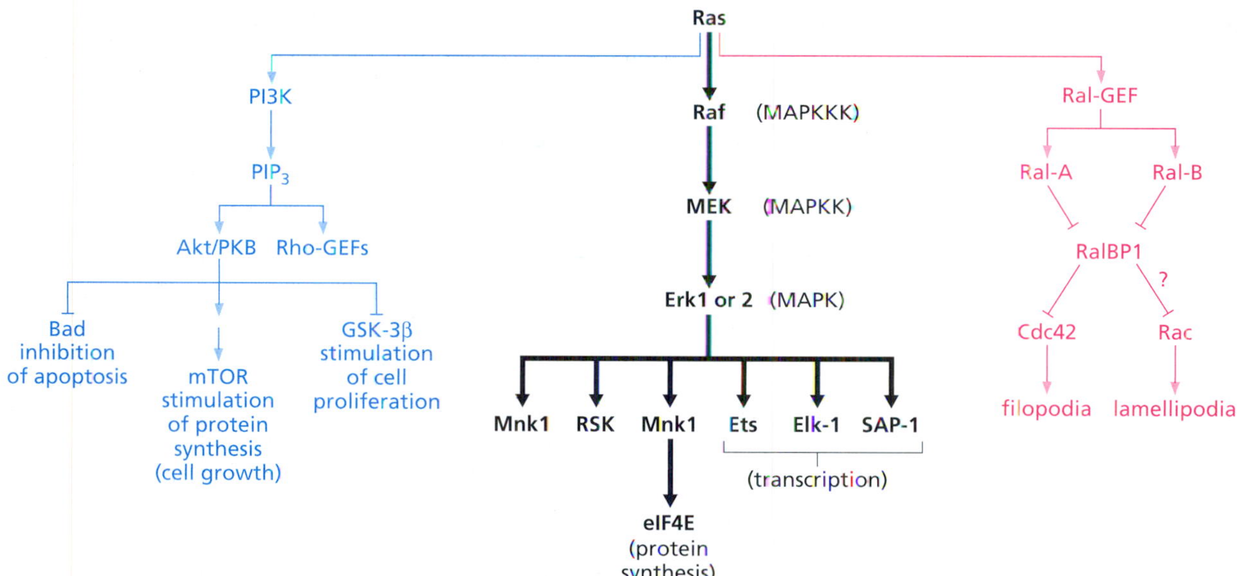

Figure 6.14 The Ras → Raf → MAP kinase pathway This signaling cascade *(black)* is one of a series of similarly organized pathways in mammalian cells that have the overall plan MAPKKK → MAPKK → MAPK. In this particular case, Ras activates the Raf kinase (a MAPKKK), the latter proceeds to phosphorylate and activate MEK (a MAPKK), and MEK then phosphorylates and activates Erk1 and Erk2 (MAPKs). The latter can then phosphorylate kinases in the cytoplasm that regulate translation as well as transcription factors in the nucleus.

serine/threonine residues as well as tyrosine residues. MEK uses these powers to phosphorylate two other kinases, the *e*xtracellular signal-*r*egulated *k*inases 1 and 2, commonly referred to as Erk1 and Erk2. Once phosphorylated and activated, each of these Erks then phosphorylates substrates that, in turn, regulate various cellular processes including transcription.

This type of kinase signaling cascade has been given the generic name of a MAPK (mitogen-activated protein kinase) pathway. In fact, there are a number of such signaling cascades in mammalian cells. Erk1 and Erk2 at the bottom of the Raf-initiated cascade are considered MAPKs. The kinase responsible for phosphorylating a MAPK is termed generically a MAPKK; in the case of the cascade studied here, this role is played by MEK. The kinase responsible for phosphorylating a MAPKK is therefore called generically a MAPKKK (Figure 6.14); in this scheme, Raf is classified as a MAPKKK.

Note that in these various kinase cascades, phosphorylation of kinase B by kinase A always results in the functional activation of kinase B. For example, when MEK is phosphorylated by Raf, MEK becomes activated as a signal-emitting kinase. Such serine/threonine phosphorylation therefore has a very different consequence from the phosphorylation of tyrosine residues carried by the cytoplasmic tails of growth factor receptors. The phosphotyrosines on these receptors attract signaling partners and are therefore responsible for the *relocalization* of partner proteins; in contrast, the phosphate groups attached to MEK by Raf cause a shift in its structure that results in its *functional activation* as a kinase.

The Ras → Raf → MEK → Erk pathway is only one of several downstream signaling cascades that are activated by the GTP-bound Ras proteins. The importance of this kinase cascade in cell transformation is indicated by the fact that in a number of cell types, the Raf protein kinase, when present in mutant, oncogenic form, can evoke most of the transformation phenotypes that are induced by the Ras oncoprotein. Hence, in such cells, this Raf pathway is responsible for the lion's share of the transforming powers of Ras oncoproteins.

Once activated, an Erk kinase at the bottom of this cascade proceeds to phosphorylate cytoplasmic substrates and can also translocate to the nucleus, where it causes the phosphorylation of transcription factors; some of the latter then initiate the immediate and delayed early gene responses. For example, as indicated in Figure 6.14, Erk can phosphorylate several transcription factors (Ets, Elk-1 and SAP-1) directly and, in addition, phosphorylate and thereby activate other kinases, which proceed to phosphorylate and activate yet other transcription factors. At the same time, these downstream kinases phosphorylate and thereby reconfigure two chromatin-associated proteins (HMG-14 and histone H3); these particular modifications place the chromatin in a configuration that is more hospitable to transcription. The Mnk1 kinase, a cytoplasmic substrate of Erk1 and Erk2, phosphorylates and thereby activates the translation initiation factor eIF4E, thereby helping to activate the cellular machinery responsible for protein synthesis.

Once the Erks can phosphorylate the Ets transcription factor, the latter can then proceed to stimulate the expression of important growth-regulating genes, such as those specifying heparin-binding EGF (HB-EGF), cyclin D1, Fos, and p21$^{\text{Waf1}}$. (The functions of the last three will be described in detail in Chapter 8.)

Especially important among the genes induced by the Raf → MEK → Erk pathway is the activation of the promoters of the two immediate early genes encoding the Fos and Jun transcription factors. Once synthesized, these two proteins can associate with one another to form AP-1, a widely acting heterodimeric transcription factor that is often found in hyperactivated form in cancer cells. The special importance and influence of Fos and Jun is indicated by the fact that each was originally identified as the product of a retrovirus-associated oncogene (Table 3.3).

This Ras → Raf → MEK → Erk cascade contributes importantly to certain cancer cell phenotypes that are induced by the Ras oncoproteins. In addition to activating a number of growth-promoting genes, as described above, this pathway confers anchorage independence and loss of contact inhibition. It also contributes to the profound change in cell shape that is associated with transformation by the *ras* oncogene.

In certain cancers, the signaling pathway lying downstream of Raf may be strongly activated without any direct involvement of Ras. Thus in 60–70% of melanomas, a close relative and functional analog of Raf, called B-Raf, is found in a mutated, constitutively activated form. It remains unclear why these melanomas develop mutant B-Raf rather than mutant Ras in order to drive their proliferation.

6.6 A second pathway downstream of Ras controls inositol lipids and the Akt/PKB kinase

A second important downstream effector of the Ras protein enables the Ras protein to evoke yet other cellular responses (Figure 6.15). In the context of cancer, the most important of these is a suppression of the cellular apoptosis (cell suicide) program. This anti-apoptotic effect is especially critical for cancer cells, since many of them are poised on the brink of activating this program, a subject covered in great detail in Chapter 9.

Early studies of the biochemistry of the plasma membrane suggested that its phospholipids serve simply as a barrier between the aqueous environments in the cell exterior and interior. The biochemistry of these phospholipids showed that they are **amphipathic**—that is, they possess a hydrophilic head, which likes to be immersed in water, and a hydrophobic tail, which prefers nonaqueous environments (Figure 6.16A). This polarity explains the structure of lipid bilayers such as the plasma membrane, in which the hydrophilic groups face and protrude into the extracellular and cytosolic aqueous environments while the hydrophobic tails are buried in the middle of the membrane, from which water is excluded.

In the 1970s, it became apparent that eukaryotic cells exploit some of the membrane-associated phospholipids for purposes that are fully unrelated to the maintenance of membrane structure. Some phospholipids contain, at their hydrophilic heads, an inositol group. Inositol is a water-soluble carbohydrate molecule (more properly termed a polyalcohol). The inositol moiety of such phospholipids can be modified by the addition of phosphate groups. The resulting phosphoinositol may then be cleaved away from the remaining, largely hydrophobic portions of a phospholipid molecule. Since it is purely hydrophilic, this phosphoinositol, termed IP_3, can then diffuse away from the membrane (Figure 6.16.B), thereby serving as an *intracellular hormone* to dispatch signals from the plasma membrane to distant parts of the cell. Such intracellular hormones are often called **second messengers**. The second product of this cleavage, termed DAG (Figure 6.16B), can serve to activate a key signaling kinase in the cell—the serine/threonine kinase known as protein kinase C (PKC). Alternatively, a phosphorylated inositol can remain attached to the remainder of the phospholipid and thus can remain embedded in the plasma membrane; there, it can serve as an anchoring point to which certain cytosolic proteins can become attached.

We now know that the inositol moiety can be modified by several distinct kinases, each of which shows specificity for phosphorylating a particular hydroxyl of inositol. Phosphatidylinositol 3-kinase (PI3K), for example, is responsible for attaching a phosphate group to the 3′ hydroxyl of the inositol moiety of membrane-embedded phosphatidylinositol (PI). While several distinct PI3 kinases have been discovered, the most important of these may well be the PI3K that phosphorylates $PI(4,5)P_2$ (also called PIP_2). Prior to modification by PI3K, PIP_2 already has phosphates attached to the 4′ and 5′ hydroxyl groups of inositol; after this modification, this inositol acquires yet another phosphate group, and PIP_2 is converted into phosphatidylinositol (3′,4′,5′)-triphosphate (PIP_3; see Figure 6.16B).

Note an important difference between the organization of this pathway and the Ras → Raf → MEK → Erk cascade we read about in the previous section. In the present case, one of the critical kinases (i.e., PI3K) attaches phosphates to a phospholipid, rather than phosphorylating a protein substrate, such as another kinase.

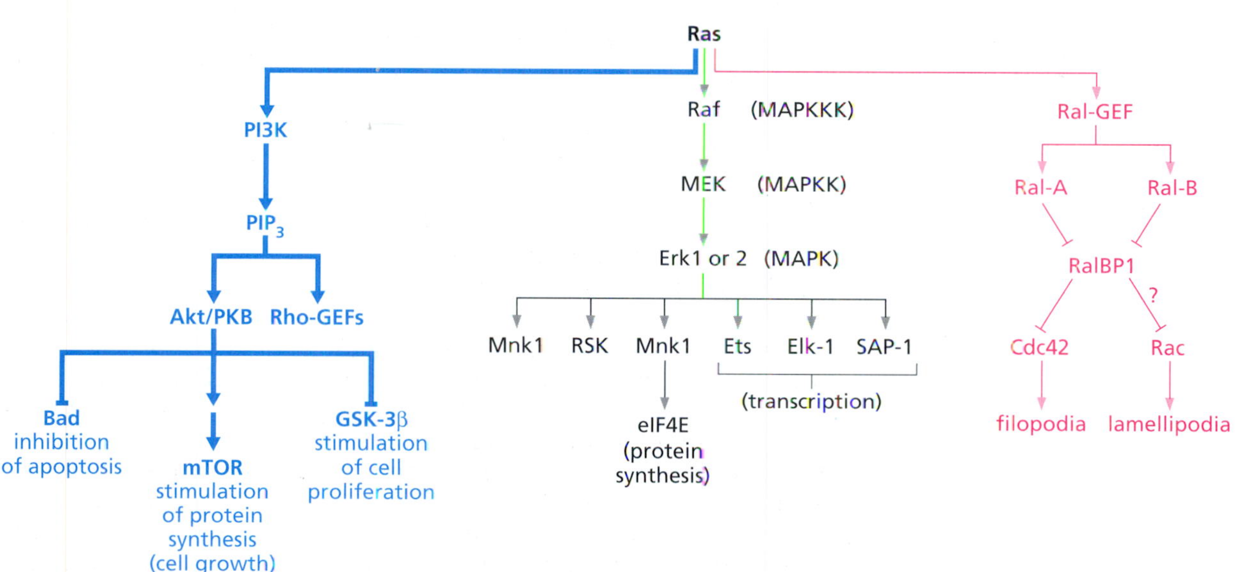

Figure 6.15 The PI3 kinase pathway A second effector pathway of Ras *(blue)* derives from its ability to associate with and activate phosphatidylinositol 3-kinase (PI3K) with the resulting formation of phosphatidylinositol (3,4,5)-triphosphate (PIP_3). This leads in turn to the tethering of PH-containing molecules, such as Akt (also called PKB) and Rho guanine nucleotide exchange proteins (Rho-GEFs), to the plasma membrane. Akt/PKB is able to inactivate Bad (suppressing apoptosis), inactivate GSK-3β, an antagonist of expression of growth-promoting proteins such as cyclin D1 and Myc, and may activate mTOR (stimulating protein synthesis and cell growth).

We first encountered PI3K, in passing, at an earlier point in this chapter (see Section 6.3). There, we described this enzyme's attraction to certain ligand-activated receptors via the SH2 group of its regulatory subunit, and the consequent phosphorylation of inositol-containing phospholipids already present in the plasma membrane. It turns out that GTP-activated Ras is also able to bind PI3K and thereby enhance its functional activation (Figure 6.17). Hence, PI3K can function as a direct downstream effector of Ras. As is the case when Raf binds Ras, the binding of PI3K to Ras causes PI3K to become closely associated with the plasma membrane, where its PI substrates are located.

Once activated by one or both of these associations, PI3K plays a central role in a number of signaling pathways (e.g., see Figure 5.7). Testimonial to this is provided by the fact that this enzyme is activated by a diverse array of signaling agents, including PDGF, nerve growth factor (NGF), insulin-like growth factor-1 (IGF-1), interleukin-3, and the extracellular matrix (ECM) attachment achieved by integrins (see Section 6.9).

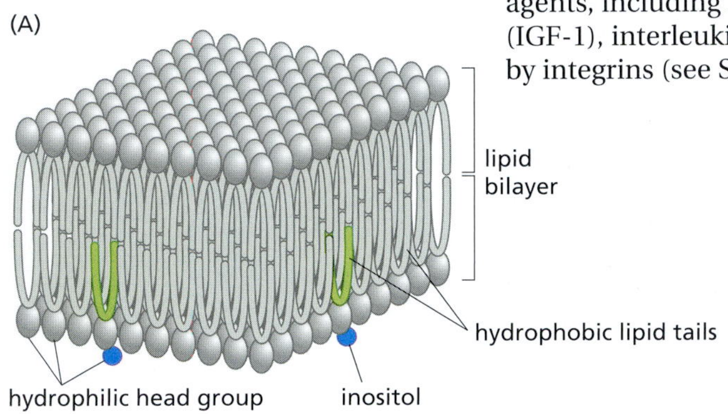

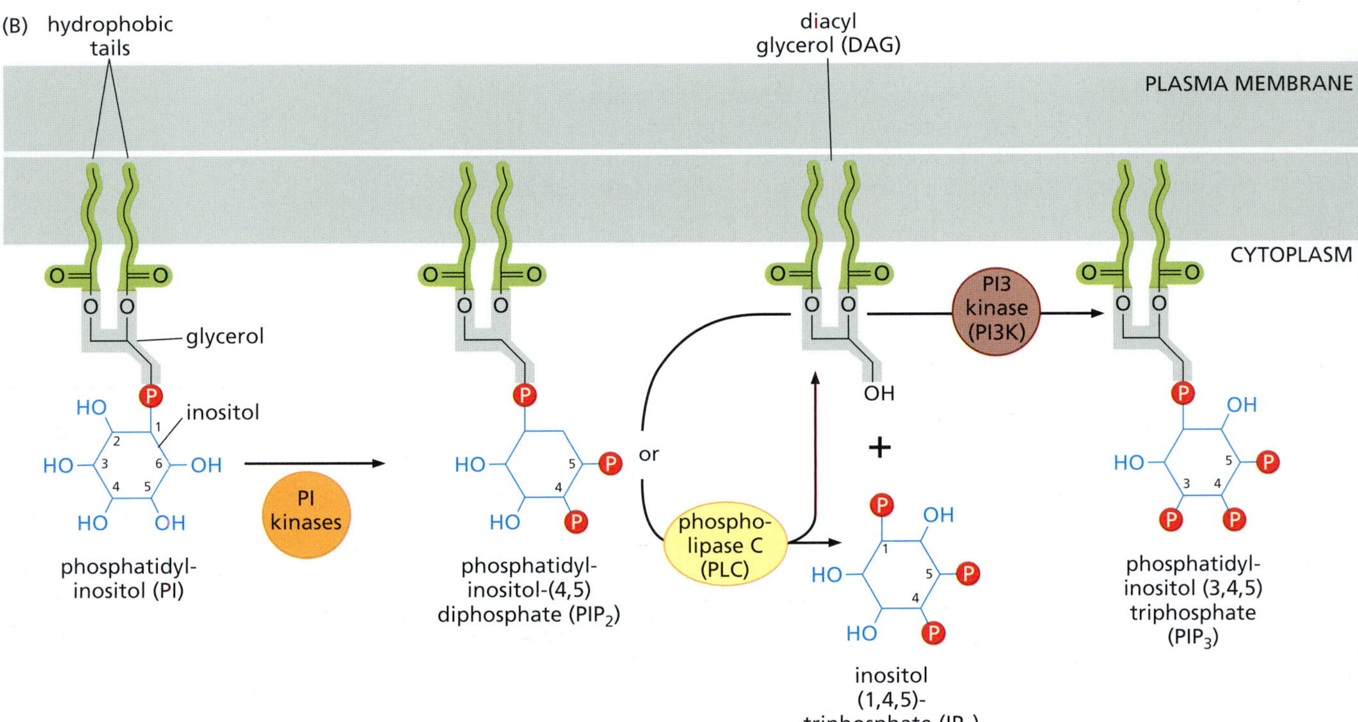

Figure 6.16 Biochemistry of lipid bilayers (A) All cellular membranes are assembled as phospholipid bilayers, in which the hydrophilic head groups *(gray ovals)* protrude into the aqueous solvent above and below the membrane, while the hydrophobic tails *(green, gray)* are hidden in the nonaqueous space within the lipid bilayer. A small minority of the head groups contain inositol sugars *(blue)*. (B) Phosphatidylinositol (PI) *(left)* is composed of three parts—two fatty acids with long hydrocarbon tails *(green)* inserted into the plasma membrane, glycerol *(gray)*, and inositol *(light blue)* attached to the glycerol via a phosphodiester linkage. PI kinases can add phosphate groups to the various hydroxyls of inositol, yielding, for example, the PI (4,5)-diphosphate (PIP_2) shown here. Cleavage of PIP_2 by phospholipase C yields diacylglycerol (DAG), which can activate protein kinase C (PKC); and inositol (1,4,5)-triphosphate (IP_3), which induces release of calcium ions from intracellular stores. Alternatively, PIP_2 can be further phosphorylated by PI3 kinase (PI3K) to yield phosphatidylinositol (3,4,5)-triphosphate (PIP_3). Once formed, the phosphorylated "head group" of PIP_3 serves to attract proteins carrying PH domains, which thereby become tethered to the inner surface of the plasma membrane.

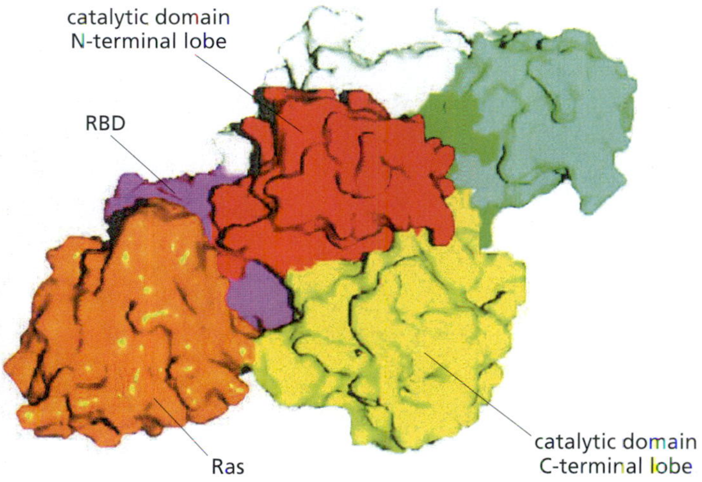

Figure 6.17 Activation of PI3K by Ras The tight binding of Ras *(orange)* via its effector domain to the Ras-binding domain (RBD; *purple*) of PI3 kinase allows Ras to aid in the activation of the PI3 kinase catalytic domain *(red, yellow)*, which can then proceed to phosphorylate the PIP_2 phospholipid, yielding PIP_3. (Other forms of PI3K are presumed to have structures very similar to the the PI3Kγ depicted here.) (From M.E. Pacold et al., *Cell* 103:931–943, 2000.)

The formation of the phosphorylated inositol head groups of phosphatidylinositol has little if any effect on the overall structure and function of the plasma membrane. (The inositol phospholipids are minor constituents of the plasma membrane.) However, a phosphorylated inositol head group protruding from the plasma membrane may be recognized and bound by certain proteins that are usually floating in the cytosol. Once anchored via a phosphoinositol to the plasma membrane, these proteins are then well positioned to release certain types of signals, as we will see below.

The most important of the phosphorylated inositol head groups appears to be PIP_3, whose inositol carries phosphates at its 3′, 4′, and 5′ positions (see Figure 6.16B). Many of the cytosolic proteins that are attracted to PIP_3 carry pleckstrin homology (PH) domains (see Table 6.2) that have strong affinity for this triply phosphorylated inositol head group (Figure 6.18). Arguably the most important PH domain–containing protein is the serine–threonine kinase named Akt, also known as protein kinase B (PKB). (Akt/PKB is yet another protein that was initially discovered because it is encoded by a retrovirus-associated oncogene.) Thus, once PIP_3 is formed by PI3K, an Akt/PKB kinase molecule can become

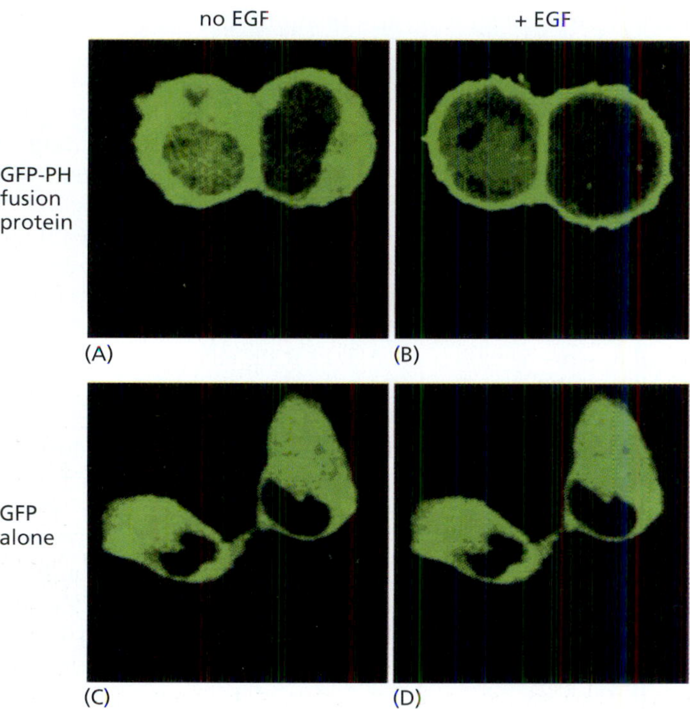

Figure 6.18 Migration of PH-containing proteins to PIP_3 in the plasma membrane After PIP_3 head groups are formed by PI3 kinase (Figure 6.16B), they become attractive docking sites for PH domain–containing proteins. This phosphorylation and homing occurs within seconds of the activation of PI3 kinase, as revealed by this micrograph, in which a fluorescing protein (green fluorescent protein; GFP) has been fused to a protein carrying a PH domain. This hybrid protein moves from a diffuse distribution throughout the cytoplasm (A) to the inner surface of the plasma membrane (B) within a minute after EGF (epidermal growth factor) has been added to cells. However, if the PH domain of the fusion protein is inactivated (C), addition of EGF has no effect on its intracellular distribution (D). (From K. Venkateswarlu et al., *J. Cell Sci.* 112:1957–1965, 1999.)

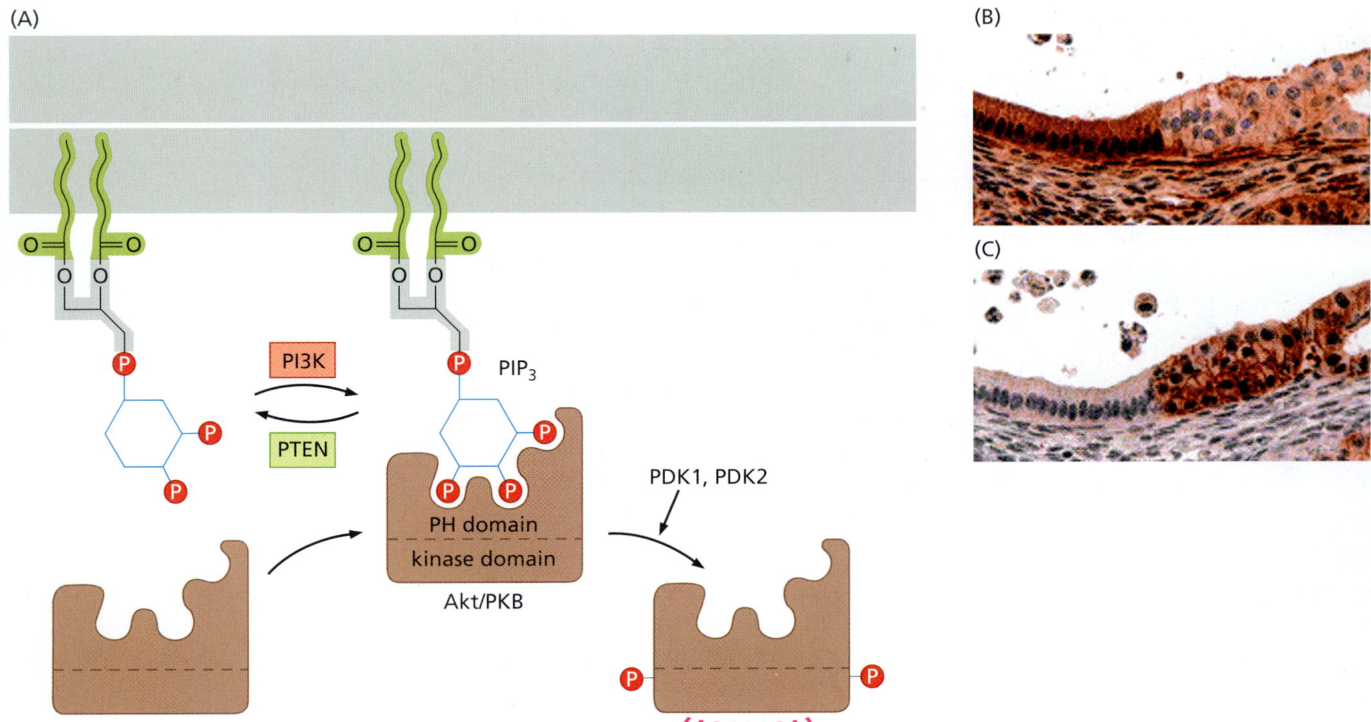

Figure 6.19 Docking of PH domains to PIP_3 (A) The formation of PIP_3 (see Figure 6.16B) by PI3K creates a docking site for cytoplasmic proteins that carry PH (pleckstrin homology) domains. The most important of these is the Akt/PKB kinase *(brown)*. Once docked to the inner surface of the plasma membrane via its PH domain, Akt/PKB becomes doubly phosphorylated by two other kinases (PDK1, PDK2) and thereby activated *(lower right)*. (Recent evidence identifies the previously poorly characterized PDK2 as mTOR, to be discussed in Chapter 16.) Akt/PKB then proceeds to phosphorylate a variety of substrates regulating proliferation, cell survival, and cell size. However, if PTEN is also present and active, it dephosphorylates PIP_3 and converts it to PIP_2, thereby depriving Akt/PKB of a docking site at the plasma membrane. (B) The dynamics of this interaction are seen in genetically altered mice having a *Pten*$^{+/-}$ genotype. Uterine epithelial cells in these mice often lose the remaining *Pten* wild-type allele, yielding *Pten*$^{-/-}$ cells, which form localized hyperplasias and cysts. When immunostained with anti-PTEN antiserum, the epithelial cells on the right have lost PTEN expression and become physically larger, indicating the effects of the PIP_3 pathway on cell size. (C) When the same lesion is stained with an anti-phospho-Akt/PKB antibody (which detects activated Akt/PKB), the cells on the right show high P-Akt activity, while those on the left show very low activity. (B and C, from K. Podsypanina et al., *Proc. Natl. Acad. Sci. USA* 98:10320–10325, 2001.)

tethered via its PH domain to the inositol head group of PIP_3 that protrudes from the plasma membrane into the cytosol (Figure 6.19).

This association of Akt/PKB with the plasma membrane (together with phosphorylations of Akt/PKB that soon follow) results in the functional activation of Akt/PKB as a kinase. Once activated by these phosphorylations, Akt/PKB proceeds to phosphorylate a series of protein substrates that have multiple effects on the cell. Three major biological effects that Akt/PKB has on cells are (1) aiding in cell survival by reducing the possibility that the cell apoptotic suicide program will become activated; (2) stimulating cell proliferation; and (3) stimulating cell growth, in the narrowest sense of the term, that is, stimulating increases in cell size (see Figure 6.19B). In addition, it also exerts an influence, still poorly understood, on cell motility.

The activation of PI3K, and thus of Akt/PKB, is under very tight control. In a quiescent, nongrowing cell lacking growth factor stimulation, the intracellular levels of PIP_3 are extremely low. Once such a cell encounters mitogens, the levels of PIP_3 in this cell increase rapidly. The normally low levels of PIP_3 and other phosphorylated PI molecules are the work of a series of phosphatases that reverse the actions of the activating kinases such as PI3K. The best-characterized of these phosphatases, PTEN, removes the 3′ phosphate group from PIP_3 that has previously been attached by PI3K (see Figure 6.19). This removal suggests two distinct mechanisms by which the Akt/PKB signaling pathway can become deregulated in cancer cells—hyperactivity of PI3K or inactivity of PTEN.

The portfolio of diverse biological functions assigned to Akt/PKB requires that this kinase be able to phosphorylate a variety of substrates, each involved in a distinct downstream signaling pathway (Table 6.3). The first of these functions—facilitating cell survival—depends on the ability of Akt/PKB to suppress any tendencies the cell may have to activate its own built-in suicide program of apoptosis. We defer detailed discussion of apoptosis until Chapter 9. For the moment, suffice it to say that phosphorylation by Akt/PKB results in the inhibition of several proteins that play prominent roles favoring the entrance of a cell into apoptosis.

The proliferative functions of Akt/PKB depend on its ability to perturb proteins that are important for regulating the advance of a cell through its cycle of growth and division, often termed the cell cycle. We will learn more about this complex program in Chapter 8.

Independent of these effects on proliferation, Akt/PKB is able to induce dramatic changes in the proteins that control the rate of protein synthesis in the cell. Akt/PKB phosphorylates and inactivates a protein termed TSC2, which otherwise acts through intermediaries to trigger inactivation of mTOR kinase; the latter then phosphorylates and inactivates 4E-BP, a potent inhibitor of translation, and, at the same time, phosphorylates and thereby activates p70S6 kinase, an activator of translation. These changes allow Akt/PKB to increase the efficiency with which the translation of a class of mRNAs is initiated; the resulting elevated rate of protein synthesis favors the accumulation of many cellular proteins and is manifested in the growth (rather than the proliferation) of cells. (We will return to the detailed organization of this pathway in Section 16.15). An illustration of the growth-inducing powers of Akt/PKB is provided by transgenic mice that express a mutant, constitutively active Akt/PKB in the β cells of the pancreas (Figure 6.20); in such mice, these particular cells grow to a size more than twice the cross-sectional area (and almost 4 times the volume) of normal β cells!

There is evidence that yet other cytoplasmic proteins use their PH domains to associate with PIP_3, and that this association also favors their functional activation. Among the PH-bearing cellular proteins are a group of guanine nucleotide exchange factors (GEFs) that function analogously to Sos, being responsible for activating various small GTPases that are distant relatives of the Ras proteins. These other GTPases belong to the Rho family of signaling proteins, which

Table 6.3 Effects of Akt/PKB on survival, proliferation, and cell growth

Biological effect	Substrate of Akt/PKB	Functional consequence
Anti-apoptotic		
	Bad (pro-apoptotic)[a]	inhibition
	caspase-9 (pro-apoptotic)[b]	inhibition
	IκB kinase (anti-apoptotic)[c]	activation
	FOXO1 TF (pro-apoptotic)[d]	inhibition
	Mdm2 (anti-apoptotic)[e]	activation
Proliferative		
	GSK-3β (anti-proliferative)[f]	inhibition
	FOXO4 (anti-proliferative)[g]	inhibition
	$p21^{Cip1}$ (anti-proliferative)[h]	inhibition
Growth		
	Tsc2 (anti-growth)[i]	inhibition

[a]Bad is an antagonist of Bcl-X; both are members of the Bcl-2 family of proteins controlling pores in the mitochondrial membrane (Section 9.13).
[b]Caspase-9 is a component of the protease cascade that effects the apoptotic program (Section 9.13).
[c]IκB kinase, usually indicated as IKK, is phosphorylated and activated by Akt/PKB (Section 6.12).
[d]Phosphorylation of the forkhead (FOXO1, previously called FKHR) TF prevents its nuclear translocation and subsequent activation of pro-apoptotic genes.
[e]Mdm2, once phosphorylated by Akt/PKB, is activated and proceeds to trigger the destruction of p53 (Section 9.7).
[f]Akt/PKB phosphorylates and inactivates glycogen synthase kinase 3β (GSK-3β) activity, which is normally responsible for phosphorylating cyclin D1 (Section 8.5), causing its degradation.
[g]FOXO4 (formerly called AFX) induces expression of the CDK-inhibitor $p27^{Kip1}$ (Section 8.4) gene and some pro-apoptotic genes; once phosphorylated by Akt/PKB, FOXO4 is exported from the nucleus.
[h]$p21^{Cip1}$ is a CDK inhibitor like $p27^{Kip1}$ (Section 8.4). Phosphorylation by Akt/PKB causes it to exit the nucleus. Once in the cytoplasm, the resulting phosphorylated $p21^{Cip1}$ has been reported to act as a caspase inhibitor, thereby acquiring anti-apoptotic functions (Section 9.13).
[i]Phosphorylation of Tsc2 by Akt/PKB causes the Tsc1/Tsc2 complex to dissociate, allowing activation of mTOR, which proceeds to up-regulate protein synthesis (Section 16.15).

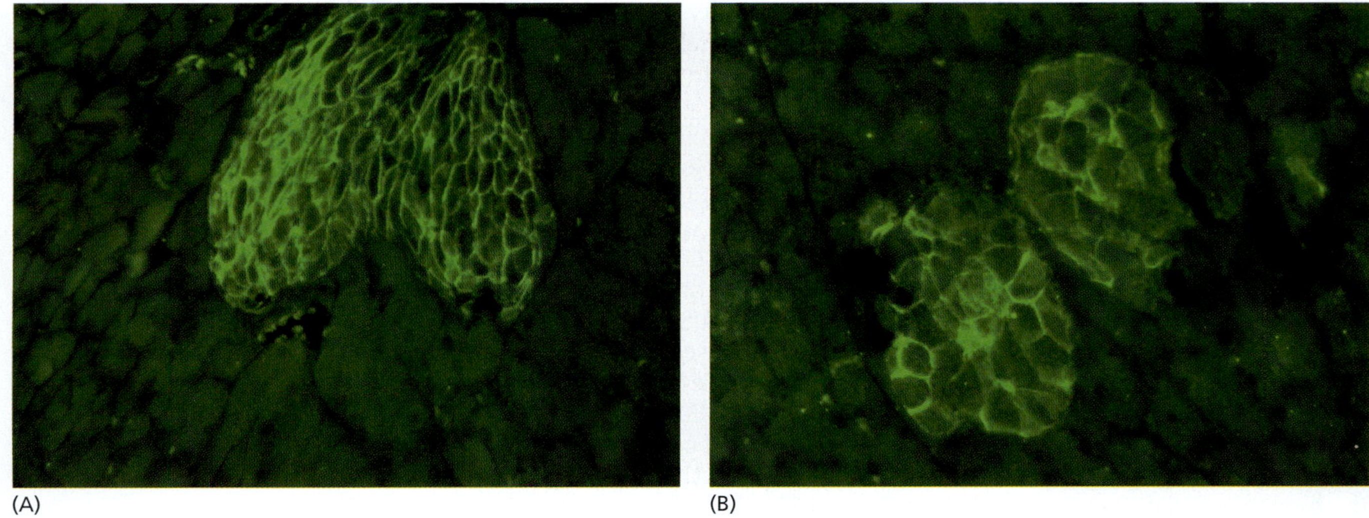

Figure 6.20 Akt/PKB and the control of cell growth The endocrine β cells in the pancreatic islets are normally responsible for secreting insulin in response to elevated concentrations of blood glucose. (A) A micrograph of a normal pancreatic islet. (B) If a constitutively active Akt/PKB kinase is expressed selectively in islet cells, they grow to volumes that are as much as fourfold larger than normal. (From R.L. Tuttle et al., *Nat. Med.* 7:1133–1137, 2001.)

includes Rho proper and its two cousins, Rac and Cdc42. Like Ras, these Rho proteins operate as binary switches, cycling between a GTP-bound actively signaling state and a GDP-bound inactive state (see Sidebar 6.4). The Rho-GEFs, once activated by binding to PIP_3, act on Rho proteins in the same way that Sos acts on Ras. In particular, the Rho-GEFs induce Rho proteins to jettison their bound GDPs, allowing GTP to jump aboard and activate the Rhos.

Once activated in this fashion, these Rho proteins have functions that differ strongly from those of Ras: they participate in reconfiguring the structure of the cytoskeleton and the attachments that the cell makes with its physical surroundings. In so doing, these Rho-like proteins control cell shape, motility, and, in the case of cancer cells, invasiveness. For example, Cdc42 is involved in reorganizing the actin cytoskeleton of the cell as well as controlling **filopodia**, small, fingerlike extensions from the plasma membrane that the cell uses to explore its environment; while Rac is involved in the formation of **lamellipodia**, broad ruffles extending from the plasma membrane that are found at the leading edges of motile cells. Because these processes are so intimately tied to the steps of tumor cell invasion and metastasis, we will defer further discussion of these Rho-like proteins and their mechanisms of action until Chapter 14.

The pathway involving PI metabolites and Akt/PKB is deregulated in a number of human cancer cell types through changes that are often quite independent of

Sidebar 6.4 Ras is the prototype of a large family of similar proteins Ras is only one of a large family of 35 similarly structured proteins that together are called "small G proteins" to distinguish them from the other class of larger, heterotrimeric guanine nucleotide–binding proteins that are activated by association with seven-membrane-spanning (serpentine) cell surface receptors (see Figure 5.25). Most of these small G proteins have been given three-letter names, each derived in some way from the initially named and discovered Ras proteins—for example, Ral, Rac, Ran, Rho, and so forth. (The Cdc42 protein is a member of the Rho family that has kept its own unique name because of its initial discovery through yeast genetics.)

Much of the sequence similarity shared between these otherwise diverse proteins is found in the cavities in each that bind and hydrolyze guanine nucleotides. Each of these small G proteins operates, as does Ras, like a binary switch, using a GTP–GDP–GTP cycle to flip back and forth between an on and an off state. Like Ras, each of these proteins has its own specialized guanine nucleotide exchange factors (GEFs) to activate it (by promoting replacement of GDP by GTP) and its own GTPase-activating proteins (GAPs) to trigger its GTPase activity.

Each of these small G proteins has been adapted to control a distinct cell physiologic or biochemical process, including regulation of the structure of the cytoskeleton and apoptosis. One is even used to regulate the transport of proteins through pores in the nuclear membrane. The only trait shared in common by these diverse cell biological processes is a need to be regulated by some type of binary switching mechanism—an operation for which the small G proteins are ideally suited.

Table 6.4 Alteration of the PI3K pathway in human tumors

Cancer type	Type of alteration
Glioblastoma (25–50%)	*PTEN* mutation
Ovarian carcinoma	*PTEN* mutation; *AKT2* amplification; *PI3K* amplification; PI3K *p85α* mutation
Breast carcinoma	increased Akt1 activity; *AKT2* amplification; *PTEN* mutation
Endometrial carcinoma (35%)	*PTEN* mutation; *PTEN* methylation[a]
Hepatocellular carcinoma	*PTEN* mutation
Melanoma	*PTEN* mutation; *PTEN* methylation[a]
Lung carcinoma	*PTEN* mutation
Renal cell carcinoma	*PTEN* mutation
Thyroid carcinoma	*PTEN* mutation; Akt/PKB overexpression
Lymphoid	*PTEN* mutation
Prostate carcinoma (40–50%)	*PTEN* mutation
Colon carcinoma (>30%)	Akt/PKB overexpression; *PI3K* mutation

[a]Methylation refers to repression of transcription of a gene through methylation of cytidines in its promoter; see Section 7.8.

Adapted from I. Vivanco and C.L. Sawyers, *Nat. Rev. Cancer* 2:489–501, 2002.

Ras oncoprotein action (Table 6.4). In some ovarian carcinomas, one form of PI3K is overexpressed and presumably highly active. In yet other tumors, such as lymphomas, cancers of the head and neck, and colon carcinomas, the Akt/PKB enzyme is overexpressed and presumably hyperactivated.

PI3K is hyperactivated in almost one-third of human colon carcinomas because of mutations affecting its regulatory subunit, called p110α. However, loss of PTEN function is far more frequently responsible for the deregulated activity of Akt/PKB, being observed in many types of human cancer cells (see Table 6.4). In a number of distinct tumor types, including breast and prostate carcinomas as well as glioblastomas, PTEN activity is lost because of chromosomal gene mutations or DNA methylation events that suppress expression of the *PTEN* gene. Altogether, PTEN activity is lost in 30 to 40% of all human cancers. The resulting accumulation of high levels of phosphorylated PIP_3 results in the downstream consequences enumerated above. This behavior of *PTEN* in suppressing cell proliferation indicates that it functions as a tumor suppressor gene, a topic that we will explore in Chapter 7.

6.7 A third Ras-regulated pathway acts through Ral, a distant cousin of Ras

The third of the three major effector pathways downstream of Ras involves a pair of Ras-like proteins termed Ral-A and Ral-B, which share 58% sequence identity with Ras (see Sidebar 6.4). As is the case with Ras, functional activation of these Ral proteins involves the replacement of bound GDP with GTP.

The communication between Ras and Ral is mediated by Ral guanine nucleotide exchange factors (Ral-GEFs), which can stimulate a Ral protein to shed its GDP and bind GTP, thereby echoing the effects of Sos on Ras. In addition to its GEF domain, each Ral-GEF has a pocket that can bind activated Ras. This association with Ras has two consequences for the Ral-GEF: Ras causes the localization of the Ral-GEFs near the inner surface of the plasma membrane; in addition, binding to Ras seems to cause a conformational shift in Ral-GEFs that activates their intrinsic guanine nucleotide exchange factor (GEF) activity.

The resulting activation of Ral-A and Ral-B proteins allows them to regulate targets further downstream in the signaling circuitry (Figure 6.21). For example, the Ral pathway can inactivate two of the Rho proteins cited in the last section—Cdc42 and Rac. Rac can also emit mitogenic signals and, by stimulating the production of reactive oxygen species, antagonize the actions of several Rho proteins. Though still poorly understood, the Ral proteins are likely to play key roles in the motility that enables cancer cells to invade and metastasize.

The functional description of Ras as a pleiotropically acting oncoprotein can now be explained at a biochemical level by the actions of the three downstream signaling cascades that radiate from Ras. It would be pleasing if each of the phenotypes that we have associated with Ras oncoprotein action (e.g., anchorage independence, survival, proliferation, biosynthetic rate, etc.) could be associated specifically with the actions of one of the three Ras effectors. In fact, things are more complicated: most Ras-induced phenotypes seem to be achieved through the collaborative actions of several of these effectors.

There are also two other dimensions of complexity. First, we have discussed Ras here as if it were a single protein, but there are indications that the four Ras proteins (one H-Ras, one N-Ras, and two K-Ras species) act in subtly different ways in distinct cytoplasmic locations. Second, our list of three Ras effectors (Raf, PI3K, and Ral-GEF) is hardly an exhaustive one. It is clear that Ras can interact with a number of other, still poorly characterized effectors that enable it to evoke yet other changes within the cell.

In any hierarchical organization, those individuals near the top exert far more power and influence than those lower down. By activating multiple control pathways concomitantly, a Ras oncoprotein (or a stimulatory signal emitted by one of its upstream controllers) induces many of the phenotypic changes acquired by a cell during the course of neoplastic transformation. In contrast, deregulation of one or another of the pathways downstream of Ras confers only a subset of these phenotypic changes and therefore yields far less growth advantage for a would-be cancer cell. This point is illustrated nicely by the mutant alleles of the gene encoding the B-Raf kinase, the close cousin of Raf, cited above, that is also activated by interaction with Ras. These mutations, which are found in many human melanomas, create oncogenic *BRAF* alleles that have transforming powers that are only about one-fiftieth those of the glycine-to-valine–substituted, activated Ras oncoprotein. Presumably, signaling components located further downstream, when altered by mutation, confer even less benefit (i.e., selective advantage) on evolving pre-neoplastic cells and therefore are rarely encountered in the genomes of human tumor cells.

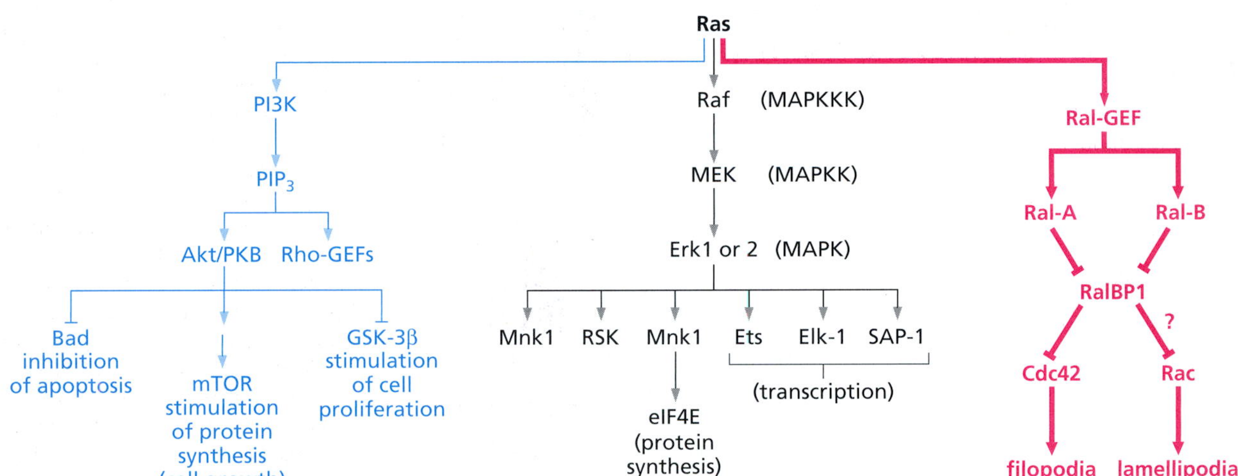

Figure 6.21 Ral and the control of the cytoskeleton The interaction of Ras with Ral-GEF *(red)* enables the latter to activate the Ral proteins Ral-A and Ral-B. The resulting GTP-bound Ral proteins proceed to activate a number of downstream pathways; prominent among them are those that activate Cdc42 and Rac, which in turn enable cellular motility through their effects on the actin cytoskeleton.

6.8 The Jak–STAT pathway allows signals to be transmitted from the plasma membrane directly to the nucleus

As mentioned in Section 5.7, study of a number of receptors for **cytokines**—growth factors that stimulate components of the hematopoietic system—has revealed receptor molecules that are constructed slightly differently from most of the growth factor receptors that were discussed in the last chapter. These particular receptor proteins do not carry covalently associated tyrosine kinase domains and instead form noncovalent complexes with tyrosine kinases of the Jak (Janus kinase) class (Figure 6.22). Included among these are the receptors for interferon (IFN), erythropoietin (EPO), and thrombopoietin (TPO). Interferon is an important mediator of immune cell function, as we will learn in Chapter 15; the other two control **erythropoiesis** (red blood cell formation) and **thrombopoiesis** (platelet formation), respectively.

Following ligand-mediated receptor dimerization, the Jak enzyme associated with each receptor molecule phosphorylates tyrosines on the cytoplasmic tails of the partner receptor molecule, much like the transphosphorylation occurring after ligand activation of receptors like the EGF-R and PDGF-R. The resulting phosphotyrosines attract and are bound by SH2-containing transcription factors termed STATs (signal transducers and activators of transcription). Once they are associated with the receptors, these STATs become phosphorylated by the Jaks. This creates individual STAT molecules that possess both SH2 groups and phosphotyrosines. Importantly, the SH2 groups displayed by the STATs have a specificity for binding the phosphotyrosine residues that have just been created on the STATs. Consequently, STAT–STAT dimers form, in which each STAT uses its SH2 group to bind to the phosphotyrosine of its partner. The resulting dimerized STATs migrate to the nucleus, where they function as transcription factors (see Figure 6.22).

The STATs activate target genes that are important for cell proliferation and cell survival. Included among the proliferation and survival genes are *myc*, the genes specifying cyclins D2 and D3 (which enable cells to advance through their

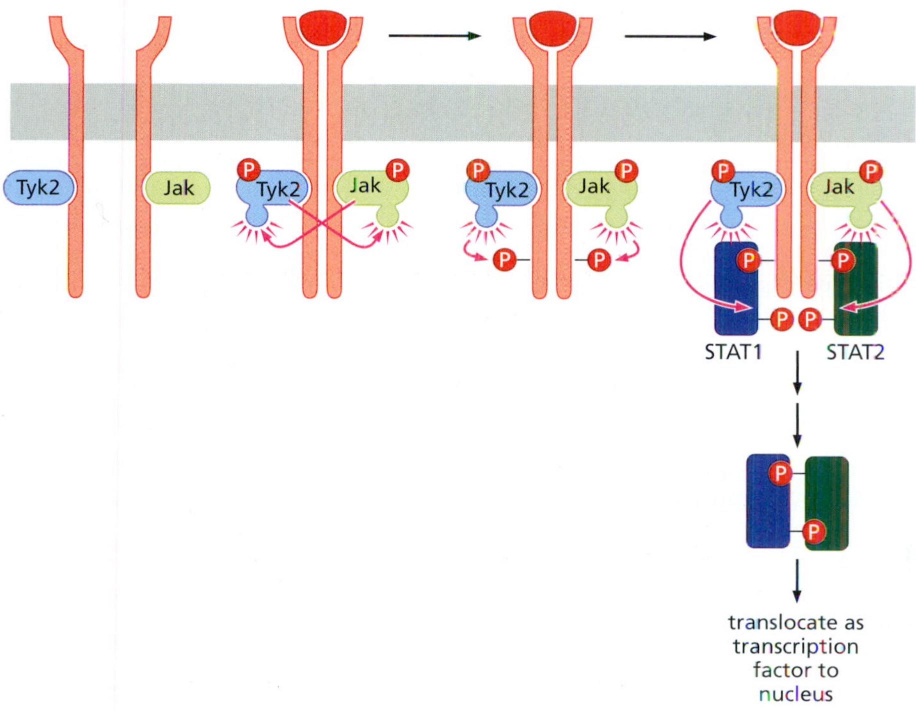

Figure 6.22 The Jak–STAT pathway The Jak–STAT pathway depends upon the actions of Jak tyrosine kinases (e.g., Jak1, Tyk2), which are attached noncovalently to a number of cytokine receptors, including those for interferons, erythropoietin (EPO), and thrombopoietin (TPO). Once a ligand has activated a receptor via dimerization, the Jaks (in this case Jak1 and Tyk 2) transphosphorylate one another as well as the C-terminal tails of the receptors. The resulting phosphotyrosines attract STAT proteins, such as the STAT1 and STAT2 shown here, which bind via their SH2 domains and become phosphorylated by the Jaks. Thereafter, the STATs dimerize, each using its SH2 domain to bind to the phosphotyrosine of its partner, and then translocate to the nucleus, where they operate as transcription factors to activate expression of key genes, e.g., the gene encoding an interferon.

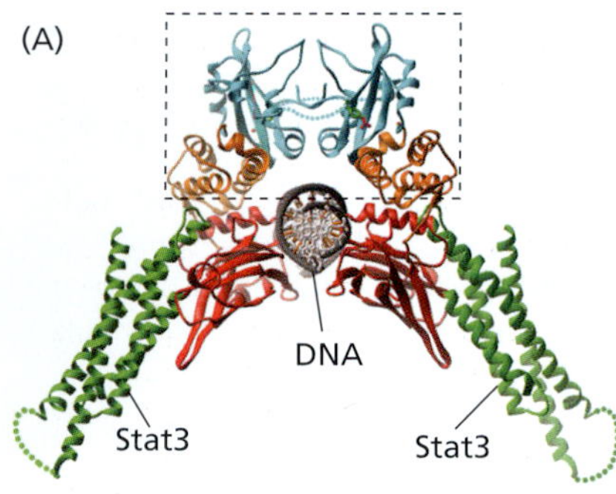

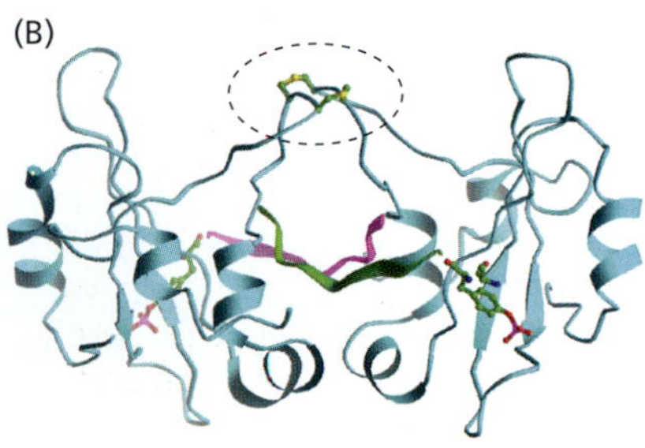

Figure 6.23 Constitutive activation of Stat3 (A) The Stat3 proteins normally dimerize through their C-terminal regions to form dimeric transcription factors that sit astride the DNA double helix (*gray circular structure, middle*). (B) Use of site-directed mutagenesis allowed the insertion of two cysteine residues in the C-terminal domain that led to the covalent dimerization of this C-terminal domain. The detailed structure of the C-terminal region seen in *aquamarine* in panel A *(dashed line box)* is shown here in more detail. The disulfide bond linking the two monomers is shown in *yellow* and *green* in the middle of the top part of this image *(dashed circle)*. (From J. Bromberg et al., *Cell* 98:295–303, 1999.)

growth-and-division cycles), and the gene encoding the strongly anti-apoptotic protein Bcl-X_L. In addition to phosphorylating STATs, the Jaks can phosphorylate substrates that activate other mitogenic pathways, including the Ras-MAPK pathway described above.

Evidence connecting STATs with cancer pathogenesis is accumulating rapidly. Perhaps the most dramatic indication of the contribution of STATs to cancer development has come from a re-engineering of the Stat3 protein through the introduction of a pair of cysteine residues, which causes the resulting mutant Stat3 to dimerize spontaneously, forming stable covalent disulfide cross-linking bonds. These stabilized Stat3 dimers are structural and functional mimics of the dimers that are formed normally when Stat3 is phosphorylated by Jaks. This mutant Stat3 protein is now constitutively active as a nuclear transcription factor and can function as an oncoprotein that is capable of transforming NIH 3T3 and other immortalized mouse cells to a tumorigenic state (Figure 6.23).

Stat3 is also known to be constitutively activated in a number of human cancers. For example, in many melanomas, its activation is apparently attributable to Src, which is also constitutively active in these cancer cells; indeed, at least four other cousins of Src, all nonreceptor tyrosine kinases, can also phosphorylate and thereby activate Stat3. This highlights the fact that STATs can be activated in the cytoplasm by tyrosine kinases other than the receptor-associated Jaks. In the case of the melanoma cells, inhibition of Src, which leads to deactivation of Stat3, triggers apoptosis, pointing to the important contribution of Stat3 in ensuring the survival of these cancer cells.

In the majority of breast cancer cells, Stat3 has been found to be constitutively activated; in some of these, Stat3 is phosphorylated by Src and Jaks acting collaboratively. As is the case with melanoma cells, reversal of Stat3 activation in these breast cancer cells leads to growth inhibition and apoptosis. In almost all head-and-neck cancers, Stat3 is also constitutively activated, possibly through the actions of the EGF receptor. Taken together, these various threads of evidence converge on the notion that STATs represent important mediators of transformation in a variety of human cancer cell types.

6.9 Cell adhesion receptors emit signals that converge with those released by growth factor receptors

In the last chapter, we read that cells continuously monitor their attachment to components of the extracellular matrix (ECM). Successful tethering to the molecules forming the extracellular matrix (ECM), achieved via integrins, causes survival signals to be released into the cytoplasm that decrease the likelihood that a cell will enter into apoptosis. At the same time, integrin activation, provoked by signals originating inside the cell, can promote cell motility by stimulating integrin molecules located at specific sites on the plasma membrane to forge new linkages with the ECM. Consequently, integrins serve the three functions of (1) physically linking cells to the ECM, (2) informing cells whether or not tethering to certain ECM components has been achieved, and (3) facilitating motility by making and breaking contacts with the ECM.

As described in the last chapter, integrins may cluster and form multiple links to the ECM in small, localized areas termed *focal adhesions* (see Figure 5.28A). Such clustering provokes activation of focal adhesion kinase (FAK), a nonreceptor tyrosine kinase like Src, which is associated with the cytoplasmic tails of the β subunits of certain integrin molecules and becomes phosphorylated, presumably by transphosphorylation, once integrins congregate in these focal adhesions. One of the resulting phosphotyrosine residues on FAK provides a docking site for Src molecules (Figure 6.24A). Src then proceeds to phosphory-

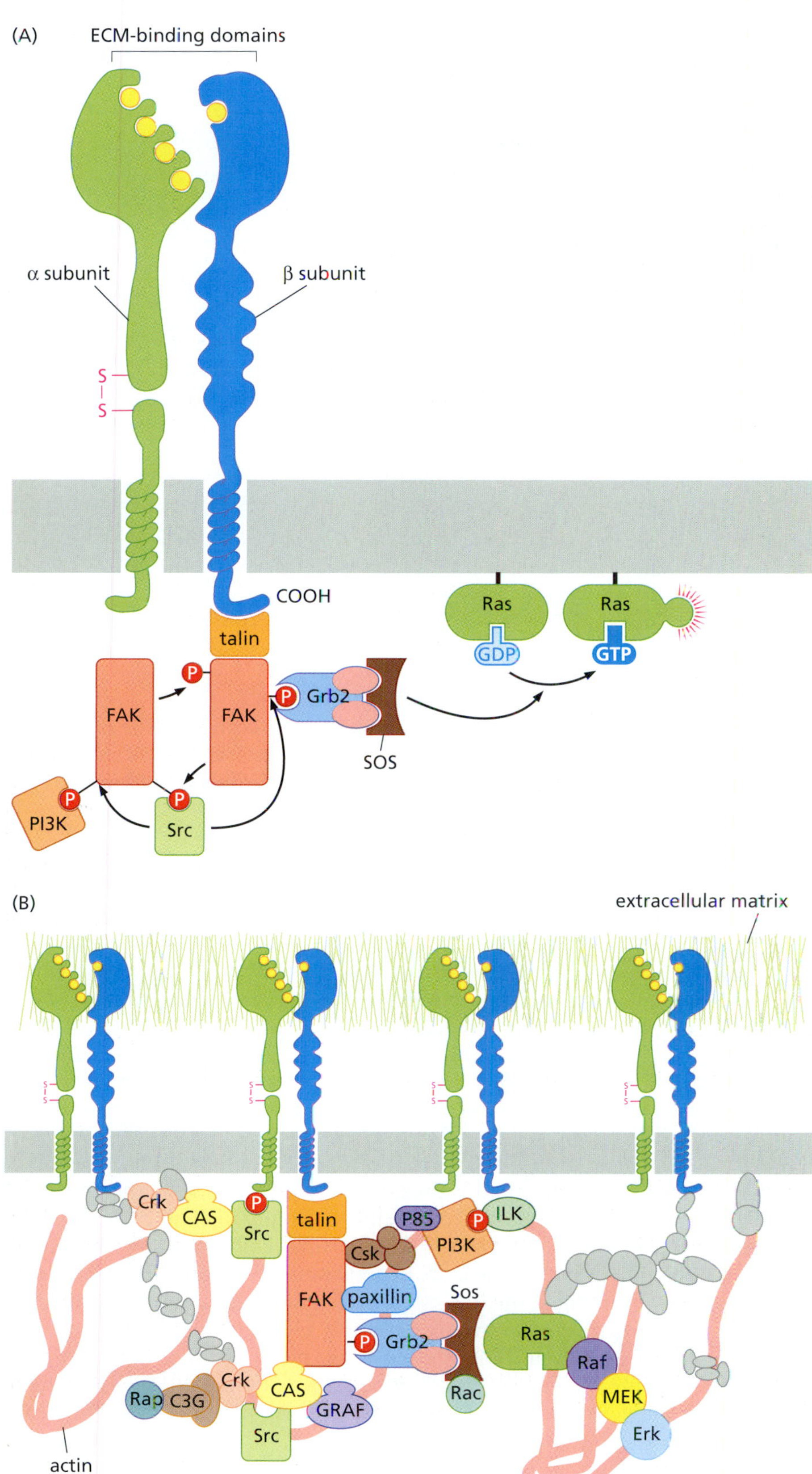

Figure 6.24 Integrin signaling (A) The integrins are assembled as α + β heterodimers. In addition to physically linking the cytoskeleton to the extracellular matrix (see Figure 5.28B), the binding by the ectodomain of the heterodimer to components of the extracellular matrix (ECM) triggers the association of a series of cytoplasmic signal-transducing proteins, such as focal adhesion kinase (FAK; *pink*), to the cytoplasmic domain of the β subunit. Resulting transphosphorylations and the attraction of SH2-containing signaling molecules release signals that activate many of the same pathways that are activated by ligand-activated growth factor receptors. (B) As indicated here, the molecules shown in panel A represent only part of a complex, still poorly understood collection of signal-transducing molecules that become physically associated with the cytoplasmic domains of various integrin β subunits following ECM binding. (Adapted from C. Miranti and J. Brugge, *Nat. Cell Biol.* 4:E83–E90, 2002.)

late additional tyrosines on FAK, and the resulting phosphotyrosines serve as docking sites for yet other SH2-containing signaling molecules, including Grb2, Shc, PI3K, and PLC-γ. There is also evidence that Grb2, once bound to

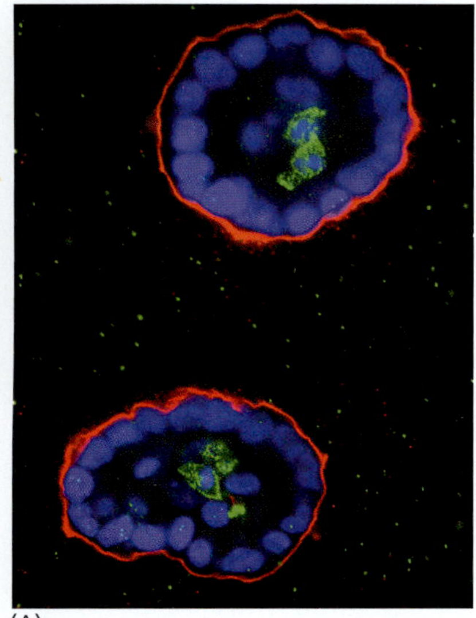

(A)

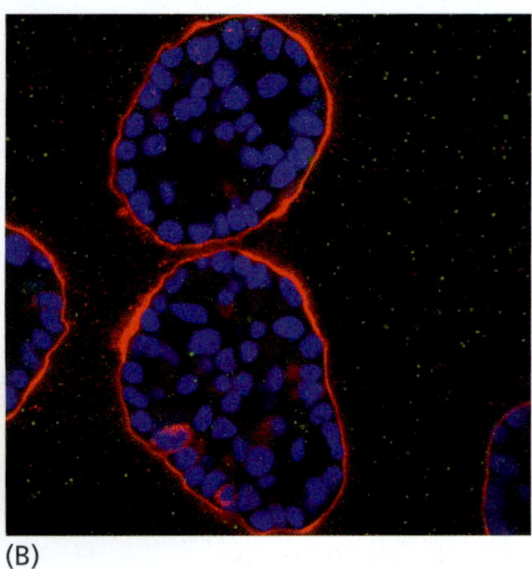

(B)

Figure 6.25 Anoikis and morphogenesis Anoikis serves during morphogenesis to excavate lumina (channels) in the middle of cylindrical aggregates of cells, thereby creating hollow ducts. When placed into suspension culture, human mammary epithelial cells (MECs) form globular clusters—acini—that mimic, at least superficially, the structures of mammary gland ducts. (A) An acinus is seen here in cross section surrounded by a layer of basement membrane (a form of extracellular matrix) containing laminin-5 *(red)* that the cells have synthesized. Cells that succeed in tethering to this basement membrane via their integrins are healthy and have large nuclei *(blue),* while those that are deprived of direct attachment to this basement membrane undergo anoikis, resulting in the formation of a protein (activated caspase-3; *green*) that is indicative of active apoptosis; these central cells also exhibit fragmented nuclei *(blue),* another sign of apoptosis. The loss of these cells will yield a lumen in this acinus. (B) In an early step of breast cancer development, normal anoikis and resulting excavation of the lumen of a mammary duct may be prevented by the actions of certain anti-apoptotic proteins. In this case, expression of the anti-apoptotic Bcl-X_L protein in MECs has resulted in the suppression of apoptosis and thus luminal filling, echoing the appearance of certain ductal hyperplasias and *in situ* carcinomas often seen in the initial steps of breast cancer development. (Courtesy of J. Debnath and J. Brugge.)

tyrosine-phosphorylated FAK, can recruit Sos into this complex. The latter proceeds to activate its normal downstream targets including Ras and PI3K. Indeed, the full complexity of signaling downstream of integrins still awaits resolution (Figure 6.24B).

The accumulated evidence suggests a pathway that has, minimally, the following components:

$$\text{ECM} \rightarrow \text{integrins} \rightarrow \text{Sos} \rightarrow \text{Ras} \rightarrow \text{Erk}$$

The parallels between integrin and tyrosine kinase signaling are striking: Tyrosine kinase receptors bind growth factor ligands in the extracellular space, while integrins bind extracellular matrix components as ligands. Once activated, these two classes of sensors activate many of the same downstream signal transduction cascades. Moreover, some integrins co-localize with certain tyrosine kinase receptors; this association appears to be required for the activation of both types of receptors.

This convergence of signaling pathways appears to explain one of the important cell-biological effects of the Ras oncoprotein—its ability to enable cells to grow in an anchorage-independent fashion. It seems that cells normally depend on integrin signaling to provide a measure of activation of the normal Ras protein. In the absence of this signaling, cells come to believe that they have failed to associate with the extracellular matrix. This may block further advance of cells through their cell growth-and-division cycle or, more drastically, may trigger entrance of cells into the form of programmed cell death termed *anoikis,* which is triggered when cells lose their attachment to solid substrates. Avoidance of such anoikis seems to represent one of the first steps in the initiation of breast cancers (Figure 6.25).

Taken together, these diverse observations suggest that oncogenic Ras promotes anchorage-independent proliferation by mimicking one of the critical downstream signals that result following extracellular matrix tethering by integrins. In effect, oncogenic Ras may delude a cell into believing that its integrins have successfully bound ECM components, thereby allowing a *ras*-transformed cell to proliferate, even when no such attachment has actually occurred.

6.10 The Wnt–β-catenin pathway contributes to cell proliferation

In most cell types, the RTK → Sos → Ras cascade appears to play a dominant role in mediating responses to extracellular mitogens, but it is hardly the only pathway having this result. Yet other, incompletely understood pathways also confer responsiveness to mitogenic signals impinging on the cell surface. Prominent among these is the pathway controlled by Wnt factors, of which more than a dozen can be found in various human tissues. In addition to transducing mitogenic signals, this pathway enables cells to remain in a relatively undifferentiated state—an important attribute of certain types of cancer cells.

As we learned in the last chapter (see Section 5.7), the molecular design of the Wnt pathway is totally different from that of the RTK → Sos → Ras → Raf → Erk pathway. Recall that Wnt factors, acting through Frizzled receptors, suppress the activity of glycogen synthase kinase-3β (GSK-3β). In the absence of Wnts, GSK-3β phosphorylates several key substrate proteins, which are thereby tagged for destruction. The most important of these substrates is β-catenin, normally a cytoplasmic protein that exists in three states. It may be bound to the cytoplasmic domain of cell–cell adhesion receptors, such as E-cadherin (Figure 6.26A). Alternatively, in a fully unrelated role, it exists in a soluble pool in the cytosol, where it turns over very rapidly, having a lifetime of less than 20 minutes. And finally, β-catenin operates in the nucleus as an important component of a transcription factor.

We focus here on the pools of β-catenin molecules that are not associated with E-cadherin. Normally, shortly after its synthesis, a β-catenin molecule will form a multiprotein complex with two other cellular proteins—Apc (the adenomatous polyposis coli protein) and axin. These proteins help to bring β-catenin together with its GSK-3β executioner (Figure 6.26B). By phosphorylating β-catenin, GSK-3β ensures that β-catenin will be tagged by **ubiquitylation**, a process that ensures the rapid destruction of β-catenin; this explains the low

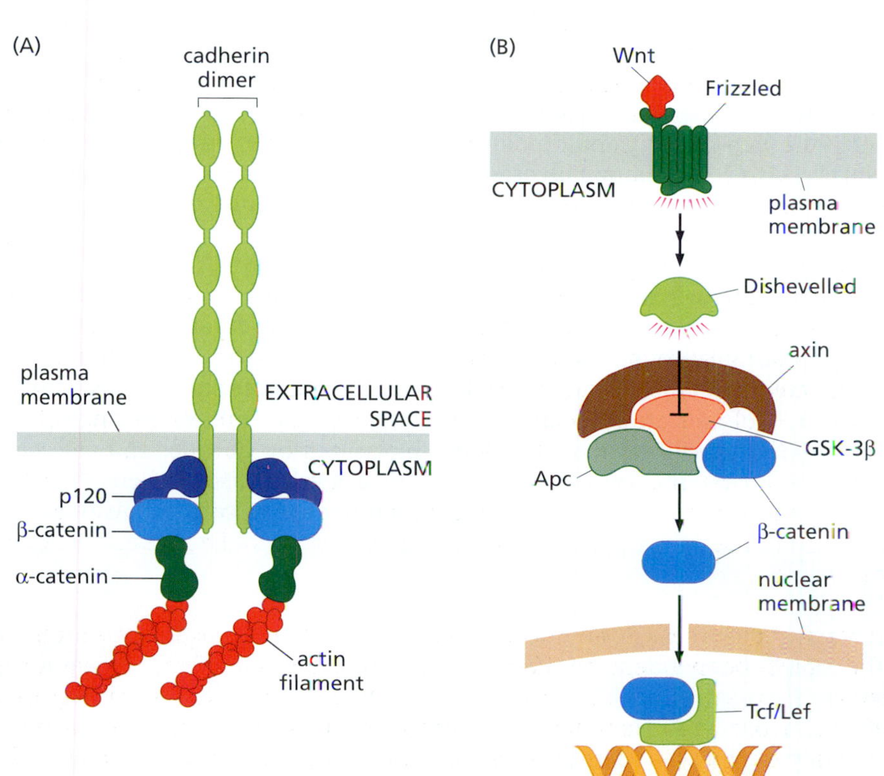

Figure 6.26 Multiple roles of β-catenin (A) Like many integrins, the cadherins *(light green)* are transmembrane proteins that form attachments in the extracellular space and become linked, via intermediary proteins, to the actin cytoskeleton *(dark red)*. (For example, the E-cadherin molecules protruding from one cell can associate with E-cadherin molecules from an adjacent cell, resulting in the formation of an *adherens junction* between them.) As seen here, β-catenin *(light blue)* serves as one of the linkers (together with α-catenin and p120) that form the mechanical linkage between cadherins and the actin cytoskeleton. (B) As indicated in Figure 5.24, the Wnt proteins *(dark red)*, by binding to their Frizzled receptors *(dark green)*, act via Dishevelled to suppress the activity of glycogen synthase kinase-3β (GSK-3β, *pink)*. This prevents GSK-3β from phosphorylating, among other substrates, β-catenin *(light blue)*, which therefore escapes degradation and accumulates in the cytoplasm and in the nucleus. Once in the nucleus, β-catenin associates with Tcf/Lef transcription factors to drive the expression of a variety of responding genes, including those involved in enabling cell proliferation.

steady-state concentrations of β-catenin normally present in the cytosol. We will encounter this degradation system later in more detail (Sidebar 7.8).

When Wnt signaling is activated, however, GSK-3β firing is blocked and β-catenin is saved from rapid destruction; its half-life increases from less than 20 minutes to 1–2 hours, and its steady-state concentrations increase proportionately. Many of the accumulated β-catenin molecules then move into the nucleus to activate transcription by binding to Tcf/Lef proteins (see Figure 6.26B). The resulting multi-subunit transcription factor complexes proceed to activate expression of a number of important target genes, including those encoding critical proteins involved in cell growth and proliferation, such as cyclin D1 and Myc, that we will study later in Chapter 8.

As mentioned, GSK-3β can phosphorylate other critical growth-regulating proteins besides β-catenin. By phosphorylating cyclin D1, GSK-3β also labels this growth-promoting protein for rapid degradation. Hence, the Wnt pathway actually modulates cyclin D1 expression at both the transcriptional and the post-translational levels. These various actions of Wnts indicate that, in addition to their originally discovered powers as morphogens, they can also act as potent mitogens, much like the ligands of many tyrosine kinase receptors. Moreover, as we will learn in Chapter 7, high levels of β-catenin in intestinal stem cells ensure that these cells remain in an undifferentiated state, rather than developing into the specialized cells that line the wall of the intestine. Such high β-catenin levels turn out to be critical to the formation of colon carcinomas.

The alternative role of β-catenin, mentioned in passing above, is to form part of the cytoplasmic complex through which cell surface adhesion receptors, notably E-cadherin, become physically linked with the cytoskeleton (see Figure 6.26A). The two unrelated roles of β-catenin represent a major puzzle. It is unclear, for example, whether the two pools of β-catenin molecules—those contributing to the formation of **adherens junctions** and those involved in regulating transcription—are in equilibrium with one another, or whether they are fully segregated in two non-equilibrating compartments in the cell. Thus, in cells that have lost E-cadherin, one might expect β-catenin to be liberated and hence available for nuclear translocation and transcriptional activation. However, in some cancer cells that have lost their E-cadherin genes through mutation, nuclear β-catenin signaling does not seem to be elevated above normal levels. Observations like these begin to suggest that the two pools of β-catenin are regulated independently and do not influence each other. Then there is the other puzzle: Why has evolution invested two such unrelated functions in a single protein? (In worms, these two functions are carried out by distinct proteins encoded by separate genes.)

In many human breast cancers, expression of certain Wnts is increased four- to tenfold above normal, suggesting the operations of autocrine and paracrine growth-stimulatory pathways. There is clear evidence of nuclear translocation of β-catenin in approximately 20% of advanced prostate carcinomas. Moreover, in 5 to 7% of prostate carcinomas, mutations of the β-catenin gene have yielded a protein that can no longer be phosphorylated by GSK-3β, and as a consequence, β-catenin now accumulates to high levels in the nucleus (see Figure 6.27). Such mutations have also been documented in carcinomas of the liver,

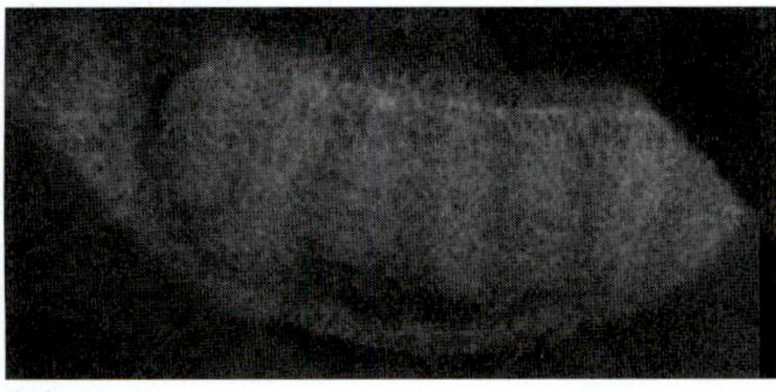
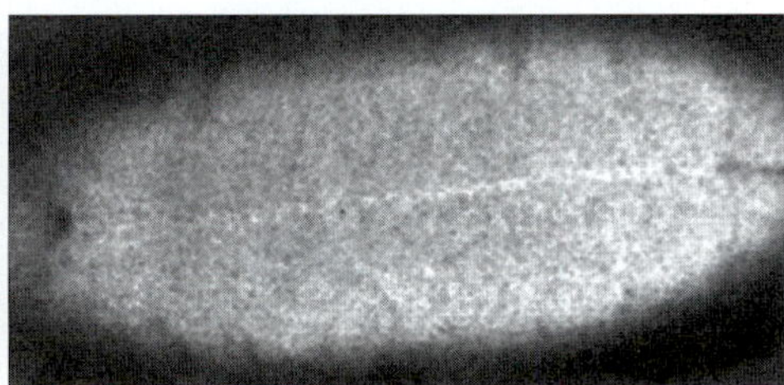

(A)

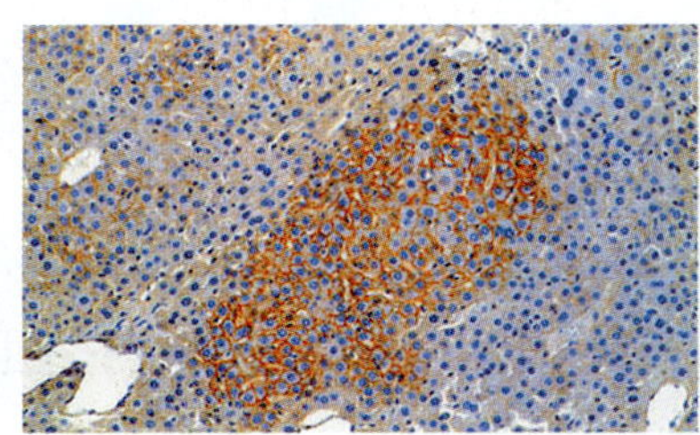

(B)

Figure 6.27 Regulation of β-catenin signaling Increases in β-catenin levels are achieved through diverse mechanisms. (A) *Drosophila* embryos usually express β-catenin in a striped pattern *(light stripes, left)*. However, when the gene encoding GSK-3β is inactivated, high levels of β-catenin accumulate throughout the embryo (*right*). (B) Mice exposed to certain liver carcinogens develop hepatocellular carcinomas (hepatomas) in which β-catenin levels rise dramatically because the β-catenin protein is mutant, enabling it to resist phosphorylation by GSK-3β and resulting degradation. The resulting cytoplasmic accumulation of β-catenin *(brown)* in this focus of pre-malignant hepatocytes is a prelude to increased levels of nuclear β-catenin (not seen here) that drive cell proliferation through association with Tcf/Lef transcription factors. (A, courtesy of E.F. Wieschaus; B, from T.R. Devereux et al., *Oncogene* 18:4726–4733, 1999.)

colon, endometrium, and ovary, as well as in melanomas. And as we will see in the next chapter, in virtually all colon carcinomas, β-catenin degradation is defective, due to defects in one of the proteins—Apc—that facilitate its degradation (see Figure 6.26B). It is likely that this pathway is also deregulated in other human cancers by mechanisms that have not yet been uncovered.

6.11 G-protein–coupled receptors can also drive normal and neoplastic proliferation

As we read in the last chapter, G-protein–coupled receptors (GPCRs) are transmembrane proteins that weave their way back and forth through the plasma membrane seven times. Upon binding their extracellular ligands, each of these "serpentine" receptors activates one or more types of cytoplasmic heterotrimeric G protein, so named because of its three distinct subunits (Gα, Gβ, and Gγ), the first of which binds either GDP or GTP. As is the case with the Ras protein, the activated state of Gα is achieved when it binds GTP.

Once stimulated by a G-protein–coupled receptor (GPCR), the Gα subunit of a heterotrimeric G protein dissociates from its two partners, Gβ and Gγ, and proceeds to activate a number of distinct cytoplasmic enzymes (Figure 6.28). Included among these are adenyl cyclase (which converts ATP into cyclic AMP) and phospholipase C-β (PLC-β), which cleaves PIP_2 to yield diacylglycerol (DAG) and inositol triphosphate (IP_3) (Figure 6.16B). The latter are potent second messengers that can function to stimulate cell proliferation.

There is even evidence that the Src kinase, the first cellular oncoprotein ever studied, can be activated by certain GTP-bound Gα subunits. At the same time, complexes of the other two subunits of the heterotrimeric G protein—Gβ and Gγ—have been found to stimulate yet other important mitogenic signaling proteins, such as one form of phosphatidylinositol 3-kinase (PI3K). The ability of GPCRs to activate mitogenic pathways suggests that deregulation of signaling by these receptors and associated G proteins may well contribute to

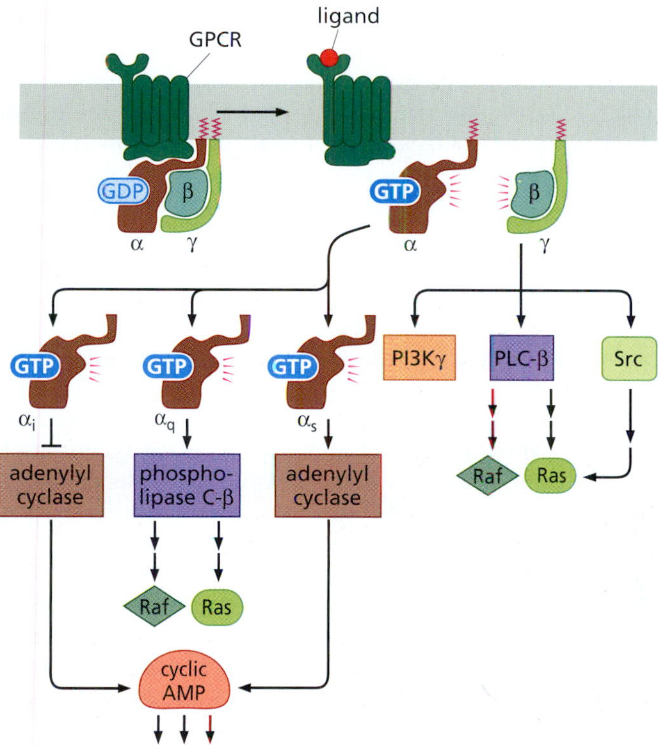

Figure 6.28 G-protein–coupled receptors The seven-membrane-spanning receptors are able to activate a variety of heterotrimeric G proteins that differ largely in the identity of their Gα subunits. Once stimulated by a GPCR, the Gα subunit separates from Gβ and Gγ and proceeds to activate or inhibit a variety of cytoplasmic enzymes, only two of which (adenylyl cyclase and phospholipase C-β) are shown here. These enzymes, in turn, can have mitogenic or anti-mitogenic influences, depending upon the cell type. At the same time, the Gβ + Gγ dimers can activate their own effectors, including phosphatidylinositol 3-kinase γ (PI3Kγ), phospholipase C-β (PLC-β), and Src. This diagram provides only a hint of the diverse array of G proteins and the effectors of Gα and Gβ + Gγ that have been discovered.

Table 6.5 G-protein–coupled receptors and G proteins involved in human cancer pathogenesis

G protein or receptor	Type of tumor
Activating mutations affecting G proteins	
$G\alpha_s$	thyroid adenomas and carcinomas, pituitary adenomas
$G\alpha_{i2}$	ovarian tumors, adrenal cortical tumors
Activating mutations affecting G-protein–coupled receptors	
Thyroid-stimulating hormone receptor	thyroid adenomas and carcinomas
Follicle-stimulating hormone receptor	ovarian tumors
Luteinizing hormone receptor	Leydig cell hyperplasias
Cholecystokinin-2 receptor	colorectal carcinomas
Ca^{2+}-sensing receptor	various neoplasms
Autocrine and paracrine activation	
Neuromedin B receptor	SCLC
Neurotensin receptor	prostate carcinomas and SCLC
Gastrin receptor	gastric carcinomas and SCLC
Cholecystokinin receptor	pancreatic hyperplasias and carcinomas gastrointestinal carcinomas and SCLC
Virus-encoded G-protein–coupled receptors	
Kaposi's sarcoma herpesvirus (HHV-8)	Kaposi's sarcoma
Herpesvirus saimiri	primate leukemias and lymphomas
Jaagsiekte sheep retrovirus	sheep pulmonary carcinomas

Adapted from M.J. Marinissen and J.S. Gutkind, *Trends Pharmacol. Sci.* 22:368–376, 2001.

cell transformation and tumorigenesis—a speculation that has been borne out by analyses of certain types of human cancer cells (see Table 6.5).

For example, the cancer cells of small-cell lung carcinomas (SCLCs), a common tumor of cigarette smokers, release a number of distinct peptide factors, some with **neuropeptide**-like properties. In some SCLCs, the tumor cells may simultaneously secrete bombesin (also known as gastrin-releasing peptide, or GRP), bradykinin, cholecystokinin (CCK), gastrin, neurotensin, and vasopressin. At the same time, these cells display the GPCRs that recognize and bind these released factors, resulting in the establishment of autocrine signaling loops.

The experimental proof that these autocrine loops are actually responsible for driving the proliferation and/or survival of the small-cell lung carcinoma (SCLC) cells is straightforward: SCLC cells can be incubated *in vitro* in the presence of an antibody that binds and neutralizes a secreted autocrine growth factor, such as gastrin-releasing peptide (GRP). In a number of SCLC cell lines, this treatment results in the rapid cessation of growth and even in apoptosis. Such a response indicates that these cancer cells depend on a GRP-based autocrine signaling loop for their survival, and it has suggested a novel therapy for SCLC patients, who have been treated with a GRP-neutralizing antibody. In a series of twelve SCLC patients treated in this way, one patient showed a complete remission, while four patients exhibited a partial shrinkage of their tumors.

In another class of neoplasias—thyroid adenomas and some thyroid carcinomas—the gene encoding the thyroid-stimulating hormone receptor (TSHR), another GPCR, is often found to carry a point mutation. This leads to constitutive, ligand-independent firing of the TSH receptor, which in turn results in the release of strong mitogenic signals into the thyroid epithelial cells (see Table 6.5). In yet other tumors of the thyroid gland, a Gα subunit has suffered a point mutation that functions much like the mutations activating Ras signaling, by depriving this Gα subunit of the ability to shut itself off through its intrinsic GTPase activity. Altogether, at least 10 of the 17 human genes encoding Gα subunits have been found to function as oncogenes in certain cell types and in various human malignancies.

The most bizarre subversion of these G-protein–coupled receptors occurs during infections by certain herpesviruses, such as human herpesvirus type 8 (HHV-8), also known as the Kaposi's sarcoma herpesvirus (KSHV; refer to Sidebar 3.10). This virus is responsible for the vascular tumors that frequently afflict AIDS patients. At one point in its evolutionary past, this virus acquired a cellular gene specifying a G-protein–coupled receptor. The viral form of this gene has been remodeled, so that the encoded receptor signals in a ligand-independent manner. Among other consequences, this signaling causes HHV-8–infected endothelial cells (lining the walls of blood vessels) to secrete vascular endothelial growth factor (VEGF); the released VEGF then creates an autocrine signaling loop by binding to its cognate receptors on the surface of these endothelial cells and driving the proliferation of these cells.

6.12 Four other signaling pathways contribute in various ways to normal and neoplastic proliferation

Four other signaling channels play important roles in the pathogenesis of some types of cancers. These pathways share in common the fact that, when activated, they control the intracellular localization of "dual-address" proteins that operate much like β-catenin: these proteins normally reside in the cytoplasm, and when activated by certain signals, they are dispatched to the nucleus, where they function as components of specific transcription factors to drive gene expression. In the last chapter, we read descriptions of how the receptors controlling these pathways operate. Now, we will briefly summarize the downstream actions of these four pathways (Figure 6.29) and cite examples of how each is deregulated in one or another type of human tumor (Figure 6.30).

Nuclear factor-κB The signaling system involving transcription factors of the nuclear factor-κB (NF-κB) class contributes to the formation of certain tumor types. The first indication of the importance of this pathway to cancer pathogenesis came from the discovery of the *rel* oncogene in a rapidly transforming turkey retrovirus responsible for reticuloendotheliosis, a lymphoma of the B-cell lineage. Later investigations into the transcription factors responsible for regulating immunoglobulin gene expression revealed Rel to be a member of a family of transcription factors that came to be called collectively NF-κBs. These proteins form homo- and heterodimers in the cytoplasm.

The most common form of NF-κB is a heterodimer composed of a p65 and a p50 subunit. Usually, NF-κB is sequestered in the cytoplasm by a third polypeptide named IκB (inhibitor of NF-κB; Figure 6.29A); while in this state, this signaling system is kept silent. However, in response to signals originating from a diverse array of sources, IκB becomes phosphorylated, and thereby tagged for rapid destruction. (Recall that β-catenin suffers the same fate following its phosphorylation in the cytoplasm.) As a result, NF-κB is liberated from the clutches of IκB, migrates into the nucleus, and proceeds to activate the expression of a cohort of at least 150 target genes.

The kinase that tags IκB for destruction (named IκB kinase or simply IKK) and thereby activates NF-κB signaling is itself stimulated by signals as diverse as tumor necrosis factor-α and interleukin-1β (extracellular factors involved in the inflammatory response of the immune system), lipopolysaccharide (a sign of bacterial infection), reactive oxygen species (ROS), anti-cancer drugs, and gamma irradiation. In the context of cancer, NF-κBs have important effects on cell survival and proliferation. Once they have arrived in the nucleus, the NF-κBs can induce expression of genes specifying a number of key anti-apoptotic proteins, such as Bcl-2 and IAP-1 and -2; we will learn more about these proteins in Chapter 9. At the same time, NF-κB functions in a mitogenic fashion by inducing expression of the *myc* and cyclin D1 genes, components of the cell-

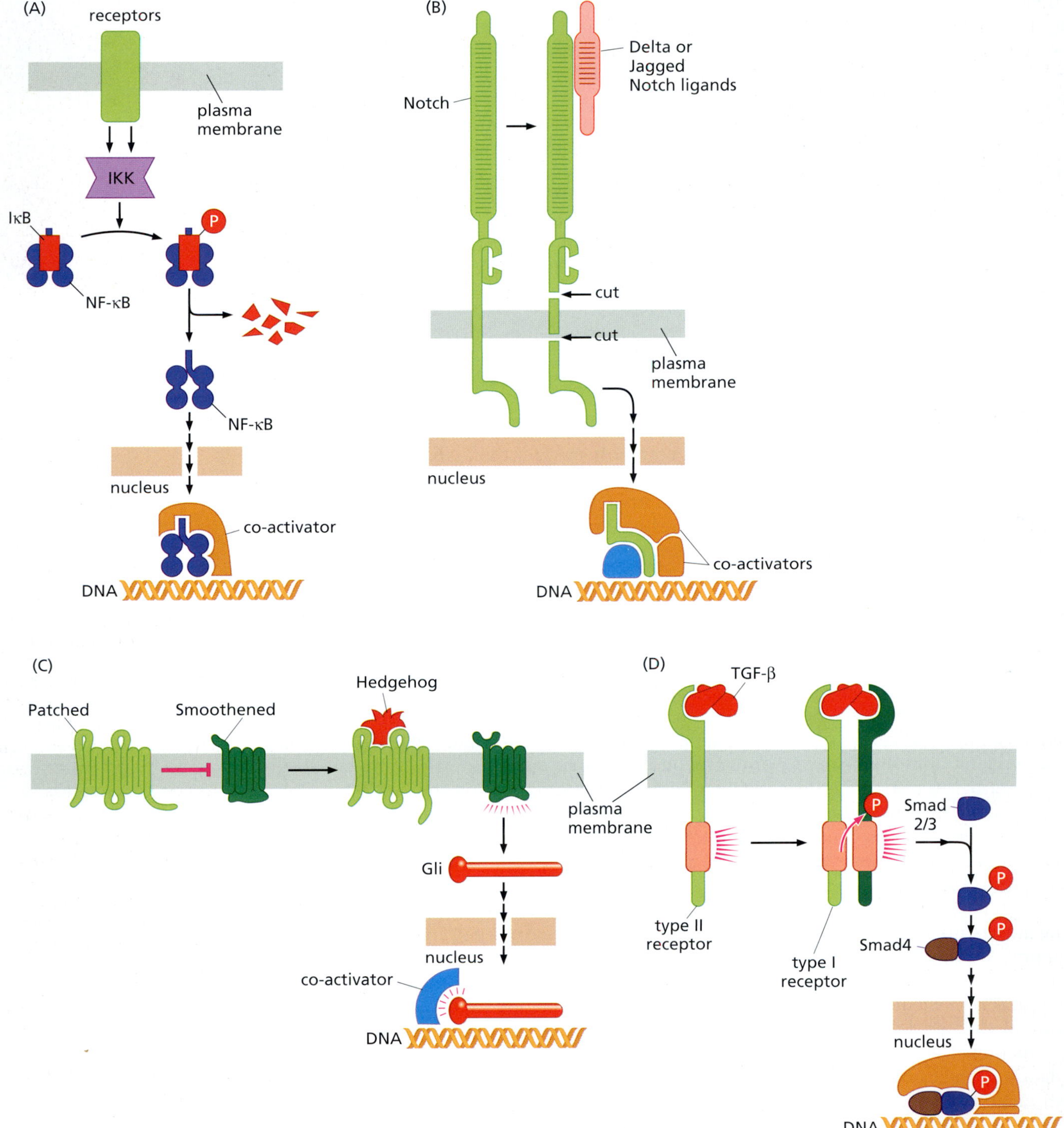

Figure 6.29 Four different "dual-address" signaling pathways (A) The NF-κB family of transcription factors, which function as heterodimers *(blue),* are sequestered in the cytoplasm by IκB. A variety of receptors and afferent signals activate IKK, which phosphorylates IκB, tagging it for proteolytic degradation. Now liberated from IκB, NF-κB can translocate to the nucleus, where it activates a broad constituency of genes, including anti-apoptotic and mitogenic genes *(not shown).* (B)The Notch receptor on the cell surface can bind ligands such as NotchL, Delta, and Jagged. Ligand binding causes two proteolytic cuts in Notch, liberating a cytoplasmic fragment of Notch that can translocate to the nucleus, where it functions as part of a transcription factor complex. (C) Binding of a Hedgehog ligand to Patched causes the latter to release Smoothened. Once free, Smoothened prevents the cleavage of a Gli protein, and the latter now translocates to the nucleus, where it helps to form functional transcription factors. (D) Binding of TGF-β ligand to the type II TGF-β receptor, which brings the type II and type I receptors together, results in the phosphorylation of the type I receptor. The latter, now activated, phosphorylates cytosolic Smad2 or Smad3 proteins, and either of these phospho-Smads then binds to Smad 4. The resulting heterodimeric Smad complexes translocate to the nucleus, where they form complexes with other transcription factors to drive gene expression.

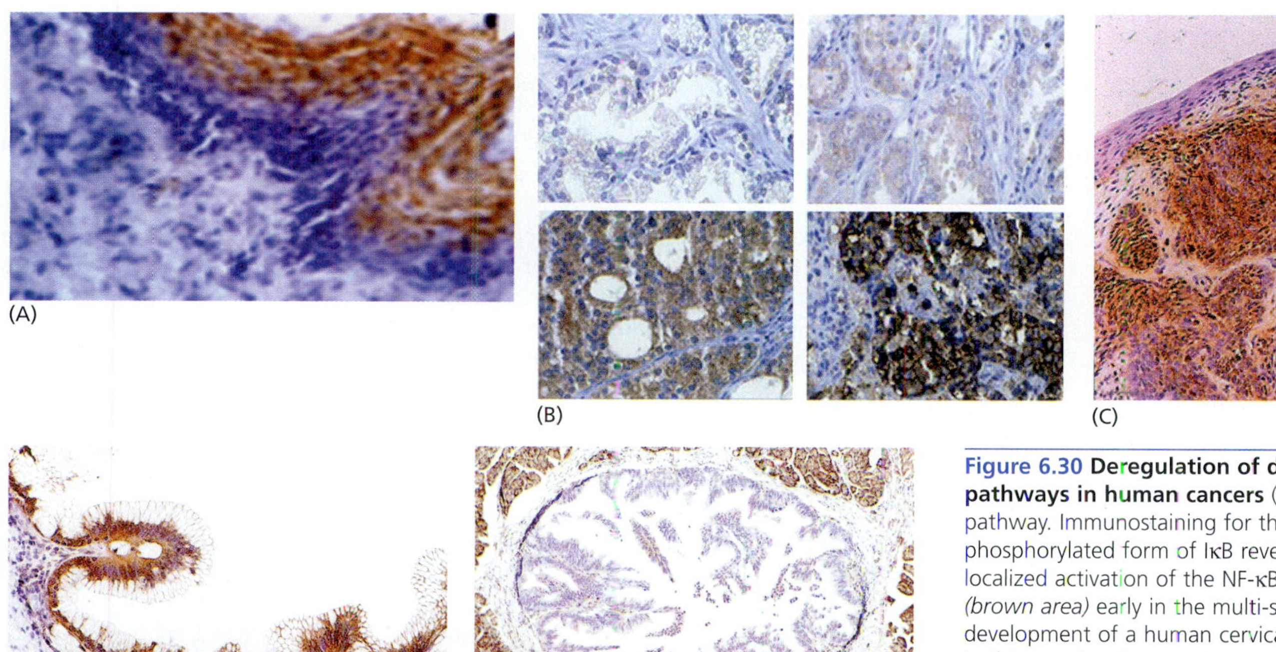

Figure 6.30 Deregulation of dual-address pathways in human cancers (A) NF-κB pathway. Immunostaining for the phosphorylated form of IκB reveals the localized activation of the NF-κB pathway *(brown area)* early in the multi-step development of a human cervical carcinoma, in this case in a low-grade squamous intraepithelial lesion. (B) Notch pathway. The Notch pathway is often deregulated through elevated expression of Notch ligands. In prostate carcinomas, increasing levels of Jagged, a ligand of the Notch receptor, are seen with increasing levels of malignancy. Seen here *(brown, left to right, top row)* is immunostaining of normal prostate and of primary carcinoma; and *(left to right, bottom row)* of a metastasis that has not been treated with anti-androgen hormone therapy and a metastasis that has become refractory to hormone therapy. Elevated Jagged expression in primary tumors correlated with increased risk of metastatic relapse. (C) Hedgehog pathway. This pathway is deregulated in the majority of basal cell skin carcinomas (BCC). Its deregulation is indicated by the elevated concentration of Gli1 protein in BCCs, detected here by immunostaining *(brown)*, which is barely detectable in normal cells *(light pink)*. Northern (RNA) blot analyses *(not shown)* indicate as much as a 40-fold elevation in Gli1 mRNA in BCCs compared with normal skin. (D) TGF-β pathway. This pathway is inactivated in a wide variety of carcinomas through several mechanisms. During pancreatic carcinoma progression, low-grade pancreatic intraepithelial neoplasias (PanIN-1) usually express Smad4, a critical transcription factor in TGF-β signaling *(brown, left panel)*, while high-grade PanIN-3 lesions often lose Smad4 expression, often because of mutation of the encoding gene *(center of right panel)*. (A, from A. Nair et al., *Oncogene* 22:50–58, 2003; B, from S. Santagata et al., *Cancer Res.* 64:6854–6857, 2004; C, from L. Ghali et al., *J. Invest. Dermatol.* 113:595–599, 1999; D, from R.E. Wilentz et al., *Cancer Res.* 60:2002–2006, 2000.)

cycle machinery that we discuss in Chapter 8. Hence, NF-κB can protect cancer cells from apoptosis (programmed cell death) and, at the same time, drive their proliferation.

While components of the NF-κB signaling cascade are rarely found in mutant form in human cancers, this pathway is frequently found to be constitutively activated. In breast cancers, for example, the pathway is often highly active, although the molecular mechanism responsible for this activation remains unclear. NF-κB seems to play its most important role in malignancies of various lymphocyte lineages. The *REL* gene, which encodes one of the subunits of NF-κB, is amplified in about one-fourth of diffuse large B-cell lymphomas, resulting in a 4- to 35-fold increase in expression of its gene product. Translocations affecting the *NFKB2* locus have been found frequently in B- and T-cell lymphomas and in myelomas (tumors of antibody-producing cells). And deregulation of this pathway can often be observed already in low-grade pre-malignant growths (Figure 6.30A).

Notch Study of another unusual signaling pathway, this one controlled by the Notch protein, traces its roots to the discovery in 1919 of an allele of a *Drosophila* gene that causes notches to form in the edges of this fly's wings. Only many decades later was it realized that Notch is a transmembrane protein; four different varieties of Notch (products of four different genes) are expressed by mammalian cells. As mentioned in the last chapter, after Notch, acting as a cell surface receptor, binds a ligand (NotchL), it undergoes two proteolytic cleavages, one in its ectodomain, the other within its transmembrane domain. The latter cleavage liberates a largely cytoplasmic protein fragment from its tethering to the plasma membrane. This fragment of Notch then migrates to the nucleus of targeted cells, where it functions, together with partner proteins, as a transcription factor (Figure 6.29B).

This signaling system operates using biochemical mechanisms that are clearly very different from those governing signaling by tyrosine kinase receptors. Each time a Notch receptor binds its ligand, the receptor undergoes an irreversible covalent alteration, specifically, a proteolytic cleavage. Hence, receptor firing occurs in direct proportion to the number of ligands encountered in the extracellular space, and each Notch receptor can presumably fire only once after it has bound its ligand. Tyrosine kinase receptors, in dramatic contrast, release multiple signals over an extended period of time following ligand binding, and therefore can greatly amplify the signal initiated by their growth factor ligands.

Truncated forms of *Notch* that specify only the cytoplasmic domain of the Notch protein are potent oncogenes for transforming cells *in vitro*. This observation suggests that altered forms of Notch contribute to human cancer pathogenesis. Indeed, overexpression of one or another of the Notch proteins is seen in the great majority of cervical carcinomas, in a subset of colon carcinomas, and in squamous carcinomas of the lung. This overexpression is often accompanied by nuclear localization of the cytoplasmic cleavage fragment of Notch, indicating that active signaling through this pathway is occurring in tumor cells. Increased expression of two ligands of Notch, termed Jagged and Delta, has also been found in some cervical and prostate carcinomas (Figure 6.30B). More dramatically, in about 10% of acute lymphocytic leukemias (ALLs), constitutively active forms of Notch are found; these result from genetic deletions of the portion of the *NOTCH-1* gene encoding the extracellular domain of the Notch-1 protein. Some experiments also suggest that Notch signaling contributes in important ways to transformation by *ras* oncogenes.

Patched As was discussed previously, binding of the Patched receptor by its ligand, Hedgehog, causes Patched to release the Smoothened protein from inhibition; Smoothened is then able to emit downstream signals. Through a poorly understood series of changes, the activated Smoothened protein alters the fate of cytoplasmic Gli protein. Normally, in the absence of intervention by Smoothened, the Gli precursor protein is cleaved into two fragments, one of which moves into the nucleus, where it functions as a transcriptional repressor. However, when Smoothened is activated, it protects the Gli precursor from being cleaved. The resulting intact Gli protein migrates to the nucleus, where it serves as an activator of transcription (Figure 6.29C).

Gli was first discovered as a highly expressed protein in glioblastomas (whence its name). When overexpressed, Gli and several closely related cousins can function as oncoproteins. Subsequent research into this pathway has revealed other instances where its malfunctioning contributes to tumor development. For example, germ-line inactivating mutations of the human patched (*PTCH*) gene cause Gorlin syndrome, an inherited cancer susceptibility syndrome, which involves increased risk of multiple basal cell carcinomas of the skin as well as other tumors, notably medulloblastomas—tumors of cells in the cerebellum. Such inactivating mutations prevent Patched from inhibiting Smoothened, giving the latter a free hand to dispatch an uninterrupted stream of active Gli protein to the nucleus.

As many as 40% of sporadic basal cell carcinomas of the skin (which occur because of somatic rather than germ-line mutations) carry mutant *PTCH* or *SMO* alleles—the latter encoding Smoothened. These basal cell carcinomas are the most common form of cancer in Western populations and fortunately are usually benign (Figure 6.30C). Somatically mutated alleles of *PTCH* have also been found in a variety of other tumors, including medulloblastomas and meningiomas, as well as breast and esophageal carcinomas. These somatic mutations, like those present in mutant germ-line alleles of *PTCH*, seem to compromise Patched function, once again permitting Smoothened to constitutively activate Gli transcription factors. Recently, mutant alleles of another downstream component of this pathway, encoded by the *SUFU* gene, have also been described in medulloblastomas. Moreover, it is now clear that there are at least three Gli proteins, two of which stimulate transcription while the third inhibits it.

These cancers, all involving mutations that affect the structure of various components of the Patched pathway, appear to represent only a small portion of the human tumors in which this signaling cascade is hyperactivated. Thus, a survey of esophageal, gastric, biliary tract, and pancreatic carcinoma cell lines and tumors has revealed that virtually all of these expressed significant levels of the Patched receptor as well as its ligands (called either Sonic Hedgehog or Indian Hedgehog), indicating the operations of an autocrine signaling loop. This notion was confirmed by demonstrating elevated nuclear levels of the Gli factor, the downstream product of the activated pathway. Moreover, treatment of tumor cells with anti-Hedgehog antibody (which sequesters both types of ligand and breaks the signaling loop) caused cessation of proliferation and/or death of tumor cells. Similar findings have been reported for small-cell lung carcinomas (SCLCs), the common malignancy of cigarette smokers. Together, these reports point to an important role of this signaling pathway in the tumors that arise in many of the tissues deriving from the embryonic gut.

TGF-β A fourth signaling pathway that involves the dispatch of cytoplasmic proteins to the nucleus is represented by the pathway leading from TGF-β receptors (Section 5.7). TGF-β and the signaling pathway that it controls appear to play major roles in the pathogenesis of many if not all carcinomas, both in their early stages, when TGF-β acts to arrest the growth of many cell types, and later in cancer progression, when it contributes, paradoxically, to the phenotype of tumor invasiveness. We will defer detailed discussions of this important pathway until Section 8.10 and Chapter 14. For the moment, suffice it to say that activation of this pathway leads to dispatch of Smad transcription factors to the nucleus, where they can collaborate with other nuclear transcription factors to activate the expression of a large constituency of genes (Figure 6.29D). In the absence of critical Smads, epithelial cancer cells can escape the growth-inhibitory actions of TGF-β and thrive—a state that is often observed in the precursors of invasive human pancreatic carcinomas (Figure 6.30D).

6.13 Synopsis and prospects

Signal transduction in cancer cells One of the hallmarks of cancer cells is their ability to generate their own mitogenic signals endogenously. These signals enable cancer cells to become liberated from dependence on external mitogenic signals, specifically those conveyed by growth factors. It is likely that the mitogenic pathway of greatest importance to human cancer pathogenesis is the one that we have discussed in detail and has the following design:

GFs → RTKs → Grb2 → Sos → Ras → Raf → MEK → ERK

As described in the last chapter and this one, a variety of molecular mechanisms cause the upstream portion of this signaling pathway to become continuously activated in cancer cells. Neoplastic cells may acquire the ability to make and release growth factors (GFs) that initiate autocrine signaling loops. Alternatively, the growth factor receptors (RTKs) may suffer significant structural alterations. In the case of the EGF receptor, the deletion of much of its ectodomain, observed in some human tumors such as glioblastomas, results in ligand-independent firing of its cytoplasmic tyrosine kinase.

Even more commonly observed are the human tumors that display elevated levels of wild-type versions of receptors such as the EGF receptor or its close relative, ErbB2/HER2/Neu. It is clear that excessively high levels of these receptor molecules, often expressed by breast, brain, and stomach cancer cells, favor increased cell proliferation. Such receptor overexpression is often achieved through amplification of the gene encoding this receptor or deregulated transcription of this gene.

The flux of signals through the RTK → Ras → Raf → MAPK pathway is also regulated by negative-feedback mechanisms that operate to attenuate signaling. For example, after the EGF-R binds ligand, the receptor–ligand complex is **internalized** into the cell via the process of endocytosis (Figure 6.31). Once sequestered in cytoplasmic membrane vesicles, the receptor may either be shuttled to lysosomes, where it is degraded, or recycled back to the surface. Defective internalization of this receptor allows it to accumulate to excessively high levels at the cell surface, yielding, in turn, greatly elevated mitogenic signaling. Defective internalization of the EGF-R and, by extension, other RTKs may play a key role in driving the proliferation of many kinds of human tumor cells.

Moving down the mitogenic pathway from the receptors, we note that few, if any, alterations in signal-transducing proteins have been documented in human tumors until we reach the Ras proteins. Recall that point mutations in the 12th, 13th, or 61st codon of the reading frames of encoding *ras* genes result in activated Ras oncoproteins. These mutant proteins fire for extended periods of time rather than the short, well-controlled bursts that characterize the behavior of normal Ras proteins. The amino acid replacements caused by these various point mutations (which are seen in more than one-fifth of all human cancers) render the Ras oncoproteins resistant to the GTPase-activating proteins (GAPs), which normally stimulate the GTPase activity of Ras. As discussed later in Chapter 7, in the disease of neurofibromatosis, one of the GAPs that usually serve to inactivate Ras is missing from certain cell types, resulting once again in an accumulation of excessively high levels of the activated, GTP-bound form of Ras proteins.

Still further down this linear signaling cascade below the level of Ras, we once again find a relative dearth of mutant proteins in human cancer cells. The B-Raf protein, a close cousin of the Raf kinase, presents one exception to this rule. It is found in mutant form in about 8% of a large panel of human tumor lines that have been examined. In addition, as many as two-thirds of melanomas possess point-mutated versions of the *BRAF* gene, whose normal product, like Raf, requires Ras stimulation before it can fire. The mutant B-Raf proteins have a greatly reduced dependence on interaction with Ras.

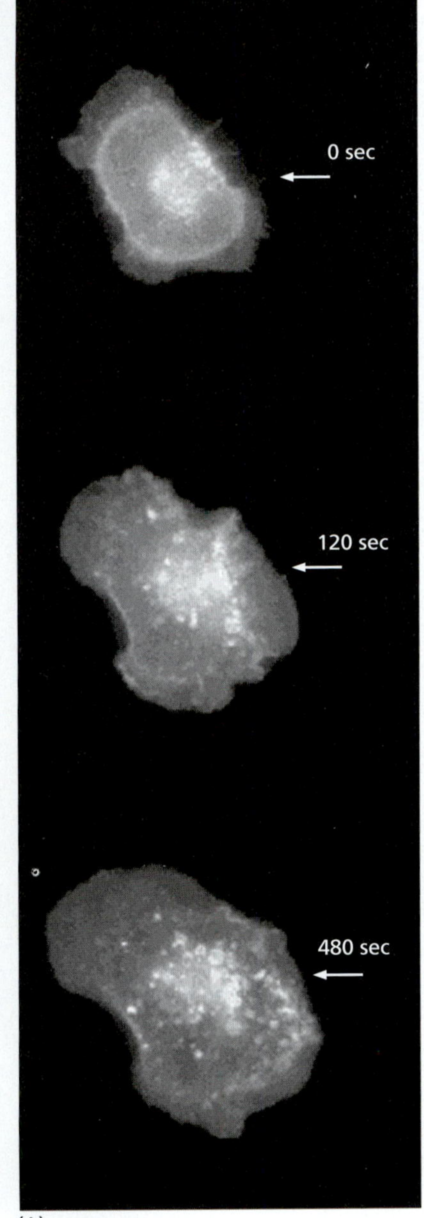

(A)

control vector
siRNA1 vector
siRNA2 vector
EGF-R
(B)

Figure 6.31 Negative-feedback control of receptor signaling Mitogenic signaling is curtailed by negative-feedback loops that ensure that cells experience physiologically appropriate doses of such signals. (A) Most of these negative-feedback loops are difficult to visualize microscopically. An exception is seen here. When a cell is exposed to epidermal growth factor (EGF), introduced into the medium through a micropipette whose location is shown *(arrowheads)*, the EGF initially causes the cell surface EGF receptor, which in this case has been linked to a fluorescent dye, to release mitogenic signals into the cell *(not shown)*. However, within 120 seconds, the fluorescing EGF-R has been internalized on the side of the cell facing the EGF source; and by 480 seconds, it is found in internalized clusters on all sides of the cell. Hence, EGF signaling is shut down because the receptor is no longer available to bind ligand at the cell surface. (B) The potent contribution of negative-feedback controls on receptor expression is revealed by an experiment in which expression of cyclin G–associated kinase (GAK) is suppressed through the actions of two interfering RNA (RNAi)-expressing vectors. In such cells, normal expression of the EGF receptor, as gauged in this immunoblot *(left channel)*, is increased 50-fold when either of the two RNAi's is expressed in these cells *(right two channels)*. GAK acts in still-unclear ways to drive the internalization of this and other cell-surface proteins. (A, from M. Bailly et al., *Mol. Biol. Cell* 11:3873–3883, 2000; B, adapted from L. Zhang et al., *Proc. Natl. Acad. Sci. USA* 101:10296–10301, 2004.)

Sidebar 6.5 The mysteries of breast cancer mitogenesis Breast cancer is a commonly occurring disease in the West that has been the subject of extensive research, yet we still have only a vague understanding at the molecular level of how the growth of the majority of these tumors is controlled. Most breast cancers express the estrogen receptor (ER), which acts through mechanisms that are very different from those of the other receptors we have studied. Like other "nuclear receptors," the ER binds its ligand, estrogen (more precisely, estradiol), and then proceeds to bind to specific DNA sequences in the promoters of certain target genes; this binding, followed by association with transcriptional co-regulators, leads to the transcription of these genes.

Estrogen acts as a critically important mitogen for ER-positive breast cancer cells. This is indicated by the fact that a number of "anti-estrogens" such as tamoxifen, which bind to the ER and block some of its transcription-activating powers, prevent the growth of ER-positive cells and tumors and can actually cause tumor regression. However, none of this seems to explain precisely how estrogen and the ER drive the proliferation of ER-positive breast cancer cells. Within minutes after estrogen is added to these cells, the Ras → MAPK signaling cascade is activated. This rapid response cannot be explained by the fact that ER acts as a nuclear transcription factor, since changes in nuclear gene transcription are not felt in the cytoplasm (in the form of proteins synthesized on newly formed mRNA templates) for at least half an hour.

One critical clue may come from the observation that some of the ligand-activated ER molecules are found to be tethered to cytoplasmic membranes rather than being localized to their usual sites of action in the nucleus. Following ligand binding, some of this cytoplasmic ER has been reported to be bound to Shc, the important SH2- and SH3-containing bridging protein that participates in SOS activation and Ras signaling. The ER–Shc association, still poorly understood, may explain how cytoplasmic mitogenic signaling cascades are activated by estrogen. This fragmentary description of a major mitogenic pathway is most unexpected, in light of the fact that breast cancer has, by now, been studied far more intensively than any other type of human malignancy.

Another exception is provided by phosphatidylinositol 3-kinase (PI3K), which also operates immediately downstream of Ras. Recall that the PI3K enzyme is activated by Ras, by growth factor receptors, or by the two acting in concert. Its product is phosphatidylinositol triphosphate (PIP_3). In certain ovarian and colon carcinomas, a subunit of PI3K may be present in mutant form. But far more commonly, inhibition of this pathway is defective: PIP_3 levels are normally held down by the actions of the PTEN phosphatase, which is frequently lost in certain types of cancers. Indeed, PTEN loss may be as effective for increasing PIP_3 levels as the hyperactivity of the PI3K that is induced by Ras oncoproteins. The increased PIP_3 levels cause activation of Akt/PKB, which in turn has widespread effects on cell survival, growth, and proliferation.

The available data suggest that this signaling pathway reaching from growth factors through Ras to the ERKs (including its side branch involving PI3K) represents the dominant mitogenic pathway in most types of non-hematopoietic cells throughout the body. This prominence is reflected in the frequent involvement of this pathway, in deregulated form, in human cancers. Still, there are other candidates waiting in the wings. Future research may reveal that other pathways, such as the four mentioned in this chapter (involving GPCRs, Patched, Notch, and NF-κB), play dominant roles in providing the mitogenic signals that drive cell proliferation in certain types of frequently occurring tumors, thereby rivaling the importance of the RTK → Ras pathway. Interestingly, the design of the main mitogenic pathway operating in the commonly occurring estrogen receptor–positive breast cancers is still poorly understood (Sidebar 6.5).

Organization of mitogenic signaling pathways Stepping back from the details of the deregulation of mitogenic signaling pathways in human cancers, we can draw some more general lessons about the design of these pathways and how this organization affects human cancer pathogenesis. Evolution might have constructed complex metazoa like ourselves by using a small number of cellular signaling pathways (or even a single one) to regulate the proliferation of individual cells in various tissues. As it happened, a quite different design plan was pasted together: each of our cells uses a number of distinct signaling pathways to control its proliferation (with one of these, involving tyrosine kinase receptors, playing an especially prominent role).

In each type of cell, the signaling pathways work in a combinatorial fashion to ensure that proliferation occurs in the right place and at the right time during development and, in the adult, during tissue maintenance and repair. Moreover, different cell types use different combinations of pathways to regulate their growth and division. This helps to explain why the biochemistry of cancer, as described in this chapter, is so complex.

This signal transduction biochemistry is organized around a small number of basic principles and a large number of idiosyncratic details. It is a patchwork quilt of jury-rigged contraptions. Many of these details represent ad hoc solutions to biological or cell-physiologic problems that were chosen during early metazoan evolution 600 or 700 million years ago; some of these solutions reach back to the origins of eukaryotes, almost a billion years earlier. Once they were cobbled together, these solutions became fixed and relatively immutable, if only because other, subsequently developed contraptions depended on them. (To be sure, the end product of all these solutions—a machinery optimized for normal development and a lifetime of health—usually works marvelously well. During an average human life span, the cells inside the body cumulatively pass through approximately 10^{16} growth-and-division cycles and, in the case of most of us, provide excellent health—something that we take for granted.)

The signaling circuitry that was described in this chapter operates largely in the cytoplasm and receives inputs from cell surface receptors—those that bind either soluble extracellular ligands, notably growth factors, or insoluble matrix components (the integrins). Having received a complex mixture of **afferent** (incoming) signals from these receptors, this cytoplasmic circuitry emits signals to molecular targets in both the cytoplasm and the nucleus.

Included among the cytoplasmic targets are the regulators of cell shape and motility, energy metabolism, protein synthesis, and the apoptotic machinery, which decides on the life or death of the cell. The remaining **efferent** (outgoing) signals are transmitted to the nucleus, where they modulate the transcription of thousands of genes involved in the programming of cell growth and differentiation as well as in cell-cycle progression. We will return to many of these themes later. Thus, the regulation of the cell cycle and differentiation is described in Chapter 8, while the regulation of programmed cell death—apoptosis—will be presented in Chapter 9. The regulators of cell motility will be described in outline in Chapter 14.

Certain major themes have recurred throughout this chapter. Most apparent are the three ways in which signals are transduced through the signaling pathways of normal and neoplastic cells. First, the *intrinsic activity* of signaling molecules may be modulated. This can be achieved by noncovalent modifications (e.g., the binding of GTP by Ras) or receptor dimerization. Alternatively, a signaling molecule may be covalently modified; the phosphorylation of MEK by Raf, the phosphorylation of PIP_2 by PI3K, and the proteolytic cleavage of Notch are three examples of this.

Second, the *concentration* of a signaling molecule may be modulated, often by orders of magnitude. Thus, the concentration of β-catenin is strongly reduced by GSK-3β phosphorylation, and the concentration of PIP_3 is regulated by both PI3K (positively) and PTEN (negatively).

Third, the intracellular *localization* of signaling molecules can be regulated, which results in moving them from a site where they are inactive to a new site where they can do their work. The most dramatic examples of these derive from the phosphorylation of receptor tyrosine kinases (RTKs) and the subsequent attraction of multiple, distinct SH2-containing molecules to the resulting phosphotyrosines. Once anchored to the receptors, many of these signal-transducing molecules are brought in close proximity with other molecules that are associated with the plasma membrane; examples of the latter are the Ras molecule (the target of Sos action) and PIP_2 (the target of PI3K action).

The translocation of cytoplasmic molecules to the nucleus is another manifestation of this third class of regulatory mechanisms. For example, in this chapter we have read of a number of distinct proteins, termed "dual-address" transcription factors, which dwell in inactive forms in the cytoplasm and may be dispatched as active transcription factors (or components thereof) to the nucleus. Examples of these include β-catenin, NF-κB, the Smads, Notch, Gli, and the STATs.

The kinetics with which signals are transduced from one site to another within a cell may also vary enormously. One very rapid mechanism involves the actions of kinase cascades. Thus, the cytoplasmic MAP kinases (MAPKs), including Erks, that lie at the bottom of these cascades are activated almost immediately (in much less than a minute) following mitogen treatment of cells, and, following activation, move into the nucleus essentially instantaneously. Once there, they proceed to phosphorylate and functionally activate a number of key transcription factors.

A slightly slower but nonetheless highly effective signaling route derives from the strategy, cited above, of activating dormant transcription factors in the cytoplasm and dispatching them in activated form to the nucleus. By far the slowest signaling mechanisms depend on modulating the concentrations of signaling proteins. This mode of regulation can occur by changes in the rates of transcription of certain genes, translation of their mRNAs, stabilization of the mRNAs, or post-translational stabilization of their protein products.

The intracellular signaling circuitry encountered in this chapter appears to be organized similarly in various cell types throughout the body. This fact has important implications for our understanding of cancer and its development. More specifically, it suggests that the biochemical lessons learned from studying the neoplastic transformation of one cell type will often prove to be applicable to a number of other cell types throughout the body. The ever-increasing body of information on the genetic aberrations of human cancer cells provides strong support for this notion.

The similar design of the intracellular growth-regulating circuitry in diverse cell types forces us to ask why different kinds of cells behave so differently in response to various external signals. Many of the answers will eventually come from understanding the spectrum of receptors that each cell type displays on its surface. In principle, knowledge of the cell type–specific display of cell surface receptors and of the largely invariant organization of the intracellular circuitry should enable us to predict the behaviors of various cell types following exposure to various mitogens and to growth-inhibitory factors like TGF-β. Similarly, by understanding the array of mutant alleles coexisting within the genome of a cancer cell, we should be able to predict how these alleles perturb behavior and how they conspire to program the neoplastic cell phenotype. In practice, we are still far from being able to do so for at least eight reasons:

1. In this chapter, we described only the bare outlines of how these circuits operate and have consciously avoided many of the details. But such details are critical. For example, individual steps in these pathways are often controlled by multiple, similarly acting proteins, each of which functions in signal transduction slightly differently from the others. Examples of this are the four structurally distinct Ras proteins, the three forms of Akt/PKB, or Raf and its close cousin, B-Raf. In most cases, we remain ignorant of the functional differences between the superficially similar proteins operating within a pathway, and therefore have an incomplete understanding of how the pathway as a whole operates.

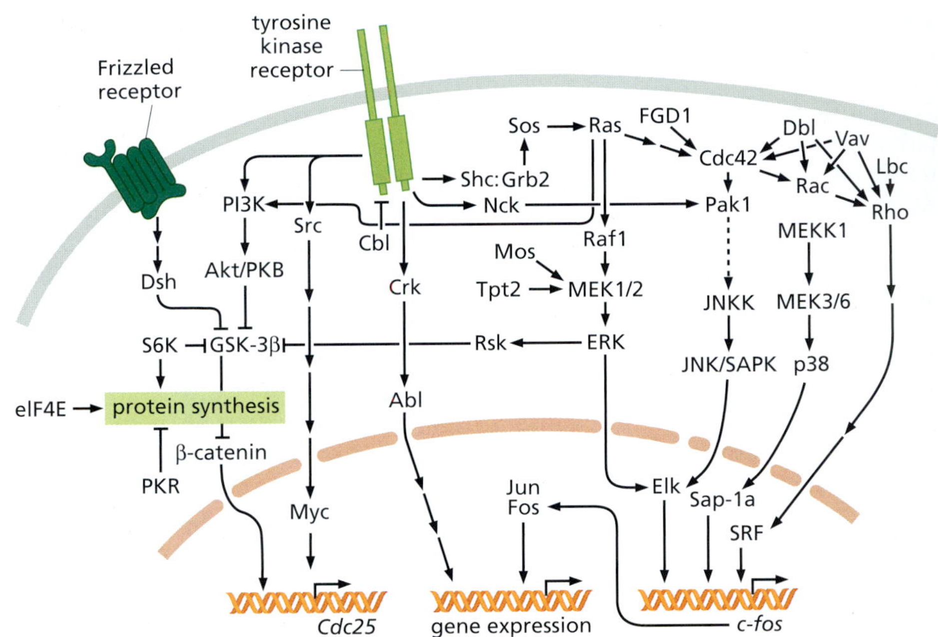

Figure 6.32 Two-dimensional signaling maps While we have depicted signaling pathways as linear cascades reaching from the plasma membrane to the nucleus, with occasional branch points, recent research reveals that these pathways interact via numerous cross-connections and points of convergence. This diagram, itself a simplification, provides some hints of these cross connections. Note, for example that the presence of GSK-3β, which plays a central role in the Wnt–β-catenin pathway, is also regulated by signaling proteins lying downstream of tyrosine kinase receptors). (Adapted from T. Hunter, *Cell* 88:333–346, 1997.)

2. Complex positive- and negative-feedback loops serve to amplify or damp the signals fluxing through each of these pathways. We touched on these only briefly. The feedback loops discovered to date are likely to represent only the tip of a much larger iceberg. For example, more than 100 phosphotyrosine phosphatase (PTP)–encoding genes have been found in the human genome. These operate to remove the phosphate groups attached by the 90 tyrosine kinases (TKs) present in our cells. We know almost nothing about how these key enzymes are regulated and what their various substrates are.

3. Each component in a circuit acts as a complex signal-processing device that is able to amplify, attenuate, and/or integrate the signals that it receives from upstream in a pathway before passing them on to downstream targets. We have not begun to understand in a quantitative way how this signal processing by individual signal-transducing proteins operates.

4. The ability of proteins to transduce signals to one another is affected by their intracellular concentrations, post-translational modifications (such as phosphorylation), and intracellular localization. We have only begun to plumb the depths of this complexity.

5. Calculations of diffusion times in cells indicate that efficient, rapid signal transduction is likely to occur only when interacting proteins are co-localized within a cell. Only very fragmentary information is available about the intracellular locations of many of these proteins and their interacting partners.

6. None of these intracellular pathways operates in isolation. Instead, each is influenced by cross connections with other pathways, some of which were also mentioned in this chapter. Understandably, the search for these cross connections has been largely postponed until the operations of each primary pathway are elucidated. Once these cross connections are added to the maps of these primary pathways, the drawings of signal transduction circuitry will likely resemble a weblike structure (e.g., Figure 6.32) rather than a series of parallel pathways that begin at the plasma membrane and extend linearly from there into the nucleus.

7. Even if these extensive cross connections did not exist, the endpoints of signaling—specific changes in cell phenotype—are the results of

combinatorial interactions between multiple, converging signaling pathways. Once again, this dimension of complexity continues to baffle us.

8. Our depiction of how signals are transmitted through a cell is likely to be fundamentally flawed. We have spoken here, time and again, about potent signaling pulses speeding down signal-transducing pathways and evoking, as endpoints, strong and clearly defined responses within a cell. In fact, each signaling cascade is likely to operate in a finely tuned, dynamic equilibrium, where positive and negative regulators continuously counterbalance one another. Accordingly, a signaling input (e.g., a mitogenic stimulus) may operate like the plucking of a fiber in one part of a spider web, which results in small reverberations at distant sites throughout the web. Here, neither our language nor our mathematical representations of signaling suffice to clarify our understanding.

Many cancer researchers would like to be able to draw a complete and accurate wiring diagram of the cell—the scheme that depicts how these pathways are interconnected. One measure of the difficulties associated with drawing the nodes and connections of this circuitry comes from census counts of the various classes of genes in the human genome. According to one enumeration, there are 518 distinct genes specifying various types of protein kinases; 40% of these genes make alternatively spliced mRNAs encoding slightly different variant structures of kinase proteins, leading to more than 1000 distinct kinase proteins that may be present in human cells. Of the 518 kinase genes, 90 encode tyrosine kinases, the remainder being serine/threonine kinases. Among the 90 tyrosine kinases, 58 function as the signaling domains of growth factor receptors. These numbers provide some measure of the complexity of the signaling circuitry that underlies cancer pathogenesis, since many of these kinases are involved in regulating the proliferation and survival of cells.

Indications of additional complexity have come from the more detailed studies of Ras signaling biochemistry. In this chapter, we emphasized the actions of Ras on three major effector pathways, specifically, Raf, PI3 kinase (PI3K), and a Ral-GEF (sometimes termed Ral-GDS); see Sections 6.5, 6.6, and 6.7. However, as the Ras proteins are subjected to more biochemical scrutiny, schemes like those shown in Figure 6.33 will emerge.

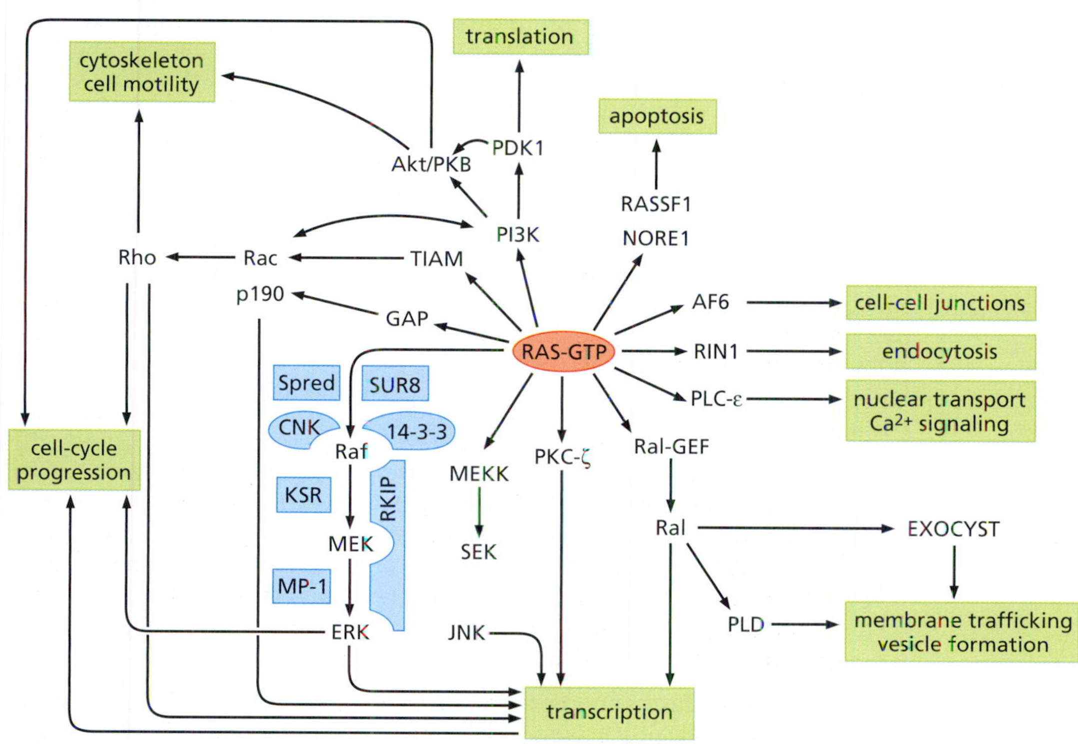

Figure 6.33 Ras effector pathways Detailed biochemical analyses of the GTP-bound, activated form of the Ras proteins has indicated that these proteins bind to far more than the three major effectors of Ras signaling—Raf, Ral-GEF, and PI3 kinase (PI3K). As indicated here, at least eight additional Ras-interacting proteins have been uncovered, most of which are known or suspected to play key roles in relaying Ras signals to specialized downstream signaling circuits involved in functions as diverse as transcription, membrane trafficking, endocytosis, and translation. (Adapted from M. Malumbres and M. Barbacid, *Nat. Rev. Cancer* 3:459–465, 2003.)

Structural biology has come of age, in that it has become almost routine to determine the detailed structures of protein molecules, including the signal-transducing proteins that play such key roles in cancer pathogenesis. Since structure determines function, the elucidation of these structures will greatly enrich our understanding of how signal-transducing pathways operate. The catalogs of these protein structures are still very incomplete. But this is changing rapidly, and in the future, the detailed structural information present in images such as Figure 6.34 will have major impact on how we understand the operations of signaling circuits.

The difficulties of explaining cell behavior in terms of the design of signal transduction circuitry underlie our current inability to understand another set of problems, these being related specifically to the genetics of cancer cell genomes: Why do various types of human tumors display highly specific types of genetic alterations? Why do 90% of pancreatic carcinomas carry a mutant K-*ras* oncogene, while this mutation is seen in only a small proportion (<10%) of breast cancers? Why are amplifications of the EGF receptor and its cousin, HER2/erbB2, encountered in more than a third of human breast cancers but seen uncommonly in certain other epithelial cancers, even though these receptors are widely displayed in many kinds of normal epithelia?

Why do approximately 50% of colon carcinomas carry a mutant K-*ras* gene, and how do the remaining 50% of these tumors acquire a comparable mitogenic signal? Some of these other tumors have mutant B-Raf or PI3K proteins. Still, this leaves many cancers with no apparent alterations affecting either the Ras → Raf → MAPK signaling cascade or the parallel PI3K → Akt/PKB pathway. Do these tumors harbor still-undiscovered genetic lesions that activate one or the other pathway? Or, alternatively, are the cells in these tumors able to proliferate uncontrollably because of mechanisms that have nothing to do with the operations of these particular pathways? And why are certain combinations of mutant or otherwise altered genes found in cancer cells, while other combinations are rarely if ever observed?

Finally, there is a major problem in clinical oncology that remains far from solution: Why do certain combinations of mutant alleles and protein expression patterns in cancer cells portend good or bad prognoses for patients? At present, we are largely limited to noting correlations between these patterns and clinical outcomes. But correlations hardly reveal the chains of causality connecting the genomes of cancer cells with the behavior of the tumors that these cells create. Many years will likely pass before these and related mysteries are solved definitively.

Key concepts

- Src has three homology domains: SH1 harbors the catalytic function; SH2 acts as an intracellular "receptor" for specific phosphotyrosines whose unique identities are determined by the particular oligopeptide sequence on their C′ side; and SH3 recognizes and binds certain proline-rich domains of substrates.
- As a consequence of ligand-induced transphosphorylation, a tyrosine kinase receptor (RTK) displays on its cytoplasmic tail an array of phosphotyrosines. Each phosphotyrosine attracts and binds a specific SH2-containing cytoplasmic protein. Such tethering of SH2-containing proteins permits them to interact with plasma membrane-associated proteins and phospholipids, thereby generating a variety of biochemical signals that can be transmitted to various downstream signaling cascades.
- Sos affects Ras in the manner of a guanine nucleotide exchange factor (GEF): it induces Ras to replace GDP with GTP, thereby activating it. Two linker proteins,

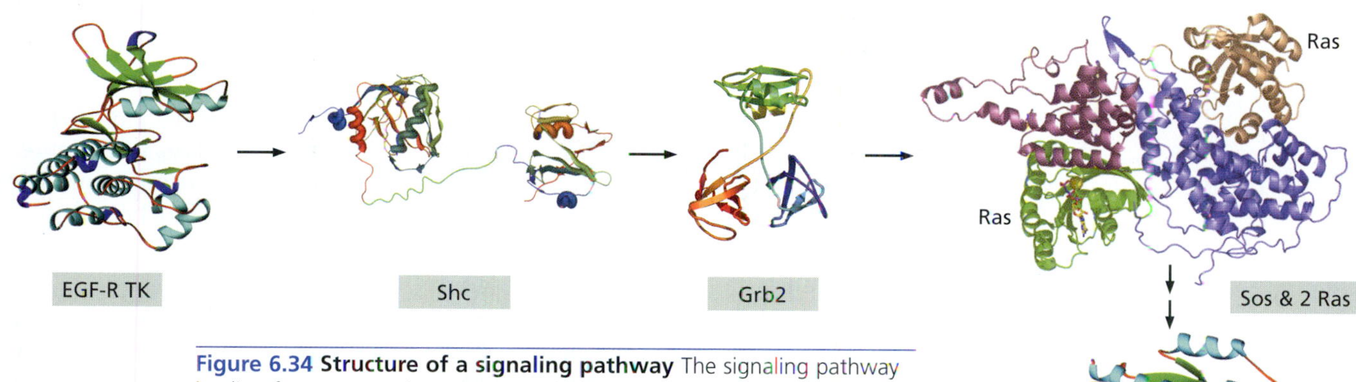

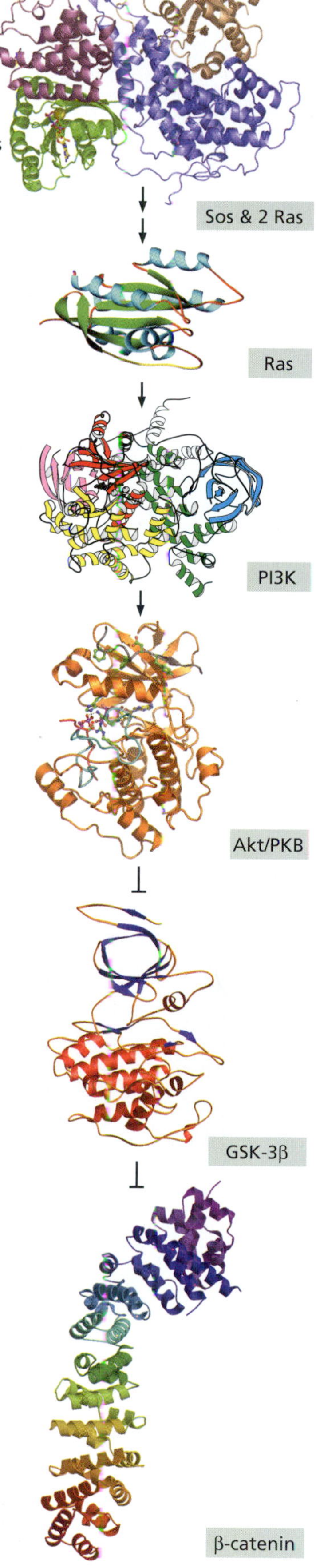

Figure 6.34 Structure of a signaling pathway The signaling pathway leading from EGF to the activation of signaling by β-catenin can now be depicted in terms of the detailed structures of all of the participating proteins, almost all of which have been determined by X-ray crystallography. The details of these structures will soon be integrated into our understanding of how signals flow through pathways like this one. Note that the structure of the central portion of Shc has not been determined and that the PI3K structure shown here reveals the structure PI3Kγ, which is closely related to, but distinct from, the PI3Ks that are activated by receptor tyrosine kinases. (EGF-R TK, courtesy of R. Latek, from J. Stamos et al., *J. Biol. Chem.* 277:46265–46272, 2002; Shc, courtesy of K.S. Ravichandran; Grb2, from S. Maignan et al., *Science* 268:291–293, 1995; Sos, from S.M. Margarit et al., *Cell* 112:685–695, 2003; Ras, from T. Schweins et al., *J. Mol. Biol.* 266:847–856, 1997; PI3K, from E.H. Walker et al., *Nature* 402:313–320, 1999; Akt/PKB, from J. Yang et al., *Nat. Struct. Biol.* 9:940–944, 2002; GSK-3β, from E. ter Haar et al., *Nat. Struct. Biol.* 8:593–596, 2001; β-catenin, courtesy of H.J. Choi and W.L. Weis.)

Shc and Grb2, form a physical bridge between a tyrosine phosphorylated receptor and Sos. This brings Sos into position to interact with Ras proteins, most of which are attached to the inner face of the plasma membrane.

- Three major downstream signaling cascades emanate from activated Ras via binding of its effector loop with downstream signaling partners—the Ras effectors; these are the Raf kinase, phosphatidylinositol 3-kinase (PI3K), and Ral-GEF.
- Raf phosphorylates residues on MEK. The resulting activated MEK phosphorylates and activates the extracellular signal-regulated kinases 1 and 2 (Erk 1 and 2), which phosphorylate yet other substrates. Erk 1 and 2 are MAPKs (mitogen-activated protein kinases), and thus the signaling cascade is often called a MAPK pathway.
- The phosphatidylinositol 3-kinase (PI3K) pathway depends on kinases phosphorylating a phospholipid and is important in suppressing apoptosis and in promoting the growth (in size) of cells. PI3K phosphorylates the membrane-embedded phosphatidylinositol (PI), by converting it to PIP_3. Akt/PKB is attracted to PIP_3 via its pleckstrin homology (PH) domain, becomes phosphorylated, and proceeds to activate a series of proteins that stimulate cell growth and proliferation and inhibit apoptosis. PIP_3 level is normally kept low in cells by phosphatases, notably PTEN. This suggests two ways in which this pathway can become deregulated in cancer cells—hyperactivation of PI3K or inactivation of PTEN.
- Among the cytoplasmic proteins that use their PH domains to associate with PIP_3 are a group of GEFs that function analogously to Sos by activating various GTPases, in this case three small G proteins belonging to the Rho family—Rho, Rac, and Cdc42. Activated Rho proteins participate in reconfiguring the cytoskeleton and the attachments cells make with their surroundings.
- In the third Ras-regulated pathway, communication between Ras and Ral is mediated by Ral-GEFs. By binding activated Ras, Ral-GEF becomes localized

near the inner surface of the plasma membrane and undergoes a conformational shift that induces Ral to shed its GDP and bind GTP. Ral then proceeds to activate downstream targets (e.g., Cdc42, Rac).

- Receptors for cytokines lack covalently associated tyrosine kinase (TK) domains and instead forms complexes with TKs of the Jak class, which phosphorylate STATs (signal transducers and activators of transcription). The STATs form dimers, and migrate to the nucleus, where they function as transcription factors of genes involved in cell growth and survival.
- While tyrosine kinase receptors bind GFs, integrins bind components of the ECM, which may lead to the formation of focal adhesions and activation of focal adhesion kinase (FAK), a nonreceptor TK associated with the cytoplasmic tails of integrins. The resulting phosphotyrosine residues on FAK provide a docking site for Src, which phosphorylates additional FAK tyrosines, providing docking sites for yet other SH2-containing molecules and activation of downstream targets, including Ras and PI3K.
- The pathway controlled by Wnt factors enables cells to remain in a relatively undifferentiated state—an important attribute of certain cancer cells. Acting through Frizzled receptors, Wnt suppresses the activity of glycogen synthase kinase-3β (GSK-3β), which otherwise would phosphorylate several key substrates, including β-catenin and cyclin D1, tagging them for destruction. The spared β-catenin moves into the nucleus and activates transcription of key growth-stimulating genes by binding to Tcf/Lef proteins.
- Upon binding their extracellular ligands, G-protein-coupled receptors (GPCRs) activate cytoplasmic heterotrimeric G proteins whose α subunit exchanges its GDP for GTP. The Gα subunit then dissociates from its two partners (Gβ + Gγ) and proceeds to activate or inhibit a number of cytoplasmic enzymes (e.g., Src, adenyl cyclase, phospholipase C) having mitogenic or antimitogenic effects, while the Gβ + Gγ dimer activates its own effectors, including PI3Kγ, PLC-β, Src, Rho-GEFs.
- The nuclear factor-κB (NF-κB) signaling system depends on the formation of NF-κB homo- and heterodimers in the cytoplasm. The inhibitor of NF-κB (IκB) usually sequesters NF-κB in the cytoplasm but in response to signaling, is tagged for destruction by IκB kinase (IKK). This leaves NF-κB free to migrate into the nucleus, where it activates expression of at least 150 genes, some of which specify key anti-apoptotic proteins. NF-κB also functions in mitogenic fashion by inducing expression of *myc* and cyclin D1 genes.
- Upon binding one of its ligands, Notch is cleaved twice, liberating a cytoplasmic fragment that migrates to the nucleus and functions as part of a transcription factor complex. The breakup of the activated receptor means that unlike RTKs, each Notch can fire only once upon binding a ligand molecule.
- Binding of the Patched receptor by a Hedgehog ligand causes Patched to release Smoothened protein from inhibition. Smoothened then emits downstream signals that protect cytoplasmic Gli protein from cleavage. Intact Gli can then migrate to the nucleus where it functions as an activator of transcription, whereas cleaved Gli yields a fragment that acts as a transcriptional repressor.
- The TGF-β signaling pathway involves the dispatch of cytoplasmic Smad transcription factors to the nucleus, where they help activate a large contingent of genes. This pathway plays a major role in the pathogenesis of many carcinomas, both in the early stages when TGF-β acts to arrest cell proliferation and later, when it contributes to tumor cell invasiveness.
- The pathways described work in a combinatorial fashion to ensure that proliferation occurs in the right place and at the right time during development; different cell types use different combinations of pathways to regulate their growth and division.
- In summary, signals are transduced through the various pathways by: (1) modulation of the intrinsic activity of signaling molecules by noncovalent

modification (e.g., GTP binding by Ras), receptor dimerization, or covalent modification (e.g., phosphorylation of MEK by Raf); (2) modulation of the concentration of a signaling molecule (e.g., increase of PIP_3 concentration by PI3K and decrease by PTEN); (3) regulation of the intracellular localization of signaling molecules (e.g., attraction of distinct SH2-containing molecules by phosphorylated RTKs; translocation of cytoplasmic molecules to the nucleus).

Thought questions

1. What molecular mechanisms have been evolved to ensure that the signals coursing down a signaling cascade reach the proper end-point targets rather than being broadcast non-specifically to "unintended" targets in the cytoplasm?
2. How many distinct molecular mechanisms can you cite that lead to the conversion of a proto-oncogene into an active oncogene?
3. Why are point mutations in *ras* oncogenes confined so narrowly to a small number of nucleotides, while point mutations in other cancer-related proteins are generally distributed far more broadly throughout the reading frames encoding these other proteins?
4. What factors determine the lifetime of the activated state of a Ras oncoprotein?
5. Treatment of cells with proteases that cleave the ectodomain of E-cadherin often result in rapid changes in gene expression. How might you rationalize this response, given what we know about this cell-surface protein?
6. What mechanisms have we encountered that ensure that a signal initiated by a growth factor receptor can be greatly amplified as the signal is transduced down a signaling cascade in the cytoplasm? Conversely, what signaling cascade(s) strongly limit the possible amplification of a signal initiated at the cell surface?
7. What quantitative parameters describing individual signal-transducing proteins will need to be determined before the behavior of the signaling cascade formed by these proteins can be predicted through mathematical modeling?

Additional reading

Alonso A, Sasin J, Bottini N et al. (2004) Protein tyrosine phosphatases in the human genome. *Cell* 117, 699–711.

Artavanis-Tsakonas S, Rand MD & Lake RJ (1999) Notch signaling: cell fate control and signal integration in development. *Science* 284, 770–776.

Blume-Jensen P & Hunter T (2001) Oncogenic kinase signalling. *Nature* 411, 355–365.

Bourne HR, Sanders DA & McCormick F (1990) The GTPase superfamily: a conserved switch for diverse cell functions. *Nature* 348, 125–132.

Bowman T, Garcia J, Turkson J & Jove R (2000) STATs in oncogenesis. *Oncogene* 19, 2474–2488.

Brazil DP & Hemmings BA (2001) Ten years of protein kinase B signalling: a hard Akt to follow. *Trends Biochem. Sci.* 26, 657–664.

Buettner R, Mora LB & Jove R (2002) Activated STAT signaling in human tumors provides novel molecular targets for therapeutic intervention. *Clin. Cancer Res.* 8, 945–954.

Campbell SL, Khosravi-Far R, Rossman KL et al. (1998) Increasing complexity of Ras signaling. *Oncogene* 17, 1395–1413.

Caron E (2003) Rac signaling: a radical view. *Nat. Cell Biol.* 5,185–187.

Chang L & Karin M (2001) Mammalian MAP kinase signalling cascades. *Nature* 410, 37–40.

Cohen P and Frame S (2001) The renaissance of GSK3. *Nat. Rev. Mol. Cell Biol.* 2, 769–775.

Dale TC (1998) Signal transduction by the Wnt family of ligands. *Biochem. J.* 329, 209–223.

Derynck R & Zhang YE (2003) Smad-dependent and Smad-independent pathways in TGF-β family signaling. *Nature* 425, 577–584.

Gilman A. (1987) G proteins: transducers of receptor-generated signals. *Annu. Rev. Biochem.* 56, 615–649.

Hazzalin CA & Mahadevan LC (2002) MAPK-regulated transcription: a continuously variable switch? *Nat. Rev. Mol. Cell Biol.* 3, 30–40.

Hunter T (1995) Protein kinases and phosphatases: the yin and yang of protein phosphorylation and signaling. *Cell* 80, 225–236.

Hunter T (1997) Oncoprotein networks. *Cell* 88, 333–346.

Hunter T (2000) Signaling—2000 and beyond. *Cell* 100, 113–127.

Hynes RO (2002) Integrins: bidirectional, allosteric signaling machines. *Cell* 110, 673–687.

Kadesch T (2004) Notch signaling: the demise of elegant simplicity. *Curr. Opin. Genet. Dev.* 14, 506–512.

Karin M, Cao Y, Greten FR & Li Z-W (2002) NF-κB in cancer: from innocent bystander to major culprit. *Nat. Rev. Cancer* 2, 301–310.

Levy DE & Darnell JE Jr (2002) STATs: transcriptional control and biological impact. *Nat. Rev. Mol. Cell Biol.* 3, 651–662.

Lewis TS, Shapiro PS & Ahn NG (1998) Signal transduction through MAP kinase cascades. In Advances in Cancer Research (GF Vande Woude, G Klein eds), pp 49–139. San Diego: Academic Press.

Lowy DR & Willumsen BM (1993) Function and regulation of Ras. *Annu. Rev. Biochem.* 62, 851–891.

Malumbres M & Barbacid M (2003) *RAS* oncogenes: the first thirty years. *Nat. Rev. Cancer* 3, 459–465.

Manning G, Whyte DB, Martinez R et al. (2002) The protein kinase complement of the human genome. *Science* 298,1912–1934.

Marinissen MJ & Gutkind JS (2001) G-protein-coupled receptors and signaling networks: emerging paradigms. *Trends Pharmacol. Sci.* 22, 368–376.

McCormick F (1999) Signalling networks that cause cancer. *Trends Cell Biol.* 9, M53–M56.

McManus EJ & Alessi DR (2002) Tsc1-Tsc2: a complex tale of PKB-mediated S6K activation. *Nat. Cell Biol.* 4, E214–E216.

Miele L & Osborne B (1999) Arbiter of differentiation and death: Notch signaling meets apoptosis. *J. Cell Physiol.* 181, 393–409.

Mumm JS & Kopan R (2000) Notch signaling: from the outside in. *Dev. Biol.* 228, 151–165.

Nelson WJ & Nusse R (2004) Convergence of Wnt, beta-catenin, and cadherin pathways. *Science* 303, 1483–1487.

Neves SR, Ram PT & Iyengar R (2002) G protein pathways. *Science* 296, 1636–1639.

Pawson T (2004) Specificity in signal transduction: from phosphotyrosine-SH2 domain interactions to complex cellular systems. *Cell* 116, 191–203.

Pawson T & Nash P (2003) Assembly of cell regulatory systems through protein interaction domains. *Science* 300, 445–452.

Pawson T & Scott JD (1997) Signaling through scaffold, anchoring and adaptor proteins. *Science* 278, 2075–2079.

Perkins ND (2004) NF-κB: tumor promoter or suppressor? *Trends Cell Biol.* 14, 64–69.

Polakis P (1999) The oncogenic activation of β-catenin. *Curr. Opin. Genet. Dev.* 9, 15–21.

Polakis P (2000) Wnt signaling and cancer. *Genes Dev.* 14, 1837–1851.

Reuther GW & Der CJ (2000) The Ras branch of small GTPases: Ras family members don't fall far from the tree. *Curr. Opin. Cell Biol.* 12, 157–165.

Richmond A. (2002) NF-κB, chemokine gene transcription and tumour growth. *Nat. Rev. Immunol.* 2, 664–674.

Scheid MP & Woodgett JR (2001) PKB/Akt: functional insights from genetic models. *Nat. Rev. Mol. Cell Biol.* 2, 760–767.

Schlaepfer DD & Hunter T (1998). Integrin signalling and tyrosine phosphorylation: just the FAKs? *Trends Cell Biol.* 8, 151–157.

Schlessinger J (2000) Cell signaling by receptor tyrosine kinases. *Cell* 103, 211–225.

Seto ES & Bellen HJ (2004) The ins and outs of Wingless signaling. *Trends Cell Biol.* 14, 45–53.

Shamji AF, Nghiem P & Schreiber SL (2003) Integration of growth factor and nutrient signaling: implications for cancer biology. *Mol. Cell* 12, 271–280.

Shi Y & Massagué J (2003) Mechanisms of TGF-β signaling from cell membrane to the nucleus. *Cell* 113, 685–700.

Shields JM, Pruitt K, McFall A et al. (2000) Understanding Ras: "it ain't over 'til it's over." *Trends Cell Biol.* 10, 147–154.

Smalley MJ & Dale TC (1999) Wnt signalling in mammalian development and cancer. *Cancer Metastasis Rev.* 18, 215–230.

Taipale J & Beachy PA (2001) The Hedgehog and Wnt signalling pathways in cancer. *Nature* 411, 349–354.

Vivanco I & Sawyers CL (2002) The phosphatidylinositol 3-kinase-AKT pathway in human cancer. *Nat. Rev. Cancer* 2, 489–501.

Vogt PK (2001) PI 3-kinase, mTOR, protein synthesis and cancer. *Trends Mol. Med.* 7, 482–484.

Weng AP & Aster JC (2004) Multiple niches for Notch in cancer: context is everything. *Curr. Opin. Genet. Dev.* 14, 48–54.

Wicking C & McGlinn E (2001) The role of hedgehog signalling in tumorigenesis. *Cancer Lett.* 173, 1–7.

Yaffe MB (2002) Phosphotyrosine-binding domains in signal transduction. *Nat. Rev. Mol. Cell Biol.* 3, 177–186.

Yaffe MB & Elia AEH (2001) Phosphoserine/threonine-binding domains. *Curr. Opin. Cell Biol.* 13, 131–138.

Yarden Y & Sliwkowski MX (2001) Untangling the ErbB signalling network. *Nat. Rev. Cancer* 2, 127–137.

Yu H & Jove R (2004) The STATs of cancer—new molecular targets come of age. *Nat. Rev. Cancer* 4, 97–105.

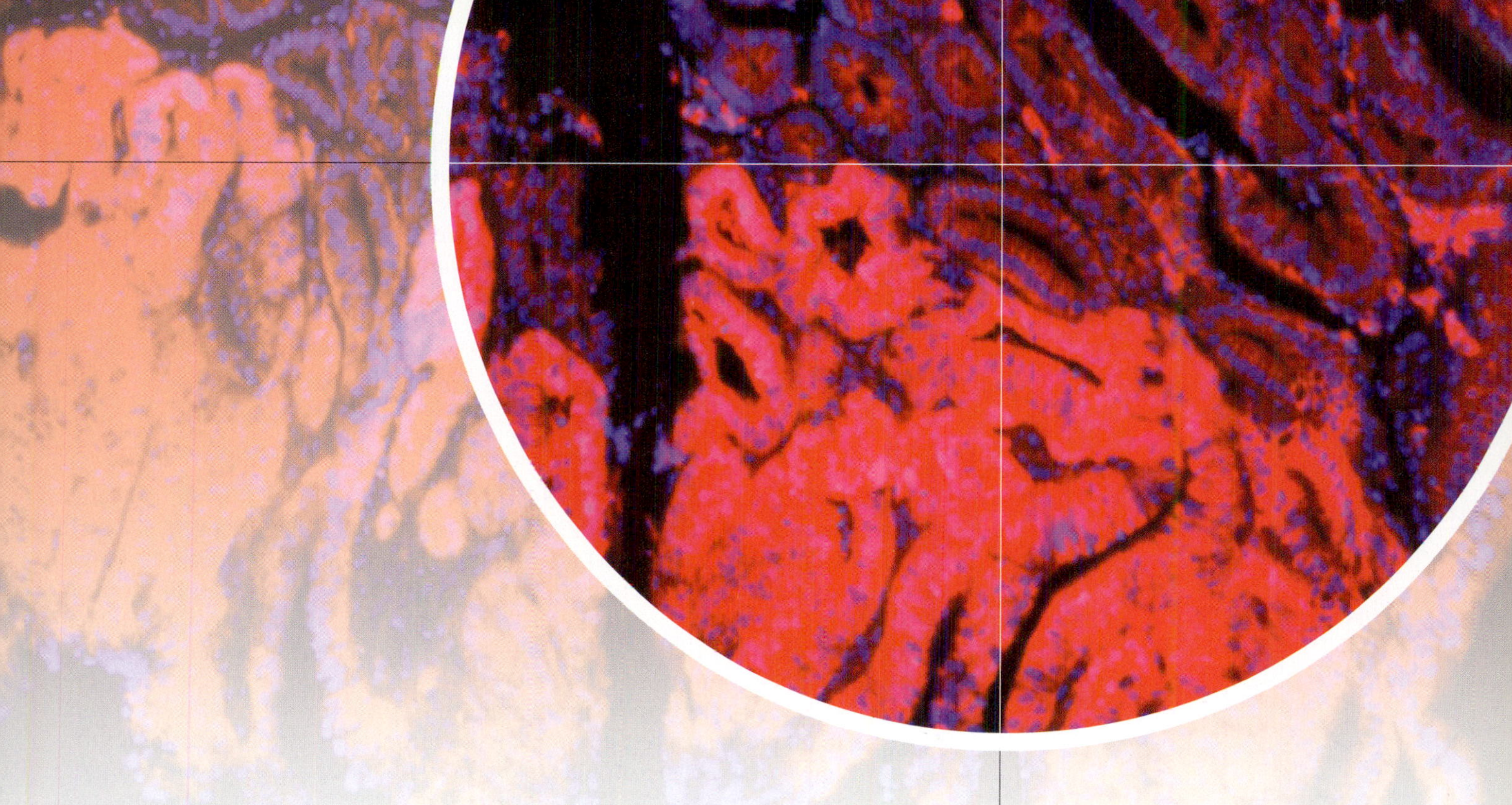

Chapter 7

Tumor Suppressor Genes

> Let me add … a consideration of the *inheritance of tumors*. … In order that a tumor may arise in such cases, the homologous elements in both series of chromosomes must be weakened in the same way.
>
> Theodor Boveri, pathologist, 1914

The discovery of proto-oncogenes and oncogenes provided a simple and powerful explanation of how the proliferation of cells is driven. The proteins encoded by proto-oncogenes participate in various ways in receiving and processing growth-stimulatory signals that originate in the extracellular environment. When these genes suffer mutation, the flow of growth-promoting signals released by these proteins becomes deregulated. Instead of emitting them in carefully controlled bursts, the oncoproteins release a steady stream of growth-stimulating signals, resulting in the unrelenting proliferation associated with cancer cells.

The logic underlying well-designed control systems dictates, however, that the components promoting a process must be counterbalanced by others that oppose this process. Biological systems seem to follow this logic as well, which leads us to conclude that the growth-promoting genes we have discussed until now provide only part of the story of cellular growth control.

In the 1970s and early 1980s, certain pieces of experimental evidence about cancer cell genetics began to accumulate that were hard to reconcile with the known properties of oncogenes. This evidence hinted at the existence of a second, fundamentally different type of growth-controlling gene—one that operates to constrain or suppress cell proliferation. The antigrowth genes came to be called **tumor suppressor genes**. Their involvement in tumor formation seemed to happen when these genes were inactivated or lost. Once a cell had shed one of these genes, this cell became liberated from its growth-suppressing effects. Now, almost three decades later, we have come to realize that the inactivation of tumor suppressor genes plays a role in cancer pathogenesis that is as important to cancer as the activation of oncogenes. Indeed, the loss of these genes from a cell's genome may be even more important than oncogene activation for the formation of many kinds of human cancer cells.

7.1 Cell fusion experiments indicate that the cancer phenotype is recessive

The study of tumor viruses in the 1970s revealed that these infectious agents carried a number of cancer-inducing genes, specifically viral oncogenes, which acted in a dominant fashion when viral genomes were introduced into previously normal cells (see Chapter 3). In particular, the introduction of a tumor virus genome into a normal cell would result in transformation of that cell. This response meant that viral oncogenes could dictate cellular behavior in spite of the continued presence and expression of opposing cellular genes within the virus-infected cell that usually functioned to ensure normal cell proliferation. Because the viral oncogenes could overrule these cellular genes, by definition the viral genes were able to induce a *dominant* phenotype—they were bringing about a cell transformation. This suggested, by extension, that cancerous cell growth was a dominant phenotype in contrast to normal (wild-type) cell growth, which was therefore considered to be *recessive* (see Section 1.1).

Still, these studies of tumor virus–induced cell transformation failed to accurately model the genetic mechanisms that seemed to be responsible for the formation of most human tumors. As was suspected at the time, and reinforced by research in the 1980s, most human cancers did not seem to arise as consequences of tumor virus infections. In the minds of many, this left the fundamental question of human cancer genetics unresolved. Thus, if human cancers were not caused by tumor viruses, then the lessons taught by the study of these tumor viruses might well be irrelevant to understanding how human tumors arise. In that case, one needed to shed all preconceptions about how cancer begins and admit to the possibility that the cancer cell phenotype was, with equal probability, a dominant or a recessive trait.

An initial hint of the importance of recessive cancer-inducing alleles came from experiments undertaken at Oxford University in Great Britain. Meaningful comparisons of two alternative alleles and specified phenotypes can occur only when both alleles are forced to coexist within the same cell or organism. In Mendelian genetics, when the allele of a gene that specifies a dominant phenotype is juxtaposed with another allele specifying a recessive trait, the dominant allele, by definition, wins out—the cell responds by expressing the phenotype of the dominant allele (Section 1.1).

The technique of *cell fusion* was well suited to force a confrontation between the alleles specifying normal growth and those directing malignant proliferation. In this procedure, cells of two different phenotypes (and often of different genotypes) are cultured together in a Petri dish (Figure 7.1). An agent is then used to induce fusion of the plasma membranes of cells that happen to be growing near one another in the cell monolayer. The fusing agent can be either a chemical, such as polyethylene glycol, or a viral glycoprotein, such as the ones displayed on the surface of certain paramyxoviruses like Sendai virus. If only two cells are close to one another and their plasma membranes are in contact, the result of treatment with a fusing agent will be a large cell with a single cytoplasm and two nuclei, often termed a **syncytium** (Figure 7.2). If a large number of cells happen to be in close proximity with one another in the cell monolayer being treated, then all of their plasma membranes may become fused to create a single syncytium—a giant multinucleated cell (a **polykaryon**) having one extremely large cytoplasm.

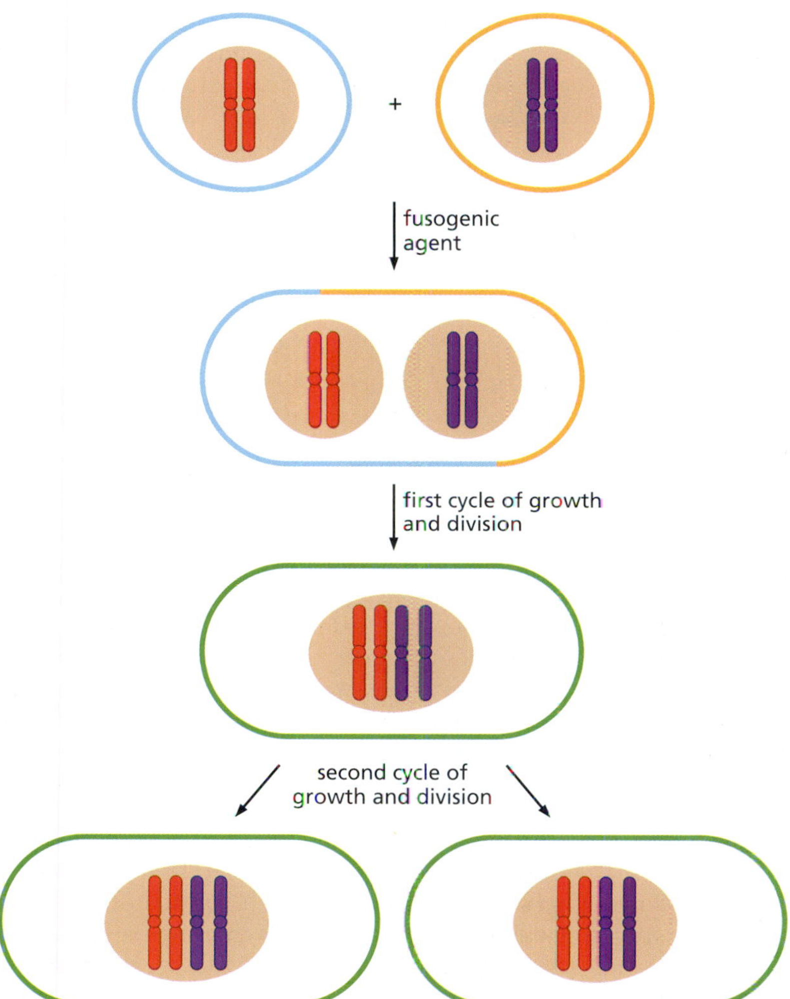

Figure 7.1 Experimental fusion of cells When cells growing adjacent to one another in monolayer culture are exposed to a *fusogenic* agent, such as inactivated Sendai virus or polyethylene glycol (PEG), they initially form a heterokaryon with multiple nuclei; formation of such a cell with two nuclei is shown here. When this cell subsequently passes through mitosis, the two sets of parental chromosomes are pooled in a single nucleus. During propagation of the resulting tetraploid cell in culture, the descendant cells often shed some of these chromosomes, thereby reducing their chromosome complement to a quasi-triploid or hyperdiploid state *(not shown)*.

These fusing agents join cells indiscriminately, so that two identical cells may be fused to one another or cells of two different types may become fused. If the two cells participating in this union happen to be of different origins, then the resulting hybrid cell is termed a **heterokaryon**, to reflect the fact that it carries two genetically distinct nuclei. Genetic tricks can then be used to select for a fused cell that has acquired two genetically distinct nuclei and, at the same time, to eliminate cells that have failed to fuse or happen to carry two identical nuclei. For example, each type of cell may carry a genetic marker that allows it to resist being killed by a particular antibiotic agent to which it would normally be sensitive. In such an instance, after the two cell populations are mixed and subjected to fusion, only heterokaryons will survive when the culture is exposed simultaneously to both antibiotics.

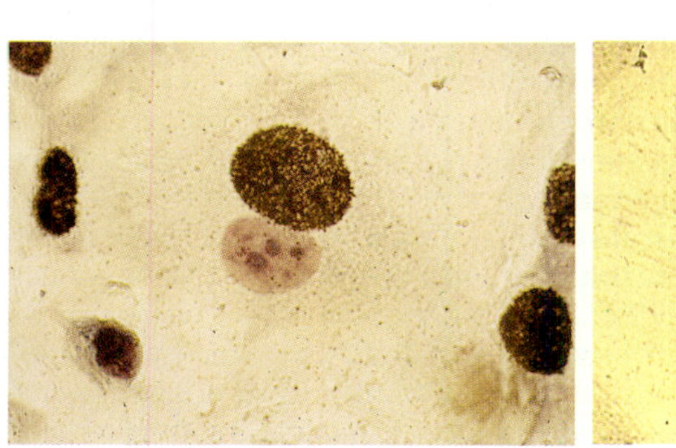
(A)

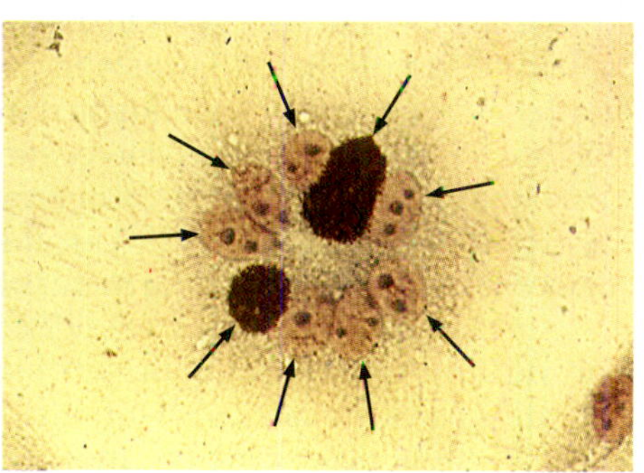
(B)

Figure 7.2 Appearance of experimentally fused cells Typical initial products of fusing cultured cells using a fusogenic agent (Figure 7.1; in this case inactivated Sendai virus) are seen here. Use of selection media and selectable marker genes can ensure the survival of a bi- or multinucleated cell carrying nuclei deriving from two distinct parental cell types. Conversely, use of these marker genes ensures the elimination of fused cells whose nuclei derive from only one or the other parental cell type. If such a hybrid cell is viable, the complements of chromosomes are intermingled in a single nucleus during the subsequent mitosis. (A) In this image, radiolabeled mouse NIH 3T3 cells have been fused with monkey kidney cells, resulting in a cell with two distinct nuclei. The (larger) mouse nucleus is identified by the silver grains that were formed during subsequent autoradiography. (B) Polykaryons may form with equal or greater frequency following such fusions but are usually unable to proliferate and spawn progeny. This polykaryon contains nine nuclei *(arrows)*. (Courtesy of S. Rozenblatt.)

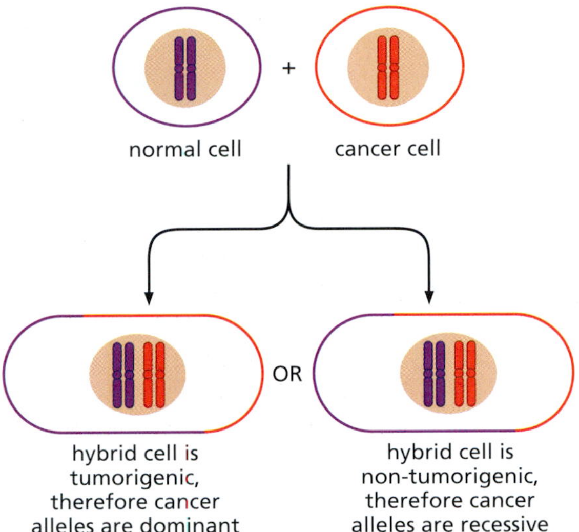

Figure 7.3 Dominance and recessiveness of the tumorigenic phenotype The use of cell fusion to test the dominance or recessiveness of the cancer cell phenotype can, in theory, yield two different outcomes—tumorigenic or non-tumorigenic hybrid cells. In fact, when cancer cells derived from most kinds of non-virus-induced human tumors (or from chemically induced rodent tumors) were used in these fusions, the hybrid cells were non-tumorigenic. In contrast, when the cancer cell derived from a virus-induced tumor, the hybrids were usually tumorigenic.

While cells having large numbers of nuclei are generally inviable, cells having only two nuclei are often viable and will proceed to grow. When they enter subsequently into mitosis, the two nuclear membranes break down, the two sets of chromosomes will flock to a single, common mitotic apparatus, and each resulting daughter cell will receive a single nucleus with chromosome complements originating from both parental cell types. We can imagine the construction of a heterokaryon formed by the fusion of a cancer cell with a wild-type cell. In this cell, the mutant, cancer-causing genes from one parental cell directly confront the wild-type alleles (governing normal cell proliferation) from the other parent, since both groups of genes now coexist within the same genome (Figure 7.3).

Some combinations of hybrid cells form genomes that are initially tetraploid, having received a complete diploid genome from each of the parent cells. Often, during the subsequent propagation of such cells in culture, they will progressively shed chromosomes and gradually approach a sub-tetraploid or even triploid chromosomal complement, since these hybrid cell lines begin their lives with far more chromosomes than they really need to grow and survive. Sometimes, the chromosomes from one parent are preferentially shed. For example, when human and mouse cells are fused, the descendant hybrid cells shed human chromosomes progressively until only a small minority of chromosomes are of human origin.

Using the cell fusion technique, hybrid cells of disparate origins, including even chicken–human hybrids, could be made. Cells from the G_1 phase of the cell cycle were fused with cells in M phase in order to see which phase of the cell cycle contained the dominantly acting controls. Undifferentiated cells were fused with more differentiated ones to determine which phenotype dominated. For us, however, the results of only one type of experiment are of special interest: How did normal + tumor hybrid cells grow?

The smart money at the time bet that cancer, a potent, dominating phenotype, would be dominant when placed in opposition to the normal cell growth phenotype (see Figure 7.3). In this instance, as is often the case, the smart money was wrong. In a number of experiments, when tumor cells were fused with normal cells, the resulting tetraploid cells or subsequent sub-tetraploid cells were discovered to have lost the ability to form tumors when these hybrid cells were injected into appropriate host animals. This meant, quite unexpectedly, that the malignant cell phenotype was recessive to the phenotype of normal, wild-type growth.

The notable exception to these observations came when the transformed parental cell in the two-cell hybrid had been transformed by tumor virus infection. On these occasions, the tumor cell phenotype (created by the acquired viral oncogenes) dominated in the cancer- plus normal-cell hybrids. So the take-home message about the cancer versus normal phenotypes was amended and narrowed slightly: if the cancer cell had originally arisen without the involvement of a tumor virus, then its malignant phenotype was recessive when this cell was fused with a normal cell.

7.2 The recessive nature of the cancer cell phenotype requires a genetic explanation

In fact, a simple genetic model could be invoked to explain these observations. It depended on the observation, made frequently in genetics, that the phenotype of a mutant, inactive allele is recessive in the presence of an intact, wild-type allele.

The hypothesis went like this. Imagine that normal cells carry genes that constrain or suppress their proliferation. During the development of a tumor, the evolving cancer cells shed or inactivate one or more of these genes. Once these growth-suppressing genes are lost, the proliferation of the cancer cells accelerates, no longer being held back by the actions of these growth-suppressing genes. As long as the cancer cell lacks these genes, it continues to proliferate in a malignant fashion. However, the moment that wild-type, intact versions of these genes operate once again within the cancer cell, after having been introduced by the technique of cell fusion, the proliferation of the cancer cell, or at least its ability to form tumors, grinds abruptly to a halt.

Since the wild-type versions of these hypothetical genes antagonize the cancer cell phenotype, these genes came to be called tumor suppressor genes (TSGs). There were arguments both in favor of and against the existence of tumor suppressor genes. In their favor was the fact that it is far easier to inactivate a gene by a variety of mutational mechanisms than it is to hyperactivate its functioning through a mutation. For example, a *ras* proto-oncogene can be (hyper)activated only by a point mutation that affects its 12th, 13th, or 61st codon (Section 5.9). In contrast, a tumor suppressor gene, or, for that matter, any other gene, can readily be inactivated by point mutations that strike at many sites in its protein-coding sequences or by random deletions that excise blocs of nucleotides from these sequences.

The logical case against the existence of tumor suppressor genes derived from the diploid state of the mammalian cell genome. If mutant, inactive alleles of tumor suppressor genes really did play a role in enabling the growth of cancer cells, and if these alleles were recessive, an incipient tumor cell would reap no benefit from inactivating only one of its two copies of a tumor suppressor gene, since the recessive mutant allele would coexist with a dominant wild-type allele in this cell. Hence, it seemed that both wild-type copies of a given tumor suppressor gene would need to be eliminated by an aspiring cancer cell before this cell would truly benefit from its inactivation.

This requirement for two separate genetic alterations seemed complex and unwieldy, indeed, too improbable to occur in a reasonably short period of time. Since the likelihood of two mutations occurring is the square of the probability of a single mutation, this made it seem highly unlikely that tumor suppressor genes could be fully inactivated during the time required for tumors to form. For example, if the probability of inactivating a single gene copy by mutation is on the order of 10^{-6} per cell generation, the probability of silencing both copies is on the order of 10^{-12} per cell generation—exceedingly unlikely, given the small size of incipient cancer cell populations, the multiple genetic alterations that are needed to make a human tumor, and the several decades of time during which multiply mutated cells expand into clinically detectable tumors.

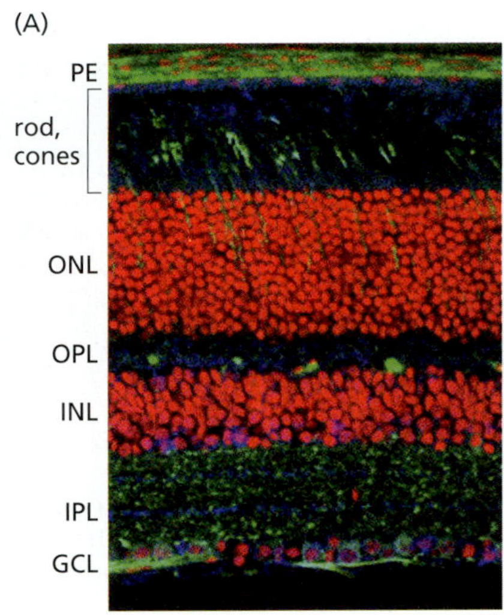

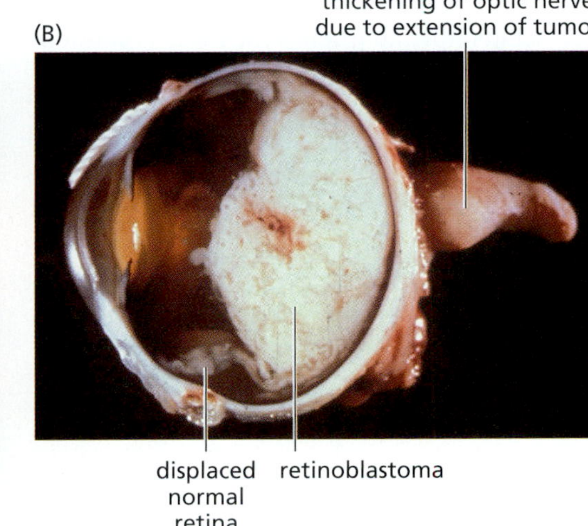

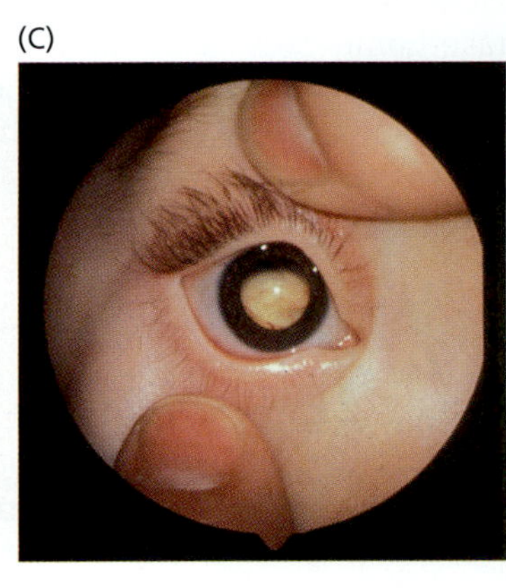

Figure 7.4 Pediatric retinoblastomas For unknown reasons, defects in the *Rb* gene lead preferentially to the formation of pediatric tumors in the retina, a small, highly specialized tissue, even though the *Rb* gene is active in governing cell proliferation in most cell types throughout the body. (A) The tumors arise from an *oligopotential* stem-cell precursor of multiple retinal cell types including rods and cones. This stem cell may be precursor of all of the cells seen here or only a subset of them. This micrograph reveals a cross-section of the retina showing, from the back of the globe toward the front of the eye (*top to bottom*), pigmented epithelium (PE), outer nuclear layer (ONL), outer plexiform layer (OPL), inner nuclear layer (INL), inner plexiform layer (IPL), and ganglion cell layer (GCL). [The nuclei of the rods and cones are located in the ONL, with the photoreceptors *(dark green)* extending upward toward the PE.] (B) A section of the globe of an eye in which a large retinoblastoma has developed. (C) A retinoblastoma may often initially present in the clinic as an opacity that obscures the retina. (A, from S.W. Wang et al., *Development* 129:467–477, 2002; B, C, courtesy of T.P. Dryja.)

7.3 The retinoblastoma tumor provides a solution to the genetic puzzle of tumor suppressor genes

The arguments for and against the involvement of tumor suppressor genes in tumor development could never be settled by the cell fusion technique. For one thing, this experimental strategy, on its own, offered little prospect of finding and isolating specific genes whose properties could be studied and used to support the tumor suppressor gene hypothesis. In the end, the important insights came from studying a rare childhood eye tumor, **retinoblastoma**.

This tumor of the retina, arising in the precursors of photoreceptor cells, is normally observed in about 1 in 20,000 children (Figure 7.4). These tumors are diagnosed anytime from birth up to the age of 6 to 8 years, after which the disease is rarely encountered. The tumor **syndrome** (i.e., a constellation of clinical traits) appears in two forms. Some children—those who are born into families with no history of retinoblastoma—present with a single tumor in one eye or the other. If this tumor is eliminated, either by radiation or by removal of the affected eye, then this child has no further risk of retinoblastoma and no risk of tumors elsewhere in the body. Because this tumor strikes in children lacking a family history, it is considered to be a manifestation of the **sporadic** form of this disease. (Since this form of the disease affects only a single eye, it is often termed *unilateral* retinoblastoma.)

The **familial** form of retinoblastoma appears in children having a parent who also suffered from and was cured of the disease early in life. In this instance, there are usually multiple foci of tumors arising in both eyes (and is therefore called *bilateral* retinoblastoma). Moreover, curing the eye tumors, which can be accomplished with radiation or surgery, does not protect these children from a greatly increased (more than 500 times above normal) risk of bone cancers (osteosarcomas) during adolescence and an elevated susceptibility to developing yet other tumors later in life (Figure 7.5A). Those who survive these tumors and grow to adulthood are usually able to reproduce; in half of their offspring, the familial form of retinoblastoma again rears its head (Figure 7.5B).

This familial form of retinoblastoma is passed from one generation to the next in a fashion that conforms to the behavior of a Mendelian dominant allele. In truth, a form of retinoblastoma that is indistinguishable from the familial disease can be seen in children with no family history. In these children, it seems as if mutations occur *de novo* during the formation of a sperm or an egg, causing the children to acquire the cancer-causing mutation in all of their cells. This creates a genetic situation that is identical to the one occurring when the mutant

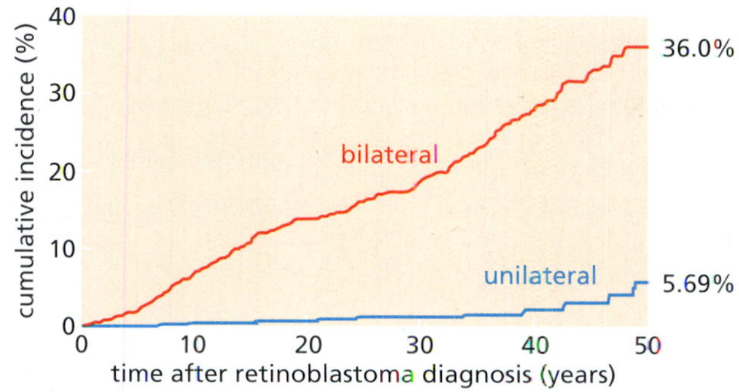

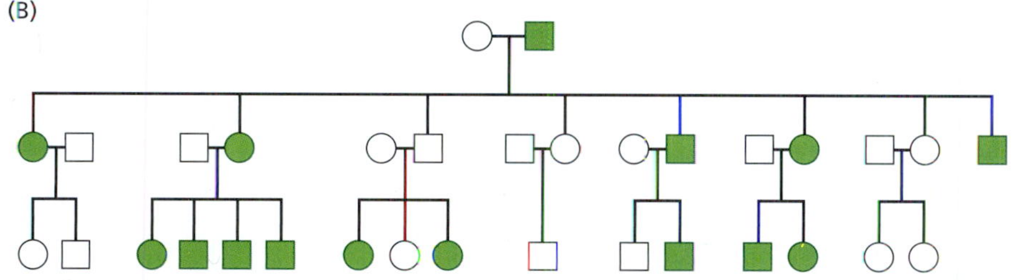

Figure 7.5 Unilateral versus bilateral retinoblastoma (A) Children with unilateral retinoblastoma and without an afflicted parent are considered to have a sporadic form of the disease, while those with bilateral retinoblastoma suffer from a familial form of this disease. In this graph, the clinical courses of a group of 1601 retinoblastoma patients who had been diagnosed between 1914 and 1984 were followed. As is apparent, those cured of bilateral tumors *(red line)* have a dramatically higher risk of developing second (and subsequently occurring tumors) in a variety of organ sites than those with unilateral tumors *(blue line)*. [A portion of this elevated risk is attributable to tumors that arose in the vicinity of the eyes because of the *radiotherapy* (i.e. treatment by X-ray irradiation) that was used to eliminate the retinoblastomas when these individuals were young.] (B) This pedigree shows multiple generations of a kindred afflicted with familial retinoblastoma, a disease that usually strikes only 1 in 20,000 children. Such multiple-generation pedigrees were rarely observed before the advent of modern medicine, which allows an affected child to be cured of the disease and therefore reach reproductive age. Males *(squares)*, females *(circles)*, affected individuals *(green filled circles, squares)*, unaffected individuals *(open circles, squares)*. (A, from R.A. Kleinerman, M.A. Tucker, R.E. Tarone et al., *J. Clin. Oncol.* 23:2272–2279, 2005; B, courtesy of T.P. Dryja.)

allele is already present in the genome of one or another afflicted parent and transmitted via sperm or egg to the offspring. (The great majority of these *de novo* mutations are of paternal origin, apparently because far more cycles of cell division precede the formation of a sperm than precede the formation of an egg; each cycle of cell growth and division presents a risk of mistakes in DNA replication and resulting mutations.)

Upon studying the kinetics with which retinal tumors appeared in children affected by either the familial or the sporadic version of the disease, Alfred Knudson concluded in 1971 that the rate of appearance of familial tumors was consistent with a single random event, while the sporadic tumors behaved as if two random events were required for their formation (Figure 7.6).

These kinetics led to a deduction, really a speculation. Let's say that there is a gene called *Rb*, the mutation of which is involved in causing childhood retinoblastoma. We can also imagine that the mutations that involve the *Rb* gene in tumor development invariably create inactive and thus recessive alleles of the *Rb* gene. If the tumor-predisposing alleles of *Rb* are indeed recessive, then both copies of the *Rb* gene must be knocked out before a retinal cell launches the uncontrolled proliferation that eventually results in a retinal tumor (Figure 7.7).

In children who inherit a genetically wild-type constitution from their parents, the formation of a retinoblastoma tumor will require two successive genetic alterations in a retinal cell lineage that operate to inactivate the two previously functional copies of the *Rb* gene—that is, two somatic mutations (see Figure 7.7). These dynamics will be quite different in half of the children born into a family in which familial retinoblastoma is present. In these children, we can imagine that one of the two required *Rb* gene mutations has been passed through the germ line from a parent to a fertilized egg This mutant *Rb* gene is therefore implanted in all the cells of the developing embryo, including all the cells in the retina. Any one of these retinal cells needs to sustain only a single (somatic) mutation knocking out the still wild-type allele in order to reach the state where it no longer harbors any functional *Rb* gene copies. Having reached this state, this cell can then launch forth to create a retinoblastoma.

Strikingly, this depiction of *Rb* gene behavior corresponded exactly to the attributes of tumor suppressor genes whose existence had been postulated from

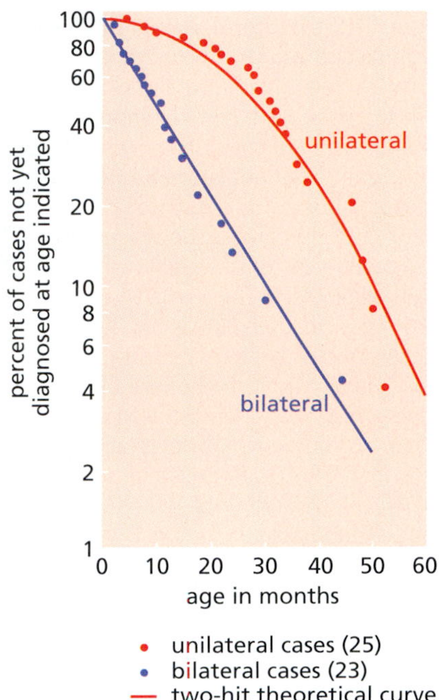

Figure 7.6 Kinetics of appearance of unilateral and bilateral retinoblastomas Alfred Knudson Jr. studied the kinetics with which bilateral and unilateral retinoblastomas appeared in children. He calculated that the bilateral cases (tumors in both eyes) arose with one-hit kinetics, whereas the unilateral tumors (affecting only one or the other eye) arose with two-hit kinetics. Each of these hits was presumed to represent a somatic mutation. The fact that the two-hit kinetics involved two copies of the *Rb* gene was realized only later. Percentage of cases not yet diagnosed *(ordinate)* at age (in months) indicated *(abscissa)*. (From A.G. Knudson Jr., *Proc. Natl. Acad. Sci. USA* 68:820–823,1971.)

results of the cell fusion studies (see also Sidebar 7.1). Still, all this remained no more than an attractive hypothesis , since the *Rb* gene had not been cloned and no molecular evidence was available showing *Rb* gene copies that had been rendered inactive by mutation.

7.4 Incipient cancer cells invent ways to eliminate wild-type copies of tumor suppressor genes

The findings with the *Rb* gene reinforced the notion that both copies of this gene needed to be eliminated, raising once again a vexing issue: How could two copies of a tumor suppressor gene possibly be eliminated, one after the other, during the formation of sporadic retinoblastomas if the probability of both required mutational events' occurring is, as calculated earlier, about 10^{-12} per cell generation? Given the relatively small target cell populations in the developing retina (Knudson guessed about 10^6 cells), it seemed highly unlikely that both gene copies could be eliminated through two, successive mutational events.

A solution to this dilemma was suggested by some geneticists. Suppose that the first of the two *Rb* gene copies was indeed inactivated by some type of mutational

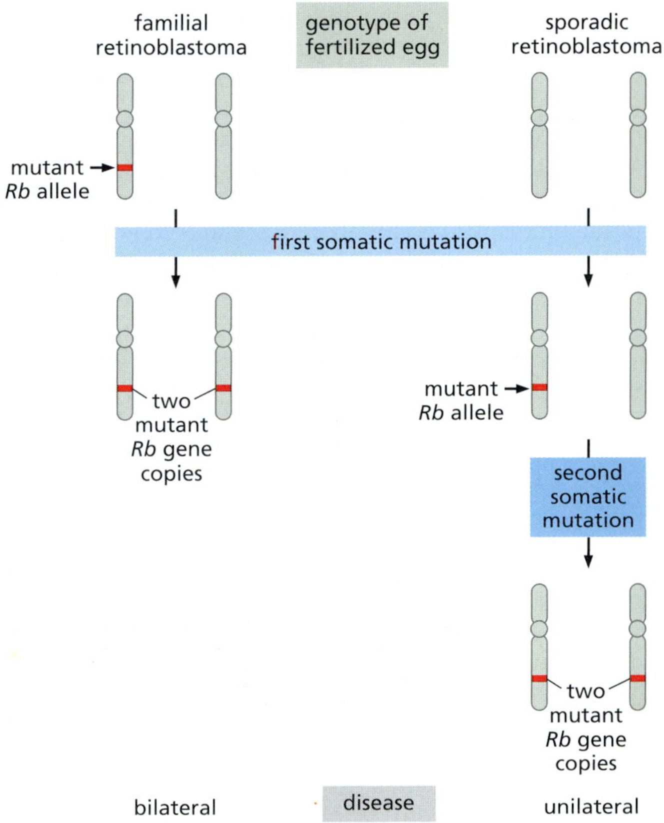

Figure 7.7 Dynamics of retinoblastoma formation The two affected genes predicted by Knudson (see Figure 7.6) are the two copies of the *Rb* gene on human Chromosome 13. In familial retinoblastoma, the zygote (fertilized egg) from which a child derives carries one defective copy of the *Rb* gene, and all retinal cells in this child will therefore carry only a single functional *Rb* gene copy. If this surviving copy of the *Rb* gene is eliminated in a retinal cell by a somatic mutation, that cell will lack all *Rb* gene function and will be poised to proliferate into a mass of tumor cells. In sporadic retinoblastoma, the zygote is genetically wild type at the *Rb* locus. In the retina of a resulting child, retinoblastoma development will require two successive somatic mutations striking the two copies of the *Rb* gene carried by a lineage of retinal precursor cells, yielding once again the same outcome—a cell poised to proliferate into a tumor mass. Because only a single somatic mutation is needed to eliminate *Rb* function in familial cases, multiple cells in both eyes are affected. However, the two somatic mutations required in sporadic disease are unlikely to affect a single cell lineage, yielding at most one tumor.

Sidebar 7.1 Mutant *Rb* genes are both dominant and recessive While the molecular evidence underlying *Rb* genetics has not yet been presented in this chapter, our discussion up to this point already carries with it an apparent internal contradiction. At the level of the whole organism, an individual who inherits a mutant, defective allele of *Rb* is almost certain to develop retinoblastoma at some point in childhood. Hence, the disease-causing *Rb* allele acts dominantly at the level of the whole organism. However, if we were able to study the behavior of the mutant allele within a cell from this individual, we would conclude that the mutant allele acts recessively, since a cell carrying a mutant and a wild-type *Rb* gene copy would behave normally. Hence, the mutant *Rb* allele acts dominantly at the *organismic* level and recessively at the *cellular* level. (A geneticist insistent on strictly proper usage of genetic terms would say that the *phenotype* of retinoblastoma behaves dominantly at the organismic level and recessively at the cellular level; we will ignore this propriety here.)

event that occurred with a frequency comparable to that associated with most mutational events—about 10^{-6} per cell generation. A cell suffering this mutation would now be in a heterozygous configuration, having one wild-type and one defective gene copy—that is, $Rb^{+/-}$. Since the mutant *Rb* allele was recessive at the cellular level, this heterozygous cell would continue to exhibit a wild-type phenotype. But what if the second, still-intact gene copy of *Rb* were inactivated by a mechanism that did not depend on its being struck directly by a second, independent mutational event? Instead, perhaps there was some exchange of genetic information between paired homologous chromosomes, one of which carried the wild-type *Rb* allele while the other carried the already-mutant, defective allele. Normally, recombination between chromosomes was known to occur almost exclusively during meiosis. What would happen if, instead, a recombination occurred between one of the chromatid arms carrying the wild-type *Rb* allele and a chromatid from the paired chromosome carrying the mutant allele (Figure 7.8)? Such recombination was thought to occur during active cell proliferation and consequently was termed **mitotic recombination** to distinguish it from the meiotic recombination events that shuffle chromosomal arms prior to the formation of sperm and egg.

In this fashion, the chromosomal arm carrying the wild-type *Rb* allele might be replaced with a chromosomal arm carrying the mutant allele derived from the paired, homologous chromosome. Since this exchange depends upon reciprocal exchange of genetic information between chromosomal arms, the chromosomes participating in this exchange would both remain full-length and, when visualized in the microscope during a subsequent metaphase, be indistinguishable from the chromosomes that existed prior to this genetic exchange. However, at the genetic and molecular level, one of the cells emerging from

Figure 7.8 Elimination of wild-type *Rb* gene copies Mitotic recombination can lead to loss of heterozygosity (LOH) of a gene such as *Rb*. Genetic material is exchanged between two homologous chromosomes (e.g., the two Chromosomes 13) through the process of genetic crossing over, occurring in the G_2 phase or, less often, the M phase of the cell cycle. Once the crossing over has occurred, the subsequent segregation of chromatids may yield a pair of daughter cells both of which retain heterozygosity at the *Rb* locus. With equal probability, this process can yield two daughter cells that have undergone LOH at the *Rb* locus (and other loci on the same chromosomal arm), one of which is homozygous mutant at the *Rb* locus while the other is homozygous wild type at this locus.

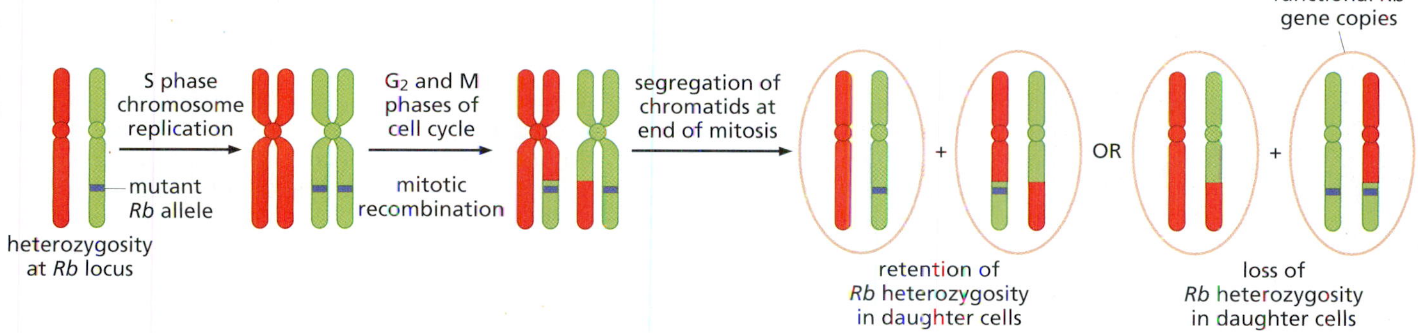

this recombination event would have shed its remaining wild-type *Rb* allele and would therefore have become $Rb^{-/-}$. Importantly, this mitotic recombination was found to occur at a frequency of 10^{-5} to 10^{-4} per cell generation, and was therefore a far easier way for a cell to rid itself of the remaining wild-type copy of the *Rb* gene than mutational inactivation of this gene copy, which, as mentioned above, was known to occur at a frequency of about 10^{-6} per cell generation.

Prior to this mitotic recombination, the two homologous chromosomes (in the case of the *Rb* gene, the two human Chromosomes 13) differ from one another in many subtle details. After all, one is of paternal, the other of maternal origin, indicating that they are heterozygous at many genetic loci. However, following the mitotic recombination that leads to homozygosity at the *Rb* locus (yielding either an $Rb^{-/-}$ or an $Rb^{+/+}$ genotype), a number of other nearby genes that are chromosomal neighbors of *Rb* on Chromosome 13 will also lose heterozygosity and become homozygous. This genetic alteration of a gene or a chromosomal region is usually termed loss of heterozygosity, or simply LOH. (An alternative term is "allelic deletion.") Interestingly, LOH events may well occur with different frequencies in different human populations (Sidebar 7.2).

Sidebar 7.2 Does endogamy lead to high rates of LOH? Experiments using laboratory and feral mice indicate that the frequency of mitotic recombination can be suppressed tenfold or more if homologous chromosomes are genetically very different from one another rather than genetically similar or identical to one another. This indicates that the pairing of homologous chromosomes that enables mitotic recombination to occur depends on extensive nucleotide sequence identity in the two DNAs.

This finding may well have important implications for tumor progression in humans. In many human populations, **endogamy** (breeding within a group) and even first-cousin marriages are quite common. Mendelian genetics dictates that in the offspring of such individuals, chromosomal regions of homozygosity are frequent. If the results of mouse genetics are instructive, then pairs of homologous chromosomes are more liable to undergo mitotic recombination in these individuals. Should one of the two chromosomes carry a mutation in a tumor suppressor gene such as *Rb*, mitotic recombination and LOH leading to loss of the surviving wild-type gene copy of this tumor suppressor gene will become proportionately more likely. Consequently, it may be that the offspring of cousin marriages in human populations may have greater susceptibility to tumors simply because their chromosomes participate more frequently in LOH. This speculation has not been tested.

There are yet other ways by which chromosomes can achieve LOH. One of these derives from the process of gene conversion (Figure 7.9). In this alternative mechanism, a DNA strand being elongated during DNA replication on one chromosome will form a hybrid with the complementary DNA strand present in the second, homologous chromosome. The strand will then be extended for some distance by a DNA polymerase using the second chromosome's strand as a template before becoming disentangled from this chromosome and once again annealing with the complementary strand of the chromosome on which its synthesis began. Thereafter, this strand will continue to be extended by using as template the DNA strand on which it originated. Accordingly, this newly synthesized strand of DNA will acquire DNA sequences from a stretch of the paired chromosome. Should this gene conversion involve copying of an already inactivated *Rb* allele, for example, then once again LOH will have occurred in this chromosomal region. Such gene conversion is known to occur even more frequently per cell generation than does mitotic recombination.

As described above, the swapping of chromosomal information between two chromosomes, occurring by either mechanism, might well involve a large number of genes and genetic markers flanking the *Rb* locus on both sides. Their behavior will also be of interest. Prior to this exchange, we can imagine that many of these genes and genetic sequences on the long arm of Chromosome 13 (where the *Rb* gene happens to be localized) will also be present in heterozygous configuration (Sidebar 1 ●). Afterward, they also will have undergone LOH.

For example, the *Rb* locus, which may originally (in an incipient cancer cell) have been in the $Rb^{+/-}$ configuration, may now become $Rb^{-/-}$ or $Rb^{+/+}$. At the same time, a genetic marker sequence in a neighboring region of the chromosome that was initially AAGCC/AAGTC (hence heterozygous) may now become either AAGCC/AAGCC or AAGTC/AAGTC (and therefore homozygous). In effect, this neighboring sequence segment has been carried "along for the ride" together with the nearby *Rb* gene during the swapping of sequences between the two homologous chromosomes.

Because LOH was thought to occur at a far higher frequency than mutational alteration of genes, this meant that the second, still-intact copy of a gene was far more likely to be lost through LOH than through a mutational event that struck this copy. Consequently, the majority of retinoblastoma tumor cells were predicted to show LOH at the *Rb* locus and at nearby genetic markers on Chromosome 13, and only a small minority were predicted to carry two distinct, mutant alleles of *Rb*, each inactivated by an independent mutational event.

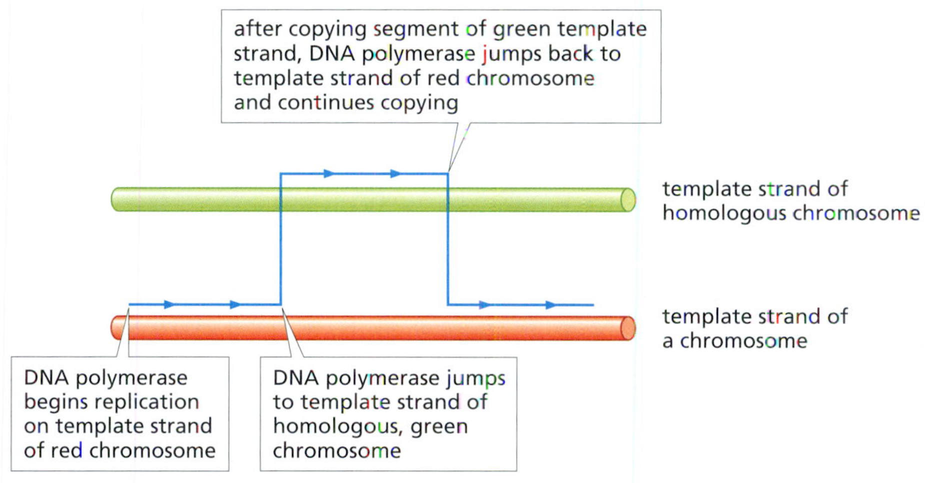

after copying segment of green template strand, DNA polymerase jumps back to template strand of red chromosome and continues copying
template strand of homologous chromosome
template strand of a chromosome
DNA polymerase begins replication on template strand of red chromosome
DNA polymerase jumps to template strand of homologous, green chromosome

Figure 7.9 Gene conversion During the process of gene conversion, DNA polymerases initially begin to use a strand on one chromosome *(red)* as a template for the synthesis of a new daughter strand of DNA *(blue)*. After advancing some distance down this template strand, the polymerase may continue replication by jumping to the homologous chromosome and using a DNA strand of this other chromosome *(green)* as a template for the continued elongation of the daughter strand. After a while, the polymerase may jump back to the originally used DNA template strand and continue replication. In this manner, a mutant tumor suppressor gene allele, such as a mutant allele of *Rb*, may be transmitted from one chromosome to its homolog, replacing the wild-type allele residing there.

7.5 The *Rb* gene often undergoes loss of heterozygosity in tumors

In 1978, the chromosomal localization of the *Rb* gene was surmised from study of the chromosomes present in retinoblastomas. In a small number of retinal tumors, careful karyotypic analysis of metaphase chromosome spreads revealed interstitial deletions (see Sidebar 1.3) within the long ("q") arm of Chromosome 13. Even though each of these deletions began and ended at different sites in this chromosomal arm, they shared in common the fact that they all caused the loss of chromosomal material in the 4th band of the 1st region of this chromosomal arm, that is, 13q14 (Figure 7.10). The interstitial deletions affecting this chromosomal band involved many hundreds, often thousands, of kilobases of DNA, indicating that a number of genes in this region had been jettisoned simultaneously by the developing retinal tumor cells. The fact that these changes involved the *loss* of genetic information provided evidence that the *Rb* gene, which was imagined to lie somewhere in this chromosomal region, had been discarded—precisely the outcome predicted by the tumor suppressor gene theory. (In the great majority of retinoblastomas, the mutations that knock out the *Rb* gene affect a far smaller number of nucleotide sequences and are therefore **submicroscopic**, i.e., invisible upon microscopic analysis of metaphase chromosome spreads.)

Through good fortune, a second gene lying in the 13q14 chromosomal region had been reasonably well characterized at the time that *Rb* was mapped. This gene, which encodes the enzyme esterase D, is represented in the human gene pool by two distinct alleles whose protein products migrate at different rates in gel electrophoresis (Figure 7.11). The esterase D locus presented geneticists with a golden opportunity to test the LOH theory. Recall that when LOH occurs, an entire chromosomal region is affected. Hence, since the esterase D locus maps closely to the *Rb* gene locus on the long arm of Chromosome 13, if the *Rb* locus suffers LOH during tumor development, then the esterase D locus should frequently suffer the same fate. In effect, the esterase D locus could act as a "surrogate marker" for the still-mysterious, uncloned *Rb* gene.

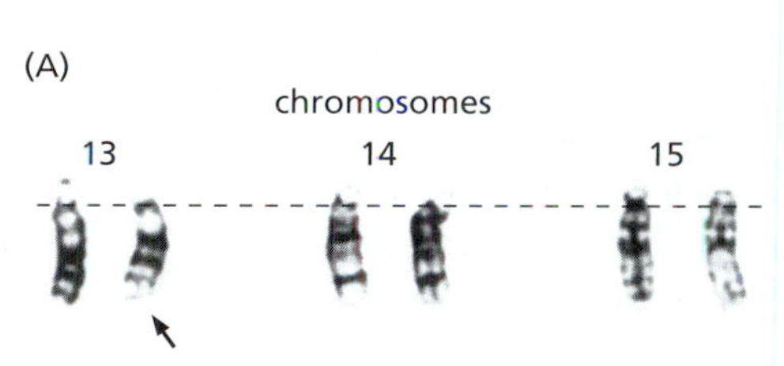

(A)
chromosomes
13
14
15

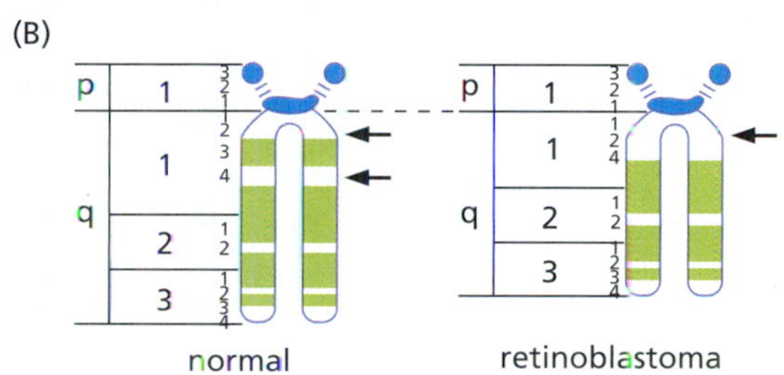

(B)
p
q
normal
retinoblastoma

Figure 7.10 Chromosomal localization of the Rb locus (A) The presence of distinct darkly staining bands in human chromosomes made it possible to delineate specific subregions in each of the chromosomes *(left)*. In this case, careful study of the karyotype of retinoblastoma cells, performed on the condensed chromosomes of late prophase or metaphase cells, revealed the presence of a deletion affecting the long arm (q) of Chromosome 13, in a 6-year-old retinoblastoma patient *(arrow, left)*, while Chromosomes 14 and 15 appeared normal. (B) Careful cytogenetic analysis revealed that the deletion was localized between bands 13q12 and q14 *(right)*, that is, between the 2nd and 4th bands of the 1st region of the long (q) arm of Chromosome 13. (A, from U. Francke, *Cytogenet. Cell Genet.* 16:131–134, 1976.)

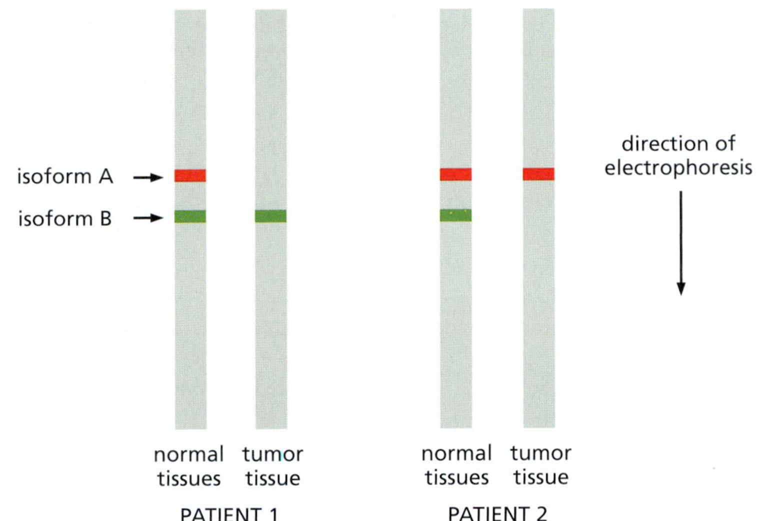

Figure 7.11 Demonstration of loss of heterozygosity at the *Rb* locus These *zymograms* represent analyses in which the two different forms (*isoforms*) of the esterase D enzyme are separated by gel electrophoresis. The presence of these enzymes can then be registered by a biochemical reaction that is carried out by the enzyme and yields a product, visualized here as a band on a gel. In one retinoblastoma patient who was heterozygous at the esterase D locus, normal tissues specified two forms of this enzyme, seen here as two distinct bands, while the tumor tissue specified only one form of the enzyme, indicating a loss of heterozygosity (LOH) at this locus in the tumor cells. Comparisons of the esterase D in the normal and tumor cells of another retinoblastoma patient also revealed LOH at the esterase D locus in his tumor cells—in that case, resulting from loss of the chromosomal region specifying the other isoform of the enzyme. (Adapted from R.S. Sparkes et al., *Science* 219:971–973, 1983.)

Indeed, when researchers scrutinized the tumor cells of several children who were born heterozygous at the esterase D locus (having inherited two distinct esterase D alleles from their parents), they found that these tumor cells had lost an esterase D allele and thus must have undergone LOH (see Figure 7.11). A similar analysis performed at the time using polymorphic DNA sequence markers reached the same conclusion—about LOH of the *Rb* gene. These 1983 discoveries provided a formal and convincing proof of the notion that the nearby *Rb* gene also frequently underwent LOH in tumors; hence its behavior conformed to the theorized behavior ascribed to a growth-suppressing gene, both of whose gene copies needed to be jettisoned before cells would grow uncontrollably.

Note, by the way, that LOH can also be achieved by simply breaking off and discarding an entire chromosomal region without replacing it with a copy duplicated from the other, homologous chromosome (see also Sidebar 7.3). This results in **hemizygosity** of this chromosomal region, where now all the genes on this chromosomal arm are present in only a single copy rather than the usual two copies per cell. Cells can survive with only one copy of some chromosomal regions, while the loss of a single copy of other chromosomal regions seems to put cells at a distinct biological disadvantage. When loss of a chromosomal segment does in fact

Sidebar 7.3 Loss of heterozygosity can occur through yet other mechanisms In many tumors, LOH seems to be achieved through the loss of an entire chromosome due to inappropriate chromosomal segregation at mitosis (the process of **nondisjunction**). The surviving chromosome, which may carry a mutant allele of a gene such as *Rb*, may subsequently be duplicated, yielding two identical chromosomes, all of whose genes are present in homozygous configuration. Alternatively, detailed examination of chromosomal complements in human colon carcinoma cells has shown that, at least in these cancers, LOH is frequently achieved via genetic alterations that change chromosomal structure and thereby affect the karyotype of cells. Many of these events appear to be translocations, in that they involve recombination between chromosomal arms on nonhomologous chromosomes. Such recombination events might be triggered by double-stranded DNA breaks followed by fusion of the resulting ends with DNA sequences originating in other chromosomes. In this fashion, some genetic regions may be duplicated and others lost altogether. Later (Section 10.4), we will discuss one specific molecular mechanism whereby such nonhomologous recombination events may occur during the development of a tumor.

occur (yielding hemizygosity), this event is registered as an LOH, since only one allelic version of a gene in this chromosomal region can be detected whereas previously two distinct alleles were present. No matter how LOH occurs, the final outcome in terms of tumorigenesis is achieved—elimination of the remaining wild-type allele of a tumor suppressor gene.

In 1986 the *Rb* gene was cloned and was found, as previously speculated, to suffer mutations that resulted in its inactivation (Figure 7.12). In some retinoblastomas, these mutations involved large deletions within the *Rb* gene and flanking DNA sequences. Southern blotting of the DNA from these tumors revealed that the resulting mutant *Rb* allele was present in homozygous configuration. This meant that after the creation of a **null** (inactive) allele of *Rb* on one chromosome, the corresponding region on the other, homologous chromosome was discarded, leading to loss of heterozygosity (LOH) at this locus. These data directly validated many of the predictions of the theoretical models that had been proposed to explain how tumor suppressor genes behave during the development of cancers.

As mentioned, children who inherit a defective *Rb* gene copy are also predisposed to osteosarcomas (bone tumors) as adolescents. With the *Rb* gene probe in hand, it became possible to demonstrate that these osteosarcomas also carried structurally altered *Rb* genes (see Figure 7.12). At the same time, these findings highlighted a puzzle that remains unsolved to this day: Why does a gene such as *Rb*, which operates in a wide variety of tissues throughout the body (as we will learn in Chapter 8), cause predominantly retinal and bone tumors when it is inherited in defective form from a parent? Why are not all tissues at equal risk?

7.6 Loss-of-heterozygosity events can be used to find tumor suppressor genes

Numerous tumor suppressor genes that operate like the *Rb* gene were presumed to lie scattered around the human genome and to play a role in the pathogenesis of many types of human tumors. In the late 1980s, researchers interested in finding these genes were confronted with an experimental quandary: How could one find genes whose existence was most apparent when they were missing from a cell's genome? The dominantly acting oncogenes, in stark contrast, could be detected far more readily through their presence in a retrovirus genome, through the transfection-focus assay, or through their presence in a chromosomal segment that repeatedly undergoes gene amplification in a number of independently arising tumors.

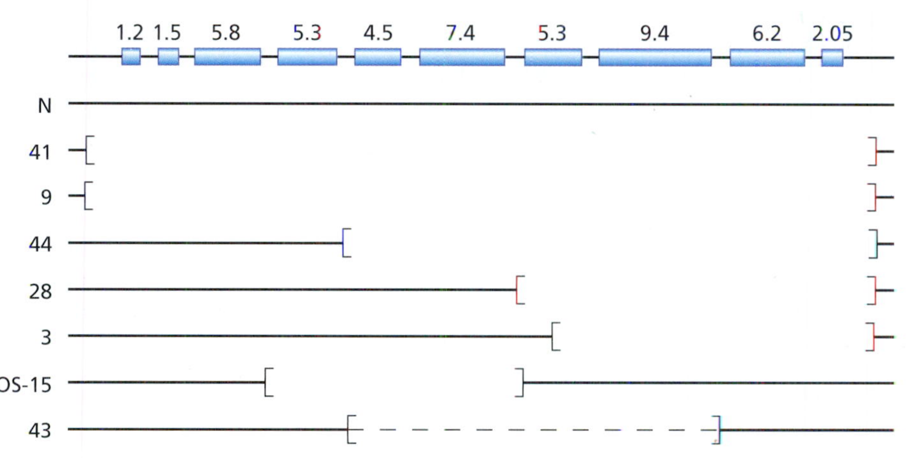

Figure 7.12 Mutations of the *Rb* gene The cloned *Rb* cDNA could be used as a probe in Southern blot analyses of genomic DNAs from a variety of retinoblastomas and an osteosarcoma. Each of the 10 exons of the *Rb* gene *(blue cylinders)* is labeled by its length in kilobases. The normal (N) version of the gene encompasses a chromosomal region of approximately 190 kilobases. However, analysis of a subset of retinoblastoma DNAs indicated that significant portions of this gene had suffered deletion. The beginning and end points of these deletions are indicated by brackets. Thus, tumors 41 and 9 lost the entire *Rb* gene, apparently together with flanking chromosomal DNA segments on both sides. Tumors 44, 28, and 3 lost, to differing extents, the right half of the gene together with rightward-lying chromosomal segments. The fact that an osteosarcoma (OS-15) and a retinoblastoma (43) lost internal portions of this gene argued strongly that this 190-kb DNA segment, and not leftward- or rightward-lying DNA segments, was the repeated target of mutational inactivation occurring during the development of these retinoblastomas and the osteosarcoma. (The dashed line indicates that the deletion in tumor 43 was present in heterozygous configuration.) (From S.H. Friend et al., *Nature* 323:643–646, 1986.)

A more general strategy was required that did not depend on the chance observation of interstitial chromosomal deletions or the presence of a known gene (e.g., esterase D) that, through good fortune, lay near a tumor suppressor gene on a chromosome. Both of these conditions greatly facilitated the isolation of the *Rb* gene. In general, however, the searches for most tumor suppressor genes were not favored by such strokes of good luck.

The tendency of tumor suppressor genes to undergo LOH during tumor development provided cancer researchers with a novel genetic strategy for tracking them down. Since the chromosomal region flanking a tumor suppressor gene seemed to undergo LOH together with the tumor suppressor gene itself, one might be able to detect the existence of a still-uncloned tumor suppressor gene simply from the fact that an anonymous genetic marker lying nearby on the chromosome repeatedly undergoes LOH during the development of a specific type of human tumor. For the sake of this argument, we can consider the esterase D gene (Section 7.5) to behave precisely like an anonymous genetic marker (even though we happen to know its role as a gene encoding a well-characterized enzyme). [While LOH is responsible for the elimination of most of the second, surviving wild-type copies of tumor suppressor genes, some of these gene copies may occasionally be lost through other mechanisms (see Sidebar 2).]

An initially exploited strategy for monitoring heterozygosity and loss thereof used certain DNA segments that were scattered throughout the genome as genetic markers. These DNA segments had no obvious affiliations with specific genes. In some individuals, such a segment of DNA could be cleaved by a restriction enzyme; in others, the same stretch of DNA resisted cleavage as a result of a single base-pair substitution in the sequence recognized by the restriction enzyme (Figure 7.13). Because such sequence variability appeared to occur as a consequence of normal genetic variability in the human gene pool, such sites were thought to represent polymorphic genetic markers (Sidebar 1). Moreover, since the allelic versions of this sequence either permit or disallow cleavage by a restriction enzyme, such a marker was termed a **restriction fragment length polymorphism** (RFLP).

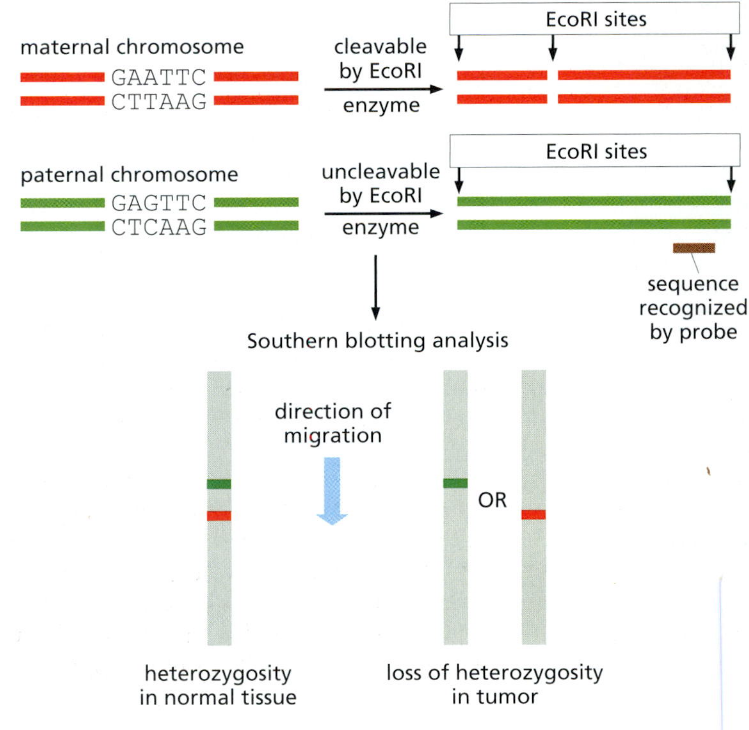

Figure 7.13 Restriction fragment length polymorphisms and localization of tumor suppressor genes The genetic analysis of chromosomal regions has been facilitated by the use of genetic markers that are defined by the presence or absence of a restriction endonuclease (i.e., restriction enzyme) cleavage site in a region of chromosomal DNA. In this illustration, the DNA segment in the chromosome of maternal origin *(red)* can be cleaved by the enzyme EcoRI, while the homologous, paternally derived segment *(green)*, because of a single base-pair substitution, resists cleavage; the presence or absence of this base substitution therefore represents a restriction fragment length polymorphism (RFLP). A radioactive probe that recognizes the right end of the paternal segment can then be used in Southern blot analysis to determine whether or not cleavage has occurred *(below)*. The presence or absence of a cleavage fragment makes it possible, in turn, to track the inheritance of DNA sequences that behave genetically like Mendelian alleles. In this example, two cancer patients' normal tissues show heterozygosity in this chromosomal region. However, loss of heterogygosity (LOH) has occurred in their tumor DNAs. In one patient, the paternal allele has been lost in the tumor DNA, while in another patient's tumor, the maternal allele has been lost.

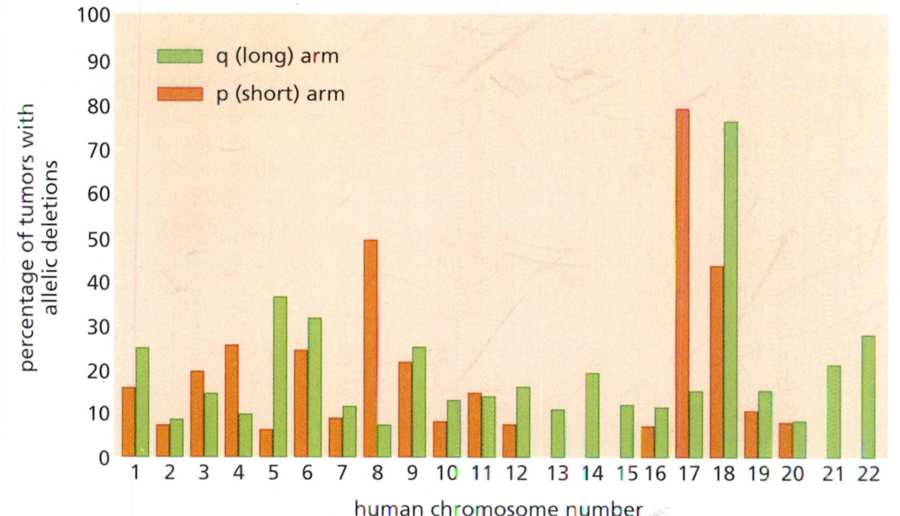

Figure 7.14 Loss of heterozygosity of chromosomal arms in colon cancers This bar graph shows the genetic behavior of the arms of human chromosomes in a series of colorectal carcinomas, which was analyzed through the use of RFLP markers (see Figure 7.13). The long (q) arms of human chromosomes are indicated by the *green bars*, while the short (p) arms are indicated by *orange bars*. Some of the 22 human chromosomes (13, 14, 15, 21, and 22) are represented only by analyses of their long (q) arms. The "allelic deletions" indicated on the ordinate are equivalent to losses of heterozygosity (LOHs). These analyses represented very imprecise measurements of the locations of critical tumor suppressor genes, since very few probes were available at the time to gauge LOH within circumscribed regions of each chromosomal arm. (In addition, the *APC* tumor suppressor gene located on Chromosome 5q, which was subsequently found to be inactivated in almost all colorectal cancers, is extremely large and therefore presents a frequent target for inactivation by mutation. As a consequence, following inactivation of the first copy of this gene, the second, still-intact copy of this gene is often lost through a second, independent mutation rather than through LOH, leading here to an underestimate of its involvement in colorectal cancer development.) These data also provide clear indication that Chromosome 8p harbors a tumor suppressor gene; this gene has never been identified. (From B. Vogelstein et al., *Science* 244:207–211,1989.)

Cancer geneticists used RFLP markers to determine whether various chromosomal regions frequently underwent LOH during the development of certain types of tumors. Figure 7.14 shows the results of using RFLPs to search for regions of LOH arising in a group of **colorectal tumors,** that is, carcinomas of the colon and rectum. In this case, the long and short arms of most chromosomes were represented by at least one RFLP marker, and the fate of the entire chromosomal arm was studied. Note that in this series of human tumors, the short (p) arm of Chromosome 17 and the long (q) arm of Chromosome 18 suffered unusually high rates of LOH. These two chromosomal arms stood out, rising far above the background level of the 15–20% LOH that affected all chromosomal arms equally in these tumor cells.

[The background level of LOH that affected all chromosomal arms in the colorectal tumors (Figure 7.14) reveals the fact that all chromosomal regions have some tendency to undergo LOH at a certain rate. However, if the LOH happens to occur in a region harboring a tumor suppressor gene, the proliferation or survival of the cells in which this has occurred may be favored, leading to high numbers of tumors carrying this particular LOH.]

The fact that specific arms of Chromosomes 17 and 18 frequently underwent LOH in tumors provided strong evidence that both arms harbored still-unknown tumor suppressor genes that also suffered LOH. Thus, this genetic localization provided the gene cloners with a clear indication of where in the genome they should search for the culprit tumor suppressor genes that seemed to be playing important roles in the development of these colorectal tumors. More recently, cancer geneticists have used other molecular strategies to detect sequence polymorphisms in the genomes of cancer cells (Sidebar 7.4).

The use of more powerful mapping techniques allowed geneticists to plant polymorphic markers more densely along the genetic maps of each chromosomal arm. Within a given chromosomal arm, some markers were found to undergo LOH far more frequently than others. Clearly, the closer these markers were to a sought-after tumor suppressor gene (i.e., the tighter the genetic linkage), the higher was the probability that such markers would undergo LOH together with the tumor suppressor gene. Conversely, markers located further away on a chromosomal arm were less likely to suffer LOH together with the tumor suppressor gene. To date, genetic analyses of DNAs prepared from various types of human tumors have revealed a large number of chromosomal regions that frequently suffer LOH. A subset of these regions have yielded to the attacks of the gene cloners, resulting in the isolation of more than 30 tumor suppressor genes (Table 7.1).

Sidebar 7.4 The polymerase chain reaction makes it possible to genetically map tumor suppressor genes rapidly In recent years, the use of RFLPs has been replaced by markers that can be detected using the polymerase chain reaction (PCR) to amplify specific chromosomal segments. In a simple form of this LOH analysis, one of the two DNA primers in a PCR reaction may be complementary to one allelic version of a DNA sequence that is polymorphic in the human gene pool; the other primer binds to a sequence on the complementary strand that is invariant in the human gene pool. The ability of the first primer to bind to one allelic version of the sequence and not to another can be used to determine whether or not the particular allele is present in a DNA preparation (Figure 7.15). The ability to perform multiple PCRs simultaneously enables geneticists to survey the configuration of many polymorphic markers at the same time; this has facilitated the discovery of thousands of these markers scattered throughout the genome. The resulting vastly increased density of markers along each chromosomal arm makes it possible to localize tumor suppressor genes far more precisely than the use of the very rough mapping strategy illustrated in Figure 7.14.

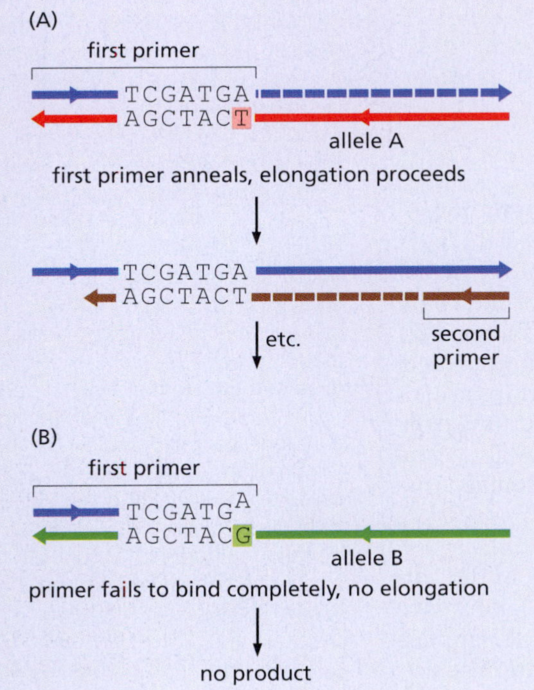

Figure 7.15 Use of the polymerase chain reaction to determine sequence polymorphisms The polymerase chain reaction (PCR) can be used to determine the presence or absence of a single-nucleotide polymorphism (SNP)—a polymorphism where the two alleles differ by only a single nucleotide. (Of the 3 million or so SNPs that have been cataloged in the human genome, about 29% are present at an allele frequency of 5% or more in the human gene pool.) (A) In one version of this procedure, the first primer *(blue)* is perfectly complementary to one allelic sequence *(red strand),* and therefore the rightward-moving primer extension reaction *(blue dashed line)* proceeds normally, thereby generating a template for the leftward-moving elongation from the second primer *(brown dashed line).* (B) However, if the first primer is not perfectly complementary to the sequence present in the other allele *(light green)*, all rightward-moving primer extensions will fail. Hence, the presence or absence of a SNP can be ascertained by the presence or absence of the resulting PCR products, which will accumulate to millions of copies in one set of reactions and will be absent in the other. The arrowheads indicate the 5′ to 3′ polarity of the DNA strands.

7.7 Many familial cancers can be explained by inheritance of mutant tumor suppressor genes

Like the *Rb* gene, most of the cloned tumor suppressor genes listed in Table 7.1 are involved in both familial and sporadic cancers. In general, inheritance of defective copies of most of these genes creates an enormously increased risk of contracting one or another specific type of cancer, often a type of tumor that is otherwise relatively rare in the human population. In some cases, mutant germline alleles of these genes lead to susceptibility to multiple cancer types, as is the case with the *Rb* gene.

Later in this chapter, we will describe in detail the mechanisms of action of some of the tumor suppressor gene–encoded proteins. However, even a cursory examination of Table 7.1 makes it clear that these genes specify a diverse array of proteins that operate in many different intracellular sites to reduce the risk of cancer. Indeed, an anti-cancer function is the only property that is shared in common by these otherwise unrelated genes. (Some of the tumor suppressor genes listed in this table are known only through their involvement in sporadic cancers; it remains unclear whether mutant alleles of these genes will eventually be found to be transmitted in the germ line and thereby predispose individuals to one or another type of cancer.)

Table 7.1 Human tumor suppressor genes that have been cloned

Name of gene	Chromosomal location	Familial cancer syndrome	Sporadic cancer	Function of protein
RUNX3	1p36	—	gastric carcinoma	TF co-factor
HRPT2	1q25–32	parathyroid tumors, jaw fibromas	parathyroid tumors	chromatin protein
FH	1q42.3	familial leiomyomatosis[a]	—	fumarate hydratase
FHIT	3p14.2	—	many types	diadenosine triphosphate hydrolase
RASSF1A	3p21.3	—	many types	multiple functions
TGFBR2	3p2.2	HNPCC	colon, gastric, pancreatic carcinomas	TGF-β receptor
VHL	3p25	von Hippel–Lindau syndrome	renal cell carcinoma	ubiquitylation of HIF
hCDC4	4q32	—	endometrial carcinoma	ubiquitin ligase
APC	5p21	familial adenomatous polyposis coli	colorectal, pancreatic, and stomach carcinomas; prostate carcinoma	β-catenin degradation
NKX3.1	8p21	—	prostate carcinoma	homeobox TF
$p16^{INK4A}$ [b]	9p21	familial melanoma	many types	CDK inhibitor
$p14^{ARF}$ [c]	9p21	—	all types	p53 stabilizer
PTC	9q22.3	nevoid basal cell carcinoma syndrome	medulloblastomas	receptor for hedgehog GF
TSC1	9q34	tuberous sclerosis	—	inhibitor of mTOR[f]
BMPR1	10q21–22	juvenile polyposis	—	BMP receptor
PTEN[d]	10q23.3	Cowden's disease, breast and gastrointestinal carcinomas	glioblastoma; prostate, breast, and thyroid carcinomas	PIP_3 phosphatase
WT1	11p13	Wilms tumor	Wilms tumor	TF
MEN1	11p13	multiple endocrine neoplasia	—	histone modification, transcriptional repressor
BWS/CDKN1C	11p15.5	Beckwith–Wiedemann syndrome	—	$p57^{Kip2}$ CDK inhibitor
SDHD	11q23	familial paraganglioma	pheochromocytoma	mitochondrial protein[e]
RB	13q14	retinoblastoma, osteosarcoma	retinoblastoma; sarcomas; bladder, breast, esophageal, and lung carcinomas	transcriptional repression; control of E2Fs
TSC2	16p13	tuberous sclerosis	—	inhibitor of mTOR[f]
CBP	16p13.3	Rubinstein–Taybi	AML[g]	TF co-activator
CYLD	16q12–13	cylindromatosis	—	deubiquitinating enzyme
CDH1	16q22.1	familial gastric carcinoma	invasive cancers	cell–cell adhesion
BHD	17p11.2	Birt–Hogg–Dube syndrome	kidney carcinomas, hamartomas	unknown
TP53	17p13.1	Li–Fraumeni syndrome	many types	TF
NF1	17q11.2	neurofibromatosis type 1	colon carcinoma, astrocytoma	Ras-GAP
BECN1	17q21.3	—	breast, ovarian, prostate	autophagy
PRKAR1A	17.q22–24	multiple endocrine neoplasia[h]	multiple endocrine tumors	subunit of PKA
DPC4[i]	18q21.1	juvenile polyposis	pancreatic and colon carcinomas	TGF-β TF
LKB1/STK11	19p13.3	Peutz–Jegher syndrome	hamartomatous colonic polyps	serine/threonine kinase
RUNX1	21q22.12	familial platelet disorder	AML	TF
SNF5[j]	22q11.2	rhabdoid predisposition syndrome	malignant rhabdoid tumors	chromosome remodeling
NF2	22q12.2	neurofibroma-position syndrome	schwannoma, meningioma; ependymoma	cytoskeleton–membrane linkage

[a]Familial leiomyomatosis includes multiple fibroids, cutaneous leiomyomas, and renal cell carcinoma. The gene product is a component of the tricarboxylic cycle.

[b]Also known as *MTS1, CDKN2,* and *p16.*

[c]The human homolog of the murine $p19^{ARF}$ gene.

[d]Also called *MMAC* or *TEP1*.

[e]*SDHS* encodes the succinate–ubiquinone oxidoreductase subunit D, a component of the mitochondrial respiratory chain complex II.

[f]mTOR is a serine/threonine kinase that controls, among other processes, the rate of translation and activation of Akt/PKB. TSC1 (hamartin) and TSC2 (tuberin) control both cell size and cell proliferation.

[g]The *CBP* gene is involved in chromosomal translocations associated with AML. These translocations may reveal a role of a segment of CBP as an oncogene rather than a tumor suppressor gene.

[h]Also termed Carney complex.

[i]Encodes the Smad4 TF associated with TGF-β signaling; also known as *MADH4* and *SMAD4*.

[j]The human SNF5 protein is a component of the large Swi/Snf complex that is responsible for remodeling chromatin in a way that leads to transcriptional repression through the actions of histone deactylases. The rhabdoid predisposition syndrome involves susceptibility to atypical teratoid/rhabdoid tumors, choroid plexus carcinomas, medulloblastomas, and extra-renal rhabdoid tumors.

Adapted in part from E.R. Fearon, *Science* 278:1043–1050, 1997; and in part from D.J. Marsh and R.T. Zori, *Cancer Lett.* 181:125–164, 2002.

Sidebar 7.5 Why are mutant tumor suppressor genes transmitted through the germ line while mutant proto-oncogenes are usually not? A number of familial cancer syndromes have been associated with the transmission of mutant germ-line alleles of tumor suppressor genes (see Table 7.1). However, with rare exception, mutant alleles of proto-oncogenes (i.e., activated oncogenes) have not been found to be responsible for inborn cancer predisposition. How can we rationalize this dramatic difference?

Mutations that yield activated oncogenes are likely to arise with some frequency during **gametogenesis**—the processes that form sperm and egg—and thus are likely to be transmitted to fertilized eggs. However, because oncogenes act at the cellular level as dominant alleles, these mutant alleles are likely to perturb the behavior of individual cells in the developing embryo and therefore to disrupt normal tissue development. Consequently, embryos carrying these mutant oncogenic alleles are unlikely to develop to term, and these mutant alleles will disappear from the germ line of a family and thus from the gene pool of the species. (For example, experiments reported in 2004 indicate that mouse embryos arising from sperm carrying a mutant, activated K-*ras* oncogene develop only to the time of mid-gestation, at which point they die because of placental and intra-embryonic developmental defects; see, however, Sidebar 5.8)

Mutant germ-line alleles of tumor suppressor genes behave much differently, however. Since these alleles are recessive at the cellular level, their presence in most cells of an embryo will not be apparent. For this reason, the presence of inherited mutant tumor suppressor genes will often be compatible with normal embryonic development, and the cancer phenotypes that they create will become apparent only in a small number of cells and after great delay, allowing an individual carrying these alleles to develop normally and, as is often the case, to survive through much of adulthood.

Still, if mutant tumor suppressor genes undergo LOH in 1 out of 10^4 or 10^5 cells, and if an adult human has many more than 10^{13} cells, why isn't a person inheriting a mutant tumor suppressor gene afflicted with tens of thousands, even millions of tumors? The response to this comes from the fact that tumorigenesis is a multi-step process (as we will see in Chapter 11). This implies that a mutant gene (whether it be an oncogene or a tumor suppressor gene) may be necessary for tumor formation but will not, on its own, be sufficient. Hence, many cells in an individual who has inherited a mutant tumor suppressor gene may well undergo LOH of this gene, but only a tiny minority of these cells will ever acquire the additional genetic changes needed to make a clinically detectable tumor.

While inheritance of a mutant tumor suppressor gene allele is likely to greatly increase cancer risk, the converse is not true: not all familial cancer syndromes can be traced back to an inherited tumor suppressor gene allele. As we will discuss later in Chapter 12, mutant germ-line alleles of a second class of genes also cause cancer predisposition. These other genes are responsible for maintaining the cellular genome, and thus act to reduce the likelihood of mutations and chromosomal abnormalities. Because cancer pathogenesis depends on the accumulation by individual cells of somatic mutations in their genomes, agents that reduce the mutation rate, such as these genome maintenance genes, are highly effective in suppressing cancer onset. Conversely, defects in genome maintenance often lead to a disastrous increase in cancer risk because they increase the mutation rate.

So we come to realize that there are really two distinct classes of familial cancer genes—the tumor suppressor genes described in this chapter, and the genome maintenance genes described in Chapter 12. We can rationalize the distinction between the two classes of genes as follows. The tumor suppressor genes function to directly control the biology of cells by affecting how they proliferate, differentiate, or die; genes functioning in this way are sometimes called **gatekeepers** to indicate their role in allowing or disallowing cells to progress through cycles of growth and division. The DNA maintenance genes affect cell biology only indirectly by controlling the rate at which cells accumulate mutant genes; these genes have been termed **caretakers** to reflect their role in the maintenance of cellular genomes. Unlike mutant gatekeeper and caretaker alleles, mutant versions of proto-oncogenes are rarely transmitted through the germ line (Sidebar 7.5).

7.8 Promoter methylation represents an important mechanism for inactivating tumor suppressor genes

As mentioned briefly in Sidebar 2, DNA molecules can be altered covalently by the attachment of methyl groups to cytosine bases. Recent research indicates that this modification of the genomic DNA is as important as mutation in shutting down tumor suppressor genes. In mammalian cells, this methylation is found only when these bases are located in a position that is 5′ to guanosines, that is, in the sequence CpG. (This MeCpG modification is often termed "methylated CpG," even though only the cytosine is methylated.) Such methylation can affect the functioning of the DNA in this region of the chromosome. More specifically, when CpG methylation occurs in the vicinity of a gene promoter, it can cause repression of transcription of the associated gene. Conversely, when methyl groups are removed from a cluster of methylated CpGs associated with the gene promoter, transcription of this gene is often de-repressed.

The means by which the methylated state of CpGs affects transcription are not entirely resolved. Nonetheless, one very important mechanism has already been uncovered. Protein complexes have been found that include one subunit that can recognize and bind methylated CpGs in DNA and a second subunit that functions as a histone deacetylase (HDAC) enzyme. Once these complexes are bound to DNA, the histone deacetylase proceeds to remove acetate groups that are attached to the side-chains of amino acid residues of histone molecules in the nearby chromatin. The resulting deacetylation of the histones initiates a sequence of events that converts the chromatin from a configuration favoring transcription to one blocking transcription.

Careful examination of methylated CpGs in cellular DNA reveals that the dinucleotide sequences in the complementary DNA strand (which are also CpG when read in a 5′ to 3′ direction on that strand) almost invariably contain MeCpGs as well. Thus, whenever a stretch of DNA is replicated in S phase, the CpGs in the newly formed daughter lying opposite the MeCpGs (in reverse orientation) in the complementary, parental strand of the DNA are initially unmethylated. This deficiency is then quickly remedied by a class of enzymes, termed **maintenance methylases**, that attach methyl groups to these recently synthesized CpGs (Figure 7.16).

The actions of maintenance methylases indicate that, in principle, the methylation state of a CpG sequence can be maintained through many rounds of DNA replication in a fashion that is as stable as the maintenance of nucleotide sequences of the DNA. Hence, the methylation state of CpGs is also a heritable property that can be passed from one cell to its daughters. Nonetheless, while heritable, DNA methylation does not alter the nucleotide sequence of the DNA and therefore is considered to be a nongenetic, that is, an **epigenetic**, mechanism for controlling gene expression.

A specific CpG sequence is often methylated in some cells of an organism while unmethylated in others. This implies that DNA methylation is reversible. Thus, enzymes may well exist that remove methyl groups from CpGs (although they have yet to be discovered). Alternatively, when DNA is replicated, the methylation of a CpG in the newly formed strand may be blocked even though the complementary CpG in the parental strand is methylated; when this newly formed

Figure 7.16 Maintenance of DNA methylation following replication When a DNA double helix that is methylated at complementary CpG sites *(green methyl groups, left)* undergoes replication, the newly synthesized daughter helices will initially lack methyl groups attached to CpGs in the recently synthesized daughter strand and will therefore be *hemi-methylated*. Shortly after their synthesis, however, a DNA maintenance methylase will detect the hemi-methylated DNA and attach methyl groups *(green)* to these CpGs, thereby regenerating the same configuration of methyl groups that existed in the parental helix prior to replication. CpG sites that are unmethylated in the parental helix *(not shown)* will be ignored by the maintenance methylase and will therefore remain so in the newly synthesized strands.

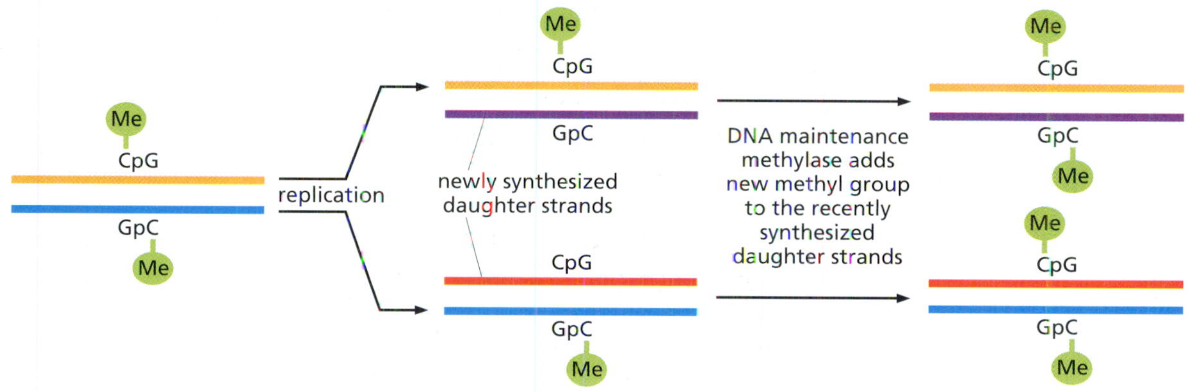

strand is copied in the succeeding round of DNA replication, the resulting double helix will now lack methyl groups at this position in both strands. Acting in the opposing direction, enzymes are known that attach methyl groups to previously unmethylated CpGs; such enzymes are termed ***de novo* methylases** to distinguish them from the maintenance methylases that perpetuate an existing pattern of methylation by attaching methyl groups to recently synthesized DNA (see Figure 7.16).

Methylation of CpGs, which is not found in all metazoa, is a clever invention. Thus, decisions concerning the transcriptional state of many genes are made early in embryogenesis. Once made, these decisions must be passed on to the descendant cells in various parts of the growing embryo. Since CpG methylation can be transmitted from cell to cell heritably, it represents a highly effective way to ensure that descendant cells, many cell generations removed from an early embryo, continue to respect and enforce the decisions made by their ancestors in the embryo.

In the genomes of cancer cells, we can imagine that the mechanisms determining whether or not DNA is properly methylated may malfunction occasionally. These important regulatory mechanisms are still poorly understood. The results of this malfunctioning are two opposing changes in the methylation state of tumor cell genomes. As the development of a tumor proceeds, the overall level of methylation throughout the genome is often found (using the technique described in Sidebar 3) to decrease progressively. This indicates that, for unknown reasons, the maintenance methylases fail to do their job effectively. Much of this "global hypomethylation" can be attributed to the loss of methyl groups attached to the DNA of highly repeated sequences in the cell genome; this loss is correlated with chromosomal instability, but it remains unclear whether it actually causes this instability.

Independent of this global hypomethylation, there are often localized regions of DNA—regions with a high density of CpGs called "CpG islands"—that become methylated inappropriately in the genomes of cancer cells. Because these CpG islands are usually affiliated with the promoters of genes (Figure 7.17), this means that the *de novo* methylases are actually silencing certain genes that should, by all rights, remain transcriptionally active.

CpG methylation is effective in shutting down the expression of a gene only if it occurs within the promoter sequences of the gene; conversely, methylation of

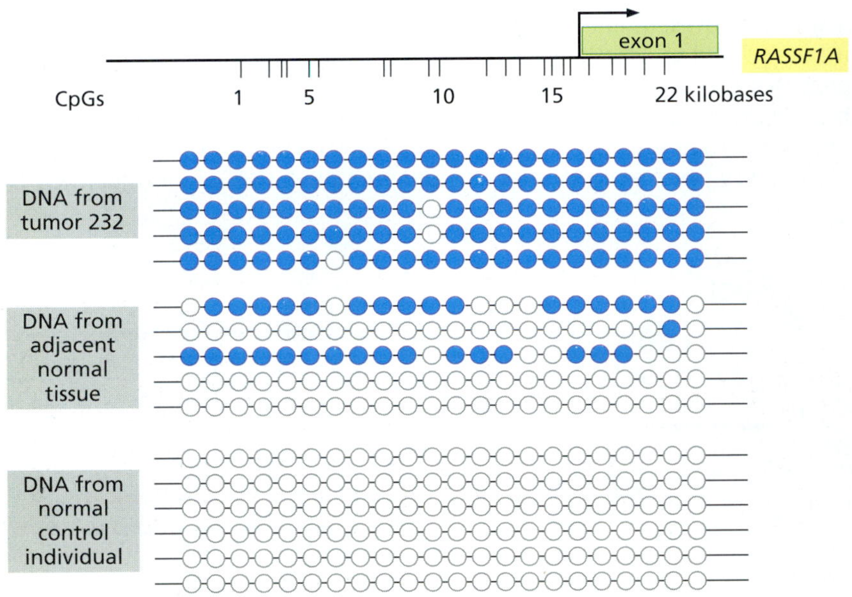

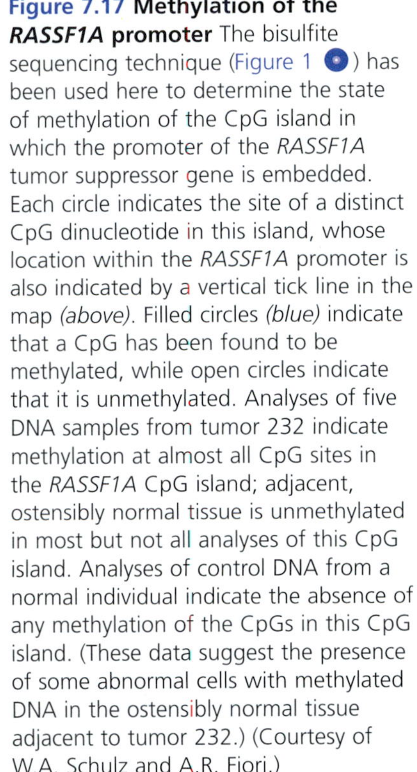
Figure 7.17 Methylation of the *RASSF1A* promoter The bisulfite sequencing technique (Figure 1) has been used here to determine the state of methylation of the CpG island in which the promoter of the *RASSF1A* tumor suppressor gene is embedded. Each circle indicates the site of a distinct CpG dinucleotide in this island, whose location within the *RASSF1A* promoter is also indicated by a vertical tick line in the map *(above)*. Filled circles *(blue)* indicate that a CpG has been found to be methylated, while open circles indicate that it is unmethylated. Analyses of five DNA samples from tumor 232 indicate methylation at almost all CpG sites in the *RASSF1A* CpG island; adjacent, ostensibly normal tissue is unmethylated in most but not all analyses of this CpG island. Analyses of control DNA from a normal individual indicate the absence of any methylation of the CpGs in this CpG island. (These data suggest the presence of some abnormal cells with methylated DNA in the ostensibly normal tissue adjacent to tumor 232.) (Courtesy of W.A. Schulz and A.R. Fiori.)

DNA sequences in the body of the gene, such as exonic sequences, seems to have little if any effect on the level of transcription. Since promoter methylation can silence a gene as effectively as a mutation of its nucleotide sequences, we might predict that methylation plays a role in the silencing of tumor suppressor genes that occurs during tumor progression.

In fact, in recent years, it has become apparent that promoter methylation is as important in shutting down tumor suppressor genes as are the various mechanisms of somatic mutation. More than half of the tumor suppressor genes that are involved in familial cancer syndromes because of germ-line mutation have been found to be silenced in sporadic cancers by promoter methylation. For example, when the *Rb* tumor suppressor gene is mutated in the germ line, it leads to familial retinoblastoma. In sporadic retinoblastomas, however, this tumor suppressor gene is inactivated either by somatic mutations or by promoter methylation. In addition, the promoters of a variety of other genes that are known or thought to inhibit tumor formation have been found in a methylated state (Table 7.2).

Some candidate tumor suppressor genes are rarely inactivated by somatic mutations in their reading frame. As an example, the *Runx3* gene, which has been implicated in stomach cancer development, is found in a methylated state in 45 to 60% of these cancers but is virtually never inactivated by mutation. Situations like this one create difficulties for researchers who would like to verify that inactivation of a candidate tumor suppressor gene contributes to the formation of one or another type of human tumor. In the past, such validation depended on finding mutant, defective copies of such a gene in tumor cell genomes. Such genetic analyses (which depend on sequencing of a tumor suppressor gene allele cloned from a tumor cell genome) can no longer be considered to be absolutely definitive, if certain tumor suppressor genes contributing to tumor progression are silenced only through promoter methylation. (In fact, the discovery of a promoter that is repeatedly methylated in the genomes of a group of tumors is difficult to interpret unambiguously; this finding may only reflect the fact that the tumors being studied arose from a normal cell type in which the gene in question is methylated as part of a normal differentiation program.)

Table 7.2 Examples of hypermethylated genes found in human tumor cell genomes

Name of gene	Nature of protein function	Type of tumor
RARβ2	nuclear receptor for differentiation	breast, lung
$p57^{Kip2}$	CDK inhibitor	gastric, pancreatic, hepatic; AML
TIMP3	inhibitor of metalloproteinases	diverse tumors
IGFBP	sequesters IGF-1 factor	diverse tumors
CDKN2A/$p16^{INK4A}$	inhibitor of CDK4/6	diverse tumors
CDKN2B/$p15^{INK4B}$	inhibitor of CDK4/6	diverse tumors
$p14^{ARF}$	inhibitor of HDM2/MDM2	colon, lymphoma
APC	inducer of β-catenin degradation	colon carcinomas
p73	aids p53 to trigger apoptosis	diverse tumors
GSTP1	mutagen inactivator	breast, liver, prostate
MGMT	DNA repair enzyme	colorectal
CDH1	cell–cell adhesion receptor	bladder, breast, colon, gastric
DAPK	kinase involved in cell death	bladder
MLH1	DNA mismatch repair enzyme	colon, endometrial, gastric
TGFBR2	TGF-β receptor	colon, gastric, small-cell lung
THBS1	angiogenesis inhibitor	colon, glioblastoma
RB	cell-cycle regulator	retinoblastoma
CASP8	apoptotic caspase	neuroblastoma, SCLC
APAF1	pro-apoptotic cascade	melanoma
CTMP	inhibitor of Akt/PKB	glioblastoma multiforme

Adapted in part from C.A. Eads et al., *Cancer Res.* 61:3410–3418, 2001.

The elimination of tumor suppressor gene function by promoter methylation might, in principle, occur through either of two routes. Both copies of a tumor suppressor gene might be methylated independently of one another. Alternatively, one copy might be methylated and the second might then be lost through a loss of heterozygosity (LOH) accompanied by a duplication of the already-methylated tumor suppressor gene copy. Actually, this second mechanism seems to operate quite frequently. For example, in a study of the normal bronchial (large airway) epithelia of the lungs of smokers, former smokers, and never-smokers, the *p16*^{INK4A} tumor suppressor gene (to be described in Section 8.4) was found to be methylated in 44% of (ostensibly normal) bronchial epithelial cells cultured from current and former smokers and not at all in the comparable cells prepared from those who had never smoked. LOH in this chromosomal region was found in 71 to 73% of the two smoking populations and in 1.5 to 1.7% of never-smokers.

The observations that methylated copies of tumor suppressor genes frequently undergo loss of heterozygosity indicate that the methylation event is relatively infrequent per cell generation, indeed, rarer than LOH events. (If methylation always occurred more frequently than LOH, tumors would show two tumor suppressor gene copies, one of paternal and the other of maternal origin, with each independently methylated—a configuration that is observed frequently in certain types of tumors and rarely in other types.) Clearly, evolving tumor cells can discard the second, still-functional tumor suppressor gene copy more readily by LOH than by a second, independent promoter methylation event.

The above-described study of morphologically normal bronchial epithelial cells also teaches us a second lesson: methylation of critical growth-controlling genes often occurs early in the complex, multi-step process of tumor formation, long before histological changes are apparent in a tissue. These populations of outwardly normal cells presumably provide a fertile soil for the eventual eruption of pre-malignant and malignant growths. This point is borne out as well in other tissues, such as the histologically normal breast tissues analyzed in Figure 7.18.

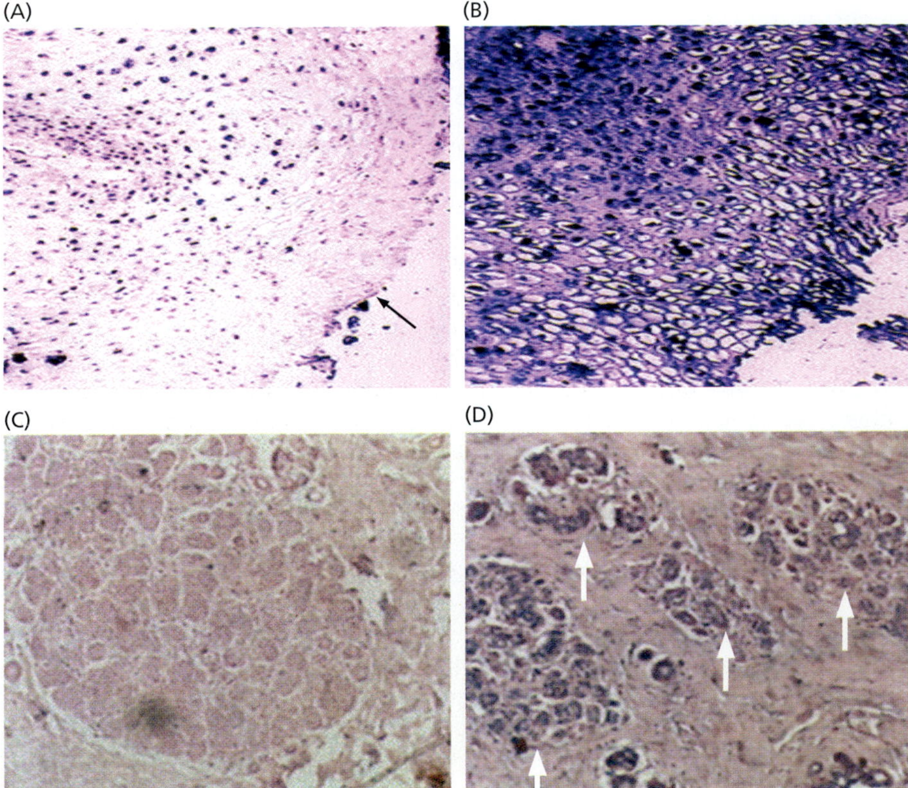

Figure 7.18 *In situ* measurements of DNA methylation The presence or absence of methylation at specific sites in the genomes of cells can be determined in tissue sections that are fixed to a microscope slide using both the methylation-specific PCR reaction (see Figure 1) and *in situ* hybridization. In these images, the methylation status of the promoter of the *p16*^{INK4A} tumor suppressor gene is analyzed with a methylation-specific probe, which yields dark staining in areas where this gene promoter is methylated. (A) In a low-grade squamous intraepithelial lesion of the cervix *(left),* nuclei of cells located some distance from the uterine surface show promoter methylation, while those near the surface *(arrow)* do not. (B) However, in an adjacent high-grade lesion, which is poised to progress to a cervical carcinoma *(right),* all the cells show promoter methylation. (C,D) In these micrographs of normal breast tissue, some histologically normal lobules show no promoter methylation (C) while others *(arrows)* show uniform promoter methylation (D). This suggests that the mammary epithelial cells in such outwardly normal lobules have already undergone a critical initiating step in cancer progression. (A,B, from G.J. Nuovo et al., *Proc. Natl. Acad. Sci. USA* 96:12754–12759, 1999; C,D, from C.R. Holst et al., *Cancer Res.* 63:1596–1601, 2003.)

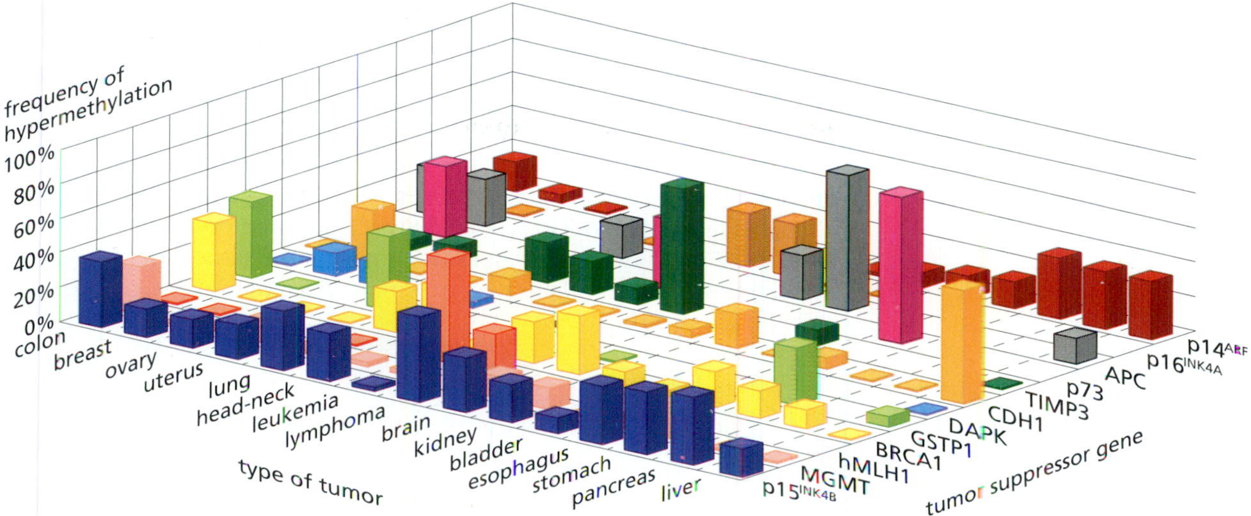

Figure 7.19 Methylation of multiple genes within tumor cell genomes This three-dimensional bar graph summarizes measurements of the methylation state of the promoters of 12 different genes (*p16*^{INK4A}, *p15*^{INK4B}, *p14*^{ARF}, *p73*, *APC*, *BRCA1*, *hMLH1*, *GSTP1*, *MGMT*, *CDH1*, *TIMP3*, and *DAPK*) that are known or presumed to play an important role in suppressing the development of human tumors. Methylation status was determined by the methylation-specific PCR technique (see Figure 1). The methylation state of each of these promoters has been studied in the DNAs of 15 different tumor types. The height of each bar indicates the proportion of tumors of a given type in which a specific promoter has undergone methylation. All of these promoters are unmethylated (or methylated to an insignificant extent) in normal tissues. (Adapted from M. Esteller et al., *Cancer Res.* 61:3225–3229, 2001.)

As tumor formation proceeds, the silencing of genes through promoter methylation can also involve the "caretaker" genes that are responsible for maintaining the integrity of the DNA sequences in the genome. An interesting example is provided by the *BRCA1* gene. Its product is responsible for maintaining the chromosomal DNA in ways that are still poorly understood. The consequence of inheritance of mutant alleles of *BRCA1* is a familial cancer syndrome involving a high lifetime risk of breast and, to a lesser extent, ovarian carcinomas (see Section 12.10). At one time, it was thought that *BRCA1* inactivation never contributes to sporadic breast cancers, since mutant alleles of this gene could not be found in this second, far more common class of mammary tumors. We now know that 10 to 15% of sporadic breast carcinomas carry inactive *BRCA1* gene copies that have been silenced through promoter methylation.

Figure 7.19 illustrates the methylation state of the promoters of 12 genes in a variety of human tumor types. This figure clearly demonstrates that the frequency of methylation of a specific gene varies dramatically from one type of tumor to the next. The data in this figure reinforce the point that tumor suppressor genes as well as caretaker genes undergo hypermethylation. Perhaps the most important lesson taught by this figure is the pervasiveness of promoter methylation during the development of a wide variety of human cancers.

Within a given tumor, multiple genes seem to be shut down by promoter methylation. For example, in an analysis of the methylation status of eight critical cancer-related genes in the genomes of 107 non-small-cell lung carcinomas (NSCLCs), 37% had at least one of these gene promoters methylated, 22% had two promoters methylated, and 2% carried five of the eight gene promoters in a methylated state. The decision to analyze these particular genes involved, by necessity, some arbitrary choices, and we can imagine that there were dozens of other hypermethylated genes in these cancer cell genomes whose inactivation contributed in various ways to cancer formation.

The cell–physiologic consequences of promoter methylation are nicely illustrated by the actions of retinoic acid on normal and neoplastic epithelial cells. In a number of epithelial cell types, retinoic acid is a potent inducer of cell cycle arrest and even differentiation. It has been used, for example, in attempts to halt the further proliferation of breast cancer cells. However, the great majority of these cancer cells are found to have silenced, by promoter methylation, the *RARβ2* gene, which encodes a critical retinoic acid receptor. Without expression of this receptor, the breast carcinoma cells are unresponsive to retinoic acid treatment, so they avoid growth arrest and continue to thrive in the presence of this agent.

As described above, the methylation of CpGs in promoters acts to attract histone deacetylase (HDAC) molecules, which proceed to reconfigure nearby chromatin proteins, placing them in a state that is incompatible with transcription. Use of an inhibitor of HDACs, termed trichostatin A (TSA), reverses this deacetylation, thereby causing reversion of the chromatin to a state that permits transcription. Hence, treatment of breast cancer cells with TSA causes reactivation of *RARβ2* gene expression and restores their responsiveness to the growth-inhibitory effects of retinoic acid. This result shows clearly that promoter methylation acts via histone deacetylation to promote cancer. At the same time, it suggests a therapy for breast carcinomas by concomitant treatment with trichostatin A and retinoic acid.

7.9 Tumor suppressor genes and proteins function in diverse ways

Earlier we read that the tumor suppressor genes and their encoded proteins act through diverse mechanisms to block the development of cancer. Indeed, the only characteristic that ties these genes and their encoded proteins together is the fact that all of them operate to reduce the likelihood of cancer development. Careful examination of the list of cloned tumor suppressor genes (see Table 7.1) reveals that some of them function to directly suppress the proliferation of cells in response to a variety of growth-inhibitory and differentiation-inducing signals. Yet others are components of the cellular control circuitry that inhibits proliferation in response to metabolic imbalances and genomic damage.

The first two tumor suppressor genes that were intensively studied, the *Rb* gene discussed earlier in this chapter and a second tumor suppressor gene, termed variously *p53, Trp53,* or *TP53,* also happen to be the two tumor suppressor genes that play major roles in human cancer pathogenesis. The protein encoded by the *Rb* gene governs the progress of a wide variety of cells through their growth-and-division cycles; and the growth control imposed by the *Rb* circuit appears to be disrupted in most and perhaps all human tumors. Because of its centrally important function, we will devote an entire chapter to discussing *Rb* and therefore defer discussion of this gene and its encoded protein until Chapter 8. The *p53* tumor suppressor gene plays an equally central role in the development of cancers, and the pathway that its product, p53, controls also appears to be disrupted in virtually all types of human cancers. This tumor suppressor gene and protein deserve an equally extended discussion, and consequently we will devote much of Chapter 9 to a detailed description of p53 function and the cell death program that it controls.

The remaining tumor suppressor genes that have been enumerated to date dispatch gene products to a variety of intracellular sites, where they operate in diverse ways to suppress cell proliferation. These proteins sit astride virtually all of the control circuits that are responsible for governing cell proliferation and survival. Several of the proteins encoded by the tumor suppressor genes listed in Table 7.1 happen to be components of the *Rb* and *p53* pathways; included in these circuits are the $p16^{INK4A}$, $p15^{INK4B}$, and $p19^{ARF}$ genes, and so discussion of these will be postponed until Chapters 8 and 9.

The mechanisms of action of some of the remaining tumor suppressor genes in Table 7.1 are reasonably well understood at present, while the actions of yet others listed in this table remain quite obscure. We will focus here in depth on three of these tumor suppressor genes in order to illustrate the highly interesting mechanisms of action of their gene products. These examples—really anecdotes—have been chosen from the large list shown in Table 7.1 because they reveal how very diverse are the mechanisms that cells deploy in order to prevent runaway proliferation. They also illustrate the difficulties encountered in determining the specific biochemical and biological actions of various tumor suppressor gene–encoded proteins. We move from the cell surface inward.

7.10 The NF1 protein acts as a negative regulator of Ras signaling

The disease of neurofibromatosis was first described by Friedrich von Recklinghausen in 1862. We now know that neurofibromatosis type 1 (sometimes called von Recklinghausen's neurofibromatosis) is a relatively common familial cancer syndrome, with 1 in 3500 individuals affected on average worldwide. The primary feature of this disease is the development of benign tumors of the cell sheaths around nerves in the peripheral nervous system. On occasion, a subclass of these **neurofibromas**, labeled plexiform, progress to malignant tumors termed **neurofibrosarcomas**. Patients suffering from neurofibromatosis type 1 also have greatly increased risk of glioblastomas (tumors of the astrocyte lineage in the brain; see Figure 2.9A), **pheochromocytomas** (arising from the adrenal glands), and myelogenous leukemias (see Figure 2.8B). These tumors involve cell types arising from diverse embryonic lineages.

Neurofibromatosis patients often suffer from additional abnormalities that involve yet other cell types. Among these are **café au lait spots**, which are areas of hyperpigmentation in the skin; subtle alterations in the morphology of the cells in the skin and long bones; cognitive deficits; and benign lesions of the iris called "Lisch nodules" (Figure 7.20). These manifestations are strongly influenced by the patient's **genetic background** (i.e., the array of all other genetic alleles in an individual's genome), since siblings inheriting the same mutant allele of the responsible gene, called *NF1*, often exhibit dramatically different disease phenotypes.

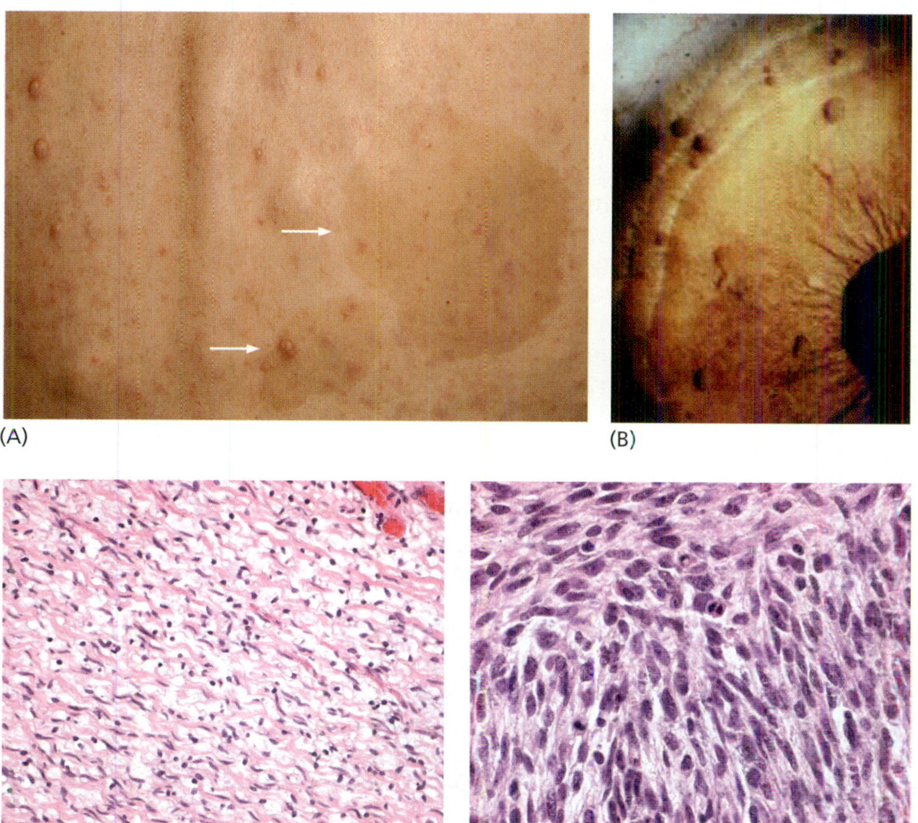

Figure 7.20 Neurofibromatosis Neurofibromatosis type 1 is considered a *syndrome* because a number of distinct conditions are associated with a single genetic state. (A) Notable among these are the numerous small, subcutaneous nodules (i.e., neurofibromas) and the light brown *café au lait spots (arrows)* seen here on the back of a patient. (B) In addition, Lisch nodules are often observed in the iris of the eyes of these patients. (C) Histological observation of afflicted tissues indicates the benign neurofibromas *(left)*, which occasionally progress to malignant peripheral nerve sheath tumors (MPNSTs, *right*). The histological complexity of the neurofibromas greatly complicated the identification of the normal precursors of these growths, which are now thought to be the Schwann cells that wrap around peripheral axons. (A and C, courtesy of B.R. Korf; B, from B.R. Korf, *Postgrad. Med.* 83:79–88, 1988.)

The *NF1* gene was cloned in 1990. (A second type of neurofibromatosis is connected with inactivation of a completely unrelated tumor suppressor gene.) The genetic behavior of the *NF1* gene parallels closely that of the *Rb* gene. Thus, mutant, inactivated alleles of the *NF1* gene transmitted through the germ line act in a dominant fashion to create disease phenotypes. At the cellular level, the originally heterozygous configuration of the gene ($NF1^{+/-}$) is converted to a homozygous state ($NF1^{-/-}$) in tumor cells through loss of heterozygosity. Finally, as many as half of neurofibromatosis patients lack a family history of the disease, indicating that the mutant allele that they carry is the consequence of a *de novo* mutation in the germ line. As is the case with the *Rb* gene, such *de novo* mutations usually occur during spermatogenesis in the fathers of afflicted patients.

Once the cloned gene was sequenced, it became possible to assign a function to neurofibromin, the *NF1*-encoded protein: it showed extensive sequence relatedness to a *Saccharomyces cerevisiae* (yeast) protein, termed IRA, that functions as a GTPase-activating protein (GAP) for yeast Ras, as well as to two mammalian Ras-GAP proteins. Like most if not all eukaryotes, yeast cells use Ras proteins to regulate important aspects of their metabolism and proliferation. Detailed genetic analyses had already shown that in yeast, the positive signaling functions of the Ras protein are countered by the IRA protein. By provoking Ras to activate its intrinsic GTPase activity, IRA forces Ras to convert itself from its activated GTP-bound form to its inactive, GDP-bound form; precisely the same function is carried out by the mammalian Ras-GAP proteins. Indeed, a Ras-GAP may ambush activated Ras before the latter has had a chance to stimulate its coterie of downstream effectors (Section 5.9, Sidebar 5.11).

This initial insight into NF1 function inspired a simple scheme of how defective forms of NF1 create disease phenotypes. NF1 is expressed widely throughout the body, with especially high levels found in the adult peripheral and central nervous systems. When cells first experience growth factor stimulation, they may degrade NF1, enabling Ras signaling to proceed without interference by NF1. However, after 60 to 90 minutes, NF1 levels return to normal, and the NF1 protein that accumulates helps to shut down further Ras signaling—a form of negative-feedback control. In neuroectodermal cells lacking NF1 function, Ras proteins are predicted to exist in their activated, GTP-bound state for longer-than-normal periods of times. In fact, in the cells of neurofibromas, which are genetically $NF1^{-/-}$, elevated levels of activated Ras and Ras effector proteins can be found (Figure 7.21). Consequently, the loss of NF1 function in a cell can mimic functionally the activated Ras proteins that are created by mutant *ras* oncogenes (see Section 5.9).

The histological complexity of neurofibromas has made it very difficult to identify the normal cell type that yields the bulk of the cells in these growths. Microscopic analyses (see Figure 7.20C) reveal that these growths are composed of a mixture of cell types, including Schwann cells (which wrap around and insulate nerve axons), neurons, perineurial cells (which seem to be of fibroblast origin and are found in the vicinity of neurons), fibroblasts, and mast cells, the latter coming from the immune system. In some of these benign tumors, Schwann cells are clearly the prevailing cell type, while in others it seems that fibroblasts or perineurial cells may dominate.

The weight of evidence indicates that the Schwann cell precursors are the primary targets of LOH in neurofibromas, and that once these cells have lost all NF1 function, they orchestrate the development of these histologically complex growths by inducing the co-proliferation of a variety of other cell types via paracrine signaling. This co-proliferation raises an interesting and still-unanswered question, however: Since these other neighboring cell types have an $NF1^{+/-}$ genotype, might they be more susceptible to growth-stimulatory signals than fully wild-type cells? Such increased susceptibility might well arise from the fact that they carry only half the normal dose of functional NF1 protein.

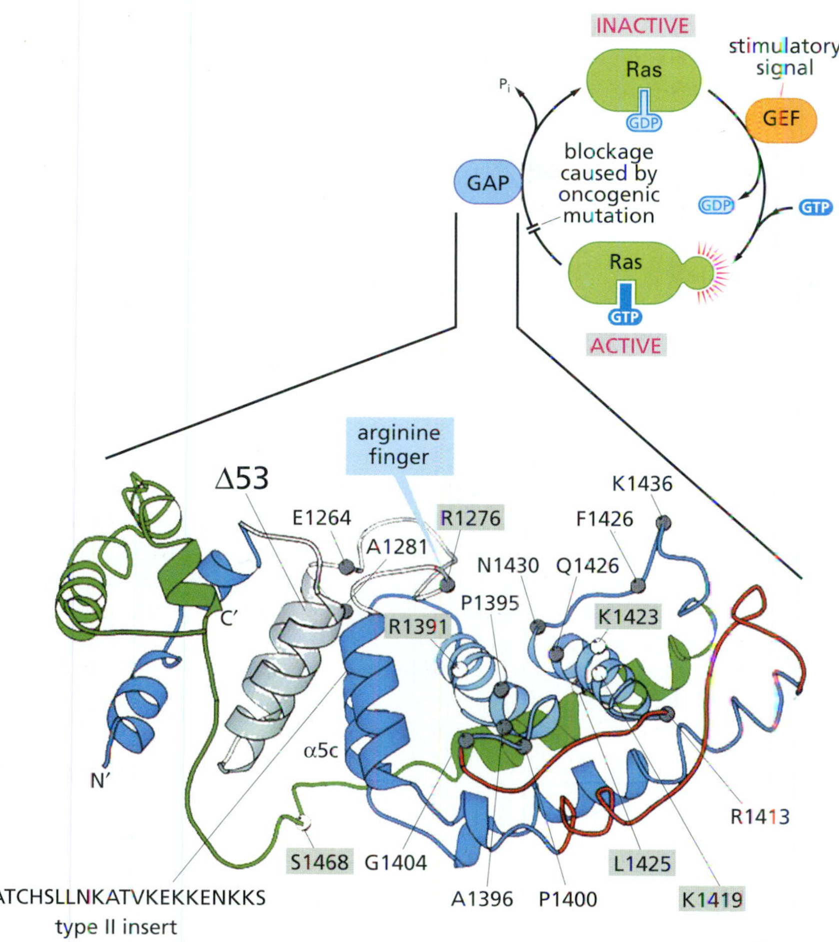

Figure 7.21 Neurofibromin and the Ras signaling cycle As illustrated in Figure 5.30, the Ras protein passes through a cycle in which it becomes activated by a guanine nucleotide exchange factor (GEF), such as Sos, and becomes inactivated by a GTPase-activating protein (a Ras-GAP). One of the main Ras-GAPs is neurofibromin (NF1), the product of the *NF1* gene. Interaction of Ras with NF1 can increase the GTPase activity of Ras more than 1000-fold. The structure of the domain of NF1 that interacts with Ras is illustrated here. A subdomain of NF1 termed the "arginine finger" *(top)* carries a critical arginine (R1276) that is inserted into the GTPase cleft of Ras and actively contributes to the hydrolysis of GTP to GDP by Ras. Mutant forms of NF1 (carrying amino acid substitutions) observed in neurofibromatosis patients are indicated by *light gray spheres* that are labeled by *gray boxes;* experimentally created mutations have also generated a number of amino acid substitutions *(dark gray spheres)* that also compromise NF1 GAP function. In addition, a large-scale deletion (Δ53) and an insertion ("Type II insert") found in patients are shown; both are also disease-associated. (From K. Scheffzek et al., *EMBO J.* 17:4313–4327, 1998.)

Here is one indication that a half-normal dose of NF1 protein can indeed affect cell and thus organismic phenotype: mast cells (a type of immune cell) in mice that express a partially defective Kit growth factor receptor are produced in abnormally low numbers, an indication of the important role of Kit in promoting the proliferation of these immune cells. However, the $Nf1^{+/-}$ genotype, when present in the precursors of mast cells that have this partially defective Kit receptor, results in a substantial reversion of the phenotype (i.e., a normalization of mast cell number), indicating (1) that NF1 operates to dampen signals from, among other proteins, the Kit receptor; and (2) that the inadequate signaling from the Kit receptor can be compensated by increased Ras signaling resulting from partial loss of NF1 function. In a larger sense, the presence of NF1 protein in half-normal concentrations in many types of cells is likely to result in a substantial increase in Ras signaling and, in this fashion, to perturb cell phenotypes.

These observations raise a more general question about tumor suppressor genes: Granted that the full phenotypic changes of tumor suppressor gene (TSG) inactivation are felt only when both gene copies are lost, might it nonetheless be the case that half-dosage of their encoded proteins (which is often observed in cells having a $TSG^{+/-}$ genotype) yields subtle but still very real changes in cell behavior—the genetic phenomenon of **haploinsufficiency** (Sidebar 7.6)?

7.11 Apc facilitates egress of cells from colonic crypts

While the great majority (>95%) of colon cancers appear to be sporadic, a small group arise as a consequence of inherited alleles that create substantial lifelong risk for this disease. The best understood of these heritable colon cancer

Sidebar 7.6 Haploinsufficiency: when half a loaf is not that much better than none The Knudson paradigm of tumor suppressor gene inactivation postulates that mutant alleles of tumor suppressor genes are recessive at the cellular level. Hence, cells that are heterozygous for a tumor suppressor gene (i.e., $TSG^{+/-}$) should be phenotypically normal. But there are perfectly good biochemical reasons to think that cells expressing half the normal level of a tumor suppressor gene protein may have a phenotype that deviates significantly from wild type, and indeed, examples of this have been observed in addition to those cited in the text.

In the case of the *Rb* gene, substantial evidence indicates that $Rb^{+/-}$ cells are phenotypically totally normal. Thus, any pathologies displayed by human heterozygotes are limited to the cell populations that have lost the remaining wild-type *Rb* gene copy, usually by LOH, and are therefore rendered $Rb^{-/-}$; the remaining $Rb^{+/-}$ tissues throughout the body seem to develop fully normally and exhibit normal function.

Observations of genetically altered mice, however, provide clear examples of haploinsufficiency. Mice carrying only one copy of the gene encoding the Smad4 transcription factor, which is used by TGF-β to inhibit cell proliferation (as we will see in Section 8.10), are predisposed to developing polyps in the stomach and small intestine, and the cells in these tumors continue to carry single copies of the wild-type allele of this gene. Mice that are heterozygous for the $p27^{Kip1}$ tumor suppressor gene are similarly tumor-prone, without indication of LOH in tumor cells; this gene encodes an important inhibitor of cell-cycle progression (Section 8.4). A third example is provided by mice that are heterozygous for the gene encoding the PTEN tumor suppressor gene (Section 6.6), where a clear acceleration of prostate cancer tumorigenesis is created by the absence of only a single copy of this TSG. Finally, deletion of one copy of the *Dmp1* gene, which encodes a transcription factor that induces expression of the $p19^{ARF}$ TSG, leads once again to increased tumor susceptibility without loss of the wild-type gene copy in tumor cells. Altogether, at least half a dozen human TSGs yield abnormal cell phenotypes when only a single wild-type gene copy is present in cell genomes.

syndromes is adenomatous polyposis coli, usually called familial adenomatous polyposis (FAP), that is, an inherited susceptibility to develop adenomatous polyps in the colon. Such polyps, while themselves nonmalignant, are prone to develop into frank carcinomas at a low but predictable frequency. This syndrome is responsible for a bit less than 1% of all colon cancers in the West.

In the colons of Western populations, in which colon cancer is relatively frequent because of still poorly understood dietary factors, polyps are often found in low numbers scattered throughout the colon. By the age of 70, as many as half of the individuals in these populations have developed at least one of these growths (see, for example, Figure 2.15A). However, in the relatively rare individuals suffering from familial adenomatous polyposis, polyps numbering in the hundreds are found to carpet the luminal surface of the colon, that is, the surface of the colon facing into the colonic cavity (Figure 7.22).

The cloning of the *APC* gene (Sidebar 7.7) led, after many years of additional research, to a reasonably clear view of how this gene and its encoded protein are able to control cell proliferation within the colon. As we discuss in greater detail in later chapters, the epithelia in the colon and duodenum are organized in a fashion that is typical of a number of epithelia throughout the body. In all cases, groups of relatively undifferentiated stem cells yield two distinct daughter cells when they divide: one daughter remains a stem cell, thereby ensuring that the number of stem cells in a tissue remains constant; however, the other daughter cell and its descendants become committed to differentiate.

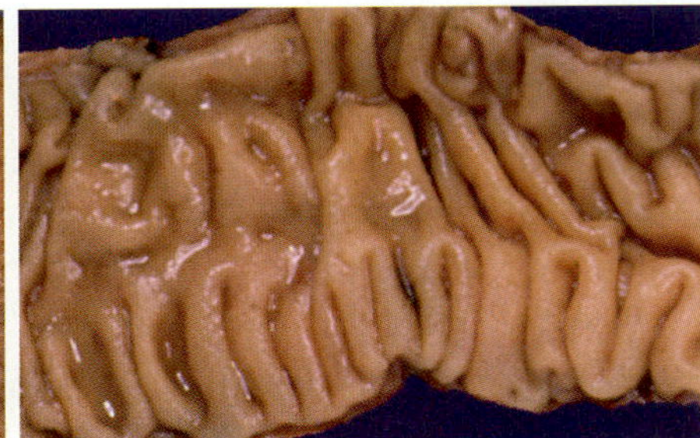

Figure 7.22 Familial adenomatous polyposis The wall of the colon from an individual afflicted with familial adenomatous polyposis (FAP) is carpeted with hundreds of small polyps *(left)*; such polyps are absent in the smooth wall of a normal colon *(right)*. The detailed structure of one type of colonic polyp is shown in Figure 2.15A. (Courtesy of A. Wyllie and M. Arends.)

Sidebar 7.7 Special human populations facilitate the detection of heritable cancer syndromes and the isolation of responsible genes The Mormons in the state of Utah represent a population that provides human geneticists with a golden opportunity to understand the genetics of various types of heritable human diseases. A substantial proportion of individuals in today's Mormon population in Utah can trace their ancestry back to a relatively small group of founding settlers who arrived in Utah in the mid-nineteenth century. Mormon couples have traditionally had large numbers of children and have reproduced relatively early in life, leading to large, multigenerational families—ideal subjects of genetic research.

Most importantly, a tenet of the Mormon religion is a belief in the retroactive baptism of individuals who are ancestors to today's Mormons. As a consequence, the Mormon church (more properly, the Church of Jesus Christ of Latter Day Saints) has encouraged its members to undertake extensive genealogical research. To facilitate such research, the Mormon church has established and maintains the world's largest genealogical archive in Salt Lake City, Utah. Finally, the Mormons in Utah have been particularly receptive to helping human geneticists trace specific disease susceptibility genes through their pedigrees.

The confluence of these factors has made it possible to assemble multigenerational pedigrees, such as that shown in Figure 7.23; these have enabled geneticists to trace with precision how the mutant allele of a gene predisposing a person to familial adenomatous polyposis is transmitted through multiple generations of a family and how this allele operates in a dominant fashion to create susceptibility to colonic polyps and resulting increased risk of colon cancer. Use of linkage analysis, in which the genetic transmission of this predisposing allele was connected to anonymous genetic markers on various human chromosomes, also revealed that this allele was repeatedly co-transmitted from one generation to the next together with genetic markers on the short arm of human Chromosome 5. This localization, together with LOH analyses, eventually made possible the molecular cloning of the *APC* gene in 1991.

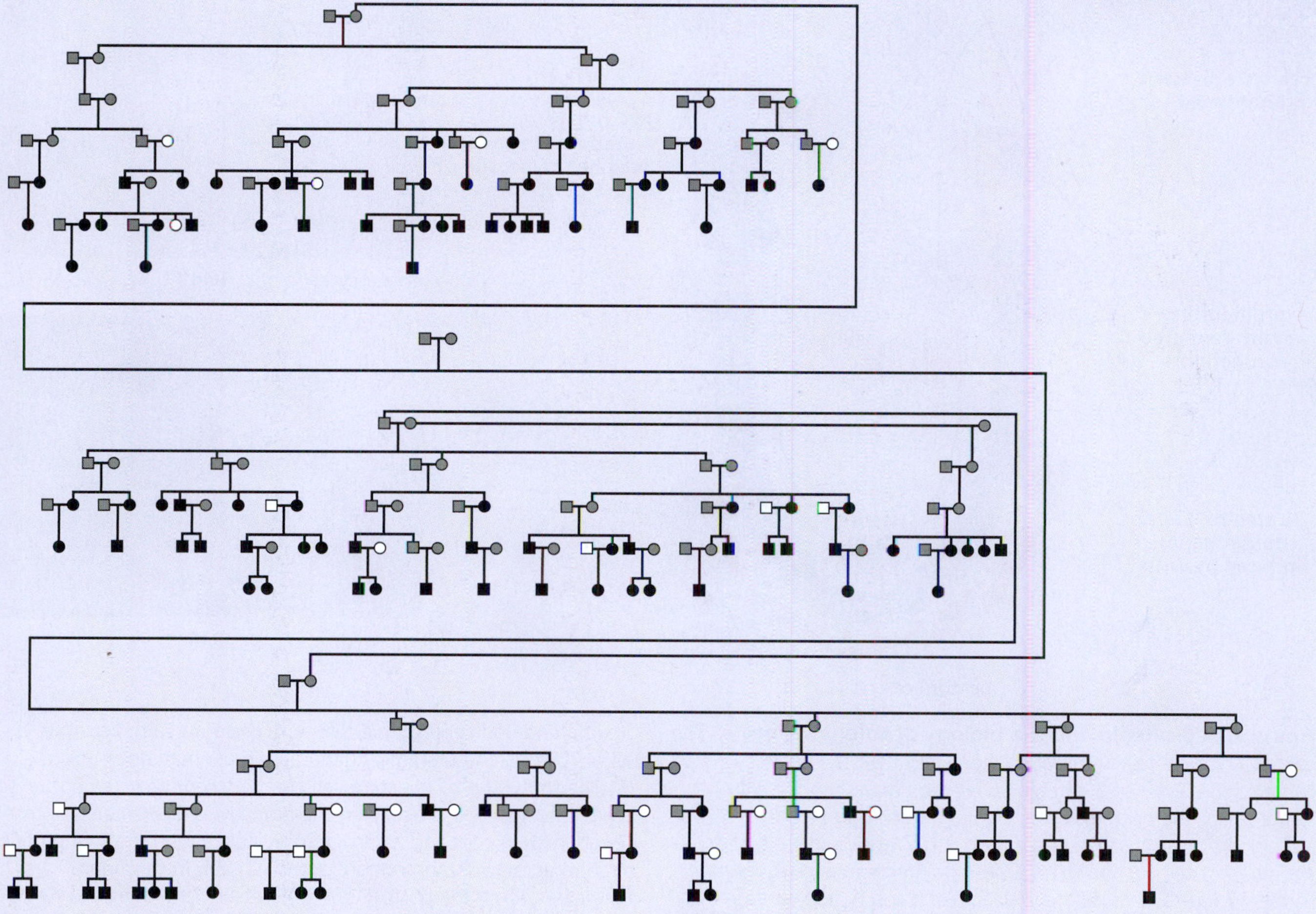

Figure 7.23 Genetic mapping using large kindreds The ability to analyze the genomes of a large number of individuals in a kindred afflicted with an inborn cancer susceptibility syndrome, such as familial adenomatous polyposis, greatly facilitates the localization of the responsible gene to a small region of a human chromosome. In the case of the *Apc* gene, the availability of a large Mormon kindred afflicted with familial adenomatous polyposis, whose relationships were traced through the extensive genealogical records kept in Salt Lake City, Utah, facilitated the localization of this gene to a small region of the short arm of human Chromosome 5. Afflicted individuals are indicated here by filled symbols. (Courtesy of R.L. White, M.F. Leppert, and R.W. Burt.)

In the small intestine, some of these differentiated epithelial cells participate in absorbing nutrients from the lumen and transferring these nutrients into the circulation; in the colon, they may absorb water from the lumen. Yet other specialized epithelial cells secrete a mucus-like material that helps to protect the colonic epithelium from the contents of the lumen. (In the gastrointestinal tract, these various epithelial cells are termed **enterocytes**.)

The locations of the stem cells and more differentiated cells in the colon are illustrated in Figure 7.24A. Note that the stem cells are protected through their locations at the bottom of the deep cavities known as **crypts**. While some of the progeny of these stem cells stay behind in the crypts in order to maintain a constant number of stem cells, most are dispatched upward and out of the crypts toward the luminal surface of the epithelium, where they will function briefly to form the epithelial lining of the gut, die by apoptosis, and be shed into the colonic lumen. This entire process of out-migration and death takes only 3 to 4 days.

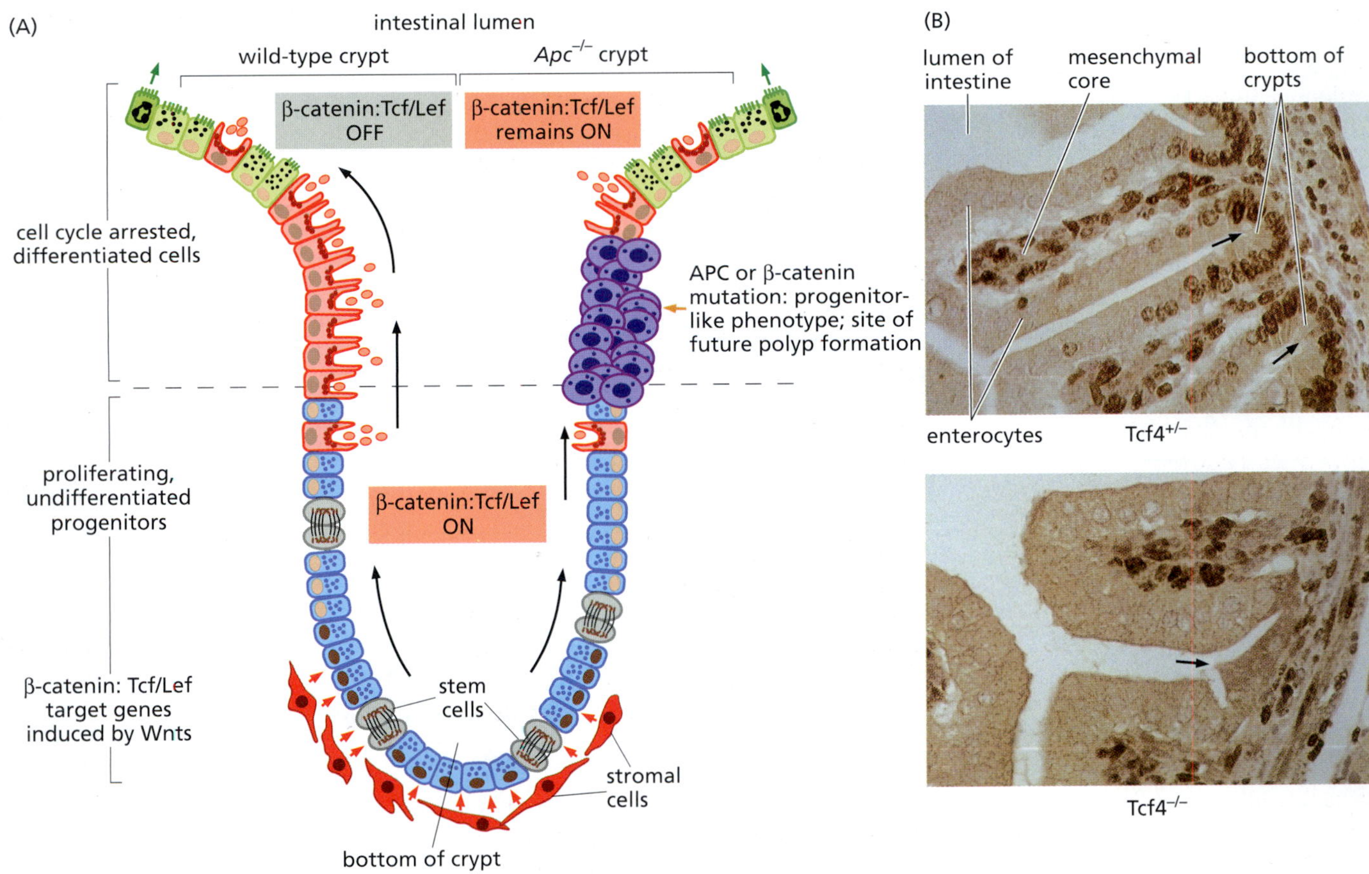

Figure 7.24 β-catenin and the biology of colonic crypts (A) The colonic crypt contains replicating stem cells near the bottom, which contain high levels of β-catenin *(small red arrows)*. In the cells located near the bottom of the crypts, intracellular β-catenin levels are high because these cells are receiving Wnt signals from the stroma *(red;* see Section 6.10). These β-catenin molecules migrate to the nucleus and associate with Tcf/Lef transcription factors; this drives increased proliferation of these cells and prevents their differentiation. In the normal intestine, many of the progeny of these stem cells migrate upward toward the lumen *(left side of crypt)*. As they do so, stimulation by Wnts decreases; this leads to increased degradation of β-catenin, which results, in turn, in cessation of proliferation and increased differentiation as these cells approach the lumen and ultimately enter apoptosis after 3 to 4 days *(small green arrows)*. In contrast, when the Apc protein is defective, β-catenin levels remain high *(right side of crypt)* even in the absence of intense Wnt signaling, and proliferating, still-undifferentiated cells *(purple)* fail to migrate upward, accumulate within crypts, and ultimately generate an adenomatous polyp. (B) The formation of the stem cell compartment near the bottom of the crypts depends critically on the high levels of β-catenin normally found in stem cells and its ability to associate with the Tcf4 protein to form an active transcription factor. As seen in this figure, epithelial cells in the bottom of the crypts in $Tcf4^{+/-}$ mice *(arrows, upper panel)* are highly proliferative, as evidenced by the staining for Ki67, a marker of cell proliferation. As they move out of the crypt up the sides of the villi, they lose proliferative potential. (The mesenchymal cells in the core of each villus are also highly proliferative, but their behavior is unrelated to that of the epithelial enterocytes.) In $Tcf4^{-/-}$ mice *(lower panel)*, in contrast, the enterocyte stem cells in the crypts are absent *(arrow)* and the entire gut therefore lacks an epithelial lining. (A, from M. van de Wetering et al., *Cell* 111:251–263, 2002; B, from V. Korinek et al., *Nat. Genet.* 19:379–383, 1998.)

The scheme depicted in Figure 7.24A represents a highly effective defense mechanism against the development of colon cancer, since almost all cells that have sustained mutations while on duty protecting the colonic wall are doomed to die within days after they have been formed. By this logic, the only type of mutations that can lead subsequently to the development of a cancer will be those mutations (and resulting mutant alleles) that block both the out-migration of colonic epithelial cells from the crypts and the cell death that follows soon thereafter. Should a colonic enterocyte acquire a mutation that causes its retention and survival in a crypt, this cell and its descendants can accumulate there, and any additional mutant alleles acquired subsequently by their progeny will similarly be retained in the crypts, rather than being rapidly lost through the apoptosis of these progeny cells. Such additional mutations might include, for example, genes that push the progeny toward a neoplastic growth state.

These dynamics focus attention on the molecular mechanisms controlling the out-migration of enterocytes from the colonic crypt. β-Catenin is the governor of much of this behavior. Recall from Section 6.10 that the levels of soluble β-catenin in the cytoplasm are controlled by Wnt growth factors. When Wnts bind cell surface receptors, β-catenin is saved from destruction, accumulates, and migrates to the nucleus, where it binds a group of DNA-binding proteins termed variously Tcf or Lef. The resulting heterodimeric transcription factors then proceed to attract yet other nuclear proteins, forming multi-protein complexes that activate expression of a series of target genes programming (in the case of enterocytes) the stem-cell phenotype.

In the context of the colonic crypt, enterocyte stem cells encounter Wnt factors released by stromal cells near the bottom of the crypt, which keep β-catenin levels high in the enterocytes (see Figure 7.24). Indeed, these cells are held in a stem cell-like state by the high levels of intracellular β-catenin. However, as some of the progeny of these stem cells begin their upward migration, these progeny no longer experience Wnt signaling, and intracellular β-catenin levels fall precipitously. As a consequence, these cells lose their stem cell phenotype, exit the cell cycle, and differentiate into functional enterocytes.

Apc, the product of the adenomatous polyposis coli gene, is responsible for negatively controlling the levels of β-catenin in the cytosol. In cells at the bottom of the normal crypts, the *APC* gene is not expressed at detectable levels and β-catenin is present at high levels. However, as cells begin their upward migration out of the crypts, the level of Apc expression in these cells increases greatly and, in the absence of Wnts, this protein drives down the intracellular levels of β-catenin.

This inverse expression is explained nicely by the known molecular mechanism of action of Apc, a large protein of 2843 amino acid residues (Figure 7.25A). Together with two scaffolding proteins termed axin and conductin, Apc forms a multiprotein complex that brings together glycogen synthase kinase-3β (GSK-3β) and β-catenin (Figure 7.25B). This association enables GSK-3β to phosphorylate four amino-terminal residues of β-catenin; the phosphorylation then leads to the degradation of β-catenin via the ubiquitin–proteasome pathway (Sidebar 7.8). In sum, Apc is essential for triggering the degradation of β-catenin, and in its absence, β-catenin levels accumulate to high levels within cells.

With this information in mind, we can place *APC* gene inactivation in the context of the biology of the colonic crypts. When the spectrum of *APC* mutations found in human colon cancers is cataloged (see Figure 7.25A), one sees many mutations that cause premature termination of translation of the Apc protein, thereby removing domains that are important for its ability to associate with β-catenin and axin and for the resulting degradation of β-catenin.

The accumulation of β-catenin is clearly the most important consequence of *APC* inactivation, which can be observed in about 90% of sporadic colon carcinomas.

One can draw this conclusion from studying the remaining minority (approximately 10%) of sporadic colon carcinomas that carry wild-type *APC* alleles. In some of these, the *APC* gene promoter is hypermethylated and rendered inactive. In others, the gene encoding β-catenin carries point mutations, and the resulting mutant β-catenin molecules lose the amino acid residues that are normally phosphorylated by GSK-3β. Since they cannot be phosphorylated, these mutant β-catenin molecules escape degradation and accumulate—precisely the outcome that is seen when Apc is missing!

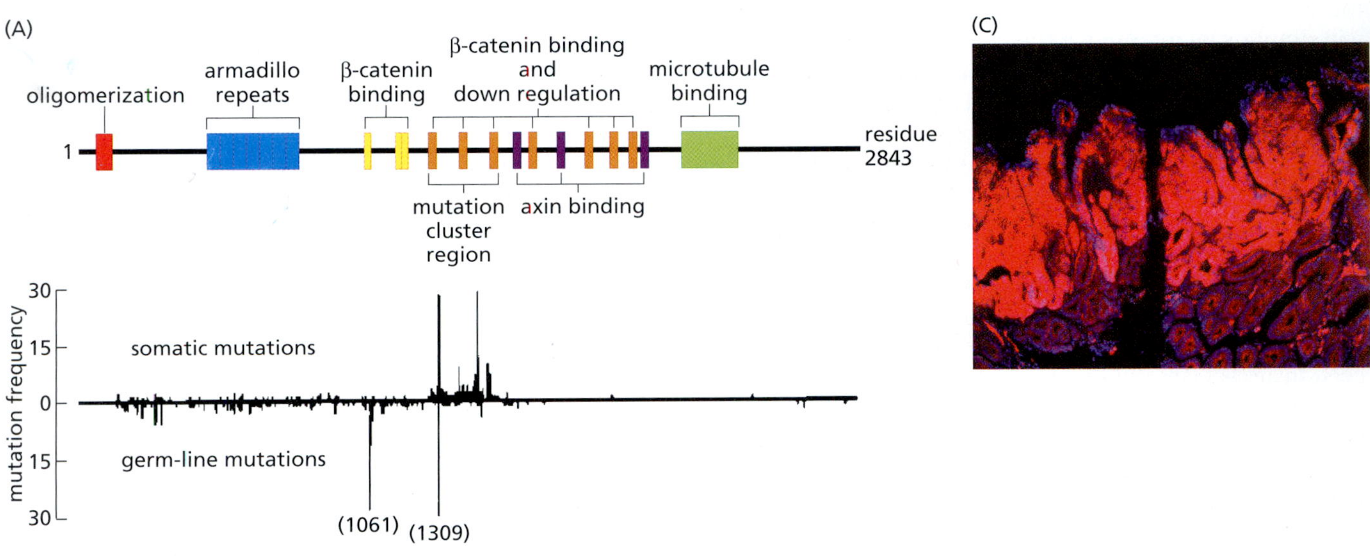

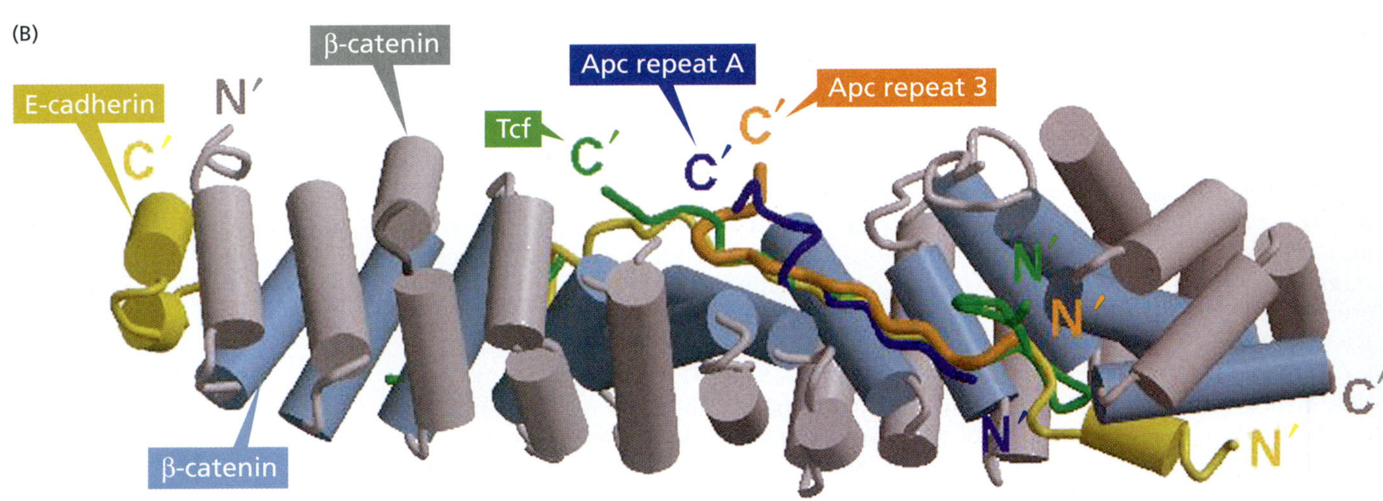

Figure 7.25 Apc, β-catenin, and Tcf/Lef (A) This diagram indicates in outline the multidomain structure of Apc *(above)* with a plot of the spectrum of *APC* mutations and where they occur *(below)*. (B) As is apparent, the β-catenin molecule *(light blue, gray cylinders)* associates with a number of alternative partner proteins during its life cycle. Following its synthesis, it may be bound in the cytosol to a domain of the Apc protein *(dark blue, orange)*, which will target it for destruction in proteasomes. If it escapes destruction, it may use the same domain to bind to the cytoplasmic domain of a cell surface receptor termed E-cadherin *(yellow)*, as we will learn in Chapters 13 and 14. Alternatively, β-catenin may migrate to the nucleus, where it will associate with a domain of a Tcf/Lef transcription factor *(green)* and activate expression of a number of target genes. As is apparent, all three β-catenin-binding partners have similarly structured domains that they use to bind β-catenin. (C) In the Min mouse model of familial polyposis, an inactivating point mutation has been introduced into the mouse *Apc* gene by the actions of a mutagenic chemical. The chromosomal region carrying this mutant allele undergoes LOH in some cells within intestinal crypts, yielding cells that lack all Apc activity. In these cells, β-catenin is no longer degraded, accumulates to high levels, and enters the cell nuclei, where it collaborates with Tcf/Lef transcription factors to drive expression of growth-promoting genes. The result is a clonal outgrowth, such as the adenoma seen here, in which the β-catenin is visualized by immunostaining *(pink, above)*, while the nuclei in all cells, including those of the normal crypts *(below)* are seen in *blue*. (A, from P. Polakis, *Biochim. Biophys. Acta* 1332:F127–F147, 1997 and adapted from R. Fodde et al., *Nat. Rev. Cancer* 1:55–67, 2001; B, courtesy of H.J. Choi and W.I. Weis, and from N.-C. Ha et al., *Mol. Cell.* 15:511–521, 2004; C, courtesy of K.M. Haigis and T. Jacks.)

When β-catenin accumulates in enterocyte precursors because of inactivation of Apc function or other mechanisms (Figure 7.25C), this causes them to retain a stem cell-like phenotype, which precludes them from migrating out of the crypts. This leads, in turn, to the accumulation of large numbers of relatively undifferentiated cells in a colonic crypt (see Figure 7.24A), which eventually form adenomatous polyps. Equally important, these accumulating cells can later sustain further mutations that enable them to form more advanced polyps and, following even more mutations, carcinomas.

This model explains the sequence of mutations that leads eventually to the formation of human colon carcinoma cells. The first of these mutations invariably involves inactivation of Apc function (or the functionally equivalent changes mentioned above). The resulting cells, now trapped in the crypts, may then suffer mutations in a number of other genes, such as K-*ras* (Section 4.4), that cause these cells to grow more aggressively. Importantly, alterations of the Apc–β-catenin pathway always come first, while the order of the subsequent genetic changes is quite variable.

Recent research has turned up another, fully unexpected consequence of Apc loss. Cells lacking Apc function have been found to exhibit a marked increase in chromosomal instability (CIN), which results in increases and decreases in chromosome number, usually because of inappropriate segregation of chromosomes during mitosis. Some Apc-negative cells even accumulate tetraploid (or nearly tetraploid) karyotypes, in which most chromosomes are represented in four copies rather than the usual two. This defect seems to derive from the fact that, in addition to Apc's cytoplasmic functions described above, Apc molecules localize to components of the microtubule arrays that form the mitotic spindle and are responsible for segregating chromosomes during the anaphase and telophase of mitosis. The aneuploidy that results from these chromosomal segregation defects alters the relative numbers of critical growth-promoting and growth-inhibiting genes. These changes, in turn, may well facilitate tumorigenesis by accelerating the rate with which pre-malignant cells acquire advantageous genotypes and thus phenotypes.

7.12 Von Hippel–Lindau disease: pVHL modulates the hypoxic response

Von Hippel–Lindau syndrome is a hereditary predisposition to the development of a variety of tumors, including clear-cell carcinomas of the kidney, pheochromocytomas (tumors of cells in the adrenal gland), and **hemangioblastomas** (blood vessel tumors) of the central nervous system and retina. Germ-line mutations of the *VHL* tumor suppressor gene have been documented in almost all of the patients suffering from the syndrome. The mutant *VHL* alleles seem to be present in the human gene pool in a frequency that results in an incidence of the disease of about 1 in 35,000 in the general population. Like mutant alleles of *Rb*, the mutant *VHL* alleles act at the organismal level in an autosomal dominant manner to create disease. And further extending the *Rb* analogy, the *VHL* locus undergoes a loss of heterozygosity (LOH) that results in a $VHL^{-/-}$ genotype in the tumor cells of patients inheriting a mutant germ-line *VHL* allele.

The *VHL* gene is also inactivated in the majority (about 70%) of sporadic (i.e., nonfamilial) kidney carcinomas. In those sporadic tumors in which mutant *VHL* alleles are not detectable, one often finds transcriptional silencing of this gene due to promoter methylation. The main, but apparently not the only, task of pVHL, the product of the *VHL* gene, is to foster destruction of a subunit of a critical transcription factor termed hypoxia-inducible factor-1 (HIF-1).

In cells experiencing normal oxygen tensions (**normoxia**), pVHL provokes the degradation of the HIF-1α subunit of HIF-1. Consequently, HIF-1α is synthesized

Sidebar 7.8 Ubiquitylation tags cellular proteins for destruction in proteasomes The concentrations of many cellular proteins must be tightly controlled in response to a variety of physiologic signals. Much of this control is achieved through the selective degradation of these proteins. Thus, under certain conditions, a protein may be long-lived and is therefore permitted to accumulate within a cell, while under other circumstances it is rapidly degraded at a rate that is regulated by certain physiologic signals. This degradation is accomplished, almost always, by the ubiquitin–proteasome system (Table 7.3).

To cite an example, many proteins are phosphorylated at critical amino acids by a kinase once their degradation is called for. The resulting phosphoamino acid, in the context of neighboring amino acid residues, attracts a complex of enzymes that covalently attaches a ubiquitin molecule to the protein—the process of **ubiquitylation** (often called ubiquitination). As an alternative mechanism, the exposure by a protein of one of its normally hidden subdomains may attract the attentions of a ubiquitylating complex even without the specific marking created by phosphorylation.

Table 7.3 Advantages of ubiquitin-regulated proteolysis

A.	Unidirectional—unlike phosphorylation and other post-translational modifications of proteins, degradation cannot be rapidly reversed
B.	Rapid—a large number of protein molecules can be eliminated in a regulated fashion in a matter of minutes
C.	Fine-tuning—provides another way to finely adjust the levels of critical regulatory proteins
D.	Localized—can be confined to specific subcellular compartments
E.	Specific—a small set of proteins can be degraded without any effect on all other proteins

Courtesy of M. Pagano.

Ubiquitin is a small (76-residue) protein whose sequence is highly conserved between single-cell eukaryotes and mammalian cells; only 3 of 76 amino acid residues differ between the yeast and human versions of this protein. (Its ubiquitous presence in the biosphere inspired its name.) One ubiquitin molecule is initially linked via its C-terminal glycine to the ε-amino side chain of a lysine present in a target protein; a second ubiquitin molecule is then linked to lysine-48 of the first ubiquitin, and the process is repeated a number of times, yielding a polyubiquitin chain (Figure 7.26). A protein molecule tagged in this fashion makes its way to a proteasome in the nucleus or cytoplasm, in which it is degraded. (An alternative version of the polyubiquitin chain, in which ubiquitin–ubiquitin linkages are formed through the lysine-63 residues of ubiquitin monomers, appears to be involved in the functional activation of tagged proteins rather than their degradation.)

The widespread importance of this protein-degradation machinery is indicated by the fact that sequence analysis of all approximately 20,000 genes in the human genome has indicated that at least 527 genes are likely to

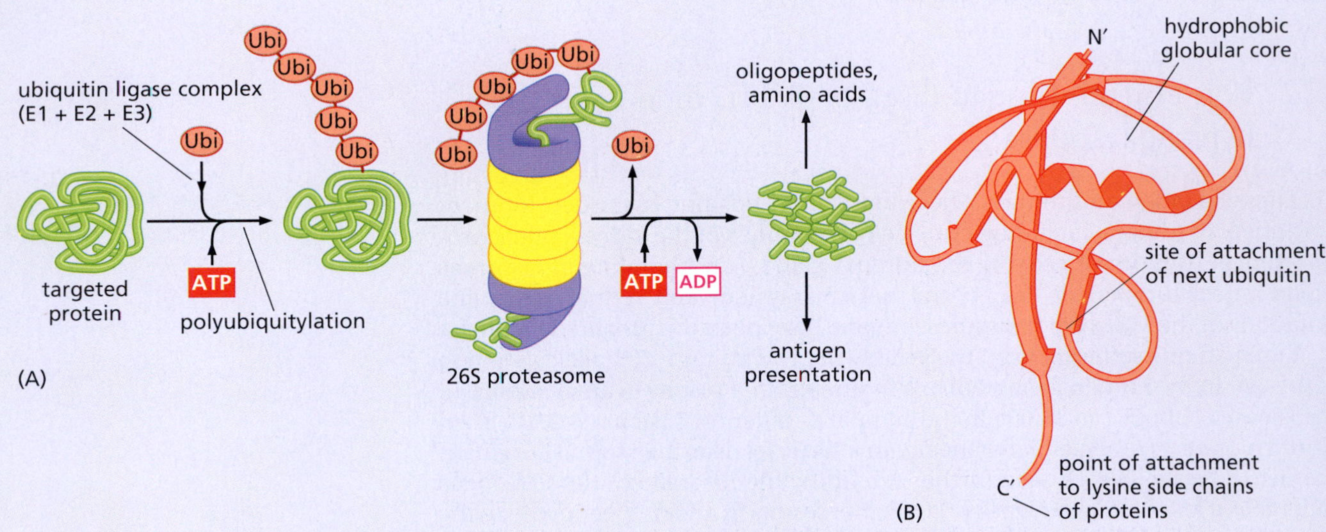

Figure 7.26 Ubiquitylation and proteasomes Much of protein degradation in the mammalian cell is carried out by the ubiquitin–proteasome pathway. (A) A complex of three proteins (E1, E2, and E3), which together constitute a ubiquitin ligase, recognizes a protein destined for degradation and tags this protein with a chain of ubiquitins (Ubi). Following polyubiquitylation, the protein is conveyed to a proteasome (see Figure 7.27), where it is de-ubiquitylated and degraded into oligopeptides that are either degraded further into amino acids or used for antigen presentation by the immune system; see Section 15.3). (In fact, a specialized proteasome is used to process proteins for antigen presentation.) (B) Ubiquitin itself is a relatively short protein of 76 amino acid residues whose amino acid sequence and structure are almost totally conserved among all eukaryotic cells. Its structure is indicated here in this ribbon diagram determined by X-ray crystallography.

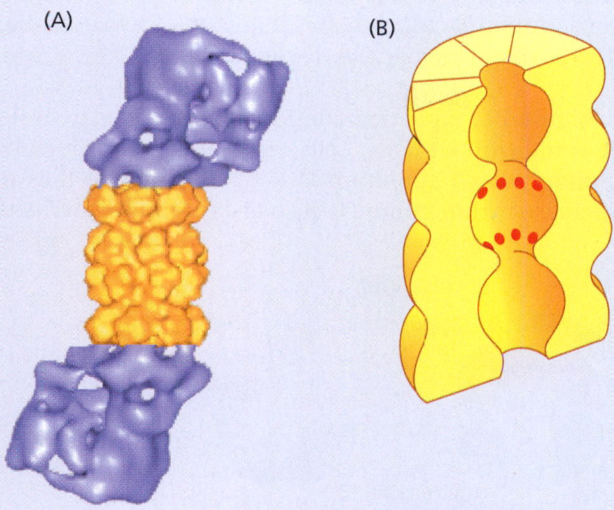

Figure 7.27 The proteasome The proteasome is a complex of about 2.5 megadaltons assembled from more than 30 distinct protein species. (A) This reconstructed image of a proteasome, determined by X-ray crystallography, indicates two "cap" regions *(purple)*, where ubiquitylated proteins are bound, de-ubiquitylated, and then introduced into the centrally located barrel region *(yellow)*, in which they are processed by proteolytic degradation into oligopeptides or amino acid residues. (B) A schematic diagram of the barrel region in cross section indicates the region *(red dots)* where the actual proteolysis takes place. (A, from W. Baumeister et al., *Cell* 92:367–380, 1998.)

encode E3 ubiquitin ligases—the proteins responsible for initially recognizing substrate proteins that are to be targeted for ubiquitylation and destruction; another analysis has identified 110 candidate de-ubiquitylating enzymes (DUBs).

Proteasomes are cellular workshops dedicated to the degradation of proteins that are presented to them. The proteasome is a large, hollow, cylindrical complex of multiple protein subunits, which uses ATP to unfold a protein prior to degrading this protein in the cylinder's interior (Figure 7.27). Polyubiquitylated proteins that are introduced into the proteasome are first de-ubiquitylated and then digested into short peptide fragments ranging in size from 3 to 25 amino acid residues.

Polyubiquitylation is known to be responsible for the degradation of many short-lived growth-regulating proteins such as Myc, p53, Jun, and certain cyclins (to be described in Chapter 8), and more than 80% of the proteins in mammalian cells are degraded in the proteasomes. While polyubiquitylation marks proteins for destruction, the function of mono-ubiquitylation (where only a single ubiquitin is attached) is complex and poorly understood. There are clear indications that mono-ubiquitylation is used in certain cellular contexts for regulating endocytosis, protein sorting, trafficking of proteins within the nucleus, and even regulating gene expression. For example, mono-ubiquitylation of the p53 tumor suppressor protein tags it for export from the nucleus; subsequent polyubiquitylation can then mark p53 for destruction in proteasomes. Mono-ubiquitylated growth factor receptors are marked for endocytosis. An additional dimension of complexity derives from the fact that ubiquitin is only one of a group of as many as 10 distinct marker proteins (e.g., the ubiquitin-like Sumo and Nedd proteins) that are used to tag various proteins, marking them for a variety of metabolic fates other than proteolysis.

and then degraded with a half-life of only 10 minutes. The result is that HIF-1α accumulates only to very low steady-state levels in cells, and therefore the functional HIF-1 transcription factor, which is made up of two essential subunits—HIF-1α and HIF-1β—remains inactive (Figure 7.28).

Such synthesis followed by rapid degradation is often called a "futile cycle." This particular cycle is interrupted when cells experience **hypoxia** (subnormal oxygen tensions), under which condition HIF-1α degradation fails to occur and HIF-1α levels increase within minutes. (Its half-life, and therefore its concentra-

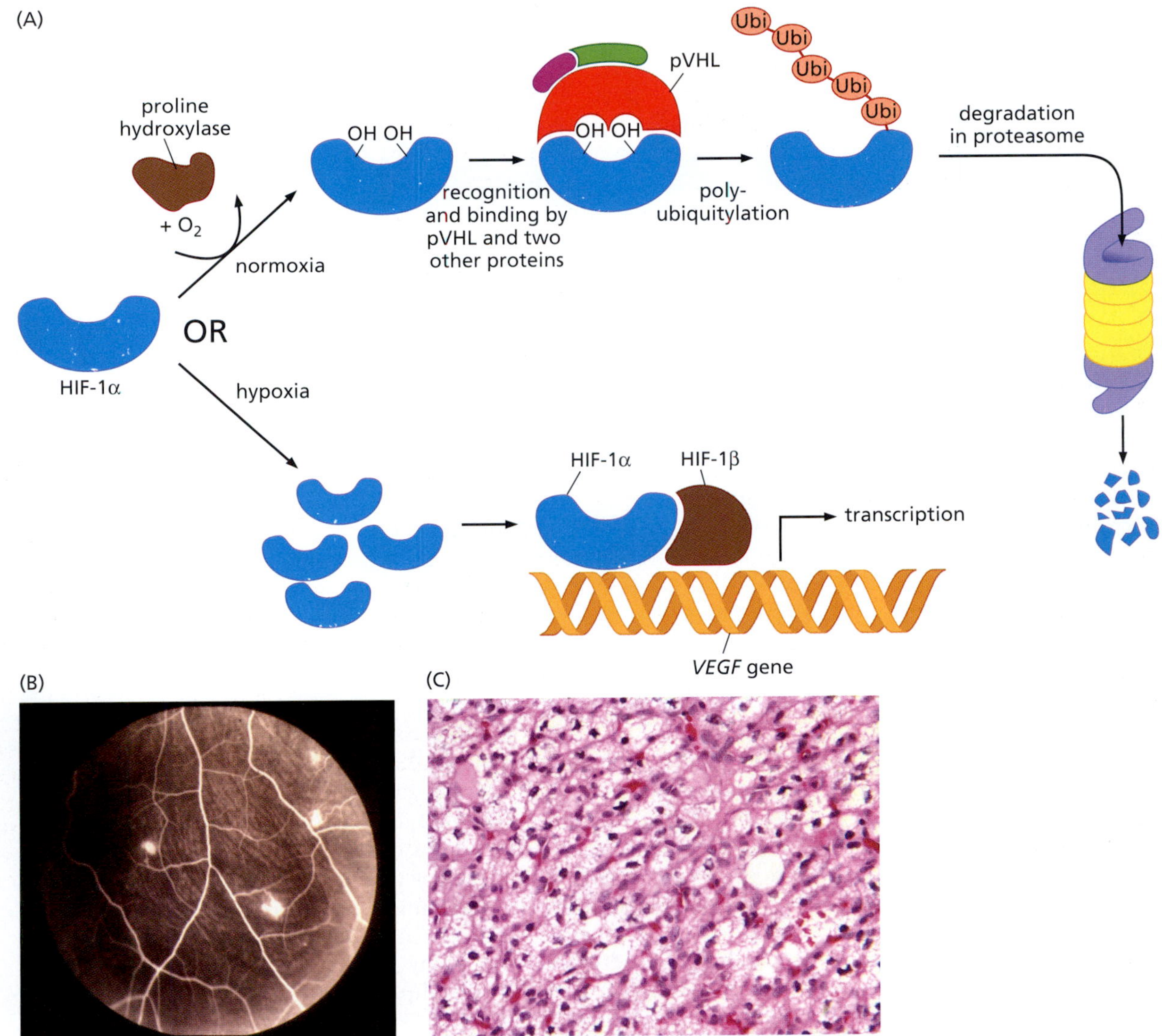

Figure 7.28 HIF-1 and its regulation by pVHL (A) The hypoxia-inducible transcription factor-1 (HIF-1) is composed of two subunits, HIF-1α and HIF-1β, both of which are required for its transcription-activating function. Under conditions of normoxia *(above)*, HIF-1α *(blue)* is synthesized at a high rate and almost immediately degraded because of the actions of pVHL *(red)*. Under these conditions, proline hydroxylase *(brown)*, an iron-containing enzyme that is responsible for oxidizing two proline residues of HIF-1α, succeeds in converting one or both of these to hydroxyproline residues *(indicated here as hydroxyl groups)*. These hydroxyprolines enable the binding of HIF-1α to pVHL, which, together with two other proteins *(purple, green)*, tag HIF-1α by ubiquitylation and thus for degradation. However, under hypoxic conditions *(below)*, proline hydroxylase fails to oxidize the two prolines of HIF-1α, pVHL fails to bind to HIF-1α, HIF-1α escapes ubiquitylation, and its levels increase rapidly. It now can form heterodimers with HIF-1β *(brown)*, and the resulting heterodimeric transcription factor can proceed to activate expression of physiologically important genes, such as the gene specifying vascular endothelial growth factor (VEGF). (As described in Chapter 13, VEGF is an important inducer of angiogenesis.) (B) One of the manifestations of von Hippel–Lindau syndrome is the presence of areas of uncontrolled vascularization in the retina, seen here as large white spots; these are ostensibly due to excessive production of VEGF in cells that have lost pVHL function. (C) A more extreme consequence of lost pVHL function is a hemangioblastoma, which arises from the precursors of endothelial cells that are stimulated by VEGF. (B and C, courtesy of W.G. Kaelin Jr.)

tion, increases more than tenfold.) The resulting formation of functional HIF-1 transcription factor complexes causes expression of a cohort of target genes whose products are involved in **angiogenesis** (generation of new blood vessels), **erythropoiesis** (formation of red blood cells), energy metabolism (specifically, glycolysis), and glucose transport into the cells. The motive here is to induce the synthesis of proteins that enable a cell to survive under hypoxic conditions in the short term and, in the longer term, to acquire access to an adequate supply of oxygen. Notable among the latter are proteins that collaborate to attract the growth of new vessels into the hypoxic area of a tissue. As we will describe later (Section 13.6), this formation of new vessels is also critical for the growth of tumors, enabling them to acquire access to oxygen and nutrients and to evacuate carbon dioxide and metabolic wastes.

The importance of HIF-1 in regulating the angiogenic response is underscored by the roster of genes whose expression it induces, among them the genes encoding vascular endothelial growth factor (VEGF), platelet-derived growth factor (PDGF), and transforming growth factor-α (TGF-α), and yet others that also promote angiogenesis. VEGF attracts and stimulates growth of the endothelial cells that construct new blood vessels; PDGF stimulates these cells as well as associated mesenchymal cells, such as pericytes and fibroblasts; and TGF-α stimulates a wide variety of cell types including epithelial cells. In the hypoxic kidney, HIF-1 mediates the induction of the gene encoding erythropoietin (EPO). This leads to rapid increases in erythropoietin in the circulation, and to an induction of red cell production in the bone marrow (see also Sidebar 6.2 and Section 6.8).

To return to the details of *VHL* function, its product, pVHL, exists in cells in a complex with several other proteins. Together, they function to acquire a molecule of ubiquitin and to link this ubiquitin molecule covalently to specific substrate proteins (Sidebar 7.8). As was discussed earlier in this chapter, once a protein has been tagged by polyubiquitylation, it is usually destined for transport to proteasome complexes in which it will be degraded. Within the multiprotein complex, pVHL is responsible for recognizing and binding HIF-1α, thereby bringing the other proteins in the complex (Figure 7.29) in close proximity to HIF-1α, which they proceed to ubiquitylate.

pVHL binds to HIF-1α only when either of two critical proline residues of HIF-1α has been oxidized to hydroxyproline; in the absence of a hydroxyproline, this binding to HIF-1α fails and HIF-1α escapes degradation. The conversion of the HIF-1α proline residues into hydroxyprolines is carried out by an enzyme that depends on oxygen for its activity. Once formed, the hydroxyproline residue(s) of HIF-1α can be inserted into a gap in the pVHL hydrophobic core. This enables

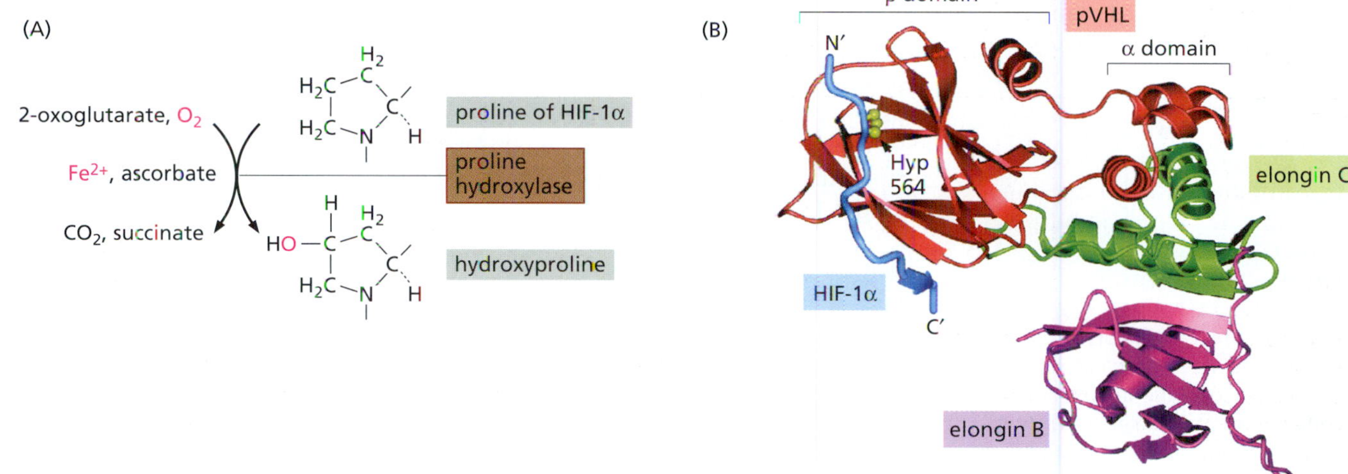

Figure 7.29 Molecular details of the HIF-1α–pVHL interaction (A) In the presence of high levels of oxygen, a proline hydroxylase enzyme will oxidize either of two proline residues in the HIF-1α transcription factor, converting it into a hydroxyproline (Hyp). (B) Once a hydroxyproline is formed on HIF-1α (e.g., Hyp at residue 564, *light green balls*), it enables a domain of HIF-1α *(light blue)* to associate with a tripartite complex consisting of pVHL *(red)* and two other proteins, elongin B *(purple)* and elongin C *(green)*. This binding, in turn, enables the tripartite complex to ubiquitylate HIF-1α, which soon leads to its destruction in proteasomes. (From J.-H. Min et al., *Science* 296:1886–1889, 2002.)

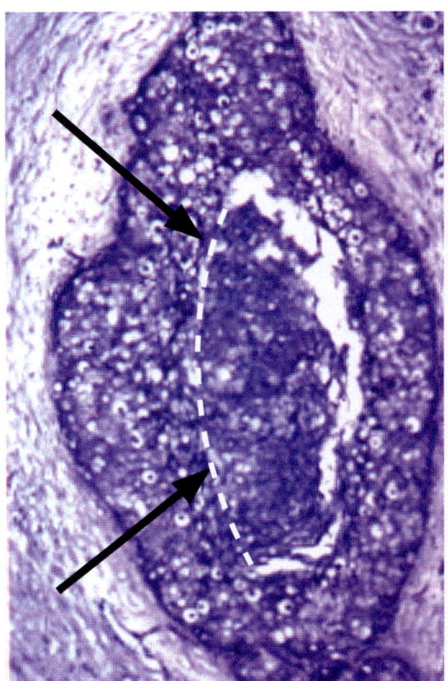
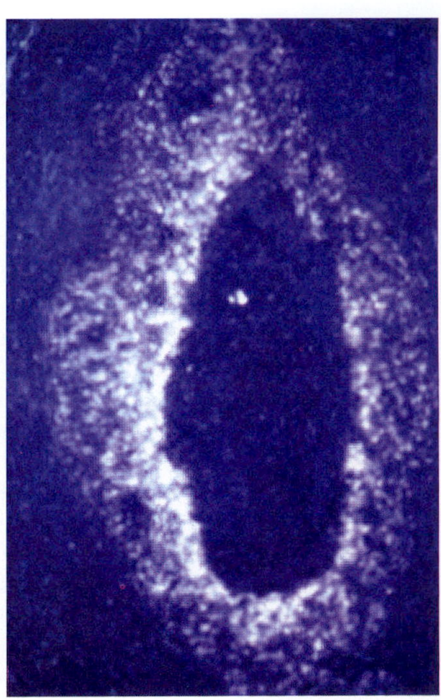

Figure 7.30 Expression of the VEGF target of HIF-1 Notable among the genes activated by HIF-1 is the gene encoding VEGF (vascular endothelial growth factor; see Figure 7.28A). Seen here *(left)* is a human *in situ* breast cancer that remains localized and thus has not yet invaded the surrounding stroma. A large area of necrotic cells has formed in the center of the tumor *(arrows, dashed line)*. Use of an *in situ* hybridization protocol *(right)* reveals that VEGF RNA *(white areas)* is expressed in a gradient increasing from the low levels at the outer rim of the tumor (where oxygen is available) to the very intense expression toward the hypoxic center of the tumor. Areas of even greater hypoxia are incompatible with cell survival, explaining the large necrotic center of the tumor, where VEGF mRNA is no longer detectable. (Courtesy of A.L. Harris.)

pVHL to bind HIF-1α and trigger its degradation, and as a consequence, the induction of gene transcription by the HIF-1 transcription factor is prevented (see Figure 7.28A).

Taken together, these facts allow us to rationalize how pVHL works in normal cells and fails to function in certain cancer cells: Under conditions of normal oxygen tension (normoxia), HIF-1α displays one or two hydroxyprolines, is recognized by pVHL, and is rapidly destroyed. Under hypoxic conditions, pVHL lacks these hydroxyprolines, cannot be bound by pVHL, and accumulates to high levels that allow the functioning of the HIF-1 transcription factor. This, in turn, allows it to activate VEGF expression in hypoxic tissues, both normal and neoplastic (Figure 7.30).

In many of the tumors associated with mutant alleles of *VHL*, the pVHL protein is undetectable in the individual cancer cells. In yet other tumors, more subtle changes may cause pVHL inactivity without affecting its levels. For example, point mutations may alter the amino acid residues in the hydrophobic pocket of pVHL that recognizes and binds the hydroxyproline residues of HIF-1α. Through either mechanism, the resulting constitutive activity of the HIF-1 transcription factor drives expression of a number of powerful growth-promoting genes, including the above-mentioned VEGF, PDGF, and TGF-α. These proceed to stimulate proliferation of a variety of cell types bearing the corresponding receptors. The result, sooner or later, is one or another type of tumor. In addition, the resulting tumor cells may cause levels of VEGF to accumulate in the circulation; VEGF can then help to stimulate erythropoiesis in the bone marrow, explaining the abnormally high red blood cell counts seen in some von Hippel–Lindau patients.

Still, these descriptions of pVHL action do not explain the full range of phenotypes that result from its defective functioning (Sidebar 7.9). Thus, it is clear that pVHL has effects that are unconnected with HIF-1α and its degradation. For example, cells lacking functional pVHL are unable to properly assemble fibronectin in their extracellular matrix, and indeed, pVHL has been reported to bind fibronectin within cells. Moreover, a substantial amount of pVHL is associated with cytoplasmic microtubules whose stability it may affect. These enigmatic observations provide hints that pVHL operates to control cellular responses that have nothing to do with the oxygen tensions experienced by cells.

Sidebar 7.9 Familial polycythemia can also result from VHL defects Curiously, one specific germ-line point mutation in the *VHL* gene, while reducing the ability of the VHL protein to associate with HIF-1α, does not create Von Hippel–Lindau syndrome. It results instead in familial **polycythemia**, in which affected individuals produce too many red blood cells because of excessive levels of erythropoietin (EPO) production, which is driven, in turn, by higher-than-normal levels of HIF-1. Hence, this particular *VHL* mutation appears to preferentially affect only one of the many known targets of HIF-1 action, indicating that VHL must influence, in still-unknown ways, the targeting of HIF-1 to specific gene promoters.

7.13 Synopsis and prospects

The tumor suppressor genes constitute a large group of genes that specify protein products mediating diverse cell–physiologic functions. As is obvious from Table 7.1, these proteins operate in all parts of the cell, and there is only one shared attribute that allows their inclusion in this gene grouping: in one way or another, each of these genes normally functions to reduce the likelihood that a clinically detectable tumor will appear in one of the body's tissues.

In the great majority of cases, both copies of a tumor suppressor gene must be inactivated before an incipient cancer cell enjoys any proliferative or survival advantages. Still, this rule is not a hard-and-fast one. In some instances, such as the case of the *Nf1* gene, it appears that loss of one copy of a tumor suppressor gene already provides a measure of growth advantage to a cell—an example of the phenomenon of haploinsufficiency. And as we will see in Chapter 9, mutant alleles of another tumor suppressor gene—*p53*—can create a partially mutant cell phenotype by actively interfering with the ongoing functions of a coexisting wild-type allele in the same cell.

The discovery of tumor suppressor genes helped to explain one of the major mysteries of human cancer biology—familial cancer syndromes. As we read in this chapter, inheritance of a defective allele of one of these genes is often compatible with normal embryonic development. The phenotypic effects of this genetic defect may only become apparent with great delay, sometimes in midlife, when its presence is revealed by the loss of the surviving wild-type allele and the outgrowth of a particular type of tumor. Elimination of these wild-type alleles often involves loss-of-heterozygosity (LOH) events; and the repeatedly observed LOH in a certain chromosomal region in a group of tumors can serve as an indication of the presence of a still-unidentified tumor suppressor gene lurking in this region. In fact, a large number of chromosomal regions of recurring LOH have been identified in tumor cell genomes, but only relatively few of these have yielded cloned tumor suppressor genes to date. This means that the roster of these genes must be far larger than is indicated by the entries in Table 7.1.

The diverse behaviors of these genes highlight an ongoing difficulty in this area of cancer research: What criteria can be used to define a tumor suppressor gene? To begin, in this book we have included in the family of tumor suppressor genes only those genes whose products operate in some dynamic fashion to constrain cell proliferation or survival. Other genes that function indirectly to prevent cancer through their abilities to maintain the genome and suppress mutations are described in Chapter 12. From the perspective of a geneticist, this division—the dichotomy between the "gatekeepers" and the "caretakers"—is an arbitrary one, since the wild-type version of both types of genes is often found to be eliminated or inactivated in the genomes of cancer cells. Moreover, the patterns of inheritance of the cancer syndromes associated with defective caretaker genes are formally identical to the mechanisms described here. Still, for those who would like to understand the biological mechanisms of cancer formation, the distinction between tumor suppressor genes (the gatekeepers) and genome maintenance genes (the caretakers) is a highly useful one and is therefore widely embraced by researchers interested in the pathogenesis of this disease.

Since the time when the first tumor suppressor genes (*p53* and *Rb*) were cloned, numerous other genes have been touted as "candidate" tumor suppressor genes because their expression was depressed or absent in cancer cells while being readily detectable in corresponding normal cells. This criterion for membership in the tumor suppressor gene family was soon realized to be flawed, in no small part because it is often impossible to identify the normal precursor of a cancer cell under study. Certain kinds of tumor cells may not express a particular gene because of a gene expression program that is played out during normal differentiation of the tissue in which these tumor cells have arisen. Hence, the absence of expression of this gene in a cancer cell may reflect only the actions of a normal differentiation program rather than a pathological loss of gene expression. So this criterion—absence of gene expression—is hardly telling.

In certain instances, expression of a candidate tumor suppressor gene may be present in the clearly identified normal precursors of a group of tumor cells and absent in the tumor cells themselves. This would appear to provide slightly stronger support for the candidacy of this gene. But even this type of evidence is not conclusive, since the absence of gene expression in a cancer cell may often be one of the myriad *consequences* of the transformation process rather than one of its root *causes*. Thus, inactivation of this gene may have played no role whatsoever in the formation of the tumorigenic cell.

Responding to these criticisms, researchers have undertaken functional tests of their favorite candidate tumor suppressor genes. In particular, they have introduced cloned wild-type versions of these genes into cancer cells that lacked any detectable expression of these genes. The goal here has been to show that once the wild-type tumor suppressor gene function is restored in these cancer cells, they revert partially or completely to a normal growth phenotype, or may even enter into apoptosis. However, interpretation of these experiments has been complicated by the fact that the **ectopic** expression of many genes—their expression in a host cell where they normally are not expressed—and their expression at unnaturally high levels often makes cells quite unhappy and causes them to stop growing or even to die. Such responses are often observed following introduction of a variety of genes that would never be considered tumor suppressor genes.

So this functional test has been made more rigorous by determining whether a candidate tumor suppressor gene, when expressed at normal, physiologic levels, halts the growth of a cancer cell lacking expression of this gene while leaving normal, wild-type cells from the same tissue unaffected. This would appear to provide strong evidence in favor of the candidacy of a putative tumor suppressor gene. Still, even these experiments yield outcomes that are not always interpretable, because of the difficulties, cited above, in identifying normal cell types that are appropriate counterparts of the cancer cells being studied.

The ambiguities of these functional tests have necessitated the use of genetic criteria to validate the candidacy of many putative tumor suppressor genes. If a gene repeatedly undergoes LOH in tumor cell genomes, then surely its candidacy becomes far more credible. But here too ambiguity reigns. After all, genes that repeatedly suffer LOH may be closely linked on a chromosome to a bona fide tumor suppressor gene that is the true target of elimination during tumor development.

These considerations have led to an even stricter genetic definition of a tumor suppressor gene: a gene can be called a tumor suppressor gene only if it undergoes LOH in many tumor cell genomes and if the resulting homozygous alleles bear clear and obvious inactivating mutations. (This latter criterion should allow an investigator to discount any bystander genes that happen to be closely linked to bona fide tumor suppressor genes on human chromosomes.)

Not surprisingly, even these quite rigid genetic criteria have proven to be flawed, since they exclude from consideration certain genes that are likely to be genuine tumor suppressor genes. Consider the fact that the activity of many tumor suppressor genes can be eliminated by promoter methylation (Section 7.8). In this event, mutant alleles might rarely be encountered in tumor cell genomes, even though the gene has been effectively silenced. Recall, for example, the behavior of the *Runx3* gene, which may well be a tumor suppressor gene. To date, its functional silencing has been associated entirely with methylation in a variety of tumor genomes. (The fact that inactivating mutations have not been reported in tumor-associated *Runx3* gene copies leaves its status ambiguous.)

The ability to inactivate ("knock out") candidate tumor suppressor genes in the mouse germ line (Sidebar 7.10) provides yet another powerful tool for validating candidate tumor suppressor genes. The biology of mice and humans differs in many respects. Nonetheless, the shared, fundamental features of mammalian biology make it possible to model many aspects of human tumor biology in the laboratory mouse.

Almost all of the genes that are listed in Table 7.1 have been knocked out in the germ line of an inbred mouse strain. For the most part, resulting heterozygotes have been found to exhibit increased susceptibility to one or another type of cancer. In many instances, the particular tissue that is affected is quite different from that observed in humans. For example, *Rb*-heterozygous mice (i.e., $Rb^{+/-}$) tend to develop pituitary tumors rather than retinoblastomas—an outcome that is hardly surprising, given the vastly different sizes and growth dynamics of target cell populations in mouse tissues compared with their human counterparts. Still, the development of any type of tumor at an elevated frequency in such genetically altered mice adds persuasive evidence to support the candidacy of a gene as a tumor suppressor gene.

All this explains why a constellation of criteria are now brought to bear when evaluating nominees for membership in this exclusive gene club. For us, perhaps the most compelling criterion is a functional one: Can the tumor-suppressing ability of a candidate tumor suppressor gene be rationalized in terms of the biochemical activities of its encoded protein and the known position of this protein in a cell's regulatory circuitry?

In the end, these many complications in validating candidate tumor suppressor genes derive from one central fact: the very existence of a tumor suppressor gene becomes apparent only when it is absent. This sets the stage for all the difficulties that have bedeviled tumor suppressor gene research, and it underscores the difficulties that will continue to impede the validation of new TSGs in the future.

With all these reservations in mind, we can nevertheless distill some generalizations about tumor suppressor genes that will likely stand the test of time. To begin, the name that we apply to these genes is, in one sense, a misnomer. Their normal role is to suppress increases in cell number, either by suppressing proliferation or by triggering apoptosis. In their absence, cells survive and proliferate at times and in places where their survival and proliferation are inappropriate (see, for example, Figure 7.32).

Another generalization is also obvious, even without knowing the identities of all TSGs: the protein products of tumor suppressor genes do not form any single, integrated signaling network. Instead, these proteins crop up here and there in the wiring diagrams of various regulatory subcircuits operating in different parts of the cell. This is explained by the simple and obvious rationale with which we began this chapter: all well-designed control systems have both positive and negative regulatory components that counterbalance one another. Thus, for every type of positive signal, such as those signals that flow through mitogenic signaling pathways, there must be negative controllers ensuring that these signaling fluxes are kept within proper limits. Perhaps cancer biologists should have deduced this from first principles, long before tumor suppressor gene research began at the laboratory bench. In hindsight, it is regrettable that they didn't consult electrical engineers and those who revel in the complexities of electronic control circuits and cybernetics. Such people could have predicted the existence of tumor suppressor genes a long time ago.

Even after another 50 tumor suppressor genes are cataloged, the pRb and p53 proteins will continue to be recognized as the products of tumor suppressor genes that are of preeminent importance in human tumor pathogenesis. The reasons for this will become apparent in the next two chapters. While all the pieces of evidence are not yet in hand, it seems increasingly likely that the two signaling pathways controlled by pRb and p53 are deregulated in the great majority of human cancers. Almost all of the remaining tumor suppressor genes (see Table 7.1) are involved in the development of circumscribed subsets of human tumors.

The fact that certain tumor suppressor genes are missing from the expressed genes within cancer cells has prompted many to propose the obvious: if only we could replace the missing tumor suppressor genes in cancer cells, these

Sidebar 7.10 Homologous recombination allows restructuring of the mouse germ line The natural process of homologous recombination can be exploited to introduce well-defined genetic changes into the mouse germ line. A cloned gene fragment that is introduced by electroporation or microinjection into a somatic cell or an embryonic stem (ES) cell is able, with a low but significant frequency, to recombine with homologous DNA sequences residing in the chromosomal DNA of this cell. Use of appropriate selection markers, in this case *neo*^{r} (for neomycin resistance gene) and *tk*^{HSV} (herpesvirus thymidine kinase gene, which renders cells sensitive to killing by the drug ganciclovir), can be exploited to select for the rare cells in which homologous recombination has taken place (Figure 7.31A). Thus, application of neomycin selects for cells that have stably integrated the donor DNA into their chromosomes, while ganciclovir causes killing of cells that have retained *tk*^{HSV} and therefore have integrated the donor DNA through nonhomologous recombination with the chromosomal DNA of the embryonic stem (ES) cell (Figure 7.31B).

ES cells that have acquired the cloned DNA via homologous recombination can then be introduced via microinjection into the **blastocoel** cavity of a mouse **blastocyst** (an early embryo), and the embryo can then be introduced into a **pseudopregnant** female (Figure 7.31C). Because ES cells are **pluripotent** (able to differentiate into all cell types in the body), the injected cells may then insert themselves into (**chimerize**) the developing tissues of the resulting embryo, creating a genetic mosaic in which some of the cells and tissues are descendants of the injected ES cells while others derive from the host blastocyst embryo. In the event that these genetically altered ES cells generate descendants that succeed in chimerizing the developing gonads, the germ cells in the gonads of resulting adults may then transmit the experimentally altered allele to descendant organisms.

The originally used donor DNA fragment can contain sequences that, following recombination with the chromosomal gene, disrupt the chromosomal gene's function, in which case this procedure is commonly termed a "gene knockout." Alternatively, other types of alterations can be introduced into this cloned donor DNA fragment prior to microinjection. Following homologous recombination, the targeted gene may retain some function and express, for example, a mutant form of the protein that it normally expresses or even a foreign protein; the introduction of such new sequences into the resident gene is often termed "gene knock-in."

Figure 7.31 Homologous recombination of mouse germ-line genes (A) A cloned fragment of a mouse gene linked using recombinant DNA procedures with the *neo*^{r} and *tk*^{HSV} drug selection markers. The *neo*^{r} marker selects for cells that have stably acquired the cloned DNA segment, while the *tk*^{HSV} marker selects against cells that have acquired the cloned DNA fragment through nonhomologous recombination. This cloned DNA fragment can then be introduced via electroporation or transfection into a mouse embryonic stem (ES) cell. (B) Drug selection can then be performed to select for those ES cells that have stably acquired the cloned DNA fragment and have done so

cells would revert partially or totally to a normal cell phenotype and the problem of cancer would be largely solved. Such "gene therapy" strategies have the additional attraction that the occasional, inadvertent introduction of a tumor suppressor gene into a normal cellshould have little if any effect if this gene is expressed at physiologic levels; this reduces the risk of undesired side-effect toxicity to normal tissues.

As attractive as such gene therapy strategies are in concept, they have been extremely difficult to implement. The viral vectors that form the core of most of

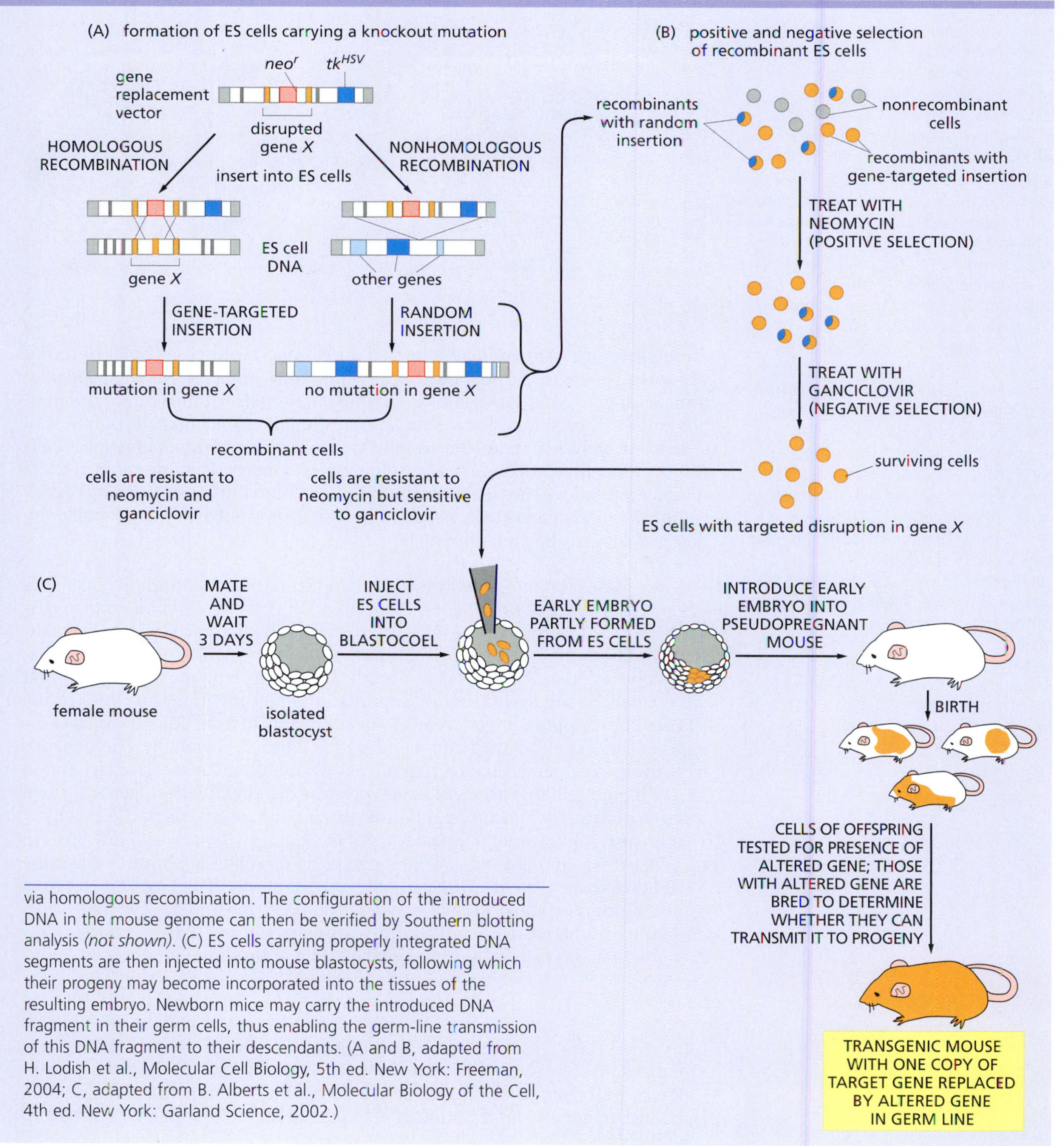

via homologous recombination. The configuration of the introduced DNA in the mouse genome can then be verified by Southern blotting analysis *(not shown)*. (C) ES cells carrying properly integrated DNA segments are then injected into mouse blastocysts, following which their progeny may become incorporated into the tissues of the resulting embryo. Newborn mice may carry the introduced DNA fragment in their germ cells, thus enabling the germ-line transmission of this DNA fragment to their descendants. (A and B, adapted from H. Lodish et al., Molecular Cell Biology, 5th ed. New York: Freeman, 2004; C, adapted from B. Alberts et al., Molecular Biology of the Cell, 4th ed. New York: Garland Science, 2002.)

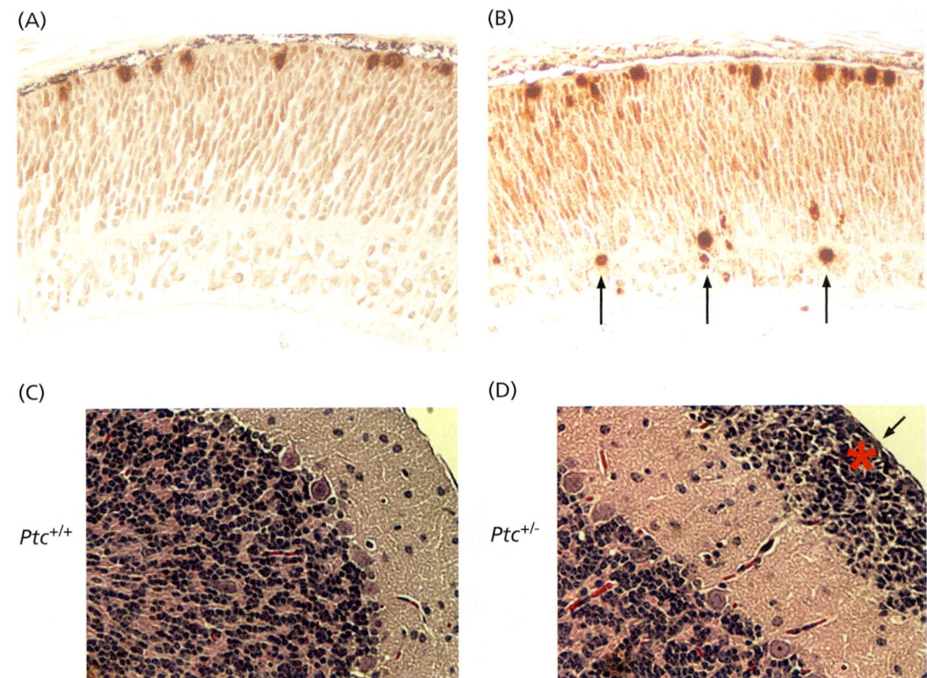

Figure 7.32 pRb and the shutdown of proliferation during differentiation The presence or absence of a functional tumor suppressor gene can have profound effects on the control of cell proliferation and thus tissue development. (A) During normal retinal development in the mouse, the proliferation of retinal precursor cells is limited to the upper layers of the developing retina *(dark brown spots)*, while the previously formed cells in the lower layers enter into post-mitotic, differentiated states. In this case, proliferation is detected through use of an antibody that detects a form of histone H3 that is present only in proliferating cells. (B) In a genetically altered mouse, however, in which both copies of the *Rb* gene have been inactivated selectively in the retina, cells in the lower levels of the developing retina, which should have become post-mitotic, continue to proliferate and fail to differentiate properly *(arrows)*. Similar cells in the human retina are presumably the precursors of retinoblastomas. (C) The granular cells *(dark purple)* of the developing cerebellum migrate from the outer layers *(right)* into inner layers of the cerebellum *(left)* during normal development of a wild-type mouse, in this case one that has two wild-type copies of the *Patched (Ptc)* tumor suppressor gene; these cells then enter in a post-mitotic state. (D) However, in a $Ptc^{+/-}$ heterozygote, many of the granular cells persist in the outer layer of the developing tissue *(arrow, red asterisk)* and continue to proliferate; such cells can become precursors of medulloblastomas, the most common pediatric brain tumor in humans. (A and B, from D. MacPherson et al., *Genes Dev.* 18:1681–1694, 2004; C and D, from T.G. Oliver et al., *Development* 132:2425–2439, 2005.)

these procedures are inefficient in delivering intact, wild-type copies of tumor suppressor genes to the neoplastic cells within tumor masses. However, efficient transfer of these wild-type gene copies into all the cells within a tumor cell population is essential for curative anti-tumor therapies: if significant numbers of cells fail to acquire a vector-borne wild-type tumor suppressor gene, these cells will serve as the progenitors of a reborn tumor mass. For this reason, while tumor suppressor genes are profoundly important to our understanding of cancer formation, in most cases, a reduction of this knowledge to therapeutic practice is still largely beyond our reach.

The exceptions to this generally discouraging scenario come from the instances where the absence of tumor suppressor function makes cells vulnerable to particular types of low-molecular-weight drugs (which are more effective in reaching cells throughout a tumor). For example, during the course of tumor progression, certain classes of tumor cells become especially dependent on signals that are created by the loss of tumor suppressor gene function. Thus, loss of the PTEN tumor suppressor gene leads to hyperactivity of Akt/PKB, on which some types of cancer cells come to depend for their continued viability. This explains why drugs that shut down firing of an essential upstream co-activator of Akt/PKB, termed mTOR, hold great promise as therapeutic agents against tumors, such as glioblastomas and prostate carcinomas, in which PTEN activity is often lost. But in general, as we will see in Chapter 16, the major advances in drug discovery and anti-cancer therapy are being made in shutting down the hyperactive oncoproteins. One can only hope that our extensive knowledge of tumor suppressor function will serve as the basis for anti-cancer treatments developed in the more distant future—a time when our skills in crafting new therapies will be vastly more sophisticated than at present.

Key concepts

- At the cellular level, the cancer phenotype is usually recessive (with the exception of viral oncogenes, which act in a dominant fashion). This indicates that the loss of genetic information is responsible for at least part of cancer cell phenotype.

- Much of the loss of functionally important genetic information is attributable to the loss of tumor suppressor genes (TSGs), which are often present in the genomes of cancer cells as inactive, null alleles.
- Consequently, tumor suppressor gene loss usually affects cell phenotype only when both copies of such a gene are lost in a cell.
- The loss of TSG function can occur either through genetic mutation or the epigenetic silencing of genes via promoter methylation.
- Inactivation (by mutation or methylation) of one copy of a TSG may be followed by other mechanisms that facilitate loss of the other gene copy; these mechanisms depend on loss of heterozygosity (LOH) at the TSG locus, and may involve mitotic recombination, loss of a chromosomal region that harbors the gene, inappropriate chromosomal segregation (nondisjunction), or gene conversion stemming from a switch in template strand during DNA replication.
- LOH events usually occur more frequently than mutations or promoter methylation, and they occur with different frequencies in different genes.
- Repeated LOH occurring in a given chromosomal region in a number of independently arising tumors often indicates the presence of a TSG in that region.
- TSGs regulate cell proliferation through many biochemical mechanisms. The only theme that unites them is the fact that the loss of each one of these increases the likelihood that a cell will undergo neoplastic transformation.
- When mutant, defective copies of a TSG are inherited in the germ line, the result is often greatly increased susceptibility to one or another specific type of cancer.
- TSGs are often called "gatekeepers" to signify their involvement in governing the dynamics of cell proliferation and to distinguish them from a second class of genes, the "caretakers," which also increase cancer risk when inherited in defective form but function entirely differently, since they work to maintain the integrity of the cell genome.
- The loss of TSGs may occur far more frequently during the development of a tumor than the activation of proto-oncogenes into oncogenes.

Thought questions

1. Why is the inheritance of mutant, activated oncogenes responsible for only a small proportion of familial cancer syndromes, while the inheritance of defective tumor suppressor genes (TSGs) is responsible for the lion's share of these diseases?
2. What factors may determine whether the inactivation of a TSG occurs at a frequency per cell generation higher than the activation of an oncogene?
3. How might the loss of TSG function yield an outcome that is, at the cell biological level, indistinguishable from the acquisition of an active oncogene?
4. Some TSGs undergo LOH in fewer than 20% of tumors of a given type. Why and how do such low rates of LOH complicate the identification and molecular isolation of such genes?
5. What criteria need to be satisfied before you would be comfortable in categorizing a gene as a TSG?
6. What factors might influence the identities of the tissues affected by an inherited, defective allele of a TSG?

Additional reading

Baylin SB & Herman JG (2000) DNA hypermethylation in tumorigenesis: epigenetics joins genetics. *Trends Genet.* 16, 168–174.

Baylin SB, Herman JG, Graff JR, et al. (1998) Alterations in DNA methylation: a fundamental aspect of neoplasia. *Adv. Cancer Res.* 72, 141–196.

Bienz M & Clevers H (2000) Linking colorectal cancer to Wnt signaling. *Cell* 103, 311–320.

Cichowski K & Jacks T (2001) NF1 tumor suppressor gene function: narrowing the GAP. *Cell* 104, 593–604.

Comings DE (1973) A general theory of carcinogenesis. *Proc. Natl. Acad. Sci. USA* 70, 3324–3328.

Fearon ER (1997) Human cancer syndromes: clues to the origin and nature of cancer. *Science* 215, 252–259.

Feinberg AP & Tycko B (2004) The history of cancer epigenetics. *Nat. Rev. Cancer* 4, 143–153.

Fodde R, Smits R & Clevers H (2001) *APC*, signal transduction and genetic instability in colorectal cancer. *Nat. Rev. Cancer* 1, 55–67.

Gregorieff A & Clevers H (2005) Wnt signaling in the intestinal epithelium: from endoderm to cancer. *Genes Dev.* 19, 877–890.

Herman JG & Baylin SB (2003) Gene silencing in cancer in association with promoter hypermethylation. *N.E. J. Med.* 349, 2042–2054.

Hershko A, Ciechanover A & Varshavsky A (2000) The ubiquitin system. *Nat. Med.* 6, 1073–1081.

Jones PA & Baylin SB (2002) The fundamental role of epigenetic events in cancer. *Nat. Rev. Genet.* 3, 415–428.

Jones PA & Takai D (2001) The role of DNA methylation in mammalian epigenetics. *Science* 293, 1068–1070.

Kern SE (2002) Whose hypothesis? Ciphering, sectorials, D lesions, freckles and the operation of Stigler's Law. *Cancer Biol. Ther.* 1, 571–581.

Koch CA, Vortmeyer AO, Zhuang Z et al. (2002) New insights into the genetics of chromaffin cell tumors. *Ann. N.Y. Acad. Sci.* 970, 11–28.

Kondo K & Kaelin WG Jr (2001) The von Hippel–Lindau tumor suppressor gene. *Exp. Cell Res.* 264, 117–125.

Laird PW (2003) The power and promise of DNA methylation markers. *Nat. Rev. Cancer* 3, 253–266.

Lustig B & Behrens J (2003) The Wnt signaling pathway and its role in tumor development. *J. Cancer Res. Clin. Oncol.* 129, 199–221.

Marsh DJ & Zori RT (2002) Genetic insights into familial cancers—update and recent discoveries. *Cancer Lett.* 181, 125–164.

Moon RT, Kohn AD, De Ferrari GV & Kaykas A. (2004) Wnt and β-catenin signaling: diseases and therapies. *Nat. Rev. Genet.* 5, 689–699.

Nelson WJ & Nusse R (2004) Convergence of Wnt, β-catenin, and cadherin pathways. *Science* 303, 1483–1487.

Ponder BA (2001) Cancer genetics. *Nature* 411, 336–341.

Pugh CW & Ratcliffe PJ (2003) The von Hippel–Lindau tumor suppressor, hypoxia-inducible factor-1 (HIF-1) degradation, and cancer pathogenesis. *Sem. Cancer Biol.* 13, 83–89.

Sherr CJ (2004) Principles of tumor suppression. *Cell* 116, 235–246.

Shtiegman K & Yarden Y (2003) The role of ubiquitylation in signaling by growth factors: implications to cancer. *Sem. Cancer Biol.* 13, 29–40.

Tischfield JA and Shao C (2003) Somatic recombination redux. *Nat. Genet.* 33, 5–6.

Vogelstein B & Kinzler KW (1998) The genetic basis of human cancer. New York: McGraw-Hill.

Voges D, Zwickl P & Baumeister W (1999) The 26S proteasome: a molecular machine designed for controlled proteolysis. *Annu. Rev. Biochem.* 68, 1015–1068.

Wijnhoven SW, Kool, HJ, van Teilingen, CM et al. (2001) Loss of heterozygosity in somatic cells of the mouse. An important step in cancer initiation? *Mutat. Res.* 473, 23–36.

Zhu Y and Parada LF (2001) Neurofibromin, a tumor suppressor in the nervous system. *Exp. Cell Res.* 264, 19–28.

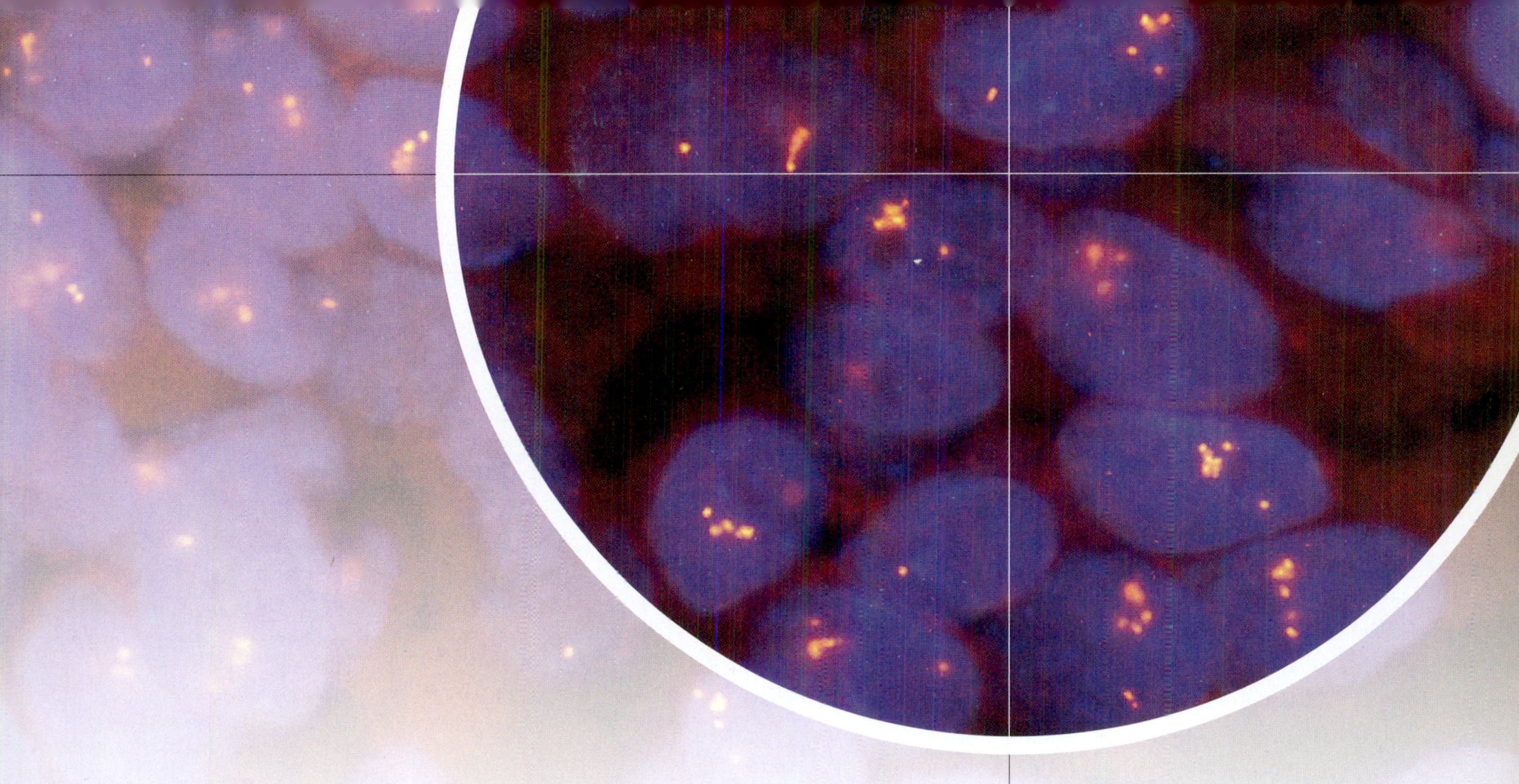

Chapter 8

pRb and Control of the Cell Cycle Clock

> This immediately leads one to ask, if the (hypothetical cellular transformation) loci can get into so much mischief, why keep them around? The logical answer is that they have some necessary function during some stage of the cell cycle, or some stage of embryogenesis.
>
> David E. Comings, geneticist, 1973

The fate of individual cells throughout the body is dictated by the signals that each receives from its surroundings—a point made repeatedly in earlier chapters. Thus, almost all types of normal cells will not proliferate unless prompted to do so by mitogenic growth factors. Yet other signaling proteins, notably transforming growth factor-β (TGF-β), may overrule the messages conveyed by mitogenic factors and force a halt to proliferation. In addition, extracellular signals may persuade a cell to enter into a **post-mitotic**, differentiated state from which it will never re-emerge and resume proliferation.

These disparate signals are collected by dozens of distinct cell surface receptors and then funneled into the complex signal-processing circuitry that operates largely in the cell cytoplasm. In some way, this mixture of signals must be processed, integrated, and ultimately distilled down to some simple, binary decisions made by the cell as to whether it should proliferate or become quiescent, and whether, as a quiescent cell, it will or will not differentiate. These behaviors suggest the existence of some centrally acting governor that operates inside the cell—a master clearinghouse that receives a wide variety of incoming signals and makes major decisions concerning the fate of the cell.

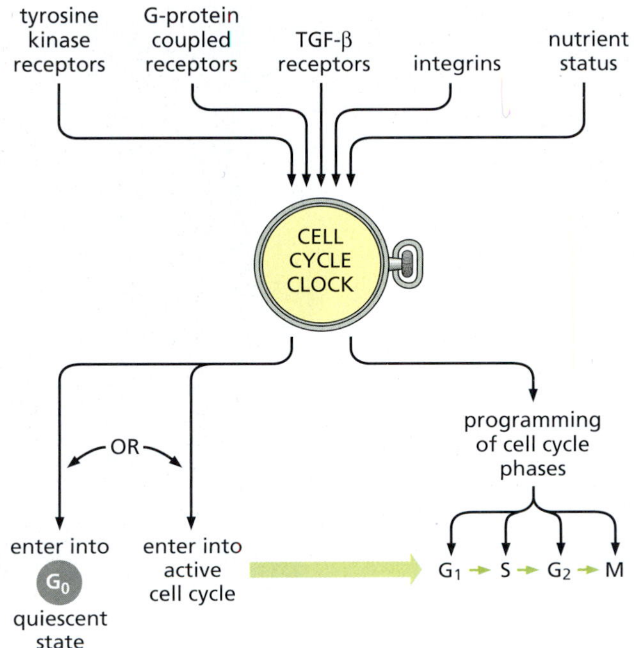

Figure 8.1 The central governor of growth and proliferation The term "cell cycle clock" denotes a molecular circuitry operating in the cell nucleus that processes and integrates a variety of afferent (incoming) signals originating from outside and inside the cell and decides whether or not the cell should enter into the active cell cycle or retreat into a nonproliferating state. In the event that active proliferation is decided upon, this circuitry proceeds to program the complex sequence of biochemical changes in a cell that enable it to double its contents and to divide into two daughter cells.

This master governor has been identified. It is the **cell cycle clock**, which operates in the cell nucleus. Its name is really a misnomer, since this clock is hardly a device for counting the passage of time. Nonetheless, we will use this term here for want of a better one. Rather than counting elapsed time, the cell cycle clock is a network of interacting proteins—a signal-processing circuit—that receives signals from various sources both outside and inside the cell, integrates them, and then decides the cell's fate. Should the cell cycle clock decide in favor of proliferation, it proceeds to orchestrate the complex transitions that together constitute the cell's cycle of growth and division. Should it decide in favor of quiescence, it will use its agents to impose this nonproliferative state on the cell (Figure 8.1).

The proliferative behavior of cancer cells indicates that the master governor of the cell's fate is influenced not only by normal proteins but also by oncogene proteins that insert themselves into various signaling pathways and disrupt normal control mechanisms. Similarly, the deletion of key tumor suppressor proteins evokes equally profound changes in the control circuitry and thus is equally influential in perturbing decision-making by the cell cycle clock. Consequently, sooner or later, the molecular actions of most oncogenes and tumor suppressor genes must be explained in terms of their effects on the cell cycle clock. To this end, we will devote the first half of this chapter to a description of how this molecular machine normally operates and then proceed to study how it is perturbed in human cancer cells.

8.1 External signals influence a cell's decision to enter into the active cell cycle

When placed into culture under conditions that encourage exponential multiplication, mammalian cells exhibit a complex cycle of growth and division that is usually referred to as the **cell cycle**. A cell that has recently been formed by the processes of cell division—**mitosis** and **cytokinesis**—must decide soon thereafter whether it will once again initiate a new round of active growth and division or retreat into the nongrowing state that we previously termed G_0. As described earlier (Sections 5.1 and 6.1), this decision is strongly influenced by mitogenic growth factors in the cell's surroundings. Their presence in sufficient concentration will encourage a cell recently formed by mitosis to remain in the active growth-and-division cycle; their absence will trigger the default decision to proceed from mitosis into the G_0, quiescent state.

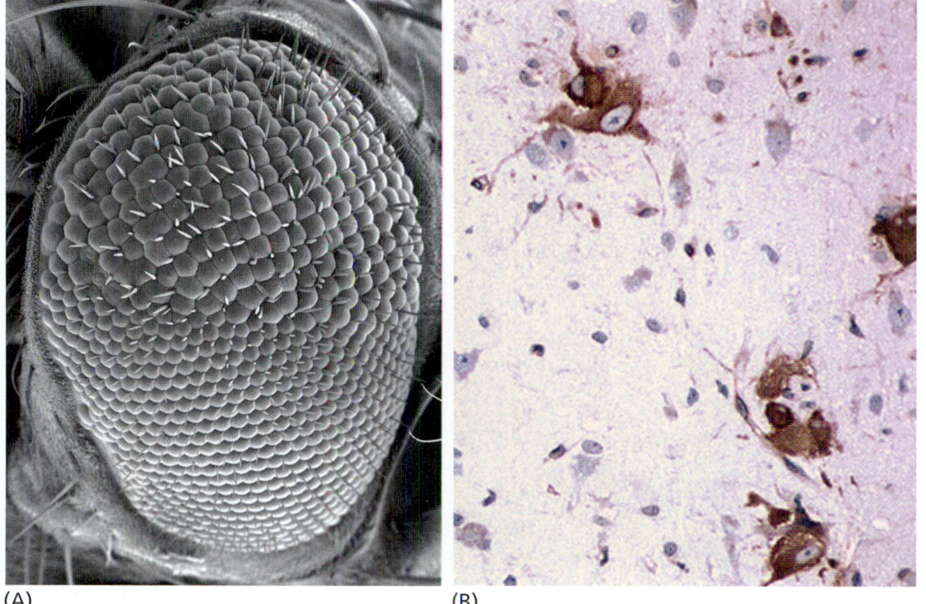

Figure 8.2 Growth versus proliferation Alterations of certain signaling proteins, such as the one encoded by the *TSC1* tumor suppressor gene, allow the processes of cell growth and division to be dissected from one another. (A) In this scanning electron micrograph of a *Drosophila* eye, the ommatidial cells in the ***upper*** portion of the eye have been deprived of the fly ortholog of *TSC1*; these cells are physically larger than the wild-type cells shown *(below)*, because they have grown more during the cell cycles that led to their formation. (B) The same behavior can be seen in the brains of patients suffering from tuberous sclerosis, in which *TSC1* function has been lost through a germ-line mutation and subsequent somatic loss of heterozygosity. Seen here are the giant cells present in a benign growth (a "tuber"). The giant cells *(brown)* are labeled with an antibody against phosphorylated S6, a ribosomal protein important in regulating protein synthesis and thus cell growth; S6 phosphorylation and functional activation is deregulated in cells lacking *TSC1* function. (A, from X. Gao and D. Pan, *Genes Dev.* 15:1383–1392, 2001; B, courtesy of J.A. Chan and D.J. Kwiatkowski.)

Withdrawal from the cell cycle may be actively encouraged by the presence of growth-inhibitory factors in the medium. Prominent among these anti-mitogenic factors is transforming growth factor-β (TGF-β). Withdrawal from the cell cycle into the G_0, quiescent state, whether due to the absence of mitogenic growth factors or the presence of anti-mitogens such as TGF-β, is often reversible, in that an encounter by a quiescent cell with mitogenic growth factors on some later occasion may induce this cell to re-enter into active growth and division. However, some cells leaving the active cell cycle may do so irreversibly, thereby giving up all option of ever re-initiating active growth and division, in which case they are said to have become *post-mitotic*. Many types of neurons in the brain, for example, are widely assumed to fall into this category.

The decision by a cell recently formed by cell division to remain in the active growth-and-division cycle requires that this cell immediately begin to prepare for the next division. Such preparations entail, among other things, the doubling of the cell's macromolecular constituents to ensure that the two daughter cells resulting from the next round of cell division will each receive an adequate endowment of them. This accumulation of cellular constituents, which drives an increase in cell size, is sometimes termed the process of cell *growth* to distinguish it from the process of cell *division*, which yields, via mitosis and cytokinesis, two daughter cells from a mother cell (see Figure 8.2). However, in the more common usage and throughout this book, the term "cell growth" implies both the accumulation of cell constituents and the subsequent cell division, that is, the two processes that together yield cell proliferation.

The accumulation of a cell's macromolecules involves, among many other molecules, the duplication of the cell's genome. In many prokaryotic cells, this duplication—the process of DNA replication—begins immediately after daughter cells are formed by cell division. But in most mammalian cells, the overall program of macromolecular synthesis is organized quite differently. While the accumulation of RNA and proteins is initiated immediately after cell division and proceeds continuously until the next cell division, the task of replicating the DNA is deferred for a number of hours (often as many as 12 to 15) after emergence of new daughter cells from mitosis and cytokinesis. During this period between the birth of a daughter cell and the subsequent onset of DNA synthesis, which is termed the G_1 (first gap) phase of the cell cycle (Figure 8.3), cells make critical decisions about growth versus quiescence, and whether, as quiescent cells, they will differentiate.

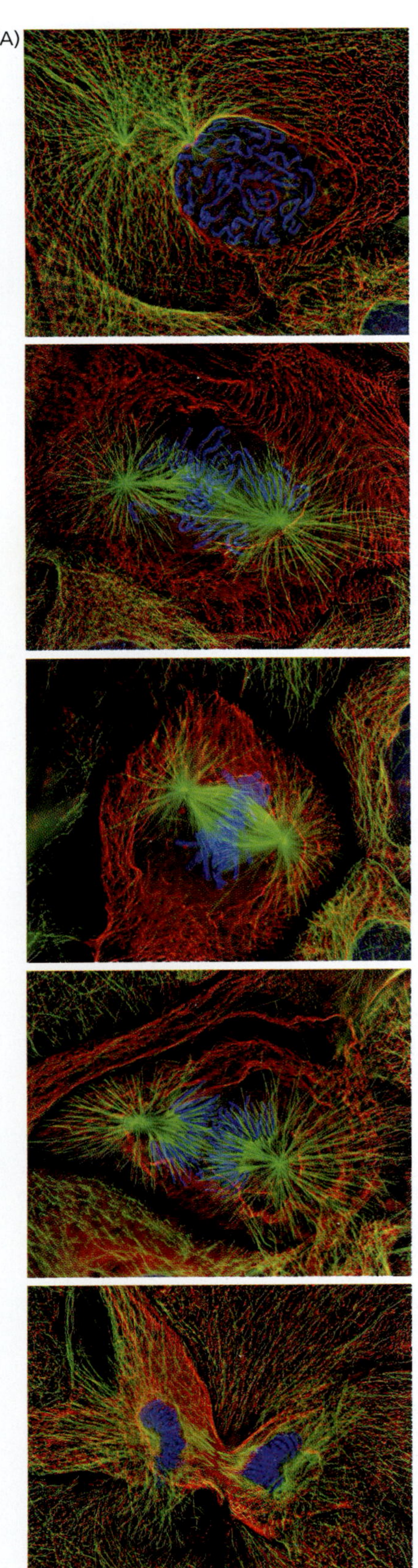

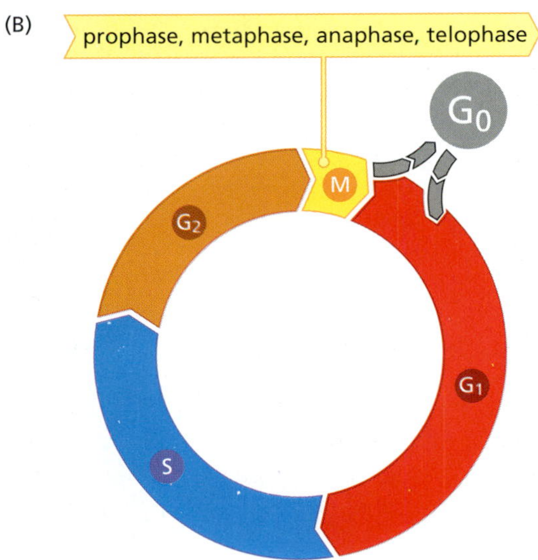

Figure 8.3 The mammalian cell cycle (A) Immunofluorescence is used here to illustrate the four distinct subphases of mitosis (M phase) in newt lung cells *(top to bottom)*. During the *prophase* of mitosis, the chromosomes (*blue*), which were invisible microscopically during *interphase* (the period encompassing G_1 through S and G_2), begin to condense and become visible under the microscope, while the *centrosomes (light green radiating bodies)* at the poles of the cell begin to assemble *(top two images)*. During *metaphase,* the chromosomes align along a plane that bisects the cell and become attached to the microtubule fibers of the mitotic spindle *(light green, third image)*. At the same time, the nuclear membrane has disappeared. During *anaphase,* the two halves of each chromosome—the chromatids—are pulled apart by the mitotic spindle (i.e., they *segregate*) to the two opposite poles of the cell (*fourth image*). During *telophase*, shortly after the chromosomes cluster into the two sets seen here (*bottom image),* the chromatids de-condense and a new nuclear membrane forms around each set of chromatids (now called chromosomes; *not shown)*. These four subphases together constitute mitosis. During the subsequent subphase—the process of *cytokinesis* (sometimes considered part of telophase, *not shown*)—the cytoplasm of the mother cell divides, yielding two daughter cells. (B) The mammalian growth-and-division cycle is divided into four phases—G_1, S (during which DNA is replicated), G_2, and M (mitosis). A fifth state, G_0 (G zero), denotes the resting, nonproliferating state of cells that have withdrawn from the active cell cycle. While exit from the active cell cycle into G_0 is depicted here as occurring in early G_1, it is unclear when during G_1 this actually occurs. (A, from Conly Rieder.)

In many types of cultured mammalian cells, the DNA synthesis that follows G_1 often requires 6 to 8 hours to reach completion. This period of DNA synthesis is termed the S (synthetic) phase, and its length is determined in part by the enormous amount of cellular DNA (~6.4×10^9 base pairs per diploid genome) that must be replicated with fidelity during this time. The actual length of S phase varies greatly among different kinds of cells, being much shorter in certain cell types, such as rapidly dividing embryonic cells and lymphocytes.

Having passed through S phase, a cell might be thought fit to enter directly into mitosis (M phase). However, most mammalian cells delay their entrance into M phase, spending 3 to 5 hours in a second gap phase, termed G_2, before they enter into M. During this G_2 period, the cell prepares itself, in some still-poorly understood fashion, for entrance into M phase and cell division. M phase itself usually encompasses an hour or so, and includes four distinct subphases—**prophase**, **metaphase**, **anaphase**, and **telophase**; this culminates in cytokinesis, the division of the cytoplasm that allows the formation of two new cells. While these times are commonly observed when studying mammalian cells in culture, they do not reflect the behavior of all cell types under all conditions. For example, actively proliferating lymphocytes may double in 5 hours, and some cells in the early embryo may do so even more rapidly.

As is the case with S phase, M phase must proceed with great precision. M phase begins with the two recently duplicated DNA helices within each chromosome; these are carried in the sister **chromatids** of the chromosome, which are aligned adjacent to one another in the nucleus. The allocation during mitosis of the duplicated chromatids to the two future daughter cells must occur flawlessly to ensure that each daughter receives exactly one diploid complement of chromatids—no more and no less. Once present within the nuclei of recently separated daughter cells, these chromatids become the chromosomes of the newly born cells.

This means that the endowment of one genome's worth of genetic material to each daughter cell depends on the precise execution of two processes—the faithful replication of a cell's genome during S phase, and the proper allocation of the resulting duplicated DNAs to daughter cells during M phase. As will be discussed later, defects in either of these processes can have disastrous consequences for the cell and the organism, one of which is the disease of cancer.

Like virtually all machinery, the machine that executes the various steps of the cell cycle is subject to malfunction. This fallibility contrasts with the stringent requirement of the cell to have the various phases of the cell cycle proceed flawlessly. For this reason, the cell deploys a series of surveillance mechanisms that monitor each step in cell cycle progression and permit the cell to proceed to the next step in the cycle only if a prerequisite step has been completed successfully. In addition, if specific steps in the execution of a process go awry, these monitors rapidly call a halt to further advance through the cell cycle until these problems have been successfully addressed. Yet other monitors ensure that once a particular step of the cell cycle has been completed, it is not repeated until the cell passes through the next cell cycle. These monitoring mechanisms are termed variously **checkpoints** or **checkpoint controls** (Figure 8.4).

One checkpoint ensures that a cell cannot advance from G_1 into S if the genome is in need of repair. Another, operating in S phase, will slow or pause DNA replication in response to DNA damage. (In mammalian cells, this may

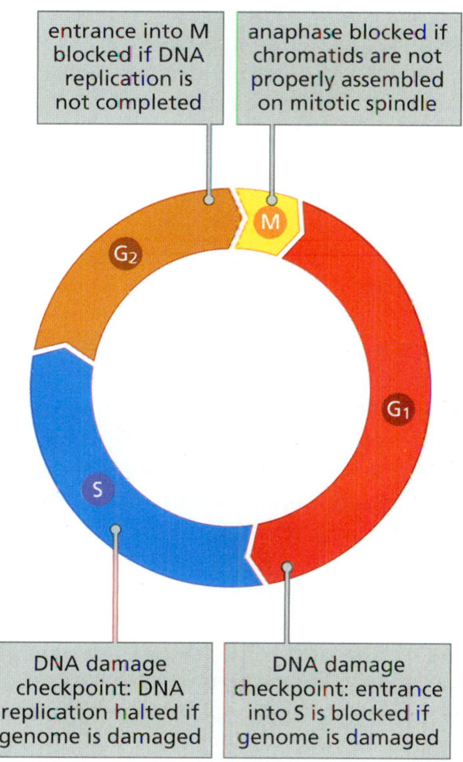
entrance into M blocked if DNA replication is not completed
anaphase blocked if chromatids are not properly assembled on mitotic spindle
M
G2
G1
S
DNA damage checkpoint: DNA replication halted if genome is damaged
DNA damage checkpoint: entrance into S is blocked if genome is damaged

Figure 8.4 Examples of checkpoints in the cell cycle Checkpoints impose quality control to ensure that a cell has properly completed all the requisite steps of one phase of the cell cycle before it is allowed to advance into the next phase. A cell will not be permitted to enter into S phase until all the steps of G_1 have been completed. It will be blocked from entering G_2 until all of its chromosomal DNA has been properly replicated. Similarly, a cell is not permitted to enter into anaphase (when the paired chromatids are pulled apart) until all of its chromosomes are properly assembled on the mitotic spindle during metaphase. In addition, a cell is not allowed to advance into S or M if its DNA has been damaged and not yet repaired. Other controls *(not shown)* ensure that once a specific step in the cell cycle has been completed, it is not repeated until the next cell cycle.

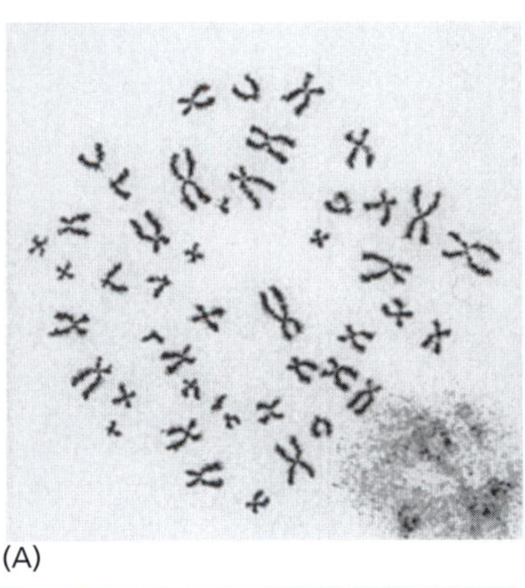

(A)

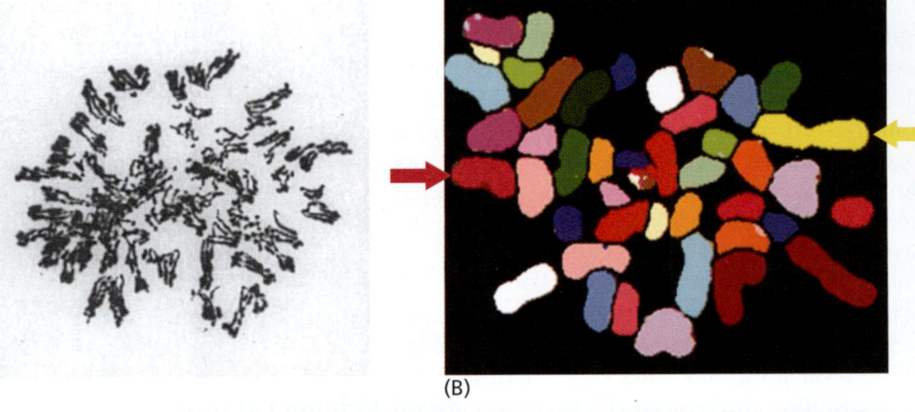

(B)

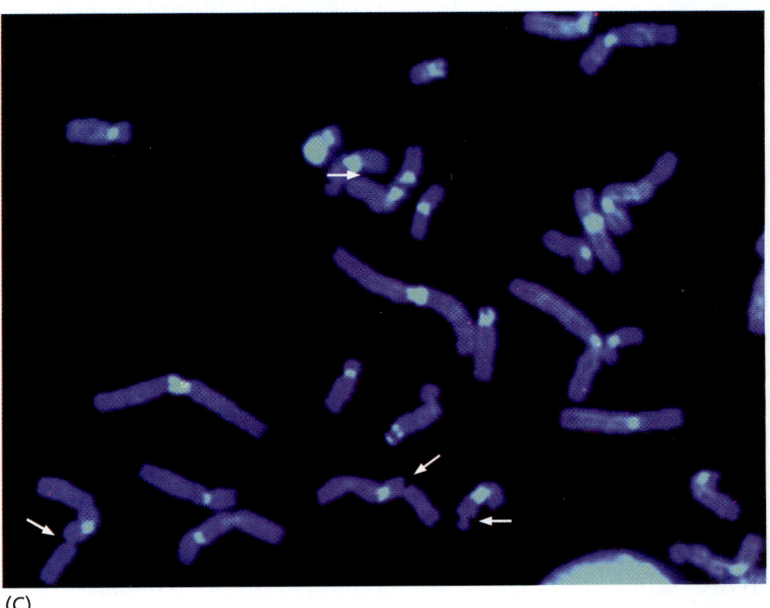

(C)

Figure 8.5 Consequences of loss of checkpoint controls Loss of critical cell cycle checkpoint control mechanisms is often manifested in the altered karyotype of cells. (A) The normal human karyotype *(left)* is contrasted with that of a cell that has been deprived of the Rad17 checkpoint protein *(right)*, which is responsible for preventing the inadvertent re-replication of already-replicated chromosomal DNA, resulting in *endoreduplication*. (B) The Bub1 protein normally prevents separation of chromosomes in the event that one or more chromosomal pairs are not properly aligned on the metaphase plate. In its absence, cells gain or lose chromosomes, as seen in this spectral karyotyping (SKY) analysis, which indicates that this human cell has only one Chromosome 1 *(yellow, arrow, upper right)* and one Chromosome 6 *(red, arrow, left middle)*. (C) The ATR (ataxia-telangiectasia and Rad3-related protein (ATR) kinase is responsible for, among other functions, halting further DNA replication until stalled replication forks are repaired. In its absence, fragile sites in the chromosome—breaks in certain sites in the chromosome—become visible upon karyotypic analysis. Here, fragile sites on human Chromosomes 3 and 16 *(white arrows)* are apparent in cells lacking ATR function. (A, from X. Wang et al., *Genes Dev.* 17:965–970, 2003; B, from A. Musio et al., *Cancer Res.* 63:2855–2863, 2003; C, from A.M. Casper et al., *Am. J. Hum. Genet.* 75:654–660, 2004.)

cause a doubling of the time required to complete DNA synthesis.) A third will not permit the cell to proceed through G_2 to M until the DNA replication of S phase has been completed. DNA damage will trigger another checkpoint control that blocks entrance into M phase. During M phase, highly efficient checkpoint controls block anaphase; these blocks are only removed once all the chromosomes have been properly attached to the mitotic spindle. Yet other checkpoint controls, not cited here, have been reported. For example, a **decatenation** checkpoint in late G_2 prevents entrance into M until the pair of DNA helices replicated in the previous S phase have been untangled from one another. Defects in some of these checkpoint controls can be observed because of their effects on cells' chromosomes (Figure 8.5).

The operations of these checkpoints also influence the formation of cancers. As tumor development (often called tumor progression, discussed in Chapter 11) proceeds, incipient cancer cells benefit from experimenting with various combinations and permutations of mutant alleles, trying out many of these in order to see which will afford them the greatest proliferative advantage. An increased mutability of their genomes accelerates the rate at which these cells can acquire advantageous combinations of alleles and thus hastens the overall pace of tumor progression. Such mutability and resulting genomic instability is incompatible with normal cell cycle progression, since checkpoint controls usually block advance of the cell through its cycle if its DNA has been damaged or its chromosomes are in disarray. So, in addition to acquiring altered growth-controlling genes (i.e., activated oncogenes and inactivated tumor suppressor genes), many types of cancer cells have inactivated one or more of their checkpoint controls. With these controls relaxed, incipient cancer cells can more rapidly accumulate the mutant genes and altered karyotypes that propel their neoplastic growth. The breakdown of the controls responsible for maintaining the cellular genome in an intact state is one of the main subjects of Chapter 12.

8.2 Cells make decisions about growth and quiescence during a specific period in the G_1 phase

As mentioned earlier, it is likely that virtually all normal cell types in the body require external signals, such as those conveyed by mitogenic growth factors, before they will undertake to grow and divide. The only exceptions to this rule appear to be very early embryonic cells, which seem able to proliferate without receiving signals that are specifically growth-stimulating (Sidebar 8.1). The rationale for this behavior of the normal cells throughout our tissues is a simple one: since these cells participate in the formation of precisely structured tissues, their proliferation must, by necessity, be coordinated with neighboring cells in those tissues. Put differently, the body cannot give each of its almost 10^{14} component cells the license to decide on its own whether to grow and divide. To do so would invite chaos.

Evidence accumulated over the past quarter century indicates that cells consult their extracellular environment and its growth-regulating signals during a discrete window of time in the active cell cycle, namely, from the onset of G_1 phase through most of G_1, ending an hour or two before the G_1-to-S transition (Figure 8.6). The operations of the G_1 decision-making machinery are indicated by the responses of cultured cells to extracellular signals. If we were to remove serum and thus growth factors from cells before they had completed 80 to 90% of G_1, they would fail to proceed further into the cell cycle and would, with great likelihood, revert to the G_0 state. However, once these cells had moved through this G_1 decision-making period and advanced into the final hours of G_1 (the remaining 10 to 20% of G_1), the removal of serum would no longer affect their progress and they would proceed through the remainder of G_1 and thereafter all the way through the S, G_2, and M phases. Similarly, anti-mitogenic factors, such as TGF-β, are able to impose their growth-inhibitory effects only during this period in the early and mid-G_1 phase. Once a cell has entered into late G_1, it seems to be oblivious to the presence of this negative factor in its surroundings.

This schedule of total dependence on extracellular signals followed by entrance in late G_1 into a state of relative independence indicates that a weighty decision must be made toward the end of G_1. Precisely at this point, a cell must make up

Sidebar 8.1 Embryonic stem (ES) cells show highly autonomous behavior Our perceptions about the behavior of normal mammalian cells have been conditioned by decades of work with a wide variety of somatic cells present in embryonic and adult tissues. However, cells in the very early embryo clearly operate under a very different set of rules. The pRb pathway and the cell cycle clock machinery to be described in this chapter appear to be operative in one form or another in virtually all types of adult cells. In contrast, a variety of experiments indicate that pRb-imposed growth control is not functional in early embryonic cells, including their cultured derivatives, embryonic stem (ES) cells. The same can be said about the p53 pathway (see Chapter 9). It seems that the mitogenic signals required to keep more differentiated cells proliferating are not required by cultured ES cells for their proliferation. For example, aside from the growth factor termed LIF (leukemia inhibitory factor), which is needed to prevent their differentiation, mouse ES cells proliferating *in vitro* appear able to drive their own proliferation through internally generated signals. (Indeed, a constitutively activated Ras-like protein, termed E-Ras, has been reported to be expressed specifically in these cells.)

ES cells seem to preserve much of the cell-autonomous behavior that we associate with the single-cell ancestors of metazoa, that is, the behavior of cells that have not yet become dependent on their neighbors for signals controlling growth and survival. The most stunning indication of their extreme autonomy is their ability to form benign tumors (**teratomas**) when introduced into many anatomical sites in an adult organism. Since these cells are genetically wild type, they represent the only example of a wild-type cell that is tumorigenic.

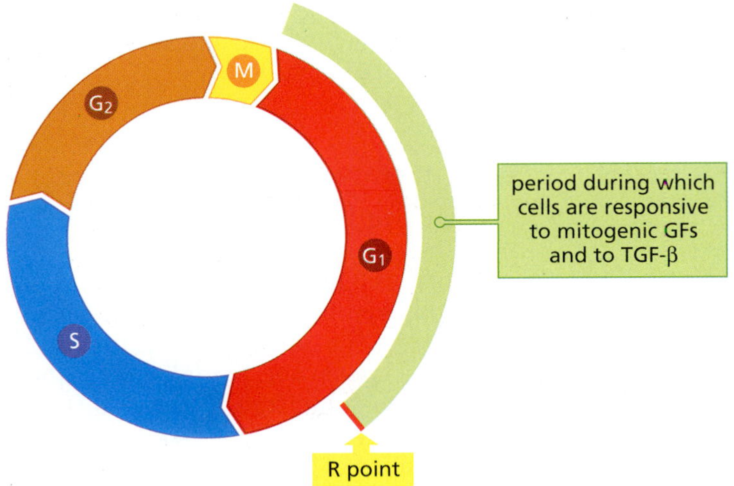

Figure 8.6 Responsiveness to extracellular signals during the cell cycle Cells respond to extracellular mitogens and inhibitory factors (such as TGF-β) only in a discrete window of time that begins at the onset of G_1 and ends just before the end of G_1. The end of this time window is designated the restriction (R) point, which denotes the point in time when the cell must make the commitment to advance through the remainder of the cell cycle through M phase, to remain in G_1, or to retreat from the active cell cycle into G_0.

its mind whether it will remain in G_1, retreat from the active cycle into G_0, or advance into late G_1 and thereafter into the remaining phases of the cell cycle. This critical decision is made at a transition that has been called the **restriction point** or **R point** (see Figure 8.6). In most mammalian cells studied to date, the R point occurs several hours before the G_1/S phase transition.

If a cell should decide at the R point to continue advancing through its growth-and-division cycle, it commits itself to proceed beyond G_1 into S phase and then to complete a rigidly programmed series of events (the entire S, G_2, and M phases) that enable it to divide into two daughter cells. This decision will be respected even if growth factors are no longer present in the extracellular space during these remaining phases of the cell cycle. We know that this series of later steps (S, G_2, and M phases) proceeds on a fixed schedule, because a cell that enters S will, in the absence of a major disaster, invariably complete S and, having done so, enter into G_2 and then advance into M phase.

For those interested specifically in understanding the deregulated proliferation of cancer cells, this fixed program holds relatively little interest, since the late $G_1 \rightarrow S \rightarrow G_2 \rightarrow M$ progression proceeds similarly in normal cells and cancer cells. Students of cancer therefore focus largely on the G_0/G_1 transition and on the one period in the life of an actively growing cell—the time window encompassing most of G_1—when a cell is given the license to make decisions about its fate.

The commitment to advance through the R point and continue all the way to M phase is, as hinted, not an absolute one. Metabolic, genetic, or physical disasters may intervene during S, G_2, or M and force the cell to call a halt, often temporary, to its further advance through the cell cycle until these conditions have been addressed. Still, in the great majority of cases, cells in living tissues seem to succeed in avoiding these various disasters. This leaves the R-point decision as the critical determinant of whether cells will grow or not. An increasingly large body of evidence indicates that deregulation of the R-point decision-making machinery accompanies the formation of most if not all types of cancer cells. Yet other decision points in late G_1 may also contribute to the deregulated proliferation of certain cancer cells (Sidebar 4).

8.3 Cyclins and cyclin-dependent kinases constitute the core components of the cell cycle clock

The existence of the R point leaves us with two major questions that we will spend much of this chapter answering. First, what is the nature of the molecular machinery that decides whether or not a cell in G_1 will continue to advance through the cell cycle or will exit into a nongrowing state? Second, how does this machinery, which we call the cell cycle clock, implement these decisions once they have been made? We begin with the second question and will return later to the first.

As described earlier (Chapters 5 and 6), when signals are broadcast from a single master control protein to many downstream responders, the signal-emitting functions are often delegated to protein kinases. These enzymes are ideally suited to this task. By phosphorylating multiple distinct targets (i.e., substrates), a kinase can create covalent modifications that serve to switch on or off various activities inherent in these substrate proteins. Indeed, the cell cycle clock uses a group of protein kinases to execute the various steps of cell cycle progression. For example, phosphorylation of **centrosome**-associated proteins at the G_1/S boundary allows their duplication in preparation for M phase. Phosphorylation of other proteins prior to S phase enables DNA replication sites along the chromosomes to be activated. Phosphorylation of histone proteins in anticipation of S and M phases places the chromatin in configurations that permit these two phases to proceed normally. And the phosphorylation of proteins forming the nuclear membrane (sometimes called the *nuclear envelope*), such as lamin and nucleoporins, causes their dissociation and the dissolution of this membrane early in M phase.

The kinases deployed by the cell cycle machinery are called collectively **cyclin-dependent kinases** (CDKs) to indicate that these enzymes never act on their own; instead, they depend on associated regulatory subunits, the **cyclin** proteins, for proper functioning. Bimolecular complexes of CDKs and their cyclin partners are responsible for sending out the signals from the cell cycle clock to dozens if not hundreds of responder molecules that carry out the actual work of moving the cell through its growth-and-division cycle.

The CDKs are serine/threonine kinases, in contrast to the tyrosine kinases that are associated with growth factor receptors and with nonreceptor kinase molecules like Src. The CDKs show about 40% amino acid sequence identity with one another and therefore are considered to form a distinct subfamily within the large (approximately 430) throng of ser/thr kinases encoded by the human genome. The cyclins associated with the CDKs activate the catalytic activity of their CDK partners (Figure 8.7). (In the well-studied example of the binding of cyclin A to CDK2, the association of the two proteins increases the enzymatic activity of CDK2 a staggering 400,000-fold!) At the same time, the cyclins serve as guide dogs for the CDKs by helping the cyclin–CDK complexes recognize appropriate protein substrates in the cell. The cyclins, for their part, also constitute a distinct family of cellular proteins that share in common an approximately 100–amino acid residue–long domain that is involved in the binding and functional activation of CDKs.

It is actually cyclin–CDK complexes that constitute the engine of the cell cycle clock machinery. During much of the G_1 phase of the cell cycle, two similarly acting CDKs—CDK4 and CDK6—are guided by and depend upon their association with a trio of related cyclins (D1, D2, and D3) that collectively are called the D-type cyclins (Figure 8.8). After the R point in late G_1 the E-type cyclins (E1 and E2) associate with CDK2 to enable the phosphorylation of appropriate substrates required for entry into S phase. As cells enter into S phase, the A-type cyclins (A1 and A2) replace E cyclins as the partners of CDK2 and thereby enable S phase to progress (see Figure 8.7). Later in S phase, the A-type cyclins switch partners, leaving CDK2 and associating instead with another CDK called either CDC2 or CDK1. (We will use CDC2 here.) As the cell moves further, into G_2 phase, the A-type cyclins are replaced as CDC2 partners by the B-type cyclins (B1 and B2). Finally, at the onset of M phase, the complexes of CDC2 with the B-type cyclins trigger many of the events of the prophase, metaphase, anaphase, and telophase that together constitute the complex program of mitosis.

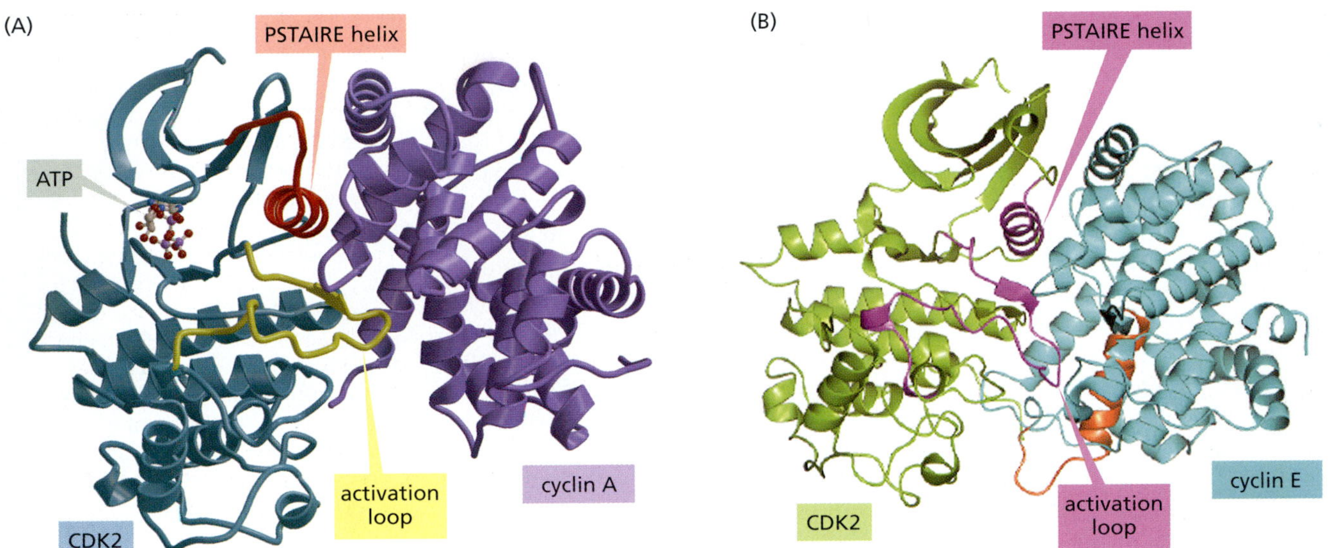

Figure 8.7 Cyclin–CDK complexes X-ray crystallographic analyses have revealed the structures of cyclin–cyclin-dependent kinase (CDK) complexes, such as these formed by CDK2 with two of its alternative partners, cyclins A and E. In each case, the cyclin activates the CDK molecule through stereochemical shifts of the CDK catalytic site and directs the resulting catalytically activated complex to appropriate substrates for phosphorylation. (A) The PSTAIRE α-helix *(red)* is present in all CDKs and is essential for binding of cyclins. The activation loop *(yellow)*, sometimes termed a T-loop, must be phosphorylated on a threonine residue by a CDK-activating kinase (CAK) in order for the catalytic function of a CDK to become activated. Cyclin A *(purple)* directs CDK2 *(aquamarine)* to substrates that must be phosphorylated in order for S phase to proceed. (B) During the late G_1 phase of the cell cycle, cyclin E directs CDK2 to substrate proteins that must be phosphorylated in preparation for entrance into S phase. The conserved PSTAIRE α-helix and the activation loop of CDK2 are shown here in *magenta*. Another segment, *(orange)* is involved, in unclear ways, with cyclin E's association with centrosomes. (A, from P.D. Jeffrey et al., *Nature* 376:313–320, 1995; B, from R. Honda et al., *EMBO J.* 24:452–463, 2005.)

As is the case with all well-regulated systems, the activities of the various cyclin–CDK complexes must be modulated in order to impose control on specific steps in the cell cycle. The most important way of achieving this regulation depends upon changing the levels and availability of cyclins during various phases of the cell cycle. In contrast, the levels of most CDKs vary only minimally.

The first insights into cyclin and CDK control came from studies of the governors of mitosis in early frog and sea urchin embryos. As these experiments

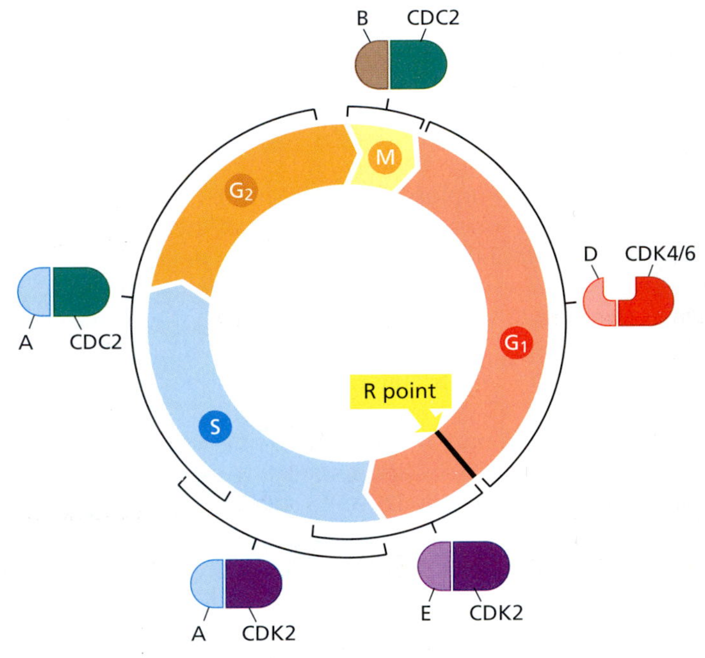

Figure 8.8 Pairing of cyclins with cyclin-dependent kinases Each type of cyclin pairs with a specific cyclin-dependent kinase (CDK) or set of CDKs. The D-type cyclins (D1, D2, and D3) bind CDK4 or CDK6, the E-type (E1 and E2) bind CDK2, the A-type cyclins (A1 and A2) bind CDK2 or CDC2, and the B-type cyclins (B1 and B2) bind CDC2. The brackets indicate the periods during the cell cycle when these various cyclin–CDK complexes are active.

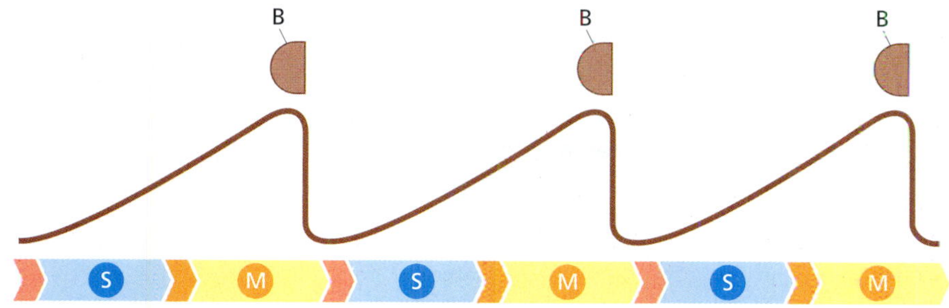

Figure 8.9 Cell cycle–dependent fluctuations in cyclin B levels The cyclic fluctuations in the levels of cyclin B in early frog and sea urchin embryos gave cyclins their name. These fluctuations were noticeable because the cell cycles in these early embryos are synchronous, i.e., all cells enter into M phase simultaneously. In these early embryos, the G_1 and G_2 phases of the cell cycle *(orange, pink chevrons)* are virtually absent and the cells, in effect, alternate between M and S phases. (Although cyclin B levels are already substantial prior to the onset of M phase, cyclin B molecules are unable to form catalytically active B–CDC2 complexes until the G_2-to-M transition.)

showed, levels of B-type cyclins increase strongly in anticipation of mitosis, allowing B cyclins and CDC2 to form complexes that initiate entrance into M phase. At the end of M phase, cyclin B levels plummet because of the scheduled degradation of this protein. Early in the next cell cycle, cyclin B is virtually undetectable in cells, and accumulates gradually later in this cycle in anticipation of the next M phase. Because the growth-and-division cycles of all of the cells in these early embryos are **synchronous** (i.e., take place coordinately), all cells in the embryo go through S and M at the same time. This results in repeated rounds of cycling of the levels of these cyclin proteins—the behavior that inspired their name (Figure 8.9).

This theme of dramatic cell cycle phase–dependent changes in the levels of cyclin B is repeated by other cyclins as well. Cyclin E levels rise abruptly after a cell has progressed through the R point and collapse as the cell enters S phase (Figure 8.10), while cyclin A increases in concert with the cell's entrance into S phase. (While there are at least two subtypes of cyclin A, cyclin B, and cyclin E, we will refer to these simply as cyclins A, B, and E, respectively, since the two subtypes of each of these cyclins appear to operate identically.)

The collapse of various cyclin species as cells advance from one cell cycle phase to the next is due to their rapid degradation, this being triggered by the actions of highly coordinated ubiquitin ligases, which attach polyubiquitin chains to these cyclins. As described in Sidebar 7.8, this polyubiquitylation leads to proteolytic breakdown in the proteasomes. The cyclins' gradual accumulation followed by their rapid destruction has an important functional consequence for the cell cycle, because it dictates that the cell cycle clock can move in only one direction, much like a ratchet. This ensures, for example, that cells that have exited one M phase cannot inadvertently slip backward into another one, but instead must advance through G_1, S, and G_2 until they have accumulated the B cyclins required for entrance, once again, into M-phase.

The sole exception to these well-programmed fluctuations in the levels of cyclins is presented by the D-type cyclins. The levels of these three, similarly structured cyclins are not found to vary dramatically as a cell advances through the various phases of its growth-and-division cycle. Instead, the levels of D-type cyclins are controlled largely by extracellular signals, specifically those conveyed

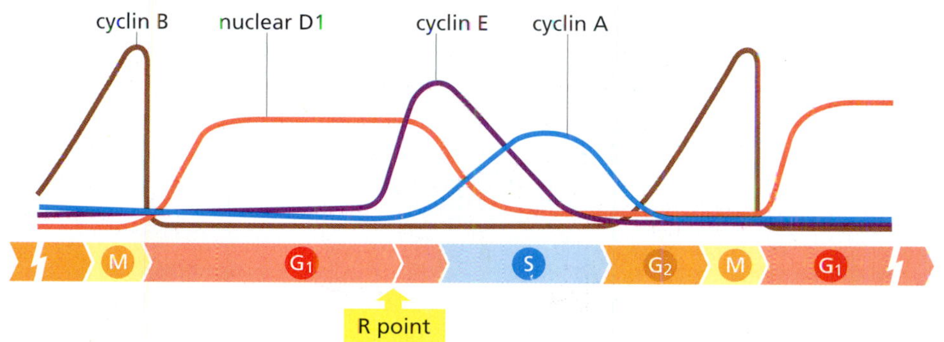

Figure 8.10 Fluctuation of cyclin levels during the cell cycle The levels of most of the mammalian cyclins fluctuate dramatically as cells progress through the various phases of the cell cycle. In the cases of most of these cyclins, these fluctuations are tightly coordinated with the schedule of advances through the various cell cycle phases. However, in the case of the D-type cyclins, extracellular signals, notably those conveyed by growth factors, strongly influence their levels. (While cyclin D1—and possibly other D-type cyclins—is present in other cell cycle phases besides G_1, following the G_1/S transition it is exported from the nucleus into the cytoplasm, where it can no longer influence cell cycle progression.)

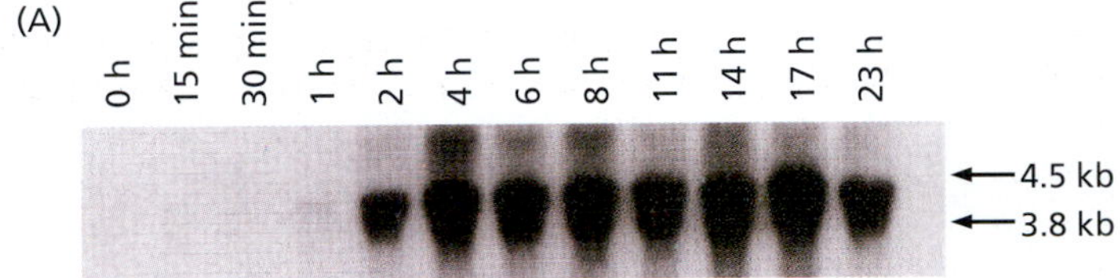

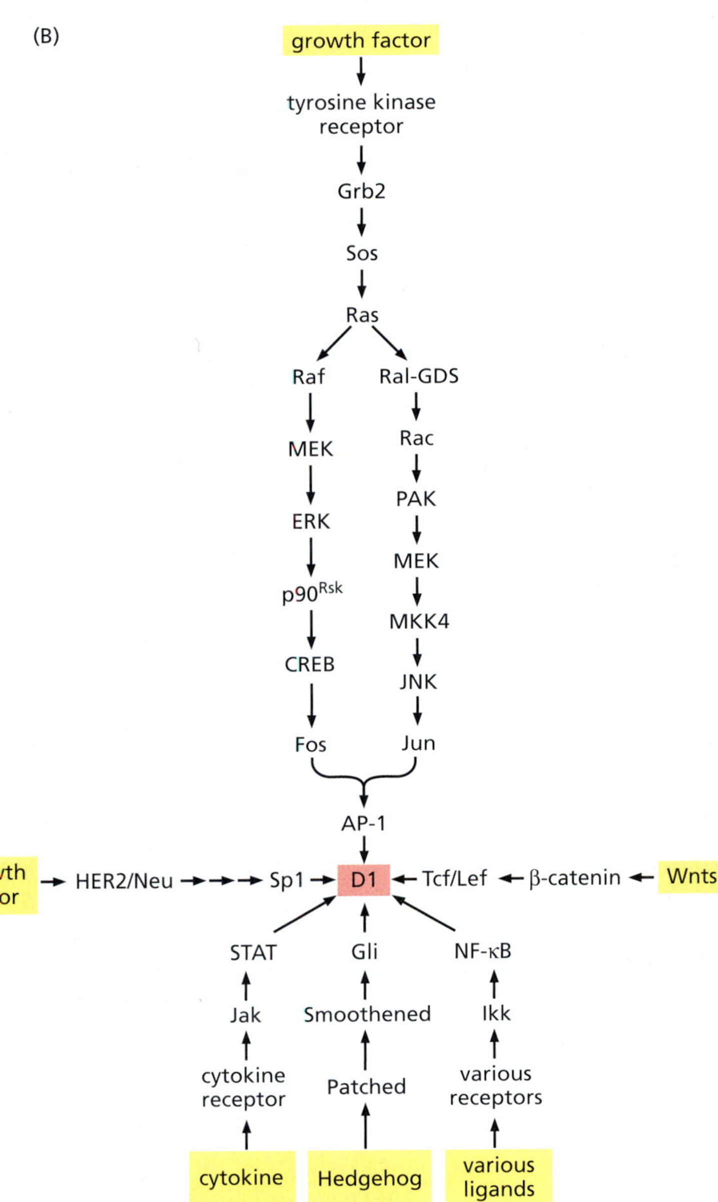

Figure 8.11 Control of cyclin D1 levels (A) Cyclin D1 was discovered as a protein whose levels are strongly induced by exposure of macrophages to the mitogen CSF-1 (colony-stimulating factor-1). In this case, a line of macrophages that had been starved of CSF-1 were exposed to fresh CSF-1 and the amounts of cyclin D1 mRNA at subsequent time periods were determined by RNA (*Northern*) blotting. (B) The control of cyclin D1 levels by extracellular mitogens can be explained, in part, by a signal-transduction cascade that leads from growth factor receptors (RTKs) to the AP-1 transcription factor, one of a number of factors that modulate the transcription of the *cyclin D1* gene in the nucleus. As indicated, a number of other signaling cascades converge on the promoter of this gene. (A, from H. Matsushime et al., *Cell* 65:701–713, 1991.)

by a variety of mitogenic growth factors. In the case of cyclin D1—the best-studied of the three D-type cyclins—growth factor activation of tyrosine kinase receptors and the resulting stimulation of several downstream signaling cascades results in rapid accumulation of cyclin D1 (Figure 8.11). Conversely, removal of growth factors from a cell's medium results in an equally rapid collapse of its cyclin D1 levels with a half-life ($T_{1/2}$) of about 30 minutes.

The distinctive behavior of the D-type cyclins has been rationalized as follows: They serve to convey signals from the extracellular environment to the cell cycle clock operating in the cell nucleus. Because the levels of D-type cyclins

Table 8.1 Induction of D-type cyclin expression by extracellular signals

Source of signal	Signaling intermediaries	Type of cyclin
RANK receptor	NF-κB pathway	D1
Prolactin receptor	Jak/STAT	D1
Estrogen receptor	AP-1 TF (?)	D1
Focal adhesion kinase		D1
HER2/Neu receptor	E2F and Sp1 TFs	D1
Wnts–Frizzled receptor	β-catenin and Tcf/Lef TFs	D1
Bcr/Abl		D2
FSH receptor	cyclic AMP	D2
Various mitogens	Myc	D2
Interleukin-4, 7 receptor		D2
Interleukin-5 receptor	STAT3/5	D3
	E2A TF	D3

Abbreviation: FSH, follicle-stimulating hormone.

fluctuate together with the levels of extracellular mitogens, D-type cyclins continuously inform the cell cycle clock of current conditions in the environment around the cell.

After D-type cyclins are synthesized in the cytoplasm and migrate to the nucleus, they assemble in complexes with their two alternative CDK partners, CDK4 and CDK6. Since these two CDKs function similarly to one another, we will refer to them hereafter as CDK4/6. The cyclin D–CDK4/6 complexes seem to have similar if not identical enzymatic activities and substrate specificities, independent of whether they contain cyclin D1, D2, or D3.

These similarities provoke the question of why mammalian cells express three apparently redundant cyclins. What appears to be a redundancy actually offers the cell refined sensory input and enhanced flexibility of response. The promoter of each of the three encoding genes is under the control of a different set of signaling pathways and thus under the control of a different set of cell surface receptors (Table 8.1). For example, the promoter of the *cyclin D1* gene (properly termed, in humans, *CCND1*) carries sites for binding by the AP-1, Tcf/Lef, and NF-κB transcription factors (see Figure 8.11B), which in turn are activated by a variety of growth factor receptors. In contrast, the *cyclin D2* promoter is responsive to activation by the Myc transcription factor and to extracellular signals that stimulate increases in the concentration of intracellular cyclic adenosine monophosphate (cAMP). The *cyclin D3* gene has been found to be responsive to STAT3 and STAT5 transcription factors, and these in turn often respond to interleukin receptors active in various types of hematopoietic cells; the E2A transcription factor, which is active in programming the differentiation of lymphocytes, also controls *cyclin D3* expression.

This arrangement enables a diverse set of extracellular signals to influence the activities of CDK4/6 by controlling the levels of its various D-type cyclin partners. In addition, more recent research has indicated that certain cyclins have functions in the cell that are apparently unrelated to their role in promoting cell cycle progression (Sidebar 8.2).

Once they are formed, the cyclin D–CDK4/6 complexes are capable of ushering a cell all the way from the beginning of the G_1 phase up to and perhaps through the R-point gate. After the cell has moved through the R point, the remaining cyclins—E, A, and B—behave in a pre-programmed fashion, executing the fixed program that begins at the R point and extends all the way to the end of M phase (Figure 8.12). Indeed, once the cell has passed through its R point, its cell cycle machinery takes on a life of its own that is quite autonomous and no longer responsive to extracellular signals.

Sidebar 8.2 D-type cyclins have other jobs besides cell cycle control Two decades of cell cycle research have created the impression that cell cycle control is the sole function of cyclins. In fact, cyclin D1 has been shown to associate with both the estrogen receptor (ER) and the transcription factor C/EBPβ. By binding the ER, cyclin D1 mimics the normal ligand of this receptor—estrogen—in stimulating the receptor's transcriptional activities. The great majority (>70%) of breast cancers express the ER, and its expression explains the mitogenic effects that estrogen has on the cells in these tumors. Since cyclin D1 is overexpressed in most of these tumors and can activate this receptor, the D1–ER complexes may also play an important role in driving the proliferation of cells in these tumors. The association of cyclin D1 with C/EBPβ results in an inhibition of this transcription factor, which is thought to play a key role in programming the differentiation of a variety of cell types. Indeed, the gene expression patterns of a variety of human breast cancers give the clear indication that cyclin D1 interactions with C/EBPβ contribute in a significant way to the transcriptional profile of these cells. These associations suggest that yet other interactions of the D-type cyclins with various nuclear proteins will be found in the future.

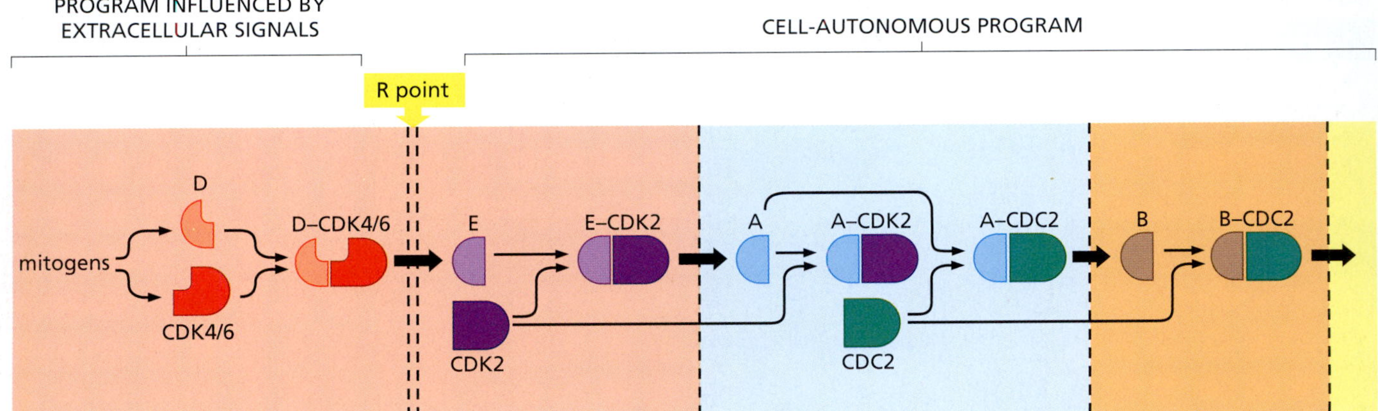

Figure 8.12 Control of cyclin levels during the cell cycle While extracellular signals strongly influence the levels of D-type cyclins during most of the G_1 phase of the cell cycle, the levels of the remaining cyclins are precisely coordinated with and control cell cycle advance. Thus, after cells pass through the R point and cyclin E–CDK2 complexes are activated, the activation of the remaining cyclin–CDK complexes occurs on a predictable schedule that appears to be independent of extracellular physiologic signals. In part, this coordination is achieved because the cyclin–CDK complexes in one phase of the cell cycle are responsible for activating those in the subsequent phase *(indicated here)* and for shutting down those that were active in the previous phase *(not shown)*.

In ways that remain poorly understood, cyclin–CDK complexes in later phases suppress the activities of the cyclin–CDK complexes that have preceded them in earlier phases of the cell cycle. For example, when cyclin A is activated by the actions of cyclin E–CDK2 complexes during the G_1/S transition, the activities of cyclin A–CDK2 result in, among other things, the inactivation of the transcription factor that served previously during the R-point transition to induce cyclin E expression. In late S and early G_2 phases, the cyclin A–CDC2 complex begins to prepare for the activation of cyclin B–CDC2 complexes that are required later for entrance into mitosis. Once the latter complexes are activated, they seem to cause a shutdown of cyclin A synthesis, and so forth.

8.4 Cyclin–CDK complexes are also regulated by CDK inhibitors

The scheme of cell cycle progression laid out above implies that physiologic signals are able to influence the activity of the cell cycle clock only through their modulation of cyclin levels. In truth, there are several other layers of control that modulate the activity of the cyclin–CDK complexes and thereby regulate advance through the cell cycle.

The most important of these additional controls is imposed by a class of proteins that are termed generically CDK inhibitors or simply CdkIs. To date, seven of these proteins have been found that are able to antagonize the activities of the cyclin–CDK complexes. A group of four of these, the INK4 proteins (named originally as **in**hibitors of CD**K4**), are targeted specifically to the CDK4 and CDK6 complexes; they have no effect on CDC2 and CDK2. These inhibitors are $p16^{INK4A}$, $p15^{INK4B}$, $p18^{INK4C}$, and $p19^{INK4D}$. The three remaining CDK inhibitors that have been discovered, $p21^{Cip1}$ (sometimes termed $p21^{Waf1}$), $p27^{Kip1}$, and $p57^{Kip2}$, are more widely acting, being able to inhibit all of the other cyclin–CDK complexes that form at later stages of the cell cycle (Figure 8.13).

The actions of these two classes of CDK inhibitors are nicely illustrated by $p15^{INK4B}$ and the pair $p21^{Cip1}$ and $p27^{Kip1}$. When TGF-β is applied to epithelial cells, it elicits a number of downstream responses from cells that antagonize cell proliferation. Among these are substantial increases in the levels of $p15^{INK4B}$, which proceeds to block the formation of cyclin D–CDK4/6 complexes (Figure 8.14) and to inhibit those that have already formed. Without active D–CDK4/6 complexes, the cell is unable to advance through early and mid-G_1 and reach the R point. Once a cell has passed through the R point, the actions of the D–CDK4/6 complexes seem to become unnecessary. This may explain why TGF-β is growth-inhibitory during early and mid-G_1 and loses most (perhaps all) of its growth-inhibitory powers once a cell has passed through the R point.

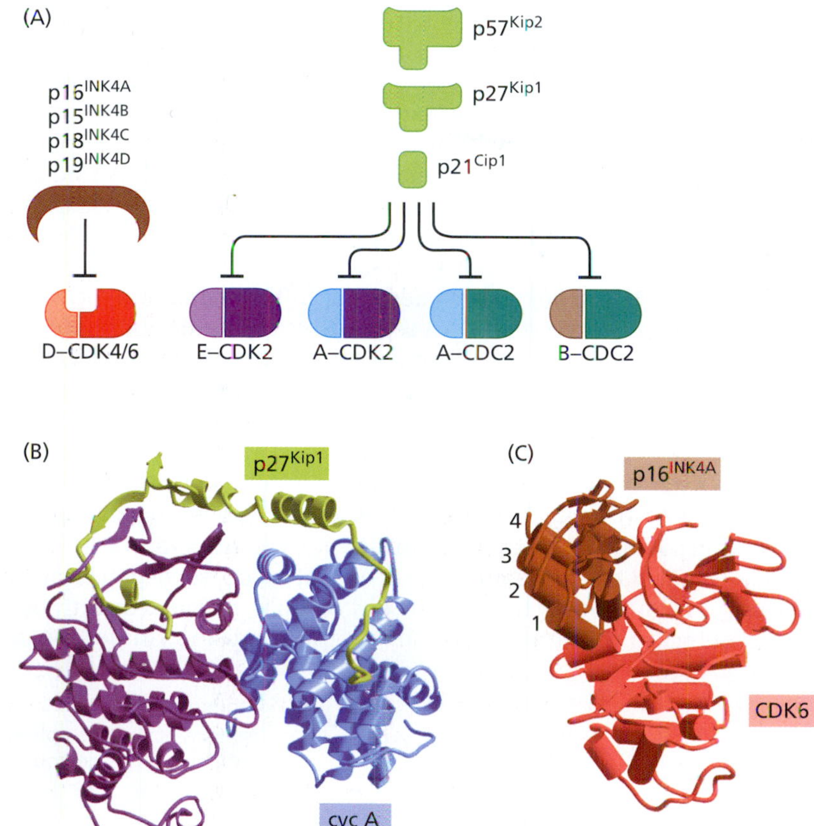

Figure 8.13 Actions of CDK inhibitors (A) The CDK inhibitors block the actions of CDKs at various points in the cell cycle. The four INK4 proteins (p16^{INK4A}, p15^{INK4B}, p18^{INK4C}, and p19^{INK4D}) are specialized to inhibit the D–CDK4 and D–CDK6 complexes that are active in early and mid-G_1. Conversely, the three Cip/Kip CdkIs (p21^{Cip1}, p27^{Kip1}, and p57^{Kip2}) can inhibit the remaining cyclin–CDK complexes that are active throughout the cell cycle. (Relatively little is known about the actions of p57^{Kip2}.) Paradoxically, p21^{Cip1} and p27^{Kip1} are known to *promote* formation of D–CDK4/6 complexes during the G_1 phase of the cell cycle (*not shown here; see* Figure 8.17). (B) This depiction of the complex between p27^{Kip1} and cyclin A–CDK2, derived by X-ray crystallography (see also Figure 8.7), illustrates how one domain of p27^{Kip1} *(light green)* blocks cyclin A–CDK2 function by obstructing the ATP-binding site in the catalytic cleft of the CDK. (C) Inhibitors of the INK4 class, such as p16^{INK4A} shown here *(reddish brown)*, bind to CDK6 *(red)* and to CDK4 *(not shown)*. These CDK inhibitors distort the cyclin-binding site of CDK6, reducing its affinity for D-type cyclins. At the same time, they distort the ATP-binding site and thereby compromise catalytic activity. Identical interactions are likely to characterize the responses of CDK4 to p16^{INK4A}. (B, from A.A. Russo et al., *Nature* 382:325–331, 1996; C, from A.A. Russo et al., *Nature* 395:237–243, 1998.)

p21^{Cip1}, a more widely acting CDK inhibitor, is also induced by TGF-β, albeit weakly. Far more important are increases in the levels of p21^{Cip1} that occur in response to various physiologic stresses (see Figure 8.14A); once present at significant levels, p21^{Cip1} can act throughout much of the cell cycle to stop a cell in its tracks. Prominent among these stresses is damage to the cell's genome. As

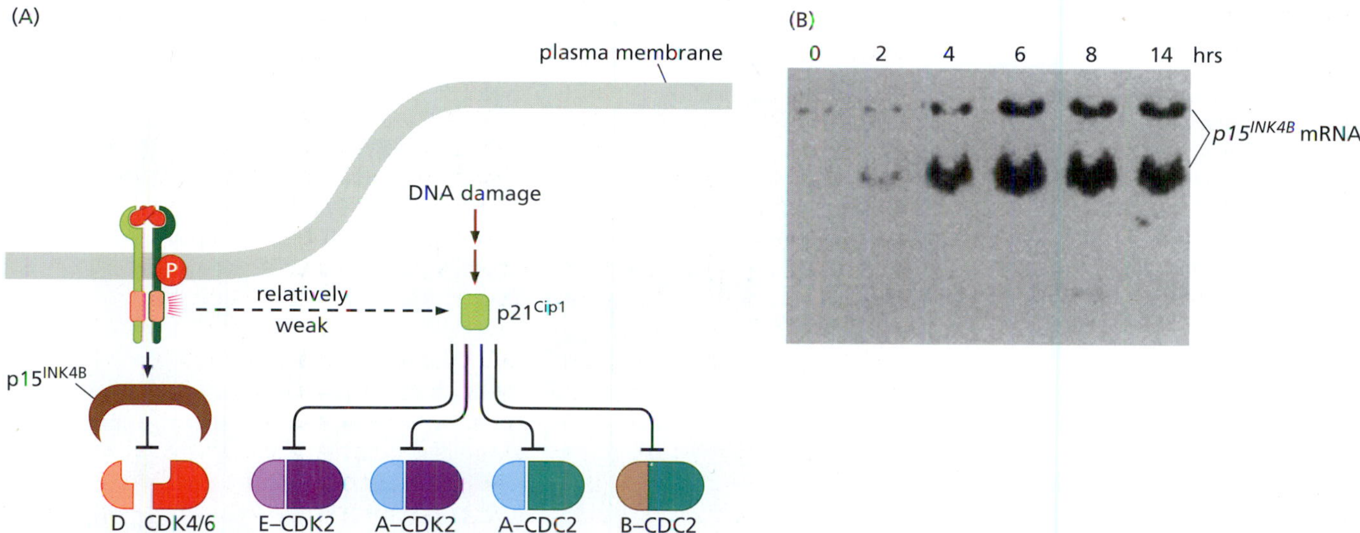

Figure 8.14 Control of cell cycle progression by TGF-β (A) TGF-β *(top left)* controls the cell cycle machinery in part through its ability to modulate the levels of CDK inhibitors. It acts to strongly induce increased expression of p15^{INK4B} and, weakly, of p21^{Cip1}. The former can block the actions of cyclin D–CDK4/6 complexes, while the latter can block the actions of the remaining cyclin–CDK complexes that are active throughout the remainder of the cell cycle. Independent of this, damage to cellular DNA causes strong, rapid increases in p21^{Cip1}, which in turn can shut down the cyclin–CDK complexes that are active in the phases of the cell cycle after the cell has passed through the R point in the late G_1 phase. (B) When TGF-β is applied to human keratinocytes, it evokes a dramatic 30-fold induction of p15^{INK4B} mRNA synthesis as demonstrated here by RNA (Northern) blotting analysis. These cells were exposed to TGF-β for the time periods (in hours) indicated *(above)*. (B, from G.J. Hannon and D. Beach, *Nature* 371:257–261, 1994.)

long as the genomic DNA remains in an unrepaired state, the p21^{Cip1} that has been induced will shut down the activity of already-formed cyclin–CDK complexes—such as E–CDK2, A–CDK2, A–CDC2, and B–CDC2—that happened to be active when this damage was first incurred; once the damage is repaired, the p21^{Cip1}-imposed block may then be relieved. Such a strategy makes special sense in G_1: if a cell's genome becomes damaged during this period through the actions of various mutagenic agents, p21^{Cip1} will block advance through the R point (by inhibiting E–CDK2 complexes) until the damage has been repaired, ensuring that the cell does not progress into S phase and inadvertently copy still-damaged DNA sequences. In addition, p21^{Cip1} can inhibit the functions of a key component of the cell's DNA replication apparatus, termed PCNA (proliferating-cell nuclear antigen); this ensures that already-initiated DNA synthesis is halted until DNA repair has been completed. We will return to the mechanisms controlling p21^{Cip1} expression in the next chapter (Section 9.9).

While DNA damage and, to a much lesser extent, TGF-β can elicit increases in the levels of p21^{Cip1} (thereby blocking cell cycle advance), mitogens act in an opposing fashion to mute the actions of this CDK inhibitor and in this way favor cell cycle advance. One mechanism by which they do so depends on the phosphatidylinositol 3-kinase (PI3K) pathway, which is activated directly or indirectly by the mitogens that stimulate many receptor tyrosine kinases (Figure 8.15A).

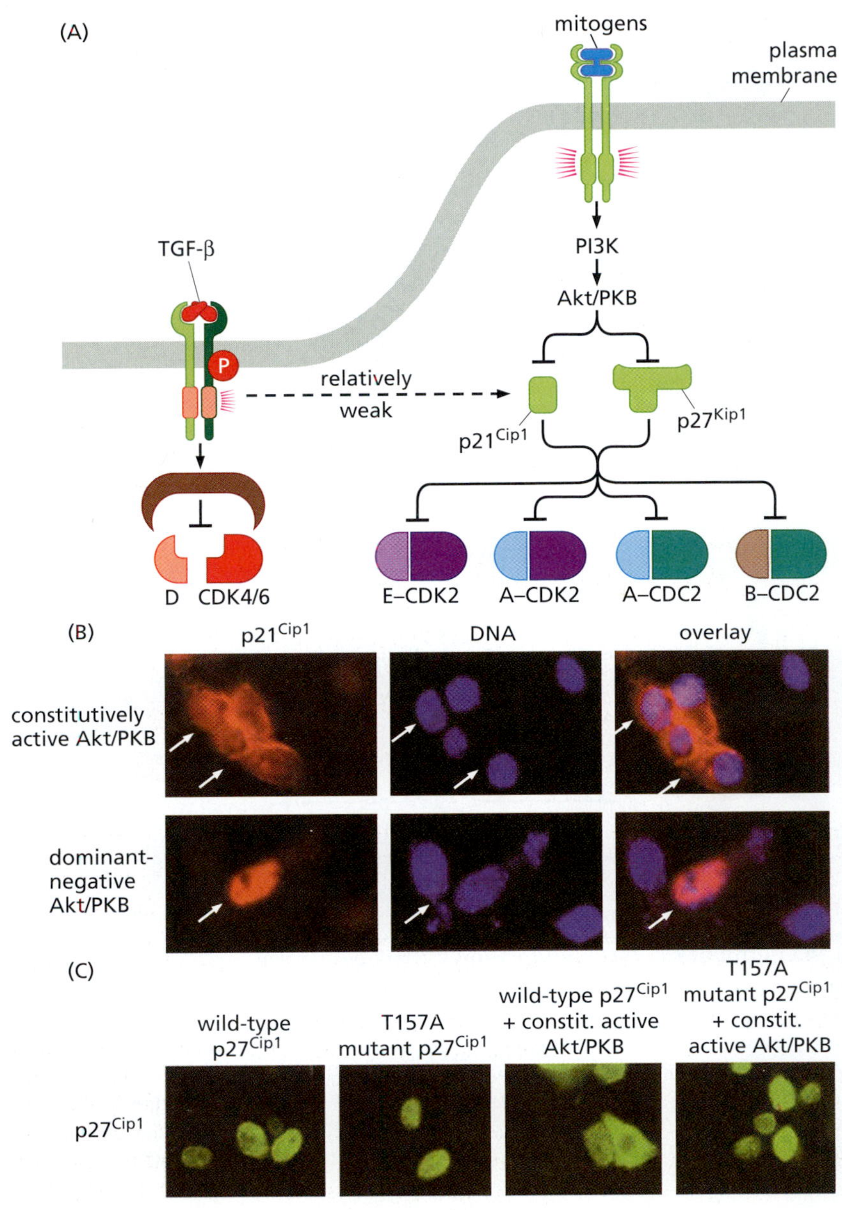

Figure 8.15 Control of cell cycle advance by extracellular signals Countervailing extracellular signals influence the cell cycle machinery, in part through their ability to control the levels and intracellular localization of CDK inhibitors. (A) During the G_1 phase of the cell cycle, TGF-β induces p15^{INK4A} and (weakly) p21^{Cip1} expression, negatively affecting the D–CDK4/6 and E–CDK2 complexes, respectively, (see Figure 8.14). Conversely, mitogens, acting through Akt/PKB, cause the phosphorylation and cytoplasmic localization of both p21^{Cip1} and p27^{Kip1}, which prevents these CDK inhibitors from entering the nucleus and blocking the activities of various cyclin–CDK complexes operating there. (B) Ectopic expression of a constitutively active Akt/PKB *(upper row)* causes p21^{Cip1} *(red, left)* to be localized largely in the cell cytoplasm. Compare its localization with that of the cell nuclei *(blue, middle)*; the overlap between these two images is also seen *(right)*. Conversely, expression of a dominant-negative Akt/PKB *(lower row)*, which interferes with ongoing Akt/PKB function, allows p21^{Cip1} to localize to the cell nucleus, where it can inhibit cell cycle progression. (C) In normal cells *(first panel)*, ectopically expressed, wild-type p27^{Kip1} *(green)* is found to be exclusively nuclear. This localization is not changed if a p27^{Kip1} mutant is expressed that lacks the threonine residue normally phosphorylated by Akt/PKB *(second panel)*. (The T157A mutant carries an alanine in place of this threonine.) If a mutant, constitutively active Akt/PKB is expressed, however, much of wild-type p27^{Kip1} is now seen in the cytoplasm *(third panel)*. But if the p27^{Kip1} mutant that cannot be phosphorylated by Akt/PKB is expressed, it resists the actions of constitutively activated Akt/PKB and remains in the nucleus *(fourth panel)*. (B, from B. Zhou et al., *Nat. Cell Biol.* 3:245–252, 2001; C, from G. Viglietto et al., *Nat. Med.* 8:1136–1144, 2002.)

Akt/PKB, the important kinase activated downstream of mitogen-activated PI3K (Section 6.6), phosphorylates p21^{Cip1} molecules located in the nucleus, thereby causing them to be exported into the cytoplasm, where they can no longer engage and inhibit cyclin–CDK complexes (Figure 8.15B). Similarly, Akt/PKB phosphorylates p27^{Kip1} (which functions much like p21^{Cip1}) and prevents its export from its cytoplasmic site of synthesis into the nucleus, where it normally does its critical work (Figure 8.15C). Taken together, these signaling responses illustrate how extracellular growth-inhibitory signals (conveyed by TGF-β) impede the advance of the cell cycle clock, while growth-promoting signals promote its forward progress.

These effects on intracellular localization appear to have clinical consequences. For example, in low-grade (i.e., less advanced) human mammary carcinomas, levels of activated Akt/PKB are low and p27^{Kip1} is able to carry out its anti-proliferative functions in the cell nucleus. In high-grade tumors, however, activated Akt/PKB is abundant and much of p27^{Kip1} is now found in the cell cytoplasm (Figure 8.16A). This intracellular localization is correlated with, and likely causally linked to, the progression of these cancers to fatal endpoints (Figure 8.16B).

One aspect of the behavior of p21^{Cip1} and p27^{Kip1} seems quite paradoxical. While they inhibit the actions of cyclin E–CDK2, cyclin A–CDC2, and cyclin B–CDC2, they actually *stimulate* the formation of cyclin D–CDK4/6 complexes (Figure 8.17A). Moreover, once a ternary (three-part) complex has formed between D–CDK4/6 and either of these CDK inhibitors, the cyclin–CDK complex can still phosphorylate its normal substrate. So the term "CDK inhibitor" is

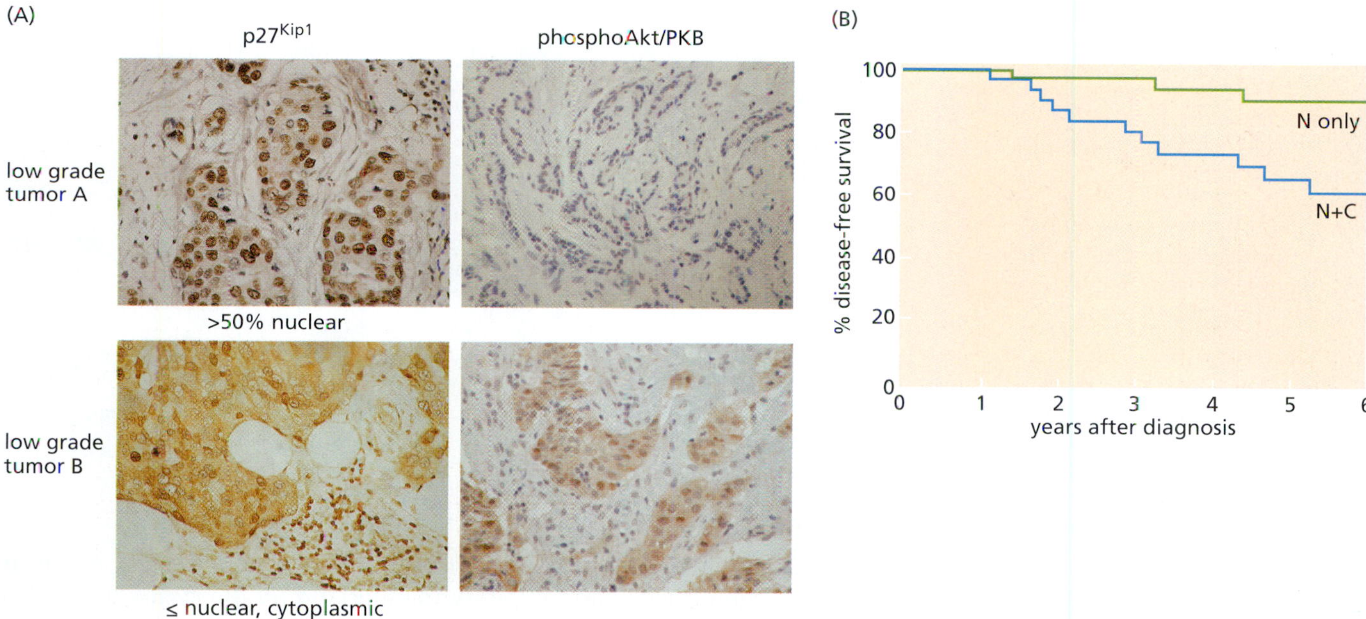

Figure 8.16 Suppression of p27^{Kip1} function by Akt/PKB in human breast cancers Various breast cancers show differing intracellular locations of p27^{Kip1} that reflect the activation state of Akt/PKB; the latter can be gauged by using an antibody that specifically recognizes the phosphorylated, functionally activated form of this enzyme. (A) In some low-grade primary breast cancers (e.g., tumor A), p27^{Kip1} is exclusively nuclear and is clearly present in more than 50% of nuclei *(brown staining, upper left,* >50% nuclear); activated phospho- Akt/PKB cannot, however, be detected in these cells *(upper right)*. Conversely, in other tumors (e.g., tumor B, *lower left panel*), p27^{Kip1} is found *(yellow brown staining)* in less than 50% of the nuclei as well as being found in significant quantities throughout the cytoplasm (≤50% nuclear, cytoplasmic) where it cannot block cell proliferation; in these tumors, activated phospho- Akt/PKB can be readily detected *(light brown staining, lower right panel)*. (B) This Kaplan–Meier graph indicates that those patients bearing primary tumors with largely nuclear p27^{Kip1} *(green curve)* have a relatively good prospect of long-term, disease-free survival over a period of 6 years following initial diagnosis and treatment. In contrast, patients whose tumors show p27^{Kip1} staining in both nuclei and cytoplasm *(blue curve)* suffer significant relapses over this time period, many of which lead to death. (Importantly, this correlation with clinical outcome does not prove causation, i.e., that p27^{Kip1} mislocalization causes clinical progression.) (A, courtesy of J. Zubovitz and J.M. Slingerland; B, from J. Liang et al., *Nat. Med.* 8:1153–1160, 2002.)

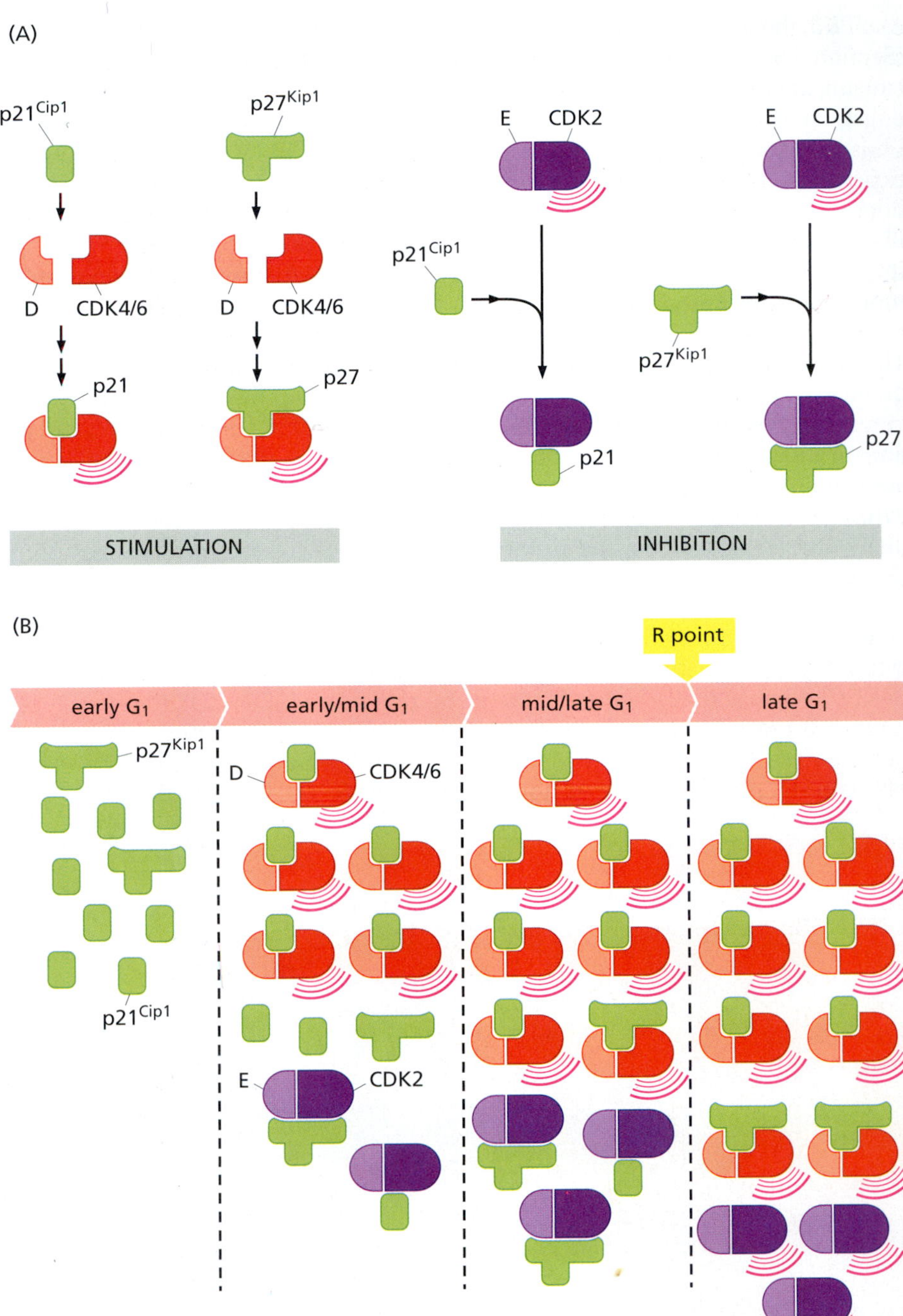

Figure 8.17 Interactions of CDK inhibitors with cyclin–CDK complexes (A) The various CDK inhibitors affect cyclin–CDK complexes in different ways. The CDK inhibitors of the $p21^{Cip1}$ and $p27^{Kip1}$ family, while inhibiting the cyclin–CDK complexes active in late G_1, S, G_2, and M phases *(only E–CDK2 is shown here; right),* actually stimulate the formation of the D–CDK4/6 complexes that are active in the early and mid-G_1 phase and permit these complexes, once formed, to be catalytically active *(left).* (B) Cells in early G_1 often have substantial concentrations of two CDK inhibitors, $p21^{Cip1}$ and $p27^{Kip1}$ molecules *(green; left).* As cyclin D–CDK4/6 complexes *(red/pink)* accumulate throughout early to mid-G_1 (being induced by extracellular mitogens), they bind and thereby sequester an ever-increasing proportion of the cellular pools of $p21^{Cip1}$ and $p27^{Kip1}$ molecules, including those that are present in soluble pools (unbound to any cyclin–CDK complexes) as well as those that are associated with cyclin E–CDK2 complexes *(light, dark purple).* The binding by D–CDK4/6 complexes of these two CDK inhibitors does not negatively affect the catalytic activity of the D–CDK4/6 complexes (see panel A). However, as this process continues, the D–CDK4/6 complexes eventually succeed in abstracting most of the $p21^{Cip1}$ and $p27^{Kip1}$ molecules away from E–CDK2 complexes in late G_1 *(right).* This finally liberates the E–CDK2 complexes *(bottom right),* enabling them to trigger the R-point transition and to initiate the phosphorylation events that drive cells into late G_1 and then S phase.

in fact a misnomer. The two proteins ($p21^{Cip1}$ and $p27^{Kip1}$) act on most cyclin–CDK complexes in an inhibitory manner and on the cyclin D–CDK4/6 complexes in a stimulatory fashion.

While $p27^{Kip1}$ functions similarly to $p21^{Cip1}$ in many ways, it has its own, highly interesting behavior. When cells are in the G_0 quiescent state, $p27^{Kip1}$ is present in high concentrations, and therefore binds to and suppresses the activity of the few cyclin E–CDK2 complexes that happen to be present in the cell. W hen cells are exposed to growth factors, cyclin D–CDK4/6 complexes accumulate, because these mitogens cause an increase in the levels of the D-type cyclins. Each time a new cyclin D–CDK4/6 complex forms, it captures and binds another molecule of $p27^{Kip1}$; consequently the pool of free $p27^{Kip1}$ molecules in the cell dwindles as they become sequestered by cyclin D–CDK4/6 complexes. After many hours, the remaining $p27^{Kip1}$ molecules are finally pulled away from cyclin E–CDK2 complexes (Figure 8.17B). Once a small portion of cyclin E–CDK2 complexes are liberated from the effects of $p27^{Kip1}$, these complexes are free to begin triggering passage of the cell through the R point.

Evidence is increasing that the post-mitotic state of many differentiated cell types in the body is imposed by various CDK inhibitors acting singly or in combination. For example, as seen here in the developing cerebellum of a mouse, the cell layers that have become post-mitotic have high levels of $p27^{Kip1}$, while those that are still in active cycle lack this CDK inhibitor (Figure 8.18A). Moreover, once this tissue is formed in the days after birth, the cells of this part of the brain are kept in a post-mitotic state by the $p27^{Kip1}$, which they express at a high and constant level throughout adulthood (Figure 8.18B).

In addition to the CDK inhibitors, yet another level of control of cyclin–CDK complexes is imposed by covalent modifications of the CDK molecules themselves (Sidebar 8.3). While the deregulated actions of the CDK inhibitors contribute importantly to cancer development, as detailed below, there is little evidence indicating that this other level of control plays a critical role in specific steps of tumor progression.

8.5 Viral oncoproteins reveal how pRb blocks advance through the cell cycle

We have viewed the cell cycle from several angles here, including its clearly delineated G_1, S, G_2, and M phases; the cyclins and CDKs that orchestrate advances through these phases; and the fact that these advances can be traced

(A)

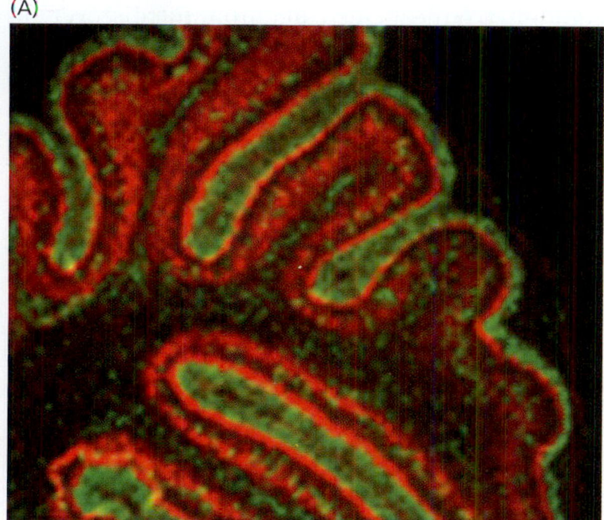

(B)

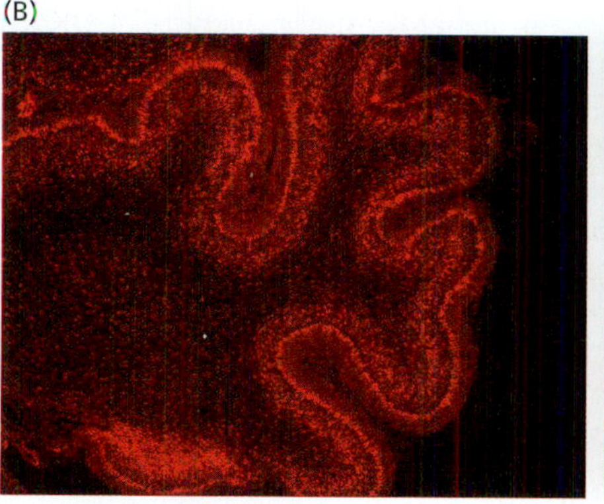

5 days

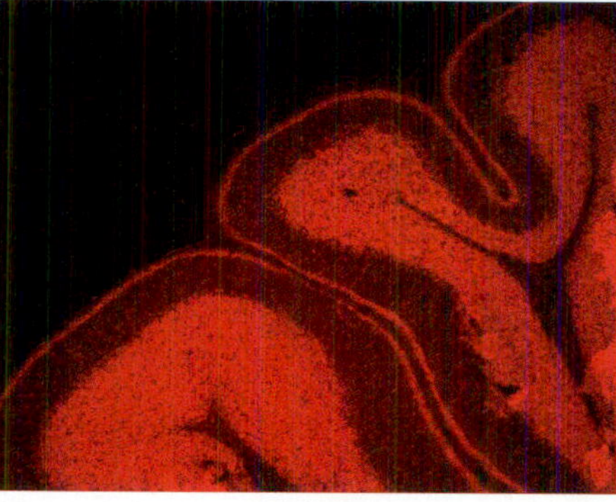

14 days

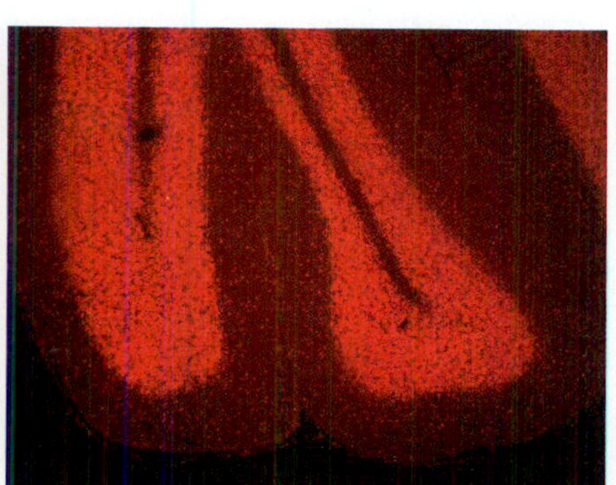

1 month

Figure 8.18 Imposition by $p27^{Kip1}$ of a post-mitotic state in cerebellar cells Much of cerebellar development in mammals occurs post-natally as cells differentiate and enter into a post-mitotic state. (A) During the course of this development in the mouse cerebellum, proliferating cells can be detected through their incorporation bromodeoxyuridine (BrdU), a thymidine analog; they are revealed here at post-natal day 7 by immunofluorescence using an antibody that recognizes BrdU-containing DNA *(green)*. The levels of $p27^{Kip1}$ have been gauged by use of a fluorescence-tagged antibody that recognizes it in the same section *(red)*. As is apparent, $p27^{Kip1}$ accumulates in cells that have ceased proliferating. (B) As the post-natal maturation of the mouse cerebellum proceeds, the levels of $p27^{Kip1}$ *(red)* increase progressively in the non-dividing granule cells of the cerebellum and are maintained at high levels throughout adulthood, ostensibly to hold these cells in a post-mitotic state. (A, from T. Uziel, F. Zindy, S. Xie et al., *Genes Dev.* 19:2656–2667, 2005;B, courtesy of A. Forget, F. Zindy, and M.F. Roussel.)

Sidebar 8.3 Phosphorylation of CDK molecules also controls their activity An additional level of control placed on the cell cycle clock machinery can be traced to the covalent modifications of the CDK molecules themselves. More specifically, CDKs must be phosphorylated at specific amino acid residues in order to be active (see Figure 8.7). At the same time, inhibitory phosphorylations of other amino acid residues must be removed. The stimulatory phosphorylations are imparted by a serine/threonine kinase termed CAK (for CDK-activating kinase), while the inhibitory phosphorylations are removed by a class of phosphatases called CDC25A, -B and -C. They are not discussed further here because there is little evidence that they are deregulated in cancer cells.

back to the extracellular signals that persuade the cell to enter into the active cell cycle and to progress through the restriction (R) point gate. These descriptions, however, fail to reveal precisely how the R-point transition is executed at the molecular level.

The solution to this puzzle was provided by the isolation of the *Rb* tumor suppressor gene, defective versions of which are involved in the pathogenesis of retinoblastomas, sarcomas, and small-cell lung carcinomas as well as other tumors (Sections 7.3–7.5). Soon after the *Rb* gene was isolated in 1986, it became apparent that it encodes a nuclear phosphoprotein of a mass of about 105 kD. This protein, called variously pRb or RB, was found to be absent, or present in a defective form, in the cells of many of the above-named tumor types.

Initial experiments with pRb revealed that it undergoes phosphorylation in concert with the advance of cells through their cell cycle. More specifically, pRb is essentially unphosphorylated when cells are in G_0; becomes weakly phosphorylated (**hypophosphorylated**) on a small number of serine and threonine residues after entrance into G_1; and becomes extensively phosphorylated (**hyperphosphorylated**) on a much larger number of serine and threonine residues in concert with advance of cells through the R point (Figure 8.19). Once cells have passed through the R point, pRb usually remains hyperphosphorylated throughout the remainder of the cell cycle. After cells finally pass through and exit mitosis, the phosphate groups on pRb are stripped off by the enzyme termed protein phosphatase type 1 (PP1). This removal of phosphate groups, in turn, sets the stage for the next cell cycle and thus for a new cycle of pRb phosphorylation.

The fact that pRb hyperphosphorylation occurs in concert with passage through the R point provided the first hint that this protein is the molecular governor of the R-point transition. An additional, critical clue came from the discovery in 1988 of physical interactions between DNA tumor virus–encoded oncoproteins and pRb. Recall that DNA tumor viruses are able to transform cells and do so through the use of virus-encoded oncoproteins. Unlike the oncoproteins of RNA

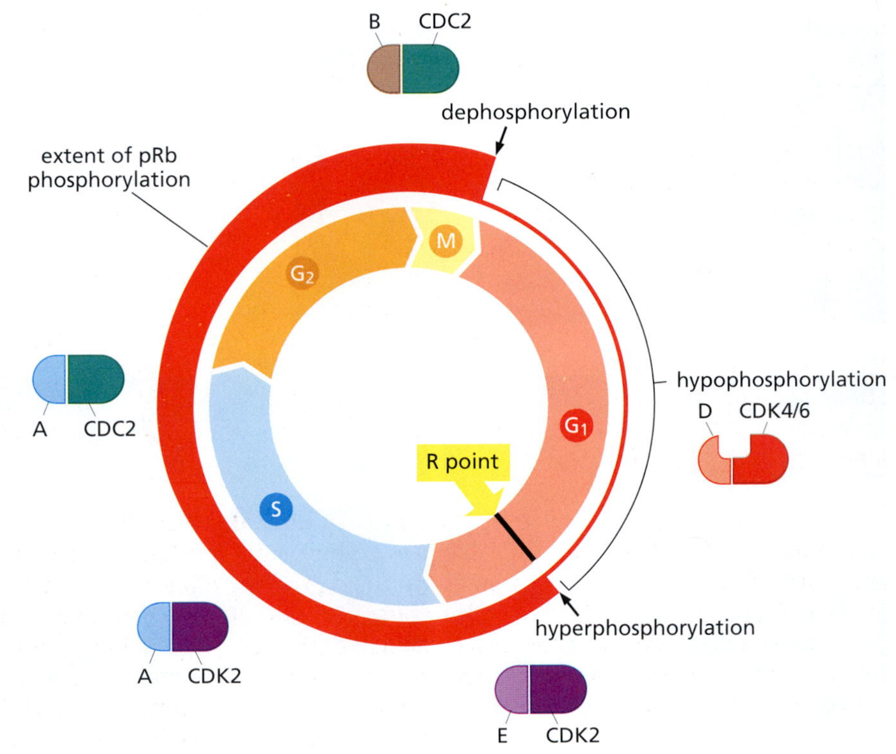

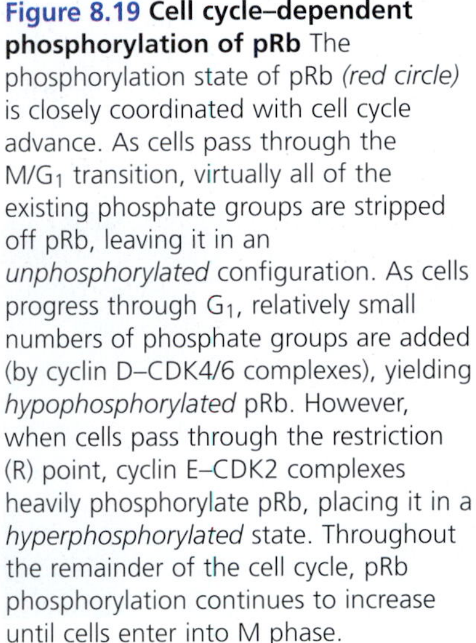

Figure 8.19 Cell cycle–dependent phosphorylation of pRb The phosphorylation state of pRb *(red circle)* is closely coordinated with cell cycle advance. As cells pass through the M/G_1 transition, virtually all of the existing phosphate groups are stripped off pRb, leaving it in an *unphosphorylated* configuration. As cells progress through G_1, relatively small numbers of phosphate groups are added (by cyclin D–CDK4/6 complexes), yielding *hypophosphorylated* pRb. However, when cells pass through the restriction (R) point, cyclin E–CDK2 complexes heavily phosphorylate pRb, placing it in a *hyperphosphorylated* state. Throughout the remainder of the cell cycle, pRb phosphorylation continues to increase until cells enter into M phase.

tumor viruses such as RSV, the DNA tumor viruses employ oncoproteins that have little if any resemblance to the proteins that exist naturally within uninfected cells (Sections 3.4 and 3.6).

The 1988 work revealed one important mechanism whereby a DNA tumor virus–encoded oncoprotein can disrupt the cellular growth-regulating circuitry: biochemical characterization of the oncoprotein made by the E1A oncogene of human adenovirus type 5 showed that in adenovirus-transformed cells, this oncoprotein is tightly bound to a number of cellular proteins, among them pRb (Figure 8.20; Sidebar 8.4). Soon thereafter, the SV40 virus-encoded large T oncoprotein and the E7 oncoprotein of certain strains of human papillomavirus (HPV) implicated in causing human cervical carcinoma were also found capable of forming physical complexes with pRb in virus-transformed cells.

These three different DNA tumor virus oncoproteins (E1A, large T antigen, and E7) are structurally unrelated to one another, yet all of them target a common cellular protein—pRb. This suggests that convergent evolution led, on at least three occasions, to the development of viral oncoproteins capable of perturbing cellular pRb. Stated differently, since evolution selects for viruses that can multiply more effectively, these discoveries suggested that sequestration or functional inactivation of pRb by the three viral oncoproteins is necessary for optimal viral replication in infected cells.

The discoveries of these viral–host protein complexes had profound effects on our understanding of how DNA tumor viruses succeed in transforming the cells that they infect. The speculation inspired by these discoveries went like this: Within a normal cell, the *Rb* gene was known to function as a tumor suppressor gene. Hence, the *Rb* gene, acting through its protein product, pRb, must inhibit cell proliferation in some fashion. DNA tumor viruses can sabotage the activities of pRb by dispatching viral oncoproteins that seek out and bind pRb, sequestering and apparently inactivating it. This sequestration removes pRb from the regulatory circuitry of the cell, yielding the same outcome as that occurring after a cell loses the two copies of its chromosomal *Rb* gene through mutations. This suggested, furthermore, that the DNA tumor viruses succeed in transforming cells through their ability to disable this key cellular growth-inhibitory protein, thereby liberating the virus-infected cell from the growth suppression that pRb normally imposes. (In fact, the HPV E7 protein goes beyond simple sequestration of pRb, since it also tags pRb for ubiquitylation and thus proteolytic degradation; see Sidebar 7.8.)

Another clue to pRb function was provided by the discovery that the oncoproteins of the DNA tumor viruses bind preferentially to the hypophosphorylated forms of pRb, that is, the forms of pRb that are present in the cell through most of the G_1 phase (see Figure 8.19). Conversely, the hyperphosphorylated pRb forms that are present in late G_1 and in the subsequent phases of the cell cycle are ignored by the viral oncoproteins. This observation led to yet another speculation: tumor virus oncoproteins focus their attentions on sequestering, and thereby inactivating, the only forms of pRb in the cell that are worthy of their attentions—the growth-inhibitory forms of pRb found in early and mid-G_1. Conversely, the viral oncoproteins ignore those forms of pRb that are already inactivated by other mechanisms.

Such reasoning indicated that the hypophosphorylated forms of pRb present in early and mid-G_1 cells are actively growth-inhibitory, while the hyperphosphorylated forms found after the R point have lost their growth-inhibitory powers, having been functionally inactivated by this phosphorylation. This meant that the R-point transition, which is defined by a physiologic criterion (acquisition of growth factor independence), is accompanied by a biochemical alteration of pRb that converts it from a growth-inhibitory protein to one that is apparently functionally inert. These speculations were soon validated by various types of experiments.

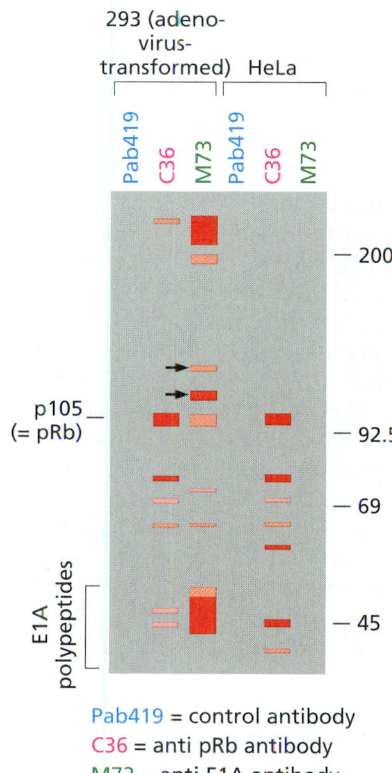

Figure 8.20 Co-precipitation of cellular proteins with adenovirus E1A The immunoprecipitation of the adenovirus E1A oncoprotein from adenovirus-transformed 293 cells, using the M73 anti-E1A monoclonal antibody, revealed a substantial group of cellular proteins that co-precipitated with E1A, including pRb (called here p105) *(third channel)*; subsequently, other co-precipitated host-cell proteins were identified, including pRb's two cousins, p107 and p130 *(third channel, arrows)*. The M73 anti-E1A monoclonal antibody failed to immunoprecipitate both E1A and the host-cell proteins from HeLa cells, which are not transformed by adenovirus *(right three channels)*. A control antibody (Pab419) failed to immunoprecipitate any of these proteins, from either cell type; and an anti-pRb antibody (C36) immunoprecipitated p105 (= pRb) from both 293 and HeLa cells. Darker and lighter bands in the autoradiogram are represented here by different shades of red. (Adapted from P. Whyte et al., *Nature* 334:124–129, 1988.)

Sidebar 8.4 The viral E1A and large T oncoproteins seek out many cellular targets As seen in Figure 8.20, an anti-E1A monoclonal antibody can immunoprecipitate not only E1A itself but also a large number of cellular proteins that are physically associated with E1A in adenovirus-transformed cells. The first of these to be identified was pRb, but soon thereafter, two other pRb-related proteins, termed p107 and p130, were identified as well; pRb and its two cousins are often called generically **pocket proteins** (Figure 8.21A), for reasons discussed below. Later, a higher-molecular weight cellular protein, termed p300, was also identified as a target of E1A binding. A number of other cellular targets of E1A binding remain to be characterized.

Of these three pocket proteins, p130 seems to be especially important in imposing quiescence on certain cell types and therefore to be involved in controlling the emergence of cells from G_0 into the G_1 phase of the cell cycle (Figure 8.21B). p107 expression, however, is induced strongly as cells pass through the R point. These two cousins of pRb (p107 and p130) play secondary roles in cancer development, since mutant versions of these two have rarely been found in spontaneously arising human tumors, in contrast to the frequently found mutant *Rb* gene copies. Later work would show that the binding of various host proteins by E1A would lead to their functional alteration, including their inactivation. Hence, this binding of E1A to multiple cellular proteins indicates that adenovirus, by deploying a single virus-encoded protein, can act to disrupt or deregulate multiple cellular control circuits concomitantly. Viral oncoproteins, such as adenovirus E1A or human papillomavirus E7, are able to bind to all three pocket proteins by associating with a shallow groove that is present in all three proteins and is highly conserved evolutionarily (Figure 8.21C).

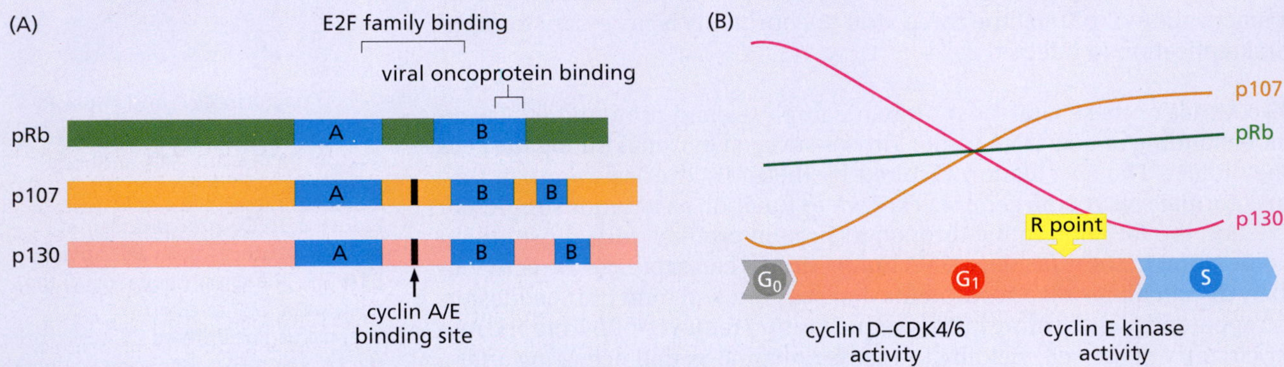

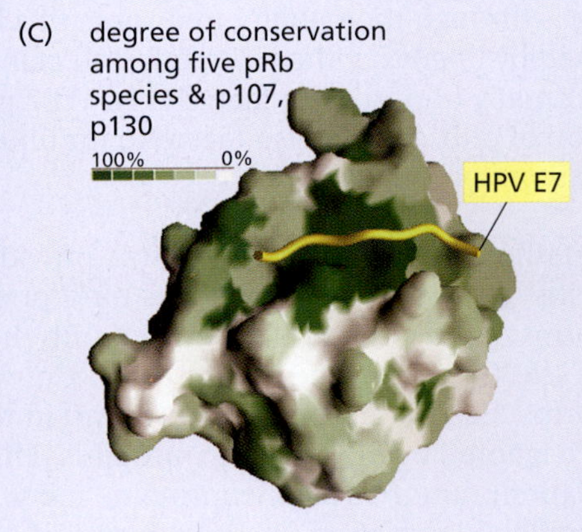

Figure 8.21 Pocket proteins (A) pRb is one of three structurally related proteins that together are often called "pocket proteins." All three are bound by viral oncoproteins, such as adenovirus E1A (Figure 8.20), SV40 large T, and human papillomavirus E7. All three pocket proteins are phosphorylated by cyclin D–CDK4/6 complexes, and the different pocket proteins bind different subsets of E2F transcription factors, as discussed later. (B) The different pocket proteins are expressed at different times as cells emerge from quiescence. p130 is involved in imposing G_0 quiescence, pRb is actively involved in controlling advance through most of G_1, and p107 is involved in controlling advance through late G_1 and S phase. (C) The shallow groove into which viral oncoproteins bind (using an LxCxE amino acid sequence carried by all of these oncoproteins, where "x" is variable) is structurally highly conserved among the pRb proteins of five mammalian species as well as the human p107 and p130 proteins, explaining the ability of the viral oncoproteins to bind all three of these cellular proteins. This groove is presented here as a space-filling model. The degree of sequence conservation is indicated by different shades of green, with the most conserved residues depicted as *dark green*. The polypeptide backbone of the pRb-binding domain of the human papillomavirus E7 is shown here in *yellow*. (B, from M. Classon and N. Dyson, *Exp. Cell Res.* 264:135–147, 2001; C, from J.O. Lee et al., *Nature* 391:859–865, 1998.)

In addition, further research has revealed that the control of pRb phosphorylation is more reversible than is indicated by the diagram in Figure 8.19. Thus, if the cell should experience serious physiologic stresses while in S phase or G_2, pRb phosphorylation can be reversed by still-unknown phosphatases, thereby returning pRb to its actively growth-inhibitory state. These stresses include, for example, hypoxia, DNA damage, and disruption of the mitotic spindle. Such reactivation of pRb-mediated growth inhibition is presumably only transient and is reversed once the physiologic stresses and/or damage have been resolved.

8.6 pRb is deployed by the cell cycle clock to serve as a guardian of the restriction-point gate

Various lines of evidence converge on the conclusion that pRb, which is growth-inhibitory through early to mid-G_1 phase, becomes inactivated by extensive phosphorylation when the cell passes through the R-point gate and thereafter is largely or completely inert as a growth suppressor. In effect, pRb serves as a guardian at this gate, holding it shut unless and until it becomes hyperphosphorylated, in which case pRb loses its growth-inhibitory powers, opens the gate, and permits the cell to enter into late G_1 and thereafter into the remaining phases of the cell cycle.

Since pRb is the ultimate arbiter of growth versus nongrowth, its phosphorylation must be carefully controlled. Not unexpectedly, its phosphorylation is governed by components of the cell cycle clock. In early and mid-G_1, D-type cyclins together with their CDK4/6 kinase partners are responsible for initiating pRb phosphorylation, leading to its hypophosphorylation. Since the levels of D-type cyclins appear to be controlled largely by extracellular signals, notably mitogenic growth factors, we can now plot out a direct line of signaling: growth factors induce the expression of D-type cyclins; D-type cyclins, collaborating with their CDK4/6 partners, initiate pRb phosphorylation (Figure 8.22).

During a normal cell cycle, this initial hypophosphorylation of pRb appears to be necessary but not sufficient for the subsequent functional inactivation of pRb at the R point, which requires its hyperphosphorylation. In fact, cyclin E levels increase dramatically at the R point. The cyclin E then associates with its CDK2 partner, and this complex drives pRb phosphorylation to completion, leaving pRb in its hyperphosphorylated, functionally inactive state (see Figure 8.22). It seems that pRb molecules that have not undergone prior hypophosphorylation by cyclin D–CDK4/6 complexes are not good substrates for completion of phosphorylation by cyclin E–CDK2 complexes. The phosphorylation and functional inactivation of pRb's cousins, p107 and p130, are also under the control of cyclin D–CDK4/6 and possibly cyclin E–CDK2 complexes.

If mitogens are withdrawn from the cell at any time in G_1 prior to the R point, then the levels of the most prominent of the D-type cyclins—cyclin D1—rapidly collapse. In the absence of cyclin D1, pRb loses its phosphate groups through the actions of a still-unidentified phosphatase and may thereby revert to a protein that is no longer a good substrate for cyclin E–CDK2 complexes. This response emphasizes the need for strong and continuous mitogenic signaling throughout the early and mid-G_1 phase up to the R point.

Figure 8.22 Control of the restriction-point transition by mitogens The levels of D-type cyclins are controlled largely by extracellular signals. Because the D-type cyclins, acting with their CDK4 and CDK6 partners, are able to drive pRb hypophosphorylation—a prerequisite to the R-point transition—this ensures that this transition is responsive to extracellular signals (see also Figure 8.15). These processes are extended by the actions of cyclin E–CDK2 complexes on hypophosphorylated pRb, which proceed to complete the work of inactivating pRb by placing it in a hyperphosphorylated state that results in its complete functional inactivation.

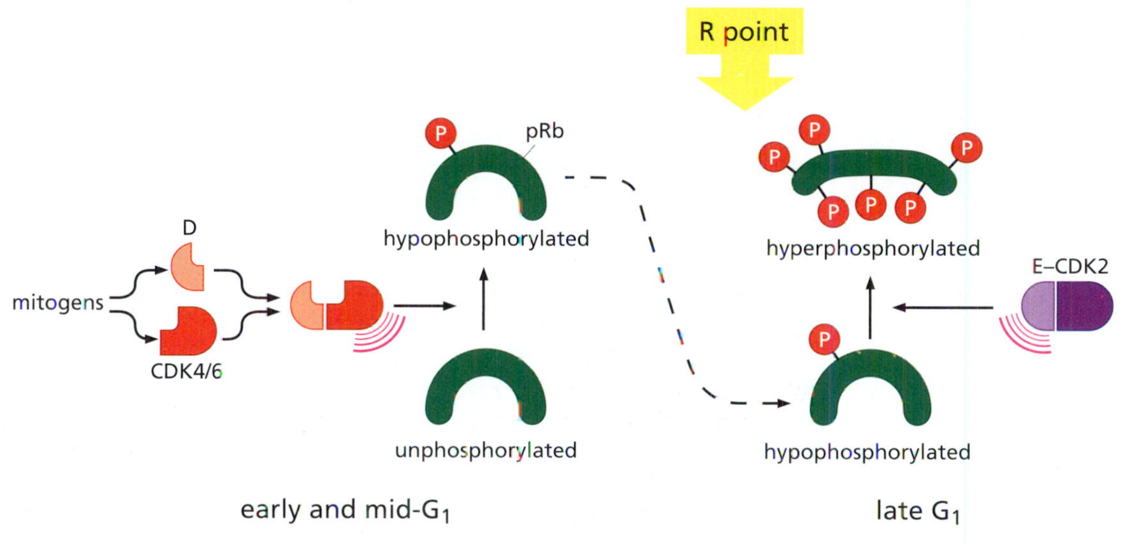

Once the cell advances through the R point, the continued hyperphosphorylation and functional inactivation of pRb is apparently maintained and increased by cyclin E-, by cyclin A-, and then by cyclin B-containing CDK complexes, none of which is responsive to extracellular signals—precisely the properties of the cell cycle clock machinery that, we imagine, operates after the R-point transition and guarantees execution of the rigidly programmed series of transitions in S, G_2, and M.

This scheme reveals why pRb is such a critical player in the regulation of cell proliferation. If its services are lost from the cell (through mutation of the chromosomal *Rb* gene copies, methylation of the *Rb* gene promoter, or the actions of DNA tumor virus oncoproteins), then this protein can no longer stand as the guardian of the R-point gate. Moreover, as we will see, in some cancer cells, pRb phosphorylation is deregulated, resulting in inappropriately phosphorylated and thus functionally inactivated pRb. In certain other cancer cells, there is evidence that the dephosphorylation (and attendant activation) of pRb, which normally happens at the M/G_1 transition through the action of the PP1 phosphatase, never occurs, leaving pRb in its hyperphosphorylated, inactivated state throughout the entire growth-and-division cycle of these cells. Without pRb on watch, cells move through G_1 into S phase without being subjected to the usual controls that are designed to ensure that cell cycle advance can proceed only when certain pre-conditions are satisfied. Indeed, as we will see, the deregulation of pRb phosphorylation is so widespread that one begins to think that this important signaling pathway is perturbed in virtually all human tumors.

The activities of pRb seem to overlap with those of its cousins, p107 and p130, raising the question why pRb is the only one of these three to play a clear and obvious role in cancer pathogenesis. A definitive answer to this puzzle is still elusive. It may come from the specific times in the cell cycle when these two pRb-related proteins are active (see Sidebar 8.4 and Figure 8.21B). Whereas p130 is specialized to suppress cell proliferation while a cell resides in G_0, p107 appears to be most active in the late G_1 and S phases. Only pRb is well positioned to control the transition at the R point from mitogen-dependent to mitogen-independent growth.

8.7 E2F transcription factors enable pRb to implement growth-versus-quiescence decisions

As described above, a number of lines of evidence indicate that hypophosphorylated pRb is active in suppressing G_1 advance and loses this ability when it becomes hyperphosphorylated. Still, this behavior provides no clue as to how pRb succeeds in imposing this control. Research conducted in the early 1990s indicated that pRb exercises much of this control through its effects on a group of transcription factors termed E2Fs.

When pRb (and its two cousins, p107 and p130) are in their unphosphorylated or hypophosphorylated state, they bind E2Fs, including E2Fs that are bound to DNA; however, when hyperphosphorylated, pRb and its cousins (the three pocket proteins; see Figure 8.21) dissociate from the E2Fs (Figure 8.23A). This suggested a simple model of how pRb is able to control cell cycle advance. In the early and middle parts of G_1, E2Fs are associated with the promoters of a number of genes under their control. At the same time, these transcription factors are bound by pocket proteins. This pocket protein association prevents the E2Fs from acting as stimulators of transcription. Accordingly, during much of the G_1 phase of the cell cycle, genes that depend on E2Fs for expression remain repressed. However, when the pocket proteins undergo hyperphosphorylation at the R point in late G_1, they release their grip on the E2Fs, permitting the E2Fs to stimulate transcription of their clientele of genes. The products of these genes, in turn, usher the cell from late G_1 into S phase. Similarly, when viral

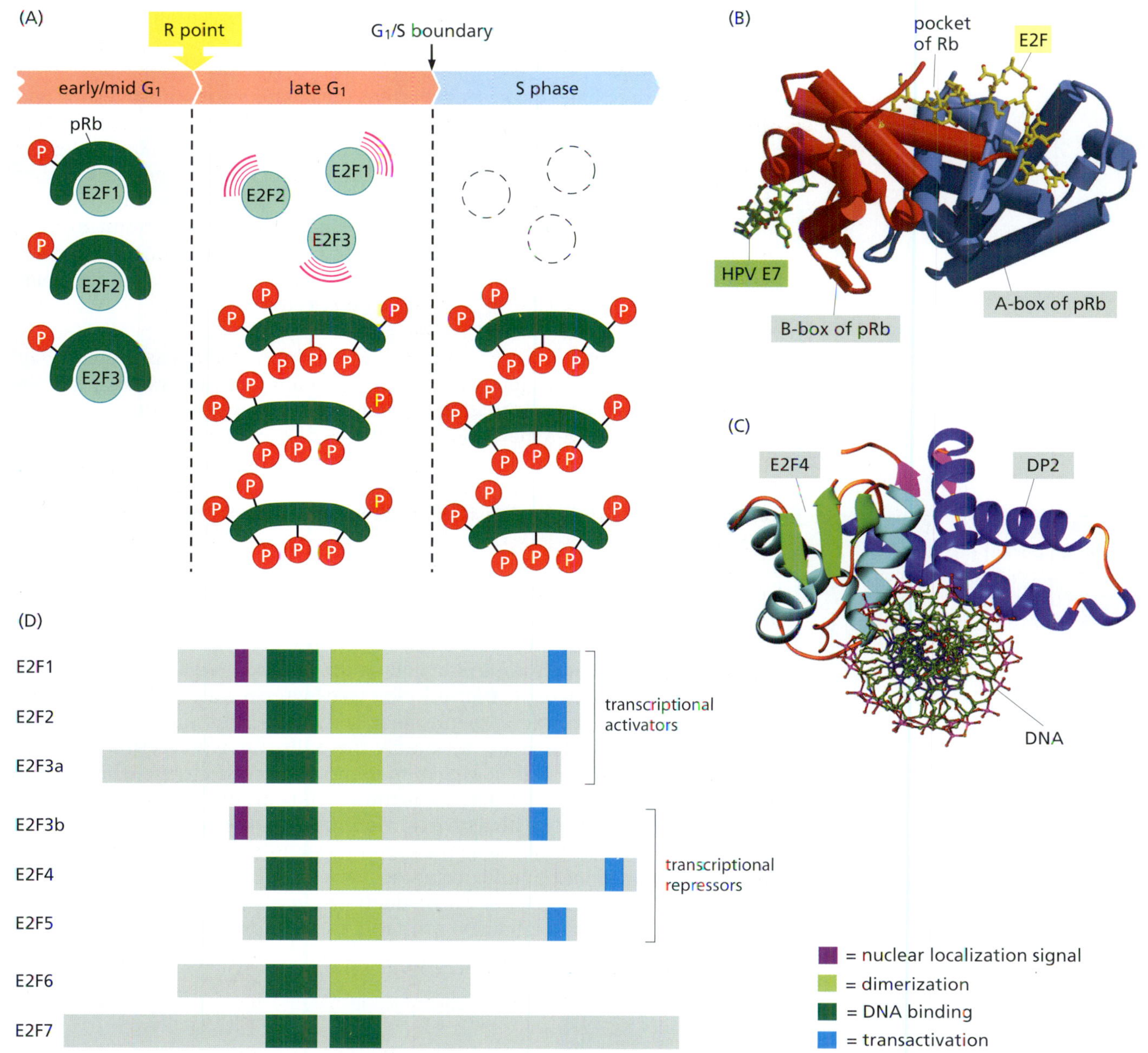

Figure 8.23 E2Fs and their interactions (A) By binding various E2F transcription factors, the three pocket proteins, pRb, p107, and p130 modulate the expression of E2F-responsive genes in various phases of the cell cycle. When pRb binds to E2Fs 1, 2, and 3 *(left)*, it blocks their transcription-activating domain; other proteins that serve to modify chromatin and repress transcription *(not shown)* are attracted to the pRb–E2F complexes. Because they are bound to the promoters of certain genes via the E2Fs, these multiprotein complexes repress the expression of such genes. (The behavior of p107 and p130, *not shown*, is similar to that of pRb.) After being hyperphosphorylated at the R point, the pocket proteins release their grip on the E2Fs, allowing the latter to activate transcription of those genes. As cells enter into S phase *(right)*, the E2Fs are inactivated and/or degraded. Hence, the E2Fs may function as active inducers of transcription only in the narrow window of time in late G_1 phase that begins at the R point and ends soon after cells enter S phase. (B) The binding of E2Fs such as E2F2 *(yellow stick figure)* by pRb *(red and blue)* is prevented by several DNA tumor virus oncoproteins. Here we see how a segment of the human papillomavirus E7 oncoprotein *(green stick figure, left)* binds to a shallow groove of pRb *(red domain, left;* also shown in Figure 8.21C*)*; this binding, together with other secondary binding sites (not yet mapped), perturbs the "pocket" formed by the B- *(red)* and A- *(blue)* boxes of pRb, thereby preventing the binding of the C-terminal, transcription-activating domain of E2F2 *(yellow)* to this pocket. This loss of E2F binding effectively neutralizes the ability of pRb to inhibit cell cycle advance. α-helices are shown here as *cylinders*. (C) The E2F transcription factors bind DNA as heterodimeric complexes with DP partners. Here the binding of an E2F4–DP2 complex to the DNA double helix *(stick figure, dark green, red)* is revealed by X-ray crystallography. The E2F4 transcription factor is seen to the left *(light blue, green)*, while DP2 is on the right *(dark blue, pink)*, both depicted as ribbon diagrams. (D) The E2Fs constitute a family of at least seven distinct proteins. E2Fs 1, 2, and 3a represent transcription-activating E2Fs, while E2F3b (arising from alternative splicing of the E2F3a pre-mRNA), E2F4, and E2F5 are involved in transcriptional repression. The functions of E2Fs 6 and 7 are poorly resolved. (B, courtesy of Y. Cho, from C. Lee et al., *Genes Dev.* 16:3199–3212, 2002; C, courtesy of R. Latek, from N. Zheng et al., *Genes Dev.* 13:666–674, 1999; D, courtesy of A. Aslanian, P.J. Iaquinta, and J.A. Lees.)

oncoproteins are present, they mimic pRb hyperphosphorylation by preventing pRb from binding E2Fs (Figure 8.23B).

While the above scheme is correct in outline, the details are more interesting and complex. The term "E2F" is now known to include a class of heterodimeric proteins composed of E2F1, 2, 3, 4, 5, and 6 subunits, all of which can bind to either a DP1 or a DP2 subunit (Figure 8.23C and D); the functions of another member of this family—E2F7—are poorly understood. (In the discussions that follow, we describe the behavior of the various E2F subunits, with association to their DP1or DP2 partner subunits being assumed.) Once assembled, E2F–DP complexes recognize and bind to a sequence motif in the promoters of various genes that seems to be TTTCCCGC or slight variations of this sequence.

Throughout this book, we have portrayed transcription factors as proteins that bind to the promoters of genes and then proceed to activate transcription. In fact, transcription factors operating like E2Fs can exert two opposing effects on transcriptional control. When bound to the promoters of genes in the absence of any associated pocket proteins (i.e., pRb, p107, or p130), E2Fs such as E2Fs 1, 2, and 3 can indeed trigger gene expression by attracting other proteins that function to remodel the nearby chromatin and recruit RNA polymerase to initiate transcription.

However, when pRb is hypophosphorylated, it physically associates with its E2F partners, which themselves are already sitting on the promoters of various genes and remain so after pRb binds to them. Once bound to these E2Fs, hypophosphorylated pRb molecules block the *transactivation* domain of the E2Fs that they use to activate transcription (see Figure 8.23D). At the same time, pRb actively recruits other proteins that actively repress transcription. One important means of doing so, but hardly the only one, is to recruit a histone deacetylase (HDAC) to this complex; by removing acetyl groups from nearby histone molecules, an HDAC remodels the nearby chromatin into a configuration that is incompatible with active transcription (Figure 8.24A).

This means that, in addition to occluding (physically blocking) the transactivation domains of E2Fs, pRb actively functions as a transcriptional repressor. E2Fs 4 and 5 seem to be involved primarily in the repression of genes and act

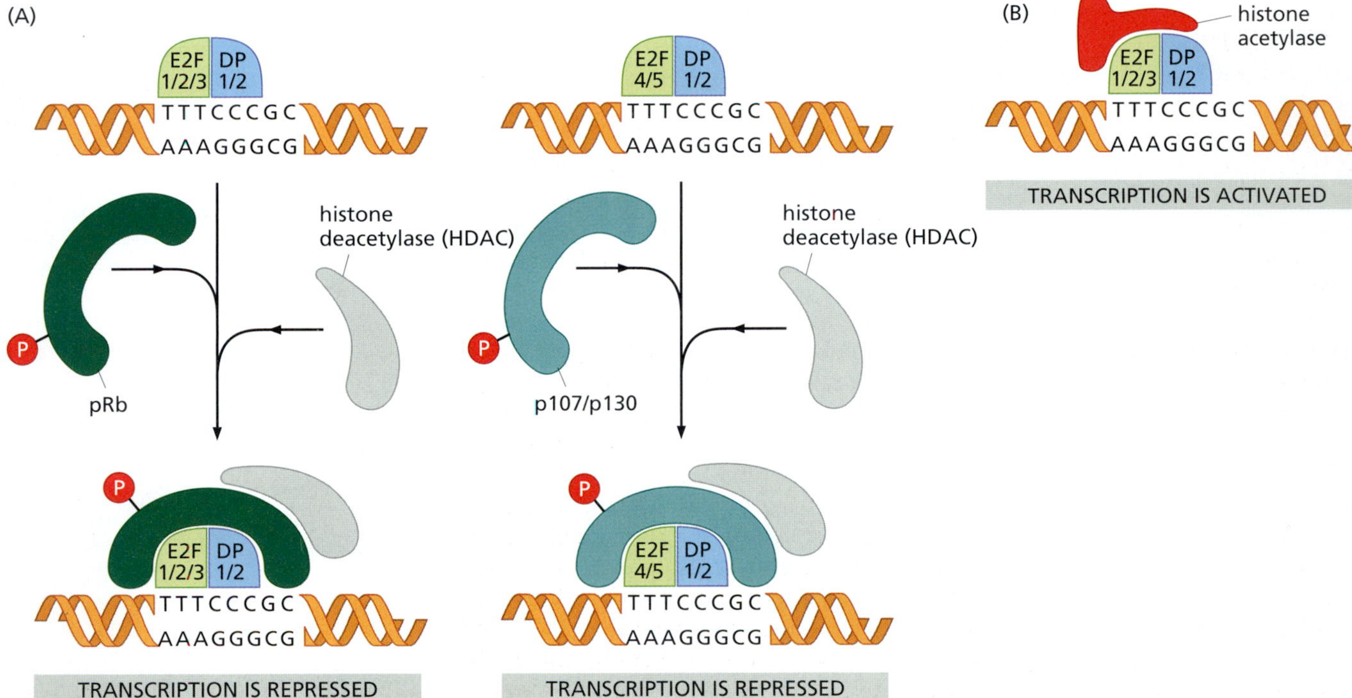

Figure 8.24 Modification of chromatin by pocket proteins (A) Hypophosphorylated pRb discourages transcription by attracting histone deacetylase (HDAC) enzymes *(gray)* to pRb–E2F complexes that are bound, in turn, to gene promoters; these enzymes prevent transcription by removing acetyl groups from nearby histones, causing the latter to shift into a configuration that is incompatible with the progression of RNA polymerase through the chromatin. (B) Conversely, when the E2Fs are not complexed with pRb (or its p107 and p130 cousins), they attract histone acetylases *(red)*, which place chromatin in a configuration that is conducive for transcription.

largely through associated p107 and p130 proteins to attract repressors of transcription to promoters and thereby shut down gene expression. E2F6, which does not associate with pocket proteins, appears to act exclusively as a transcriptional repressor.

In quiescent G_0 cells, for example, E2F4 and E2F5 are present in abundance (and are associated largely with p130), whereas E2Fs 1, 2, and 3 are hardly present at all, being expressed largely in proliferating cells. Moreover, cells that have been genetically deprived of E2F4 and E2F5 have lost responsiveness to the growth-inhibitory effects of the $p16^{INK4A}$ CDK inhibitor. This implies that the lion's share of the downstream effects that result from inhibiting the cyclin D–CDK4/6 complexes and blocking the phosphorylation of pocket proteins is mediated via the binding of p107 and p130 to these two E2Fs; ultimately, this binding results in the repression of a constituency of genes that carry TTTCCCGC sequences in their promoters.

Once hyperphosphorylated, pRb releases its grip on E2F complexes, and the E2Fs can then attract transcription-activating proteins, such as histone acetylases, which modify the histones and place them in a configuration that encourages transcription (Figure 8.24B). A series of genes that are expressed specifically in late G_1 and are known or assumed to be important for S-phase entry have been found to contain binding sites in their promoters for E2Fs. Included among these are some genes encoding proteins involved in synthesizing DNA precursor nucleotides (such as dihydrofolate synthetase and thymidine kinase), as well other genes involved directly in DNA replication.

Prominent among these E2F-activated genes is the gene encoding cyclin E. Consequently, the levels of both cyclin E mRNA and protein rise quickly after passage through the R point. Recall that cyclin E, for its part, is responsible, together with CDK2, for driving the hyperphosphorylation of pRb. Hence pRb inactivation leads to increases in cyclin E levels, and cyclin E, once formed, drives further pRb inactivation (Figure 8.25A). These relationships enable the operations of a powerful, self-reinforcing positive-feedback loop that is triggered as cells pass through the R point. (Such regulation can also be termed a "feed-forward loop.")

A second positive-feedback loop is activated simultaneously at the R point through the following mechanism: Cyclin E–CDK2 complexes phosphorylate $p27^{Kip1}$, and the latter, once phosphorylated, is ubiquitylated and rapidly degraded (Figure 8.25B). The destruction of $p27^{Kip1}$ molecules, which normally inhibit E–CDK2 firing, liberates additional cyclin E–CDK2 complexes that proceed to phosphorylate and thereby inactivate additional $p27^{Kip1}$ molecules. Such positive-feedback mechanisms operate in many control circuits to guarantee the rapid execution of a decision once that decision has been made. Equally important, they ensure that the decision, once made, is essentially irreversible—a situation that corresponds precisely with the behavior of cells advancing through the R point.

The period of active transcriptional promotion by E2Fs 1, 2, and 3 appears to be rather short-lived (see Figure 8.23A). This time window begins at the R point when pRb becomes hyperphosphorylated and liberates the E2Fs. These three E2Fs then proceed to induce expression of critical late G_1 genes, the products of which are required to prepare the cell for entrance into S phase, as described above. Soon thereafter, as the cell traverses the G_1/S transition into S phase, cyclin A becomes activated and, acting together with its CDK2 partner, phosphorylates both the E2F and DP subunits of these heterodimeric transcription factors; this results in the dissociation of the E2F–DP complexes and in the attendant loss of their transcription-activating abilities. At the same time, E2F1 (and possibly other E2Fs) are tagged for degradation by ubiquitylation, and E2F7, which seems to function largely to antagonize E2F-mediated gene activation, is

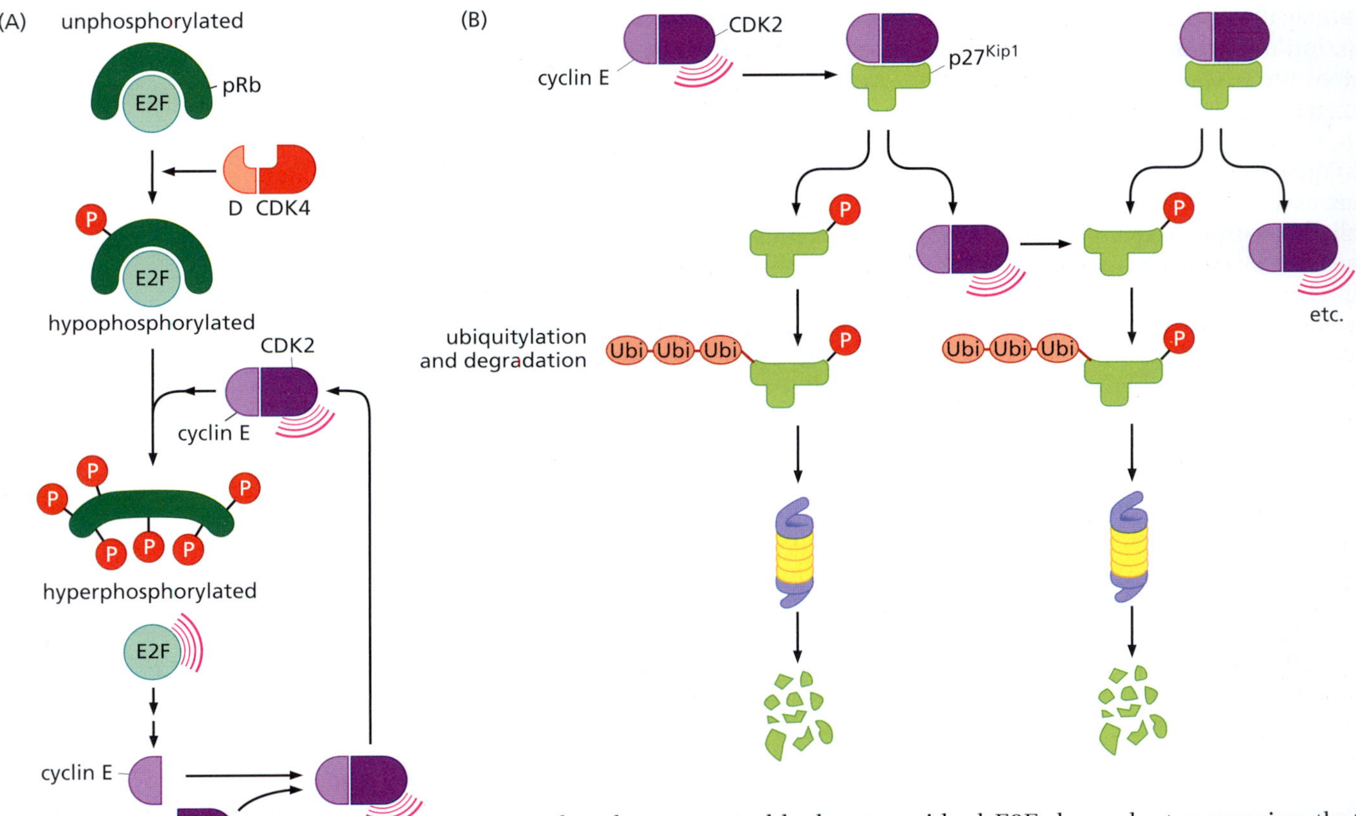

Figure 8.25 Positive-feedback loops and the irreversibility of cell cycle advance The irreversibility of certain key steps in cell cycle progression and the rapidity of their execution is ensured, in part, by the activation of certain positive-feedback loops. (A) When E–CDK2 complexes drive pRb hyperphosphorylation, this liberates E2F transcription factors from pRb control, enabling the E2Fs to trigger increased transcription of the cyclin E genes; this leads to the synthesis of more cyclin E protein and the formation of more E–CDK2 complexes, which function, in turn, to drive additional pRb phosphorylation. (B) The activation of a small number of E–CDK2 complexes enables the latter to phosphorylate $p27^{Kip1}$, marking it for ubiquitylation. This $p27^{Kip1}$ is then degraded, liberating more E–CDK2 complexes from inhibition by $p27^{Kip1}$, allowing this process to repeat itself and amplify its effects.

expressed and appears to block any residual E2F-dependent expression that persists after cells have entered in S phase. Together, these changes effectively shut down further transcriptional activation by the E2F factors, whose brief moment of glory as active transcription factors will come again only during the end of the G_1 phase of the next cell cycle.

pRb has been reported to bind to a variety of transcription factors, and these might, in principle, be controlled by pRb and its state of phosphorylation. In the great majority of cases, however, detailed functional analyses of their interactions with pRb have not been undertaken. There is a compelling reason to believe that, among all of these transcription factor clients of pRb, the E2Fs are of central, preeminent importance in controlling cell cycle advance: when the E2F1 protein is micro-injected into certain types of serum-starved, quiescent cells (which are therefore in G_0), this transcription factor, acting on its own, is able to induce these cells to enter G_1 and advance all the way into S phase; E2F2 and E2F3 have similar powers.

8.8 A variety of mitogenic signaling pathways control the phosphorylation state of pRb

Throughout this chapter we have emphasized that a number of distinct signaling pathways converge on the cell cycle clock and modulate its functioning. Acting as a central clearinghouse, the clock machinery collects and processes these various *afferent* signals and then, acting via its cyclin–CDK complexes, emits the signals that energize cell cycle advance (see Figure 8.1). It is worthwhile at this point to step back and summarize the ways in which these various signals regulate distinct components of the cell cycle clock. Prominent among the signals impinging on this nuclear regulatory machinery are those conveyed by the mitogenic signaling pathways. Because excessive signals flow through these pathways in virtually all types of cancer cells, the connections between these pathways and the cell cycle clock are critical to our understanding of cell transformation and tumor pathogenesis.

Mitogenic signals transduced via Ras result in the hypo- and eventual hyperphosphorylation of pRb. This phosphorylation is initiated by inducing the expression of genes encoding the D-type cyclins, notably cyclin D1. When Ras signaling is blocked (through the introduction of a **dominant-negative** mutant Ras protein), cell proliferation is blocked in wild-type cells exposed to serum, while mutant $Rb^{-/-}$ cells lacking Ras function enter S phase quite normally. This demonstrates that the main function of Ras during the G_1 phase is to ensure that the mitogenic signals released by ligand-activated growth factor receptors succeed in causing the phosphorylation and functional inactivation of pRb. Moreover, the block to S-phase entrance imposed by the dominant-negative Ras protein can be circumvented by ectopically expressing either cyclin E or E2F1 genes in cells. Together, this suggests a linear chain of command of the following sort:

growth factors → growth factor receptors → Ras → cyclins D1 and E → inactivation of pRb → activation of E2Fs → S-phase entrance

The ability of mitogenic growth factors to increase the levels of D-type cyclins depends on several distinct signaling pathways (Figure 8.26; see also Section 8.3 and Table 8.1). The best studied of these signals flow via the Ras → Raf → MAPK pathway (Section 6.5). For example, members of the Fos family of transcription factors, which lie at the bottom of this signaling cascade (Table 6.1), create the heterodimeric AP-1 transcription factors by forming complexes with Jun proteins; these AP-1 complexes are known to be powerful activators of cyclin D1 gene transcription. In addition, a variety of cytoplasmic signal-transducing proteins impinge on the cyclin D1 promoter in ways that are poorly understood (see

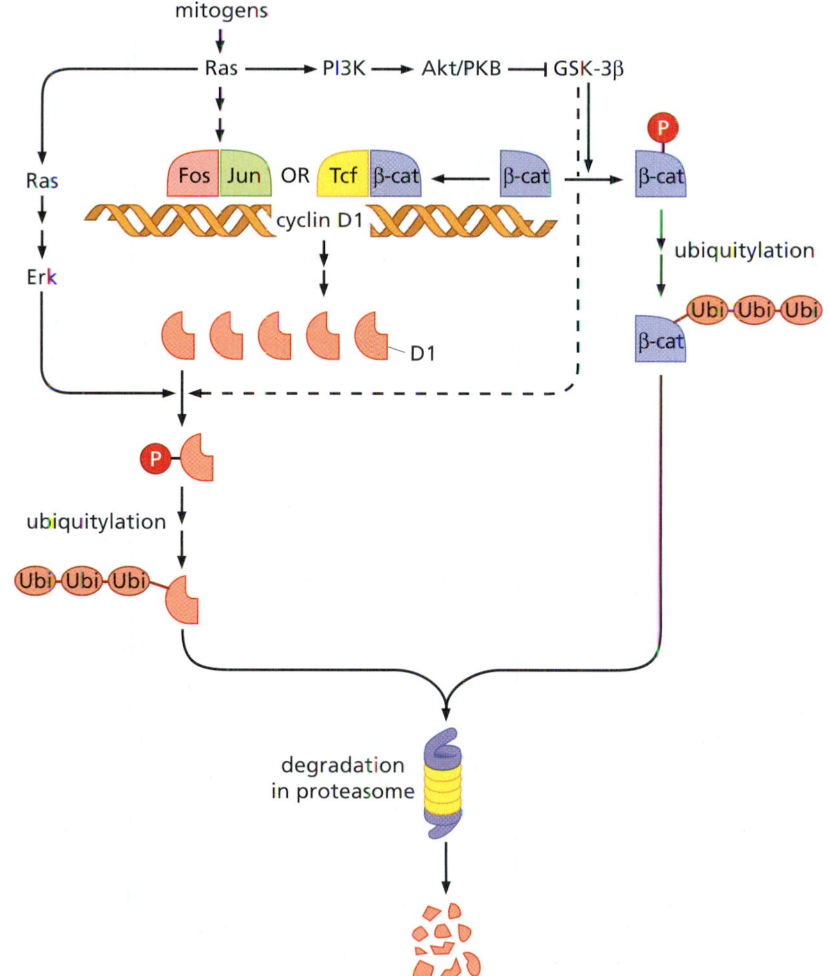

Figure 8.26 Countervailing controls on cyclin D1 levels A variety of signaling pathways transducing mitogenic signals modulate the levels of cyclin D1. For example, the Ras signaling pathway influences cyclin D1 levels in at least three ways. First, the AP-1 transcription factor acts on the *cyclin D1* promoter, yielding increased levels of the cyclin D1 mRNA; AP-1 is composed of Fos–Jun heterodimers, both of which are increased in amount and/or functionally activated by mitogens *(top middle*; see also Figure 8.11). Second, by activating PI3 kinase and thus Akt/PKB, Ras is able to inhibit the actions of GSK-3β. This, in turn, spares β-catenin (β-cat) from phosphorylation, ubiquitylation, and degradation, allowing it to partner with the Tcf transcription factor to induce *cyclin D1* transcription. Third, operating in the opposing direction is Erk, which lies at the bottom of the Ras → Raf → MEK → Erk kinase cascade (Section 6.5). Cyclin D1 is phosphorylated by Erk *(left)* and thereby marked for ubiquitylation and degradation in the proteasomes. A fourth connection has also been reported, which involves the phosphorylation of cyclin D1 by GSK-3β, which also results in its ubiquitylation and degradation *(dashed line)*; once again, Ras action favors cyclin D1 accumulation. (In general, the powers of the Ras pathway to increase cyclin D1 levels seem to greatly exceed its effects in causing cyclin D1 degradation.)

Table 8.1). As mentioned (Section 8.3), the cyclin D2 and D3 promoters have their own array of upstream regulators that confer responsiveness to certain extracellular signals. Many of the signaling pathways converging on these two gene promoters are also deregulated in human cancer cells, leading in turn to excessive expression of one or the other of these D-type cyclins, inappropriate phosphorylation and inactivation of pRb, and deregulated proliferation.

Activation of the Ras signaling pathway also leads, via phosphatidylinositol 3-kinase (PI3K), to activation of the Akt/PKB kinase; the latter proceeds to phosphorylate and inactivate glycogen synthase kinase-3β (GSK-3β, Section 6.10; see Figure 8.26). This has several benefits for growth promotion. Normally, GSK-3β targets β-catenin molecules for destruction by phosphorylating them; this phosphorylation results in the ubiquitylation of β-catenin molecules and their destruction in proteasomes. However, when GSK-3β is inactivated by PKB/Akt phosphorylation, β-catenin can accumulate and migrate to the nucleus, where it forms transcription factor complexes with Tcf/Lef that stimulate cyclin D1 transcription. It is also possible that the Wnt growth factors (Section 6.10), by suppressing GSK-3β activity, have similar effects on the cell cycle clock.

Adding to this complexity are other mitogenic signals that affect specific components of the cell cycle clock. For example, in ways that are poorly understood, continuous exposure to serum growth factors causes a degradation of the cellular pool of the $p27^{Kip1}$ CDK inhibitor, resulting in the progressive decline in the overall levels of this protein as the G_1 phase of the cell cycle proceeds. [This loss of $p27^{Kip1}$ molecules from the cell seems to cooperate with the sequestration of $p27^{Kip1}$ achieved by cyclin D1–CDK4/6 complexes (see Figure 8.17) to liberate some cyclin E–CDK2 from inhibition by $p27^{Kip1}$, an event in late G_1 that triggers the explosive onset of the R-point transition.]

There are yet other mechanisms, still poorly characterized, that enable mitogenic growth factors to energize the cell cycle apparatus. For instance, the physical association of D-type cyclins with CDK4 is also dependent, for unknown reasons, on mitogenic signaling. The elucidation of this and other connections between growth-promoting signals and the cell cycle clock will surely come from future research.

8.9 The Myc oncoprotein perturbs the decision to phosphorylate pRb and thereby deregulates control of cell cycle progression

In the previous section, we described how the excessive mitogenic signals flowing through cancer cells lead to elevated levels of D-type cyclins and the resulting phosphorylation and functional inactivation of pRb. Prior to that discussion, we already had read about two other ways in which pRb function is compromised in neoplastic human cells: pRb function can be lost through mutation of the *Rb* gene, as is seen in retinoblastomas, osteosarcomas, and small-cell lung carcinomas. A functionally similar outcome can be observed in cells that have been infected and transformed by human papillomavirus (HPV); pRb binding by the viral E7 oncoprotein (see Figures 8.21C and 8.23B) plays a key role in the development of the great majority (>99.7%) of cervical carcinomas. As we will now see, human cancer cells devise a variety of other strategies to derail control of cell cycle progression.

We begin with the Myc protein, which, when expressed in a deregulated fashion, operates as an oncoprotein. *Myc* gene expression is deregulated in 15 to 30% of human cancers, resulting in elevated levels of the Myc protein. The Myc oncoprotein functions very differently from most of the oncoproteins that we encountered in Chapters 4, 5, and 6. These others, notably Ras and Src, operate in close proximity to cytoplasmic membranes and trigger complex signaling cascades that activate cytoplasmic signal-transducing proteins and, ultimately, nuclear transcription factors. Myc, for its part, is found in the nucleus, where it functions as a growth-promoting transcription factor.

Myc is a member of the large family of bHLH transcription factors. They were named after their shared three-dimensional structures, which include a **b**asic DNA-binding domain followed by amino acid sequences forming an α-**h**elix, a **l**oop, and a second α-**h**elix. Members of this transcription factor family form homo- and heterodimers with themselves and with other members of the family. Such dimeric complexes then associate with specific regulatory sequences, termed E-boxes (composed of the sequence CACGTG), which are found in the promoters of target genes that they regulate (Figure 8.27).

In the particular case of Myc, its actions are determined by its own levels as well as by its associations with partner bHLH proteins that either enhance or suppress its function as a transcription-activating factor. Phosphorylation of the Myc protein also modulates its functioning and stability. When Myc associates with Max, one of its bHLH partners that enhance its transcription-activating powers, the resulting Myc–Max heterodimeric transcription factor drives the expression of a large cohort of target genes. The products of many of these genes, in turn, have potent effects on the cell cycle, favoring cell proliferation.

(While Myc levels are strongly influenced by mitogenic signals, the levels of Max are kept relatively constant within cells. Accordingly, when normal cells are cultured in the presence of serum mitogens, Myc accumulates to substantial levels; conversely, Myc levels collapse when serum mitogens are withdrawn. This means that the levels of the Myc-Max heterodimer are continuously controlled by the flux of mitogenic signals that normal cells are receiving.)

As described later in this chapter, some bHLH proteins are also involved in orchestrating certain tissue-specific differentiation programs. Acting through intermediaries, Myc can prevent these other bHLH transcription factors from executing various differentiation programs. Consequently, Myc can simultaneously promote cell proliferation and block cell differentiation; this is precisely the biological behavior that is associated with actively growing cells that have not yet entered into post-mitotic, differentiated states. However, as cells slow their proliferation and become differentiated, the Myc–Max complexes disappear, since increasing levels of Mad protein (another bHLH partner) displace Myc from existing Myc–Max complexes (see Figure 8.27). The resulting Mad–Max complexes then function as transcriptional repressors, enabling cells in many human tissues to enter into post-mitotic differentiated states.

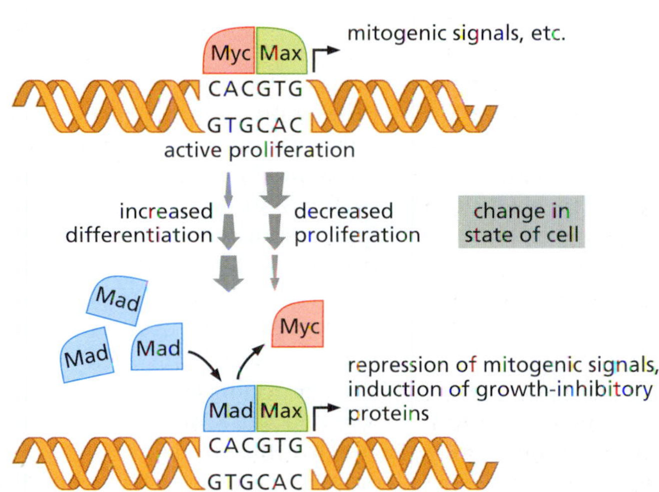

Figure 8.27 The Myc transcription factor Myc belongs to a family of bHLH (basic helix–loop–helix) transcription factors that act as heterodimers to modulate the transcription of a large cohort of target genes possessing E-box sequences. Myc–Max complexes act to promote transcription, while Mad–Max complexes act to repress transcription of most target genes. As cells differentiate, Mad levels increase progressively and Myc is displaced by Mad, resulting in the disappearance of Myc–Max complexes, which otherwise would block differentiation.

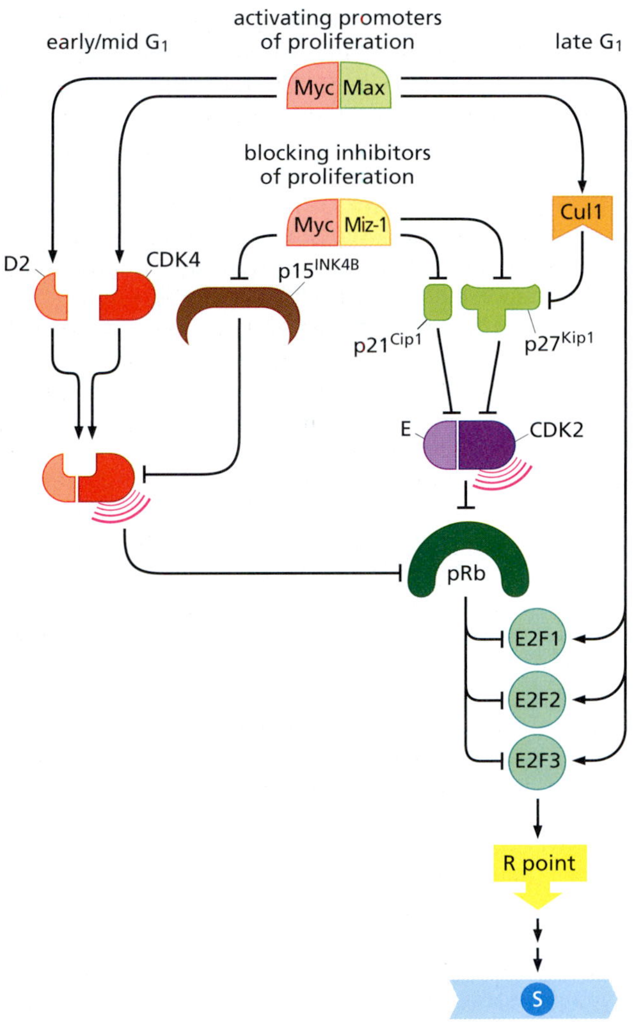

Figure 8.28 Actions of Myc on the cell cycle clock The Myc protein, acting with its Max partner *(top)*, is able to modulate the actions of a number of positive and negative regulators of cell cycle advance. For example, the Myc–Max heterodimer is able to induce expression of the growth-promoting proteins cyclin D2 and CDK4 *(left)*, which can promote advance through early G_1. At the same time, by increasing the expression of *Cul1* (which is responsible for degrading the $p27^{Kip1}$ CDK inhibitor; *upper right*) as well as E2Fs 1, 2, and 3 *(right, below)*, Myc favors advance into S phase. In addition, Myc, acting with its Miz-1 partner, is able to repress expression of the $p15^{INK4B}$, $p21^{Cip1}$, and $p27^{Kip1}$ CDK inhibitors; once again, effects are felt both in early/mid and in late G_1.

Myc interacts closely with the cell cycle machinery. Perhaps the first indication of this came from observations in which pairs of oncogenes were found to collaborate with one another in transforming rodent cells to a tumorigenic state. We will revisit oncogene collaboration later in Chapter 11. For the moment, suffice it to say that the *ras* and *myc* oncogenes were found to be highly effective collaborators in cell transformation, implying that each made its own unique contribution to this process. Similarly, a *ras* oncogene was found to collaborate with the adenovirus E1A oncogene in cell transformation. Together, these observations indicated that the *myc* and E1A oncogenes acted analogously in this experimental setting. Precisely how they did so was a mystery.

The fact that the E1A oncoprotein binds and inactivates pRb together with its p107 and p130 cousins (Section 8.5) suggested that the Myc protein could have similar effects on these vital cellular proteins, especially pRb. However, a direct association between Myc and pRb was ruled out. Instead, we have come to realize that Myc regulates the expression of a number of other critical components of the cell cycle clock. When expression of these components is driven by the abnormally high levels of Myc present in cells carrying *myc* oncogenes, the result is a physiologic state similar to that seen in cells lacking pRb function. Both changes deprive a cell of the normal control of progression through the G_1 phase of its cell cycle and deregulate passage through the R point.

Among the many targets of Myc is the cyclin D2 gene (Figure 8.28), whose elevated expression leads, in turn, to the hypophosphorylation of pRb. Myc also drives expression of the CDK4 gene, and the increased levels of CDK4 enable the formation of the cyclin D–CDK4 complexes that hypophosphorylate pRb and, at the same time, sequester the $p27^{Kip1}$ CDK inhibitor, thereby liberating cyclin E–CDK2 complexes from inhibition (see Figure 8.17). Myc also drives expression of the Cul1 protein, which plays a central role in the degradation of the $p27^{Kip1}$ CDK inhibitor through ubiquitylation. In one way or another, these Myc-driven gene expression changes serve to push the cell through the G_1 phase of its cell cycle.

By associating with a second transcription factor named Miz-1, Myc can also function as a transcriptional repressor (see Figure 8.28). In this role, Myc can repress expression of the genes encoding the $p15^{INK4B}$ and $p21^{Cip1}$ CDK inhibitors, which shut down the actions of CDK4/6 and CDK2, respectively. In fact, as we will read later, TGF-β, uses these two CDK inhibitors to block progression through the G_1 phase of the cell cycle. Hence, by preventing the expression of these two CDK inhibitors, Myc confers resistance to the growth-inhibitory actions of TGF-β. This represents an important means by which cancer cells can continue to proliferate under conditions (the presence of TGF-β in their surroundings) that would normally preclude their proliferation.

Finally, Myc is able to induce the expression of the genes encoding the E2F1, E2F2, and E2F3 transcription factor proteins. As we read earlier, these E2F transcription factors are negatively regulated by pRb and its two cousins. By causing these growth-promoting transcription factors to accumulate in a cell, Myc once again tips the balance in favor of cell proliferation.

Now, we begin to understand how the Myc oncoprotein can function so effectively to deregulate cell proliferation. It reaches in and pulls various regulatory levers within the cell cycle clock. The resulting changes in levels of key proteins cause a shift in the balance of growth-promoting and growth-inhibiting mechanisms, strongly favoring pRb inactivation and cell growth. In addition, Myc can also affect the long-term proliferation of cell lineages that are unrelated to its effects on the cell cycle apparatus (Sidebar 8.5).

Sidebar 8.5 Myc can also exert long-term effects on cell proliferation As discussed here, the Myc oncoprotein can act dynamically to deregulate the cell cycle clock machinery, thereby facilitating the transit of cells through their growth-and-division cycles. Myc can also affect cell proliferation in an unrelated way. Later, in Chapter 10, we will learn that normal cell populations are usually able to pass through only a limited number of cell cycles before their proliferation is halted, an effect that can be reversed by the actions of the telomerase enzyme. By binding to E-boxes in the promoter of the gene encoding a key subunit of human telomerase, Myc can increase expression of this enzyme in certain cell types, thereby enabling them to pass through many more growth-and-division cycles than would normally be allotted to them.

Many of Myc's actions in perturbing the cell cycle apparatus, as described in this section, provide a satisfying explanation of its ability to deregulate cell proliferation and, in this way, to contribute so importantly to cancer pathogenesis. The actions of the Myc protein illustrate a more general principle of cell physiology that affects many of our discussions throughout this text. Repeatedly, we make reference to the fact that a protein like Myc turns its downstream targets on or off. Such wording represents a simplification of what actually occurs inside living cells. In reality, most regulators do not operate by switching targets totally on or off. Instead, they increase or decrease the levels of their downstream targets, often incrementally. In effect, they *reset the regulatory dials*, thereby shifting the balance of signals in ways that favor or disfavor a certain cellular response.

Perhaps the most graphic demonstration of the potent mitogenic powers of Myc has come from an experiment in which Myc is engineered to be expressed at high, constant levels but as a fusion protein that is formed together with the estrogen receptor (ER). In the absence of an estrogen receptor ligand, such as estrogen or tamoxifen, the Myc–ER fusion protein is sequestered in a functionally inactive state in the cytoplasm. However, upon addition of ligand, this fusion protein is liberated from sequestration, rushes into the nucleus, and operates like the normal Myc transcription factor. When quiescent, serum-starved cells (in the G_0 state) expressing the Myc–ER fusion protein are treated with either estrogen or tamoxifen (but left in serum-deficient medium), these cells are induced to enter into the G_1 phase and progress to S phase (Figure 8.29). This illustrates the fact that Myc, acting on its own and in the absence of serum-associated mitogens, is able to relieve all of the constraints on proliferation that have held these cells in G_0 and is able to shepherd them all the way through G_1—an advance that usually requires substantial, extended stimulation by growth factors. In addition to Myc, a number of other transcription factors display the wide-ranging powers of oncoproteins (Sidebar 8.6).

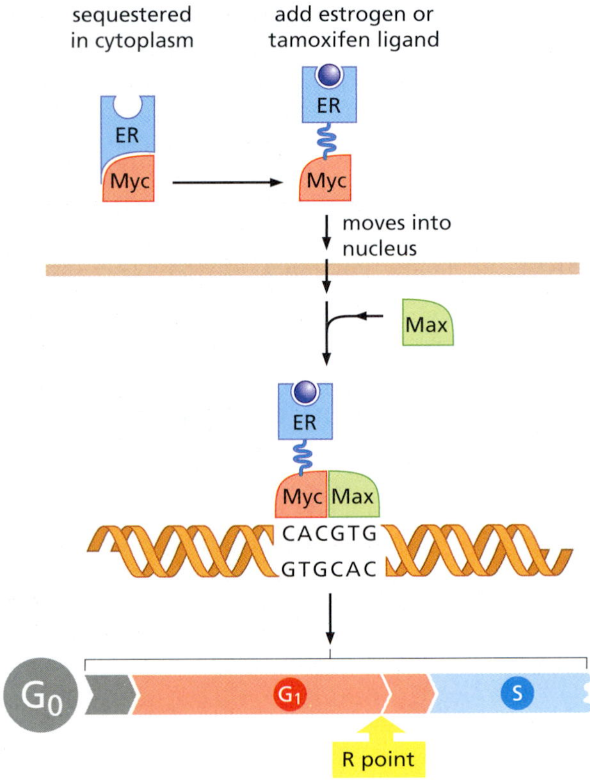

Figure 8.29 Powers of the Myc oncoprotein The wide-ranging effects of the Myc protein are illustrated by an experiment in which the Myc protein has been fused to the estrogen receptor (ER) protein *(blue)*. In the absence of ER ligands, such as estrogen or tamoxifen, the Myc–ER protein is trapped in the cytoplasm (through association with heat shock proteins, *not shown*). When estrogen or tamoxifen ligands of the ER *(small purple ball)* are added to cells, the Myc–ER protein migrates into the nucleus, associates with Max, and activates Myc target genes within minutes. Such activation, when induced in serum-starved cells in the G_0 phase, enables them to enter the active cell cycle and advance all the way through G_1 into the S phase.

As it turns out, Myc also acts on many other target genes that operate outside of the cell cycle machinery (Figure 8.30A). Analyses of both *Drosophila* and mammalian cells indicate that Myc associates with 12 to 15% of the genes present in the genomes of each of these cell types. (It is unclear how many of these genes are actually *regulated* by Myc, since, by some estimates, only about 1000 molecules of Myc are present in a normal mammalian cell.) In the case of *Drosophila*, examination of some of these genes indicates that Myc acts widely within the cell to promote cell growth in the narrow sense of the word, that is, increases in cell size. This may have implications for mammalian cells as well: if a human cell is unable to proceed through the entire cell cycle unless it grows above a certain minimum size, Myc may facilitate cell proliferation by promoting increases in cell size. Such an effect may act synergistically with Myc's ability, described above, to regulate many of the key proteins that govern a cell's passage through the R point. In the context of cancer pathogenesis, Myc seems to have yet other cell biological effects, as indicated by its influence on tumorigenesis by embryonic stem (ES) cells (Figure 8.30B).

8.10 TGF-β prevents phosphorylation of pRb and thereby blocks cell cycle progression

TGF-β represents a major growth-inhibitory signal that normal cells, especially epithelial cells, must learn to evade in order to become cancer cells. As we will learn later, TGF-β also exerts other, quite distinct effects on cells, by forcing them to change their differentiation programs. Often the resulting changes in cell phenotype actually favor tumor progression, as some of these phenotypic changes enable cancer cells to become anchorage-independent (Sidebar 5), angiogenic, and even invasive.

These two major effects of TGF-β—one antagonizing, the other favoring tumor progression—are clearly in direct conflict with one another. Cancer cells often resolve this dilemma by learning how to evade the cytostatic effects of TGF-β

Sidebar 8.6 Many oncoproteins function as transcription factors Myc is only one of a large group of transcription factors that can function as oncoproteins (Table 8.2). We first encountered oncoproteins of this type in our discussion of the immediate early genes *myc*, *fos*, and *jun* (Section 6.1), each of which encodes a transcription factor that can function as an oncoprotein. Later, we read about STAT proteins (Section 6.8), NF-κB, Notch, and Gli (Section 6.12), which also can play these dual roles. These various transcription factors act on a variety of target genes to induce a subset of the phenotypes that we associate with transformed cells.

The constituency of target genes that are acted upon by each of these transcription factors remains poorly defined. This lack of information is explained largely by the fact that at the experimental level, it has been extremely difficult to determine the identities of gene promoters to which a transcription factor of interest binds. The recent development of the chromatin immunoprecipitation (ChIP) technique holds the promise of addressing this problem. In the ChIP procedure, chromatin (containing both DNA and associated proteins that have been chemically cross-linked to the DNA) is mechanically sheared into small fragments. An antibody that reacts with a specific transcription factor is then used to immunoprecipitate those few chromatin fragments that have the transcription factor of interest associated with them. DNA molecules are then extracted from the precipitate and their sequences determined using PCR-based amplification. The resulting sequences are then matched with the sequences of all of the known genes present in the human or mouse genome, resulting in the identification of specific genes whose promoters are bound by this transcription factor and may consequently be activated or repressed through this association. Application of this ChIP technique to a wide variety of transcription factors should soon help to reveal the identities of the cohorts of genes that are targets of transcriptional induction or repression by each of these oncogenic transcription factors.

Table 8.2 Representatives of the classes of genes encoding TFs that act as oncoproteins[a]

Transcription factor	Representative genes
Leucine zipper + basic DNA-binding domain	*fos* and *jun*
Helix + loop + helix (bHLH)	*myc, N-myc, L-myc, tal,* and *sci*
Zinc finger	*myl/RARα, erbA, evi-1,* and *gli-1*
Homeobox	*pbx* and *Hox2.4*
Others	*myb, rel, ets-1, ets-2, fli-1, spi-1, ski*

[a]Genes are grouped according to the structural features of the encoded proteins.

Adapted from T. Hunter, *Cell* 64:249–270, 1991.

while leaving intact the other responses, notably those favoring tumor cell invasiveness. Not unexpectedly, the cytostatic actions of TGF-β derive from its direct and indirect effects on pRb, the central controller that determines whether or not a cell will proliferate.

We previously learned that TGF-β has its own, quite unique receptor, which uses serine/threonine kinases in its cytoplasmic domains (rather than tyrosine kinases) to emit signals. The primary targets of these signals documented to date are proteins of the Smad transcription factor family. Once phosphorylated by a TGF-β receptor, Smad2 (or Smad3) protein molecules then associate with Smad4 proteins (which are not substrates for phosphorylation by the receptor), and the resulting heterodimeric protein complex migrates to the nucleus, where it functions as a transcription factor (see Figure 6.29D).

As we read earlier, in the context of cell cycle control, the most important targets of action by TGF-β are the genes encoding the two CDK inhibitors $p15^{INK4B}$ and, to a lesser extent, $p21^{Cip1}$ (see Figure 8.14A). Each has a CAGAC sequence in its promoter that attracts a heterodimeric Smad3–Smad4 transcription factor complex. On its own, such a Smad complex cannot activate transcription of these genes. Instead, it collaborates with yet another transcription factor—the Miz-1 factor mentioned in the last section—that binds to adjacent DNA sequences in the promoters of these two genes in order to activate their transcription (Figure 8.31, left).

We also noted earlier that the Myc oncoprotein, acting in the opposite direction, can block the induction of these two CDK inhibitors, doing so through its ability to associate with Miz-1. By repressing expression of the genes encoding

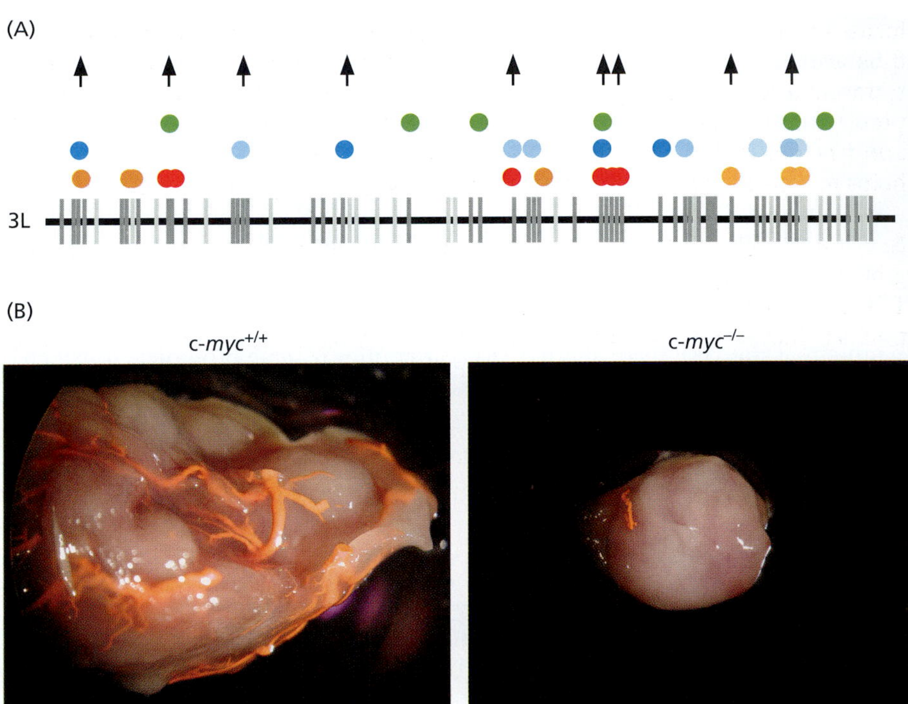

Figure 8.30 Wide-ranging actions of Myc Myc acts in diverse ways on cells beyond its effects on the cell cycle. (A) A systematic study of the target genes acted upon by the Myc, Max, and Mad-related Mnt transcription factors in the cells of *Drosophila melanogaster* indicated that they are present in large numbers and are scattered throughout the *Drosophila* genome on its four chromosomes. While the entire chromosomal array was surveyed for Myc binding sites, only the long arm of chromosome 3 (i.e., 3L) is shown here. This survey analyzed the ability of *Drosophila* Myc *(green circles)*, the Mad-like Mnt *(red circles)*, and Max *(blue circles)* to associate with specific chromosomal DNA segments. Altogether, this survey indicated that about 2000 *Drosophila* gene promoters are associated with one or more of these bHLH proteins; a majority of these promoters are affiliated with genes that can be shown to be up- or down-regulated by pairs of these bHLH transcription factors. A number of genes that are orthologs of mammalian genes and known to be targets of Myc action in mammalian cells are indicated with *black arrowheads*. (B) Other indications of the pleiotropic actions of Myc come from biological experiments, such as this one, in which the tumorigenicity of mouse embryonic stem (ES) cells has been gauged. Wild-type ES cells *(left,* c-$myc^{+/+}$) form large, highly *vascularized* tumors *(orange vessels)*, while cells lacking Myc function *(right,* c-$myc^{-/-}$) fail to grow robustly, in large part because of a failure to develop an adequate blood supply; the two cell populations grow with equal rates *in vitro*. (A, adapted from A. Orian et al., *Genes Dev.* 17:1101–1114, 2003; B, from T.A. Baudino et al., *Genes Dev.* 19:2530–2543, 2002.)

$p15^{INK4B}$ and $p21^{Cip1}$, Myc removes two major obstacles standing in the way of cell cycle progression (see Figure 8.28). Stated differently, the constitutively high levels of Myc protein made by *myc* oncogenes ensure that expression of these two CDK inhibitors is strongly repressed, thereby paving the way for vigorous cell cycle advance.

In normal cells, TGF-β needs to have the last word in determining whether or not these cells proliferate. In order to do so, TGF-β must overrule any conflicting signals that Myc might release. To guarantee success in this endeavor, TGF-β must ensure that Myc does not thwart its scheme to activate expression of the $p15^{INK4B}$ and $p21^{Cip1}$ CDK inhibitors. So, in normal cells, TGF-β keeps Myc away from the promoters of these genes by shutting down expression of the *myc* proto-oncogene (Figure 8.31, right). The *myc* gene promoter has a sequence to which a TGF-β–activated Smad3 transcription factor can bind. Adjacent to this sequence is another that allows binding of either E2F4 or E2F5, the two E2F transcription factors that favor transcriptional repression (Section 8.7). By forming a tripartite complex with E2F4/5 plus p107, Smad3 ensures the shutdown of *myc* transcription, thereby eliminating the growth-promoting effects of the Myc protein from the cell's regulatory circuitry. Once TGF-β has succeeded in removing Myc from the scene—from the promoters of the $p15^{INK4B}$ and $p21^{Cip1}$ CDK inhibitors—TGF-β can then use its Smad3–Smad4 complexes, working together with Miz-1, to activate these two promoters, thereby inducing expression of these critically important CDK inhibitors (Figure 8.31, left).

Many types of cancer cells must evade TGF-β-imposed growth inhibition if they are to thrive. More specifically, such cancer cells depend on high levels of the Myc transcription factor to drive their proliferation and therefore must liberate *myc* transcription from the repressive actions of TGF-β. This explains why, for example, in a series of 12 human breast carcinomas, *myc* expression in all 12 of these tumors was no longer responsive to TGF-β–imposed shutdown even though many TGF-β–induced responses were found to be intact in 11 of these tumors. (Since the *myc* gene itself was apparently in a wild-type configuration in these tumors, this suggested some deregulation of the transcription factors responsible for expression of this gene.) Similarly, when *myc* oncogenes are formed by mutation, these genes acquire promoters that fire constitutively and thus are no longer responsive to TGF-β–induced repression. Once protected from the threat of this repression, the *myc* oncogene, more specifically, its Myc product, proceeds to hold expression of the two CDK inhibitors (i.e., $p15^{INK4B}$ and $p21^{Cip1}$) at a very low level and does so indefinitely. As noted earlier, this helps to create a condition in the cell permitting rapid proliferation.

In a more general sense, cancer cells place a high premium on escaping the growth-inhibitory influences of TGF-β while leaving other, potentially useful TGF-β responses intact. Often this evasion of TGF-β–mediated growth inhibition is achieved through inactivation of the pRb signaling pathway. In effect, in normal cells pRb operates as the brake lining that calls a halt to cell proliferation in response to TGF-β-initiated signals. Accordingly, if the pRb protein is eliminated from the regulatory circuitry through one of the various mechanisms described earlier, then the ability of TGF-β to impose growth arrest is greatly compromised, since the $p15^{INK4B}$ induced by TGF-β now fails to effectively block cell cycle advance.

A partial inactivation of the TGF-β responses can also be achieved, for example, through mutations of the gene encoding the Smad2 protein; such mutations are frequently observed in colon carcinomas. A similar partial blunting of TGF-β–induced anti-mitogenic responses seems to be achieved by the Ski

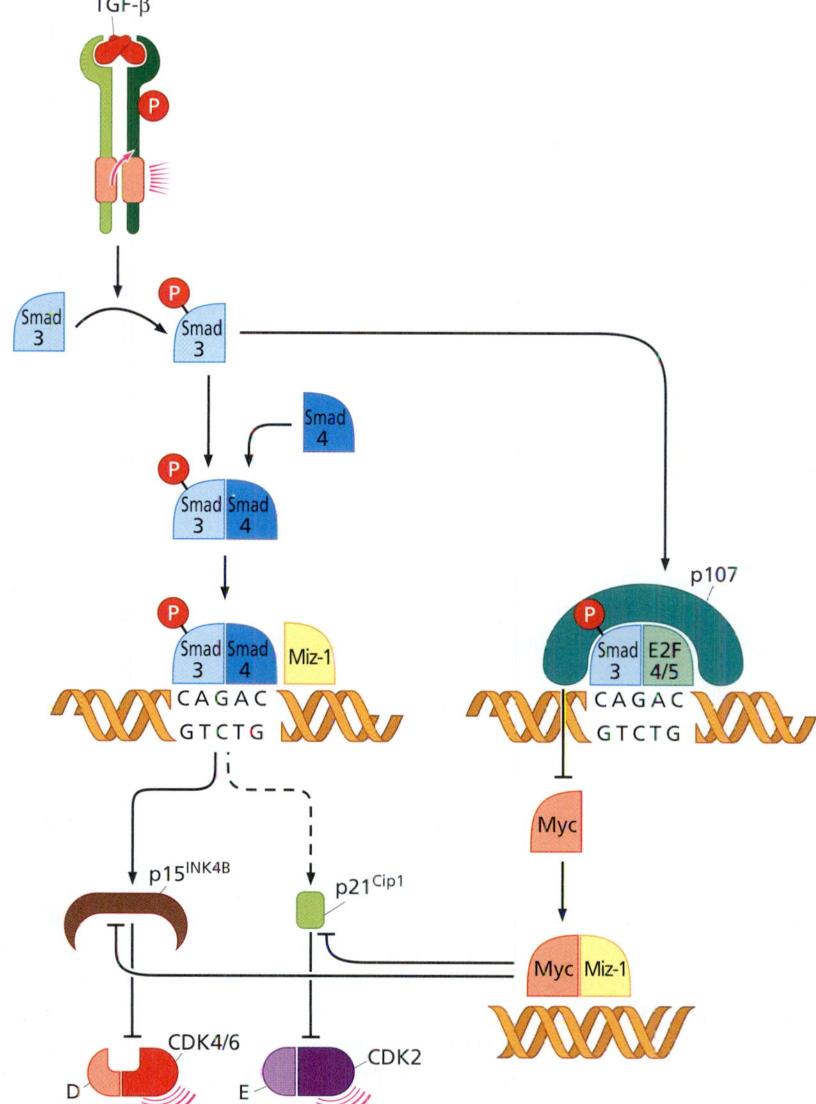

Figure 8.31 Countervailing actions of TGF-β and Myc TGF-β, acting through its receptor, causes phosphorylation of several Smad proteins, such as Smad3 shown here *(left top)*. (Smad2, *not shown,* acts equivalently to Smad3.) Smad3 then forms a heterodimeric complex with Smad4, which migrates to the nucleus, where it teams up with Miz-1 to induce expression of the $p15^{INK4A}$ and (weakly) $p21^{Cip1}$ CDK inhibitors. Myc, for its part, is capable of teaming up with Miz-1 to repress expression of these CDK inhibitors *(lower right)*. However, this action of Myc can be preemptively blocked by TGF-β, which dispatches Smad 3 to form a complex with E2F4 or E2F5 plus p107 (pRb's cousin; *right*) that represses expression of the *myc* gene, thereby causing Myc levels to collapse. This ensures that TGF-β succeeds in inducing expression of the two CDK inhibitors—$p15^{INK4A}$ and $p21^{Cip1}$—and thereby shuts down cell cycle progression in the early/mid G_1 phase of the cell cycle.

oncoprotein (Table 3.3). Both this protein and its cousin, termed Sno, are able to bind to the Smad3–Smad4 transcription factor complex and block its ability to repress *myc* transcription.

Half of all pancreatic carcinomas and more than a quarter of all colon cancers carry mutant, inactivated Smad4 proteins. Without the presence of Smad4, neither Smad2–Smad4 nor Smad3–Smad4 complexes can form. These two complexes are the chief agents dispatched by the TGF-β receptor to the nucleus with the important assignment to shut down proliferation.

Some types of cancer cells undertake a far more drastic evasive maneuver, however: they jettison all responsiveness to TGF-β by inactivating the genes encoding the TGF-β receptor. For example, the great majority of colon cancers that suffer "microsatellite instability" (a state leading to high rates of mutation; see Section 12.9) carry mutant, inactivated TGF-βII receptors. While such cells acquire the ability to evade TGF-β–imposed growth inhibition, they also forgo potential benefits that might be conferred by TGF-β later in tumor progression, when TGF-β helps cancer cells to acquire malignant phenotypes such as invasiveness.

8.11 pRb function and the controls of differentiation are closely linked

Cell differentiation is a process of crucial importance to cancer pathogenesis and, as such, is mentioned repeatedly in this text. For the moment, we can imagine, a bit simplistically, that cells throughout the body can exist in either of two alternative growth states. They may be found in a relatively undifferentiated state in which they retain the option to divide in the event that mitogenic signals call for their proliferation; this is essentially the behavior of stem cells. The alternative is that cells leave this state and enter into a more differentiated state, whereupon they give up the option of ever proliferating again and thus become post-mitotic. The decisions that govern these fundamental changes in cell biology must be explainable in terms of the molecular controls that determine whether a cell remains in the active growth-and-division cycle, exits reversibly into G_0, or exits irreversibly from this cycle into a post-mitotic, differentiated state.

The opposition between cell proliferation and differentiation is seen most clearly in the formation of most types of cancer cells in which differentiation is partially or completely blocked (see, for example, Sidebar 8.7). Knowing this, we might ask whether two independent changes in the cellular control circuitry must be made during the formation of a cancer cell—one that deregulates control of proliferation and another that blocks differentiation. The alternative is simpler: might certain alterations, in a single stroke, deregulate proliferation and prevent differentiation?

Evidence supporting the latter idea, which implies a close coupling between the mechanisms controlling these two processes, has been forthcoming from studies of muscle cell differentiation *in vitro*. In this experimental system, relatively undifferentiated **myoblasts** can be induced to differentiate into **myocytes** (muscle cells). An enforced block in the G_1 phase of the myoblast cell cycle, achieved experimentally by introducing genes encoding elevated levels of either the $p16^{INK4A}$ or the $p21^{Cip1}$ CDK inhibitor, causes the myoblasts to differentiate under conditions (e.g., high levels of growth factors in the medium) in which such differentiation would otherwise not occur (Figure 8.33). Genetically altered mouse myoblasts that lack both *Rb* gene copies, and therefore express no pRb, are unresponsive to the differentiation-inducing influences of the introduced 16^{INK4A}- and $p21^{Cip1}$-encoding genes. Additionally, overexpression of cyclin D1 in myoblasts (which drives pRb phosphorylation and inactivation) blocks the differentiation that is normally seen when growth factors are removed from their culture medium. These experiments indicate that hypophosphorylated pRb is needed for two ostensibly unrelated functions—halting the proliferation of myoblasts, and facilitating the differentiation of these cells into myocytes.

Yet other molecular connections between cell cycle control and differentiation have been uncovered. Thus, enforced expression of CDK inhibitors, such as $p21^{Cip1}$ or $p27^{Kip1}$, is able to induce differentiation of neuroblastoma cells and myelomonocytic leukemia cells, while enforced expression of E2F4 (a repressive E2F; Section 8.7) induces the differentiation of neuronal precursor cells. Mice that are genetically deprived of this transcription factor (and thus have an $E2f4^{-/-}$ genotype) die in gestation because of defective differentiation of their erythrocytes. The C/EBP transcription factor, which induces the differentiation of adipocytes (fat cells), associates with E2Fs, and an osteoblast-specific transcription factor (CBFA1) has been found to bind pRb. Finally, retinoic acid induces the differentiation of acute promyelocytic leukemia (APL) cells, in part through its ability to trigger degradation of cyclin D1.

These diverse experimental observations indicate that the machinery controlling the phosphorylation state of pRb and its cousins is also an active participant

Sidebar 8.7 Blocked differentiation can accompany the advance of tumor progression Some hematopoietic malignancies, specifically, leukemias, come in two forms—chronic and acute. The chronic diseases, such as chronic myelogenous leukemia (CML), are composed of more mature, differentiated cells and can proceed for years without becoming life-threatening (Figure 8.32). In the case of CML, after a period of 3 to 5 years, suddenly there is an eruption of a more malignant, aggressive form of the disease that is termed "blast crisis." The cells present in blast crisis clearly originate from the same cell clone that initiated the chronic phase of the disease. However, they are less differentiated and multiply ceaselessly, creating a life-threatening disease that is difficult to treat. It seems that the cells in the chronic phase are derived from a mutant self-renewing stem cell, but this disease remains relatively benign, because the differentiation of these mutant cells into post-mitotic neutrophils proceeds quite normally. The moment that such differentiation is blocked [by some still-unknown genetic change(s)], cells in this population are trapped in a less differentiated stem cell compartment (of either erythroid or myeloid type) and proliferate without a compensating exit of their progeny into a more differentiated, post-mitotic compartment.

Figure 8.32 Disease progression and dedifferentiation (A) The disease of chronic myelogenous leukemia (CML) leads to the accumulation in the blood of elevated numbers of cells that are almost fully differentiated neutrophils *(purple cells, complex nuclei)*. The *pink* cells are erythrocytes. (B) After several years, when the chronic phase of the disease progresses to an "accelerated" phase, these more differentiated neutrophils are increasingly replaced by less differentiated blast cells *(arrows)*. (C) Ultimately, this accelerated phase progresses to "blast crisis," when the peripheral blood carries large numbers of undifferentiated blast cells. This last phase appears to reflect the behavior of leukemic stem cells that were already present in the chronic phase of the disease but are now prevented from differentiating. (A and B, courtesy of P.G. Maslak; C, courtesy of the American Society of Hematology Image Bank.)

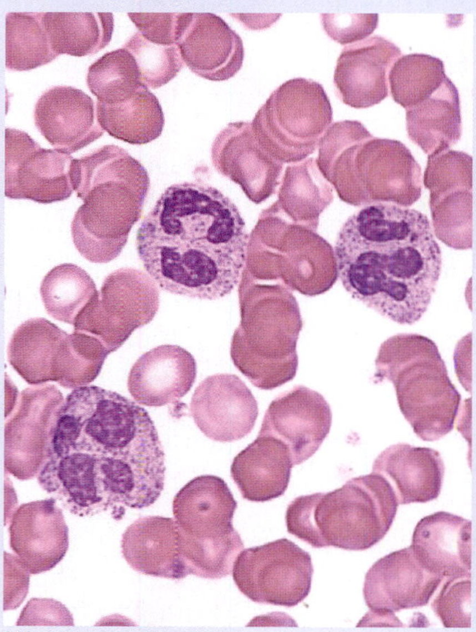

(A) chronic phase

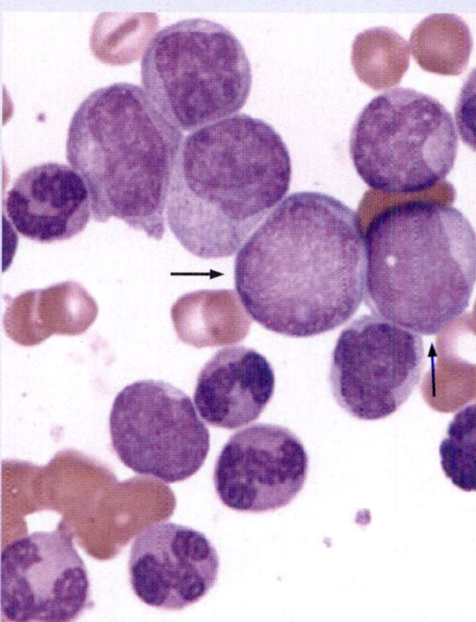

(B) accelerated phase

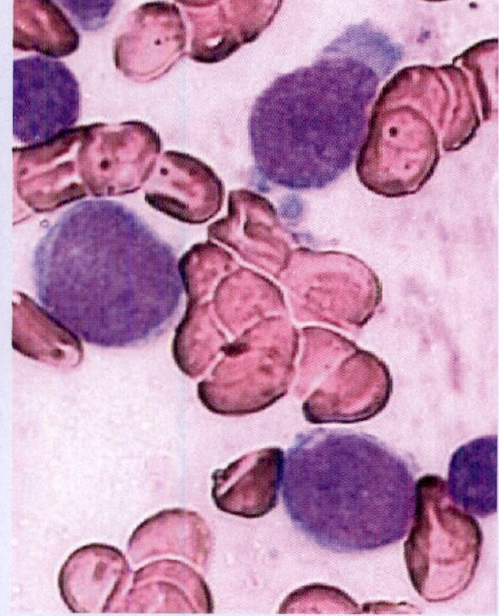

(C) blast crisis

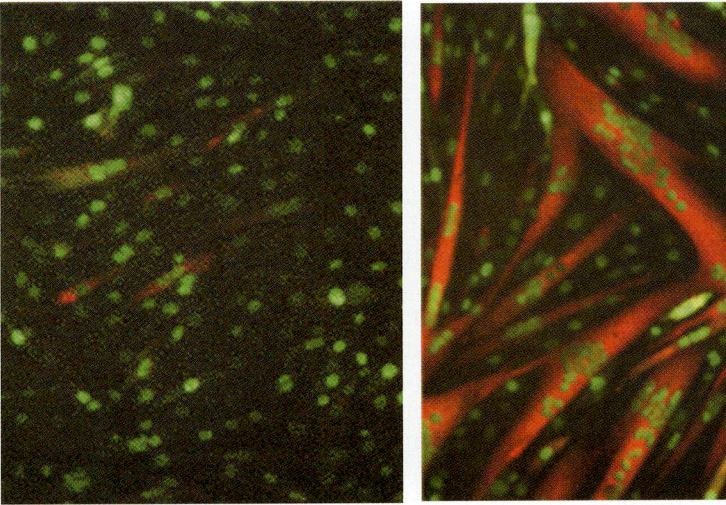

Figure 8.33 Differentiation of myoblasts into myocytes Myoblasts, the less differentiated precursors of muscle cells (myocytes), can be cultured *in vitro* and remain in the undifferentiated state *(left panel)*. However, various types of physiologic signals can induce them to differentiate *in vitro* into myocytes, whereupon they fuse to form muscle fibers *(red, right panel)*. For example, removal of serum (and its associated growth factors from the culture medium leads to such differentiation). Numerous experiments have shown that differentiation can be prevented by forcing pRb phosphorylation and inactivation. Conversely, signals that prevent pRb phosphorylation favor and often induce differentiation. (From E.M. Wilson et al., *Mol. Biol. Cell* 15:497–505, 2003.)

in the differentiation process, at least in the well-studied biological systems described above. More specifically, signals favoring the hyperphosphorylation (and thus functional inactivation) of the three pocket proteins (pRb, p107, and p130) operate to block differentiation, while those that prevent pRb hyperphosphorylation favor differentiation. This mechanistic linkage, still poorly documented for most cell types, suggests another idea that remains largely speculative: when a cell decides to leave the active cell cycle in order to differentiate, it exits the cycle sometime during the G_1 phase, when pRb is active in controlling the fate of the cell.

Yet another line of evidence indicating a coupling between differentiation and the cell cycle machinery comes from research on the Myc protein. As mentioned earlier (Section 8.9), the Myc protein shifts the balance between proliferation and post-mitotic differentiation in favor of proliferation. In order to understand these actions, we need to know that many of the transcription factors controlling differentiation programs are, like Myc, members of the bHLH family of transcription factors. However, these bHLH proteins function very differently from Myc: they orchestrate complex, tissue-specific differentiation programs, while Myc acts in the opposite direction to block differentiation and promote proliferation. Differentiation-associated bHLH transcription factors have been most extensively studied in the embryonic development of several distinct cell lineages, including those leading to the formation of muscles, the nervous system, the pancreas, and the immune system. For example, four distinct bHLH proteins—termed MyoD, Myf5, myogenin, and MRF4—operate in myoblast-to-myocyte differentiation (Figure 8.33) by controlling various phases of the muscle-specific differentiation program.

Myc works to increase production of the Id2 protein, and the latter acts as an antagonist of the bHLH transcription factors that program differentiation. Id2 is one of a group of related proteins (Id1 to Id4) that are also members of the bHLH family of transcription factors. The Id proteins operate, however, as dominant-negative inhibitors of the other bHLH transcription factors (Figure 8.34). More specifically, the Ids can form heterodimers with bHLH transcription factors but lack the ability to bind to DNA (because they lack the basic domain involved in DNA recognition). This explains how the Id proteins such as Id2 can act as inhibitors of the bHLH differentiation-inducing transcription factors, thereby blocking differentiation.

Id proteins are present at high levels in many types of actively growing cells, and this in itself reduces the likelihood that these cells will differentiate. Thus, by associating with MyoD, an Id protein can prevent MyoD from programming muscle differentiation in actively growing myoblasts. During the normal course

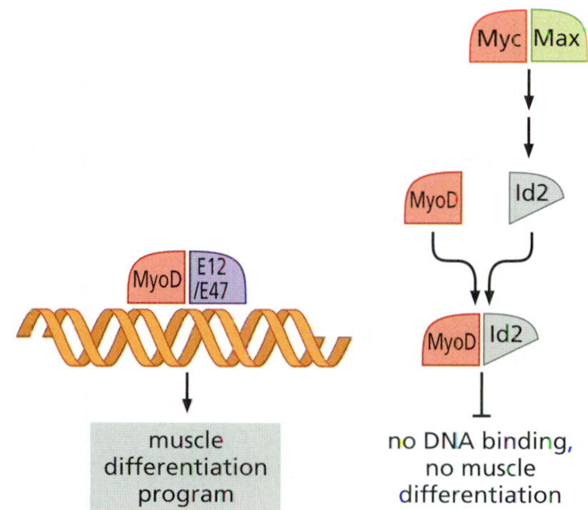

Figure 8.34 Id transcription factors and inhibition of differentiation Lineage-specific bHLH transcription factors, such as MyoD, shown here *(pink)*, form heterodimers with widely expressed E12 or E47 bHLH partners *(purple)*. The resulting heterodimeric transcription factors orchestrate differentiation programs in a variety of tissues, including the muscle differentiation indicated here *(left)*. The formation of these functional heterodimeric transcription factor complexes can, however, be blocked by Id proteins *(light gray)*, which form heterodimers with the lineage-specific bHLH proteins, thereby preventing association of the latter with E12 or E47. Because the Id proteins lack a DNA-binding domain, they act as natural "dominant-negative" inhibitors of the lineage-specific bHLH proteins. Expression of certain Id proteins, such as the Id2 protein shown here, can be induced by Myc (acting with its partner Max; *top right)*; this helps to explain the observed ability of the Myc oncoprotein to block differentiation of various cell types.

of differentiation, however, the levels of Id proteins sink to undetectably low levels, and MyoD, now free of interference by Ids, is able to dimerize with its bHLH partners (called E12 and E47) in order to activate the muscle-specific differentiation program (see Figure 8.34). Depressed synthesis of the Id1 protein is known to be required for cell cycle withdrawal and differentiation in numerous cell lineages: muscle, pancreas, mammary epithelial cells, myeloid cells, erythroid cells, myocardial cells, B cells, T cells, and osteoblasts.

As might be expected from the above, the Id proteins also have been associated with cancer pathogenesis. In many types of normal cells, the molecules of Id2 are bound and sequestered by the far more abundant molecules of pRb. However, in neuroblastomas, a relatively common pediatric tumor, Id2 is often overexpressed because its expression is driven by extra copies of the N-Myc protein, a *myc* cousin that acts on the same targets as *myc* in cells (Section 4.5). Now the tables are turned: in neuroblastoma cells, Id2 accumulates to such high levels that it is in great (>10×) molar excess of pRb. Consequently, pRb can no longer sequester and regulate these cells' Id2 proteins. They are then free to block the actions of differentiation-inducing bHLH transcription factors.

The theme of mutual antagonism between growth-promoting components of the cell cycle clock and the differentiation machinery is further illustrated by other cross-connections between these two regulatory circuits. For example, during active proliferation, cyclin E–CDK2 and cyclin A–CDC2 phosphorylate MyoD in myoblasts, thereby triggering its degradation and eliminating its differentiation-inducing effects. Hints of yet other cross connections continue to be uncovered. One research report indicates that the cyclin D1 protein associates with and inhibits a transcription factor that is responsible for expression of the differentiation-associated proteins in the epithelial cells of the duodenum.

Antagonism operating in the opposite direction is exhibited by key components of the differentiation machinery. Thus, the MyoD protein, which is so important in programming muscle differentiation, reaches back into the cell cycle clock and inhibits proliferation by causing increased production of pRb and of the $p21^{Cip1}$ CDK inhibitor. MyoD is also reported to bind directly to CDK4 and inhibit its activity.

Previously, we learned that when a cell receives signals that encourage differentiation, Myc's partner, Max, is kidnapped by Mad, the levels of which increase progressively during differentiation (see Figure 8.27). This creates a state that is permissive for cell cycle arrest and differentiation. Thus, the Mad–Max complexes cause increased expression of the two cousins of pRb, p107 and p130. While this is happening, Myc, deprived of help from its Max partner, becomes sidelined from the molecular circuitry governing the fate of the cell, and thus loses its ability to block differentiation.

A dramatic demonstration of the opposition between the Myc oncoprotein and cell differentiation has come from a mouse model of liver cancer pathogenesis, which depends on the targeted expression of a *myc* transgene in hepatocytes. Large hepatocellular carcinomas form, and these regress when the *myc* transgene is shut down. At the same time, many of the carcinoma cells, which previously lacked most of the traits of normal hepatocytes, rapidly differentiate into normally appearing liver cells that regenerate many of the histological features of the normal liver.

These various controls on cell differentiation, involving pRb, Myc, Ids, and other regulatory proteins, clearly have major implications for cancer pathogenesis, since tumors formed by more differentiated cells are usually less aggressive while those composed of poorly differentiated cells tend to be much more aggressive and carry a worse prognosis for the patient.

8.12 Control of pRb function is perturbed in most if not all human cancers

Deregulation of the pRb pathway yields an outcome that is an integral part of the cancer cell phenotype—unconstrained proliferation. This explains why normal regulation of the R-point transition, as embodied in pRb phosphorylation, is likely to be disrupted in most if not all types of human tumor cells (Tables 8.3

Table 8.3 Molecular changes in human cancers leading to deregulation of the cell cycle clock

Specific alteration	Clinical result
Alterations of pRb	
Inactivation of the *Rb* gene by mutation	retinoblastoma, osteosarcoma, small-cell lung carcinoma
Methylation of *Rb* gene promoter	brain tumors, diverse others
Sequestration of pRb by Id1, Id2	diverse carcinomas, neuroblastoma, melanoma
Sequestration of pRb by the HPV E7 viral oncoprotein	cervical carcinoma
Alteration of cyclins	
Cyclin D1 overexpression through amplification of *cyclin D1* gene	breast carcinoma, leukemias
Cyclin D1 overexpression caused by hyperactivity of *cyclin D1* gene promoter driven by upstream mitogenic pathways	diverse tumors
Cyclin D1 overexpression due to reduced degradation of cyclin D1 because of depressed activity of GSK-3β	diverse tumors
Cyclin D3 overexpression caused by hyperactivity of *cyclin D3* gene	hematopoietic malignancies
Cyclin E overexpression	breast carcinoma
Defective degradation of cyclin E protein due to loss of hCDC4	endometrial, breast, and ovarian carcinomas
Alteration of cyclin-dependent kinases	
CDK4 structural mutation	melanoma
Alteration of CDK inhibitors	
Deletion of $\mathit{15}^{INK4B}$ gene	diverse tumors
Deletion of $\mathit{16}^{INK4A}$ gene	diverse tumors
Methylation of $\mathit{p16}^{INK4A}$ gene promoter	melanoma, diverse tumors
Decreased transcription of $\mathit{p27}^{Kip1}$ gene because of action of Akt/PKB on Forkhead transcription factor	diverse tumors
Increased degradation of $\mathrm{p27}^{\mathrm{Kip1}}$ protein due to Skp2 overexpression	breast, colorectal, and lung carcinomas, and lymphomas
Cytoplasmic localization of $\mathrm{p27}^{\mathrm{Kip1}}$ protein due to Akt/PKB action	breast, esophagus, colon, thyroid carcinomas
Cytoplasmic localization of $\mathrm{p21}^{\mathrm{Cip1}}$ protein due to Akt/PKB action	diverse tumors
Multiple concomitant alterations by Myc, N-myc or L-myc	
Increased expression of Id1, Id2 leading to pRb sequestration	diverse tumors
Increased expression of cyclin D2 leading to pRb phosphorylation	diverse tumors
Increased expression of E2F1, E2F2 E2F3 leading to expression of cyclin E	diverse tumor
Increased expression of CDK4 leading to pRb phosphorylation	diverse tumors
Increased expression of Cul1 leading to $\mathrm{p27}^{\mathrm{Kip1}}$ degradation	diverse tumors
Repression of $\mathrm{p15}^{\mathrm{INK4B}}$ and $\mathrm{p21}^{\mathrm{Cip1}}$ expression allowing pRb phosphorylation	diverse tumors

Table 8.4 Alteration of the cell cycle clock in human tumors A plus sign indicates that this gene or gene product is altered in at least 10% of tumors analyzed. Alteration of gene product can include abnormal absence or overexpression. Alteration of gene can include mutation and promoter methylation. More than one of the indicated alterations may be found in a given tumor.

Tumor type	Gene product or gene						% of tumors with 1 or more changes
	Rb	Cyclin E1	Cyclin D1	p16^{INK4A}	p27^{Kip1}	CDK4/6	
Glioblastoma	+	+		+	+	+/+	>80
Mammary carcinoma	+	+	+	+	+	+/	>80
Lung carcinoma	+	+	+	+	+	+/	>90
Pancreatic carcinoma			[a]		+		>80
Gastrointestinal carcinoma	+	+	+[b]	+	+	+/[e]	>80
Endometrial carcinoma	+	+	+	+	+	+/	>80
Bladder carcinoma	+	+	+	+	+		>70
Leukemia	+	+	+	+[c]	+	+/	>90
Head and neck	+		+	+	+	+/	>90
Lymphoma	+	+	+[d]	+[c]	+	/+	>90
Melanoma		+	+	+	+	+/	>20
Hepatoma	+	+	+	+[c]	+	+/[e]	>90
Prostate carcinoma	+	+	+	+	+		>70
Testis/ovary	+	+	+[b]	+	+	+/	>90
Osteosarcoma		+		+		+/	>80
Other sarcomas		+	+	+	+	/+	>90

[a]Cyclin D3 is up-regulated in some tumors.
[b]Cyclin D2 is up-regulated in some tumors.
[c]p15^{INK4B} also found to be absent in some tumors.
[d]Cyclin D2 and D3 also found up-regulated in some lymphomas.
[e]CDK2 also found to be up-regulated in some tumors.

Adapted from M. Malumbres and M. Barbacid, *Nat. Rev. Cancer* 1:222–231, 2001.

and 8.4). These disruptive mechanisms are summarized in Figure 8.35. We are already familiar with the most direct mechanism for deregulating advance through the R point—inactivation of the *Rb gene* through mutation. In some tumors, an equivalent outcome is achieved through methylation of the *Rb* gene promoter. In others, pRb, though synthesized in normal amounts, may be functionally inactivated by viral oncoproteins, such as the HPV E7 protein, which prevent pRb from binding and regulating E2Fs.

Yet another strategy used by cancer cells to inactivate pRb function is indicated by the presence of very high levels of cyclin D1 in a variety of human tumor cells. This is most widely documented in breast cancer cells, in which as many as half of tumors have been reported to show elevated levels of this protein. In these and other carcinomas, the overexpression is sometimes achieved by increases in

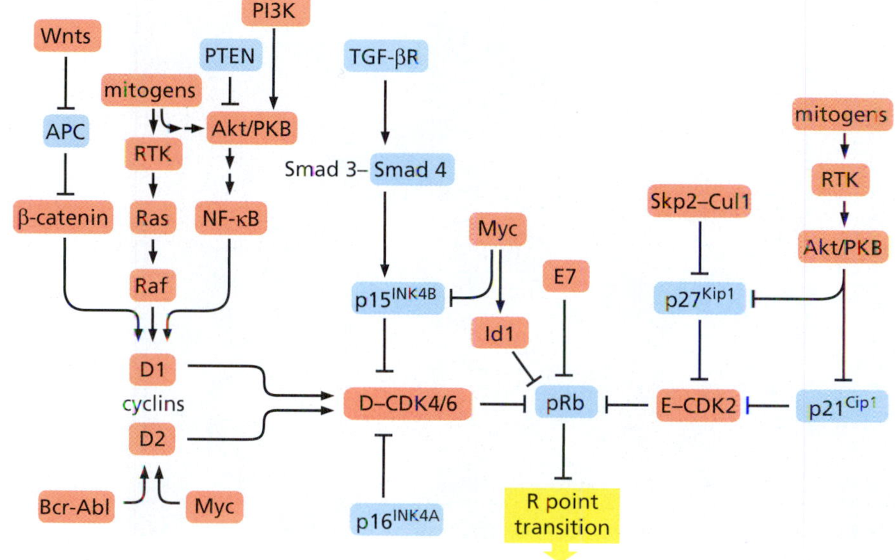

Figure 8.35 Perturbation of the R-point transition in human tumors The decision to advance through the R-point transition *(yellow, middle, bottom)* can be perturbed in a variety of ways in human tumors. Elements that favor advance through the R point are drawn in *red*, while those that undertake to block this advance are shown in *blue*. Almost all human tumors show either a hyperactivity of one or more of the agents favoring this advance *(red)* or an inactivation of the agents blocking this advance *(blue)*.

the copy number of the cyclin D1 genes (i.e., gene amplification; Figure 8.36). More frequently, however, breast cancer cells acquire excessive cyclin D1 by altering the upstream signaling pathways (Section 8.8) that are normally responsible for controlling expression of the cyclin D1 gene.

A more devious ploy is frequently exploited by cancer cells to disable the pRb machinery: they shut down expression of their $p16^{INK4A}$ tumor suppressor protein. Recall that the $p16^{INK4A}$ protein, like its $p15^{INK4B}$ cousin, inhibits the cyclin D–CDK4/6 responsible for initiating pRb phosphorylation. In the absence of the $p16^{INK4A}$ protein, pRb phosphorylation operates without an important braking mechanism, resulting in excessive cyclin D–CDK4/6 kinase activity, deregulated pRb phosphorylation, and inappropriate inactivation of pRb (see Figure 8.35). Individuals suffering from one form of familial melanoma inherit defective versions of the $p16^{INK4A}$ gene. It is unclear why loss of this particular CDK inhibitor, which seems to operate in all cell types throughout the body, should affect specifically the melanocytes in the skin that are the normal precursors of melanoma cells. In sporadic (i.e., nonfamilial) tumors of various sorts, cancer cells resort far more frequently to another strategy to shed $p16^{INK4A}$ function—they methylate the CpG sequences present in the promoter of the $p16^{INK4A}$ gene.

Evidence of an even more cunning strategy for destabilizing this control circuit has been found in the genomes of a small number of both sporadic and familial melanomas. In these cancers, point mutations in the CDK4 gene (the R24C mutation) create CDK4 molecules that are no longer susceptible to inhibition by the family of INK4 molecules (i.e., p15, p16, p18, and p19). While these various CDK inhibitors may be perfectly intact and functional in such tumor cells, their normally responsive CDK4 target now eludes them. Once again, this permits CDK4, together with its cyclin D partners, to drive the initial steps of pRb phosphorylation in a deregulated fashion. [Since the R24C mutation creates a dominant allele of CDK4 (at the cellular level), only one of the two copies of the gene encoding this CDK needs to be mutated in order for a cancer cell to derive proliferative benefit.]

The most critical CDK inhibitor involved in cancer pathogenesis may well be $p27^{Kip1}$. As mentioned earlier, $p27^{Kip1}$ is involved largely in inhibiting the activity of cyclin E–CDK2 complexes (e.g., see Figure 8.15). As cells exit the cell cycle into the G_0 quiescent state, $p27^{Kip1}$ levels rise (Figure 8.37A). Conversely, as cells re-enter the cell cycle and advance through its G_1 phase, the levels of $p27^{Kip1}$ are reduced progressively throughout early and mid-G_1 and then are made to fall

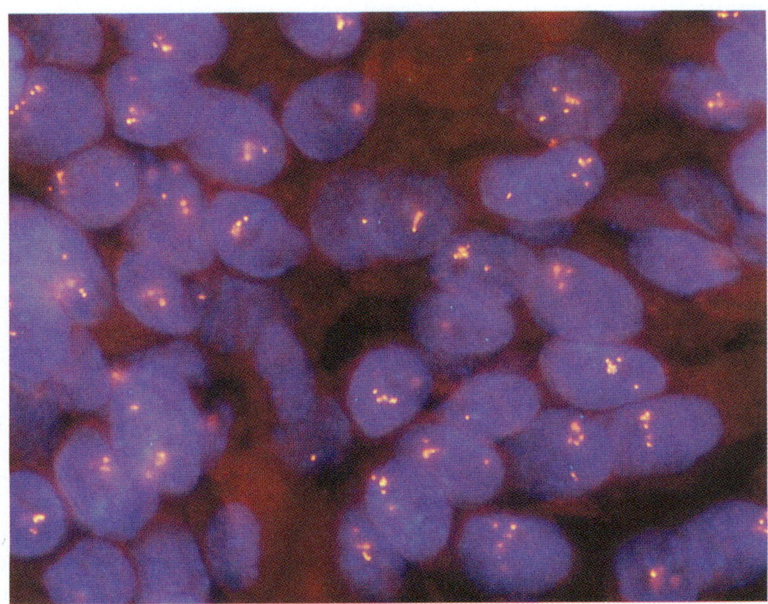

Figure 8.36 Amplification of the cyclin D1 gene The use of fluorescent *in situ* hybridization (FISH) makes it possible to detect the copy number of specific genes in histological sections. Here is the result of using a probe for *CCND1*, the human gene encoding cyclin D1, to analyze the copy number of this gene in the cells of a head-and-neck squamous cell carcinoma (HNSCC). Each bright spot represents a single copy of this gene. Individual nuclei are stained *purple*. Hence, *CCND1* is amplified to various extents here, being present in three to five copies per cell in this tumor. In general, CCND1 is found to be amplified in about one-third of all of these tumors, leading to corresponding increases in cyclin D1 expression and resulting loss of proper control of pRb phosphorylation. Similar observations *(not shown)* can be made with human breast cancer cells. (From K. Freier et al., *Cancer Res.* 61:1179–1182, 2003.)

precipitously during late G_1 phase by the actions of cyclin E–CDK2 complexes. These low levels are not created by reduced transcription of the gene encoding $p27^{Kip1}$. Instead, it seems $p27^{Kip1}$ levels are reduced by the actions of the Skp2 protein, which acts together with Cul1 and several other proteins (Figure 8.37B) to recognize $p27^{Kip1}$ and ubiquitylate it, thereby tagging it for destruction in proteasomes. Indeed, the declining levels of Skp2 explain the increase in $p27^{Kip1}$ as cells enter into G_0 (see Figure 8.37A and C).

The inverse relationship between $p27^{Kip1}$ levels and those of Skp2 is especially apparent in various human cancers. Thus, in mammary and oral carcinomas, as well as lymphomas, the levels of $p27^{Kip1}$ are inversely correlated with those of Skp2 (see, for example, Figure 8.37D), with higher levels of Skp2 portending

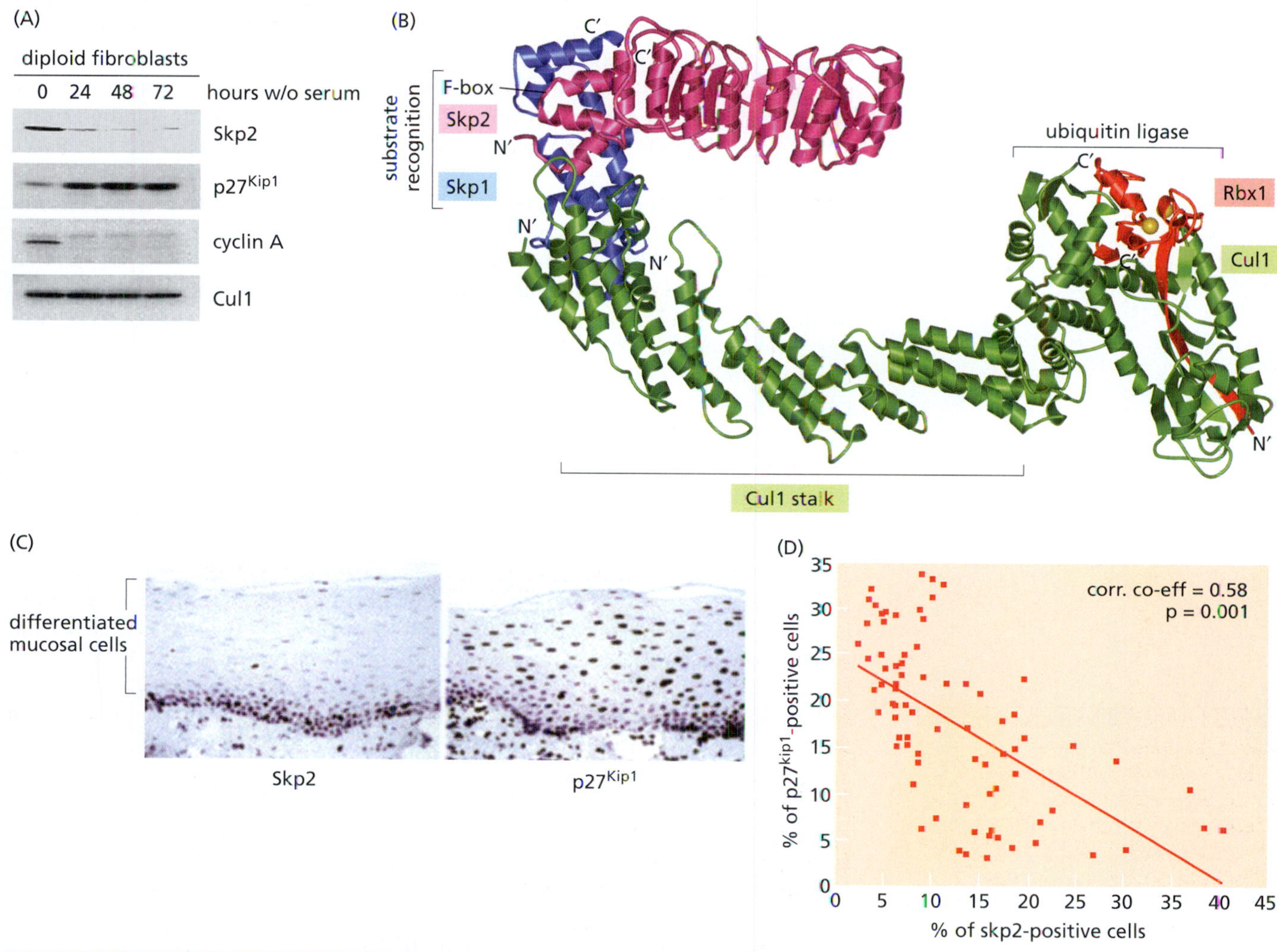

Figure 8.37 Suppression of $p27^{Kip1}$ levels by Skp2 Levels of $p27^{Kip1}$ are controlled by, among other factors, the levels of the SCF^{Skp2} complex, which constitutes an E3 ubiquitin ligase (Sidebar 7.8) that ubiquitylates $p27^{Kip1}$, thereby tagging it for degradation in proteasomes. (A) As normal human cells enter into the G_0, quiescent state, in this case because they are deprived of serum mitogens, Skp2 levels fall, allowing the accumulation of $p27^{Kip1}$; the latter, in turn, is responsible for blocking the actions of CDK2, which is critical for advance through the late G_1 phase of the cell cycle. (B) Skp2 *(pink)*, which is responsible for recognizing the $p27^{Kip1}$ substrate, assembles with several other proteins to form this complex. Among these are Skp1 *(blue)*, the Cul1 protein *(green)*, whose expression is induced by Myc (Section 8.9, Figure 8.28), and the Rbx1 protein *(red)*; the latter binds to an E2 ubiquitin ligase, which is responsible for transferring a ubiquitin moiety to $p27^{Kip1}$. The Cul1 stalk holds the two ends of the complex 100 Ångstroms apart, which is important for ubiquitylation of the substrate. (C) As seen here, in the normal oral mucosa, when Skp2 levels are low *(left)*, levels of $p27^{Kip1}$ are high *(right)* and help to hold cells in a post-mitotic state; the opposite situation is observed in many cancers *(not shown)*. (D) The inverse relationship between Skp2 levels and those of $p27^{Kip1}$ is plotted here for a large number of oral epithelial dysplasias and carcinomas. (Similar relationships, *not shown*, are seen in lymphomas and colorectal carcinomas.) (A, from T. Bashir et al., *Nature* 428:190–193, 2004; B, from B. Schulman et al., *Nature* 408:381–386, 2000, and N. Zheng et al., *Nature* 416:703–709, 2002; C and D, from M. Gstaiger, *Proc. Natl. Acad. Sci. USA* 98:5043–5048, 2001.)

shorter patient survival. Moreover, when the levels of $p27^{Kip1}$ are measured in the cells of human esophageal, breast, colorectal, and lung carcinomas, poor patient survival is correlated with low levels of this CDK inhibitor. In all of these tumors, Skp2 seems to act as a growth-promoting oncoprotein by forcing the degradation of the critical $p27^{Kip1}$ proliferation inhibitor. One clue to the ultimate source of elevated Skp2 levels comes from research indicating that the Notch protein (Section 6.12), which is hyperactive in many types of human cancer, increases transcription of the gene encoding Skp2.

We have read repeatedly about the hyperactive state of the Akt/PKB kinase in many human tumors, which is caused by a variety of molecular defects, including hyperactive growth factor receptors, loss of the PTEN tumor suppressor protein, and mutations in PI3K. It, too, has effects on the cell cycle clock in cancer cells, by reducing the effective levels of important CDK inhibitors (see Figure 8.15). By phosphorylating $p21^{Cip1}$ and $p27^{Kip1}$, Akt/PKB ensures the cytoplasmic localization of these two critical antagonists of cell cycle advance, thereby marginalizing them. At the same time, Akt/PKB can suppress expression of the gene encoding $p27^{Kip1}$ by phosphorylating a transcription factor of the Forkhead family, which serves to further reduce the overall concentrations of $p27^{Kip1}$ in the cell.

In Section 8.9, we learned of the various ways in which the Myc protein, operating in normal cells, extends its reach into the cell cycle clock and pulls many key regulatory levers. The actions of the Myc oncoprotein are likely to be qualitatively similar. The difference between the two is largely, and possibly entirely, a matter of protein level: In normal cells, levels of Myc protein are highly dependent upon extracellular mitogenic growth factors. In many cancer cells, however, the Myc protein is produced constitutively, independent of mitogenic signals coming into the cell.

With this idea in mind, we can deduce that the many perturbations wrought by the Myc protein are magnified in the many types of human cancer cells that carry either *myc* oncogenes or the oncogenic versions of its cousins, N-*myc* and L-*myc*. Among the many products induced by Myc, the Id proteins may well have the greatest physiologic importance. On the one hand, they can act as dominant-negative inhibitors of bHLH transcription factors. On the other, they can bind to pRb and seem to inhibit its functioning in a way that is reminiscent of the actions of viral oncoproteins such as large T, E7, and E1A.

In fact, ectopically expressed Id1 protein can replace the pRb-binding actions of SV40 large T antigen in experiments designed to reverse the quiescent growth state of cells. This particular Id protein has been reported to be overproduced in diverse tumors including endometrial, head-and-neck, breast, pancreatic, esophageal, and cervical carcinomas, as well as melanomas and neuroblastomas. Its expression is positively correlated with the degree of malignancy and invasiveness in many of these tumor types.

The diverse genetic and biochemical strategies shown in Figure 8.35 are all focused on one common goal—that of overwhelming and deregulating pRb function, thereby destroying the tight control that it normally imposes on the R-point transition. Darwinian selection, occurring in the microcosm of living tissues, favors the outgrowth of cells that, by hook or by crook, have succeeded in inactivating the critical pRb braking system and thus deregulating the R-point transition.

8.13 Synopsis and prospects

All of the physiologic signals and signaling pathways affecting cell proliferation must, sooner or later, be connected in some fashion to the operations of the cell cycle clock. It represents the brain of the cell—the central signal processor that receives afferent signals from diverse sources, integrates them, and makes the final decisions concerning growth versus quiescence, and in the latter case, whether or not the exit from the active cell cycle will be reversible. Connected with the latter decision are the mechanisms governing entrance into tissue-specific differentiation programs.

The core components of the cell cycle clock are already present in clearly recognizable form in single-cell eukaryotes, including the much-studied baker's yeast, *Saccharomyces cerevisiae*. Single-cell organisms such as this one respond to a far smaller range of external signals than do metazoan cells residing within complex tissues. These simple organisms lack the hundred and more distinct types of growth factor receptors that vertebrate cells display on their surfaces, as well as other receptors, such as integrins, that our cells use to sense and control attachment to the extracellular matrix. Yeast cells also lack the growth-inhibitory receptors, such as the TGF-β receptors, that play such a critical role in the economy of mammalian tissues.

All this explains why the peripheral wiring that regulates the core cell cycle machinery of animal cells has been added relatively recently in the history of life on this planet—perhaps 600 million years ago when metazoa may first have appeared. The need to respond to a wide variety of afferent signals explains why so many distinct layers of regulation have been imposed on the core machinery. Without these additional regulators, notably the CDK inhibitors, the core machinery could not be made responsive to the diverse array of signals that impinge on individual metazoan cells and modulate their proliferation.

While these connections between the cell exterior and the cell cycle clock were being forged, other critical regulators became integrated into this complex circuitry. Actually, the invention of the key governors of G_1 progression—pRb and its two cousins, p107 and p130—seems to have occurred well before the rise of metazoa: a pRb–E2F signaling pathway involved in cell cycle control is already present in *Chlamydomonas reinhardtii*, a single-cell alga that is related to the ancestors of land plants. Similarly constructed pRb- and E2F-like proteins are present in worms and flies, indicating their presence in early metazoans.

pRb may not have been the first of these three pocket proteins to have evolved, but during the ascendance of mammals, it gained the upper hand in governing the critical decision made at the R point. These three proteins preside over various aspects of the growth-versus-quiescence decision of the cell and in this sense lie *upstream* of the cell cycle clock. At the same time, by affecting gene transcription, these proteins create a coupling between the cell cycle clock and the *downstream* circuits that must be activated in order for cells to execute the complex biochemical changes that enable them to enter into S phase.

Some of the neoplastic growth state can be explained by the workings of the cell cycle clock. We can explain the deregulated proliferation of cancer cells in terms of the operations of pRb and the molecules that control its state of phosphorylation. Without pRb at the helm, the requirement for the growth-promoting actions of oncoproteins such as Ras is greatly reduced, and cells advance through G_1 without fulfilling many of the prerequisites that normally determine whether or not the R-point transition will proceed. The coupling between proliferation and blocked differentiation, still incompletely understood, seems to be traceable to the operations of proteins such as Myc, which simultaneously drive the cell cycle clock forward through G_1 and antagonize some of the master regulators of differentiation programs.

Perhaps surprisingly, one important aspect of the proliferative control of cells operates independently of the cell cycle clock. As we will learn in detail in Chapter 10, normal cells can replicate only a limited number of times before they lose the ability to proliferate further; cancer cells have an unlimited proliferative capacity—the phenotype of **replicative immortality**. The molecular devices within cells that tally the number of replicative generations through which a cell lineage has passed are embedded in the chromosomal DNA, and these devices do not seem to be controlled by the cyclin–CDK complexes regulating cell cycle advance.

Yet other aspects of the malignant growth program are not controlled by the cell cycle clock. Cells respond to severe, essentially irreparable genomic damage by activating their cell suicide program—apoptosis; this response does not seem to be connected directly to the cell cycle machinery. This program will be the subject of much of our discussions in the next chapter.

Many other peculiarities of the cancer cell phenotype are largely the purview of cytoplasmic oncoproteins such as Ras. Included here are phenotypes of cell motility, changes in cell shape, anchorage independence, alterations in energy metabolism, and invasiveness. These behaviors are also not controlled by the cell cycle clock, which does its work in the nucleus.

In spite of significant deregulation that the cell cycle clock suffers in cancer cells, we recognize that only a small portion of the cell cycle machinery is actually perturbed in these cells—specifically, those components that govern advance through the G_1 phase of the cell cycle. We can understand this by noting that the R-point transition occurring toward the end of G_1 represents a critical decision point in the life of a cell; this decision must be deregulated if cancer cells are to gain proliferative advantage. However, once a cell has moved past the R point and reached the G_1/S transition, the remaining steps of the cell cycle proceed in an essentially automatic, pre-programmed fashion. Accordingly, the S, G_2, and M phases of cancer cells, which together represent an extraordinarily complex program of biochemical and cell biological steps, closely resemble the comparable cell cycle phases of normal cells. This helps to explain an often-noted aspect of cancer cells: their growth-and-division cycles are not necessarily shorter than those of many normal cells in the body. Instead, cancer cells continue to enter into these cycles and thus continue to proliferate under conditions that would force normal cells to halt proliferation (see, for example, Figure 7.32).

In addition to the critical regulators of G_1 progression, the only other components of the cell cycle clock that seem to be affected in the malignant growth state are the small cohort of proteins that serve as checkpoint controls during the S and M phases, notably those functioning at the G_1/S transition, during S phase, and during M phase. Their inactivation, which occurs in some cancers, is not directed toward the immediate goal of deregulating proliferation. Rather, the inactivation of these checkpoint controls serves to destabilize the cellular genome, enabling incipient cancer cells to generate a wide variety of permutations of the normal human genome. These cells and their descendants then test out the resulting novel genetic configurations, searching for those that are particularly useful in enabling neoplastic growth. We will return to the theme of genomic destabilization in Chapter 12.

Large gaps in this scenario remain to be filled in. We still do not really understand why pRb inactivation is so important for the creation of cancer cells while the inactivation of its two cousins, p107 and p130, seems rarely, if ever, to be a priority during the course of tumorigenesis. We do not understand how the proliferation-versus-differentiation decision is made in the great majority of the body's cell types. Relevant here may be observations indicating that pRb interacts with a number of other transcription factors in addition to the intensively studied E2Fs. Some of these other transcription factors may well govern the expression of genes that contribute to certain differentiation programs. And we still do not understand how many oncoproteins that function as transcription factors (Table 8.2) succeed in perturbing the workings of the cell cycle clock. Moreover, many of the currently held preconceptions about the operations of the cell cycle clock, as described in this chapter, may one day require substantial revision (Sidebar 8.8).

Sidebar 8.8 We know less about the cell cycle clock than we thought Studies of genetically altered mice, reported in the first years of the new millennium, indicate that the depictions of the cell cycle machinery, as presented in this chapter, will eventually require revision. For example, mutant mouse embryos that have been deprived of both copies of each of the three D-type cyclin genes (i.e., with a *D1*$^{-/-}$ *D2*$^{-/-}$ *D3*$^{-/-}$ genotype) are able to pass through most stages of embryonic development, some dying as late as embryonic day 16. The discussions in this chapter provide no clue how the cells of these embryos succeed in advancing through many cycles of growth and division without D-type cyclins driving their advance through the G_1 phase of the cell cycle. Similarly, mouse embryos lacking both copies of each of the two cyclin E genes (i.e., with an *E1*$^{-/-}$ *E2*$^{-/-}$ genotype) develop until mid-gestation, at which point they die because of placental defects; if these are circumvented, then the embryos can develop to term, whereupon they die. Once again, the existing models provide no insight how their cells are able to complete the last steps of the G_1 phase and initiate the S phase of the cell cycle. These startling results might suggest the operations of a normally latent cell cycle clock, possibly inherited from our protozoan ancestors, that is able to rise to the occasion and take over when the more modern, metazoan cell cycle machinery fails to do its job.

While we have read about the details of cell cycle control and the various ways by which it is disrupted in many types of cancer cells, the ultimate motivation behind these discussions is a need to understand clinical disease: How do these changes actually affect tumor progression and patient outcome? In fact, losses of pRb function can have particularly striking effects on the behavior of cancer cells and thus tumors. For example, as indicated in Figure 8.38, abnormally high levels of cyclin E in the cancer cells of breast carcinoma patients are strongly predictive of aggressive malignancy and poor patient outcome, while low levels indicate long-term, disease-free survival. In this case, expression of the cyclin E mRNA may be elevated, and the degradative mechanisms that are normally responsible for reducing cyclin E levels are likely to be compromised. The result-

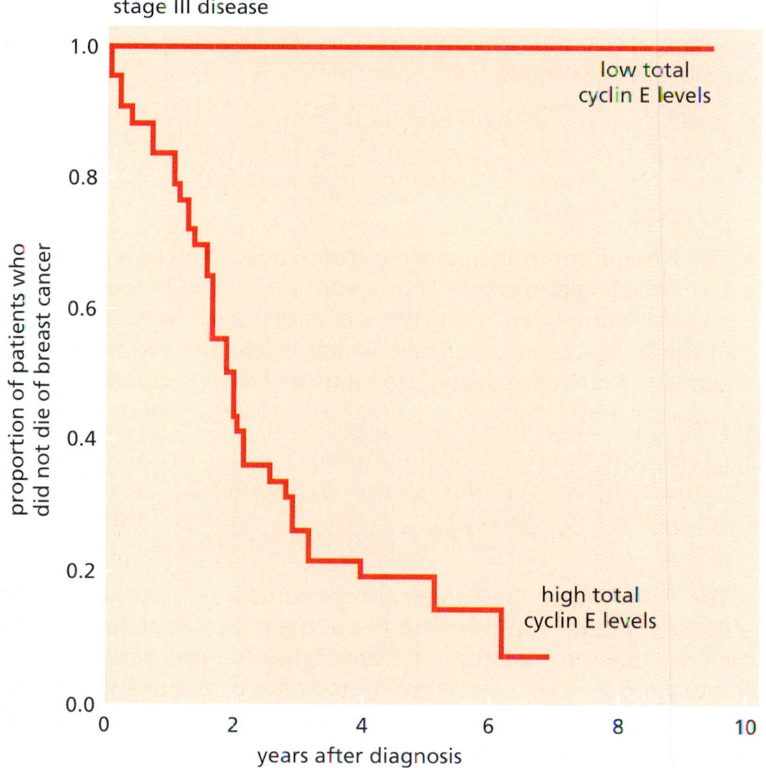

Figure 8.38 Cyclin E and breast cancer progression This Kaplan–Meier plot presents the clinical progression of disease of women with stage III breast cancer, that is, those having relatively large primary tumors and cancer cells in regional lymph nodes but lacking observable metastases at distant anatomical sites. Plotted is the fraction of patients *(ordinate)* who are still alive at the indicated times after initial diagnosis *(abscissa)*. (From K. Keyomarsi et al., *N. Engl. J. Med.* 347:1566–1575, 2002.)

ing excessively high levels of cyclin E drive deregulated pRb phosphorylation and inactivation. Observations like these provide compelling indications that the cell cycle clock is, in its normal state, an important deterrent to cancer development and, when deranged by various lesions, a potent agent for promoting cancer progression.

Our discussions of pRb and cell cycle control would suggest the preeminent importance of this protein among all the many products of tumor suppressor genes. In fact, pRb shares this position with a second protein, p53, which plays an equally important role in normal cell physiology and in cancer development. In essence, the pRb circuitry deals with the relations between the cell and the outside world. The p53 circuitry has a very different function, since it monitors the internal well-being of the cell and permits cell proliferation and cell survival only if all the vital operating systems within the cell are functioning properly. As we will see in the next chapter, the inactivation of this p53 signaling pathway is as important to developing cancer cells as the deregulation of the controls governing pRb and the R-point transition.

Key concepts

- The cell cycle is a precisely programmed series of events that enables a cell to duplicate its contents and to generate two daughter cells. This series of events is controlled by the machinery that is often termed the cell cycle clock.
- The machinery that constitutes the cell cycle clock seems to operate similarly in all cell types throughout the body.
- The cell cycle clock uses a subfamily of serine/threonine protein kinases—the cyclin-dependent kinases (CDKs)—to execute the various steps of the cell cycle.
- Specific steps of the cell cycle are controlled by changing the levels and availability of cyclins, which function by activating the catalytic function of their CDK partners.
- The D-type cyclins convey information about extracellular mitogenic signals to the core cell cycle machinery, while the remaining cyclins operate on a preordained schedule, once the decision to advance past the G_1 phase of the cell cycle has been made.
- While the levels of the D cyclins are controlled primarily through extracellular signals, the other cyclins' gradual accumulation followed by their rapid destruction (via ubiquitylation) dictates that the cell cycle clock can move in only one direction.
- Checkpoint controls operating throughout the cell cycle ensure that a new step in cell cycle progression is not undertaken before the preceding step is properly completed. Many types of cancer cells have inactivated one or more of these checkpoint controls, which helps them to accumulate the mutant genes and altered karyotypes that propel their neoplastic growth.
- The critical decisions concerning growth versus quiescence are made in the G_1 phase of the cell cycle. In normal cells, the decision to grow and replicate requires signals from the external environment (hence the dependence of D-cyclin levels on mitogenic signals). Thereafter, advance through the other phases of the cell cycle is relatively independent of external signals.
- The restriction point (R point) represents a point at which the cell commits itself, essentially irrevocably, to complete the remainder of the cell cycle or, alternatively, to remain in G_1 and possibly retreat from the active cell cycle into the G_0, quiescent state. Deregulation of the R-point decision-making machinery accompanies the formation of most types of cancer cells, since it leads to unconstrained cell proliferation.

- The decision concerning growth versus quiescence is governed by the state of phosphorylation of the retinoblastoma protein, pRb. The D cyclins and cyclin E control the degree to which pRb is phosphorylated. Hypophosphorylated pRb blocks passage through the R point, while hyperphosphorylated pRb permits this passage.
- pRb controls passage through the R point by binding or releasing E2F transcription factors associated with promoters of genes that usher the cell from G_1 to S phase. Hypophosphorylated pRb binds E2Fs, while hyperphosphorylated pRb releases them. When viral oncoproteins are present, they mimic pRb hyperphosphorylation by preventing pRB from binding E2Fs.
- In cancer cells, a number of alternative mechanisms operate to ensure that cell proliferation is not constrained by pRb. Many of these work to cause pRb hyperphosphorylation and resulting functional inactivation. This permits, in turn, the deregulated proliferation of these cells.
- pRb function can be lost in a variety of ways, including: excessive mitogenic signals (since these lead to elevated levels of D cyclins, which then initiate pRB inactivation via phosphorylation); mutation of the *Rb* gene; binding of pRb by a viral oncoprotein (e.g., HPV E7); and the actions of cellular oncoproteins (e.g., Myc) that deregulate pRb phosphorylation or directly affect pRb activity.
- The control of cell differentiation is coupled to the regulation of cell cycle progression. Hypophosphorylated pRB is needed to halt proliferation of cells and to facilitate their differentiation. Conversely, other regulatory proteins, such as Myc and the Ids, work to inhibit cell differentiation.
- In most types of cancer, differentiation is partially or completely blocked. In general, the more differentiated the cells are that form a tumor, the less aggressive is the disease of cancer.

Thought questions

1. Why is pRb function compromised in human tumors through mutations of its encoding gene while the genes encoding its two cousins, p107 and p130, have virtually never been found to suffer mutations in the genomes of cancer cells?
2. Why have DNA tumor viruses evolved the ability to inactivate pRb function?
3. How might loss of a CDK inhibitor function affect the control of cell cycle advance?
4. What molecular mechanisms operate to ensure that once the decision to advance through the restriction point has been made, this leads to an essentially irreversible commitment to complete the remaining phases of the cell cycle through M phase?
5. How are the decisions of cell growth versus quiescence coupled mechanistically with the decisions governing cell differentiation? Why must these two processes be tightly coupled?
6. How do cells ensure that the transcription-activating functions of the E2F transcription factors are limited to a narrow window of time in the cell cycle? What might occur if E2F function were allowed to continue throughout the cell cycle?
7. In what ways does the Myc oncoprotein deregulate cell proliferation and differentiation?
8. Why is it important that the cell cycle clock never runs backwards? And what mechanisms are in place to ensure that this does not happen?

Additional reading

Bashir T & Pagano M (2003) Aberrant ubiquitin-mediated proteolysis of cell cycle regulatory proteins and oncogenesis. *Adv. Cancer Res.* 65, 101–144.

Blagosklonny MV & Pardee AB (2001) The restriction point of the cell cycle. In Cell cycle checkpoints and cancer (MV Blagosklonny ed), pp 52–64. Georgetown, TX: Landes Bioscience.

Blais A & Dynlacht BD (2004) Hitting their targets: an emerging picture of E2F and cell cycle control. *Curr. Opin. Genet. Dev.* 14, 527–532.

Bloom J & Pagano M (2003) Deregulated degradation of the cdk inhibitor p27 and malignant transformation. *Semin. Cancer Biol.* 13, 41–47.

Classon M & Dyson N (2001) p107 and p130: versatile proteins with interesting pockets. *Exp. Cell Res.* 264, 135–147.

Classon M & Harlow E (2002) The retinoblastoma tumour suppressor in development and cancer. *Nat. Rev. Cancer* 2, 910–917.

DeSalle LM and Pagano M (2001) Regulation of the G1 to S transition by the ubiquitin pathway. *FEBS Lett.* 490, 179–189.

Dyson N (1998) The regulation of E2F by pRB-family proteins. *Genes Dev.* 12, 2245–2262.

Eisenman RN (2001) Deconstructing Myc. *Genes Dev.* 15, 2023–2030.

Harbour JW & Dean DC (2000) The Rb/E2F pathway: expanding roles and emerging paradigms. *Genes Dev.* 14, 2293–2409.

Hasskarl J & Münger K (2002) Id proteins—tumor markers or oncogenes? *Cancer Biol. Ther.* 1, 91–96.

Hunter T & Pines J (1994) Cyclins and cancer II: cyclin D and CDK inhibitors come of age. *Cell* 79, 573–582.

Levens D (2002) Disentangling the MYC web. *Proc. Natl. Acad. Sci. USA* 99, 5757–5759.

Liu H et al. (2004) New roles for the RB tumor suppressor protein. *Curr. Opin. Genet. Dev.* 14, 55–64.

Look AT (1997) Oncogenic transcription factors in the human acute leukemias. *Science* 278, 1059–1064.

Lutz W, Leon J & Eilers M (2002) Contributions of Myc to tumorigenesis. *Biochim. Biophys. Acta Revs. Cancer* 1602, 61–71.

Malumbres M & Barbacid M (2001) To cycle or not to cycle: a critical decision in cancer. *Nat. Rev. Cancer* 1, 222–231.

Marcu KB, Bossone SA & Patel AJ (1992) *myc* function and regulation. *Annu. Rev. Biochem.* 61, 809–860.

Massagué J (2000) How cells read TGF-β signals. *Nat. Rev. Mol. Cell Biol.* 1, 169–178.

Massagué J (2004) G1 cell-cycle control and cancer. *Nature* 432, 298–307.

Massari ME & Murre C (2000) Helix-loop-helix proteins: Regulators of transcription in eucaryotic organisms. *Mol. Cell Biol.* 20, 429–440.

Mulligan G & Jacks T (1998) The retinoblastoma gene family: cousins with overlapping interests. *Trends Genet.* 14, 223–229.

Nevins JR (1992) E2F: a link between the Rb tumor suppressor protein and viral oncoproteins. *Science* 258, 424–429.

Nevins JR (2001) The Rb/E2F pathway and cancer. *Hum. Mol. Genet.* 10, 699–703.

Perk J, Iavarone A & Benezra R (2005) Id family of helix-loop-helix proteins in cancer. *Nat. Rev. Cancer* 5,603–614.

Reed S (2003) Ratchets and clocks: the cell cycle, ubiquitylation and protein turnover. *Nat. Rev. Mol. Cell Biol.* 4, 855–864.

Reiss M (1999) TGF-β and cancer. *Microbes Infect.* 1, 1327–1347.

Sherr CJ (1996) Cancer cell cycles. *Science* 274, 1672–1677.

Sherr CJ (1998) Tumor surveillance via the ARF-p53 pathway. *Genes Dev.* 12, 2984–2991.

Sherr CJ (2001) The INK4a/ARF network in tumour suppression. *Nat. Rev. Mol. Cell Biol.* 2, 731–737.

Sherr CJ & McCormick F (2002) The RB and p53 pathways in cancer. *Cancer Cell* 2, 103–112.

Sherr CJ & Roberts JM (1999) CDK inhibitors: positive and negative regulators of G1-phase progression. *Genes Dev.* 13, 1501–1512.

Siegel PM & Massagué J (2003) Cytostatic and apoptotic actions of TGF-β in homeostasis and cancer. *Nat. Rev. Cancer* 3, 807–820.

Slingerland J & Pagano M (2000) Regulation of the cdk inhibitor p27 and its deregulation in cancer. *J. Cell Physiol.* 183, 10–17.

Tapon N, Moberg KH & Hariharan IK (2001) The coupling of cell growth to the cell cycle. *Curr. Opin. Cell Biol.* 13, 731–737.

Trimarchi JM & Lees JA (2002) Sibling rivalry in the E2F family. *Nat. Rev. Mol. Cell Biol.* 3, 11–20.

Yamasaki L & Pagano M (2004) Cell cycle, proteolysis and cancer. *Curr. Opin. Cell Biol.* 16, 623–628.

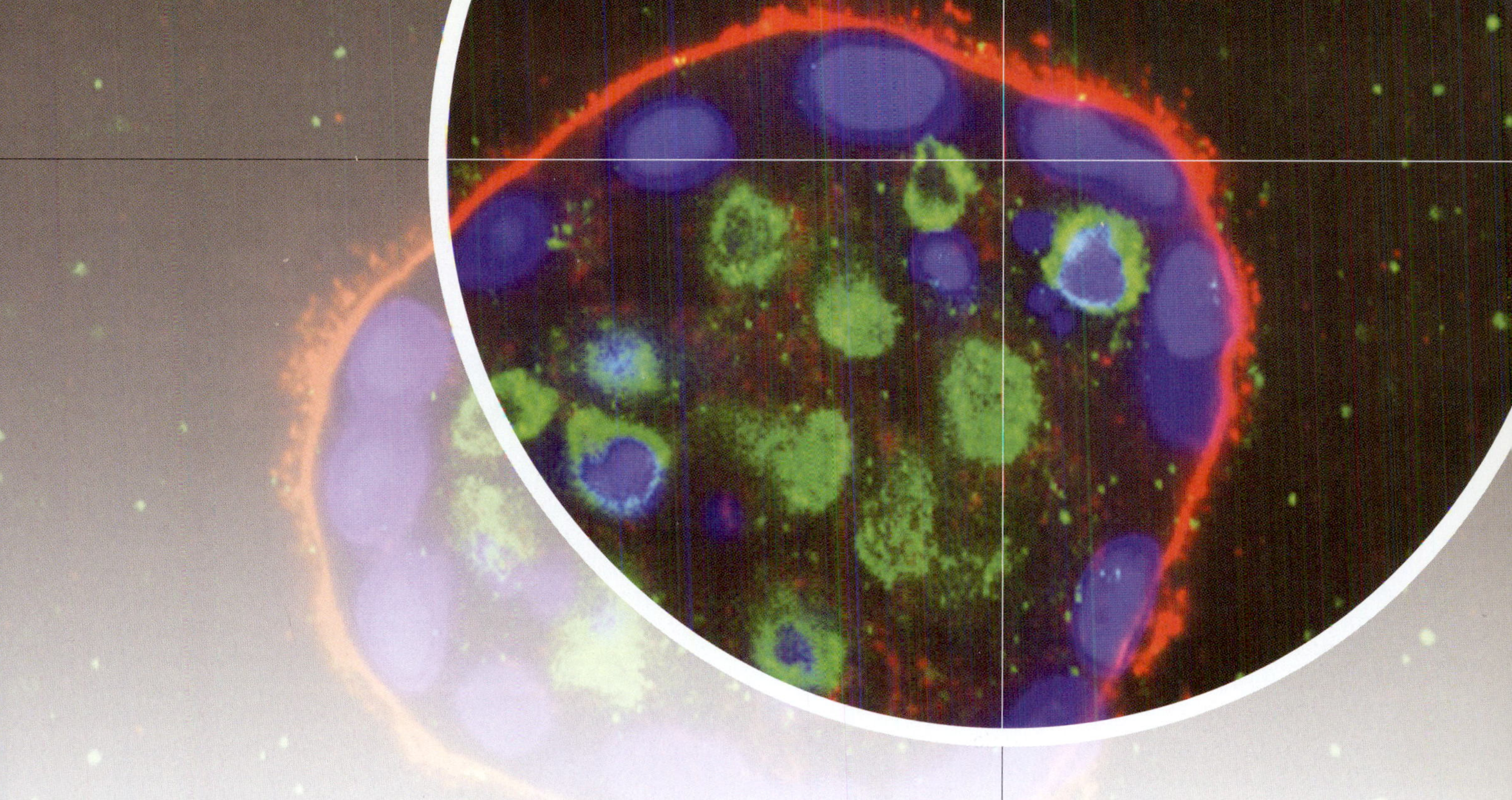

Chapter 9

p53 and Apoptosis: Master Guardian and Executioner

> To examine the causes of life, we must first have recourse to death.
> Mary Shelley, *Frankenstein*, 1831

> There cannot however be the least doubt, that the higher organisms, as they are now constructed, contain within themselves the germs of death.
> August Weissmann, philosopher of biology, 1889

Metazoan organisms have a vital interest in eliminating defective or malfunctioning cells from their tissues. Responding to this need, mammals have implanted a loyal watchman in their cells. Within almost all cells in mammalian tissues, the p53 protein serves as the local representative of the organism's interests. p53 is present on-site to ensure that the cell keeps its household in order.

If p53 receives information about metabolic disorder or genetic damage within a cell, it may arrest the advance of the cell through its growth-and-division cycle and, at the same time, orchestrate localized responses in that cell to facilitate the repair of damage. If p53 learns that metabolic derangement or damage to the genome is too severe to be cured, it may decide to emit signals that awaken the cell's normally latent suicide program—**apoptosis**. The consequence is the rapid death of the cell. This results in the elimination of a cell whose continued growth and division might otherwise pose a threat to the organism's health and viability.

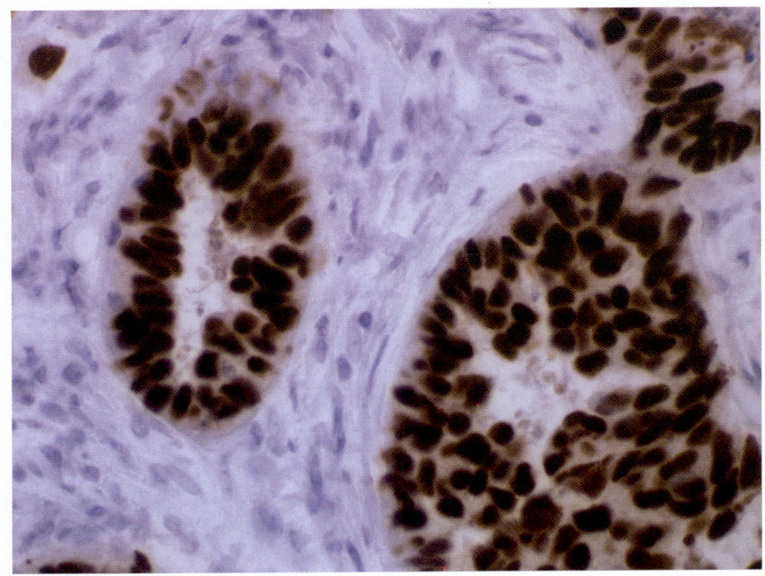

Figure 9.1 Large T antigen in SV40-transformed cells Antibodies that bind the SV40 large T (LT) antigen can be used to detect LT in the nuclei of SV40-transformed tumor cells. In the present case, such antibodies have been used to stain human mammary epithelial cells (MECs) that were transformed by introduction of the SV40 early region plus two other genes. A similar image would be seen if such antibodies were used to stain SV40-transformed mouse cells. LT was detected by linking these antibody molecules to peroxidase enzyme, which generated the *dark brown* spots. In this image of a tumor xenograft, the transformed MECs form ducts (seen in cross section), which are surrounded by normal stromal cells *(light blue nuclei)*. (Courtesy of T.A. Ince.)

The apoptotic program that may be activated by p53 is built into the control circuitry of most cells throughout the body. Apoptosis consists of a series of distinctive cellular changes that function to ensure the disappearance of all traces of a cell, often within an hour of its initial activation. The continued presence of a latent but intact apoptotic machinery represents an ongoing threat to an incipient cancer cell, since this machinery is poised to eliminate cells that are en route to becoming neoplastic. This explains why p53 function must be disabled before a clone of pre-malignant cells gains a sure and stable foothold within a tissue. Without a clear description of p53 function and apoptosis, we have no hope of understanding a fundamental component of the process that leads to the creation of virtually all types of human tumors.

9.1 Papovaviruses lead to the discovery of p53

When murine cells that have been transformed by the SV40 DNA tumor virus are injected into a mouse of identical genetic background (i.e., a *syngeneic* host), the immune system of the host reacts by mounting a strong response; antibodies are made that react with a nuclear protein that is present in the virus-transformed cells and is otherwise undetectable in normal mouse cells (Figure 9.1). This protein, the large tumor (large T, LT) antigen, is encoded by a region of the viral genome that is also expressed when this virus infects and multiplies within monkey kidney cells—host cells that permit a full infectious (lytic) cycle to proceed to completion (Section 3.4).

Large T is a multifunctional protein that SV40 virus uses to perturb a number of distinct regulatory circuits within infected and transformed cells. Indeed, large T was cited in the previous chapter because of its ability to bind and thus functionally inactivate pRb (Sidebar 8.4). Anti-large T sera harvested from mice and hamsters bearing SV40-induced tumors were used in 1979 to analyze the proteins present in SV40-transformed cells. The resulting immunoprecipitates contained both large T and an associated protein that exhibited an apparent molecular weight of 53 to 54 kilodaltons (Figure 9.2). Antisera reactive with the p53 protein were found to detect this protein in mouse embryonal carcinoma cells and, later on, in a variety of human and rodent tumor cells that had never been infected by SV40. However, monoclonal antibodies that recognized only large T immunoprecipitated the 53- to 54- kD protein in virus-infected but not in uninfected cells.

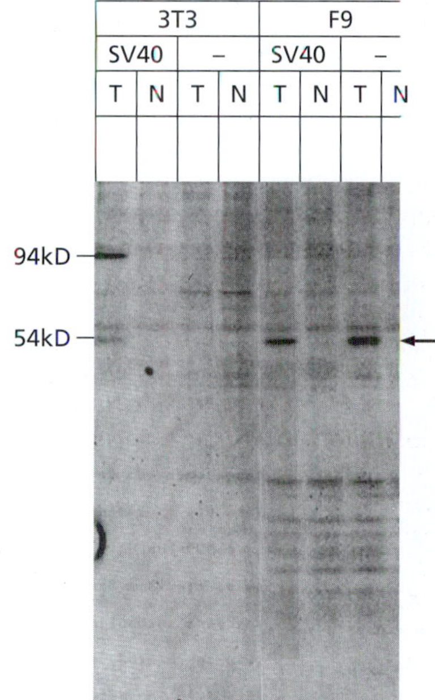

Figure 9.2 The discovery of p53 Normal BALB/c 3T3 mouse fibroblasts (3T3) transformed by SV40 and F9 mouse embryonal carcinoma (EC) cells were exposed to ^{35}S-methionine, and resulting lysates were incubated with either normal hamster serum (N) or hamster antiserum reactive with SV40-transformed hamster cells (T). The anti-tumor serum immunoprecipitated a protein of 94 kD from virus-infected but not uninfected 3T3 cells. In addition, a second protein running slightly ahead of the 54-kD marker was immunoprecipitated from SV40-transformed 3T3 cells but not from normal 3T3 cells. Moreover, this same protein could be immunoprecipitated from F9 cells, whether or not they had been exposed to SV40 *(arrow)*. [These particular data, on their own, did not prove a physical association of SV40 large T (the 94-kD protein) with p53, but they did show that p53 was a cellular protein that was present in elevated amounts in two types of transformed cells.] (From D.I. Linzer and A.J. Levine, *Cell* 17:43–52, 1979.)

Taken together, these observations indicated that the large T protein expressed in SV40-transformed cells was tightly bound to a novel protein, which came to be called p53. Antisera that reacted with both large T and p53 detected p53 in certain uninfected cells, notably tumor cells that were transformed by non-viral mechanisms, such as the F9 embryonal carcinoma (EC) cells analyzed in Figure 9.2. The latter observations indicated that p53 was of cellular rather than viral origin, a conclusion that was reinforced by the report in the same year that mouse cells transformed by exposure to a chemical carcinogen also expressed p53.

These various lines of evidence suggested that the large T oncoprotein functions, at least in part, by targeting host-cell proteins for binding. (The discovery that large T antigen is also able to bind pRb, the retinoblastoma protein, came seven years later.) In the years since these 1979 discoveries, a number of other DNA viruses and at least one RNA virus have been found to specify oncoproteins that associate with p53 or perturb its function (Table 9.1). (As we will discuss later in this chapter, and as is apparent from this table, these viruses also target pRb and undertake to block apoptosis.)

Table 9.1 Tumor viruses that perturb pRb, p53, and/or apoptotic function

Virus	Viral protein targeting pRb	Viral protein targeting p53	Viral protein targeting apoptosis
SV40	large T (LT)	large T (LT)	
Adenovirus	E1A	E1B55K	E1B19K[a]
HPV	E7	E6	
Polyomavirus	large T	large T?	middle T (MT)[b]
Herpesvirus saimiri	V cyclin[c]		v-Bcl-2[d]
HHV-8 (KSHV)	K cyclin[c]	LANA-2	v-Bcl-2,[d] v-FLIP[e]
HCMV	IE72[f]	IE86	vICA,[g] pUL37[h]
HTLV-I	Tax[i]		
Epstein–Barr		EBNA-1[j]	LMP1[j]

[a]Functions like Bcl-2 to block apoptosis.
[b]Activates PI3K and thus Akt/PKB.
[c]Related to D-type cyclins.
[d]Related to cellular Bcl-2 anti-apoptotic protein.
[e]Viral caspase-8 (FLICE) inhibitory protein; blocks an early step in the extrinsic apoptotic cascade.
[f]Interacts with and inhibits p107 and possibly p130; may also target pRb for degradation in proteasomes.
[g]Binds and inhibits procaspase-8.
[h]Inhibits the apoptotic pathway below caspase-8 and before cytochrome *c* release.
[i]Induces synthesis of cyclin D2 and binds and inactivates p16^{INK4A}.
[j]LMP1 facilitates p52 NF-κB activation and thereby induces expression of Bcl-2; EBNA-1 acts via a cellular protein, USP7/HAUSP, to reduce p53 levels.

9.2 *p53* is discovered to be a tumor suppressor gene

Transfection of a *p53* cDNA clone into rat embryo fibroblasts (REFs) revealed that this DNA could collaborate with a co-introduced *ras* oncogene in the transformation of these rodent cells. Such activity suggested that the *p53* gene (which is sometimes termed *Trp53* in mice and *TP53* in humans) might operate as an oncogene, much like the *myc* oncogene, which had previously been found capable of collaborating with the *ras* oncogene in rodent cell transformation (see Section 11.10). Like *myc*, the introduced *p53* cDNA seemed to contribute certain growth-inducing signals that resulted in cell transformation in the presence of a concomitantly expressed *ras* oncogene.

But appearances deceived. As later became apparent, this *p53* cDNA had originally been synthesized using the mRNA extracted from tumor cells as a template. Subsequent manipulation of a *p53* cDNA cloned instead from the mRNA of normal cells revealed that this *p53* cDNA clone, rather than favoring cell transformation, actually suppressed it (Figure 9.3). Comparison of the sequences of the two cDNAs revealed that the two differed by a single base substitution—a point mutation—that caused an amino acid substitution in the p53 protein. Hence, the initially used clone encoded a mutant p53 protein with altered function.

These results indicated that the wild-type allele of *p53* really functions to suppress cell proliferation, and that *p53* acquires growth-promoting powers only when it sustains a point mutation in its reading frame. Because of this discovery, the *p53* gene was eventually categorized as a tumor suppressor gene.

By 1987 it became apparent that such point-mutated alleles of *p53* are common in the genomes of a wide variety of human tumor cells. Data accumulated by 2002 indicated that the *p53* gene is mutated in 30 to 50% of commonly occurring human cancers (Figure 9.4). Indeed, among all the genes examined to date in human cancer cell genomes, *p53* is the gene found to be most frequently mutated, being present in mutant form in the genomes of almost half of all human tumors.

Further functional analyses of *p53*, conducted much later, made it clear, however, that *p53* is not a typical tumor suppressor gene. In the case of most tumor suppressor genes, when the gene was inactivated (i.e., "knocked out") homozygously in the mouse germ line (using the strategy of targeted gene inactivation described in Sidebar 7.10), the result was, almost invariably, a disruption of embryonic development due to deregulated morphogenesis in one or more tissues. These tumor suppressor genes seemed to function as negative regulators of proliferation in a variety of cell types; their deletion from the regulatory circuitry of cells led, consequently, to inappropriate proliferation of certain cells and thus to disruption of normal development.

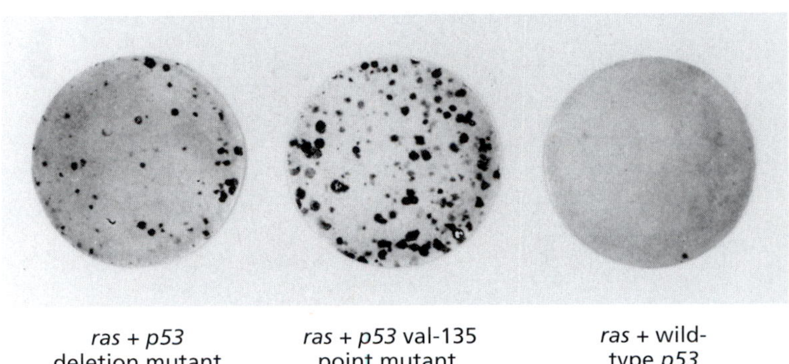

Figure 9.3 Effects of p53 on cell transformation A cDNA encoding a *ras* oncogene was co-transfected with several alternative forms of a *p53* cDNA into rat embryo fibroblasts (REFs). In the presence of a *p53 dl* mutant vector, which contains an almost complete deletion of the *p53* reading frame *(left)*, a small number of foci were formed. In the presence of a *p53* point mutant *(middle)*, a large number of robust foci were formed. However, in the presence of a *p53* wild-type cDNA clone *(right)*, no foci were formed. (From D. Eliyahu et al., *Nature* 312:646–649, 1984.)

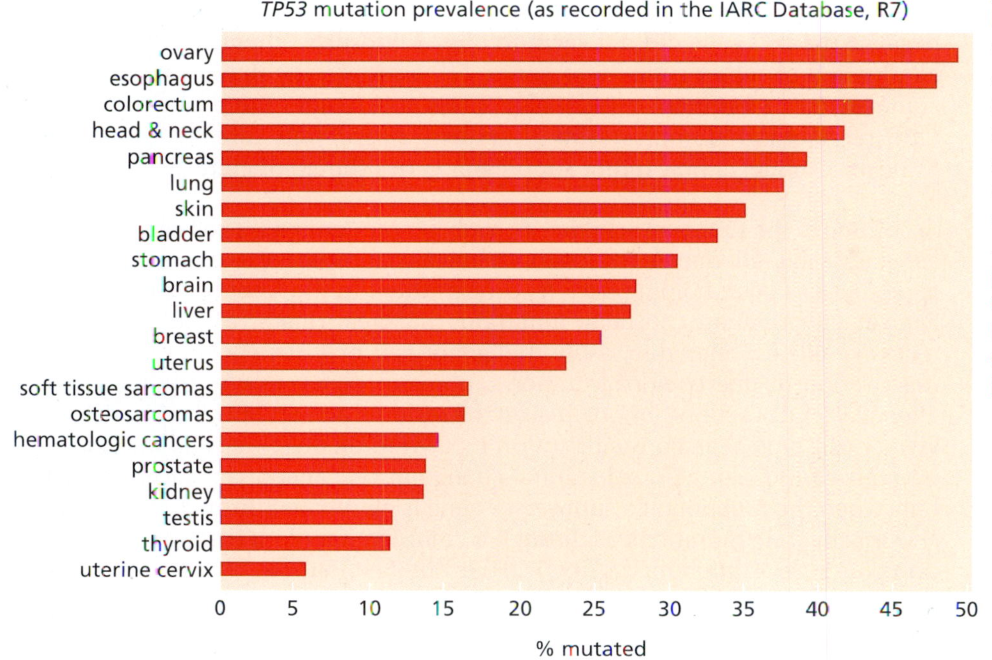

Figure 9.4 Frequency of mutant *p53* alleles in human tumor cell genomes As indicated in this bar graph, mutant alleles of *p53* are found frequently in commonly occurring human tumors. This dataset includes 17,689 somatic mutations of *p53* and 225 germ-line mutations that had been reported by June 2002. The bars indicate the percentage of each tumor type found to carry a mutant *p53* allele. (From International Agency for Research on Cancer, TP53 genetic variations in human cancer, IARC release R7, 2002.)

In stark contrast, deletion of both *p53* gene copies from the mouse germ line had no apparent effect on the development of the great majority of $p53^{-/-}$ embryos. Therefore, *p53* could not be considered to be a simple negative regulator of cell proliferation during normal development. Still, *p53* was clearly a tumor suppressor gene, since mice lacking both germ-line copies of the *p53* gene had a short life span (about 5 months), dying most often from lymphomas and sarcomas (Figure 9.5). This behavior provided the first hints that the p53 protein does not operate to transduce the proliferative and anti-proliferative signals that continuously impinge on cells and regulate their proliferation. Instead, *p53* seemed to be specialized to prevent the appearance of abnormal cells, specifically, those cells that were capable of spawning tumors.

9.3 Mutant versions of p53 interfere with normal p53 function

The observations of frequent mutation of the *p53* gene in tumor cell genomes suggested that many incipient cancer cells must perturb or eliminate p53 function before they can thrive. This notion raised the question of precisely how

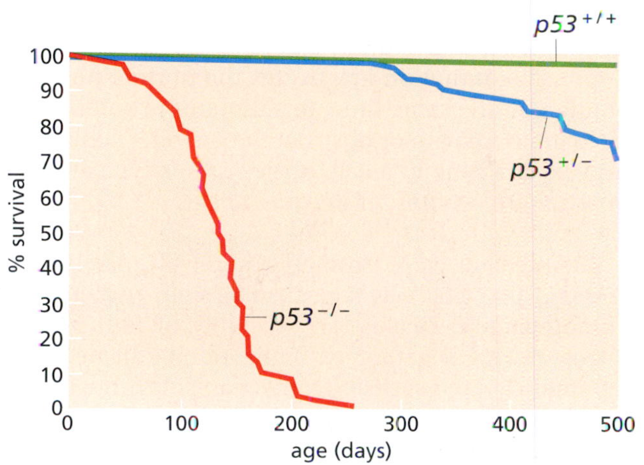

Figure 9.5 Effects of mutant *p53* alleles in the mouse germ line This Kaplan–Meier plot indicates the percent of mice of the indicated phenotype that have survived *(ordinate)* as a function of elapsed lifetime in days *(abscissa)*. While the absence of p53 function in the $p53^{-/-}$ mice had little effect on their embryologic development and viability at birth, it resulted in a greatly increased mortality relatively early in life, deriving largely from the development of tumors such as sarcomas and leukemias. All $p53^{-/-}$ homozygotes succumbed to malignancies by about 250 days of age *(red line)*, and even $p53^{+/-}$ heterozygotes *(blue line)* began to develop tumors at this time, while wild-type *($p53^{+/+}$)* mice *(green line)* showed virtually no mortality until almost 500 days of age. (Adapted from T. Jacks et al., *Curr. Biol.* 4:1–7, 1994.)

these cells succeed in shedding p53 function. Here, another anomaly arose, because the *p53* gene did not seem to obey Knudson's scheme for the two-hit elimination of tumor suppressor genes. For example, the finding that a cDNA clone encoding a mutant version of p53 was able to alter the behavior of wild-type rat embryo fibroblasts (as described above) ran directly counter to Knudson's model of how tumor suppressor genes should operate (Section 7.3).

According to the Knudson scheme, an evolving pre-malignant cell can only reap substantial benefit once it has lost both functional copies of a tumor suppressor gene that has been holding back its proliferation. In the Knudson model, such gene inactivations are caused by mutations that create inactive ("null") and thus recessive alleles. Therefore, a pre-malignant cell may benefit minimally from inactivation of one copy of a tumor suppressor gene—due to the halving of effective gene function—or not at all, if the residual activity specified by the surviving wild-type gene copy suffices on its own to mediate normal function. As we learned in Chapter 7, substantial change in cell phenotype usually occurs only when the function of a suppressor gene is eliminated through two successive inactivating mutations or through a combination of an inactivating mutation plus a loss-of-heterozygosity (LOH) event (Section 7.4).

Knudson's model was hard to reconcile with the observed behavior of the mutant *p53* cDNA introduced into rat embryo fibroblasts. The mutant *p53* cDNAs clearly altered cell phenotype, even though these embryo fibroblast cells continued to harbor their own pair of wild-type *p53* gene copies. This meant that the introduced mutant *p53* cDNA could not be functioning as an inactive, recessive allele. It seemed, instead, that the point-mutated *p53* allele was actively exerting some type of *dominant* function when introduced into these rat embryo cells.

Another clue came from sequence analysis of mutant *p53* alleles in various human tumor cell genomes. These analyses indicated that the great majority of tumor-associated, mutant *p53* alleles carry point mutations in their reading frames that create **missense** codons (resulting in amino acid substitutions) rather than **nonsense** codons (which cause premature termination of the growing polypeptide chain). To date, more than 15,000 tumor-associated *p53* alleles originating in human tumor cell genomes have been sequenced, 75% of which have been found to carry such missense mutations (Figure 9.6). Furthermore, deletions of sequences within the reading frame of the *p53* gene are relatively uncommon. Consequently, researchers came to the inescapable conclusion that tumor cells can benefit from the presence of a slightly altered p53 protein rather than from its complete absence, as would occur following the creation of **null alleles** by nonsense mutations or the outright deletion of significant portions of the *p53* gene.

A solution to the puzzle of how mutant p53 protein might foster tumor cell formation arose from two lines of research. First, studies in the area of yeast genetics indicated that mutant alleles of certain genes can be found in which the responsible mutation inactivates the normal functioning of the encoded gene product. At the same time, this mutation confers on the mutant allele the ability to interfere with or obstruct the ongoing activities of the surviving wild-type copy of this gene in a cell. Alleles of this type are termed variously *dominant-interfering* or *dominant-negative* alleles.

A second clue came from biochemical analyses of the p53 protein, which revealed that p53 was a nuclear protein that normally exists in the cell as a **homotetramer**, that is, an assembly of four identical polypeptide subunits (Figure 9.7A). Together with the dominant-negative concept, this observed tetrameric state suggested a mechanism through which a mutant allele of *p53* could actively interfere with the continued functioning of a wild-type *p53* allele being expressed in the same cell.

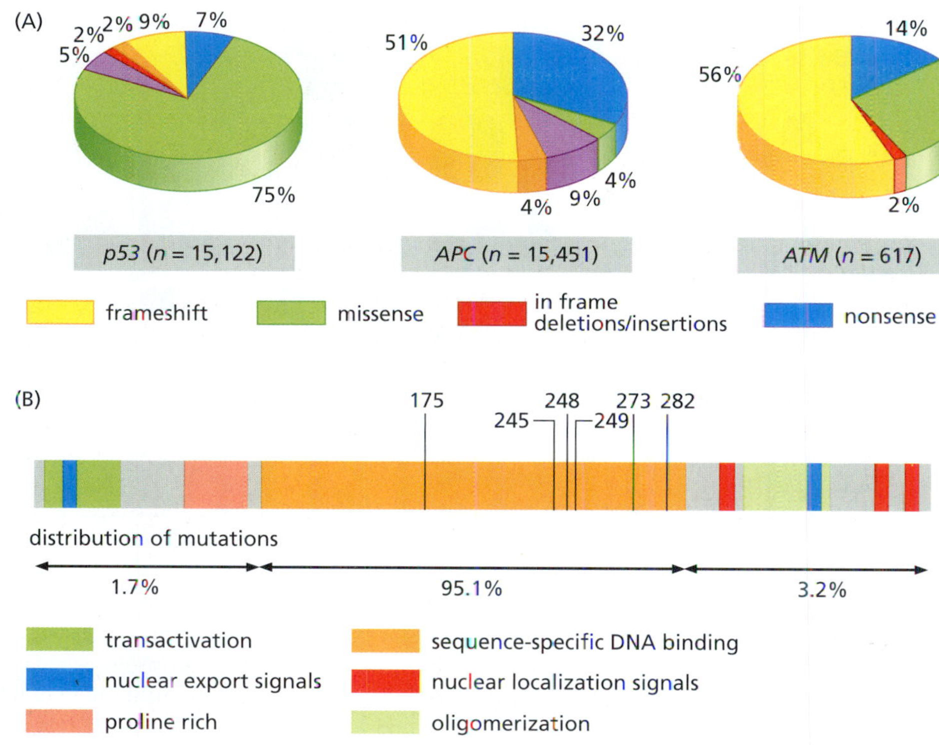

Figure 9.6 Nature of *p53* mutations (A) As indicated in these pie charts, point-mutated alleles of *p53* (leading almost always to amino acid substitutions; *green*) represent the great majority of the mutant *p53* alleles found in human tumors, while other types of mutations are seen relatively infrequently. In contrast, the mutations striking other tumor suppressor genes (*APC*) or "caretaker" genes involved in maintenance of the genome (*ATM*, *BRCA1*) represent reading-frame shifts (*yellow*) or nonsense codons (*blue*) in the majority of cases, both of which disrupt protein structure, usually by creating truncated versions of proteins; such defective forms of these proteins are often degraded rapidly in cells. (B) More than 15,000 mutant alleles of *p53* have been sequenced in human tumors, most of which are point mutations. The locations of these point mutations across the *p53* reading frame are plotted here. As is apparent, the great majority of *p53* mutations (95.1%) affect the DNA-binding domain of the p53 protein. The numbers above the figure indicate the amino acid residue number. (A, from A.I. Robles et al., *Oncogene* 21:6898–6907, 2002; B, from K.H. Vousden and X. Lu, *Nat. Rev. Cancer* 2:594–604, 2002.)

Assume that a mutant *p53* allele found in a human cancer cell encodes a form of the p53 protein that has lost most normal function but has retained the ability to participate in tetramer formation. If one such mutant allele were to coexist with a wild-type allele in this cell, the p53 tetramers assembled in such a cell would contain mixtures of mutant and wild-type p53 proteins in various proportions. The presence of only a single mutant p53 protein in a tetramer might well interfere with the functioning of the tetramer as a whole. Figure 9.7B illustrates the fact that 15 out of the 16 equally possible combinations of mutant and wild-type p53 monomers would contain at least one mutant p53 subunit and might therefore lack some or all of the activity associated with a fully wild-type p53 tetramer. Consequently, only 1/16 of the p53 tetramers assembled in this heterozygous cell (which carries one mutant and one wild-type *p53* gene copy) would be formed purely from wild-type p53 subunits and retain full wild-type function.

In an experimental situation in which a mutant *p53* cDNA clone is introduced by gene transfer (transfection) into cells carrying a pair of wild-type *p53* alleles, the expression of this introduced allele is usually driven by a highly active transcriptional promoter, indeed, a promoter that is far more active than the gene promoter controlling expression of the native *p53* gene copies in the cell. As a consequence, in such transfected cells, the amount of mutant p53 protein expressed by the introduced gene will be vastly higher than the amount of normal protein produced by the cells' endogenous wild-type *p53* gene copies. Therefore, far fewer than 1/16 of the p53 tetramers in such cells will be formed purely from wild-type p53 subunits. This explains how an introduced mutant *p53* allele can be highly effective in compromising virtually all p53 function in such cells.

The above logic might suggest that many human tumor cells, which seem to gain some advantage by shedding p53 function, should carry one wild-type and one mutant *p53* allele. Actually, in the great majority of human tumor cells that are mutant at the *p53* locus, the *p53* locus is found to have undergone a loss of heterozygosity (LOH), in which the wild-type allele has been discarded, yielding a cell with two mutant p53 alleles. Thus, in such a cell, one copy of the *p53* gene

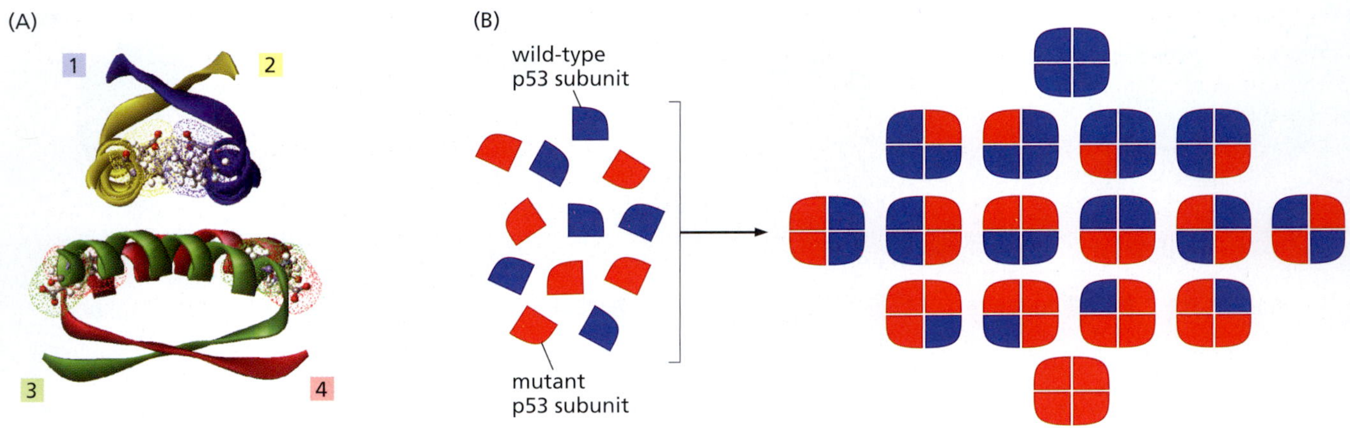

Figure 9.7 Mechanism of *p53* dominant-negative mutations (A) The tetramerization domain of p53, composed of an α-helix, is revealed here by use of X-ray crystallography. This domain usually remains intact in the mutant p53 proteins found in tumor cells, enabling these mutant proteins to form tetramers with other mutant p53 proteins or with wild-type p53 molecules. This domain also allows p53 to form mixed tetramers with related proteins (Sidebar 10 ●); and an amino acid substitution at the end of the helix, which destabilizes tetramers, is responsible for a rare familial tumor predisposition (Sidebar 9.5). The four helical domains illustrated here, each in a different color, are assembled in two pairs, the pairs assembling at right angles to one another. (B) The p53 protein normally functions as a homotetrameric transcription factor. However, in cells bearing a single mutant *p53* allele that encodes a structurally altered protein, the mutant protein may retain its ability to form tetramers but may lose its ability to exert normal p53 function. Consequently, mixed tetramers composed of differing proportions of wild-type *(blue)* and mutant *(red)* p53 subunits may form, and the presence of even a single mutant protein subunit may compromise the functioning of the entire tetramer. Therefore, in a cell that is heterozygous at the *p53* locus, 15/16 of the subunits may lack fully normal function. (A, from E.L. DiGiammarino et al., *Nat. Struct. Biol.* 9:12–16, 2002.)

is initially mutated followed by elimination of the surviving wild-type copy achieved through some type of loss-of-heterozygosity mechanism.

It is clear that an initial mutation resulting in a mutant, dominant-negative (dn) allele is far more useful for the incipient tumor cell than one resulting in a null allele, which causes total loss of an encoded p53 protein. The dominant-negative allele may well cause loss of 15/16 of p53 function, while the null allele will result, at best, in elimination of one-half of p53 function. (Actually, if the levels of p53 protein in the cell are carefully regulated, as they happen to be, then this null allele will have no effect whatsoever on a cell's overall p53 concentration, since the surviving wild-type allele will compensate by making more of the wild-type protein.)

Why, then, is elimination of the surviving wild-type *p53* allele even necessary? The answer seems to lie in the residual 1/16 of fully normal *p53* gene function; even this little bit seems to be more than most tumor cells care to live with. So, being most opportunistic, they jettison the remaining wild-type *p53* allele in order to proliferate even better. Observation of genetically altered embryonic stem cells provides further evidence for p53's dominant-negative mode of action (Sidebar 9.1).

9.4 p53 protein molecules usually have short lifetimes

The nuclear localization of the p53 protein in many normal and neoplastic cells suggested that this protein might function as a transcription factor (TF). At least three mechanisms were known to regulate the activity of transcription factors. (1) Levels of the transcription factor in the nucleus are modulated. (2) Levels of a transcription factor in the nucleus are held constant, but the intrinsic activity of the factor is boosted by some type of covalent modification. (3) Levels of certain collaborating transcription factors may be modulated. In some instances, all three mechanisms cooperate. In the case of p53, the first mechanism—changes in the level of the p53 protein—was initially implicated. Measurement of p53 protein levels indicated that they could vary drastically from one cell type to another and, provocatively, would increase rapidly when cells were exposed to certain types of physiologic stress.

These observations raised the question of how p53 protein levels are modulated by the cell. Many cellular protein molecules, once synthesized, persist for tens or hundreds of hours. (Some cellular proteins, such as those forming the ribosomal subunits in exponentially growing cells, seem to persist for many days.) Yet other cellular proteins are metabolically highly unstable and are degraded almost as soon as they are assembled. One way to distinguish between these alternatives is to treat cells with cycloheximide, a drug that

Sidebar 9.1 Mutant embryonic stem cells demonstrate the dominant-negative action of mutant *p53* alleles The most compelling evidence that mutant *p53* alleles act in a dominant-negative fashion comes from genetic manipulation of embryonic stem (ES) cells in which point mutations have been introduced into the endogenous *p53* gene copies through a homologous recombination technique (see Sidebar 7.10). In one such experiment, two alternative single-nucleotide changes causing amino acid substitutions in the DNA-binding domain of p53 were introduced into the genomes of two sets of these cells; these substitutions mirrored two mutant alleles encountered frequently in human tumor genomes. In both cases, the mutant alleles were present in heterozygous configuration.

These mutant cells showed substantial reductions in p53 function. In contrast, yet other ES cells lacking one of the two *p53* gene copies (i.e., carrying a null allele of *p53*) responded almost like wild-type cells to certain physiologic stimuli that depend on p53 function. Evidence like this lends great weight to the argument that many mutant versions of p53 are able to actively interfere with the functions of the co-expressed wild-type p53 protein.

blocks further protein synthesis. When such an experiment was performed in cells with wild-type *p53* alleles, the p53 protein disappeared with a half-life of only 20 minutes. This led to the conclusion that p53 is usually a highly unstable protein, being broken down by proteolysis soon after it is synthesized.

This pattern of synthesis followed by rapid degradation might appear to be a "futile cycle," which would be highly wasteful for the cell. Why should a cell invest substantial energy and synthetic capacity in making a protein molecule, only to destroy it almost as soon as it has been created? Similar behaviors have been associated with other cellular proteins such as Myc (Section 6.1).

The rationale underlying this ostensibly wasteful scheme of rapid protein turnover is a simple one: a cell may need to rapidly increase or decrease the level of a protein in response to certain physiologic signals. In principle, such modulation could be achieved by regulating the level of its encoding mRNA or the rate with which this mRNA is being translated. However, far more rapid changes in the levels of a critical protein can be achieved simply by stabilizing or destabilizing the protein itself. For example, in the case of p53, a cell can double the concentration of p53 protein in 20 minutes simply by blocking its degradation.

Under normal conditions, a cell will continuously synthesize p53 molecules at a high rate and rapidly degrade them at an equal rate. The net result of this is a very low "steady-state" level of the protein within this cell. In response to certain physiologic signals, however, the degradation of p53 is blocked, resulting in a rapid increase of p53 levels in the cell. This finding led to the further question of why a normal cell would wish to rapidly modulate p53 levels, and what types of signals would cause a cell to halt p53 degradation, resulting in rapidly increasing levels of this protein.

9.5 A variety of signals cause p53 induction

During the early 1990s, a variety of agents were found to be capable of inducing rapid increases in p53 protein levels. These included X-rays, ultraviolet (UV) radiation, certain chemotherapeutic drugs that damage DNA, inhibitors of DNA synthesis, and agents that disrupt the microtubule components of the cytoskeleton. Within minutes of exposing cells to some of these agents, p53 was readily detected in substantial amounts in cells that previously had shown only minimal levels of this protein. This rapid induction occurred in the absence of any marked changes in *p53* mRNA levels and hence was not due to increased transcription of the *p53* gene. Instead, it soon became apparent that the elevated protein levels were due entirely to the post-translational stabilization of the normally **labile** p53 protein.

In the years that followed, an even greater diversity of cell-physiologic signals were found capable of provoking increases in p53 levels. Among these were low oxygen tension (hypoxia), which is experienced by cells, normal and malignant, that lack adequate access to the circulation and thus to oxygen borne by the blood. Still later, introduction of either the *myc* or adenovirus *E1A* oncogene (Sections 8.5 and 8.9) into cells was also found to be capable of causing increases in p53 levels.

By now, the list of stimuli that provoke increases in p53 levels has grown even longer. Expression of higher-than-normal levels of the E2F1 transcription factor, widespread demethylation of chromosomal DNA, and a deficit in the nucleotide precursors of DNA all trigger p53 accumulation. Exposure of cells to nitrous oxide or to an acidified growth medium, depletion of the intracellular pool of ribonucleotides, and blockage of either RNA or DNA synthesis also increase p53 levels.

These various observations made it clear that a diverse array of sensors are responsible for monitoring the integrity and functioning of various cellular systems. When these sensors detect damage or aberrant functioning, they send signals to p53 and its regulators, resulting in a rapid increase in p53 levels within a cell (Figure 9.8).

The same **genotoxic** (i.e., DNA-damaging) agents and physiologic signals that provoked p53 increases were already known from other work to act under certain conditions in a **cytostatic** fashion, forcing cells to halt their advance through the cell cycle, a response often called "growth arrest." In other situations, some of these stressful signals might trigger activation of the apoptotic (cell suicide) program. These observations, when taken together, showed a striking parallel: toxic agents that induced growth arrest or apoptosis were also capable of inducing increases in p53 levels. Because such observations were initially only correlations, they hardly proved that p53 was involved in some fashion in *causing* cells to enter into growth arrest or apoptosis following exposure to toxic or stressful stimuli.

The definitive demonstrations of causality came from detailed examinations of p53 functions. For example, when genotoxic agents, such as X-rays, evoked an increase in cellular p53 levels, the levels of the $p21^{Cip1}$ protein (Section 8.4) increased subsequently; this induction was absent in cells expressing mutant p53 protein. This suggested that p53 could halt cell cycle advance by inducing expression of this widely acting CDK inhibitor (Figure 9.9). Indeed, the long-term biological responses to irradiation were often affected by the state of a cell's *p53* gene. Thus, cells carrying mutant *p53* alleles showed a greatly decreased tendency to enter into growth arrest or apoptosis when compared with wild-type cells that were exposed in parallel to this stressor (Figure 9.10).

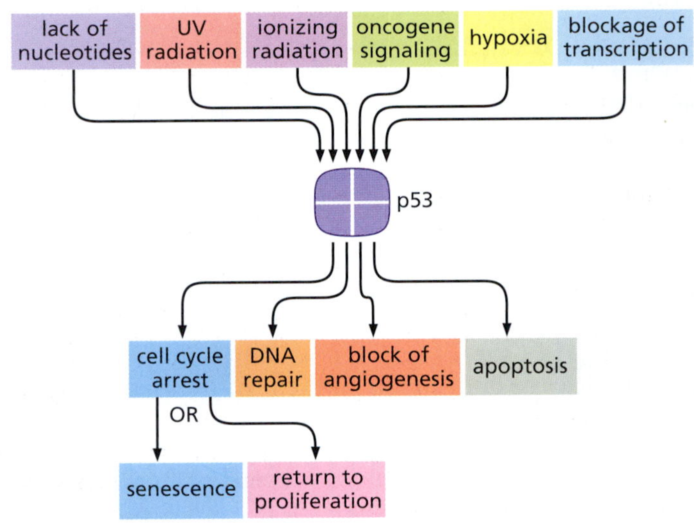

Figure 9.8 p53-activating signals and p53's downstream effects Studies of p53 function have revealed that a variety of cell-physiologic stresses can cause a rapid increase in p53 levels. The resulting accumulated p53 protein then undergoes certain post-translational modifications (to be discussed later) and proceeds to induce a number of responses. A cytostatic response ("cell cycle arrest", often called "growth arrest") can either be irreversible ("senescence") or reversible ("return to proliferation"). DNA repair proteins may be mobilized as well as proteins that antagonize blood vessel formation ("block of angiogenesis"). As an alternative, in certain circumstances, p53 may trigger apoptosis.

Taken together, these various observations could be incorporated into a simple, unifying mechanistic model: p53 continuously receives signals from a diverse array of surveillance systems. If p53 receives specific alarm signals from these monitors, it calls a halt to cell proliferation or triggers the apoptotic suicide program (see Figure 9.8).

In fact, these cytostatic and pro-apoptotic powers of p53 represent a major threat to incipient cancer cells that are advancing toward the malignant growth state. A number of stresses, including hypoxia, genomic damage, and imbalances in the signaling pathways governing cell proliferation, are commonly experienced by cancer cells during many stages of tumor development. In the presence of any one of these stresses, an intact, functional p53 alarm system threatens the viability of would-be cancer cells. Consequently, p53 activity must be blunted or even fully eliminated in these cells if they are to survive and prosper.

This explains why most and perhaps all human tumor cells have partially or totally inactivated their p53 alarm response. Without p53 on duty, cancer cells are able to tolerate hypoxia, extensive damage to their genomes, and profound dysregulation of their growth-controlling circuitry. Once a cell acquires resistance to these normally debilitating factors, the road is paved for it and its descendants to continue their march toward a highly malignant growth state. In the same vein, normal cells must also avoid excessive p53 activity, since it threatens to end their lives and thereby cause depletion of the cells needed to maintain normal bodily functions (Sidebar 6).

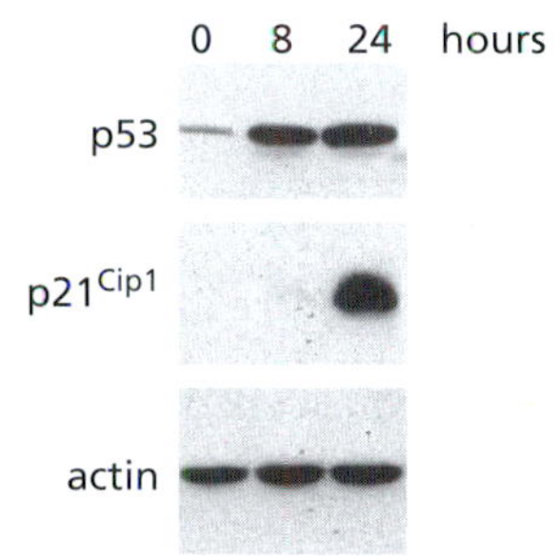

Figure 9.9 Induction of p53 and p21^{Cip1} following DNA damage Exposure of cells to X-rays serves to strongly increase p53 levels. Once it is present in higher concentrations (8, 24 hours) and functionally activated *(not measured here)*, p53 induces expression of the p21^{Cip1} protein (Section 8.4). p21^{Cip1} acts as a potent CDK inhibitor of the cyclin–CDK complexes that are active in late G_1, S, G_2, and M phases and can thereby act to halt further cell proliferation at any of these phases of the cell cycle. The actin protein present in all three samples is included as a "loading control" to ensure that equal amounts of protein were added to the three gel channels prior to electrophoresis. (Courtesy of K.H. Vousden.)

9.6 DNA damage and deregulated growth signals cause p53 stabilization

Three well-studied monitoring systems have been found to send alarm signals to p53 in the event that they detect damage or signaling imbalances. The first of these responds to double-strand (ds) breaks in chromosomal DNA, notably those that are created by ionizing radiation such as X-rays. Indeed, a single dsDNA break occurring anywhere in the genome seems sufficient to induce a measurable increase in p53 levels, While the identities of the proteins that detect such breaks remain unclear, it is known that they transfer signals to the ATM kinase. (As described later in Chapter 12, a deficiency of ATM leads to the disease of ataxia telangiectasia and to hypersensitivity of cells to X-irradiation.) ATM, in turn, transfers its signals on to the ATR (ATM-related) kinase, which is able to phosphorylate p53 itself; ATM also appears able to directly phosphorylate p53. This phosphorylation of p53 protects it from destruction.

A second signaling pathway is activated by a wide variety of DNA-damaging agents, including certain chemotherapeutic drugs and UV radiation; certain inhibitors of protein kinases also stimulate this pathway. It depends on the ATR kinase (see above) to pass signals on to casein kinase II (CKII), which in turn phosphorylates p53.

A third pathway leading to p53 activation is triggered by aberrant growth signals, notably those that result in deregulation of the pRb–E2F cell cycle control

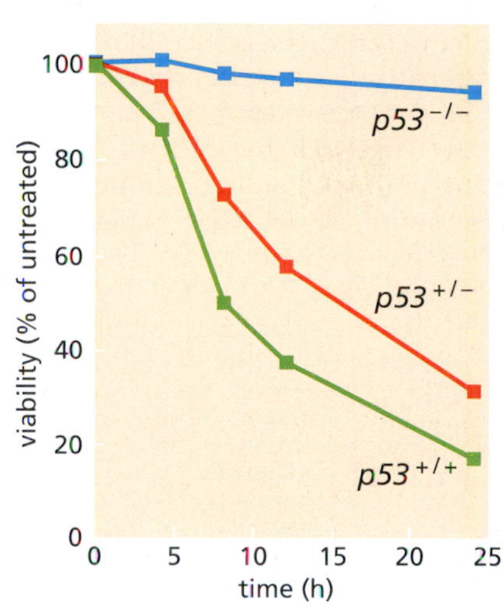

Figure 9.10 *p53* genotype and cellular responses to irradiation Thymocytes (leukocytes derived from the thymus) of wild-type mice show an 80% loss of viability during the 25 hours following X-irradiation *(green)*, while thymocytes from *p53*$^{+/-}$ heterozygous mice (with one wild-type and one null allele) show almost as much loss of viability *(red)*. In contrast, thymocytes prepared from *p53*$^{-/-}$ homozygous mutant mice exhibit less than a 5% loss of viability during this time period *(blue)*. In all cases, the loss of viability was attributable to apoptosis. (From S.W. Lowe et al., *Nature* 362:847–849, 1993.)

pathway. This pathway does not depend upon kinase intermediates to induce increases in p53 levels and signaling. The mechanisms by which other physiologic stresses or imbalances, such as hypoxia, trigger increases in p53 levels remain poorly understood.

These converging signaling pathways reveal a profound vulnerability of the mammalian cell. It has entrusted a single protein—p53—with the task of receiving signals from lookouts that are charged with monitoring a wide variety of important physiologic and biochemical intracellular systems (Figure 9.8). The funneling of these diverse signals to a single protein would seem to represent an elegant and economic design of the cellular signaling circuitry. But it also puts cells at a major disadvantage, since loss of this single protein from a cell's regulatory circuitry results in a catastrophic loss of the cell's ability to monitor its own well-being and respond with appropriate countermeasures in the event that certain operating systems malfunction.

In one stroke (actually, the two strokes that cause successive inactivation of the two *p53* gene copies), the cell becomes blind to many of its own defects. It thereby gains the ability to continue active proliferation under circumstances that would normally cause it to call a halt to proliferation or to enter into apoptotic death. In addition, as we will learn shortly, loss of the DNA repair and genome-stabilizing functions promoted by p53 will make descendants of a *p53*$^{-/-}$ cell more likely to acquire further mutations and advance more rapidly down the road of malignancy (see Sidebar 9.2 for an example).

Sidebar 9.2 Sunlight, p53, and skin cancer The p53 protein stands as an important guardian against skin cancer induced by sunlight. In the event that the genome of a **keratinocyte** in the skin has suffered extensive damage from ultraviolet-B (UV-B) radiation, p53 will rapidly trigger its apoptotic death. One manifestation of this is the extensive scaling of skin several days after a sunburn. At the same time, UV-B exposure may cause the mutation and functional inactivation of a *p53* gene within a keratinocyte. This is indicated by the fact that mutant *p53* alleles found in human squamous cell carcinomas of the skin often occur at dipyrimidine sites—precisely the sites at which UV-B rays induce the formation of pyrimidine–pyrimidine cross-links (see Section 12.6). Such mutant *p53* alleles can also be found in outwardly normal skin that has suffered chronic sun damage. Once p53 function is compromised by these mutations, keratinocytes may be able to survive subsequent exposures to UV-B irradiation, because apoptosis will no longer be triggered by their p53 protein. Such *p53*-mutant cells may then acquire additional mutant alleles that together enable them to form a squamous cell carcinoma. Interestingly, mice that lack functional *p53* gene copies in all cells also respond to UV-B exposure by developing ocular melanomas—tumors of pigmented cells of the eye; similar tumors are suspected to be caused in humans by UV exposure.

9.7 Mdm2 and ARF battle over the fate of p53

The diverse alarm signals that impinge on p53 have a common effect—causing a rapid increase in the levels of the p53 protein. Researchers have begun to understand how this dramatic change is achieved. Like a wide array of other cellular proteins, p53 protein molecules are degraded by the ubiquitin–proteasome system. As was described in Sidebar 7.8, proteins that are destined to be degraded by this system are initially tagged by the covalent attachment of polyubiquitin side chains, which causes them to be transported to proteasomes, in which they are digested into oligopeptides. The critical control point in this process is the initial tagging process.

The degradation of p53 in normal, unperturbed cells is regulated by a protein termed Mdm2 (in mouse cells) and Hdm2 (in human cells). This protein recognizes p53 as a target that should be ubiquitylated shortly after its synthesis and therefore marked for destruction (Figure 9.11). Mdm2 was initially identified as a protein encoded by double-minute chromosomes present in murine sarcoma cells (hence **m**ouse **d**ouble **m**inutes). Subsequently, the human homolog of the *mdm2* gene) was discovered to be frequently amplified in sarcomas. In many human lung tumors, Mdm2 (as we will call it) is overexpressed through mechanisms that remain unclear.

As is the case with other oncogenes, it seemed at first that amplification of the *mdm2* gene (indicated by the presence of many double-minute chromosomal particles in tumor cells; see Figure 1.12) afforded tumor cells some direct, immediate proliferative advantage. Only long after the Mdm2 protein was first identified did its role as the agent of p53 destruction become apparent. In fact, the detailed effects exerted by Mdm2 on p53 are slightly more complex than indicated above.

As we will learn below, p53 operates by acting as a transcription factor; Mdm2 binding to p53 immediately blocks the ability of p53 to function in this role. Thereafter, Mdm2 directs the attachment of a ubiquitin moiety to p53 and the export of p53 from the nucleus (where p53 does most of its work) to the cytoplasm; subsequent polyubiquitylation of p53 ensures its rapid degradation in

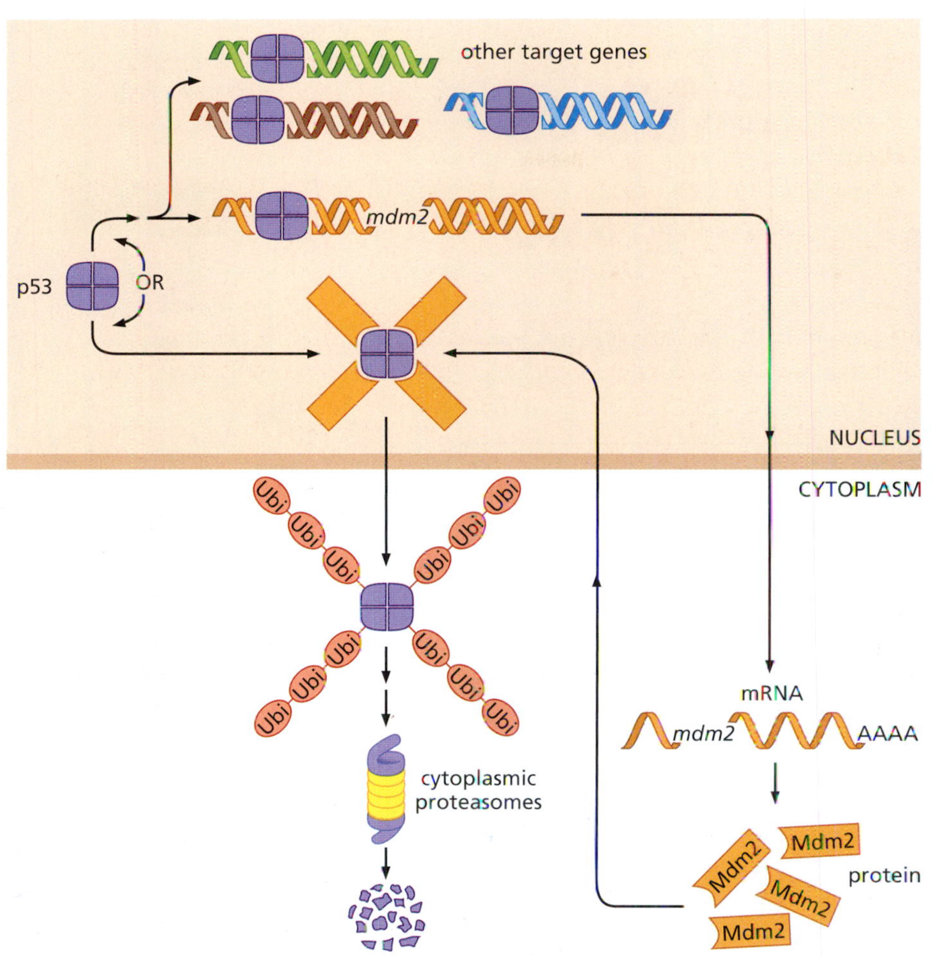

Figure 9.11 Control of p53 levels by Mdm2 After p53 concentrations increase in response to certain physiologic signals *(not shown)*, the p53 tetramers bind to the promoters of a large constituency of target genes whose transcription they induce *(above)*. Among the induced genes is *mdm2*; this results in a large increase in *mdm2* mRNA and Mdm2 protein *(right)*. Once synthesized, Mdm2 molecules bind to the p53 protein subunits and trigger their ubiquitylation and export to the cytoplasm, where they are degraded in proteasomes. This negative-feedback loop ensures that p53 levels eventually sink back to a low level and, in undisturbed cells, helps to keep p53 levels very low.

cytoplasmic proteasomes (see Figure 9.11). The continuous, highly efficient actions of Mdm2 ensure the short, 20-minute half-life of p53 in normal, unstressed cells.

In some circumstances—specifically, when cells are suffering certain types of stress or damage—p53 protein molecules must be protected from their Mdm2 executioner so that they can accumulate to functionally significant levels in the cell. This protection is often achieved by phosphorylation of p53, which blocks the ability of Mdm2 to bind p53 and trigger its ubiquitylation. More specifically, phosphorylation of p53 on amino acid residues in its N-terminal domain (Figure 9.12)

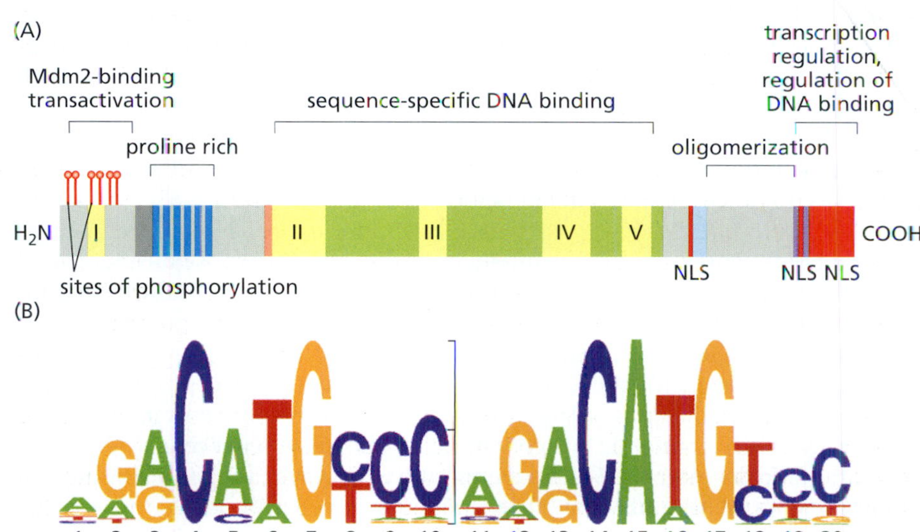

Figure 9.12 Specialized domains of p53 (A) The p53 protein binds to DNA via its central DNA-binding domain *(yellow, green)*. It is bound by Mdm2 (see Figure 9.11) in a small domain near its N-terminus, where its transactivation domain (involved in the activation of transcription) is also located. The phosphorylation of p53 amino acid residues in this region *(red lollipops; not all are indicated)* blocks Mdm2 binding and thus saves p53 from ubiquitylation and degradation. Toward its C-terminus, an oligomerization domain (*gray*; see Figure 9.7A) allows it to form tetramers. Near the C-terminus are also nuclear localization signals (NLS, *red*), which allow its import into the nucleus, as well as amino acid sequences that regulate its DNA binding. The proline-rich domain *(blue)* near the N-terminus contributes to p53's pro-apoptotic functions. (B) Use of anti-p53 antibodies allows the immunoprecipitation of p53–DNA complexes present in the chromatin of human cells (the ChIP procedure). Sophisticated sequence analyses of DNA fragments in the precipitates led to the identification of 452 sites in the human genome to which p53 binds. The consensus DNA sequence to which p53 binds is shown here, where the relative sizes of each letter indicates how frequently a DNA base was found at the indicated position in the binding site. (A, from D.E. Fisher, ed. Tumor Suppressor Genes in Human Cancer. Totowa, NJ:Humana Press, 2000; B, from C.L. Wei, *Cell* 124:207–219, 2006.)

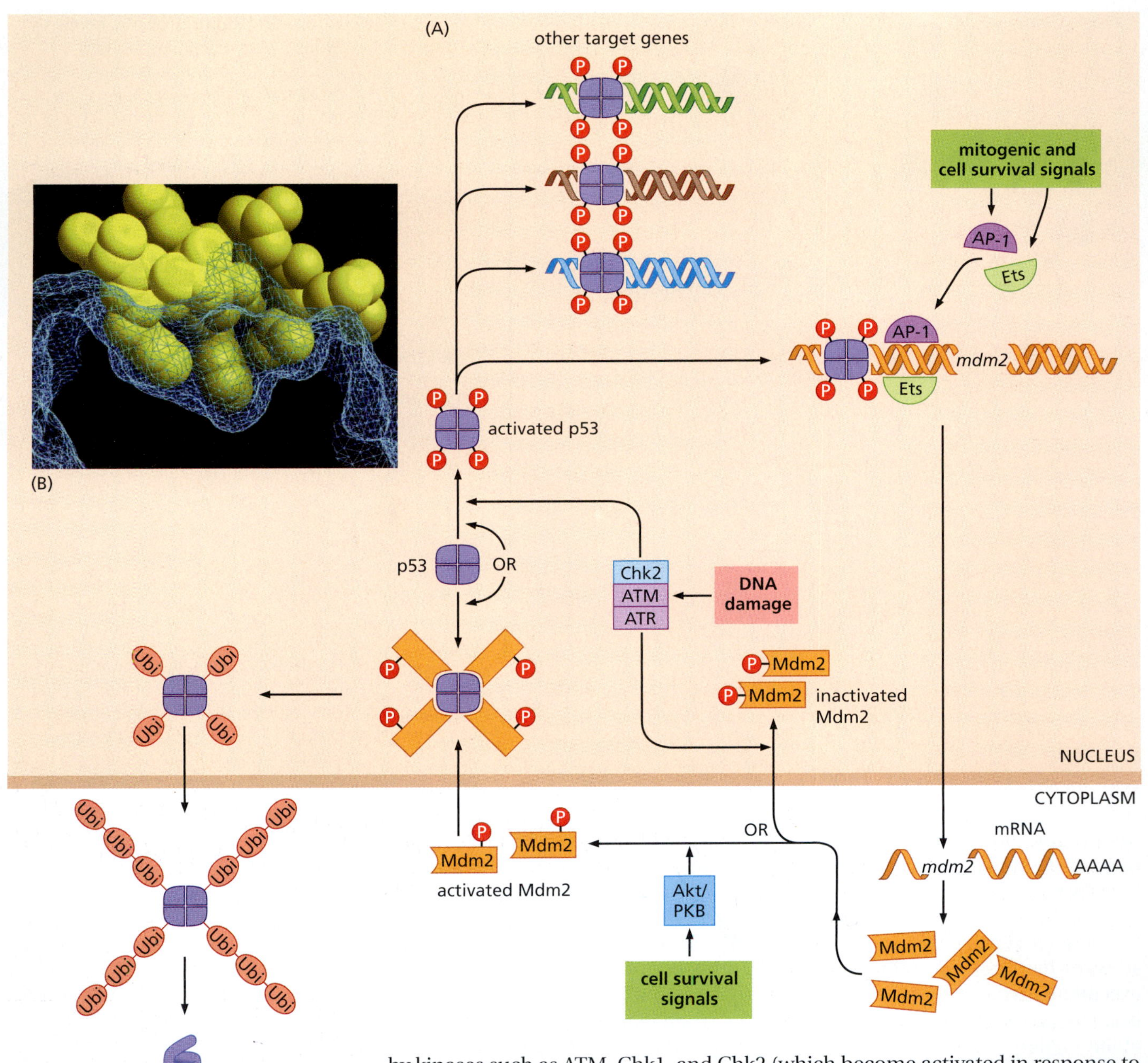

by kinases such as ATM, Chk1, and Chk2 (which become activated in response to DNA damage) alters the domain of p53 that is normally recognized and bound by Mdm2, and in this way prevents the association of Mdm2 with p53 (Figure 9.13). At the same time, the ATM kinase can phosphorylate Mdm2 in a way that causes its functional inactivation. As a consequence of this phosphorylation of both p53 and Mdm2, Mdm2 fails to initiate ubiquitylation of p53, p53 escapes destruction, and p53 concentrations in the cell increase rapidly. Once present in substantial amounts, p53 is then poised to evoke a series of downstream responses, to be discussed in detail below.

Note that Mdm2 operates here as an oncoprotein, but one whose mechanism of action is very different from those of the various oncoproteins that we encountered in Chapters 4, 5, and 6. The latter function as components of mitogenic signal cascades and thereby induce cell proliferation by mimicking the signals normally triggered by the binding of growth factors to their receptors. Mdm2, in contrast, operates by antagonizing p53 and thereby prevents entrance of a cell into cell cycle arrest, into the nongrowing state known as **senescence**, or into the apoptotic suicide program. The final outcome is, however, the same: the actions of both oncoproteins and Mdm2 favor increases in cell number.

Figure 9.13 Control of p53 levels by various kinases (A) The cycle of p53 synthesis and destruction indicated in Figure 9.11 can be modulated by a series of regulators. DNA damage-sensing kinases, such as ATM and ATR, acting directly or indirectly via Chk2, can phosphorylate p53 *(center)* in its N-terminal domain (see Figure 9.12), which prevents the binding of Mdm2 *(gold, left center)*. At the same time, phosphorylation of Mdm2 molecules by these kinases blocks their ability to associate with p53 *(center)*. These alterations save p53 from Mdm2-mediated binding, ubiquitylation, and destruction in proteasomes *(lower left)* that would otherwise occur. Acting in an opposing manner, certain survival signals (such as those conveyed by mitogenic growth factors), acting through the AP-1 and Ets transcription factors, collaborate with p53 to promote expression of the *Mdm2* gene *(center right)*, resulting in increases in Mdm2 mRNA and protein synthesis *(lower right)*. These survival signals also activate the Akt/PKB kinase, which phosphorylates and activates the already synthesized Mdm2 molecules at another site *(bottom)*. The resulting activated Mdm2 then proceeds to bind p53 and trigger its ubiquitylation and proteasome-mediated destruction *(lower left)*. (Recent evidence suggests that p53 becomes monoubiquitylated in the nucleus and polyubiquitylated thereafter in the cytoplasm.) (B) The structure of the interface where p53 and Mdm2 interact has been revealed by X-ray crystallography. The interacting domain of p53 is shown as a yellow space-filling model that includes p53 residues 18 through 27, while the surface of the complementary pocket of Mdm2 is shown as a blue wire mesh. Phosphorylation of p53 by kinases such as Chk2, ATM, and ATR prevent this interaction, sparing p53 from Mdm2-mediated ubiquitylation and subsequent degradation. (B, from P.H. Kussie et al., *Science* 274:948–953, 1996.)

The activity and levels of the Mdm2 protein are affected by yet other positive and negative signals. The signaling pathway that favors cell survival through activation of the PI3 kinase (PI3K) pathway leads, via the Akt/PKB kinase, to Mdm2 phosphorylation (at a site different from that altered by the ATM kinase described above) and to the resulting translocation of Mdm2 from the cytoplasm to the nucleus, where it is poised to attack p53 (see Figure 9.13). Because PI3K itself is activated by Ras and growth factor receptors, we come to realize that the mitogenic signaling pathway does indeed influence Mdm2 and thereby p53, albeit indirectly. At the same time, activation of the mitogenic Ras → Raf → MAPK signaling pathway leads, via the Ets and AP-1 (Fos + Jun) transcription factors, to greatly increased transcription of the *mdm2* gene, yielding higher levels of *mdm2* mRNA and protein (see Figure 9.13). These elevated levels of Mdm2 protein amplify the phosphorylation-induced activation of Mdm2 achieved by the PI3K → Akt/PKB signaling pathway. Ultimately, all these effects converge on suppressing p53 protein levels.

Yet another mechanism that affects Mdm2 has been revealed through the discovery of an Mdm2 antagonist, which is termed $p19^{ARF}$ in mouse cells and $p14^{ARF}$ in human cells. Astute sequence analysis led to the discovery of ARF, as we will call it hereafter. Its encoding gene was originally uncovered in mouse cells as a gene whose sequences are intertwined with those specifying $p16^{INK4A}$, the important inhibitor of the CDK4 and CDK6 kinases that initiate pRb phosphorylation (Section 8.4).

Through use of a transcriptional promoter located 13 kilobases upstream of the $p16^{INK4A}$ promoter and an alternative splicing program, an mRNA is assembled that encodes in an alternative reading frame, the structure of the ARF protein (Figure 9.14). Forced expression of an ARF-encoding cDNA in wild-type rodent

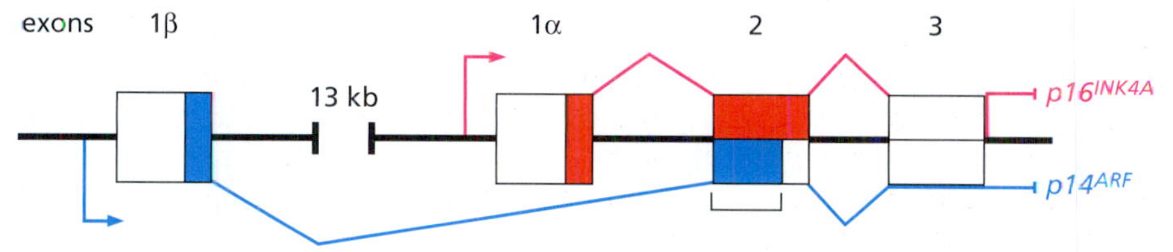

Figure 9.14 The gene encoding $p16^{INK4A}$ and $p14/19^{ARF}$ Analysis of the $p16^{INK4A}$ gene *(red)* has revealed that it shares its second exon with a second gene encoding a 19-kD protein in mice and a 14-kD protein in humans. The gene encoding this p14/p19 protein uses an alternative transcriptional promoter *(blue arrow, left)* located more than 13 kilobases upstream of that used by $p16^{INK4A}$ *(red arrow, center)*; because translation of its mRNA uses an alternative reading frame *(bracket)* present in this exon 2 *(red, blue)*, the resulting protein and thus gene came to be called $p19^{ARF}$ (or in humans $p14^{ARF}$). The patterns of RNA splicing are indicated by the carets connecting the various exons of the two intertwined genes. The boxes indicate exons, while the filled areas within each exon indicate reading frames. (From C. Sherr, *Genes Dev.* 12:2984–2991, 1998.)

cells was found to cause a strong inhibition of their proliferation. However, this inhibition was not observed when the ARF cDNA was expressed in cells that lacked wild-type p53 function. This indicated that the growth-inhibitory powers of ARF depend absolutely on the presence of functional p53 in these cells.

Further investigation revealed that in wild-type cells, the expression of ARF causes a rapid increase in p53 levels. We now understand the molecular mechanisms that explain how this response works. ARF binds to Mdm2 and inhibits its action, apparently by sequestering Mdm2 in the nucleolus—the nuclear structure that is largely devoted to manufacturing ribosomal subunits (Figure 9.15A).

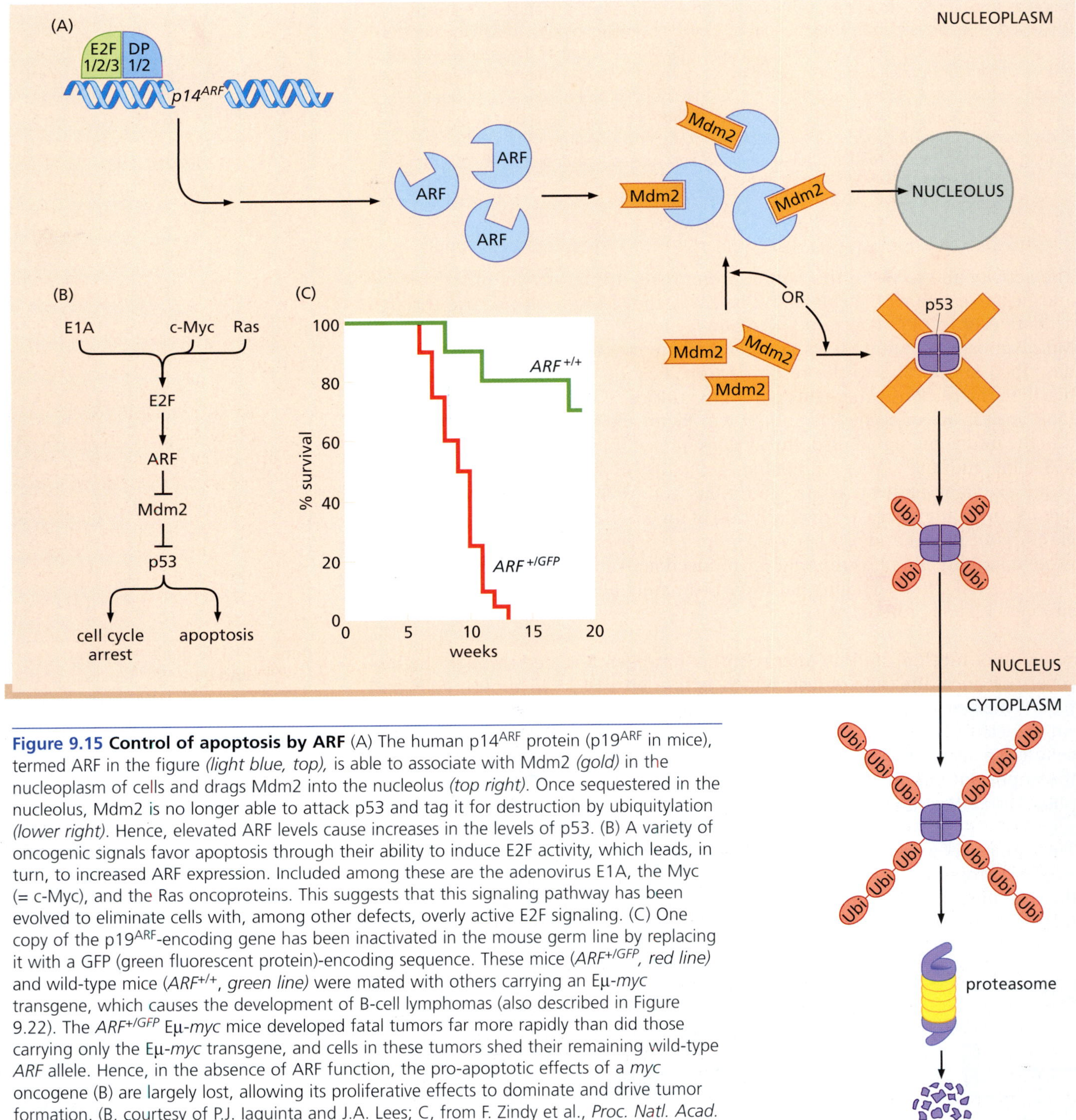

Figure 9.15 Control of apoptosis by ARF (A) The human p14^{ARF} protein (p19^{ARF} in mice), termed ARF in the figure *(light blue, top),* is able to associate with Mdm2 *(gold)* in the nucleoplasm of cells and drags Mdm2 into the nucleolus *(top right)*. Once sequestered in the nucleolus, Mdm2 is no longer able to attack p53 and tag it for destruction by ubiquitylation *(lower right)*. Hence, elevated ARF levels cause increases in the levels of p53. (B) A variety of oncogenic signals favor apoptosis through their ability to induce E2F activity, which leads, in turn, to increased ARF expression. Included among these are the adenovirus E1A, the Myc (= c-Myc), and the Ras oncoproteins. This suggests that this signaling pathway has been evolved to eliminate cells with, among other defects, overly active E2F signaling. (C) One copy of the p19^{ARF}-encoding gene has been inactivated in the mouse germ line by replacing it with a GFP (green fluorescent protein)-encoding sequence. These mice (*ARF*$^{+/GFP}$, *red line)* and wild-type mice (*ARF*$^{+/+}$, *green line)* were mated with others carrying an Eμ-*myc* transgene, which causes the development of B-cell lymphomas (also described in Figure 9.22). The *ARF*$^{+/GFP}$ Eμ-*myc* mice developed fatal tumors far more rapidly than did those carrying only the Eμ-*myc* transgene, and cells in these tumors shed their remaining wild-type *ARF* allele. Hence, in the absence of ARF function, the pro-apoptotic effects of a *myc* oncogene (B) are largely lost, allowing its proliferative effects to dominate and drive tumor formation. (B, courtesy of P.J. Iaquinta and J.A. Lees; C, from F. Zindy et al., *Proc. Natl. Acad. Sci. USA* 100:15930–15935, 2003.)

Once Mdm2 is diverted away from its interactions with p53, the latter escapes Mdm2-mediated ubiquitylation and resulting destruction and therefore accumulates rapidly to high levels in the cell. The enemy of an enemy is a friend: ARF can induce rapid increases in p53 levels because it kidnaps and inhibits p53's destroyer, Mdm2.

Importantly, in normal, unstressed cells, Mdm2 must be allowed its normal role of keeping p53 levels very low. Otherwise, p53 will build up to intolerably high levels and inappropriately shut down cell proliferation or induce apoptosis (Sidebar 9.3).

The series of mutual antagonisms indicated in Figure 9.15 makes ARF an ally of p53 and, like p53, a tumor suppressor protein. In many human tumors, inactivation of the $p16^{INK4A}/\ p14^{ARF}$ locus by genetic mutation or epigenetic promoter methylation can be demonstrated. Once a cell has lost ARF activity, it loses the ability to block Mdm2 function. As a consequence, Mdm2 is given a free hand to drive p53 degradation, and the cell is deprived of the services of p53 because the latter can never accumulate to functionally significant levels. Since ARF has a central role in increasing p53 levels in many cellular contexts, this means that the $p14^{ARF}$ gene, like the gene encoding its p53 target, is an extremely important tumor suppressor gene. Moreover, it seems likely that many of the human cancer cells that retain wild-type *p53* gene copies have eliminated p53 function by inactivating their two copies of the gene encoding ARF. Finally, we should note that the co-localization of the $p16^{INK4A}$ and $p14^{ARF}$ genes (see Figure 9.14) represents yet another concentration of power that creates additional vulnerability for normal cells (Sidebar 9.4).

Sidebar 9.3 Mdm2 and p53 are locked in a death grip The antagonistic actions of Mdm2 and p53 are highlighted by the results of inactivating both *mdm2* gene copies in the genomes of mouse embryos. These die very early in embryogenesis, ostensibly because p53 levels increase in embryonic cells to physiologically intolerable levels, preventing the normal proliferation of embryonic cells or causing them to die. As was discussed earlier (Section 9.2), inactivation of both *p53* gene copies has virtually no effect on embryonic development. However, when both copies of both genes are inactivated in a mouse embryo (yielding the $Mdm2^{-/-}\ p53^{-/-}$ genotype), development occurs perfectly normally. Taken together, these results indicate that the profoundly disruptive effects of *Mdm2* gene inactivation can be attributed totally to runaway p53 activity and can be fully reversed by removing p53 from the scene. Such a clean and unambiguous outcome is rarely observed in experiments of this sort, since most genes and proteins exert multiple effects that make dramatically simple experimental outcomes like this one quite uncommon.

9.8 ARF and p53-mediated apoptosis protect against cancer by monitoring intracellular signaling

The influential role of ARF in increasing p53 levels raises the question of how ARF itself is regulated. In this instance, we learn something highly relevant from our discussion in Chapter 8 of the pRb pathway, and from the fact that mammalian cells are very sensitive to higher-than-normal levels of E2F1 activity. In fact, a cell seems to monitor the activity level of this particular transcription

Sidebar 9.4 Have mammalian cells placed too many eggs in one basket? The discovery of the $p16^{INK4A}/p14^{ARF}$ genetic locus, which is inactivated through one mechanism or another in about half of all human tumors, raises a provocative question: Why have mammalian cells invested a single chromosomal locus with the power to encode two proteins regulating the two most important tumor suppressor pathways, those of pRb and p53? Deletion of this single locus results in the simultaneous loss of normal regulation of both pathways. As was the case with *p53* itself, enormous power has been concentrated in the hands of a single genetic locus.

Placing two such vital eggs in a single genetic basket seems foolhardy for the mammalian cell, as it causes the cell to be vulnerable to two types of deregulation through loss of a single gene. To make matters even worse, the gene encoding $p15^{INK4B}$, another important regulator of pRb phosphorylation (Section 8.4), is closely linked to the $p16^{INK4A}/p14^{ARF}$ locus, indeed so close that all of these genetic elements are often lost through the deletion of only about 40 kb of chromosomal DNA. We have yet to discern the underlying rationale of this genetic arrangement. Maybe there is none, and perhaps mammalian evolution has produced a less-than-optimal design of one part of the cellular regulatory machinery.

factor (perhaps together with those of E2F2 and E2F3) as an indication of whether its pRb circuitry is functioning properly; excessively high levels of active E2F transcription factors provide a telltale sign that pRb function has gone awry.

Evolution has created several ways to eliminate cells that carry too much E2F activity and, by implication, have lost proper pRb control (Figure 9.16). Runaway E2F1 activity drives expression of a number of genes encoding proteins that directly participate in the apoptotic program. Included among these are genes encoding **caspases** (types 3, 7, 8, and 9), pro-apoptotic Bcl-2 related proteins (Bim, Noxa, PUMA), Apaf-1, and p53's cousin, p73; these proteins collaborate to drive cells into apoptosis. We will learn more about them later.

In addition, the p53-dependent apoptotic program is often triggered by elevated E2F activity. It turns out that the *p14*^{ARF} gene carries an E2F recognition sequence in its promoter. In a way that is still incompletely understood, unusually high levels of E2F1 or E2F2 activity induce transcription of *p14*^{ARF} mRNA. The ARF protein soon appears on the scene and blocks Mdm2 action. p53 then accumulates and triggers, in turn, apoptosis (see Figure 9.15A,B), leading to a signaling cascade configured like this:

$$\text{pRb} \dashv \text{E2F} \rightarrow \text{ARF} \dashv \text{Mdm2} \dashv \text{p53} \rightarrow \text{apoptosis}$$

This pathway, working together with the several p53-independent pro-apoptotic signals cited above, accomplishes the goal of eliminating cells that lack proper pRb function. Such E2F-initiated apoptosis seems to explain why mouse embryos that have been deprived of both copies of their *Rb* gene die in mid-gestation due to the excessive proliferation and concomitant apoptosis of certain critical cell types, including those involved in erythropoiesis (formation of red blood cells) and in placental function.

The discovery of the critical role of ARF in the control of p53 function suggested the possibility that ARF function is eliminated by a variety of molecular strategies during tumor formation. Such elimination may well confer on cancer cells the same benefits as those resulting from mutation of the *p53* gene itself. In fact, regulation of transcription of the *p14*^{ARF} gene is quite complex and therefore susceptible to disruption through a variety of alterations (Sidebar 7).

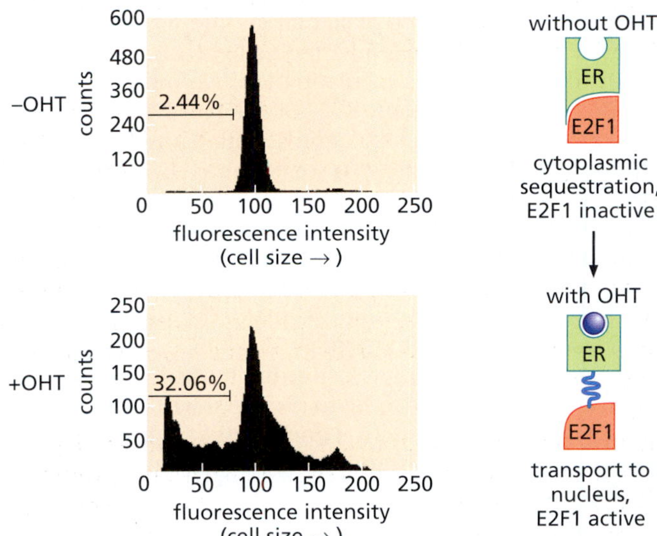

Figure 9.16 E2F1-mediated induction of apoptosis The apoptotic state of cells can be monitored by fluorescence-activated cell sorting (FACS), which in this case is used to measure the size of individual cells or subcellular fragments *(abscissa)* and the number of cells of a given size *(ordinate)*. In this experiment, the E2F1 transcription factor *(E2F, green)* has been fused to the estrogen receptor (ER) protein *(red)*, making E2F1 activity dependent on the presence of tamoxifen (OHT), a ligand of the ER. In the absence of tamoxifen *(upper panel)*, the E2F1 factor is sequestered in the cytoplasm; under this condition almost all cells in such a population have a size of roughly 100 (arbitrary) units (with 2.44% having a smaller size). However, when tamoxifen is added to these cells *(lower panel, purple ball)*, the NLS is exposed and the E2F1-containing fusion protein is imported into the nucleus and becomes activated, resulting in the expression of a cohort of genes, among them those that have pro-apoptotic effects. As a result, a significant proportion (32.06%) of the cells now show a size smaller than that of normal healthy cells, indicative of their having fragmented during the process of apoptosis. (From T. Hershko and D. Ginsberg, *J. Biol. Chem.* 279:8627–8634, 2004.)

In sum, because loss of pRb control within a cell represents a grave danger to the surrounding tissue, cells are poised to trigger apoptosis whenever E2F1 deregulation occurs. These connections between E2F1 activity and apoptosis suggest another idea, still speculative: the great majority of cells that suffer loss of pRb control never succeed in generating clones of pre-neoplastic or neoplastic descendants, because these cells suffer apoptosis as soon as they lose this important control mechanism.

Consistent with this logic are some of the known properties of the *E1A* and *myc* oncogenes. Both deregulate pRb control, and both are highly effective in inducing apoptosis. Recall that the adenovirus E1A oncoprotein binds and effectively sequesters pRb and its cousins. Myc, for its part, pulls regulatory levers in the cell cycle clock that ensure that pRb is inactivated through phosphorylation (Sections 8.5 and 8.9).

Many studies of *myc* oncogene function indicate that this gene exerts both potent mitogenic and pro-apoptotic functions. Indeed, the pro-apoptotic effects of the *myc* oncogene are so strong that it is highly likely that most cells that happen to acquire a *myc* oncogene are rapidly eliminated through apoptosis. On occasion, the apoptotic program may be blunted or inactivated, and only then can the mitogenic actions of *myc* become apparent.

As an example, when a *myc* oncogene becomes activated in the lymphoid tissues of a mouse, it prompts a substantial increase in cell proliferation. However, there is no net increase in cell number, since the newly formed cells are rapidly lost through apoptosis. If one of the *myc* oncogene–bearing cells happens subsequently to inactivate its *p53* gene copies, then *myc*-induced apoptosis is diminished and the cell proliferation driven by *myc* leads to a net increase in the pool of mutant lymphoid cells. As might be expected from the organization of the pathway drawn above, a similar effect operates in mice that carry only a single functional $p19^{ARF}$ gene (see Figure 9.15C).

These discussions suggest that E2F-induced apoptosis functions solely as an anti-cancer mechanism designed to eliminate unwanted, pre-neoplastic cells. However, research with genetically altered mice provides evidence that normal physiologic mechanisms also depend on E2F-induced apoptosis to weed out extra cells that are not required for the development of a normal immune system (Sidebar 8).

9.9 p53 functions as a transcription factor that halts cell cycle advance in response to DNA damage and attempts to aid in the repair process

The p53 protein has a DNA-binding domain (see Figure 9.12A) with an affinity for binding a sequence motif composed of the sequence Pu-Pu-Pu-C-A/t-T/a-G-Py-Py-Py repeated twice in tandem (where Pu represents the purine nucleotides A or G while Py represents the pyrimidine nucleotides C or T; A/t represents a site at which A occurs more frequently than T; and T/a denotes a site where T occurs more frequently than A). Between 0 and 13 nucleotides of random sequence are found to separate these two tandemly arrayed recognition sequences (see Figure 9.12B). This sequence motif is present in the promoters or initial introns of a number of the downstream target genes whose expression is induced (or repressed) by p53.

Actually, the transcription-activating powers of the p53 tetramer depend on more than simply recognizing and binding this sequence within a promoter. In addition, a complex array of covalent modifications of p53 must occur, many affecting the C-terminal domain of p53. These include acetylation, glycosylation, phosphorylation, ribosylation, and sumoylation (involving respectively attachment of acetyl, sugar, phosphate, ribose, and sumo groups, the latter being a ubiquitin-like peptide that appears to target proteins for localization to specific intracellular sites, often in the nucleus; see Figure 9.35). These modifications are likely to affect the ability of p53 to interact physically with other transcription factors that modulate its transcriptional powers. Indeed, it seems likely that combinatorial interactions of p53 with these other transcription factors determine the identities of the specific target genes that are activated in various circumstances by p53.

Significantly, as described in Figure 9.6B, the great majority (>90%) of the mutant *p53* alleles found in human tumor cell genomes encode amino acid substitutions in the DNA-binding domain of p53. The resulting defective p53 proteins, being unable to bind to the promoters of downstream target genes, have therefore lost the ability to mediate most if not all of p53's multiple functions.

As suggested earlier, one highly important target of the p53 transcription factor is the *Mdm2* gene. Consequently, when active as a transcription factor, p53 encourages the synthesis of Mdm2—the agent of its own destruction (see Figure 9.11). This creates a negative-feedback loop that usually functions to ensure that p53 molecules are degraded soon after their synthesis, resulting in the very low steady-state levels of p53 protein observed in normal, unperturbed cells.

The operations of this p53–Mdm2 feedback loop explain a bizarre aspect of p53 behavior. In human cancer cells that carry mutant, defective *p53* alleles, the p53 protein is almost invariably present in high concentrations (for example, see Figure 9.17), in contrast to its virtual absence from normal cells. At first glance, this might appear paradoxical, since high levels of a growth-suppressing protein like p53 would seem to be incompatible with malignant cell proliferation.

The paradox is resolved by the fact, mentioned above, that the great majority of the mutations affecting the *p53* gene cause the p53 protein to lose its transcription-activating powers. As a direct consequence, p53 is unable to induce *Mdm2* transcription and thus Mdm2 protein synthesis. In the absence of Mdm2, p53 escapes degradation and accumulates to very high levels. This means that many types of human cancer cells accumulate high concentrations of essentially inert p53 molecules.

This logic explains why the presence of readily detectable p53 in a population of tumor cells, usually revealed by **immunostaining** (see Figure 9.17), is a telltale

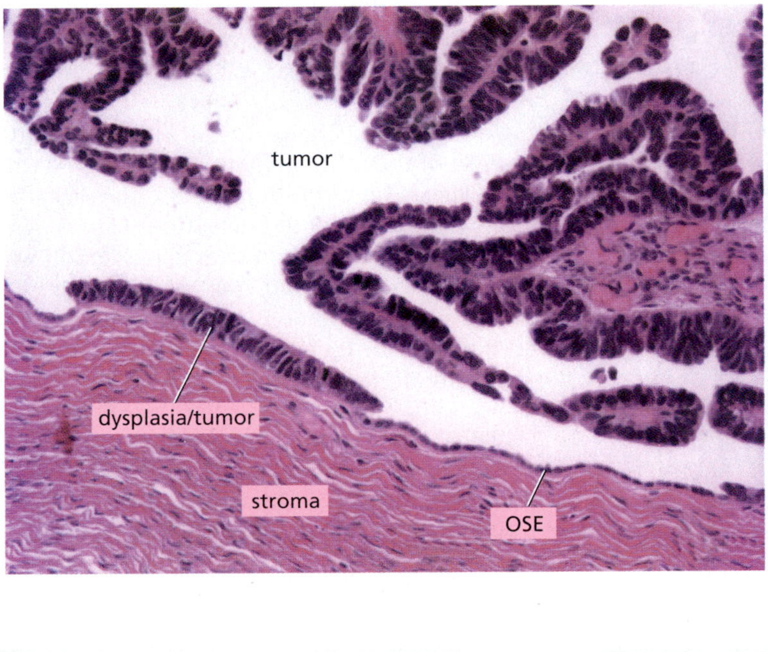

Figure 9.17 Accumulation of p53 in *p53*-mutant cells This microscope section of ovarian tissue has been stained with an anti-p53 antibody, which is coupled to the peroxidase enzyme, resulting in the blackened nuclei seen here. Large patches of epithelial cells in an ovarian carcinoma *(above)* are composed of cells that have high levels of p53; a patch of dysplasia *(left middle)* is also p53-positive. Stromal cells *(small black nuclei, pink matrix, below)* are unstained, as is a patch of normal ovarian surface epithelium (OSE, *below, right)*. (Courtesy of R. Drapkin and D.M. Livingston.)

Table 9.2 Examples of p53 target genes according to function The expression of genes in this table is induced by p53 unless otherwise indicated.

Class of genes	Name of gene	Function of gene product
p53 antagonist	*MDM2/HDM2*	induces p53 ubiquitylation
Growth arrest genes	*p21^{Cip1}*	inhibitor of CDKs, DNA polymerase
	Siah-1	aids β-catenin degradation
	14-3-3σ	sequesters cyclin B–CDC2 in cytoplasm
	Reprimo	G_2 arrest
DNA repair genes	*p53R2*	ribonucleotide reductase—biosynthesis of DNA precursors
	XPE/DDB2	global NER
	XPC	global NER
	XPG	global NER, TCR
	GADD45	global NER ?
	DNA pol κ	error-prone DNA polymerase
Regulators of apoptosis	*BAX*	mitochondrial pore protein
	PUMA	BH3-only mitochondrial pore protein
	NOXA	BH3-only mitochondrial pore protein
	p53AIP1	dissipates mitochondrial membrane potential
	Killer/DR5	cell surface death receptor
	PIDD	death domain protein
	PERP	pro-apoptotic transmembrane protein
	APAF1	activator of caspase-9
	NF-κB	transcription factor, mediator of TNF signaling
	Fas/APO1	death receptor
	PIG3	mitochondrial oxidation/reduction control
	PTEN	reduces levels of the anti-apoptotic PIP_3
	Bcl-2	(repression of) its expression
	IGF-1R	(repression of) its expression
	IGFBP-3	IGF-1–sequestering protein
Anti-angiogenic proteins	*TSP-1* (thrombospondin)	antagonist of angiogenesis

sign of the presence of a mutant *p53* allele in the genome of these cells. (Such a conclusion cannot be drawn, however, from analyzing human tissue that has recently been irradiated, since radiation can also evoke the widespread expression of p53 throughout a tissue for days, even weeks after radiotherapy.) The identical logic explains the large amounts of p53 protein in SV40-infected or SV40-transformed cells, in which sequestration of p53 by the viral large T (LT) antigen prevents p53-induced expression of the *Mdm2* gene expression and resulting p53 degradation (see Figure 9.1).

Mdm2 is only one of a large cohort of genes whose expression is induced by p53 (Table 9.2). As earlier discussion implied, another highly important target gene is *p21^{Cip1}* (sometimes called *Cdkn1a* in the mouse), which we encountered previously as a widely acting CDK inhibitor (Section 8.4). Its induction explains the cytostatic (rather than pro-apoptotic) actions of p53. In fact, the gene encoding p21^{Cip1} was originally discovered by a molecular search strategy designed to uncover genes whose expression is increased by p53. Soon after its discovery, it became apparent that p21^{Cip1} functions as an important inhibitor of a number of the cyclin-dependent kinases (CDKs). Thus, the ability of p21^{Cip1} to inhibit two CDKs—CDK2 and CDC2—that are active in the late G_1, S, G_2, and M phases of the cell cycle explains how p53 is able to block forward progress at many points in this cycle.

This information also provides us with insight into the physiologic roles played by p53 in the life of a cell. For example, if the chromosomal DNA of a cell should suffer some damage during the G_1 phase of the cell cycle, p53 will become activated, both by rapid increases in its concentration and by post-translational modifications that enable it to function effectively as a transcription factor. p53 will then induce p21^{Cip1} synthesis, and p21^{Cip1}, in turn, will halt further cell proliferation.

At the same time, components of the cellular DNA repair machinery will be mobilized to repair the damage. Some of these are directly induced by p53. This is suggested by observations that certain DNA repair proteins are mobilized far more effectively in cells carrying wild-type *p53* alleles than in those with mutant *p53* alleles. For example, cells lacking functional p53 are unable to efficiently repair the DNA lesions caused by benzo[*a*]pyrene (a potent carcinogen present in tars) and the cyclobutane pyrimidine dimers caused by ultraviolet (UV) radiation. In addition, DNA polymerase β, which plays a critical role in reconstructing DNA strands after chemically altered bases have been excised by DNA repair proteins, is much less active in p53-negative cells than in their wild-type counterparts (see Table 9.2). We will return to these DNA repair proteins and their mechanisms of action in Chapter 12.

In the event that the DNA is successfully repaired, the signals that have protected p53 from destruction will disappear. The consequence is that the levels of p53 collapse and $p21^{Cip1}$ follows suit. This allows cell cycle progression to resume, enabling cells to enter S phase, where DNA replication now proceeds.

The rationale for this series of steps is a simple one: by halting cell cycle progression in G_1, p53 prevents a cell from entering S phase and inadvertently copying still-unrepaired DNA. Such copying, if it occurred, would cause a cell to pass mutant DNA sequences on to one or both of its daughters. The importance of these cytostatic actions of $p21^{Cip1}$ can be seen from the phenotype of genetically altered mice in which both germ-line copies of the $p21^{Cip1}$ gene have been inactivated. Although not as tumor-prone as p53-null mice, they show an increased incidence of tumors late in life. This milder phenotype is what we might expect, since $p21^{Cip1}$ mediates some, but not all, of the tumor-suppressing activities of p53.

If a cell suffering DNA damage has already advanced into S phase and is therefore in the midst of actively replicating its DNA, the $p21^{Cip1}$ induced by p53 can engage the DNA polymerase machinery at the replication fork and halt its further advance down DNA template molecules. (It does so by preventing activation of the centrally important DNA polymerase δ.) Once again, the goal here is to hold DNA replication in abeyance until DNA damage has been successfully repaired.

The p53 protein uses yet other genes and proteins to impose a halt to further cell cycle advance. For example, Siah, the product of another p53-induced gene, participates in the degradation of β-catenin; the latter helps to induce cyclin D1 synthesis and thus progression through most of the G_1 phase of the cell cycle (Section 8.3). The loss of β-catenin may also cause a decrease in transcription of the *myc* gene, which in turn may slow progression through several phases of the cell cycle in addition to its effects on G_1 advance (Section 8.9).

Two other genes that are activated by p53 encode the 14-3-3σ and Reprimo proteins (see Table 9.2), which help to govern the G_2/M transition. The 14-3-3σ protein, for its part, sequesters the cyclin B–CDC2 complex in the cytoplasm, thereby preventing it from moving into the nucleus, where its actions are needed to drive the cell into mitosis. This mechanism holds mitosis in abeyance until the chromosomal DNA is in good repair.

These various actions of p53 have caused some to portray this protein as the "guardian of the genome." By preventing cell cycle advance and DNA replication while chromosomal DNA is damaged and by inducing expression of DNA repair enzymes, p53 can reduce the rate at which mutations accumulate in cellular genomes. Conversely, cells that have lost p53 function may proceed to replicate damaged, still-unrepaired DNA, and this can cause them, in turn, to exhibit relatively mutable genomes, that is, genomes that accumulate mutations at an abnormally high rate per cell generation. And in the event that severe DNA damage has been sustained (i.e., damage that taxes a cell's DNA repair functions), p53 may trigger apoptosis.

In one particularly illustrative experiment, pregnant *p53*$^{+/-}$ mice that had been bred with *p53*$^{+/-}$ males were treated with the highly mutagenic carcinogen ethylnitrosourea (ENU). In all, 168 offspring were born. Of these, 70% of the *p53*$^{-/-}$ pups (which had been exposed *in utero* to this carcinogen) developed brain tumors, 3.6% of the *p53*$^{+/-}$ pups did so, and none of the *p53*$^{+/+}$ offspring gave evidence of brain tumor formation. Hence, in the absence of p53 function, fetal cells that have been mutated by ENU can survive and spawn the progeny forming these lethal tumors.

The absence of p53 results in the accumulation of genomic alterations far more profound than the point mutations caused by ENU. For example, when mouse fibroblasts are deprived of p53 function, they show greatly increased rates of chromosomal loss and duplication (ascribable, at least in part, to the loss of G_2/M checkpoints) and also exhibit increased numbers of *interstitial deletions*, that is, deletions involving the loss of a microscopically visible segment from within the arm of a chromosome.

9.10 p53 often ushers in the apoptotic death program

As indicated at several places in this chapter, p53 can opt, under certain conditions, to provoke a response that is far more drastic than the reversible halting of cell cycle advance. In response to massive, essentially irreparable genomic damage, **anoxia** (extreme oxygen deprivation), or severe signaling imbalances, p53 will trigger apoptosis. We now begin to explore the apoptotic program in more detail.

The cellular changes that constitute the apoptotic program proceed according to a precisely coordinated schedule. Within minutes, patches of the plasma membrane herniate to form structures known as **blebs**; indeed, in time-lapse movies, the cell surface appears to be boiling (Figure 9.18A). The nucleus collapses into a dense structure—the state termed **pyknosis** (Figure 9.18B)—and fragments (Figure 9.18E) as the chromosomal DNA is cleaved into small segments (Figure 9.18C). Yet other changes in other parts of the cell can also be observed (Figure 9.18D and E). Ultimately, usually within an hour, the apoptotic cell breaks up into small fragments, sometimes called *apoptotic bodies*, which are rapidly ingested by neighboring cells in the tissue or by itinerant macrophages, thereby removing all traces of what had recently been a living cell (Figure 9.18F). A variety of procedures can be used to detect apoptotic cells within a tissue or in culture (Sidebar 9 ●).

Apoptosis is exploited routinely during normal morphogenesis in order to discard unneeded cells. It serves to chisel away unwanted cell populations during the sculpting that results in well-formed, functional tissues and organs (Figure 9.19). Mice that have been genetically deprived of various key components of the apoptotic machinery show a characteristic set of developmental defects, including excess neurons in the brain, facial abnormalities, delayed destruction of the webbing between fingers, and abnormalities in the palate and lens.

Apoptosis also plays an important role in normal tissue physiology. In the small intestine, for example, epithelial cells are continually being eliminated by apoptosis after a four- or five-day journey from the bottom of intestinal crypts to the tips of the villi that protrude into the lumen (see Section 7.11). During the development of red blood cells (the process of erythropoiesis), more than 95% of the erythroblasts—the precursors of mature red cells—are eliminated as part of the routine operations of the bone marrow. However, in the event that the rate of oxygen transport by the blood falls below a certain threshold level, either because of hemorrhage, various types of anemia, or low oxygen tension in the surrounding air, levels of the blood-forming hormone erythropoietin (EPO; Sidebar 5.4) rise rapidly and block apoptosis of these erythrocyte precursors, enabling their maturation into functional red blood cells. This yields a rapid increase in the concentration of these cells in the blood.

A particularly dramatic example of the contribution of apoptosis to normal physiology is provided by the regression of the cells in the mammary gland following the weaning of offspring. As many as 90% of the epithelial cells in this gland, which have accumulated in large numbers during pregnancy in order to produce milk for the newborn, die via apoptosis during this regression, which is usually termed **involution**.

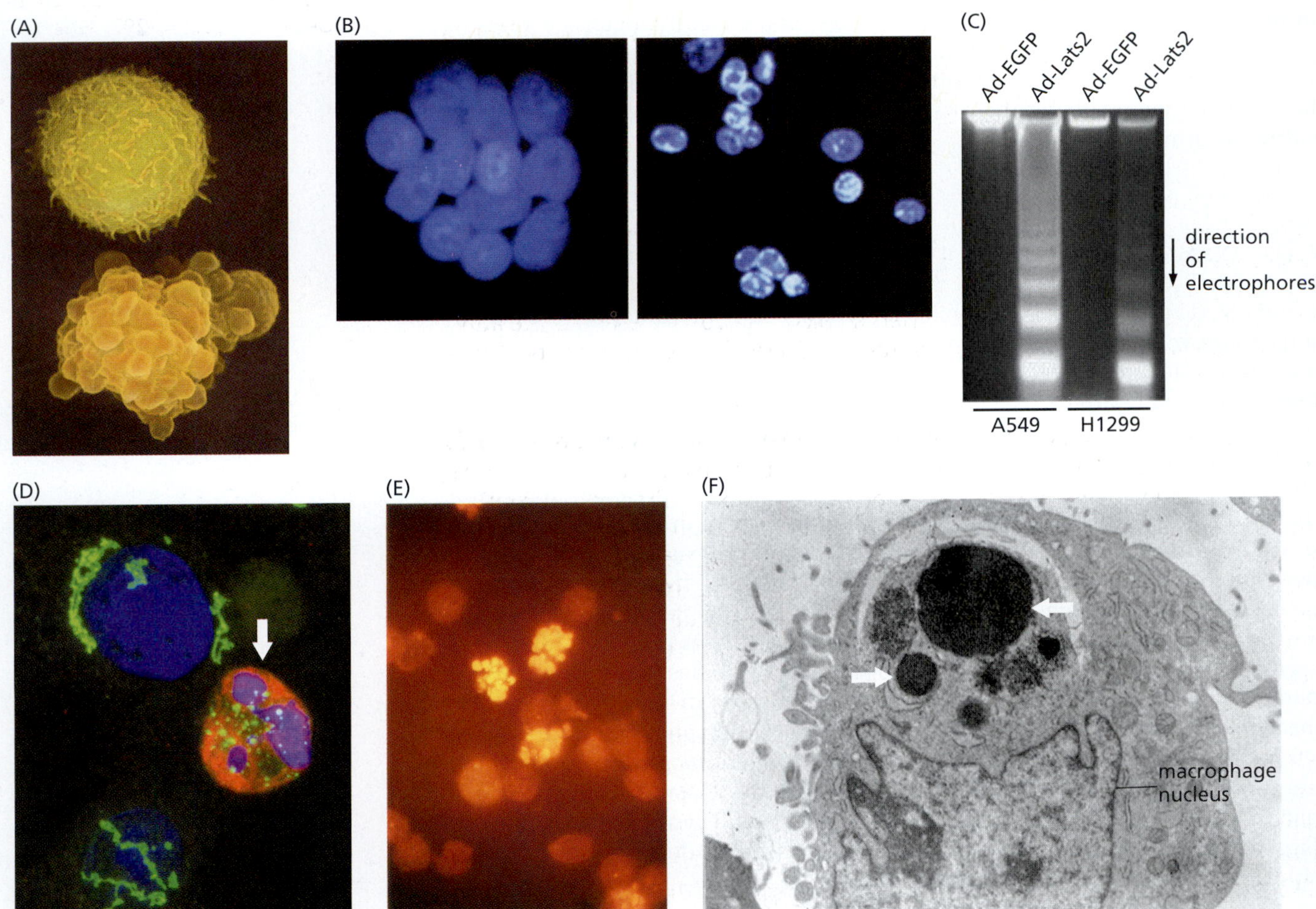

Figure 9.18 Diverse manifestations of the apoptotic program (A) The upper of these two lymphocytes, visualized by scanning electron microscopy (SEM), is healthy, while the lower one has entered into apoptosis, resulting in the numerous blebs protruding from its surface. (B) HeLa cells—a line of human cervical carcinoma cells—have been fused, resulting in the formation of a syncytium *(left)*. Prior to fusion, some of these cells were treated with staurosporine, an apoptosis-inducing drug. As a consequence, these cells' nuclei, which normally are quite large *(left)*, have undergone *pyknosis*, which involves a condensation of their chromatin and a collapse of nuclear structure that accompanies apoptosis *(right)*. Fragmentation of nuclei follows soon thereafter. (C) When apoptosis is induced, in this case through the expression of the pro-apoptotic Lats2 protein *(2nd, 4th channels)*, the normally high–molecular-weight DNA of A549 and H1299 cells *(1st, 3rd channels)* is cleaved into low–molecular-weight fragments that run rapidly upon gel electrophoresis, forming a "DNA ladder." (D) The Golgi bodies *(green)* are usually found in peri-nuclear locations in a normal cell *(upper left)*, while in an apoptotic cell *(arrow)*, the Golgi bodies have become fragmented. Chromatin is stained in blue, while cleaved PARP, a nuclear protein that is cleaved during apoptosis, is immunostained in red. (E) The wide-ranging effects of the apoptotic program are illustrated here by use of an antibody that specifically reacts with histone 2B molecules (in the chromatin) that are phosphorylated on their serine 14 residues. This antibody stains apoptotic nuclei (which are already undergoing fragmentation). The precise consequences of this phosphorylation on chromatin structure and function remain to be elucidated. (F) The end result of apoptosis is the phagocytosis of apoptotic bodies—the fragmented remains of apoptotic cells—by neighbors or by macrophages. In this image, the pyknotic nuclear fragments of a phagocytosed apoptotic cell are seen above *(white arrows)* and contrast with the normal nucleus of the phagocytosing macrophage *(below)*. (A, courtesy of K.G. Murti; B, from K. Andreau et al., *J. Cell Sci.* 117:5643–5653, 2004; C, from H. Ke et al., *Exp. Cell Res.* 298:329–338, 2004; D, from J.D. Lane et al., *J. Cell Biol.* 156:495–509, 2002; E, from W.L. Cheung et al., *Cell* 113:507–517, 2003; F, courtesy of G.I. Evan.)

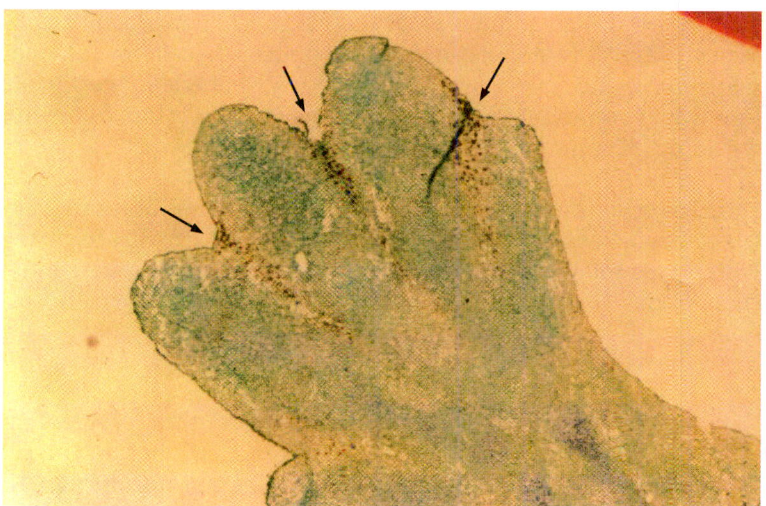

Figure 9.19 Apoptosis and normal morphogenesis The webs of tissue between the future fingers of a mouse paw are still in evidence in this embryonic paw, which are preferentially labeled using the TUNEL assay (Figure 3B), as indicated by the numerous dark dots in these webs *(arrows)*. The apoptosis of cells forming the webs ultimately results in the formation of fingers that are joined by webs only near the palm. (From Z. Zakeri et al., *Dev. Biol.* 165:294–297, 1994.)

In a more general sense, apoptosis is used to maintain appropriate numbers of different cell types in a wide variety of human tissues. The importance of this process is indicated by the fact that each year of our lives, the turnover of cells—the number that are newly formed and the equal number that are eliminated—approximates the total number of cells ($\sim 3 \times 10^{14}$) present in the adult body at any one time. The majority of these discarded cells appear to be eliminated by apoptosis.

The apoptosis that is used for routine tissue maintenance does not seem to depend on p53 function and is, instead, triggered by other mechanisms that we will discuss later. Thus, the actions of p53 seem to be limited to nonroutine, emergency situations that occasionally threaten cells and thus tissues. In the specific context of cancer pathogenesis, as noted earlier, the organism uses p53-triggered apoptosis as a means of weeding out cells that have the potential to become neoplastic, including some cells that have sustained certain types of growth-deregulating mutations and others that have suffered widespread damage to their genomes.

p53 initiates apoptosis in part through its ability to promote expression of several downstream target genes that specify components of the apoptotic machinery. Among these are the genes encoding a diverse group of pro-apoptotic proteins (see Table 9.2). At the same time, p53 represses expression of genes specifying anti-apoptotic proteins. We will return to the biochemical details of the apoptotic program and their relevance to cancer pathogenesis in Section 9.13.

To summarize, the various observations cited here indicate that the biological actions of p53 fall into two major categories. In certain circumstances, p53 acts in a cytostatic fashion to halt cell cycle advance. In other situations, p53 activates a cell's previously latent apoptotic machinery, thereby ensuring cell death. The choice made between these alternative modes of action seems to depend on the type of physiologic stress or genetic damage, the severity of the stress or damage, the cell type, and the presence of other pro- and anti-apoptotic signals operating in a cell. At the biochemical level, it remains unclear how p53 decides between imposing cell cycle arrest and triggering apoptosis.

9.11 p53 inactivation provides advantage to incipient cancer cells at a number of steps in tumor progression

As we will learn in Chapter 11, the formation of a malignant human cell involves more than half a dozen distinct steps that usually occur over many years' time. An early step in the formation of a cancer cell may involve activation of an oncogene through some type of mutation. This oncogene activation may put the cell at great risk of p53-induced apoptosis. Recall, for example, the fact that a *myc* oncogene can, on its own, trigger p53-dependent apoptosis. Hence, cells that have acquired such an oncogene accrue additional growth advantage by shedding p53 function.

Later in tumor development, a growing population of tumor cells may experience anoxia because they lack an adequate network of vessels to provide them with access to blood-borne oxygen. While normal cells would die in the face of such oxygen deprivation, tumor cells may survive because their ancestors managed to inactivate the *p53* gene during an earlier stage of tumor development (having done so for a quite different reason).

During much of this long, multi-step process of tumor progression, the absence of p53-triggered responses to genetic damage will permit the survival of cells that are accumulating mutations at a greater-than-normal rate. Such increased mutability increases the rate at which oncogenes become activated and tumor suppressor genes become inactivated; the overall rate of evolution of pre-malignant cells to a malignant state is thereby accelerated. Telomere collapse, another danger faced by evolving, pre-malignant cells (see Chapter 10), also selects for the outgrowth of those cells that have lost their p53-dependent DNA damage response.

The advantages to the incipient tumor cell of shedding p53 function do not stop there. One of the important target genes whose expression is increased by p53 is the *TSP-1* gene, which specifies thrombospondin-1. As we will see later in Chapter 13, Tsp-1 is a secreted protein that functions in the extracellular space to block the development of blood vessels. Consequently, a reduction of Tsp-1 expression following p53 loss removes an important obstacle that otherwise would prevent clusters of cancer cells from developing an adequate blood supply during the early stages of tumor development.

Together, these diverse consequences of p53 inactivation illustrate dramatically how the malfunctioning of a single component of the alarm response circuitry permits cancer cells to acquire multiple alterations and survive under conditions that usually lead to the death of normal cells. These multiple benefits accruing to cancer cells explain why the p53 pathway is disrupted in most if not all types of human tumors.

In almost half of these tumors, the p53 protein itself is damaged by reading-frame mutations in the *p53* gene (see Figure 9.4). In many of the remaining tumors, the ARF protein is missing (see Sidebar 7) or the Mdm2 protein is overexpressed. In addition, p53 function may be compromised by defects in the complex signaling network in which p53 and its antagonist, Mdm2, are embedded (Sidebar 10). There are reasons to suspect that yet other, still-undiscovered genetic mechanisms subvert p53 function. While the organism as a whole benefits greatly from the p53 watchman stationed in its myriad cells, it suffers grievously once p53 function is lost in some of them, because the resulting p53-negative cells are now free to begin the long march toward malignancy.

9.12 Inherited mutant alleles affecting the p53 pathway predispose one to a variety of tumors

In 1982, a group of families was identified that showed a greatly increased susceptibility to a variety of different tumors, including glioblastoma; leukemias; carcinomas of the breast, lung and pancreas; Wilms tumor; and soft-tissue sarcomas (Figure 9.20). In some kindreds, as many as half the members were

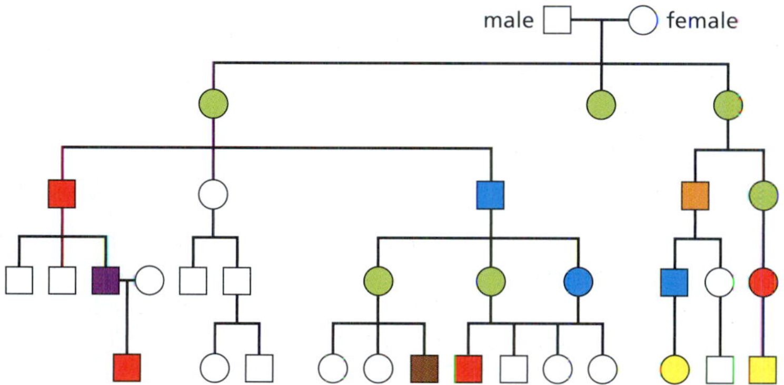

Figure 9.20 Familial cancer susceptibility due to mutant *p53* germ-line alleles This pedigree of a family suffering Li–Fraumeni syndrome reveals a variety of distinct cancers arising at elevated incidence levels. The diversity of cancer types suffered by members of Li–Fraumeni families contrasts with most familial cancer syndromes, in which a single type of cancer is usually seen to occur at elevated frequency. Family members developed the following malignancies: breast cancer (*green*), glioblastoma (*yellow*), leukemia (*purple*), lung cancer (*blue*), pancreatic carcinoma (*orange*), sarcoma (*red*); Wilms tumor (*brown*); males (*squares*), females (*circles*). (From F.P. Li and J.F. Fraumeni, *J. Am. Med. Assoc.* 247:2692–2694, 1982.)

afflicted with one or another of these cancers, and two-thirds of these developed some type of cancer by the time they reached age 22. Some family members were even afflicted with several types of cancer concurrently.

This familial cancer syndrome, termed Li–Fraumeni after the two human geneticists who first identified and characterized it, is most unusual, in that it causes susceptibility to a wide variety of cancers. Recall the starkly contrasting behavior of the other familial cancer syndromes that we encountered in Chapter 7. Mutant germ-line alleles of most tumor suppressor genes typically increase susceptibility to a narrow range of cancer types (see Table 7.1).

In 1990, eight years after the Li–Fraumeni syndrome was first described, researchers discovered that many of the cases were due to a mutant allele at a locus on human Chromosome 17p13—precisely where the *p53* gene is located. In about 70% of these multicancer families, mutant alleles of *p53* were found to be transmitted in a Mendelian fashion. Family members who inherited a mutant *p53* allele had a high probability of developing some form of malignancy, often early in life. The age of onset of these various malignancies was found to be quite variable: about 5 years of age for adrenocortical carcinomas, 16 years for sarcomas, 25 years for brain tumors, 37 years for breast cancer, and almost 50 years for lung cancer. In light of the various roles that p53 plays in suppressing cancer risk in cells throughout the body, it seems reasonable that mutant germ-line alleles of this gene should predispose a person to such a diverse group of malignancies. (An intriguing exception is provided by rare *p53* germ-line mutations that confer a predisposition to only a narrow set of cancer types; see Sidebar 9.5).

We now know that the mutant *p53* alleles that are transmitted through the germ lines of Li–Fraumeni families carry a variety of point mutations that are scattered across the *p53* reading frame, with a distribution reminiscent of that shown by the somatic mutations that have been documented in more than 15,000 tumor genomes (see Figure 9.6B). Analysis of the spectrum of germ-line mutations has shown a predominance of G:C to A:T transitions at CpG sites—precisely those that would occur if a 5-methylcytosine underwent spontaneous deamination, causing it to be replaced by a thymidine.

Provocatively, a familial cancer syndrome that behaves very much like the Li–Fraumeni syndrome has been discovered in families that inherit mutant, defective forms of the Chk2 kinase, which is responsible for phosphorylating p53, saving it from destruction, and thereby activating the p53 alarm machinery (see Figure 9.13). A mutant germ-line allele of the *Chk2* gene is present in 1.1 to 1.4% of Western populations, and in higher percentages (as much as 5%) in cohorts of breast cancer patients. This suggests that inheritance of a mutant *Chk2* allele confers a measurably increased risk of developing this malignancy. Yet another related cancer predisposition syndrome derives from an alteration in the promoter of the *MDM2* gene (Sidebar 11).

Sidebar 9.5 Mutant *p53* alleles cause highly specific tumors Some mutant germ-line alleles of *p53* have effects that are quite different from those observed typically in Li–Fraumeni families. One such unusual example has been reported in southern Brazil, where pediatric adrenal cortical carcinomas (affecting the cortex of the adrenal glands, which sit above the kidneys) are encountered at rates that are 10 to 15 times higher than those found elsewhere in the world. Remarkably, of 36 patients from this region who were examined, 35 showed an identical germ-line mutation in the *p53* gene, which caused an arginine-to-histidine substitution in amino acid residue 337. This finding is difficult to reconcile with the current preconception that inherited *p53* alleles should affect a variety of tissues throughout the body. However, recent efforts to characterize the effects of this amino acid substitution on p53 structure have shown that the tetramerization domain of the mutant protein (see Figure 9.7A) is less stable than that of wild-type p53 and is sensitive to disruption at acidic pH—precisely the environment encountered within the adrenal gland. This may account for the peculiar, tissue-specific effects of this mutant germ-line allele.

9.13 Apoptosis is a complex program that often depends on mitochondria

Many consider the loss of a fully functional apoptotic program to be one of the hallmarks of all types of malignant human cells. Until now, however, our descriptions of apoptosis have not done justice to this suicide program and its role in tumor pathogenesis. Accordingly, we now revisit apoptosis and explore it in greater depth.

We have already learned that the initiation of apoptosis by p53 represents an important mechanism by which tissues can eliminate aberrantly functioning or irreparably damaged cells. Importantly, as we read earlier, apoptosis can also be initiated through a variety of signaling channels that do not depend on the actions of p53. A particularly dramatic example of this is seen in Figure 9.21: the loss by a cell of anchorage to extracellular matrix triggers anoikis, the specialized form of apoptosis that occurs without the intervention of p53. This hints at a larger theme—that p53 is only one of many players in the apoptotic program, and inactivation of p53 function is only one means by which cancer cells evade apoptosis.

The first indication of contributions of other proteins to the regulation of apoptosis came from an exploration of the functioning of the *bcl-2* (B-cell lymphoma gene-2) oncogene. Like many cancer-associated genes found in the genomes of human hematopoietic tumors, the oncogenic version of the *bcl-2* gene is formed through a reciprocal chromosomal translocation, in which portions of the arms of human Chromosomes 14 and 18 are exchanged. At the breakpoint where the translocated arms are joined in human follicular B-cell lymphoma cells, the reading frame of the *bcl-2* gene is placed under the control of a promoter that drives its high, constitutive expression.

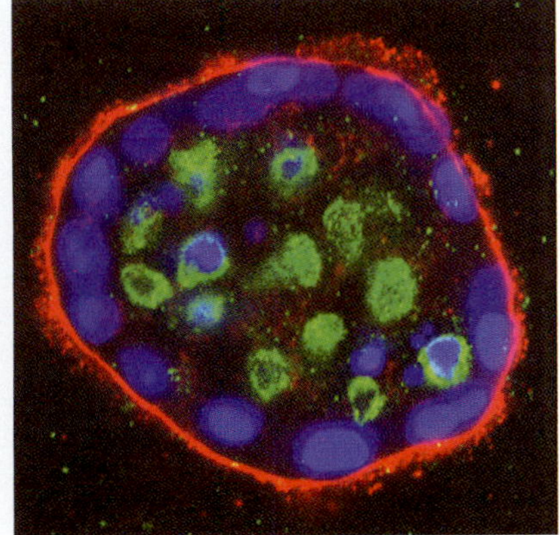

Figure 9.21 Anoikis triggered by loss of anchorage to extracellular matrix Suspension of immortalized but non-tumorigenic human mammary epithelial cells (MECs) in a medium of extracellular matrix proteins forces these cells to grow in an anchorage-independent fashion, that is, without attachment to a solid substrate. These MECs *(large blue nuclei)* succeed in attaching themselves via integrins *(red)* to the extracellular matrix (ECM) proteins present in the surrounding medium; this attachment enables these cells to avoid the apoptotic program termed anoikis. However, cells in the middle of this spherical cell colony are unable to attach themselves to these ECM molecules and therefore activate anoikis, as indicated through use of the assay described in Figure 3C —their staining *(green)* with an antibody specific for one of the enzymes (caspase 3) that participate in the apoptotic program. (Courtesy of J. Debnath and J. Brugge.)

When such a *bcl-2* oncogene was inserted as a transgene into the germ-line of mice under conditions that ensured its expression in lymphocyte precursor cells, there was no observable effect on the long-term survival of these transgenic mice (Figure 9.22A,B). On the other hand, expression of an oncogenic *myc* transgene in these cells led to lymphomas and to the death of half the mice within two months of birth. The concomitant expression of the two transgenes, however (achieved by breeding *bcl-2* transgenic mice with *myc* transgenic mice), led to offspring having an even more rapid death rate, with virtually all of the mice being dead less than two months after birth (see Figure 9.22B).

The inability of *bcl-2,* on its own, to trigger tumor formation argued against its acting as a typical oncogene, for example, an oncogene like *myc* that emits potent growth-promoting signals. Careful study of the lymphocyte populations in mice carrying only the *bcl-2* transgene indicated that the effects of this gene on cells were actually quite different from those of *myc* or *ras*: *bcl-2* prolonged the lives of lymphocytes that were otherwise destined to die rapidly. Indeed, when B lymphoid cells from these transgenic mice were cultured *in vitro,* they showed a remarkable extension of life span. *In vivo,* the lymphoid cells in which *bcl-2* was being expressed accumulated in amounts several-fold above normal, but significantly, these cells were not actively proliferating, explaining the absence of a hematopoietic tumor in mice carrying only the *bcl-2* transgene.

The *myc* oncogene, when acting on its own, acted as a potent mitogen, but its growth-stimulatory powers were attenuated by its death-inducing effects, which we described earlier in this chapter (Section 9.7). However, *myc* and *bcl-2,* when acting collaboratively, created an aggressive malignancy of the B-lymphocyte lineage; *myc* would drive rapid cell proliferation, and its accompanying death-inducing effects were neutralized by the life-prolonging actions of *bcl-2.* Since these early experiments, even more dramatic examples of synergy between *myc* and *bcl-2*-like oncogenes have been reported (e.g., Figure 9.22C). (Note, by the way, that these dynamics parallel one that we discussed in Sections 9.7 and 9.8, where the death-inducing effects of *myc* were blunted by inactivation of the *p53* gene.)

This work on *bcl-2* soon converged with research on apoptosis. Although the apoptotic program had been recognized as a normal biological phenomenon in animal tissues by nineteenth-century histologists, it was rediscovered and described with far greater precision in 1972. Prior to this rediscovery, cells in metazoan tissues were thought to be eliminated solely by **necrosis**. As indicated in Table 9.3, these two death processes are actually quite different. By the late-1980s, research on the genetics of worm development (the species *Caenorhabditis elegans*) revealed that apoptosis is exploited to eliminate various cell types as part of the normal developmental program of these tiny animals (see also Figure 10.1). This led to the recognition that apoptosis is a basic biological process that is common to all metazoa.

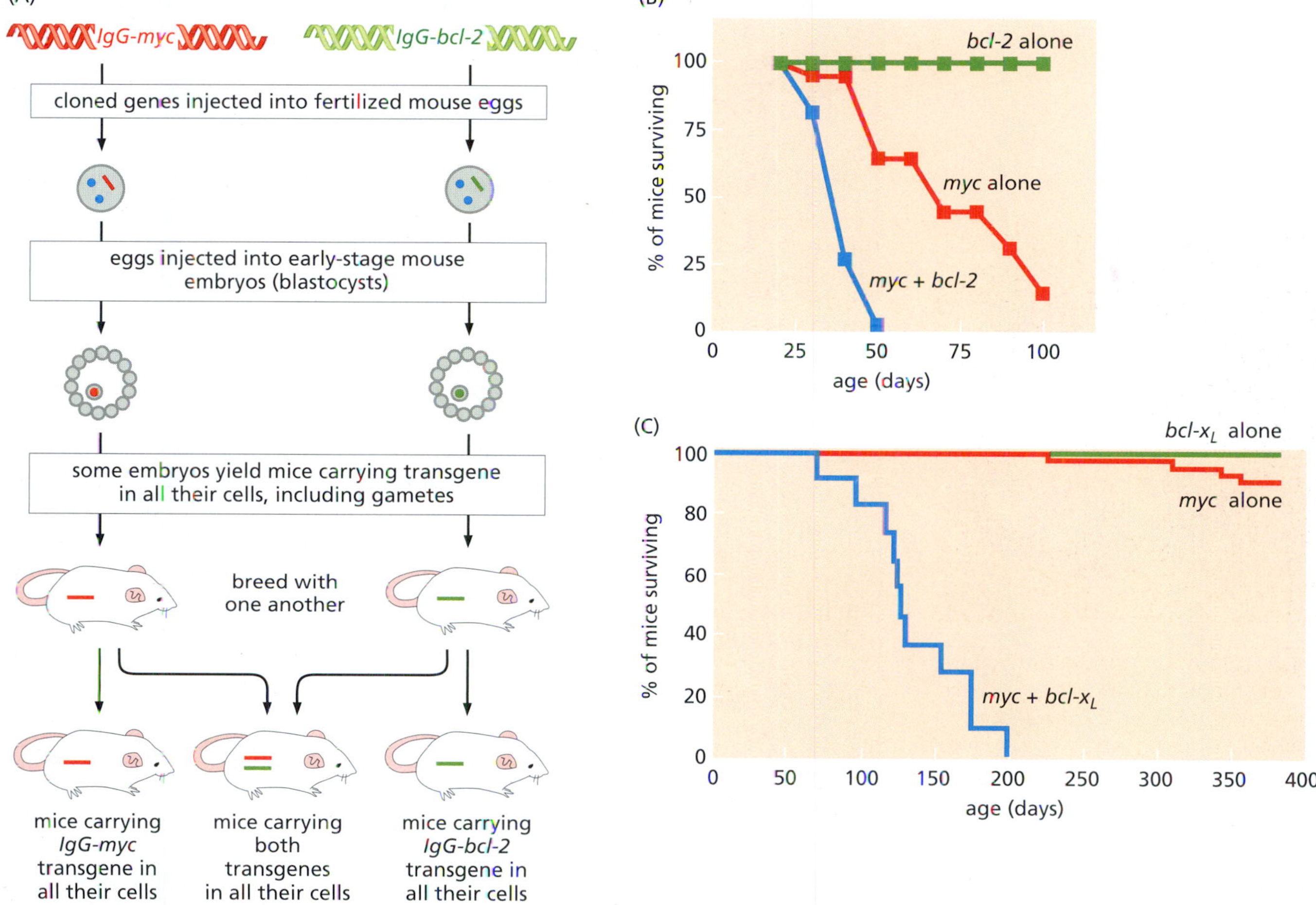

Figure 9.22 Actions of the anti-apoptotic Bcl-2-like genes (A) Clones of the *myc* and the *bcl-2* genes were constructed in which each gene was placed under the control of an antibody gene promoter *(IgG)*. This promoter ensured that both genes were expressed specifically in cells of the B lymphocyte lineage. Each gene was then introduced into the germ line of a mouse as shown, yielding in each case a *transgene*. (B) Mice bearing the *IgG-bcl-2* transgene experienced no increased mortality *(green squares)*. In contrast, the *myc* transgene led to a greatly increased mortality from lymphomas, with almost all of the mice dying from this disease by 100 days of age *(red squares)*. However, when both *bcl-2* and *myc* transgenes were present in the germ line of mice, lymphomagenesis was greatly accelerated, with virtually all mice succumbing to lymphomas by 50 days of age *(blue squares)*. (C) A far more dramatic outcome was observed years later when two strains of transgenic mice were developed that expressed the *myc* or the *bcl-x_L* oncogenes specifically in the plasma cell (antibody-secreting) lymphocytes of the immune system. (*bcl-x_L* is a cousin of *bcl-2* that functions in a similar fashion to *bcl-2*.) Normal mice and transgenic mice carrying only the *bcl-x_L* transgene all survived for more than a year without any tumor development *(green line)*; similarly, transgenic mice expressing a *myc* transgene were largely healthy until about 10 months of age, when a small number of them developed plasma cell tumors. However, when the two strains of mice were bred together, the double-transgenic strain developed plasma cell tumors rapidly beginning at about 75 days of age, and all died from these tumors by 200 days of age. (B, from A. Strasser et al., *Nature* 348:331–333, 1990; C, from W.C. Cheung et al., *J. Clin. Invest.* 113:1763–1773, 2004.)

Table 9.3 Apoptosis versus necrosis

	Apoptosis	Necrosis
Provoking stimuli		
	programmed tissue remodeling	metabolic stresses
	maintenance of cell pool size	absence of nutrients
	genomic damage	changes in pH, temperature
	metabolic derangement	hypoxia, anoxia
	hypoxia	
	imbalances in signaling pathways	
Morphological changes		
Affected cells	individual cells	groups of cells
Cell volume	decreased	increased
Chromatin	condensed	fragmented
Lysosomes	unaffected	abnormal
Mitochondria	morphologically normal initially	morphologically aberrant
Inflammatory response	none	marked
Cell fate	apoptotic bodies consumed by neighboring cells	lysis
Molecular changes		
Gene activity	required for program	not needed
Chromosomal DNA	cleaved at specific sites	random cleavage
Intracellular calcium	increased	unaffected
Ion pumps	continue to function	lost

Adapted from R.J.B. King, Cancer Biology, 2nd ed. Harlow, UK: Pearson Education, 2000.

Once the apoptotic program was clearly understood, the role of the Bcl-2 oncoprotein in cancer pathogenesis became apparent. It operates as an anti-apoptotic agent rather than as a mitogen. Hence, Bcl-2 acts in a fashion precisely opposite to that of p53. Bcl-2 blocks apoptosis, while p53 promotes it. Stated in genetic terms, *bcl-2* oncogene activation and *p53* inactivation confer similar (but hardly identical) benefits on cancer cells, in that both reduce the likelihood of activation of the apoptotic program.

Careful biochemical and cell-biological sleuthing eventually revealed the intracellular site of Bcl-2 action. It was found to operate in a most unexpected intracellular locale—the outer membrane of the mitochondrion (Figure 9.23). At first, this discovery made little sense, since mitochondria were thought to be specialized for one task only, namely, that of generating energy in the form of ATP for the cell. As a by-product of this energy production, mitochondria were also known to make metabolites of glucose that are used in the biosynthesis of some of the cell's diverse biochemical species.

Soon, however, the role of mitochondria in the apoptotic program was clarified. Cytochrome *c*, the central actor in this process, normally resides in the space between the inner and outer mitochondrial membranes (see Figure 9.23B), where it functions to transfer electrons as part of oxidative phosphorylation. However, when certain signals trigger the initiation of apoptosis, the outer mitochondrial membrane becomes depolarized (i.e., loses its normal voltage gradient), and cytochrome *c* spills out of the mitochondrion into the surrounding cytosol (Figure 9.24). Once present in the cytosol, cytochrome *c* associates with other proteins to trigger a cascade of events that together yield apoptotic death. Therefore, the workshop in which the bulk of the cell's energy is produced was co-opted at some point during eukaryotic cell evolution for a fully unrelated function—to harbor and release a biochemical messenger that triggers the changes leading to cell death.

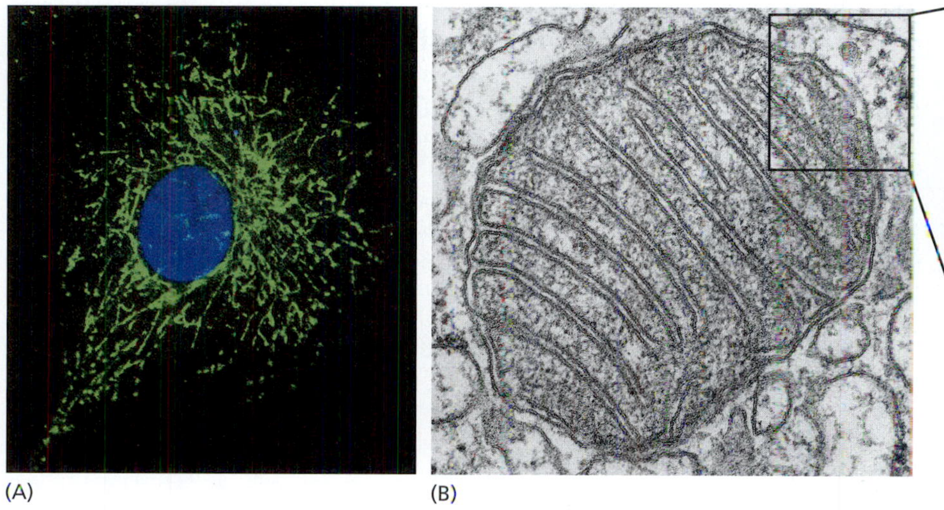

Figure 9.23 Mitochondria
(A) Mitochondria stained here in the cytoplasm of a human liver cell *(green)* were once regarded only as the sites of oxidative phosphorylation as well as the sites of biosynthesis of certain key metabolites. The discovery of their ability to regulate apoptosis through cytochrome c release forced a major rethinking of their role in the life of a cell. (B) Cytochrome c is stored in the space between the outermost mitochondrial membrane and the inner membranes, where it normally plays a key role in electron transport. The channels in the outer membrane that allow its release into the cytosol during apoptosis are not visualized here.
(A, courtesy of L.B. Chen; B, courtesy of D.S. Friend.)

We now know that the Bcl-2 protein is a member of a large and complex family of proteins (Figure 9.25) that control the flow of cytochrome *c* through specialized channels in the outer membrane of the mitochondrion. More specifically, these channels regulate whether or not cytochrome *c* and several other proteins (which normally reside in the space between the inner and outer mitochondrial membranes) are released from the mitochondrion into the cytosol. Some members of the Bcl-2 protein family, including Bcl-2 itself and Bcl-X_L, work to keep these channels closed and thereby keep cytochrome *c* penned up in the mitochondrion.

Yet other relatives of Bcl-2, such as Bax, Bad, Bak, and Bid (see Figure 9.25), function oppositely by striving to pry open this channel. Bax, for one, is encoded by a gene whose transcription is known to be activated by p53; its actions begin to explain how p53 succeeds in inducing apoptosis. Bad can be phosphorylated by Akt/PKB, which decreases Bad's ability to hold the mitochondrial channel in an open configuration; this explains some of the anti-apoptotic effect of the Akt/PKB kinase.

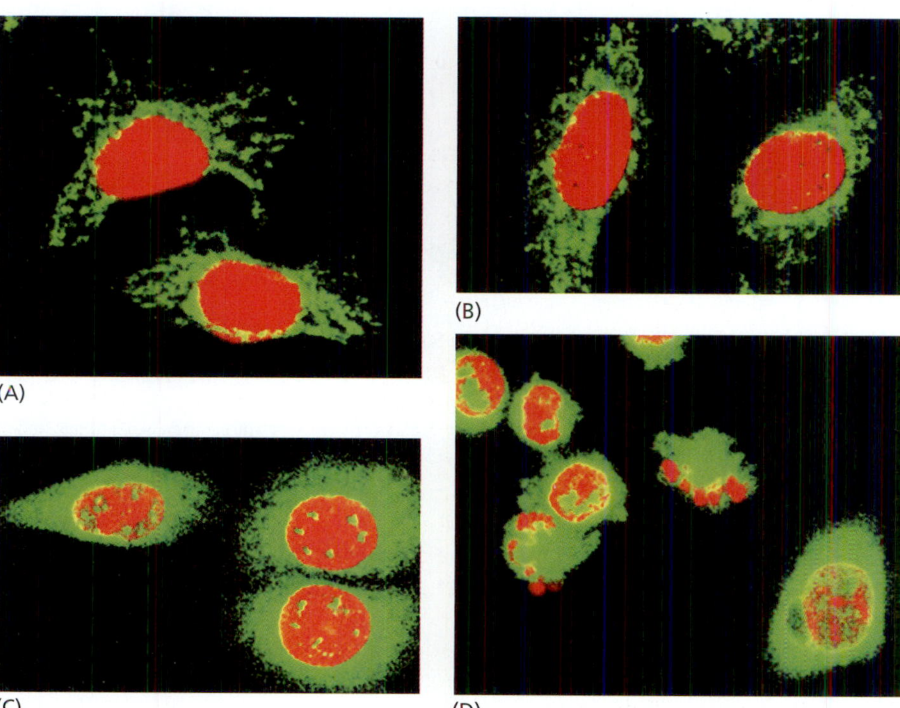

Figure 9.24 Release of cytochrome c from mitochondria into the cytosol
The presence of cytochrome c can be detected by staining with a specific fluorescence-labeled antibody *(green)*, which contrasts with the dye-labeled nuclei *(orange)*. (A) Initially, the distribution of cytochrome c coincides with the distribution of mitochondria in the cytoplasm. As apoptosis proceeds, however (B,C,D), cytochrome c staining becomes increasingly uniform in the cytoplasm as it is released from the mitochondria. At the same time, some nuclei give evidence of the fragmentation that is associated with apoptosis (C,D). (From E. Bossy-Wetzel et al., *EMBO J.* 17:37–49, 1998.)

As apoptosis proceeds, large complexes of Bax and Bak proteins congregate at the outer surface of the mitochondrion, where they participate, in ways that are still unclear, in causing fragmentation of this organelle (Figure 9.25C). This seems to result in the further release of pro-apoptotic proteins into the cytosol and a total collapse of a cell's primary ATP-generating machinery. Altogether, the human genome is known to encode twenty-four Bcl-2–related proteins; six of these are anti-apoptotic, while the remaining eighteen are pro-apoptotic.

The relative levels of the pro- and anti-apoptotic proteins within each channel determine whether cytochrome *c* will be retained in the mitochondrion or spilled out. In this fashion, the mitochondrial membrane channel determines the life and death of a cell. For example, mice that have been deprived of both germ-line copies of the anti-apoptotic *Bcl-2* gene suffer from kidney failure and immune

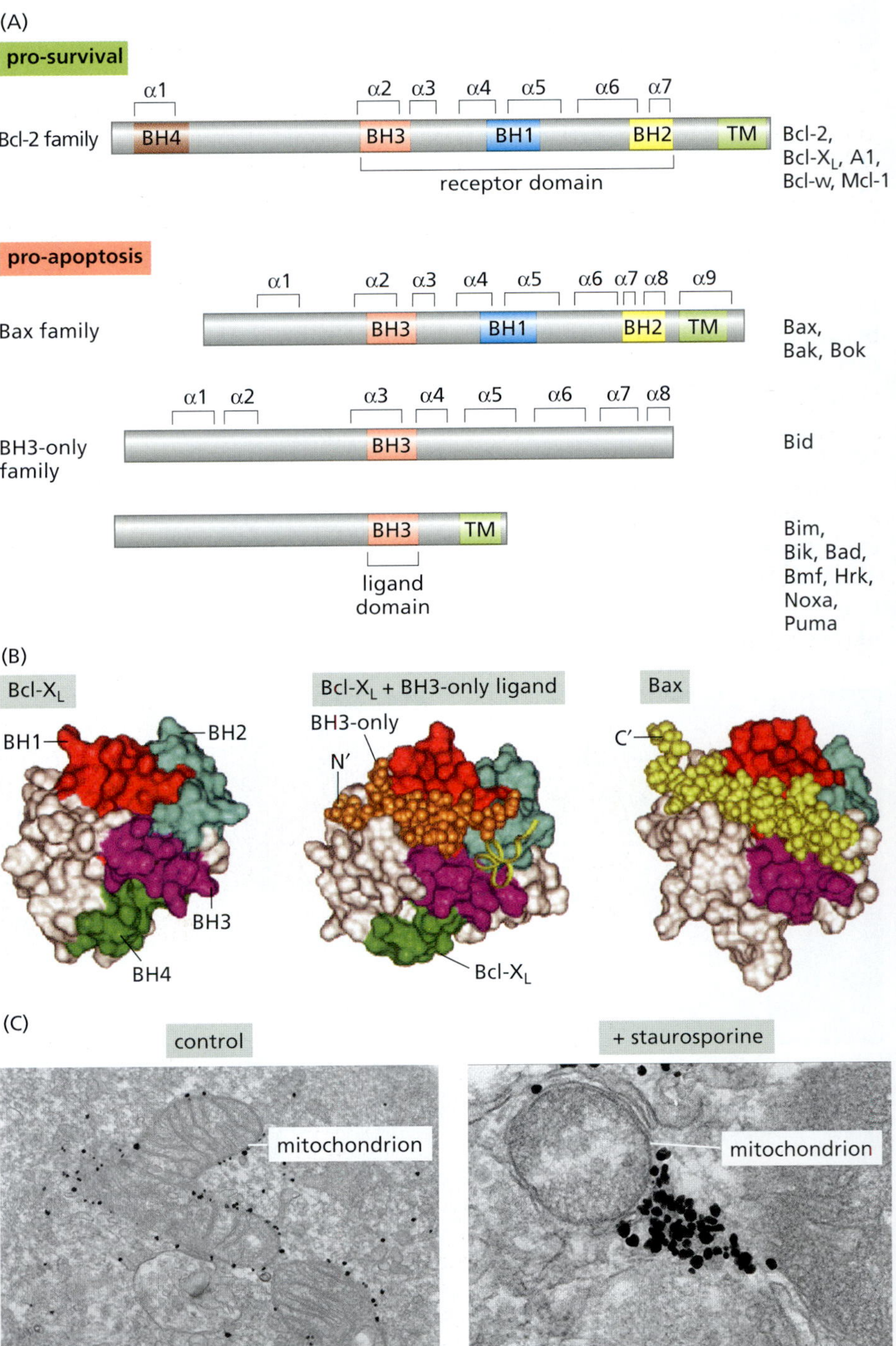

Figure 9.25 Bcl-2 and related proteins (A) The Bcl-2 protein is a member of a large family of proteins that play various roles in preventing apoptosis *(top)* or in promoting it *(bottom)*. These proteins share in common a BH3 domain *(pink)*, with some, like Bcl-2 itself, having the additional BH1 and BH2 domains indicated here *(blue, yellow)*. Several also have transmembrane (TM) domains *(light green)* for membrane anchoring and BH4 domains *(brown)* of unknown function. The pro-apoptotic proteins are subgrouped into those of the "Bax" family, which have several domains that are homologous to domains of Bcl-2; while the "BH3-only" family of proteins possess only the BH3 domain that is present in Bcl-2. Members of the Bax family normally reside in inactive form in the outer mitochondrial membrane or in the cytosol, whereas inactive members of the BH3-only family normally are limited to the cytosol; both are activated and/or translocated to the mitochondria by pro-apoptotic signals. (B) The Bcl-X$_L$ anti-apoptotic protein *(left)* is shown with its BH1 *(red)*, BH2 *(aquamarine)*, BH3 *(purple)*, and BH4 domains *(green)*. When a pro-apoptotic, BH3-only protein *(orange, middle)* binds to Bcl-X$_L$, an α helix of this BH3 protein binds in the groove of Bcl-X$_L$ lying between its BH1 and BH3 domains, thereby inhibiting Bcl-X$_L$. Bax, another pro-apoptotic protein *(right)*, tucks its C-terminal, α-helical tail *(yellow)* into its own comparable groove (i.e., forming an intramolecular complex). (Note that the N′ to C′ orientation of this tail is opposite to that of the BH3-only protein bound to Bcl-X$_L$.) The precise mechanisms by which these various molecules regulate opening of the mitochondrial membrane channel remain to be elucidated. (C) Bax is normally found in the cytosol *(not visible)*, whereas Bak is attached to the outer mitochondrial membrane *(black spots, left panel)*. Upon initiation of apoptosis following treatment of cells with the drug staurosporine (STS), thousands of molecules of Bak *(black spots)* coalesce together with Bax *(not shown)* in punctate foci on the surface of mitochondria *(right panel)*. These clusters participate in the mitochondrial fragmentation observed during apoptosis, and this fragmentation appears to contribute to the release of cytochrome *c* from the spaces between the inner and outer mitochondrial membranes. (A and B, from S. Cory and J.M. Adams, *Nat. Rev. Cancer* 2:647–656, 2002; C, from A. Nechushtan et al., *J. Cell Biol.* 153:1265–1276, 2001.)

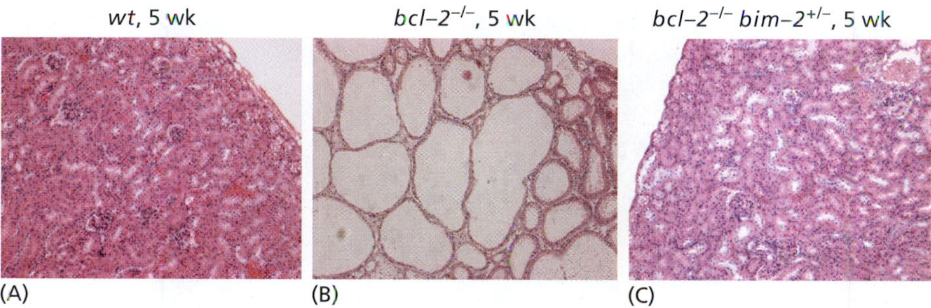

Figure 9.26 The delicate balance between pro- and anti-apoptotic proteins (A) The wild-type mouse kidney is a well-consolidated tissue at 5 weeks of age. (B) However, deletion of both germ-line copies of the anti-apoptotic *bcl-2* gene (i.e., *bcl-2*$^{-/-}$ genotype) results in widespread apoptosis and polycystic kidney disease at this age. (C) These effects can be fully reversed if one copy of the pro-apoptotic *bim* gene is deleted from the germ line (i.e., *bcl-2*$^{-/-}$ *bim*$^{+/-}$ genotype). (C, from P. Bouillet et al., *Dev. Cell* 1:645–654, 2001.)

collapse due to widespread apoptotic death of cells in these tissues; these lethal phenotypes can be avoided through the additional deletion from their germ line of just one copy of the pro-apoptotic cousin gene of *Bcl-2* called *bim* (Figure 9.26). This illustrates how, in tissues like these, the fate of individual cells is governed by the delicate balance between pro- and anti-apoptotic Bcl-2-like proteins.

These accounts provide no indication of why mammalian cells express so many distinct Bcl-2-like proteins. In this instance, important clues have come from studies indicating that various cell-physiologic stimuli act through distinct Bcl-2–related proteins to activate pro-apoptotic proteins, which proceed to antagonize Bcl-2, thereby favoring opening of the outer mitochondrial membrane and the initiation of apoptosis (Figure 9.27A). Moreover, a large body of other observations, not illustrated here, indicate that the actions of the BH3-only proteins

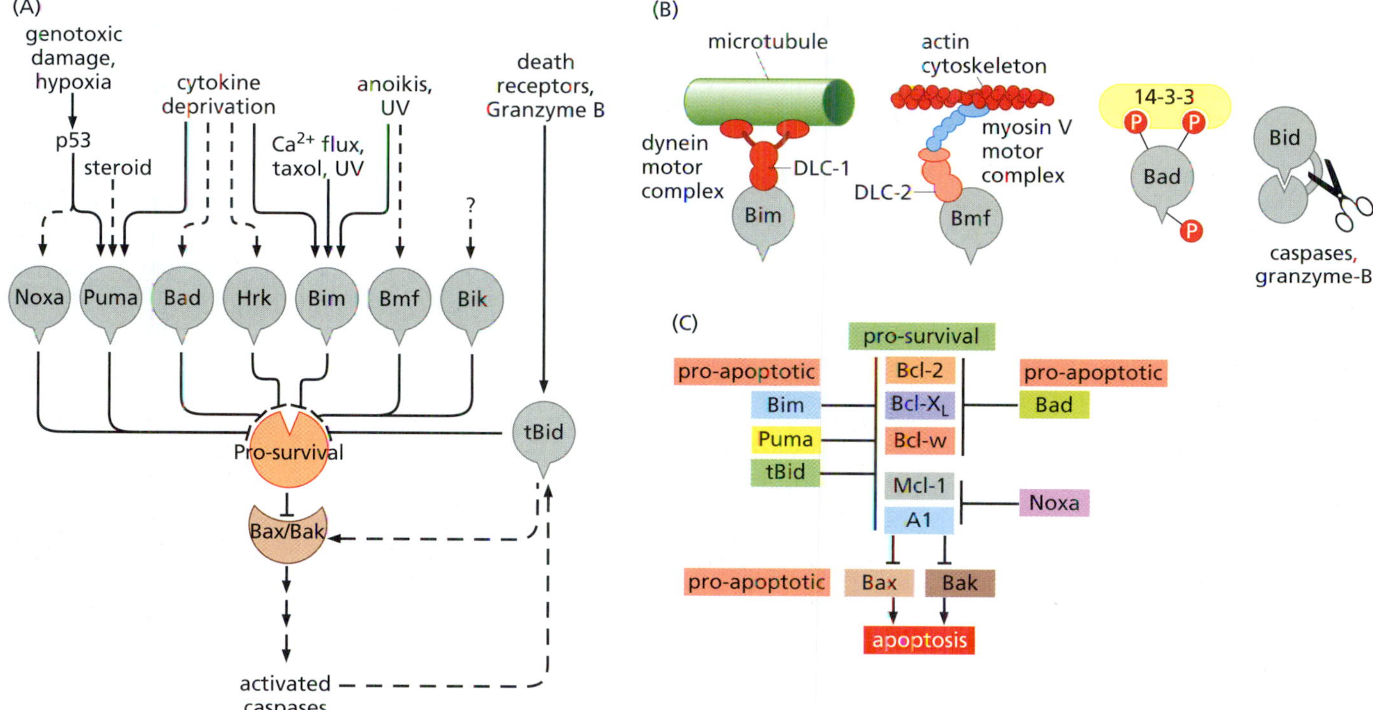

Figure 9.27 Pro-apoptotic signals acting through various Bcl-2–related proteins (A) The rationale for the multiplicity of Bcl-2-like proteins is suggested here by the fact that various cell-physiologic stresses operate through different pro-apoptotic proteins to antagonize the anti-apoptotic Bcl-2 protein. The Bcl-2 protein is thereby prevented from neutralizing Bax, the dominant pro-apoptotic protein that functions to release cytochrome *c* from the mitochondrial membrane. (B) The multiple ways of activating pro-apoptotic BH3 proteins depend, in part, on their being tethered in inactive form to various cytoplasmic structures or complexed with inhibitory domains. (C) Each member of the Bcl-2 family of proteins has its own set of opposing partners. At the center of this network lie the anti-apoptotic proteins Bcl-2, Bcl-X$_L$, Bcl-w, and Mcl-1. Each has its own set of antagonistic partners. For example, Bcl-2, Bcl-X$_L$, and Bcl-w are opposed by the pro-apoptotic Bad protein. At the bottom of this network lie the pro-apoptotic Bax and Bak proteins; if unopposed, they will promote apoptosis by opening the mitochondrial membrane channel. Not all members of the Bcl-2 family of proteins (see Figure 9.25A) are indicated here. (A, courtesy of J. Adams; adapted from S. Cory et al., *Oncogene* 22:8590–8607, 2003; B, from S. Cory and J.M. Adams, *Nat. Rev. Cancer* 2:647–656, 2002; C, courtesy of D. Huang.)

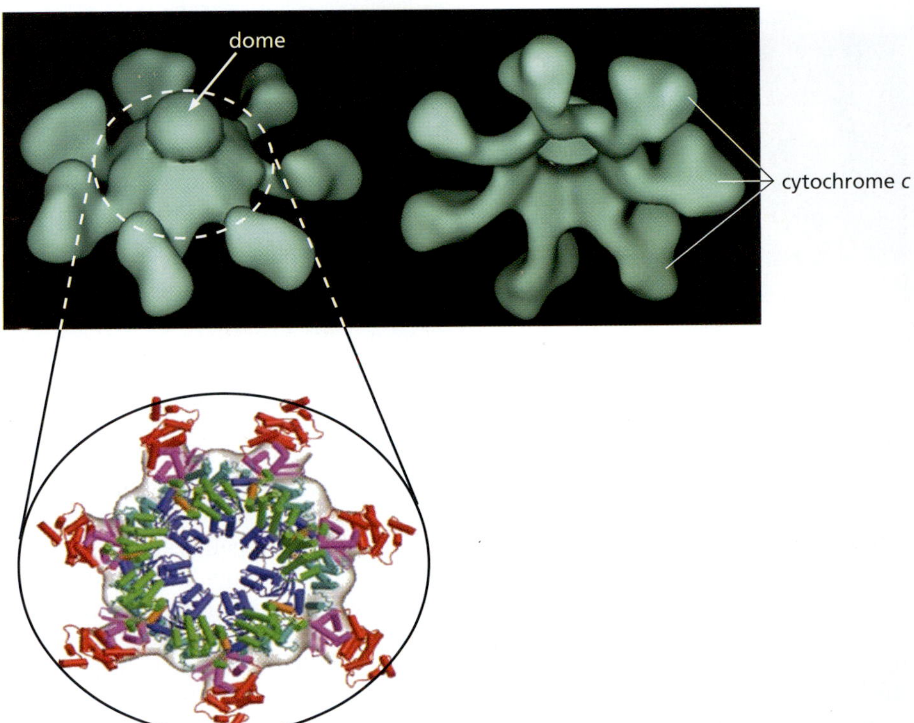

Figure 9.28 The wheel of death The apoptosome—the wheel of death—is assembled in the cytosol when cytochrome *c* molecules are released from mitochondria into the cytosol and associate with Apaf-1. This causes the assembly of a seven-spoked wheel, in which Apaf-1 forms the spokes and cytochrome *c* molecules form the tips of the spokes *(above)*. Once assembled, this attracts procaspase 9 to the hub of the wheel *("dome")*, which causes conversion of procaspase 9 into active caspase 9. The detailed structure of the inner circle of the apoptosome which is formed from seven Apaf-1 subunits, is illustrated in the molecular model *(below)*. The blue helices in the center are responsible for interacting with and activating procaspase 9. The resulting caspase 9 proceeds, in turn, to cleave and activate other caspase molecules, thereby triggering the apoptotic cascade. (Top, from D. Acehan et al., *Mol. Cell* 9:423–432, 2002; bottom, from S.J. Riedl et al., *Nature* 434:926–933, 2005.)

(see Figure 9.25A) can be controlled by modulating transcription of their respective genes, by their intracellular localization, by proteolytic cleavage, or by phosphorylation (e.g., Figure 9.27B). The pro- and anti-apoptotic members of the Bcl-2 family of proteins operate in direct opposition to one another, with each member having its own opposing partner or partners (Figure 9.27C).

Once they have leaked out of mitochondria into the cytoplasm, cytochrome *c* molecules associate with the Apaf-1 protein and form a structure that has been called the **apoptosome** (Figure 9.28). The resulting apoptosome complexes then proceed to activate a normally latent cytoplasmic protease termed procaspase 9, converting it into active caspase 9. Caspase 9 is one member of a family of **c**ysteine **asp**artyl-specific prote**ases**; genes encoding 11 of these proteases have been mapped in the human genome.

Having been converted into an active protease by the apoptosome, caspase 9 then cleaves procaspase 3 and thereby activates yet another related protease (see Figure 3C), which proceeds to cleave and activate yet another procaspase, and so on down the line (Figure 9.29). This sequence of cleavages constitutes a signaling cascade in which one protease activates the next one in the series by cleaving it. Such behavior is reminiscent of the organization of kinase cascades that we read about earlier (Section 6.5), where a group of ser/thr kinases activate one another in a linear sequence. (The fact that each of these caspases acts catalytically means that a relatively minor initiating signal at the top of the cascade can be amplified to yield a large number of activated caspases at the bottom.)

While cytochrome *c* in the cytosol serves to activate the caspases, Smac/DIABLO—another protein that is released from mitochondria together with cytochrome *c*—inactivates a group of anti-apoptotic proteins termed IAPs (inhibitors of apoptosis; see Figure 9.29). These IAPs normally operate to block caspase action in two ways: they bind directly and thereby inhibit caspase proteolytic activity (Figure 9.30), and in certain cases, they can mark caspases for ubiquitylation and degradation. Without the continued influence of IAPs, caspases are free to initiate the proteolytic cleavages that result ultimately in apoptosis.

So, the apoptotic signal is activated in two ways following the opening of the mitochondrial channels. Cytochrome *c* is released and activates caspase 9 in the

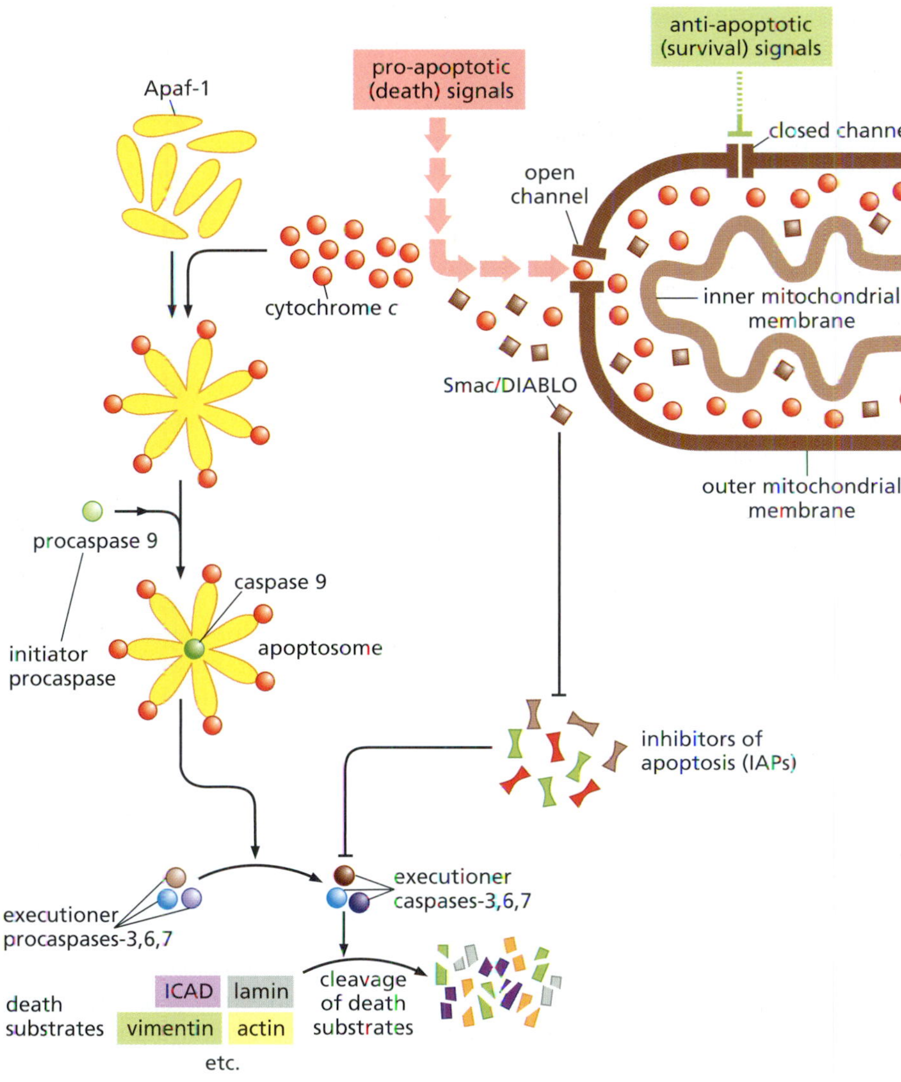

Figure 9.29 The apoptotic caspase cascade The outer mitochondrial membrane *(dark brown)* creates a critical gate that determines the life or death of the cell. Various anti-apoptotic signals work to hold channels in the outer mitochondrial membrane closed; conversely, pro-apoptotic death signals strive to open these channels *(top right)*. Important among the molecules released into the cytosol are cytochrome *c* molecules *(red circles)* and the Smac/DIABLO molecules *(brown squares)*. Cytochrome *c* molecules proceed to aggregate with Apaf-1 molecules to form the seven-spoked apoptosome *(left;* see Figure 7.29*)*. The latter attracts and activates procaspase 9 into active caspase 9 *(green circle)*, which in turn activates procaspase 3 *(orange circle, left)*. The resulting active caspase 3 activates, in turn, a series of other procaspases (executioner caspases), which, once active, cleave various "death substrates" whose cleavage creates the apoptotic cell phenotype. Normally, a number of inhibitors of apoptosis (IAPs; *lower right)* attach to and inactivate caspases. However, the Smac/DIABLO that is also released from mitochondria antagonizes these IAPs *(right)*, thereby protecting caspases from IAP inhibition.

apoptosome. At the same time, Smac/DIABLO, also released through the channel, marginalizes the IAPs, thereby liberating caspases from IAP inhibition. (As an aside, one of the functionally important proteins that are phosphorylated by the anti-apoptotic Akt/PKB kinase is caspase 9. The phosphorylation of caspase

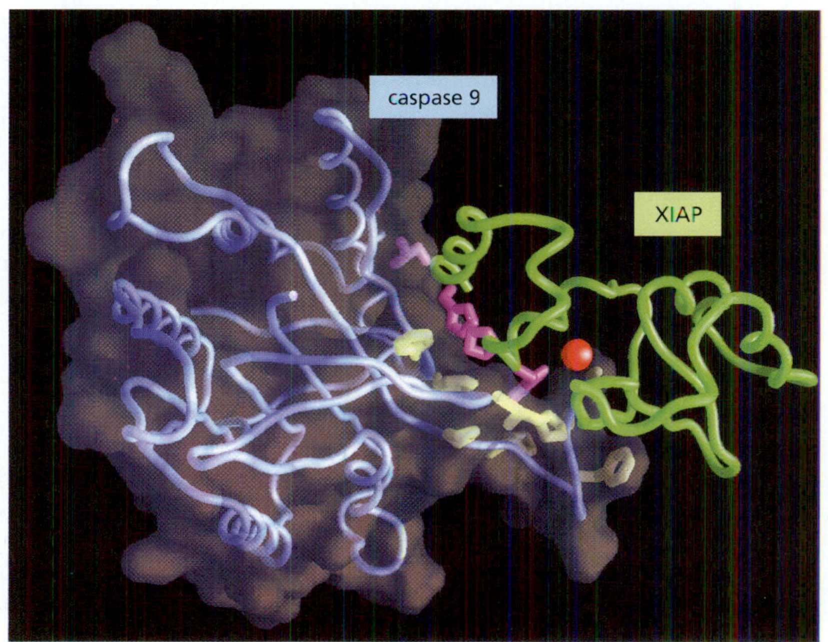

Figure 9.30 Structure of an inhibitor of apoptosis The large family of IAPs (inhibitors of apoptosis) function by binding and inhibiting various caspase molecules, thereby preventing inadvertent activation of the caspase cascade and the triggering of apoptosis. Shown here is the molecular structure of the complex formed by the BIR3 domain of an IAP molecule termed XIAP *(green)* with caspase 9 *(light blue)*. (Only the peptide backbone of XIAP is shown, while both the peptide backbone and a space-filling model of caspase 9 are shown.) A zinc atom at the center of this complex is shown in *red*, while critical residues in the interaction domains of XIAP and caspase 9 are shown in *pink* and *yellow*, respectively. This binding by XIAP prevents the homodimerization of caspase 9, which is required for its activity. (From E.N. Shiozaki et al., *Mol. Cell* 11:519–527, 2003.)

Sidebar 9.6 Amplification of initial pro-apoptotic signals ensures the efficient execution of apoptosis Once the decision to activate the pro-apoptotic caspase cascade has been made, the execution of this decision is aided by control mechanisms that amplify and reinforce some of the initial steps. (Such positive-feedback mechanisms are sometimes termed "feed-forward" controls.) Thus, one of the proteins that is cleaved by caspase 3 is a Rel subunit of the NF-κB transcription factor. As noted earlier (Section 6.12), this transcription factor is an important activator of important anti-apoptotic genes, such as *bcl-2*. By destroying a critical subunit of NF-κB, caspase 3 shifts the regulatory balance further in favor of apoptosis. Provocatively, an oncogenic derivative of Rel (an important component of the NF-κB complex) is the v-Rel oncoprotein made by the avian reticuloendotheliosis retrovirus. During its evolution in the retroviral genome, the viral *rel* oncogene has acquired a deletion and two point mutations that render the encoded v-Rel oncoprotein resistant to cleavage by caspase 3, thereby greatly enhancing its anti-apoptotic powers and resulting oncogenic potency.

9 inhibits this protease, as might be anticipated from the known anti-apoptotic actions of Akt/PKB.)

The cascade of caspase activations thus initiated proceeds until the final caspases in the cascade cleave "death substrates," that is, proteins whose degradation creates the cellular changes that one associates with the apoptotic death program (see Figure 9.29). The specialized roles of these various caspases has caused some to classify them into two functional groups—the *initiator caspases*, which trigger the onset of apoptosis by activating the caspase cascade; and the downstream *executioner caspases*, which undertake the actual work of destroying critical components of the cell.

A number of key cellular components are cleaved by these executioner caspases—caspases 3, 6 and 7. Their degradation causes the profound morphological transformations that accompany the death throes of apoptotic cells. Cleavage of lamins on the inner surface of the nuclear membrane is involved in some fashion in the observed chromatin condensation and nuclear shrinkage (pyknosis) that is characteristic of the apoptotic program. Cleavage of the inhibitor of caspase-activated DNase (ICAD) liberates this DNase, which then fragments the chromosomal DNA. Cleavage of cytoskeletal proteins such as actin, plectin, vimentin, and gelsolin leads to collapse of the cytoskeleton, the formation of the blebs protruding from the plasma membrane, and the formation of apoptotic bodies—the condensed hulks of cells that remain after all this has occurred (see Figure 9.18).

The executioner caspases even extend their reach back into the mitochondria, where the apoptotic program is initiated: one of their substrates is a protein that is part of the electron transport chain within mitochondria. Its cleavage by caspases leads to disruption of electron transport, loss of ATP production, release of reactive oxygen species (ROS) from the mitochondria, and the loss of mitochondrial structural integrity. The efficient execution of many of these steps of the apoptotic program is enabled, in part, by certain changes that further amplify the signals initiating apoptosis (Sidebar 9.6). Some researchers estimate that 400 to 1000 distinct cellular proteins undergo cleavage during apoptosis, but it remains unclear how many of these are active participants in the apoptotic program and how many are victims of the collateral damage that occurs as cells undergo disintegration.

The caspase cascade forms the central signaling pathway that determines whether or not apoptosis will occur and, in the event that apoptosis is decided upon, how the various death-inducing steps are activated within cells. The importance of careful regulation of these caspases is underscored by findings concerning E2F1 activity and apoptosis. Earlier, we noted that excessive E2F activity warns the cell of a malfunctioning pRb pathway and induces apoptosis through its ability to activate expression of the gene encoding ARF. It has also become clear that E2Fs, such as E2F1, are able to increase transcription of several of the key caspase genes five- to fifteenfold. These increases, on their own, do not trigger apoptosis. Instead, they sensitize the cell to specific pro-apoptotic stimuli, such as those emitted by p53, thereby tilting the playing field in favor of apoptosis. (The E2Fs can compound these effects by inducing the expression of several genes encoding pro-apoptotic relatives of Bcl-2.)

9.14 Two distinct signaling pathways can trigger apoptosis

The apoptotic program described above can be initiated by p53 in response to a series of signals, including those indicating substantial genomic damage, anoxia, and imbalances in the growth-regulating signals coursing through a cell's intracellular circuits. To review, once p53 becomes activated and placed in an apoptosis-inducing configuration, it can drive the expression of genes encoding proteins such as Bax (see Table 9.2). Bax then works to open the mitochondrial channels, whereupon cytochrome *c* and other pro-apoptotic proteins leak out of the mitochondria and apoptosis is initiated through the activation of caspases.

This series of events is sometimes termed the *intrinsic apoptotic* program, since the signals triggering it originate from within the cell; others have termed it the *stress-activated* apoptotic pathway. In ways that remain poorly understood, yet other types of signals can also cause the mitochondria to release cytochrome *c* through mechanisms that do not seem to involve p53. Included here are stresses such as excessive calcium within the cell, excessive oxidants, certain types of DNA-damaging agents, and yet other agents that disrupt the microtubules—key components of the cytoskeleton and mitotic spindle. Infections by various tumor viruses may also trigger apoptosis. Interestingly, these viruses take great pains to ensure that the intrinsic apoptotic program is not initiated, lest their cellular hosts die before their infectious cycles reach completion (Sidebar 12 ●).

Importantly, apoptosis can also be triggered through an alternative route. This one is initiated outside the cell and activates pro-apoptotic cell surface receptors. These are transmembrane proteins that are often termed *death receptors* to indicate their ability to activate the apoptotic program (Figure 9.31). After binding their cognate ligands in the extracellular space, the death receptors activate a cytoplasmic caspase cascade that converges rapidly on the intrinsic apoptotic pathway described above, thereby triggering an apoptotic response identical to the one seen following activation of the intrinsic apoptotic program (Figure 9.32). Because the signals that activate the death receptors originate from outside the cell, the apoptotic program initiated by these receptors has been called either the *extrinsic apoptotic* program or the *receptor-activated* apoptotic pathway.

The ligands of the death receptors are members of the tumor necrosis factor (TNF) family of proteins, which includes TNF-α, TRAIL, and Fas Ligand (FasL) (Table 9.4). TNF-α was first identified and studied because of its ability to cause the death of cancer cells. Subsequently, this group of protein ligands was found capable of causing the death of a wide variety of normal cell types that display appropriate receptors on their surfaces. Similar to growth factors and their receptors, each of these TNF-like proteins, including FasL and TRAIL/Apo2L, binds to its own cognate death receptor (Figure 9.31). Members of this family of receptors—there are as many as 30 of them encoded in the human genome—share in common a cytoplasmic "death domain." An understanding of these receptors and their action is confounded by the multiple names that many of these receptors and ligands bear (see Table 9.4).

Once activated by ligand binding, the death domains of these receptors bind and activate an associated protein termed FADD (Fas-associated death domain protein) in the cytoplasm. The resulting complex is termed a DISC (death-inducing signaling complex) and summons the inactive, pro-enzyme forms of caspase 8 and, less commonly, caspase 10, which are among the initiator caspases. The DISC then triggers the self-cleavage of these caspases and their conversion into active proteases. Caspases 8 and 10 (the initiator caspases of this pathway) then activate the executioner caspases 3, 6, and 7, thereby converging on the signaling pathway through which the intrinsic apoptotic program operates. In addition, caspase 3 can cleave and activate a Bcl-2–related protein in the cytosol termed Bid, which then migrates to the mitochondrial channel and undertakes to pry it open (see Figure 9.32). In doing so, Bid succeeds in amplifying pro-apoptotic signals of the extrinsic apoptotic program by recruiting elements of the intrinsic program into the process as well.

Some cell types in the body are able to utilize both apoptotic pathways to trigger their own death, while others rely solely on either the extrinsic or the intrinsic program. For example, a cell can trigger its own death via the extrinsic program by secreting a ligand for one of the death receptors that it displays on its surface; this ligand can then act in an autocrine fashion to initiate apoptosis. The use by a cell of the intrinsic or extrinsic program has implications for the actions of anti-apoptotic proteins such as Bcl-2. Thus, in cells that can activate the extrinsic program, the overexpression of Bcl-2 may not confer much benefit, since the death receptors can circumvent the mitochondrion-based intrinsic apoptotic program by communicating directly with the caspase cascade.

In fact, there is actually a third way of triggering apoptosis, in effect another extrinsic apoptotic program. This one is triggered by cytotoxic T lymphocytes

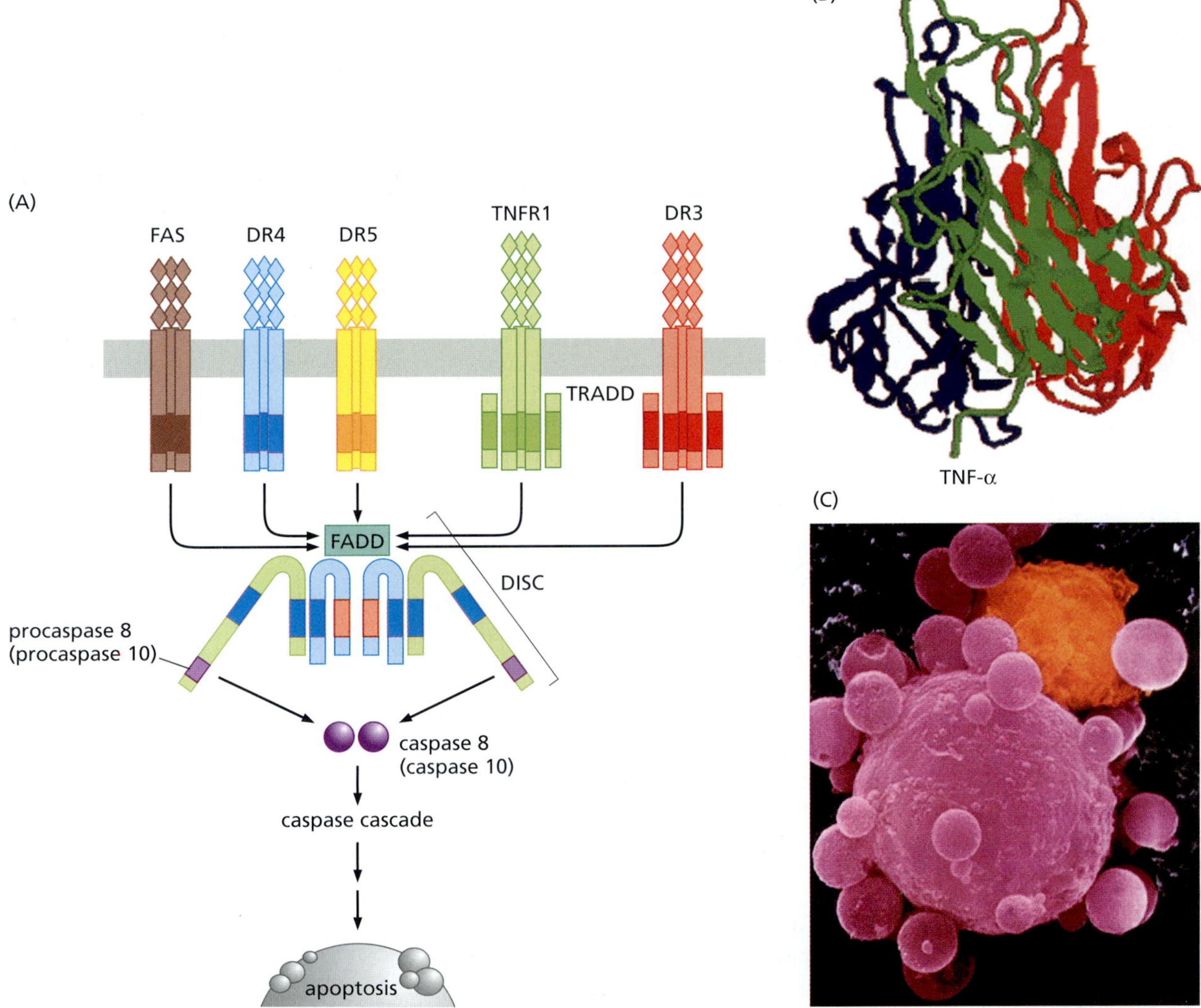

Figure 9.31 Death receptors (A) Five families of "death receptor" proteins are displayed on the surfaces of various types of mammalian cells. For example, the FASL and APO2L/TRAIL ligands bind to the FAS, DR4, and DR5 death receptors *(top left)*. The cytoplasmic tails of these receptors act via the FAS-associated death domain (FADD) protein to assemble a death-inducing signaling complex (DISC), and the latter proceeds to convert procaspases 8 and 10 into their respective active caspases. These then converge on the caspase cascade that is activated through the intrinsic apoptotic program. A similar sequence of events occurs following the activation of the TNFR1 and DR3 receptors *(top right)*. These death receptors differ from receptor tyrosine kinases (and their ligand-induced dimerization) in that three-subunit ligand complexes cause receptor trimerization. (B) The ligands of these receptors, sometimes called "death ligands," are homotrimers and appear to work by causing the trimerization of their cognate receptors. [These can be compared with the ligands of growth factor receptors (Figure 5.16), many of which are homodimers.] Shown here is the molecular structure of tumor necrosis factor-α (TNF-α), the ligand of the TNF-R1, with each monomer depicted in a distinct color. (C) Death receptor ligands, notably FasL, are frequently used by cytotoxic T lymphocytes (CTLs) to kill cancer cells, which display the FAS receptor on their surfaces. (In addition, granzymes may be injected by CTLs into target cells; see Figure 9.32.) In this colorized scanning electron micrograph, a CTL *(orange)* is seen to be attacking a cancer cell, which is in the throes of apoptosis, as evidenced by the numerous blebs on its surface. (A, adapted from A. Ashkenazi, *Nat. Rev. Cancer* 2:420–430, 2002; B, from M.J. Eck and S.R.Sprang, *J. Biol. Chem.* 264:17595–17605, 1989; C, from Andrejs Liepin/Science Photo Library.)

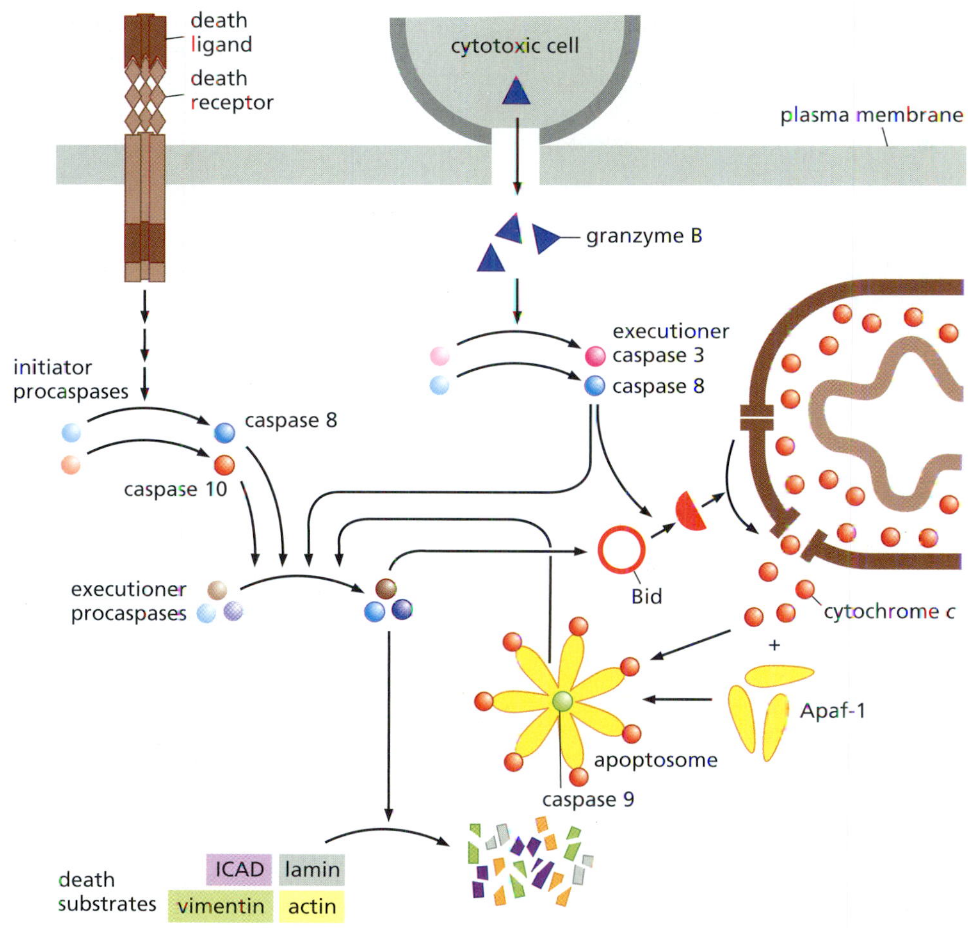

Figure 9.32 Convergence of intrinsic and extrinsic apoptotic pathways After a death receptor (Figure 9.34) activates procaspase 8 and/or 10 into the corresponding active caspases *(top left)*, the latter converge on the intrinsic apoptotic cascade by cleaving and activating the executioner caspases *(middle left)*. In addition, after procaspase 3 has been activated into caspase 3, the latter cleaves and activates Bid (see Figure 9.27B), a Bcl-2–related protein *(center, red circle)*, which moves from its cytosolic location to the mitochondrion, where it initiates opening of the outer mitochondrial membrane channel, further activating the apoptotic cascade by forming new, activated apoptosomes *(right center, lower right)*. A second, alternative form of the extrinsic apoptotic pathway is initiated by cytotoxic cells that are dispatched by the immune system *(gray, top middle)* and attach to the surfaces of targeted cells, into which they introduce granzyme B molecules; once present in the cytosol, they cleave and thereby activate procaspases 3, 8 and 9.

and natural killer (NK) cells that undertake to kill target cells. These two types of killer cells—some of the frontline troops of the immune system—can activate death receptors such as Fas displayed on the surfaces of cells that they have chosen for destruction. But in addition, the killer cells can attach a protease, termed granzyme B, to the surfaces of targeted cells; once internalized by these cells, granzyme B cleaves and activates procaspases 3, 8, and 9 (see Figure 9.32). At this point, there is immediate convergence with the other apoptotic pathways described above.

This mapping of the intrinsic and extrinsic apoptotic cascades enriches our understanding of how p53 succeeds in evoking apoptosis (Figure 9.33). p53 promotes expression of the genes encoding the Bax, Fas receptor, and IGFBP-3 proteins (see Table 9.2). Bax antagonizes Bcl-2 function and thereby works to open the mitochondrial channels that release cytochrome *c*. The Fas receptor, once expressed at the cell surface, increases a cell's responsiveness to extracellular death ligands, specifically FasL. IGFBP-3 is released from the cell, whereupon it binds and sequesters IGF-1 (insulin-like growth factor-1), a growth factor that operates through its own cell surface receptor to produce **trophic**

Table 9.4 Death receptors and their ligands

Alternative names of receptors	Alternative names of ligands
Fas/APO-1/CD95	FasL/CD95L
TNFR1	TNF-α
DR3/APO-3/SWL-1/TRAMP	APO3L
DR4/TRAIL-R1	APO2L/TRAIL
DR5/TRAIL-R2/KILLER	APO2L/TRAIL

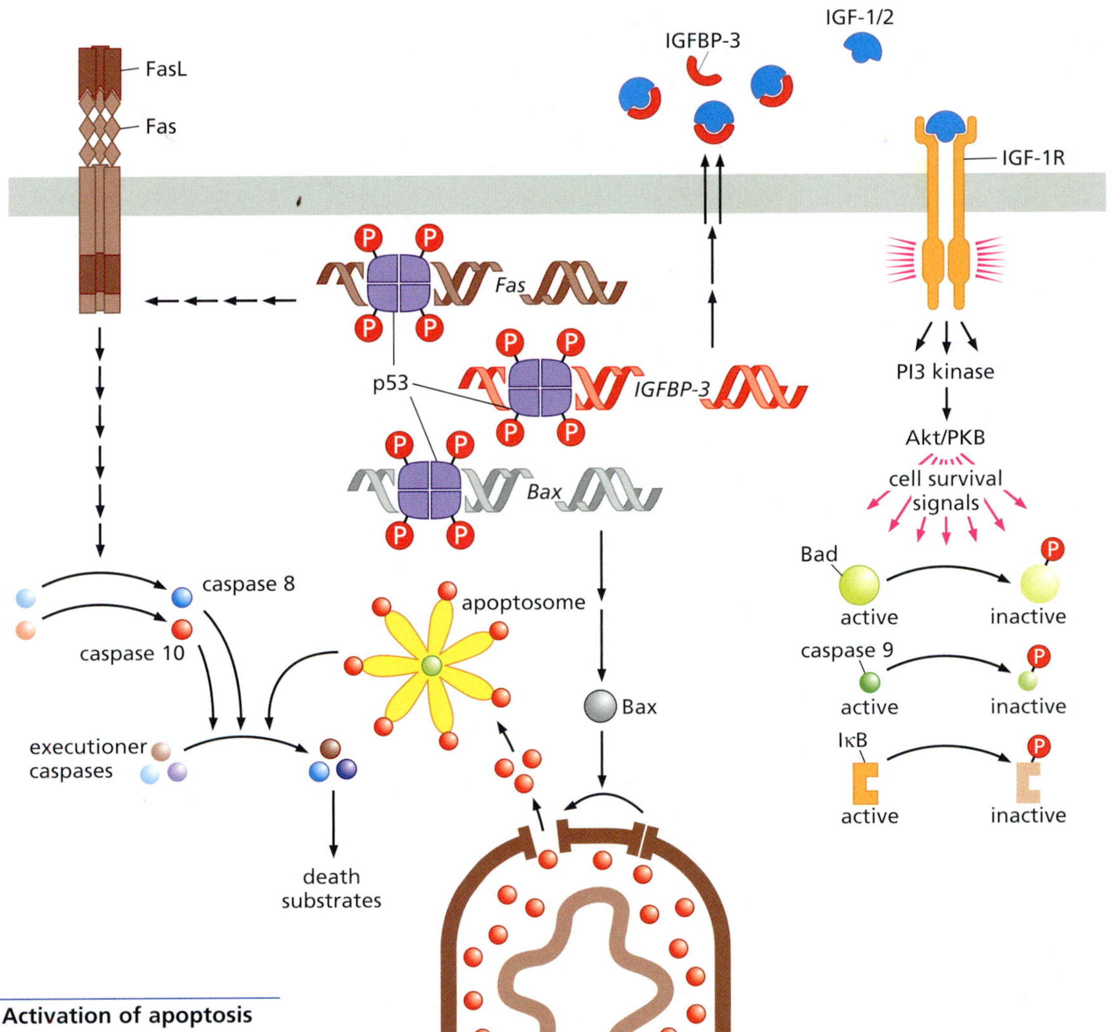

Figure 9.33 Activation of apoptosis by p53 p53 utilizes multiple signaling pathways to activate the apoptotic program. By inducing expression of the gene encoding the Fas receptor, it causes display of this death receptor at the cell surface *(brown, top left),* thereby sensitizing the cell to any Fas ligand (FasL) that may be present in the extracellular space. By inducing expression of IGF-binding protein-3 (IGFBP-3; *red),* p53 causes release of this protein in the extracellular space, where it binds and sequesters IGF-1 and IGF-2 *(blue, top center),* the pro-survival, anti-apoptotic ligands of the IGF-1 receptor (IGF-1R; *top right).* In the absence of IGFBP-3, IGF-1 would bind to its receptor and cause release of anti-apoptotic signals in the cell, including those leading to the inactivation of the pro-apoptotic proteins Bad (related to Bcl-2), caspase 9, and IκB (antagonist of NF-κB). In addition, p53 drives expression of Bax *(center),* the pro-apoptotic, Bcl-2–related protein, which causes the release of cytochrome *c* and other proteins from the mitochondria *(bottom center).* p53 also inactivates several other anti-apoptotic agents, such as Smac/DIABLO (see Figure 9.29).

(survival) signals in the cell; in many cells, the resulting reduction of IGF-1–initiated survival signals places them in grave danger of succumbing to apoptosis (Sidebar 9.7).

We will return to the extrinsic apoptotic program in Chapter 15, because it can play two important roles in the complex interactions between cancer cells and the immune system. On the one hand, cells of the immune system may release death ligands (such as FasL) in order to trigger apoptosis in cancer cells that they have targeted for elimination. On the other, cancer cells may display or release death ligands that they deploy to kill immune cells that have approached too closely and threaten their survival.

The complexity of the signals flowing through apoptotic pathways underscores the fact that the decision for or against apoptosis is achieved by altering the *balance* of complex arrays of pro- and anti-apoptotic proteins that influence the ultimate decision whether or not to flip the "apoptotic master switch."

9.15 Cancer cells invent numerous ways to inactivate some or all of the apoptotic machinery

As emphasized earlier, cells advancing down the long road toward neoplasia encounter a number of physiologic stressors that threaten their existence, since each of these stressors threatens to trigger apoptosis. Included among these are anoxia, deregulation of the pRb pathway, activation of the *myc* oncogene, and various forms of DNA damage. These dangers explain why most and likely all types of cancer cells have inactivated key components of their apoptosis-inducing machinery (Table 9.5; Figure 9.34). We can also view these changes from the perspective of the organism as a whole: the fact that virtually all human cancer cells employ anti-apoptotic strategies suggests that apoptosis is a vital anti-neoplastic mechanism that is implanted in all normal cells in order to obstruct tumor progression.

One anti-apoptotic strategy widely exploited by cancer cells is, as already noted, inactivation of the p53 pathway. The *p53* gene is altered in almost half of all human cancer cell genomes, usually by point mutations that lead to amino acid substitutions in the p53 DNA-binding domain. In a significant percentage of other tumors, ARF is no longer expressed. This is often achieved through outright deletion of the encoding gene or by promoter methylation, the latter resulting in repression of transcription. Other mechanisms shutting down ARF expression may also intervene, as was indicated in Sidebar 7. A small percentage of human tumors (mostly sarcomas) will overexpress Mdm2. All of these mechanisms affecting ARF or Mdm2 expression serve to drive down levels of p53 protein in cells. Finally, in some human tumor cells, p53 appears to be mislocalized, being sequestered in the cytoplasm, where it cannot carry out its main task—transcriptional activation.

Still, this elimination of p53 function does not seem to suffice for many types of human tumors, which additionally strive to alter specific components of the apoptotic machinery. For example, many melanoma cells exhibit methylation and thus functional inactivation of the promoter of the *APAF1* gene, which encodes the cytosolic protein that assembles with released cytochrome *c* to form the apoptosome and activate caspase 9. The pro-apoptotic *Bax* gene (see Table 9.5) is inactivated by mutation in more than half the colon cancers that show microsatellite instability (Chapter 12). Deregulated expression of *BCL2* (the human version of *bcl-2*) is found in follicular B-cell lymphomas, and *BCL2* expression is found to be elevated in large numbers of human tumors of diverse tissue origins; by some estimates, at least half of all human tumors show increased *BCL2* expression.

Yet another highly effective strategy for acquiring resistance to apoptosis derives from hyperactivation of the PI3K → Akt/PKB pathway (Section 6.6). Activation of this pathway can be achieved through the collaborative actions of signal-

Sidebar 9.7 Extracellular factors also favor cell survival We have read about a variety of alterations of the intracellular signaling circuitry that are used by cancer cells to decrease the likelihood that their apoptotic programs will become triggered. However, the most widely used pro-survival strategy appears to involve an extracellular signaling pathway that plays a role in normal tissue physiology: many normal cells seem to depend on continuous exposure to trophic (i.e., survival) signals, such as those conveyed by insulin-like growth factors-1 and -2 (IGF-1, IGF-2), in order to remain viable. Inadequate levels of these factors in the cellular environment may trigger apoptosis, a process that has been called "death by neglect."

Many cancer cells secrete abnormally high levels of IGF-1 or IGF-2. Once released by cancer cells, both growth factors bind and activate the IGF-1 receptor (IGF-1R), which is displayed by the great majority of cell types throughout the body. This autocrine receptor activation, in turn, causes the release of intracellular signals that are strongly anti-apoptotic, notably through their ability to activate the PI3K–Akt/PKB pathway (Figure 9.34). Some human tumors have been reported to overexpress IGF-2 mRNA by as much as several hundredfold, although this degree of overexpression is quite unusual.

Another common strategy involves reduction in the effective levels of an IGF-binding protein (IGFBP), which normally reduces the effective levels of active IGFs in the extracellular space by binding and sequestering them (see Figure 9.33). (At least six distinct IGFBPs have been described.) In some tumors, methylation of the promoter of an *IGFBP* gene reduces the expression of its encoded protein and thus the levels of this protein in the immediate vicinity of a cancer cell. In other tumors, cancer cells may increase the synthesis and secretion of certain matrix metalloproteinases (MMPs), which proceed to cleave and thus inactivate IGFBP molecules in the surrounding extracellular space. By reducing the levels of functional IGFBPs, both mechanisms cause increases in the effective concentrations of IGFs, which in turn favor cancer cell survival.

Table 9.5 Examples of anti-apoptotic alterations found in human tumor cells

Alteration	Mechanism of anti-apoptotic action	Types of tumors
CASP8 promoter methylation	inactivation of extrinsic cascade	SCLC, pediatric tumors
CASP3 repression	inactivation of executioner caspase	breast carcinomas
Survivin overexpression[a]	caspase inhibitor	mesotheliomas, melanomas, many carcinomas
ERK activation	repression of caspase-8 expression	many types
ERK activation	protection of Bcl-2 from degradation	many types
Raf activation	sequestration of Bad by 14-3-3 proteins	many types
PI3K mutation/activation	activation of Akt/PKB	gastrointestinal
NF-κB constitutive activation[b]	induction of anti-apoptotic genes	many types
p53 mutation	loss of ability to induce pro-apoptotic genes	many types
$p14^{ARF}$ gene inactivation	suppression of p53 levels	many types
Mdm2 overexpression	suppression of p53 levels	sarcomas
IAP-1 gene amplification	antagonist of caspases-3 and 7	esophageal, cervical
APAF1 methylation	loss of caspase-9 activation by cytochrome *c*	melanomas
BAX mutation	loss of pro-apoptotic protein	colon carcinomas
Bcl-2 overexpression	closes mitochondrial channel	~½ of human tumors
PTEN inactivation	hyperactivity of Akt/PKB kinase	glioblastoma, prostate carcinoma, endometrial carcinoma
IGF-1/2 overexpression	activates PI3K	many types
IGFBP repression	loss of anti-apoptotic IGF-1/2 antagonist	many types
Casein kinase II	activation of NF-κB	many types
TNFR1 methylation	repressed expression of death receptor	Wilms tumor
FLIP overexpression	inhibition of caspase-8 activation by death receptors	melanomas, many others
Akt/PKB activation	phosphorylation and inactivation of pro-apoptotic Bcl-2-like proteins	many types
Stat3 activation	induces expression of $Bcl\text{-}X_L$	several types
TRAIL-R1 repression	loss of responsiveness to death ligand	small-cell lung carcinoma
FAP-1 overexpression	inhibition of Fas receptor signaling	pancreatic carcinoma
XAF1 methylation[c]	loss of inhibition of anti-apoptotic XIAP	gastric carcinoma
Wip1 overexpression[d]	suppression of p53 activation	breast and ovarian carcinomas, neuroblastoma

[a]Survivin is an inhibitor of apoptosis (IAP) in gastric, lung, and bladder cancer and melanoma in addition to the mesotheliomas indicated here. The expression of a number of IAP genes is directly induced by the NF-κB TFs.
[b]Induces synthesis of c-IAPs, XIAP, $Bcl\text{-}X_L$, and other anti-apoptotic proteins.
[c]XAF1 (XIAP-associated factor 1) normally binds and blocks the anti-apoptotic actions of XIAP, the most potent of the IAPs.
[d]Wip1 is a phosphatase that inactivates p38 MAPK, which otherwise would phosphorylate and stimulate the pro-apoptotic actions of p53.

emitting tyrosine kinase receptors and the Ras oncoprotein, both of which activate PI3K; the result is an increased level of PIP_3 and activation of Akt/PKB. Inactivation of the gene encoding PTEN, the phosphatase that breaks down PIP_3, has a similar effect. Once Akt/PKB becomes activated by the PIP_3 that accumulates, it can phosphorylate and thereby inhibit pro-apoptotic proteins such as Bad, caspase 9, and IκB, and at the same time phosphorylate and activate Mdm2, the key antagonist of p53 (Table 6.3).

A diverse collection of cancer cell types, ranging from leukemias to colon carcinomas, have been found to resort to yet another mechanism to protect themselves against apoptosis: they activate expression of the NF-κB transcription factor. As we read earlier, NF-κB is usually sequestered in inactive form in the cytoplasm and, in response to certain physiologic signals, is liberated and translocates to the nucleus, where it activates a large constituency of target genes (Section 6.12). A number of these genes are involved in pro-inflammatory and angiogenic functions. In addition, a significant cohort of NF-κB target genes function as apoptosis antagonists; included among these are *IAP-1, IAP-2, XIAP, IEX-1L, TRAF-1,* and *TRAF-2,* which serve to block both the intrinsic and extrinsic apoptotic programs. This list of NF-κB targets is expanding rapidly, and it is likely that the overexpression of some of these anti-apoptotic genes (and encoded proteins) found in many types of human cancer cells will be traced directly to the actions of hyperactivated NF-κB within these cells.

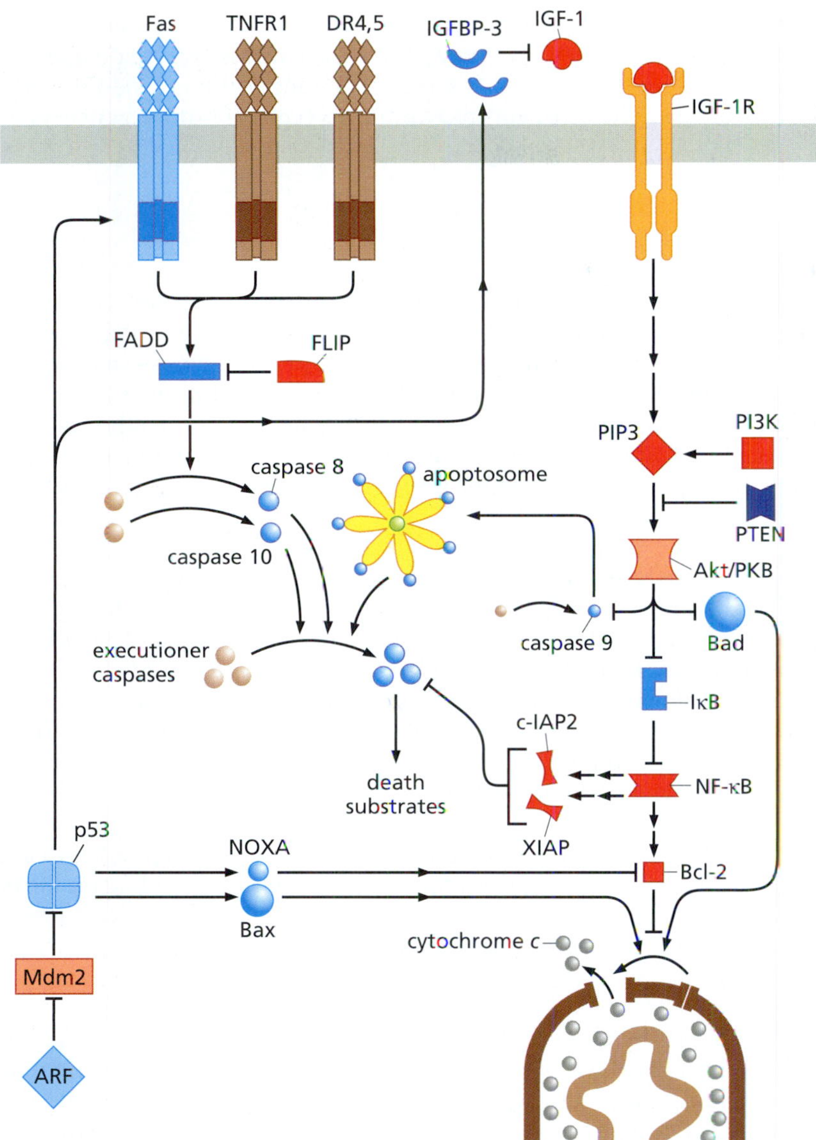

Figure 9.34 Anti-apoptotic strategies used by cancer cells Cancer cells resort to numerous strategies in order to decrease the likelihood of apoptosis. This diagram indicates that in various cancer cell types, the levels or activity of important pro-apoptotic proteins are decreased *(blue)*. Conversely, the levels or activity of certain anti-apoptotic proteins may be increased *(red)*. (Some of these, because they are downstream of controllers that are known to be altered in cancer cells, are affected by the documented changes in their upstream controllers.)

Yet another strategy for avoiding apoptosis is evident in childhood neuroblastomas, in which amplification of the N-*myc* gene normally indicates aggressive tumor growth and short life expectancy. On its own, this amplification actually threatens the proliferation of neuroblastoma cells since, as mentioned earlier, the actions of the *myc* oncogene (and those of its N-*myc* cousin) represent a double-edged sword. These oncogenes release potent mitogenic signals, but at the same time are strongly pro-apoptotic. Many neuroblastoma cells that have overexpressed N-*myc* through gene amplification seem to solve this problem, at least in part, by inactivating the gene encoding caspase 8, which is either methylated or deleted.

The actions of the extrinsic apoptosis-inducing pathway are also relevant for cancer cell evolution, as evidenced by the fact that this pathway is inactivated in certain types of human tumors. Some cancer cells deploy a protein termed FLIP (FLICE-inhibitory protein) which can interfere with the activation of the extrinsic apoptotic cascade by binding to the death domains of death receptors, thereby blocking the processing and autocatalytic activation of the initiator caspase-8 (see Figure 9.32). Yet other tumor cells suppress expression of their TNF receptors by methylating the promoter of the encoding *TNFR1* gene.

These maneuvers by cancer cells provide clear evidence that they can gain survival advantage by inactivating the extrinsic pathway. This leads, in turn, to the speculation that cells of the immune system often besiege cancer cells and attempt to kill them through the release of TNF-like factors and resulting initiation of the extrinsic apoptotic program. Given the complex wiring diagram of the circuitry controlling apoptosis, it is virtually certain that many additional anti-apoptotic strategies will be uncovered in various types of human cancer cells.

9.16 Synopsis and prospects

In the early chapters of this book, we concentrated on the molecular mechanisms that propel cell proliferation forward or hold it back. The regulators—the products of (proto-) oncogenes and tumor suppressor genes—act to process extracellular mitogenic and anti-mitogenic signals that impinge upon the cell cycle machinery. These signals converge on the final decisions that determine whether cells advance through the active growth-and-division cycle, retreat reversibly into quiescence, or enter irreversibly into a post-mitotic, differentiated state.

In this chapter, we have focused on entirely different cell-physiologic processes, most of which are responsible for monitoring the internal well-being of cells. The monitoring apparatus conducts continuous surveillance of vital cell systems, including access to oxygen, the physical state of the genome, and the balance of signals flowing through growth-regulatory circuits. If cellular monitors determine that a vital system is damaged or malfunctioning in a cell, an alarm system is sounded and growth is halted or, more drastically, the cell is induced to activate its previously latent apoptotic program.

Sidebar 9.8 p53 acts directly on mitochondria p53 has been widely viewed as a protein that exerts all of its pro-apoptotic effects through its ability to activate transcription of key pro-apoptotic genes in the nucleus. However, several scientific reports indicate that p53 can contribute to apoptosis even if it is constrained to remain in the cytoplasm. These reports demonstrate that p53 is capable of liberating the pro-apoptotic Bak protein from the clutches of the anti-apoptotic Bcl-X_L protein. Moreover, p53 can associate directly with Bax, another key pro-apoptotic protein, and functionally activate it, enabling it to open mitochondrial membrane channels, leading in turn to the release of cytochrome *c* into the cytosol and the triggering of apoptosis. Consequently, p53 appears to be far more directly involved in activating the pro-apoptotic machinery than previously suspected. For reasons that remain poorly understood, mutant p53 forms from human tumors that lack transcription-activating activity have also lost the ability to activate Bax.

The p53 protein lies at the heart of this signaling machinery. Remarkably, a wide variety of alarm signals converge on this single protein. Depending on the gravity of cellular stresses and the ability of cells to repair defects that have been detected, p53 will temporarily arrest cell cycle advance or activate the apoptotic program.

The *p53* gene and encoded protein have been the subject of more than 30,000 research reports that have appeared in the scientific literature in the quarter century following the discovery of the p53 protein in 1979. This extraordinary concentration of research on a single gene and protein reflects both the central role that this protein plays in cancer pathogenesis and the frequent presence of mutant *p53* alleles in the genomes of human cancer cells. In spite of the enormous corpus of scientific literature about p53, research on this protein is continually turning up new surprises (see Sidebar 9.8).

Its vastness notwithstanding, this body of research on p53 has failed to provide a clear answer to the most perplexing puzzle created by the protein's existence: Why do mammalian cells entrust so many vital alarm functions to a single protein? Once a cell has lost p53 function, it becomes oblivious, in a single stroke, to a whole variety of conditions that would normally call a halt to its proliferation or trigger entrance into the apoptotic program.

Unlike the other tumor suppressor genes that we have encountered, this gene is usually not deleted from cancer cell genomes. Instead, mutant alleles encoding single amino acid replacements are usually observed—a testimonial to the dominant-negative powers of the resulting mutant p53 proteins. As argued here, such slightly defective alleles are far more effective in disrupting surviving wild-type *p53* function than are null alleles that would arise from deletion of part or all of *p53*'s coding sequences.

As we will see in Chapter 11, tumor development is a complex, multi-step process involving changes in multiple cellular control circuits. Because p53 receives signals from so many distinct sensors of damage or physiologic distress, the loss of p53 function plays a role in many of these steps. For example, early in the development of bladder, breast, lung, and colon carcinomas, when premalignant growths are present, DNA damage signaling is already active, as indicated by the presence of phosphorylated, functionally active ATM and Chk2 kinases. In addition, the p53 substrate of the ATM and Chk2 kinases is also phosphorylated and apparently active, since the *p53* gene is still in its wild-type configuration at this stage of tumor progression. This indicates that DNA lesions are widespread in the cells of these growths, even at this early stage of tumorigenesis, and suggests that p53 is striving to constrain the proliferation of the cells in these growths and may even induce their apoptosis with some frequency. Consequently, these premalignant cells stand to benefit greatly by eliminating p53 function, thereby liberating themselves from its cytostatic and pro-apoptotic effects.

At this stage of tumor development, cells that succeed in shedding p53 function may also gain special advantage because they have a better chance of avoiding the apoptosis that is triggered by p53 in response to severe hypoxia. (In general, early premalignant growths have not yet developed an adequate access to the circulation and the oxygen that it brings.) In addition, because p53 induces expression of the potently anti-angiogenic molecule thrombospondin (Tsp-1), the loss of p53 can accelerate the development of nearby capillary networks that can cure the hypoxia that incipient cancer cells have been suffering.

Still later in tumor progression, when malignant carcinomas develop, the absence of p53 function will compromise some of the DNA repair systems of the cell and, at the same time, allow cells carrying damaged genomes greater survival advantage; the result is an increased rate of genomic alteration and associated acceleration in the rate of tumor development. During these later stages of tumor development, the absence of p53 function may confer additional benefit by enabling cells to survive the pro-apoptotic side effects of oncogenes such as *myc*. This short list reveals how truly catastrophic the loss of p53 function can be for a tissue and, ultimately, for the organism as a whole.

The disastrous effects of p53 inactivation may be compounded later should *p53*-mutant tumor cells become clinically apparent and become targeted for chemotherapy and radiation. These treatments are often directed toward damaging the genomes of these cells, thereby provoking their death by apoptosis. Indeed, the pro-apoptotic effects of most existing anti-cancer therapies are the central reason why these therapies succeed, in many instances, in halting the further growth of tumors. Unfortunately, it seems that the loss of p53 function, which is seen so often in human tumors, renders these neoplasias far less responsive to many of these therapeutic strategies.

Our understanding of the p53 protein is still quite superficial, in spite of the vast research literature that describes it. One indication of how much more needs to be learned has come from attempts at systematically cataloging the list of p53-regulated genes in the human genome; more than 120 have been identified, only a small proportion of which are cited in Table 9.2. Yet another indication derives from analyses of the post-translational modifications of p53. As we have read, these modifications affect p53's stability and thus intracellular concentration. In addition, they are also likely to affect the interactions of p53 with other nuclear proteins, notably transcription factors, that influence the genes that are targeted by p53 action. The extent of the complexity of this regulation is suggested by Figure 9.35, in which are indicated the sites of post-translational modification of p53 and the nature of these modifications. At least six distinct types of modification, involving phosphate, acetyl, ubiquityl, sumoyl, neddyl, and methyl groups, decorate the p53 molecules present in human cells. (The neddyl and sumoyl groups are similar to ubiquityl groups but have very different effects on the proteins to which they become linked.)

Three additional dimensions of complexity are indicated in Figure 9.35. First, certain amino acid residues, such as serine 15 (S15), can be phosphorylated by multiple protein kinases; at least five alternative kinases are known to phosphorylate this particular residue. Second, other residues, such as lysine 372 (L372), can be modified by multiple alternative means; this particular residue can undergo methylation, acetylation, ubiquitylation, or neddylation. Third the number of enzymes that modify p53—at least 28 are indicated here—is likely to reflect the number of distinct signaling pathways that impinge on p53 and perturb its functioning.

Inactivation of p53 is only one of many strategies that cancer cells employ to protect themselves against the danger of apoptosis. These maneuvers compromise the apoptotic response at many other levels, including perturbing the expression of death receptors, their ability to activate caspases, and the activities of the caspases themselves. Yet other defensive maneuvers undertaken by cancer cells help to prevent opening the outer membrane pores of mitochondria (see Figure 9.34). These diverse strategies illustrate the fact that the apoptotic machinery, like all well-regulated machines, operates through the balanced actions of mutually antagonistic elements. Cancer cells exploit preexisting anti-apoptotic components of this machinery by increasing their levels or activities, thereby nudging the balance away from apoptosis. These changes in specific components of the apoptotic machinery may compound the effects of p53 loss, further increasing the resistance of cancer cells to a variety of signals that usually succeed in provoking cell death. At the same time, the levels of insulin-like growth factor-1 (IGF-1) in the tissues surrounding clones of incipient cancer cells may also have profound effects on tumor development by reducing the likelihood of apoptosis (see Sidebar 13).

The ability of certain growth-promoting transcription factors, notably Myc and the E2Fs, to provoke apoptosis creates something of a paradox. Ectopic expression of either E2F1 or Myc can, on its own, force mitogen-starved cells to leave G_0 quiescence, enter into the active growth-and-division cycle, traverse G_1, and enter into S phase. Why do these transcription factors not drive actively growing cells into apoptosis long before they have succeeded in completing even a single cell cycle?

One important answer derives from the fact that as cells pass through G_1 and activate their pro-apoptotic functions, they concomitantly activate anti-apoptotic signals; the balance between these two determines the ultimate fate of the

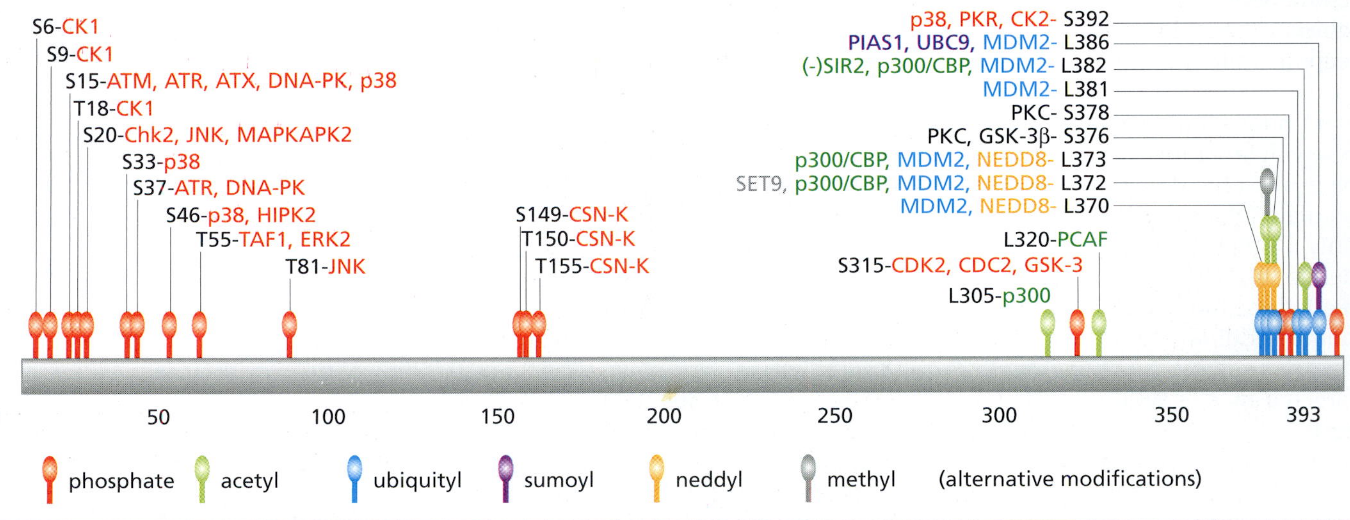

Figure 9.35 Multiple types of post-translational modifications of p53 The phosphorylation of p53 by various kinases (see Figure 9.12) represents only one type of modification that appears to be important to its functioning. In addition to the attachment of phosphate groups *(red)*, other enzymes modify it by attaching acetyl *(green)*, ubiquityl *(blue)*, sumoyl *(purple)*, neddyl *(orange)*, and methyl *(gray)* at the indicated amino acid residues and residue numbers *(black type)*. The identities of the modifying enzymes are indicated *(various font colors)* to the right or the left of the amino acid residue numbers. SIR2 acts to deacetylate p53. (Courtesy of K.P. Olive and T. Jacks.)

cell. For example, many of the mitogenic growth factors that stimulate Myc and E2F1 function in G_1 also activate a range of anti-apoptotic proteins, and it is the latter that usually predominate and allow the cell to survive and proliferate.

Such thinking raises yet another question: Why do mitogen-stimulated cells entering into the active growth cycle bother at all to synthesize a variety of pro-apoptotic proteins, whose very presence places these cells close to the edge of apoptotic abyss? According to an attractive but still-speculative model, whenever cells enter the growth-and-division cycle, they run the risk of unforeseen disasters, notably those occurring during its S and M phases. Some of these accidents can lead to dangerous changes in the genome, and these in turn may trigger the outgrowth of malignant cell clones. So all cells, upon entering the active cell cycle, place their pro-apoptotic machinery in a state of high alert, sensitizing it to signals that can rapidly activate the apoptotic program in the event that these cells stumble at one point or another as they attempt to execute the complex program of growth and division.

As additional features of cell death have been uncovered, it has become increasingly apparent that apoptosis, as we have described it in this chapter, is not the only form of programmed cell death. A second death program, known as **autophagy**, seems increasingly likely to play a key role in the elimination of undesired cells in various tissues throughout the body. This program is often activated when cells suffer nutrient starvation and, in response, digest their own intracellular organelles in cytoplasmic lysosomes in order to recycle and scavenge various chemical species that may aid in their survival (Figure 9.36).

Autophagy may also be used as a means of eliminating cancer cells. Thus, the gene encoding Beclin-1, a key autophagy-promoting protein, is often found in reduced copy numbers in various types of human cancers, and the deletion of one of the two copies of this gene in the mouse germ line leads to greatly increased tumor incidence. Together, these lines of evidence indicate the opening of an entirely new area of cell death research that has clear implications for our understanding of cancer pathogenesis.

Later, in Chapter 16, we will learn about some recent attempts at developing novel forms of therapy that are designed to mobilize components of the pro-apoptotic machinery that remain intact in cancer cells. These therapies are crude beginnings, and we have only begun to recognize the difficulties and the opportunities that the apoptotic pathway provides to those who would like to cure the disparate collection of diseases that we call cancer. The hope is that at

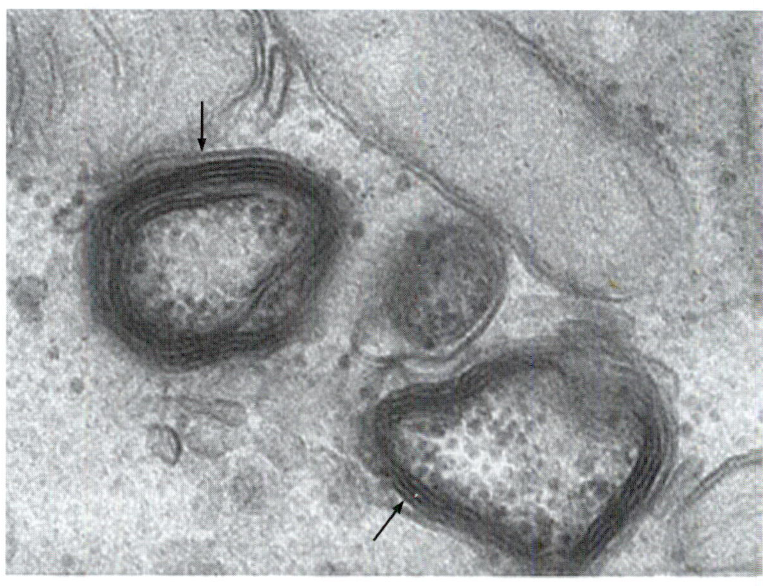

Figure 9.36 Autophagy This electron micrograph reveals the multilayered vesicles (*arrows*) that form in the cytoplasms of cells undergoing autophagy. Various cytoplasmic organelles become incorporated into lysosomes, where their contents are broken down to be used later to supply nutrients to the cell. (Courtesy of J. Debnath and J.S. Brugge.)

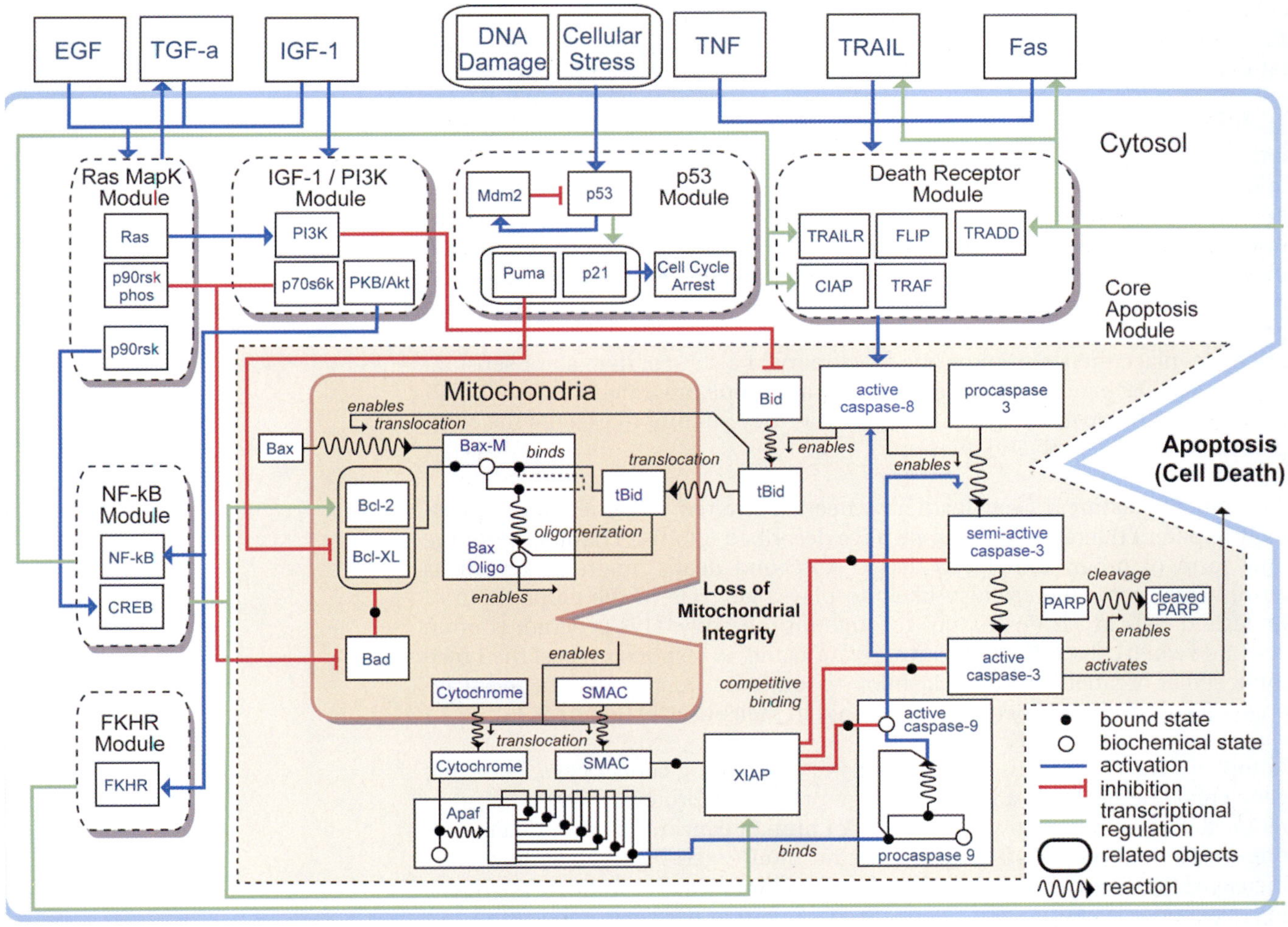

Figure 9.37 The apoptotic circuit board The full array of components governing apoptosis is not yet known. One of the goals of apoptosis research is to understand in a quantitative way how these components interact within a cell to influence the decision whether or not to initiate apoptosis. Such work should eventually yield mathematical models that predict, with some accuracy, how the apoptotic circuitry will respond to diverse extra- and intracellular signals. An initial attempt at modeling this signaling system is shown here. (Courtesy of I. Khalil and Gene Network Sciences, Inc.)

some time in the not-too-distant future, the circuitry governing apoptosis will be understood in a quantitative way (perhaps through a schematic diagram such as Figure 9.37), and that we will be able to predict its behavior with precision and manipulate it at will within cancer cells.

Key concepts

- Organisms attempt to block the development of cancer through the actions of the p53 alarm protein, which can cause cells to enter quiescence or apoptosis in the event that the machinery regulating cell proliferation is malfunctioning or the cell is exposed to various types of physiologic stress.
- p53 normally turns over rapidly. This turnover is blocked when a variety of signals indicate cell-physiologic stress or DNA damage.
- The mutant p53 found in human tumors usually carries amino acid substitutions. This allows the mutant p53 to interfere with wild-type p53, with which it can form tetrameric complexes.

- p53, which functions as a transcription factor, can impose cell cycle arrest through its ability to induce expression of $p21^{Cip1}$, and apoptosis through its ability to induce the expression of a number of pro-apoptotic proteins.
- A variety of cell-physiologic stresses can lead to the activation of p53; included among these are anoxia, damage to the genome, and signaling imbalances in the intracellular growth-regulating machinery.
- p53 becomes functionally activated when its normally rapid degradation is blocked. In addition, covalent modifications of the resulting accumulated p53 protein potentiate its activity as a transcription factor.
- p53 levels are controlled by two critical upstream regulators, Mdm2 and $p19^{ARF}$.
- Excessive activity of E2Fs, which is triggered by deregulation of the pRb pathway, results in activation of ARF and thus p53.
- Apoptosis is a complex cellular response program that can be activated by p53 as well as signals impinging on the cell from the outside.
- Apoptosis involves the activation of a cascade of caspases that result in the destruction of a cell, usually within an hour.
- Activation of the apoptotic caspase cascade can be triggered through the opening of a channel in the outer membrane of mitochondria, which releases several pro-apoptotic proteins, notably cytochrome *c*.
- Opening of the mitochondrial membrane channel is determined by the relative levels of Bcl-2–related anti-apoptotic and pro-apoptotic proteins.
- The apoptotic caspase cascade can be activated by cell surface death receptors as well.
- Loss of apoptotic functions allows cancer cells to survive a variety of cell-physiologic stresses, including anoxia, signaling imbalances, DNA damage, and loss of anchorage.
- Cancer cells invent numerous ways to inactivate the apoptotic machinery in order to survive and thrive. Included among these are the activation of Akt/PKB firing, the increase in the levels of anti-apoptotic Bcl-2-related proteins, the inactivation of p53 through changes in the *p53* gene or the upstream regulators of p53, methylation of the promoters of a variety of pro-apoptotic genes, interference with cytochrome *c* release from mitochondria, and inhibition of caspases.

Thought questions

1. In light of the fact that DNA tumor viruses must suppress the apoptosis of infected cells in order to multiply, what are the range of molecular strategies that are available to them to do so?
2. What types of factors influence the decision of p53 to act in a cytostatic versus a pro-apoptotic fashion?
3. How can anti-cancer therapeutics be successful in treating cancer cells that have inactivated components of their apoptotic machinery? Given the physiologic stresses that are known to activate p53-induced apoptosis, what types of anti-cancer therapeutic drugs might be created to treat cancer cells?
4. What side-effects would you predict could result from a general inhibition of apoptosis in all tissues of the body?
5. How might the loss of components of the apoptosis machinery render cancer cells more susceptible than normal cells to certain types of cell death?
6. Can you enumerate the range of apoptosis-inducing physiologic stresses that cancer cells must confront and circumvent during the course of tumor development?

Additional reading

Ashkenazi A & Dixit VM (1998) Death receptors: signaling and modulation. *Science* 281, 1305–1308.

Baehrecke EH (2002) How death shapes life during development. *Nat. Rev. Mol. Cell Biol.* 3, 779–787.

Bartek J & Lukas J (2003) Chk1 and Chk2 kinase in checkpoint control and cancer. *Cancer Cell* 3, 421–429.

Baserga R, Peruzzi F & Reiss K (2003) The IGF-1 receptor in cancer biology. *Int. J. Cancer* 107, 873–877.

Bursch W (2001) The autophagosomal-lysosomal compartment in programmed cell death. *Cell Death Differ.* 8, 569–581.

Calle EE & Kakas R (2004) Overweight, obesity, and cancer: epidemiological evidence and proposed mechanisms. *Nat. Rev. Cancer* 4, 579–591.

Calle EE, Rodriguez C, Walker-Thurmond K & Thun MJ (2003) Overweight, obesity, and mortality from cancer in a prospectively studied cohort of U.S. adults. *N. Engl. J. Med.* 348, 1625–1638.

Cory S & Adams JM (2002) The Bcl-2 family: regulators of the cellular life-or-death switch. *Nat. Rev. Cancer* 2, 647–656.

Danial NN & Korsmeyer SJ (2004) Cell death: critical control points. *Cell* 116, 205–219.

Downward J (1998) Ras signalling and apoptosis. *Curr. Opin. Genet. Dev.* 8, 49–54.

El-Deiry WS (2003) The role of p53 in chemosensitivity and radiosensitivity. *Oncogene* 22, 7486–7495.

Fürstenberger G & Senn H-G (2002) Insulin-like growth factors and cancer. *Lancet Oncol.* 3, 298–302.

Giaccia AJ & Kastan MB (1998) The complexity of p53 modulation: emerging patterns from divergent signals. *Genes Dev.* 12, 2973–2983.

Ginsberg D (2002) E2F1 pathways to apoptosis. *FEBS Lett.* 529, 122–125.

Giovannucci E (2003) Nutrition, insulin, insulin-like growth factors and cancer. *Horm. Metab. Res.* 35, 694–704.

Green DR & Evan GI (2002) A matter of life and death. *Cancer Cell* 1, 19–30.

Harris CC (1996) p53 tumor suppressor gene: from the basic research laboratory to the clinic—an abridged historical perspective. *Carcinogenesis* 17, 1187–1198.

Hengartner MO (2000) The biochemistry of apoptosis. *Nature* 407, 770–776.

Hursting SD et al. (2003) Calorie restriction, aging and cancer prevention: mechanisms of action and applicability to humans. *Annu. Rev. Med.* 54,131–152.

Igney FH & Krammer PH (2002) Death and anti-death: tumour resistance to apoptosis. *Nat. Rev. Cancer* 2, 277–288.

Karin M & Lin A (2002) NF-κB at the crossroads of life and death. *Nat. Immunol.* 3, 221–227.

Levine AJ (1997) p53, the cellular gatekeeper for growth and division. *Cell* 88, 323–331.

Levine B & Klionsky DJ (2004) Development by self-digestion: molecular mechanisms and biological functions of autophagy. *Dev. Cell* 6, 463–477.

Lowe SW, Cepero E & Evan G (2004) Intrinsic tumour suppression. *Nature* 342, 307–315.

Martinou JC & Green DR (2001) Breaking the mitochondrial barrier. *Nat. Rev. Mol. Cell Biol.* 2, 63–67.

Melino G, De Laurenzi V & Vousden KH (2002) p73: friend or foe in tumorigenesis. *Nat. Rev. Cancer* 2, 605–615.

Michael D & Oren M (2003) The p53-Mdm2 module and the ubiquitin system. *Semin. Cancer Biol.* 13, 49–59.

Nevins JR & Vogt PK (1996) Cell transformation by viruses. In Fields Virology, 3rd ed (BN Fields, DM Knipe, PM Howley eds), pp 301–343. Philadelphia: Lippincott–Raven.

Pollak MN, Schernhammer ES & Hankinson SE (2004) Insulin-like growth factors and neoplasia. *Nat. Rev. Cancer* 4, 505–518.

Reed JC (1999) Dysregulation of apoptosis in cancer. *J. Clin. Oncol.* 17, 2941–2953.

Reed JC (2000) Mechanisms of apoptosis. *Am. J. Pathol.* 157, 1415–1530.

Riddle DL and Gorski SM (2003) Shaping and stretching life by autophagy. *Dev. Cell* 5, 364–365.

Robles AI, Linke SP & Harris CC (2002) The p53 network in lung carcinogenesis. *Oncogene* 21, 6898–6907.

Salvesen GS & Duckett CS (2002) IAP proteins: blocking the road to death's door. *Nat. Rev. Mol. Cell Biol.* 3, 401–410.

Sharpless NE & DePinho RA (1999) The *Ink4A/Arf* locus and its two gene products. *Curr. Opin. Genet. Dev.* 9, 22–30.

Sherr CJ (2001) The *Ink4a/ARF* network in tumour suppression. *Nat. Rev. Mol. Cell Biol.* 2, 731–737.

Sherr CJ (2004) Principles of tumor suppression. *Cell* 116, 235–246.

Sherr CJ & McCormick F (2002) The RB and p53 pathways in cancer. *Cancer Cell* 2, 103–112.

Sherr CJ & Weber JD (2000) The ARF/p53 pathway. *Curr. Opin. Genet. Dev.* 10, 94–99.

Smith GD, Gunnell D & Holly J (2000) Cancer and insulin-like growth factor-I. *BMJ* 321, 847–848.

Soussi T & Béroud C (2001) Assessing *TP53* status in human tumors to evaluate clinical outcome. *Nat. Rev. Cancer* 1, 233–240.

Varley JM (2003) Germline TP53 mutations and Li-Fraumeni syndrome. *Hum. Mutat.* 21, 313–320.

Vaux DL, Cory S & Adams JM (1988) *Bcl-2* gene promotes haematopoietic cell survival and cooperates with c-*myc* to immortalize pre-B cells. *Nature* 335, 440–442.

Vogelstein B, Lane B & Levin AJ (2000) Surfing the p53 network. *Nature* 408, 307–310.

Vousden KH & Lu X (2002) Live or let die: the cell's response to p53. *Nat. Rev. Cancer* 2, 595–604.

Zahir N & Weaver VM (2004) Death in the third dimension: apoptosis deregulation and tissue architecture. *Curr. Opin. Genet. Dev.* 14, 71–80.

Zamzami N & Kroemer G (2001) The mitochondrion in apoptosis: how Pandora's box opens. *Nat. Rev. Mol. Cell Biol.* 2, 67–71.

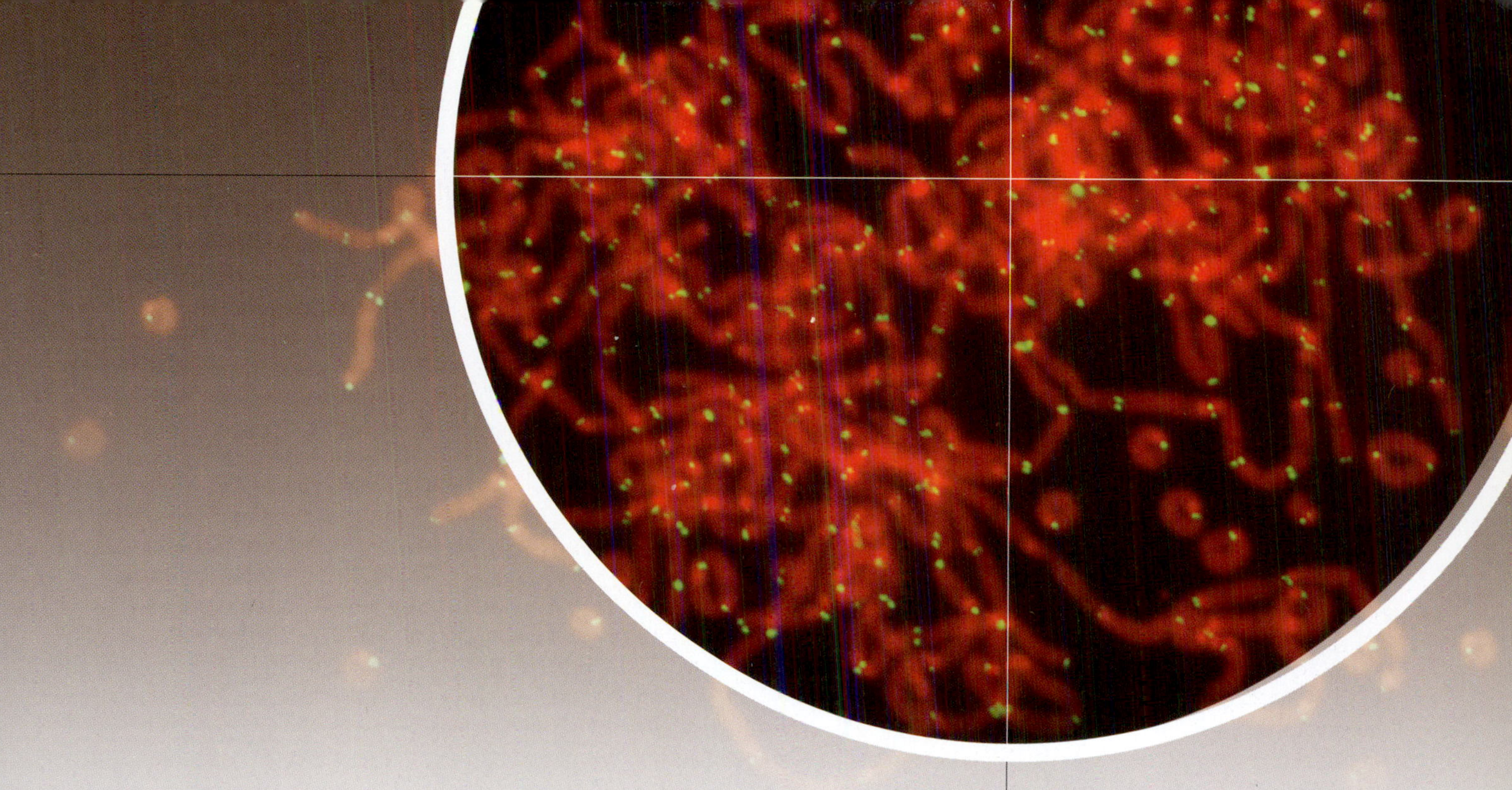

Chapter 10

Eternal Life: Cell Immortalization and Tumorigenesis

> Death takes place because a worn-out tissue cannot forever renew itself, and because a capacity for increase by means of cell division is not everlasting but finite.
>
> August Weissmann, biologist, 1881

In previous chapters, we read about a number of distinct traits displayed by cancer cells. In some instances, these traits are acquired through the actions of activated oncogenes; in others, cancer cell–specific traits can be traced back to the loss of tumor suppressor genes. As we will discuss in Chapter 11, the acquisition by human cells of these neoplastic traits (and thus the development of a clinically apparent human tumor) usually requires several decades' time.

During this extended period of development, populations of human cells pass through a long succession of growth-and-division cycles as they evolve toward the neoplastic growth state. Such extensive proliferation, however, conflicts with a fundamental property of normal human cells: they are endowed with an ability to replicate only a far smaller number of times. Once normal human cell populations have exhausted their allotment of allowed doublings, the cells in these populations cease proliferating and may even enter apoptosis.

These facts lead us to a simple, inescapable conclusion: in order to form tumors, incipient cancer cells must breach the barrier that normally limits their proliferative potential. Somehow, they must acquire the ability to multiply for

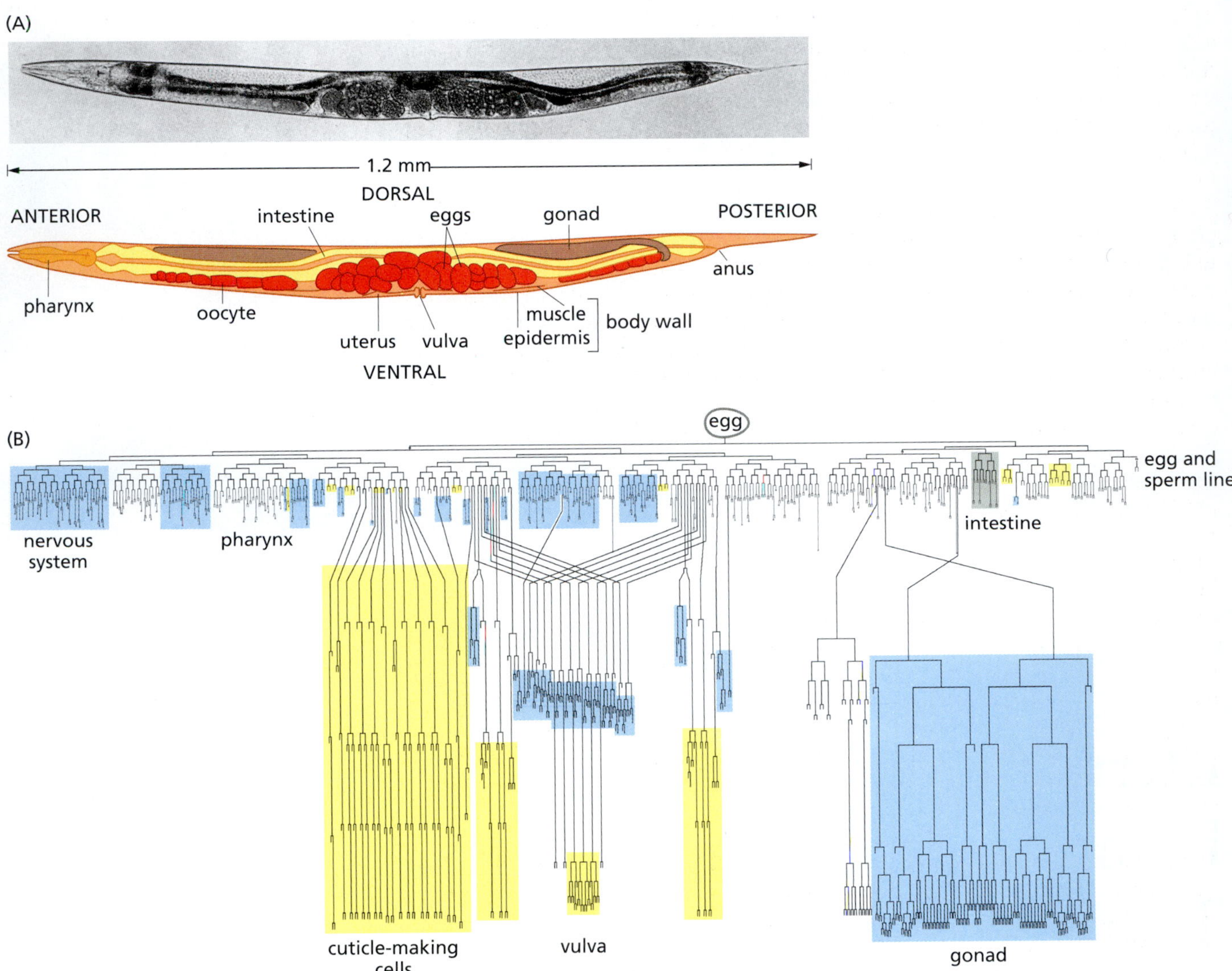

Figure 10.1 The pedigree of cells in the body of a worm (A) The adult worm *Caenorhabditis elegans* is composed of 959 somatic cells, and the lines of descent of all cells in this worm have been traced with precision back to the fertilized egg. (B) As is apparent, many of the cell lineages end abruptly because cells stop proliferating or are discarded (usually by apoptosis). The cells in each of the major branches are destined to form a specific tissue. For example, cells in the major, multigenerational branch on the right are destined to form the gonads, while those in a small branch at the far left will form the nervous system. (A, from J.E. Sulston and H.R. Horvitz, *Dev. Biol.* 56:110–156, 1977. © Academic Press; B, University of Texas Southwestern Medical Center.)

an abnormally large number of growth-and-division cycles, so that they can successfully complete the multiple steps of tumor development.

In this chapter, we explore the nature of the regulatory machinery that limits cell proliferation and how it must be neutralized in order for cells to become fully neoplastic and form clinically detectable tumors. By neutralizing this machinery, cells gain the ability to proliferate indefinitely—the phenotype of cell immortality. This immortality is a critical component of the neoplastic growth program.

10.1 Normal cell populations register the number of cell generations separating them from their ancestors in the early embryo

In multicellular (metazoan) animals such as ourselves, the origin of each cell can, in principle, be traced through multiple cell generations back to a single ancestor—the fertilized egg. Looking in the other direction, the sequence of cell divisions that stretches from an ancestral cell that existed in the embryo to a descendant cell that exists many cell generations later is often termed a **cell lineage**. Indeed, in a relatively simple metazoan—the worm *Caenorhabditis elegans*—the lineage of all 959 somatic cells in the adult body has been traced and can be depicted as a pedigree (Figure 10.1).

In large, complex mammals, however, the assembling of a comparable pedigree will never happen, because the total number of cell divisions is astronomical: the adult human body, for example, comprises almost 10^{14} cells, and the organism as a whole undergoes as many as 10^{16} cell divisions in a lifetime. Still, we can imagine that for each human body such a cell pedigree must exist, if only as a theoretical construct.

As is the case with *C. elegans*, during the course of human development, early embryonic cells become the founders of specific cell lineages that are committed to assuming various tissue-specific cellular phenotypes (Section 8.11). Indeed, the science of developmental biology focuses much of its attention on how individual cells in various cell lineages acquire the information from their surroundings that causes them to enter into one or another program of differentiation. However, developmental biologists do not address a question of great relevance to tumor development: are there specific controls that determine the number of cell generations through which a particular cell lineage can pass during the lifetime of an organism? Can each branch and twig of the cell pedigree grow indefinitely, or is the number of replicative generations in each cell lineage predetermined and limited?

Currently available techniques do not allow us to determine with any accuracy how many times specific cell lineages within the human or mouse body pass through successive growth-and-division cycles. Still, a crude measure of the replicative capacity of a cell lineage can be undertaken by culturing cells of interest *in vitro*. For example, one can prepare fibroblasts from living tissue, introduce them into a Petri dish, and determine how many times these cells will double. (In practice, such experiments require **serial passaging**, in which a portion of the cells that have filled one dish are removed and introduced into a second dish and allowed to proliferate, after which some of their number are introduced into a third dish, and so forth.)

As first demonstrated in the early 1960s, cells taken from rodent or human embryos exhibit a limited number of replicative cycles in culture. The work of Leonard Hayflick showed that cells would stop growing after a certain, apparently predetermined number of divisions and enter into the state that came to be called **replicative senescence** or simply **senescence** (Figure 10.2). Senescent cells remain metabolically active but seem to have lost irreversibly the ability to re-enter into the active cell cycle. Such cells will spread out in monolayer culture, acquire a large cytoplasm, and persist for weeks if not months, as long as they are given adequate nutrients and growth factors; such cells are often described as taking on the appearance of a fried egg (Figure 10.3). The growth factors help to sustain the viability of the senescent cells, but they are unable to elicit the usual proliferative response observed when these factors are applied to

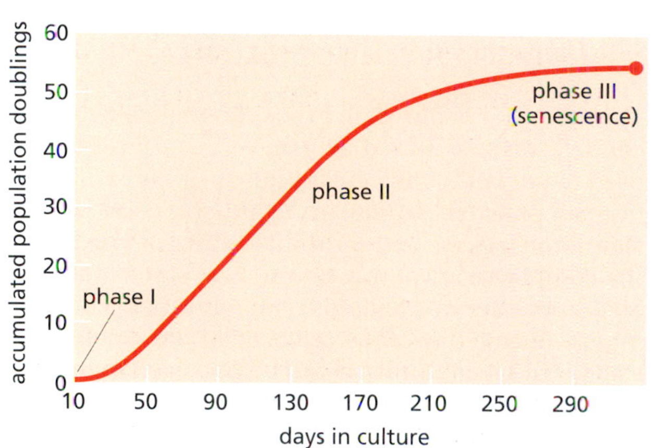

Figure 10.2 The proliferative capacity of cells passaged extensively in culture The ability of human fibroblasts to proliferate in culture was gauged in Leonard Hayflick's work by counting the number of times that the population of cells had doubled *(ordinate)*. As is apparent, these cells, beginning soon after explantation from living tissue into culture (phase I), were able to proliferate robustly for about 60 doublings (phase II) before entering into senescence (phase III), in which state they could remain viable but nonproliferating for as long as a year. (From J.W. Shay and W.E. Wright, *Nat. Rev. Mol. Cell Biol.* 1:72–76, 2000.)

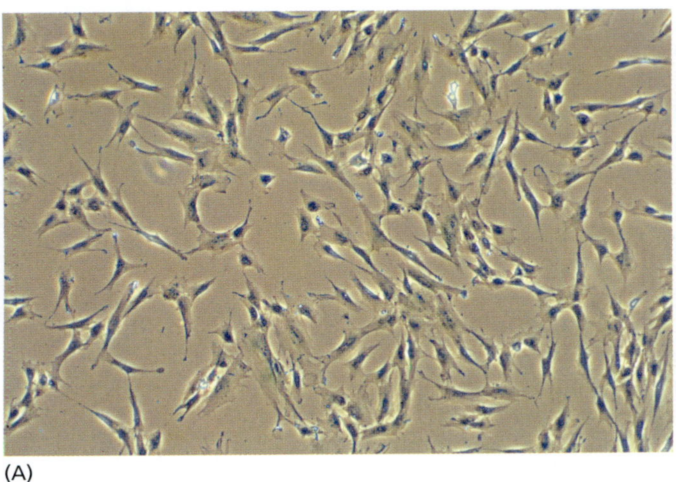

(A)

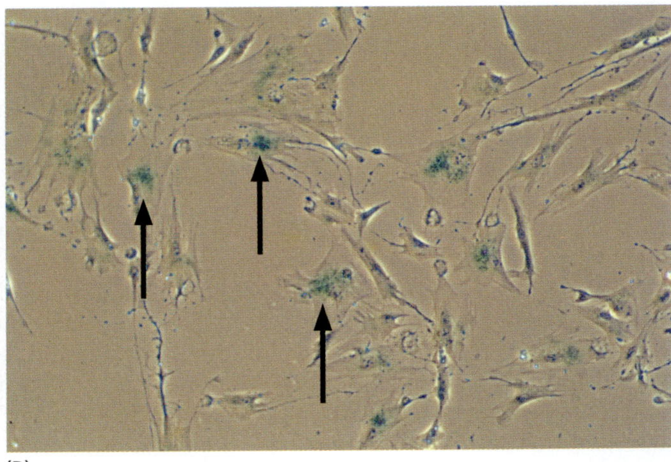

(B)

Figure 10.3 Senescent cells *in vitro* and *in vivo* (A) When viewed with the phase-contrast microscope, pre-senescent human fibroblasts are still vigorously growing and retain the appearance of early-passage cells. (B) However, once cells enter into senescence, they cease proliferating but remain viable. Many of their number develop extremely large cytoplasms, giving them a "fried egg" appearance. (These two micrographs were produced at the same magnification.) In addition, senescent cells characteristically express the senescence-associated, acidic β-galactosidase enzyme, which can be detected by supplying them with a substrate that turns *blue* upon cleavage by this enzyme *(arrows)*; in contrast, early-passage cells seen in (A) show very faint staining. (Courtesy of C. Scheel.)

healthy, nonsenescent cells. Like actively proliferating cells, the senescent cells display growth factor receptors, but the downstream signaling pathways have been inactivated through still poorly understood mechanisms.

The precise number of replicative doublings exhibited by cultured cells before they reach senescence is dependent on the species from which the cells were prepared, on the tissue of origin, and on the age of the donor organism. Some experiments with human cells indicate that cells prepared from embryos or newborns are able to double in culture a larger number of times (e.g., 50 to 60 population doublings, or PDs) than comparable cells taken from middle-aged or elderly adults (Figure 10.4). Such behavior suggests, but hardly proves (Sidebar 10.1), that cells from older individuals have already used up part of their allotment of replicative doublings prior to being introduced into tissue culture.

A contrasting behavior is shown by embryonal stem (ES) cells, which are prepared from very early embryos and retain the ability, under the proper conditions, to seed all the differentiated lineages in the body (Sidebar 8.1). When provided with the proper nutrients, these cells show unlimited replicative potential in culture and are thus said to be **immortal**. (The term is a bit misleading, since it is really a *lineage* of ES cells that is immortal rather than individual ES cells.)

Taken together, these various observations convey the notion that very early in embryogenesis, cells have an unlimited replicative capacity. However, as specific lineages of cells in the organism (e.g., dermal fibroblasts, neurons, mammary epithelial cells) are formed, each seems to be allocated a predetermined number of postembryonic doublings. The replicative behavior of cancer cells resembles, at least superficially, that of ES cells. When many types of cancer cells are propagated in culture, they seem able to proliferate forever when provided with proper *in vitro* culture conditions.

This behavior is illustrated most dramatically by HeLa cells. Over the past half century, these cultured cells have been the human cell type most frequently used to study the molecular biology of human cells. They were derived in 1951 from an unusual, particularly aggressive cervical adenocarcinoma discovered in Henrietta Lacks, a young woman in Baltimore, Maryland, who soon died from the complications of this tumor. Ever since that time, these cells have proliferated in culture in hundreds of laboratories across the world, dividing approximately once a day. HeLa cells constitute a **cell line**, in that they have become established in culture and can be passaged indefinitely, in contrast to many cell populations that have a limited replicative ability after being removed from living tissue.

(A)

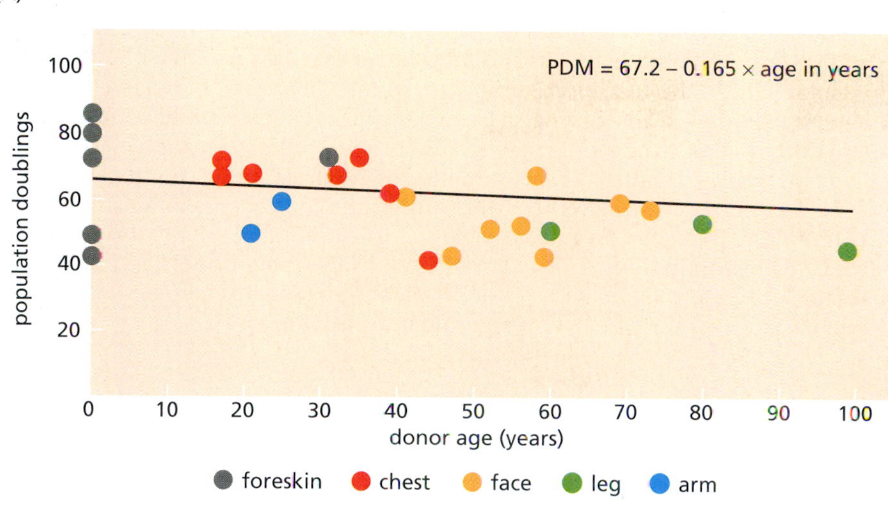

(B)

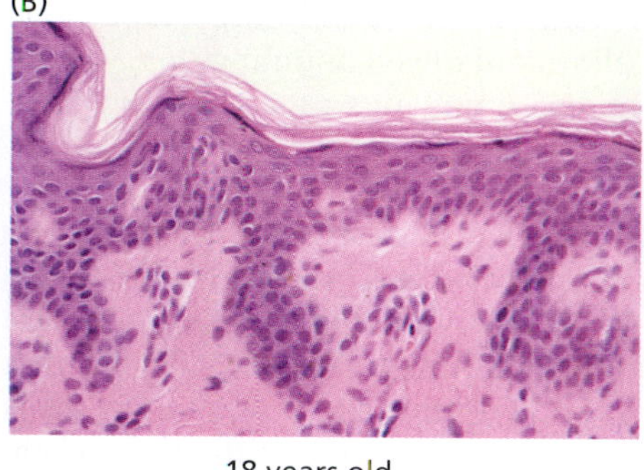

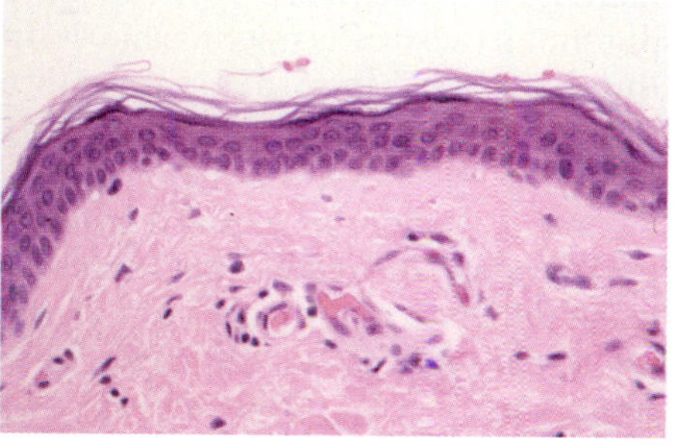

Figure 10.4 Loss of proliferative capacity with age (A) The proliferative capacity of cells can be gauged by determining the number of times cultures of such cells will double in cell number *in vitro* (mean population doublings, PDM). Here, doublings of dermal fibroblasts prepared from various anatomical sites have been measured as a function of the age of the donors. (Because of inter-individual variability in cell proliferative capacity, data points from such measurements show substantial scatter around the mean at each age.) (B) As the keratinocyte stem cells in the skin lose proliferative capacity with increasing age, the overall ability of the skin to regenerate itself declines, leading to a thinning of the keratinocyte layer of the skin *(dark pink)* and a loss of the ridge architecture seen in the younger skin. The sun-protected skin of an 18-year-old female *(left)* is compared here with that of a 76-year-old female *(right)*. (A, courtesy of J.G. Rheinwald and T.M. O'Connell-Willstaedt; B, courtesy of T. Brenn.)

10.2 Cancer cells need to become immortal in order to form tumors

The observation that cancer cells, once adapted to growth in tissue culture (Sidebar 10.2), are often found to be immortal strongly suggests that immortalization is an integral component of the cancer cells' transformation to a neoplastic growth state, that is, that cancer cells are immortal because they need to be if they are to succeed in forming a tumor.

Why, conversely, do lineages of normal cells lack immortalized growth properties? Perhaps the body endows its normal cells with only a limited number of

Sidebar 10.1 Why do cultured cells from older people proliferate less than those from younger people? The simplest explanation for the observation (see Figure 10.4) that cells from the elderly stop proliferating *in vitro* sooner than the cells of juveniles is that the older cell populations have already passed through more cell division cycles *in vivo* and thereby have exhausted a portion of their allowed doublings. In truth, other explanations are equally plausible. For example, nondividing cells within a living tissue may sustain damage over an extended period of time; this accumulated damage may be caused, for example, by long-term exposure to reactive oxygen species (ROS), which are capable of damaging many cellular components. In this case, the subsequently observed loss of proliferative capacity *in vitro* may be proportional to elapsed time since these cells were first formed and may be unconnected with the number of cell generations that their ancestors passed through during the lifetime of the organism.

Sidebar 10.2 Cancer cells often fail to adapt to culture Cells that are extracted from a variety of normal tissues and introduced into tissue culture often do not adapt readily to the conditions of *in vitro* culture. This applies to many kinds of cancer cells as well. Like normal cells, most types of cancer cells depend on several other cell types to support their viability and proliferation, as we will learn in great detail in Chapter 13. This complicates attempts at determining whether certain populations of cancer cells are immortalized, since the initial attempts to propagate these cells in culture usually fail. Consequently, conclusions about cancer cell immortality are based on the relatively small proportion of human tumors carrying cells that have readily adapted to *in vitro* culture conditions and have been found, thereafter, to proliferate indefinitely.

replicative generations as an anti-cancer defense mechanism. For example, if one or another cell in the body were to accidentally acquire certain oncogenes and shed critical tumor suppressor genes, its descendants would begin to proliferate uncontrollably, and the population of these tumor cells might well increase exponentially. However, if endowed with only a limited replicative potential, these cells might exhaust their allotment of cell doublings long before they succeeded in forming a life-threatening tumor mass; as a consequence, tumor development would grind to a halt.

The credibility of this model depends on some critical numbers. Specifically, we need to know how many successive cell generations are required to make a clinically detectable human tumor (Figure 10.5A) and how many generations are granted to normal cell lineages throughout the body. We know that human tumors are clonal, in the sense that all the neoplastic cells in the tumor mass descend from a common ancestral cell that underwent transformation at one point in time (Section 2.5). With this fact in mind, we can ask how many cell generations separate the tumor cells in a large human tumor from their common progenitor.

The arithmetic works out like this. The volume of a cubic centimeter (cm^3) within a tumor cell mass contains about 10^9 cells, and a life-threatening tumor has a size, say, of 10^3 cm^3. We can calculate that these 10^{12} cells seem to have arisen following 40 cycles of exponential growth and division (Figure 10.5B)—that is, 40 cell generations separate the founding cell from its descendants in the end-stage, highly aggressive tumor. ($10^3 \cong 2^{10}$; hence $10^{12} \cong 2^{40}$).

However, as mentioned above, some types of normal human cells are known to pass through 50 or 60 cycles of growth and division in culture before they become senescent and stop growing. According to the arithmetic above, these 50 or 60 cell generations of exponential growth are far more than are required in order for a founding cell to spawn enough descendants to constitute a life-threatening tumor mass. Indeed, 60 cell doublings is enough to create a tumor mass of about 10^{18} cells $\cong 10^9$ cm^3 $\cong 10^6$ kilograms. Something is drastically wrong with these numbers!

The error in our calculations lies in a flawed premise: they assume an exponential expansion of populations of cancer cells (Figure 10.5A and B). The biological reality of tumor growth is much different. Thus, a number of defense mechanisms built into the body's tissues make life very difficult for incipient cancer cells, indeed, so difficult that in each cell generation, a significant number of these cells die off (Figure 10.5C). Early in tumor development, the defense mechanisms deployed by the tissue include depriving tumor cells of growth factors, of adequate oxygen, and of the ability to eliminate metabolic wastes via the vasculature. Moreover, a number of the anti-tumor cell defense mechanisms in the hard-wired regulatory circuitry of cells operate to weed out aberrantly behaving, pre-malignant cells (Chapter 9). As a consequence of the resulting

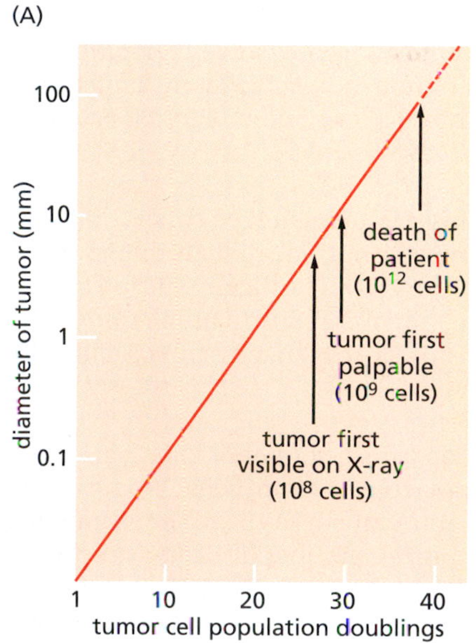

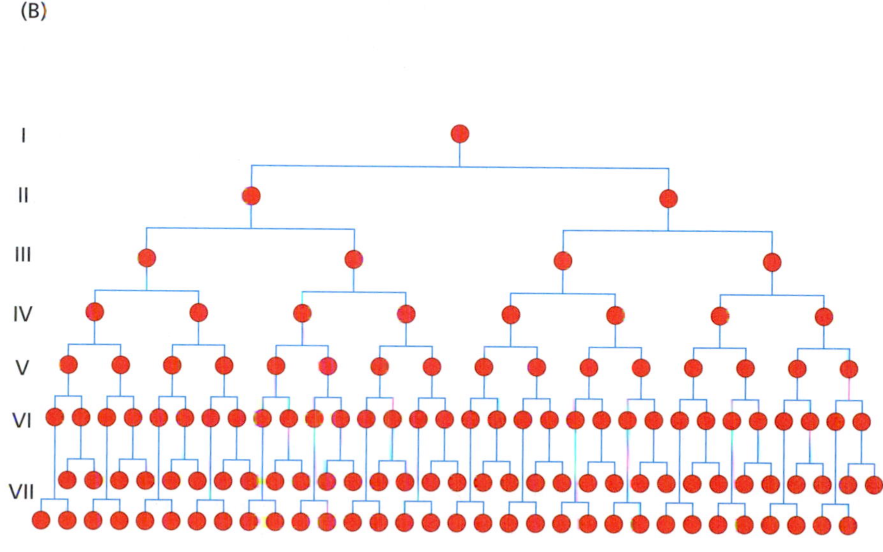

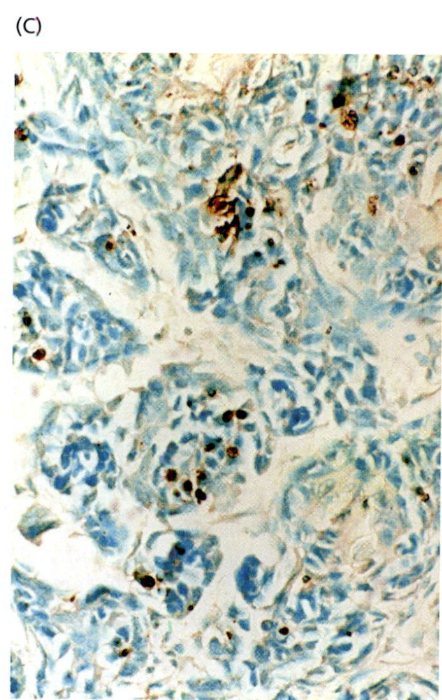

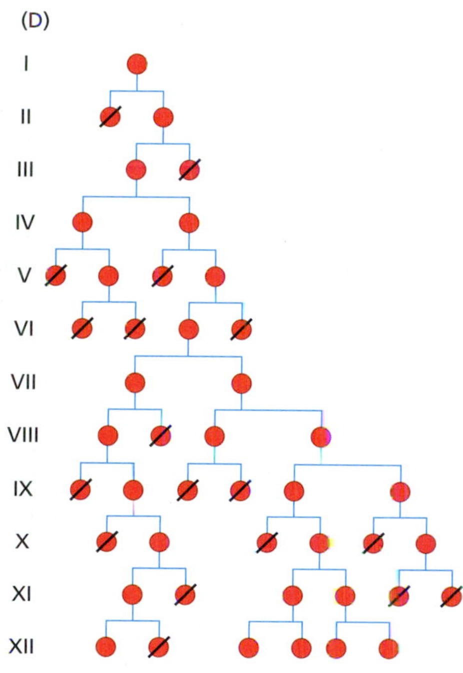

Figure 10.5 Generations of cells forming a tumor mass (A) The growth of a tumor can be related to its size when it is first detectable by X-rays, when it becomes palpable, and when it reaches a life-threatening size. (B) In principle, each time the cells in a tumor divide, this cell population, as a whole, should double in size. (Only six populations are shown here.) Consequently, the number of successive population doublings theoretically required to generate a lethal tumor from a single founding ancestral cell can be calculated from the graph in panel A to be about to be about 40. (C) In reality, however, cell populations that are evolving toward the neoplastic state and those that are already neoplastic experience substantial attrition during each cell generation. The TUNEL stain *(dark brown spots*; see Figure 3B) is used here to detect apoptotic cells in a mouse mammary tumor that arose in a mouse carrying an oncogenic *Wnt-1* transgene in its germ line. In fact, this assay greatly underestimates the rate of apoptosis in each cell generation, since apoptotic cells persist only for an hour before they are consumed by neighboring cells and by macrophages. (D) The high rate of attrition leads to loss of many cells in each cell generation *(diagonal slashes)*. Accordingly, an ancestral founder cell that should have generated 2^{11} descendants leaves instead only five descendants in the 12th generation. The number of successive cell generations required to generate a tumor of life-threatening size is therefore far larger and, in the absence of precise knowledge of attrition rates, incalculable. (A, from B. Alberts et al., Molecular Biology of the Cell, 4th ed. New York: Garland Science, 2002; C, courtesy of L.D. Attardi and T. Jacks.)

attrition in each cell generation, the pedigree of cells in a tumor mass actually looks quite different (Figure 10.5D). Many branches of the tree are continually being pruned by the high death rate of tumor cells in each generation.

This ongoing attrition means that the number of cell generations required to form a tumor mass of a given size is far greater than would be predicted by simple exponential growth kinetics. For example, a clonal population of 10^3 tumor cells might be thought, on the basis of its size, to have gone through 10 cycles of exponential growth and division since its founding by an ancestral cell; in reality, 20 or 30 or more cell generations may have been required to accumulate this many cells.

These revised calculations provide credible support for the notion that the human body endows its cell lineages with only a limited number of growth-and-division cycles in order to protect itself against the development of tumors. For example, if cell populations must pass through 100 replicative generations in order to form a clinically detectable human tumor, it is likely that most incipient tumor cell populations will use up their normal allotment of 50 or 60 cell divisions long before they succeed in creating such a mass.

Having accepted, at least for the moment, this argument and its conclusions, we are now left with some major puzzles: How can normal cells throughout the body possibly remember their replicative history? And how can aspiring cancer cells erase the memory of this history and acquire the ability to proliferate indefinitely, or at least as long as is required to form a macroscopic, clinically detectable tumor?

A solution to the problem of replicative history must, sooner or later, be spelled out in terms of the actions of specific molecules within cells. Whatever its nature, the solution is hardly an obvious one. Organisms as complex as humans possess no innate biological clock for counting the number of (organismic) generations that separate each of us from ancestors who lived 100 or 1000 years ago. When we humans wish to learn the number of these generations over extended periods of time, we usually hire genealogists to chart them for us. No biological counting device inside our bodies can provide us with such answers. How, then, can far simpler biological entities—individual human cells—keep track of their generations?

In addition, this generational counting mechanism is likely to be, in the language of developmental biologists, **cell-autonomous**; that is, it must be intrinsic to a cell and not influenced by the ongoing interactions of the cell with its neighbors and with the body as a whole. Recall the contrasting situation of oncogenes and tumor suppressor genes. The products of most of these genes perturb the pathways responsible for processing the signals received by cells from their surroundings—non-cell-autonomous processes. Finally, the hypothesized generation-counting device must be relatively stable biochemically, because it needs to store a record of the past history of a cell lineage in a form that survives over extended periods of time, often many decades.

In principle, the counting processes of this "generational clock" (which measures elapsed cell generations rather than elapsed time) might depend on the concentrations of some soluble intracellular molecule that (1) is synthesized early in development and not thereafter, (2) is present in high concentrations in the embryonic cells that are the ancestors to various lineages, and (3) undergoes progressive dilution by a factor of 2 each time a cell in the lineage divides. Senescence might then be triggered after the levels of this compound fall below some threshold level. Though plausible to mathematicians, this arrangement cannot work in the real world of biology. Biochemistry dictates that no molecule can be present in a cell over a concentration range of 2^{50} or 2^{60}.

This realization forced researchers to look elsewhere for the molecular embodiments of the generational clock. They came upon two regulatory mechanisms that govern the replicative capacity of cells growing *in vitro* and possibly *in vivo*. The first of these appears to measure the cumulative physiologic stress that lineages of cells experience over extended periods of time and halts further proliferation once that damage exceeds a certain threshold; this causes cells to enter into the state, described above, termed senescence.

The second regulator measures how many replicative generations a cell lineage has passed through and sounds an even more drastic alarm once an allowed quota of replicative doublings has been used up; this leads to the state termed **crisis**, which results in the apoptotic death of most cells in a population. We will address the mechanisms inducing senescence first and later describe those governing crisis.

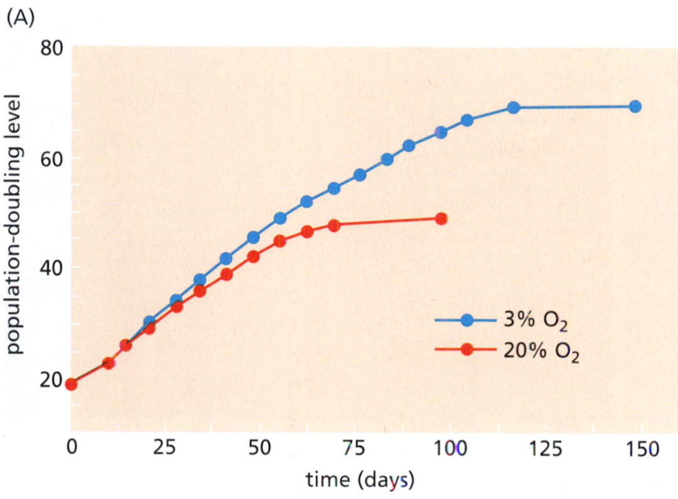

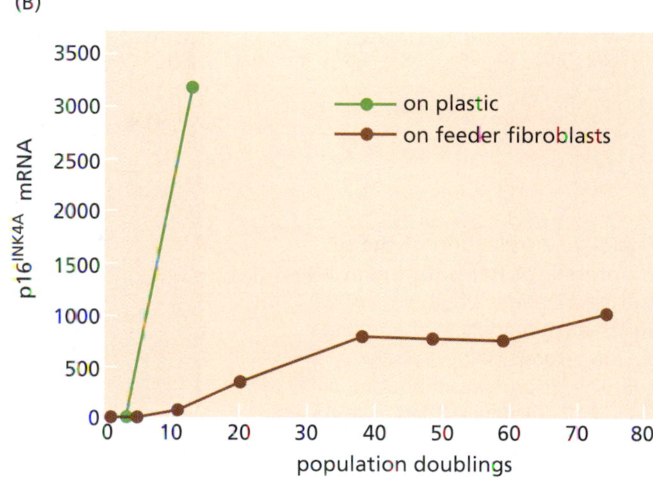

Figure 10.6 Influence of culture conditions on the onset of senescence (A) The proliferative capacity of cell populations (population doubling level, *ordinate*) has been measured here when they are exposed in culture to either 3% or 20% oxygen. The 3% oxygen concentration is far more physiologic, since it reflects the oxygen concentration that cells actually experience in living tissues. (B) The dependence of normal epithelial cells on biological support from the stroma creates additional stresses when such cells are propagated *in vitro*. When grown as pure cultures, human foreskin keratinocytes rapidly induce expression of the $p16^{INK4A}$ tumor suppressor protein (*green line*; see Section 8.4), which soon imposes growth arrest and senescence on these cells. However, when these epithelial cells are grown above a feeder layer of fibroblasts (*brown line*), they can proliferate for an extended period of time without experiencing a strong induction of $p16^{INK4A}$ and resulting senescence. (A, from Q. Chen, A. Fischer, J.D. Reagan et al., *Proc. Natl. Acad. Sci. USA* 92:4337–4341, 1995; B, from B. Fu, J. Quintero, and C.C. Baker, *Cancer Res.* 63:7815–7824, 2003.)

10.3 Cell-physiologic stresses impose a limitation on replication

Evidence pointing to the causes of replicative senescence comes from experiments in which the conditions of *in vitro* culture are varied. The most dramatic observations derive from experiments where the oxygen tension to which cultured cells are exposed has been reduced from 20% to 1 to 3%; this results in substantial increase in the replicative life span *in vitro*. In one set of experiments, populations of human diploid fibroblasts had a more than 20% longer life span (i.e., went through 20% more doublings) in culture when grown in 1% oxygen than in 20% oxygen; in another, cells went through 50% more doublings when cultured in 3% oxygen rather than 20% oxygen (Figure 10.6A). The lower oxygen tensions more closely reflect the oxygen tensions that most cells experience in living tissues, rather than the tension that is most conveniently provided to cells growing in a tissue culture incubator. In addition, in the case of epithelial cells, the presence of a stromal feeder layer has profound effects on the ability of these cells to survive and proliferate in culture (Figure 10.6B).

The contribution of cumulative oxidative damage to senescence is suggested by other types of evidence as well. For example, senescent cultured cells have been found to produce oxidized forms of guanine at four times the rate of young cells, suggesting a progressive breakdown of mitochondrial function during *in vitro* culture, a resulting leakage of reactive oxygen species (ROS) from the mitochondria, and the accumulation of large numbers of oxidation-induced lesions in these cells. Many of the lesions that induce senescence are found in the chromosomal DNA of cells.

Provocatively, the levels of two important growth-suppressing CDK inhibitors—$p16^{INK4A}$ and $p21^{Cip1}$—increase progressively in a variety of human cell types during extended culture *in vitro* (e.g., Figure 10.7A). Moreover, variants of human keratinocytes that succeed in escaping from early senescence *in vitro* often are found to carry inactivated copies of the gene encoding the $p16^{INK4A}$ tumor suppressor protein. As we read earlier, once present above a certain level, $p16^{INK4A}$ blocks phosphorylation of pRb by D-type cyclins and CDK4/6, and the resulting hypophosphorylated pRb can act to halt further proliferation (Section 8.4). Similarly, $p21^{Cip1}$ can halt cell cycle progression but has a more wide-ranging action, since it can block signaling by all of the cyclin-dependent kinases involved in progression through the late G_1, S, G_2, and M phases of the cell cycle.

The ectopic expression of $p16^{INK4A}$ (Figure 10.7B) in cells causes them to develop many of the attributes of replicative senescence. These and other observations indicate that cells respond to the stresses of *in vitro* culture by inducing

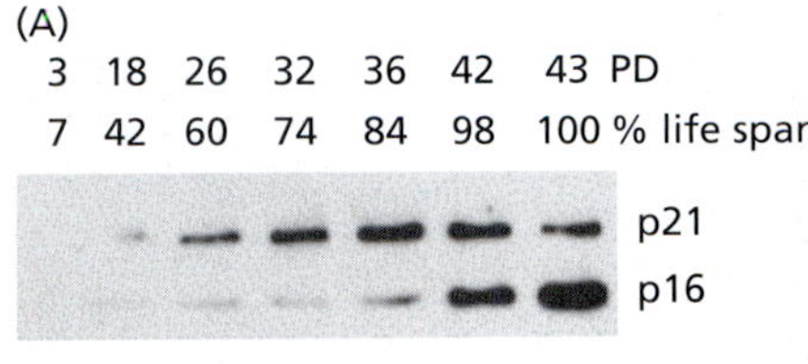

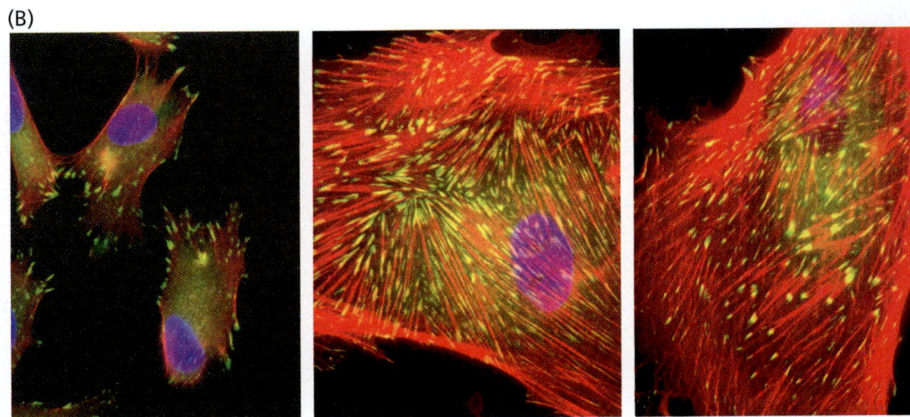

Figure 10.7 Induction of tumor suppressor proteins during *in vitro* culture (A) When adult human endometrial fibroblasts are propagated *in vitro,* expression of both the $p16^{INK4A}$ and $p21^{Cip1}$ proteins is induced, albeit with distinct kinetics. These cells have an expected life span of 40 to 43 population doublings (PD) in culture, whereupon they enter into senescence. (B) Normal cells *(left)* are relatively small and have few focal contacts with the substrate *(yellow)* and relatively few actin stress fibers *(orange),* in contrast to senescent cells *(middle and right).* Ectopic expression of $p16^{INK4A}$ is able to induce a cell phenotype *(middle)* that is indistinguishable from that of cells that have entered replicative senescence *(right).* This effect may depend on its ability to inhibit CDK4 and CDK6, thereby blocking phosphorylation of pRb. (A, from S. Brookes, J. Rowe, A. Gutierrez del Arroyo et al., *Exp. Cell Res.* 298:549–559, 2004; B, courtesy of I. Ben-Porath.)

expression of key CDK inhibitors; once induced, these proteins force cells to assume many of the attributes associated with the senescent state. Significantly, the early senescence of cultured human keratinocytes can be circumvented by ectopic expression in these cells of high levels of CDK4; this cyclin-dependent kinase serves to sequester $p16^{INK4A}$ (see Figure 8.13), thereby preventing it from halting cell proliferation and inducing senescence.

As indicated in Figure 10.7A, the induction of $p21^{Cip1}$ expression can also occur as cells approach senescence. In such senescing cells, the induction of $p21^{Cip1}$ is mediated by increased levels and activity of the p53 tumor suppressor protein. In some cells, such as mouse embryo fibroblasts (MEFs), the levels of p53 increase 10- to 40-fold as cells approach senescence. Once induced, p53 seems to act in a cytostatic fashion (via $p21^{Cip1}$) to arrest further cell proliferation; like $p16^{INK4A}$ (see Figure 10.7B), ectopic expression of $p21^{Cip1}$ has been reported to induce the senescent cell phenotype.

We can imagine that the environment experienced by pre-neoplastic and neoplastic cells within tissues forces them to adapt to a variety of physiologic stresses similar to those experienced by cells in culture. Indeed, the same signaling pathways—pRb and p53—that must be inactivated in order to avoid senescence *in vitro* (Figure 10.8) are also found to be inactivated in the great majority

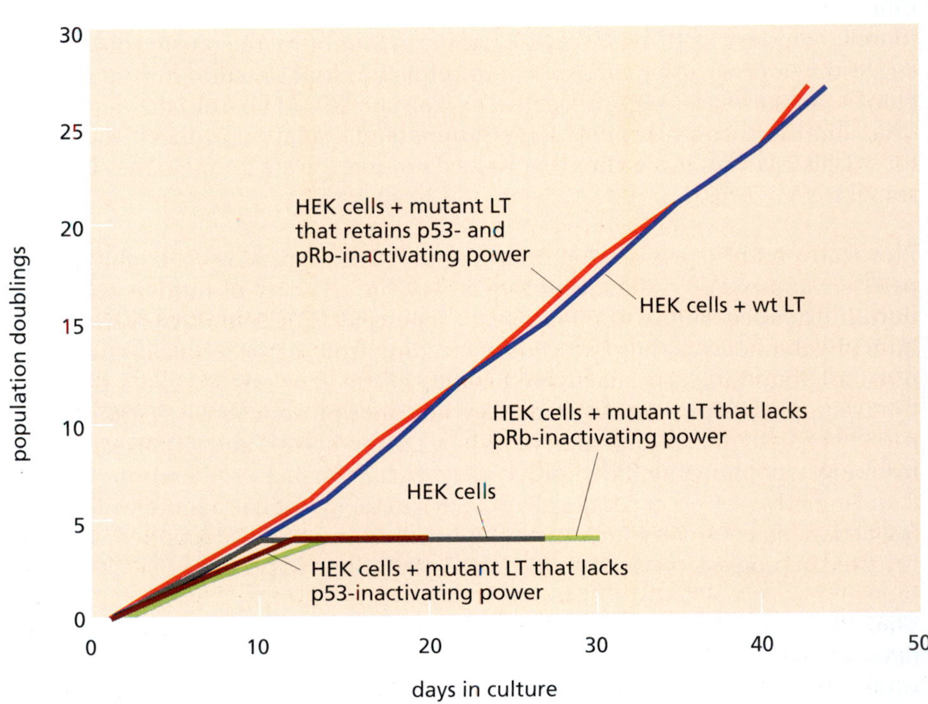

Figure 10.8 Role of large T antigen in circumventing senescence Experiments with a variety of human cell types indicate that inactivation of both pRb and p53 is needed to ensure that these cells do not senesce in culture. This can be achieved through the expression of the SV40 large T antigen (LT) in these cells. As seen here, human embryonic kidney (HEK) cells senesce after 10 to 12 population doublings (PDs) in culture *(gray line).* However, if they express either the wild-type LT protein *(dark blue line)* or a mutant of LT that retains the ability to bind and sequester both p53 and pRb *(red line),* they circumvent senescence and continue to proliferate for an extended period of time. Mutants of LT that have lost the ability to sequester either p53 *(brown line)* or pRb *(light green line)* fail to prevent senescence. (From W.C. Hahn, S.K. Dessain, M.W. Brooks et al., *Mol. Cell. Biol.* 22:2111–2123, 2002.)

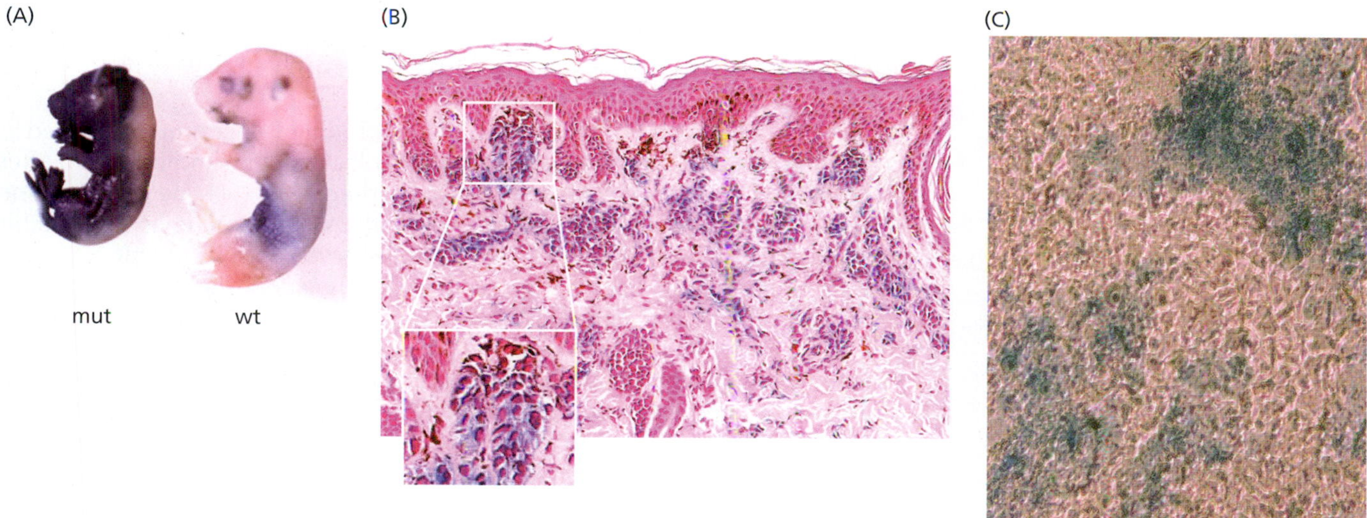

Figure 10.9 Evidence of senescent cells in living tissues The existence of senescent cells *in vivo* is suggested by several lines of evidence. (A) Mouse embryos have been produced that are homozygous mutants at the *Brca1* locus, which is involved in maintaining genomic integrity (Section 12.10). As a consequence, cells throughout these embryos (mut, *left*) suffer extensive genomic damage, causing their cells to register as senescent *(blue)* using the assay for the senescence-associated β-galactosidase (SA-β-gal) enzyme (see Figure 10.3), in contrast to wild-type embryos (wt, *right*), where such senescent cells are not apparent. (B) The presence of senescent human melanocytes within dysplastic nevi (benign precursors of melanomas) can also be demonstrated by staining for SA-β-gal *(blue; see inset)*. (C) Treatment of tumors with chemotherapeutic drugs also appears to induce senescence in tumor cells, as gauged by blue staining with the SA-β-gal senescence-specific stain. Seen here is a portion of a lung carcinoma from a patient who had been treated with the drugs carboplatin and taxol prior to surgical excision of the tumor. Normal lung tissue showed scant traces of SA-β-gal–staining cells. (A, from L. Cao , W. Li, S. Kim, S.G.Brodie and C.-X.Deng, *Genes Dev.* 17:201–213, 2003; B, courtesy of D. Peeper; C, from R.S. Roberson, S.J. Kussick, E. Vallieres et al., *Cancer Res.* 65:2795–2803, 2005.)

of human tumors, as we read in Chapters 8 and 9. Consequently, when cancer cells are extracted from tumors and placed into tissue culture, the previous inactivation of their *Rb* and *p53* pathways may help these cells to resist many of the stresses imposed by *in vitro* culture conditions and thereby to avoid senescence.

In line with this thinking is the observation that the senescence of cultured, early-passage human fibroblasts and epithelial cells can be avoided if the large T oncoprotein of SV40 is expressed in these cells (see Figure 10.8). By sequestering both the p53 and pRb tumor suppressor proteins (see Sidebar 12), the large T protein is able to neutralize the key cellular regulatory circuits that impose senescence in response to cell-physiologic stress.

As researchers have devised cell culture conditions that more closely approach the conditions experienced by cells within tissues, the life-span of cell lineages in culture has increased progressively and senescence *in vitro* can be delayed and, in certain cases, avoided altogether. This raises the issue of whether cell senescence ever occurs *in vivo*. Resolving this issue is complicated by the fact that relatively few biochemical markers are available that can be used to identify senescent cells within tissues that reside amid large populations of cells that are in the nongrowing, G_0 state. One widely used marker of senescence is the enzyme acidic β-galactosidase, which is expressed in senescent cells (see Figure 10.3). Use of this marker has provided evidence, still fragmentary, that cell senescence does indeed occur *in vivo* (Figure 10.9). Yet other biochemical markers of senescence are being discovered (Sidebar 14).

The likelihood that senescence is a physiologic process (rather than simply an artifact of experimentation with cultured cells) is increased by observations indicating that a variety of other cell-physiologic stresses besides extended *in vitro* propagation are also able to induce a cellular state that is indistinguishable from replicative senescence. Included among these are hyperoxia (i.e., oxidative stress), DNA damage, and the aberrant signaling by certain oncoproteins. For example, elevated signaling by the Ras protein can induce senescence in certain cell types (see Figure 6B).

A definitive proof that senescence is an *in vivo* phenomenon has become critical to our understanding of cancer development. Thus, if cell senescence is indeed a cell-physiologic process occurring in cells within living tissues, we can imagine that it can serve as an important barrier to the development of spontaneously arising tumors. Conversely, if cell senescence is found to be largely an artifact of tissue culture, its role in cancer pathogenesis becomes less plausible.

10.4 The proliferation of cultured cells is also limited by the telomeres of their chromosomes

While cultured human fibroblasts expressing the large T oncoprotein succeed in bypassing senescence (see Figure 10.8) , they still will not be immortalized. After an additional number of cell generations—perhaps 10 to 20—beyond the time when they would usually senesce, cell populations enter into crisis and exhibit widespread apoptosis. The SV40 large T antigen, even with its potent p53-inactivating ability, clearly does not protect against the apoptotic death associated with crisis.

Senescence represents a halt in cell proliferation with retention of cell viability over extended periods of time, while crisis involves death by apoptosis (Figure 10.10). Senescent cells seem to have a reasonably (but not totally) stable karyotype, while cells in crisis show widespread karyotypic instability.

The timing of crisis and the appearance of cells that have entered into this state suggest the involvement of a second mechanism operating independently of the mechanism(s) triggering senescence. The workings of this second mechanism have been traced back to the chromosomal DNA of cells. Unlike the mechanism(s) leading to senescence, the molecular apparatus that initiates crisis is truly a functional counting device that tallies how many successive growth-and-division cycles cell lineages have passed through since their founding in the early embryo.

At first glance, the chromosomal DNA would seem to be an unlikely site for mammalian cells to construct such a "generational clock." We know that the structure of chromosomal DNA is highly stable and therefore should be unchanged from one cell generation to the next. A generational clock, in contrast, must depend on some progressive molecular change that is noted and recorded during each cell generation. How can chromosomal DNA molecules, which seem to be so immutable, register an additional cell generation each time a cell passes through a growth-and-division cycle?

Each mammalian chromosome carries a single, extremely long DNA molecule. As it turns out, the two ends of this DNA molecule create a serious problem for the cell—a problem dramatically revealed by the experimental technique of DNA transfection (see Section 4.2). After entering cells, transfected linear DNA molecules are rapidly fused end-to-end through the actions of a variety of nucleases and DNA ligases that are active in most if not all mammalian cell types. Hence, linear DNAs are intrinsically unstable in our cells, yet the linear DNA molecules within chromosomes clearly persist.

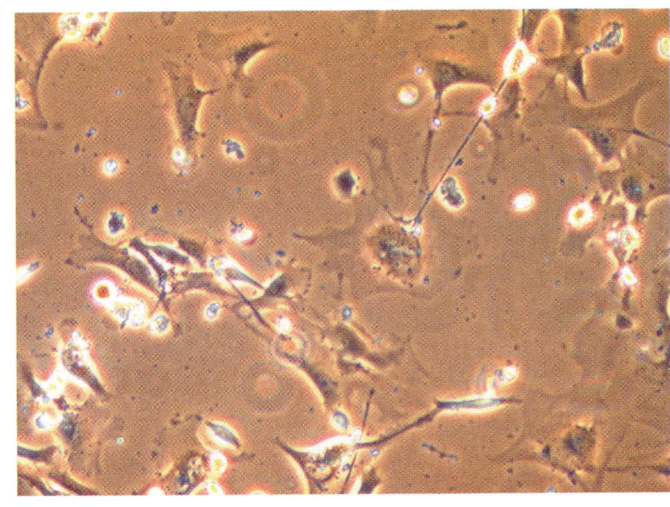

Figure 10.10 Apoptosis associated with cell populations in crisis Cell populations in crisis show widespread apoptosis, with a high percentage of cells in the midst of disintegrating and numerous cell fragments *(white refractile spots)*. Because populations of cells enter into crisis asynchronously, some of the cells in such populations still appear relatively normal *(top right)*. (Courtesy of S.A. Stewart.)

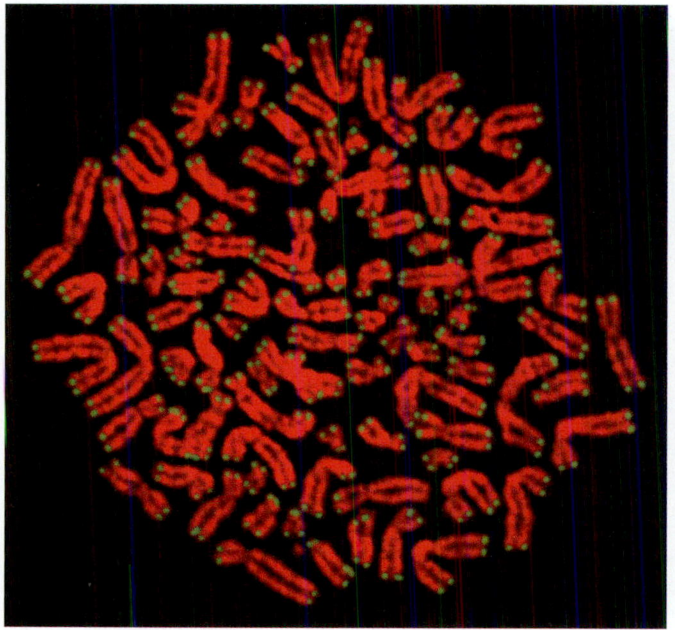

(A)

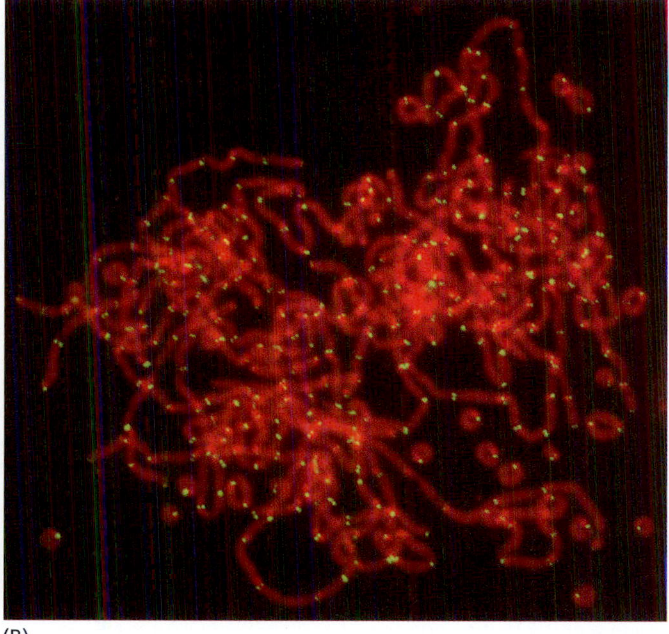

(B)

Figure 10.11 Telomeres detected by fluorescence *in situ* hybridization Telomeres can be detected by the technique of fluorescence *in situ* hybridization (FISH), in which a DNA probe that recognizes the repeated sequence present in telomeric DNA is coupled with a fluorescent label. (A) In this micrograph, human cells have been trapped in metaphase using a microtubule antagonist. The telomeres of the resulting condensed chromosomes *(red)* are then visualized through a probe that labels the telomeric DNA *(yellow green dots)*. Each chromatid carries telomeres at both its ends. Because these cells are in metaphase, the paired, recently replicated chromatids have not yet separated. (B) The karyotype seen in panel A can be compared with that of cells that have been deprived of TRF2, a key protein in maintaining normal telomere structure (see Figure 10.19). In these cells, the telomeres lose their protective function, resulting in massive end-to-end fusion of chromosomes that often results in fusing of all the chromosomes of a cell into one huge chromosome. (A, courtesy of J. Karlseder and T. de Lange; B, from G.B. Celli and T. de Lange, *Nat. Cell Biol.* 7:712–718, 2005.)

The **telomeres** located at the ends of the chromosomes (Figure 10.11) explain how these linear DNA molecules can stably coexist with the cell's various DNA-modifying enzymes. Telomeres act to prevent the end-to-end fusion of chromosomal DNA molecules and, hence, to prevent the fusion of chromosomes with one another. In effect, telomeres serve as protective shields for the chromosomal ends, much like the aglets safeguarding the tips of shoelaces. As we will see, the catastrophic events of crisis are triggered when cells lose functional telomeres from their chromosomes.

Discoveries reported in 1941 by Barbara McClintock (Figure 10.12) in her studies of the chromosomes of corn first revealed that chromosomes that have lost functional telomeres at their ends soon fuse, end-to-end, with one another. The resulting mega-chromosomes possess two or more **centromeres**, the specialized chromosomal structures that become attached during mitosis to the fibers of the mitotic spindle. An extreme form of this fusion is shown in Figure 10.11B, in which virtually all the chromosomes of a cell have fused into one giant chromosome.

Figure 10.12 Barbara McClintock Barbara McClintock's detailed studies of the chromosomes of corn (maize) revealed specialized structures at the ends of chromosomes—the telomeres—that protected them from end-to-end fusions. This work, and her demonstration of movable genetic elements in the corn genome, later called transposons, earned her the Nobel Prize in Physiology or Medicine in 1983. (Courtesy of American Philosophical Society.)

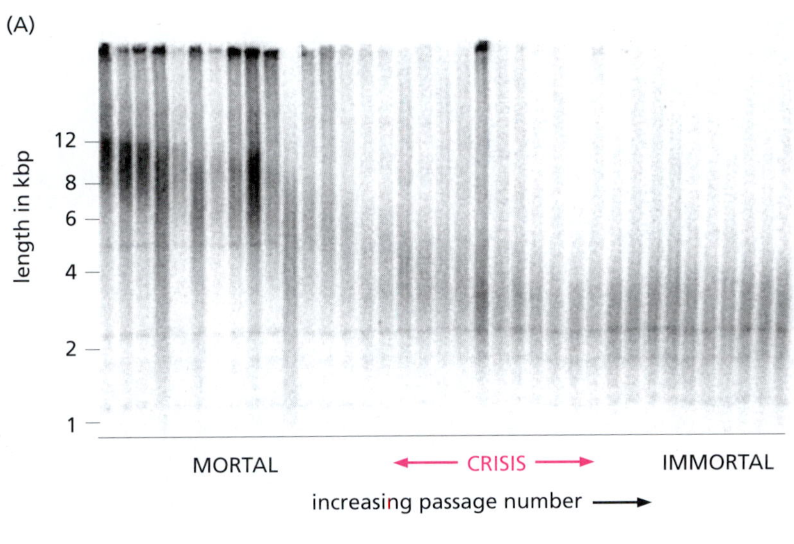

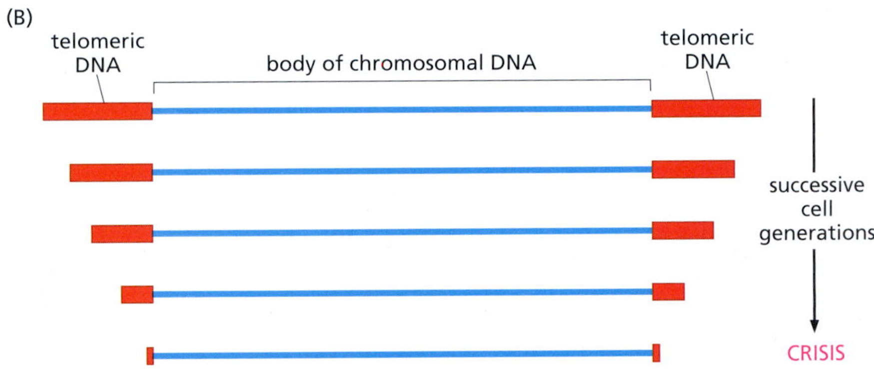

Figure 10.13 Shortening of telomeric DNA in concert with cell proliferation (A) The length of telomeric DNA was measured in a population of non-immortalized human lymphocytes. Each time the culture was passaged, part was set aside and the cells' telomeric DNA was analyzed. Each passage represented three to six population doublings (PDs). The length of this DNA can be measured by treating genomic DNA with several restriction enzymes that do not cleave within the repeating TTAGGG sequence that constitutes telomeric DNA. The only high–molecular-weight genomic DNA fragments left behind are those with the repeating hexanucleotide sequence that constitutes telomeric DNA. The length of the resulting telomeric restriction fragments (TRFs) can then be determined by gel electrophoresis followed by Southern blotting analysis using a probe that recognizes this sequence. Since the telomeres within a given cell are of heterogeneous length, and since the extent of telomeric DNA shortening differs from one cell to another, the size distribution of telomeric DNA in each sample is quite heterogeneous. In the analysis shown, the TRFs of mortal (i.e., non-immortalized) human lymphocytes grew shorter with every successive passaging of these cells (usually calculated to be between 50 and 100 base pairs per cell generation). When the TRFs became less than about 3 kbp long, cells entered crisis. Later, when cells emerged spontaneously from crisis (and became immortalized), they maintained their telomeric DNA at sizes that were slightly longer than those seen in the cells in crisis. (B) Observations like these have led to the model shown, in which telomeric DNA *(red)* shortens progressively each time a cell passes through a growth-and-division cycle until the telomeres are so eroded that they can no longer protect the ends of chromosomal DNA. This telomere collapse provokes a cell to enter into crisis. (In fact, it is not known precisely how short telomeric DNA must become before crisis is triggered.) Note that this drawing is not to scale: telomeric DNA is usually 5 to 10 kilobase pairs long, while the body of chromosomal DNA *(blue)* is often many tens of megabase pairs long. (A, from C.M. Counter, F.M. Botelho, P. Wang et al., *J. Virol.* 68:3410–3414, 1994.)

As discussed in greater detail later, the DNA component of each telomere in our cells is composed of the 5′-TTAGGG-3′ hexanucleotide sequence, which is tandemly repeated thousands of times; these repeated sequences, together with associated proteins, form the functional telomere. The telomeric DNA (and thus the telomeres) of normal human cells proliferating in culture shorten progressively during each growth-and-division cycle, until they become so short that they can no longer effectively protect the ends of chromosomes (Figure 10.13). At this point, crisis occurs, chromosomes fuse, and widespread apoptotic death is observable. Hence, in such cells, it is telomere shortening that functions to register the number of cell generations through which cell populations have passed since their origin in the early embryo.

In human cells entering crisis, the initial end-to-end fusion events usually occur between the eroded telomeric ends of the two sister chromatids that form the two halves of the same chromosome (Figure 10.14A). Recall that paired chromatids exist during the G_2 phase of the cell cycle—a period after S phase has created two chromatids from a parental chromosome and before M phase, when these two sister chromatids are destined to be separated from one another (see Figure 8.3). Such fusions between the ends of sister chromatids (rather than between the ends of two unrelated chromosomes) are favored for at least two reasons. First, the two chromatids, and thus their ends, are held in close proximity through their joined centromeres. Second, for unknown reasons, telomeres shorten at different rates on different chromosomal arms. Consequently, the two homologous telomeric DNAs (e.g., the pair of telomeres at the ends of the long arms of the chromatids of human Chromosome 9), having just been generated anew by DNA replication, possess virtually identical molecular structures. Therefore, if the telomeric end of one chromatid has become shortened, frayed, and vulnerable to fusion, its counterpart on the other paired chromatid is likely to be in the same state.

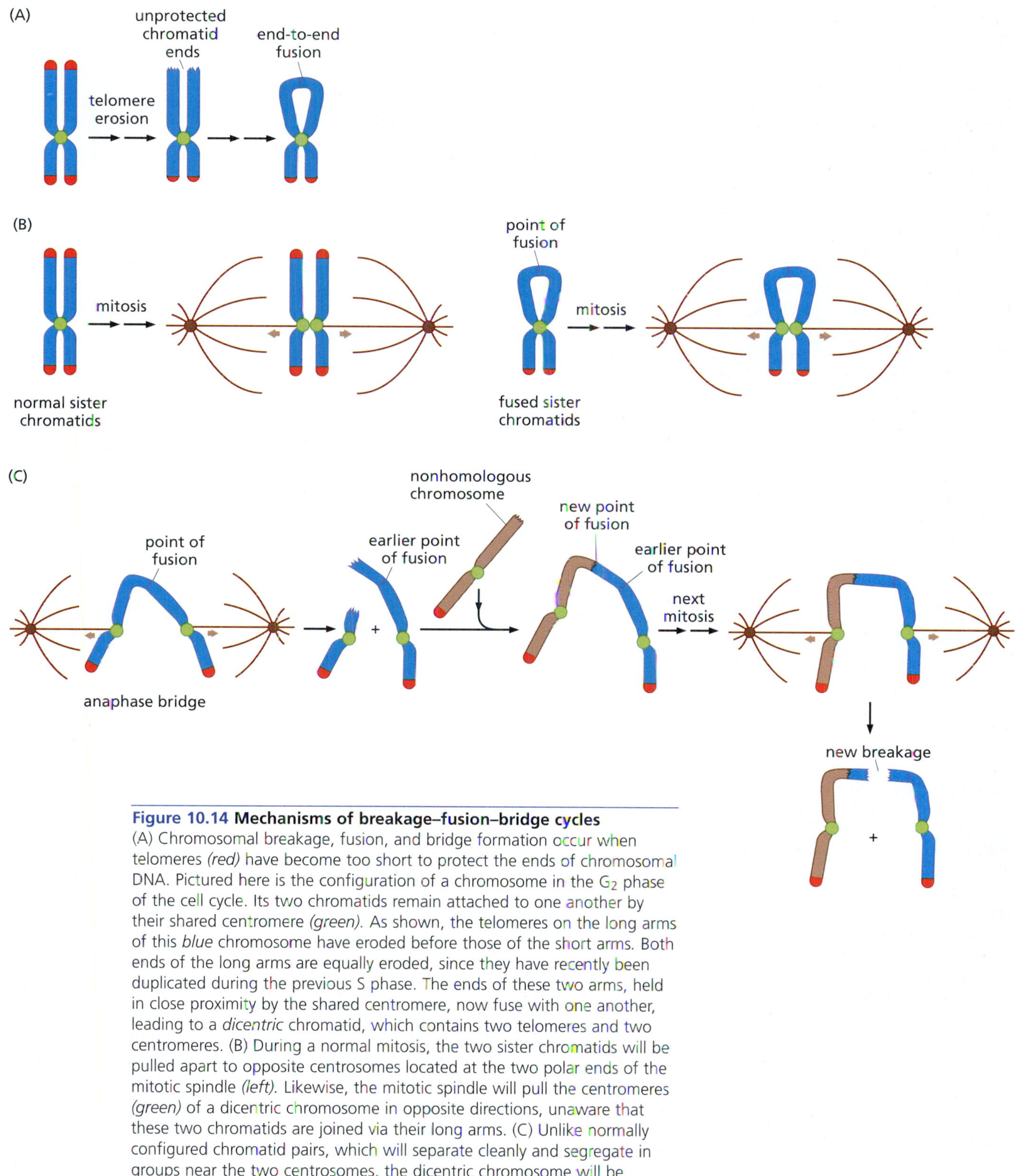

Figure 10.14 Mechanisms of breakage–fusion–bridge cycles (A) Chromosomal breakage, fusion, and bridge formation occur when telomeres *(red)* have become too short to protect the ends of chromosomal DNA. Pictured here is the configuration of a chromosome in the G_2 phase of the cell cycle. Its two chromatids remain attached to one another by their shared centromere *(green)*. As shown, the telomeres on the long arms of this *blue* chromosome have eroded before those of the short arms. Both ends of the long arms are equally eroded, since they have recently been duplicated during the previous S phase. The ends of these two arms, held in close proximity by the shared centromere, now fuse with one another, leading to a *dicentric* chromatid, which contains two telomeres and two centromeres. (B) During a normal mitosis, the two sister chromatids will be pulled apart to opposite centrosomes located at the two polar ends of the mitotic spindle *(left)*. Likewise, the mitotic spindle will pull the centromeres *(green)* of a dicentric chromosome in opposite directions, unaware that these two chromatids are joined via their long arms. (C) Unlike normally configured chromatid pairs, which will separate cleanly and segregate in groups near the two centrosomes, the dicentric chromosome will be unable to do so and will instead create a bridge between the two poles of the mitotic spindle during the anaphase of mitosis. Eventually, the dicentric chromosome will be ripped apart at some weak point. During the next cell cycle, the larger fragment lacking a telomere at one end *(blue)* may fuse with another atelomeric chromosome *(beige)*, creating a new dicentric chromosome, which itself will be pulled apart during a subsequent mitosis, resulting once again in a breakage–fusion–bridge cycle. (The fate of the shorter chromatid fragment generated by the first cycle of breakage is not shown.)

(A)

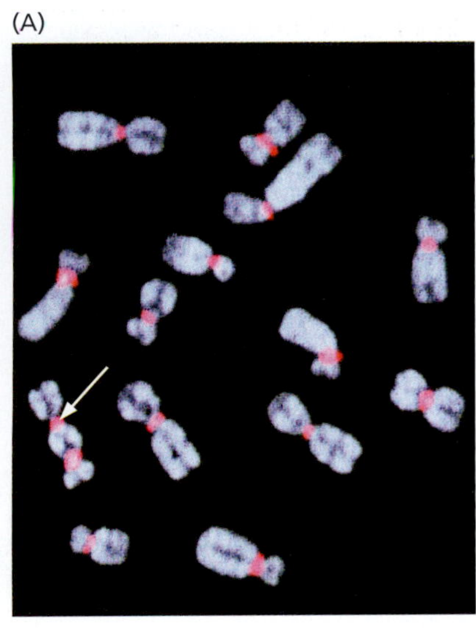

(B)

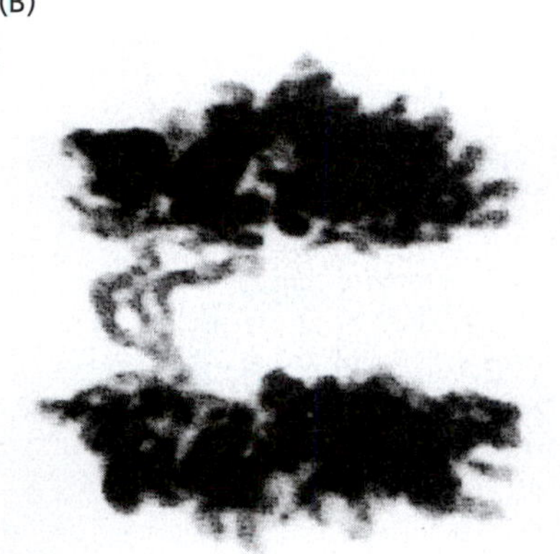

(C)

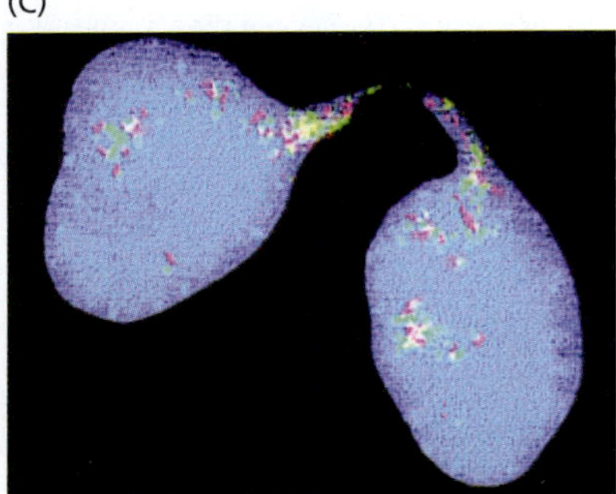

Figure 10.15 Dicentric chromosomes, anaphase bridges, and internuclear bridges (A) A dicentric chromosome is seen *(white arrow)* in a metaphase spread of cells from a human pancreatic carcinoma cell line. Centromeres have been stained *(pink)*. (B) Such dicentric chromosomes often result in the formation of anaphase bridges, several of which are seen in a cell of a human malignant fibrous histiocytoma. The two opposing groups of chromosomes are unable to separate properly from one another because they are still connected by multiple anaphase bridges. Observation of such anaphase bridges (see Figure 10.14) provides strong evidence that chromosomes within such cells are participating in breakage–fusion–bridge cycles because they no longer possess adequate telomeres. (C) While most anaphase bridges are broken during telophase or cytokinesis, some may persist in the resulting interphase cells, causing, for example, the nuclei of two daughter cells of a human lipomatous tumor cell to be linked by a chromatin bridge. (From D. Gisselsson, L. Pettersson, M. Hoglund et al., *Proc. Natl. Acad. Sci. USA* 97:5357–5362, 2000.)

When these fused chromatids participate in the mitosis that follows, their two centromeres will be pulled in opposite directions by the mitotic spindle, creating the dramatic *anaphase bridges* that are often seen following extensive telomere erosion (Figure 10.14B). Sooner or later, the mitotic apparatus that is pulling on such a **dicentric** chromatid (i.e., one with two centromeres) will succeed in ripping it apart at some random site between the two centromeres (Figure 10.14C). This yields two new chromosomal ends, neither of which possesses a telomere. Such unprotected ends may fuse with one another or with the unprotected ends of other, nonhomologous chromosomes.

In the event that the resulting defective, nontelomeric end of a chromosome fuses with the end of another (nonhomologous) chromosome, the fate of the resulting new, dicentric chromatid is more ambiguous (see Figure 10.14C). Half the time, on average, the two centromeres will be pulled in the same direction toward the centrosome that nucleates one of the two soon-to-be-born daughter cells, and no breakage will occur. Half the time, the two centromeres of a dicentric chromatid will become attached to the two opposing spindle bodies and thus to the two centrosomes located at opposite sides of the mitotic cell. When chromatids are separated during the anaphase of mitosis, this second configuration (like the earlier one between sister chromatids) will once again prove to be disastrous, since it will involve the tearing apart of a chromatid and the generation of ends that are, as before, unprotected by telomeres.

These newly created chromosomal ends will once again attempt to fuse with yet other chromosomes, yielding more dicentric chromosomes and a new cycle of chromosome breakage. This sequence of events is termed the *breakage–fusion–bridge* (BFB) *cycle,* since it involves the initial *breakage* of dicentric chromatids in anaphase, the subsequent *fusion* of the resulting nontelomeric DNA ends with yet other chromatids, and the formation once again of anaphase *bridges* by the new dicentric chromosomes resulting from these fusions (Figure 10.15). While they are occurring, these BFB cycles create karyotypic chaos that has the potential to affect many chromosomes within a cell, as proposed in 1941 by McClintock.

The relevance of telomeres to the state of crisis is indicated by the fact that cells that have entered crisis show precisely the type of karyotypic disarray that is observed when chromosomes lose their telomeres. Moreover, the chromosomes that are fusing within these cells possess especially short telomeres, or none at all. These disparate observations provide clear indication that the molecular machinery triggering crisis resides in the telomeres.

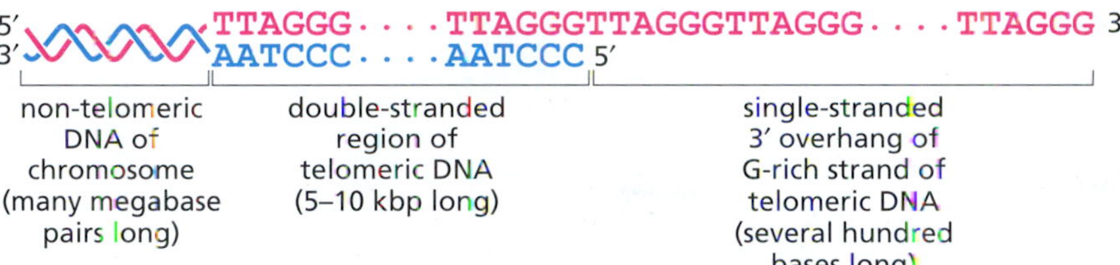

Figure 10.16 The 3′ overhang of telomeric DNA The G-rich strand of telomeric DNA *(pink)* extends beyond the C-rich strand *(blue)*. This creates a 3′ overhang *(right)* that is often several hundred nucleotides long. This overhang is far shorter than the double-stranded portion of telomeric DNA *(left)*, which is often 5 to 10 kilobase pairs long and is not drawn here to scale.

10.5 Telomeres are complex molecular structures that are not easily replicated

Research conducted since the mid-1980s has revealed the molecular structure of telomeres and their DNA. To begin, and as cited earlier, the telomeric DNA of mammalian cells (as well as the cells of many other metazoa) is formed from the repeating hexanucleotide sequence 5′-TTAGGG-3′ in one strand (the "G-rich" strand) and the complementary 5′-CCCTAA-3′ in the other (the "C-rich" strand). In normal human cells, telomeric DNA is formed from several thousand of these hexanucleotide sequences, resulting in 5- to 10-kilobase pair-long stretches of such repeating sequences at the ends of all chromosomes.

The telomeric DNA of mammalian cells, and possibly the cells of all eukaryotes, possesses an additional, distinctive feature: its G-rich strand is longer by one hundred to several hundred nucleotides, resulting in a long 3′ single-strand overhang (Figure 10.16). This overhanging strand is often found in a most unusual molecular configuration termed the **T-loop**. It was discovered in the late 1990s, when telomeric DNA analyzed in the electron microscope often was found to be configured in a loop, in effect a lasso (Figure 10.17A). This structure has been interpreted to depend on the formation of a three-stranded complex of DNA (Figure 10.17B and C). The T-loop may be present at the ends of all telomeres, although it has been observed in only a subset of those viewed, probably because of the technical difficulties associated with preserving and visualizing this structure in the electron microscope. The T-loop may help to protect the ends of linear DNA molecules, because the end of the single-stranded overhanging region is tucked into a double-stranded region, out of harm's way.

Both the relatively long double-stranded telomeric DNA and the short overhanging end are bound by specific proteins. As might be expected, some of the proteins possess domains that specifically recognize and bind to the hexanucleotide sequence present in the double- and the single-stranded regions of

Figure 10.17 Structure of the T-loop (A) Purification of telomeric DNA from mouse cells (following stabilization of double-stranded structures through chemical cross-linking) reveals, upon electron microscopy, lassos at the ends of chromosomal DNA, which have been called *T-loops*. (B) A schematic drawing of the T-loop indicates that the 3′ overhanging end of the G-rich strand *(pink)* is annealed to a small region of the C-rich strand *(blue)*, causing the formation of a displacement loop *(pink strand)*. The 5′-to-3′ polarities of the two strands are indicated by the arrowheads. (C) The T-loop is illustrated once again, on this occasion with the double helices drawn and the 3′ overhanging end emphasized in a wider line *(pink)*. (A, from J.D. Griffith, L. Comeau, S. Rosenfield et al., *Cell* 97: 503–514, 1999.)

(A)

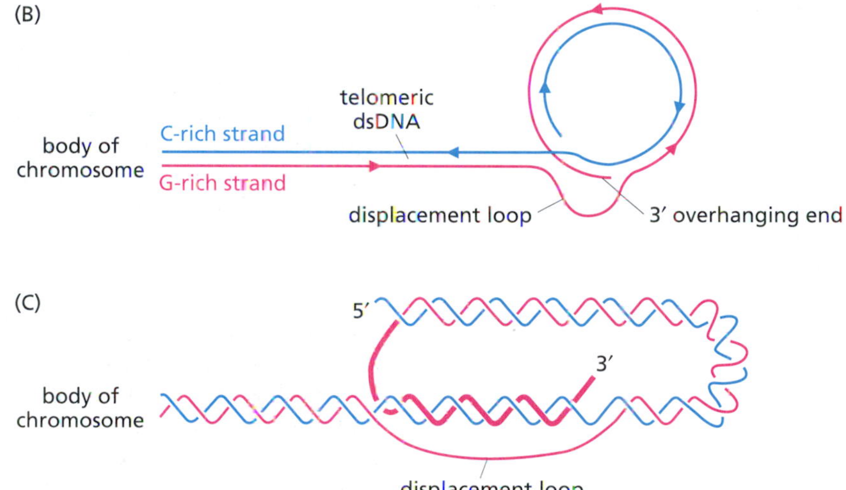

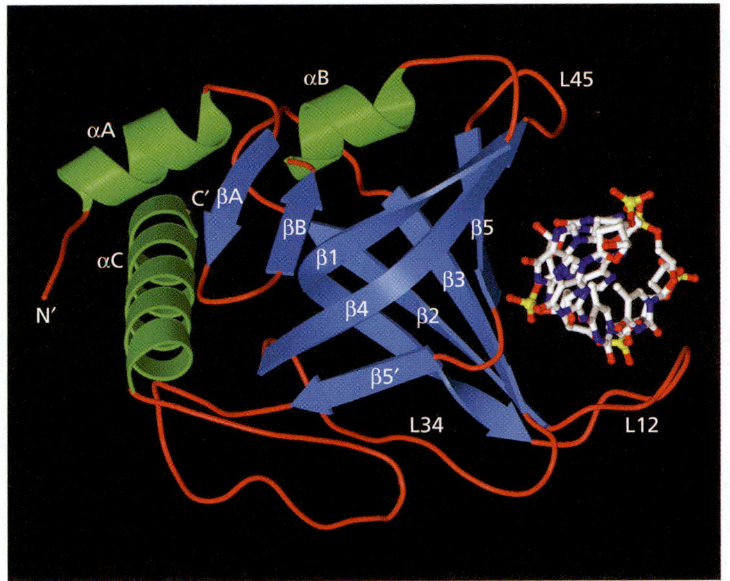

Figure 10.18 Structure of a telomere-associated protein In addition to its complex structure (Figure 10.17), the telomeric DNA is protected from degradation by a number of associated, telomere-binding proteins. This is the structure of one of them, termed Pot1, depicted as a ribbon diagram with its α-helices, β-pleated sheets, and loops numbered. Pot1 binds both the single- and double-stranded telomeric DNA in ways that help to stabilize its structure. Seen to the right is a stick figure of the single-stranded telomeric DNA that is bound by Pot1. (Courtesy of M. Lei.)

telomeric DNA (Figure 10.18). Together, these telomere-binding proteins and the telomeric DNA form the nucleoprotein complexes that we call telomeres (Figure 10.19).

While the replication machinery that operates during the S phase of the cell cycle is highly effective at copying the sequences in the middle of linear DNA molecules, such as those in the body of each of our chromosomes, this machinery has great difficulty copying sequences at the very ends of these molecules. The difficulty can be traced to the requirement that the synthesis of all DNA strands during DNA replication must be initiated at the 3′-hydroxyl end of an existing DNA strand, which serves as a **primer** to nucleate DNA strand elongation; alternatively, in the absence of an available DNA primer, the 3′ end of an RNA molecule can serve as a primer for DNA synthesis (Figure 10.20).

If the **primase** enzyme, which is responsible for laying down short RNA primers, happens to deposit a primer at some distance from the 3′ terminus of a template strand on which "leading strand synthesis" is occurring (see Figure 10.20), a DNA polymerase will initiate synthesis of a new DNA strand that lacks

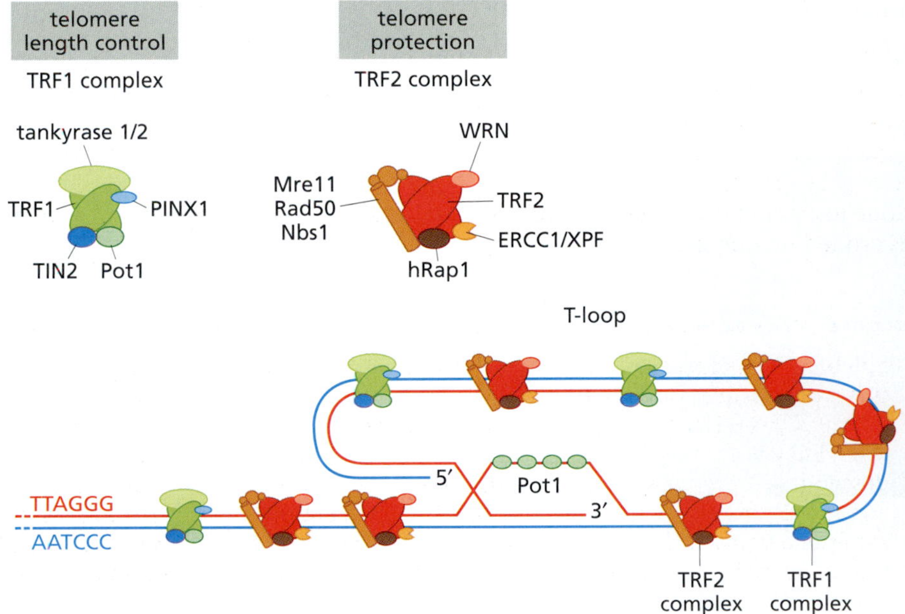

Figure 10.19 Multiple telomere-specific proteins bound to telomeric DNA Pot1 (Figure 10.18) is only one of a large array of telomere-specific binding proteins that have been uncovered. Their precise contributions to telomere stabilization remain to be elucidated. At least two distinct complexes of proteins have been found, both of which bind to the double-stranded portion of telomeric DNA. Provocatively, the TRF2-containing complexes carry a number of proteins that are known to participate in the repair of genomic damage, some of which are described in Chapter 12. Pot1 can also bind to the loop of the G-rich single-strand that is displaced by formation of the T-loop *(bottom center)*. (From T. de Lange, *Nat. Rev. Mol. Cell Biol.* 5:323–329, 2004.)

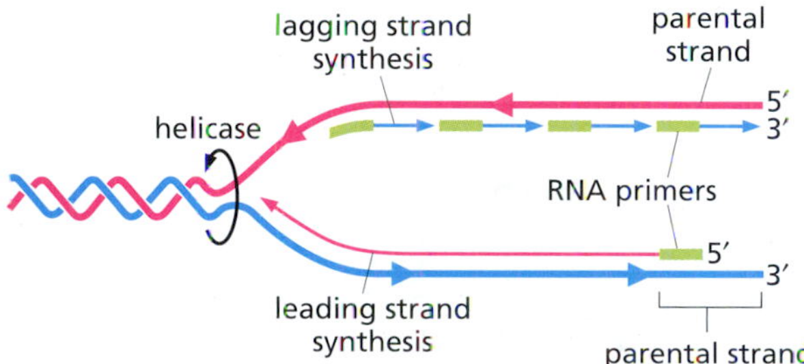

Figure 10.20 Primers and the initiation of DNA synthesis During DNA replication, the parental DNA double-helix *(left)* is unwound by a helicase enzyme, allowing the replication process as a whole to progress in a leftward direction. The synthesis of the new "lagging strand" *(above)* is made possible by the presence of short RNA primer segments *(green rectangles)* that are laid down at intervals of several hundred nucleotides by the primase enzyme. The 3′-hydroxyl ends of these RNA molecules serve as sites of initiation of newly made daughter strands *(light blue)*. Because this synthesis (like all DNA synthesis) occurs in a 5′-to-3′ direction, these lagging strand segments grow in a direction opposite to the direction of advance of the replication fork. The copying of the other parental strand *(below)*, which is also initiated by an RNA primer *(green)*, can proceed continuously, as this "leading" strand can grow at its 3′ end as the parental DNA unwinds. However, the RNA primer that is responsible for initiating "leading-strand" synthesis may sit at a site many nucleotides away from *(to the left of)* the 3′ end of the blue parental strand; in addition, the RNA primer itself will be lost when it is degraded subsequently during the maturation of the recently synthesized DNA. For these reasons, the leading-strand synthesis will lead to under-replication of one of the parental *(blue)* strands of DNA *(bottom right)*.

a substantial number of the bases complementary to the very 3′ end of this template strand. (Even if the primase "sits down" and constructs an RNA primer at the very end of the template strand, the sequences corresponding to the approximately 10 nucleotides of the RNA primer will not be present in the newly synthesized daughter strand, since the RNA primer will be degraded after it has served its purpose of initiating DNA strand elongation.)

This *end-replication problem* provides a molecular explanation for the observed shortening of telomeric DNA each time a normal human cell passes through a cell cycle. In addition to the under-replication of telomeric DNA ends, there appear to be exonucleases within cells that slowly chew on the ends of telomeric DNA and may ultimately contribute far more to telomere erosion. For whatever reason, in many types of normal human cells, telomeres lose 50 to 100 base pairs of DNA during each cell generation (see Figure 10.13). This progressive erosion of telomeric DNA represents a simple molecular device that limits how many generations of descendant progeny a cell can spawn.

We can imagine, for example, that in human embryonic cells, the telomeric DNA begins rather long, perhaps 8 to 10 kb in length. As various lineages of descendant cells throughout the developing body proceed through their repeated cycles of growth and division, the telomeres in these cells grow progressively shorter. Ultimately, in some cells, the telomeric DNA erodes down to a size that is so short that it can no longer perform its intended function of protecting the ends of the chromosomal DNA, the result being the breakage–fusion–bridge cycles and chromosomal translocations that are illustrated in Figure 10.14. Indeed, it is plausible that the aging of certain tissues derives from the loss by individual cells of replicative potential, and that this loss is attributable, in turn, to telomere erosion. However, we have not yet discovered how to use telomere length to predict the onset of BFB cycles and crisis (Sidebar 10.3).

While the supporting evidence is still indirect, it is highly likely that telomere shortening, by limiting the replicative potential of cell lineages, creates an obstacle to the accumulation of large populations of cancer cells. Thus, as argued earlier, should a clone of cells acquire oncogenic mutations (involving oncogene activation and tumor suppressor gene inactivation), the ability of this clone to expand to a large, clinically detectable size is likely to be constrained by its eroded telomeres.

Sidebar 10.3 Measurements of telomeric DNA length do not provide accurate prediction of future replicative potential Even if the average length of telomeric DNA in a cell is known, the number of replicative generations that its descendants can pass through before the onset of crisis cannot be predicted with any precision. As implied above, some telomeres within a cell shorten more rapidly than others. Accordingly, the length of the shortest telomeres in a cell may determine the replicative potential of its progeny, since these telomeres will determine the timing of the earliest breakage–fusion–bridge cycles and thus the onset of crisis. Moreover, we don't really know how short telomeric DNA must become before it begins to lose its protective functions. In some cases, it seems as if telomeric DNA that is still several kilobases long has already lost its ability to prevent end-to-end chromosomal DNA fusions.

10.6 Incipient cancer cells can escape crisis by expressing telomerase

The detailed behavior of pre-neoplastic cell clones *in vivo* is difficult to study, and so we are largely forced to extrapolate from observations of cultured cells, such as human fibroblasts. As mentioned earlier, if these cells are allowed to circumvent senescence through the expression of the SV40 large T oncoprotein, they will continue to replicate another 10 to 20 cell generations and then enter crisis. On rare occasion, a small group of cells will emerge spontaneously from the vast throng of cells in the midst of crisis. These variant cell clones—arising perhaps from a single cell among 10 million cells in crisis—proceed to proliferate and continue to do so indefinitely (e.g., see Figure 10.13A). They have become immortalized, seemingly as the consequence of some random event.

This transition, observed with cultured cells *in vitro*, seems to recapitulate the behavior of pre-malignant cell populations *in vivo*. Thus, both cell populations enter crisis sooner or later, having passed through large numbers of replicative generations and suffered extensive erosion of their telomeres, and both are capable of spawning rare immortalized variants that have apparently solved the problem of telomere collapse.

How can these cell populations in crisis address and solve the problem of telomere collapse? At first glance, crisis and telomere collapse would appear to be irreversible processes from which cells can never escape. As it turns out, the route to immortality is simple and, in retrospect, obvious: cells can emerge from crisis by regenerating their telomeres, thereby erasing the molecular record (their shortened telomeres) that previously blocked their proliferation and drove them into crisis.

Telomere regeneration can be accomplished through the actions of the **telomerase** enzyme, which functions specifically to elongate telomeric DNA. A striking finding is that telomerase activity is clearly detectable in 85 to 90% of human tumor cell samples, while being present at very low levels in the lysates of most types of normal human cells, as measured by the TRAP assay (Figure 10.21). These low levels of telomerase activity, while they may enable some type of minimal maintenance of the ends of telomeric DNA (see Sidebar 10.4), are clearly unable to prevent the progressive telomere erosion that accompanies passage of normal human cells through each cell cycle.

In adult humans, there are actually several known exceptions to the generally observed low levels of telomerase activity in normal cells. For example, substantial enzyme activity is present in the germ cells of the testes, and lymphocytes express a burst of telomerase activity when they become functionally activated.

The available evidence indicates that strong expression of the telomerase enzyme is present early in embryogenesis and is largely lost during the cellular differentiation that produces the great majority of the body's tissues. Thus, during the course of early mouse and cow embryogenesis, telomerase is strongly expressed in embryonic cells in the time between the blastocyst and morula stages and seems to disappear thereafter; a similar control of telomerase expression is likely to operate during human embryonic development. Accordingly, the great majority of normal human somatic cells, while carrying the full complement of genes specifying the telomerase enzyme, are denied the services of this enzyme because they do not *express* these gene(s) at significant levels.

Among the large populations of cultured cells in crisis, however, rare variants find a way to de-repress the gene or genes encoding telomerase and thereby acquire high levels of constitutively expressed enzyme. This enables them to

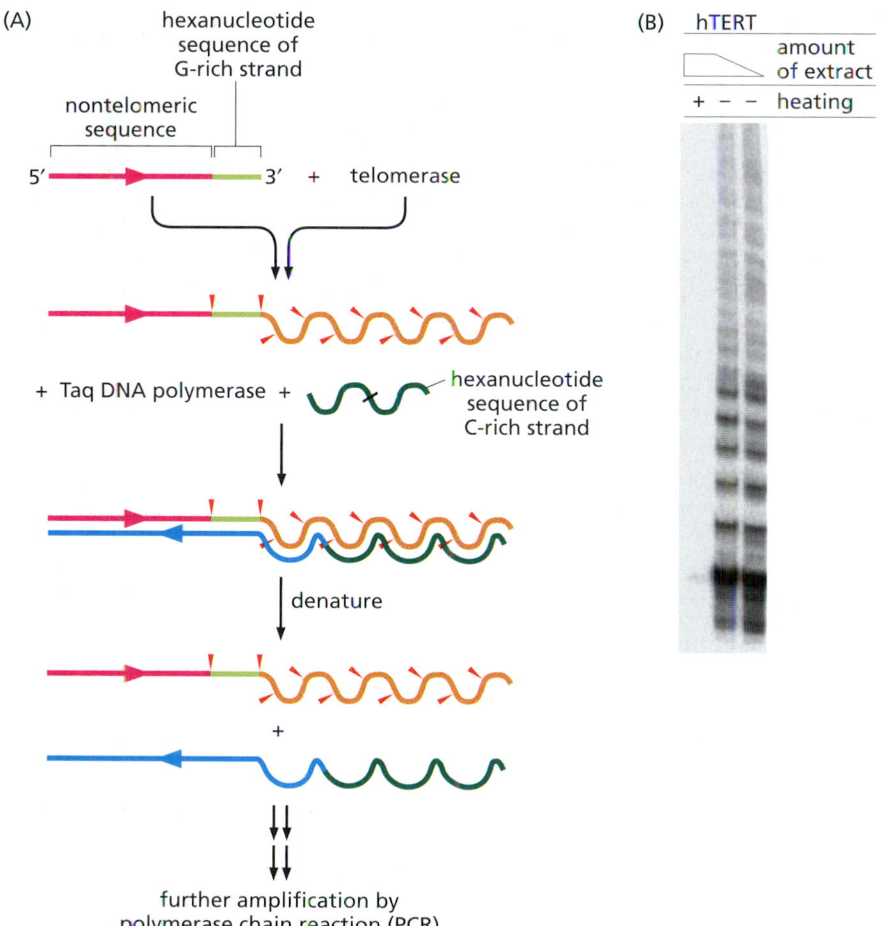

Figure 10.21 The TRAP assay (A) The telomeric repeat amplification protocol (TRAP) assay permits detection of minute levels of telomerase activity in cell lysates by relying on the polymerase chain reaction (PCR) to amplify the products of the telomerase enzyme. A primer consisting of nontelomeric sequences *(pink)* and telomeric hexanucleotide sequences (from the G-rich strand; *light green*) is added to a cell lysate in the presence of deoxyribonucleotide triphosphates. This primer is extended *(light orange)* by any telomerase that may be present in a cell lysate. The thermostable Taq polymerase is then added together with a primer from the C-rich strand *(dark green)*, and the second strand is elongated *(light blue)*. These two DNA strands are then denatured and recopied repeatedly by the PCR in the presence of proper primers. (B) The products of the TRAP reaction are analyzed by gel electrophoresis to resolve DNA molecules differing in size by hexanucleotide increments. As a negative control, brief heat treatment of a portion of the lysate is used at the beginning of the reaction to denature and inactivate any telomerase that may be present *(1st channel)*; this is done to determine whether any heat-resistant DNA polymerases are present in the lysate in addition to the telomerase itself. In this experiment, pre-crisis cells have been infected with an expression vector that specifies hTERT, the catalytic subunit of the telomerase holoenzyme, thereby inducing high levels of telomerase activity that are readily detectable when both high *(2nd channel)* and low levels *(3rd channel)* of cell extract are tested. (From M. Meyerson, C.M. Counter, E.N. Eaton et al., *Cell* 90:785–795, 1997.)

extend their telomeric DNA to a length that permits further proliferation. Indeed, as long as their descendants continue to express this enzyme, such cells will continue to proliferate and thus will be considered to be immortalized. A similar sequence of events is presumed to occur *in vivo* when populations of pre-neoplastic cells enter into crisis and, on rare occasion, generate immortalized variants that become the progenitors of large neoplastic cell populations.

The telomerase enzyme was initially characterized in baker's yeast using genetic analyses and in ciliates through biochemical purification (Sidebar 15). It is a complex enzyme composed of a number of distinct subunits, not all of which have been characterized. At the core of the mammalian telomerase **holoenzyme** are two subunits. One subunit is a DNA polymerase, more specifically a reverse transcriptase, which functions, like the enzymes made by retroviruses (Section 3.7) and a variety of other viruses and transposable elements, to synthesize DNA from an RNA template (Figure 10.22). However, unlike these other reverse transcriptases, the telomerase holoenzyme cleverly packs it own RNA template—the

Sidebar 10.4 Telomerase is actually expressed in normal human cells, albeit at low levels At one time, it was thought that the great majority of human cancer cells express high levels of the telomerase enzyme while normal cells express none (see Figure 10.21B). In fact, use of a monoclonal antibody to immunoprecipitate the telomerase enzyme prior to TRAP analysis has revealed that in normal human cells, a small amount of enzyme activity is detectable transiently as cells enter into S phase and disappears as cells advance into the G_2 phase of their cell cycles. This small amount of enzyme activity may suffice to repair or regenerate the T-loops at the ends of telomeres (Figure 10.17) but clearly does little to maintain or extend the overall length of telomeric DNA, which is determined largely by its double-stranded region. Unanswered by these experiments is the question whether stem cells in certain normal tissues express significant levels of telomerase activity while their more differentiated progeny do not.

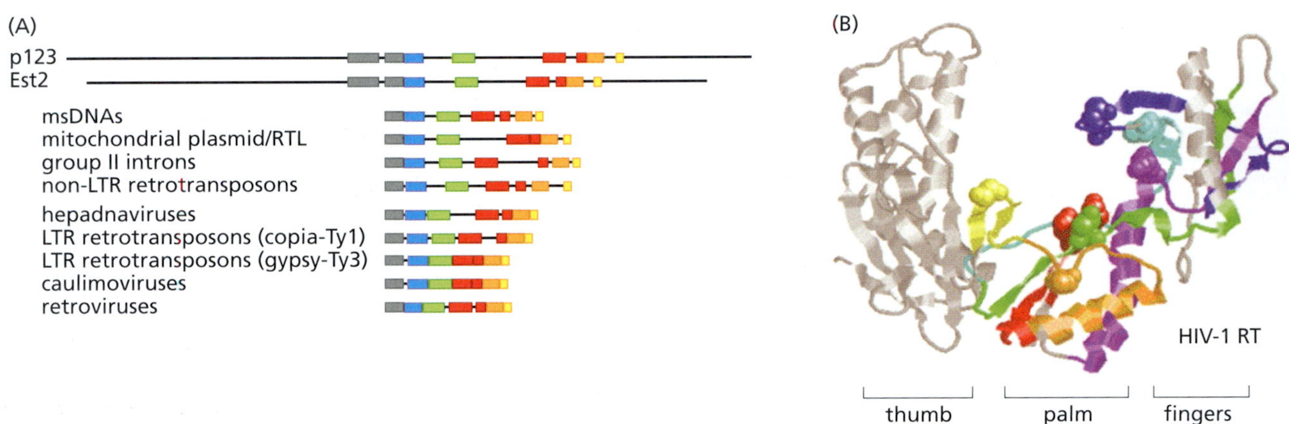

Figure 10.22 The catalytic subunit of telomerase (A) Determination of the amino acid sequence of the catalytic subunit of the telomerase holoenzyme of *E. aediculatus* (Figure 7) enabled the cloning of the encoding gene, termed initially p123. This gene was found to be homologous to the catalytic subunit of the *Saccharomyces cerevisiae* (yeast) telomerase holo-enzyme, termed Est2 (ever-shorter telomeres). Detailed sequence analysis revealed extensive sequence relatedness of these two catalytic subunits to the catalytic clefts of reverse transcriptases (RTs) specified by a variety of retro-transposons as well as viruses, including retroviruses such as human immuno-deficiency virus. (B) The regions of homology are mapped on the three-dimensional structure of the human immunodeficiency virus reverse transcriptase (HIV-1 RT). (From T.M. Nakamura, G.B. Morin, K.B. Chapman et al., *Science* 277:955–959, 1997.)

second essential subunit. A short segment of this 451-nucleotide-long RNA molecule serves as the template that instructs the reverse transcriptase activity of the holoenzyme.

Isolation of substantial quantities of telomerase protein from the ciliate *Euplotes aediculatus* allowed determination of the amino acid sequence of its catalytic subunit and, in turn, to cloning of the gene encoding the ciliate enzyme as well as the homologous human gene. In human cells, this catalytic subunit, termed hTERT (for **h**uman **te**lomerase **r**everse **t**ranscriptase), synthesizes a DNA molecule that is complementary to six nucleotides present in the telomerase-associated RNA molecule (hTR; its encoding gene is sometimes called *TERC*), attaching these nucleotides to the G-rich 3′ overhanging end of the preexisting telomeric DNA (Figure 10.23). The complementary strand of the telomeric DNA is then presumably synthesized by conventional DNA polymerases.

As would be predicted from the scenario described earlier, before entering crisis, human cells are essentially telomerase-negative and do not express appreciable levels of the *hTERT* mRNA. However, should a rare immortalized variant emerge from a population of cells in crisis, its descendants usually express significant levels of *hTERT* mRNA and exhibit substantial levels of telomerase enzyme activity (Figure 10.24). This observation demonstrates directly that derepression of *hTERT* gene expression accompanies the escape of these cells from crisis.

The sudden acquisition of telomerase activity might be only a *correlate* of escape from crisis rather than a *cause*. This ambiguity can be resolved by a simple experiment: a cDNA version of the *hTERT* gene can be introduced into cells just before they are destined to enter crisis. The introduced *hTERT* cDNA confers telomerase activity on these cells, causes elongation of their greatly shortened telomeres, prevents entrance into crisis, and enables such cells to grow indefinitely (Figure 10.25).

The outcomes of this experiment and similar experiments actually prove three related points. First, while the telomerase holoenzyme may be composed of multiple, distinct subunits, the only subunit missing from normal, pre-crisis human cells is the catalytic subunit encoded by the *hTERT* gene; the remaining subunits of the telomerase holoenzyme, including the *hTR* telomerase-associated RNA molecule, seem to be present in adequate amounts in pre-crisis cells. Second, because expression of telomerase (rather than another enzyme) allows cells to avoid crisis, and because telomerase acts specifically on telomeric DNA, these observations demonstrate that one cause of crisis must indeed be critically shortened telomeres. Third, this experiment shows that the acquisition of telomerase activity, which can be achieved spontaneously by rare variant cells amid a population of cells in crisis, is sufficient to enable cells to escape crisis and generate descendants that can grow in an immortalized fashion.

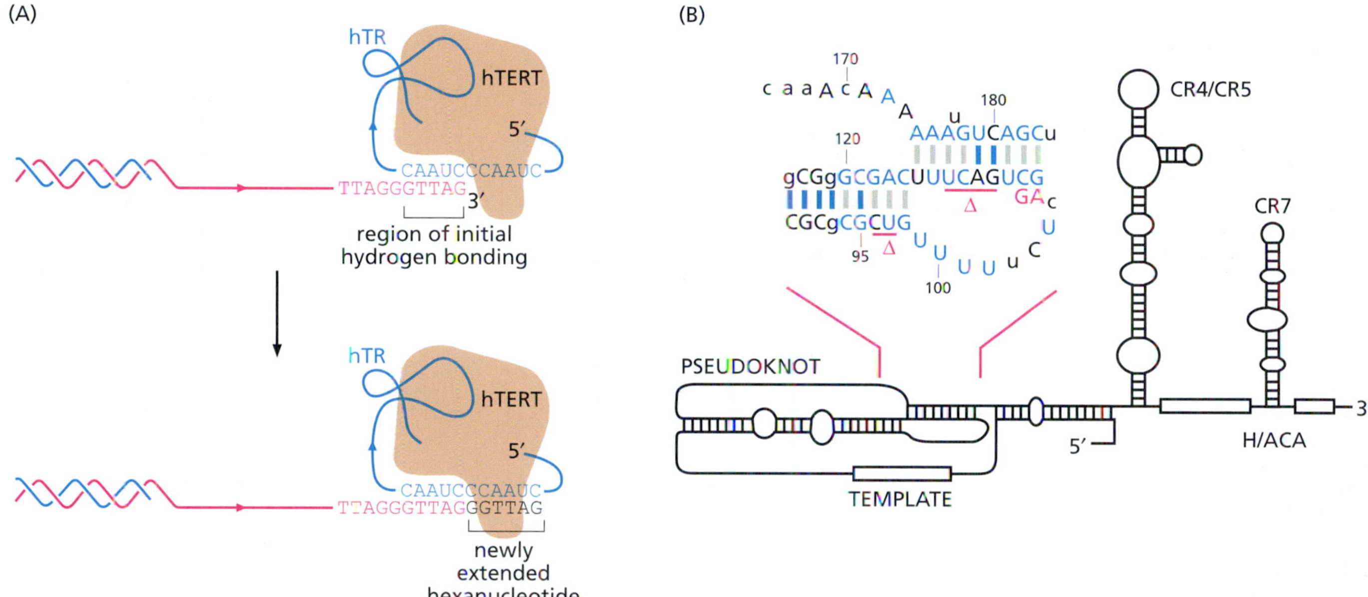

Figure 10.23 Structure of the human telomerase holoenzyme (A) The human telomerase holoenzyme is composed of at least two essential subunits, the hTERT catalytic subunit *(light brown)* and the associated *hTR* RNA subunit *(blue)*. (At least eight other subunits may exist in the holoenzyme but have not been characterized.) The holoenzyme attaches to the 3′ end of the G-rich strand overhang *(pink)*, doing so in part through the hydrogen bonding of *hTR* to the last five nucleotides of the G-rich strand. Subsequently, by reverse transcription of sequences present in the hTR subunit, hTERT is able to extend the G-rich strand by six nucleotides *(black)*. By repeating this process, the enzyme can extend the G-rich strand by hundreds, even thousands of nucleotides. Arrowheads indicate the 5′-to-3′ orientation of the nucleic acids. (B) The hTR molecule is formed from only 451 ribonucleotides. In its center is a "pseudoknot," a topologically complex structure *(blue letters)* that has, among its other features here, a three-stranded helix. Pathological human conditions arise from mutations *(pink)* in the sequence, involving either base substitutions or deletions (Δ), all of which compromise telomerase function. (B, from C.A. Theimer, C.A. Blois, and J. Feigon, *Mol. Cell* 17:671–682, 2005.)

We can conclude further that telomerase succeeds in allowing cells to circumvent crisis because it subverts the operations of the generational clock. When telomerase is expressed at significant levels, telomeres are maintained at lengths that are compatible with unlimited further replication. In effect, the generational clock that depends on progressive telomere shortening is rendered inoperative. The molecular strategies used by cancer cells to de-repress *hTERT* expression and thereby acquire high levels of telomerase activity are poorly understood (Sidebar 10.5).

With these insights in mind, the normal biological roles of telomerase now become clearer. In single-cell eukaryotic species, such as ciliates and yeast, the exponential growth of cells requires the continuous presence of high levels of telomerase activity to ensure that telomeres are maintained indefinitely at a length that is compatible with chromosomal stability. By elongating telomeric

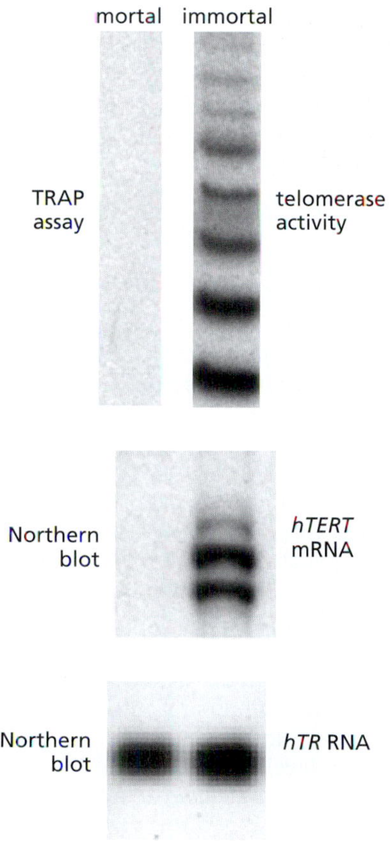

Figure 10.24 Activation of telomerase activity following escape from crisis Individual B lymphocytes escape at low frequency from a cell population in crisis, begin to grow robustly, and are now immortalized (see Figure 10.13A). As seen here *(top)*, the pre-crisis "mortal" cells lack telomerase activity, as measured by the TRAP assay (Figure 10.21), while the spontaneously immortalized ("immortal") cells show abundant activity. The *hTERT* mRNA, which encodes the catalytic subunit of the telomerase holoenzyme, is essentially absent in the pre-crisis cells but clearly present in the immortalized cells *(middle)*. However, levels of the *hTR* RNA, which functions as template for hTERT during telomeric DNA elongation (Figure 10.23B), are increased only slightly *(bottom)*, suggesting that levels of *hTERT* expression determine the levels of enzyme activity in the cell. (From M. Meyerson, C.M. Counter, E.N. Eaton et al., *Cell* 90:785–795, 1997.)

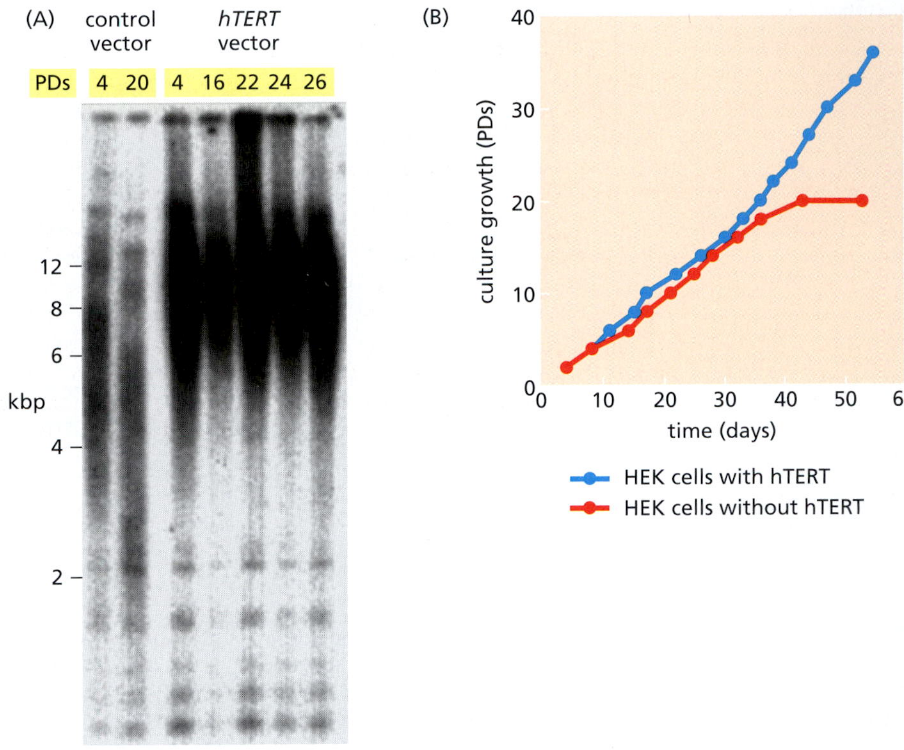

Figure 10.25 Prevention of crisis by expression of telomerase (A) This Southern blot analyzes the lengths of telomeric DNAs that have been generated by cleavage of genomic DNA with a restriction enzyme that does not cleave within the repeating hexanucleotide sequence of the telomeres (see Figure 10.13A). These DNAs are then probed with a DNA probe that recognizes this hexanucleotide sequence. As seen, telomerase-negative human embryonic kidney (HEK) cells infected with a control retroviral vector that specifies no protein carry telomeres that decrease in size between 4 and 20 population doublings (PDs; *left two channels*). However, infection of such cells with a retroviral vector specifying hTERT causes the telomeres to substantially increase in length, maintaining their sizes through 26 population doublings *(right channels)*. (Because these telomeres are much longer, they generate a stronger signal upon Southern blotting.) The size markers are given in kilobase pairs. (B) This population of human embryonic kidney cells was destined to enter crisis by 40 days, which is the equivalent of roughly 20 population doublings (PDs; *red line*). However, when hTERT was ectopically expressed in these cells *(blue line)*, they gained the ability to proliferate indefinitely. (Courtesy of C.M. Counter and R.L. Beijersbergen.)

DNA, the telomerase in these protozoan cells is able to compensate for the continued erosion of telomeric DNA due to the under-replication of DNA termini by the bulk DNA replication machinery. Being single-celled, these eukaryotes have no need to fear cancer.

Expression of telomerase activity is programmed differently in complex metazoa, notably humans. Because telomerase expression is largely repressed in postembryonic cell lineages, these lineages are granted only limited postembryonic replicative potential before they enter crisis. This limitation seems to represent a key component of the human body's anti-cancer defenses.

This mechanistic model also provides a compelling explanation of how human cells can escape from crisis and become immortalized. It fails, however, to address another long-standing mystery: How does expression of telomerase enable certain types of cultured cells to avoid senescence (Sidebar 10.6)?

Sidebar 10.5 Oncoproteins and tumor suppressor proteins play critical roles in governing *hTERT* expression The mechanisms that lead to the de-repression of *hTERT* transcription during tumor progression in humans are complex and still quite obscure. Multiple transcription factors appear to collaborate to activate the *hTERT* promoter. Once expressed, the resulting *hTERT* transcripts allow the synthesis of substantial amounts of hTERT protein; the latter, by complexing with other subunits of the telomerase holoenzyme, suffices to generate high levels of telomerase activity. The Myc protein, which acts as a transcription factor that deregulates the cell cycle clock (Section 8.9), also contributes to *hTERT* transcription. There are at least two recognition sequences for bHLH transcription factors like Myc in the promoters of the mouse and human *TERT* genes.

In addition, the connection between Myc and hTERT expression can be shown by biological experiments. For example, a human melanoma cell line has been isolated that expresses high levels of Myc and is tumorigenic. Three clones of variants of this cell line have been isolated that exhibit as much as an eightfold reduction in Myc expression and have lost tumorigenicity. Their tumorigenic growth can be restored if they are supplied with an *hTERT*-expressing gene construct. Therefore, in these cells, the loss of tumorigenicity following reduction of Myc expression levels can be attributed largely, if not exclusively, to an associated loss of *hTERT* gene expression. In addition to Myc, Menin—the product of the *MEN1* tumor suppressor gene (Table 7.1)—associates with the *hTERT* promoter, but it represses *hTERT* transcription. A similar repressive role has been attributed to the Sp1 transcription factor, whose expression is induced by TGF-β.

Sidebar 10.6 Telomerase also affects replicative senescence It is reasonably clear that (1) cumulative physiologic stress, including the stress of tissue culture, dictates the timing of onset of replicative senescence, and (2) this senescence occurs in cells that still possess quite long telomeres. Nevertheless, telomeres do seem to play a key role in governing its onset. This is indicated by the fact that in certain cell types, such as cultured human fibroblasts, senescence can be postponed by expressing hTERT prior to the expected time for entering replicative senescence.

This puzzle may be resolved by observations indicating that when some cells enter replicative senescence, they lose many of the single-strand overhangs at the ends of their telomeric DNA; this loss occurs ostensibly in response to some type of cell-physiologic stress. The resulting degraded telomeric ends, which are likely to lack a T-loop, may cause the release of a DNA damage signal, thereby provoking a p53-mediated halt in cell proliferation that is manifested as the senescent growth state (Figure 10.26). By protecting the single-strand overhang from this degradation, hTERT avoids the triggering of senescence. This model, yet to be validated, implies that hTERT is involved both in the maintenance of this single-strand telomeric DNA overhang and in the maintenance of the far longer double-stranded region of telomeric DNA.

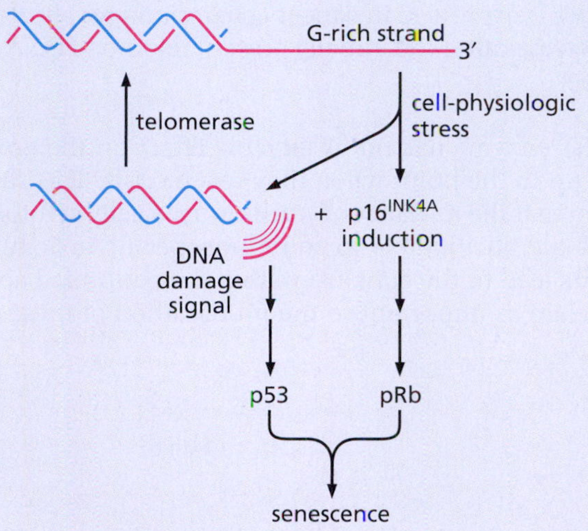

Figure 10.26 Replicative senescence and the actions of telomerase This diagram presents a still-speculative mechanistic model of how and why telomerase expression (achieved by the ectopic expression of the hTERT catalytic subunit) can prevent human cells from entering into replicative senescence. The T-loop that normally exists at the end of the telomeric DNA (Figure 10.17) is not shown here. The G-rich overhanging strand of telomeric DNA *(pink, top)* has been found to be largely degraded in cells that have entered replicative senescence. As indicated here, this loss may be caused by certain cell-physiologic stresses. The resulting blunted telomeric DNA may then emit a DNA damage signal that activates p53. Together with p16^{INK4A} expression (Section 8.4), which is activated by cell-physiologic stresses, this may impose a senescent growth state. By reversing this loss of the G-rich overhang, telomerase is able to prevent certain cell types, such as human fibroblasts, from entering into senescence.

10.7 Telomerase plays a key role in the proliferation of human cancer cells

The experiments described above using pre-crisis human cells in culture suggest that the telomerase activity detectable in the great majority of human cancer cells plays a causal role in their immortalization, and that such immortalization is a key component of the neoplastic growth state. This mechanistic model has been tested experimentally by suppressing the telomerase activity in cancer cells and following their subsequent responses. Telomerase activity can be reduced through the expression of antisense RNA in the telomerase-positive cells. For example, an RNA that is complementary in sequence to the *hTR* RNA subunit of the telomerase holoenzyme (see Figure 10.23B) can be introduced experimentally into telomerase-positive human cancer cells. Such an antisense molecule is presumed to anneal to the *hTR* molecule, forming an RNA–RNA double helix, thereby blocking the ability of the *hTR* subunit to participate in the synthesis of telomeric DNA. It is difficult to achieve total inhibition of hTR function using this strategy. Nonetheless, such an antisense experiment performed with HeLa cells caused them to stop growing 23 to 26 days after they were initially exposed to the antisense RNA.

An alternative experimental strategy that is more effective derives from the use of a mutant hTERT enzyme carrying amino-acid substitutions in its catalytic cleft; these alterations yield a catalytically inactive enzyme. When overexpressed in telomerase-positive cells, the mutant hTERT protein can act in a dominant-negative fashion (dn; see Section 9.3) to interfere with the existing endogenous telomerase activity of these cells. The dn hTERT, which can be expressed at levels vastly higher than the endogenous hTERT protein, is likely to associate with and thereby monopolize the other subunits that normally assemble to form the telomerase holoenzyme. Of course, when these other molecules associate with the dn hTERT protein, they become recruited into unproductive holoenzyme complexes.

Expression of the dn hTERT subunit in a number of different telomerase-positive human tumor cell lines causes them to lose all detectable telomerase activity and, with some delay, to enter crisis. The crisis occurs with a lag time from 5 to 25 days, depending on the length of the telomeric DNA in these cells at the time when the dn hTERT subunit was introduced. Cells with telomeres that initially were 2 to 3 kb in length enter crisis almost immediately, while those with telomeres that were initially 4 to 5 kb in length require 30 days of further passaging in culture before entering crisis (Figure 10.27). These delayed reactions suggest that after the dn hTERT is expressed in cancer cells, telomeres shorten progressively, and that crisis ensues after the initially present telomeric DNA erodes to some threshold length.

Importantly, the dn hTERT enzyme has no observable effect on the growth of telomerase-positive cells up to the point when they reach crisis. This rules out an alternative explanation: that the mutant enzyme is intrinsically cytotoxic and that its effects on the cell are attributable to some nonspecific toxicity. Taken together, such experiments lead to the conclusion that the continued activities of the telomerase enzyme are as important to the proliferation of these cancer

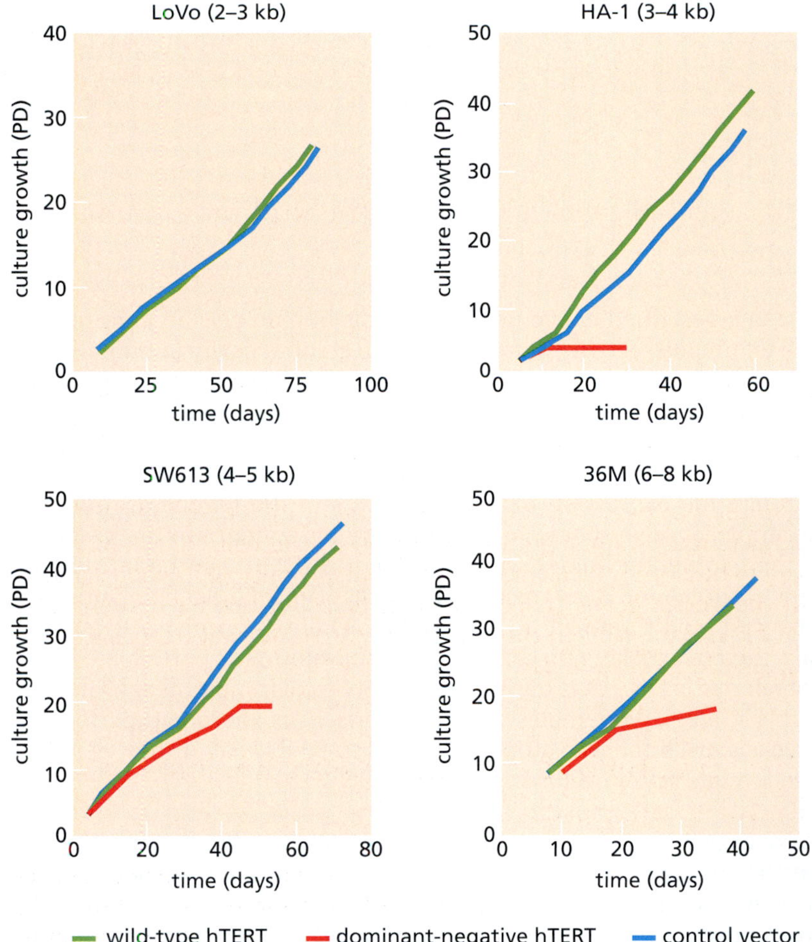

Figure 10.27 Suppression of telomerase activity and resulting loss of the neoplastic growth program Expression vectors specifying a mutant dominant-negative (dn) hTERT protein *(red lines)*, a wild-type enzyme *(green lines)*, or an empty control vector *(blue lines)* have dramatically different effects on four different human cancer cell lines. The respective lengths of their telomeric DNAs, as determined by TRF Southern blots (see Figure 10.13) at the onset of the experiment, are indicated in kilobases (kb). The empty control vector and the hTERT-expressing vector permit the continued proliferation of these cells, as measured in population doublings (PD). However, introduction of the dn hTERT enzyme, which differs from the wild-type enzyme by a single amino-acid substitution in its catalytic cleft, causes these cells to stop growing, doing so with various lag times (in days) following introduction of the dn vector. Microscopic examination revealed that the cells that ceased proliferating entered into crisis and showed widespread apoptosis. Cells of the LoVo cell line *(top left)* entered crisis immediately after expressing the dn hTERT enzyme, and therefore no cells were available to count. (From W.C. Hahn, S.A. Stewart, M.W. Brooks et al., *Nat. Med.* 5:1164–1170, 1999.)

Sidebar 10.7 Telomerase fuels the growth of some pediatric tumors Neuroblastomas are tumors of cells in the peripheral (sympathetic) nervous system and are usually encountered in very young children. These tumors have highly variable outcomes, with some regressing spontaneously and others progressing into invasive, metastatic tumors that ultimately prove fatal. As described earlier (Section 4.5), the N-*myc* gene often undergoes amplification in these tumors, and greater copy numbers of N-*myc* indicate a worse prognosis for the patient. Telomerase activity is an even more useful prognostic indicator of the eventual outcome of the disease. As shown in Figure 10.28A, children whose neuroblastomas are telomerase-negative (as judged by the TRAP assay) do very well in response to therapy, while those whose tumors are telomerase-positive do poorly. Expression of these two genes—N-*myc* and *hTERT*—is likely to be functionally linked, since N-*myc*'s cousin, *myc*, is already known to be a strong inducer of *hTERT* transcription (Sidebar 10.5). (Similar dynamics operate in another childhood tumor—Ewing's sarcoma—which may also arise from cells of the peripheral nervous system; Figure 10.28B.)

Some pathologists argue that neuroblastomas are extremely common in very young children, and almost all of these regress spontaneously and never become clinically apparent. The causes of this regression and the benign outcomes of many clinically detected neuroblastomas that respond well to treatment may now have found a molecular explanation: without significant hTERT expression, neuroblastoma cells lose telomeres and progress into crisis, from which they fail to emerge.

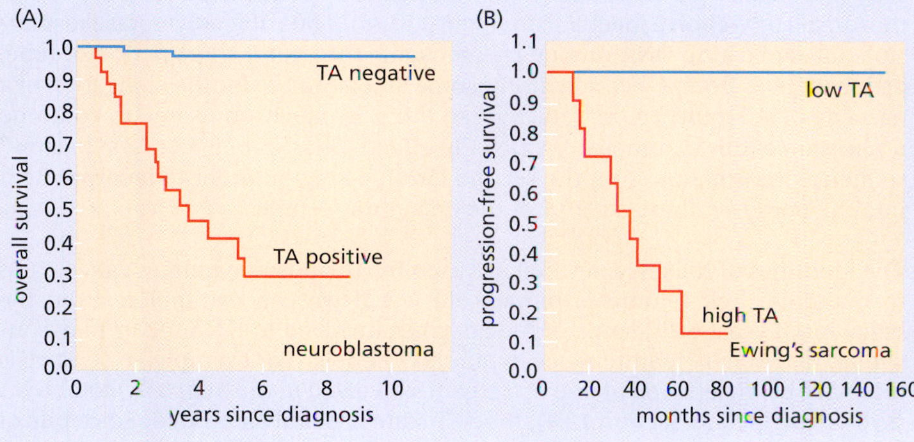

Figure 10.28 Telomerase activity and the prognosis of pediatric tumors The Kaplan–Meier plot has been used here to illustrate the *event-free survival* (or *progression—free survival*) of a group of patients as a function of the time after diagnosis or treatment. (A) In this case, pediatric neuroblastoma patients who had received no cytotoxic chemotherapy treatment were followed for many months *(abscissa)* after initial clinical presentation. These patients were segregated into two groups—those whose tumors were telomerase activity (TA)–positive *(red)*, and those whose tumors appeared to lack this activity *(blue)*; other criteria of tumor staging were not included in establishing these groupings. The ordinate indicates overall survival, i.e., the proportion of patients who remained alive at various times after initial treatment. (B) A similar study of children with Ewing's sarcoma was undertaken, in which the patients were segregated into two groups, depending on whether their tumors had high *(red)* or low *(blue)* telomerase activity. In this case, the ordinate indicates the proportion of children who experienced *progression-free survival,* i.e., those whose tumors had not advanced beyond the stage initially encountered in the clinic. (A, from C. Poremba, C. Scheel, B. Hero et al., *J. Clin. Oncol.* 18:2582–2592, 2000; B, from A. Ohali, S. Avigad, I.J. Cohen et al., *J. Clin. Oncol.* 21:3836–3843, 2003.)

cells as the actions of oncogenes and the inaction of tumor suppressor genes. This conclusion is extended and strengthened by the dramatic outcomes of studies of certain human pediatric tumors (Sidebar 10.7).

Responding to these various discoveries, some researchers have used an automotive analogy to illustrate the growth of human cancer cells: activated oncogenes are said to be akin to accelerator pedals that are stuck to the floor; inactivated tumor suppressor genes are compared to defective braking systems; and telomerase is likened to an agent that ensures that the runaway car has an endless, self-replenishing supply of gasoline in its tank.

10.8 Some immortalized cells can maintain telomeres without telomerase

As noted earlier, 85 to 90% of human tumors have been found to be telomerase-positive. The remaining 10 to 15% lack detectable enzyme activity, yet the cells within this second group of tumors are presumably faced with the need to maintain their telomeres above some minimum length in order to proliferate indefinitely. In fact, these cells have learned to maintain their telomeric DNA using a mechanism that does not depend on the actions of telomerase.

This non-telomerase-based mechanism was first discovered in the yeast *Saccharomyces cerevisiae* following inactivation of one of the several yeast genes encoding subunits of the telomerase holoenzyme. Resulting mutant yeast cells soon entered into a state that is analogous to crisis in mammalian cells. The vast majority of the cells died, but rare variants emerged from these populations of dying cells that used the ALT (alternative lengthening of telomeres) mechanism, which is telomerase-independent, to construct and maintain their telomeres. This ALT mechanism is also used by the minority of human tumor cells that lack significant telomerase activity.

Details of the molecular mechanisms used by ALT-positive human cells to maintain their telomeres are poorly understood. An important clue, however, comes from an experiment in which a traceable molecular marker—a neomycin resistance gene—was introduced into the midst of the telomere repeat sequence of one chromosome carried by a mammalian cell in the ALT state. When the telomeric DNAs were examined in the descendants of this cell, copies of this molecular marker were found in a number of other telomeres as well (Figure 10.29). Hence, sequence information appears to be exchanged between the telomeres of cells in the ALT state.

This exchange of sequence information seems to depend on a type of interchromosomal **copy-choice** mechanism (Figure 10.30). Thus, the polymerase responsible for replicating DNA on one chromosome may, for a short period of time, use sequences from a second chromosome as a template for the elongation of a nascent DNA strand before returning to the original chromosome to continue replication of this chromosome's DNA. In effect, the DNA polymerases "borrow" sequence information from the second chromosome in order to incorporate it into the newly synthesized copy of the first chromosome.

The identities of the enzymes that are used by ALT-positive human cancer cells to maintain their telomeres remain elusive. If we can extrapolate from the behavior of yeast cells, some of the enzymes involved in DNA repair may contribute to telomere maintenance in human ALT cells. For example, in yeast cells that have been deprived of one of the genes involved in *mismatch repair* of DNA (to be described in Section 12.4), the ALT state is readily activated. Since one of the functions of mismatch repair is to suppress recombination between imperfectly homologous DNA sequences, this finding supports the notion that interchromosomal recombination is an important mechanistic component of the ALT state.

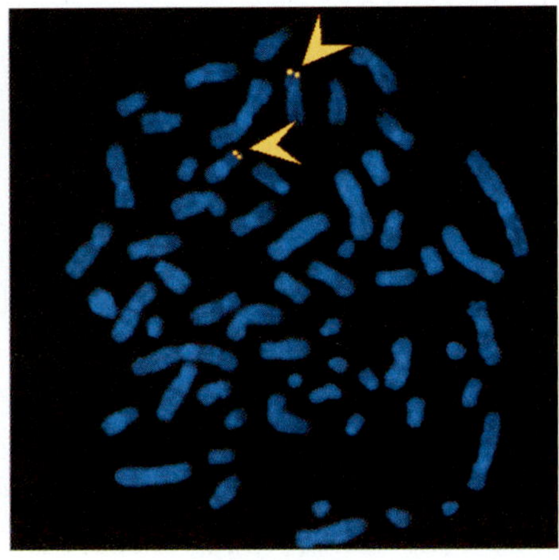

early passage cells

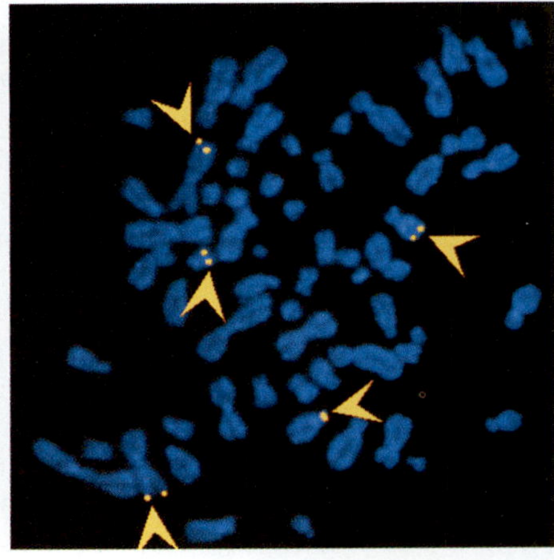

40 population doublings later

Figure 10.29 The ALT mechanism The alternative lengthening of telomeres (ALT) mechanism can be shown to depend, at least in certain cells, on the exchange of sequence information between the telomeres of different chromosomes. As seen here, two of the telomeres of human GM847 cells were labeled via the targeted insertion (using homologous recombination) of a genetic marker into one of their telomeres. By the time of the analysis, the marker was already present in two telomeres *(arrows, left panel)*. Forty population doublings later, this sequence could be found to be associated with five distinct chromosomes *(arrows, right panel)*. In both cases, the fluorescence *in situ* hybridization (FISH) technique was used to detect these marker sequences in metaphase chromosomal preparations. (From M.A. Dunham, A.A. Neumann, C.L. Fasching, and R.R. Reddel, *Nat. Genet.* 26:447–450, 2000.)

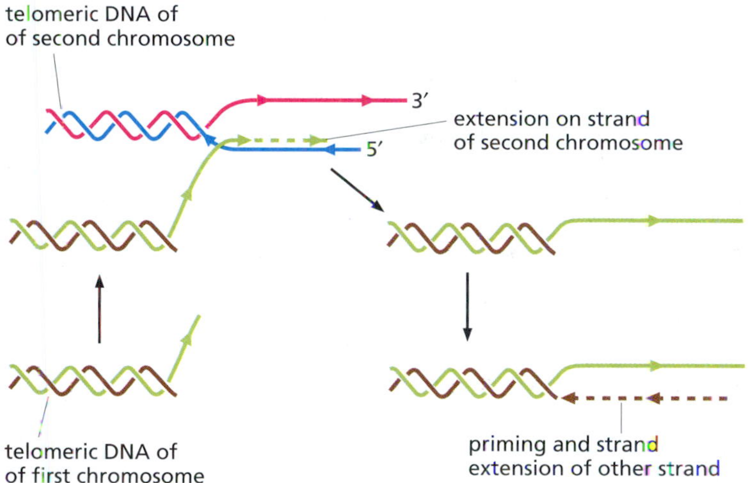

Figure 10.30 Copy choice The exchange of sequence information between telomeres that occurs in ALT (Figure 10.29) may have its basis in DNA polymerases using more than one chromosome as a template during chromosome replication. According to one plausible mechanism, one telomere *(brown and green, lower left)* extends its 3′ overhanging end *(green)*, which displaces a strand of the same polarity *(pink)* in the telomere of another chromosome, allowing this 3′ overhanging end to anneal to the complementary strand *(blue)* of this other telomere. Once it is associated with this other telomere, conventional DNA polymerases can extend this strand *(broken green line)* by using the complementary *(blue)* strand as template *(top)*. Subsequently, the resulting elongated G-rich strand can disengage from the complementary *(blue)* strand *(right)*. This recently elongated G-rich strand can be converted by other conventional DNA polymerases into double-stranded form *(broken brown line, lower right)*. This can be repeated dozens of times, resulting in the transfer of sequence information from one telomere to another and the lengthening of telomeres by many kilobases without the involvement of the telomerase enzyme. (Note that telomeres with 3′ overhanging ends can be created in the absence of telomerase by the actions of exonucleases on the C-rich strand.) The arrowheads indicate the 5′-to-3′ orientation of the DNA strands.

The existence of the ALT mechanism emphasizes the fact that all types of human cancer cells must develop ways to maintain their telomeres above a certain threshold length in order to proliferate in an immortalized fashion. Most human cancer cells de-repress expression of the *hTERT* gene, and a small minority of incipient cancer cells find a way to activate the ALT mechanism.

The ALT state is associated preferentially with a specific subset of the tumors encountered in oncology clinics. These include perhaps half of osteosarcomas and soft-tissue sarcomas as well as many glioblastomas. Perhaps these associations may one day be explained by the fact that the spectrum of proteins—ostensibly DNA repair proteins—that are expressed in these particular types of cancer cell types allows the ALT state to be activated far more readily than in most other kinds of cancer cells.

However, at the moment, the precise reasons why one mechanism (involving hTERT) is usually favored over the other (ALT) remain obscure. Equally unclear are the mechanisms that cause the telomeric DNA of human cells in the ALT state to grow to lengths (>30 kb) far greater than those usually seen in telomerase-positive cells (5–10 kb). Importantly, the dn hTERT enzyme (Section 10.7), which is able to induce crisis in a number of telomerase-positive human cancer cell lines, fails to do so in a human ALT-positive tumor cell line, providing further support of the notions (1) that ALT-positive cells don't depend on telomerase for their growth, and (2) that the dn hTERT enzyme is not intrinsically cytotoxic.

The hTERT enzyme represents a very attractive target for those researchers who are intent on developing novel types of anti-cancer therapeutics. The ALT state, however, holds important implications for such development. As we will learn in Chapter 16, the catalytic clefts of enzymes like hTERT can often be blocked by highly specific therapeutic drug molecules. Significantly, hTERT is a distant relative of the reverse transcriptase of human immunodeficiency virus (HIV; see Figure 10.22), against which effective drug inhibitors have been successfully developed. These existing successes greatly increase the likelihood that anti-hTERT compounds will one day be produced as well. In addition, this enzyme is expressed in the great majority of human cancers, while not being present at significant levels in normal tissue. This offers the prospect of drug selectivity—being able to affect cancer cells while leaving cells in normal tissues untouched.

Imagine that an inhibitor of hTERT is indeed developed by a drug company—a goal that has proven elusive until now. It is already clear that at least 10% of human tumors will never respond to such an anti-telomerase drug, because they are ALT-positive and therefore do not depend on this enzyme for their continued proliferation. While the remaining approximately 90% of human tumors, which do indeed exhibit robust telomerase activity, may initially be susceptible to the killing effects of this drug, sooner or later, rare variant cancer cells may arise that activate the ALT mechanism. This will allow the variants to evade the killing actions of the anti-telomerase drug. Consequently, even an anti-telomerase drug that might initially seem to be a "silver bullet" (because of its ability to selectively kill a diverse array of human tumor cell types) may ultimately be outmaneuvered by the ever-changing genotypes and phenotypes of cancer cells.

10.9 Telomeres play different roles in the cells of laboratory mice and in human cells

Our descriptions of telomerase expression repeatedly refer to its actions in human cells, with good reason. Rodent cells, specifically those of the laboratory mouse strains, control telomerase expression in a totally different way. While human cells repress hTERT expression to low levels in almost all postembryonic cell lineages, the cells in comparable tissues of the laboratory mouse continue to express significant levels of telomerase throughout life. This has important consequences for the role of telomeres in murine cell biology and, in turn, for the relative susceptibility of human and mouse cells to malignant transformation.

The expression of telomerase throughout life in murine tissues is likely to be the cause of the relatively long telomeres normally found in the cells of laboratory mice; the double-stranded region of their telomeric DNA is as much as 30 to 40 kb long—about five times longer than corresponding human telomeric DNA. In fact, mouse telomeres are so long that they are never in danger of eroding down to critically short lengths during the lifetime of a mouse, even when expression of mouse telomerase is suppressed experimentally.

The lengths of mouse telomeres permit cell lineages to pass through a far larger number of replicative generations than are required for tumor formation. This indicates that laboratory mice do not rely on telomere length to limit the replicative capacity of their normal cell lineages and that telomere erosion cannot serve as a mechanism for constraining tumor development in these rodents. Nevertheless, it seems that the significant level of telomerase activity normally present in mouse somatic cells and their long telomeres are still not sufficient for the robust growth of tumor cells in these animals (Sidebar 10.8).

To the extent that cells derived from laboratory mice show limited replicative ability *in vitro*, that ability is never determined by telomere shortening. In stark contrast to the behavior of human cells, mouse cells can be immortalized relatively easily following extended propagation in culture (which selects for cells that are spontaneously immortalized *in vitro*). Human cells require, instead, the introduction of both the SV40 *large T* oncogene (to avoid senescence) and the *hTERT* gene (to avoid crisis).

These differences suggest that one of the central mechanisms governing human cell biology—telomere shortening—is irrelevant to the biology of mouse cells, at least those of laboratory mice. The biological rationale behind these interspecies differences in the organization of an important cellular control circuit remains unclear. One attractive speculation is that humans, whose cells pass through approximately 10^{16} mitoses in an average lifetime, have a far greater risk of cancer than do mice, which experience only about 10^{11} cell divisions (i.e., mice have only about 0.1% as many cells as humans and live on average only about 1% of a human lifespan). This 10^5-fold larger number of cell divisions in humans would seem to create a proportionately greater lifetime cancer risk, dictating the need for large, long-lived mammals, such as humans, to develop additional anti-tumor mechanisms beyond those operating in small, short-lived

Sidebar 10.8 Long telomeres do not suffice for tumor formation Experiments with several strains of transgenic mice show that the mouse homolog of hTERT, termed mTERT, provides some function that contributes to tumorigenesis even though the mouse cells in which this enzyme acts already possess very long (>30 kb) telomeres. For example, one strain of transgenic mice engineered to overexpress mTERT in the basal keratinocytes of the skin show a twofold increase in the rate of skin cancer, while another strain of mice that overexpress mTERT in a wide variety of tissues show greatly increased risk of mammary carcinomas. It is also true that mTERT activity, which is already present at considerable levels in normal mouse cells, increases progressively during tumor development in the mouse. Clearly, the mTERT enzyme aids tumorigenesis through mechanisms other than simple telomere extension. Since the TERT enzyme is likely to act largely on telomeres, it is possible that this additional function involves protection of the single-strand overhang of the G-rich strand at the ends of telomeres (see Sidebar 10.6).

mammals, such as mice. (Attempts at drawing broad, generalizable conclusions from these interspecies differences are complicated by the fact that certain strains of wild-type mice control their telomerase and telomeres much like humans.)

The role of telomeres in the lives of mice has been revealed dramatically by researchers who have inactivated (see Sidebar 7.10) the gene encoding the *mTR* RNA subunit of telomerase in the mouse germ line. As expected, the homozygous mutant offspring of initially created heterozygous mice carrying the mutant *mTR* allele exhibit no telomerase activity in any of their cells. During their lifetimes, the telomeres of these telomerase-negative mice may shorten from approximately 30 kb down to about 25 kb; the latter length is still far longer than is required to protect chromosomal ends.

These telomerase-negative mice are indistinguishable phenotypically from wild-type mice. This absence of any observable mutant phenotype provides compelling evidence that the telomerase enzyme plays no essential role in the tissues of these mammals beyond its task of maintaining telomeres.

The homozygous telomerase-negative (i.e., $mTR^{-/-}$) mice can be bred with one another through at least three more organismic generations without showing any discernible phenotype. Finally, however, in the fifth mouse generation, telomerase-negative mice begin to show distinctive phenotypes—an indication that after five organismic generations without telomerase, their telomeres have eroded down to dangerously short lengths (Figure 10.31). These fifth-generation $mTR^{-/-}$ mice and, even more so, their progeny in the sixth generation are sickly and show a diminished capacity to heal wounds, indicative of the inability of their cells to respond properly to mitogenic signals. The sixth-generation $mTR^{-/-}$ mice suffer also from substantially reduced fertility.

The $mTR^{-/-}$ mice that are born in this sixth generation already exhibit very short telomeres at birth, and their inability to subsequently maintain these telomeres, particularly in mitotically active tissues, results in widespread cell death and loss of tissue function. For example, highly proliferative tissues, such as the gastrointestinal epithelium, the hematopoietic system, and the testes, show substantial **atrophy** (loss of cells). These observations are striking and without parallel in the field of mouse genetics, because they demonstrate organismic phenotypes that only become manifest five or six organismic generations after a mutation has been introduced into the germ line of these animals.

Moreover, such observations dramatically demonstrate the differences between the telomeres of human cells and those of these mutant laboratory mice.

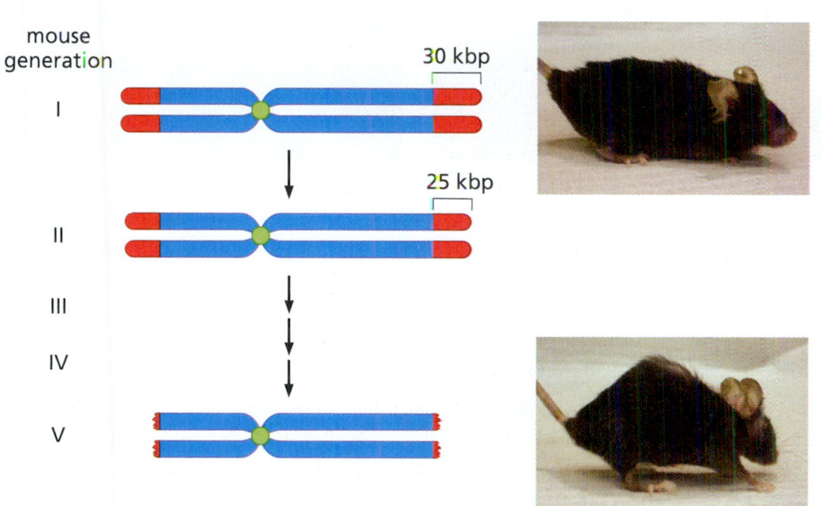

Figure 10.31 Erosion of telomeres over multiple generations in populations of $mTR^{-/-}$ mice During the lifetimes of the first $mTR^{-/-}$ mouse generation, their telomeres, which are initially 30 kbp long *(red)*, are reduced in size by about 5 kbp *(left)*. These mice *(top, right)* are phenotypically normal, as are their descendants in the second and third generations. However, by the fourth and fifth organismic generations, the telomeres in their cells have eroded to the extent that they can no longer protect the ends of chromosomes; the tissues of these mice begin to lose their ability to renew themselves and heal wounds, and these mice begin to show symptoms of premature aging, including wasting away of muscle tissues and a hunched back *(bottom, right)*. The relative sizes of the telomeres in these images have been exaggerated 1000-fold for the sake of illustration. (Courtesy of R.S. Maser and R.A. DePinho.)

Human telomeres begin relatively short, and loss of telomerase activity can lead, already in the first organismic generation, to severe organismic phenotypes (Sidebar 10.9).

10.10 Telomerase-negative mice show both decreased and increased cancer susceptibility

Laboratory mice are susceptible to spontaneous cancers, largely lymphomas and leukemias. This background of susceptibility can be increased by experimentally introducing specific mutations (Sidebar 7.10) into the mouse germ line that cause inactivation of tumor suppressor genes or activation of proto-oncogenes. To what extent do the critically short telomeres in fifth- and sixth-generation telomerase-negative mice affect the cancer rate in these animals?

Sidebar 10.9 Defective telomerase function explains a rare human familial syndrome Humans suffering from the inherited condition known as dyskeratosis congenita exhibit defects in highly regenerative tissues, such as skin and bone marrow, as well as reduced chromosomal stability. In the X-linked form of the disease, a protein known as dyskerin is absent or malfunctioning. Dyskerin associates with the telomerase-associated RNA subunit (*hTR*) and helps *hTR* to assemble in the telomerase holoenzyme complex (see Figure 10.23). An autosomal dominant version of the same disease is caused by the inheritance of a defective *hTR*-encoding gene. Cells in proliferating tissues of affected individuals exhibit dicentric chromosomes, which are indicative of interchromosomal fusions. Some affected individuals succumb to a collapse of their hematopoietic system due to the loss of regenerative capacity of the stem cells in their bone marrow (Figure 10.32).

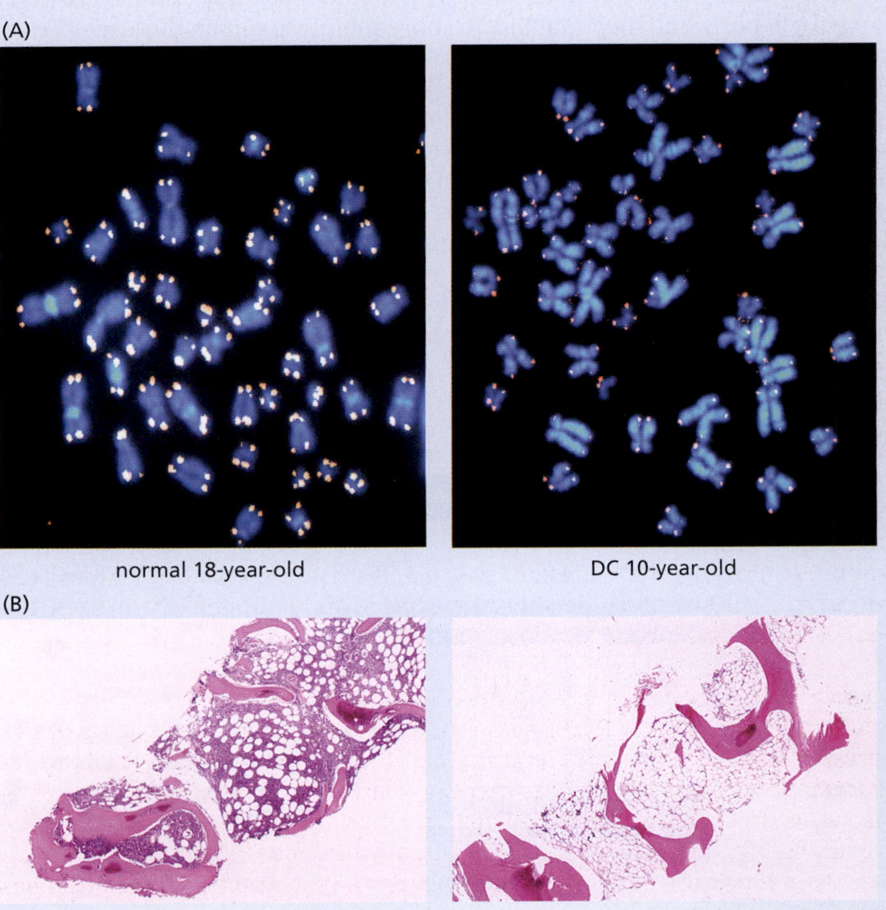

Figure 10.32 Telomere and bone marrow collapse of dyskeratosis congenita patients Humans suffering from dyskeratosis congenita (DC) experience atrophy of a number of tissues, which is traceable to the absence of functional telomerase and to the catastrophic collapse of their telomeres. (A) Fluorescence *in situ* hybridization (FISH) has been used to detect the telomeres in a metaphase spread of chromosomes in a cell of a normal 18-year-old *(left)* as well as in a cell of a 10-year-old suffering from DC. Many of the chromosomes in the latter cell have telomeres that are so short that they are no longer detectable by FISH. (B) The most life-threatening tissue atrophy of DC patients occurs among the rapidly proliferating cells of the bone marrow. As indicated here, in normal bone *(left panel)*, regions between mineralized bone *(pink)* are filled with hematopoietic cells of the marrow *(purple)*. However, cells in the marrow of dyskeratosis patients suffering germ-line mutations in the *hTR* gene *(right panel*; see also Figure 10.23B*)* are almost totally absent because the stem cells in these patients have lost the ability to maintain the various lineages of hematopoietic cells. (A, courtesy of S. Kulkarni and M. Bessler; B, courtesy of M. Bessler.)

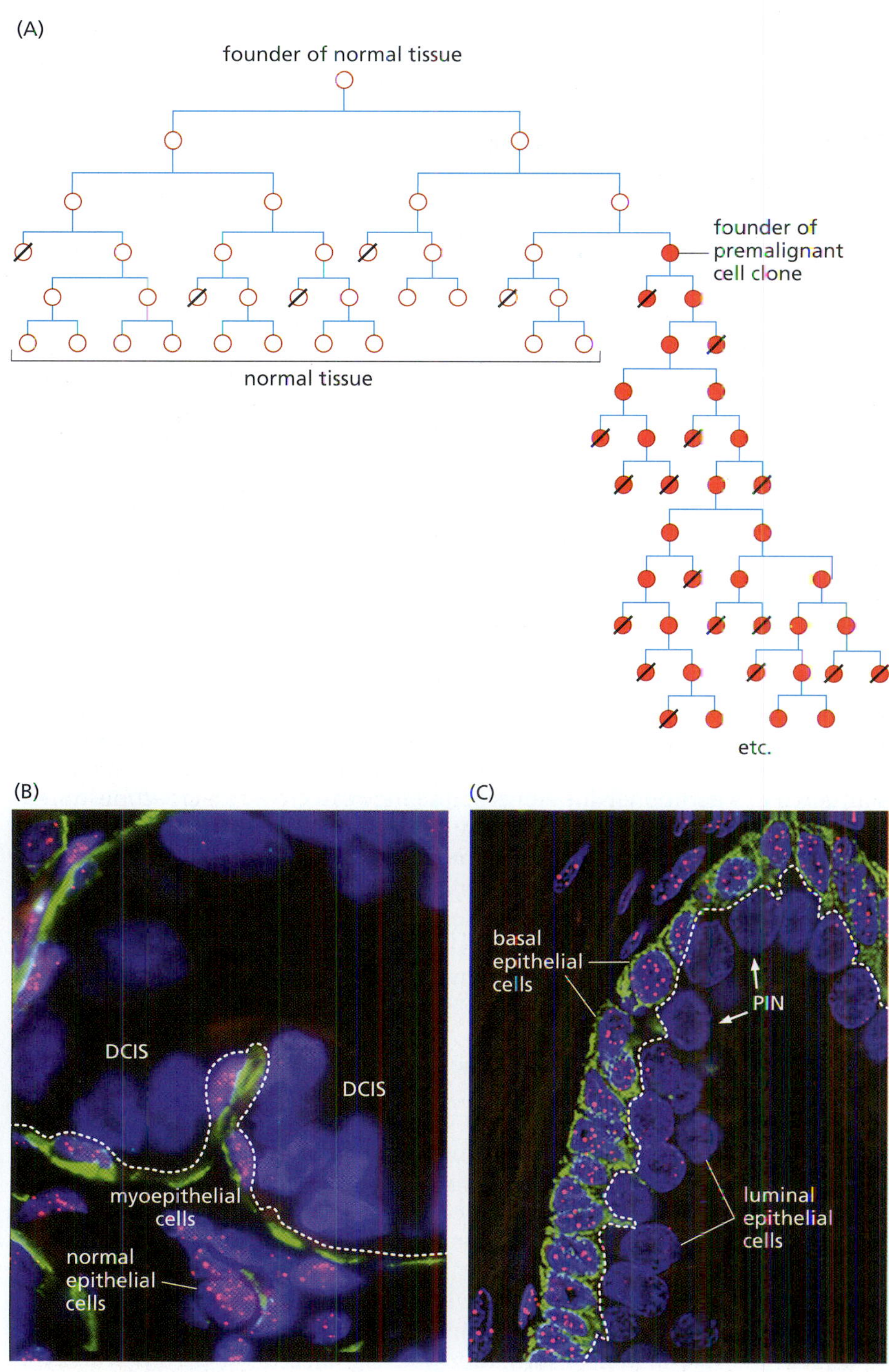

Figure 10.33 Normal and neoplastic cell lineages in mTR$^{-/-}$ mice and in humans (A) This diagram indicates the pedigree of normal cells *(open circles)* in a small patch of tissue. A founder cell in the embryo becomes the ancestor of all the cells in this tissue, e.g., all the epithelial cells. At one point, one of these epithelial cells suffers changes that make it the ancestor of the cells *(closed red circles)* that develop into a pre-cancerous cell clone. This clone of cells must pass through many more replicative generations in order to eventually form a tumor mass. Slashes indicate the substantial loss of cells through apoptosis or necrosis that occurs in almost every cell generation. (B) The large number of cell generations through which pre-malignant cells must pass en route to neoplasia results in considerable telomere erosion. In these images, telomeres are revealed by fluorescence *in situ* hybridization (FISH, *pink dots).* In normal human mammary ducts, telomeres are apparent in both the luminal epithelial cells *(below)* and the myoepithelial cells *(not shown).* However, in ductal carcinoma *in situ* (DCIS), the luminal epithelial cells, which constitute the neoplastic cell population *(above dotted line)*, have lost all detectable telomeric DNA, while the nearby normal myoepithelial cells *(below dotted line)*, which have been stained for α-smooth muscle actin *(green)*, still carry strongly staining telomeres. (In all cases, cell nuclei are stained *blue.*) (C) Similar dynamics apply in the case of prostatic intraepithelial neoplasia (PIN)—a precursor lesion to prostatic carcinoma—in which many of the luminal epithelial cells lining ducts *(below and right of dotted line)* have lost telomeric DNA staining, while the underlying keratin-positive *(green, above and left of dotted line)* basal epithelial cells *(above and left)* continue to display a strong telomeric DNA signal. (B and C, courtesy of A.K. Meeker; see also A.K. Meeker, J.L. Hicks, E.A. Platz et al., *Cancer Res.* 62:6405–6409, 2002.)

The answer to this question would seem to be straightforward. We can imagine that the deterioration of cells in these *mTR*$^{-/-}$ mice leaves a barely adequate number of healthy cells to sustain tissue and organismic viability. Stem cells charged with the task of replenishing the pools of differentiated cells in these tissues will likely have great difficulty doing so.

Since the normal cells in these tissues are barely able to maintain themselves, incipient cancer cells should, by all rights, have an even more difficult time. The process of tumor development requires that clones of pre-malignant cells evolving toward malignancy must pass through a large number of growth-and-division cycles—a number substantially greater than those experienced by nearby normal cells (Figure 10.33A). Consequently, any tumor development that is successfully initiated by these cell clones is likely to be aborted long before it has reached completion.

In order to test these predictions, mice were first made cancer-prone by inactivating tumor suppressor genes in their germ line (Sidebar 7.10), in this case the locus encoding both the 16^{INK4A} and $p19^{ARF}$ tumor suppressor proteins (Sections 8.4 and 9.7). Such mice, as might be expected, are highly susceptible to cancer; indeed, they frequently develop lymphomas and fibrosarcomas relatively early in life. This susceptibility is manifested even more dramatically when they are exposed to carcinogens. In one set of experiments, sequential exposure to dimethylbenzathracene (DMBA), a potent carcinogen, followed by repeated exposures to ultraviolet-B (UV-B) radiation, led after 20 weeks to a 90% tumor incidence in the ($16^{INK4A}/p19^{ARF}$-negative) mutant mice, while a control group of similarly treated wild-type mice had no tumors at this time.

This experiment was extended by introducing the $16^{INK4A}/p19^{ARF}$ germ-line inactivation into mice that also lacked *mTR*, the gene encoding the RNA subunit of the mouse telomerase holoenzyme (see Figure 10.31). The results were that 64% of control, telomerase-positive, $16^{INK4A}/p19^{ARF}$-negative mice contracted tumors, while only 31% of the generation 4 and 5 $mTR^{-/-}$ $16^{INK4A}/p19^{ARF}$-negative mice developed tumors. These results were even more dramatic when survival of mice after 16 weeks was measured: 88% of the telomerase-positive mice had been lost to cancer while only 46% of the fifth-generation telomerase-negative mice had been lost to this disease.

This reduced rate of cancer in telomerase-negative mice fulfills the prediction that the normal tissues of $mTR^{-/-}$ mice have exhausted most of their endowment of replicative generations even before tumorigenesis has begun. Once tumors are initiated in these mice, incipient tumor cell populations must pass through many additional doublings before they can create macroscopic tumors (see Figure 10.33A). However, relatively early during the course of tumor formation, these aspiring cancer cells will be driven into crisis by telomere collapse and their agenda of forming tumors will be aborted. Indeed, in human tissues, pre-malignant cells that are poised to become fully neoplastic already show drastically truncated telomeres; this truncation is ostensibly due to the fact that these cell populations have passed through many more successive cell cycles than nearby normal cells (Figure 10.33B and C).

A very different and ultimately far more interesting outcome was seen when mice were used that had been deprived of both *p53* gene copies in their germ line (rather than inactivating their $16INK4A/p19^{ARF}$ locus, as described above.) Ordinarily, germ-line inactivation of the *p53* gene in mice leads, on its own, to an increased tumor incidence and resulting mortality, mirroring aspects of the Li–Fraumeni syndrome of humans (Section 9.12). The mutant *mTR* locus was then introduced into the $p53^{-/-}$ genetic background of these mice.

When the $p53^{-/-}$ genotype was present in the genomes of fifth- and sixth-generation telomerase-negative $mTR^{-/-}$ mice, something quite unexpected was observed: the rate of cancer formation was significantly *increased* above that created by the $p53^{-/-}$ genotype alone. In addition, the spectrum of tumors, namely lymphomas and angiosarcomas, that are commonly seen in $p53^{-/-}$ mice was shifted in favor of carcinomas—just the types of tumors commonly seen in humans. These trends became even more apparent in the seventh and eight generation of telomerase-negative mice (Figure 10.34).

How can these intriguing results be explained? We can imagine that as fifth- and sixth-generation mice exhaust their telomeres, their cells will begin to experience chromosomal breakage–fusion–bridge (BFB) cycles (see Figure 10.14) and thus crisis. We already know that double-strand DNA breaks, which are formed in BFB cycles, can provoke apoptosis through a p53-dependent pathway (Section 9.6). Hence, cells that carry functional *p53* genes and experience BFB cycles are likely to be rapidly eliminated from tissues.

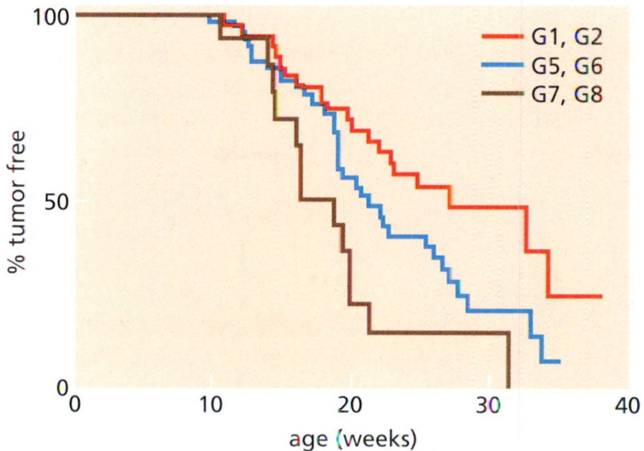

Figure 10.34 Rate of tumor formation in cancer-prone *mTR*$^{-/-}$ *p53*$^{-/-}$ mice These Kaplan–Meier plots reveal that tumor incidence actually increases with increasing organismic generation in mice that lack both the p53- and mTR-encoding genes in their germ line. In generations 1 and 2 (G1, G2, *red*), about 50% of the mice had readily detectable tumors by the age of 24 weeks, while in generations 5 and 6 (G5, G6, *blue*), the mice developed tumors more rapidly. In generation 7 or 8 (G7, G8, *brown*), about 50% of the mice already exhibited tumors by the age of 17 weeks. (From L. Chin, S.E. Artandi, Q. Shen et al., *Cell* 97:527–538, 1999.)

As cells of fifth- and sixth-generation *mTR*$^{-/-}$ *p53*$^{-/-}$ mice experience BFB cycles, they may well struggle to stay alive because of the repeated breakage of their chromosomal DNA. However, many of these cells may manage to survive because a key component of their pro-apoptotic response machinery—p53—is missing. These cells will now limp through a number of growth-and-division cycles in spite of the ongoing karyotypic chaos that afflicts their genomes. All the while, their chromosomes will participate in numerous BFB cycles.

These cycles may continue for many cell generations, since p53 will not be on watch to trigger apoptosis and eliminate these cells. As a consequence, the genomes of such cells will become increasingly scrambled by the nonreciprocal chromosomal translocations generated by the BFB cycles. In addition, and perhaps even more important, there is evidence that dsDNA breaks lead to the amplification or deletion of the chromosomal regions adjacent to these breaks.

While these BFB cycles will not prove to be immediately fatal for the p53-negative cells, the cycles will surely compromise their proliferative ability. This means that individual cells in these populations will be under great selective pressure to escape these BFB cycles and to reacquire karyotypic stability; once they do so, they will be able to grow more rapidly. In these cells, however, the acquisition of telomerase activity is not an option (because of the germ-line knockout of *mTR* gene copies). For this reason, these cells will resort to activating the ALT telomere maintenance system. Having done so, these cells will stabilize whatever aberrant karyotypes arose since the time when their ancestors first lost functional telomeres and will begin to grow robustly again.

Clearly, the scrambled karyotypes that these cells have acquired generate novel combinations of genes (including many translocated, deleted, and amplified genes), and among the novel genotypes that are created, there will likely be some that *favor* neoplastic proliferation. Stated differently, the period of genetic instability appears to *increase* the probability that cancer-promoting genetic configurations will be produced. Such reasoning explains how collapsing telomeres in a p53-negative background can actually promote the development of tumors.

These *mTR*$^{-/-}$ mice have human counterparts: patients suffering the dyskeratosis congenita syndrome (see Sidebar 10.9) often go on to develop cancer once their collapsing hematopoietic systems have been successfully reconstructed by bone marrow transplantation. Some develop hematopoietic malignancies (myelodysplasia and acute myelogenous leukemia) or carcinomas of the gastrointestinal tract. These neoplasias arise in those tissues that undergo constant, intense proliferation—precisely the sites where telomeres might become rapidly eroded in the absence of telomerase activity, resulting in repeated breakage–fusion–bridge cycles.

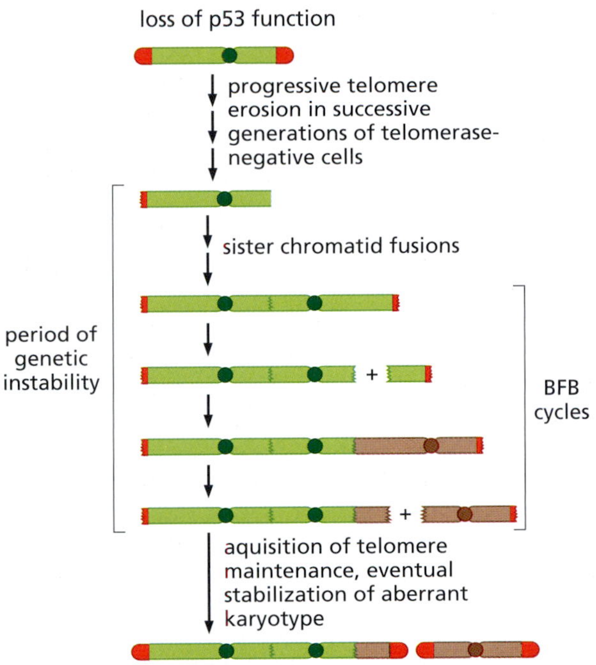

Figure 10.35 A mechanistic model of how BFB cycles promote human carcinoma formation During the early steps of tumor progression, inactivation of the *p53* tumor suppressor gene is often favored, because it allows incipient cancer cells to escape apoptotic death due to oncogene activation, inadequate levels of trophic signals, anoxia, and poisoning by metabolic wastes due to inadequate vascularization. Later, as tumor progression proceeds, the telomeric DNA of evolving pre-malignant cell populations will erode below a level needed to protect chromosomal ends. Breakage–fusion–bridge (BFB) cycles (see Figure 10.14) will ensue, which will lead to increasing chromosomal rearrangements and the amplification and deletion of chromosomal segments adjacent to breakpoints *(not shown)*.

Cells undergoing BFB cycles would normally be eliminated by p53-induced apoptosis. However, in the absence of p53 function, such cells may survive, albeit with scrambled karyotypes that will slow down their proliferation but will not trigger their apoptotic death. Emerging from such cell populations will be variants that have learned how to regenerate telomeres and thereby stabilize their karyotype; such cells can once again grow rapidly. In cells of the mutant $mTR^{-/-}$ mice (Figure 10.34), this stabilization may be achieved through the activation of the ALT telomere maintenance system, while in wild-type organisms (including human cancer patients), de-repression of hTERT expression allows reconstruction of telomeres to proper length *(red)*. Once telomeres have been regenerated at the ends of all chromosomes, the further generation of karyotypic disorder through BFB cycles will be halted. However, any karyotypic disorder that was generated previously during BFB cycles will be perpetuated in descendant cell populations.

10.11 The mechanisms underlying cancer pathogenesis in telomerase-negative mice may also operate during the development of human tumors

These observations of genetically altered mice, when taken together, lead us to consider a most interesting mechanistic model (Figure 10.35) that explains why aneuploidy is seen in the great majority of human carcinomas. Imagine that relatively early in the long process of human tumor development, cells in pre-malignant cell populations are able to jettison their *p53* gene copies. The selective pressure favoring this genetic loss might derive, for example, from the fact that cells in the early pre-malignant cell populations experience anoxia and, in turn, suffer from p53-induced apoptosis. Alternatively, *p53* gene copies may have been shed by incipient cancer cells in order to reduce the likelihood of oncogene-triggered apoptosis (see Section 9.6).

Some time after the loss of p53 function, as tumor progression proceeds, cells in these pre-malignant populations will begin to experience substantial telomere erosion and eventually telomere collapse, since they lack significant levels of telomerase activity. Repeated BFB cycles will ensue and result in aneuploidy. However, because these cells also lack functional p53, they will survive and even continue to proliferate, albeit slowly.

Sooner or later, in this population of cells suffering BFB cycles, a cell will emerge that has acquired the ability to de-repress *hTERT* gene expression and, having done so, has gained the ability to stabilize its karyotype and prevent further BFB cycles. With telomerase on hand, we can imagine that these cells will repair the ends of their chromosomal DNA molecules and once again enjoy robust growth. Of course, the belated appearance of telomerase on the scene cannot unscramble the considerable aneuploidy that accumulated during the time window that began with telomere collapse and ended with the acquisition of *hTERT* expression. (The same arguments will apply in the event that the ALT mechanism, rather than telomerase activation, is used to protect the genome from further BFB cycles.)

Observations of human pancreatic cancer progression provide support for this model. Relatively early in the long, multi-step process of tumorigenesis in the pancreas, the mitotic cells in low-grade adenomas have few of the anaphase bridges that are characteristic of BFB cycles. However, more advanced, highly dysplastic adenomas show large numbers of these bridges. Still later, the even more advanced, *in situ* carcinomas once again exhibit lower levels of anaphase bridges. This behavior reinforces the model, described above, that proposes that BFB cycles and associated chromosomal instability are found only in a defined window of time during the course of multi-step tumor progression.

Additional support for this model comes from a study of human carcinomas of the esophagus, colon, and breast. In each case, tumors possessing relatively short telomeres are strongly associated with poor long-term prognosis, while tumors carrying long telomeres are associated with a far better prognosis, including greater long-term survival. Here, once again, we can imagine that the chromosomes of cancer cells with sub-optimal telomere lengths undergo relatively frequent breakage–fusion–bridge cycles, perhaps because they express hTERT at levels that are barely adequate to maintain telomeric DNA. These BFB cycles, in turn, create an ongoing chromosomal instability that continues throughout the life of the tumor to generate novel, scrambled genotypes, some of which may favor phenotypes such as more rapid proliferation and increasing aggressiveness. Certain human chronic inflammatory conditions, which are associated with increased risks of cancer, may also be attributable to eroded telomeres and BFB cycles (Sidebar 10.10).

This model provides an attractive mechanism that explains how many types of human tumor cells acquire highly aneuploid karyotypes. Although not directly demonstrated, it is widely assumed that the resulting aneuploid genomes confer growth advantages on these cells by creating novel oncogenes through translocations, by increasing the dosage of growth-promoting proto-oncogenes, and by eliminating tumor suppressor genes that have been holding back cell proliferation. Hence, these BFB cycles may be instrumental in accelerating tumor progression because they increase genomic mutability and allow evolving, pre-malignant cells to explore a multitude of novel, possibly advantageous genotypes.

These speculations still do not address an issue raised by some observations cited earlier: the repeated BFB cycles in *mTR*$^{-/-}$ *p53*$^{-/-}$ mice affect the *types* of tumors that these mice exhibit, causing them to develop carcinomas, which are common in humans, rather than hematopoietic and mesenchymal malignancies, which are frequently seen in mice. How does telomere biology possibly help to explain why cancers tend to arise in some tissues and not others?

10.12 Synopsis and prospects

When Barbara McClintock first described telomeres at the ends of maize chromosomes, she could not have anticipated the dramatic convergence of her discoveries with the molecular models of cancer pathogenesis. Telomeres are now known to be major determinants of the ability of cells to multiply for a limited number of growth-and-division cycles before halting proliferation and entering into the state that we call crisis. This circumscribed proliferative potential of normal human cells appears to operate as an important barrier to the development of cancers by limiting the proliferation of pre-neoplastic cell clones.

This mechanistic model of cell immortalization and cancer pathogenesis is elegant, if only because it is so simple. Still, the simplicity of this model should not obscure the fact that much of telomere structure and function remains poorly understood, and some of the conclusions described in this chapter may one day require substantial revision.

Sidebar 10.10 Telomere collapse may contribute to cancer in organs affected by chronic inflammation In Chapter 11, we will read about a number of human tumors arising in tissues that suffer continuous loss of cells due to chronic infections or inflammation. Such ongoing attrition of cells, which may occur over a period of decades, occurs in tissues such as those affected by ulcerative colitis, Barrett's esophagus, and hepatitis B or C virus infection. In response to losses of differentiated cells in the affected tissues, the stem cells in these tissues are continually producing replacements to ensure maintenance of tissue functions. These stem cells are therefore forced to pass through many more cycles of growth and division than are the corresponding stem cells of normal tissues. The resulting repeated cell cycles may tax the regenerative powers of the stem cell pools and lead eventually to telomere collapse, to the triggering of breakage–fusion–bridge cycles, and to the generation of karyotypes that favor neoplastic cell proliferation.

Such a mechanism may well account for the observed high rates of cancer in individuals suffering from these various inflammatory and infectious diseases. In fact, in the intestinal epithelial cells of individuals suffering from ulcerative colitis, one can often find the anaphase bridges that are telltale signs of BFB cycles and hence of telomere collapse—precisely what we would expect if telomere collapse were contributing to the appearance of the colon carcinomas seen in these patients (Figure 10.36). Similarly, anaphase bridges have been documented in the cells of Barrett's esophagus (Figure 2.13), a condition in which the reflux of stomach acid causes a high turnover of esophageal epithelial cells that leads, with significant frequency, to esophageal carcinomas.

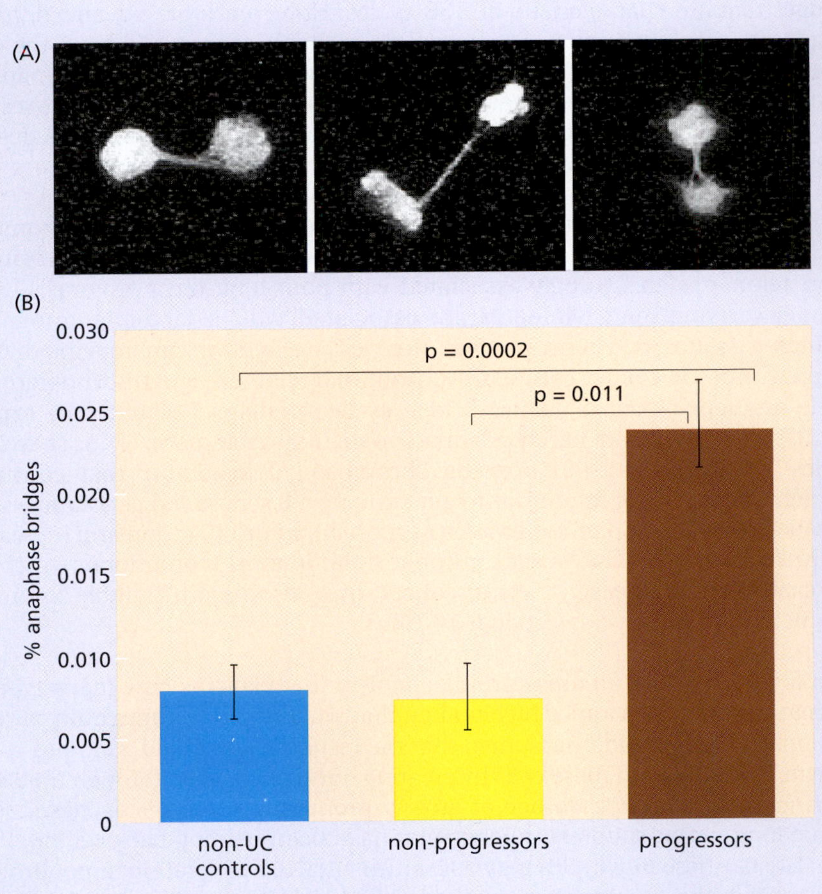

Figure 10.36 Ulcerative colitis and anaphase bridges Ulcerative colitis (UC) involves chronic irritation and inflammation of the colonic epithelium. These provoke ongoing turnover of colonic epithelial cells that can continue over the course of several decades. This condition leads, with significant frequency, to the development of colon carcinomas. (A) UC patients can be separated into "progressors," whose colitis has led to the appearance of one or more carcinomas, and "non-progressors," who show no evidence of intestinal tumors. Among the colonic epithelial cells of progressors, the anaphase bridges characteristic of BFB cycles are frequent (see also Figures 10.14 and 10.15). In addition, their colonic epithelial cells exhibit extensive shortening of telomeres and numerous karyotypic aberrations *(not shown)*. (B) Colon tissue samples from normal individuals and UC non-progressors show relatively few anaphase bridges, while much greater numbers are found in the inflamed epithelium of the progressors. This indicates a correlation between the occurrence of anaphase bridges and the onset of tumor development. The p values indicate the probability of these differing measurements arising by chance. (From J.N. O'Sullivan, M.P. Bronner, T.A. Brentnall et al., *Nat. Genet.* 32:280–284, 2002.)

Among the many unresolved issues are the connections between the findings described in this chapter and another major problem of biomedical research: Is organismic aging, as it occurs in human beings and other long-lived mammals, attributable to progressive telomere erosion, the resulting loss of the ability of cells in certain tissues to continue proliferating, and ultimately, the depletion of cell populations needed to maintain tissue integrity and function? While attractive in concept, there are surely alternative mechanisms of aging that we need to consider. For example, the cells in aging tissues may lose their robustness due to a variety of accumulated chemical and genetic insults that they suffer over many decades' time, and this progressive loss of viability may have nothing to do with telomere shortening. Intertwined with these issues is another: Does age-related telomere shortening contribute to the dramatically increased rates of cancer in the elderly (Sidebar 10.11)?

Much of our understanding of how cell proliferation is controlled has been influenced by observations of cells propagated *in vitro*. Tissue culture experiments tell us that the great majority of normal human cell types in culture stop proliferating sooner or later and enter into the viable, albeit nongrowing, state termed senescence. Cells that somehow circumvent senescence are able to proliferate for a limited period thereafter before entering into crisis, at which stage telomere collapse triggers devastating chromosomal changes.

Senescence remains a hotly debated issue in cancer biology, since there are indications that a number of cell types do not senesce because of eroded telomeres at the ends of their chromosomes. Instead, physiologic stresses brought on by sub-optimal conditions of culture seem to represent the major provoking causes. These observations lead to the question, still not definitively answered, whether processes comparable to replicative senescence affect cells within living tissues. Moreover, the basic biochemical mechanisms governing cell senescence continue to be uncovered (Sidebar 10.12), indicating that our understanding of this cellular state will undergo considerable revision in the years to come.

The collapse of telomeres during crisis represents a double-edged sword from the point of view of cancer progression. On the one hand, this collapse and the crisis that it provokes appear to represent an effective way for the body to limit

Sidebar 10.11 Do shortened telomeres contribute to increased cancer rates among the aged? The incidence of many types of human tumors increases dramatically with age. The precise reasons for this age dependence of cancer incidence are clearly complex, and are likely to include accumulated damage to the genome, age-related declines in immune function, and the fact that tumor progression is a protracted, multi-step process that often encompasses many decades' time. Progressive telomere shortening may contribute as well. The data from measurements such as those given in Figure 10.37 yield three conclusions: (1) In some (and perhaps most) human tissues, telomeres shorten progressively with increasing age. (2) The rate of shortening differs in different tissues, ostensibly because of differing mitotic activity in the cell lineages being studied. (3) The scatter on each curve indicates substantial inter-individual variability in the lengths of telomeric DNA early in life and/or rates of telomere shortening during a lifetime. (In addition, not discussed here, it is known that the telomeric DNAs at the ends of various chromosomal arms within a cell exhibit differing, chromosome-specific lengths, and that these vary from one individual to another.)

Telomeres that erode below the threshold length are no longer capable of protecting the ends of chromosomal DNA. In some cells, such critically shortened telomeres may limit the subsequent proliferative powers of a cell lineage, thereby acting in an anti-neoplastic fashion. At the same time, critically short telomeres may stimulate end-to-end chromatid fusions, and the latter may trigger, in turn, breakage–fusion–bridge cycles that can contribute to karyotypic instability and cancer development. Telomere collapse, then, may well be an important factor in the increased cancer incidence rates of the elderly—a notion that has yet to be tested experimentally.

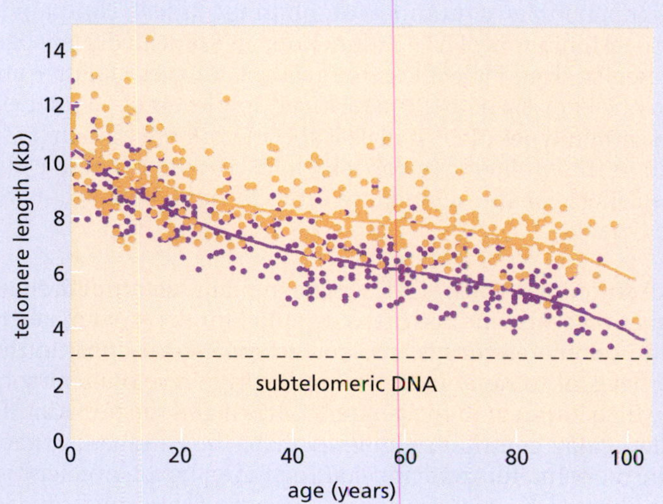

Figure 10.37 Telomere shortening, aging, and cancer Telomeric DNA lengths were measured in the lymphocytes and granulocytes of 400 individuals. As is apparent, the length of telomeres decreased in both cell types progressively with increased age. Inaccuracy in measuring telomere lengths is ± 0.5 kb. The horizontal line *(dashed)* represents the threshold level below which telomeric DNA no longer protects chromosomal ends. (Courtesy of G.M. Baerlocher and P. Lansdorp.)

the outgrowth of clones of neoplastic cells. On the other, some cells may coexist with collapsed telomeres if their response to damaged DNA is defective (e.g., because of loss of p53 function). The resulting multiple breakage–fusion–bridge (BFB) cycles may create altered genotypes (Figure 10.38) that strongly favor malignant proliferation. Subsequent acquisition of telomerase activity can then enable such genetically altered cells to take advantage of their aberrant genotypes by launching into a highly aggressive growth phase that yields, sooner or later, malignant tumors.

The actions of telomerase itself are also quite contradictory. Since expression of hTERT in normal human cells confers substantial enzyme activity, these cells appear to contain all the requisite components of the telomerase holoenzyme save the hTERT catalytic subunit. Once expressed, hTERT seems to provide cells with full telomerase function and to contribute to their subsequent immortalization. This enhanced proliferative capacity has been of great interest to gerontologists, who agonize about the prevention of aging and the restoration of robust tissues in the aged. As mentioned above, some believe that telomere collapse is the cause of aging. Such researchers tout hTERT as an obvious solution to the flagging proliferative powers of aging tissues. But here the double-edged sword of telomerase action becomes apparent. By supplying aged tissues with telomerase, one may inadvertently immortalize pre-malignant cell clones that are undoubtedly scattered about these tissues and, in so doing, contribute to the appearance of tumors that otherwise would never rear their heads.

Arguably the most direct and compelling proof of the importance of telomeres and telomerase in cancer pathogenesis comes from experiments that will be described in detail in the next chapter: a clone of the *hTERT* gene has been found to be an essential ingredient in the cocktail of introduced genes that are used experimentally to transform normal human cells into tumorigenic derivatives. Without the presence of *hTERT,* human cell transformation fails. (Alternative protocols for transforming human cells arrive at the same endpoint, by introducing genes, such as *myc,* that induce expression of cells' endogenous *hTERT* gene.)

The central role of telomerase in fueling the growth of cancer cells suggests a novel mode of anti-cancer therapy. Telomerase inhibitors should prove to be potent in killing cancer cells that carry relatively short telomeres—the situation observed in many telomerase-positive human tumor cells. Such drugs, if developed, should at the same time have little if any effect on normal cells throughout the body, which express relatively low levels of telomerase and then only transiently during S phase.

While attractive in concept, this plan may be derailed by certain realities. First, research programs launched by a number of pharmaceutical companies designed to develop novel inhibitors of hTERT function have repeatedly failed to yield potent, highly specific drug molecules. The reasons for these failures are unknown. Second, no one knows whether the relatively low levels of telomerase enzyme detectable in certain types of normal cells throughout the human body play a significant role in the proliferative capacity and viability of these cells. Consequently, the side effects of anti-telomerase drugs, should these agents be developed, are unpredictable.

In spite of these initial difficulties, the hTERT enzyme remains an attractive target for drug development. Because its activity is essential for the great majority of human cancers, anti-telomerase drugs are likely to show great utility in the treatment of diverse types of tumors. Indeed, if such drug development succeeds, this research, which began with the genetics of corn and the peculiar life cycles of ciliates, will one day contribute significantly to the conquest of neoplastic disease. For the moment, this remains nothing more than a fond wish.

Sidebar 10.12 Is cell senescence contagious? The descriptions of cell senescence provided in this chapter have depicted this state as a cell-autonomous process, in which intracellular signaling mechanisms trigger entrance into this state and maintain this state once it is established. However, an experimental screen for cellular genes that are responsible for maintaining senescence has turned up an unlikely suspect—the gene encoding plasminogen activator inhibitor-1 (*PAI-1*).

Plasminogen activator is an extracellular protease that converts plasminogen to plasmin; the latter, also a protease, functions to cleave a variety of substrates in the extracellular space. Among these substrates are the inactive precursors of growth factors, which become active mitogens and survival factors following plasmin-mediated cleavage. PAI-1 can block the activation of plasmin and thereby prevent growth factor activation in the extracellular space.

The *PAI-1* gene is one of the many targets of p53 transcriptional activation (Table 9.2). After p53 induces *PAI-1* transcription and thus PAI-1 release from a cell, proteolytic activation of growth factors is blocked. If *PAI-1* activation by p53 is prevented experimentally, then p53 loses its senescence-inducing properties. In addition, ectopic expression of PAI-1 causes robustly growing cells to senesce. Hence, PAI-1 expression is both necessary and sufficient for p53-mediated cell senescence. Since PAI-1, once released by a cell into the extracellular space, can block growth factor activation by neighboring cells, they too may be induced to enter senescence. Consequently, senescence can be governed by extracellular agents (e.g., PAI-1) and is, from this perspective, a cell-nonautonomous process.

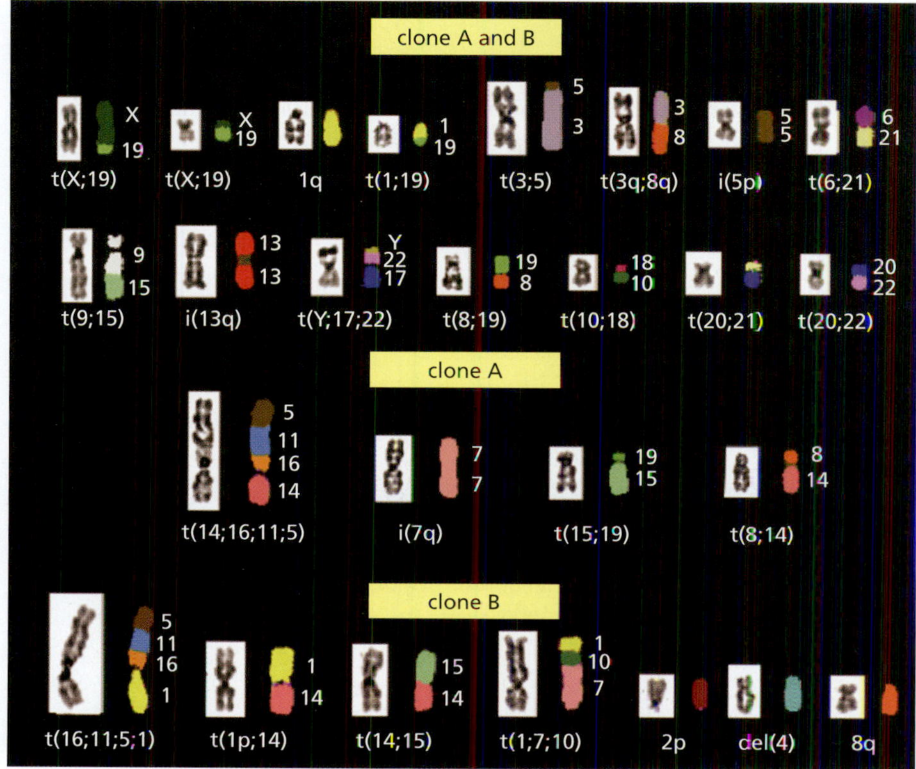

Figure 10.38 Karyotypic chaos The karyotype of cancer cells can be revealed by using the SKY technique to "paint" each chromosome with its own characteristic color. A large number of chromosomes have participated in translocations in the cells belonging to two subclones (A and B) of a human bladder carcinoma cell line. This is indicated by the fact that many chromosomes carry segments originating from two normal human chromosomes and some carry segments deriving from as many as four distinct normal human chromosomes. Each translocation is indicated by a "t" followed by the numbers of the parental chromosomes from which these arms derive. An "i" indicates an isochromosome, in which a single arm of a chromosome has been duplicated. "p" and "q" indicate the short and long arms of normal human chromosomes, respectively. "del" indicates that a normally present chromosomal segment has been deleted. "Clone A and B" indicates aberrant chromosomes shared by the two subclones, while "Clone A" and "Clone B" indicate chromosomes that are present uniquely in one or another of these subclones. The fact that the two subclones differ from one another in their karyotypes indicates ongoing chromosomal instability while in culture. Comparable degrees of aneuploidy can be observed in cancer cells that have recently been introduced into culture. (From H.M. Padilla-Nash, W.G. Nash, G.M. Padilla et al., *Genes Chromosomes Cancer* 25:53–59, 1999.)

Key concepts

- Two barriers prevent cultured cells from replicating indefinitely in culture—senescence and crisis.
- Senescence involves the long-term residence of cells in a non-growing but viable state; crisis involves the apoptotic death of cells.
- Senescence is provoked by physiologic stresses that cells experience *in vitro*, but its role in confining cell proliferation *in vivo* remains to be demonstrated.
- Crisis is provoked by the erosion of telomeres, which results in widespread end-to-end chromosomal fusions, karyotypic chaos, and cell death.
- Most pre-malignant cells escape from crisis by activating expression of hTERT, the telomerase enzyme, which is specialized to elongate telomeric DNA by extending it in hexanucleotide increments.
- Some cancer cells escape crisis by regenerating their telomeric DNA through the ALT mechanism.
- Cells that have stabilized their telomeres through the actions of telomerase or the ALT mechanism can then proliferate indefinitely and are therefore said to be immortalized.
- Cell immortalization is a step that appears to govern the development of all human cancers.
- The end-to-end chromosomal fusions that accompany crisis lead to repeated breakage–fusion–bridge cycles, which appear, in turn, to be responsible for much of the aneuploidy associated with the karyotypes of many kinds of solid human tumors.

Thought questions

1. What advantages do breakage–fusion–bridge cycles confer on evolving populations of pre-malignant cells?
2. Why is the acquisition of an immortalized proliferative potential so important for human tumors?
3. What types of evidence connect telomeres and telomerase to entrance into the senescent growth state?
4. What complications and side effects might result from the shutdown of telomerase activity by (still-hypothetical) anti-cancer drugs that are specific inhibitors of this enzyme?
5. How does the molecular configuration of the T-loop protect the ends of telomeric DNA?
6. How does the telomerase-associated *hTR* RNA molecule facilitate the maintenance of telomeric DNA by the hTERT enzyme?

Additional reading

Aisner DL, Wright WE & Shay JW (2002) Telomerase regulation: not just flipping the switch. *Curr. Opin. Genet. Dev.* 12, 80–85.

Artandi SE & DePinho RA (2000) Mice without telomerase: what can they teach us about human cancer? *Nat. Med.* 6, 852–855.

Bérubé NG, Smith JR & Pereira-Smith OM (1998) Insights from model systems: the genetics of cellular senescence. *Am. J. Hum. Genet.* 62, 1015–1019.

Blasco MA (2002) Telomerase beyond telomeres. *Nat. Rev. Cancer* 2, 627–633.

Blasco MA (2005) Telomeres and human disease: ageing, cancer and beyond. *Nat. Rev. Genet.* 6, 611–622.

Blasco MA & Hahn WC (2003) Evolving views of telomerase and cancer. *Trends Cell Biol.* 13, 389–394.

Campisi J (2005) Senescent cells, tumor suppression, and organismal aging: good citizens, bad neighbors. *Cell* 120, 513–522.

Cong Y-S, Wright WE & Shay JW (2002) Human telomerase and its regulation. *Microbiol. Mol. Biol. Rev.* 66, 407–425.

Cristofalo VJ, Lorenzini A, Allen RG et al. (2004) Replicative senescence: a critical review. *Mech. Ageing Dev.* 125, 827–848.

de Lange T. (2004) T-loops and the origins of telomeres. *Nat. Rev. Mol. Cell Biol.* 5, 323–329.

DePinho RA (2000) The age of cancer. *Nature* 408, 248–254.

Hanahan D (2000) Cancer: benefits of bad telomeres. *Nature* 406, 573–574.

Harrington L (2004) Those dam-aged telomeres! *Curr. Opin. Genet. Dev.* 14, 22–28.

Hayflick L & Moorhead PS (1961) The serial cultivation of human diploid cell strains. *Exp. Cell Res.* 25, 585–621.

Lundblad V (2001) Genome instability: McClintock revisited. *Curr. Biol.* 11, R957–R960.

Mathon NF & Lloyd A (2001) Cell senescence and cancer. *Nat. Rev. Cancer* 1, 203–213.

Neumann AA & Reddel RR (2002) Telomere maintenance and cancer—look, no telomerase. *Nat. Rev. Cancer* 2, 879–884.

Shay JW & Wright WE (2000) Hayflick, his limit, and cellular ageing. *Nat. Rev. Mol. Cell Biol.* 1, 72–76.

Shay JW, Zou Y, Hiyama E & Wright WE (2001) Telomerase and cancer. *Hum. Mol. Genet.* 10, 677–685.

Stewart SA & Weinberg RA (2000) Telomerase and human tumorigenesis. *Semin. Cancer Biol.* 10, 399–406.

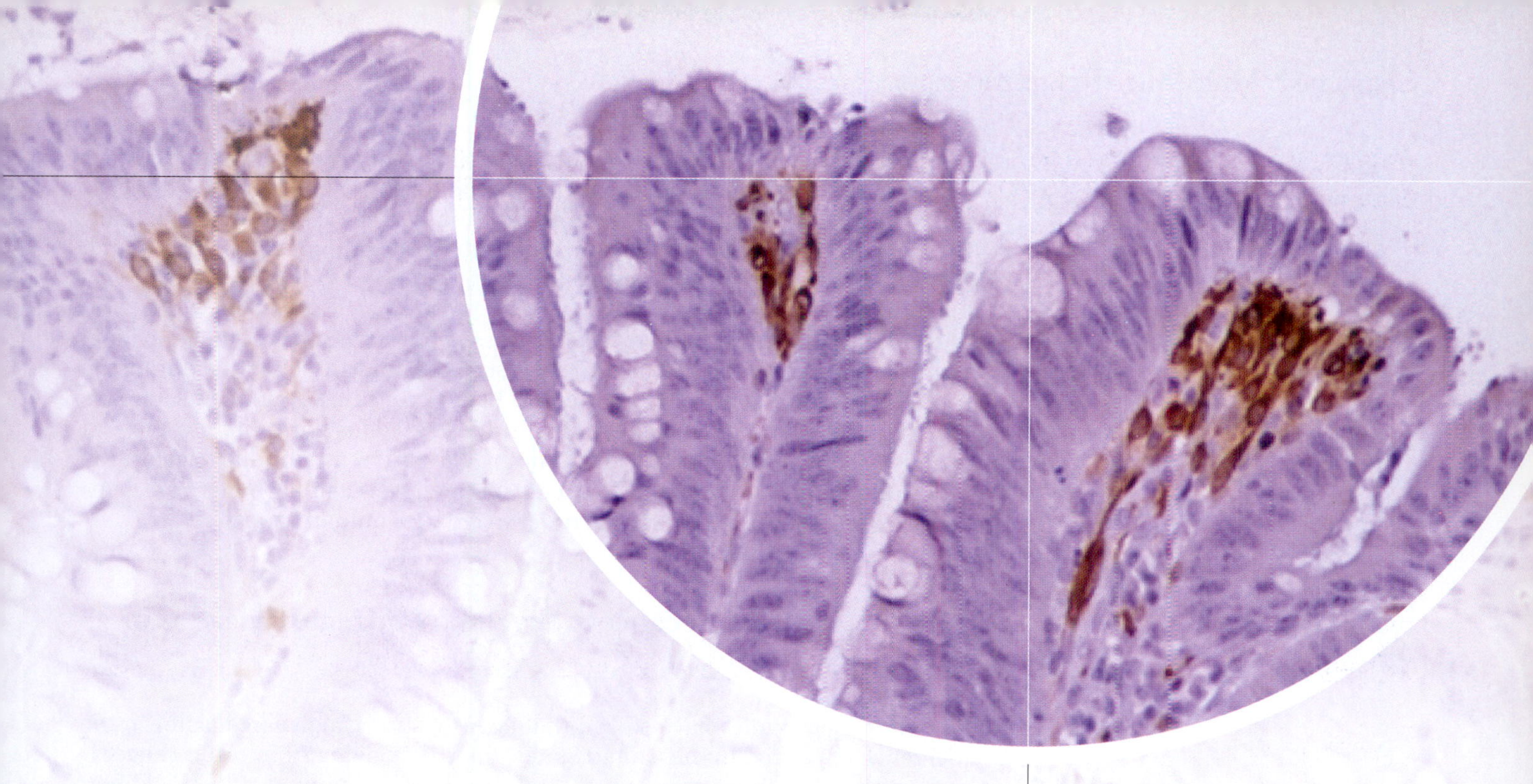

Chapter 11

Multi-Step Tumorigenesis

> In the survival of favoured individuals and races, during the constantly-recurring struggle for existence, we see a powerful and ever-acting form of selection.
>
> Charles Darwin, biologist, 1859

The formation of a tumor is a complex process that usually proceeds over a period of decades. Normal cells evolve into cells with increasingly neoplastic phenotypes through a process termed **tumor progression**. This process takes place at myriad sites throughout the normal human body, advancing further and further as we get older. Rarely does it proceed far enough at any single site to make us aware of its end product, a clinically detectable tumor mass.

Tumor progression is driven by a sequence of randomly occurring mutations and epigenetic alterations of DNA that affect the genes controlling cell proliferation, survival, and other traits associated with the malignant cell phenotype. The complexity of this process reflects the work of evolution, which has erected a series of barriers between normal cells and their highly neoplastic derivatives. Accordingly, completion of each step in tumor progression can be viewed as the successful breaching of yet another barrier that has been impeding the progress of a clone of pre-malignant cells.

One might think that these barriers are the handiwork of relatively recent evolutionary processes. Perhaps, we might imagine, the forces of evolution first worked to design the architecture and physiology of complex metazoan bodies and, having completed this task, then proceeded to tinker with these plans in order to reduce the risk of cancer.

An alternative scenario is far more likely, however: the risk of uncontrolled cell proliferation has been a constant companion of metazoans from their very beginnings, roughly 650 million years ago. By granting individual cells within their tissues the license to proliferate, even simple metazoans ran the risk that one or another of their constituent cells would turn into a renegade and trigger the disruptive, runaway cell multiplication that we call cancer. Consequently, the erection of defenses against cancer must have accompanied, hand-in-hand, the evolution of organismic complexity.

Most of the preceding chapters have addressed one or another of the individual cellular control systems that defend against cancer and are subverted during the process of tumorigenesis. Now, we will begin tying these individual threads together and examine how alterations in these systems contribute to the end product—the formation of a primary tumor. We start by attempting to gauge the scope of the problem at hand: How many different sequential changes are actually required in cells and tissues in order to create a human cancer?

11.1 Most human cancers develop over many decades of time

Epidemiologic studies have shown that age is a surprisingly large factor in the incidence of cancer. In the United States, the risk of dying from colon cancer is as much as 1000 times greater in a 70-year-old man than in a 10-year-old boy. This fact, on its own, suggests that this type of cancer and, by extension, many other cancers common in adults (Figure 11.1) require years if not decades to develop. The late onset of most cancers also has an important and unexpected public health implication: curing all cancers will have a relatively small effect on expected life span (Sidebar 11.1).

This late age of onset indicates that the development of most cancers requires an extended period of time. A more direct measure of this is provided by the incidence of lung cancers among males in the United States. Cigarette smoking was relatively uncommon among this group until World War II, when large numbers of men first acquired the habit, encouraged in part by the cigarettes they received as part of their rations while serving in the armed forces during this war. Thirty years later, in the mid-1970s, the rate of lung cancer began to climb steeply (Figure 11.2). At the same time, cigarette smoking spread throughout the world and peaked in 1990, leading to estimates that global lung cancer mortality, which currently exceeds 1.1 million deaths annually, will peak only sometime in the decade after 2020. In these other areas of the world outside the United States, the approximately 35-year lag between marked increases in smoking and the onset of large numbers of lung cancers seems to apply as well.

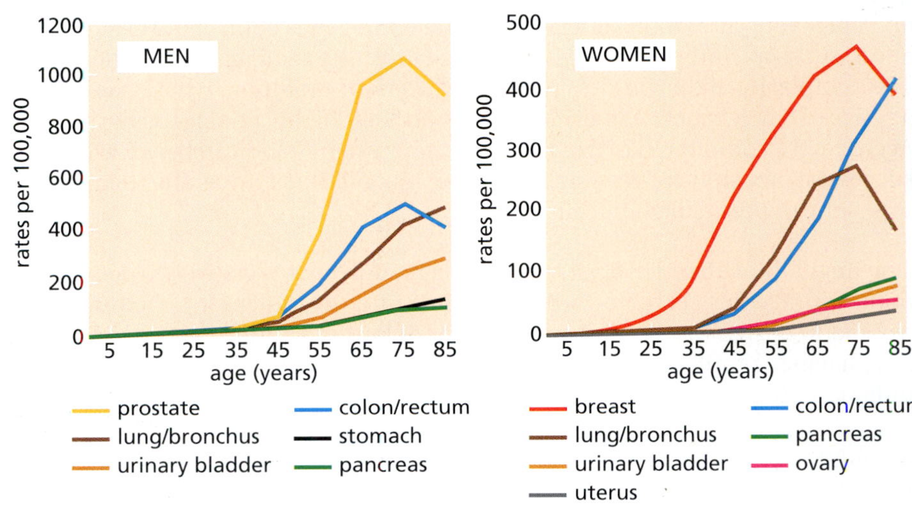

Figure 11.1 Cancer incidence at various ages This graph of diagnosis of various types of epithelial cancers shows a steeply rising incidence with increasing age, indicating that the process of tumor formation generally requires decades to reach completion. (Courtesy of W.K. Hong, compiled from *SEER Cancer Statistics Review*.)

Sidebar 11.1 A life without cancer would not lead to extreme changes in longevity The curves of cancer incidence as a function of age illustrated in Figure 11.1 leave us with a surprising and, for some, depressing conclusion. Even though cancer-associated deaths are so frequent in Western populations, being responsible for some 20% of all mortality, curing all forms of cancer would extend life expectancy by only a few years. A calculation of cancer-related deaths occurring in the Netherlands in 1990 indicates that if all cancer-related deaths were eliminated, this would yield an extension of life expectancy of 3.83 years for men and 3.38 years for women. (In less industrialized countries, where infectious diseases take a toll in midlife, effects on life expectancy would be far less.)

Because the great majority of cancers strike late in life, successful prevention or curing of this late-occurring disease will have relatively little effect on the life span of a person who has already lived eight or more decades. Conversely, even though infant mortality is extremely low in the West, further reductions in early deaths caused by various conditions will have profound quantitative effects in increasing overall life expectancy. In the middle, between these two sources of mortality, are maladies such as heart disease, which often strike down individuals a decade or more before they would be likely to contract cancer.

Epidemiologists have developed formulas that predict the frequency of various cancers as a function of age. These formulas indicate that, depending upon the cancer type, disease **incidence** (the rate at which the disease is diagnosed) and mortality rates occur as a function of a^4 to a^7, where a represents the age of patients at initial diagnosis. For epithelial cancers as a whole, the risk of death from cancers increases approximately as the fifth or sixth power of elapsed lifetime (Figure 11.3). Algebraic functions like these are interpreted quite simply. If the probability of an outcome is indicated by a^n, this means that $n + 1$ independent events, each occurring randomly and with comparable probability per unit of time, must take place before the ultimate outcome—in this case a diagnosed tumor—is achieved. The fact that the incidence of a disease like colon cancer begins to shoot upward only in the seventh and eighth decades of life indicates that each of these events occurs with a very low probability each year. More specifically, each event is likely to occur on average once every 10 to 15 years, and the entire succession of events may usually require 40 to 60 years to reach completion.

These calculations provide us with only a very rough estimate of the complexity of tumor progression. In fact, some critical steps may occur far more rapidly than others. Because these calculations of the kinetics of tumor progression reflect the influence of the slowest, "rate-limiting" steps, the more rapid changes will not be registered in these calculations. For this reason, it seems likely that the actual number of events required to form most tumors is actually higher than is predicted by use of this a^n formula.

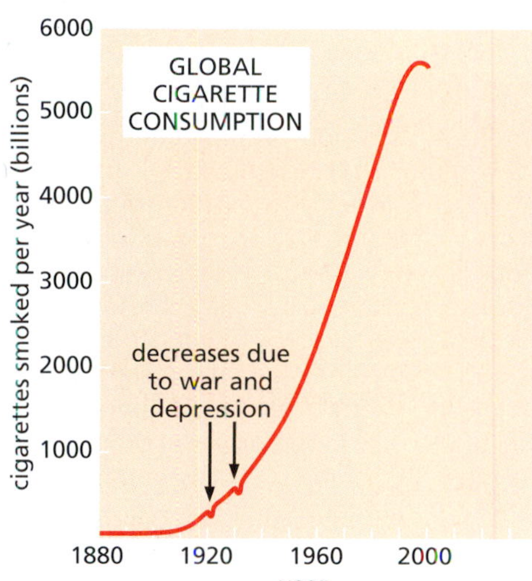

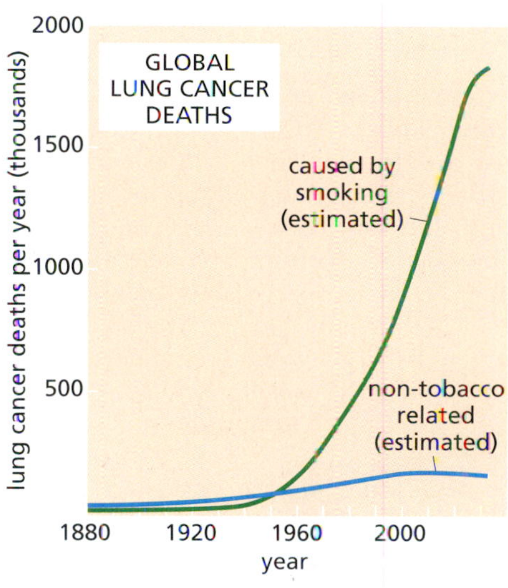

Figure 11.2 Cigarette consumption and lung cancer These curves compare the annual global consumption of cigarettes (billions smoked per annum, *red curve, left panel*) with the recorded and predicted annual worldwide mortality from lung cancer (thousands of deaths per annum; *green, blue curves, right panel*). Annual mortality from tobacco-induced lung cancer is estimated to peak sometime in the fourth or fifth decade of the twenty-first century. During the twentieth century, there was an increase in what is judged to be "non-tobacco–related" lung cancer mortality *(blue curve)*. The precise number of these cases is unclear, and there is some debate about how many of these cases are attributable to secondhand tobacco smoke. (From R.N. Proctor, *Nat. Rev. Cancer* 1:82–86, 2001.)

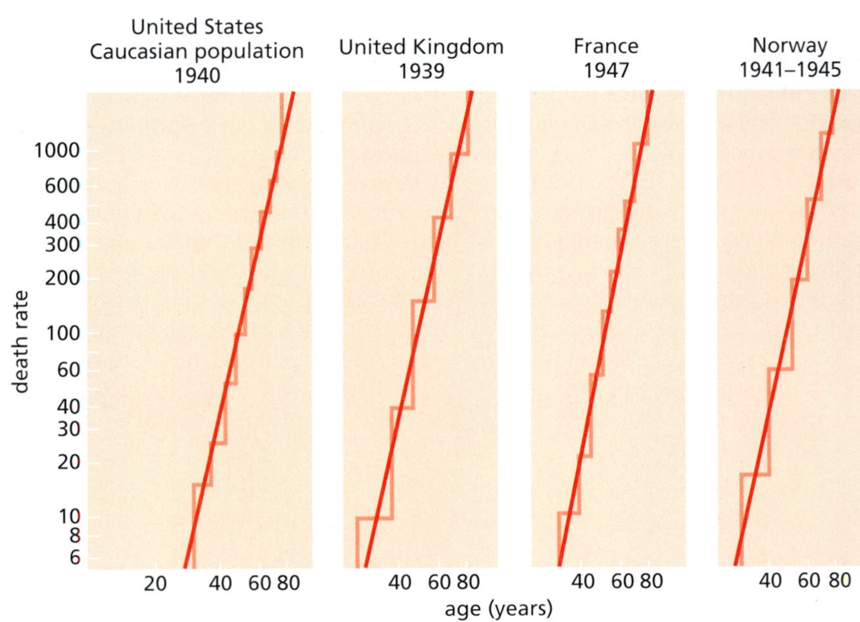

Figure 11.3 Age at death from various epithelial cancers The graphs indicate the general mortality from cancer from 1939 to 1947 in four countries where public health statistics were kept with some precision. These log–log plots of male cancer death rates (deaths per 100,000 population, *ordinate*) at different ages *(abscissa)* have a slope indicating that completion of five or six rate-limiting events is required to produce a lethal cancer. For example, a slope of 5 indicates six rate-limiting events, whose nature is not revealed by these analyses. Interestingly, these slopes are remarkably similar between different countries. The slopes differ slightly with different types of cancer (*not shown*). (From A.G. Knudson, *Nat. Rev. Cancer* 1:157–162, 2001, reproduced from C.E. Nordling, *Brit. J. Cancer* 6:68–72, 1953.)

From the kinetics of tumor progression we can safely conclude that tumorigenesis is a complex, multi-step process. It turns out that these multi-step kinetics complicate calculations of the true rates with which cancers strike human populations (Sidebar 11.2).

The conclusion that tumorigenesis is a multi-step process hints at another interesting idea. Assume that (1) a sequence of unlikely events is required in order for a tumor to appear and (2) many of these events happen at comparable frequencies in all of us. Together, these assumptions indicate that as we grow older, virtually all of us will carry populations of cells in many locations throughout the body that have completed some but not all of the steps of tumor progression. Since most of us will not live long enough for the full schedule of requisite events to be completed (because we will succeed in dying from another disease), we will never realize that any of these tumor progressions had been initiated in our bodies. Viewed from this perspective, cancer is an inevitability; if we succeeded in avoiding the death traps set by all the other usual diseases, sooner or later most of us would become victims of cancer.

Of course, some—though not most—of us will actually contract a neoplastic disease such as colon cancer. This suggests something else: while the a^5 or a^6 expression may predict colon cancer incidence averaged over a large human

Sidebar 11.2 Multi-step tumorigenesis complicates cancer epidemiology Calculating the true incidence rates of cancers or the death rates due to these cancers in various human populations is hardly a simple and straightforward exercise. For example, presenting the colon cancer incidence in different populations as the number of diagnosed cases per hundred thousand people per year is meaningless, since some populations may contain a disproportionate number of young people, while others may be heavily weighted in favor of oldsters. The latter populations will, of course, have far higher numbers of colon cancers per unit of total population.

The strong age dependence of the incidence of almost all cancers therefore forces epidemiologists to calculate *age-adjusted incidence,* which corrects for the fact that the age distributions of different human populations differ markedly. A calculated age-adjusted incidence can tell us, for instance, what the probability is of a 60-year-old woman in the United States contracting breast cancer during the course of a year compared with the risk experienced by a woman of the same age living in Egypt, Kazakhstan, or Portugal. At the same time, age-adjusted incidence measurements enable us to make meaningful comparisons in a specific population over an extended period of time—for example, the age-adjusted colon cancer risk of the U.S. population in 1930 compared with that of the successor population in 1990, where effects of differences in the age distributions of the two populations are once again eliminated.

population, the probability of the rate-limiting pathogenic events occurring per unit of time varies dramatically from one individual to another, being affected by hereditary predisposition, diet, lifestyle, and all the other variables that are known to strongly influence colon cancer incidence in various human populations.

Epidemiology provides us with another important insight into the multi-step nature of tumorigenesis. If we examine the frequencies of mesothelioma in humans (caused largely by asbestos exposure and smoking) and skin cancer in mice (induced by repeated benzo[*a*]pyrene painting), it becomes apparent that the formation of each of these tumors requires an extended period of repeated exposure to carcinogens and that it is the *duration* of this exposure (rather than the absolute *age* of exposed individuals or the age when exposure began) that determines the timing of the onset of detectable disease (Figure 11.4). In these cases, tumors are created by the actions of exogenous carcinogens rather than occurring spontaneously within the body; these carcinogens increase the rate of tumor progression, often by many orders of magnitude above the spontaneous background rate, and the pathogenesis of each of these tumors seems to involve a predetermined exposure schedule before it reaches completion.

11.2 Histopathology provides evidence of multi-step tumor formation

Pathologists are skilled in the art of examining normal and diseased tissues under the microscope and, in the case of cancer, rendering diagnoses about the tissue of origin of a tumor and its stage of development. As we discussed briefly in Section 2.4, histopathological analyses have provided ample support for the idea that most types of tumors arise as the end products of a complex sequence of events. We revisit these analyses here in greater detail, because they form the core of this chapter.

The notion of human tumor development as a multi-step process has been documented most clearly in the epithelia of the intestine. The intestinal epithelial

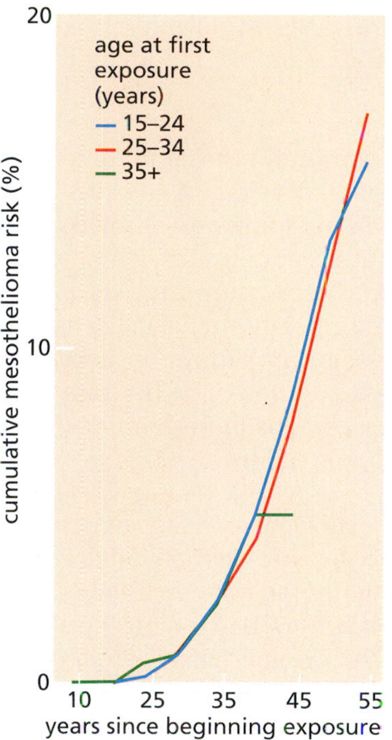

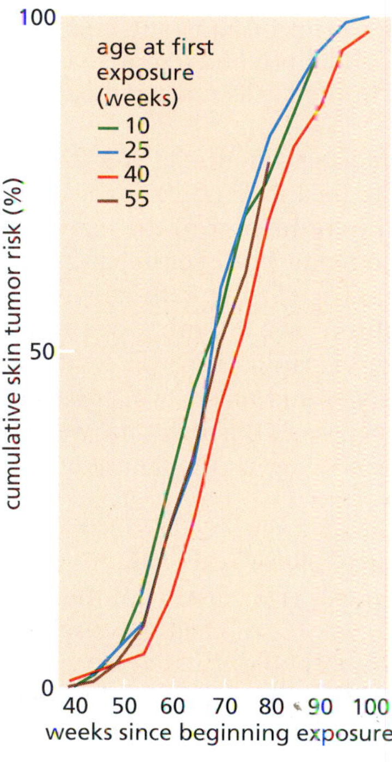

Figure 11.4 Cancer incidence and carcinogen exposure These two graphs both indicate that cumulative exposure to a carcinogenic stimulus, rather than the age at which this exposure began, determines the likelihood of developing a detectable tumor. The left graph presents the cumulative risk of developing mesothelioma (a tumor of the mesodermal lining of the abdominal organs and the lungs) among insulation workers in the United States, many of whom were occupationally exposed to asbestos. The right graph presents the cumulative risk of developing a skin tumor among mice treated in an experimental protocol for inducing squamous cell carcinomas of the skin. (From J. Peto, *Nature* 411:390–395, 2001.)

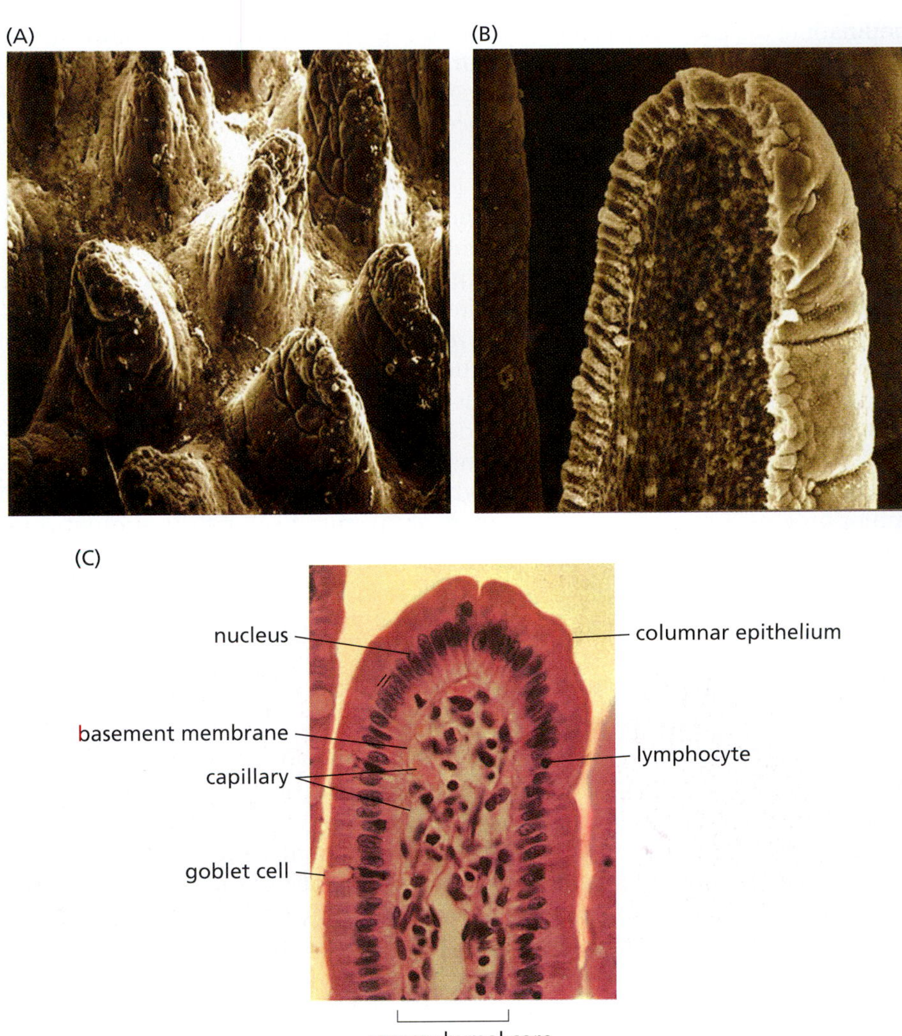

Figure 11.5 Microanatomy of the normal intestinal wall (A) This scanning electron micrograph of intestinal epithelium (in this case that of the small intestine) shows villi (fingers) of the intestinal mucosa extending into the lumen at regular intervals. (B) Each villus is covered with a layer of epithelial cells (which are separated from the core by a basement membrane). (C) The core of each villus, which is separated from the overlying epithelium by a basement membrane, is composed of various types of mesenchymal cells, including fibroblasts, endothelial cells, pericytes, and various cells of the immune system *(not indicated)*. (A and B, from S. Canan; C, from University of Iowa Virtual Hospital, Atlas of Microscopic Anatomy, Plate 194.)

cells, which face the interior cavity (lumen) of the gastrointestinal tract, form a layer that is only one cell thick in many places (Figure 11.5). These epithelial cell populations are in constant flux. Each minute, 20 to 50 million cells in the human duodenum and a tenth as many in the colon die and an equal number of newly minted cells replace them!

Underlying these epithelia is a basement membrane (basal lamina) to which these cells are anchored. As is the case in other epithelial organs, this basement membrane forms part of the extracellular matrix and is assembled from proteins secreted by both epithelial cells above and stromal cells lying beneath the membrane (see Figure 2.3). The mesenchymal cells composing the stroma are largely fibroblasts; but other cell types, including endothelial cells, which form the walls of capillaries and lymphatic vessels, and immune cells, such as macrophages and mast cells, are also scattered about. Lying beneath this layer of stromal cells in the intestinal wall is a thick layer of smooth muscle responsible for moving along the contents of the colonic lumen through periodic contractions.

The epithelial layer is the site of most of the **pathological** (i.e., disease-associated) changes associated with the development of colon carcinomas. Analyses of human colonic biopsies have revealed a variety of tissue states, with degrees of abnormality that range from mildly deviant tissue, which is barely distinguishable from the structure of the normal intestinal **mucosa** (the lining of the colonic lumen), to the chaotic jumble of cells that form highly malignant tissue

(Figure 11.6A). Like the normal intestinal lining, these growths are composed of a variety of distinct cell types, indeed almost all of the cell types found in the normal tissue.

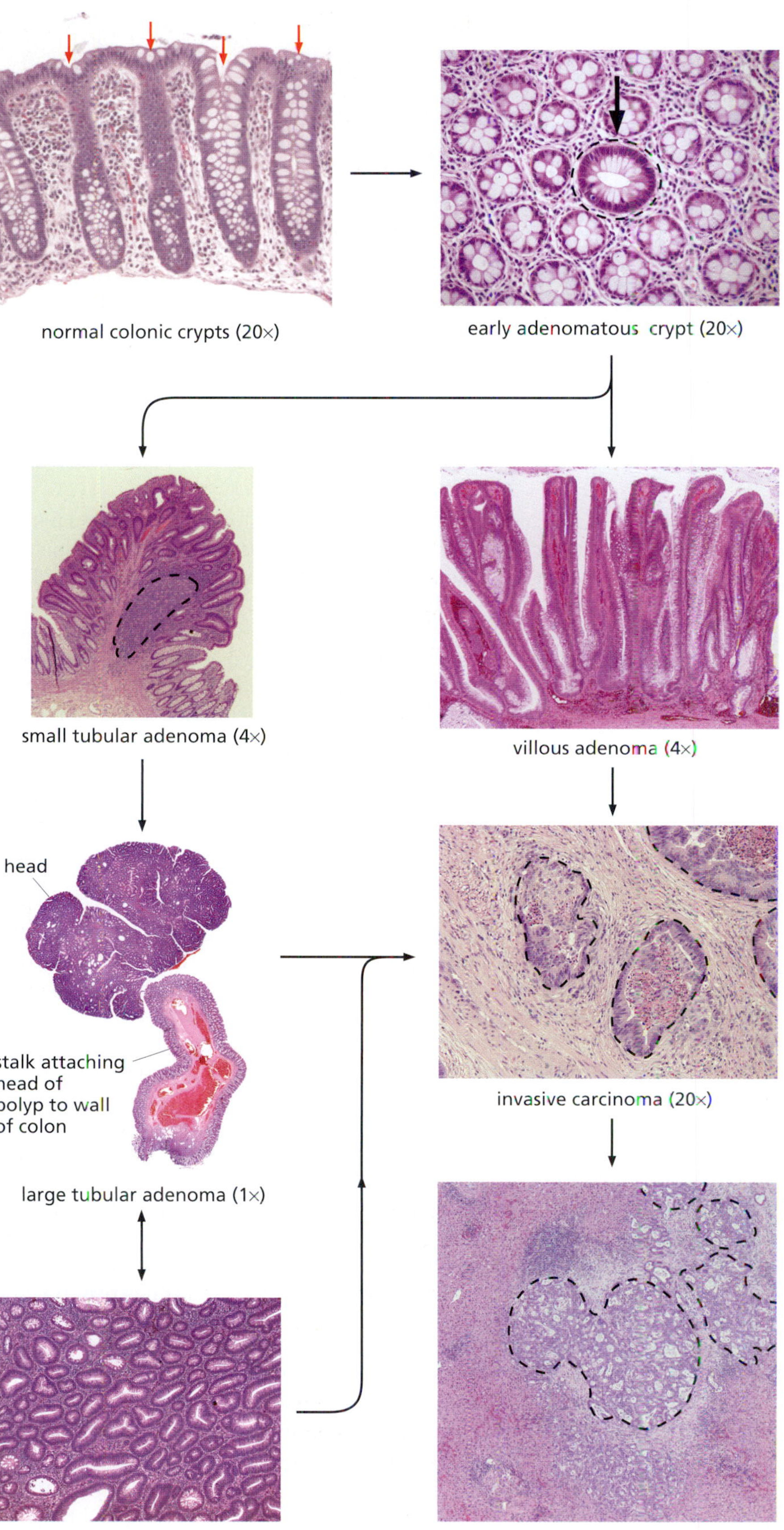

Figure 11.6 Histopathological alterations of the human colon The various types of abnormal tissues revealed by histopathological analyses of the human colon can be arrayed in a succession of ever-increasing abnormality. In fact, more detailed successions can be drawn, since adenomas can be further subdivided into various subtypes. The normal colonic crypts are seen here in longitudinal section *(top left)*. A small adenomatous crypt *(arrow, circled)* is shown, together with normal crypts, in cross section *(top right)*. The boundary of a small tubular adenoma is circled *(left, middle)*. The larger tubular adenoma *(below)* is sometimes termed *pedunculated,* indicating its attachment via a stalk to the colonic wall. The locally invasive carcinoma is seen here as small islands of carcinoma cells *(circled)* surrounded by extensive stroma *(right, lower middle)*. Metastases to the liver *(circled)* are surrounded by layers of recruited stromal cells; the normal liver tissue is seen to the *left*. Such a histopathological progression, indicated here schematically, would seem to be the most logical way by which normal tissue, in this case the colonic epithelium, is transformed through a series of intermediate steps, into carcinomas and ultimately spawns metastatic growths. However the evidence for most of these precursor-product relationships is actually quite fragmentary. (Courtesy of C. Iacobuzio-Donahue and B. Vogelstein.)

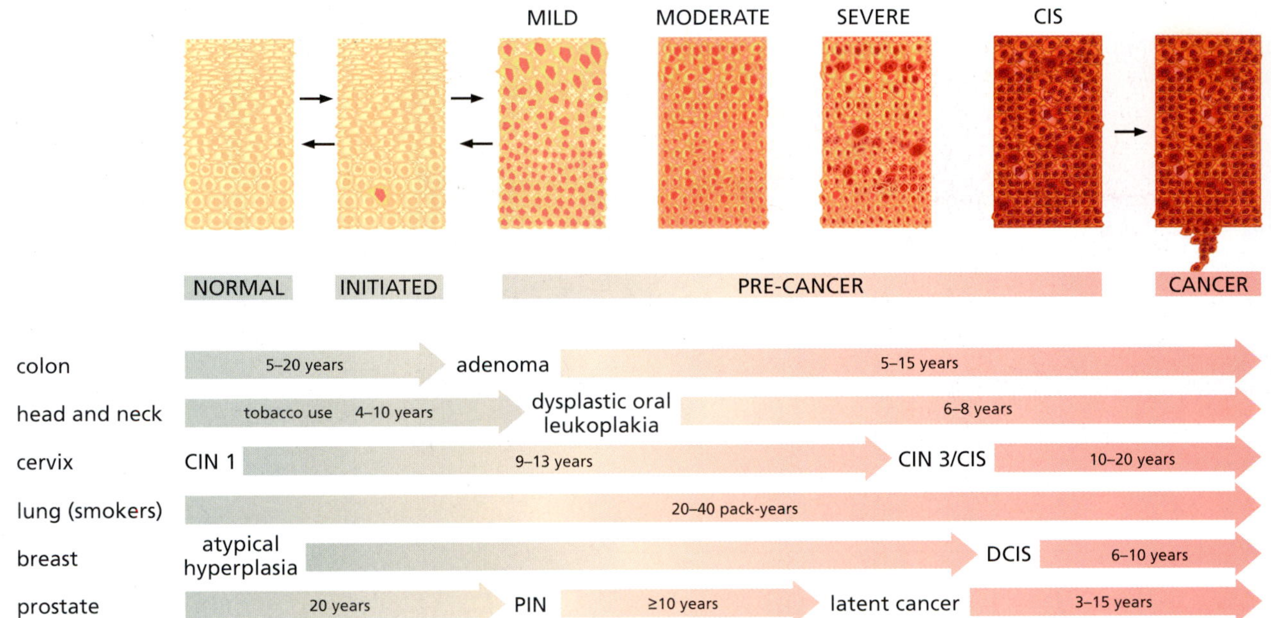

Figure 11.7 Multi-step tumorigenesis in a variety of organ sites The pathogenesis of carcinomas is thought to be governed by very similar biological mechanisms operating in a variety of epithelial tissues. Accordingly, multi-step tumorigenesis involving similar histological entities has been proposed to progress along parallel paths in these various organ sites. These similarities are obscured by the fact that the nomenclature is quite variable from one tissue to another. CIS, carcinoma *in situ*; CIN, cervical intraepithelial neoplasia; DCIS, ductal carcinoma *in situ*; PIN, prostatic intraepithelial neoplasia. (Courtesy of W.K. Hong.)

Some growths that are classified as *hyperplastic* exhibit almost normal histology, in that the individual cells within these growths have a normal appearance. However, it is clear that in these areas of hyperplasia, the rate of epithelial cell division is unusually high, yielding thicker-than-normal epithelia. Yet other growths show abnormal histologies, with the epithelial cells no longer forming the well-ordered cell layer of the normal colonic mucosa and the morphology of individual cells deviating in subtle ways from that of normal cells; these growths are said to be *dysplastic* (see Figure 2.14). A much larger and more deviant growth that has dysplastic cells and marked thickening is termed a *polyp* or an *adenoma* (see Figures 2.15 and 11.6). In the colon, several distinct types of polyps are encountered; some are attached to the wall of the colon, while others are tethered to the colonic wall by a stalk. Importantly, all these growths are considered benign, in that none has broken through the basement membrane and invaded underlying stromal tissues.

The more abnormal growths that have broken through the basement membrane and beyond and are considered to be malignant. There are distinctions among these more aggressive colon carcinomas and associated cancer cells, depending on whether they have penetrated deeply into the stromal layers and smooth muscle and whether they have migrated—metastasized—to anatomically distant sites in the body, where they may have succeeded in founding new tumor cell colonies.

Having arrayed these growths in a succession of tissue phenotypes that advance from the normal to the aggressively malignant (see Figure 11.6), we might imagine that this succession depicts with some accuracy the course of actual tumor development as it occurs in the human colon. In truth, the evidence supporting this scheme is quite indirect. Some tumors may well develop through a series of intermediate growths, most of which are arrayed here. Alternatively, it is possible that some of the tissue types depicted as intermediates in this sequence represent dead ends rather than stepping stones to more advanced tumors. In certain cases of colon cancer, it is also possible that the development of the tumor depended on the ability of early growths to leapfrog over intermediate steps, allowing them to arrive at highly malignant endpoints far more quickly than is suggested by this succession. Similar successions have been proposed for a variety of other epithelial cancers (Figure 11.7).

At least three types of evidence strongly support the precursor–product relationship between colon adenomas and carcinomas. First, on rare occasion, one can

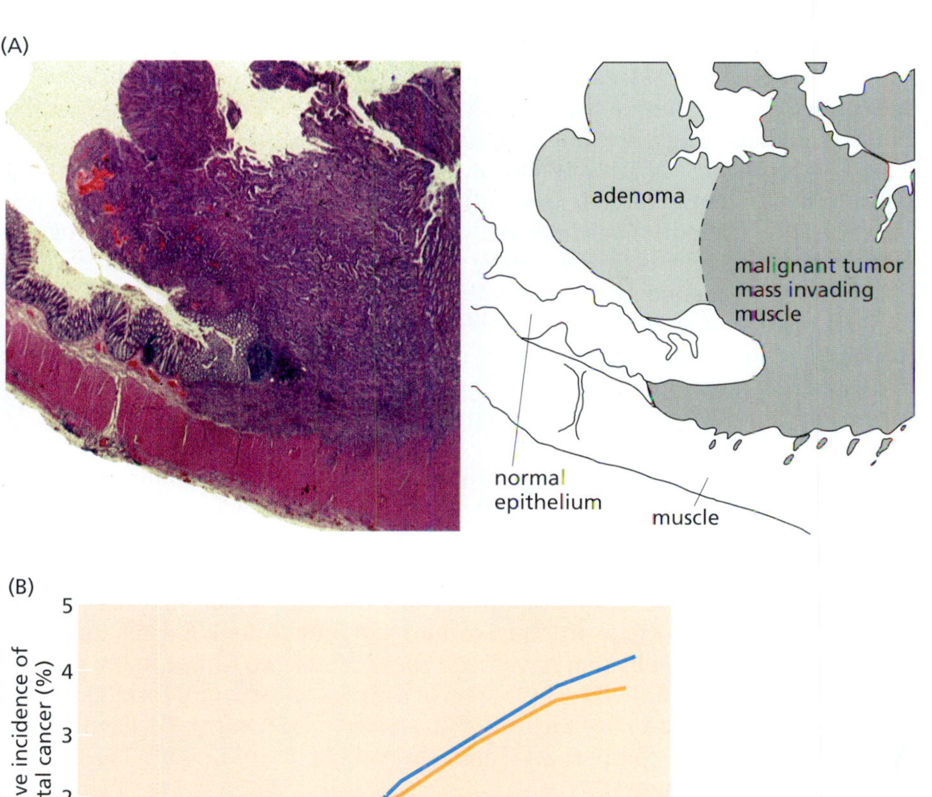

Figure 11.8 Evidence for adenoma-to-carcinoma progression The evidence for precursor–product relationships between intestinal adenomas and carcinomas comes from at least two types of clinical observations. (A) On occasion, carcinomas are observed to be growing directly out of adenomas. As seen here, the demarcation between the two can sometimes be drawn with some precision. (B) A second line of evidence derives from clinical studies in which colonoscopy is used to screen large cohorts of patients. Any polyps that are discovered are removed by the procedure of *polypectomy* (surgical removal of polyps). As seen in the graph here, two independent studies *(yellow-orange, blue lines)* predict a certain expected number of carcinomas in such cohorts on the basis of historical experience. However, in patients who have undergone polypectomy, the observed numbers of colorectal cancers diagnosed in subsequent years have been reduced by more than 80% *(red line)*. (A, courtesy of Paul Edwards; B, courtesy of W.K. Hong, data from S. Winawer, A.G. Zauber, M.N. Ho et al., *N. Engl. J. Med.* 329:1977–1981, 1993.)

actually observe a carcinoma growing directly out from an adenomatous polyp (Figure 11.8A). We can surmise that outgrowths like these occur routinely during the development of virtually all colon carcinomas and that, more often than not, the rapid expansion of the carcinoma soon overgrows and obliterates the adenoma from which it arose.

Second, clinical studies have been performed of large cohorts of patients who have undergone colonoscopy, which is performed routinely to survey the colon for occasional adenomatous polyps and to remove any that are detected. In one such study, those patients whose polyps were removed experienced, in subsequent years, about an 80% reduction in the incidence of colon carcinomas (Figure 11.8B). This indicates that in this patient population, at least 80% of colon carcinomas derive from preexisting, readily detectable adenomas; because colonoscopy may have missed some polyps, the actual proportion may be higher. (These observations still do not prove that every single colon carcinoma arising in humans must arise from a preexisting adenoma.)

The third support for the adenoma-to-carcinoma progression comes from the disease termed familial adenomatous polyposis (FAP; Section 7.11), in which individuals inheriting a mutant form of the *APC* tumor suppressor gene are prone to develop anywhere from dozens to more than a thousand polyps in the intestine (Figure 7.22). With a certain low but measurable frequency, one or another of these polyps will progress spontaneously into a carcinoma. This explains the high risk of colon cancer in these individuals.

The development of carcinomas in other organ sites throughout the body is thought to resemble, at least in outline, the multi-step progression observed in the colon (see, for example, Figure 11.7). Many of these other tissues, such as the breast, stomach, lungs, prostate, and pancreas, also exhibit growths that can be called hyperplastic, dysplastic, and adenomatous, and these growths would seem to be the benign precursors of the carcinomas that arise in these organs. However, the histopathological evidence supporting multi-step tumor progression in these tissues is, for the most part, less well developed than for the colon—a consequence of the accessibility of the colon through colonoscopy and the relative inaccessibility of these other tissues. In the case of nonepithelial tissues, including components of the nervous system, the connective tissues, and the hematopoietic system, the histopathological evidence supporting multi-step tumor progression is even more fragmentary.

In fact, the altered histopathology of pre-malignant and malignant growths is largely a reflection of changes that have occurred in a subset of the cells forming each of these masses. In the case of carcinomas, the progressive alteration of epithelial cells is assumed to drive the process of tumor progression and the associated changes in histopathology. The other cell types in these tumor masses, specifically those in the stroma, are often depicted as innocent bystanders—normal cells that have been recruited to the tumor mass and co-opted by abnormal epithelial cells to help in the process of tumor formation. But as we will see in Chapter 13, the role of these ostensibly normal cells in tumor formation is far more complex and interesting.

11.3 Colonic growths accumulate genetic alterations as tumor progression proceeds

The forces driving tumor progression in various tissues are hardly obvious from the histopathological descriptions cited above. It is possible that epigenetic alterations contribute to the progressive changes in cell and tissue phenotype as a tumor develops. For instance, some of the steps in tumor progression might recapitulate specific changes in cell behavior that occur normally during embryogenesis. We know that virtually all of the steps in embryological development depend on changes in programs of gene expression rather than alterations in the genome itself. (The cloning of Dolly the sheep proved this point dramatically, since it demonstrated that the genome of a highly differentiated mammary epithelial cell is essentially identical to that of a recently fertilized egg.)

According to the alternative model, many of the steps of tumor progression are driven by genetic alterations accumulated in the genomes of developing tumor cells. This second scenario is supported by extensive evidence garnered over the past several decades, which documents the accumulation of increasing numbers of mutant genes in cells as they evolve from a more benign to a more malignant growth state. We have encountered many of these mutant genes in earlier chapters.

This parallel between genetic evolution and phenotypic progression is best documented for human colon carcinomas. We know so much about the genetic basis of tumor progression in this organ because the colonic epithelium is, as mentioned, relatively accessible through colonoscopy; because colon cancer is a common disease in the West, making it relatively easy to collect and study large numbers of samples of pre-malignant and malignant growths; and because of the energies of researchers at the Johns Hopkins Medical School in Baltimore, Maryland, whose interests were focused on understanding these tumors and their genetic origins.

These researchers recognized the extensive evidence that mutant alleles of genes such as *ras* and *p53* can contribute to cell transformation under experimental

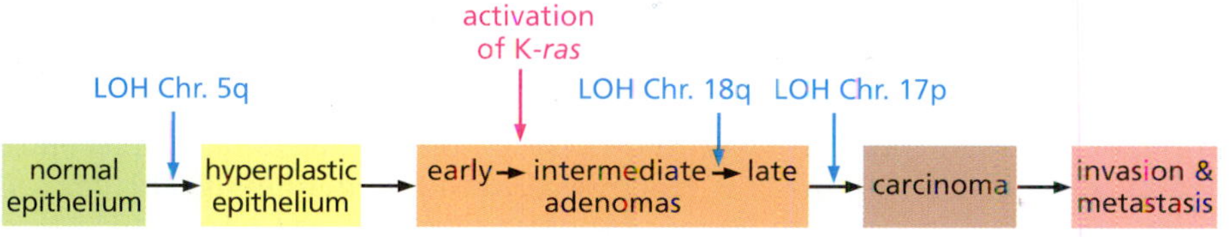

Figure 11.9 Colon tumor progression and loss of heterozygosity in various chromosomal arms DNA from tissue samples representing various stages of colon cancer progression was analyzed for loss of heterozygosity (LOH) by examining the behavior of chromosomal markers present on the long and short arms of most chromosomes, as described in Figure 7.14. In general, each chromosomal arm was represented by one or more genetic loci that existed in heterozygous form in the normal tissue of a patient. As tumor progression proceeded, the involved colonic epithelial cells exhibited increasing numbers of chromosomal arms that had lost heterozygosity. Certain chromosomal regions *(blue lettering)* suffered especially high rates of LOH, suggesting that they carried tumor suppressor genes that were being inactivated, in part, through the LOH mechanism. (Adapted from B. Vogelstein, E.R. Fearon, S.E. Kern et al., *Science* 244:207–211, 1989.)

conditions *in vitro*. They therefore sought to determine whether they could find *in vivo* correlates by examining the genomes of sizable groups of small colonic adenomas, mid-sized adenomas, large adenomas, and frank carcinomas. It was plausible that as colonic tissues advanced progressively from normalcy to a high-grade malignancy, the epithelial cells in these various tissues would accumulate increasing numbers of mutations in various genes.

This is just what these scientists discovered in the late 1980s. The genes that they examined included the K-*ras* oncogene and a number of tumor suppressor genes. In fact, the identities of the tumor suppressor genes that participate in colon cancer pathogenesis were not known when they began their work, so they searched instead for chromosomal regions that suffered loss of heterozygosity (LOH) during tumor progression. Recall that a high rate of LOH in a particular chromosomal region provides strong indication that this region harbors a tumor suppressor gene, and that developing cancer cells exploit the LOH mechanism as a means of shedding the still-intact (i.e., wild-type) alleles of such a tumor suppressor gene (Section 7.4).

In landmark work, this research discovered that early-stage adenomas often showed loss of heterozygosity in the long arm of Chromosome 5 (i.e., 5q). Almost half of slightly larger adenomas carried, in addition, a mutant K-*ras* oncogene. Even larger adenomas tended to also have high rates of LOH on the long arm of Chromosome 18 (i.e., 18q); and about half of all carcinomas showed, in addition, an LOH on the short arm of Chromosome 17 (i.e., 17p; Figure 11.9; see also Figure 7.14).

These observations provided strong support for the idea that as epithelial cells acquire increasingly neoplastic phenotypes during the course of tumor progression, their genomes show a corresponding increase in the number of altered genetic loci. Equally important, these changes involve both the activation of a proto-oncogene into an oncogene (K-*ras*) and the apparent inactivation of at least three distinct tumor suppressor genes. This was the first suggestion of a phenomenon that is now recognized to be quite common: the number of inactivated tumor suppressor genes greatly exceeds the number of activated oncogenes present in human tumor cell genomes.

The identities of the genetic loci on Chromosomes 5 and 17 were revealed soon after this genetic progression was laid out (Figure 11.10). The Chromosome 5q21 gene that is often the target of LOH was found to be the *APC* (adenomatous polyposis coli) tumor suppressor gene (Section 7.11), while the Chromosome 17p13 gene was identified as the *p53* tumor suppressor gene (Chapter 9). The identity of the gene or genes on Chromosome 18q that are inactivated during colon carcinoma pathogenesis remains unclear. While this chromosomal region suffers loss of heterozygosity in more than 60% of human colon cancers, the best

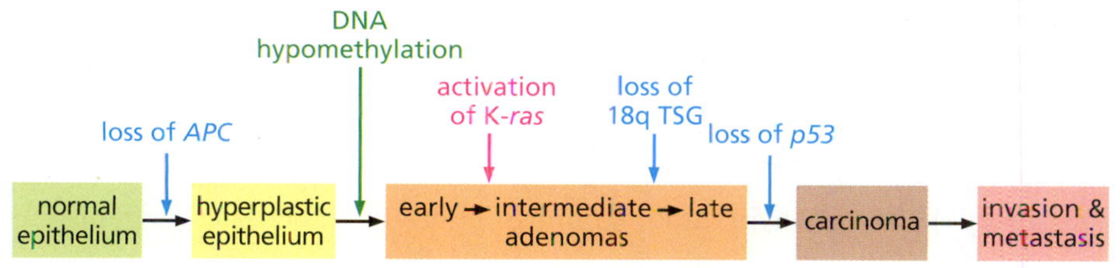

Figure 11.10 Tumor suppressor genes and colon carcinoma progression Each of the chromosomal regions that underwent loss of heterozygosity (LOH; Figure 11.9) was considered to harbor a tumor suppressor gene (TSG) whose loss provided growth advantage to evolving pre-neoplastic colonic epithelial cells. Eventually two TSGs were identified *(blue lettering)*—*APC* on Chromosome 5q, and *p53* on Chromosome 17p. The identity of the inactivated TSG or TSGs on Chromosome 18q remains unclear. In addition, as indicated here, about half of colon carcinomas were found to acquire mutant, activated alleles of the K-*ras* gene *(pink lettering)*, and the genomes of most evolving, pre-neoplastic growths were found to suffer hypomethylation (loss of methylated CpGs; *green lettering*). The precise contribution of hypomethylation to tumor progression remains unclear; some evidence suggests that it creates chromosomal instability.

candidate tumor suppressor gene here is *DPC4/MADH4*, which specifies Smad4, the protein that relays growth-inhibitory signals from the TGF-β receptor to the cell nucleus (Section 6.12). However, the copies of *DPC4/MADH4* are present in mutant form in fewer than 15% of colon cancers, and a second, nearby gene specifying Smad2 is inactivated even less frequently. This leaves unanswered the identity of the tumor suppressor gene(s) on Chromosome 18q that are the targets of inactivation and LOH in the majority of these cancers.

The most obvious way to rationalize the steps in colon cancer development involves an ordered succession of genetic changes that strike the genomes of colonic epithelial cells as they evolve progressively toward malignancy, as illustrated in Figure 11.10. In reality, the specific sequence of genetic changes depicted in this figure presents the history of only a very small proportion of all colonic tumors. This number can be calculated from the known rates of genetic alteration of the various genes or chromosomal regions known to participate in colon cancer pathogenesis. While the great majority (~90%) of colon carcinomas suffer inactivation of the *APC* gene on Chromosome 5q as an early step in this process, only about 40 to 50% acquire a K-*ras* mutation, 50 to 70% show an LOH of Chromosome 17p involving *p53*, and about 60% show an LOH on Chromosome 18q. In addition, as many as 12% of colonic tumors have mutations that lead to functional inactivation of the type 2 TGF-β receptor (Section 5.7). A further complication is created by the observation that tumors bearing K-*ras* oncogenes rarely have mutant *p53* alleles, and vice versa.

Most colon cancers will therefore begin with a Chromosome 5 alteration, but then will take alternative genetic paths on the road toward full-fledged malignancy (Figure 11.11A). These other paths will presumably involve a comparable number of genetic alterations. It seems plausible, for example, that those colon carcinomas that do not acquire a mutant K-*ras* oncogene will sustain alterations in other components of the Ras signaling pathway. Presumably, these alternative mutations confer a cell-physiologic advantage on colon cancer cells similar to that resulting from the formation of a mutant, activated K-*ras* oncogene.

This suspicion has been borne out by the systematic sequencing of 90 distinct kinase genes present in the genomes of a group of 35 colorectal carcinomas. These analyses indicated that about 15% of these tumors carry mutant, presumably activated tyrosine kinase enzymes; another genetic study reported mutant forms of the 110-kD subunit of PI3 kinase (PI3K; Section 6.6) in about one-third of colorectal tumors analyzed. Yet another analysis, this time of colonic polyps, revealed that those growths lacking activated K-*ras* oncogenes often carried mutant, oncogenic alleles of the B-*raf* gene, which encodes a serine/threonine kinase (Section 6.5); significantly, the mutant *ras* and *raf* oncogenes were mutually exclusive, never being found together in the same polyps. These various mutant kinases are excellent candidates for the agents that serve as functional alternatives to the K-*ras* oncoprotein during colon carcinoma progression.

It is also true that the four alterations documented by the Johns Hopkins researchers, even when they define the history of a tumor, do not always occur in the order described in Figure 11.10. The loss of heterozygosity on Chromosome 5q is almost always the first in the progression, but the precise order of the subsequent changes may vary from tumor to tumor. This unique role of the *APC* locus as the site of the first genetic step in colon cancer pathogenesis seems to be dictated by the unique biological effects of *APC* inactivation (Sidebar 11.3).

Finally, we should note that these various genetic changes do not represent an upper limit to the number of alterations that contribute in essential ways to colon cancer progression. For example, the type of genetic analysis used to identify this series of four chromosomal regions registered only those loss-of-heterozygosity (LOH) events in tumor suppressor genes that occur at a significant

(A)

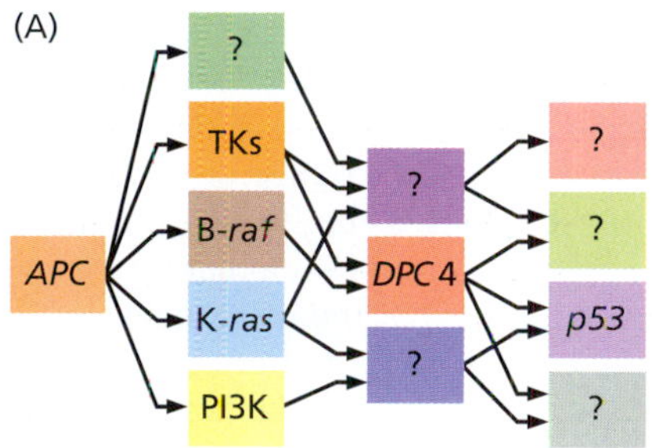

(B)

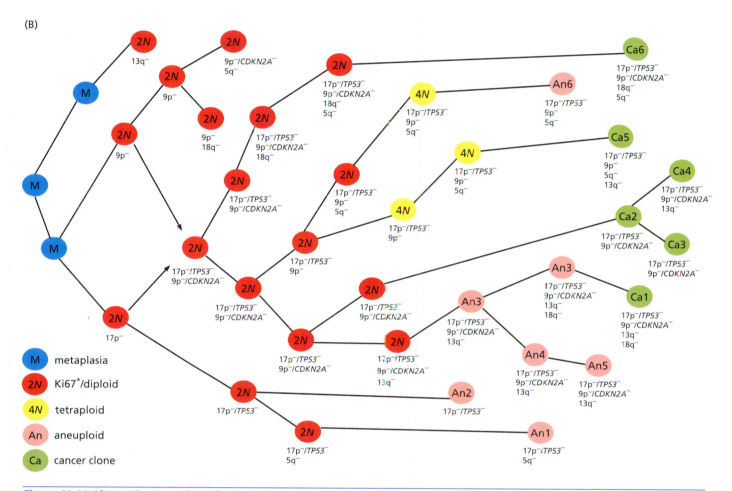

Figure 11.11 Alternative genetic paths during cancer progression (A) The series of genetic alterations shown in Figure 11.10 does not represent an invariant sequence that defines the genetic paths followed by all colon carcinomas. Loss of APC function (or the functionally equivalent gain of β-catenin function) represents a starting point that is common to almost all human colon carcinomas. However, the identities of the genes altered in subsequent steps are variable, and the precise order of these changes also seems to vary. This makes possible, in theory, the multiple paths depicted here. Moreover, the exact number of genetic steps occurring during human colon cancer pathogenesis is not known and may be far larger than the four depicted here. (B) Detailed study of human Barrett's esophagus (see Figure 2.13), which is a precursor lesion to esophageal carcinoma, has led to the discovery of a number of alternative genetic paths depicted here between the initial metaplasia (in which cells of gastric origin displace the normally present squamous cells of the esophagus; *left*) through dysplasia to the cancer clones *(right)* that may eventually arise. Note that the changes listed here include changes in chromosome number (including aneuploidy and tetraploidy) as well as loss of heterozygosity (LOH) and apparent loss of tumor suppressor gene expression through promoter methylation. Ki67$^+$, positive for the S-phase Ki67 marker and therefore actively proliferating; *CDKN2A*$^-$, mutation or promoter methylation of *p16^{INK4A}* gene; *TP53*$^-$, mutation of *p53* gene; 9p$^-$, LOH of markers on the short arm of Chromosome 9, etc.; 2*N*, diploid; 4*N*, tetraploid. (B, from M.T. Barrett, C.A. Sanchez, L.J. Prevo et al., *Nat. Genet.* 22:106–109, 1999.)

frequency above the general background rate of LOH associated with all chromosomal arms in advanced tumor cells (see Figure 7.14). Accordingly, LOH events that occur with a relatively low frequency in a collection of tumor samples (i.e., those present in 20% or fewer of the tumors analyzed) would not be

Sidebar 11.3 The organization of colonic crypts explains why *APC* loss occurs early in colon cancer progression As described earlier (Section 7.11), once normal colonic enterocytes are formed, they migrate out of the colonic crypts, differentiate, and die within 3 to 4 days by apoptosis. Consequently, most mutations that strike the genomes of these cells (e.g., a point mutation creating a K-*ras* oncogene) will soon be lost from this tissue because of the rapid death of such cells within days of their initial formation. Loss of *APC* function, however, results in *trapping* of cells within the colonic crypts and thus in the accumulation of *APC*-negative cell populations in these crypts. (A minority of colorectal tumors express a mutant β-catenin that resists degradation, yielding a biochemical state similar to that resulting from loss of Apc.) Because emigration of enterocytes from the crypts is blocked, when these cells and their descendants sustain additional mutations, the resulting mutant cells will also be retained in the crypts rather than being lost rapidly through emigration and apoptosis. We cannot, however, make similarly compelling arguments to rationalize why the mutation of other genes involved in colon cancer progression (*ras*, *p53*, and *DPC4*/*MADH4*) should occur at a particular time or in a particular sequence during the course of colon tumor progression. Indeed, the sequence of mutation of these other genes appears to be quite variable from one colonic tumor to another.

registered in such an analysis, even though these events might lead to the elimination from the cell genome of functionally important tumor suppressor genes.

Epigenetic events, including the repression of some genes through promoter methylation (Section 7.8) and the de-repression of others through demethylation, may also contribute importantly to tumor progression. Recently, evidence has accumulated that hypomethylation (i.e., demethylation of normally methylated sequences), such as that observed in early adenomas (see Figure 11.10), has an effect independent of its possible effects on gene expression: operating through unknown mechanisms, hypomethylation contributes to chromosomal instability, and the latter state favors acceleration of the rate of tumor progression.

The publication of this "genetic biography" of colon cancer tumorigenesis, as depicted in Figure 11.10, should ideally have been followed by similar descriptions of a wide variety of other tumor types, each biography involving its own particular set of oncogenes and tumor suppressor genes. However, only a handful of such descriptions (e.g., of bladder, pancreatic, and esophageal carcinomas; see Figure 11.11B) have actually been reported. This means that, at present, we cannot cite lists of genetic alterations in tumor cell genomes to illustrate the multi-step nature of cancer progression in most organ sites throughout the body. With the development of more sophisticated and sensitive tools for analyzing tumor cell genomes, the genetic biographies of many types of tumors should be forthcoming.

11.4 Multi-step tumor progression helps to explain familial polyposis and field cancerization

The genetic pathway laid out above (Figure 11.10) and modifications thereof depict the genetic events that occur during the pathogenesis of many sporadic colonic tumors. These tumors arise in the colons of individuals whose genotypes are, as far as we know, genetically wild type at the moment of conception. During the course of colon tumor development, somatic mutations (and methylation events) progressively alter the initially pristine genomes of colonic epithelial cells, yielding the corrupted genomes found in highly malignant colon cancer cells.

All this adds weight to the notion that the complexity of this multi-step process reflects the existence of an equally complex set of defense mechanisms that block the appearance of tumors in normal tissues. Each of these defense mechanisms must be thwarted or disabled, one after another, in order for an evolving population of cells to reach the final endpoint of full malignancy.

Given the vast number of colonic epithelial cells that are formed during a human lifetime ($>10^{14}$), these defense mechanisms are highly successful, and sporadic colon carcinomas occur at remarkably low rates. Still, the development of colon cancer is virtually ensured in some individuals, as discussed in Section 7.11 and again above. Recall that a mutant germ-line allele of the *APC* tumor suppressor gene predisposes an individual to the development of dozens, hundreds, or even a thousand colonic polyps—the syndrome of familial adenomatous polyposis (FAP).

The first step in the development of almost all *sporadic* colon carcinomas involves the inactivation of the *APC* gene—precisely the same gene that is inherited in mutant form by individuals suffering from the *familial* polyposis syndrome. Now, in the context of multi-step tumorigenesis, we can understand why inheritance of a mutant *APC* allele results in polyposis and colon cancer: the first step in colon cancer progression, which involves the inactivation of an *APC* gene copy, has already occurred in all of the colonic epithelial cells of an individual suffering from familial polyposis. That is, each of their cells, including those in the colon, is $APC^{+/-}$ rather than $APC^{+/+}$.

Since loss of heterozygosity is a relatively frequent event per cell generation, hundreds if not thousands of individual colonic epithelial cells may lose all APC function relatively early in the lives of individuals suffering from familial polyposis, doing so by advancing to the $APC^{-/-}$ state. Each of these Apc-negative cells can spawn, with substantial probability, an adenomatous polyp; and such polyps, once formed, have a significant probability (perhaps one per several hundred) of progressing into a carcinoma. Consequently, a common mechanism of inborn susceptibility to cancer involves an acceleration of multi-step tumor progression, since one of its critical, rate-limiting steps is no longer dependent on infrequently occurring somatic mutations (because it has already occurred in the germ line).

Organs affected by sporadic tumors occasionally sprout multiple, apparently independently arising tumors—the phenomenon that is called **field cancerization**. These multiple growths seem to resemble, at least superficially, processes of inborn cancer susceptibility such as familial polyposis. In fact, a unique mutation (or set of mutations) appears to be shared by these multiple, ostensibly independent growths, but in this case, the shared genetic alteration is the product of a *somatic* mutation mechanism (Sidebar 16).

11.5 Cancer development seems to follow the rules of Darwinian evolution

The observations about colon cancer made at Johns Hopkins University demonstrated that the histopathological changes occurring during tumor progression were correlated with genetic changes striking the genomes of cells in the colonic mucosa. More important, it became plausible that these genetic changes were actually *causing* the phenotypic evolution of these cells and the tissues they form.

Years earlier, others had speculated that tumor development could be understood in terms of a biological process that resembles Darwinian evolution. The results of the genetic analyses of human colon cancer progression provided further evidence supporting this model. (While Darwin himself knew virtually nothing about genes and genetics, the "modern synthesis" of Darwinian theory introduces Mendelian and population genetics into the evolutionary processes that Darwin first postulated.)

In the case of cancer development, the evolving units are individual cells competing with one another in a population of cells, rather than individual organisms competing with one another within a species. Like the modern depiction of Darwinian evolution, random mutations are presumed to create genetic variability in a cell population. Once a genetically heterogeneous population has arisen, the forces of selection may then favor the outgrowth of individual cells (and their descendants) that happen to be endowed with mutant alleles conferring advantageous traits, notably traits that favor the proliferation and survival of these cells in the microenvironment of a living tissue.

Combining Darwinian theory with the assumptions of multi-step tumor progression, researchers could now depict tumorigenesis as a succession of *clonal expansions*. The scheme goes like this: A random mutation creates a cell having particularly advantageous growth and/or survival traits. This cell and its descendants then proliferate more effectively than their neighbors, eventually yielding a large clonal population that dominates the tissue and crowds out genetically less favored neighbors. Sooner or later, this cell clone will reach a large enough size (e.g., 10^6 cells) that another advantageous mutation, which strikes randomly with a probability of about 1 per 10^6 cell generations, may now plausibly occur in one or another cell within this clonal population (Figure 11.12).

The resulting doubly mutated cell, which will proliferate (or survive) even more effectively than its 10^6 clonal brethren, will spawn a new subclone that will expand and eventually dominate the local tissue environment, overshadowing and possibly obliterating the precursor population from which it arose. After this doubly mutated cell clone reaches a large size, as before, a third mutation may now strike, and the process of clonal expansion and succession will repeat itself. Quite possibly, a sequence of four to six such clonal successions, each triggered by a specific mutation, suffices to explain how cancer progression occurs at the cellular and genetic level.

To be sure, this Darwinian model of cancer progression is simplistic. For example, it must be amended to respond to the discovery that epigenetic alterations of genes, specifically, promoter methylation (see Section 7.8), play an important role in eliminating the activities of tumor suppressor genes. (Here, we encounter

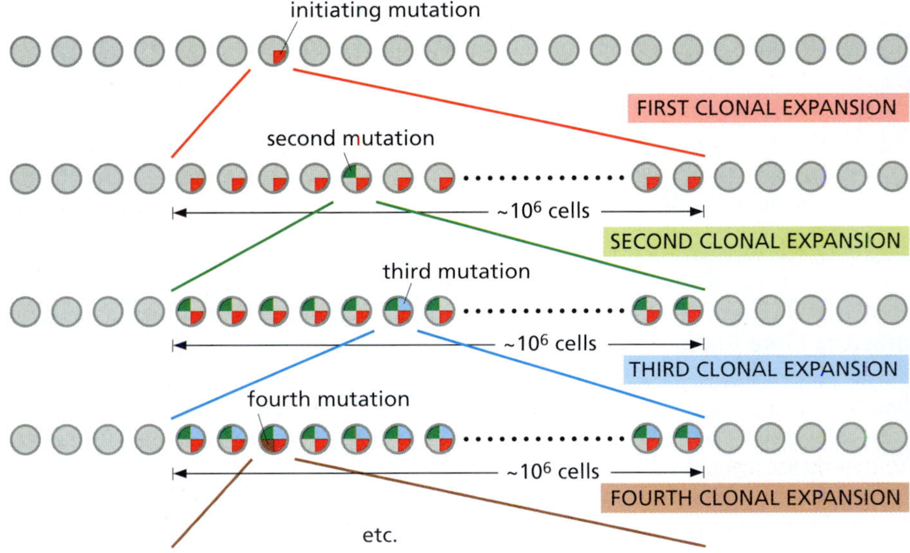

Figure 11.12 Darwinian evolution and clonal successions Darwinian evolution involves expansions of organisms that are endowed with advantageous genotypes and thus phenotypes; a similar scheme seems to describe how tumor progression occurs. One cell amid a large cell population sustains an initiating mutation *(red sector, top)* that confers on it a proliferative and/or survival advantage compared with those cells lacking this mutation. Eventually, the clonal descendants of this mutant cell dominate in a localized area by displacing the cells that lack this mutation, resulting in the first clonal expansion. When this clone expands to a large enough size (e.g., 10^6 cells), the occurrence of a second mutation that strikes with a frequency of 10^{-6} per cell generation may occur *(green sector)*, resulting in a doubly mutated cell that has even greater proliferative and/or survival advantage. The process of clonal expansion then repeats itself, and the newly mutated population displaces ("succeeds") the previously formed one. This results once again in a large descendant population, in which a third mutation *(blue sector)* occurs, and so forth.

a major discordance between tumor progression and Darwinian evolution, since heritable epigenetic alterations, such as DNA methylation events, have never been shown to drive the evolution of species.)

This scheme is simplistic in other respects as well. Thus, the number of distinct steps in tumor progression may be underrepresented by counting the number of genetic loci that are altered during this process. As discussed in Chapter 7, the inactivation of a tumor suppressor gene is, almost always, a two-step process. First, one gene copy is mutated (or methylated) to inactivity. Thereafter, the surviving, still-intact gene copy is eliminated, usually through the process of loss of heterozygosity (LOH). Knowing this and the fact that elimination of tumor suppressor genes often represents the majority of the genomic changes occurring during tumor progression, we conclude that the number of distinct alterations occurring during tumor progression may be almost twice as many as the number of loci that are involved.

Since each of these clonal expansions is triggered by an infrequently occurring genetic or epigenetic alteration (or pair of alterations, in the case of tumor suppressor genes), these expansions are likely to be spaced far apart in time. During the development of sporadic colon cancers, a decade or more may separate one critical genetic alteration from the next, and in many individuals, the process as a whole may stretch over a century. (For example, one of the key steps—the evolution from polyp to invasive cancer—has been estimated to take from 4 to 11 years, depending on the degree of dysplasia present in the polyp.) Still, it is clear that some people develop sporadic colon carcinomas in far less time, and so this schedule must be compressed in the colons of these individuals.

We know almost nothing about the processes that govern the rates of tumor progression in most tissues. In the case of colon cancer, certain constituents of diet may greatly increase the rate at which the genomes of their colonic epithelial cells accumulate mutations (Sidebar 11.4). Such increased mutation rates can, in turn, compress the time between clonal successions. Perhaps in individuals who consume certain foodstuffs, 5 or fewer years intervene between successive clonal expansions rather than the usual 10 or 20. As a consequence, a disease process that usually requires a century to reach completion may reach its neoplastic endpoint in 30 or 40 years, well within the human life span.

Sidebar 11.4 What carcinogens affect the human colon? Colon cancer rates vary by as much as twentyfold between countries. These dramatic differences are due to environmental differences rather than different degrees of genetic susceptibility, since populations migrating from one country to another exhibit cancer rates typical of their new country within a generation (e.g., see Figure 2.20). The "environment" in this case is, without doubt, foodstuffs in the diet.

The determination of carcinogens in the human colon represents an undertaking of daunting complexity. The foodstuffs that we consume contain, by some estimates, between 5000 and 10,000 biochemical species that exist naturally in plants and animals. Cooking further increases the number of these chemical species. In addition, the several hundred species of bacteria living in the human colon metabolize the breakdown products of our food in complex ways, resulting in further increases in the number of distinct chemical species that exist in our gastrointestinal tracts. The resulting enormous chemical complexity means that at present, and likely for the foreseeable future, our insights into the connections between diet and multi-step colon cancer progression will be derived largely from epidemiology rather than from analyses of colonic contents and their biochemical and biological mechanisms of action.

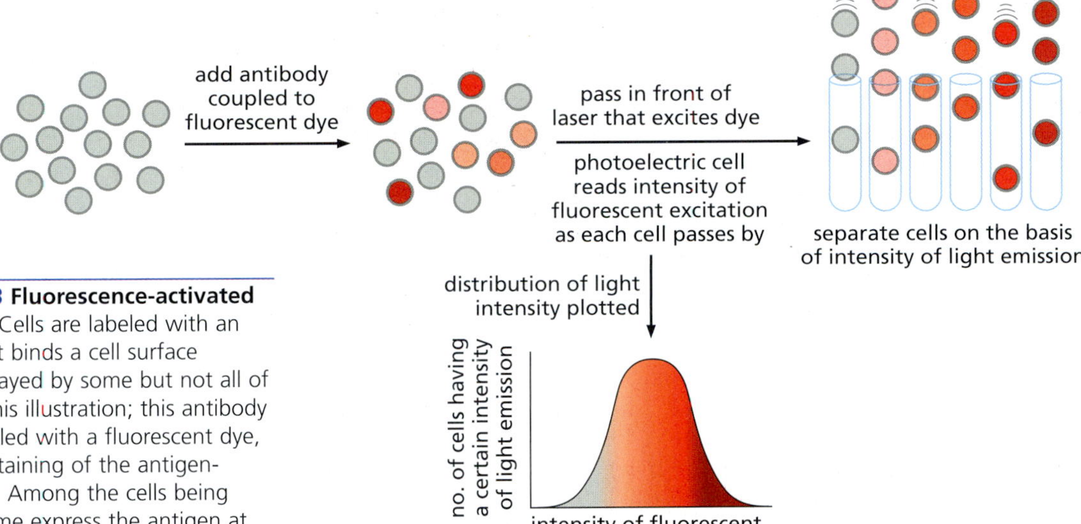
add antibody coupled to fluorescent dye
pass in front of laser that excites dye
photoelectric cell reads intensity of fluorescent excitation as each cell passes by
separate cells on the basis of intensity of light emission
distribution of light intensity plotted
no. of cells having a certain intensity of light emission
intensity of fluorescent emissions

Figure 11.13 Fluorescence-activated cell sorting Cells are labeled with an antibody that binds a cell surface antigen displayed by some but not all of the cells in this illustration; this antibody may be coupled with a fluorescent dye, resulting in staining of the antigen-positive cells. Among the cells being analyzed, some express the antigen at higher levels and therefore are stained more darkly by the antibody *(darker red)*, while others are stained weakly by the antibody *(pink)* or not at all *(gray)*. This mixture of cells, placed in suspension, is then passed in single file past a laser beam that excites the dye (and causes it to emit light) and a photoelectric cell that measures the intensity of fluorescence emission by the dye molecules staining each cell. These measurements may then be integrated and plotted as a distribution of signals, in which case this procedure is often termed "flow cytometry" *(below)*. Alternatively, if the labeling procedure has not damaged these cells, they can be separated from one another on the basis of their fluorescence intensity (e.g., through use of an electrical field), which allows the biological properties of these cells to be gauged through other tests *(right)*. Often, the intensity of staining by two different antibodies, each bearing a distinct fluorescent dye, allows the simultaneous measurement of intensities of two different antigens, in which case a different graphing convention is used, as shown in Figure 11.14. Measurements of fluorescence intensity obtained by in this way are usually plotted on a logarithimic scale.

Certain alterations within pre-malignant cells may conspire with these exogenous foodstuff carcinogens to accelerate the pace of colon tumor progression. Thus, increased mutability may be caused by defects in the complex cellular machinery that is dedicated to maintaining and repairing cellular DNA. For example, as discussed in Chapter 12, some individuals inherit mutations that compromise the function of the cellular DNA mismatch repair apparatus. Because of the resulting defective DNA maintenance, greatly increased mutation rates are seen in their colonic epithelial cells, and these lead, in turn, to a greatly accelerated rate of formation of pre-malignant and malignant growths in the colon—the syndrome of hereditary non-polyposis colon cancer (HNPCC). More commonly, however, defects in components of the DNA repair apparatus result from somatic mutation (or promoter methylation) occurring early in tumor progression; the resulting crippling of the repair apparatus and increased genome mutability guarantee that subsequent mutational steps in colon cancer progression will occur relatively rapidly. These multiple factors influencing colon carcinoma formation greatly complicate attempts to delineate the clonal successions predicted by the Darwinian model of tumor progression.

11.6 Tumor stem cells further complicate the Darwinian model of clonal succession and tumor progression

The clonal succession model proposes that a mutant cell spawns a large flock of descendants and that among these numerous descendants, a new mutational event will trigger yet another wave of clonal expansion (see Figure 11.12). However, certain experiments have cast serious doubt on the notion that all of the cells within a pre-neoplastic (or neoplastic) cell clone are biologically equivalent and therefore equally capable of becoming ancestors of a new successor clone of cells. In these experiments, the cancer cells within a human tumor were separated into distinct subclasses. These separations took advantage of cell surface proteins displayed by the different subpopulations. In particular, the technique of fluorescence-activated cell sorting (FACS) was used to separate living cancer cells after labeling them (via their surface proteins) with antibodies linked to fluorescent dyes (Figure 11.13). Cells separated by this procedure can be recovered in viable form and used in biological experiments, including *in vivo* tests of their ability to seed tumors following injection into immunocompromised host mice.

Use of the FACS technique initially enabled researchers to segregate populations of acute myelogenous leukemia (AML) cells into majority and minority populations; in one such experiment, the latter represented fewer than 1% of the neoplastic cells in the tumor mass. As few as 5000 cells in the minority subpopulation were able to produce new tumors upon injection into host mice and were therefore deemed to be "tumorigenic"; in contrast, as many as 500,000 AML cells from the majority subpopulation were unable to seed a tumor. Importantly, the cells in this majority subpopulation exhibited many of the attributes of differentiated granulocytes or monocytes, and these cells had limited ability to proliferate. These observations provided compelling evidence that the AML tumors were composed of small populations of self-renewing, tumorigenic cells and large populations of more differentiated cells that had little, if any, ability to proliferate *in vivo*.

Subsequent experiments extended these results to human breast cancer cells prepared directly from tumors. In these later experiments, the minority tumorigenic cell population within a tumor represented only about 2% of the overall neoplastic cell population. Several hundred of these minority cells seeded a new tumor when injected into a host mouse, while as many as 20,000 cells from the majority cell population failed to do so (Figure 11.14). Importantly, the majority

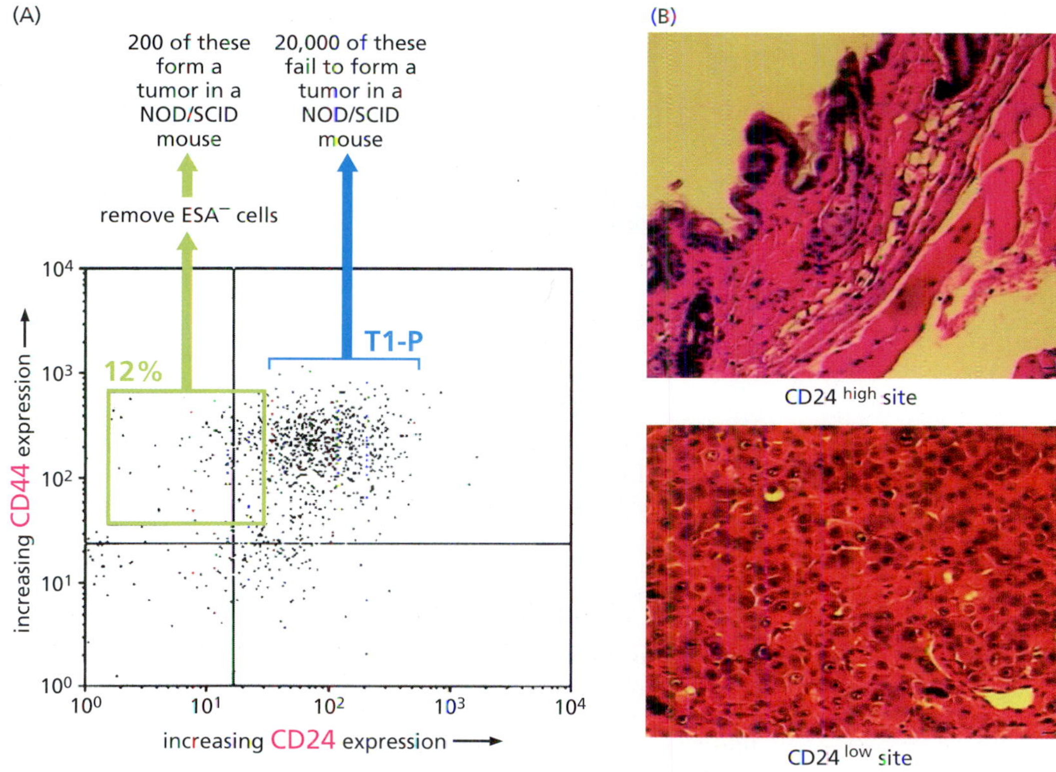

Figure 11.14 Enrichment of breast cancer stem cells
(A) Metastatic human breast carcinoma cells were freed of contaminating noncancerous (stromal) cells and separated from one another in a fluorescence-activated cell sorter (FACS; see Figure 11.13). The expression of two distinct cell surface antigens—CD24 and CD44—was gauged simultaneously, each being detected with a specific monoclonal antibody linked to a distinct fluorescent dye. The intensity of staining is plotted logarithmically on each axis. Each black dot in the graph represents the detection of a single cell. In this experiment, a 12% subpopulation of cells that expressed low CD24 and high CD44 antigens *(green box)* was separated from cells (T1-P) that showed high CD24 expression and high CD44 expression *(blue bracket)*. [The cells in the 12% minority population were further enriched by sorting for those that expressed epithelial surface antigen (ESA), which resulted in the elimination of contaminating cells that did not express this antigen.] (B) Two hundred of the resulting enriched $CD24^{low}$ $CD44^{high}$ ESA^{+} cells were able to form a tumor following injection into a NOD/SCID immunocompromised mouse ($CD24^{low}$ site, *below*), while 20,000 of the $CD24^{high}$ $CD44^{high}$ cells failed to do so ($CD24^{high}$ site, *above*). The *upper* image shows a section through the subcutaneous site of cell implantation, in which relatively normal skin and underlying muscle wall are apparent; the *lower* image shows a section through the tumor that formed. (From M. Al-Hajj, M.S. Wicha, A. Benito-Hernandez et al., *Proc. Natl. Acad. Sci. USA* 100:3983–3988, 2003.)

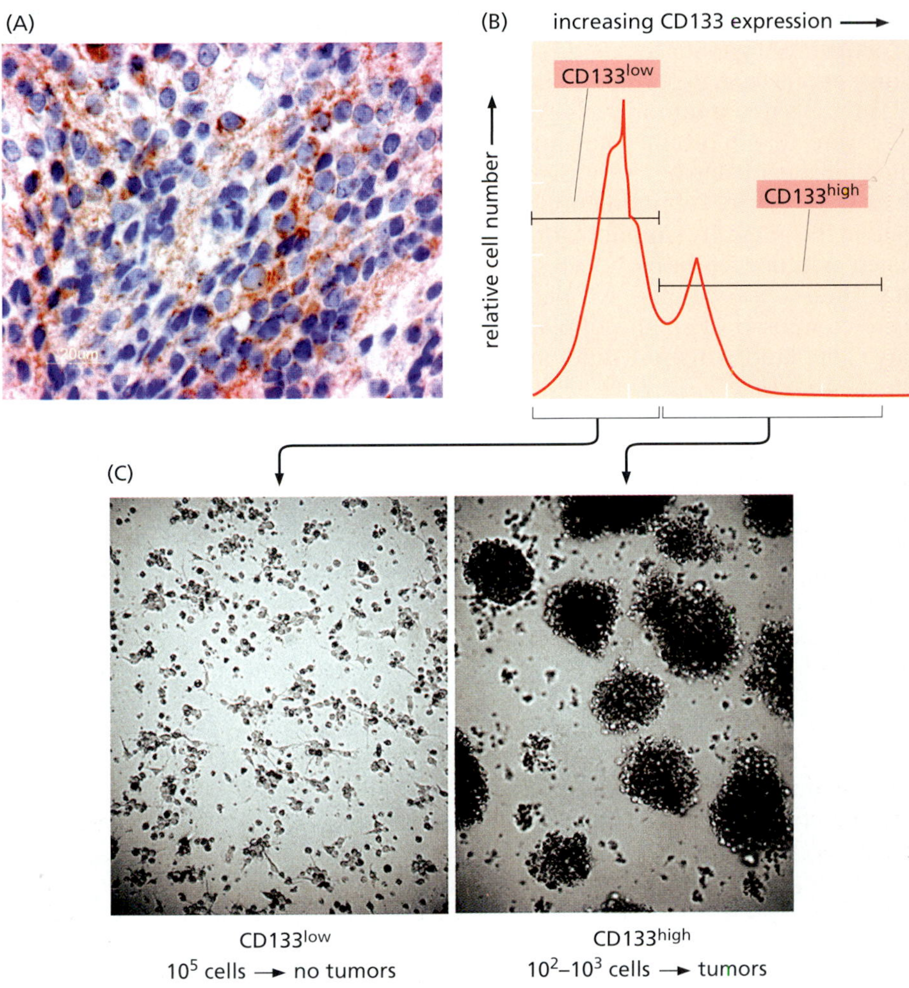

Figure 11.15 Enrichment of brain tumor stem cells Brain tumor stem cells can be detected because of their expression of CD133, a neural cell surface stem cell antigen. (A) In this immunofluorescence-stained section of a medulloblastoma (a tumor of cerebellar cells), tumor cells expressing high levels of CD133 *(dark red)* lie scattered among other tumor cells that have low CD133 expression. Cell nuclei are stained here in *blue*. (B) Cell populations from a human medulloblastoma were sorted by FACS (Figure 11.13) into those that express greater or lesser levels of CD133 *(increasing levels to the right along the abscissa)*. The number of cancer cells *(ordinate)* expressing low levels of CD133 *(left peak)* represents 80–85% of this tumor, while the CD133 high expressors *(right peak)* are in the minority. (Note that the intensity of staining is presented on a logarithmic scale on the abscissa.) (C) CD133^{low} cells from the left peak show limited proliferative ability in suspension culture *(left panel)*, and in some experiments, as many as 10^5 of these cells fail to seed tumors in host mice. In contrast, CD133^{high} cells from the right peak formed numerous colonies in suspension culture *(right panel),* and injection of as few as 10^2 (more typically 10^3) of these cells into an immunocompromised mouse host resulted in the formation of a tumor *(not shown)*. (From S.K. Singh, I.D. Clarke, M. Terasaki et al., *Cancer Res.* 63:5821–5828, 2003.)

and minority breast cancer cell subpopulations both contained equivalent proportions of cells in the active growth cycle, and both subpopulations were purified away from nonmalignant cells (such as stromal cells) that were present in the original tumor masses.

The tumors that eventually arose from the injected minority cells were once again composed of minority and majority cell populations that showed, as before, vast differences in their ability to seed new tumors, that is, in their tumorigenicity. Comparable results have since been obtained with brain tumor cells (Figure 11.15).

Taken together, these experiments indicate that the neoplastic cell populations in breast and brain cancers and, with some likelihood, solid tumors of many other organs are organized much like normal epithelial tissues, in which relatively small pools of self-renewing stem cells are able to spawn large numbers of descendant cells that have only a limited proliferative potential *in vivo*. In each of the experiments shown here (Figures 11.14 and 11.15), the tumorigenic minority cells expressed a set of antigenic markers on their surfaces that were distinct from those exhibited by cells in the majority populations, indicating that the two groups of cells were in different states of differentiation.

As was discussed in Section 8.11, in many normal tissues, stem cells are usually less differentiated, and their non-stem cell descendants usually enter into states of increased differentiation. In addition, stem cells appear to have an essentially unlimited ability to proliferate and, because some of their progeny remain as

Figure 11.16 Stem cells and their progeny (A) The simplest organization of stem cell behavior involves an asymmetric cell division of a stem cell *(blue),* in which one of the daughter cells becomes a stem cell like its mother, while the other daughter *(red)* differentiates and loses the ability to divide again. This can be depicted in two graphic conventions, both of which are shown here. (B) In many tissues, a more complex scheme appears to operate. As in panel A, one of the daughters of a dividing stem cell becomes a stem cell. However, the other becomes a "transit-amplifying cell" (sometimes termed a "progenitor cell"; *gray),* which is committed to enter a differentiation pathway but does not yet participate in end-stage differentiation. Instead, this cell and its progeny undergo a series of symmetric cell divisions before its descendants eventually enter into a fully differentiated state *(red).*

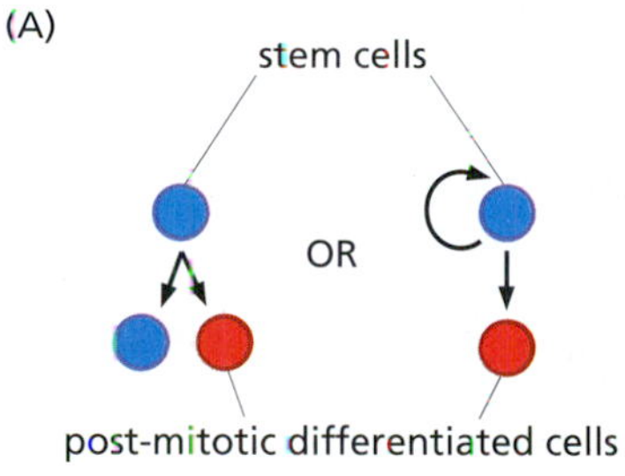

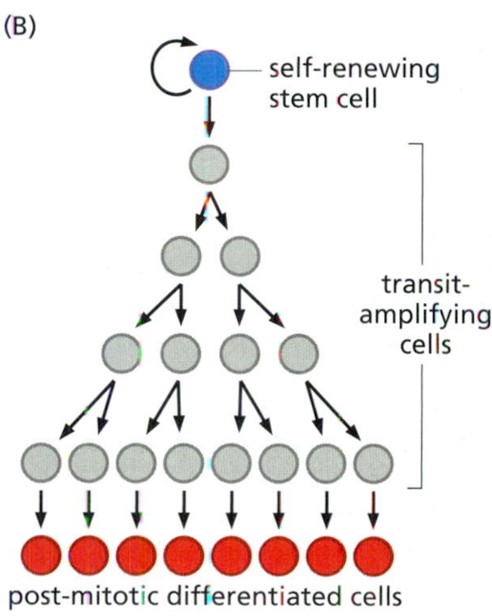

stem cells, are said to be "self-renewing." In contrast, the more differentiated progeny of a stem cell often enter a post-mitotic state, from which they will never emerge to re-enter into the active growth-and-division cycle (Figure 11.16).

It is therefore tempting to think that the same organization of cell behavior operates in these human tumors, but definitive evidence of such a relationship is still missing. Moreover, the cancer cells in the majority populations behave much like **transit-amplifying cells**, also called **progenitor cells** (see Figure 11.16), that are present in many normal tissues and represent intermediates between stem cells and their fully differentiated descendants (Sidebar 11.5).

The outcomes of these various experiments force us to reconsider how multi-step tumor progression occurs. In a revised scheme, the ancestors of each new successor population are drawn from a small minority of the predecessor cancer cell population—the tumorigenic cancer stem cells—rather than from the much larger population of neoplastic transit-amplifying cells. This means that mutations striking the cancer stem cell population (Figure 11.17) can be transmitted to descendant cells in the population, while mutations that strike the (far more numerous) non-tumorigenic transit-amplifying cells cannot, because these cells have only a limited proliferative potential. Consequently, the genetic evolution that we associate with multi-step tumor progression may occur in relatively small subpopulations of cancer cells—the minority tumor stem cells.

(This behavior is analogous to the behavior of germ-line mutations versus somatic mutations in metazoa. Germ-line mutations can, in principle, be transmitted to an unlimited number of germ cells and somatic cells in subsequent organismic generations, whereas somatic mutations cannot, because they occur in cells that leave only a limited number of descendants, all of which disappear whenever an individual organism dies. Following this analogy, the cancer stem cells represent the "germ line" of a tumor.)

Sidebar 11.5 Transit-amplifying cells enable small numbers of stem cells to generate large numbers of differentiated progeny The simplest scheme for organizing a continuously renewing tissue requires one recently formed daughter of a stem cell to retain the stem cell phenotype of the mother while the other daughter immediately proceeds into a highly differentiated, post-mitotic state, that is, becomes a cell that will never again divide (Figure 11.16A). In reality, a slightly more complex scheme seems to operate in most tissues (Figure 11.16B). As before, one daughter cell retains the stem cell phenotype of the mother, while the other becomes the progenitor of a large flock of differentiated, post-mitotic cells. This second daughter can do so because it and its immediate progeny—the transit-amplifying cells—undertake a program of exponential growth for a limited period of time before the descendant cells differentiate and become post-mitotic. Consequently, this single daughter cell can spawn dozens if not hundreds of differentiated progeny. This means that (1) the stem cell needs to divide only once in order to generate large numbers of differentiated descendants; (2) the stem cell may therefore divide only periodically rather than continually, even in a tissue in which differentiated cells are continually being lost and replaced; and (3) the great majority of cell divisions within a tissue can be ascribed to the exponentially growing transit-amplifying cells.

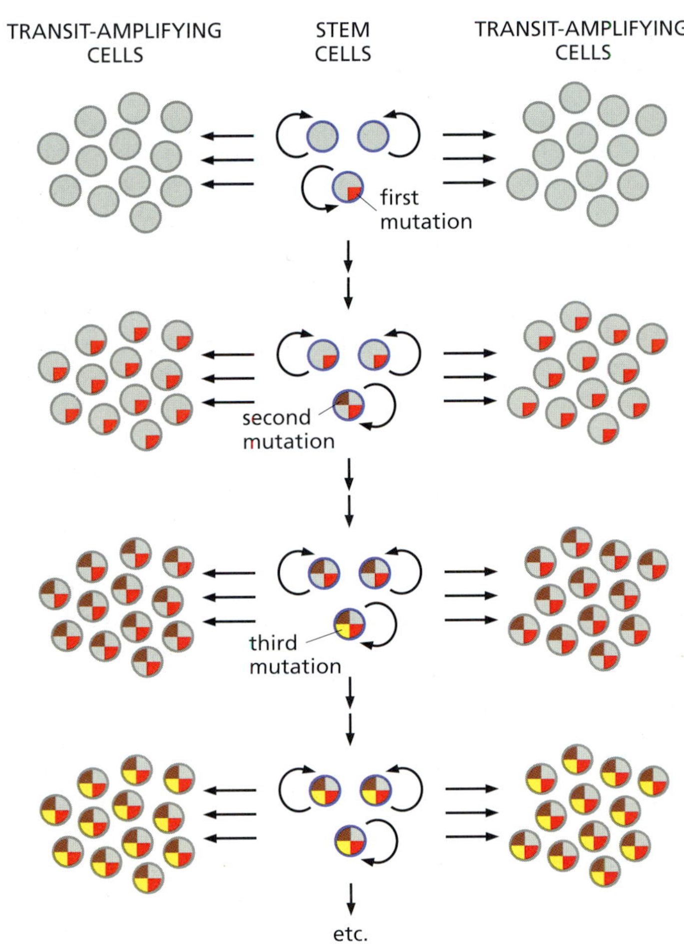

Figure 11.17 Cancer stem cells and clonal succession The existence of cancer stem cells has important implications for how tumor progression and clonal succession occur. In the scheme depicted here, less differentiated cells (i.e., cancer stem cells) can differentiate into transit-amplifying cells, but the reverse process (dedifferentiation of transit-amplifying cells into cancer stem cells) does not occur. Moreover, the cancer stem cells represent a minority of the neoplastic cells in tumor masses, while the transit-amplifying cells represent the majority of neoplastic cells. Therefore, mutations that strike the genomes of tumor stem cells can be transmitted to descendant tumor stem cells, which can then launch new clonal successions (see Figure 11.12). Conversely, mutations that strike the genomes of transit-amplifying cells cannot be transmitted further, because these cells have only limited replicative ability. (An exception to this scheme may occur if a mutation striking a transit-amplifying cell converts it into a stem cell.)

The revised scenario also has implications for the mutational mechanisms that propel tumor progression forward. During the formation of some tumors, such as the breast cancer analyzed in Figure 11.14, the mutations responsible for initiating new rounds of clonal succession must occur in cell populations that number in the tens of thousands rather than the millions, and the mutation rates (i.e., the number of mutations occurring per cell division) required to trigger these clonal successions must be vastly higher than one previously dared to imagine. In other tumors, however, in which the proportion of cancer stem cells is far higher (see Figure 11.15), the size of the target population would be predicted to be much larger, requiring a lower mutation rate in order for tumor progression to proceed.

11.7 A linear path of clonal succession oversimplifies the reality of cancer

The two schemes of clonal successions proposed until now (Figures 11.12 and 11.17) suggest that all of the cells within a tumor mass that participate in a particular clonal expansion are genetically identical to one another and that tumor formation occurs as a consequence of a linear series of these clonal successions. According to these schemes, if we were to examine the cells within a pre-malignant or malignant cell mass, we would almost always find that a single, genetically homogeneous clone of cells dominates in this mass, since it would have outgrown and largely displaced the preceding cell clone from which it arose. (Note that the stem cells and transit-amplifying cells within a clonal population are genetically identical to one another and differ from one another only phenotypically, having distinct gene expression programs.)

However, the actual course of tumor progression is complicated by yet another factor that we must take into account: as tumor progression advances, tumor genomes often become increasingly unstable, and the rate at which mutations are acquired during each cell generation soars. The rate of genetic change and resulting genetic diversification soon outpaces the rate at which Darwinian selection (and the elimination of less-fit subclones of cells) can occur. As a consequence, rather than looking like a linear series of clonal successions, actual tumor progression in many tumor masses is likely to resemble the highly branched scheme shown in Figure 11.18, in which a number of genetically distinct subclones of cells coexist within a single tumor mass.

Demonstrations of the genetic diversification of cells within tumor masses can be obtained by tracking the state of a single gene of interest within various cells of a primary tumor. One vivid (and extreme) example of such diversification is shown in Figure 11.19, which presents an analysis—accomplished by the FISH (fluorescence *in situ* hybridization) technique—to determine the copy number of Chromosomes 11 and 17 in individual cells within a human breast carcinoma. As is apparent, within this single tumor mass, the copy number of these chromosomes varies enormously from one cell to another. Such variability implies that the tumor cell genomes being analyzed are quite plastic and are continually changing the number of *CCND1* gene copies that they carry, doing so almost randomly, and a rate that greatly exceeds the ability of (Darwinian) selection to eliminate less fit variants.

Analyses such as these only scratch the surface of the genetic diversity that accumulates as cancer cells progress to ever more malignant growth states. Another, more systematic measure of this widespread genetic diversification has come from sequence analyses of randomly chosen DNA segments present in the genomes of a series of 58 human carcinomas. (Analysis of each DNA segment depended on the use of the polymerase chain reaction, or PCR, which enabled its selective amplification prior to determining its nucleotide sequence.) By extrapolating from the relatively small number of segments analyzed,

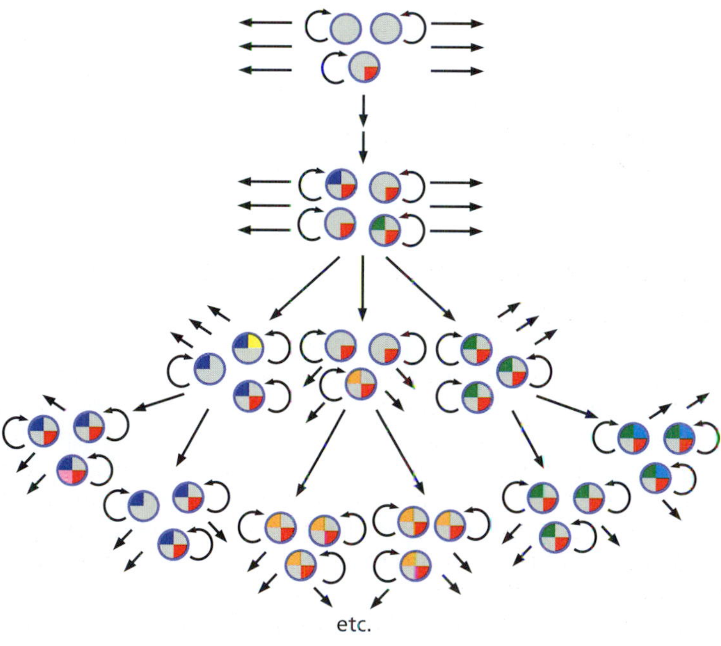

Figure 11.18 Clonal diversification due to high mutation rates As tumor progression proceeds, the genomes of populations of tumor cells often become increasingly unstable. As this occurs, the rate of generation of new mutant alleles may exceed the rate at which Darwinian selection eliminates less-fit clones. Consequently, the tumor mass becomes composed of an increasing number of distinct sectors, each dominated by a genetically distinct subclone. In this diagram, only the populations of tumor stem cells are depicted, while the arrows pointing outward (or downward) indicate the production of far larger numbers of transit-amplifying neoplastic cells *(not shown)*.

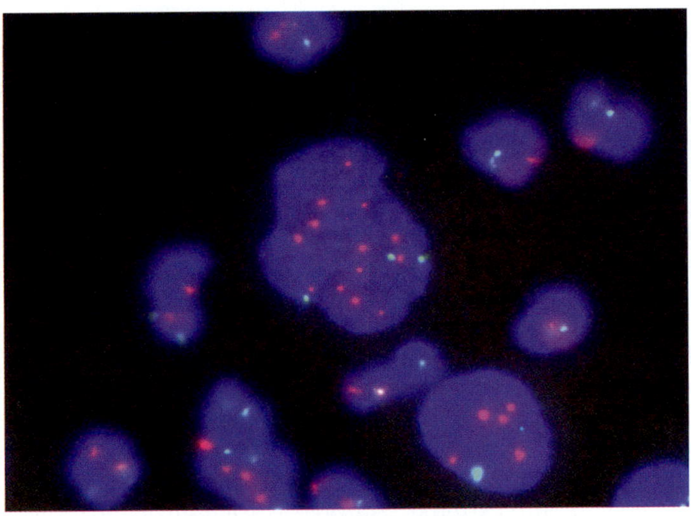

Figure 11.19 Genetic diversification within a tumor cell population In *high-grade* (i.e., highly malignant) tumors, the numbers of chromosomes often fluctuate wildly from one cell to another, indicative of great intra-tumor genetic heterogeneity. In this image, the copy numbers of Chromosomes 11 *(green)* and 17 *(pink)* have been revealed by fluorescence *in situ* hybridization (FISH) in cells from a *pleural effusion* present in a non-small-cell lung carcinoma patient. In addition to fluctuations in chromosome number, highly polyploid giant nuclei are apparent. (Courtesy of M. Fiegl.)

researchers estimated that at least 10^4 sequence alterations, including myriad single-nucleotide changes, were present in the genome of each of these sporadic tumors. Strikingly, a comparable number of alterations were found in the genomes of 11 sporadic adenomas. This latter observation indicates that widespread destabilization of cell genomes and extensive genetic diversification already occur quite early in multi-step tumor progression.

Use of the procedure of comparative genomic hybridization (CGH; Figure 11.20) reveals a further dimension of genetic instability. CGH is used to gauge increases and decreases in the copy number of various chromosomal DNA segments in tumor cell genomes and yields a plethora of genetic data (Figure 11.21). Some of

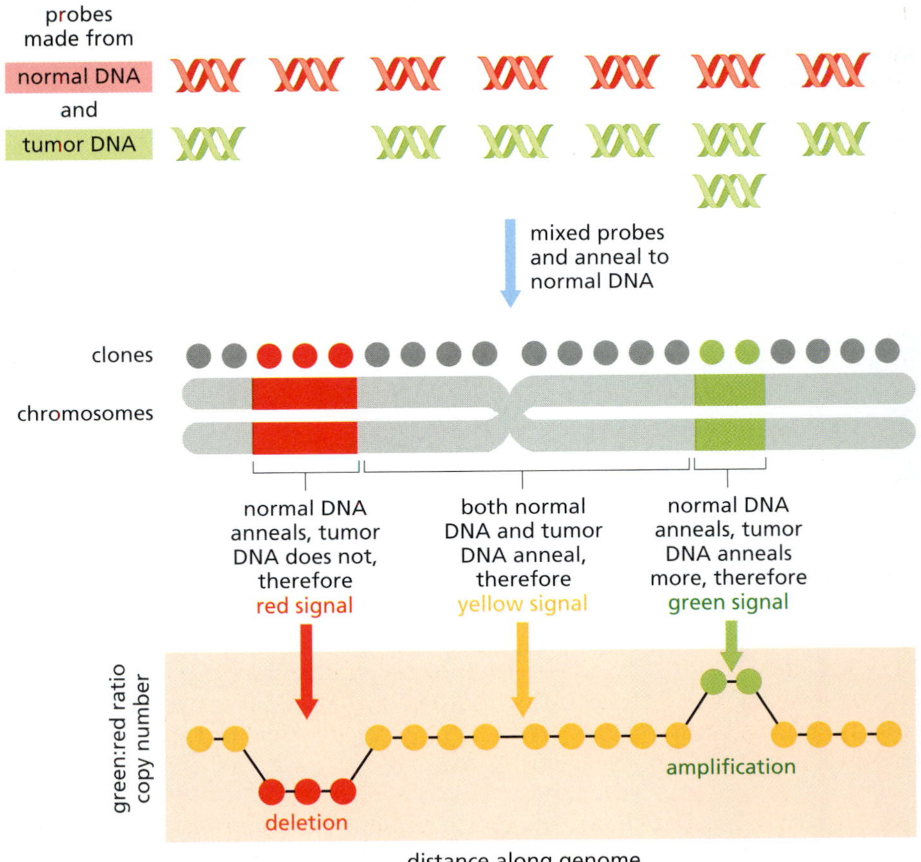

Figure 11.20 Comparative genomic hybridization CGH enables an investigator to determine whether the genome of a tumor cell contains segments that are present in greater or lesser copy numbers than exist in the normal diploid genome. As shown here, fragments of normal DNA *(red)* and tumor DNA *(green)*, each labeled with distinctly colored dye molecules, can be hybridized to DNA segments derived from defined chromosomal regions throughout the normal cell genome. Those chromosomal segments that anneal with only normal DNA will be registered as *red*, indicating that the corresponding DNA is missing in the tumor DNA. Conversely, those segments that are labeled in *green* will indicate that an elevated copy number of the segment (e.g., as a consequence of gene amplification) is present in the tumor DNA. DNA segments that are present in equal number in normal and tumor DNA will be registered as *yellow*. (Courtesy of J.W. Gray.)

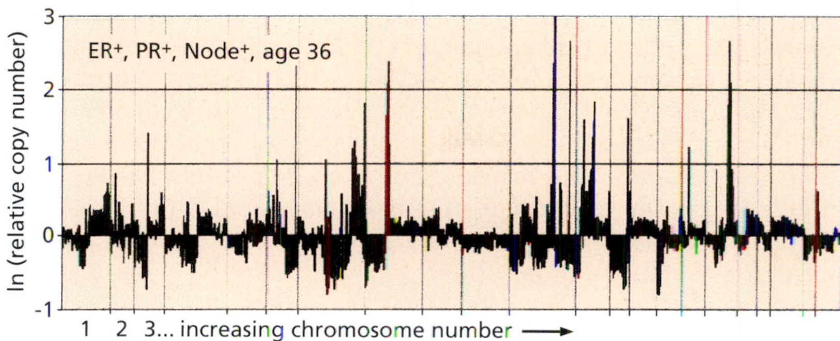

Figure 11.21 CGH analysis of a breast cancer genome The genomic DNA of an ER^+ (estrogen receptor–positive), PR^+ (progesterone receptor–positive), $Node^+$ (draining lymph nodes containing metastatic cancer cells), stage 4 (advanced) human breast cancer was analyzed here using comparative genomic hybridization analysis (Figure 11.20). All 22 human autosomes were represented by probes, beginning from Chromosome 1 *(left)* to Chromosome 22 *(right)*, each fitted into one of the areas delineated by the vertical lines. The degree of amplification or deletion is indicated by excursions of the line from the normal copy number, which is indicated on the ordinate by "0," since the data are plotted logarithmically, with amplifications shown above the line and deletions shown below the line. For example, one segment of Chromosome 8 was found to be present in this tumor in a copy number that is more than 2 natural-logarithmic units (i.e., a factor > e^2) above the normal copy number. Such an analysis is unable to resolve between those segments whose altered copy number played a critical role in tumor development and other segments whose altered copy number reflects the widespread chromosomal instability known to operate in many advanced human tumors. This analysis indicates rampant genomic instability and resulting genetic diversification of subclones of cancer cells within this tumor. (Courtesy of J.W. Gray.)

these fluctuations change the dosage of key tumor-inducing and tumor-suppressing genes and thereby affect the proliferation of subclones of neoplastic cells within tumors. The number of changes in the copy number of chromosomal segments greatly exceeds the number of clonal successions occurring during tumor progression, pointing once again to extensive genetic diversification within single tumor masses.

The consequences of this accumulated genetic heterogeneity are likely to be manifested at two levels. *Within* a given tumor mass, different subclones will carry distinct sets of genetic alterations, as implied in Figure 11.18. This heterogeneity also affects comparisons *between* tumors of the same type that arise in different patients. The genotypes of these tumors (e.g., colon carcinomas at the same histopathological stage of tumor progression arising in 20 different patients) are likely to be markedly different from one another.

11.8 The Darwinian model of tumor development is difficult to validate experimentally

While the Darwinian model of tumor development, as depicted schematically in Figures 11.12, 11.17, and 11.18, is attractive in concept, it remains little more than a theoretical construct. The outlines of this model are undoubtedly true, but its details are very difficult to validate for a number of reasons. To begin, a convincing validation of the Darwinian model would require an identification of the key genetic and epigenetic alterations of cell genomes that are responsible for each clonal expansion and thus for each step of multi-step tumorigenesis. However, the vast number of these alterations accumulated in the genomes of tumors (Section 11.7) greatly exceeds the number of clonal successions that drive tumor progression. Consequently, sequence analyses of the genomes of cells at different states of tumor progression are unlikely to converge on the critical genetic changes that are responsible for many clonal successions.

Such sequence analyses are encumbered by another problem: as we have read, inactivation of key tumor suppressor genes is often caused by the epigenetic process of gene silencing via promoter methylation (Section 7.8). Genes functionally inactivated in this way will appear as wild-type alleles upon DNA sequencing, and their silencing can be determined only by analyses of the methylation state of their promoters or by screening for their transcripts within tumor cells. The latter screens, often termed **functional genomics**, may also not be particularly useful, since they are unlikely to distinguish between the gene silencing that is a consequence of a normal differentiation program and the gene silencing that results from the pathological process of promoter methylation.

A clear view of this Darwinian model also requires some knowledge of the *kinetics* of each step of the multi-step progression—that is, how long does each step

Sidebar 11.6 Lung cancer epidemiology provides a clue to the final step in multi-step tumor progression The kinetics with which individual steps of cancer progression occur are extremely difficult to measure. However, one peculiar aspect of lung cancer epidemiology provides us with a measure of the final step in tumor progression. After longtime smokers cease smoking, the rate at which they contract lung cancer remains relatively constant per year for each year of their remaining lifetime. This suggests that all the steps in tumor progression leading to a lung cancer except the final step have already occurred in smokers' lungs by the time they stop their habit. The final required step, which cannot be triggered directly by smoking, then strikes with a constant probability per unit time for the rest of their lives. Once this final, rate-limiting step occurs, formation of a clinically detectable tumor seems to occur rapidly. Unfortunately, the biological and genetic nature of this final step remains elusive.

take? Some of these steps, such as the point mutations that activate *ras* oncogenes, may occur with a frequency of 10^{-6} to 10^{-7} per cell generation, while other critical steps, such as loss of heterozygosity (LOH), seem to occur with frequencies 100- to 1000-fold higher. We still do not know the frequency per cell generation of other critical events, including promoter methylation, gene amplification, gene deletion, and losses of entire chromosomes. Hence, some steps of tumor progression may occur so rapidly (compared with others) that they will never be registered as "rate-limiting" steps, that is, steps that hold up the forward march of tumor progression.

Some of these processes may be influenced by the mutagen environment of the cell or, in the case of chromosomal instability, may occur only episodically during a narrow window of time over the course of tumor progression (Section 10.11). Then there is another moving target: as mentioned earlier, tumor cell genomes often become more mutable as tumor progression advances because of the breakdown of one or another component of the DNA repair apparatus.

With rare exception (see Sidebar 11.6), these facts make it almost impossible to measure the kinetics of individual steps of tumor progression; this in turn makes an enumeration of all the individual steps virtually impossible. When taken together, these various considerations indicate that we are still far from being able to understand with any precision how any single human tumor arises.

11.9 Multiple lines of evidence reveal that normal cells are resistant to transformation by a single mutated gene

The difficulties in cataloging the key steps in tumor progression, as enumerated above, indicate that we cannot rely on observations of naturally arising human tumors as the sole source of our insights into the biology of carcinogenesis. Far more definitive lessons about cancer development may well be learned by actively intervening in the process of tumorigenesis, that is, by reconstructing it in detail in the laboratory. In particular, the introduction of well-defined genetic alterations into previously normal (i.e., wild-type) cells offers the prospect of elucidating with precision how specific changes in genotype collaborate to create the cancer cell phenotype.

The roots of this experimental strategy can be traced back to Temin's experiments (Section 3.2) and, later on, to those of others in which chicken and mammalian cells propagated in culture were exposed to a variety of oncogenes by infecting tumor viruses. Subsequently, analogous experiments exploited the technique of DNA transfection to introduce oncogenes into cultured cells (Section 4.2). In all these experiments, successful transformation was gauged by the appearance of foci of morphologically transformed cells in culture dishes. Additional tests of anchorage-independent growth and the ability to form tumors in suitable animal hosts provided further validation of the transformed state of such genetically modified cells.

Some of these experiments seemed to indicate that the genetic rules governing the transformation of mammalian cells are actually extremely simple. Recall, for example, the experiment in which a mutant, activated H-*ras* oncogene from a human bladder carcinoma was introduced via transfection into previously normal NIH 3T3 mouse fibroblasts. Having acquired the mutant *ras* oncogene, these cells became fully transformed, to the point that they were capable of seeding tumors in appropriate host mice (Chapter 4).

This behavior of the *ras*-transformed NIH 3T3 cells indicated that the requirements for transforming these cells were quite minimal. A single genetic alteration of these cells—the acquisition of a *ras* oncogene—sufficed to convert them to a transformed, tumorigenic state. Moreover, the mutation that originally created the *ras* oncogene was itself a simple point mutation. This suggested that a point mutation affecting one of the NIH 3T3 cells' own native H-*ras* proto-oncogenes would yield the identical outcome—full transformation to a neoplastic state. In one stroke, a point mutation should transform a normal cell into a tumor cell.

We know from our earlier discussions of human tumor cell genetics that this conclusion is demonstrably wrong. A single point mutation—indeed, a single mutational event of any sort—cannot, on its own, generate a cancer cell from a preexisting normal cell. We can verify this conclusion with a simple calculation. Given the rate at which specific point mutations occur randomly in the human genome and the number of cells in the human body ($>3 \times 10^{13}$), some have estimated that several thousand new, point-mutated *ras* oncogenes are created every day throughout the human body and that the total body burden of cells carrying *ras* oncogenes must number in the millions. Clearly, human beings are not afflicted with a comparable number of new tumors each day.

Something has gone terribly wrong here, either in these calculations or in the transfection experiments that we relied upon to gauge the genetic complexity of the transformation process. The natural place to search for problems is in the design of the experiments used to inform our thinking, specifically in the cells that were used in the transformation assay. The NIH 3T3 cells, as it turns out, are not truly normal, since they constitute a cell line—a population of cells that has been adapted to grow in culture and can be propagated indefinitely (Chapter 10). This implies that these cells at some point underwent one or more genetic or epigenetic alterations that enabled them to grow in culture and to proliferate in an immortalized fashion.

Knowing this, investigators began in the early 1980s to examine the consequences of introducing a *ras* oncogene into truly normal cells—those from rat, mouse, or hamster embryos that had been recently explanted from living tissues and propagated *in vitro* for only a short period of time before being used in gene transfer experiments. Such cells—sometimes called **primary cells**—were unlikely to have undergone the alterations that apparently affected NIH 3T3 cells during their many-months-long adaptation to tissue culture and attendant immortalization.

The results obtained with primary rat and hamster cells were very different from those observed previously with NIH 3T3 cells. These primary cells were not susceptible to *ras*-induced transformation. Control experiments left no doubt that these cells had indeed acquired the transfected oncogene and were able to express the encoded Ras oncoprotein, but somehow they did not respond by undergoing transformation. This provided the first evidence that the act of adapting rodent cells to culture conditions and selecting for those that have undergone immortalization yields cells that have become responsive to transformation by an introduced *ras* oncogene.

The further implications of these observations are clear. Immortalized cells are not truly normal, even though they exhibit many normal traits, such as contact inhibition and anchorage dependence. Indeed, because their abnormal state renders them susceptible to *ras*-induced transformation, we might consider them to have undergone some type of pre-malignant genetic (or epigenetic) change long before they are confronted with this introduced oncogene.

It is clear that the selective pressures *in vitro* that yield immortalized cell lines are quite different from those that evolving pre-malignant cells experience within living tissues. Nonetheless, the biological traits and, quite possibly, the underlying mutant genes acquired during propagation *in vitro* may be identical to many of those arising during tumor progression *in vivo*. (In fact, our discussions in the last chapter revealed the **same** regulatory pathways—those controlled by p53 and pRb proteins—that are altered during cell immortalization are also found to be disrupted in a wide variety of cancer cell genomes, including those of human tumors.)

These experiments with primary cells have been extended by introducing activated *ras* oncogenes into the colonic epithelial cells of mice. Once activated, the resulting *ras* oncogene-expressing cells create nothing more than hyperplastic epithelia, that is, essentially normal cells that are present in excessive numbers but are, in other respects, essentially normal (Figure 11.22A).

Certain experiments of nature also support the notion that single mutations are not sufficient for the development of cancers. For example, some individuals are born carrying a germ-line mutation of the gene encoding the Kit growth factor receptor; such mutations create a constitutively active, ligand-independent Kit receptor, which functions as a potent oncoprotein. These individuals are at high risk for developing gastrointestinal stromal tumors (GISTs), but these tumors

(A)

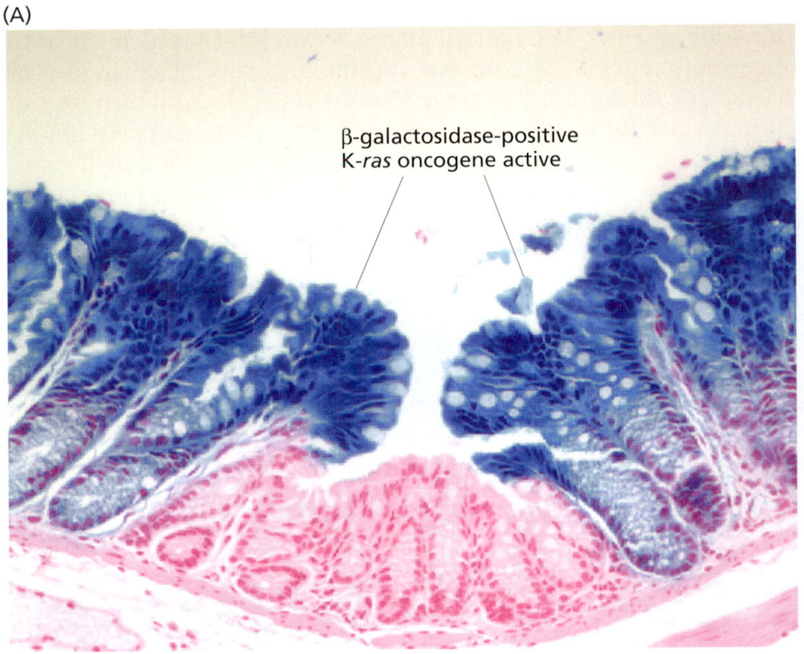

(B)

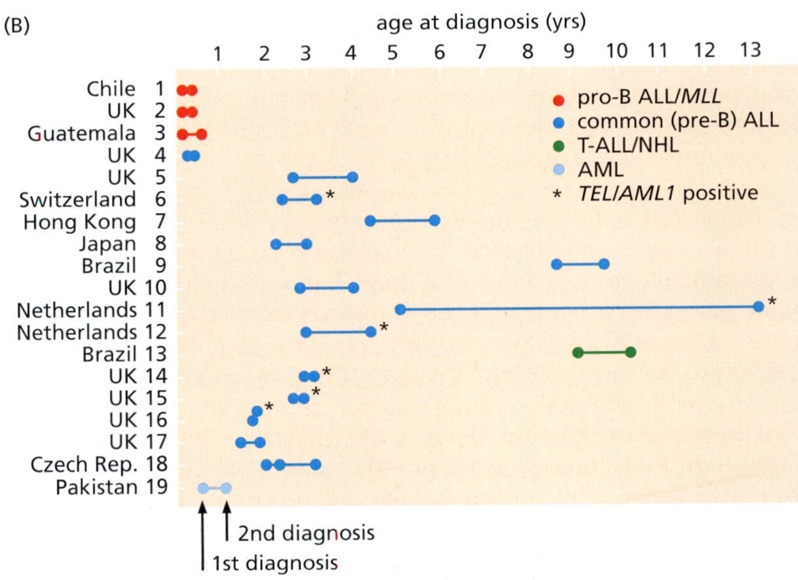

Figure 11.22 Single genetic lesions and tumor initiation: laboratory experiments and experiments of nature (A) The mouse germ line has been re-engineered to create colonic epithelial cells in which both a mutant K-*ras* oncogene and the β-galactosidase gene can be activated in scattered cells by an infecting adenovirus. As indicated here, colonic epithelial cells in which both β-galactosidase *(blue)* and the K-*ras* oncogene *(not visible)* have become expressed create localized regions of hyperplasia in which the epithelial cells are otherwise normal, indicating that the *ras* oncogene on its own does not suffice to transform these cells into tumorigenic state. (B) A number of *monozygotic* (identical) twin pairs have been documented throughout the world in which both twins develop the same type of leukemia. The leukemias invariably share a common chromosomal marker or mutation, indicating that they derive from the same clone of initiated cells. The fact that many of these leukemias are diagnosed at quite different postnatal ages *(dots)* indicates that these initiating somatic mutations, which occurred *in utero,* were not, on their own, sufficient to trigger the formation of a clinically apparent leukemia. The labels *(top right)* and the associated colors denote different subtypes of leukemia identified by distinctive gene markings. (A, courtesy of K.M. Haigis and T. Jacks; B, from M.G. Greaves, A.T. Maia, J.L. Wiemels and A.M. Ford , *Blood* 102:2321–2333, 2003.)

Sidebar 11.7 Identical twins provide striking evidence that single mutations are necessary but not sufficient for transformation Beginning in 1882, a series of medical reports has described more than 70 sets of **monozygotic** (i.e., identical) twins both of whom developed the same type of leukemia, often in infancy. Over the past two decades, a group of 19 such twin sets has been genetically characterized to validate (1) their common origin, in each case, from the same fertilized egg and (2), in the great majority of cases, the presence of identical chromosomal translocations in the leukemic cells of both children within each twin set. Some of these translocations have been subjected to gene cloning and detailed sequence analysis to verify their identity, removing all doubt about the common, **prenatal** origin of these mutations. In each case, the initiating chromosomal translocation must have arisen in one of the twins *in utero*. Thereafter, cells from the resulting mutant cell clone are passed during gestation to the other twin through a shared placental circulation. (The non-leukemic cells in both twins of a set are always genetically normal, indicating that the initiating mutation is of somatic rather than germ-line origin.)

In addition to the mutation that is shared in common between the two leukemic cell populations, each leukemia often shows its own distinctive DNA and/or chromosomal abnormalities, indicating that further mutations occurred during postnatal tumor progression. As seen in Figure 11.22B, the twins sometimes develop leukemia asynchronously, in that months to years separate the time of diagnosis of leukemia in the two twins within a set. This experiment of nature indicates that a single genetic event, while necessary for the triggering of leukemia, is not sufficient, and that other rate-limiting events must intervene before these tumors become clinically apparent.

only become apparent several decades after birth, even though a constitutively active Kit oncoprotein has been active in many of their cells since birth (Sidebar 5.8). Similarly, some individuals have been documented that carry mutant H-*ras* alleles in their germ lines yet usually develop tumors only after several decades' time. Early childhood leukemias in twins provide equally dramatic examples of the inability of single mutations, acting on their own, to create clinically apparent tumors (Sidebar 11.7). Taken together, these diverse observations teach us that multiple changes seem to be required in order for a cell to reach a tumorigenic state.

11.10 Transformation usually requires collaboration between two or more mutant genes

The resistance of fully normal rodent cells to *ras*-induced transformation led to an interesting question: Were there yet other oncogenes that could immortalize embryo cells and, at the same time, render these cells susceptible to transformation by *ras*? In the early 1980s, research on DNA tumor viruses indicated that some of these agents carried multiple oncogenes in their genomes (Sidebar 17). Polyomavirus, for example, bears two oncogenes, termed *middle T* and *large T*; in 1982, these two oncogenes were found to collaborate with one another to transform rodent cells. The large T oncoprotein seemed to aid in the adaptation of cells to tissue culture conditions and to facilitate their immortalization, while the middle T protein elicited many of the phenotypes associated with the *ras* oncogene—rounding up of cells, loss of contact inhibition, and acquisition of anchorage-independent growth. Soon, a number of other DNA tumor viruses were found to employ similar genetic strategies for cell transformation.

The genetics of transformation by DNA tumor viruses suggested that mutant cellular genes might also collaborate in cell transformation. In fact, a line of human promyelocytic leukemia cells was discovered to carry both an activated N-*ras* and an activated *myc* oncogene. This suggested the possibility that these two cellular oncogenes were cooperating to create the malignant phenotype of the leukemia cells. This notion was soon borne out by a simple experiment: when a *myc* oncogene was introduced together with an H-*ras* oncogene into rat embryo fibroblasts (REFs), the cells responded by becoming morphologically transformed (Figure 11.23) and, more important, tumorigenic; neither of these oncogenes, on its own, could create such transformed cells.

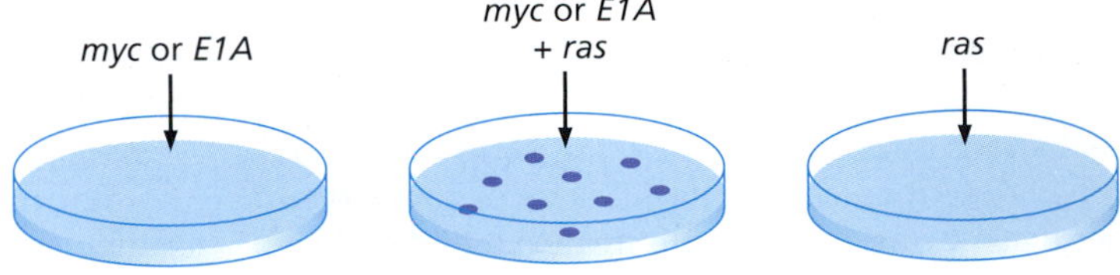

Figure 11.23 Oncogene collaboration in rodent cells *in vitro* Cultures of early-passage rat embryo fibroblasts (REFs) or baby hamster kidney cells (BHKs) were prepared and exposed to cloned DNAs via the calcium phosphate gene transfection procedure. Introduction of a *myc* or adenovirus *E1A* oncogene into these cells *(left),* on its own, did not yield foci of transformants, although these introduced oncogenes did facilitate the establishment of these early-passage cells in long-term culture. Introduction of an H-*ras* oncogene via transfection *(right)* also did not yield foci of transformed cells, although it did allow significant numbers of these cells to form anchorage-independent colonies when they were introduced into a semi-solid medium such as dilute agar. However, the simultaneous introduction of *ras* + *myc* or, alternatively, *ras* + *E1A* did generate foci of transformed cells that were found to be able to form tumors when these cells were injected into syngeneic or immunocompromised hosts.

This result yielded several interesting conclusions. These two cellular oncogenes clearly affected cell phenotype in quite different ways, since they were able to complement one another in eliciting cell transformation. Each seemed to be specialized to evoke a subset of the cellular phenotypes associated with the transformed state. For example, *ras* was able to elicit anchorage independence, a rounded, refractile appearance in the phase microscope, and loss of contact inhibition; *myc* helped the cells to become immortalized and reduced somewhat their dependence on growth factors. Similar results were found in experiments in which the *E1A* oncogene of human adenovirus 5 was used as the collaborating partner of a *ras* oncogene. Once again, the two collaborating oncogenes were found to have complementary effects on cell phenotype.

Soon, a number of other pairs of oncogenes were discovered to be capable of collaborating with one another to induce transformation of cells *in vitro* and tumorigenesis *in vivo* (Table 11.1). The *ras* oncogene, as an example, could also collaborate with the SV40 *large T* oncogene, with the polyoma *large T* oncogene, or with a mutant *p53* gene in cell transformation. Conversely, *myc* could collaborate also with the polyoma *middle T* oncogene, with *src,* or with the *raf* oncogene to transform cells.

In most cases, the genes within a collaborating pair could be placed in two functional groups—those with *ras*-like and those with *myc*-like properties. In fact, not all *ras*-like oncogenes elicit identical effects in cells; the same could be said of the members of the *myc*-class. Provocatively, the *ras*-like oncogenes encode largely cytoplasmic oncoproteins, while the *myc*-like oncogenes encode products that tend to be nuclear (Table 11.2). We now know that the Ras-like oncoproteins are components of the cytoplasmic mitogenic signaling cascade (Chapter 6), while the Myc-like oncoproteins perturb in various ways the cell cycle control machinery, which operates in the nucleus (Chapter 8).

Table 11.1 Examples of collaborating oncogenes *in vitro* and *in vivo*

"*ras*-like" oncogene[a]	"*myc*-like" oncogene[a]	Target cell or organ
***In vitro* transformation**		
ras	*myc*	transfected rat embryo fibroblasts (REFs)
ras	*E1A*	transfected rat kidney cells
ras	SV40 *large T*	transfected REFs
Notch -1	*E1A*	transfected rat kidney cells
***In vivo* tumorigenesis**		
middle T	*large T*	polyomavirus-induced murine tumors
mil (= raf)	*myc*	MH2 avian leukemia virus chicken tumors
erbB	*erbA*	avian erythroblastosis virus chicken tumors
pim1	*myc*	mouse leukemia virus tumors
abl	*myc*	mouse leukemia virus tumors
Notch-1/2	*myc*	thymomas in transgenic mice
bcl-2	*myc*	follicular lymphomas in transgenic mice

[a]The terms "*ras*-like" and "*myc*-like" refer to functional classes rather than genes encoding components of a common signaling pathway. "*ras*-like" oncogenes tend to encode components of cytoplasmic signaling cascades, while "*myc*-like" oncogenes tend to encode nuclear proteins.

Table 11.2 Physiologic mechanisms of oncogene collaboration[a]

Oncogene pair	Cell type	Mechanisms of action
ras + SV40 *large T*	rat Schwann cells	*ras*: proliferation + proliferation arrest *large T*: prevents proliferation arrest and reduces mitogen requirement
ras + *E1A*	mouse embryo fibroblasts	*ras*: proliferation and senescence *E1A*: prevents senescence
erbB + *erbA*	chicken erythroblasts	*erbB*: induces GF-independent proliferation *erbA*: blocks differentiation
TGF-α + *myc*	mouse mammary epithelial cells	*TGF-α*: induces proliferation and blocks apoptosis *myc*: induces proliferation and apoptosis
v-*sea* + v-*ski*	avian erythroblasts	v-*sea*: induces proliferation v-*ski*: blocks differentiation
bcl-2 + *myc*	rat fibroblasts	*bcl-2*: blocks apoptosis *myc*: induces proliferation and apoptosis
ras + *myc*	rat fibroblasts	*ras*: induces anchorage independence *myc*: induces immortalization
raf + *myc*	chicken macrophages	*raf*: induces growth factor secretion *myc*: stimulates proliferation
src + *myc*	rat adrenocortical cells	*src*: induces anchorage and serum independence *myc*: prolongs proliferation

[a]In each pair, the first oncogene encodes a cytoplasmic oncoprotein while the second oncogene encodes a nuclear oncoprotein.

These oncogene collaboration experiments provided a crude *in vitro* model of multi-step transformation *in vivo* and suggested a rationale for the complex genetic steps that accompany and cause tumor formation in human beings: each of the genetic changes provides the nascent tumor cell with one or more of the phenotypes that it needs in order to become tumorigenic (see Table 11.2). These unique contributions seem to derive from the ability of each of these oncogenes to perturb a specific subset of regulatory circuits within a cell. Moreover, these experiments suggested that cell proliferation and cell survival are governed by a number (two or more) of distinct regulatory circuits, all of which must be perturbed before the cell will become tumorigenic.

Previously, we noted that oncogenes act in a pleiotropic fashion on cell phenotype, in that each of these genes is able to concomitantly induce a number of distinct changes in cell phenotype. Accepting this, we must also recognize that, as multi-talented as oncogenes are, none of them seems able, on its own, to evoke all of the changes that are required for a normal cell to become transformed into a tumorigenic state. In the case of cellular oncogenes, there would seem to be an obvious evolutionary rationale for this, which we touched on earlier in this chapter: a mammalian cell cannot tolerate the presence of a proto-oncogene in its genome that could, through a single mutational event, yield an oncogene capable of transforming this cell into a full-fledged tumor cell. Such a proto-oncogene would place each cell in the body only a single, small step away from malignancy and create too much of a liability for the organism as a whole. This represents another version of the argument that cells and tissues must place multiple obstacles in the path of normal cells in order to prevent them from becoming tumorigenic. Interestingly, under certain experimental conditions, researchers can thwart these defense mechanisms and succeed in transforming cells with a single genetic element (Sidebar 11.8).

11.11 Transgenic mice provide models of oncogene collaboration and multi-step cell transformation

In many rodent models of cancer pathogenesis, tumors can be triggered by exposure of an animal to a mutagenic carcinogen, which acts in a random (sometimes termed **stochastic**) fashion to generate the mutant cellular alleles

Sidebar 11.8 The rules of multi-step transformation can be circumvented There are a number of experimental situations in which a single oncogene or genetic agent can, on its own, create a transformed cell. For example, a chicken embryo fibroblast (CEF) infected *in vitro* with a Rous sarcoma virus (RSV) particle appears to be transformed to tumorigenicity in this single step. However, *in vivo*, it seems that RSV-infected cells form tumors only at sites of wounding, including those sites along the needle track formed when RSV is injected into muscle; hence, the changes occurring in fibroblasts during wound healing seem to be required to help the RSV *src* gene transform these cells into tumor cells.

Experiments with cultured rat embryo fibroblasts (REFs) indicate that when a *ras*-transformed cell is isolated from other cells in the Petri dish (and therefore is not surrounded by normal, nontransformed cells), it can proliferate to form a colony of tumorigenic cells, even without the aid of a second collaborating oncogene like *myc*. However, when such a cell is surrounded by normal neighbors (as might occur *in vivo* during the initial step of tumor progression), it is unable to proliferate to generate a focus. Accordingly, single-step transformation experiments can sometimes succeed because they fail to recapitulate certain anti-cancer mechanisms operating in living tissues, which normally require additional alterations within a cell or tissue before tumor progression can proceed.

leading to cancer. An alternative to such experimental protocols can be achieved through the insertion of an already mutant, activated oncogene into the germ line of a laboratory mouse, thereby guaranteeing expression of this gene in some of its tissues. In practice, the expression of this oncogenic allele must be confined to a small subset of tissues in the mouse. (If its expression were allowed in all tissues, including those of the developing embryo, chances are that embryogenesis would be so profoundly disrupted that the developing fetus would die long before the end of gestation.)

An early version of this strategy for creating cancer-prone, **transgenic** mice involved the insertion of oncogenic alleles of the *ras* or *myc* genes into the mouse germ line (Figure 11.24). In one influential set of experiments, expression of the *ras* oncogene and the *myc* oncogene was placed under the control of the transcriptional promoter of mouse mammary tumor virus (MMTV), a retrovirus that specifically targets mammary tissues. The viral promoter is expressed at significant levels only in mammary glands and, to a lesser extent, in salivary glands.

As anticipated, the presence of either one of these oncogenic **transgenes** in the mouse germ line predisposed mice to breast cancer and, to a much lesser extent, to salivary gland tumors. In spite of the expression of either a *ras* or a *myc* oncogene in most if not all of the mammary epithelial cells of these mice, their mammary glands showed either minimal morphologic changes (in the case of the *myc* transgene) or hyperplasia (in the case of the *ras* transgene). Moreover, breast tumors were observed only beginning at four weeks of age—a significantly long latency period (see Figure 11.24). This proved conclusively that the presence of a single oncogene within a normal cell in living tissue is not, on its own, sufficient to transform this cell into a tumor cell. Instead, the kinetics of breast cancer formation in these mice pointed to the necessary involvement of one or more additional stochastic events that needed to happen before these *ras* or *myc* oncogene–bearing mammary cells progressed to a tumorigenic state (see also Figure 11.22A).

Double-transgenic mice that carried both MMTV-*ras* and MMTV-*myc* transgenes were created through mating between the two transgenic strains described above. These double-transgenic mice contracted tumors at a greatly accelerated rate and at high frequency compared with mice inheriting only one of these transgenes (see Figure 11.24). Therefore, the two oncogenic transgenes could collaborate *in vivo* to generate tumors, corroborating the conclusions of the *in vitro* experiments described earlier (Section 11.10).

Interestingly, even with two mutant oncogenes expressed in the great majority of mammary cells from early in development, tumors did not appear in these mice soon after birth, but instead were seen with great delay. Hence, the concomitant expression of two powerful oncogenes was still not sufficient to fully transform mouse mammary epithelial cells (MECs); instead, these cells clearly required at least one additional stochastic event, ostensibly a somatic mutation, before they would proliferate like full-blown cancer cells. (A hint about the identity of this third, stochastic event has come from careful analysis of rat cells that have been transformed *in vitro* by the *ras* + *myc* protocol; sooner or later, such cells usually acquire a mutation or methylation event that leads to inactivation of the p53 tumor suppressor pathway; see Chapter 9.)

Note that the collaborative actions of transgenic oncogenes were already mentioned earlier, when we read of the synergistic actions of *myc* and *bcl-2* transgenic oncogenes in promoting lymphomagenesis (see Figure 9.22). In this case, the benefit of *bcl-2* (and its *bcl-*x_L cousin) derives largely from its anti-apoptotic effects. This illustrates the fact that oncogenes can collaborate through a variety of cell-physiologic mechanisms to promote tumor formation, a point made in Table 11.2.

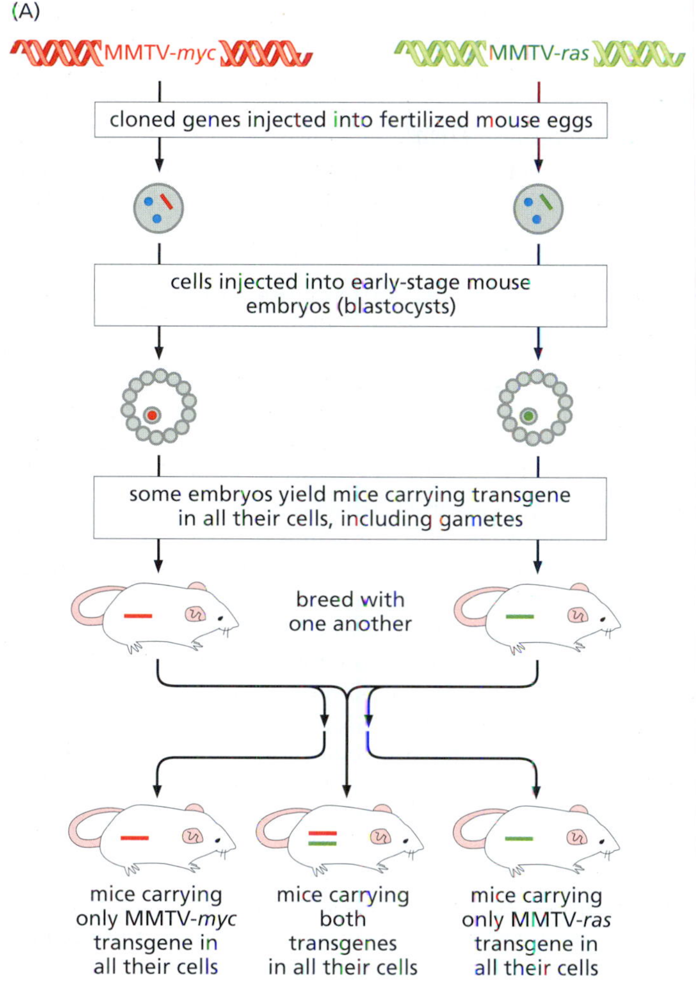

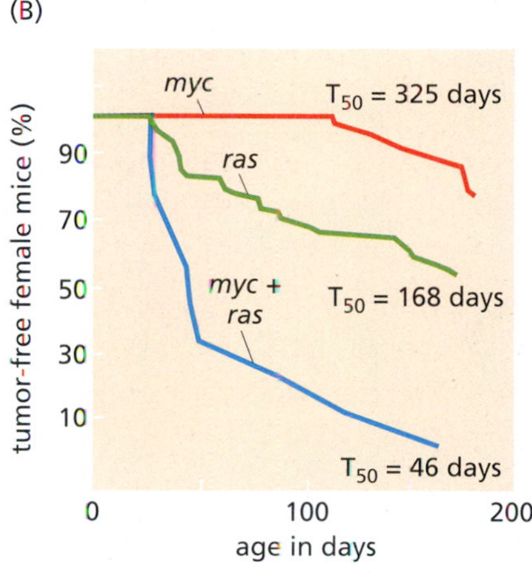

Figure 11.24 Oncogene collaboration in transgenic, cancer-prone mice The ability to create transgenic mice (see Figure 9.22A) made it possible to determine whether oncogenes are able to collaborate *in vivo* as well as *in vitro*. (A) Mice were created that bore either the MMTV-*ras* or the MMTV-*myc* transgene in their germ line. The MMTV (mouse mammary tumor virus) transcriptional promoter ensured expression of the transgene largely in the mammary glands. Mice of these two transgenic strains were then bred to create double-transgenic mice carrying both transgenes *(bottom)*. (B) The incidence of mammary carcinomas in mice carrying either the MMTV-*myc* (*red curve*) or MMTV-*ras* (*green curve*) transgenes in their germ lines was followed over many months. In addition, the two transgenic mouse strains were bred with one another to create double-transgenic mice, and the effects of both transgenes on tumor incidence *(blue curve)* were also tracked. The percentage of tumor-free mice *(ordinate)* is plotted versus the age in days of the various strains of mice. "T_{50}" indicates the number of days required for one-half of the mice of a particular genotype to develop detectable mammary carcinomas. (From E. Sinn, W. Muller, P. Pattengale et al., *Cell* 49:465–475, 1987.)

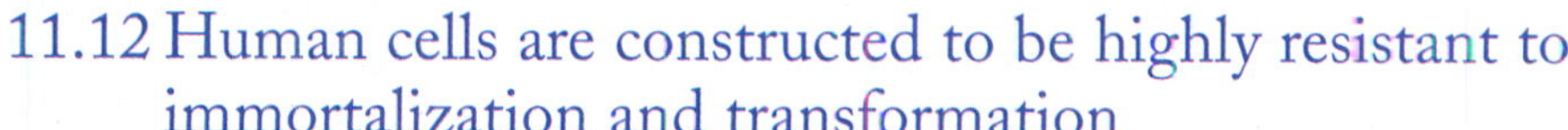

11.12 Human cells are constructed to be highly resistant to immortalization and transformation

The biological lessons derived from studying mice and rats are usually directly transferable to understanding various aspects of human biology. Even though 80 million years may separate us from the most recent common ancestor we share with rodents, the great majority of biological and biochemical attributes of these distant mammalian cousins are present in very similar, if not identical, form in humans. The genomes of humans and rodents also seem very similar: essentially all of the roughly 20,000 genes discovered in the human genome have been found to have mouse orthologs. It stands to reason that the biological processes of immortalization and neoplastic transformation should also be essentially identical in rodent and human cells.

The biological reality is, however, quite different. It is easy to immortalize rodent cells simply by propagating them through a relatively small number of passages *in vitro*. Spontaneously immortalized cells arise frequently and become the progenitors of cell lines, such as the NIH 3T3 cells discussed earlier. In contrast, human cells rarely, if ever, become immortalized following extended serial passaging in culture (see Chapter 10). Eventually, cultured human cells stop growing and become senescent, and spontaneously immortalized cell clones do not emerge.

Attempts to experimentally transform cells have shown comparable inter-species differences in cell behavior. Primary rodent cells become transformed *in vitro* following the introduction of pairs of oncogenes (such as *ras* and *myc*;

Section 11.10), while such pairs of introduced oncogenes consistently fail to yield tumorigenic human cells. In fact, the human cells emerging from such co-transfections are not even immortalized and therefore senesce sooner or later.

These repeated failures at cell transformation prevented researchers from addressing a simple yet fundamental problem in human cancer biology: How many intracellular regulatory circuits need to be perturbed in order to transform a normal human cell into a cancer cell? Sequence analyses of human cancer cell genomes were of little help here. As argued earlier (Section 11.7), cancer cells derived from human tumors possess a plethora of genetic alterations, far more than the relatively small number that play causal roles in tumorigenesis. This forced researchers to consider experimental cell transformation as an alternative way of addressing this problem. Thus, they asked precisely how many genetic changes must be introduced experimentally into human cells in order to transform them.

The general strategy was inspired by the experience with cultured rodent cells, which indicated that once cells were immortalized in culture, they became responsive to transformation by a *ras* oncogene. The fact that the telomere biology of rodent and human cells differs so starkly (Section 10.9) seemed to explain at least part of the difficulty of immortalizing human cells, and thus their very different responses to introduced oncogenes. Recall that the cells of laboratory mice usually carry extremely long telomeric DNA (as long as 40 kilobases) and express readily detectable levels of telomerase enzyme activity. In marked contrast, normal human cells have far shorter telomeres, and most human cell types lack significant telomerase activity. Accordingly, immortalization of human cells might be facilitated by adding the *hTERT* gene to other immortalizing oncogenes introduced into these cells.

In fact, introduction of an *hTERT* gene in addition to the SV40 *large T* oncogene (whose product inactivates both pRb and p53 tumor suppressor proteins) did indeed yield immortalized human cells. (Alternative means of inactivating pRb and p53, such as introduction of human papillomavirus *E6* and *E7* oncogenes, succeeded as well.) And once immortalization was achieved through these changes, the resulting human cells could then be transformed morphologically in the culture dish by introduction of an activated *ras* oncogene.

These morphologically transformed human cells were still not fully transformed, however, as indicated by their inability to form tumors when implanted in immunocompromised host mice. (The faulty immune systems of such mice ensure that tissues of foreign origin are not eliminated by immunological attack.) These cells still required one more alteration, this one achieved by introduction of the gene encoding the SV40 small T oncoprotein. Small T perturbs a subset of the functions of the abundant cellular enzyme termed protein phosphatase 2A (PP2A; Sidebar 18).

Taken together, these experiments demonstrated that five distinct cellular regulatory circuits need to be altered experimentally before human cells can grow as tumor cells in immunocompromised mice (Figure 11.25). These changes involve (1) The mitogenic signaling pathway controlled by Ras (Chapter 6). (2) The cell

pathway	Ras	pRb	p53	telomeres	PP2A
genes/agents used to deregulate pathway	*ras*	*CDK4 + D1* SV40 *LT* HPV *E7*	DN *p53* SV40 *LT* HPV *E6*	*hTERT* *myc + SV40 LT*	SV40 *sT* sometimes: *myc* *Akt/PKB+Rac1* *PI3K* *B56* shRNA

Figure 11.25 Intracellular pathways involved in human cell transformation The experimental transformation of human cells has been achieved through the insertion of various combinations of cloned genes into cells. Initially, a combination of three genes encoding the SV40 large T oncoprotein (LT), the hTERT telomerase, and the SV40 small T oncoprotein (sT) was found to suffice for the transformation of a variety of normal human cell types to a tumorigenic state. These genes were found to deregulate five distinct regulatory pathways involving (1) Ras mitogenic signaling, (2) pRb-mediated G_1 cell cycle control, (3) p53, (4) telomere maintenance, and (5) protein phosphatase 2A (PP2A). Subsequent work has found that other combinations of cloned genes suffice as well. For example, the disruption of pRb function can be achieved by a combination of ectopically expressed, CDK inhibitor–resistant CDK4 + cyclin D1; p53 can be disrupted by an introduced dominant-negative *p53* allele; telomeres can be maintained through a combination of SV40 LT + *myc;* and PP2A function can be disrupted by an shRNA construct that inhibits the synthesis of the B56 subunit of PP2A. It is unknown whether these five pathways are required for the experimental transformation of all human cell types, and whether deregulation of all of these five pathways occurs in spontaneously arising human tumors.

Sidebar 11.9 Must the same set of regulatory pathways be disrupted to create all types of human tumors? A wide variety of normal adult human cell types can be transformed experimentally by perturbing the five pathways listed in Figure 11.25. Included among these are fibroblasts; kidney cells; mammary, prostate, ovarian, and small-airway (lung) epithelial cells; and astrocytes. These identical requirements for transformation suggest that a common set of biochemical pathways must be deregulated in a wide variety of adult human cell types in order for transformation to succeed. (Each of these pathways can presumably be perturbed through several alternative genetic and epigenetic mechanisms in various types of cancers.)

Nonetheless, it is plausible that certain normal human cells require a greater or lesser number of changes before they will become transformed. For example, a number of pediatric cancers occur so early in life that it is difficult to imagine how the cells in these tumors could have had sufficient time to accumulate the cohort of mutations (and epigenetic changes) that seem to be required for the formation of many adult malignancies.

This suggests the possibility that some pediatric cancers arise directly from certain embryonic cell types (e.g., stem cells), and that such embryonic cells may be more readily transformed (through a smaller number of alterations) than the cells that serve as the precursors of adult tumors. The extreme case of an embryonic cell type is provided by embryonic stem (ES) cells that can be extracted from very early embryos. ES cells are, by all measurements, genetically wild type, yet are tumorigenic (yielding teratomas) when implanted in syngeneic hosts. Indeed, they seem to be the only example of wild-type cells that are tumorigenic. Perhaps certain cells within later embryos require, for their transformation, a number of genetic changes that is intermediate between the number needed for the transformation of ES cells (zero) and the number required for the experimental transformation of adult human cells (five).

cycle checkpoint controlled by pRb (Chapter 8). (3) The alarm pathway controlled by p53 (Chapter 9). (4) The telomere maintenance pathway controlled by hTERT (Chapter 10). (5) The signaling pathway controlled by protein phosphatase 2A.

Experiments like these provide clear indications of why human cells are highly resistant to transformation. At the same time, it remains unclear whether the steps required to experimentally transform human cells *in vitro* accurately reflect the changes that normal cells must undergo within human tissues before they succeed in proliferating like cancer cells. Clearly, four of these changes (involving Ras and hTERT activation, pRb and p53 inactivation) are commonly seen in the cells of human cancers. However, it remains unclear whether the fifth alteration—deregulation of a subset of the actions of PP2A—occurs during the formation of spontaneously arising human cancers. In addition, yet other genetic changes, not revealed by these experiments, may be required before the cells within certain human tissues are able to generate clinically detectable tumors. These experiments also leave another issue unsettled: Are the genetic and biochemical rules governing human transformation identical in all human cancers (Sidebar 11.9)?

The stark differences in the behavior of human versus mouse cells require some type of biological rationale. Indeed, we have already discussed one of them (Section 10.9): the cells in a mouse pass through about 10^{11} mitoses in a mouse lifetime, while those in a human body pass through about 10^{16} cell cycles in a human lifetime. In response to the ever-present dangers created by passage through each cell cycle, our cells and tissues have been hard-wired by evolution to be far more resistant to cell transformation. This notion—really a speculation—still requires some validation and generalization. For example, do the cells of a bumblebee bat or Etruscan shrew (both of approximately 2 g body weight) and those of the blue whale (of approximately 1.3×10^8 g body weight; Figure 11.26) require proportionally fewer and more hits, respectively, before they become cancerous? (The difference in cumulative mitoses in a lifetime is likely to be even greater, since large mammals often live almost 100 times longer than tiny ones!)

The various experimental demonstrations of oncogene collaboration in mouse and human cells may well serve as good models of how multi-step tumorigenesis actually occurs within the human body. Thus, each mutation (or gene

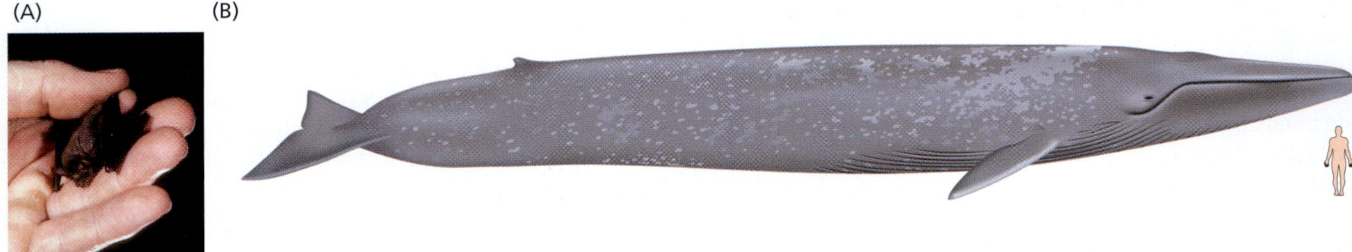

Figure 11.26 Mammalian body size and relative risk of cell transformation While the sizes of individual cells are quite comparable in various mammalian species, the overall body mass and thus cell number varies enormously. Since passage through each cell cycle creates the danger of genome alterations, this suggests that the risks of cancer can vary enormously from one species to the next. (A) The bumblebee bat of Thailand, said to be the smallest mammal, was discovered in 1973; it weighs 1.5 g and has a wingspan of 15 cm. (B) The blue whale has a body weight of about 1.3×10^8 g and a life span of about 80 years. The eight-orders-of-magnitude difference in mass (and therefore cell number) between the bat and the whale together with the approximately 1.5 orders of magnitude difference of life span indicates a difference of a factor of some 10^9 in the number of cell divisions that the two organisms experience in a lifetime. (Since the metabolic rate of this shrew may be as much as 10^3 times higher than that of the whale, and since much of the mutational burden of the cell genome derives from byproducts of oxidative metabolism, the blue whale may experience only a 10^6-fold higher risk of cancer than the bat.) (A, courtesy of Merlin D. Tuttle, Bat Conservation International; B, courtesy of Uko Gorter.)

methylation) sustained by a population of cells perturbs or deregulates yet another intracellular signaling pathway, until all the key control circuits have been disrupted. Once this is accomplished, the cells in this population may be fully transformed and therefore capable of generating a vigorously growing tumor. In fact, analyses of human tumor cell genomes reveal far more confounding results that are difficult to reconcile with such a simple and satisfying conceptual scheme (Sidebar 11.10).

Significantly, the experimental manipulations used to transform human cells to a tumorigenic state usually yield cells that form localized primary tumor masses having little if any tendency (1) to extend beyond their boundaries and invade nearby tissues and (2) to seed distant metastases. Therefore, analyses of the genetic changes within cells that are needed to make them tumorigenic do not address the identities of the genes and proteins that program the phenotypes of advanced, highly aggressive malignancies, an issue that we confront only later, in Chapter 14. Moreover, the "5-hit" scenario suggested by the experimental transformation of human cells sidesteps a critical issue that we will discuss in the next chapter: the mutability of human cell genomes is normally very low, making it highly unlikely that cell populations within our tissues can acquire all of the genetic changes needed to complete tumor progression within a human life span.

Sidebar 11.10 The genetics of actual human tumors confounds our understanding of how cancer progression occurs The simplest scheme of multi-step tumor progression states that each successive step involves the disruption or deregulation of yet another key cellular signaling pathway. Hence, each of the mutant (or methylated) genes found in the genome of a human cancer cell should affect a distinct regulatory pathway, and the mutations accumulated by the end of tumor progression should collaborate with one another to program neoplastic growth. In fact, the actual genetic evidence often conflicts with such thinking.

Many human colorectal carcinoma genomes carry mutations that lead to the activation of both PI3 kinase and B-Raf, which makes sense, since these two mutations affect distinct, complementary pathways that lie downstream of Ras (Chapter 6). However, many other tumors of this type have mutations that activate both Ras and PI3 kinase, which makes no sense, since a Ras oncoprotein is thought to be capable of directly activating PI3 kinase; these two mutations therefore seem to be functionally redundant rather than complementary.

Conversely, mutations that are expected to collaborate with one another, such as those affecting the *ras* and *p53* genes, are often mutually exclusive. Thus, among human colon carcinomas, some bear *ras* mutations, while others carry *p53* mutations, and tumors carrying both are quite uncommon (contrary to the initial depiction of the genetic pathway leading to these cancers; see Section 11.3). Similarly, among human bladder carcinomas, those tumors bearing mutations that activate the fibroblast growth factor receptor-3 (yielding an effect similar to *ras* oncogene activation) rarely carry *p53* mutations and vice versa. Observations like these are difficult to reconcile with our current understanding of how these genes and encoded proteins operate—which only says that our perceptions about these issues will, sooner or later, require substantial revision.

11.13 Nonmutagenic agents, including those favoring cell proliferation, make important contributions to tumorigenesis

The clinical observations and experimental results that we have read about in this chapter provide us with a crude picture of the genetic and epigenetic changes required to generate a cancer cell. They fail, however, to reveal *how* these changes are actually acquired during tumor progression. So now, we turn to these issues—the processes occurring *in vivo* that enable cells to accumulate the large number of alterations needed for tumor formation.

The schemes described here dictate that a succession of genetic changes provide the major impetus for tumor progression. Since many of these changes are caused by the actions of mutagens, this implies that cancer progression is fueled largely by the genetic hits inflicted by mutagenic carcinogens. Of course, we need to revise this scenario to accommodate the clearly important role of epigenetic changes, specifically those caused by methylation of gene promoters (see Section 7.8). (At present, it is unclear whether these methylation events are actively provoked by external agents or occur spontaneously because of occasional random mistakes made by the cellular proteins responsible for regulating methylation.)

In addition to the clearly documented contributions of mutagenic carcinogens to cancer induction (Section 2.9), extensive evidence points to a wide variety of *nonmutagenic* agents that participate in tumor formation. Indications of the importance of nonmutagenic (sometimes termed **nongenotoxic**) carcinogens first came from attempts in the early 1940s to develop effective methods for inducing skin cancers in mice. The experimental model used in such research depended on exposing mouse skin to highly carcinogenic tar constituents, such as benzo[*a*]pyrene (BP), 7,12-dimethylbenz[*a*]anthracene (DMBA), or 3-methylcholanthrene (3-MC; Figure 2.22). For example, mice subjected to daily painting of DMBA on a patch of skin would develop skin carcinomas after several months of this treatment.

But another experimental protocol proved to be even more revealing about the mechanisms of skin cancer induction. Following a single painting with an agent such as DMBA, the same area of skin could be treated on a weekly basis with a second agent, termed TPA (12-*O*-tetradecanoylphorbol-13-acetate; Figure 11.27), a skin irritant prepared from the seeds of the croton plant. (Another often-used term for TPA is PMA, for phorbol-12-myristate-13-acetate.) Repeated painting of the DMBA-exposed area with TPA resulted in the appearance of papillomas after 4 to 8 weeks, depending on the strain of mice being used (Figure 11.28A–C). (These papillomas are in many ways analogous to the adenomas observed in early-stage colon cancer progression.)

At first, the survival and growth of these skin papillomas depended upon continued TPA paintings, since cessation of TPA treatments caused the papillomas to regress (Figure 11.28E). However, if TPA painting was continued for many weeks, TPA-independent papillomas eventually emerged, which would not regress after cessation of TPA painting and instead persisted for extended periods of time (Figure 11.28F). Some of these TPA-independent papillomas might, with low probability, evolve further into malignant squamous cell carcinomas of the skin after about six months.

In the absence of initial DMBA treatment, however, repeated painting with TPA failed to provoke either papillomas or carcinomas (Figure 11.28B). Even more interesting, an area of skin could be treated once with DMBA and then left to rest for a year. If this patch of skin was then treated with a series of TPA paintings (as in Figure 11.28C), it would "remember" that it had been exposed previously to DMBA and respond by forming a papilloma.

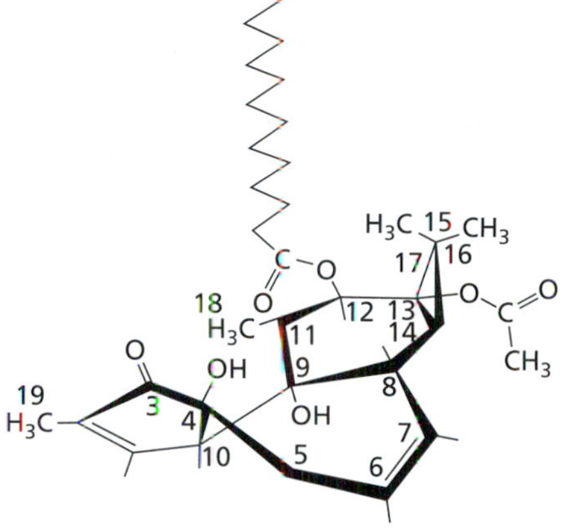

Figure 11.27 TPA, an important promoter of skin tumorigenesis The stereochemical structure of 12-*O*-tetradecanoylphorbol-13-acetate (TPA), also known as phorbol-12-myristate-13-acetate (PMA), is shown here. TPA is extracted from the croton plant (*Croton tiglium*). Its target in cells is protein kinase Cα (PKCα).

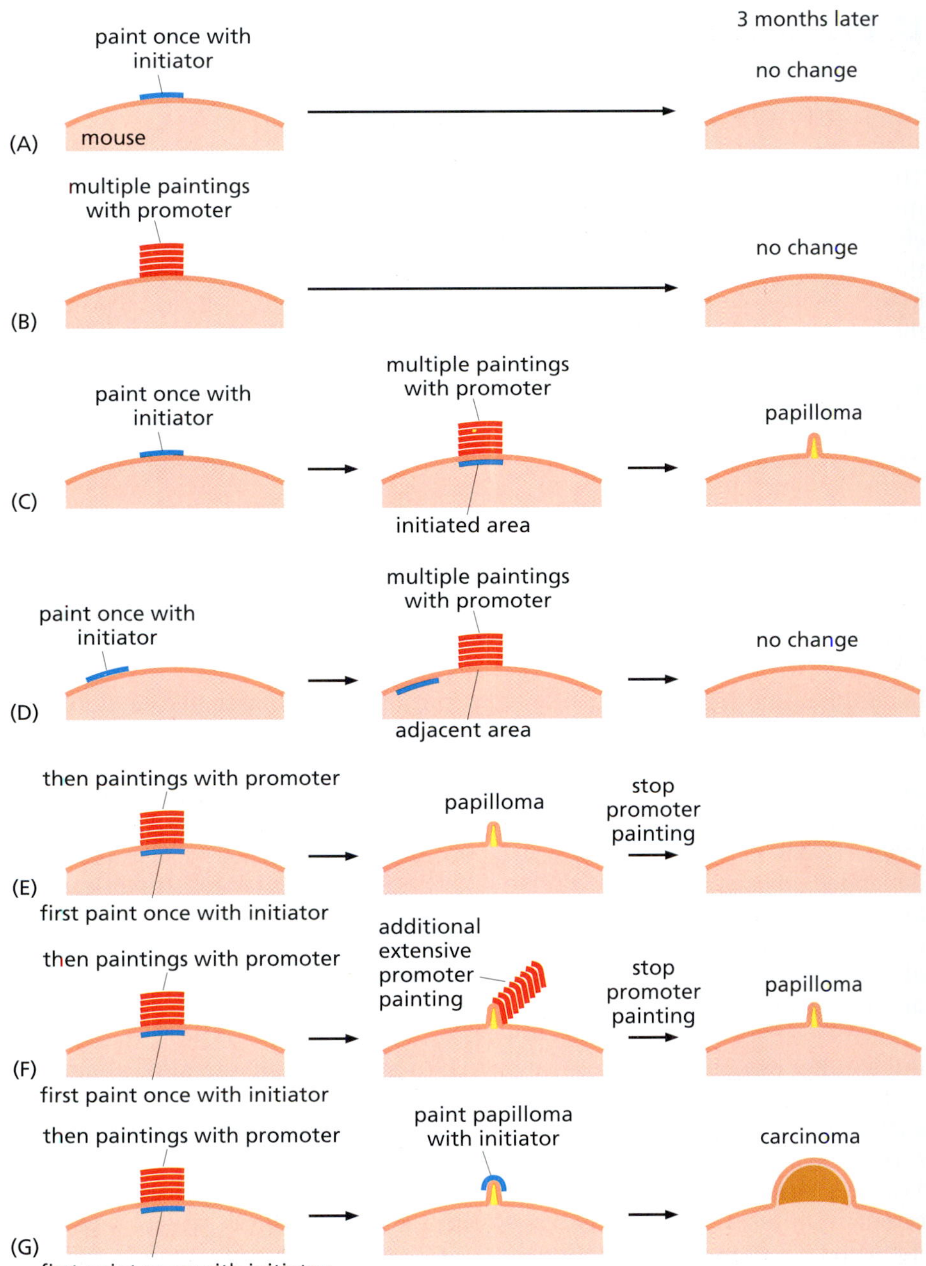

Figure 11.28 Protocols for inducing skin carcinomas in mice The induction of skin carcinomas by painting carcinogens on the backs of mice requires certain combinations of treatments with initiators and promoters. (A) A single treatment with an initiating carcinogen, such as DMBA (dimethylbenz[*a*]anthracene), leads to no skin carcinomas observed 3 months later. (B) Multiple treatments with promoting agents, such as TPA (Figure 11.27), also do not lead to significant numbers of tumors. (C) If an area of skin is painted once with an initiating agent followed by repeated paintings with a promoting agent, a papilloma will often appear several months later. (D) If an area of skin is painted once with an initiating agent and a promoting agent is then used to repeatedly paint another nearby but non-overlapping patch of skin, no papillomas will be seen 3 months later. (E) In a variation of the protocol depicted in panel C, an initiator such as DMBA is applied followed by repeated TPA treatments, which lead to papillomas. However, if repeated painting with the TPA tumor-promoting agent is halted soon after the papillomas appear, they then regress, indicating that they are dependent on ongoing promoter stimulation. (F) In another variation of the protocol depicted in panel C, TPA promoter painting can be continued for several months after papillomas first appear. Thereafter, TPA painting can be halted. Under these conditions, some of the papillomas will persist, indicating that they have become independent of continued promoter stimulation. (G) If a papilloma is produced by the protocols of panels C or F and the papilloma is then treated with an initiating agent, a carcinoma may appear, even in the absence of further promoter treatment.

These phenomena were rationalized as follows (Figure 11.29). A single treatment by an **initiating agent** (or **initiator**) like DMBA left a stable, long-lived mark on a cell or cluster of cells; this mark was apparently some type of genetic alteration. Subsequent repeated exposures of these "initiated" cells to TPA (termed the **promoting agent** or simply the **promoter**) allowed these cells to proliferate vigorously while having no apparent effect on nearby uninitiated cells. (Note that use of the word "promoter" in this context is unconnected with its other meaning—namely, the DNA sequences controlling the transcription of a gene.) The localized proliferation of initiated cells that was encouraged by the promoter would eventually produce a papilloma. However, as mentioned, if TPA painting were halted, the papilloma would disappear. Accordingly, the effects of the promoter were reversible, suggesting that it exerted a *nongenetic* effect on the cells in the papilloma. Clearly, this nongenetic effect, whatever its nature, could collaborate with the apparent genetic alteration created by the initiator to drive the proliferation of cells.

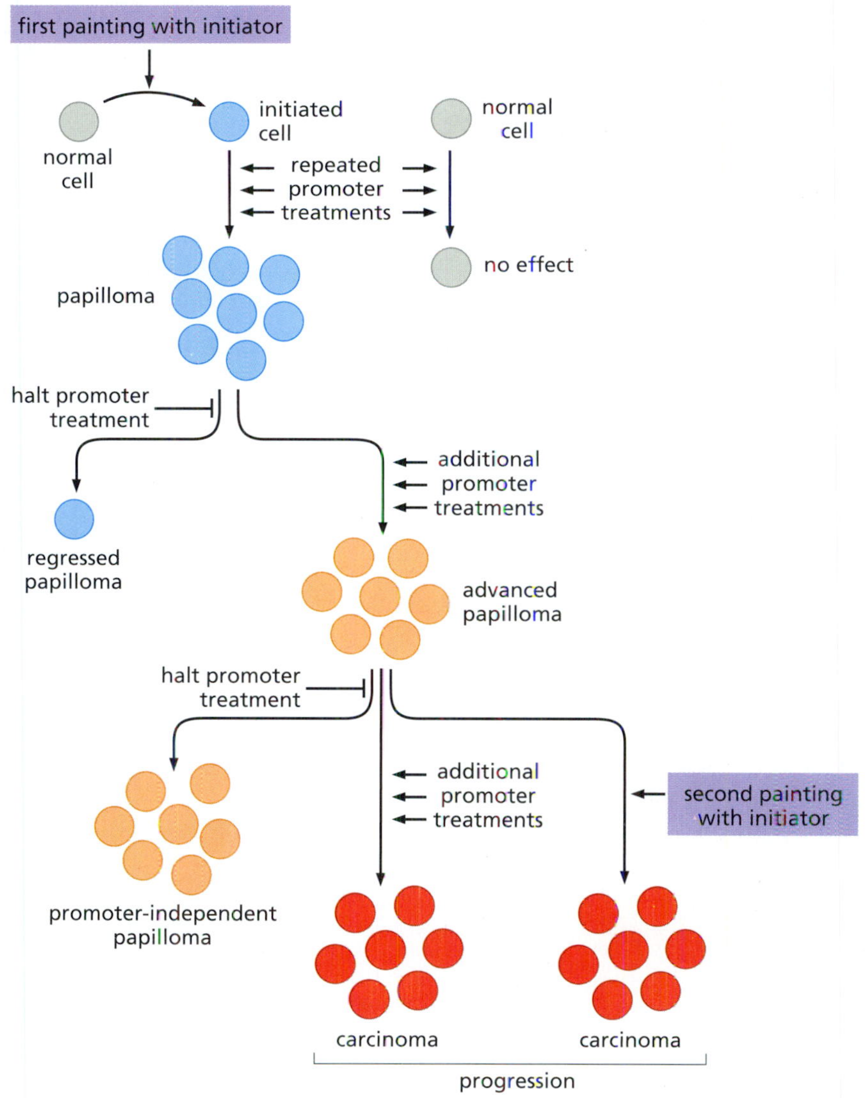

Figure 11.29 Scheme of initiation and promotion of epidermal carcinomas in mice The observations of Figure 11.28 can be rationalized as depicted here. The initiating agent converts a normal cell *(gray, top left)* into a mutant, initiated cell *(blue)*. Repeated treatment of the initiated cells with the TPA promoter generates a papilloma *(cluster of blue cells)*, while TPA treatment of normal, adjacent cells *(gray, top right)* has no effect. Further treatment of the initially formed papilloma can be halted *(middle left)*, in which case the papilloma regresses. Alternatively, further repeated treatment of the initially formed papilloma can yield a more progressed papilloma *(orange cells)*, which persists even after promoter treatment is halted *(bottom left)*; further repeated treatment of this more progressed papilloma with TPA eventually yields, with low frequency, a carcinoma *(bottom middle, red cells)*. Alternatively, exposure of the initially formed papilloma to a second treatment by the initiating agent yields doubly mutant cells that also form a carcinoma *(bottom right, red cells)*.

As we read above, if the initiated cells were treated with the TPA promoter for many months' time, eventually some papillomas would evolve to become TPA-independent; in this case, even after TPA withdrawal, the papillomas continued to increase in size and some eventually developed into skin carcinomas. This permanent change in cell behavior seemed to reflect the actions of a second, independent genetic alteration. Indeed, this evolution to a carcinomatous state could be strongly accelerated by treating a papilloma with a second dose of the initiating agent, already suspected to be a mutagen (see Figure 11.29). This third step in tumorigenesis (coming after initiation and promotion) is termed **progression**; the term is used more generally, throughout this book and elsewhere, to indicate the evolution of cells to an increasingly malignant growth state.

Four decades after the mouse skin cancer induction protocol was first developed, the identities of the genes and proteins that are the main actors in this skin tumorigenesis were discovered (Figure 11.30). As long suspected, the DMBA used as the initiating agent is indeed a potent mutagen in the context of skin carcinogenesis. (For support of this conclusion, see Sidebar 11.11.) Since it is a randomly acting mutagen, DMBA creates a wide variety of mutations in the genomes of exposed cells. However, the skin tumors that emerge invariably bear point-mutated H-*ras* oncogenes, indicating that this particular mutant allele confers some strong selective advantage on cells in the skin.

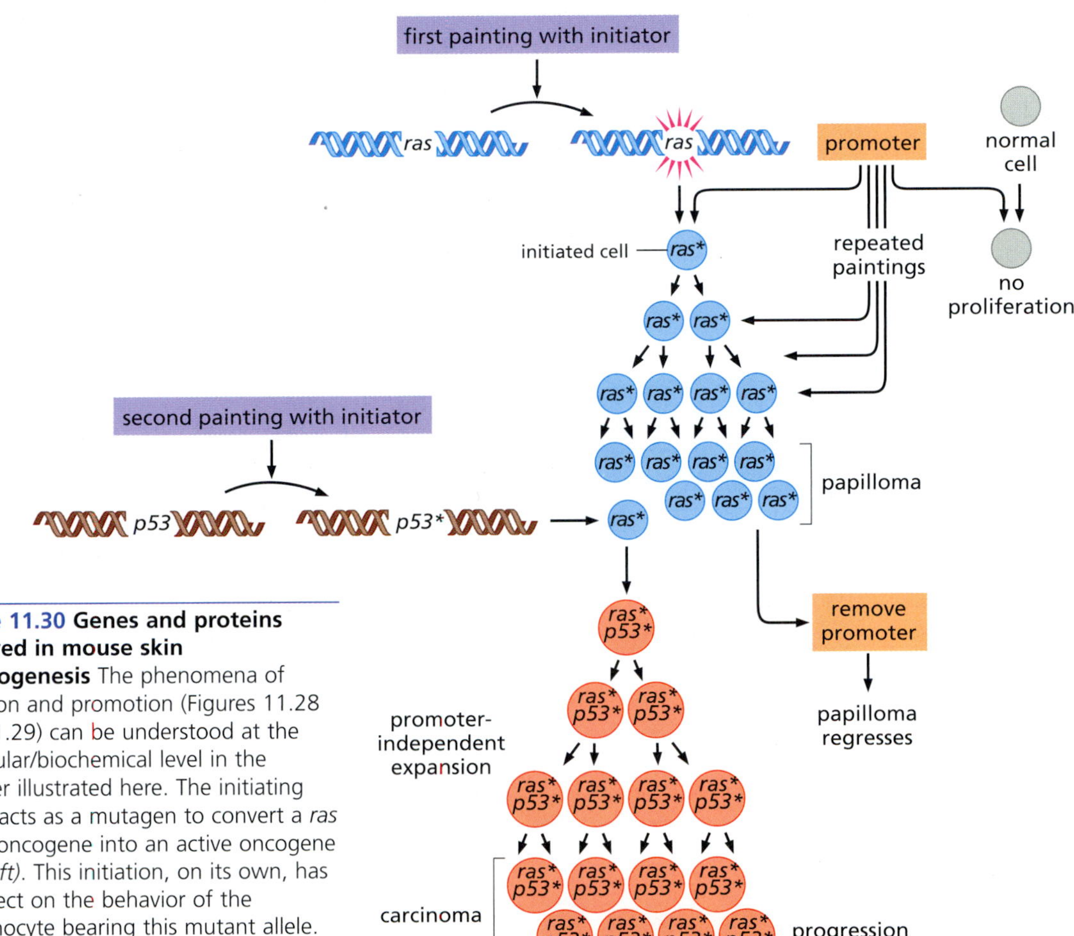

Figure 11.30 Genes and proteins involved in mouse skin carcinogenesis The phenomena of initiation and promotion (Figures 11.28 and 11.29) can be understood at the molecular/biochemical level in the manner illustrated here. The initiating agent acts as a mutagen to convert a *ras* proto-oncogene into an active oncogene *(top left)*. This initiation, on its own, has no effect on the behavior of the keratinocyte bearing this mutant allele. However, in the presence of repeated stimulation by a promoting agent *(top right),* the *ras*-bearing cell is induced to pass through repeated growth-and-division cycles, leading to the formation of a papilloma *(blue cells)*. Conversely, a cell lacking a *ras* oncogene *(gray, top right)* is not stimulated by the promoting agent and thus does not divide in response to repeated exposure to this agent. If repeated treatment by the promoter is halted *(lower right),* the papilloma regresses. However, if the papilloma is exposed a second time to a mutagenic initiating agent *(left),* a second genetic lesion, often involving the mutation of the *p53* tumor suppressor gene, is created. This mutant *p53* allele collaborates with the *ras* oncogene to create a population of cells *(light orange)* that are no longer dependent on the promoter and are capable of forming a carcinoma.

The subsequent repeated treatments with TPA, the promoting agent, act synergistically with the activated H-*ras* oncogenes to drive the proliferation of oncogene-bearing cells, yielding a papilloma. Treatment of the papilloma with TPA over an extended period of time may generate a papilloma that can persist, even after TPA treatment is halted. Alternatively, if cells in a papilloma are exposed, once again, to an initiating agent like DMBA, this papilloma may progress into a carcinoma, whose cells now carry, in addition to the H-*ras* oncogene, a mutant *p53* gene.

If we were to describe these phenomena at the level of signal-transduction biochemistry (Figure 11.31), we would say that the TPA promoter functions as a potent stimulator of cell proliferation through its ability to activate the cellular serine/threonine kinase known as protein kinase Cα (PKC-α), which we encountered earlier in the context of cytoplasmic signal transduction. More specifically, TPA acts as a functional mimic of diacylglycerol (DAG), the molecule that is generated endogenously by cells as a means of activating protein kinase C (PKC; Figure 6.16). We will encounter some of the downstream effectors of PKC-α later in this chapter. Somehow, the downstream effectors of PKCα collaborate with an H-*ras* oncogene, in still unknown ways, to drive proliferation of initiated keratinocytes whose descendants form papillomas and, on rare occasion, progress to forming carcinomas. These diverse observations of mouse skin carcinogenesis (Figures 11.28, 11.29, and 11.30) leave us with the conclusion that tumor promoters like TPA, which do not directly affect the genomes of cells, can nevertheless function as important agents in driving forward multi-step tumorigenesis.

Sidebar 11.11 Molecular analyses show that initiation is caused directly by the mutagenic actions of an initiating carcinogen The experimental convenience of the mouse skin carcinogenesis model has made it possible to directly validate some of the mechanistic speculations made here. For example, introduction of an H-*ras* oncogene into mouse skin cells through use of a retrovirus vector creates cells that closely mimic the behavior of DMBA-initiated cells, in that they are responsive to the effects of subsequently applied TPA promoter. This outcome demonstrates that the creation of an H-*ras* oncogene by an initiating agent is sufficient, on its own, to yield an initiated cell.

The mouse skin carcinogenesis model raises yet other questions: Do initiating agents, such as the DMBA or 3-methylcholanthrene (3-MC) carcinogens, act directly on the H-*ras* proto-oncogene to create an oncogene? Or do they do so indirectly, by stimulating the actions of some intermediary molecules that are then responsible for reacting with DNA and creating the critical initiating mutations?

When mouse skin tumors were initiated by exposure to either of two alkylating carcinogens (Chapter 12), *N*-methyl-*N'*-nitro-*N*-nitrosoguanidine (MNNG) or methylnitrosourea (MNU), the resulting skin tumors showed only G-to-A **transitions** (i.e., purine–purine or pyrimidine–pyrimidine substitutions) at codon 12 of the H-*ras* gene. In contrast, when mouse skin tumors were initiated by painting with 3-methylcholanthrene (3-MC), the resulting papilloma and carcinoma cells were found to carry predominantly G-to-T **transversions** (purine–pyrimidine substitutions or vice versa) in codon 13 and A-to-T transversions in codon 61 of the H-*ras* gene. Dimethyl benzanthracene (DMBA), which reacts with A's, induced tumors with A-to-T transversions in this gene. These nucleotide substitutions conform to the known mutagenic activities of these carcinogens.

Because the specific base substitutions reflect the chemical identities of the initiating agents, these observations provide strong evidence of the direct chemical interaction between the initiating carcinogen molecules and specific bases present in the H-*ras* proto-oncogene of mouse skin cells. Moreover, when taken together with the observed effects of the H-*ras* retrovirus vector, they indicate that the main mutational contribution of these agents to tumor initiation derives from their ability to mutate H-*ras* proto-oncogenes.

11.14 Toxic and mitogenic agents can act as human tumor promoters

The experimental model of mouse skin carcinogenesis is useful for illustrating the principles of tumor initiation, promotion, and progression. However, it tells us almost nothing about how analogous mechanisms operate in the human body to create cancer. In fact, a diverse array of biochemical and biological mechanisms appear to be responsible for the tumor promotion leading to human cancers. Among these are mechanisms that act in either toxic or mitogenic fashion on various human tissues.

A striking example of cytotoxic mechanisms in human tumor promotion is provided by cancers of the mouth and throat (often called head-and-neck cancers). These carcinomas are often encountered in cigarette smokers who are also consumers of distilled alcoholic drinks. In fact, a serious cigarette habit together with frequent consumption of distilled alcohol leads to as much as a 100-fold increased risk for certain types of head and neck cancers.

Cigarette smoke is rich in a variety of mutagenic carcinogens, including 3-methylcholanthrene (3-MC). Ethanol, in contrast, has weak, if any, mutagenic powers. Instead, the contribution of distilled alcoholic drinks to tumor induction seems to derive from their toxic effects on the epithelial cells lining the mouth and throat. After exposure to a drink containing a high percentage of ethanol, many of these cells die and slough off. Stem cells underlying the epithelium respond by dividing in order to regenerate epithelial cell layers within the mouse and throat. While these stem cells may normally divide at a low and steady rate, their mitotic rate increases substantially after widespread **denuding** (stripping) of an epithelium by ethanol.

The cells in the mouth and throat whose proliferation is stimulated by alcohol may already carry mutant alleles induced by tobacco tar. The promoting effect of alcohol then causes the clonal expansion of these initiated cells and may thereby enable their descendants to acquire yet other mutations that lead ultimately to the clinically aggressive head-and-neck cancers. This represents a dramatic illustration of a toxic agent acting as a tumor promoter.

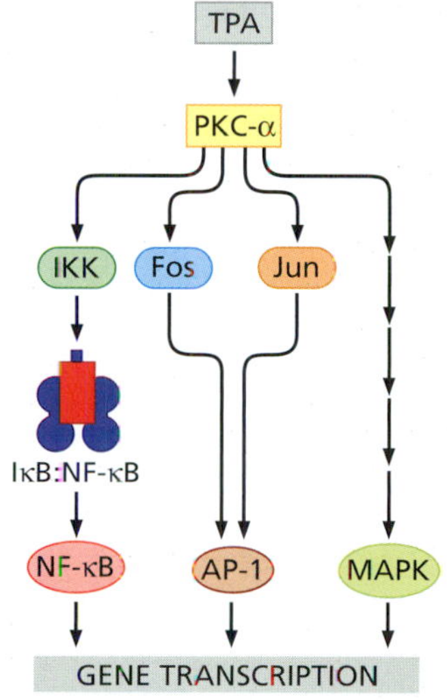

Figure 11.31 Activation of Fos, Jun, and NF-κB by PKC-α The precise mechanism by which an activated *ras* oncogene acts synergistically with TPA-activated protein kinase C (PKC-α) to drive the proliferation of keratinocytes is poorly understood. It is clear that once PKC-α is activated by binding TPA, it is able to stimulate transcription of several distinct signaling pathways, including those involving the NF-κB and AP-1 transcription factors and the ERK/MAPK enzyme. These biochemical changes have not yet yielded a clear explanation why TPA stimulates the proliferation of initiated keratinocytes bearing a *ras* oncogene while having minimal effect on nearby wild-type keratinocytes.

Sidebar 11.12 A rare cancer syndrome illustrates the importance of cell proliferation in tumor promotion Kostmann syndrome is a rare, heritable syndrome characterized by the almost complete absence of **neutrophils**, the cells in the blood that are responsible for killing bacterial and fungal infectious agents. In the autosomal dominant form of this disease, the genetic defect inducing the syndrome causes the synthesis of a mutant, neutrophil-specific elastase (a protease)—an enzyme that is normally expressed at increasing levels in neutrophils as they differentiate. The mutant elastase present in Kostmann patients is cytotoxic for their neutrophils, and once it becomes expressed at significant levels in these cells, it causes their depletion through apoptosis (Figure 11.32). Progenitor myeloid stem cells in the marrow respond to the resulting **neutropenia** (depressed numbers of neutrophils in the blood) by proliferating and attempting to replenish the pool of differentiated neutrophils. This leads to endless futile cycles of stem-cell proliferation, since all attempts at neutrophil production by these progenitors are frustrated by the rapid, elastase-induced death of these cells once they differentiate. In some patients, as a consequence of the continuous, excessive proliferation of these stem cells over many years' time, acute myelogenous leukemias arise from this myeloid stem-cell population. This syndrome provides a dramatic example of how a cytotoxic agent (in this case one of endogenous origin) functions directly as a tumor promoter.

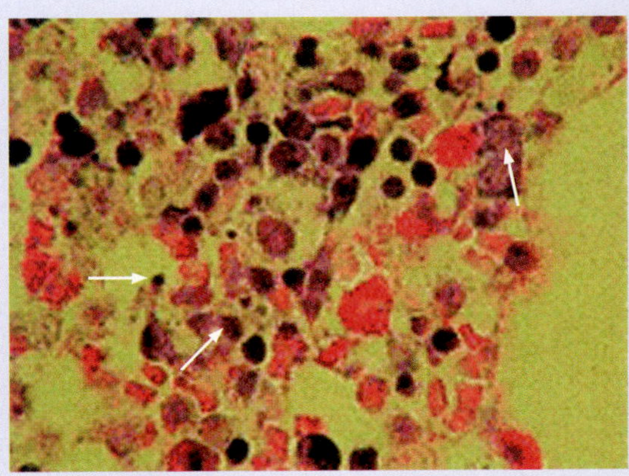

Figure 11.32 Kostmann syndrome This disease is manifested by ongoing loss of differentiating neutrophil precursor cells from the bone marrow, as revealed here by the *white arrows*, which point to unusually large numbers of apoptotic bodies, i.e., fragments of cells that result from the triggering of their apoptotic program. This apoptosis is attributable, in turn, to their synthesis of a mutant, toxic variant of an elastase enzyme. (From G. Carlsson. A.A. Aprikyan, R. Tehranchi et al., *Blood* 103:3355–3361, 2004.)

Imagine, in a more general sense, compounds that are highly toxic for certain cell populations within a tissue. These cytotoxic agents can function as tumor promoters simply by causing the proliferation of the cells that have survived the toxic effects of these agents. Certain types of human cancer nicely illustrate the role of cytotoxicity in tumor progression (see Sidebar 11.12).

In fact, mitogenic agents can also function as tumor promoters. Prominent among these are the steroid hormones—estrogen, progesterone, and testosterone. In the female body, for example, estrogen and progesterone are involved in programming the proliferation of cells in reproductive tissues. The monthly menstrual cycles of women between **menarche** and menopause result in the proliferation and then regression of the cells forming the epithelia of the ducts in the mammary gland (Figure 11.33). The endometrial lining of the uterus undergoes similar cycles of proliferation and regression.

Epidemiology makes it clear that the more menstrual cycles a woman experiences in a lifetime, the higher the risk of breast cancer. By one estimate, lifetime breast cancer risk decreases by 20% for each year that menarche is delayed during adolescence. (The most compelling illustration of the importance of the timing of menarche has come from studies of identical, or monozygotic, twins both of whom eventually developed breast cancer; the twin whose menarche began earlier had a 5.4-fold greater risk of being the first to be diagnosed with this disease.) Women who stop menstruating before age 45 have only about one-half the risk of breast cancer of those who continue to menstruate to age 55 or beyond.

Removal of the ovaries, the prime source of estrogen in the female body, causes breast cancer risk to plummet. Reinforcing this observation are the results of a Dutch study, which has shown that women who enter menopause before the age of 36, due to the side effects of chemotherapy for Hodgkin's lymphoma, have

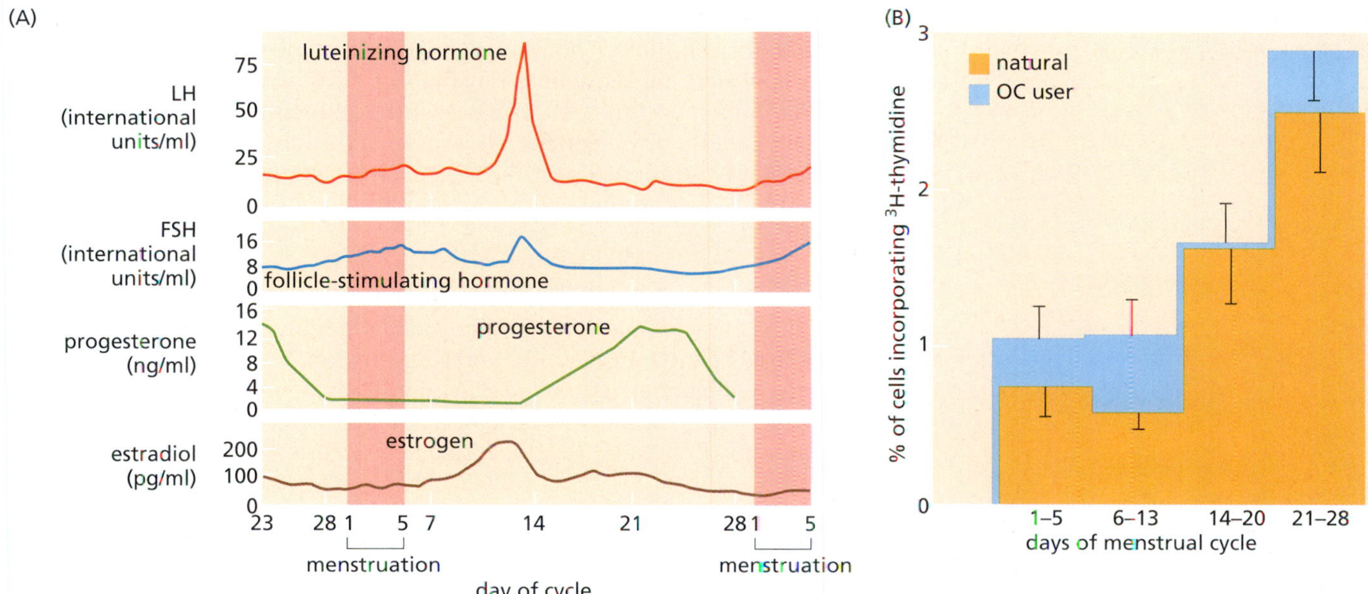

Figure 11.33 Fluctuations of hormone levels, cell proliferation, and mammary gland morphology during the menstrual cycle The periodic stimulation of mammary epithelial cell proliferation in the human breast under the control of hormones such as progesterone, prolactin, and estrogen (estradiol) appears to be coupled with increased risk of developing breast cancer. (A) As seen here, the levels of several of these hormones vary dramatically throughout the menstrual cycle. (B) The periodic monthly cycling leads to strong fluctuations in the rate of proliferation of cells. Here the percentage of cells in a tissue biopsy incorporating tritiated thymidine is plotted as a function of the days of the menstrual cycle. Use of oral contraceptives (OC) had minimal effect on DNA synthesis rates. (A, from R.A. Rhoades and R.H. Pflanzer, Human Physiology. Philadelphia: W.B. Saunders, 1996; B, from T.J. Anderson, S Battersby, R.J. King et al., *Hum. Pathol.* 20:1139–1144, 1989.)

a 90% decreased risk of subsequently developing breast cancer. Conversely, postmenopausal women who contract breast cancer have on average a 15% higher level of circulating estrogen than unaffected women.

The effects of estrogen on breast cancer are surely complex, and it appears that this hormone acts on other cells in the mammary gland besides the epithelial cells. Still, it is evident that estrogen, and perhaps other hormones such as progesterone and even prolactin, periodically induce cell proliferation in a way that enables the progression of initiated mammary epithelial cells (MECs) into the MECs found in the various types of breast cancers. (Some have argued that metabolites of estrogen are mutagenic, and that these metabolites contribute to breast cancer development; if so, estrogen's mutagenic effects on breast cancer development are surely overshadowed by its power to promote the proliferation of MECs.)

In this example, we confront a tumor promoter that is endogenous to the mammalian body rather than being an agent of foreign origin. Nonetheless, the actions of this hormone adhere closely to the properties of classic tumor promoters.

11.15 Chronic inflammation often serves to promote tumor progression in mice and humans

Relatively few human tumor promoters act through purely cytotoxic or mitogenic mechanisms, such as those described in Section 11.14. Instead, the great majority seem to drive clonal expansion through mechanisms involving inflammation. Hints of this come from the model of mouse skin tumor initiation and promotion (Section 11.13). This experimental model is clearly artificial, in that it involves a promoter (TPA) that is rarely, if ever, involved in skin tumor development in a mammal. Nonetheless, it teaches an important, generalizable lesson about tumor promotion: TPA was initially chosen because it is an irritant of mouse skin and thus an inducer of localized inflammation.

A diverse set of other observations lend weight to the notion that inflammation is commonly involved in tumor promotion. For example, when cells of a human colonic adenoma cell line were implanted subcutaneously into nude mice, they were found to be non-tumorigenic. However, when they were introduced into host mice together with a fragment of plastic (to which they were attached), localized stromal inflammation was induced by the plastic and the adenoma cells grew to form tumors. The tumorigenic phenotype of these cells persisted even after they were subsequently transplanted (without the plastic) to another host animal, indicating that their neoplastic proliferation was now driven by some stable genetic or epigenetic alteration. Yet other observations of tumor development in the mouse gastrointestinal tract support the argument that inflammation is a centrally important component of colonic carcinogenesis (Sidebar 11.13).

Chronic inflammation also plays a clear role in the pathogenesis of human carcinomas. For example, those arising in the gallbladder are usually associated with a decades-long history of gallstones and resulting inflammation of the epithelial lining of the gallbladder (Figure 11.35A). Similarly, hepatocellular carcinomas (HCCs), which are common in East Asia (Sidebar 11.14), are associated with chronic hepatitis B virus (HBV) infections and accompanying inflammation of

Sidebar 11.13 Inflammation contributes to carcinogenesis in the gastrointestinal tract A dramatic example of the role of inflammation in colon carcinoma formation has come from the study of mutant mice that lack the ability to make TGF-β1. Such mice tend to develop an autoimmune disease, which results in their death after several weeks. In order to spare these mice, their immune systems were crippled by germ-line inactivation (see Sidebar 7.10) of their *Rag2* gene copies, a loss that prevents the formation of the antigen-specific lymphocytes that trigger autoimmune disease. The doubly mutant mice ($TGF\text{-}\beta 1^{-/-}$ $Rag2^{-/-}$) now survived but developed areas of inflammation in the colon, as well as colonic adenomas and adenocarcinomas, between 3 and 6 months of age. However, if these doubly mutant mice were reared in a germ-free environment (which yields colons free of the usual bacterial populations), no colonic inflammation was seen and neither adenomas nor adenocarcinomas developed.

If such germ-free mice were introduced into animal quarters in which *Helicobacter hepaticus* (a bacterium that commonly inhabits the mouse colon) was present, once again the polyps and carcinomas formed in their colons (Figure 11.34A,B). Such observations indicate that bacterial flora in the gut contribute importantly to inflammation, both in these mice and, one imagines, in humans suffering from ulcerative colitis, an inflammatory condition that predisposes to colon cancer (e.g., see Figure 10.36). Moreover, results like these suggest that localized areas of colonic inflammation in otherwise normal individuals can also contribute to increased risk of forming adenomatous polyps that can eventually progress to carcinomas.

(A)

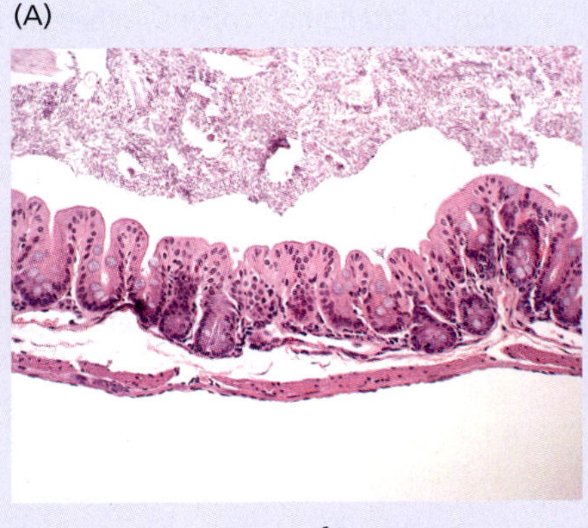

germ free

(B)

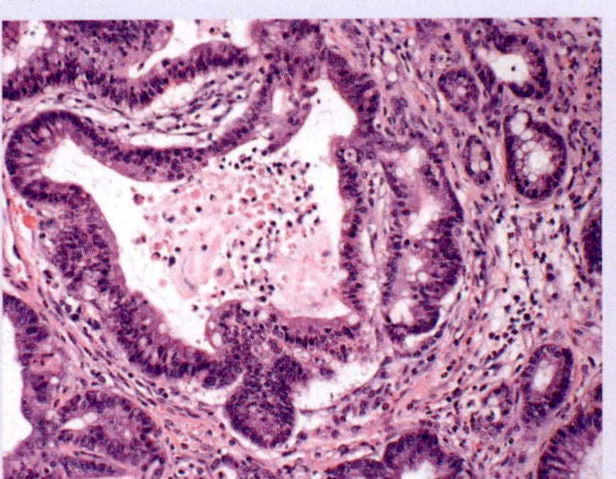

infected with *Helicobacter hepaticus*

(C)

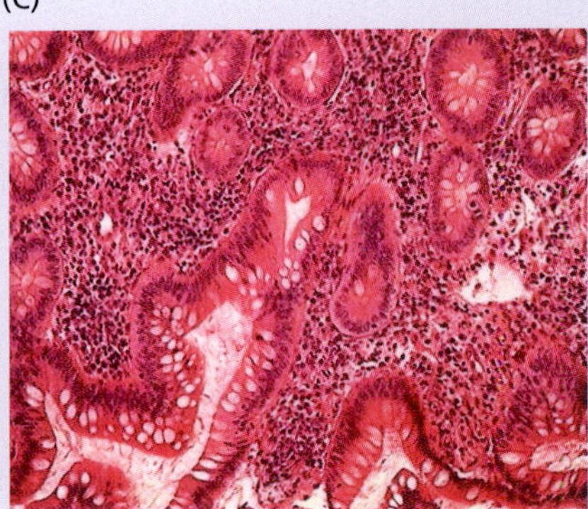

Figure 11.34 Colonic inflammation and tumor promotion The development of adenomas and carcinomas in the colon is strongly dependent on chronic inflammation occurring in this organ. Thus, genetically altered mice that lack the ability to make TGF-β1 succumb early in life to overwhelming autoimmune disease. (A) However, if their immune system is crippled through the inactivation in their germ line of the gene encoding the Rag-2 enzyme, which is required for the generation of functional antibody and T-cell receptor genes, these mice survive when they are housed in a germ-free facility. Under such conditions, their intestinal epithelium is quite normal histologically. (B) In contrast, if these doubly mutant mice are housed in a facility in which *Helicobacter hepaticus* bacteria are present, their colons become infected with these bacteria and they develop adenomas and carcinomas. This indicates that the chronic inflammation created by the presence of these intestinal bacteria is essential to the formation of these colonic lesions. (C) Human ulcerative colitis involves chronic inflammation of regions of the colonic mucosa, which leads after many years to substantially increased risk of colon carcinoma. In this condition, the stromal areas between the colonic crypts are infiltrated with large numbers of lymphocytes *(small dark purple nuclei)*. (A and B, from S.J. Engle, I. Ormsby, S. Pawlowski et al., *Cancer Res.* 62:6362–6366, 2002; C, courtesy of D. Lamarque.)

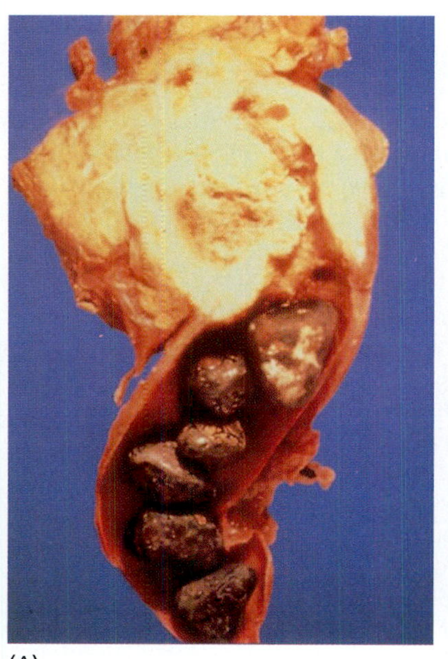

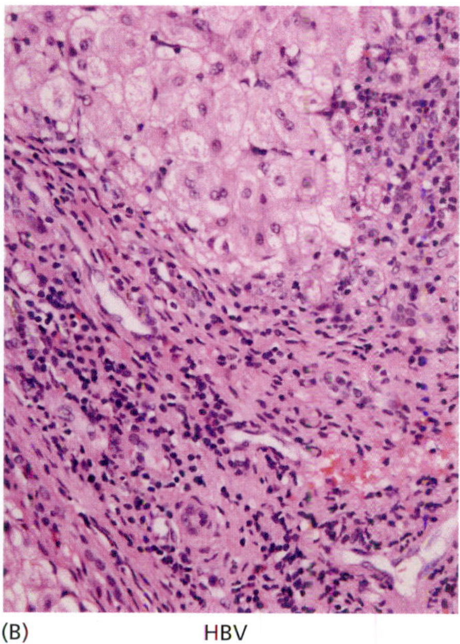

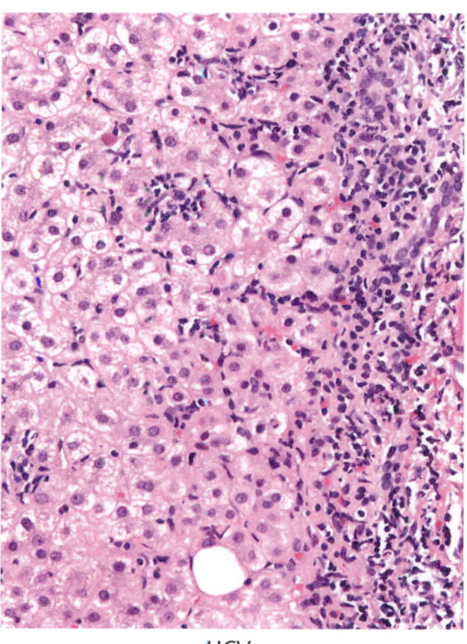

Figure 11.35 Chronic inflammation leading to cancer (A) A graphic demonstration of chronic inflammation leading to cancer is provided by carcinomas of the epithelial lining of the gallbladder *(white mass, above)*, which are commonly associated with the formation of gallstones arising as precipitates from the bile *(brown masses, below)*. (B) Chronic hepatitis B virus (HBV) infection creates continual cell death of hepatocytes together with chronic inflammation *(numerous small dense nuclei, left panel)*. Over a period of decades, this can lead to an almost 100-fold increased risk of liver cancer. The inflammation in the liver of someone suffering from chronic hepatitis C virus (HCV) infection is strikingly similar *(right panel)*. The inflammatory cells are in the right part of this micrograph. The fact that the two inflammatory conditions are so similar and lead to comparably increased risks of hepatocellular carcinoma (HCC) suggests that the inflammatory states, rather than some specific aspect of viral function, are responsible for the appearance of HCC in patients infected with either of these viruses. (A, from A.T. Skarin, Atlas of Diagnostic Oncology, 3rd ed. Philadelphia: Elsevier Science Ltd., 2003; B, left—courtesy of A. Perez-Atayde, right—courtesy of A.K. Bhan.)

the liver (Figure 11.35B, left panel). In many infected individuals, HBV infection is well established in the liver early in life and continues in a chronically active form for decades. The resulting hepatitis may have relatively few outward effects on the individual, since the continual HBV-induced killing of hepatocytes (the cells forming the bulk of the liver) is compensated by an equal proliferation of surviving cells.

The HBV genome does carry a gene, termed *HBX*, that shows weak oncogenic and pro-apoptotic powers, but this gene, on its own, can hardly explain HBV's ability to induce hepatic carcinomas after decades of chronic infection. This led to a search for other carcinogenic mechanisms. Thus, HBV might act as a liver carcinogen through a mechanism that echoes the carcinogenic strategy of non-oncogene-bearing retroviruses. Recall that these viruses, notably avian leukosis virus and murine leukemia virus, induce cancer slowly and inefficiently. When, after many months, they finally succeed in doing so, this occurs through insertional mutagenesis: the chance integration of a provirus next to a critical cellular growth-controlling gene—a proto-oncogene (Section 3.11). The resulting deregulation of expression of the proto-oncogene effectively converts this gene into an oncogene, paving the way for cancer formation.

In the case of HBV-induced liver cancer, however, the situation is quite different. Extensive molecular analyses have failed to demonstrate that the genomes of virus-associated human liver cancers carry HBV genomes integrated next to critical cellular growth-controlling genes, such as the *myc* proto-oncogene. So, HBV is unlikely to act directly as a mutagen in infected liver cells.

This leaves open two plausible explanations for the virus-induced cancer. HBV creates liver cancer through its ability to cause continuous cell proliferation in an organ that normally experiences hardly any at all; this proliferation is required to replace hepatocytes that are continually being killed by HBV infectious cycles. Alternatively, HBV infection causes cells of the immune system to attempt to eliminate virus-infected cells, yielding a chronic inflammatory state in the liver (see Figure 11.35B). It is likely that both of these mechanisms conspire in the pathogenesis of the hepatocellular carcinomas that so frequently afflict individuals with chronic HBV infections.

In the years since these studies were undertaken, it has become clear that chronic hepatitis C virus (HCV) infections act in a similar way to increase liver

Sidebar 11.14 Hepatitis B virus infections lead to hepatomas in Taiwanese government workers Epidemiological studies usually yield outcomes that make it difficult to discern a strong correlation between occasional exposure to certain environmental factors and a subsequent moderately increased risk of cancer. The incidence of liver cancer, however, varies dramatically throughout the world, making it possible to strongly associate this disease with a causative factor. In certain parts of the world, including much of Asia—especially China—and sub-Saharan Africa, hepatocellular carcinomas (HCCs, also known as hepatomas) are one of the leading causes of death due to cancer; in the United States, liver cancer, in stark contrast, ranks 25th as a cause of cancer-related deaths.

A stunning correlation can be made between hepatitis B virus (HBV) infections and susceptibility to liver cancer. A prospective epidemiologic study was undertaken in 1975 among 22,707 men who worked in government service in Taiwan. Their HBV status was determined at the beginning of the study by measuring, among other things, the presence of viral antigens in the blood. The causes of deaths in this cohort were then chronicled over the next decade. By 1986, 152 of the 3454 men who were initially positive for HBV viral antigen had died of hepatocellular carcinomas. By comparison, 9 men of the 19,253 who were negative for this antigen in their blood had died of this disease. The relative risk (RR) of dying from this disease if one carried viral antigen in the circulation (which is indicative of an active HBV infection in the liver) was 98.4. This means that an individual afflicted with chronic, active HBV infection experienced an almost 100-fold increased risk of contracting and dying from this cancer compared with someone who lacked viral antigen in his liver and was apparently uninfected. These numbers contrast strongly with most epidemiologic correlations made between specific exposures of patients to suspected carcinogens and disease incidence, where relative risk of disease is often only two- to threefold—often hovering on the borderline of statistical significance.

cancer rates. While the two viruses are totally unrelated to one another with respect to genome structure and replication cycles, they evoke very similar biological outcomes through their shared ability to create chronic infections, cytotoxicity, and inflammation in the liver (Figure 11.35B, right panel). (Significantly, a variety of other types of chronic liver injury, including that inflicted by alcoholism, are also associated with increased incidence of hepatocellular carcinoma, although the relative risks are vastly less than that associated with lifelong HBV infection.)

HBV, acting as a tumor promoter, can also function synergistically with aflatoxin-B1, a highly mutagenic compound that is made by *Aspergillus* fungi that proliferate on peanuts, nuts, and corn stored under conditions of high humidity (see Figure 2.25). This combination of infection plus aflatoxin exposure proves to be deadly. In one, relatively small prospective epidemiologic study carried out in Shanghai, infection with HBV increased the risk of hepatocellular carcinoma about 7-fold while exposure to aflatoxin-B1–contaminated food yielded about a 3-fold increased risk. When an individual experienced both agents, the risk of liver cancer increased about 60-fold. The parallel between the pathogenesis of these human liver carcinomas and the actions of initiators and promoters of mouse skin cancer is striking.

Yet other indications of infectious agents inducing inflammation and, in turn, human malignancies come from individuals afflicted with lymphomas arising in gastric mucosa-associated lymphoid tissue (MALT). Seventy-five percent of MALT lymphomas can be cured if patients are treated with antibiotics that eradicate the *Helicobacter pylori* bacterial populations in the stomach. These MALT lymphomas are clearly dependent on continued promoter stimulation—in this case the presence of *H. pylori*. The 25% of lymphomas that do not respond to this treatment have evolved beyond this dependence to become "promoter-independent," possibly because they have sustained the Chromosome 11 to 18 translocation that is frequently observed in these lymphomas.

11.16 Inflammation-dependent tumor promotion operates through defined signaling pathways

Evidence supporting the role of inflammation in cancer development also comes from a large number of epidemiologic observations demonstrating that

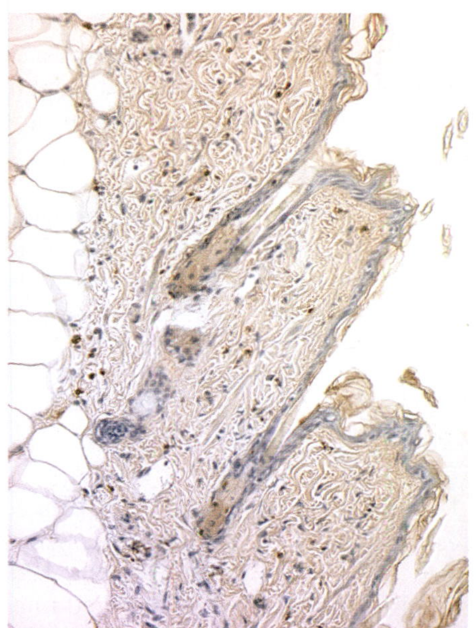

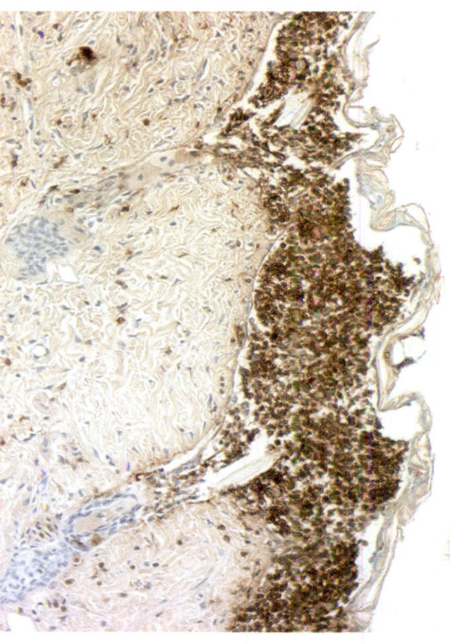

Figure 11.36 TPA and mouse skin induction The inflammatory powers of TPA are revealed in exaggerated form in transgenic mice that overexpress PKC-α, the intracellular target of TPA, specifically in their keratinocytes. (This is achieved by constructing a transgene in which PKC-α expression is driven by a keratin 5 transcriptional promoter, which is specific to the skin.) Eighteen hours after wild-type or transgenic mice have been treated with TPA, the skin of wild-type mice *(left)* shows relatively few neutrophils *(dark spots)*, which are important immune mediators of inflammation. In contrast, the skin of the transgenic mice *(right)* shows extensive neutrophil infiltration. Neutrophils were detected through use of an antibody that recognizes a neutrophil-specific antigen. (From C. Cataisson. E. Joseloff, R. Murillas et al., *J. Immunol.* 171:2703–2713, 2003.)

anti-inflammatory drugs, such as aspirin and sulindac, function to reduce the incidence of a variety of carcinomas in humans. For example, one study showed that those who took low doses of a nonsteroidal anti-inflammatory drug (NSAID)—an aspirin tablet once every day or two over a period of 15 years—had a lung cancer rate of 0.68, a breast cancer rate (in women) of 0.70, and, in younger men, a colorectal cancer rate of 0.35 compared with the rates of these cancers in corresponding control groups. Another prospective epidemiologic study of a large cohort of women over a seven-year period found that those regularly taking aspirin had a risk of pancreatic cancer that was about one-half that of a control group. Two studies have shown that regular use of aspirin (or another NSAID) resulted in about a 40% reduction of stomach cancer (in the region of the stomach outside of the cardia) in individuals who were infected with *Helicobacter pylori*, a bacterium that often inhabits the human stomach; stomach cancer rates in uninfected individuals were not reduced by aspirin usage. Yet other studies indicate reduced death rates associated with ovarian, bladder, and prostate carcinomas.

(These outcomes might suggest that NSAIDs, which can also reduce deaths from certain types of cardiovascular disease, should be used routinely as disease preventives in the general population. However, a caution comes from another public health statistic: in 1998, 16,550 deaths in the United States were attributed to NSAID-induced gastrointestinal complications.)

The epidemiologic observations linking long-term NSAID use with reduced cancer incidence, together with the experiments in mouse skin and liver carcinogenesis, lead us to the biochemical and cell-biological mechanisms that are responsible for tumor promotion in many tissues. One important clue has come from the identification of the downstream cellular target of TPA—protein kinase C-α (PKC-α). A dramatic demonstration of the key role of this enzyme in mediating TPA-induced skin inflammation is provided by mouse skin keratinocytes that are forced to overexpress it (Figure 11.36).

In TPA-treated keratinocytes, PKC-α signals, in part, by activating IKK (IκB kinase); the latter phosphorylates IκB (inhibitor of NF-κB), tagging it for destruction. NF-κB is thereby liberated (see Figure 6.29A) and migrates from the cytoplasm to the nucleus, where it induces transcription of a large constituency of genes. Among these are genes that block apoptosis and favor cell proliferation. Importantly, NF-κB also induces expression of the gene encoding tumor necrosis factor-α (TNF-α), a potent attractant of immune cells (i.e., a **cytokine**) that triggers localized inflammation in mouse skin and in many other epithelial tissues as well. (As its name implies, TNF-α was originally discovered as an inducer of the death of cancer cells. However, when its actions were explored in greater detail, its role as a major intermediary in tissue inflammation became apparent.)

Mice lacking functional copies of the TNF-α gene respond to a skin carcinogenesis protocol (involving the DMBA initiator and the TPA promoter) by developing 5 to 10% as many skin carcinomas as wild-type mice. In the latter animals, TPA treatment elicited TNF-α production in the **epidermal** keratinocytes, which then induced inflammation in the underlying stromal cells of the **dermis**. When taken together, these various observations suggest the following pathway:

$$\text{TPA} \rightarrow \text{PKC-}\alpha \rightarrow\rightarrow \text{IKK} \rightarrow\rightarrow \text{NF-}\kappa\text{B} \rightarrow \text{TNF-}\alpha \rightarrow \text{inflammation}$$

In another illustrative mouse tumor model, liver carcinogenesis was provoked through deletion of the *Mdr* (multi-drug resistance) gene in the mouse germ line (see Sidebar 7.10); loss of this gene and its encoded gene product leads to accumulation of bile acids and resulting chronic liver inflammation. Affected mice develop nodules of dysplastic hepatocytes, localized hepatocellular carcinomas (HCCs), and eventually metastatic HCCs.

In the livers of these mice, TNF-α is initially produced by inflamed endothelial cells as well as infiltrating inflammatory immune cells in the stroma, such as neutrophils and macrophages. The released TNF-α acts in a paracrine manner on nearby hepatocytes, which display receptors for TNF-α. Like PKC-α in keratinocytes, the ligand-activated TNF-α receptor of these hepatocytes funnels signals through the NF-κB pathway (Section 6.12). As described above, NF-κB is dispatched to the nucleus, where it activates anti-apoptotic genes, genes favoring proliferation, and the TNF-α gene. The resulting TNF-α, once released by hepatocytes, attracts more inflammatory cells through *paracrine* signaling and amplifies hepatocyte NF-κB signaling through *autocrine* signaling.

When NF-κB signaling was blocked in the hepatocytes of the *Mdr*$^{-/-}$ mice, tumor incidence was strongly suppressed (Figure 11.37). This inhibition of signaling could be achieved through the introduction of an anti-TNF-α antibody into the *Mdr*$^{-/-}$ mice (which blocked the paracrine signaling between inflammatory stromal cells and hepatocytes) or by expressing a nondegradable IκB in the hepatocytes (which inhibited NF-κB activation). In both cases, loss of NF-κB signaling resulted in greatly increased rates of apoptosis in the pre-neoplastic hepatocytes. Indeed, prevention of apoptosis was likely to explain much of the tumor-promoting effect of inflammation in this mouse model. Interestingly, shutdown of NF-κB signaling in hepatocytes did not prevent the early steps of liver tumor progression—hepatitis, tumor initiation, and the development of dysplasia—but the subsequent progression of dysplastic tissue to hepatocellular carcinoma was blocked. Hence, in this experimental model, TNF-α and NF-κB are involved in promotion and not in the earlier stages of initiation (see Figure 11.37).

A similar conclusion was reached in studies of a mouse model of colitis-associated colon cancer: inactivation of IKK signaling led to greatly increased rates of apoptosis in pre-cancerous enterocytes—the epithelial cells lining the colon. Without NF-κB signaling, these cells failed to express elevated levels of the potently anti-apoptotic Bcl-X_L (Figure 9.25) and instead produced high levels of the pro-apoptotic Bax and Bak proteins. Once again, some of the tumor-promoting effects of NF-κB could be traced to its ability to protect initiated epithelial cells from apoptosis.

These descriptions indicate that NF-κB signaling is highly active in the epithelial cells of inflamed tissues (e.g., hepatocytes in the liver, enterocytes in the colon). The inflammation is created by inflammatory cells of the immune system (notably macrophages, neutrophils, eosinophils, mast cells, and lymphocytes) that are recruited into the stromal compartments of these tissues, where they release pro-inflammatory signals. Importantly, the NF-κB pathway *also* operates within these stromal immune cells, enabling them to release pro-inflammatory signals such as TNF-α. For example, in the above-cited model of colitis and colorectal cancer, shutdown of IKK signaling in the inflammatory immune cells of the intestinal stroma also led to a significant suppression of colorectal tumor formation.

(A)

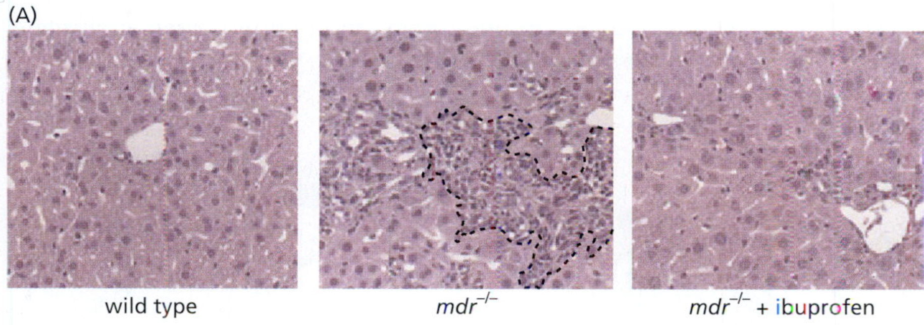

(B)

Figure 11.37 Chronic liver inflammation acts via NF-κB to induce hepatocellular carcinomas In this mouse model of liver carcinogenesis, deletion of the *mdr* (multi-drug resistance) gene causes chronic liver inflammation, which generates dysplastic foci among the hepatocytes, some of which progress to form hepatocellular carcinomas. (A) An infiltrate of inflammatory cells in the liver of an $mdr^{-/-}$ mouse *(outlined, center panel)* contrasts with the normal liver *(left panel)* or with the liver of an $mdr^{-/-}$ mouse that has been treated with a potent anti-inflammatory drug, ibuprofen *(right panel)*. These areas of inflammatory infiltrate appear to be critical to the formation of dysplastic foci, some of which eventually progress to hepatocellular carcinomas. (B) These various steps can be summarized in this flow diagram. A state of chronic inflammation provoked by liver bile attracts inflammatory cells to the liver stroma that release TNF-α, which impinges on the hepatocytes located within the epithelial compartment of the liver. Once activated, the TNF receptor of the hepatocytes acts via IKK to activate NF-κB signaling in these cells, which results in the activation of anti-apoptotic genes, such as Bcl-X_L, proliferative genes such as that encoding cyclin D1, and the gene specifying TNF-α. Together, these proteins function in dysplastic nodules of hepatocytes to facilitate their progression to hepatocellular carcinomas. This progression can be blocked by antibodies reactive with TNF-α as well as a dominant-negative, non-degradable IκB. (The actions of anti-inflammatory drugs are not explained in this diagram.) (A, from E. Pikarsky, R.M. Porat, I.Stein et al., *Nature* 431:461–466, 2004.)

None of this explains how NSAIDs, such as aspirin, ibuprofen, and sulindac, succeed in blocking tumor promotion and thus tumor progression. These drugs, and a host of other NSAIDs, have as their common target the enzyme cyclooxygenase-2 (COX-2; Figure 11.38). In fact, the *Cox-2* gene is yet another gene that is strongly induced by NF-κB. The critical role of this enzyme in mediating the inflammatory responses that lead to epithelial malignancies is illustrated by an experiment with genetically altered mice that were predisposed, because of a germ-line mutation of the *Apc* gene, to develop hundreds of intestinal polyps (see Section 7.11) : the number of polyps was reduced by a factor of 7 when the *Cox-2* gene was inactivated in the germ line of these mice. (Since COX-2, like NF-κB, operates in both stromal inflammatory cells and nearby epithelial cells, loss of this enzyme is likely to have affected both types of cells.) Similarly, when COX-2 enzyme activity was shut down pharmacologically (through use of an NSAID), the development of mammary carcinomas in tumor-prone transgenic mice was strongly suppressed (Figure 11.39).

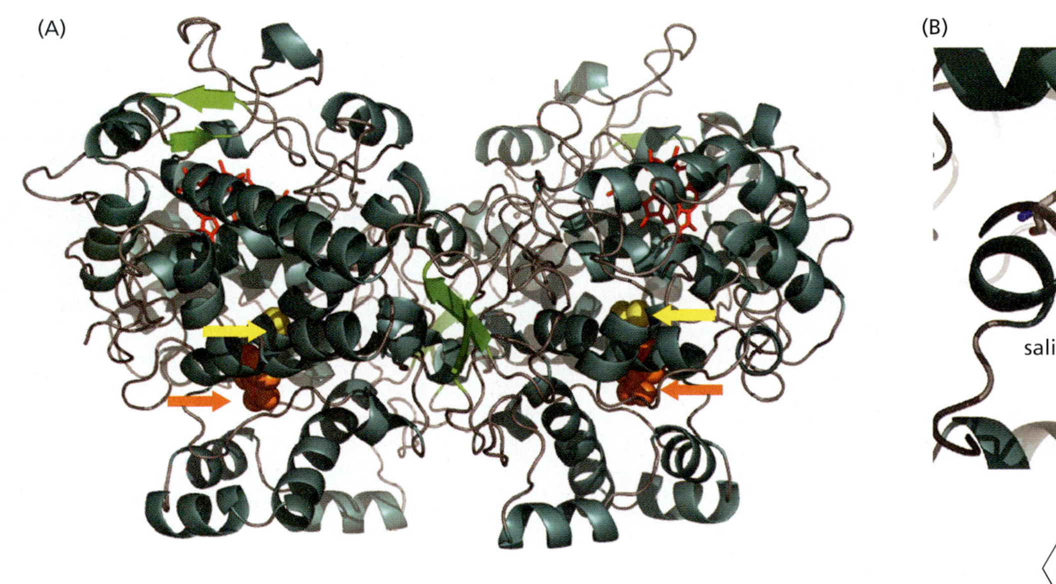

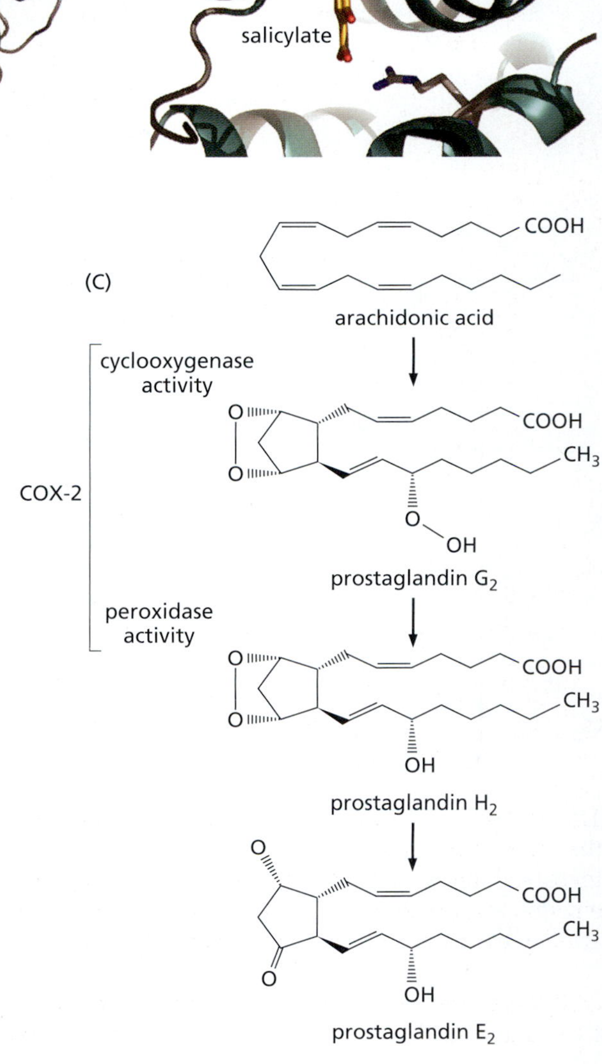

Figure 11.38 The actions of the COX-2 enzyme and their inhibition by aspirin The profound effects of aspirin in suppressing the incidence of a variety of commonly occurring cancers appears to be due to its ability to act on the catalytic activity of the cyclooxygenase-2 (COX-2) enzyme, which is sometimes called prostaglandin H_2 synthase. COX-2 is inducible in many tissues in response to inflammatory stimuli, unlike the very similar COX-1 enzyme, which is expressed constitutively. (A) COX-2 is a homodimeric enzyme. Its active sites that convert arachidonic acid to prostaglandin G_2 are largely inactivated by aspirin—acetylsalicylic acid—which is able to acetylate *(yellow balls, arrows)* serine 530 of COX-2, leaving the salicylate moiety of the aspirin molecule, which remains weakly bound in the catalytic cleft *(orange balls, arrows).* α helices *(teal),* loops *(gray),* β pleated sheets *(light green),* heme group *(red).* (B) In more detail, the actions of aspirin can be traced with great precision to the catalytic cleft of COX-2, in which the acetylation of serine 530 obstructs the enzymatic activity that is responsible for converting arachidonic acid to the precursor of all prostanoids, prostaglandin G_2; the latter is converted further to prostaglandin H_2 by a second site in the COX-2 enzyme. (C) The two distinct catalytic activities of COX-2, result in the conversion of arachidonic acid into prostaglandin H_2. The latter is then metabolized further to prostaglandin E2, which evokes many of the tumor-promoting responses described here. (A and B, courtesy of R.M. Garavito.)

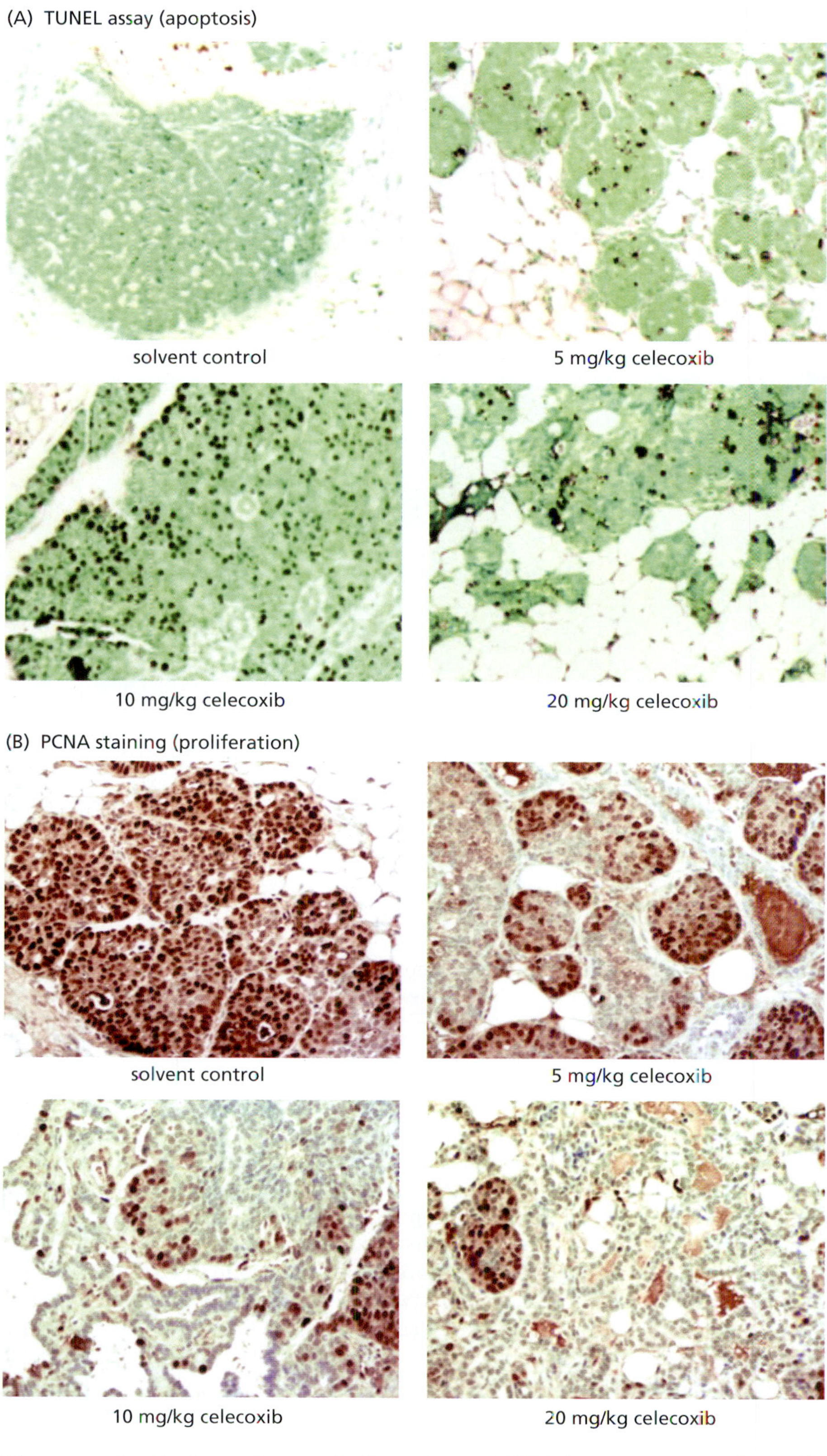

Figure 11.39 COX-2 inhibition in mammary carcinomas of tumor-prone transgenic mice In a transgenic mouse model of mammary carcinoma pathogenesis, the development of tumors, driven by a polyomavirus middle T oncoprotein, is strongly reduced by treatment with celecoxib, an NSAID that is a potent and selective inhibitor of cyclooxygenase-2 (COX-2). (A) As indicated here, increasing doses of this drug generate progressively higher numbers of tumor cells *(light green)* that stain with the TUNEL assay (*dark brown spots*); see Figure 3B ●), a marker of apoptosis. (B) Conversely, increasing concentrations of celecoxib lead to decreasing cell proliferation, as revealed by staining *(dark red)* for PCNA (proliferating cell nuclear antigen), a marker of cell-cycle progression. In all cases, the indicated doses of celecoxib are presented as milligrams of drug per kilogram of body weight. (From G.D. Basu, L.B. Pathangey, T.L. Tinder et al., *Mol. Cancer Res.* 2:632–642, 2004.)

The opposite effect was observed when elevated COX-2 expression was forced in the mammary tissue of transgenic mice: these mice showed significantly increased rates of breast cancers. An analogous experiment, in which this enzyme was driven by a mouse transgene causing its elevated expression in keratinocytes, led to skin hyperplasia and dysplasia.

COX-2 is expressed in the stromal compartment early in tumorigenesis in some tissues and in the epithelial compartment of others (Figure 11.40). In certain

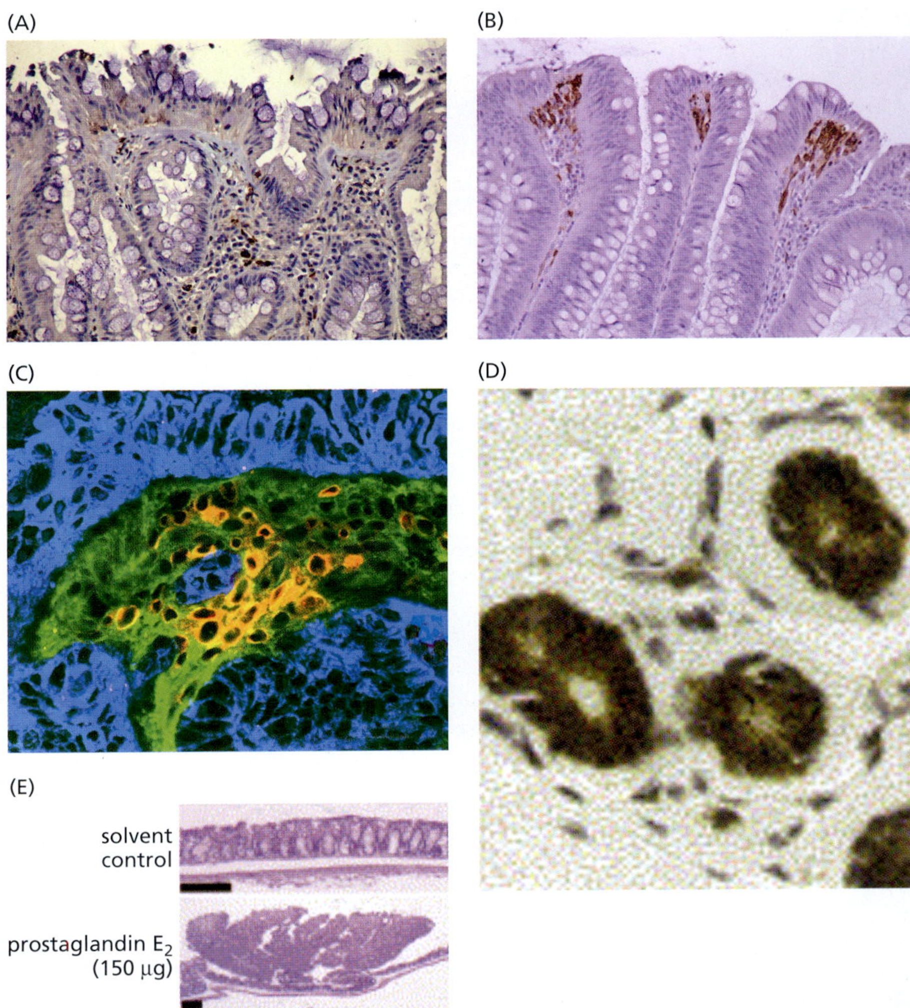

Figure 11.40 Expression of COX-2 and the effects of its PGE_2 downstream product The cyclooxygenase-2 (COX-2) enzyme is an important mediator of inflammation through its ability to convert arachidonic acid into prostaglandins, which yield, among other products, prostaglandin E_2 (PGE_2). Once produced, PGE_2 may act in a paracrine manner to induce pre-neoplastic changes in nearby cells, or in an autocrine manner on the cell that produced it. COX-2 function is likely be important in the early stages of carcinoma progression in a variety of tissue sites. However, its cell type–specific expression differs from one tissue to another. (A) In a human colorectal hyperplastic polyp, the COX-2 enzyme can be seen by immunostaining *(brown spots)* to be expressed by endothelial cells and macrophages. (B) In a more advanced adenomatous polyp, the expression is found to be expressed increasingly in myofibroblasts, another type of stromal cell *(brown spots).* (C) In this intestinal adenomatous polyp of a mouse that is genetically predisposed to polyposis because of a germ-line mutation in the *Apc* gene (Section 7.11), the epithelial cells are immunostained *blue* with an anti-cytokeratin antibody, and the stromal cells (85% fibroblasts, 5% endothelial cells) are stained *green* with an anti-vimentin antibody. Areas expressing both vimentin and COX-2 are stained yellow. (D) In the human breast, however, COX-2 expression *(dark brown immunostaining)* is found to be high in early pre-malignant growths in the epithelial cells rather than the surrounding stromal cells. (E) Feeding PGE_2 (a downstream product of COX-2 function) to mice of the *Apc mutant* strain causes a dramatic increase in the number of colonic polyps *(below)* compared with those seen in control *Apc mutant* mice *(above).* (A and B, from P.A. Adegboyega, O. Ololade, J. Saada et al., *Clin. Cancer Res.* 10:5870–5879, 2004; C, courtesy of M. Sonoshita and M.M. Taketo; D, from Y.G. Crawford, M.L. Gauthier, A. Joubel et al., *Cancer Cell* 5:263–273, 2004; E, from D. Wang, H. Wang, Q. Shi et al., *Cancer Cell* 6:285–295, 2004.)

epithelial tissues, COX-2 expression increases in the epithelial compartment as tumor progression advances until its expression is elevated tenfold or more above normal levels.

The COX-2 enzyme expressed by stromal and epithelial cells produces a series of prostaglandins from arachidonic acid. Prostaglandin E_2 (PGE_2) appears to be the most important of these (see Figure 11.38C). Application of this prostaglandin to normal intestinal epithelial cells in culture causes them to exhibit many of the traits associated with cell transformation—loss of contact inhibition, increased anchorage-independent growth, down-modulation of expression of the cell surface protein E-cadherin, reduced apoptosis, and increased rate of proliferation. Even more important, when mice that are prone to developing gastrointestinal polyps were fed PGE_2, the number of these growths in the colon increased dramatically (Figure 11.40E).

Since PGE_2 can diffuse from cell to cell, it is likely that early in tumor progression, COX-2 expression in stromal inflammatory cells results in release of this prostaglandin, which acts in a *paracrine* manner to induce the various transformation-associated traits in nearby epithelial cells (e.g., the enterocytes of the colon). However, as tumor progression advances, the rising level of COX-2 in the enterocytes enables them to make their own PGE_2, which stimulates their proliferation in an *autocrine* fashion and, once again, allows them to assume many of the traits associated with cell transformation.

Early-stage colonic polyps appear to depend on the continued inflammation induced by prostaglandins such as PGE_2 for their maintenance. However, later, as these growths evolve toward a higher degree of neoplasia, they no longer depend on prostaglandins for their maintenance and expansion. This echoes the experimental model of mouse skin carcinogenesis, in which early-stage papillomas depend on tumor promoters, such as TPA, for their maintenance, while more advanced papillomas become promoter-independent.

Taken together, these observations lead to an attractive model (Figure 11.41) of how inflammation in a variety of epithelial tissues functions as a tumor-promoting mechanism and ultimately leads to carcinomas. This scheme is likely to change, since many of the indicated steps continue to be subjects of intensive investigation. Perhaps the greatest uncertainty lies in the mechanism(s) used by NSAIDs to inhibit tumor progression: while COX-2 is clearly a key target of their action, it is likely that other cellular enzymes are also affected by these drugs, and that inhibition of these other NSAID targets also contributes to slowing down or blocking multi-step tumorigenesis.

The details of the scheme depicted in Figure 11.41 also provide us with another way of viewing tumor promotion. Many of the cell phenotypes conferred by the inflammation-associated prostaglandins are uncannily similar to those conferred by the oncogenes that we read about earlier. Included here are traits such as loss of contact inhibition, as well as a gain of anchorage-independent growth and the ability to proliferate more rapidly. In effect, tumor promotion can create a **phenocopy** of the actions of an oncogene, that is, a biological state that closely resembles one created by an oncogene but arises through very different mechanisms.

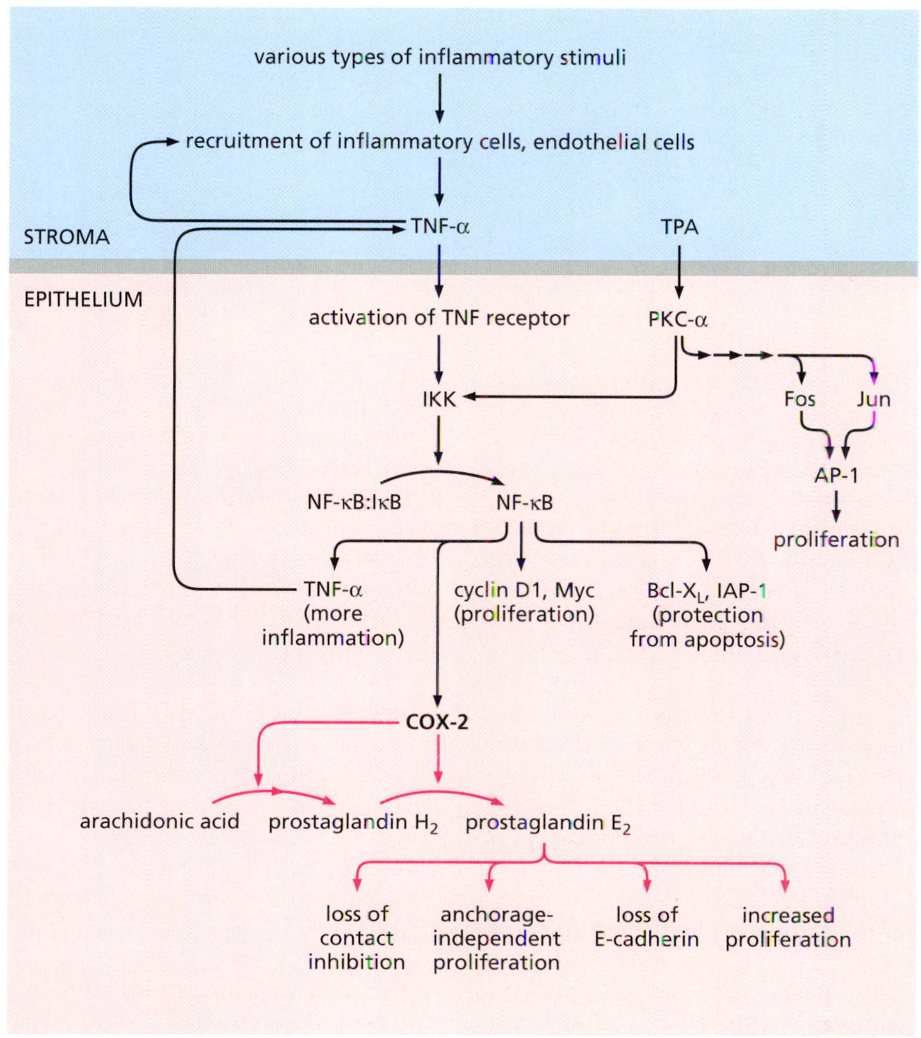

Figure 11.41 A model of epithelial inflammation and tumor promotion The information from skin and liver tumor promotion, as illustrated in Figures 11.31 and 11.37, can be integrated into a more general model. Inflammatory stimuli, including chronic infections, result in recruitment of inflammatory cells into areas of inflammation. These cells, as well as endothelial cells in the vicinity, release TNF-α, which activates the NF-κB pathway in epithelial cells, causing the production of more TNF-α, anti-apoptotic proteins (e.g., Bcl-X_L, IAP-1 and -2), and mitogenic proteins (Myc, cyclin D1), as well as COX-2. Prostaglandins synthesized by COX-2 *(pink arrows)* induce multiple cancer cell phenotypes, including loss of contact inhibition, acquisition of increased proliferation rate, and increased anchorage-independent growth. The TPA skin tumor promoter converges on this pathway through the activation of PKC-α; in addition, TPA-activated PKC-α can induce mitogenic signals by activating the Fos, Jun, and thus AP-1 transcription factors (Section 6.5). Together, these actions of inflammatory cells and of TPA create phenotypic states that closely resemble those induced by many oncogenes, enabling an initiated cell to launch a clonal expansion that eventually results in the acquisition of additional mutant alleles by its descendants.

This suggests that a tumor-promoting mechanism can collaborate with an oncogene in a way that resembles the collaboration between two oncogenes (Section 11.10). Thus, initiated tumor cells may rely on this form of collaboration until the time when their descendants acquire additional oncogenes and therefore no longer need to depend on the readily reversible effects of tumor promoters for their continued proliferation and survival.

11.17 Tumor promotion is likely to be a critical determinant of the rate of tumor progression in many human tissues

A wide variety of agents have, by now, been classified as human tumor promoters (Table 11.3). They share in common an ability to promote expansion of initiated clones. The key role of this promotion in multi-step tumorigenesis can best be understood in the context of the clonal succession models depicted in Figures 11.12, 11.17, and 11.18. In order for an initiated cell to acquire an additional mutation, its clonal descendants must become so numerous that a second, low-probability mutational event is likely to strike one or another cell of the now-expanded clonal population. Without such clonal expansion, the new, secondary mutation is unlikely to strike even a single descendant cell (because it is a rare event per cell generation), and tumor progression will halt.

Table 11.3 Known or suspected human tumor promoters and their sites of action

Agent or process	Cancer site
Hormones	
Estrogen	endometrium
Estrogen and progesterone	breast
Ovulation	ovary
Testosterone	prostate
Drugs	
Oral contraceptives, anabolic steroids	liver
Analgesics	renal pelvis
Diuretics	kidney
Infectious agents	
Hepatitis B/C viruses	liver
Schistosoma haematobium—blood fluke	bladder
Schistosoma japonicum—blood fluke	colon
Clonorchis sinensis—liver fluke	biliary tract
Helicobacter pylori—bacterium	stomach
Malarial parasites	B cell
Tuberculosis bacillus	lung
Chemical agents	
Betel nut, lime	oral cavity
Chewing tobacco	oral cavity
Bile	small intestine
Salt	stomach
Acid reflux	esophagus
Physical or mechanical trauma	
Asbestos	mesothelium, lung
Gallstones	gallbladder
Coarsely ground corn	stomach
Head injury	meninges
Chronic irritation/inflammation	
Tropical ulcers	skin
Chronic ulcerative colitis	colon
Chronic cystitis	bladder
Chronic pancreatitis	pancreas

Adapted in part from S. Preston-Martin, M.C. Pike, R.K. Ross et al., *Cancer Res.* 50:7415–7421, 1990.

Actually, these various mechanisms of tumor promotion can contribute to tumor progression in at least three ways. First, as just mentioned, promoters can stimulate the clonal expansions that yield the large cell populations in which otherwise improbable events become possible.

Second, since cell proliferation requires DNA replication, and since DNA replication generates miscopied and thus mutant DNA sequences at a low but significant frequency, agents favoring cell proliferation are indirectly mutagenic. And repeated cell divisions hold yet other dangers for cellular genomes, since mitotic recombination and faulty chromosomal segregation yield loss-of-heterozygosity (LOH) events that enable mutant tumor suppressor genes to contribute to tumor progression.

Third, as was discussed in the last chapter, repeated cycles of growth and division lead to progressive shortening of telomeric DNA in stem cells. Ultimately, the ensuing telomere collapse and breakage–fusion–bridge (BFB) cycles result in karyotypic disarray and hence mutagenesis. And as described in Section 10.10, BFB cycles occurring in the absence of functional p53 are likely to lead directly to increases in cancer incidence.

In fact, one type of tumor promotion—that involving inflammation—is likely to have an additional effect on tumor progression: the inflammatory cells that are recruited into a tissue often release reactive oxygen species (ROS); these may attack and mutate the DNA of nearby epithelial cells, thereby adding to the mutagenic effects generated by these cells' own endogenous processes.

When taken together, these diverse observations indicate that the carcinogen = mutagen equation, which we previously distilled from the work of Bruce Ames (Chapter 2), is simplistic. It is now obvious that many carcinogens succeed in doing their work through means that do not depend on an ability to directly damage DNA. As mentioned, such nonmutagenic agents cannot be detected by the Ames test. This means that a positive result in the Ames test, which demonstrates the genotoxicity of a chemical agent, is likely (but not guaranteed) to predict its carcinogenic powers in rodents and humans. However, a negative result in this test does not exclude the possibility that the compound being tested can contribute to the formation of human cancers by acting as a nongenotoxic tumor promoter.

Some carcinogenic agents, such as the benzo[*a*]pyrene (BP) described earlier, when applied to laboratory animals as single agents, are able on their own to induce tumor formation after repeated applications. They thereby function as both initiators and promoters of tumorigenesis and consequently have been termed **complete carcinogens**. We can imagine, for example, that certain mutagenic agents are also cytotoxic at high concentrations and thus able to act as tumor promoters by killing cells and inducing compensatory proliferation of surviving cells in a target tissue. But such agents are likely to play minor roles in human carcinogenesis. Most genotoxic carcinogens enter into human tissues at concentrations that are far too low to evoke cytotoxic effects. Hence, the tumors incited by these agents seem to depend heavily for their formation on other substances that function as pure tumor promoters.

11.18 Synopsis and prospects

We began this chapter with the mindset that genetic changes are responsible for fueling many of the steps of cancer progression and that each of these mutations is, in principle, traceable to the actions of specific mutagenic agents. As our discussions of multi-step tumorigenesis proceeded, however, it became increasingly clear that agents other than mutagens contribute to the pathogenesis of human cancer. More specifically, promoters seem to play an equally weighty role in driving forward the development of tumors, including many, if not all, human cancers. This realization both clarifies and confounds our attempts to ferret out the root causes of these malignancies.

It is possible, for example, that most human carcinogens act as tumor promoters (rather than initiators), and that many types of human tumors arise entirely without the contributions of exogenous genotoxic agents, that is, mutagens originating from outside the human body. In such tumors, the genetic damage would, by necessity, be generated entirely by endogenous processes.

This speculation comes from two sources. First, the origins of most human cancers have not been associated with exposure to specific mutagenic agents. The major exceptions here are the combustion products of tobacco and the products of cooking meat at high temperature, as we will see in Chapter 12.

Second, the genome within a single human cell sustains as many as 10,000 chemical modifications each day, according to some estimates. These modifications are created by chemical species that have been generated endogenously by various metabolic reactions. Notable among these endogenous mutagens are several types of reactive oxygen species (ROS) that arise as by-products of oxidative metabolism in the mitochondria.

Clearly, the great majority of these chemical lesions are removed by the highly effective DNA repair systems operating in most types of cells (Chapter 12). Some of these lesions, however, escape detection and subsequent repair, become fixed in the genomes of the cells in which they initially were formed, and are then transmitted as mutations to descendant cells. In most human tissues, the number of endogenously generated mutations is likely to dwarf those of exogenous origin.

This information allows us to entertain the following scenario: Endogenously generated mutant alleles, acting in concert with tumor promoters of exogenous origin, drive the progression that leads to the appearance of many kinds of human cancers. These exogenous promoters include foodstuffs, infections, and even tobacco smoke, which contains many tumor-promoting chemicals. If so, searches for the exogenous mutagenic carcinogens that are suspected to be etiologic (causative) agents of most human tumors are doomed to failure.

The existence of such tumor-promoting substances, whatever their nature, greatly complicates attempts to devise laboratory-based screens for human carcinogens. On the one hand, the frequently used *in vitro* tests for mutagenicity (Section 2.9), such as the Ames test, lead one astray because they do not register the presence of tumor promoters. On the other, if used incautiously, certain candidate human carcinogens being screened in the laboratory may register as tumor promoters even if this is not their normal role (Sidebar 11.15).

These arguments indicate that arcane debates about cancer initiation versus promotion have enormous public health implications. And misinterpretations of biological processes such as tumor promotion lead, in turn, to major changes in the substances that we consume and those that we have been told to avoid (see, for example, Sidebar 11.16).

The recognition that chronic inflammation (Table 11.4) and infections are two important sources of tumor promotion also has profound implications for reducing cancer incidence. Inflammation can be suppressed using anti-inflammatory drugs, such as NSAIDs, and there is great hope that some of these will be developed in the coming years that can be taken chronically without unacceptable side effects.

On a worldwide scale, it is estimated that 9% of all cancer deaths arise because chronic infections of the stomach by the bacterium *Helicobacter pylori* lead to gastric carcinomas, and that 6% of all cancer deaths are associated with liver cancers, most of which are caused by chronic infections by hepatitis B and hepatitis C viruses. The chronic infection of the cervical epithelium caused by human papillomaviruses, notably strains 16, 18, and 45, leads in some women to the cervical carcinomas that account for 5% of worldwide cancer mortality. (In the latter case, the chronic inflammatory effects of a viral infection are compounded by the two HPV-borne oncogenes, *E6* and *E7*.) While these figures are daunting, in fact they represent an enormous opportunity for reducing worldwide cancer incidence and mortality: infectious diseases are ultimately far easier to control and prevent (through immunization) than the effects of foodstuffs

Sidebar 11.15 Are rodent carcinogen tests reliable indicators of danger to humans? Bruce Ames, whose *Salmonella* test has had such a profound effect on our testing of potential human carcinogens (Section 2.9), has questioned the meaningfulness of many of the laboratory tests currently performed to assess the potential carcinogenicity of various chemical compounds. These tests involve exposing rats and mice to high doses of a test compound for an extended period of time, followed by histopathological surveys to detect tumors that may have arisen in the exposed animals.

Many of these rodent tests are conducted at doses that are close to or at the maximum tolerated doses (MTDs) of these compounds; above these doses, the test compounds begin to induce obvious damage in various organ systems of the exposed animals and therefore generate obvious effects on the animals' health. [These MTD tests usually involve exposure to concentrations of agents that are orders of magnitude higher than those experienced by human populations. For example, human exposure to heterocyclic amines, a product of cooking meat at high temperature (see Section 12.6), has been estimated to be under 1 μg per kg body weight per day, although most laboratory animal studies have been conducted at doses in excess of 10 mg/kg body weight per day—more than 10 thousand times higher. The motive for testing at these high dose levels is a desire to detect biological effects in a population of several hundred laboratory animals that might normally only be observed in several individuals among a human population size of one million.]

Ames argues that a compound tested at the MTD may well be exerting its effect by killing cells in specific tissues, thereby provoking compensatory proliferation of the surviving cells in those tissues. Hence, a compound introduced into a rodent at concentrations near the MTD may be acting as a tissue-specific mitogen and thus as a tumor promoter, whether or not it also has mutagenic powers (Section 11.13). Alternatively, certain chemicals tested at the MTD may favor inflammation, once again fostering tumor progression.

The toxicity and secondary mitogenic effects of compounds, both natural and synthetic, are usually apparent only above a certain threshold dose, below which these effects are not detectable. Thus, unlike mutagens, whose powers for creating mutations in the genome are likely to be a linear function of dosage, compounds that act like tumor promoters (including cytotoxic agents) generally have nonlinear dose response curves (Figure 11.42). If actual human exposures to a compound being tested involve effective doses that are thousands of times lower than the MTD of this agent, and if the carcinogenic effect being scored in rodents derives largely from tumor promotion (because of toxicity at the MTD), then the outcomes of such high-dose rodent tests may have no implications whatsoever for human health risks. Ames's critique of MTD carcinogen testing remains hotly debated, and the issue is largely unresolved.

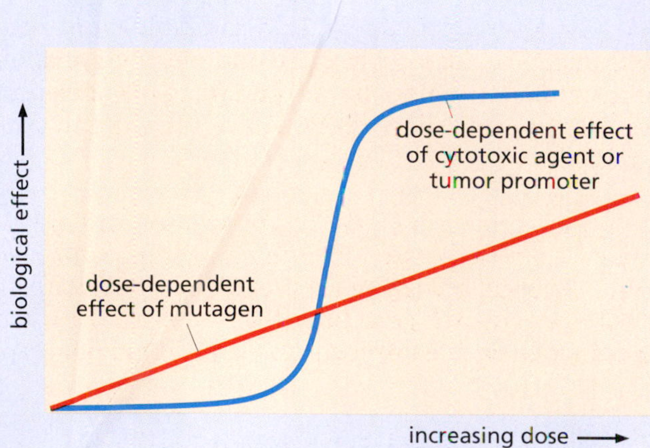

Figure 11.42 Dose response curves of genotoxic and nongenotoxic agents Numerous studies have shown that the mutational burden inflicted by mutagens is, up to a certain concentration, linearly proportional to the cumulative dose of administered mutagen *(red line)*. The risk of cancer, in turn, is likely to be linearly proportional to the burden of inflicted mutations. For example, 1% of a full dose of a mutagenic carcinogen is likely to yield 1% as much risk of carcinogenesis. In contrast, many tumor-promoting agents, including cytotoxic agents, operate differently *(blue curve)*. Because their effects depend on the binding affinities of promoters to protein targets within cells, the biological effects that are elicited by promoters usually have a sigmoid curve of the sort indicated. In this type of dose response, once the agent falls significantly below a certain threshold, it has virtually no biological effect. Conversely, once it rises above a certain threshold, its biological effects are maximal and cannot increase further. Accordingly, 1% of the dose of such a promoting agent may elicit far less than 1% of a biological effect.

Sidebar 11.16 Does saccharin cause cancer? A scientific controversy has roiled the community of chemical carcinogenesis researchers for almost four decades. In 1970, researchers discovered that saccharin pellets implanted in the bladders of male rats resulted in a markedly increased rate of bladder cancer. This resulted in a widespread fear that the millions of obese and diabetic individuals who were using this artificial sweetener were being exposed to a carcinogen in their daily diet.

As it turned out, the evidence pointing to saccharin's carcinogenicity was most peculiar. It never registered as a mutagen in the Ames test. Indeed, since the saccharin ion present in solution is negatively charged, it has little affinity for DNA. Moreover, the carcinogenic effects of the implanted pellets were seen long after the saccharin had been leached out, leaving behind the vehicle used to construct the pellets—cholesterol. This suggested that mechanical irritation of the wall of the bladder created by these pellets contributed to the observed tumorigenesis. Tumors were observed only when saccharin constituted 2.5% or 5.0% of the total diet of male rats (this being a dose equivalent to that deriving from a human's drinking 750 cups of coffee each day, each with a saccharin pill in it). Bladder cancer was not observed in any other mammalian species exposed to saccharin, including monkeys who consumed substantial doses for more than 20 years. Mice exposed to a diet of 7.5% saccharin showed no sign of bladder cancer.

Male (but not female) rats excrete into their urine large amounts of a protein called "major urinary protein" (MUP); the concentration of MUP in rat urine has been estimated to be 100 to 1000 times higher than in human urine. MUP, acting together with saccharin ions, triggered the formation of co-precipitates and co-crystals, which accumulated in high concentrations in the male rat bladder. (These crystals were also formed by salts of other ionized acids, which were also effective bladder tumor promoters in male rats.) These precipitates acted to irritate and inflame cells of the male rat urothelium—the specialized epithelium lining the bladder—yielding a tumor-promoting effect.

Experiments like these illustrate the perils of carcinogen testing in rodent models. It is a widespread but still not universal consensus in the cancer research community that saccharin is totally harmless. Thirty years after this controversy began, there was still no epidemiologic evidence that the vast numbers of diabetics experiencing chronic exposure to saccharin over many years' time suffered any increased risk of bladder carcinomas. (The sweetener remains banned in Canada and California!) So, tumor promoters work in strange and complex ways, and interspecies comparisons designed to detect them are fraught with danger.

that we ingest and the tobacco products that we inhale. By some estimates, 450 million people worldwide are infected chronically with HBV, while 200 million are long-term HCV carriers. These numbers indicate the enormous benefit that may derive from preventing these viral illnesses.

Inflammatory processes are also implicated in the promotion of a number of nonviral human cancer types. In these cases, opportunities for the development of future anti-cancer drugs derive from the recent advances in elucidating the biochemical details of inflammation and its specific contributions to tumor promotion. Cyclooxygenase-2 (COX-2) is clearly at the center of this process, and a number of anti-inflammatory agents (i.e., NSAIDs) targeting this enzyme are known to be effective in decreasing the incidence of various types of cancer. Unfortunately, however, long-term use of some of these agents

Table 11.4 Links between inflammation and cancer pathogenesis

Many inflammatory conditions predispose to cancer
Cancers arise at sites of chronic inflammation
Functional polymorphisms of cytokine genes are associated with cancer susceptibility and severity
Distinct populations of inflammatory cells are detected in many cancers
Extent of tumor-associated macrophage infiltrate correlates with prognosis
Inflammatory cytokines are detected in many cancers; high levels are associated with poor prognosis
Chemokines are detected in many cancers; they are associated with inflammatory infiltrate and cell motility
Deletion of cytokines and chemokines protects against carcinogens, experimental metastases, and lymphoproliferative syndrome
Inflammatory cytokines are implicated in the action of nongenotoxic liver carcinogens
The inflammatory cytokine tumor necrosis factor is directly transforming *in vitro*
Long-term NSAID use decreases mortality from colorectal cancer

Courtesy of F. Balkwill; from F. Balkwill and A. Mantovani, *Lancet* 357:539–545, 2001.

induces unacceptable side effects in some individuals, including fatal cardiovascular complications. Presumably these side effects are due to the fact that COX-2 generates a number of distinct prostaglandins, and each of these, in turn, elicits multiple downstream cellular responses.

Nonetheless, the prospects are bright for the development of potent anti-cancer agents having **prophylactic** (preventative) and therapeutic activity that affect the COX-2 pathway but do not create the side effects associated with long-term NSAID use. Thus, the key product of COX-2 action in tumor promotion, prostaglandin E_2 (PGE_2), is known to bind and activate at least eight different cell surface serpentine receptors (see Sections 5.7 and 6.11), each of which presumably evokes its own subset of downstream responses. Accordingly, the development of highly specific drugs designed to inhibit only one or another of these receptors is likely to yield agents having significant effects in reducing inflammation-associated tumor promotion while having few of the side effects associated with the broadly acting COX-2 antagonists.

Estrogen, progesterone, and androgens loom large in most discussions about human cancer because they play such critical roles in the pathogenesis of breast, endometrial, and prostatic carcinomas. Together, these tumors account for about 9% of all mortality from neoplasias in the West. These hormones act to stimulate the proliferation of epithelial cells in the corresponding responsive tissues, and so we can easily incorporate them into our conceptual schemes of tumor initiation and promotion. Another systemic hormone—IGF-1 (insulin-like growth factor-1)—also seems to function as an important tumor promoter (Sidebar 13), because it protects pre-malignant cells from apoptosis and may also spur their proliferation.

Significantly, these hormones continue to play important roles in tumorigenesis once multi-step tumor progression has yielded full-blown malignancies. The proliferation of cancer cells in many established breast tumors and almost all prostatic tumors depends upon the continued presence of estrogen and androgens, respectively. The continued presence of IGF-1 also seems to be required by many already-formed tumors. This ongoing dependence highlights a more general question that remains relatively unexplored by cancer researchers: To what extent do a variety of malignancies, once they are formed, continue to depend on the tumor promoters that helped to create them in the first place?

We can also pose a different but related question: Do full-blown tumors continue to depend upon the mutant alleles that were created decades earlier by initiating mutagens? Or do these early mutations, which occur as the initial steps of tumor progression, become irrelevant later, when subsequently acquired mutant alleles take over the job of programming cancer cell proliferation?

The most direct answers to this question come from experiments with mouse models of cancer development, in particular transgenic mice in which certain oncogenes are activated on a tissue-specific basis; these gene activations lead eventually to tumors in those tissues. In some of these mouse models, the initiating transgenes, such as *myc* and *ras*, have been inactivated by a variety of experimental tricks long after substantial tumors have formed. The effects on tumor growth that have been observed to date are conflicting.

Usually, these tumors collapse rapidly when deprived of the oncogenes that led originally to their formation. Hence, in these cases, the mutant oncogenic alleles were important both for the *initiation* of the cancerous state as well as the *maintenance* of this state long afterwards. Stated differently, the mutant alleles that are accumulated in later steps of multi-step tumorigenesis do not render the earlier ones unnecessary. However, in several research reports, shutdown of the initiating *neu*, *myc* or *wnt* oncogenes after tumors had formed led, after short remissions, to the regrowth of these tumors. In these cases, the initiating oncogenes were no longer required for the expansion of the relapsing tumors.

We can assume, for the moment, that these various mouse models of cancer formation reflect mechanisms operating in human tumors as well. If so, these observed effects of shutting down initiating oncogenes hold profound implications for the development of new types of anti-cancer therapeutics, a topic that we will return to in Chapter 16. For example, if a mutant *ras* allele is found in a human tumor cell genome, can this tumor be cured by drugs targeted to the Ras oncoprotein, or has this oncoprotein, which may have played a key role in the initiation of tumor formation, become irrelevant, later on, to the continued proliferation and survival of the cells in this tumor?

The vast numbers of genetic alterations present in human cancer cell genomes greatly complicate one major goal of current cancer research: descriptions of the "genetic biographies" of various types of human tumors—the successions of genetic alterations that program the neoplastic phenotype of the cells within these tumors. The number of altered DNA sequences in the genomes of most kinds of human cancer cells clearly dwarfs the small number of genetic alterations (and promoter methylations) that play causal roles by driving tumor progression forward. At present, we possess only very crude tools for distinguishing the wheat from the chaff—the small number of biologically important genetic changes from the large throng of functionally irrelevant genetic changes that are present in the genomes of almost all human cancer cells.

In spite of these truly daunting problems, there is optimism that we may soon be able to formulate some basic organizing principles that place all types of human tumors under a common conceptual roof. Once these principles are well established, we should be able to use them to explain why various human tumors acquire certain combinations of mutant (and methylated) genes. Specifically, examinations of the biological phenotypes of a variety of human tumor types have led to the proposition that all highly advanced human cancer cells share in common a number of essential attributes that they have acquired en route to the malignant state. These are (1) a reduced dependence on exogenous mitogenic growth factors (Chapters 5 and 6); (2) an acquired resistance to growth-inhibitory signals, such as those conveyed by TGF-β (Chapter 8); (3) the ability to multiply indefinitely, that is, immortalized cell proliferation (Chapter 10); (4) a reduced susceptibility to apoptosis (Chapter 9); (5) the ability to generate new blood vessels—angiogenesis (Chapter 13); (6) the acquisition of invasiveness and metastatic ability (Chapter 14); and a seventh that remains less well documented: (7) the ability to evade elimination by the immune system (Chapter 15).

Actually, some of the observations described in this chapter argue strongly for yet another shared attribute. Our discussion of clonal evolution and succession concluded that in order for these events to occur at reasonable frequency, mutation rates within tumor stem cell populations must be abnormally high. In the absence of such increased mutability, clones of cells may not accumulate the multiple genetic alterations during a human life span that are required for cancer formation. Together, these speculations and observations converge on an eighth attribute that may well be shared in common by almost all types of human cancer cell genomes: (8) acquisition of genomic instability (Chapter 12).

If we embrace this list, at least tentatively, we need to relate the acquisition of each of these traits to the specific steps of multi-step tumor progression. For example, we might assume that each step in the progression of a human tumor is demarcated by the alteration of a distinct gene, achieved either through genetic mutation or epigenetic methylation. Moreover, we could imagine that each of the steps in tumor progression yields one of the seven or eight attributes listed above. This would create a one gene–one phenotype scenario that would greatly simplify our thinking about human tumor progression.

While attractive, such thinking is simplistic, since there is no one-to-one mapping between specific gene alterations and corresponding changes in cell phenotype. Instead, many of the phenotypes of cancer cells are achieved collaboratively through the actions of several genes or genetic alterations (Figure 11.43). For example, the acquired resistance to apoptosis shown by human cancer cells often results from the activation of the Ras signaling pathways (which energizes the anti-apoptotic PKB/Akt kinase) as well as the inactivation of the p53 signaling pathway.

Conversely, the mutation of certain pleiotropically acting cancer genes may confer several distinct phenotypic benefits simultaneously. Thus, the formation of a *myc* oncogene may concomitantly deregulate the pRb signaling pathway (which normally enables cells to respond to growth-inhibitory signals), aid in the de-repression of the *hTERT* gene (which enables cell immortalization), and reduce the mitogen dependence of a cell. So the hope of a simple scheme is frustrated by the biological realities of how each of these genes and encoded proteins actually operates.

Knowing this complexity, we might still wish to attempt another type of mapping. We could imagine that each of the biological phenotypes of cancer cells is the direct result of an alteration in one of the regulatory subcircuits that govern the life of the cell (Figure 11.44). If so, perhaps we can explain tumor progression as the progressive deregulation of a number of distinct subcircuits within the cell.

An examination of Figure 11.44 gives us some encouragement. For example, the acquisition of mitogen independence is achieved largely by the activation of the receptor tyrosine kinase (RTK) → Ras → MAPK pathway, while the resistance to apoptosis is acquired through lesions in the subcircuit that governs programmed cell death. Yet other such acquired attributes can also be related to distinct portions of this large circuit diagram. Still, even here there is no neat and simple compartmentalization, because there are numerous cross connections between the various subcircuits operating within cells. In addition, several subcircuits often collaborate to create a distinct cancer cell phenotype.

Nonetheless, with all these reservations in mind, Figure 11.44 and others like it inspire a hope that may well be realized sometime during the first decades of the new millennium: At some point, we will truly understand in detail how each of these subcircuits operates to regulate cell phenotype. We will be able to model the operations of each mathematically. And we will be able to rationalize the behavior of the cancer cell as a whole in terms of the interactions of specific molecular defects in each of these regulatory circuits.

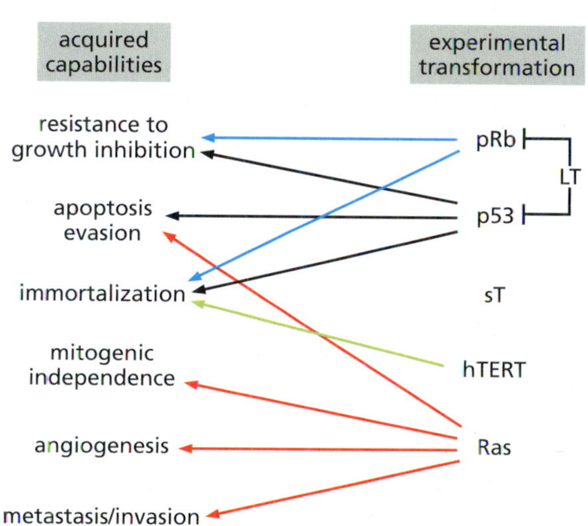

Figure 11.43 Cancer cell genotypes versus phenotypes During the course of tumor progression, human cells acquire a number of distinct cancer-associated phenotypes. Independent of this, laboratory experiments reveal that a number of distinct genes must be introduced into human cells in order to transform them to a tumorigenic state. This raises the question how each of these introduced genes contributes to the cell phenotypes associated with tumorigenicity. As indicated here, genes such as that encoding hTERT, the catalytic subunit of telomerase, affect only the phenotype of immortalization, while other genes, such as the gene encoding p53, affect at least three distinct phenotypes—resistance to growth inhibition, evasion of apoptosis, and immortalization. The most widely acting protein is likely to be Ras, which affects susceptibility to apoptosis, dependence on exogenous mitogens, angiogenesis, and invasiveness/metastatic ability. Hence, a one-to-one mapping between genes and cancer-associated phenotypes is not possible.

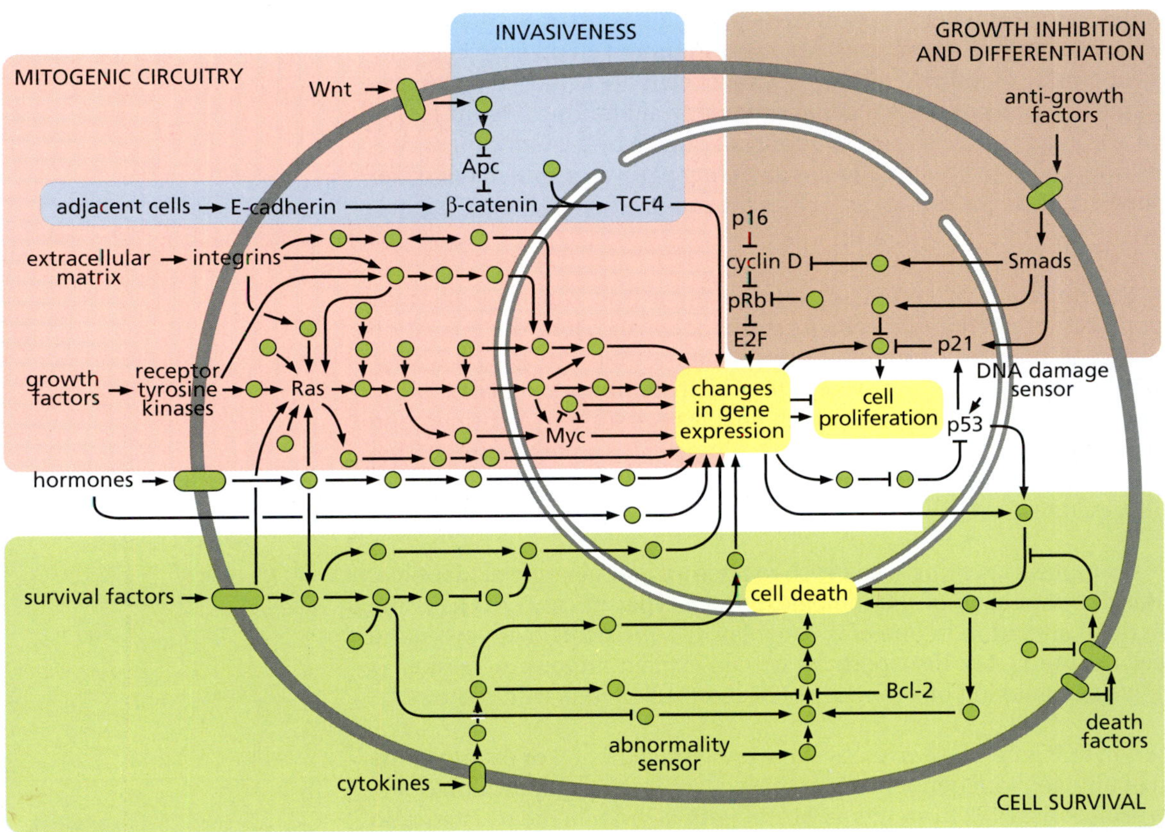

Figure 11.44 The intracellular signaling circuitry and collaboration between cancer-associated genes The design of the signal-transduction circuitry of mammalian cells has been uncovered, piece by piece, over the past quarter century. The diagram here indicates only a portion of the proteins that play critical roles in modulating the flow of signals through the various circuits operating within our cells. As indicated here by the various shadings, different subcircuits are involved in regulating distinct cell physiologic processes. Thus, the growth-promoting, mitogenic circuit *(light red),* the circuit governing growth-inhibitory signals *(light brown),* the circuit governing apoptosis *(light green),* and the circuit governing invasiveness and metastasis *(light blue)* can be assigned to distinct regions in the map of the master circuitry of the mammalian cell. (Note that the circuit governing mitogenesis overlaps in part with that governing cancer cell invasiveness, indicative of the fact that a common set of proteins mediates both biological responses.) Presumably, mathematical modeling of these various circuits will one day be able provide a mechanistic rationale of why tumorigenesis is a multi-step process in mammals. (Adapted from D. Hanahan and R.A. Weinberg, *Cell* 100:57–70, 2000.)

We close this chapter by recalling the eighth attribute that was ascribed to cancer cells—their acquisition of mutable genomes. Given the low probability of each individual step of tumor progression, completion of the process as a whole becomes extremely improbable, yet cancers occur with substantial frequency in the human population. The next chapter is focused on the attempts to resolve this dilemma and reveal how a mathematically impossible disease process becomes, most unfortunately, quite commonplace and the cause of 20% of human mortality.

Key concepts

- The process of tumor formation is a complex one of multiple steps involving multiple alterations of cells and their physiologic control mechanisms.
- The complexity of this process is reflected in the long time periods required for most human cancers to develop.
- These changes involve both the activation of oncogenes and the inactivation of tumor suppressor genes.

- The number of steps required to experimentally transform human cells is larger than is needed to transform cells of laboratory mice.
- These alterations affect multiple distinct regulatory circuits within cells and function in a complementary fashion to create the neoplastic cell phenotype
- Some of these changes occur as the direct result of the actions of exogenous mutagens and exposure to such mutagens may represent a "rate-limiting" determinant of tumor progression.
- In many instances, however, the rate of tumor progression may be governed by the actions of nonmutagenic promoting agents, which may determine the rate of expansion of mutant cell clones.
- In many human cancers, these critical nonmutagenic, tumor-promoting stimuli include chronic mitogenic stimulation and inflammation.
- The multiplicity of steps required for human cancers to arise is not known, in part because certain changes may occur rapidly and therefore may not be "rate-limiting," while others may require a decade or more to complete.
- Multi-step tumor progression can be depicted as a form of Darwinian evolution occurring within tissues. However, because some of the critical changes occurring during tumorigenesis are epigenetic, and because the rate of genetic diversification can occur very rapidly, the classic depictions of Darwinian evolution must be modified.
- In most, but not all, transgenic models of tumorigenesis, the initiating changes continue to be required for tumor progression, long after this process has reached completion.
- The number of genetic changes found in the genomes of human cancer cells vastly exceeds the number required for tumorigenesis to reach completion, complicating identification of the critical changes that are causally important in tumor formation.
- The discovery of cancer stem cells greatly changes our concepts about the mechanisms of multi-step tumorigenesis, since these self-renewing cells, rather than the bulk populations of cancer cells, may be the objects of genetic alteration and clonal selection.

Thought questions

1. Knowing the various genetic and epigenetic changes that occur during multi-step tumorigenesis, which of these would you say are likely to be readily uncovered and which may be difficult to identify? Describe the reasons for these assignments.
2. Some tumor suppressor genes inactivated during multi-step tumorigenesis may be readily identified because of LOH in the chromosomal region carrying them, while others may be difficult to identify in this way. Describe the factors that allow or complicate this identification.
3. What arguments favor the notion that all of us carry myriad clones of initiated pre-malignant cells throughout the body?
4. What arguments can be mustered that favor the notion that the bulk of human carcinogens act as promoters rather than initiators of tumorigenesis?
5. What different approaches can be used to estimate the number of steps in multi-step tumor progression, and how is each of these approaches flawed?
6. How does the current available information about multi-step tumor progression provide insights into strategies for the prevention of clinically detectable cancers?
7. What mechanisms enable chronic viral infections to exert a carcinogenic influence on a tissue?
8. Describe the various mechanisms of tumor promotion and the features that they share in common and those that distinguish them from one another.

Additional reading

Adams JM & Cory S (1992) Oncogene co-operation in leukaemogenesis. *Cancer Surveys* 15, *Oncogenes in the Development of Leukaemia*, 119–141.

Aggarwal BB (2004) Nuclear factor-κB. The enemy within. *Cancer Cell* 6, 203–208.

Al-Hajj M, Wicha MS, Benito-Hernandez A et al. (2003) Prospective identification of tumorigenic breast cancer cells. *Proc. Natl. Acad. Sci. USA* 100, 3983–3988.

Al-Hajj M & Clarke ME (2004) Self-renewal and solid tumor stem cells. *Oncogene* 23:7274–7282.

Armitage P & Doll R (1954) The age distribution of cancer and a multistage theory of carcinogenesis. *Brit. J. Cancer* 8, 1–12.

Balkwill F (2004) Cancer and the chemokine network. *Nat. Rev. Cancer* 4, 540–550.

Boland CR & Ricciardiello L (1999) How many mutations does it take to make a tumor? *Proc. Natl. Acad. Sci. USA* 96, 14675–14677.

Braakhuis BJM, Tabor MP, Kummer JA et al (2003) A genetic explanation of Slaughter's concept of field cancerization: evidence and clinical implications. *Cancer Res.* 63, 1727–1730.

Buendia MA (1992) Hepatitis B virus and hepatocellular carcinoma. *Adv. Cancer Res.* 59, 167–226.

Cairns J (1975) Mutation, selection and the natural history of cancer. *Nature* 255, 197–200.

Clevers H (2004) At the crossroads of inflammation and cancer. *Cell* 118,671–674.

Cohen SM & Ellwein LB (1990) Cell proliferation in carcinogenesis. *Science* 249, 1007–1011.

Coussens LM & Werb Z (2001) Inflammatory cells and cancer: think different! *J. Exp. Med.* 193, F23–F26.

Coussens LM & Werb Z (2002) Inflammation and cancer. *Nature* 420, 860–867.

Dick JE (2003) Breast cancer stem cells revealed. *Proc. Natl. Acad. Sci. USA* 100, 3547–3549.

Glick AB & Yuspa SH (2005) Tissue homeostasis and the control of the neoplastic phenotype in epithelial cancers. *Semin. Cancer Biol.* 15, 75–83.

Gold LS, Ames BN & Slone TH (2002) Misconceptions about the causes of cancer. In Human and Environmental Risk Assessment: Theory and Practice (D Paustenbach ed), pp 1415–1460. New York: John Wiley & Sons.

Grady WM & Markowitz SD (2002) Genetic and epigenetic alterations in colon cancer. *Ann. Rev. Genomics Hum. Genet.* 3, 101–128.

Gupta RA & DuBois RN (2001) Colorectal cancer prevention and treatment by inhibition of cyclooxygenase-2. *Nat. Rev. Cancer* 1, 11–19.

Hahn WC & Weinberg RA (2002) Rules for making human tumor cells. *N. Engl. J. Med.* 347, 1593–1603.

Hahn WC & Weinberg RA (2002) Modelling the molecular circuitry of cancer. *Nat. Rev. Cancer* 2, 331–341.

Hanahan D (1989) Transgenic mice as probes into complex systems. *Science* 246, 1265–1275.

Hanahan D & Weinberg RA (2000) The hallmarks of cancer. *Cell* 100, 57–70.

Hansel DE, Kern SE & Hruban RH (2003) Molecular pathogenesis of pancreatic cancer. *Ann. Rev. Genomics Hum. Genet.* 4, 237–256.

Henderson BE & Feigelson HS (2000) Hormonal carcinogenesis. *Carcinogenesis* 21, 427–433.

Hunter T (1991) Cooperation between oncogenes. *Cell* 64, 249–270.

Huntly BJP & Gilliland DG (2004) Blasts from the past: new lessons in stem cell biology from chronic myelogenous leukemia. *Cancer Cell* 6, 199–201.

Hussain SP, Hofseth LJ & Harris CC (2003) Radical causes of cancer. *Nat. Rev. Cancer* 3, 276–285.

Janssens V, Goris J & Van Hoof C (2005) PP2A: the expected tumor suppressor. *Curr. Opin. Genet. Dev.* 15, 34–41.

Kelly LM & Gilliland DG 2002 Genetics of myeloid leukemias *Ann. Rev. Genomics Hum. Genet.* 3, 179–198.

Kern S , Hruban RH, Hidalgo M & Yeo CJ (2002) An introduction to pancreatic adenocarcinoma genetics, pathology and therapy. *Cancer Biol. Therapy* 1, 607–613.

Knudson AG (2001) Two genetic hits (more or less) to cancer. *Nat. Rev. Cancer* 1, 157–162.

Marx J. (2003) Mutant stem cells may seed cancer. *Science* 301, 1308–1310.

Miller EC (1978) Some current perspectives on chemical carcinogenesis in humans and experimental animals: presidential address. *Cancer Res.* 38, 1479–1496.

Moore MA & Tsuda H (1998) Chronically elevated proliferation as a risk factor for neoplasia. *Eur. J. Cancer Prev.* 7, 353–385.

Nowell PC (1976). The clonal evolution of tumor cell populations. *Science* 194, 23–28.

Pardal R, Clarke MF & Morrison SJ (2003) Applying the principles of stem-cell biology to cancer. *Nat. Rev. Cancer* 3, 895–902.

Parsonnet J (ed) (1999) Microbes and Malignancy: Infection as a Cause of Human Cancers. Oxford, UK: Oxford University Press, New York.

Renan MJ (1997) How many mutations are required for tumorigenesis? Implications from human cancer data. *Mol. Carcinogenesis* 7, 139–146.

Reya T, Morrison SJ, Clarke MF & Weissman IL (2001) Stem cells, cancer and cancer stem cells. *Nature* 414, 105–111.

Thun MJ, Henley SJ & Calle EE (2002) Tobacco use and cancer: an epidemiologic perspective for geneticists. *Oncogene* 21, 7307–7325.

Vogelstein B, Fearon ER, Hamilton SR et al. (1988) Genetic alterations during colorectal-tumor development. *N. Engl. J. Med.* 319, 525–532.

Vogelstein B & Kinzler KW (eds) (1998) The Genetic Basis of Human Cancer. New York: McGraw-Hill.

Yuspa SH (1994) The pathogenesis of squamous cell cancer: lessons learned from studies of skin carcinogenesis. *Cancer Res.* 54, 1178–1189.

Zha S (2004) Cyclooxygenases in cancer: progress and perspective. *Cancer Lett.* 215, 1–20.

Zhao JJ, Roberts TM & Hahn WC (2004) Functional genetics and experimental models of human cancer. *Trends Mol. Med.* 10, 344–350.

zur Hausen H (1991) Viruses in human cancers. *Science* 254, 1167–1173.

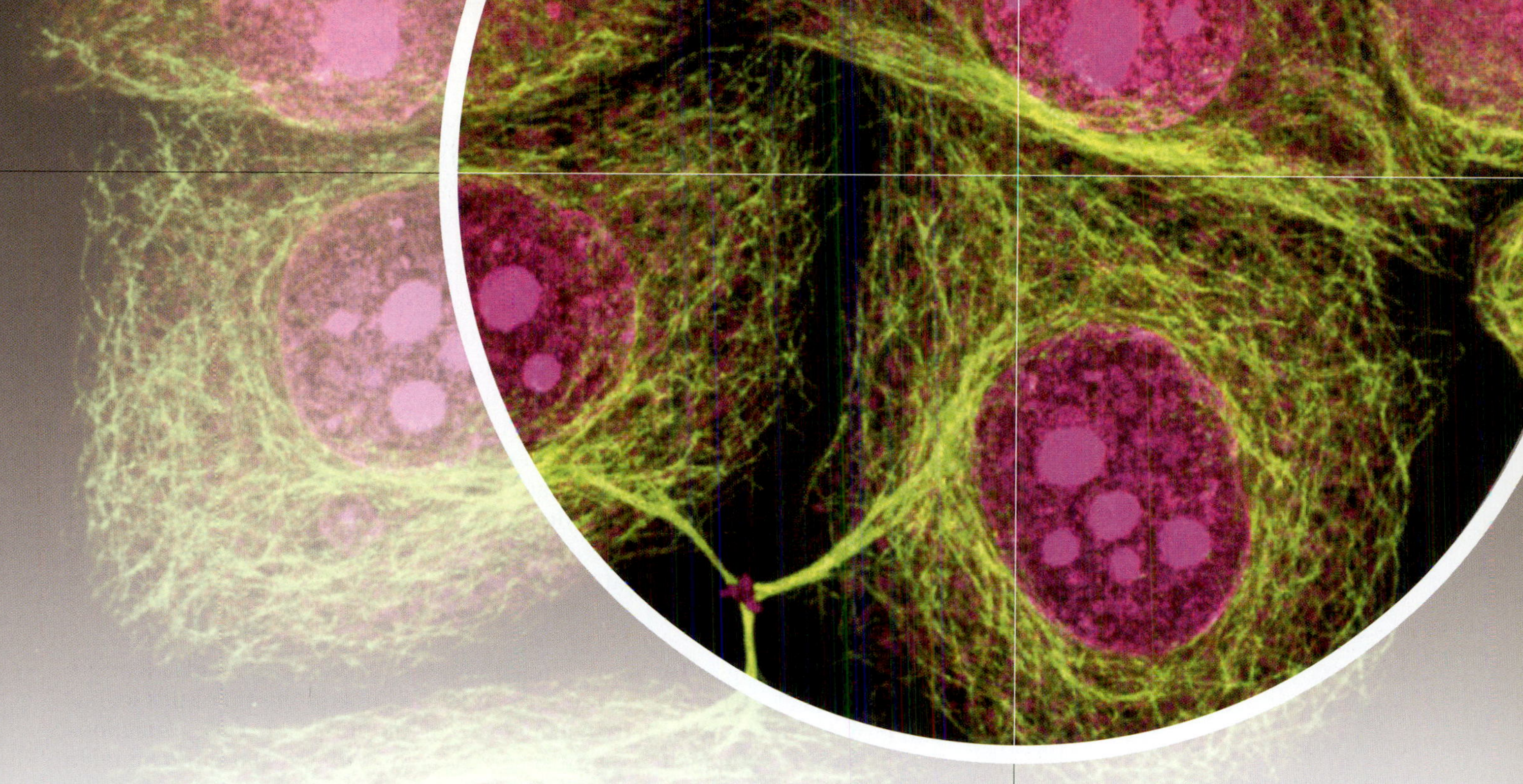

Chapter 12

Maintenance of Genomic Integrity and the Development of Cancer

> When first interpreting the ramifications of DNA and the genetic code... We totally missed the possible role of enzymes in repair...I later came to realise that DNA is so precious that probably many distinct repair mechanisms could exist.
>
> Francis H.C. Crick, molecular biologist, 1974

> The capacity to blunder slightly is the real marvel of DNA. Without this special attribute, we would still be anaerobic bacteria and there would be no music.
>
> Lewis Thomas, biologist, 1979

The fact that human tumor formation is a complex, multi-step process reflects the multiple lines of defense against cancer that have been established within our cells, each maintained by the hard-wiring of a complex regulatory circuit. The human body—actually, its individual cells—must entrust the maintenance of these anti-cancer defenses to their most stable, reliable constituents: DNA molecules. Over extended periods of time, DNA sequences are the most fixed, unchangeable components of a cell; most of its other parts are in constant flux, being created and broken down continuously.

Following this logic, it is really the stability of DNA molecules that underpins the most robust defenses against cancer. Because there are multiple cellular lines of defense that depend on DNA stability and because the breaching of each defense usually requires a rare mutational event, the probability of cell populations' advancing all the way to the neoplastic state must be astronomically small.

So the cancerphobe can rest easy at night, reassured by the multiplicity of cellular and tissue defense mechanisms that evolution has assembled to protect us from neoplasia. But there is a troubling inconsistency here: if the number of anti-cancer defense mechanisms were truly as great as depicted in this text, and if the breaching of each of these defenses were usually dependent on rare mutational events, then cancers should never strike human populations. Yet they do. In Western populations, in which deaths from infectious diseases are relatively infrequent, about 1 person in 5 is destined to die from one or another form of cancer. So, cancer cell populations accomplish what seems to be the impossible—acquiring a substantial array of mutant (and methylated) alleles over a period of several decades.

Researchers working in Seattle, Washington, attempted to resolve this inconsistency as far back as 1974. They proposed that the only resolution of this logical quandary must depend on a drastic increase in mutation rate: cell populations en route to becoming malignant must carry genomes that are far more mutable than the genomes of normal human cells—a condition sometimes termed the *mutator phenotype*. Such speculation has received increasing support in recent years, as different types of genetic instability have been documented in the genomes of cancer cells.

In this chapter, we will direct much of our attention to two major issues. First, how do normal human cells and tissues manage to keep the mutation rate so low? And second, how are the strategies for suppressing mutations thwarted during human cancer pathogenesis?

12.1 Tissues are organized to minimize the progressive accumulation of mutations

On a number of occasions throughout this text, we have described the effects of carcinogens and tumor promoters on target cells throughout the body. However, the specific biological identities of these target cells have never been spelled out. As it turns out, knowledge of the nature of these cells is critical to understanding how genome integrity is maintained. To explore this issue, we need to delve deeply into the organization of tissues and the various types of cells that form tissues. Their biological behavior furnishes us with insights into the strategies exploited by tissues and cells to minimize the accumulation of genetic lesions.

As described earlier (Section 11.6), a common scheme seems to explain the construction and maintenance of many tissues throughout the body. Within each tissue, a relatively small number of cells populate its stem cell **compartment**. These self-renewing cells may constitute a minute fraction of the entire cell population within a tissue, sometimes as few as 0.1 to 1% of the total. In truth, in most tissues, these numbers represent nothing more than poorly informed guesses. Because stem cells are present in very small numbers, have appearances that are not particularly distinctive, and are often scattered among other cell types within tissues, they are difficult to identify and study. Consequently, much of what is described below rests on inference rather than on direct observation of stem cells and their properties.

We will briefly review the discussions of Chapter 11 concerning stem cells. Here, however, we focus on the *normal* stem cells within tissues rather than the pool

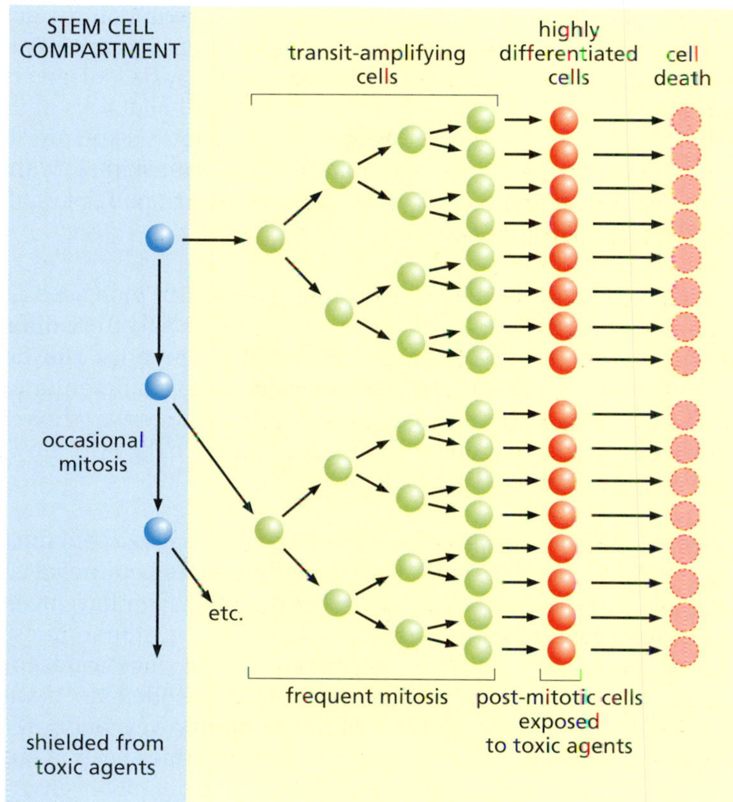

Figure 12.1 Tissue organization and protection of the stem cell genome The organization of many epithelial tissues seems to conform to the scheme shown here. As indicated, each stem cell *(blue)* divides only occasionally in an asymmetric fashion to generate a new stem cell daughter and a transit-amplifying daughter. These stem cells are often shielded anatomically from toxic agents. The transit-amplifying cells *(green)* undergo repeated rounds of growth and division, expanding exponentially. Eventually, the products of these divisions undergo further differentiation into post-mitotic, highly differentiated cells *(red)*. The highly differentiated cells, which are often in direct contact with various toxic agents, are shed with some frequency; hence, any mutant alleles that arise in these cells will be lost, sooner or later, from the tissue. This means that the genomes of stem cells are protected through two mechanisms: stem cells rarely divide, and they are protected anatomically from noxious, potentially mutagenic influences.

of *cancer* stem cells that reside within a tumor. As is the case in tumors, the stem cells in a normal tissue are self-renewing, since at least one of the two daughters of a dividing stem cell will retain the phenotype exhibited by the mother cell prior to cell division. In many tissues, the second daughter cell and its transit-amplifying descendants will pass through a substantial number of cell divisions before entering into a post-mitotic, highly differentiated state. These actively dividing cells, which serve as intermediates between a stem cell and its differentiated descendants, may thereby generate dozens, possibly even hundreds, of differentiated descendants of the second daughter cell (Figure 12.1).

The exponential increase in the number of transit-amplifying cells means that a stem cell needs to divide only on rare occasion in order to maintain a large pool of end-stage, highly differentiated cells in a tissue. Therefore, while one might think that stem cells participate in continual cycles of growth and division, the reality is usually much different: the transit-amplifying cells create the great bulk of mitotic activity in many tissues. Since the DNA replication occurring during each cell cycle is inherently prone to making errors, this scheme reduces the risk that mutations will accumulate in the genomes of the stem cells within a tissue.

In many epithelial tissues, the differentiated epithelial cells are especially vulnerable to damage, since they form cell sheets that line the walls of various ducts and cavities containing toxic material. In the cases of the colon and the bile duct, the epithelial cells confront fecal contents and highly corrosive bile, respectively. The cells lining the alveoli in the lungs cope every day with particulates and pollutants in the air. The keratinocytes in our skin are exposed directly to the outside world and hence are liable to sustain several types of damage, including that inflicted by ultraviolet radiation.

The differentiated end-stage cells (see Figure 12.1) in these and other tissues have a finite lifetime and are discarded sooner or later. Some cell types may simply age and lose their viability, being worn out from carrying on the active business of the tissue. For example, red blood cells have an average lifetime of approximately 120 days, after which they are scavenged by the spleen and broken down and their contents recycled or excreted. The epithelial cells in the colon live for 5 to 7 days before they are induced to enter apoptosis and are sloughed off into the lumen of the intestine. The keratinocytes in our skin die within 20 to 30 days of being formed, and they are shed continually in small flakes of dead skin (see, for example, Figure 2.6A).

Hence, the transit-amplifying cells may well run an increased risk of sustaining mutations because of their high mitotic activity, and their differentiated progeny may often be located in mutagenic environments. However, any genetic damage that the transit-amplifying cells and their differentiated progeny have sustained will have little consequence for the tissue as a whole: sooner or later, these cells are flushed out of the tissue, and once they die, any mutations they may have accumulated disappear with them.

The dynamics of stem cells and their progeny are illustrated most graphically by the stem cells and the *enterocytes* (the differentiated epithelial cells) of the small intestine and the colon. These cells and their behavior have been described earlier in the context of our discussions of the *Apc* tumor suppressor gene and β-catenin (Section 7.11). Here, we return to them once again to illustrate other principles. Recall that the stem cells are embedded deep within the crypts (Figure 12.2A). There they are well out of harm's way, being shielded from the mutagenic contents of the intestinal lumen by a thick layer of mucus secreted by cells in the crypt. This mucus, which is formed from highly glycosylated proteins termed mucins, creates a jelly-like barrier that prevents the contents of the intestinal lumen from penetrating deep into the crypt (e.g., see Figure 12.2B) and illustrates yet another strategy by which mutations can be minimized in the genomes of stem cells: evolution has created mechanisms by which stem cells are anatomically shielded from the actions of toxins, including carcinogens. (Thus, mice that have been genetically deprived of the gene encoding Muc2, the most abundant gastrointestinal mucin, are prone to develop adenomas in the small intestine, many of which progress to adenocarcinomas.) This confers further protection on the genomes of stem cells, complementing the mechanism mentioned earlier, in which the descendants of these cells, which may have sustained mutations, are driven out of the crypts and eliminated after a period of 5 to 7 days (Figure 12.2C).

In theory, the stem cell compartment within a tissue has an inexhaustible ability to generate differentiated progeny without ever suffering depletion. However, almost inevitably, a stem cell will be lost from this compartment through one or another mishap. This gap in the ranks must be filled by other stem cells. More specifically, both daughters of a surviving stem cell will need to retain the phenotype of their mother, which therefore undergoes a symmetrical division (Figure 12.3). This may also have implications for genome maintenance, as we will see below.

12.2 Stem cells are the likely targets of the mutagenesis that leads to cancer

The properties of the intestinal cells (see Figure 12.2) also provide important clues about the identities of the cells that are the likely targets of carcinogenesis. Having excluded the more differentiated cells (because they are rapidly discarded), we must shift our focus to the long-lived cells and cell lineages within epithelial tissues. Actually, we already have come across evidence that points to the central role of long-lived cells in the process of carcinogenesis. In Section 11.13, we read about experimental protocols used to induce skin cancer in mice.

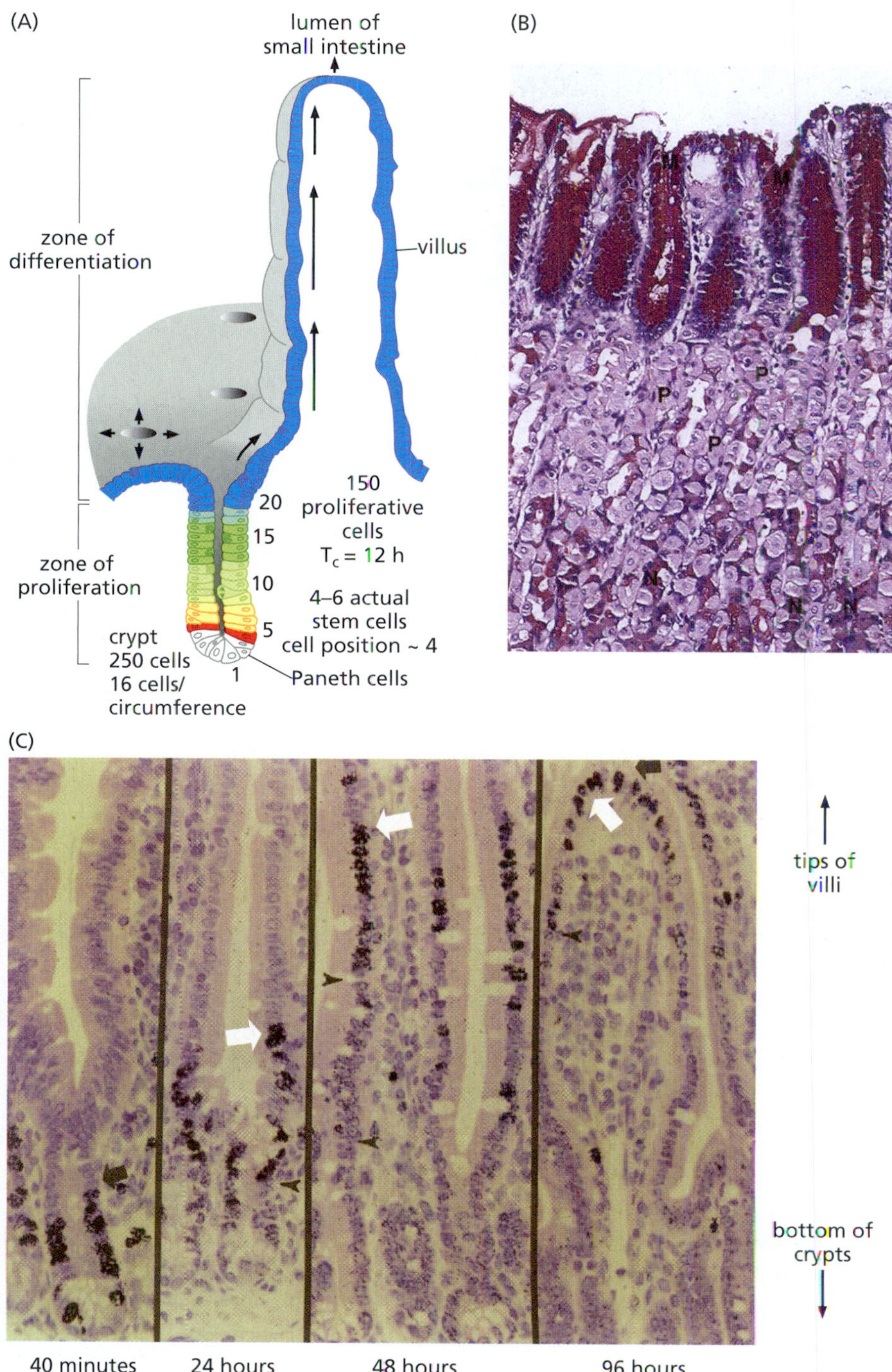

Figure 12.2 Stem cells and the organization of gastrointestinal crypts (A) The scheme of tissue organization depicted in Figure 12.1 is illustrated nicely by the organization of the epithelial cells—enterocytes—in the small intestine (see also Figure 11.5). The 4 to 6 stem cells located near the bottom of the crypts *(red)* are shielded from the contents of the small intestine by their location and by mucus that prevents the entrance of fluids from the intestinal lumen into the crypt (see panel B). The stem cells spawn a large number (~150) of highly proliferative transit-amplifying cells *(yellow, green)*, which divide every 12 hours or so. Their division yields approximately 3500 enterocytes *(blue)*, which cover the villus—the fingerlike structure that projects into the small intestine. The enterocytes are continuously migrating toward the tip of the villus, where they undergo apoptosis and are shed into the lumen of the small intestine. The numbers to the right of the crypt indicate the cell position number from the bottom of the crypt. (B) Copious amounts of protective mucus *(dark purple)* are secreted by the cells lining comparable pits in the wall of the stomach; similar mucus, composed of highly glycosylated proteins termed mucins, is found in the crypts of the small and large intestine. (C) The emigration of transit-amplifying cells from the crypts of the small intestine can be tracked by injecting a dose of ^{3}H-thymidine (i.e., radioactive thymidine carrying a tritium atom) into a mouse and following the resulting incorporation of radiolabel into DNA by *autoradiography*; radioactive decay is indicated by *dark silver grains*. (Incorporation occurs only during a brief period of time.) Seen here are the cells in the crypts of the duodenum of the mouse at the indicated times after injection of the ^{3}H-thymidine. Cells that multiplied only a small number of times after initial incorporation of ^{3}H-thymidine remain heavily labeled *(broad arrows)*, while the great majority underwent multiple additional divisions after labeling (during the chase period) and therefore exhibit diluted radiolabeling. After four days, virtually all of the cell genomes that were synthesized at the beginning of the experiment have been carried out of the crypts to the tips of the villi. (A, courtesy of C.S. Potten; B, from B. Young and J.W. Heath et al., Wheater's Functional Histology, 4th ed. Edinburgh: Churchill Livingstone, 2003; C, from C.S. Potten, *Phil. Trans. Royal Soc. London B* 353:821–830, 1998.)

One such protocol involved painting a patch of skin with an initiating agent, allowing the patch to remain untouched for some months, and then painting it repeatedly with TPA, a potent skin tumor promoter. Cells that had been exposed to the initiating carcinogen "remembered" that exposure one year later by undergoing proliferation and forming a skin papilloma in the presence of the promoter. In the skin, as in many other epithelial tissues, the long-lived cells are those in the stem cell compartment.

Provocatively, the number of skin papillomas and carcinomas induced by the mouse skin carcinogenesis protocol (see Figure 11.28) is not reduced if the mouse skin is treated with 5-fluorouracil (5-FU) shortly after being exposed to a mutagenic initiating agent. Since 5-FU selectively kills actively cycling cells, this indicates that the cell targeted by carcinogenic mutagens during initiation is not

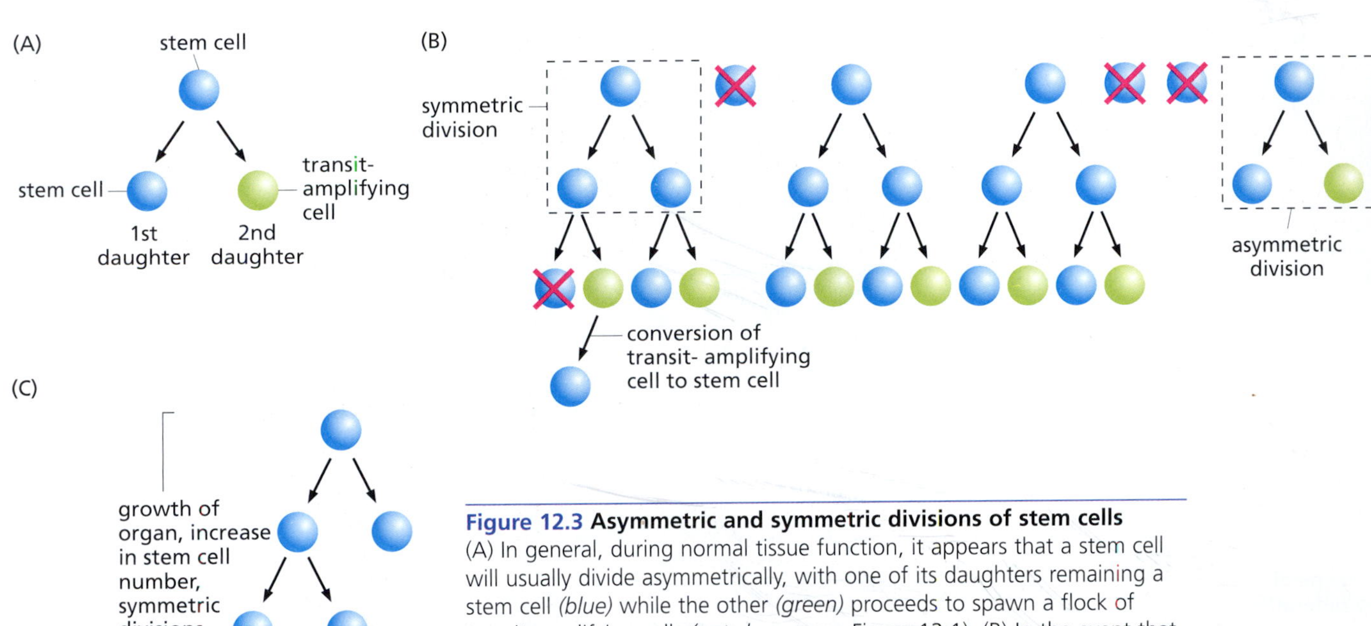

Figure 12.3 Asymmetric and symmetric divisions of stem cells (A) In general, during normal tissue function, it appears that a stem cell will usually divide asymmetrically, with one of its daughters remaining a stem cell *(blue)* while the other *(green)* proceeds to spawn a flock of transit-amplifying cells *(not shown*; see Figure 12.1). (B) In the event that several stem cells in a tissue are lost, some of the surviving stem cells may divide symmetrically in order to re-populate the stem cell pool with the proper number of cells. As seen here, three stem cells have been lost *(red crosses, top row)* from a pool of seven stem cells. The subsequent symmetric divisions undertaken by the surviving stem cells make possible a regeneration of the original population size of the stem cell pool. Alternatively, the loss of a stem cell *(red cross, third row)* may cause its transit-amplifying sister to revert back to a stem cell *(bottom)*. (C) Similarly, when an organ is growing, the number of stem cells must increase proportionately, requiring some stem cells to undergo symmetric divisions.

in the active cell cycle at the time of initiation and shortly thereafter, lending weight to the notion that the target for initiation is a cell type that divides only occasionally.

Analyses of several types of leukemia suggest that the initial targets of carcinogenesis in the hematopoietic system are also stem cells. The most dramatic example is provided by chronic myelogenous leukemia (CML). As described earlier, the Philadelphia (Ph^1) chromosome, which results from a reciprocal chromosomal translocation that fuses the *bcr* and *abl* genes (see Section 4.6), is observed in almost all cases of this disease. Extensive evidence points to this particular translocation as the genetic lesion that initiates this disease.

A number of distinct hematopoietic cell types within a CML patient may carry the Ph^1 chromosome. Included are lymphoid cells (both B and T lymphocytes), as well as cells of the myeloid lineage (including neutrophils, granulocytes, the megakaryocyte precursors of platelets, and erythrocytes). This observation provides persuasive evidence that the cell type in which the translocation originally occurred was the common progenitor of all of these hematopoietic cell lineages—the **pluripotent** stem cell that serves as the precursor for many types of hematopoietic cells (Figure 12.4). Like a variety of other stem cells, this hematopoietic stem cell (HSC) is thought to have a very long lifetime in the hematopoietic system, more specifically in the bone marrow. In the particular case of CML, a stem cell that has suffered a critical mutation—formation of the Ph^1 chromosome—retains the option to dispatch its progeny into a number of distinct hematopoietic cell lineages. Yet other indications of the role of stem cells as targets for tumor formation come from other types of hematopoietic disorders (Sidebar 12.1).

Highly compelling observations of stem cells' role in cancer derive from transgenic mice in which the expression of an activated *ras* oncogene is limited to either the keratinocyte stem cells in the skin (which in this case are located in hair follicles) or the keratinocytes that have begun to enter into a terminally differentiated state. When the transgene directs expression of the *ras* oncogene in the stem cells, the mice develop malignant carcinomas. In contrast, when the same oncogene is expressed in the differentiating keratinocytes, benign papillomas are formed, and these tend to regress.

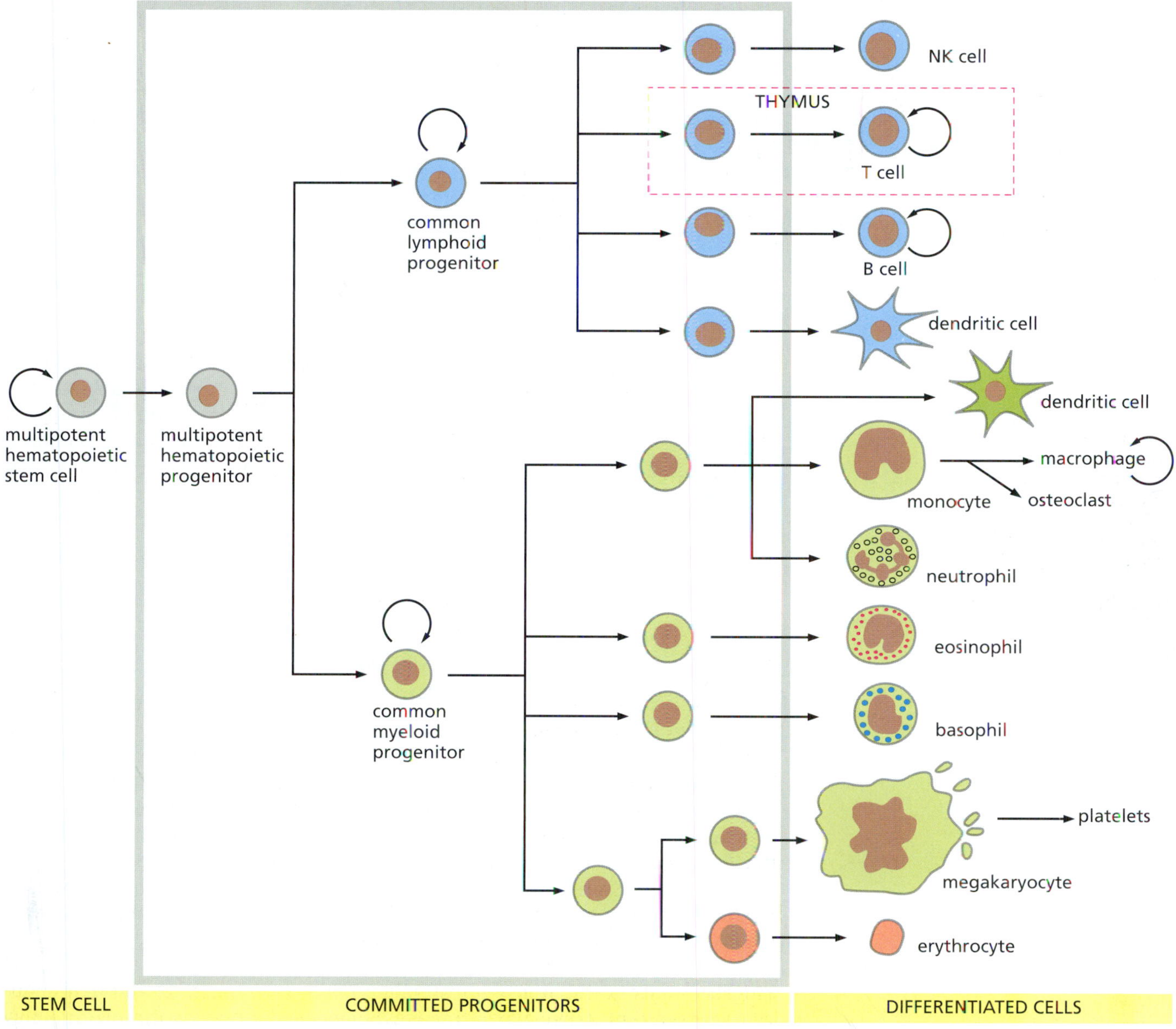

Figure 12.4 Hematopoietic differentiation Our current understanding of hematopoietic cell differentiation teaches a number of lessons. (1) It indicates that a single cell type—the multipotent hematopoietic stem cell (HSC; *left)*—is capable of generating virtually all of the cell types in the blood and in the immune system. (2) It shows that a single stem cell type can spawn multiple types of "committed" stem progenitor cells, in this case, the two stem cell types that are committed to generating lymphoid and myeloid cell types. (3) It shows that self-renewal ability *(curved arrows)* is not confined to a single stem cell type in a tissue; instead, in certain tissues such as this one, "committed progenitors" (i.e., the lymphoid and myeloid stem cells shown here) as well as some of their descendants have self-renewal capability. The fact that a patient suffering from CML (chronic myelogenous leukemia) often exhibits several distinct differentiated lymphoid and myeloid cell types carrying the Ph[1] chromosome (and a *BCR-ABL* translocation) provides strong indication that this abnormal chromosome was initially formed in some multipotent HSC or progenitor. (From B. Alberts et al., Molecular Biology of the Cell, 4th ed. New York: Garland Science, 2002.)

Sidebar 12.1 Blocked differentiation is a frequent theme in the development of hematopoietic malignancies There are dozens of examples of malignancies in animals and in humans where inhibition of differentiation favors the appearance of neoplasias. Possibly the first of these situations to be defined genetically involved the avian erythroblastosis virus, a retrovirus that encodes two oncoproteins: its *erbB* oncogene specifies a constitutively active version of the epidermal growth factor (EGF) receptor (see Section 5.4), which drives the proliferation of erythroblasts (precursors of red blood cells); while its *erbA* oncogene encodes a nuclear receptor (a homolog of the thyroid hormone receptor), which inhibits differentiation of the hyperproliferating erythroblasts created by *erbB*. Similarly, in human acute myelogenous leukemia (AML), a large variety of genetic lesions found in the leukemic cells have been assigned to two functional classes: those that are required to drive the proliferation of the myeloid precursor cells, and others that are required in the same cells to block subsequent differentiation.

In the megakaryoblastic leukemias (a malignancy of platelet precursor cells) encountered with some frequency in Down syndrome patients, the gene encoding the GATA1 transcription factor has been found to be frequently mutated, preventing the proper maturation and differentiation of these precursors of platelets. These few examples point to the notion that the exit of cells from stem cell compartments must be impeded in order for tumorigenesis to succeed.

Not addressed by these observations are the precise identities of the stem cell targets of transformation. In many cases, the target is not likely to be the pluripotent hematopoietic stem cell, but instead one of its derivatives that is already committed to one or another lineage of differentiation. Such "committed progenitors" (see Figure 5.4) normally may have significant (but limited) self-renewal capacity and are not yet fully differentiated, and thereby can be considered stem cells. Their transformation from normal to tumor stem cells involves, among other changes, an acquisition of unlimited self-renewal capability.

These various strands of evidence, obtained from several types of tissue, converge on the conclusion that self-renewing cells of various types are the targets of the genetic changes that lead, sooner or later, to the formation of tumors. In some instances, the target cells may be stem cells with unlimited self-renewal capacity; in others, committed progenitors, which normally have only a limited ability to renew themselves, may acquire unlimited self-renewal capability during the course of tumorigenesis. This idea, in turn, may explain some of the complex epidemiology of certain types of human cancers, such as breast cancer (Sidebar 19 ●).

12.3 Apoptosis, drug pumps, and DNA replication mechanisms offer tissues a way to minimize the accumulation of mutant stem cells

The apparently prominent role played by normal stem cells as targets for transformation indicates that the genomes of these cells must be protected by whatever biological and biochemical strategies these cells and the tissues around them can muster. We have already come across two such strategies: the relatively infrequent replication of stem cell DNA and the placement of stem cells in anatomically protected sites. Still, these mechanisms do not seem to suffice, so the organism has developed yet other strategies.

The stem cells in the mouse intestinal crypts (see Figure 12.2) and mammary glands represent especially attractive objects for study of these protective strategies. In the case of the crypts, the need for additional protective mechanisms is clear: the enterocyte stem cell lineages in the crypts of the mouse small intestine have been estimated to pass through a succession of 1000 growth-and-division cycles during a lifetime, and each of these cycles exposes the stem cells to various types of genetic damage. Similarly, in the human gut, the number of cell divisions occurring each year greatly exceeds the total number of cells residing at any time within the entire body; this enormous mitotic activity, most of which involves transit-amplifying cells, must also depend on many successive stem cell divisions, although in the human case the approximate number is not known.

One protective mechanism is suggested by the responses of stem cells in the crypts to massive genetic damage. In the intestinal crypts of the mouse, stem cells that have suffered genetic damage inflicted by X-rays will rapidly initiate apoptosis rather than halt their proliferation and attempt to repair the damage. The motive here seems to be associated with the error-prone nature of DNA repair. As we will learn later, the DNA repair apparatus is highly efficient but hardly perfect, and therefore often leaves a residue of unrepaired or incorrectly repaired lesions in the chromosomal DNA. If such lesions are encountered by the DNA replication machinery, they may cause mutant DNA sequences to be copied and passed on to daughter cells, including those that will themselves become stem cells. So, rather than risk this outcome, stem cells in the mouse crypts are primed to activate apoptosis in response to DNA damage. It is unclear whether stem cells in other tissues are similarly poised to enter apoptosis.

Yet another mechanism is suggested by a commonly used technique for separating stem cells from the bulk of cells in a tissue via fluorescence-activated cell sorting (FACS; see Figure 11.13). Stem cells efficiently pump out certain fluorescence dye molecules, while these cells' differentiated derivatives do so much less actively. As a consequence, after exposure of cell populations to such dyes, the stem cells fluoresce much more weakly than all other cells in these populations.

The active excretion of these fluorescent dye molecules is due to the actions of a plasma membrane protein termed Mdr1 (multi-drug resistance 1), which was first discovered because it is exploited by many cancer cells to pump out, and therefore acquire resistance to, chemotherapeutic drug molecules. The unusually high levels of Mdr1 expressed by many types of stem cells seem to represent a strategy that they use to protect their genomes from potentially mutagenic compounds that may have entered into their cytoplasms from outside.

The mechanism of asymmetric DNA strand allocation may also play an important role in preventing the stem cells in certain tissues from accumulating genetic damage. The experimental observations supporting this proposed mechanism are still fragmentary. Nonetheless, it is presented here, because of its interest and potential importance to understanding cancer pathogenesis.

The rationale behind this strategy, first proposed in 1975, derives from the molecular details of the DNA replication occurring in stem cells. We have just revisited the model that when stem cells divide, the division is usually asymmetric, in that one daughter cell remains a stem cell and the other enters into a differentiation pathway by producing transit-amplifying cells (see Figure 12.1). Ideally, the genome that is donated to the daughter that remains a stem cell should be afforded *more protection* than the genome that is passed on to the daughter destined for differentiation, because descendants of the latter are destined to be discarded sooner or later. As illustrated in Figure 12.5A, the asymmetric allocation of DNA strands can help to accomplish this aim.

The idea here is based, once again, on the fact that DNA replication is inherently error-prone. By some estimates, each time a cell passes through S phase and replicates its DNA, several nucleotide substitutions occur per cell genome because DNA polymerases make mistakes that escape subsequent detection and repair. (In truth, this number may greatly underestimate the errors in DNA replication.) Consequently, DNA strands that were not synthesized during the most recent cycle of DNA replication—the "conserved" strands—are more likely to retain wild-type sequences than are those "nonconserved" strands that are indeed the products of this DNA synthesis.

This suggests that in a well-designed tissue, the DNA strands that have not been created by recent DNA replication should be retained by the daughter cell that remains in the stem cell compartment, while those DNA strands that are the

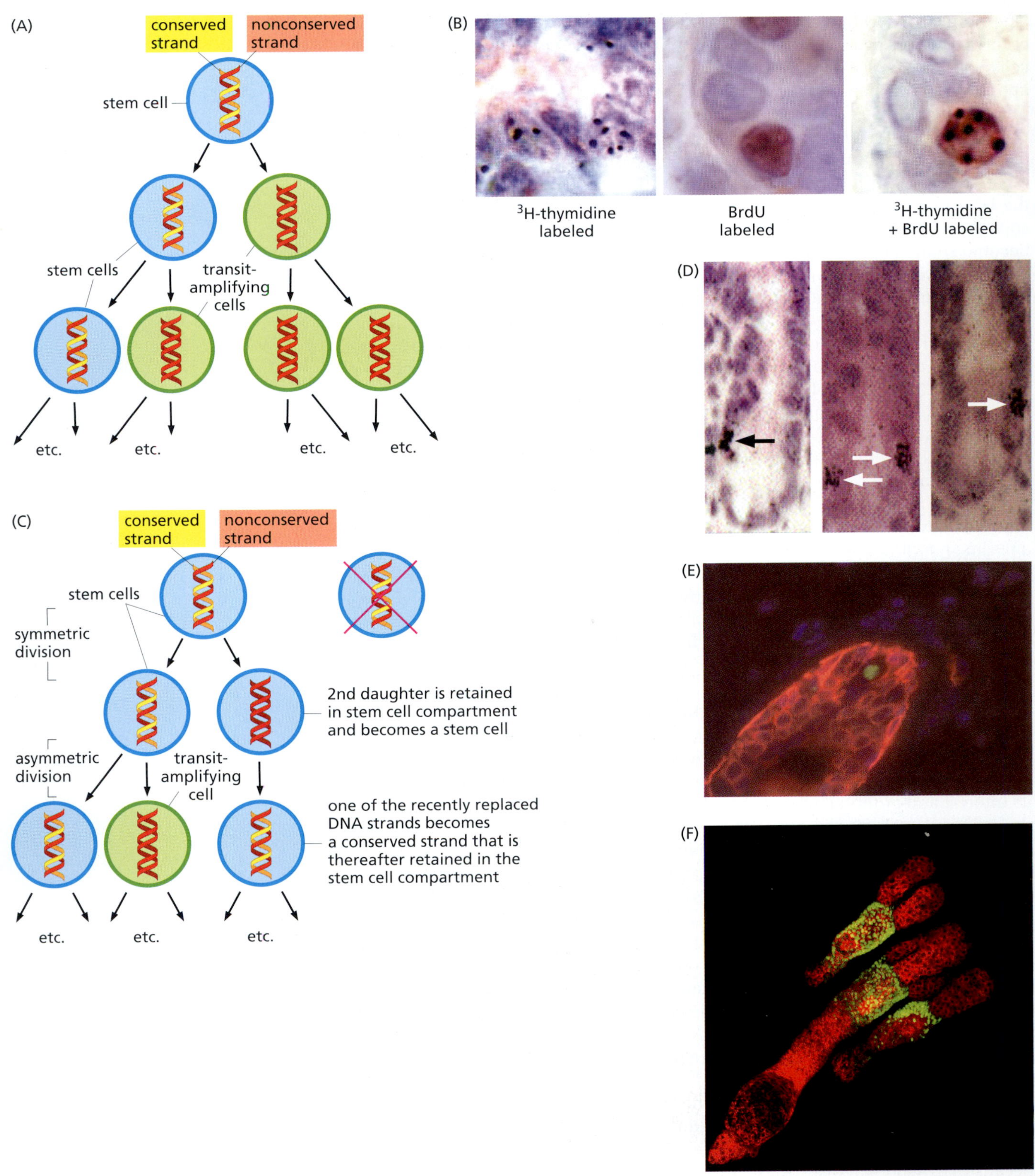

products of DNA replication should be allocated to the daughter cell whose descendants are destined to differentiate and eventually die. As Figure 12.5A makes clear, one DNA strand (the conserved, "immortal" strand) can, in principle, be transmitted indefinitely through a lineage of stem cells by such asymmetric segregation of DNA molecules. Stated differently, stem cells may carry DNA strands that have repeatedly served as *templates* for DNA replication but are only infrequently synthesized as *products* of replication (at least since the

Figure 12.5 Conserved DNA strands and the stem cell genome (A) The "immortal-strand" model depicts one conserved DNA strand *(yellow)* of the chromosomal DNA of a stem cell *(light blue)* that is donated to its daughter cell that will remain a stem cell and is therefore retained in the stem cell compartment; this conserved DNA strand is not the product of recent DNA replication. Conversely, the "nonconserved strand" *(red)* that is indeed the product of recent DNA replication will be allocated preferentially to the daughter cell that spawns transit-amplifying cells *(light green)* and therefore exits the stem cell compartment; the new round of DNA replication adds a new *red* strand to the nonconserved parental *red* strand. This model predicts that one DNA strand can persist indefinitely within the stem cell compartment without undergoing replication. (B) This prediction is fulfilled in the mouse mammary gland. Mice can be exposed to a brief pulse of 3H-thymidine at a time during puberty when the gland is still growing and the number of mammary epithelial stem cells is continuously increasing, necessitating symmetrical divisions in which both daughters of a stem cell become stem cells (see Figure 12.3C) and therefore, hypothetically, both strands of DNA are retained as conserved DNA strands. Hence, label incorporated during this period may be retained indefinitely in conserved DNA strands. DNA that has incorporated the ^{3}H-thymidine tradiolabel is detected by incubating tissue slices with a photographic emulsion *(dark grains)*—the procedure of *autoradiography*. Five weeks after initial exposure to ^{3}H-thymidine, radiolabel is retained in only about 2% of the mammary epithelial cells *(left)*. At that time, mice can be exposed to a pulse of bromodeoxyuridine (BrdU), a thymidine analog whose incorporation into DNA can be detected by a specific antibody that recognizes BrdU-containing DNA *(red staining nucleus, middle)*. Following this pulse, BrdU can be detected in the majority of the cells that retained radiolabel from the exposure to ^{3}H-thymidine 5 weeks earlier *(right panel)*. Hence, these label-retaining cells are actively proliferating 5 weeks later yet they retain a conserved strand that was synthesized 5 weeks earlier and is not lost from these cells by the repeated rounds of growth and division that they are undergoing. (C) When a stem cell is lost *(top right)* in an adult (in which the size of the stem cell pool should be constant), a surviving stem cell will divide symmetrically, so that both of its daughters will remain as stem cells, thereby reconstituting the population of stem cells in the pool (see Figure 12.3B) In this daughter *(right)*, a DNA strand that was previously nonconserved and thus the product of recent replication *(red)* will be retained in the stem cell compartment and become an immortal, conserved strand *(yellow)*. Hence, the killing of stem cells in an adult should make it possible to label stem cell DNA in such a fashion that this label is retained indefinitely in the stem cell compartment. (D) This prediction is fulfilled by the behavior of cells in the duodenum of the mouse. In the crypts, the enterocytes are normally replenished by the continual multiplication of the stem cells located near the bottom of the crypts (see Figure 12.2A). As in panel (B), the DNA molecules in proliferating cells can be radiolabeled by a brief exposure to ^{3}H-thymidine and detected subsequently by autoradiography. Normally, all of the radiolabeled DNA that is initially synthesized in the crypts moves out of the crypts together with the differentiating transit-amplifying cells and their enterocyte progeny and hence is lost from the crypts after several days (see Figure 12.2C). However, if the duodenum is exposed to 8 grays (Gy) of X-irradiation (which kills some of the stem cells) before radiolabelling, cells in the crypts can be found to retain label even 8 days after a brief pulse of radioactive thymidine. Four examples of these label-retaining cells (LRCs), which are found precisely in the location of stem cells in the crypts, are shown here *(arrows)*. (E) In this mammary duct, the mammary epithelial cells (MECs) are stained for cytokeratin expression *(red)*. An LRC that incorporated BrdU nine weeks earlier is immunostained in *green*. (F) LRCs can be found in the "bulge" region of the hair follicles of the mouse, where keratinocyte stem cells are known to reside. These cells, which were briefly induced to express a stable form of green fluorescent protein (GFP, *green*) at four weeks of age, continue to express GFP four weeks later. The epithelial cells are labeled here in *red*. (B, from G.H. Smith, *Development* 132:681–687, 2005; D, courtesy of C.S. Potten; E, courtesy of B. Welm and M.A. Goodell; F, courtesy of T. Tumbar, V. Greco, and E. Fuchs.)

adult tissue was first formed). While Figure 12.5A illustrates the behavior of a short stretch of DNA, we can imagine that the entire chromosomal DNA of stem cells behaves in this fashion as well.

This model of asymmetric strand allocation (also called the "conserved–strand" model) can be tested experimentally. In fact, we have already seen one manifestation of this behavior in the experiment illustrated in Figure 12.2C. In that case, stem cells were allowed to incorporate ^{3}H-thymidine for a brief period of time. The radioactive precursor became incorporated into the DNA strands being replicated during this short period of time, after which further incorporation ceased. (Such an experimental protocol is often termed "pulse-chase" labeling.) The fate of the radiolabeled DNA molecules was then followed through the technique of **autoradiography**, in which a photographic emulsion is placed on a slice of tissue; this emulsion yields a readily visualized dark grain whenever a radioactive atom such as a **tritium** atom decays. If the allocation of DNA strands were symmetrical, then we would expect some of the radiolabel would remain behind in the stem cell compartment and some would be distributed to the differentiating cells that had left the stem cell compartment.

As Figure 12.2C demonstrated, virtually all of the radiolabeled DNA strands migrated out of the stem cell compartment together with the transit-amplifying cells that had begun to differentiate. This supports the notion that the newly synthesized strands (i.e., those that were synthesized during the ^{3}H-thymidine pulse) were preferentially donated to the daughter cells that spawned transit-amplifying cells and their more differentiated descendants that migrated out of the crypts. Conversely, we discover that it is extremely difficult to label the DNA strands that are retained in the stem cell compartment within the crypts.

Actually, there is an alternative interpretation to this observation: the radiolabel leaves the stem cell compartment because it is rapidly diluted by repeated cycles of growth and division in the stem cell compartment. This notion can be tested critically by exposing stem cells to ^{3}H-thymidine at a time when the stem cell compartment is expanding; under these conditions, stem cells must undergo symmetric division in order to increase their number (see Figure 12.3C), and *both* radiolabeled DNA strands should therefore be retained in the stem cell compartment. This is just what is seen when the mammary epithelial stem cells of the mouse are allowed to incorporate ^{3}H-thymidine during puberty, when the mammary gland is growing rapidly (Figure 12.5B). Under these conditions, the radiolabel is retained many weeks later in the stem cell compartment, even though these "label-retaining" cells (LRCs) can be shown to be actively dividing at this later time. (The radiolabeled strand inherited from their ancestors many cell generations earlier retains its radiolabel in spite of repeated intervening cycles of cell growth and division.)

Another test of the conserved-strand model comes from experiments in which some of the stem cells are killed by exposure to X-rays. The remaining cells in the stem cell compartment will attempt to replace the lost stem cells through symmetric cell divisions in which both daughters remain as stem cells (see Figure 12.3B). Consequently, a newly made DNA strand, which would normally be allocated to the differentiating daughter cell, will now be converted into a conserved strand and retained in the stem cell compartment (Figure 12.5C).

If we expose the stem cell compartment of mouse intestinal crypts to ^{3}H-thymidine during the period of time when the lost stem cells are being replaced, we can indeed label DNA molecules that subsequently remain within the stem cell compartment for an indefinite period of time; that is, the labeled molecules are not "chased" out when the stem cells are exposed subsequently to non-radiolabeled thymidine precursors (Figure 12.5D). Hence, the only time in an adult animal that we can introduce long-lived radiolabel into the stem cell compartment seems to be when we perturb this compartment by killing some of its cells. Under these conditions, both the immortal DNA strand and the recently synthesized, non-immortal strand are retained in cells that become stably ensconced in the crypts of the small intestine.

Label-retaining cells are also found in other epithelial tissues (Figure 12.5E and F). In spite of this and other evidence, most of which has been gathered in the mouse, the "immortal-strand" model remains largely a matter of speculation for

Sidebar 12.2 Some carcinogens may produce mutant cells through their ability to be cytotoxic The "immortal-strand" model makes predictions about how certain carcinogens work. We can imagine, for example, that some carcinogens, rather than being directly mutagenic, act through their ability to be cytotoxic and thus kill cells within a tissue, including some of its stem cells. In the event that such a carcinogen kills a stem cell recently formed by mitosis, the sister of that stem cell, formed by the same mitosis, may be retained in the stem cell compartment rather than being dispatched into the pool of differentiating cells, as would normally be its fate (see Figure 12.3B). However, this sister cell may happen to carry a mutant sequence due to a DNA replication error that occurred during the most recent S phase. Once this sister cell is recruited into the stem cell compartment, the DNA strand carrying this mutant sequence may then be chosen to become an "immortal" DNA strand, thereby ensuring that this mutation is now permanently established in the stem cell compartment.

We have encountered nongenotoxic carcinogens previously in the context of the discussion of tumor promoters (Section 11.14). There, we argued that some tumor promoters, such as ethanol, act through their ability to cause the death of cells in a target tissue, resulting in a compensatory proliferation by the surviving cells in the tissue. Now, as we view these promoters in the context of stem cell biology, we can speculate that many tumor promoters may act through their ability to kill stem cells. An even more dangerous agent would be a "complete" carcinogen (Section 11.17)—one that is able to act both as an initiator through its mutagenic actions on stem cell genomes and as a promoter through its cytotoxic effects on stem cells.

most tissues and requires far more experimental validation before we can accept it as a well-established fact. The immortal-strand theory, if further validated, also holds important implications for the process of carcinogenesis (Sidebar 12.2).

12.4 Cell genomes are threatened by errors made during DNA replication

The design of stem cell compartments and the behavior of individual stem cells illustrate several biological strategies used by tissues to reduce the burden of accumulated somatic mutations. These mechanisms serve to protect stem cell genomes, which constitute, in effect, the "germ lines" of tissues. Importantly, these strategies represent only the first line of defense against genomic damage. The next line of defense is a biochemical one that depends on the ability of various proteins to recognize and repair damaged DNA molecules within cells.

In fact, DNA molecules are under constant attack by a variety of agents and processes. For the sake of simplicity, we can place these mutagenic processes in three categories. First, as mentioned above, the replication of DNA sequences by DNA polymerases during the S phase of the cell cycle is subject to a low but nonetheless significant level of error. Included among these errors are those generated when chemically altered nucleotide precursors are inadvertently incorporated into DNA in place of their normal counterparts. Second, even in the absence of attack by mutagenic agents, the nucleotides within DNA molecules undergo chemical changes spontaneously; these changes often alter the base sequence and thus the information content of the DNA. Finally, DNA molecules may be attacked by various mutagenic agents, including those molecules generated endogenously by normal cell metabolism as well as agents of exogenous origin—chemical species and physical mutagens (X-rays and UV rays) that are introduced into the body from outside. We will return to the latter two processes in the next sections.

The molecular machinery that is responsible for replicating almost all chromosomal DNA sequences has a remarkably low rate of error. The basic replication machinery in the cell nucleus is powered by the actions of three polymerases, pol-α, pol-δ, and pol-ε. (In all, 15 distinct DNA polymerase genes have been cataloged in the human genome, and more are likely to be found; as will be apparent later, many of these are not involved in DNA replication per se but rather in the repair of damaged DNA molecules.)

A cell has two major strategies for detecting and removing the miscopied nucleotides arising during DNA replication. The first strategy lies in the hands of the DNA polymerases themselves, which are structurally complex aggregates assembled from a number of distinct protein subunits. While they are advancing down single-strand DNA templates and extending nascent DNA strands in a 5′-to-3′ direction, DNA polymerases such as pol-δ continuously look backward, "over their shoulder," scanning the stretch of DNA that they have just polymerized; such monitoring is often called **proofreading**. Should a polymerase detect a copying error, it will use its 3′-to-5′ exonuclease activity to move backward and digest the DNA segment that it has just synthesized and then copy this segment once again, with the hope for a better outcome the second time (Figure 12.6).

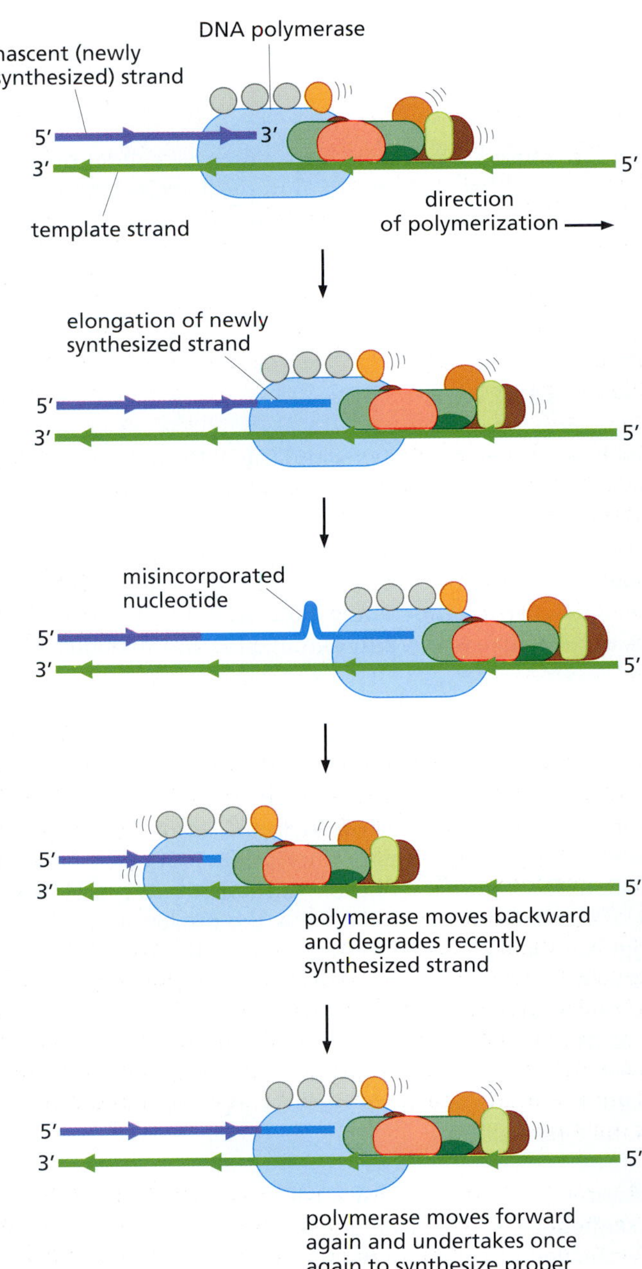

Figure 12.6 Proofreading by DNA polymerases A number of DNA polymerases have a proofreading ability that allows them to minimize the number of bases that are misincorporated and retained in the recently synthesized strand. Thus, as a DNA polymerase extends a nascent strand *(dark blue)* in a 5′-to-3′ direction *(moving rightward)*, it will use the existing 3′-OH of the nascent strand as the primer for further elongation *(light blue)*. However, if a base has been misincorporated *(third drawing)*, the DNA polymerase, which is continuously looking backward to check whether it has incorporated the correct bases in the growing DNA strand, can degrade in a 3′-to-5′ *(leftward)* direction the recently elongated strand *(fourth drawing)* and undertake once again to synthesize this stretch of nascent strand *(bottom drawing)*.

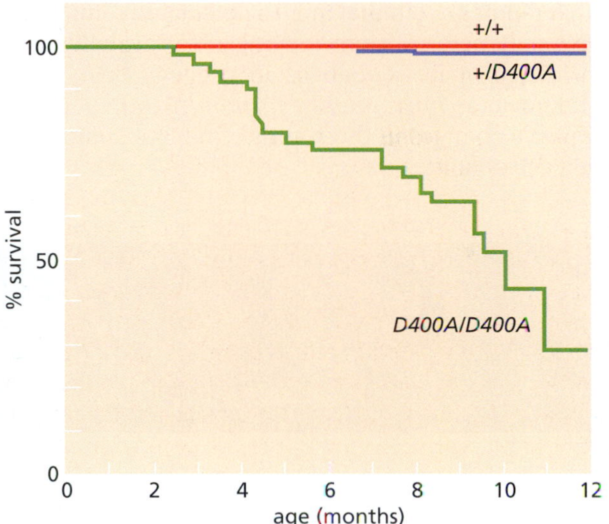

Figure 12.7 Proofreading by DNA polymerase and cancer incidence A point mutation has been introduced into the germ-line copy of the mouse gene encoding DNA polymerase δ, the mammalian DNA polymerase that is responsible for the bulk of leading and lagging strand synthesis. This mutation, termed *D400A*, alters the amino acid sequence in the proofreading domain of the polymerase by specifying the replacement of an aspartic acid by an alanine at residue position 400 of the polymerase molecule. Shown here is the fate of 53 wild-type mice (+/+), 97 heterozygotes (+/*D400A*), and 49 homozygous mutants (*D400A*/*D400A*). Deaths of the mutant homozygotes were all due to malignancies; these included lymphomas, squamous cell carcinomas of the skin, and several other types of cancer that occurred relatively infrequently. Two of the heterozygotes died from causes that were unrelated to cancer, while the homozygous wild-type mice all survived to the age of one year. Their survival curves are shown here in this Kaplan–Meier plot. (From R.E. Goldsby, N.A. Lawrence, L.E. Hays et al., *Nat. Med.* 7:638–639, 2001.)

The importance of this proofreading mechanism for the prevention of cancer has been illustrated dramatically by the creation of a mouse strain whose germ-line pol-δ –encoding gene has been subtly altered (by a single amino acid substitution). The resulting mutant pol-δ retains its polymerizing activity but has lost its 3′-to-5′ exonuclease activity; this loss eliminates its proofreading function. In a cohort of 49 mice carrying the mutant pol-δ allele in a homozygous configuration, 23 developed tumors by one year of age while no tumors developed in a group consisting of twice as many heterozygous mice (Figure 12.7). This experiment provides a dramatic demonstration that the maintenance of wild-type genomic sequences, in this case by a DNA polymerase, represents a critical defense against the onset of cancer. Moreover, for us, this observation represents the first of many indications that the mutations leading to cancer may arise through endogenous processes rather than being triggered exclusively by invading foreign carcinogenic agents.

Following close on the heels of the DNA polymerases and their proofreading activities are a complex set of **mismatch repair** (MMR) enzymes. These enzymes monitor recently synthesized DNA in order to detect miscopied DNA sequences that have been overlooked by the proofreading mechanisms of the DNA polymerases.

The actions of the mismatch repair system become especially critical in regions of the DNA that carry repeated sequences. These sequence blocks include simple mononucleotide repeats (such as AAAAAAA), dinucleotide repeats (such as AGAGAGAG), and repeats of greater sequence complexity. Because of strand slippage, which occurs when the parental and nascent strands slip out of proper alignment, DNA polymerases appear to occasionally "stutter" while copying these repeats, resulting in incorporation of higher or lower copy numbers of the repeat sequence into the newly formed daughter strands (Figure 12.8). Thus, the sequence AAAAAAA, that is, A_7, might well cause a polymerase to synthesize a T_6 or T_8 sequence in the complementary strand. The resulting insertions or deletions may elude detection by the proofreading components of the DNA polymerases and are therefore prime targets for recognition and repair by the mismatch repair machinery.

For historical reasons, highly repeated sequences in the genome, often carrying 100 or more nucleotides per repeat unit, have been called "satellite" sequences. Because the simple, shorter sequences discussed here are also found in many places in the genome, they have been named **microsatellites**. A defective mismatch repair system that fails to detect and remove stuttering mistakes made by DNA polymerases when copying a microsatellite will result in the expansion or shrinkage of its sequences in progeny cells. This creates the genetic condition known as *microsatellite instability* (MIN; Figure 12.9), which may ultimately involve changes in thousands of microsatellite sequences scattered throughout a cell genome.

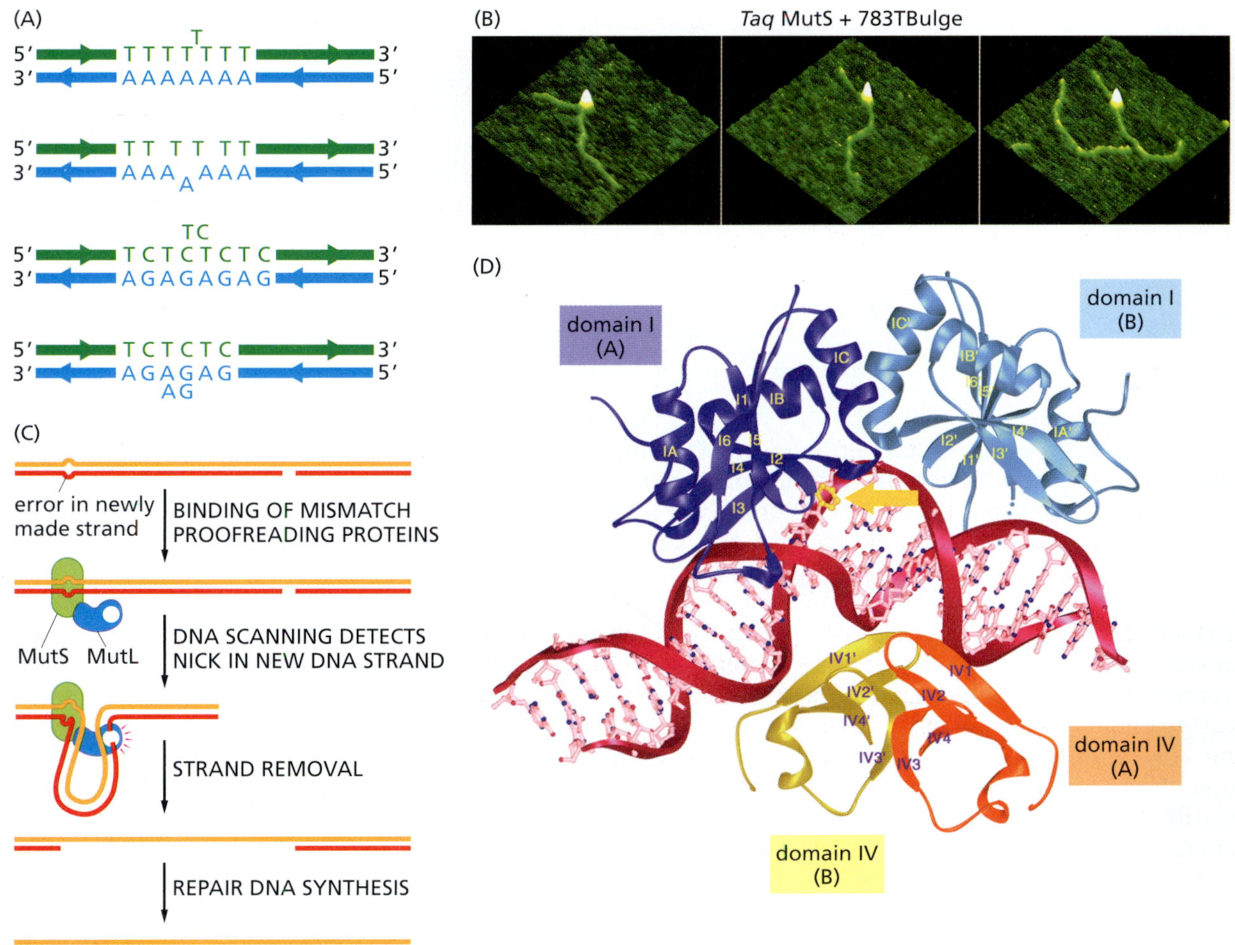

Figure 12.8 DNA polymerase errors and mismatch repair (A) The DNA polymerases, notably pol-δ, occasionally "stutter," or skip a base within a repeating sequence of DNA (e.g., a microsatellite sequence) present in the template strand *(blue)* indicated here. As a consequence, the newly synthesized strand *(green)* may acquire an extra base that increases the length of the repeating sequence or may lack a base *(top two images)*. Identical dynamics may cause similar changes in microsatellite sequences where the repeat unit is a TC dinucleotide segment *(bottom two images)*, or a more complex repeating sequence *(not shown)*. (B) Mismatch repair (MMR) proteins function to recognize and repair the mistakes made by DNA polymerases, including misincorporated bases and inaccurate replication of microsatellite sequences. One powerful technique to visualize the functions of individual MMR proteins uses atomic-force microscopy. Here, the MutS MMR protein of the bacterium *Thermus aquaticus,* a homolog of a number of the mammalian MMR proteins, has been visualized binding to a DNA fragment into which a mismatch has been introduced at a specific nucleotide site. MutS kinks the DNA double helix as it scans for and ultimately finds regions of mismatch, where it binds in a stable fashion, seen here as a *white pyramid.* (C) In eukaryotic cells, two components of the MMR apparatus, MutS and MutL, collaborate to remove mismatched DNA. As illustrated in panel B, MutS *(green)* scans the DNA for mismatches. MutL then scans the DNA for single-strand nicks, which identify the strand *(red)* that has recently been synthesized; the under-methylation of the recently synthesized strand may also aid in this identification. MutL then triggers degradation of this strand back through the detected mismatch, allowing for repair DNA synthesis to follow and generate a properly matched DNA strand. It is unclear whether MutL also uses other clues to determine the recently synthesized DNA strand. (D) The function of the *Thermus aquaticus* MutS MMR protein is revealed in even more detail by X-ray crystallography. Part of the structure the *T. aquaticus* MutS homodimeric protein in complex with a mismatched helix *(red)* is shown here. Domains I and IV of subunit A are in *dark blue* and *orange,* while the corresponding domains of subunit B are in *light blue* and *yellow.* An arrow *(yellow)* indicates where phenylalanine residue 39 of subunit I is associated with an unpaired thymidine in one of the two strands. Defects in the human homolog of this protein play a critical role in triggering hereditary non-polyposis colon cancer (HNPCC), discussed in Section 12.9. (B, from H. Wang, Y. Yang, M. J. Schofield et al., *Proc. Natl. Acad. Sci. USA* 100:14822–14827, 2003; C, from B. Alberts et al., Molecular Biology of the Cell, 4th ed. New York: Garland Science, 2002; D, from G. Obmolova, C.Ban, P. Hsieh and W. Yang, *Nature* 407:703–710, 2000.)

Yet other, more subtle copying mistakes made by a DNA polymerase, such as the incorporation of an inappropriate base in a nonrepeating sequence, may also be detected and erased by mismatch repair proteins, which are highly sensitive to bulges and loops in the double helix caused by inappropriately incorporated nucleotides. The mismatch repair machinery must be able to distinguish the recently synthesized DNA strand from the complementary "parental" strand that served as the template; this enables the MMR apparatus to direct its attention to removing and then repairing the recently synthesized and therefore defective DNA strand (see Figure 12.8C). Mismatch repair involves the excision of the nucleotides that have created the mismatch and a new attempt at synthesis of this strand.

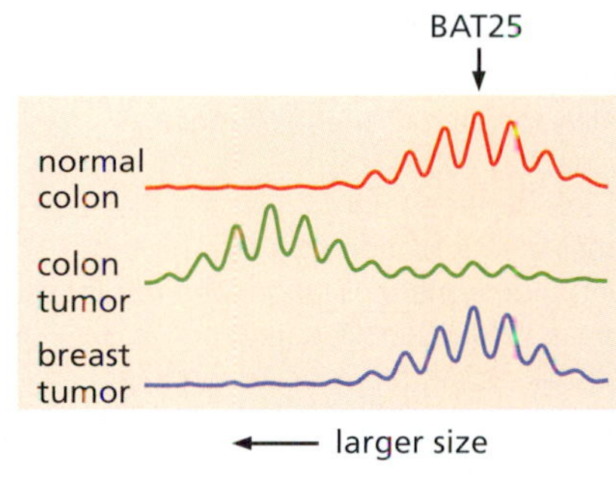

Figure 12.9 Detection of microsatellite instability Microsatellite instability (MIN) often causes an expansion or contraction of the size of a microsatellite repeat sequence. In the analysis shown here, the size of a mononucleotide repeat is revealed using a PCR (polymerase chain reaction), in which the primers bind to sequences flanking the repeat on both sides. The BAT25 sequence, which is located on human Chromosome 4q12, consists of the sequence TTTTxTxTTTTxT7xxT25, where "x" indicates a nucleotide other than T. Because of errors made by the polymerase used in the PCR reaction, the products of a reaction show a Gaussian distribution of lengths grouped around a PCR product that is the actual length of the genomic DNA segment being amplified. This analysis shows the lengths of a microsatellite repeat in a woman suffering from HNPCC (hereditary non-polyposis colon cancer), who has been diagnosed with both colon and breast carcinomas; the DNA of normal tissue adjacent to the colon carcinoma graph has also been analyzed. This analysis reveals a clear increase in size of the microsatellite repeat in the colon carcinoma *(leftward shift)*, while the breast tumor exhibits a microsatellite repeat that is precisely the same as normal, control DNA. (This observation strongly suggests that the breast carcinoma, unlike the colon carcinoma, is unlikely to have been caused by MIN.) (From A. Müller, T.B. Edmonston, D.A. Corao et al., *Cancer Res.* 62:1014–1019, 2002.)

Working together, these various error-correcting mechanisms yield extremely low rates of miscopied bases that survive to become mutant DNA sequences. To begin, DNA polymerases make copying mistakes in only about 1 out of 10^5 polymerized nucleotides. The $3' \rightarrow 5'$ proofreading by the polymerases overlooks about 1 out of every 10^2 nucleotides initially miscopied by the polymerase, thereby reducing the error rate to about 1 in 10^7 nucleotides. After the DNA polymerase has passed through a stretch of DNA, the mismatch repair proteins check the recently synthesized DNA strand a second time. The mismatch repair enzymes fail to correct only about 1 miscopied base out of 100 that have escaped the attentions of the proofreading carried out by the DNA polymerase. Together, this yields a stunningly low mutation rate of only about 1 nucleotide per 10^9 that have been synthesized during DNA replication. As we will see, defects in these error-correcting mechanisms can lead to both familial and sporadic human cancers.

Finally, DNA replication holds yet other dangers for the genome. Some measurements indicate that as many 10 double-strand (ds) DNA breaks occur per cell genome each time a cell passes through S phase. These breaks appear to occur near replication forks, ostensibly because the single-strand DNA at the unwound but not-yet-replicated portion of the parental DNA is susceptible to inadvertent breakage (Figure 12.10). Cells have well-developed mechanisms for dealing with such dsDNA breaks, as we will see later. Failure to repair such breaks properly can lead to disastrous consequences, including chromosomal breaks and translocations.

12.5 Cell genomes are under constant attack from endogenous biochemical processes

Most accounts of the origins of contemporary cancer research contain a strong emphasis on the actions of carcinogenic agents that enter the body through various routes, attack DNA molecules within cells, and create mutant cell genomes that occasionally cause the formation of cancer cells. Unrecognized by these models of cancer pathogenesis are the mutagens and mutagenic mechanisms of endogenous origin. In recent decades, however, analytical techniques of greatly improved sensitivity have allowed researchers to detect altered bases and nucleotides in the DNA of normal cells that have not been exposed to exogenous mutagens. The results of these analyses have caused a profound shift in thinking about the origins of most mutant genes present in the genomes of human cells, because they have shown that endogenous biochemical processes usually make far greater contributions to genome mutation than do exogenous mutagens. Since mutagenic events, independent of their origin, are potentially carcinogenic, this has forced a rethinking of how many human cancers arise.

The structure of the DNA double helix, with its bases facing inward, offers a measure of protection from all types of chemical attack by shielding its potentially reactive chemical groups, notably the amine side chains of the bases, from

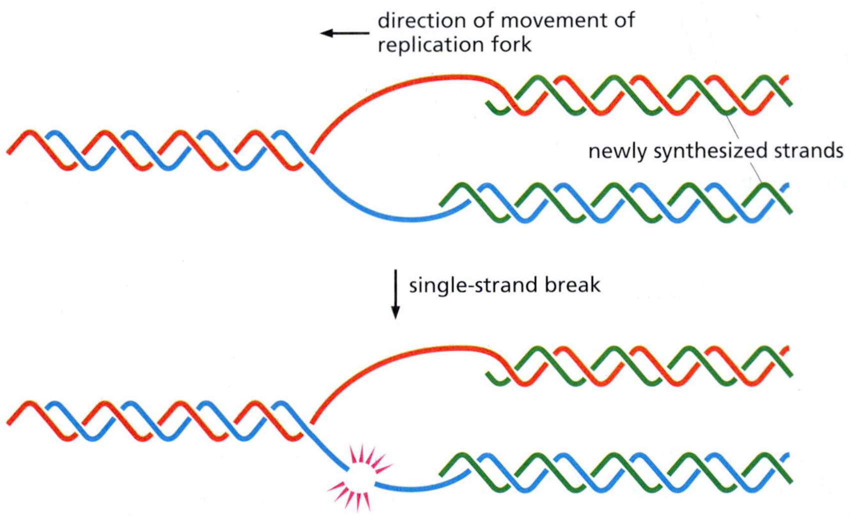

Figure 12.10 Double-strand DNA breaks at replication forks During DNA replication, the DNA molecules are especially vulnerable to breakage in the single-stranded portions of the molecule near the replication fork that have not yet undergone replication. The resulting break is functionally equivalent to a double-strand break occurring in an already-formed double helix, in that the break leaves two helices unconnected by either strand.

various mutagenic agents. In spite of this clever design, DNA molecules are subject to chemical alteration and physical damage. Some of this damage appears to occur through the actions of hydrogen and hydroxyl ions that are present at low concentration ($\sim 10^{-7}$ M) at neutral pH. Often cited in this context is the process of **depurination**, in which the chemical bond linking a purine base (adenine or guanine) to deoxyribose breaks spontaneously (Figure 12.11A). By some estimates, as many as 10,000 purine bases are lost by depurination each day in a mammalian cell. (This amounts to more than 10^{17} chemically altered nucleotides generated each day in the human body!) **Depyrimidination** occurs at a 20- to 100-fold lower rate, but still results in as many as 500 cytosine and thymine bases lost per cell per day. Estimates of the steady-state level of base-free nucleotides present in a single human genome range from 4000 to 50,000.

At the same time, **deamination** may occur, in which the amine groups that protrude from guanine, adenine, and cytosine rings of the bases are lost. This deamination leads respectively to xanthine, hypoxanthine, and uracil (Figure 12.11B). The uracil, for instance, may subsequently be read as a thymine during subsequent DNA replication, thereby causing a C–T point mutation, known as a **transition** mutation, in which one pyrimidine replaces another. The bases generated by deamination are all foreign to normal DNA, and consequently can be recognized as such and removed by DNA repair enzymes. However, any such altered bases that escape detection and removal by these repair enzymes represent potential sources of point mutations.

Spontaneous deamination of 5-methylcytosine—the methylated form of cytosine that we encountered earlier (Section 7.8)—occurs even more frequently, yielding thymine (see Figure 12.11B). This creates a serious problem for the DNA repair apparatus, since thymine (unlike the other three products of deamination described above) is a component of normal DNA, and the T:G basepair may therefore escape detection, survive, and ultimately serve as template during a subsequent cycle of DNA replication, leading to a C-to-T point mutation.

In fact, this deamination of 5-methylcytosine represents a major source of point mutations in human DNA. By one estimate, 63% of the point mutations in the genomes of tumors of internal organs (i.e., in those tissues shielded from UV radiation) arise in CpG sequences. Among mutant *p53* alleles, about 30% seem to arise from CpG sequences present in the wild-type *p53* allele. [To be accurate, this percentage is inflated somewhat by the fact that during lung carcinogenesis, methylated CpG sequences are also favored targets for attack by chemically activated forms of benzo [*a*] pyrene (see Section 12.6), a **polycyclic aromatic hydrocarbon** (PAH) present in tobacco smoke. Hence, not all mutations arising at CpG sites derive from deamination events.]

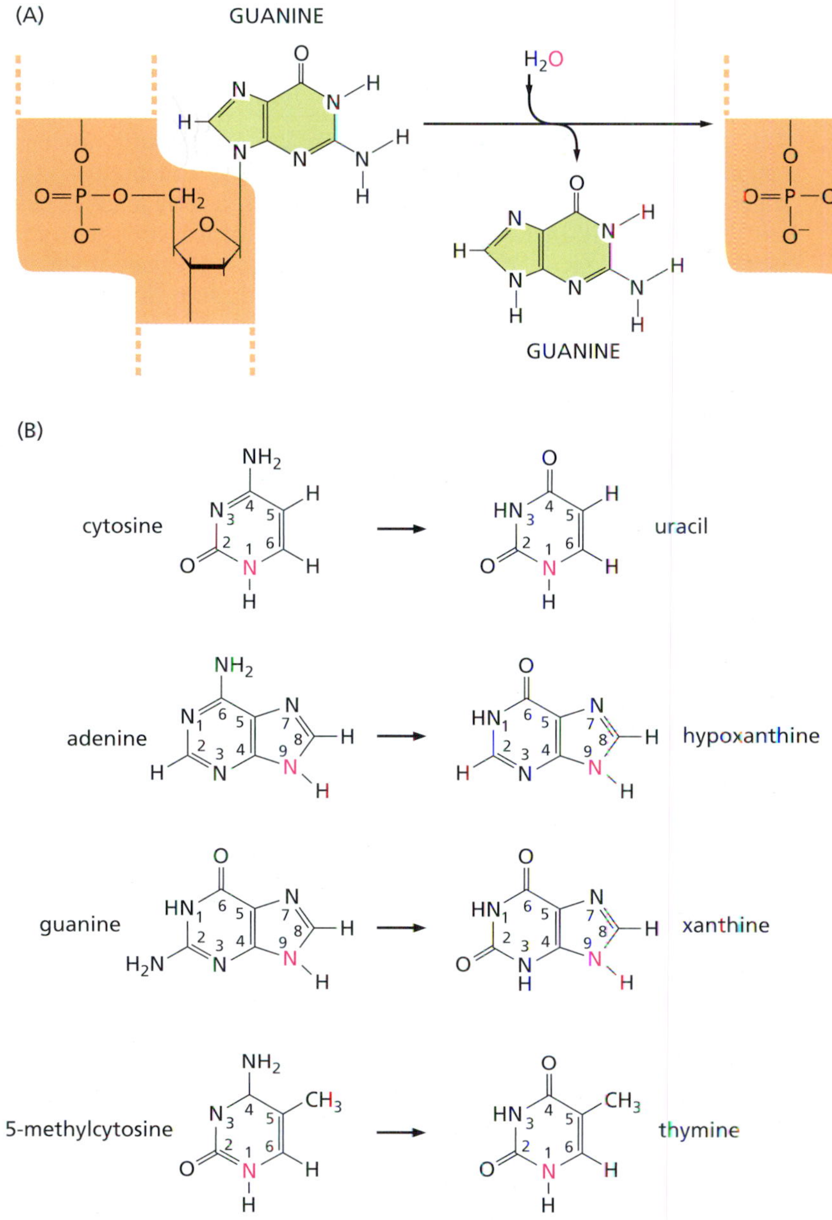

Figure 12.11 Depurination and base deamination (A) Spontaneous depurination frequently affects guanine within DNA, leaving behind a deoxyribose. (B) The deamination reactions affecting purine and pyrimidine bases, which occur spontaneously at various rates at neutral pH, lead to changes in nucleotide sequences unless they are repaired. In each case, the nitrogen atom in red participates in the formation of a covalent glycosidic bond with the 1-carbon of deoxyribose. The deamination of 5-methylcytosine yields thymine *(bottom)*; because this base is naturally present in DNA, it is not always recognized as being aberrant by the repair machinery, explaining the frequent mutations at sites bearing this methylated base. (Adapted from B. Alberts et al., Molecular Biology of the Cell, 4th ed. New York: Garland Science, 2002.)

The intracellular environment holds yet other dangers for the chromosomal DNA. The greatest of these comes from the processes of oxidation, which may inflict far more damage on DNA than the reactions mentioned above. Most important here are the reactions that occur in the mitochondria and generate a variety of intermediates as oxygen is progressively reduced to water:

$$O_2 + e^- \rightarrow \underset{\text{superoxide ion}}{O_2^{\bullet -}} + e^- \rightarrow \underset{\text{hydrogen peroxide}}{H_2O_2} + e^- \rightarrow \underset{\text{hydroxyl radical}}{{}^{\bullet}OH} + e^- \rightarrow H_2O$$

Some of these intermediates, the so-called reactive oxygen species (ROS), may leak out of the mitochondria into the cytosol and thence into the rest of the cell. Included among these are the superoxide ion, hydrogen peroxide, and the hydroxyl radical—the intermediates in the reactions listed above. Yet other oxidants arise as by-products of various oxygen-utilizing enzymes, including those in **peroxisomes** (cytoplasmic bodies that are involved in the oxidation of various cellular constituents, notably lipids), and from spontaneous oxidation of lipids, which results in their peroxidation. Inflammation also provides an important source of the oxidants that favor mutagenesis and therefore carcinogenesis (Sidebar 12.3).

The highly reactive molecules produced by these various processes proceed, usually within seconds, to form covalent bonds with many other molecular species in the cell. Among the many targets of attack by reactive oxygen species are the bases within DNA, including both purines and pyrimidines (Figure 12.12). In addition, reactive oxygen species can induce single- and double-stranded DNA breaks, **apurinic** and **apyrimidinic** sites (together, known as **abasic** sites, in which bases are cleaved away from deoxyribose; e.g., see Figure 12.11A), as well as DNA–protein cross-links. As described below, many of the resulting altered bases are recognized by a repair machinery that proceeds to excise them from the DNA. Some of the excised bases, including thymine glycol, which derives from deoxythymidine glycol, and 8-oxoguanine, which derives from 8-oxo-deoxyguanosine (8-oxo-dG), can be detected and quantified in the urine of mammals, providing some indication of the rate at which they are produced throughout the body (Sidebar 12.4).

Some experiments have shown that the yield of these compounds is directly proportional to the rate of oxidative metabolism in various species (Figure 12.13). The formation of 8-oxo-dG creates a danger of mutation, as one conformation of this altered base can readily pair with A. This mispairing of bases during DNA replication can lead, in turn, to the replacement of a G:C base pair, via G:A pairing, to a T:A base pair (see Figure 12.12B). Such a G $\rightarrow$ T replacement of a purine by a pyrimidine (or the opposite) is often termed a **transversion**. Yet other damage can occur through the methylation of bases triggered by reaction with S-adenosylmethionine, a common metabolic intermediate in cells that carries a highly reactive methyl group. Taken together, the continuing hail of damage from oxidation, depurination, deamination, and methylation, which together may alter thousands of bases per cell genome each day, greatly exceeds the amount of damage created by exogenous mutagenic agents in most tissues.

Sidebar 12.3 Inflammation can have both mitogenic and mutagenic consequences Chronic inflammation of tissues is often provoked by certain infectious agents, such as hepatitis B and C virus infections of the liver, *Helicobacter* infection of the gastric epithelium, and human papillomavirus (HPV) infection of the cervical epithelium (Section 11.15). These infections lead to cell death and resulting compensatory proliferation of the surviving cells—a type of tumor promotion discussed in the last chapter.

These inflammatory responses can also have direct mutagenic consequences. Here's how. The immune system dispatches cells, notably macrophages and neutrophils, to sites of infection; these **phagocytic** cells destroy and consume infected cells within infected tissues. Importantly, such phagocytes kill infected cells in part by releasing bursts of a powerful mix of oxidants, which include nitric oxide (NO), superoxide ion ($O_2^{\bullet -}$), hydrogen peroxide (H_2O_2), and hypochlorite (OCl^-). These highly reactive chemical species kill the intended target cells but, in addition, may cause collateral damage by leaving behind significant numbers of survivors whose genomes they have also damaged. Like the by-products of normal oxidative metabolism, these oxidants act as mutagens on the genomes of nearby bystander cells through their ability to generate chemically modified bases via nitration, oxidation, deamination, and halogenation. Indeed, the DNAs of inflamed and neoplastic tissues have been found to carry substantially increased concentrations of 8-oxo-dG (see Figure 12.12), one of the primary products of DNA oxidation. The increased mutation rate in the cells bearing such oxidized bases helps to explain why chronic inflammation in many tissues favors tumor progression.

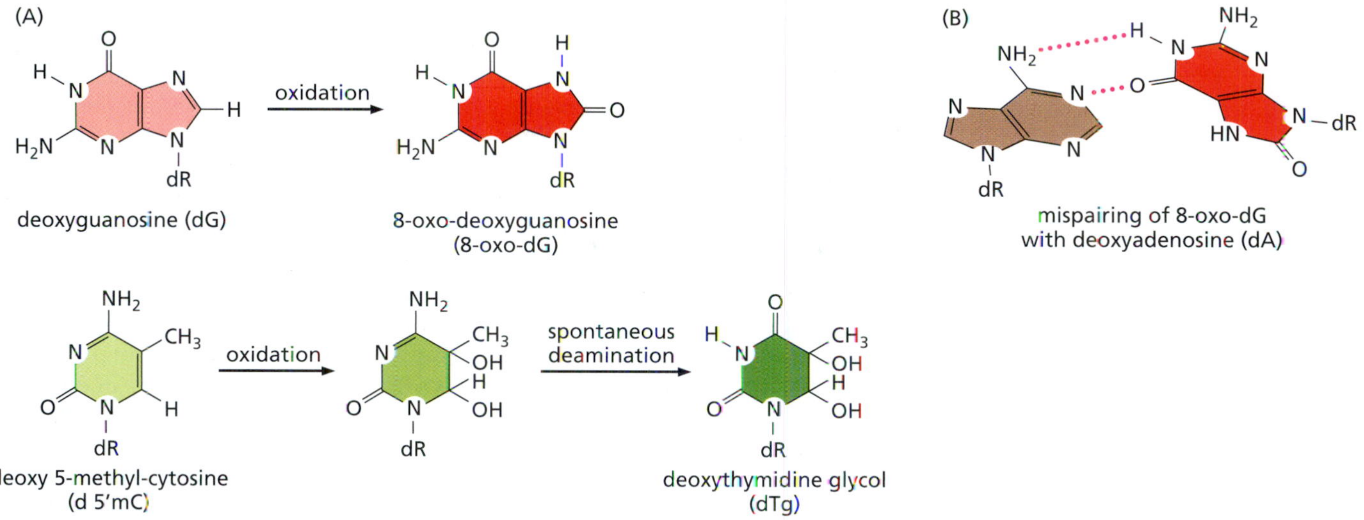

(A)

deoxyguanosine (dG) —oxidation→ 8-oxo-deoxyguanosine (8-oxo-dG)

deoxy 5-methyl-cytosine (d 5′mC) —oxidation→ —spontaneous deamination→ deoxythymidine glycol (dTg)

(B)

mispairing of 8-oxo-dG with deoxyadenosine (dA)

Figure 12.12 Oxidation of bases in the DNA The oxidation of DNA bases, which often results from the actions of reactive oxygen species (ROS), can be mutagenic in the absence of subsequent DNA repair reactions. (A) Two frequent oxidation reactions involve deoxyguanosine (dG), which is oxidized to 8-oxo-deoxyguanosine (8-oxo-dG); and deoxy-5-methylcytosine (d-5′-mC), the nucleotide that is present in methylated CpG sequences. Upon oxidation, the latter initially forms an unstable base that rapidly deaminates, yielding deoxythymidine glycol (dTg). (B) The 8-oxo-dG, which is formed by the oxidation of dG, can mispair with deoxyadenosine (dA) rather than forming a normal base pair with deoxycytosine (dC). Hence, if 8-oxo-dG is not removed from a double helix, the DNA replication machinery may inappropriately incorporate a dA rather than a dC opposite it, resulting in a C–A point mutation. "dR" signifies deoxyribose in all cases. The purines are shown in various shades of *red* and *brown*, while the pyrimidines are shown in various shades of *green*.

12.6 Cell genomes are under occasional attack from exogenous mutagens and their metabolites

As we have seen repeatedly in this text, cellular genomes are also damaged by exogenous carcinogens, including various types of radiation as well as molecules

Sidebar 12.4 Oxidation products in urine provide an estimate of the rate of ongoing damage to the cellular genome By recent estimates, the genomes of some human cells suffer as many as 10^3 oxidative hits a day, about 10-fold less than the rate at which depurination of bases occurs. The resulting oxidized bases are largely but not totally removed and replaced with the appropriate normal bases. Rats cells suffer about 10-fold more oxidative hits per cell per day in their genomes than do human cells because they have about a 7-fold greater metabolic rate (Figure 12.13). Any unrepaired oxidative lesions will accumulate with time, especially in the genomes of cells that are not mitotically active.

8-oxo-dG is the most frequently observed nucleotide product of oxidative damage. It seems that 1 to 2% of these oxidized nucleotides fail to be removed by the DNA repair apparatus. Oxidants may oxidize the nucleotide precursor of dG prior to its incorporation into DNA; the oxidized nucleoside triphosphate may then be incorporated instead of dGTP into the DNA. Alternatively, oxidants may attack the guanine base after its incorporation into DNA. The importance of the oxidized dGTP (i.e., 8-oxo-dG triphosphate) is indicated by the fact that a special enzyme—MTH1—is used by mammalian cells to degrade this oxidized DNA precursor; mice lacking MTH1 develop tumors at a 3 to 4 times higher rate than their wild-type counterparts. (Yet another highly specialized enzyme, called MUTYH, excises adenines that have been misincorporated opposite 8-oxo-dG bases in the DNA.) The 8-oxo-dG excised from DNA is largely excreted in the urine.

Unfortunately, the analyses of oxidation products of DNA have been subject to a number of artifacts, including notably the inadvertent oxidation of DNA and nucleosides *in vitro*. On one occasion, aliquots of one DNA preparation were sent to 21 laboratories in Europe for measurement of 8-oxo-dG content; the resulting analyses yielded estimates ranging over a factor of more than 200. The estimates of the numbers of oxidized bases in cell genomes have fallen dramatically in recent years. Nonetheless, the newer, more conservative estimates place the steady-state level of 8-oxo-dG residues in the DNA isolated from an average human cell at about 3000. These steady-state levels are comparable to the level of chemically altered bases that are formed in the DNA of target tissues of laboratory animals that have been exposed to high, carcinogenic doses of compounds such as aflatoxin and heterocyclic amines.

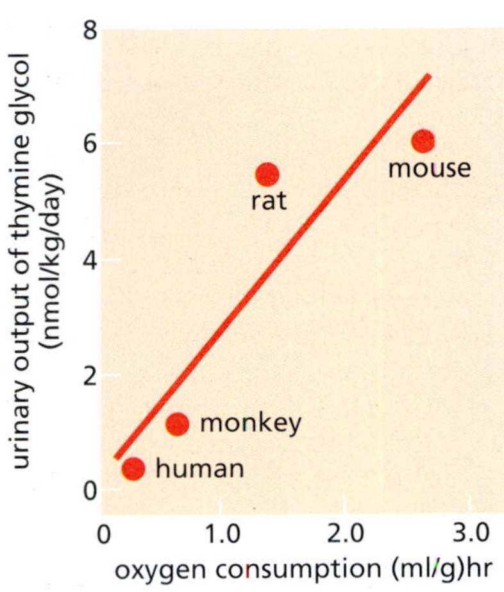

Figure 12.13 Metabolic rate and DNA oxidation The metabolic rate, and thus the rate of oxygen consumption, varies inversely with body size in mammals, being about one order of magnitude higher in rodents than in humans. As indicated here, higher oxygen consumption (ml of O_2 per g of body weight per hour, *abscissa*) correlated with an increased rate of base oxidation of DNA, ostensibly created by ROS (reactive oxygen species) that are the by-products of oxidative phosphorylation in the mitochondria. Thymine glycol *(ordinate)* is the product of excising from DNA the pyrimidine base present in deoxythymidine glycol (see Figure 12.12), one of the common oxidation products of DNA; this base is eventually excreted in the urine. The 6 nmol per kg of body weight per day excretion rate measured in mice and rats corresponds to approximately 3000 thymidines oxidized per cell per day. (From B.N. Ames, *Free Radic. Res. Commun.* 7:121–128, 1989; and from B.N. Ames and L.S. Gold, *Mutat. Res.* 250:3–16, 1991.)

that enter into the body via the food we eat and the air we breathe. Among the best studied of the exogenous carcinogens are X-rays, often termed "ionizing radiation" because of the ionized, chemically reactive molecules that this form of electromagnetic energy creates within cells. As much as 80% of the energy deposited in cells by X-rays is thought to be expended in stripping electrons from water molecules. The resulting free radicals proceed to generate reactive oxygen species (ROS). X-rays can alter the DNA, because they often generate free radicals that create single- and double-strand breaks in the double helix. As discussed later, these double-strand breaks are often difficult to repair and may, on occasion, generate breaks in a chromosome that are visible microscopically during metaphase.

Ultraviolet (UV) radiation from the sun is a far more common source of environmental radiation than X-rays. Living organisms have had to contend with UV radiation since life first formed on this planet some 3.5 billion years ago. Once oxygen accumulated to high levels in the atmosphere about 0.6 billion years ago, the ozone formed from atmospheric oxygen provided a protective shield that significantly attenuated the flux of UV radiation striking the Earth's surface. Nonetheless, a significant amount of UV still succeeds in penetrating the ozone shield and reaching the biosphere.

Should UV photons strike a DNA molecule in one of our skin cells, a frequent outcome is the formation of pyrimidine dimers—that is, covalent bonds form between two adjacent pyrimidines in the same strand of DNA. In principle, these can form between two adjacent C's, two adjacent T's, or a C and an adjacent T. In mammals, where the percentages of A's, C's, G's, and T's are similar, more than 60% of the pyrimidine dimers are TT and perhaps 30% are CT dimers, with the remaining dipyrimidines being CC dimers. As seen in Figure 12.14, a pair of covalent bonds are formed between adjacent pyrimidines, resulting in the creation of a four-carbon (cyclobutane) ring. Another, less common class of DNA photoproducts, termed pyrimidine (6-4) pyrimidinone, also involves covalent linkage between two adjacent pyrimidines. Once formed, pyrimidine dimers are very stable and can persist for extended periods of time in the DNA unless they are recognized and removed by certain DNA repair enzymes.

The fact that these pyrimidine dimers are mutagenic is demonstrated dramatically by the spectrum of *p53* mutations found in the DNAs of **keratoses** (benign skin lesions) and basal cell carcinomas of the skin. In these growths, many of the mutant *p53* alleles carry a dipyrimidine substitution. While the TT dimer is the one most frequently formed by UV radiation, it is only weakly mutagenic, because various DNA repair and replication enzymes, to be discussed later, are able to deal with it effectively. This explains why CC → TT substitutions, which arise from CC (rather than TT) dimers, are the most common consequences of UV light mutagenesis. Since UV photons characteristically cause this mutation, these observations provide further support for the notion that UV rays are directly mutagenic and carcinogenic for the human skin. (Other evidence for a direct causal role of UV radiation is epidemiologic: the incidence of squamous cell skin carcinomas doubles with each 10° decline in latitude, reaching its peak at the equator, where cumulative UV exposure is highest.) As might be expected, this particular type of *p53* mutation is otherwise extremely rare in the genomes of the many types of human tumors that arise in internal organs and are therefore shielded from UV radiation.

A variety of chemical species can enter the body from outside, undergo chemical modification, and then proceed to react with the macromolecules within cells, among them the DNA. Many of these modified chemical species are **electrophilic**, that is, they are molecules that seek out and attack electron-rich regions of target molecules. Among the most potent mutagens are a class of chemicals, termed **alkylating** agents, that are capable of attaching alkyl groups covalently to the DNA bases (Figure 12.14C).

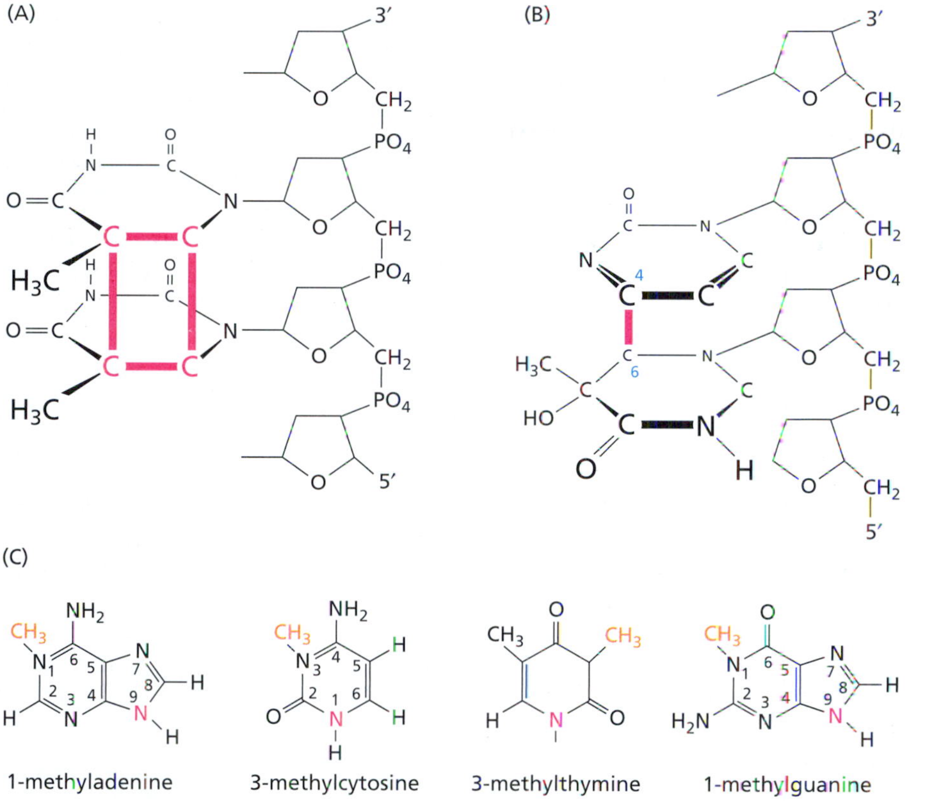

Figure 12.14 Products of UV irradiation and methylation of DNA Ultraviolet (UV) radiation produces covalent cross-links between adjacent pyrimidine bases in the DNA. When purified DNA is irradiated with 254-nm photons, 71% of the resulting photoproducts formed are the cyclobutane pyrimidine dimers (CPD; A), while 24% are the pyrimidine (6-4) pyrimidinone (6-4 PP) photoproducts (B). The cyclobutane ring of a CPD is highlighted in *red* in panel A, as is the bond linking the 6-position of one pyrimidine to the 4-position of the adjacent pyrimidine in panel B. These structures are relatively stable and must be removed by transcription-coupled repair and global genomic repair (to be described in Section 12.8). (C) Exogenous alkylating agents can covalently alter DNA bases by the attachment of alkyl groups, such as the methyl groups *(orange)* shown here. Many of these methyl groups may also be generated endogenously by the inadvertent actions of S-adenosyl methionine, which carries a highly reactive methyl that plays a key role in many normal biosynthetic reactions. The nitrogens that form glycosidic linkages with deoxyribose are shown in *pink*.

The alkylation of a base may destabilize its covalent bond to deoxyribose, resulting in the loss of the purine or pyrimidine base from the DNA. Alternatively, the alkylated bases may be misread by the DNA polymerase machinery during DNA replication. Because of their potent mutagenicity, alkylating agents are often used experimentally to induce various types of tumors in laboratory animals. (Moreover, because certain alkylating agents that are used clinically as anti-cancer chemotherapeutics are also potent mutagens, a delayed outcome of chemotherapy may be the appearance, at a second anatomical site, of a new, therapy-induced tumor.)

A number of potent mutagens are formed when ingested or inhaled compounds become altered by metabolic processes occurring inside our cells. Take, as an example, benzo[*a*]pyrene (BP), a potent carcinogen that falls in the class of polycyclic aromatic hydrocarbons (PAHs), that is, molecules carrying multiple benzene rings fused together in various combinations (see Figure 2.22). Experiments conducted in Britain in the late 1920s indicated that this compound is a prominent carcinogen found amid the complex mixture of compounds in coal tar.

An elaborate array of enzymes belonging to the class of cytochrome P-450 enzymes (Cyp's) are dispatched by the cell to oxidize polycyclic hydrocarbons. (The genes for 46 distinct P-450s have been uncovered in the human genome.) The goal of the cell is to detoxify these foreign chemical species and convert them into molecules that are soluble and can be readily excreted (Figure 12.15A). However, an inadvertent outcome of this detoxification is often the creation of chemical species that are highly reactive with the DNA and are therefore actively mutagenic (Figure 12.15B). As a consequence, chemically inert, unreactive *procarcinogens* are converted into highly reactive *ultimate carcinogens* that can attack DNA molecules directly through their ability to form covalent bonds with various bases. The chemical entity formed after reaction of a carcinogen with a DNA base is often termed a **DNA adduct** (Figure 12.16).

(A)

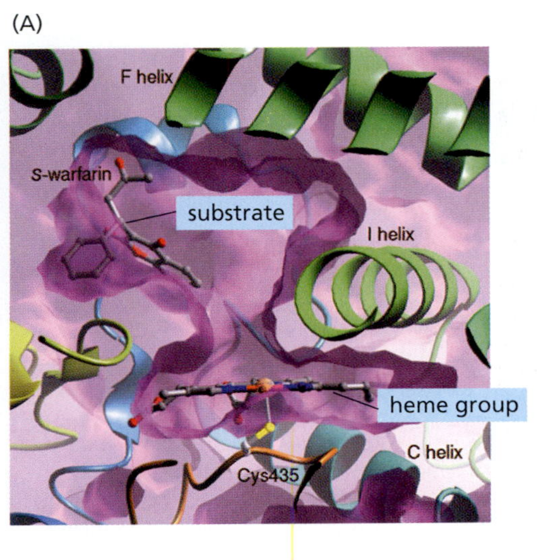

Figure 12.15 Actions of cytochromes on procarcinogens (A) Cytochrome P-450s (Cyp's) are involved in the biosynthesis of a variety of metabolites such as steroid hormones, cholesterol, and bile acids as well as the degradation of compounds such as fatty acids and steroids. In addition, they are used to aid in the oxidation and associated detoxification of a variety of *xenobiotics* (compounds originating outside the body), such as drugs and carcinogens. The substrate-binding cavity of human CYP2C9, shown here, carries a molecule of a xenobiotic substrate, in this case warfarin (used both as an anticoagulant drug and as a rat poison; substrate, *above*). The warfarin molecule is juxtaposed to the heme ring *(below)*, which functions to oxidize the warfarin, thereby detoxifying it. The large substrate-binding cavities of Cyp's allow them to accommodate a wide range of substrates, most of which are quite hydrophobic. (B) Among the xenobiotic compounds entering the body are a variety of polycyclic aromatic hydrocarbons (PAHs) that derive from tobacco smoke, broiled foods, and polluted environments. A common PAH is benzo[*a*]pyrene (BP, *far left*), which, following two successive oxidation reactions mediated by cytochrome P-450 enzymes (largely CYP1A1), is converted to benzo[*a*]pyrenediolepoxide (BPDE) *(far right)*. This highly reactive molecule is termed an *ultimate carcinogen* (Section 2.9) because, unlike its BP precursor, it is able to directly attack and form covalent adducts with DNA bases, which often generate oncogenic mutations. (A, from P.A. Williams, J. Cosme, A. Ward et al., *Nature* 424:464–468, 2003; B, from E.C. Miller, *Cancer Res.* 38:1479–1496, 1978.)

(B)

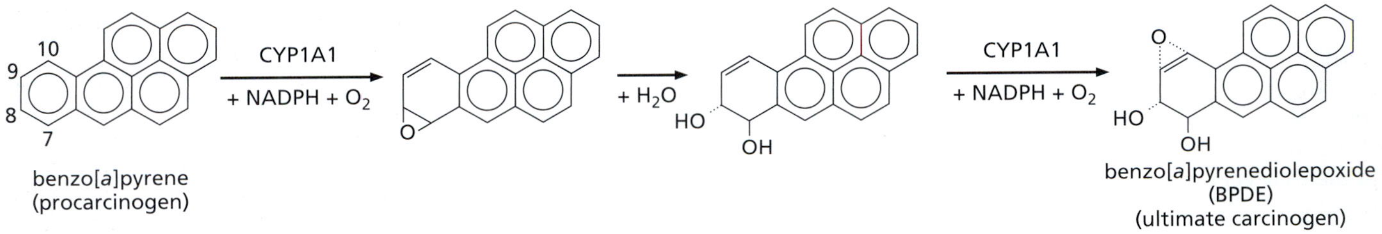

10 9 8 7 benzo[*a*]pyrene (procarcinogen) → CYP1A1 + NADPH + O_2 → + H_2O → CYP1A1 + NADPH + O_2 → benzo[*a*]pyrenediolepoxide (BPDE) (ultimate carcinogen)

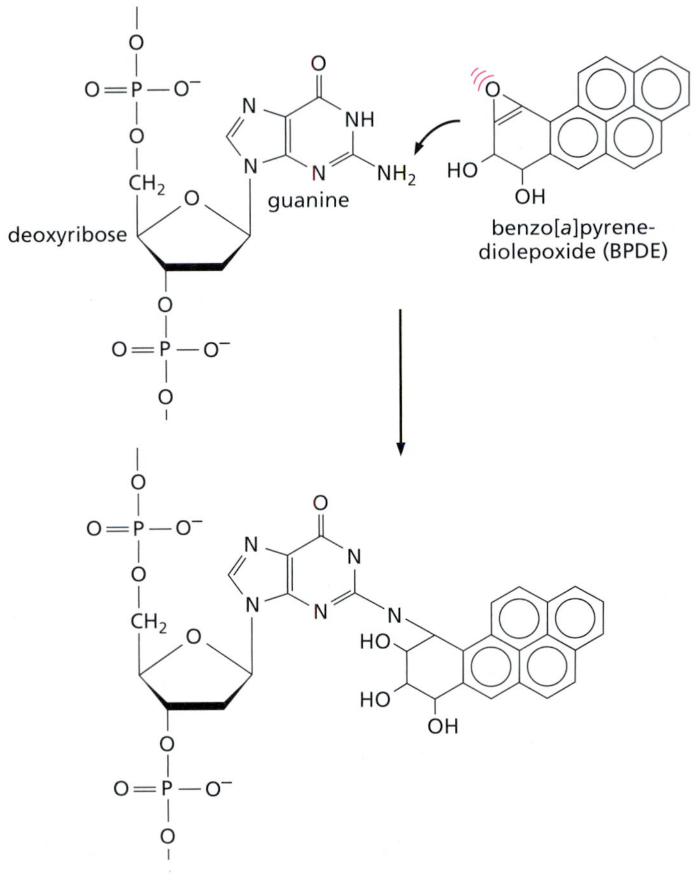

deoxyribose guanine benzo[*a*]pyrene-diolepoxide (BPDE)

Figure 12.16 DNA adducts The chemically reactive epoxide group of benzo[*a*]pyrenediolepoxide (BPDE; see Figure 12.15B) can attack a number of chemical sites in DNA, including the extracyclic amine of guanine *(shown here)*, as well as the two ring nitrogens and the O[6] of this base *(not shown)*. Because the cell may remove these various adducts with different efficiencies, the O[6] adduct of BP may be more potently mutagenic than the more frequently formed adduct shown here. Even though polycyclic aromatic hydrocarbons and BP have been studied for more than half a century, their precise contributions to human cancer development remain a matter of debate, although indirect evidence provides strong suggestion of an important role (see Figure 12.17).

In most cases, chemically reactive, ultimate carcinogens attack other molecules almost immediately after these mutagens are formed. Consequently, they have lifetimes as free molecules that are measured in seconds; this dictates that many of the genetic lesions created by the activated carcinogens arise in the same cells where these molecules underwent initial metabolic activation. For example, benzo[*a*]pyrene (BP), cited above as an important carcinogenic component of coal tar, is also a prominent carcinogenic component of tobacco smoke. Consequently, BP is often activated in the first cells that it enters—the epithelial cells in the lungs of smokers. Once formed, the activated derivative, benzo[*a*]pyrenediolepoxide (BPDE), proceeds directly to form adducts with the guanosine residues in the DNA of these epithelial cells (see Figure 12.16), as suggested by some of the characteristic base substitutions found in the mutant *p53* alleles of resulting lung cancers (Figure 12.17).

Among the most potent of exogenous carcinogens is aflatoxin B1 (AFB1), which is produced by molds belonging to the *Aspergillus* genus. These molds grow on peanuts and grains that have not been stored properly. Those people living in areas where AFB1 exposure is high run a 3-fold elevated risk of hepatocellular carcinoma (HCC), while those carrying a chronic hepatitis B viral infection (Section 11.15) have a 7-fold increased risk of this disease. In some areas of the world, individuals who suffer this viral infection and are also exposed to AFB1-tainted food run a 60-fold increased risk of contracting liver cancer (Figure 12.18A).

Once AFB1 is activated by Cyp's in the liver, the resulting metabolite can attack guanine and form a DNA adduct by becoming covalently linked to this base (Figure 12.18B). AFB1 causes a characteristic G-to-T mutation in DNA. Such

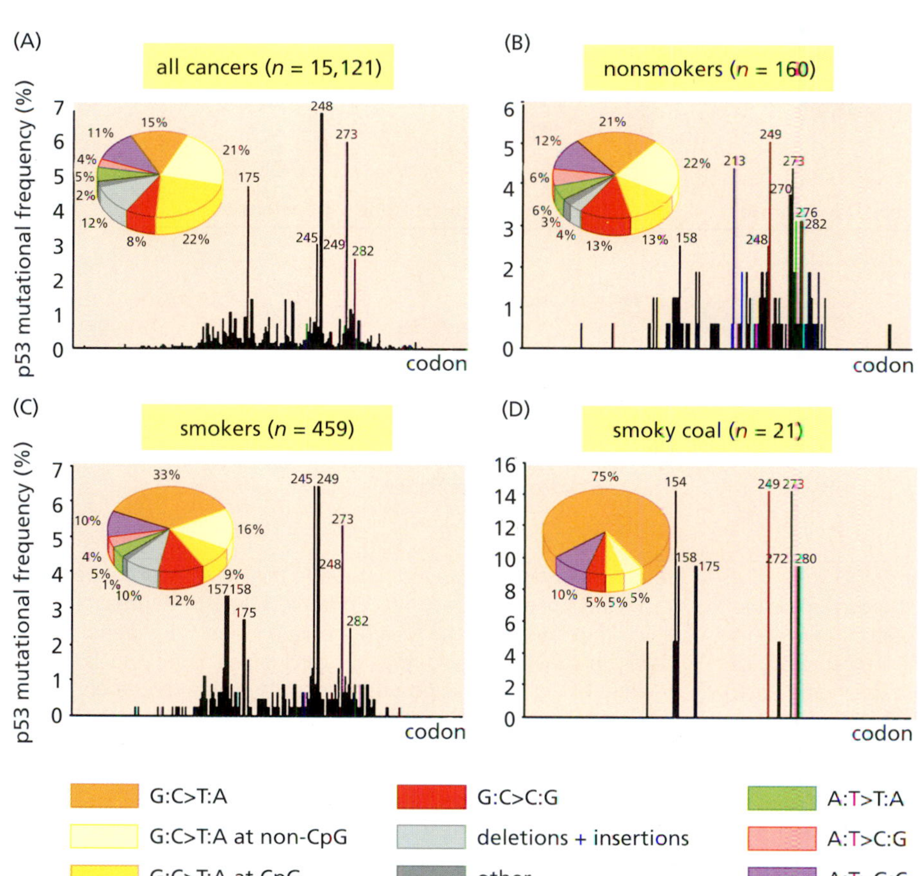

Figure 12.17 *p53* point mutations caused by mutagens The point mutations found in the *p53* alleles carried by human cancer cells provide clues about the identities of the responsible mutagenic agents. (A) All types of cancer; (B) lung cancers in nonsmokers; (C) lung cancers in smokers; (D) lung cancers in nonsmokers exposed to smoky coal emissions. In each case the number of tumors being analyzed is indicated by *n*.

Pie charts: G:C-to-T:A (i.e., G-to-T) transversions have been found in 15% of a group of more than 15,000 mutant *p53* alleles associated with a variety of human tumors. However, in lung carcinomas of nonsmokers, 21% of mutant *p53* alleles carried this transversion, while 33% of the mutant *p53* alleles of lung cancers arising in cigarette smokers showed this transversion, which was present in 75% of the lung cancers of nonsmokers who had a history of repeated exposure to smoky coal emissions. This G-to-T transversion has been found experimentally to be the mutation usually induced by benzo[*a*]pyrene (BP), which is known to be present in the combustion products of fossil fuels as well as tobacco. *Bar graphs:* Additional clues about the mutagenic process, whose meaning is not yet clear, appear to be provided by the locations in the *p53* gene of the various point mutations seen in the bar graphs. The number above each bar in a bar graph designates the codon in the human *p53* reading frame that was found to be affected by a point mutation. The ordinate indicates the percentage of the tumors studied within a group that carried mutations in each of the codons indicated along the abscissa. (From A.I. Robles, S.P. Linke and C.C. Harris, *Oncogene* 21:6898–6907, 2002.)

point mutations, where the sequence AGG has been converted to AGT, are found at codon 249 of the *p53* tumor suppressor gene in about half of the hepatocellular carcinomas occurring in individuals exposed to this carcinogen. These characteristic changes in the DNA provide compelling evidence of the direct interaction of this mutagenic carcinogen with bases in the DNA.

Another widely studied example of carcinogens of exogenous origin involves the heterocyclic amines (HCAs), a class of molecules that are formed in large amounts when meats of various sorts are cooked at high temperatures (Figure 12.19A). These compounds arise through the reactions that take place between naturally occurring molecular species in cells, notably creatine, glucose phosphates, dipeptides, and free amino acids. The HCAs are undoubtedly carcinogenic. For example, the most abundant of these compounds in meats cooked at high temperature—2-amino-1-methyl-6-phenylimidazo[4,5-b]pyridine (PhIP)—is capable of inducing colon and breast carcinomas in rats and lymphomas in mice. PhIP is recognized as being the principal HCA in the human diet (Sidebar 20).

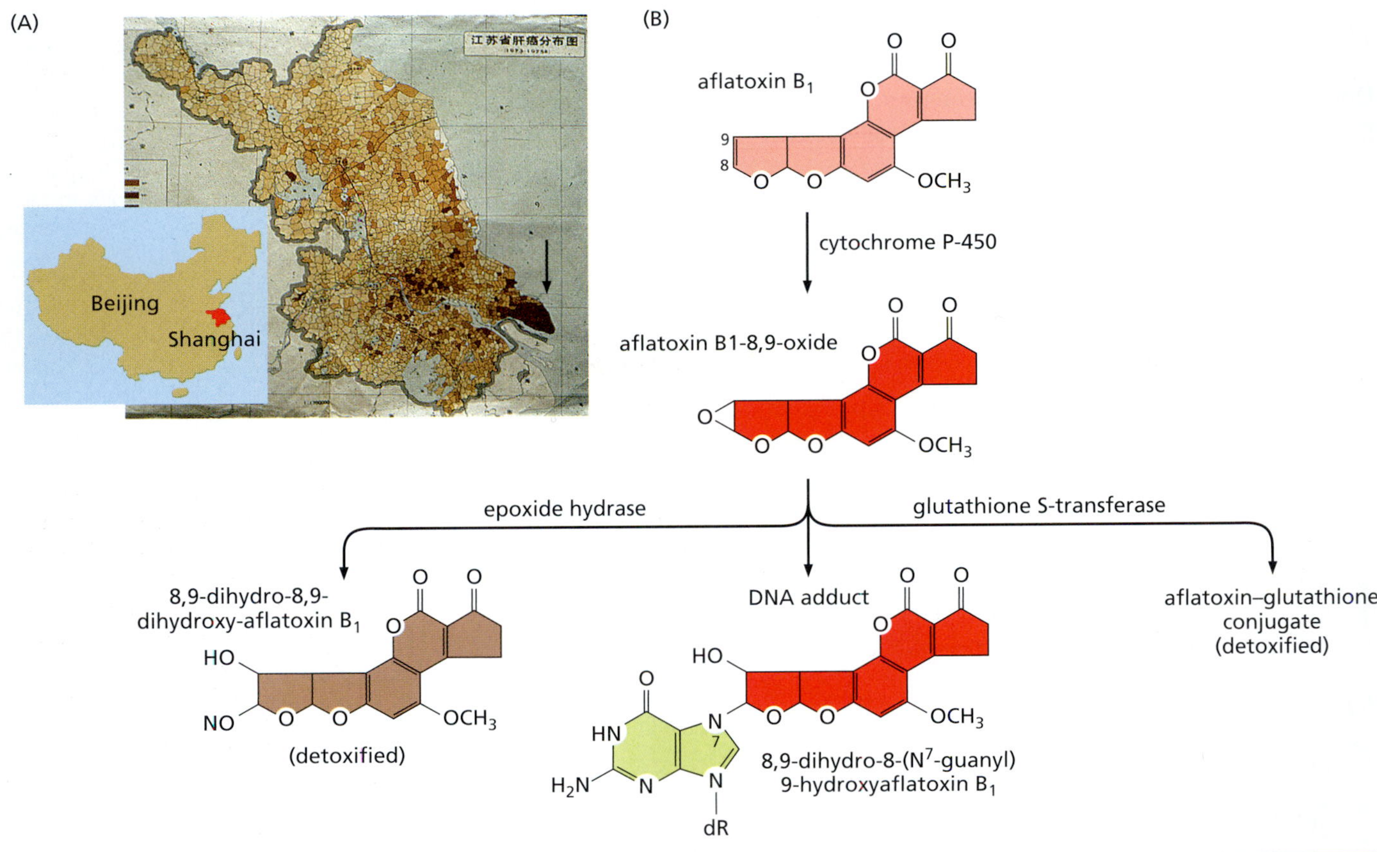

Figure 12.18 Aflatoxin and liver carcinogenesis (A) The fungal toxin aflatoxin B1 (AFB1) is made by a mold that grows on peanuts and grains that have been stored improperly, notably those stored in areas of high rainfall and humidity. Within the Jiangsu province of eastern China, the incidence of hepatocellular carcinoma (HCC) cases *(brown dots)* is 8-fold higher in the very humid southeastern Qidong peninsula *(arrow)* than in the northwestern parts of the province. The incidence of hepatitis B virus (HBV) infection—a critical co-carcinogen in HCC pathogenesis (Section 11.15)—is relatively constant across the province. (B) Activation of AFB1 *(pink)* by cytochrome P-450s results in the 8,9-oxide form *(red)* that is highly reactive. This may be detoxified through several side reactions *(bottom left, right)*. However, this activated form of AFB1 can, with some frequency, react directly with DNA, forming a covalent adduct with the N^7 atom of guanosine *(light green)* that is highly mutagenic. Indeed, the liver carcinomas of individuals living in areas of high AFB1 exposure often carry mutant *p53* alleles with a characteristic G-to-T transversion in codon 249—precisely the type of mutation that would be expected from the known reactivity of AFB1. (A, from T.W. Kensler, G.S. Qian, J.G. Chen and J.D. Groopman, *Nat. Rev. Cancer* 3:321–329, 2003; B, from J.D. Groopman and L.G. Cain, in C.S. Cooper and P.L Grover, eds., Interactions of fungal and plant toxins with DNA in chemical carcinogenesis and mutagenesis. Berlin: Springer Verlag, 1990.)

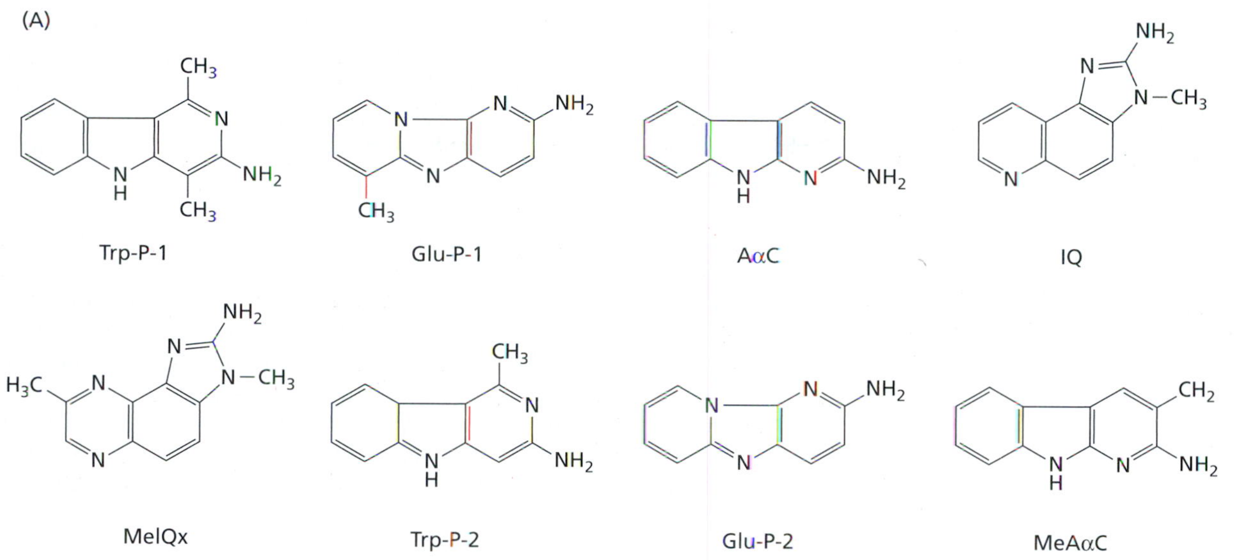

(A)

Trp-P-1 Glu-P-1 AαC IQ

MeIQx Trp-P-2 Glu-P-2 MeAαC

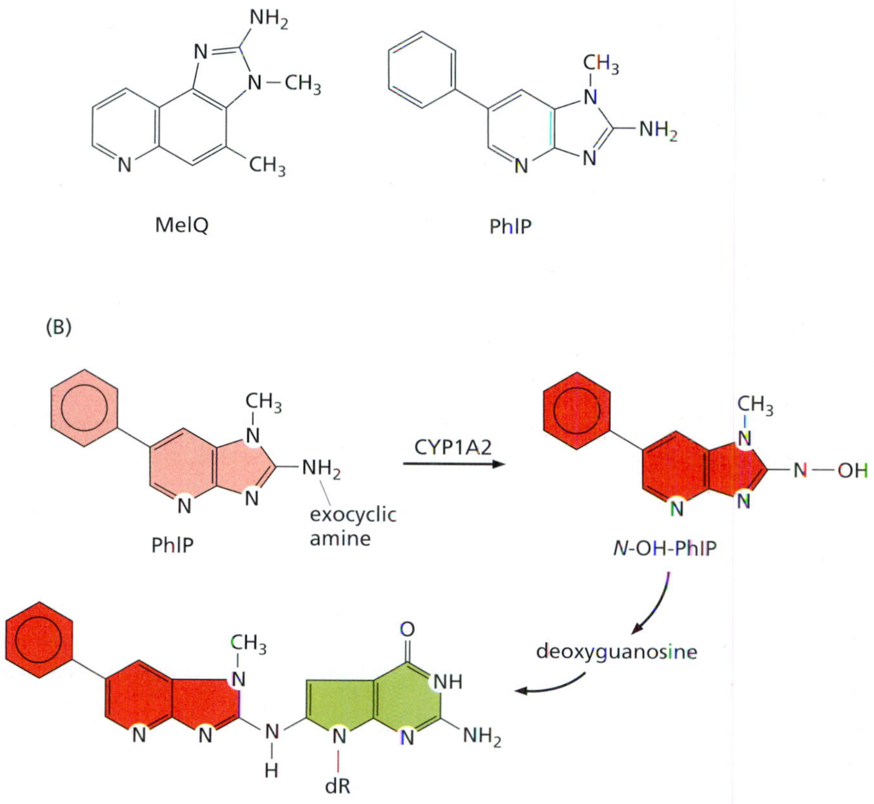

MeIQ PhIP

(B)

PhIP exocyclic amine CYP1A2 N-OH-PhIP deoxyguanosine

Figure 12.19 Heterocyclic amines (A) The 10 heterocyclic amines (HCAs) whose structures are shown here are the most common HCAs to which humans are exposed. This class of compounds derives its name from the fact that each of them is composed of multiple fused rings, which are generally formed from both carbon and nitrogen that bear one or more *exocyclic* amine groups protruding from the ring structure. HCAs arise through cooking various foods, notably red meats, at high temperature. PhIP has been estimated to constitute two-thirds of the total dietary intake of HCAs among Americans. (B) The oxidation of the exocyclic amine of PhIP (2-amino-1-methyl-6-phenylimidazo [4,5-b] pyridine; *pink*) by CYP1A2, a cytochrome P-450, leads to the highly reactive compound N-OH-PhIP (*red*). It can react with the 8-C of deoxyguanosine *(green)* to form a mutagenic adduct (*lower left*). (A, from T. Sugimura, *Carcinogenesis* 21:387–395, 2000; B, from M. Nagao and T. Sugimura, eds., Food-borne Carcinogens. New York: John Wiley & Sons, 2000.)

Once heterocyclic amines have entered into cells, Cyp's are used by the cells to oxidize these molecules. Some Cyp's will oxidize the rings of heterocyclic amines, while others will oxidize the **exocyclic** amine groups, that is, those that protrude from the rings. Ring oxidation by Cyp's leads to successful detoxification; amine group oxidation, however, leads to the formation of highly reactive compounds that can readily form covalent bonds with proteins and DNA (Figure 12.19B). While these and other chemical conversions of HCAs are largely achieved in the liver, the resulting reactive molecules often survive long enough to be released by the liver and to pass via the circulation into other organs where they may exercise their mutagenic activity by attacking DNA. (For example, it is suspected that heterocyclic amines derived from the cooking of red meats contribute to the high incidence of prostate cancer in Western populations.) Together, these various examples, only a few among many that could be cited here, illustrate how the detoxifying enzymes in our cells, often present in high concentration in liver cells, yield genotoxic compounds rather than the end-products intended by evolution—harmless, readily excretable chemical species.

The notion that exogenous and endogenous mutagens (the latter including DNA replication errors, spontaneous depurinations, and the actions of endogenously generated chemical species) constitute distinct, separable causes of human cancer is suggested by the analyses of the specific point mutations present in the mutant *p53* alleles carried by lung carcinomas of smokers and nonsmokers (see Figure 12.17). For example, G-to-T transversions, in which a pyrimidine base (T) replaces a purine (G), are found in one-third of the mutant *p53* alleles in the tumors of smokers. This base substitution conforms to the known mutagenic actions of the polycyclic aromatic hydrocarbons, notably benzo[*a*]pyrene (BP), that are present in high concentrations in cigarette smoke. These genetic lesions are found less frequently in the mutant *p53* alleles in other kinds of tumors (except liver cancers). Moreover, only about 21% of the mutant *p53* alleles in the lung tumors of nonsmokers show these transversions, and a significant fraction of the tumors in this subgroup may well have arisen in passive smokers, that is, those living in close contact with smokers. The remaining mutant *p53* alleles of the nonsmokers' tumors carry mutations that are more typical of the spontaneous alterations in DNA described in Section 12.5.

Yet another perspective on the apparent contribution of benzo[*a*]pyrene to human cancers comes from study of mammary carcinomas. In one study, those women who were actively smoking at the time of breast cancer diagnosis had more than a 10-fold greater likelihood of carrying G-to-T transversions in their tumor-associated *p53* alleles than did nonsmokers.

The successes in identifying electrophilic compounds and alkylating agents that are potent mutagens and thus carcinogens has led to the widespread assumption that other, similarly acting chemical species that enter into the body through food, water, or air are also important in provoking many types of human cancers. However, the fraction of human malignancies that are traceable to the actions of specific mutagenic carcinogens in our environment or food supply remains a matter of great contention (Section 11.18). And it is plausible that, with the exception of a small number of exogenous mutagens, including UV radiation, tobacco combustion products, aflatoxin, and heterocyclic amines, relatively few mutagenic agents enter our bodies, create genetic damage, and thereby cause cancer.

12.7 Cells deploy a variety of defenses to protect DNA molecules from attack by mutagens

The most effective way for a cell to defend its genome from disruption by mutagenic agents is to *physically* shield its DNA molecules from direct attack. In the case of ultraviolet rays from the sun, these penetrate poorly into the body's tissues, leaving the cells of the skin and pigmented cells in the retina as the only vulnerable tissues. The skin shields itself from UV radiation using the **melanin** pigment that is donated by **melanocytes** to keratinocytes located in the basal region of the epidermis (the epithelial layer of the skin) (Figure 12.20).

Skin color in humans is determined by the amount of melanin that is transferred from the melanocytes to the keratinocytes. The role of this pigmentation in cancer pathogenesis is highlighted by the oft-cited case of skin cancers in Australia. There, a high flux of UV radiation (because of proximity to the equator) and a lightly pigmented population (deriving until recently largely from the British Isles) combine to create the world's highest incidence of these diseases. In Africa, in contrast, the darkly pigmented human populations living at similar latitudes rarely experience skin cancers. (Among those few who do contract melanoma in Central Africa, tumors of the unpigmented area of the sole of the foot are common.) In the case of X-rays and cosmic radiation, there is no effective physical shielding that can be erected by the body, since these types of radiation can penetrate easily through all biological substances.

These limited options for protection against physical carcinogens contrast with the large number of mechanisms that cells can deploy to intercept *chemical* carcinogens before they have had the opportunity to damage the cellular genome. The ambushing of reactive oxygen species (ROS) and free radicals is the job of a variety of enzymes, including the enzymes superoxide dismutase and catalase; they collaborate to detoxify an ROS into unreactive forms of oxygen. The ROS may also be intercepted by a series of free-radical scavengers, including vitamin C, α-tocopherol (vitamin E), bilirubin, and urate. These molecules will chemically react with the ROSs, thereby detoxifying them.

Yet another important line of defense is erected by enzymes of the class termed glutathione-S-transferases (GSTs), which function to link electrophilic compounds, and thus many carcinogens, with glutathione, thereby detoxifying these compounds and preparing them for further metabolism and secretion. For example, one type of GST, glutathione-S-transferase-π (GST-π), is used to detoxify a wide range of carcinogens (Figure 12.21A,B). Significantly, as many as 90% of human prostate adenocarcinomas show a shutdown of GST-π expression due to methylation of the promoter of the *GSTP* gene (Figure 12.21C)—the same mechanism that is often used by cancer cells to shut down expression of a variety of tumor suppressor genes (Section 7.8). Frequent inactivation of this gene has been reported as well in a number of other human carcinomas. This loss of GST-π expression, which often occurs relatively early in tumor progression, suggests that pre-malignant prostate cancer cells acquire a distinct growth advantage by inactivating this gene. Thus, without this enzyme to defuse certain carcinogens, the genomes of these prostate epithelial cells are attacked more often by actively mutagenic carcinogens. The resulting increased rate of mutagenesis is likely to accelerate the forward march of tumor progression.

A connection between the glutathione-S-transferase enzyme and cancer susceptibility is also suggested by epidemiologic studies. In one such study, the allelic configurations of two separate GST-encoding genes, termed *GSTT1* and

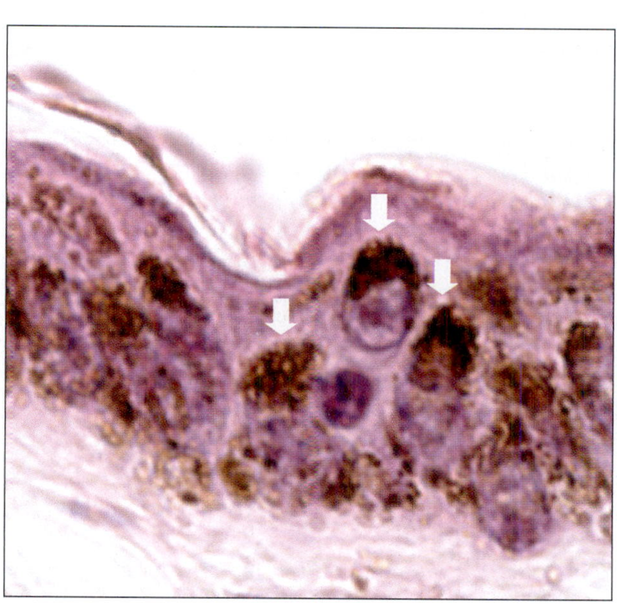

Figure 12.20 Physical shielding of keratinocyte nuclei from ultraviolet radiation The main protection that keratinocytes in the skin have from ultraviolet (UV) radiation, notably UVB photons, derives from the melanosomes—vesicles that carry melanin pigment that have been transferred from melanocytes into keratinocytes located in the basal layers of the epidermis. As seen here, once the melanosomes are acquired by the keratinocytes, they are assembled into tiny parasols/sun umbrellas (sometimes called supranuclear caps) that sit above keratinocyte nuclei *(arrows)* and shield these nuclei from visible and, more importantly, UVB radiation. Keratinocyte nuclei that lack these umbrellas sustain as much as four-fold more UV-induced DNA damage than those that carry them. (Courtesy of D.E. Fisher.)

(A)
glutathione (GSH)

$$\mathrm{HOOC{-}CH_2{-}NH{-}C(=O){-}CH(CH_2SH){-}NH{-}C(=O){-}CH_2CH_2{-}CH(NH_2){-}COOH}$$

glycine | cysteine | glutamic acid

(B)

$$\mathrm{GSH + R{-}CH{-}CH_2\ (epoxide,\ O) \xrightarrow{GST} R{-}C(SG){-}CH_2(OH) \xrightarrow{GST} R{-}CH(OH){-}CH_2(OH)}$$

(C)

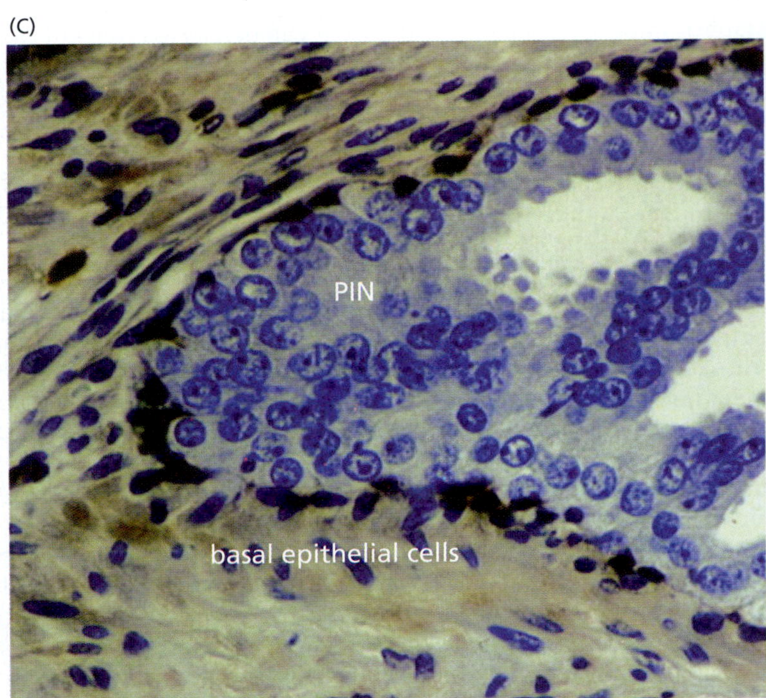

Figure 12.21 Glutathione-S-transferase and its loss from tumors
(A) Glutathione (GSH) is an unusual tripeptide in which a glutamic acid is attached to the amine group of cysteine via the carboxyl side chain of the glutamic acid. (B) Glutathione-S-transferase (GST) enzymes use the sulfhydryl (SH; *pink*) group of the cysteine residue of glutathione to detoxify a number of reactive compounds before they are able to react with cellular target molecules, such as DNA. Shown here is a typical reaction mediated by a GST in which the SH group of glutathione is used to disrupt the highly reactive epoxide group of a compound that has suffered oxidation (where R can be any of a variety of chemical groups). (C) As many as 90% of prostate carcinomas show the loss of expression of an important glutathione-S-transferase, GST-π; this deprives the tumor cells of the means to detoxify many electrophilic mutagens. In many of these cases, loss of GST-π expression is traceable to methylation of the *GSTP1* promoter. As seen here, immunohistochemical analysis using an antibody reactive with GST-π demonstrates that in a pre-malignant PIN (prostatic intraepithelial neoplasia) lesion—a benign precursor to prostate carcinoma—the presence of GST-π is detectable in the basal epithelial cells *(dark brown, left side),* while the luminal epithelial cells *(light blue, right side)* have already lost all GST-π expression. The presence of luminal epithelial cells is revealed here by use of a stain, DAPI, that is specific for cell nuclei *(blue).* (C, from C. Jeronimo, G. Varzim, R. Henrique et al., *Cancer Epidemiol. Biomarkers Prev.* 11:445–450, 2002.)

GSTM1, were examined in a normal control population and in individuals suffering from **myelodysplastic syndrome** (MDS), a hyperproliferative disorder of the bone marrow that often progresses to acute myelogenous leukemia (AML). Of the patients suffering from MDS, 46% carried two null alleles (which encode no enzyme) of *GSTT1;* this genetic state was present in only 16% of the general population used as controls. (In the case of the related *GSTM1* gene, homozygosity of the null allele was found in comparable proportions of the two populations.) Calculations indicated that individuals inheriting two null alleles of the *GSTT1* gene run more than a 4-fold increased risk of myelodysplastic syndrome over those who carry at least one functional allele of the gene. These observations suggest that the T1 isoenzyme of glutathione-S-transferase is involved in some way in detoxifying the compounds that provoke MDS. (An alternative interpretation, which is less likely but nonetheless difficult to exclude at present, is that the *GSTT1* gene is closely linked on the chromosome to a second gene that predisposes an individual to MDS.) Yet other epidemiologic studies hint at connections between carcinogen metabolism and individual cancer risk (Sidebar 12.5).

Still, these elaborate chemical defenses against attacking mutagens fail to protect the genome with a certain frequency. The failure rate depends on the chemical nature of the attacking mutagen and the levels of the molecules deployed by cells to intercept these attackers. As a consequence, altered bases are indeed created in substantial numbers every day in the life of a cell.

These discussions of carcinogen activation lead inevitably to another question: If many of the activating enzymes (such as cytochromes) are designed by nature

Sidebar 12.5 Inter-individual differences in carcinogen activation seem to contribute to cancer risk and responses to therapy Cells use a broad spectrum of enzymes to modify potential carcinogens in a variety of chemical ways, including the attachment of acetyl, glucuronic acid, glutathione, and sulfate groups; many of these chemical modifications aid in the detoxification and eventual excretion of potentially toxic compounds including mutagens. Because of genetic heterogeneity in the human gene pool, the level of expression of many of the enzymes responsible for these detoxification reactions may vary greatly between different individuals. These differences may, in turn, strongly influence the biological responses of an individual to potential carcinogens. For example, a study of 416 lung cancer patients and 446 healthy control patients determined that individuals of a certain genotype had a twofold increased risk of lung cancer. These susceptible individuals carried a particular allele of the cytochrome-encoding *Cyp1A1* gene and null alleles of the gene specifying glutathione-S-transferase M1 (*GSTM1*).

Another study focused on breast adenomas among women who consumed more than 27 ng per day of a certain heterocyclic amine (2-amino-3,8-dimethylimidazo[4,5-f] quinoxaline—MeIQx; see Figure 12.19A) that was present in their diet, apparently from eating large quantities of burnt meat. Women who expressed high levels of *N*-acetyltransferase 1 (NAT1, an enzyme that can help to convert heterocyclic amines into active mutagens) were reported to experience a sixfold increased rate of adenomas, while those with much lower levels of NAT1 showed only a twofold increased risk. In both cases, adenoma frequency was compared with the incidence of adenomas in a third group of women who had low levels of MeIQx in their diet. These correlations suggest that enzymes can influence cancer incidence in two ways: some enzymes affect the rate at which a number of potentially mutagenic compounds are detoxified, while others (inadvertently) convert otherwise nonreactive compounds into chemically reactive mutagens.

In fact, these enzymes intersect with the disease of cancer in yet another way, because they can also function to detoxify chemotherapeutic drugs, thereby blunting the effects of treatment. A particularly dramatic effect of these enzymes on the success of cancer therapy was observed when examining the clinical responses of breast cancer patients who carried functional alleles of the *GSTM1* and *GSTT1* genes; their responses were compared with those of a group of women who had only null alleles at both genetic loci. All of these women were treated with a combination chemotherapy regimen (involving cyclophosphamide, Adriamycin, and 5-fluorouracil) plus radiation. Of those women who had functional *GSTM1* and *GSTT1* alleles, almost all succumbed to their disease within 6 years of treatment. In contrast, of those women with a double-null genotype at both genetic loci, about two-thirds were still alive 8 years after treatment.

In all of the cases cited in this sidebar, these differential responses were correlated with specific alleles. It is important to remember, however, that these are only correlations and not proofs of causality. In each case, it is possible, in principle, that the alleles being studied are closely linked to yet other alleles that are the true causes of the observed increased or decreased responses of patients.

to neutralize and detoxify xenobiotics, what are the origins and the daily burdens of these compounds that our various tissues must routinely contend with? And among these xenobiotics, do man-made carcinogens contribute substantially to this burden (Sidebar 12.6)?

12.8 Repair enzymes fix DNA that has been altered by mutagens

If genotoxic chemicals are not intercepted before they attack DNA, mammalian cells have a backup strategy for minimizing the genetic damage caused by these potential carcinogens. An elaborate DNA repair system exists to continuously monitor the integrity of the genome, to remove inappropriate bases created by chemical or physical attack, and to replace them with those bases that existed prior to the attack. Components of this system also work to stitch together double helices that have been broken by genotoxic agents or accidentally during replication. It seems likely that a mammalian cell deploys several hundred distinct proteins to ensure that damage to DNA is unlikely to result in a mutation being transmitted to daughter cells. Some of these DNA repair proteins figure large in the process of human carcinogenesis, since defects in these proteins result in increased rates of mutation, thereby accelerating the rate of tumor progression.

Cells deploy a wide variety of enzymes to accomplish the very challenging task of restoring normal DNA structure. Importantly, these functions are different from the mismatch repair (MMR) enzymes described above (Section 12.4), since

Sidebar 12.6 Do man-made xenobiotic mutagens ever cause cancer? The heated debate about the man-made carcinogens in the air and human food chain—specifically, the products of synthetic organic chemistry—has continued unabated for half a century. Much of this debate has been focused on trace contaminants in the food chain, notably pesticides, and on the possibility that they may become metabolically activated into potent mutagens and thus carcinogens once they have entered our bodies. Bruce Ames, of the Ames test, has estimated that, by eating naturally occurring foodstuffs, humans are exposed on a daily basis to between 5000 and 10,000 distinct natural chemical compounds and their metabolic breakdown products. Included among these are about 2000 milligrams (mg) of burnt material (the products of cooking various foodstuffs at high temperatures) and 1500 mg of naturally occurring pesticides (used by plants to protect themselves against insect predators). In contrast, the average daily exposure to all synthetic pesticide residues contaminating the food chain is about 0.1 mg. About half of the naturally occurring plant pesticides are found to be carcinogenic when tested in laboratory rodents using standard testing protocols. Since (1) synthetic pesticides are as likely to register as carcinogens in rodent tests as are randomly chosen compounds of natural (i.e., plant) origin; since (2) plant-derived compounds, such as those in the vegetables we eat, are generally presumed to be safe; and since (3) concentrations of synthetic pollutants in the food chain are many orders of magnitude below the natural (and equivalently carcinogenic) plant compounds, this raises the question whether synthetic pesticides are ever responsible for significant numbers of human cancers in Western populations. It may well be that the role of synthetic chemical species in creating human cancers (with the exception of tobacco combustion products and the products of cooking food at high temperature) is limited largely to those chemicals that are encountered repeatedly and at very high concentrations in certain occupations, such as agricultural workers who handle large quantities of pesticides routinely.

the MMR enzymes are largely focused on detecting nucleotides of normal structure that have been incorporated into the wrong positions, while the repair mechanisms discussed here detect nucleotides of abnormal chemical structure.

The simplest strategy for restoring the structure of chemically altered DNA involves an enzyme-catalyzed reversal of the chemical reaction that initially created the altered base. For example, one type of DNA alkyltransferase removes methyl and ethyl adducts from the O^6 position of guanine, thereby restoring the structure of the normal base (Figure 12.22A and B). The importance of this enzyme [O^6-methylguanine DNA methyltransferase (MGMT), often referred to simply as DNA alkyltransferase] in the development of certain kinds of human tumors is suggested by observations that the *MGMT* gene is silenced by promoter methylation in as many as 40% of gliomas and colorectal tumors, and in about 25% of non-small-cell carcinomas, lymphomas, and head and neck carcinomas. (In contrast, methylation of its promoter is not detected in a large cohort of other tumor types.) As was the case with detoxifying genes, such as the glutathione-S-transferase (*GST*) alleles discussed earlier, we can imagine that the loss of this DNA repair function in certain tissues favors increased rates of mutation and hence accelerated tumor progression. (Conversely, when the *MGMT* gene, in the form of a transgene, is overexpressed in the mouse mammary gland or thymus, such expression renders these glands quite resistant to the otherwise potently carcinogenic effects of methylnitrosourea, a widely used alkylating mutagen; see, for example, Figure 12.22C.)

Provocatively, in certain animal models of cancer, and quite possibly in humans, the *types* of tumors caused by certain carcinogens can be influenced by the specific types of DNA repair defenses that have been deployed in various tissues (Sidebar 12.7). Moreover, as was the case with the glutathione-S-transferase enzyme (Sidebar 12.5), these DNA repair enzymes can influence responses to therapy. In one instance, after a group of glioblastoma patients expressing high levels of the MGMT enzyme were treated with an alkylating chemotherapeutic agent, they survived for another 12 months; other patients in the same cohort, whose cells expressed only low levels of MGMT, survived for 22 months after therapy—almost twice as long. This suggests that high levels of the MGMT enzyme are very effective in removing the alkyl groups created by the chemotherapeutic agent, thereby neutralizing its effects.

The MGMT system described above is only one way by which cells deal with methylated bases. Another, involving homologues of the bacterial AlkB DNA repair protein, works by oxidizing methyl groups that have become attached to bases, which are then shed as formaldehyde from the rings of all four DNA bases; similarly, AlkB enzymes cause the larger ethyl group to be released as acetaldehyde. (Aficionados of DNA repair portray these enzymes as "burning off" the unwanted methyl groups!) As mentioned earlier, methylation of DNA bases may occur frequently during the lives of cells through the actions of S-adenosylmethionine, the methyl donor that participates in the biosynthesis of many molecules in the cell; its highly reactive methyl group may accidentally become diverted to methylate a variety of cellular macromolecules, including the bases of DNA.

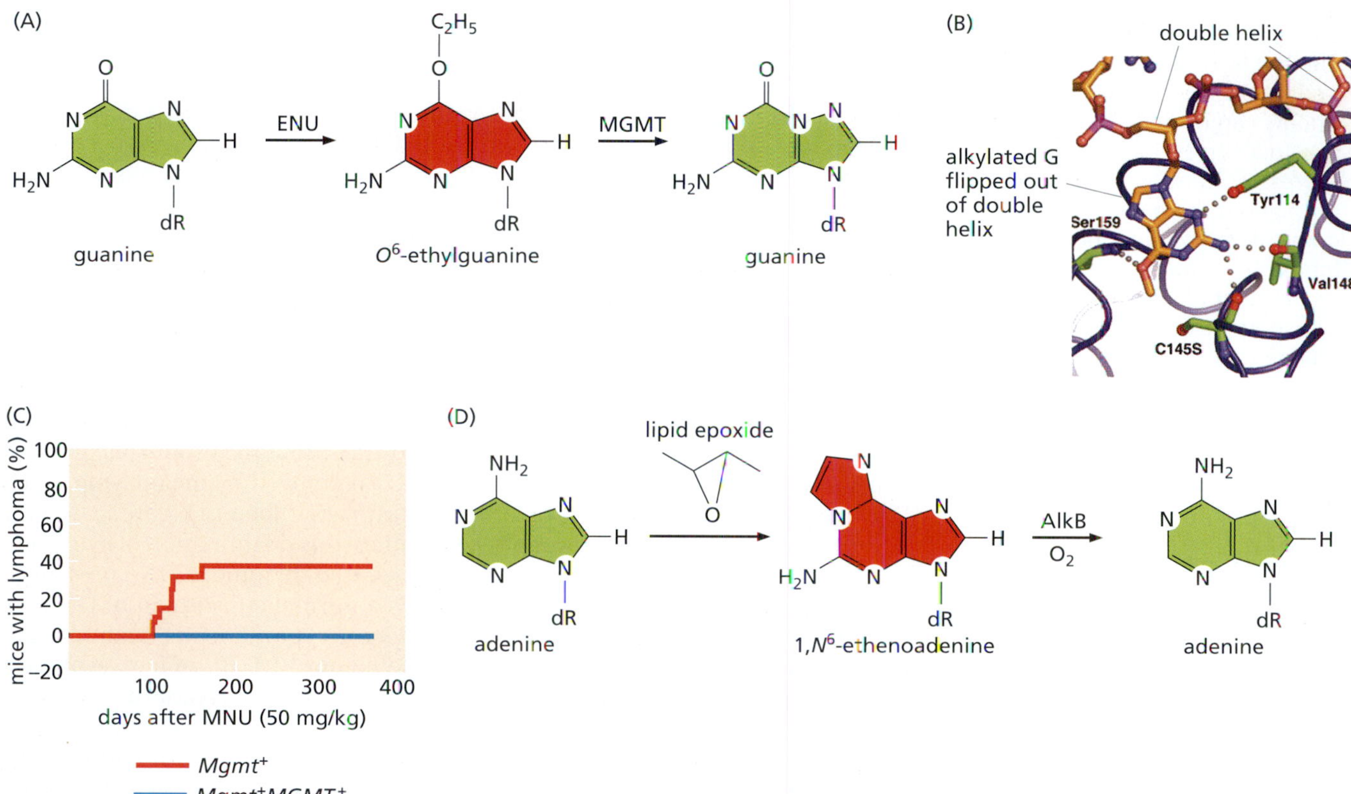

Figure 12.22 Restoration of normal base structure by dealkylating repair enzymes (A) The O^6 position of guanosine is especially vulnerable to alkylation by agents such as ethylnitrosourea (ENU). Unlike many DNA repair enzymes, which respond to altered bases by excising them or the entire nucleotide containing them from the DNA, the enzyme O^6-methylguanine-DNA methyltransferase [MGMT; also known as O^6-alkylguanine DNA alkyltransferase (AGT)] restores an altered guanosine to its normal structure. It does so by removing the alkyl group from the O^6 atom of guanine. In the absence of such repair, the alkylated guanosine often leads to a G-to-A transition mutation. (B) Structural analyses of DNA repair proteins have revealed much about how they function. In the case of the MGMT protein, these analyses, together with enzymological studies, indicate that this enzyme *(below, light and dark blue)* works by flipping the damaged base *(blue, orange)* out of the double helix *(above, orange and light purple)* before removal of the alkyl group. Moreover, the reaction between enzyme and substrate is stoichiometric, in that the cysteine 145 (C145S) residue *(red, green, below)* in the active site becomes irreversibly alkylated following restoration of normal guanosine structure; hence, each enzyme molecule is able to dealkylate only a single alkylated deoxyguanosine. (C) The effects of MGMT ectopic expression are seen here in mice that are exposed to the alkylating carcinogen methylnitrosourea (MNU). Wild-type mice (*Mgmt*⁺) are highly susceptible to the induction of thymic lymphomas *(red line)*, while transgenic mice that overexpress the MGMT enzyme because of a transgene in their germ line (*Mgmt*⁺ *MGMT*⁺, *blue curve*) are largely protected from these effects. (D) Highly reactive lipid peroxides, which are common in inflamed tissues, can attack and modify adenine *(shown here)* as well as other DNA bases *(not shown)*. The AlkB enzyme of bacteria can remove the resulting adducts as well as simpler methyl adducts, such as those shown in Figure 12.14C; homologues of AlkB are assumed to act similarly in human cells. (A, from S.L. Gerson, *Nat. Rev. Cancer* 4:296–307, 2004; B, from D.S. Daniels, T.T. Woo, K.X. Loo et al., *Nat. Struct. Mol. Biol.* 11:714–720, 2004; C, from L.L. Dumenco, E. Allay, K. Norton and S.L. Gerson, *Science* 259:219–222, 1993.)

Sidebar 12.7 Expression patterns of repair enzymes explain certain tissue-specific susceptibilities to cancer Mammalian cells seem to express only a single MGMT enzyme. The importance of this activity in influencing carcinogenesis is indicated by experiments in which pregnant rats are exposed to the carcinogen *N*-ethylnitrosourea (ENU) during the 15th day of gestation. Virtually all the rat pups that are born succumb to neuroectodermal tumors arising in the central nervous system several months after birth. The peculiar ability of ENU to preferentially induce these tumors can be explained by the fact that the MGMT enzyme is expressed at significant levels throughout the bodies of developing embryos and newborns but is only minimally expressed in the central nervous system. Consequently, alkylated guanine residues that are formed in the cells of the nervous system persist rather than being quickly removed and are ultimately able to generate the point mutations that are re-sponsible for the creation of the oncogenes in the resulting tumors. In one experiment conducted with newborn rats that had been exposed *in utero* to ENU or the related alkylating agent MNU (*N*-methylnitro-sourea), the levels of O^6-alkylguanine adducts surviving in unrepaired form a week after exposure to these carcinogens were 20-fold (ENU) and 90-fold (MNU) higher in brain DNA than in liver DNA.

Bacterial AlkB (and quite possibly its mammalian homologues) has also been found to be capable of removing more complex base adducts. For example, the inadvertent oxidation of unsaturated lipids, yielding lipid peroxides, occurs at high rates in inflamed tissues; these highly reactive peroxides can generate complex adducts with DNA bases (Figure 12.22D) that are highly mutagenic. Indeed, such adducts have been found in tissues of ulcerative colitis patients, which are known to progress, with significant frequency, to carcinomas (see Sidebar 11.13). A role in cancer pathogenesis of the human homologues of AlkB, termed hABH2 and hABH3, has not yet been directly demonstrated.

Far more important, however, than these dealkylating enzymes are the numerous cellular enzymes that recognize chemically altered bases in the DNA and respond in two other ways, depending on the specific modification of the DNA. In some cases, specialized enzymes will cleave the bond linking a modified base to the deoxyribose sugar, the process of **base-excision repair** (BER; Figure 12.23A). In other cases, the entire nucleotide containing both the base and associated deoxyribose will be cut out, this being the process termed **nucleotide-excision repair** (NER; Figure 12.23B).

Base-excision repair (BER) tends to repair lesions in the DNA that derive from endogenous sources, such as those attributed to the reactive oxygen species and depurination events described earlier (Section 12.5). Nucleotide-excision repair (NER), in contrast, focuses its energies largely on repairing the lesions created by exogenous agents, such as UV photons and chemical carcinogens (e.g., see Figures 12.16 and 12.18). BER seems to concentrate on fixing lesions that do not create structural distortions of the DNA double helix, while NER directs its attention to helix-distorting alterations.

BER is initiated by a group of DNA **glycosylases**, each specialized to recognize an abnormal base and cleave its covalent bond to deoxyribose. For example, a uracil base in the DNA is recognized readily by the proteins responsible for BER because U is not normally present in DNA. U is removed by the enzyme uracil DNA-glycosylase and soon replaced, usually with a C. (Refer to Figure 12.11B for how uracil can arise in DNA through the spontaneous deamination of cytosine.) However, the presence of an inappropriately located thymine in DNA presents a quandary for these repair enzymes, since T is a normal constituent of DNA. As we have read, 5-methyl-C occasionally undergoes spontaneous deamination, leading to a T, and thus to TG base pairs (see Figure 12.11B). In fact, evolution has responded to this problem by implanting a T:G glycosylase in our cells, which is designed specifically to excise T's that happen to arise opposite G's. Nonetheless, it is clear that the T:G base pairs formed by this deamination occasionally escape detection by this enzyme and persist to yield point mutations.

After an aberrant base is removed by a DNA glycosylase, the base-free sugar that results is then cleaved by a second enzyme, an endonuclease named APE (apurinic/apyrimidinic endonuclease) that is specialized to cut the strand carrying the base-free deoxyribose, doing so on the 5′ side of this sugar; a second enzyme, termed an AP lyase, then cleaves on the 3′ side, liberating the base-free sugar. The resulting single-strand gap in the DNA is repaired by a DNA polymerase, often polymerase β. The single-strand nick that results is finally closed by a DNA **ligase**, which rejoins adjacent nucleotides through the formation of phosphodiester bonds between them, thereby reconstructing the normal chemical structure of DNA. (An occasionally used variant form of BER, termed "long patch repair," involves the excision of 4 to 7 nucleotides around the damaged base followed by a filling of the resulting gap.)

Nucleotide-excision repair (NER; see Figure 12.23B) is accomplished by a large multiprotein complex composed of almost two dozen subunits. This complex seems to require two distinct changes in DNA before it will initiate repair: significant distortion of the normal Watson–Crick structure of the double helix plus

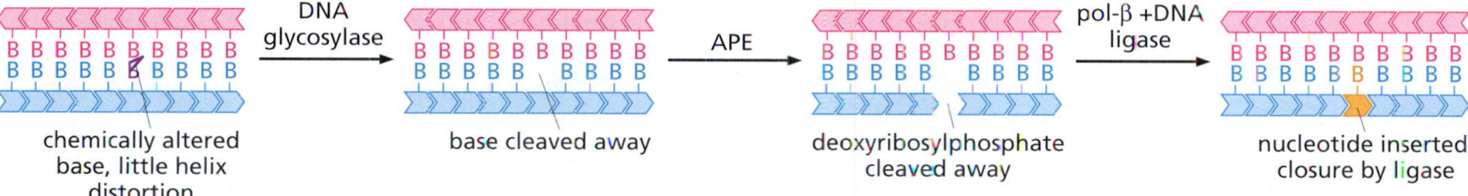

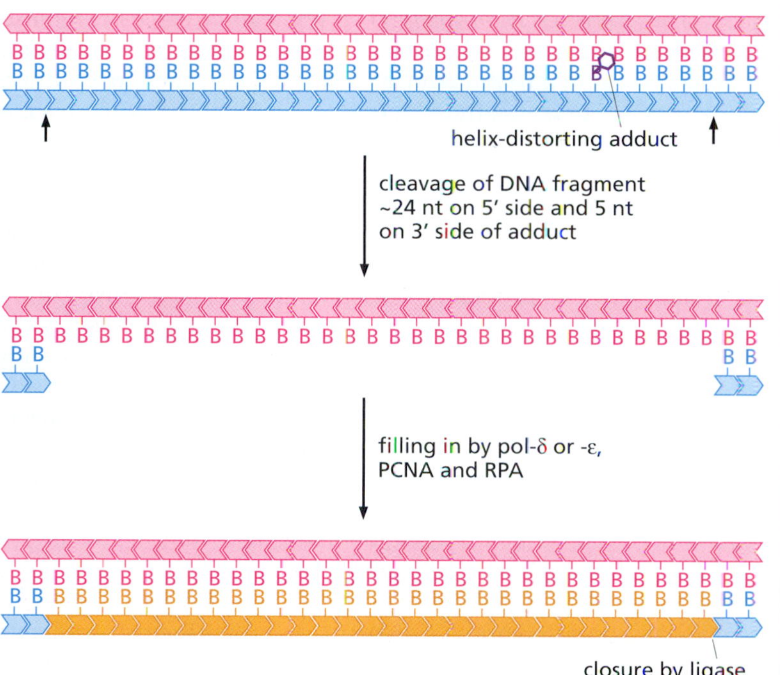

Figure 12.23 Base- and nucleotide-excision repair (A) Base-excision repair (BER) is achieved by enzymes that recognize chemically altered bases having minimal helix-distorting effect. These DNA glycosylase enzymes cleave the glycosyl bond linking the altered base *(purple)* and the deoxyribose. The base-free deoxyribosylphosphate is then excised by an enzyme—apurinic/apyrimidinic endonuclease (APE)—specialized to remove base-free sugars. The resulting single nucleotide gap is then filled by DNA polymerase β and sealed by a DNA ligase. (B) Nucleotide-excision repair (NER) is accomplished by enzymes that recognize bulky, helix-distorting lesions and cleave the flanking oligonucleotide sequences at sites approximately 24 nucleotides (nt) on the 5′ side and about 5 nucleotides on the 3′ side. The resulting approximately 29-nt single-strand gap in the DNA is then filled by DNA polymerase δ or ε, acting together with PCNA (proliferating-cell nuclear antigen) and RPA (single-strand DNA-binding protein), and finally is sealed by a DNA ligase. The chevrons represent deoxyribose nucleotides; in all cases they are pointing in a 5′-to-3′ direction.

the presence of a chemically altered base. Once this large complex recognizes the problem, it proceeds to cleave the damaged strand upstream and downstream of the damage, yielding a single-strand fragment of 25 to 30 nucleotides in length, which is then removed. DNA polymerases that are specialized to fill in the resulting gap in the DNA (using the complementary, undamaged strand as a template) then take over, followed by a DNA ligase, which erases the final trace of the damage.

Included among the NER enzymes are those that can recognize and remove structures resulting from the formation of bulky base adducts (i.e., those composed of complex molecular structures covalently bound to bases) created by certain exogenous mutagens, such as polycyclic hydrocarbons, heterocyclic amines, and aflatoxin B1, as well as the pyrimidine dimers formed by UV radiation. For example, following exposure to UV radiation, cultured human cells can repair approximately 80% of their pyrimidine dimers in 24 hours. The NER apparatus active here will remove 5 nucleotides on the 3′ side of the photoproduct (the pyrimidine dimer) and 24 nucleotides on the 5′ side.

The various reactions that constitute NER can actually be divided into two subtypes. The first of these is focused specifically on the template strand of actively transcribed genes and is coupled to the actions of RNA polymerase molecules that are proceeding down these template strands during transcription; these actions are termed transcription-coupled repair (TCR). The second subtype of NER addresses the remainder of the genome, including the nontemplate strand of transcribed genes as well as the nontranscribed regions of the genome. This type of NER is sometimes termed *global genomic repair* (GGR). The p53 tumor suppressor protein activates expression of several genes encoding NER proteins involved in GGR (Table 9.2), explaining the defectiveness of GGR in *p53* mutant cells; in contrast, transcription-coupled repair is intact in these cells. This distinction between TCR and GGR holds profound implications for the maintenance of cell genomes in the half of all human tumors in which the *p53* gene is mutant (Chapter 9). Many of the remaining cancers, in which p53 function is compromised in other ways, may also have defects in global genomic repair.

An alternative strategy for the cell to cope with damaged DNA—actually an act of desperation—involves DNA replication of a still-unrepaired stretch of template-strand DNA. This is termed *error-prone* DNA replication, since the replication apparatus involved here must often "guess" which of the four nucleotides is appropriate for incorporation into the growing DNA strand when it encounters a still-damaged base or set of bases; these guesses are not always correct, leading quite frequently to misincorporated bases (Figure 12.24).

To date, at least nine distinct mammalian error-prone human DNA polymerases have been discovered. Some of these can add a nucleotide to a growing strand even when a base in the complementary strand is missing. Yet others can extend a nascent DNA strand, using as primer a nucleotide that has been misincorporated by another DNA polymerase. A third type can incorporate a base when the corresponding base in the complementary strand carries a bulky, covalently attached DNA adduct that has not yet been removed by nucleotide-excision repair. One of these enzymes, encoded by the *XPV* gene, is highly specialized, being able to recognize the T–T thymine dimers created by UV radiation and insert two A's on the opposite strand (see Figure 12.24). While the DNA polymerases responsible for the bulk of DNA synthesis in a cell have error rates as low as 10^{-5}, these error-prone polymerases generally have error rates as high as 1 misincorporated base per 100 bases replicated.

The mistakes made by the error-prone polymerases would seem to generate unacceptably high rates of mutation in cell genomes. Still, the price paid for accumulating such mutations should be balanced against the alternative: the risk of imminent death confronted by a cell whose DNA replication forks are stalled because of difficult-to-copy lesions in its DNA.

Perhaps the best studied of these error-prone polymerases is DNA polymerase β (pol-β), which is usually involved in replacing the nucleotides that have been removed because of BER. This relatively small polymerase molecule lacks the proofreading capabilities of the larger polymerase enzymes (Section 12.4), and this absence may explain much of its error-prone DNA replication activities. In a variety of ovarian carcinoma cell lines, this enzyme has been found to be overexpressed by as much as a factor of 10. The overexpression of the error-prone DNA polymerase β may represent an effective strategy used by these cancer cells to increase the mutability of their genes and hence accelerate the rate of tumor progression. In support of this idea, the forced overexpression of polymerase β in cultured human fibroblasts has been found to encourage microsatellite instability and to increase overall mutation rates as much as threefold.

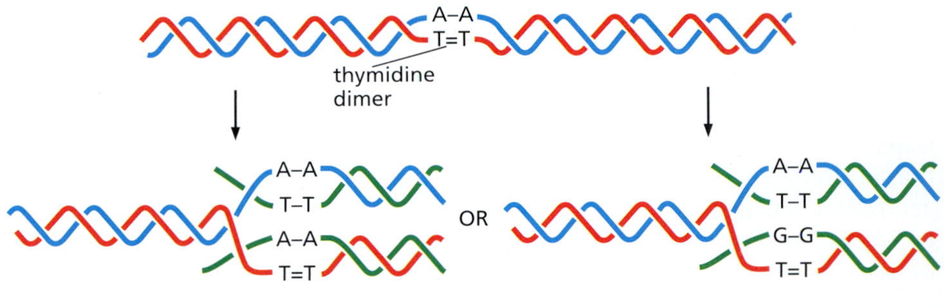

Figure 12.24 Error-prone repair Error-prone DNA synthesis occurs when a DNA replication fork is advancing during replication and encounters a still-unrepaired DNA lesion, such as the thymidine dimer shown here. In the great majority of cases *(left)*, the error-prone DNA polymerase (sometimes called a *by-pass polymerase)* that is called in and used to respond to this lesion in the damaged template strand *(red)* inserts the appropriate bases (in this instance, an A–A dinucleotide) in the growing DNA strand *(dark green)*. However, in several percent of these encounters *(right)*, the error-prone polymerase will fail to properly "guess" the structure of the lesion in the template strand and will instead incorporate a G–G dinucleotide opposite the thymidine dimer.

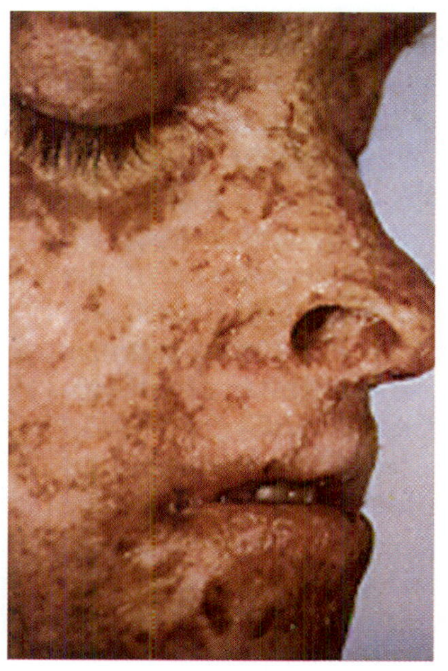

Figure 12.25 A xeroderma pigmentosum patient A patient suffering from xeroderma pigmentosum (XP) has severe and extensive lesions in all areas of sun-exposed skin. These lesions can develop into squamous and basal cell carcinomas as well as melanomas. The tumors develop at rates that are as much as 1000-fold higher than in the general population. (Courtesy of K.H. Kraemer.)

The deployment of error-prone polymerases by a cell represents a situation in which this cell is making the best of a desperate situation: it gambles that misincorporated bases are an acceptable compromise to avoid the death that would inevitably ensue from a failure to complete DNA replication. In fact, there is at least one enzyme encoded by the mammalian genome that *purposely* inserts mutations into the genome. Such an enzyme may also, quite inadvertently, contribute to cancer development (Sidebar 21).

12.9 Inherited defects in nucleotide-excision repair, base-excision repair, and mismatch repair lead to specific cancer susceptibility syndromes

In 1874, two Austro–Hungarian physicians, Ferdinand Hebra and Moritz Kaposi, described an unusual syndrome that involved high rates of the development of squamous and basal cell carcinomas of the skin. (Kaposi subsequently described the unusual sarcoma that bears his name.) As became apparent later, affected individuals have extreme sensitivity to UV radiation, and infants will often suffer severe burning of the skin after only minimal exposure to sunlight (Sidebar 22). These individuals show dry, parchment-like skin (xeroderma) and many freckles ("pigmentosum"; Figure 12.25). In aggregate, individuals suffering from the *xeroderma pigmentosum* (XP) syndrome have a 1000-fold increased risk of skin cancer compared with the general population and about a 100,000-fold increased risk of squamous cell carcinoma of the tip of the tongue. Skin cancers appear in children with a median age of 8 years, compared with 50 years before such cancers are encountered in the general population (Figure 12.26).

Inherited defects in any one of eight distinct genes can lead to xeroderma pigmentosum. The genetic complexity of this syndrome was first recognized through the use of somatic cell genetics. Cells from two different XP patients were fused together in culture in order to determine the repair phenotype of the resulting hybrid cells (Figure 12.27). On many occasions, the hybrids were found to repair DNA normally, indicating that the two parental cells carried defects in DNA repair that were associated with two distinct genes. For example, using nomenclature developed later, cells from an individual carrying a mutant *XPA*

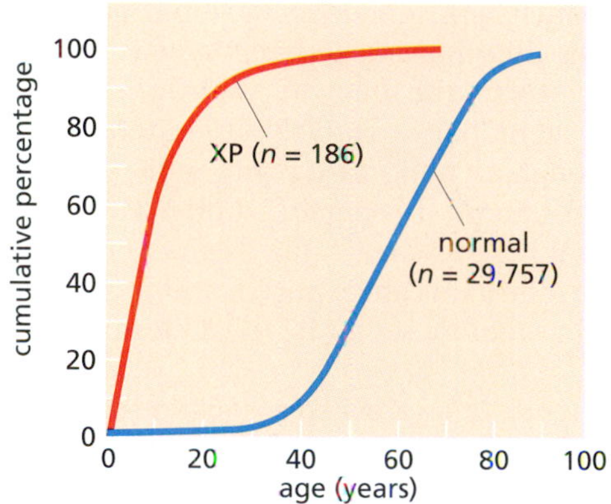

Figure 12.26 Epidemiology of XP patients Patients suffering from xeroderma pigmentosum (XP) exhibit skin cancers far earlier than the general population (*normal*). In the general population, these skin cancers appear with a median age of onset of about 60 years, in contrast to the XP population, in which skin cancers are diagnosed with a median age of about 10 years. The percentages on the ordinate are calculated by dividing the number of patients who have already been diagnosed with one or more skin cancers by a given age by the total number of patients in this population who will eventually be diagnosed with these cancers. Hence, in the XP population, virtually everyone who will develop skin cancers has already done so by around the age of 25. *n* is the number of individuals in each population studied. (From J.E. Clever and K.H. Kraemer, in C.R. Scriver et al. (eds.), The Metabolic Basis of Inherited Disease, 6th ed. New York: McGraw-Hill, 1989, pp. 2949–2971.)

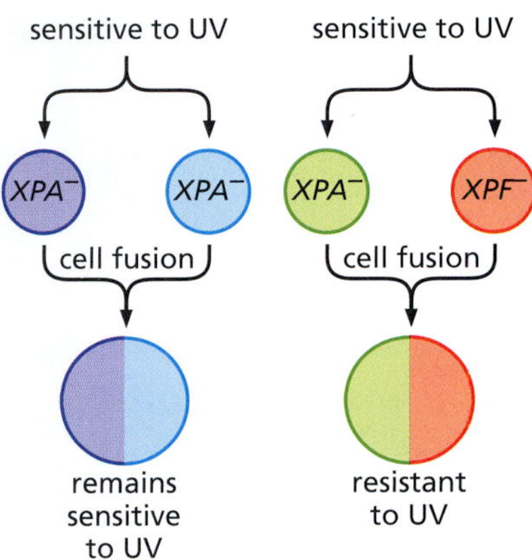

Figure 12.27 Discovery of XP complementation groups Cell fusion experiments using cultured cells from xeroderma pigmentosum (XP) patients have revealed that certain combinations of fibroblasts, each derived from a different patient *(dark blue, light blue, left)*, when fused together, yield hybrid tetraploid cells that are as UV-sensitive as the two parental cell populations. On other occasions, however *(right)*, combinations of cells from two patients *(light green, light red)*, yield cell hybrids after cell fusion that are as resistant to UV-mediated killing as the cells from normal individuals. Such findings caused the pair of cells that did not complement one another's DNA repair defect *(dark blue, light blue)* to be assigned to the *same* complementation group, ostensibly because the two populations carried mutations in the same gene. Conversely, successful complementation by other pairs of cells *(light green, light red)* allowed the respective parental cell types to be assigned to two *distinct* complementation groups, indicating that mutant alleles of at least two distinct genes were involved in predisposing to XP. In this way, eight distinct XP complementation groups were eventually delineated. Years later the respective genes, each one representing a complementation group, were isolated by gene cloning.

gene (and having a wild-type *XPC* gene) were able to repair DNA normally after being fused to cells from an individual carrying a mutant *XPC* gene (and having a wild-type *XPA* gene). Such collaboration, or "genetic **complementation**," led to the classification of XP-associated mutant alleles into eight distinct complementation groups, each ostensibly defined by the identity of a responsible gene. Only years later were the responsible genes isolated by molecular cloning. Almost always, it has been possible to show that an affected individual has inherited two mutant, null alleles of a gene representing one or another XP complementation group.

Seven of the eight XP-associated genes, named *XPA* through *XPG*, encode proteins that function as components of the large, multiprotein nucleotide-excision repair (NER) complex. The eighth gene, *XPV*, specifies the error-prone DNA polymerase pol-η that many cells seem to use when their regular DNA polymerases (e.g., pol-δ) are unable to copy over unrepaired DNA lesions such as pyrimidine dimers. As mentioned in the last section, error-prone polymerases (sometimes called **by-pass polymerases**) are able to copy a template strand of DNA containing still-unrepaired T–T dimers, usually synthesizing two A's in the complementary strand. In general, pol-η is thought to be so accurate that it incorporates AA nucleotides in the growing DNA strand opposite a TT dimer 95% of the time.

Individuals afflicted with XP also have some increased risk of other diseases, notably neurological problems, which are observed in about 18% of these patients. And mice that have been deprived of one of several XP genes suffer markedly increased susceptibility to tumors following exposure to chemical carcinogens. These two observations provide evidence that components of the nucleotide-excision repair system encoded by some of the XP genes are, not unexpectedly, responsible for repairing genetic damage created by other agents besides UV radiation. This raises the following question: Why does a human who lacks one or another XP gene have relatively little increased risk of cancers in internal organs, even though an important component of the NER machinery is missing from all cells throughout this person's body? The simplest and possibly correct explanation is that UV rays are, by far, the most important environmental mutagen to which most humans are exposed, and thus the source of the great majority of the lesions that require repair by the NER machinery.

Several other inherited syndromes are also associated with defects in NER. For example, individuals suffering from Cockayne syndrome (CS) appear to be defective in one of two genes that are involved in transcription-coupled NER.

Their cells have increased photosensitivity like those of XP patients. The median age of death of patients from this disease is 12 years of age. Those few who survive through adolescence do not show increased rates of cancer. Still unaddressed is the possibility that individuals who are *heterozygous* for one of the XP- or Cockayne syndrome–associated mutant alleles have an elevated risk of developing certain types of cancer.

XP was only the first of many human cancer susceptibility syndromes that have been discovered to be caused by inherited defects in various types of DNA repair (Table 12.1). We will explore another one here—hereditary non-polyposis colon cancer (HNPCC). HNPCC is a familial cancer syndrome that represents a quite common cause of inherited predisposition to colon cancer, being responsible for 2 to 3% of all colon cancer cases. HNPCC is, as its name implies, distinct from the other type of hereditary colon cancer predisposition that we encountered previously—adenomatous polyposis coli (Section 7.11). A subclass of HNPCC patients have increased susceptibility to endometrial, stomach, ovarian, and urinary tract carcinomas in addition to colon carcinomas.

The increased cancer susceptibility of the HNPCC patients can be traced back to the accelerated rate with which tumor progression proceeds in their colons: while the adenoma-to-carcinoma progression is estimated to require 8 to 10 years in the general population, the genetic instability afflicting the cells of HNPCC patients allows this step to occur in only 2 to 3 years. Indeed, because their adenomas progress so quickly to carcinomas, these premalignant growths have a relatively short lifetime and are therefore not found in significant numbers in the colons of these patients.

The majority of HNPCC cases result from germ-line mutations in the genes encoding two important mismatch repair proteins, MSH2 and MLH1 (Sidebar 12.8). Mutant germ-line alleles of two other MMR genes, *MSH6* and *PMS2*, are

Table 12.1 Human familial cancer syndromes due to inherited defects in DNA repair

Name of syndrome	Name of gene	Cancer phenotype	Enzyme or process affected
HNPCC	(4–5 genes)[a]	colonic polyposis	mismatch repair enzymes
XP[b]	(8 genes)[b]	UV-induced skin cancers	nucleotide-excision repair
AT[c]	*ATM*	leukemia, lymphoma	response to dsDNA breaks
AT-like disorder[c]	*MRE11*	not yet determined	dsDNA repair by NHEJ
Familial breast, ovarian cancer	*BRCA1, BRCA2*[d]	breast and ovarian carcinomas	homology-directed repair of dsDNA breaks
Werner	*WRN*	several cancers	exonuclease and DNA helicase[e], replication
Bloom	*BLM*	solid tumors	DNA helicase, replication
Fanconi anemia	(9 genes)[f]	AML, HNSCC	repair of DNA cross-links and ds breaks
Nijmegen break[g]	*NBS*	mostly lymphomas	processing of dsDNA breaks, NHEJ
Li–Fraumeni	*TP53*	multiple cancers	DNA damage alarm protein
Li–Fraumeni	*CHK2*	colon, breast	kinase signaling DNA damage

[a]Five distinct MMR genes are transmitted as mutant alleles in the human germ line. Two MMR genes—*MSH2* and *MLH1*—are commonly involved in HNPCC; two other MMR genes—*MSH6* and *PMS2*—are involved in a small number of cases; a fifth gene, *PMS1*, may also be involved in a small number of cases.

[b]Xeroderma pigmentosum, at least eight distinct genes, seven of which are involved in NER. The seven genes are named *XPA* through *XPG*. An eighth gene, *XPV*, encodes DNA polymerase η.

[c]Ataxia telangiectasia, small number of cases.

[d]Mutant germ-line alleles of *BRCA1* and *BRCA2* together may account for 10–20% of identifiable human familial breast cancers.

[e]An exonuclease digests DNA or RNA from one end inward; a DNA helicase unwinds double-stranded DNA molecules.

[f]Nine genes have been cloned and at least eleven complementation groups have been demonstrated. Complementation group J encodes the BACH1 protein, the partner of BRCA1.

[g]The NBS1 protein (termed nibrin) forms a physical complex with the Rad50 and Mre11 proteins, all of which are involved in repair of dsDNA breaks. The phenotypes of patients with Nijmegen break syndrome are similar but not identical to those suffering from AT.

Adapted in part from B. Alberts et al., Molecular Biology of the Cell, 4th ed. New York: Garland Science, 2002; and from E.R. Fearon, *Science* 278:1043–1050, 1997.

Sidebar 12.8 A convergence of bacterial, yeast, and human genetics led directly to the discovery of hereditary non-polyposis colon cancer genes The route through which the genetic basis of HNPCC was discovered is quite fascinating. By following the co-inheritance of disease susceptibility in a number of families suffering from this syndrome and the inheritance of a set of chromosomal markers scattered throughout the human genome, geneticists discovered two distinct genetic regions that seemed to be associated with disease susceptibility. In some families, disease susceptibility co-segregated with genetic markers from human Chromosome 2p16, while in a second set of families, this susceptibility seemed to be inherited together with markers from chromosomal region 3p21.

At the same time, astute observation of the types of genetic aberrations found in a subset of sporadic colon carcinomas noted that stretches of A's (e.g., A_7) often were expanded or compressed (i.e., were replaced by A_6 or A_8) in the genomes of colon carcinoma cells; thereafter, more complex microsatellite sequences [e.g., $(CA)_8$] were also found to undergo expansion or shrinkage (see Section 12.4). This microsatellite instability (MIN) was found to be rampant in the cancer cells of individuals suffering from HNPCC.

Yet other researchers noted that these changes in DNA sequence were strikingly similar to the alterations that accumulated rapidly in *E. coli* that bore mutant versions of either of two genes—*mutS* or *mutL*—involved in mismatch repair (MMR; Section 12.4); yeast carrying mutations in the homologous genes showed identical genomewide aberrations. This similarity led some scientists to ascertain whether human homologs of these or other known mismatch repair genes were affected in HNPCC patients, leading rapidly to the confirmation that, indeed, a human *mutS* homolog (named *hMSH2*) mapped to human Chromosome 2p16, while a human *mutL* homolog (*hMLH1*) mapped to 3p21. Eventually, members of one set of HNPCC families were found to transmit mutant alleles of *hMSH2*, while those belonging to a second set of HNPCC kindreds carried mutant alleles of *hMLH1*.

involved in a small proportion of these cases; however, two other MMR genes (*PMS1, MSH3*), which have been found to play equally important roles in DNA repair, are rarely if ever transmitted as mutant alleles in the human germ line. Similar to the genetics of most tumor suppressor genes, patients inherit one defective allele of an MMR gene; the genomes in any tumor cells that arise have, almost always, undergone a loss of heterozygosity (LOH) that results in the discarding of the surviving wild-type gene copy of the MMR gene in question.

The resulting inability to properly detect and repair sequence mismatches leads to, among other consequences, high rates of mutations in genes that have microsatellite repeats nested in their sequences. A dramatic and early illustration of the consequences of this repair defect came from study of a group of 11 colorectal cancer cell lines that showed microsatellite instability. In nine of these cells lines, the gene encoding the type II TGF-β receptor (TGF-βRII) was found to be mutant. More specifically, the wild-type reading frame of this gene carries a stretch of ten A's in a row (Figure 12.28). However, in these nine tumor cell

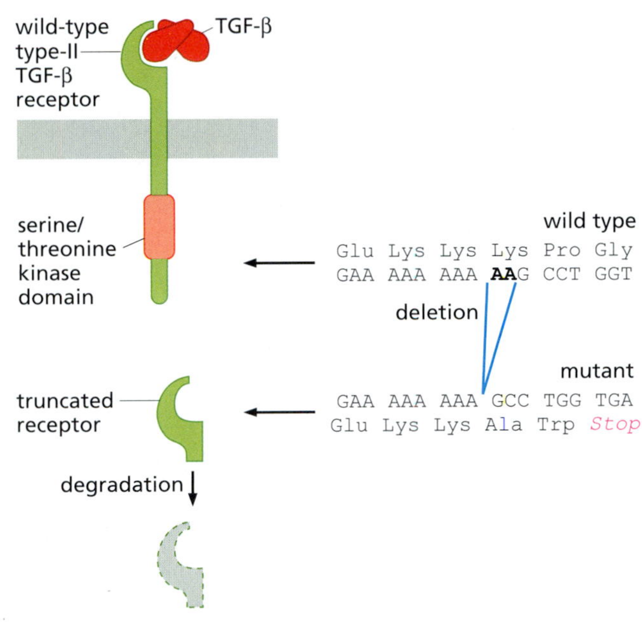

Figure 12.28 A TGF-β receptor gene affected by microsatellite instability The type II TGF-β receptor (TGF-βRII) is frequently inactivated in human colon cancers exhibiting microsatellite instability and therefore carrying defects in mismatch repair (MMR) genes. In the particular colon cancer whose DNA was analyzed here, the last two of ten adenines *(boldface)* were deleted from the reading frame of the receptor-encoding gene, ostensibly because of the MMR defect. This particular deletion ($A_{10} \rightarrow A_8$) resulted in a nonsense mutation that caused premature termination of translation of the nascent TGF-βRII protein and hence loss of functionally critical signaling domains in the C-terminus of the receptor. This loss, in turn, allowed the progenitors of the colon carcinoma cells to become resistant to the growth-inhibitory effects of TGF-β. Alterations of this tract of A's in TGF-βRII gene were subsequently found in 100 of 111 colorectal carcinomas showing defective MMR, in which they caused translation of the TGF-βRII mRNA to yield polypeptides of either 129 or 161 amino acid residues (depending on how many A's were deleted) rather than the 565 amino acid residue-long present in the wild-type receptor. (From C. Lengauer, K.W. Kinzler and B. Vogelstein, *Nature* 396: 643–649, 1998.)

lines, the TGF-βRII gene was found to have lost one or two A's of the normally present **homopolymeric** stretch of ten A's. These sequence changes forced the coding sequence of the TGF-βRII gene out of its normal reading frame and resulted in nonfunctional TGF-βRII proteins.

We can imagine that once tumor cell precursors no longer express functional TGF-βRII, they can escape the growth-inhibitory effects of this anti-mitogenic factor (Section 8.4)—a highly advantageous trait if it is acquired early in tumor progression by epithelial cells. In a subsequent study of a series of 110 colon carcinomas exhibiting microsatellite instability, 100 were found to carry mutant, defective alleles of the TGF-βRII gene, with almost all mutant alleles being present in homozygous configuration. Hence, once one of the receptor-encoding genes suffers an inactivating mutation, the surviving wild-type allele is discarded through loss of heterozygosity.

Later, yet other genes were found to have suffered similar mutations in mismatch repair–defective cancer cells (Table 12.2). In the great majority of cases listed in Table 12.2, the MMR defect and resulting mutant alleles were discovered in sporadic (rather than familial) cancers. These observations point to the fact that in nonfamilial tumors, MMR genes, like tumor suppressor genes, can be rendered defective either by somatic mutation or by promoter methylation and resulting transcriptional silencing (Section 7.8). In fact, the second mechanism is responsible for the lion's share of defective MMR in these tumors: about 15% of sporadic gastric, colorectal, and endometrial tumors show defective MMR, and in almost all of these, the observed microsatellite instability can be

Table 12.2 Genes and proteins that have been inactivated in human cancer cell genomes because of mismatch repair defects

Gene	Function of encoded protein	Wild-type coding sequence	Colon	Stomach	Endometrium
ACTRII	GF receptor	A_8	X		
AIM2	interferon-inducible	A_{10}	X		
APAF1	pro-apoptotic factor	A_8	X	X	
AXIN-2	Wnt signaling	A_6, G_7, C_6	X		
BAX	pro-apoptotic factor	G_8	X	X	X
BCL-10	pro-apoptotic factor	A_8	X	X	X
BLM	DNA damage response	A_9	X	X	X
Caspase-5	pro-apoptotic factor	A_{10}	X	X	X
CDX2	homeobox TF	G_7	X		
CHK1	DNA damage response	A_9	X		X
FAS	pro-apoptotic factor	T_7	X		X
GRB-14	signal transduction	A_9	X	X	
hG4-1	cell cycle	A_8	X		
IFRIIR	decoy GF receptor	G_8	X	X	X
KIAA097	unknown	T_9	X		
MLH3	MMR	A_9	X		X
MSH3	MMR	A_8	X	X	X
MSH6	MMR	C_8	X	X	X
NADH-UO8	electron transport	T_9	X		
OGT	glycosylation	T_{10}	X		
PTEN	pro-apoptotic	A_6	X		X
RAD50	DNA damage response	A_9	X	X	
RHAMM	cell motility	A_9	X		
RIZ	pro-apoptotic factor	A_8, A_9	X	X	X
SEC63	protein translocation into endoplasmic reticulum	A_{10}, A_9	X		
SLC23A1	transporter	C_9	X		
TCF-4	transcription factor	A_{10}	X	X	X
TGF-βRII	*TGF-β receptor*	A_{10}	X	X	X
WISP-3	growth factor	A_9	X		

From A. Duval and R. Hamelin, *Cancer Res.* 62:2447–2454, 2002.

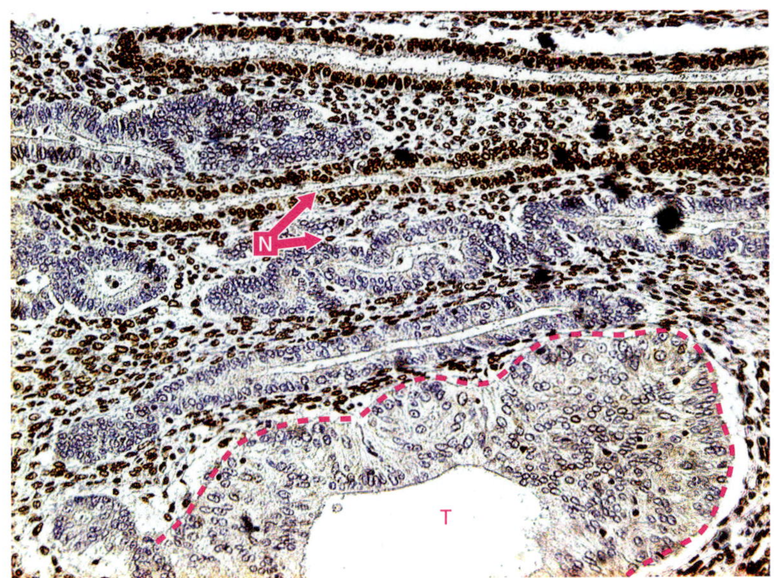

Figure 12.29 Suppression of MLH1 activity in endometrial tissue Anti-hMLH1 antibody has been used in immunohistochemistry, in which the antibody has been coupled to a peroxidase enzyme to generate a *dark brown* spot wherever it binds MLH1 antigen. As indicated here, the normal endometrial tissue *(pink N)* of an endometrial cancer patient shows areas of intense, dark brown staining, indicating high hMLH1 expression *(N, upper pink arrow)* as well as very weak staining *(N, lower pink arrow)*, where cell nuclei *(light blue)*, stained with a DNA-specific stain (DAPI), are apparent but no MLH1 staining is seen. The endometrial carcinoma tissue *(dashed pink line,* T) is virtually devoid of hMLH1 staining, which therefore reveals only the weakly staining of cell nuclei by DAPI. Molecular analysis *(not shown)* indicated that the promoter of the *hMLH1* gene in the carcinoma cells was strongly methylated. The fact that hMLH1 expression was absent in some of the histologically normal tissue adjacent to the tumor indicates that loss of hMLH1 expression occurred relatively early in tumor progression and preceded the histopathological alterations that led to the formation of carcinoma. (From T. Kanaya, S. Kyo, Y. Maida et al., *Oncogene* 22:2352–2360, 2003.)

traced to the methylation and resulting silencing of the *MLH1* gene. Interestingly, in the histologically normal endometrial tissue adjacent to tumors with defective MMR, the *MLH1* gene is often found to be methylated, suggesting that this methylation is one of the earliest events of tumor progression in this tissue (Figure 12.29).

In addition to losing mismatch repair function, cells that have lost MLH1 or MSH2 expression also become resistant to alkylating mutagens that would normally activate a G_2/M cell cycle checkpoint or induce apoptosis; such cells continue to advance into G_2/M and succeed in avoiding apoptosis following exposure to these DNA-damaging agents. Moreover, mutant versions of MSH2 can be created that selectively inactivate its MMR function without affecting its ability to trigger apoptosis in response to certain types of DNA damage. This suggests that in this particular MMR protein, distinct domains are involved in detecting damaged DNA, repairing this damage, and emitting alarm signals, including those leading to apoptosis.

The repeated observation of methylated mismatch repair genes provides direct evidence that a somatically acquired (i.e., non-inherited) defect in a DNA repair function confers replicative advantage on evolving, pre-malignant cells during the course of tumor progression. In many of the tumor cell genomes showing microsatellite instability, there are hundreds and likely thousands of genes that are concomitantly mutated; the genes shown in Table 12.2 represent only a small proportion of this group.

The data in Table 12.2 are biased by the fact that only an arbitrary set of genes was examined, and only the sequences associated with homopolymeric microsatellites within these genes were sequenced. Still, this list is most interesting. It shows that the *BAX* gene, which encodes an important pro-apoptotic protein (Section 9.9), can be silenced through mutations provoked by a stretch of eight Gs in its normal reading frame and an MMR defect. Yet other pro-apoptotic genes have also been found to have undergone mutations directly traceable to changes in the number of bases in one of their homopolymeric sequences. Even the genes that encode MMR proteins are themselves inactivated by MMR defects!

Future research will reveal how the inactivation of some of the genes listed in Table 12.2 results in replicative advantage for tumor cells. Still, successes in such research will leave another major question unanswered: Why are MMR defects and resulting microsatellite instability associated preferentially with carcinomas of the colon, stomach, ovary, and endometrium and much less often with tumors arising elsewhere in the body?

12.10 A variety of other DNA repair defects confer increased cancer susceptibility through poorly understood mechanisms

By far the most notorious genes associated with cancer, at least in the mind of the public, are *BRCA1* and *BRCA2*. Mutant germ-line alleles of either of these genes confer an inborn susceptibility to breast and ovarian carcinomas. For example, almost half of all identified familial breast cancers involve germ-line transmission of a mutant *BRCA1* or *BRCA2* allele; by some estimates, 70 to 80% of all familial ovarian cancers are due to mutant germline alleles of *BRCA1* or *BRCA2*. When these two genes were first discovered, it seemed that they should be included among the tumor suppressor genes, which are known to be involved in regulating the dynamics of cell proliferation, survival, and differentiation (Chapter 7). But diverse lines of evidence built an increasingly persuasive case that these two genes are actually involved in the maintenance of genomic integrity, and that they therefore should be considered "caretakers" (DNA repair proteins) rather than "gatekeepers" (tumor suppressor proteins).

The case for their participation in genomic maintenance can be argued using guilt-by-association evidence. The BRCA1 and BRCA2 proteins are found in large physical complexes with one another and with a large number of other proteins in the cell nucleus. These massive complexes carry, among other components, both the RAD50/Mre11 and the RAD51 proteins—homologs of two proteins in yeast that were initially discovered because of the important roles they play in repairing DNA breaks caused by ionizing radiation (i.e., X-rays). Mismatch repair proteins have been found in these complexes as well.

Quite dramatically, treatment of cells with hydroxyurea, which results in a stalling of replication forks during S phase, causes BRCA1 molecules to change their localization in the nucleus. Many of these stalled forks are thought to be sites of dsDNA breakage, which results from accidental breaks in the, still-unreplicated single-strand strands at the forks (see Figure 12.10); the resulting breaks are usually fixed by homology-directed repair (HDR). As visualized by immunofluorescent microscopy, BRCA1 molecules are normally distributed in a large number of tiny dots throughout the nucleus; hydroxyurea causes the BRCA1 molecules to leave these dots and flock together in a far smaller number of large, discrete spots, in which the proliferating-cell nuclear antigen (PCNA)—known to be localized to replication forks—is also found (Figure 12.30A). These spots have also been found to contain a number of other known DNA repair proteins, including Rad50 and Rad51. The BRCA2 protein is also found in these spots, providing additional presumptive evidence of its collaboration in DNA repair processes. Moreover, when dsDNA breaks are intentionally created in discrete areas within cell nuclei using a narrow laser beam, the BRCA1 protein co-localizes in these areas together with γ-H2AX, a phosphorylated histone that is present in the chromatin flanking sites of dsDNA damage (Figure 12.30B; see also Sidebar 14 ●). Altogether, 12 distinct cellular proteins, most known to be involved in DNA repair, have been found to be recruited to these areas of damage.

Mice that have been deprived genetically of all BRCA1 function die during early embryogenesis, but mutant germ-line alleles of BRCA2 that cause only partial loss of function result in susceptibility to lymphoid malignancies and unusual chromosomal aberrations. These aberrations have structures that suggest high rates of illegitimate recombination, that is, recombination events (or fusions) between two chromosomal arms that are nonhomologous to one another (Figure 12.31A). Such chromosomal structures result characteristically from improper repair of dsDNA breaks, many of which may arise accidentally at replication forks during a typical S phase of the cell cycle (see Figure 12.10). Moreover, the two daughters of a mother cell lacking full BRCA2 function are unable to separate from one another during the cytokinesis that follows mitosis; the precise effects that this **abscission** defect (Figure 12.31B) has on karyotype still require elucidation. This abscission defect indicates that BRCA2 has functions that are not directly connected with its main role in organizing the repair of dsDNA breaks. Yet other indications of defective dsDNA repair in *BRCA1* and *BRCA2* mutant cells come from experiments that test the ability of cultured cells to recover from double-strand (ds) breaks introduced into their chromosomal DNA by X-rays. Cultured cells lacking the bulk of BRCA1 function show greatly increased sensitivity to killing by X-rays.

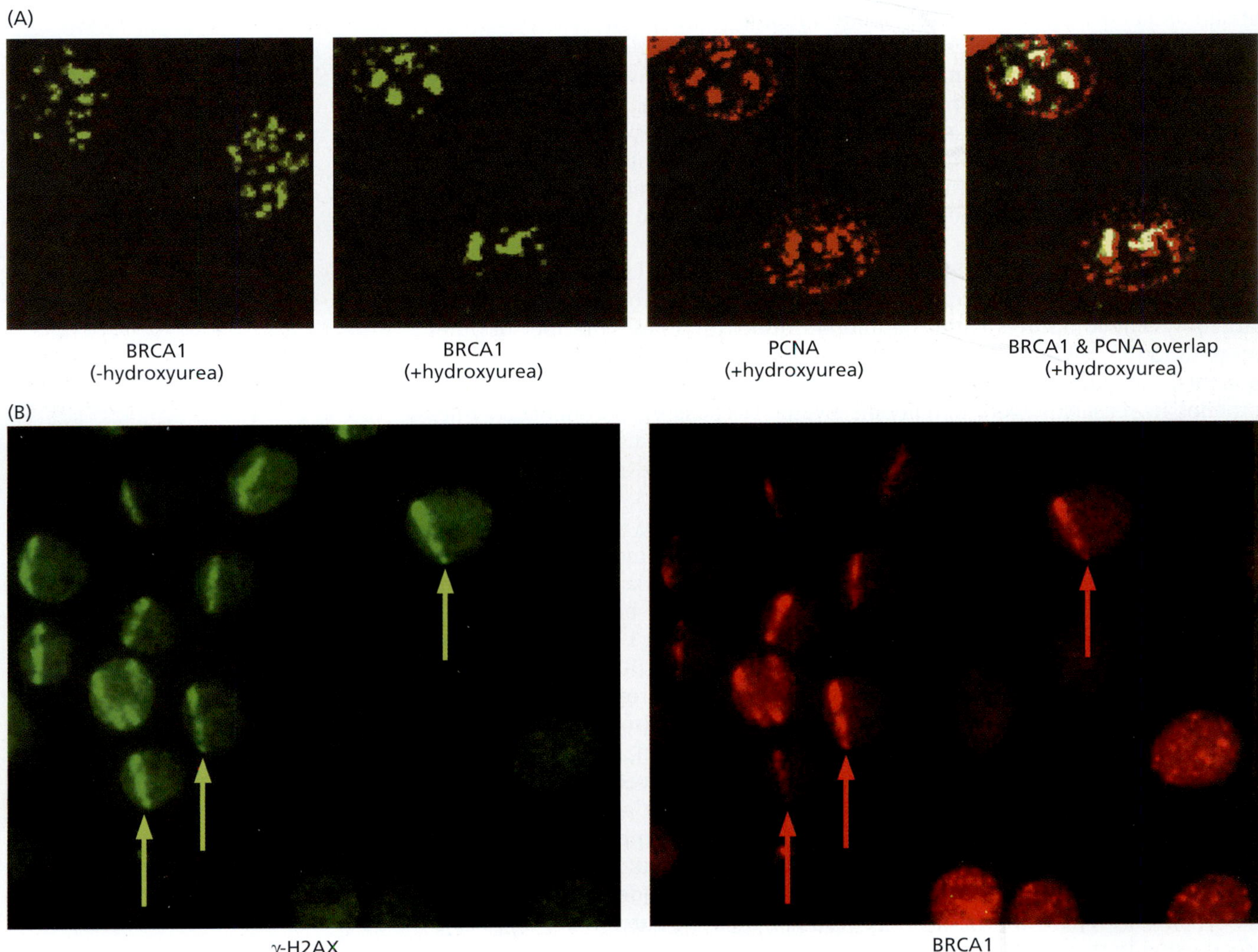

Figure 12.30 BRCA1 and the response to DNA damage (A) The BRCA1 protein, which can be detected by fluorescence-labeled antibodies, is normally found during S phase in a large number of discrete, small dots throughout the nucleus *(green, 1st panel)*. However, when cells in S phase are treated with hydroxyurea (HU), which stalls replication forks, the BRCA1 protein molecules leave these dots and congregate in a small number of quite large spots *(green, 2nd panel)*. A similar relocalization pattern can be observed with the proliferating-cell nuclear antigen (PCNA), which is known to be associated with replication forks *(red, 3rd panel)*. Substantial co-localization of the BRCA1 and PCNA is indicated by the yellow spots *(4th panel)*. These stalled replication forks often are sites of double-strand (ds) DNA breaks caused by the accidental breakage of the still-unreplicated (and thus fragile) single-strand DNA at the replication forks (see Figure 12.10); these observations suggests that BRCA1 is recruited to sites of dsDNA breaks. (B) A 355nm UV laser was used to paint narrow stripes across individual nuclei, which were then analyzed by immunostaining with antibodies reactive with either γ-H2AX (a phosphorylated histone that is known to localize to chromatin flanking dsDNA breaks; *green)* or BRCA1 *(red)*. This co-localization indicates that BRCA1 is attracted to areas of dsDNA breaks. (A, from R. Scully, J. Chen, R.L. Ochs et al., *Cell* 90:425–435, 1997; B, from R.A. Greenberg, B. Sobhian, S. Pathania et al., *Genes Dev.* 20:34–46, 2006.)

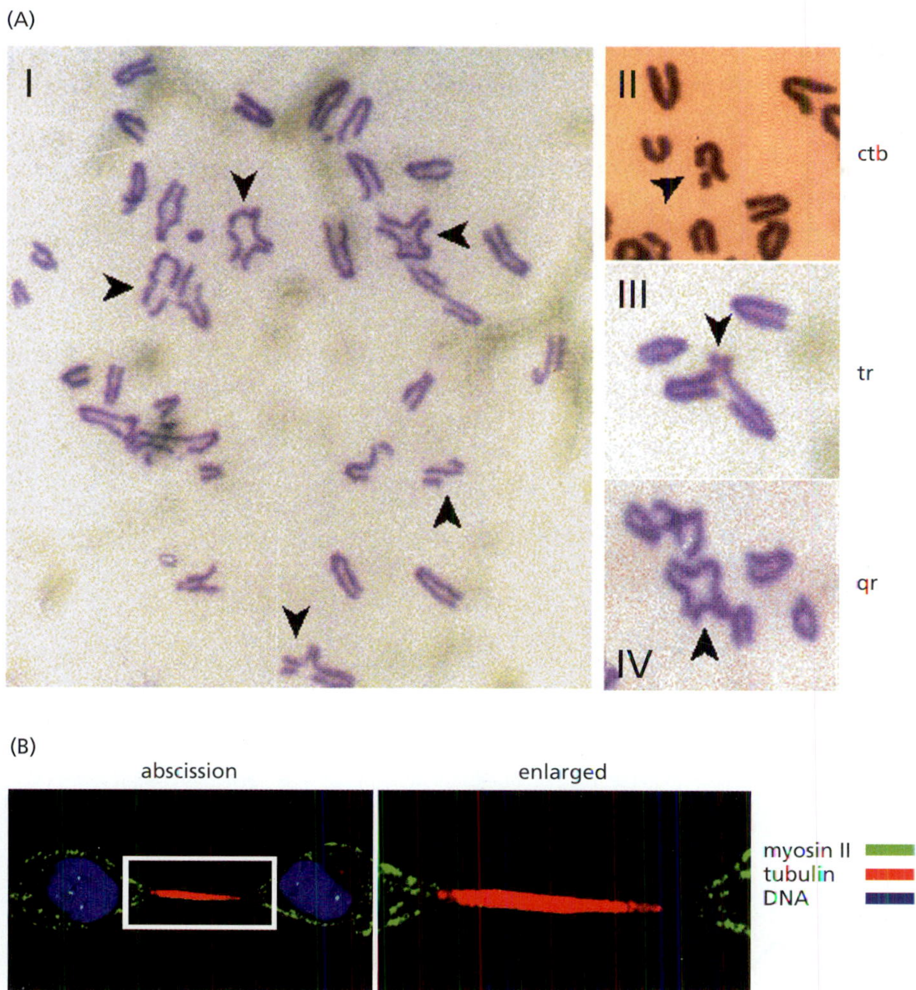

Figure 12.31 Karyotypic alterations due to partial loss of BRCA2 function (A) Various karyotypic abnormalities have been observed in cultured fibroblasts prepared from a mouse embryo that was homozygous for an allele of the *Brca2* gene encoding a truncated Brca2 protein. These include fusions between chromosomal arms, resulting in chromosomal translocations that are often manifested by presence of aberrant chromatid pairings at the metaphase of mitosis. These fusions are often caused by unrepaired or improperly repaired dsDNA breaks. When the hybrid chromatids resulting from such fusions pair with sister chromatids that have not been affected by such translocations, structures such as those seen here are observed (panel I). Among the aberrant chromosomal structures resulting from dsDNA breaks are the chromatid breaks (panel II), triradial chromosomes (panel III), and quadriradial chromosomes (panel IV) in these metaphase chromosomal spreads. Chromatid pairings like these are rarely observed in wild-type cells. (B) In the absence of proper BRCA2 function, cell growth and division may be disrupted through a mechanism that seems to be unconnected with defective DNA repair. Thus, the last stage of cytokinesis (the time after mitosis when daughter cells separate from one another) involves the process of *abscission*, which depends in turn on the cleavage of the remaining spindle microtubules that still connect the two presumptive daughter cells. In *Brca2* mutant cells, this component of the abscission program is also defective. (A, from K.J. Patel, V.P. Yu, H. Lee et al., *Mol. Cell* 1:347–357, 1998; B, from M.J. Daniels, Y. Wang, M. Lee, and A.R. Venkitaraman, *Science* 306:876–879, 2004.)

The repair of a dsDNA break in one chromatid often depends on the ability of the repair apparatus to consult the undamaged, homologous DNA sequences present in a sister chromatid, using the sequences of the still-intact DNA segment to instruct the repair apparatus as to how the broken double helix should be reconstructed (Figure 12.32). Thus, such homology-directed repair (HDR) occurs largely during the late S and the G_2 phases of the cell cycle, when the double helix in a sister chromatid can provide the sequence information for repairing the damaged chromatid. (Recall that during S phase, DNA replication results in the production of two identical chromatids that remain associated as part of a common chromosome until they are separated during the next mitosis.) HDR is also used if inter-strand covalent cross-links within a double helix should arise.

All types of homology-directed repair (HDR) are compromised in cells lacking either BRCA1 or BRCA2 function. This may be explained, in part, by the behavior of the RAD51 protein, with which BRCA1's partner, BRCA2, associates directly. RAD51 is known to bind single-strand DNA molecules, enabling them to invade (and thereby unwind) homologous double-strand helices, a process essential to initiating HDR (see Figure 12.32). (The fact that BRCA2 has eight BRC domains, each of which can, in principle, bind a RAD51 molecule, suggests that BRCA2 may actually assemble strings of RAD51 molecules to bind coordinately to a ssDNA.) In the absence of BRCA1 or BRCA2, RAD51 may not be properly recruited to sites of dsDNA breaks, and the subsequent steps of HDR may not be able to occur correctly. The triradial and quadriradial chromosomes that are encountered in the metaphase of *Brca2* mutant cells (see Figure 12.31A) are manifestations of the inability of the repair apparatus in these cells to use HDR to fix dsDNA breaks.

Homology-directed repair is also defective in patients suffering from Nijmegen break syndrome (see Table 12.1). Their cells, which lack the Nbs1 protein, fail to execute the initial steps of HDR. Lacking an ability to exploit HDR, the cells of these patients will resort to a fusion of two dsDNA ends via the process termed **nonhomologous end joining** (NHEJ; Figure 12.33).

NHEJ is inevitably an error-prone process, simply because the alignment between the two DNA segments being fused is not informed by the wild-type DNA sequences present in a sister chromatid. Consequently, the resulting end-to-end fusions generate mutant sequences at the site of joining, explaining the high rates of hematopoietic malignancies in patients suffering from Nijmegen break syndrome. Mice that are defective for one or another component of the NHEJ machinery and also lack p53 function develop lymphomas at extremely high rates.

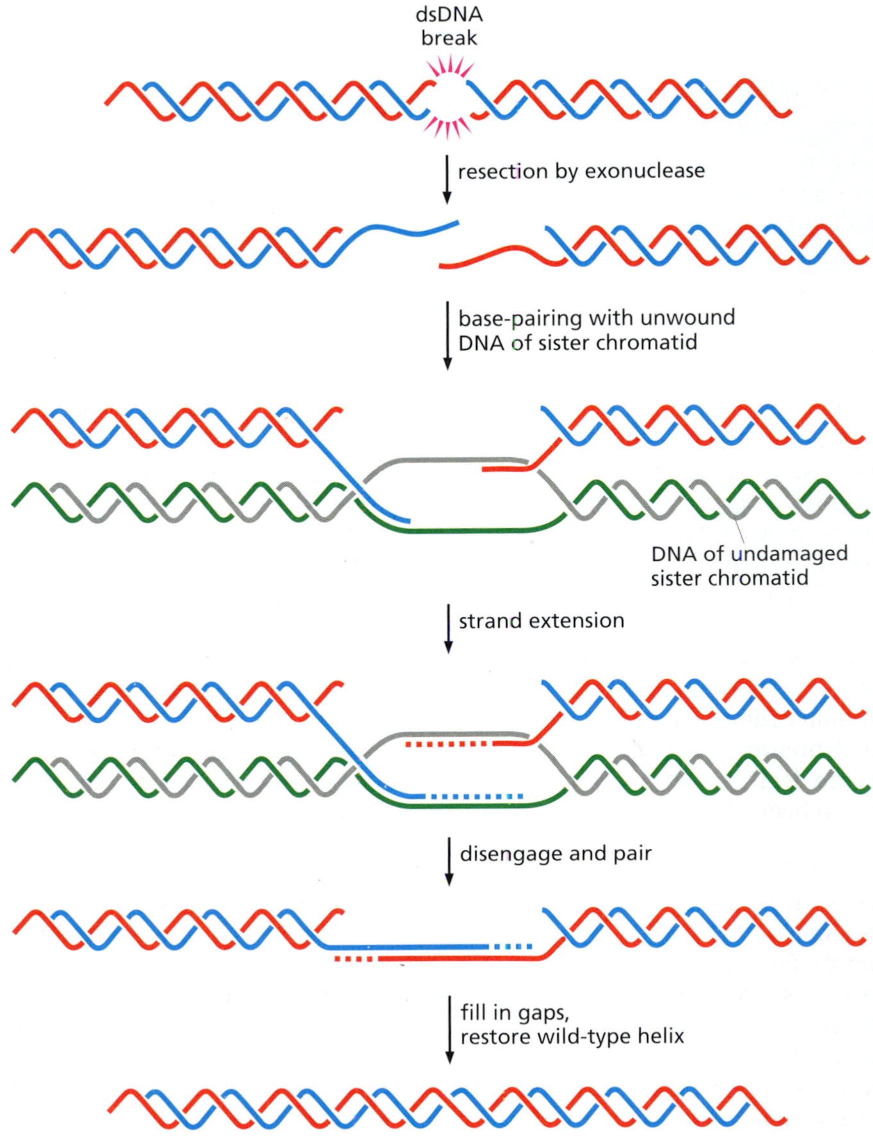

Figure 12.32 Homology-directed repair The repair of dsDNA breaks during the late S phase and G_2 phase of the cell cycle often depends on the ability of the repair apparatus to consult the sequences present in the undamaged sister chromatid that was previously formed, together with the damaged chromatid, during the most recent S phase. Such homology-directed repair (HDR) begins *(top)* with the *resection* (removal) by an exonuclease of one of the two DNA strands at each of the ends formed by a dsDNA break. Each of the resulting ssDNA strands *(blue, red)* is then forced to invade the double helix formed by the two strands *(green, gray)* of the undamaged sister chromatid; this undamaged sister chromatid has been unwound by the repair apparatus in order to accommodate the pairing of the invading ssDNA strands with complementary sequences present in the undamaged sister chromatid. The ssDNA strands from the damaged chromatid are then elongated in a 5′-to-3′ direction by a DNA polymerase, using the strands of the sister chromatid's DNA as templates. Thereafter, the resulting extended ssDNA strands are released from the sister chromatid and caused to pair with one another, allowing further elongation by a DNA polymerase and a ligase, which together reconstruct a double helix possessing wild-type DNA sequences. Included among the DNA repair proteins known or thought to facilitate these complex steps of HDR are RAD51, BRCA1, and BRCA2.

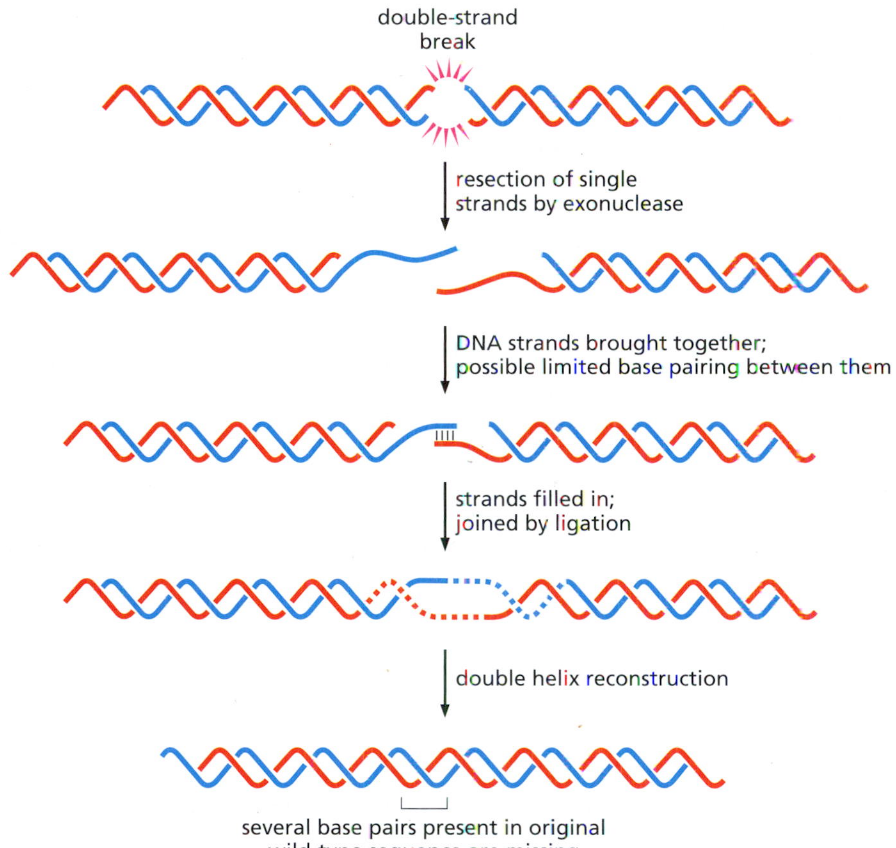

Figure 12.33 Nonhomologous end joining NHEJ is used to restore a DNA double helix following a double-strand break in the event that the nucleotide sequences present in a sister chromatid are not available to instruct the repair apparatus how these ends should be properly joined (see Figure 12.32). In NHEJ, the resection of single strands from both broken ends results in ssDNAs that can then be joined to one another, possibly by a limited degree of base pairing between them; four base pairs are shown here. The subsequent filling in of single-strand gaps and the ligation of any remaining ssDNA breaks results in the reconstruction of a double helix that lacks some of the base pairs that were present in the original undamaged DNA helix. (Adapted from M.R. Lieber, Y. Ma, U. Pannicke and K Schwarz, *DNA Repair* 3:817–826, 2004.)

Interestingly, NHEJ occurs largely in the G_1 phase of the cell cycle, when sister chromatids are not available to allow homology-directed repair. NHEJ is virtually unique among the DNA repair processes, because it plays a role in a normal, physiologic function unrelated to repairing DNA damage (Sidebar 12.9).

It is still not clear precisely how BRCA1 and -2 contribute to the maintenance of normal DNA and chromosomal structure and thereby ward off cancer. The fact that BRCA1 and BRCA2 can bind to so many distinct nuclear proteins, many involved in DNA repair (Figure 12.34A), suggests that they act, at least in part, as molecular scaffolding that helps to assemble large complexes of these other proteins and coordinate their actions. Once assembled, these various repair proteins presumably collaborate in fixing lesions, largely double-strand breaks, in the DNA. Moreover, as Figure 12.34B illustrates, loss of BRCA1 can cripple certain cell cycle checkpoint controls that normally respond to damaged DNA.

We do not know why inheritance of mutant alleles of the *BRCA1* and *BRCA2* genes leads preferentially to cancers of the breast and ovary, or why somatic mutation of *BRCA2* is occasionally associated with prostate and colon carcinomas. Moreover, the **penetrance** of *BRCA1* and *BRCA2* mutant germ-line alleles (i.e., the degree to which each allele exerts an observable effect on phenotype) has also been difficult to quantify (Sidebar 23).

As BRCA1 function is studied in ever-increasing detail, its mechanisms of action become increasingly confusing, if only because additional distinct biochemical functions are ascribed to this very large protein and its physically associated partners. For example, BRCA1 associates with a number of transcriptional regulators. It is also clear that female cells lacking BRCA1 function are unable to properly inactivate one of the two X chromosomes in their somatic cells (Sidebar 1.2). How this intersects with its DNA repair functions and the tendency of mutant germ-line *BRCA1* alleles to generate almost exclusively female cancers remains obscure.

Sidebar 12.9 NHEJ and the development of immune function The NHEJ machinery plays an integral role in the normal process of gene rearrangement that leads to the formation of functional antibodies and T-cell receptors. For example, the formation of the DNA sequences encoding antigen-binding sites depends on the rearrangement and fusion of chromosomal V, D, and J segments. In the absence of the full complement of repair proteins needed for NHEJ, such DNA segment fusions cannot occur. This results in the inability to make proper immunoglobulin (antibody) molecules and T-cell receptors, compromising both the humoral and cellular arms of the immune system (to be described in Section 15.1) and creating the syndrome of severe combined immunodeficiency (SCID). Similarly, NHEJ is needed in order to allow most types of **class switching** to occur (see also Sidebar 21); this process normally enables fused VDJ segments, which encode the antigen-binding portions of immunoglobulins, to join with alternative constant-region immunoglobulin gene segments to generate various classes of antibody molecules.

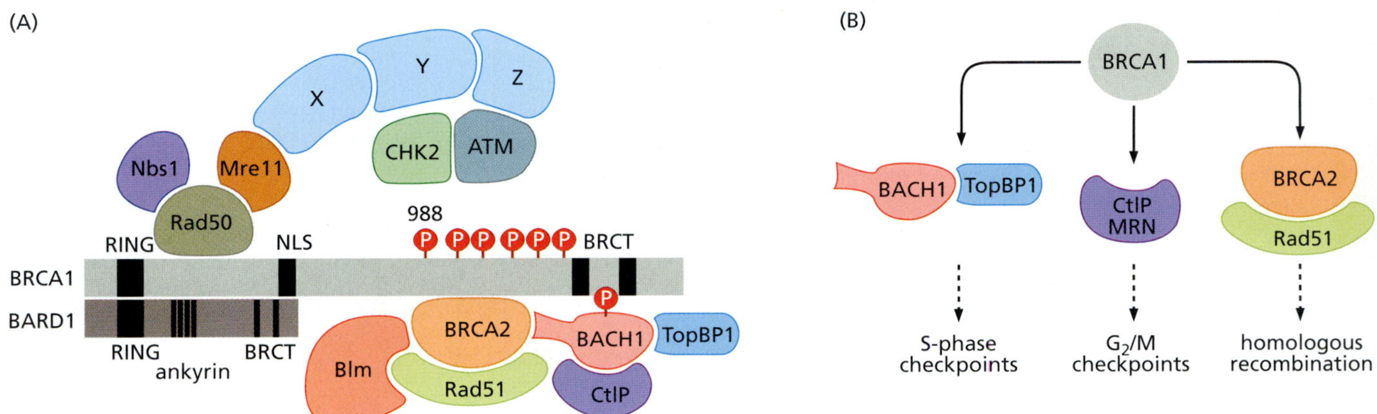

Figure 12.34 BRCA1, BRCA2, and their partners (A) The BRCA1 and BRCA2 proteins act, at least in part, as scaffolds to assemble a cohort of other DNA repair proteins into large physical complexes. Once assembled, these multiprotein complexes aid in the repair of dsDNA breaks, usually via homology-directed repair (HDR). For example, one exon of the *BRCA2* gene encodes eight copies of a "BRC domain" *(not shown);* these aid in the recruitment of multiple RAD51 molecules, which form filaments and coat ssDNA strands as part of the HDR process illustrated in Figure 12.32. The MRN complex, composed of MRE11, Rad 50, and Nbs1, appears able to recognize the end created by a dsDNA break and to activate ATM kinase function in response. (B) The loss of different partners of BRCA1 specifically affects different checkpoint controls in the cell cycle in addition to compromising the processes of homologous recombination and homology-directed repair. This illustrates the central role that BRCA1 plays as a scaffolding for a diverse array of proteins involved in a variety of processes involving DNA function. (A and B, courtesy of R.A. Greenberg and D.M. Livingston.)

12.11 The karyotype of cancer cells is often changed through alterations in chromosome structure

Long before the subtleties of DNA damage and its repair were recognized, aberrant karyotypes were known to be present in cancer cells. Indeed, for almost a century, cancer cells have been known to often carry chromosomes that appear abnormal. The triradial and quadriradial metaphase chromosomes seen in cells lacking BRCA1 or BRCA2 function are examples of these aberrations (see Figure 12.31A). Stepping back for a moment from these particular aberrations, we can recognize that two distinct classes of karyotypic abnormalities can be seen in cancer cells: changes in the *structures* of individual chromosomes, and changes in chromosome *number* that have no effect on chromosome structure.

One frequent deviation from the normal diploid karyotype involves an increase or decrease in the number of specific chromosomes. On occasion, through various accidents occurring during mitosis, cancer cells may acquire **polyploid** genomes, where an additional haploid complement of chromosomes is acquired (leading to a **triploid** state) or even an extra diploid complement of chromosomes is acquired (leading to a **tetraploid** state). Alternatively, extra copies of individual chromosomes may be present or, less commonly, a chromosome copy may be missing.

The term **aneuploidy** is usually reserved for the types of deviation from a normal (or **euploid**) karyotype that involve changes in chromosome *number.* In recent years, however, use of the term aneuploid has occasionally been broadened to include changes in the *structures* of individual chromosomes, which are prevalent in cells of the great majority (>85%) of solid tumors; a more specific term is "chromosomal aberration," which we will use here. These two major types of karyotypic alteration arise through fundamentally different mechanisms.

We will postpone discussions of changes in chromosome number until the next section. For the moment, we review the mechanisms responsible for changes in chromosome structure, some of which we have already encountered at various points in this text. For example, as we saw in the previous section, unrepaired dsDNA breaks, many of which occur accidentally at DNA replication forks, are thought to be a major source of chromosomal translocations.

In addition, much earlier (Chapter 4) we learned of a class of cancer-associated chromosomal alterations as part of a discussion of the mechanisms leading to the creation of the *myc* and *bcr-abl* oncogenes. Recall the well-studied case of the translocations that fuse the *myc* proto-oncogene to promoter sequences deriving from one of three alternative immunoglobulin genes (Section 4.5). In these and other lymphomas, it is likely that the complex machinery dedicated to rearranging the immunoglobulin and T-cell receptor (TCR) genes (see Sidebar 12.9) misfires. Instead of rearranging the immunoglobulin or TCR gene sequences, this machinery inadvertently catalyzes inappropriate interchromosomal recombination events that join the immunoglobulin genes promiscuously with sequences scattered throughout the genome, the *myc* gene being only one of them. Those rare translocations that happen to involve the *myc* proto-oncogene and deregulate its transcription seem to confer special proliferative advantage on cells, resulting in the appearance of cell clones and ultimately in lymphomas that carry these very characteristic karyotypic alterations.

In fact, many other highly specific translocations have been documented in a variety of hematopoietic malignancies (see, for example, Tables 4.4 and 4.5). The molecular mechanisms that lead to these various alterations in protein structure remain unclear. However, one highly attractive possibility was presented in Chapter 10, where we read about telomere collapse resulting in breakage– fusion–bridge (BFB) cycles (Figure 10.14). These BFB cycles create large-scale aberrations in the structures of individual chromosomes, apparently striking all chromosomes with comparable frequency. On occasion, translocations may occur that provide growth advantage to the cells carrying them, which one imagines results in clonal outgrowth of the cells carrying such aberrant chromosomes.

Superficially similar structural aberrations are seen in a variety of hematopoietic tumors. Many of these are termed *recurrent* because they have been seen on several occasions in a series of independently arising human tumors. By now, hundreds of these recurrent translocations have been cataloged. Because recurrent translocations map to highly specific chromosomal sites, it would seem that the molecular mechanisms creating them are distinct from the breakage–fusion–bridge cycles described above.

Still, the molecular mechanisms that yield these translocations remain obscure, since they occur in hematopoietic cell types in which the enzymes involved in the rearrangement of immunoglobulin genes and the similarly configured T-cell receptor genes are not expressed. Sequence analyses of the DNA flanking translocation breakpoints have revealed duplications, deletions, and inversions of sequence blocs, findings that suggest but hardly prove the involvement of some type of "error-prone" DNA repair mechanism, such as the nonhomologous end joining (NHEJ) discussed earlier.

To conclude, in the case of chromosomal translocations, we are left with two mysteries: (1) Which components of the DNA repair machinery are normally on guard to prevent the formation of these structural aberrations? (2) How do these chromosomal abnormalities, once formed, contribute to cancer?

12.12 The karyotype of cancer cells is often changed through alterations in chromosome number

As stated earlier, some types of genetic instability affect karyotype by altering the number of individual chromosomes without affecting their structure. These changes create aneuploidy. While the term "mutation" is often reserved for changes in DNA sequence (and thus includes changes in chromosome structure), we should recognize the fact that alterations in chromosome number also represent significant changes in a genome that can have equally profound effects on cell behavior and are, strictly speaking, also a type of mutation.

Changes in chromosome number are often created in cancer cells that are afflicted by the condition termed *chromosomal instability* (CIN). When CIN-positive cancer cells are removed from patients and propagated *in vitro*, the consequences of CIN become evident, for these cells continue to reshuffle their complement of chromosomes during propagation in the Petri dish. Such cancer cells were already quite aneuploid when they were removed from the patient, and their karyotypic instability seen *in vitro* is presumably only an extension of comparable instability that existed *in vivo* during the course of tumorigenesis (Figure 12.35).

The aneuploid karyotypes of cancer cells can be interpreted in two ways. One point of view portrays aneuploidy as a *consequence* of the general chaos that progressively envelops cancer cells as they advance toward highly malignant states. The other point of view ascribes a *causal* importance to aneuploidy, arguing that it is an essential component of tumorigenesis. Thus, some contend that most cancer cells require chromosomal instability during their development in order to scramble their genomes and arrive at chromosomal configurations that are more favorable for neoplastic growth. According to this second line of thinking, in the absence of the increased mutability associated with aneuploidy, most clones of incipient cells could never succeed in acquiring all of the genetic alterations needed to complete multi-step tumorigenesis.

An important observation that will help settle this debate has come from the study of a series of human colon and rectal carcinomas. The few tumors that exhibit microsatellite instability (MIN) show relatively little aneuploidy and virtually no chromosomal instability (CIN). Conversely, the far more numerous tumors that have CIN are not prone to show nucleotide sequence alterations that are characteristic of MIN (Figure 12.36). Together, these observations provide a suggestion that, at least in the case of colorectal carcinomas, tumor cells must acquire increased mutability of their genomes, and that either one or the other of these mechanisms suffices to provide such mutability. This conclusion begins to persuade one that CIN, like MIN, is an effective mechanism for remodeling the cellular genome in a way that favors evolution toward neoplasia. Whether this logic pertains as well to the genetic mechanisms creating hematopoietic malignancies remains unclear (Sidebar 12.10).

The changes in chromosome number that characterize chromosomal instability are usually (and perhaps always) the consequences of mis-segregation of chromosomes during mitosis. During the normal M phase of the cell cycle, the chromosomes line up in a plane, the *metaphase plate*, and associate with spindle

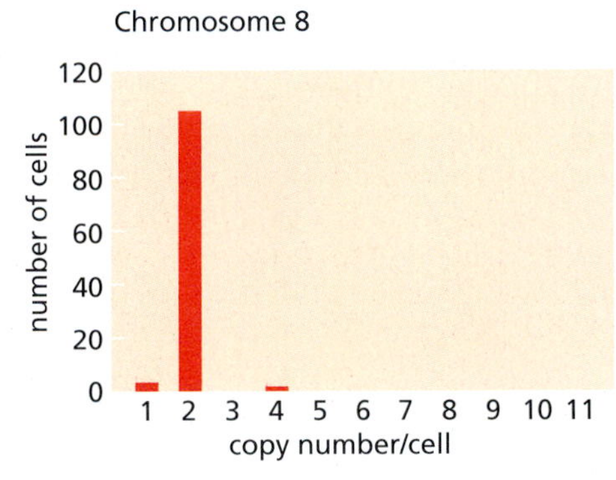

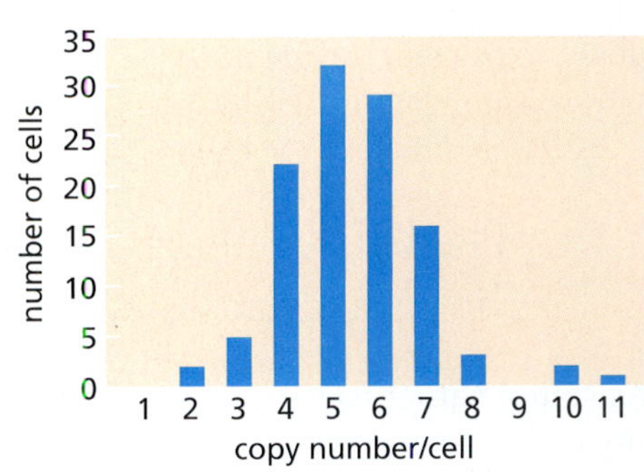

Figure 12.35 Chromosome instability in cultured cancer cells The copy number of Chromosome 8 was measured here using fluorescence *in situ* hybridization (FISH) in normal cells *(left)* and also in cultured breast cancer cells *(right)* that were found to suffer from chromosomal instability (CIN). As indicated, almost all of the normal cells had two copies of Chromosome 8, while the copy number of Chromosome 8 varied extensively in the cells afflicted with CIN. (An essentially identical distribution of chromosome copy number was observed upon study of a second, arbitrarily chosen chromosome.) This great cell-to-cell variability of chromosome number indicates that fluctuations of this number continue to occur frequently as these cancer cells are propagated in culture. (From G.A. Pihan, A. Purohit, J. Wallace et al., *Cancer Res.* 58:3974–3985, 1998.)

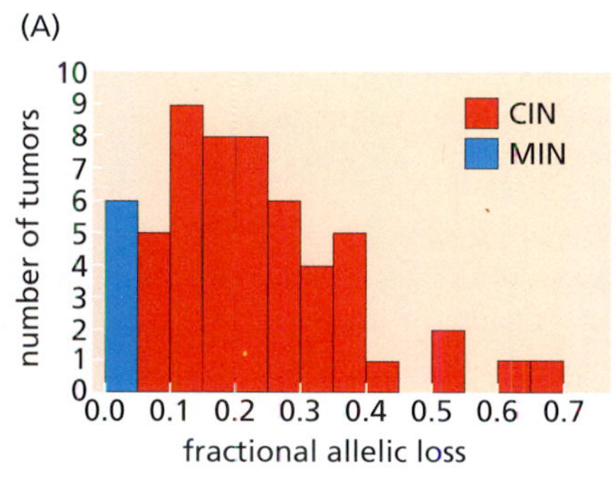

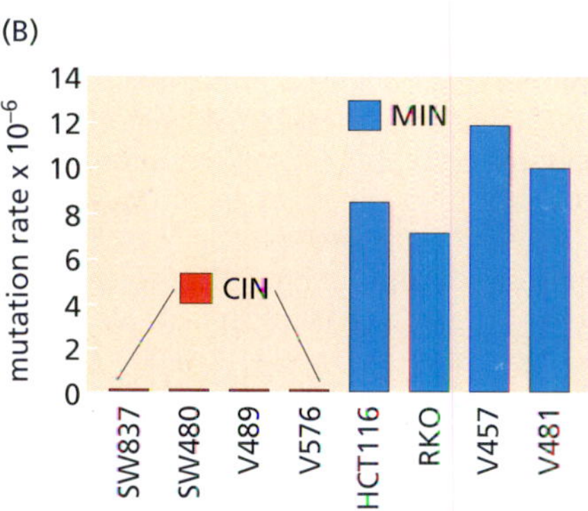

Figure 12.36 Chromosomal instability vs. gene mutation The presence of chromosome instability (CIN) can be gauged by measuring the loss of alleles from chromosomal arms. (A) In the colorectal carcinomas studied here, analyses of a large number of tumors have revealed that many tumors have lost heterozygosity (LOH; see Section 7.4) at a substantial number of chromosomal loci. On the abscissa, 0.3 allelic loss, for example, refers to tumors in which 30% of the loci that were previously heterozygous, as revealed by analyses of chromosomal markers, no longer exhibit heterozygosity *(red bars)*. Most of this LOH is attributable to the loss of whole chromosomes. In contrast, among the tumors afflicted with microsatellite instability (MIN; *blue bar*), the loss of alleles and hence the loss of entire chromosomes is negligible. (B) In colorectal tumor cells lines that exhibit CIN, as gauged by the loss of chromosomal markers (see panel A), the rate of inactivation of the *HPRT* (hypoxanthine phosphoribosyltrans-ferase) gene is virtually zero *(first four bars, red)*. In contrast, in those that exhibit MIN, the rate of mutation of this gene is significant and is occasionally 100-fold higher than in CIN tumor cell lines *(last four bars, blue)*. (A, from C. Lengauer, K.W. Kinzler and B. Vogelstein, *Nature* 396:643–649, 1998, and B. Vogelstein, E.R. Fearon, S.E. Kern et al., *Science* 244:207–211, 1989; B, from C. Lengauer, K.W. Kinzler and B. Vogelstein, *Nature* 396:643–649, 1998, and J.R. Eshleman, E.Z. Lang, G.K. Bowerfind et al., *Oncogene* 10:33–37, 1995.)

fibers constructed of microtubule proteins. The fibers together form a metaphase spindle. The metaphase spindle, in turn, is a bipolar structure in which each half spindle is constituted of microtubule fibers, many of which extend from the **kinetochores** on the chromosomes (the nucleoprotein bodies associated with the centromeric DNA of the chromosome) back to the centrosomes; the latter are responsible for organizing the entire metaphase spindle structure. When this apparatus is working properly, the spindle fibers pull sister chromatid pairs apart, so that each chromatid moves toward one of the two centrosomes. This ensures that the two daughter cells that will eventually arise after cell division receive precisely equal allotments of chromosomes (see Figure 8.3B).

This complex process of chromosome segregation is monitored by a series of checkpoint controls, which ensure initially that precisely two centrosomes and two half spindles form; that each chromatid in a pair associates with its own, distinct half spindle; and that chromatid separation is not allowed to proceed unless and until all pairs of chromatids are properly aligned on the metaphase plate. When these checkpoint mechanisms fail to impose quality control on chromosomal segregation, both sister chromatids in a pair may be pulled to one or the other centrosome (the process of **nondisjunction**). As a consequence, one of the subsequently arising daughter cells may become haploid for this chromosome and the other triploid. Alternatively, a chromatid may fail to attach to a spindle fiber and may simply be lost from the genomes of descendant cells.

More widespread karyotypic chaos may occur if the spindles themselves are not properly assembled. Aberrant mitoses, which result from inappropriate spindle organization, were noticed as early as 1890 and, in retrospect, represented the first clue that cancer cells are genetically abnormal. In normal **interphase** cells, a single centrosome can be visualized in the cytoplasm (Figure 12.37A); during

Sidebar 12.10 Widespread chromosomal aberrations are not present in all types of human cancer cells The karyotypes of carcinoma cells and of hematopoietic tumor cells show a striking discrepancy: the epithelial cancer cells almost invariably exhibit widespread karyotypic chaos, including a variety of nonreciprocal translocations, deletions of chromosomal arms, and duplications of others. In contrast, the karyotypes of hematopoietic tumor cells are often diploid, with the exception of one or two reciprocal translocations that seem to be responsible for initiating the cancer or triggering a specific step of tumor progression (e.g., the one creating the *BCR-ABL* oncogene). Therefore, chaotic karyotypes are not required for the formation of all types of human malignancies.

It is highly likely that the small number of observable karyotypic alterations found in most hematopoietic cancer cells do not, on their own, suffice to enable full neoplastic proliferation. (In one case—that of chronic myelogenous leukemia—the acquisition of the *BCR-ABL* oncogene is often followed during blast crisis relapse by the loss of p53 function; point mutations cause this loss, and they are, of course, karyotypically invisible.) Moreover, hematopoietic tumor cells have not been reported to suffer from microsatellite instability. It therefore remains unclear what genetic mechanisms enable hematopoietic cells to acquire the entire ensemble of mutant alleles needed in order for them to proliferate as fully neoplastic cells. Indeed, we do not even know whether the formation of hematopoietic tumors requires as many genetic changes as those needed for the formation of solid tumors (see Section 11.12).

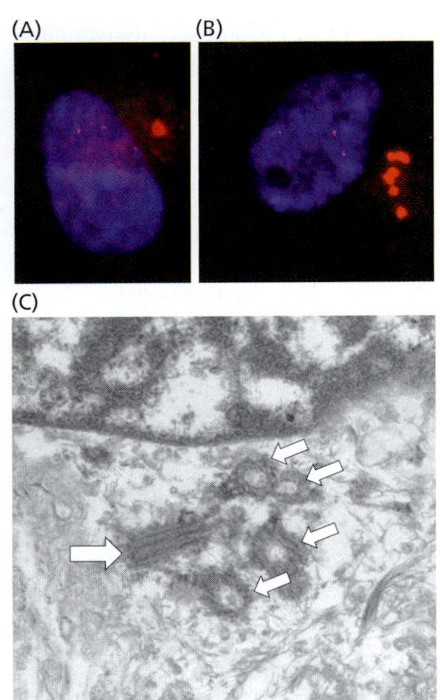

Figure 12.37 Centrosomes and the organization of the mitotic spindle Centrosomes are responsible for organizing the microtubule spindle fibers at mitosis. (A) In immortalized but nonmalignant interphase cells, the presence of a single centrosome can be detected in the cytoplasm through use of an antibody that detects pericentrin, a centrosome-associated protein *(red).* This centrosome is normally duplicated at the G_1/S transition to generate the two centrosomes found at the poles of the mitotic spindle. (B) In contrast, during interphase of human breast cancer cells, multiple centrosomes can often be observed. These will often create multipolar spindles (see Figure 12.38) when such cells enter mitosis. (C) The pair of centrioles that forms the core of each centrosome can be best seen using transmission electronic microscopy (TEM), in this case of a human colon carcinoma cell. Four centrioles are seen here in cross section *(small arrows),* while a side view of a fifth *(large arrow)* is apparent, indicating a deregulation of centriole number. The nuclear membrane is seen above. (A and B, from G.A. Pihan, A. Purohit, J. Wallace et al., *Cancer Res.* 58:3974–3985, 1998; C, courtesy of M.J. Difilippantonio and T. Ried.)

mitosis, two centrosomes are arrayed at opposite poles within the cell. Cancer cells, however, often show marked defects in this organization, including multiple centrosomes at interphase (Figure 12.37B and C). The result may be mitotic spindles that have multiple poles rather than the two seen in normal cells (Figure 12.38A and B) and the division of the normal chromosomal array of chromosomes among three or more daughter cells (Figure 12.38C). As a further consequence, the resulting mis-segregation of chromosomes into daughter cells may lead to wild fluctuations in chromosome number and overall karyotype. In one survey of a series of 87 different tumors, 81 of these showed abnormalities in centrosome number or in the microstructure of individual centrosomes; such defects were never encountered in normal cells used as controls in this study.

It seems that once the complex apparatus designed to ensure proper chromosomal segregation has been damaged, such damage is irreversible. For example, as was seen in Figure 12.35, the enormous cell-to-cell variability in the number of Chromosome 8 copies in certain breast cancer cells indicated that chromosome instability (CIN) persisted in these cells long after tumor progression had reached completion. In this respect, CIN differs from the breakage–fusion–bridge (BFB) cycles described earlier (Section 10.4), which seem to plague the genomes of cancer cells for a limited window of time during tumor progression and then cease once cells succeed in acquiring telomerase and thereby stabilize their karyotypes.

In recent years, some of the molecular defects that contribute to various types of chromosomal instability have come to light. Not surprisingly, the duplication of centrosomes is closely coordinated with cell cycle advance; it seems to occur at or near the G_1/S transition. More specifically, an increasing body of evidence indicates that centrosome duplication is coordinated in some way by cyclin E– and A–containing cyclin-dependent kinase (CDK) complexes (Section 8.3). Provocatively, primary human cells (i.e., those that have been in culture for only a short period of time) in which the human papillomavirus (HPV) E7 oncoprotein is expressed show a deregulation of centrosome number, often leading to **supernumerary** (extra) centrosomes beyond the two that are normally present in cells poised to enter mitosis.

Recall that the E7 oncoprotein binds to cellular pRb (the retinoblastoma protein), inactivates it functionally, and accelerates its degradation (Section 8.5). Significantly, a mutant form of the HPV E7 oncoprotein that fails to bind to pRb also fails to induce the appearance of extra centrosomes. Together, these strands of evidence suggest that one of the consequences of loss of pRb function is a deregulation of centrosome duplication. Since the centrosomes play a central

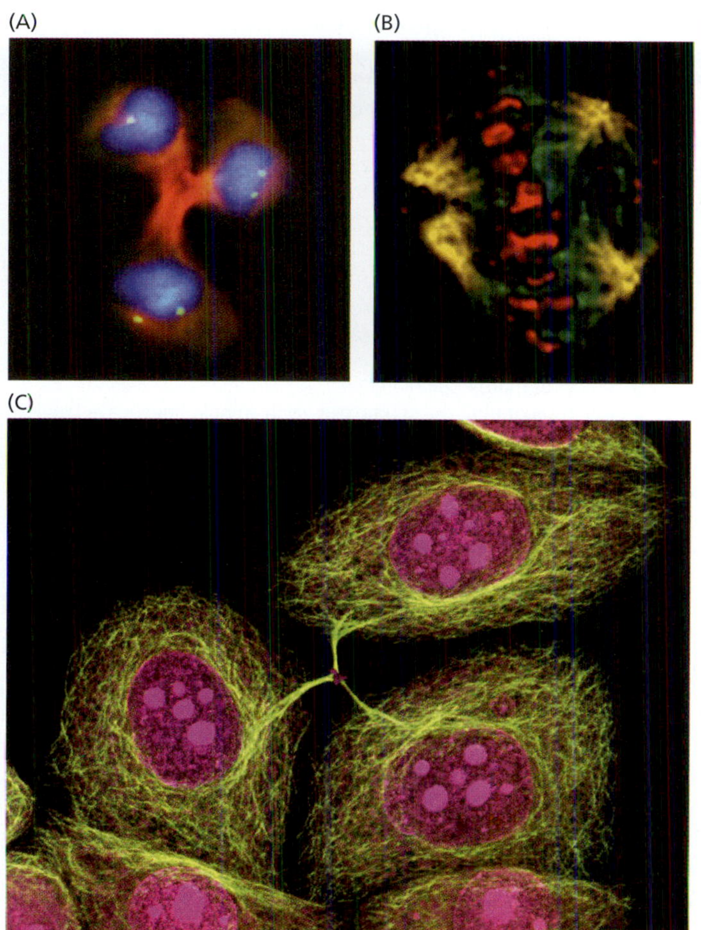

Figure 12.38 Multipolar mitotic apparatuses In a normal mitotic metaphase of cultured cells, the mitotic spindle fibers, which are composed of microtubules, reach from the two centrosomes at the mitotic poles to the kinetochores of the chromosomes—the proteinaceous structures associated with the centromeric chromosomal DNA. (A) In this micrograph of *p53*$^{-/-}$ mouse embryo fibroblasts, the formation of three centrosomes has resulted in the formation of a triradial mitotic spindle array, in which the spindle fibers have been immunostained *red* with an antibody reactive with α- and β-tubulin while the chromosomal clusters have been stained *blue* with the DAPI, a DNA dye. The three centrosomes are seen as small *yellow* spots that have been immunostained with an antibody reactive with γ-tubulin. Observations like these implicate p53 in the regulation of centrosome number. (B) The presence of multipolar spindles is evident in this human oral carcinoma cells in which four centrosomes are active. The spindle fibers have been immunostained with an anti-tubulin antibody *(yellow)*. The kinetochores, immunostained in *red*, have become aligned in one major and one minor metaphase plate. Regions of overlap between the chromosomal DNA (stained blue with the DAPI dye, *not visible)* and the spindle fibers appear *green*. (C) Here at the cytokinesis following telophase, three daughter cells of the human HeLa cervical carcinoma cell line are seen to have been formed through mitosis of a mother cell whose chromosomal complement they have divided up. Microtubules, including those of the spindle fiber, have been stained *yellow*; DNA is stained *purple*. The abscission of these cells at the end of cytokinesis is not proceeding properly. (A, from P. Tarapore and K. Fukasawa, *Oncogene* 21:6234–6240, 2002; B, from W.S. Saunders, M. Shuster, X. Huang et al., *Proc. Natl. Acad. Sci. USA* 97:303–308, 2000; C, from K.G. Murti, *BioTechniques*, Oct. 2004 cover picture.)

role in organizing the mitotic spindles, this deregulation may soon lead to dramatic fluctuations in chromosome number. Indeed, precisely such chromosomal instability is observed in lines of human cervical carcinoma cells, almost all of which express the early region of the HPV genome, which carries the genes encoding the viral E6 and E7 oncoproteins.

An equal contribution to aneuploidy may be made by the HPV E6 protein (Sidebar 12 ●) and other agents that disrupt p53 function, since cells lacking the services of this important tumor suppressor protein seem particularly susceptible to acquiring aneuploid genomes. Thus, the HPV E7 oncoprotein destabilizes centrosome number, while the E6 oncoprotein, through its actions on p53, causes cells to tolerate any centrosome abnormalities that may eventually arise. And in a large variety of other human cancers that are not caused by viral infection, cells that have lost p53 function through other mechanisms also show relatively high rates of instability in their chromosome number. The precise signaling connections between p53 function, mitotic checkpoint controls, and the monitoring of chromosomal number remain obscure.

Inactivation of pRb regulation may confer immediate proliferative advantage on an incipient, HPV-infected cancer cell by inactivating the control that governs passage through the R-point transition (Section 8.2). In the longer term, additional advantages may accrue to the descendants of this cell, since centrosome number will become destabilized, leading in turn to derangement of mitotic spindles and resulting destabilization of karyotype. The changes in chromosome number that ensue may yield constellations of growth-promoting and growth-retarding genes in proportions that expedite tumor progression.

Table 12.3 Mutated, methylated, and overexpressed genes in cancer cells that perturb chromosomal stability

Gene	Function of gene product	Consequence of alteration in cancer cells
BUB1	spindle assembly checkpoint	progress through mitosis, even in the presence of microtubule inhibitors[a]
MAD1[b]	spindle assembly checkpoint	large-scale aneuploidy
MAD2[b,c]	spindle assembly checkpoint	premature entrance into anaphase[d], aneuploidy
Securin	attachment of sister chromatids	nondisjunction of chromosomes[e]
ATM	chromosome segregation	defective metaphase-anaphase transition
Aurora-A,-B,-C	separation of chromatids at anaphase	premature entrance into anaphase[d]
CHFR	spindle assembly checkpoint	nondisjunction, chromosome loss[e]
14-3-3σ	DNA damage checkpoint	segregation of unrepaired chromosomes

[a]Microtubule inhibitors such as colchicine and nocodazole block the assembly of mitotic spindle fibers.
[b]Mad1 and Mad2 form complexes at the kinetochore that prevent chromatid separation until complexes with spindle fibers have been properly formed.
[c]The *MAD2* gene is transcriptionally repressed in a number of solid tumors and is frequently mutated in gastric carcinomas. Mice that are heterozygous at the *Mad2* locus (i.e., are $Mad2^{+/-}$) develop lung cancers as adults.
[d]Premature entrance into anaphase can lead to loss of entire chromosomes.
[e]Nondisjunction is the failure of sister chromatids to separate at anaphase.

The great complexity of mitosis raises the question of how prone this process is error, and how many regulatory proteins, including checkpoint control proteins, are in place to monitor the progression of the various steps of M phase. Yeast genetic analyses have revealed as many as 100 distinct genes and proteins that are involved in the various steps of spindle assembly and dynamics, spindle attachment, and the separation of chromosomes during mitosis; mutation of many of these genes results in chromosome instability in yeast. Many of these proteins are highly conserved evolutionarily, and their homologs are likely to be components of the mammalian mitotic machinery.

To date, only a small proportion of human tumors have been found to carry mutations in the human homologs of yeast genes known to be involved in chromatid assembly and separation during mitosis. Some examples are provided in Table 12.3. The functions of some of these critical mitosis-regulating genes may be lost through methylation of their promoters (Section 7.8), but this remains largely undocumented. In fact, only a small proportion of the large cohort of human genes involved in mitotic functions have been explored to determine whether the functioning of these genes is compromised during the course of tumor progression.

One example of such mitotic control genes is the *CHFR* gene, which encodes a checkpoint protein that normally prevents advance from prophase into metaphase in the event that the spindle microtubules and centrosomes are not properly arrayed. The gene encoding this checkpoint protein, which is normally expressed ubiquitously in human cells, has been found to be fully repressed due to promoter methylation in 3 of 8 randomly chosen human cancer cell lines, and in 7 out of 37 lung cancer biopsies.

Another key regulator of entrance into M phase is the 14-3-3σ protein. When genomic DNA is damaged, p53 induces synthesis of 14-3-3σ, which proceeds to trap cyclin B–Cdc2 complexes in the cytoplasm (Section 9.9); by sequestering these cyclin–CDK complexes, 14-3-3σ succeeds in blocking entrance into M phase, thereby holding mitosis in abeyance until damaged chromosomal DNA has been repaired. The gene encoding 14-3-3σ has also been found to be frequently methylated in common cancers, including those of the lung, breast, stomach, and liver. The resulting loss of 14-3-3σ must surely contribute to the aneuploidy observed in these kinds of cancer cells.

These two examples, involving the CHFR and 14-3-3σ proteins, only begin to plumb the depths. There are clearly many other "caretaker" genes to be discovered that play critical roles in stabilizing DNA sequences or karyotype and in this way determine whether or not tumor progression will proceed slowly or move ahead in leaps and bounds (Sidebar 24).

12.13 Synopsis and prospects

Genome instability has been inherent in life since the first cells appeared 3.5 billion years ago. In the intervening time, living organisms have continually struck a balance between too little and too much instability in their genomes. If they went too far in suppressing the rate at which mutations accumulated, Darwinian evolution, which depends directly on the continued generation of genetic diversity, would have ground to a halt. Conversely, if they allowed mutation rates to increase too much, their ability to reproduce and even their viability would have been seriously compromised. The relatively low level of genomic instability that operates in our cells—specifically those carrying the germ line—represents a compromise between these two conflicting needs.

This balance between too little and too much genetic instability does not need to be struck in the individual cells of our somatic tissues. In these, any genetic instability—mutability, as we have called it—is undesirable, since it opens the door to neoplasia. This would seem to explain why evolution has worked hard to ensure that the genomes of our somatic cells are so stable.

Multiple layers of defense mechanisms operate to hold somatic mutation rates to extremely low levels. At the biological level, they are embodied in the organization of stem cells and their differentiated progeny. At the biochemical level, an array of enzymes and a variety of low-molecular-weight biochemical species are deployed to confront and neutralize mutagens before they succeed in striking the genome. And should damage be inflicted, either because mutagens have slipped through the outer defenses or because of errors in DNA replication, then a large group of DNA repair enzymes—the caretakers—lie in wait, always alert to structural aberrations in the double helix and its nucleotides. More often than not, these enzymes deal very effectively with incurred damage and restore the DNA to its pristine state, erasing any sign that damage was ever sustained. In addition to these, a complex array of proteins ensures that mitosis and meiosis occur only when the chromosomes are aligned properly at the metaphase plate, thereby sustaining the euploid karyotype.

Our perception of DNA and its much-touted stability is changed by an understanding of these caretakers and their multiple roles in maintaining the genome. Previously, we depicted DNA as a rock-solid, unchanging entity within the cell, a unique island of stability sitting amid countless other molecules that are constantly forming and being degraded. Now, we realize that this was simplistic and an illusion. Like all other molecules in the cell, DNA is vulnerable to many types of damage. Its apparent stability reflects nothing more than a dynamic equilibrium, an ongoing battle between the forces of order and chaos. Any stability that chromosomal DNA does exhibit, and it is considerable, represents a stunning testimonial to the elaborate array of caretakers that are always on watch, ready to fix even the most minor lesion in the double helix.

The implications of this for cancer are simple and clear: if a breakdown of genomic integrity is an essential ingredient in forming human tumors, this can derive most readily from weakening the ever-vigilant repair machinery and its controllers.

Our initial encounter with the breakdown of genomic stability came in Chapter 9, where we learned that the p53 tumor suppressor protein is occasionally called the "guardian of the genome," because cells lacking p53 function acquire a variety of genetic defects at an elevated rate. In large part, this increased mutability, which includes alterations in DNA sequence as well as changes in karyotype, does not reflect p53's role in directly maintaining the genome. Instead, the loss of p53 function creates an environment that is permissive for the survival of mutant cells. In this chapter, we changed our focus by posing a different question, one that goes beyond p53 inactivation: If p53 loss permits mutant (and highly mutable) cells to survive, how, precisely, do the mutations acquired by these cells arise in the first place?

These mutations occur frequently during tumor progression, and elevated mutability is increasingly accepted as an essential element of cancer pathogenesis. As noted at the beginning of this chapter, the proposal was made in 1974 that a departure from DNA's highly stable state is essential for the formation of human cancers. This proposal arose from calculations of the rates at which mutations accumulate in normal cells and an estimate of the number of genetic alterations that are needed in order for tumor progression to reach completion.

Without such increased mutation rates—so the thinking went—the time intervals between clonal successions (see Section 11.5) would be far too long. Actually, the readings from the last chapter and this one suggest at least two alternative ways by which clonal successions can be accelerated during multi-step tumor progression. Tumor promoters (including endogenous processes such as inflammation) can compress the time between clonal successions; alternatively, acceleration can be achieved by the destabilization of the genome, as described here. Because the two processes usually work hand-in-hand, mathematical modeling of tumor progression becomes very difficult.

We now realize that the 1974 mathematical analyses depended on so many quantitative assumptions that their major prediction represented little more than an inspired speculation. As is almost always the case in biology, observations of living systems speak more loudly than theorizing: During the three decades that followed, a diverse array of experimental observations provided strong support, indeed validation, of this speculation. We now know that genetic instability of one sort or another is pervasive in human cancers. The accumulated evidence indicates that destabilization of the genome is an essential ingredient in the creation of almost all tumor cells that arise in the human body. In fact, genetic instability is so common in human tumors that we can include it among the hallmarks of cancer—the roster of cellular traits that together define virtually all types of malignancies.

Years after genetic instability was predicted to play a critical role in tumor progression, a biological discovery was made that is likely to provide additional support for this idea. In the last chapter, we read at length about the role of stem cell-like populations within breast tumors that function, like stem cells in normal tissues, to populate tumor masses with neoplastic cells of a limited replicative potential (Section 11.6). Other work has extended this paradigm to include human brain and gastrointestinal tumors as well, and it appears likely that these hierarchical organizations of stem cells and transit-amplifying progeny will be present in most types of human tumors.

In most tumors, these stem cell-like cells constitute only a small portion of the overall population of cancer cells, yet they appear to be entrusted with transmitting the tumor cell genome—the tumor's germ line—to future generations of cancer cells. Consequently, the mutations that drive tumor progression in the breast and brain, and likely elsewhere, must strike the genomes of these relatively rare cells. (Conversely, mutations that strike the non-stem cells in the tumor mass, notably the transit-amplifying cells, will generally not be productive for tumor progression, because these cells have only a limited replicative potential. A possible exception here may be those mutations, if they exist, that strike transit-amplifying cells and confer on them an unlimited self-renewal capability.)

If most of the mutations contributing to cancer development are limited to those striking stem cells, this dictates that the effective mutation rate (mutations accumulated per cell division) during tumor progression must be far higher than was imagined previously. Recall that earlier calculations assumed that all the pre-malignant cells in a tumor mass are equally qualified, once mutated, to become the progenitors of new, more highly evolved clones of malignant cells (see Figure 11.12).

One crude measure of the actual extent of genome destabilization in cancer cells is provided by observations that, in addition to its normal role in eliminating mutant cells, p53 does play a direct role in supporting genome maintenance: p53-negative cells seem to have defective global nucleotide-excision repair (NER). Given the widespread occurrence of *p53* gene inactivation in almost half of all human tumors and deregulation of the p53 pathway in most of the others, this suggests that a high percentage of human neoplasms lack at least one type of NER. Importantly, this partial loss of repair capacity is likely to represent only one of several defects in genome maintenance that *p53*-negative cells acquire during their long march toward malignancy. (Moreover, accumulating evidence indicates that p53's two cousins, p63 and p73, which frequently collaborate with p53 to form multi-subunit transcription factor complexes, are also responsible for inducing expression of genes encoding certain DNA repair proteins.)

The submicroscopic changes in genome structure created by defective BER and NER—far too small to affect the karyotype of cancer cells—are compounded by the widespread scrambling of chromosome structure, and thus the karyotype, found in cells from the great majority of solid tumors. In Section 10.4, we read that telomere erosion is likely to be responsible for much of this instability through its ability to trigger breakage–fusion–bridge cycles. At the same time, we learned that this chromosomal chaos also contributes to driving tumor progression forward.

A quantitative measure of the extent of this chaos is provided by the technique of comparative genomic hybridization (CGH; Figure 11.20), which measures the copy number of different chromosomal DNA segments and thus chromosome arms. As was shown in Figure 11.21, the genome of a single human breast cancer shows substantial fluctuations in the copy number of individual chromosomal segments; these CGH profiles vary dramatically from one tumor to the next. However, these excursions from normal copy number are not entirely random: when the profiles of a large number of independently analyzed breast cancers are averaged, then certain recurring changes in chromosomal-segment copy number become apparent (Figure 12.39). Since these changes are seen in a number of independently arising tumors, they are presumably advantageous for cancer cell proliferation, but we are still far from knowing precisely why.

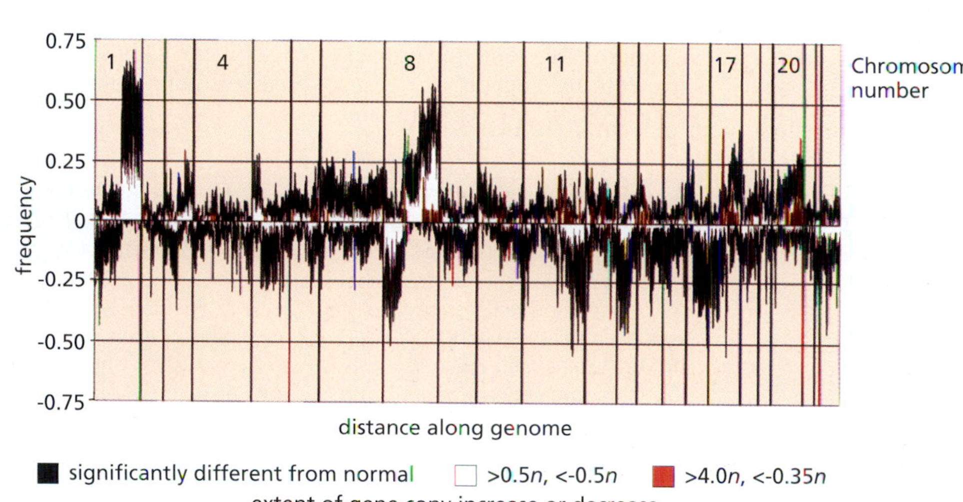

Figure 12.39 Comparative genomic hybridization analysis of breast cancer genomes The genome of a human breast cancer cell can exhibit wide fluctuations in the copy number of various chromosomal segments and associated genes. The technique of comparative genomic hybridization (CGH) makes it possible to compare the copy number of the chromosomal segments in a tumor with the copy number of the related segments present in normal human DNA (see Figure 11.20). CGH analysis of each human breast cancer genome yields a distinct profile of segment gains and losses; however, an averaging of the CGH profile patterns of a large number of breast cancers, as shown here, reveals that gains and losses of particular chromosomal segments are present in multiple tumors and are therefore recurrent. The proportion of tumors showing various degrees of amplification or loss *(ordinate)* is plotted as a function of distance along the genome *(abscissa)* moving *rightward* from Chromosome 1. The proportions of breast cancers showing abnormally high copy number abnormalities are plotted as positive values, while the proportions showing abnormally low copy numbers are plotted as negative values. Vertical lines indicate chromosome boundaries. Values plotted in *black* show the proportions of tumors in which the copy number of a chromosomal region is significantly different than the average. Values plotted in *white* show the proportions of tumors in which the copy number is 50% greater than average or 50% lower than average. Values plotted in *red* show frequencies of tumors for which copy number is more than that are four times the average (i.e., highly amplified) or less than 35% of average (i.e., more than one copy of the tumor genome has been lost). In this case, the "average" is the expected copy number of a gene, given the overall ploidy of the breast cancer genome being studied. Thus, segments of Chromosomes 1 and 8 are repeatedly found to be amplified in these cancers, while segments from Chromosomes 8, 11, 13, 16, and 17 are often deleted. (Courtesy of K. Chen and J.W. Gray).

The acquired mutability of cancer cell genomes is a double-edged sword, since it places cancer cells at some risk: the resulting frequent errors in DNA replication and in chromosomal segregation make these cells especially vulnerable to death because of lethal, unrepaired defects in their genomes. Indeed, it is possible that much of the attrition of cancer cells during each cell generation (see Figure 10.5) occurs because of frequent genomic catastrophes that are incompatible with continued cell survival. Moreover, cancer cells may be exposed to a second danger because of their defective DNA repair apparatus: they may be especially susceptible to certain forms of chemotherapy (Sidebar 12.11).

The fact that diploid human cancer cells can be created experimentally through the introduction of a specific set of cloned genes (Section 11.12) indicates that the myriad chromosomal changes present in many cancer cells are not essential for neoplastic proliferation if a cell is provided with the proper combination of genetic elements, including mutant proto-oncogenes and tumor suppressor genes. This must mean that only a small proportion of the hundreds or thousands of the alterations present in many human cancer cell genomes are actually playing causal roles in the process of tumorigenesis. So, we might speculate as follows: mutation of a small number of critical control genes is one way to acquire a neoplastic cell phenotype; genome scrambling, as revealed by comparative genome hybridization, is yet another. A combination of the two strategies appears to be the route chosen by most human cancer cells during multi-step tumor progression.

We still know rather little about the time during multi-step tumor progression when increased mutability is first acquired. Genomic destabilization may occur early in the development of some cancers and quite late during the develop-

Sidebar 12.11 Certain chemotherapeutic drugs may have selective toxicity for cancer cells lacking fully competent DNA repair systems A number of anti-cancer treatments under development take advantage of the inability of many types of cancer cells to properly repair or maintain their genomes. The PARP1 [poly(ADP-ribose) polymerase] enzyme facilitates DNA repair by binding to DNA breaks, thereby attracting other repair proteins. *Parp*$^{-/-}$ mice are viable and fertile, indicating redundant DNA repair systems; this explains why **pharmacologic** inhibitors of PARP have little if any effect on normal tissues.

Cells lacking BRCA1 or BRCA2 function appear to survive by depending on PARP and its collaborators as a back-up system to maintain their genomes at some minimal level of integrity through HDR. This type of repair seems to be particularly critical in restoring the genome following the accidental breaks that occur at replication forks during S phase (see Figure 12.10). As a consequence, cells lacking BRCA1 or BRCA2 function have been found to be highly sensitive to killing by pharmacologic inhibitors of PARP1 (Figure 12.40). Because these inhibitors have relatively little effect on normal cells, chemical derivatives of these PARP1 inhibitors may one day prove to be useful as highly selective cytotoxic agents for treating breast tumors that lack either BRCA1 or BRCA2 function.

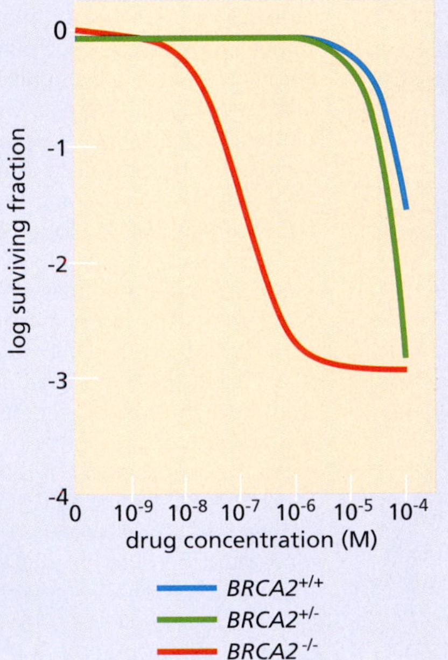

Figure 12.40 Killing of BRCA2-negative cells by an anti-PARP drug The reliance of BRCA1- and BRCA2-negative human breast cancer cells on a redundant HDR pathway mediated by PARP may make these cells especially vulnerable to killing by an anti-PARP drug. Here, mouse ES (embryonal stem) cells of three genotypes have been treated with KU0058948, a PARP inhibitor. While the *BRCA2*$^{+/+}$ *(blue)* and *BRCA2*$^{+/-}$ *(green)* ES cells were killed only by very high drug concentrations *(abscissa)*, the *BRCA2*$^{-/-}$ ES cells were killed by the drug at concentrations that were almost 1000-fold lower *(red)*. (From H. Farmer, N. McCabe, C.J. Lord et al., *Nature* 434:917–921, 2005.)

ment of others. For example, chromosome instability is often detectable early in colon cancer progression, being already present in many small adenomatous polyps. The immediate consequence, we can imagine, is an acceleration of many of the subsequent steps of colon carcinoma progression.

This scenario of acquired genetic instability seems to create a logical quandary. Almost all of the genetic alterations that occur during tumor progression appear to confer some immediate growth or survival benefit on the cells that acquire them. Many of the earlier chapters in this book documented the specific growth advantages resulting from each of these alterations. These dynamics caused us to speculate that tumor progression is a process that is analogous to Darwinian evolution (Section 11.5). However, unlike many traits acquired by cancer cells, the trait of genomic instability, once acquired by a cell, does not provide an immediate pay-off—a marked advantage in proliferation or survival. Instead, acquisition of this trait represents a "long-term investment", that is, the benefit will only be realized by the distant descendants of this cell.

We can understand this long-term advantage in the following way. Cell clones that have acquired some of the initial genetic alterations leading to cancer seem to arise with great frequency throughout the body's tissues. The vast majority of them remain in a dormant, pre-malignant state for an entire human lifetime, unobtrusive and unthreatening, because their stable genomes preclude the acquisition of the additional mutations that would render these cell clones truly dangerous. On rare occasion, however, a cell in one or another of these already-initiated clones acquires an alteration in one of its caretaker genes. Now, for the first time, this cell and its descendants have the opportunity to tinker with their genomes by testing new combinations of genetic elements and new sequences, some of which will allow them to advance steadily down the long road of tumor progression.

To be sure, additional agents, including tumor promoters and exogenous mutagens, further enhance the rate at which this forward march occurs. Still, acquired genomic destabilization seems to play a critical role in the development of most human tumors by opening a floodgate of genetic change. In sum, clones of would-be cancer cells that acquire genomic instability are destined, over the long term, to prosper, while their brethren, lacking this instability, are likely to remain unchanged and indolent for decades.

As we learned in this chapter, a variety of familial cancer syndromes are attributable to the inheritance of mutant forms of genes specifying important components of the DNA repair apparatus. Genomic instability or a tendency toward genetic instability is already implanted in all the cells of such genetically afflicted individuals. In these cases, genetic instability is not an acquired attribute—unlike the situation operating in the great majority of human cancers.

On some occasions, such as in the many subtypes of xeroderma pigmentosum (XP), both copies of a critical DNA repair gene are inherited in defective form. This nullizygous state leaves all the cells in the skin without the means to cope with the UV-induced formation of pyrimidine dimers. In other cases, such as hereditary non-polyposis colon cancer (HNPCC), an individual inherits a defective gene copy of a mismatch repair gene, and the remaining copy is then discarded in various somatic cells through a loss-of-heterozygosity (LOH) event.

The familial cancer syndromes that have been identified to date, including those involving heritable defects in DNA repair, reflect the actions of strongly penetrant germ-line alleles. Such alleles ensure that individuals carrying them will, with high likelihood, manifest obvious disease at some point in their lifetimes. However, the known familial cancer syndromes may well represent only the small tip of a very large iceberg. Thus, a variety of less penetrant, mutant alleles of the genes involved in genome maintenance may be widespread in the human gene pool. Each of these alleles may confer only a slightly increased risk of cancer, but one that is not readily apparent when studying individual families and their susceptibilities to various types of cancer. Consequently, a significant portion of many commonly occurring types of human tumors, such as breast cancer (see, for example, Sidebar 24) may be associated with inheritance of these still-unknown germ-line alleles.

Our thinking about the family cancer syndromes associated with inheritance of defective caretaker genes is confounded by one major mystery: the majority of these diseases involve only a very narrow subset of tissues in the body, even though we have every reason to believe that the services of affected caretakers are required ubiquitously. Why, for example, do inherited mismatch defects have such a strong preference for causing tumors in the intestinal tract? The puzzle is drawn even more starkly in the case of breast cancer. By one account, inherited defects in at least seven genes that are involved, directly or indirectly, in the maintenance of genomic integrity generate substantially increased risks of mammary tumors in humans. These include the *BRCA1* and *BRCA2* genes, two genes encoding BRCA1-associated proteins (*BARD1* and *BACH1*), *CHK2* (Sidebar 24), *p53*, and *ATM* (which specifies one of the kinase sensors that activate the p53 alarm following DNA damage; Section 9.6).

Of course, this is not the first time that we have encountered a puzzle of this sort. Recall, for example, the behavior of inherited mutant alleles of the retinoblastoma gene, *Rb*. This gene specifies a protein, pRb, that appears to function in almost every cell type in the body as the critical controller of advance through the cell cycle, yet children who inherit defective *Rb* alleles are predisposed peculiarly to a rare eye tumor when they are very young and to bone tumors as adolescents (Sections 7.3–7.5).

To date, our perceptions about the genes that are responsible for mammalian genome maintenance have been shaped largely by bacterial and yeast genetics. These disciplines have yielded a wealth of genes that are essential for genome maintenance in these microbes and, by extension, in our own genomes as well. For example, the resonance between yeast genetics and its harvest of mismatch repair genes and the HNPCC syndrome led to the rapid enumeration of six human genes and proteins that are responsible for this type of genome maintenance (see Sidebar 12.8).

These two lines of microbial genetics may have netted only a portion of the genes in the human genome that are responsible for maintenance of its integrity. How large, then, is the universe of human caretaker genes and proteins? The sequencing of the yeast genome begins to give us a feeling for this number. Recent estimates indicate that this genome consists of approximately 5600 distinct genes. Of these, 153 are classified as being involved in DNA replication and repair, while 88 seem to be involved exclusively with repair. The fact that many of the genes encountered in this chapter are metazoan inventions suggests that the size of the cohort of human genes involved in these processes is likely to be much larger. Indeed, by 2001, more than 130 DNA repair genes had already been identified in the human genome; by 2004, the number approached 150. Genes and proteins that play critical roles in the maintenance of the genome and in the suppression of cancer are being discovered continually, some in very unexpected places (Sidebar 12.12).

Defects in the machinery that protects the genome from damage will one day be exploited by oncologists intent on developing new and highly effective cancer chemotherapies. For example, in Sidebar 12.5, we read that levels of certain detoxifying enzymes, such as a glutathione-S-transferase (GST), are likely to influence the responsiveness of a patient's tumor to various chemotherapeutic treatments by detoxifying the chemotherapeutic molecules before they have had a chance to attack their intended targets.

Sidebar 12.12 Histones function to reduce cancer risk For many decades, histones and the nucleosomes they form were depicted as static structural components of chromatin that serve to package and compact the chromosomal DNA. Beginning in the 1990s, it became clear that covalent modification of histones is critical to make regions of chromatin more or less hospitable for transcription. Even more recently, histones have been found to play a dynamic role in DNA repair. H2AX is a variant of the abundant H2A histone that is present in virtually all nucleosomes. H2AX constitutes some 15% of the H2A-like histones in cells and substitutes for H2A in nucleosomes scattered here and there throughout the chromatin (see Figure 6 ⦿). When a double-strand DNA break is sustained in chromosomal DNA, H2AX molecules (but not the other histones) become phosphorylated, primarily by the ATM and ATR kinases, on a specific serine residue located four amino acid residues from the carboxy-terminal end of these molecules; such phosphorylated H2AX is observed in a large chromosomal region (involving as much as 2 megabases of DNA) flanking the break (Figure 12.41). (These two kinases are also responsible for phosphorylating and thereby mobilizing p53; Figure 9.13.) The resulting phosphorylated H2AX (sometimes termed γ-H2AX) serves to attract DNA repair proteins, such as BRCA1 and NBS1, as well as at least four others that aid in the task of rejoining the DNA ends (see also Figure 12.30).

Mice that have lost the H2AX gene (because of inactivation of this gene in the mouse germ line) are viable but stunted in growth. Their cells are unable to execute homology-directed repair (Section 12.10) and are prone to accumulate structurally abnormal chromosomes. Mice lacking both the H2AX and p53 proteins are highly prone to both hematopoietic and solid tumors. Even mice lacking one copy of the H2AX gene in the context of p53 deficiency show significantly increased rates of lymphomas.

These responses illustrate how highly complex the DNA repair process is, and how defects in any of its individual components—many still unidentified—open the door to the appearance of cancer. Provocatively, the human *H2AX* gene maps to a chromosomal region (11q23) that frequently undergoes loss of heterozygosity and/or deletion, creating the possibility that this gene and encoded histone are frequently shed by human cells en route to malignancy.

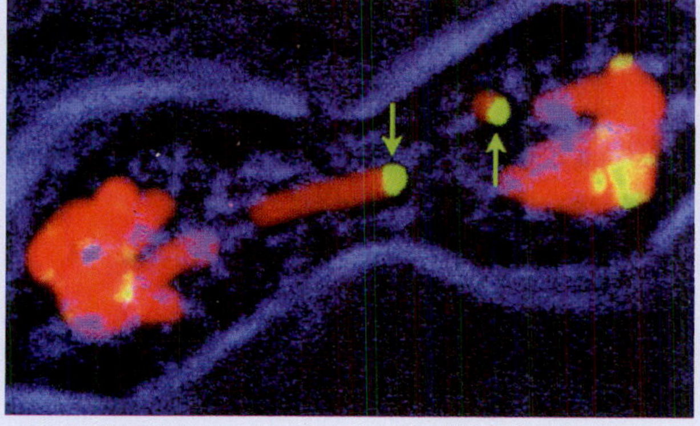

Figure 12.41 γ-H2AX and double-strand DNA breaks The creation of dsDNA breaks by various mechanisms results in the phosphorylation of H2AX histone (a variant form of histone H2A), yielding γ-H2AX. Like H2A, H2AX participates in the formation of nucleosomes. Regions of a chromosome that flank a dsDNA break and may encompass as much as a megabase of DNA are found to carry γ-H2AX (see also Figure 12.30B). In the absence of H2AX, critical DNA repair proteins, including Nbs1 and Brca1, fail to be recruited to the site of dsDNA breaks. In the image shown here, cells of the Indian muntjac, used because of their small number of chromosomes, have been irradiated with a low dose (0.6 Gy) of X-rays. Late anaphase cells were then stained for γ-H2AX *(yellow green, arrows)* and for chromosomal DNA *(red)*. The outlines of the dividing cell are seen in *blue*. Note the staining of γ-H2AX at the broken ends of a chromosome. (From E.P. Rogakou, C. Boon, C. Redon and W.M. Bonner, *J. Cell Biol.* 146:905–916, 1999.)

DNA repair enzymes are also likely to influence the efficacy of chemotherapeutic regimens. Take, for example, the enzyme DNA alkyltransferase (MGMT; Section 12.8). As we read earlier, colorectal tumors and gliomas frequently use promoter methylation to shut down expression of this enzyme; its loss would seem to allow their genomes to gain increased mutability, since alkylation events may occur frequently in the life of a cell as by-products of the actions of molecules like S-adenosylmethionine (SAM). Many chemotherapeutic agents also act through their ability to alkylate the guanine bases of DNA—precisely the modification that is reversed by the MGMT enzyme. Indeed, mice genetically deprived of this enzyme are hypersensitive to even very low levels of these chemotherapeutic alkylating agents and either die or age prematurely following exposure; conversely, mice overexpressing MGMT are resistant to the effects of these drugs. This suggests that in the future, oncologists intent on treating tumors like these gliomas and colorectal cancers will assess the levels of MGMT expressed by the tumor cells in order to choose the most effective means of chemotherapy.

This area of cancer research is also having a profound effect on our understanding of another, seemingly unrelated biological process—aging. At least ten distinct premature aging syndromes—each a form of **progeria**—have been traced to inherited defects in one or another component of the DNA repair system. Individuals suffering from these syndromes often show many of the phenotypes of the aged by the time they reach adolescence. These clinical effects of defective DNA repair provide compelling indications that much of the normal aging process will one day be traced to genetic damage that we accumulate in our cells throughout life. Consequently, cancer and aging may be found to share a common root—the progressive deterioration of our genomes as we get older. And both cancer development and aging may one day be forestalled by treatments and lifestyles that protect our genomes from the ongoing attacks that they suffer, decade after decade, deep within our cells.

Key concepts

- The structural integrity and thus low mutability of DNA depends on a large and complex set of biological and biochemical mechanisms that work to ensure that somatic mutations accumulate in tissues at very low rates.
- Some of the mechanisms depend on the organization of tissues, in which the long-lived cells (the stem cells) are protected from genetic damage, while the short-lived cells (the transit-amplifying and differentiated cells) are vulnerable to sustaining such damage but are soon discarded.
- Misincorporated bases generated by errors in DNA replication can contribute to the burden of accumulated mutations. The numbers of these alterations are held down by the low error rates of DNA polymerases together with an array of error-correcting proteins, such as those involved in mismatch repair. Inherited defects in mismatch repair proteins can lead to increased susceptibility to certain types of cancer, notably hereditary non-polyposis colon cancer.
- Cell genomes are under continuous attack by a variety of chemically reactive molecules, many of them deriving from the cellular process of oxidative phosphorylation and the resulting reactive oxygen species that are generated as by-products of this process. Cell genomes may also suffer spontaneous chemical alterations, which affect DNA bases at a low but significant rate.
- In addition, the genomic DNA of cells can be attacked by mutagenic molecules of foreign origin. Such xenobiotics and their chemically reactive derivatives may come from pollutants and, to a far greater extent, from commonly consumed foodstuffs.
- Cells attempt to detoxify many of these compounds before they can attack genomic DNA. However, the side products of these reactions may actually be more reactive and mutagenic than the initially introduced molecular species.
- If the attacking mutagenic agents succeeds in damaging DNA, an elaborate array of proteins involved in base-excision and nucleotide-excision repair lies in wait in order to remove the vast majority of damaged bases. Inherited defects in base-excision repair proteins can lead to cancer susceptibility—specifically, xeroderma pigmentosum.
- Other types of DNA damage include double-strand DNA breaks, which can be created by X-rays or, more commonly, by DNA breakage at replication forks.
- Double-strand DNA breaks can be repaired in the G_1 phase of the cell cycle by nonhomologous end joining or in the S and G_2 phases by homology-dependent repair. Inherited defects in double-strand DNA repair can explain the breast and ovarian cancer susceptibility among patients inheriting mutant *BRCA1* or *BRCA2* germ-line alleles.
- Genomes may also be scrambled by mechanisms that affect the karyotype of cells. One class of such alterations includes those that affect chromosomal

structure, including the translocations that are created by the fusions of unrelated chromosomal arms to one another. Such fusions seem to be commonly triggered by eroded telomeres or unrepaired double-strand DNA breaks.

- Changes in chromosome number are also common in cancer cell genomes and appear to facilitate the accumulation of genes in proportions that favor the proliferation and survival of these neoplastic cells. Many of these changes derive from defects in the mitotic apparatus and its regulators, notably the centrosomes.
- Without a breakdown of the various mechanisms responsible for maintaining the integrity of the genome, it seems likely that the mutation rate would be too slow to allow cell genomes to accumulate the ensemble of genetic changes required for tumor progression to reach completion.

Thought questions

1. What types of evidence suggest that karyotypic alterations of cell genomes are not absolutely essential for neoplastic transformation?
2. When calculating the rates of mutation required in order for multi-step tumor progression to reach completion, what parameters must one know in order for such a calculation to accurately describe the actual biological process?
3. How does our understanding of defective DNA repair processes in tumor cells make possible the development of new anti-cancer therapeutic strategies?
4. In which ways do the defectiveness of p53 function and resulting defects in apoptosis and DNA repair facilitate the forward march of tumor progression?
5. What types of tumor promotion, as described in Chapter 11, favor the genetic evolution of premalignant cell clones?
6. What evidence implicates mutagenic chemicals originating outside the body in the pathogenesis of human cancers? How can one gauge their contribution to human carcinogenesis compared with that of mutagens and mutagenic processes of endogenous origin?
7. How do defects in various cell cycle checkpoints allow for accelerated rates of the accumulation of mutations?
8. How do the biological properties of stem cells help to reduce the rate at which tissues accumulate mutant genes?
9. How does the existence of cancer stem cells affect the calculations of the rate at which mutations must be accumulated in order to allow multi-step tumor progression to advance?
10. How does the genetic heterogeneity in the human gene pool affect the functioning of various types of biological defenses that have been erected to prevent the accumulation of mutant alleles in human somatic cells?

Additional reading

Ames BN (1989) Endogenous DNA damage as related to cancer and aging. *Mutat. Res.* 214, 41–46.

Bartek J, Lukas C & Lukas J (2004) Checking on DNA damage in S phase. *Nat. Rev. Mol. Cell Biol.* 5, 792–804.

Bartek J & Lukas J (2003) Chk1 and Chk2 kinases in checkpoint control and cancer. *Cancer Cell* 3, 421–429.

Burgers PMJ (1998) Eukaryotic DNA polymerases in DNA replication and DNA repair. *Chromosoma* 107, 218–227.

Cahill DP, Kinzler KW, Vogelstein B & Lengauer C (1999) Genetic instability and darwinian selection in tumours. *Trends Cell Biol.* 9, M57–M60.

D'Assoro AB, Lingle WL & Salisbury JL (2002) Centrosome amplification and the development of cancer. *Oncogene* 21, 6146–6153.

de Boer J & Hoeijmakers JHJ (2000) Nucleotide excision repair and human syndromes. *Carcinogenesis* 21, 453–460.

Downs JA & Jackson SP (2003) Cancer: protective packaging for DNA. *Nature* 424, 732–734.

Duensing S & Münger K (2002) Human papillomavirus and centrosome duplication errors: modeling the origins of chromosomal stability. *Oncogene* 21, 6241–6248.

Fearon ER (1997) Human cancer syndromes: clues to the origin and nature of cancer. *Science* 278, 1043–1050.

Friedberg EC (2001) How nucleotide excision repair protects against cancer. *Nat. Rev. Cancer* 1, 22–33.

Friedberg EC, McDaniel LD & Schultz RA (2004) The role of endogenous and exogenous DNA damage and mutagenesis. *Curr. Opin. Genet. Dev.* 14, 5–10.

Friedberg EC, Walker GC & Siede W (1995) DNA Repair and Mutagenesis. Washington, DC: ASM Press.

Gold LS, Ames BN & Slone TH (2002) Misconceptions about the causes of cancer. In Human and Environmental Risk Assessment: Theory and Practice (D Paustenbach ed), pp 1415–1460. New York: John Wiley & Sons.

Grady WM (2004) Genomic instability and colon cancer. *Cancer Metast. Rev.* 23, 11–27.

Hartwell L (1992) Defects in a cell cycle checkpoint may be responsible for the genomic instability of cancer cells. *Cell* 71, 543–546.

Heinen C, Schmutte C & Fishel R (2002) DNA repair and tumorigenesis. *Cancer Biol. Ther.* 1, 477–485.

Hoeijmakers JH (2001) Genome maintenance mechanisms for preventing cancer. *Nature* 411, 366–374.

Jackson SP (2002) Sensing and repairing DNA double-strand breaks. *Carcinogenesis* 23, 687–696.

Jallepalli PV & Lengauer C (2001) Chromosome segregation and cancer: cutting through the mystery. *Nat. Rev. Cancer* 1, 109–117.

Jiricny J & Nyström-Lahti M (2000) Mismatch repair defects in cancer. *Curr. Opin. Genet. Dev.* 10, 157–161.

Kastan MB & Bartek J (2004) Cell cycle checkpoints and cancer. *Nature* 432, 316–323.

Kinzler K & Vogelstein B (eds) (1998) The Genetic Basis of Human Cancer. New York: McGraw-Hill.

Lengauer C, Kinzler KW & Vogelstein B (1998) Genetic instabilities in human cancers. *Nature* 396, 643–649.

Lindahl T & Wood RD (1999) Quality control by DNA repair. *Science* 286, 1897–1905.

Loeb KR & Loeb LA (2000) Significance of multiple mutations in cancer. *Carcinogenesis* 21, 379–385.

Loeb LA (2001) A mutator phenotype in cancer. *Cancer Res.* 61, 3230–3239.

Loeb LA, Loeb KR & Anderson JP (2003) Multiple mutations and cancer. *Proc. Natl. Acad. Sci. USA* 100, 776–781.

Lynch HT & de la Chapelle A (2003) Hereditary colorectal cancer. *N. Engl. J. Med.* 348, 919–932.

Marnett LJ (2000) Oxyradicals and DNA damage. *Carcinogenesis* 21, 261–270.

Marnett LJ & Plastaras JP (2001) Endogenous DNA damage and mutation. *Trends Genet.* 17, 214–221.

Masuda A & Takahashi T (2002) Chromosome instability in human lung cancers: possible underlying mechanisms and potential consequences in the pathogenesis. *Oncogene* 21, 6884–6897.

Modrich P & Lahue R (1996) Mismatch repair in replication fidelity, genetic recombination, and cancer biology. *Annu. Rev. Biochem.* 65, 101–133.

Nakabeppu Y, Tsuchimoto D, Ichinoe A et al. (2004) Biological significance of the defense mechanisms against oxidative damage in nucleic acids caused by reactive oxygen species: from mitochondria to nuclei. *Ann. N.Y. Acad. Sci.* 1011, 101–111.

Nowell PC (1976) The clonal evolution of tumor cell populations. *Science* 194, 23–28.

Perucho M (2003) Tumors with microsatellite instability: many mutations, targets and paradoxes. *Oncogene* 22, 2223–2225.

Pierce GB, Shikes R & Fink LM (1978) Cancer: A Problem of Developmental Biology, Englewood Cliff, NJ: Prentice-Hall.

Pihan GA (1998) Centrosome defects and genetic instability in malignant tumors. *Cancer Res.* 58, 3974–3985.

Rahman N & Stratton MR (1998) The genetics of breast cancer susceptibility. *Ann. Rev. Genet.* 32, 95–121.

Rajagopalan H & Lengauer C (2004) Aneuploidy and cancer. *Nature* 432, 338–341.

Scully R & Livingston DM (2000) In search of the tumour-suppressor functions of BRCA1 and BRCA2. *Nature* 408, 429–432.

Sedgwick B (2004) Repairing DNA-methylation damage. *Nat. Rev. Mol. Cell Biol.* 5, 148–157.

Sieber O, Heinemann K & Tomlinson I (2005) Genomic stability and tumorigenesis. *Semin. Cancer Biol.* 15, 61–66.

Shiloh Y (1997) Ataxia-telangiectasia and the Nijmegen breakage syndrome: related disorders but genes apart. *Ann. Rev. Genet.* 31, 635–662.

Shiloh Y (2004) ATM and related protein kinases: safeguarding genome integrity. *Nat. Rev. Cancer* 3, 155–168.

Smith ML & Seo YR (2002) p53 regulation of DNA excision repair pathways. *Mutagenesis* 17, 149–156.

Starita LM & Parvin JD (2003) The multiple nuclear functions of BRCA1: transcription, ubiquitination and DNA repair. *Curr. Opin. Cell Biol.* 15, 345–350.

Sugimura T (2000) Nutrition and dietary carcinogens. *Carcinogenesis* 21, 387–395.

Tainer JA & Friedberg EC (2000) Dancing with the elephants: the structural biology of DNA pathways. *Mutat. Res.* 460, 139–141.

Venkitaraman AR (2002) Cancer susceptibility and the functions of BRCA1 and BRCA2. *Cell* 108, 171–182.

Wilson JH & Elledge SJ (2002) *BRCA2* enters the fray. *Science* 297, 1822–1823.

Wogan GN, Hecht SS, Felton JS et al. (2004) Environmental and chemical carcinogenesis. *Semin. Cancer Biol.* 14, 473–486.

Wood RD, Mitchell M, Sgouros J & Lindahl T (2001) Human DNA repair genes. *Science* 291, 1284–1289.

Zhou BB & Elledge SJ (2000) The DNA damage response: putting checkpoints in perspective. *Nature* 408, 433–438.

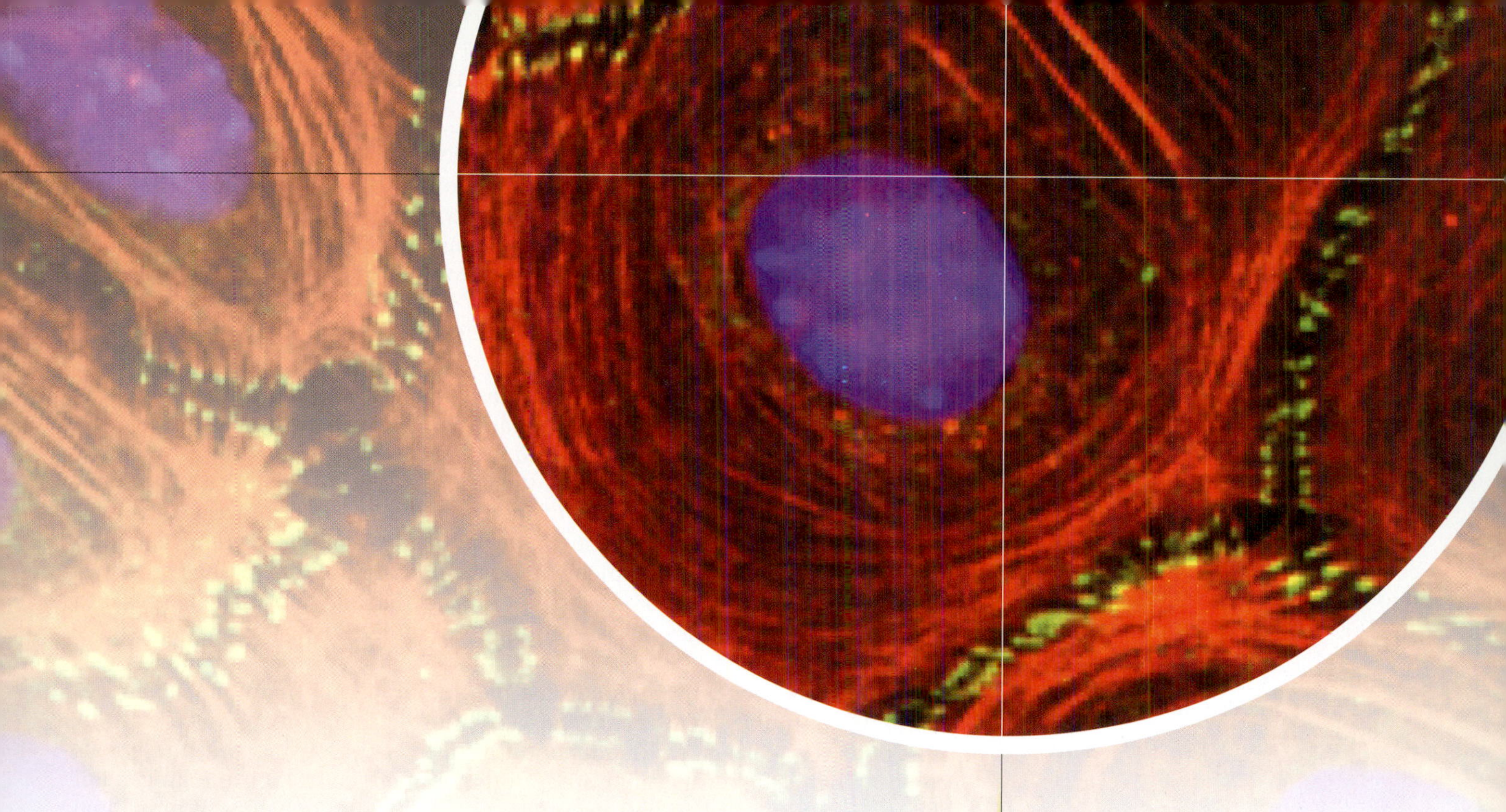

Chapter 13

Dialogue Replaces Monologue: Heterotypic Interactions and the Biology of Angiogenesis

Simple ideas are wrong. Complicated ideas are unattainable.
Paul Valery, poet, 1942

A simple and powerful conceptual paradigm pervades most of the previous discussions in this book: cancer is a disease of cells, and the phenotypes of cancer cells can be understood by examining the genes and proteins within them. The origins of this idea are clear, being traceable directly back to bacterial and yeast genetics. These two research specialties thrived because they were well served by the postulate that cell genotype determines all aspects of cell phenotype. Indeed, virtually all the attributes of individual bacterial and yeast cells could be shown to derive directly from the genes that these microorganisms carry in their genomes.

Applying this concept to metazoans and their tissues has had obvious advantages for biologists. Metazoan organisms are complex almost beyond measure, and this complexity has frequently prevented researchers from extracting simple and irrefutable truths about organismic function. In response, many researchers, notably molecular biologists, cell biologists, and biochemists, embraced the credo of **reductionist** science: when working with complex systems, the best way to arrive at solid and rigorous conclusions is to take apart these systems into simpler, more tractable components and study each separately. While the lessons learned may relate only to small parts of very large systems, at least these lessons are solid and definitive and will not require substantial revision each time a new generation of researchers revisits these complex systems and their component parts.

Such reductionism has allowed many areas of cancer research to thrive. Witness the progress described in the early chapters of this textbook: when the first proto-oncogene was discovered three-quarters of the way through the twentieth century (Section 3.9), almost nothing was known about the genetics and molecular biology of cancer. By the end of the century, information was available in abundance about how cancers begin and progress to highly aggressive, malignant states. Much of this avalanche of information came from taking the reductionist paradigm to its limits—disassembling normal and neoplastic tissues into component cells and the cells into component molecules.

As mentioned above, the reductionist pact embraced by many in the cancer research community treated cancer as a cell-autonomous process: all the attributes of cancer cells can be understood in terms of the genes that these cells carry. Actually, there was yet another notion embedded in their thinking: all of the traits of a tumor can be traced directly back to the behavior of individual cancer cells within the tumor mass.

By the end of the twentieth century, it became increasingly clear that many of the traits of tumors could not in fact be traced directly back to individual cancer cells and the genes they carry. The grand simplifications agreed upon a generation earlier by cancer researchers had begun to lose their utility. Increasingly, evidence turned up that cancer is actually a disease of tissues, in particular, the complex tissues that we call tumors.

The reductionists' way of thinking had, all along, denied certain realities that clinical oncologists and pathologists confronted on a daily basis, namely, that carcinomas—which constitute more than 80% of the human cancer burden—derive from epithelial tissues of very complex microscopic structure. Histopathological characterizations of these epithelial tumors reveal that they are composed of a number of distinct cell types. In fact, even cursory examinations under the microscope indicate that most carcinomas are as complex histologically as the normal epithelial tissues from which they have arisen; examples of this complexity can be seen in Figure 13.1. In some cases of commonly occurring carcinomas, such as those of the breast, colon, stomach, and pancreas, the non-neoplastic cells, which constitute the tumor stroma, may account for as many as 90% of the cells within the tumor mass. An extreme version of this is presented by the cells of another type of solid tumor—Hodgkin's disease, a lymphoma in which non-neoplastic cells account for 99% of the cells in a tumor mass (Figure 13.2).

One's first instinct is to dismiss these non-neoplastic cells as distracting contaminants that confound rather than enlighten attempts at understanding the biology of tumors. To do so now seems increasingly unwise. Recent research results make it clear that non-neoplastic cells, notably the stromal cells of carcinomas, are active, indeed essential collaborators of the neoplastic epithelial cells within tumor masses, having been recruited and then exploited by the cancer cells. The biology of these recruited cells now appears to be almost as important as that of the neoplastic cells in enabling growth of tumors.

This means that the disease of cancer is far more than cancer cells talking to themselves in endless monologues. We know, instead, that in most tumors, the neoplastic cells are in continuous communication with their non-neoplastic

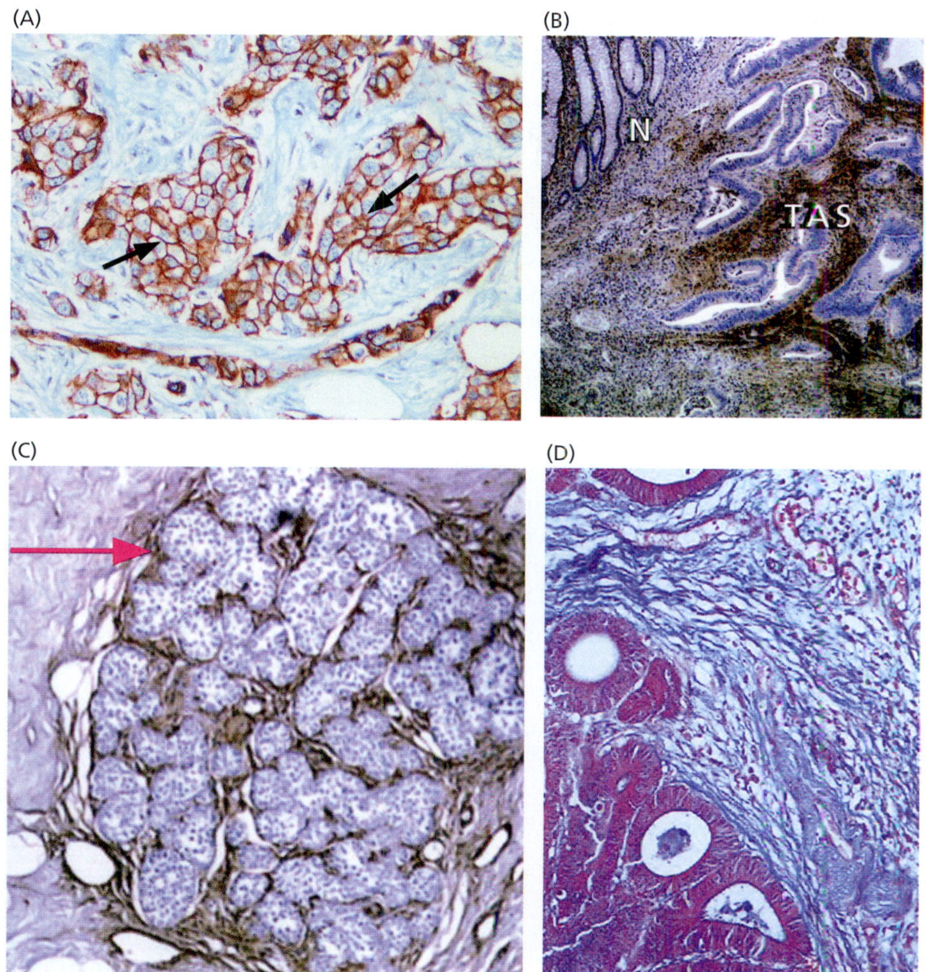

Figure 13.1 Stromal components of several commonly occurring carcinomas A variety of common carcinomas contain significant proportions of both epithelial cancer cells and recruited stromal cells. This is apparent in the carcinomas shown here. (A) A high-grade invasive ductal carcinoma of the breast, in which the membranes *(arrows)* of the epithelial cells are stained *brown,* while the nuclei of the stromal cells are *light blue.* (B) A colon carcinoma *(right side)* and normal colonic mucosa (N, *left side),* in which the tumor-associated stroma (TAS, *right)* is immunostained *dark brown* for PINCH (a protein associated with the cytoplasmic tails of integrins), while the nuclei of epithelial cancer cells lining ducts are *light blue.* (C) A lobular carcinoma *in situ* of the breast showing small clusters of cancer cells *(light blue, arrow)* surrounded by thin layers of stromal fibroblasts that have been immunostained for PINCH antigen *(dark brown).* (D) An adenocarcinoma of the stomach in which the cancer cells *(purple, lower left)* are adjacent to extensive stroma, which has been stained for collagen *(blue).* (A, from A. Gupta, C.G. Deshpande and S. Badve, *Cancer* 97:2341–2347, 2003; B and C, from J. Wang-Rodriguez, A.D. Dreilinger, G.M. Alsharabi and A. Rearden, *Cancer* 95:1387–1395, 2002; D, from S. Ohno, M. Tachibana, T. Fujii et al., *Int. J. Cancer* 97: 770–774, 2002.)

neighbors. In this chapter, we focus our discussions specifically on carcinomas, in which the interactions of neoplastic and non-neoplastic cells within tumors are far better understood than in the many other types of cancer that afflict humans.

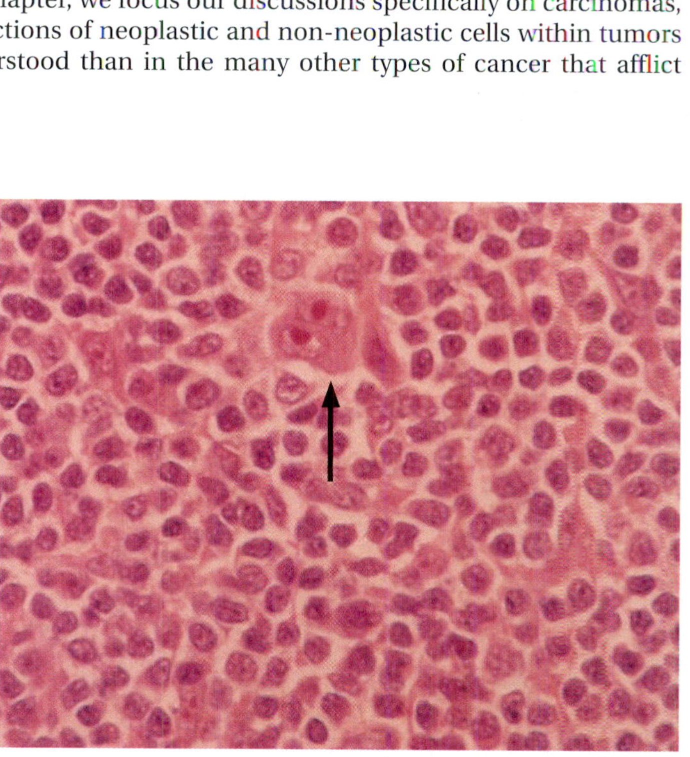

Figure 13.2 Rare neoplastic cells in Hodgkin's lymphoma Normal, non-neoplastic cells may constitute the majority of living cells in many tumor masses. An extreme form of this is illustrated here by Hodgkin's disease, a lymphoma in which as many as 99% of the cells are recruited normal lymphocytes, which surround a rare neoplastic Reed–Sternberg cell *(arrow).* (From A.T. Skarin, Atlas of Diagnostic Oncology, 3rd ed. Philadelphia: Elsevier Science Ltd., 2003.)

13.1 Normal and neoplastic epithelial tissues are formed from interdependent cell types

The true complexity of the stroma of epithelial cancers becomes apparent only at high magnification under the microscope. In addition to the neoplastic epithelial (i.e., carcinoma) cells, a number of stromal cell types populate the tumor. These stromal cells include fibroblasts, myofibroblasts, endothelial cells, pericytes, smooth muscle cells, adipocytes, macrophages, lymphocytes, and mast cells (Figure 13.3; see also Section 13.3). Because some of these cell types constitute only a small proportion of the stroma of certain tumors, their presence becomes apparent only through use of immunostaining with antibodies that bind cell type–specific antigens.

The diverse stromal cell types within tumors are all members of several mesenchymal cell lineages that generate both connective tissue and various types of immune cells in the blood and immune tissues, and therefore are biologically very different from the epithelial cells whose transformation drives the growth of carcinomas. (In the case of non-epithelial tumors such as sarcomas, where the cancer cells are themselves of mesenchymal origin, the boundaries between the malignant cell compartment and the nonmalignant cells within the tumor mass are less easily defined.)

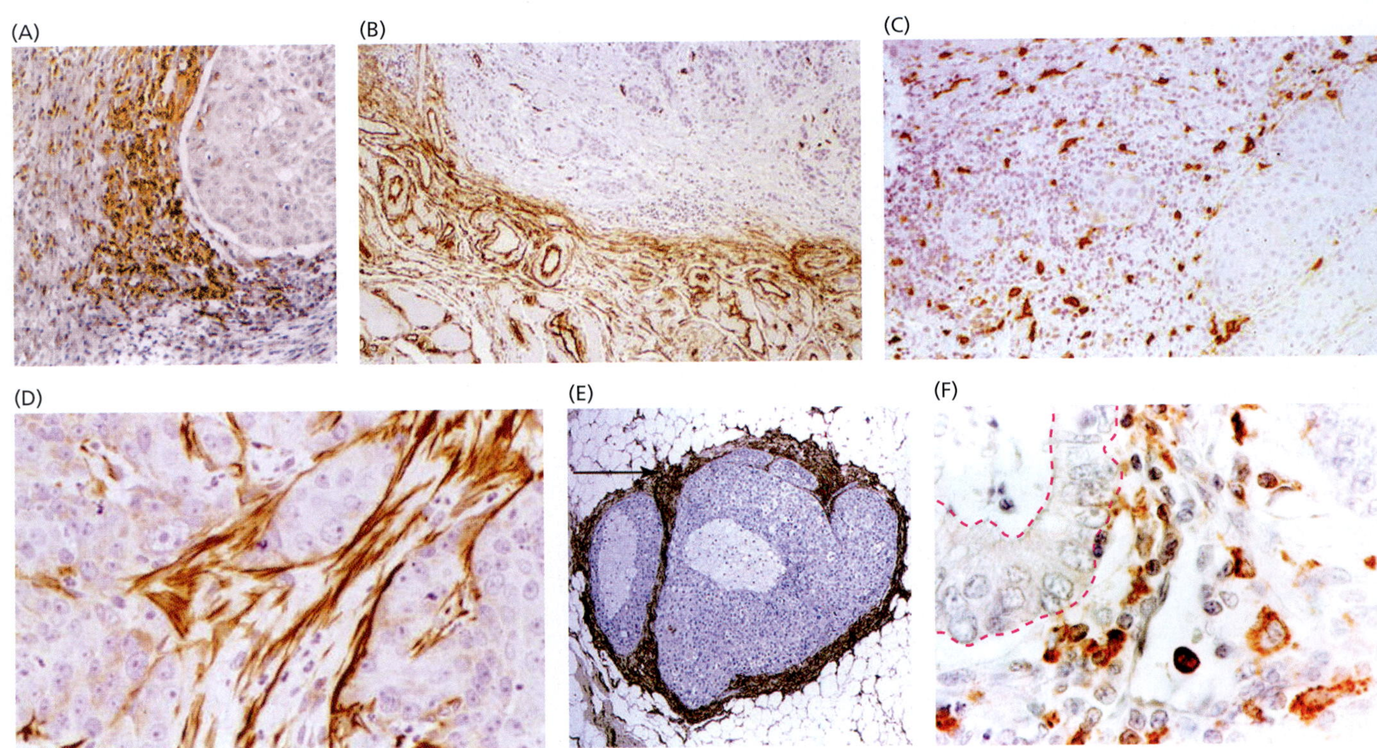

Figure 13.3 A variety of distinct cell types in the stroma of carcinomas Various antibodies can be used to immunostain specific mesenchymal cell types in the stroma of carcinomas. (A) CD4 antigen–positive T lymphocytes *(brown)* are apparent in the stroma of a non-small-cell lung carcinoma (NSCLC). The stroma of squamous cell carcinomas of the oral cavity, pharynx, and larynx may contain (B) CD34 antigen–positive fibrocytes *(brown),* (C) CD117 antigen–positive mast cells *(brown),* and (D) α-smooth muscle actin–positive myofibroblasts *(brown).* (E) The stroma of an *in situ* ductal carcinoma of the breast (DCIS) shows PINCH antigen–positive fibroblasts *(dark brown*; see Figure 13.1B*)* arrayed in a ring immediately around the carcinoma *(arrow)*, which is surrounded, in turn, by adipocytes *(light blue rims).* (F) Monocytes, the precursors of macrophages, can be detected through their display of the CD11b+ antigen in this human colorectal adenocarcinoma. The epithelial cancer cells *(light blue)* form a duct outlined in the *upper left.* (A, from O. Wakabayashi, K. Yamazaki, S. Oizumi et al., *Cancer Sci.* 94:1003–1009, 2003; B, C, and D, from P.J. Barth, T. Schenck zu Schweinsberg, A. Ramaswamy and R. Moll, *Virchows Arch.* 444:231–234, 2004; E, from J. Wang-Rodriguez, A.D. Dreilinger, G.M. Alsharabi and A. Rearden, *Cancer* 95:1387–1395, 2002; F, from E. Barbera-Guillem, J.K. Nyhus, C.C. Wolford et al., *Cancer Res.* 62:7042–7049, 2002.)

In the absence of other information, we might explain the presence of these various stromal cell types within a carcinoma in two ways. They might represent the remnants of cells that resided in the stroma of a tissue before tumor development began. During the subsequent expansion of the tumor cell population, groups of cancer cells may have inserted themselves between these preexisting normal stromal cell layers, thereby generating the richly textured tissues that are seen when most carcinomas are examined microscopically.

The alternative explanation for the presence of these numerous stromal cells is more intriguing and is likely to be the correct one for the great majority of tumors. The rationale behind this second model depends on our current understanding of how normal, architecturally complex tissues maintain their structure and function. In such tissues, the proper proportions of the various component cells must depend on the continuous exchange of signals among them. These interactions involve multiple distinct cell types, each following its own particular differentiation program. Such communication between dissimilar cell types is termed **heterotypic** signaling and is used by each cell type to encourage or limit the proliferation of the other types of cells nearby.

Extending this thinking, we might speculate that many of the heterotypic interactions operating in normal tissues continue to play important roles in the biology of the tumors arising in these tissues. This would suggest that, like normal epithelial cells, carcinoma cells continue to control the populations of stromal cells near them, perhaps by recruiting the latter from adjacent normal tissues and then encouraging their proliferation. Operating in the other direction, stromal cells may also influence epithelial cell proliferation and survival in the tumor mass.

In normal tissues, these heterotypic signaling channels depend in large part on the exchange of (1) mitogenic growth factors, such as hepatocyte growth factor (HGF), transforming growth factor-α (TGF-α), and platelet-derived growth factor (PDGF); (2) growth-inhibitory signals, such as transforming growth factor-β (TGF-β); and (3) trophic factors, such as insulin-like growth factor-1 and -2 (IGF-1 and -2), which favor cell survival (Figure 13.4).

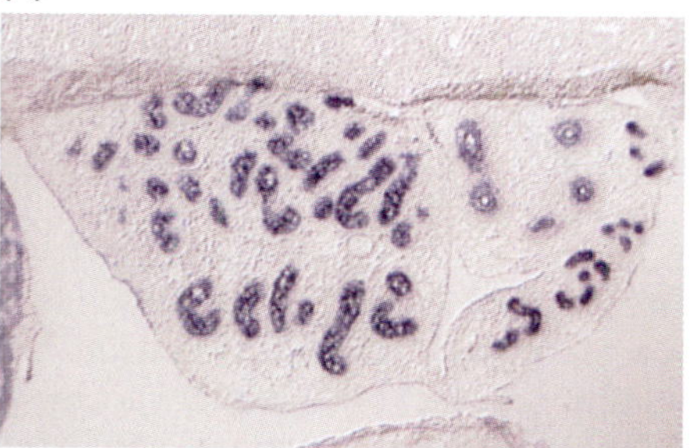

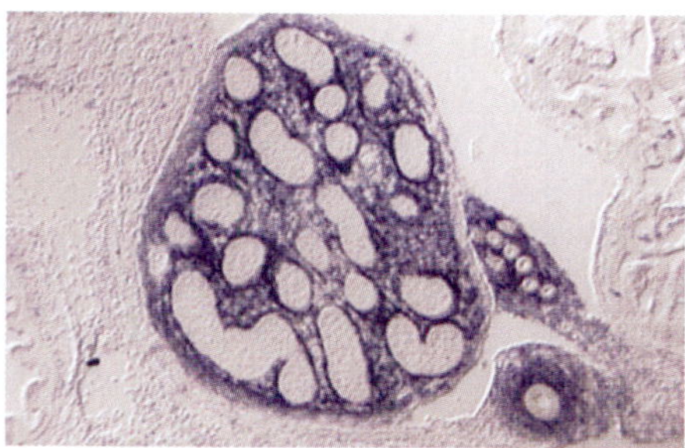

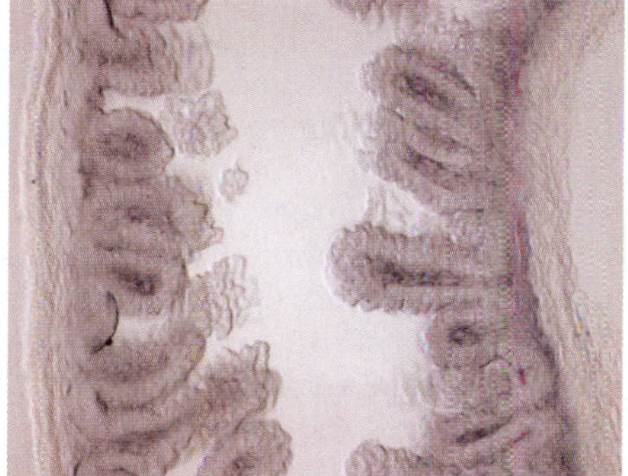

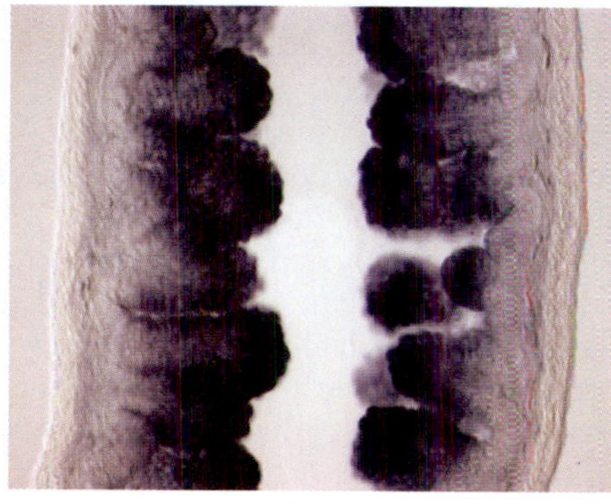

Figure 13.4 Heterotypic ligand–receptor signaling (A) In epithelial tissues, the synthesis of PDGF-A and its receptor (PDGF-Rα) is usually confined to distinct cell types. In the testis, as in many bona fide epithelial tissues, expression of PDGF is confined to the cells of the tubular epithelium *(above)*, while expression of its receptor is confined to the surrounding mesenchymal cells *(below)*. (B) The reciprocal expression of the Met receptor and its ligand, termed variously hepatocyte growth factor (HGF) or scatter factor (SF), is shown here in the embryonic mouse ileum (the lower section of the small intestine). In each fingerlike villus, seen in longitudinal section, HGF/SF mRNA, detected by *in situ* hybridization, is made by the mesenchymal cells in the core of the villus *(above)* while Met is expressed by the epithelial cells coating the surface of the villus *(below)*. (A, from L. Gnessi, S. Basciani, S. Mariani et al., *J. Cell Biol.* 149:1019–1026, 2000; B, courtesy of S. Britsch and C. Birchmeier.)

During the initial formation of tumors, heterotypic interactions play an important role in driving many types of tumor promotion. Thus, stromal cells, such as macrophages, neutrophils, and lymphocytes, participate in an inflammatory response that involves the release of TNF-α and prostaglandins; these molecules then stimulate the proliferation of nearby epithelial cells (Section 11.16).

Many heterotypic interactions continue to operate after tumors are formed. To cite some examples, epithelial cells within a carcinoma often release PDGF, for which stromal cells—notably fibroblasts, myofibroblasts, and macrophages—possess receptors; the stromal cells reciprocate by releasing IGF-1 (insulin-like growth factor-1), which benefits the growth and survival of the nearby cancer cells. Similarly, the neoplastic cells within melanomas release PDGF, which elicits IGF-2 production from nearby stromal fibroblasts; this IGF-2 helps to maintain the viability of the melanoma cells. Stromal cells in breast cancers release the factor CXCL12 (a *chemokine;* see Section 13.5) and the HGF/SF growth factor, which stimulate the proliferation and survival of nearby epithelial cancer cells. By regulating each other's numbers and positions, epithelial and stromal cells can ensure the optimal representation and localization of each cell type within normal and neoplastic tissues.

The stromal and epithelial cells in normal epithelial tissues collaborate in the construction of the specialized extracellular matrix (ECM) that lies between them, called variously the basement membrane or basal lamina (Figure 13.5; see also Figure 2.3). For example, the epithelial cells of the skin, the keratinocytes, express genes specifying many of the major protein components of the basement membrane, including type IV collagen and laminins; the stromal cells also contribute to its construction, doing so in ways that have yet to be documented in detail. Certain **proteoglycans** in the basement membrane, such as perlecan, provide it with increased hydration and with sites to which growth factors can be attached for long-term storage. Continuous tethering to the basement membrane, mediated largely by integrins and the hemidesmosomes that they construct (see Figure 13.5A), is essential for the survival of many kinds of epithelial cells, and loss of this tethering often provokes anoikis, the form of apoptosis resulting from loss of anchorage to a solid substrate (see, for example, Figure 9.21).

The **endothelial** cells, which assemble to form the linings of the walls of capillaries and larger blood vessels (Figure 13.6A) as well as **lymphatic** ducts, represent a vital component of the normal and neoplastic stroma. As discussed in greater detail later, their proliferation is encouraged by other cells in both the epithelium and stroma in order to guarantee access by all of these cells to an adequate blood supply. Once capillaries are assembled and become functional, they provide essential nutrients and oxygen to nearby cells. We were introduced to these interactions earlier, when we read that cells lacking adequate access to oxygen release **angiogenic factors** that stimulate the ingrowth of capillaries (Section 7.12).

While capillary formation is proceeding, the endothelial cells secrete growth factors that stimulate the proliferation of nearby nonendothelial cell types. Most important, endothelial cells release PDGF and HB-EGF (heparin-binding EGF), which enables them to attract the peri-endothelial cells called **pericytes** and the vascular smooth muscle cells that together create the outer cell layers of capillaries, sometimes called the **mural** cells to distinguish them from the luminal endothelial cells (Sidebar 13.1). Once in place, the pericytes (which closely resemble smooth muscle cells) reciprocate by releasing vascular endothelial growth factor (VEGF) and angiopoietin-1 (Ang-1), which provide important survival signals to the endothelial cells that recruited them in the first place. In addition, these mural cells provide the endothelial tubes with structural stability and an ability to resist the forces exerted by the pressure of blood (Figure 13.6).

(A)

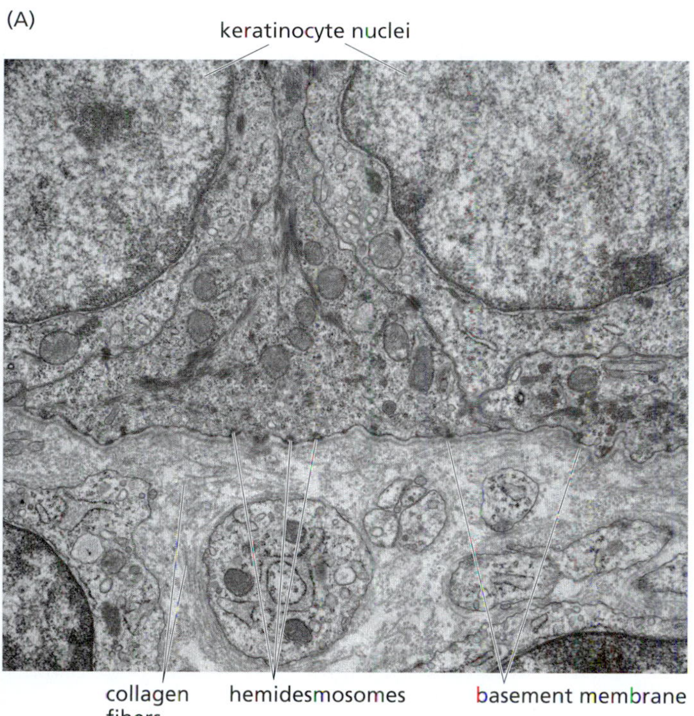

Figure 13.5 The basement membrane (A) In epithelial tissues, the specialized extracellular matrix (ECM) termed the basement membrane (BM) separates the epithelial cells *(above)* from the stroma *(below)*. As shown in this transmission electron micrograph of the skin of a newborn mouse, the epithelial cells of the skin (termed keratinocytes) are tethered to one side *(above)* of the BM, being anchored in part through structures termed *hemidesmosomes*, which are composed of clusters of integrins that bind BM proteins, notably laminin; the keratinocyte nuclei are labeled here. Below the basement membrane are the collagen fibers and portions of mesenchymal cells constituting the stroma of the skin, i.e., the dermis. (B) This schematic drawing indicates that the basement membrane is composed largely of four major ECM proteins, namely, laminin, collagen type IV, perlecan, and nidogen. This highly permeable molecular meshwork allows a variety of molecules to pass through in both directions. A specialized basement membrane, not shown here, underlies endothelial cells in capillaries. (A, courtesy of H.A. Pasolli and E. Fuchs; B, from B. Alberts et al., Molecular Biology of the Cell, 4th ed. New York: Garland Science, 2002, based on H. Colognato and P.D. Yurchenko, *Dev. Dynamics* 218:213–234, 2000.)

(B)

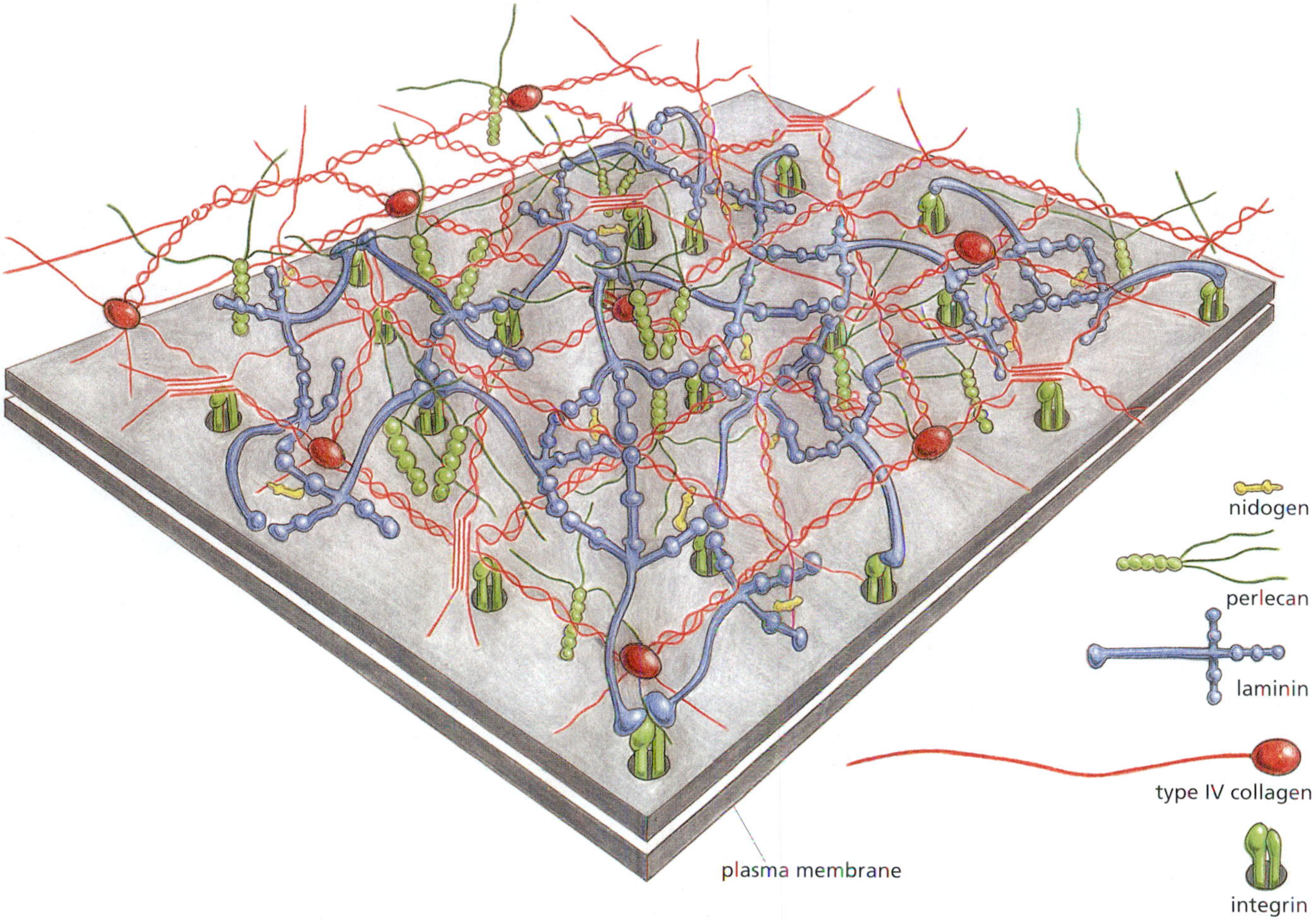

nidogen
perlecan
laminin
type IV collagen
integrin
plasma membrane

The locations and large populations of stromal cells within carcinomas provide clear indication that as the epithelial components of these tumors expand, this growth is accompanied by coordinated proliferation of stromal cells. In fact, stromal cells are even found to be layered between carcinoma cells in most **metastases**—the secondary tumors that become established in anatomical sites

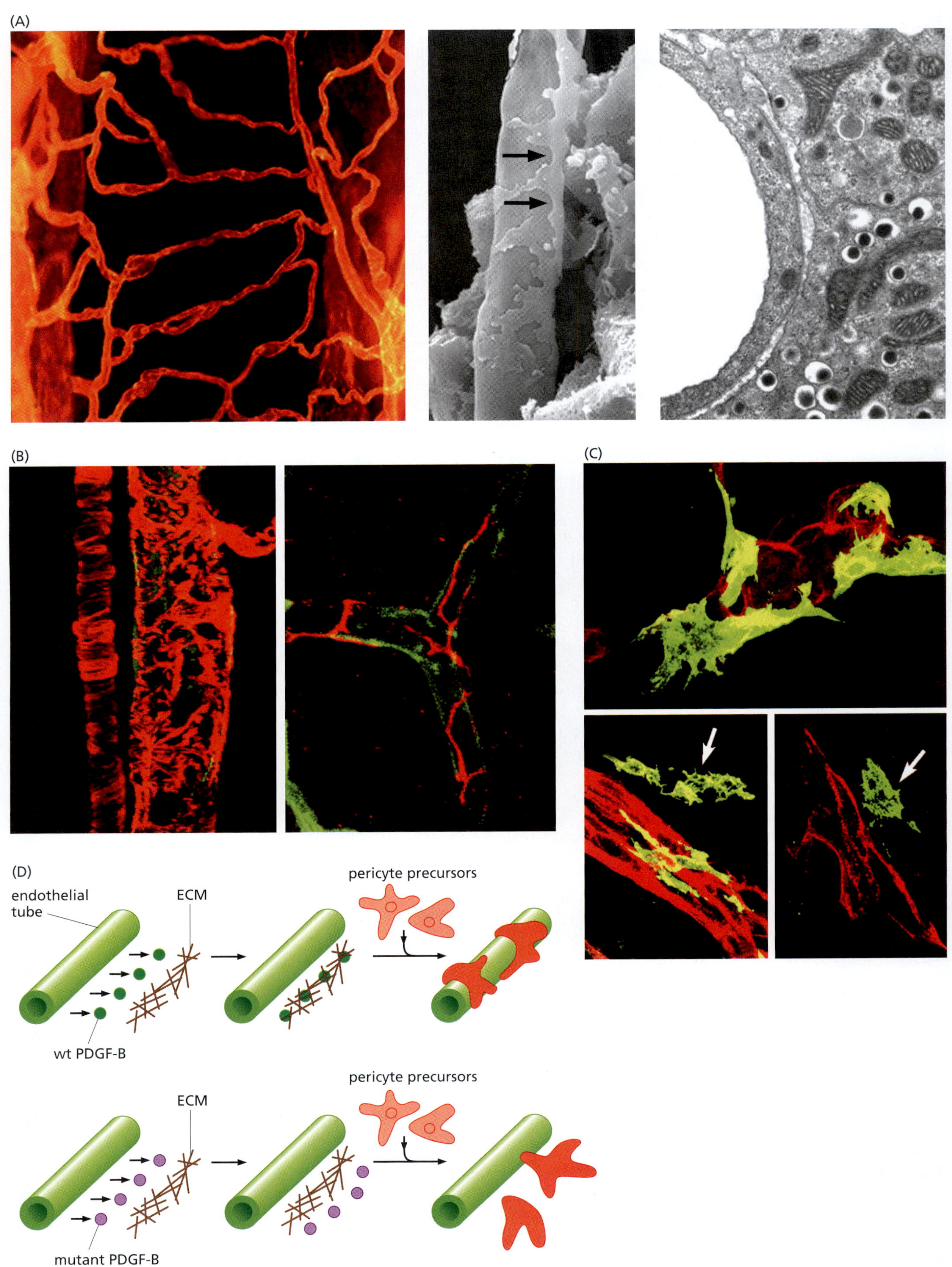
(A)
(B)
(C)
(D)
endothelial tube
ECM
pericyte precursors
wt PDGF-B
ECM
pericyte precursors
mutant PDGF-B

Figure 13.6 Pericytes, smooth muscle cells, and endothelial cells of the microvessels (A) Microvessels can be studied using at least three types of microscopy. Immunofluorescent microscopy *(left)* of the capillaries in a normal mouse trachea (using an antibody that recognizes the endothelial cell–specific CD31 antigen) reveals a well-organized network of vessels. At the level of scanning electron microscopy *(middle)*, a capillary is seen to resemble a smooth-walled cylindrical tube with pericytes attached here and there to its surface *(arrows)*. In the transmission micrograph *(right)*, a section of a capillary reveals that its wall is constructed by the folded cytoplasm of an endothelial cell. (B) This immunofluorescence micrograph resolves the pericytes and endothelial cells. Normally, the pericytes and smooth muscle cells *(both red orange)* coat the outside of the tubes of endothelial cells *(green)*, forming normally structured venules and arterioles *(left panel)*. In these vessels, coverage of the endothelial cells by pericytes and smooth muscle cells is often so complete that it is difficult to visualize the endothelial cells lying under the pericytes. In contrast, in capillaries *(right panel)*, the pericytes are more sparsely disposed, but are nonetheless tightly attached to the endothelial cells forming the capillaries. (C) Normally, the pericytes *(yellow green)* are tightly attached to capillaries *(red, upper panel)*. However, mice have been genetically altered to make mutant PDGF-B molecules that cannot attach to the extracellular matrix and therefore diffuse away from the endothelial cells making this growth factor. As a consequence, the pericytes are no longer attracted to the endothelial cells and lack intimate contact with the capillary tubes *(arrows, lower panels)*. (D) This close juxtaposition of pericytes with the capillaries depends on the localization of PDGF-B, which is secreted by endothelial cells *(lighter green)*. Normally *(above)*, this PDGF-B *(dark green)* is trapped immediately adjacent to the endothelial cells in the extracellular matrix (ECM, *brown*), where it serves to attract pericyte precursors *(pink)* that attach themselves *(red)* tightly to the capillary tubes formed by the endothelial cells. However *(below)*, if PDGF-B is deprived of the amino acid sequences that normally tether it to the ECM *(purple)*, it diffuses away from the capillary tubes, and recruited pericytes are scattered at some distance from the capillaries. (A, from T. Inai, M. Mancuso, H. Hashizume et al., *Am. J. Pathol.* 165:35–52, 2004; B, from S. Morikawa, P. Baluk, T. Kaidoh et al., *Am. J. Pathol.* 160:985–1000, 2002; C, from A. Abramsson, P. Lindblom, and C. Betsholtz, *J. Clin. Invest.* 112:1142–1151, 2003.)

far away from the initially formed, primary tumors. Consequently, even those cancer cells that are independent enough to leave the primary tumor and to found new tumors at distant sites usually find it necessary to recruit stromal cells and encourage their proliferation (Figure 13.7A), ostensibly to aid in the construction of these new tumor colonies. On occasion, metastatic cells will insinuate themselves into the existing stroma of the tissue in which they have landed (Figure 13.7B), once again demonstrating their need to secure stromal support.

In the extreme version of this scenario, we might imagine that all of the heterotypic interactions needed to maintain normal tissue function continue to operate within carcinomas, being required for the neoplastic cells to thrive and multiply within these tumors. However, such a depiction cannot be literally true, since many of the acquired traits that we associate with cancer cells (Chapter 2),

Sidebar 13.1 Localization of growth factors is important for proper heterotypic interactions Highly localized concentrations of PDGF are needed to ensure that the pericytes recruited to the outer layers of capillaries become directly **apposed** to the endothelial cells rather than dispersed randomly some distance away (Figure 13.6C and D). Like many other growth factors, the PDGF-B molecule—the specific form of PDGF that operates here—contains a "retention motif," which ensures its binding to proteoglycans in the extracellular matrix (ECM). Consequently, soon after PDGF-B molecules are secreted by endothelial cells into the space around capillaries, these molecules are trapped in the nearby ECM and therefore accumulate in high concentrations immediately next to the endothelial cells. Once liberated from this tethering by proteases, the resulting high local concentrations of activated PDGF-B molecules ensure the recruitment of pericytes to locations directly adjacent to the endothelial cells. However, in mice expressing a mutant PDGF-B that lacks this retention motif, the secreted PDGF-B molecules diffuse away from endothelial cells, causing the recruitment of pericytes to the general area of the endothelial cells but not closely apposed to them. These endothelial cells therefore lack the structural support afforded by the pericytes, as well as the VEGF supplied by the pericytes that is required for long-term endothelial cell survival.

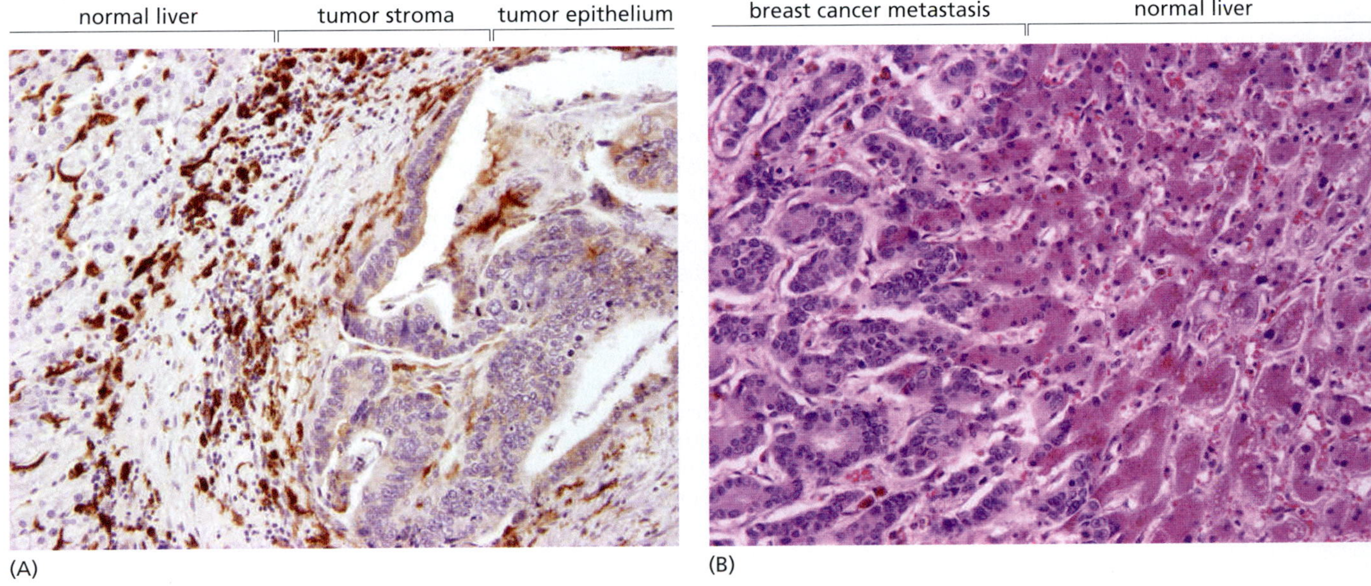

Figure 13.7 Metastasis and dependence on stroma (A) This micrograph of a metastasis originating from a primary colon carcinoma reveals that the metastasis is as complex histologically as the primary tumors seen in Figure 13.1. Here, we see that the carcinoma cells that have formed a metastasis *(light purple, right)* have constructed the same ductal structures as are seen in primary adenocarcinomas of the colon (see, for example, Figure 13.1B). In addition, the carcinoma cells have recruited substantial stroma *(middle)*, which includes macrophages *(dark brown)*. (B) Most breast cancer metastases to the liver *(left)* behave quite differently from colon carcinoma metastases to this organ, in that the mammary carcinoma cells infiltrate into normal liver tissue *(right)* and displace resident hepatocytes, thereby taking advantage of the existing stroma, including its vasculature. These metastatic cells appear to be as dependent on stromal support as the colorectal carcinoma cells of (A) but acquire it in a totally different way. (From F. Stessels, G. van den Eynden, I. van der Auwera et al., *Brit. J. Cancer* 90:1429–1436, 2004.)

including a decreased dependence on mitogens, an increased resistance to apoptosis, and acquisition of anchorage independence, are certain to affect the interactions between epithelial and stromal compartments and to lessen their interdependence. So, we conclude that these acquired traits *reduce* the dependence of epithelial cancer cells on stroma but do not seem to *eliminate* it.

13.2 The cells forming cancer cell lines develop without heterotypic interactions and deviate from the behavior of cells within human tumors

The continued dependence of carcinoma cells on stromal support explains the enormous difficulties that researchers have experienced when attempting to adapt tumor cells to *in vitro* culture conditions. Their goal here has been to create cancer cell lines—populations of cancer cells that can be propagated in tissue culture indefinitely. To do so, cells are prepared directly from tumor biopsies and placed into culture, with the hope that colonies of vigorously growing cancer cells will eventually emerge in culture dishes. In most cases, however, such cell colonies do not appear, foiling attempts to generate carcinoma cell lines. More often than not, the cells that do thrive in culture are of stromal origin—specifically the fibroblasts whose growth is favored by the platelet-derived growth factor (PDGF) that is present in the serum component of standard tissue culture medium.

It is clear that human carcinoma cells propagated under standard conditions of tissue culture are being forced to proliferate in an environment very different from the one experienced by their ancestors that resided in the tumors of cancer patients. For example, the high concentration of serum present in most culture media is growth-inhibitory for many types of normal and neoplastic epithelial cells, which rarely experience large concentrations of the serum-associated factors *in vivo* except acutely after wounding. Even more important, carcinoma cells are being selected for their ability to proliferate *in vitro* without intimate contact with stroma, since the proper balance of epithelial and stromal cells present in tumors cannot be maintained in the tissue culture dish.

On rare occasion, vigorously proliferating colonies of carcinoma cell populations do indeed emerge following extended culture of tumor fragments in the Petri dish. Because they have been selected *in vitro* for their ability to grow

autonomously, and therefore independently of stromal support, such cancer cells have evolved beyond the stage of tumor progression that is reached *in vivo* by the neoplastic cells of most human tumors. In fact, many of the successes in establishing human tumor cell lines have depended on culturing cells from the few tumors that had already progressed *in vivo* to a stage where they no longer depended on stroma for their survival and proliferation (Sidebar 25 ●).

Such human cancer cell lines—there are several dozen in common use—are standard reagents used in many types of cancer research. Cells from virtually all of these can be implanted into immunocompromised host mice, where they proliferate, often vigorously, and form tumors. These tumors are often termed **xenografts**, since they result from grafting the tissues of one species into a host animal of another. Mice of the Nude, NOD/SCID, and RAG1/2 strains, lacking functional immune systems, are used in these experiments, because they tolerate the growth of introduced, genetically foreign cells. The resulting tumor xenografts growing in these mice can then be tested for their responsiveness to anti-cancer drugs under development (Sidebar 13.2). The outcome of such experiments is not necessarily predictive of eventual clinical responses to these drugs.

13.3 Tumors resemble wounded tissues that do not heal

As argued above, heterotypic signaling governs much of the biology of carcinomas, and experimental models of cancer that ignore this important process generate tumors that are biologically very different from those found in cancer patients. We now proceed to examine in greater detail the nature of this signaling within tumors and the cell types that participate in it. Actually, the heterotypic interactions operating within tumors are highly complex and involve the exchange of dozens of distinct molecular species that mediate cell-to-cell signaling between the various cell types that together form these growths. Importantly, many of these signals are not routinely exchanged by the cell types constituting normal, undamaged tissue and therefore seem, at first glance, to be unique to tumors.

This complexity raises some fundamental questions: How do cancer cells learn to release and respond to an array of diverse heterotypic signals? Are the complex signaling programs and resulting biological responses invented anew, being cobbled together piece by piece, each time normal cells evolve progressively into cancer cells? Or do cancer cells exploit preexisting biological programs that are normally used by tissues for other purposes?

One attractive solution to these questions has come from an insight gained in 1986. A researcher studying the histological appearance of both tumors and wounds noted the striking resemblance between many of the signaling processes involved in tumor progression and those that occur during wound healing. If the similarities between these two biological programs could indeed be documented, this would explain much of the cleverness of tumor cells: they simply activate a complex, normal physiologic program—wound healing—that is already encoded in their genomes. By accessing and exploiting this preexisting biological program, cancer cells are spared the task of re-inventing it anew each time a tumor arises.

Wound healing has been studied most extensively in the context of the skin. Following the formation of a superficial wound to the skin and underlying tissues, blood platelets aggregate and release granules containing, among other factors, platelet-derived growth factor (PDGF) and transforming growth factor-β (TGF-β; see Figure 5.3). Wounding also causes release of **vasoactive** factors, which increase the permeability of blood vessels near the wound. This helps the wound site to acquire fibrinogen molecules from the blood plasma, which,

Sidebar 13.2 The development of anti-cancer therapies has been imperfectly served by the use of existing human cancer cell lines Human cancer cell lines, when grown as tumor xenografts in immunocompromised mice, often respond to the anti-proliferative effects of both untargeted cytotoxic drugs (which affect a wide spectrum of normal and neoplastic cell types) and drugs that are designed to affect only cancer cells. However, the ability of these xenograft models to predict the responses of patients to these drugs is limited. A 2001 retrospective study of 39 different anti-cancer drugs showed a weak correlation between the responses of xenografted tumors to these drugs and patient responses observed in the oncology clinic. For example, candidate drugs that were effective in halting the growth of more than one-third of a disparate collection of mouse tumor xenografts had only a 50% likelihood of showing any therapeutic activity (including halting tumor growth) in some human tumor types treated in the clinic. And drugs that affected smaller proportions of the mouse xenograft models had essentially no meaningful activity in the clinic.

These disappointing outcomes are not surprising, in light of the fact that the human tumor cell lines commonly used in such drug testing experiments bear little resemblance to cells in the tumors frequently encountered in the oncology clinic. Thus, the tumors that these cell lines form often look quite different histologically from those routinely encountered in a clinical pathology laboratory (Figure 13.8).

Another difficulty arises from the fact that cells of various human tumor cell lines are usually implanted in mice subcutaneously (beneath the skin). While this location affords researchers the ability to readily monitor the growth of these tumors, it almost always corresponds to an **ectopic** site within the host, that is, an anatomical location quite different from the one in which the cancer cells initially arose. For example, pancreatic carcinoma cells may behave differently when growing under the skin than when they are implanted into the host animal's pancreas, their natural, **orthotopic** site, presumably because the stromal environments in these two locations are so different. Similarly, breast cancer cells may generate tumors having more histological resemblance to clinical breast tumors if these cells are implanted into the mammary stromal fat pad instead of other locations.

Research into cancer stem cells suggests yet another disparity between cancer cell lines and the neoplastic cell populations within human tumors, this one not related to heterotypic interactions: if commonly occurring tumors are organized in a hierarchical fashion, involving cancer stem cells and transit-amplifying cells with limited self-renewal capacity, this organization is unlikely to be reestablished by cancer cell lines that are propagated *in vitro* (see Section 11.6).

One experimental response to these difficulties involves attempts at introducing fragments of tumors directly into immunocompromised mice without intervening propagation in tissue culture. These implanted tumor grafts contain both epithelial and stromal cells, which may co-proliferate to form histologically complex tumors in host mice that resemble the original tumors from which they derive (see Figure 13.8). Unfortunately, however, like the cancer cells in actual human tumors, the cells in these xenografts usually proliferate very slowly, and their propagation involves labor-intensive transplantation from one host animal to another. These properties preclude their routine use in the testing of anti-cancer drugs.

The inadequacies of currently available human tumor models and the resulting inability to accurately predict how clinically effective anti-cancer drugs will be (before they are actually used in humans) cost the pharmaceutical industry hundreds of millions of dollars annually and therefore represent one of the major impediments to the development of new anti-cancer drugs.

when converted to fibrin, create the scaffolding of the blood clot (Figure 13.9). The resulting fibrin bundles, which become entwined around clumps of platelets, help to stanch further bleeding.

The PDGF released by the platelets attracts fibroblasts and stimulates their proliferation (Figure 13.10). Thereafter, the platelet-derived TGF-β activates these fibroblasts and induces them to release a class of secreted proteases termed **matrix metalloproteinases** (MMPs; Table 13.1). Unlike most secreted proteases, which have a serine in their catalytic clefts, MMPs carry zinc ions to aid in catalysis; indeed, these ions inspired their name. Activated fibroblasts also secrete mitogens, such as various fibroblast growth factors (FGFs) that can stimulate the proliferation of certain epithelial cells.

Once released, the MMPs begin to degrade specific components of the extracellular matrix (ECM; Table 13.1). This degradation has two major consequences. On the one hand, it allows for structural remodeling of the ECM, thereby carving out space for new cells. On the other, it results in the release of a variety of growth factors that have been tethered in an inactive form to the proteoglycans of the ECM and now become solubilized and activated. Included among these

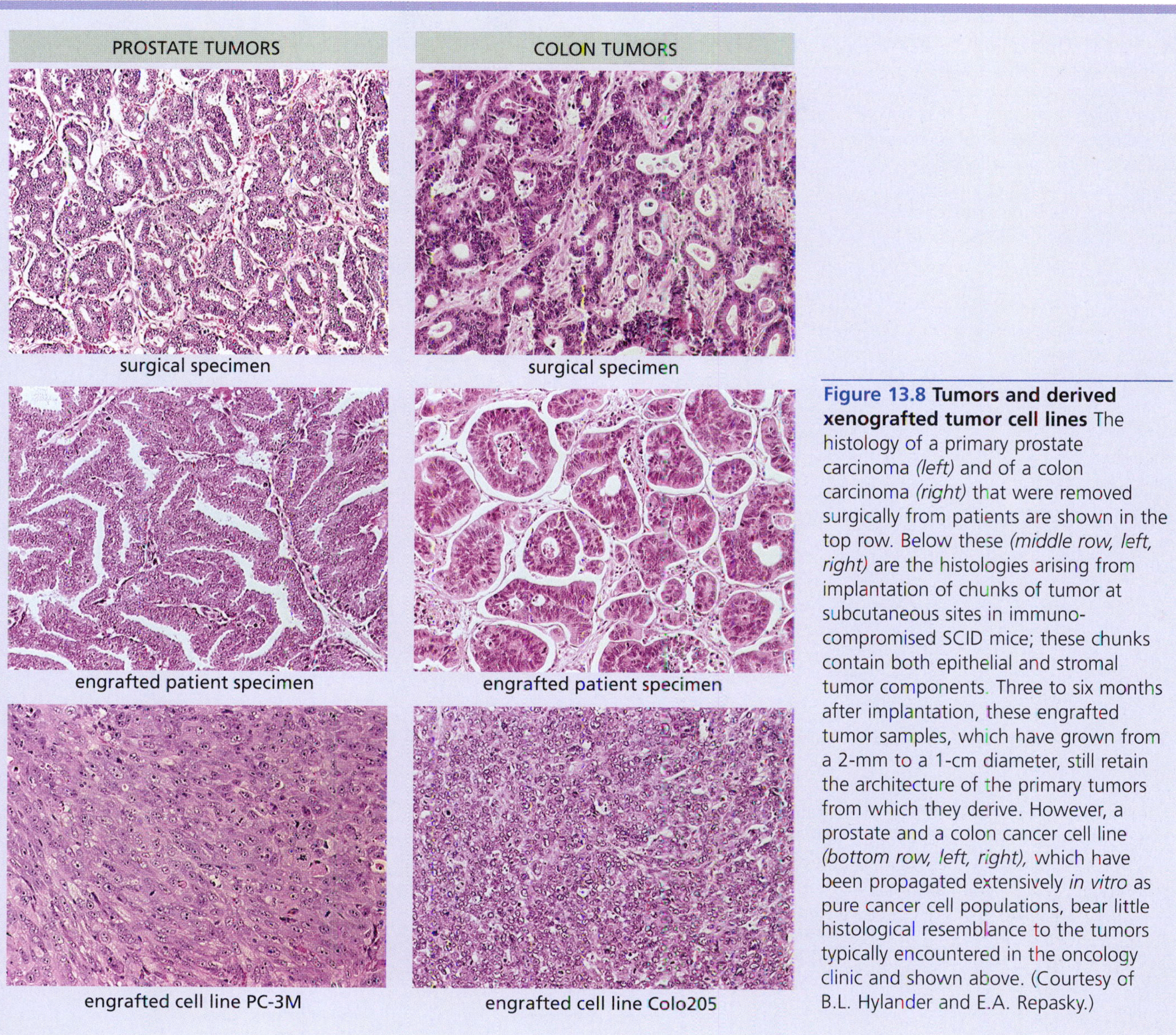

Figure 13.8 Tumors and derived xenografted tumor cell lines The histology of a primary prostate carcinoma *(left)* and of a colon carcinoma *(right)* that were removed surgically from patients are shown in the top row. Below these *(middle row, left, right)* are the histologies arising from implantation of chunks of tumor at subcutaneous sites in immunocompromised SCID mice; these chunks contain both epithelial and stromal tumor components. Three to six months after implantation, these engrafted tumor samples, which have grown from a 2-mm to a 1-cm diameter, still retain the architecture of the primary tumors from which they derive. However, a prostate and a colon cancer cell line *(bottom row, left, right),* which have been propagated extensively *in vitro* as pure cancer cell populations, bear little histological resemblance to the tumors typically encountered in the oncology clinic and shown above. (Courtesy of B.L. Hylander and E.A. Repasky.)

are basic fibroblast growth factor (bFGF), TGF-β1, PDGF, several EGF-related factors, and interferon-γ (IFN-γ). (In addition, virtually every MMP has been found to act on a variety of non-ECM proteins in the extracellular space; among these substrates are the latent **proenzyme** forms of other proteases, which are converted into active enzymes following cleavage by MMPs.)

The growth factors released by platelets and those mobilized from the ECM then attract **monocytes**, macrophages, and another class of phagocytes, termed **neutrophils**, which infiltrate the wound site. (Yet other types of immune cells, including **eosinophils**, **mast cells**, and **lymphocytes**, are also recruited to this site.) These recruited cells scavenge and remove foreign matter, bacteria, and tissue debris from the wound site; at the same time, they release and activate mitogenic factors, such as fibroblast growth factors (FGFs) and vascular endothelial growth factor (VEGF; see also Section 7.12). Such factors proceed to stimulate endothelial cells in the vicinity to multiply and to construct new capillaries—the process of **angiogenesis** (sometimes termed **neoangiogenesis**).

While all this is occurring, largely in the stromal region of a wound site, the epithelial cells around the edges of the wound have been undergoing their own

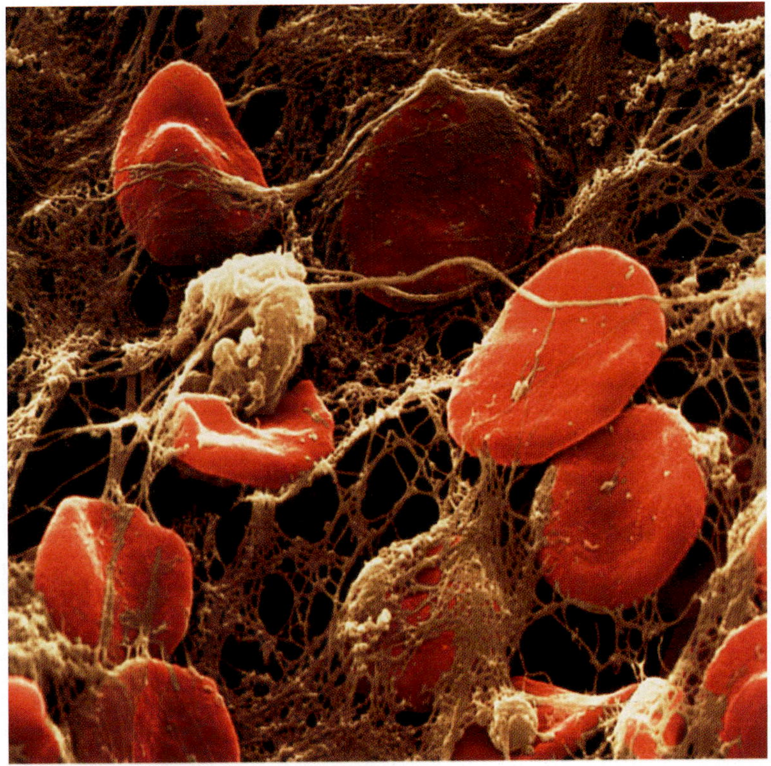

Figure 13.9 Scanning EM of a blood clot This scanning electron micrograph, which has been colorized, shows that clots are composed of dense networks of fibrin fibers *(light brown)* that have trapped platelets *(red)*. Prior to clotting, platelets undergo activation that leads to the release of the granules seen in Figure 5.3. (CNRI/Photoresearchers; courtesy of J. Weisel, Philadelphia.)

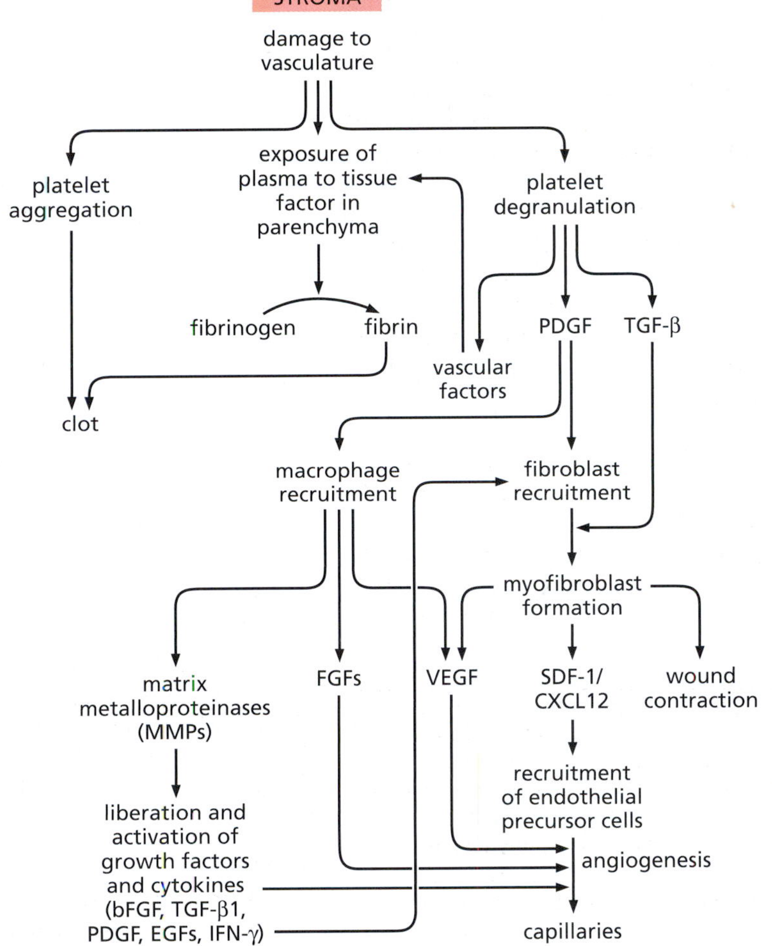

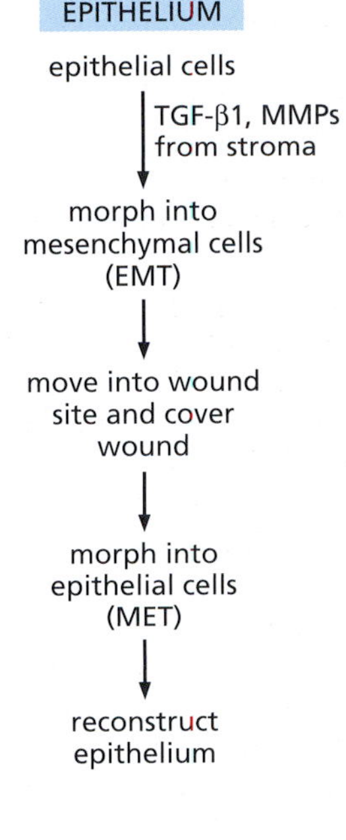

Figure 13.10 Flowchart of wound healing These flowcharts indicate many of the major events occurring after an epithelial tissue is wounded. Many of the changes occur in the stroma of the wounded tissue *(left)* in order to reconstruct it and thereby provide a foundation for reconstructing the epithelial cell layer *(right)*.

Table 13.1 Some matrix metalloproteinases and their extracellular matrix substrates

Name of MMP	Alternative name of MMP	ECM substrates[a]
MMP-1	collagenase-1	various collagens, gelatin, entactin, aggrecan, tenascin
MMP-2	gelatinase A	elastin, fibronectin, various collagens, laminin, aggrecan, vitronectin
MMP-3	stromelysin-1	proteoglycans, laminin, fibronectin, gelatin, various collagens, fibrinogen, entactin, tenascin, vitronectin
MMP-7	matrilysin	same as MMP-3
MMP-9	gelatinase B	same as MMP-2
MMP-11	stromelysin-3	inactive serpin

[a]In addition to ECM substrates, most of the MMPs also cleave other substrates present in the extracellular space. For example, MMP-3 has been reported to cleave pro-HB-EGF, pro-IL-1β, plasminogen, E-cadherin, IGFBP-3, α1-anti-chymotrypsin, α1-proteinase inhibitor, pro-MMPs-1, -3, -7, -8, -9, and -13, and pro-TNF-α.

Adapted from L.J. McCawley and L.M. Matrisian, *Curr. Opin. Cell Biol.* 13:534–540, 2001.

alterations. Their goal is to reconstruct the epithelial sheet that existed prior to wounding. To do so, epithelial cells reduce their adhesion to the ECM, especially to the basement membrane (Section 2.2 and Figure 2.3) that separates them from the stromal compartment (Figures 13.5 and 13.11). By severing these connections, the epithelial cells gain increased mobility.

The epithelial cells also sever their attachments to one another. These side-by-side associations are stabilized in part by **adherens junctions** (Figure 13.12), which are assembled as associations of E-cadherin molecules that are displayed by adjacent epithelial cells and tether the cells' apposed plasma membranes to one another. Accordingly, E-cadherin expression is suppressed in the epithelial cells that are situated around the edges of a wound site and is often replaced by N-cadherin, another cell–cell adhesion molecule that is normally displayed by mesenchymal cells, notably fibroblasts. (These N-cadherin molecules, though related structurally to E-cadherin, do not reestablish adherens junctions between the epithelial cells, since they bind much more weakly to one another.)

While shifting their display of cell-surface cadherins, the epithelial cells at the edge of a wound undergo a major change in phenotype, which causes them to assume, at least superficially, a fibroblastic appearance (Figure 13.13). This profound shift is termed the **epithelial–mesenchymal transition** (EMT) and enables the epithelial cells to become motile and invasive. The traits acquired during the EMT allow these cells to move into the wound site, where they fill in the gap in the epithelium created by the wounding.

Stromal cells in the wound site help to trigger the EMT of nearby epithelial cells. The matrix metalloproteinases (MMPs) secreted by stromal cells liberate and activate latent growth factors, such as TGF-β1, that have been stored in the extracellular matrix. In ways that are poorly understood, these released factors seem to stimulate and reinforce expression of the EMT program in epithelial cells. Importantly, the EMT is only a temporary shift in cell phenotype. After the

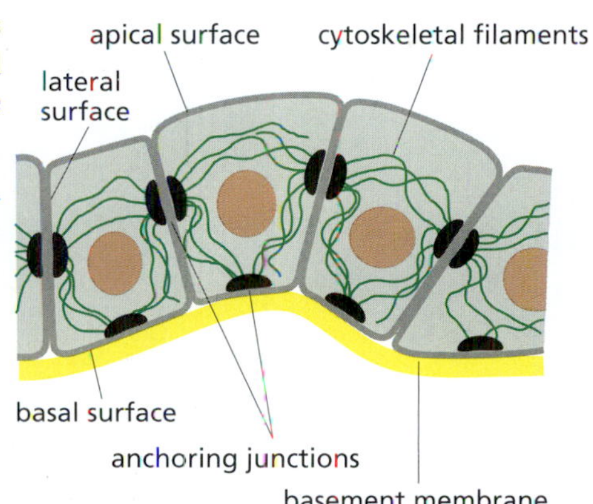

Figure 13.11 Attachments of epithelial cells to the structures around them Epithelial cells are anchored via their basal surfaces to underlying basement membrane *(yellow)* and to adjacent epithelial cells via their lateral surfaces, which form several distinct types of connections (anchoring junctions, *black*) that physically tie the cytoskeletons of neighboring cells *(green lines)* to one another; these cytoskeleton attachments involve both actin filament and microtubule components. (From B. Alberts et al., Molecular Biology of the Cell, 4th ed. New York: Garland Science, 2002.)

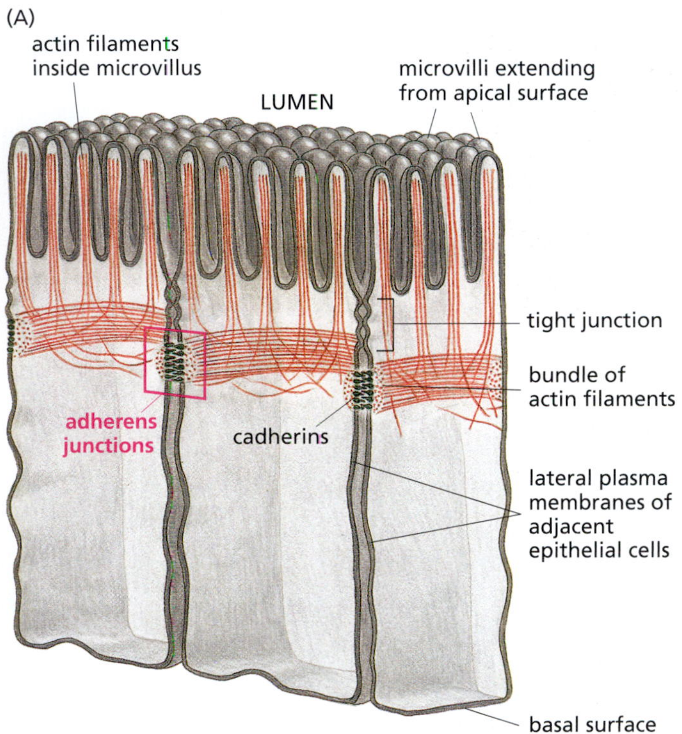

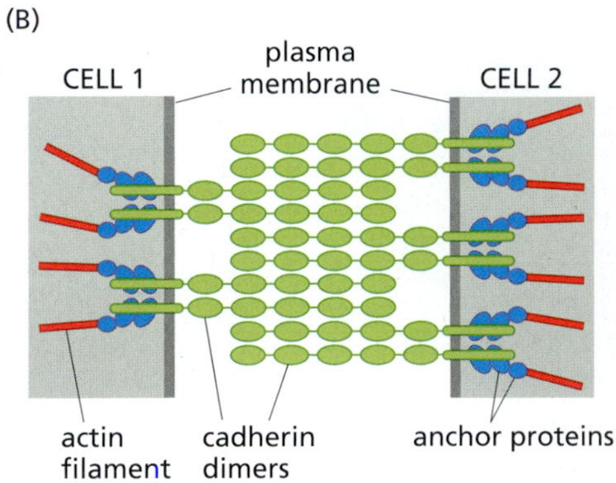

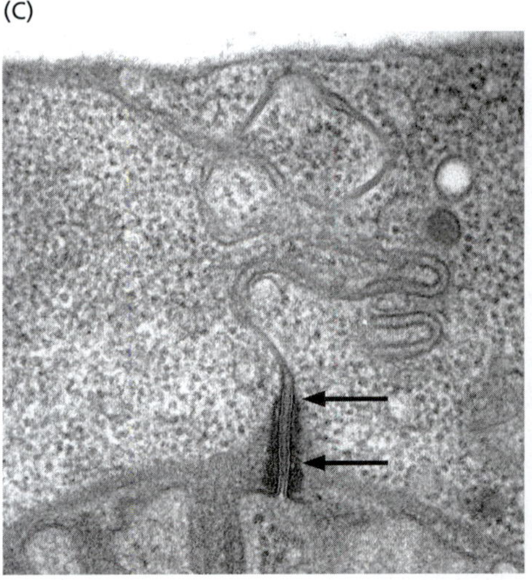

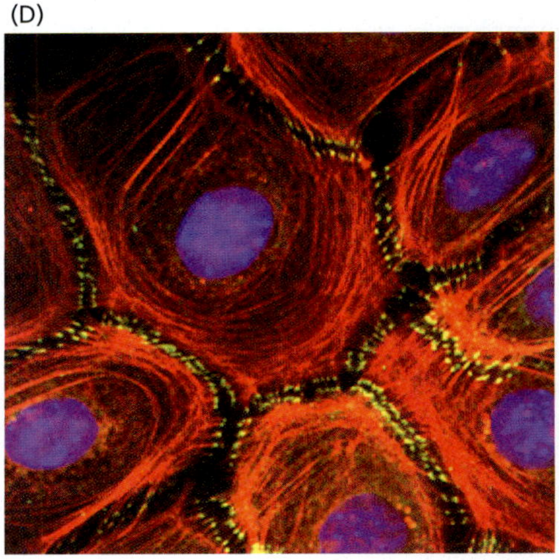

Figure 13.12 Adherens junctions (A) This drawing illustrates the adhesion belt that is formed by adherens junctions between the lateral surfaces of adjacent epithelial cells. E-cadherin molecules are responsible for forming these adherens junctions in many epithelial cell types. (B) In more detail, these cell–cell connections depend on oligomerization between cadherin molecules displayed by two adjacent cells. E-cadherin *(light green)* plays a dominant role in most epithelia in forming the resulting adherens junctions. (C) As visualized in this transmission electron micrograph, the adherens junctions *(black arrows)* between two closely apposed plasma membranes form between the lateral surfaces of adjacent epithelial cells, in this instance those creating the intestinal lining of the worm *Caenorhabditis elegans.* Virtually identical structures are seen in mammalian cells. (D) This immunofluorescence micrograph illustrates the interactions between the lateral surfaces of neighboring keratinocytes. E-cadherin "zippers" *(yellow)* form adherens junctions between neighboring cells and are attached via their cytoplasmic domains and other intermediary proteins to the actin cytoskeletons of these cells *(red).* Nuclei are stained in *blue*. (A and B, from B. Alberts et al., Molecular Biology of the Cell, 4th ed. New York: Garland Science, 2002; C, courtesy of D. Hall; D, from M. Perez-Moreno, C. Jamora and E. Fuchs *Cell* 112:535–548, 2003.)

migrating cells have moved into position and covered the wound site, they reconstruct the epithelium by reverting to an epithelial state via the program termed the **mesenchymal–epithelial transition** (MET). As a consequence, once wound healing is complete, the cells in the reconstituted epithelium show no trace of having passed transiently through a mesenchymal state.

If we compare these processes (Figure 13.14) with the interactions of epithelial cancer cells and their stromal neighbors, we find striking parallels between wound healing and tumorigenesis. One clear similarity between the two processes derives from the presence of clumps of fibrin in the tumor-associated stroma. In the case of tumors, this does not derive from traumatic damage to

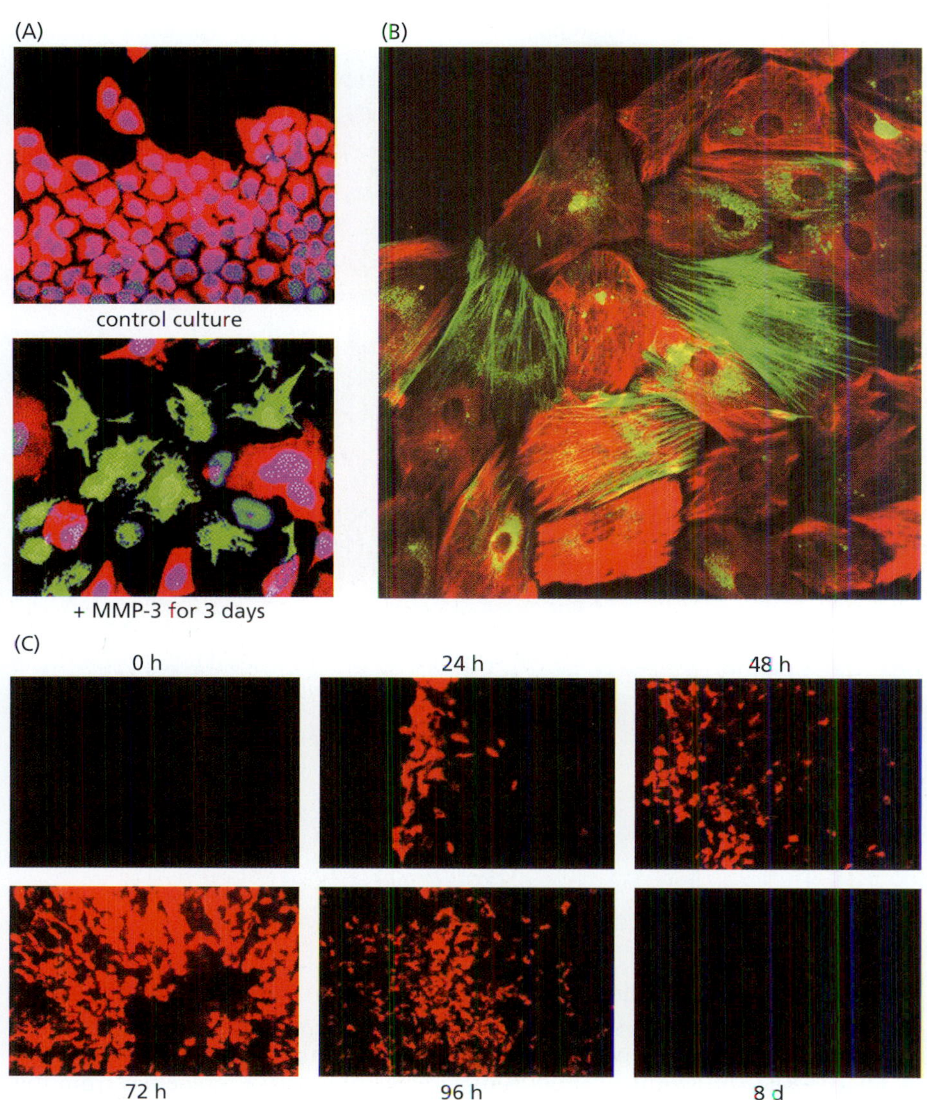

Figure 13.13 The epithelial–mesenchymal transition (A) When epithelial cells in monolayer culture undergo an epithelial–mesenchymal transition (EMT), they shut down expression of many typical epithelial cell markers, such as cytokeratins *(red)* and E-cadherin, and induce expression of mesenchymal proteins, such as vimentin *(green)*, fibronectin, and N-cadherin. In addition, they change their typically polygonal shape *(top)* to a fibroblastic shape *(bottom)* and often acquire motility and invasiveness. In this case, the EMT was provoked by a 3-day-long exposure to matrix metalloproteinase-3 (MMP-3), which may initiate the EMT through its ability to degrade E-cadherin. (B) In some cell types in culture, individual cells may spontaneously shift from an epithelial phenotype, indicated by cytokeratin expression *(red)*, to a mesenchymal phenotype, indicated by α-smooth muscle actin expression *(green)*. This mirrors a plasticity that is apparent in certain cells during early embryogenesis. (C) A confluent monolayer of MCF10A cells, a line of nontransformed, immortalized human mammary epithelial cells (MECs), has been disturbed by removal of a patch of cells from the monolayer *(removed from the left of each image)*. Initially, no vimentin is expressed by these cells. However, soon after the monolayer has been wounded, the epithelial cells at the edge of the wound undergo a partial EMT, express vimentin (characteristic of mesenchymal cells; *red*) and migrate into and fill the wound site. After 8 days, at which time the cell monolayer has been restored, these cells revert to a fully epithelial phenotype and shut down vimentin expression (indicating a mesenchymal–epithelial transition, an MET). (A, from M.D. Sternlicht, A. Lochter, C.J. Sympson et al., *Cell* 98:137–146, 1999; B, from O.W. Petersen, H.L. Nielsen, T. Gudjonsson et al., *Am. J. Pathol.* 162:391–402, 2003; C, from C. Gilles, M. Polette, J.M. Zahm et al., *J. Cell Sci.* 112:4615–4625, 1999.)

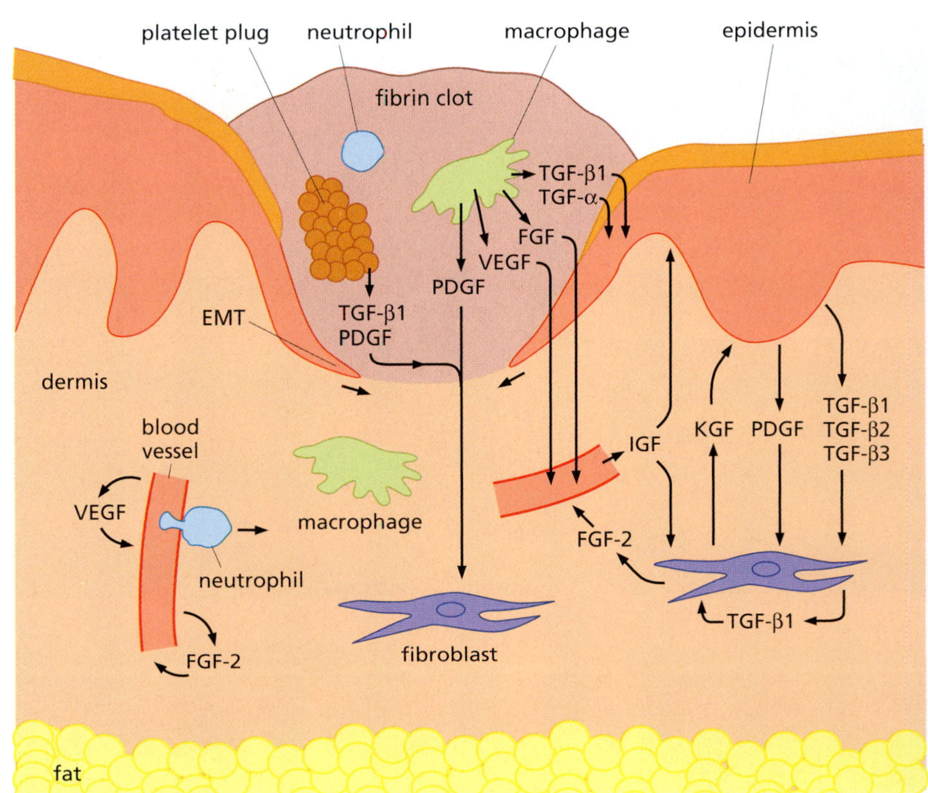

Figure 13.14 The program of wound healing This drawing depicts schematically the heterotypic signaling occurring after the skin, a typical epithelial tissue, is wounded. In the site of the wound, a fibrin clot that carries platelets has initially filled the wound site. Day 3 of the wound-healing process is sometimes termed its inflammatory phase, because of the involvement of a number of cell types that are often associated with inflammation, notably neutrophils and monocytes; they are involved in removing cell debris, the clot, and invading bacteria. In addition, macrophages, which derive from differentiation of monocytes, release TGF-β1 and TGF-α; these factors stimulate the proliferation of nearby epithelial cells from the epidermis and induce them to undergo an epithelial–mesenchymal transition (EMT). Tongues of these cells, now exhibiting mesenchymal behavior, are invading under the clot from both sides in order to cover the wound site. At the same time, TGF-β1 and PDGF-A and -B, which have been released by the platelets during initial clot formation, are stimulating the proliferation of fibroblasts in the underlying dermal stroma and inducing them to convert into myofibroblasts *(not shown)*; because of their contractility, myofibroblasts will pull the sides of the wound together. Myofibroblasts also release angiogenic factors, notably VEGFs and FGF-2 (fibroblast growth factor-2), which stimulate the formation of new blood vessels in the stroma underlying the wound site. Finally, after the wounded stroma is covered, the cells that had previously undergone an EMT revert, via a mesenchymal–epithelial transition, (MET) to an epithelial state, thereby reconstituting the epithelium *(not shown)*. (From A.J. Singer and R.A.F. Clark, *N. Engl. J. Med.* 341:738–746, 1999.)

blood vessels. Instead, the capillaries within tumors are constitutively leaky, unlike those in normal tissues (Figure 13.15A). Later, we will discuss the causes of this leakiness. For the moment, suffice it to say that the permeability of the capillary walls and **venules** (small veins) allows fibrinogen molecules from the plasma to come in direct contact with cancer cells; this provokes, through a series of intermediate reactions, the conversion of fibrinogen to fibrin and the formation of large bundles of fibrin strands (Figures 13.15B and C).

Many kinds of cancer cells, including those forming carcinomas of the breast, prostate, colon, and lung, continuously release significant levels of PDGF. This contrasts to the situation in wounds, in which PDGF is released in a brief burst by platelets as they form the initial blood clot. As is the case in wound healing, the targets of the PDGF produced by cancer cells are mesenchymal cells in the stroma that display PDGF receptors, including smooth muscle cells, fibroblasts, and macrophages. PDGF functions here both as an attractant and as a mitogen for these stromal cells and appears to be the most important signaling molecule used by carcinoma cells to recruit and stimulate the proliferation of stromal cells. In breast cancers, for example, the level of PDGF expression generally increases with increased tumor progression, which may explain the high degree of **stromalization** of many advanced tumors.

The PDGF released by carcinoma cells initially succeeds in recruiting fibroblasts into the fibrin matrix. At the same time, the fibrin matrix (see Figure 13.15B and C) provides an important scaffolding that helps these recruited mesenchymal cells to attach, via integrins, and migrate. The invading stromal cells remodel this "provisional matrix" by degrading many of the initially formed fibrin molecules and replacing them with a more permanent matrix that is assembled from the collagen secreted by the fibroblasts. This closely parallels the sequence of steps occurring during wound healing.

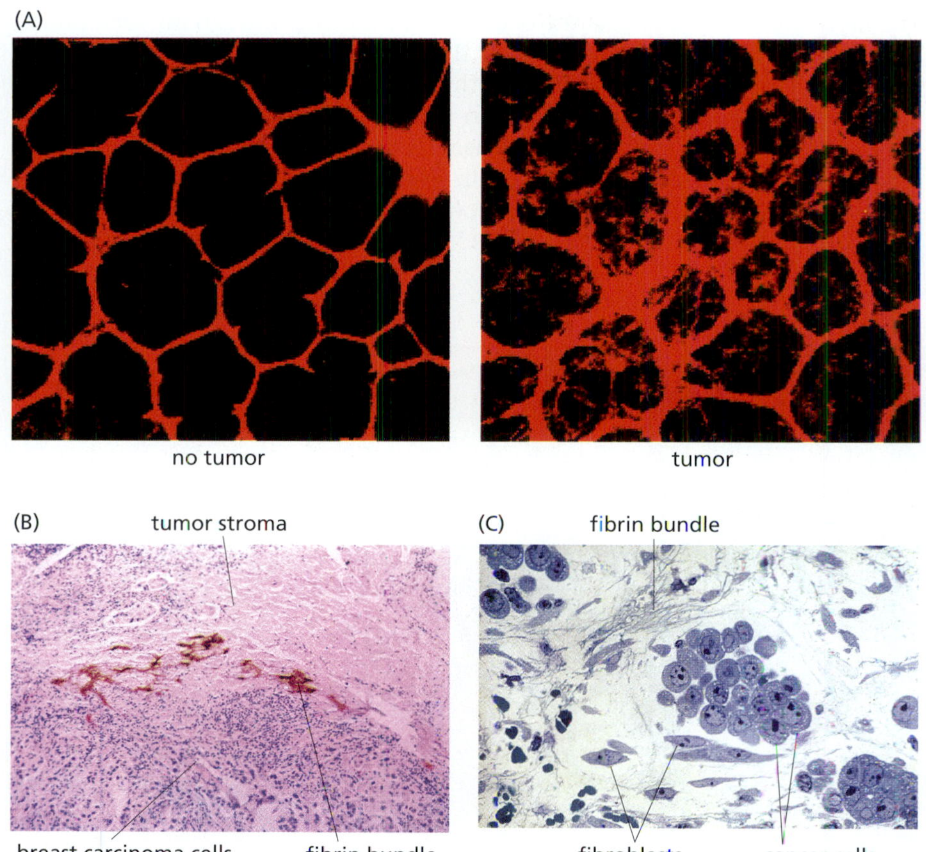

Figure 13.15 Capillary leakiness and deposition of fibrin bundles in the tumor-associated stroma (A) Injection of red dextran dye into normal microvasculature reveals that the walls of associated capillaries have relatively low permeability, as indicated by the sharply delineated profiles of these vessels *(left)*. However, when injected into tumor-associated vasculature, the dye leaks out of the capillaries and diffuses into the nearby parenchyma, yielding diffusely staining outlines of vessels *(right)*. (B) The leakiness of tumor-associated microvasculature results in the continuous leaking of thrombin and fibrinogen molecules from the plasma into the parenchyma surrounding blood vessels. *Tissue factor* is a protein displayed on the surface of cancer cells and many other cell types. Thrombin from the plasma is activated upon contact with tissue factor and converts fibrinogen into fibrin. This results in the extensive network of fibrin bundles *(dark red)* seen here at the border between actively growing breast cancer cells *(below)* and the tumor stroma *(above)*. (C) The fibrin bundles *(above and to left of cluster of cancer cells)* form an extracellular matrix to which migrating cells, such as fibroblasts *(below cluster of cancer cells)* as well as endothelial cells *(not seen)*, can attach, using it as a substrate on which they can move forward. In many tumors, such as this guinea pig liver carcinoma of bile duct origin, the fibrin matrix may eventually be dissolved by the process of *fibrinolysis* and replaced by a collagenous matrix. (A, courtesy of R. Muschel; B, from C.G. Colpaert, P.B. Vermeulen, P. van Beest et al., *Histopathology* 42:530–540, 2003; C, from H.F. Dvorak, *Am. J. Pathol.* 162:1747–1757, 2003.)

One of the tasks of the stromal cells in wound healing involves the physical contraction of the wound site in order to close wounds. This contraction is mediated by the specialized class of fibroblasts termed **myofibroblasts**, which express α-smooth muscle actin (α-SMA) and are capable of using the actin–myosin contractile system (related to the one operating in muscle cells) to generate the mechanical tension needed for wound closure (Figure 13.16A). Myofibroblasts are also present in sites of chronic wounding, that is, continuously inflamed tissues (Figure 13.16B). Provocatively, essentially identical myofibroblasts are prominent components of the stroma present in the majority of advanced carcinomas (Figure13.16C–E).

Some of the tumor-associated myofibroblasts seem to arise from normal stromal fibroblasts. In fact, myofibroblasts can be produced *in vitro* simply by exposing normal fibroblasts to TGF-β1. This suggests that the TGF-β1 released by many types of carcinoma cells, especially those that have progressed to higher levels of malignancy, is a major factor responsible for the formation of myofibroblasts in the tumor-associated stroma. In the context of wounds, the major source of TGF-β1 is likely to be the platelets; as mentioned earlier, they release large amounts of this factor when they participate in clot formation. The origins of many stromal myofibroblasts and fibroblasts are unclear (Sidebar 13.3).

The tumor stroma formed by myofibroblasts differs substantially in appearance from the stroma present in normal epithelial tissues. This distinctive appearance has caused pathologists to label it a "reactive" or **desmoplastic** stroma. The latter term refers to the *hardness* of the tumor mass as a whole, which results from the deposition of extensive extracellular matrix (ECM) by the myofibroblasts. As the progression of a carcinoma advances to a higher, more aggressive grade, the proportion of stroma that is desmoplastic often increases in parallel (Figure 13.17).

The myofibroblasts construct the desmoplastic stroma through the secretion of large amounts of collagen types I and III, fibronectin, proteoglycans, and glycosaminoglycans (GAGs), which together give this stroma its characteristic appearance at the microscopic level. In addition, these cells secrete urokinase plasminogen activator (uPA, a protease) and a series of matrix metalloproteinases (MMPs), which help to mobilize growth factors that were previously sequestered in the ECM. As the desmoplastic stroma matures, many of the initially present stromal cells disappear, being replaced progressively by the dense,

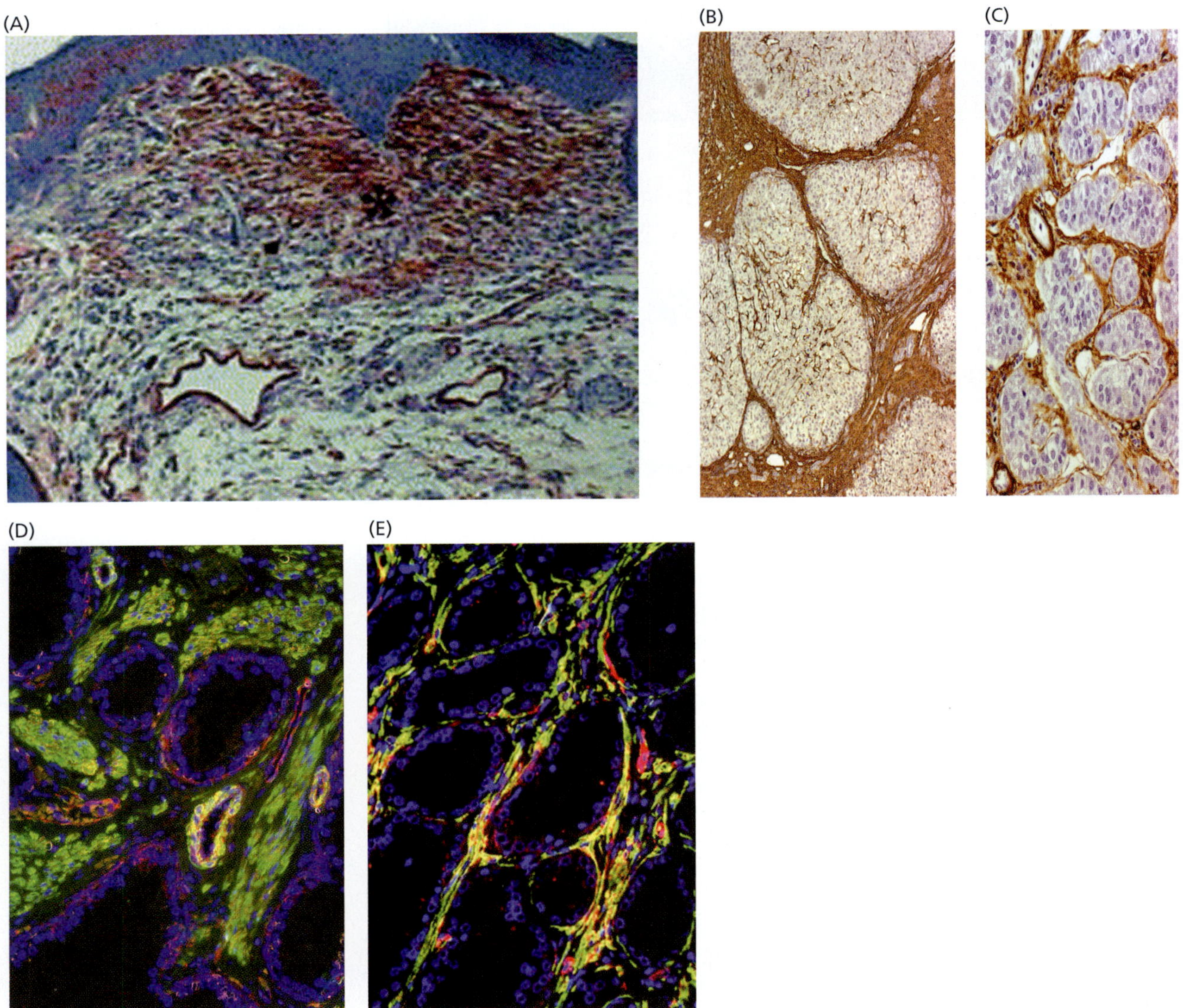

Figure 13.16 Myofibroblasts Myofibroblasts arise in sites of wound healing and in the tumor-associated stroma through the transdifferentiation of stromal fibroblasts and through the recruitment of myofibroblast precursors, possibly fibrocytes, from the circulation. Their presence is revealed typically by their expression of α-smooth muscle actin (α-SMA), the antigen that is being stained here. (A) α-SMA-positive myofibroblasts are present in abundance 3 days after the skin of a mouse has suffered a wound *(reddish brown)*. (B) Chronically inflamed tissues acquire a fibrotic stroma, such as the cirrhotic liver seen here stained with anti-α-SMA antibody, which reveals the presence of numerous myofibroblasts *(brown)*. (C) A section of a hepatocellular carcinoma stained with anti-α-SMA antibody *(brown)*. This reveals the clear resemblance of the stroma of a chronically inflamed tissue to the stroma of a carcinoma arising in that tissue. (D) α-SMA *(green)* is expressed by the numerous smooth muscle cells present in the stroma of the normal human prostate gland. Unlike myofibroblasts, smooth muscle cells are known to lack significant vimentin staining *(red)*. The few areas of overlapping expression of these two proteins are indicated in *yellow*; these represent the occasional pericytes surrounding a blood vessel. The nuclei of epithelial cells lining the ducts are in *blue*. (E) In prostate carcinomas, in contrast, the stroma is filled with myofibroblasts that express both vimentin *(red)* and α-SMA *(green)*. The overlap between these two is indicated by the *yellow*. Epithelial cell nuclei are indicated in *blue*. (A, from P. Martin, D. D'Souza, J. Martin et al., *Curr. Biol.* 13:1122–1128, 2003; B and C, from A. Desmouliere, A. Guyot and G. Gabbiani, *Int. J. Dev. Biol.* 48:509–517; D and E, from J.A. Tuxhorn, G.E. Ayala, M.J. Smith et al., *Clin. Cancer Res.* 8:2912–2923, 2002.)

Sidebar 13.3 Fibroblasts and myofibroblasts may be recruited from afar It remains unresolved precisely where most stromal fibroblasts and myofibroblasts originate. During wound healing, fibroblasts may initially be recruited from the stroma of adjacent normal tissues; myofibroblasts may then be produced through the transdifferentiation of some of these recruited fibroblasts. However, an intriguing alternative, still to be rigorously documented, is that many myofibroblasts and fibroblasts within wounds derive from recruited **fibrocytes**—less differentiated, collagen-producing precursor cells that are present in the circulation. Fibrocytes, which originate in the bone marrow, are known to be capable of homing to areas of tissue damage, where they settle and seem to differentiate into myofibroblasts and, quite possibly, fibroblasts, thereby contributing to the rapid reconstruction of stroma. This organization ensures that large numbers of cells can be mobilized rapidly from sources outside a wound site in order to reconstruct damaged stroma. The strong parallels between wound healing and tumorigenesis, described here, suggest that fibrocyte recruitment also plays a significant role in generating tumor-associated stroma.

acellular, collagenous ECM that is the hallmark of this type of tumor-associated stroma. The processes leading to the formation of the tumor-associated desmoplastic stroma seem to be similar to those operating in sites of wound healing. However, in the tumor, these processes operate continuously over many months and years, rather than transiently over periods of several days, as happens during wound healing.

Another hint of parallels between wound healing and tumorigenesis comes from several studies indicating that agents known to inhibit tumor-associated angiogenesis also antagonize wound healing. Included among these antagonists are certain soluble inhibitors of the growth factor receptors that play key roles in angiogenesis, as well as an ECM component, termed thrombospondin-1 (Tsp-1), which is a potent anti-angiogenic molecule. These common mechanisms shared by wound healing and by the epithelial–stromal interactions in tumors also have implications for clinical practice (Sidebar 13.4).

The term "carcinoma-associated fibroblasts" (CAFs) is sometimes used to describe the mixed populations of fibroblasts and myofibroblasts that are present in the stroma of epithelial tumors. Analyses of the gene expression patterns of CAFs underscore the strong similarities between these cells and the fibroblasts present in wound sites. In addition, these studies provide an indication of the role of these cells as key players in tumor progression.

These analyses initially determined the gene expression pattern of serum-stimulated fibroblasts propagated *in vitro,* that is, fibroblasts that had been starved of serum and then exposed to high concentrations of fresh serum. This exposure re-creates the environment of fibroblasts during the initial stages of wound healing (a time when the various serum factors released by activated platelets act on stromal fibroblasts at the wound site, converting them, at least transiently, into myofibroblasts). A relatively small cohort of serum-induced and -repressed genes was extracted from these data and used to represent the characteristic "signature" of serum-stimulated fibroblasts (Figure 13.18A).

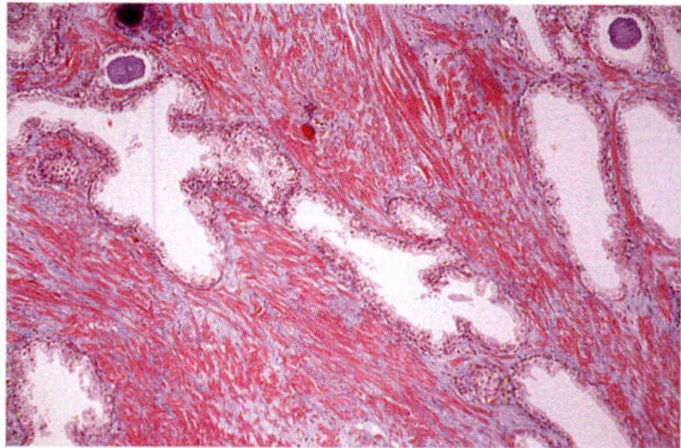

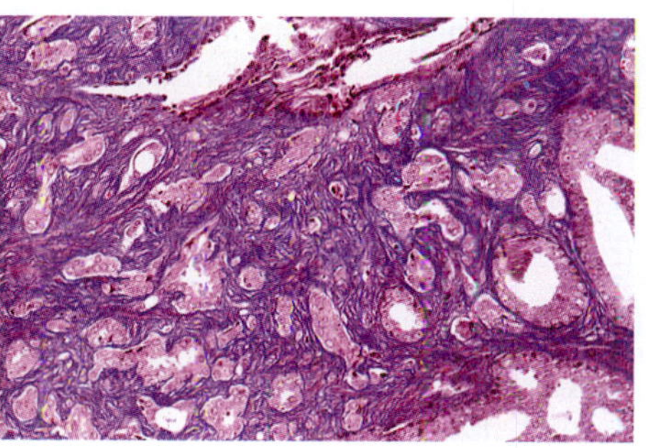

Figure 13.17 Micrograph of normal stroma and desmoplastic stroma The histologically complex stroma of normal tissue may eventually be replaced by the desmoplastic stroma of an advanced carcinoma. Here we see normal human prostate tissue *(left panel)* stained with Masson's trichrome stain, which reveals the extensive smooth muscle cells in the stroma as *pink;* normal ducts are scattered throughout this stroma. In contrast, an advanced prostate carcinoma *(right panel)* stained identically and viewed at the same magnification reveals the extensive desmoplastic stroma *(blue purple),* which is rich in extracellular matrix, notably collagen I. Islands of prostate carcinoma cells forming small ducts *(pink)* are scattered throughout this desmoplastic stroma, which lacks significant numbers of viable cells such as myofibroblasts and fibroblasts. (From G. Ayala, J.A. Tuxhorn, T.M. Wheeler et al., *Clin. Cancer Res.* 9:4792–4801, 2003.)

Sidebar 13.4 Breast cancer surgery may lead to the stimulation of tumor growth Epidemiologic studies of breast cancer patients undergoing surgery (i.e., partial or total mastectomy) for removal of their primary tumors have revealed a peak of recurrence of breast cancers at the primary site as well as in distant anatomical sites occurring about 3 years after surgery. The timing of these relapses suggests that the surgery is responsible for stimulating their formation. Examination of the fluids draining from the wound sites in the breast created by the surgery has demonstrated the presence of potent mitogenic factors, which are associated with the wound-healing process. Among these are mitogens for breast cancer cells, especially those cancer cells that over-express the HER2/Neu protein. In addition, a burst of vascular endothelial growth factor (VEGF) synthesis occurs following surgery; as discussed later in this chapter, VEGF is a potent stimulant of tumor angiogenesis. It is therefore plausible that surgery stimulates the proliferation of residual **micrometastases** (small deposits of metastatic cells) that are not detected and removed by the surgeons when they excise primary tumors. Indeed, some have argued that the clinical relapses stimulated by surgery nullify much of the therapeutic benefit that should, by all rights, be achieved by the removal of primary tumors and nearby lymph nodes. The importance of such surgery-stimulated tumor progression remains a matter of great contention.

Another clinical observation may eventually be found to bear on this discussion: when Herceptin (a therapeutic monoclonal antibody that blocks the action of the HER2/Neu receptor; see Section 15.19) is applied postoperatively in women who have borne relatively small, low-grade tumors, the rate of clinical relapse is reduced by as much as 50%. This stunning therapeutic success may one day be traced to the ability of Herceptin to block stimulation of residual cancer cells by the growth factors elaborated during healing of the surgical wound site.

The RNA expression patterns of a large group of human carcinomas were then analyzed. In each case, these analyses determined whether a tumor expressed the signature of serum-stimulated fibroblasts. (Since all the genes being analyzed are associated specifically with fibroblasts, these analyses, by necessity, reflected the RNA expression patterns of the CAFs in each tumor.) Many carcinomas were indeed found to express the signature of serum-stimulated fibroblasts. Significantly, the carcinomas that expressed this signature more intensely were associated with a grimmer clinical prognosis (Figure 13.18B and C). This correlation suggests that the activated, myofibroblast-rich stroma represents a force that can drive aggressive tumor progression.

13.4 Stromal cells are active contributors to tumorigenesis

Several biological experiments provide even more direct demonstrations of the profound influence that recruited stromal cells exert on epithelial cell tumorigenesis. In one study, previously non-tumorigenic, immortalized keratinocytes were forced to secrete PDGF at high levels (achieved through the introduction of a PDGF expression vector). The released growth factor had no effect on the proliferation of these epithelial cells *in vitro,* because they do not display PDGF receptors on their surface. However, when these cells were implanted into host mice, they acquired the ability to form robustly growing tumors, clearly derived from the ability of the released PDGF to recruit and activate stromal cells. The stromal cells then reciprocated by driving the proliferation of the PDGF-secreting keratinocytes, eventually causing the latter to undergo malignant transformation.

In a complementary experiment, genetically engineered alterations of the mouse germ line enabled investigators to selectively inactivate the TGF-β type II receptor in the stromal fibroblasts in a variety of tissues. As a consequence, these stromal cells were no longer susceptible to TGF-β–mediated growth inhibition, and stromal hyperplasia occurred in many of these tissues. In some of these, the hyperproliferating fibroblasts drove nearby epithelial cell layers to proliferate and, eventually, to develop into carcinomas (Figure 13.19). This demonstrates, once again, the powers of stromal cells to stimulate epithelial cell proliferation, doing so in ways that can lead to neoplastic transformation of the epithelial cells.

The important role of stromal fibroblasts in supporting tumor growth can be demonstrated by yet another type of experiment: transformed, weakly tumorigenic human mammary epithelial cells (MECs) were found to require more than two months to form a tumor after being introduced into immunocompromised host mice (Figure 13.20). However, if these cancer cells were mixed with human mammary stromal fibroblasts (from normal breast tissue) prior to injection into

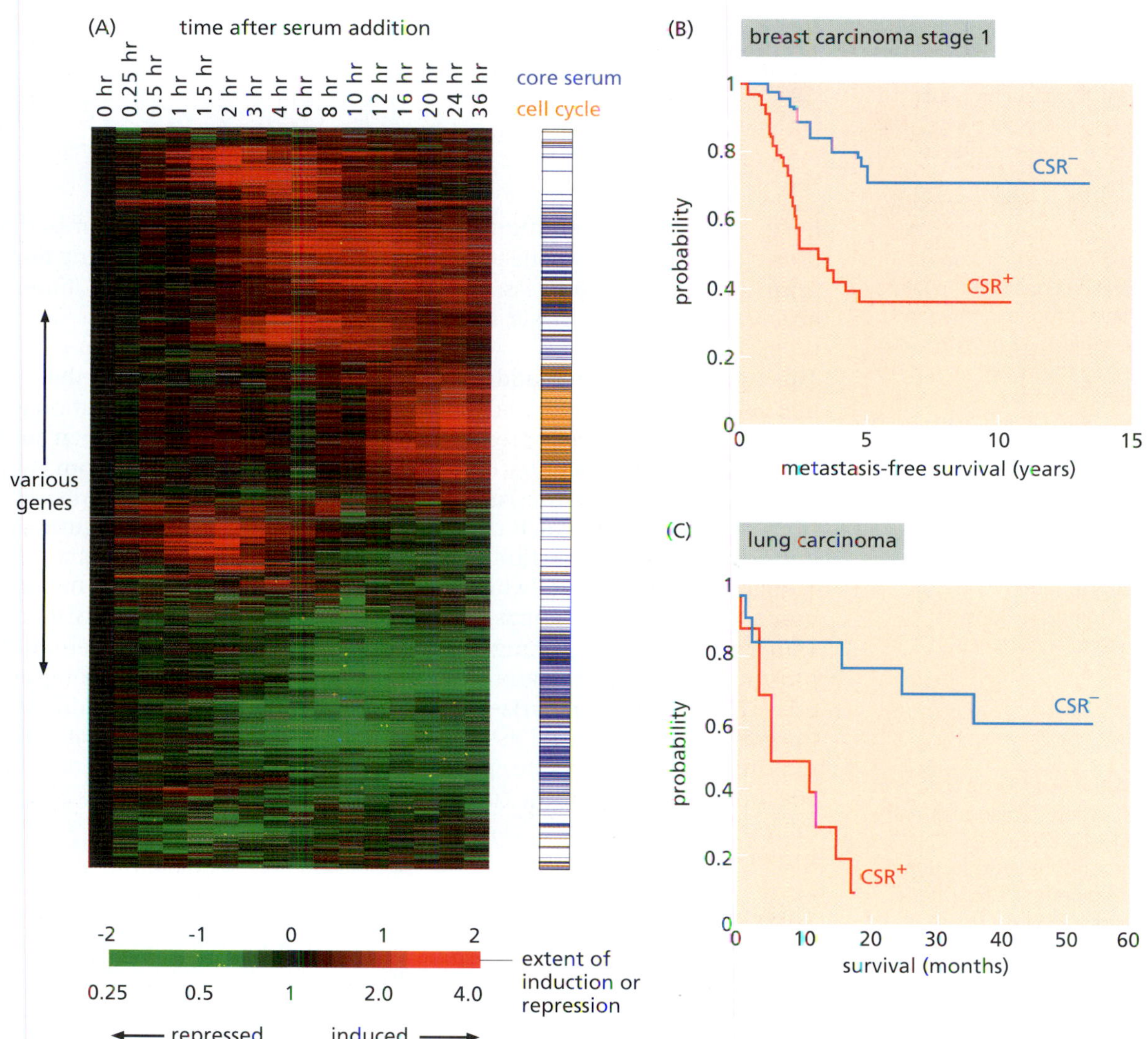

Figure 13.18 Gene expression arrays of tumor-associated myofibroblasts and serum-activated fibroblasts The analysis of the spectrum of genes expressed in a cell population or a tissue is often called *functional genomics*. (A) In this gene expression analysis, the expression of cellular mRNAs was analyzed following the addition of fresh serum to previously quiescent, serum-starved human fibroblasts; this caused the entrance by these cells into the active cell cycle and the induction or repression of the expression of a large cohort of genes. Gene expression was measured at various times *(above)* after addition of serum. The cohort of genes analyzed *(not named)* is arrayed vertically from *top to bottom*. Genes that were induced or repressed late after the addition of serum were classified as "cell cycle genes" *(orange lines, right vertical bar)*, whereas genes that were induced or repressed early and did not fluctuate with cell cycle phase were classified as "core serum response" (CSR) genes *(blue lines, right vertical bar)*. The horizontal bar *(below)* is a key indicating the degree of induction *(red)* or repression *(green)* following serum addition, this being indicated logarithmically above the bar and in absolute numbers below the bar. (Genes that are induced more than 4-fold above or repressed more than 4-fold below these levels are simply registered here as being induced 4-fold or repressed 4-fold.) (B) The expression pattern of the CSR genes was analyzed in a series of tumors in women presenting with stage 1 (i.e., early-stage) breast cancer. As indicated here, those tumors expressing the CSR gene pattern *(red line)* showed a much greater probability for developing metastases in the years following initial treatment compared with those whose tumors did not show this gene expression pattern *(blue)*. (C) Similarly, those patients whose lung adenocarcinomas (including all stages of tumor progression at the time of diagnosis) showed the CSR gene expression signature *(red line)* suffered dramatically higher mortality rates compared with those whose tumors did not show this gene expression signature *(blue line)*. (From H.Y. Chang, J.B. Sneddon, A.A. Alizadeh et al., *PLoS Biol.* 2:E7, 2004.)

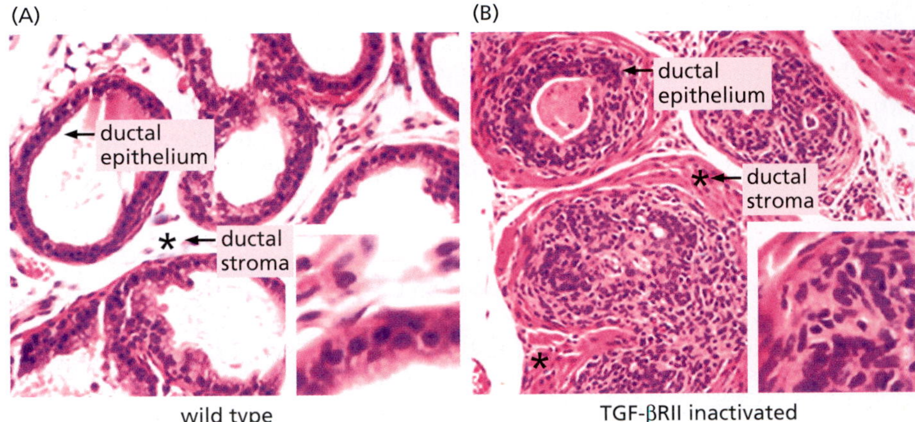

Figure 13.19 Prostatic tumors from mice with genetically altered stromal fibroblasts The gene encoding the TGF-β type II receptor has been inactivated selectively in the fibroblasts present in a variety of epithelial tissues of the mouse. (A) The prostates of male mice normally exhibit a thin layer of epithelial cells lining the lumina of the ducts and a relatively thin layer of stromal cells *(asterisk)* outside the epithelial cell layers. (B) As a consequence of TGF-β type II receptor inactivation in the stromal fibroblasts, both cell layers became hyperplastic. The epithelial cells soon created tissue that closely resembled the prostatic intraepithelial neoplasia (PIN) commonly seen in humans (see Figure 13.40). This suggests that the hyperplastic stroma released proliferative signals to the nearby epithelium. Indeed, production by the stromal cells of hepatocyte growth factor (HGF)—a potent epithelial cell mitogen—increased threefold after they lost their responsiveness to the inhibitory effects of TGF-β. In the stomachs of these mice, in which the TGF-β type II receptor was inactivated in the gastric stroma, the hyperplastic epithelia that formed became dysplastic and soon progressed to form carcinomas. Insets at higher magnification. (From N.A. Bhowmick, A Chytil, D. Plieth et al., *Science* 303:848–851, 2004.)

hosts, the cells formed tumors in one-third of the time. These admixed fibroblasts clearly obviated the need of the MECs to spend time recruiting fibroblasts from host mice—a process that usually requires many weeks' time; in the absence of these fibroblasts, tumor growth could not take off.

This experiment did not address a related issue, specifically, whether the stromal cells of normal epithelial tissues can accelerate tumor formation as effectively as the stromal cells present in carcinomas. This issue has been addressed by comparing the actions of stromal fibroblasts extracted from a normal epithelial tissue with the carcinoma-associated fibroblasts (CAFs) prepared from carcinomas (in which myofibroblasts are abundant; see Figure 13.16). In one set of experiments, fibroblasts were purified from the stroma of normal human prostate glands, while the CAFs were prepared from the stroma of human prostate carcinomas. Each of these cell populations was then mixed with otherwise non-tumorigenic, SV40 large T antigen–immortalized human prostate epithelial cells and then implanted in immunocompromised Nude mice. The results, summarized in Figure 13.21, showed dramatic differences in the growth of these mixed tissue grafts. In particular, the grafts containing CAFs plus immortalized prostate epithelial cells formed tumors that were as much as 500 times larger than those containing normal prostate fibroblasts plus immortalized prostate epithelial cells. (When injected alone, the CAFs formed no tumors at all.)

This experiment demonstrated that these CAFs were functionally very different from the stromal fibroblasts present in normal prostatic tissue. Stated differently, during the course of tumor progression, stromal cells become increasingly

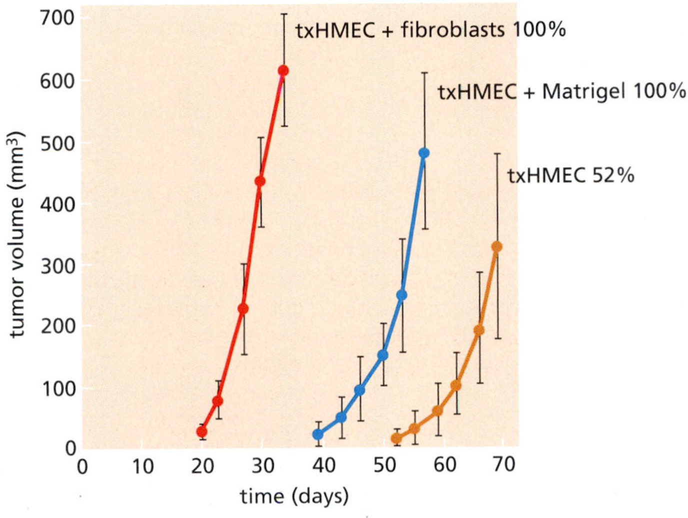

Figure 13.20 Admixed fibroblasts and tumor growth When human mammary epithelial cells (HMECs) were transformed through the introduction of the SV40 early region, the *hTERT* gene, and a weakly expressed *ras* oncogene (Section 11.12), the resulting transformed HMECs (txHMECs) formed tumors, albeit with a long lag time after injection, and then only in about half of the mice into which they had been injected *(orange curve)*. This slow development was accelerated a bit and tumor-forming efficiency was doubled when a preparation of extracellular matrix (Matrigel) produced by fibroblasts was mixed with these cells prior to introduction into host mice *(blue curve)*. However, when mammary stromal fibroblasts from a normal human breast were admixed to the transformed human MECs prior to injection, the cancer cells formed tumors rapidly and with 100% efficiency *(red curve)*. This illustrates that the recruitment of stromal fibroblasts is an important rate-limiting step in tumor formation. (From B. Elenbaas, L Spirio, F. Koerner et al., *Genes Dev.* 15:50–65, 2001.)

adept at helping their epithelial neighbors to survive and proliferate. Similar observations have since been made with CAFs extracted from human breast cancers.

Still, these experiments do not reveal precisely *how* the myofibroblast-rich CAF populations accelerate tumor growth. Once established within the tumor-associated stroma, the myofibroblasts are likely to confer multiple benefits on nearby epithelial cancer cells. Possibly the most important of these benefits is angiogenesis. Myofibroblasts aid tumorigenesis through their ability to release stroma-derived factor-1 (SDF-1), also called CXCL12; this factor (a chemokine) serves to recruit circulating endothelial precursor cells (EPCs) into the tumor stroma (Figure 13.22). The VEGF secreted by myofibroblasts helps to induce these recruits to differentiate into the endothelial cells that form the tumor **neovasculature**. Because angiogenesis is usually a rate-limiting step in tumor formation, the tumor-stimulating effects of admixed CAFs may be largely due to their ability to accelerate tumor angiogenesis.

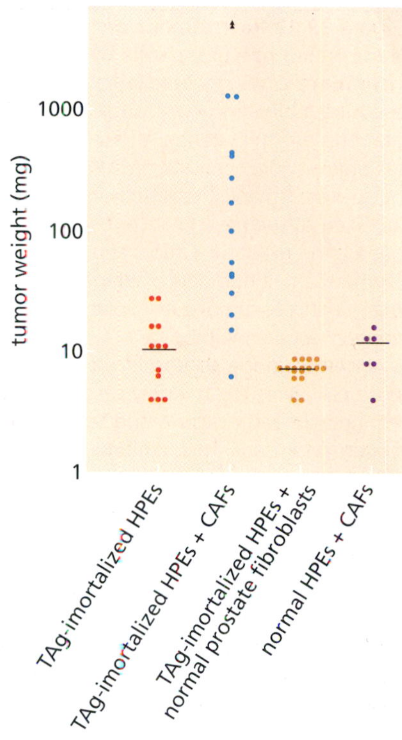

Figure 13.21 Effects of prostatic stromal cells on immortalized prostatic epithelial cells When normal human prostate epithelial cells (HPEs) that had been immortalized by SV40 T antigen (Tag-immortalized HPEs) were introduced into Nude mice, they formed growths of ~10 mg after many weeks, not significantly larger than the initial inoculum *(red dots)*. However, if these same cells were mixed with carcinoma-associated fibroblasts (CAFs) prepared from a human prostate carcinoma, tumors arose and the median weight of these growths was 10 times larger (Tag-immortalized HPEs + CAFs), *blue dots*). In contrast, when these T antigen–immortalized epithelial cells were mixed with stromal fibroblasts from a normal human prostate, the median weight of the resulting growth was ~7 mg, possibly even less than the starting inoculum *(orange dots)*. And if CAFs were mixed with normal human prostate epithelial cells prior to inoculation, the median weight after many weeks was, once again, only ~10 mg (*purple dots*). (From A.F. Olumi, G.D. Grossfeld, S.W. Hayward et al., *Cancer Res.* 59:5002–5011, 1999.)

In some tumors, the functionally altered stroma may be traced to a mechanism quite different from the one described above: stromal cells in advanced carcinomas, having coexisted with epithelial cancer cells for many years, may change their genotype and acquire traits that genetically normal stromal cells cannot achieve. This suggests that stromal cells co-evolve with their neoplastic neighbors during these long periods of tumor development by altering their genomes in order to adapt to the physiologic stresses present within tumors.

For example, analyses of a large group of human breast cancers for mutations in the *PTEN* and *TP53* tumor suppressor genes have shown the presence of somatically mutated alleles of these genes in stromal cells isolated from the tumors. These experiments exploited the procedure of *laser capture microdissection* (LCM; Figure 13.23) to isolate small patches of epithelial or stromal cells from these tumors. In some tumors, distinct mutant *TP53* alleles were found in the epithelial and in the stromal compartments. In others, mutant alleles were found in one cell population but not the other. These experiments have opened the door to the possibility that the well-documented genetic evolution of neoplastic epithelial cells (Chapter 11) is often accompanied by changes in the genomes of nearby stromal cells.

13.5 Macrophages represent important participants in activating the tumor-associated stroma

The active recruitment of macrophages into tumor masses would seem, at first glance, to be counterproductive for tumor formation, since macrophages are usually deployed by the immune system to scavenge and destroy infectious agents and abnormal cells, as we will explore in more detail in Chapter 15. However, increasing evidence indicates that these immune cells also play important roles in *furthering* tumor development.

In more detail, monocytes from the myeloid lineage in the bone marrow enter into the general circulation, from which they are recruited by cancer cells into a tumor; once ensconced there, the monocytes are induced to differentiate into macrophages (see Figure 13.14). This recruitment depends on attraction signals that are conveyed by **chemotactic** factors. By definition, these factors provide directional cues to motile cells rather than mitogenic stimulation. In the case of **leukocytes** (white blood cells such as monocytes), the relevant chemotactic factors are often called **chemokines**, that is, chemotactic cytokines.

The chemokine known as monocyte chemotactic protein-1 (sometimes called macrophage chemoattractant protein; MCP-1) is expressed in significant quantities by a wide range of neuroectodermal and epithelial cancer cell types. It

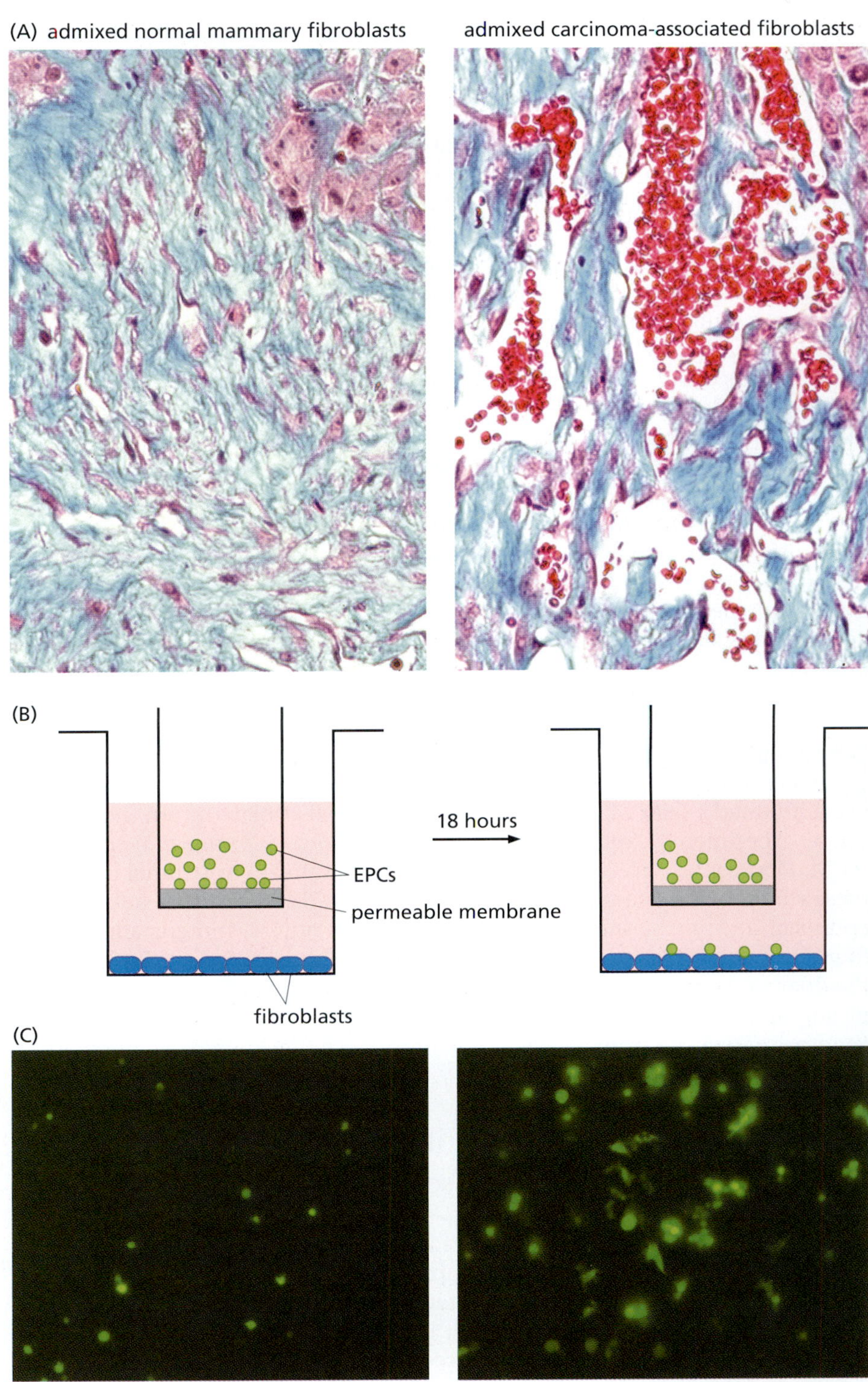

Figure 13.22 Recruitment of endothelial precursor cells by mammary CAFs (A) Tumors have been assembled by admixing stromal fibroblasts to cells of the human MCF7-ras breast cancer cell line. When normal human mammary stromal fibroblasts are admixed to the breast cancer cells *(left panel)*, the resulting tumors show relatively small numbers of blood vessels *(red)* amid the tumor-associated stroma *(blue)*. However, admixture of carcinoma-associated fibroblasts (CAFs; *right panel)* results in highly vascularized tumors with large vessels and associated erythrocytes *(red)*. The resulting access to the circulation is likely to greatly facilitate tumor growth. (B) The ability of the CAFs, which are essentially myofibroblasts, to attract endothelial cells could be demonstrated by an *in vitro* experiment, in which green fluorescent protein (GFP)–labeled bone marrow cells expressing surface antigens characteristic of endothelial precursor cells (EPCs) were placed above a permeable membrane in a Boyden chamber. Either normal mammary stromal fibroblasts or CAFs from a breast cancer were placed on the bottom surface of the lower chamber. After 18 hours, the number of EPCs that had been recruited to the bottom of the lower chamber was gauged by fluorescence microscopy. (C) Analysis of the cells attached to the lower chamber indicated that the CAFs *(right panel)* are able to attract far more GFP-labeled EPCs than the normal mammary stromal fibroblasts *(left panel)*. This recruitment could be reduced by 60% by placing antiserum that neutralized the SDF-1/CXCL12 chemokine released by the CAFs in these chambers, indicating its important role in mediating this recruitment. (From A. Orimo, P.B. Gupta, D.C. Sgroi et al., *Cell* 121:1–14, 2005.)

appears to be a critical signal for attracting monocytes to some tumors and inducing their differentiation into macrophages. In other tumors, vascular endothelial growth factor (VEGF), colony-stimulating factor-1 (CSF-1; often called M-CSF, macrophage-colony–stimulating factor), and the PDGF released by tumor cells also seem to help in this recruitment, while CSF-1 aids in stimulating the monocyte-to-macrophage differentiation (Figure 13.24).

Once established in the tumor stroma, macrophages play important roles in stimulating angiogenesis. Thus, in some cancers, such as breast carcinomas, there is a direct correlation between the level of MCP-1 produced, the number of macrophages present, and the level of angiogenesis induced by the various

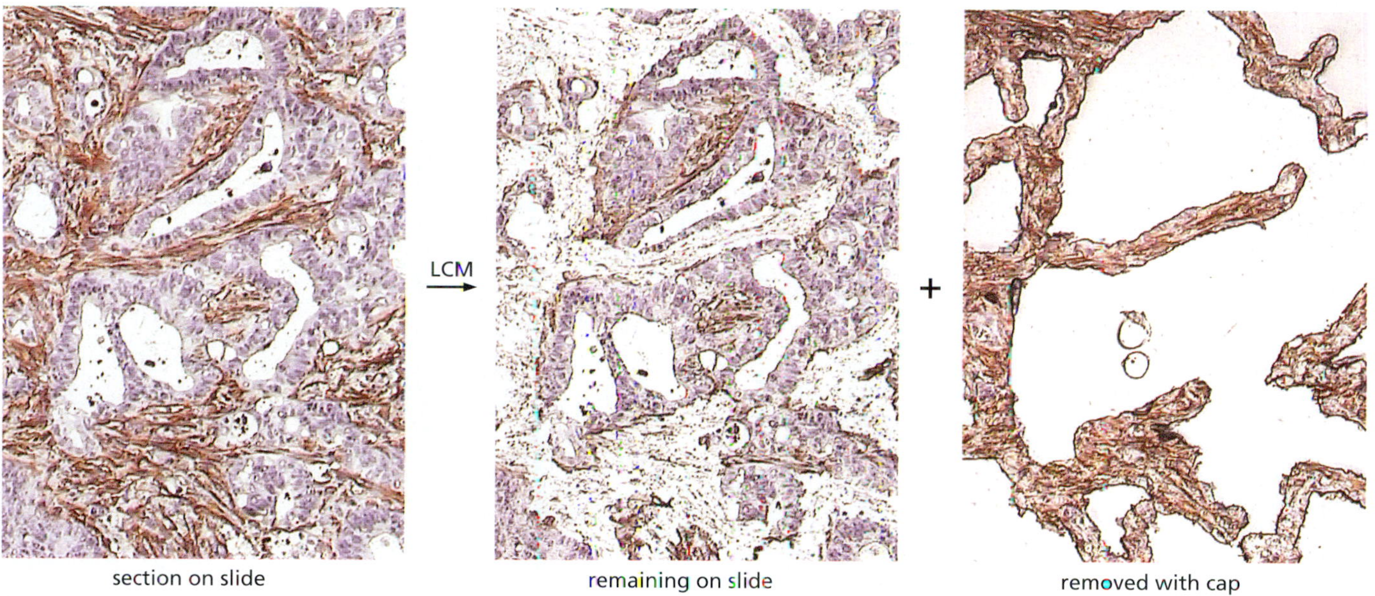

Figure 13.23 Laser capture microdissection The procedure of laser capture microdissection (LCM) allows an investigator to use a laser beam to microdissect tissue sections that have been affixed to a microscope slide. As part of the LCM procedure, a transparent material, termed a "cap," is layered over a microscope slide to which a tissue section has been affixed. A motor-driven laser beam then irradiates chosen areas, which adhere to the cap; the cap can then be lifted with the adherent cells, which can then be analyzed biochemically. As seen here, the section of a human colon carcinoma shown in the *left panel* contains both epithelial regions *(light blue)* and stromal regions *(brown)*. In this instance, the LCM procedure caused the epithelial areas to be left behind on the slide *(middle panel)* while the stromal regions were lifted up *(right panel)*. Both groups of cells could then be studied by various analytical procedures, including gene expression arrays. (Courtesy of H. Dolznig.)

tumors; in other tumors, the density of infiltrating tumor-associated macrophages is correlated with microvessel density (see, for example, Figure 13.25A, B, and C; this density can be gauged by counting the number of capillaries per microscopic field in a tumor section.) A major role of macrophages in driving tumor progression is suggested by numerous reports demonstrating a direct correlation between the presence of a high density of macrophages in tumor masses and a poor prognosis for cancer patients carrying such tumors. By now, this correlation has been documented in gliomas and in carcinomas of the breast, ovary, prostate, cervix, bladder, and lung.

Such evidence, however, is only *correlative*. More compelling evidence of the *causal* role of macrophages, at least in tumor angiogenesis, is supported by experiments in which cancer cells are forced experimentally to express higher levels of MCP-1. The expression of this chemokine allows the cancer cells to attract more macrophages, which proceed to secrete important angiogenic factors, notably VEGF and interleukin-8 (IL-8), resulting in marked increases in angiogenic activity and the formation of more extensive tumor-associated vasculature. In addition, when lung cancer cells are exposed *in vitro* to factors that

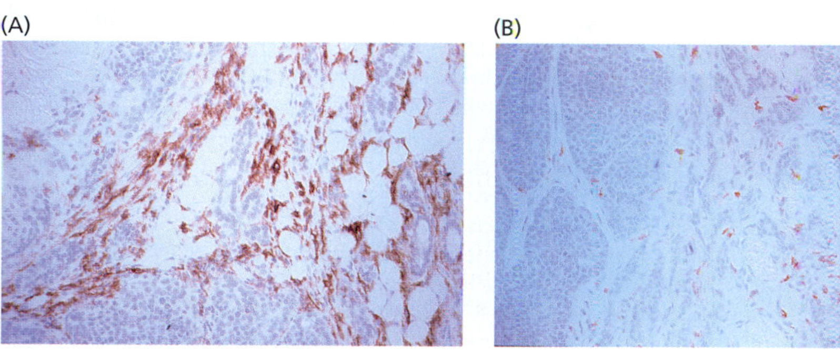

Figure 13.24 Recruitment of macrophages by CSF-1 In some tumors, colony-stimulating factor-1 (CSF-1) released by tumor cells serves as the key attractant for monocytes and macrophages. In the transgenic mice studied here, mammary tumorigenesis was driven by a transgene, which was constructed from an MMTV (mouse mammary tumor virus) transcriptional promoter that drives expression of the polyoma middle T (PyMT) oncogene. (A) This tumor arose in a transgenic PyMT mouse that was heterozygous at the *Csf1* locus, since it carried one null allele of the CSF-1–encoding gene, termed $Csf1^{op}$. The presence of macrophages in the stroma of a tumor is revealed by a monoclonal antibody that detects macrophages *(reddish brown)*. (B) This tumor arose in a transgenic mouse that was genetically $Csf1^{op/op}$ and therefore lacked all CSF-1 activity. Consequently, the tumor cells in the mice also were unable to release this chemoattractant. In this mammary tumor, the macrophages are barely detectable. The absence of macrophages did not affect the growth of the primary tumor but strongly suppressed progression to a metastatic state. (From E.Y. Lin, A.V. Nguyen, R.G. Russell et al., *J. Exp. Med.* 193:727–740, 2001.)

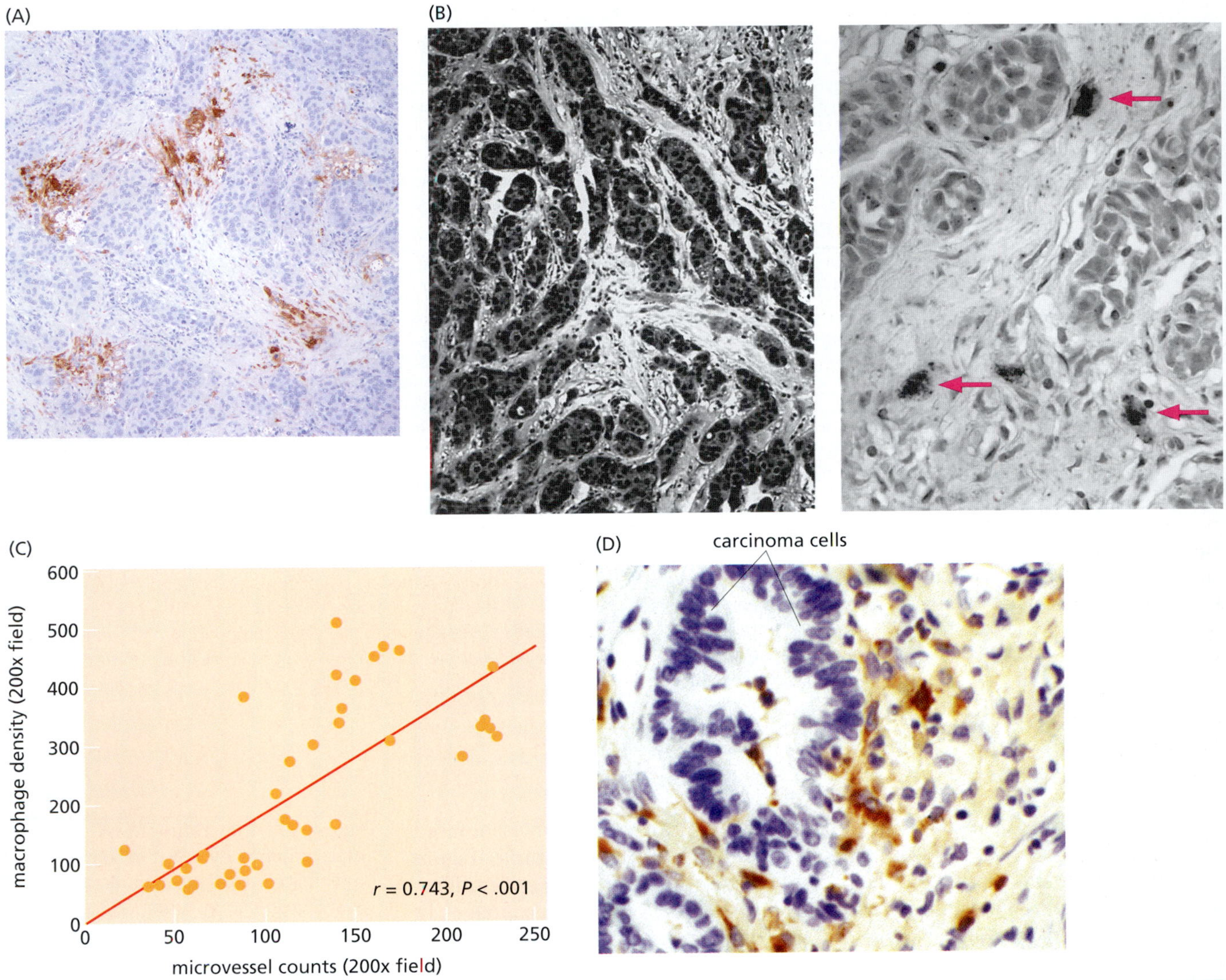

(A)

(B)

(C)

macrophage density (200x field)

600
500
400
300
200
100
0

0 50 100 150 200 250

microvessel counts (200x field)

$r = 0.743$, $P < .001$

(D)

carcinoma cells

(E)

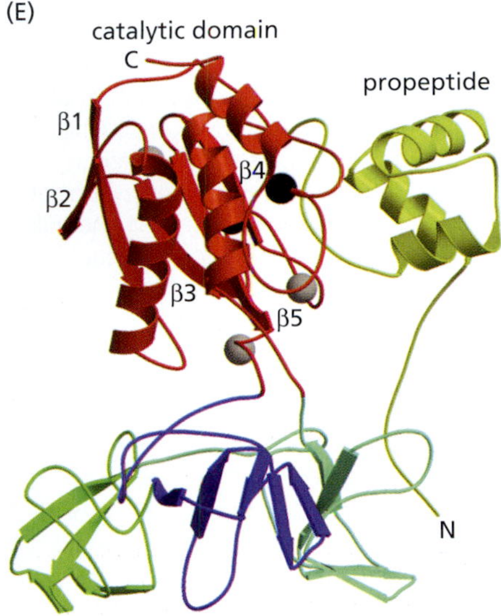

Figure 13.25 Macrophage proteins and angiogenesis in the tumor stroma (A) The presence of tumor-associated macrophages (TAMs; *red*) can be demonstrated, as is seen here in this section of a human breast cancer, through the use of an antibody that specifically recognizes the HIF-2α protein displayed by these cells in the stroma of breast cancers. Like its cousin, HIF-1α (see Section 7.12), HIF-2α is involved in the up-regulation of angiogenic proteins in hypoxic cells. (B) In certain breast cancers *(left panel)*, the cancer cells synthesize the potent angiogenic factor VEGF (vascular endothelial growth factor); as seen here, the islands of carcinoma cells stain positively *(dark areas)* with an anti-VEGF monoclonal antibody. However, in other breast tumors *(right panel)*, VEGF production originates in isolated macrophages within the tumor-associated stroma *(arrows)*. (C) In a series of human non-small-cell lung carcinoma (NSCLC) specimens, each represented here by a point on the graph, the density of TAMs per microscope field has been plotted versus the density of microvessels (i.e., capillaries) per microscope field. Such correlations add further weight to the notion that TAMs play a key role in fostering tumor-associated angiogenesis. (D) Macrophages in the stroma of a human colorectal adenocarcinoma produce matrix metalloproteinase-9 (MMP-9; *brown spots*), a key enzyme in angiogenesis and invasiveness. MMP-9 can promote angiogenesis by liberating angiogenic factors from the extracellular matrix. (E) MMP-9 is synthesized as a pro-enzyme that becomes activated through cleavage by another protease, which results in release of a propeptide *(yellow green, right)*. The active MMP-9 is then able to cleave a diverse set of substrates in the intercellular space. The zinc ion, which gives the MMPs their name, is shown as a *black sphere*; calcium ions are seen as *gray spheres*. (A, from R.D. Leek and A.L. Harris, *J. Mammary Gland Biol. Neoplasia* 7:177–189, 2002; B, from R.D. Leek, N.C. Hunt et al., *J. Pathol.* 190:430–436, 2000; C, from J.J.W. Chen, Y. C. Lin, P.L. Yao et al., *J. Clin. Oncol.* 22:953–964, 2005; D, from B.S. Nielsen, S. Timshel, L. Kjeldsen et al., *Int. J. Cancer* 65:57–62, 1996; E, from P.A. Elkins, Y.S. Ho, W.W. Smith et al., *Acta Crystal. Series D* 58:1182–1192, 2002.)

have been secreted by cultured macrophages, the cancer cells respond by producing IL-8 and a number of other proteins that promote angiogenesis and cell invasiveness.

Hypoxic areas within tumors attract macrophages, which appear to tolerate hypoxia quite well. Once established in hypoxic regions of tumors, these inflammatory cells begin to secrete significant amounts of VEGF (see Figure 13.25B and Section 7.12), which reduces the hypoxia by bringing in endothelial cells, capillaries, and thus oxygen-rich blood. The fact that hypoxic areas of tumors often remain poorly vascularized indicates that the recruited macrophages, on their own, are unable to fully cure local defects in angiogenesis. Thus, in rapidly growing tumors, the macrophages and the vasculature that they induce cannot keep pace with the rate of tumor expansion, and large areas within a tumor mass eventually become *necrotic* (i.e., filled with dead and dying cells) as a consequence.

Like myofibroblasts (Section 13.3), macrophages are adept at secreting matrix metalloproteinases (MMPs; see Table 13.1). MMP-9, prominently involved in cancer progression, is produced by tumor-associated macrophages (TAMs; Figure 13.25D) and, once activated, proceeds to cleave a number of important protein substrates in the extracellular space. In certain invasive carcinomas, including those of the breast, bladder, and ovary, TAMs have proven to be the major source of this enzyme. It contributes to tumor progression by enhancing angiogenesis, by disrupting existing tissue structure and thereby carving out space for expanding tumor masses, and by liberating critical mitogens that have been immobilized through tethering to proteoglycans of the ECM. MMP-9 can also cleave IGFBPs (insulin-like growth factor-binding proteins; Sidebar 9.7), which normally sequester IGF molecules in the extracellular space. This liberates the IGFs, notably IGF-1, which then provide survival signals to nearby cells, including cancer cells (Figure 13.26). In advanced breast cancers, tumor-

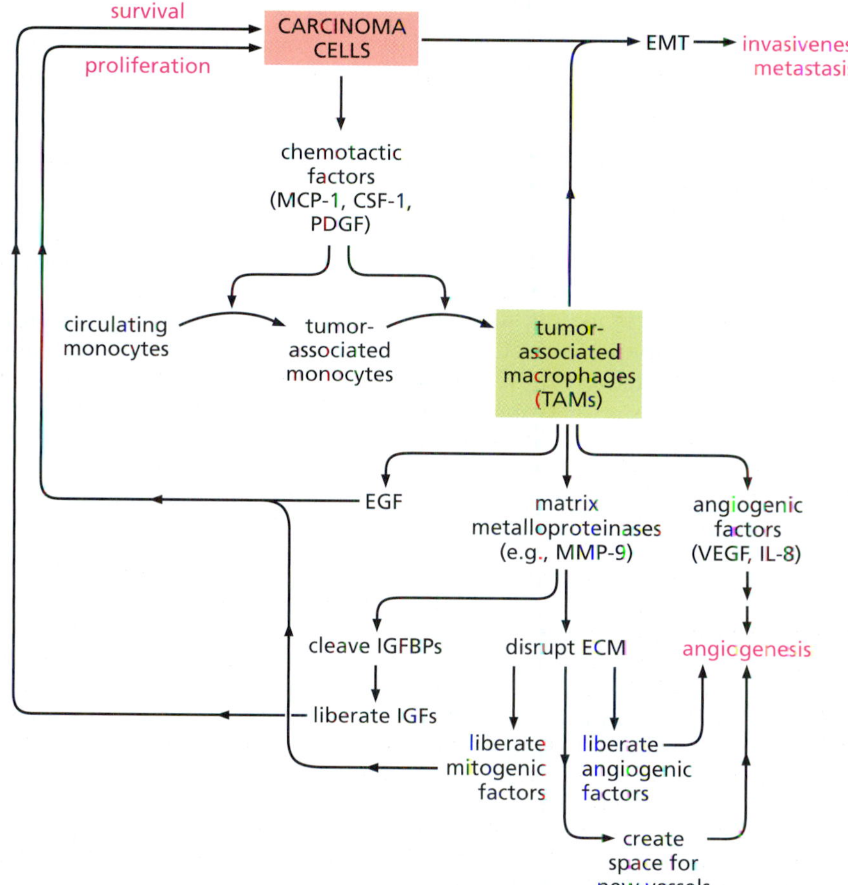

Figure 13.26 Contribution of macrophages to tumorigenesis Macrophages play a major role in releasing mitogenic factors for carcinoma cells as well as reorganizing the tumor stroma in order to facilitate angiogenesis and, in some tumors, carcinoma cell invasiveness.

associated macrophages are also able to help the carcinoma cells directly, since they are the major source of the epidermal growth factor (EGF) that drives the proliferation of EGF-R–expressing carcinoma cells.

While there is abundant evidence implicating some macrophages as active collaborators in tumor progression, it is also clear that another, morphologically distinct subset of macrophages, by acting as deputies of the immune system, can detect and kill cancer cells. Yet they clearly fail to do so in many types of tumors. It may be that some cancer cells acquire the ability to inactivate or blunt the **tumoricidal** (cancer-killing) actions of this second class of macrophages, while leaving intact the functions of the first type that are involved in aiding tumor progression. Alternatively, cancer cells may engage in evasive maneuvers that render them invisible to the macrophages that have been dispatched by the immune system to destroy them. We will return to the role of macrophages and other types of immune cells in blocking tumor development in Chapter 15.

13.6 Endothelial cells and the vessels that they form ensure tumors adequate access to the circulation

The most obvious stromal support required by tumors has already been cited repeatedly in this chapter: like normal tissues, tumors require access to the circulation in order to grow and survive. As early as the mid-1950s, pathologists noted that cancer cells grew preferentially around blood vessels. Those tumor cells that were located more than about 0.2 mm away from blood vessels were found to be nongrowing, while others even farther away were seen to be dying (Figure 13.27A).

We now realize that this threshold of approximately 0.2 mm represents the distance that oxygen can effectively diffuse through living tissues. Cells located within this radius from a blood vessel can rely on diffusion to guarantee them oxygen; those situated further away suffer from moderate or severe hypoxia (Figure 13.27B and C). Tissues suffering from hypoxia are in danger of becoming necrotic, as discussed above and illustrated in Figures 13.27 and 13.28. p53-triggered apoptotic death also threatens hypoxic cells (Chapter 9), and the inactivation of the p53 signaling system often enables cancer cells to survive beyond the small perimeter surrounding each capillary.

Hypoxia is only one price that is paid by cells lacking close proximity to the circulatory system. Cells also require effective interactions with the vasculature in order to acquire nutrients and to shed metabolic waste products and carbon dioxide. The capillary networks threading their way through many normal tissues are arrayed so densely that virtually all cells in a tissue are no more than several cell diameters away from the nearest capillary (Figure 13.29). In some normal tissues with an especially high metabolic activity, most cells enjoy direct contact with at least one capillary. This intimate association means that their access to oxygen and critical nutrients is not dependent on the diffusion of these molecules over large distances and through densely packed cell layers.

These observations illustrate the central importance of the vasculature to the growth and survival of all types of tissue, normal and neoplastic. The process of developing this vasculature through angiogenesis can be observed during embryonic development, implantation of the placenta, wound healing, and certain pathological conditions such as diabetic retinopathy, psoriasis, rheumatoid arthritis, and, of course, tumorigenesis.

During embryonic development, the overall size and architecture of an organ or tissue is genetically programmed, but the precise locations of its individual component cells are not. This applies as well to the vasculature: the genome determines the overall layout of the major vessels and delegates to the individual cells within the tissue the task of designing and assembling local capillary networks and slightly larger vessels. Heterotypic interactions operating over short distances determine the routing of individual capillaries constituting the microvasculature. More specifically, capillaries appear to form wherever they are needed by the nonvascular (i.e., **parenchymal**) cells in a tissue.

A variant of this plan operates during tumorigenesis. The overall architecture of tumors cannot be determined by the genome of an organism, since each tumor

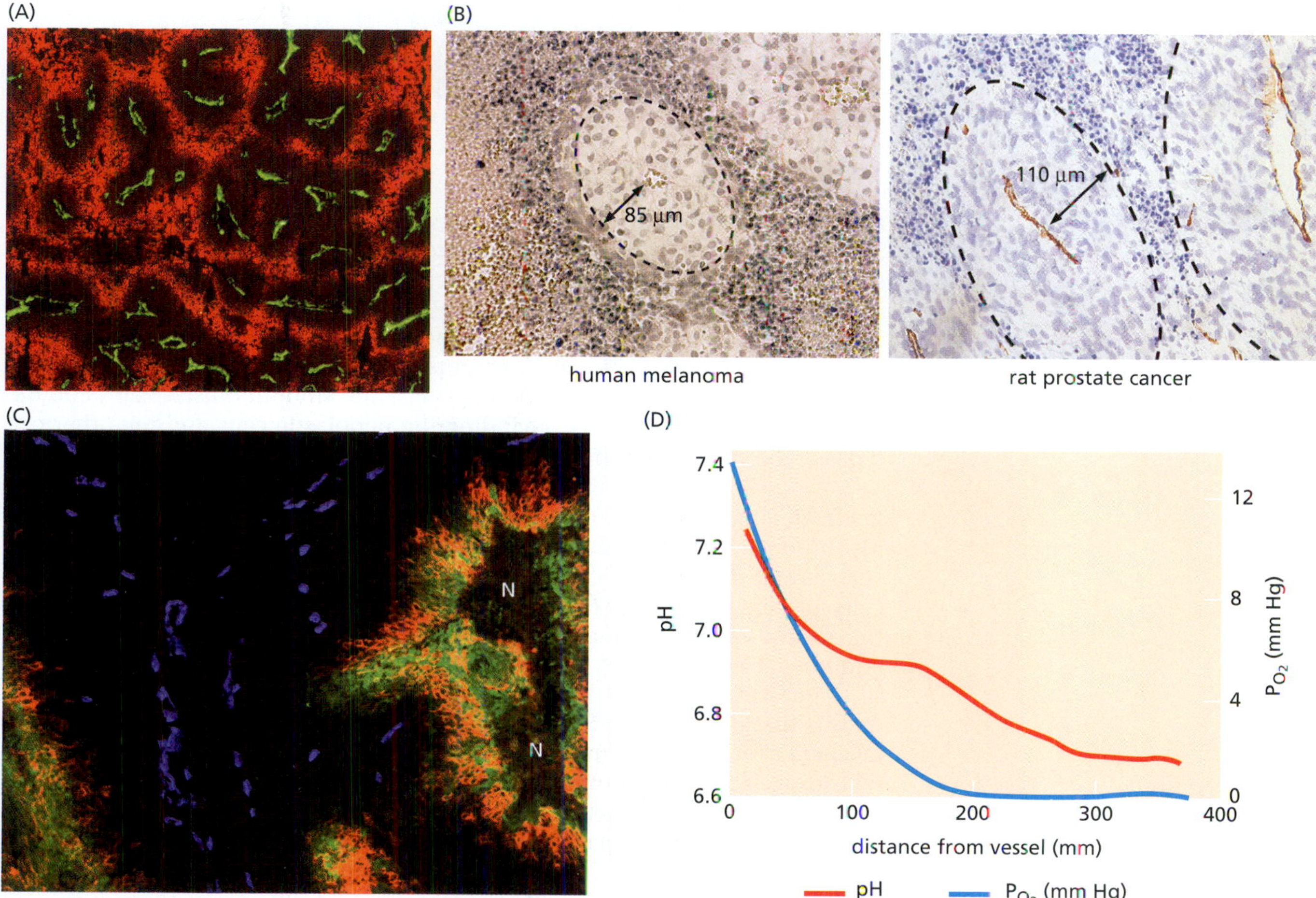

Figure 13.27 Hypoxia and necrosis of cells in poorly vascularized sections of tissues Various techniques have been used to demonstrate hypoxia and necrosis in regions of tissues surrounding capillaries. (A) This micrographic image reveals capillaries *(green)* as well as the degrees of oxygenation in the surrounding tumor parenchyma. Immunostaining using an antibody reactive with EF5, a molecule that localizes to hypoxic regions of tissues, reveals areas that are hypoxic *(red)* at some distance from the capillaries, while those cells that are well oxygenated are unstained and therefore appear *dark brown*. (B) These sections of a human melanoma *(left panel)* and rat prostate carcinoma *(right panel)* have been immunostained with an anti-CD31 antibody, which reveals capillaries *(brown)*. Immediately around each capillary, a region of healthy cells is apparent. However, beyond a perimeter *(dashed line)* surrounding each capillary a region of necrosis *(granular area)* is apparent. The necrotic region begins as close as 85 µm from the melanoma capillary *(left)* and 110 µm from the prostate carcinoma capillary *(right)*. This necrosis reveals the limitations of diffusion in conveying oxygen and nutrients from capillaries to cells in the tumor parenchyma. (C) The dynamics of oxygenation on a larger scale are apparent here in this human squamous cell head and neck tumor. Blood vessels *(blue spots)* provide good oxygenation of the tumor, which appears *black*. However, in areas of poor vascularization and moderate hypoxia, a carbonic anhydrase enzyme is expressed *(red)*, while in areas of extreme hypoxia, the pimonidazole dye is detectable *(green)*. Overlap between these two markers appears *orange*. Areas of necrosis, located even further away from the tumor vasculature, are indicated by "N." (D) Use of indicator dyes and intra-vital microscopy makes it possible to measure pH and P_{O_2} (partial oxygen tension) in microenvironments located at various distances from a blood vessel. These measures explain the prime causes of necrosis in poorly vascularized areas: oxygen pressure falls to zero at a distance of more than 0.2 mm, while pH drops to levels that are incompatible with normal cellular metabolism, largely a consequence of lactic acid production occurring under anaerobic conditions. (A, courtesy of B.M. Fenton; B, left, courtesy of N. Almog and M.J. Folkman; B, right, from L. Hltaky, P. Hahnfeldt and J. Folkman *J. Natl. Cancer Inst.* 94:883–893, 2002; C, from J.H. Kaanders, K.J. Wijffels, H.A. Marres et al., *Cancer Res.* 62:7066–7074, 2002; D, from G. Helmlinger, F. Yuan, M. Dellian and R.K. Jain, *Nat. Med.* 3:177–182, 1997.)

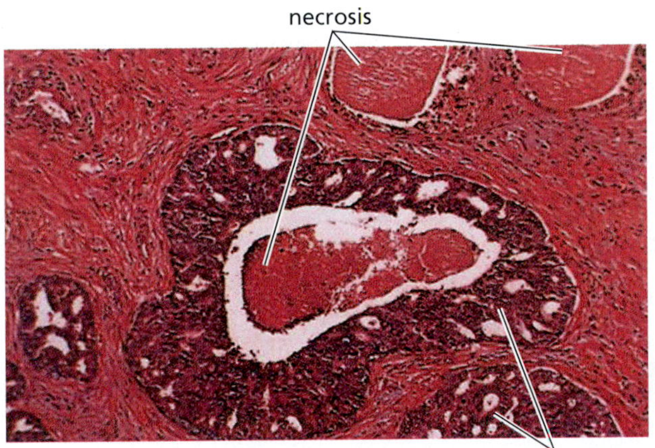

Figure 13.28 Large-scale necrosis within a tumor The inability of large sectors of a tumor to gain access to the vasculature present in the stroma is demonstrated by this high-grade adenocarcinoma of the prostate. As seen most clearly in the growth in the *center*, a nest of adenocarcinoma cells *(dark red)* is surrounded by extensive stroma *(light red)*. In the middle of the nest of adenocarcinoma cells is a large clump of necrotic tissue *(light red)* that has shrunk away from the viable adenocarcinoma cells that surround it; the latter thrive because they have better access to the stroma and associated vasculature. (From A.T. Skarin, Atlas of Diagnostic Oncology, 3rd ed. Philadelphia: Elsevier Science Ltd., 2003.)

that arises is a novel invention whose design is not anticipated by the genome. This has implications for tumor angiogenesis as well: tumors cannot rely on some genetic blueprint to aid them in configuring their blood supply and must, instead, design the layout of their own vasculature, doing so step-by-step as they grow. In particular, when assembling their own blood supply, tumors depend totally on the localized heterotypic interactions between the cells of the vasculature (including endothelial cells, pericytes, and smooth muscle cells) and the nonvascular cells in tumors (including the neoplastic cells and the other cell types of the supporting stroma).

We have already learned of two ways by which tumors assemble vasculature. Myofibroblasts in the tumor-associated stroma can release chemotactic signals, such as SDF-1/CXCL12, which helps to recruit circulating endothelial precursor cells into the stroma (see Figure 13.22). This recruitment is likely to be aided by the release of VEGF, which also helps these cells to mature into functional endothelial cells.

As was first described in Section 7.12, production of VEGF is governed by the availability of oxygen. Thus, many types of cells take stock of their own intracellular oxygen tension through the actions of the VHL protein and its partners. Under conditions of hypoxia, this complex of proteins allows functional HIF-1 transcription factor to accumulate, which in turn drives the expression of a

(A)

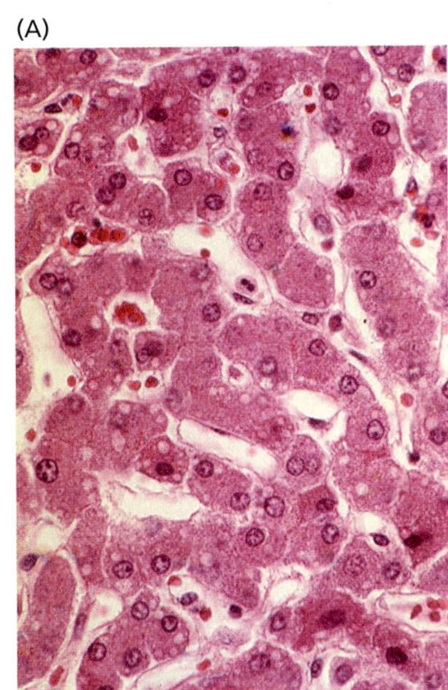

(B)

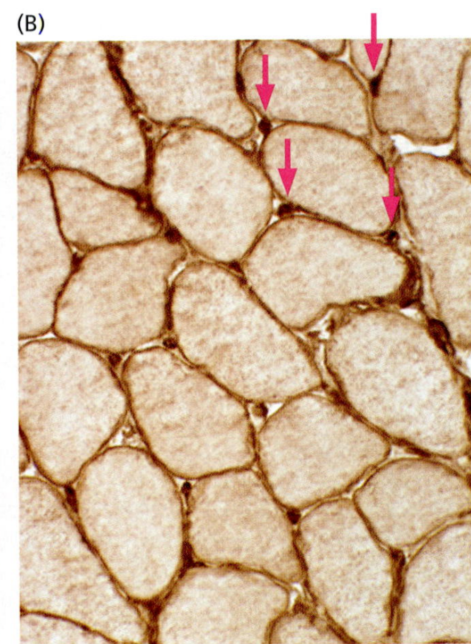

Figure 13.29 Dense microvasculature in liver and muscle Because of high metabolic rates, certain tissues develop an extremely high density of capillaries, ensuring almost every cell direct contact with endothelial cells and the capillaries that they form. (A) Virtually every hepatocyte in the liver is adjacent to a space, termed a *sinusoid*, that is lined with endothelial cells and serves as a capillary. (B) Each muscle fiber, immunostained with an anti-laminin antibody, is a single syncytial cell, formed by the fusion of a number of myoblasts. Each muscle fiber lies adjacent to one or more capillaries *(arrows, dark brown spots)*. (A, from Loyola University Medical Education Network; B, from Washington University of St. Louis Neuromuscular Disease Center.)

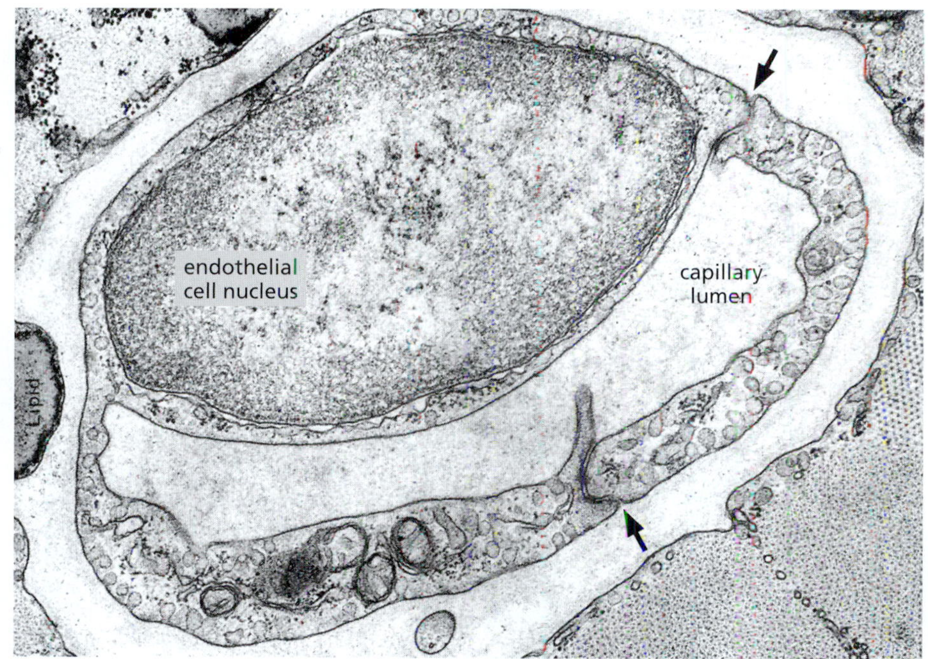

Figure 13.30 Endothelial cells and the formation of capillaries This transmission electron micrograph reveals the cross section through a capillary in the midst of cardiac muscle and shows how the cytoplasms of two endothelial cells have become joined *(arrows)* to form the lumen of a capillary. The tight sealing of the plasma membranes of adjacent endothelial cells to one another results in capillary tubes whose walls are continuous and usually without gaps. The nucleus of one of the endothelial cells is also seen here. (From D.W. Fawcett, *J. Histochem. Cytochem.* 13:75–91, 1965.)

number of genes whose products encourage angiogenesis. Prominent among these is VEGF. Production of this key angiogenic factor has been associated with tumor cells, macrophages, and myofibroblasts, depending on the tumor type and its stage of progression. (In fact, there are two structurally related, functionally similar proteins termed VEGF-A and VEGF-B. We will continue here to use the generic term VEGF to refer to both of these.)

Like many other growth factors, VEGF functions as a ligand of tyrosine kinase receptors—in this case, VEGF receptor-1 (also termed Flt-1) and VEGF receptor-2 (known also as Flk-1/KDR). Both are displayed on the surfaces of endothelial cells. Similarly, basic fibroblast growth factor (bFGF), another important angiogenic factor, binds to its own cognate receptors displayed by endothelial cells. Once stimulated by these angiogenic factors, the endothelial cells proliferate and contort their cytoplasms to construct the cylindrical walls of capillaries (Figure 13.30). These capillaries also penetrate through existing tissue layers, moving toward the highest localized concentration of angiogenic factors.

Similar mechanisms appear to operate during the formation of lymphatic vessels (Sidebar 13.5). As we will see later in this chapter, lymph ducts are important regulators of fluid balance in the tumor stroma. In the next chapter, they assume great importance, because they provide avenues for cancer cells to escape primary tumors and disseminate to distant sites in the body.

The existence and powers of angiogenic factors were first revealed by implanting small chunks of tumor on the cornea or the ears of laboratory animals such as rabbits. Within a matter of days, dense networks of capillaries and larger vessels were seen to emerge from preexisting capillary beds and to converge on the implanted tumor chunks. Images like these (Figure 13.32A) have strongly influenced our thinking about the behavior of tumors and their vasculature, because they demonstrate so vividly that tumors actively recruit blood vessels into their midst.

The formation of blood capillaries is actually far more complicated than is suggested by these descriptions of angiogenesis. On the one hand, this complex morphogenetic process involves a number of other factors in addition to the VEGFs, which clearly play a major role in attracting blood vessels to tumors (Figure 13.32B). These other factors include several forms of TGF-β, basic

Sidebar 13.5 Endothelial cells also construct lymph ducts The lymph ducts have two major functions in normal physiology. They drain fluid from the interstices between cells and empty this fluid into the venous circulation. In addition, they allow antigen-presenting cells of the immune system to convey antigens from various tissues to the lymph nodes, where immune responses are often initiated (Chapter 15).

Interestingly, lymph ducts are assembled from endothelial cells originating in the same embryonic stem cell population that yields the endothelial cells of capillaries and larger blood vessels. During embryonic development, lymphatic vessels can often be observed to bud from developing capillaries before they separate and construct their own parallel network of interconnecting vessels (Figure 13.31). In addition, the factors that stimulate **lymphangiogenesis**—vascular endothelial growth factors C and D (VEGF-C and VEGF-D)—are homologous to VEGF-A and -B, which play a major role in stimulating the angiogenesis that creates the blood vasculature.

As might be expected, the receptor for VEGF-C and VEGF-D displayed by lymphatic endothelial cells—VEGF receptor 3 (VEGF-R3)—is structurally related to the dominant VEGF receptor of blood capillaries, VEGF-R2. In addition, there is clear evidence that VEGF-D, which is mainly responsible for driving lymphangiogenesis, may also help stimulate angiogenesis by binding and activating the VEGF-R2. So these two systems—the blood and lymphatic networks—derive from common evolutionary roots, develop from common precursors in the embryo, and continue to interact with one another in complex ways within adult tissues.

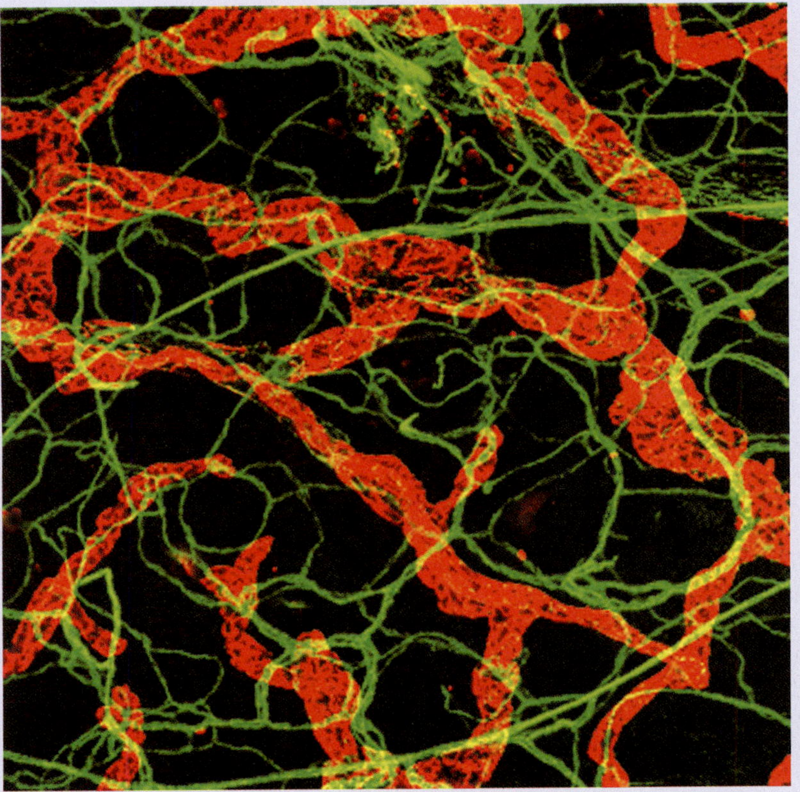

Figure 13.31 Networks of lymphatic vessels and capillaries Most normal tissues are interlaced with networks of capillaries *(green)* and lymphatic vessels *(red orange)*. As is apparent here, the diameters of lymphatic vessels are far larger than those of capillaries. Unlike capillaries, however, the endothelial cells of lymphatic vessels are not supported by underlying mural cells—pericytes and smooth muscle cells. (Courtesy of T. Tammela and K. Alitalo.)

fibroblast growth factors (bFGFs), interleukin-8 (IL-8), angiopoietin, angiogenin, and PDGF. Moreover, several distinct cell types in addition to endothelial cells contribute to the construction of capillaries and larger vessels. As mentioned, the endothelial cells form the lumen of a capillary; these cells, in turn, are surrounded by the mural pericytes and vascular smooth muscle cells (see Figure 13.6).

The systematic covering of capillaries by pericytes seen in normal tissues (see 13.6B) can be contrasted with their chaotic dispersion near tumor-associated capillaries (Figure 13.33). Capillaries in tumors typically have diameters that are three times greater than their normal counterparts. In addition, the overall layout of blood vessels around and within tumor masses is quite chaotic (Figure 13.34). Often vessels stop abruptly in dead-end pouches or circle back and attach to themselves. At the submicroscopic level, it is also apparent that the capillaries in tumor masses are assembled haphazardly. This is because the plasma membranes of adjacent endothelial cells do not contact one another to form a seamless lining around the capillary lumen, but instead leave gaps, often several microns wide (Figure 13.35), which allow direct access of the blood plasma to the cells surrounding the capillary. The resulting leakage seems to be responsible for the deposition of fibrin in the tumor parenchyma described earlier.

Quantitative measures indicate that the walls of capillaries in tumors are about 10 times more permeable than those of normal capillaries. Some of this leakiness is attributable to the defective assembly of capillary walls noted here. But this, on its own, does not suffice to explain all of this leakiness. Instead, most of seems to be due to the deregulated production of VEGF within tumors (Sidebar 13.6).

The precise reasons why capillaries and larger vessels within tumors are so haphazardly constructed are unclear. One possible factor may lie in the balance between two mutually antagonistic growth factors, angiopoietin-1 and -2. While VEGF is responsible for initiating the growth of capillaries by attracting and stimulating endothelial cells, a mix of angiopoietin-1 and -2 induces endothelial cells to recruit the pericytes and smooth muscle cells that enable newly formed capillaries to mature into well-constructed vessels containing appropriate proportions of these three cell types. An imbalance of these two antagonistic

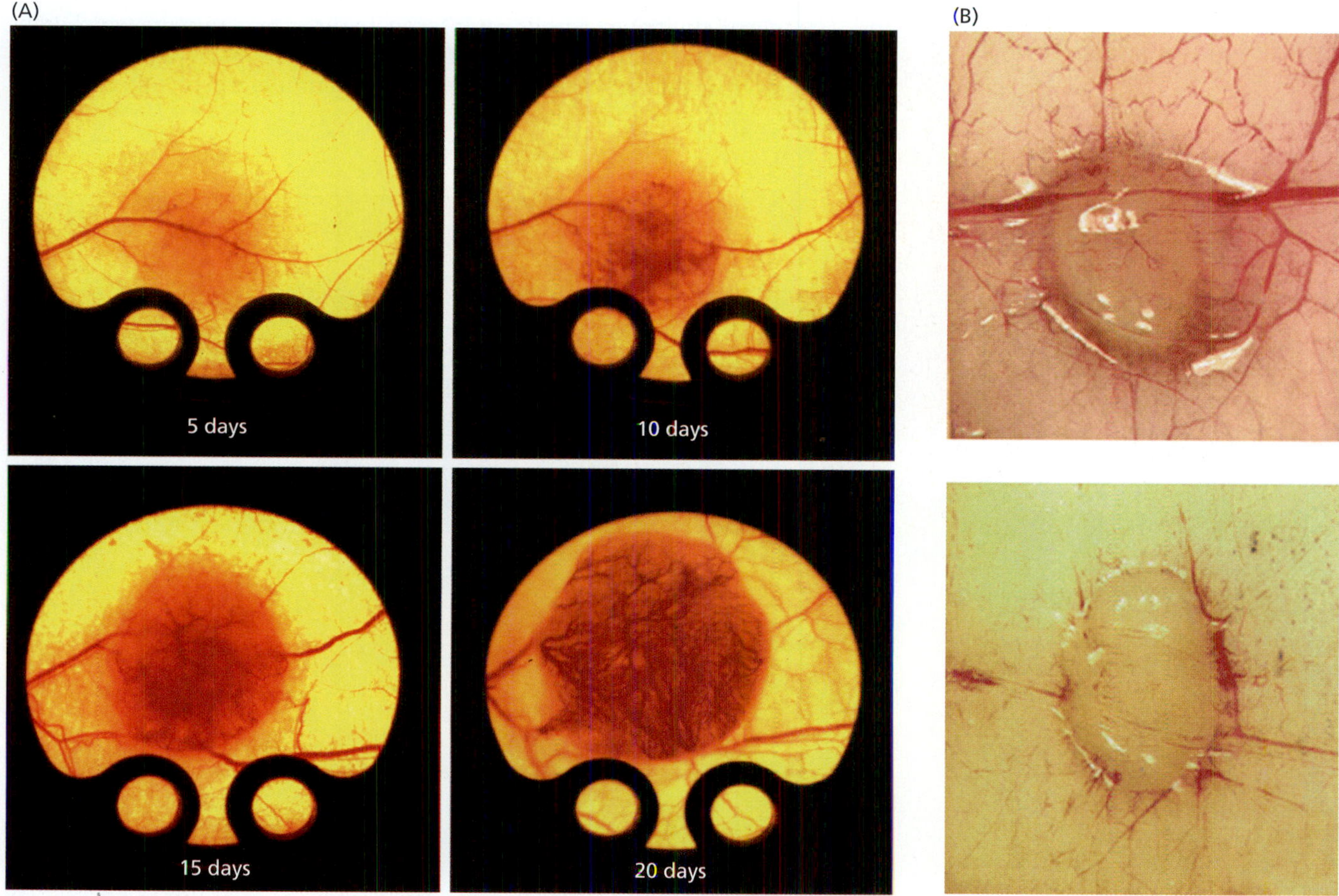

Figure 13.32 Recruitment of capillaries by an implanted tumor (A) These images show the growth of a small group of subcutaneously implanted human colorectal adenocarcinoma cells over a period of 20 days. The growth of the tumor-associated vessels is observed through a window that has been inserted in the skin of the mouse above the tumor. (B) Such vascularization can be suppressed by antagonists of the VEGF receptor. Seen here are the effects of ZD6474, an inhibitor of the VEGF receptor, in mice bearing human adenocarcinoma cell xenografts. The widespread pink background in the untreated control mouse, which is indicative of numerous capillaries, and the number of major vessels entering into a tumor mass *(top) are* strongly decreased in the presence of the drug *(bottom)*. (A, from M. Leunig, F. Yuan, M.D. Menger et al., *Cancer Res.* 52:6553–6560, 1992; B, from S.R. Wedge, D.J. Ogilvie, M. Dukes et al., *Cancer Res.* 62:4645–4655, 2002.)

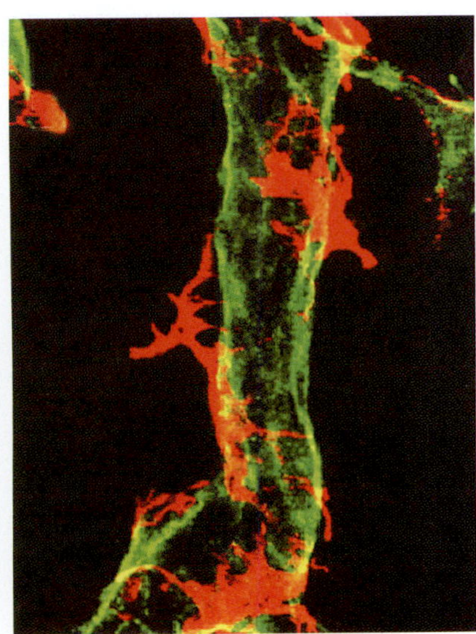

Figure 13.33 Tumor-associated pericytes This fluorescent micrograph reveals the structure of a tumor-associated microvessel, in this case a vessel formed in a mouse by Lewis lung carcinoma (LLC) cells. The endothelial cells forming the lumen of this vessel *(green)* are only partially overlain by pericytes and smooth muscle cells *(red orange)*. The loose attachment of pericytes to the capillaries can be contrasted to their tight attachment seen in Figure 13.6B. (From S. Morikawa, P. Baluk, T. Kaidoh et al., *Am. J. Pathol.* 160:985–1000, 2002.)

angiopoietins, together with the greatly elevated levels of VEGF within tumors, is suspected to be responsible for much of the defective construction of the capillaries and larger vessels within neoplasms.

The leakiness of tumor-associated capillaries leads to the accumulation of substantial amounts of fluid in the parenchymal spaces within a tumor. In normal tissues, these fluids are drained by the lymphatic vessels, which eventually empty their contents into the venous circulation (Sidebar 13.5). However, within solid tumors, the ongoing expansion of cancer cell populations exerts pressure on those few lymphatic vessels that do succeed in forming, causing their collapse (Figure 13.36). (Blood capillaries are more capable of resisting this pressure, because of their own significant internal hydrostatic pressure, which lymphatic vessels lack.) The resulting defective lymphatic drainage within the cores of solid tumors further exacerbates the elevated accumulation of fluid caused by

(A)

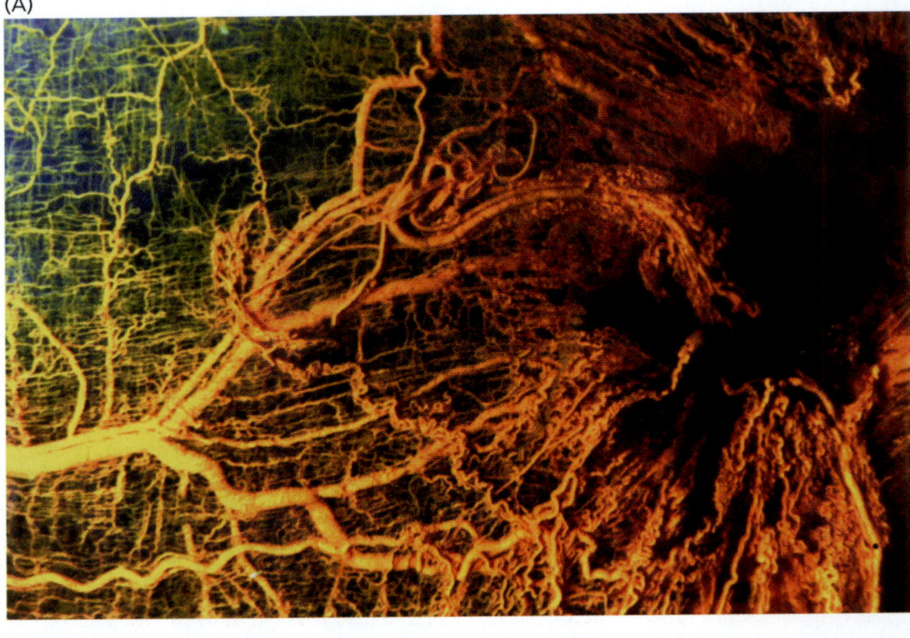

(B)

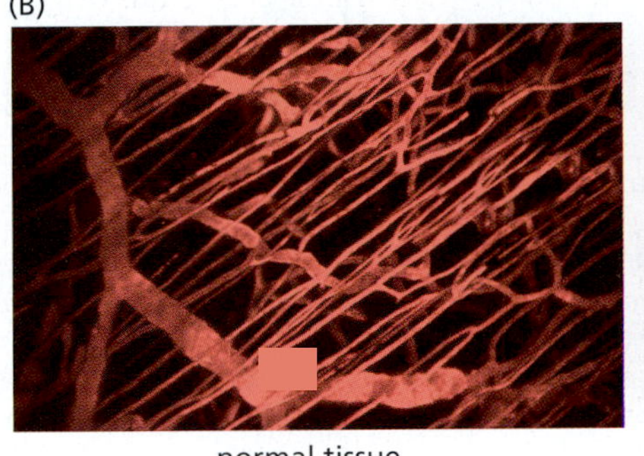

normal tissue

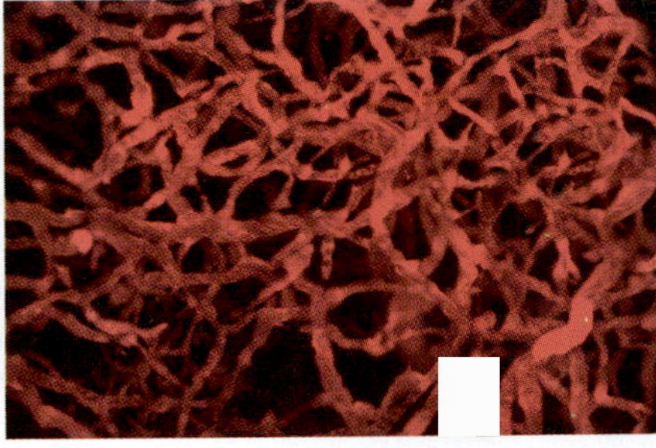

tumor

Figure 13.34 The chaotic organization of tumor-associated vasculature (A) This image, obtained by injecting a resin into blood vessels prior to tissue fixation, shows that the vasculature feeding a mass of tumor cells *(black area, far right)* is tortuous and poorly organized, in contrast to the well-organized vasculature in normal tissue *(turquoise area, top left)*. (B) Techniques of imaging living tissues (*intravital* imaging) have made it possible to visualize the capillary beds of normal and neoplastic tissues. This technique reveals that the capillary networks in a normal tissue are well organized *(left)*, while those in a tumor are chaotic *(right)*. (A, courtesy of L. Heiser and R. Ackland, University of Louisville; B, from R.K. Jain, *Nat. Med.* 9: 685–693, 2003.)

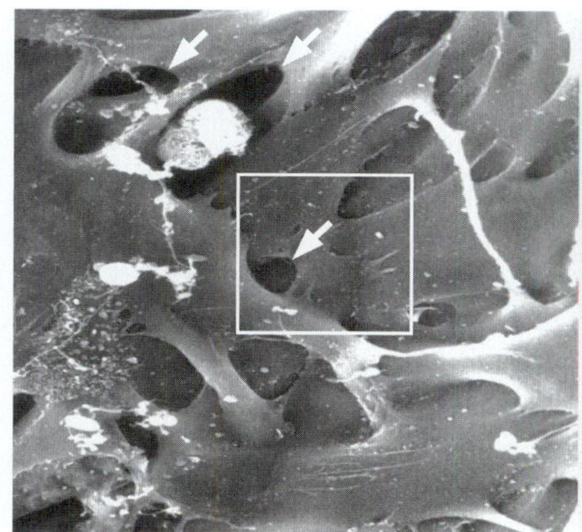

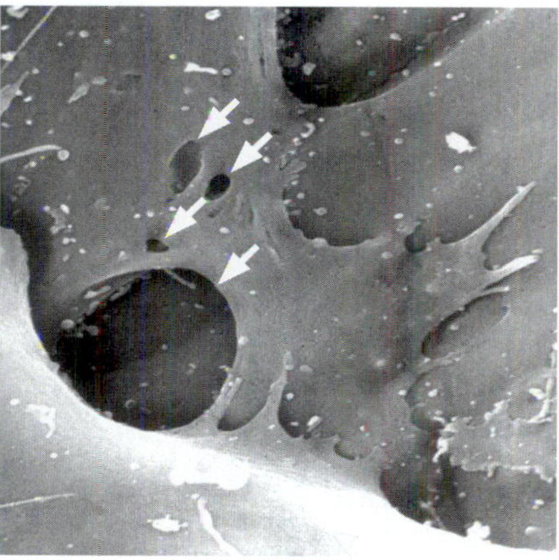

Figure 13.35 Scanning EM of gaps in tumor vasculature These scanning electron micrographs reveal that the sheets of endothelial cells, which form a continuous, uninterrupted surface around the walls of a normal capillary *(not shown)*, fail to do so in the capillary within a tumor. Instead, the tumor-associated endothelial cells overlap one another and, as indicated here *(arrows)*, show gaps of significant size between them. The box in the *left panel* is shown at higher magnification in the *right panel*. Such gaps permit the seepage of plasma fluids into the interstitial spaces between the cancer cells, contributing to high hydrostatic pressure in these spaces. (From H. Hashizume, P. Baluk, S. Morikawa et al., *Am. J. Pathol.* 156:1363–1380, 2000.)

capillary leakage, generating relatively high fluid pressure in the nonvascular parts of tumors. This pressure, in turn, greatly complicates the administration of anti-cancer therapeutic drugs (Sidebar 26).

13.7 Tripping the angiogenic switch is essential for tumor expansion

The descriptions of angiogenesis given above, as well as those in an earlier chapter (see Section 7.12), would seem to suggest that the release of angiogenic factors is virtually automatic: whenever groups of cells, including cancer cells, suffer hypoxia, they release angiogenic factors and thereby provoke the growth of capillaries into their midst. This cures the hypoxia and results in an appropriate density of capillaries in the tissue harboring these cells.

In fact, the ability to attract blood vessels seems to be a trait that many tumor cell populations initially lack and must acquire as tumor progression proceeds. This idea was initially suggested by the observation, cited above, that in certain tumors, cancer cells thrive near capillaries, but those that are located further than 0.2 mm from capillaries stop growing and may enter apoptosis or become

Sidebar 13.6 Vascular endothelial growth factor (VEGF) causes capillary permeability Two independent lines of research led to the initial cloning of the gene encoding VEGF-A. One research project characterized a factor secreted by cancer cells that caused blood vessels to leak fluid and was therefore termed vascular permeability factor (VPF). The other work focused on a factor released by tumors that functioned as an attractant and mitogen for the endothelial cells that assemble to form the luminal walls of capillaries; this agent was called vascular endothelial growth factor (VEGF)—the name that prevails today. The two research efforts culminated in 1989 in the independent isolations of the VPF- and VEGF-encoding genes and the discovery that they are one and the same. Cancer cells, as well as macrophages and fibroblasts in the tumor-associated stroma, continuously release significant amounts of VEGF, which are responsible for much of the elevated permeability of the tumor-associated capillaries and venules.

The precise mechanisms responsible for VEGF-induced permeability remain unclear. VEGF may cause the plasma membranes of adjacent endothelial cells to separate slightly, generating gaps between them. In addition, it may stimulate endothelial cells to transport fluid and solutes through their cytoplasm from the luminal to the **abluminal** surface.

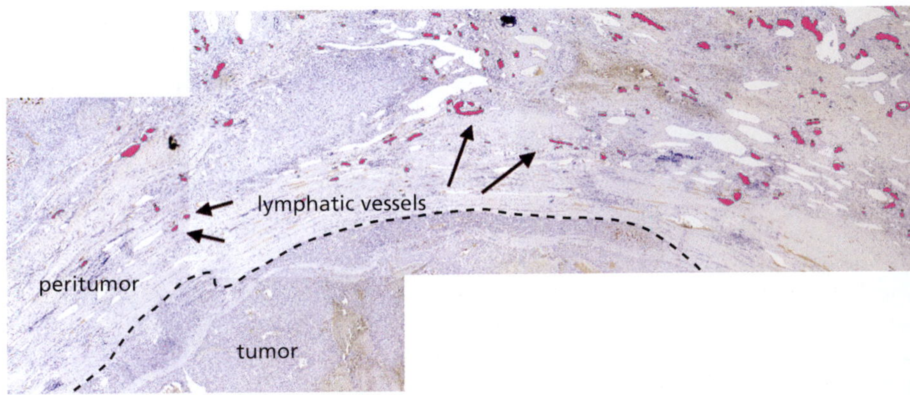

Figure 13.36 Absence of lymphatic vessels within solid tumors Analysis of a section of a hepatocellular carcinoma (HCC; liver cancer), using an antibody to specifically stain lymph ducts *(dark red)*, reveals that these ducts are present in the normal tissue above the tumor margin *(dotted line)* but are absent within the tumor mass itself. This absence may be attributed (1) to the fact that these ducts were not generated initially during tumor formation and/or (2) to the collapse and degeneration of these ducts because of the high hydrostatic pressure within solid tumors. (From C. Mouta Carreira, S.M. Nasser, E. di Tomaso et al., *Cancer Res.* 61:8079–8084, 2001.)

necrotic (Section 13.6). So, even though these cancer cells experience hypoxia, they lack the ability to induce the formation of nearby capillaries.

We know much about these dynamics from detailed study of experimental models of tumorigenesis. The most informative of these has been the Rip-Tag transgenic mouse. It carries a transgene in its germ line, in which the expression of the SV40 large T antigen (Sections 8.5 and 9.1) and small T antigen (Sidebar 18) is driven by the promoter of the insulin gene. This promoter ensures expression of these viral oncoproteins in the β cells that form the islets of Langerhans in the pancreas (in which insulin is normally produced).

Tumor progression in the 400 or so islets of the mouse pancreas can be easily followed, because these islets can readily be distinguished from the surrounding tissue of the **exocrine** pancreas, which is involved in manufacturing and secreting digestive enzymes. As many as half of these islets in a Rip-Tag mouse form hyperplastic nodules by 10 weeks of age, and 8 to 12% of the hyperplastic islets eventually progress to become angiogenic, that is, they acquire the ability to recruit new blood vessels into their midst. By 12 to 14 weeks, about 3% of the initially formed hyperplastic islets finally progress to form carcinomas (Figure 13.37).

Early in tumor progression in the Rip-Tag mice, the hyperplastic pancreatic islets begin to expand slightly to a small diameter of about 0.1–0.2 mm and then halt their forward march (see Figure 13.37), at least for a while. In these small nests of tumor cells, cell division continues unabated, being driven by the oncogenic transgene. However, the overall size of the tumor cell population within each islet remains constant, because of a compensating attrition of cells occurring through apoptosis. This mouse model suggests that in humans, small tumor nests may also remain in this dynamic but nongrowing state for many years, unable to break through the barrier that is holding them back.

In principle, the barrier to expansion of the tumor cell nests might be a physical one—lack of adequate space within the tissue for these cancer cells to multiply. But detailed histological analysis of these small nests of cancer cells reveals

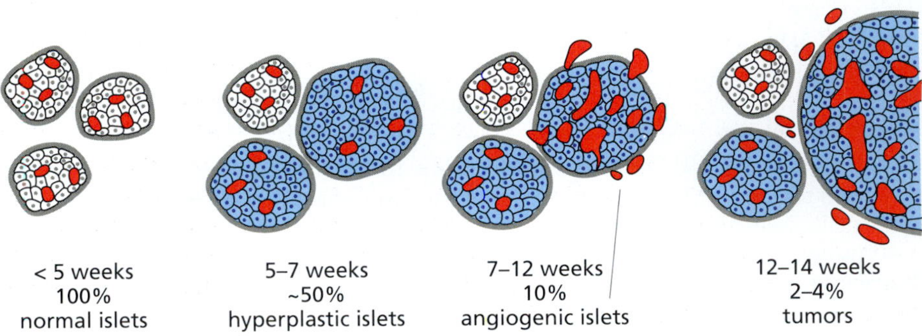

Figure 13.37 The Rip-Tag model of islet cell tumor progression Multi-step tumorigenesis in the Rip-Tag transgenic mouse model proceeds in distinct stages. A normal pancreatic islet (also known as an islet of Langerhans; *left*) carries a small number of capillaries to support the β cells. By 5–7 weeks of age, about half of these islets become hyperplastic, but the density of blood vessels is not increased. During a subsequent period from 7 to 12 weeks, about 10% of the hyperplastic islets become angiogenic, as indicated by the greatly increased density of capillaries in the surrounding, nearby exocrine pancreas. Finally, by 12–14 weeks of life, 2–4% of the initially formed hyperplastic islets have become invasive carcinomas that grow rapidly and invade the surrounding exocrine pancreas. The exocrine pancreas, which surrounds the islets, is not shown here. (From D. Hanahan and J. Folkman, *Cell* 86:353–364, 1996.)

something quite different. Because these cells have not yet become angiogenic, they lack vasculature. The resulting hypoxia that they experience triggers p53-dependent apoptosis, which explains their high rate of attrition. (It is possible that other sub-optimal conditions within these poorly vascularized cell nests, including inadequate nutrient supply, high levels of carbon dioxide and metabolic wastes, and low pH caused by lactic acid accumulation, also contribute to the death of these cells.)

At some point in time, however, small clusters of these pre-neoplastic islet cells suddenly acquire the ability to provoke neoangiogenesis (see Figure 13.37). Once these cells learn how to induce capillaries to form nearby, they and their descendants seem to be liberated from the major constraint that has been holding back their multiplication. This sudden, dramatic change in the behavior of the small tumor masses has been termed the "**angiogenic switch**."

These phenomena suggest an interesting idea, really a speculation: the body purposefully denies its cells the ability to readily induce angiogenesis. By doing so, the body erects yet another impediment to block the development of large neoplasms. According to such thinking, the angiogenic switch—a clearly important step in tumor progression—represents the successful breaching of this defensive barrier and the acquisition by cancer cells of a forbidden fruit: the ability to induce blood vessel growth at will.

One might conclude that the angiogenic switch in these transgenic mice is driven by the pre-malignant β cells' suddenly acquiring the ability to express and release VEGFs. Actually, these cells make large amounts of VEGFs long before the angiogenic switch occurs, as do fully normal pancreatic islets. However, the VEGF molecules secreted by these β cells are efficiently sequestered by the surrounding extracellular matrix (ECM). As a consequence, the VEGF molecules are unable to stimulate angiogenesis.

This sequestered state of the VEGF explains why the angiogenic switch in the Rip-Tag pancreatic islets is accompanied by the sudden appearance of substantial amounts of matrix metalloproteinase-9 (MMP-9; Section 13.5). MMP-9 acts in a targeted fashion to cleave specific structural components of the ECM, thereby releasing and mobilizing VEGF for active signaling to nearby endothelial cells. This MMP-9 is synthesized and released by inflammatory cells—mast cells and, quite possibly, macrophages—that have been attracted to the pre-malignant islets (Sidebar 13.7). Hence, in this particular tissue, tripping of the angiogenic switch ultimately depends on an acquired ability to recruit inflammatory cells.

If the gene encoding VEGF is selectively deleted from the islet cells through genetic engineering, the islets survive and the early steps of tumor progression still proceed normally, but angiogenic switching never occurs. This reinforces the conclusion that VEGF molecules of islet cell origin play a critical role in triggering the onset of angiogenesis, and that recruited stromal cells cannot compensate for this absence of VEGF by bringing in some of their own.

This scenario (Figure 13.38) involves heterotypic interactions among three distinct cell types: (1) the release of still-unidentified signals from the pre-malignant islet cells that recruit mast cells and, quite possibly, macrophages; (2) the release of MMP-9 by these inflammatory cells to activate previously latent VEGF made by the pre-malignant islet cells; and (3) the proliferative response of endothelial cells to the activated VEGF. In fact, yet other cell types are likely to be partners in islet cell angiogenesis. Thus, many of the endothelial cells may derive from recruited endothelial precursor cells in the circulation (EPCs, sometimes called circulating endothelial progenitors, or CEPs); and the capillaries that arise are eventually covered, albeit haphazardly, with another cell type—pericytes—whose origins are unclear.

Sidebar 13.7 Mast cells from the bone marrow can play a key role in the angiogenic switch Rip-Tag mice carrying the insulin-SV40 transgene in their germ line can be bred with others that lack the Kit growth factor receptor (see Sidebar 5.7). Kit serves as the receptor for stem cell factor (SCF), an important growth factor that triggers the development of certain subclasses of hematopoietic cells. Among other deficits, *Kit*$^{-/-}$ mice lack the ability to form mast cells. In Rip-Tag mice lacking Kit receptor, islet cell tumors are initiated at rates routinely observed in standard Rip-Tag mice. However, these tumors never succeed in becoming angiogenic, and the rate of cell proliferation in these small growths is compensated by an equal rate of apoptotic death. Consequently, these tumors remain at a very small size (0.1–0.2 mm diameter).

If these mice are provided with a bone marrow transplant containing wild-type hematopoietic precursor cells, which cures their mast cell deficit, angiogenesis is initiated in the pancreatic islets, tumor cells in the islets are no longer lost through apoptosis, and large, life-threatening neoplasms appear soon thereafter. This proves that the ability to provoke angiogenesis depends on the actions of at least one nonendothelial component of the tumor-associated stroma—the mast cells originating in the bone marrow. The main contribution of these mast cells to the angiogenic switch seems to be the release of MMP-9, which proceeds to mobilize latent VEGF, thereby provoking angiogenesis (see Figure 13.38).

Importantly, the Rip-Tag model does not typify the angiogenic mechanisms occurring during the formation of all types of tumors. For example, in some tumors, angiogenesis appears to increase progressively, as if a controlling rheostat is gradually being turned. This contrasts with the behavior of Rip-Tag islets, in which a binary, On-Off switch seems to be tripped.

Other tumors may depend on different angiogenic factors to provoke angiogenesis. For example, when transformed mouse embryonic stem (ES) cells are deprived of both copies of the VEGF-A–encoding gene, they lose almost all their power to make malignant teratomas. In contrast, transformed adult mouse dermal fibroblasts remain highly tumorigenic after they have been deprived of both copies of this gene. And the sarcomas generated by these transformed fibroblasts continue to grow even when the mice bearing them have been treated with an antibody that binds and inactivates VEGF-R2 (the primary endothelial cell receptor that confers responsiveness to both VEGF-A and VEGF-B). This behavior is likely to be explained by the fact that transformed dermal fibroblasts can make a complex mixture of angiogenic factors—including VEGF-B, acidic and basic fibroblast growth factors (aFGFs and bFGFs), and transforming growth factor-α (TGF-α). This deployment of multiple angiogenic factors (see Table

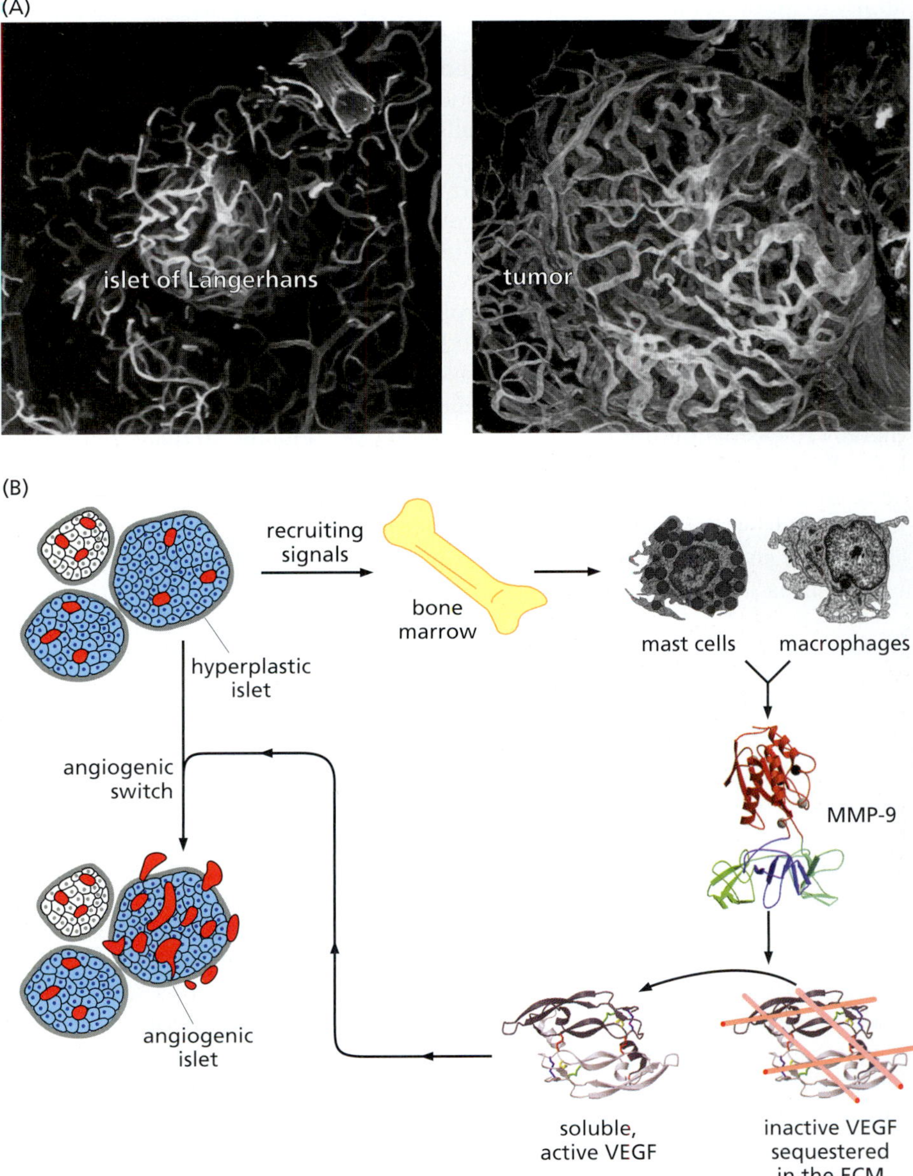

Figure 13.38 The angiogenic switch and recruitment of inflammatory cells (A) The normal islet of Langerhans *(circular area, center of left panel)* is poorly vascularized and is sustained largely through diffusion from the microvessels surrounding it. Following tripping of the angiogenic switch, there is an increase in the number of cells in the islet, due to tumor expansion, and, as seen here, a dramatic induction of vessel formation *(right panel)*. Imaging was achieved by wholemount microscopy of lectin-perfused vessels. (B) According to this scheme, a pre-angiogenic, hyperplastic islet sends signals to the bone marrow and/or circulation *(not illustrated)* that lead to the recruitment of mast cells and macrophages. Once in the vicinity of the islet, these release metalloproteinases, notably MMP-9 (see Figure 13.25E). MMP-9 then cleaves extracellular matrix (ECM) components, liberating vascular endothelial growth factor (VEGF) from its sequestered state. VEGF proceeds to induce angiogenesis around the islet, thereby tripping the angiogenic switch. (A, from G. Bergers, D. Hanahan and L.M. Coussens, *Int. J. Dev. Biol.* 42:995–1002, 1998; B, adapted from data in G. Bergers, R. Brekken, G. McMahon et al., *Nat. Cell Biol.* 2:737–744, 2000; VEGF: Y.A. Muller, B. Li, H.W. Christinger et al., *Proc. Natl. Acad. Sci. USA* 94:7192–7197, 1997.)

Table 13.2 Important angiogenic factors

Name	Mol. wt. (kD)
Vascular endothelial GF (VEGF)	40–45
Basic fibroblast growth factor (bFGF)	18
Acidic fibroblast growth factor (aFGF)	16.4
Angiogenin	14.1
Transforming growth factor-α (TGF-α)	5.5
Transforming growth factor-β (TGF-β)	25
Tumor necrosis factor-α (TNF-α)	17
Platelet-derived growth factor (PDGF)	45
Granulocyte-colony–stimulating factor	17
Placental growth factor	25
Interleukin-8 (IL-8)	40
Hepatocyte growth factor (HGF)	92
Proliferin	35
Angiopoietin	70
Leptin	16

13.2), often observed in advanced human cancers, complicates the development of anti-angiogenesis cancer therapies, as we will see later.

13.8 The angiogenic switch initiates a highly complex process

Angiogenesis begins in the stroma surrounding the Rip-Tag tumors long before the basement membrane has been broken down. This behavior typifies that of many tumors in both mice and humans (Figure 13.39). Somehow, angiogenic

(A)

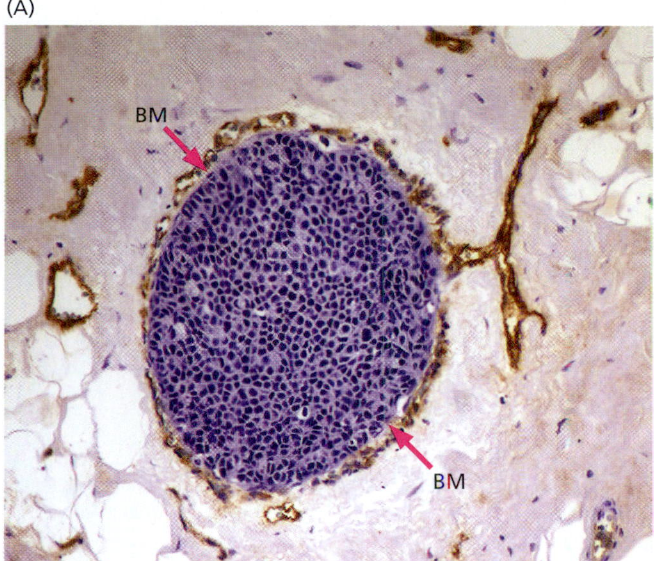

(B)

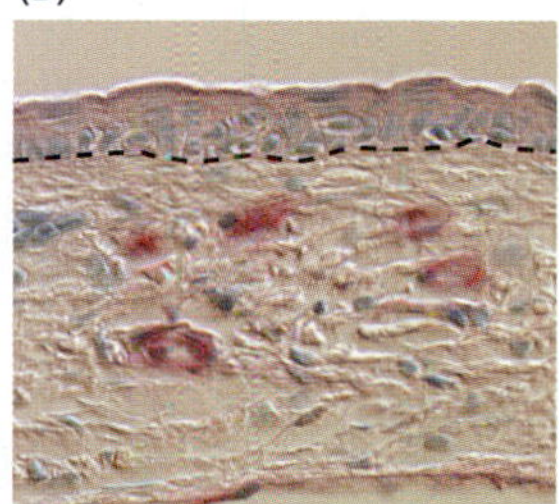

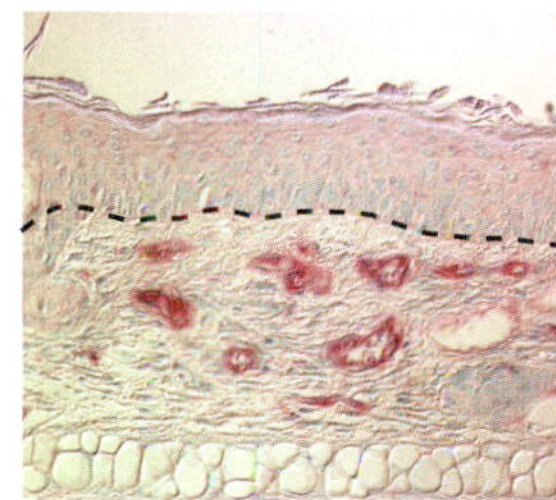

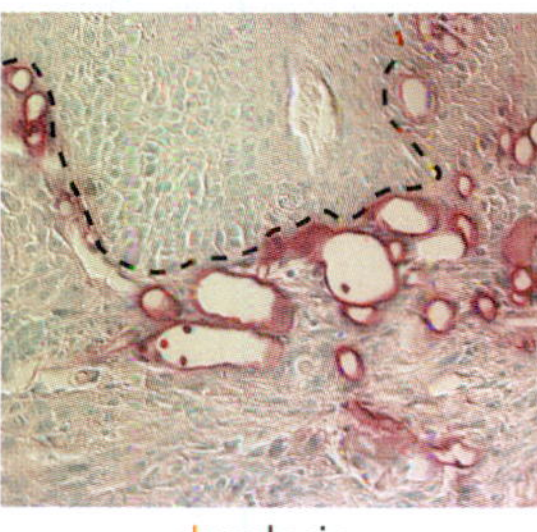

Figure 13.39 Signaling through the basement membrane early in tumor progression (A) A human ductal *in situ* breast carcinoma (DCIS) contains a collection of carcinoma cells *(purple)* that are noninvasive and therefore have not yet breached the basement membrane (BM) that surrounds them and separates them from the mammary stroma. As is apparent, this DCIS has nonetheless succeeded in transmitting angiogenic signals through the BM into the nearby stroma that have resulted in the growth of a small vessel *(dark brown)* that surrounds the tumor mass but does not penetrate into the tumor itself because of the continued integrity of the BM. (B) In this transgenic mouse model of skin carcinogenesis, the human papillomavirus (HPV) 16 early region is expressed under the control of a keratin-14 gene promoter. In the early hyperplastic stage of tumor progression, the noninvasive carcinoma cells on the epithelial side *(above)* are able to transmit signals through the BM *(dashed line)* that provoke angiogenesis on the stromal side, as evidenced by the increased density of capillaries *(red)*, detected in this case through their display of the CD31 antigen. Dysplastic tissue, in which the BM *(dashed line)* has not yet been breached, shows even more intensive angiogenesis in the nearby stroma. (A, courtesy of A.L. Harris; B, from L.M. Coussens, W.W. Raymond, G. Bergers et al., *Genes Dev.* 13:1382–1397, 1999.)

(A)

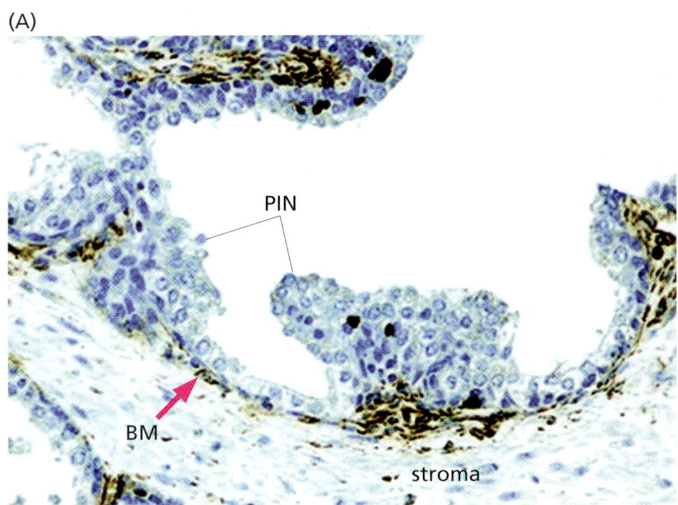

(B)

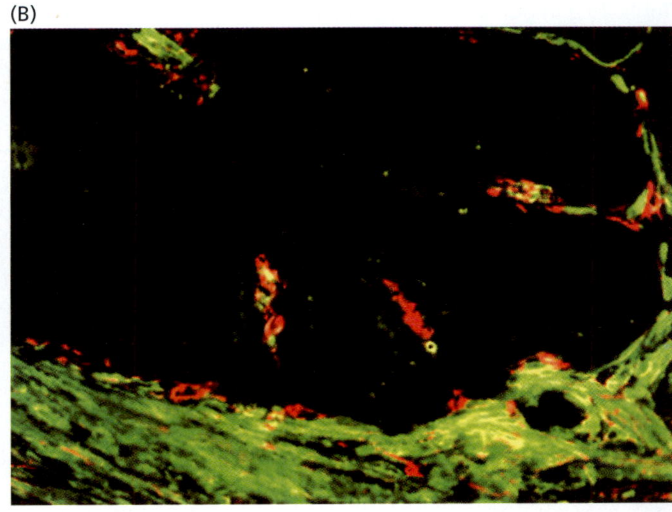

Figure 13.40 Recruitment of myofibroblasts on the stromal side of the basement membrane Myofibroblasts also foster angiogenesis, in part through their ability to recruit endothelial precursor cells (EPCs) to the tumor stroma. (A) In human prostate intraepithelial neoplasia (PIN)—a counterpart of ductal carcinoma *in situ* (DCIS) in the breast—the carcinoma cells *(darker blue nuclei)* remain on the epithelial side of the still-intact basement membrane (BM), yet they have stimulated the accumulation of stromal fibroblasts beyond the BM, as indicated by their display of vimentin *(dark brown)*. (B) The identification of many of these stromal cells more specifically as myofibroblasts is indicated by the fact that they can be immunostained with both an anti-vimentin antibody *(red)* and α-smooth muscle actin *(green)*; cells co-expressing both markers, and therefore identified as myofibroblasts, are seen in *yellow*. These cells can also be identified as myofibroblasts through their expression of collagen I (*not shown*). (From J. Tuxhorn, G.E. Ayala, M.J. Smith et al., *Clin. Cancer Res.* 8,2912–2923, 2002.)

signals are dispatched by benign cancer cells through the porous basement membrane (BM) in order to encourage increased angiogenesis on the stromal side of this membrane. Yet other signals transmitted through the basement membrane recruit myofibroblasts to the nearby stroma (Figure 13.40). As we read earlier, these myofibroblasts can also help to foster angiogenesis.

Nevertheless, this early angiogenesis is circumscribed, and it is clear that intense angiogenesis can begin only when cancer cells become invasive, penetrate the basement membrane, and acquire direct, intimate contact with stromal cells (Figure 13.41). This suggests that tumor invasiveness and intense

(A)

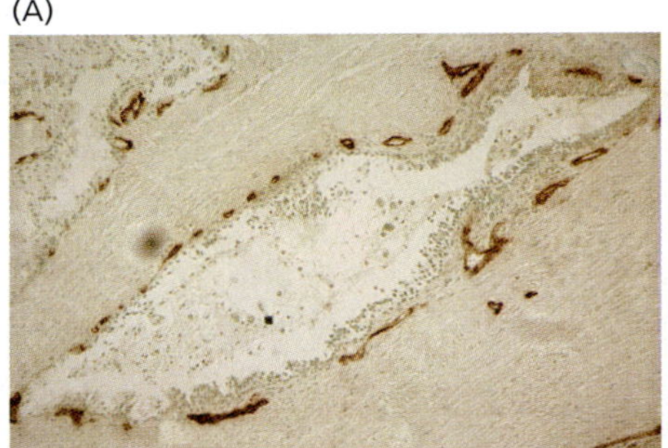

prostate cancer (PIN; *in situ*)

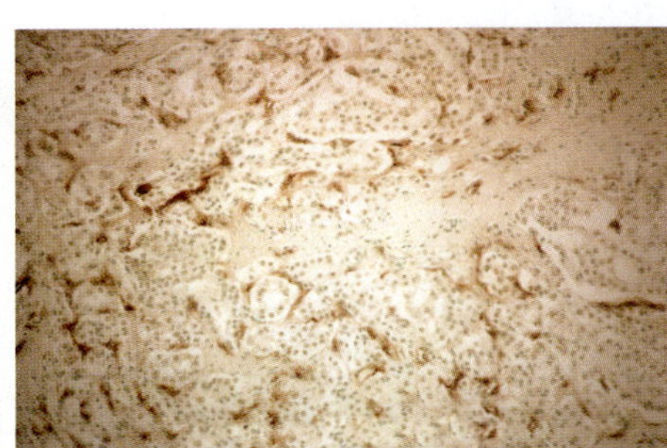

invasive prostate cancer

(B)

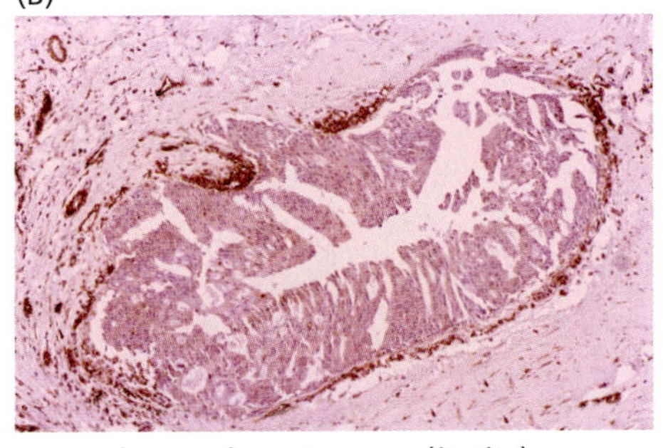

human breast cancer (*in situ*)

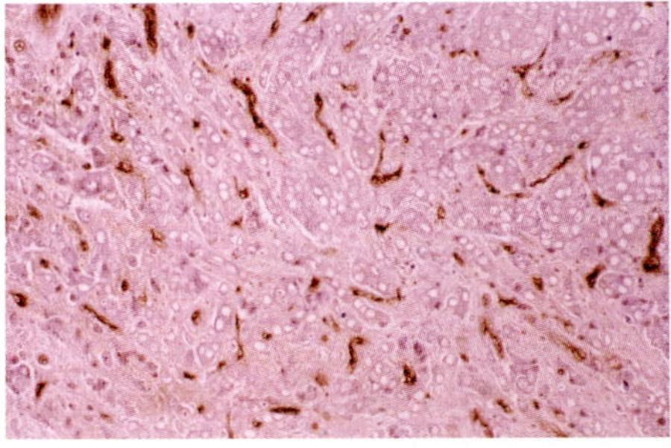

invasive human breast cancer

Figure 13.41 Angiogenesis before and after acquisition of invasiveness Tumor angiogenesis is circumscribed as long as human carcinomas remain benign. However, once they become invasive, the intensity of angiogenesis increases, leading to a higher density of capillaries *(brown)* threading their way through tumors. (A) The capillaries ringing a benign prostatic intraepithelial neoplasia (PIN) lesion *(left)* are far fewer than those in an invasive prostate carcinoma *(right)*. (B) Similarly, those in a benign ductal *in situ* breast carcinoma (DCIS; *left)* are far fewer than those in an invasive ductal carcinoma *(right)*. (A, courtesy of J. Folkman; B, from N. Weidner, J.P. Semple, W.R. Welch and J. Folkman, *N. Engl. J. Med.* 324: 1–8, 1991.)

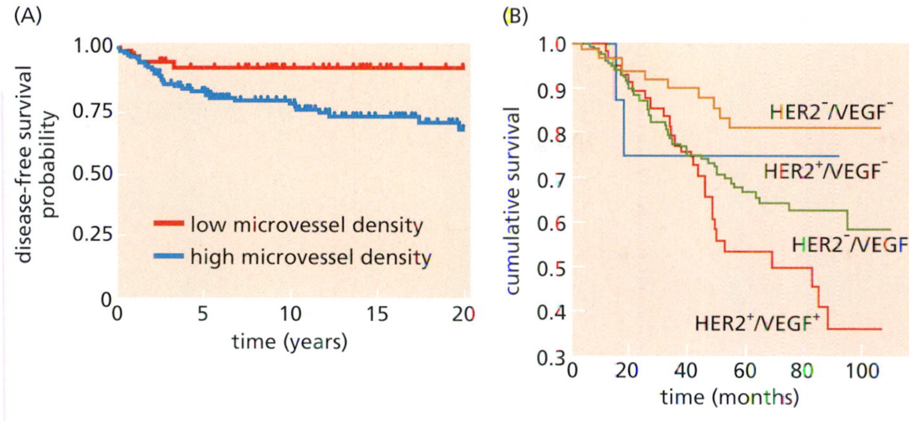

Figure 13.42 Clinical outcomes and the intensity of angiogenesis (A) Breast carcinomas were analyzed for the density of capillaries (microvessel count), which was determined as the number of capillaries per microscopic field. This Kaplan–Meier graph demonstrates that those patients whose tumors had a higher microvessel count in their tumors *(blue curve)* had a markedly lower probability of disease-free survival in the 20 years following initial diagnosis than did those whose tumors had a lower microvessel count *(red curve)*. (B) As indicated earlier (see Figure 4.6), breast cancer patients whose tumors overexpress HER2/Neu have a markedly poorer prognosis than those who do not. In the group analyzed here, all of the patients showed metastatic cancer cells in one or more of the lymph nodes draining the breast. The differences in survival are even more dramatic when the levels of VEGF produced by their tumors are also measured. In this relatively small clinical study, more than 80% of patients whose tumors expressed low, basal levels of both HER2/Neu and VEGF were alive eight years after diagnosis. In contrast, only about 35% of the patients whose tumors expressed elevated levels of both HER2/Neu and VEGF were still alive at this time. (A, from R. Heimann and S. Hellman. *J. Clin. Oncol.* 16:2686–2692, 1998; B, from G.E. Konecny, Y.G. Meng, M. Untch et al., *Clin. Cancer Res.* 10:1706–1716, 2004.)

angiogenesis are often tightly coupled processes. We will study tumor invasiveness in detail in the next chapter.

In many human tumor types, the density of capillaries per microscope field increases in lockstep with increasing degrees of malignancy. For example, among human breast carcinomas that have already grown to a considerable size, those that have managed to attract dense networks of capillaries into their midst are indicative, on average, of a far worse prognosis than those that are poorly vascularized (Figure 13.42A). Moreover, patients with breast tumors that express large amounts of VEGF (in addition to HER2) also fare badly following initial diagnosis and treatment (Figure 13.42B). Altogether, this suggests that the angiogenic switch is only the first of many gradations that enable tumors to become progressively more angiogenic and hence increasingly vascularized.

These striking correlations are actually susceptible to two alternative interpretations. It is possible that intense vascularization enables cancer cells to grow more aggressively, thereby leading to poor clinical outcomes. Alternatively, intense angiogenesis is only a marker of an underlying aggressive phenotype but is not causally involved in driving high-grade malignancy. Available clinical data do not allow a clear resolution between these alternatives.

We have spoken of the angiogenic switch as if small numbers of cells within a tumor undergo this shift and proliferate to ultimately dominate within the tumor mass, thereby imparting the angiogenic phenotype to the tumor as a whole. In fact, analyses of cells isolated from explanted human tumors indicate great heterogeneity in the angiogenic powers of different subpopulations of cancer cells within a given tumor, with some cancer cells being highly angiogenic while others are poorly so. Even within tumor cell lines, individual subcloned cell populations show greatly differing angiogenic powers when engrafted *in vivo* (Figure 13.43). This introduces yet another idea: that within a tumor mass, long after the tripping of the angiogenic switch, the weakly angiogenic cancer cells rely on help from their friends—their strongly angiogenic neighbors—in order to acquire adequate vasculature.

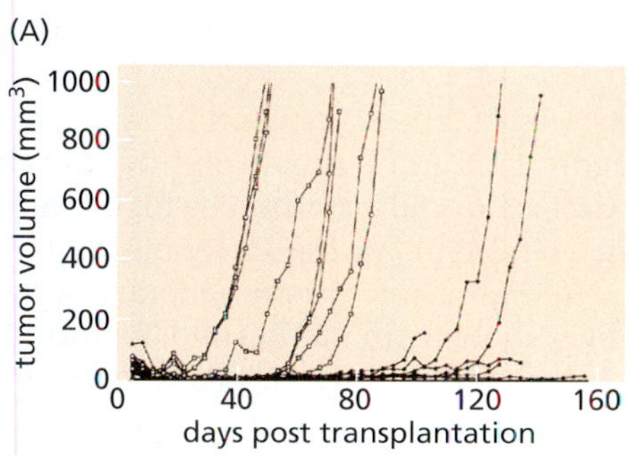

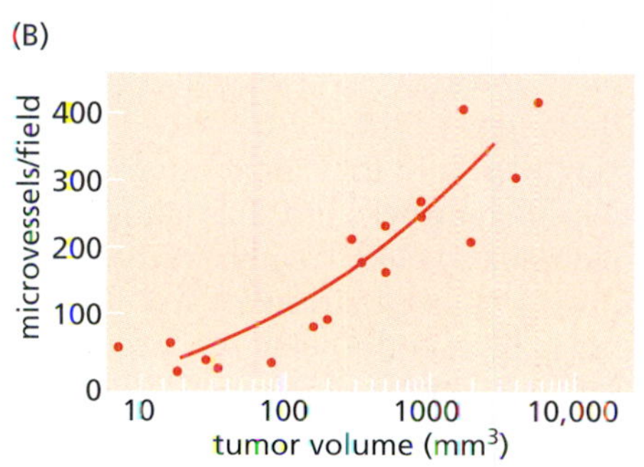

Figure 13.43 Heterogeneous degrees of vascularization within a tumor cell population (A) When a population of cells in a human liposarcoma cell line is subjected to single-cell cloning (in which the cells in each resulting cell clone all derive from a common ancestor), the various cloned cell populations show greatly differing abilities to form tumors *(individual curves)* when implanted into immuno-compromised mice. (B) When the resulting tumors are analyzed for microvessel density (microvessels per microscopic field) and tumor volume, plotted logarithmically here, it becomes clear that they have greatly differing angiogenic capabilities and that angiogenesis is likely to be the rate-limiting determinant of the ability to form tumors. (From E.G. Achilles, A. Fernandez, E.N. Allred et al., *J. Natl. Cancer Inst.* 93:1075–1081, 2001.)

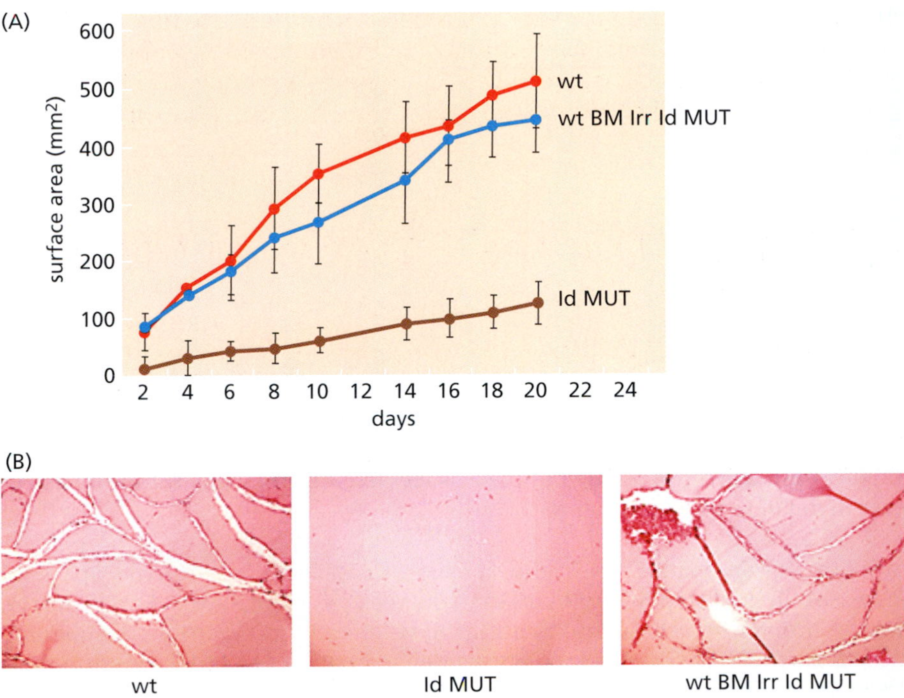

Figure 13.44 Defective tumor angiogenesis in Id1$^{+/-}$ Id3$^{-/-}$ mice Mutant mice (Id MUT), which lack three of four *Id* gene copies (and have an *Id1$^{+/-}$ Id3$^{-/-}$* genotype), show impaired mobilization of endothelial precursor cells (EPCs) from the bone marrow and thus impaired recruitment of circulating EPCs into tumors that they may carry. (A) Wild-type (wt) mice bearing Lewis lung carcinoma (LLC) cells develop rapidly growing tumors *(red curve)*, while those of the mutant strain (Id MUT) are unable to support vigorous tumor growth *(brown curve)*. This defect is essentially reversed if the bone marrow (BM) of the mutant mice is eliminated by irradiation and replaced with transplanted wild-type bone marrow *(blue curve,* wt BM Irr Id MUT*)*, indicating that the recruitment of bone marrow–derived cells is defective in the mutant mice and is responsible for the inability of the tumor to grow in them. (B) This defective recruitment can be further localized by the use of plugs of Matrigel (an extracellular matrix material) that have been impregnated with VEGF and are then implanted subcutaneously. These plugs are able to recruit neovasculature when implanted in wild-type mice *(left)* but not in the mutant mice *(center)*. However, if the mutant mice receive a graft of wild-type bone marrow cells, the defect is cured and now angiogenesis within these plugs is comparable to that seen in wild-type mice *(right)*. (From D. Lyden, K. Hattori, S. Dias et al., *Nat. Med.* 7:1194–1201, 2001.)

The angiogenic switch may have effects that extend far beyond the immediate vicinity of a tumor. At first, the tumor-associated blood vessels that are formed may sprout directly from existing vessels in adjacent normal tissues. At this stage, it is likely that the endothelial cells forming these new capillaries arise through the proliferation of endothelial cell precursors residing in these already-formed vessels. However, as tumor progression proceeds, neovascularization is likely to rely increasingly on the recruitment of endothelial precursor cells (EPCs) that originate in the marrow and travel via the circulation to the tumor mass.

VEGF released into the circulation by a tumor stimulates the production of EPCs in the bone marrow and helps to attract EPCs to the tumor mass. As described earlier (Section 13.4), stroma-derived factor-1 (SDF-1, also called CXCL12), which is liberated by stromal myofibroblasts, can also help in this recruitment. Once they have arrived in the tumor mass, the EPCs are induced to differentiate into functional endothelial cells and construct the tumor-associated vasculature. Detailed analyses have shown that the proportion of endothelial cells derived from circulating EPCs (versus those arising from nearby capillaries) varies greatly from one type of tumor to another.

These dynamics of EPC recruitment are dramatically illustrated by the behavior of mice that lack one copy of the *Id1* gene and both copies of the *Id3* gene. Recall that these *Id* genes encode transcription factors regulating differentiation of a variety of cell types (Section 8.11). Many of these mutant mice survive to adulthood and show a very peculiar phenotype: they are defective in neoangiogenesis and will not permit several types of engrafted tumors to grow (Figure 13.44A).

The defect in these *Id1$^{+/-}$ Id3$^{-/-}$* mice can be cured by transplanting wild-type stem cells into their bone marrow (see Figure 13.44A). Alternatively, the subset of bone marrow stem cells whose production is normally spurred by high levels of circulating VEGF can be harvested from a wild-type mouse and transplanted into the bone marrow of *Id1$^{+/-}$ Id3$^{-/-}$* mice. These mice soon generate large numbers of endothelial precursor cells (EPCs) in their circulation and, following implantation of cancer cells, develop rapidly growing tumors. Taken together, these observations indicate that these tumors rely on VEGF to mobilize EPCs from the bone marrow (Figure 13.44B) and to recruit the resulting circulating cells into the tumor stroma. These findings, as well as others demonstrating the recruitment of myofibroblast precursors into tumor stroma (Sidebar 13.3), show that primary tumors can extend their reach throughout the body long before they metastasize.

13.9 Angiogenesis is normally suppressed by physiologic inhibitors

In finely tuned physiologic processes, the actions of positive effectors must be counterbalanced by negative regulators. We have read much about the positive effectors of angiogenesis, such as VEGF and bFGF, but their antagonists have remained offstage until now. They turn out to be as interesting and important as the angiogenic factors whose actions they antagonize.

During the process of wound healing, for instance, the burst of angiogenesis that is required to repair the wound site must be shut down once the newly formed capillaries have reached a density that suffices to support normal tissue function. This shutdown is achieved, at least in part, by suppressing formation of the HIF-1 transcription factor. Its assembly is induced under hypoxic conditions and is reversed once normal oxygenation in the wound site has been restored (see Section 7.12).

In addition, a number of the components of the extracellular matrix are used by tissues to actively block excessive angiogenesis (Table 13.3). The best characterized of these is the thrombospondin-1 (Tsp-1) protein, which is secreted by many cell types into the surrounding extracellular space, where it forms homotrimers and carries out several distinct functions. Most important for our discussion, Tsp-1 associates with a receptor (termed CD36) that is displayed on the surfaces of endothelial cells and halts their proliferation. In addition, research indicates that Tsp-1 treatment of endothelial cells causes them to release Fas ligand (FasL), the pro-apoptotic signaling protein that acts by binding to the Fas death receptor. Recall that the latter, once it has bound its ligand, activates an intracellular caspase cascade that triggers apoptosis (Section 9.14). Therefore, once Tsp-1 causes endothelial cells to release FasL, the latter may act in an autocrine fashion to trigger the death of these cells in the event that they also display the Fas receptor.

Interestingly, the Fas receptor is displayed on endothelial cells that are actively proliferating or have recently ceased proliferation, but its display is suppressed once these cells have successfully formed mature capillaries and retreated into quiescence (Figure 13.45A). This seems to explain a most intriguing aspect of Tsp-1 behavior: it selectively inhibits and causes regression of newly formed and still-growing capillaries but has little if any effect on already-formed, mature capillaries. In fact, a number of other natural anti-angiogenic factors also seem to exhibit such selectivity. The actions of these other anti-angiogenic agents seem, once again, to depend upon activation of the pro-apoptotic caspase cascade in endothelial cells, since compounds that inhibit the caspase enzymes also protect endothelial cells from the anti-angiogenic effects of Tsp-1 and the other natural blockers of angiogenesis.

Transcription of the *TSP1* gene is strongly induced by p53, ostensibly as part of the p53-mediated emergency response that leads to a generalized shutdown of cell proliferation and tissue growth. Conversely, the loss of p53 function, which is seen in almost all human tumors (Chapter 9), leads to a substantial decrease in Tsp-1 levels. This permits angiogenesis to be induced by cells that normally would have been prevented from doing so by the high Tsp-1 concentrations in the surrounding extracellular matrix.

Table 13.3 Endogenous inhibitors of angiogenesis

Inhibitor	Description
A. Derived from extracellular matrix	
Arresten	fragment of type IV collagen α_1 chain of vascular basement membrane
Canstatin	fragment of type IV collagen α_2 chain of vascular basement membrane
EFC-XV	fragment of type XV collagen
Endorepellin	fragment of perlecan
Endostatin	fragment of collagen type XVIII
Anastellin	fragment of fibronectin
Fibulin	fragment of basement membrane protein
Thrombospondin-1 and -2	ECM glycoproteins
Tumstatin	fragment of type IV collagen α_3 chain
Chondromodulin-I	component of cartilage ECM
Troponin I	component of cartilage ECM
B. Non-matrix–derived	
Growth factors and cytokines	
Interferon-α (IFN-α)	cytokine
Interleukins (IL-1β, -12, -18)	cytokines
Pigment epithelium-derived factor (PEDF)	growth factor
Platelet factor-4	released by platelets during degranulation
Other types	
Angiostatin	fragment of plasminogen
Antithrombin III	fragment of antithrombin III
2-Methoxyestradiol	endogenous metabolite of estrogen
PEX	fragment of MMP-2
Plasminogen kringle 5	fragment of angiostatin
Prolactin fragments	specific cleavage fragment
Prothrombin kringle 2	fragment of prothrombin
sFlt-1	soluble form of VEGF-R1 (= Flt-1)
TIMP-2	inhibitor of metalloproteinase -2
TrpRS	fragment of tryptophan yl-tRNA synthetase
Vasostatin	fragment of calreticulin

Adapted from P. Nyberg, L. Xie and R. Kalluri, *Cancer Res.* 65:3967–3979, 2005.

The Ras oncoprotein, acting through a complex signaling cascade, acts in the opposite fashion, since it causes shutdown of *TSP1* gene expression. The resulting absence of significant levels of Tsp-1 can also contribute substantially to the elevated angiogenic powers of *ras*-transformed cells compared with their normal neighbors. In one study of melanomas arising in transgenic mice, expression of the H-*ras* oncogene, which had been used to initiate these tumors, was shut down after they had grown to a considerable size. These tumors collapsed rapidly thereafter, and this collapse was directly traceable to their loss of functional vasculature, which began to disintegrate within 6 hours after the loss of Ras function (Figure 13.45B). The observed rapid entrance into apoptosis of endothelial cells forming the tumor vasculature preceded any decline in the levels of VEGF (which provides survival signals for endothelial cells). While not demonstrated directly in these experiments, it seems likely that this apoptosis derived from the rapid re-expression of Tsp-1 that follows close on the heels of Ras shutdown.

Tsp-1 is surely a major governor of angiogenesis, but as hinted above, it is only one of a large cohort of natural inhibitors of angiogenesis that are found in the spaces between cells (see Table 13.3). The existence of several other anti-angiogenic molecules was first suggested by observations of the behavior of certain primary tumors and their derived metastases. In some mouse models of tumorigenesis, metastases were found to remain small in size as long as the primary

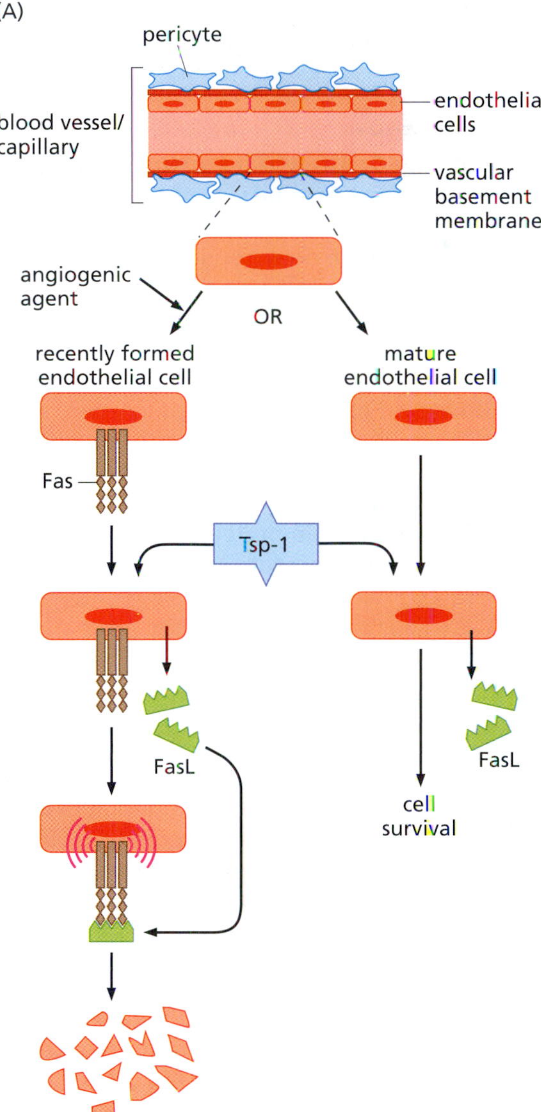

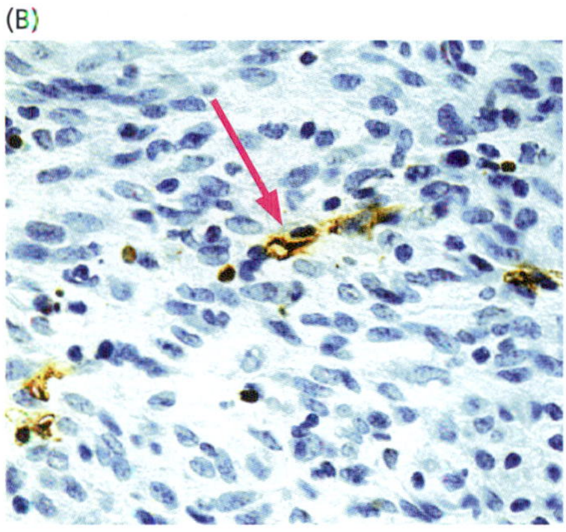

Figure 13.45 Thrombospondin, endothelial cell survival, and tumorigenesis (A) The stimulation of resting endothelial cells with angiogenic agents, such as VEGF or bFGF, creates activated, growing endothelial cells *(left branch),* which soon express the Fas death receptor on their surface. Subsequent treatment of mature and recently formed endothelial cells with Tsp-1 *(blue)* causes both groups of cells to secrete FasL *(green),* the ligand of Fas. However, because only the activated endothelial cell displays the Fas receptor, it is preferentially induced to enter into apoptosis by Tsp-1. This seems to explain why Tsp-1 can block new angiogenesis but has relatively minimal effects on already-constructed vasculature, whose endothelial cells are only rarely involved in active growth. (B) The loss of *ras* oncogene expression can lead to a rapid increase in Tsp-1 expression. This may explain why, in this transgenic mouse model of melanoma development, shutdown of *ras* oncogene expression in already-formed tumors leads to rapid collapse of the tumor-associated vasculature and tumor regression. Here, the TUNEL assay (Figure 3B) for apoptosis reveals apoptotic endothelial cells within a capillary *(arrow)* of a regressing melanoma. (A, adapted from O.V. Volpert, T. Zaichuk, W. Zhou et al., *Nat. Med.* 8:349–357, 2002; B, from L. Chin, A. Tam, J. Pomerantz et al., *Nature* 400:468–472, 1999.)

tumor that spawned them continued to grow. However, the moment the primary tumor was surgically removed, the metastases began to grow vigorously. This behavior is echoed by anecdotal reports of cancer surgeons, who have observed that after the successful surgical removal of a primary tumor, substantial numbers of metastases may suddenly sprout and flourish, doing so within several months' time.

Such observations suggested that some type of inhibitory substance released by the primary tumor acted, via the circulation, to suppress the proliferation of distant nests of metastatic cells. More specifically, these inhibitory factors, whatever their nature, seemed to block angiogenesis in these secondary growths, which failed to expand to a diameter of more than several tenths of a millimeter. Once the primary tumors were excised, the hypothetical inhibitory factor(s) disappeared from the circulation, removing some constraint on the growth of already-seeded metastases.

The subsequent isolation of these circulating factors, initially from the urine of tumor-bearing animals, yielded the intensively studied angiostatin and endostatin molecules. Determination of the amino acid sequences of these two protein species indicated that they arise as cleavage products of familiar proteins of the extracellular matrix (ECM) or plasma. With the passage of time, yet other anti-angiogenic substances have been isolated (see Table 13.3), many of which are also formed by the proteolysis of extracellular proteins. Taken together, these discoveries suggest that as angiogenesis proceeds in normal tissues during development and wound healing, this process is eventually curtailed by the accumulation of anti-angiogenic protein fragments in the extracellular space. These naturally occurring anti-angiogenic substances can therefore be depicted as important components of negative-feedback loops operating to ensure that excessive vascularization of a tissue does not occur.

In more detail, the specialized basement membrane surrounding most capillaries is the source of several proteins that have potent anti-angiogenic powers. Proteolytic cleavage of collagen XVIII yields the 20-kD C-terminal fragment that was discovered as endostatin. Yet another, tumstatin, derives from cleavage of one of the chains of collagen IV (which constitutes as much as half of the capillary basement membrane), yielding a 28-kD fragment. Collagen IV cleavage also yields another anti-angiogenic fragment termed arresten. And cleavage of plasminogen (the precursor of the plasmin protease involved in activating coagulation) generates the 38-kD internal fragment known as angiostatin.

Another important class of natural anti-angiogenic proteins function as antagonists of the matrix metalloproteinases (MMPs). VEGF stimulates the localized production of MMPs 1 to 4, which enable elongating capillaries to invade through the extracellular matrix between cells. A class of secreted proteins, termed tissue inhibitors of metalloproteinases (TIMPs), can prevent this elongation by blocking the actions of these and other MMPs. For example, by forcing the ectopic expression of TIMP-2 in tumor cells, researchers have blocked the angiogenic and thus tumorigenic powers of these cells. However, the precise mechanisms by which TIMPs block angiogenesis are still not clearly resolved (Sidebar 13.8).

Sidebar 13.8 TIMPs suppress angiogenesis in multiple ways Accumulating evidence indicates that the TIMPs can act on other molecular targets besides the MMPs. For example, the VEGF and fibroblast growth factor (FGF) receptors, which are required for endothelial cell responses to these two critical angiogenic growth factors, are also inhibited by TIMP-2. Evidence indicates that this receptor antagonism proceeds through a quite indirect route: TIMP-2 binds to a cell surface integrin, and the latter, acting via its cytoplasmic tail, activates an intracellular phosphatase that proceeds to dephosphorylate and thereby shut down signaling by the VEGF and FGF receptors. These diverse antineoplastic activities of TIMPs have made them attractive agents for development as anticancer therapeutics. However, in spite of numerous research-and-development projects, none of the TIMPs or TIMP analogs has yet proven to be a useful antagonist of tumor development when used in the clinic.

Arguably the most bizarre natural angiogenesis inhibitor is a variant form of tryptophanyl-tRNA synthetase. This enzyme is normally responsible for charging a tRNA with the amino acid tryptophan, and therefore is one of the core components of the protein synthesis apparatus. However, an alternatively spliced version of the mRNA encoding this enzyme leads to a truncated form of this protein, called mini TrpRS, which is secreted from cells; its production is inducible by interferon treatment of cells. This protein, which can specifically trigger the apoptosis of endothelial cells, is an example of the opportunistic use by evolution of any substance at hand to solve certain biological problems.

When integrated, these disparate observations about the physiologic regulators of angiogenesis reinforce the idea, cited earlier, that angiogenesis is not a binary state—either on or off. Instead, different types of tumor cells acquire greater or lesser angiogenic powers, and even within a given tumor, the tumor cells are likely to show differing abilities to attract vasculature. Such behavior is likely to be explained by a scheme (Figure 13.46) in which the balance between pro- and anti-angiogenic factors determines whether neoangiogenesis will proceed, and if so, how intensively regions within tumors will become vascularized.

13.10 Certain anti-angiogenesis therapies hold great promise for treating cancer

The more complex a system becomes, the more vulnerable it is to various types of disruption. The process of angiogenesis, as described here, clearly falls in the class of highly complex systems, as shown by its dependence on multiple cell types and signaling molecules. Indeed, while we have featured several important angiogenic factors, by some counts there are at least a dozen involved in regulating various steps of vascular morphogenesis (see Table 13.2).

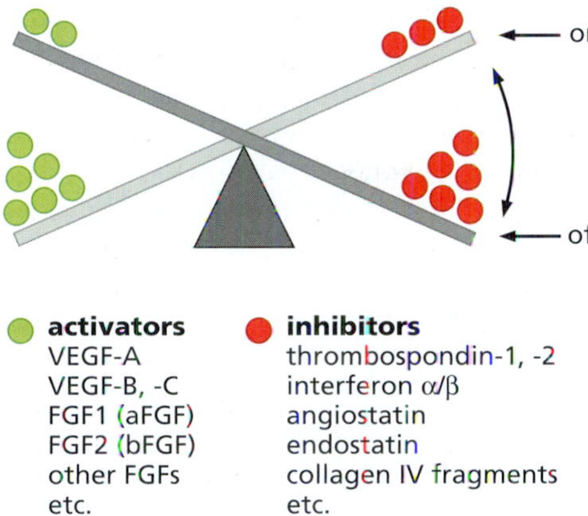

Figure 13.46 Balancing the angiogenic switch This diagram presents the major physiologic regulators that work to promote or inhibit angiogenesis within tissues and indicates that it is the balance between these two countervailing groups of regulators that determine whether or not angiogenesis proceeds. While the angiogenic switch is depicted here as an essentially on–off binary decision, in fact the vasculature that grows within a tumor mass after the switch has been activated shows various gradations of development. (From D. Hanahan and J. Folkman, *Cell* 86:353–364, 1996, and courtesy of D. Hanahan.)

For those researchers intent on developing new types of anti-cancer therapeutics, this complexity offers multiple targets for intervention. In particular, highly targeted therapies may be devised to inhibit the several cell types that participate in angiogenesis as well as the multiple signaling channels through which they intercommunicate. Since tumors depend absolutely on angiogenesis to grow above a certain size (~0.2 mm diameter), any successes in blocking angiogenesis or in undoing the products of angiogenesis should represent a highly effective strategy for treating cancer. Microscopic tumors should be prevented from growing larger, while larger tumors should collapse once their already-established blood supply disintegrates.

In principle, anti-angiogenesis therapies have a major advantage over those directed at the neoplastic cells within tumors. As we will learn in Chapter 16, one of the great frustrations of anti-cancer drug development comes from the fact that, sooner or later, tumors that have responded initially to a drug treatment will relapse and become **refractory** (resistant) to further treatment by the drug. Almost always, these relapses can be traced to the emergence of drug-resistant variants within tumor cell populations; these variants arise at an almost-predictable frequency and proceed to proliferate and regenerate aggressively growing tumor masses. The emergence of these drug-resistant variants seems to be one of the consequences of the highly unstable genomes of cancer cells (see Sections 12.11 and 12.12) and their resulting ability to spawn mutants at high frequency.

Many anti-angiogenesis therapies, in stark contrast, are directed at killing the genetically *normal* cells that have been recruited into tumor masses and co-opted by the cancer cells to do their bidding. There is every reason to believe that the endothelial cells within tumors possess normal, stable genomes and are therefore stable phenotypically. Hence, drug therapies directed against these cells are not likely to select for the outgrowth of drug-resistant variants, and tumors should not, in theory, become refractory to anti-angiogenic drug therapy (see, however, Sidebar 13.9).

This interest in treating tumor-associated endothelial cells is further heightened by the peculiar biology of these cells. They are continually being formed and lost within tumor masses, with lifetimes measured as short as a week, while their counterparts that line normal blood vessels elsewhere in the body rarely divide and have lifetimes that are measured in hundreds of days, some being as long as seven years. Cycling cells (i.e., cells racing around the active cell cycle) are, almost always, far more sensitive to drug-induced killing than are quiescent cells. For this reason, cytotoxic therapies directed against endothelial cells should have drastic effects on the tumor-associated vasculature, while leaving blood vessels located elsewhere in the body unscathed (but for the caveat cited in Sidebar 13.9).

Sidebar 13.9 Tumors may outsmart even the best anti-angiogenic therapies Much of the allure of anti-angiogenic therapies comes from the likelihood that the cells being targeted, notably the endothelial cells, are unlikely to generate drug-resistant variants. The complexity of heterotypic signaling may, however, allow tumors to circumvent even the cleverest anti-angiogenic therapies. Imagine, for example, that we devise a strategy to block the VEGF signaling that is so critical for the formation of new vessels in a tumor and their subsequent maintenance (since VEGF is required for endothelial cell survival). A tumor treated in this way should rapidly collapse, since its capillary beds will disintegrate. Indeed, just such a response has been observed in Rip-Tag mice that developed pancreatic islet tumors (see Figure 13.37): When treated with an anti-VEGF-R2 antibody, their tumors regressed by more than 50%. However, the residual, surviving tumor cells responded by spawning variants that acquired the ability to produce elevated levels of fibroblast growth factors (FGFs), which are also potent angiogenic factors. These growth factors then supplanted VEGF as the main conveyor of signals from the islet tumor to the endothelial cells and succeeded in triggering the regeneration of vasculature and, in turn, the rebirth of a vigorously growing tumor. Accordingly, truly effective anti-angiogenesis strategies will require the inhibition of multiple angiogenic pathways.

Anti-angiogenesis therapies are also attractive because of their potential selectivity for killing the proliferating endothelial cells within tumors, since many blood vessels in these growths seem to be in a constant state of formation and collapse, and are therefore susceptible to cytotoxic drugs that may leave the quiescent endothelial cells in normal tissues untouched. For example, dramatically contrasting growth states of normal and tumor-associated blood vessels can be seen in many mouse models of cancer, in which engrafted tumors expand rapidly. However, it is possible that many of the vessels in slowly growing human tumors may have existed for years and therefore may have had ample time to consolidate and mature into robust, well-structured vessels in which the endothelial cells turn over slowly. These tumor-associated vessels may then be as resistant to anti-angiogenic drugs as the normal blood vessels elsewhere in the body.

We will learn in great detail about the general principles of anti-cancer therapy later in Chapter 16 but will take the opportunity here to describe the details of emerging therapies that are directed specifically against the tumor-associated vasculature. Some of the first efforts in this area have come from experiments in which the natural anti-angiogenesis inhibitors mentioned above have been used to treat tumors borne by mice.

Since these agents, such as angiostatin and endostatin, are native to the body, they have the advantage of being reasonably well tolerated without toxic side effects. They have virtually no effect on the *in vitro* proliferation of a variety of cells, and instead are biologically active only on endothelial cells that are actively participating in neoangiogenesis *in vivo*. Moreover, these proteins can persist for some time in the circulation, thereby increasing the exposure of tumor vasculature to their anti-angiogenic effects. On the negative side, their manufacture in large quantities, like that of all proteins, is costly and challenging, and their precise mechanisms of action have been elusive.

Both angiostatin and endostatin had only modest effects in blocking the angiogenic switch in the pancreas of Rip-Tag mice (Section 13.7). But when applied to already-vascularized small tumors, endostatin reduced by more than 80% their subsequent growth, while angiostatin reduced this size by 50%. When large, well-established tumors were treated with either of these agents, relatively little effect was seen, but the two introduced together caused a 75% reduction in the mass of such tumors. Long-term endostatin treatment of mice bearing tumors formed from a lung carcinoma cell line led to regression of the tumors. This success might suggest that endostatin may be a highly effective therapeutic in the oncology clinic. However, responses like these in mice are rarely predictive of similar efficacy in the oncology clinic. Some dramatic responses have been reported following endostatin treatment in a small number of patients treated in early clinical trials.

Yet another class of natural angiogenesis antagonists (see Table 13.3) are the interferons, which are usually studied in the context of their ability to modulate the activities of various cell types operating in the immune system. Interferon-α and -β have proven to be potent suppressors of the synthesis of basic fibroblast growth factor (bFGF) and of interleukin-8 (IL-8), both of which are strong angiogenic agents. And administration of interferon-α has proven to be useful in causing the regression of some hemangiomas (endothelial cell tumors) as well as Kaposi's sarcomas, which are also of endothelial cell origin (possibly the endothelial cells forming lymph ducts). In both cases, the tumor regressions have been attributed to the anti-angiogenic effects of the interferon.

An experiment reported in 1991 provided one of the earliest indications of the promise of another type of anti-angiogenic therapy. In this work, cancer cells were engineered to release a modified basic fibroblast growth factor (bFGF), which greatly increased their tumorigenicity in mice, simply because this bFGF strongly enhanced the angiogenic powers of these cells. Use of a monoclonal antibody that specifically bound and neutralized this bFGF (but had no effect on the endogenous bFGF of the mouse) blocked the angiogenicity of these tumor cells and led to a dramatic reduction in tumor volume. Within two years, a similar experiment was performed with an anti-human-VEGF monoclonal antibody. It succeeded in blocking the proliferation of two human sarcoma cell lines as well as a glioblastoma in Nude (immunocompromised) mouse hosts.

A more modern version of this therapy came a decade later with the use of a monoclonal antibody that binds and neutralizes VEGF-A. This antibody, termed variously Avastin and bevacizumab, showed significant efficacy in large-scale clinical trials. For example, patients with metastatic colon carcinoma who were treated with this antibody plus chemotherapy (the drug 5-fluorouracil) survived, on average, four months longer than patients treated with chemotherapy alone, and the addition of Avastin to conventional chemotherapy extended the survival of patients with non-small-cell lung carcinoma (NSCLC) by about two months. Similarly, Avastin could retard the progression of renal cell carcinomas in patients, but in the end had no effect on their long-term survival. It is plausible that the synergistic effects of Avastin with conventional chemotherapeutic drugs derive directly from the ability of this VEGF-A inhibitor to normalize tumor-associated vasculature (see Figure 14 ◉), thereby greatly facilitating the delivery of drugs to the tumor parenchyma.

Some synthetic low–molecular-weight anti-angiogenic compounds have been discovered or developed that are directed against various molecular targets in the angiogenic program. The first of these was fumagillin, an anti-angiogenic compound of fungal origin, and its chemical derivative, TNP-470 (Table 13.4C). These two compounds were found to inhibit the proliferation of endothelial cells both *in vitro* and *in vivo*; this suggests that their anti-angiogenic effects *in vivo* derive from their ability to prevent the growth of new capillaries, which depends on endothelial cell proliferation. Importantly, TNP-470 had no effect on the proliferation of tumor cells *in vitro* but strongly blocked their tumorigenicity in mice. For example, this drug reduced by 70 to 80% the sizes of pancreatic islet tumors arising in the Rip-Tag mice. The mechanism of action of TNP-470 remains obscure, but is under active investigation. Its proven ability to inhibit the methionine aminopeptidase-2 enzyme in endothelial cells may one day help to explain its preferential effects on these cells while it has almost no effects on other cultured cell types.

The most informative studies of angiogenic inhibitors have come using synthetic receptor inhibitors (see Table 13.4B) in the transgenic Rip-Tag model of pancreatic islet tumorigenesis (see Section 13.7). Two types of low–molecular-weight synthetic compounds have been utilized in attempts to block various stages of islet tumor progression. One of these drugs is directed against VEGF-R2, which is the main receptor for angiogenesis in this tumor model. By inhibiting the tyrosine kinase of this receptor, the agent termed SU5416 should mimic the effects of Avastin (the anti-VEGF monoclonal antibody described earlier), that is, both should be able to shut down VEGF signaling. Drugs of a second class, such as an agent termed SU6668, are directed primarily against the tyrosine kinase of the PDGF receptor. Our discussions here about the role of PDGF

Table 13.4 Angiogenesis inhibitors and their development and use in clinical trials

Name	Status	Responses
A. Endogenous inhibitors of angiogenesis		
Endostatin	in clinical trial	scattered responses
Interferons-α and -β	effective in treating hemangioblastomas	Kaposi's sarcomas; limited efficacy against most other types of tumors
B. Agents that block VEGF and VEGF-R signaling		
Avastin anti-VEGF MoAb	in clinical trial	delayed progression 1–3 months in lung, 3–4 months in colon
SU5416 inhibitor of VEGF-R2 (Flk-1)	trial abandoned	severe vascular toxicities
ZD6474 inhibitor of VEGF-R2	under clinical test	
CP547, 632 inhibitor of VEGF-R2	in trial	
C. Miscellaneous other drugs		
Thalidomide	in trial	inhibits bFGF- and VEGF-dependent angiogenesis
Squalamine sterol from shark liver	in trial	strong anti-angiogenic activity
Celecoxib anti-inflammatory drug	in trial	multiple anti-neoplastic effects
ZD6126	in trial	antagonist of tubulin in endothelial cell cytoskeleton
Fumagillin and TNP-470	in trial; slowed tumor growth	antagonist of methionine aminopeptidase in endothelial cells
D. Inhibitors of ECM breakdown—MMP inhibitors		
Marimastat	in clinical trial	no delay of tumor progression
Prinomastat	in clinical trial	no slowing of tumor progression
BMS275291	in clinical trial	
BAY12-9566	in clinical trial	
Neovastat (shark cartilage MMPI)	in clinical trial	

in recruiting pericytes and smooth muscle cells to growing capillaries (Sidebar 13.1) indicated that these "mural" cells are highly important for consolidating and strengthening recently formed capillaries.

SU5416, the anti-VEGF-R agent, was able to block 90% of the early-stage, dysplastic islets from undergoing the angiogenic switch, thereby holding them to a small size and a noninvasive state (Figure 13.47C). However, SU5416 had no effect on late-stage, well-established tumors, which continued to progress in spite of its presence in equally high concentrations. Hence, in the early stages of angiogenesis, VEGF signaling plays a critical role, while later on, this process seems to become increasingly independent of VEGF.

The anti-PDGF receptor agent, SU6668, had a far weaker effect on preventing dysplastic islets from undergoing the angiogenic switch, reducing by about half the islets that did so (see Figure 13.47C). But it was far more potent than the anti-VEGF-R drug in treating the late-stage, advanced tumors, reducing their size by about half and substantially reducing their vascularity (Figure 13.47D). Importantly, the only cells expressing PDGF-R in and near these tumors were the capillary-associated pericytes and related smooth muscle cells (Figure 13.47B), indicating that these mural cells were the targets of SU6668 action. Indeed, microscopic examination confirmed that SU6668 had prevented these mural cells from associating with and reinforcing the capillary tubes formed by endothelial cells. Together, these experiments showed that the initial steps of angiogenesis could proceed reasonably well without PDGF receptor function, but that later in tumor progression, PDGF signaling, and thus the involvement of mural cells, became increasingly important to angiogenesis and growth of the tumor masses.

Combination therapy using the two agents proved to be a highly potent way of intervening at various stages of tumor formation (see Figure 13.47C and D). Thus, simultaneous inhibition of both VEGF and PDGF receptors prevented

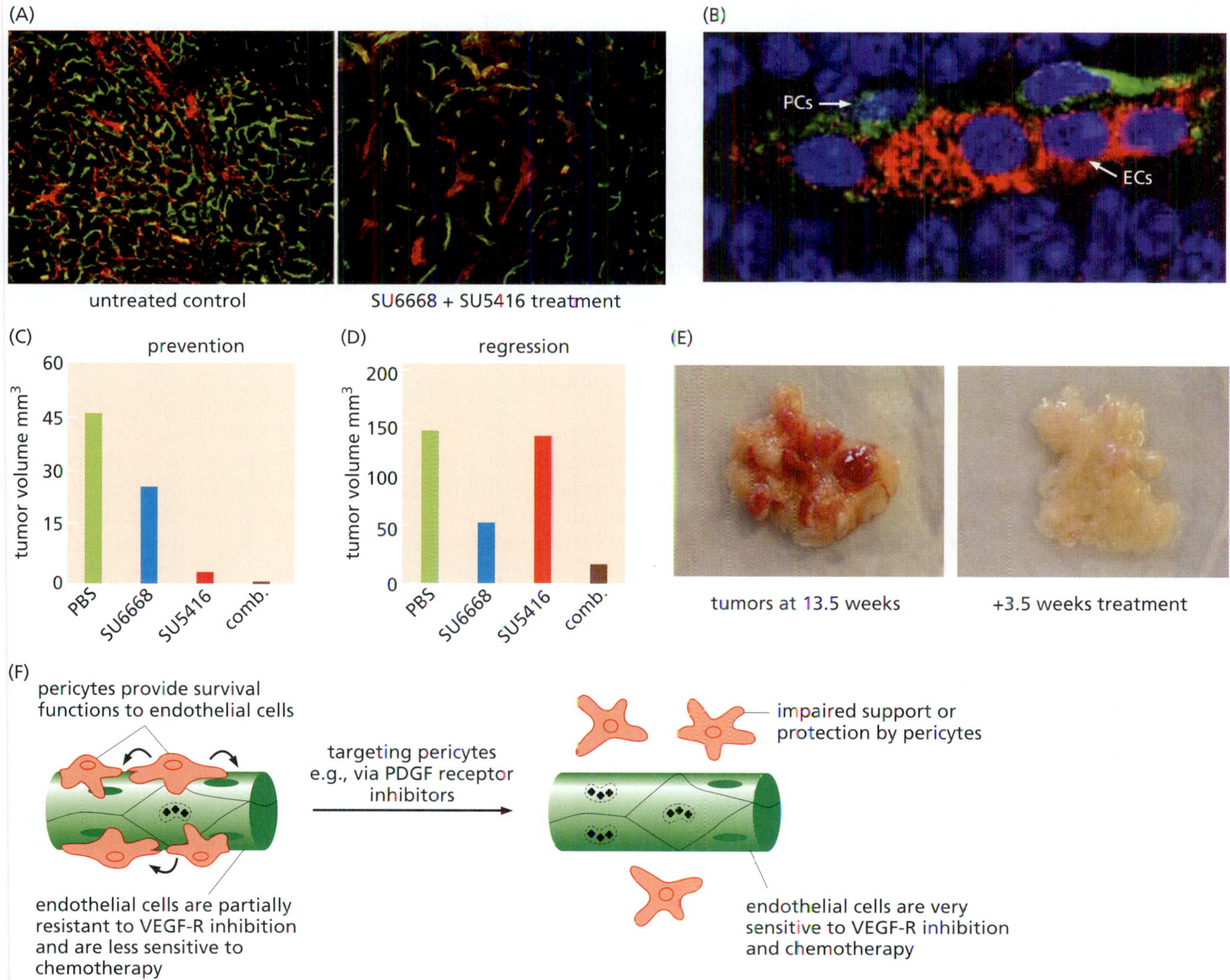

Figure 13.47 Angiogenesis inhibitors as treatments of islet cell carcinogenesis The Rip-Tag transgenic mouse model of pancreatic islet cell carcinogenesis makes possible the testing of anti-angiogenic pharmacologic inhibitors. (A) The treatment of mice bearing established angiogenic islet tumors *(left)* for 4 weeks with the PDGF receptor inhibitor SU6668 plus the VEGF-R inhibitor SU5416 results in the regression of the vasculature *(right)*. The lumina of the capillaries are labeled *yellow* while the associated pericytes are labeled *red*. (B) The actions of the SU6668 PDGF-R inhibitor could be traced specifically to its effects on the pericytes that act as support cells for the endothelial cells. In this image, an antibody reactive with the PDGF-R reveals that the only cells within the islets that express this receptor are the pericytes (PCs, *green)* that are closely associated with the endothelial cells (ECs, *red)*. Nuclei are stained *blue*. (C) While neither the SU6668 PDGF-R inhibitor *(blue bar)* nor the SU5416 VEGF receptor inhibitor *(red bar)* was able, on its own, to fully prevent the formation of such tumors, the two applied in concert (combined, *purple bar)* succeeded in doing so. PBS, phosphate-buffered saline control *(green)*. (D) In an attempt to cause regression of already-formed tumors, SU5416 and SU6668 were introduced at 12 weeks of age into Rip-Tag mice either singly or in combination and the size of tumors was measured 4 weeks later. While the SU5416 VEGF-R antagonist on its own *(red bar)* showed only minimal reduction of tumor volume, the SU6668 PDGF-R antagonist showed greater effects *(blue bar)*, and the two together *(brown bar)* worked synergistically to reduce overall tumor volume by approximately 85%. This indicates, once again, that antagonists of the endothelial cells (which depend on the VEGF-R) together with antagonists of the supporting pericytes (which depend on the PDGF-R) can act synergistically in anti-angiogenic therapy. (E) The combined effects of these two drugs can be seen: when 13.5-week-old mice that have developed substantial pancreatic islet tumors *(red growths, left)* are treated for 3.5 weeks with this combination therapy, their tumors largely regress *(right)*. (F) A schematic summary of these and other observations indicates that endothelial cells depend on closely apposed pericytes for various types of biological support. Inhibition of PDGF signaling, as achieved by certain inhibitors, causes dissociation of pericytes from endothelial cells and renders the latter sensitive to various types of subsequent therapy, including inhibition of VEGF-R function. This emphasizes the fact that anti-angiogenesis therapy is most effective when two synergistically acting treatments are applied. (A, B, and E, from G. Bergers, S. Song, N. Meyer-Morse et al., *J. Clin. Invest.* 111:1287–1295, 2003; C and D, courtesy of D. Hanahan; F, from K. Pietras and D. Hanahan, *J. Clin. Oncol.* 23:939–952, 2005.)

angiogenic switching, holding virtually all (>98%) islets at a pre-angiogenic stage. This drug combination blocked the further expansion of small, already-established angiogenic tumors by 90% and caused an approximately 80% regression of large tumors, as did another PDGF-R inhibitor, the drug Gleevec, which we will read about in great detail in Chapter 16. In some mice, these combination drug therapies held already-formed tumors to a small size for as long as two months.

Significantly, these combinations of drugs had virtually no toxic effects on normal pancreatic tissue adjacent to neoplastic islets, confirming that recently formed capillaries within a tumor are far more vulnerable to disruption than the well-established vessels within normal tissues. In addition, it seemed that PDGF played a major role in initially attracting pericytes to the capillary tubes formed by endothelial cells, but in normal tissues, continued PDGF signaling was not required to maintain this association. These observations indicate that the most effective ways of inhibiting angiogenesis and thus blocking tumor progression are likely to depend on targeting several of the cell types that construct the tumor-associated vasculature.

The results of these targeted anti-angiogenesis treatments illustrate the highly promising nature of this general therapeutic strategy. In fact, this type of therapy may have been in use for many years: unbeknownst to the community of clinical oncologists, much of the efficacy of some widely used, traditional anti-tumor therapies may also derive from their effects on tumor-associated microvasculature. One striking example of this, which sheds light on how radiation therapy succeeds in destroying certain tumors, has come from studies of genetically altered mice that lack the genes encoding either the Bax protein or the acid sphingomyelinase (asmase) enzyme. Both of these proteins are important pro-apoptotic regulators in a variety of cell types including, importantly, endothelial cells. Consequently, the endothelial cells from these genetically altered mice are far more resistant to toxic agents, including X-rays, than their wild-type counterparts.

In the experiments shown here, mouse tumor cells—from either a fibrosarcoma or a melanoma cell line—were introduced into both wild-type and *asmase*$^{-/-}$ mice. After tumors had grown to a substantial size, these animals were exposed to a dose of 15 grays (Gy) of X-rays (Figure 13.48A and B). This dose of radiation usually causes a significant reduction in the tumor burden, and indeed it succeeded in doing so in the wild-type mice. However, the identically sized tumors grown in the *asmase*$^{-/-}$ host mice responded quite differently, in that their growth continued unabated. When, as a further experiment, the bone marrow of *asmase*$^{+/+}$ mice was replaced with transplanted mutant bone marrow cells, the tumors became highly resistant to X-ray–induced killing (Figure 13.48C).

These experiments demonstrate that the radiosensitivity of these tumors was not intrinsic to the tumor cells themselves. Instead, it was governed by host cells, specifically by cells recruited into engrafted tumors from the host bone marrow. These recruited host cells were indeed endothelial cells, as evidenced by the fact that (1) the apoptosis of the endothelial cells in tumor fragments irradiated *in vitro* directly paralleled and predicted the *in vivo* behavior of the tumors, and (2) the only cells that showed significant apoptosis in the tumors shortly after irradiation were associated with capillaries (Figure 13.48B).

Observations like these indicate that anti-angiogenesis therapy has played a far greater role in a conventional anti-tumor **radiotherapy** than anyone dared to imagine; similarly, clinical responses to certain types of conventional chemotherapy may also be strongly influenced by the sensitivity of the tumor-associated microvasculature to these agents. This suggests that in the future, treatments with many anti-cancer chemotherapeutics may be optimized by gauging their effects on the tumor-associated microvessels rather than on the tumor cells themselves.

13.11 Synopsis and prospects

Metazoan tissues are organized as condominiums of various cell types that are continuously communicating with one another. To the developmental biologist, the need for this organizational plan is self-evident: only through such interactions can the proper numbers and locations of each of these cell types be ensured. These heterotypic interactions continue to operate after embryogenesis has been completed in order to support the maintenance and repair of already-formed tissues.

As we have learned in this chapter, this organizational plan confers another, quite distinct benefit on the organism. By making its cells so interdependent, none is easily able to extricate itself from its complex web of interactions and go off on its own. Interdependence imposes the regimentation that wards off the chaos of neoplasia.

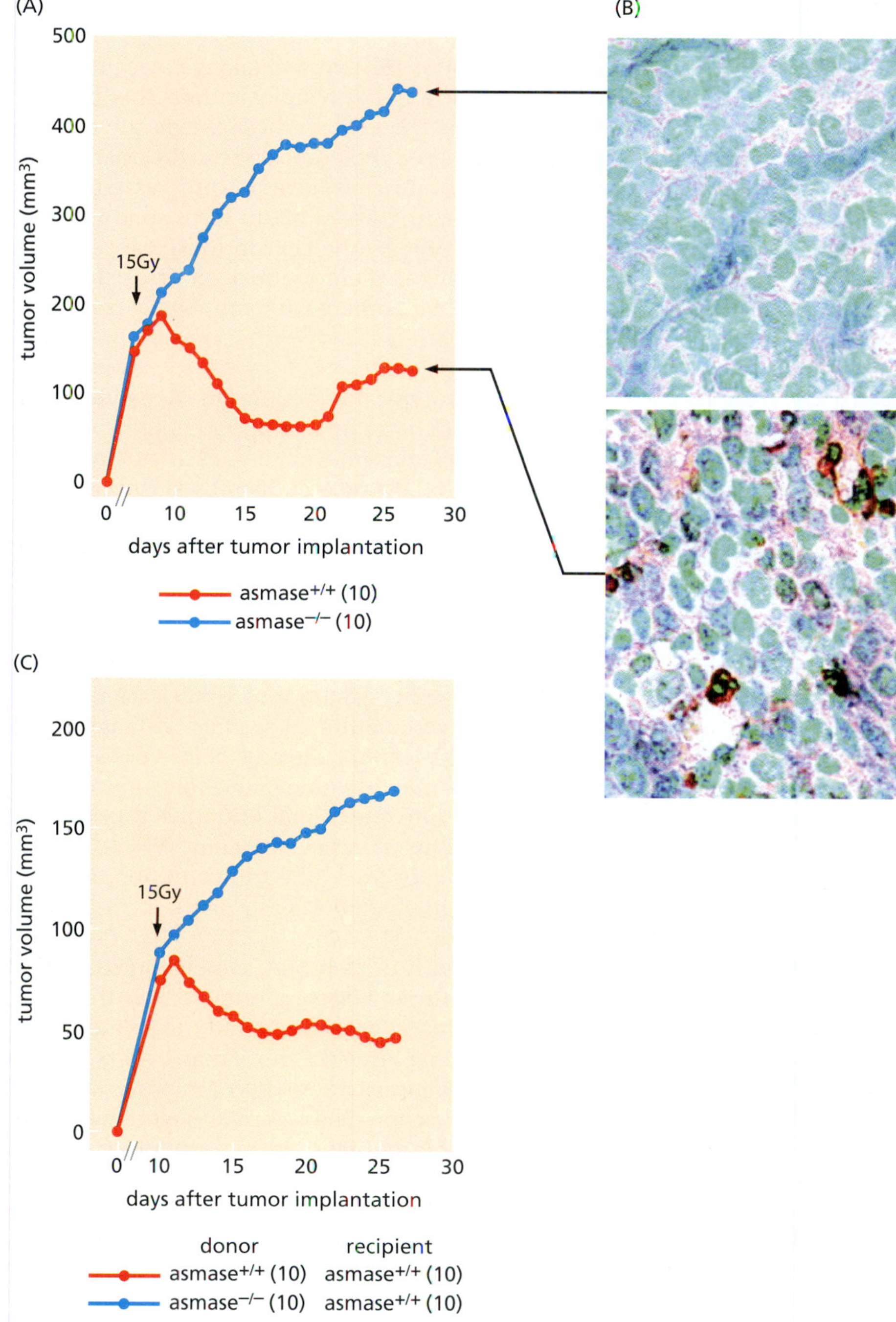

Figure 13.48 Survival of wild-type and asmase-negative tumor-bearing mice The responses of tumors to radiotherapy may often be determined by the radiosensitivity of the endothelial cells that form their vasculature. Mice were bred to be either wild type or homozygous null for the gene encoding the pro-apoptotic enzyme acid sphingomyelinase (asmase). When mouse fibrosarcoma cells were implanted in the wild-type and the *asmase*$^{-/-}$ hosts, the tumors grew twice as fast in the mutant mice as in the wild-type mice, suggesting that a host factor, such as recruited endothelial cells, was governing the rate of tumor growth. (A) When these tumors were given a therapeutic dose (15 Gy) of radiation *(arrow)*, the tumors in wild-type mice regressed *(red filled circles)* and, after a while, began to grow again. In contrast, the tumors in the *asmase*$^{-/-}$ hosts *(blue filled circles)* continued to grow unabated. (B) When these tumors were examined microscopically following irradiation, the endothelial cells in tumors carried by wild-type mice *(below)*, identified by an endothelial cell–specific immunostain, were apoptotic, as indicated by the TUNEL staining *(brown spots)*, while the endothelial cells in tumors borne by the *asmase*$^{-/-}$ hosts showed no signs of apoptosis *(above)*. (C) When the bone marrow of wild-type hosts was replaced through transplantation of either wild-type or *asmase*$^{-/-}$ donor marrow cells, the tumors implanted in mouse hosts with engrafted mutant marrow *(blue curve)* continued to grow following 15 Gy of radiation *(arrow)*, while tumors implanted in mouse hosts engrafted with wild-type marrow were stopped by the radiation *(red curve)*. This demonstrated that the radiosensitivity of the tumor was determined by cells of host bone marrow origin, not by the tumor cells themselves. (From M. Garcia-Barros, F. Paris, C. Cordon-Cardo et al., *Science* 300:1155–1159, 2003.)

Sidebar 13.10 Kaposi's sarcoma cells hold the record for the number of documented heterotypic signals they receive Kaposi's sarcoma (KS) seems to derive from the cells related to those generating the endothelial linings of lymphatic vessels. The causal agent of KS—human herpesvirus-8 (HHV-8)—flourishes when immune defenses have been compromised, explaining the high frequency of this tumor in AIDS patients (see Sidebar 3.10). Gene expression analyses from several laboratories have documented the increased expression of mRNAs encoding 15 distinct receptors in KS cells; the respective ligands for these receptors appear not to be produced by the KS cells. This suggests that the elevated expression of these receptors allows KS cells to take advantage of ligands that are supplied, via *paracrine* signaling, by other nearby cell types. The list of receptors includes those binding the following ligands: TNF-α, INF-γ, oncostatin M, bFGF, VEGF-C and -D, insulin, thromboxane A2, IL-2, IL-6, IL-10, IL-13, thrombospondin, VEGF 165, CXCL12, and CSF-1. (The dependence of KS cells on all of these signals has not yet been demonstrated experimentally.) This complexity hints that other histologically complex tumors, like KS, may rely on multiple heterotypic signals to sustain their proliferation and survival.

Earlier, we viewed multi-step cancer progression as the hurdling of successive barriers placed in the path of developing cancer cells (Chapter 11). Each successfully completed step, whether achieved by genetic or epigenetic changes, removes one of these obstacles and places the cancer cell incrementally closer to full-fledged malignancy. Now we can conceptualize multi-step tumor progression in a quite different way: as pre-malignant cells evolve toward malignancy, they progressively sever their ties with their neighbors and their dependence on neighborly support.

Perhaps the biggest surprise is how dependent most cancer cells remain on stromal support in spite of having completed multiple steps of tumor progression. The epithelial cells residing in many carcinomas continue to rely on many of the physiologic signals that sustained their precursors in normal tissues (see, for example, Sidebar 13.10). This conservatism is evident under the microscope. Thus, a well-trained pathologist can recognize the origins of perhaps 95% of the tumor samples viewed under the microscope, because most of the heterotypic interactions that govern normal morphology are still operative in the great majority of cancers.

The remaining approximately 5% of tumors present a challenge, because they are anaplastic and therefore have lost most of the histologic traits that make it possible to identify their tissue of origin. The cells in these anaplastic tumors have shed most forms of dependence that tied their precursors to normal neighboring cells. However, most anaplastic tumor cells have not progressed all the way to total independence, because they still assemble to form solid tumors. The ultimate independence is achieved only by the cells in those cancers that have advanced so far that they can grow as pleural effusions or ascites (see Sidebar 25) and therefore have no direct contact with supporting cells and, apparently, with an extracellular matrix (ECM).

Even without detailed knowledge of heterotypic interactions, the dependence of most carcinoma cells on at least one form of stromal support could have been predicted from the known physiology of mammalian tissues: virtually all of them depend on a functional blood supply. Less predictable was the mechanism by which tumors acquire their vasculature. Rather than invading normal tissue and expropriating existing capillary beds, tumors actively recruit endothelial cells that proceed to construct capillaries and larger vessels within the tumor masses.

The sources of these endothelial cells could also not be deduced from first principles. The proliferation of endothelial cells in neighboring tissues seems to be the main mechanism for acquiring neovasculature soon after the angiogenic switch has been tripped. However, in many tumors, the bulk of the vessels made thereafter seem to be constructed by cells that originate in the bone marrow and then differentiate in the tumor stroma into functional endothelial cells. The contribution of these circulating endothelial precursor cells (EPCs) to the tumor-associated vasculature seems to vary from one tumor to another, but clearly represents a significant source in many neoplasias.

Major aspects of angiogenesis are still poorly understood, even paradoxical. For example, it seems apparent that once tumors become angiogenic, they can launch into a prolonged phase of growth and expansion, and that tumors that are more angiogenic can grow even more rapidly than those that are less able to attract new vasculature. Indeed, as we read, measurements of microvessel density—the number of capillaries per microscope field—correlate quite well with the likelihood that a primary tumor, such as a breast cancer, will progress to a highly malignant endpoint.

The paradox comes from the frequent observations that patients bearing highly hypoxic tumors also confront a poor prognosis. This makes no sense, given that hypoxia starves tumor cells and often leads to their death through apoptosis and, on a larger scale, to extensive necrotic regions within tumors. The paradox may one day be resolved by invoking the actions of the HIF-1 transcription factor, which becomes activated in hypoxic cells (Section 7.12) and induces production of a large number of other proteins besides VEGF. Indeed, a high level of HIF-1 expression is also an indicator of poor clinical prognosis. Included among the genes activated by HIF-1 are those specifying PDGF, TGF-α, TGF-β, and several matrix metalloproteinases (MMPs) that are responsible for remodeling the extracellular matrix (ECM). As we have learned, several of these secreted proteins act as potent mitogens that drive the proliferation of both epithelial cells and their stromal neighbors.

An additional HIF-1–induced gene encodes the Met protein, which functions as the receptor for hepatocyte growth factor (HGF), also known as scatter factor (SF). Once the Met receptor is activated by binding its HGF ligand, it activates a diverse set of responses in epithelial cells, including the epithelial–mesenchymal transition (EMT), increased motility, invasiveness, and proliferation. HGF seems to be widely available in many human tumors of both epithelial and mesenchymal origin. Consequently, HIF-1–mediated increases in Met expression can sensitize tumor cells to HGF molecules that are present in their surroundings, such as the HGF that has been released by stromal cells. These diverse products of HIF-1 action may therefore help explain the poor prognosis attached to hypoxic tumors, which thrive in spite of great adversity and actually turn out to be more aggressive than their well-oxygenated counterparts.

The recently discovered dynamic interactions between tumors and the bone marrow have been a surprise to most cancer researchers. In the absence of metastasis, most tumors have traditionally been considered to be localized diseases confined to one or another corner of the body. But as we learn more about the cells of the stromal compartment, the more we come to realize that even localized tumors extend their reach far and wide throughout the body in order to recruit the cells that they need to support their own survival and proliferation programs. Besides endothelial precursor cells (EPCs), cited above, carcinomas recruit mast cells and monocytes from the marrow, the latter differentiating on-site into macrophages. Even some stromal fibroblasts and myofibroblasts may have bone marrow origins, since they may often arise from circulating precursor cells termed fibrocytes.

The physiology of the carcinomas, on which we have focused in this chapter, is far more complex than bidirectional exchanges of signals between epithelium and stroma. The latter is composed of more than half a dozen distinct cell types (fibroblasts, myofibroblasts, endothelial cells, pericytes, smooth muscle cells, mast cells, macrophages, lymphocytes, and, in some tissues, adipocytes), each of which signals to the neoplastic epithelial cells as well as to the other components of the stromal compartment. Physicists have struggled, so far unsuccessfully, with solutions to the three-body problem. Here, we confront a world of as many as eight or more distinct cell types, each of which is sending a complex mixture of signals to other types of cells within tumors. We have only begun to scratch the surface of the complex epithelial–stromal interactions operating in neoplastic tissues and in their normal counterparts.

For more than half a century, the mindset of oncologists has been focused on eradicating the bulk of the cancer cells within solid tumors, with the hope of achieving the impossible—cures of common carcinomas that have traditionally been incurable. As we learned, first in Chapter 11 and now in this one, this focus needs to be radically redirected. First, the true targets of anti-cancer therapies can no longer be the bulk of neoplastic cells in a tumor, because they have limited self-renewal capacity; instead, we are likely to find that truly durable cures can come only from eradicating the still-elusive tumor stem cells that hide out, here and there, throughout tumor masses and represent their engines of self-renewal. Second, many of the useful anti-cancer therapies to be developed in the future will not come from targeting the cancer cells themselves. Instead, it

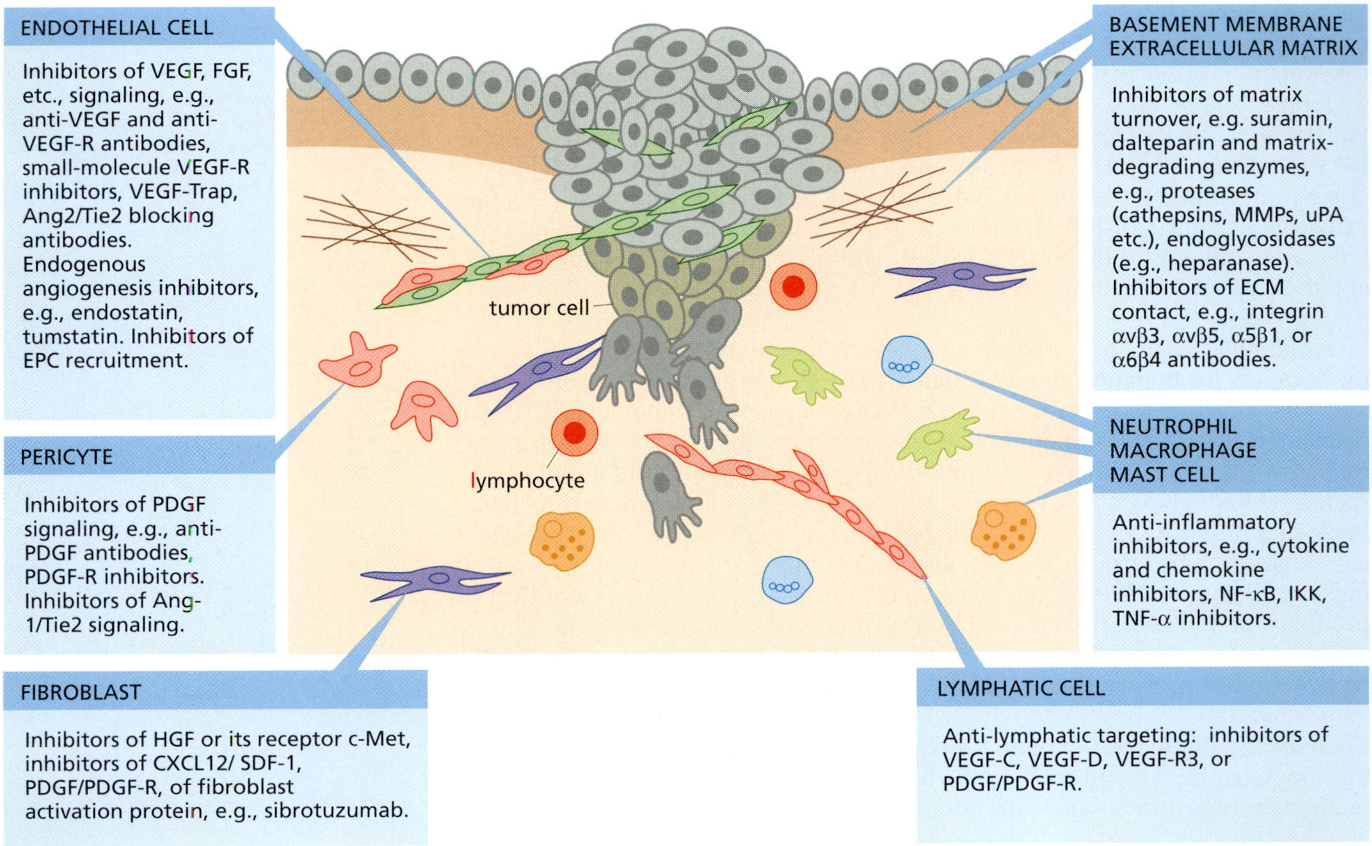

ENDOTHELIAL CELL

Inhibitors of VEGF, FGF, etc., signaling, e.g., anti-VEGF and anti-VEGF-R antibodies, small-molecule VEGF-R inhibitors, VEGF-Trap, Ang2/Tie2 blocking antibodies. Endogenous angiogenesis inhibitors, e.g., endostatin, tumstatin. Inhibitors of EPC recruitment.

PERICYTE

Inhibitors of PDGF signaling, e.g., anti-PDGF antibodies, PDGF-R inhibitors. Inhibitors of Ang-1/Tie2 signaling.

FIBROBLAST

Inhibitors of HGF or its receptor c-Met, inhibitors of CXCL12/ SDF-1, PDGF/PDGF-R, of fibroblast activation protein, e.g., sibrotuzumab.

tumor cell

lymphocyte

BASEMENT MEMBRANE
EXTRACELLULAR MATRIX

Inhibitors of matrix turnover, e.g. suramin, dalteparin and matrix-degrading enzymes, e.g., proteases (cathepsins, MMPs, uPA etc.), endoglycosidases (e.g., heparanase). Inhibitors of ECM contact, e.g., integrin $\alpha v\beta 3$, $\alpha v\beta 5$, $\alpha 5\beta 1$, or $\alpha 6\beta 4$ antibodies.

NEUTROPHIL
MACROPHAGE
MAST CELL

Anti-inflammatory inhibitors, e.g., cytokine and chemokine inhibitors, NF-κB, IKK, TNF-α inhibitors.

LYMPHATIC CELL

Anti-lymphatic targeting: inhibitors of VEGF-C, VEGF-D, VEGF-R3, or PDGF/PDGF-R.

Figure 13.49 Heterotypic interactions as targets for therapeutic intervention As described in this chapter, cancer cells are dependent on the nearby stromal microenvironment for a variety of cell-physiologic supports. This dependence on heterotypic interactions has inspired a new type of cancer therapy, some of which has been featured in this chapter. Instead of focusing on the intracellular signaling defects within cancer cells, this new type of therapy is directed toward interrupting heterotypic signaling, thereby depriving cancer cells of essential stromal support. This scheme indicates some of the anti-tumor therapies that are being developed or under consideration. (From J.A. Joyce, *Cancer Cell* 7:513–520, 2005.)

may often be far more profitable to attack the cells that provide them with vital physiologic support; by undermining the elaborate stromal support network on which most cancer cells depend, truly dramatic regressions of solid tumors may one day be achieved (Figure 13.49).

The insight that tumors are "wounds that do not heal" extends and echoes our earlier discussions (Section 11.15), in which the critical role of chronic inflammation in promoting tumor formation was discussed in great depth. Inflammation and wound healing are intertwined processes, and the mechanisms of inflammation-driven tumor promotion, which lead to the initial formation of cancers, are extended and elaborated by the chronic wound healing that seems to best characterize the biology of the stroma of well-established tumors.

Here, too, there has been another surprise. Inflammatory cells, notably macrophages, have traditionally been depicted as the front-line soldiers of the immune response that deal effectively with infectious agents, such as bacteria, by phagocytosing them and help guide the long-term immune response through antigen presentation, as we will see later in Chapter 15. Now we learn that macrophages are also critical agents in producing mitogenic growth factors, liberating angiogenic factors, and remodeling the extracellular matrix (ECM); the latter process is also critical for tumor invasion and metastasis (Chapter 14).

So, the traditional job assignments of various differentiated cell types are being extended and blurred. Cells of the immune system, which are purportedly dispatched to protect us from infection and even cancer, are often active collaborators in tumor development. And the deletion of one or another cell type from the immune system, achieved in mice through germ-line re-engineering, often creates a host organism that is, paradoxically, less able to support tumorigenesis.

Normal, quite elaborate morphogenetic programs, such as wound healing and the epithelial–mesenchymal transition (EMT), are likely to explain how carcinoma cells are clever enough to acquire the complex cell phenotypes that they need in order to execute the later stages of malignant progression. In hindsight, this notion is not so surprising, since the more we learn about cancer cells, the more we realize how opportunistic they are in co-opting and exploiting normal biological processes in order to further their own ends. This leaves us with the last question of this chapter: Have we begun to truly understand the mechanistic complexity of heterotypic interactions, or is there an entire universe of undiscovered signaling pathways and behavioral programs lurking within tumors, waiting, like intergalactic dark matter, to surprise us?

Key concepts

- Tumors are complex tissues that depend on intercommunication between various cell types. Indeed, most tumors are as complex histologically as the normal tissues in which they arise.
- In carcinomas, these cell types can be separated into the neoplastic epithelial cells and recruited stromal cells, which include fibroblasts, myofibroblasts, and macrophages, as well as the various cell types that participate in the construction of the tumor-associated vasculature, specifically endothelial cells, pericytes, and smooth muscle cells.
- Most carcinomas depend absolutely on recruited stromal cells for various types of physiologic support. This dependence is lost only in the small subset of tumors that progress to an extremely malignant state, notably the tumor cells growing in ascites and pleural fluid.
- At the biochemical level, this interdependence is manifested by the exchange of various types of mitogenic and trophic factors. For example, carcinoma cells may release PDGF to recruit and activate stromal cells, while the latter respond by releasing IGFs that sustain the carcinoma cells.
- The formation of tumor-associated vasculature, formed by the process of neoangiogenesis, is a critical, rate-limiting determinant of the growth of all tumors larger in size than approximately 0.2 mm.
- In the case of carcinomas, the acquisition of tumor-associated stroma closely resembles the process of healing in wounded epithelial tissues. The genesis of stroma therefore relies on the same gene expression programs that are activated during wound healing.
- As tumor progression proceeds, the fibroblast-rich stroma is increasingly replaced by myofibroblasts, which eventually generate collagen-rich, desmoplastic stroma.
- The recruitment of the cells that participate directly in the construction of the neovasculature of tumors involves the release of factors, such as VEGF, by both the tumor cells and inflammatory cells, notably macrophages.
- Neoangiogenesis represents an attractive target for the development of novel anti-cancer agents, in that the targeted cells are the various normal stromal cell types participating in angiogenesis rather than the ever-changing cancer cells.

Thought questions

1. What diverse lines of evidence prove directly that most carcinoma cells depend on stromal cell types for various types of physiologic support?
2. How might anti-angiogenesis therapies improve (in some cases) or neutralize (in other cases) the efficacy of conventional chemotherapeutic agents?
3. How might macrophages facilitate or antagonize tumorigenesis?
4. Which lines of evidence persuade you that the generation of tumor-associated stroma depends on the same biological programs that are activated during wound healing?
5. Which biological forces cause the tumor-associated vasculature to be defective in so many respects?
6. Which types of anti-cancer therapeutic agents would you deploy in order to encourage the collapse of an established tumor by depriving it of vasculature support?

7. What biochemical strategies can tumor cells use to lessen their dependence on stromal support?
8. Can you cite examples of how oncoproteins perturb the interactions between carcinoma cells and the nearby tumor-associated stroma?
9. How might you determine what proportion of endothelial cells in the vasculature of a tumor derive from the expansion of adjacent vasculature and what proportion of these cells arise from circulating endothelial precursor cells?

Additional reading

Balkwill F & Mantovani A (2001) Inflammation and cancer: back to Virchow? *Lancet* 357, 539–545.

Bergers G & Benjamin LE (2003) Tumorigenesis and the angiogenic switch. *Nat. Rev. Cancer* 3, 401–410.

Bhowmick NA, Neilson EG & Moses HL (2004) Stromal fibroblasts in cancer initiation and progression. *Nature* 432, 332–337.

Bingle L, Brown NJ & Lewis CE (2002) The role of tumour-associated macrophages in tumour progression: implications for new anticancer therapies. *J. Pathol.* 196, 254–265.

Bissell MJ & Radisky D (2001) Putting tumours in context. *Nat. Rev. Cancer* 1, 46–54.

Bouck N, Stellmach V & Hsu SC (1996) How tumors become angiogenic. *Adv. Cancer Res.* 69, 135–174.

Carmeliet P (2000) Mechanisms of angiogenesis and arteriogenesis. *Nat. Med.* 6, 389–395.

Conway EM & Carmeliet P (2003) Cardiovascular biology: signalling silenced. *Nature* 425, 139–141.

Coussens LM & Werb Z (2002) Inflammation and cancer. *Nature* 420, 860–867.

Doljanski F (2004) The sculpturing role of fibroblast-like cells in morphogenesis. *Perspect. Biol. Med.* 47, 339–356.

Dvorak HF (1986) Tumors: wounds that do not heal. *N. Engl. J. Med.* 315, 1650–1689.

Dvorak HF (2002) Vascular permeability factor/vascular endothelial growth factor: a critical cytokine in tumor angiogenesis and a potential target for diagnosis and therapy. *J. Clin. Oncol.* 20, 4368–4380.

Dvorak HF (2003) How tumors make bad blood vessels and stroma. *Am. J. Pathol.* 162, 1747–1757.

Elenbaas B & Weinberg RA (2001) Heterotypic signaling between epithelial tumor cells and fibroblasts in carcinoma formation. *Exp. Cell Res.* 264, 169–184.

Ferrara N (2002) VEGF and the quest for tumour angiogenesis factors. *Nat. Rev. Cancer* 2, 795–803.

Ferrara N, Gerber HP & LeCouter J (2003) The biology of VEGF and its receptors. *Nat. Med.* 9, 669–676.

Folkman J & Kalluri R (2003) Tumor angiogenesis. In Cancer Medicine, 6th ed (DW Kufe, RE Pollock, RR Weichselbaum et al. eds), pp. 161–194. Hamilton, Ontario: BC Decker.

Gold LI (1999) The role for transforming growth factor-beta (TGF-beta) in human cancer. *Crit. Rev. Oncol.* 10, 303–360.

Hanahan D & Folkman J (1996) Patterns and emerging mechanisms of the angiogenic switch during tumorigenesis. *Cell 86*, 353–364.

Harris AL (2002) Hypoxia—a key regulatory factor in tumour growth. *Nat. Rev. Cancer* 2, 38–47.

Jain RK (1994) Barriers to drug delivery in solid tumors. *Sci. Am.* 271(1), 58–65.

Jain RK (2003) Molecular regulation of vessel maturation. *Nat. Med.* 9, 685–693.

Jain RK (2005) Normalization of tumor vasculature: an emerging concept in antiangiogenic therapy. *Science* 307, 58–62.

Jain RK & Booth MF (2003) What brings pericytes to tumor vessels? *J. Clin. Invest.* 112, 1134–1136.

Kalluri R (2003) Basement membranes: structure, assembly and role in tumour angiogenesis. *Nat. Rev. Cancer* 3, 422–433.

Kalluri R & Neilson EG (2003) Epithelial-mesenchymal transition and its implications for fibrosis. *J. Clin. Invest.* 112, 1776–1784.

Kerbel RS (2000) Tumor angiogenesis: past, present and the near future. *Carcinogenesis* 21, 505–515.

Kerbel R & Folkman J (2002) Clinical translation of angiogenesis inhibitors. *Nat. Rev. Cancer* 2, 727–739.

Leek RD & Harris AL (2002) Tumor-associated macrophages in breast cancer. *J. Mammary Gland Biol. Neoplasia* 7, 177–189.

Littlepage LE, Egeblad M and Werb Z (2005) Evolution of cancer and stromal cellular responses. *Cancer Cell* 7,499–500.

McCawley LJ and Matrisian LM (2001) Matrix metalloproteinases: they're not just for the matrix anymore! *Curr. Opin. Cell Biol.* 13, 534–540.

McDonald DM & Baluk P. (2002) Significance of blood vessel leakiness in cancer. *Cancer Res.* 62, 5381–5385.

McDonald DM & Foss AJ (2001) Endothelial cells of tumor vessels: abnormal but not absent. *Cancer Metastasis Rev.* 19, 109–120.

Mueller MM & Fusenig N (2004) Friends or foes—bipolar effects of the tumour stroma in cancer. *Nat. Rev. Cancer* 4, 839–849.

Nyberg P, Xie L & Kalluri R (2005) Endogenous inhibitors of angiogenesis. *Cancer Res.* 65, 3967–3979.

Pepper MS (2001) Lymphangiogenesis and tumor metastasis: myth or reality? *Clin. Cancer Res.* 7, 462–468.

Rabbani SY, Heissig B, Hattori K & Rafii S (2003) Molecular pathways regulating mobilization of marrow-derived stem cells for tissue revascularization. *Trends Mol. Med.* 9, 109–117.

Radisky D, Hagios C & Bissell MJ (2001) Tumors are unique organs defined by abnormal signaling and context. *Semin. Cancer Biol.* 11, 87–95.

Rafii S, Lyden D, Benezra R et al. (2002) Vascular and hematopoietic stem cells: novel targets for anti-angiogenesis therapy? *Nat. Rev. Cancer* 2, 826–835.

Savagner P (2001) Leaving the neighborhood: molecular mechanisms involved during epithelial-mesenchymal transition. *BioEssays* 23, 912–923.

Serini G & Gabbiani G (1999) Mechanisms of myofibroblast activity and phenotypic modulation. *Exp. Cell Res.* 250, 273–283.

Sleeman JP (2000) The lymph node as a bridgehead in the metastatic dissemination of tumors. *Recent Results Cancer Res.* 157, 55–81.

Thiery JP (2003) Epithelial-mesenchymal transitions in development and pathologies. *Curr. Opin. Cell. Biol.* 15, 740–746.

Tlsty T (2001) Stromal cells can contribute oncogenic signals. *Semin. Cancer Biol.* 11, 97–104.

Walker RA (2001) The complexities of breast cancer desmoplasia. *Breast Cancer Res.* 3, 143–145.

Wiseman BS & Werb Z (2002) Stromal effects on mammary gland development and breast cancer. *Science* 296, 1046–1049.

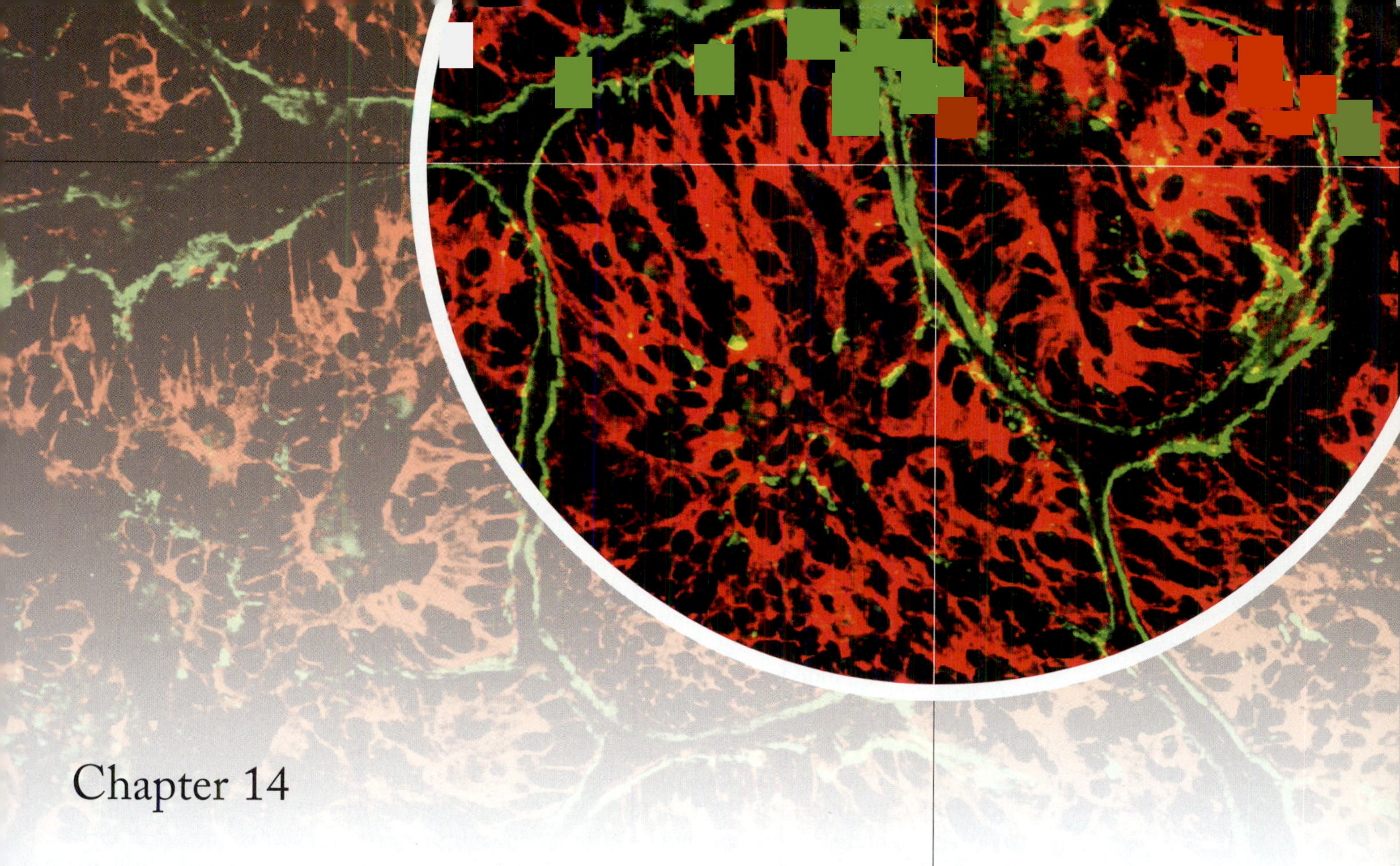

Chapter 14

Moving Out: Invasion and Metastasis

> The fact of cells identical with those of the cancer itself being seen in the blood may tend to throw some light upon the mode of origin of multiple tumors existing in the same person.
>
> T.R. Ashworth, physician, 1869

> It is not birth, marriage, or death, but gastrulation, which is truly the most important time in your life.
>
> Lewis Wolpert, embryologist, 1986

In the early phases of multi-step tumor progression, cancer cells multiply near the site where their ancestors first began uncontrolled proliferation. The result, usually apparent only after many years' time, is a **primary tumor** mass. Given the fact that a cubic centimeter of tissue may contain as many as 10^9 cells, we can easily imagine that tumors may often reach a size of 10^{10} or 10^{11} cells before they make themselves apparent to the individual carrying them or to the clinician in search of them.

Primary tumors in some organ sites—specifically those arising within the peritoneal or pleural space—may well expand without causing any discomfort to the patient, simply because these cavities are expansible and their contents are quite plastic; in other sites, such as the brain, the presence of a tumor is often apparent when it is still relatively small. Sooner or later, however, in all sites throughout the body, tumors of substantial size compromise the functioning of the organs in which they have arisen and begin to evoke symptoms.

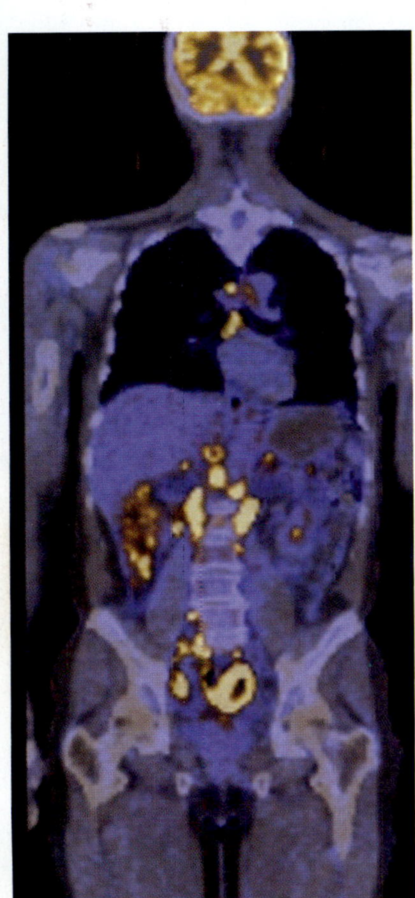

Figure 14.1 Disseminated tumors The diagnosis of metastatic disease often represents a death sentence for a cancer patient, yet the mechanisms by which cancer cells metastasize from a primary tumor to distant sites in the body remain poorly understood. Seen here is a whole-body scan of a patient with metastatic non-Hodgkin's lymphoma (NHL). This is a fusion image of a CT (computed X-ray tomography) scan of the body's tissues (*gray, blue*) and a PET (positron-emission tomography) scan in which the uptake of radioactively labeled fluorodeoxyglucose (FDG) in various tissues *(yellow)* has been detected. FDG uptake indicates regions of highly active cell metabolism throughout the body. The activity associated with the brain is normal. However, the presence of numerous yellow spots in the abdominal regions indicates multiple NHL metastases. (Courtesy of S.S. Gambhir.)

In many cases, the effects on normal tissue function come from the physical pressure exerted by the expanding tumor masses. In others, cells from the primary tumor mass invade adjacent normal tissues and, in so doing, begin to compromise vital functions. Large tumors in the colon may obstruct passage of digestion products through the lumen, and in tissues such as the liver and pancreas, cancer cells may obstruct the flow of bile through critical ducts. In the lungs, airways may be compromised.

As insidious and corrosive as these primary tumors are, they ultimately are responsible for only about 10% of deaths from cancer. The remaining approximately 90% of patients are struck down by cancerous growths that are discovered at sites far removed from the locations in their bodies where their primary tumors first arose (Figure 14.1; see also Figure 2.2). These **metastases** are formed by cancer cells that have left the primary tumor mass and traveled by the body's highways—blood and lymphatic vessels—to seek out new sites throughout the body where they may found new colonies (Figure 14.2). Breast cancers often spawn metastatic colonies promiscuously in many tissues throughout the body, including the brain, liver, bones, and lungs. Prostate tumors are most often seeded to the bones, while colon carcinomas preferentially form new colonies in the liver.

Such wandering cancer cells are the most dangerous manifestations of the cancer process. When they succeed in founding colonies in distant sites, they often wreak great havoc. The female body can dispense with its mammary glands without losing vital physiologic functions, and so almost all primary breast carcinomas do not compromise survival while they are confined to the breast. However, the metastatic colonies that breast cancer cells initiate in the bone can cause localized erosion of bone tissue, resulting in agonizing pain and skeletal collapse. Metastases in the brain may rapidly compromise central nervous system function, while those in the lung or liver are similarly threatening to life because of the vital functions of these organs.

For reasons that remain obscure, tumors in certain tissues have a high probability of metastasizing, while those arising in other tissues almost never do so. After primary melanomas penetrate a certain distance downward into the tissue underlying the skin, the presence of metastases at distant sites in the body is almost a certainty. In contrast, squamous cell carcinomas of the skin and astrocytomas—primary tumors of the glial cells in the brain—rarely spawn metastases.

In a variety of human tumor types, the dissemination of cancer cells throughout the body has already occurred by the time a primary tumor is first detected; at the time of initial diagnosis, these scattered cells will be inapparent because they only form minute tumor colonies—*micrometastases*. Such behavior provokes a question that we will confront in this chapter and again in Chapter 16: Do the properties of a primary tumor (Sidebar 14.1) reveal whether it has broadcast

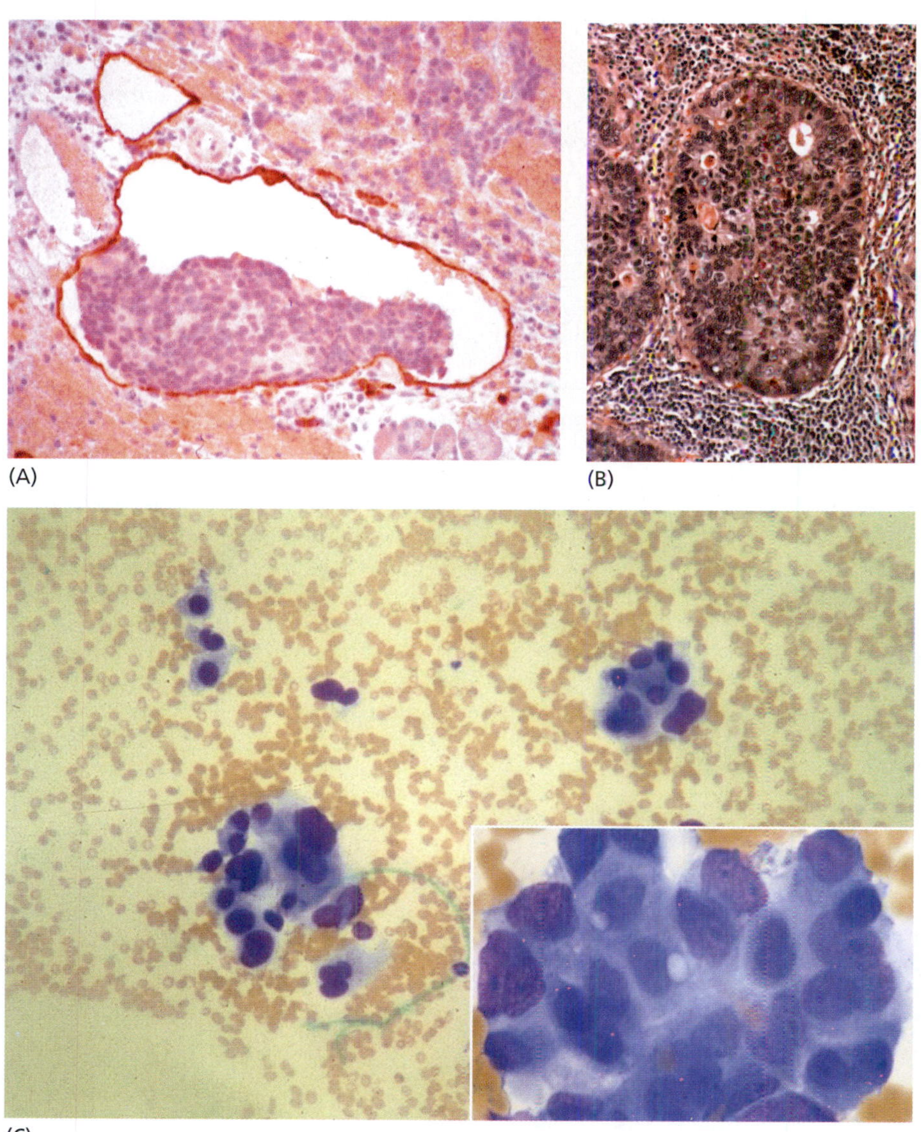

Figure 14.2 Histology of metastases in various tissues throughout the body (A) In the Rip-Tag transgenic mouse model of pancreatic islet cell tumorigenesis (see Figure 13.37), metastasis via the lymph nodes can be encouraged through the forced expression of VEGF-C—a lymphangiogenic factor—in the islet tumor cells. Seen here is a small metastasis of islet cells *(dark pink)* to a lymphatic vessel that is lined with endothelial cells *(brown)*. (B) A small metastasis of a human breast cancer is seen growing within a lymph node associated with one of the lymphatic ducts draining the breast. Note the fact that this metastasis exhibits the detailed structure, including ducts and stroma, that is characteristic of many primary tumors of the breast, as well as the numerous lymphocytes in the surrounding lymph duct *(dark nuclei)*. (C) The presence of metastatic carcinoma cells *(blue)* in the bone marrow can be revealed using specific immunohistochemistry to detect cells displaying epithelial markers, which sets them apart from the mesenchymal cells naturally present in the marrow; or, as seen here, through use of the Wright–Giemsa stain, which has rendered them blue. (A, from S.J. Mandriota, L. Jussila, M. Jeltsch et al., *EMBO J.* 20:672–682, 2001; B, courtesy of T.A. Ince; C, from A.T. Skarin, Atlas of Diagnostic Oncology, 3rd ed. Philadelphia: Elsevier Science Ltd., 2003.)

cancer cells throughout the body that will eventually create life-threatening metastatic disease long after the primary tumor has been surgically removed?

In this chapter, we confront the processes that create these most aggressive products of tumor progression. These processes depend on complex biochemical and biological changes in cancer cells and in the associated stroma. Most of the steps of cancer formation, as described in earlier chapters, are understood in considerable detail. In contrast, our understanding of invasion and metastasis is still quite incomplete, explaining why these late steps of tumor progression represent the major unsolved problems of cancer pathogenesis.

14.1 Travel of cancer cells from a primary tumor to a site of potential metastasis depends on a series of complex biological steps

The great majority (>80%) of life-threatening cancers occur in epithelial tissues, yielding carcinomas. Consequently, most of our discussions in this chapter, as in the last, will refer to this class of tumors, with the understanding that cancers

Sidebar 14.1 The connection between tumor size and disease prognosis is unclear The crude measurement of primary tumor size can often be correlated with the likelihood that the tumor has already seeded metastases in distant sites in the body. For example, as indicated in Figure 14.3, in one study 22% of women with primary breast cancers of less than 1 cm diameter (at initial diagnosis) eventually developed metastatic disease. In contrast, 77% of those women whose primary tumors were more than 8 cm in diameter progressed to metastatic disease. Independent of these measurements are others indicating that, in general, larger tumors have passed through more of the genetic (and epigenetic) steps of tumor progression. Thus, 4% of breast cancers of less than 1 cm diameter have been found to carry mutant *p53* alleles, while 42% of tumors greater than 3 cm bear cells with these mutations. We might conclude from such observations that the ability to metastasize is acquired by cancer cells as a relatively late step in tumor progression when a primary tumor is expanding from a small to a large size. Moreover, acquisition of this ability might well be due to the loss of p53 function.

Actually, the true meaning of these statistics is very difficult to assess and can be explained by a number of alternative mechanisms: (1) The trait of metastasis may be acquired only relatively late in the growth of the primary tumor, thereby limiting this ability to cells present in the largest primary tumors. (2) Individual cancer cells in small and large tumors may be equally capable of metastasizing. However, the larger tumors may dispatch proportionately greater numbers of metastasizing cells (per unit of time) to the rest of the body simply because they contain more cells. (3) At some point during multi-step tumor progression, those cells showing an enhanced ability to proliferate in the primary tumor also exhibit a closely associated trait—the ability to metastasize. Such cancer cells generate larger primary tumors more rapidly than other cells that proliferate slowly, and these rapidly multiplying cells also serve as sources of widely disseminated metastases.

Similar arguments apply to mutant genes present in advanced tumors: The greater frequency of mutant *p53* alleles in large tumors may not mean that this mutant allele is acquired late in primary tumor progression. Instead, tumor cell populations that acquire mutant *p53* alleles relatively early in multi-step tumor progression may be able to grow more rapidly and generate tumors of a larger size by the time of diagnosis.

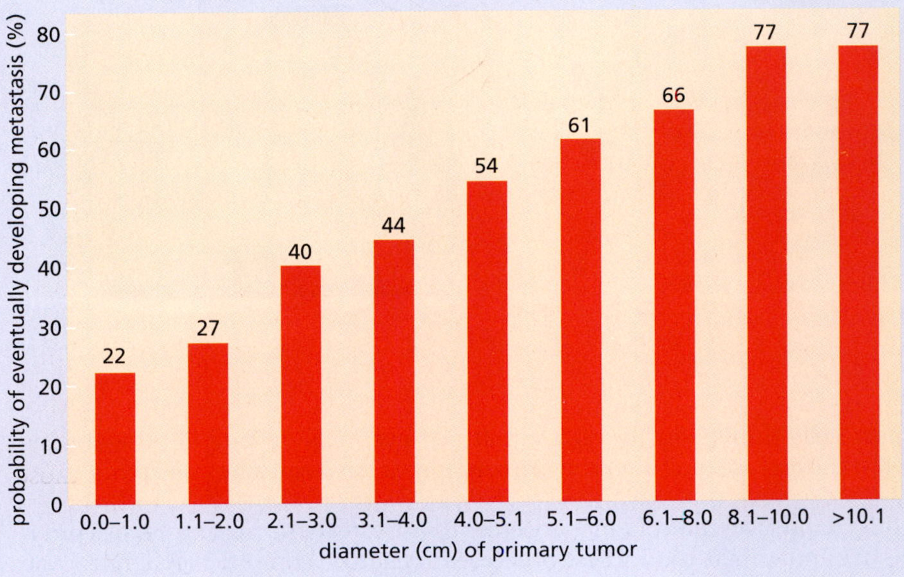

Figure 14.3 Primary tumor size and risk of metastasis This bar graph reveals that as the diameter of an initially diagnosed primary breast cancer increases, the probability increases that distant metastases will arise in a patient's body. This is indicated here as the percentage of women bearing primary tumors of a given size who eventually develop distant macroscopic metastases. In this study, 1589 breast cancer patients were followed for periods as long as 46 years after initial diagnosis and treatment. (From R. Heimann and S. Hellman, *J. Clin. Oncol.* 16:2686–2692, 1998.)

arising in other tissue types, such as connective and nervous tissues, often follow similar paths when they become invasive and metastatic. Even certain hematopoietic tumors, notably lymphomas, often have an early, localized phase and a later phase during which they become disseminated to distant anatomical sites (see Figure 14.1). This complex sequence of steps, sometimes called "the *invasion–metastasis cascade,*" is illustrated in Figure 14.4.

Our focus on carcinomas requires us to draw from earlier discussions of these tumors and the epithelial tissues in which they arise (see, for example, Sections 2.2 and 13.1). To recap briefly, the great majority of epithelial tissues are constructed according to a common set of architectural principles; in most cases,

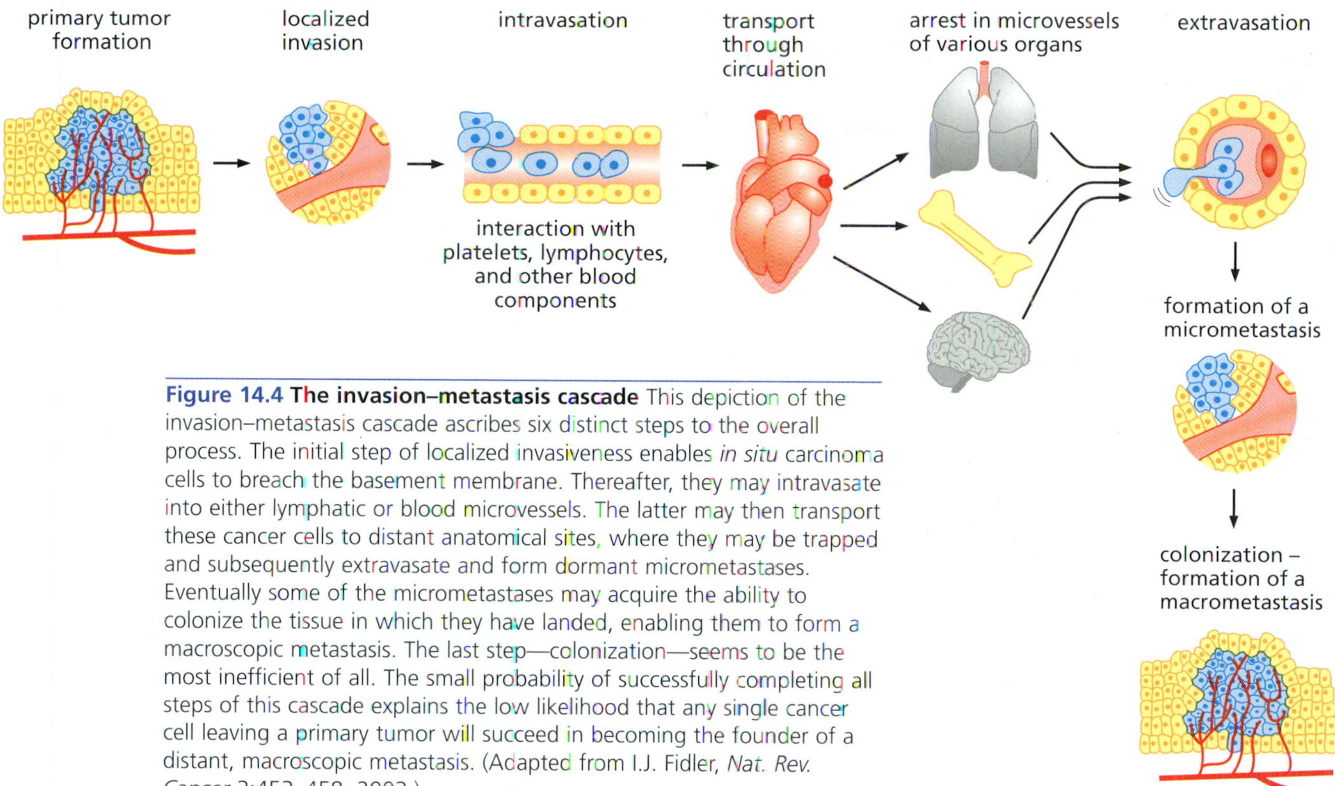

Figure 14.4 The invasion–metastasis cascade This depiction of the invasion–metastasis cascade ascribes six distinct steps to the overall process. The initial step of localized invasiveness enables *in situ* carcinoma cells to breach the basement membrane. Thereafter, they may intravasate into either lymphatic or blood microvessels. The latter may then transport these cancer cells to distant anatomical sites, where they may be trapped and subsequently extravasate and form dormant micrometastases. Eventually some of the micrometastases may acquire the ability to colonize the tissue in which they have landed, enabling them to form a macroscopic metastasis. The last step—colonization—seems to be the most inefficient of all. The small probability of successfully completing all steps of this cascade explains the low likelihood that any single cancer cell leaving a primary tumor will succeed in becoming the founder of a distant, macroscopic metastasis. (Adapted from I.J. Fidler, *Nat. Rev. Cancer* 3:453–458, 2003.)

relatively thin sheets of epithelial cells sit atop deep, complex layers of stroma. Separating the two is the specialized type of extracellular matrix (ECM) known as the basement membrane (see Figures 2.3 and 13.5). This proteinaceous meshwork is constructed collaboratively by proteins secreted by both epithelial and stromal cells.

By definition, carcinomas begin on the epithelial side of the basement membrane and are considered to be *benign* as long as the cells forming them remain on this side. Sooner or later, however, many carcinomas acquire the ability to breach the basement membrane, and individual cancer cells or groups of cancer cells begin to invade the nearby stroma (Figure 14.5). This mass of neoplastic cells is now reclassified as *malignant*. In fact, as mentioned in an earlier chapter, many pathologists and surgeons reserve the word "cancer" for those epithelial tumors that have acquired this invasive ability. To be sure, this dissolution of the basement membrane by invading carcinoma cells removes an important physical barrier to the further expansion of tumor cell populations. But in addition, as we learned in the last chapter, by degrading various components of the basement membrane, invasive cells harvest growth and survival factors that have been sequestered by attachment to this specialized extracellular matrix.

Recall that even before carcinoma cells breach the basement membrane, they often succeed in stimulating angiogenesis on the stromal side of the membrane, apparently by dispatching angiogenic factors through this porous barrier to endothelial cells within the stroma (see Figure 13.39). However, invasion through the basement membrane places them in a far better position for executing subsequent steps of the invasion–metastasis cascade. Once present in the stromal compartment, carcinoma cells can gain direct access to the blood and lymphatic vessels (see Figure 13.41), which are normally found only on the stromal side of the basement membrane (see, however, Sidebar 14.2). Close contact with the capillaries affords tumor cells improved access to the nutrients and

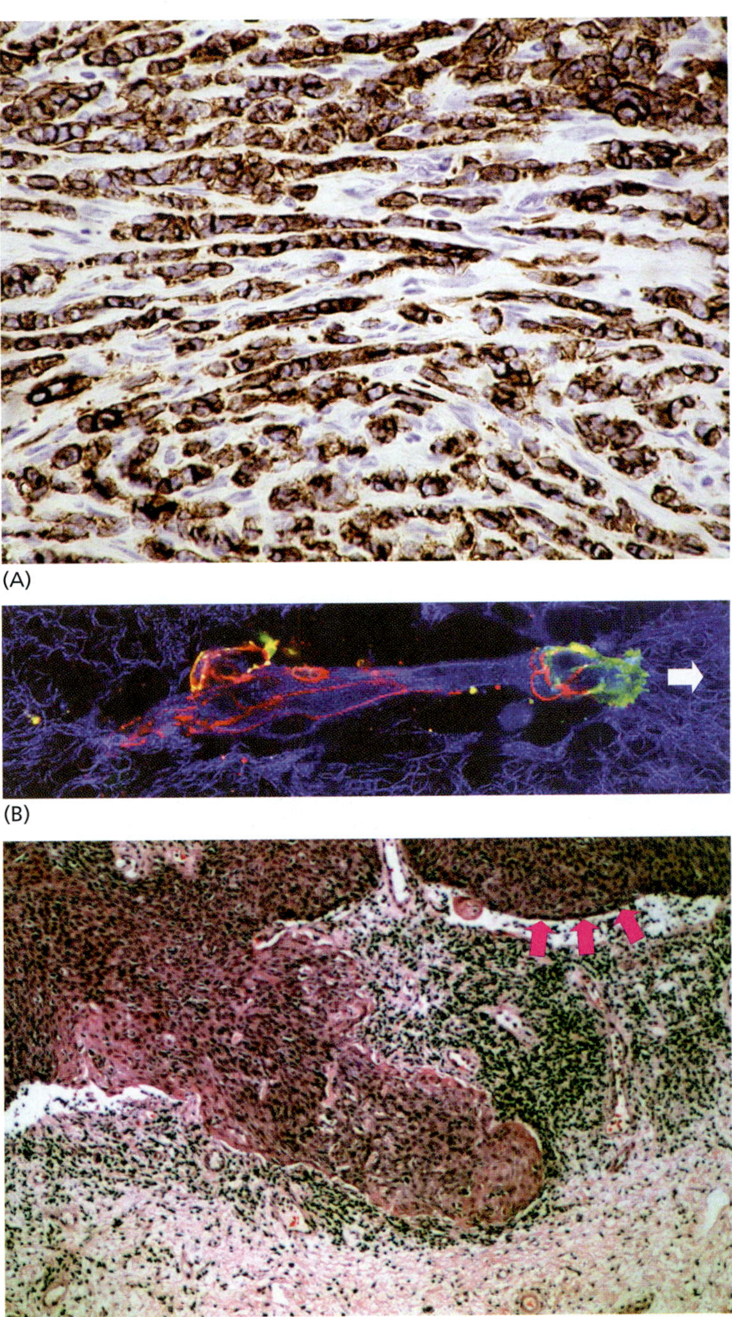

Figure 14.5 Patterns of invasion (A) These invasive lobular mammary carcinoma cells *(brown)* have left the primary tumor *(not shown to the left)* and are proceeding rightward, one-by-one in single file, through channels they have carved in the adjacent stroma. (B) However, far more typical is the behavior of this cohort of 5–10 melanoma cells, viewed in a confocal micrograph, which are moving together through a collagen matrix *(blue)*; like normal melanocytes, they continue to adhere to one another through adherens junctions that are formed by E-cadherin molecules *(red)*. Large gaps in the matrix *(black)* indicate areas that have been degraded by the advancing cancer cells. At the leading invasive edge *(white arrow)*, the melanoma cells are displaying β1 integrins *(green)*, which enable them to attach to the extracellular matrix ahead of them. (C) On a far larger scale, such coordinated invasion is reflected by the cells in this squamous cell carcinoma of the cervix, in which a large finger or tongue of many hundreds of cancer cells *(pink, brown)* has breached the basement membrane and is invading the stroma. The latter being characterized by both stromal fibroblasts and inflammatory cells *(dark green)*. The basement membrane is seen here as a dark brown, horizontal line *(pink arrows, top right)* that separates the bulk of the carcinoma cells *(above)* from the stromal cells *(below)* and is uninterrupted except for a capillary and the tongue of invasive cancer cells. (A, courtesy of J. Jonkers; B, from P. Friedl, Y. Hegerfeldt and M. Tusch, *Int. J. Dev. Biol.* 48:441–449, 2004; C, courtesy of T.A. Ince.)

oxygen carried by the blood. In addition, their invasive properties enable these cancer cells to move through the walls and into the lumina (i.e., the bores) of blood and lymphatic vessels. This invasion into vessels is often termed **intravasation**.

Once they have arrived within the lumen of a blood or lymphatic vessel, individual cancer cells may travel with the blood or lymph to other areas in the body. These long-range migrations are fraught with great danger for the wanderers. Like normal cells, the cancer cells may continue to depend on anchorage to solid substrates; without such attachment, the migrating cells may die rapidly from anoikis, the form of apoptosis that is triggered by detachment of a cell from a solid substrate such as an extracellular matrix (Sections 5.8 and 9.13). Also, like their forebears in the primary tumor, these pioneers may depend on various types of stromal support, which will be lacking the moment they leave the primary tumor mass. Recall that the stroma can benefit carcinoma cells in multiple ways by supplying both mitogenic and trophic (survival) factors.

The blood represents an actively hostile environment for metastasizing cancer cells. Hydrodynamic shear forces in the circulation, which are often substantial in smaller vessels, may tear the wandering cancer cells apart. Some experimental models of metastasis in the mouse indicate that survival of metastasizing cancer cells in the general circulation is greatly enhanced if they can attract an entourage of blood platelets to escort them through the rapids into safe pools within tissues (Sidebar 27).

If metastasizing cells survive these initial dangers and gain access to the larger vessels in the venous system, they will travel with the blood through the heart and then lodge, with high probability, in the first set of capillaries that they encounter after their initial passage through the heart—the capillary beds of the lungs (Figure 14.6). Unlike red and white blood cells, carcinoma cells are ill suited for passage through most capillaries, whose internal diameters are far too small to accommodate them. Capillaries usually have internal diameters in the range of 3 to 8 μm, and the blood cells that must pass through them are well adapted to do so. Erythrocytes, for example, are only about 7 μm in diameter and are easily deformed, facilitating their passage through capillaries (Figure 14.7). Most cancer cells, in contrast, are more than 20 μm in diameter and are not especially deformable. (Moreover, if cancer cells in the blood are coated with platelets, as discussed in Sidebar 27, their effective diameters are greatly increased, causing them to be trapped in vessels far larger than capillaries, e.g., the small arteries known as **arterioles**.)

Once trapped within the lungs, some metastasizing cancer cells may attempt to found metastases there. However, the metastases of many types of human tumors are often found elsewhere in the body, indicating that cancer cells frequently succeed in escaping from the lungs and travel further to other sites in the body. How they do so is unclear. In some experiments, cancer cells trapped in capillaries have been observed to pinch off large amounts of cytoplasm, leaving

Sidebar 14.2 Cancer cells usually invade as a unified phalanx The language used here and elsewhere to describe cancer cell invasiveness attributes this property to individual cells that venture outside the perimeter of a primary tumor, make their way into nearby stroma, and eventually intravasate. In fact, there are such cancers—for example, invasive lobular carcinomas of the breast (see Figure 14.5A)—in which individual cancer cells leave the primary tumor mass, one by one, and proceed in single file (sometimes called "Indian file") through channels that they have carved in the nearby stroma. However, a far more typical behavior of invasive carcinoma cells is shown in Figure 14.5C, in which a *phalanx* (i.e., a well-organized cohort) of these cells invades nearby stroma. As we will see later in this chapter, the carcinoma cells in direct contact with the stroma may pave the way for others to follow behind them. Sooner or later, however, as these invasive tongues approach nearby vasculature, any intravasation that occurs must depend on the ability of individual cancer cells or small clumps of these cells to break away from their neoplastic neighbors and enter on their own into the circulation.

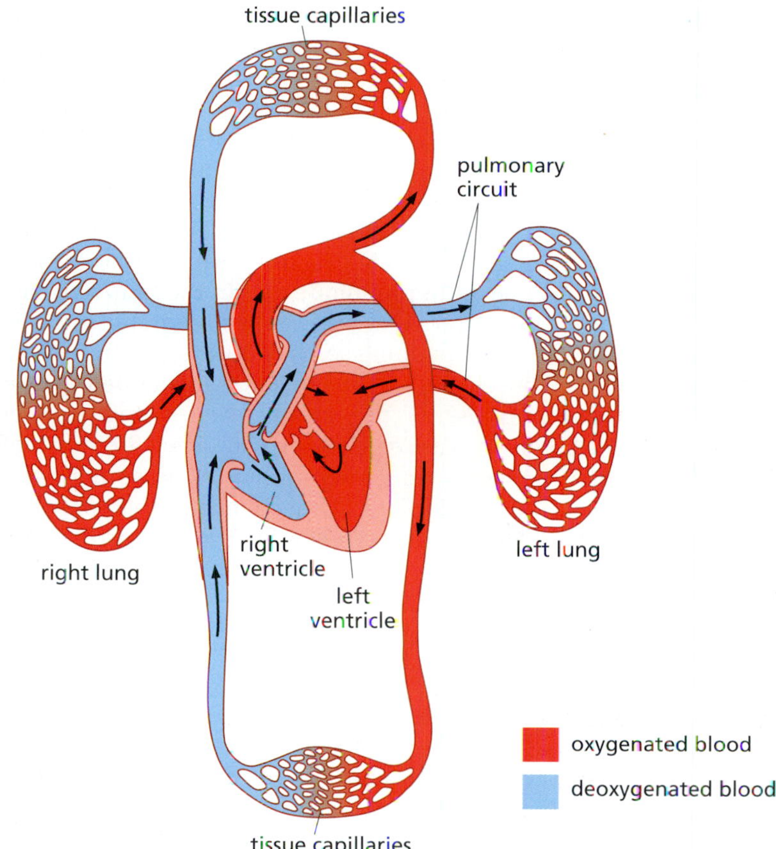

Figure 14.6 Major routes of blood circulation through the body This schematic diagram of the mammalian circulation indicates that venous blood *(blue)* leaving a tissue (and thus cancer cells that have escaped from a primary tumor and intravasated) must first pass through the right ventricle of the heart and thence through the lungs before it passes through the left ventricle and is pumped into the general arterial circulation. Since passage through the pulmonary circulation of the lung requires passage through its capillaries, almost all metastasizing cells entering into the venous circulation are rapidly trapped in the pulmonary capillary beds. (From P.H. Raven, G.B. Johnson, S. Singer et al., Biology, 7th ed. New York: McGraw-Hill, 2005.)

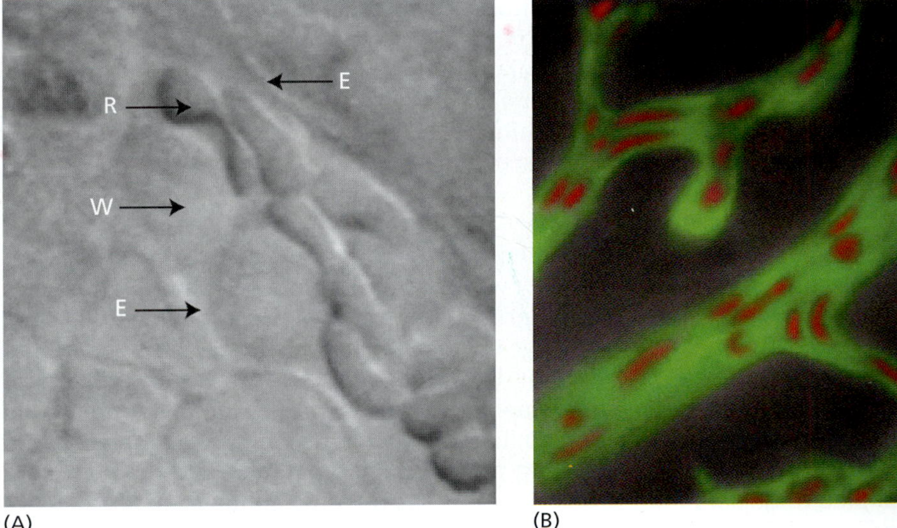

Figure 14.7 Passage through capillaries (A) *Intra-vital* microscopy reveals the two endothelial walls of a capillary (E), erythrocytes (R), and some leukocytes (W). The small size and deformability of these cells allow them to pass through capillaries without becoming trapped. (B) The vessels in this intra-vital fluorescence micrograph are slightly larger than capillaries. The plasma has been stained with a green dye, while the erythrocytes have been colorized red. The high deformability of these red blood cells is apparent. Since most cancer cells have more than twice the diameter of erythrocytes and are not deformable, they are unable to negotiate narrow passages, such as the lumina of capillaries. (From I.C. MacDonald, A.C. Groom and A.F Chambers, *BioEssays* 24:885–893, 2002.)

behind cells that, while greatly reduced in size, seem to be viable; once they have undergone this amputation, such slenderized cells may succeed in negotiating passage through the narrow straits of the lung capillaries. A more plausible explanation is that wandering cancer cells may avoid being trapped altogether: in many organs, including the lung, metastasizing cells can bypass capillaries by traveling through arterial–venous shunts, which form large-bore, direct connections between the two parts of the circulatory system.

Having snaked their way through the lungs and arrived in the general arterial circulation (see Figure 14.6), roaming cancer cells can then scatter to all tissues in the body. Some experiments suggest that cancer cells use specific cell surface receptors, such as integrins, to initially adhere to the luminal walls of arterioles and capillaries in certain tissues. However, far more extensive evidence indicates that simple physical trapping within small vessels, as discussed above in the context of the lung, provides most wandering cancer cells with their first foothold within a tissue.

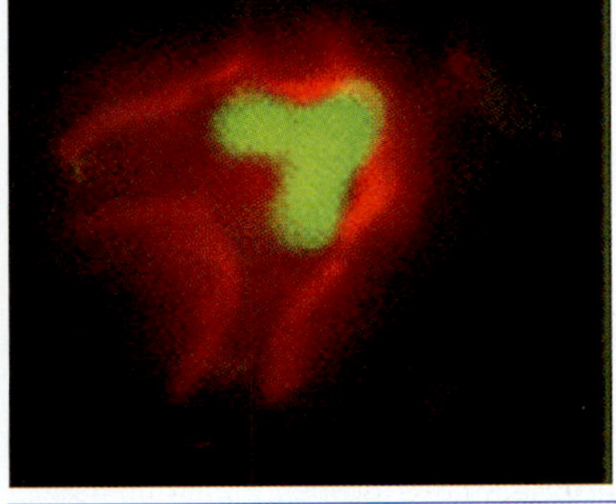

Figure 14.8 Cancer cells growing in the lumen of a microvessel Confocal microscopy reveals a colony of rat fibrosarcoma cells, which express green fluorescent protein (GFP), proliferating within an arteriole whose walls *(red)* are labeled through use of a dye bound to low-density lipoprotein (LDL). (LDL binds to the LDL receptor displayed on the luminal surfaces of endothelial cells.) These cells were visualized 5 days after single cells were introduced into the venous circulation of a mouse. (Courtesy of R.J. Muschel.)

Once lodged in the blood vessels of various tissues, cancer cells must escape from the lumina of these vessels and penetrate into the surrounding tissue—the step termed **extravasation**. The process of extravasation depends on complex interactions between cancer cells and the walls of the vessels in which they have become trapped. Cancer cells can use several alternative strategies to extravasate. They may begin to proliferate within the lumen of the vessel, creating a small tumor that grows and eventually obliterates the adjacent vessel wall (Figures 14.8 and 14.9). In doing so, they push aside endothelial cells, pericytes, and smooth muscle cells that previously separated the vessel lumen from the surrounding tissue, the latter often being called the tissue *parenchyma*. Alternatively, cancer cells may proceed immediately to elbow their way through the vessel wall. Their ability to do so may depend on the same biochemical and cell-biological mechanisms that previously enabled them or their immediate ancestors to invade from the primary tumor and to intravasate (see Sidebar 14.3).

14.2 Colonization represents the most complex and challenging step of the invasion–metastasis cascade

Once they have arrived within the parenchyma of a tissue, metastasizing cancer cells may begin forming a tumor mass in their newfound homes, the process often termed **colonization**. This step is also a challenging one—perhaps the

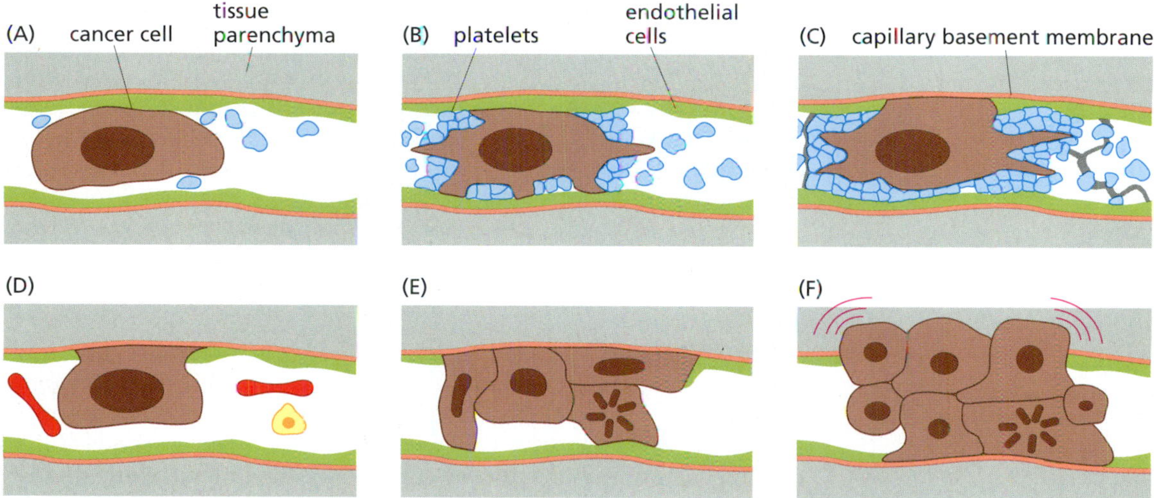

Figure 14.9 Steps leading to extravasation Electron-microscopic observations of metastasizing lung cancer cells injected into the venous circulation of mice have suggested that the process of extravasation often proceeds through the following sequence of steps. (A) A metastasizing cell *(light brown cytoplasm, dark brown nucleus)* is trapped physically in a capillary. (B) Within minutes, a large number of platelets *(blue)* become attached to the cancer cell, forming a microthrombus. Some of these have not yet degranulated and released growth factors, proteases, etc. (C) The cancer cell pushes aside an endothelial cell *(green)* on one wall of the capillary, thereby achieving direct contact with the underlying capillary basement membrane *(orange)*. (D) Within a day, the microthrombus is dissolved by the proteases in the blood that are usually responsible for dissolving clots. (E) The cancer cell begins to proliferate in the lumen of the capillary. (F) Within several days, sometimes earlier, the cancer cells break through the capillary basement membrane and invade the surrounding tissue parenchyma *(gray area)*. (Note that in this scenario, the microthrombus forms only *after* the cancer cell has become trapped in a capillary.) (From J.D. Crissman, J.S. Hatfield, D.G. Menter et al., *Cancer Res.* 48:4065–4072, 1988.)

most difficult step of all, ostensibly because the foreign tissue environments do not provide cancer cells with the collection of familiar growth and survival factors that allowed their progenitors to thrive in the primary tumor. Without these various types of physiologic support, the metastasizing cells may rapidly die or, at best, survive for extended periods of time as single cells or small clumps of cancer cells—so-called micrometastases—that can only be detected microscopically and rarely increase beyond this size. In general, the number of micrometastases in the body of a cancer patient vastly exceeds those that will eventually grow large enough (several millimeters or more in diameter) to be clinically detectable. These micrometastases may be widely disseminated throughout the tissues of a cancer patient, occasionally leading to disastrous outcomes (Sidebar 28 ●).

Sidebar 14.3 Cancer cells are clumsy escape artists The complex task of escaping from the circulation into the surrounding tissue parenchyma is accomplished routinely by leukocytes, which must be able to enter into the parenchyma in response to certain inflammatory stimuli, including the presence of infectious agents. Through a sequence of steps known as **diapedesis**, leukocytes are able to induce endothelial cells in post-capillary venules to retract and provide a portal into the underlying tissue. The entire process from attachment to the endothelial wall to entrance into the tissue parenchyma takes less than a minute and involves an elaborately choreographed program of biochemical and cell-biological changes!

In contrast, the vast majority of metastasizing cancer cells are not endowed with the receptors and biochemical response mechanisms required to execute diapedesis. Accordingly, if neoplastic cells do succeed in penetrating through the wall of a capillary or slightly larger vessel, they seem to do so by brute force, perhaps by degrading patches of endothelium in a process that requires many hours, or even a day, rather than a minute to complete. Of additional interest here is the fact that the thrombin produced during the formation of microthrombi (see Sidebar 27 ●) is quite effective in cleaving the various proteins used by endothelial cells to attach to the underlying vascular basement membrane; this may cause endothelial cells to retract from microemboli, thereby exposing the capillary basement membrane to direct attack by invasive cancer cells and the proteases that they generate.

Antibodies reactive with **cytokeratins** are useful for detecting the micrometastases that primary carcinomas spawn in the bone marrow and blood, while an antibody against the epithelial cell adhesion molecule (EpCAM) is often used to detect micrometastases in the lymph nodes. In all these cases, the presence of isolated cytokeratin-positive (and thus epithelial) cells in otherwise fully mesenchymal tissues represents a clear sign that metastatic seeding has taken place. Current microscopic techniques using cytokeratin-specific antibodies make it possible to detect a single-cell micrometastasis among 10^5 or even 10^6 surrounding mesenchymal cells in the blood, bone marrow, or lymph node (Figure 14.10).

Many of the steps proceed inefficiently, and the probability of an individual cell successfully completing all of them is very low. For example, in some mice carrying primary tumors of about 1 gram mass (~10^9 cells), as many as a million cells may be seeded into the circulation each day, yet the visible metastases formed in such animals may be counted on the fingers of one hand.

This low rate of success in forming metastases is sometimes termed *metastatic inefficiency* and is the end product of the sequence of inefficient steps that together make up the invasion–metastasis cascade. Some experiments indicate that the earlier steps in this cascade are executed quite efficiently by metastasizing cells, while the last step, involving colonization, succeeds only rarely and therefore is the *rate-limiting* determinant of the process as a whole. Consequently, vast numbers of micrometastases may be seeded throughout the body and many may persist for extended periods of time before one or another

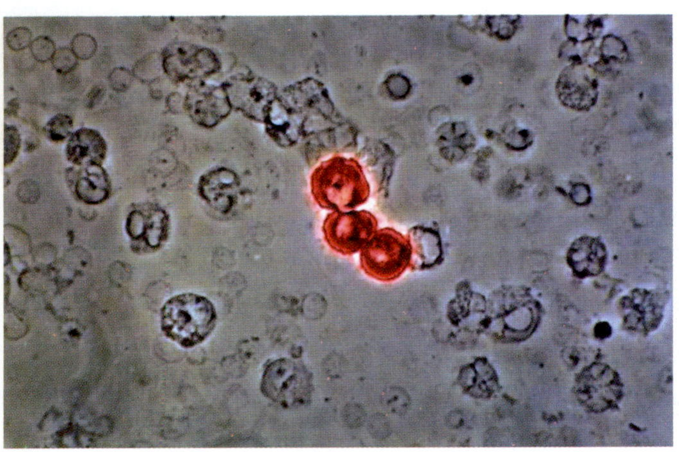

(A)

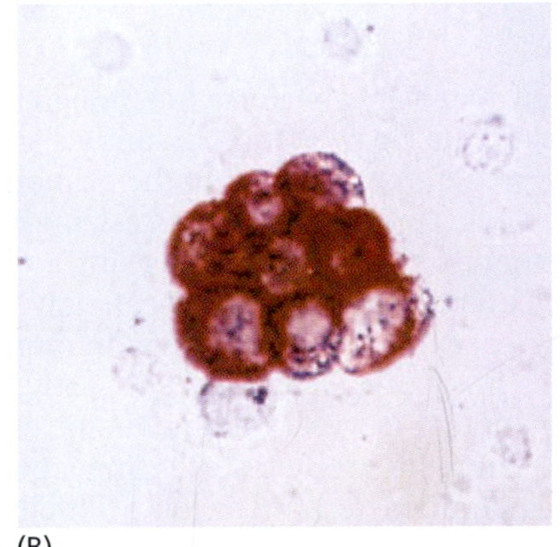

(B)

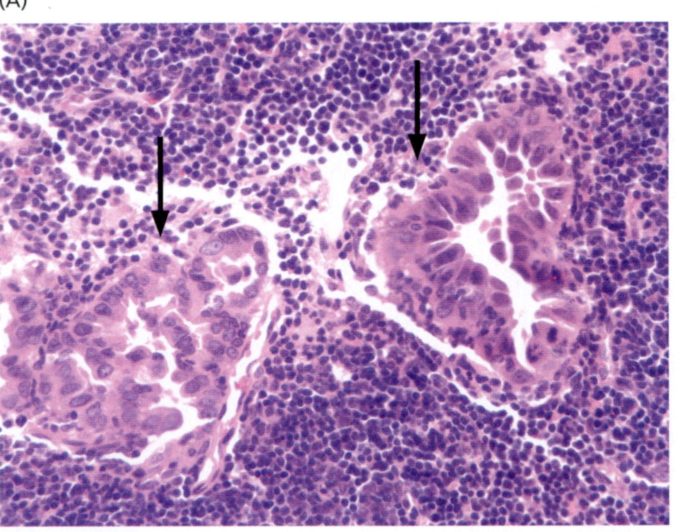

(C)

Figure 14.10 Detection of a micro-metastasis in a lymph node or in the marrow (A) The presence of metastatic cancer cells in the bone marrow is usually analyzed by withdrawing marrow from the iliac crest of the pelvis. In this micrograph, the presence of a micrometastasis containing several cancer cells in the bone marrow of a colon cancer patient has been detected by staining with an anti-cytokeratin antibody. (B) Essentially identical images can be detected in the bone marrow extracted from breast cancer patients. In this case, a cluster of eight mammary carcinoma cells, viewed at slightly lower magnification, form a micrometastasis. (C) The presence of micrometastases in a lymph node is indicated by their appearance, which contrasts strongly with that of the surrounding lymphocytes. Here two metastases of a mouse lung adenocarcinoma *(arrows)* to a lymph node are seen amid a sea of lymphocytes. Note the formation of a ductlike structure by the right micrometastasis. (A, courtesy of I. Funke and G. Riethmüller; B, from S. Braun, K. Pantel, P. Muller et al., *N. Engl. J. Med.* 342:525–533, 2000; C, courtesy of K. P. Olive and T. Jacks.)

of these migrants or its lineal descendants finally acquires the ability to grow into a clinically detectable mass (Sidebar 14.4).

Support for the existence of dormant micrometastases, which persist in a non-growing state for extended periods of time, comes from experiments in which living cancer cells were initially marked by brief exposure to fluorescence label–containing particles, which persist for extended periods of time within cells but do not affect their viability. Following such labeling, the intracellular concentration—and thus the fluorescence intensity of the dye particles—decreases by a factor of 2 each time a cell divides; hence, the residual fluorescence intensity in cells after an extended period of time allows the experimenter to estimate how many cell divisions occurred since the time when the cells were initially marked.

Such dye-labeled cancer cells were introduced via the portal circulation into mouse livers, in which they formed large numbers of single-cell micrometastases. Eleven weeks later, cancer cells were recovered from these livers, and many of these still possessed full fluorescence intensity (Figure 14.12), indicating that they had not divided even once since their arrival in the liver. Importantly, these recovered cancer cells remained capable of proliferating *in vitro* and were able to generate new tumors when injected subcutaneously into other host mice.

The above experiment shows dramatically that metastatic cancer cells can remain viable for extended periods of time in a nondividing, dormant state within foreign tissue sites. (Note that a different type of micrometastasis arises when disseminated cancer cells succeed in proliferating and forming colonies of a very small size within a foreign tissue; however, these micrometastases never increase in size, since the rate of cell proliferation in these clumps is counterbalanced by an equal rate of apoptosis, perhaps because of the failure of these cells to execute the angiogenic switch. See Section 13.7.) Whatever their nature, micrometastases represent an imminent threat, since they are often present in vast numbers throughout the body and may erupt years after the cancer has been judged to be cured.

In summary, the multiple steps of the invasion–metastasis cascade encompass as many distinct biological changes as all the steps that preceded them during the course of primary tumor formation. The complexity of this cascade, as outlined in these two sections, raises questions that will motivate many of our discussions in the rest of this chapter: How do cancer cells learn to become metastatic? Does each of the steps in this cascade require the actions of a specific gene that becomes altered during tumor progression? Or are many of the individual steps of invasion and metastasis orchestrated by a single master control gene or a small group of such genes?

We will also touch on another simple but profoundly important issue: Do highly malignant cells carry genes that in mutant form are specialized to induce invasiveness or metastasis? Or do these late steps in tumor progression depend on the actions of familiar actors, specifically the oncogenes and tumor suppressor genes that we have encountered repeatedly throughout this book?

14.3 The epithelial–mesenchymal transition and associated loss of E-cadherin expression enable carcinoma cells to become invasive

The first of the many steps leading to metastasis—the acquisition of local invasiveness—involves major changes in the phenotype of cancer cells within the primary tumor. As before, we will focus this discussion on epithelial tissues and

Sidebar 14.4 Genetic analyses suggest that the evolution of metastatic ability can occur outside of the primary tumor In perhaps 30% of breast, prostate, and colon cancer patients whose primary carcinomas have been surgically removed, one can still detect micrometastases in the marrow, lymph nodes, or blood; these patients are considered to have "minimal residual disease." At this stage, genetic analysis of the micrometastases indicates that they are genetically heterogeneous. However, years later, when a patient develops disease relapse and manifests readily detectable metastatic masses, the patient's single-cell micrometastases are now much more similar to one another genetically (Figure 14.11A).

This suggests a stage of tumor progression in which the ability to colonize is acquired separately from the ability to disseminate to distant organ sites. Initially, genetically diverse cancer cells are seeded by a primary tumor throughout the body, but none of these succeeds in establishing a macroscopic metastasis simply because none is capable of doing so. After a period of time, however, genetic evolution occurring in a micrometastasis somewhere in the body yields a clone of cells with the newly acquired ability to colonize efficiently. As this clone expands, it also begins to seed cancer cells throughout the body, and therefore generates a new, second wave of metastatic dissemination. The individual cancer cells that are released by this clone soon constitute the majority of the single-cell micrometastases in the marrow of the patient, and these micrometastases are genetically very similar to one another because of their shared descent from the same clonal cell population. Importantly, because the cells in these new micrometastases all have inherited the ability to colonize, many grow rapidly into macroscopic metastases, creating a life-threatening burden of disseminated cancer cells in the patient. This model (Figure 14.11B) suggests that the evolution toward advanced malignancy often occurs at anatomical sites far removed from the primary tumor. While suggested by these observations, the scheme of Figure 14.11B is not yet validated by large, statistically robust sets of observations.

Figure 14.11 Genetic heterogeneity of micrometastases and the evolution of colonizing ability (A) Single-cell micrometastases of primary carcinomas can be identified in bone marrow biopsies because of their display of epithelial cell markers, such as cytokeratins or EpCAM (see Figure 14.10), and then isolated using a micropipette. (B) Adaptation of the comparative genomic hybridization (CGH; see Figure 11.20) procedure is used to analyze individual micrometastatic cells for the gain or loss of various chromosomal arms. This, in turn, has made it possible to relate the resulting "genetic profile" of each micrometastasis with the profiles of several other micrometastases isolated from the same patient. Those profiles that are similar to one another are placed closely to one another on a branch of a tree termed a *dendrogram*; conversely, those that have very different genetic profiles are located far away from one another on the tree. In the event that multiple micrometastases from a single patient are clustered next to one another on a branch of the dendrogram (and thus are closely related to one another genetically), this is indicated by a *blue horizontal bar* associated with this pair or with a larger group of micrometastases. Micrometastases in the bones of patients (identified by *labels)* carrying breast, prostate, and colorectal tumors were obtained at the time of surgical removal of their primary tumors, a stage termed "minimal residual disease." As seen in the upper dendrogram, only a minority of micrometastases in patients with minimal residual disease are clustered together on the dendrogram, indicating substantial genetic heterogeneity of micrometastases within each patient at this stage of disease. However, months or years later, when disease relapse with macroscopic metastases occurs, the several micrometastases detected in almost every patient are located close to one another on the lower dendrogram, indicating close genetic relationship to one another. (C) The analyses shown in panel (B) suggest the following model. An initially formed, genetically heterogeneous primary tumor cell population (see Figure 11.18) seeds equally heterogeneous micrometastases throughout the body of the cancer patient. The primary tumor is then removed surgically, leaving behind only the micrometastases and creating the state of minimal residual disease. Over a period of years, in one or another site in the body, one of these micrometastatic cell clones *(aquamarine)* acquires the ability to colonize, i.e., to grow into a macroscopic metastasis. The latter now acts as a source of cells that generate a new cascade of metastatic dissemination throughout the body. Because these newly dispersed cells all share a common clonal origin, the micrometastases that they form are genetically very similar to one another. Moreover, because the cells in each of these secondary micrometastases are endowed with the ability to colonize (since they all derive from a cell clone in a macrometastasis that has previously acquired this ability), many of these micrometastases can rapidly grow into macroscopic, clinically detectable metastases that result in disease relapse. (A and B, from C.A. Klein, T.J. Blankenstein, O. Schmidt-Kittler et al., *Lancet* 360:683–689, 2002.)

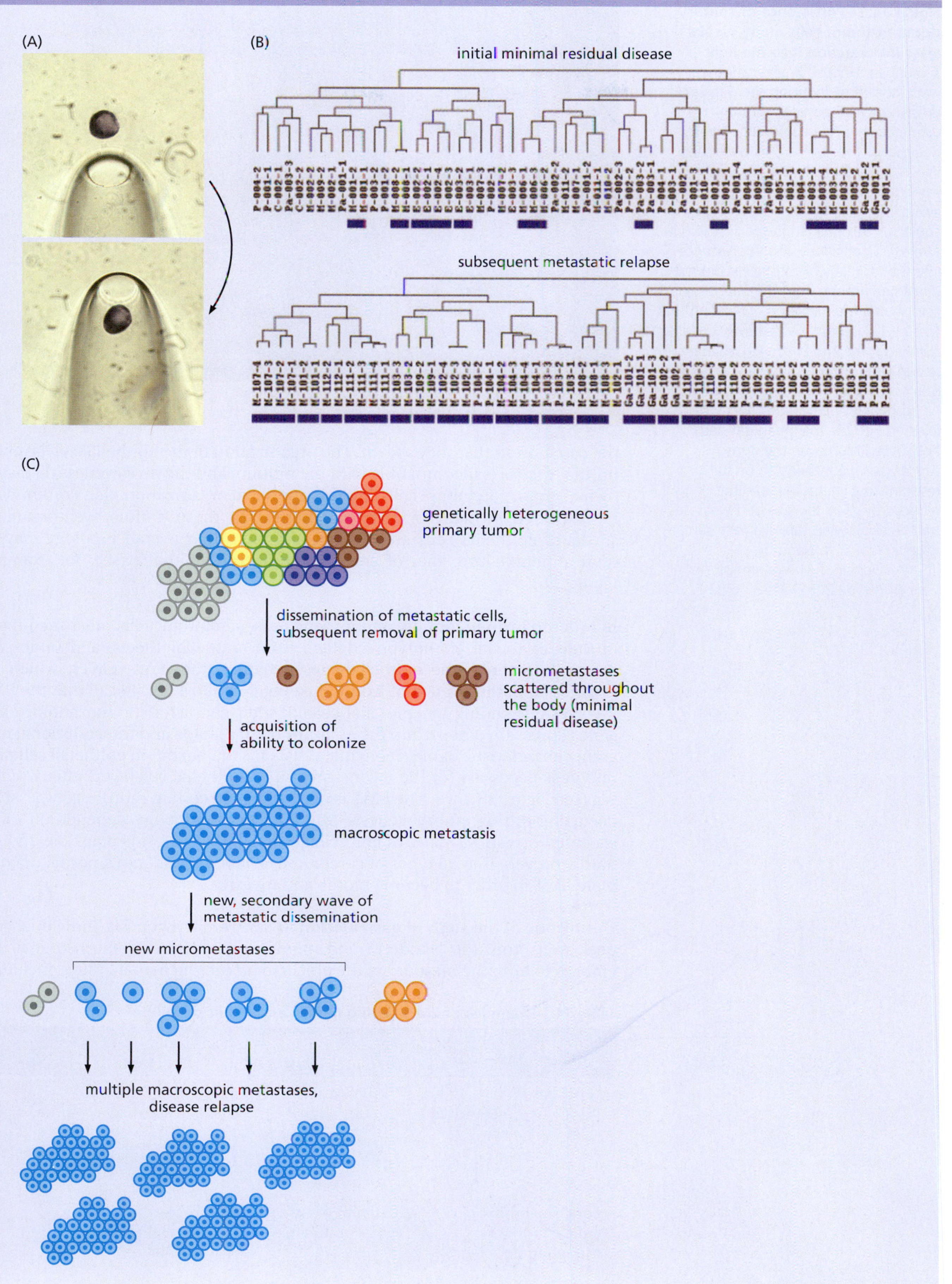
(A)
(B)
initial minimal residual disease
subsequent metastatic relapse
(C)
genetically heterogeneous primary tumor
dissemination of metastatic cells, subsequent removal of primary tumor
micrometastases scattered throughout the body (minimal residual disease)
acquisition of ability to colonize
macroscopic metastasis
new, secondary wave of metastatic dissemination
new micrometastases
multiple macroscopic metastases, disease relapse

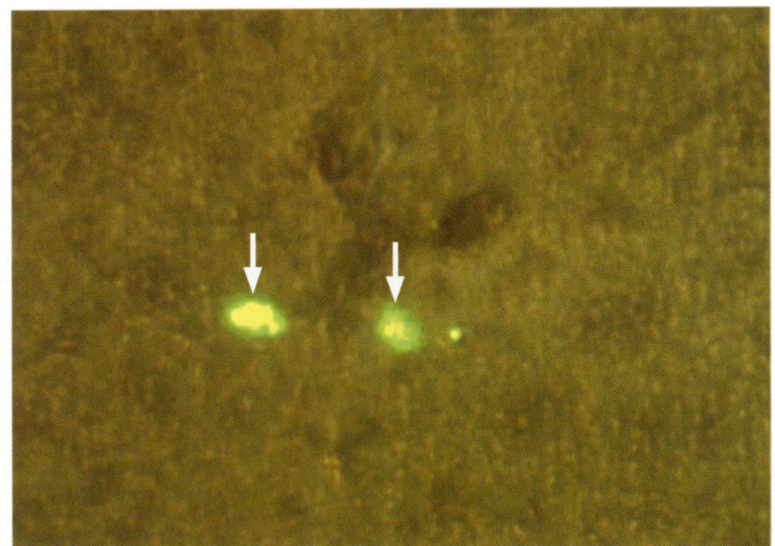

Figure 14.12 Persistence of solitary dormant tumor cells many weeks after introduction into the liver Tumorigenic mouse breast cancer cells were labeled by inducing them to take up styrene nanoparticles (48 nm diameter) that had been tagged with a fluorescent dye. These cells were then injected into a mesenteric vein, which carried them via the portal vein into the liver. Eleven weeks later, tumor cells could still be detected in the liver *(white arrows)*. Significantly, the fluorescence intensity of many of these cells did not differ significantly from the intensity of cells shortly after labeling, indicating that these cells had not divided even once following labeling. Following isolation and *in vitro* culturing, the descendants of many of these cells were tumorigenic, i.e., capable of forming tumors when injected into the mammary fat pads of host mice. (Moreover, the dormant cancer cells were found to be fully resistant to a chemotherapy that reduced by 75% the size of metastases that were growing rapidly in the same mice.) (From G.N. Naumov, I.C. MacDonald, P.M. Weinmeister et al., *Cancer Res.* 62:2162–2168, 2002.)

the carcinomas that they spawn. The organization of the epithelial cell layers in normal tissues is incompatible with the motility and the invasiveness displayed by malignant carcinoma cells, yet this epithelial organization plan continues to be respected in many primary carcinomas. In these tumors, well-organized sheets of epithelial cells are present, although their overall topology may be quite different from that of comparable normal epithelia (see, for example, Figure 2.6).

In order to acquire motility and invasiveness, carcinoma cells must shed many of their epithelial phenotypes, detach from epithelial sheets, and undergo a drastic alteration—the epithelial–mesenchymal transition (EMT), which we mentioned in the context of wound healing (Section 13.3). Recall that the EMT involves a shedding by epithelial cells of their characteristic morphology and gene expression pattern and the assumption of a shape and transcriptional program characteristic of mesenchymal cells. This major shift in epithelial cell phenotype is necessary for the reconstruction of epithelial cell layers after wounding (see Figure 13.13C). The EMT is used widely in certain morphogenetic steps occurring during embryogenesis, when tissue remodeling depends on EMTs executed by various types of epithelial cells (Table 14.1). It is plausible, though hardly proven, that all types of carcinoma cells must undergo a partial or complete EMT in order to become motile and invasive.

During one of the steps of **gastrulation** in early embryogenesis, individual cells peel away from the ectoderm and migrate inward toward the center of the embryo to form the mesoderm, the precursor of mesenchymal tissues, including

Table 14.1 Examples of EMTs during mouse embryonic development

Process	Transition	
	From	To
Gastrulation	epiblast	mmesoderm
Prevalvular mesenchyme in the heart	endothelium	atrial and ventricular septum
Neural crest cells	neural plate	neural crest cells, which can yield bone, muscle, peripheral nervous system
Somitogenesis	somite walls	sclerotome
Palate formation	oral epithelium	mesenchymal cells
Müllerian duct regression	Müllerian tract	mesenchymal cells

Adapted from P. Savagner, *BioEssays* 23:912–923, 2001.

(A)

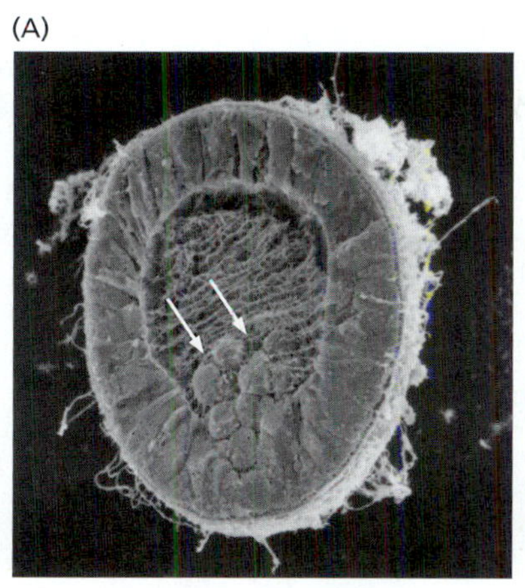

(B)

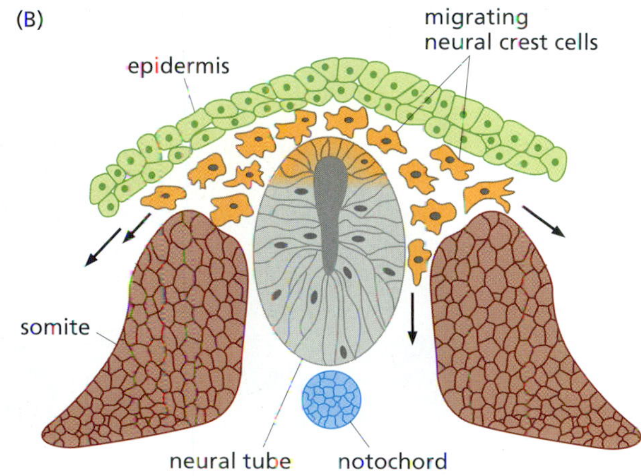

Figure 14.13 Embryogenesis and the epithelial–mesenchymal transition (A) This scanning electron micrograph shows the delamination of cells from the primitive ectoderm of a sea urchin embryo *(white arrows)* and their migration into the interior of the embryo. The cells have become round and acquired motility in anticipation of their forming the rudiments of the mesoderm—changes associated with an epithelial–mesenchymal transition (EMT). They are migrating along strands of extracellular matrix in the lumen of this early embryo. (B) The delamination of neuroepithelial cells from the neural tube also requires cells to undergo an EMT. Cells of the embryonic neural crest *(orange)*, which derive from the upper region of the initially epithelial neural tube (*gray*), delaminate from this epithelium, acquire motility and invasiveness, and disperse throughout the embryo and eventually throughout the body of the resulting organism, where they form melanocytes, much of the peripheral nervous system, and much of the skeleton of the face. (A, courtesy of G. Cherr; B, from A.E. Vernon and C. LaBonne, *Curr. Biol.* 14:R719–R721, 2004.)

fibroblasts and hematopoietic cells (in chordates). This conversion of ectodermal cells, which at this stage are arrayed in an epithelial cell layer, to those having a mesodermal phenotype involves an EMT (Figure 14.13A). At the same time, these cells undergoing an EMT acquire the ability to translocate from one location (the outer cell layer) to another (the interior) within the embryo.

The migration of neuroepithelial cells from the neural crest into the mesenchyme of early vertebrate embryos also depends on a transformation of cell phenotype that can best be described as an EMT (Figure 14.13B). Similarly, the migration of myogenic precursor cells (the progenitors of muscle cells) from the dermomyotome of the early embryo to the limb buds depends on an EMT-like transformation of cell phenotype. All of these processes, in turn, bear a striking resemblance to the EMT undertaken by the cells at the edge of a wound within an epithelium; these cells must undergo a transient EMT in order to migrate into the wound site and close the gaps in the epithelial cell sheet that were created by the wounding process (see Figure 13.14).

An EMT can also be seen at the edges of carcinomas that are invading adjacent tissues (Figure 14.14). This pathological process is strikingly similar to the EMTs occurring during early embryogenesis and wound healing. Once again, epithelial cells cease expressing epithelial protein markers and express mesenchymal ones in their stead; at the same time, these cells lose their epithelial morphology and take on the appearance of fibroblasts (see Figure 13.13).

The strong resemblance between the pathological process of tumor invasiveness and normal steps of embryogenesis and wound healing suggests a plausible mechanistic model: according to this model, which is supported by extensive evidence gathered in recent years, the complex program of cellular reorganization exhibited by invasive carcinoma cells depends on the reactivation of latent behavioral programs whose expression is usually confined to early embryogenesis and to damaged adult tissues. According to this thinking, once carcinoma cells acquire access to the EMT program, they can exploit it to profoundly change their own morphology, motility, and ability to invade nearby cell layers. This model implies that the multiple changes in cell phenotype associated with invasiveness, some of which are described below, need not be acquired piecemeal by carcinoma cells. Instead, these cells simply activate a morphogenetic program that is already encoded in their genomes. (This logic echoes our earlier discussion of epithelial–stromal interactions in Section 13.3, where we argued that cancer cells exploit the wound-healing program in order to acquire an activated stroma.)

The normal and pathological versions of the EMT involve, in addition to changes in shape and the acquisition of motility, fundamental alterations in the gene expression profiles of cells (Table 14.2). Expression of E-cadherin and cytokeratins—hallmarks of epithelial cell protein expression—is repressed, while the expression of **vimentin**, an intermediate filament component of the mesenchymal cell cytoskeleton, is induced. Epithelial cells that have undergone an EMT often begin to make fibronectin, an extracellular matrix protein that is normally secreted only by mesenchymal cells such as fibroblasts. At the same time, expression of a typical fibroblastic marker—N-cadherin—is often acquired in place of E-cadherin (Figure 14.15).

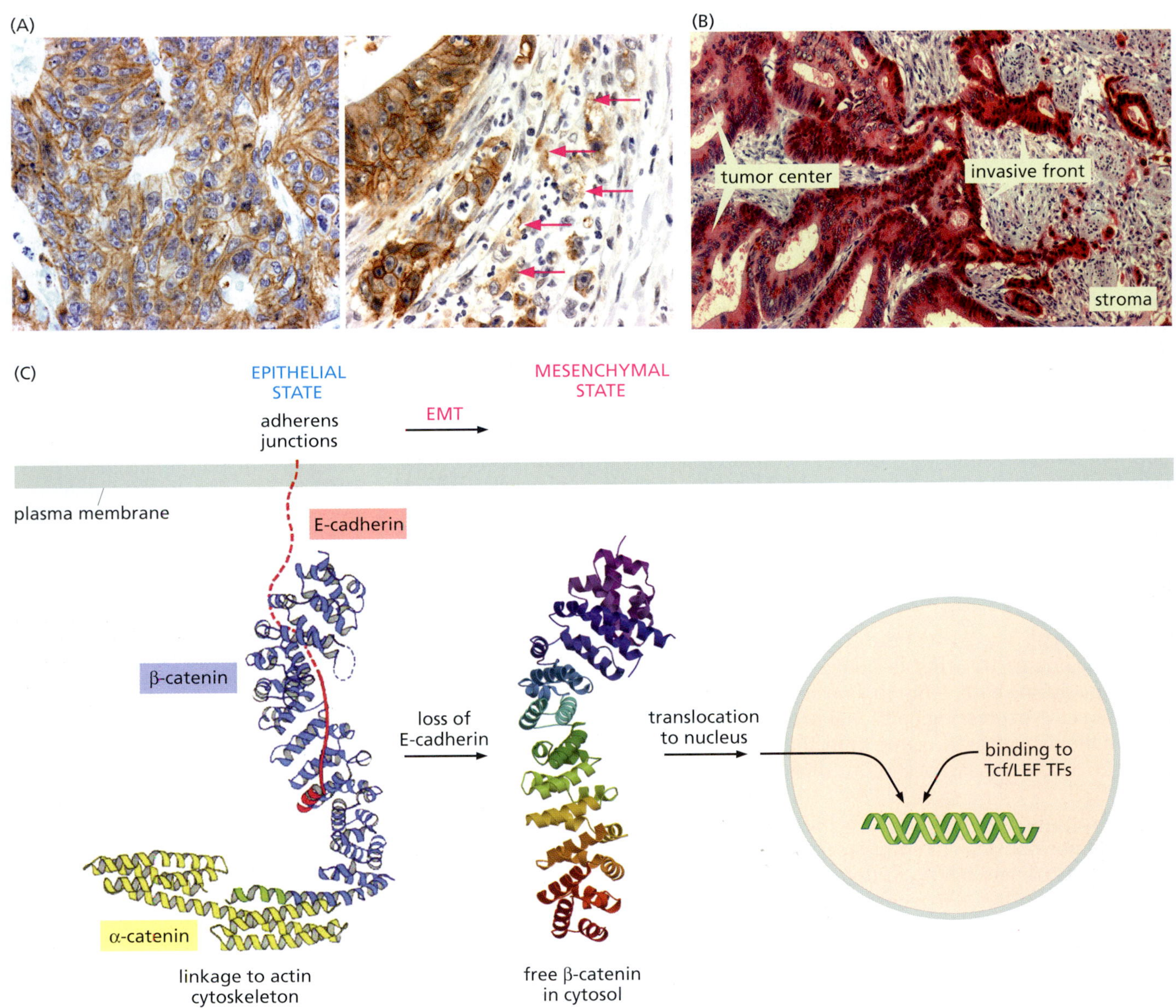

Figure 14.14 Epithelial–mesenchymal transition at the invasive edge of a tumor Colon carcinoma cells at the invasive edge of a primary human tumor undergo changes in gene expression and the localization of certain proteins. (A) While E-cadherin *(brown)* is strongly expressed on the plasma membranes of cells in the core of a primary tumor, where it forms adherens junctions *(left panel)*, its expression decreases substantially in individual invasive cells at the edge of this tumor *(red arrows, right panel)* and is no longer localized to their plasma membranes. (B) At the same time, cells in the core of this tumor express β-catenin *(dark red, left side)* under their plasma membranes and diffusely throughout the cytoplasm, while tumor cells at the invasive edge of this tumor show intense staining β-catenin in their nuclei *(right side)*. (C) In normal epithelial cells, β-catenin serves to link the cytoplasmic tail of E-cadherin, which forms adherens junctions with neighboring cells (see Figure 13.12), to the actin cytoskeleton. It also functions in the cytoplasm as a key intermediary in the Wnt signaling pathway (Section 6.10). Loss of E-cadherin from the plasma membrane liberates β-catenin molecules, which may then migrate to the nucleus and associate with Tcf/Lef transcription factors, thereby inducing expression of genes orchestrating the EMT program. (A, from T. Brabletz, A. Jung, S. Reu et al., *Proc. Natl. Acad. Sci. USA* 98:10356–10361, 2001; B, courtesy of T. Brabletz and T. Kirchner; C, from T.A. Graham, C. Weaver, F. Mao et al., *Cell* 103:885–896, 2000, courtesy of H.-J. Choi and W.L. Weis.)

Table 14.2 Cellular changes associated with the epithelial–mesenchymal transition

Loss of
Cytokeratin (intermediate filament) expression
Epithelial adherens junction protein (E-cadherin)
Epithelial cell polarity
Acquisition of
Fibroblast-like shape
Motility
Invasiveness
Mesenchymal gene expression program
Mesenchymal adherens junction protein (N-cadherin)
Protease secretion (MMP-2, MMP-9)
Vimentin (intermediate filament) expression
Fibronectin secretion
PDGF receptor expression
$\alpha v\beta 6$ integrin expression

Of all these proteins, the transmembrane E-cadherin molecule plays the dominant role in influencing epithelial versus mesenchymal cell phenotypes. Recall our earlier encounters with E-cadherin and its role in enabling epithelial cells to adhere to one another (see Figures 6.26A and 13.12). In normal epithelia, the ectodomains of E-cadherin molecules extend from the plasma membrane of one epithelial cell to form complexes with other E-cadherin molecules protruding from the surface of an adjacent epithelial cell. This enables homodimeric (and higher-order) bridges to be built between adjacent cells in an epithelial cell layer, resulting in the adherens junctions that are so important to the structural integrity of epithelial cell sheets.

The cytoplasmic domains of individual E-cadherin molecules are tethered to the actin fibers of the cytoskeleton via a complex of α- and β-catenins (see Figures 13.12 and 14.14) and other ancillary proteins. The actin cytoskeleton, for

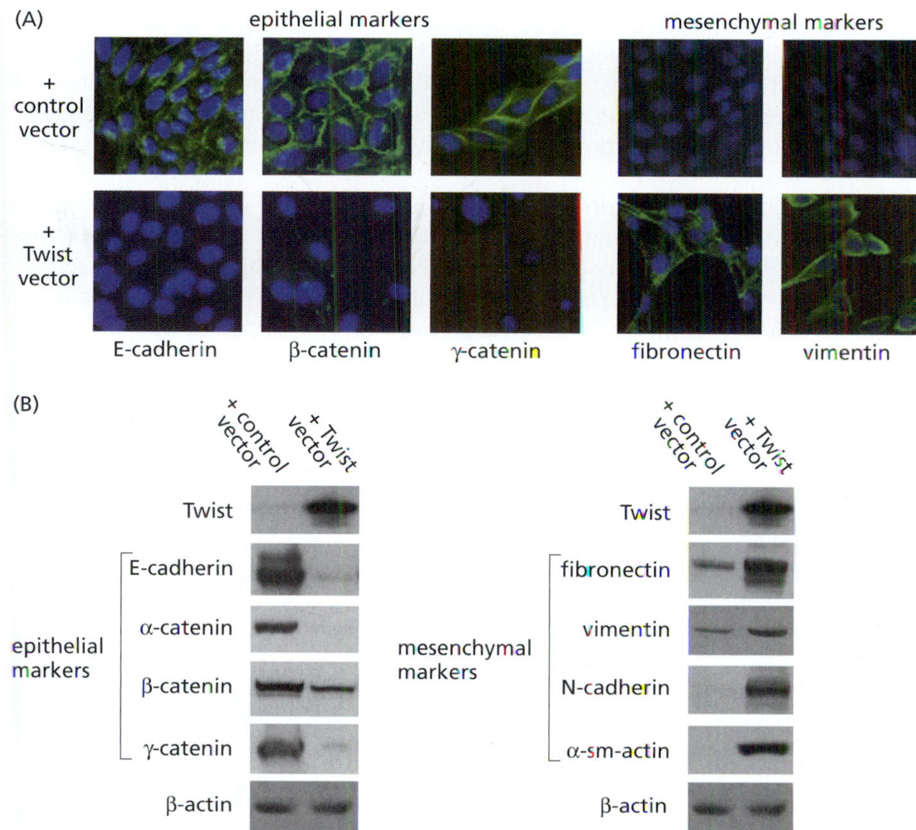

Figure 14.15 Biochemical changes accompanying the EMT As discussed later, the EMT can be induced by several transcription factors. Shown here are the effects of expressing the Twist transcription factor in MDCK (Maden–Darby canine kidney) cells, which are widely used to study epithelial cell biology. (A) These immunofluorescence analyses indicate that expression of epithelial markers, specifically E-cadherin, β-catenin, and γ-catenin, is depressed, while expression of mesenchymal markers, specifically vimentin and fibronectin, is induced by ectopic expression of Twist. (B) Immunoblots confirm the results of immunofluorescence, but in a more quantitative fashion. Lysates of control cells are analyzed in the *left channels,* while lysates of MDCK cells forced to express Twist are analyzed in the *right channels.* β-actin, whose expression is unaffected by the EMT, is used here as a control to ensure that equal amounts of cell lysate have been analyzed in all cases. sm-actin, α-smooth muscle actin. (A and B, from J. Yang, S.A. Mani, J.L. Donaher et al., *Cell* 117:927–939, 2004.)

its part, provides tensile strength to the cell. Hence, by knitting together the actin cytoskeletons of adjacent cells, E-cadherin molecules help an epithelial cell sheet resist mechanical forces that might otherwise tear it apart. Once E-cadherin expression is suppressed, many of the other cell-physiologic changes associated with the EMT seem to follow suit. Some experiments indicate that simply by suppressing the expression of the E-cadherin protein, cells acquire a mesenchymal morphology and increased motility.

The pivotal role of E-cadherin in the acquisition of malignant cell phenotypes is further supported by observations indicating that the *CDH1* gene, which specifies E-cadherin, is repressed by promoter methylation in many types of invasive human carcinomas (see Table 7.2) and in others by certain transcriptional repressors; this gene can also be inactivated by reading-frame mutations. For example, an analysis of 26 human breast cancer cell lines indicated that 8 had mutations that led to inactivation of E-cadherin gene expression, 5 had truncating mutations in the E-cadherin reading frame, while 3 had in-frame deletions resulting in the expression of mutant E-cadherin molecules at the cell surface. By now, loss of E-cadherin expression or expression of mutant E-cadherin proteins has been documented in advanced carcinomas of the breast, colon, prostate, stomach, liver, esophagus, skin, kidney, and lung. And mutant germline alleles of the *CDH1* gene result in familial gastric cancer (see Table 7.1).

Additionally, in studies of several types of carcinoma cells that had lost E-cadherin expression, re-expression of this protein (achieved experimentally by introduction of an E-cadherin expression vector) strongly suppressed the invasiveness and metastatic dissemination of these cancer cells. Together, these diverse observations indicate that E-cadherin levels are key determinants of the biological behavior of epithelial cancer cells and that the cell-to-cell contacts constructed by E-cadherins impede invasiveness and hence metastasis.

The replacement of E-cadherin by N-cadherin during the EMT (see Figure 14.15B) is also seen during gastrulation in early embryogenesis. Moreover, hepatocyte growth factor (HGF) promotes the E- to N-cadherin switch in cultured epiblast cells (which derive from the embryonic ectodermal cell layer); in this way, it induces an EMT and enables emigration of muscle and dermal precursor cells from the primitive dermomyotome—the collection of epithelial-like cells located in the somites of early vertebrate embryos. In the context of cancer pathogenesis, HGF potently induces motile and invasive behavior in carcinoma cells.

Like E-cadherin, the N-cadherin that is produced in its stead participates in homophilic interactions, that is, binds to other molecules of the same type displayed by nearby cells. Consequently, the N-cadherin molecules expressed on the surface of a carcinoma cell that has undergone an EMT increase the affinity of this cancer cell for the stromal cells that normally display N-cadherin, notably the fibroblasts in the stroma underlying the epithelial cell layer. This association seems to help invading carcinoma cells insert themselves amid stromal cell populations. Precisely the same dynamics have been proposed to explain how melanomas develop: normal melanocytes express E-cadherin, which binds them to the keratinocytes around them; melanoma cells—the transformed derivatives of melanocytes—express N-cadherin, which facilitates their invasion of the dermal stroma of the skin and their association with its fibroblasts and endothelial cells (Figure 14.16).

Importantly, the acquisition of N-cadherin expression does not result in the assembly of large sheets of cancer cells that might, in principle, be created by the formation of cell–cell N-cadherin bridges. It seems that the intermolecular bonds formed between pairs of N-cadherin molecules are far weaker than those formed by E-cadherin homodimers. This helps to explain why cell-surface N-cadherin molecules actively favor cell motility, and thus behave very differently from their E-cadherin cousins, which function to immobilize cells within

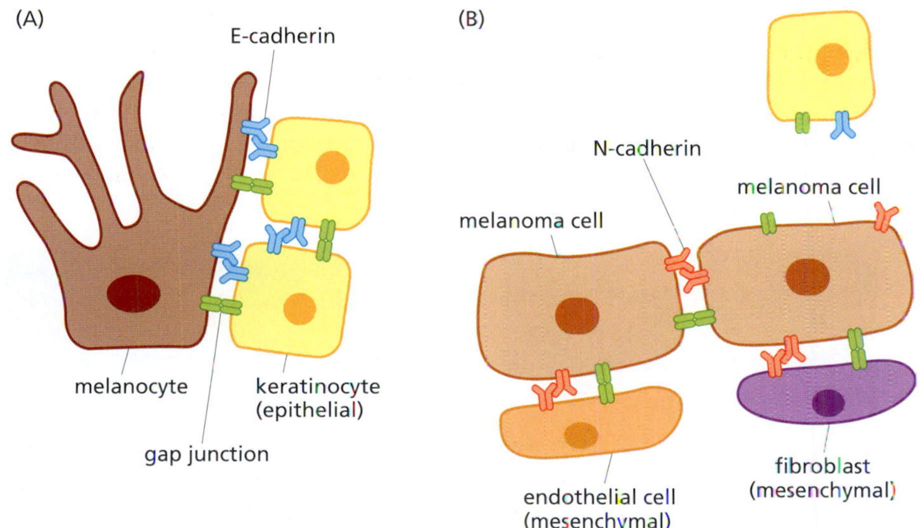

Figure 14.16 Cadherin shifts and melanoma cell invasiveness Melanomas are among the most malignant tumors because of their tendency to metastasize widely once they reach a certain stage of tumor progression. This behavior is attributable, in part, to the reactivation of a cell-biological program that enabled the migratory behavior of their neural crest ancestors. The shift from E- to N-cadherin expression, which occurs when melanocytes (A) become transformed into melanoma cells (B), has been proposed to facilitate invasion of the stroma, since the shutdown of E-cadherin expression *(blue)* enables these cancer cells to extricate themselves from their keratinocyte neighbors in the epidermis, while its replacement by expression of N-cadherin *(red)* allows these tumor cells to form homotypic interactions with various types of mesenchymal cells, such as fibroblasts and endothelial cells, that reside in the stroma of the skin (i.e., the *dermis*). (From N.K Haass, K.S.M Smalley and M. Herlyn, *J. Mol. Histol.* 35:309–318, 2004.)

epithelial cell layers. This point is driven home by experiments in which expression vectors are used to force high levels of N-cadherin expression in otherwise normal cultured epithelial cells. Such ectopic expression causes the epithelial cells to acquire motility and invasiveness, as indicated by their ability to break through reconstructed extracellular matrix (ECM) introduced into a culture dish.

14.4 The epithelial–mesenchymal transition is often induced by stromal signals

As described above, the epithelial–mesenchymal transition (EMT) seems to be an irreversible change that carcinoma cells acquire as they advance down the road toward a highly malignant growth state. Actually, there are reasons to believe that during the development of many carcinomas, the EMT phenotype is acquired *reversibly*, and that once carcinoma cells have completed the multiple steps of invasion and metastasis, they often revert back to a more epithelial phenotype by passing through the mesenchymal–epithelial transition (MET) mentioned in the last chapter.

This reversion suggests that the EMT is often triggered by signals that cancer cells experience in one environment but no longer experience in another. Thus, carcinoma cells at the invasive front of a primary tumor may receive specific signals from the nearby reactive stroma (Section 13.3) that has developed during the formation of this tumor. However, once these cancer cells have left the primary tumor and settled at a distant site, they may experience a stromal environment that does not release EMT-inducing signals. In the absence of these contextual signals, some of which are described below, the carcinoma cells may then undergo a mesenchymal–epithelial transition (MET) and revert to the phenotype of their ancestors in the heart of the primary tumor (Figure 14.17; Sidebar 14.5).

In fact, traditional histopathological techniques have failed to demonstrate the EMT at the invasive edges of primary carcinomas for a simple reason: once tumor cells undergo a full EMT (i.e., shed all epithelial traits and acquire mesenchymal ones instead), they are essentially indistinguishable from the mesenchymal cells in the surrounding stroma. For this reason, demonstrations of the EMT at the invasive edges of tumors has required the use of antibodies and cellular reagents that are not normally used in diagnostic pathology laboratories. The use of a fluorescent antibody to detect intracellular β-catenin localization (see Figure 14.14B) is one example of such a reagent. Another demonstration comes from xenografted human tumor cells, in which expression of the $\alpha_v\beta_6$ integrin has been detected, once again through immunostaining (Figure 14.19A). In this instance, $\alpha_v\beta_6$ expression, which is another marker of the EMT, is seen only in a thin outer layer of tumor cells that are in contact with the tumor-associated stroma.

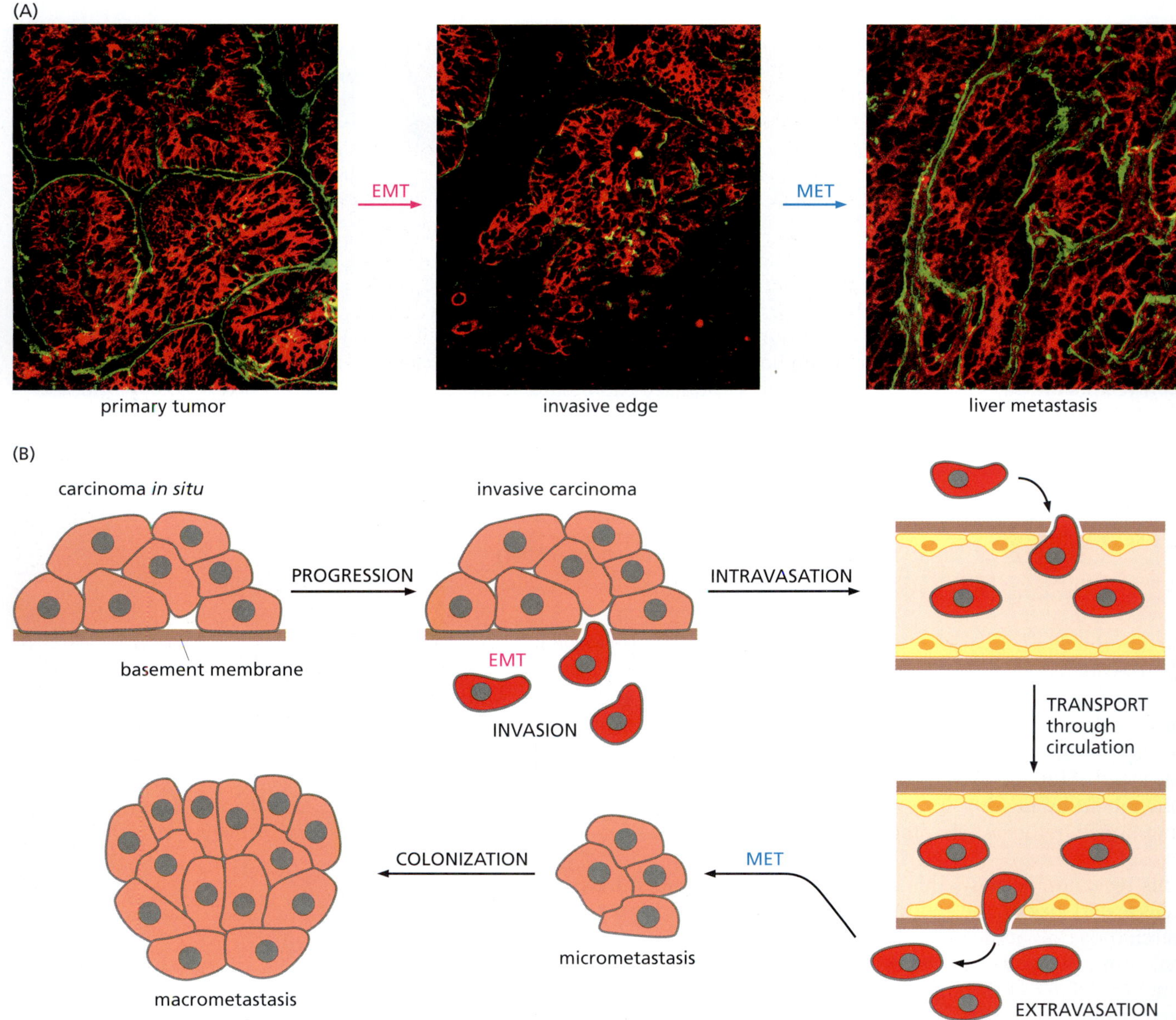

Figure 14.17 Reversibility of EMT While cells at the invasive edge of a primary carcinoma often give evidence of an EMT, derived metastases may exhibit a histology typical of the center of the primary tumor. (A) Release of degradative enzymes, notably matrix metalloproteinases (MMPs), is one of the many manifestations of the EMT. These cells in a primary colorectal carcinoma show expression of both cytokeratin 18 *(red)* and a basement membrane protein *(green)*. However, at the invasive edge of this tumor, the cells have undergone a partial EMT, in that they have degraded the adjacent basement membranes while still expressing cytokeratin 18, a key epithelial marker. In a subsequently arising metastasis in this patient, which presumably descends from cells that acquired invasiveness *en route* to metastatic dissemination, the cancer cells form a growth having, once again, the histological appearance of cells in the heart of the primary tumor. (B) Observations such as those of panel A have suggested the scheme depicted here. Thus, epithelial cancer cells at the edge of a primary carcinoma *(pink)* undergo an EMT as they invade into the stroma and become mesenchymal *(red cells)*. This change would seem to be triggered by signals that these carcinoma cells receive from the tumor-associated stroma. The newly acquired mesenchymal state enables these cells to invade locally, intravasate, and subsequently to extravasate into the parenchyma of a distant tissue. Once they have become established in this tissue, these cells find themselves in a new type of stromal environment that lacks the signals that previously induced their ancestors to undergo an EMT. This allows these cells to revert to an epithelial phenotype via a mesenchymal-epithelial transition (MET). (The regeneration of a basement membrane is not indicated here.) (A, courtesy of T. Brabletz; B, adapted from J.P. Thiery, *Nat. Rev. Cancer* 2:442–454, 2002.)

Yet another marker of the EMT is provided by laminin 2γ. Normally, this protein is one of three subunits of laminin-5, an important constituent of the basement membrane to which epithelial cells attach via specific integrins. However, cells that have undergone an EMT and become invasive now release only laminin 2γ (and not the other two subunits of laminin-5), and cleavage of laminin-2γ by matrix metalloproteinases (Section 14.6) releases a fragment that functions as an EGF-receptor ligand, facilitating cell survival and motility (Figure 14.19B).

Possibly the most vivid demonstration of the phenotypic conversion of carcinoma cells at the invasive edge of a tumor has come from the use of a human-specific anti-vimentin antibody, which demonstrates the EMT at the invasive edge of a tumor formed by experimentally transformed human mammary epithelial cells growing as a xenograft in an immunocompromised mouse

Sidebar 14.5 The reversibility of the EMT explains one peculiarity of many metastases All carcinoma cells that become invasive and metastatic may need to undergo an EMT in order to acquire these complex phenotypes. (Such cells may undergo a full EMT, during which they shed all epithelial characteristics, or, alternatively, may enter only partially into an EMT, where some epithelial characteristics are retained together with newly acquired mesenchymal traits.) This behavior would seem to be incompatible with one frequently observed aspect of human metastases: these secondary growths often closely resemble, at the histopathological level, the primary tumors from which they originated (see, for example, Figures 14.17A and 14.18). Indeed, the cells in such metastases seem to be as epithelial in their behavior as the bulk of the cells in the primary tumor, yet they are descended from invasive cells that supposedly underwent an EMT in order to initiate metastatic spread. This inconsistency is resolved if one assumes that the EMT is fully reversible, and that lineages of highly malignant cells often pass only transiently through a mesenchymal state while traveling from the primary tumor to the site of metastasis formation.

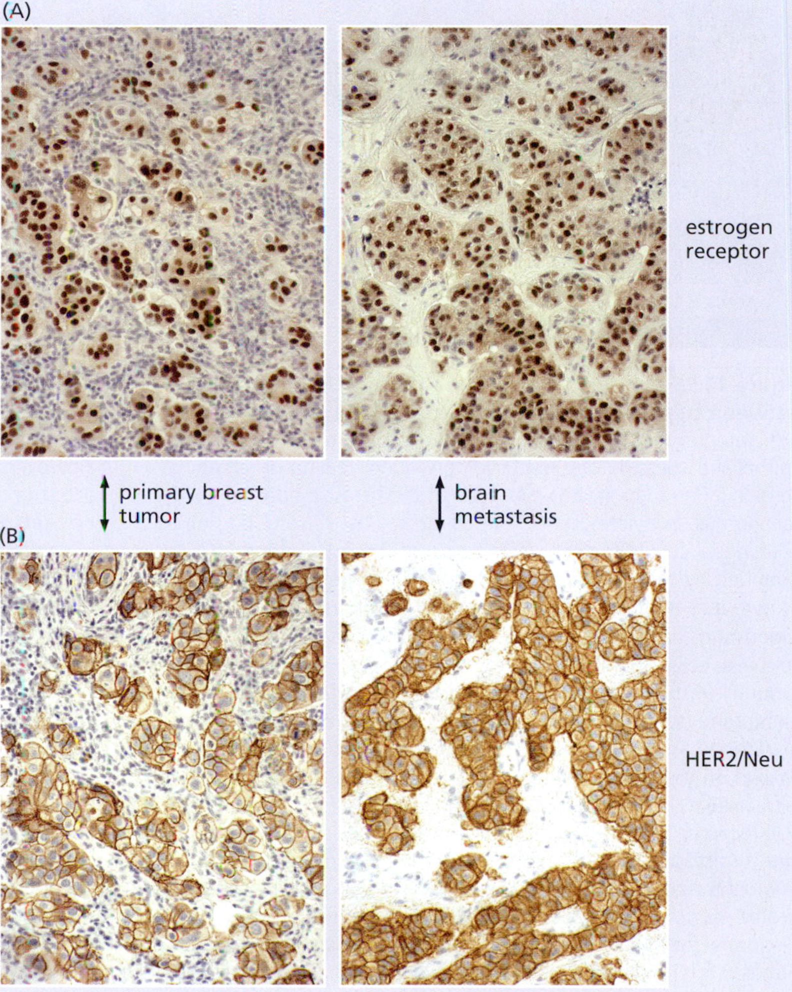

Figure 14.18 Appearance of a primary tumor and derived metastasis These micrographs illustrate the striking similarity in the appearance and expression patterns of a resected primary breast tumor *(left images)* and a brain metastasis that was detected two years later *(right images)*. (A) Immunohistochemistry using an anti-estrogen receptor (ER) antibody reveals the presence of this protein *(brown)* in the nuclei of both primary tumor cells and the cells of the derived metastasis. (B) Immunohistochemistry using an anti-HER2/Neu antibody reveals the expression of this receptor at the surfaces of tumor cells in both the primary tumor *(rimmed in brown)* and the derived brain metastasis. In both cases, the appearance of the epithelial cells and that of the recruited stromal cells are remarkably similar. Indeed, such resemblances often help a pathologist to identify the primary tumor origin from which a metastasis derives and provide support for the model depicted in Figure 14.17B. (Courtesy of T.A. Ince.)

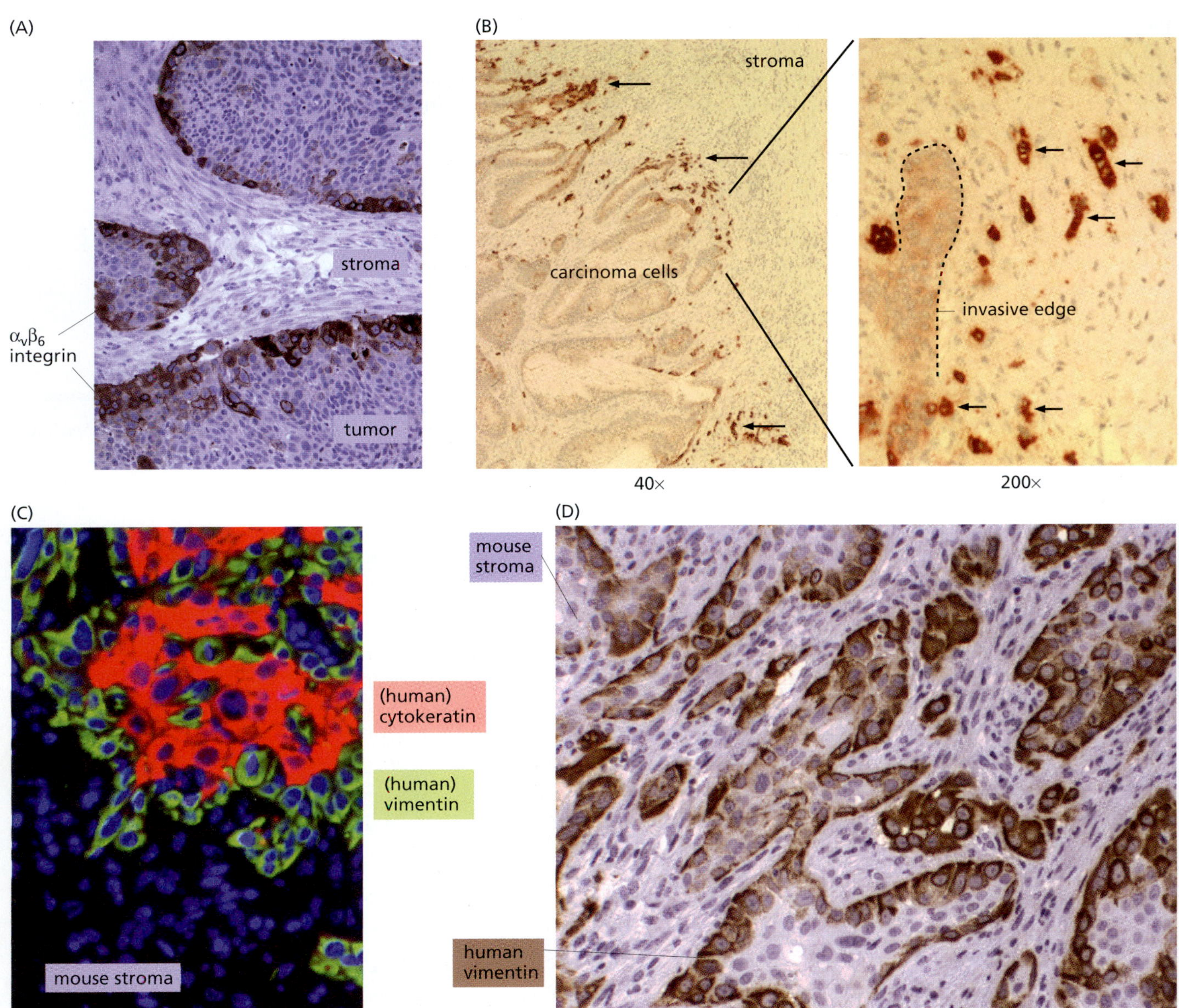

Figure 14.19 Manifestations of the EMT at the interface between tumor epithelium and stroma (A) The expression of the $\alpha_v\beta_6$ integrin is associated with the EMT. This integrin is expressed in epithelial tissues that are undergoing wound healing or suffering chronic inflammation; it is also seen at the invasive edge of carcinomas. In a xenografted tumor formed by Detroit 562 human pharyngeal carcinoma cells, expression of the $\alpha_v\beta_6$ integrin is exhibited by carcinoma cells at the invasive edge of the tumor *(dark brown)* that are in direct contact with the tumor-associated stroma, suggesting that stromal signals are responsible for its expression in epithelial cells. (B) Laminin 2γ normally serves as as one of the subunits of the heterotrimeric laminin-5 molecule of the basement membrane (see Figure 13.5). However, carcinoma cells that have undergone an EMT at the invasive edge of the tumor secrete this protein on its own, whereupon it becomes cleaved in the extracellular space by a matrix metalloproteinase (MT1-MMP) into an EGF receptor ligand that encourages cancer cell survival and motility. In this micrograph, cells from the invasive edge of a human colorectal carcinoma *(dotted line, left)* have begun to invade into the stroma *(right)*. As indicated by immunohistochemistry, individual carcinoma cells that have already invaded deep into the stroma express high levels of laminin 2γ *(brown, arrows)*. (C) Experimentally transformed human mammary epithelial cells (MECs) were implanted in an immunocompromised mouse host. The cytokeratin-positive human carcinoma cells *(red)* toward the center of the tumor mass are not in direct contact with the surrounding mouse stromal cells, whose presence is indicated only by their DAPI-stained nuclei *(blue)*. However, many of the human MECs that are in direct contact with the stroma have undergone an EMT, as indicated by their loss of cytokeratin staining and their display instead of human-specific vimentin *(green)*. [The use of antibody that specifically recognizes human (and not mouse) vimentin ensures that the green cells at the invasive edge derive from the engrafted human cells rather than from the mouse host.] Moreover, some of these cancer cells at the invasive edge have lost the cuboidal shape of the epithelial cancer cells and have assumed, instead, a more elongated, fibroblastic shape. (D) The preferential display of vimentin at the edge of these carcinoma cell islands is revealed here. A tumor formed by the same strain of transformed human MECs described in panel C is seen here at lower magnification. This image makes it apparent that the carcinoma cells that are in direct contact with the surrounding stroma have undergone an EMT, as judged by the criterion of human vimentin display *(brown)*, while tumor cells at the interior of these islands do not display human vimentin and have presumably remained in an epithelial state. (A, courtesy of D.R, Leone, B.M. Dolinski, and S.M. Violette, BiogenIdec; B, from E. Shinto, H. Tsuda, H. Ueno et al., *Lab. Invest.* 85:257–266, 2005; C, courtesy of K. Hartwell and T.A. Ince; D, courtesy of T.A. Ince.)

(Figure 14.19C and D). (The use of a human-specific antibody guarantees, in this instance, that the vimentin-expressing cells derive from keratin-positive human carcinoma cells rather than from the surrounding stroma produced by the mouse host.)

These observations and others like them indicate the involvement of certain heterotypic signals that originate in the reactive stroma of primary carcinomas, impinge on neoplastic cells located at the outer edges of the epithelial cell mass, and induce these cells to undergo an EMT. Abundant evidence indicates that TGF-β is an important agent for conveying these stromal signals. Other observations implicate a variety of other factors, including TNF-α (tumor necrosis factor-α), epidermal growth factor (EGF), HGF (hepatocyte growth factor), and IGF-1 (insulin-like growth factor-1). It appears likely that these stromal signals act in various combinations with one another and with mutant alleles, such as *ras* oncogenes, that reside in transformed epithelial cells in order to activate their EMT program. (Presumably the diversity of EMT-inducing factors reflects the multitude of steps in normal embryogenesis during which EMTs occur, these being induced at various sites throughout the developing embryo by diverse heterotypic signals.)

In one set of influential experiments, exposure of *ras*-transformed EpRas mouse mammary epithelial cells (MECs) to TGF-β resulted in the progressive loss of epithelial morphology and a reduction of epithelial markers, including cytokeratins and E-cadherin. At the same time, these transformed cells acquired mesenchymal protein markers, such as vimentin, and assumed a morphology resembling that of fibroblasts—all the hallmarks of an EMT. Provocatively, once these *ras*-transformed cells underwent an EMT, they began to produce their own TGF-β1; this TGF-β1, acting via an autocrine signaling loop, allowed them to maintain their mesenchymal phenotype for extended periods of time, long after the inciting TGF-β was withdrawn from their culture medium (Figure 14.20A–D). These studies suggest that TGF-β signaling can conspire with a *ras* oncogene to cause epithelial cancer cells to undergo an EMT. Similarly, maintenance of TGF-β signaling through a positive-feedback loop may play an important role in maintaining $\alpha_v\beta_6$ integrin expression and the EMT in human carcinoma cells. This figure also provides indication that TGF-β may often be produced in abundance by the tumor-associated stroma (Figure 14.20E).

Two of the downstream effectors of Ras function (see Sections 6.5 and 6.6) seem to be responsible for the collaboration between Ras and TGF-β signaling. The Raf oncoprotein, which functions immediately downstream of Ras, can elicit an EMT on its own: it causes transformed EpRas cells to secrete TGF-β, which then acts in an autocrine fashion to induce the EMT in these cells. PI3K, another effector of the Ras oncoprotein, protects Ras-transformed cells from the cytostatic and pro-apoptotic effects of TGF-β.

A related mechanism also seems to explain the progression of mouse skin tumors that have been initiated by *ras* oncogene activation. The cells in these tumors evolve from having a highly differentiated, squamous cell phenotype (see Figure 2.6A,B) into cancer cells that are motile, spindle-shaped, and metastatic. (Spindle-shaped cells exhibit a morphology similar to that of transformed fibroblasts.) This progression to a highly malignant state is accompanied by and likely caused by progressively increasing levels of the Ras oncoprotein as well as increasing nuclear localization of the Smad2 transcription factor, the latter indicating intense signaling through the TGF-β pathway (see Figure 6.29D).

TGF-β's role in actively promoting the aggressiveness of malignant cancer cells contrasts starkly with our earlier discussions of its anti-proliferative effects (see Sidebar 29 ●). Strong support that TGF-β can favor malignant cell behavior is provided by numerous studies in which the levels of tumor-associated TGF-β (often TGF-β1) were found to rise in parallel with increasing degrees of tumor

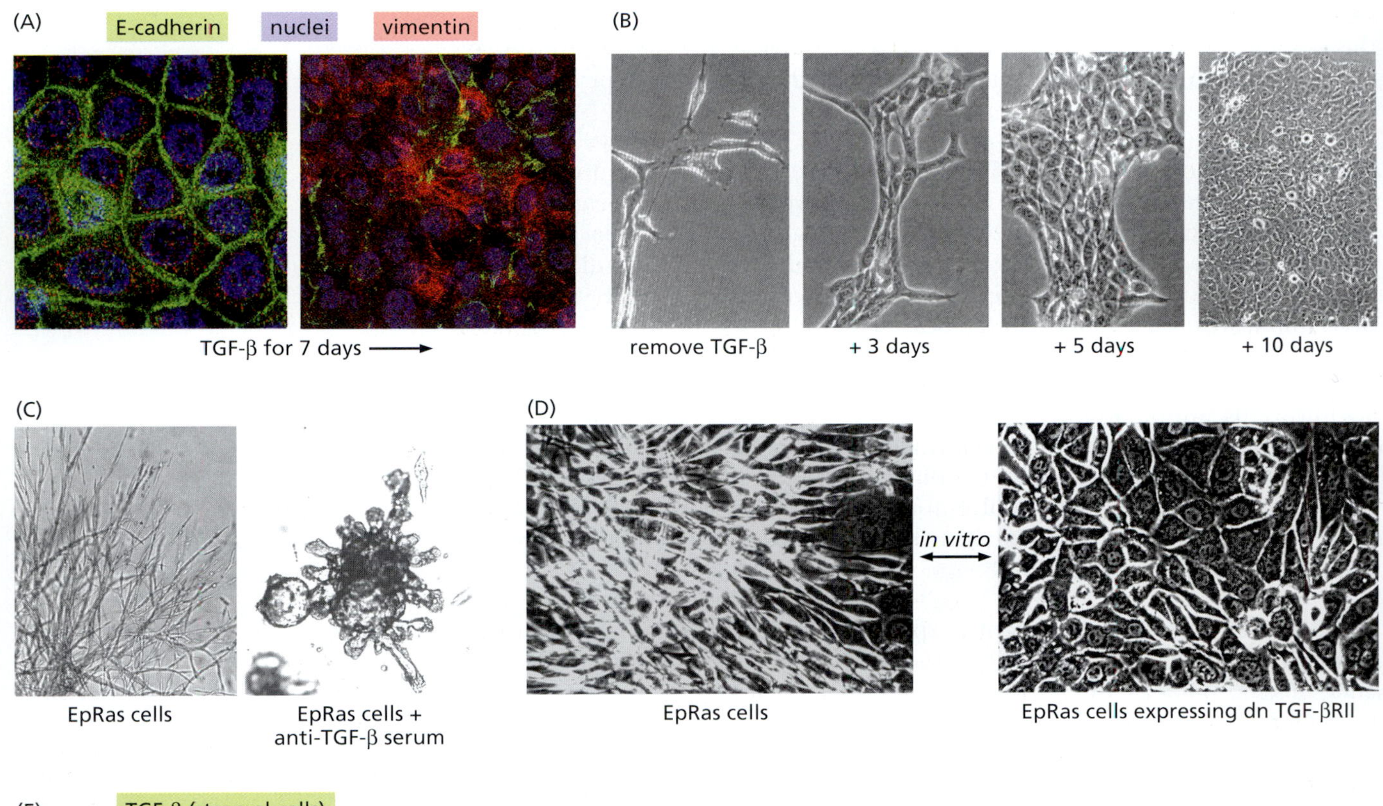

(E) TGF-β (stromal cells)

$\alpha_v\beta_6$ integrin (epithelial cells)

Figure 14.20 Control of the EMT by TGF-β and its effects on tumorigenic cells Mouse mammary epithelial cells (MECs) of the EpH4 cell line were transformed into tumor cells through the introduction of a *ras* oncogene, yielding EpRas cells. (A) EpRas cells *(left)* usually have an epithelial, cobblestone-like appearance and express E-cadherin *(green)* at their cell–cell junctions. However, when they are cultured for 7 days in the presence of TGF-β1 *(right)*, they undergo an EMT and assume an elongated fibroblastic appearance (not visible here). In addition, they suppress expression of E-cadherin and express instead vimentin *(red)*, thereby shifting from an epithelial to a mesenchymal gene expression program. Nuclei are stained blue with the DAPI dye. (B) After the EpRas cells undergo an EMT (panel A), they maintain the resulting mesenchymal, fibroblast-like state through their own production of and response to TGF-β1 (i.e., via TGF-β1 autocrine signaling). However, when these cells are cultured in a medium that lacks added TGF-β1 and their growth medium is changed on a daily basis (to remove any TGF-β that they may have secreted into the medium), their appearance gradually reverts to an epithelial cobblestone phenotype, as shown after 3, 5, and 10 days of culture *(left to right)*, indicating that they have undergone a mesenchymal-epithelial transition (MET). (C) When EpRas cells that have been growing in a tumor in the presence of high concentrations of autocrine TGF-β1 are placed into collagen gels *in vitro*, they form the highly invasive structures seen here *(left panel)*. However, if these cells are propagated *in vitro* under identical conditions in the presence of antiserum that neutralizes TGF-β1 (to sequester any TGF-β produced by these cells), they revert to an epithelial phenotype and now form ductal structures in collagen gels that are typical of those formed by cultured normal MECs *(right panel)*. (D) Use of a dominant-negative (dn) type II TGF-β receptor (which effectively blocks autocrine TGF-β signaling) provides further proof that autocrine TGF-β signaling by EpRas cells is required to maintain their mesenchymal state. When this signaling is blocked by expression of this receptor, the mesenchymal appearance of the EpRas cells *(left)* disappears and they assume an epithelial appearance *(right)*, indicating, as before, that they have undergone a mesenchymal–epithelial transition (MET). (E) Detroit 562 carcinoma cells growing in a tumor express the $\alpha_v\beta_6$ integrin *(red*, see Figure 14.19A), which is indicative of their having undergone an EMT; while TGF-β *(green)* is being produced by cells in the nearby tumor-associated stroma. The $\alpha_v\beta_6$ displayed by the tumor cells can activate the latent form of TGF-β produced by the stromal cells, thereby creating a factor that is a strong inducer of the EMT and additional $\alpha_v\beta_6$ integrin expression by epithelial cells, yielding a self-sustaining, positive-feedback loop. (A, from E. Janda, K. Lehmann, I. Killisch et al., *J. Cell Biol.* 156:299–314, 2002; B and C, from M. Oft, J. Peli, C. Rudaz et al., *Genes Dev.* 10:2462–2477, 1996; D and E, courtesy of D.R. Leone, B.M. Dolinski and S.M. Violette, BiogenIdec.)

invasiveness and general aggressiveness. Indeed, high levels of TGF-β, both in the tumor mass and in the general circulation, augur poorly for the long-term survival of the cancer patient.

To summarize these and other observations, TGF-β can contribute to cancer cell invasiveness for at least four reasons. First, most human carcinomas arising outside the intestine retain at least some functional TGF-β receptor signaling, allowing them to continue to respond to TGF-β during the entire course of tumor progression. (This contrasts with the situation observed in a subset of colon carcinomas in which receptor function may be lost entirely through mutations in the receptor-encoding genes; see Section 12.9.) Second, the inactivation of the pRb pathway, which occurs in most if not all human cancers, causes malignant cells to lose their responsiveness to the cytostatic effects of TGF-β (see Section 8.10); this loss allows these cells to respond to other types of downstream signals that are released by ligand-activated TGF-β receptors. Third, in the absence of the cytostatic effects of TGF-β, exposure of cancer cells to this factor may actually *favor* their proliferation. For example, glioblastoma and osteosarcoma cells that are treated with TGF-β respond by producing and secreting PDGF; once released, the latter acts in an autocrine fashion to stimulate the proliferation of these cancer cells via the PDGF receptors that they display. Finally, exposure of breast cancer cells to TGF-β causes them to release other factors that accelerate the breakdown of mineralized bone—a critical step in the formation of *osteolytic* metastases; as we will learn later, this breakdown liberates additional mitogens that drive cancer cell proliferation.

TNF-α, acting on its own or in concert with TGF-β, also appears to be an important agent for inducing the EMT. Early in tumor progression, TNF-α is often produced by inflammatory cells, such as macrophages (see Section 11.16). At this stage, it functions via its receptor to activate the NF-κB signaling pathway in epithelial cells. TGF-β also activates the NF-κB pathway in epithelial cells, such as the immortalized mouse mammary epithelial cells discussed above. In various tumors, TNF-α and TGF-β may contribute, to differing extents, to the long-term maintenance of active NF-κB signaling. This signaling seems to be critical for the induction and maintenance of the EMT, as indicated by work with the EpRas mouse mammary epithelial cells. Thus, blockage of NF-κB signaling prevents expression of their EMT program (Figure 14.21).

The influence of stromal macrophages on the invasive and metastatic behavior of primary cancer cells can be demonstrated by studying genetically altered mice that lack the ability to make colony-stimulating factor-1 (CSF-1). As was discussed in the last chapter, mammary carcinomas arising in cancer-prone

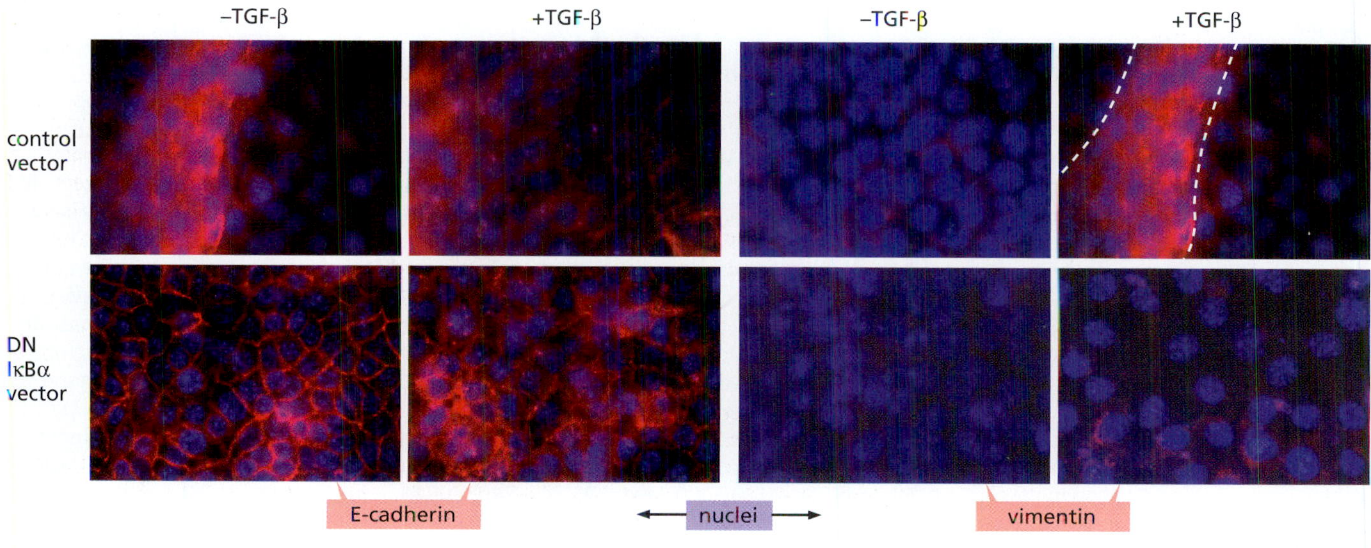

Figure 14.21 The NF-κB signaling pathway and induction of the EMT Cells of the EpRas line of transformed mouse mammary epithelial cells (see Figure 14.20) were treated with TGF-β *(2nd and 4th micrographs, upper row)*. This caused a suppression of E-cadherin expression *(pink, 1st and 2nd micrographs, upper row)* as well as an induction of vimentin *(pink, 3rd and 4th micrographs, upper row)*. However, when a dominant-negative form of IκBα, which blocks NF-κB signaling (see Figure 6.29A), was expressed in these cells *(lower row)*, TGF-β treatment failed to suppress E-cadherin expression *(pink, 1st and 2nd micrographs)* and failed to induce vimentin expression *(pink, 3rd and 4th micrographs)*. This indicated that NF-κB signaling was necessary for induction of the EMT in these cells. Other experiments (not shown) indicated that constitutive activation of NF-κB signaling was sufficient to induce the EMT in these cells. (From M.A. Huber, N. Azoitei, B. Baumann et al., *J. Clin. Invest.* 114:569–581, 2004.)

transgenic mice usually recruit large numbers of tumor-associated macrophages (TAMs). However, when the tumor cells in such mice lack the ability to make CSF-1, TAMs are virtually absent (see Figure 13.24). The absence of CSF-1 and TAMs has no effect on primary tumor growth (Figure 14.22A), but such tumors show a benign, noninvasive behavior, in contrast to tumors that succeed in recruiting TAMs (Figure 14.22B). The influence of these macrophages on metastatic behavior is striking: without TAMs, these breast tumors fail to seed metastases to the lungs (Figure 14.22C).

This experiment provides compelling evidence that the invasive and metastatic behavior of these mouse breast carcinoma cells is strongly influenced by signals that these carcinoma cells receive from stromal cells, in this case macrophages. It fails, however, to reveal the precise nature of these signals. Macrophage-derived TNF-α, as argued above, is likely to contribute to induction of the EMT by cancer cells, and therefore to the invasive and metastatic behavior described in Figure 14.22. Another key macrophage-derived signal is likely to be conveyed by EGF.

Some of the evidence favoring EGF as a key inducer of cancer cell invasiveness comes from studies of mouse breast cancer cells both *in vivo* and *in vitro*. Like most epithelial cells, these carcinoma cells express the EGF receptor, and activation of this receptor by EGF causes them to acquire both motility and invasiveness

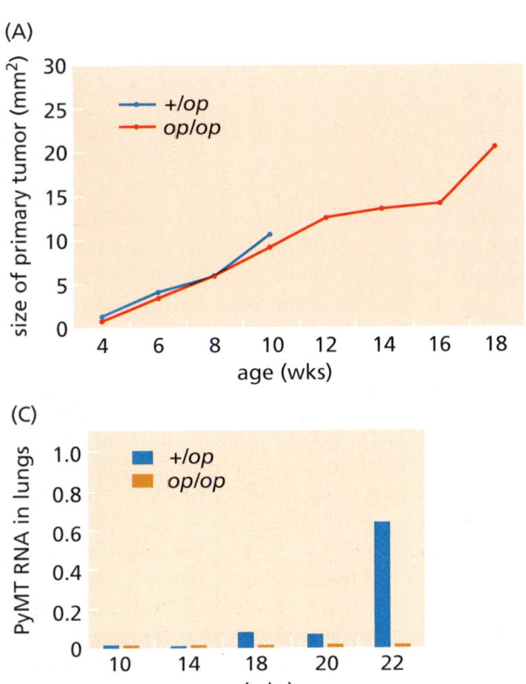

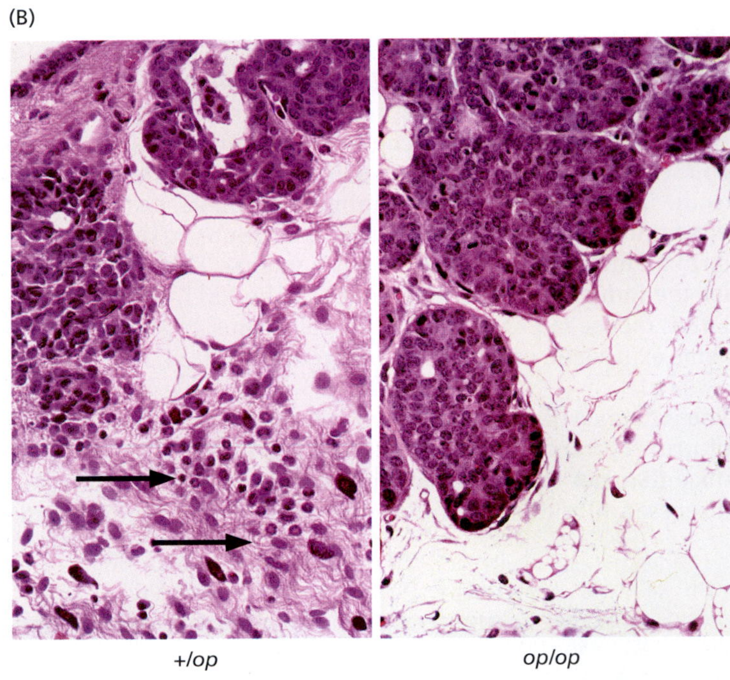

Figure 14.22 Effects of macrophages on invasion and metastasis Mice of a transgenic strain develop mammary carcinomas through the expression of a transgene in which the MMTV promoter drives expression of the polyoma middle T *(PyMT)* oncogene. This transgene has been introduced, through breeding, into mice that can ($Csf^{+/op}$) or cannot ($Csf^{op/op}$) make colony-stimulating factor-1 (CSF-1), which is needed to recruit macrophages to the tumor mass (see Figure 13.24). (A) The presence (in $Csf^{+/op}$ mice) or absence (in $Csf^{op/op}$ mice) of recruited tumor-associated macrophages (TAMs) has no effect on the ability of the primary breast tumors to grow in these transgenic mice. (B) Such mammary tumors arising in $Csf^{+/op}$ mice (whose tumors contain many TAMs, *not shown*) develop a highly invasive phenotype, in which individual carcinoma cells invade the nearby stroma in large numbers *(left panel, arrows)*. However, if tumors develop in $Csf^{op/op}$ mice, in which macrophages cannot be recruited into the tumor-associated stroma *(right panel)*, the tumor cells do not break through the basement membrane and the tumor as a whole remains encapsulated and benign, indicating that the macrophages contribute in essential ways to tumor invasiveness. (C) In the $Csf^{+/op}$ mice, metastases in the lungs begin to appear at 18 weeks of age and increase progressively thereafter *(blue bars)*, as gauged by the amount of polyomavirus middle T RNA (expressed in tumor cells) present in the lungs *(ordinate)*. However, in the $Csf^{op/op}$ TAM-negative mice *(orange bars)*, metastases are virtually absent. (From E.Y. Lin, A.V. Nguyen, R.G. Russell and J.W. Pollard, *J. Exp. Med.* 193:727–740, 2001.)

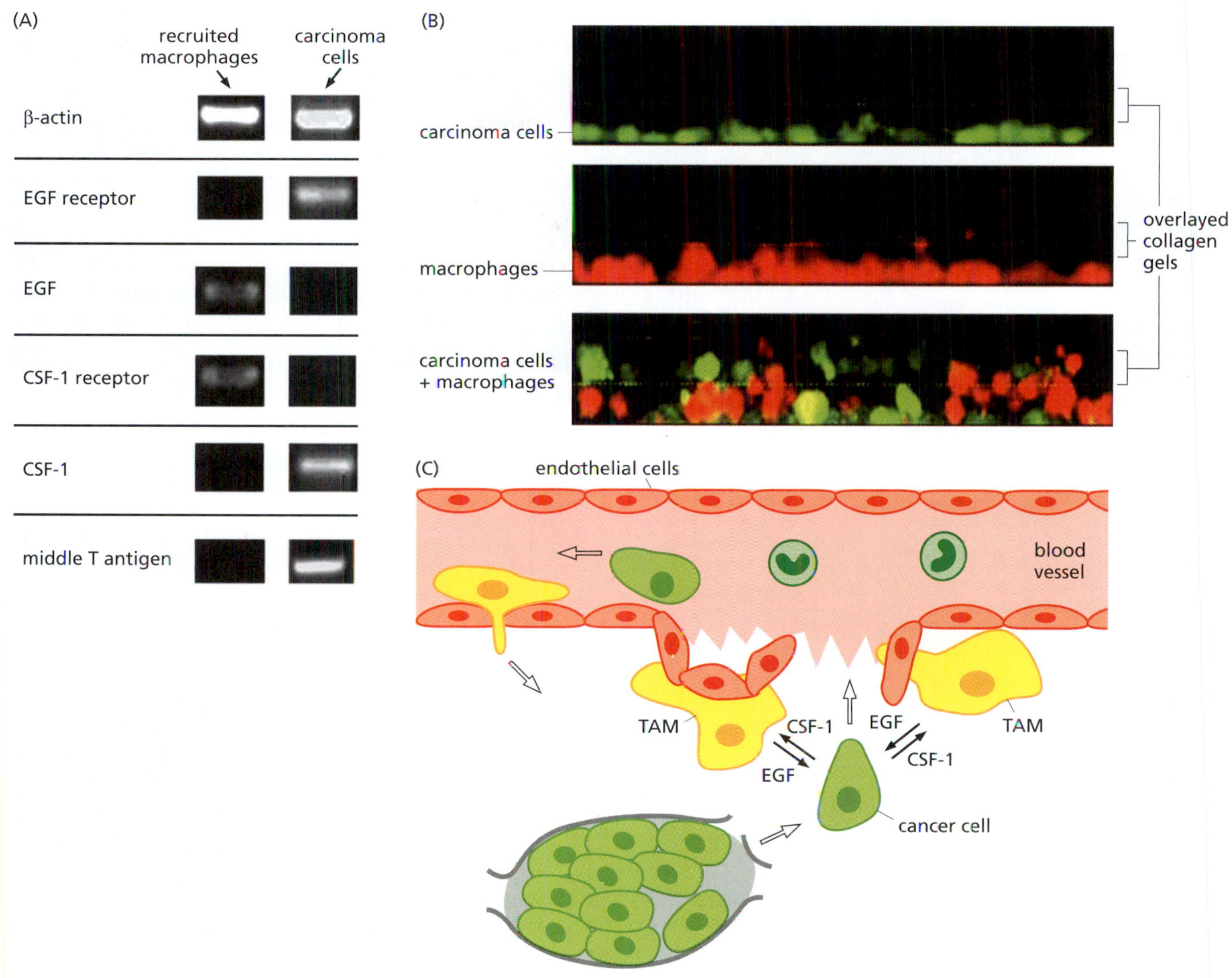

Figure 14.23 Reciprocal stimulation by breast cancer cells and macrophages A variety of experiments indicate that macrophages are the major source of EGF in breast cancers. EGF is known to be able to stimulate epithelial cancer cells to proliferate and invade through extracellular matrix. In addition, EGF exposure causes breast cancer cells to release CSF-1, which allows them to recruit macrophages and stimulates production by the macrophages of more EGF, resulting in a positive-feedback loop between these two cell types. (A) Using PCR analysis, the mRNA levels of these two growth factors and their receptors are found to be reciprocally expressed in mammary carcinoma cells arising in tumor-prone transgenic mice (see Figure 14.22) and in recruited stromal macrophages. (B) Breast cancer cells (labeled here with *green* fluorescent protein, GFP) were placed at the bottom of a Petri dish *(seen here in side view)* below a layer of collagen gel *(top image)*, where they remained. Similarly, macrophages *(red)* also remained where they were initially placed at the bottom of the Petri dish below a collagen gel *(middle image)*. However, when the two populations were co-cultured at this location, the breast cancer cells *(green)* were now induced to move upward and invade the overlying collagen gel *(lower image)*. (C) The reciprocal interactions between breast cancer cells and macrophages are illustrated schematically here. Because macrophages are often found in close proximity to microvessels, the stimulation by tumor-associated macrophages (TAMs) of breast cancer cell motility and invasiveness may also contribute to cancer cell intravasation, as depicted here. (A, from J. Wyckoff, W. Wang, E.Y. Lin et al., *Cancer Res.* 64:7022–7029, 2004; B, from S. Goswami, E. Sahai, J.B. Wyckoff et al., *Cancer Res.* 65:5278–5283, 2005; C, from W. Wang, S. Goswami, E. Sahai et al., *Trends Cell Biol.* 15:138–145, 2005.)

and to secrete CSF-1, the attractant and stimulant of macrophages (Figure 14.23). Macrophages respond to CSF-1 by proliferating and releasing EGF, which activates the cancer cells. These effects all proceed through paracrine rather than autocrine signaling, since the breast cancer cells do not express the CSF-1 receptor and the macrophages do not express the EGF receptor (see Figure 14.23A). Accordingly, these two cell types collaborate by reciprocally stimulating one another, yielding another type of positive-feedback loop. While cancer cell motility and invasiveness are clearly demonstrated in these experiments, the

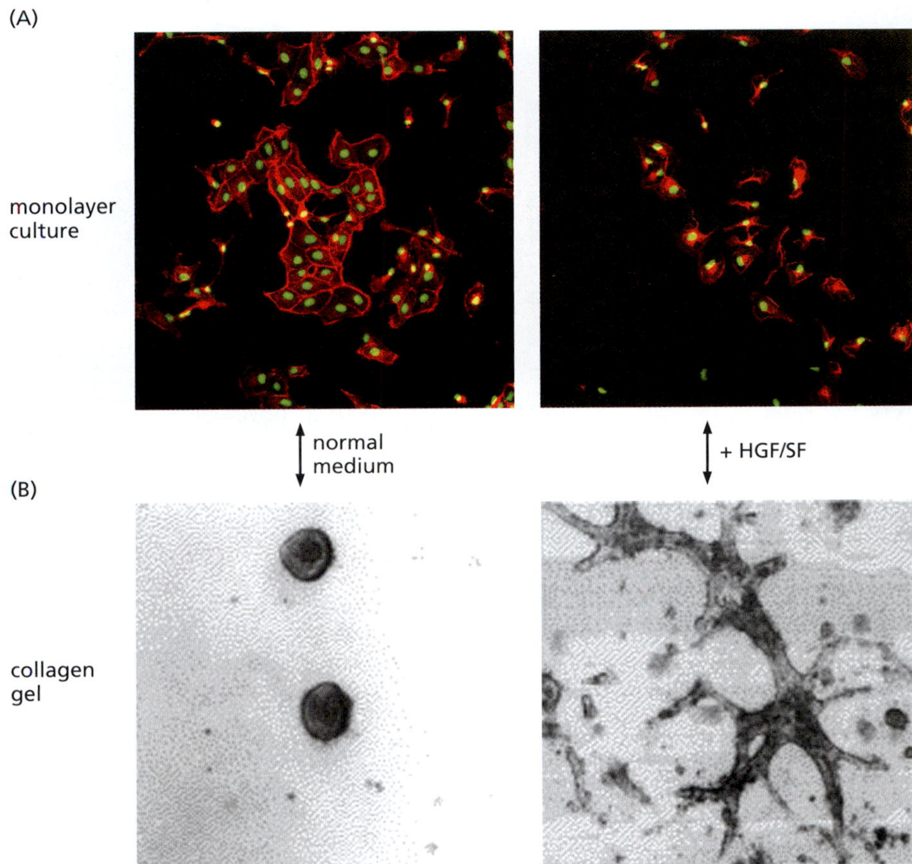

Figure 14.24 Cell scattering and invasive behavior induced by HGF Hepatocyte growth factor (HGF), also known as scatter factor (SF), is produced by a variety of stromal cell types. It has profound effects on epithelial cells that display its cognate receptor, Met. (A) Shown here are MDCK (Maden–Darby canine kidney) cells *(red cytoplasms),* which are used widely to study epithelial cell biology. The cells in the left panel were grown in a normal medium, while those in the right panel were grown in medium to which HGF/SF was added. In monolayer culture, these epithelial cells normally form clusters of cobblestone-like cells. However, after treatment by HGF/SF, they become motile and scatter in many directions. (B) When introduced into collagen gels, MDCK cells normally form small spherical clumps *(left panel).* However, following exposure to HGF/SF, these cells grow in long processes that invade the surrounding collagen gel *(right panel).* (A, courtesy of J.H. Resau; B, courtesy of U. Schaeper, from M. Rosario and W. Birchmeier, *Trends Cell Biol.* 13:328–335, 2003.)

induction of an EMT in the cancer cells can only be inferred from their acquisition of motile, invasive behavior. HGF, another ligand of stromal origin, is also capable of inducing many of the attributes of the EMT in epithelial cells, which generally display Met, its cognate receptor, on their surface (Figure 14.24).

These diverse lines of evidence strongly suggest that the acquisition of malignant traits by cancer cells, including induction of the EMT, is not governed solely by the genomes of these cells. Instead, these profound shifts in cell phenotype are often initiated by a collaboration between specific mutant alleles harbored in cancer cell genomes (e.g., a *ras* oncogene) and the signals that these cancer cells receive in some tissue microenvironments, specifically at the boundary between a tumor epithelium and reactive stroma. In many tumors, these contextual signals are conveyed by certain factors, such as TGF-β and TNF-α, that are released by cells in the reactive stroma (Figure 14.25). Yet other stromal signals, such as those carried by EGF and HGF, may also help to elicit many of the changes that we associate with cancer cell invasiveness and the EMT. While this scenario depicts the behavior of many carcinomas, it does not accurately describe all of them (Sidebar 30).

Throughout much of this text, our thinking has been driven by the notion that the phenotypes of cancer cells are dictated by their genotype and that tumorigenic growth is essentially a cell-autonomous phenomenon. Our encounters with heterotypic interactions (Chapter 13) revised this notion slightly, by indicating that cancer cells show a surprising degree of dependence on normal neighbors for various types of sustenance and support. Now, we must come to terms with the idea that the microenvironment of the cancer cell can also fundamentally reshape that cell's phenotype, specifically by inducing the profound changes in cell behavior that comprise the EMT.

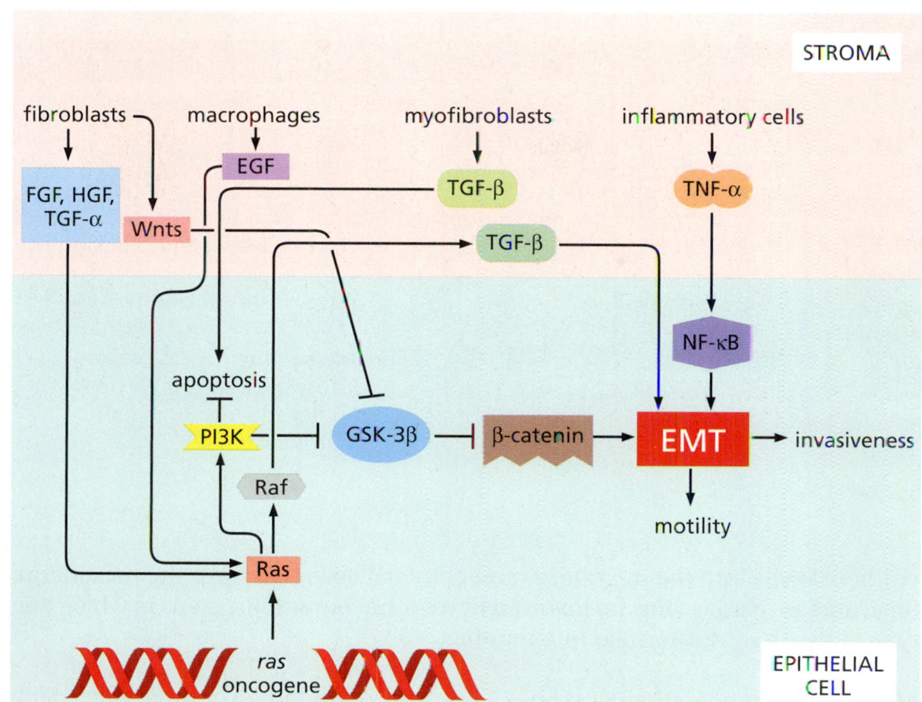

Figure 14.25 Signals that trigger the EMT This diagram presents a highly simplified view of the signaling channels that originate in the stroma and influence epithelial cancer cells to undergo a partial or complete EMT. It is likely that the EMT is normally triggered in response to a mixture of signals that carcinoma cells receive from the stroma together with intracellular signals, such as those released by a *ras* oncogene, as indicated here. The precise identities of these stromal signals and their combinatorial mechanisms of action remain to be elucidated.

14.5 EMTs are programmed by transcription factors that orchestrate key steps of embryogenesis

Execution of the EMT program depends on changes in the expression of dozens, possibly hundreds of distinct genes. These changes affect many aspects of cell biology, not all of which have been enumerated here. The changes include the organization of a cell's intermediate filament cytoskeleton, its motility, its association with neighboring cells, its release of proteases, and even its display of cell surface integrins and growth factor receptors (see Table 14.2). While extensive evidence implicates stromal signals as key elements in triggering the EMT of carcinoma cells, none of this evidence, on its own, reveals how the complex EMT program is actually coordinated within the responding epithelial cells.

The genetics of early development has provided much of the answer to this question. A number of genes that specify pleiotropically acting, EMT-inducing transcription factors have been identified, largely in *Drosophila melanogaster*. Many of these genes and the transcription factors that they encode are conserved in chordates and have been found to control key steps in early embryogenesis in frog and mouse embryos; these steps involve various types of EMT. (The strong conservation of these genes indicates that the EMT and key steps of early embryogenesis were developed early in metazoan evolution, long before the radiation of the various metazoan phyla.) By activating these transcription factors, cancer cells gain access to the complex, multi-component EMT programs that they orchestrate.

More than half a dozen such transcription factors have been described, each of which is capable of inducing an EMT when ectopically expressed in certain epithelial cells (Table 14.3). For example, Snail is a transcription factor that was first described in *Drosophila* (Figure 14.26). Since its initial description, Snail has been discovered in a wide range of metazoa, including vertebrates, insects, worms, and mollusks. In early vertebrate embryos, Snail is first expressed in the portion of the ectoderm that is destined to become mesoderm following gastrulation. When operating during embryogenesis, Snail, Slug, and Twist convert

Table 14.3 Transcription factors orchestrating the EMT

Name	Where first identified	Type of transcription factor	Cancer association
E47/E2A	associated with E-cadherin promoter	bHLH	
FOXC2	mesenchyme formation	winged helix/forkhead	basal-like breast cancer
Goosecoid	gastrulation in frog	paired homeodomain	various carcinomas
SIP1	neurogenesis	2-handed zinc finger/homeodomain	ovarian, breast, liver carcinomas
Slug	delamination of the neural crest and early mesoderm in chicken	C2H2-type zinc finger	breast cancer cell lines, melanoma
Snail	mesoderm induction in *Drosophila;* neural crest migration in vertebrates	C2H2-type zinc finger	invasive ductal carcinoma
Twist	mesoderm induction in *Drosophila;* emigration from neural crest	bHLH	invasive lobular breast cancer, diffuse-type gastric carcinoma, high-grade melanoma and neuroblastoma

epithelial cells into the migratory mesenchymal cells that form the mesoderm. Snail and its relative Slug are involved in yet other embryonic steps in which one type of tissue is transformed into another.

Moreover, when epithelial monolayers are wounded experimentally, Slug expression is induced in the surviving epithelial cells at the edge of the wound in order to enable these cells to acquire motility and migrate into the wound site (Figure 14.27). Expression of Slug helps to explain how epithelial cells at the edges of wound sites undergo a transient EMT in order to reconstruct epithelial cell sheets (see Figure 13.13C). Observations like these broaden our perspective on these transcription factors (TFs) and their normal biological roles: in addition to programming key steps in early embryogenesis, the expression of some of these TFs may be resurrected transiently in adults in order to reconstruct damaged tissues.

Snail and Slug are members of the C2H2-type zinc finger TFs. The Snail–Slug TFs seem to operate largely as repressors of transcription. Thus, both have been found to be able to repress transcription of the E-cadherin gene. As we read earlier, the loss of E-cadherin expression can, on its own, cause epithelial cells to assume many of the phenotypic changes associated with the EMT.

The Snail TF has been found to be expressed in the invasive fronts of chemically induced mouse skin carcinomas, and its expression is associated with the degree of lymph node metastasis of human breast cancers. Moreover, embryonic expression of Snail, the related Slug TF, and Goosecoid is induced by contextual signals, such as TGF-β and Wnts, that are known to be responsible for inducing the EMT conversion of mouse tumor cells. Twist is expressed during the gastrulation of *Drosophila* embryos (see Figure 14.26B) and the out-migration of neuroepithelial cells from the neural crest of chordate embryos. Its expression is also induced by exposure to TGF-β.

Rapidly accumulating evidence associates some of these transcription factors with various types of human malignancy. For example, Snail has been found to be expressed in islands of human mammary ductal carcinoma cells that lack E-cadherin expression. Slug has also been implicated in the repression of E-cadherin expression in human breast cancers (Figure 14.28A). Twist is found to be expressed at elevated levels in many invasive mammary lobular carcinomas and, to a much lesser extent, in invasive ductal carcinoma cells (Figure 14.28B). Similarly, Twist is expressed preferentially in the "diffuse" subtype of gastric carcinomas and to a lesser extent in the "intestinal" subtype of gastric carcinomas, in which expression of Sip1—another EMT-inducing TF—is elevated (Figure 14.28C). And significantly, both Twist and Slug enable cells to resist apoptosis and anoikis, and therefore can protect metastasizing cells from some of the physiologic stresses that would normally cause their death long before they succeed in reaching distant tissue sites and forming micrometastases.

These associations have not been studied systematically, and in any event, even when they are observed, they do not prove definitively that the various TFs are causally involved in programming the invasive and metastatic traits of human

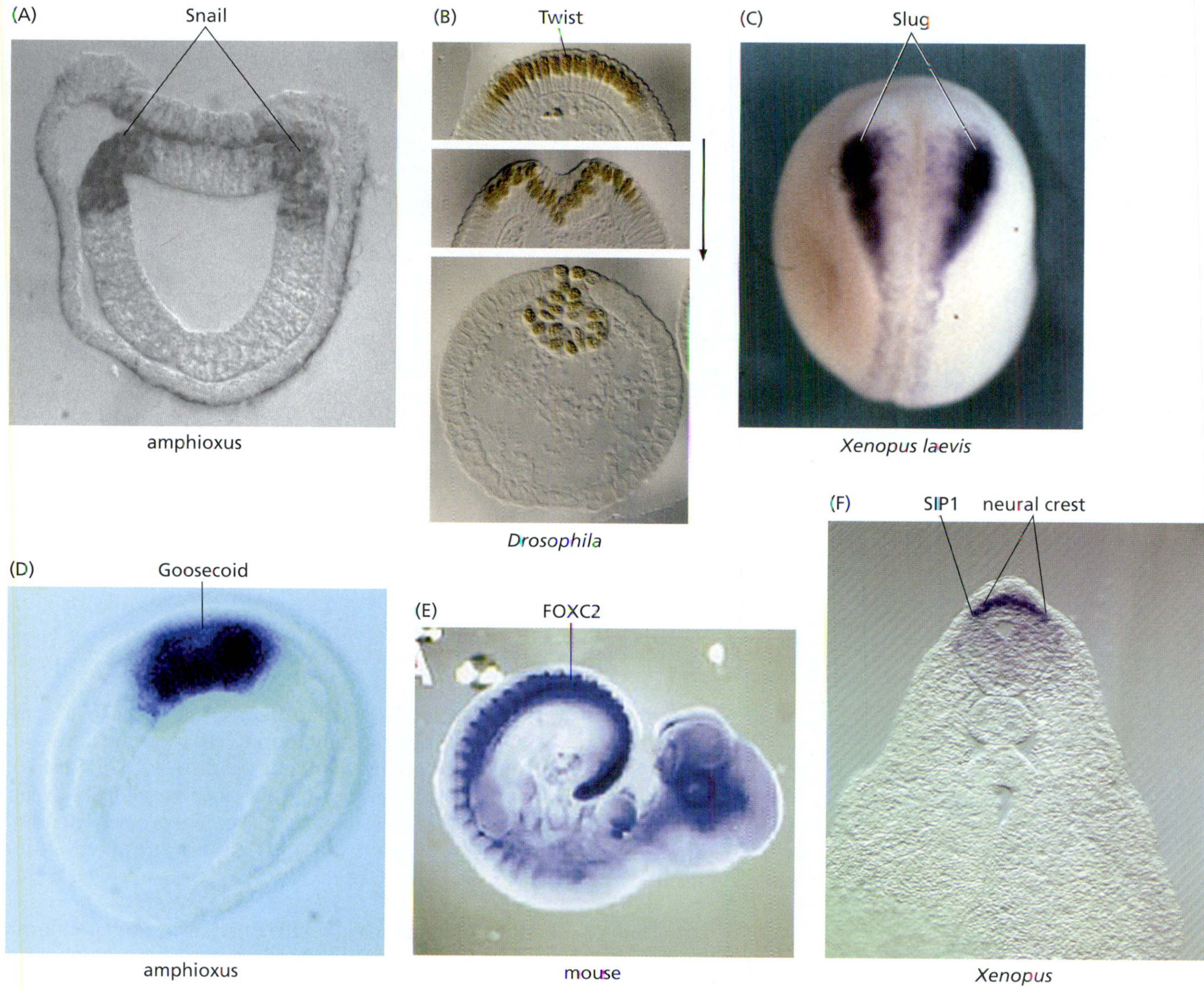

Figure 14.26 Embryonic transcription factors programming epithelial–mesenchymal transitions (A) The Snail transcription factor (TF), which can program the EMT, is shown being expressed in *Amphioxus,* a primitive chordate, in the cells *(dark areas)* that are counterparts of the cells in higher chordates that will form the neural crest. (B) The Twist TF, which also can program an EMT, is shown here *(brown)* in an early *Drosophila* embryo at the site of gastrulation, which it helps to orchestrate. (C) The Slug TF, a close relative of Snail, is also expressed in the embryonic neural crest. Here its expression is visualized *(dark blue)* in the neural crest of an embryo of *Xenopus laevis*, the African clawed toad. (D) The Goosecoid TF is expressed at the blastopore lip in gastrulating chordate embryos. Here its expression, which is inducible by the TGF-β signaling pathway, is shown adjacent to the blastopore 8 hours after fertilization of an amphioxus egg. (E) The FOXC2 TF, previously known as Mesenchyme Forkhead-1, is expressed in important mesodermal structures in this day 9.5 mouse embryo, including the mesoderm around the developing spinal column as well as the somites, which are precursors to many of the body's muscles. (F) In *Xenopus laevis,* expression of the SIP1 TF is seen in the neural crest and neural tube, where it appears to be responsible for the cell movements that lead to closure of the neural tube and emigration of cells from the neural crest to other parts of the body. (A, courtesy of J. Langeland; B, from M. Leptin, J. Casal, B. Grunewald and R. Reuter *Dev. Suppl.* 22–31, 1992. © The Company of Biologists; C, courtesy of C. LaBonne; D, from A.H. Neidert, G. Panopoulou and J.A. Langeland, *Evol. Dev.* 2:303–310, 2000; E, from T. Furumoto, N. Miura, T. Akasaka et al., *Dev. Biol.* 210:15–29,1999; F, from L. van Grunsven, C. Papin, B. Avalosse et al., *Mech. Dev.* 94:189–193, 2000.)

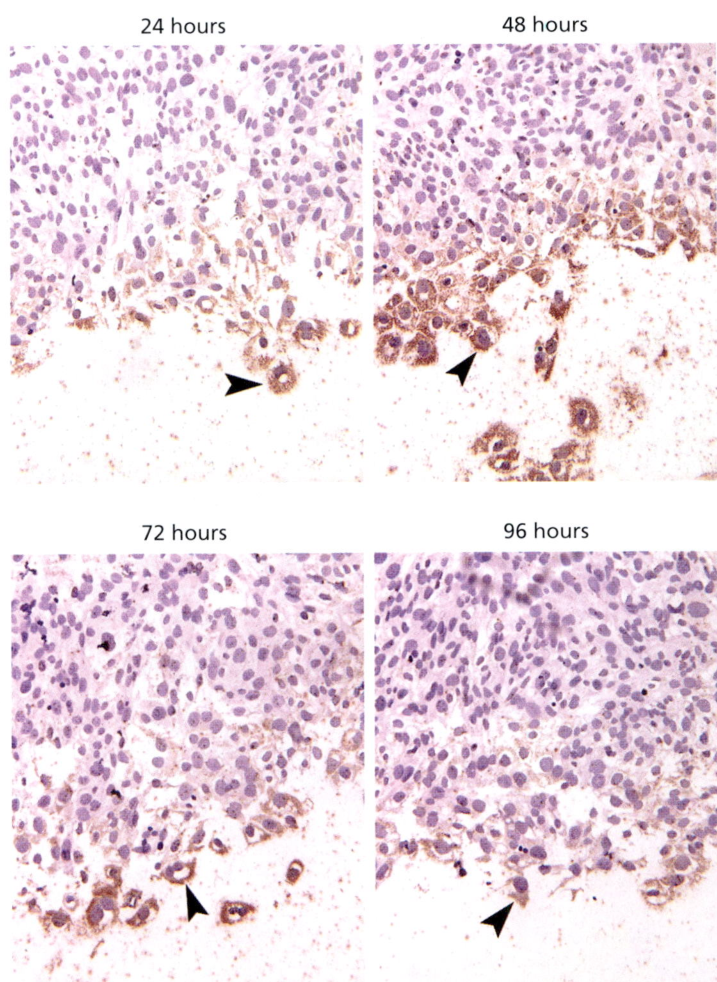

Figure 14.27 Transient expression of an EMT-inducing transcription factor in wound healing Expression of the Slug transcription factor is induced transiently in a monolayer of keratinocytes that have been wounded by scraping away a swath of cells. As seen here, 48 hours after wounding, keratinocytes at the edge of the wound induce expression of Slug *(dark brown)* as they separate from the monolayer and begin to make their way into the wound site *(bottom of each panel)* in order to reconstruct an intact monolayer. By 96 hours, most of these cells cease expressing Slug and become integrated into a continuous monolayer. (Courtesy of P. Savagner, L. Hudson, and D. Kusewitt.)

tumor cells. Still, it seems increasingly likely that once some of these TFs are expressed in cancer cells, they act singly or collaboratively to orchestrate the complex cellular changes associated with invasion and metastasis (Figure 14.29; see also Figure 14.15). Finally, we should note that the close parallels between embryonic EMTs and those contributing to cancer pathogenesis are extended by the similarities in the signaling pathways that are active in these two situations (Figure 14.30).

These descriptions of the molecules that contribute in key ways to carcinoma cell invasiveness are reminiscent of our discussion in the previous chapter about the activated stroma and its contributions to the tumorigenicity of carcinoma cells. In both places, for example, we encounter cadherins, the EMT, and TGF-β. Such connections hint at an interesting but still speculative idea: perhaps the formation of primary carcinomas and the acquisition of invasiveness are not as separate and distinct as most descriptions of cancer would suggest. Although it is convenient to place them in separate conceptual boxes, the biological reality may be quite different. Quite possibly, cancer cell invasiveness is a natural extension—an exaggerated form—of the cell transformation processes that lead initially to the formation of many types of primary tumors. Because transformation and invasiveness depend on many of the same regulatory circuits and effector proteins, they may lie on a continuum in which one process blends seamlessly into the next.

To summarize, the discovery of the EMT-inducing TFs suggests at least three important ideas about malignant progression. First, many malignant cell pheno-

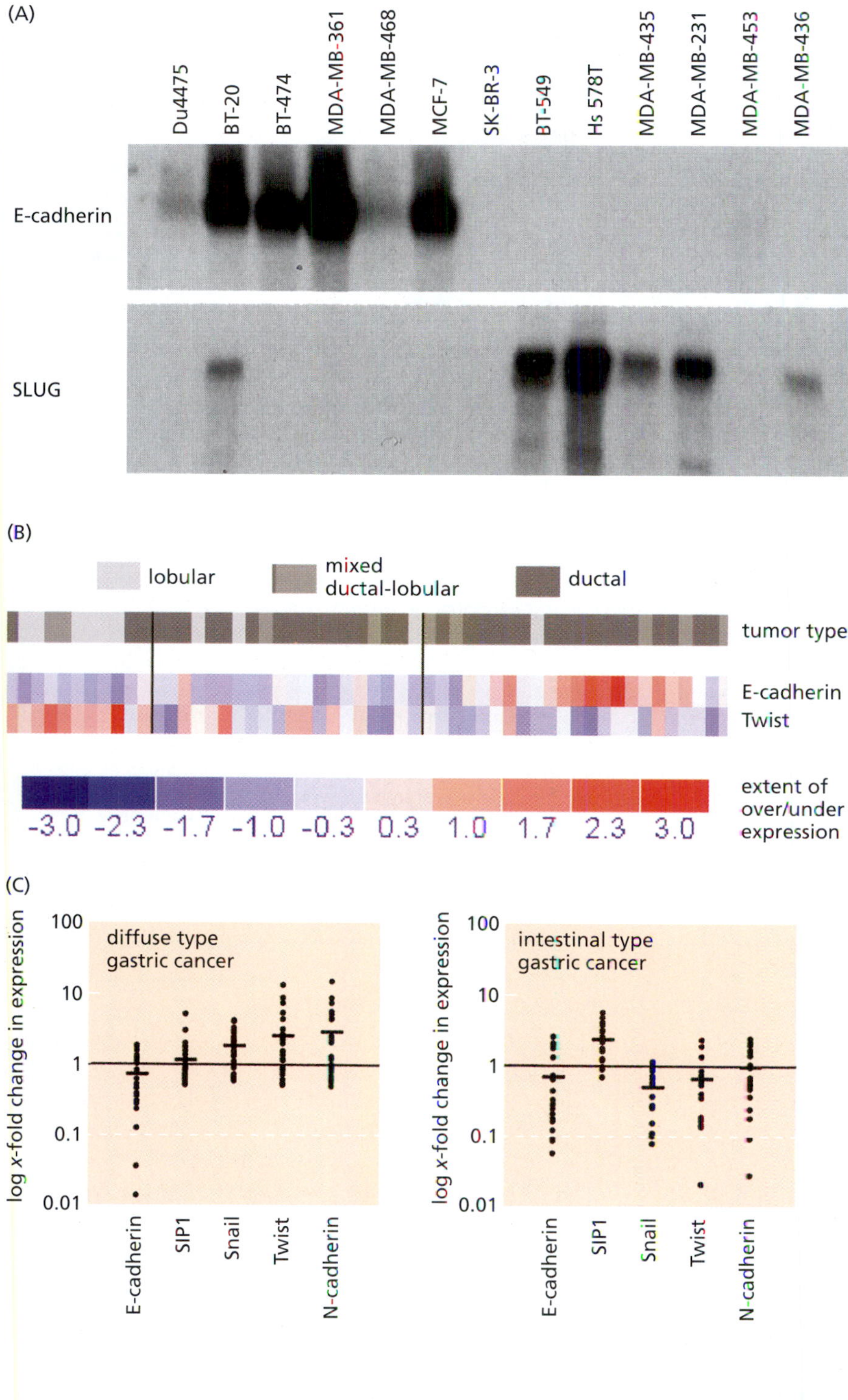

Figure 14.28 Expression of EMT-inducing embryonic transcription factors in human tumors Still-fragmentary evidence implicates a number of embryonic transcription factors (TFs) in the induction of the EMT (and associated loss of E-cadherin) in human cancer cells. (A) This analysis of the mRNA transcripts expressed in a panel of human breast cancer lines indicates that Slug is present in tumor cell lines lacking E-cadherin expression; conversely, E-cadherin is generally present in tumor cell lines lacking significant expression of this TF. These inverse patterns of expression are likely due to the ability of Slug to bind an E-box sequence in the E-cadherin promoter and to potently repress transcription of this key epithelial gene. (B) Expression array analysis demonstrates that Twist expression varies inversely with E-cadherin expression in these human breast carcinomas. Since Twist can repress expression of the encoding *CDH1* gene 250-fold in cultured cells, Twist is likely to be responsible for the loss of the E-cadherin mRNA, which is lost preferentially in the more aggressive invasive lobular carcinomas of the breast but is present at significant levels in the less aggressive invasive ductal carcinomas. (C) Analysis of mRNA levels of three EMT-inducing TFs—SIP1, Snail, and Twist—reveals that in the "diffuse" subtype of gastric cancer, expression of Snail and Twist is significantly elevated, as is expression of N-cadherin, a mesenchymal marker. In contrast, in the "intestinal" subtype of gastric cancer (which is very different in its etiology and biological behavior), SIP1 expression is elevated and the mRNAs encoding the other TFs are either repressed or expressed at the basal level seen in the normal gastric mucosa. Observations like these indicate that EMT-inducing TFs act in various combinations in different types of human cancers to program invasive growth. Note the logarithmic scale of the ordinate. (A, from K.M. Hajra, D.Y. Chen and E.R. Fearon, *Cancer Res.* 62:1613–1618, 2002; B, from J. Yang, S.A. Mani, J.L. Donaher et al., *Cell* 117:927–939, 2004; C, from E. Rosivatz, I. Becker, K. Specht et al., *Am. J. Pathol.* 161:1881–1891, 2002.)

types may be induced by non-genetic changes—heterotypic signals of stromal origin—rather than genetic changes occurring within carcinoma cells. Second, cancer cells do not need to cobble together all of the phenotypes associated with highly malignant cells; instead, a multitude of malignancy-associated traits may be acquired concomitantly because these TFs, once expressed, act in a highly pleiotropic fashion. Third, because expression of these TFs and the resulting EMT is dependent on heterotypic signaling from the stroma of the primary tumors, carcinoma cells may revert from the mesenchymal state to an epithelial state once they have left the primary tumor and landed in new stromal microenvironments, such as those present in sites of metastasis.

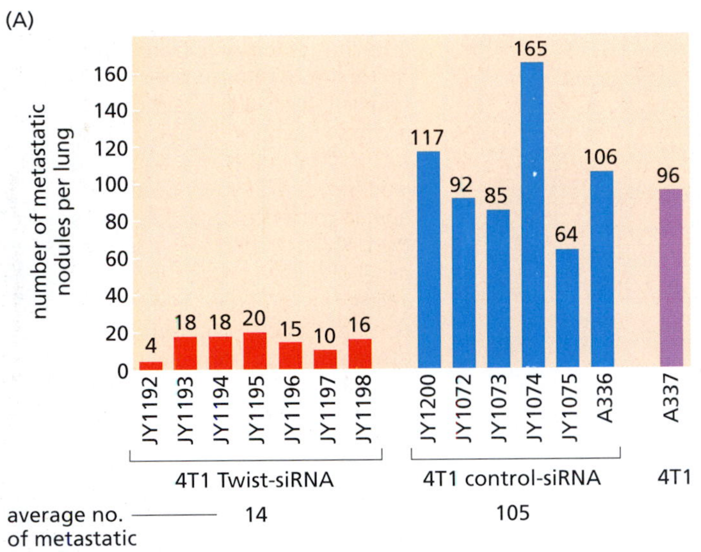

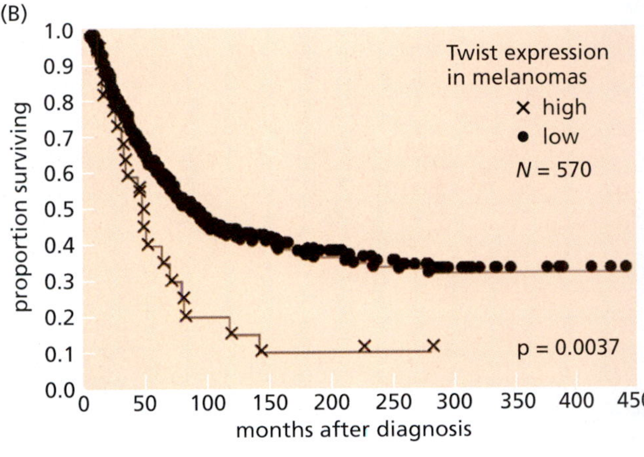

Figure 14.29 Embryonic transcription factors and tumor progression The evidence that the various embryonic transcription factors act to cause malignant progression in humans is still fragmentary and indirect. (A) Mouse 4T1 breast cancer cells, when implanted subcutaneously in a mouse host, generate large numbers of metastases in the lungs *(purple bar, far right)*. This number is unaffected when these cancer cells are infected by a control retrovirus vector *(blue bars)*. However, when the cancer cells were deprived of Twist expression through infection with an siRNA-expressing viral vector (which causes degradation of *Twist* mRNA), growth of the resulting primary tumors was unaffected, but the number of metastases that they generated was strongly (~85%) reduced *(red bars)*. Significantly, those few metastases that did form *(red bars)* all continued to express high levels of Twist, indicating that they derived from primary tumor cells in which Twist mRNA levels had never been properly suppressed. Each bar represents one tumor-bearing mouse. Each bar represents the metastases counted in a single tumor-bearing mouse. (B) Twist expression was gauged in a large group of melanomas by immunohistochemistry. As seen in this Kaplan–Meier plot, those patients whose tumors expressed elevated levels of Twist fared far worse than those whose tumors showed low, basal levels of Twist. (A, from J. Yang, S.A. Mani, J.L. Donaher et al., *Cell* 117:927–939, 2004; B, from K. Hoek, D.L. Rimm, K.R. Williams et al., *Cancer Res.* 64:5270–5282, 2004.)

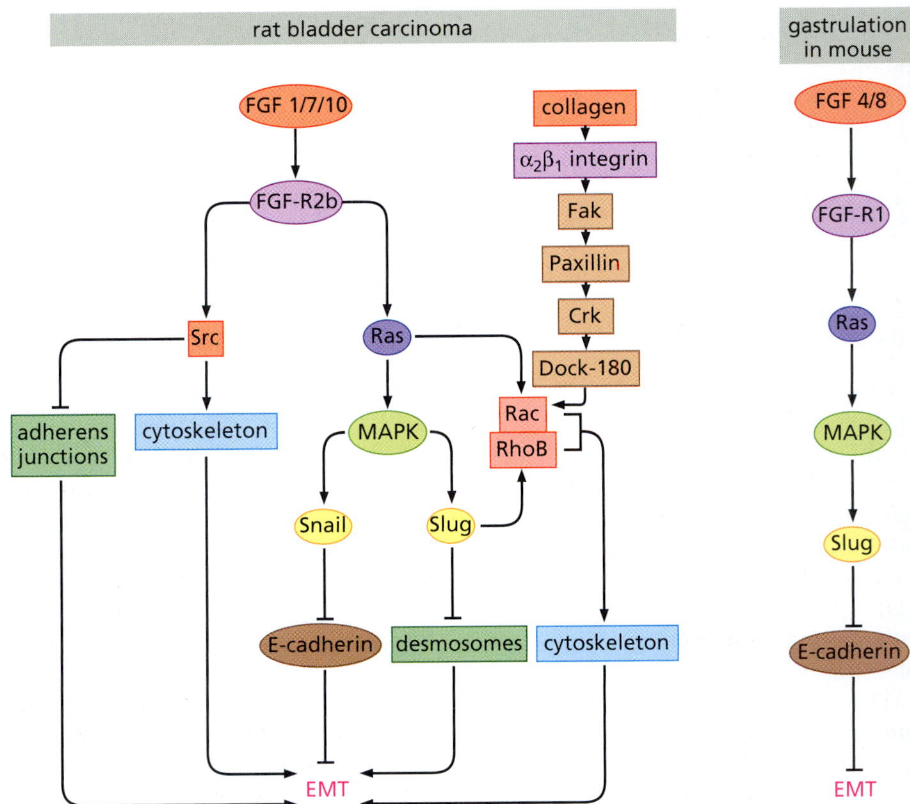

Figure 14.30 Similarities between EMT signaling during embryogenesis and tumor progression The signal transduction cascades that are responsible for activating the epithelial–mesenchymal transition (EMT) in a rat bladder carcinoma model *(left)* and during gastrulation early in mouse embryogenesis *(right)* have striking parallels. These similarities provide further support for the notion that the EMT program expressed by invasive carcinoma cells represents a reactivation of latent cell-biological programs, many of which are normally active in early mammalian embryonic development. (From J.P. Thiery, *Nat. Rev. Cancer* 2:442–454, 2002.)

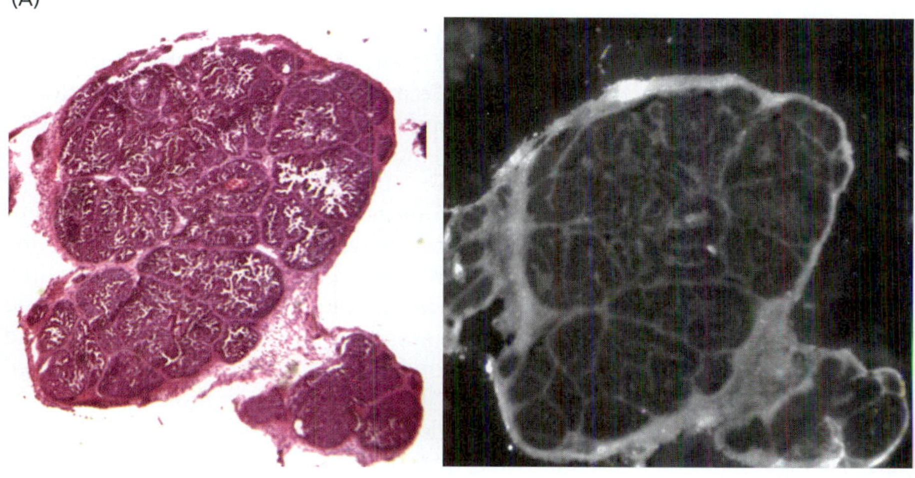

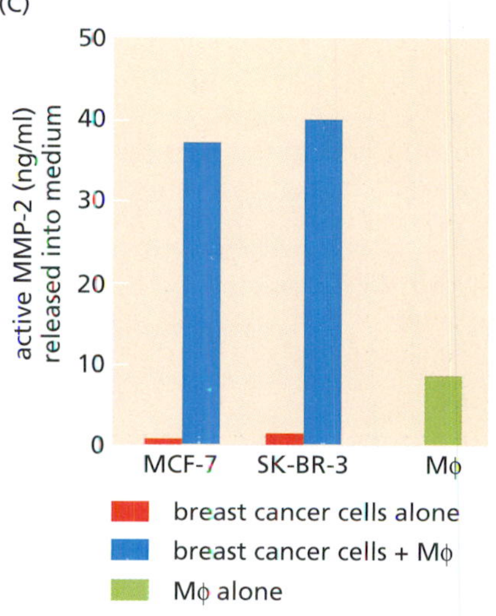

Figure 14.31 Matrix metalloproteinases produced by tumor-associated cells MMPs are produced largely by inflammatory cells and fibroblasts in the stroma. (A) In this mammary carcinoma arising in a transgenic MMTV–polyoma middle T mouse, the presence of MMPs can be detected by their ability to cleave a synthetic substrate, which releases a polycationic fluorescent tag that migrates into nearby cells, generating a fluorescent signal. The tumor stained with hematoxylin–eosin *(left panel)* is seen to generate a halo of proteolysis *(right panel)*, suggesting the involvement of surrounding stromal cells in protease production. (B) The ability of tumors to degrade collagen IV, a major component of the basement membrane (see Figure 13.5), can be measured by generating a modified collagen IV substrate that creates a fluorescent green color upon cleavage. In this experiment, both human mammary carcinoma cells *(not seen)* and human mammary fibroblasts *(red)* showed relatively weak ability to degrade the collagen IV substrate. However, when these two cell populations were co-cultured, regions of collagen IV cleavage *(green)* were evident, often in areas where fibroblasts were also present *(yellow: overlap of green and red)*. This cleavage was essentially eliminated in the presence of MMP inhibitors. (C) An even more important source of MMPs is the populations of macrophages (Mϕs) that are recruited into the tumor stroma. In this *in vitro* culture experiment, the presence of MMP-2 was measured in the culture medium of Mϕs that were either cultured alone *(right)* or co-cultured in the presence of two human breast cancer cell lines—MCF7 or SK-BR-3 *(left, middle)*. Neither of these cancer cell types made significant levels of MMP-2 on its own, but in the presence of Mϕs, both caused the Mϕs to increase MMP-2 production 4- to 5-fold; the increase could be traced to the induction of MMP-2 mRNA expression by the Mϕs *(not shown)*. The released MMP-2 imparted increased invasiveness to these breast cancer cells *(not shown)*. (A, courtesy of E. Olson, T. Jiang, L. Ellies, and R. Tsien; B, from M. Sameni, J. Dosescu, K. Moin and B.F. Sloane *Mol. Imaging* 2:159–175, 2003; C, from T. Hagemann, S.C.Robinson, M. Schulz et al., *Carcinogenesis* 25:1543–1549, 2004.)

14.6 Extracellular proteases play key roles in invasiveness

The epithelial–mesenchymal transition (EMT) represents a complex biological program that enables cancer cells to acquire the attributes of invasiveness and motility. In order to properly appreciate the processes that together constitute the EMT, we need to examine the roles of some of its key *effectors*—the proteins that work to create the phenotypes associated with the EMT. To begin, we examine the most obvious trait of malignant cells—their ability to invade adjacent cell layers. This burrowing requires that cancer cells remodel the nearby tissue environment by excavating passageways through the extracellular matrix (ECM) and pushing aside any cells that stand in their path.

The most important effectors of these complex changes are the matrix metalloproteinases (MMPs; see Table 13.1). In carcinomas, the great bulk of these proteases are secreted by recruited stromal cells, notably macrophages, mast cells, and fibroblasts, rather than by the neoplastic epithelial cells (Figure 14.31). By dissolving the dense thickets of ECM molecules that surround and confine individual cells within tissues, MMPs create spaces for these cells to move. Included among the ECM components that are cleaved by MMPs are fibronectin, tenascin, laminin, collagens, and proteoglycans. During the course of degrading ECM components, MMPs also mobilize and activate certain growth factors that have been tethered in inactive form to the ECM or to the surfaces of cells.

MT1-MMP (membrane type-1 MMP) is one of six MMP types that are membrane-anchored and therefore limited to cleaving substrate proteins in the immediate vicinity of the cells that have produced them. Of the 187 distinct metalloproteinases known to be encoded in the human genome, 28 are secreted MMPs. In contrast to MT1-MMP, which acts as a plasma membrane– bound enzyme, most of the MMPs function as soluble enzymes in the spaces between cells. MT1-MMP may attack and cleave cell surface adhesion molecules, such as cadherins and integrins, as well as growth factor receptors and chemokines. It can also cleave inactive **pro-enzymes**, such as pro-MMP-2, into enzymatically active MMPs.

These secreted proteases clearly play important roles in normal cell survival and proliferation. After all, each time a cell within a normal tissue goes through a cycle of growth and division, space within the ECM must be carved out for its future daughters, and once these two are formed, each of these cells must, in turn, reconstruct new ECM around itself. Hence, the remodeling of the ECM takes place continuously in mitotically active tissues. Consequently, rather than being aberrations of invasive cancer cells, the activities of MMPs and other extracellular proteases, such as urokinase plasminogen activator (uPA), accompany and participate in normal cell proliferation. Of relevance here are clinical trials of certain MMP-inhibitory drugs, which have been terminated due to the effects of these inhibitors on a variety of normal tissues; because these agents suppress the normal turnover of cartilage and other joint components, they created unacceptable levels of joint stiffness and pain.

Each type of MMP usually acts on a well-defined set of substrates (see Table 13.1), doing so in a highly regulated and localized fashion. It is likely that these enzymes continue to show such substrate specificity during the process of cancer cell invasion. However, in this instance, the proteolysis seems to proceed continuously rather than in the brief spurts that accompany normal cell growth and division.

One of the consequences of the EMT programmed by several well-studied embryonic transcription factors (Section 14.5) is the induced synthesis and release by carcinoma cells of MMPs, notably MMP-2 and –9. However, it is clear that the bulk of the MMPs found in tumors originate in the stroma. For example, the best-studied of the matrix metalloproteinases, MMP-9, is expressed largely by macrophages (Section 13.5), neutrophils, and fibroblasts at the invasive fronts of tumors. MMP-9 expression at these fronts correlates positively with the metastatic ability of a primary tumor, suggesting that MMPs like this one can act at several stages of the invasion–metastasis cascade, including local invasion of the primary tumor stroma, intravasation, and extravasation. *In vitro* assays indicate that MMP-9 can degrade collagens that are prominent components of the ECM including basement membranes, specifically collagen types IV, V, XI, and XIV. Other targets of MMP-9 include laminin (another important constituent of the basement membrane; see Figure 13.5), chemokines, fibrinogen, and latent TGF-β. In the case of the latter two, cleavage by MMPs converts them from latent into activated forms.

These widely ranging functions of MMPs indicate that their enzymatic activity must be tightly controlled, at least in normal tissues. Reflecting this requirement is the fact that the soluble MMPs, such as MMP-9, are all initially synthesized as inactive pro-enzymes that can only function, like the caspases (Section 9.13),

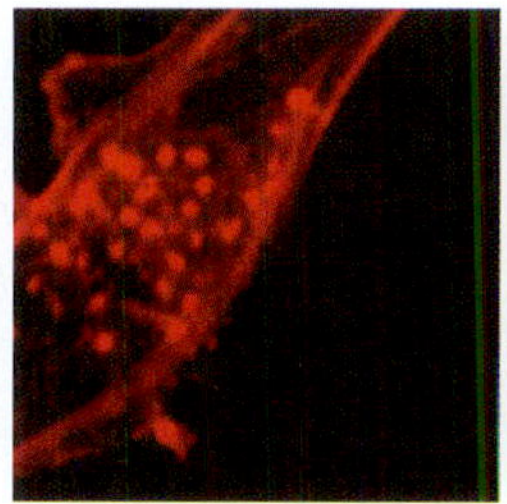
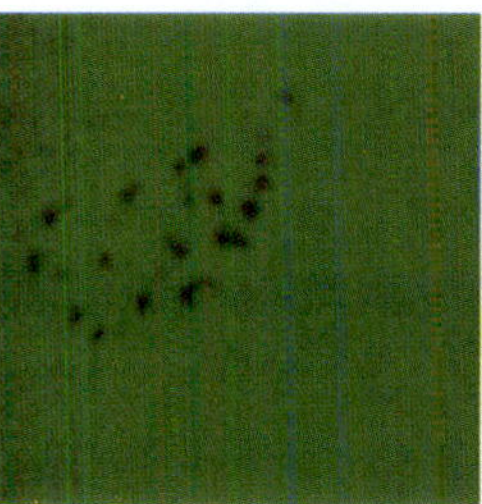

Figure 14.32 Podosomes and focal degradation of the extracellular matrix Podosomes are small, focal protrusions from the cell surface that are displayed by many cell types and used by these cells to degrade highly localized areas of extracellular matrix (ECM) in their immediate vicinity. Podosomes are thought to be used by invasive cancer cells as well, to direct controlled degradation of the ECM near the leading edge of an invading cell; when operating in this context, they are sometimes called "invadosomes." Here, cells were stained for actin fibers with the dye phalloidin *(red, left micrograph)*, revealing a number of discrete clusters of actin that are associated with the podosomes, which are located on the ventral (underside) surface of these *src*-transformed rat fibroblasts. These cells have been growing on a layer of the ECM protein fibronectin coupled to the FITC dye *(green, middle)*, revealing that directly below the podosomes, discrete holes *(black)* have been eroded in the fibronectin sheet, ostensibly by the actions of podosome-associated proteases, such as MT1-MMP. Direct overlap between the actin clusters of the podosomes and the eroded holes in the fibronectin are indicated by *white arrowheads* in the right panel. (From K. Mizutani, H. Miki, H. He et al., *Cancer Res.* 62:669–674, 2002.)

following activation by other proteases. Negative regulation is also provided by a class of proteins termed tissue inhibitors of metalloproteinases (TIMPs), which bind MMPs and put them in an inactive configuration (Section 13.9). Moreover, the activities of some of these extracellular proteases seem to be further confined through their concentration at discrete foci, termed **podosomes**, where membrane-tethered MMPs, notably MT1-MMP, are active in creating highly localized areas of proteolysis (Figure 14.32).

While MMPs have been depicted as the direct effectors of specific steps in invasion and metastasis, it is clear that the deregulation of MMPs can, through unknown mechanisms, drive the progression of cells through all of the stages of multi-step tumorigenesis including completion of the invasion–metastasis cascade. Thus, when expression of MMP-3 (also known as stromelysin-1) is forced in the mammary gland of transgenic mice, these mice initially develop mammary hyperplasias (Figure 14.33). Some of these growths progress to carcinomas that eventually become invasive and metastatic. These mice reveal how critical the regulation of MMP action is and why it must be kept under control in normal tissues.

The complexity of the regulatory network governing MMP activation is further illustrated by the behavior of urokinase plasminogen activator (uPA), a non-MMP protease. uPA is secreted by stromal cells as an inactive pro-enzyme. This form of uPA proceeds to bind to its own cell surface receptor (termed uPAR), which is displayed by epithelial cells, including malignant ones, and thereby becomes catalytically active (Figure 14.34). The tethered, activated uPA can then be wielded by the epithelial cells to cleave a variety of extracellular substrates in their immediate vicinity. Prominent among these are a series of pro-enzymes of other extracellular proteases that are activated by this cleavage. For example, by cleaving **plasminogen** into plasmin, uPA creates an activated protease that, in turn, proceeds to cleave and thereby activate the pro-enzyme forms of yet other extracellular proteases, notably, the matrix metalloproteinase (MMP) types 1, 2, 3, 9, and 14. Alternatively, uPA may cleave and activate some MMPs directly. Not surprisingly, inhibitors of the uPA–uPAR complex have been found to block both tumor growth and metastasis in animal models of cancer pathogenesis. Moreover, high levels of solubilized uPAR in the serum represent a bad prognosis for cancer patients.

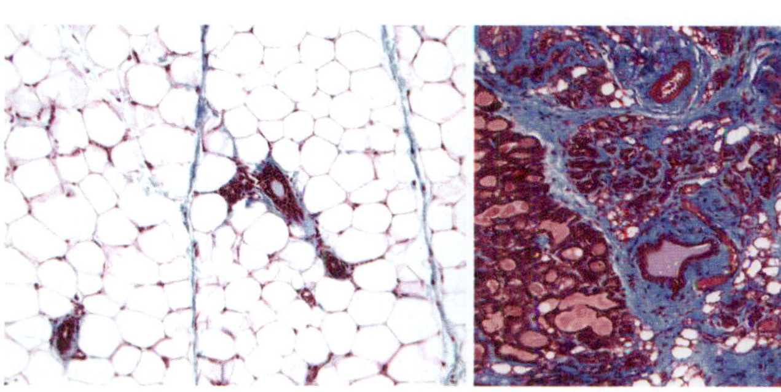

Figure 14.33 Ectopic expression of MMP-3 and mammary tumor progression The normal mouse mammary gland *(left panel)* is composed of resting ducts *(purple)* and abundant adipose tissue *(white, lipid-filled cells)*, as well as collagen *(light blue)*. However, when the gene encoding MMP-3 (also known as stromelysin-1) is expressed constitutively as a transgene that directs its expression to the mammary epithelium, the mice develop abundant hyperplasia *(right panel)*, including extensive islands of hyperplastic epithelial cells *(purple)* forming ducts, as well as a fibrotic, collagen-rich stroma *(light blue)* and abnormal adipocytes *(white oval structures, lower right)*. Many of these areas subsequently progress to invasive, metastatic tumors *(not shown)*. (From M.D. Sternlicht, A. Lochter, C.J. Sympson et al., *Cell* 98:137–146, 1999.)

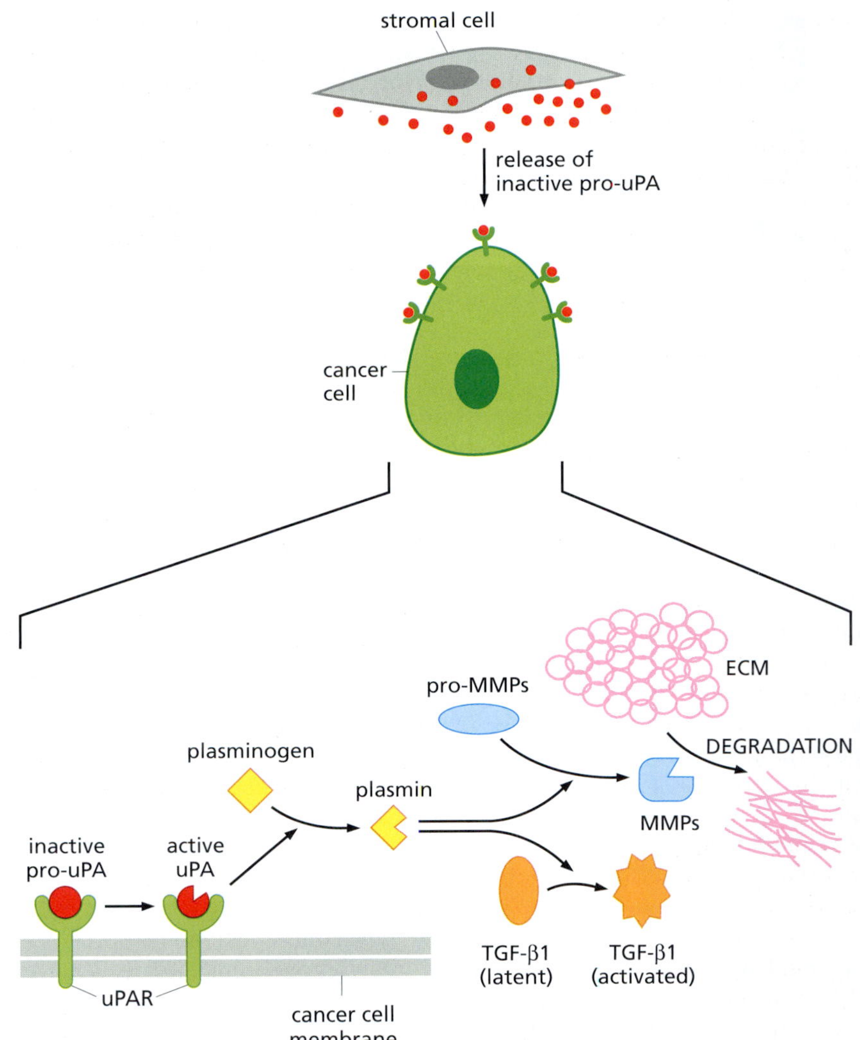

Figure 14.34 uPA, uPAR, and the activation of extracellular proteins The inactive, pro-enzyme form of urokinase plasminogen activator (pro-uPA) is released by stromal cells. Once released, it binds to its cognate receptor, uPAR, which is displayed on the surface of epithelial cells; this binding converts uPA into a catalytically active protease. Active, receptor-bound uPA can then convert inactive, soluble plasminogen to the activate plasmin form; the latter functions as a protease to cleave pro-enzyme forms of matrix metalloproteinases (pro-MMPs) into active MMPs and latent TGF-β1 into its active form. At the same time, there is evidence that uPA can also act directly on pro-MMPs to convert them into active proteases. (From F. Blasi and P. Carmeliet, *Nat. Rev. Mol. Cell Biol.* 3:932–943, 2002.)

These brief vignettes of proteases and their contributions to cancer cell invasiveness describe only small parts of what are surely highly complex networks of interacting proteases and substrates. The total number of proteases made by mammalian cells is vast and rivals the number of proteins that form the highly complex intracellular signal-processing circuits described in Chapter 6. To date, the actions of only a small proportion of these enzymes have been studied in the context of cancer pathogenesis (Sidebar 31).

14.7 Small Ras-like GTPases control cellular processes including adhesion, cell shape, and cell motility

The actions of extracellular proteases, notably the MMPs, explain at the biochemical level how paths are cleared for the advance of invasive cancer cells through the extracellular matrix and thus through tissues. They fail, however, to tell us how individual cancer cells take advantage of these cleared paths to move ahead—the trait of cell motility. The motile behavior of cells has been studied extensively with cultured cells, and it is presumed that their crawling on solid substrates *in vitro* reflects the *in vivo* behavior of cancer cells as they invade nearby cell layers and intravasate. Such motility is also presumed to be important for cancer cells' escape from blood vessels or lymph ducts—the process of extravasation.

Motile behavior can be induced in cultured cells by exposing them to a variety of growth factors. (The ability of these factors to induce such locomotion is sometimes recognized by designating them as being **motogenic** in addition to being mitogenic.) In the case of epithelial cells, the best inducer of motility is usually hepatocyte growth factor (HGF); this protein is also called scatter factor (SF) in recognition of its ability to induce multidirectional movement of cells in monolayer culture. Most and perhaps all types of epithelial cells express Met, the receptor for HGF, and many of these cell types have been found to acquire motility in response to HGF treatment (see Figure 14.24A). Similarly, EGF is clearly able to induce motility of breast cancer cells (see Figure 14.23B).

The cellular machinery that responds to motogenic signals and operates as the engine of motility is extraordinarily complex at the molecular level. Cell motility involves continuous restructuring of the actin cytoskeleton in different parts of a cell (Figure 14.35), as well as the making and breaking of attachments between the migrating cell and the extracellular matrix (ECM). (In the case of cultured cells, the ECM in question is the network of proteins that has previously been laid down by these cells on the surface of the Petri dish.)

The process of cellular movement can be broken down into several distinct steps. To begin, a cell will extend its cytoplasm in the direction of intended movement. This extension involves the protrusion from the cell surface of **lamellipodia**—broad, flat, sheetlike structures that may be tens of microns in width but only 0.1 to 0.2 μm thick (Figure 14.36). At the same time, cell surface proteases, such as those described earlier, are used to selectively degrade ECM proteins that stand in the way of the "leading edge" of the migrating cell. While this is going on, the cell deploys integrins to construct new points of attachment between the lamellipodia and the ECM at its leading edge and breaks such adhesions at its "trailing edge," thereby liberating cytoplasm and plasma membrane for redeployment to the leading edge.

Protruding from the lamellipodia are spikelike structures termed *filopodia* that are thought to enable an advancing cell to explore the territory that lies ahead (Figure 14.37). Like the lamellipodia, these are assembled through the

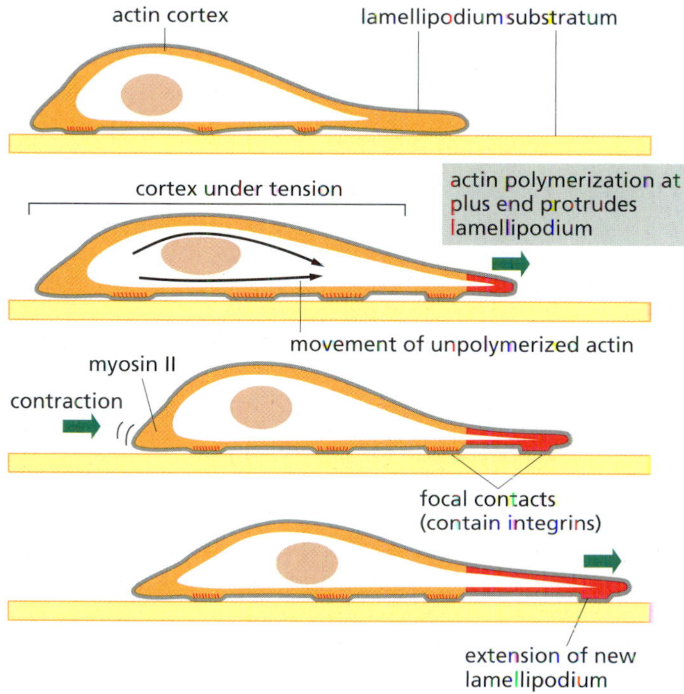

Figure 14.35 Locomotion of cells on solid substrates The locomotion of a cultured cell depends on the coordination of a complex series of changes in the cytoskeleton as well as the making and breaking of focal contacts with the underlying solid substrate. The cell organizes actin fibers in order to extend lamellipodia at its advancing/leading edge and to establish new focal contacts. At the same time, stress fibers, also consisting of actin, are used to contract the trailing edge of the cell, where focal contacts are being broken. The making and breaking of the focal contacts *(red)* depends on localized modulation of the affinities of various integrins for extracellular matrix (ECM) components, represented here by the *yellow* substratum. (From B. Alberts et al., Molecular Biology of the Cell, 4th ed. New York: Garland Science, 2002.)

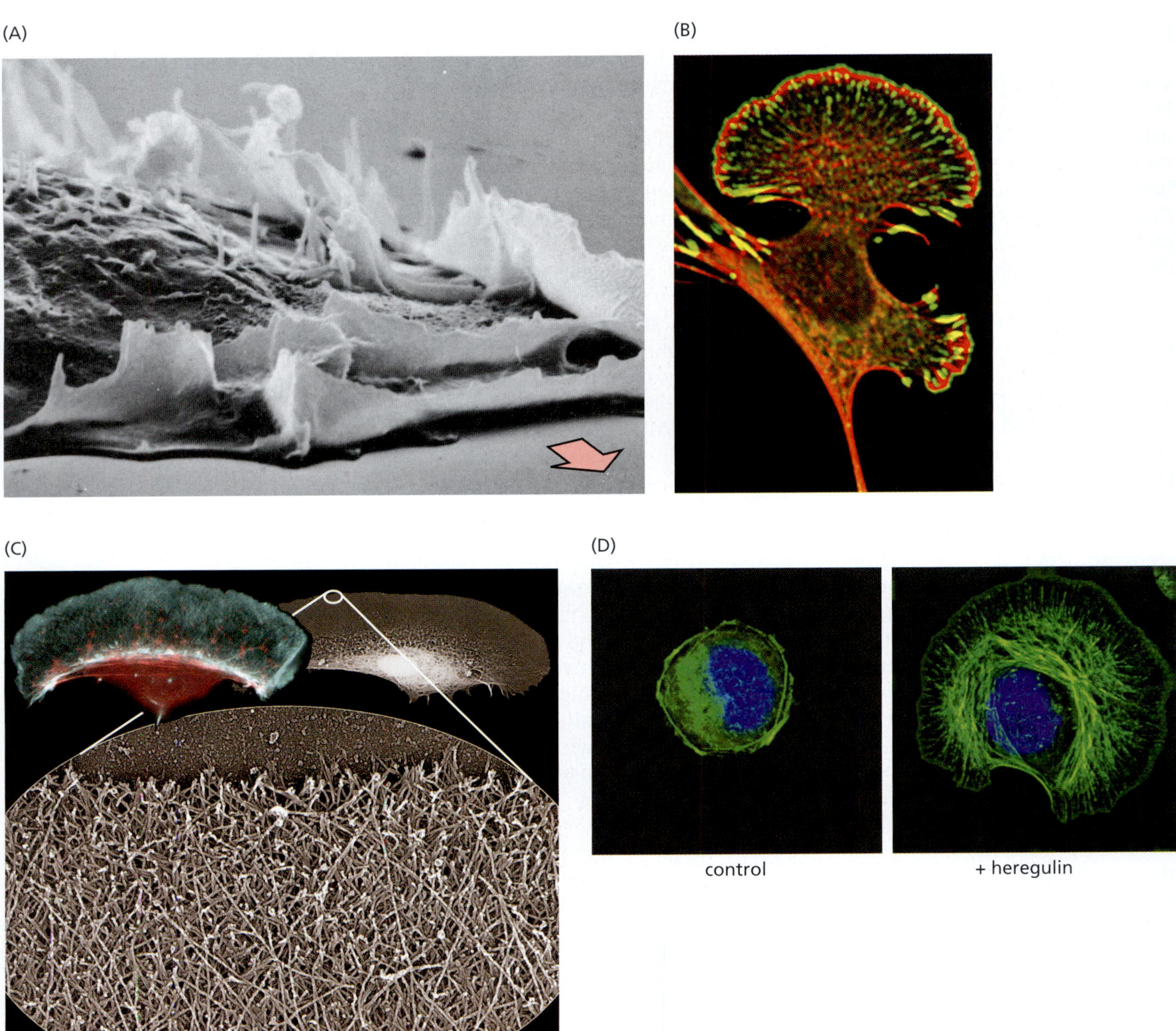

Figure 14.36 Lamellipodia (A) This scanning electron micrograph (SEM) of a spontaneously transformed rat liver cell shows the elaborate ruffles—lamellipodia—that migrating cells extend at their leading edge during forward locomotion *(arrow)*. Lamellipodia are presumed to play a key role in the advance of invasive cancer cells *in vivo,* but such a role has not yet been directly demonstrated. Lamellipodia that are extended on the leading edge of the cell but fail to attach to the substrate seem to be swept back as ruffles along the dorsal (top) side of the cell; their component parts are then reintegrated into the larger plasma membrane and cytoskeleton. (B) This fluorescence micrograph image shows a lamellipodium being extended by a fibroblast in anticipation of locomotion. As seen here, the actin fibers are labeled with phalloidin *(red),* while the Ena protein, which programs the advance of the lamellipodium by organizing its focal adhesions and its outer edge, is labeled here with green fluorescent protein (GFP), to which it has been fused. (C) Phase microscopy of a fish keratocyte, studied because of its prominent lamellipodia, indicates the presence of substantial actin in this protrusion *(cyan, top left figure)*. Electron microscopy of such a cell *(top right image)* and its cytoskeleton *(bottom image)* at much higher magnification indicates a densely woven network of actin filaments that are organized in order to extend the lamellipodium in the direction of cell movement. (D) Lamellipodium formation and resulting cell motility are strongly stimulated by a number of growth factors and their cognate tyrosine kinase receptors. Seen here are the effects of adding heregulin, a ligand of the erbB2/erbB3 family of receptors, to a human breast cancer cell. An untreated cancer cell is to the *left,* while a cell exposed to heregulin for 20 minutes is to the *right*. The actin cytoskeletons have been stained *green* (using phalloidin coupled to a fluor), while the nuclei are stained *blue*. (Because the heregulin is present uniformly in the surrounding medium, this cell has been induced to develop a lamellipodium that faces in all diretions rather than toward a single, localized source of this motogen.) Elevated signaling by erbB2 (= HER2/Neu) is correlated with increased metastatic progression of human breast cancer cells, which may be explained in part by the receptor-mediated induction of lamellipodium formation and associated cell motility. (A, courtesy of Julian Heath; B, from J.J. Loureiro, D.A. Rubinson, J.E. Bear et al., *Mol. Biol. Cell* 13:2533–2546, 2002; C, from T.M. Svitkina, A.B.Verkhovsky, K.M. McQuade and G.G. Borisy *J. Cell Biol.* 139:397–415, 1997; D, courtesy of A. Badache and N.E. Hynes.)

(A)

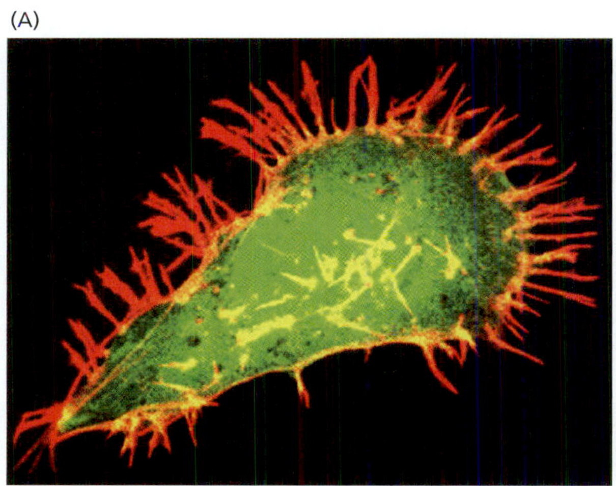

(B)

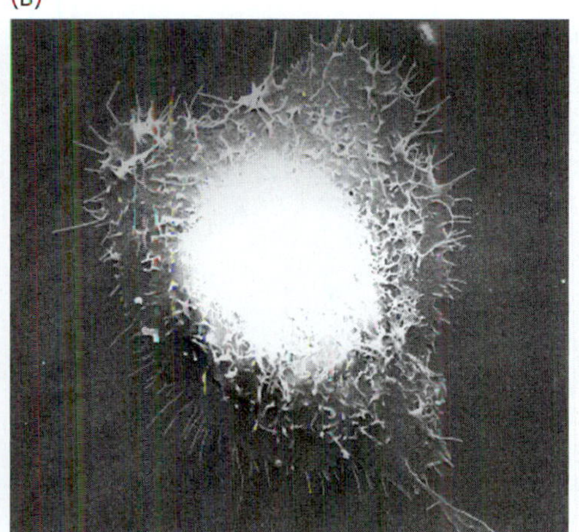

(C)

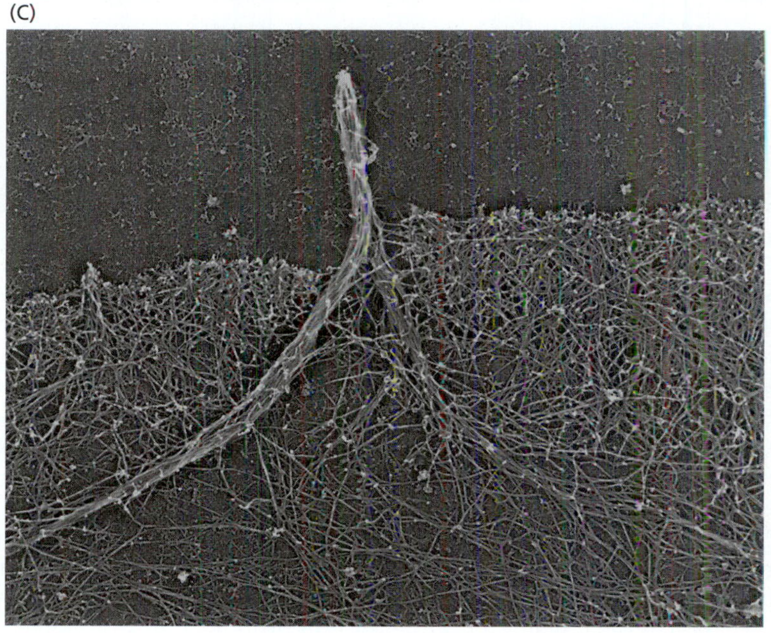

Figure 14.37 Filopodia (A) Filopodia *(orange spikes)* extend from the cell surface and are presumed to allow an invading cell to explore its extracellular environment. In this case the actin fibers that form within filopodia have been induced to assemble through expression of a Cdc42-interacting protein termed Zizimin. (B) The leading edges of lamellipodia are often interspersed with filopodia, which serve, in some poorly understood fashion, as sensors for the advancing cell. The cell shown here has extended lamellipodia and filopodia in all directions and therefore does not have a "leading edge." (C) The actin fibers within a filopodium are bundled tightly together and extend forward in the direction of cell migration in order to enable protrusion of the filopodium from the leading edge of a lamellipodium. (A, courtesy of W.B. Kiosses and M.A. Schwartz; B, courtesy of M. Cayer, L. Lim and C.A. Heckman; C, courtesy of E. Bulanova; see also T.M, Svitkina, E.A. Bulanova, O.Y. Chaga et al., *J. Cell Biol.* 160:409–421, 2003.)

reorganization of actin fibers, in this case fibers that are tightly bundled together beneath the plasma membrane of each filopodium. The precise role played by filopodia in cancer cell invasiveness remains to be elucidated.

The detailed management of cell shape and motility is under the control of members of a group of Ras-related proteins belonging to the Rho family. As discussed briefly in Chapter 6, the Rho proteins, like Ras, operate as binary switches, being in a functionally active state while binding GTP and in an inactive state once they hydrolyze their bound GTP to GDP. More than 20 members of the Rho family of proteins have been discovered in human cells. They are divided into three subfamilies—the Rho proteins proper, the Rac proteins, and Cdc42. Like the Ras proteins, most members of the Rho protein family bear lipid groups at their C-termini that enable anchoring to intracellular membranes. Each of these has specialized functions in reorganizing cell shape and enabling cell motility (Figure 14.38).

Figure 14.38 actually misrepresents the actions of these various Rho-like proteins in one important respect: it implies that each of them acts globally throughout the cell to organize certain changes in the configuration of the actin cytoskeleton. In reality, they act quite differently. The complex program of cell motility depends on the *localized* activation of each of these proteins in very

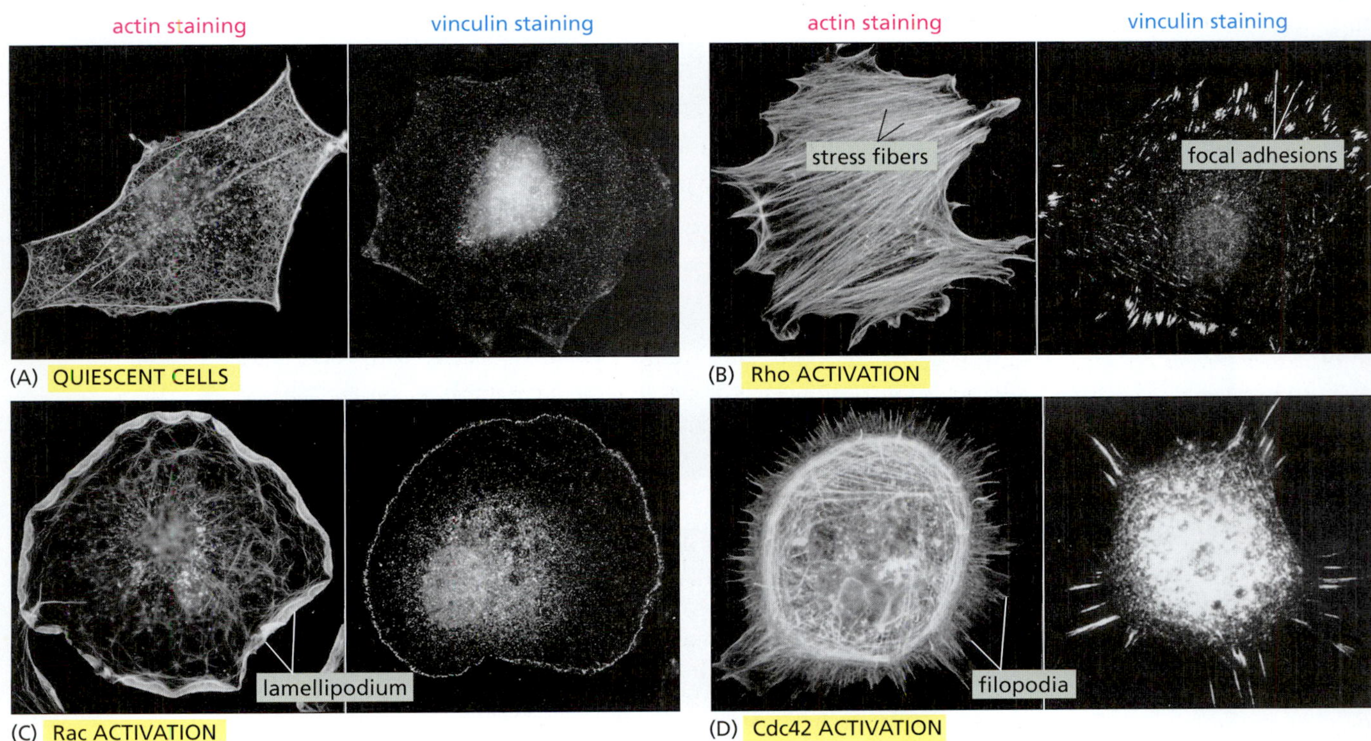

Figure 14.38 Effects of Rho-like proteins on the actin cytoskeleton and cell adhesion Members of the Rho family of small GTPases, which consists of the Rho, Rac, and Cdc42 subfamilies, control both the actin cytoskeleton and the formation of focal adhesions, seen here through staining with an antibody reactive with vinculin, one of the proteins that link integrins at focal adhesions to this cytoskeleton (see Figure 5.28). The actin fibers were labeled by fluorescent phalloidin, a compound with high affinity for actin. (A) Quiescent, serum-starved 3T3 fibroblasts, which serve as controls for the panels that follow. (B) Exposure of cells to lysophosphatidic acid, which specifically activates Rho subfamily proteins, causes a cell to construct large numbers of stress fibers and focal adhesions. (C) Micro-injection of a constitutively activated form of a Rac protein into a cell causes it to construct a single enormous lamellipodium around its entire circumference. (In contrast, a focal source of a Rac-activating signal is likely to induce lamellipodia only on the side of the cell facing this source.) (D) Micro-injection of a guanine nucleotide exchange factor (GEF) of Cdc42 into these cells causes a cell to extend hundreds of filopodia in all directions. (From A. Hall, *Science* 279:509–514, 1998.)

small domains of the cytoplasm, which in turn enables the cell as a whole to move in one direction or another. The alternative—global activation—would lead to attempts by a cell to move simultaneously in all directions, a scenario suggested by the lamellipodia of Figures 14.36D and 14.38C, which form a continuous ring around the entire perimeter of the cytoplasm. (The global activation of Rac function in this cell is an artifact of introducing mutant, constitutively activated Rac protein into the cell by micro-injection.)

Growth factor activation of tyrosine kinase receptors leads to the activation of many members of the Rho family of G proteins. For example, treatment of cultured fibroblasts with platelet-derived growth factor (PDGF), a potent mitogen for these cells, activates a number of Rho proteins and stimulates these fibroblasts to move across the bottom of a Petri dish. Alternatively, when fibroblasts are placed in three-dimensional culture by being suspended in a collagen gel, PDGF induces them to invade through this gel. All three subfamilies of Rho proteins (i.e., Rho, Rac, and Cdc42) appear to contribute to this invasion, while only Rac may be needed for the movement of fibroblasts across a solid substrate in culture. These various behaviors also illustrate an important distinction between the Ras proteins and their distant Rho family cousins: in cancer cells, a Ras protein is often activated by alterations in its structure (more specifically, amino acid substitutions), while the various Rho proteins are functionally activated by their upstream physiologic regulators.

All of the signaling connections between the PDGF receptor and these Rho family proteins are not known. However, it is clear that this receptor, by activating Ras, stimulates at least three downstream signaling pathways involving the Raf, Ral-GEF, and PI3K (phosphatidylinositol-3 kinase) effectors (see Sections 6.5, 6.6, and 6.7). In addition, activated Ras binds and appears to activate Tiam1, which functions as a guanine nucleotide exchange factor (GEF) for Rac. (Recall that GEFs are responsible for causing small G proteins, such as Ras and Rho, to jettison bound GDP and take on GTP, thereby activating signaling by these G proteins.) Hence, Tiam1 should also be considered to be another effector of Ras.

In fact, from the perspective of cell motility, PI3K is clearly the most important of the Ras effectors. By generating PIP_3 [phosphatidylinositol (3,4,5) triphosphate], the PI3K enzyme creates a chemical structure on the cytoplasmic face of the plasma membrane to which a variety of cytosolic proteins can attach via their PH domains (Section 6.6). Among these proteins are a number of guanine nucleotide exchange factors (GEFs) that are responsible for activating members of the Rho family of G proteins. These Rho-GEFs become activated following their tethering to the plasma membrane.

(The overarching role of PI3K and PIP_3 in choreographing cell motility is illustrated by studies of the motility of the slime mold *Dictyostelium discoideum*. PI3K, and thus its product, PIP_3, is localized at the leading edge of an advancing slime mold cell. Conversely, PTEN, the enzyme that destroys PIP_3 and thereby antagonizes PI3K, is localized to the sides and the rear, lagging edge of such a cell. This introduces another element into our thinking: while growth factor receptors, such as the PDGF-R, may release signals encouraging PI3K activation throughout the cell, the actual signaling by this enzyme may also be influenced by its localization within a cell.)

Tiam1 was originally identified as the product of a **T**-cell lymphoma **i**nvasion **a**nd **m**etastasis gene, indicating the importance of its encoded protein to these late steps of tumor progression. Tiam1 function appears to be stimulated both by its association with GTP-bound, active Ras and by its binding to PIP_3. By activating Rac proteins, the Tiam1 GEF encourages the localized polymerization of actin at the leading edge of migrating cells, thereby yielding the lamellipodia that are so critical to cell locomotion (see Figures 14.36C and 14.38C).

The other Rho-like proteins that are activated by Rho-GEFs are responsible for a number of distinct components of the cell motility program. For example, Rho proteins like RhoA and RhoB, acting in concert with Rac proteins, promote the establishment of new points of adhesion between the leading edge of the cell and the extracellular matrix. The reverse is also true: the forging of new focal adhesions (see Figure 14.38B) also encourages Rac activation, suggesting the operation of some self-sustaining, positive-feedback loop that ensures the continuity of forward motion. Rac and Cdc42 proteins also appear able to induce expression of certain secreted proteases, notably the matrix metalloproteinases described in the last section. By doing so, they may coordinate localized remodeling of the extracellular matrix with extension of lamellipodia at the leading edge of a motile cell.

The contraction of the cell body (which helps to pull the lagging edge of the cell forward toward the leading edge) is equally important for a cell's directed movement. This contraction is also governed largely by members of the Rho subfamily of proteins. By encouraging the formation of actin bundles in the cytoplasm, Rho proteins are able to create the structures known as "stress fibers" (see Figures 14.35 and 14.38B) and thus contribute to the regulation of the contractility of the cytoplasm. As many as 22 distinct guanine nucleotide exchange factors (GEFs) for the Rho subfamily of proteins have been discovered. Much of this complexity is likely attributable to the need to activate certain Rho proteins in microdomains of the cytoplasm in response to specific cell-physiologic signals. Stated differently, the activation of these Rho proteins needs to be tightly coordinated in time and space in order for the cell motility program to perform properly.

Cdc42, which represents the third subfamily of Rho-like proteins, has its own specialized function: it is able to induce the extension of the finger-like filopodia (see Figures 14.37 and 14.38D). Precisely how filopodia contribute to cell motility remains unclear. In addition, activated Cdc42 is able to stimulate generalized cell motility, independent of its specific effects on filopodia.

To complicate things even more, the actions of Rho, Rac, and Cdc42 differ in different cell types. For example, in normal epithelial cells (rather than the fibroblasts discussed above), the Rac and Rho subfamily proteins are responsible for

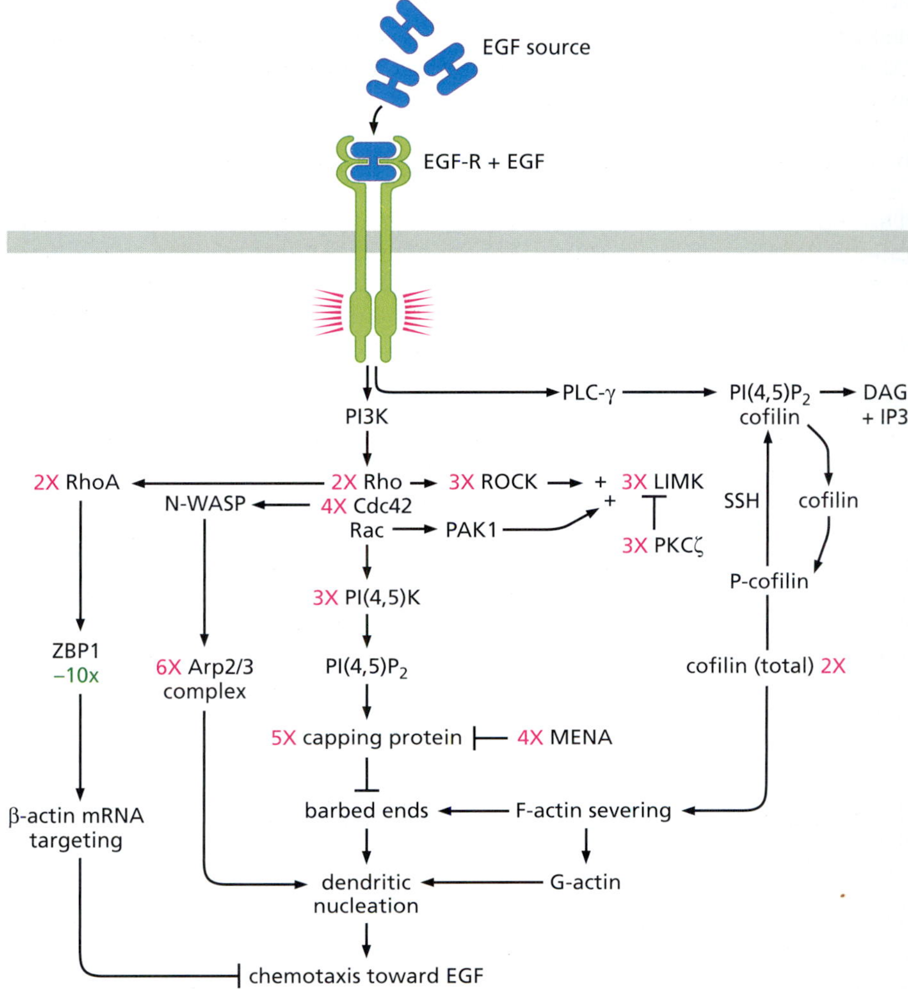

Figure 14.39 The circuitry mediating EGF-induced cell motility Many of the signal-transducing proteins that lie downstream of the EGF receptor and are responsible for mediating EGF-induced cell motility are indicated here. The relative proportions of these proteins change as breast cancer cells acquire increased motility and invasiveness: the extent of overexpression of certain proteins in the cancer cells is indicated by the *pink* numbers, while the extent of reduced expression is indicated by the *green* number. Some proteins, such as cofilin, sever existing actin fibers in certain regions of the cell, while capping protein prevents further extension of existing fibers. The actions of some of these signaling proteins are discussed in Section 6.6. (From W. Wang, S. Goswami, E. Sahai et al., *Trends Cell Biol.* 15:138–145, 2005.)

maintaining the E-cadherin–dependent cell–cell adherens junctions; as we have read, these junctions are vital for preserving the epithelial cell sheet and therefore immobilize participating epithelial cells. However, in transformed epithelial cells, such as colon carcinoma cells that have undergone a partial or complete EMT, Rac clearly contributes to increased motility. In nonmotile cells, Tiam-1 (the Rac exchange factor) is found in these adherens junctions, while in migrating cells, Tiam-1 localizes to lamellipodia and related membrane ruffles.

The task of integrating these disparate observations into a single scheme has only begun. An initial attempt is seen in Figure 14.39, which depicts part of the signaling circuitry that lies downstream of the EGF receptor and is responsible for coordinating EGF-stimulated cell motility. Virtually all of the subcircuits depicted in this scheme are likely to participate in organizing the motility stimulated by other motogenic growth factors, such as HGF and PDGF, as well.

The relevance of the Rho family proteins to cancer metastasis has been highlighted by searches for genes that are specifically expressed in metastatic cells but are expressed to a much lesser extent in nonmetastatic cells. In one set of experiments, strongly metastatic variants of mouse and human melanoma tumor cell lines were selected and the gene expression patterns of the cell lines were then compared with weakly metastatic cells. Prominent among the genes whose expression was elevated in the metastatic variants was the gene encoding the RhoC protein (one of the Rho subfamily). Indeed, introduction of a RhoC-expression vector into the poorly metastatic melanoma cells caused them to become highly metastatic, while ectopic expression of a dominant-interfering form of RhoC reduced the metastatic powers of the usually metastatic cells (Figure 14.40). RhoC has also been found to be strongly expressed in cells of inflammatory breast cancers, a particularly aggressive form of this disease, and in pancreatic carcinomas, which are almost always highly aggressive.

Forced expression of the RhoA gene in normally noninvasive rat hepatoma cells causes them to exhibit invasive behavior; it also enhances the metastatic behavior of transformed NIH 3T3 cells and of poorly metastatic melanoma cells. And a comparison of the gene expression patterns of aggressive (i.e., invasive and metastatic) and nonaggressive variants of the human T24 bladder carcinoma cell line indicates that one of the genes whose expression is most diminished in the aggressive cells specifies the RhoGDI-2 protein, a known inhibitor of Rho activation.

We still do not know how accurately the studies of the motility of cancer cells in culture reflect their behavior in living tissues. Nonetheless, it seems likely that both the *in vitro* and *in vivo* motile behaviors of cells are coordinated by a network of Rho-like proteins that modulate cell shape, adhesion, and the localized proteolysis of nearby extracellular matrix. These cellular functions are clearly critical to the invasive and metastatic behaviors of malignant cells. Perhaps most disconcerting is the potential complexity of this signaling machinery: analyses of the human genome sequence indicate that the approximately 20 Rho family proteins are regulated by as many as 80 distinct guanine nucleotide exchange factors (i.e., Rho-GEFs).

14.8 Metastasizing cells can use lymphatic vessels to disperse from the primary tumor

After invasive, motile cells enter into the vessels of blood or lymphatic systems—the process of intravasation—they disperse and, should they survive the rigors of the voyage, eventually settle in a tissue site that lies at some distance from the primary tumor. Travel via the blood circulation is often called **hematogenous**

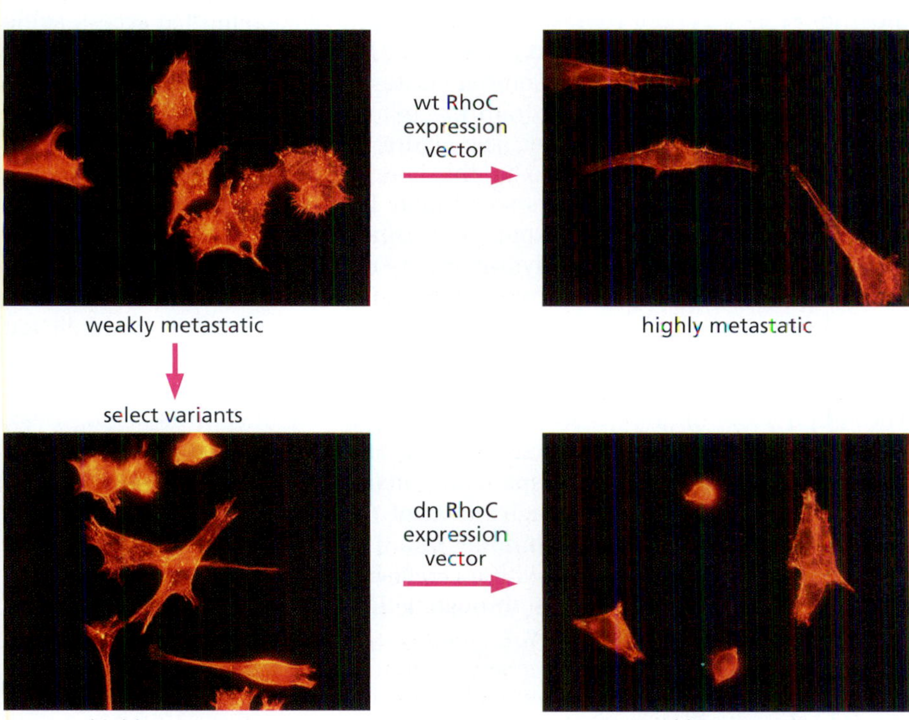

Figure 14.40 Influence of RhoC on metastasis Variants of a human melanoma cell line were derived that were either weakly or potently metastatic; the latter were far more motile and invasive *in vitro*. Metastatic ability was measured by injecting tumor cells into the mouse tail vein and counting resulting lung metastases; this assay gauges some of the steps of the invasion–metastasis cascade (e.g., extravasation) but not all of them. The weakly metastatic cells *(top left)* showed an epithelial morphology, while the highly metastatic cells *(bottom left)* appeared fibroblastic. Introduction of a retrovirus vector expressing high levels of wild-type (wt) RhoC into the weakly metastatic cells yielded cells *(top right)* that showed an elongated, fibroblastic appearance and were highly metastatic. Conversely, introduction of a dominant-negative (dn) RhoC protein into the potently metastatic cells yielded cells that were more epithelial in appearance *(bottom right)* and lost almost all of their metastatic ability as well as their motility and invasiveness *in vitro*. (From E.A. Clark, T.R. Golub, E.S. Lander and R.O.Hynes *Nature* 406:532–535, 2000.)

spread, and it depends on prior successful angiogenesis by the tumor. This emphasizes the fact that angiogenesis benefits cancer cells in two distinct ways. On the one hand, it supports the metabolic activity required in order for these cells to survive and proliferate. On the other, it provides tumor cells with direct access to avenues through which they can disperse throughout the body.

The extended discussion of hematogenous spread in Section 14.1 reflects the clearly important role of the blood circulation in metastatic dissemination. The contribution of the lymphatic vessels to the dispersion of cancer cells is, however, less obvious. Almost all tissues in the body carry networks of lymphatic vessels that are responsible for continuously draining the **interstitial** fluid that accumulates in the spaces between cells (see Sidebars 13.5 and 26). Most of these vessels converge on a major abdominal vessel that empties its lymph into the left subclavian vein near the heart and thence into the general circulation. Consequently, cancer cells present in lymphatic vessels may occasionally enter through this cross connection into the general circulation.

Tumor cells and recruited stromal companions may secrete VEGF-C, which drives lymphangiogenesis—the formation of new lymphatic vessels (Section 13.6). Moreover, experimental tumors forced to secrete increased levels of VEGF-C will seed larger numbers of metastatic cells in nearby "draining" lymph nodes—the lymph nodes associated with the lymphatic ducts that drain the tissue in which the tumor lies (Figure 14.41A). However, detailed histological analyses of spontaneously arising tumors indicate that functional lymphatic vessels are rarely found throughout tumor masses. Instead, they are largely present in a zone at the periphery of solid tumors. Those few lymphatic vessels discovered in the central regions of tumor masses are usually collapsed (see Figure 13.36). As discussed in the previous chapter, it seems that the expanding masses of cancer cells within a tumor press on these vessels; because the lymphatic ducts have little internal hydrostatic pressure, they cannot resist these forces and collapse.

The absence of functional lymphatic vessels within tumor masses must influence the paths used by metastasizing cancer cells to leave the primary tumor. Without ready access to lymph ducts, most motile cancer cells are forced to use the far more numerous, functional capillaries, which are threaded throughout the tumor mass, as their routes of escape. In spite of such limited access, some cancer cells do indeed succeed in entering the lymphatic system. In the specific case of mammary carcinomas, some metastasizing cancer cells enter into the lymphatic vessels that directly drain the mammary gland and collect in the nearby downstream lymph nodes (see Figure 14.41A). These wandering carcinoma cells are readily detected in the lymphatic ducts and lymph nodes, because their appearance differs so strongly from the surrounding lymphoid cells (Figure 14.41C) and they express epithelial proteins, such as cytokeratins, that are otherwise absent from lymphatic tissues (Figure 14.41D). Histological examination of draining lymph nodes is routinely used to determine whether a primary breast cancer has begun to dispatch metastatic pioneer cells to distant sites in the body (Sidebar 14.6).

The lymph nodes draining a primary tumor might well function as staging areas. Thus, once cancer cells multiply and form small metastases within these nodes, they may disperse further by dispatching metastatic pioneers to more distant sites in the body. In fact, through much of the twentieth century, surgeons believed that the draining lymph nodes of a tissue function as filters, and that once these nodes become filled with metastasizing cancer cells, these cells spill over into other lymphatic vessels, through which they travel in order to disseminate widely throughout the body.

The alternative notion is that draining lymph nodes represent dead ends for disseminated cancer cells, that is, that those cancer cells that proliferate within these nodes rarely move on to more distant sites in the body. Indeed, studies of patients carrying breast, head and neck, gastric, and colorectal carcinomas indicate that surgical removal of draining lymph nodes has no effect on long-term patient survival. Such observations suggest that metastasis through the lymph and through the blood operate in parallel, and that the cancer cells that arrive in lymph nodes usually venture no farther. Accordingly, lymph nodes represent useful "**surrogate markers**" of metastasis by providing useful diagnostic and prognostic data without being directly involved in the processes that lead to widespread cancer cell dissemination and metastatic disease.

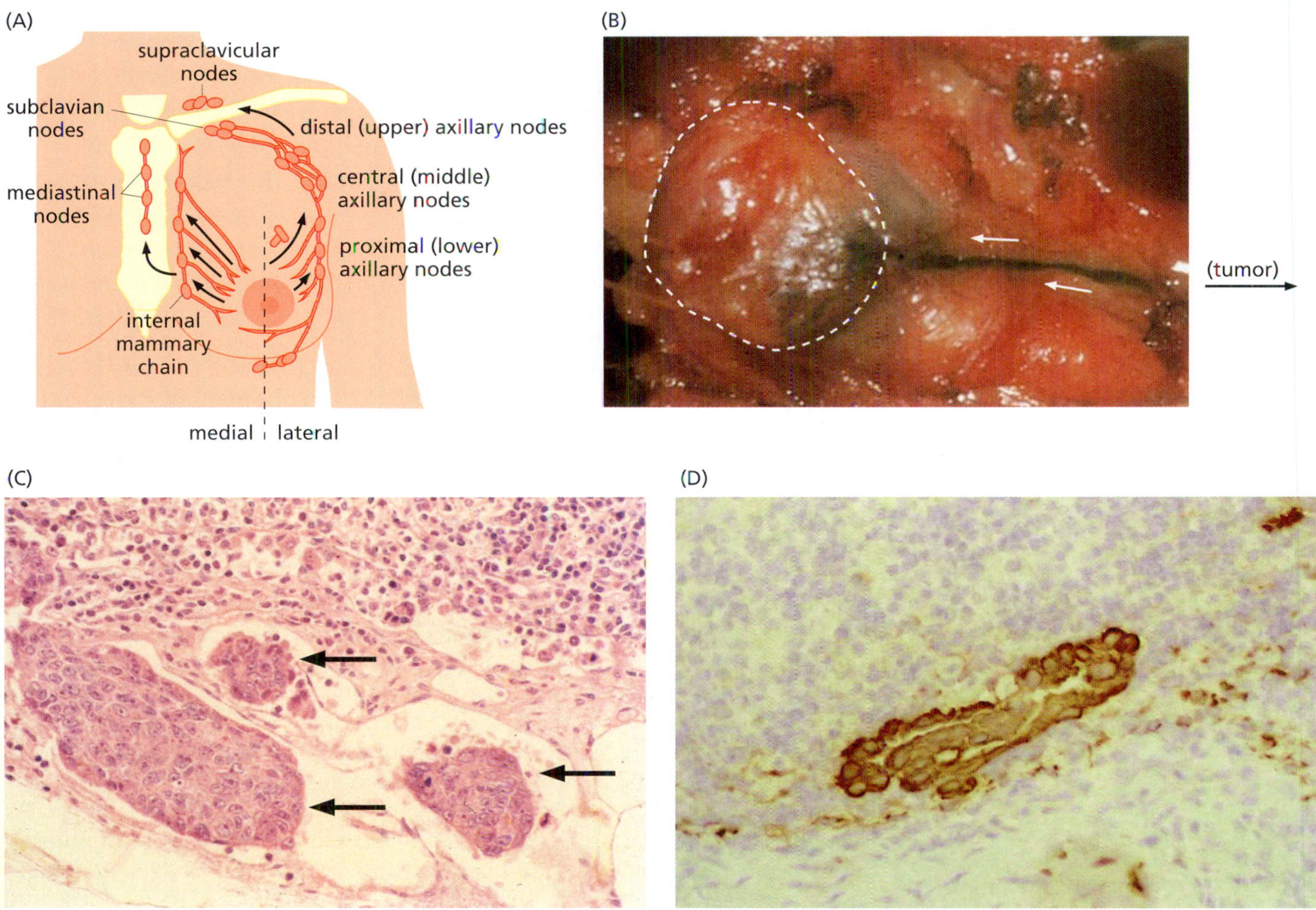

Figure 14.41 Draining lymph nodes of the mammary gland (A) The lymphatic ducts *(red)* and the lymph nodes draining the breast *(swellings along ducts)* are initial sites of metastatic spread, being carried there by the flow of lymph *(arrows)* leaving various sectors of the breast. Carcinomas arising in these different sectors tend to deposit metastatic cells in different sets of draining lymph nodes. Discovery of carcinoma cells in these lymph nodes, which is observed in more than 30% of human breast carcinomas at the time of initial diagnosis, suggests the possibility of deposits of metastatic cells in more distant sites in the body, particularly if large numbers of draining nodes are found to carry breast cancer cells. (B) The lymph node that serves as the sentinel node of a tumor can usually be identified among all of the lymph nodes draining the breast (panel A) by injecting a blue dye into the tumor *(outside photographic field to right)* and following the trail of the dye via the lymphatic duct *(arrows)* to the draining node *(outlined in dashed line, left)*. (C) Hematoxylin–eosin (H&E) staining of a section of an *axillary* lymph node reveals that three micrometastases arising from a primary breast tumor *(arrows)* have grown in the space between the capsule surrounding this node *(not seen, below)* and the mass of lymphocytes within the node *(small cells, dark nuclei, above)*, displacing the latter upward. (D) Immunohistochemistry using an antibody specific for cytokeratins *(brown)* reveals this small micrometastasis in a sentinel node. This procedure is far more sensitive than the H&E staining of panel C in detecting small micrometastases, since the mesenchymal cells of the lymph node do not expressed cytokeratin, which is made by epithelial cells. (A, B, and C, from A.T. Skarin, Atlas of Diagnostic Oncology, 3rd ed. Philadelphia: Elsevier Science Ltd., 2003; D, from J.P. Leikola, T.S. Toivonen, L.A. Krogerus et al., *Cancer* 104:14–19, 2005.)

Sidebar 14.6 Lymph nodes are sentinels that carry important prognostic information The lymph nodes associated with the lymphatic ducts that drain various tissues operate as collection points for subcellular debris and cells shed by these tissues; in addition, cells of the immune system become activated in response to antigens that they first confront in these nodes. This explains why draining nodes associated with a variety of organs (e.g., the mammary gland; Figure 14.41A) are routinely examined to determine whether cancer cells have been released by primary tumors in those organs. For example, upon initial diagnosis, about one-third of breast, colorectal, cervical, and oral carcinoma patients have metastasized cancer cells in the lymph nodes near their primary tumors (see Figure 14.41C and D).

These draining lymph nodes serve as proverbial "canaries in the mine," by providing early warning of the presence of metastasizing cells in the body. Among these regional lymph nodes, the single node that directly drains the primary tumor is often termed the "sentinel" node (Figure 14.41B). Patients with small numbers of affected nodes often have only localized spread of the breast cancer and may never develop metastatic disease, while those with many affected nodes are far more likely to harbor other deposits of metastatic cells in distant sites in the body. For example, in one study 90% of long-term survivors of a variety of carcinomas had one, two, or occasionally three "positive" lymph nodes at diagnosis. Conversely, fewer than 5% of patients with more than five positive lymph nodes when their primary tumors were removed enjoyed long-term, disease-free survival.

14.9 A variety of factors govern the organ sites in which disseminated cancer cells form metastases

The descriptions in the previous sections of the mechanisms of invasion and metastatic dissemination seem to explain, at least in outline, how most of the steps of the invasion–metastasis cascade proceed. Moreover, it is plausible that the dispersion strategies used by a wide variety of invasive, metastatic cancer cell types will one day be found to be governed by a common set of mechanistic principles, such as those discussed here. Importantly, however, our discussions did not address the last step of the invasion–metastasis cascade—colonization.

The growth of micrometastases (<2 mm diameter) into macrometastases (>2 mm diameter) is clearly the key step in determining whether or not metastatic disease will ever develop. For example, 30% of the women diagnosed with primary breast carcinomas have thousands of micrometastases in their bone marrow, many composed of single cells or tiny clusters of cells (see, for example, Figure 14.2C). Yet only half of these women will ever suffer a disease relapse triggered by the appearance of macroscopic metastases. Clearly colonization is an extremely inefficient process, and the vast majority of cells that end up forming small micrometastases never succeed in properly adapting to the tissue in which they have landed by spawning a macrometastasis.

In addition, while a variety of cancer cell types may execute the earlier steps of the invasion–metastasis cascade in a very similar fashion, it is likely that colonization by each type of cancer cell proceeds quite differently. Thus, successful adaptation of metastasized breast cancer cells to the bone marrow (which, by definition, enables these cells to colonize the marrow) is likely to involve a quite different set of cellular changes from those required for successful bone marrow colonization by prostate cancer cells. In addition, the changes required for a breast cancer cell to colonize the bone marrow are likely to be quite different from those needed for it to succeed in brain or lung colonization.

Abundant evidence supports the notion that metastatic cancer cells that have colonized a certain target organ have become *highly specialized* to do so. Here are some indications of this. (1) 75% of young patients with papillary thyroid carcinomas have significant lymph node metastases, but only 3% will ever develop distant metastases. Hence, adaptation to the lymph nodes by metastasizing thyroid carcinoma cells does not allow them to colonize other tissues in the body. (2) Similarly, duodenal carcinoid tumors greater than 1 cm in diameter (containing $>10^9$ cells) have a high rate of lymph node metastasis, yet they rarely metastasize

Sidebar 14.7 Cancer cells must "learn" to colonize sites of metastasis Implicit in these descriptions of colonization is the notion that the neoplastic cells within a primary tumor do not gain the ability to colonize one or another target organ the moment they become tumorigenic. Instead, for most kinds of cancer cells, colonizing ability is a trait that seems to be acquired independently of the acquisition of tumorigenicity. This makes sense, since the biological forces operating within the primary tumor site select for the ability of variant cells to proliferate in that particular microenvironment and not in the microenvironment of a distant, biologically unrelated tissue.

Unanswered by these arguments is the question of *where* and *when* tumor cells learn to colonize a particular target organ. Some subclones of cells within the primary tumor mass may, through happenstance, acquire this ability as an unselected side product of the forces that led initially to their development of tumorigenic potential. More frequently, however, the acquisition of colonizing ability is likely to occur at the site of eventual metastasis formation, where large numbers of micrometastases reside for extended periods of time until one of the dispersed cancer cells happens to acquire colonizing ability. In this case, the Darwin-ian selection favoring such a variant cell is clear and obvious, since its descendants will thrive and proliferate strongly in this distant site, while all the other micrometastatic neighbors in this tissue will stagnate and eventually disappear.

to the liver, which is the common site of metastasis of the tumors that arise in the nearby colon. (3) Cancer cells isolated from human lymph node metastases have been found, after injection into the venous system of mice, to grow preferentially in the lymph nodes of their mouse hosts rather than other possible sites of colonization. (4) Surgical removal of isolated, relatively large colorectal carcinoma metastases present in the liver or lung often results in disease-free survival for a number of years, in spite of the fact that the circulation of patients with these metastases clearly carries large numbers of metastasizing cells, including some that already possess colonizing ability in one or several organs. (5) Mouse melanoma cells can be selected that metastasize preferentially to lungs, or breast cancer cells that metastasize to the lungs or, alternatively, to the bone. These disparate observations reinforce the notion that the ability to colonize a certain organ represents an acquired specialization (Sidebar 14.7).

Yet another factor affects these dynamics: different types of cancer cells acquire the ability to colonize a given tissue more or less readily. Thus, the ability of metastasizing prostate cancer cells to colonize the bone marrow seems to be much more readily acquired than their ability to colonize the liver or the pancreas. This suggests that the differentiation program of normal prostatic epithelial cells exerts a strong influence on the ability of derived carcinoma cells to metastasize to specific organs. If we were to place prostate carcinoma cells and potential target organs on a map that depicts metastatic tendencies (Figure 14.42), we would indicate that the prostatic cells have relatively easy "access" to the bone marrow, implying that they need to undergo fewer changes in order to adapt to this site. Conversely, they have more limited access to other organs, such as the liver or pancreas, in which they rarely form macroscopic metastases.

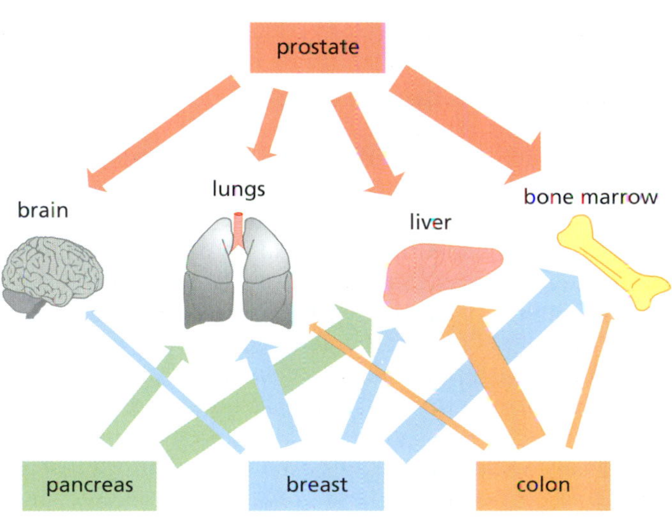

Figure 14.42 Primary tumors and their metastatic tropisms In this diagram, the relative width of each arrow indicates the relative proportion of clinically apparent metastases that are generated by a primary adenocarcinoma. Four types are indicated here: prostate, breast, pancreas, and colon. In some cases, the tendencies of a tumor to spawn metastases in one or another tumor reflect the abilities of the cancer cells from the primary tumor cells to adapt to (and thus colonize) the microenvironment of distant tissues; this is likely to explain the strong tendencies of prostate and breast cancers to generate metastases in the bone marrow. In other cases, the layout of the circulation may strongly influence the site of metastasis. For example, the high proportion of liver metastases deriving from primary colon cancers may reflect the drainage via the portal vein of blood from the colon directly into the liver (see Figure 14.44).

Figure 14.43 Stephen Paget The British physician Stephen Paget was the first to enunciate the "seed and soil" hypothesis, which states that the ability of a disseminated cancer cell to successfully found a metastasis depends on whether a distant tissue offers it a hospitable environment to survive and proliferate. (From I.J. Fidler, *Nat. Rev. Cancer* 3:453–458, 2003.)

This predilection to form macrometastases in one or another organ site was noted as early as 1889 by the British pathologist Stephen Paget (Figure 14.43). He proposed the "seed and soil" hypothesis, in which he analogized the seeding of cancer cells to the dispersal of the seeds of plants. After studying the clinical course of 735 breast cancer patients, Paget concluded that the patterns of metastasis formation in these patients could not be explained either by random scattering throughout the body or by the patterns of dispersal from the breast through the general circulation. He therefore proposed that the metastasizing cancer cells (the seed) find a compatible home only in certain especially hospitable tissues (the soil). He wrote, "a plant goes to seed, its seeds are carried in all directions; but they can only live and grow if they fall on congenial soil." This ability to form macroscopic metastases in some sites but not others has been highlighted by certain clinical procedures (Sidebar 32).

The seed and soil model states quite explicitly that disseminating cancer cells are dispersed "in all directions," that is, throughout the body. This phrase injects another idea into our thinking: the reason why many kinds of cancer cells form metastases in a specific target organ is not attributable to their directed migration or homing to this organ. Instead, they are scattered randomly, and only those cancer cells that happen to land in a reasonably hospitable tissue succeed in surviving, forming micrometastases, and occasionally, having learned how to colonize, forming macroscopic metastases. Conversely, essentially identical disseminating cells may land, with an even higher frequency, in other organs where they perish immediately or survive as micrometastases without ever succeeding to colonize these sites.

The seed and soil hypothesis cannot, however, explain the metastatic patterns of *all* types of human cancers (see Sidebar 14.8). Instead, in certain cases, the predilection to metastasize to a certain target organ is likely to be dictated by the layout of the vessels connecting the site of a primary tumor and the site of metastasis. For example, the strong tendency of colon carcinoma cells to metastasize to the liver may simply reflect the fact that these cancer cells leave the gut via the portal vein (which drains the lower gastrointestinal tract and the spleen) and, after a very brief trip, almost inevitably become lodged in the capillary beds of the liver that are fed by this vein (Figure 14.44). Even if individual metastasizing colon cancer cells colonize the liver with an extremely low efficiency, the sheer numbers of the cancer cells trapped in the liver guarantee that, with the passage of enough time, substantial numbers of metastases will arise in this target organ.

Sidebar 14.8 Contralateral metastases are relatively rare Possibly the greatest embarrassment for the seed and soil hypothesis comes from its failure to explain the rarity of contralateral metastases. For example, cancer cells disseminated from a primary tumor in one breast should find that the **contralateral** (i.e., opposite) breast provides the most hospitable environment for colonization. In fact, only about 2% of breast cancer cases result in contralateral metastases, comparable to the frequency of tumors in the breast that arise as metastases of primary tumors located outside of the breast. Similarly, primary kidney cancers metastasize infrequently to contralateral kidneys. These behaviors are clearly incompatible with the seed and soil hypothesis and still require explanation.

The same logic may explain why breast cancer cells often form metastases in the lungs. As is the case with metastasizing colorectal carcinoma cells, wandering mammary carcinoma cells may not find that the lungs provide them with an especially hospitable environment, and individual cancer cells will have a low probability of successfully colonizing the lungs. Nonetheless, some metastases will eventually form there, simply because so many of these cells become physically trapped in this tissue (see Figure 14.44). This logic suggests that, in general, the frequency of metastases to an organ is governed by two parameters—the frequency with which metastasizing cells are physically trapped in an organ, and the ease with which they can adapt to the microenvironment of that organ, thereby colonizing it.

There are also indications that tissues that are normally not hospitable sites for colonization may become so through specific pathological processes, such as localized wounding (Sidebar 33). This suggests that areas of chronic inflammation within the body of a cancer patient may occasionally become congenial environments for metastasizing cancer cells, simply because they provide a spectrum of mitogenic and trophic signals, as discussed in Chapter 13.

Yet other mechanisms have been proposed to explain the tissue **tropisms** of metastasizing cells. For example, target organs may release specific chemical

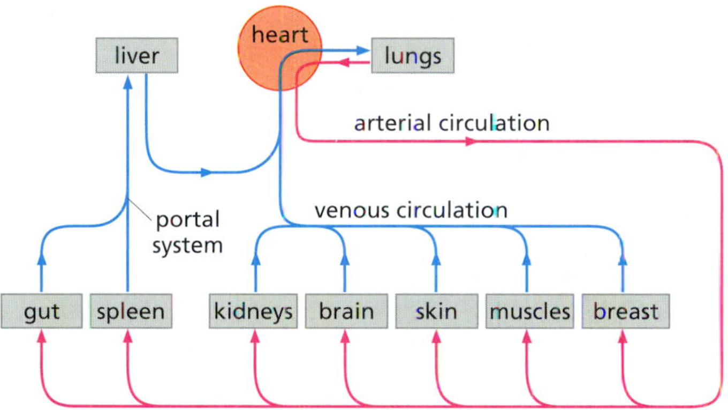

Figure 14.44 Portal circulation and liver metastasis While the venous systems of most tissues drain to the right side of the heart and thereafter into the capillary beds of the lungs, the veins draining the spleen and gut are organized differently, in that they empty directly into the liver via the portal circulation. (After passing through the liver, venous blood is only then dispatched back to the heart.) Consequently, vast numbers of metastasizing colorectal carcinoma cells are trapped in capillary beds of the liver within seconds of leaving the colon. (Adapted from I.C. MacDonald, G.C. Groom and A.F. Chambers, *BioEssays* 24:885–893, 2002.)

messages—the chemoattractants sometimes termed chemokines—that might actively recruit wandering cancer cells to enter into them from the circulation. Such chemoattraction clearly operates to ensure the homing of a variety of circulating immune cells to specific tissues as part of the normal operations of the immune system. In one study, when B16 mouse melanoma cells were forced to express the CXCR4 chemokine receptor, their metastases to the lung increased by a factor of 10. However, when an expression vector specifying the CXCR7 receptor was introduced into these melanoma cells, they then showed substantially increased metastasis to the lymph nodes, thereby appropriating a mechanism normally used by lymphocytes for homing to these nodes. (In truth, since these chemokine-activated receptors often provide mitogenic and survival signals to cancer cells, it is difficult to know whether these receptors, when activated by their ligands, induce metastasizing cells to migrate into a tissue or simply encourage the survival and proliferation of these cells after they have landed in the tissue.)

According to another mechanistic model of metastatic tropism, the capillaries forming the *vascular beds* (i.e., the networks of blood vessels) in various tissues express tissue-specific molecules on their luminal surfaces. These molecules may offer specialized docking sites for cancer cells that express certain adhesion molecules, such as integrins, on their surfaces. This model is sometimes termed the "vascular **ZIP code**" theory, because it implies that the luminal surfaces of vessels in different tissues carry, in chemical form, specific homing addresses, much like those used by a postal system. This model fails to take into account the fact that cancer cells are often surrounded by clouds of platelets that are capable of blocking direct association between the cancer cell and the luminal surfaces of endothelial cells.

One detailed study of the behavior of metastasizing human cancers calculated that 66% of metastases could be explained simply by the blood flow patterns between the primary tumor and the sites of observed metastases. In 20% of the cases, the specialized microenvironments of target tissues (rather than blood flow patterns) seemed to provide the explanation of the tendency of certain cancer types to form macroscopic metastases in these tissues. And finally, in 14% of cases, negative interactions (in which tissues seemed to actively repel wandering cancer cells) seemed to explain smaller-than-expected numbers of metastases predicted by blood flow patterns.

To summarize, these diverse observations suggest that metastasizing cells disperse to many organ sites in the body and that their dispersion is affected by the layout of the vasculature. Once arrived in these various sites, the cancer cells will usually survive and eventually colonize only those tissues that provide them with specific chemokines, trophic factors, and mitogens. On occasion, however, these cells may succeed in founding macroscopic metastases in relatively inhospitable organ sites, only because the routing of the blood circulation introduces these cells in vast numbers into such sites.

14.10 Metastasis to bone requires the subversion of osteoblasts and osteoclasts

The development of bony metastases represents one instance in which we understand in some detail the biochemical and biological mechanisms that permit metastasized cancer cells to thrive in a specific tissue microenvironment. This fact, on its own, justifies a detailed discussion of **osteotropic** metastasis. In addition, and as mentioned repeatedly, several of the most common types of cancer occurring in the Western world—carcinomas of the lung, breast, and prostate—show a strong tendency to metastasize to the bone. In fact, patients with advanced breast and prostate cancer almost always develop bone metastases. And in those patients who succumb to these cancers, the bulk of the tumor cells in their bodies at the time of death are usually found among the metastases scattered throughout their bones.

We usually think of bone as being a static tissue which, once formed, retains its structure throughout life. The truth is far more interesting. In mammals, about 10% of skeletal bone mass is replaced each year, resulting in an essentially complete replacement over the course of a decade. This continuous remodeling enables the bones to respond to mechanical stresses by compensatory reinforcing of stressed regions. For example, the bones of the legs are continuously being remodeled in response to the weight-bearing signals that different portions of each leg bone receives.

The turnover of bone is the work of **osteoclasts**, which break down mineralized bone, and of **osteoblasts**, which reconstruct it. The osteoclasts function first to **demineralize** the bone (by dissolving its calcium phosphate crystals) and then to degrade the now-exposed extracellular matrix, which previously formed the organic scaffolding for the calcium phosphate crystals (Figure 14.45). Osteoblasts move in soon after to carry out reconstruction, which involves both the assembly of new ECM and the deposition of calcium phosphate crystals in the interstices of this matrix. As can be deduced from this description, the two cell types normally work in close coordination.

Most kinds of metastasizing cancer cells are, on their own, incapable of remodeling bone structure. Instead, they manipulate and exploit these two types of cells present in the bone in order to change its shape. Thus, breast cancer cells activate the osteoclasts, resulting in **osteolytic** metastases—literally, metastases that dissolve bone. Prostate cancer cells tend, on the other hand, to activate osteoblasts, yielding **osteoblastic** lesions, in which mineralized bone actually accumulates in the vicinity of the metastases (Figure 14.46).

In fact, these two behaviors represent the extremes of a continuum, since both types of cancers activate both osteoblasts and osteoclasts to a greater or lesser extent. For example, while osteolytic metastases predominate in advanced breast cancer patients, as many as one-quarter of these women also have clearly defined osteoblastic lesions in their bones. Similarly, prostate carcinomas also generate occasional osteolytic metastases scattered among the many osteoblastic growths spawned by these tumors. One exception to this rule of a mingling of both types of bone metastases is provided by myeloma cells (tumors of the B-cell, antibody-secreting lineage), which create exclusively osteolytic lesions in bone.

The close coordination between osteoblasts and osteoclasts is mediated, at least in part, by the exchange of growth factor signals. An important inducer of osteoclast differentiation is RANK (receptor activator of NF-κB) ligand, or simply RANKL. RANKL is produced by and displayed on the surface of osteoblasts. When an osteoclast precursor displaying the RANK receptor comes into contact with an osteoblast and its cell surface RANKL molecules, this results in activation of the RANK receptors of the osteoclast precursor and its maturation into a

(A)

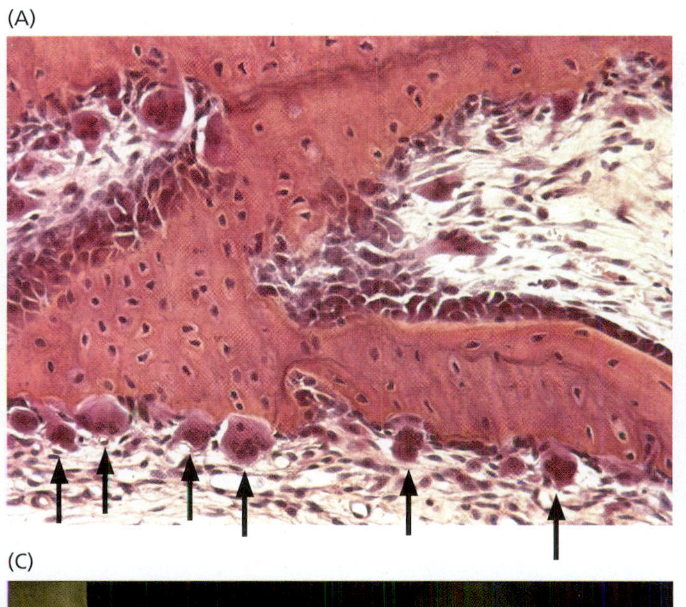

(B)

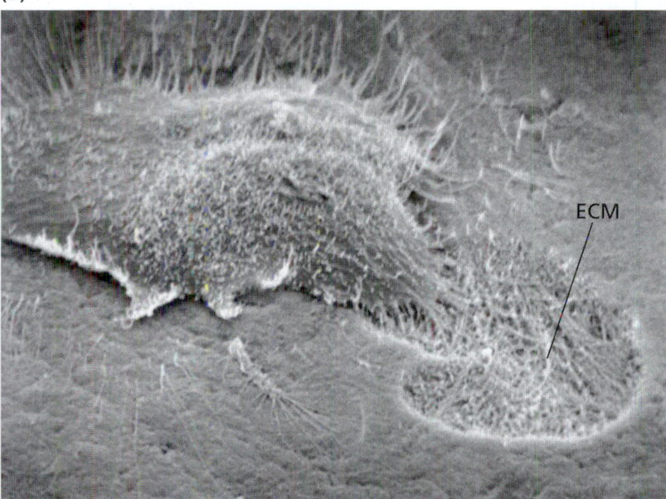

(C)

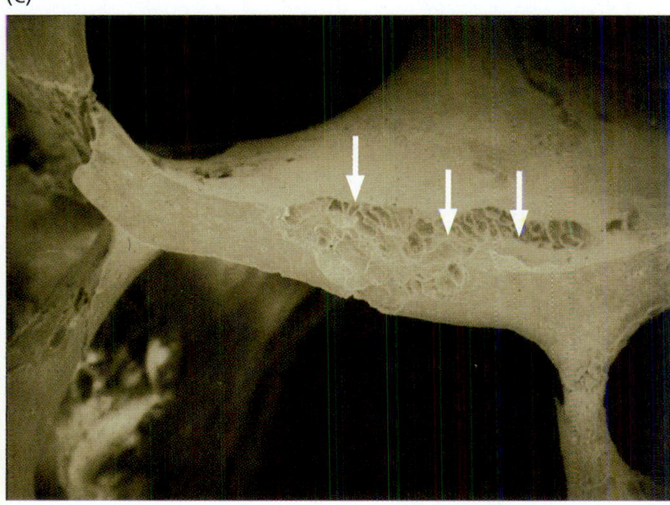

Figure 14.45 Bone degradation by osteoclasts (A) This micrograph shows osteoclasts *(purple, arrows)* excavating small pits in the surface of a mouse jawbone *(pink)*. (B) At far higher magnification, this scanning electron micrograph shows a rat osteoclast that has excavated a shallow pit in the surface of mineralized bone. The calcium apatite crystals in the bone have been dissolved away by acid secreted by the osteoclast, revealing the complex meshwork of collagen-rich extracellular matrix (ECM) at the bottom of the pit. Associated with this ECM are mitogens and survival factors that become available to cancer cells after osteoclasts break down the ECM. (C) This scanning electron micrograph reveals how devastating the osteolytic lesions *(arrows)* can be in terms of compromising bone structure in a patient with metastatic osteolytic lesions. (A and B, from T.R. Arnett and D.W. Dempster, *Endocrinol.* 119:199–124, 1986; C, courtesy of G.R. Mundy.)

functional osteoclast (Figure 14.47). At the same time, osteoblasts produce a soluble decoy receptor, termed osteoprotegerin (OPG), which can bind RANKL and ambush it before it succeeds in activating the RANK receptor on the surface of osteoclast precursors. The result is a blockage of the RANKL–RANK signaling and the inhibition of osteoclast maturation. Hence, the balance between the RANKL (stimulatory) and OPG (inhibitory) signals determines the state of activation of osteoclasts.

This dynamic interaction of osteoblasts and osteoclasts provides the background for the actions of cancer cells that metastasize to bones. Their attraction to the bone derives ultimately from the nonmineralized, collagenous extracellular matrix (ECM) that forms the organic scaffolding in which calcium phosphate crystals are deposited (see Figure 14.45B). As it happens, bone ECM is an unusually rich source of the mitogenic and trophic factors that allow several types of carcinoma cells to thrive. Consequently, by provoking the demineralization of bone, cancer cells gain access to the storehouse of factors sequestered in the bone ECM and use them to support their own proliferation and survival.

Metastasizing cancer cells reach the bone through the vessels feeding the marrow. Once there, they adhere to specialized stromal cells coating the surfaces of the bone facing the marrow. Metastasizing breast cancer cells, in particular, upon arrival in bone, revert to a behavior characteristic of their normal precursors (mammary epithelial cells, or MECs). During lactation, when producing

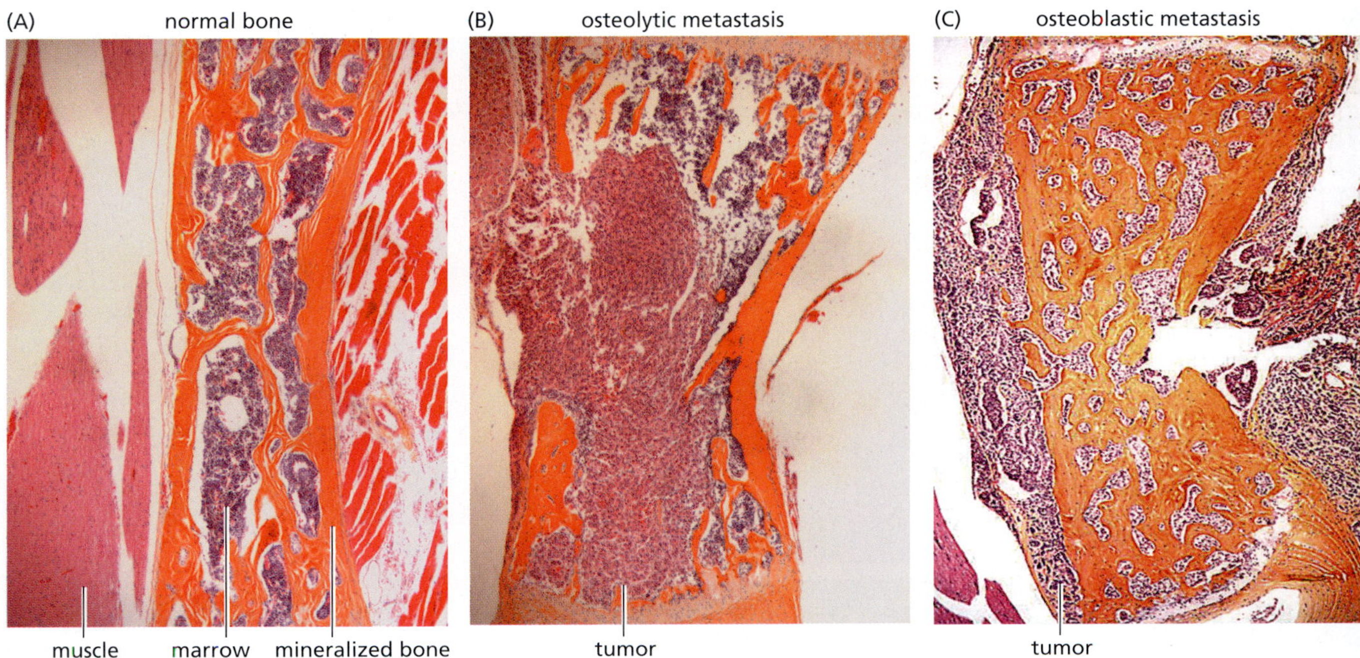

Figure 14.46 Osteolytic and osteoblastic metastases These micrographs present sections of vertebrae and femurs in which the mineralized bone *(orange)*, surrounding muscle *(bright red)*, and bone marrow *(dark purple)* are clearly delineated. (A) This vertebra of a control mouse injected only with buffer is seen to be composed of extensive marrow with ribbons of mineralized bone running through the marrow. (B) In a mouse bearing a human breast cancer cell line (MDA-MB-231) that creates osteolytic lesions, much of the mineralized bone is seen to be missing, and the marrow has been displaced by tumor cells *(dark red)*. (C) In a mouse bearing a human breast cancer cell line (ZR-75-1) that creates osteoblastic lesions, much of the marrow space is now filled with mineralized bone *(orange)* with tumor masses evident to the left and right. (From J.J. Yin, K.S. Mohammad, S.M. Kakonen et al., *Proc. Natl. Acad. Sci. USA* 100:10954–10959, 2003.)

milk, MECs forming the small sacs (**alveoli**) of the mammary gland release parathyroid hormone–related peptide (PTHrP). PTHrP then travels through the circulation to the bones, where it triggers a chain of events that encourages the dissolution of bone minerals by osteoclasts. This results in the mobilization of calcium ions, which travel back via the circulation to the mammary gland, where they are incorporated into the milk by the MECs.

This normal calcium-mobilizing mechanism is subverted by metastasizing breast cancer cells that become established in bones (Figure 14.48). Having attached to the stromal cells covering the surfaces of mineralized bone, the breast cancer cells, reverting to the habit of normal MECs, release PTHrP. The PTHrP, in turn, impinges directly on its receptors displayed by osteoblasts, causing these cells to release RANKL. RANKL then induces the differentiation of osteoclast precursors into active osteoclasts. The activated osteoclasts degrade nearby mineralized bone, thereby liberating the rich supply of growth factors attached to the extracellular matrix of the bone.

The growth factors liberated from the bone ECM, including PDGF, bone morphogenetic proteins (BMPs), fibroblast growth factors (FGFs), insulin-like growth factor-1 (IGF-1), and TGF-β, fuel the further growth of the breast cancer cells, inducing them to secrete more PTHrP. This PTHrP engenders more osteolysis by the osteoclasts, leading to a self-perpetuating signaling system that has been called a "vicious cycle" (see Figure 14.48) in which TGF-β also plays a key role (Sidebar 14.9).

This cycle suggests possible points of therapeutic intervention. Most promising are drug compounds such as **bisphosphonates**, which are taken orally and become adsorbed to the apatite crystals that constitute the mineral portion of bone; the drug molecules can persist there for extended periods of time as long as a decade or more. When bisphosphonate-containing bone is later dissolved by osteoclasts, the latter are poisoned by the liberated bisphosphonates, leading to their apoptosis. Hence, bisphosphonates are useful for reducing the burden of osteolytic lesions in patients with various types of metastatic cancer.

When immunocompromised mice carrying human breast cancer cells are treated with bisphosphonates, the number of osteolytic lesions is reduced and, at the same time, the total burden of tumor cells in these animals is decreased. This observation provides additional indication that late in tumor progression,

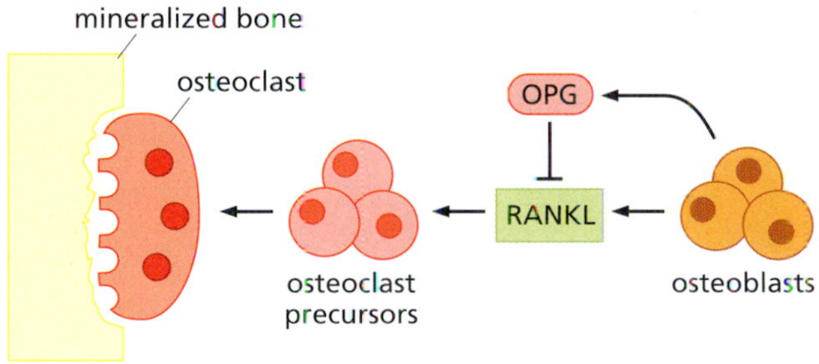

Figure 14.47 Osteoblasts versus osteoclasts The physiologic balance between bone formation and resorption is created by signaling between osteoblasts, which assemble bone, and osteoclasts, which dissolve it. In an ongoing cycle, osteoclasts remove mineralized bone by covering and sealing off a section of bone and secreting digestive acid into the bone below them; this is followed by osteoblastic filling of resulting cavities with new bone. The osteoblasts release RANKL, which acts via the RANK receptor *(not shown)* displayed by osteoclast precursors to induce the latter to mature into functional osteoclasts. The osteoblasts may also secrete osteoprotegerin (OPG), which acts as a decoy receptor to ambush RANKL before it has had a chance to activate osteoclast precursors. Hence, the balance between RANKL and OPG determines the net rate of bone growth/loss.

the proliferation of these breast cancer cells depends greatly on osteolysis and the resulting liberation of growth factors from dissolved bone. Moreover, bisphosphonate therapy can provide additional benefits to patients suffering from metastatic breast cancer (see Sidebar 14.10). Recently, derivatives of osteoprotegerin (OPG), which should also block the "vicious cycle," have also been found, in early-phase clinical trials, to substantially reduce the rate of bone dissolution in patients with myelomas and metastatic breast cancers.

As might be predicted, osteoblastic lesions depend on other signals—ones that activate osteoblasts rather than osteoclasts. In this case, the release by metastatic cancer cells of the growth factor termed "endothelin-1" (ET-1) plays a dominant role in stimulating osteoblasts and, at the same time, suppressing osteoclast activity. Thus, prostate cancer cells in primary tumors release endothelin; since its cognate receptor is also expressed by these cancer cells, an autocrine growth-stimulatory loop results. However, when these cancer cells arrive in the marrow, the endothelin that they release also acts via heterotypic signaling to stimulate osteoblasts, creating the osteoblastic lesions characteristic of this malignancy. (Precisely how osteoblast activation benefits the prostate cancer cells is less well understood. It is plausible that activated osteoblasts secrete large amounts of growth factors during the construction of mineralized bone, and that some of these factors are diverted by the cancer cells in osteoblastic metastases.)

So, Paget's seed and soil metaphor is useful, but it does not go far enough. Like seeds, metastatic cells are cast in many directions, but once they fall on certain ground, they can hardly be portrayed as being passive participants in their

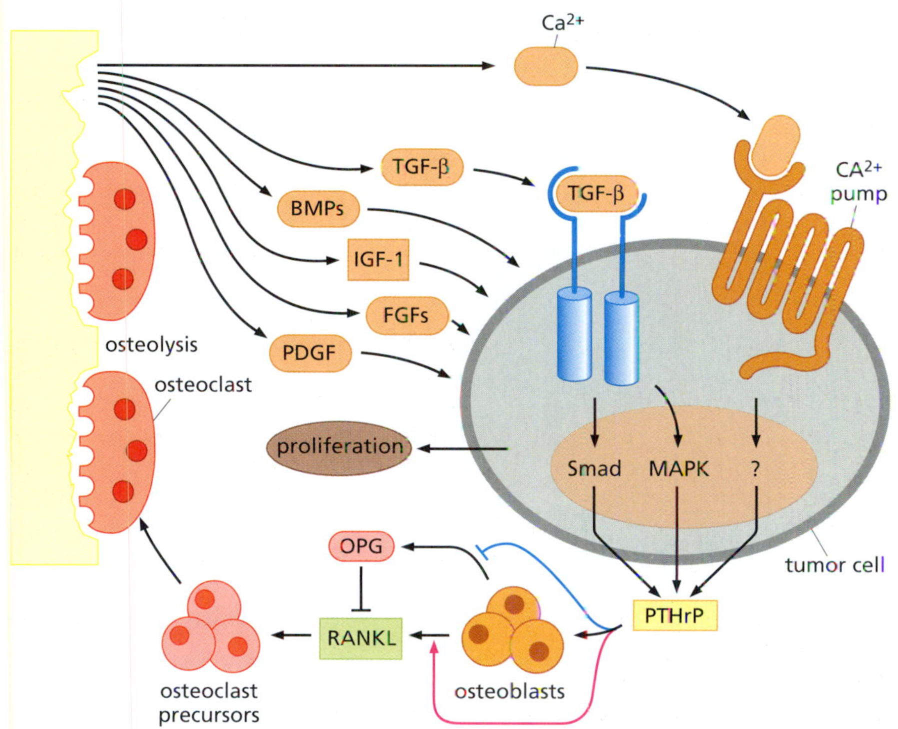

Figure 14.48 The vicious cycle of osteolytic metastasis Release by a breast cancer cell *(right, gray)* of PTHrP (parathyroid hormone–related peptide) causes nearby osteoblasts to change the mix of signals that they release: they increase RANKL synthesis *(red arrow)* and decrease OPG (osteoprotegerin) synthesis *(blue line)*. RANKL induces osteoclast precursors to mature into functional osteoclasts (see Figure 14.47). The latter undertake osteolysis, which causes bone demineralization, exposes the extracellular matrix within the bone (Figure 14.45B), and results in liberation of TGF-β, Ca^{2+}, and IGF-1 *(upper left and middle)*. IGF-1 and Ca^{2+} cause cancer cell proliferation and survival, and the additional presence of TGF-β induces the cancer cell to release more PTHrP, resulting in a self-sustaining positive-feedback loop that has been termed the "vicious cycle" of osteolytic metastasis. (From G.R. Mundy, *Nat. Rev. Cancer* 2:584–593, 2002.)

future fate. Instead, these cancer cells begin to actively till the soil in which they have landed, cultivating it so that it is guaranteed to become fertile ground for their own proliferation and that of their descendants.

14.11 Metastasis suppressor genes contribute to regulating the metastatic phenotype

We have read here of a number of genes that actively promote some of the steps in the invasion–metastasis cascade. Many of these encode familiar growth factors, growth factor receptors, or signal-transducing proteins that we encountered

Sidebar 14.9 TGF-β and PTHrP play pivotal roles in the vicious cycle of breast cancer osteolytic metastases Breast cancer cells that have metastasized to the bone produce far more PTHrP than do others in the same animal that have not—a reflection of the fact that certain growth factors liberated from the bone ECM stimulate PTHrP production by the metastatic cancer cells. The most important of these bone-derived factors is TGF-β, as illustrated by some simple experiments. In one of these, a dominant-negative TGF-β receptor (which blocks a cell's ability to respond to TGF-β) has been expressed in human breast cancer cells. Such cells now cease producing PTHrP and lose the ability to efficiently produce osteolytic metastases in the bone (Figure 14.49). In another experiment, breast cancer cells that usually lack the ability to metastasize to bone and fail to secrete TGF-β can be forced (through the use of an expression vector) to secrete TGF-β. The latter then acts in an autocrine fashion to stimulate these cells to produce their own PTHrP, allowing them to form large numbers of bone metastases. Finally, antibodies that bind and neutralize PTHrP are able to block the ability of human breast cancer cells to generate osteolytic lesions in mice. These are some of the disparate observations that have inspired the "vicious cycle" model depicted in Figure 14.48.

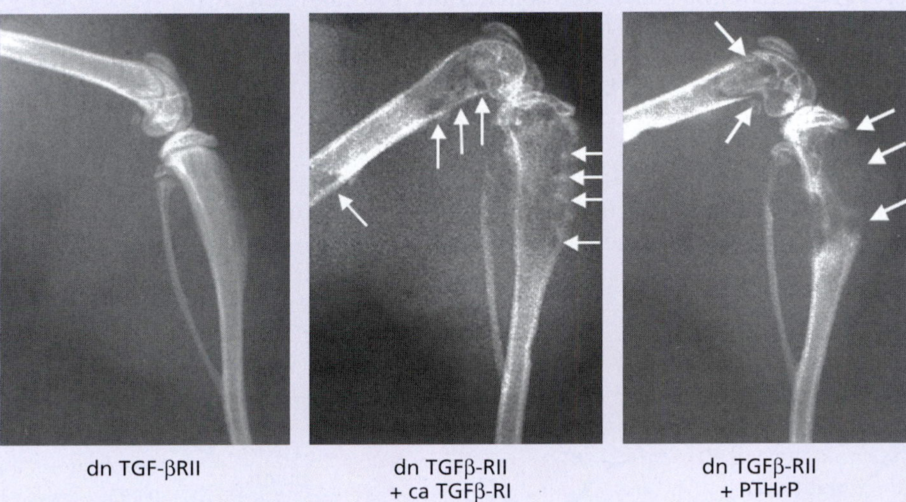

Figure 14.49 TGF-β and the formation of osteolytic metastases The evidence supporting the "vicious cycle" model of osteolytic metastasis (Figure 14.48) comes in part from experiments involving the use of MDA-MB-231 cells, a line of human breast cancer cells that usually show a high tendency to produce osteolytic metastases (see Figure 14.46B). The ability of these cells to do so is gauged here by X-ray analyses of the hind limbs of mice that have borne MDA-MB-231 tumor xenografts. When a dominant-negative type II TGF-β receptor (dn TGF-βRII) expression construct is introduced into these cancer cells, this mutant protein blocks their ability to respond to TGF-β, specifically the TGF-β that they would otherwise be liberated from the extracellular matrix of the osteolytic lesions that they may have induced. Without TGF-β stimulation, these cancer cells fail to form osteolytic metastases *(left panel)*. However, if this inability to respond to TGF-β is overridden by introducing additionally into these cells an expression construct specifying a constitutively active type I TGF-β receptor (ca TGF-βRI), then the powers of these breast cancer cells to induce osteolytic lesions are restored *(arrows, center panel)*. This observation, on its own, does not indicate precisely how the ca TGF-βRI succeeds in restoring the osteolytic activity to these cells. The explanation comes from an experiment in which a vector causing PTHrP expression (instead of ca TGF-βRI expression) is introduced into the dn TGF-βRII. Now, they regain the ability to form osteolytic metastases *(arrows, right panel)*. This demonstrates that TGF-β stimulates osteolytic activity by forcing the breast cancer cells to release PTHrP, which proceeds to activate osteoblasts and, in turn, osteoclasts (see Figure 14.48). (From J.J. Yin, K. Selander, J.M. Chirgwin et al., *J. Clin Invest.* 103:197–206, 1999.)

in our earlier discussions of oncogenes and their mechanisms of action. When introduced into a variety of epithelial cells, these genes are able to encourage changes such as the epithelial–mesenchymal transition (EMT), the acquisition of cell motility, and even invasiveness. Indeed, it seems increasingly likely that deregulated versions of these genes are the primary forces driving many of the steps of invasion and metastasis.

Importantly, the protein products of these various genes operate as components of the complex regulatory circuits that govern many aspects of cell physiology. And like all well-designed circuits, these have both positive regulators and counterbalancing negative regulators in order to ensure finely tuned outputs. This logic leads to the conclusion that there must be a number of control elements operating in cells that counteract and balance the invasive and metastatic actions of the positive effectors of advanced malignancy. Such negative regulators, in analogy with the tumor suppressors, have been called *metastasis suppressor* genes.

As one might anticipate, these metastasis suppressors operate at various levels in regulating the steps of invasion and metastasis, ranging from master, pleiotropically acting regulators and signal-transducing proteins to the ultimate effectors of the various biochemical changes (Table 14.4). These genes have been identified through a variety of experimental strategies. Quite often, their expression in primary tumors and their far lower expression in derived metastases have suggested important roles in blocking the late steps of malignant progression.

Such observations, being only correlations, do not prove *causal* roles in preventing metastasis, which can only be demonstrated through other types of experiments. For example, the role of a candidate gene as a bona fide metastasis suppressor gene can be tested by a simple operational criterion: when the gene's expression is forced in the cells of a primary tumor, does this expression permit the continued expansion of this tumor mass while, at the same time, block the appearance of distant metastases that are usually seeded by this tumor and others like it? Some of these genes have passed such a test, while others act in a less specific way by inhibiting proliferation by all types of cells, including some that lack invasive and metastatic properties. Yet other candidate tumor suppressor genes have been found able to suppress metastasis in only a small subset of malignant tumor types

The first of these genes to be discovered—*NM23*—was reported in 1988 to reduce the metastatic powers of mouse melanoma cells. Subsequent research indicates that it suppresses the motility and invasiveness of cells *in vitro*. Its ability to suppress the metastatic powers of experimental tumors has also been reported. The biochemical mechanism of action of the NM23 protein may be related to its ability to form physical complexes with the KSR protein, which serves as a scaffolding protein that brings together several of the kinases in the ERK–MAPK cascade

Sidebar 14.10 Osteolytic lesions wreak havoc in other ways In addition to the corrosive defects of osteolytic lesions on the skeleton, including severe bone pain and fractures, osteolysis has another, more subtle effect: The large-scale degradation of bone mineral leads to **hypercalcemia**—elevated concentrations of calcium in the circulation. At the same time, the PTHrP (parathyroid hormone–related peptide) released into the circulation by bone metastases causes kidneys to secrete less calcium into the urine, further increasing calcium levels in the blood. Such hypercalcemia, which usually signals the final stages of malignant disease, causes a wide variety of symptoms, including gastrointestinal, urinary tract, cardiovascular, and neuropsychiatric problems. Hypercalcemia can be substantially reduced by treating patients with bisphosphonates (to reduce osteolysis) and with antibody to PTHrP (to increase calcium excretion by the kidneys).

Table 14.4 Candidate metastasis suppressor genes

Name	Cellular location	Mechanism of action
BRMS1	nuclear protein	involved in chromatin remodeling
CRSP3	nuclear protein	transcription factor
KAI1/CD82	transmembrane protein	cell–cell association (?)
KISS1	secreted protein	ligand of G-protein–coupled receptor
NM23	cytoplasmic kinase	regulator of MAPK cascade (?)
RhoGDI-2	cytoplasmic protein	negative regulator of Rho action
SseCKs	cytoplasm	cytoskeleton-associated protein
VDUP1	cytoplasm	regulator of MAPK cascade (?)
CDH1 (=E-cadherin)	cell surface adhesion protein	favors formation of epithelial cell sheets
TIMPs	secreted protein	inhibitor of metalloproteinases
MKK4	cytoplasm	protein kinase component of MAPK cascade

Adapted in part from P.S. Steeg, *Nat. Rev. Cancer* 3:55–63, 2003.

(Section 6.5) in order to facilitate rapid and efficient transfer of signals between them. Hence, NM23 may affect the metastatic powers of cells through its ability to modulate the flux of signals passing through this critical survival and mitogenic pathway. NM23 has also been reported to function variously as a nucleoside diphosphate kinase (which converts nucleoside diphosphates to triphosphates), a histidine kinase, or a serine kinase. Studies of the *Drosophila* homolog of *NM23,* termed *awd,* indicate that it acts during development to suppress unwanted cell migration. Its candidacy as a metastasis suppressor is complicated by the fact that it has been reported to be overexpressed in advanced ovarian, gastric, and colon carcinomas, as well as in sarcomas.

The E-cadherin molecule, about which much has been said in this chapter, is also considered to be the product of a metastasis suppressor gene. Its role in stabilizing cell–cell contacts in epithelial sheets and in preventing the epithelial–mesenchymal transition (Section 14.3) clearly places it among the major molecular obstacles that block acquisition of the invasive cell phenotype. As we have read, its expression is lost in invasive carcinomas through a variety of mechanisms. Similarly, we can easily imagine how tissue inhibitors of metalloproteinases (TIMPs), which bind and inactivate MMPs in the intercellular space, might block a number of steps of invasion and metastasis.

Another tumor suppressor gene encodes the KAI1/CD82 protein, which weaves its way back and forth four times through the plasma membrane. Its expression has been found to be substantially repressed in many advanced lung, pancreatic, prostate, colon, and gastric carcinomas. A poor prognosis for breast cancer patients is associated with low KAI1 expression in their cancer cells. In cultured cells, KAI1 suppresses migration and invasiveness and, at the same time, enhances their aggregation with one another. Its location near adherens junctions is compatible with its playing a role in cell–cell adhesion. KAI1 has also been reported to act as an antagonist of EGF receptor signaling.

Another gene of interest encodes the KISS1 protein, which has been identified tentatively as a ligand of a cell-surface G-protein–coupled receptor (GPCR; see Section 5.7). Ectopic expression of the *KISS1* gene in tumor cells suppressed their metastatic tendencies without affecting the growth of these cells in primary tumors. Like several others in this class of genes, its precise biochemical role in metastasis suppression is poorly understood.

There are yet more candidates for suppressors of metastasis. The Rho guanine nucleotide dissociation inhibitor-2 (RhoGDI-2) is a negative regulator of Rho proteins, which acts by sequestering the GDP-bound forms of these proteins in the cytoplasm, thereby preventing them from remodeling the actin cytoskeleton (Section 14.7). The expression of its encoding gene was found to be correlated inversely with the invasive tendencies of a large group (105) of bladder carcinomas. Given the critical roles of Rho-like proteins in cell motility and invasiveness, RhoGDI-2 becomes an attractive candidate for being an inhibitor of cancer cell invasiveness and metastasis.

The breast cancer metastasis suppressor-1 (*BRMS-1*) gene was identified because of its decreased expression in breast cancer metastases. Its ectopic expression in breast carcinoma and melanoma cells suppressed their metastatic tendencies without affecting their tumorigenicity. It has been reported to increase the gap-junctional communication between cells, which involves channels that allow adjacent cells to exchange molecules of molecular weight less than about 10^3. At the same time, the BRMS-1 protein has been found in the nucleus as part of a complex of proteins involved in chromatin remodeling. Clearly, these disparate roles will need to be reconciled in the future.

Research on metastasis suppressor genes is still in its infancy, and in most cases, clear and definitive molecular mechanisms have yet to emerge. Some of the genes in this category, including those specifying E-cadherin, RhoGDI-2, and

TIMPs, produce proteins that are part of the known biochemical mechanisms of invasion and metastasis. The biochemical connections between many of the other candidate metastasis suppressor proteins and malignant cell phenotypes are less apparent. Until these genes have been found to be inactivated in human tumor cell genomes, either by mutation or promoter methylation, their involvement in regulating the malignant phenotypes of these cells will remain unclear.

14.12 Occult micrometastases threaten the long-term survival of cancer patients

Throughout this chapter, we have read repeatedly about the extraordinary inefficiency with which metastases are produced. Some of this metastatic inefficiency is created by the profound difficulties that cancer cells experience as they undertake the initial steps of the invasion–metastasis cascade. Most of those that do manage to reach distant sites and survive in their newfound homes fail to form clinically detectable metastases. The result is the presence of myriad micrometastases seeded throughout the tissues of many cancer patients.

While micrometastases are, with rare exception, unable to expand to form clinically detectable metastases, they do provide clear indication that a primary tumor has seeded cells throughout the body. These micrometastases represent a threat to the long-term survival of cancer patients, if only because some of them may erupt into full-fledged, clinically significant macroscopic metastases years after they become implanted in some distant tissue site. Breast cancers are notorious for yielding relapses one and even two decades after the primary tumor has been removed and the patient has been declared to be free of cancer.

In one study of breast cancer patients, micrometastases were detected by sampling the bone marrow of the iliac crest of the pelvis. About 1% of a population of patients suffering from nonmalignant conditions showed cytokeratin-positive cells (i.e., epithelial cells) in their marrow. In contrast, 36% of breast cancer patients carrying tumors of stages I, II, or III had such micrometastases in their marrow. The presence of these micrometastases in the marrow proved to be a highly useful prognostic marker for the risk of relapsing with clinically detectable metastasis (Figure 14.50A). Thus, within four years, one-quarter of the marrow-positive patients had died from cancer, while only 6% of those lacking cancer cells in their marrow had died from this disease. Overall, the presence of micrometastases in these patients represented about a 4-fold increased risk of eventual relapse or death from this disease. Another study found a more than 10-fold increased risk of death from breast cancer among those whose marrow carried micrometastases composed of single cells or small clumps of cancer cells.

Colon cancer patients who have undergone **resection** (surgical excision) of their primary tumor will often appear in the cancer clinic a year or two later with a small number of metastases in their liver but none elsewhere; these can then be removed surgically, often with significant clinical benefit. Once again, micrometastases in the marrow of the pelvis can be scored. About 90% of those who lack these micrometastases are still alive 15 months later, while only 30% of those who carry such micrometastases survive to this point (Figure 14.50B).

A procedure used to treat cancer of the esophagus provides yet another insight into metastatic spread. These tumors are often treated surgically, which necessitates the removal of one or more rib segments, from which marrow can be easily flushed. Two independent studies reported that 79% and 88% of these patients, respectively, harbored carcinoma micrometastases in their rib marrow at the time of their surgery. These numbers, which contrast with the approximately 30% of initially diagnosed breast cancer patients with micrometastases, correlate with the far grimmer prognosis for the patient suffering this type of cancer.

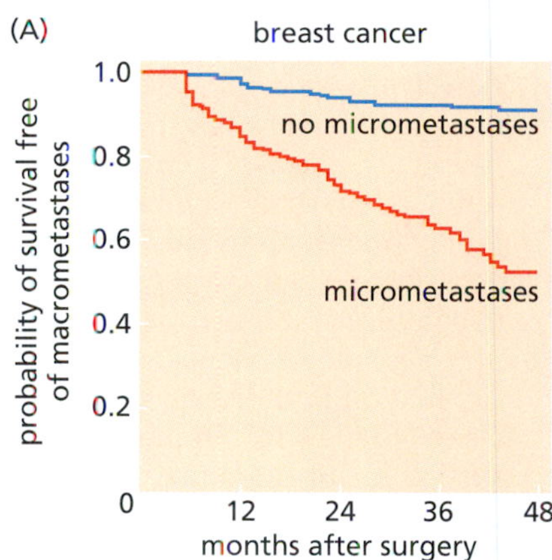

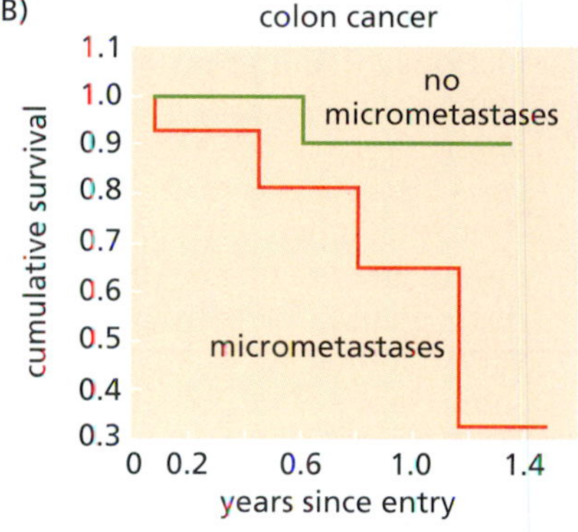

Figure 14.50 Micrometastases and clinical prognosis (A) This Kaplan–Meier plot presents the proportion of breast cancer patients who survive free of distant macroscopic metastases as a function of the elapsed time following initial surgical treatment of their primary tumors. The patients with micrometastases in the bone marrow at the time of surgery *(red line)* suffered from a far higher relapse rate than those who lacked such micrometastases *(blue line).* (B) Strikingly similar patterns characterize the probability of survival of a group of 54 colon cancer patients, all of whom exhibited no apparent metastases outside of the liver at the time of preoperative diagnosis; all were treated surgically to remove large metastases from the liver. Those whose bone marrow showed no cytokeratin-positive metastatic cells in the marrow had a far more favorable clinical course *(green curve)* than those whose marrow did indeed carry such cancer cells *(red curve).* In this case the fraction of patients surviving *(ordinate)* is plotted versus the time in years since they were initially treated and entered into a clinical study. (A, from S. Braun, K. Pantel, P. Muller et al., *N. Engl. J. Med.* 342:525–533, 2000; B, courtesy of R.A. Tollenaar.)

Sidebar 14.11 Are all micrometastases truly dormant? The dormancy of micrometastatic disease may often be an illusion. Thus, we can imagine that in many patients with "minimal residual disease" following removal of their primary tumors, micrometastatic clones occasionally acquire the ability to colonize and thereby create a new cascade of metastatic dissemination and disease relapse (see Figure 14.11). Importantly, such a major alteration of cell phenotype (and possibly an underlying change of genotype) has a low probability per cell generation of occurring in populations of nongrowing, dormant cells. Instead, extensive observations made over many years' time indicate that such changes happen spontaneously only in proliferating cell populations. Hence, in many patients with minimal residual disease, some clones of micrometastatic cancer cells must be passing through repeated growth-and-division cycles and occasionally spawning variants that have, through some random accident, acquired colonizing ability. (These micrometastases may remain clinically inapparent for many years simply because the rate of cell proliferation in these growths is balanced by an equal rate of cell attrition by apoptosis.)

The melanoma literature provides equally dramatic testimony of the long-term dangers posed by **occult**, dormant micrometastases (i.e., those that are hidden and apparently not growing). In one particularly well documented case, kidneys were prepared for organ transplantation from the cadaver of a patient who had undergone resection of a small melanoma 16 years earlier. The patient had been followed closely for 15 years after removal of this small primary tumor and had remained symptom-free. However, soon after transplantation, the two recipients of his kidneys developed aggressive melanomas that were directly traceable to this donor.

The mechanisms that prevent micrometastases from erupting into clinically threatening growths are poorly understood. In some instances, one can observe micrometastases growing as cuffs around small vessels; this suggests that they lack their own angiogenic capabilities but are nonetheless able to take limited advantage of host capillaries that happen to be nearby. In the great majority of micrometastases found in the marrow, the involved cells lack any indication of cell proliferation markers and thus are in a nongrowing, G_0-like state for extended periods of time (see Figure 14.12), perhaps for months and even years (see, however, Sidebar 14.11). (Because such cells are nongrowing, they may be especially resistant to chemotherapeutic treatment designed to eliminate the residual disease that persists following surgical removal of a primary tumor.)

Immune mechanisms may also contribute to suppressing the growth of micrometastases, thereby preventing metastatic disease relapse. This is suggested by the occasionally observed explosive growth of aggressive metastatic tumors in immunosuppressed organ transplant recipients. In addition, the phenomenon of tumor stem cells may help to explain the inability of the great majority of initially seeded micrometastases to generate macrometastases (Sidebar 34 ●). Beyond this, relatively little is known about the mechanisms that preclude most micrometastases from successfully colonizing the tissues in which they have landed.

14.13 Synopsis and prospects

Like all other biological phenotypes, those contributing to invasion and metastasis must be directed by the actions of genes. Several major issues have complicated the search for the genetic determinants of these aggressive phenotypes of cancer: Are these phenotypes programmed by a small number of pleiotropically acting, master control genes, or do the actions of multiple genes collaborate to create each of these phenotypes? Do these genes undergo mutation during tumor progression, or do they become involved in the late steps of tumor progression through epigenetic mechanisms that control their expression? Are there master control genes that are specialized to program the phenotypes of invasion and metastasis, or are these behaviors the by-products of the actions of familiar oncogenes and tumor suppressor genes?

While the genetic mechanisms involved in metastasis remain unclear, some progress is being made in solving another puzzle: Are the cells within a primary human tumor that undertake invasion and metastasis rare variants (among the larger population of tumor cells) that have, through some genetic or epigenetic accident, acquired the ability to execute these steps? Or are all the cancer cells within certain primary tumors equally capable of invading and metastasizing (albeit with extraordinarily low efficiency), while the great majority of the cancer cells in other tumors lack these abilities? Recent analyses of the gene expression patterns of various human tumors have lent some weight to the second mechanism. Still, this issue remains a bone of much contention.

One such study indicates that the tendency to metastasize is associated with a particular pattern of gene expression in some but not other human breast cancers. Moreover, this expression pattern is manifested by the bulk of the cells in

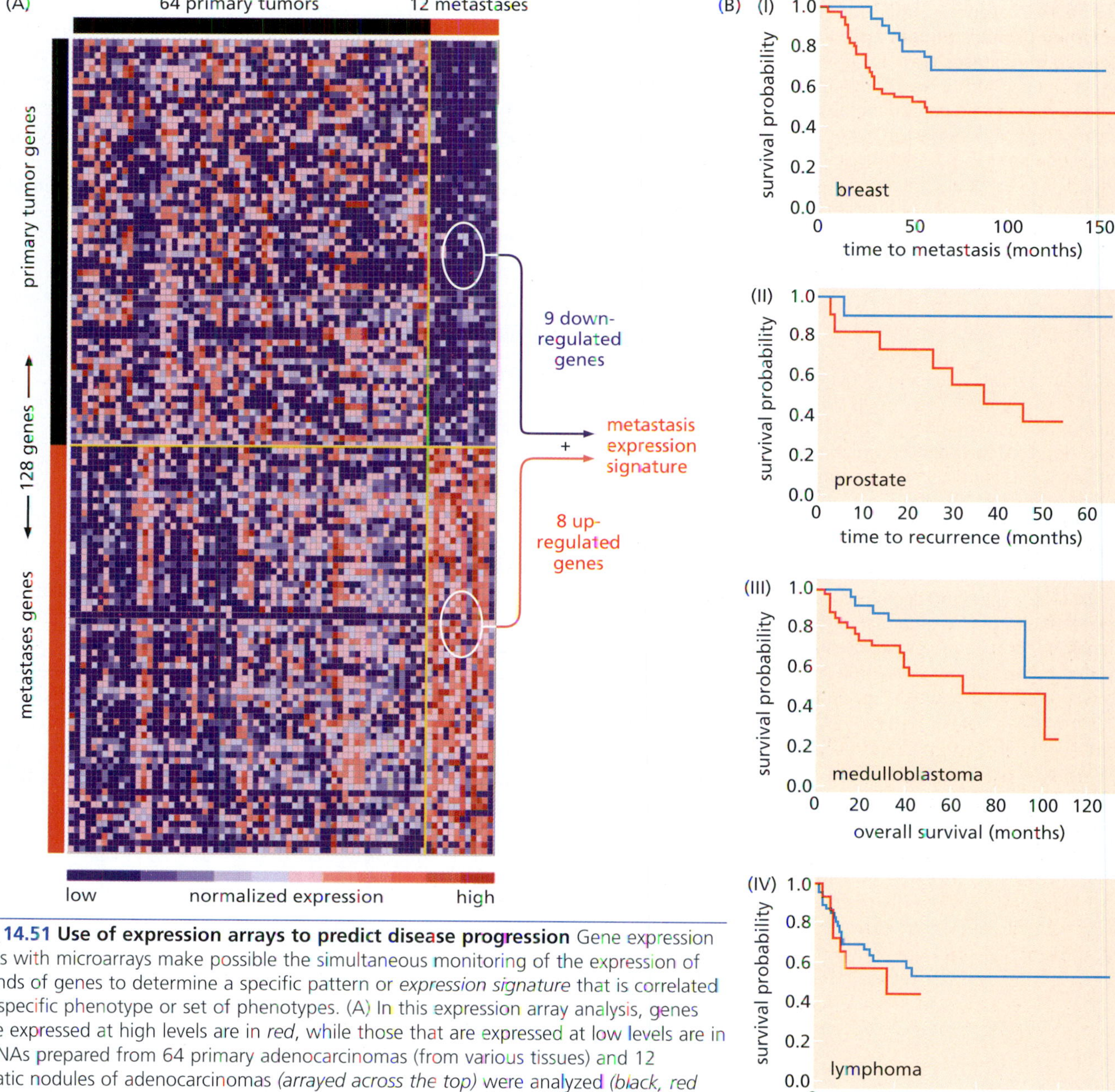

Figure 14.51 Use of expression arrays to predict disease progression Gene expression analyses with microarrays make possible the simultaneous monitoring of the expression of thousands of genes to determine a specific pattern or *expression signature* that is correlated with a specific phenotype or set of phenotypes. (A) In this expression array analysis, genes that are expressed at high levels are in *red*, while those that are expressed at low levels are in *blue*. RNAs prepared from 64 primary adenocarcinomas (from various tissues) and 12 metastatic nodules of adenocarcinomas *(arrayed across the top)* were analyzed *(black, red bars, respectively)*. Of the thousands of genes analyzed in an initial gene expression array *(not shown)*, the expression of 128 genes *(arrayed vertically here)* was found to be associated—because of over- or under-expression—with metastasis *(vertical red, black bars, respectively)*. Further distillation of the data yielded a set of 17 genes whose expression was as useful as that of the larger 128 gene set in distinguishing metastases from primary tumors. Importantly, this metastasis-specific expression signature was found to be exhibited by a small subset of the initially analyzed primary tumors, suggesting that it could be used to predict the metastatic tendencies of other groups of human tumors. (B) When researchers used the metastasis expression signature of panel A to analyze the gene expression patterns of other groups of primary tumors, they were able to separate the patients bearing adenocarcinomas of the breast (I) and prostate (II) as well as medulloblastomas (III) into two groups *(blue, red lines)* having markedly different times to clinical progression or relapse following initial surgery. However, the clinical progression of lymphomas (IV) was not predicted using the metastasis-specific expression signature, suggesting that a common set of genes is involved in mediating metastasis by solid tumors and other, unknown genes mediate malignant progression and metastasis by hematopoietic tumors. (From S. Ramaswamy, K.N. Ross, E.S. Lander and T.R. Golub, *Nat. Genet.* 33:49–54, 2002.)

each of a group of primary tumors, rather than by a small subset of cells within each tumor (Figure 14.51). This suggests that the proclivity to metastasize was developed during the course of the multi-step progression that culminated in

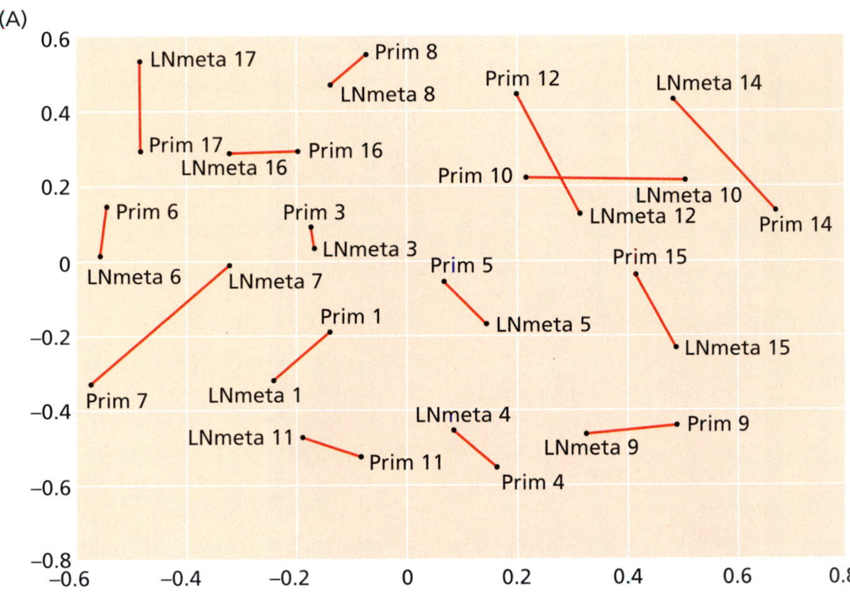

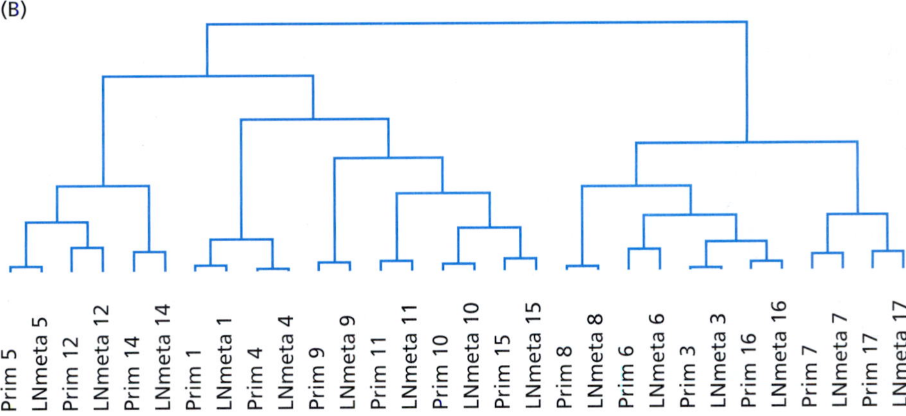

Figure 14.52 Genetic similarity between primary tumors and derived metastases Gene expression arrays can be used to classify different primary tumors and derived metastases according to their respective gene expression profiles (e.g., see Figure 14.51). If this is done, the degree of similarity between pairs of biopsies can be calculated using statistical methods. (A) As seen here, the biopsies of primary tumors and derived lymph nodes from a group of patients (each patient identified by a number) have been placed on a two-dimensional map, in which proximity indicates similarity in gene expression patterns. This reveals that the great majority of primary tumors (Prim) map closely to their derived lymph node metastases (LNmeta), indicating similarity in gene expression patterns. (B) Alternatively, gene expression patterns can be used to map the relatedness of primaries and derived metastases, where the degree of relationship is placed on a *dendrogram*—i.e., the most closely related tumor samples are placed near one another on the same or neighboring branches. Once again, the gene expression pattern of a metastasis is, almost always, most closely related to the parental primary tumor from which it arose. (Courtesy of B. Weigelt and L.J. van't Veer.)

primary tumor formation, not afterward by a small, specialized subpopulation of cancer cells within a primary tumor. (If the ability to metastasize were limited to only a small minority of cancer cells in the primary tumor, their gene expression pattern would not significantly influence the expression pattern of the tumor cell population as a whole, and this larger population would therefore not manifest a metastasis-prone gene expression signature.)

Another analysis indicates strong resemblance between the gene expression pattern of the bulk of the cells in a primary tumor and that present in its derived metastases; this held true for 15 pairs of primary tumors and associated metastases (Figure 14.52). If further validated, this means that cancer cells in the primary tumor do not need to undergo major changes in gene expression in order to metastasize and colonize distant sites.

The idea that the great bulk of cells in certain primary tumors are capable of metastasizing receives further support from the phenomenon of "cancer of unknown primary" (CUP), which constitutes as much as 5% of total cases in the oncology clinic. In these cases, metastatic growths represent the first clinical indication that a patient is harboring cancer. In about 30% of these patients, the primary tumors are never detected, while in the remaining 70%, these primaries are found only upon autopsy.

In the great majority of these CUP cases, the primary tumors are likely to have been very small when they began seeding metastatic cells. Given the profound inefficiency of metastasis (possibly as low as a 10^{-6} to 10^{-8} probability of success) and the small sizes of these primary tumor cell populations, it seems plausible that the great bulk of cells in these primaries were capable of seeding metastatic progeny. And, as before, it seems likely that the genetic evolution that yielded the tumorigenic cells in these primary tumors simultaneously created cell populations, all of whose members had the ability to disseminate their progeny to distant sites in the body.

While none of these studies represents an unassailable proof, they provide support for the argument that metastatic cells are drawn from the general population of cells in a primary tumor rather than from small, specialized, genetically unrepresentative subclones of cells. If this thinking is eventually validated, it would suggest that the reason why relatively few macroscopic metastases are formed is not because there are few cells within primary tumors that are intrinsically competent or qualified to metastasize. Instead, the majority of cells in certain primary tumors may be endowed with metastatic ability, and many of these cells may journey to distant anatomical sites. However, the actual number of macroscopic metastases that are eventually formed may be severely limited by metastatic inefficiency. Some speculations about the genetic evolution of metastatic competence further reinforce this notion (see Sidebar 35).

Contrary results derive, however, from experiments in mice in which primary tumors have been allowed to metastasize, for example, to the lungs. As described in this chapter, the cells isolated from these pulmonary metastases have been found, time and again, to have greater powers to metastasize to the lungs than do the bulk of cells in the primary tumor. Such observations argue strongly that these metastatic cells are at least marginally different from most cells in the primary tumor and that their genotypes or gene expression programs facilitate greater metastatic ability. This suggests, in turn, that these metastatic cells are indeed representative of only a minority subpopulation in the primary tumor and are likely to be variants that arise relatively late in tumor progression.

Whether one or the other model of metastasis is ultimately validated, it is clear that the identities of the genes that are specifically involved in programming metastasis have been elusive. In addition, it has been difficult to learn precisely how certain genes become functionally activated during the multi-step cascade of invasion and metastasis. Experimental resolution of these problems is confounded by complications at every level.

1. The first dimension of difficulty derives from the complexity of the invasive–metastatic process. Given the distinct steps that are involved in this process, are there a comparable number of genes involved? Or do metastasizing cancer cells activate and exploit latent, preexisting developmental programs (Section 14.3), such as the epithelial–mesenchymal transition (EMT), which simultaneously impart an ability to execute multiple steps, including breaching of the basement membrane, invasion into the stroma, intravasation, and extravasation (Sidebar 14.12)?
2. Experimental analyses are complicated by the inefficiencies of the metastatic process when compared with other steps in tumor pathogenesis. Even when cancer cells have purportedly acquired the genotype and phenotype enabling metastasis, they succeed in metastasizing with extraordinarily low efficiency. Such a weak connection between genotype and measurable phenotype derails most currently available experimental strategies.
3. Yet another complexity arises from the apparent collaboration of genetic and epigenetic factors in creating the metastatic trait. Recall, for example, that in certain experimental models of cancer, the epithelial–mesenchymal transition (EMT) is achieved when *ras*-transformed cells are exposed to TGF-β (Section 14.4). This transition, which may operate in many

Sidebar 14.12 Does the invasion–metastasis cascade need to be reconfigured? The invasion–metastasis cascade is often portrayed as a succession of six distinct steps (localized invasiveness, intravasation, translocation, extravasation, micrometastasis formation, and colonization; see Figure 14.4). This scheme might imply that cancer cells must acquire the ability to execute each of these steps through a comparable number of genetic or epigenetic alterations of their genomes. However, the pleiotropic powers of EMT-inducing transcription factors (TFs) and evidence gathered on the genetic evolution of micrometastases (see, for example, Sidebar 14.4) may soon reshape the depiction of the invasion–metastasis cascade. In many tumors, a three-step process may be a far more appropriate description. The *first step* would involve the activation of expression of one or several of these TFs and the concomitant acquisition of an ability to invade, intravasate, translocate to a distant site, extravasate, and form a micrometastasis. Thus, many primary tumors (e.g., 30% of human breast cancers) may be able to seed micrometastases even when these tumors are quite small, because the great majority of the cells in these tumors express one or more of the EMT-inducing TFs. The *second step* would involve acquisition of colonizing ability, which may be achieved only by rare micrometastases, often years after they were seeded in various tissue sites throughout the body. The *third step* would occur when such a micrometastasis, now grown into a macrometastasis, showers the body with metastatic cells that are capable of colonization, thereby generating clinical relapse with widely disseminated metastatic disease.

human carcinomas and enable their invasiveness, can be triggered by specific signals that genetically altered cells encounter in some tissue microenvironments but not in others. Hence, in these cases, invasion and subsequent metastasis can hardly be portrayed as genetically templated traits and, for this reason, are not readily studied by commonly used experimental techniques.

4. In many tumors, the genes and proteins that participate directly in programming invasion and metastasis may be expressed only at the invasive edges of primary tumors (see Figure 14.19), and the cancer cells in these invasive edges may represent only a tiny fraction of the neoplastic cell populations in these tumors. This enormously complicates experiments designed to reveal the biochemical and genetic bases of invasiveness and metastatic ability, which often rely on analyzing bulk populations of cancer cells prepared from large chunks of surgically resected tumors.
5. Carcinomas constitute the most common class of human cancers, and the neoplastic epithelial cells within these tumors may need to undergo an EMT in order to become invasive and metastatic. However, if invading carcinoma cells pass through a complete EMT and shed all epithelial traits, they become the proverbial "wolves in sheep's clothing," since most commonly used histological analyses are unable to distinguish these cells from the non-neoplastic mesenchymal cells of the tumor-associated stroma. (Indeed, this difficulty explains why many tumor pathologists deny the very existence of the EMT as the key process in the development of carcinoma invasiveness.)
6. Metastatic dormancy creates another experimental problem. In the case of breast cancers, for example, metastases may suddenly appear as long as 20 years after the initial primary tumor has been removed. Because of this long latency period and the sheer number of micrometastases carried by many patients, it has been difficult to learn how only a few of them suddenly acquire the ability to mushroom into macroscopic, life-threatening tumors.

These experimental difficulties have greatly retarded the progress of invasion–metastasis research, leaving many simple yet fundamental questions unanswered. For example, are there really genes that are specialized to impart an invasive or metastatic phenotype to cancer cells (see Sidebar 35)? And in the same vein, are there specialized metastasis suppressor genes (Section 14.11) that must be inactivated before a population of tumor cells can acquire invasive and/or metastatic ability? Or do the genes and proteins that affect metastasis operate as components of the regulatory circuits that we have repeatedly encountered throughout this book, namely, the circuits governed by the products of oncogenes and tumor suppressor genes?

The tissue tropisms of metastasizing cancer cells—their tendencies to colonize some but not other organs—represent another class of unsolved problems. Some insights have been gained from the substantial advances in understanding the detailed mechanisms of osteotropic metastasis, as described in Section 14.10. However, this class of metastases represents a rare exception, in that, in general, we know almost nothing about the functionally important interactions of disseminated cancer cells with the other tissues in which they settle and colonize. This is beginning to change (Sidebar 14.13).

The existence of micrometastases represents a major challenge for clinicians who would like to prevent disease relapse years after the primary tumor has been eliminated. Micrometastases of less than 0.2 mm diameter may carry several hundred to several thousand cells, and their detection in an organism carrying approximately 5×10^{13} cells represents a daunting undertaking. Without eradication, these micrometastases represent an ongoing threat, since some of them may erupt at an unpredictable future time into a lethal growth.

Sidebar 14.13 Some expressed genes within tumor cells facilitate specific types of metastasis One strategy for discovering the genes and proteins responsible for specific organ tropisms involves the isolation of tumor cells that show preference for colonizing one target organ but not another. By retrieving resulting metastatic cells from that organ, propagating them *in vitro*, and re-injecting them into host mice, followed, once again, by isolating metastatic cells from that organ, it is possible to select clones of cancer cells that stably express a highly specific tropism for that organ.

Alternatively, single-cell clones (i.e., clonal cell populations each descended from an isolated cell) can be prepared from a heterogeneous population of cells present in a human cancer cell line. The gene expression profile of each can then be analyzed, and independently, its tendency to form metastases in one or another target organ can be determined. This can lead to the identification of certain genes whose expression in a cancer cell is correlated with the metastatic tropism of that cell and may even contribute causally to this behavior (Figure 14.53). Indeed, ectopic expression of a group of such genes in otherwise poorly metastatic clonal cell populations can induce these cells to exhibit potent osteotropic metastasis. Such experiments also indicate that within a heterogeneous tumor cell population, various preexisting gene expression patterns can influence individual cells to exhibit a variety of metastatic behaviors.

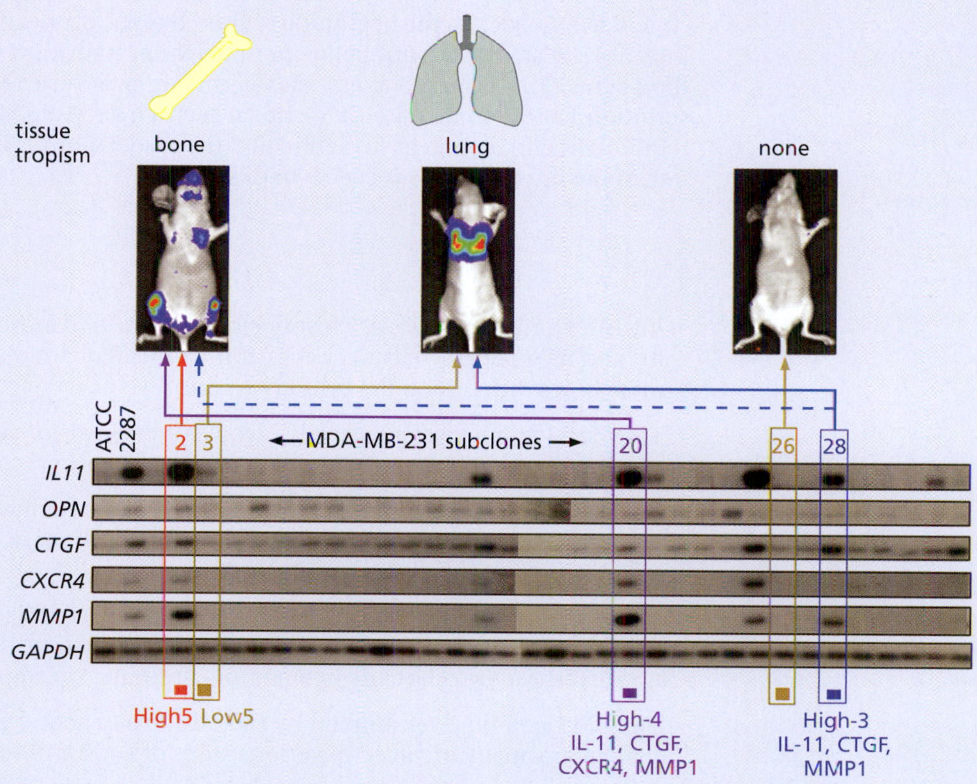

Figure 14.53 Gene expression patterns and metastatic tropism Thirty-three single cells were picked from a large population of human MDA-MB-231 cells (see Figure 14.49), and each was expanded into a clonal population in culture. The mRNA expression pattern of each subclone was analyzed *(columns, arrayed left to right)* using probes for the mRNAs of a series of five genes—*IL11* (interleukin-11), *OPN* (osteopontin), *CTGF* (connective tissue growth factor), *CXCR4* (chemokine receptor 4), and *MMP1* (matrix metalloproteinase-1)—and, a sixth, as loading control, *GAPDH* (glyceraldehyde-3-phosphate dehydrogenase) mRNA. In addition, the expression patterns of the original tumor cell population (ATCC, *left column*) and a subcloned cancer cell population termed 2287 (which was selected for its ability to generate osteolytic metastases; *2nd column)* were analyzed. The five experimental genes were chosen because of their overexpression in osteotropic metastatic cells and their known biological properties in promoting osteolytic metastases. As seen here, clone 2 cells *(red box)*, when injected into the arterial circulation, showed a tendency to produce osteotropic metastases, as indicated by *in vivo* imaging of the tumors developed by these mice; these cells expressed high levels of all five experimental mRNAs. Clone 3 cells *(yellow box)*, in contrast, expressed low levels of all five mRNAs and preferentially formed lung metastases. And clone 26 genes *(yellow box)*, which expressed almost none of these mRNAs, formed no metastases at all. Moreover, when otherwise poorly metastatic cells were forced to express combinations of three of these genes, they acquired the ability to form bone metastases efficiently *(not shown)*, pointing to their causal role in forming these metastases. Metastases were visualized through the presence of a luciferase gene in the tumor cells, which causes these cells to release a bioluminescent signal. (From Y. Kang, P.M. Siegel, W. Shu et al., *Cancer Cell* 3:537–549, 2003.)

This issue leads directly to another: Can the therapies used to treat primary tumors also be used to treat their metastatic derivatives? Or are metastatic cells so different from their progenitors in the primary tumor that they require their own customized therapies? In fact, the expression array analyses indicating substantial similarity between the gene expression profiles of primary tumors and their metastatic offshoots (see Figure 14.52) provide some hope. Such similarities make it plausible that metastatic cells will often be found to respond to the same therapies that succeed in attacking and destroying the primary tumors from which they derive.

A deep understanding of the processes leading to the formation of metastases is surely critical for the future development of cancer cures. For the moment, however, the questions and unresolved issues listed above give us pause, because they show how little we understand about the details of the metastatic cascade and why, in the eyes of many, invasion and metastasis represent the remaining major challenges of basic cancer research.

To end, we go back to the beginning of this chapter: if, as experimental evidence increasingly shows, the epithelial–mesenchymal transition is a critical event in the acquisition of invasiveness, and if cancer cells resurrect embryonic transcription factors to acquire these traits, then Lewis Wolpert's statement might require revision, in that gastrulation and the associated EMT might well loom as one of the most *dangerous* events in our lives!

Key concepts

- Invasion and metastasis are responsible for 90% of cancer-associated mortality, and the majority of cancer cells at the time of death may often be found in metastases rather than the primary tumor.
- The invasion–metastasis cascade involves local invasion, intravasation, transport, extravasation, formation of micrometastases, and colonization.
- The sequence of steps in this cascade is completed only infrequently, resulting in metastatic inefficiency. The least efficient of these steps appears to be colonization.
- Many of these steps can be executed by carcinoma cells that activate a cell-biological program called the epithelial–mesenchymal transition (EMT), which is normally used by cells early in embryogenesis and during wound healing.
- The EMT can be programmed by pleiotropically acting transcription factors that are normally involved in various steps of early embryogenesis.
- Signals released by the stromal microenvironment of a cancer cell, operating together with genetic and epigenetic alterations of the cancer cell genome, are often responsible for inducing expression of the EMT-inducing transcription factors in the cancer cell and thus the EMT.
- The EMT involves loss of an epithelial cell gene expression program and acquisition of mesenchymal gene expression. The latter enables cells to acquire invasiveness and motility.
- Cell motility is regulated by a series of small G proteins of the Rho family that are activated by cytoplasmic signal-transducing pathways and control the assembly of the actin cytoskeleton.
- Cell invasiveness is controlled in large part by various matrix metalloproteinases (MMPs) that function to degrade components of the extracellular matrix. These enzymes are often manufactured by cells within the tumor-associated stroma.
- Metastatic cancer cells may travel via the lymph ducts to nodes. However, their spread via the blood circulation is responsible for the great majority of distant metastases.

- Many cancer cells that are carried through the circulation form microthrombi that lodge in the arterioles and capillaries of various tissues.
- The ability of cancer cells to extravasate may depend on many of the same activities that were used earlier to execute invasiveness and intravasation.
- While the great majority of earlier steps of the invasion–metastasis cascade are likely to be similar in various types of human tumors, the last step—colonization—is likely to depend on complex interactions between various types of metastasizing cells and the microenvironments of the host tissues in which they land.
- The details of colonization are well understood only in the context of osteotropic metastases, especially the osteolytic metastases initiated by breast cancer cells.
- In some cases, the metastatic tropisms of cancer cells can be explained by the organization of the circulation between the primary tumor site and the target site of metastasis. However, in many other cases, the reasons why cancer cells metastasize from primary tumors to certain target organs are poorly understood.
- The acquisition of invasive and metastatic powers does not appear to involve major changes in the genotype of cancer cells within the primary tumor.

Thought questions

1. What arguments can be mustered for or against the notion that invasion and metastasis are likely to be orchestrated by specific mutant alleles that are acquired by cancer cells late in tumor progression?
2. What explanations can be offered for the inefficiency of colonization by the cells within micrometastases?
3. What arguments suggest that the ability to metastasize is expressed either by the bulk of cancer cells in a primary tumor or only by a minority of cells that are specialized to do so?
4. What evidence suggests that genetic and phenotypic evolution of cancer cells can occur in sites within the body that are far removed from the primary tumor?
5. What specific types of physiologic support may be supplied by tissues that are frequently the sites of successful cancer cell metastases? In what way do these supports affect the ultimate success of the colonization procedure?
6. How might primary tumors exhibit metastatic powers as soon as they form?
7. Would the ability to prevent metastasis have demonstrable effects on the clinical course of some human tumors but not others?
8. What evidence supports the involvement of the EMT in human tumor pathogenesis, and what evidence argues against it?
9. How might the ability to accurately determine the prognosis of a diagnosed prostate or mammary tumor lead to dramatic changes in the practice of clinical oncology?

Additional reading

Akhurst RJ & Balmain A (1999) Genetic events and the role of TGF-β in epithelial tumor progression. *J. Pathol.* 187, 82–90.

Barrallo-Gimeno A & Nieto MA (2005) The Snail genes as inducers of cell movement and survival: implications in development and cancer. *Development* 132,3151–3161.

Birchmeier C, Birchmeier W, Gherardi E & Vande Woude GF (2003) Met, metastasis, motility and more. *Nat. Rev. Mol. Cell Biol.* 4, 915–925.

Blasi F & Carmeliet P (2002) uPAR: a versatile signalling orchestrator. *Nat. Rev. Mol. Cell Biol.* 3, 932–943.

Brabletz B, Jung A, Spaderna S et al., (2005) Migrating cancer stem cells–an integrated concept of malignant tumour progression. *Nat. Rev. Cancer* 5,744–749.

Braun S, Pantel K, Muller P et al. (2000) Cytokeratin-positive cells in the bone marrow and survival of patients with Stage I, II or III breast cancer. *N. Engl. J. Med.* 342, 525–533.

Brinkerhoff CD & Matrisian LM (2002) Matrix metalloproteinases: a tail of a frog that became a prince. *Nat. Rev. Mol. Cell Biol.* 3, 207–214.

Bray D (2002) Cell Movements. New York: Garland Science.

Burridge K & Wennerberg K (2004) Rho and Rac take center stage. *Cell* 116, 167–179.

Cavallaro U & Christofori G (2001) Cell adhesion in tumor invasion and metastasis: loss of the glue is not enough. *Biochim. Biophys. Acta/Revs. Cancer* 1552, 39–45.

Cavallaro U & Christofori G (2004) Cell adhesion and signaling by cadherins and Ig-CAMs in cancer. *Nat. Rev. Cancer* 4, 118–132.

Chambers AF, Groom AC & MacDonald IC (2002) Dissemination and growth of cancer cells in metastatic sites. *Nat. Rev. Cancer* 2, 563–572.

Comoglio PM & Trusolino L (2002) Invasive growth: from development to metastasis. *J. Clin. Invest.* 109, 857–862.

Condeelis JS (2001) Lamellipodia in invasion. *Semin. Cancer Biol.* 11, 119–128.

Conway EM & Carmeliet P (2003) Cardiovascular biology: signalling silenced. *Nature* 425, 139–141.

Coussens LM, Fingleton B & Matrisian LM (2002) Matrix metalloproteinase inhibitors and cancer: trials and tribulations. *Science* 295, 2387–2393.

Coussens L & Werb Z (2002) Inflammation and cancer. *Nature* 420, 860–867.

deCaestecker MP, Piek E & Roberts AB (2000) Role of transforming growth factor-beta signaling in cancer. *J. Natl. Cancer Inst.* 92, 1388–1402.

Derynck R, Akhurst RJ & Balmain A (2001) TGF-beta signaling in tumor suppression and cancer progression. *Nat. Genet.* 29, 117–129.

Egeblad M & Werb Z (2002) New functions for the matrix metalloproteinase in cancer progression. *Nat. Rev. Cancer* 2, 161–174.

Feldner J & Brandt BH (2002) Cancer cell motility—on the road from c-erbB-2 receptor steered signaling to actin reorganization. *Exp. Cell Res.* 272, 93–108.

Fidler IJ (2001) in Cancer: Principles and Practice of Oncology (V.T. DeVita, jr., S. Hellman, SA Rosenberg eds), pp 135–152. Philadelphia: Lippincott-Raven.

Fidler IJ (2001) Seed and soil revisited: contribution of the organ microenvironment to cancer metastasis. *Cancer Metastasis: Biol. Clin. Aspects* 10, 257–269.

Fidler IJ (2003) The pathogenesis of cancer metastasis: the "seed and soil" hypothesis revisited. *Nat. Rev. Cancer* 3, 453–458.

Hay ED (2005) The mesenchymal cell, its role in the embryo, and the remarkable signaling mechanisms that create it. *Dev. Dynamics* 3, 706–720.

Jaffe AB & Hall A (2002) Rho GTPases in transformation and metastasis. *Adv. Cancer Res.* 84, 57–80.

Jechlinger M, Grünert S & Beug H (2003) Mechanisms in epithelial plasticity and metastasis: insights from 3D cultures and expression profiling. *J. Mammary Gland Biol. Neoplasia* 7, 415–432.

Karkkainen MJ, Mäkinen T & Alitalo K (2002) Lymphatic endothelium: a new frontier of metastasis research. *Nat. Cell Biol.* 4, E2–E5.

Karpanen T & Alitalo K (2001) Lymphatic vessels as targets of tumor therapy? *J. Exp. Med.* 194, F37–F42.

Klein CA (2003) The systemic progression of human cancer: a focus on the individual disseminated cancer cell—the unit of selection. *Adv. Cancer Res.* 89, 35–67.

Linder S & Aepfelbacher M (2003) Podosomes: adhesion hot-spots of invasive cells. *Trends Cell Biol.* 13, 376–385.

MacDonald IC, Groom AC & Chambers AF (2002) Cancer spread and micrometastasis development: quantitative approaches for in vivo models. *Bioessays* 24, 885–893.

Mareel M & Leroy A (2003) Clinical, cellular, and molecular aspects of cancer invasion. *Physiol. Rev.* 83, 337–376.

Mundy GR (2002) Metastasis to bone: causes, consequences, and therapeutic opportunities. *Nat. Rev. Cancer* 2, 584–593.

Murphy PM (2001) Chemokines and the molecular basis of cancer metastasis. *N. Engl. J. Med.* 345, 833–835.

Nieto MA (2002) The Snail superfamily of zinc-finger transcription factors. *Nat. Rev. Mol. Cell Biol.* 3, 155–166.

Overall CM & Lopez-Otin C (2002) Strategies for MMP inhibition in cancer: innovations for the post-trial era. *Nat. Rev. Cancer* 2, 657–672.

Paget S (1889) The distribution of secondary growths in cancer of the breast (republication of 1889 *Lancet* article). *Cancer Metast. Rev.* 8, 98–101.

Pollard JW (2004) Tumour-educated macrophages promote tumour progression and metastasis. *Nat. Rev. Cancer* 4, 71–78.

Pollard TD & Borisy GG (2003) Cellular motility driven by assembly and disassembly of actin filaments. *Cell* 112, 453–465.

Ridley A (2000) Cancer: molecular switches in metastasis. *Nature* 406, 466–467.

Ridley AJ (2001) Rho family proteins: coordinating cell responses. *Trends Cell Biol.* 11, 471–477.

Ridley AJ (2001) Rho GTPases and cell migration. *J. Cell Sci.* 114, 2713–2722.

Ridley AJ, Schwartz MA, Burridge K et al. (2003) Cell migration: integrating signals from front to back. *Science* 302, 1704–1709.

Riethmüller G & Klein CA (2001) Early cancer cell dissemination and late metastatic relapse: clinical reflections and biological approaches to the dormancy problem in patients. *Semin. Cancer Biol.* 11, 307–311.

Rosário M & Birchmeier W (2003) How to make tubes: signaling by the Met receptor tyrosine kinase. *Trends Cell Biol.* 13, 328–335.

Rossi D & Zlotnik A (2000) The biology of chemokines and their receptors. *Annu. Rev. Immunol.* 18, 217–242.

Sahai E & Marshall CJ (2002) RHO-GTPases and cancer. *Nat. Rev. Cancer* 2, 133–142.

Savagner P (2001) Leaving the neighborhood: molecular mechanisms involved during epithelial-mesenchymal transition. *Bioessays* 23, 912–923.

Schmidt A & Hall A (2002) Guanine nucleotide exchange factors for Rho GTPases: turning on the switch. *Genes Dev.* 13, 1587–1609.

Shevde LA & Welch DR (2003) Metastasis suppressor pathways—an evolving paradigm. *Cancer Lett.* 198, 1–20.

Siegel PM & Massagué J (2003) Cytostatic and apoptotic actions of TGF-β in homeostasis and cancer. *Nat. Rev. Cancer* 3, 807–820.

Stacker SA, Achen MG, Jussila L et al. (2002) Lymphangiogenesis and cancer metastasis. *Nat. Rev. Cancer* 2, 573–583.

Steeg PS (2003) Metastasis suppressors alter the signal transduction of cancer cells. *Nat. Rev. Cancer* 3, 55–63.

Thiery JP (2002) Epithelial-mesenchymal transitions in tumour progression. *Nat. Rev. Cancer* 2, 442–454.

Thiery JP (2003) Epithelial-mesenchymal transitions in development and pathologies. *Curr. Opin. Cell Biol.* 15, 740–746.

Vernon AE & LaBonne C (2004) Tumor metastasis: a new twist on epithelial-mesenchymal transitions. *Curr. Biol.* 14, R719–R721.

Wakefield LM & Roberts AB (2002) TGF-β signaling: positive and negative effects on tumorigenesis. *Curr. Opin. Genet. Dev.* 12, 22–29.

Weigelt B, Peterse JL & van't Veer LJ (2005) Breast cancer metastasis: markers and models. *Nat. Rev. Cancer* 5, 591–602.

Weiss L (1990) Metastatic inefficiency. *Adv. Cancer Res.* 54, 159–211.

Yoshida BA, Sokoloff MM, Welch DR, & Rinker-Schaeffer CW (2000) Metastasis-suppressor genes: a review and perspective on an emerging field. *J. Natl. Cancer Inst.* 92, 1717–1730.

Chapter 15

Crowd Control: Tumor Immunology and Immunotherapy

> It is by no means inconceivable that small accumulations of tumour cells may develop and, because of their possession of new antigenic potentialities, provoke an effective immunological reaction with regression of the tumour and no clinical hint of its existence.
>
> Macfarlane Burnet, immunologist, 1957

Throughout this text, we have studied various defenses that the body erects against the appearance of cancerous growths. Many of these defenses are inherent in cells, more specifically in their hard-wired regulatory circuitry. The most obvious of these are the controls imposed on cells by the apoptotic machinery, which is poised to trigger the death of cells that are misbehaving or suffering certain types of damage or physiologic stress. The pRb circuit and the DNA repair apparatus are similarly configured to frustrate the designs of incipient cancer cells.

The organization of tissues also places constraints on how incipient cancer cells can proliferate. For example, normal epithelial cells that lose their tethering to the basement membrane activate the form of apoptosis that is called anoikis. This mechanism limits the ability of epithelial cells to stray from their normal locations within tissues and grow in ectopic (i.e., abnormal) sites. At the same time, the special status afforded to stem cells and their genomes (Section 12.3) also reduces the probability of mutant cancer cells' gaining a foothold within a tissue.

Beyond these cell- and tissue-specific mechanisms, mammals may have another line of defense—the immune system. The immune system is highly effective in detecting and eliminating foreign infectious agents, including viruses, bacteria, and fungi, from our tissues. One of the major questions in cancer research over the last half century has been whether the immune system can also recognize cancer cells as foreigners and proceed to kill them.

Actually, evidence is rapidly accumulating that the immune system does indeed contribute to the body's multilayered defenses against tumors. The difficulties associated with establishing this type of anti-cancer defense are apparent from the outset: the immune system is organized to recognize and eliminate foreign agents from the body while leaving the body's own tissues unmolested. Cancer cells, however, are native to the body and are, in many respects, indistinguishable from the body's normal cells. How can cancer cells be recognized by the immune system as being different and, therefore, appropriate targets of immune-mediated killing? We will wrestle with this problem and its ramifications repeatedly throughout this chapter.

The field of tumor immunology, more than any other area of cancer research, remains in great flux, with basic concepts still a matter of great debate. Consequently, in this chapter, you will find many observations and conclusions to be much more tentatively stated than elsewhere in this book and subject, no doubt, to future revision. Still, this is an area of cancer biology that is well worth our time and study, since it holds great promise for new insights into cancer pathogenesis and new ways of treating human tumors.

Research conducted on mammals over the past three decades has revealed an immune system of great complexity and subtlety. Before we enter into discussions of its anti-tumor functions, we need to take an excursion into the workings of the general immune system. An understanding of its mechanisms of action, at least in outline, is a prerequisite for engaging the three major questions that will occupy us in this chapter. First, what specific molecular and cellular mechanisms enable the immune system to recognize and attack incipient cancer cells? Second, do these immune mechanisms represent effective defenses that prevent the appearance of tumors? Third, how can the immune system be mobilized by oncologists to attack tumors once they have formed? (An introduction to immunology will occupy our attention in Sections 15.1 through 15.6; an overview will be provided in Figure 15.14.)

15.1 The immune system functions in complex ways to destroy foreign invaders and abnormal cells in the body's tissues

The mammalian immune system launches several types of attack against foreign infectious agents and the body's own cells that happen to be infected with such agents. It identifies its targets by recognizing specific molecular entities—**antigens**—that are made by these agents. Having done so, the immune system undertakes to **neutralize** or destroy the infectious particles (bacterial and fungal cells, virus particles), as well as infected cells displaying these antigens. To the extent that the immune system also functions to ward off cancer, one assumes that it exploits many of the same mechanisms that it uses to eliminate foreign infectious agents.

The most familiar of the immunological defense strategies involves the **humoral immune response**—the arm of the immune system that generates soluble **antibody** molecules capable of specifically recognizing and binding antigens (Figure 15.1). Thus, a virus particle or bacterium displaying antigens on its surface may rapidly become coated with antibody molecules, which may result in the neutralization of these pathogens (Figure 15.2). Similarly, an infected cell

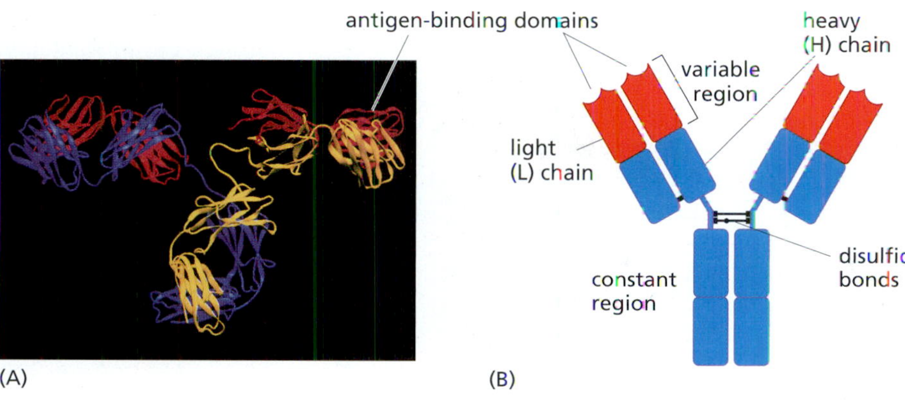

Figure 15.1 Structure of antibody molecules and their binding to antigens The most abundant antibody molecule in the plasma is the immunoglobulin γ (IgG) molecule. (A) X-ray crystallography of an IgG molecule reveals the symmetry that allows the two antigen-binding domains *(top left, top right)* to bind two antigen molecules simultaneously. (B) IgG molecules are divided into two functional regions. One region is designed to recognize and bind antigen molecules. Because the IgG molecules in plasma can recognize an essentially unlimited number of antigens, IgG molecules have a comparable diversity of structures in their antigen-binding portions, which are called their *variable* domains *(red)*, to recognize this diversity. The remainder of the IgG molecule is termed its *constant* region (*blue*) and is invariant among all IgG molecules of a given subclass, e.g., all IgG1 molecules. An IgG molecule as a whole is a heterotetramer composed of two light (L) chains and two heavy (H) chains. Two separate antigen-recognizing and binding pockets are displayed *(top left, top right)*, each composed of an H- and an L-chain N-terminal domain *(concave shapes)*. (C) The detailed structure of an antigen–antibody complex is seen here in this space-filling molecular model in which the antigen-binding domains of the heavy chain *(purple)* and light chain *(yellow)* are seen to contact the antigenic molecule, in this case the chicken egg-white lysozyme molecule *(light blue)*. Only parts of the variable regions of the heavy and light chains are shown here. (A glutamine residue, *red*, is indicated that is important for the hydrogen bonding of the antigen to the antibody molecule.) (From C.A. Janeway Jr. et al., Immunobiology, 6th ed. New York: Garland Science, 2005.)

may display on its surface the antigens made by the agents that have infected it and its surface may become coated with antibodies that recognize and bind these antigens. Once a mammalian cell or an infectious agent is coated by antibody molecules, it may be recognized, engulfed, and destroyed by phagocytic

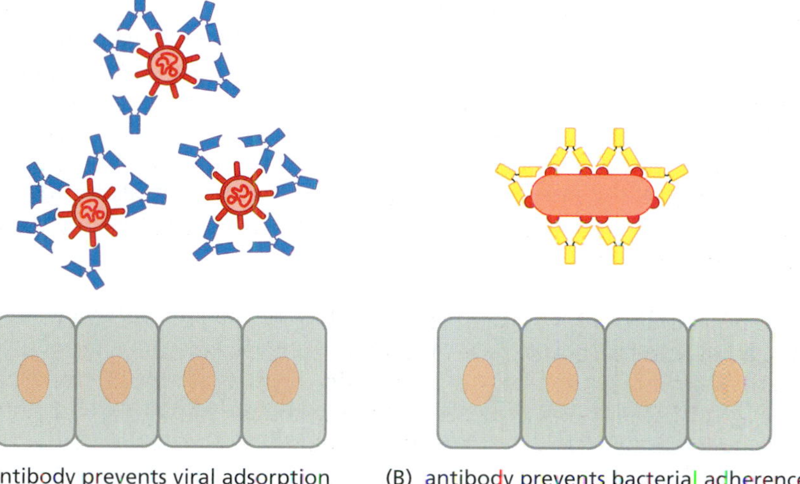

Figure 15.2 Neutralization by antibody molecules (A) Virus particles *(red)* can become coated by antibody molecules *(blue)* developed by the immune system of an infected host. This coating *neutralizes* (inactivates) the infectivity of the particles by blocking their adsorption to host cells. (B) Similarly, a bacterium displaying certain surface antigens *(red)* can also be prevented from adhering to host cells by bound antibody molecules.

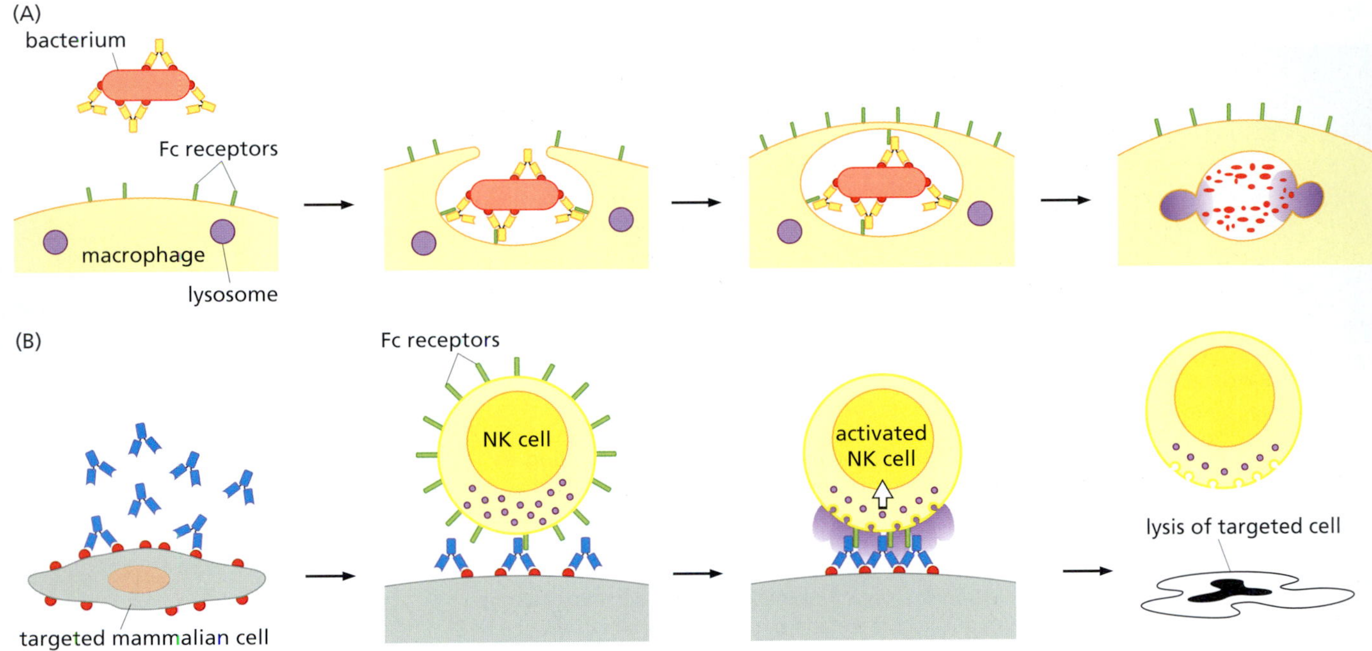

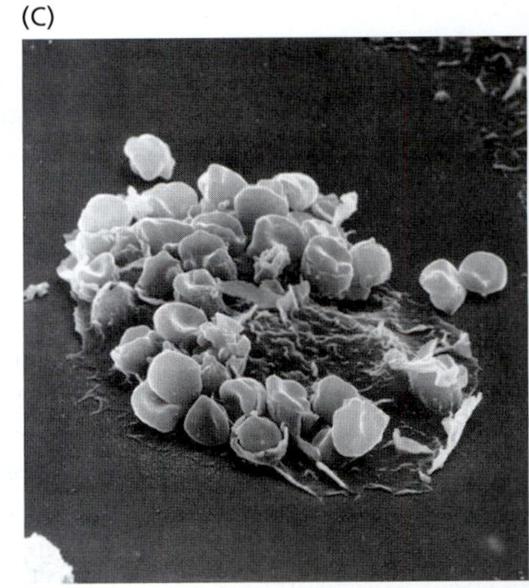

Figure 15.3 Coating of pathogens by antibody molecules and their elimination by effector cells of the immune system The coating of viruses, bacteria, and mammalian cells by antibody molecules is often the prelude to their being phagocytosed (engulfed) or destroyed by cytotoxic cells of the immune system. (A) The coating of a bacterium *(red)* by antibody molecules *(yellow)* may provoke a macrophage to use specialized receptors on its surface, termed Fc receptors *(green)*, to recognize and bind the constant regions of the antibody molecules (which are not involved in antigen recognition; see Figure 15.1). This often results in the phagocytosis of the antibody-coated bacterium and its eventual destruction in lysosomes within the cytoplasm of the macrophage. (B) A mammalian cell *(gray)* becomes coated by antibody molecules *(blue)* that recognize and bind antigens *(red)* on its surface. A type of lymphocyte termed a natural killer (NK) cell then uses its Fc cell surface receptors *(green)* to bind the constant regions of the coating antibody molecules. This binding results in activation of the NK cell, which proceeds to destroy the targeted cell, using cytotoxic granules *(purple dots)*, whose contents it introduces into the targeted cell, to do so. (C) Sheep red blood cells were treated with an antibody that recognizes an antigen displayed on their surface. As seen in this scanning electron micrograph, a large number of them have become adsorbed to a macrophage via the Fc receptors on the surface of the latter. (A and B, from C.A. Janeway Jr. et al., Immunobiology, 6th ed. New York: Garland Science, 2005; C, from J. Swanson, University of Michigan.)

cells, such as macrophages, or killed by cytotoxic cells, such as natural killer (NK) cells (Figure 15.3). Importantly, these immune cells do not, on their own, have the ability to recognize specific foreign antigens. Instead, the antibody molecules that have bound to antigens on the surfaces of viruses, bacteria, or mammalian cells alert these immune cells to the presence of targets that should be destroyed.

The other arm of the immune system involves the **cellular immune response**. This response is mounted when specialized cytotoxic cells are developed by the immune system that can, on their own, recognize and directly attack other cells displaying certain antigens on their surface. In this case, soluble antibodies are not required as intermediaries to recognize antigens displayed by targeted cells, since cytotoxic cells of the T-lymphocyte lineage (CTLs) have developed their own antigen-recognizing machinery in the form of T-cell receptors (TCRs), which they use to identify cells bearing particular antigens; such cells are then targeted for destruction by the cytotoxic T lymphocytes (Figure 15.4).

We can also depict the immune system in another dimension: many of the responses of the immune system to an infectious agent (e.g., a specific strain of virus) and its antigens depend on a previous encounter with this agent. The immune system has been "educated" through the initial encounter to recognize certain antigens displayed by this agent and to mount a vigorous counterattack against it in the event that it encounters this agent a second time; this represents the **adaptive immune response.** At the same time, other cellular components of the immune system are *naturally endowed* with the ability to recognize certain infectious agents or abnormal cells and thus do not require prior exposure and education; this inborn ability is termed the **innate immune response.** For example, the natural killer (NK) cells cited above have the ability to recognize specific cell-surface molecules displayed by aberrant cells, even without having encountered such cells previously.

15.2 The adaptive immune response leads to antibody production

Adaptive immune responses begin when infectious particles or abnormal cells are engulfed by specialized phagocytic cells of the immune system, notably macrophages and **dendritic cells (DCs)** (Figure 15.5). Having ingested these objects or fragments thereof, the phagocytic cells are then charged with the task of presenting the ingested contents to other cellular components of the immune system. More specifically, these cells must inform the immune system of the set of antigens that were associated with the particles that they have ingested. This presentation of ingested antigens by phagocytic cells takes place in the lymph nodes, to which these cells migrate following uptake of antigen. (As discussed in detail below, the antigens are presented in the lymph nodes to various types of T cells.)

In order to educate the immune system, these **antigen-presenting cells (APCs)** first digest the particles that they have phagocytosed (i.e., ingested outright) or endocytosed (i.e., bound via cell surface receptors and then internalized). This digestion, which is carried out in specialized cytoplasmic vesicles, snips internalized proteins into small oligopeptides that are 18–22 amino acid residues long. The oligopeptides are then loaded onto the **major histocompatibility complex** (MHC) class II molecules as the latter are making their way to the surface of APCs

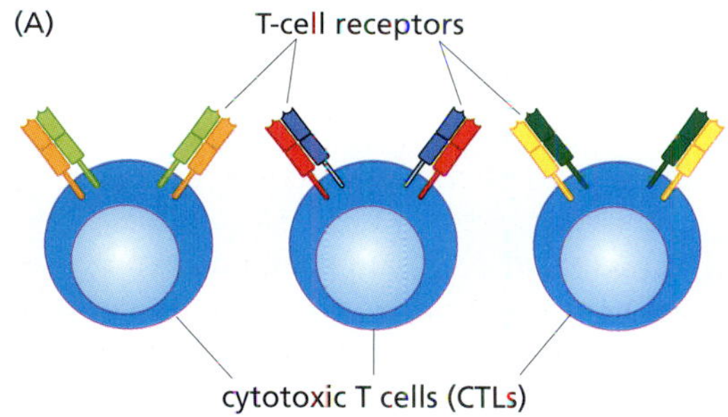

Figure 15.4 Cytotoxic T lymphocytes The cellular arm of the immune response results in the formation of cytotoxic cells, such as cytotoxic T cells (T_C's, CTLs) that are able to recognize and kill other cells displaying certain antigens on their surface. (A) CTLs develop antibody-like molecules on their surface termed T-cell receptors (TCRs). A diverse array of TCRs are developed during the development of the immune system, paralleling the development of a diverse repertoire of soluble antibodies. Each CTL displays a particular antigen-recognizing TCR. (B) Seen here is a CTL *(upper right, 1st panel)* that has already used its TCR to recognize and bind to a target cell *(diagonally below it to the left).* The cytotoxic granules within this CTL *(red spot)* begin over a period of minutes to migrate toward the point of contact between the killer and its victim. By 40 minutes, the contents of these granules (such as granzymes; see Section 9.14) have been introduced into the target cell, which has already advanced into apoptosis and begun to disintegrate. (From C.A. Janeway Jr. et al., Immunobiology, 6th ed. New York: Garland Science, 2005.)

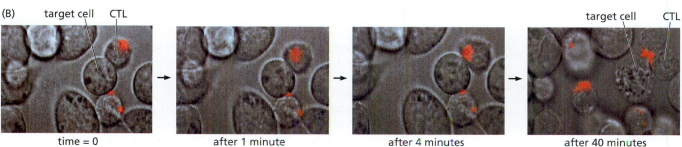

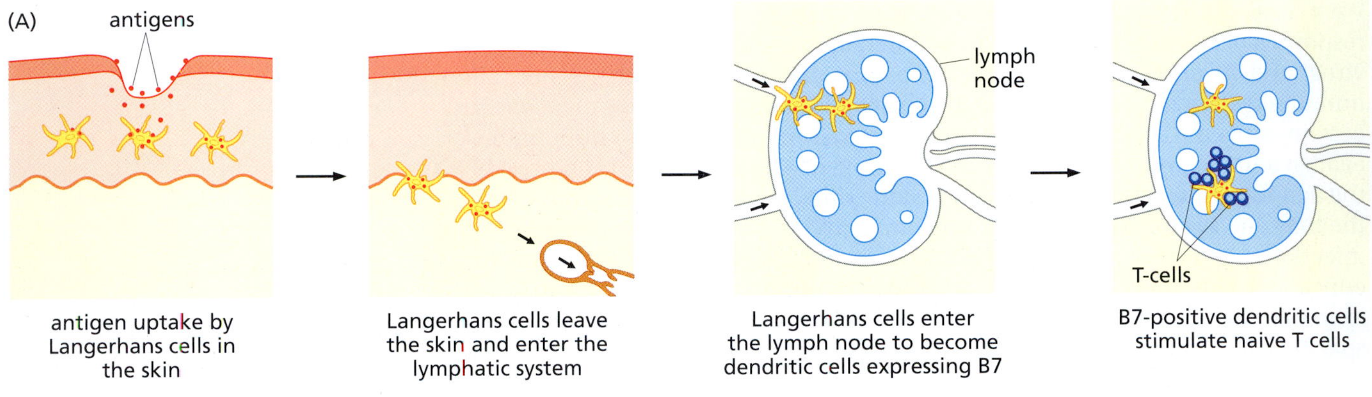

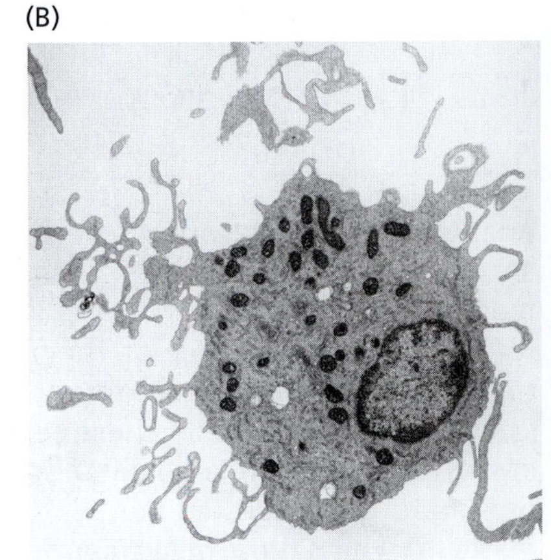

Figure 15.5 Antigen presentation by dendritic cells The immune system becomes aware of infectious agents and their antigens largely through the actions of antigen-presenting cells (APCs), notably dendritic cells. (A) Here we see a drawing of specialized phagocytic cells (i.e., Langerhans cells, *yellow)* residing in the skin, which take up antigens *(red dots)* by phagocytosis and then migrate to the lymph nodes *(light blue)*, where they mature into dendritic cells. In the lymph nodes, these cells confront T cells *(dark blue circles)*, to which they present antigens; this results in the functional activation of the T cells and the subsequent mounting of a specific immune response against cells and viruses that display these antigens. (B) The dendritic cells take their name from their multiple arms extending out from the cell body. (From C.A. Janeway Jr. et al., Immunobiology, 6th ed. New York: Garland Science, 2005.)

(Figure 15.6). More specifically, the oligopeptide fragments become attached to the specialized antigen-presenting domains (Figure 15.7A) of MHC class II molecules. (In humans, the MHC molecules are often termed HLA, or human leukocyte antigen, molecules, but we will use the more generic term, MHC, throughout this chapter to refer to both human and murine molecules of this type.)

The class II MHC molecules function much like a street hawker's hands used for displaying wares to passers-by. In this case, the wares are oligopeptide antigens captured by the APCs and the intended customers are other cells of the immune system, specifically a class of lymphocytes termed helper T cells (T_H cells), often called $CD4^+$ cells to reflect a specific cell surface antigen that they display

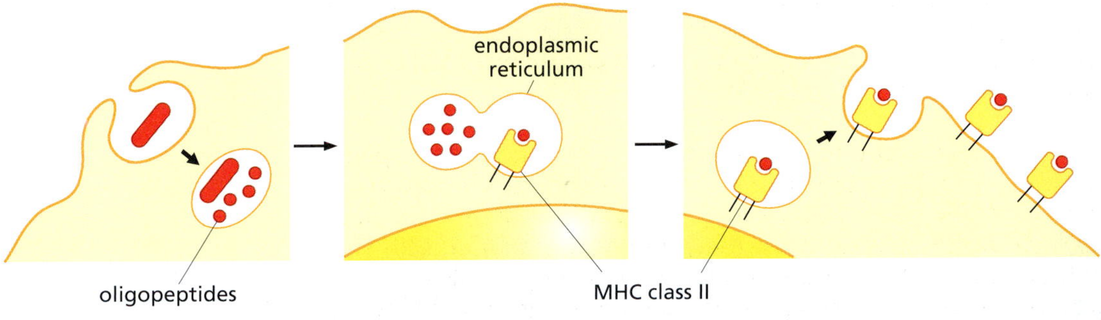

Figure 15.6 Antigen processing by antigen-presenting cells After phagocytes, notably dendritic cells and macrophages, have internalized potential antigenic particles *(red oblongs)*, these are fragmented into oligopeptides *(red dots)* by proteolysis. The resulting oligopeptides are then loaded onto MHC class II molecules in the endoplasmic reticulum, which then move to the cell surface, allowing the MHC class II molecules to display the oligopeptide fragments on their surface and present the oligopeptide fragments to T cells in the lymph nodes. (From C.A. Janeway Jr. et al., Immunobiology, 6th ed. New York: Garland Science, 2005.)

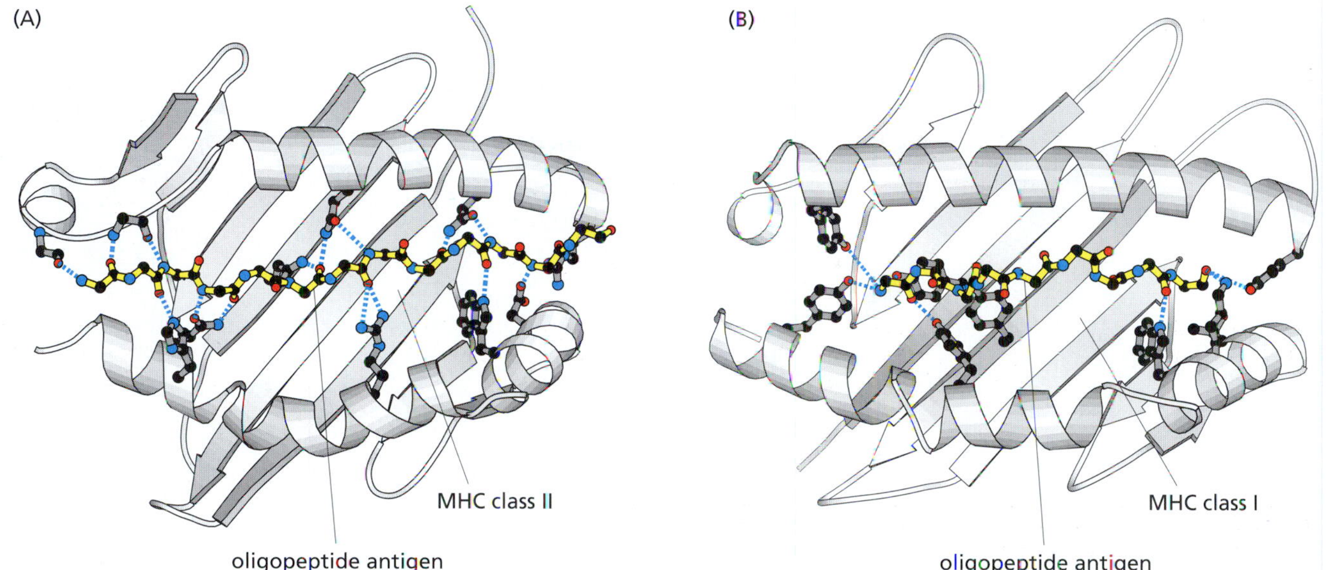

Figure 15.7 Antigen presentation by MHC molecules (A) The structure of the antigen-presenting groove of an MHC class II molecule is shown here as determined by X-ray crystallography. The oligopeptide antigen *(stick figure, in color)* that is bound via hydrogen bonds to the "palm" of the MHC molecule *(ribbon diagram)* is shown with its N-terminus to the left and C-terminus to the right. The oligopeptide antigen together with the nearby amino acid residues of the MHC molecule form the molecular structure that is recognized by other immune cells, which may, for example, use T-cell receptors to do so. (B) A very similar arrangement characterizes the structure of the antigen-presenting domain of MHC class I molecules. (From C.A. Janeway Jr. et al., Immunobiology, 6th ed. New York: Garland Science, 2005.)

(Figure 15.8). Because macrophages and dendritic cells are specialized to use their MHC class II molecules to present antigens scavenged from their environment, immunologists sometimes call them "professional" APCs, to distinguish them from other types of cells that are not specialized for this type of antigen presentation.

Note that it is the combined molecular structures formed by the class II ectodomains (the "hands") and their bound oligopeptide antigens (the "wares") that are presented to helper T (T_H) cells (see Figure 15.7). Antigen presentation to certain helper T cells provokes the latter to activate, in turn, the B cells that can manufacture immunoglobulin (antibody) molecules that specifically recognize and bind the particular antigen (see Figure 15.8). The subsequent maturation of these B cells yields a population of cells (called **plasma cells**) that actively secrete this particular antibody species into the circulation, that is, antibody molecules that are specialized to recognize and bind the particular antigen that originally triggered this series of responses. (Dendritic cells, once again functioning as APCs, can also activate cytotoxic T cells—not indicated in Figure 15.8.)

This system works well when confronting infectious agents such as virus particles, bacteria, and fungi in the extracellular spaces. Thus, these infectious agents can be internalized by the professional antigen-presenting cells, and the peptides deriving from the ingested agents can be presented again to the outside world. The antibody molecules that are eventually formed by B cells and their descendants as a result of this antigen presentation can recognize and bind the infectious particles and thereby neutralize them (see Figure 15.2). By the same token, we can imagine that cancer cells displaying certain distinctive antigenic proteins on their surfaces might also provoke an antibody response by the immune system and become coated by antibody molecules bound to these cell surface antigenic molecules.

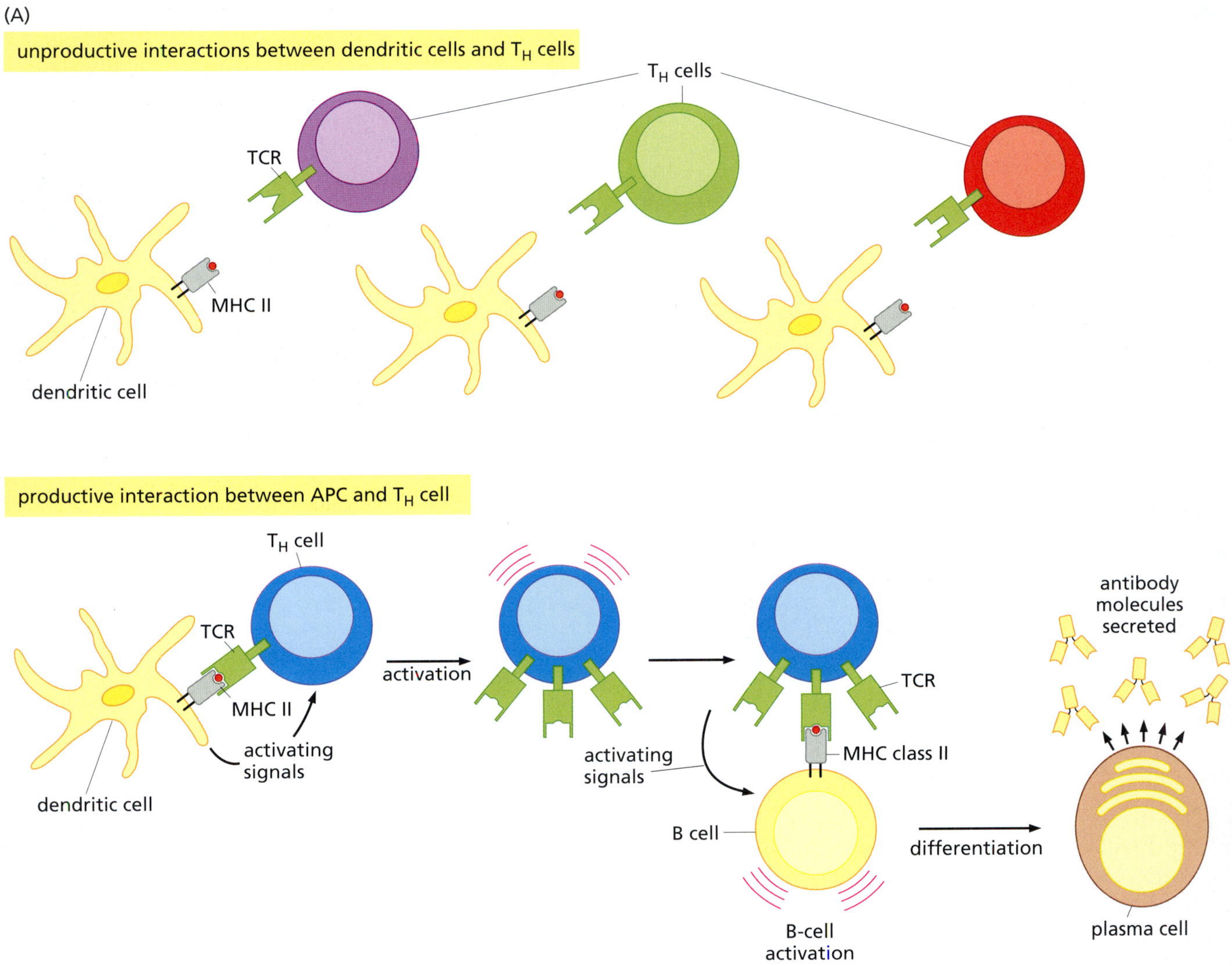

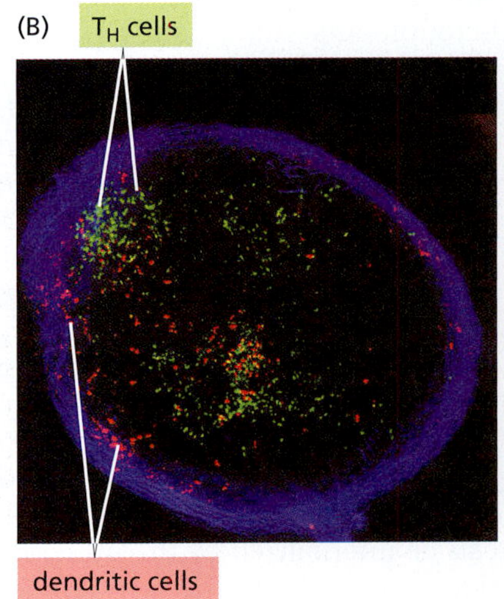

Figure 15.8 Immunocyte encounters within lymph nodes Dendritic cells interact directly with helper T cells and present antigen to them in the lymph nodes. (A) Dendritic cells engulf, process, and present antigenic oligopeptide fragments *(red dots;* see Figure 15.6) on their surface to T cells in the lymph nodes, using their MHC class II molecules *(gray)* to do so. Here, a dendritic cell meets a number of T cells *(above)*, known hereafter as helper T (T_H) cells. Each of them displays its own distinct T-cell receptor (TCR; *green)* on its surface. However, in the first three encounters, none of the T_H cells' T-cell receptors recognizes and binds the antigen being presented by the MHC II molecule of the dendritic cell. Nonetheless, on occasion, the dendritic cell succeeds in finding a T_H cell whose T-cell receptor does indeed recognize the oligopeptide antigen being presented by the dendritic cell's MHC class II molecules *(below).* This causes the T_H to become activated; the T_H cell leaves the dendritic cell and proceeds to search for B cells that also display on their surface the same antigen in the context of MHC II. When and if the T_H finds such a B cell *(light yellow, 2nd diagram from right)*, it activates the B cell, which proliferates and, having differentiated into a plasma cell *(light brown)*, begins to release antibody molecules that are capable of recognizing this oligopeptide antigen. (B) Multiphoton microscopy reveals the capsule of a mouse lymph node *(blue)* and a number of recently arrived, dye-labeled dendritic cells *(red dots)* as well as dye-labeled T cells *(green dots)* to which antigen will be presented by the dendritic cells. The two cell types are largely segregated from one another within the lymph node, and their mechanisms of trafficking and interaction within the lymph node remain poorly understood. The T_H cells have arrived in the lymph node from the venous circulation and have extravasated via diapedesis (Sidebar 14.3) in order to take up residence in the node. (B, from T.R. Mempel, S.E. Henrickson and U.H. Von Andrian, *Nature* 427:154–159, 2004.)

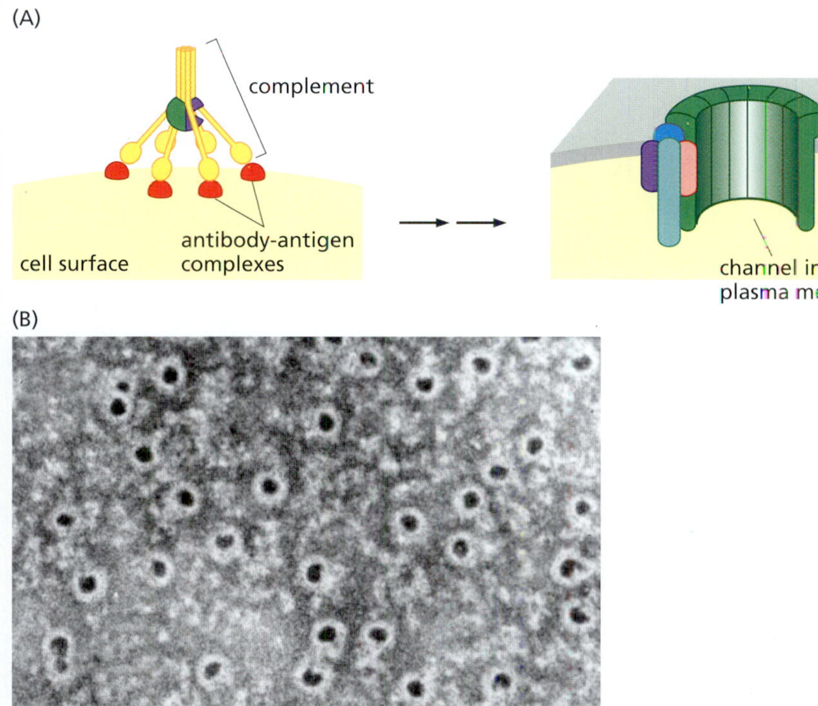

Figure 15.9 Complement-mediated killing (A) Antigen–antibody complexes *(red spheres)* formed by the binding of antibody molecules to cell surface antigens *(left)* can attract complement proteins present in the plasma *(yellow, green, purple)* and induce them to form complexes that lead, through a series of steps, to the formation by other complement proteins of channels in the plasma membrane of the cell *(right)* at a site adjacent to where the antigen–antibody complexes initially formed. (B) The resulting channels, seen here in this electron micrograph, destroy the integrity of the barrier functions of the plasma membrane and lead rapidly to cell death. (From C.A. Janeway Jr. et al., Immunobiology, 6th ed. New York: Garland Science, 2005.)

The antibodies coating a cell or infectious agent may elicit an alternative type of immune attack: a set of proteins in the plasma, termed **complement**, will recognize the constant regions of antibody molecules tethered to the surface of a cell (including bacterial, fungal, and mammalian cells), bind to these antibody molecules, and proceed to punch holes in the adjacent plasma membrane, thereby killing the cell (Figure 15.9).

This series of steps leading to adaptive humoral responses tells us something important about the molecular structure of the antigens that are **immunogenic**, that is, that elicit immune responses: they are not intact proteins, but instead are oligopeptide fragments derived from the cleavage of much larger proteins (see Figures 15.6 and 15.7). (The major exceptions to this generality are certain complex carbohydrate chains and linked side chains that may, under some circumstances, also be immunogenic.)

15.3 Another adaptive immune response leads to the formation of cytotoxic cells

The type of immunologic response described above fails to deal effectively with infectious agents that have entered into cells and are therefore shielded by the plasma membrane from scrutiny. Similarly, in the case of cancer cells, the humoral response system will fail to recognize aberrant cellular proteins that are hiding deep within these cells. In principle, such shielding should create a serious problem for the immune system, which needs to monitor what is going on *inside* cells in addition to its task of monitoring the contents of the extracellular spaces and the surfaces of cells.

The problem is solved by an antigen-presenting mechanism that echoes the one used by the professional antigen-presenting cells (APCs) described above. Actually, this other antigen-presenting mechanism is the more widespread of the two, since it is used by the great majority of cell types throughout the body.

It works like this (Figure 15.10): rather than being used for their normally designated functions, a portion of the proteins synthesized within cells (by some

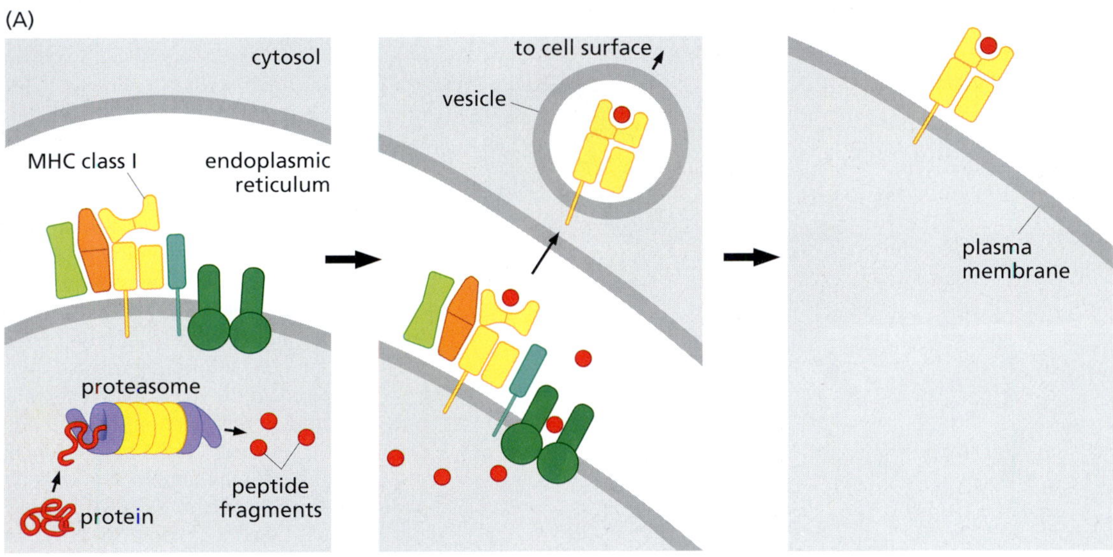

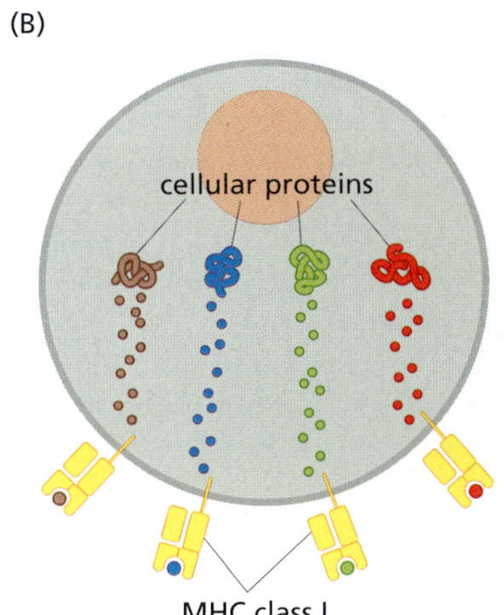

Figure 15.10 Display of intracellular antigens by MHC class I molecules Almost all cell types throughout the body, including cancer cells, routinely divert a portion of their recently synthesized proteins to the antigen-presenting machinery. (A) Some of the recently synthesized proteins in the cytosol are diverted to proteasomes *(purple, yellow)*, in which they are broken down into oligopeptides *(red dots)*; resulting oligopeptides are then introduced into the endoplasmic reticulum, where they may encounter MHC class I molecules *(yellow)* that bind them relatively tightly (see Figure 15.7B). This will cause the multi-protein complexes to move via membranous vesicles to the cell surface, where these protein complexes serve to display to the immune system fragments of the proteins that are being synthesized within the cell. The overall process of displaying these antigens is similar to that undertaken by MHC class II molecules (Figure 15.6); however, MHC class II antigen presentation is the speciality of "professional antigen-presenting cells", such as macrophages, dendritic cells, and B cells, while MHC class I presentation is undertaken routinely by almost all cell types in the body. (B) A broad spectrum of oligopeptide fragments deriving from a large number of cellular proteins (here represented as four distinct protein species) are displayed simultaneously by cells using their MHC class I proteins. (A, from C.A. Janeway Jr. et al., Immunobiology, 6th ed. New York: Garland Science, 2005.)

accounts, as much as one-third in certain cells) is routinely diverted to specialized proteasomes. There these proteins are cleaved into oligopeptides. These cleavage products, of 8 to 11 amino acid residues in length, are then attached to MHC molecules en route to the cell surface and displayed on the outside of cells by the other major class of antigen-presenting molecules—the MHC class I molecules (see Figure 15.7B).

Included among the intracellular peptides displayed by the MHC class I molecules are those synthesized normally by cells as well as those made by foreign infectious agents within the cell, such as viruses and bacteria. This external presentation of internal antigens occurs routinely and continuously, whether or not foreign proteins happen to be present within a cell.

The display by a cell of certain oligopeptide antigens on its surface (via its MHC class I molecules) may attract the attention of cytotoxic T cells (T_C's, also called cytotoxic T lymphocytes, CTLs, or $CD8^+$ cells), which proceed to kill this cell (see Figure 15.4). The origins of this killing can be traced back to the actions of helper T cells. Recall that some helper T cells are able to activate the humoral immune response by interacting with and stimulating antibody-producing B cells (see

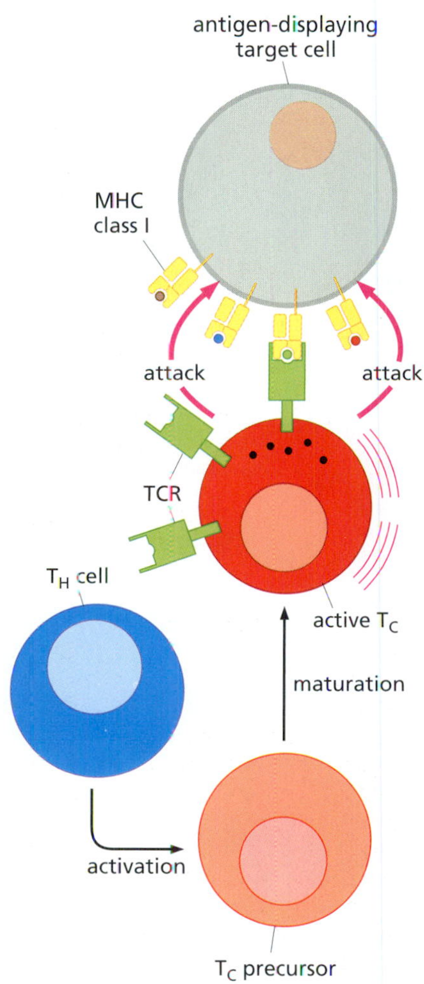

Figure 15.11 Activation of cytotoxic T cells by helper T cells In addition to inducing B cells to make antibody molecules (Figure 15.8), helper T cells (T_H) of a second subtype *(blue)* can activate the precursors of cytotoxic T cells *(light red, bottom)* to become active cytotoxic T cells (termed T_C's or CTLs, *red*) that can use their T-cell receptors (TCRs) to recognize and bind antigens presented on the surfaces of many cell types throughout the body by MHC class I molecules. This recognition results in attack on the antigen-displaying cell *(gray, top)*, as shown by the micrographs of Figure 15.4B. The T_C's often use cytotoxic granules *(black dots)* containing perforin and granzymes to kill targeted cells (Figure 15.12).

Figure 15.8). Now, we encounter a second, independent function of helper T cells: some of them can contribute to the activation of cytotoxic T cells, which are specialized to recognize and kill target cells displaying the particular oligopeptide antigen that initially provoked an immune response (Figure 15.11). This attack on antigen-displaying cells by cytotoxic lymphocytes represents the *cellular arm* of the adaptive immune response.

The capacity of helper T cells to facilitate development of both humoral and cellular immune responses reflects the ability of distinct subpopulations of T_H cells to produce and release the soluble immune factors known as cytokines: T_H's that promote humoral immunity (by stimulating B cells) produce interleukin-4 (IL-4), while T_H's that promote cell-mediated immunity (by stimulating cytotoxic T cells) secrete interferon-γ (IFN-γ).

Cytotoxic T cells can kill their cellular victims through two separate mechanisms. They can expose their intended victims to certain toxic proteins (Figure 15.12A and B). One of these, **perforin**, punches holes in the plasma membrane of a targeted cell. These holes then enable granzymes released by the cytotoxic cells to enter into the cytoplasm of the victim. As described earlier (Section 9.14), once in the cytoplasm of the targeted cell, granzymes cleave and thereby activate pro-apoptotic caspases.

The second killing mechanism, also discussed in Section 9.14, involves the Fas death receptor, which is displayed on many cell types throughout the body. Cytotoxic T cells can present the ligand of the Fas receptor, termed FasL, to their intended victims. FasL then activates the Fas death receptors on the surfaces of the targeted cells, thereby activating their extrinsic apoptotic pathway (Figure 15.12C).

Killing by cytotoxic T cells can play an important role in limiting the infectious spread of viruses. For example, a recently infected cell in which a virus is actively replicating will use its MHC class I molecules to display oligopeptide antigens derived from cleaved viral proteins. This antigen display will warn the immune system that abnormal proteins are being produced deep within the cell. If the immune system is functioning well, its cytotoxic T cells will recognize the viral oligopeptide antigens displayed by the cell's MHC class I molecules and kill this cell long before the virus has had a chance to multiply and release progeny virus particles.

This means that the immune system actually uses two arms of the adaptive immune response to limit viral infections: the cellular response is used to kill virus-infected cells, while the humoral response is used to neutralize virus particles that have been released into extracellular spaces, including the circulation, by coating these particles with antibody molecules (see Figure 15.2A). As we will see, the anti-viral responses are important means by which the immune system blocks the appearance of virus-induced human tumors.

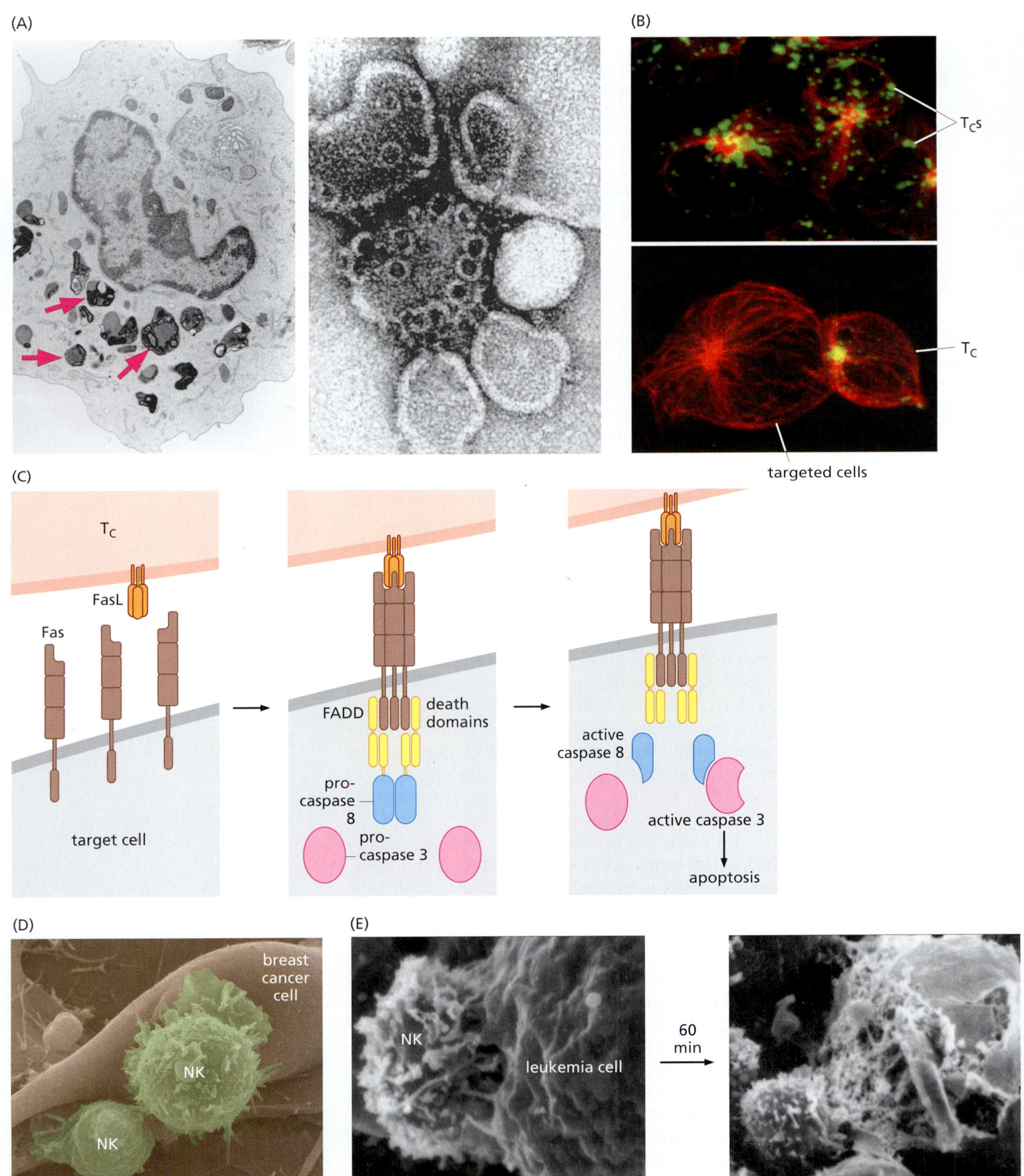

15.4 The innate immune response does not require prior sensitization

Ninety-nine percent of the animal species on the planet do not possess adaptive immune responses to protect them from attack by pathogens. These organisms rely on innate immunological responses for such protection. Importantly, this ancient, widespread innate immunity system has been conserved during the

Figure 15.12 Mechanisms of cell killing by cytotoxic lymphocytes (A) This electron micrograph of a cytotoxic T lymphocyte (T_C, CTL) reveals a series of lytic granules in its cytoplasm *(pink arrows, left panel)*. When contact is made with a targeted cell (which was initially recognized by the T-cell receptors borne by the T_C), these granules release perforin, which forms cylindrical channels in the plasma membrane of the target *(center cell, right panel)*; pro-apoptotic proteins such as granzymes (see Section 9.14), which are also carried in these granules, are then introduced through these channels into the cytoplasm of the targeted cell, where they initiate the apoptotic cascade by cleaving procaspases. (B) In the absence of a cellular target, the lytic granules *(green, yellow)*, which contain perforin and granzymes, are scattered throughout the cytoplasm of cytotoxic T lymphocytes (T_Cs; *upper panel)*. In the lower panel, a *synapse* has been formed with a targeted cell *(left)*, and the lytic granules have congregated at the synapse in preparation for killing the targeted cell. (C) An alternative mechanism of killing cells that have been targeted for destruction depends on the display of FasL *(orange)* by the T_C *(top, pink)*. FasL, which is a trimer, then engages the Fas receptor *(brown)* displayed by the targeted cell *(bottom cell, gray)* and triggers receptor trimerization and resulting activation of the extrinsic apoptotic cascade in the targeted cell via the sequential activation of caspases 8 and 3 (see also Figure 9.31). (D) NK cells are programmed to recognize and kill other cells, including cancer cells, that do not display normal levels of MHC class I molecules on their surface. This scanning electron micrograph (SEM) reveals that NK cells *(colorized green)*, one of which has spread a portion of its cytoplasm across the surface of a human ductal breast carcinoma cell in the initial stage of such an attack. (E) This SEM reveals the initial attack of an NK cell *(left panel, smaller cell)* on a leukemia cell. Sixty minutes later, the NK cell has caused extensive damage to the plasma membrane of the leukemia cell, which has fragmented and rolled up its plasma membrane in response to this attack *(right panel)*. (A, C, and E, from C.A. Janeway Jr. et al., Immunobiology, 6th ed. New York: Garland Science, 2005; B, from R.H. Clark, J.C. Stinchcombe, A. Day et al., *Nat. Immunol.* 4:1111–1120, 2003; D, courtesy of S.C. Watkins and R. Herberman.)

evolution of mammals and continues to play a critical role in various immunological responses.

The cellular components of the innate immune response are able to recognize and attack foreign particles and aberrant cells without having been "educated" through prior exposure to these agents. Thus, these **immunocytes** (cells of the immune system) "instinctively" recognize aberrant cells, such as cancer cells, in the body's tissues and target these cells for attack and destruction. Instead of recognizing specific antigens, the cells mediating innate immunity recognize characteristic molecular patterns that are present on the surfaces of infectious agents (or transformed cells) but are not displayed by normal cells.

A major component of the innate response is the natural killer (NK) cell. It is likely that many initial encounters of the immune system with cancer cells are made by NK cells. As we will discuss in greater detail later, the NK cells recognize configurations of cell surface proteins displayed by a wide variety of cancer cell types. Hence, NK cells are "pre-programmed" to recognize cancer cells and to eliminate them from the body's tissues. In addition to NK cells, yet other cellular components of the innate immune system, including macrophages and neutrophils, are likely to contribute to mounting innate immune responses against cancer cells.

After an NK cell has initiated the innate immune response by recognizing and attacking a target cell (Figure 15.12D and E), it sends out cytokine signals, notably interferon-γ (IFN-γ), in order to recruit yet other immune cells, including macrophages, to the site of attack. The actions of this second wave of immunocytes will often enable the immune system to mount more specific and ultimately more effective responses, in particular, adaptive humoral and cellular responses. For example, large numbers of cytotoxic T cells can be mobilized by the adaptive immune response to efficiently kill cancer cells.

15.5 The need to distinguish self from non-self results in immune tolerance

The immune system is finely tuned and highly specific. Most critically, it must be able to distinguish foreign proteins (e.g., those made by invading infectious agents) from those proteins that are normally made by the body's own cells. As a consequence, if the oligopeptides displayed by one of the normal cells in the body are similar or identical to those routinely encountered by the immune system, this cell will remain unmolested by the various arms of the immune system—one of the manifestations of immune **tolerance**. In fact, immune tolerance represents the major puzzle of current immunological research: How does the immune system learn to discriminate foreign proteins and peptides from the body's normal repertoire of proteins? Immunologists often refer to this behavior as the ability of the immune system to discriminate between "non-self" and "self."

A variety of mechanisms operating during the development of the immune system ensure that any T cells and B cells that happen to recognize self-antigens are eliminated; alternatively, if such cells escape elimination, their actions will be strongly suppressed. Failure to delete such **self-reactive** or *auto-reactive* lymphocytes from the large pool of lymphocytes in the body results in the survival of immune cells that may target the body's own normal tissues. Should such auto-reactive cells actually do so, this breakdown of tolerance may lead to autoimmune diseases, such as rheumatoid arthritis, ulcerative colitis, and lupus erythematosus, in which the immune system dispatches antibodies and cytotoxic cells to attack normal cells and tissues (Figure 15.13).

(A) pancreatic islet

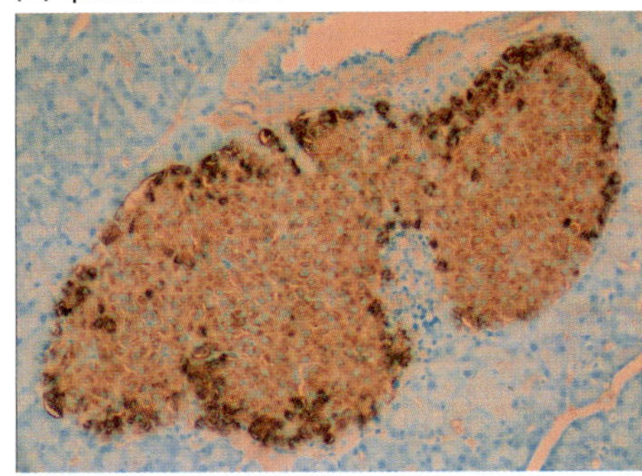

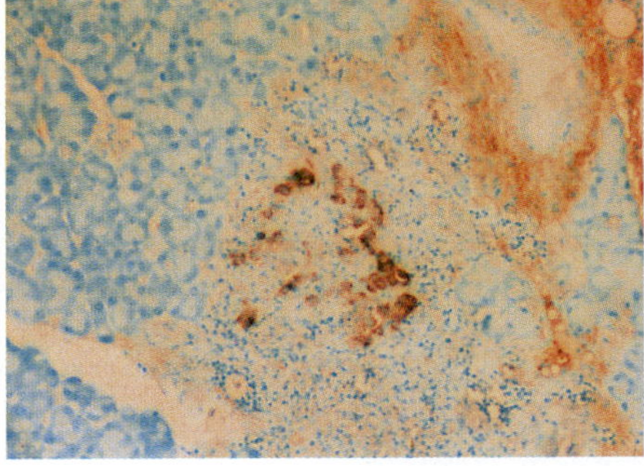

(B) kidney glomerulus

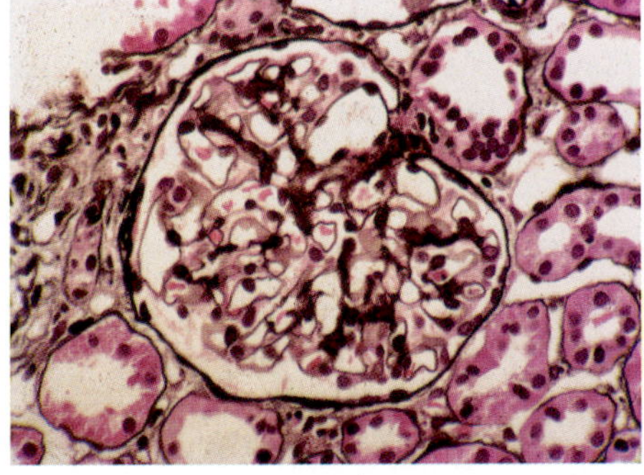

normal

autoimmune destruction

Figure 15.13 Destruction of normal tissues by autoimmune attack Extensive tissue damage can be wrought by an immune system that has been provoked to attack the body's normal tissues. In principle, the same immune mechanisms that are at work here can also operate to attack and destroy malignant tissues. (A) A normal pancreatic islet (i.e., islet of Langerhans) in a mouse *(left panel)* is composed largely of insulin-secreting β cells *(light brown)* with a small number of α cells at its periphery *(dark brown)*. The pancreas of a mouse suffering from diabetes resulting from autoimmune attack on β cells is seen to have lost almost all of them *(right panel)*. (B) A normal glomerulus in the kidney *(center of left panel)*, which contains a complex network of tubules, is responsible for the filtering of plasma and the production of urine. In the disease of systemic lupus erythematosus (SLE), an autoimmune attack on the basement membrane located beneath the epithelial cells of the glomerulus results in the accumulation of antibody protein and the invasion of a variety of inflammatory cells; together, these eventually destroy the architecture of the glomeruli *(right)* and thus kidney function. (A, from C.A. Janeway Jr. et al., Immunobiology, 6th ed. New York: Garland Science, 2005; B, courtesy of A.B. Fogo.)

Immune tolerance raises a simple and obvious point that will dominate the discussions that follow: How does the immune system, which is designed to be tolerant of the body's own cells, recognize and attack cancer cells, which are, to a great extent, very similar at the biochemical level to cells that are normally present in the body? And if it does undertake attacks against cancer cells, including those transformed by tumor viruses, how might these cells evade and thwart the attacks launched by various arms of the immune system (see Sidebar 36)?

15.6 Regulatory T cells are able to suppress major components of the adaptive immune response

Research beginning in the 1990s has described an entirely new class of T cells that have come to be called regulatory T cells (T_{reg} cells or simply T_{reg}s). Indirect evidence suggesting their existence came from the observation that in normal individuals, a significant proportion of cytotoxic T cells (CTLs) recognize normal tissue antigens presented by these individuals' MHC class I molecules—a situation that should lead directly to extensive immune attack on normal tissues and resulting autoimmune disease. However, such attacks do not occur, apparently because of suppression of these cells' actions by some unknown agents.

The discovery of T_{reg} cells seems to have largely solved this problem, since these cells are able to block the actions of the cytotoxic T cells that are scattered throughout our tissues. Indeed, in genetically altered mice lacking T_{reg} cells, lethal autoimmune disease develops; a comparably aggressive, ultimately fatal autoimmune disease has also been documented in humans who are unable to make T_{reg}s.

Like T helper (T_H) cells, the T_{reg}s display the CD4 antigen on their surface. However, the T_{reg}s are distinguished by their additional display of the CD25 surface antigen and their expression of a transcription factor, termed FOXP3, that programs their development. Because T_{reg}s express antigen-specific T-cell receptors (TCRs; see Figure 15.4), they can specifically block the actions of those cytotoxic T lymphocytes whose TCRs recognize the same antigens. In addition, when located in the lymph nodes, the T_{reg}s can prevent the activation of T_H cells by dendritic cells. It appears that the T_{reg}s must be in close proximity with the T_H and T_C cells that they suppress, and that the release of TGF-β and interleukin-10 (IL-10) by the T_{reg}s is often used to inhibit or kill these other types of T lymphocytes.

Research on T_{reg}s is still in its infancy. However, it is possible that their behavior holds the key to understanding the pathogenesis of a number of autoimmune diseases. At the same time, the actions of T_{reg}s may explain how many types of tumor cells can thrive in the presence of large numbers of CTLs that should, by all rights, be able to eliminate them—a topic pursued later in this chapter.

An overview of the various components of the immune system that we have covered until now is provided in Figure 15.14.

15.7 The immunosurveillance theory is born and then suffers major setbacks

As suggested by the quotation at the beginning of this chapter, the notion that the immune system is able to defend us against cancer is an old one. Burnet's 1957 speculation about the immune system's role in monitoring tissues for the presence of tumors, together with other speculations made by Lewis Thomas, represented the first instance that the notion of the **immunosurveillance** of cancer was clearly articulated.

At the time, infecting microorganisms, specifically, bacteria, viruses, and fungi, were known to be strongly immunogenic, in that they usually provoke an immune response that leads to their total eradication by various arms of the immune system. By analogy, it was plausible that the immune system continuously monitors its tissues for the presence of cancer cells. Having identified them—so this thinking went—the immune system would treat these cancer cells as foreign invaders and eliminate them long before they had a chance to proliferate and form life-threatening tumors.

Early attempts in the 1950s to test this model were not definitive. When tumors were removed from some mice and implanted in others, the tumors were rapidly destroyed in a way that gave clear indication of the actions of vigorous host immune responses. Soon it became clear, however, that this rejection had nothing to do with the *neoplastic* nature of the tumor cells. Instead, their elimination was a consequence of what came to be called **allograft rejection**. Thus, cells and tissues from one strain of mice are invariably recognized as being foreign when implanted in mice of a second strain. This is a consequence of the fact that the

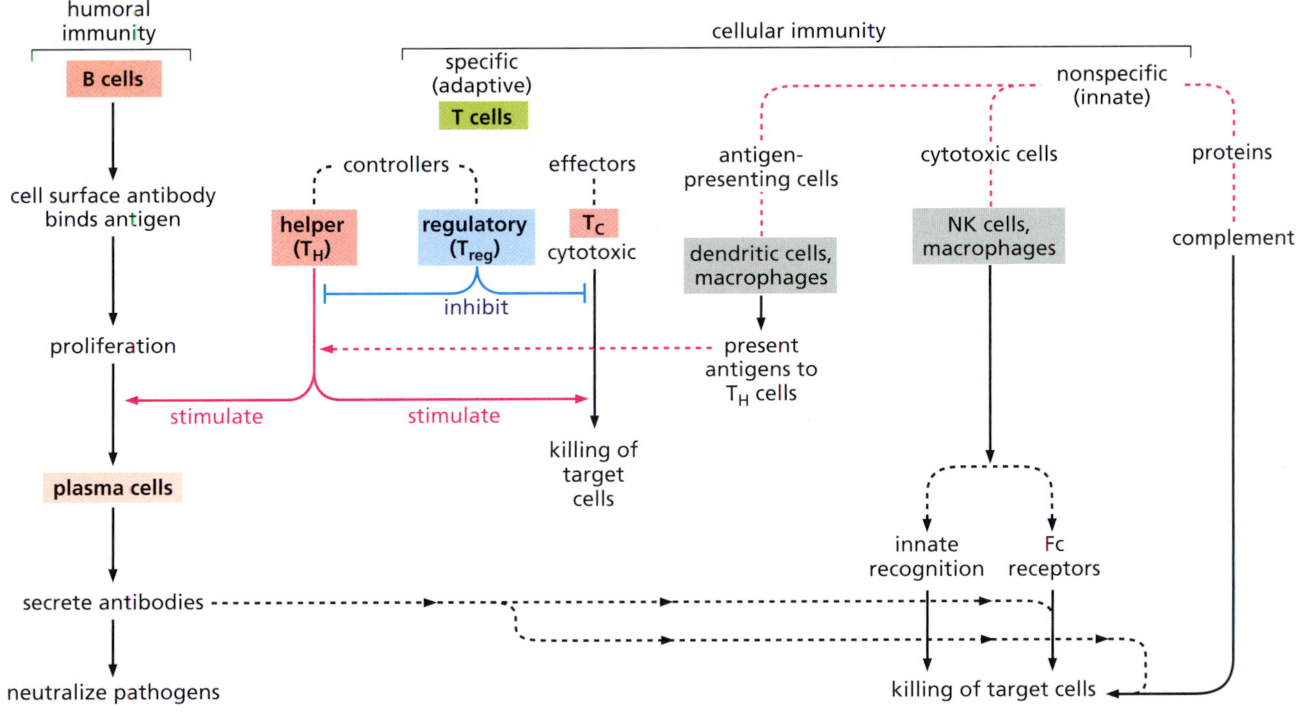

Figure 15.14 Overview of the humoral and cellular arms of the immune system The humoral immune system *(left)* is driven by the actions of B cells that develop millions of distinct antigen-specific antibody molecules through the rearrangement of antibody-encoding genes and the diversification of the sequences encoding the antigen-combining sites in the V regions of antibody molecules. This humoral response depends on activation by helper T (T_H) cells, which depends, in turn, on their prior activation in the lymph nodes by antigen-presenting cells, largely dendritic cells. The latter process proteins into oligopeptides that are recognized by the T-cell receptors of T_H cells, which proceed to activate B cells that have developed, by chance, antibodies that recognize the oligopeptide antigens.

T-cell receptors (TCRs) are used as well by cytotoxic T cells (T_C) cells (also termed CTLs), which rely on these receptors to recognize and kill target cells displaying cognate antigens. The activation of the T_C cells also depends on prior stimulation by T_H cells. A third class of T cells that also expresses antigen-specific T-cell receptors are the regulatory T cells, often called T_{reg}s. These play important roles in suppressing the actions of both T_C and T_H cells and thereby prevent inappropriate activation of immune responses that might otherwise lead to a breakdown of tolerance and resulting autoimmune disease.

These various manifestations of adaptive immunity are augmented by arms of the innate immune system *(right)*, specifically cell types that can aid in the elimination of pathogens and cancer cells without any prior "education" through previous exposure to these entities. Thus, natural killer (NK) cells are primed to kill many types of cancer cells because of the abnormal configuration of cell-surface molecules displayed by these cells; macrophages are also capable of recognizing and killing many cellular pathogens without any prior exposure to these agents. Although macrophages and NK cells cannot themselves recognize most cell-surface antigens, the coating of potential target cells by antibody molecules (produced by the adaptive immune response) will attract macrophages and NK cells, which will use their F_c receptors to bind to the constant (C) regions of antibody molecules and proceed to kill the antibody-coated cells. Similarly, the complex group of plasma proteins known as complement may also recognize antibody molecules bound to the surface of a cell and then kill this cell by inserting channels in its plasma membrane.

cells of various strains of mice display distinct, genetically templated major histocompatibility (MHC) molecules on their surfaces. (In this instance, however, it is not the bound oligopeptide antigens that evoke an immune response but the MHC molecules themselves, which vary slightly in structure from one strain of mouse to another.)

For example, engrafted cancer cells from BALB/c mice were recognized as being of foreign origin (and were therefore histo*incompatible*) when introduced into C57/BL6 mice, and vice versa (Figure 15.15). These graft rejections from dissimilar, **allogeneic** (i.e., genetically distinct) mouse strains were not observed when tumor cells of BALB/c origin were grafted into BALB/c hosts, that is, into **syngeneic** hosts that, by definition, shared an identical genetic background and identical histocompatibility antigens with the engrafted cells. [In fact, the term **histocompatibility** derives from the observation that tissue ("histo-") from mice of one inbred strain can be grafted and established in the bodies of other members of the same genetic strain and are in this sense "compatible."]

The observed rejections of allogeneic tumors represented a detour for the young field of tumor immunology, since they shed no light on how the immune system of a mouse or human host would recognize cancer cells that arise in its own tissues. Still, this early work did make one profoundly important point: in addition to eliminating microbes and various types of viruses, the immune system is capable of destroying mammalian cells that it recognizes as foreign or, quite possibly, as otherwise abnormal. As an additional corollary, these observations of immune function led to the conclusion that cancer cells could never be transmitted from one individual to another (but see Sidebar 37, and Figure 15.16).

An alternative strategy was then embraced for studying the immunosurveillance problem. If the immune system were indeed responsible for suppressing the appearance of tumors, animals with compromised immune systems should suffer increased rates of cancer. Such cancers, which originated within their own bodies—so-called **autochthonous** tumors—were, of course, of the same histocompatibility type as the remaining tissues in these animals. In these situations, the issue of histocompatibility (and -incompatibility) was rendered irrelevant.

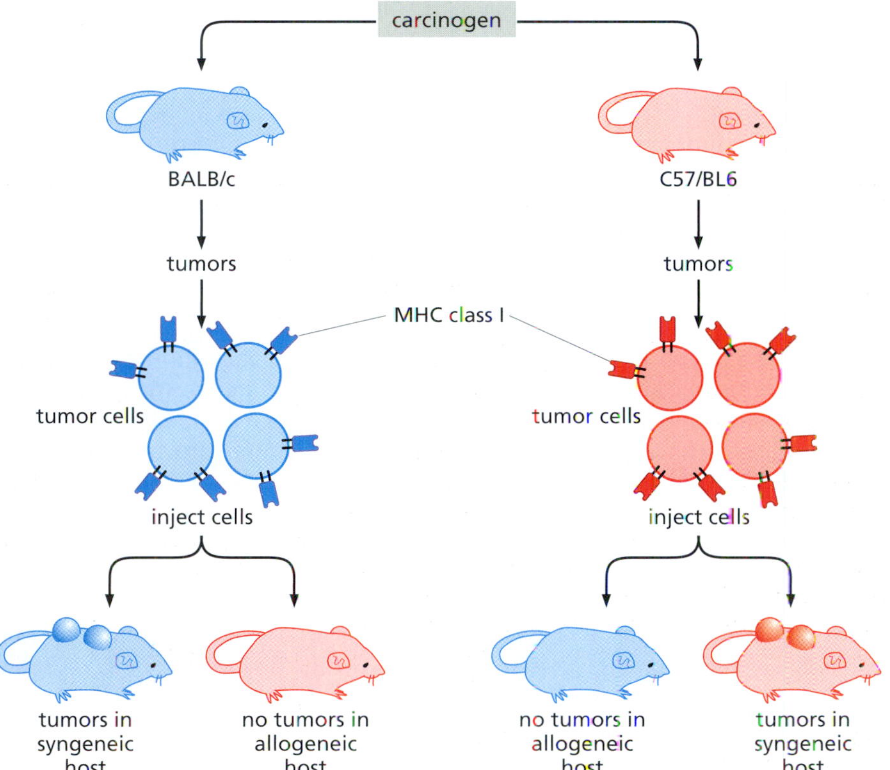

Figure 15.15 Syngeneic mice and MHC variability The use of inbred strains of mice has revealed that major determinants of the immunogenicity of the cells of these mice (and of mammals in general) are the MHC class I molecules. These molecules are highly polymorphic within a species, indicating that one individual (or one inbred strain of mice) almost always has a different set of MHC class I molecules from another *(red, blue cell surface molecules)*. Therefore, if a tumor arises within a BALB/c mouse, it is often transplantable into a syngeneic host, i.e., another BALB/c mouse, but not into an allogeneic host, such as a C57/BL6 mouse. The converse is true for tumors arising in C57/BL6 mice.

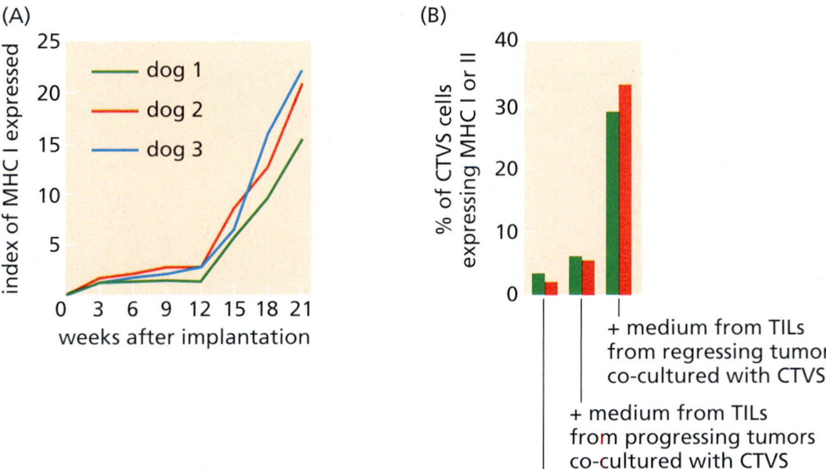

Figure 15.16 Regression of CTVS and re-expression of MHC antigens One exception to the rule of the non-transmissibility of cancer from one organism to another comes from canine transmissible venereal sarcoma; its cells are transferred from one dog to another via sexual intercourse. The transferred cells initially form a vigorously growing tumor, which is, however, rejected after several months (see also Sidebar 37). (A) While the canine transmissible venereal sarcoma (CTVS) cells initially express very low levels of MHC class I proteins, after 12 weeks of tumor growth these proteins begin to be re-expressed, as seen here in the tumors borne by three dogs. (B) This re-expression results, at least in part, from signals released by tumor-infiltrating lymphocytes (TILs). Fresh culture medium has little effect on the expression of MHC class I *(green)* or class II *(red)* proteins by CTVS cells *(left)*. Factors released by TILs isolated from progressing tumors (1st 12 weeks) also have little effect *(middle)*. However, factors released by TILs from regressing tumors (12–21 weeks) potently induce MHC protein expression *(right)* by cultured CTVS cells. Such re-expression appears to be responsible for tumor regression *in vivo*. (From Y.W. Hsiao, K.W. Liao, S.W. Hung, and R.M. Chu, *J. Immunol.* 172:1508–1514, 2004.)

In the late 1960s, immunocompromised mice of the Nude strain first became available to cancer researchers. These mice lack a functional thymus—the tissue in which the T lymphocytes of the immune system initially develop. Their lack of hair, another distinct phenotype of this strain, gave them their name (see Figure 3.13).

The research that followed in the early and mid-1970s revealed that these mice are no more susceptible to spontaneously arising or chemically induced autochthonous tumors than are their normal, wild-type littermates. So, the immunosurveillance theory suffered a major setback, having failed a major critical test of its validity. It lost credibility and retreated from the main arena of cancer research for two decades.

But this rejection was premature. Only years later did it become apparent that mice of the Nude strain, while lacking many of their normal T lymphocytes, retain other components of their immune system in an intact form. For example, some types of T cells may be able to develop outside the thymus, the normal site of maturation of these cells. In addition, a very important type of immune cell—the natural killer (NK) cell—is able to develop totally outside the thymus, and thus NK cells are present in large numbers in Nude mice.

In the 1980s, researchers began to accumulate evidence that NK cells are actually very important in recognizing and killing a variety of abnormal cells, including cancer cells. So in the end, the lessons taught by the low cancer rates of Nude mice were of limited value, since these mice did indeed continue to harbor functionally important components of the immune system. Still, Nude mice, as well as other types of immunocompromised mice, have proven to be of great value in cancer research (Sidebar 38).

Evidence also began to accumulate that certain chemically induced tumors in mice were antigenic and could be recognized and eliminated by the immune system. For example, in one set of experiments, cells from a 3-methylcholanthrene (3MC)–induced tumor were irradiated prior to injection into mouse hosts in order to prevent the proliferation of these cells in the hosts (Figure 15.17). (The chemically induced tumor had been induced in the same strain of mice as these hosts.) Subsequently, the mice received a second injection of live tumor cells originating from the same tumor or from a second 3MC-induced tumor; the cells originating from the same tumor did not grow, while the cells from the second tumor did grow and form a new tumor. This indicated that the two tumors were antigenically different and that the initial exposure to dead cancer cells had immunized the mice against live cells originating in the same tumor. Hence, tumor cells could have distinctive antigens, and under certain conditions, these antigens could provoke the immune system to attack and kill such cells.

15.8 Use of genetically altered mice leads to a resurrection of the immunosurveillance theory

In the mid-1990s, several lines of research gave new life to the long-discredited immunosurveillance theory. These experiments derived from the then recently gained ability to create genetically altered strains of mice at will. This technology (see Sidebar 7.10) was exploited to create mice whose genomes lacked one or more of the genes known to play critical roles in the functioning of the immune system.

One group of experiments used mice that were rendered incapable of expressing the receptor for interferon-γ (IFN-γ) through targeted inactivation of the responsible gene in their germ line. Like growth factors, IFN-γ is a diffusible protein factor that conveys signals from one cell to another and induces responses in cells by binding and activating its cognate cell surface receptor. Importantly, IFN-γ has not been found to be released by cells other than those of the immune system. Consequently, any changes observed following deletion of the IFN-γ receptor gene from the mouse genome could be attributed to defects associated with immune cells and their interactions with the remaining cells in the body. Strikingly, mice that lack the IFN-γ receptor in all of their cells were found to be 10 to 20 times more susceptible to tumor induction by the chemical carcinogen 3-methylcholanthrene.

In another set of experiments, tumor cells were forced to express a dominant-negative IFN-γ receptor, rendering them unresponsive to the IFN-γ released by various types of immunocytes. These cells were then injected into wild-type mice and found to be more tumorigenic than tumor cells carrying the corresponding wild-type receptor. This particular experiment suggested that the IFN-γ receptor displayed by cancer cells enables them to respond to IFN-γ released by immunocytes, and that this response usually prevents or retards the growth of tumors formed by these cells.

These striking effects of IFN-γ could be associated, at least in part, with the actions of the natural killer cells. The NK cells were discovered and named because of their innate ability to recognize tumor cells as abnormal and to eliminate them. Once NK cells identify cancer cells as targets for elimination, they release IFN-γ in the vicinity of the targeted cells. The released IFN-γ, in turn, elicits several distinct responses.

As mentioned earlier, IFN-γ enables the NK cells to call in other types of immune cells to assist in killing targeted cancer cells, thereby amplifying the immune system's response. Among the responding immune cells are macrophages, which aid not only by killing the cancer cells directly but also indirectly, by functioning as professional antigen-presenting cells (APCs) that process and display antigenic molecules derived from the corpses of their victims (see Figure 15.6).

At the same time, IFN-γ stimulates targeted cancer cells to display on their surfaces increased levels of class I MHC molecules that may carry oligopeptide antigens capable of provoking further, highly specific adaptive immune responses. This helps to explain why transformed cells lacking the IFN-γ receptor are more tumorigenic than counterpart cells that do display this receptor. All of these responses seemed to be defective in genetically altered mice lacking the IFN-γ receptor; such mice were also found to have an increased susceptibility to certain types of spontaneously arising tumors. When taken together, these experiments provided compelling validation of the idea that immune surveillance plays a critical role in tumorigenesis, at least in chemically induced tumors of mice.

Further support of the immunosurveillance theory came from mice that had been deprived of the gene encoding perforin, the protein used by lymphocytes

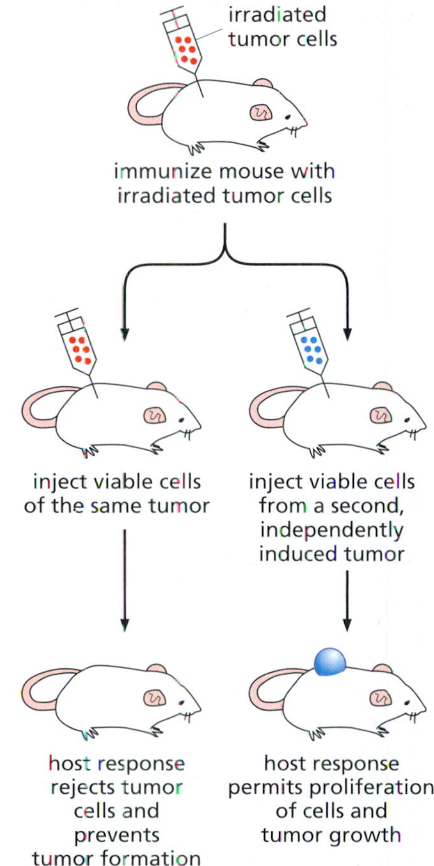

Figure 15.17 Immunization of mice by exposure to killed cancer cells Mice were initially injected with irradiated, killed cancer cells *(red)* deriving from one chemically induced tumor. When these mice were subsequently injected with live cells from the same tumor, the cells failed to grow *(lower left)*. However, when these mice were injected with live cells from a second tumor *(blue)*, the cells proliferated and formed a tumor mass *(lower right)*. The reciprocal experiment *(not shown)* yields the opposite results, i.e., injection with killed blue cancer cells rendered mice immune to the blue tumor but not to the red tumor.

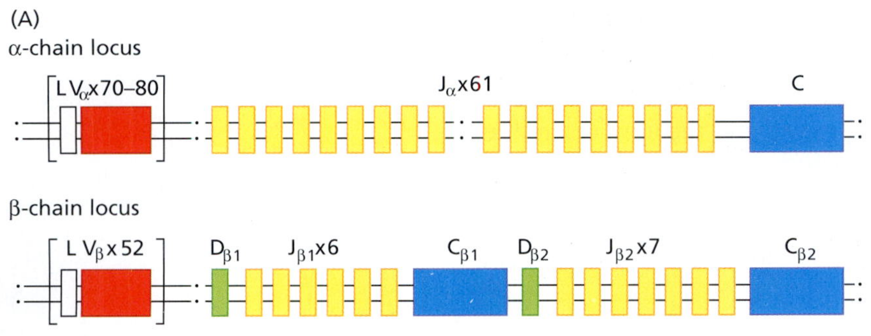

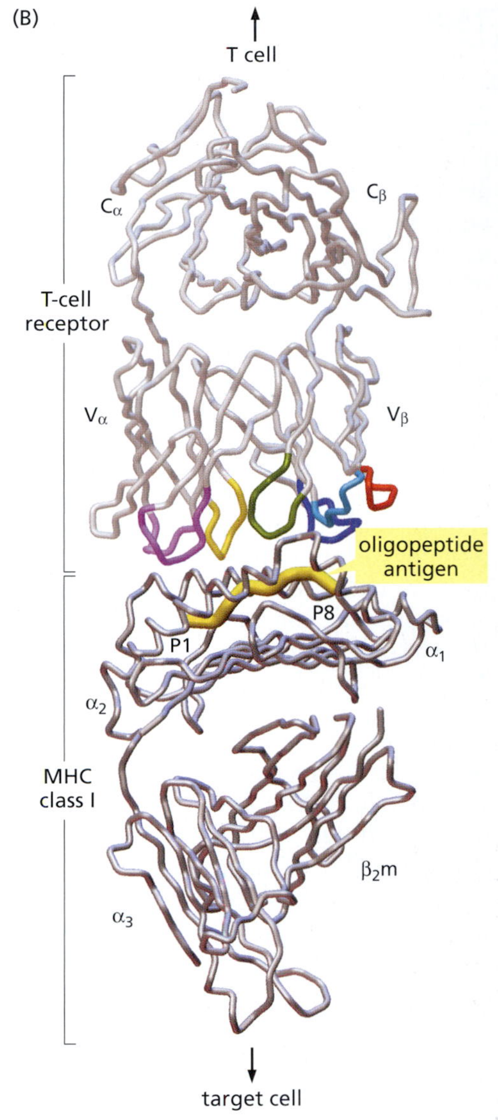

Figure 15.18 RAG proteins and TCR gene rearrangement The RAG-1 and RAG-2 proteins are responsible for the rearrangement of DNA segments that leads to the formation of both antibody molecules and T-cell receptors (TCRs). (A) This diagram illustrates the organization of the unrearranged genes encoding the α and β chains of the TCR. The RAG-1 and RAG-2 proteins rearrange the germ-line versions of these two genes through the deletion of intragenic segments and the attendant fusion of previously distantly linked DNA segments within the α and within the β gene. Rearrangement within the α gene is achieved when the RAG proteins juxtapose an L segment encoding a leader sequence (L) at the N-terminus of the α chain with one of the 70 to 80 V_α gene segments *(red)* and one of the 61 J_α segments *(yellow);* since the choice of individual V_α and J_α segments to be fused is essentially random, this results in a large number of combinations of joined V_α– J_α segments and a comparably large number of distinct antigen-binding pockets. A similar set of rearrangements occurring independently on a different chromosome results in the formation of the β chain-encoding segments of the TCR. Since the antigen-recognition domains of the TCR are created cooperatively by the amino acid sequences encoded by the α and β chains, these RAG-1/2-mediated rearrangements of the TCR-encoding genes are able to create a vast number of distinct antigen-binding domains. (B) The TCR *(above)* on the surface of a cytotoxic T cell (T_C, CTL) enables the T cell to recognize a specific oligopeptide antigen *(yellow)* displayed by an MHC class I molecule *(below;* see also Figure 15.7B*)* on the surface of a potential target cell; such recognition and binding by the TCR is achieved by its V_α and V_β domains *(colored loops)* whose generation is described in panel A. (Confusingly, the V_α domain of the TCR is created by juxtaposition of V_α and J_α DNA segments of the α-chain locus, while the V_β domain is created by fusion of the V_β, D_β, and J_β DNA segments of the β chain locus, all illustrated in panel A.) Once such recognition has occurred, it may result in the killing of the target cell (see Figure 15.12). Since TCRs are also used by T_H and T_{reg}s for other immune functions (*not shown*; see Figure 15.14), loss of TCRs caused by inactivation of a *RAG* gene leads to crippling of many components of the multi-faceted cellular immune response as well as the inactivation of the humoral response, which dependends on RAG1/2-mediated rearrangement of antibody genes. (From C.A. Janeway Jr. et al., Immunobiology, 6th ed. New York: Garland Science, 2005.)

and NK cells to mediate killing of targeted cells. Recall that perforin is used by cytotoxic cells to create channels in the plasma membrane of their victims, allowing the entrance of apoptosis-inducing granzymes (see Figure 15.12A). Mutant mice lacking the ability to make perforin showed an elevated incidence of spontaneous tumors and were also more susceptible to developing tumors following exposure to 3-methylcholanthrene.

Similarly, increased cancer susceptibility was registered in genetically altered mice that lacked the RAG-1 or RAG-2 proteins; these two proteins are responsible for rearranging the genes encoding soluble antibody molecules as well as those encoding the antigen-recognizing T-cell receptors (TCRs) displayed on the surfaces of T cells (Figure 15.18). Such RAG-1 or -2–negative mice lack T lymphocytes, B lymphocytes, γδ T cells (not discussed further in this chapter), and a subclass of NK cells called NKT cells. As a consequence, these mice have severely compromised adaptive immune responses.

For example, in one experiment, 3-MC treatment caused 30 of 52 RAG-$2^{-/-}$ mice to develop sarcomas, while only 11 of 57 wild-type mice of the same genetic background and treated in parallel formed these tumors. The mutant mice were also found to be far more susceptible to spontaneously arising cancers. Thus, 50% of older (18-month-old) RAG-2–negative mice developed spontaneous gastrointestinal malignancies—a tumor that is otherwise rare in wild-type mice of this age.

Arguably the most persuasive evidence supporting the role of immunosurveillance in cancer prevention comes from detailed studies of the 3-MC–induced sarcomas growing in either *RAG2*$^{-/-}$ or wild-type mice. When tumor cells prepared from these two groups of sarcomas were grafted into new *RAG2*$^{-/-}$ hosts, both groups of sarcomas seeded tumors in these new hosts with high efficiency (Figure 15.19).

A very different outcome was observed, however, when tumor cells were transplanted into syngeneic wild-type (and thus immunocompetent) hosts. Cells from 17 tumors that had previously been induced in wild-type mice all succeeded in generating tumors in their new hosts. In contrast, cells from 8 of 20 tumors that had previously been induced by 3-MC in RAG$2^{-/-}$ mice failed to form tumors, being rejected by the immune systems of these wild-type hosts (see Figure 15.19).

These observations open our eyes to an entirely new dimension of tumor immunology. They suggest that when 3-MC–transformed cells arise in an immunocompetent host, those that happen to be *strongly immunogenic* (and thus capable of provoking some type of immune response) are effectively eliminated by the host, resulting in the survival and outgrowth of only those cancer cells that happen to be *weakly immunogenic*. The latter then multiply and form tumors in their original hosts and, later on, succeed in doing so when transplanted into other immunocompetent hosts. Hence, these tumors represent a subset of those that originally arose in the primary hosts. The missing, strongly immunogenic tumors are apparently eliminated early in tumor progression by host immune systems and therefore never see the light of day (see Figure 15.19).

In contrast, when 3-MC–transformed cells arise in an immunocompromised host (see Figure 15.19A), two classes of tumors are initially formed, as before—those that are strongly immunogenic and those that are weakly immunogenic; both types of tumor cells survive in an immunodeficient host. Later, when these tumors are transplanted into immunocompetent hosts, those that are strongly immunogenic fail to form tumors, while those that are weakly immunogenic succeed in doing so. We conclude that in wild-type mice, a functional immune system plays an important and effective role in eliminating a significant fraction of the tumors that are initially induced by 3-MC.

These observations indicate that the immune system of these mice plays an active role in determining the identities of tumors that arise and the antigens that they express. This active intervention in the phenotype of tumors has been termed **immunoediting**, to indicate the weeding out of some tumors and the tolerance of others. Immunoediting can be thought of as a type of Darwinian selection, in which the selective force is created by the directed attacks of the immune system on incipient tumors.

15.9 The human immune system plays a critical role in warding off various types of human cancer

Because the biology of mice and humans differs in so many respects, we need to interpret the results described above with caution when attempting to understand the role of the human immune system in defending us against cancer. In addition, the chemical carcinogens used in the experiments described above may well create tumors in mice that are far more antigenic or immunogenic than spontaneously arising human tumors (to be discussed in Section 15.12).

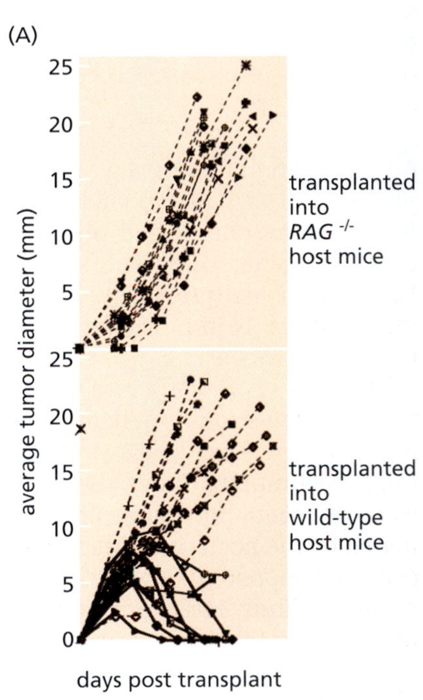

Figure 15.19 Effects of immune function on the development of anti-tumor immune responses Both wild-type (wt) and $RAG2^{-/-}$ immunocompromised mice were exposed to the potent carcinogen 3-methylcholanthrene (3-MC). (A) When the tumors induced in the $RAG2^{-/-}$ mice were transplanted back into $RAG2^{-/-}$ hosts, they all formed tumors *(above)*. However, when the tumors induced in the $RAG2^{-/-}$ mice were transplanted back into wild-type hosts, 8 of 20 tumors failed to form *(below)*. Each line presents the kinetics of growth of a single implanted tumor. (B) This experiment and other experiments using tumors induced in wt mice *(not shown)* are summarized here. Following exposure to 3-MC, the wt mice developed fewer tumors *(blue)* than did the $RAG2^{-/-}$ mutants *(blue and red)*. The tumors from the two groups of mice were excised and cells from each were converted to a cell line that could be propagated *in vitro*. Cells from each of these cell lines were then transplanted back into either wt mice or $RAG2^{-/-}$ mutant mice. Cells from all of the tumors that appeared initially in the wt mice *(blue)* were able to form new tumors in both wt and $RAG2^{-/-}$ hosts *(left)*. However, cells from all of the tumors that arose and grew initially in the $RAG2^{-/-}$ mice were able to form new tumors in $RAG2^{-/-}$ mice *(red, blue)*, but only some of these *(blue)* were able to form new tumors in the wt mice *(right)*. These experiments suggested that 3-MC initially induced two types of tumor cells in all of the mice: strongly immunogenic *(red)* and weakly immunogenic *(blue)*. Both red and blue cells formed tumors in the $RAG2^{-/-}$ mice, but only blue cells formed tumors in the wt mice, since any initially formed red tumor cells were eliminated by the functional immune systems of these mice. This meant that the tumors that did arise in wt mice were already selected for being weakly immunogenic and thus capable of forming new tumors in other wt mice. (A, from V. Shankaran, H. Ikeda, A.T. Bruce et al., *Nature* 410:1107–1111, 2001.)

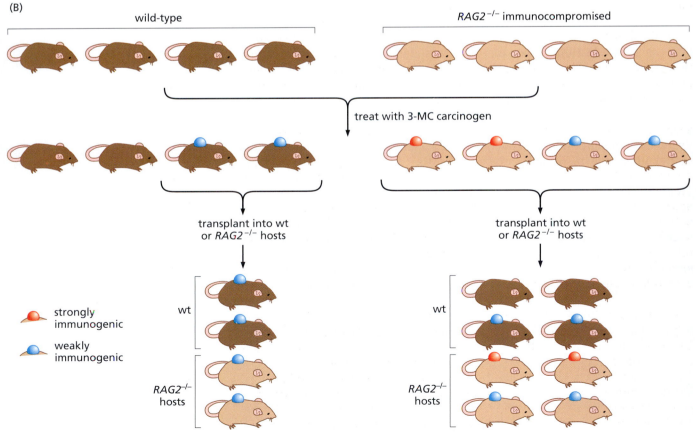

(A)

tissue transplanted	no. of grafts in USA (2002)
kidney	15,680
liver	5594
heart	2231
pancreas	1492
lung	1077
cornea	~40,000
bone marrow	15,000

(B)

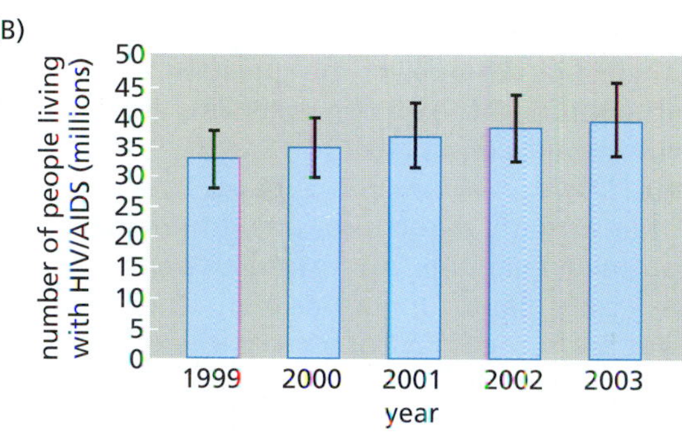

Figure 15.20 Immunosuppressed populations (A) The number of organ transplants performed has increased progressively over the past two decades. This table lists the number of organ transplants of various sorts that were performed in the United States in the year 2002. (By the middle of 2003, more than 66,000 heart transplants and 17,000 lung transplants had been performed worldwide since these procedures were introduced into the clinic.) (B) The numbers of individuals who are immunosuppressed because they are recipients of organs from allogeneic donors is dwarfed by the number of people who have been infected by human immunodeficiency virus and who are in various stages of developing AIDS. (A, from C.A. Janeway Jr. et al., Immunobiology, 6th ed. New York: Garland Science, 2005; B, from World Health Organization AIDS epidemic update, 2003.)

Nevertheless, evidence compiled in the 1990s provides clear indications that the human immune system does indeed play an important role in warding off cancer. The bulk of this evidence comes from observations that immunocompromised humans are far more susceptible than the general population to certain types of cancer.

Actually, immunocompromised humans were uncommon until recently. Those who were born with dysfunctional immune systems died early in life, while others whose immune systems deteriorated later in life died as their defenses against infectious agents declined. However, over the past few decades, the number of individuals who live for extended periods of time with compromised immune systems has increased dramatically, for three reasons.

First, organ transplants involving kidneys, hearts, and livers have become common throughout the developed world (Figure 15.20A). Because these organs derive from donors who are, almost always, genetically different from the recipients, the donor cells in the transplanted organs are recognized as foreign (i.e., allogeneic and thus histoincompatible) by the immune systems of the graft recipients, which proceed to eliminate them. This unwanted reaction of the host immune system against the engrafted organs is controlled by long-term treatment with several types of immunosuppressive drugs.

Second, and independent of these clinically induced immunodeficiencies, are the cases of more than 40 million people throughout the world who have been infected by human immunodeficiency virus (HIV); as a consequence, increasing numbers of these individuals suffer from acquired immunodeficiency syndrome (AIDS; Figure 15.20B). Third, the long-term survival of immunocompromised individuals, including organ graft recipients and many HIV-infected patients, has been enabled by the development and use of a diverse group of antibiotic (i.e., antibacterial), antifungal, and antiviral drugs. These extended survival periods of immunocompromised patients represent time spans that are long enough for pre-malignant growths that were previously latent in these individuals to progress to a state where they become clinically apparent.

Thousands of patients bearing transplanted organs have developed a variety of solid tumors and hematopoietic malignancies over the past two decades. As discussed earlier (Sidebar 28), a very small proportion of these tumors have derived from occasional metastatic cancer cells that were hiding in the bodies of organ donors, escaped detection during the transplantation procedure, and began to multiply aggressively once they were introduced into the bodies of graft recipients. These tumors can be shown to be of allogeneic (i.e., of organ donor) origin by analysis of the genetic markers that they bear. However, the eruption of tumors triggered by transplanted cells provides no insights that might help us answer the question whether tumors of *endogenous origin* (i.e., autochthonous tumors) arise with greater-than-normal frequency in immunocompromised patients.

In fact, many autochthonous tumors have been documented in immunocompromised patients. An Australian study followed patients who had received kidney transplants for periods of up to 24 years after the transplantation procedure was performed; 72% of these patients had developed at least one type of cancer. A similar study in the United States found a three- to fivefold increased risk of cancer in transplant recipients. Patients who have undergone liver transplants commonly develop new tumors within five years of transplantation and initiation of immunosuppression. The transplant registry in Ohio (Sidebar 28) has cataloged more than 15,000 such malignancies in transplant patients.

The most common cancers among transplant patients are those that are known to be caused by viral infections, including a 400- to 500-fold increased risk of Kaposi's sarcoma, a malignancy of the hemangioblast precursors of endothelial cells (caused by human herpesvirus 8, HHV-8); a 28- to 49-fold increased risk of lymphoproliferative diseases, including Hodgkin's disease (most of which appear to be associated with Epstein–Barr virus, or EBV, infections); a 100-fold increased risk of squamous cell vulvar and anal carcinomas (triggered by human papillomaviruses, HPV); a 20- to 38-fold increased risk of hepatocellular carcinomas (usually caused by hepatitis B and C viruses—HBV and HCV); and a 14- to 16-fold increased risk of *in situ* uterine carcinomas (also associated with HPV infections).

Comparably increased rates of virus-induced malignancies are seen in patients who are congenitally immunodeficient or have become so because of HIV infection and subsequent AIDS. Together, these disparate observations clearly show that the immune system represents an important defensive bulwark against the 20% and more of human tumors arising worldwide whose development is traceable, directly or indirectly, to the viruses mentioned above.

The extraordinary effectiveness of the normal immune system in defending us against virus-induced malignancies is illustrated by the fact that some 90% of adults in the West are infected by the potently oncogenic Epstein–Barr virus yet EBV-induced malignancies are relatively uncommon in the general population. To cite another example, the introduction of multi-drug therapy to suppress HIV replication has resulted in the regeneration of immune competence in large numbers of HIV-infected patients; as a consequence, the incidence of new cases of Kaposi's sarcoma has decreased as much as 40-fold in some AIDS clinics.

Unanswered by these studies is the precise mechanism by which virus-induced cancers are normally controlled by the immune system. We could entertain two plausible models that can explain this control. (1) The immune system is normally responsible for protecting us against all types of viral infections, independent of whether certain viruses are bent on inducing cancer. In the absence of fully functional immune systems, many viruses, including those initiated by oncogenic viruses, are able to persist and proliferate for extended periods of time within the body, in contrast to the usual fate of infecting viruses, which are cleared rapidly from the body by the immune system. (2) Alternatively, the normal immune system is responsible for recognizing and eliminating virus-transformed cells. In immunocompromised individuals, however, such cells may be able to survive indefinitely.

Either or both mechanisms may explain the greatly increased rates of virus-induced cancers in immunocompromised people. In AIDS patients, for example, high levels of circulating Epstein–Barr virus are not commonly observed, while the levels of actively proliferating EBV-infected lymphoid cells often increase dramatically, yielding, in turn, the virus-induced lymphomas. Consequently, in the condition of AIDS, the elimination of virus-infected cells rather than the virus itself seems to be defective. (In the blood of healthy carriers of EBV infection, who represent 90% of the general population, as many as 5% of the long-lived cytotoxic T cells have been found to react with a certain EBV antigen, indicating that the cellular arm of the immune system normally devotes an astounding proportion of its operations to controlling a single infectious agent–EBV–in this case by eliminating cells that carry actively proliferating virus.)

Still, virus-induced malignancies represent only a minority of the tumors routinely treated in oncology clinics. This causes us to ask whether a competent immune system also erects defenses against the great majority of human tumors (~80%) that are of nonviral origin. In fact, a two- to fourfold increased risk of melanoma has been found among adult organ transplant recipients, while non-Kaposi's sarcomas were found at 3 times increased rates above the general population. A population of heart transplant recipients has been found to experience a 25-fold increased risk of carcinomas of the lung. And registries of transplant patients in Australia, New Zealand, and Scandinavia have documented elevated rates of carcinomas of the colon, lung, bladder, and kidney as well as tumors of the endocrine system (Table 15.1). Attempts to associate these cancers with tumor virus infections have consistently failed.

Another, fully independent line of evidence supports the role of immune defenses in protecting us from cancer. This one derives from histopathological observations that human tumors often have substantial numbers of lymphocytes that have infiltrated into the tumor mass (Figure 15.21). These tumor-infiltrating lymphocytes (TILs) might represent yet another type of stromal cell that has been recruited into the tumor mass by neoplastic cells in order to support the expansion of the tumor as a whole, as argued in Chapter 13. However, an alternative explanation is even more intriguing: these TILs may have been dispatched by the immune system in order to eliminate cancer cells.

Observations of a group of ovarian carcinoma patients have provided some of the most dramatic testimony supporting the second alternative. This evidence comes from correlating the clinical course of these ovarian carcinoma patients with the presence or absence of substantial numbers of tumor-infiltrating lymphocytes in their cancers. In one group of patients, who had been treated initially by surgical removal of the bulk of their tumors followed by chemotherapy, 74% were alive five years later if their initial tumors carried large numbers of these TILs. In contrast, among those patients whose ovarian tumors lacked significant populations of TILs, only 12% were still alive.

Similar outcomes are associated with patients carrying malignant melanomas that are infiltrated with large numbers of TILs; these patients live 1.5 to 3 times longer following diagnosis than do those patients whose tumors lack large numbers of the tumor-infiltrating lymphocytes. Yet other correlations have been made between the presence of these infiltrating lymphocytes and the survival of patients bearing carcinomas of the breast, bladder, colon, prostate, and rectum. Still, such observations, while dramatic, are only correlative and therefore do not prove definitively that these lymphocytes are important agents responsible for holding back tumor progression.

Moreover, these studies do not address the role of the humoral immune response in defending us against various types of cancer. In fact, a number of research reports have demonstrated the presence of anti-tumor antibodies in the blood of patients suffering from various types of cancer; the presence of these antibodies clearly suggests some type of immunosurveillance. However, once again, it remains unclear whether these antibodies actively contribute to eliminating tumor cells from the body.

Table 15.1 Cancer incidence in immunosuppressed transplant patients[a]

Site of cancer	No. of cases observed	No. of cases expected[b]	Ratio observed/expected
Non-melanoma skin	127	5.1	24.7
Thyroid, other endocrine	30	2.1	14.3
Mouth, tongue, lip	22	1.6	13.8
Cervix, vulva, vagina	39	3.6	10.8
Non-Hodgkin's lymphoma	25	2.4	10.3
Kidney, ureter	32	3.5	9.1
Bladder	26	4.7	5.5
Colorectal	38	10.5	3.6
Lung	30	12.5	2.4
Brain	10	4.1	2.4
Prostate	11	5.2	2.1
Melanoma	7	4.1	1.7
Breast	15	13.6	1.1

[a]Data from S.A Birkeland, H.H. Storm, L.U. Lamm et al., *Int. J. Cancer* 60:183–189, 1995, as adapted by J. Peto, *Nature* 411:390–395, 2001.
[b]These numbers represent the numbers of cases of the various cancers expected to occur in an age-matched control population over the same period of time.

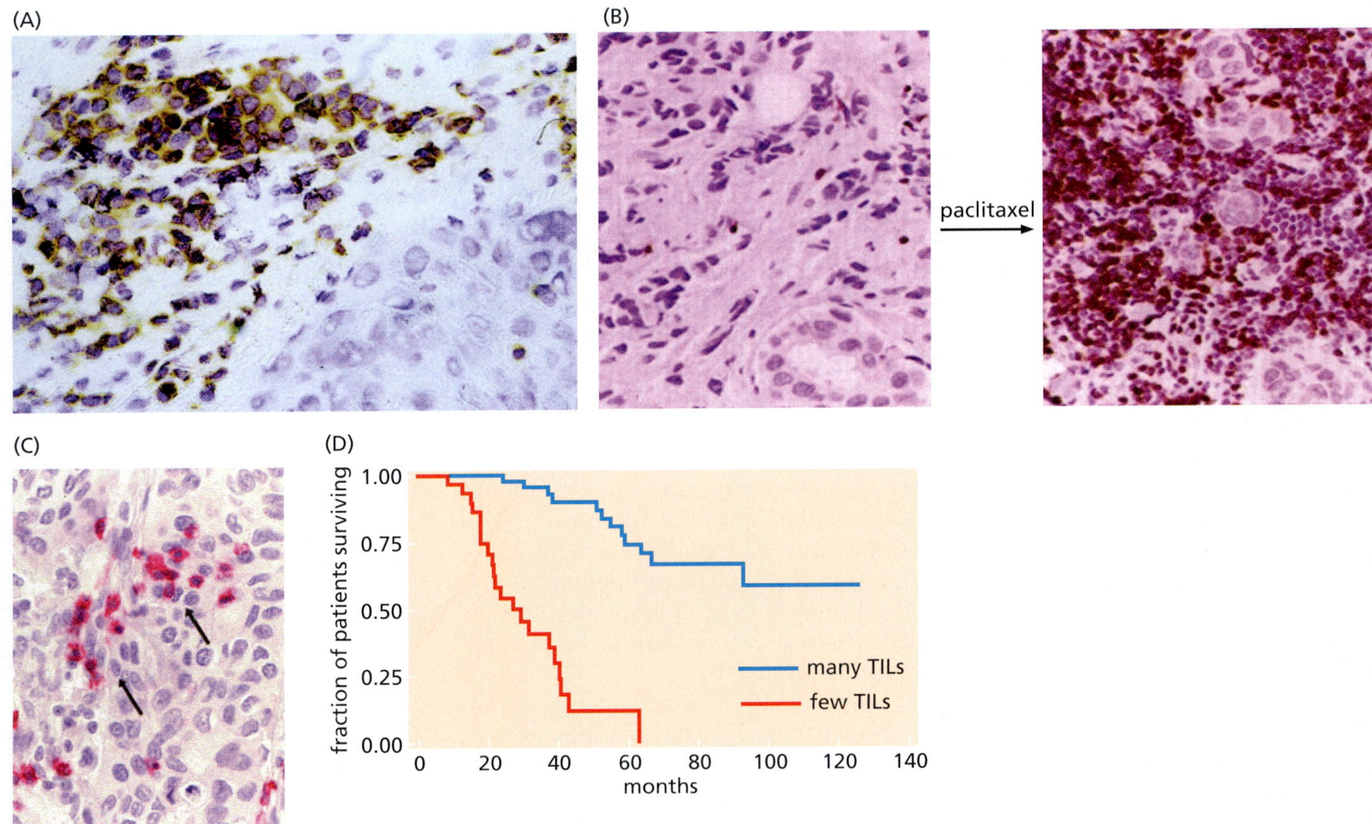

Figure 15.21 Tumor-infiltrating lymphocytes (A) This immunostaining of an oral carcinoma reveals tumor-infiltrating lymphocytes (TILs; *brown)* that are abundant in certain areas of the tumor and have been detected with an antibody that recognizes the CD3 antigen expressed by T lymphocytes. More detailed characterizations revealed several T-lymphocyte subtypes among these cells *(not shown)*. (B) This immunostaining demonstrates that TILs, detected once again with an anti-CD3 antibody *(dark purple)*, are relatively rare in an untreated breast tumor *(left)* but become abundant in areas of the tumor following chemotherapy with the chemotherapeutic drug paclitaxel *(right)*. (C) TILs are also frequently found in invasive non-small-cell lung carcinomas (NSCLCs; *arrows)*. The expression here of the CD8 antigen *(dark pink)* indicates that these cells are largely cytotoxic lymphocytes (CTL, T_C). (D) The clinical prognosis of a set of ovarian carcinoma patients was dramatically affected by whether or not their tumors carried significant concentrations of TILs. As shown by this Kaplan–Meier plot, the proportion of patients surviving after initial diagnosis *(ordinate)* is plotted versus the months of survival *(abscissa)*. Those patients whose tumors had high levels of TILs *(blue line)* fared significantly better than did those whose tumors lacked significant concentrations of TILs *(red line)*. (A, from T.E. Reichert, L. Strauss, E.M. Wagner et al., *Clin. Cancer Res.* 8:3137–3145, 2002; B, from S. Demaria, M.D. Volm, R.L. Shapiro et al., *Clin. Cancer Res.* 7:3025–3030, 2001; C, from A. Trojan, M. Urosevic, R. Dummer et al., *Lung Cancer* 44:143–147, 2004; D, adapted from L. Zhang, J.R. Conejo-Garcia, D. Katsaros et al., *N. Engl. J. Med.* 348:203–213, 2003.)

Finally, a large body of evidence supporting the immunosurveillance theory comes from observations, some already cited (see Sidebar 36), that cancer cells employ various strategies for evading detection and elimination by the immune system. We will defer detailed discussion of these various mechanisms of *immunoevasion* until later. For the moment, suffice it to say that cancer cells make extensive efforts to lower their immunological profiles, enabling them to "fly under the immunological radar." While also correlative, these observations are so frequent as to represent persuasive evidence that immunosurveillance and resulting immunoevasion represent important dynamics in tumor progression.

15.10 Subtle differences between normal and neoplastic tissues may allow the immune system to distinguish between them

The successes of the immune defenses against infectious agents depend on the ability to recognize these intruders as foreigners in the human body. Infectious agents invariably display molecules that betray their alien origins, provoking attack by specialized cells of the innate and adaptive arms of the immune system. Typically, these foreign molecules contain oligopeptide sequences that are recognizably different from the sequences present in the body's own repertoire of native proteins.

At the same time, a well-functioning immune system turns a blind eye to the proteins and thus amino acid sequences that are native to the body—a reflection of the immunological tolerance discussed earlier. The phenomenon of immunological tolerance greatly complicates our attempts at understanding how the immune system defends the body against spontaneously arising tumors. Ultimately, the success of these defenses depends on a critical issue: can the immune system recognize cancer cells arising in a person's own tissues as being foreign, even though these cells are truly native to the body?

The answer to this question is hardly obvious. The vast majority of proteins expressed by tumor cells are clearly normal, both in their structures and levels of expression. Nevertheless, among the 20,000 or so distinct protein species (and variants thereof) made by one or another type of cancer cell, there are certainly a small number that are expressed by some cancer cells but are not present in normal tissues. Such structurally novel, and in this sense "foreign," antigens might well provoke a vigorous immune response.

An obvious example of a novel cancer antigen is provided by the Ras oncoproteins, which are created by amino acid substitutions in residues 12, 13, or 61 of the four subtypes of Ras proteins seen in normal cells. These oncoproteins clearly exhibit altered chemical structures, and the sequence of amino acids in a Ras oncoprotein that surrounds and includes an altered residue may indeed constitute an oligopeptide antigen that evokes an immune response. Similarly, the numerous mutant alleles of the *p53* tumor suppressor gene also specify amino acid substitutions that might cause the altered versions of this protein to be immunogenic in the almost 50% of common human tumors that carry altered p53 proteins (see, however, Sidebar 39).

Yet other examples of cancer-specific proteins derive from the numerous types of chromosomal translocations that specify fusion proteins encoded by pairs of previously unlinked genes. Such translocations are found frequently in hematopoietic malignancies. Recall, for example, the Bcr-Abl fusion protein, which is found in chronic myelogenous leukemia (CML) cells (Section 4.6). While the bulk of the amino acid sequences of the fusion protein are identical in structure to the sequences of the two parental proteins (Bcr and Abl) found in normal cells, the short region where these two proteins are joined constitutes a novel amino acid sequence that has the potential to be recognized as foreign and therefore to be immunogenic.

Because of the genetic instability of human tumor cell genomes (Chapter 12) and the large number of successive cell cycles through which cell lineages pass during the course of multi-step tumor progression (see Figure 10.5), mutant alleles encoding structurally altered proteins may be present at elevated frequency in these genomes. The great majority of these alleles may have nothing to do with accelerating tumor cell proliferation, being incidental by-products of this genomic instability, but the proteins made by these mutant alleles may happen to be quite immunogenic—an issue to which we will return later.

A well-studied melanoma provides a clear example of an antigen that arises as a by-product of tumor progression. This tumor was found to express a mutant triosephosphate isomerase (a key enzyme in glycolysis). Given the normal function of this protein, its mutant structure is unlikely to have played a role in tumor pathogenesis. Intracellular processing of this protein yielded a mutant oligopeptide antigen that was able to bind to MHC class II proteins with an affinity 5 orders of magnitude greater than that of the corresponding wild-type oligopeptide. This binding greatly enhanced antigen presentation and the overall immunogenicity of the melanoma.

In general, attempts at predicting the immunogenicity of various proteins and their relative abilities to provoke robust immune responses is highly challenging. Many proteins having novel structures may be overlooked by the immune system because of their overall structural similarity to normal cellular proteins. Others may be tolerated because, while structurally distinct from normal cellular proteins, they may be present in such low concentrations that they are effectively invisible to the immune system. Ras exemplifies both circumstances: the mutant Ras oncoproteins found in many human tumors have an almost-normal three-dimensional structure and, like normal Ras proteins, are present in relatively low concentrations in cancer cells. Moreover, because of their amino acid sequences, certain mutant oligopeptides may not be readily bound by the antigen-displaying hands of a patient's MHC molecules (see Figure 15.7).

Predictions of immunogenicity are also complicated by the fact that we do not fully understand the rules governing the establishment of tolerance in the developing immune system. Much of this tolerance is achieved in the thymus gland and bone marrow during embryonic and early postnatal development, where the B and T lymphocytes that happen to have developed immunological reactivity to the body's normal proteins are eliminated or functionally inactivated. Later, lymphocytes circulating throughout the body develop "peripheral tolerance" to proteins that they have encountered in tissues distant from the thymus and bone marrow. Once again, tolerance is probably achieved by eliminating lymphocytes that recognize normal tissue antigens. In addition, the regulatory T cells described in Section 15.6 are known to suppress the actions of lymphocytes that have escaped elimination by these other mechanisms.

What would happen if the various populations of developing lymphocytes were never exposed to certain normal proteins? Some proteins may be expressed only transiently at specific stages of early embryonic development, long before immunological tolerance is developed. Other proteins, as suggested above, may be present in very low concentrations, so that developing lymphocyte populations rarely experience them and hence do not delete the few cells that happen to recognize these weakly expressed proteins. Yet other proteins may be expressed in "immune-privileged" tissues such as the brain, where effective surveillance by the immune system is usually blocked by the complex mechanisms constituting the "blood–brain barrier." Such shielding may also operate in the testes, which seem to be protected from routine monitoring by cells of the immune system.

In sum, a number of distinct mechanisms can prevent the development of tolerance toward certain normal cellular proteins. However, if one of these proteins happens to be displayed at abnormally high levels by cancer cells, these gaps in immunological tolerance may enable immune recognition of these cells. For example, the catalytic subunit of the telomerase holoenzyme, hTERT, is not expressed at readily detectable levels in normal human cells, while it is found to be expressed at significant levels in 85–90% of human tumors (Section 10.6); a significant percentage of liver cancer patients studied in Japan show detectable levels of antibody against hTERT, which increase steadily during tumor progression. (It is unclear whether these anti-hTERT antibodies have any effect on the course of tumor progression in these patients.)

Similarly, in many breast carcinomas, the HER2/Neu receptor is often expressed at levels far higher (10- to 20-fold) than are encountered in normal epithelial tissues (Figure 4.6); once again, immune reactivity may be mounted against this protein. Some human melanomas overexpress a class of cell surface carbohydrates, termed gangliosides, that can also provoke an immune response. Expression of one of these, termed GD3, is sometimes vastly higher in melanomas than in their normal precursors, the melanocytes. In each of these cases, the overexpressed proteins or carbohydrate moieties attract the attentions of an immune system that is normally oblivious to their existence.

Imagine that a cancer cell expresses an embryonic protein that is never found in normal adult tissues—a situation commonly seen in many kinds of human tumors. These embryonic proteins might therefore represent potent antigens when expressed in adult tissues. Through yet another mechanism—the alternative splicing of mRNA precursors— structurally distinct versions of normal adult proteins may be displayed by tumor cells that are rarely, if ever, experienced by the adult immune system.

Taken together, these examples indicate that vigorous immunological attacks may well be launched against tumor cells expressing proteins in unnatural (ectopic) anatomical locations, at stages of development where they are not usually encountered, or at abnormally high levels. Occasionally, these immunological responses against tumors, which result from the breakdown of tolerance, are actually counterproductive for cancer patients, because they lead to certain types of bizarre autoimmune disorders (Sidebar 40 ●).

15.11 Immune recognition of tumors may be delayed until relatively late in tumor progression

When does the anti-tumor immune response become activated during the course of multi-step tumor progression? Some cancer immunologists argue that early in tumor progression, nests of tumor cells may be too small to be recognized by the immune system. Accordingly, they propose that immune recognition may occur only much later, after tumors have become large and have acquired the ability to invade adjacent tissues.

The evidence favoring a close connection between tumor cell invasiveness and the initiation of immune recognition of these cells is not definitive. As we read in the previous two chapters, the process of invasion involves, among other things, the recruitment of a number of immune cell types that together constitute an inflammatory response. This inflammatory response is, in turn, intimately connected with the processes that normally contribute to wound healing.

We can imagine that wound healing actually involves two distinct biological processes. First, damaged tissue must be reconstructed; this involves the activation of stromal fibroblasts, reversible epithelial–mesenchymal transitions

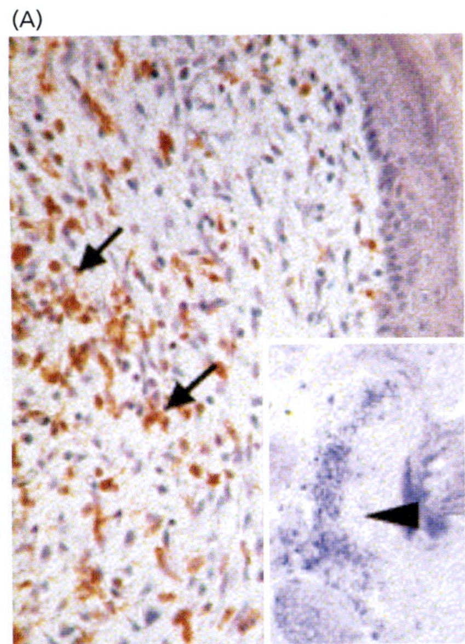

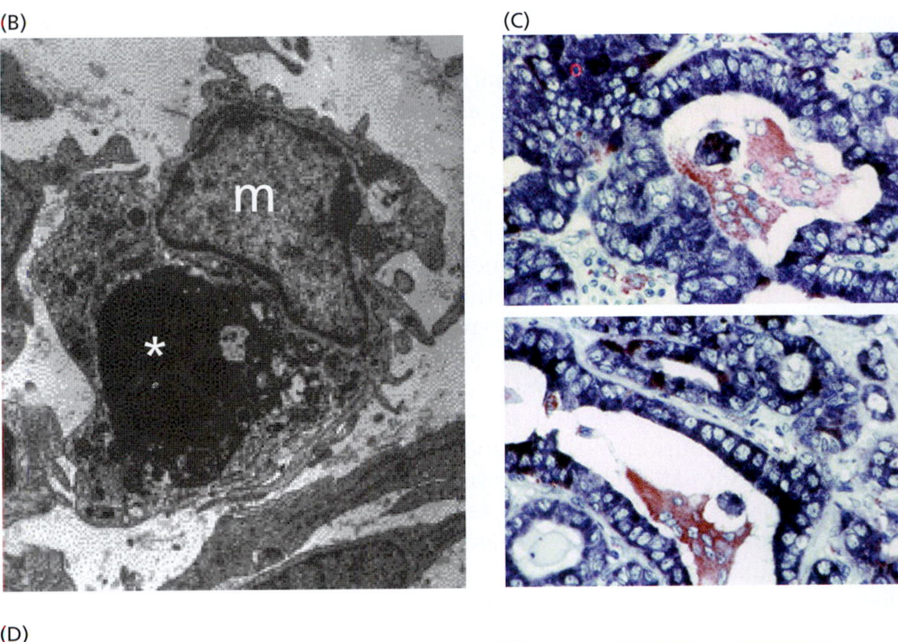

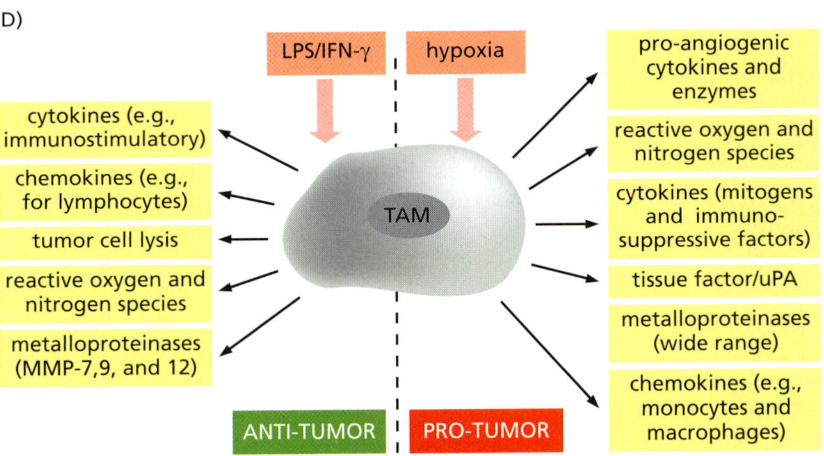

Figure 15.22 Multiple roles of macrophages Macrophages play critical roles in promoting wound healing and tumor progression; at the same time, they act as agents of the immune system to present tumor antigens and to consume tumor cells. It seems that these differing functions are carried out by distinct subtypes of macrophages. (A) In this site of healing of a skin wound in a mouse, macrophages *(brown)* are detected in abundance *(arrows)* through an antibody that recognizes the F4/80 macrophage-specific antigen and the c-Fms receptor marker *(arrowhead, inset)*. (B) Transmission electron microscopy reveals a macrophage (m) that is engorged with the phagocytosed corpses of apoptotic cells *(asterisk, dark body)* in the middle of a wound-healing site. (C) Two macrophages *(pink, top, bottom panels)* are seen as they begin to phagocytose tumor cells in the lumina of the ducts of a papillary thyroid carcinoma. (D) Macrophages can be activated by diverse signals, such as interferon-γ (IFN-γ) and bacterial lipopolysaccharide (LPS), the latter being used as an *adjuvant* to potentiate the immune response. Once activated, they can function to present antigen to helper T (T_H) cells and to trigger tumor cell killing by antibody-dependent cellular cytotoxicity (ADCC). Acting in the opposite direction, hypoxia as well as a variety of physiologic signals released by tumors (see Chapters 13 and 14) activate the macrophages and cause them to further tumor progression. (A and B, from P. Martin, D. D'Souza, J. Martin et al., *Curr. Biol.* 13:1122–1128, 2003; C, from A. Fiumara, A. Belfiore, G. Russo et al., *J. Clin. Endocrinol. Metab.* 82:1615–1620, 1997; D, from L. Bingle, N.J. Brown and C.E. Lewis, *J. Pathol.* 196:254–265, 2002.)

(EMTs) of epithelial cells, and the complex signaling mechanisms governing angiogenesis (Chapter 13). Second, wound healing must include the mobilization of an immune defense in order to ensure that wounded areas, which lack the normally present physical barriers to infection, are kept free of pathogens, such as bacteria and fungi.

Over the course of evolution, certain classes of immune cells have been endowed with abilities to mediate both types of biological functions. Thus, macrophages are responsible for releasing proteases that facilitate tissue remodeling and releasing factors that provoke angiogenesis—two aspects of tissue reconstruction (Section 13.5). They are also responsible for removing cellular debris in sites of tissue damage (Figure 15.22A and B). At the same time, macrophages are capable of engulfing bacteria and fungi; and by playing the role of professional antigen-presenting cells (APCs), they can process and present oligopeptide antigens to helper T (T_H) cells. Similarly, tumor-associated macrophages (TAMs) may phagocytose (ingest) tumor cells or fragments thereof (Figure 15.22C). Some studies indicate that those macrophages within a tumor that have ingested tumor-derived antigens produce especially high levels of vascular endothelial growth factor (VEGF), the key governor of angiogenesis.

The fact that these diverse activities reside in a single cell type (or group of closely related cells) persuades one that tumor invasiveness, inflammation, angiogenesis, and the immune response are often interconnected processes. These interconnections may also have an anatomical explanation: as long as neoplastic epithelial cells remain on their own side of the basement membrane, they are largely shielded from the immune cells that are continuously patrolling the stromal compartment on the other side of this membrane. However, once they have burst through the basement membrane, invasive cancer cells may attract the attention of immunocytes, initially natural killer (NK) cells; the NK cells may then sound alarms that trigger a second wave of immune attack mediated by macrophages and by lymphocytes of the adaptive immune response.

While plausible, this sequence of events remains to be demonstrated experimentally. Moreover, there are indications that a certain degree of immune surveillance may already occur before epithelial cells have become invasive and breached the basement membrane. For example, in some epithelial tissues, such as the gut, there is evidence of intraepithelial lymphocytes (IELs); these immunocytes are responsible for monitoring epithelial cell populations for the presence of invading pathogens. Such IELs might also serve as early-warning sentinels, allowing the immune system to detect neoplastic cells long before they have become invasive. In addition, as seen in Figure 15.22C, macrophages can ingest cancer cells on the epithelial side of a basement membrane (in this case in the lumen of a duct formed by an adenocarcinoma). It remains unclear whether such macrophages, having ingested cancer cells, migrate to lymph nodes and present cancer cell–associated antigens to the helper T cells in these nodes, thereby initiating adaptive humoral and cellular immune responses.

15.12 Tumor-specific transplantation antigens often provoke a potent immune response

To the extent that cancer cells do succeed in provoking an adaptive immune response, this response must be traceable to specific antigens displayed by these cells. Tumor immunologists have placed these antigenic proteins in two major categories—tumor-specific transplantation antigens (TSTAs), and tumor-associated transplantation antigens (TATAs; Table 15.2). (Because tumor transplantation experiments cannot be undertaken in humans, the analogous human proteins are called TSAs and TAAs.)

TSTAs are said to be *specific* to a tumor or a type of tumor and are therefore not present among the repertoire of proteins and oligopeptides normally expressed within the body's tissues. TSTAs may be encoded, for example, by viral genomes or by the somatically mutated alleles (such as those of *ras*, *p53*, or *bcr-abl*) arising during tumor progression. Because they are structurally novel, these proteins are unlikely to have induced tolerance during the normal development of the immune system.

TATAs, in contrast, are only *associated* with tumor cells, and their expression is not limited to malignant tissues. TATAs represent the large class of normal proteins

Table 15.2 Some tumor-associated and tumor-specific antigens and the antigenic peptides recognized by human T cells

Human tumor	Protein	Antigenic peptide
Melanoma, esophageal carcinoma, liver carcinoma, NSCLC	MAGE	EADPTGHSY, SAYGEPRKL
Melanoma	tyrosinase	MLLAVLYCL, YMNGTMSQV
Colon carcinoma	carcinoembryonic antigen (CEA)	YLSGANLNL
Breast and ovarian carcinomas	HER2/Neu	KIFGSLAFL
Head and neck carcinoma	caspase 8	FPSDWCYF
Chronic myelogenous leukemia (CML)	Bcr-Abl	ATGFKQSSKALQRPVAS
Prostate carcinoma	prostate-specific antigen (PSA)	FLTPKKKLQCV, VISNDVCAQV

From R.A. Goldsby et al., Immunology, 5th ed. New York: Freeman, 2002; and B.J. Van den Eynde and P. van der Bruggen, *Curr. Opin. Immunol.* 9:684–693, 1997.

that, for one reason or another, have failed to elicit complete tolerance and, when expressed by tumor cells, attract the attention of the immune system. The fact that TATAs are normal cellular proteins explains the observation that a certain TATA may be displayed by many independently arising tumors of a specific type, for example, a group of melanomas.

As their names imply, the existence of these two classes of antigens can be demonstrated by tumor transplantation experiments. For example, to revisit an experiment described earlier (see Figure 15.17), cells of a sarcoma arising from exposure of a mouse to the chemical carcinogen 3-methylcholanthrene can be introduced into a second, syngeneic host mouse. The tumor cells are then allowed to multiply for several weeks before being surgically removed. If this mouse is then challenged by being re-inoculated with tumor cells from the original sarcoma, the mouse will often reject the tumor and thereby block tumor formation. However, if cells from another 3-MC–induced sarcoma are injected into this mouse, these cells will indeed succeed in forming a tumor.

Such behavior indicates that the cells in the two independently induced sarcomas are antigenically distinct, and that cells of the first sarcoma carry one or more unique antigenic determinants that evoke an immune response rendering a mouse resistant to subsequently introduced cells derived from the same sarcoma. This suggests that the mutagenic carcinogen (3-MC) that originally induced these tumors also created one or more mutant cellular genes in the tumor cells whose products functioned as TSTAs. Moreover, it seems that each time 3-MC caused a tumor to form, it generated a distinct TSTA or set of TSTAs (Figure 15.23).

The actual detection and identification of proteins that function as TSTAs are often challenging experimentally. Imagine that a 3-MC–induced sarcoma expresses a potently antigenic TSTA. As a consequence, the immune system of a tumor-bearing mouse will respond vigorously to this TSTA, attempting to eliminate those cells within the sarcoma that display high levels of this TSTA while sparing those cells in the tumor that display only very low levels of this antigen. The resulting Darwinian selection will dictate that only those tumor cells and

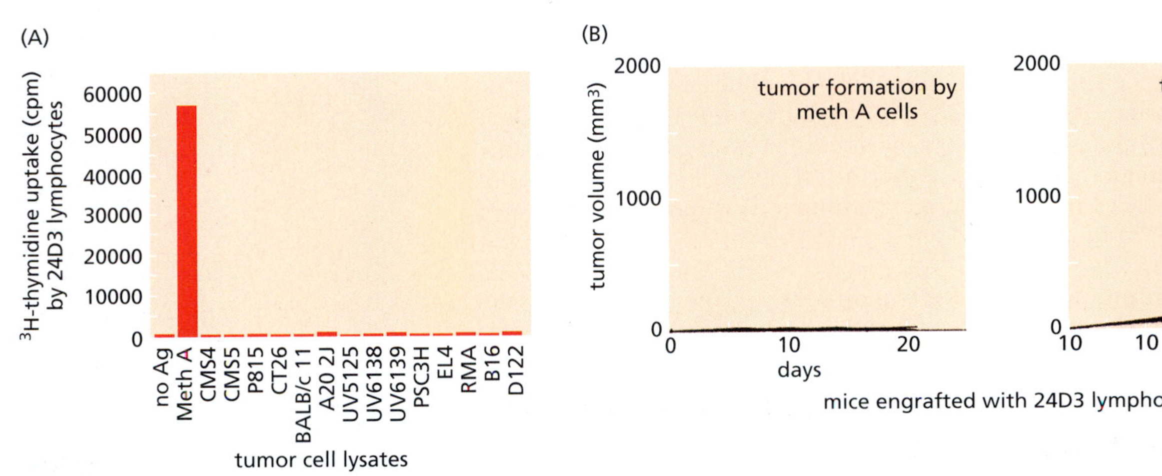

Figure 15.23 Specificity of the antigen display by a chemically induced fibrosarcoma Mice of the BALB/c strain were immunized with lysates of cells of the 3-methylcholanthrene–induced Meth A fibrosarcoma line. A line of antigen-presenting lymphocytes, termed 24D3, was developed from these mice. (A) As gauged by their incorporation of ^{3}H-thymidine, proliferation of these lymphocytes in culture was stimulated by addition of Meth A cell lysates *(2nd bar)*. However, lysates prepared from 14 other tumor cell lines, including other methylcholanthrene-induced sarcomas, UV-induced squamous cell skin carcinomas, lymphomas, a melanoma, and a lung carcinoma, failed to stimulate proliferation of these lymphocytes *(remaining channels)*. In the absence of antigen *(1st channel)* no proliferation was seen. (B) When a clonal population of these 24D3 lymphocytes was introduced into BALB/c hosts, the formation of tumors by subsequently injected Meth A fibrosarcoma cells was fully blocked *(left panel)*. However, the formation of tumors by an unrelated fibrosarcoma line, termed CMS5, was unaffected by the presence of these lymphocytes *(right panel)*. (From T. Matsutake and P.K. Srivastava, *Proc. Natl. Acad. Sci. USA* 98:3992–3997, 2001.)

their descendants expressing low levels of TSTA will survive long enough to be studied by an experimenter, greatly complicating the biochemical isolation and identification of the TSTA protein.

In recent years, several of the genes encoding 3-MC–induced TSTAs have nevertheless been isolated by gene cloning procedures. In one cloning strategy, the oligopeptides that were bound to MHC class I molecules on the surfaces of 3-MC–transformed cells and served as targets for immune recognition were eluted from the MHC molecules, purified, and subjected to amino acid sequencing. The resulting amino acid sequences were then used to predict the nucleotide sequences of the encoding genes, which made possible the cloning of these genes. Sequence analyses of the TSTA-encoding genes cloned from these tumors showed that the genes were all point-mutated alleles of normal cellular genes encoding various cellular proteins, none involved in any obvious way in the transformation of these cells (Sidebar 41 ●).

These observations suggest that during the course of chemical carcinogenesis, the 3-MC carcinogen, a known point mutagen (Section 12.6), mutates both a proto-oncogene (often the K-*ras* gene) in target cells and additional genes that, as mutant alleles, specify TSTAs; the latter genes are struck at random—innocent bystanders that play no causal role in tumorigenesis but happen to have been damaged by the large doses of mutagenic carcinogen used to provoke tumor formation.

Importantly, the behavior of these chemically induced TSTAs is quite different from that of the TSTAs resulting from tumor virus infection. For example, SV40 virus can be used to induce a sarcoma in a mouse. Subsequent removal of this SV40-induced sarcoma will result in immunization of the mouse against subsequently inoculated tumor cells that derive from this particular SV40-induced sarcoma as well as any other tumors that have been induced by SV40 virus. In this instance, there is indeed a **cross-immunity** established, in that all the SV40-induced tumor cells seem to share a common TSTA or set of TSTAs. It happens that the dominant TSTA responsible for this cross-immunity is a familiar protein: it is the virus-encoded large T oncoprotein, which is expressed at significant levels in all SV40 virus–transformed cells. This contrasts with the behavior of a group of 3-MC–induced cancers, where each tumor expresses its own unique TSTA or set of TSTAs.

Observations like these raise the question whether similar mechanisms operate during human tumorigenesis. Thus, do the highly mutable genomes of cancer cells (Chapter 12) generate mutant, antigenic proteins as inadvertent by-products of the mutagenesis that drives tumor progression (see Sidebar 15.1)? Or are the 3-MC–induced TSTAs artifacts of the high doses of mutagenic carcinogens used in many mouse tumorigenesis experiments that do not accurately reflect the mutagenic processes that create human tumors?

15.13 Tumor-associated transplantation antigens may also evoke anti-tumor immunity

As noted above, tumor-associated transplantation antigens (TATAs) represent normal cellular proteins that, for one reason or another, have failed to induce tolerance. When these normal proteins are expressed by tumors they evoke a measurable immune response, often involving both the humoral and cellular arms of the immune system.

For a variety of reasons, the antigenicity of melanomas has been more intensively studied than that of all other human tumors (Sidebar 42 ●). Much of their antigenicity stems from their display of certain TATAs. Melanoma cells may

Sidebar 15.1 **Microsatellite instability often leads to more immunogenic tumors** As described in Section 12.4, defects in the DNA mismatch repair machinery create the condition of microsatellite instability (MIN), which leads to mutations accumulating in hundreds, possibly thousands of cellular genes within tumor cell genomes. Among other consequences, these mutations generate shifts in the reading frames of many of these genes. The resulting mutant alleles often encode novel amino acid sequences, sometimes termed "frameshift peptides," some of which may function as potent tumor-specific antigens.

This logic predicts that the 15% of human colorectal cancers that exhibit microsatellite instability should interact with the host immune system differently from the majority of colorectal cancers that show no MIN and instead exhibit chromosomal instability (CIN). In fact, the MIN tumors show a markedly higher degree of tumor-infiltrating lymphocytes (TILs) and a lower degree of metastasis. Moreover, antigen presentation by their MHC class I proteins is compromised far more frequently than in colorectal tumors with chromosomal instability (60% vs. 30%), suggesting that the MIN tumors are under greater pressure to evade killing by various arms of the immune system, and that they undertake the immunoevasive maneuver of blocking antigen presentation by their cell-surface MHC class I proteins. Taken together, these observations suggest that the MIN that enables some colorectal tumors to evolve more rapidly can exact a price in the markedly higher immunogenicity of these tumors.

overexpress certain proteins that are present in their normal melanocyte precursors, albeit at lower levels. Such lineage-specific proteins are sometimes called *differentiation antigens*, implying that their display is a vestige of the differentiation program that previously governed the behavior of the normal cellular precursors of tumor cells. Included among the melanoma TATAs are transferrin, tyrosinase (Figure 15.24A), gp100, Melan-A/MART-1, and gp75.

The display of these differentiation antigens by melanoma cells often provokes a vigorous response by the immune system, which results in a very peculiar form of autoimmune disease—**vitiligo**—the depigmentation of large areas of skin seen in some melanoma patients (Figure 15.25). This depigmentation is a specific response to the presence of a melanoma. For example, when 104 renal carcinoma patients were treated with the cytokine interleukin-2 (IL-2) in order to enhance their anti-tumor immune responses, none developed vitiligo; in contrast, of 74 melanoma patients who were treated similarly, 11 developed vitiligo.

In these melanoma patients, it is clear that the immune response provoked by the melanoma TATAs leads, as a by-product, to attack and destruction of normal melanocytes, which also express these antigens. This type of vitiligo is formally analogous to the paraneoplastic syndromes discussed earlier (Sidebar 40), in which the display by tumors of cellular proteins results in the destruction of normal tissues that also happen to express these proteins. Significantly, melanoma patients showing vitiligo usually survive for longer periods than those who don't—suggesting that their immune systems are effective in controlling the melanomas, at least for a period of time. (For example, in a large population of melanoma patients described in 1987, 75% were still alive five years after initial diagnosis; among the subgroup of these patients who exhibited concomitant vitiligo, 86% survived for this period of time.)

The antigenicity of human melanoma cells may also derive from their display of the other major subclass of TATAs, the **oncofetal** antigens—literally those antigens that are displayed during embryogenesis and once again by tumors. Included among these are the antigens called either cancer germ-line or cancer-testis (CT) antigens, to reflect their normal expression in the germ cells of the testis and the fetal ovary. The genes for a number of these antigens, such as

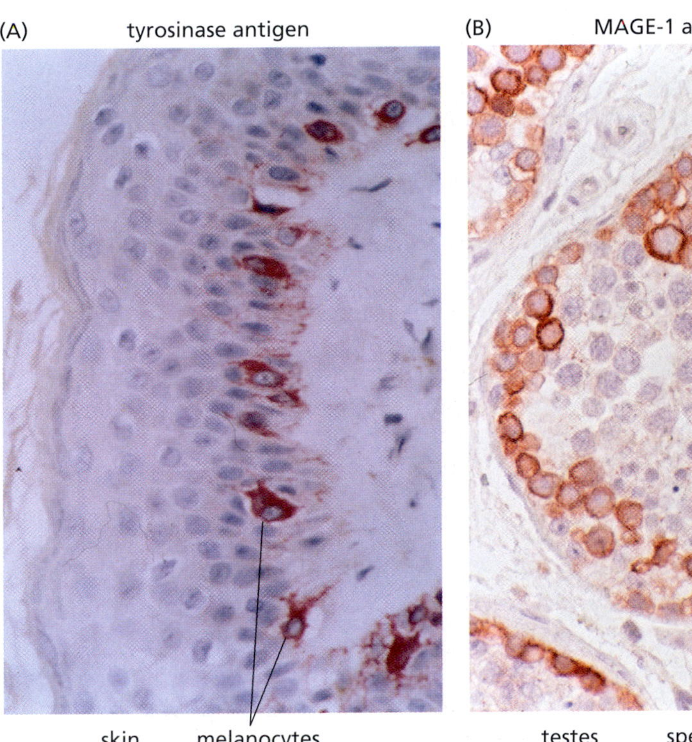

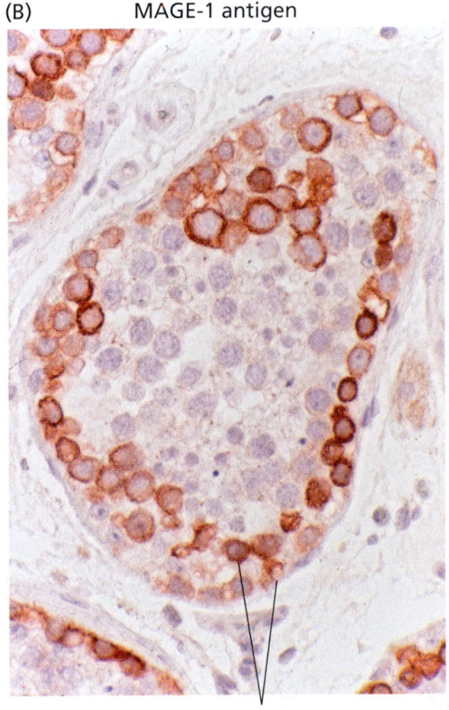

Figure 15.24 Normal proteins displayed as tumor-associated antigens on melanoma cells (A) The tyrosinase enzyme, which is involved in pigment production in melanocytes and melanomas, is detected here in the melanocytes of the skin (*red,* located just above the basement membrane in the skin) through the use of a monoclonal antibody; it is not detectable in other normal tissues. Its expression by melanoma cells can cause them to become immunogenic and the target of killing by cytotoxic lymphocytes. (B) The spermatogonia in the testis have been stained here with a monoclonal antibody against the MAGE-1 antigen; normally this antigen is seen only in the placenta. Its expression has been detected in a variety of human tumor types and has been studied in detail in melanomas because it is often immunogenic when expressed by these tumors. (A, from Y.T. Chen, E. Stockert, S. Tsang et al., *Proc. Natl. Acad. Sci. USA* 92:8125–8129, 1995; B, from J.C. Cheville and P.C. Roche, *Mod. Pathol.* 12:974–978, 1999.)

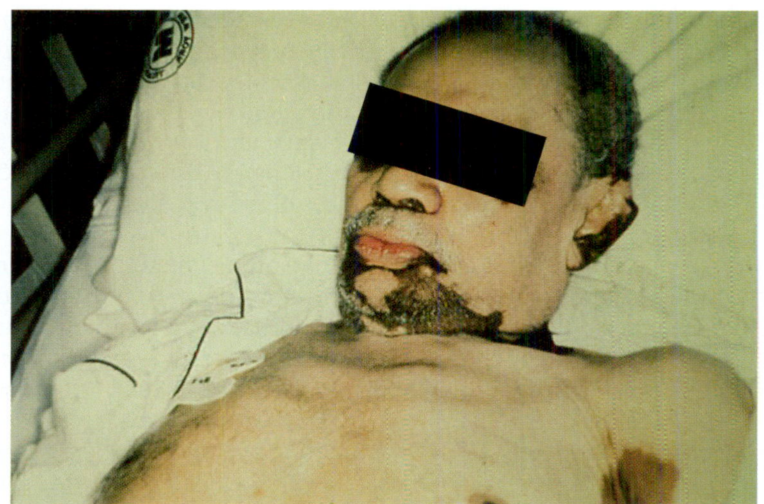

Figure 15.25 Autoimmune depigmentation provoked by melanomas The melanoma patient shown here, who was dark-skinned prior to the onset of melanoma, has lost almost all of his skin pigment except for several isolated areas (face, armpit) due to the autoimmune attack incited by the melanoma cells. The condition of pigment loss, known as *vitiligo,* is often correlated with a longer survival of melanoma patients. (Courtesy of A.N. Houghton.)

MAGE-1 (Figure 15.24B), MAGE-3, BAGE, GAGE-1, and GAGE-2, have been cloned (Sidebar 43). By 2003, 44 genes or gene families encoding a total of 89 distinct cancer-testis antigens had been identified.

As an aside, it seems that the absence of immune responses against these antigens in males is likely due to the fact that several of the cell types in the testes do not express MHC class I molecules and are thereby prevented from presenting their internal contents to the immune system. (Of course, females may never express these proteins in any of their tissues.) In addition, the expression by cells within the testes of significant levels of FasL, the ligand of the Fas death receptor (Section 9.14), may also be used by these cells to keep wandering T lymphocytes and other immunocytes at a safe distance, further ensuring that the proteins expressed by these testicular cells are tolerated within these immunologically privileged sites and do not provoke immune responses. All this may help to explain why the display of these germ-cell proteins in an ectopic site in the body frequently provokes a vigorous immune response.

The conclusion that tolerance for many of these melanoma proteins can be readily circumvented ("broken," in the language of immunologists) leads to a simple and obvious strategy for anti-cancer therapy, in which these antigens are treated as if they were the products of invading foreign agents, such as viruses or bacteria. Such thinking has led to attempts to immunize mice or humans with a vaccine consisting of one or another of these antigenic proteins.

For example, some melanoma cells will overexpress by a factor of 100 the normal receptor for transferrin, a protein that is involved in iron uptake and metabolism in many cell types throughout the body. Mice can mount an immune response against this protein following injection with purified murine transferrin receptor, indicating that any tolerance they may have had toward this protein can be readily broken through exposure to it at high levels. These vaccinated mice will reject any subsequently introduced melanomas that happen to overexpress the mouse transferrin receptor, suggesting a more general strategy for causing cancer patients to develop potent immunity against the tumors that they carry. We will return later in this chapter to strategies that can be used to mobilize the immune system to attack human tumors.

15.14 Cancer cells can evade immune detection by suppressing cell surface display of tumor antigens

The descriptions of immune function cited earlier might suggest that attempts by the immune system to erect defenses against the outgrowth of most tumors are relatively rare, in large part because tolerance causes the immune system to be blind to the antigens displayed by most types of cancer cells. In fact, relatively high titers (concentrations) of antibody molecules that bind tumor cell-surface antigens are often encountered in the blood of cancer patients; they may even be present in most patients. Both helper and cytotoxic T cells that recognize tumor-associated transplantation antigens (TATAs) are also readily detectable in the blood, lymph nodes, and tumors of many patients.

These observations indicate that in spite of the phenomenon of immunological tolerance, the human immune system succeeds in mounting attacks on many (and perhaps all) types of cancer cells. That some of these attacks do indeed stem tumor development is indicated by the elevated cancer incidence in immunocompromised individuals (Section 15.9).

Still, the development of many human cancers is clearly not blocked successfully by immunological defense mechanisms that are, by some measures, extraordinarily efficient. For example, immunologists have found that the display (by MHC class I molecules) of as few as 10 copies of a TATA- or TSTA-derived oligopeptide on the surface of a tumor cell suffices to attract cytotoxic T cells expressing the appropriate antigen-recognizing T-cell receptor, which then proceed to kill this tumor cell.

The repeated failures of the human immune system to use these potent lymphocytes and other means of immunological attack to effectively block tumor development require some explanation. The most attractive scenario is suggested by the behavior of the genetically altered mice described earlier (Section 15.8): some human tumors arise that are strongly antigenic, and these are efficiently eliminated by normal immune systems; such tumors are rarely if ever detected within the human body. Other tumors are formed that are from the outset poorly antigenic because they express only proteins to which the immune system is tolerant; these tumors thrive and become clinically apparent.

Another possibility is that some antigenic cancers initially may suffer severe attrition because of attack by one or another arm of the immune system but find ways to escape elimination—the strategies of **immunoevasion** (Table 15.3). Such cells may then flourish, and their progeny may go on to create large, life-threatening tumors. This last scenario may well describe the history of most human cancers, as suggested by frequently observed changes in human cancer cells that can only be interpreted as maneuvers undertaken by these cells in order to evade immune attack.

The most obvious immunoevasive maneuver that a tumor cell can undertake is to stop displaying a tumor-associated (TATA) or tumor-specific (TSTA) transplanta-

Table 15.3 Immunoevasive strategies used by cancer cells

Strategy	Mechanism	Agent being evaded
Hide identity	repress tumor antigens (TATA or TSTA), repress MHC class I proteins	cytotoxic T lymphocytes
Hide stress	repress NKG2D ligands (e.g., MICA)	NK cells
Inactivate immunocytes	destroy immunocyte receptors	NK cells; cytotoxic T lymphocytes
	saturate immunocyte receptors with adenosine, MICA	NK cells; variety of immunocytes
Avoid apoptosis	inhibit caspase cascade by increasing IAPs, acquire resistance to FasL-mediated apoptosis	
Induce immunocyte apoptosis	release soluble FasL	cytotoxic T lymphocytes
	release cytokines (IL-10, TGF-β)	cytotoxic T lymphocytes, dendritic cells, macrophages
Neutralize intracellular toxins	enzymatic detoxification of H_2O_2, prostaglandin E_2	macrophages, NK cells
Neutralize complement	overexpress mCRPs	complement system

tion antigen that has attracted the attention of the immune system and its cytotoxic lymphocytes. Importantly, the great majority of these antigens represent molecules that do not participate *causally* in neoplastic transformation, but instead only reflect tissue-specific differentiation antigens, often those typically expressed in the tissues in which tumors have arisen (see Figure 15.24). Expression of these antigens can often be repressed by cancer cells with impunity, that is, without compromising the continued survival and proliferation of these cells. In fact, tumor cell populations often harbor variants in their midst that have suppressed expression of antigen-encoding genes by promoter methylation (Section 7.8). These antigen-negative variants may therefore be able to escape an immune attack that has been mounted against the bulk of the cells in a tumor mass, and their descendants may eventually emerge as the dominant cell population in this mass.

In one illustrative case, a melanoma patient was vaccinated with tyrosinase protein in order to induce immune reactivity against this protein, which was being expressed by his melanoma cells. (The tyrosinase enzyme participates in pigment synthesis by melanocytes and is therefore not involved in causing neoplastic transformation; see Figure 15.24A.) At first, the resulting immune response caused his melanoma metastases to regress. Soon, however, some tyrosinase-negative variant cells emerged among the melanoma metastases. While his tyrosinase-positive melanoma cells continued to regress, the tyrosinase-negative cells began to proliferate rapidly and eventually killed the patient, a phenomenon sometimes called "immune escape" by tumor immunologists. However, in another case of melanoma, this type of immunoevasion did not guarantee tumor cells long-term protection from immune attack (Sidebar 15.2).

In many tumors, the cancer cells cannot resort to the simple expedient of repressing expression of a tumor antigen, simply because the continued expression of a TSTA or TATA antigen may be essential for their continued neoplastic proliferation. For example, the overexpressed but otherwise normal HER2/Neu protein—the growth factor receptor displayed by approximately 30% of breast cancers (see Figure 4.6)—may well stimulate an immune attack. However, the neoplastic cells in these breast carcinomas cannot afford to shut down the display of this protein, because its continued expression at high levels is critical for their proliferation and their ability to avoid apoptosis.

Consequently, cancer cells that cannot down-modulate expression of a TATA or TSTA because of its essential contribution to neoplastic growth may resort to alternative strategies in order to avoid immune killing. An important and widely used immunoevasive strategy derives from the ability of cancer cells to repress expression of their MHC class I proteins that are responsible for antigen presentation (see Sidebar 36 ◉). This is often achieved through repression of MHC class I gene transcription (Sidebar 15.3). Actually, many types of human cancer cells have been found to lack normal levels of these MHC molecules, thereby preventing oligopeptide antigens from appearing on the surfaces of these cells (Figure 15.27).

Loss of MHC class I protein expression is often associated with more invasive and metastatic tumors. For example, in more than half of advanced breast cancers, the display of these critical antigen-presenting molecules has been totally lost, and such loss correlates with poor prognosis. In addition, the carcinoma cells that form micrometastases in the bone marrow (Figure 14.10) often show little if any MHC class I expression, possibly because of intensive surveillance by the immune cells operating in this tissue environment.

Cancer cells can also use post-translational mechanisms to reduce MHC class I–mediated antigen presentation. For example, the migration of MHC class I molecules from the endoplasmic reticulum to the cell surface depends on their

Sidebar 15.2 Immunoevasion may offer tumor cells only a temporary reprieve from immune attack Five years after resection of a primary melanoma, a melanoma patient returned with multiple metastases in lymph nodes, which were removed surgically (Figure 15.26). On this occasion, the tumor-infiltrating cytotoxic lymphocytes (CTLs) recognized MART-1, one of the well-studied melanoma-associated tumor antigens. The patient was free of symptoms for another six years, when he developed a recurrence of melanoma in a regional lymph node, which was also removed. Now, the MHC class I–mediated presentation of MART-1 oligopeptides was absent from the tumor cells, and the CTLs present in the tumor that could recognize MART-1 were almost gone. (MART-1 synthesis by the tumor cells continued, however.) Instead, the tumor-associated CTLs recognized a second melanoma-associated antigen—tyrosinase—which was now being presented by the tumor cells' MHC class I molecules.

This behavior indicated that the patient's immune system was able to adapt dynamically to the initial evasive maneuver of the melanoma cells (shutdown of MART-1 antigen presentation) by redirecting its energies to mount an attack on another tumor-associated antigen (tyrosinase). However, by the time of the third surgery, the patient's melanoma cells had begun to lose almost all MHC class I expression. This loss should have eventually allowed the disseminated tumor cells that remained after the third surgery to escape all CTL attack and to multiply rapidly into life-threatening metastases. Nonetheless, the patient survived for another five years and died of causes unrelated to his melanoma. His ability to survive a usually highly aggressive malignancy for another five years suggested that his immune system was able to keep any residual melanoma cells in check through mechanisms that remain unclear. Quite possibly, NK cells, which are known to attack cells lacking MHC class I expression (to be discussed shortly), were responsible for the last five, symptom-free years of this patient's life.

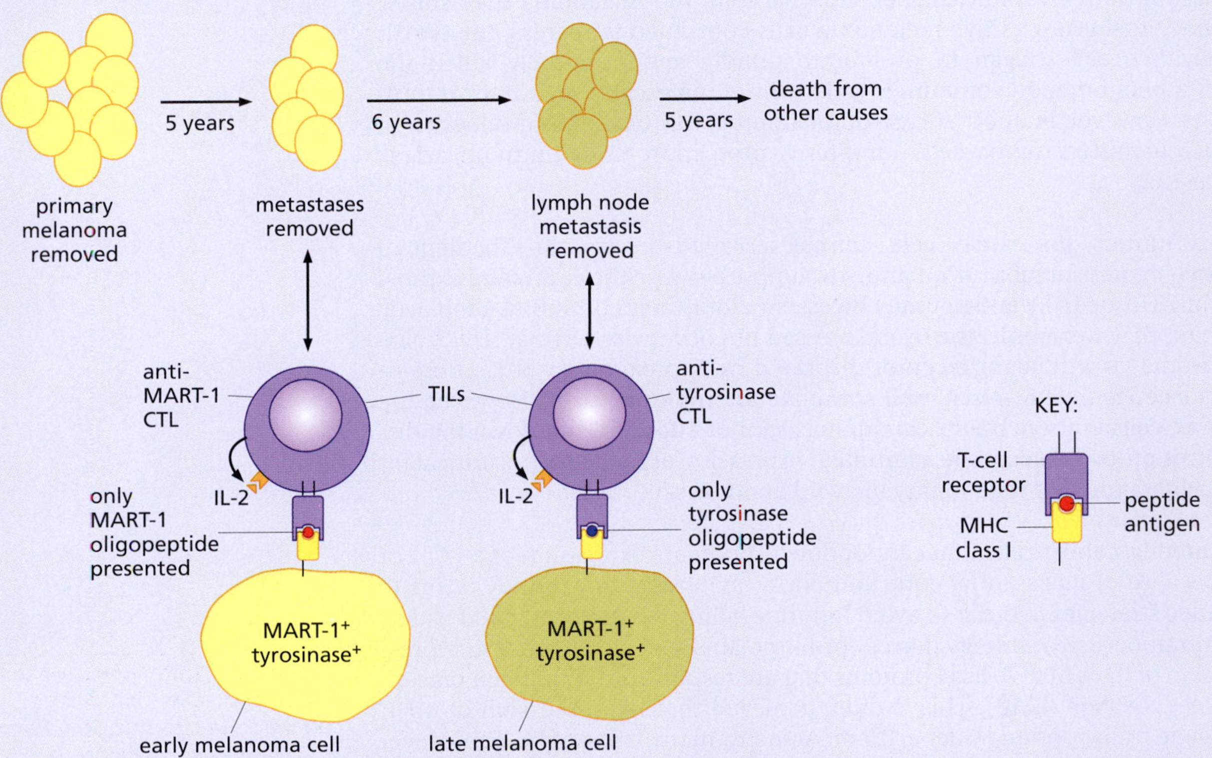

Figure 15.26 Dynamic adaptation by the immune system to shifting tumor cell antigen presentation A melanoma tumor was removed, and 5 years later, when the metastases appeared and were removed, the tumor cells presented only MART-1 oligopeptide antigen on their MHC class I receptors *(red dot)* even though these cells synthesized both MART-1 and tyrosinase proteins; at this time, the tumor-infiltrating cytotoxic lymphocytes (CTLs) recognized MART-1 antigen via their T-cell receptors. Six years later, when a lymph node metastasis was discovered and removed from this patient, the tumor cells continued to make both MART-1 and tyrosine, but now their MHC class I molecules presented only tyrosinase oligopeptide antigen *(blue dot)* and the tumor-infiltrating CTLs recognized only the tyrosinase antigen. Hence, the immune system was able to co-evolve with the tumor cells, generating CTLs of novel specificity as required in order to combat the melanoma cells.

association with β_2-microglobulin protein (β_2m). Normally, β_2m escorts the MHC molecules together with their oligopeptide cargo to the cell surface (Figure 15.28A). In certain high-grade tumors, however, the lack of β_2m synthesis prevents the oligopeptide-loaded MHC class I molecules from reaching the

Sidebar 15.3 Oncoproteins can help tumors to evade immune attack In addition to their effects in driving cancer cell proliferation, some oncoproteins can help these cells to evade immune detection. Thus, in some cancers, the actions of Myc and its close cousins, N-Myc and L-Myc, are responsible for reducing MHC class I expression. In childhood neuroblastomas, this loss is often attributable to overexpression of the N-*myc* gene (Section 4.5). In melanomas, loss of HLA-B (one type of MHC class I molecule) expression is correlated with *myc* overexpression. And in small-cell lung cancers (SCLCs), overexpression of L-myc correlates with loss of MHC class I expression by tumor cells.

In many of these tumors, the loss of MHC class I expression seems to be due to a complex sequence of events: the overexpressed N-Myc and L-Myc proteins block expression of another transcription factor, termed CIITA, that is responsible for interferon-γ–mediated induction of MHC class I and class II gene expression. In many commonly occurring tumors, however, such as melanomas and carcinomas of the colon and prostate, suppression of MHC class I expression is achieved through poorly understood molecular mechanisms.

cell surface (Figure 15.28B). An earlier step in antigen presentation may also be compromised: in some tumors, defects in the TAP1 or TAP2 (transporter associated with antigen presentation) proteins can be found (Figure 15.28C). These two proteins are needed to transport oligopeptides generated by the proteasomes in the cytosol to the MHC class I molecules in the lumen of the endoplasmic reticulum. Without both TAP proteins, antigen presentation by the MHC molecules also fails.

Because they are only correlative, these observations of crippled MHC-mediated antigen presentation do not prove that these defects are *causally* involved in enabling cancer cells to evade immune attack. Evidence of a direct link between defective MHC function and immunoevasion is limited and anecdotal. For example, in one well-documented case of a human melanoma that was being treated by anti-tumor immunotherapy, cells emerged that had down-regulated MHC class I expression and were resistant to therapy. In other cases, such as many neuroblastomas and small-cell lung carcinomas (SCLCs), the loss of MHC class I expression is correlated with the virtual absence of tumor-infiltrating lymphocytes (TILs; Section 15.9), which are thought to be key elements in leading the immune attack on tumors.

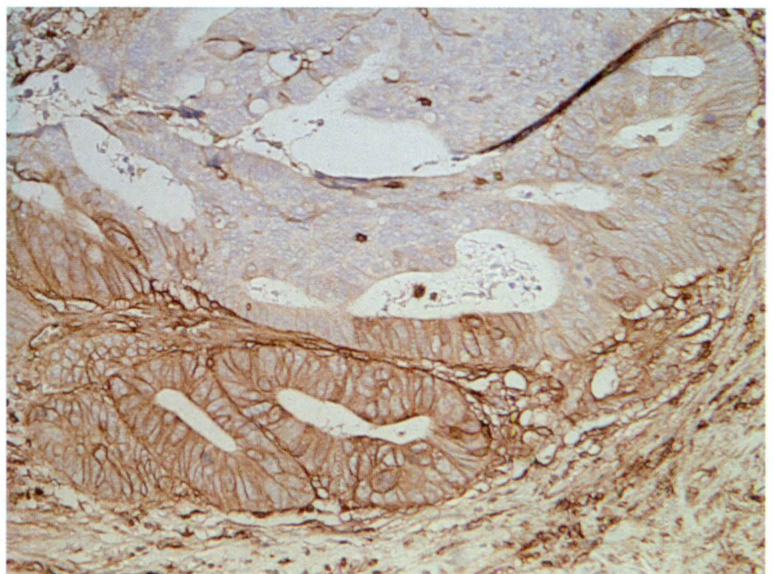

Figure 15.27 Immunoevasion through suppression of MHC class I expression Cancer cells will often down-regulate expression of the MHC class I molecules, ostensibly in order to avoid recognition and attack by cellular components of the adaptive immune response. In this immunohistochemical staining, cells of a human colorectal tumor are visualized. They have been immunostained with an antibody that recognizes a human MHC class I molecule—HLA-A. As seen here, the cells in the lower part of the tumor strongly express HLA-A, while the cells in the upper part of the tumor have partially or totally lost HLA-A expression. (From A.G. Menon, H. Morreau, R.A. Tollenaar et al., *Lab. Invest.* 82:1725–1733, 2002.)

In fact, the total absence of MHC class I molecules invites an attack by NK cells, which continuously patrol the body's tissues looking for cells that have lost these key proteins from their surface. This explains why some tumor cells will selectively suppress expression of only one of the six key MHC class I molecules that are normally expressed concomitantly by cells throughout the body. This suppression may block presentation of a particular tumor antigen, thereby allowing the tumor cell to survive free of attack by antigen-specific cytotoxic lymphocytes. However, because only a small proportion of the total population of cell surface MHC class I molecules has been lost, these cells may not provoke an attack by NK cells.

Taken together, these diverse observations suggest that by attacking cancer cells displaying MHC class I molecules and associated oligopeptide antigens, the immune system creates great selective pressure for the outgrowth of variant cancer cells that no longer display these antigen-presenting molecules on their

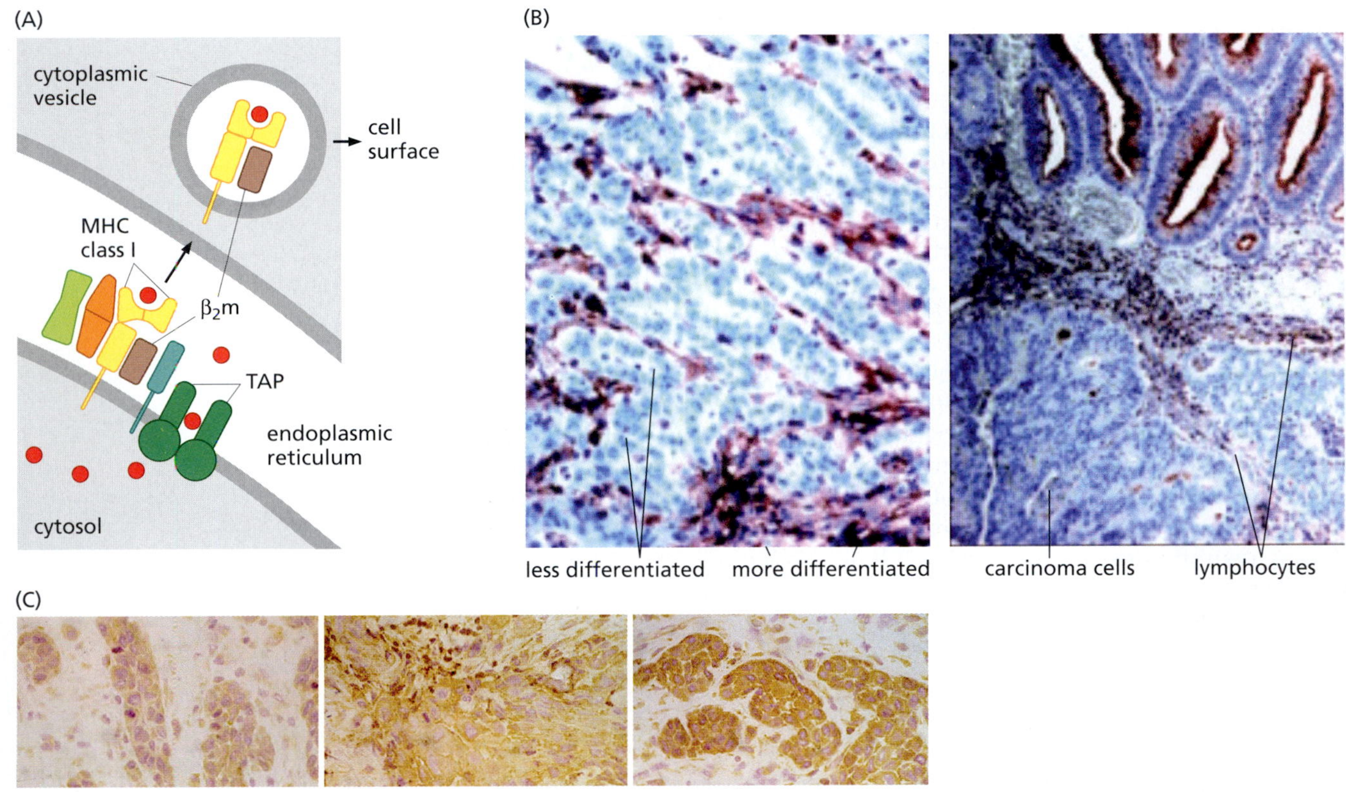

Figure 15.28 Failure of antigen presentation due to loss of β_2-microglobulin or TAP1 (A) A more detailed depiction of Figure 15.10A reveals that the TAP complex (a heterodimer of the TAP1 and TAP2 proteins (**t**ransporters associated with **a**ntigen **p**rocessing; *green*) is involved in transporting oligopeptides *(red dots)* from the cytosol into the lumen of the endoplasmic reticulum, where the oligopeptides can associate with MHC class I molecules *(yellow)*. Independent of this, the β_2-microglobulin (β_2m, *brown*) molecule associates with MHC class I molecules and helps to shepherd them, together with their captured oligopeptide antigens, to the cell surface. The MHC class I molecules require both a bound oligopeptide antigen and an associated β_2m molecule in order to reach the cell surface (see also Figure 15.18B). Therefore, defects in either the β_2m or a TAP protein preclude presentation of oligopeptide antigens on the cell surface. (B) In one human colorectal carcinoma *(left panel)*, β_2m staining *(reddish brown)* is present in the cancer cells forming the more differentiated tubular structures *(below)*, while the less differentiated carcinoma cells *(light blue, above)* fail to express any β_2m. In a second colorectal carcinoma *(right panel)*, β_2m is present in infiltrating lymphocytes *(dark red)* but is totally absent in many of the carcinoma cells *(light blue, bottom half)*. In 5 of 17 microsatellite-unstable (MIN) colorectal tumors (see Sidebar 15.1), the *β_2m* gene was found to have suffered inactivating mutations, while this gene was transcriptionally silenced through unknown mechanisms in a number of other tumors. (C) Defective antigen processing is frequently seen in head-and-neck squamous cell carcinomas (HNSCCs), in which expression of TAP1 or TAP2 is often lost. The immunohistochemical staining shown here reveals the expression of TAP1 in a group of HNSCCs to be variable, being absent in some tumors *(left)*, present in patches in a some others *(middle)*, and normally expressed in yet others *(right)*. (A, from C.A. Janeway Jr. et al., Immunobiology, 6th ed. New York: Garland Science, 2005; B, from M. Kloor, C. Becker, A. Benner et al., *Cancer Res.* 65:6418–6424, 2005; C, from M. Meissner, T.E. Reichert, M. Kunkel et al., *Clin. Cancer Res.* 11:2552–2560, 2005.)

surface. Since the display of these cell surface molecules tends to be lost in invasive and metastatic cells, it may be, as speculated earlier, that the immune system only launches a full-scale attack on cancer cells once they have invaded the stroma, where immune cells are present in substantial numbers.

15.15 Cancer cells protect themselves from NK-mediated attack

As mentioned several times above, the evasive tactic of suppressing MHC class I expression, clever as it may be, carries its own dangers: the immune system anticipates this trick and uses its natural killer (NK) cells to attack cells that lack adequate numbers of MHC class I molecules on their surface. This NK response can also foil the plans of tumor viruses that attempt to elude immune detection by preventing infected cells from displaying MHC class I molecules and thus viral oligopeptide antigens on their surface.

A receptor normally displayed on the surface of NK cells recognizes MHC class I molecules displayed by potential target cells (Figure 15.29). Binding of the target cell MHC class I molecules to this NK cell receptor (called killer inhibitory receptor, or KIR) causes this receptor to release signals into the NK cytoplasm indicating that any attempts to proceed with the killing of the MHC class I–positive cell are inappropriate and should be curtailed immediately. Conversely, these suppressive signals are missing when an NK cell encounters a potential target that lacks MHC class I molecules on its surface, in which case target cell killing is permitted to proceed. (This elimination of MHC class I–negative cells by NK cells may explain why, in certain classes of human tumors, the absence of class I expression actually correlates with a better clinical outcome—a behavior opposite to that discussed earlier.)

Yet another innate interaction of NK cells with cancer cells appears to be equally important. This particular interaction depends on the fact that many types of human cells are programmed to display specific proteins on their surface whenever these cells suffer certain physiologic stresses, including the stresses resulting from viral infections and neoplastic transformation. These stress proteins have names such as MICA, MICB, ULBP4, and so forth (Figure 15.30A). NK cells, for their part, display a complementary cell surface receptor, called NKG2D, that

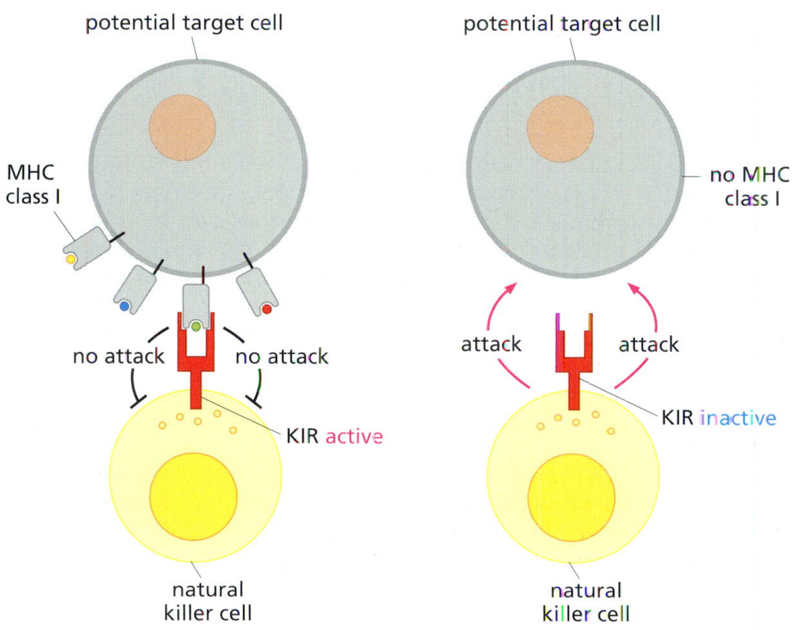

Figure 15.29 Control of natural killer cell–mediated killing by the KIR receptor Natural killer (NK) cells *(yellow, bottom)* display a killer inhibitory receptor (KIR, *red*) that can recognize MHC class I molecules being displayed on the surface of potential target cells *(gray, above)*. In the event that such recognition occurs, the KIR releases inhibitory signals that prevent attack by the NK cell *(left)*. Conversely, in the absence of such recognition and the functional silencing of KIR, attack by the NK cell is permitted *(right)*. (Note that KIR recognizes and binds a constant structural feature of MHC class I molecules and conversely does not recognize an oligopeptide antigen ensconced in the antigen-binding pocket of the MHC class I molecule.)

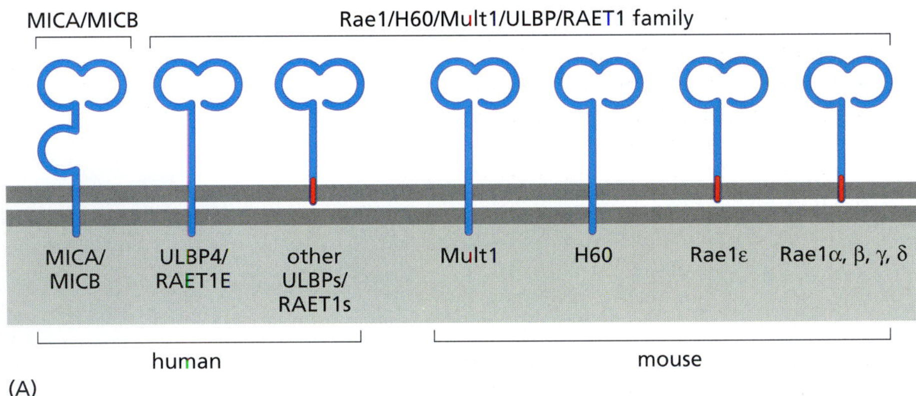

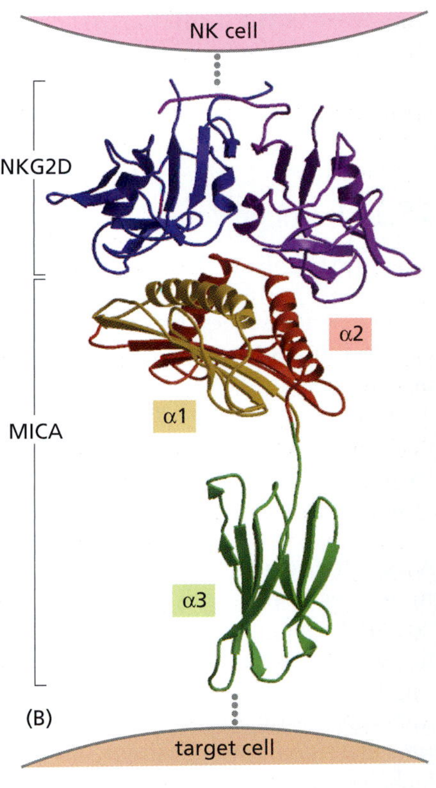

Figure 15.30 Regulation of NK function by the NKG2D receptor (A) The NKG2D receptor displayed by NK cells binds a number of ligands that are expressed by other cell types suffering certain physiologic stresses, notably the stresses caused by viral infection and neoplastic transformation. Shown here is an assembly of various alternative ligands of NKG2D that have been identified in human and murine cells. For example, MICA and MICB are normally expressed only by certain epithelial cells suffering certain types of cell-physiologic stress. (B) The molecular structure of the complex between NKG2D and its MICA ligand is shown here. The two subunits of the NKG2D receptor are shown above *(blue, purple)* as they might be displayed on the surface of an NK cell. The structure of the three subdomains of the ectodomain of MICA *(beige, red, green)* is shown as it might be displayed on the surface of a cancer cell. (The domains that anchor these proteins to the surfaces of the NK and the cancer cell are not shown.) (A, from D. Raulet, *Nat. Rev. Immunol.* 3:781–790, 2003; B, from P. Li, D.L. Morris, B.E. Willcox et al., *Nat. Immunol.* 2:443–451, 2001.)

is specialized to recognize these stress-associated cell proteins on the surfaces of potential target cells (Figure 15.30B). Binding of these proteins to the NKG2D receptor results in strong activation of an NK cell's cytotoxic response and is therefore followed rapidly by killing of cells that express these stress-associated proteins on their surfaces.

As one demonstration of these dynamics, a gene encoding the mouse homolog of the human MICA antigen has been introduced into mouse lymphoma cells, causing them to express significant levels of this antigen on their surface. These cells thereupon lost their tumorigenicity when injected into host mice. However, if these mice were first deprived of their NK cells, then the lymphoma cells expressing the MICA homolog once again became tumorigenic.

In summary, NK cells instinctively recognize cancer cells in two ways: they sense either the absence of MHC class I proteins, or the presence of one or more stress-associated proteins (e.g., MICA) on the surfaces of cancer cells. When both conditions are satisfied, the killing of cancer cells is far more efficient than the killing of cells that satisfy only one or the other of these conditions. Indeed, some immunologists believe that both conditions must be satisfied (i.e., low levels of MHC class I and high levels of stress-associated NKG2D ligand such as MICA) before NK killing proceeds.

In the event that expression of NKG2D ligands threatens cancer cells, their immunoevasive response is straightforward: they suppress expression of these ligands, such as MICA, on their surface, thereby depriving NK cells of an important means of recognizing them. Precisely how this suppression is usually achieved is not well understood, although promoter methylation has been demonstrated in a small number of cases.

Actually, many types of human carcinoma cells and melanoma cells disable this NKG2D signaling pathway through an alternative immunoevasive strategy: they continue to synthesize significant amounts of MICA but shed much of this protein into the medium around them (rather than retaining it attached to their cell surface). The released, soluble MICA can be detected in the serum of many cancer patients and acts as a decoy ligand that distracts the NKG2D receptors of their NK cells. Thus, soluble MICA binds the NKG2D receptors on their NK cells (and cytotoxic T cells) and causes the endocytosis and degradation of these receptors, thereby rendering these immunocytes incapable of responding to the MICA molecules that continue to be displayed on the surfaces of patients' cancer cells.

To conclude our discussion of tumor cells and their interactions with NK cells, it is also worthwhile noting the peculiar way in which metastasizing cancer cells in the circulation avoid being ambushed by NK cells. As was described in Sidebar 27, cancer cells that have intravasated and thus come into direct contact with the blood rapidly acquire a cloak of adhering platelets that, together with the cancer cells, form microthrombi. When the coagulation mechanisms that create the microthrombi are rendered defective through one or another experimental strategy, the success rates of the metastasizing cells in founding new tumor colonies plummet to several percent of what is otherwise observed. The attrition is due to attacks by NK cells, which are normally prevented from recognizing and attacking their neoplastic cell targets by the platelet cloaks. This particular immunoevasive maneuver is not developed through the selection of rare variant cancer cells within a tumor, but instead is almost inadvertent, being deployed routinely by all cancer cells that have managed to enter into the circulation and begun their journeys to distant tissue sites throughout the body.

15.16 Tumor cells launch counterattacks on immunocytes

Earlier, we read that tumor cells thrive in immunodepressed environments, such as the bodies of transplant recipients whose immune systems have been compromised by immunosuppressive drugs as well as those suffering from AIDS. In fact, the great majority of human tumors may develop in immunodeficient environments. More precisely, tumors may create localized microenvironments in which immune function is compromised. By keeping functional cytotoxic cells at some distance, tumors can ensure safety zones for themselves at various sites in the body.

One strategy for doing so is suggested by one of the mechanisms, described above, that is normally used by cytotoxic lymphocytes to kill their victims: these lymphocytes wield Fas ligand (FasL) molecules on their surface, which bind and activate the Fas death receptor displayed by other cell types throughout the body. Binding of the FasL ligand to this death receptor leads to activation of the extrinsic apoptotic pathway (Section 9.14).

Many cancer cells, however, subvert this signaling system in a two-step process. First, they develop resistance to FasL-mediated killing through mechanisms that are not well understood. For example, by acquiring resistance to multiple forms of apoptosis (see Table 9.5), cancer cells can protect themselves from activation of the extrinsic apoptotic cascade that is triggered by the FasL-activated Fas receptor. Clearly, this resistance, on its own, provides these cells one means whereby they can avoid destruction by cytotoxic cells, such as cytotoxic T cells.

In a second step, many kinds of cancer cells turn this FasL–Fas system on its head, by acquiring the ability to produce and release soluble forms of FasL themselves (Figure 15.31A and B). This FasL does not affect these cancer cells, which have already become resistant to its effects. However, a number of studies indicate that this ligand can activate Fas death receptors displayed on the surfaces of several types of lymphocytes, causing their death. By launching such counterattacks on nearby immune cells, cancer cells can hold them at a safe distance, once again reducing the risk of being killed by them.

In many cancer patients, there is evidence of depressed levels of certain types of circulating lymphocytes, indicating a systemic defect in immune function. In oral cancer patients, for example, membranous vesicles displaying membrane-bound, biologically active FasL can be found in the general circulation (Figure 15.31C); these vesicles seem to be released by the cancer cells as a means of depressing lymphocyte function throughout the body. The importance of this particular immunosuppressive mechanism for the growth of these and other types of human tumors still requires extensive documentation.

Alternative forms of counterattack are used by the many types of human cancer cells that have been found to release either interleukin-10 (IL-10) or TGF-β. Both of these secreted proteins are potently immunosuppressive and act through their ability to induce T lymphocytes to enter into apoptosis. In addition, TGF-β can induce apoptosis of dendritic cells and macrophages—the two key antigen-presenting cells of the immune system.

Interestingly, the release of IL-10 by human cancer cells is mimicked by Epstein–Barr virus (EBV), which has acquired and remodeled a cellular IL-10 gene. As we have read (Section 3.4), this virus is an important etiologic agent of Burkitt's lymphomas, other lymphomas of the B-cell lineage, and nasopharyngeal carcinomas. By forcing virus-infected cells to release an IL-10-like cytokine, Epstein–Barr virus protects these cells from direct attack by cytotoxic immune cells, thereby ensuring a safe cellular haven for itself over extended periods of time.

One set of experiments is particularly illustrative of the immunosuppressive powers of TGF-β and demonstrates the truth of the old adage that the best defense is the launching of a preemptive attack (Figure 15.32A). In these experiments, a group of 10 mice were injected with a strain of mouse melanoma cells that happen to secrete high levels of TGF-β; all of these mice succumbed to cancer within 45 days. In a complementary experiment, the lymphocytes of mice were first rendered resistant to TGF-β–mediated killing by grafting into their bone marrow genetically altered hematopoietic stem cells that displayed a dominant-negative (dn) TGF-β type II receptor (which blocks normal receptor function). When cells of the same melanoma cell line were injected into this second set of mice, 7 out of 10 mice were still alive 45 days later, and examination of their lungs revealed a dramatic reduction in melanoma metastases (see Figure 15.32B). Similar outcomes were observed when prostate carcinoma cells were engrafted into normal mice or into mice whose marrow cells had been modified in this way.

These observations indicate that the release of TGF-β by the melanoma and prostate cancer cells enabled the tumor cells to defend themselves against killing by immune cells. However, once the host immune cells were made resistant to TGF-β–induced apoptosis, these immunocytes showed themselves to be perfectly capable of eliminating the cancer cells. Moreover, in the tumor-bearing mice whose hematopoietic cells displayed the dn TGF-βRII, cytotoxic lymphocytes showing an ability to kill the melanoma cells *in vitro* could be recovered from the spleen; such splenic lymphocytes were virtually absent in control animals whose lymphocytes lacked this mutant receptor.

(These experiments might suggest to you the outlines of a novel anti-tumor therapy, in which the bone marrow cells of a cancer patient are rendered resistant to TGF-β–induced apoptosis, making these cells especially effective in attacking the many human tumors that release TGF-β. Unfortunately, once

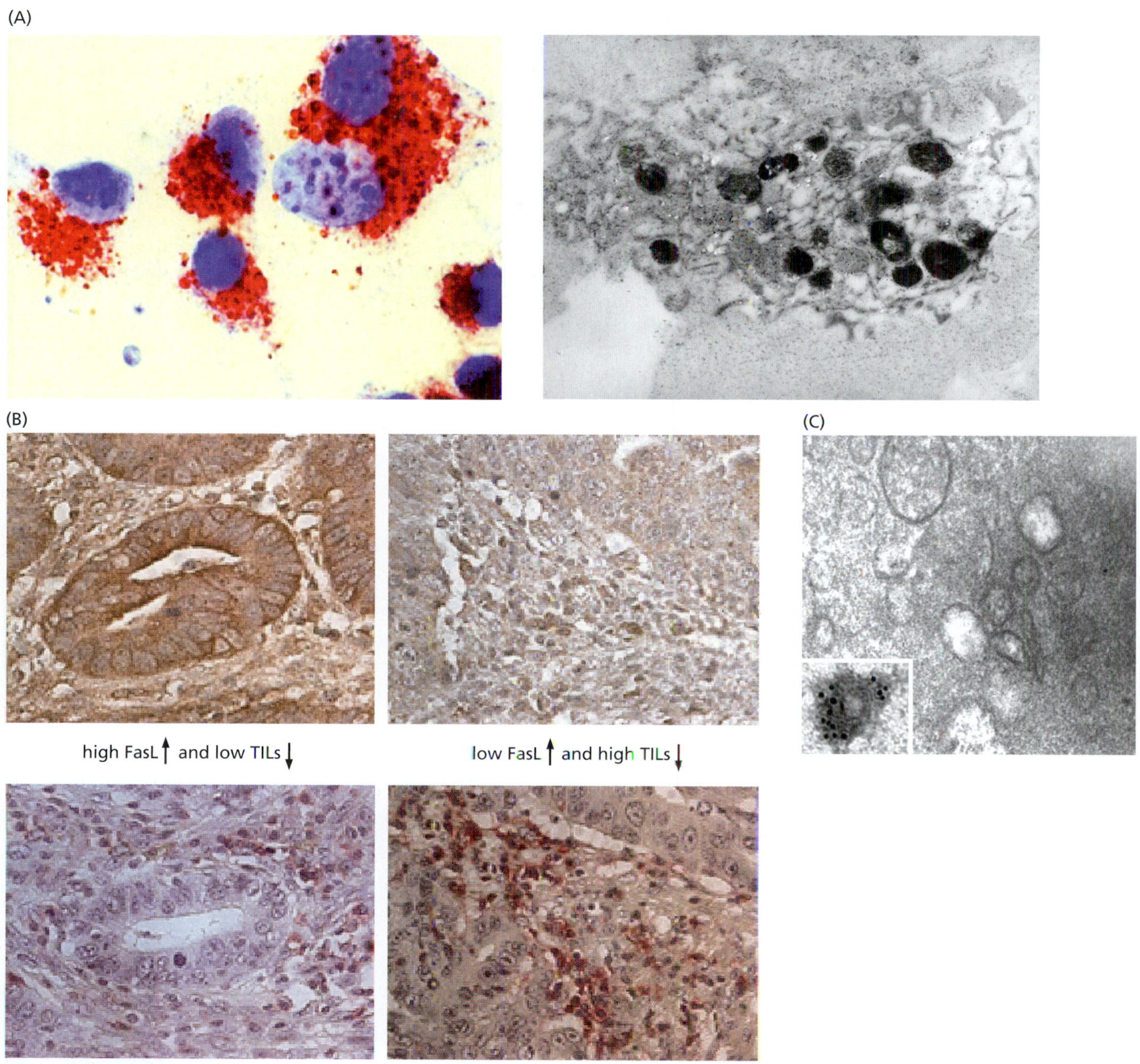

Figure 15.31 Contributions of FasL to immunoevasion Some cancer cells are able to protect themselves from killing by T lymphocytes through a two-step process. First, they acquire resistance to killing by FasL, the ligand of the Fas death receptor. Second, they acquire the ability to produce and release FasL, which allows them to kill lymphocytes and other cells that may stray too close to them. (A) When melanoma cells are stained with an anti-FasL antibody *(left panel)*, significant amounts of FasL are seen to be concentrated in vesicles *(red)* in the cytoplasm; nuclei are in *blue*. Far higher resolution is obtained by use of immunoelectron microscopy *(right panel)*, which reveals that the FasL in melanoma cells *(large black spots)* is actually localized to melanosomes (cytoplasmic bodies that carry the dark melanin pigment molecules of normal melanocytes); melanoma cells can release these melanosomes into the extracellular space, where the FasL acquires the ability to trigger the death of Fas-expressing cells, such as nearby lymphocytes. (B) In an adenocarcinoma of the colon, those areas of the tumor in which high levels of FasL *(brown)* were evident *(top left)* showed very low levels (in the next tissue slice) of tumor-infiltrating lymphocytes (TILs; *red, bottom left*). Conversely, those areas that showed low levels of FasL *(brown, top right)* showed high levels of TILs *(red, bottom right)*. TILs that were near areas of high FasL showed high rates of apoptosis *(not shown)*. (C) One explanation for the frequently observed suppression of functional circulating T lymphocytes in oral carcinoma patients may come from the discovery that the great majority of these patients (but not normal controls) have membranous microvesicles in their circulation that display membrane-bound FasL, which is a particularly potent apoptosis-inducing form of this death receptor ligand. Such microvesicles have been purified from the serum of such patients and found by immunoelectron microscopy (see panel A) to contain membrane-bound FasL *(black dots, inset)*. (A, left, from G. Andreola, L. Rivoltini, C. Castelli et al., *J. Exp. Med.* 195:1303–1316, 2002; A, right, from L. Rivoltini, M. Carrabba, V. Huber et al., *Immunol. Rev.* 188:97–113, 2002; B, from A. Houston, M.W. Bennett, G.C. O'Sullivan et al., *Brit. J. Cancer* 89:1345–1351, 2003; C, from J.W. Kim, E. Wieckowski, D.D. Taylor et al., *Clin. Cancer Res.* 11:1010–1020, 2005.)

immune cells are rendered unresponsive to TGF-β, they often launch devastating autoimmune attacks on tissues throughout the body, yielding a condition that can be far more debilitating and rapidly lethal than a neoplastic disease.)

Complementary results come from experiments in which the ability of cancer cells to release TGF-β has been greatly reduced by inserting antisense constructs into these cells. Such cancer cells lose much of their tumorigenic powers, in large part because they are now surrounded by flocks of cytotoxic lymphocytes that are capable of killing them.

Taken together, these various experiments indicate that even when cytotoxic T cells succeed in acquiring adaptive immunoreactivity against tumor cells, these T cells and their NK collaborators are often prevented from eliminating tumor

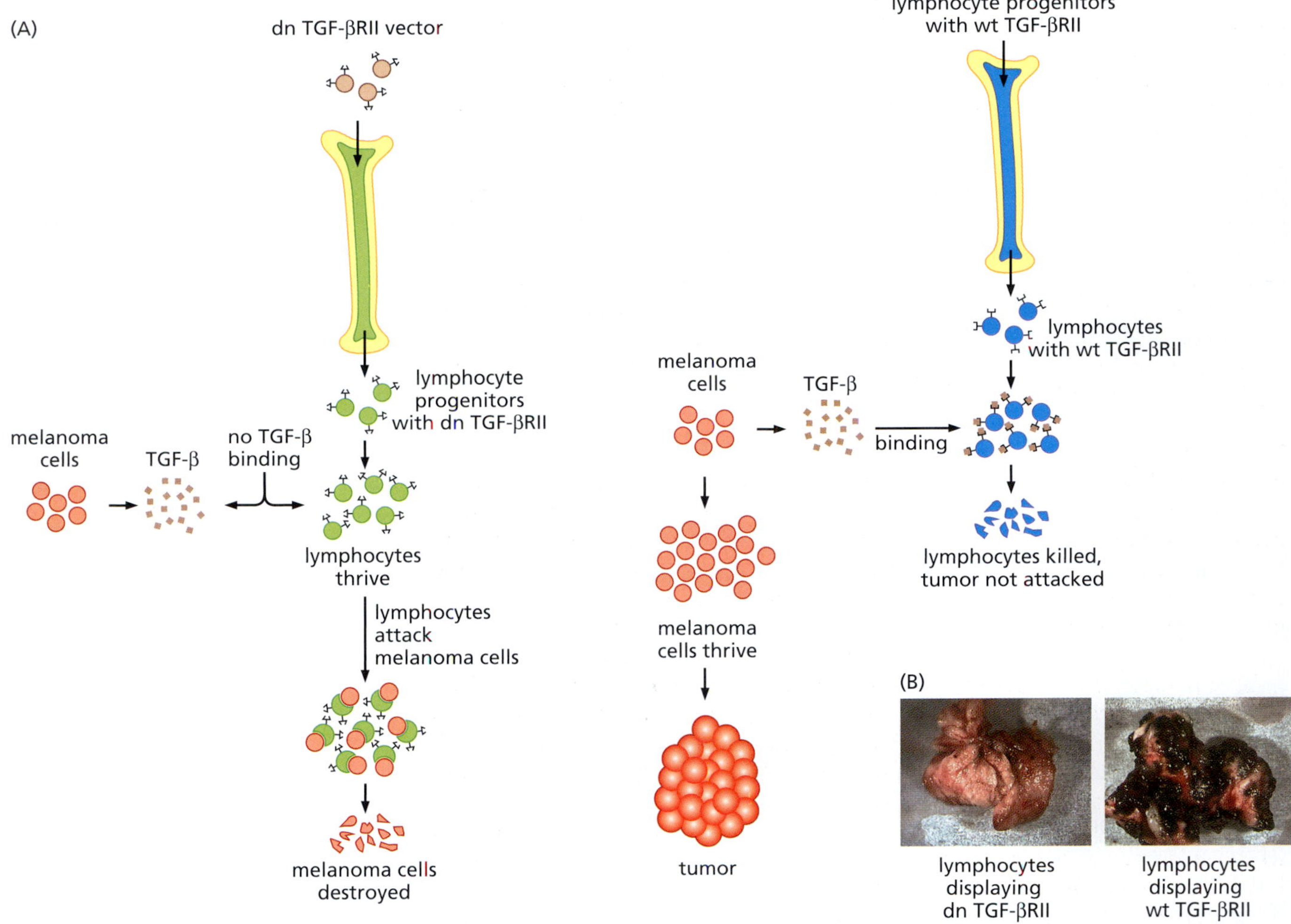

Figure 15.32 Role of TGF-β in controlling immune responses (A) If lymphocyte progenitor cells bearing a dominant-negative TGF-β II receptor (dn TGF-βRII, *top)* are introduced into the marrow of a mouse, differentiated lymphocytes *(green)* are formed that do not show responsiveness to the killing effects of TGF-β *(left diagram).* If melanoma cells *(pink, left)* are then introduced into a mouse carrying such lymphocytes, even though these melanoma cells secrete copious amounts of TGF-β (which is normally toxic for lymphocytes; *beige squares)*, the lymphocytes thrive and proceed to kill the melanoma cells, thereby blocking tumor formation. In contrast, when the same melanoma cells are introduced into a host mouse that has been engrafted with lymphocyte progenitors expressing wild-type (wt) TGF-βII receptor *(right drawing)*, the TGF-β released by the melanoma cells induces the lymphocytes *(blue)*, which display fully functional TGF-β receptors, to enter apoptosis *(bottom)*. This proves that TGF-β secretion is a highly effective means by which these melanoma cells can defend themselves against immune attack. (B) The dramatic differences in tumor-forming ability are revealed by the relative abilities of the pigmented melanoma cells to form lung metastases. Wild-type C57/BL6 mice rapidly developed many such metastases *(black spots, right)*, while those expressing the dominant-negative TGF-β type II receptor (dn TGF-βRII) in their T cells efficiently blocked metastasis formation *(left)*. The lungs were observed here 21 days after B16 melanoma cells were injected intravenously. (From data of L. Gorelik and R.A. Flavell, *Nat. Med.* 7:1118–1122, 2001; B, from A. Shah, W.B. Tabayoyong, S.D. Kundu et al., *Cancer Res.* 62:7135–7138, 2002.)

cells by a variety of defenses thrown up by the latter. At least three agents—TGF-β, IL-10, and FasL—have been proposed as weapons used by cancer cells to launch counterattacks on the various immunocytes that threaten them. Since the killing of antibody-coated cancer cells is often carried out by cytotoxic immunocytes (e.g., NK cells, macrophages; see Figure 15.3), the secretion of these immunosuppressive agents may also protect tumor cells from elimination in those patients who have high levels of circulating anti-tumor antibodies.

15.17 Cancer cells become intrinsically resistant to various forms of killing used by the immune system

Cancer cells may also alter their own biochemistry to make themselves intrinsically less responsive to attacks launched by the immune system. One example of this general strategy was already mentioned in passing above: some cancer cells become resistant to the FasL released by several kinds of cytotoxic immunocytes. By altering the signaling pathways downstream of the Fas death receptor, these cells may become relatively insensitive to the FasL released by various immunocytes.

A related defensive maneuver responds to the other major mechanism used by cytotoxic T lymphocytes and NK cells to kill targeted cells including tumor cells. Recall that these immunocytes introduce a protease—a granzyme—into targeted cells (Sections 9.14 and 15.3), which functions to induce apoptosis in the latter by cleaving and activating a caspase pro-enzyme. Cancer cells can escape this killing mechanism simply by increasing their levels of certain inhibitor-of-apoptosis proteins (IAPs), which operate by sequestering and thereby inactivating key pro-apoptotic caspases (Section 9.13).

These two strategies for avoiding killing are extensions of mechanisms that we encountered earlier in Chapter 9, where acquisition of resistance to apoptosis was described as one of the hallmarks of cancer. Accordingly, immune-initiated cytotoxicity can now be added to the other physiologic stressors, including hypoxia, intracellular signaling imbalances, and loss of anchorage, that force cancer cells to disable their pro-apoptotic signaling pathways during the course of tumor progression.

Cancer patients often have significant levels of anti-tumor antibodies in their circulation, indicating that their tumor cells are likely to carry a coating of bound antibody molecules. As we learned from Figure 15.9, such cells are vulnerable to killing by the cohort of plasma proteins that together constitute complement. When participating in the process of complement-dependent cytotoxicity (CDC), complement molecules associate with antibody molecules bound to a cell surface and punch holes in the nearby plasma membrane, leading quickly to cell death. Normal cells protect themselves from inadvertent killing by complement by expressing on their plasma membranes one or several anti-complement proteins, called membrane-bound complement regulatory proteins (mCRPs). The most important of these are CD46, CD55, and CD59.

These mCRPs have been found to be overexpressed on the surfaces of a variety of human cancer cell types. Such overexpression affords the cancer cells a measure of protection from complement-dependent cytotoxicity (Figure 15.33). The repeated observations of elevated mCRP expression in a diverse array of cancer cell types suggest that the progenitors of these cancer cells were threatened, at some point in their development, with complement-dependent cytotoxicity, and that, in response, variant cancer cells were selected that could resist such killing through overexpression of one or another mCRP protein. (Interestingly, herpesvirus saimiri, a virulent herpesvirus that causes lymphomas and leukemias in New World monkeys, expresses a protein closely related to human

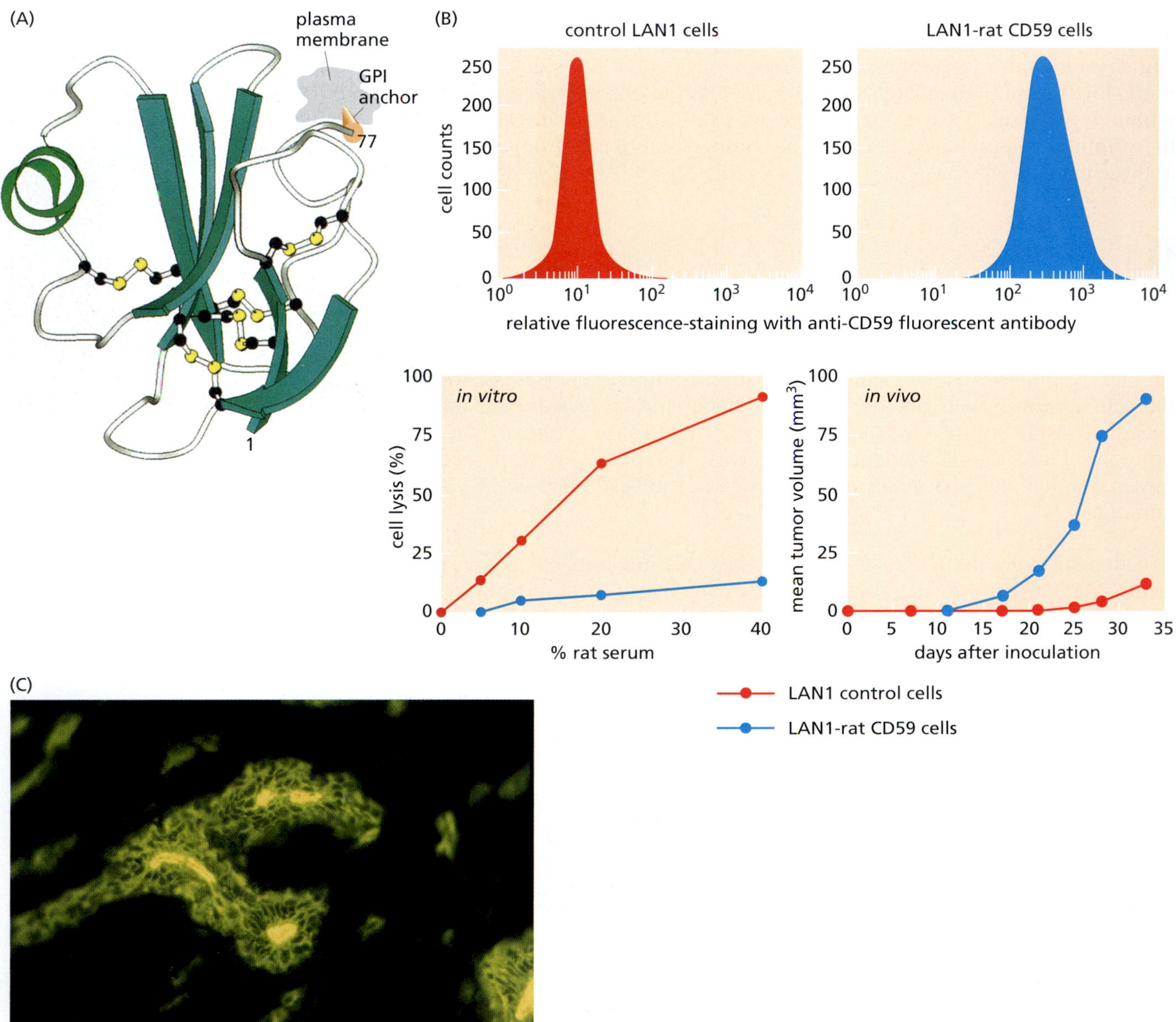

Figure 15.33 Regulation of the complement cascade by mCRPs The mCRPs (CD46, CD55, and CD59) are membrane-bound inhibitors of the complement cascade that protect cells from complement-mediated cell lysis and are frequently over-expressed by human tumor cells. (A) CD59 is a glycoprotein that is tethered to the plasma membrane of cells *(behind molecule in the plane of the page)* by a phosphatidyl inositol (GPI) anchoring tail *(brown)*. The structure of CD59 shown here was determined by nuclear magnetic resonance (NMR). CD59 binds certain components of complement and thereby prevents full assembly of the membrane attack complex (MAC), which mediates the cytolytic activity of complement. CD59 is structurally related to the uPAR receptor (see Figure 14.34) and to several snake venom neurotoxins. Carbon and sulfur residues of cysteines *(black, yellow)*, α-helix *(green)*, β-pleated sheets *(aquamarine)* are highlighted. (B) Human LAN1 neuroblastoma cells, which are normally very sensitive to lysis by rat complement, were transfected with an expression vector that forced them to overexpress the rat CD59 mCRP protein. As seen in the fluorescence-activated cell sorting (FACS) analysis *(top left and right)*, this caused the normal LAN1 cells *(left)* to express 20- to 30-fold higher levels of CD59 *(right)*. When these two cell populations were incubated *in vitro* in medium containing rat serum plus an antibody that activates the complement present in this serum *(bottom left)*, the cells overexpressing CD59 were protected from complement-mediated lysis *(blue curve)*, in contrast to the unmanipulated LAN1 cells, which were highly sensitive to this lysis *(red curve)*. When these two cell populations were implanted in immunocompromised Nude rats *(bottom right)*, the LAN1 cells formed tumors only slowly *(red curve)*, while those overexpressing CD59 formed tumors rapidly *(blue curve)*. This indicated that complement-mediated killing usually impedes the growth of LAN1 tumors in these rats. (C) This immunofluorescence micrograph of section of a human ductal carcinoma of the breast shows intense expression of CD59 in the duct-forming carcinoma cells *(green)* with virtually no staining in the surrounding stromal tissue. (Because CD59 is loosely tethered via a GPI anchor to luminal surfaces of the ductal epithelial cells, it is released in large amounts into the lumina of the ducts, generating the intense yellow color.) (A, from C.M. Fletcher, R.A. Harrison, P.J. Lachmann and D. Neuhaus, *Structure* 2:185–199, 1994; B, from S. Chen, T. Caragine, N.-K. V. Cheung, and S. Tomlinson, *Cancer Res.* 60:3013–3018, 2000; C, from J. Hakulinen and S. Meri, *Lab. Invest.* 71:820–827, 1994.)

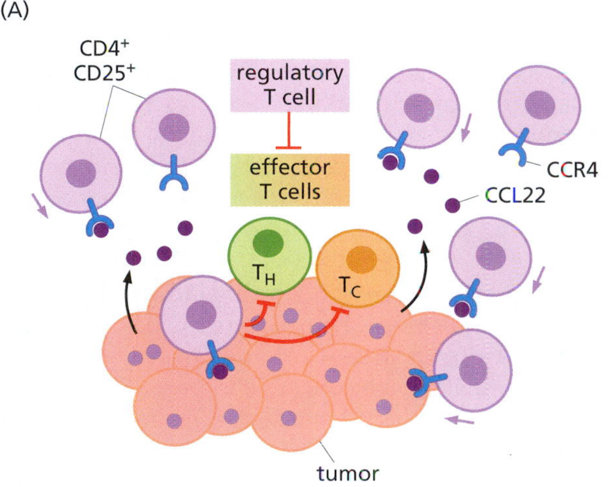

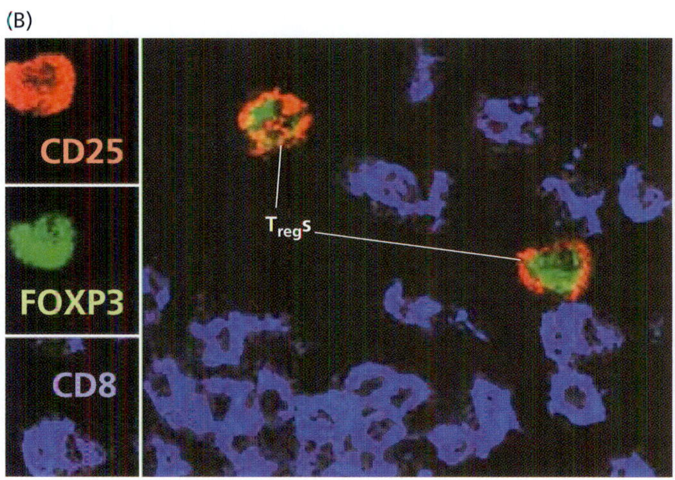

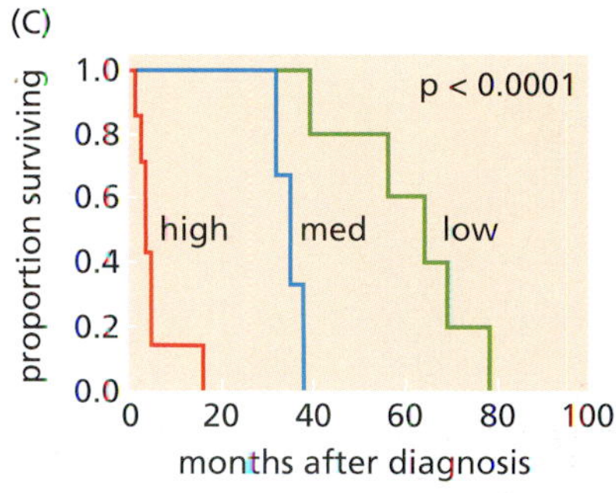

Figure 15.34 Regulatory T cells and tumor immunoevasion (A) Tumor cells *(pink)* release the CCL22 chemokine *(small purple circles)*, which binds the CCR4 chemokine receptor *(blue)* displayed by the T_{reg}s and in this way attracts them into the tumor. Once they are recruited, the T_{reg}s, which express the CD4 and CD25 cell-surface antigens, are able to inhibit two types of "effector T cells"—the $CD4^+CD25^-$ helper T (T_H) lymphocytes *(light green)* and the $CD8^+CD25^-$ cytotoxic (T_C) lymphocytes *(light orange)*. Such actions can cripple major components of the host immune response to a tumor. (B) Immunofluorescence staining reveals the presence of T_{reg}s through their expression of the CD25 surface antigen *(red)* and the FOXP3 transcription factor *(green)*. They are seen here amid CD8-positive cytotoxic T cells *(blue)*, whose actions they are ostensibly inhibiting; these lymphocytes were present in the ascites of a patient suffering from ovarian cancer. (C) Among advanced (stage IV) ovarian cancer patients, the concentration of tumor-infiltrating T_{reg}s in tumor sections is a strong predictor of long-term survival, as indicated by this Kaplan–Meier plot. (A, from E.M. Shevach, *Nat. Med.* 10:900–901, 2004; B and C, from T. Curiel, G. Coukos, L. Zou et al., *Nat. Med.* 10:942–949, 2004.)

CD59 mCRP, which it apparently uses to protect virus-infected cells from rapid killing by host complement.) Several other mechanisms that protect cancer cells against complement-dependent cytotoxicity have been described but remain poorly studied.

15.18 Cancer cells attract regulatory T cells to fend off attacks by other lymphocytes

Regulatory T cells (T_{reg}s)—a more recently characterized type of lymphocyte (Section 15.6)—seem also to play a major role in immunoevasion by cancer cells. Recall that a T_{reg} can directly inhibit and even kill both cytotoxic and helper T lymphocytes that recognize the same antigen as the one recognized by the T_{reg}. [In all cases, antigen recognition is achieved by the T-cell receptors (TCRs) that these various T lymphocytes display.]

In normal individuals, the T_{reg}s represent only 5 to 10% of the population of $CD4^+$ lymphocytes, the remainder being helper T cells. In cancer patients, however, this number may increase to 25 to 30%. Moreover, T_{reg}s have been found, often in large numbers, among the tumor-infiltrating lymphocytes (TILs) present in lung, ovarian, breast, and pancreatic carcinomas as well as in tumor ascites (Figure 15.34A and B). Together, these various observations suggest that T_{reg}s play an important role in influencing anti-tumor immunity.

Tumors release the chemokine CCL22 in order to recruit T_{reg}s; the latter display the cognate receptor, termed CCR4, on their surface. Once present within a tumor mass, the T_{reg}s can suppress the actions of helper T cells that are instrumental in mobilizing both the humoral and cellular arms of the adaptive immune response as well as cytotoxic T cells that are otherwise fully competent to attack and kill tumor cells (Section 15.1). Hence, the ability to carry out this immunoevasive maneuver can be traced to the ability of tumor cells to produce and secrete CCL22.

The existence of T_{reg}s clearly complicates many conclusions drawn in recent years concerning the role of tumor-infiltrating lymphocytes (TILs) in tumor pathogenesis. Such lymphocytes have been widely assumed to be cytotoxic T cells that are actively involved in eliminating the cancer cells around them (see, for example, Figure 15.21D). However, if T_{regs} constitute a significant proportion of the TILs within some tumors, then the significance of the total number of TILs within such tumors is unclear. This means that assessments of the *relative proportions* of cytotoxic T cells and T_{reg}s must be made in order to understand the real dynamics of a tumor's interactions with the host's cellular immune response. Observations such as those presented in Figure 15.34C suggest that the dysfunctional state of many tumor-associated cytotoxic T lymphocytes (CTLs) can be explained by the presence of many T_{reg}s in their midst and that T_{reg}s may therefore be critical determinants of whether or not the immune system can keep tumors under control.

15.19 Passive immunization with Herceptin can be used to kill breast cancer cells

Until now, we have grappled largely with the question of how effective the human immune system is in defending us against spontaneously arising tumors. The answers to this question are surely complex and continue to provoke vigorous debate. However, the eventual resolution of this debate will not pre-ordain the answers to a second question: Can the immune system of a cancer patient be manipulated in ways that enable it to kill cancers that have already formed at various sites throughout the body?

Two types of manipulation could be entertained. Following one major strategy, we might wish to activate or enhance the powers of patients' immune systems to mount an attack against tumors that they carry. This could involve the use of certain *immunostimulatory* factors that can incite the development and proliferation of immunocytes capable of launching an effective attack. Such enhancement of anti-tumor immune function might also be achieved by exposing patients to TATAs or TSTAs displayed by their tumors, in effect immunizing the patient against the tumor in a way that is analogous to vaccination against viral or bacterial infection.

The alternative therapeutic strategy involves various forms of **passive immunization**. Use of this class of strategies implies that a patient's own immune system is incapable of mounting an effective immune defense, even after immunostimulatory therapies are applied, and involves supplying the patient with immune products (e.g., antibodies) or cells originating in another organism's immune system. (Immunologists reserve the term "passive immunization" for procedures involving the introduction of antibodies into a patient, but we will use the term more broadly here.) When cells are supplied from the immune system of another individual, this procedure is often termed **adoptive transfer**. We first describe various types of passive immunization and will then return to the immunostimulatory strategies in the next section.

The best known of the passive immunization treatments involves the monoclonal antibody termed **Herceptin**, also called trastuzumab by its developers.

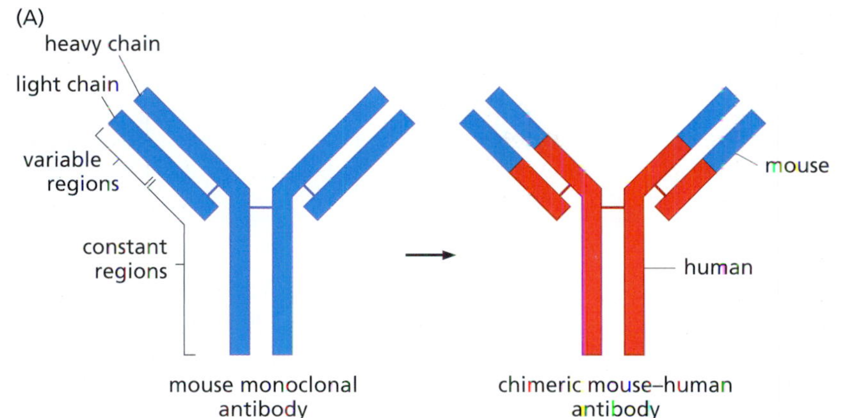

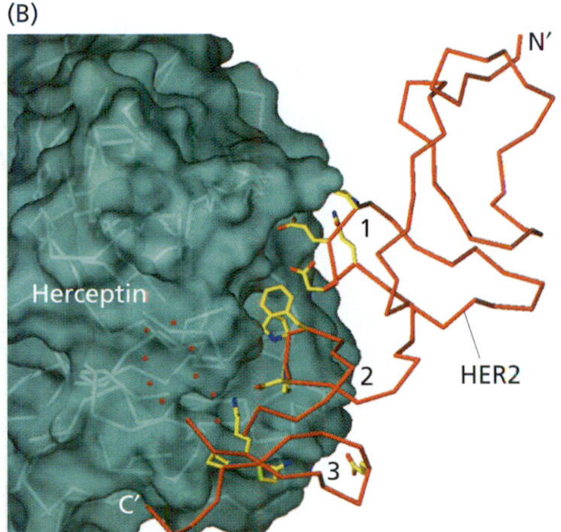

Figure 15.35 Humanizing of therapeutic antibodies (A) The monoclonal antibodies (MoAbs) that are produced in mice often cannot be used therapeutically in humans over extended periods of time because the constant regions of the heavy and light chains of the mouse antibody *(left figure)* are themselves immunogenic when introduced into humans, thereby provoking the formation of an anti-MoAb serum response in treated patients that can neutralize the actions of the introduced MoAb. To deal with this, the cDNAs encoding the heavy and light chains of the mouse MoAb *(blue)* are cloned, and the DNA segments encoding the antigen-combining domains of the mouse MoAb, which are located in the variable regions of the heavy and light chains, are fused with the genes encoding the constant regions of the human antibody *(red)*, yielding a chimeric antibody that is not immunogenic or, at worst, weakly so. (B) This X-ray crystallography study resolves the detailed structure of the complex between the Herceptin and HER2 molecules at a resolution of 2.5 Å. The surface of the antigen–binding domain of Herceptin (seen here as the Fab proteolytic fragment of the antibody; *aquamarine*) recognizes and binds an epitope located in one of the four subdomains of the ectodomain of HER2 *(red, yellow stick figure)*, termed subdomain IV *(right)*; the side chains of HER2 residues that contact Herceptin are shown as yellow stick figures. This subdomain of HER2 is immediately adjacent to the transmembrane domain of the receptor where it passes through the plasma membrane (see Figure 15.38B). (A, adapted from R.A. Goldsby et al., Immunology, 5th ed. New York: Freeman, 2002; B, from H.S. Cho, K. Mason, K.X. Ramyar et al., *Nature* 421:756–760, 2003.)

Herceptin derives from a mouse monoclonal antibody that reacts strongly with the EGF receptor–related protein that is called variously HER2, erbB2, or Neu (Sections 4.3 and 5.6; Table 5.2); as we read earlier, this receptor is overexpressed in 30% of breast cancers diagnosed in the West.

While the HER2 protein is not itself a tumor-specific antigen, its display at abnormally high levels—often 10 to 100 times above normal—may create a target cell that is preferentially affected by Herceptin. Such **selectivity**—preferential killing of cancer cells—derives directly from differences between normal and cancer cells, in this case the overexpression of HER2 by breast cancer cells. This overexpression, which is often due to amplification of the receptor-encoding gene, generally represents a poor prognosis, often involving progressive disease in the years after the initial diagnosis is made (see Figure 4.6).

In order to make Herceptin, a mouse monoclonal antibody (MoAb) against HER2 was initially produced. However, this antibody, like all murine antibodies, could not be used directly for anti-tumor therapy in humans, simply because the constant region of the antibody (see Figure 15.1), being of mouse origin, constitutes a potent antigen on its own and therefore provokes the human immune system to produce high levels of human antibodies that bind and neutralize the mouse antibody molecules. Moreover, these human anti-mouse antibodies (HAMAs) occasionally induce anaphylactic shock in a patient re-treated with a mouse antibody.

Consequently, the cDNA encoding the mouse anti-HER2 antibody was re-engineered so that the constant (C) regions of the encoded antibody molecule were composed of human rather than mouse sequences; as hoped, the resulting "**chimerized**" anti-HER2 MoAb (Figure 15.35) was usually not immunogenic when injected into patients. Importantly, following its injection, the chimerized Herceptin molecules were found to remain in the circulation at significant levels for as long as a month, indicating the absence of a significant anti-Herceptin host immune response and a potential for long-term therapeutic effects. [A further modification of this antibody, which also substitutes human for mouse sequences in the portions of the variable (V) region of the antibody that are outside the antigen-combining site, results in a **humanized antibody**; such a humanized Herceptin has even fewer mouse antigenic determinants than the chimeric version.]

Use of the Herceptin antibody has resulted in extension of the life span of breast cancer patients whose tumors overexpress HER2 protein. Herceptin is rarely used on its own, but instead is applied in combination with established chemotherapeutic treatments of breast cancer. In one large clinical study, the addition of Herceptin to standard chemotherapy treatment of women with advanced breast cancer resulted in a longer time to disease progression (7.4 versus 4.6 months with chemotherapy alone), a lower rate of death at 1 year (22 versus 33%), and a longer overall survival (25 versus 20 months).

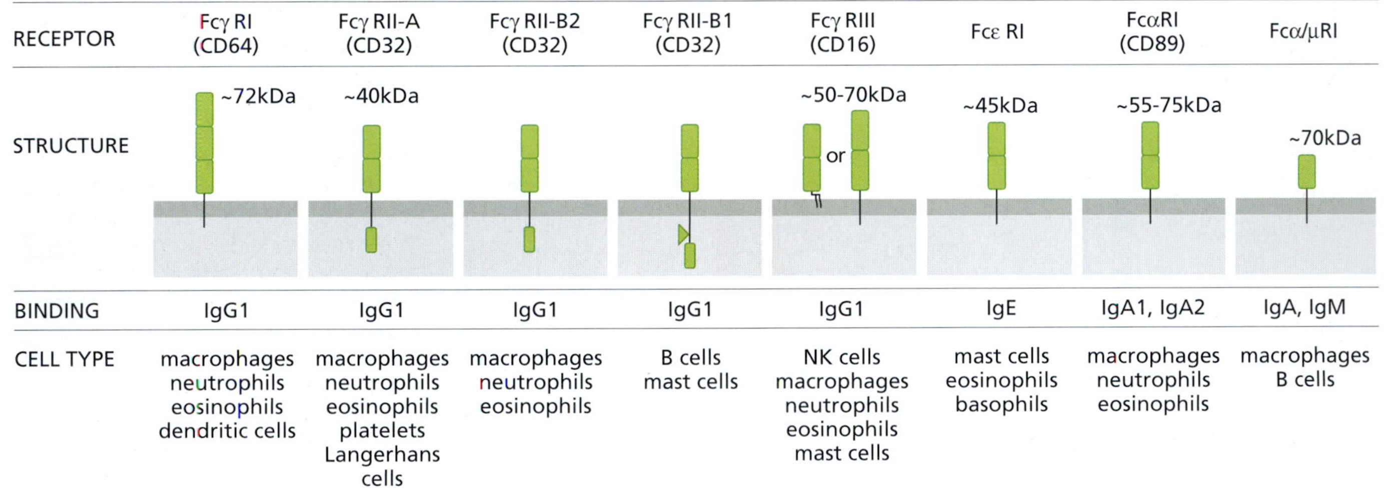

RECEPTOR
FcγRI (CD64)
FcγRII-A (CD32)
FcγRII-B2 (CD32)
FcγRII-B1 (CD32)
FcγRIII (CD16)
FcεRI
FcαRI (CD89)
Fcα/μRI
STRUCTURE
~72kDa
~40kDa
~50-70kDa
or
~45kDa
~55-75kDa
~70kDa
BINDING
IgG1
IgG1
IgG1
IgG1
IgG1
IgE
IgA1, IgA2
IgA, IgM
CELL TYPE
macrophages neutrophils eosinophils dendritic cells
macrophages neutrophils eosinophils platelets Langerhans cells
macrophages neutrophils eosinophils
B cells mast cells
NK cells macrophages neutrophils eosinophils mast cells
mast cells eosinophils basophils
macrophages neutrophils eosinophils
macrophages B cells

Figure 15.36 Fc receptors displayed by immunocytes Fc cell surface receptors *(green)* are used by various types of immunocytes to attach to the constant regions (see, for example, Figure 15.3) of antibody molecules. Once antibody molecules are bound to the surface antigens of a potential target cell, the Fc receptors then enable various immunocytes to attach to the constant regions of the bound antibodies, triggering activation of the immunocytes and attack on the target cell. As seen here, a variety of Fc receptor variants have been evolved by the immune system, with the subclass of Fcγ receptors, for example, being specialized to bind the constant regions of IgG antibody molecules. (From C.A. Janeway Jr. et al., Immunobiology, 6th ed. New York: Garland Science, 2005.)

Even more impressive responses were announced in 2005: Women with **operable**, early-stage HER2-overexpressing breast tumors were treated postoperatively either with chemotherapy alone or with chemotherapy plus Herceptin; most of these women carried micrometastases in draining lymph nodes (see Figure 14.41). After four years, only 15% of the women who received the double treatment had **relapsed**, while 33% of the women who had received chemotherapy alone showed disease relapse. It is plausible, though unproven, that much of Herceptin's therapeutic benefit observed in this study derived from its ability to block the proliferation of residual cancer cells that were left behind after removal of the primary tumors; in the absence of Herceptin, such cells might be responsive to mitogens produced by the wound-healing process that followed surgery (see Sidebar 13.4).

The precise mechanisms by which Herceptin antibodies kill HER2-overexpressing breast cancer cells remain matters of controversy. One important mechanism of cancer cell destruction derives from the presence of Fcγ receptors displayed on the surface of a variety of cytotoxic and phagocytic cells, including, most importantly, NK cells and macrophages (Figure 15.36; see also Figure 15.3). These Fcγ receptors bind the constant regions of immunoglobulin γ (IgG) antibody molecules that are coating the surface of other cells, such as Herceptin-treated breast cancer cells. Such antibody coating informs the Fcγ receptor–expressing cytotoxic cell of the presence of a cell that should be targeted for elimination. Accordingly, NK cells and macrophages become tethered to IgG antibody–coated cancer cells via their Fcγ receptors and proceed to kill the cancer cells—the process termed antibody-dependent cellular cytotoxicity (ADCC).

The germ lines of immunocompromised mice have been modified by genetic deletion of the gene specifying a critical Fcγ receptor. Such mice have a greatly reduced ability for Herceptin-dependent killing of engrafted human breast cancer cells, providing direct evidence of ADCC as a key mechanism of Herceptin-initiated cell killing. Interestingly, the efficacy of Herceptin therapy seems to be compromised in certain cancer patients, simply because they lack normal levels of the NK cells that are responsible for much of the ADCC that occurs during Herceptin therapy. Taken together, these various observations lend weight to the

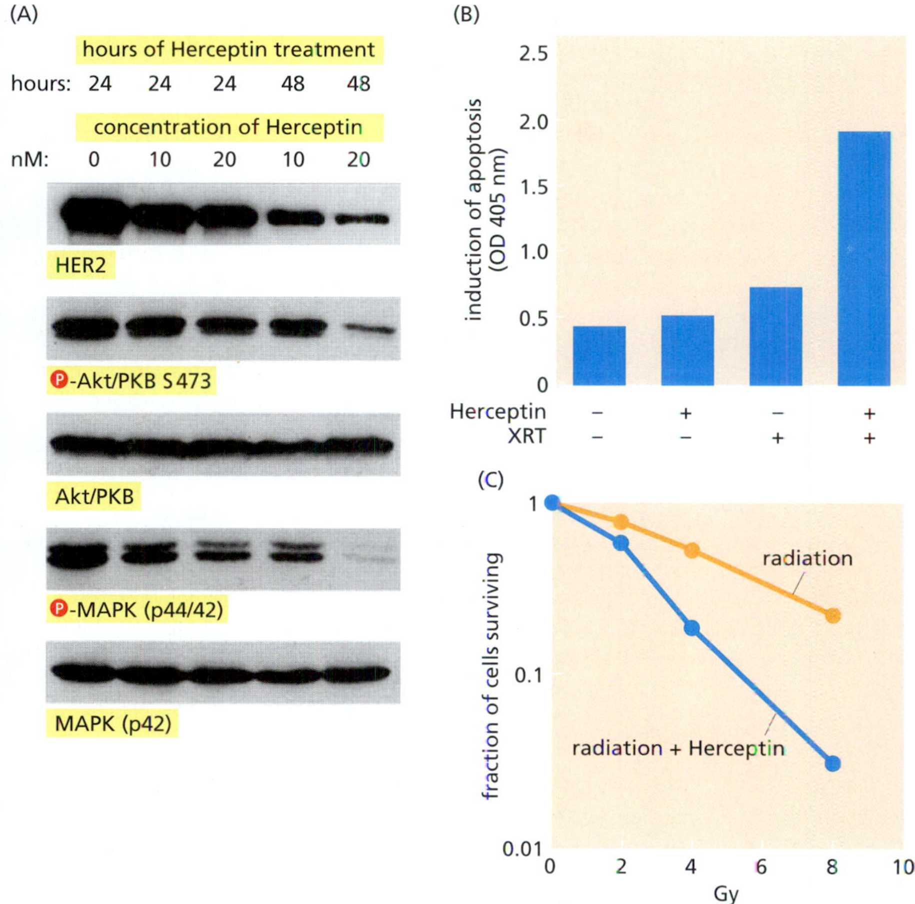

Figure 15.37 Herceptin treatment of breast cancer cells Herceptin has multiple effects on HER2/Neu-overexpressing breast cancer cells. In the work shown here, MCF7 human breast cancer cells were forced to overexpress the HER2/Neu receptor protein. (A) As a consequence of extended incubation with Herceptin, the levels of HER2/Neu declined substantially, doing so in proportion to Herceptin concentration and duration of treatment *in vitro*. In parallel, the levels of the activated, phosphorylated anti-apoptotic Akt/PKB and mitogenic MAP kinases (Sections 6.5 and 6.6) declined precipitously (while the overall levels of these two proteins did not change). (B) Treatment of cancer cells with Herceptin on its own had minimal effect on the extent of apoptosis by these cells. X-ray treatment (XRT) also had minimal effects. However, the two acted synergistically to induce apoptosis as indicated on the ordinate. (C) The collaborative effects of Herceptin with radiation are shown here, in which cells treated with the antibody have considerably more radiation sensitivity than do untreated cells. The doses of radiation are given here in grays (Gy). (Note that the ordinate here is logarithmic.) (From K. Liang, Y. Lu, W. Jin et al., *Mol. Cancer Ther.* 2:1113–1120, 2003.)

notion that much of the killing of tumor cells achieved by Herceptin depends on NK cells and their Fcγ receptors. This killing can be enhanced by further re-engineering of the Herceptin molecule (Sidebar 44).

Yet other mechanisms may contribute to Herceptin-induced killing. For example, when many cell surface proteins are exposed to reactive antibodies, these proteins are internalized and degraded. The same process leads to the Herceptin-induced decrease of cell surface HER2 displayed by breast cancer cells. This loss deprives the breast cancer cells of the high levels of HER2-activated PI3 kinase and Akt/PKB that have protected them from apoptosis, and thereby renders these cancer cells far more vulnerable to radiation- and chemotherapy-induced killing (Figure 15.37).

It is also clear that in many breast tumors, the ectodomain of the HER2 protein, once it arrives at the cell surface, is cleaved away by extracellular proteases. The residual protein, which contains the transmembrane and the cytoplasmic tyrosine kinase domains of HER2, exhibits constitutively activated kinase function and is therefore a very potent oncoprotein (e.g., see Figure 5.11). The post-translational cleavage of HER2 that produces this deregulated receptor protein is known to be blocked by Herceptin binding.

Yet another possible mechanism contributing to the anti-tumor actions of Herceptin may be associated with its anti-angiogenic effects: ligands of the HER2 receptor are known to be released by endothelial cells in order to aid in the recruitment of the pericytes and smooth muscle cells needed for the construction of well-formed capillaries. As we read, such recruitment is critical to the maturation of blood vessels, including those formed within tumors (Section 13.1). The reduction by Herceptin of this recruitment may explain how it is able to exert anti-angiogenic effects in many tumors. Another monoclonal antibody that binds HER2 and an antibody reactive with the EGF receptor (HER1) have

Sidebar 15.4 Herceptin is only the first of many clinically useful anti-receptor antibodies The EGF receptor (EGF-R; HER1) appears to be overexpressed in about one-third of all human carcinomas and in almost half of glioblastomas as well. As is the case with HER2, its overexpression is likely to play a key role in providing mitogenic and anti-apoptotic signals to tumor cells. Moreover, since HER2 and the EGF-R heterodimerize, the latter may play a key role in driving the proliferation of HER2-overexpressing breast cancer cells.

For various historical reasons, the development of anti-EGF-R monoclonal antibodies, notably Erbitux (also called cetuximab) lagged behind that of Herceptin by about three years. Its clinical use for treating advanced colorectal cancers was approved by the U.S. Food and Drug Administration (FDA) in 2004. In the long run, antibodies such as Erbitux may have greater utility than Herceptin, if only because the EGF-R seems to participate in the pathogenesis of a far larger number of tumors than does HER2.

Figure 15.38 Multiple mechanisms of receptor inhibition by monoclonal antibodies (A) Unlike other anti-receptor antibodies, Erbitux (Fab, *yellow, orange, left panel)* binds to a region of the EGF receptor *(circled area)* that is also involved in ligand binding by the EGF ligand *(aquamarine, right panel),* thereby blocking ligand-stimulated receptor activation. (The monoclonal antibody binds 50-fold more tightly to the EGF-R ectodomain than does the EGF ligand.) The antigen-combining domains of the light and heavy chains of the antibody molecule are indicated by V_L and V_H, respectively. Only a portion of the antibody (including the antigen-combining domain), termed the Fab fragment, was analyzed in this X-ray crystallography study. (B) When three distinct antibody–receptor complexes are viewed from the side and at greater distance, it becomes clear that the antibodies act on the receptors in very different ways. The specific binding sites of the heavy *(aquamarine)* and light *(yellow)* chains of the various anti-receptor monoclonal antibodies have been mapped, as in panel A, by X-ray crystallography. Erbitux *(left panel),* whose binding to domain III of the EGF-R/HER1 ectodomain was mapped in finer detail in panel (A), can be seen to block the binding site of the EGF ligand (compare Figure 5.16C). In contrast, Herceptin *(middle panel)* binds to the HER2/Neu receptor in a way that allows the latter to continue to homo- or heterodimerize with members of the HER family of receptors (HER1, HER2, HER3, HER4; see also Figure 15.35B), while a third antibody, termed Omnitarg or pertuzumab *(right panel),* binds to the rightward-pointing finger of the HER2 domain II ectodomain that is critical for such dimerization, thereby blocking the ability of HER2 to heterodimerize with other related receptors. (Like Herceptin, Omnitarg can trigger antibody-dependent cellular cytotoxicity.) (A, from S. Li, K.R. Schmitz, P.D. Jeffrey et al., *Cancer Cell* 7:301–311, 2005; B, from S.R. Hubbard, *Cancer Cell* 7:287–288, 2005.)

also been developed. They function differently from Herceptin and may eventually show greater utility (Sidebar 15.4).

15.20 Passive immunization with antibody can be used to treat B-cell tumors

A quite different form of passive anti-cancer immunization has been inspired by observations that many lymphoma cells arising from the B-cell lineage display on their surface high levels of the specific antibody molecules that their normal B-cell ancestors developed through rearrangement and mutation of immunoglobulin genes (Figure 15.39A). Since these tumors are clonal outgrowths, all of the B cells forming a given tumor will display identical antibody molecules. While most of the peptide sequences of these cell surface antibody molecules derive from the constant regions of these molecules (and are therefore

As indicated in Figure 15.38, Erbitux functions quite differently from Herceptin, in that it appears to inhibit receptor activation by blocking ligand binding, including binding by EGF, TGF-α, amphiregulin, and other ligands of the EGF-R. A third antibody, called Omnitarg or pertuzumab, acts on HER2 by blocking its ability to heterodimerize with other members of the EGF-R family.

These various anti-receptor antibodies are more specific than low–molecular-weight drugs in shutting down receptor signaling, in that they are unlikely to directly affect unrelated receptor tyrosine kinases (RTKs). However, they may not be able to penetrate into the interstices of tumors as effectively as low–molecular-weight RTK antagonists (to be discussed in Chapter 16). Even more importantly, many human tumors express truncated versions of the EGF-R that lack the ectodomain and fire constitutively (in a ligand-independent fashion); these proteins have lost the antigenic determinants that are recognized by anti-receptor monoclonal antibodies and therefore elude inhibition by these antibodies.

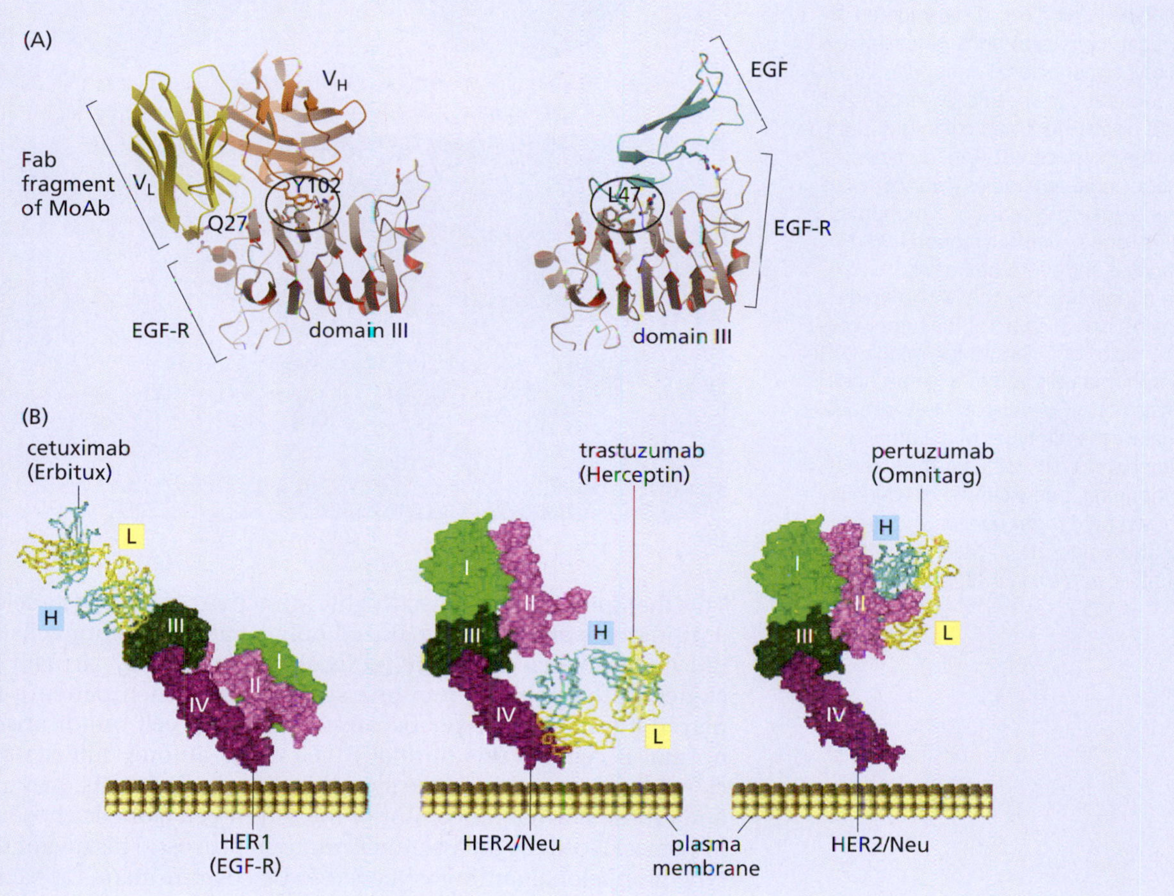

shared commonly with all other antibody molecules of the same class), the amino acid sequences in the antigen-combining domains represent unique antigens in their own right; such antigens are termed **idiotypes**.

Accordingly, anti-idiotype antibodies can be produced that specifically recognize the antibody molecules displayed on the surface of a clone of malignant B cells. This inspired the creation of anti-idiotype antibody molecules and their exploitation in treating B-cell tumors. In one widely heralded case, a mouse anti-idiotype monoclonal antibody was made against the antibody molecule displayed on the surface of a patient's lymphoma cells. A small number of injections of the monoclonal antibody resulted in a long-term remission in this patient, who had already developed advanced disease and had a very short life expectancy. The strategy of using anti-idiotype antibodies against a variety of tumors of the B-cell lineage eventually resulted in many positive responses among the approximately 50 patients who were eventually treated.

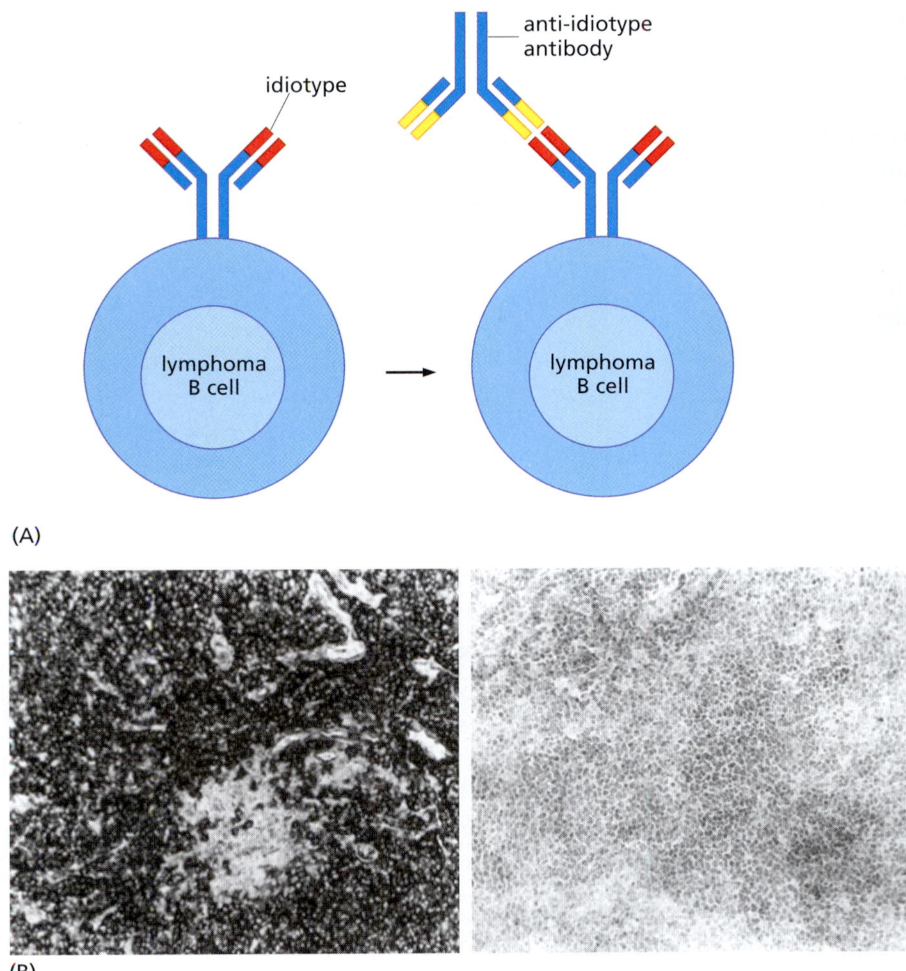

Figure 15.39 Anti-idiotype therapy of B-cell lymphomas (A) B-cell lymphomas derive from B cells that have developed the ability to produce antibody molecules with highly specific antigen-binding sites in their variable regions *(red)*. (The cell surface immunoglobulin molecule that these cells display is of the IgM class and is expressed, at this stage of B-cell development, as a transmembrane cell-surface molecule, being therefore configured much like a growth factor receptor.) The amino acid sequences in the antigen-combining domain of the receptor constitute an *idiotype*, which can be considered an antigen in its own right, since anti-idiotype monoclonal antibodies can be made that can specifically recognize B cells displaying a particular IgM idiotype on their surface. (B) Anti-idiotype monoclonal antibodies (MoAbs) were made against a series of lymphomas. Here, one patient's lymphoma *(left panel)* is shown to be heavily immunostained by the MoAb made against the idiotype of his tumor cells. (An island of T cells in the middle of the lymphoma cells within a lymph node is unstained.) Conversely, MoAbs made against the idiotypes of 10 other lymphomas did not stain this patient's lymphoma, one example of which is shown here *(right panel)*. (B, from K. Thielemans, D.G. Maloney, T. Meeker et al., *J. Immunol.* 133:495–501, 1984.)

This therapeutic strategy was highly attractive because of its selectivity, since an anti-idiotype antibody recognized only a patient's tumor cells and, conversely, did not recognize antibody molecules displayed by virtually all other B-cell clones in the patient's immune system. Thus, each patient's tumor cells displayed a unique idiotype, because (1) every B-cell tumor arises from a single normal B cell; (2) this normal B cell exists among millions of distinct B-cell clones in a normal immune system, each expressing its own unique antibody; and (3) the normal progenitor of the tumor cell clone is chosen from the ranks of normal B cells by an essentially random process. This meant that an anti-idiotype monoclonal antibody needed to be custom-made for each patient suffering a B-cell tumor (Figure 15.39B). These facts help explain why, in the end, this strategy was not economically sustainable, since enormous cost was associated with the production of antibodies against each patient's tumor.

An alternative therapeutic strategy was therefore undertaken: a monoclonal antibody (MoAb) was developed that binds **CD20**, a widely expressed cell surface antigen. CD20 is displayed by cells at various stages of B-cell differentiation and in B-cell–derived tumors (Figure 15.40). The monoclonal antibody, called variously **Rituxan** or rituximab, was humanized in order to reduce the likelihood that treated patients would develop an immunological reaction against injected Rituxan molecules. Rituxan has proven to be useful for treating, among other tumors, many non-Hodgkin's lymphomas (NHLs), which constitute the fifth and sixth most common cause of cancer-related deaths in American males and females, respectively. (More than 90% of B-cell NHLs express the CD20 antigen.)

Treatment of patients with weekly doses of Rituxan has been effective in treating about half of all patients with *relapsed* follicular non-Hodgkin's lymphoma

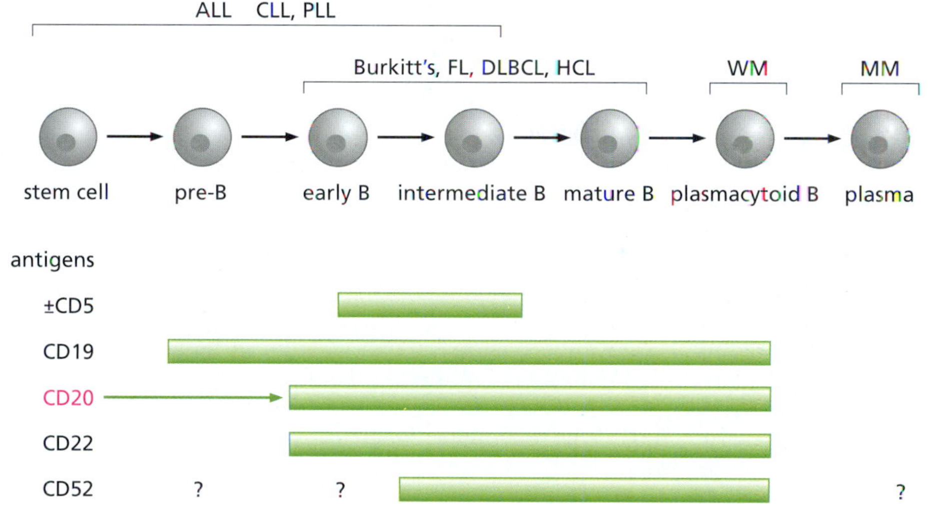

Figure 15.40 Rituxan and the treatment of B-cell tumors The utility of the Rituxan/rituximab monoclonal antibody derives in part from the fact that the CD20 antigen that it recognizes is displayed as a cell surface transmembrane protein by B cells at various stages of differentiation, beginning with early B cells and continuing through the formation of plasmacytoid B cells, the immediate precursors of the antibody-secreting plasma cells. The *green bars* indicate the stages of B-cell differentiation that express the indicated antigen. For example, CD20 is expressed by early B cells and all subsequent cell types up to and including plasmacytoid B cells. As is indicated here, a variety of tumors of the B-cell lineage, including Burkitt's lymphoma, follicular lymphoma (FL), diffuse large B-cell lymphoma (DLBCL), hairy-cell leukemia (HCL), and Waldenström's macroglobulinemia (WM), which phenocopy the traits of B cells at various discrete stages of differentiation, also express this antigen, making Rituxan a highly useful agent for treating these various leukemias. Significantly, most multiple myelomas (MMs) no longer express the CD20 antigen, because they mimic the gene expression pattern of highly differentiated plasma cells, which explains why Rituxan is ineffective in treating these tumors. The fact that several other cell surface antigens (e.g., CD19) have equally broad if not broader expression suggests that antibodies against these other antigens may one day also prove therapeutically useful. (Courtesy of R. Levy.)

(i.e., tumors that have regrown after earlier, initially successful treatments) or **refractory** cases of these tumors (i.e., tumors that have become unresponsive to other available treatment regimens). Interestingly, the populations of normal B cells in treated patients are also eliminated by this MoAb treatment and rebound only after 6 to 9 months. Nonetheless, this loss of normal B-cell function and the associated immunosuppression, together with the other, more transient consequences of rituximab treatment, represent acceptable, relatively minor side effects of this treatment.

By 2003, Rituxan had been used to treat more than half a million patients worldwide afflicted with various malignancies of the B-cell lineage. In many cases, it has been found to be an extremely useful adjunct to existing treatment. For example, addition of Rituxan to the standard chemotherapeutic treatment of diffuse large B-cell lymphoma (DLBCL)—a protocol employing a cocktail of four drugs termed CHOP—resulted in a 41% decrease in the risk of disease progression or mortality. A similar modification of the CHOP protocol for treating follicular lymphoma led to a 66% reduction in treatment failure.

While this monoclonal antibody treatment has been found capable of stabilizing disease and thereby prolonging survival, it has not been curative, and virtually all patients relapse within several years. The tumors in some of the relapsed patients may respond to a second round of Rituxan treatment, while others may have developed resistance to Rituxan. To date, the precise mechanisms of acquired resistance are not well understood; among the mechanisms proposed are alterations in CD20 expression, elevated resistance to apoptosis, loss of complement activity, and loss of immune cells capable of mediating cytotoxicity. Provocatively, the genetic background of the patients under treatment also has a strong influence on the success of Rituxan treatment (Sidebar 15.5).

Rituxan and Herceptin—both widely used in the clinic—represent the vanguard of a far larger group of therapeutic antibodies that are in various stages of research and development (see, for example, Figure 15.38). In general, when Herceptin or Rituxan is used on its own, it succeeds in extending the life span of patients by several months, occasionally by several years. This fact, together with rapidly accumulating clinical evidence, makes it apparent that monoclonal antibody therapies like these will be most effective when used in conjunction with other anti-cancer therapies with which they may be highly synergistic in inducing durable remissions and, quite possibly in the future, cures.

Research into anti-tumor monoclonal antibodies has also taken another direction: many experiments have explored the possibility of enhancing the cytotoxic effects of these antibody molecules. Until now, we have discussed the fact that

Sidebar 15.5 Genetic and nongenetic factors are strong determinants of the effectiveness of Rituxan treatment As discussed earlier, the antibody-dependent cellular cytotoxicity (ADCC)–mediated killing of targeted cancer cells depends on the ability of cytotoxic cells, such as NK cells, to use their Fcγ receptors to attach to cancer cells whose surfaces have been coated with immunoglobulin γ (IgG antibody) molecules (see Figures 15.3 and 15.36). An important Fcγ receptor of human NK cells, termed FcγRIIIa, is found in two polymorphic variant forms, which carry either a valine (V) or a phenylalanine (F) at amino acid residue position 158. The V variant FcγRIIIa shows a significantly higher affinity for binding human IgG molecules than does the F variant, and indeed, cytotoxic cells bearing the V variant FcγRIIIa are far more effective in ADCC than are those bearing the F variant.

A patient's clinical responses to Rituxan treatment are dramatically affected by the type of Fcγ receptor displayed by his or her NK cells. Thus, B-cell lymphoma patients who were homozygous for the V allele (VV genotype) and were treated with rituximab showed a median time to progression (time between initial treatment and recurrence of disease) of 534 days, while those who were homozygous or heterozygous for the F allele (VF or FF genotype) had a time to progression of 170 days. Polymorphic variants of a second Fcγ receptor, called FcγRIIa, also have a strong effect on therapeutic response. Hence, in addition to the influences of tumor cell phenotype, the outcomes of immunotherapy are often strongly affected by the genetic constitutions of the patients under treatment, in this case a gene that determines the structure of an Fc receptor.

Yet other factors predict whether or how patients will respond to Rituxan therapy. For example, when the serum levels of Rituxan are measured in patients under treatment, those who maintain higher levels for longer periods are more likely to show favorable clinical response (Figure 15.41). This highlights a key constraint of monoclonal antibody–based treatments and an unsolved issue in the development of immunotherapeutics: we still do not understand the biological mechanisms that determine the lifetimes of monoclonal antibodies in the circulation.

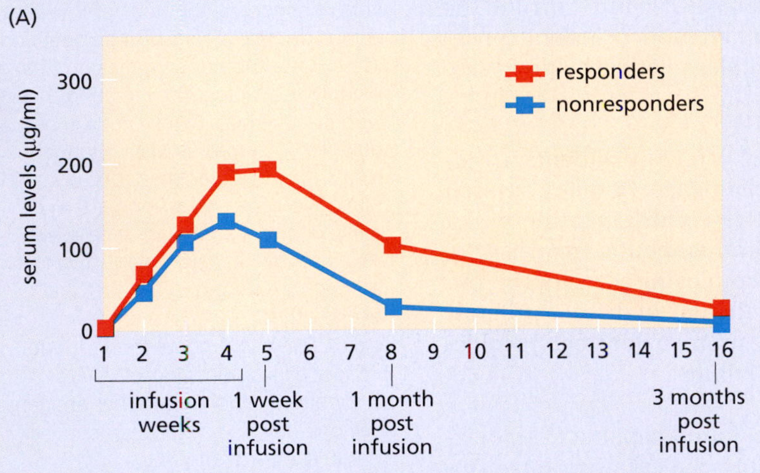

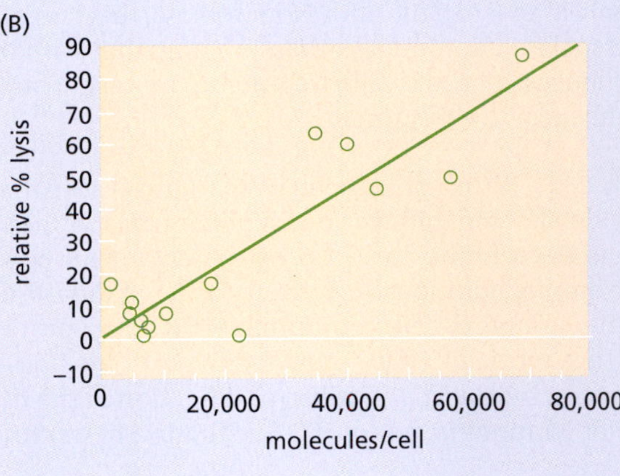

Figure 15.41 Determinants of success of Rituxan treatment (A) The time course of the persistence of a therapeutic agent in the circulation—its pharmacokinetics (PK)—often determines its therapeutic effects. As shown here, Rituxan was injected into a group of leukemia and lymphoma patients at regular intervals over a period of four weeks and the serum levels of Rituxan were followed during this time and for the next three months. The PK of Rituxan was markedly different among those who did or did not show clear clinical responses to this monoclonal antibody treatment. This behavior suggests that some type of "antigen sink" exists in certain patients that absorbs available Rituxan, thereby accelerating its disappearance from the circulation and compromising its therapeutic effectiveness. (B) Another important parameter governing the success of Rituxan therapy appears to be the levels of CD20 that are displayed by various tumor types (see Figure 15.40). Cell-surface expression levels of CD20 were measured in a collection of 14 freshly isolated leukemia samples (prolymphocytic and chronic lymphocytic leukemias, both of the B-cell lineage). As shown here, the susceptibility of these various tumors to lysis in the presence of Rituxan plus complement varies dramatically, depending on the number of CD20 molecules that are displayed per cell by each of the tumor samples. (A, from N.L. Berinstein, A.J. Grillo-Lopez, C.A. White et al., *Ann. Oncol.* 9:995–1001, 1998; B, from J. Golay, M. Lazzari, V. Facchinetti et al., *Blood* 98:3383–3389, 2001.)

an antibody molecule may attach to the surface of a tumor cell and achieve subsequent cell killing through the actions of complement or cytotoxic cells bearing Fc receptors. An attractive alternative is to link antibody molecules to toxic agents, thereby creating **immunotoxins** that are guided like "smart bombs" to the tumor, where they present the toxin in high concentrations to the targeted cells. For example, antibody molecules can be linked *in vitro* to highly toxic biological substances, such as the ricin A chain or a toxin made by *Pseudomonas* bacteria, and used to convey these toxins to the tumor.

A second approach links radioactive molecules to the antibody molecules prior to injecting them into cancer patients. The hope here is that radioactive decay will kill nearby cancer cells. A variation of this involves conjugating highly potent chemotherapeutic drug molecules, such as adriamycin, to antibody molecules.

In a third strategy, specific enzymes are linked to antibody molecules; these enzymes are capable of converting nontoxic **pro-drugs** into actively toxic drugs. Once the antibody and linked activating enzyme are concentrated in the tumor mass, the pro-drug can be injected into the patient and become activated by the enzyme in the vicinity of the tumor. The advantage of this approach derives from the fact that the enzyme can generate hundreds, possibly thousands of toxic drug molecules near targeted cancer cells, thereby amplifying the toxic effects of a single bound monoclonal antibody molecule. These various uses of monoclonal antibody molecules as tumor-targeting vectors are the subjects of active, ongoing research and development.

15.21 Passive immunization can be achieved by transfer of immunocytes from one individual to another

A quite different kind of passive immunization involves bone marrow transplantation (BMT). The original rationale for this treatment came from the discovery that the immune system of a mouse or human can be **ablated** (eliminated) through drug treatments or X-irradiation. The subsequently introduced donor marrow graft, because it contains hematopoietic stem cells (HSCs), can repopulate the recipient's bone marrow and regenerate all of the cell lineages required for normal hematopoiesis and immune function.

In the case of many hematopoietic malignancies, notably lymphomas and leukemias, the intent of bone marrow transplantation was to rid the body of the neoplastic stem cells that were present throughout the body, most importantly in the marrow itself. According to this thinking, such bone marrow transplantation (BMT) would prevent the tumor from ever regenerating itself because the tumor stem cells would be eliminated from patients' marrow.

When this BMT treatment strategy was first employed, the most effective bone marrow donor was thought to be one whose histocompatibility antigens closely matched those of the marrow recipient. This would allow the most effective repopulation of the bone marrow and minimize the likelihood of attack on a recipient's tissues by immune cells arising from the graft. However, with the passage of time, it became apparent that a minimum level of histo*incompatibility* between donor and recipient is actually desirable.

As it turned out, much of the therapeutic effect of bone marrow transplantations in treating hematopoietic malignancies derives from the graft-versus-tumor (GvT) response, in which donor immunocytes identify and attack residual tumor cells—those that have survived the radiation and chemotherapy used to ablate a patient's bone marrow. This attack is presumably provoked because the recipient's tumor cells express antigens that are unfamiliar to the engrafted donor immune cells. At present this GvT response is the only truly effective mechanism for achieving durable cures of chronic myelogenous leukemia (CML). (For reasons that remain unclear, the GvT reaction is usually not accompanied by a graft-versus-host (GvH) attack of comparable severity, which would, if it occurred, lead to widespread inflammation and the destruction of a variety of the recipient's normal tissues by the engrafted immune cells.)

Here, once again, we see an instance where the inability of a cancer patient's own immune system to mount an effective anti-tumor response is addressed by introducing the products of a foreign immune system, in this case immunocytes that are often capable of mounting a potent attack on the patient's cancer cells. Unfortunately, for multiple reasons, bone marrow transplantation has not proven to be an effective strategy for treating patients with solid (rather than hematopoietic) tumors.

15.22 Patients' immune systems can be mobilized to attack their tumors

The other major class of cancer immunotherapies depends on mobilizing and enhancing the endogenous immune defenses of cancer patients. Implicit in these approaches is the notion that their immune systems are intrinsically capable of attacking and eliminating tumors and that aggressive anti-tumor immune responses can be elicited by increasing the numbers and activity of various cytotoxic immune cells.

(These strategies are often said to constitute novel types of anti-cancer "vaccines." The term is unfortunate, because vaccines have traditionally represented substances that act to *prevent* disease rather than to *treat* existing disease. This explains why many hearing this term believe that the immunotherapies under development hold the promise of eventually serving as preventives—a goal that is far from the minds of those who are currently developing anti-tumor immunotherapies.)

The ongoing, intensive efforts at developing new immunotherapeutic protocols depend heavily on our rapidly increasing insights into immune function at the molecular and cellular level. Many of these projects exploit our knowledge of the signaling molecules that are used naturally by the immune system to regulate its various arms. Curiously, one highly successful form of immune mobilization has been used for many years to successfully treat early-stage bladder carcinomas, without a clear understanding of how it works (Sidebar 15.6).

An important strategy for mobilizing an anti-tumor response depends on activating the dendritic cells (Section 15.2). These antigen-presenting cells (APCs) are normally charged with the task of ingesting infectious agents and other antigen-bearing particles (including tumor cells) throughout the body's tissues and then rushing back to nearby draining lymph nodes, where they use their MHC class II molecules to present oligopeptide fragments of the consumed material to helper T cells (see Figure 15.8). Dendritic cells are known to be functionally activated by exposure to the growth factor termed GM-CSF (granulocyte macrophage-colony–stimulating factor).

With these interactions in mind, some researchers are culturing tumor cells from patients ***ex vivo*** (i.e., outside of the body) and forcing these cells to express GM-CSF through introduction of an expression vector. The resulting modified cells are then re-introduced into the patient, with the hope that they will attract and activate dendritic cells that happen to pass nearby. Ideally, the activated dendritic cells should then scurry back to regional lymph nodes and activate helper T cells that can then launch a potent anti-tumor immune response.

As an alternative manipulation, the precursors of dendritic cells can be harvested from the blood and induced to differentiate into functional dendritic cells by incubation with a cocktail of growth factors, specifically, GM-CSF, TNF (tumor necrosis factor), and IL-4 (interleukin-4). The intent here is to incubate the resulting large populations of dendritic cells together with tumor fragments *ex vivo* and then introduce these dendritic cells into the circulation of the cancer patient, with the hope that they will home to lymph nodes, present their acquired tumor antigens to helper T cells, and in this way mobilize an effective anti-tumor immune response.

Sidebar 15.6 Bacteria can be used to treat bladder cancer The use of bacteria to treat tumors reaches back to the end of the nineteenth century, when William Coley, a New York surgeon, noticed that cancers often regressed in patients who had experienced (and recovered from) bacterial skin infections. This led to a many-decades-long research program that undertook to use live or killed bacteria as agents for inciting an immune response, against both themselves and any tumor cells that happened to reside in the body. These attempts have fallen by the wayside with one exception, involving Bacillus Calmette–Guérin (BCG), an *attenuated* (weakened) strain of mycobacterium that was originally developed as a vaccine against tuberculosis.

BCG is frequently injected into the bladders of patients suffering from early-stage bladder carcinomas. It is usually effective in halting or delaying the progression of these tumors, doing so through a treatment that is far less traumatic than surgery or radiation (Figure 15.42). The BCG treatment clearly functions through its ability to attract a variety of immunocytes—including $CD4^+$ T_H and $CD8^+$ T_C lymphocytes as well as macrophages and NK cells—to the bladder, where these cells create localized inflammatory responses. However, the precise mechanisms of BCG antitumor action remain elusive.

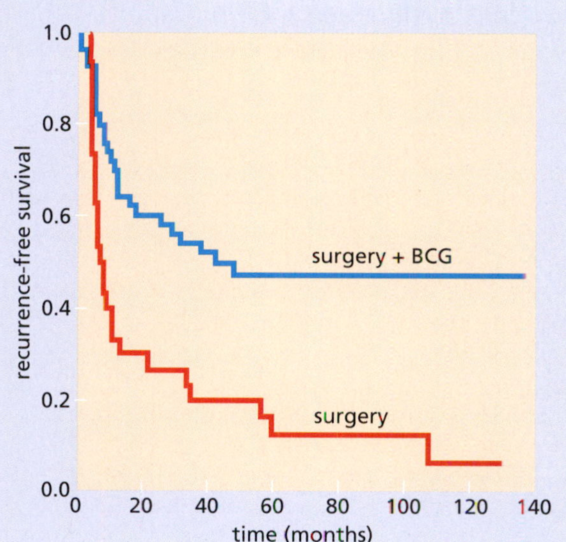

Figure 15.42 BCG enhancement of the immune response The use of Bacillus Calmette–Guérin (BCG), an attenuated mycobacterium, as an immunostimulant has clear effects in preventing the progression of human bladder carcinomas. Among patients who were treated only surgically *(red line)* with the procedure known as transurethral resection, the proportion of recurrence-free patients *(ordinate)* declined precipitously in the months after treatment *(abscissa)*, as indicated in this Kaplan–Meier plot. In contrast, among those patients who were surgically treated and were additionally exposed to BCG, which was injected into their bladders *(blue line)*, the frequency of recurrence was much less, and some were still recurrence-free after 5 years. (From J. Patard, S. Moudouni, F. Saint et al., *Urol.* 58:551–556, 2001.)

An even more direct strategy is to force the dendritic cells of a patient to express or process tumor-associated antigen proteins. The hope in this case is that these dendritic cells, following re-introduction into the body of a cancer patient, will proceed to stimulate the patient's T_H and T_C lymphocytes and thereby directly energize anti-tumor immunological responses. A diverse array of experimental protocols have been employed in order to achieve these ends. Included among these are the transfection of tumor cell mRNAs into the dendritic cells, as well as exposure of dendritic cells to certain tumor proteins or antigenic peptides, to apoptotic bodies derived from tumor cells, to tumor cell lysates, and even to irradiated tumor cell populations. In some instances, use of one or another of these strategies has resulted in significant extension of the life span of advanced cancer patients, notably those with advanced (stage-IV) melanomas. It is likely that the success rates of future cancer immunotherapy protocols will be greatly enhanced by the use of mature dendritic cells rather than the immature DCs that have been used in most of the clinical trials to date.

The responses of T lymphocytes, specifically T_H and T_C cells, to the oligopeptide antigens being displayed by antigen-presenting dendritic cells (APCs) can also be enhanced in various ways. In detail, the responses of these T lymphocytes to antigen presentation by dendritic cells usually proceed through two phases.

During the first phase (Figure 15.43A), the initial encounter of T lymphocytes with APCs induces various lymphocyte responses, these being mediated by the CD28 cell surface receptor of the T lymphocytes; CD28 participates in the antigen-presenting encounter in conjunction with the T-cell receptor (TCR), the latter being responsible for antigen recognition. Subsequently, during the second phase (Figure 15.43B), another cell surface T lymphocyte receptor, termed

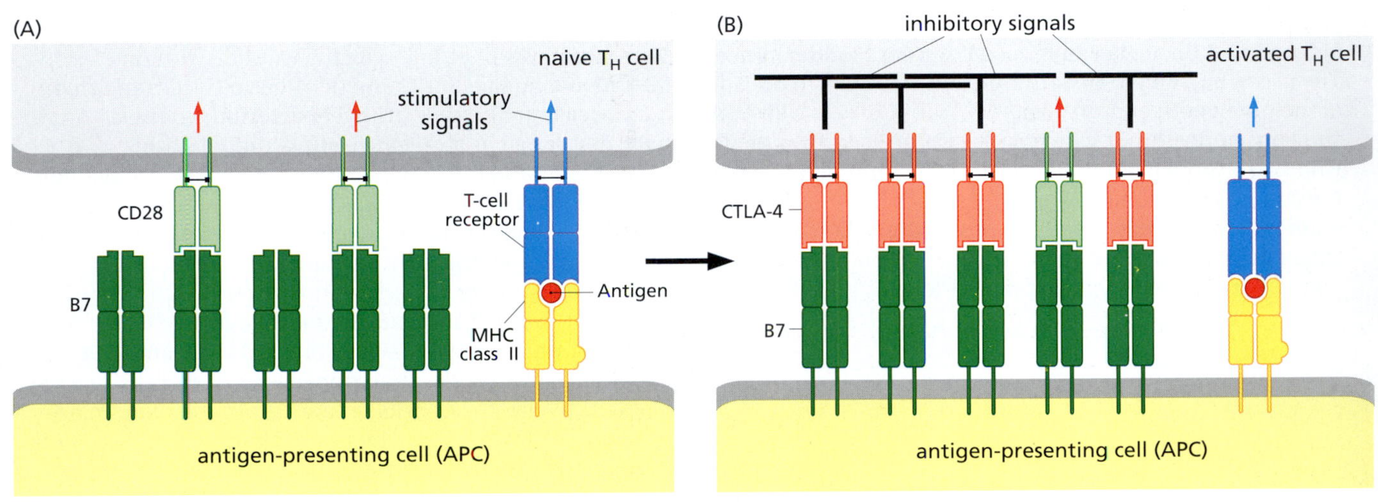

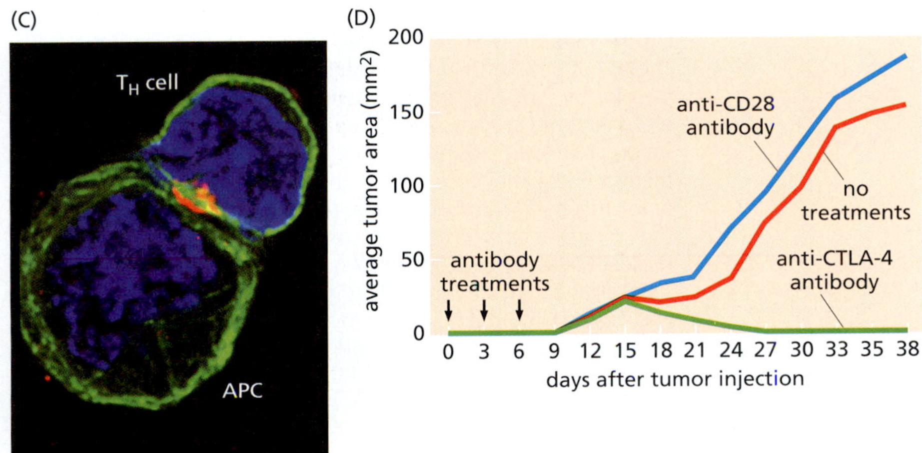

(E)

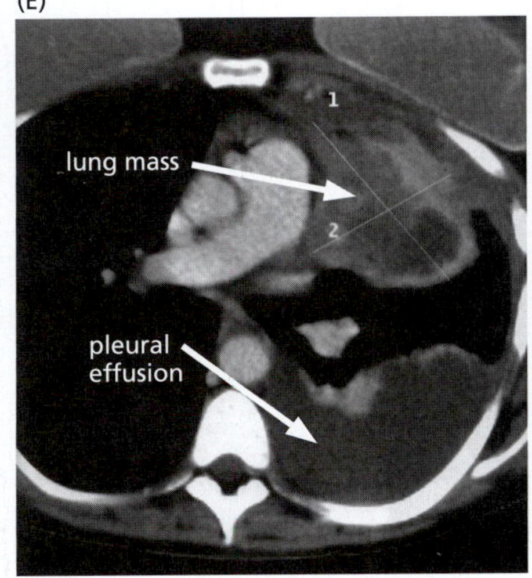

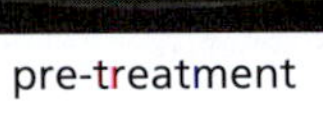
pre-treatment

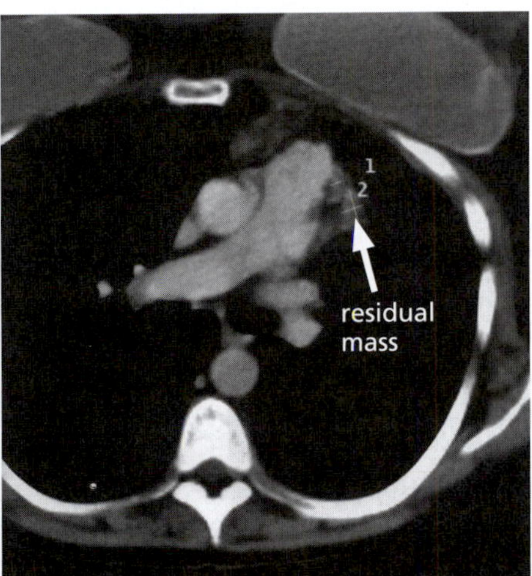

5 months post-treatment

CTLA-4 (cytotoxic T-lymphocyte antigen-4), is produced by the T cells, and it too moves into close proximity to the T-cell receptor (Figure 15.43C) and strongly shuts down further stimulation of the T lymphocyte by antigen-presenting dendritic cells. Hence, the CTLA-4 receptor mediates a critical negative-feedback control loop in order to ensure that T-lymphocyte activation is only transient and that any resulting immune responses are limited.

Monoclonal antibodies that bind and neutralize the CTLA-4 inhibitory receptor have been found to greatly potentiate the immune responses achieved by T lymphocytes, ostensibly because these antibodies prolong and intensify the

Figure 15.43 CTLA-4 modulation of the cellular immune response The interaction of a T_H lymphocyte *(above)* with an antigen-presenting cell (APC, *below*), such as a dendritic cell, proceeds in two phases. (A) Initially, the CD28 surface protein *(light green)* displayed by the T_H lymphocyte engages complementary B7 proteins *(dark green)* on the surface of the APC. This causes the CD28 molecules to release co-stimulatory signals *(red arrows)* that act together with the signals *(blue arrow)* released by the antigen-binding T-cell receptor *(blue)* to induce activation of the T_H cell. (B) In the second phase, the T cell, as part of its activation program, begins to synthesize CTLA-4 molecules *(pink)*. These CTLA-4 molecules bind the B7 molecules of the APC with higher affinity than the CD28 molecules, thereby displacing CD28 from this association and shutting down further T-cell activation. (The term "B7" subsumes two similarly acting molecules, termed B7-1 and B7-2.) (C) As seen here, when an APC and a T lymphocyte interact, CTLA-4 *(red orange)* molecules become concentrated at the *synapse* (interface) between them. (D) Modulation of CTLA-4 activity can have substantial effects on the immune response against an implanted tumor. In the absence of any intervention, syngeneic carcinoma cells grow vigorously beginning 9 days after implantation of cancer cells into a host mouse *(red line)*. Anti-CD28 serum, which should interfere with the activation of T lymphocytes by APCs (see panel A), has at best a minor effect in potentiating the growth of these tumors *(blue vs. red line)*. However, when tumor-bearing mice are exposed to anti-CTLA-4 antibody at the indicated times *(arrows)*, the normally occurring shutdown of T lymphocyte function is apparently prevented and the resulting hyperactivated T lymphocytes proceed to eliminate the tumors *(green line)*. This suggests a strategy for immunotherapy of cancer in which CTLA-4 function is suppressed through the use of an anti-CTLA-4 antibody. (E) Occasionally, the clinical use of an anti-CTLA-4 monoclonal antibody, termed MDX-010, yields dramatic responses, such as that seen here in a patient with metastatic melanoma, whose major lung mass and pleural effusion *(left panel)* largely disappeared five months after initiation of immunotherapy *(right panel)*, as revealed by this computerized tomography (CT) X-ray scan. This patient was first exposed to repeated injections of dendritic cells whose MHC class II receptors had been loaded *ex vivo* with an antigenic oligopeptide derived from the MART-1 melanoma antigen (see Figure 15.26); this treatment was followed later by injection of the anti-CTLA-4 antibody. Dramatic responses like this one indicate the potential of this type of immunotherapy, which has yet to be realized because it succeeds to this degree in only a small proportion of treated patients. (A and B, from C.A. Janeway Jr. et al., Immunobiology, 6th ed. New York: Garland Science, 2005; C, courtesy of J.P. Allison; D, from D.R. Leach, M.F. Krummel and J.P. Allison, *Science* 271:1734–1736, 1996; E, courtesy of A. Ribas, see A. Ribas, J.A.Glaspy, Y. Lee et al., *J. Immunother.* 27:354–367, 2004.)

lymphocyte activation triggered by antigen-presenting dendritic cells. In the context of cancer, treatment of tumor-bearing mice with anti-CTLA-4 monoclonal antibody has resulted in strong stimulation of the immune response against tumors, leading in some cases to tumor regression under circumstances in which these tumors would otherwise progress (Figure 15.43D). As an undesired side effect, anti-CTLA-4 antibodies induce or exacerbate autoimmune reactions in both mice and humans, an indication that this particular signaling system is one of the primary mechanisms used by the immune system to prevent the inadvertent destruction of normal tissues.

Initial clinical trials in humans carrying advanced melanomas and ovarian carcinomas indicate that the efficacy of several types of anti-tumor immunotherapy can be strongly enhanced by the injection of the anti-CTLA-4 monoclonal antibody termed MDX-010 without serious concomitant autoimmune reactions (Figure 15.43E). On some occasions, this anti-CTLA-4 antibody has been administered to patients who had previously not experienced any immunotherapy. At other times, the anti-tumor immune response of patients was first stimulated by either (1) vaccination with purified melanoma-associated antigens like those listed earlier or (2) injection with **autologous** (i.e., their own) tumor cells that had been forced *ex vivo* to express GM-CSF and were then irradiated (to prevent the proliferation of these cells after they were injected back into the patients). As mentioned earlier, this GM-CSF expression serves to attract dendritic cells to tumor cells. These patients were then treated with anti-CTLA-4 antibody.

The second strategy worked best: immunization with autologous GM-CSF–expressing tumor cells followed, sometimes months later, by injection with anti-CTLA-4 antibody. In one initial trial, 5 of 5 patients showed clear responses to this particular treatment protocol, with their tumors showing areas of necrosis and extensive damage to the tumor-associated vasculature.

Some cancer patients have been treated with interferons. The three interferons (IFNs)—IFN-α, -β, and -γ—can induce expression of MHC class I molecules on the surfaces of many cell types, including cancer cells. Moreover, as described earlier, IFN-γ can also induce MHC class II expression by professional antigen-presenting cells and can mobilize various types of immune cells, notably macrophages, cytotoxic T cells, and natural killer cells. Thus, interferons may counteract the tendency of tumor cells to suppress their expression of MHC class I molecules in order to evade immune surveillance.

In fact, a number of patients with hematopoietic malignancies, including myelomas, lymphomas, and leukemias, have shown a partial or complete regression of their tumors in response to various interferon treatments. Similarly, some melanomas, Kaposi's sarcomas, and breast and renal carcinomas have also shown such responses.

The demonstrated ability of interleukin-2 (IL-2) to activate lymphocytes has also been exploited by adding this cytokine (also called a **lymphokine**) to mixtures of killed tumor cells and lymphocytes *in vitro*. This results in the functional activation of the lymphocytes and increased killing of tumor cells in tumor-bearing mice. These *lymphokine-activated killer* (LAK) cells are almost entirely NK cells, which, as we have read, are specialized to kill target cells lacking normal amounts of MHC class I molecules and expressing certain stress- or transformation-associated proteins on their surface (see Figures 15.29 and 15.30). In some clinical trials, the resulting LAK cells have been co-injected into patients together with additional IL-2, and in one well-described trial, 16 of 222 cancer patients showed complete regression of their tumors. However, undesirable side effects of IL-2 precluded further development of this treatment as a strategy that could be used routinely in the oncology clinic.

A related strategy involves the preparation of lymphocytes that are already present in a patient's tumor, manipulating these cells *ex vivo*, and then re-introducing these cells back into the patient—the procedure of adoptive cell transfer (ACT). Once again, the bulk populations of tumor-infiltrating lymphocytes (TILs; see Figure 15.21) that are prepared from surgically removed tumor samples are NK cells. However, a functionally critical subpopulation of the cells in these TIL preparations consists of cytotoxic T lymphocytes that have acquired specific reactivity against the antigens displayed by the tumor from which they were isolated. Importantly, when these "educated" cytotoxic T cells are functionally activated *ex vivo*, they require only one-hundredth the concentration of IL-2 needed to activate the larger populations of NK cells. Injection of these tumor-infiltrating CTLs into patients (after increasing their numbers by extended propagation *in vitro*) has caused partial regressions of tumors in about one-quarter of treated melanoma and renal cancer patients. Attempts at generating TILs having specific anti-tumor cytotoxic activity from other types of tumors have not met with comparable success to date.

The responses to the TIL treatment described above have generally been short-lived, which is explained in part by the inability of the introduced populations of TILs to establish themselves stably in the bodies of cancer patients under treatment. This observation has led to attempts to stabilize and expand the populations of anti-tumor lymphocytes *in vivo* by treating melanoma patients with chemotherapy prior to injection of *ex vivo*–expanded TILs (Figure 15.44A). In this case, the motive of chemotherapeutic treatment was not to kill cancer cells. Instead, it was employed in order to reduce the populations of lymphocytes and lymphocyte precursors in patients' bone marrow, thereby "making room" for clones of subsequently introduced TILs. The latter

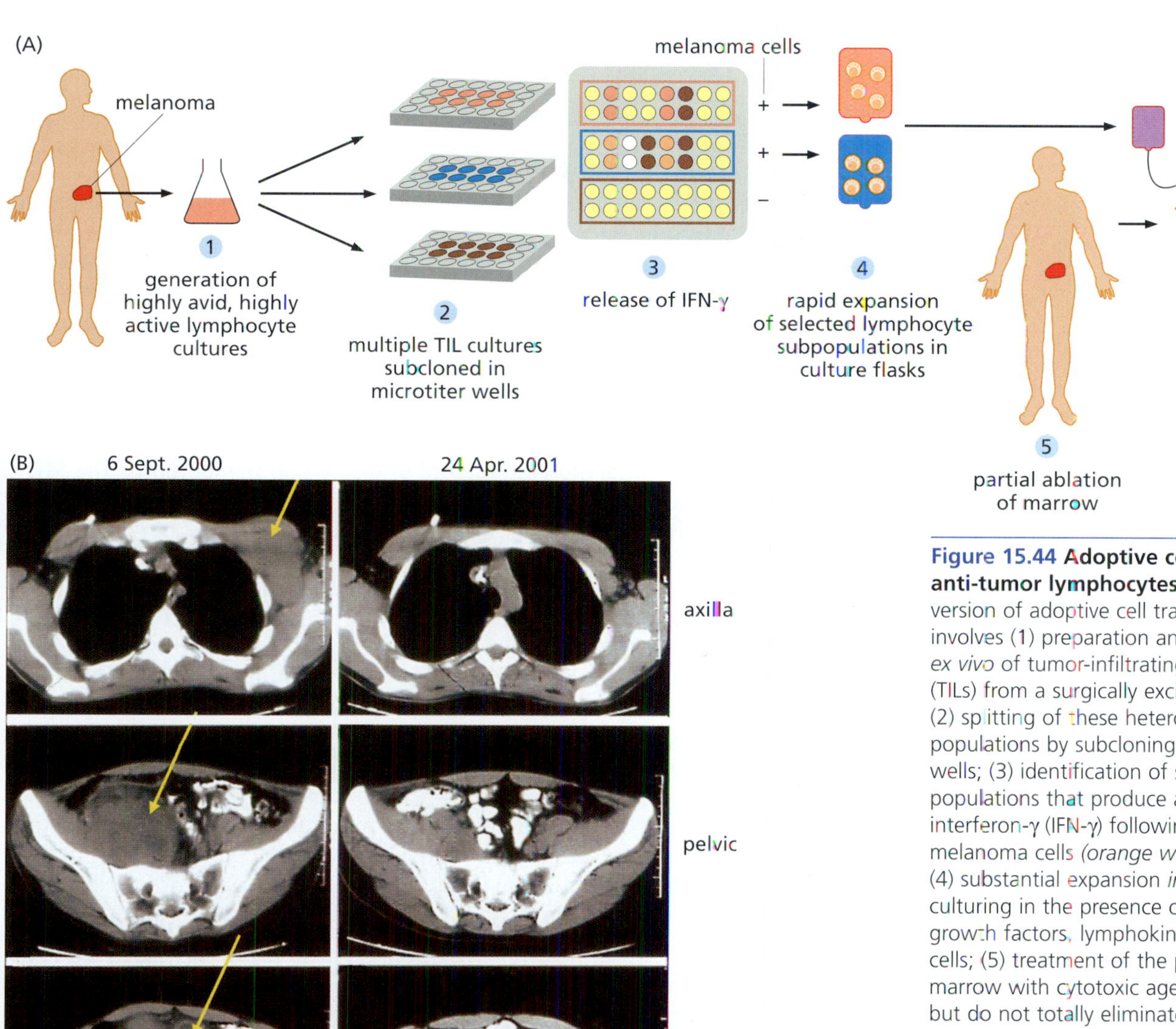

Figure 15.44 Adoptive cell transfer of anti-tumor lymphocytes (A) One version of adoptive cell transfer (ACT) involves (1) preparation and propagation *ex vivo* of tumor-infiltrating lymphocytes (TILs) from a surgically excised melanoma; (2) splitting of these heterogeneous TIL populations by subcloning in microtiter wells; (3) identification of subcloned populations that produce abundant interferon-γ (IFN-γ) following exposure to melanoma cells *(orange wells)*; (4) substantial expansion *in vitro* by culturing in the presence of appropriate growth factors, lymphokines, and feeder cells; (5) treatment of the patient's bone marrow with cytotoxic agents that reduce but do not totally eliminate the resident lymphocytes, thereby creating a biological niche for the lymphocytes that have been expanded *in vitro*; and (6) infusion of the *in vitro*–expanded lymphocytes of TIL origin. These introduced cells, which may become stably established in the marrow of the patients, can then launch attacks on the tumor cells. (B) The use of ACT has occasionally yielded dramatic responses, such as those seen in this patient who was treated for metastatic melanoma. This patient's metastases *(yellow arrows)* in an axilla (armpit; *top)*, pelvic region *(middle)*, and mesenteric lymph nodes (associated with the intestine; *bottom)*, as analyzed by computerized tomography (CT) X-ray scanning on September 6, 2000 *(left panels)*, had shrunk dramatically by April 24, 2001 *(right panels)*. Such responses, while not typical, suggest that the immune systems of cancer patients may one day be manipulated to generate outcomes like these reproducibly. (A, from M.E. Dudley and S.A. Rosenberg, *Nat. Rev. Cancer* 3:666–675, 2003; B, from S.A. Rosenberg and M.E. Dudley, *Proc. Natl. Acad. Sci. USA* 101:14639–14645, 2004.)

could then establish themselves in the marrows of patients and could persist and even expand within the patients during the course of immunotherapy. In some patients, some clonal populations of introduced anti-tumor CTLs ultimately formed the *majority* of all the cytotoxic T cells in their immune systems. These cytotoxic T cells, whose growth *in vivo* was sustained by injecting these patients with interleukin-2 (IL-2), showed high potency for killing the melanoma cells borne by the patients. In several of such patients, dramatic regressions of melanoma tumors were observed (Figure 15.44B).

Such responses, compelling as they may be, are only anecdotal, and none of these immunotherapy protocols has advanced to a stage where it can reproducibly yield robust therapeutic responses in a substantial proportion of treated patients. It seems evident that we are only beginning to learn how to manipulate the immune systems of cancer patients in ways that will cause tumor regression. This explains why the protocols described here are only first steps, and that they will soon be superseded by far more effective ways of energizing the immune responses against tumors.

15.23 Synopsis and prospects

The interactions between the immune system and tumors are surely very complex, and the precise roles that immune cells play in the suppression of most types of human tumors remain poorly understood. Studies of immunosuppressed patients provide clear testimony to the fact that immune surveillance is responsible for helping to prevent the appearance of a variety of virus-induced tumors. Humoral immune responses are likely to be responsible for suppressing the infectious spread of virus through tissues, thereby minimizing the number of virus-infected cells, some of which may eventually progress to a tumorigenic state.

Even more important, the continued expression of viral proteins by virus-transformed tumor cells creates clearly recognizable foreign antigens that can trigger a highly effective immune attack, much of it mediated by the cellular arm of the immune response. Antigenic viral oligopeptides, when presented by the MHC class I proteins of the tumor cells, attract the attention of cytotoxic T cells, which proceed to kill a virus-infected or virus-transformed cell.

The contributions of the immune response to protecting us against the many types of tumors of *nonviral* origin remain more ambiguous. It is difficult to gauge the extent to which immune surveillance plays a role in eliminating or retarding the formation of these tumors. Some measure of the importance of these mechanisms derives from studying immunocompromised patients, in whom the rates of certain solid tumors of nonviral etiology are two or three times higher than in the general population.

Some tumor cells may succeed in eluding annihilation by the immune system because they present few antigens that are clearly foreign, and thus they benefit from the tolerance that the normal immune system develops toward "self" proteins. Yet other tumor cells may express proteins and thus potential antigens in aberrant amounts or anatomical sites; such ectopic expression may alert the immune surveillance system, which will begin to hunt down and eliminate the responsible tumor cells.

Once under attack, tumor cells often defend themselves with countermeasures. They may down-modulate expression of the antigen that initially attracted the interest of the immune system, often using the same strategy—promoter methylation—that they use to rid themselves of unwanted tumor suppressor gene activity. Alternatively, they may suppress the display of the MHC class I molecules that enable the immune system to detect the presence of the antigen on their surface. The fact that the MHC class I molecules are often absent from metastatic cells feeds the suspicion that the migration of cancer cells through the body's tissues and circulation represents a dangerous passage that is often cut short by immune cells lying in ambush along the route.

Should these various immunoevasive maneuvers not suffice, cancer cells may deploy an even more effective defense by driving away or killing potential attackers. For example, release of potent, pro-apoptotic factors such as TGF-β and FasL can often ensure elimination of any immunocytes that venture too close to neoplastic cells.

In the end, the most effective defense mounted by tumor cells may derive from the body's main cellular systems for establishing and maintaining immune tolerance—the mechanisms that normally prevent the development of autoimmune attacks on the body's own tissues. Tolerance toward many antigens seems to depend on the actions of the regulatory T cells (T_{reg}s), which operate to ensure that both the humoral and cellular arms of the immune response do not destroy our normal tissues. The ability of human tumors to release the chemotactic factor CCL22 (which serves to attract T_{reg}s) may ultimately benefit them more than all the other immunoevasive maneuvers cited here.

When taken together, these various well-documented maneuvers of tumor cells persuade us that escape from immune attack is an important step in the progression of most, and perhaps all, tumors toward the highly malignant growth state. Indeed, immunoevasion may come to be recognized as a hallmark of neoplasia that is as fundamental as the half dozen or so others that were enumerated earlier (Section 11.18).

Previously, we depicted each of the steps of tumor progression as the successful breaching of an important anti-cancer defense mechanism in our cells and tissues (Chapter 11). If this is so, we might ask about the relative contributions of these various defenses to preventing the onset of cancer. Most of these mechanisms represent the actions of the regulatory circuitry that is hard-wired in all of our cells. How much of our anti-cancer defense can be attribute to these cell-autonomous mechanisms, that is, mechanisms that operate within individual cells en route to neoplasia? And how much of our defenses against cancer derives from the organism as whole and the actions of its immune system? Definitive answers to these questions are still elusive, and another decade may pass before they are in hand.

The frequent failures of the normal immune system to erect effective defenses against most tumors do not preclude the eventual development of truly effective immunotherapeutic strategies against cancer. By learning how various components of the immune system are regulated, immunologists are gaining the ability to manipulate it and to empower these components to attack cancer cells. Only a few of the many immune-mobilizing strategies currently under investigation (Table 15.4, Figure 15.45) have been described in this chapter.

To date, more than one hundred cytokines (regulatory factors of the immune system including interleukins and interferons) and their cognate receptors have been identified. Some of these have been functionally characterized, while most have not. They are almost evenly divided into those factors that potentiate immune function (e.g., interferons, interleukin-2) and those that inhibit it (e.g., IL-10, TGF-β). Such a long list of cytokines is indicative of the complexity of immune regulation. Indeed, we have only begun to understand the physiology of normal immune function and the possibilities of manipulating it to strengthen anti-cancer defenses. The fact that a fundamentally important type of immunocyte—the regulatory T cell (T_{reg})—has only recently come into clear view indicates that the field of immunology continues to revise many of its fundamental concepts.

Table 15.4 Examples of anti-tumor immunotherapy strategies

Passive immunization
Infuse tumor-specific monoclonal antibodies (e.g., Herceptin, Rituxan)
Engraft histoincompatible marrow
Active immunization
Infuse activated tumor-infiltrating lymphocytes (TILs)
Infuse dendritic cells loaded with tumor-specific oligopeptide antigen
Add B7 co-activating receptor to introduced tumor-specific antigen
Block CTLA-4 function
Inhibit regulatory T cells

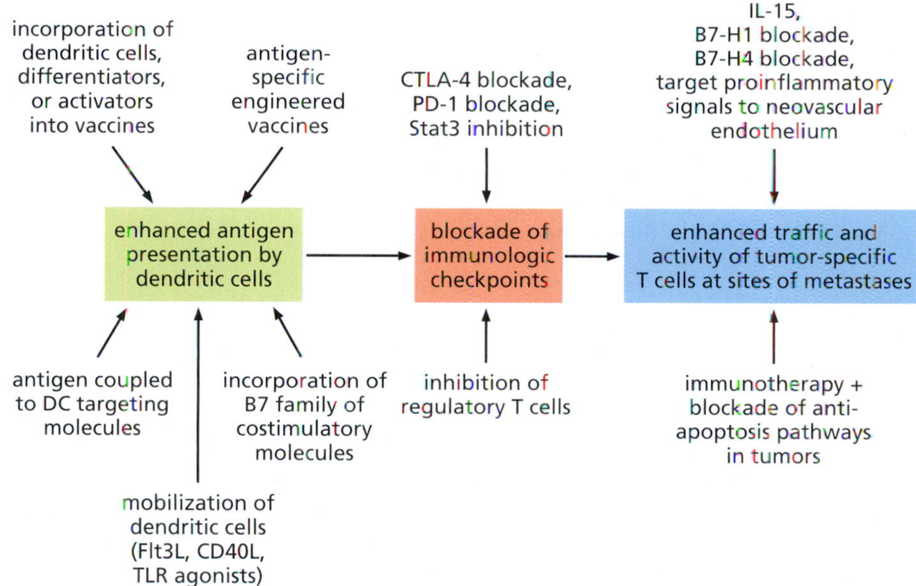

Figure 15.45 Strategies for anti-cancer immunotherapy A number of distinct strategies are currently being employed in order to create more effective anti-cancer immunotherapies. As indicated here, these focus on three major steps of immunological function: enhancing antigen presentation, suppressing immunological checkpoints, and increasing the migration of activated T cells to tumor masses. Many of the strategies indicated in this figure have not been discussed in this chapter. B7-H1 and B7-H4, B7-related proteins; DC, dendritic cell; Flt3L, ligand of Fms-like cytokine receptor; PD-1, CTLA-4-related protein; TLR, toll-like receptor. (From D. Pardoll and J. Allison, *Nat. Med.* 10:887–892, 2004.)

It remains unclear precisely how and when during multi-step tumor progression the immune system is first alerted to the presence of cancer cells and begins its initial, often unsuccessful attempts at eliminating them. This may occur when carcinoma cells invade the stroma and directly confront immunocytes. An attractive alternative model, still unproven, places these initial encounters earlier during the inflammatory phases of tumor progression, when macrophages and other leukocytes are initially recruited into the tumor stroma.

At present, it is impossible to say which of the dozens of immunotherapy strategies under development will become the precursors of anti-cancer treatments that will prove to be vastly more effective than those developed to date and will generate robust, durable cures for the majority of patients under treatment. Less optimistically, the development of truly curative immunotherapies may always be thwarted by the inventiveness and plasticity of cancer cells. Perhaps fundamental aspects of neoplastic disease create obstacles to cancer immunotherapy that can never be overcome by the cleverness of immunologists and cancer biologists. We wrestle with these issues once again in the next chapter, where the biological discoveries reported throughout this book are applied to the development of new types of low–molecular-weight drug molecules designed to bring down the elusive quarry—the ever-changing cancer cell.

Key concepts

- The immune system launches two types of attack against infectious agents or cells that it has targeted for destruction or neutralization; these involve humoral and cellular immunity.
- Some types of immune cells have an innate ability to recognize cells that should be destroyed; macrophages and NK cells are the two most prominent of these cell types.
- T lymphocytes develop the ability to recognize antigenic targets through the display of T-cell receptors, which are key elements of adaptive immunity and are created through the rearrangement of gene segments similar to those leading to the formation of soluble antibodies. These T lymphocytes are largely cytotoxic T cells (CTLs, or T_C's), helper T cells (T_H's), and regulatory T cells (T_{reg}s).
- The helper T cells aid B cells to develop antibodies, thereby creating humoral immunity, and aid cytotoxic T cells to develop and recognize cells that should be destroyed. The regulatory T cells suppress the actions of helper T cells and cytotoxic T cells.
- Normal cells throughout the body routinely present oligopeptide fragments of their proteins on the cell surface, using MHC class I molecules to do so. Professional antigen-presenting cells, such as macrophages and dendritic cells, use MHC class II molecules to present oligopeptide fragments of proteins that they have scavenged from tissue environments. These various oligopeptide fragments represent the antigens that are recognized by T-cell receptors of helper and cytotoxic T cells.
- Antibodies that recognize antigens on the surface of a cell can direct the killing of the cell through two alternative mechanisms: (1) cells that express Fc receptors, specifically macrophages and NK cells, can attach to the antibody-coated cells and kill these cells; (2) complement can attach to the surface-bound antibody and kill the cell by inserting channels into its plasma membrane.
- The immune system has the intrinsic ability to develop immune recognition of both normal tissue antigens and those expressed by foreign elements, specifically, infectious agents. However, the immune system uses a number of distinct mechanisms to suppress the reactivity of its various arms against normal tissue antigens, thereby developing tolerance of them.

- Many of the actions of the immune system directed toward recognizing and eliminating infectious agents may also be used to launch attacks against cancer cells. However, its tolerance-inducing mechanisms complicate these attacks, since the great majority of tumor cell antigens are components of normal cellular proteins.
- Some tumor-associated antigens may nevertheless attract the attention of the immune system, because they are normally displayed only in embryos or in immunologically privileged sites, such as the testes and brain, where development of tolerance toward cellular antigens does not develop. Other tumor-associated antigens may provoke immune recognition and attack because they are expressed at elevated levels.
- The fact that immunocompromised individuals experience elevated levels of various cancers provides strong suggestion that the immune system is continuously monitoring the body's tissues for the presence of tumors and attempting to eliminate them—the process of immunosurveillance.
- Some cancers may thrive in the body in spite of immune surveillance, simply because these cancers, for one or another reason, are weakly antigenic. Others may have originally been strongly antigenic and may have generated weakly antigenic variants. This represents one of many immunoevasive strategies used by tumor cells.
- Tumor-associated antigens (TAAs) are expressed by tumor cells and often reflect the differentiation programs of the tissues in which these tumor cells arose. Suppression of TAA expression may allow the tumor cells to escape immune surveillance and, at the same time, can often occur without compromising the ability of these cells to proliferate.
- Another immunoevasive strategy involves the release of factors, such as IL-10, FasL, and TGF-β, that are capable of eliminating immune cells that have ventured too close to tumor cells.
- Cancer cells may also attract regulatory T cells, which can inactivate any cytotoxic T cells that have entered into tumor masses.
- Because the immune system often fails to block the development of tumors, researchers have devised a number of treatment strategies that supplement or strengthen the existing immune response. An important strategy is to provide cancer patients with monoclonal antibodies (MoAbs) that can bind their tumor cells, leading to the killing of these cells through a variety of mechanisms. Included among these MoAbs are Herceptin and Rituxan.
- Alternatively, some attempts at developing anti-cancer immunotherapeutic protocols involve perturbing the signaling agents that normally regulate the activities of various immune cell types. For example, GM-CSF can be used to activate dendritic cells, while anti-CTLA-4 antibody can be used to enhance the interactions between antigen-presenting cells and helper T cells.
- The responses of the immune system to tumors are still imperfectly understood, and the multiplicity of immune regulators creates the opportunity to activate anti-tumor responses in many ways, most of which have not yet been attempted in the oncology clinic.

Thought questions

1. How do viral genes help us understand how cancer cells escape immune surveillance?
2. In what way might it be possible to force cancer cells to become more immunogenic?
3. How might measurements of immune functions provide insight into the presence of tumors in the body?
4. How might one determine whether tumor-associated macrophages (TAMs) are working to support tumor growth or are working as agents of the immune system to eliminate a tumor?

5. In what diverse ways might oncoproteins render cancer cells more resistant to killing by various components of the immune system?
6. What do you believe to be the most compelling evidence that the immune system plays a significant role in suppressing the appearance of many commonly occurring solid tumors?
7. How would you evaluate the relative importance of innate versus adaptive immunity in suppressing the appearance of clinically apparent tumors?

Additional reading

Alegre M-L, Frauwirth KA & Thompson CB (2001) T-cell regulation by CD28 and CTLA-4. *Nat. Rev. Immunol.* 1, 220–228.

Bingle L, Brown NJ & Lewis CE (2002) The role of tumour-associated macrophages in tumour progression: implications for new anticancer therapies. *J. Pathol.* 196, 254–265.

Boye J, Elter T & Engert A (2003) An overview of the current clinical use of the anti-CD20 monoclonal antibody rituximab. *Ann. Oncol.* 14, 520–535.

Cheson BD (2002) Rituximab: clinical development and future directions. *Expert Opin. Biol. Ther.* 2, 97–110.

Darnell R (2004) Tumor immunity in small-cell lung cancer. *J. Clin. Oncol.* 22, 762–764.

Dillman RO (2001) Monoclonal antibody therapy for lymphoma: an update. *Cancer Pract.* 9, 71–80.

Dranoff G (2002) Tumour immunology: immune recognition and tumor protection. *Curr. Opin. Immunol.* 14, 161–164.

Dranoff G (2004) Cytokines in cancer pathogenesis and cancer therapy. *Nat. Rev. Cancer* 4, 11–22.

Dudley ME & Rosenberg SA (2003) Adoptive-cell-transfer therapy for the treatment of patients with cancer. *Nat. Rev. Cancer* 3, 666–675.

Dunn GP, Bruce AT, Ikeda H et al. (2002) Cancer immunoediting: from immunosurveillance to tumor escape. *Nat. Immunol.* 3, 991–998.

Dunn GP, Old LJ & Schreiber RD (2004) The immunobiology of cancer immunosurveillance and immunoediting. *Immunity* 21, 1–20.

Figdor CG, de Vries IJ, Lesterhuis WJ & Melief CJ (2004) Dendritic cell immunotherapy: mapping the way. *Nat. Med.* 10, 475–480.

Finn OJ (2003) Cancer vaccines: between the idea and reality. *Nat. Rev. Immunol.* 3, 630–641.

Fishelson Z, Donin N, Zell S et al. (2003) Obstacles to cancer immunotherapy: expression of membrane complement regulatory proteins (mCRPs) in tumors. *Mol. Immunol.* 40, 109–123.

Gelderman KA, Tomlinson S, Ross GD & Gorter A (2004) Complement function in mAb-mediated cancer immunotherapy. *Trends Immunol.* 24, 158–164.

Greenwald RJ, Freeman GJ & Sharpe AH (2005) The B7 family revisited. *Annu. Rev. Immunol.* 23, 515–548.

Harris M (2004) Monoclonal antibodies as therapeutic agents for cancer. *Lancet Oncol.* 5, 292–302.

Houghton AN, Gold JS & Blachere NE (2001) Immunity against cancer: lessons learned from melanoma. *Curr. Opin. Immunol.* 13, 134–140.

Khong HT & Restifo NP (2002) Natural selection of tumor variants in the generation of "tumor escape" phenotypes. *Nat. Immunol.* 3, 999–1005.

Korman A, Yellin M & Keler T (2005) Tumor immunotherapy: preclinical and clinical activity of anti-CTLA4 antibodies. *Curr. Opin. Invest. Drugs* 6, 582–591.

Lieberman FS & Schold SC (2002) Distant effects of cancer on the nervous system. *Oncology* 16, 1539–1551.

Melief CJM, Van Der Burg Sh, Toes RE et al. (2002) Effective therapeutic anticancer vaccines based on precision guiding of cytolytic T lymphocytes. *Immunol. Rev.* 188, 177–182.

Old LJ (2001) Cancer/testis (CT) antigens—a new link between gametogenesis and cancer. *Cancer Immun.* 1, 1.

Old L & Chen Y-T (1998) New paths in human cancer serology. *J. Exp. Med.* 187, 1163–1167.

Pardoll DM (2002) Spinning molecular immunology into successful immunotherapy. *Nat. Rev. Immunol.* 2, 227–238.

Pardoll D (2003) Does the immune system see tumors as foreign or self? *Annu. Rev. Immunol.* 21, 807–839.

Pardoll D & Allison J (2004) Cancer immunotherapy: breaking the barriers to harvest the crop. *Nat. Med.* 10, 887–892.

Penn I (1994) Depressed immunity and the development of cancer. *Cancer Detect. Prevent.* 18, 241–252.

Rathmell JC & Thompson CB (1999) The central effectors of cell death in the immune system. *Annu. Rev. Immunol.* 17, 781–828.

Raulet D (2003) Roles of the NKG2D immunoreceptor and its ligands. *Nat. Rev. Immunol.* 3, 781–790.

Riddell SR (2001) Progress in cancer vaccines by enhanced self-presentation. *Proc. Natl. Acad. Sci. USA* 98, 8933–8935.

Sakaguchi S (2005) Naturally arising Foxp3-expressing $CD25^+CD4^+$ regulatory T cells in immunological tolerance to self and non-self. *Nat. Immunol.* 6, 345–352.

Scanlan MJ, Gure AO, Jungbluth AA et al. (2002) Cancer/testis antigens: an expanding family of targets for cancer immunotherapy. *Immunol. Rev.* 188, 22–32.

Schwartz RH (2005) Natural regulatory T cells and self-tolerance. *Nat. Immunol.* 6, 327–330.

Slamon DJ, Leyland-Jones B, Shak S et al. (2001) Use of chemotherapy plus a monoclonal antibody against HER2 for metastatic breast cancers that overexpress HER2. *N. Engl. J. Med.* 344, 783–792.

Smith MR (2003) Rituximab (monoclonal anti-CD20 antibody): mechanisms of action and resistance. *Oncogene* 22, 7359–7368.

Smyth MJ, Godfrey DI & Trapani JA (2001) A fresh look at tumor immunosurveillance and immunotherapy. *Nat. Immunol.* 2, 293–299.

Smyth MJ, Hayakawa Y, Takeda K & Yagita H (2002) New aspects of natural-killer cell surveillance and therapy of cancer. *Nat. Rev. Cancer* 2, 850–861.

Turk MJ, Wolchok JD, Guevara-Patino JA et al. (2002) Multiple pathways to tumor immunity and concomitant autoimmunity. *Immunol. Rev.* 188, 122–135.

Wang E, Panelli MC & Marincola FM (2005) Gene profiling of immune responses against tumors. *Curr. Opin. Immunol.* 17, 423–427.

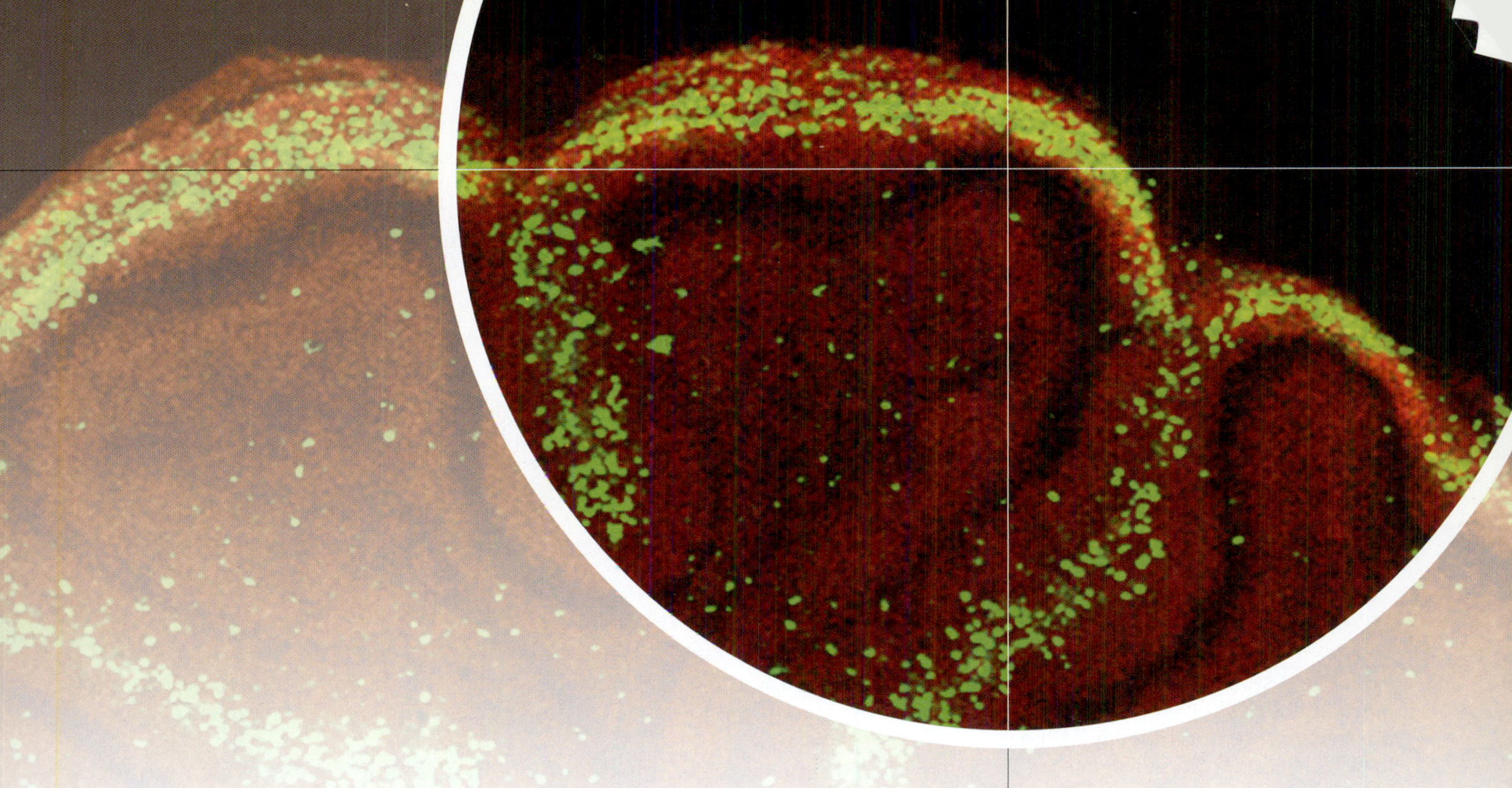

Chapter 16

The Rational Treatment of Cancer

> All substances are poisonous, there is none that is not a poison; the right dose differentiates a poison from a remedy.
>
> Paracelsus (Auroleus Phillipus Theostratus Bombastus von Hohenheim), alchemist and physician, 1538

> Doctors are men who prescribe medicines of which they know little, to cure diseases of which they know less, in human beings of whom they know nothing.
>
> Voltaire (François-Marie Arouet), author and philosopher, 1760

The research described throughout this book represents a revolution in our understanding of cancer pathogenesis. In 1975, there were virtually no insights into the molecular alterations within human cells that lead to the appearance of malignancies. One generation later, we possess this knowledge in abundance. Indeed, the available information and concepts about cancer's origins can truly be said to constitute a science with a logical and coherent conceptual structure.

In spite of these extraordinary leaps forward, relatively little progress has been made in exploiting these insights into etiology (i.e., the causative mechanisms of disease) to prevent the disease and, equally important, to treat it. Most of the anticancer treatments in widespread use today were developed prior to 1975, at a time when the development of therapeutics was not yet informed by the genetic and biochemical mechanisms of cancer pathogenesis. This explains the widely felt

(A)

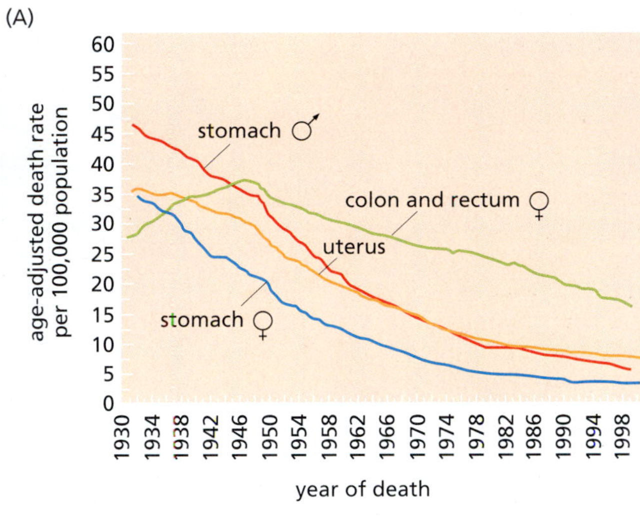

(B)

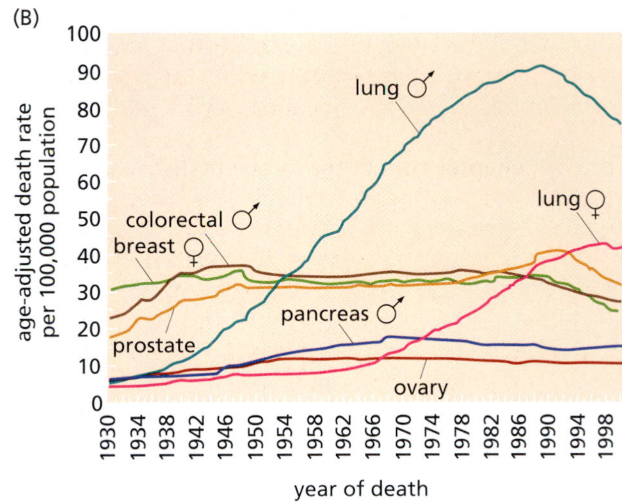

Figure 16.1 Statistics of cancer mortality over the past three-quarters of a century: the lay of the land The statistics compiled in the United States on the age-adjusted death rates from various types of cancer reveal two different long-term trends. (A) Mortality from several major killers has declined significantly since 1930. This is due to changes in food storage practices and possibly *Helicobacter pylori* infection rates, in the case of stomach cancer, and to screening, in the cases of cervical and colorectal cancers. (B) A number of major sources of cancer-related death have proven resistant to most forms of traditional therapy, especially when these tumors progress to a highly malignant, metastatic stage. (From A. Jemal, T. Murray, E. Ward et al., *CA Cancer J. Clin.* 55:10–30, 2005.)

frustration among molecular oncologists that the potential of their research for contributing to new anti-cancer therapeutics has not yet been realized.

This frustration is fueled most strongly by the slow pace at which advances have been made in the treatment of common solid tumors. For example, in 1970 in the United States, 7% of the patients diagnosed with lung cancer were still alive 5 years after their initial diagnosis. Three decades later, this number had risen to only 14%, a relatively minor improvement. And even this degree of therapeutic success may be illusory, since modern diagnostic techniques often detect tumors far earlier in their clinical course, creating a greater time span between initial diagnosis and ultimate progression to end-stage disease. Death rates for colon cancer have begun to fall, because of early detection and surgical removal of growths that have advanced through only the early stages of tumor progression (Figure 16.1A; see also Figure 11.8C). However, mortality caused by the more advanced colorectal tumors has changed little in recent decades—a testimonial to the failures of chemotherapy and radiation to eliminate these malignancies once they have invaded and begun to metastasize (Sidebar 45).

Surely the now-abundant information on the molecular and cellular mechanisms of cancer pathogenesis will inspire new ways to effectively treat the disease. In this chapter, we explore a number of strategies of therapy under development or recently introduced into the clinic. The goal here is not to survey the full range of current research in these areas. That would be unreachable: a 2003 compilation of anti-cancer therapies in pre-clinical development or in clinical testing listed more than 1300 research and development projects that were being pursued by various large pharmaceutical companies and smaller biotechnology firms. The therapeutic agents under development included small–molecular-weight drugs, proteins, monoclonal antibodies, and gene therapy strategies including viral vectors.

Rather than being encyclopedic, we will concentrate here on a small number of recently developed therapies that illustrate how discoveries described in the previous chapters have inspired novel strategies for treating cancer, and how molecular diagnosis will increasingly play a part in the development and clinical introduction of novel therapies.

At the same time, we will pass over discussions of how molecular biology is changing cancer prevention strategies. Thus, we will not examine the major advances that have been made in developing vaccines that protect against hepatitis B virus (HBV) and human papillomavirus (HPV) infections; these vaccines should be highly effective in reducing the incidence of hepatomas and cervical carcinomas, which are rampant in certain parts of the world. (If the past history of public health is any guide, the prevention of cancer will ultimately yield far greater reductions in disease-related mortality than will therapies of the sort discussed in this chapter; see Sidebar 45 .)

The present chapter concentrates on small-molecule therapeutics and on their intracellular targets that have been identified through research on the signaling pathways within cancer cells. These strategies all hold great promise, and invariably their true potential is yet to be realized. The anecdotes surrounding the development of each of these drugs are interesting and provocative, because they teach important lessons about the triumphs and pitfalls of developing novel anti-cancer treatments. Note that a number of monoclonal antibody–based therapeutic strategies have already been discussed in some detail in previous chapters, as have several therapies focused on preventing or blocking tumor angiogenesis. Note also that the conventional means of treating cancer, most of which have been in use for decades, are not discussed here, because their development was not informed by the more recent discoveries described in this book.

Almost all of the research findings described throughout this textbook will stand the test of time and be considered credible and correct (though perhaps not that interesting) a generation from now. However, those who love certainty and eternal truths will find the stories that follow to be unsatisfying for a very simple reason: the work reported is in great flux and is therefore far more time-sensitive.

Many of these therapies will seem quaint and anachronistic a decade after this chapter is written. The campaign to convert insights about cancer's molecular causes into new ways of curing disease has just begun. And so we encounter its first few stumbling steps.

16.1 The development and clinical use of effective therapies will depend on accurate diagnosis of disease

In previous chapters, we repeatedly categorized cancers in terms of their tissues of origin and their stages of clinical progression. Almost always, these assignments have been dictated by the appearance of normal and malignant tissues under the microscope. On some occasions, to be sure, we have refined these classifications by describing certain molecular markers (e.g., expression of HER2 in breast cancers) and the implications that they hold for prognosis. But in general, histopathology has reigned supreme in our discussions, as it has in the practice of clinical oncology for more than half a century.

Even without insights into the molecular origins of human cancers, it has become increasingly clear that the traditional ways of classifying cancers have limited utility. Truly useful diagnoses must inform the clinician about the underlying nature of diseases and, more important, how each disease entity will respond to various types of therapy. As we have learned more about human cancers, we have come to realize that many human cancers that have traditionally been lumped together as examples of a single disease entity should, in fact, be separated into several, distinct disease subcategories. This helps to explain why many existing anti-cancer therapeutic strategies used over the past three decades have had such low overall success rates. These response rates also have important implications for the development of new drugs (Sidebar 46).

For patients bearing the subtype of a tumor that responds to a particular treatment, this therapy may prove to be a boon, extending life and even offering cures on occasion. For the remaining patients, a uniformly applied therapy yields no positive clinical effect and may divert these patients from receiving other treatments that might prove to be truly beneficial. Worse yet, anti-cancer treatments often incur numerous side effects and some may actually increase the incidence of second-site cancers arising years later. For example, in the early 1980s, breast cancer patients receiving the then-standard dose of cyclophosphamide (a chemotherapeutic drug that is also an alkylating agent) experienced a 5.7-fold increased risk of subsequently developing acute myelogenous leukemia (AML). (Current treatment protocols use lowered drug dosage and result in greatly decreased incidence of such second-site cancers.) All this points to the great need for more refined diagnostic tools—ones that can accurately predict responsiveness to various anti-tumor therapies and avoid use of therapies when they are not needed (see, for example, Sidebar 47).

In 2003, about 192,000 invasive breast cancers and 46,000 *in situ* breast carcinomas were detected in the United States; the disease was predicted to claim about 40,000 lives that year. The great majority of the patients with invasive mammary carcinomas were treated aggressively with chemotherapy. Since the age-adjusted death rate from breast cancer in the United States did not change significantly throughout most of the twentieth century (Figure 16.2) during a period when therapies changed dramatically, this suggests that a large proportion of invasive breast carcinomas currently being diagnosed—possibly more than three-quarters—are not likely to cause the death of the women carrying them, even without therapeutic intervention, much like the prostate cancers that are diagnosed in vast numbers in the West (see Sidebar 47). As screening for breast cancer increases and the power to detect small, previously overlooked tumors improves (Figure 16.3), this disparity between breast cancer incidence and mortality is likely to grow.

Statistics like these underscore the desperate need to develop molecular markers that enable oncologists to distinguish between those tumors that truly require aggressive treatment and those that can be ignored or monitored periodically for signs of progression. In the case of other types of cancer, equally important distinctions must also be made, but of a far grimmer sort—between those cancers that are likely to show some response to therapy and those that will not, in which case compassionate care dictates that the disease should be allowed to run its natural course.

Gene expression arrays, of the type first described in Figure 13.18, show great promise by allowing clinicians to **stratify** cancers—to classify them in subgroups having distinct biological properties and prognoses. These gene expression arrays, often referred to as the key analytical tools of the science of *functional genomics*, allow a researcher to survey the expression levels of 10,000 or

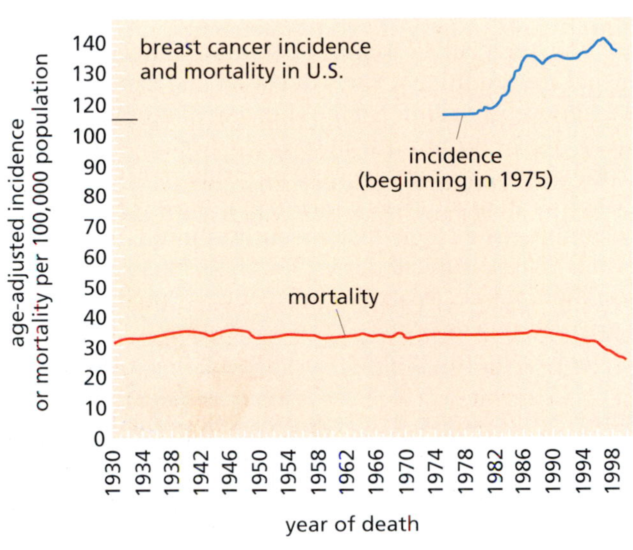

Figure 16.2 Breast cancer incidence vs. mortality in the United States The age-adjusted *incidence* of breast cancer has been increasing steadily over the past several decades, while *mortality* from this disease was quite constant until the end of the twentieth century, when it began to decline. Most of the increase in incidence appears to be attributable to increasing screening, but a small proportion of it may be due to real changes in the rate at which the disease strikes because of changes in reproductive practices, nutrition, and so forth. (From A. Jemal, T. Murray, E. Ward et al., *CA Cancer J. Clin.* 55:10–30, 2005.)

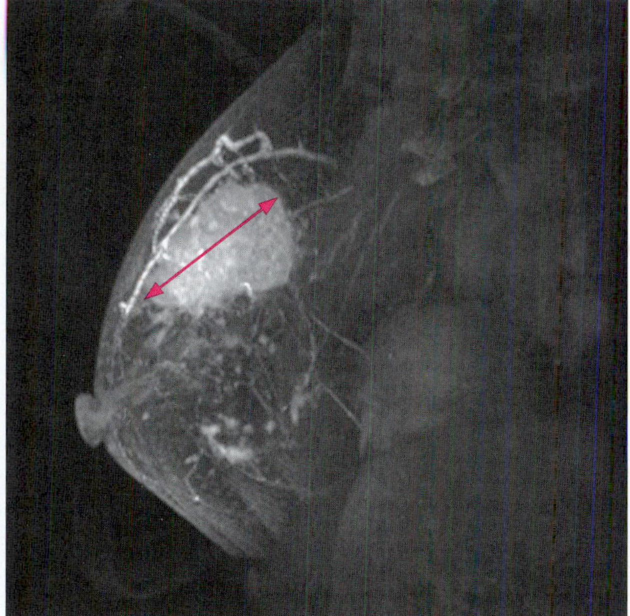

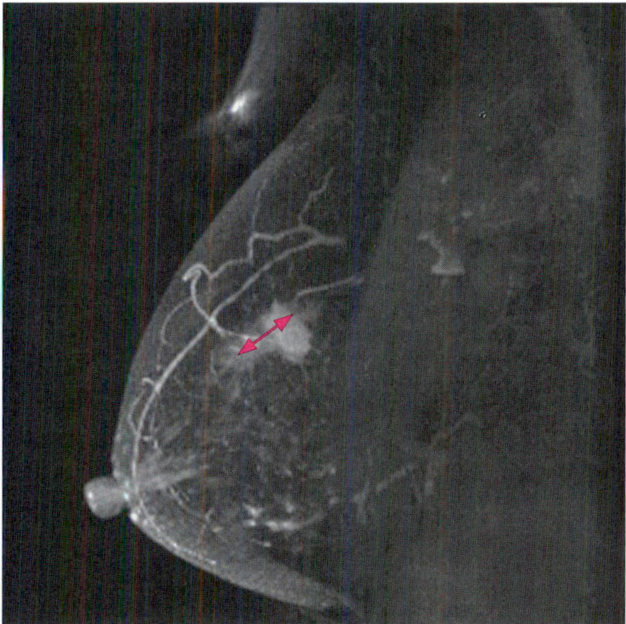

Figure 16.3 High-resolution noninvasive imaging of human tissues The development of magnetic resonance imaging (MRI) has enabled increasingly higher-resolution, *noninvasive* visualization of living tissues. MRI now allows breast tumors of very small size (several mm diameter) to be detected and, as shown here, makes it possible to view the progress of anti-tumor therapy—in this case chemotherapy with an anthracycline cytotoxic agent—in exquisite detail. Widespread use of such highly sensitive imaging techniques is likely to result in further increases in the incidence rate of breast cancer. (Courtesy of N.M. Hylton and L.J. Esserman.)

even 20,000 distinct genes in a tissue preparation. Subsequent computerized analyses of these expression arrays using **bioinformatics** make it possible to identify a small subset of these genes whose expression (at characteristically high or low levels) correlates with a specific biological phenotype, drug responsiveness, or prognosis. For example, the expression of a cohort of several dozen genes by a tumor may suffice to serve as a strong predictor of its degree of progression or its association with one or another subtype of cancer.

In the case of breast cancers, there has been a crying need to distinguish those primary tumors that are likely to become metastatic from those that will remain indolent and are therefore not likely to spread during the lifetime of the patient. Traditionally, the main prognostic parameters that have been used to predict the course of tumor development have been patient age, tumor size, status of axillary lymph nodes, histologic type of the tumor, pathological grade, and hormone-receptor status (i.e., the expression of estrogen and progesterone receptors). Because these factors, when used singly or in combination, do not yield prognoses with an extremely high degree of accuracy, the great majority of patients diagnosed with primary breast cancers in the United States have been treated aggressively, even though only some 15% of such patients will ever develop metastatic, life-threatening disease.

The use of gene expression arrays and bioinformatics has made it possible to predict the clinical course of breast cancer progression with more than 90% accuracy (Figure 16.4), and further improvements in predictive power are being developed. Such highly accurate prognostic information holds the promise of sparing many women exposure to unnecessary chemotherapy. And in the future, the details of an expression array analysis may also inform the oncologist about the treatment protocol that is most likely to yield a durable clinical response or even a cure.

Analyses like the one shown in Figure 16.4 are only the beginning steps in a large-scale effort to analyze a variety of human cancer types by means of expression arrays, to stratify the types into subtypes, and, on the basis of the information the arrays yield, to devise therapies tailored to each specific subtype. For example, B-cell lymphomas have presented a quandary to the oncologist because their outcomes are so variable in the clinic, with some patients dying within four weeks of diagnosis while others are being cured, or are at least achieving 10-year survival without clinical symptoms. At the same time, all these tumors have a very similar appearance under the microscope (Figure 16.5A, top). However, use of gene expression arrays (below in Figure 16.5A) has allowed these tumors to be segregated into three distinct diseases with quite different clinical outcomes—primary mediastinal B-cell lymphomas, germinal-center B-cell-like lymphomas, and activated B-cell-like lymphomas (Figure 16.5B).

Of these three, both the activated B-cell-like lymphomas (ABCs) and the primary mediastinal B-cell lymphomas (PMBLs) exhibit constitutively high levels of NF-κB activity (Figure 16.5C); this transcription factor (Section 6.12) appears to be driving their proliferation and, at the same time, protecting them from apoptosis. Accordingly, drugs that target the NF-κB pathway, specifically its upstream activator, IKK, have been used in attempts to affect these two subtypes of DLBCL cells propagated in culture, and indeed both groups of cells are killed once they lose IKK activity (Figure 16.5D). Cultured cells from the third lymphoma subtype, germinal-center B-cell-like, do not show high NF-κB activity and are essentially unaffected by such treatment.

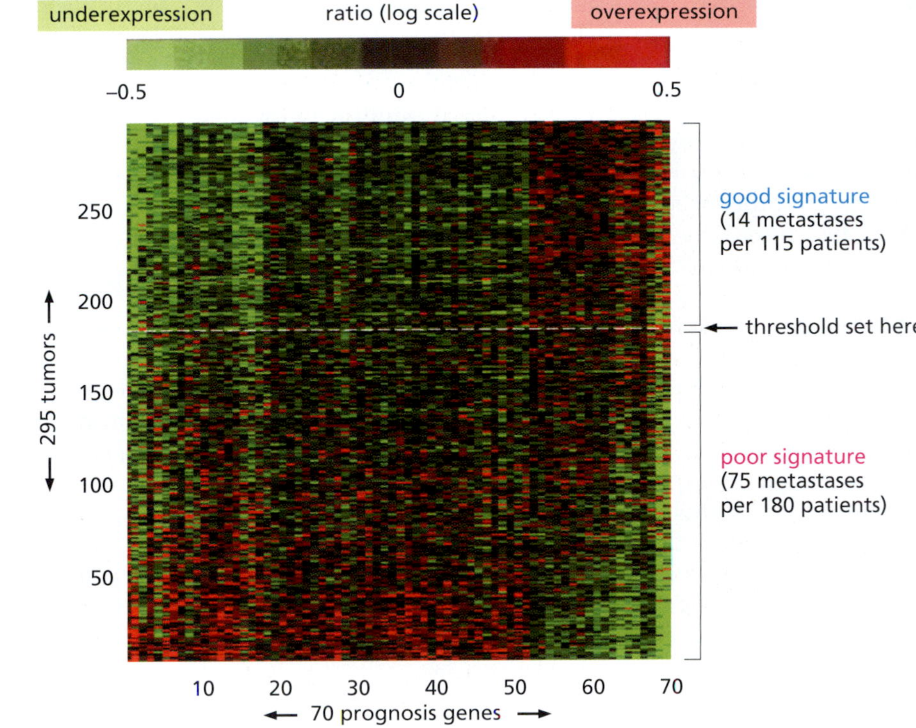

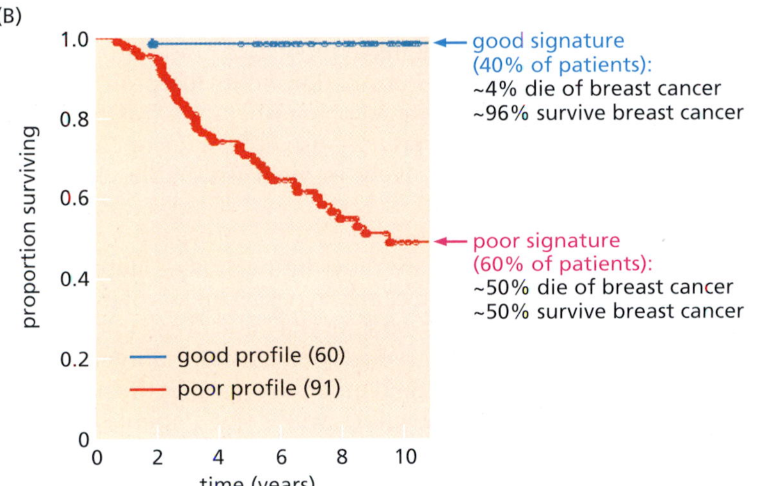

Figure 16.4 Stratifying breast cancers using functional genomics
(A) Expression arrays were used to analyze the gene expression of a group of 295 primary breast cancers diagnosed in women of less than 53 years of age. The group included patients with metastatic cells in their axillary lymph nodes as well as patients whose lymph nodes were free of cancer cells. Bioinformatics analyses of these tumors were then employed to choose a set of 70 "prognosis genes" whose expression could be used to stratify these breast cancer patients *(arrayed along vertical axis)*, whose clinical course had been followed for a mean time of 7 years. The expression levels of these 70 genes *(arrayed along horizontal axis, names not given)* together with information about the patients' clinical history was then used to set a threshold that separated tumors that had a "good expression signature" from tumors that had a "poor expression signature." (B) This Kaplan–Meier plot reveals the stratification of a group of 151 breast cancer patients whose survival had been followed for 10 years following initial diagnosis. Using the criteria of panel (A), they could be separated into two groups with dramatically different clinical courses. Taken together with other factors (such as the efficacy of chemotherapy), calculations indicate that women whose tumors carry a good expression signature derive virtually no benefit from adjuvant chemotherapy. (From M.J. van de Vijver, Y.D. He, L.J. Van't Veer et al., *N. Engl. J. Med.* 347:1999–2009, 2002.)

Interpretation of the gene expression patterns of tumors of complex histology, such as carcinomas composed of both epithelial and stromal cell types, is often confounded by the fact that the RNA transcripts being measured are a mixture deriving from multiple cell types. The technique of *laser capture microdissection* (LCM; see Figure 13.23) now makes it possible to physically isolate the epithelial from the stromal cells present in a carcinoma sample that has been mounted on a microscope slide. This allows the gene expression pattern of the two groups of cells to be analyzed separately, enabling further refinement of these analyses and, potentially, even greater accuracy in the stratification of tumor samples.

Beyond these gene expression analyses stands a generation of novel diagnostic tools involving the science of **proteomics**, in which the spectrum of proteins

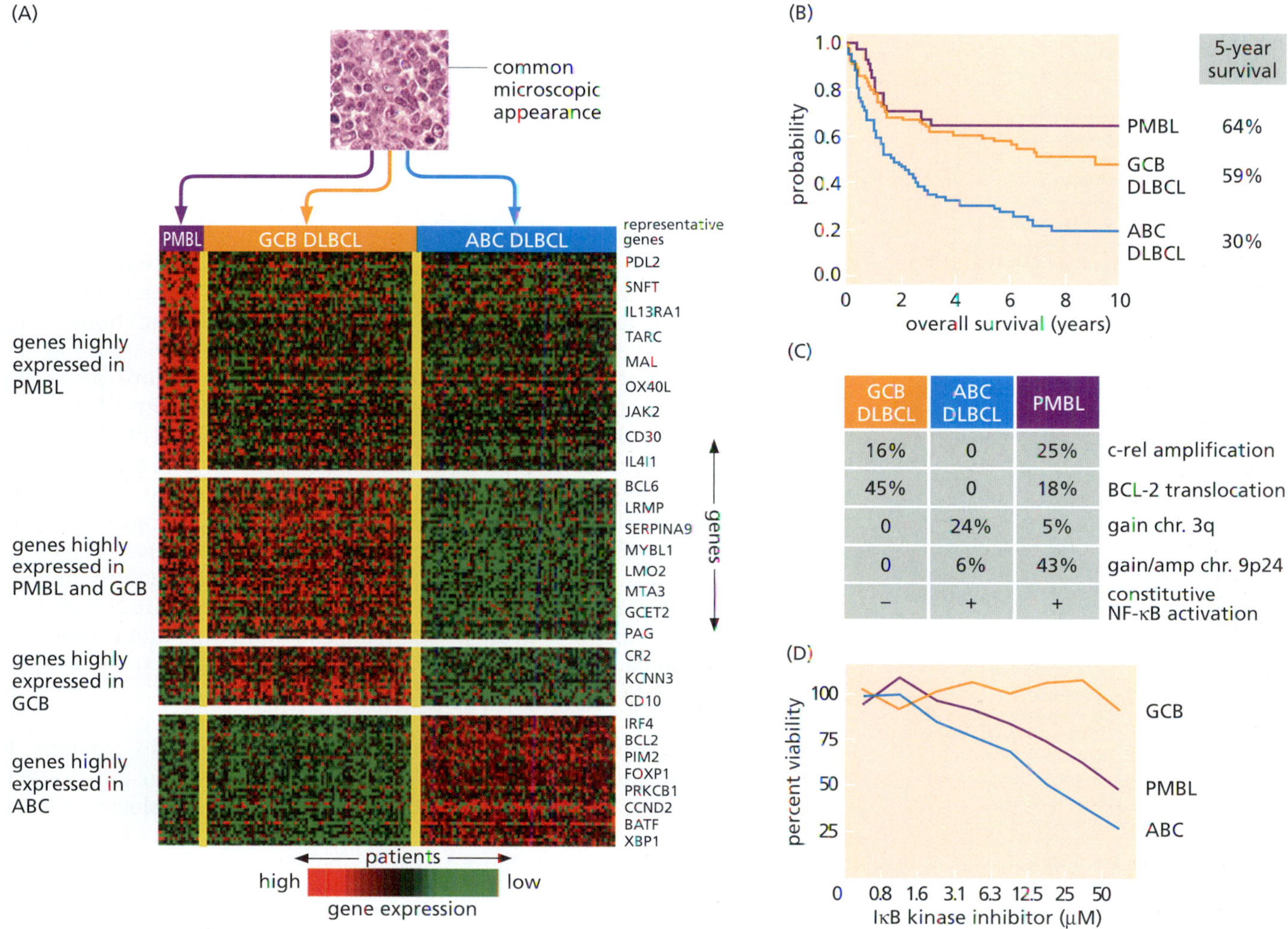

GCB DLBCL	ABC DLBCL	PMBL	
16%	0	25%	c-rel amplification
45%	0	18%	BCL-2 translocation
0	24%	5%	gain chr. 3q
0	6%	43%	gain/amp chr. 9p24
–	+	+	constitutive NF-κB activation

Figure 16.5 Stratification of diffuse large B-cell lymphomas (A) DLBCLs represent a group of several subtypes of B-cell neoplasia that are essentially indistinguishable in the microscope from one another and from primary mediastinal B-cell lymphomas *(above)*. However, use of gene expression arrays *(below)* allows these tumors to be stratified into three subtypes, termed primary mediastinal B-cell lymphomas (PMBLs), germinal-center B-cell (GCB) DLBCLs, and activated B-cell-like (ABC) DLBCLs. In this expression array, a set of genes whose expression levels have been found to be useful in making this stratification is plotted along the vertical axis, while a set of patient tumors *(unlabeled)* is plotted along the horizontal axis. (B) This Kaplan–Meier curve illustrates the greatly differing disease courses that patients with the three subtypes experience. (C) The tumors that are classified through the expression array analyses shown in panel A exhibit distinct karyotypic and biochemical alterations. (D) The fact that the PMBLs and the ABCs show high constitutive levels of NF-κB activity suggests that they may be particularly susceptible to disruption of this signaling pathway by inhibition of the upstream activator of NF-κB, IκB kinase (IKK; see Figure 6.29). This is borne out by experiments in which MLX105, a pharmacologic inhibitor of IKK, was applied to cell lines derived from the three types of lymphoma and indeed showed differential effects on these three cell populations growing *in vitro*. (A and C, from L.M. Staudt and S. Dave, *Adv. Immunol.* 87:163–208, 2005; B, from A. Rosenwald, G. Wright, K. Leroy et al., *J. Exp. Med.* 198:851–862, 2003; D, from L.T. Lam, R.E. Davis, J. Pierce et al., *Clin. Cancer Res.* 11:28–40, 2005.)

expressed in a patient's tumor or serum provides critical diagnostic information. The long-term goal of all of these analytic techniques—both functional genomics and proteomics—is to assign each patient's tumor to a specific subtype of disease and to apply drug therapies that are proven to be effective for treating a particular subtype of cancer but not for other, superficially similar tumors, for which treatment may not be effective. Such use of tailor-made drug therapies holds the promise of yielding high response rates in narrowly defined patient populations.

16.2 Successful anti-cancer drugs can elicit several responses from tumor cells

In principle, several distinct biological strategies might prove to be equally successful in eliminating established tumors or holding their growth in check. The most obvious of these are designed to induce the death of cancer cells, usually via apoptosis. Indeed, almost all of the existing nonsurgical strategies for eliminating cancer cells lead in one way or another to activation of their apoptotic pathways. However, an alternative therapeutic strategy relies on inducing differentiation, and we will consider this one first, if only briefly.

As described in Chapter 8, the acquisition of the malignant phenotype is usually accompanied by defective differentiation and associated entrance into a post-mitotic state. Recall that as tumor cell populations evolve to greater degrees of malignancy, they usually shed more and more markers of differentiation.

These behaviors suggest an attractive strategy for treating tumors: persuade cancer cells to differentiate and thereby enter into post-mitotic states. While we have learned much about the connections between cell cycle control and the regulation of differentiation programs (see Section 8.11), most of this information has not yet been translatable into effective forms of therapy. To date, the prominent exception has been one form of treatment of acute promyelocytic leukemia (APL). In the course of this therapy, the undifferentiated leukemic **blast** cells of this disease can be induced to differentiate into neutrophils by treatment with all-*trans*-retinoic acid (ATRA; Figure 16.6A). Treatment of APL patients with ATRA together with concomitant chemotherapy often results in complete **remissions**, with 5-year survival rates approaching 75 to 85%—suggestive of complete cures.

During the initial development of APL, the normal differentiation program of certain hematopoietic cells is blocked by the actions of the fusion protein termed PML-RAR, which results from a 15;17 chromosomal translocation seen in the leukemic cells of almost all APL patients (Figure 16.6B). This hybrid protein is composed of the PML (promyelocytic leukemia) protein, of unknown normal function, fused to the nuclear retinoic acid receptor (RAR) protein. The latter, upon binding its retinoic acid ligand, is normally able to induce the expression of genes that program cell differentiation in a variety of cell types throughout the body.

In fact, the precise mechanism by which the PML-RAR protein prevents the differentiation of promyelocytic leukemia cells is not clear. An attractive possibility is that the PML-RAR fusion protein found in leukemic cells interferes with the differentiation-inducing powers of the normal retinoic acid receptor (RAR) in bone marrow cells, thereby causing accumulation of large numbers of these cells in a stem cell-like state. (While the ligand-activated RAR may act to induce expression of certain target genes associated with the differentiation of hematopoietic precursor cells, the PML-RAR fusion protein may act as a repressor, or at least an antagonist, of RAR.) Much of the therapeutic effect of all-*trans*-retinoic acid treatment appears to derive from its ability to induce ubiquitylation and proteasome-mediated degradation of the PML-RAR fusion protein, resulting in relief of the block to differentiation.

A similar mechanism likely explains the successes of therapy using 13-*cis*-retinoic acid (13cRA; Figure 16.6C), related chemically to vitamin A and the all-*trans*-retinoic acid described above. It has been used with great success to cause the regression of pre-malignant lesions in the mouth and throat, thereby preventing or delaying their progression to head-and-neck cancers. Interestingly, loss of RARβ expression is often observed in these pre-malignant growths as well as in a variety of other human carcinomas, ostensibly because it helps the cells in these various lesions to avoid entrance into a differentiated, post-mitotic state. Moreover, reduction of RARβ receptor expression in mice, achieved through the use of an antisense transgene, results in the formation of large numbers of lung carcinomas by 18 months of age, providing further support for the notion that evasion of retinoic acid–induced differentiation serves as an important mechanism of carcinoma pathogenesis.

Beyond these two striking examples, differentiation-inducing strategies have had limited success in treating established cancers. For this reason, many cancer therapies that are under development are directed toward activating pro-apoptotic signals within cancer cells. At first glance, attempts at awakening the

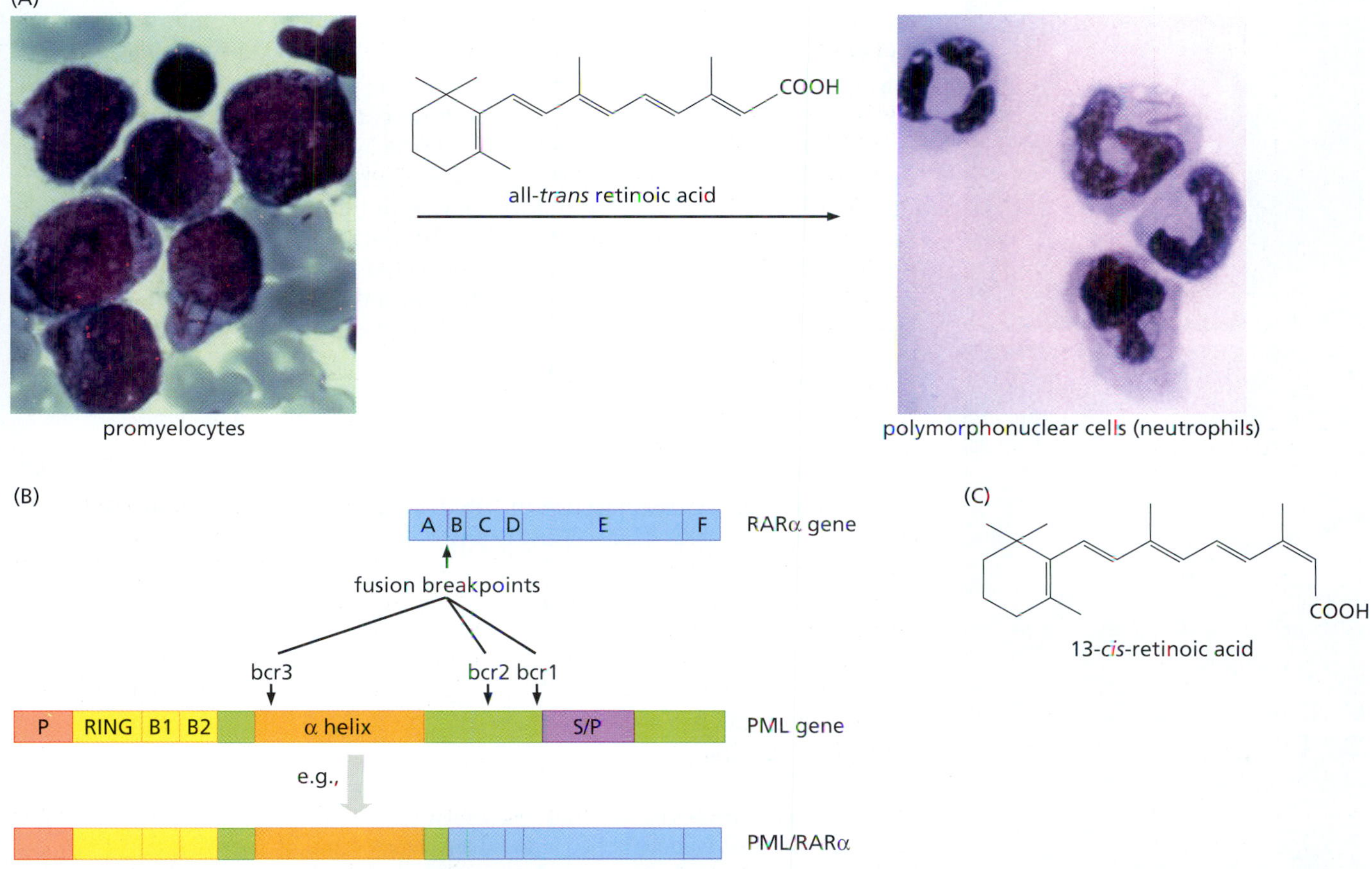

Figure 16.6 Induced differentiation of acute promyelocytic leukemia cells (A) In acute promyelocytic leukemia (APL), large numbers of promyelocytes carrying many granules in their cytoplasms are apparent in the circulation *(left)*. However, following all-*trans*-retinoic acid (ATRA) treatment, these immature promyelocytes disappear and are replaced by differentiated myelocytes, specifically the polymorphonuclear (PMN) neutrophils *(right)*. (B) In 99% of APL cases, a translocation involving Chromosomes 17 and 11 results in the fusion of the gene encoding the retinoic acid receptor (RAR) with the promyelocytic leukemia (PML) gene. Once formed, the resulting fusion protein appears to block RAR-triggered differentiation of promyelocytes. (C) 13-*cis*-retinoic acid has been used to cause regression of the pre-malignant precursors of head-and-neck carcinomas. (A, courtesy of P.P. Pandolfi; B, adapted from S. Kalantry, L. Delva, M. Gaboli et al., *J. Cell Physiol.* 173:288–296, 1997.)

apoptotic response in cancer cells might seem to represent a futile undertaking, since we read earlier of the numerous ways in which cancer cells disable their apoptotic machinery (Section 9.15). But the complexity and functional redundancies of the apoptotic circuitry dictate that, almost inevitably, important components of this circuitry remain intact even in the most aggressive tumors. It is these still-functional components that can be targeted for activation, directly or indirectly, in order to eliminate tumor cells from the bodies of cancer patients. We are just beginning to learn how to predict the responsiveness of tumors to certain apoptosis-inducing therapies (see, for example, Sidebar 48).

Many of the therapeutic strategies under development are designed to kill cancer cells by depriving them of the anti-apoptotic signals that sustain them. As we read in Chapter 9, cancer cells often depend on hyperactive growth factor signaling to generate intracellular anti-apoptotic signals (e.g., those released by Akt/PKB) that suppress the actions of the pro-apoptotic circuitry. Effective cancer therapies may be devised by interfering with this signaling at one or another step in the upstream signaling cascades that regulate Akt/PKB activity.

An alternative set of therapeutic strategies take advantage of the vulnerabilities that cancer cells have once they have discarded critical checkpoint controls operating during the normal cell cycle. For example, some cancer cells lack the checkpoint control that normally blocks entrance into mitosis (M phase) from the G_2 phase until significant damage to the genomic DNA or the chromosomes has been repaired. Consequently, tumor tissue may be treated by inflicting genomic damage through chemotherapeutics or radiation. While normal cells will tarry and repair this damage before advancing into M phase, cancer cells may ignore such damage and, as a consequence, will proceed blithely into mitosis, where they may stumble into a "mitotic catastrophe" that threatens their continued viability when they attempt, later in M phase, to segregate their still-damaged chromosomes (Figure 16.7). This damage may be so overwhelming that it succeeds in triggering the residual apoptotic responses that these cells possess. Indeed, many of the traditionally used cancer therapeutics are suspected to take advantage of these defects in checkpoint controls to destroy cancer cells, but hard evidence to sustain this point is not yet in hand. In the discussions that follow, however, we will focus on agents that target critical proteins rather than the genomes of cancer cells.

16.3 Functional considerations dictate that only a subset of the defective proteins in cancer cells are attractive targets for drug development

Researchers interested in developing novel, highly specific anti-cancer drugs that are directed to treating certain types of cancer are confronted by the fact that, with rare exception, drugs—usually low-molecular-weight organic compounds—

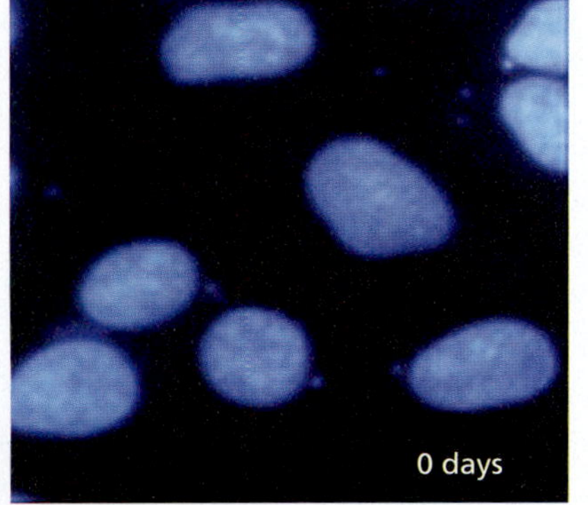

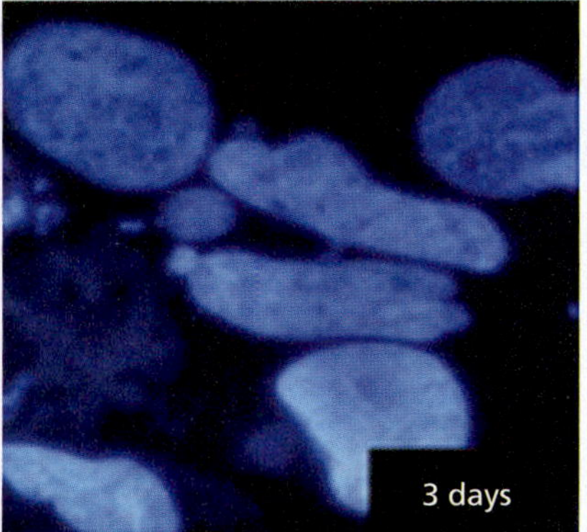

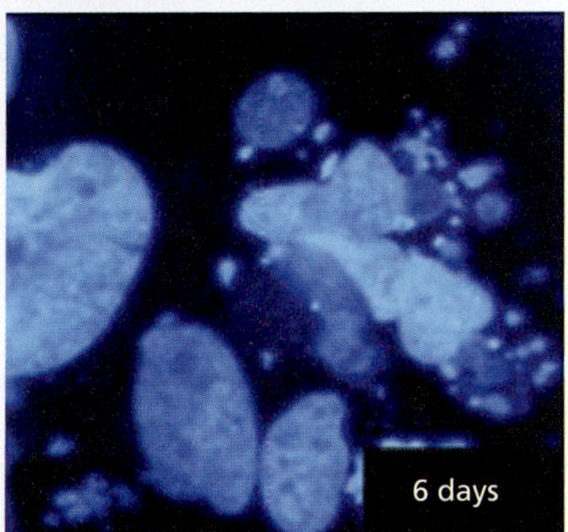

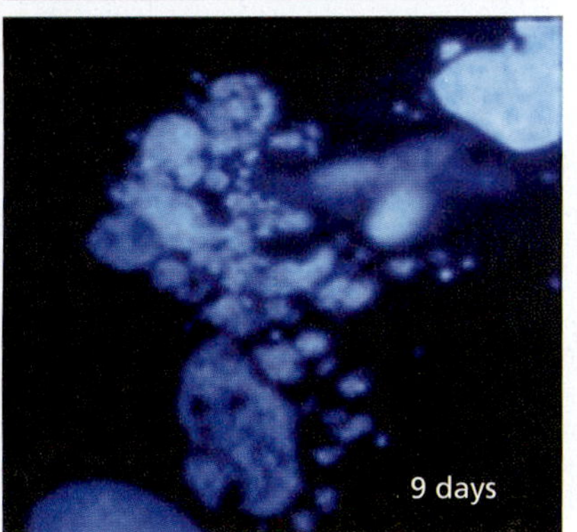

Figure 16.7 Chemotherapy and mitotic catastrophe Many chemotherapeutic drugs may damage the chromosomes of cancer cells. Because the cancer cells lack key G_2/M checkpoint controls, they may advance into mitosis without having repaired the chromosomal damage. This may cause them to enter into "mitotic catastrophe" that results in aneuploidy, polyploidy, formation of micronuclei, and the eventual death of these cells. Seen here are the effects of low doses (50 ng/ml) of doxorubicin, a widely used chemotherapeutic drug, on Huh-7 human hepatoma cells. Over a period of 9 days, the nuclei of these cells grow larger or smaller and many eventually fragment into micronuclei, each of which carries a small number of chromosomes; this leads eventually to cell death, often by apoptosis. (From Y.W Eom, M.A. Kim, S.S. Park et al., *Oncogene* 24:4765–4777, 2005.)

inhibit biochemical functions rather than enhance them. This simple fact drastically narrows the options for the development of anti-cancer drugs.

As we saw in Chapter 7, the protein products of tumor suppressor genes—the so-called gatekeepers—contribute to cancer development through their absence, and attempts at developing low–molecular-weight compounds to replace or replicate these missing functions are implausible at present and may remain so forever. The few successes here represent relatively minor victories. For example, certain compounds can restore some p53 function by shifting mutant forms of the p53 protein from their functionally defective stereochemical configurations back into a wild-type configuration.

Precisely the same arguments apply to the proteins responsible for maintaining the cellular genome—the caretakers (Chapter 12). Once again, their functions, often missing from cancer cells, cannot be restored by even the most complex drug molecules. And even if they were, little utility would derive from such successes. After all, if tumor progression has been driven by defective DNA repair and resulting accumulation of mutant alleles, restoration of the missing repair function will have no effect on the many mutant sequences that have already accumulated in a cancer cell genome.

Once gatekeepers and caretakers are removed from consideration, such logic leaves oncoproteins—hyperactive forms of normal cellular growth- or survival-promoting proteins—as the most attractive targets for the development of anti-cancer therapies. These are molecules that, in principle, can be inhibited by drugs, resulting in reduction of their activity and, with luck, collapse of the neoplastic growth program. In fact, the signal-transducing proteins immediately downstream of hyperactive oncoproteins are also attractive targets, since most of these are also important positive effectors of signaling (see, for example, Figure 16.8).

Certain genetic considerations may further narrow the range of molecules that are attractive targets for anti-tumor drug development. Earlier, we learned that as cancer progression proceeds, cell populations acquire a succession of genetic and thus biochemical alterations that ultimately lead these cells to the neoplastic growth state (Chapter 11). This scenario raises a provocative question: Do the changes that were responsible for the *early* steps of multi-step

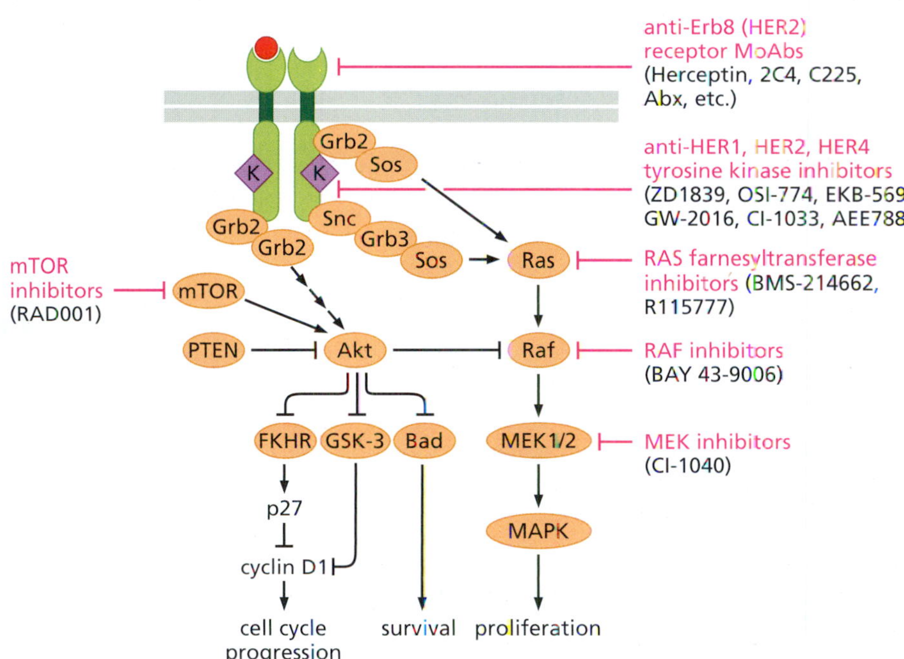

Figure 16.8 Inhibition of tumor growth by targeting downstream signaling elements As indicated in this diagram, signaling from receptors such as the EGF (HER1) and HER2/Neu receptors can be blocked in a number of ways. The ectodomains of the receptors can be targeted by monoclonal antibodies such as Herceptin. Moreover, the tyrosine kinase signal-emitting domains of these receptors can be targeted by a variety of low–molecular-weight compounds. In addition, however, a number of drugs have been developed that target proteins functioning as components of downstream signaling pathways, including those that inhibit Ras (through inhibition of its post-translational maturation involving farnesylation), as well as Raf and MEK (through inhibition of their serine/threonine kinase catalytic functions), and mTOR (through inhibition of the formation of functional signaling complexes between mTOR and associated partner proteins). (Courtesy of J. Baselga.)

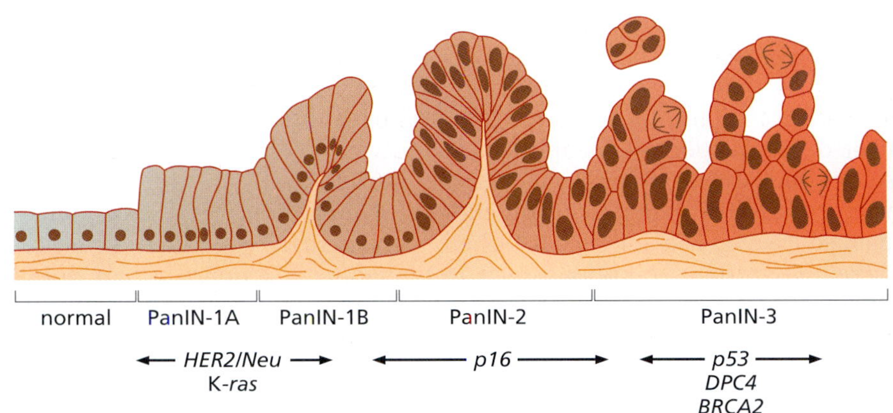

Figure 16.9 Pancreatic cancer progression A key question in identifying targets for therapeutic intervention is whether genetic alterations and mutant proteins that arise early in multi-step tumor progression continue to be required much later when a full-blown tumor appears. For example, a series of histological entities termed PanINs (pancreatic intraepithelial neoplasias) have been defined as discrete stages in early pancreatic tumor progression. In PanIN-1A, the K-*ras* oncogene is often found in a mutant activated state, raising the question whether its continued activity is required much later when invasive and metastatic pancreatic carcinomas *(not shown)* finally arise. (From R.E. Wilentz, C.A. Iacobuzio-Donahue, P. Argani et al., *Cancer Res.* 60:2002–2006, 2000.)

tumor progression continue to play critical roles far *later,* when the full-blown malignant phenotype has finally been acquired? For example, if the initial step in the development of a tumor involved the formation of a *ras* oncogene, are the continued actions of this oncogene still required later by the highly malignant descendant cells? Or have some of the changes occurring later during tumor progression rendered the continued services of a Ras oncoprotein unnecessary?

Take the case of pancreatic carcinomas, in which the K-*ras* oncogene is found in the great majority (~90%) of tumors. The acquisition of this oncogene occurs relatively early in tumor progression, since mutant K-*ras* oncogenes are often found in pancreatic intraepithelial neoplasias (PanINs; Figure 16.9), the benign precursors of frank carcinomas. Do some of the subsequently acquired changes in, for example, the *Smad4/DPC4* and $p16^{INK4A}$ genes make the K-Ras4B oncoprotein superfluous? If so, drugs designed to block K-Ras signaling will never prove useful for treating this class of cancers.

Here, we can take some encouragement from a number of mouse models of cancer development (Table 16.1). The oncogene responsible for initiating tumor progression in these transgenic mice can be shut off experimentally many weeks later in the tumors that eventually form. Such experiments have exploited the H-*ras* oncogene to create melanomas, the K-*ras* oncogene to induce lung adenocarcinomas, the *bcr-abl* oncogene to create leukemias, and the *myc* oncogene to make islet cell tumors of the pancreas as well as leukemias and lymphomas.

Table 16.1 Effects of shutting down expression of an initiating oncogenic transgene in tumor-prone mice[a]

Transgenic oncogene	Response of tumors
Permanent regression after shutdown of transgene	
H-*ras*	melanoma collapsed
K-*ras*	lung adenocarcinoma regressed
bcr-abl	B-cell leukemia regressed
myc	T-cell lymphoma, acute myelogenous leukemia regressed
fgf-7	lung epithelial hyperplasia regressed
SV40 large T	salivary gland hyperplasia regressed if transgene expressed < 4 months
Persistence or relapse after shutdown of transgene	
SV40 large T	salivary gland hyperplasia relapsed if transgene expressed > 4 months
neu	mammary adenocarcinoma relapsed
myc	mammary adenocarcinoma persisted
wnt	mammary adenocarcinoma persisted

[a]Adapted in part from D.W. Felsher, *Curr. Opin. Genet. Dev.* 14: 37–42, 2004.

In all of these cases, the tumor cells that arise continue to be dependent on the initiating oncogenes, as indicated by the regression of these tumors once expression of the initiating oncogenes is shut down. The behavior of certain human tumor cells also indicates the continuing contributions of initiating genetic elements (see, for example, Sidebar 16.1).

However, experiments with another mouse strain carrying a transgenic *myc* oncogene have yielded an equally dramatic but quite different outcome: shutdown of *myc* expression initially caused a regression of the transgene-induced lymphomas, but the tumors relapsed in about 20% of these mice; the relapsing lymphoma cells invariably showed additional karyotypic alterations beyond those that were present in the initially formed tumors. This suggests that the *myc* transgene, in addition to initially triggering lymphoma formation, encouraged the formation of genetic changes in the tumor cells that made its continued actions unnecessary later. Observations like this one clearly complicate targeting certain tumor-initiating oncoproteins for inactivation by anti-cancer drugs, since these proteins may no longer be playing critical roles, years later, in maintaining tumor cell viability and growth.

Sidebar 16.1 HeLa cells provide the most dramatic example of the ongoing importance of initiating genetic lesions Arguably the most extreme example of the continued influence of initiating genetic lesions is provided by the behavior of the cells of the human HeLa cervical cancer cell line. Recall that these cells were derived in 1951 from a highly aggressive cervical carcinoma (Sidebar 25 ●) and, like almost all (>99.7%) cervical carcinomas, the initiation of this tumor is traceable to a human papillomavirus (HPV) infection, in this case HPV type 18 and its two encoded oncoproteins, E6 and E7. Half a century later, during which time cultured HeLa cells passed through many thousands of growth-and-division cycles, genetic strategies were used to shut down the HPV18 oncogenes being expressed in these cells. The shutdown of E7 expression led to reactivation of pRb function and cell senescence, while shutdown of E6 led to the reappearance of p53 and subsequent senescence or apoptosis in these cells. Therefore, in this case, the initiating genetic change (acquisition of an HPV genome) continued to be absolutely essential for the maintenance of cancer cell proliferation and viability thousands of cell generations later.

16.4 The biochemistry of proteins also determines whether they are attractive targets for intervention

The biochemical subtleties of the proteins that have been chosen as attractive targets for drug intervention further complicate attempts at developing novel anti-cancer drugs. These drugs are, almost invariably, low–molecular-weight organic compounds, since (1) in general, such molecules are synthesized far more readily than molecules of higher molecular weight; and (2) small molecules are more likely to penetrate into the interstices of a tumor, thereby exerting therapeutic effects on all of its component cells.

Target molecules, for their part, must have domains within their structures that are capable of strong and specific interactions with small drug molecules. Pharmaceutical chemists intent on sending drugs into cancer cells to attack specific targets place these potential molecular targets (e.g., oncoproteins) into two major categories—those that are **druggable** and those that are not. "Druggability" implies that the target molecule has a structure indicating that it should be vulnerable to attack and inhibition by low–molecular-weight compounds. Given these and other constraints, target molecules are always proteins of various sorts.

A protein is considered druggable if it carries an identifiable enzymatic function and a well-defined catalytic cleft that it uses to carry out this function. Such clefts are attractive for drug developers, because they usually represent relatively small cavities that can bind small organic molecules in a highly specific manner. Thus, these cavities often make it possible for a low–molecular-weight compound to form noncovalent bonds simultaneously with multiple amino acid residues lining their walls (Figure 16.10). Such multiple independent contacts enable a drug molecule to bind to the targeted protein with great specificity and avidity. Equally important, such binding has a high likelihood of perturbing protein function, since the drug molecule occupies a functionally critical domain of the protein.

Proteins lacking such catalytic clefts are often dismissed as being "undruggable." Transcription factors, for example, are widely thought (rightly or wrongly) to be undruggable, because they usually lack catalytic clefts and thus the much-sought drug-binding pockets. These considerations place transcription factor oncoproteins, such as Myc and Fos, in the category of undruggable targets and, at the same time, allow the many kinases involved in cancer formation to be placed in

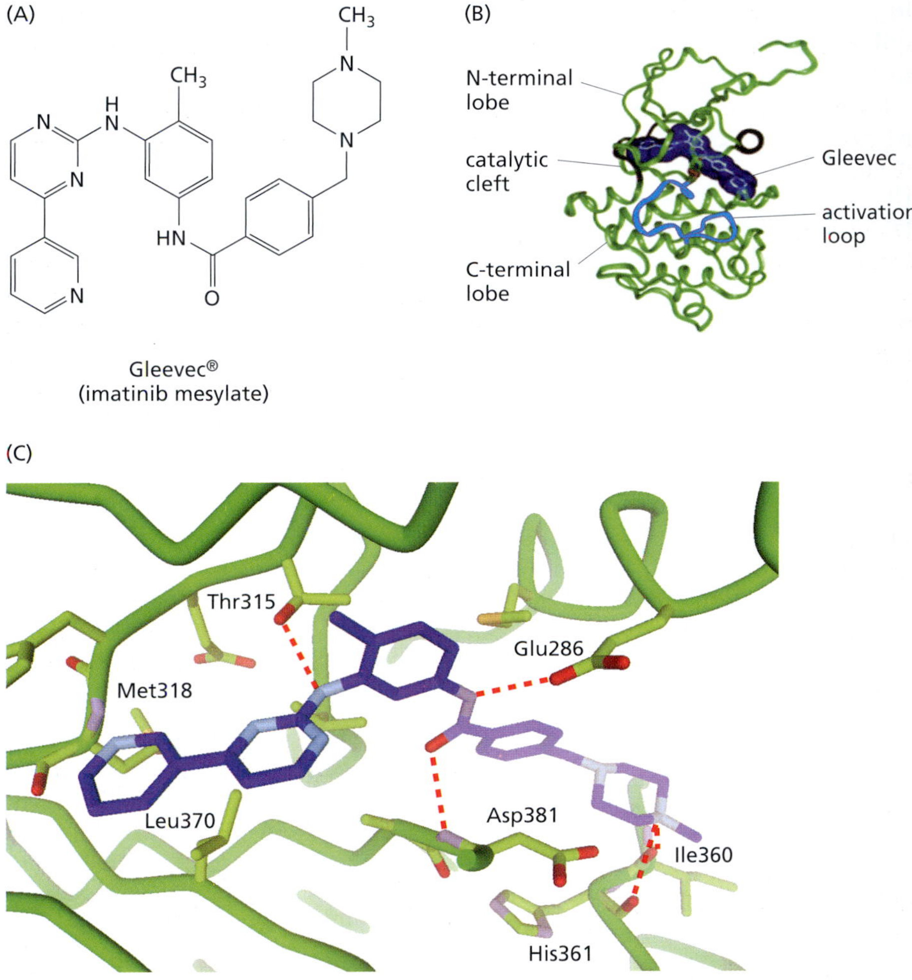

Figure 16.10 Multiple contacts between drugs and their targets (A) The chemical structure of Gleevec, which was developed to inhibit the tyrosine kinase activity of the Bcr-Abl fusion protein active in chronic myelogenous leukemia (CML), was the result of optimizing the structure of a precursor compound to which certain side chains were added while others were removed in order to improve drug binding to the catalytic cleft of the Abl tyrosine kinase domain. (B) The catalytic cleft of the Abl kinase is found between its N- and C-terminal lobes, shown here as *green* ribbon structures. A space-filling model (with van der Waals radii) of a Gleevec molecule is shown in *dark blue*, while the "activation loop" of Abl, which normally blocks access to substrates by the catalytic cleft, is shown in *light blue*. (This activation loop swings out of the way when the kinase shifts into its active configuration.) (C) The avid and specific association of Gleevec *(magenta)* with the catalytic cleft of Abl depends on the formation of multiple hydrogen bonds *(red dashed lines)*, as well as weaker van der Waals interactions *(not shown)*. These bonds increase the binding affinity of the drug for the protein; at the same time, they explain the specificity of association, since the various pairs of proton donors and acceptors, which participate as partners in hydrogen bond formation, must be precisely positioned in three-dimensional space. (A and B, from B. Nagar, W.G. Bornmann, P. Pellicena et al., *Cancer Res.* 62:4236–4243, 2002; C, courtesy of E. Buchdunger, S.W. Cowan-Jacob, G. Fendrich, J. Liebetanz, D. Fabbro and P.W. Manley, Novartis Pharmaceuticals.)

the camp of druggable target molecules. (The major exceptions to the lack of druggability of transcription factors are created by the nuclear hormone receptors, such as the estrogen and progesterone receptors. Because they have hormone-binding domains, these receptor proteins are, in principle, vulnerable to disruption by pseudo-ligands, such as tamoxifen, which binds and antagonizes certain functions of the estrogen receptor.) On average, pharmaceutical chemists judge about 1 in 5 cellular proteins to be druggable.

The presence of an identifiable catalytic function and apparent druggability does not, on its own, guarantee that an attractive target has been identified. Consider, for example, the case of the Ras oncoprotein, which has a clearly identifiable catalytic activity—its GTPase function. This enzymatic activity in Ras-expressing cells has never been the object of drug development, because the Ras GTPase, as we learned in Section 5.9, functions as a *negative* regulator of Ras signaling. Its inhibition would only augment the already-disastrous effects of the amino acid substitutions that create Ras oncoproteins in the first place. The same can be said of some tyrosine phosphatases, whose designated roles are to reverse the effects of growth-promoting tyrosine kinases. In response to the difficulty in attacking the Ras protein itself, a number of drug development strategies have focused instead on the enzymes that modify this protein and thereby enable it to become functional (Sidebar 49 ●).

The notion that molecular cavities provide attractive targets for drug development might also suggest that many types of protein–protein interactions represent druggable targets. After all, the confined space between two physically apposed proteins would seem to create a highly specific drug-binding pocket, and insertion of a drug into such a cavity might therefore destabilize or block the protein–protein interaction. Obvious candidates for such inhibition are the several types of cyclin–Cdk pairs, whose actions drive the proliferation of all cancer cells (Chapter 8).

Unfortunately, most attempts at preventing these and other protein–protein associations through custom-made drug molecules have been unsuccessful. The numerous failures have been rationalized as follows: the association of two proteins with one another involves multiple points of binding between their interacting faces. These points of contact extend over molecular domains that greatly exceed the dimensions of drug molecules, which typically have a rather low molecular weight (e.g., $<10^3$). Consequently, only a small fraction of the contact points between two associating proteins can be blocked by any such drug, and the association as a whole remains essentially unperturbed.

An exception to this widely accepted lore was announced in late 2003, when a low–molecular-weight compound that inhibits formation of the Mdm2–p53 complex (Section 9.7) was described (Figure 16.11A). This development means that, in principle, the absence of p53 protein observed in many types of human tumors—often resulting from loss of $p14^{ARF}$ expression or overexpression of Mdm2—can be reversed by drug treatment.

Within months, more rule breakers were reported. Naturally occurring compounds isolated from green and black tea were found to block the binding and neutralization of pro-apoptotic BH3 proteins by the anti-apoptotic Bcl-2 and Bcl-X_L proteins, doing so at relatively low concentrations (Figure 16.11B). And a compound, ICG-001, identified through high-throughput screening (HTS; see Sidebar 50) was found to be capable of inhibiting the association of β-catenin with CBP (cyclic AMP response element–binding protein), a widely acting transcriptional co-activator that works with β-catenin to induce expression of key genes, including the gene encoding survivin (see Table 9.5 and Section 9.13), an important anti-apoptotic IAP protein (Figure 16.11C), as well as cyclin D1. A year later, yet another inhibitor of the Bcl-2, Bcl-X_L, and Bcl-w anti-apoptotic proteins was announced. This one, also the product of synthetic organic chemistry (Figure 16.11D), associates with these anti-apoptotic proteins with an affinity of approximately 1 nM—more than two orders of magnitude more avid than the tea compound.

Agents that function like the β-catenin inhibitor hold the promise of being useful in treating colon carcinomas. As we have read, deregulated accumulation of β-catenin plays a key role in the pathogenesis of almost all of these cancers, resulting in the association of this protein with Tcf/Lef transcription factors in the nucleus and an excessive expression of a constituency of key target genes, such as those encoding cyclin D1 and Myc (Section 7.11). Once assembled at gene promoters in the nucleus, the β-catenin–Tcf/Lef complexes recruit several additional proteins, such as the CBP transcriptional co-activator; the latter aids in the remodeling of the chromatin around promoter sites. The drug ICG-001 inhibits the association of β-catenin–Tcf with CBP, resulting in loss of transcription-inducing activity by this complex. In a more general sense, successes like the four listed here are sure to encourage attempts by others at discovering pharmacological inhibitors of key protein–protein interactions.

These successes notwithstanding, a major lesson, learned time and again, is that kinases are among the few classes of cellular molecules that are attractive, druggable targets for anti-cancer therapy. As we have seen, many of these kinases function as oncoproteins that act to drive neoplastic proliferation and, at the same time, are enzymes possessing well-defined catalytic clefts. Recall that at least 518 distinct kinase-encoding genes have been enumerated in the human

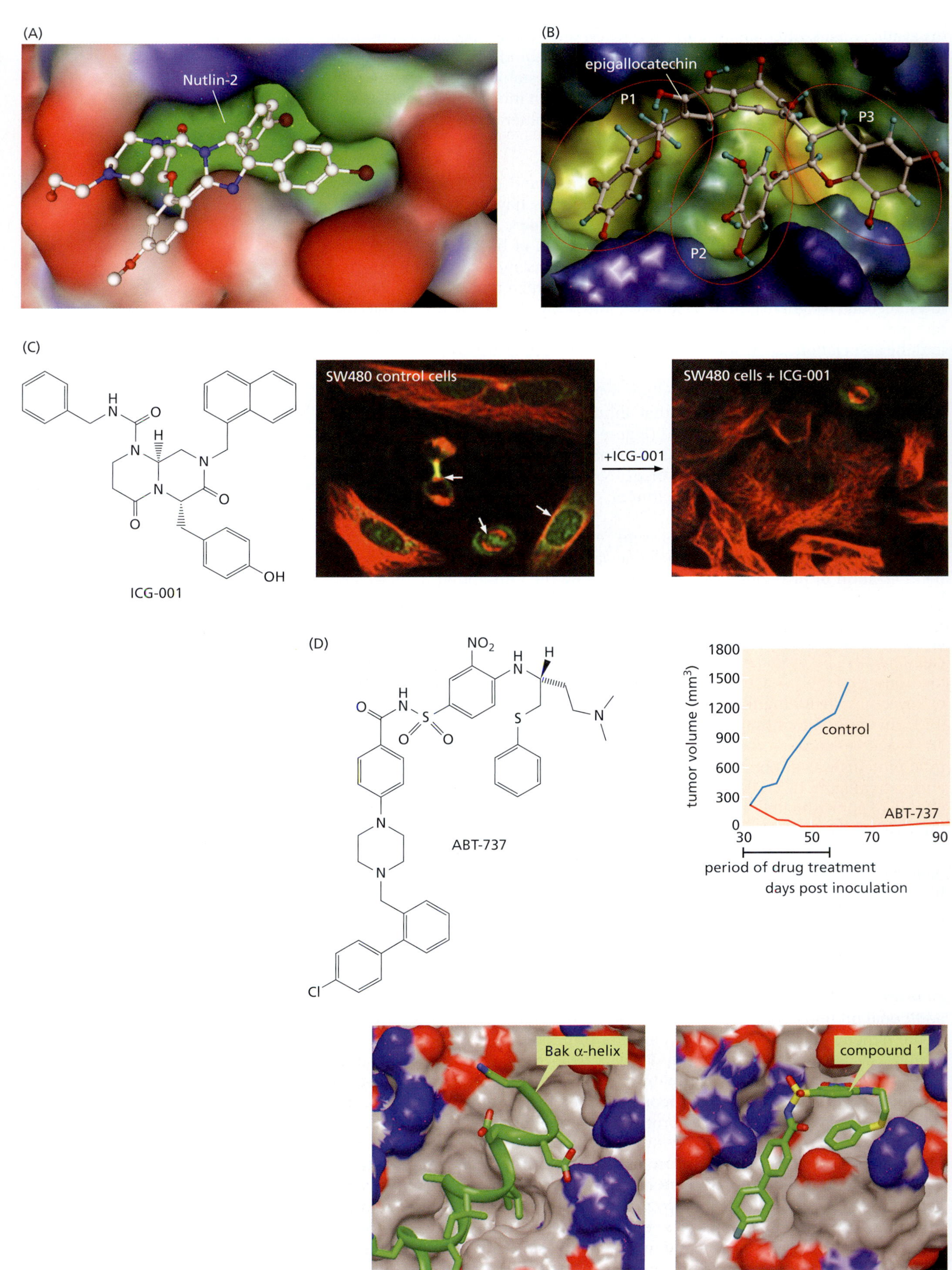
(A)
Nutlin-2
(B)
epigallocatechin
P1
P2
P3
(C)
ICG-001
SW480 control cells
+ICG-001
SW480 cells + ICG-001
(D)
ABT-737
tumor volume (mm³)
1800
1500
1200
900
600
300
0
30
50
70
90
control
ABT-737
period of drug treatment
days post inoculation
Bak α-helix
compound 1

Figure 16.11 Inhibitors of protein–protein interactions (A) A search for compounds that inhibit Mdm2–p53 binding resulted in the discovery of Nutlin-2, which associates with the p53-binding pocket of Mdm2, whose surface is shown here. The bottom of the pocket is hydrophobic and is therefore *green,* while patches on the side of the pocket that are hydrophilic are *blue.* Those amino acid residues that are exposed to solvent are *red.* The Nutlin-2 compound is shown here as a stick figure with carbon atoms *(white),* nitrogen *(blue)*, oxygen *(red)*, and bromine *(brown)*. This and closely related compounds prevent Mdm2-mediated p53 degradation and induce apoptosis of cancer cells, achieving both effects in the low micromolar concentration range. (B) The healthy effects claimed to come from drinking black and green teas have been ascribed to a number of the constituents of the tea leaves, some of which induce apoptosis in tumor cells at very low concentrations. Epigallocatechin gallate (EGCG; *stick figure*) extracted from green tea binds the important anti-apoptotic protein Bcl-X_L with an inhibitory constant (K_i) of 490 nM and Bcl-2 with a K_i of 335 nM. (The values of these constants reflect the concentrations at which 50% of the activity of these proteins is inhibited; values in the submicromolar range are indicative of high potency.) A combination of three technologies—nuclear magnetic resonance (NMR) binding assays, fluorescence polarization assays, and computational docking—was used to derive this image, which shows the docking of EGCG in three adjacent hydrophobic pockets of Bcl-X_L *(yellow and green)*, whose molecular surfaces are shown here and labeled P1, P2, and P3. These pockets together constitute a hydrophobic domain that Bcl-X_L, which is overexpressed in many cancer cells, uses to bind and neutralize pro-apoptotic BH3-only proteins (see Figures 9.25 and 9.27). (C) The small molecule ICG-001 *(left)* inhibits the association of the β-catenin–Tcf/Lef complex with the transcriptional co-activator CBP. The transcription factor complex is responsible for inducing, among other genes, expression of the gene encoding survivin, a key inhibitor of apoptosis (IAP; see Section 9.13). As seen here, expression of survivin *(green, yellow immunofluorescence, both right panels)* that is normally present in SW480 human colon carcinoma cells *(arrows)* is strongly reduced by the presence of ICG-001; tubulin *(red)* is used in both cases as a *counterstain.* The concentration of ICG-001 used in this experiment causes an approximately 6-fold reduction in the activity of the survivin gene promoter *(not shown)*. (D) A compound termed ABT-737 *(1st panel)* has been developed, using nuclear magnetic resonance (NMR)–based screening, parallel synthesis, and structure-based design to visualize the interacting structures of a pro-apoptotic protein (Bak) and candidate drug molecules (compound 1). Thus, the α helix of the pro-apoptotic Bak protein *(green helix, 3rd panel*; see Figure 9.25*)* that binds and inhibits the anti-apoptotic Bcl-X_L protein *(red, white, blue surface)* is also occupied by a close chemical relative of ABT-737, termed compound 1 *(green stick figure, 4th panel)*. In the *2nd panel*, the effects of ABT-737 on the growth of a human small-cell lung carcinoma (SCLC) xenograft are seen. The *black bar* indicates the time window during which ABT-737 was applied. The drug does not appear to induce apoptosis on its own, but instead makes cancer cells vulnerable to apoptosis induced by other agents. (A, from L.T. Vassilev, B.T. Vu, B. Graves et al., *Science* 303:844–848, 2004; B, from M. Leone, D. Zhai, S. Sareth et al., *Cancer Res.* 63:8118–8121, 2003; C, left, from K.H. Emami, C. Nguyen, H. Ma et al., *Proc. Natl. Acad. Sci. USA* 101:12682–12687, 2004; C, right, from H. Ma, C. Nguyen, K.S. Lee and M. Kahn, *Oncogene* 24:3619–3631, 2005; D, from T. Oltersdorf, S.W. Elmore, A.R. Shoemaker et al., *Nature* 435:677–681, 2005.)

genome, of which 90 encode tyrosine kinases, the latter being major players in many kinds of human cancer.

For some cancer researchers, this multiplicity of potentially druggable cancer targets represents an embarrassment of riches. However, for pharmaceutical chemists, numbers like these create a nightmare. Because almost all protein kinases are evolutionarily related to one another (Figures 16.12 and 16.13A), their catalytic clefts are structurally quite similar. The similarity is even more striking among the clefts of the more closely related tyrosine kinases (Figure 16.13B), which are involved in cancer pathogenesis. How can one possibly develop agents that affect the actions of certain cancer-associated kinases, while leaving untouched the kinases required for normal cell proliferation and survival? Rational drug design and high-throughput screening (HTS), both described below, attempt to address these issues.

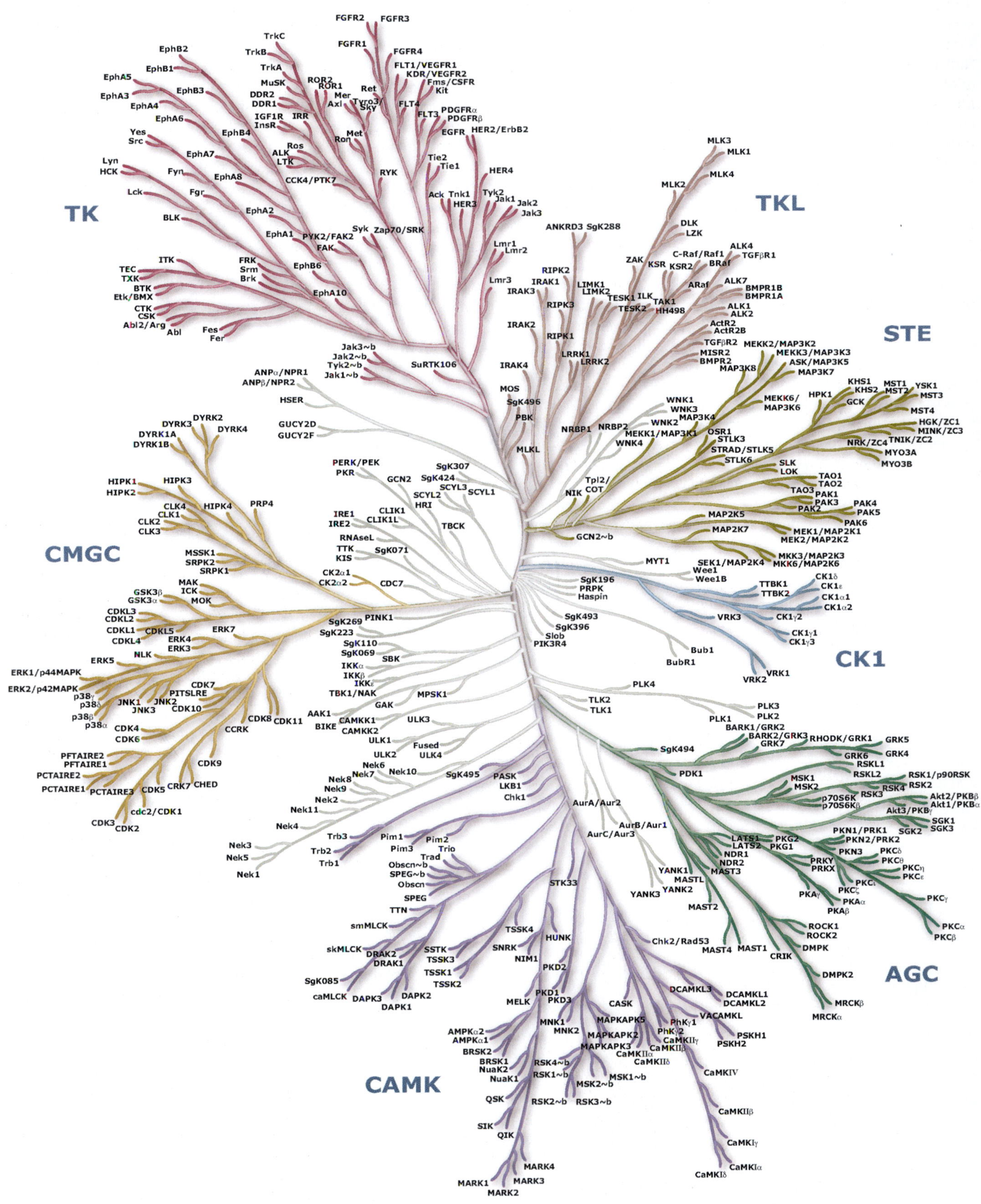
TK
TKL
STE
CK1
AGC
CAMK
CMGC
FGFR2
FGFR3
FGFR1
FGFR4
TrkC
TrkB
TrkA
EphB2
EphB1
EphA5
EphA3
EphA4
EphA6
EphB3
EphB4
Yes
Src
Lyn
HCK
Fyn
Lck
Fgr
BLK
EphA7
EphA8
EphA2
EphA1
MuSK
DDR2
DDR1
IGF1R
InsR
IRR
ROR2
ROR1
Mer
Axl
Tyro3/
Sky
Ret
FLT1/VEGFR1
KDR/VEGFR2
Fms/CSFR
Kit
FLT4
FLT3
PDGFRα
PDGFRβ
EGFR
HER2/ErbB2
Met
Ron
Ros
ALK
LTK
CCK4/PTK7
RYK
Tie2
Tie1
HER4
Ack
Tnk1
HER3
Tyk2
Jak1
Jak2
Jak3
Syk
Zap70/SRK
PYK2/FAK2
FAK
EphB6
EphA10
Lmr1
Lmr2
Lmr3
ITK
TEC
TXK
BTK
Etk/BMX
CTK
CSK
Abl2/Arg
Abl
Fes
Fer
FRK
Srm
Brk
Jak3~b
Jak2~b
Tyk2~b
Jak1~b
SuRTK106
ANPα/NPR1
ANPβ/NPR2
HSER
GUCY2D
GUCY2F
MLK3
MLK1
MLK4
MLK2
DLK
LZK
ALK4
TGFβR1
C-Raf/Raf1
BRaf
ZAK
KSR
KSR2
ARaf
ALK7
BMPR1B
BMPR1A
ANKRD3
SgK288
RIPK2
IRAK1
IRAK3
LIMK1
LIMK2
TESK1
TESK2
ILK
TAK1
HH498
ALK1
ALK2
RIPK3
ActR2
ActR2B
IRAK2
RIPK1
TGFβR2
MISR2
BMPR2
LRRK1
LRRK2
IRAK4
MOS
SgK496
PBK
MLKL
MEKK2/MAP3K2
MEKK3/MAP3K3
ASK/MAP3K5
MAP3K7
MAP3K8
KHS1
KHS2
MST1
MST2
YSK1
MST3
HPK1
GCK
MST4
HGK/ZC1
MINK/ZC3
TNIK/ZC2
NRK/ZC4
MYO3A
MYO3B
MEKK6/
MAP3K6
WNK1
WNK3
WNK2
MAP3K4
MEKK1/MAP3K1
WNK4
NRBP1
NRBP2
OSR1
STLK3
STRAD/STLK5
STLK6
SLK
LOK
TAO1
TAO2
TAO3
PAK1
PAK3
PAK2
PAK4
PAK5
PAK6
Tpl2/
COT
NIK
MAP2K5
MAP2K7
MEK1/MAP2K1
MEK2/MAP2K2
GCN2~b
MKK3/MAP2K3
MKK6/MAP2K6
SEK1/MAP2K4
MYT1
Wee1
Wee1B
SgK196
PRPK
Haspin
TTBK1
TTBK2
CK1δ
CK1ε
CK1α1
CK1α2
VRK3
CK1γ2
CK1γ1
CK1γ3
SgK493
SgK396
Slob
PIK3R4
Bub1
BubR1
VRK1
VRK2
DYRK2
DYRK3
DYRK4
DYRK1A
DYRK1B
HIPK1
HIPK2
HIPK3
HIPK4
PRP4
CLK4
CLK1
CLK2
CLK3
MSSK1
SRPK2
SRPK1
PERK/PEK
PKR
GCN2
SgK307
SgK424
SCYL2
SCYL3
SCYL1
HRI
IRE1
IRE2
CLIK1
CLIK1L
TBCK
RNAseL
TTK
KIS
SgK071
CK2α1
CK2α2
CDC7
MAK
ICK
MOK
GSK3β
GSK3α
CDKL3
CDKL2
CDKL1
CDKL5
CDKL4
ERK7
ERK4
ERK3
NLK
ERK5
ERK1/p44MAPK
ERK2/p42MAPK
p38γ
p38δ
p38β
p38α
JNK1
JNK2
JNK3
CDK7
PITSLRE
CDK10
CDK4
CDK6
CDK8
CDK11
CCRK
PFTAIRE2
PFTAIRE1
PCTAIRE2
PCTAIRE1
PCTAIRE3
CDK5
CRK7
CHED
CDK9
cdc2/CDK1
CDK3
CDK2
PINK1
SgK269
SgK223
SgK110
SgK069
SBK
IKKα
IKKβ
IKKε
TBK1/NAK
MPSK1
GAK
AAK1
BIKE
CAMKK1
CAMKK2
ULK3
ULK1
Fused
ULK2
ULK4
Nek6
Nek7
Nek8
Nek9
Nek10
Nek2
Nek11
Nek4
Nek3
Nek5
Nek1
SgK495
PASK
LKB1
Chk1
PLK4
TLK2
TLK1
PLK3
PLK2
PLK1
BARK1/GRK2
BARK2/GRK3
RHODK/GRK1
GRK7
GRK5
GRK6
GRK4
SgK494
PDK1
RSKL1
RSKL2
MSK1
MSK2
RSK1/p90RSK
RSK2
RSK3
RSK4
p70S6K
p70S6Kβ
Akt2/PKBβ
Akt1/PKBα
Akt3/PKBγ
SGK1
SGK2
SGK3
AurA/Aur2
AurB/Aur1
AurC/Aur3
LATS1
LATS2
PKG2
PKG1
NDR1
NDR2
MAST3
PKN1/PRK1
PKN2/PRK2
PKN3
PKCδ
PKCθ
PKCη
PKCε
PRKY
PRKX
PKCι
PKCζ
PKAγ
PKAα
PKAβ
PKCγ
YANK1
MASTL
YANK2
YANK3
MAST2
PKCα
PKCβ
ROCK1
ROCK2
DMPK
CRIK
MAST4
MAST1
DMPK2
MRCKβ
MRCKα
Trb3
Trb2
Trb1
Pim1
Pim2
Pim3
Trio
Trad
Obscn~b
SPEG~b
Obscn
SPEG
TTN
STK33
smMLCK
TSSK4
HUNK
skMLCK
DRAK2
DRAK1
SSTK
TSSK3
TSSK1
TSSK2
SNRK
NIM1
PKD2
Chk2/Rad53
SgK085
caMLCK
DAPK3
DAPK2
DAPK1
MELK
PKD1
PKD3
CASK
DCAMKL3
DCAMKL1
DCAMKL2
VACAMKL
MNK1
MNK2
MAPKAPK5
MAPKAPK2
MAPKAPK3
PhKγ1
PhKγ2
CaMKIIγ
CaMKIIβ
CaMKIIα
CaMKIIδ
PSKH1
PSKH2
AMPKα2
AMPKα1
BRSK2
BRSK1
NuaK2
NuaK1
RSK4~b
RSK1~b
MSK2~b
MSK1~b
CaMKIV
QSK
RSK2~b
RSK3~b
CaMKIIβ
SIK
QIK
CaMKIγ
MARK4
CaMKIα
MARK1
MARK3
MARK2
CaMKIδ

Figure 16.12 *(left)* The human kinome As illustrated here, the great majority of the protein kinases of mammalian cells, including serine/threonine and tyrosine kinases, share substantial structural similarity, indicating that they all descend from a primitive protein kinase that existed long before the radiation of eukaryotic life forms. The serine/threonine and tyrosine kinases diverged from one another relatively early and have been further diversified over the past approximately 10^9 years.

Altogether, the human genome sequence reveals 518 distinct genes encoding protein kinases, which, as a group, have been called the "human *kinome*." Of these, 90 phosphorylate tyrosine residues, while the remainder phosphorylate serine or threonine residues of substrate proteins. All of the tyrosine kinases (TKs) and 318 of the serine/threonine kinases show clear structural relatedness to one another and can be arrayed on an evolutionary tree that depicts how they appear to have evolved from one another through repeated gene duplications followed by diversification. (A small number of "atypical" protein kinases *(not shown)* appear to represent independent evolutionary inventions.) The shared origins of the kinases located on this tree dictate that many of these enzymes are structurally similar to one another, which complicates the creation of drugs that interfere selectively with only a few members of this large family of enzymes.

The TKs *(top left)* represent relatively recent evolutionary inventions, as they are absent in prokaryotes and are present in only very small numbers (e.g., 3) in the genomes of single-cell eukaryotic protozoa sequenced to date. Their great diversification and specialization appear to have been critical to the evolution of anatomically complex metazoa. The remaining groupings on this tree are TKL, tyrosine kinase-like; STE, homologs of yeast sterile 7, 11, and 20 kinases; CK1, casein kinase-1; AGC, members of protein kinase A, kinase G, and kinase C families; CAMK, calcium/calmodulin-dependent protein kinases; and CMGC, containing CDK, MAPK, GSK-3, and CLK families. (Courtesy of Cell Signaling Technology, Inc.)

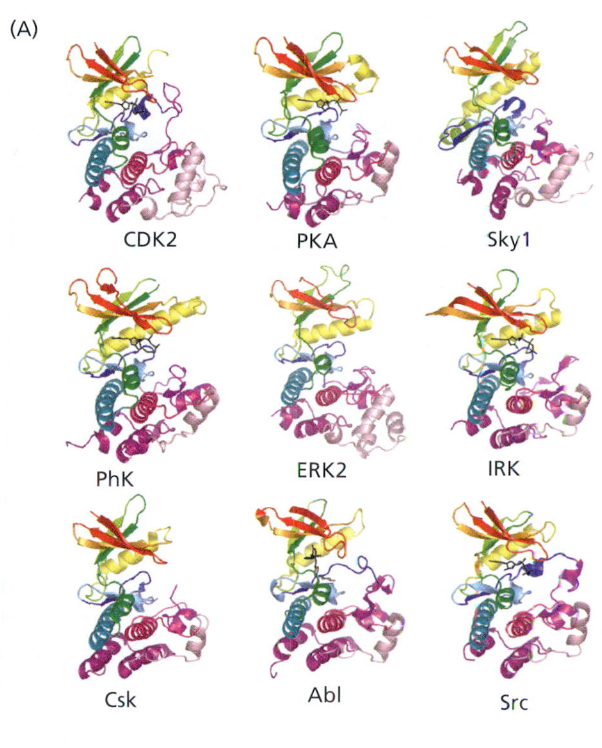

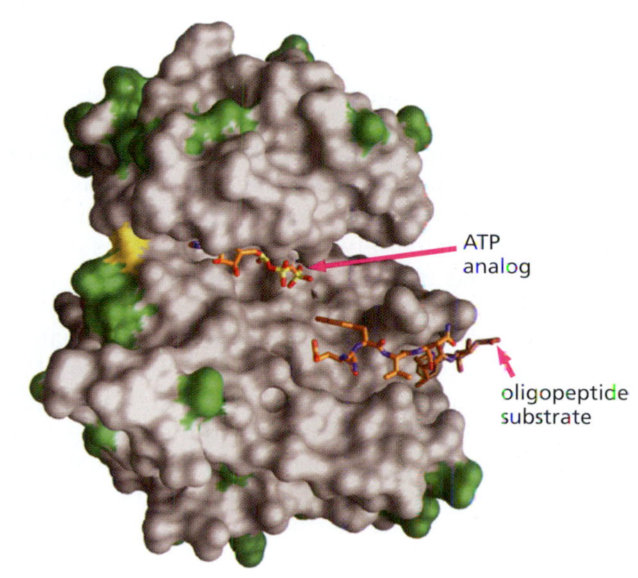

Figure 16.13 Similar structures of kinases (A) The difficulty of producing a specific kinase inhibitor is suggested by the striking similarities in structure of a variety of serine/threonine as well as tyrosine kinases. Depicted here as ribbon diagrams are the kinase domains of five serine/threonine kinases: CDK2 (see Chapter 8), PKA (cyclic AMP-regulated protein kinase A), Sky1 (SR-protein-specific kinase of budding yeast) a yeast serine kinase involved in nuclear RNA processing, PhK (phosphorylase kinase involved in glycogen metabolism) and ERK2 (extracellular-regulated kinase of the MAPK casacade, see Section 6.5);as well as four tyrosine kinase (TK) domains: IRK (insulin receptor kinase), Csk (C-terminal Src kinase; a tyrosine kinase), Abl (see Section 16.11), and Src (see Chapter 5). In all cases, the catalytic clefts of these kinases are sandwiched between the two major lobes (N- and C-terminal, *above and below, respectively*) of these proteins. (B) An extreme example of the structural similarities between related kinases is illustrated by this surface diagram in which the catalytic clefts and adjacent amino acid residues of the tyrosine kinase domains of the insulin receptor (IR; see the IRK of panel A) and the insulin-like growth factor receptor (IGF-1R) are compared. Identical amino acid residues are in *gray* while dissimilar ones are in *green.* This shows how very similar the catalytic regions of the two TK domains are and explains why it has been so difficult to find inhibitors of one tyrosine kinase receptor that do not affect the other. A threonine in the peptide linking the two lobes of the kinases is shown in *yellow,* while stick figures (*orange, blue, red, yellow)* of an ATP analog *(left)* and an oligopeptide substrate *(right)* are also shown. Almost all TK antagonist drugs bind in the ATP-binding sites of the kinases that they inhibit. (A, courtesy of N.M. Haste, S.S. Taylor and the Protein Kinase Resource; B, from S. Favelyukis, J.H. Till, S.R. Hubbard and W.T. Miller, *Nat. Struct. Biol.* 8:1058–1063, 2001.)

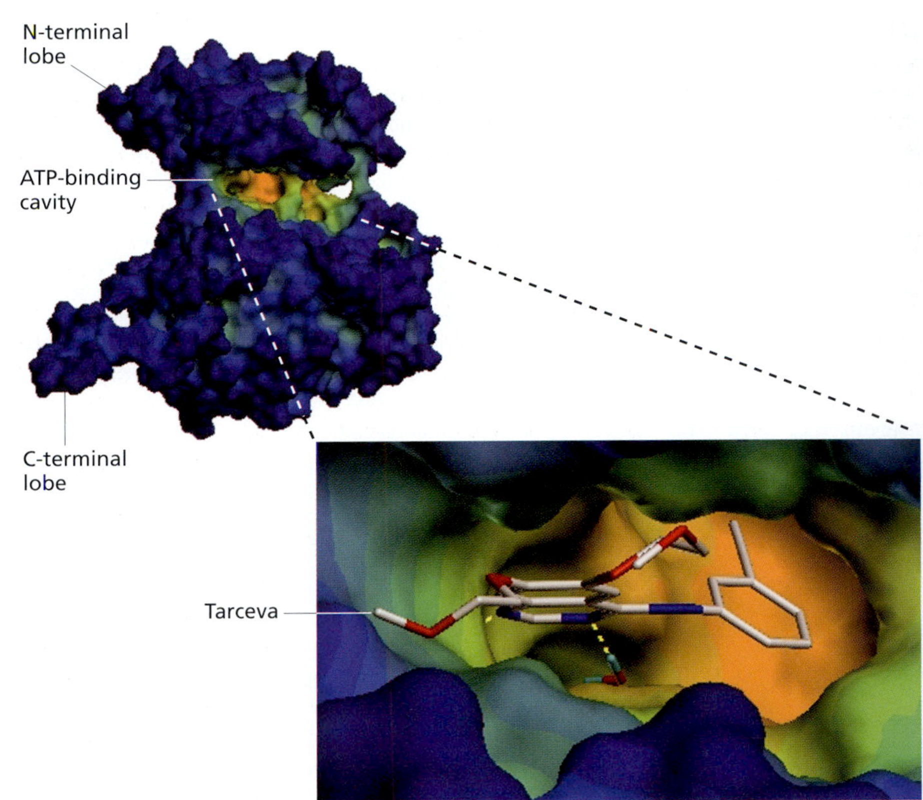

Figure 16.14 Tarceva and the inhibition of the EGF receptor kinase domain This space-filling model of the catalytic cleft of the EGF receptor tyrosine kinase (TK) domain, derived by X-ray crystallography, shows how the drug Tarceva *(stick figure)* fits snugly within the ATP-binding cavity of the cleft and in this way inhibits receptor signaling. The three-dimensional complementarity between drug and targeted protein is necessary for drug binding but not sufficient, as specific bonding, largely achieved through the formation of hydrogen bonds *(not shown)*, must occur between the drug molecule and amino acids lining the drug-binding pocket (see, for example, Figure 16.10C). A water molecule *(behind, below)* that is hydrogen-bonded to the Tarceva molecule is also shown. (Courtesy of C. Sambrook-Smith and A. Castelanho, OSI Pharmaceuticals Inc.; see also J. Stamos, M.X. Sliwkowski and C. Eigenbrot, *J. Biol. Chem.* 277:46265–46272, 2002.)

16.5 Pharmaceutical chemists can generate and explore the biochemical properties of a wide array of potential drugs

The ideology of "rational drug design," as it is often called, embraces the notions that (1) drugs should be targeted against specific proteins known to be malfunctioning within cells, thereby contributing to a disease state; (2) the candidacy of these proteins as attractive targets for therapeutic intervention should be further determined by their predicted druggability; and (3) the detailed molecular structures of such target proteins should inform the design of the chemical structures of the drugs that are to be developed. More specifically, chemical species must be synthesized whose detailed three-dimensional structures (i.e., whose *stereochemistry*) enable them to fit, in a key-in-lock fashion, into specific pockets or sites within the far larger proteins that they are supposed to attack and disable (see, for example, Figure 16.14).

In principle, knowledge of the detailed structure of a potential drug-binding cavity in a targeted protein should allow a skilled organic chemist to design and synthesize a molecule that fits snugly into this cavity and forms multiple noncovalent bonds with the amino acids lining its walls. However, this purely theoretical route for designing a novel drug structure has not, in practice, yielded many useful products. For this reason, current drug discovery relies on more empirical ways of finding useful molecular structures (Sidebar 50 ●).

The specifications noted in Sidebar 50 ● do not address the many biochemical and biological problems associated with the development of anti-cancer drugs. The major challenge here is to predict *clinical* (i.e., patient) responses from the *pre-clinical* research conducted at the laboratory bench and on animals. Imagine, for example, that high-throughput screening has yielded a drug molecule that inhibits the activity of a targeted protein in living cells, doing so at 10 or 100 micromolar concentrations (that is, concentrations in this range are required for 50% inhibition of the activity of the targeted protein). Further development of this particular drug becomes unrealistic, given the massive amounts of this agent that would need to be delivered into a patient's body in order to achieve a therapeutic effect. The chemical properties of this molecular species may or may not allow *derivatization* (the synthesis of modified derivatives of this compound) that yields a molecule with a potency in the nanomolar concentration range.

The pre-clinical testing that follows involves measurements of a drug's *relative* effects on its intended target compared with its **off-target** effects on other, similar proteins in the cell. The goal here is determine whether the drug acts selectively by inhibiting the targeted protein at drug concentrations that are substantially below (10- to 100-fold) those affecting other, similar proteins in the cell (Figure 16.15). (In truth, given the 20,000 or more distinct protein structures present in mammalian cells, these measurements do not preclude possible effects on structurally unrelated proteins that may, through happenstance, be affected by an agent under development.)

In the case of tyrosine kinase inhibitors, which are the focus of much current drug development, attempts at identifying all of the kinases that might be affected by a drug have, until recently, involved assays of only a small proportion of the large cohort of protein kinases known to be present in human cells. Consequently, certain off-target effects are likely to have eluded drug developers. This has begun to change with the advent of more systematic screening of a far larger portion of the kinases that might be affected by these inhibitors.

For example, one biotechnology company has developed an assay (Figure 16.16) for measuring the binding affinities of a test drug for 156 distinct kinases that are located on various branches of the kinome tree (see Figure 16.12). Two EGF receptor inhibitors that we will discuss in greater detail later—Iressa and Tarceva—are indeed found to bind preferentially the EGF-R tyrosine kinase, while the ability of staurosporine, which is thought to inhibit a wide spectrum of protein kinases of all types, is confirmed to exhibit a wide-ranging kinase-binding ability. (The *binding affinity* of a test drug for a kinase, as measured in this assay, has been found to predict the ability of this drug to *inhibit the activity* of the kinase.)

As we will see later, discovering off-target activities of a drug, which is enabled by screens such as this one, is actually useful in two ways. (1) It may explain toxicities of a drug—undesired side effects of the drug in tissues other than the targeted tumor. (2) It may reveal new clinical applications for the drug, since the drug may be found to inhibit an enzyme, such as a kinase, that is active in a type of tumor that was not targeted during the initial drug development.

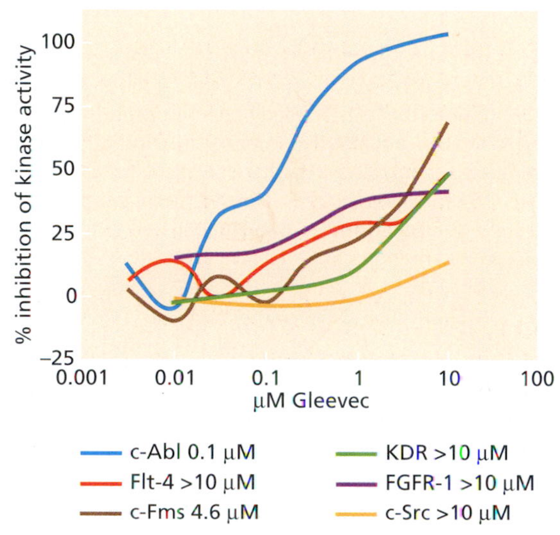

Figure 16.15 Dose response curves of Gleevec The ability to inhibit a targeted protein without affecting other, related cellular proteins is critical to the success of therapy, making possible therapeutic responses without undesired side-effect toxicities. Here we see the responses of a group of six tyrosine kinase (TK) enzymes to the drug Gleevec (see Figure 16.10). In all cases, the kinase activity of the purified enzymes was measured. Note that the graph was constructed using the log of Gleevec concentration on the abscissa while the percent of inhibition of catalytic activity is plotted linearly on the ordinate. At a concentration of approximately 0.1 μM, about 50% of c-Abl enzyme activity was inhibited, whereas comparable inhibition of the c-Fms TK only occurred at a concentration as high as 4.6 μM. Under these conditions, the c-Src TK was hardly inhibited at all. (Courtesy of E. Buchdunger, Novartis Pharmaceuticals.)

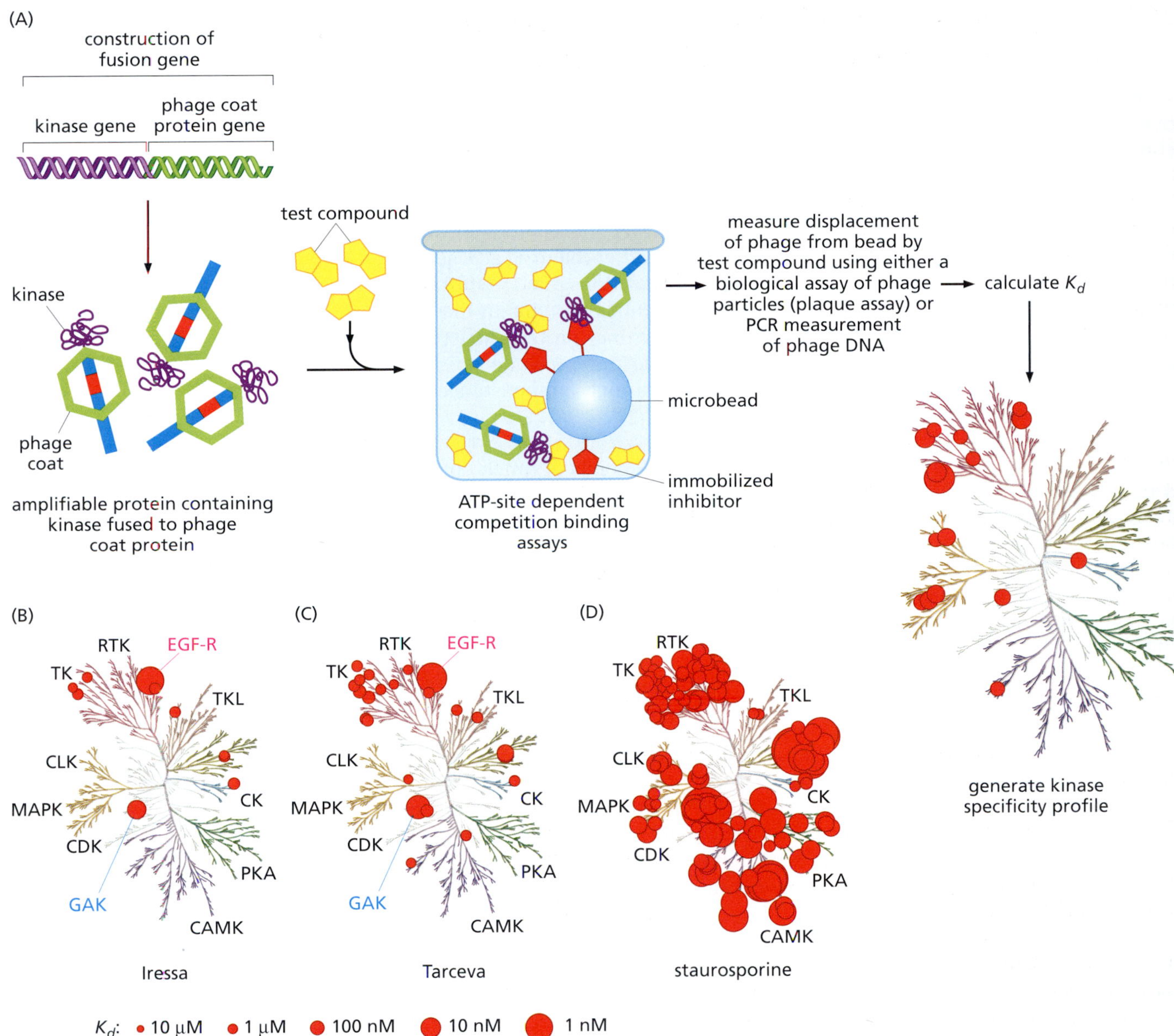

Figure 16.16 Effects of kinase inhibitors on a wide spectrum of protein kinases The individual responses to various kinase inhibitors of an array of 156 distinct tyrosine and serine/threonine kinases have been tested by measuring the *binding affinities* of each inhibitor for each of these kinases. This test depends on the fact that almost all kinase inhibitors bind to the ATP-binding sites of targeted enzymes. (A) A cDNA encoding the kinase domain of a protein is cloned into a bacteriophage vector so that some of the phage capsid (coat) proteins *(green)* are synthesized as fusion proteins with the kinase *(purple)*. An ATP analog that is known to bind the ATP-binding sites of many kinases *(red)* is then immobilized by being linked to microbeads *(light blue)*. This allows phage particles to become linked via the fusion proteins in their capsids to the ATP analog on the beads. The assay then measures the ability of a test compound (e.g., a candidate tyrosine kinase inhibitor, *yellow)* to compete with the immobilized *(red)* inhibitor, thereby blocking association of the phage to the bead. A reduction in binding of the phage to the beads, as revealed by the number of phage particles released from the beads (using a phage plaque assay or a polymerase chain reaction for the phage DNA) then indicates the binding affinity of the test compound for the kinase present in the fusion protein, as represented by its displacement constant, K_d, i.e., the concentration at which 50% of the phage is displaced from the beads.

In the remaining panels, each kinase that showed a K_d of less than 1 µM is indicated by a red circle whose diameter varies inversely with the K_d (binding constant affinity) and whose position is dictated by the location of the kinase on the kinome tree (see Figure 16.12). The effects of two inhibitors of the EGF receptor (EGF-R) that have already been licensed for clinical use, Iressa and Tarceva, are analyzed in panels (B) and (C), respectively. Both, reassuringly, show greater specificity for the EGF-R than for the 155 other kinases tested. However, Iressa also binds GAK (cyclin G–associated kinase; see Sidebar 5.3) at about a 10-fold higher concentration, while Tarceva affects GAK at an even lower concentration, barely higher than that required to inhibit the EGF-R itself. (D) In contrast, staurosporine, a widely used experimental reagent that is thought to inhibit many kinases, is seen here to indeed bind a large number of these enzymes, some while it is present at subnanomolar concentrations. (A, from J.D. Griffin, *Nat. Biotechnol.* 23:308–309, 2005; B, C, and D, courtesy of P.R. Zarrinkar and D.J. Lockhart, Ambit Biosciences; see also M.A. Fabian, W.H. Biggs 3rd, D.K. Treiber et al., *Nat. Biotechnol.* 23:329–336, 2005.)

16.6 Drug candidates must be tested on cell models as an initial measurement of their utility in whole organisms

The demonstrated ability of a drug to inhibit an isolated target protein in solution is usually followed by tests of its effects on cultured cells. Take the case of Gleevec, the compound (see Figure 16.10A) found initially to inhibit the tyrosine kinase activity of the isolated Bcr-Abl fusion protein; this protein was known to be responsible for driving the proliferation and survival of the cancer cells of chronic myelogenous leukemia (CML). Having established its effects on the isolated Bcr-Abl protein (see Figure 16.15), drug developers could then proceed to the next step, which involved the use of cultured cells whose proliferation or survival *in vitro* depended on the continued actions of this fusion protein.

Figure 16.17 shows an example of such a cell-based test that happens to have been conducted long after Gleevec was developed. This test used the cells of a murine pre-B-lymphocyte line that normally depend for their *in vitro* survival and proliferation on the presence of interleukin-3 (IL-3) in their culture medium. These cells could be rendered independent of IL-3 if a Bcr-Abl oncoprotein was ectopically expressed in them. The modified cells were then cultured in the absence of IL-3 (making them totally dependent on continued Bcr-Abl firing) and exposed to various drugs that were candidate antagonists of the Bcr-Abl oncoprotein; the proliferation and/or survival of these cells was then gauged (Figure 16.17B).

Cell-based tests like these are designed to determine whether the drug being examined induces apoptosis in treated cells, or cytostasis (i.e., a halt in cell proliferation), or has no effect whatsoever. And if the drug does evoke a desired response, does it do so at a reasonably low concentration?

The outcomes of such cell-based tests are rarely obvious beforehand. Many compounds that are highly hydrophobic may be excluded from these tests from the outset because they are poorly soluble and therefore cannot be placed on cultured cells in significant concentrations (see Table 1). Their more hydrophilic chemical relatives may be highly soluble and may work well on purified Bcr-Abl protein, but may not be readily transported through the plasma membranes of cells; these chemical species are likely to prove useless, simply because they cannot accumulate within cells at concentrations that would allow them to be effective.

Imagine that these hurdles have been successfully surmounted and that the proliferation of Bcr-Abl–dependent cells is indeed inhibited at nanomolar concentrations of a candidate therapeutic agent. The fact that the candidate drug acts on these cells does not exclude the possibility that it also affects dozens of other kinases in these and other cells, some of which may be essential for normal cell metabolism—the property of biological **selectivity**. (Its *biochemical* selectivity is likely to have been determined previously by tests like those illustrated in Figure 16.15.)

So next, it becomes necessary to determine whether cancer cells whose growth is driven by other tyrosine kinases are equally sensitive to the actions of an identified anti-Bcr-Abl agent like Gleevec. And how are fully normal cultured cells affected by a candidate drug like Gleevec? With luck, one may begin to see a high **therapeutic index** emerge; for example, Bcr-Abl–dependent cells may be killed by drug concentrations that have little discernible effect on comparable cells grown in the presence of IL-3 or on a variety of other cancer cells whose growth is driven by other tyrosine kinase oncoproteins. This will provide hope that *in vivo* the drug may perturb the tumor without having unacceptable side effects on normal tissues. Good outcomes in these tests will then encourage the drug developers to proceed to the next steps, in which the biological effects of drugs at the cellular and tissue level are evaluated *in vivo*, as we learn below.

16.7 Studies of a drug's action in laboratory animals are an essential part of pre-clinical testing

Once a candidate anti-tumor agent has been found to have potent killing effects on cultured cancer cells *in vitro*, drug development inevitably moves to the next step – testing whether it will kill cancer cells proliferating within tumor masses *in vivo*. Ideally, the *in vitro* behavior of a drug should predict its actions *in vivo*.

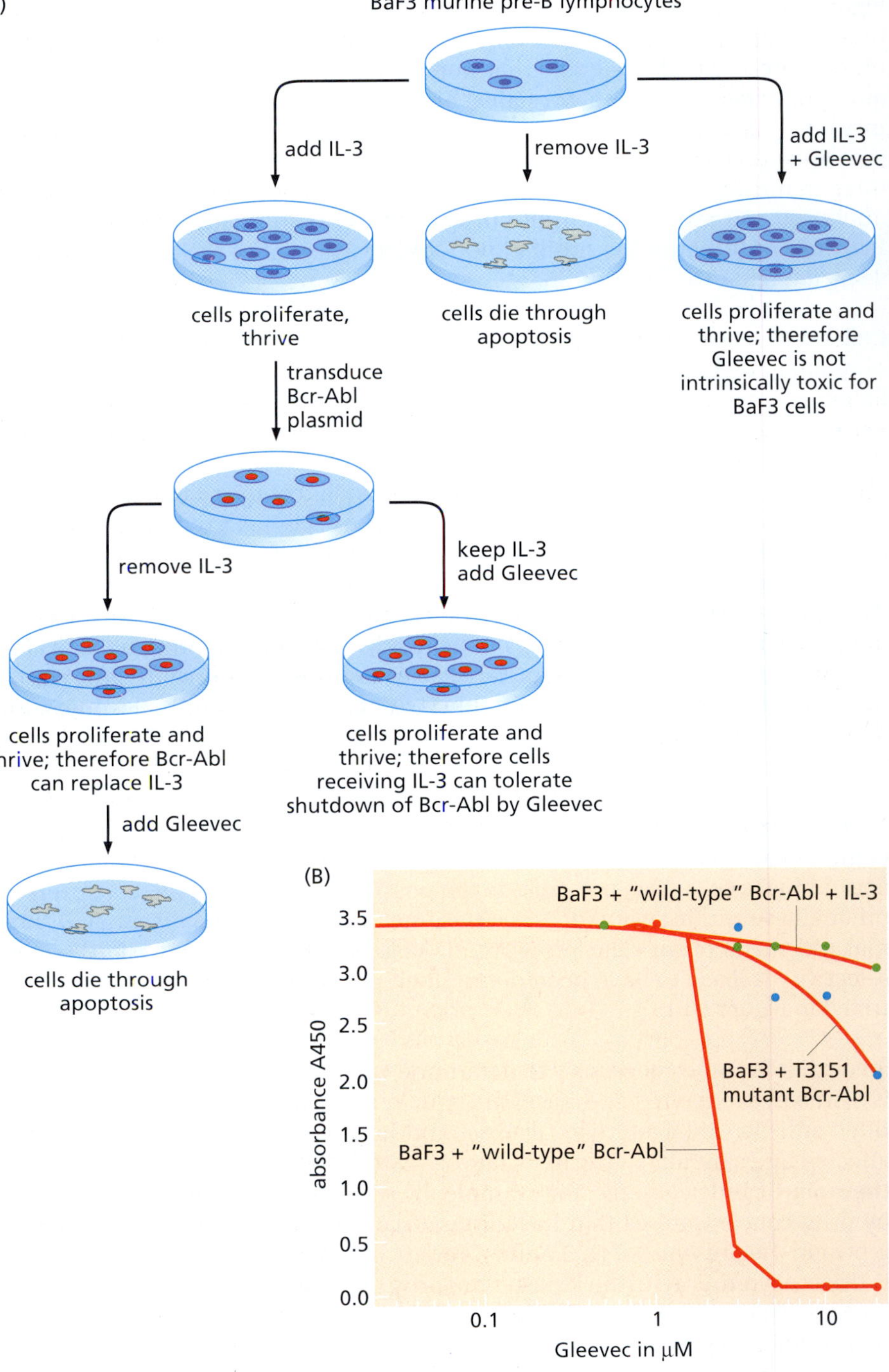

Figure 16.17 Testing of Gleevec in cell culture (A) BaF3 cells, a line of murine pre-B lymphocytes, are normally dependent on the addition of interleukin-3 (IL-3) for their proliferation and survival *(top left)*. When Gleevec is added together with IL-3, these cells continue to thrive *(top right)*, indicating that the IL-3–based survival mechanism is not sensitive to Gleevec inhibition. An expression plasmid specifying the Bcr-Abl oncoprotein can be introduced into these cells, and under these conditions, the BaF3 cells continue to proliferate, even after IL-3 is withdrawn *(left side)*, indicating that Bcr-Abl can replace IL-3 and sustain these cells on its own. However, because these cells are now dependent on Bcr-Abl for their survival, the addition of Gleevec at doses that inhibit the Abl kinase will cause them to die *(lower left)*, while addition of Gleevec to Bcr-Abl–expressing cells that continue to receive IL-3 does not affect their survival, (indicating that Gleevec is not toxic to BaF3 cells that express Bcr-Abl as long as they continue to receive IL-3 stimulation). Therefore, in the absence of IL-3, the Bcr-Abl–expressing BaF3 cells can serve as highly sensitive and specific indicators of the actions of Gleevec and similarly acting drugs on the Bcr-Abl oncoprotein. (B) The information in panel A can be used to develop an assay system, in which the number of BaF3 cells surviving after certain treatments is indicated by the optical density of BaF3 cell suspensions *(ordinate)*. In the presence of IL-3, Gleevec has almost no effect on BaF3 cell survival whether or not the Bcr-Abl oncoprotein is being expressed in BaF3 cells *(green dots)*. In the absence of IL-3, however, survival of Bcr-Abl–expressing cells is strongly suppressed above about 2 μM Gleevec concentration *(red dots)*. If instead of the "wild-type" Bcr-Abl protein, cloned from a patient's CML cells at the beginning of Gleevec treatment, a highly drug-resistant mutant version of Bcr-Abl (termed T3151) that arose in a CML patient during the course of Gleevec treatment is expressed in the BaF3 cells, far higher drug concentrations are required to kill these cells *(blue dots)*; this last curve is measured, once again, in the absence of IL-3 in the growth medium. (Courtesy of M. Azam and G.Q. Daley.)

Here, further complications arise. One is suggested by experimental results that we encountered in Section 13.10. There we read that the sensitivity of tumors to radiation may be determined by the radiosensitivity of the endothelial cells in their vasculature, rather than by the responses of the neoplastic cells in these masses; some drugs may act similarly, by affecting the supporting stromal cells of a tumor (which are not studied *in vitro*) rather than the cancer cells themselves. Independent of this, stromal cells may provide certain types of anti-apoptotic survival signals, such as IGF-1, that are not available in comparable amounts to cancer cells in culture. In a more general sense, the complexities of tumor biology created by heterotypic interactions with the tumor-associated stroma often dictate that the drug responses of pure populations of cancer cells proliferating *in vitro* fail to predict their responses within growing tumors *in vivo*.

Because rodent and human cells differ so substantially in their biology (Section 11.12), the *in vivo* testing of candidate anti-cancer drugs involves, almost always, human (rather than murine) cancer cells grown in mouse hosts. The presumption is that the human tumor xenografts formed in immunocompromised mice will behave much like the tumors encountered by oncologists in human patients.

Once again, there are highly challenging complications. The human tumor cells that are used to form these xenografts are propagated as tumor cell lines—cancer cells that have been propagated in culture as pure populations for many years, often decades. A set of 60 of these human cancer cell lines has been established by the National Cancer Institute as standard reagents to be used in the United States for gauging the efficacy of candidate anti-cancer agents. Many of these cell lines are not representative of neoplasms encountered routinely in the cancer clinic, because they derive from particularly aggressive human tumors that yielded cells that were especially adaptable to propagation in tissue culture (see, for example, Figure 16.18). Other cancer cell lines have, almost inevitably, evolved in culture far beyond the ancestral cells that were originally removed from actual human tumors; indeed, the cells in such lines have been selected for optimal proliferation under *in vitro* conditions that differ dramatically from those in living tissues.

These facts help to explain why human tumor xenograft models are relatively unpredictive of the responses of the actual tumors borne by patients in the cancer clinic. Indeed, in some cases, it is questionable whether cancer cells that are purportedly from, for example, a pancreatic carcinoma continue to reflect pancreatic behavior, or whether they have been inadvertently contaminated by colon or breast carcinoma cells at some point over the previous several decades of *in vitro* passage in one or another laboratory. Still, these highly imperfect xenograft models are the best reagents available and are unlikely to be supplanted in the near future by improved animal models of human cancer.

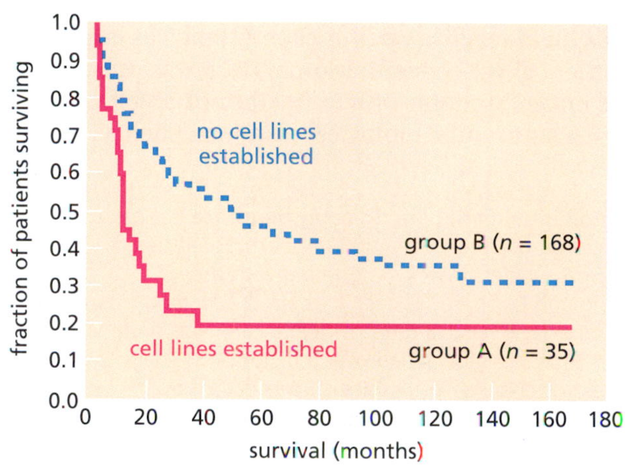

Figure 16.18 Cancer cell lines as representatives of human tumors Many researchers have attempted to create cancer cell lines by extracting cells from human tumors and adapting them to culture. Their experience, largely anecdotal, is that only the most malignant cancer cells can be propagated *in vitro,* yielding cancer cell lines. This notion has finally been tested systematically in a 12-year-long study in which the esophageal carcinoma cells of 203 patients were introduced into culture. Of these, only 35 cell lines (from about 17% of the tumors) became established in culture. The patients whose tumors were in this group (group A) experienced a far worse clinical progression *(red line)* than did those whose cells failed to adapt to culture (group B; *blue line*). This illustrates graphically why tumor xenografts produced from established cancer cell lines usually fail to recapitulate the properties of the tumors typically encountered in a cancer clinic (since the cancer cell lines usually derive from tumors at the far end of the spectrum–the most aggressive subset). (From Y. Shimada, M. Maeda, G. Watanabe et al., *Clin. Cancer Res.* 9:243–249, 2003.)

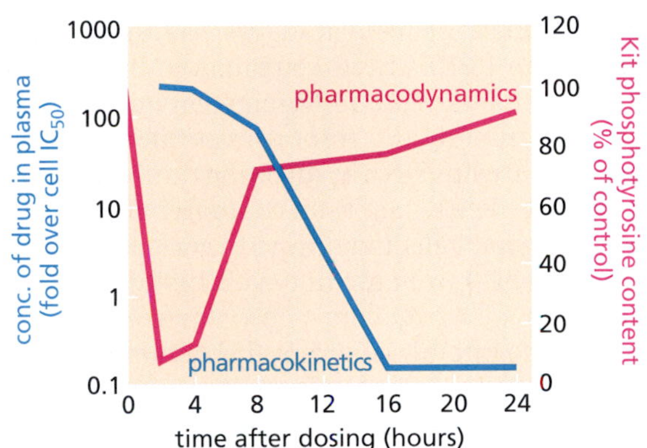

Figure 16.19 Pharmacokinetics and pharmacodynamics of Gleevec The pharmacokinetics of a drug represent the kinetics of its accumulation in and disappearance from the plasma, which in turn are presumed to provide a good indication of the drug concentrations that tumor cells experience in a laboratory animal or a patient under therapy. The plasma level of the drug Gleevec, plotted on a logarithmic scale *(left ordinate)*, fluctuates dramatically following injection of the drug into a mouse *(blue curve)*. Its concentration is indicated here as a multiple of the drug concentration known to inhibit the firing of the tyrosine kinase of the Kit receptor by 50% (i.e., the IC_{50} of this agent). (The tyrosine kinase domain of the Kit growth factor receptor is also a target of inhibition by Gleevec.) As seen here, the amount of phosphotyrosine associated with the Kit receptor (a reflection of Kit tyrosine kinase activity) expressed by engrafted human mast cell leukemia cells *(red curve)*, which was initially set as 100%, is reduced to <<1% of preexisting levels within an hour after drug injection but rebounds within 8 hours as the concentration of the drug declines in the plasma. (Courtesy of D.L. Emerson, OSI Pharmaceuticals Inc.)

The **pharmacokinetics** (PK) of a drug represent a key determinant of its efficacy *in vivo*: Does it accumulate to significant levels in the plasma or tissues for an extended period of time? Or is it present in the body only transiently, being excreted by the kidneys within minutes of entering into the circulation? Is it resistant to rapid degradation, or do certain drug-metabolizing systems, such as the cytochrome P-450s (Cyps) that we encountered in Chapter 12, rapidly convert it into an innocuous molecular species (Figure 16.19)? (A key pharmacokinetic parameter that is often measured is the "area under the curve," or AUC, calculated by integrating the concentration of a drug in the plasma as a function of time; the AUC is thought to reflect the cumulative drug dose experienced by cells in a tumor.) And can it be administered orally rather than requiring injection?

Laboratory animals give some rough indication of a drug's pharmacokinetics, but are by no means accurate predictors of how humans metabolize and excrete various agents. Moreover, as we read earlier (Sidebar 12.5), the rates at which various compounds, including drugs, are metabolized or excreted can even vary dramatically from one person to another (for example, see Figure 16.20). (In some pharmaceutical companies, the pharmacokinetics of candidate compounds may be measured even *before* any tests of therapeutic efficacy against xenografted tumors; those showing poor pharmacokinetics in laboratory animals are often eliminated from further testing. Discarding such drugs may occasionally be premature, given the dramatically differing rates of drug metabolism and excretion between rodents and humans.)

In fact, Figure 16.19 reveals a second attribute of a drug—its **pharmacodynamics** (PD), in this case those of Gleevec. Pharmacodynamics gauge the ability of a drug to affect a targeted biochemical function in a tumor under treatment. In the PD presented in this figure, as is often the practice, a **surrogate marker** of the targeted Bcr-Abl function was measured—the behavior of the Kit receptor. As we will read in more detail later, Kit is one of several tyrosine kinases affected by Gleevec, and its responses to the drug presumably parallel those of Bcr-Abl. Figure 16.19 reveals that Kit activity in this experiment was inhibited only briefly at the time when the highest concentration of drug was present in the circulation. Such a transient inhibition—only a fraction of a cell cycle—is generally insufficient to elicit a substantial biological response, such as tumor cell killing.

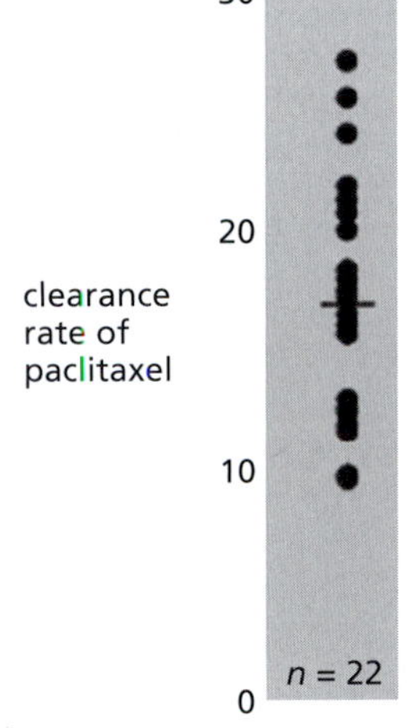

Figure 16.20 Inter-individual variability in drug clearance Paclitaxel is a chemotherapeutic drug used to treat a number of malignancies; it works by stabilizing microtubules, thereby interfering with the progression of cells through M phase. As seen here, in this study of 22 ovarian cancer patients, the relative rates of clearance of this drug from the plasma after initial injection varied over a factor of 3. These rates may be influenced by changes in the rate of metabolism by enzymes such as cytochrome c and by excretion in the kidneys. (From M. Nakajima, Y. Fujiki, S. Kyo et al., *J. Clin. Pharm.* 45:674–682, 2005.)

During the course of animal testing, information may surface about the toxic side-effects that the drug elicits in whole organisms. They represent the bane of almost all existing cancer treatments. Quite frequently, various normal organ systems, including the liver, kidneys, gastrointestinal tract, and the hematopoietic system, show toxic effects of a drug when it is used at the concentrations required to kill tumor cells. These toxicities are rarely predicted by *in vitro* tissue culture tests, and toxicities detected in laboratory animals, including dogs, monkeys, mice, and rats, may or may not be predictive of human responses.

Such observations direct our attentions once again to the therapeutic index of an agent—the efficiency with which it affects cancerous tissues compared with its toxic effects in normal tissues. Clearly, ideal cancer treatments should have high therapeutic indices, wreaking havoc on cancer cells while leaving normal tissues relatively untouched. The fundamental obstacle to achieving such selectivity is suggested by the fact that the vast majority of the 20,000 or so genes expressed in cancer cells are also being expressed by their normal counterparts.

The failure of animal models to predict the toxic side effects of a drug in humans creates serious problems. The roughly 80 million years of independent evolution that separate us from our rodent cousins have led to substantial differences in metabolism; we may react to certain drugs much differently from mice or rats, or even the more closely related Old World monkeys, which may eventually be exposed to a candidate drug in order to obtain a marginally more accurate prediction of toxicities in humans. In the event that a drug passes these various tests without raising too many warning flags, it may be promoted to a candidate for testing in humans.

16.8 Promising candidate drugs must be subjected to rigorous and extensive clinical trials in Phase I trials in humans

The discussions above explain why the first true tests of a drug's tolerability usually come in initial patient exposures, which are termed Phase I trials in the United States. Here, candidate drugs are tested at various doses, including the presumed therapeutic doses, to gauge toxic side-effects. The usual practice is to begin these trials at drug dosages that are likely to be far below the level of any overt **toxicity** (e.g., one-tenth the drug concentration that created toxicity in laboratory animals) and then, in a series of patients, increase the dosages incrementally until drug levels are reached that begin to induce unacceptable toxicities. This "dose escalation" yields a value—the maximum tolerated dose (MTD)—that is then used to guide further treatment protocols. Certain side effects, such as a skin rash or transient nausea, may be tolerable and not scuttle further drug development, while others, such as massive diarrhea or bone marrow depletion, may be so burdensome or life-threatening that they cause rapid abandonment of all further development of a drug.

During these Phase I trials, pharmacokinetic measurements, like those made previously in animals, will also be taken in order to ascertain whether the drug is reaching tumor cells at a sufficient concentration and for an extended period of time. Still, these measurements give no indication whether the cancer cells are responding in any way—the property of pharmacodynamics. For example, in Figure 16.21, we see the pharmacodynamic responses to treatments with EGF receptor antagonists (which happen in this case to include both a monoclonal antibody and a low–molecular-weight tyrosine kinase inhibitor). The oncologists who undertook this particular clinical trial wished to obtain some measure of the effects of therapies on the EGF-R in patients' tumors. To do so, they chose to use, as a surrogate marker, the EGF-R of patients' skin cells, which were far more easily monitored, simply by taking small skin biopsies from patients in treatment.

As seen in Figure 16.21A, patient exposure to an EGF-R tyrosine kinase inhibitor resulted in a strong suppression of EGF-R signaling in the skin. In addition, the activity of MAP kinase, which functions as an important downstream transducer of EGF-R signaling (Section 6.5), was also suppressed, indicating successful inhibition of downstream mitogenic signaling.

Similar results were observed in biopsies taken from a colon cancer patient's tumor following treatment with an anti-EGF-R monoclonal antibody (Figure 16.21B). Pharmacodynamic measurements like these provide reassurance that the administered treatment (in this case a monoclonal antibody) is reaching its intended target at concentrations that suffice to shut down much of the activity of its intended target.

Interestingly, a number of the signal-transducing proteins operating downstream of the EGF-R, including Akt/PKB, were only minimally suppressed in the colonic tumor (Figure 16.21B), indicating that the tumor cells had acquired alternative means for activating these signaling molecules. Hence, pharmacodynamics measurements ensure that one precondition of therapeutic success—delivery of the therapeutic agent to the targeted cells and molecules—has been satisfied, but do not, on their own, guarantee that the therapy will succeed, as other factors may thwart it.

When taken together, the measurements of maximum tolerated dose (MTD), pharmacokinetics (PK), and pharmacodynamics (PD) define the **therapeutic window**—the range of concentrations that are higher than that needed to elicit a therapeutic effect and lower than the maximum tolerated dose (Figure 16.21C). Ideally, a therapeutic window of a drug should be broad so as to allow clinicians some flexibility in administering the drug, adjusting dosage to the patient and the condition being treated. As the therapeutic window narrows, the likelihood that a candidate drug will prove clinically useful diminishes.

Occasionally, Phase I clinical trials, which are usually undertaken with very small groups of patient volunteers who have failed other available therapies, may reveal some favorable responses in terms of tumor regression or halting of further tumor growth, doing so at acceptably low levels of toxicity. However, even if there are hints of clinical efficacy, the positive results observed in Phase I trials are never statistically significant and thus not regarded as definitive. Instead, these trials are really undertaken to discover unanticipated toxicities and tolerable levels of drug dosage.

16.9 Phase II and III trials provide credible indications of clinical efficacy

Acceptably low levels of toxicity in a Phase I trial will encourage testing a candidate drug's efficacy in a Phase II trial, in which larger groups of cancer patients are involved. Now, for the first time, critical decisions must be made about the **indications** for enlisting specific patients in the trial—that is, which type of tumor or what stage of tumor progression will justify enrolling patients in such a trial?

Sometimes the clinical indications are obvious. For example, as we noted earlier, the effects of an agent targeted against the Bcr-Abl oncoprotein should be tested in patients diagnosed with chronic myelogenous leukemia (CML). Another drug directed against the HER2/Neu receptor molecule should be tested in the approximately 30% of breast cancer patients whose tumor cells overexpress this protein. Yet another agent—an inhibitor of Raf kinases—can be tried in patients with advanced melanomas, in which the B-Raf kinase molecule is often (70% of cases) mutant and constitutively activated. (Interestingly, in the last case, a B-Raf inhibitor failed to effectively stop further proliferation of metastatic melanomas, while its use in combination with a conventional chemotherapeutic drug has yielded dramatic, albeit only anecdotal responses.)

But more often than not, the choice of indications is neither rational nor optimal. Which class of cancer patients should be treated, for example, with a drug

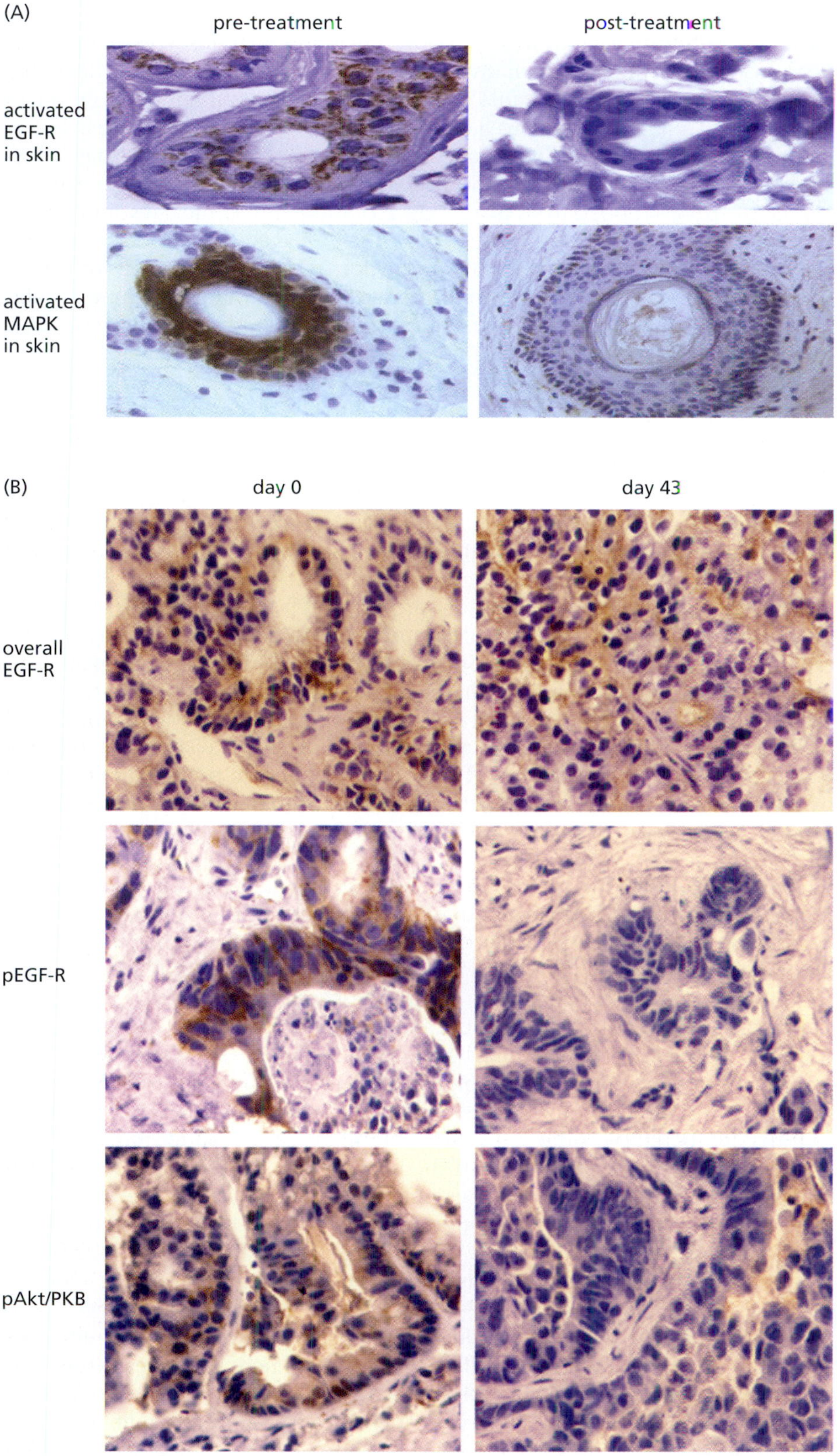

(A)

pre-treatment | post-treatment

activated EGF-R in skin

activated MAPK in skin

(B)

day 0 | day 43

overall EGF-R

pEGF-R

pAkt/PKB

Figure 16.21 Measurements of pharmacodynamics and determination of the therapeutic window The extent of inhibition of the EGF-R in a tumor can, in principle, be gauged by measuring effects of drug treatment on the EGF-R in the skin; the latter is readily assessed through small skin biopsies. In the cases illustrated here, patients under treatment were suffering from a variety of tumors, including carcinomas of the ovary, lung, colon, prostate, and head-and-neck. (A) Shown here are the effects of treating a cancer patient with Iressa, a low–molecular-weight EGF-R tyrosine kinase inhibitor (see Figure 16.31). The upper panels show immunohistochemistry using an antibody against phospho-EGF-R *(brown)*, i.e., the activated form of the receptor. The lower panels used an antibody against phospho-MAPK, the activated form of this kinase. Both measurements depended on the normally intense signaling occurring in keratinocytes present in the hair follicles. (B) The effects of an anti-EGF receptor (EGF-R) monoclonal antibody (termed EMD7200) were gauged by immunohistochemical staining of a colon carcinoma biopsy. In this case, long-term treatment resulted in a minimal reduction in the overall level of the EGF-R *(brown)* and a strong reduction in the level of phosphorylated (and therefore activated) receptor *(brown;* pEGF-R*)*. The reduction in the level of phosphorylated, activated Akt/PKB *(brown;* pAkt/PKB*)* was slight and the patient showed only a partial response to this antibody therapy, which may have reflected this minimal reduction of Akt/PKB activity in tumor cells. (C) Measurements of pharmacodynamics such as these, taken together with studies of pharmacokinetics and toxicity, define the *therapeutic window* in which a drug should be given—the range of concentrations that are efficacious without creating an unacceptable level of toxic side effects. (A and B, courtesy of J. Baselga.)

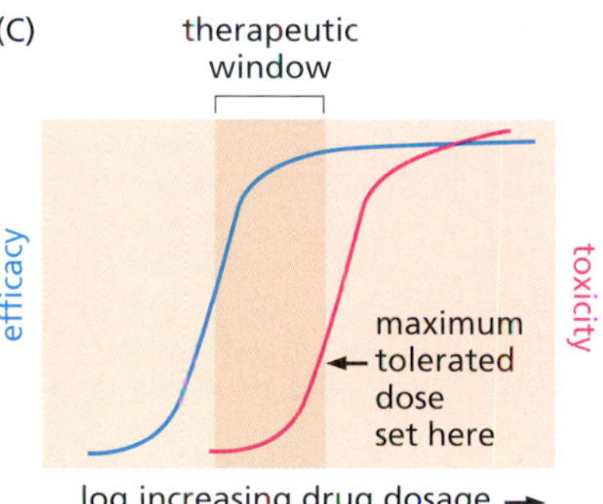

that acts as a general inducer of apoptosis in many types of cancer cells? How should a drug directed against the anti-apoptotic Akt/PKB kinase be used in the clinic? Will an anti-EGF receptor drug prove useful in carcinomas that express elevated levels of this receptor protein (Sidebar 51)? As we will see later in this chapter, certain types of cancer that would never be identified by genetics or molecular biology as attractive targets for drug treatment turn out, on occasion, to be highly susceptible to certain drugs under development. In these cases, the therapeutic utility of such drugs is discovered only by chance.

(Given the arbitrary ways in which tumor indications are chosen in many Phase II trials, we can wonder how many truly useful candidate drugs have been discarded in the past, simply because good luck did not favor them in the design of these trials. Thus, a drug may well have spectacular efficacy against gastric carcinomas, but this effect is never realized, since it is tested in Phase II trials for its effects on pancreatic or lung carcinomas, where it fails to show any useful effects and is therefore dropped from further development and clinical testing.)

If Phase II trials yield clear signs of efficacy for treating certain types of cancer with a drug candidate, Phase III trials, undertaken in far larger patient populations, will be launched. These trials are very costly but are ultimately critical, if only because they may show, for the first time, whether any clinical responses ascribed to a drug are statistically significant. The results of these trials usually become compelling only if control experiments are performed by treating equally large populations of patients with another therapy in parallel, usually one that is already licensed and in widespread use. Importantly, the licensing of a candidate drug for a specific disease indication (in the United States by the Food and Drug Administration, FDA) usually depends on whether it yields a therapeutic benefit that is measurably greater than the existing standard of care.

Patients in Phase III trials usually have gone through several previous rounds of chemotherapy with various types of cytotoxic agents, each ending with a relapse and the appearance of tumors that are *refractory* (nonresponsive, insensitive) to established therapies. Moreover, these tumors are often highly aggressive. This helps to explain why the bar is not set too high for FDA approval of a new drug or drug combination, since the drugs in Phase III trials are dispatched to attack the most difficult types of cancer. Thus, improvements in patient quality of life or temporary shrinkage of a tumor may suffice even without improvement in long-term survival.

As one illustration of this, we can cite the development of current treatments for pancreatic cancer. This disease is an extreme example, to be sure, in that the 5-year survival of this disease (from the time of initial diagnosis) has consistently been found to be less than 4%. Gemcitabine (difluorodeoxycytidine), which is widely employed as a therapy for pancreatic carcinoma, received initial FDA approval for treating this tumor because in some patients it resulted in an improvement in symptoms, weight gain, and a temporary stabilization in tumor growth, although it offered only a modest increase in survival time after disease diagnosis: patients treated with gemcitabine had a median survival time of 5.65 months compared with those given the standard care—5-fluorouracil (5-FU), which afforded them a 4.41-month median survival time (Figure 16.22). This and similar anecdotes reveal how desperate is the need for truly effective means of treating solid tumors. It also illustrates the fact that the FDA's requirements for approving anti-cancer agents are far less stringent than for other disease states, where much greater efficacy is required to gain licensing of new drugs.

Nonetheless, even with these relatively modest regulatory requirements, the other complications of drug development described here keep the current success rate for anti-cancer drug development extremely low. Perhaps one drug in a hundred is advanced all the way through the drug development "pipeline" from initial *in vitro* testing through a Phase III trial that culminates in some clear

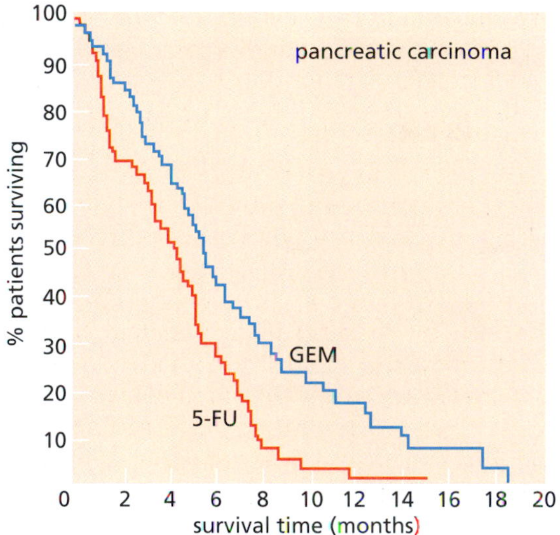

Figure 16.22 Gemcitabine as a treatment for pancreatic cancer This Kaplan–Meier plot illustrates the high mortality exacted by pancreatic cancer. Patients treated with gemcitabine (GEM) lived slightly longer than did those treated with 5-fluorouracil (5-FU)—the standard of treatment in the 1990s. Both of these agents are pyrimidine derivatives whose cytotoxicity derives from their ability to inhibit DNA synthesis, in part through misincorporation into the DNA. (5-FU also interferes with pyrimidine biosynthesis.) As is apparent, gemcitabine treatment offered only a modest increase in patient survival in this study reported in 1997, but nevertheless this effect sufficed to allow its approval by the U.S. Food and Drug Administration. (From H.A. Burris 3rd, M.J. Moore, J. Andersen et al., *J. Clin. Oncol.* 15:2403–2413, 1997.)

improvement in patient outcome and licensing by the FDA. (After licensing has occurred, a Phase IV trial may be conducted to determine how a newly introduced drug compares with other drugs used for similar indications, how certain subgroups of patients respond to the drug, and whether concerns about a drug's safety eventually emerge from its use in very large patient populations.)

16.10 Tumors often develop resistance to initially effective therapy

A complication that dogs all anti-cancer drugs is illustrated by the behavior of *HER2/neu* transgenic mice, in which the mutant, oncogenic transgene has been programmed to induce mammary tumors on a predictable schedule and can be shut down thereafter. While transgene-induced primary breast tumors and metastases all collapsed when the *HER2/neu* transgene was shut down, new tumors recurred in most of these mice between 1 and 9 months later (see Table 16.1). These tumors clearly represented variants of the initially observed ones that had developed alternative means of propelling their growth—that is, had become independent of *HER2/neu* oncogene expression.

As we saw in Chapter 12, the unstable, mutable genomes of cancer cells continually generate new alleles and novel genetic configurations. Evolving cancer cells can pick and choose among these genetic variations, searching for combinations that improve their ability to survive and proliferate. In this *HER2/neu* example, the relatively small number of cancer cells that survived oncogene shutdown seem to have spent months thereafter trolling for newly arising oncogenes (or other cancer-causing alleles) in their genomes that might enable them to re-launch their program of aggressive proliferation. The rare cells that happened to acquire such a novel genetic/epigenetic change then began clonal expansions that led to relapsing tumors.

A similar dynamic complicates almost all types of cancer therapies, where initial clinical successes in reducing tumor cell populations are usually followed by the re-emergence in patients of tumor cell populations that have, through one means or another, developed resistance to the initial treatment, thereby acquiring a new lease on life.

For example, a variety of commonly occurring human tumors, including breast, small-cell lung, and ovarian carcinomas, are initially quite responsive to cytotoxic drugs that are currently used in chemotherapy, but after a while, these

Table 16.2 Examples of Drug Resistance Mechanisms Developed by Tumors

General mechanism	Example
(A) Tumor cell–based mechanisms	
Decreased intracellular drug accumulation	overexpression of P-glycoprotein multi-drug efflux pump
Decreased drug activation	decreased metabolic activation of pro-drug by loss of expression of a cytochrome P-450
Increased inactivation of drug or toxic intermediate	increase in nucleoside deaminase that inactivates nucleoside analogs, increase in glutathione-*S*-transferase (GST) detoxification
Increased repair of drug-induced damage	increased repair of DNA cross-linking induced by cisplatin, a chemotherapeutic drug
Increased resistance to drug-induced toxicity	inactivation of apoptotic responses by mutation of p53 or overexpression of Bcl-2
Drug targets altered (quantitatively or qualitatively)	overexpression of Bcr-Abl, alteration of Bcr-Abl catalytic cleft
(B) Host-dependent mechanisms	
Anatomic drug barriers (tumor sanctuaries)	tumor growth in brain behind blood–brain barrier, or in testes
Host–drug interactions: increased drug inactivation by normal tissues	detoxification of cyclophosphamide chemotherapeutic drug in liver
Host–drug interactions: decreased drug activation by normal tissues	decreased cyclophosphamide activation in liver

Adapted from J. Moscow, C.S. Morrow and C.H. Cowan, in J. Holland and E. Frei (eds.), Drug Resistance and Its Clinical Circumvention in Cancer Medicine, 6th ed. Hamilton, ON: B.C. Decker, 2003.

tumors become refractory to drug treatment and therefore relapse, as suggested earlier. Much of this acquired resistance is attributable to the genetic and therefore phenotypic plasticity of cancer cell populations.

The acquired mechanisms of drug resistance are quite variable and illustrate the ingenuity of cancer cells in devising various maneuvers for eluding killing by chemotherapeutic drugs. As indicated in Table 16.2, some of these mechanisms involve loss of the ability to import drug molecules through the plasma membrane or an acquired ability to pump drug molecules out through this membrane. Yet others depend on an acquired ability to metabolize drug molecules, in some cases using the same classes of enzymes that are important in detoxifying other types of toxic compounds that have entered into the cell (Section 12.6). Cells may also neutralize components of their apoptotic machinery or may acquire an increased ability to repair DNA molecules damaged by chemotherapeutics or radiation.

These behaviors represent a general challenge to all types of anti-tumor therapy. The only clear solution to the evasive maneuvers undertaken by cancer cells derives from the fact that most resistance mechanisms are acquired at a relatively low probability per cell generation. Accordingly, by applying two unrelated drugs simultaneously, the probability of tumor cell populations spawning variants that can survive this double onslaught is roughly equal to the square of the probability of acquiring resistance to a single agent, and the probability of surviving triple therapy should be the cube of this low probability, etc.

However, even these multi-drug therapy strategies are often foiled by cancer cells, which develop powerful strategies for evading killing, such as the acquisition of multi-drug resistance (MDR). For example, high-level expression of the *MDR1* gene, which encodes a transmembrane drug efflux pump, enables cancer cells to efficiently excrete a variety of chemically unrelated drugs, thereby lowering intracellular drug concentrations to sub-toxic levels (Figure 16.23). Similarly, inactivation of certain parts of the apoptotic machinery may also confer concomitant resistance to a number of distinct cytotoxic agents. In spite of these complications, it is a widespread consensus among drug developers that monotherapies involving either low–molecular-weight drugs or biological molecules are unlikely to cure most types of cancer, and that effective multi-agent therapies must be devised if definitive, durable cures are to be achieved in the future.

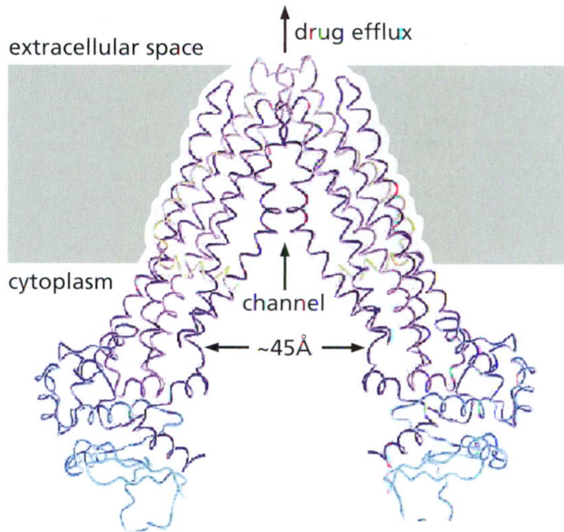

Figure 16.23 Multi-drug resistance and the P-glycoprotein The *MDR1* gene encodes P-glycoprotein, a protein that is often present at elevated levels in cancer cells being treated with various types of chemotherapy. P-glycoprotein, whose structure is shown here, is a 170-kD ATP-dependent transmembrane protein that can pump a wide variety of drug molecules out of cells. It is a member of a large family of mammalian transporter molecules, of which 49 have been discovered until now. Its actions appear to be responsible for the acquisition of drug resistance by a variety of human tumor types under treatment with low-molecular-weight chemotherapeutic drugs. (From D. Mahadevan and A.F. List, *Blood* 104:1940–1951, 2004.)

With these considerations in mind, we will read about a series of illustrative anecdotes in the sections that follow. Each concerns a type of drug and its targets within cancer cells. The stories are arranged in an order, beginning with a well-established therapy and ending with a speculative one that holds great promise but is still far from clinical validation. In some cases, the specific therapy that has been developed was inspired by discoveries of malfunctioning proteins within cancer cells; these discoveries allowed drug development to be pursued logically and methodically. In other cases, strokes of good luck or intuitive leaps enabled the development of highly active compounds. Inevitably, these anecdotes represent arbitrary choices and draw from a vast pool of agents currently under investigation or development. They represent the forerunners of a large flock of such drugs that will be developed and licensed for clinical use in the years to come.

16.11 Gleevec development has paved the way for the development of many other highly targeted compounds

In the previous sections, we made repeated reference to the Bcr-Abl oncoprotein and to experimental strategies for antagonizing it. Now, we backtrack and review some history of how the Bcr-Abl oncoprotein was discovered and validated as an attractive drug target and finally used as an object of rational drug design. This story is valuable, if only because it illustrates the long course through which drug development passes from initial discovery at the laboratory bench to the oncology clinic.

This particular story begins in 1914, when the German cytologist Theodor Boveri proposed that chromosomal defects might cause a cell to proliferate abnormally, resulting ultimately in the formation of some type of cancer. Almost half a century passed before Boveri's idea received some validation. In 1960, two cytologists working in Philadelphia noted that an abnormal, unusually small Chromosome 22 was characteristically present in the great majority of cells of chronic myelogenous leukemia (CML); since that time, it has been called the Philadelphia chromosome or simply Ph^1. It took another dozen years before a researcher in Chicago demonstrated that a reciprocal translocation between Chromosomes 9 and 22 was responsible for creating the Ph^1 chromosome (initially described in Section 4.6). (Since a larger chunk of chromosome 22 is donated to the tip of chromosome 9 than is received from this chromosome, this

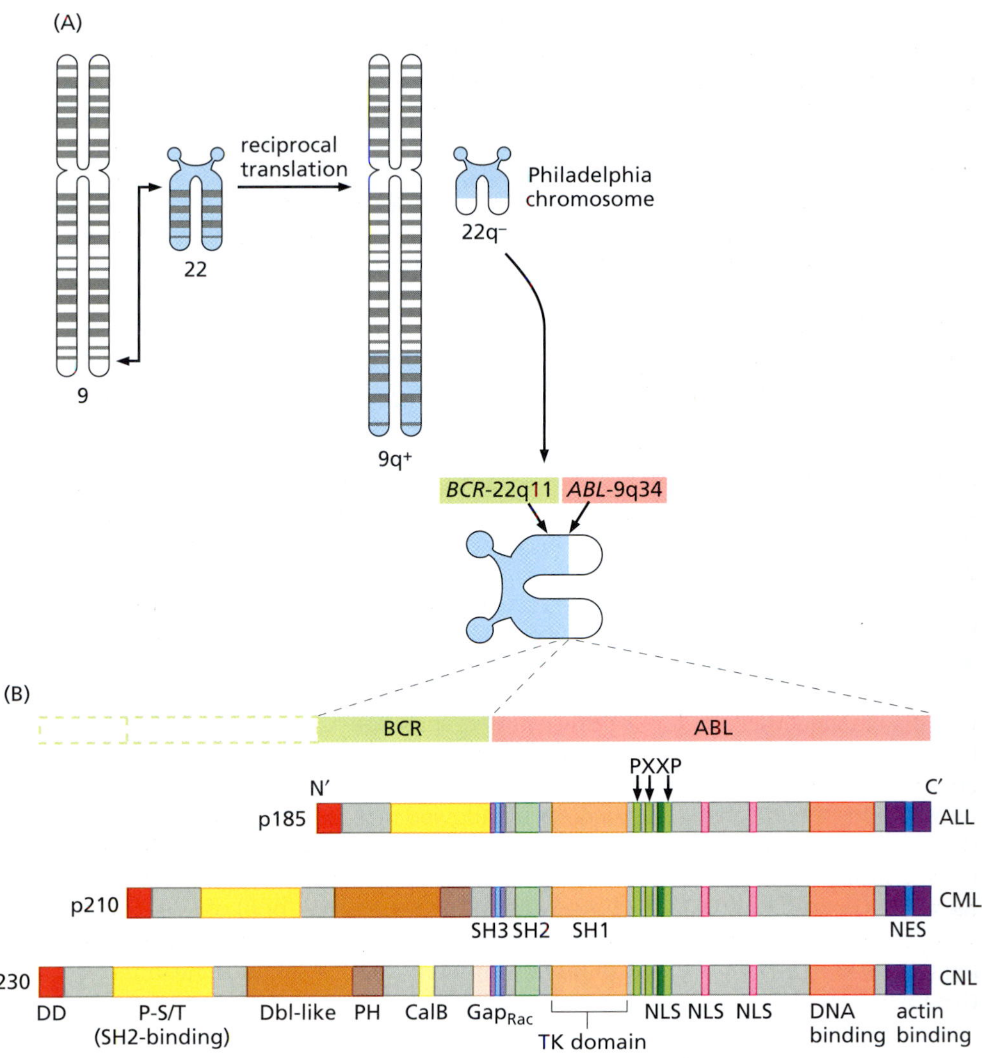

Figure 16.24 Origin and structure of the Bcr-Abl protein (A) More than 95% of cases of chronic myelogenous leukemia (CML) exhibit the Philadelphia chromosome, which results from a reciprocal translocation between chromosomes 9 and 22. The q34 region of chromosome 9 carrying most of the *ABL* gene is transferred to the q11 region of chromosome 22, replacing a larger segment of chromosome 22 that is translocated reciprocally to chromosome 9. The net result is a truncated chromosome 22 (i.e., 22q–) that is often termed the Philadelphia chromosome (Ph^1) and a fusion of the 5′ portion of the *ABL* gene with a 3′-proximal portion of the *BCR* gene, which normally resides at 22q11. (B) Depending on the precise location of the breakpoint in *BCR,* three distinct Bcr-Abl fusion proteins may be formed which are found in ALL (acute lymphoblastic leukemia), CML, and CNL (chronic neutrophilic leukemia). Each of these *BCR-ABL* fusion genes encodes a multidomain (and thus multifunctional) protein. (From A.S. Advani and A.M. Pendergast, *Leuk. Res.* 26:713–720, 2002.)

leaves the already-tiny chromosome 22 even more diminished in size; this remnant of 22 plus the small translocated segment is Ph^1; see Figure 16.24A) The chromosomal aberration—clearly the consequence of a somatic mutation—was proposed as a potential cause of this malignancy. As mentioned earlier, we now know that this particular translocation is present in more than 95% of cases of CML.

The genes that were fused through this translocation remained unknown for another decade. Finally, in 1982, molecular biologists discovered that *ABL,* the human homolog of the mouse c-*abl* proto-oncogene, participates directly in these chromosomal translocations, becoming fused with a second, still unknown gene. The breakpoints of this other gene (the chromosomal sites at which it becomes fused to the *ABL* gene) were soon found to be scattered over many kilobases of DNA, yielding the name "**b**reakpoint **c**luster **r**egion" or simply *BCR.* In fact, three distinct fusion proteins arise through the inclusion of variously sized Bcr proteins at the N-termini of the fusion proteins with almost the entire Abl protein at the C-termini (Figure 16.24). As indicated in the figure, the different fusion proteins tend to be associated with distinct types of leukemia.

Within two years of its discovery, the Bcr-Abl protein was found to function as a constitutively activated tyrosine kinase. In this respect, it functions like the Abl oncoprotein of Abelson mouse leukemia virus. The genome of this retrovirus carries an *abl* oncogene derived from the corresponding proto-oncogene residing in the normal mouse genome.

By 1990, a cDNA encoding the Bcr-Abl fusion protein had been introduced into a retrovirus vector, and the resulting virus was then found to induce a leukemia in mice that closely resembled human CML. Like the human disease, this leukemia involved large numbers of fully differentiated granulocytes in the blood. Under certain conditions, the mouse leukemia, like its human counterpart, progressed to a "blast crisis," involving the accumulation of immature cells of the lymphoid or myeloid lineages (see Figure 12.4). These observations in mice represented the first formal proof that the Bcr-Abl fusion protein operates as the central motive force of leukemogenesis in CML.

Unfortunately, this demonstration of the critical role of Bcr-Abl revealed nothing about the mechanisms by which it functions. The bewildering complexity of Bcr-Abl signaling is indicated by the diverse array of structural and functional domains in the two contributing proteins (see Figure 16.24B). Altogether, the domains present in this fusion protein enable it to activate the Ras pathway, the PI3 kinase–Akt/PKB pathway, the Jak–STAT pathway, and transcription factors including Jun, Myc, and NF-κB. In addition, the Ras-like Rac protein, which regulates activities as diverse as cellular migration, survival, and proliferation, is activated, as are two nonreceptor tyrosine kinases, Hck and Fes (not shown). These various associations enable the Bcr-Abl protein to extend its reach into almost all of the regulatory circuits governing cell proliferation and survival.

In spite of this complexity, the tyrosine kinase domain of Bcr-Abl, derived from the Abl proto-oncogene protein, was found to be the key element in leukemogenesis. For example, subtle alterations of the Bcr-Abl protein that inactivated its tyrosine kinase catalytic activity led to total loss of its transforming function. In the early 1990s, a research program was begun to develop low-molecular-weight antagonists of the Bcr-Abl tyrosine kinase activity. A drug emerged, termed variously imatinib mesylate, STI-571, Glivec, and Gleevec (see Figure 16.10A), which was able to bind the catalytic cleft of the Bcr-Abl tyrosine kinase. As is the case with all other kinases of this family, the cleft is located between the two major structural lobes of the kinase protein (see Figure 16.10B).

Even though the Abl kinase domain shares roughly 42% amino acid identity with a large number of other tyrosine kinases, the inhibitory effects of Gleevec on Bcr-Abl were found to be relatively specific (see Figure 16.15). Subsequently, four other tyrosine kinases—those belonging to the PDGF (α and β) and Kit receptors as well as the Arg (Abelson-related gene) protein—were also found to be inhibited by Gleevec. Accordingly, this drug, when used at therapeutic concentrations, appears to target only 4 of the 90 or so human tyrosine kinases. Like most other kinase inhibitors, the Gleevec molecule associates with the ATP-binding pocket of the Abl kinase domain (see Figure 16.16). While other kinase inhibitors block ATP binding in this cleft, Gleevec works differently: it binds and stabilizes a catalytically inactive conformation of this enzyme.

This success with Gleevec encouraged many other attempts at creating low–molecular-weight kinase antagonists, which were realized to have certain therapeutic advantages when compared with anti-receptor monoclonal antibodies (Table 16.3). In addition, it emboldened pharmaceutical chemists to try to make narrowly targeted tyrosine kinase inhibitors, some of which have demonstrated extraordinary specificity (Sidebar 52 ●).

By 1996, Gleevec had been found able to inhibit the growth of CML cells *in vitro* while having no effect on normal bone marrow cells. More specifically, the proliferation of Bcr-Abl–dependent cells could be inhibited at drug concentrations as low as 40 nM, indicating a high affinity of Gleevec for the catalytic cleft of the tyrosine kinase domain. (Cells that depend on Bcr-Abl for survival can be forced into apoptosis by Gleevec's inhibition of Abl kinase function.) The initial clinical trials, begun in 1998, revealed remissions from disease in all of the 31 treated

Table 16.3 Strengths and weaknesses of anti-receptor antibodies versus low–molecular-weight tyrosine kinase inhibitors as anti-cancer agents[a]

	Small molecule	Antibody
Target	tyrosine kinase domain	receptor ectodomain
Specificity	+++	++++
Binding	most are rapidly reversible	receptor internalized, only slowly regenerated
Dosing	oral daily	intravenous, ≤ weekly
Distribution in tissues	more complete	less complete
Toxicity	rash, diarrhea, pulmonary	rash, allergy
Antibody-dependent cellular cytotoxicity	no	possibly

[a]Courtesy of N.J. Meropol and from N. Damjanov and N. Meropol, *Oncol. (Huntington)* 18:479–488, 2004.

CML patients, with only minimal side effects registered, even when taken daily for many years. Four years later, 6000 patients had already been entered into Gleevec clinical trials.

Treatment of early-stage (chronic) CML with Gleevec leads to a hematologic response in 90% of cases: microscopic analysis of blood smears reveal a profound shift in the cellular composition of the blood (Figure 16.25A), and PCR analysis reveals an extraordinary decline in the levels of the *BCR-ABL* mRNA in blood cells (Figure 16.25B). In 50% of these cases, the translocated Philadelphia chromosome is no longer detectable by karyotypic analyses of patients' white cells. About 60% of the patients who have already progressed to blast crisis respond to Gleevec, but they generally relapse after a period of some months.

Figure 16.25 Measuring responses to Gleevec treatment (A) The successes of Gleevec in treating chronic myelogenous leukemia (CML) patients can be gauged from cytological analyses of the patients' blood. As seen here, treatment with Gleevec converted the blood smear from a state in which many leukemia cells *(large, dark nuclei, above)* appear to one in which only normal granulocytes are visible *(below)* among the red blood cells. (B) A more sensitive and quantitative measure of therapeutic success comes from use of quantitative polymerase chain reaction (qPCR) measurements of the level of Bcr-Abl mRNA (which is initially reverse-transcribed prior to PCR amplification). In an untreated patient *(red curve)*, 50% of maximum *(red arrow)* PCR-mediated gene amplification is observed at about the 29th cycle of gene amplification (in which each cycle results in the doubling of the amplified sequence). However, following Gleevec treatment *(blue curve)*, a comparable degree of amplification is only achieved at about the 39th cycle *(blue arrow)*, indicating that the Bcr-Abl RNA–expressing cells are present at a level that has been reduced by approximately a factor of about 2^{10}. PCR-based assays can detect as few as one CML cell amid 10^5 to 10^6 normal blood cells. (A and B, courtesy of B.J. Druker.)

(A)

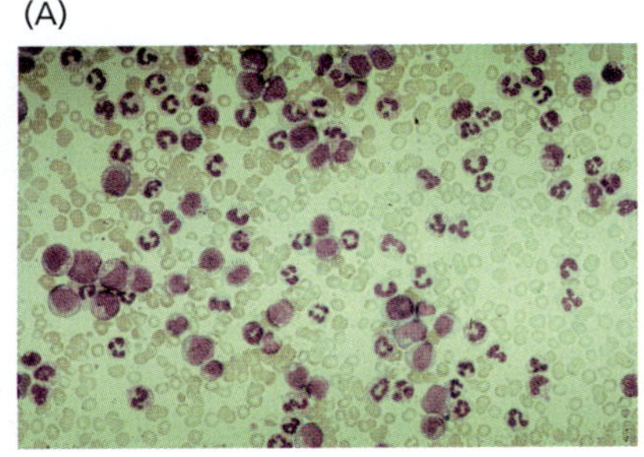

before treatment

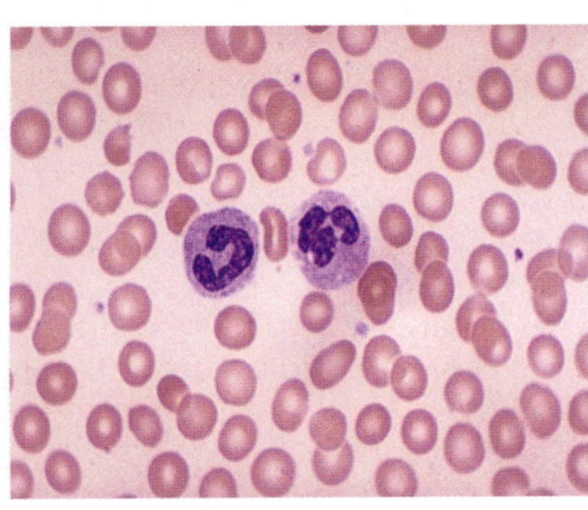

after treatment

(B)

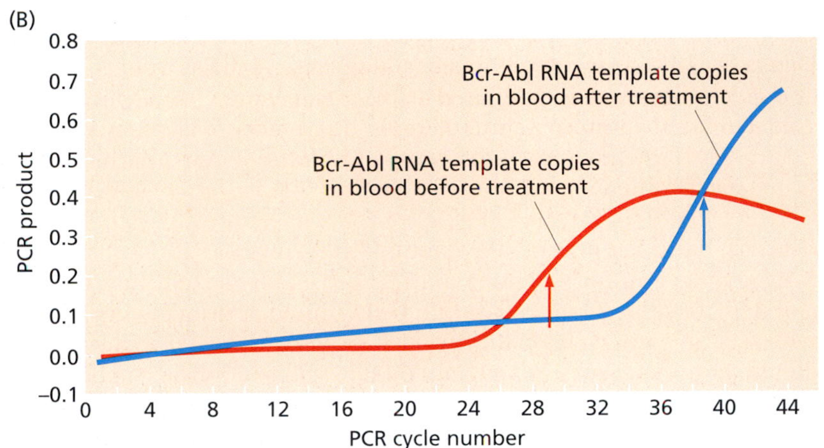

Initial studies have indicated that the average chronic-phase patient under treatment runs a risk of about 10% per year of relapsing by progressing into the crisis phase of CML; the relapse rates of recently diagnosed patients, who are generally at an early stage of disease progression, may be as low as 5% per year.

The molecular mechanisms that allow tumor cells to eventually escape Gleevec inhibition are interesting, in that they shed further light on the Bcr-Abl oncoprotein and its actions and, more generally, they reveal how cancer cells can acquire resistance to highly targeted drugs. Analyses of *BCR-ABL* sequences in the tumors of patients with Gleevec-resistant, relapsed disease revealed that 29 of 32 tumors harbored mutations in the *BCR-ABL* gene; altogether these yielded substitutions of 13 distinct amino acid residues in the kinase domain. (Another dozen have been cataloged in subsequent studies.)

Some of these mutations prevent Gleevec from binding to the catalytic cleft, either by directly interfering with its binding or, less directly, by creating a stereochemical shift in the oncoprotein (Figure 16.26A and B). In a minority of patients, Gleevec resistance was achieved through amplification of the *BCR-ABL* gene in their leukemia cells, yielding increased levels of the encoded

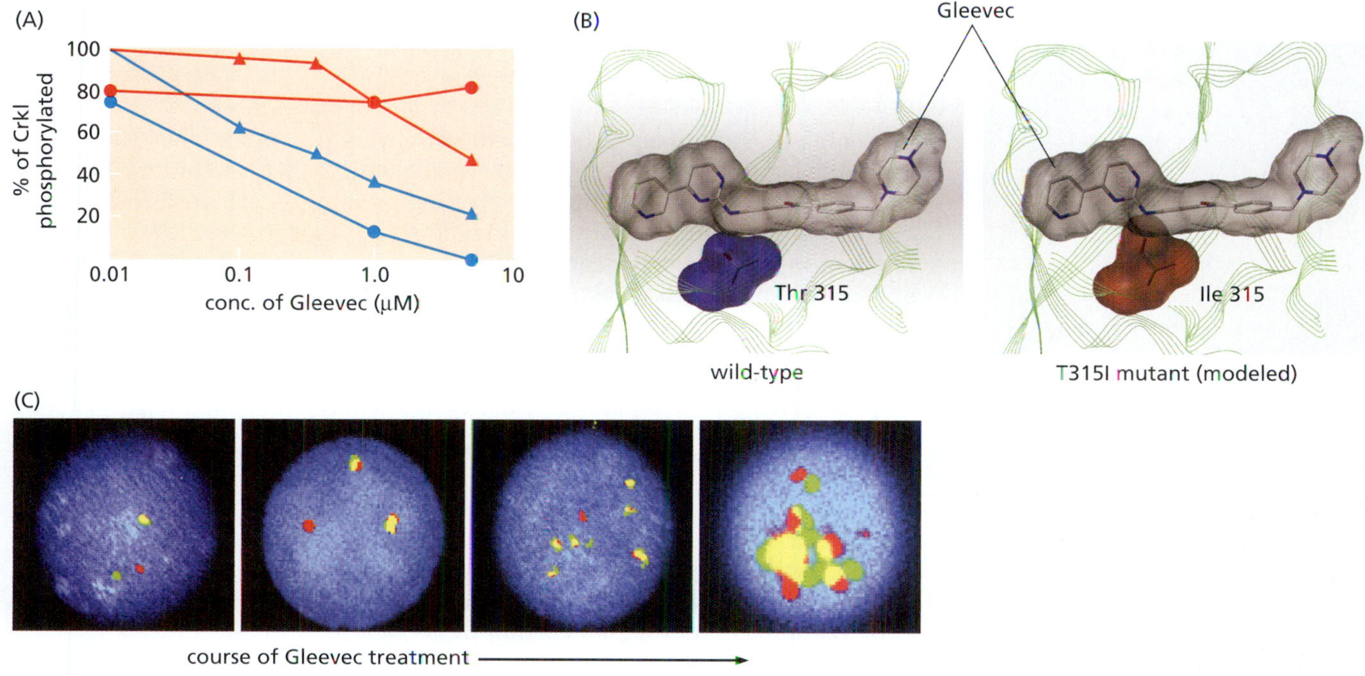

Figure 16.26 Acquisition by CML cells of resistance to Gleevec The ability of Gleevec to inhibit Bcr-Abl kinase activity changes dramatically following relapse and acquired resistance to drug treatment. (A) In this case, the kinase activity was gauged in isolated leukemia cells by the degree of phosphorylation of Crkl, a protein that is a good substrate for phosphorylation by Bcr-Abl. At the onset of therapy, the Bcr-Abl kinase (in cultured leukemia cells) suffered about a 50% inhibition in the presence of approximately 0.1 μM Gleevec in two patients *(blue triangles, blue circles)*. However, after Gleevec resistance developed in these patients, about an 8-μM concentration of the drug was required to inhibit one patient's Bcr-Abl kinase *(red triangles)*, while the other patient's was totally resistant to the drug *(red circles)*. (B) The Gleevec molecule is able to nestle tightly in a molecular cavity created in part by a threonine residue *(blue)* in position 315 of the wild-type Bcr-Abl oncoprotein *(left;* see also Figure 16.10C*)*. However, in a mutant Bcr-Abl found in the leukemic cells of a Gleevec-resistant patient *(right)*, this threonine residue was replaced by an isoleucine *(brown)*, which protrudes into the drug-binding cavity and interferes with insertion of Gleevec into the cavity. (C) The number of *BCR-ABL* gene copies in a patient's leukemic cells has been gauged here using fluorescent *in situ* hybridization (FISH). Nuclei are visualized here in *blue*, *ABL* sequences in *red*, and *BCR* sequences in *green*. *Yellow* indicates an overlap of *ABL* and *BCR* sequences, i.e., sites of the fused gene created by the chromosomal translocation. The levels of the fused gene *(yellow)* at the beginning of therapy *(left)* were quite low, but as treatment proceeded *(rightward)*, the levels of the fused gene (and therefore Bcr-Abl fusion protein) increased progressively until the patient's leukemia became resistant to Gleevec treatment. In this particular patient, Gleevec resistance was acquired by the tumor cells because the fusion protein became overexpressed, thereby exceeding the ability of the normal therapeutic concentration of Gleevec to bind and inactivate it. (From M.E. Gorre, M. Mohammed, K. Ellwood et al., *Science* 293:876–880, 2001.)

Table 16.4 Changed responsiveness to Gleevec of chronic myelogenous leukemia cells following extended treatment

	Number of CML patients analyzed	Mean Gleevec IC_{50} of Crkl phosphorylation (μM)	Range of Gleevec IC_{50} (μM)
Pre-treatment	6	0.45±0.33	0.21–1.1
After relapse	5	7.5±5.2	2.5–14.5

From M.E. Gorre, M. Mohammed, K. Ellwood et al., *Science* 293:876–880, 2001.

oncoprotein that apparently could no longer be inhibited by the concentrations of drug used to treat patients (Figure 16.26C).

These demonstrations that acquired resistance to Gleevec (Table 16.4) is usually accompanied by structural alterations of the Bcr-Abl protein or overexpression of this protein provide compelling proof that Gleevec's ability to evoke therapeutic responses can be attributed directly to its effects on the Bcr-Abl protein. These insights have been taken a step further by introducing random mutations into a vector encoding the Bcr-Abl protein and then determining which of the resulting mutant forms of this protein are able to resist inhibition by Gleevec (Figure 16.27). Such an experimental strategy, which uses cultured cells whose growth and viability are dependent on Bcr-Abl (see Figure 16.17), can in principle reveal the full spectrum of structural alterations of Bcr-Abl that are capable of rendering it resistant to Gleevec inhibition. The results of this screen should prove valuable in the future for understanding the molecular mechanisms of acquired drug resistance.

Subsequent research has also led to the discovery that other kinase inhibitors, including a drug originally developed as a Src antagonist, are quite effective at low concentrations (<10 nM) in inactivating mutant, Gleevec-resistant Bcr-Abl oncoproteins (Figure 16.28). This provides hope that many Gleevec-resistant tumors can be effectively treated in the future with other kinase inhibitors, and that in the long term, concomitant treatment of CML patients with several inhibitors will greatly reduce the emergence of drug-resistant tumors.

The ability of Gleevec to also inhibit the platelet-derived growth factor receptor (PDGF-R) suggests that it may one day be useful in treating other types of malignancies as well. For example, mutations in the genes encoding the PDGF-Rα and -Rβ have been found in a number of chronic **myeloproliferative** diseases, that is, conditions involving elevated levels in the circulation of one or another cell type arising from the myeloid lineage of hematopoiesis (see Figure 12.4). Indeed, patients suffering from one of these diseases—hypereosinophilic syndrome—have shown a **complete response** following Gleevec treatment, with virtual disappearance of their eosinophiles.

The PDGF-Rβ is overexpressed in some 85% of metastatic medulloblastomas but not in nonmetastatic tumors of this type. The cells in these more aggressive growths appear to rely on a PDGF–PDGF-R autocrine loop to drive their motility and proliferation. This type of cancer commonly affects children, and the therapeutic means for treating it have largely been confined to **radiotherapy**, which has serious longterm neurological side effects. More generally, angiogenesis in a wide variety of tumor types depends in part on the PDGF-mediated recruitment of pericytes by their endothelial cell neighbors (see Figure 13.47). Accordingly, future therapeutic protocols for treating these solid tumors may involve Gleevec or similarly acting agents.

Gleevec's effects on a third tyrosine kinase—the Kit receptor—also make it an attractive agent for attacking gastrointestinal stromal tumors (GISTs), a relatively uncommon tumor for which few therapeutic options have been available. The Kit receptor is typically mutated in these cancers and seems to represent the primary mitogenic force in the tumor cells (see Figure 5.18). In one study, responses involving clear regression of the tumor were observed in almost 70% of treated patients (see Figure 16.29). By 2005, SU11248—a second inhibitor of Kit tyrosine kinase function—was approved by the FDA for the treatment of

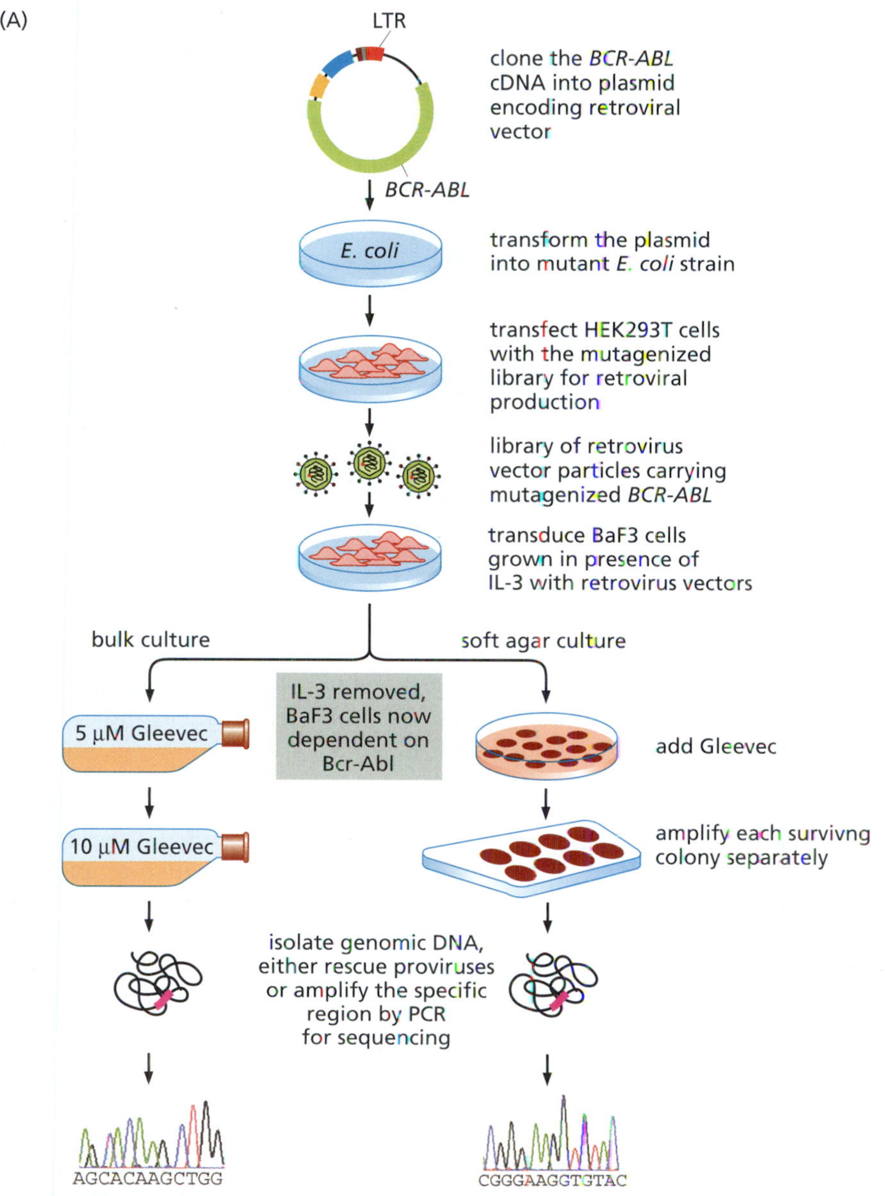

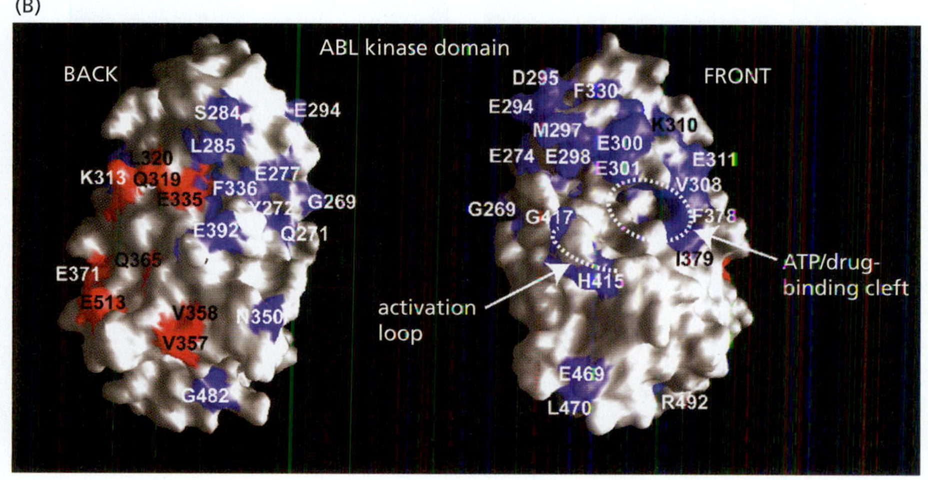

Figure 16.27 Screening *in vitro* for Gleevec-resistant mutant forms of Bcr-Abl One strategy to detect drug-resistant variants of Bcr-Abl involves cultured cells, such as the BaF3 cells described in Figure 16.17, whose continued survival can be made dependent on the presence of a functionally active Bcr-Abl oncoprotein. When such Bcr-Abl–expressing cells are treated with Gleevec, they are killed because of their dependence on continued Bcr-Abl signaling. (A) A cDNA clone expressing the "wild-type" BCR-ABL protein (i.e., the direct product of the chromosomal translocation) can be mutated by passage through *E. coli* bacteria that are highly error-prone in DNA replication and therefore generate mutant variants of the introduced, plasmid-borne *BCR-ABL* sequence. The resulting collection of randomly mutated Bcr-Abl–expressing clones is then introduced, via a retrovirus vector, into BaF3 cells, which are then exposed to Gleevec. The rare cells that resist being killed by Gleevec are then isolated, either in bulk or as soft agar colonies, and the sequence of the mutant BCR-ABL protein that conferred Gleevec resistance is determined. (B) When the resulting Gleevec-resistant mutant BCR-ABL proteins are analyzed, many are found to have single amino acid substitutions of residues located throughout the Abl kinase domain of the Bcr-Abl oncoprotein. The "front" and "back" of the ABL kinase domain are shown here, together with the sites of these mutant residues and the identities of the normally present amino acid residues. Surprisingly, the individual mutations, each of which confers Gleevec resistance, alter residues at many sites in the ABL domain, indicating that CML cells have multiple options for developing drug resistance. Many of these mutant residues are found on the side of Abl opposite the catalytic cleft *(left)*; some of these residues *(red)* participate in interactions between the kinase (i.e., SH1) domain and the SH2 and SH3 domains of Abl *(not shown)*. Many other mutant sites *(blue)* found elsewhere in the kinase domain produce drug resistance through poorly understood mechanisms. This *in vitro* screen for Gleevec-resistant Bcr-Abl mutants revealed most of those discovered in patients plus a number of others that have not been documented in patients to date. (From M. Azam, R.R. Latek and G.Q. Daley, *Cell* 112:831–843, 2003.)

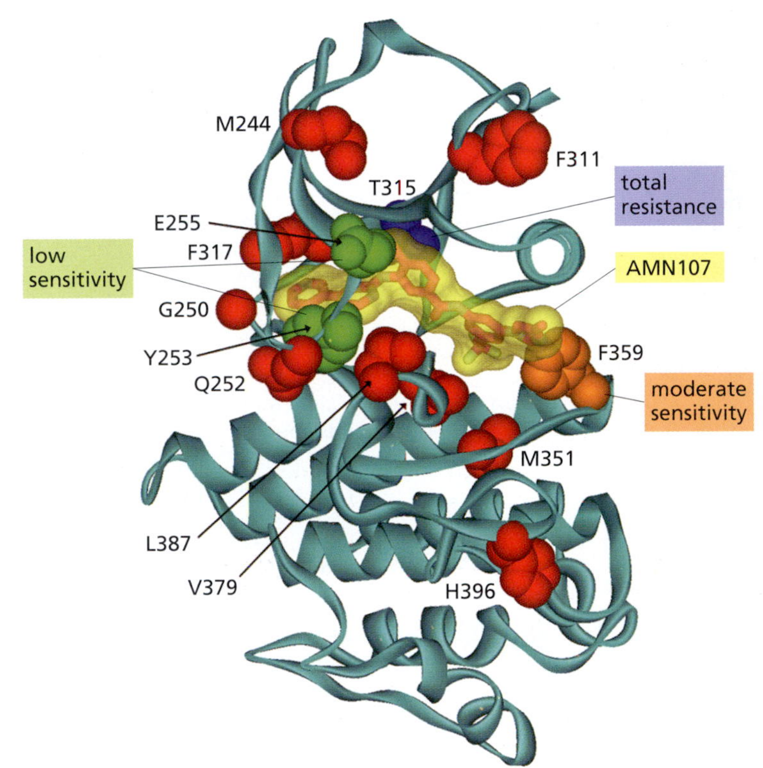

Figure 16.28 Backup inhibitors of Bcr-Abl for patients with Gleevec-resistant tumors The fact that patients in the acute (blast crisis) phase of CML often develop resistance to Gleevec (see, for example, Figure 16.26) has stimulated the development of alternative inhibitors of the Abl tyrosine kinase. One of these inhibitors, AMN107, is shown here *(yellow space-filling model, orange stick figure)* in complex with the tyrosine kinase domain of Bcr-Abl, on which are also indicated the sites of a number of amino acid substitutions found in the mutant forms of Bcr-Abl discovered in Gleevec-resistant tumors of patients. (The number of *colored spheres* at a site indicates the number of atoms present in the side-chain of the substituted amino acid.) Mutant, Gleevec-resistant forms of Bcr-Abl having amino acid substitutions areas with the *red spheres* are highly sensitive to inhibition by AMN107. A mutant of Bcr-Abl with an amino acid substitution at the site indicated with *orange spheres* (i.e., F359) shows moderate sensitivity to AMN107 inhibition, while variants showing low sensitivity to AMN107 inhibition carry amino acid substitutions shown in *light green* (e.g., Y253). One Gleevec-resistant mutant form of Bcr-Abl is also totally resistant to AMN107 *(blue spheres,* residue T315). ("M244" indicates that the normally present residue at position 244, which happens to be valine, *not shown,* was replaced by a methionine; etc.) (From T. O'Hare, D.K. Walters, E.P. Stoffregen et al., *Cancer Res.* 65:4500–4505, 2005.)

GISTs, including those that had developed a resistance to the anti-tumor effects of Gleevec. Moreover, quite unexpectedly, the utility of Gleevec (and thus other, similarly acting kinase inhibitors) may occasionally extend far beyond the treatment of the malignancies described here (Sidebar 53).

The clear successes of Gleevec represented the first validation that rational drug design can succeed in producing agents that are highly useful for treating various types of human cancer. The fact that Gleevec interferes with multiple tyrosine kinases was initially viewed as a disadvantage of this drug, since it was feared that this broader activity would lead to unacceptable side-effects. However, with the passage of time, it is becoming increasingly clear that such multi-target effects may actually prove useful in treating certain malignancies. Thus, the viability and proliferation of many tumors depend on the coordinate actions of multiple tyrosine kinases, and the ability to strike at several of these simultaneously may one day be found to confer great therapeutic advantage.

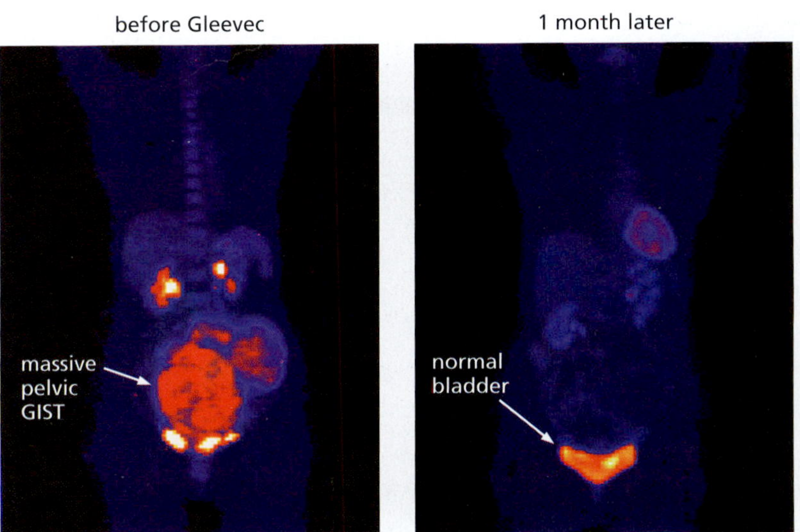

Figure 16.29 Use of Gleevec to treat gastrointestinal stromal tumors The fact that Gleevec also shows inhibitory activity against the tyrosine kinase function of the Kit receptor suggested that it might prove useful against gastrointestinal stromal tumors (GISTs), in which mutant, constitutively active Kit receptors are commonly found. As seen here, this patient's GIST *(red mass, pelvic region, left image),* which was visualized because of its uptake of a labeled glucose analog, responded dramatically to Gleevec treatment *(right image)*. (The residual labeling following treatment reflects the accumulation of the labeled dye in the patient's bladder.) Unfortunately, over time, most GISTs develop resistance to Gleevec, so that 2.5 years after starting treatment, about 75% of tumors no longer respond well to Gleevec treatment. (Courtesy of G.D. Demetri.)

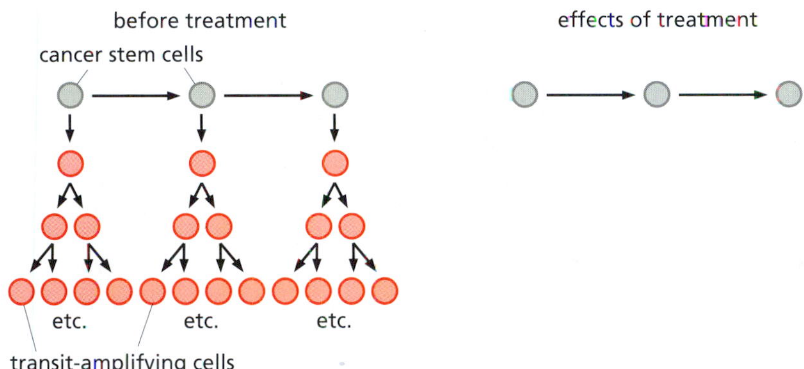

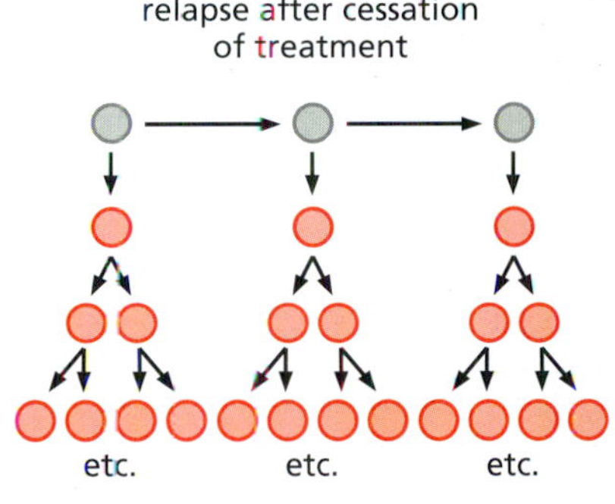

Figure 16.30 Role of tumor stem cells in the response to anti-cancer treatments While the evidence is still fragmentary, it appears that in many tumors, a small proportion of the neoplastic cell population is composed of self-renewing tumor stem cells *(gray)*. These spawn the bulk of the cancer cells in tumors *(pink)*, which have many of the properties of normal progenitor or transit-amplifying cells. If, as is the case with Gleevec, an anti-cancer therapy results in the depletion of the neoplastic transit-amplifying cells without eliminating the tumor stem cells, then the latter can regenerate the tumor soon after that therapy is halted.

Unfortunately, the existence of tumor stem cells limits the utility of Gleevec. Recall that research on human cancers, including hematopoietic tumors, breast carcinomas, and brain tumors, has revealed that these tumor stem cells often constitute only a small proportion (<<5%) of the neoplastic cells in these tumors, and that their existence can only be revealed by the biological test of their tumor-forming ability or by the use of fluorescence-activated cell sorting (FACS; Section 11.6). As it happens, Gleevec is quite potent in killing actively cycling leukemia cells (i.e., the "transit-amplifying" or "progenitor" cells). However, most of the cells in the neoplastic stem cell population, which are outside the active cell cycle at any single point in time, have proven to be quite resistant to drug treatment. Consequently, following the cessation of treatment, such cancer stem cells may re-enter into the growth-and-division cycle and regenerate transit-amplifying progeny, leading to regrowth of a tumor and clinical relapse (Figure 16.30). This seems to explain why Gleevec treatment needs to be chronic and why, in the future, drug development needs to be focused on agents that strike at the core of tumors by destroying their stem cells (Sidebar 16.2).

16.12 EGF receptor antagonists may be useful for treating a wide variety of tumor types

The proposal to develop Gleevec initially met with considerable resistance in the pharmaceutical company where it originated, simply because the market for this drug was judged to be too small to justify the high costs of its development

Sidebar 16.2 Cancer stem cells greatly complicate the evaluation of anti-cancer therapies The existence of cancer stem cells in many solid tumors has profound implications for the evaluation of many types of anti-cancer treatments. For example, if a candidate drug is able to eliminate the stem cells in a tumor while leaving the bulk cancer cell population intact, the tumor mass may at first appear to be unaffected by the treatment, and will begin to shrink slowly only when these transit-amplifying cells gradually senesce and die in the weeks and months that follow. Although such a drug may have succeeded in delivering a death blow to the tumor, it may be judged to be unworthy of further development because of these initially minimal responses to treatment.

In fact, rituximab, the anti-CD20 monoclonal antibody that is used to treat B-cell tumors (Section 15.20), shows just this behavior: it eliminates the tumor stem cells of multiple myelomas but not the more differentiated, far more abundant antibody-secreting cells that form the bulk of the tumor. (Fortunately, early in its development, rituximab was recognized to have efficacy for treating a wide range of B-lymphocyte–lineage tumors and was therefore approved for further development and introduction into the clinic.)

Conversely, candidate drugs that kill only the transit-amplifying cells create the illusion of success: while a tumor will shrink substantially in response to treatment, it will rebound quickly once treatment is halted (see Figure 16.30). This is sometimes called the "dandelion effect," referring to the rapid re-emergence of weeds in a lawn following mowing, which cuts off their leaves but leaves their roots intact.

Nonetheless, if a drug such as Gleevec succeeds in generating clinical remissions that are durable over many years' time, its inability to kill the stem cells of a CML may represent an acceptable limitation. In any case, Gleevec represents a major triumph of anti-cancer drug development, because it is vastly superior to all alternative treatments of this otherwise inexorably progressive disease.

and testing in the clinic. The same could not be said of the class of drugs designed to inhibit the epidermal growth factor receptor (EGF-R). Carcinomas are common tumors, and this receptor is believed to play a key role in the development of as many as one-third of them, being frequently overexpressed.

At least six distinct EGF-related ligands, including EGF itself, have been found to bind and activate the EGF-R. This means that even in those carcinomas in which the EGF-R is present but not overexpressed, it may nonetheless emit critical oncogenic signals through the actions of active autocrine or paracrine signaling loops driven by the presence of one or more of its ligands. Moreover, in breast carcinomas that overexpress the HER2/Neu receptor, the oncogenic actions of this protein may depend on its ability to form heterodimers with its cousin, the EGF-R; in such heterodimers, the EGF-R can phosphorylate HER2/Neu, thereby activating signaling by the latter.

The best-characterized inhibitors of the EGF-R tyrosine kinase are the drugs Iressa, also known as gefitinib and ZD1839, and Tarceva, also called erlotinib and OSI-774 (Figure 16.31A). The two drugs have very similar but not identical properties, and both have been mentioned in passing in several earlier sections in this chapter. These two drugs act by blocking the ATP-binding site of the receptor-associated kinase (Figure 16.31B; see also Figure 16.14).

Once cancer cells are deprived of receptor signaling through inhibition of the EGF-R, they should lose the benefit of the strong mitogenic and anti-apoptotic signals emitted by this receptor. For example, in many types of epithelial cells, the continuous firing of the EGF-R sustains expression of $Bcl-X_L$ (the potently anti-apoptotic cousin of Bcl-2) and, acting via MAPK, drives phosphorylation and attendant functional inactivation of the pro-apoptotic Bad protein.

Because Iressa and Tarceva target a cell-surface receptor, their therapeutic utility must be compared with that of the monoclonal antibodies that also affect this receptor (Sidebar 15.4). In principle, these low–molecular-weight compounds should be able to penetrate into all the interstices of a solid tumor, including those where the far larger antibody molecules may have trouble gaining access (Table 16.3). Also, it is generally far easier and less expensive to produce low–molecular-weight compounds on an industrial scale than it is to generate large amounts of monoclonal antibodies.

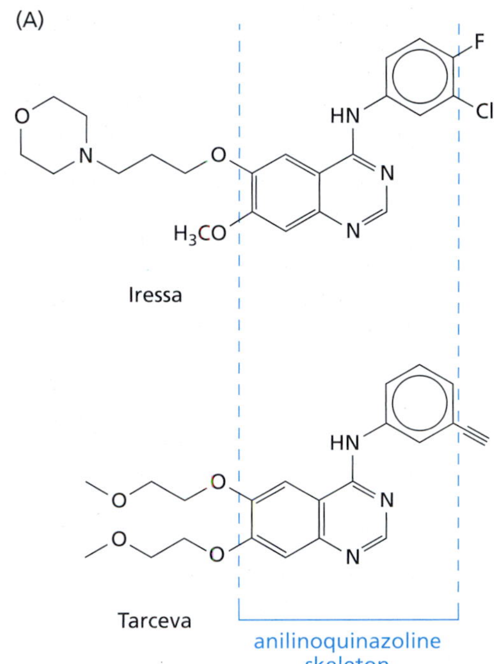

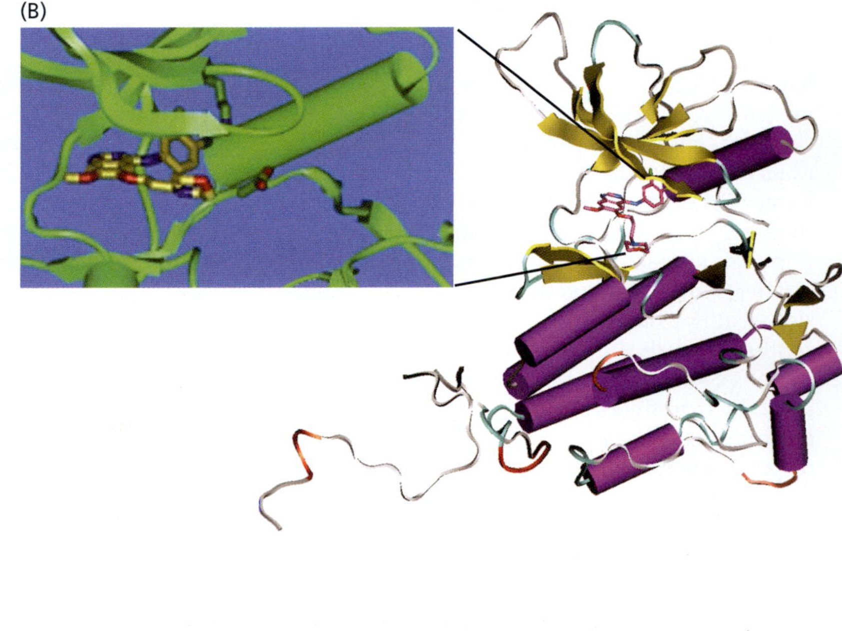

Figure 16.31 Iressa and Tarceva (A) The two epidermal growth factor receptor (EGF-R) antagonists are constructed from a common anilinoquinazoline skeleton, which confers on them an affinity for the ATP-binding site of the receptor tyrosine kinase. The chemical side groups that are attached to this skeleton have biological effects, since the two drugs have differing efficacies in treating, for example, non-small-cell lung carcinomas (NSCLCs). (B) Iressa, also termed ZD1839, binds to a very similar region of the EGF-R tyrosine kinase as does Tarceva (see Figure 16.14). A blown-up view *(left)* of the drug-binding site of the EGF-R tyrosine kinase domain *(right)* is shown here, with the drug molecule shown as a colored stick figure. This binding is so strong that 50% inhibition of TK enzyme activity is achieved at a concentration of about 0.030 μM. (B, courtesy of A.C. Kay, AstraZeneca.)

There are yet other possible advantages of low-molecular-weight tyrosine kinase inhibitors. For example, as we have read, in many human carcinomas, truncated forms of the EGF-R are expressed that lack the normally present ectodomain; these mutant EGF-Rs may signal in a ligand-independent, constitutive fashion and therefore can function as potent oncoproteins (Figure 5.11). Similarly, about half of high-grade (i.e., advanced) gliomas, termed glioblastoma multiforme (GBM), exhibit overexpressed EGF-R, and of these, about 40% display a form of the receptor that lacks the ectodomains specified by exons 2 through 7 of the EGF-R coding sequence. Such decapitated receptors cannot be bound by the monoclonal antibodies (MoAbs) that have been developed to recognize antigenic **epitopes** present in the ectodomain of the normal receptor protein. However, these aberrations should not derail the low-molecular-weight tyrosine kinase inhibitors, which target the cytoplasmic, signal-emitting domain of the receptor.

Weighing against these drugs are their pharmacokinetic properties: compounds like Iressa may have lifetimes in the circulation that are often measured in hours to days, whereas therapeutic monoclonal antibodies may persist for weeks in the circulation. As is the case with all such agents, extensive measurements of the steady-state concentrations of Iressa in the circulation were required in order to ensure adequate dosing of tumor cells *in vivo* (Figure 16.32). If taken daily, effective concentrations were found to be maintained with half-lives of the drug as long as 40 hours or more.

Iressa has approximately a 50-fold more potent activity against the EGF-R–associated tyrosine kinase than against a number of other tyrosine kinases that have been tested (see, for example, Figure 16.16B), and its initial use in the oncology clinic was reasonably encouraging. In the first clinical trials, 10% of patients with non-small-cell lung carcinomas (NSCLCs) showed **partial responses** to the drug including disease stabilization of tumor growth; these patients tended to be women, nonsmokers and those with the bronchioalveolar subtype of lung cancers. A parallel study in Japan found a far higher rate (27%) of partial responses to Iressa; this difference, which continues to be observed, appears to represent a difference in the genetic constitutions of the Japanese and Caucasian populations. (The tumors classified as NSCLCs have represented difficult diseases to treat, as fewer than 15% of patients survive for five years following initial diagnosis.)

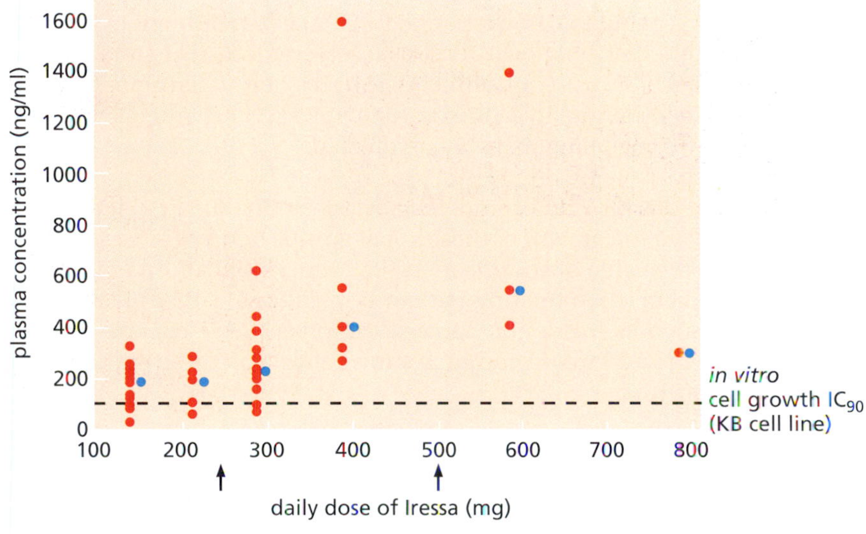

Figure 16.32 Pharmacokinetics of Iressa In a Phase I trial of Iressa, steady-state concentrations of the drug in the plasma were achieved after 7 days of daily dosing. There was substantial inter-patient variability in the steady-state concentrations achieved by different daily doses *(red circles)*; the geometric mean concentration among this group is indicated by *blue circles*. The concentration of Iressa needed to inhibit 90% of proliferation of KB human epidermoid carcinoma cells (i.e., IC_{90}) growing *in vitro* is given by the dashed line (~100 ng/ml). Hence, even the smallest administered clinical dose produced higher mean concentrations in the plasma. Daily doses of 250 and 500 mg/day *(arrows)* were ultimately chosen for the subsequent Phase II trial. (From R.S. Herbst, A.M. Maddox, M.L. Rothenberg et al., *J. Clin. Oncol.* 20:3815–3825, 2002.)

These outcomes were gratifying, if only because they represented clear responses in patients who otherwise had few if any other treatment alternatives. However, the hoped-for synergistic actions of Iressa with standard chemotherapeutic agents did not provide any survival advantage over standard chemotherapy alone for the treatment of NSCLC tumors, which constitute almost 80% of lung cancer cases in the United States. When used on its own, Tarceva (but not Iressa) increased the overall survival time of patients whose NSCLCs had become refractory to treatment by standard chemotherapeutic drugs.

Some valuable lessons were learned from these initial trials that may improve the responses in subsequent clinical trials of these and similar drugs: First, the specific contribution of the EGF-R to the growth of the tumors under treatment was not documented. Hence, far greater response rates might have resulted from a stratification of the NSCLC patients and limiting the use of Iressa only to those tumors having specific molecular signatures.

Second, the possible contribution of other mutant proteins to mitogenic and anti-apoptotic signaling was not assessed. There is evidence, for example, that PTEN-negative tumors (which have a hyperactive PI3 kinase pathway; see Section 6.6) do not respond to Iressa, and that inhibitors of Akt/PKB (the downstream beneficiary of PTEN inactivation) can act synergistically with Iressa to halt tumor growth. Third, relatively few pre-clinical studies were undertaken in order to optimize the dosage and the schedule of treatments with this drug.

In 2004, four years after results of the initial clinical trials with Iressa were first reported, two research collaborations in Boston independently provided a molecular explanation for the observed responses to Iressa. Previously, the status of the EGF receptor in NSCLC cells was assessed by determining whether it was overexpressed and whether it was present in truncated, constitutively active form, as is the case in human glioblastomas. In the 2004 studies, however, investigators undertook detailed sequencing of the reading frames of the EGF-R–encoding gene in the NSCLC patients who had been treated with Iressa.

Quite dramatically, they found that almost all of the small group (~10% of the total) of NSCLC patients who had responded well to Iressa treatment (Figure 16.33A) bore tumor cells displaying structurally altered EGF-Rs. Such mutant receptors were not found among the tumors that failed to respond to Iressa, including those that expressed elevated levels of this receptor (see, for example, Sidebar 51 ◉). The responsible mutations created amino acid substitutions and small deletions in the kinase domain (Figure 16.33B) rather than the major deletions of the receptor ectodomain typically found in glioblastomas. Similarly, the tumors in 5 of 7 patients who responded to Tarceva treatment expressed such mutant receptors. For unknown reasons, these mutant receptors showed distinctive patterns of tyrosine phosphorylation of their C-terminal tails (Section 6.3) and selectively stimulated the downstream Akt/PKB and STAT5 pathways, leaving the MAPK signaling pathway unaffected.

These observations provided compelling evidence that the EGF-R played a central role in driving the growth of these small groups of tumors. In addition, they demonstrated the value of stratification (i.e., subclassification) of tumors using molecular markers when treating patient populations with targeted molecular therapeutics, such as Iressa and Tarceva. However, these experiments did not reveal why Iressa and Tarceva had such strong effects on these particular tumors (but see Sidebar 16.3).

Still, virtually all of these successes have been short-lived, and most patients relapsed in 6 to 18 months, having developed a resistance to drug treatment. This underscores, once again, the need for alternative agents to treat drug-resistant receptors, and for multi-drug therapy, in which several drugs with synergistic effects are applied simultaneously. Indeed, several drug molecules that are structurally distinct from the anilinoquinazolines (Figure 16.31A) are able to shut down Iressa- or Tarceva-resistant EGF-Rs, providing hope for those patients who have relapsed following extended treatment with one or another of these two drugs.

16.13 Proteasome inhibitors yield unexpected therapeutic benefit

Serendipity plays an unusually prominent role in the world of drug discovery. On occasion, the development of an anti-cancer drug is launched as part of a rational drug design program and ultimately yields an agent that turns out to be highly useful, albeit for reasons quite unrelated to those that inspired its

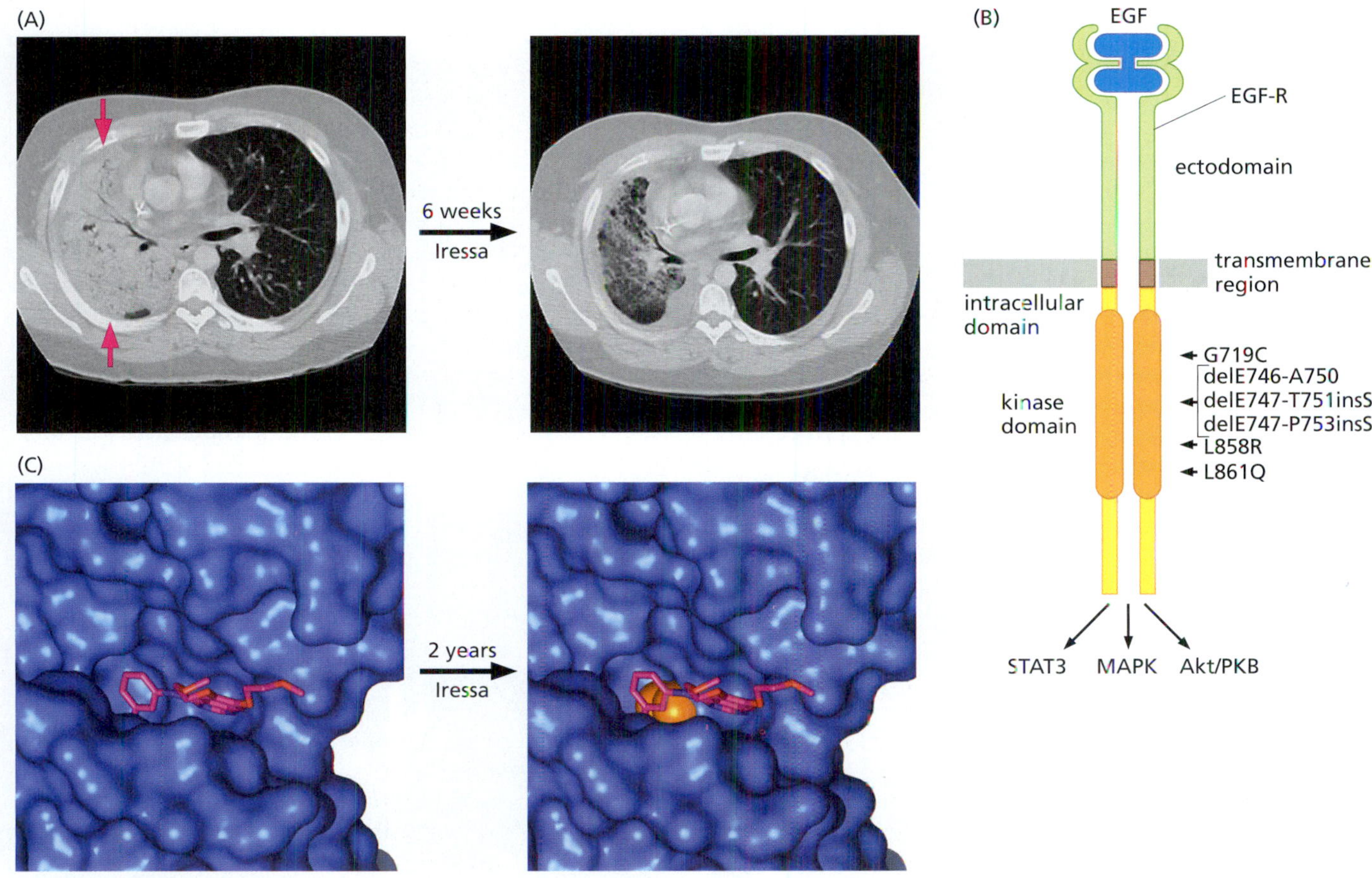

Figure 16.33 NSCLCs responsive to Iressa treatment
(A) A minority of patients with refractory non-small-cell lung cancers (NSCLCs)—tumors that have failed to respond or no longer respond to standard chemotherapy—show dramatic responses to treatment by Iressa. These computed tomographic X-ray images reveal a large mass in the right lung *(left)* of a patient that underwent dramatic regression following six weeks of Iressa treatment *(right)*. (B) A substantial proportion of NSCLCs that respond to Iressa treatment have been found to carry mutations in the gene encoding the EGF-R that affect the cytoplasmic domain of the receptor and include both deletions ("del") and point mutations. These alterations in EGF-R structure deregulate and activate the tyrosine kinase function of the receptor, thereby stimulating the downstream Akt/PKB and STAT signaling pathways, which protect these tumor cells from apoptosis. (C) Eventually, patients with some of the indicated mutations relapse from Iressa or Tarceva therapy. As is the case with acquired resistance to Gleevec (see Figure 16.26B), the EGF-Rs in these NSCLC patients often acquire structural changes that block drug binding. Here a patient, whose tumor-associated EGF-R showed a delE747-P753insS mutation (B), enjoyed a remission achieved by Iressa therapy; after two-years, however, his tumor regrew. Sequencing of the EGF-R gene in the relapsed tumor showed that the binding site present in the wild-type receptor *(left)* was now partially occluded by a threonine-to-methionine substitution *(right)*, causing the bulkier methionine side-chain *(orange sphere)* to block binding of Iressa. (A and B, from T.J. Lynch, D.W. Bell, R. Sordella et al., *N. Engl. J. Med.* 350:2129–2139, 2004; C, from S. Kobayashi, T.J. Boggon, T. Dayaram et al., *N. Engl. J. Med.* 352:786–792, 2005.)

Sidebar 16.3 Oncogene addiction may explain how Iressa and Tarceva succeed in killing NSCLCs The mutant EGF-Rs that are found in certain NSCLCs cause the cells from these particular tumors to be approximately 100 times more sensitive to Iressa than tumors expressing the wild-type receptors (Figure 16.34A). Moreover, the actual drug concentrations in the plasma of patients being treated fall in the range that allows such selective inhibition to operate.

The mechanism of "oncogene addiction" may explain the selective effects of both EGF-R inhibitors on tumors expressing mutant EGF-Rs. Oncogene addiction refers to the fact that certain cancer cells appear to be particularly dependent on a certain oncogene or oncoprotein for their growth and survival, while other tumors can lose this gene or protein without suffering any obvious consequences.

To explain this behavior, we can imagine that some oncogenes are generally deleterious when expressed in wild-type cells but are actually beneficial on cells

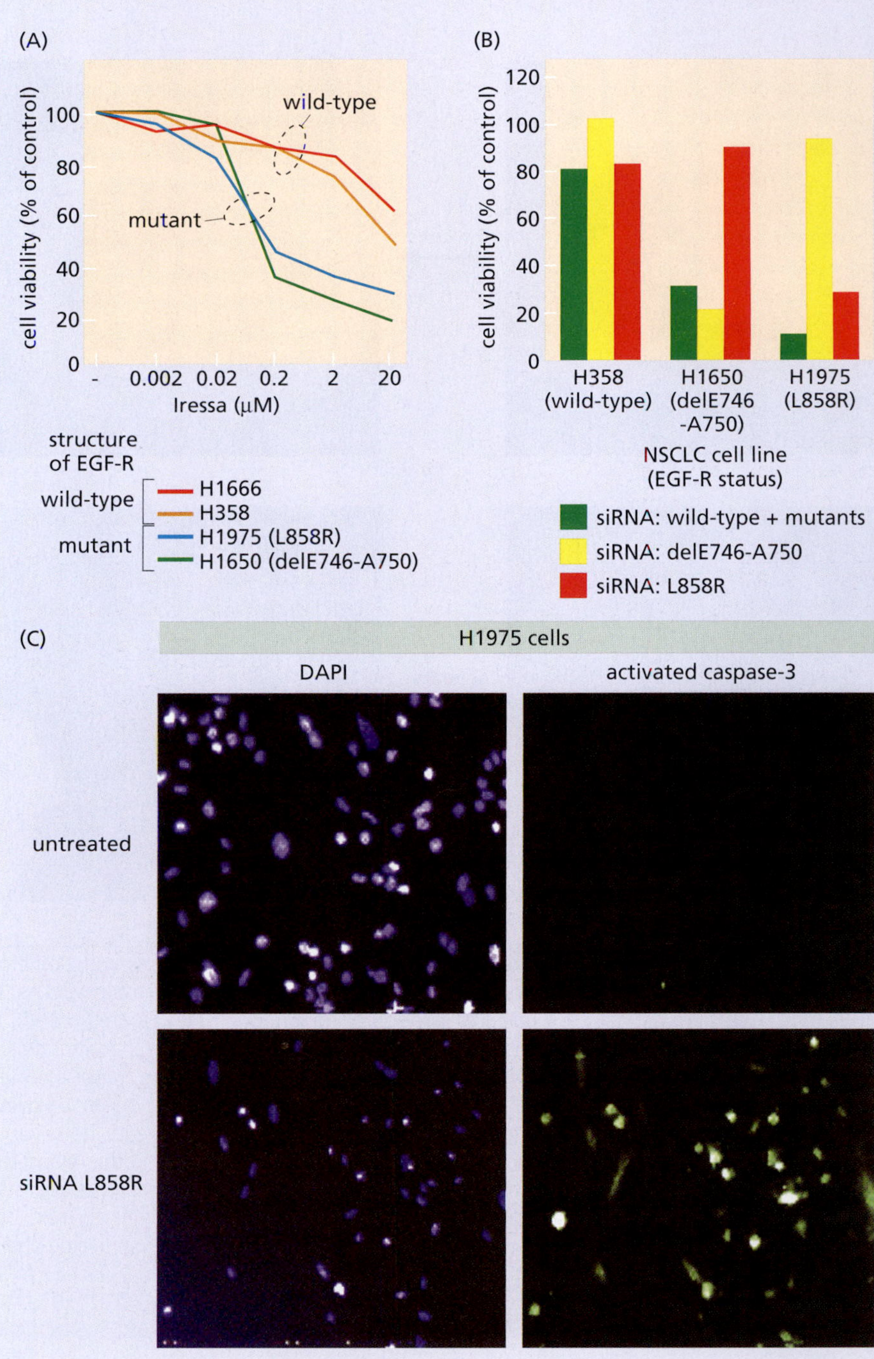

that previously acquired certain mutant alleles. A good example is provided by the *myc* oncogene, which has pro-apoptotic effects on cells unless they have been protected from apoptosis by some other, previously acquired anti-apoptotic allele (e.g., a *ras* oncogene); in the presence of the anti-apoptotic mutation, the strongly mitogenic effects of the *myc* oncogene then become apparent. Hence, tumor cells that carry both the *ras* and *myc* oncogenes would behave as if they were "addicted" to the *ras* expression, since they would die quickly by apoptosis if they were deprived of the *ras* oncogene.

Accordingly, early in tumor progression, the acquisition of a certain oncogene, such as a mutant EGF-R gene, might create a cellular environment that permits acquisition of other oncogenes (or losses of tumor suppressor genes) that would otherwise be highly deleterious for tumor cells. If the mutant receptor is now lost, then the deleterious effects of these other oncogenes, notably those favoring apoptosis, would become apparent and result in rapid loss of cell viability.

In the case of NSCLC, those tumors bearing mutant receptors may have come to depend on the firing by their mutant EGF-Rs in order to survive and proliferate; i.e., they are "addicted" to the mutant receptors. Conversely, the far more numerous NSCLCs expressing the wild-type EGF-R may have developed alternative means of securing mitogenic and survival signals, as is indeed suggested by the observation of receptor-independent firing of the MAPK and PI3K pathways in some lung cancers.

This scenario is further supported by experiments using siRNAs to inhibit the expression of wild-type or mutant receptors: NSCLC cells with mutant EGF-R die quickly, while those displaying wild-type receptor are only slightly affected (Figure 16.34B and C). Consequently, the death of cancer cells with mutant EGF-Rs is not due to some unknown, off-target effect of Iressa or Tarceva, but instead is caused directly by the loss of beneficial signals released by these receptors. Moreover, experiments like these suggest that EGF-R inhibitors may have far greater effects on NSCLCs expressing wild-type receptors if they are applied together with a second drug that inhibits another, functionally redundant signaling pathway, such as the one controlled by PI3K.

Figure 16.34 Effects of siRNAs to suppress EGF-R expression It is unclear why non-small-cell lung carcinoma (NSCLC) cells bearing structurally altered EGF receptors (see Figure 16.33) are especially responsive to Iressa or Tarceva therapy. (A) When cultured *in vitro,* two NSCLC cell lines that overexpress structurally normal (i.e., wild-type) EGF-R are relatively resistant to Iressa treatment *(orange, red),* while two NSCLC lines expressing either an amino acid–substituted *(blue)* or a partially deleted *(green)* receptor protein were approximately 100 times more sensitive to killing. (B) The biological mechanism of these differences could be examined by depriving NSCLC cell lines of EGF-R by expressing siRNAs in these cells; these siRNAs cause degradation of the EGF-R mRNA. H358 cells *(left 3 bars)* expressing wild-type EGF-R were relatively unaffected by siRNAs directed either against all forms of the receptor *(green)* or against the two mutant forms *(yellow, red).* However, an siRNA directed specifically against the mRNA encoding a deleted form of the receptor (DelE746; *yellow)* caused loss of viability in some 80% of the NSCLC cells expressing this mutant receptor (*middle bars*), while having no effect on cells with an amino acid–substituted receptor *(red, middle bars).* Conversely, loss of viability was observed when an siRNA directed against the amino acid–substituted receptor *(red)* was expressed in cells that express this particular mutant receptor *(right bars)* but not when the siRNA directed against the deletion mutant was used. In the case of both mutant cell lines, the siRNA directed against all forms of the receptor *(green bars)* also caused widespread cell death. Hence, the two NSCLC cell lines with mutant receptors were dependent on ("addicted to") EGF-R function, while the NSCLC cells with wild-type receptor showed virtually no dependence on continued EGF-R function. (C) The loss of viable cells after siRNA treatment seen in panel B is due specifically to an induction of apoptosis, as revealed by immunostaining of fixed cells with an antibody reactive with cleaved, activated caspase-3. NSCLC cells were alternatively stained with 4′,6′-diamidino-2-phenylindole (DAPI) to reveal nuclei and thus cell number. (From R. Sordella, D.W. Bell, D.A. Haber and J. Settleman, *Science* 305:1163–1167, 2004.)

development in the first place. This best describes the development of the drug known as Velcade, also called PS-341 and bortezomib (Figure 16.35A).

On many occasions throughout this book, we have seen how the levels of key cellular regulatory proteins are determined by the balance between their synthesis and their degradation. Much of this degradation is mediated by the ubiquitin–proteasome system (Sidebar 7.8). Recall that the tagging of a protein by polyubiquitylation results in its transport to proteasomes and its degradation in these intracellular machines.

The phenomenon of cancer-associated **cachexia** initially stimulated interest in inhibitors of proteasome function. Cachexia occurs late in tumor progression and represents a progressive wasting of the cancer patient's tissues through mechanisms that remain poorly understood. Use of a proteasome inhibitor was speculated to be useful in retarding the widespread degradation of proteins occurring in the tissues of cachectic patients. While at least five distinct classes of proteasome inhibitors have been developed, most of these have been abandoned because of metabolic instability, lack of specificity, or irreversible binding and inactivation of proteasomes. Velcade, one of these proteasome inhibitors, is a boronic acid dipeptide that was designed as a specific inhibitor of the **peptidase** (peptide-cleaving) activity present in the 20S core of the proteasome (Figure 16.35B). It has extraordinary potency, since it is able to inhibit 50% of the proteasome's chymotryptic activity at a concentration (i.e., its K_i) of only 0.6 nM. Functioning as a competitive inhibitor of this enzyme activity, Velcade slows down the flux of substrates through proteasomes, which soon become clogged and dysfunctional.

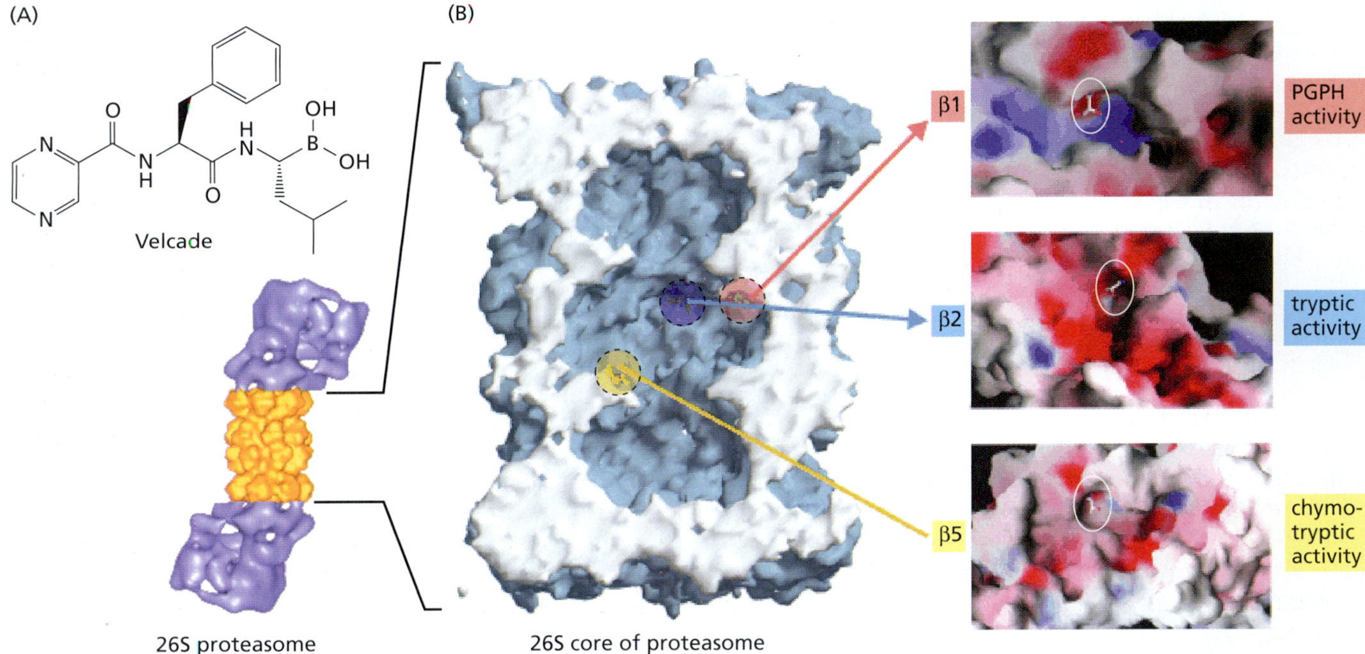

Figure 16.35 Velcade and its effects on proteasomes (A) The chemical structure of Velcade reveals the presence, unusual among drugs, of a boron atom. Peptide boronic acids like Velcade were known to bind to the active site of serine proteases of the chymotrypsin class (which cleave substrate proteins adjacent to phenylalanine and tyrosine residues) by mimicking the normal substrates of these enzymes. This suggested that such compounds might inhibit the chymotrypsin-like active site in the 20S core of the proteasome. (B) A cross section (*center*) through the central (20S) core of the yeast proteasome (*left;* see also Figure 7.27) reveals the locations of three distinct catalytic sites, involving the β1, β2, and β5 subunits, which are responsible for its PGPH (peptidyl-glutamyl-peptide hydrolyzing; *pink*), tryptic *(light blue)*, and chymotryptic *(light yellow)* proteolytic activities, respectively. Velcade shows a strong preference for inhibiting the β5 chymotryptic activity, whose detailed structure is seen here *(lower right);* a weak interaction with the β2 tryptic activity *(middle right);* and no interaction with the β1 PGPH activity *(top right)*. The key nucleophilic threonine residue present in each of these catalytic sites is shown as a stick figure *(white inside white ovals)*; basic amino acids in the catalytic clefts *(right)* are shown in *blue,* acidic amino acids are shown *red,* and hydrophobic residues are shown in *white.* (B, from M. Groll, M. Bochtler, H. Brandstetter et al., *Chembiochem* 6:222–256, 2005.)

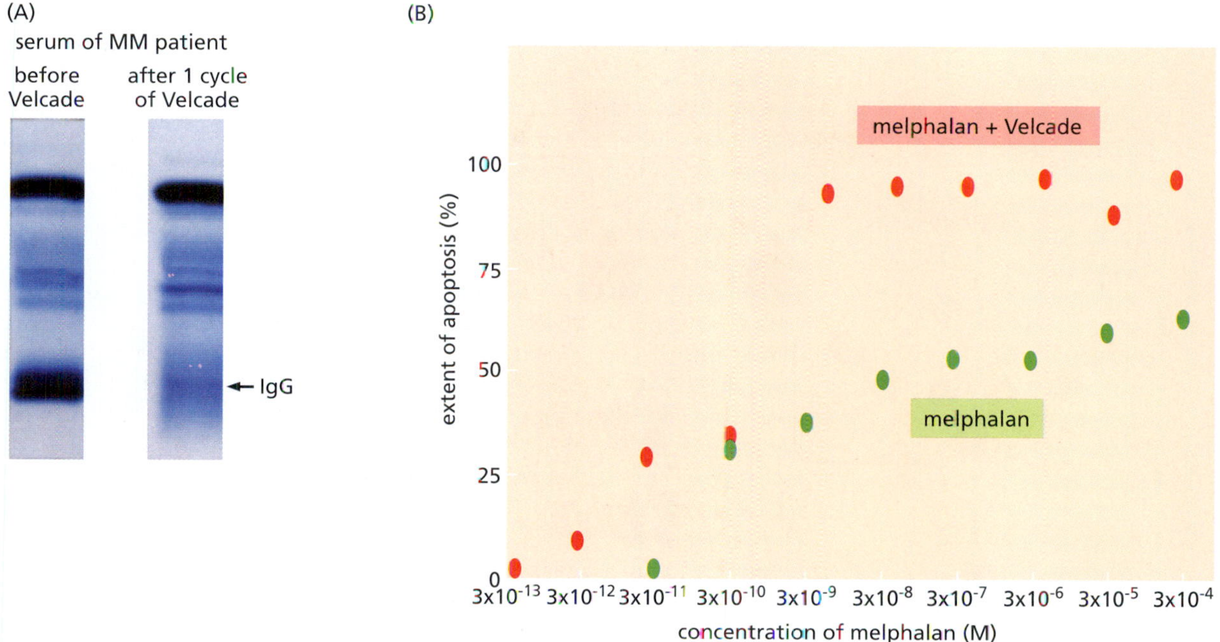

Figure 16.36 Multiple myeloma and the biological effects of Velcade (A) Velcade can have profound effects on the cellular contents of the marrow and thus the composition of antibody molecules in the blood. In one multiple myeloma (MM) patient, after one cycle of eight doses of Velcade, the neoplastic plasma cells in the marrow declined from 41% of total cells to 1%. At the same time, there was a precipitous decline in the level of the single species of immunoglobulin γ (IgG) made by these myeloma cells. As seen here, upon gel electrophoresis, the small number of IgG species present before treatment (indicative of a monoclonal tumor, *left*) resolved into the heterogeneously migrating, polyclonal pattern of IgGs that are present in the circulation of a healthy individual (*right*; see also Figure 2.19A). (B) Melphalan, an alkylating chemotherapeutic drug used routinely to treat MM, was added at various concentrations to an MM cell line *in vitro*, either on its own *(green)* or in the presence of a noncytotoxic dose of Velcade *(red)*. In the presence of Velcade, melphalan was able to induce widepsread apoptosis at approximately 3 nM concentration, while when applied on its own, melphalan is unable to induce this degree of apoptosis, even at vastly higher concentrations. (A, from R.Z. Orlowski, T.E.Stinchcombe, B.S. Mitchell et al., *J. Clin. Oncol.* 20:4420–4427, 2002; B, from M.H. Ma, H.H. Yang, K. Parker et al., *Clin. Cancer Res.* 9:1136–1144, 2003.)

Proteasome-mediated degradation was subsequently found to play a critical role in regulating a number of key cellular signaling pathways, broadening the horizons of Velcade's developers. Because other proteasome inhibitors had been found to be especially potent in killing a variety of cultured cancer cells, Velcade was used in early-phase (i.e., Phase I) trials to treat cancer patients who had failed other available therapies. Those with solid tumors showed few striking responses. However, among a group of patients with hematologic malignancies was one suffering from multiple myeloma (MM), a malignancy of the B-cell lineage in which a single clone of antibody-producing plasma cells dominates the bone marrow (see Figure 2.19A). The myeloma cells create osteolytic bone lesions leading to fractures and ultimately crowd out the remaining cellular components of the marrow, resulting in severe immune depression and, typically, death from overwhelming infection. Survival after initial diagnosis is usually three to five years. The myeloma carried by this initially treated patient showed a dramatic regression (see, for example, Figure 16.36A), which soon led to inclusion of other myeloma patients in this Phase I trial and eventually to large-scale clinical trials in which myeloma patients were treated with Velcade.

In a subsequent Phase I clinical trial with a group of multiple myeloma patients suffering from rapidly progressing disease, Velcade showed clear "objective responses" in slowing down disease progression in 55% of the patients and halted progression in another 25%. In a Phase II clinical trial, half of patients were given Velcade while the other half, who served as controls, were given dexamethasone, a standard treatment for multiple myeloma. The great majority of these patients had already failed the chemotherapies commonly used for myeloma. This trial was stopped prematurely in 2003 because the disease showed either a "complete response (CR)" in a small number of patients (i.e., myeloma cells disappeared completely from the blood for a period of at least 6 weeks) or a "partial response (PR)" (at least 50% reduction in myeloma cell–secreted antibody in blood and 90% reduction of this protein in urine over the same time period) in 35% of the Velcade-treated patients, thereby showing clear superiority over the existing treatments. As a consequence, the control patients were then allowed to take the drug as well. In a subsequent trial, the progression of myeloma to a higher stage of disease occurred with a median time of 7 months in Velcade-treated patients compared with 3 months in a control group studied in parallel. Moreover, pre-clinical studies indicate that relatively low doses of Velcade can sensitize myeloma cells to chemotherapeutic drugs, making the latter far more effective (see Figure 16.36B).

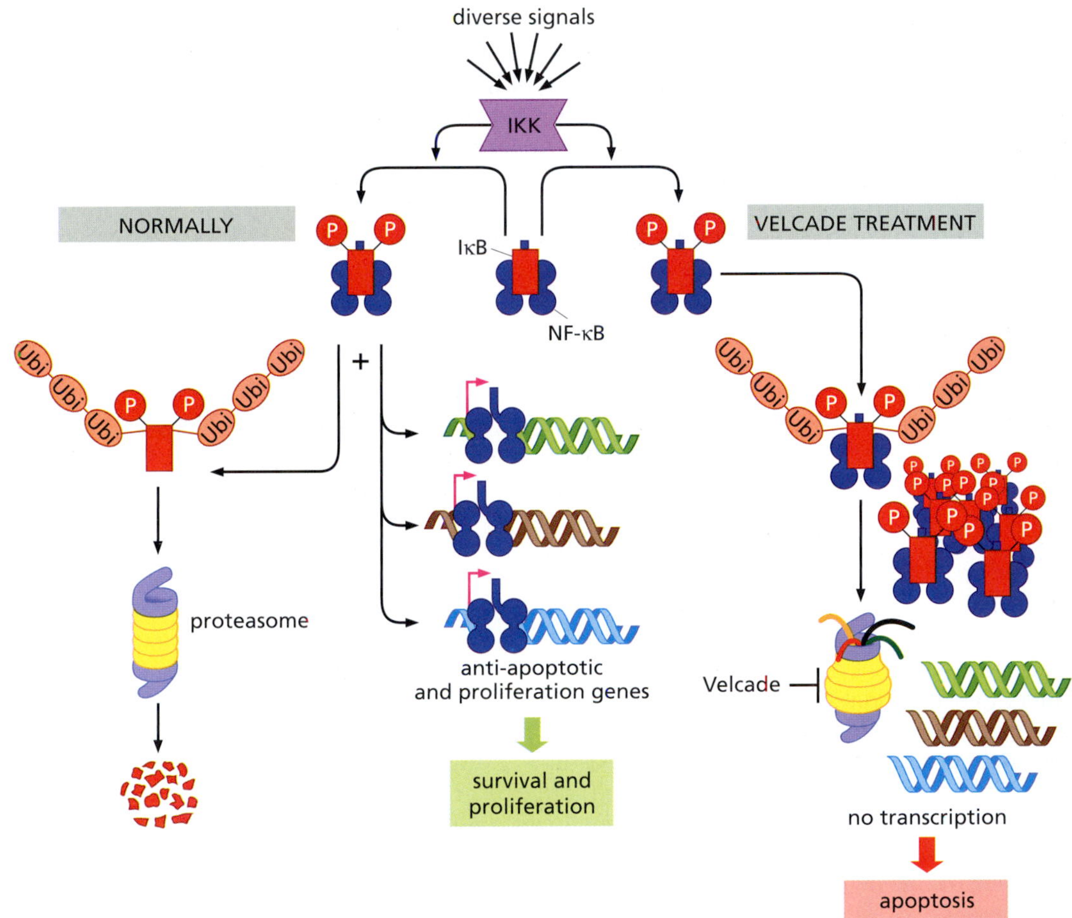

Figure 16.37 Mechanism of Velcade action The inhibition of signaling by the NF-κB transcription factor contributes significantly to Velcade's ability to induce the apoptosis of myeloma cells within the bone marrow. In normal cells *(left)*, a variety of stress, mitogenic, and trophic (survival) signals activate IκB kinase (IKK; *purple*); similar signaling operates in myeloma cells, as well as in a variety of other cancer cell types, in which various mechanisms are responsible for the constitutive activation of IKK. Once active, IKK proceeds to phosphorylate IκB (*red*), the inhibitor of NF-κB. This phosphorylation causes IκB to become ubiquitylated *(left)* and degraded in proteasomes *(lower left)*. In the absence of IκB (*center*), NF-κB *(blue)* is free to move into the nucleus, where it activates the expression of a large constituency of proliferation and anti-apoptotic genes. In the presence of Velcade *(right)*, the phosphorylation and ubiquitylation of IκB proceed normally, but the ubiquitylated IκB cannot be degraded in the proteasomes because the latter have become engorged with unprocessed polypeptides. This leads to an accumulation of IκB in the cytoplasm and to the continued sequestration of NF-κB by the IκB molecules that have built up *(right)*. As a consequence, NF-κB is prevented from moving into the nucleus and activating expression of its constituency of key anti-apoptotic genes. This tips the regulatory balance in myeloma cells in favor of apoptosis, resulting in the death of these cells. Velcade treatment may additionally lead to accumulation of p53 molecules, which are also suspected to contribute to the apoptosis of treated myeloma cells *(not shown)*.

In truth, inclusion of a myeloma patient in the initial clinical trial was hardly accidental. Multiple myeloma was thought to be an attractive target for treatment by a proteasome inhibitor because of the known elevated activity of the NF-κB signaling pathway in the myeloma cells and its physiologic importance in driving the survival and proliferation of these cells. In Section 6.12, we noted that NF-κB transcription factors are normally sequestered in the cytoplasm by a class of inhibitors termed IκBs (inhibitors of NF-κB). When these IκBs are phosphorylated by a group of specialized kinases termed IκB kinases, or simply IKKs, the IκBs undergo polyubiquitylation and resulting degradation; this liberates the NF-κBs, allowing them to migrate into the nucleus, where they activate a number of anti-apoptotic genes as well as growth-promoting genes (Figures 16.37 and 16.38; see also Figures 6.29A and 11.37B).

Like many other polyubiquitylated proteins, the IκBs end up being degraded in proteasomes. Hence, by inhibiting proteasome action, IκBs should be protected from degradation, survive in the cytoplasm, and continue to sequester the NF-κBs, thereby blocking NF-κB nuclear translocation and activation of transcription. Nuclear, functionally active NF-κB was known to be important for inducing the expression of IL-4 and IL-6, two interleukins that operate as important autocrine factors required for the growth and survival of myeloma cells. In addition, as was learned later, NF-κB plays a prominent role in anti-apoptotic signaling in a number of cancer cell types; hence, loss of active NF-κB might well tilt the signaling balance within these cells toward apoptosis. More specifically, once cancer cells lose the potently anti-apoptotic Bcl-2, cIAP-2, and XIAP proteins (all of whose expression is induced by NF-κB), they are in grave danger of slipping into the apoptotic abyss.

All this does not explain, however, why Velcade is far more potent against myelomas than other tumors that rely on NF-κB signaling to protect them from apoptosis. A possible clue comes from observations that the growth and viability of myeloma cells is highly dependent on their ability to synthesize VEGF (vascular endothelial growth factor; Section 13.1) and adhesion molecules; the latter enable myeloma cells to attach to bone marrow stem cells (BMSCs), with which the myeloma cells establish critically important heterotypic interactions. The genes encoding these various proteins are all under NF-κB control.

In addition, the malignant plasma cells that form the myeloma tumor synthesize and secrete large amounts of protein—antibody molecules. A certain portion of these molecules may be routinely degraded in the proteasomes because of misfolding or other mishaps occurring during their assembly. Consequently, the plasma cells may be especially sensitive to inhibitors of protein degradation and may rapidly become engorged with such defective protein molecules.

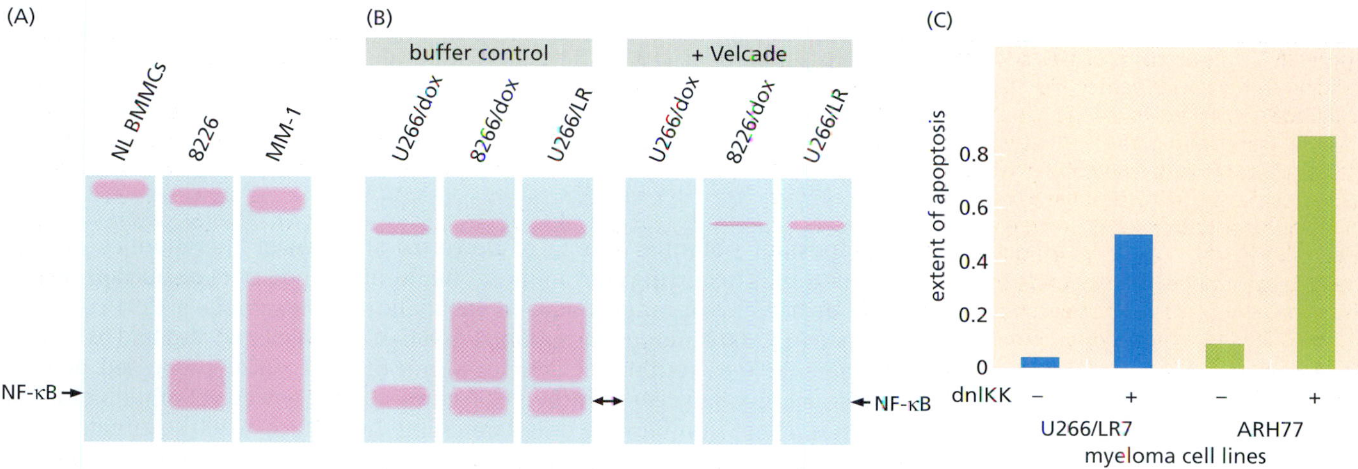

Figure 16.38 Evidence supporting the importance of NF-κB signaling in Velcade-induced apoptosis The scheme presented in Figure 16.37 is supported by a number of lines of evidence. (A) In an electrophoretic mobility shift assay (EMSA; also called a gel retardation assay), the presence and concentration of a functional, DNA-binding transcription factor (TF) are assessed by mixing an extract of nuclear proteins with a radiolabeled dsDNA oligonucleotide that carries a binding site for the TF. The presence of the DNA-binding TF is reflected by the amount of oligonucleotide that has formed a nucleoprotein complex with the TF; the large mass of protein associated with the oligonucleotide retards its migration during electrophoresis, causing it to migrate to a characteristic position in the gel. The arrow indicates the expected location of a complex containing the NF-κB transcription factor and the radiolabeled oligonucleotide, in this case one derived from the promoter of the TNF-α gene, a target of NF-κB activation. The assay indicates little if any detectable NF-κB activity in normal bone marrow mononuclear cells (NL BMMCs) and considerable activity in a multiple myeloma (MM) cell line (8226) and an enormous amount of NF-κB activity in the bone marrow cells prepared directly from a multiple myeloma patient (MM-1). (B) An EMSA has been used, as in panel A, to measure the level of functional NF-κB transcription factor in three MM cell lines treated with a control buffer *(left three channels)* or with Velcade *(right three channels)*. Velcade is able to eliminate essentially all NF-κB activity in these cells. (C) The importance of ongoing NF-κB signaling to the survival of MM cells is demonstrated by this experiment, in which a vector expressing a dominant-negative IKK (dnIKK) has been introduced into two different MM cell lines. If NF-κB signaling were critical to the action of Velcade, then the dnIKK should mimic the effects of Velcade by inducing MM cell apoptosis (see Figure 16.37)—the outcome that is indeed observed here. A vector that does not express dnIKK was used as control here. (From M.H. Ma, H.H. Yang, K. Parker et al., *Clin. Cancer Res.* 9:1136–1144, 2003.)

These consequences of suppressing proteasome function may well explain much of Velcade's potent effect against myeloma cells, some but not all of which depends on perturbation of NF-κB signaling. Still, when viewed in another light, the rationale of using Velcade to inhibit proteasome function makes no sense as a therapeutic strategy: since proteasomes are used by cells to degrade a diverse array of cellular proteins—likely many thousands of distinct protein species—inhibition of proteasome function should be highly toxic for all types of cells throughout the body. Hence, agents that function as proteasome inhibitors are unlikely to have a significant therapeutic index, in that they are apt to harm normal cells as much as cancer cells.

In spite of this logic, low concentrations of Velcade exhibit great potency in killing cancer cells *in vitro*, while having minimum or tolerable toxicity against cells from various normal tissues. This drug can kill a variety of cultured human cancer cells when applied in concentrations ranging from tens to hundreds of nanomolar. Still, its potency in killing myeloma cells—doing so at concentrations as low as 1 nM—is truly amazing.

NF-κB antagonists are likely to have utility for treating a number of other kinds of cancer. Recall that NF-κB plays a key role in the development of a variety of carcinomas (Section 11.16) and may be required for the maintenance of these tumors once they are formed. In addition, the use of gene expression arrays (Section 16.1) has revealed that diffuse large B-cell lymphomas (DLBCLs), which appear under the microscope to constitute a single, homogeneous type of tumor, can actually be classified into three distinct subgroups (see Figure 16.5). Tumor cells belonging to the activated B-cell and mediastinal lymphoma subgroups have a constitutively activated IKK. Consequently, tumors belonging to these two DLBCL subgroups, as well as a variety of other tumors with hyperactive NF-κB signaling, have become attractive targets for treatment by either Velcade or a number of IKK inhibitors that have been developed by pharmaceutical companies.

16.14 A sheep teratogen may be useful as a highly potent anti-cancer drug

An important potential source of powerful anti-cancer therapeutics derives from naturally occurring compounds. The number of distinct **natural products** made by bacteria or fungi is beyond calculation. For example, a 1994 compilation listed 11,900 different compounds that had been isolated and had exhibited antibacterial (i.e., antibiotic) activity, some of which also possessed activity against mammalian cells. Another 3000 compounds showed yet other biological activities. These numbers are only scratching the surface: a 2001 estimate of the number of distinct, biologically active compounds made by the *Streptomyces* genus of bacteria ran into the hundreds of thousands. A portion of these are likely to possess cytostatic or cytotoxic powers against mammalian cells including cancer cells. The universe of biologically active, plant-derived compounds is even less well explored.

In all these cases, the forces of evolution, rather than the cleverness of synthetic organic chemists, have generated molecular species that are potent and highly specific pharmacologic agents. Many of these molecules seem to be used by the organisms that make them to attack competitors or to defend against predators. Since the number of these naturally occurring agents is beyond reckoning, they are likely to be the sources of novel anti-cancer agents for decades to come.

An illustrative example of such a natural product comes from the discovery of cyclopamine, a natural product of plant origin. This particular story starts with the observation that sheep flocks grazing in highland areas of the Western

United States occasionally showed epidemics of congenital malformations in lambs, many of which were stillborn. The most extreme of these malformations were those involving **cyclopia**—a single central eye. (The term comes from Cyclops, the one-eyed mythical giant, vanquished, along with many other enemies, by Ulysses.)

Veterinary detective work begun in the 1950s revealed that newborn lamb cyclopia was seen if pregnant ewes grazed on false hellebore, *Veratrum californicum* (Figure 16.39A), during day 14 of gestation. Yet other malformations, including cleft palate and shortened legs, were evident if the grazing occurred at earlier or later stages of pregnancy. By 1968, the **teratogenic** (malformation-inducing) effects of the false hellebore were traced to an alkaloid that came to be called cyclopamine (Figure 16.39B), which can induce cyclopia in a wide variety of organisms (Figure 16.39C).

Many of these deformities resembled a condition in humans termed holoprosencephaly, in which the development of bilaterally symmetrical structures in the embryonic head fails to proceed normally. Some human fetuses afflicted with this condition were found to carry inherited germ-line mutations in either the *PTC* (*patched*) receptor gene or the *SHH* (*sonic hedgehog*) gene, which encodes its ligand. As many as 23 distinct *SHH* mutations and 3 *PTC* mutations have been associated with this condition. (More generally, cyclopia is associated with about 1 in 250 spontaneously aborted human fetuses.) This provided that first clue that cyclopamine—a potent teratogenic agent—perturbs the Hedgehog-activated signaling pathway.

In this signaling pathway, a precursor of the Gli transcription factor is usually cleaved in the cytoplasm, enabling a cleavage product to move into the nucleus, where it acts as a transcriptional repressor (Figure 16.40; see also Sections 5.7 and 6.12). The plasma membrane protein Smoothened, which can protect the Gli precursor protein from this cleavage, is normally prevented from doing so because of poorly understood inhibition by the Patched (Ptc) receptor protein,

(A)

(B)

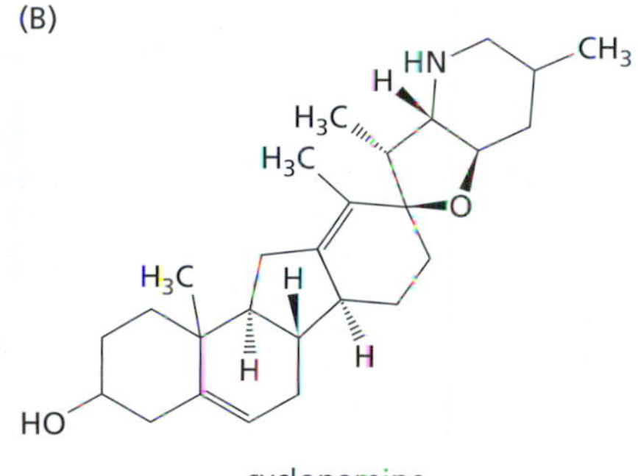

(C)

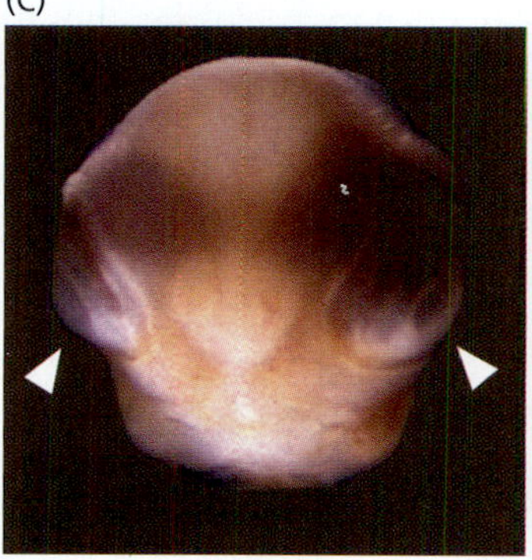

normal

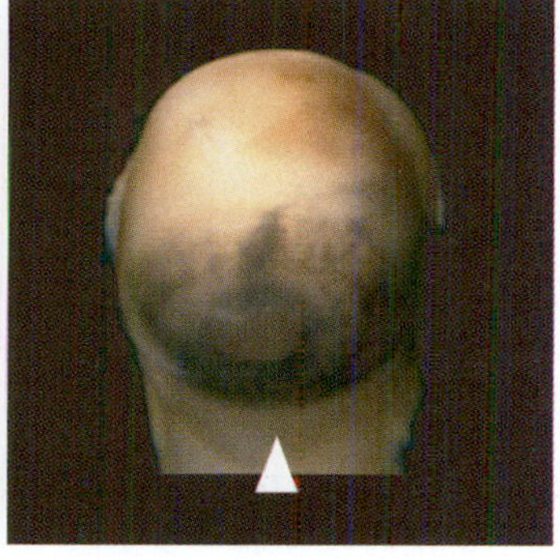

+ cyclopamine

Figure 16.39 False hellebore and its teratogenic product, cyclopamine (A) The plant *Veratrum californicum*, commonly called either the false hellebore or corn lily, grows in highland meadows in the American West. Pregnant sheep that graze on this plant often give birth to stillborn lambs with major morphogenetic defects in the head region, indicating the presence of a potent teratogen in this plant. (B) The teratogen has been found to be cyclopamine and is produced in a complex series of biosynthetic steps within the cells of the false hellebore. Its structure resembles that of a steroid, such as estrogen or progesterone. The complexity of its structure means that creation of this compound by using the techniques of synthetic organic chemistry is very challenging, effectively precluding the manufacture of this drug on an industrial scale. (C) Treatment of a chicken embryo with cyclopamine has dramatic effects on the formation of the head, as it does in all vertebrate embryos tested. As seen here, the normally present two eyes *(white arrowheads, left)* are replaced by the rudiment of a single central eye (cyclopia) following exposure to this drug *(white arrowhead, right)*. (A, from Henriette Kress; C, courtesy of J.P. Incardona and H. Roelink.)

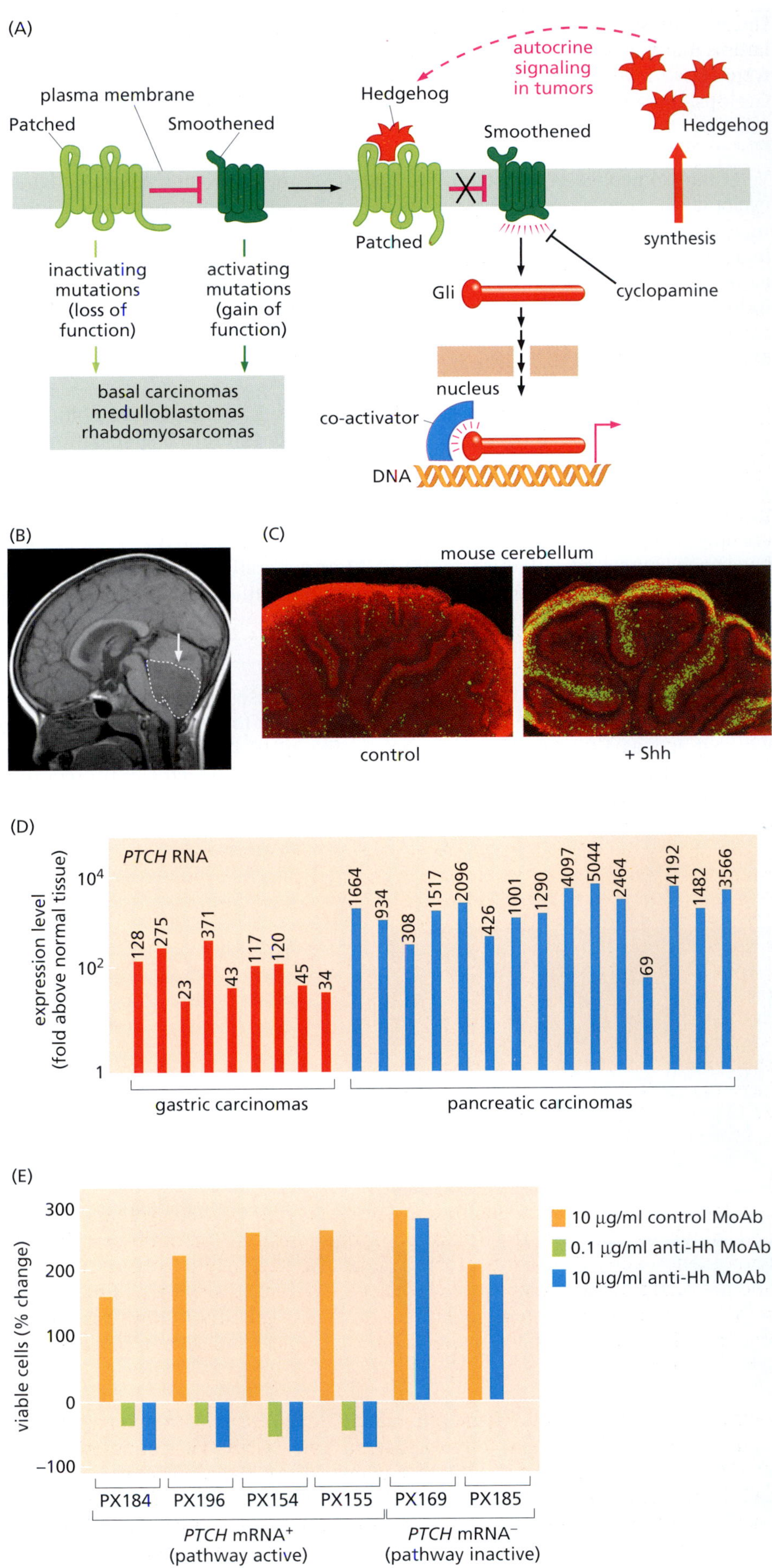
(A)
autocrine signaling in tumors
plasma membrane
Patched
Smoothened
Hedgehog
Smoothened
Hedgehog
Patched
synthesis
inactivating mutations (loss of function)
activating mutations (gain of function)
Gli
cyclopamine
basal carcinomas
medulloblastomas
rhabdomyosarcomas
nucleus
co-activator
DNA
(B)
(C)
mouse cerebellum
control
+ Shh
(D)
PTCH RNA
expression level (fold above normal tissue)
10^4
10^2
1
128
275
23
371
43
117
120
45
34
1664
934
308
1517
2096
426
1001
1290
4097
5044
2464
69
4192
1482
3566
gastric carcinomas
pancreatic carcinomas
(E)
viable cells (% change)
300
200
100
0
–100
10 μg/ml control MoAb
0.1 μg/ml anti-Hh MoAb
10 μg/ml anti-Hh MoAb
PX184
PX196
PX154
PX155
PX169
PX185
PTCH mRNA+ (pathway active)
PTCH mRNA– (pathway inactive)

Figure 16.40 Perturbations of the Patched–Smoothened signaling pathway (A) Normally, Patched inhibits Smoothened activity *(top left)*. However, when Patched binds Hedgehog (or a related ligand; *top right*), it no longer interacts with Smoothened, and thus allows the latter to activate Gli signaling, which results in the movement of Gli into the nucleus, where it can act as a growth-promoting transcription factor. The consequences of Hedgehog binding can be mimicked by mutations causing loss of Patched function or gain of Smoothened function—alterations often seen in the common basal cell carcinomas of the skin *(lower left)*. Many carcinomas of endodermal origin release a Hedgehog-type ligand *(top right*; e.g., Indian Hedgehog, Sonic Hedgehog) that acts in an autocrine fashion to drive their proliferation and survival. Moreover, many medulloblastomas trace their origins to dysregulation of this signaling pathway. Recent evidence implicates cyclopamine as an antagonist of Smoothened that prevents the latter from activating Gli (see Figure 16.41). (B) Medulloblastomas arise from the granular cell precursors of the cerebellum. Such a tumor, revealed here by computer tomography (CT) scanning *(arrow)*, is the most common brain tumor of children and often has devastating clinical outcomes. (C) The granular cells of the normal cerebellum respond strongly to the mitogenic actions of an added Hedgehog ligand, in this case Sonic Hedgehog (Shh). As seen here in these slices of the cerebellum of a young mouse, the proliferative activity of the granulosa cells at the surfaces of the cerebellar folds, which in the control slice *(left)* is minimal at this stage of development, is stimulated strongly by the addition of Shh *(right)*. In this case, proliferative activity *(green)* is measured by incorporation into DNA of bromodeoxyuridine (BrdU), a deoxythymidine analog. (D) The level of Patched provides a direct measure of the activity of the Hedgehog pathway (since Gli induces transcription of the *PTCH* gene). Here the levels of *PTCH* mRNA were measured by real-time PCR (RT-PCR) in a series of surgically resected human stomach and pancreatic carcinomas, and in each case, the level of *PTCH* mRNA expression was compared with that of corresponding normal tissue. The extent of overexpression (*number above each bar*) indicates a striking activation of Gli signaling (and thus of the Hedgehog pathway) in these tumors. (E) The importance of an autocrine signaling loop is indicated by the responses of six early-passage pancreatic carcinoma cell cultures, four of which showed active Gli signaling (being *PTCH* mRNA–positive, *left*) and two of which did not *(right)*. When added to cultures of these cells, 5E1, a monoclonal antibody that neutralizes both the Indian and Sonic Hedgehog proteins, exerted a strong effect in suppressing growth and viability of the carcinoma cells in which Gli signaling was high but had no effect on the carcinoma cells that had no Gli activity. This indicated the importance of an extracellular autocrine signaling loop, involving Hedgehog proteins, in driving the growth of these cells. (B, courtesy of St. Jude Children's Research Hospital; C, from R.J. Wechsler-Reya and M.P. Scott, *Neuron* 22:103–114, 1999; D and E, from D.M. Berman, S.S. Karhadkar, A. Maitra et al., *Nature* 425:846–851, 2003.)

also located in the plasma membrane. However, when the Hedgehog (Hh) ligand binds to Patched, the latter no longer inhibits Smoothened (Smo); Smoothened now rushes to protect Gli from the normally occurring cleavage, and the intact Gli can move into the nucleus, where it acts as a zinc finger transcription factor to induce gene expression. This signaling pathway has been implicated in a wide variety of morphogenetic steps in both *Drosophila* and vertebrate embryos.

Germ-line mutations in the gene encoding Patched have been detected in the skin condition termed basal cell nevus syndrome (BCNS). Loss of heterozygosity (LOH) at the *PTC* locus in skin cells allows these nevi to develop into basal cell carcinomas (BCCs; see Section 6.12). Moreover, at least 40% of sporadic basal cell carcinomas—extremely common skin tumors caused by UV radiation—carry inactivating mutations of the *PTCH* gene or activating mutations of the *SMO* gene. (Projections of future disease incidence indicate that 28% of Caucasians born in the United States after 1994 will develop at least one BCC during their lifetime.) Fortunately, these skin tumors are relatively innocuous and easily treated. But other types of tumors associated with mutant alleles of these two genes, notably muscle cell and cerebellar tumors (i.e., medulloblastomas; Figure 16.40B), are not.

An important extension of these findings came from the discovery of another sort of deregulation of this pathway. A diverse group of cultured human cancer cells were found to express unusually high levels of one of the two major ligands of the Smoothened receptor, specifically Indian Hedgehog or Sonic Hedgehog. These cells derived from carcinomas arising in the epithelia lining the esophagus, stomach, bile duct, lung, and colon. (Intriguingly, these diverse tumors all derive from organs arising out of the embryonic endoderm.) Later, prostate carcinomas, which are not of endodermal origin, were added to this list. In the case of pancreatic carcinomas, overexpression of a Hedgehog-type ligand has been found in more than 70% of tumor samples tested. (In normal epithelia, the Hedgehog ligand and the downstream signaling pathway that it activates are thought to be responsible for the maintenance of self-renewing stem cells.)

The secretion of high levels of Hedgehog ligand by these various tumor cells is presumed to activate an autocrine signaling loop that results in the constitutive activation of signaling and, therefore, in the continuous dispatching of intact, transcription-activating Gli to the nuclei of cancer cells (see Figure 16.40D). The key role of Hedgehog in driving the proliferation of some of these tumor cell types was confirmed by adding neutralizing Hedgehog antibody to their culture medium, which stopped their proliferation (see Figure 16.40E). Conversely, Hedgehog added to their growth medium was found to be potently mitogenic (Figure 16.40C). As might be expected, these cytostatic and even cytotoxic effects of anti-Hedgehog antibody were seen only in tumor cells that also showed expression of the Smoothened receptor protein.

In 2000, cyclopamine was found to directly inhibit the Smoothened protein (Figure 16.41). Moreover, this interaction blocked the abnormal signaling resulting from excessive Hedgehog synthesis or mutations in the *SMO* gene. This suggested that the teratogenic effects of cyclopamine derive directly from its ability to block Hedgehog signaling at critical junctures in embryonic development. Moreover, this lack of adequate Hedgehog signaling during development contrasted with the excessive activity of this pathway in a variety of malignancies.

The discovery of the cyclopamine–Smoothened association led, in turn, to treatment of a variety of Hedgehog-positive human tumor cell lines with cyclopamine, which resulted in a 75 to 95% inhibition of cell proliferation. For example, cultured human medulloblastoma cells, in which the Hedgehog signaling pathway has also been found to be hyperactivated, responded to cyclopamine by stopping growth and rapidly losing viability, whereas cells from two other kinds of brain tumors (glioblastomas and ependymomas) were unaffected by cyclopamine treatment. This treatment had no effect on yet other tumor cell lines in which the Hedgehog signaling pathway was not activated, demonstrating that cyclopamine was not simply a nonspecific, widely acting cytotoxic agent. In addition, treatment of mice in which a gallbladder carcinoma (another Hedgehog-secreting endodermal tumor) had been implanted showed total blockage of tumor-forming ability (Figure 16.42A).

The presence of high levels of Hedgehog in approximately 70% of human pancreatic carcinomas suggests that activation of the Hh–Smo–Ptc–Gli pathway is an integral part of the neoplastic growth programs of these tumors, while this signaling pathway appears to play little if any role in many other types of cancers and in the maintenance of many normal tissues. In addition, long-term exposure of adult mice to therapeutic levels of cyclopamine has, to date, not yielded any indications of toxicity. All this should augur well for the candidacy of cyclopamine as a highly useful agent for the treatment of the subsets of human cancers that exhibit hyperactivated Hedgehog signaling pathways.

In fact, the candidacy of cyclopamine as a useful anti-cancer therapeutic agent has three strikes against it. Like other natural products, cyclopamine is the end result of a complex series of enzymatic reactions that are difficult to recapitulate in the synthetic organic chemistry laboratory. Second, harvesting significant amounts from *V. californicum* is not practical. Finally, in spite of the above-mentioned results with mice, cyclopamine is considered to be too toxic to be used in humans.

So, alternative Smoothened antagonists have been developed that are likely to prove as potent as cyclopamine in interrupting the Hedgehog pathway but lack cyclopamine's toxicity. Smoothened is a seven-membrane-spanning cell surface receptor, and therefore has an overall structure that closely resembles that of the many G-protein–coupled receptors (GPCRs; Section 5.7) made by mammalian cells. The development of low–molecular-weight, highly specific compounds that target GPCRs has proven, in many cases, to be relatively easy. Accordingly, a number of pharmaceutical companies have developed drugs that target Smoothened with high specificity and show minimal effects on other cellular receptors.

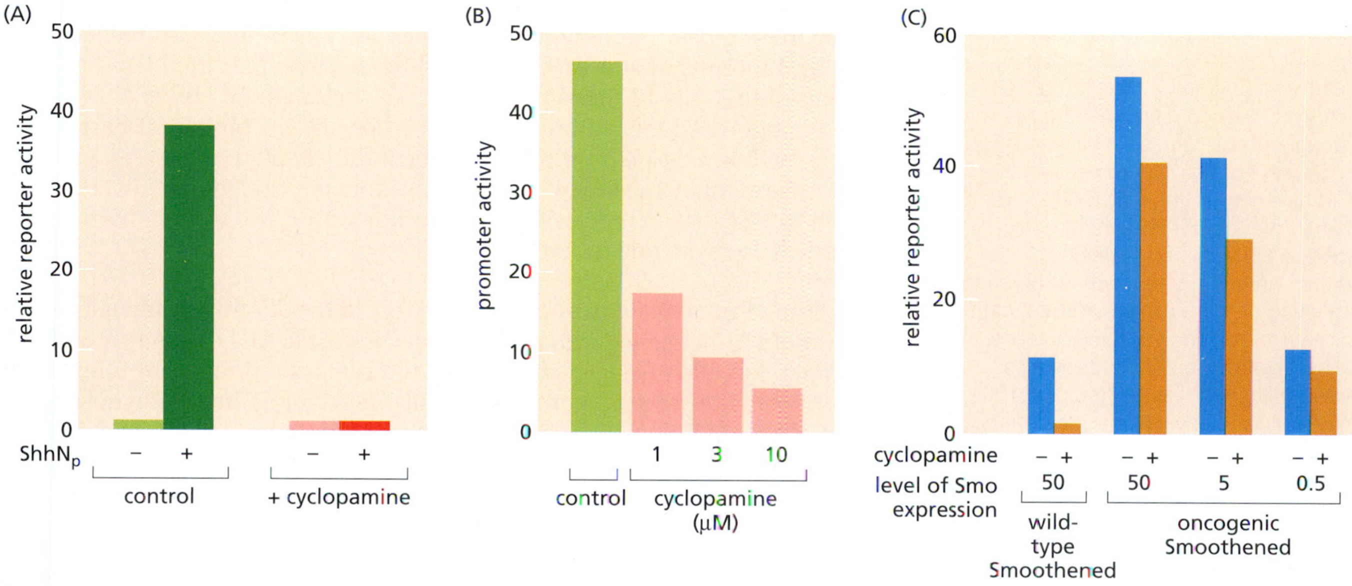

Figure 16.41 Actions of cyclopamine on the Patched–Smoothened pathway (A) In the experiment shown here, the activity of Smoothened was gauged indirectly by measuring the activity of a reporter gene whose transcription is driven by Gli in mouse NIH 3T3 cells. In the absence of added Sonic hedgehog ligand (ShhN$_P$), a Hedgehog variant that is also a ligand of the Patched receptor, there was no activity of Gli *(light green bar)*. In the presence of ShhN$_P$, Gli activity was strongly stimulated *(dark green bar)*, and this induction was reversed by cyclopamine treatment *(pink, red bars)*. This demonstrated that cyclopamine counteracts the effects of Hedgehog ligand and is therefore likely to lie *downstream* of the Patched receptor in the signaling pathway. (B) The target of cyclopamine action was further localized by this experiment, in which the activity of Gli (measured as in panel A) was measured in *PTCH*$^{-/-}$ cells. Gli activity was, as before, suppressed by cyclopamine, confirming that this drug is likely to interfere with a step downstream of and independent of Patched. (C) When wild-type Smoothened (Smo) was expressed at high levels in NIH 3T3 cells, its activity was, once again, suppressed by addition of cyclopamine, as indicated by the activity of the Gli-regulated reporter gene *(blue, orange bars, left)*. However, when a mutant, dominantly acting, oncogenic Smoothened was expressed at the same or lower levels *(right bars)* signaling was quite resistant to cyclopamine inhibition. This indicated that Smoothened was either downstream of or a direct target of cyclopamine action. Subsequent studies generated a series of mutant, constitutively active Smoothened proteins that were all resistant to cyclopamine inhibition, reinforcing the notion that cyclopamine interacts directly with Smoothened (see Figure 16.40A). Biochemical analyses then demonstrated the direct binding of the cyclopamine molecule to Smoothened *(not shown)*. (From J. Taipale, J.K. Chen, M.K. Cooper et al., *Nature* 406:1005–1009, 2000.)

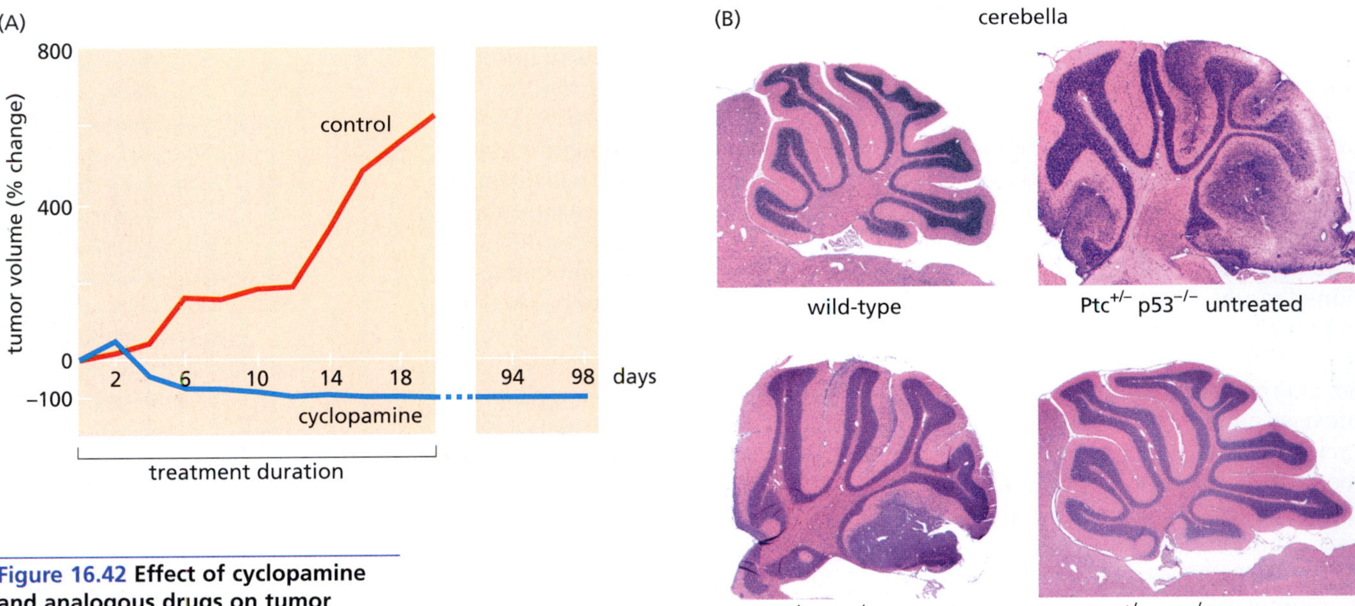

Figure 16.42 Effect of cyclopamine and analogous drugs on tumor growth (A) Human cholangiosarcoma (bile duct tumor) cells formed tumor xenografts in mice of 180 mm^3 volume and either were left untreated *(red line)* or were then treated for 22 days with cyclopamine *(blue line)*. In the latter case, the tumor shrank and did not reappear in the 76 days that followed in the absence of further cyclopamine treatment. (B) Mice with a $Ptc^{+/-}p53^{-/-}$ genotype develop medulloblastomas throughout their cerebella early in life. By 5 weeks of age, the cerebellum in a wild-type mouse (*top left*) is far smaller than in the tumor-prone mutant (*top right*). A Smoothened antagonist, termed HhAntag, was identified through screening of a drug library. If the mutant mice were treated with the drug twice daily between the third and the fifth week of life, with either 20 mg or 100 mg per kg of body weight, the tumors regressed partially or completely (*bottom left and right*). Subsequent treatments of 8- and 10-week-old mutant mice with far larger tumors have yielded comparable therapeutic responses *(not shown)*. (A, from D.M. Berman, S.S. Karhadkar, A. Maitra et al., *Nature* 425:846–851, 2003; B, from J.T. Romer, H. Kimura, S. Magdaleno et al., *Cancer Cell* 6:229–240, 2004.)

In order to test these new compounds, a mouse model of human medulloblastoma has been created that depends on inactivation (see Sidebar 7.10) of one copy of the *Ptc* gene and both copies of the *p53* gene in the mouse germ line, yielding a $Ptc^{+/-}p53^{-/-}$ genotype; virtually all such mice develop medulloblastomas by 3 months of age. An inhibitor of Smoothened, termed HhAntag, was synthesized that has 10 times the potency of cyclopamine and is able to pass easily through the blood–brain barrier, the specialized biological barrier that protects the brain tissue from the contents of the circulation. As seen in Figure 16.42B, treatment of 3-week-old mutant mice that developed medulloblastomas with HhAntag causes a regression of the tumor within two weeks; this occurred with little if any systemic toxicity.

In the case of pancreatic cancer, the prospect of developing a clinically useful inhibitor of the Hedgehog signaling pathway is an exciting one. At present, this carcinoma, in which Hedgehog signaling often plays a prominent role, has an almost inevitable fatal outcome: once this cancer has been diagnosed in a patient, the probability of surviving for another five years is less than 4%. This contrasts with the five-year survival in 1998 of American patients diagnosed with breast cancer (86%) and prostate cancer (97%).

Medulloblastomas, largely pediatric tumors, occur about one-tenth as often as pancreatic carcinomas; at present, almost two-thirds of patients are cured of this tumor through a combination of surgery, radiation, and chemotherapy; these treatments can, however, leave survivors with significant neurological impairment, including compromised cognitive functions. Ironically, however, the major economic incentive for developing cyclopamine mimetics is likely to derive from the need to treat the most benign but also the most common human cancer type—basal cell carcinomas of the skin.

16.15 mTOR, a master regulator of cell physiology, represents an attractive target for anti-cancer therapy

The final anecdote is the shortest of all, if only because it describes a regulatory circuit that is still incompletely understood and has yielded few clinical successes to date. Nonetheless, this circuit has all the attributes of generating

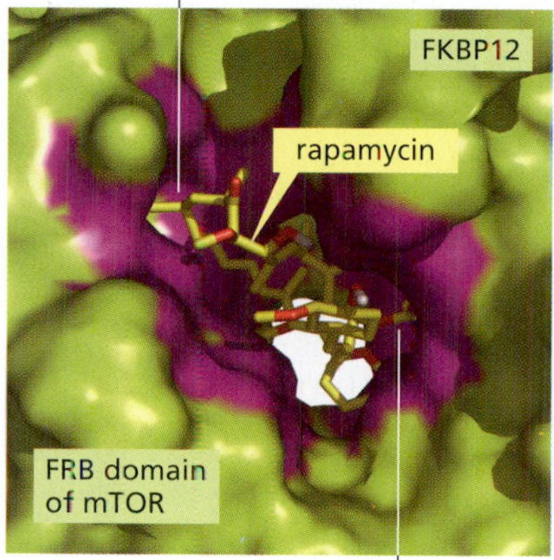

Figure 16.43 Rapamycin, FKBP12 and mTOR (A) Rapamycin is described chemically as a macrocyclic lactone and biologically as a macrolide antibiotic, one of many that are made by bacteria belonging to the *Streptomyces* genus. Rapamycin and its chemical derivatives act as potent immunosuppressants without inducing severe side effects in other organ systems in the body. Some of its effects are due to its ability to inhibit mTOR signaling. (B) The binding of rapamycin *(green, red stick figure)* to FKBP12 *(blue ribbon and space-filling model, right)* occurs with high affinity, the dissociation constant *(K_d)* being in the range of 0.2 to 0.4 nM. This bimolecular complex forms a molecular surface that can then associate with mTOR *(red ribbon and space-filling model, left)* and prevent the latter from functioning as a serine/threonine kinase. In this image, only the FRB (FKBP12-rapamycin-binding) domain of mTOR is shown. (C) The details of the interface between rapamycin *(yellow, red stick figure)* and the surfaces of the two proteins are shown here. Areas of high stereochemical complementarity between rapamycin and the proteins are highlighted in *purple*. Some of the high affinity association depends on the insertion of chemical groups of rapamycin into deep cavities within FKBP12 *(right)* and the FRB domain of mTOR *(left)*. (B and C, courtesy of Y. Mao and J. Clardy, and from J. Choi, J. Chen, S.L. Schreiber and J. Clardy, *Science* 273:239–242, 1996.)

therapies that will rival and even eclipse some of those that have been described earlier in this chapter.

This story also starts with a natural product—rapamycin—that was isolated in the 1960s from *Streptomyces hygroscopicus* bacteria growing in the soil of Rapa Nui, known to us as Easter Island, in the middle of the Pacific. In the early 1970s, it was re-isolated by a drug company, which developed it as an antifungal agent. In the decades that followed, it became clear that rapamycin (Figure 16.43A) can act to halt the growth of an extraordinarily wide spectrum of eukaryotic cells, ranging from those of yeast to mammals.

Rapamycin was also found to have powerful immunosuppressive powers, even when used at low concentrations. In 1999, it was approved by the U.S. Food and Drug Administration (FDA) to prevent immune rejection of transplanted organs, largely kidneys. This drug, also called sirolimus, functions synergistically with other immunosuppressants, specifically cyclosporine and steroids, to ensure long-term engraftment without causing major side effects in transplant recipients. The reasons for its selective actions in preferentially affecting the immune system are not fully understood. [Intriguingly, immunosuppression by cyclosporine in organ transplant recipients leads to increased risk of malignancies (see Section 15.9), while rapamycin-induced immunosuppression in these patients actually decreases the risk of post-transplantation lymphoproliferative disorders. Hence the notion that immunosuppression always leads to increased cancer risk needs to be refined, since some types of immunosuppression yield increased tumor incidence while other types do not.]

Biochemical analyses show that rapamycin binds directly to a low–molecular-weight protein, called FKBP12 (FK506-binding protein of 12 kD), originally discovered because it is also bound by FK506, a similarly acting drug. Once formed, the rapamycin–FKBP12 complex (Figure 16.43B) associates with a protein that was identified in 1994, termed mTOR (mammalian target of rapamycin), and shuts it down. mTOR is a large (289 kD) protein that functions as a serine/threonine kinase; its kinase domain resembles that of PI3 kinase and related enzymes.

mTOR is of special interest, because it operates as a critical node in the control circuitry of mammalian cells (Figure 16.44A). Thus, mTOR integrates a variety of afferent (i.e., incoming) signals, including nutrient availability and mitogens, and, having done so, acts to control glucose import and protein synthesis. More specifically, mTOR phosphorylates two key governors of translation: p70S6 kinase (S6K1) and 4E-BP1. This phosphorylation activates S6K1, which then proceeds to phosphorylate the S6 protein of the small (40-S) ribosomal subunit,

Figure 16.44 The mTOR circuit and tumor responses to mTOR inhibitors (A) mTOR sits in the middle of a complex regulatory circuit that integrates incoming signals about nutrient availability, oxygen tension, ATP levels, and mitogenic signals and, in response, releases signals that govern ribosome biogenesis, protein synthesis, cell proliferation, protection from apoptosis, angiogenesis, and even cell motility. mTOR exists in two alternative complexes with its Rictor *(left)* and Raptor *(right)* partners; the two complexes intercommunicate in still-obscure ways. The mTOR–Rictor complex governs the activity of Akt/PKB by adding a critical second phosphate to the latter and thereby gains control over Akt/PKB's multiple downstream effectors. Exposure to rapamycin *(lower right)* rapidly inhibits the mTOR–Raptor complex and, after extended periods, causes a progressive shutdown of the mTOR–Rictor complex. (B) BALB/c mice bearing injected cells of a syngeneic colon adenocarcinoma cell line develop large, well-vascularized tumors *(left)* by 35 days after injection. However, if the tumors are allowed to grow for a week after which the mice receive continuous treatment with doses of rapamycin comparable to those used in humans for immunosuppression, the tumors are much smaller *(right)* and the density of microvessels in these tumors is less than half of that seen in the control tumors *(not shown)*. (C) Osteosarcomas generally respond poorly to various types of chemotherapy. However, in the case of a 23-year-old osteosarcoma patient, treatment with AP23573, an analog of rapamycin, yielded a more than 50% decrease in the maximum standard uptake value (SUV_{max}) of radiolabeled glucose by a metastasis within 5 days of treatment, and a more than 85% decrease by 54 days of treatment *(white arrows)*. While such responses are not typical, they indicate the potential of this type of treatment and the possibility that, in the future, conditions will be found to enable similar responses in a significant proportion of such patients. Each of these images is a fusion of two images initially obtained by CT (computerized X-ray tomography) and PET (positron-emission tomography); the latter measures the extent of uptake of radiolabeled glucose, which is generally elevated in neoplastic tissue. (A, from D.A. Guertin and D.M. Sabatini, *Trends Mol. Med.* 11:353–361, 2005; B, from M. Guba, von Breitenbuch, M. Steinbauer et al., *Nat. Med.* 8:128–135, 2002; C, courtesy of S.P. Chawla and K.K. Sankhala, Century City Doctors' Hospital and John Wayne Cancer Institute, and of C.L. Bedrosian, Ariad Pharmaceuticals, Inc.)

enabling this subunit to participate in ribosome formation (by associating with the large ribosomal subunit) and thus in protein synthesis.

In addition, by phosphorylating 4E-BP1, mTOR causes 4E-BP1 to release its grip on the key translational initiation factor eIF4E (eukaryotic initiation factor 4E); once liberated, eIF4E forms complexes with several other initiation factors, and the resulting complexes enable ribosomes to initiate translation of certain

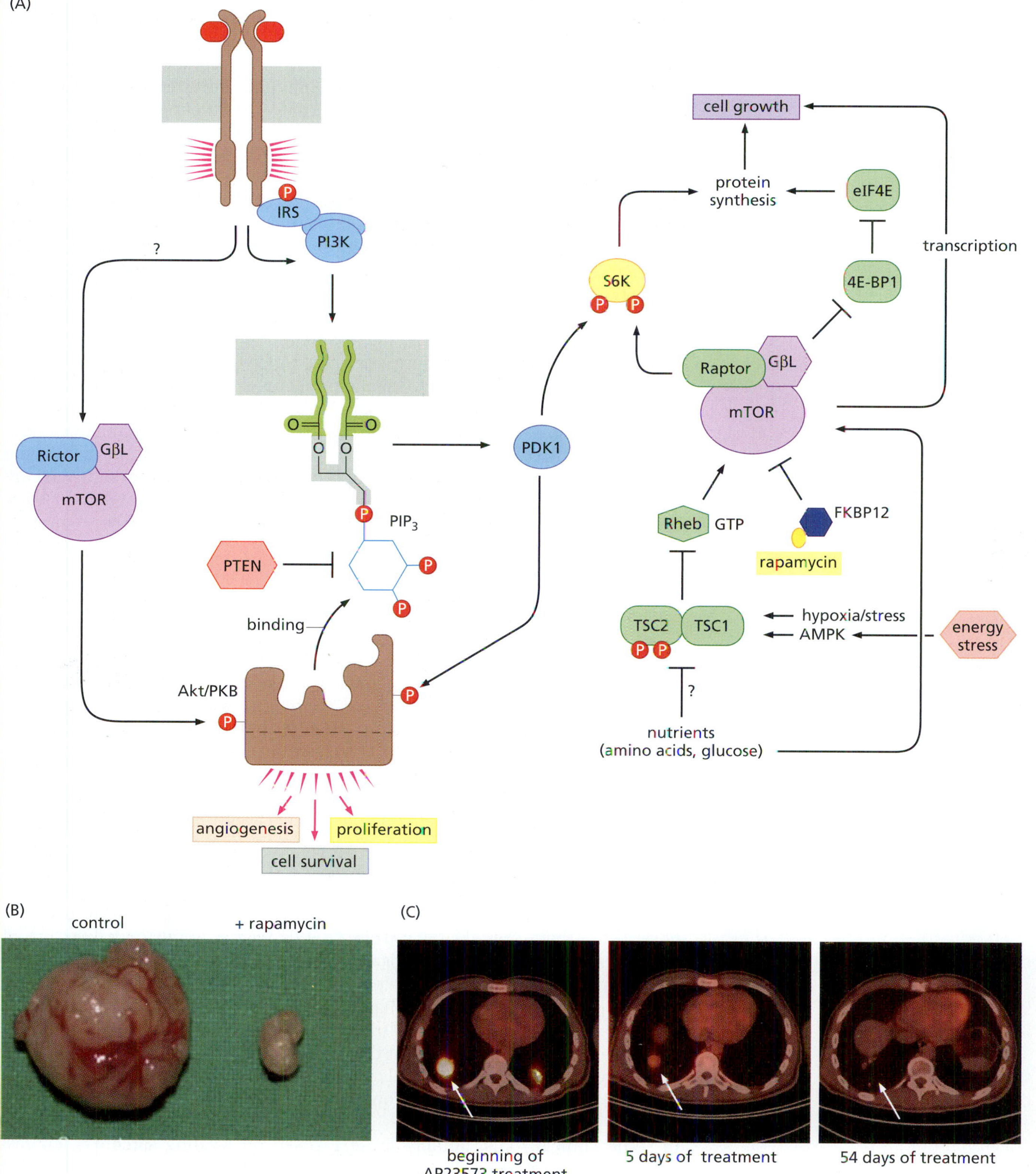

mRNAs, specifically those with oligopyrimidine tracts in their 5′, untranslated regions. Together, these various actions allow mTOR to be a key governor of cell growth (rather than cell proliferation; see Figure 8.2).

Until recently, mTOR was thought to be one of the multiple downstream substrates of Akt/PKB, specifically the one allowing Akt/PKB to regulate cell growth by controlling protein synthesis. But the tables have been turned: mTOR is now realized to be a key *upstream activator* of Akt/PKB (Sidebar 54). This shift puts mTOR in a far more powerful position in the cell. By controlling Akt/PKB, mTOR can regulate apoptosis and proliferation in addition to its known ability to regulate cell growth.

In fact, mTOR appears in two places in the circuitry depicted in Figure 16.44A, since it is able to associate with two alternative partners, called Raptor and Rictor. The mTOR–Rictor complex (together with a third protein, GβL) is regulated in unknown ways by growth factors and is responsible for activating Akt/PKB. The mTOR–Raptor complex (+ GβL), about which more is known, is responsible for activating protein synthesis (by phosphorylating S6K1 and 4E-BP1). Acting together with FKBP12, rapamycin directly interacts with the mTOR–Raptor complex, which is rapidly inhibited after this drug is applied to cells. If, however, rapamycin treatment is continued for many hours, eventually the mTOR–Rictor complex is also shut down, resulting in the inhibition of Akt/PKB. The mechanism by which rapamycin succeeds in inhibiting the mTOR–Rictor complex is poorly understood.

This inhibitory effect on Akt/PKB signaling seems to be responsible for much of rapamycin's effect on cancer cells that exhibit a hyperactivated PI3K or loss of PTEN expression. It is plausible that such cells, much like the small-cell lung carcinoma cells with mutant EGF receptors (Sidebar 16.3), have become "addicted" to Akt/PKB signals and lurch into apoptosis the moment they are deprived of these signals by the actions of rapamycin and related drugs. However, the precise rules that determine sensitivity to rapamycin treatment are yet to be worked out.

The regulatory circuit shown in Figure 16.44A intersects in additional ways with cancer pathogenesis. For example, TSC1 and TSC2 (also called hamartin and tuberin) have already appeared in this book in the context of their role as tumor suppressor proteins. Loss of either of these proteins leads to tuberous sclerosis (Table 7.1); and, as seen in Figure 8.2, loss of TSC1 results in the formation of giant cells in both flies and humans. TSC2 acts as a GAP (GTPase-activating protein; see, for example, Sidebar 5.11) for Rheb, a small Ras-like protein. As long as it remains in its GTP-bound state, Rheb contributes in unknown ways to stimulating the mTOR–Raptor–GβL complex; however, once TSC2 has induced Rheb to hydrolyze its GTP, Rheb loses this stimulatory activity. Yet other signaling connections between the mTOR circuit and critical growth-inducing and mitogenic proteins are being forged by ongoing research.

Various derivatives of rapamycin have been produced, and three are in early-phase clinical trial. Their development has been encouraged, in part, by the observation that drugs like rapamycin can be tolerated for extended periods of time by transplant recipients, indicating a tolerably low level of side-effect toxicity. In pre-clinical experiments, rapamycin given to mice at levels that are used for chronic immunosuppression has strong effects in suppressing tumor-associated neoangiogenesis and thus tumor growth (Figure 16.44B), an effect that may be explained by the fact that one of the three Akt/PKB isozymes, Akt1, is critical to the ability of endothelial cells and their precursors to respond to stimulation by vascular endothelial growth factor (VEGF).

In some clinical trials, notably those focused on treating sarcomas, clinical responses have occasionally been observed that are nothing short of remarkable (Figure 16.44C). And in 2006, rapamycin was reported to induce regression of astrocytomas associated with tuberous sclerosis (see Figure 8.2B). Indeed, it is clinical responses like these that have motivated discussion of the mTOR circuit in this chapter. They provide tantalizing hints of how this circuit may one day be manipulated to induce cancer cell death, yielding substantial improvements in the therapy of solid tumors. These advances are likely to come as oncologists learn which types of cancer cells are particularly sensitive to rapamycin analogs, often in the presence of other collaborating therapeutic drugs.

16.16 Synopsis and prospects: Challenges and opportunities on the road ahead

"When is cancer going to be cured?" This is the simple and reasonable question posed most often to cancer researchers by those who are not directly involved in this area of biomedical research. In their minds are the histories of other public health measures. Infectious diseases, such as polio and smallpox, can be prevented, and bacterial infections are, almost invariably, cured. Heart disease is, in the eyes of many, well on its way to being prevented (Sidebar 55). Why should cancer be any different?

The information in this book provides some insights into the answers to these questions. As much as we have invoked unifying concepts to portray cancer as a single disease, the reality—at least in the eyes of clinical oncologists—is far different. Cancer is really a collection of more than 100 diseases, each affecting a distinct cell or tissue type in the body.

Pathological analyses have led us to embrace this number, or one a bit larger. (For example, there are at least eight distinct histopathological categories of breast cancer.) However, even the expanded number, large as it may be, represents an illusion: the current use of molecular diagnostics, specifically gene expression arrays, is leading to an explosion of subcategories, so that by the second decade of the new millennium, several hundred distinct neoplastic disease entities are likely to be recognized, each following its own, reasonably predictable clinical course and exhibiting its own responsiveness to specific forms of therapy. With the passage of time, cancer diagnoses will increasingly be made using bioinformatics rather than the trained eyes of a pathologist.

So the initial response to questions about "the cure" is that there won't be a single major breakthrough that will cure all cancers—a decisive battlefield victory—simply because cancer is not a single disease. Instead, there will be many small skirmishes that will steadily reduce the overall death rates from various types of cancer. And because certain molecular defects and pathological processes (e.g., angiogenesis) are shared by multiple human cancers, there will be occasions when therapeutic advances on a number of fronts will be made concomitantly.

Before we speculate on the future of cancer therapy, it is worthwhile to step back and assess the scope of the challenge: (1) How large is the problem of cancer and, in the future, how desperate will the need be to cure various types of neoplastic disease? (2) How well are we doing now in curing the major solid tumors?

Epidemiology and demographics provide some answers to the first question. They yield sobering assessments of the road ahead. The statistics in Figure 16.45 demonstrate that cancer is largely a disease of the elderly, whose numbers are growing rapidly and will continue to do so, generating progressive increases in the numbers of cancer-related deaths (mortality) over the coming decades.

Equally important, we still have only very imperfect ways of measuring incidence—how often the disease strikes. This greatly complicates assessments of the effectiveness of current therapies and future needs for therapy. As indicated

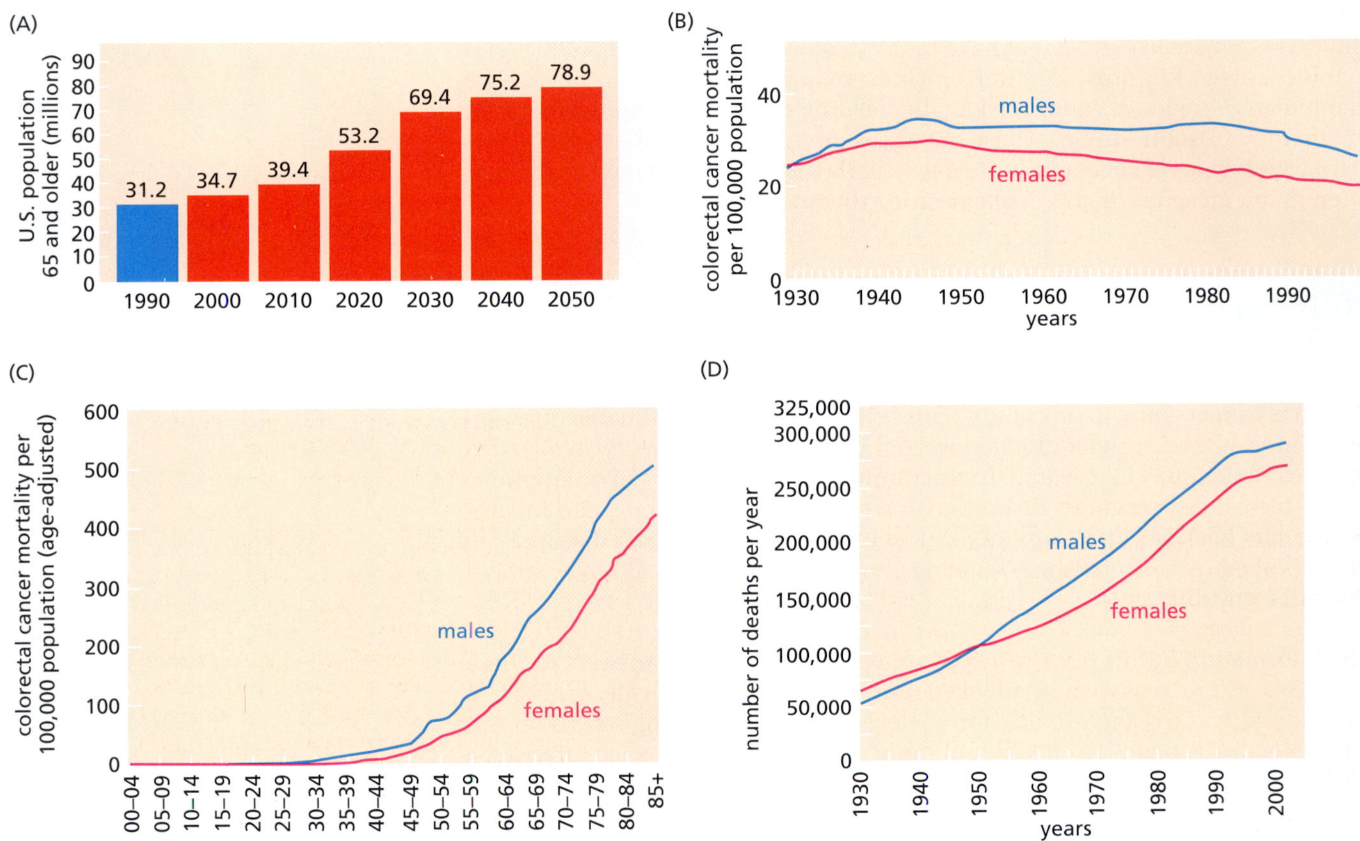

Figure 16.45 The demographics and epidemiology of cancer (A) Because of dramatic decreases in midlife mortality, populations in industrialized countries are aging rapidly. In the United States, the number of individuals over 65 years of age has increased 11-fold since 1900, while the number of those under 65 has increased by a factor of 3. Comparable increases in the aged population are likely to occur worldwide over the next generation. (B) The age-adjusted death rate from colorectal cancer has changed only slightly over the past several decades. Other major cancers show similar curves. (C) Like a number of other diseases, cancer is uncommon during early and midlife and then increases rapidly. Shown here is the age-dependent death rate of colorectal cancer. (D) Because (1) the number of elderly has increased steadily over the past century and will continue to do so (see panel A), (2) cancer is a disease of the elderly (panel C), and (3) the age-adjusted death rate of most cancers has been relatively constant for many decades (as suggested in panel B), the absolute number of annual deaths due to cancer has increased dramatically over the past three-quarters of a century in the United States. Since these trends are likely to continue, the burden of cancer cases in industrialized societies will continue to climb for many decades. (A, courtesy of D. Singer and R. Hodes, from the U.S. Bureau of the Census Projections of 1996; B, courtesy of M.J. Thun, from Cancer Statistics, 2003, American Cancer Society; C, from A. Jemal, T. Murray, E. Ward et al., *CA Cancer J. Clin.* 55:10–30, 2005; D, from A. Jemal, E.M. Ward and M.J. Thun, in V. DeVita et al. (eds.), Cancer: Principles and Practice of Oncology, 7th ed. Philadelphia: Lippincott/Williams & Wilkins, 2004.)

in Figure 16.46, perceptions of the incidence of certain types of neoplastic disease are strongly influenced by diagnostic practices.

For many types of cancer, the more one looks, the more one finds. Statistics like those in Figure 16.46 suggest that in the past many cancers remained undiagnosed and asymptomatic, and that these tumors are contributing the lion's share to perceived increases in disease incidence, notably of common tumors, such as those arising in the breast and prostate. (The major exceptions here are the cancers related to tobacco use, whose increased incidence is real and beyond dispute, since the incidence rates are closely paralleled by the rates of mortality.) Such statistics indicate that for many types of tumors, we have only a poor appreciation of how large is the number of tumors that truly require treatment (see Sidebar 16.4).

Data like those in Figure 16.46 also undermine the notion (Chapter 11), deeply embedded in the thinking of many cancer biologists and clinical oncologists, that benign growths are in danger of becoming, sooner or later, highly malignant

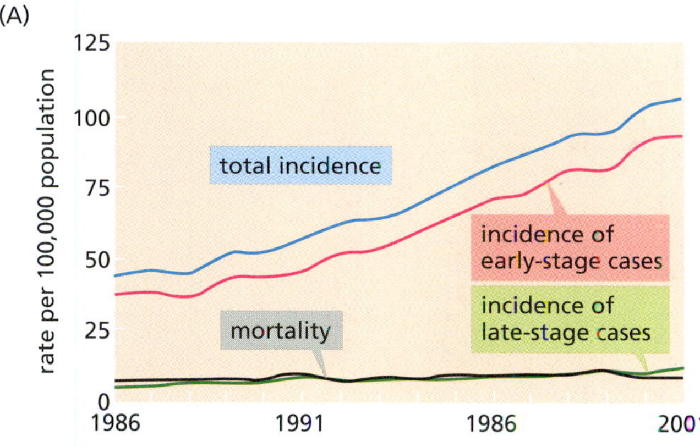

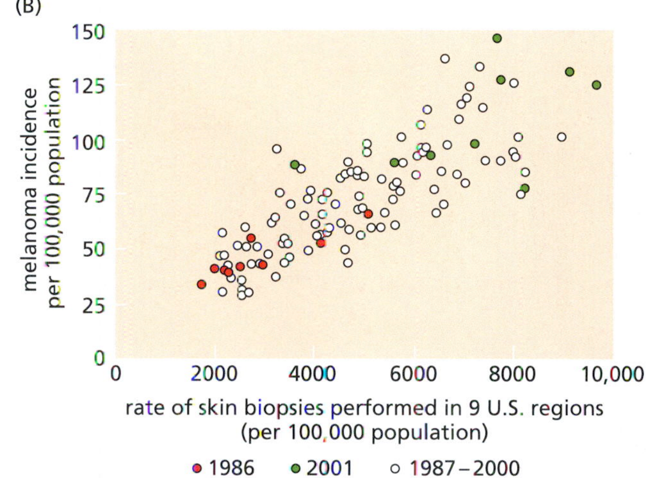

Figure 16.46 Melanoma incidence and mortality Melanoma incidence in the United States has increased by a factor of six over the past half century, raising the question of whether this trend has created a comparable increase in the number of cases that require aggressive clinical treatment. (A) The breakdown of melanoma incidence (into early-stage localized and late-stage disseminated disease categories, both diagnosed through skin biopsy) in nine areas of the United States among individuals over 65 years of age indicates a dramatic increase over the past two decades in the incidence of early-stage disease and a relatively constant incidence of late-stage disease. Age-adjusted mortality has been relatively constant during these years. (Melanoma is a disease whose incidence rises progressively with age, making this age bracket the one with the highest incidence and mortality from this disease.) These data raise the question of whether the real incidence of the early-stage disease has been increasing over this time period, or whether the real incidence has been relatively constant and the registered incidence of this disease has increased due to changes in screening practices. (B) If the incidence rates, such as those of panel A, are plotted against the rates of screening for melanoma in nine areas of the United States, as registered during several time periods, the resulting scatter plot reveals a close correlation between the two. This provides strong indication that the incidence of the disease is strongly influenced by diagnostic practices. It is therefore possible that (1) the true incidence of life-threatening melanomas (panel A) has not changed significantly over the past two decades; or (2) the true incidence of these tumors has increased, but intensification of screening has held the mortality rate at levels observed two decades ago by allowing the removal of early-stage tumors before they progress to become invasive and metastatic. (From H.G. Welch, S. Woloshin and L.M. Schwartz, *BMJ* 331:481, 2005.)

Sidebar 16.4 What is the true incidence of life-threatening cancers? To some, data like those shown in Figure 16.52 might suggest that the age-adjusted incidence of life-threatening melanomas has not changed over the past two decades, and that the screening for melanomas, which is increasingly practiced in the industrialized countries, has only led to the discovery of growths that, in earlier times would have remained unnoticed and would have never progressed to a highly malignant state. This would suggest that screening for melanoma, and by extension for other early-stage malignancies, has little beneficial effect in terms of reducing mortality.

But there is another, equally plausible view: that the true incidence of melanomas with the potential to become life-threatening has indeed been increasing steadily in recent decades (e.g., as a consequence of increased sun exposure), and that screening has prevented an associated increase in mortality, since it often leads to the removal of such growths long before they have had time to progress to a highly malignant state. In parts of Australia, for example, where melanoma screening has been widely practiced for three decades, the age-adjusted mortality from melanoma has actually decreased in recent years. Significantly, mortality has decreased in cohorts of people who were born after 1950, when recognition of the danger of UV radiation first became widespread. But even this statistic is subject to alternative interpretation: the decrease may be due to prevention (in the form of sun screen lotion and greater skin coverage outdoors) rather than screening.

In either case, it is clear that existing metrics of disease incidence for this and a number of other relatively common neoplasms are highly inaccurate, which greatly complicates the development of strategies for reducing cancer-related mortality. Thus, as the technologies of detection improve (see Figure 16.3), the proportion of tumors discovered that are likely to become highly malignant is likely to decrease, and an increasing proportion of tumors will be subjected to therapy when none is justified by the natural course of these more benign growths. These trends highlight the twin dilemmas facing those who are involved in the development of many types of cancer therapy: the absolute numbers of cases that require treatment are unknown, and it is difficult to distinguish with any certainty between those growths that require aggressive treatment and those that can be ignored or, at most, subjected to "watchful waiting."

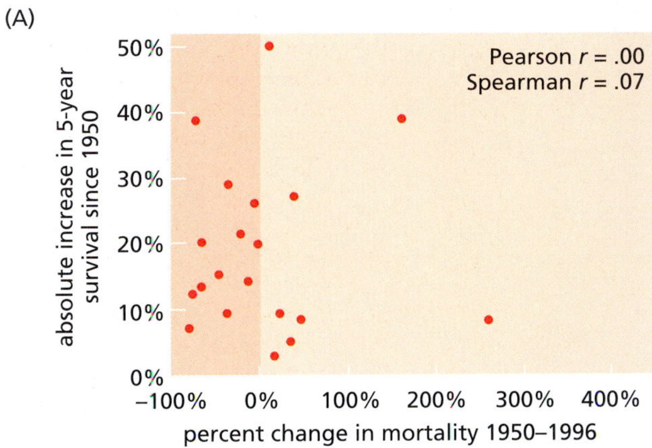

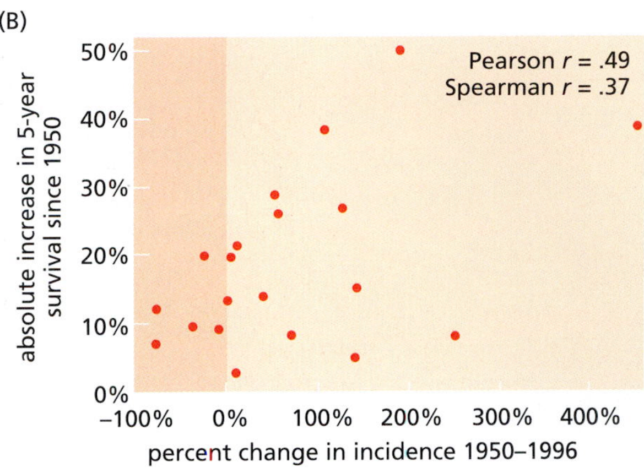

Figure 16.47 Cancer incidence and survival since 1950 in the United States One way of gauging therapeutic success in treating a certain type of tumor is to determine the percentage of patients who survive for 5 years following initial diagnosis. (A) If extension of patient survival time following diagnosis were a reflection of increased therapeutic efficacy, this extension should be correlated with a decrease over the past half-century in the age-adjusted mortality from this type of tumor. In fact, as indicated here, there is no correlation between these two parameters when changes in 5-year survival rates of a number of solid tumors are plotted versus changes in mortality. (Note the low correlation coefficients, *upper right*). (B) Instead, there is a significant correlation between changes in 5-year survival and changes in disease incidence. If the latter is strongly affected by diagnostic bias (see Figure 16.46), then increases in 5-year survival become very difficult to interpret, since they may largely reflect the detection of a disease at an earlier stage of its natural clinical course. (From H.G. Welch, L.M. Schwartz and S. Woloshin, *J. Am. Med. Assoc.* 282:2976–2978, 2000.)

ones. Cancer epidemiology now makes us confront the alternative possibility: many kinds of early-stage tumors, including those associated with commonly occurring types of cancer, are unlikely to progress to high-grade malignancy during an average human life span. Unfortunately, we are only beginning to learn how to segregate those tumors that are truly deserving of aggressive treatment from those that are not (see, for example, Figure 16.4).

As to the second question, which deals with the effectiveness of current cancer therapies, our perceptions are strongly influenced by the fact that people are living longer with their cancers. This provides a measure of reassurance that progress is being made. However, some of these perceived improvements in therapy may, once again, be artifacts of increased screening and more sensitive detection techniques that increasingly uncover tumors relatively early in their development, giving the patient additional years of survival before tumor progression advances through its natural course, whether or not treatments are applied (Figure 16.47). This logic forces the conclusion that the efficacy of therapies can be accurately gauged only by well-controlled experiments: comparisons of several patient populations that are afflicted by the same malignancy and exposed in parallel to different agents or treatment protocols. Such side-by-side comparisons have, until recently, yielded only incremental gains in treating most solid tumors (see, for example, Figure 16.22), but this is beginning to change, as new drugs and monoclonal antibodies are introduced into the clinic.

Gleevec, Avastin, and Rituxan (see Sections 13.10 and 15.20) come to mind here. The major hope is that agents such as these are the forerunners of dozens and eventually hundreds of targeted, highly efficacious drugs. Efforts to develop such highly targeted drugs, including antibodies, will surely yield steadily increasing numbers of approved anti-cancer agents with every passing year. The use of computer-driven procedures, such as high-throughput screening (HTS), automated determinations of the structures of target molecules, and computer-aided design of drug molecules (Figure 16.48), has already begun to put this trend into motion.

Traditionally, new drugs have been evaluated as single agents during pre-clinical development and Phase I clinical trials. This practice contrasts with the growing belief of cancer researchers that most monotherapies are unlikely to yield curative treatments and that, with rare exception, truly successful clinical outcomes will depend on the use of combinations of anti-cancer drugs. Part of their rationale derives from the known genetic plasticity of tumor cell populations. As we learned in Chapter 12, tumor cell genomes are genetically unstable, and the large numbers of neoplastic cells within tumors are continually producing variant subclones. Some of these may, through happenstance, have acquired the means to resist killing by a therapeutic agent, and once confronted with such an agent, these variants will thrive even as the bulk of the cells in a tumor are being eliminated.

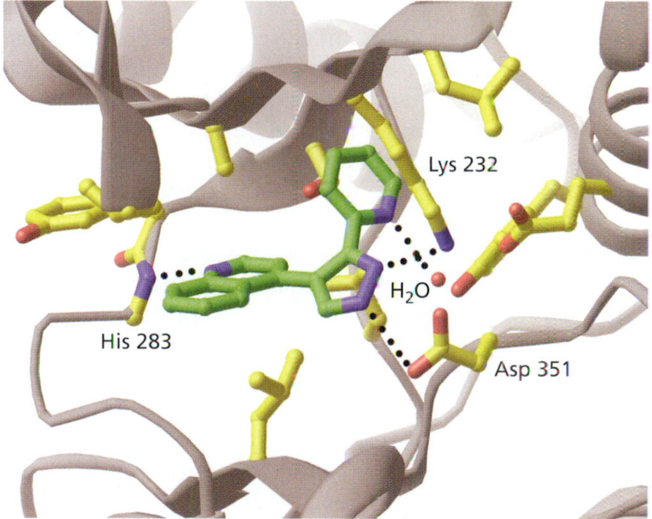

Figure 15.48 Virtual screening—designing drugs in the desktop computer High-throughput screening has yielded a large number of highly attractive drug candidates, many of which have advanced into clinical trial. In the future, however, much of this screening, which is very expensive, may be obviated by the development of powerful algorithms that enable pharmacologists to design drugs using the known structures of target proteins. The development of the drug illustrated here began with a compound that showed very weak affinity (IC_{50} of 30 μM) for the ATP-binding site of the TGF-β type I receptor kinase. Researchers then queried a database of 200,000 known compounds for compounds that shared some chemical features with the starting compound and conformed to the constraints imposed by the known structure of the ATP-binding site. This yielded the structures of 87 chemical species that satisfied these criteria and were then screened using conventional biochemical techniques. One of these, named HTS466284, shown here ensconced in the ATP-binding site of the TGF-β type I receptor kinase, exhibits an IC_{50} of 27 nM and thus functions as a potent inhibitor of TGF-β signaling. Significantly, another research group, working independently, arrived at the identical inhibitor molecule using conventional high-throughput screening. Hydrogen bonds are indicated by *dotted lines*. (From J. Singh, C.E. Chuaqui, P.A. Boriack-Sjodin et al., *Bioorg. Med. Chem. Lett.* 13:4355–4359, 2003.)

Such thinking, originally inspired by the behavior of bacterial populations treated with antibiotics, calls for therapies involving simultaneous treatment with two or more agents having very different mechanisms of action. If the probability of acquired resistance to killing by a single agent is small (e.g., 1 in 10^5), the probability of cell clones acquiring resistance to two agents is far less, perhaps the square of this probability (1 in 10^{10}). The latter number may be larger than the total number of cells in a tumor, making it unlikely that any single cell will be able to simultaneously resist killing by both agents. (The existence of multi-drug resistance weakens this argument somewhat.)

The logic of multi-drug therapy is further supported by the accumulated information about multi-step tumorigenesis, in which certain cancer cell phenotypes can be achieved through the combined actions of several genetic and/or epigenetic alterations (see Figure 11.43). One example is provided by tumors in which the Ras signaling pathway has been activated and the p53 and PTEN tumor suppressor pathways have been inactivated. Each of these changes, in its own way, reduces the likelihood that a cancer cell can be pushed to the brink of apoptosis. This overlap of anti-apoptotic function may require that a treatment target all three signaling pathways in order to destroy the cancer cell. (Interestingly, low–molecular-weight tyrosine kinase inhibitors can act synergistically with monoclonal antibodies on a common target, the EGF receptor, indicating that therapeutic benefits may often come from surprising combinations of agents.)

At present, the choice of drugs to be used singly or in combination is inspired by biological intuition or poorly informed guesses. Increasingly over the coming decade, strategies for organizing multi-drug treatment protocols will be influenced by our rapidly evolving understanding of the design of the signaling circuitry within human cells and by molecular diagnostics that tell us how certain signaling pathways have been perturbed in some tumors and not in others. Functional genomics using gene expression arrays has already begun to change this, as indicated by the analyses of diffuse large B-cell lymphomas described here (Figure 16.5).

What if the real powers of a candidate drug are only realizable when it is used in combination with several others? In theory, many anti-cancer agents should fall into this category, suggesting that many truly useful drug candidates have been discarded in the past and that many others will suffer this fate in the future, simply because their true utility as combination agents will never be tested. Here again, one hopes that our increasing understanding of the subcellular signaling circuitry will ameliorate this situation (but see Sidebar 16.5).

Sidebar 16.5 Legal and financial disincentives decrease the likelihood of drugs being tested in combination At present, the biological difficulties of testing candidate drugs in combination with others are compounded by economic forces that often create disincentives for pharmaceutical companies to test their own proprietary drugs in combination with those produced by their competitors. Patent regulations have also discouraged certain uses of patented compounds by firms that are in direct competition with the patent holders.

In the past, the difficulties of organizing multi-drug early clinical trials were further compounded in the United States by the Food and Drug Admini-stration's (FDA) insistence that the clinical efficacy of drug candidates be demonstrated when they were used as monotherapies. But this has begun to change. For example, Erbitux, the anti-EGF-receptor monoclonal antibody (Sidebar 15.4), was initially approved for use together with irinotecan, because the two used in concert yielded far better responses than did irinotecan on its own. (Irinotecan is a more traditional cytotoxic agent that functions as a topoisomerase I inhibitor.)

The discovery of tumor stem cells, as described in Chapter 11, creates another major challenge for anti-cancer drug development. Recall that the self-renewing tumor stem cells can seed new tumors, while their far more numerous transit-amplifying progeny, with limited self-renewal capacity, cannot. This organizational scheme was demonstrated initially in leukemias, breast carcinomas, and brain tumors, but in the longer term is likely to be found to describe the organization of most other kinds of human tumors as well.

Such a hierarchical organization operating within tumors holds important implications for the development of anti-cancer drugs. Traditionally, clinical validation of the therapeutic efficacy of these drugs has depended on demonstrations of their ability to halt further growth of tumors or to cause significant decreases in tumor size. Many anti-tumor agents being tested in clinical trials may shrink tumor masses by eliminating populations of transit-amplifying cells, which constitute the great bulk of the neoplastic cells in these growths. However, if the self-renewing tumor stem cells are left untouched by these drugs (see Figure 16.30), the tumor has a high probability of regrowing, leading sooner or later to clinical relapse.

This organization of neoplastic cells within a tumor means that durable remissions and cures can only derive from therapies that strike at the heart of the tumor— the self-renewing tumor stem cells. Currently, assays for the presence of these cells in most tumor types are essentially nonexistent. Hence, drug development efforts are hobbled because they lack a key analytical tool needed for developing truly efficacious therapies.

A major, still-unsolved problem concerns the biological models of human tumors that are used in pre-clinical drug development. Some xenograft models of human cancer are useful in predicting the behavior of tumors encountered in the oncology clinic, but many are not (see Sidebar 13.2). The development of predictive pre-clinical models of human cancer, if it occurs, will surely reduce the expense of drug development and, quite possibly, may even obviate certain early-phase clinical trials. For the moment, the development of truly useful animal tumor models is little more than a fond hope.

The non-neoplastic stromal cells within a tumor may be major determinants of responsiveness to most drug therapies, yet their contributions are not recognized in the design of many pre-clinical models of human cancer. In Chapter 13, we learned, for example, that the most radiosensitive cells in some tumors are likely to be the endothelial cells forming their neovasculature (see Figure 13.48). In addition, it seems increasingly likely that many widely used anti-cancer chemotherapeutics have strong effects on the tumor-associated endothelial cells that were never suspected in the past. In fact, some researchers are redesigning chemotherapeutic treatment protocols in order to optimize their toxic effects on the tumor-associated neovasculature. Recognition of these two classes of cells—cancer stem cells and tumor stromal cells—as critically important biological targets of chemotherapy will surely change the entire landscape of the drug development field.

The ultimate test of many candidate anti-cancer drugs comes only when these drugs are tested in substantial numbers of cancer patients. Here, the drug developer is often confronted by the dilemma of not knowing which types of tumors are likely to respond. Should a candidate drug be tested on patients suffering from pancreatic carcinomas or neuroblastomas? The molecular lesions discovered in these and other types of cancer cells would seem to be highly useful indicators that can be used to inform this decision. But quite often, certain types of tumors respond for reasons that cannot be predicted by the mutant genes and deregulated signaling pathways known to be present in these growths, and so the choices of patients recruited into clinical trials are arbitrary and sub-optimal. Once again, we can only hope that our increasing insights into the molecular etiologies of various cancers will provide truly useful guidelines for oncologists to follow.

Ultimately, the biggest challenge of current drug development is to demonstrate long-term efficacy: Does a drug being tested have significant effects on extending the life expectancy of cancer patients, and do we dare to hope that it can achieve durable cures? Some populations of cancer cells may develop drug evasion mechanisms by increasing their anti-apoptotic defenses. Yet others may lose checkpoint controls that previously made them sensitive to certain types of drug treatment. Tumors formed from cancer cells whose killing was previously achieved through antibody-mediated attack may simply down-regulate expression of the antigen recognized by the therapeutic antibody. A significant number of cancer cells may develop resistance to a drug through strategies that prevent the intracellular accumulation of drugs. At present, it remains unclear whether we will, one day, be able to devise treatment strategies that anticipate the plasticity and evasiveness of cancer cells, allowing us to develop definitive, life-long cures of malignancies that have long been incurable.

Independent of the challenges of cancer drug development and testing are the more transcending problems created by the complex biology of human cancers: Will different types of drugs need to be developed for different classes of cancers, or will a small number of treatments find wide applicability? Will different tumors within a given class (e.g., colon carcinomas) require distinct, tailor-made treatments based on their particular genotypes and phenotypes? And will we one day be able to provide "personalized molecular medicine" at affordable cost, in which the detailed characteristics of each patient's tumor and genetic constitution inform the design of a customized therapy?

Will anti-neoplastic drugs ever be developed that have lethal effects on malignant growths while having minimal side effects on normal tissues? And should drug designers undertake to develop anti-cancer drugs that keep tumors under control rather than attempting to wipe them out? The goal of totally curing many kinds of tumors may well be an unreachable one, and for these tumors, reducing cancer to a chronic but bearable disease may be a more realizable goal. (Such is the thinking of researchers developing new types of anti-HIV therapies.)

A 2004 census of genes mutated in human cancer cell genomes netted 291 distinct genes—almost 1.5% of the genes present in the human genome (Figure 16.49). Of these, 228 become involved in cancer only through somatic mutation, while 32 become involved through mutant alleles that are acquired both somatically and through the germ line. This list will undoubtedly grow; for example, genes whose involvement in cancer is only evident from the methylation of their promoters were not even included in this census.

This census provides many druggable targets for researchers involved in the development of new types of anti-cancer therapeutics. At the same time, it represents bewildering complexity. Some of these genes are mutated only in rare cancers, and the costs of developing therapeutic drugs directed against their protein products are unlikely to ever be recouped through sales. Most of the encoded mutant proteins contribute in still-obscure ways to the neoplastic growth of various types of human cancer cells. How will we ever ascribe key roles to each of the roughly 300 mutant genes that are already known or suspected to be involved in the pathogenesis of many cancers? Our current methods of assimilating and interpreting data on human cancer cell genomes and signaling pathways are not up to the task.

Many cancer researchers would like to understand the entirety of a biological system, such as a living cancer cell, rather than its individual functional components. In their eyes, reductionist biology, which focuses on the individual, isolatable components of complex systems, has had its day, and the time has come for

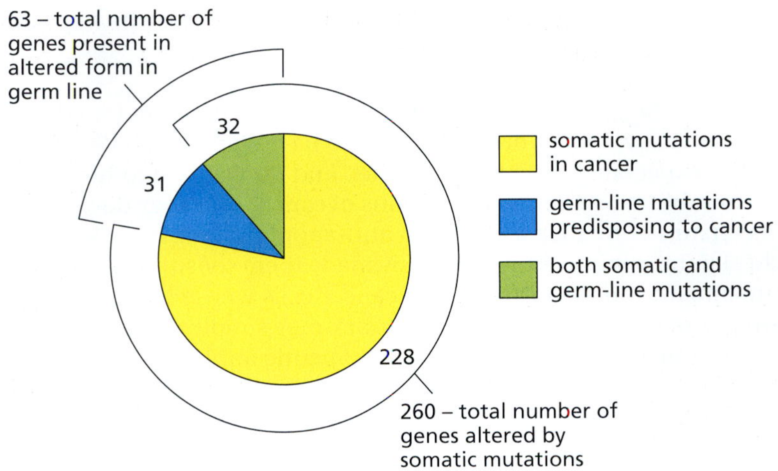

Figure 16.49 Targets for future cancer drug development A survey of the cancer literature conducted in 2004 revealed at least 291 genes that have been implicated as contributing causally to human cancers. Of these, 260 are known to be present as somatically mutated alleles in human cancer cell genomes. However, the great bulk of the latter are somatically mutated in the genomes of mesenchymal tumors (leukemias, lymphomas, and sarcomas), leaving only 77 genes that are mutant in the far more common tumors of the remaining types. Given the number of kinase genes present in the human genome, 6 kinases were expected to be among the 291 cancer-associated genes, whereas 27 were actually present in this list. Conversely, the more numerous G-protein–coupled receptors (GPCRs) were expected to yield at least 8 cancer-associated genes, but only 1 was present. (Courtesy of M.R. Stratton.)

the vast amounts of information known about these components to be integrated into complex interacting *systems* whose behavior can be predicted by bioinformatics.

Successes in these efforts, involving the new discipline of "systems biology," will surely benefit cancer research. Imagine a day—still years away—when the biological responses of various human cells, normal and malignant, can be predicted by mathematical models of these cells and their internal control circuits. Such advances will render many current practices in experimental biology, including many steps of drug development, unnecessary. If this ever becomes possible, drug development will be more a matter of dry bioinformatics than wet biology at the laboratory bench.

But for the moment, most of this remains a pipe dream, still far in the future. For now, at least, we need to wrestle with the grim realities of drug development, the inadequate animal models, our ignorance of the behavior of cellular regulatory circuitry, and the confounding biological complexities of human cancer.

Key concepts

- The science of molecular oncology has revealed dozens of proteins whose malfunction contributes to the formation and maintenance of tumors.
- Among these proteins are those whose molecular properties make them attractive targets for novel anti-cancer therapeutic agents, such as monoclonal antibodies or low–molecular-weight drugs.
- Proteins that are attractive targets for attack by antibodies are invariably located at the cell surface or in the extracellular space.
- Most proteins that are attractive targets for attack by low–molecular-weight compounds are enzymes that possess druggable catalytic clefts.
- Recent advances have expanded the range of druggable targets to include certain protein–protein interactions that can also be inhibited by low–molecular-weight drugs.
- Protein targets should be chosen whose inactivation is predicted to lead to the cessation of proliferation of tumor cells or to their death by apoptosis.
- The most successful anti-cancer drugs developed to date have been those that interfere with the functioning of various growth- and survival-promoting kinases, specifically, receptor-associated tyrosine kinases.
- Successful drugs must have a high therapeutic index, appropriate pharmacokinetics and pharmacodynamics, and minimal side effects on major organ systems.

- Studies of drugs in Phase I, II, and III trials are essential because pre-clinical studies of drug efficacy and tolerability are poorly predictive of the behavior of drugs in humans.
- The indications for the clinical use of certain drugs may be based on the known behavior of the targeted proteins in cancer cells (as in the case of Gleevec) or instead depend primarily on empirical tests of how various types of human tumors respond to treatment (as with Velcade).
- Stratification of outwardly similar tumors into narrow subclasses greatly helps researchers and clinicians to match drugs to the specific tumor cell types they can most effectively treat.
- The greatest benefit of certain drugs, such as Gleevec, may eventually prove to derive from their broad target specificity, allowing them to be used in a wide range of cancers.

Thought questions

1. What are the therapeutic advantages and disadvantages of using a drug that affects a broad range of molecular targets?
2. Given the large and heterogeneous collection of signaling molecules that have been portrayed in this book as playing key roles in the pathogenesis of various cancers, which classes of molecules do you think might become the targets for the development of a new range of anti-cancer therapeutics besides the much-studied kinases?
3. How might tumors that were initiated by the formation of a certain oncogene become independent of this oncogene later in tumor progression?
4. What strategies might you implement upon finding that cells within a tumor have become resistant to a drug therapy after extended treatment?
5. Having concluded that natural products represent a rich resource of potential anti-cancer drugs, what obstacles might limit the search and testing of such drugs?
6. What obstacles stand in the way of developing drugs for tumors that represent only a very small proportion of the total cancer burden in a population?
7. In the oncology clinic of the future, what types of information might be included in the assembly of anti-cancer therapies that are tailor-made specifically to respond to an individual patient's tumor?
8. What strategies would you pursue to develop pre-clinical models of human tumors that are highly useful in predicting patient responses to candidate drugs?

Additional reading

Adams J (2004) The proteasome: a suitable antineoplastic target. *Nat. Rev. Cancer* 4, 349–360.

Al-Hajj M, Becker MW, Wicha M et al. (2004) Therapeutic implications of cancer stem cells. *Curr. Opin. Genet. Dev.* 14, 43–47.

Baselga J & Arteaga CL (2005) Critical update and emerging trends in epidermal growth factor receptor targeting in cancer. *J. Clin. Oncol.* 23, 2445–2459.

Bernardi R & Pandolfi PP (2003) Role of PML and the PML-nuclear body in the control of programmed cell death. *Oncogene* 22, 9048–9057.

Brunner RB, Hahn SM, Gupta AK et al. (2003) Farnesyltransferase inhibitors: an overview of the results of preclinical and clinical investigations. *Cancer Res.* 63, 5656–5668.

Chabner BA & Roberts TG Jr (2005) Chemotherapy and the war on cancer. *Nat. Rev. Cancer* 5, 65–72.

Chen ZJ (2005) Ubiquitin signalling in the NF-κB pathway. *Nat. Cell Biol.* 7, 758–765.

Cleator S & Ashworth A (2004) Molecular profiling of breast cancer: clinical implications. *Brit. J. Cancer* 90, 1120–1124.

Courtneidge SA (2003) Escape from inhibition. *Nature* 422, 827–828.

Downward J (2003) Targeting Ras signalling pathways in cancer therapy. *Nat. Rev. Cancer* 3, 11–22.

Druker BJ, Talpaz M, Resta DJ et al. (2001) Efficacy and safety of a specific inhibitor of the BCR-ABL tyrosine kinase in chronic myeloid leukemia. *N. Engl. J. Med.* 344, 1031–1037.

Duensing A, Medeiros F, McConarty B et al. (2004) Mechanisms of oncogenic KIT signal transduction in primary gastrointestinal stromal tumors (GISTs). *Oncogene* 23, 3999–4006.

Elrick LJ, Jorgensen HG, Mountford JC and Holyoake TL, (2005) Punish the parent not the progeny. *Blood* 105, 1862–1866.

Felsher DW (2004) Reversibility of oncogene-induced cancer. *Curr. Opin. Genet. Dev.* 14, 37–42.

Futreal PA, Coin L, Marshall M et al. (2004) A census of human cancer genes. *Nat. Rev. Cancer* 4, 177–183.

Gibbs JB (2000) Mechanism-based target identification and drug discovery in cancer research. *Science* 287, 1969–1973.

Gills JJ, Granville CA & Dennis PA (2004) Targeting aberrant signal transduction pathways in lung cancer. *Cancer Biol. Ther.* 3, 147–155.

Gingras A-C, Raught B & Sonenberg N (2001) Regulation of translation initiation by FRAP/mTOR. *Genes Dev.* 15, 807–826.

Guertin DA & Sabatini DM (2005) An expanding role for mTOR in cancer. *Trends Mol. Med.* 11, 353–361.

Hideshima T (2002) NF-kappa B as a therapeutic target in multiple myeloma. *J. Biol. Chem.* 277, 16639–16647.

Hingorani SR & Tuveson DA (2003) Targeting oncogene dependence and resistance. *Cancer Cell* 3, 414–417.

Ince TA & Weinberg RA (2002) Functional genomics and the breast cancer problem. *Cancer Cell* 1, 15–17.

Kamb A. (2005) What's wrong with our cancer models? *Nat. Rev. Drug Discov.* 4,161–165.

Klein S, McCormick F & Levitzki A (2005) Killing time for cancer cells. *Nat. Rev. Cancer* 5, 573–580.

Kloog Y & Cox AD (2004) Prenyl-binding domains: potential targets for Ras inhibitors and anti-cancer drugs. *Semin. Cancer Biol.* 14, 253–261.

Lynch TJ, Bell DW, Sordella R et al. (2004) Activating mutations in the epidermal growth factor receptor underlying responsiveness of non-small-cell lung cancer to Gefitinib. *N. Engl. J. Med.* 350, 2129–2139.

McCormick F (2001) Cancer gene therapy: fringe or cutting edge? *Nat. Rev. Cancer* 1, 130–141.

Mendelsohn J & Baselga J (2003) Status of epidermal growth factor receptor antagonists in the biology and treatment of cancer. *J. Clin. Oncol.* 21, 2787–2799.

Nahta R, Hortobágyi GN & Esteva FJ (2003) Growth factor receptors in breast cancer: potential for therapeutic intervention. *Oncologist* 8, 5–17.

Piazza F, Gurrieri C & Pandolfi PP (2001) The theory of APL. *Oncogene* 20, 7216–7222.

Ramaswamy S & Golub TR (2002) DNA microarrays in clinical oncology. *J. Clin. Oncol.* 20, 1932–1941.

Rothenberg ML, Carbone DP & Johnson DH (2003) Improving the evaluation of new cancer treatments: challenges and opportunities. *Nat. Rev. Cancer* 3, 303–309.

Salesse S & Verfaillie CM (2002) BCR/ABL: from molecular mechanisms of leukemia induction to treatment of chronic myelogenous leukemia. *Oncogene* 21, 8547–8559.

Sawyers CL (1999) Chronic myeloid leukemia. *N. Engl. J. Med.* 340:1330–1340.

Sawyers CL (2003) Opportunities and challenges in the development of kinase inhibitor therapy for cancer. *Genes Dev.* 17, 2998–3010.

Sawyers C (2004) Targeted cancer therapy. *Nature* 432, 204–207.

Schlessinger J (2005) SU11248: Genesis of a new cancer drug. *The Scientist* 19, 17–18.

Scott MP (2003) A twist in hedgehog's tale. *Nature* 425, 780–782.

Sebolt-Leopold JS & Herrera R (2004) Targeting the mitogen-activated protein kinase cascade to treat cancer. *Nat. Rev. Cancer* 4, 937–947.

Sebti S & Der CJ (2003) Searching for the elusive targets of farnesyltransferase inhibitors. *Nat. Rev. Cancer* 3, 945–951.

Segal E, Friedman N, Kaminski N et al. (2005) From signatures to models: understanding cancer using microarrays. *Nat. Genet.* 37:S38-S45.

Shah NP & Sawyers CL (2003) Mechanisms of resistance to STI571 in Philadelphia chromosome-associated leukemias. *Oncogene* 22, 7389–7395.

Shawver LK, Slamon D & Ullrich A (2002) Smart drugs: tyrosine kinase inhibitors in cancer therapy. *Cancer Cell* 1, 117–123.

Staudt LM (2003) Molecular diagnosis of the hematologic cancers. *N. Engl. J. Med.* 348, 1777–1785.

Strausberg RL, Simpson AJ, Old LJ & Riggins GJ (2004) Oncogenomics and the development of new cancer therapies. *Nature* 429, 469–474.

Weigelt B, Peterse JL & van't Veer LJ (2005) Breast cancer metastasis: markers and models. *Nat. Rev. Cancer* 5, 591–602.

Vande Woude GF, Kelloff GJ, Ruddon RW et al. (2004) Reanalysis of cancer drugs: old drugs, new tricks. *Clin. Cancer Res.* Voorhees PM, Dees EC, O'Neil B & Orlowski RZ (2003) The proteasome as a target for cancer therapy.

Voorhees PM, Dees EC, O'Neil B & Orlowski RZ (2003) The proteasome as a target for cancer therapy. *Clin. Cancer Res.* 9, 6316–6325.

Zwick E, Bange J & Ullrich A (2002) Receptor tyrosine kinases as targets for anticancer drugs. *Trends Mol. Med.* 8, 16–23.

Abbreviations

A (1) adenine; (2) adenosine
ABC activated B-cell-like subtype of DLBCL
ABH alkB homolog
Abl Abelson leukemia virus oncoprotein
ACT adoptive cell transfer
ACTH adrenocorticotropic hormone
ADCC antibody-dependent cellular cytotoxicity
AEV avian erythroblastosis virus
AFB1 aflatoxin B1
Ag antigen
AID activation-induced cytidine deaminase
AIDS acquired immunodeficiency syndrome
ALL acute lymphocytic (or lymphoblastic) leukemia
ALT alternative lengthening of telomeres
ALV avian leukosis virus
AML acute myelogenous leukemia
AMP adenosine monophosphate
AMV avian myelocytomatosis virus
Ang angiopoietin
AP (1) apurinic; (2) apyrimidinic
Apaf-1 apoptotic protease-activating factor-1
APC (1) antigen-presenting cell; (2) anaphase-promoting complex; (3) adenomatous polyposis coli
APE (1) apurinic endonuclease; (2) apyrimidinic endonuclease
APL acute promyelocytic leukemia
ARF alternative reading frame (protein)
asmase acid sphingomyelinase
AT ataxia telangiectasia (syndrome)
ATM mutated in ataxia telangiectasia
ATP adenosine triphosphate
ATR ATM-related kinase
ATRA all-*trans*-retinoic acid
AUC area under the curve
BCC basal cell carcinoma
BCG Bacillus Calmette–Guérin (mycobacterium)
Bcl-2 B-cell lymphoma gene-2
BCNS basal cell nevus syndrome
BCR breakpoint cluster region
BER base-excision DNA repair
β-gal β-galactosidase
β_2m β_2 microglobulin
BFB breakage–fusion–bridge (cycle of chromosomes)
BH Bcl-2–homologous (domain)
BHK baby hamster kidney (cells)
bHLH basic helix–loop–helix (transcription factor)
BL Burkitt's lymphoma
BMI body-mass index
BMP bone morphogenetic protein
BMPR1 BMP receptor-1
BMSC bone marrow stem cell
BMT bone marrow transplantation
BP benzo[*a*]pyrene
bp base pair
BPDE benzo[*a*]pyrenediolepoxide
BPV bovine papillomavirus
BrdU bromodeoxyuridine
BTK Bruton's TK
C (1) carboxy (terminus of protein); (2) constant region (of an antibody molecule); (3) cytidine; (4) cytosine
C′ C (carboxy) terminus; C-terminal (adj.)
CA constitutively active
CAD caspase-activated DNase
CAF carcinoma-associated fibroblast
CAK CDK-activating kinase
CalB calcium–phospholipid binding domain
cAMP cyclic adenosine monophosphate
CASP caspase-encoding gene
CBP cyclic AMP response element–binding protein
CCK cholecystokinin
CDC complement-dependent cytotoxicity
CDH1 E-cadherin
CDK cyclin-dependent kinase
CdkI CDK inhibitor
cDNA complementary DNA copy of mRNA
CEA carcinoembryonic antigen
CEF chicken embryo fibroblast
CEP circulating endothelial progenitor (cell)
CGH comparative genomic hybridization
ChIP chromatin immunoprecipitation
CHOP cyclophosphamide, doxorubicin, vincristine, and prednisone (chemotherapy protocol)
CI confidence interval
CIITA class II transactivator (transcription factor)
CIN (1) cervical intraepithelial neoplasia; (2) chromosomal instability
CIS carcinoma *in situ*
CKII casein kinase II
CML chronic myelogenous leukemia
CMV cytomegalovirus
CNL chronic neutrophilic leukemia
CNS central nervous system
COX-2 cyclooxygenase-2
CPD cyclobutane pyrimidine dimer
CPE cytopathic effect
CR complete response
13cRA 13-*cis*-retinoic acid
Crkl Crk-like (*also* **CrkL**)
CS Cockayne syndrome
CSF-1 colony-stimulating factor-1
Csk C-terminal Src kinase
CT (1) cancer-testis (antigen); (2) computed (X-ray) tomography; (3) chemotherapy
CTGF connective tissue growth factor
CTL cytotoxic T lymphocyte
CTLA-4 cytotoxic T lymphocyte–associated antigen-4
CTVS canine transmissible venereal sarcoma
CUP (metastatic) cancer of unknown primary
CXCL ligand of chemokine receptor
CXCR chemokine receptor
Cyp cytochrome P-450 enzyme; *also* **CYP**
d5′-mC deoxy-5′-methylcytosine
D/D dimerization domain
dA deoxyadenosine
DAG diacylglycerol
DAPI 4′,6′-diamidino-2-phenylindole
DC (1) dendritic cell; (2) dyskeratosis congenita
DCIS ductal carcinoma *in situ* (of the breast)
DFS disease-free survival
dG deoxyguanosine
Diablo direct IAP-binding protein with low pI
DISC death-inducing signaling complex
DLBCL diffuse large B-cell lymphoma
DM double minute chromosome
DMBA dimethylbenz[*a*]anthracene
DN dominant-negative (allele)
DP differentiation-related transcription factor protein
DPC4 deleted in pancreatic cancer-4

dR deoxyribose
ds double-stranded (DNA or RNA)
dTg deoxythymidine glycol
DUB de-ubiquitylating enzyme
dUTP deoxyuridine triphosphate
4E-BP eIF4E-binding protein
E2F transcription factor activating adenovirus E2 gene
EBV Epstein–Barr virus
EC embryonal carcinoma (cell)
ECM extracellular matrix
EFC-XV endostatin-like fragment of type XV collagen
EFS event-free survival
EGF epidermal growth factor
EGF-R EGF receptor
eIF eukaryotic (translation) initiation factor
eIF4E eukaryotic (translation) initiation factor 4E
EM electron micrograph (or electron microscope)
EMSA electrophoretic mobility shift assay
EMT epithelial–mesenchymal transition
ENU *N*-ethylnitrosourea
env retrovirus envelope glycoprotein
EPC endothelial precursor cell; endothelial progenitor cell
EpCAM epithelial cell adhesion molecule
EPO erythropoietin
ER estrogen receptor
ERK extracellular signal–regulated kinase
ERV endogenous retrovirus
ES embryonic stem (cells)
ESA epithelial surface antigen
ET-1 endothelin-1
FA Fanconi anemia (syndrome)
FACS fluorescence-activated cell sorting
FADD Fas-associated death domain (protein)
FAK focal adhesion kinase
FAP (1) familial adenomatous polyposis; (2) fibroblast-activating protein
FasL ligand of the Fas death receptor
Fc constant region of an IgG antibody molecule
FDA U.S. Food and Drug Administration
FDG 2-fluoro-2-deoxy-D-glucose
FGF fibroblast growth factor
FISH fluorescence *in situ* hybridization
FITC fluorescein isothiocyanate (dye)
FKBP12 FK506-binding protein of 12 kD
FL (1) follicular (B-cell) lymphoma; (2) ligand of the Flt-3 receptor
FLICE FADD-like interleukin-1β–converting enzyme
FLIP FLICE-inhibitory protein
Flt Fms-like tyrosine kinase
FLV Friend murine leukemia virus
FPT farnesyl protein transferase
FRB FKBP12 + rapamycin–binding (domain of mTOR)
Frz Frizzled
FSH follicle-stimulating hormone
FTI farnesyltransferase inhibitor
5-FU 5-fluorouracil
G (1) guanine; (2) guanine nucleotide; (3) guanosine
G_1 gap 1 phase (of cell cycle)
G_2 gap 2 phase (of cell cycle)
G6PD glucose-6-phosphate dehydrogenase
GAG glycosaminoglycan
gag group-specific antigen (retrovirus capsid protein)
GAK cyclin G–associated kinase
GAP GTPase-activating protein
GAPDH glyceraldehyde-3-phosphate dehydrogenase
GBM glioblastoma multiforme
GCB germinal-center B-cell lymphoma
GDNF glial cell–derived neurotrophic growth factor
GDP guanosine diphosphate
GEF guanine nucleotide exchange factor
GF growth factor
GFP green fluorescent protein
GGR global genomic repair
GH growth hormone
GIST gastrointestinal stromal tumor
GM-CSF granulocyte macrophage-colony–stimulating factor
gp glycoprotein
GPCR G protein–coupled receptor
GRP gastrin-releasing peptide
GSK-3β glycogen synthase kinase-3β
GSTP1 glutathione-S-transferase π
GTP guanosine triphosphate
GTPase enzyme that cleaves GTP, usually generating GDP
GvH graft versus host (immunological reaction)
GvT graft versus tumor (immunological reaction)
Gy gray (radiation dose unit)
^{3}H tritium
h prefix referring to the human form of a gene or protein
H&E hematoxylin–eosin (tissue stain)
HAMA human anti-mouse antibody
HB-EGF heparin-binding EGF
HBV hepatitis B virus
HCA heterocyclic amine
HCC hepatocellular carcinoma
hCG human chorionic gonadotropin
HCL hairy-cell leukemia
HCMV human cytomegalovirus
HCV hepatitis C virus
HDAC histone deacetylase
HDM2 human homolog of MDM2
HDR homology-directed repair
HEK human embryonic kidney (cells)
HERV human endogenous retrovirus
HGF hepatocyte growth factor; *also called* **scatter factor (SF)**
HGPRT hypoxanthine guanine phosphoribosyltransferase
Hh Hedgehog
HHV-8 human herpesvirus type 8; *also called* **Kaposi's sarcoma herpesvirus (KSHV)**
HIF hypoxia-inducible (transcription) factor
HIP1 huntingtin-interacting protein-1
HIV human immunodeficiency virus
HLA human leukocyte antigen (*equivalent to* **MHC**)
HMEC human mammary epithelial cell
HNPCC hereditary non-polyposis colon cancer
hnRNA heterogeneous nuclear RNA
HNSCC head and neck squamous cell carcinoma
HPE human prostate epithelial (cell)
HPRT hypoxanthine phosphoribosyltransferase
HPV human papillomavirus
HSA human serum albumin
HSC hematopoietic stem cell
HSIL high-grade squamous intraepithelial lesion
HSR homogeneously staining region
HSV-1 herpes simplex virus type 1
HTLV-I human T-cell lymphotropic virus type I (or human T-cell leukemia virus)
HTS high-throughput screening
HU hydroxyurea
IAP inhibitor of apoptosis
IC_{50} concentration required to obtain 50% inhibition
ICAD inhibitor of caspase-activated DNase
IEG immediate early gene
IEL intraepithelial lymphocyte
IFN interferon
IFN-R interferon receptor
IGF-1 insulin-like growth factor-1
IGF-1R IGF-1 receptor
IGFBP insulin-like growth factor–binding protein
IκB inhibitor of NF-κB
IKK IκB kinase
IL interleukin
ILK integrin-linked kinase
IP_3 inositol (1,4,5)-triphosphate
IR insulin receptor

IRK insulin receptor kinase
Jak Janus kinase
JCV JC virus, a close relative of SV40
JM juxtamembrane
kb kilobase
kbp kilobase pair
kD kilodalton
KGF keratinocyte growth factor
KIR killer inhibitory receptor
KO knockout; knocked out (gene)
KSHV *see* **HHV-8**
LAK lymphokine-activated killer (cell)
LCM laser capture microdissection
LDL low-density lipoprotein
LEF lymphoid enhancer factor
LH luteinizing hormone
LIF leukemia inhibitory factor
LLC Lewis lung carcinoma
LOH loss of heterozygosity
LRC label-retaining cell
LT large T antigen (of SV40 or polyomavirus)
LTR long terminal repeat (of a provirus)
M mitosis
m prefix referring to the murine form of a gene or protein
M-CSF macrophage-colony–stimulating factor
MAC membrane attack complex
MALT mucosa-associated lymphoid tissue
MAPK mitogen-activated protein kinase; *also* **MAP-kinase**
MAPKK kinase that phosphorylates a MAPK
MAPKKK kinase that phosphorylates a MAPKK
3-MC 3-methylcholanthrene
MCP-1 monocyte chemotactic protein-1
mCRP membrane-bound complement regulatory protein
MDCK Maden–Darby canine kidney (cells)
MDM2 mouse double minute chromosome 2
MDR multi-drug resistance (phenotype)
MDS myelodysplastic syndrome
MEC mammary epithelial cell
MEF mouse embryo fibroblast
MeIQx 2-amino-3,8-dimethylimidazo[4,5-f]-quinoxaline
MEK MAPK/Erk kinase
MEN multiple endocrine neoplasia
MET mesenchymal–epithelial transition
MGMT O^6-methylguanine DNA methyltransferase
MHC major histocompatibility complex (*see also* **HLA**)
MIN microsatellite instability
MKP MAP kinase phosphatase
MLV murine leukemia virus
mM millimolar (10^{-3} molar)
MM multiple myeloma
MMP matrix metalloproteinase
MMR mismatch repair (of DNA)
mms methylmethane sulfonate
MMTV mouse mammary tumor virus
MNNG *N*-methyl-*N′*-nitro-*N*-nitrosoguanidine
MNU *N*-methylnitrosourea
MoAb monoclonal antibody
moca 4,4′-methylenebis(2-chloroaniline)
MPNST malignant peripheral nerve sheath tumor
MRI magnetic resonance imaging
mRNA messenger RNA
MSP methylation-specific polymerase chain reaction
MT middle T antigen of polyomavirus
mt mutant type
MT1-MMP membrane type MMP
MTD maximum tolerated dose
MTH1 *mutT* homolog-1
mTOR mammalian target of rapamycin
μM micromolar (10^{-6} molar)
MUP major urinary protein
N′ N (amino) terminus; N-terminal (adj.)
NAT *N*-acetyltransferase
NBCCS nevoid basal cell carcinoma syndrome
NBS Nijmegen break syndrome
NER nucleotide-excision DNA repair
NES nuclear export signal
Nf/NF neurofibromatosis
NF-κB nuclear factor-κB
NGF nerve growth factor
NHEJ nonhomologous end joining
NHL non-Hodgkin's lymphoma
NHPF normal human prostate fibroblast
NK natural killer (lymphocyte)
NKT natural killer lymphocyte expressing T-cell receptor
NLS nuclear localization (import) sequence
nM nanomolar (10^{-9} molar)
NMR nuclear magnetic resonance
NOD non-obese diabetic
NOD/SCID non-obese diabetic/severe combined immunodeficient
NotchL Notch ligand
NRG neuregulin
NSAID nonsteroidal anti-inflammatory drug
NSCLC non-small-cell lung carcinoma
NTD N-terminal domain
OD optical density
OHT 4-hydroxy tamoxifen
OPG osteoprotegerin ligand
OPN osteopontin
OSE ovarian surface epithelium
8-oxo-dG 8-oxo-deoxyguanosine
p (1) short arm of a chromosome; (2) probability of an event
PAH polycyclic aromatic hydrocarbon
PAI-1 plasminogen activator inhibitor-1
PanIN pancreatic intraepithelial neoplasia
PARP poly (ADP-ribose) polymerase
PBS phosphate-buffered saline
PCNA proliferating-cell nuclear antigen
PCR polymerase chain reaction
PD (1) pharmacodynamics; (2) population doubling
PDGF platelet-derived growth factor
PDGF-R PDGF receptor
PDK1 phosphoinositide-dependent kinase 1
PEG polyethylene glycol
PET positron-emission tomography
PFS progression-free survival
PGE_2 prostaglandin E_2
PH pleckstrin homology (PIP_3 binding) (domain)
PhIP 2-amino-1-methyl-6-phenylimidazo [4,5-b] pyridine
PhK phosphorylase kinase
PI (1) phosphatidylinositol; (2) propidium iodide
$PI(3,4,5)P_3$ *equivalent to* **PIP_3**
PI3K phosphatidylinositol 3-kinase
PIN prostatic intraepithelial neoplasia
PINCH particularly interesting new cysteine-histidine-rich (protein)
PIP_2 phosphatidylinositol (4,5)-diphosphate
PIP_3 phosphatidylinositol (3,4,5)-triphosphate
PK pharmacokinetics
PKA protein kinase A
PKB protein kinase B (=Akt)
PKC protein kinase C
PLC phospholipase C
PLD phospholipase D
PMA phorbol-12-myristate-13-acetate (=TPA)
PMBL primary mediastinal B-cell lymphoma
PML (1) promyelocytic leukemia; (2) progressive multifocal leukoencephalopathy
PMN polymorphonuclear leukocytes/neutrophils
PND paraneoplastic neurological degeneration
pol (1) polymerase; (2) retrovirus reverse transcriptase
pol II RNA polymerase II
6-4 PP pyrimidine (6-4) pyrimidinone photoproduct
PP1 protein phosphatase 1

PP2A	protein phosphatase 2A
PR	partial response
pRb	retinoblastoma protein
PRL	prolactin
pro-B ALL	pro-B-cell acute lymphocytic leukemia
PS	phosphatidylserine
PSA	prostate-specific antigen
PTB	phosphotyrosine-binding (domain)
PTC	(1) papillary thyroid carcinoma; (2) Patched
PTEN	phosphatase and tensin homolog deleted on Chromosome 10
PTH	parathyroid hormone
PTHrP	parathyroid hormone–related peptide
PTP	phosphotyrosine phosphatase
PyMT	polyomavirus middle T oncoprotein
q	long arm of a chromosome
qPCR	quantitative polymerase chain reaction
R	restriction point
RA	retinoic acid
RAG	recombination-activating gene
RANK	receptor activator of NF-κB
RANKL	ligand of RANK (receptor)
RAR	retinoic acid receptor
Rb	retinoblastoma protein
REF	rat embryo fibroblast
RFLP	restriction fragment length polymorphism
Rheb	Ras homolog enriched in brain
ROS	reactive oxygen species
RPA	replication protein A
RR	relative risk
RSV	Rous sarcoma virus
RT	reverse transcriptase
RTK	tyrosine kinase receptor
S	(1) DNA synthesis phase (of cell cycle); (2) svedberg (unit of sedimentation in centrifuge)
S1P	sphingosine-1-phosphate
S6	protein 6 of the small ribosomal subunit
S6K1	p70 S6 kinase-1
SA-β-gal	senescence-associated β-galactosidase
SAHF	senescence-associated heterochromatic foci
SCE	sister chromatid exchange
SCF	stem cell factor
SCID	severe combined immunodeficiency (syndrome)
SCLC	small-cell lung carcinoma
SDF-1	stroma-derived factor-1
SEM	scanning electron micrograph/microscope/microscopy
Ser	serine
SF	scatter factor; *see* **HGF**
SH1	Src homology 1 domain (tyrosine kinase)
SH2	Src homology 2 domain (phosphotyrosine binding)
SH3	Src homology 3 domain (proline-rich binding)
SHP	SH2-containing phosphatase
SIP1	Smad-interacting protein 1
SKY	multicolor spectral karyotyping
Sky1	SR-protein–specific kinase of budding yeast
SLE	systemic lupus erythematosus
SMA	smooth muscle actin
Smac	second mitochondria-derived activator of caspase
Smo	Smoothened
SNP	single-nucleotide polymorphism
SOD	superoxide dismutase
Sos	son of sevenless
Srf	serum response factor
ss	single-stranded (DNA or RNA)
STAT	signal transducer and activator of transcription
Str-1	stromelysin-1
STS	staurosporine
SUMO	small ubiquitin-like modifier
SV	simian virus
T	(1) thymine; (2) thymidine
t	referring to a chromosomal translocation
$T_{1/2}$	half-life
T_C	cytotoxic T cell
T_H	helper T cell
T_m	melting temperature
T_{reg}	regulatory T cell
TAA	tumor-associated antigen
T-ALL	T-cell acute lymphocytic leukemia
TAM	tumor-associated macrophage
TAP1/2	transporter associated with antigen processing
TATA	tumor-associated transplantation antigen
TCF	T-cell factor
TCR	(1) T-cell receptor; (2) transcription-coupled DNA repair
TdT	terminal deoxynucleotidyl transferase
TEL/AML1	chromosomal translocation
TEM	transmission electron microscopy
TERT	telomerase reverse transcriptase
TF	transcription factor
TGF	transforming growth factor
TGF-α	transforming growth factor-α
TGF-β	transforming growth factor-β
TGF-βR	TGF-β receptor
Thr	threonine
TIAM-1	T-cell lymphoma invasion and metastasis gene-1
TIL	tumor-infiltrating lymphocyte
TIMP	tissue inhibitor of metalloproteinase
TK	tyrosine kinase
TM	transmembrane (domain of a protein)
TNF	tumor necrosis factor
TOR	target of rapamycin
TPA	12-*O*-tetradecanoylphorbol-13-acetate (=PMA)
TPO	thrombopoietin
TRAIL	TNF-related apoptosis-inducing ligand
TRAP	telomeric repeat amplification protocol
TRF	telomeric restriction fragment
tRNA	transfer RNA
TrpRS	tryptophanyl-tRNA synthetase
ts	temperature-sensitive
TSA	(1) trichostatin A; (2) tumor-specific antigen
TSC	tuberous sclerosis
TSG	tumor suppressor gene
TSHR	thyroid-stimulating hormone receptor
Tsp-1	thrombospondin-1
TSTA	tumor-specific transplantation antigen
TUNEL	TdT-mediated dUTP nick end labeling
U	(1) uracil; (2) uridine
UC	ulcerative colitis
uPA	urokinase (type) plasminogen activator
uPAR	receptor of uPA
UV	ultraviolet
UV-B	ultraviolet-B radiation
V	variable region (of an antibody molecule).
VEGF	vascular endothelial growth factor (=VPF)
VHL	von Hippel–Lindau
VPF	vascular permeability factor (=VEGF)
VSMC	vascular smooth muscle cell
WM	Waldenström's macroglobulinemia
wt	wild type
XAF1	XIAP-associated factor 1
XIAP	X-chromosome–linked inhibitor of apoptosis
XP	xeroderma pigmentosum

Glossary

abasic
Referring to a nucleotide that has lost its purine or pyrimidine base. *See also* **apurinic**; **apyrimidinic**.

ablate
To eliminate.

abluminal
Located away from the lumen of a duct or other hollow structure.

abscissa
Horizontal or *x* axis of a Cartesian graph.

abscission
Final step of cytokinesis, when the remaining linkages, including those created by microtubules, are severed, allowing complete separation of the two daughter cells.

acellular
Lacking or deprived of cells.

acetylation
Covalent attachment of an acetyl group to a second molecule such as a protein.

acromegaly
Pathological condition of excessive growth of certain tissues, usually due to the elaboration of excessive growth hormone by a pituitary tumor.

adaptive immune response
An immune system response that is acquired or learned following exposure of an organism to an antigen or antigen-bearing agent. *See also* **innate immune response**.

adduct
The novel molecular structure arising after covalent linkage of a mutagen with one or another portion of a DNA molecule.

adenocarcinoma
Tumor derived from secretory epithelial cells.

adenoma (adj., **-omatous)**
Any of a series of pre-malignant, noninvasive growths in various epithelial tissues, many of which have the potential to progress further to carcinomas. *See also* **polyp**.

adherens junctions
Lateral junctions between adjacent cells, notably epithelial cells, that serve to mechanically link their cytoplasmic actin cytoskeletons.

adipocytes
Specialized cells of the mesenchymal lineage, closely related to fibroblasts, that create fat and store it in large globules in the cytoplasm; the dominant cell type in fatty tissues.

adipogenic
Causing differentiation into adipocytes.

adjuvant
(1) A substance that is a powerful stimulant of the immune response while not being antigenic itself. (2) A substance that is used in concert with a therapy whose efficacy it enhances.

adoptive transfer
Procedure in which immune cells are transferred from a donor to a recipient, undertaken with the hope that the donor's cells will be able to mediate an immune function lacking in the recipient.

adrenal
Referring to the secretory glands that sit above the kidneys

afferent
Referring to incoming signals.

agonist
Activating agent; opposite of antagonist.

alkylating
Capable of attaching an alkyl group or similarly structured chemical group to a substrate such as a DNA base.

allele
One alternative among different versions of a gene that may be defined by the phenotype that it creates, by the protein that it specifies, or by its nucleotide sequence.

allogeneic
(1) Referring to two genetically distinct members of the same species. (2) Describing the relationship between two sets of cells or tissues deriving from distinct genetic backgrounds.

allograft
Implantation of cells of an animal of one genetic background into a host animal of another genetic background but from the same species.

allograft rejection
Situation arising when a donor tissue graft is rejected by the immune system of a recipient because the donor and recipient, although members of the same species, are genetically distinct.

alternative splicing
Process whereby a pre-mRNA may be spliced in several alternative ways, resulting in mRNAs composed of different combinations of exons.

***Alu* repeat**
Sequence block of about 300 bp that is found in 500,000 to 1 million copies scattered throughout the human genome.

alveolus (pl., **alveoli)**
A small sac within a tissue that is connected to a duct, such as the sacs seen in the lung and in the mammary gland during pregnancy.

amphipathic
Describing a molecule containing distinct hydrophobic and hydrophilic domains.

amplicon
A defined stretch of (chromosomal) DNA that undergoes amplification.

amplification
Genetic mechanism by which the copy number of a gene is increased above its normal level in the diploid genome.

anaphase
Third subphase of mitosis, during which the paired chromatids are segregated to the two opposite poles of the cell.

anaplastic
(Referring to a tumor) having a tissue and cellular architecture lacking the differentiated characteristics of the tissue of origin.

anastomosis
End-to-end connection formed between two ductal structures, such as a connection formed directly between an artery and a vein.

anchorage dependence
Requirement of normal cells for tethering to a solid substrate before they will grow.

anchorage independence
Ability of a cell to proliferate without attachment to a solid substrate.

aneuploid
(1) Describing a karyotype that deviates from diploid because of increases or decreases in the numbers of certain chromosomes. (2) Less commonly, describing a karyotype that carries structurally abnormal chromosomes.

angiogenesis
Process by which new blood vessels are formed; *also called* **neoangiogenesis** in certain circumstances.

angiogenic factor
Type of growth factor that is specialized to induce angiogenesis.

angiogenic switch
The shift by a clump of tumor cells from a state in which they are unable to induce neovascularization to one in which they exhibit this ability.

angiosarcoma
Tumor of the cells that are precursors to endothelial cells.

anoikis
Form of apoptosis that is triggered by the failure of a cell to establish anchorage to a solid substrate, such as the extracellular matrix, or by loss of such anchorage.

anoxia
State or environment in which oxygen is essentially absent.

antibody
A soluble protein produced by plasma cells of the immune system that is capable of recognizing and binding particular antigens with high specificity. *See also* **immunoglobulin**.

antigen
A molecule or portion of a molecule, often an oligopeptide, that can be specifically recognized and bound by an antibody or a T-cell receptor or that provokes the production of an antibody.

antigen-presenting cells
A class of cells—often termed professional antigen-presenting cells and including dendritic cells, macrophages, and B cells—that present oligopeptide antigens via class II MHC molecules to other immunocytes, notably helper and cytotoxic T cells.

antiserum
Serum that is produced by an animal exposed to a specific antigen and is able to recognize and bind that antigen.

apatite
Mineral component of bone composed of calcium phosphate crystals.

apoptosis
Complex program of cellular self-destruction, triggered by a variety of stimuli and involving the activation of caspase enzymes, that results in quick fragmentation and phagocytosis of a cell.

apoptosome
Multiprotein complex that consists of cytochrome *c* molecules and Apaf-1 and helps to initiate apoptosis by activating procaspase 9 into caspase 9.

apposed
Placed directly next to something.

apurinic
Referring to the product of depurination, in which the glycosidic bond linking a deoxyribose or ribose to a purine base is broken, leaving behind only the deoxyribose or ribose in the DNA or RNA, respectively.

apyrimidinic
Referring to the product of depyrimidination, in which the glycosidic bond linking a deoxyribose or ribose to a pyrimidine base is broken, leaving behind only the deoxyribose or ribose in the DNA or RNA, respectively.

aromatic
Referring to an organic molecule that contains one or more benzene rings.

arteriole
A small artery that empties into capillaries.

ascites
The fluid that accumulates in the peritoneal cavity of some cancer patients, often containing malignant cells.

astrocytoma
A tumor of the astrocytes, a type of nonneuronal support cell in the brain.

asynchronous
Referring to a population of cells that are dispersed throughout the cell cycle during any point of time and therefore do not execute specific cell cycle steps in a coordinated or synchronous manner.

ataxia
Loss of muscle coordination, often due to cerebellar dysfunction.

atrophy
Shrinkage of a tissue, often due to loss of viability of its component cells or to loss of normal cell numbers.

autochthonous
(1) Of native origin. (2) Referring to a tumor that arises within an organism (rather than from implanted cells or tumor fragments).

autocrine
Referring to the signaling path of a hormone or factor that is released by a cell and proceeds to act upon the same cell (or same cell type) that has released it.

autoimmune
Referring to a process or disease in which the immune system attacks an organism's own normal cells and tissues.

autologous
Referring to biological material, usually cells or tissues, that originates in a patient's own body (and may be reintroduced into that patient following some manipulation *ex vivo*).

autophagy
Program of cellular responses to nutrient deprivation involving the digestion of a cell's organelles within its own lysosomes.

autophosphorylation
Phosphorylation of a protein molecule by its own associated kinase activity.

autoradiography
Procedure for detecting radiolabeled molecules by placing them (or the samples carrying them) adjacent to a radiographic emulsion, which responds to radioactive decay by producing silver granules.

autosome
A chromosome that is not a sex chromosome, i.e., neither an X nor a Y chromosome.

axillary
Referring to the armpit.

Barr body
The condensed, inactive X chromosome found in each of the cells of females of placental mammals.

Barrett's esophagus
Metaplasia in which squamous epithelium of the esophagus is replaced by secretory epithelial cells of a type normally found in the stomach.

basal
(1) Referring to a lower physical location. (2) Referring to a (low) rate of activity or function observed in the absence of any activating stimulus.

basal lamina
See basement membrane.

base-excision repair
A form of DNA repair that initially involves cleavage by a repair enzyme of the glycosidic bond between a base and a deoxyribose, leaving behind an abasic nucleotide.

basement membrane
A specialized extracellular matrix that forms a sheet separating epithelial from stromal cells or endothelial cells from pericytes; *sometimes called* **basal lamina.**

benign
(1) Describing a growth that is confined to a specific site within a tissue and gives no evidence of invading adjacent tissue. (2) Referring to an epithelial growth that has not penetrated through the basement membrane.

bioinformatics
The science of using computational methods for analyzing biological information, notably complex sets of biological data.

bi-specific
Able to specifically recognize two objects simultaneously.

bisphosphonates
A class of drugs, characterized by a chemical backbone with the structure P–C–P, that are incorporated into bone apatite and subsequently become available to poison osteoclasts that might later dissolve the bone.

blast
Term, often used as prefix or suffix, indicating a relatively undifferentiated or embryonic precursor cell.

blastocoel
Inner cavity present in a blastocyst.

blastocyst
An early stage of vertebrate embryogenesis in which the embryo consists largely of an outer layer of cells enclosing an inner cavity.

bleb
A small, bubble-like herniation through a membrane such as the plasma membrane.

body-mass index
Weight in kilograms divided by the square of height in meters.

breakpoint
Location in a chromosomal region or gene at which it becomes fused through chromosomal translocation with another chromosomal region or gene.

bronchial
Referring to a major airway of the lung.

buccal
Referring to the tissues of the oral cavity, specifically the epithelial lining of the cheek.

by-pass polymerase
A DNA polymerase that will copy over an unrepaired lesion in the template strand of the DNA, "guessing" the nucleotides that should be incorporated into the nascent complementary DNA strand in order to avoid inappropriate incorporation and resulting mutation.

cachexia
Physiologic state, often seen late in cancer development, in which the patient loses appetite and suffers wasting of tissues throughout the body.

café au lait spots
Coffee-with-milk–colored spots on the skin that are seen characteristically in the neurofibromatosis type 1 syndrome.

capsid
Protein coat of a virus particle that envelops and protects the viral genome.

carcinogen
An agent that contributes to the formation of a tumor.

carcinogenic
Capable of causing or contributing to the causation of cancer.

carcinoma (adj., **-omatous)**
A cancer arising from epithelial cells.

caretaker
A gene that encodes a protein responsible for maintaining the integrity of the genome and thereby preventing the appearance of neoplastic cells.

caspase
A cysteine aspartyl–specific protease.

CD4
A cell surface protein displayed by helper T cells that enables them to recognize MHC class II proteins on the surface of professional antigen-presenting cells.

CD8
A cell surface protein displayed by cytotoxic T cells that enables them to recognize MHC class I proteins on the surface of cells that they may target for destruction.

CD20
An antigen that is displayed on the surface of many cell types of the B-cell lineage.

cell-autonomous
Referring to a trait or behavior of a cell that is governed by its own genome and internal physiology and not by its ongoing interactions with other cells.

cell cycle
The sequence of changes in a cell from the moment when it is created by cell division, continuing through a period in which its contents including chromosomal DNA are doubled, and ending with the subsequent cell division and formation of daughter cells.

cell cycle clock
Network of signaling proteins in the nucleus that regulate and orchestrate progression of the cell through the cell cycle.

cell line
A strain of cells that has been adapted to grow indefinitely in culture.

cellular immune response
The arm of the immune system that depends on the ability of specific cell types, such as cytotoxic T cells, natural killer cells, and macrophages, to recognize and destroy specific entities, including abnormal cells and infectious agents.

centriole
Component of centrosome from which spindle fibers radiate during mitosis and meiosis.

centromere
Region of a chromosome that holds the two chromatids together and that binds, via a kinetochore, with mitotic or meiotic spindle fibers.

centrosome
A body in the cytoplasm containing a centriole and ancillary proteins that functions to organize one-half of a mitotic spindle.

checkpoint
Control mechanism that ensures that the next step in the cell cycle does not proceed until a series of preconditions have been fulfilled including the completion of all previous steps.

chemokine
A chemical message, often a polypeptide, that serves as an attractant for motile cells, notably leukocytes.

chemotaxis (adj., **-tactic**)
Directed migration of a cell toward high concentrations of an attractive factor, such as a chemokine.

chimera
Organism in which different cells or tissues derive from genetically distinct parents or genetically distinct cells.

chimerization
Process whereby a donor cell introduced into a host embryo is able to insert itself into that embryo and participate in the formation of some of the subsequently arising tissues.

chimerized
(1) Referring to the protein product of a re-engineered gene in which two portions of the protein derive from two distinct sources, notably, two distinct species. (2) An antibody molecule whose constant (C) region amino acid sequences have been replaced by the homologous sequences from another species, e.g., one in which a mouse C region is replaced by a human C region.

chromatid
A half-chromosome that exists after S phase and before M phase; paired chromatids are separated at M phase, whereupon each becomes a chromosome.

chromatin
Complex of DNA, RNA, and proteins that constitutes a chromosome.

class switching
The gene rearrangement occurring in an immunoglobulin gene in which an already-formed antigen-binding variable region sequence is switched from its juxtaposition with one constant region to another constant region sequence, resulting in a change in the coding of the constant region of an antibody molecule by the immunoglobulin mRNA.

clone (adj., **clonal**)
(1) Copy of a gene that has been isolated by recombinant DNA procedures and amplified into a large number of identical copies. (2) Population of cells, all of which descend from a common progenitor cell. (3) Offspring of a procedure of asexual reproduction in which the genome of a somatic cell of one organism is used to form a cell that functions equivalently to a fertilized egg that may then itself develop into another organism.

co-carcinogen
An agent or substance that, while not carcinogenic on its own, collaborates with another agent to enable carcinogenesis to proceed.

colonization
Proliferation of cells within a micrometastasis that leads to the formation of a macroscopic metastasis.

colony
(1) A cluster of cells, usually of clonal origin. (2) A cluster of cells that is able to proliferate in the absence of anchorage to a solid substrate.

colorectal
Referring to the lower gastrointestinal tract including the colon and rectum.

compartment
Physical or virtual space that contains all cells of a given type within a tissue, e.g., a stem cell compartment.

complement
A group of collaborating plasma proteins that can associate with antibody molecules that are bound to cell-surface antigens, including those displayed by bacterial, yeast, or mammalian cells; once attached via these antibody molecules to a cell surface, complement can kill the cell by introducing pores into nearby plasma membrane.

complementation
Ability of two mutant genotypes, when coexisting in the same cell or organism, to compensate for each other's defects and thereby create a wild-type phenotype, indicating that the two genotypes carry changes in distinct genes.

complete carcinogen
An agent that can act as both initiator and promoter of tumor progression.

complete response
Elimination of all detectable tumor mass following anti-cancer therapy.

confluence
State reached when cells in monolayer culture proliferate until they fill all available space at the bottom surface of a Petri dish.

congenital
Referring to a condition that is already present at birth and may persist thereafter.

constitutive
Describing a state of activity that occurs at a constant level and is therefore not responsive to modulation by physiologic regulators, or a type of control that yields such a constant output.

contact inhibition
A behavior exhibited by cells propagated in monolayer culture: the cells will continue to proliferate until they touch one another, after which they stop proliferating.

contralateral
Referring to the opposite side.

copy choice
A molecular genetic mechanism whereby a growing DNA strand can select the template for its further elongation from among alternative complementary DNA strands.

counterstain
Histological procedure in which the microscopic entity of interest is viewed in the context of or is contrasted with other entities by staining the latter with a different dye or other substance.

crisis
State arising when cells lose telomeres of adequate length, resulting in the end-to-end fusion of chromosomes, karyotypic chaos, and widespread cell death by apoptosis.

cross-immunity
Immunity that has initially arisen against a particular antigen or infectious agent that is subsequently found to confer immunity against another distinct antigen or infectious agent.

crypt
A deep cavity in the wall of the small or large intestine in which enterocyte stem cell proliferation and initial differentiation occur.

cyclin
A protein that associates with a cyclin-dependent kinase and serves as a regulatory subunit of this kinase by activating its catalytic activity and directing it to appropriate substrates.

cyclin-dependent kinase
Type of serine/threonine kinase deployed by the cell cycle machinery that depends on an associated cyclin protein for proper functioning.

cycloheximide
A drug able to prevent the movement of ribosomes down an mRNA template, thereby blocking protein synthesis.

cyclopia
Malformation of the head resulting in embryos with only a single, centrally located eye.

-cyte
Suffix indicating a type of cell.

cytoarchitecture
Physical structure of a cell.

cytocidal
Referring to an effect or influence that causes cell killing.

cytokeratin
Forms of the intermediate filament protein keratin that constitute part of the cytoskeleton of an epithelial cell; term used to distinguish it from the keratins that constitute hair, nails, and feathers.

cytokines
(1) Growth factors that stimulate one or several of the cell types constituting the hematopoietic system. (2) Regulatory factors of the immune system, including interferons and interleukins, that, like mitogenic growth factors, convey signals between cells.

cytokinesis
Last step of M phase, often considered part of telophase, during which the cytoplasm divides and daughter cells separate.

cytology (adj., **-logical**)
(1) Analysis of subcellular structure under the microscope. (2) The microscopic appearance of a cell.

cytopathic
Causing damage or death to a cell.

cytoskeleton
Network of proteins in the cell, largely in the cytoplasm, that provides it with structure and rigidity and enables it to exhibit motile behavior.

cytosol
Portion of the cytoplasm that contains soluble material untethered to either cytoskeleton or membranes.

cytostatic
Referring to an influence or a force that inhibits cell proliferation without necessarily having any effect on cell viability.

de novo
(1) Arising or formed anew. (2) Occurring for the first time.

***de novo* methylase**
An enzyme that attaches a methyl group to the C of a CpG dinucleotide in the absence of methylation of the complementary CpG.

deamination
Loss of an amine group from a larger molecule such as a DNA base.

decatenation
Untangling and separation of DNA helices from one another, often accomplished by the topoisomerase II enzyme.

dedifferentiation
Reversion of a differentiated cell to the phenotype of a less differentiated cell, such as its precursor.

degranulation
Discharge of the contents of cytoplasmic granules into the extracellular space following a physiologic stimulus.

delayed early
Referring to genes whose expression depends on *de novo* protein synthesis and is induced with some delay following growth factor stimulation of a cell.

demineralize
To dissolve the inorganic apatite (i.e., calcium phosphate) component (of bone).

denaturation
Process that causes a molecule, such as a macromolecule (DNA, RNA, or protein), to lose its natural three-dimensional structure.

dendritic cell
An immune cell that phagocytoses fragments of cells or infectious agents and then presents oligopeptides derived from these phagocytosed particles to several types of helper T cells in the lymph nodes.

dendrogram
Diagram in which closely related entities, such as genes, cells, or organisms, are placed close to one another on the branches of a multibranched tree.

density inhibition
See contact inhibition.

denuding
Removal or stripping away of a tissue or population of cells, such as those forming an epithelium.

depurination
Breakage of the glycosidic bond that links purine bases to the deoxyribose or ribose of DNA or RNA, respectively.

depyrimidination
Breakage of the glycosidic bond that links pyrimidine to the deoxyribose or ribose of DNA or RNA, respectively.

dermal
Referring to a thick layer of stromal cells, largely fibroblasts, found beneath the epidermal keratinocyte layer of the skin.

dermis
The stromal layer of the skin, composed largely of fibroblasts.

desmoplastic
Referring to the hard, collagenous stroma that is formed by fibroblasts and myofibroblasts within carcinomas and certain chronically inflamed tissues.

detoxify
To render a previously toxic substance harmless.

diagnostic bias
Tendency of an observation or conclusion to be influenced by the properties or applications of a diagnostic technique rather than to accurately reflect an intrinsic property of a disease state.

diapedesis
Complex sequence of steps enabling leukocytes to extravasate from the lumen of a blood vessel into the tissue parenchyma.

dicentric
Referring to a chromosome or chromatid that bears two distinct centromeres.

differentiation
Process whereby a cell acquires a specialized phenotype, such as a phenotype characteristic of cells in a particular tissue.

dimer
Molecular complex composed of two subunits.

diploid
Describing a genome in which all chromosomes are present in pairs, one of each pair being inherited from a father and the other from a mother, with the exception of the sex chromosomes, which in placental mammals are paired in either the XX or the XY configuration.

disseminated
Spread or seeded widely.

dominant
(1) Referring to one of several alternative traits (phenotypes) that can be specified by a genetic locus; when the locus is heterozygous and carries information specifying two distinct traits, the dominant trait will be the one actually exhibited. (2) Describing an allele of a gene that determines phenotype in spite of the presence of a second gene allele that specifies a different phenotype. *See also* **recessive**.

dominant-interfering
See **dominant-negative**.

dominant-negative
Referring to a mutant allele of a gene that, when co-expressed with the wild-type allele of the gene, is able to interfere with the functioning of the latter.

double minute
A chromosomal segment that becomes separated from the chromosome in which it normally resides and is able to perpetuate itself as an extrachromosomal particle unlinked to a centromere.

druggable
Referring to a molecular species, such as a protein, that has the structural and functional properties suggesting that low-molecular-weight therapeutic compounds can be developed that specifically interact with and perturb its functioning.

dysplasia (adj., **-plastic**)
A pre-malignant tissue composed of abnormally appearing cells forming a tissue architecture that deviates from normal.

ectoderm
Outer layer of cells in an early embryo.

ectodomain
Portion of a cell surface protein that protrudes from the plasma membrane into the extracellular space.

ectopic
(1) Referring to the expression of a gene in a place or physiologic situation where it normally would not be expressed. (2) Referring to the presence of cells or tissues in an anatomical site where they would not naturally be found in the body. *See also* **orthotopic**.

effector
An agent (such as a protein) that carries out the actual work of a biological process rather than just regulating it.

effector loop
The region of the Ras protein that participates in direct physical interaction with one of its effector proteins; such interaction usually leads to the functional activation of the effector protein.

efferent
Referring to outgoing signals.

efficacy
Ability of a therapeutic agent to elicit a desired clinical response.

electrophilic
Referring to a molecule that seeks out and reacts with electron-rich substrates.

embolization
Process of forming an embolus.

embolus
A blood clot that can travel through the circulation and eventually lodge somewhere in the body.

-emia
Suffix denoting an excess of a substance or a cell type in the blood.

encapsidation
Process of packaging a viral genome in a capsid

endocrine
(1) Referring to a gland that secretes fluids into the general circulation. (2) Referring to the signaling path of a hormone or factor that is made by cells in one tissue, passes through the blood, and affects the behavior of cells in another tissue at a distant site. *See also* **exocrine**.

endocytosis
Process by which patches of plasma membrane and associated proteins are internalized into the cell cytoplasm, resulting in their forming cytoplasmic vesicles. *Compare* **exocytosis**.

endoderm
Innermost layer of cells in an early embryo, which is precursor to the gastrointestinal tract and associated tissues

endogamy
The practice of marrying within one's own ethnic group, tribe, or clan.

endogenous
Originating from within a cell, tissue, or organism.

endogenous provirus
The provirus of a retrovirus that is transmitted through the germ line of a species.

endometrium
Epithelial lining of the uterus.

endoplasmic reticulum
Elaborate network of membranous structures in the cytoplasm in which glycoproteins are assembled.

endoreduplication
Process whereby a genome or portions thereof are replicated without subsequent segregation of the replicated copies, leading to their duplication.

endosomes
Membranous vesicles lying beneath the plasma membrane that are formed after patches of the plasma membrane invaginate into (fold into) the cytoplasm and pinch off from the plasma membrane.

endothelial cells
(1) Mesenchymal cells that form the walls of capillaries or lymph ducts by assuming a tubelike shape. (2) Mesenchymal cells lining the luminal walls of larger blood vessels or lymph ducts.

enhancer
A relatively short sequence of nucleotides near or within a gene to which a transcription factor may bind and in turn influence the transcription of this gene.

enterocyte
Epithelial cell lining the luminal wall of the gastrointestinal tract.

eosinophil
A motile phagocytic granulocyte that can migrate from blood into tissue spaces, displays cell surface IgE receptors, and is thought to play a role in eliminating parasitic organisms.

epidermis (adj., **-dermal**)
The epithelial layer of the skin, composed largely of keratinocytes at various stages of differentiation.

epigenetic
Referring to changes in the behavior of a cell or in the activity state of its chromatin that do not depend on alterations of DNA nucleotide sequence.

epithelial–mesenchymal transition
Acquisition by epithelial cells of the phenotype of mesenchymal cells such as fibroblasts.

epithelium
A layer of cells that forms the lining of a cavity or duct; included here is the specialized epithelium that forms the skin.

epitope
A specific chemical structure—usually a short oligopeptide segment of a protein antigen—that is recognized and bound by an antibody molecule.

erythroblastosis
Malignancy of the precursors of red blood cells, usually referring to a condition of birds.

erythrocyte
Red blood cell.

erythroleukemia
A leukemia of the nonpigmented precursors of red blood cells.

erythropoiesis
Process by which red blood cells are formed.

erythropoietin
Growth factor that stimulates the production of red blood cells, often in response to inadequate oxygen transport by the blood.

etiology (adj., **-ologic)**
(1) Mechanism or agent that is responsible for causing a specific pathological state. (2) The study of causative mechanisms of pathology.

euchromatin
Chromatin that contains transcriptionally active genes and is therefore relatively expanded (rather than being condensed) and stains lightly. *See also* **heterochromatin**.

eukaryotic
Referring to the large, complex, nucleated cells of metazoa, metaphyta, and many protozoa.

euploid
(1) Referring to a collection of chromosomes that corresponds precisely in number (usually diploid) and structure to the array present in normal, wild-type cells. (2) Describing a karyotype having such a complement of chromosomes.

event-free survival
Period after initial treatment during which no additional clinical episodes of disease are registered.

ex vivo
Occurring outside of a living body or organism.

exocrine
Referring to a gland that secretes fluids via a duct into the gastrointestinal tract or to the surface of the skin. *See also* **endocrine**.

exocyclic
Referring to a chemical group that protrudes from the ring of a molecule such as a DNA base.

exocytosis
Process by which cells secrete products by storing them in cytoplasmic membrane vesicles that are caused to fuse with the plasma membrane, allowing the products carried in these vesicles to be released into the extracellular space. *Compare* **endocytosis**.

exogenous
Originating from outside a cell, tissue, or organism.

exon
The portion of a primary RNA transcript that is retained in the RNA product of splicing.

expression
(1) Transcription of an active gene or synthesis of a protein from it. (2) Release of interstitial fluid from a tissue caused by contraction of cells in that tissue.

expression program
The coordinated expression of a series of genes.

expression signature
A constellation of up-regulated and down-regulated genes that can be correlated with a defined biological phenotype.

extracellular matrix
Mesh of secreted proteins, largely glycoproteins and proteoglycans, that surrounds most cells within tissues and creates structure in the intercellular space.

extravasation
Process of leaving a blood or lymphatic vessel and invading the surrounding tissue

familial
Referring to a trait or a syndrome that is heritable and therefore found in clusters in certain families.

fibrin
Protein that is formed by the cleavage of plasma fibrinogen and assembles to form the fibers binding the platelets together in clots.

fibrinogen
See **fibrin**.

fibrinolysis
Dissolution of fibrin bundles achieved by certain proteases.

fibroblast
Mesenchymal cell type that is common in connective tissue and in the stromal compartment of epithelial tissues and is characterized by its secretion of collagen.

fibrocyte
A relatively undifferentiated circulating cell of bone marrow origin that expresses collagen, adheres tightly to substrates when cultured *in vitro*, and appears to serve as the precursor of fibroblasts and/or myofibroblasts in tissues to which it has been recruited.

fibrosis
Development within a tissue, often following chronic inflammation, of dense fibrous stroma that replaces normally present epithelium, resulting in loss of function of that tissue.

field cancerization
Process in which a region of an organ produces multiple, ostensibly independently arising pre-malignant growths or neoplasms.

filopodium
Filamentous, spikelike protrusion from the surface of a cell, usually extending from the leading edge of a lamellipodium, that may be used by the cell to explore territory that lies ahead in its path.

focus (pl., **foci**)
A cluster of transformed cells growing amid a surrounding monolayer of normal cells in culture.

functional genomics
(1) Technology in which cell phenotypes are gauged by measuring the expression level of multiple genes, usually numbering in the thousands, in the cells. (2) Such analysis performed comparatively by examining various cell types or tissues that exist in different physiologic, pathological, or developmental conditions.

fusogenic
Capable of causing the fusion of membranes, such as the plasma membranes of two adjacent cells.

gametogenesis
Processes that yield gametes, i.e., sperm and eggs.

gastric
Referring to the stomach.

gastrulation
An early stage of embryogenesis in which the embryo, initially composed of an outer, ectodermal cell layer, develops an internal cavity that serves as the precursor of endodermal and mesodermal cell layers.

gatekeeper
A gene that operates to hinder cell multiplication or to further cell differentiation or cell death and in this way succeeds in preventing the appearance of neoplastic cells.

gene amplification
Increase in the number of copies of a gene normally present in the diploid genome.

gene family
Group of genes all of which are descended evolutionarily from a common ancestral gene. The members of a gene family often encode distinct, structurally related proteins.

gene pool
Collection of genes present among the genomes of all members of a species.

genetic background
The entire array of alleles carried in a genome with the exception of a small number of genes that are the subject of study.

genetic polymorphism
A variant sequence element in an organism's genome that has no effect on phenotype yet is transmitted genetically as a Mendelian determinant.

genomic clone
The product of a procedure that generates many copies of a specific segment of an organism's genome (e.g., its chromosomal DNA) through isolation and amplification.

genotoxic
Referring to an agent that is capable of damaging the genome, i.e., is mutagenic.

genotype
(1) Genetic constitution of an organism. (2) Genetic basis of a trait of an organism.

germ line
(1) The collection of genes that is transmitted from one organismic generation to the next. (2) The cells within a multicellular organism that are responsible for carrying and transmitting genes from one organismic generation to its offspring.

glioblastoma
Tumor of the nonneuronal glial cells of the brain.

glioma
See **glioblastoma**.

glycoprotein
A protein that has been modified post-translationally through the addition of carbohydrate side chains.

glycosaminoglycan
A charged polysaccharide of the extracellular matrix, such as chondroitin sulfate, hyaluronic acid, or heparin, that is attached covalently to a protein core and is composed of repeating monosaccharides, some of which are amino sugars.

glycosylase
An enzyme that cleaves a glycosidic bond, such as that linking a purine or pyrimidine base to a ribose or deoxyribose sugar.

glycosylation
Covalent attachment of a sugar side chain, including a covalently linked network of sugars, to a second molecule, e.g., to the asparagine residue of a protein.

granulocyte
Leukocyte that contains cytoplasmic granules, such as a basophil, an eosinophil, or a neutrophil.

gray (Gy)
Unit of radiation exposure equal to the absorption of 1 joule of energy per kilogram of exposed tissue.

growth factor
Protein that is able to stimulate the growth and/or proliferation of a cell by binding to a specific cell surface receptor displayed by that cell.

half-life
(1) Time during which half of a population of metabolically unstable molecules decays or is eliminated or the time required for half of a physiologic signal to decrease. (2) Time in which 50% of the atoms of a quantity of a radioactive isotope decay.

hamartoma
Benign overgrowth of tissue that normally involves mesenchymal cells; for example, gastrointestinal hamartomas are characterized by a benign expansion of stromal cells, often with concomitant hyperplasia of adjacent epithelial cells

haploid
Describing a genome in which all chromosomes are present in a single copy.

haploinsufficiency
State in which the presence of only a single functional copy of a gene yields a mutant or partially mutant phenotype.

hemangioblastoma
Tumor of the precursors of the endothelial cells forming blood vessels.

hematogenous
Depending upon or facilitated by circulating blood.

hematopoiesis (adj., **-poietic**)
The formation of all of the cells in the blood, including its red and white cells including various cells of the immune system.

hemidesmosome
A cluster of integrin molecules displayed on the basal surface of epithelial cells and used to anchor these cells to the underlying basement membrane.

hemi-methylated
Referring to a DNA molecule in which only one of the two complementary strands of a particular DNA segment is methylated.

hemizygosity
Presence of only one copy of an autosomal gene per cell.

heparin
An extracellular matrix glycosaminoglycan.

hepatocellular carcinoma
See **hepatoma**.

hepatocyte
Epithelial cell type that forms the bulk of the liver and is responsible for virtually all of its metabolic activities.

hepatoma
Tumor of the liver, also known as **hepatocellular carcinoma**.

Herceptin
Chimeric anti-HER2/Neu monoclonal antibody bearing murine antigen-combining (variable) domains and a human constant domain (*also called* **trastuzumab**).

heterochromatin
Chromatin that contains transcriptionally inactive genes or no genes at all and is condensed and stains darkly. *See also* **euchromatin**.

heterodimer
Molecular complex composed of two distinct subunits.

heterokaryon
Cell carrying two (or more) genetically distinct nuclei.

heterotrimer
A molecule that is composed of three distinct subunits such as three distinct protein chains.

heterotypic
Referring to interactions between two or more distinct cell types. *See also* **homotypic**.

heterozygous
Referring to the configuration of a genetic locus in which the two copies of the associated gene carry different versions (alleles) of the gene.

high-grade
Referring to a tumor that has progressed through many steps of multi-step tumorigenesis and become highly malignant.

histo-
Referring to tissue.

histocompatibility
Ability of tissues or cells to be tolerated by the immune system of the host organism into which they are engrafted.

histocompatibility antigen
Cell surface protein that determines whether or not an engrafted cell or tissue will be tolerated by the immune system of a host organism. *See also* **major histocompatibility antigen**.

histology
Study of tissue structure at the microscopic level. *See also* **histopathology**.

histopathology
Study of tissue structure at the microscopic level, often with reference to abnormal tissue. *See also* **pathology**.

holoenzyme
Enzyme that is assembled from multiple subunits that collaborate to mediate and regulate enzymatic activity.

homodimer
Molecular complex composed of two identical subunits.

homogeneously staining region
A region of a chromosome consisting of amplified copies of a chromosomal segment that have become fused end-to-end.

homolog
A gene that is related to another because of evolutionary descent from a common ancestral gene.

homophilic
Describing a molecule that binds preferentially to one or more additional molecules of the same type.

homopolymer
A polymer that is assembled from monomers of a single type, such as one or another of the four possible deoxyribonucleotides.

homotetramer
An assembly of four identical subunits, usually referring to proteins.

homotypic
(1) Referring to interactions between two cells of the same type; *see also* **heterotypic**. (2) Referring to interactions between molecules of the same type.

homozygous
Referring to the configuration of a genetic locus in which the two copies of the gene carry identical versions (alleles) of the gene.

housekeeping gene
A gene that is used virtually universally in all cells throughout the body independent of their differentiated state and is widely assumed to be essential for cellular viability.

humanize
To impart human properties to something, e.g., an antibody molecule.

humanized antibody
An antibody molecule of one species whose constant (C) region and variable (V) region amino acid sequences outside of the antigen-combining site have been replaced by the homologous sequences of human origin, leaving only the antigen-combining sequences unmodified.

humoral
Referring to a soluble substance or a fluid.

humoral immune response
The arm of the immune system mediated by the antibodies that it produces.

hydrocarbon
A molecule composed of hydrogen and carbon atoms.

hydrophilic
Referring to a chemical moiety or environment that prefers association with water. *Compare* **hydrophobic**.

hydrophobic
Referring to a chemical moiety or environment that avoids direct interaction with water. *Compare* **hydrophilic**.

hypercalcemia
Presence of elevated concentrations of calcium ions in the blood.

hyperinsulinemia
Elevated level of circulating insulin.

hyperoxia
State of oxygen tension that is elevated above physiologic levels.

hyperphosphorylated
Referring to the elevated phosphorylation (of a protein).

hyperplasia (adj., **-plastic)**
Accumulation of excessive numbers of normal-appearing cells within a tissue.

hypophosphorylated
Referring to relatively weak phosphorylation (e.g., of a protein).

hypoxia
State of lower-than-normal oxygen tension.

idiotype
The antigenic structure that is created by the amino acid sequence forming the antigen-combining pocket of an antibody molecule.

immediate early
Referring to the group of genes whose expression is induced within half an hour of growth factor stimulation of a cell, even when protein synthesis is inhibited.

immortality
Trait of a cell or population of cells to proliferate indefinitely.

immortalization
Process whereby a cell population normally having limited replicative potential acquires the ability to multiply indefinitely.

immunocompetent
Referring to an organism whose immune system is fully functional.

immunocompromised
Describing an organism lacking a fully functional immune system.

immunocyte
A cell associated with one of the functional arms of the immune system.

immunodeficient
See **immunocompromised**.

immunoediting
Process whereby the immune system permits some weakly immunogenic tumors to survive while selectively eliminating those that are strongly immunogenic.

immunoevasion
Any biological strategy that enables an abnormal cell or an infectious agent to evade detection and/or elimination by the immune system.

immunofluorescence
Use of antibodies linked directly or indirectly to fluorescent dyes in order to stain tissue sections displaying antigens that are specifically recognized by such antibodies.

immunogen (adj., **-genic)**
A chemical structure that is capable of provoking a specific immune response, e.g., an antigen that can provoke the synthesis of antibody molecules capable of recognizing and binding it.

immunoglobulin
Protein molecule that assembles to form a functional antibody molecule.

immunohistochemistry
Procedure in which an antigen is located in a histological section through the use of an antibody that has been coupled to an enzyme (e.g., peroxidase) capable of generating a product that is visible in the light microscope.

immunoprecipitation
Process of precipitating a molecule using an antibody that specifically recognizes and binds such molecule.

immunoproteasome
A proteasome, usually found in professional antigen-presenting cells, that is specialized to generate oligopeptides for presentation by MHC molecules at the cell surface

immunostaining
The use of antibodies to stain specific cells or subcellular structures present in histological sections.

immunosurveillance
Process by which the immune system is continuously monitoring tissues for the presence of aberrant cells, including cancer cells.

immunotoxin
Toxin that is targeted to certain tissues or cells because it has

been coupled to an antigen-specific antibody, usually a monoclonal antibody.

in situ
(1) Occurring in the site of origin. (2) In the case of carcinomas, confined to the epithelial side of the basement membrane.

in utero
Occurring in the womb during embryonic or fetal development

in vitro
(1) Occurring in tissue culture, or in cell lysates or in purified reaction systems in the test tube. (2) Referring to the propagation of living cells in a vessel (e.g., a Petri dish) rather than in living tissues.

in vivo
(1) Occurring in a living organism. (2) Occurring in a living, intact cell.

inbreeding
Breeding of a strain of organisms, such as a strain of mice or rats, with one another in order to achieve genetic identity among all individuals of the strain (with the exception of male/female genetic differences).

incidence
Frequency with which a condition or a disease occurs or is diagnosed in a population.

indication
A call for the treatment of a disease (or a tumor) with a certain type of therapy as suggested by a set of diagnostic parameters.

initiation
(1) Process of changing a cell, usually in a stable fashion, so that it is able to respond subsequently to the growth-stimulatory actions of a promoting agent. (2) Such a process, with the implication that the change involves a mutation. (3) The first step in multi-step tumorigenesis.

initiator
Agent that triggers the first step in multi-step tumorigenesis. *See also* **initiation**.

innate immune response
An immune system response toward an antigen or an agent bearing an antigen that occurs in the absence of prior exposure of an organism to this particular agent or cell. *See also* **adaptive immune response**.

inoculum
Inoculated material.

insertional mutagenesis
Process whereby a gene becomes mutated through the insertion or integration of another genetic element (such as a retroviral provirus) nearby in the chromosomal DNA.

integration
Insertion of a fragment of foreign DNA (e.g., the DNA genome of an infecting virus) into chromosomal DNA so that the viral DNA becomes covalently linked to the chromosomal DNA segments flanking it on both sides.

integrin
A heterodimeric cell surface receptor that binds components of the extracellular matrix and transmits information about this binding to the cell interior; the cytoplasmic domain of an integrin may also be coupled with components of the cytoskeleton, thereby linking the extracellular matrix to the cytoskeleton.

interleukin
A growth and differentiation factor that stimulates various cellular components of the immune system.

intermediary metabolism
The collection of biochemical reactions within a cell that allows the interconversion of various molecular species into one another.

internalization
Process by which proteins and other molecules are imported into a body, usually referring to importation of molecules into cells.

interphase
The portion of the cell cycle outside of mitosis.

interstitial
Referring to the space within a tissue that lies between cells.

interstitial deletion
Genetic alteration, often observed through analysis of karyotype, that causes deletion of a segment of chromosomal DNA and thus chromatin from the midst of a chromosomal arm without affecting material at the ends of the arm, with subsequent fusion of the segments on either side of the deletion.

intraepithelial
(1) Referring to a change or an attribute within an epithelium. (2) Referring to a neoplasia that remains on the epithelial side of a basement membrane.

intravasation
Process of invading a blood or lymphatic vessel from the surrounding tissue.

intra-vital
Referring to a process that occurs in living tissue, such as imaging a process as it occurs in living tissue.

intron
Portion of a primary RNA transcript that is deleted during the process of splicing.

invasion
(1) Process by which cancer cells or groups thereof move from a primary tumor into adjacent normal tissue. (2) In the case of carcinomas, a movement that involves breaching of the basement membrane.

involution
Regression or disappearance of a tissue, notably, the regression of the mammary epithelium upon weaning.

isoform
Protein that is functionally and structurally similar to another protein.

junk DNA
Genomic DNA that cannot be associated with any biological function .

juxtamembrane
Located adjacent to a membrane.

Kaplan–Meier plot
A convention for graphing various clinical observations in which the percentage of surviving patients (or another clinical parameter such as disease-free or progression-free survival) is plotted on the ordinate while the time course after initial diagnosis or treatment is plotted (usually in increments of months or years) on the abscissa.

karyotype
The array of chromosomes carried by a cell, as determined by detailed examination of these chromosomes, usually performed with condensed chromosomes at metaphase. (2) Image of the metaphase chromosomes of a cell arrayed systematically by homologous pairs from the largest to the smallest pair.

keratinocyte
Epithelial cell type found in tissues such as the skin.

keratosis
A benign lesion of the keratinocyte lineage in the skin usually caused by exposure to ultraviolet radiation.

kinase
Enzyme that covalently attaches phosphate groups to substrate molecules, often proteins.

kinetochore
Nucleoprotein complex that is associated with the centromeric DNA of a chromosome and is responsible during mitosis (or meiosis) for forming a physical connection between the chromosome and the microtubules of the spindle fibers.

kinome
The complete repertoire of kinases encoded by a genome, such as the human genome.

labile
Highly susceptible to change, including alteration or destruction.

lamellipodium
A broad, sheetlike ruffle extending from the plasma membrane into the extracellular space that is typically found at the leading edge of a motile cell.

laser capture microdissection
Procedure in which a laser beam is used to dissect a patch of cells away from other cells present in a tissue section that has been mounted on a microscope slide.

leiomyoma
Benign tumor of the mesenchymal cells forming the wall of the uterus.

leukemia
Malignancy of any of a variety of hematopoietic cell types, including the lineages leading to lymphocytes and granulocytes, in which the tumor cells are nonpigmented and dispersed throughout the circulation.

leukemogenesis (adj., **-genic)**
Process that creates a leukemia.

leukocyte
A nonpigmented white blood cell such as a lymphocyte, monocyte, macrophage, neutrophil, or mast cell.

leukosis
A leukemia-like disease of chickens.

library
(1) Referring to a collection of gene clones derived from an organism's genome, of which each component clone ideally derives from a distinct element in this genome. (2) Referring to other collections of DNA clones, notably cDNA clones, each of which derives from reverse transcription of a distinct mRNA expressed by a certain cell type.

ligand
Molecule that binds specifically to a receptor and activates its signaling powers.

ligase
An enzyme that covalently joins the ends of two molecules together; in the context of DNA, ligases join the 3′ end of one ssDNA to the 5′ end of the other via a phosphodiester linkage.

lineage
Linear succession of cells extending from an ancestral cell to its descendants.

locus
Chromosome site that can be studied genetically and is presumed to be associated with a specific gene.

low-grade
Referring to a tumor that has progressed minimally and is still relatively benign.

lumen (pl., **lumina)**
The bore of a hollow, tubelike structure, such as the gut, a bronchiole in the lung, a blood vessel, or a duct in a secretory organ.

luminal
Referring to the cells that line and face into the lumen of a tubelike structure.

lymph
The interstitial fluid between cells that is drained via the lymph nodes to larger lymphatic vessels that eventually empty into the venous circulation.

lymphangiogenesis
The process of forming new lymphatic vessels.

lymphocytes
Class of leukocytes that mediate humoral or cellular immunity, encompassing B cells, T cells, and NK cells and derivatives thereof.

lymphoid
(1) Referring to the lymphatic system. (2) Referring to the lineage of hematopoietic cells that yields B and T lymphocytes as well as natural killer cells.

lymphokine
A cytokine or growth factor specialized to attract and/or activate lymphocytes.

lymphoma
Solid tumor of lymphoid cells .

lysate
Product of dissolving the structure of a tissue or a population of cells, usually created in order to liberate the internal contents of the tissue or cells.

lysosome
A cytoplasmic lipid vesicle that contains degradative enzymes in a solution of low pH, allowing it to degrade various molecules that are introduced into it.

lytic
Dissolving a cell or tissue; often associated with the potent cytopathic effects of certain viruses on specific host cells.

lytic cycle
Cycle of viral infection and replication that results ultimately in the death of the infected host cell.

macronucleus
Larger of the two nuclei in many ciliate cells, which carries multiple copies of each gene and is used for the production of mRNAs.

maintenance methylase
An enzyme that attaches a methyl group to an unmethylated CpG that is complementary to a methylated CpG in a DNA double helix.

major histocompatibility antigen
One of a group of cell surface proteins that are responsible for the presentation of oligopeptide antigens to responding cells of the immune system. *See also* **histocompatibility antigen**.

malignant
Describing a growth that shows evidence of being locally invasive and possibly even metastatic.

mammary
Referring to the breast and its milk-producing glands.

mast cell
A cell of bone marrow origin that bears Fc receptors for IgE and undergoes IgE-mediated degranulation following encounter with certain antigens.

medical
Referring to the use of drugs, rather than surgery, to treat disease.

medulloblastoma
Tumor of the primitive precursors of neurons in the cerebellum.

megakaryocyte
Large cell of the hematopoietic system that produces platelets by pinching off fragments of its cytoplasm.

melanin
A brown or black pigment that is synthesized by melanocytes and, in the skin, transferred into basal keratinocytes, thereby creating pigmentation of the skin.

melanocyte
Cell of neural crest origin that creates pigmentation of the skin and iris.

melanoma
Tumor arising from melanocytes, the pigmented cells of the skin and iris.

melanosome
Melanin-containing body in the cytoplasm of a melanocyte that is transferred to a keratinocyte in order to impart pigmentation to the latter.

menarche
Time in life when menstrual cycling begins.

mesenchymal
(1) Referring to tissue composed of cells of mesodermal origin, including fibroblasts, smooth muscle cells, endothelial cells, and immunocytes. (2) Referring to an individual cell type belonging to this class of cells.

mesenchymal–epithelial transition
The reversal of the epithelial–mesenchymal transition (*compare* **epithelial–mesenchymal transition**).

mesoderm
Middle layer of cells in an early embryo lying between the ectoderm and endoderm, which is the precursor of mesenchymal tissues including connective tissues and the hematopoietic system.

metabolite
The chemical species resulting from the metabolic conversion by enzymes of a precursor chemical species.

metalloproteinase
A protease that contains a metal atom, usually zinc, in its catalytic site.

metaphase
Second subphase of mitosis, during which chromosomes complete condensation and attach to the mitotic spindle as the nuclear membrane disappears; chromosomes are now readily seen in the light microscope.

metaphyta
Plants composed of many cells.

metaplasia
Replacement in a tissue of cells of one differentiation lineage by cells of another lineage.

metastasis
Malignant growth forming at one site in the body, the cells of which derive from a malignancy located elsewhere in the body.

metazoa
Animals composed of many cells.

micrometastasis
A metastasis that is composed of a single cell or a small clump of cells and is only apparent through microscopy.

micronucleus
(1) A small fragment of a nucleus that has its own nuclear membrane and results from certain aberrations in cell division or from damage inflicted on a cell. (2) Smaller of the two nuclei normally present in many ciliate cells, which is used to carry and transmit the ciliate genome to progeny cells.

microsatellite
A DNA sequence block that consists of a succession of repeating units of identical or similar nucleotide sequence.

microthrombus
See **embolus**

mismatch repair
The class of DNA repair processes that depend on scanning a recently synthesized segment of DNA and removing any misincorporated bases.

missense codon
A triplet codon in the genetic code that specifies an amino acid residue different from that specified by the codon that it replaces.

missense mutation
Mutation causing an amino acid substitution.

mitogen (adj., **-genic**)
An agent that provokes cell proliferation.

mitosis
(1) Cell division, composed of the four steps of prophase, metaphase, anaphase, and telophase. (2) More properly, process by which a single cell separates its complement of chromosomes into two equal sets in preparation for the division into two daughter cells.

mitotic recombination
Recombination between homologous chromosomal arms occurring during somatic cell proliferation, often during the G_2 phase of the cell cycle (rather than mitosis).

monoclonal
Describing a population of cells, all of which derive by direct descent from a common ancestral cell.

monoclonal antibody
An antibody that is made by a population of cells that has been produced through the immortalization of a single antibody-producing B lymphocyte, so that all antibody molecules in a preparation are identical to one another and show identical antigen specificity.

monocyte
A phagocytic leukocyte that circulates briefly in the blood before migrating into tissues where it differentiates into a macrophage.

monolayer
A population of cells growing as a layer one cell thick.

monomer
Molecule composed of a single unit or a single subunit of a kind that is capable of forming higher–molecular-weight complexes by association with or covalent linkage to identical or similar units.

monotherapy
A therapy in which only a single drug is used at one time.

monozygotic
Derived from a single fertilized egg, thereby referring to identical twins.

morphogen
A substance that induces cells to construct a tissue of a certain shape and form.

morphogenesis
Process whereby shape is created, usually referring to the creation of shape of various structures during embryonic development.

morphology
Shape and form of a cell, tissue, or organism.

mortal
Referring to a cell population having limited proliferative capacity.

mortality
The rate or frequency of death from a particular disease condition.

motility
Tendency for movement, usually of individual cells, from one location to another.

motogenic
Referring to an agent or signal that stimulates cell movement or motility.

mucosa
Epithelial cell layer that secretes a mucus-like substance that forms a protective layer above the secreting cells.

multiparous
Referring to a woman who has given birth multiple times.

multipotent
Referring to the ability of a stem cell to differentiate into multiple distinct cell types.

mural
Referring to the outer cell layers of blood vessels, which are composed of pericytes and smooth muscle cells and surround the luminal endothelial cells.

murine
Referring to the subgroup of rodents that includes mice and rats.

mutagen (adj., **-genic**)
An agent that induces a mutation.

mutation
Change in the genotype of a species that may involve an alteration in the nucleotide sequence of a DNA segment, the arrangement of a segment within a chromosome, the number of copies of a segment, the physical structure of a chromosome, or even the number of copies of a structurally normal chromosome.

myelocytomatosis
A malignancy of avian bone marrow cells.

myelodysplastic syndrome
A hyperproliferative condition of cells of the myeloid lineage in the bone marrow which often progresses to acute myelogenous leukemia.

myelogenous
Originating in the bone marrow. *See also* **myeloid**.

myeloid
(1) Referring to the lineage of hematopoietic cells that yields granulocytes, macrophages, and mast cells. (2) Pertaining to or resembling bone marrow; often used as a synonym for **myelogenous**.

myeloma
A malignancy of the antibody-producing cells of the bone marrow, often called **multiple myeloma** to indicate the large number of osteolytic lesions that are encountered in patients with advanced disease.

myeloproliferative
Referring to excessive proliferation and resulting elevated levels of one of the several cell types generated by the myeloid branch of the hematopoietic system.

myoblasts
The undifferentiated precursors of the myocytes present in differentiated muscle.

myocytes
The cells constituting functional muscles.

myofibroblast
A type of fibroblast that is normally involved in wound healing and inflammation and is often defined by its expression of α-smooth muscle actin, which, together with myosin, imparts to it contractile powers.

natal
Referring to birth.

natural product
A chemical species that is a product of the natural metabolism of an organism, such as a bacterium, fungus, or plant.

necrosis
Process of cell death involving the breakdown of a cell and its constituents through steps that are distinct from those in the apoptotic death program.

neoangiogenesis
See **angiogenesis**.

neoplasia (adj., **-plastic**)
(1) The state of cancerous growth. (2) Benign or malignant tumor composed of cells having an abnormal appearance and abnormal proliferation pattern.

neoplasm
A tumor. *See also* **neoplasia.**

neovasculature
Vasculature that is newly developed.

neural crest
Region of the early embryo that serves as precursor of various specialized tissues and cell types, including certain cells of the peripheral nervous system, bones of the face, melanocytes, and several types of neurosecretory cells.

neuroblastoma
Tumor of primitive neuronal precursor cells of the peripheral nervous system and adrenal medulla.

neuroectodermal
Referring to the components of the nervous system, which derive from the embryonic ectoderm.

neurofibroma
Benign tumor of cells forming the sheath around nerve axons.

neurofibrosarcoma
Malignant tumor of cells forming the sheath around nerve axons.

neuropeptides
Oligopeptides that are released by certain cells of the central nervous system or by neuroendocrinal cells and impinge upon other cells inside and outside of the central nervous system.

neurosecretory
Referring to a cell type that will secrete a substance, such as a hormone, in response to neuronal signals.

neutral mutation
Change in DNA sequence that has no effect on phenotype, including mutations that have no effect on protein structure.

neutralization
Inactivation of a biological activity, such as inactivation of viral infectivity by antibody molecules.

neutropenia
Deficiency of neutrophils in the marrow and circulation.

neutrophil
An abundant granulocyte in the circulation that expresses Fc receptors and is responsible for recognizing and engulfing various types of infectious agents, notably bacteria.

niche (stem cell)
A functional locale that supports the proliferation of stem cells and prevents them from differentiating.

nondisjunction
(1) Failure of two chromatids to separate from one another during mitosis or two homologous chromosomes to separate from one another during meiosis. (2) State created by this failure of separation.

nonhomologous end joining
Type of DNA repair consisting of fusion of two dsDNA ends, in which the joining of the two ends is not informed or directed by sequences in a sister chromatid or homologous chromosome.

noninvasive
Referring to procedures that allow diagnosis or treatment without the need to enter into the body with diagnostic instruments or surgery.

nonsense codon
A triplet codon in the genetic code that specifies termination of a growing amino acid chain.

nonsense mutation
Mutation causing premature termination of a growing protein chain.

non-small-cell lung carcinoma
Any of several types of lung cancers with the exception of small-cell lung carcinoma

normoxia
A level of oxygen that corresponds to that normally experienced by cells in a specific tissue environment.

Northern blot
Adaptation of the Southern blotting procedure in which RNA (rather than DNA) is resolved by gel electrophoresis and transferred to a filter that is subsequently incubated with a sequence-specific, radiolabeled DNA probe.

nucleophilic
Referring to a molecule that seeks out and reacts with electron-poor substrates.

nucleosome
Protein octamer, composed of two each of histones H2A, H2B, H3, and H4, around which DNA is wrapped in chromatin.

nucleotide-excision repair
A type of DNA repair in which the initial step involves the excision of DNA bases together with associated nucleotides.

null allele
An allele of a gene that eliminates all normal function of the gene.

nulliparous
Referring to a female who has never given birth.

occult
Hidden, inapparent.

off-target effects
Effects of a drug on molecules other than the intended, targeted molecule, such as drug effects on proteins other than a targeted protein.

oligomer
A polymer of more than two (but not a large number of) subunits.

oligopeptide
A short protein polymer consisting of a relatively small number of amino acid residues.

oligopotent
Referring to the ability of a less differentiated cell to differentiate into several distinct cell types.

-oma
Denoting a benign or malignant growth.

oncofetal
Referring to an antigen that is normally expressed during embryonic development and is also expressed by some tumor cells.

oncogene (adj., **-genic**)
(1) A cancer-inducing gene. (2) A gene that can transform cells.

oncologist
A physician who treats cancer, usually through medical rather than surgical means.

oncoprotein
A protein specified by an oncogene.

operable
Capable of being treated successfully by surgery.

ordinate
The vertical or *y* axis of a Cartesian graph.

ortholog
A gene in one species that is the closest relative of a gene in another species; usually orthologs represent direct counterparts of one another in the genomes of two species.

orthotopic
Referring to an anatomically proper or native site. *See also* **ectopic**.

osteoblast
Mesenchymal cell type related to fibroblasts that constructs mineralized bone through the deposition of a collagenous matrix and apatite crystals.

osteoblastic
Referring to a class of bone lesions that involve localized increases in the amount of mineralized bone. *See also* **osteolytic**.

osteoclast
Cell type of the macrophage lineage that functions to degrade and demineralize already-assembled bone.

osteolytic
Referring to a bone lesion that involves localized dissolution of mineralized bone. *See also* **osteoblastic**.

osteotropism
The trait of cells, notably metastasizing cancer cells, to migrate to and colonize bone.

overexpression
Expression of an RNA or a protein at higher-than-normal levels.

papilloma
A benign, adenomatous proliferation of epithelial cells; term often used to describe benign lesions of the skin.

papovaviruses
The class of viruses that includes SV40, polyomavirus, and papillomaviruses.

paracrine
Referring to the signaling path of a hormone or factor that is released by one cell and acts on a nearby cell.

paraneoplastic
Referring to a biological effect evoked in the body by a tumor at a site in the body that is located some distance from the tumor itself and is apparently not directly involved in the pathogenesis of the tumor.

parenchyma
The portion of a tissue that lies outside the circulatory system and often is responsible for carrying out the specialized functions of the tissue.

parity
(1) The condition of having given birth. (2) The number of times that a female has given birth.

parous
Referring to a woman who has given birth at least once.

partial response
A 50% or greater reduction in tumor mass following anti-cancer therapy.

passaging
Practice of transferring cells from one culture vessel to another, often performed because the cell population has filled up the first vessel. *See also* **serial passaging**.

passive immunization
Procedure in which the immune responses of an organism are supplemented or strengthened through the introduction of immunological agents, usually antibodies, of foreign origin.

pathogenesis
Process that leads to the creation of a disease state.

pathological (n., **pathology**)
(1) Diseased or associated with a disease. (2) Referring to the study of a disease process, often at the level of light microscopy.

pathologist
Physician who examines tissues microscopically to study and classify disease.

pediatric
Referring to an attribute or condition of children.

penetrance
Degree to which or frequency with which an allele of a gene can influence phenotype, e.g., the likelihood that a germ-line allele will induce a clinical phenotype in a carrier of this allele.

peptidase
Peptide-cleaving enzyme.

perforin
A protein made by cytotoxic immune cells and inserted by them into the plasma membrane of a targeted cell; once inserted, perforin creates a channel through the membrane that causes the death of the cell, often by allowing the introduction of pro-apoptotic proteins into the cell.

pericytes
Cells closely related to smooth muscle cells that surround capillaries and provide the capillary walls formed by endothelial cells with tensile strength and contractility.

peritoneal
Referring to the cavity in the abdomen that is limited by an enclosing membrane and includes the lower gastrointestinal tract and associated organs, including pancreas and liver.

permissive
(1) Describing a temperature that allows growth of a temperature-sensitive virus. (2) Describing a type of host cell that allows viruses to proliferate.

peroxisome
A cytoplasmic organelle that is involved in the oxidation of various substrates, notably lipids.

phagocyte
A cell of the immune system—e.g., a macrophage or dendritic cell—that is specialized to engulf and destroy other cells, cellular fragments, and other debris.

phagocytosis
The process by which a cell, usually a component of the immune system, engulfs a particle (which may be another cell), internalizes this particle, and usually proceeds to degrade it.

pharmacodynamics
Time course of responses within a tissue or its cells that are induced by a drug.

pharmacokinetics
Kinetics describing the rise and fall in the concentration of a drug in the body, usually measured in the serum.

pharmacologic
Referring to a drug, usually a low–molecular-weight compound.

phenocopy
Phenotypic state created by one gene or regulatory mechanism that closely resembles the state created by a different gene or regulatory mechanism.

phenotype
(1) A measurable or observable trait of an organism. (2) The sum of all such traits of an organism.

pheochromocytoma
Tumor of the neuroectodermal cells of the adrenal glands.

phosphatase
An enzyme that removes phosphate groups from phosphorylated substrates, such as the phosphoamino acid residues in a protein or the phosphorylated inositol of a phospholipid.

phosphoprotein
A protein to which one or more phosphate groups have been covalently attached.

phosphorylation
Covalent attachment of a phosphate group to a substrate, often a protein.

physiology
(1) Biological functioning of cells, tissues, organs, and organisms. (2) The study thereof.

pimonidazole
1-[(2-hydroxy-3-piperidinyl)propyl]-2-nitroimidazole hydrochloride, a chemical used to detect regions of hypoxia within a tissue.

plasma cells
Cells of the B-cell lineage that secrete antibodies into the blood plasma.

plasma membrane
Lipid bilayer membrane that surrounds a eukaryotic cell and separates the aqueous environment of the cytoplasm from that in the extracellular space.

plasminogen
The inactive pro-enzyme that is converted into the active plasmin protease through proteolytic cleavage.

pleiotropy
Ability of certain genes or proteins to concomitantly evoke a series of distinct downstream responses within a cell or organism.

pleural
Referring to the space between the lungs and the membrane surrounding the lungs.

pleural effusion
Accumulation of cancer cells and fluid in the space between the lungs and the surrounding pleural membrane.

pluripotent
Referring to the ability of a stem cell to seed progeny that can participate in the formation of all of the tissues of an embryo except the extraembryonic membranes.

pocket protein
Term referring to pRb and two related proteins, p107 and p130.

podosome
An organized focal area on the surface of a cell in which plasma-membrane–bound proteases degrade nearby extracellular matrix.

point mutation
Substitution of a single base for another in a DNA sequence.

polyclonal
Describing a population of cells that trace their origins to two or more founding ancestral cells (*see also* **monoclonal**).

polycyclic
Referring to molecules with a structure that contains multiple, covalently closed rings.

polycyclic aromatic hydrocarbon
A hydrocarbon that carries multiple benzene rings.

polycythemia
Condition involving higher-than-normal levels of circulating red blood cells.

polykaryon
Cell carrying multiple nuclei in a single cytoplasm.

polyp
An adenomatous growth; often such a growth in the intestine.

polypectomy
Surgical removal of a (colonic) polyp.

pool
A population or collection of similar entities, e.g., a gene pool or a pool of stem cells.

post-mitotic
Describing a cell that has given up the option of ever entering into an active growth-and-division cycle again.

post-translational modification
Covalent alteration of a protein occurring after the initial polymerization of the polypeptide backbone of the protein.

prenatal
Occurring before birth. *See also* **natal**.

primary cells
(1) Literally, cells that have been recently explanted from living tissue into culture dishes and have not been propagated thereafter in culture. (2) More commonly, cells that have been explanted from living tissue into culture dishes and have been subjected to only a small number of successive *in vitro* passages thereafter.

primary tumor
Tumor growing at the anatomical site where tumor progression began and proceeded to yield this mass.

primase
An enzyme that initiates DNA synthesis by laying down a short RNA segment on the template strand; the 3′-hydroxyl end of this RNA primer then serves as the site for attachment of the initial deoxyribonucleotide by a DNA polymerase. *See also* **primer**.

primer
A DNA or RNA molecule whose 3′ end serves as the initiation point of DNA synthesis by a DNA polymerase.

primitive
(1) Referring to the relatively undifferentiated phenotype of a cell. (2) An embryonic cell.

probe
An RNA or DNA, usually radiolabeled, that anneals specifically with a complementary nucleic acid being analyzed, enabling the detection of the targeted nucleic acid sequence.

procarcinogen
A chemical compound that, while itself relatively nonreactive chemically, can be converted into a highly reactive carcinogen, usually through metabolic processes.

pro-drug
An inactive precursor of a biologically active drug.

pro-enzyme
Catalytically inactive form of an enzyme that requires some type of alteration (e.g., proteolytic cleavage) in order to become catalytically active.

professional antigen-presenting cell
See **antigen-presenting cells**.

progenitor cells
See **transit-amplifying cells**.

progeria
Syndrome in which an individual undergoes premature or accelerated aging.

prognosis
A prediction about the future clinical course of a disease, often influenced by detailed analyses of its existing attributes, such as histopathology and biochemical markers.

programmed cell death
See **apoptosis**.

progression
See **tumor progression**.

progression-free survival
Period between the initiation of anti-tumor therapy and subsequent clinical relapse or clear worsening of clinical condition.

prokaryotic
Referring to the relatively small, nonnucleated cells of bacteria and related organisms.

promoter
(1) An agent that furthers the progression of multi-step tumorigenesis by nongenetic mechanisms, notably those involving inflammation and/or mitogenesis. (2) The sequence within a gene that controls its transcription.

promotion
Process that stimulates or accelerates tumor progression, usually presumed to do so without directly damaging the genome.

proofreading
Process by which a DNA polymerase scans the deoxyribonucleotide segment that it has just synthesized in order to ensure that the sequence of this segment is precisely complementary to that of the template strand.

prophase
First subphase of mitosis, in which chromosomes begin to condense and centrosomes begin to assemble.

prophylactic
Preventative.

protease
An enzyme that cleaves protein substrates

proteoglycan
Molecule with one or more glycosaminoglycan chains attached covalently to a protein core.

proteolysis (adj., **-lytic**)
Process, usually mediated by proteases, of cleaving a polypeptide to lower–molecular-weight fragments including individual amino acids.

proteomics
Technology by which systematic surveys are made of the expression of large numbers of distinct protein species in a biological sample, such as a cell lysate or a biological fluid.

proto-oncogene
A normal cellular gene that, upon alteration by DNA-damaging agents or viral genomes, can acquire the ability to function as an oncogene.

provirus
The dsDNA copy of a retroviral genome that is the product of reverse transcription; it can exist transiently, as an episomal (nonchromosomal) plasmid, or stably, following its integration into the chromosomal DNA of an infected host cell.

pseudopregnant
Referring to a female that has been placed in a physiological state that closely resembles that of pregnancy through exposure to certain hormones.

pulmonary
Referring to the lungs.

pyknosis (also spelled **pycnosis)**
Collapse of nuclei into densely staining structures.

quartile
Any one of four equal groups into which a large sample of individuals or test objects has been subdivided through the use of one or another measurement technique.

rad
Unit of radiation corresponding to 0.01 joule of absorbed radiation or 0.01 gray.

radioautography
See **autoradiography**.

radiosensitive
Describing cells or tissues that are sensitive to killing by radiation, including radiation therapies.

radiotherapy
Treatment of a disease, notably cancer, through X-irradiation.

rate-limiting
Referring to a step in a multi-step process that governs the over-

all rate at which the process reaches completion because this step is kinetically among the slowest to occur.

reading frame
(1) The base sequence within a gene that encodes the amino acid sequence of a protein. (2) The registration of triplet codons within this sequence that enables the proper translation of this protein sequence.

receptor
Protein found on the plasma membrane or within a cell that is capable of specifically binding a signaling molecule (its ligand). Most types of receptors emit signals, such as those inducing cell proliferation, in response to such binding.

recessive
(1) Referring to one of several alternative traits that can be specified by a genetic locus; when the locus is heterozygous and carries information specifying two distinct traits, the dominant trait will be exhibited by the organism and the recessive will not. (2) Referring to an allele of a gene that is unable to dictate phenotype when in the presence of a second allele that acts dominantly. *See also* **dominant**.

reciprocal translocation
Exchange of chromosomal segments between two chromosomes from different chromosome pairs, resulting in the conservation of all participating chromosomal segments.

reductionism
A scientific research strategy that involves the study of individual, relatively simple components of complex systems rather than the systems as a whole.

refractory
Unresponsive to some type of signal or therapeutic agent.

rejection
Process whereby the immune system of an organism prevents the growth of engrafted cells, usually by killing these cells. *See* **allograft rejection**.

relapse
(1) (n.) Reoccurrence of a disease state, such as the reappearance of a tumor, after treatment with an initial, ostensibly successful therapy. (2) (v.) To sustain such a reoccurrence.

remission
Retreat or disappearance of a disease state with the implied possibility of its eventual reappearance or worsening.

renal
Referring to the kidney.

replicative immortality
See **immortality**.

repressed
Referring to a gene that is not being expressed and therefore is not being transcribed.

repression
Regulatory mechanism that causes shutdown of the expression of a gene.

resection
Removal by surgical excision.

restriction fragment length polymorphism
Variation in DNA sequence that can be detected through its effect of allowing or preventing cleavage of a chromosomal DNA segment by a restriction enzyme.

restriction point
Decision-making point in the late G_1 phase of the cell cycle at which a cell commits itself to completing the remaining phases of the cell cycle, remaining in G_1, or exiting the active cell cycle and entering into G_0.

retinoblastoma
Tumor of the oligopotential stem cells of the retina.

retrovirus
A class of viruses that uses a reverse transcriptase enzyme to copy its genomic RNA into DNA.

reverse transcriptase
Enzyme capable of making a DNA complementary copy of an RNA molecule using the RNA molecule as template.

reverse transcription
Enzymatic reaction whereby an enzyme, such as reverse transcriptase, copies an RNA template into a complementary DNA copy.

ribosomes
Bodies in the cell that synthesize proteins.

Rituxan
The chimeric anti-CD20 monoclonal antibody bearing a murine antigen-combining (variable) domain and a human constant domain (*also called* **rituximab**).

sarcoma
Tumor derived from mesenchymal cells, usually those constituting various connective tissue cell types, including fibroblasts, osteoblasts, endothelial cell precursors, and chondrocytes.

schwannoma
Tumor of the nonneuronal Schwann cells forming sheaths around the axons of neurons.

second messenger
A low–molecular-weight molecule that is able to act as an intracellular hormone, conveying signals from one part of the cell to another.

section
A slice through a tissue.

segregation
(1) Separation of chromosomes at the end of mitosis. (2) Separation of alleles during meiosis.

selectivity
Relative ability of a therapy to affect targeted cells or tissue compared with its (side) effects on normal cells or tissue. *See also* **therapeutic index**.

self-reactive
Referring to the ability of certain components of the immune system of an organism to recognize and react with the normal tissue and normal cellular antigens of that organism.

seminoma
A tumor of the germ cells in the testes.

senescence
A nongrowing state of cells in which they exhibit distinctive cell phenotypes and remain viable for extended periods of time but are unable to proliferate again. Often arises after extended passaging *in vitro*.

sequence motif
(1) Short oligonucleotide sequence that is characteristically associated with one or another biological function. (2) Amino acid sequence that is characteristically associated with a structural or functional aspect of a protein.

serial passaging
The practice of transferring a cell population from one culture vessel to the next and subsequently taking an aliquot of the resulting cell population and transferring it to yet another culture vessel, repeating this cycle for an extended period of time.

serpentine
Referring to the class of G-protein–coupled receptors that wend their way back and forth (in a snakelike pattern) through the plasma membrane a total of seven times.

serum (pl., **sera**)
The fluid left behind when blood clots.

sinusoid
A capillary-like channel located between liver hepatocytes that is lined with endothelial cells but lacks mural cells as well as a capillary basement membrane.

small-cell lung carcinoma
A lung cancer of specialized cells having neurosecretory properties.

soma
All the tissues in the body outside of the germ cells (sperm and egg) and the immediate precursors of the germ cells.

somatic mutation
Mutation that strikes the genome of a cell outside of the germ line; such a mutation cannot, by definition, be transmitted to the next organismic generation.

Southern blot
Procedure in which DNA molecules, usually produced by restriction enzyme cleavage, are resolved by gel electrophoresis and transferred to a filter to which they adsorb; the filter is subsequently incubated with a sequence-specific radiolabeled DNA probe to reveal, upon subsequent autoradiography, the sizes of the DNA fragments recognized by the probe.

splicing
(1) Process that causes the deletion of a defined segment of a primary RNA transcript and the fusion of the two RNA segments flanking the deleted RNA segment. (2) Process occurring in the nucleus whereby a pre-mRNA precursor is converted into an mRNA through the deletion of introns and the fusion of remaining exons.

sporadic
Describing a disease or condition that occurs randomly in a large population without any apparent predisposition, such as one caused by a heritable genetic susceptibility.

squamous
Referring to epithelial cells that line a duct or the skin and lack secretory function.

stem cell
Cell type within a tissue that is capable of self-renewal and is also capable of generating daughter cells that develop new phenotypes, including those that are more differentiated than the phenotype of the stem cell.

stereochemistry
Description of the three-dimensional structure of a molecule (such as a protein) and the influence that this structure has on the chemical behavior or biochemical function of the molecule.

stochastic
Referring to an event that occurs randomly with a certain probability rather than in a precisely predetermined fashion.

stoichiometric
A relationship or a reaction between two or more molecular species in which the relative molarities of the participating species are precisely specified.

stratify
To classify superficially similar entities (e.g., tumors) into several distinct categories or subclasses.

stressor
An agent that causes some type of physiologic stress.

stroma (pl., **stromata)**
The mesenchymal components of epithelial and hematopoietic tissues and tumors, which may include fibroblasts, adipocytes, endothelial cells, and various immunocytes as well as associated extracellular matrix.

stromalization
Referring to the process by which stroma is generated in a normal or neoplastic tissue.

subcutaneous
Beneath the skin.

submicroscopic
Too small to be seen through the light microscope.

substrate
A molecule that is acted upon by an enzyme.

supernumerary
Referring to a greater-than-normal number of some object.

surrogate marker
A measurable parameter, often a diagnostic parameter, that serves to indicate the behavior of another process whose behavior it parallels and reflects.

synapse
Physical connection formed between two interacting immune cells or between a cytotoxic lymphocyte and a targeted cell that facilitates exchange of signals between them and, in the case of cytotoxic cells, the transfer of cytotoxic granules from the cytotoxic cell to the targeted cell.

synchronous
(1) Occurring at the same time; occurring in a temporally coordinated fashion. (2) Referring to a population of cells that enter a specific phase of the cell cycle at the same time.

syncytium
Cell formed when the plasma membranes of two or more cells are fused together.

syndrome
Collection of symptoms that together define a specific disease condition.

syngeneic
(1) Referring to two organisms that share the identical genetic background, such as two members of an inbred strain of mice. (2) Describing the relationship between two sets of cells or tissues deriving from identical genetic backgrounds, or between a set of cells and an organism.

T-loop
Lasso-shaped configuration of the end of telomeric DNA that involves the insertion of its 3′ overhanging strand into its double-stranded region.

tamoxifen
A synthetic analog of estrogen that can antagonize certain types of estrogen action upon binding the estrogen receptor.

telomerase
An enzyme specialized to extend telomeric DNA; those telomerases characterized to date carry an RNA subunit and a reverse transcriptase-like catalytic subunit.

telomere
Protective nucleoprotein structure at the end of a eukaryotic chromosome that protects this end from degradation and from fusion with other chromosomes.

telophase
Fourth subphase of mitosis, during which chromosomes decondense and the nuclear membrane reassembles.

temperate
(Of an infectious agent, such as a virus) creating minimal damage in an infected host cell or organism. *See also* **virulent**.

temperature-sensitive
Describing a phenotype that is apparent when cells or viruses grow at one temperature but not at another.

teratogen
An agent that causes malformations by perturbing embryonic morphogenesis.

teratoma
Benign tumor formed by embryonic stem cells in which a wide variety of differentiated cell types are formed.

tetraploid
Karyotype having precisely four haploid complements (or two diploid complements) of chromosomes.

therapeutic index
(1) A measurement of the extent to which a treatment affects a targeted diseased tissue, such as a tumor, compared with its effects on untargeted, normal tissues. (2) The ratio of these two effects.

therapeutic window
Range of concentrations of a drug that are higher than that needed to elicit a therapeutic effect and lower than the maximum tolerated dose.

thrombin
A plasma protease that is activated following wounding and triggers blood coagulation by activating platelets and cleaving fibrinogen to fibrin.

thromboembolus
See **embolus**.

thrombopoiesis
Process leading to the formation of blood platelets.

thrombopoietin
A growth factor that stimulates the production of megakaryocytes, and thus of blood platelets.

thrombus (pl., **thrombi**)
A blood clot.

tissue culture
Procedure of propagating cells outside of living tissues in various types of flasks and dishes.

tissue factor
A cell-surface glycoprotein expressed by many cell types in the body that interacts with clotting factors in the plasma, thereby initiating the coagulation cascade.

tissue-specific gene
Gene that is expressed only in cells of certain individual tissue types.

tolerance
State in which the immune system shows a lack of reactivity toward certain antigens, notably those that are expressed by normal cells and tissues.

tomography
A computerized image-processing technique that integrates images obtained by X-rays, ultrasound, or other imaging procedures to generate a cross-sectional image of an object, such as the human body, at a given depth.

topoinhibition
See **contact inhibition**.

totipotent
Referring to the ability of a stem cell to generate all the differentiated cell lineages existing in the embryo as well as the extraembryonic membranes.

toxicity
The undesired side effect(s) of a drug on normal tissues and normal metabolism.

transcription
Copying of DNA sequences into RNA molecules.

transcription factor
Protein that is involved in regulating the transcription of a gene, often by associating with sequences in the promoter region of the gene.

transdifferentiation
Acquisition by a cell from one differentiation lineage of a phenotype characteristic of cells from another, distinct differentiation lineage.

transduction
(1) Process whereby a signaling element, such as a protein, receives a signal and, in response, emits another signal. (2) Process by which a gene is introduced into a cell, usually by a vector such as a viral vector.

transfection
Procedure of introducing DNA into mammalian cells, often achieved by calcium phosphate co-precipitation.

transferase
An enzyme that attaches a complex molecule, such as glutathione, to its substrate.

transformant
A transformed cell.

transformation
(1) Process of converting a normal cell into a cell having some or many of the attributes of a cancer cell. (2) Alteration of a cell through the introduction of a genetic element.

transgene
(1) A cloned gene that has been inserted experimentally into the germ line of an animal. (2) Less commonly, any experimentally altered gene in the germ line.

transgenic
(1) Referring to an animal or breed of animal whose germ line has been experimentally altered, usually through the insertion of a cloned gene. (2) Less commonly, referring to an animal or breed of animal whose germ line has been altered through any of a variety of genetic manipulations, including the addition of a cloned gene or the alteration of a resident gene through homologous recombination.

transit-amplifying cells
Relatively undifferentiated cells that are initially generated by division of a stem cell and are capable of exponential proliferation for a limited number of cell generations before spawning highly differentiated progeny, which in many tissues are post-mitotic.

transition
Point mutation in which one purine base replaces the other, or in which one pyrimidine base replaces the other.

translation
Synthesis of proteins according to the base sequences of RNA molecules.

translocation
(1) Rearrangement of chromosomes that results in the fusion of two chromosomal segments that are not normally attached to one another, often resulting in a microscopically visible alteration of karyotype. (2) Movement of a physical entity from one part of the cell to another. (3) Movement of a ribosome down an mRNA being translated.

transmembrane
Referring to the domain of a protein that is threaded through a membrane and therefore exists in the hydrophobic environment of a lipid bilayer.

transphosphorylation
Phosphorylation of one protein molecule by another, such as the phosphorylation of one receptor subunit by the kinase carried by another.

transposon
A genetic element or DNA segment that is able to move from one chromosomal integration site to another within a cell.

transversion
Point mutation in which a purine base replaces a pyrimidine or vice versa.

triploid
Describing a karyotype having precisely three haploid complements of chromosomes.

tritium
A radioactive isotope of hydrogen.

-trophic
Aiding in or supporting survival.

-tropic
Referring to a tendency of a cell or an organism to move toward or turn toward some object or source or to direct its actions toward that source.

tropism
(1) Tendency of a cell to face or move toward a specific location or signaling source. (2) Tendency of a cell to migrate in a specific direction or, in the case of metastatic cancer, to appear to home to a specific tissue site in the body.

tumor progression
(1) Process of multi-step evolution of a normal cell into a tumor cell. (2) Evolution of a benign into a malignant cancer cell. (3) Evolution of a pre-malignant cell from a promoter-dependent to a promoter-independent state.

tumor rejection
Process by which an organism prevents the formation of a tumor (including tumor formation by engrafted cells), often achieved through the action of its immune system.

tumor suppressor gene
(1) A gene whose partial or complete inactivation, occurring in either the germ line or the genome of a somatic cell, leads to an increased likelihood of cancer development. (2) Such a gene that is responsible for constraining cell proliferation. *See also* **gatekeeper**.

tumoricidal
Able to kill cancer cells and/or destroy a tumor.

tumorigenic
(1) referring to the ability of cells to form tumors when introduced into appropriate animal hosts. (2) Less commonly, pertaining to an agent such as a tumor virus that imparts this ability to cells.

ubiquitylation
Process by which one or more ubiquitin molecules are attached to a protein substrate molecule, which often results in the degradation of the tagged protein.

ultimate carcinogen
A chemical compound that is able to directly contribute to the induction of cancer without prior (or further) chemical modification, usually by direct chemical interaction with DNA, thereby altering the structure of the latter.

urothelium
The specialized epithelial cell lining of the bladder.

vacuoles
Small, fluid-filled, bubble-like structures, often seen in the cytoplasm of cells that are under physiologic stress and in cells infected by certain viruses.

vascularized
Referring to the presence of blood vessels in a tissue such as a tumor.

vasculature
Network of blood vessels.

vasoactive
Referring to a regulator of vascular function, such as a regulator of vascular permeability or constriction.

vector
(1) Agent, often a virus, that is able to carry a gene from one cell to another. (2) An infected organism that serves to transmit and distribute an infectious agent to other organisms.

vehicle
The solvent that is used to deliver a drug.

venule
A small vein that conducts blood from capillaries to larger veins.

villus
Fingerlike structure of epithelium that protrudes from the wall of the small intestine into the lumen.

vimentin
An intermediate filament protein of the cytoskeleton of mesenchymal cells such as fibroblasts.

viremia
Presence of high concentrations of virus in the bloodstream.

virion
Virus particle including a capsid (coat) and the viral genome.

virulent
(Of an infectious agent, such as a virus) creating damage such as cell or tissue destruction in an infected host cell or organism. *See also* **temperate**.

virus stock
A solution of virus particles used experimentally to infect cells or organisms.

vital dye
A dye that can be used to stain cells or tissues and is retained for extended periods of time in these objects without compromising viability.

vitiligo
A skin disorder, often of autoimmune origin, that leads to loss of patches of melanocytes from the epidermis and resulting loss of pigmentation.

wild type
The allele of a gene that is commonly present in the great majority of individuals in a species.

xenobiotic
A biologically active compound that originates outside of the body and is foreign to its normal metabolism.

xenograft
A normal or neoplastic tissue derived from one species that has been grafted into a host animal from another species.

xenotropic
Referring to a class of retroviruses from one species that can infect and replicate in cells of another species.

ZIP code
Zone improvement plan, a numerical scheme devised by the U.S. Postal Service that designates each area of delivery with its own five-digit number; by analogy, hypothesized homing address that disseminating cancer cells seek on the luminal surfaces of capillaries in specific tissues.

zymogen
An inactive precursor form of an active enzyme.

zymogram
Analytic technique in which the migration rates of various proteins upon gel electrophoresis are gauged by their localized enzymatic activity following such electrophoresis.

Index

Notes: Page references followed by the suffixes F, S and T refer to figures, sidebars and tables respectively, those in *italics* refer to material on the companion CD. Please note that where both human and nonhuman homologous genes/proteins are mentioned in the text, that the gene/protein is listed using the nonhuman nomenclature (e.g., *Ras* as opposed to *RAS*).

A

B

C

D

E

I

J

K

L

M

S

T

U

V

W

X

Y

Z